Comparative Animal Physiology

Philip C. Withers

Senior Lecturer, Department of Zoology
The University of Western Australia

SAUNDERS COLLEGE PUBLISHING
Harcourt Brace Jovanovich College Publishers

Fort Worth   Philadelphia   San Diego   New York
Orlando   Austin   San Antonio   Toronto
Montreal   London   Sydney   Tokyo

Text Typeface: Times Roman
Compositor: Monotype Composition
Acquisitions Editor: Julie Levin Alexander
Developmental Editor: Gabe Goodman
Managing Editor: Carol Field
Project Editor: Nancy Lubars
Copy Editor: Zanae Rodrigo
Manager of Art and Design: Carol Bleistine
Art Director: Christine Schueler
Art and Design Coordinator: Caroline McGowan
Text Designer: Rebecca Lemna
Cover Designer: Doris Bruey and Christine Schueler
Text Artwork: J&R Art Services Inc.
Layout Artwork: Dorothy Chattin
Director of EDP: Tim Frelick
Production Manager: Charlene Squibb
Senior Marketing Manager: Marjorie Waldron

Cover Credit: Kangaroo in "X-ray" style; Australian aboriginal rock art. (© R. A. Clevenger/Westlight, Los Angeles.)

Printed in the United States of America.

COMPARATIVE ANIMAL PHYSIOLOGY

ISBN 0-03-012847-1

Library of Congress Catalog Card Number: 91-050762

89   032   9876

Comparative Animal Physiology

Dedicated to Adelaide Lindsay Withers

About the Author

Dr. Philip Carew Withers received his B.Sc. (Hons.) degree in Zoology from the University of Adelaide in 1972; his honors project investigated the physiological adaptations of hairy-nosed wombats. After visiting the Zoology Department at Monash University, he completed his Ph.D. degree in the Department of Biology at the University of California, Los Angeles, on the physiology of burrowing and hibernating mammals and birds. He was a postdoctoral fellow first in the Department of Zoology at the University of Cape Town, where he studied the physiological adaptations of small mammals, including those on rocky outcrops in the Namib desert, then at the Percy Fitzpatrick Institute for African Ornithology, University of Cape Town, where he studied the flight of storm petrels. Later as a visiting scientist at the Department of Zoology, Duke University, Dr. Withers studied the aerodynamics of bird wings and the energetics of activity in salamanders. He became a professor in the Department of Biology at Portland State University and is currently a senior lecturer in the Department of Zoology, The University of Western Australia.

Dr. Withers retains his research interests in the physiology of burrowing and hibernating mammals, but his broad research activities include the physiology and ecology of arid-adapted terrestrial vertebrates, specifically, the energetics, water balance, and osmoregulation of mammals, birds, reptiles, and amphibians.

Dr. Withers teaches a variety of topics related to animal physiology to first, second and third year students, and supervises honors and postgraduate students.

Dr. Withers is a member of a number of scientific societies, including the American Society of Icthyologists and Herpetologists, the American Society of Zoologists, the Australian Mammal Society, the Royal Society of Western Australia, and the Zoological Society of Southern Africa. He is also a member of the Naturalists Club of Western Australia.

Preface

How animals function has long fascinated zoologists. Despite being made of the same atomic, molecular, subcellular and cellular "nuts and bolts," different species of animals have evolved distinct, complex modes of functioning, which enable them to survive in their individual environments. Comparing and contrasting these diverse mechanisms provides a cohesive understanding of animal function that is the foundation for the discipline of comparative animal physiology.

This discipline is now in a critical and exciting period. The rapid pace of scientific and technological advances have separated the different aspects of comparative animal physiology into narrower, more specialized, and more technical subdisciplines, each with its own methodology, jargon, and journals. Comparative studies are becoming increasingly difficult for animal physiologists because the wealth of new information becoming available every day limits their capacity to follow the intricacies of even a few subdisciplines simultaneously. Nevertheless, I believe that there is still an important role to be played by the comparative study of animal physiology, particularly by integrating the advances made in specific subdisciplines with the broader picture of how differently animals function. Animal physiology as a whole is advanced more rapidly by a comparative approach rather than by isolated advances in discrete subdisciplines. For example, advances in one subdiscipline, such as buoyancy regulation in crustaceans, can be equally applicable to buoyancy regulation in other animals, such as sharks. It is easier and quicker to transfer a conceptual advance from one area of research to another, than it is to make the same conceptual advance independently to each area. Such information can also be just as instructive if it proves to be inapplicable to another research area. The "why doesn't it work" questions can be as important as the "it works the same way" answers!

Consequently, I hope that this book serves to sustain the comparative approach to animal physiology by convincing the next generation of animal physiologists of the usefulness of this method, and to remind students in isolated, narrow subdisciplines that there are useful things to be learned from the progress of others, no matter how unrelated and distant they may appear to be.

Comparative Animal Physiology is intended for use by students who are majoring in biology or zoology and already have a general biology background, for courses such as General Physiology, Principles of Physiology, Comparative Animal Physiology, Vertebrate Physiology, or Invertebrate Physiology.

Instructional Features

Comparative Approach with Accessible Writing Style. This book presents an up to date and extremely broad overview of animal physiology with a thoroughly comparative approach. In general, at least equal coverage is given to invertebrate and vertebrate animals using a quantitative perspective. A physico-chemical background is provided so that students can appreciate that all animals obey the same physical and chemical constraints as machines. Equations are used to illustrate quantitative relationships. I have endeavored to maintain a writing style that is direct and to the point, rather than cumbersome or encyclopedic, encouraging students to read and use the book as an integrated work. For this reason, it is not possible or desirable to discuss thoroughly every invertebrate taxonomic group. Many entire phyla of animals are not mentioned in specific chapters because they are "lesser phyla" and either have not been well studied or their physiology resembles that of other phyla that have been discussed. Nevertheless, most of the major phyla, or groups of phyla, are included in each chapter.

Illustrations and Tables. Based on the belief that structure is irretrievably related to function, and that it is necessary to appreciate both structure and function, each chapter has numerous illustrations. A good illustration can more succinctly explain structure and function than could a thousand words. Where possible, illustrations of physiological relationships have been taken or adapted from original sources so that students can see not just the summarizing line or curve of best fit, but actual data. The illustrations use color to emphasize important aspects. Photographs are frequently used to show what animals actually look like, and photomicrographs are used to show organs, tissues, and cell structures. Tables summarize and compare actual values for physiological parameters that illustrate important concepts, and provide in-depth analysis and comparison of different taxomomic groups. Most figures and tables also include a reference to the original source of the material.

Chapter Supplements. Supplements at the end of each chapter provide a more detailed discussion of topics that merit further attention for the advanced student, without distracting readers from the overall flow of the chapter material.

Appendices. Appendix A provides a taxonomic overview of invertebrates and vertebrates to the level at which reference might be made in the text, e.g., phylum for the "lesser phyla," class and order for most invertebrates, and family for vertebrates. Appendix B summarizes the SI system of fundamental and derived units, the unitary system to which this book adheres. Appendix C tabulates a variety of constants and coefficients that are referred to in the text and used particularly in equations.

References. A comprehensive list of over 1,300 references is provided at the end of the book.

Glossary. A glossary is provided at the end of the book to define the majority of biological terms used, particularly those that might not be generally known to students.

Indexes. A comprehensive subject index is provided to the text, illustrations, and tables. The subject index is organized both by topic and by taxon. A separate genus index directs the reader to each text, illustration, and table reference for that genus name.

Organization

The general organization of chapters in this book proceeds from the molecular, subcellular, and cellular levels to the more organismal levels, but is intentionally designed to be flexible. Each chapter is self-contained and cross-referenced, allowing instructors to assign readings in almost any order according to their individual orientation. Each chapter begins with a list of major section headings indicating the overall organization of the chapter, and closes with a summary section. A list of recommended readings directs the student to important review articles or books. Reference is made throughout the text to material that has historical importance, provides a good survey of the topic, or presents new material.

There is considerable variation in the manner in which the taxonomic diversity and physiological diversity are presented in each chapter. The invertebrates are often covered taxonomically because they represent very different types of animals with correspondingly diverse physiologies, whereas the vertebrates are taxonomically more related animals in which evolutionary progression is evident. Nevertheless, the particular approach in each chapter, and the order in which invertebrates and vertebrates are discussed, varies to suit the topic.

The first two chapters in the book, the **Introduction** and **External and Internal Environments**, are essentially background material to the remaining chapters. Each of the remaining chapters in the book is organized into introductory background concepts followed by a thorough comparative discussion of the various animal groups with an overall evolutionary perspective.

The third chapter, **Cellular Energetics**, discusses cellular metabolism from a physico-chemical and biochemical perspective, whereas Chapter 4, **Animal Energetics**, discusses metabolism from the perspective of the whole organism.

Temperature has a profound effect on almost every physiological function, and so Chapter 5, **Temperature**, provides a thorough analysis of the physico-chemical effects of temperature, the avenues for heat exchange between animals and their environment, and patterns of thermoregulation in various groups of animals.

The structure and function of membranes described in Chapter 6, **Membrane Physiology**, provides essential background to understanding solute regulation by cells and the electrophysiology of cells. The physiology of excitable cells, described in Chapters 7 and 8, **Sensory Physiology** and **Nervous Systems**, describes how cells and animals react to their environment. The ways that animal cells move by ciliary, flagellar, or muscular contraction are described in Chapter 9, **Cell Movement**, and the various methods of animal locomotion in water, on land, or in the air are described in Chapter 10, **Animal Movement**.

The important role of endocrine systems in motor control, especially by the nervous system, is described in Chapter 11, **Endocrinology**. This chapter describes how both neurohormonal and classic hormonal systems are used to control a variety of body functions such as growth and regeneration, reproduction, iono- and osmoregulation, cellular metabolism, and color.

Chapters 12 and 13, **Aquatic Respiration** and **Aerial Respiration**, describe the physical principles involved in gas exchange, and explain how these principles apply to the wide variety of different types of gas exchange structures used by animals, such as the body surface, gills, lungs, and tracheal systems. The structure and function of circulatory systems, which generally transport respiratory gases between the respiratory surfaces and tissues, is described in Chapter 14, **Circulation**, and the role of the blood, or hemolymph, is described in Chapter 15, **Blood.**

Chapter 16, **Water and Solutes**, describes the important physiological roles of water and solute regulation in animals, both aquatic and terrestrial, and the structures and functioning of the specific excretory organs (protonephridia, nephridia, coelomoducts, nephrons, Malpighian tubules, etc.) are discussed in Chapter 17, **Excretion**.

The structure and function of digestive tracts, described in Chapter 18, **Digestion**, integrates aspects of animal behavior and structure (feeding mechanisms), cell secretion and movement (secretion and peristalsis by the gut tube), cellular biochemistry (digestive enzymes and the biochemistry of hydrolysis of various organic substrates), and nutrition (the role of vitamins).

Acknowledgments

The basic concepts of this book were developed as lecture material for a *Comparative Animal Physiology* course which I offered at Portland State University. Many additional topics have been added during the preparation of the book and thus it may appear to have little resemblance to the original course. Nevertheless, I thank all my students for their participation in that course, and for thereby unknowingly contributing to the preparation of this book.

A great number of colleagues have assisted in the preparation of this book, principally by critiquing chapters. I especially thank Warren Burggren of the University of Nevada, Las Vegas, for his heroic efforts in reviewing and re-reviewing the entire book. Despite the efforts of these colleagues in correcting my errors of fact or omission, I am of course, responsible for any remaining errors. My thanks also to the following reviewers for their valuable chapter critiques:

Dr. Gary Bell, Biological Sciences Center, Boston University

Dr. Don Bradshaw, Department of Zoology, University of Western Australia

Dr. Warren Burggen, Department of Zoology, University of Nevada

Dr. Stan Hillman, Department of Biology, Portland State University

Dr. Delbert Kilgore, Division of Biological Sciences, University of Montana

Dr. John Knesel, Department of Biology, Northeast Louisiana University

Dr. John Minnich, Department of Biological Sciences, University of Wisconsin–Milwaukee

Dr. Jamie O'Shea, Department of Zoology, University of Western Australia

Dr. Bela Piacsek, Department of Biology, Marquette University

Dr. Gerald Robinson, Department of Biological Sciences, Towson State University

Dr. Dick Rowe, Department of Biology, Greensboro College

Dr. William Stickle, Department of Zoology and Physiology, Louisiana State University

Dr. Richard Walker, Department of Biological Sciences, University of Calgary

Dr. Donald Whitmore, Department of Biology, University of Texas at Arlington

Dr. Paul Yancey, Biology Department, Whitman College

Dr. Randy Zelick, Department of Biology, Portland State University.

I also thank the editorial staff of Saunders College Publishing for their continued support and encouragement. Ed Murphy and Julie Levin Alexander provided invaluable assistance and encouragement as Acquisition Editors, and Gabe Goodman provided outstanding support as Developmental Editor. The Project Editor, Nancy Lubars, was extremely patient in coaxing, cajoling, and guiding me through the various stages of production. I also thank the Art Director, Christine Schueler, and the artists for their fine efforts at deciphering my diagrams and producing high quality artwork.

I especially thank Stan Hillman, my friend and colleague, for his continued encouragement with this book. I am grateful to Stan Hopwood and Tom Stewart for photographic and histological assistance with figures for the book. Maureen McAllister, Yvonne Patterson, and Sue Beardman provided invaluable library help. I thank the many hundreds of authors, and their publishers, who without remuneration allowed me to reproduce their artwork in original or modified form.

Finally, I thank my family for their patience and support during the many years of writing this book. My daughter Adelaide showed maturity beyond her years in understanding why I spent so much time working on the book rather than the other things that fathers should do. I am especially indebted to my wife Patricia O'Neill for her continued support, help, and encouragement.

P. C. WITHERS

Department of Zoology
The University of Western Australia
Nedlands 6009, Western Australia

MARCH 1992

Contents Overview

Contents

Comparative Animal Physiology

Chapter 1

Introduction

How animals work is exceedingly more complex than how even the most complicated man-made machine works. *(Department of Energy/Photo Researchers, Inc./Science Source)*

W hat is physiology? A typical dictionary definition of physiology (Henderson and Henderson 1975) is

• • • • • •

physi·ol'o·gy (fizioloji) *n*. [Gk. physis, nature; logos, discourse] That part of biology dealing with functions and activities of organisms.

A physiologist's definition might be more detailed and precise, but it would have the same basic theme—that physiology is the study of the biological processes that enable life to exist and function.

Physiology includes the study of many **levels of organization** of matter, from molecules to organisms. This is readily apparent if we examine the hierarchial fashion in which matter is organized in living tissues (Figure 1–1). Each level of organization is a specialized research area that tends to have its own techniques, jargon, and literature (Bartholomew 1986). Individual researchers tend to be restricted to one, or a few, adjacent levels of organization. However, it is necessary to look down the hierarchy to lower levels (e.g., towards macromolecules, molecules, atoms) for an analysis of how biological mechanisms function, i.e., how things work. We look up the table to higher levels (e.g., organisms, populations)

for a better understanding of effects and adaptive significance, i.e., why things work that way.

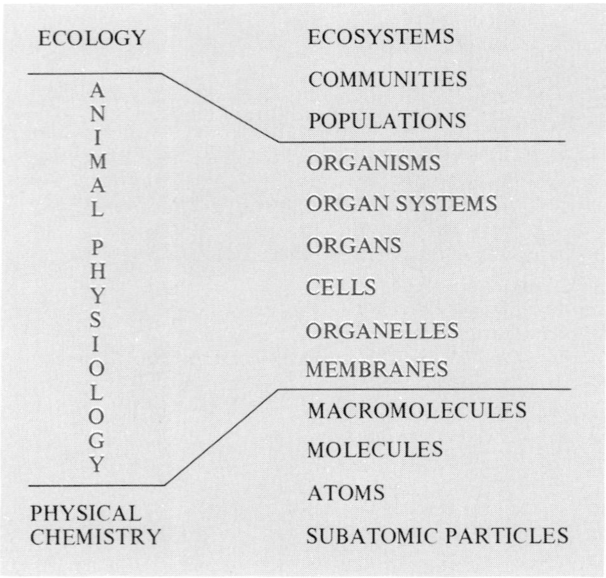

Physiology is a broad field of science, encompassing a number of subdisciplines. One criterion for defining subdisciplines is taxonomy. The primary topic of this book is **animal physiology**. Aspects of viral, microbial, fungal, and plant physiology are almost totally absent from this text, although the physiological principles of these organisms are frequently identical with, or at least very similar to, principles of animal physiology. Occasional reference will be made where it is heuristically useful to microbial physiology (e.g., internal environment, bioluminescence, osmotic and ionic regulation) or plant physiology (e.g., osmotic regulation). The physiology of protozoans (phylum Protista) will often be used as an introduction to the more complex physiology of multicellular animals.

Comparative Animal Physiology

Comparative animal physiology, defined in a broad context, is the comparing and contrasting of physiological mechanisms, processes, or responses of different species of animals, or of a single species, under differing conditions. The most interesting comparisons and contrasts are often the physiological convergences observed for distantly related animals adapted to similar environments (e.g., desert-adapted insects and mammals) because this illustrates the independent evolution of solutions to common problems, or the physiological divergence observed for closely related animals adapted to different environments (e.g., fresh water, brackish water, and seawater teleost fish) because this illustrates the modification of originally similar physiological processes to cope with different problems, or for the physiological responses of individual animals under differing conditions (e.g., at different temperatures) because this illustrates the intraindividual plasticity of an animal's physiology. Physiological adaptation is often most evident at high taxonomic levels (e.g., comparing classes, orders, families) but is also evident at lower levels (genus, species, and even subspecies).

It is often interesting and instructive to relate physiological adaptation with evolutionary progression, but it should be remembered that the phylogenetic tree is not an evolutionary progression of functional complexity from "lower organisms" to "higher animals" (e.g., from protozoans to mammals) to the "highest animal," man. Rather, all

ECOLOGY	ECOSYSTEMS
A N I M A L P H Y S I O L O G Y	COMMUNITIES
	POPULATIONS
	ORGANISMS
	ORGAN SYSTEMS
	ORGANS
	CELLS
	ORGANELLES
	MEMBRANES
	MACROMOLECULES
	MOLECULES
	ATOMS
PHYSICAL CHEMISTRY	SUBATOMIC PARTICLES

FIGURE 1–1 The hierarchy of organization in a biological system from subatomic particles to ecosystems showing the range of levels that comparative animal physiologists study (although they tend to concentrate on organs and organ systems).

branches of the phylogenetic tree represent evolutionary change over the same geologic time scale, and the physiological mechanisms of the lower animals are often as advanced and specialized as those of the higher animals.

Comparative animal physiology can be further subdivided into numerous (and nonexclusive) sections, such as general, medical, clinical, cellular, neural, sensory, reproductive, endocrinological, respiratory, biophysical, renal, cardiovascular, gastrointestinal, thermal, etc. The subdisciplines of physiological ecology and environmental physiology are at the interface of physiology with other disciplines, such as ecology and environmental science. This diversity within animal physiology is well reflected by the names of the scientific journals specializing in these areas, as can be readily observed by perusal of the journal holdings of university libraries. Table 1–1 provides a selected list of journals and abstract journals covering the type of material contained in this book. The original research articles and review articles found in such journals are the ultimate source of detailed and current information concerning animal physiology; books, particularly textbooks, are less detailed and more synthetic than journal articles and are often less current.

This book primarily examines the physical, chemical, and biochemical principles common to physiological systems, such as respiration, excretion, and metabolism—that is, it presents a **systems approach**. It then examines the variability of physiological processes for the wide diversity of animal groups, all of which operate within basic physical and chemical constraints but may differ markedly in approach and operation.

Objectives

The first major objective of this book is to illustrate the great diversity of physiological processes employed by different taxa of animals and to examine how these adaptations are related to the physical environment. The greatest taxonomic diversity is found among the invertebrates (about 30 phyla) compared to the vertebrates (a subphylum). Even amongst the invertebrates, the arthropods greatly outnumber the other invertebrates in numbers of species (Figure 1–2). A summary of animal classification is provided in Appendix 1 so that students can place those animals discussed in this book in their taxonomic/phylogenetic context. The amount of knowledge, particularly physiological knowledge, is not proportional to the numbers of species in a

TABLE 1–1

Selected list of journals and abstract journals (in boldface ital) that publish research or review papers relevant to the field of comparative animal physiology; the list is not exhaustive and further journal titles can be found in the Reference sections of this book and in biological sciences listings of libraries.

Acta Physiologica Scandinavica
American Journal of Physiology
American Zoologist
Annual Review of Physiology
Auk
Australian Journal of Zoology
Bioabstracts
Biochimie Biophysica Acta
Biological Bulletin
Biological Reviews
Canadian Journal of Zoology
Comparative Biochemistry and Physiology
Comptes Rendus Academie Sciences
Comptes Rendus Société Biologie
Condor
Copeia
Doklady Akademia Nauk
Endocrinology
Endocrinology Abstracts
Entomology Abstracts
Experientia
Forma et Functio
International Reviews of Biochemistry
International Reviews of Physiology
Japanese Journal of Physiology
Journal de Physiologie
Journal of Applied Physiology
Journal of Comparative Physiology
Journal of Endocrinology
Journal of Experimental Biology
Journal of Experimental Marine Biology and Ecology
Journal of Experimental Zoology
Journal of General and Comparative Endocrinology
Journal of General Physiology
Journal of Insect Physiology
Journal of Mammalogy
Journal of the Marine Biological Association, UK
Journal of Neurophysiology
Journal of Physiology
Journal of Theoretical Biology
Journal of Thermal Biology
Marine Biology
Nature
Physiological Reviews
Physiological Zoölogy
Pflügers Archiv
Quarterly Reviews of Biology
Respiration Physiology
Science
Science Citation Index
South African Journal of Zoology
Zoological Record

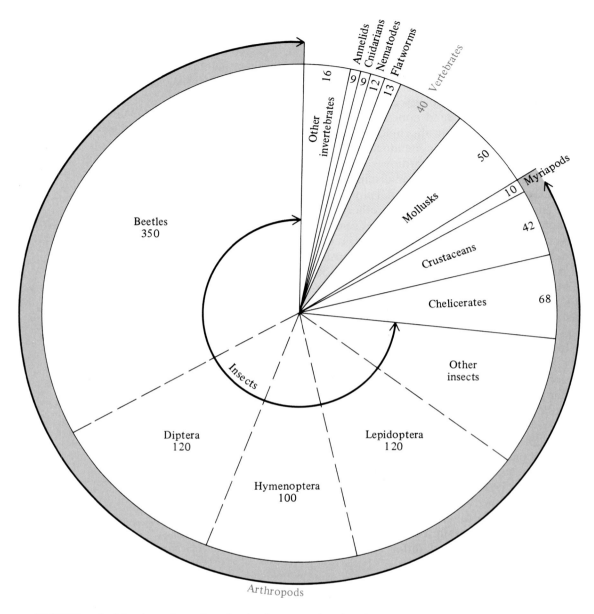

FIGURE 1–2   Number of species of animal groups showing the great predominance of the invertebrates (96.3% of the total) and, especially, of the arthropods (86% of the total). The numbers represent thousands of species.

taxon, but disproportionately favors the vertebrate animals, with mammals being the best studied of vertebrates. This book gives about equal coverage to vertebrates and the immensely more diverse invertebrate animals.

The second major objective of this book is to provide a physicochemical perspective to animal physiology. Physiological principles almost invariably reduce to some basic physical and chemical laws, e.g., the first and second laws of thermody-namics, Fick's laws of diffusion, the ideal gas law, and the Nernst equation. Animals are no less constrained by these, and other, physical laws than are nonliving physical and chemical reactions, although even the simplest of animals is immensely more complex in operation than any machine. Energy is the common denominator to all physiological processes, from membrane function to movement, biochemistry to bioluminescence, and thermoregulation to thinking. The ultimate definition of animal life would at least include the utilization by organisms of externally derived energy for their self-perpetuation. All of the proximate physiological processes responsible for maintaining life require energy, usually in the form of adenosine triphosphate (ATP). Consequently, energy and processes of energy exchange and flow (i.e., thermodynamics) are an important aspect of the physicochemical perspective that will be continually emphasized throughout the book.

Having established the physical, chemical, and energetic constraints on animal physiological systems, the third objective is to examine the processes by which homeostasis (constancy) and regulation of physiological processes are achieved by animals in varying and variable environments. Constancy of the internal environment is a characteristic of many animals. For many, the constancy is a consequence of regulatory mechanisms. A common regulatory mechanism involves a sensor to determine the value of a variable, a comparator to relate this value with a specified setpoint value, and an effector system to alter the variable towards the setpoint; this is **negative-feedback regulation.**

A final objective of this book is to present a quantitative approach to the study of physiology because all scientific disciplines including physiology rely on the collection, testing, and interpretation of data. Any scientific theory or concept is only as good as the data upon which it is based. It is therefore important for students to learn to analyze and interpret scientific data. Furthermore, information clearly presented in a figure or table can convey much more detail than can be accomplished in words. Consequently, extensive use will be made of figures and tabular material to supplement the text material. Graphical and tabular material will, wherever possible, reflect data in its original form, rather than hypothetical figures or numbers.

Some systematization of units is required if we are to have a quantitative approach to comparative animal physiology, particularly in figures and tables. This book uses the International System of Units (Système international d'unités, or SI units), which has seven fundamental units, two supplementary units, and a number of units derived from these (see Appendix 2). Some SI units are already used even in "non-SI" countries (e.g., meter, kilogram, degrees Kelvin) but many SI units are unfamiliar, disconcerting, and confusing in comparison with the older unit systems (e.g., pascal, newton, joule). SI units are used not only because of the convenience of having common and standardized units but, being metric, they provide a greater ease of calculation. Many scientific journals require strict adherence to the SI unit system. The nomenclature and values for prefixes to the SI units (e.g., kilo [k] = 10^3; micro [μ] = 10^{-6}) are also included in Appendix 2, along with conversion factors between units of many systems. Many conversion factors are included, perhaps redundantly, at strategic locations throughout the text, figures, and tables for the more unfamiliar SI units. The degrees Celsius (centigrade) temperature scale is not, strictly speaking, an SI unit for temperature; the SI unit is degrees Kelvin. Nevertheless, °C will be used in this book because it is in extremely common usage, and the difference is not in scale but in absolute value i.e., °C = °K − 273.

Summary

This introductory chapter defines the field of Comparative Animal Physiology, and describes the four major objectives of the book:

1. to describe the immense diversity of physiological processes and systems;
2. to use a physicochemical perspective to understand how physiological processes work;
3. to examine animals in relation to their environment and investigate the basic principles of regulation by animals of their internal environment;
4. to use a quantitative rather than a descriptive approach

Chapter 2

External and Internal Environments

· ·

The stark contrast between an oasis and the Saharan desert illustrates the great diversity in the nature of the external environment in which animals live. *(M. Thonig/H. Armstrong Roberts, Inc.)*

The **external environment** of an animal is a complex combination of many physical variables (the abiotic environment) and the presence of living organisms, e.g., plants and animals (the biotic environment). Some of the more important abiotic variables are temperature, water availability, air humidity, water salinity, light intensity, photoperiod, ambient pressure, the gaseous composition of air, and the partial pressures of gases in solution. The climate is the particular combination of these abiotic variables over relatively large geographic areas, whereas the microclimate is the particular combination of variables over a small, often highly localized, region. For example, the climate of a desert environment has high daytime air temperatures, low and erratic rainfall, and low ambient humidity, but a variety of often strikingly different microclimates can be found in deserts. Subterranean burrows are one example of such a microclimate. Burrow temperatures are more moderate and have a lower daily variation than the macroclimatic temperatures; the relative humidity of burrow air is much higher than that of the ambient air, often approaching saturation.

A basic understanding of the various abiotic climatic variables defining the external environment provides a framework for the investigation of the **internal environment** (body fluids) of animals, and the processes of physiological regulation. Changes in the external environment can either change the composition of the internal environment of animals or require regulatory mechanisms to maintain a relative constancy of the internal environment. **Homeostasis** is the constancy of the internal environment. It may be accomplished, even in the face of an altered external environmental change, by mechanisms that **regulate** a constant internal environment. For example, many animals maintain a constant body temperature, or blood pH, pO_2, or solute concentration. The basic principles of homeostatic regulatory systems will be described in more detail after we briefly examine the nature of the external and internal environment.

The External Environment

For convenience, we can separate a discussion of the external environment into three sections: the atmosphere, aquatic environments, and terrestrial environments.

The Atmosphere

The **fractional composition** of normal, dry atmospheric air is 0.2095 for oxygen and 0.0003 for carbon dioxide; 0.781 is nitrogen and the balance is other inert gases. The standard pressure exerted by the atmosphere at sea level is about 101 kiloPascals; this is equivalent to 1 atmosphere, or 760 torr (1 torr = 1 mm mercury). The partial pressure of any particular constituent of air is that fraction of the total pressure that it exerts; it is equal to the gas fractional composition times the total pressure. For example, the partial pressure of oxygen at sea level (pO_2) is 0.2095 × 101 = 21.2 kPa (160 torr). Atmospheric air also contains a variable amount of water vapor, which reduces the fractional content and partial pressure of the other gases in air. The partial pressure of water vapor varies widely depending on the air temperature and relative humidity, but generally is from 0.2 to 2.0 kPa (2 to 20 torr).

The fractional composition of atmospheric air is relatively constant over the surface of the earth and is unaffected by altitude (at least below 10000 m where most animals are found). The pressure exerted by the atmosphere is, however, profoundly influenced by altitude; it declines by about ½ for every 5500 meters increase in altitude. This is because the thickness (hence mass) of the atmosphere above that altitude is less. Consequently, the partial pressure exerted by each of the constituent atmospheric gases also decreases with altitude. For example, the normal pO_2 is reduced from 21.1 kPa (160 torr) to 10.6 kPa (80 torr) at 5500 m altitude.

Air temperature, as we know, varies widely on both geographic and seasonal scales. It also varies markedly with altitude, decreasing by about 6 to 10° C per 1000 meter increase in altitude. This dependence of air temperature on altitude is called the **lapse rate**.

Aquatic Environments

Water is the ambient medium of aquatic environments, e.g., oceans, lakes, rivers, streams, and ponds. One of the most important characteristics of an aquatic environment is its **salinity**, the content of dissolved inorganic material (grams of total solids per kg solution). Salinity is usually measured in units of grams per kilogram (i.e., parts per thousand, or ‰). Related measures are osmotic concentration, chlorinity, and freezing point depression. The salinity of aquatic environments varies markedly, from almost zero for raindrops and fog condensate, to saturation levels in hypersaline lakes and ponds (Table 2–1).

The Venice system provides a classification of water bodies based on their salinity. Fresh water (e.g., rain and the water found in most lakes and rivers) has a salinity of 0 to 0.5 ‰. Brackish waters of intermediate salinity are oligohaline (0.5 to 5 ‰), mesohaline (5 to 18 ‰), or polyhaline (18 to 30 ‰). Euhaline waters, such as seawater, are 30 to 40 ‰. Hyperhaline waters, salt lakes for example, have a salinity >40 ‰.

The specific **ionic composition** of the inorganic material dissolved in water is quite variable. The major ions are generally Na^+ and Cl^-, and occasionally Ca^{2+}, Mg^{2+}, and SO_4^{2-}. Seawater is mostly sodium chloride with small amounts of many other ions (Table 2–1). Many dilute waters are formed from seawater by dilution with fresh water (e.g., estuaries) and some hypersaline waters are formed by the concentration of seawater through evaporation (e.g., hypersaline estuaries). The ionic composition of these waters is proportional to that of seawater. Most hyperhaline waters have a strikingly different ionic composition from seawater due to salts leached from the earth.

The **dissolved gas content** of aquatic environments (usually measured in ml gas $liter^{-1}$ of water, or parts per million) depends on the partial pressure of the gas and its solubility coefficient (α; $ml\ liter^{-1}\ kPa^{-1}$). The solubility coefficient in turn depends on the water temperature and salinity. The oxygen content of the oceans is generally from 1 to 6 ml $O_2\ liter^{-1}$; the content is higher at the surface where there is

rapid exchange between atmospheric air and the water, and is low in the deep ocean where there is little exchange between the atmospheric air and the water, and utilization of O_2 by marine organisms depletes the O_2. The dissolved carbon dioxide content of water, like the oxygen content, depends on the solubility coefficient and the partial pressure. The dissolved CO_2 content of water is generally significant, despite its low ambient partial pressure, because CO_2 is more soluble than O_2. The total CO_2 content is greater than the dissolved CO_2 content because of its chemical reaction with water to form bicarbonate ions (HCO_3^-) and hydrogen ions (H^+).

$$CO_2 + H_2O \longleftrightarrow H^+ + HCO_3^- \qquad (2.1)$$

This reaction is often catalyzed in biological fluids by an enzyme, **carbonic anhydrase**. The slightly alkaline pH of natural waters buffers the H^+ and facilitates the formation of HCO_3^-. Even very pure water, such as raindrops and fog condensate, has a measurable salinity because of the HCO_3^- and H^+ formed from dissolved CO_2 (as well as from traces of inorganic ions, such as Na^+ and Cl^-).

The **temperature** can vary markedly for water bodies, from the freezing point (0° C for fresh water) to the boiling point (100° C for fresh water). Dissolved solutes lower the freezing point (e.g., $-1.9°$ C for seawater) and raise the boiling point (e.g., 100.55° C for seawater). The boiling point can be well over 100° C for subterranean and submarine water because of the high pressures. There generally is a vertical stratification in water temperature,

TABLE 2–1

Ionic concentrations (mEq l^{-1}), salinity (‰), and osmotic concentration (mOsm) of cloud water, "world average" river water, seawater and hypersaline water bodies, the Great Salt Lake (US), and Lake Koombekine (Aust).

	Cloud	River	Seawater	Great Salt Lake	Lake Koombekine
Sodium	0.09	0.27	468	3000	4070
Potassium	—	0.06	10	90	17
Magnesium	0.01	0.34	107	230	109
Calcium	0.002	0.75	20	9	32
Chloride	0.10	0.22	547	3100	1670
Sulfate	0.006	0.24	56	150	72
Salinity	0.01	0.18	35	210	263
Osmotic Conc	<1	≈1	1100	6000	7500

Source: Data from Hutchinson, G.E. 1957 *A treatise on limnology.* Vol. I, Part 2. New York: Wiley-Interscience; Weyl, P.K. 1970. *Oceanography: an introduction to the marine environment.* New York: Wiley & Sons; and Geddes, M.C., P. DeDekker, W.D. Williams, D.W. Morton, and M. Topping. 1981. *Hydrobiologica* 82:201–222.

which can range from 30° C at the surface to 4° C at the bottom of shallow, temperate waters. Throughout the year, surface temperatures of large water bodies (oceans, lakes) fluctuate little in the tropics (<2° C variation), more in polar regions (4° C variation), and most at midlatitudes (8° C variation). Water temperatures at great depths and pressures are generally 0 to 4° C. Small bodies of water, such as ponds and tide pools, tend to have greater diurnal temperature fluctuations than larger bodies of water because of their lower thermal inertia. For example, the temperature of a small tide pool may exceed 35 to 40° C in the middle of a hot day, whereas the ocean water only a few feet away might be 10 to 20° C. Shallow desert ponds may fluctuate in temperature from near 40° C during the day to 20° C or less at night.

Hydrostatic pressure alters markedly with depth below the water surface because the weight of a 10-meter-high water column is equivalent to the weight of the entire air column (101 kPa) above the earth's surface! Consequently, the normal atmospheric pressure is increased to 202 kPa at 10 m depth, 303 kPa at 20 m, etc. The mean depth of the oceans is 3800 m, which is equivalent to 38000 kPa or 38 megapascals, and some deep-sea trenches extend to 10000 m, which is 100 MPa pressure. Such high pressures can have significant effects on animals both directly (e.g., compression of air spaces within the body) and indirectly (e.g., altering the ionization of water, protein structure, and enzyme function). For example, the transition point for biological membranes between the normal liquid-crystalline state (a high membrane fluidity) and the gel state (low fluidity) is determined by both temperature and pressure. For a typical biological membrane, a pressure rise of 1 MPa is equivalent to a temperature decrease of 0.2° C, i.e., $\Delta T/\Delta P = -0.2°$ C MPa^{-1}, or -0.02 °C atm^{-1} (MacDonald and Cossins 1985). Thus, a water temperature of 0° C for deep-sea conditions (10000 m depth, 100 MPa pressure) has an equivalent physiological effect as a temperature of $-20°$ C at normal atmospheric pressure!

Penetration of light through water depends on its clarity. For typical seawater, 50% of the light that is not reflected at the surface is absorbed within 2 meters depth, 75% in 4 meters, and 90% in 8 meters. In the exceptionally clear water of the Sargasso Sea, 99% of the light is not absorbed until a considerable depth, about 80 m, whereas 99% of the light is absorbed at 8 m in the murky waters of Woods Hole, Massachusetts. Not all wavelengths of light penetrate the water equally well; red light is absorbed the most, and blue light the least.

Terrestrial Environments

Many classification schemes have been used to categorize the climate of terrestrial environments. Rainfall and temperature are two variables commonly used to define the major terrestrial climatic zones, as defined by their predominant community types, i.e., biomes. For example, the various biomes, such as desert, woodland, tundra, and rainforest, can be defined within a triangular figure (Figure 2–1). On one side of the triangle we have mean annual precipitation; on a second side the ratio of potential evapotranspiration (PET; evaporation from a water surface + transpirational evaporation by plants) to annual precipitation; and on a third side humidity, which is a combination of annual precipitation and the ratio of PET to annual precipitation (Holdridge 1947). Consequently, deserts, dry tundra, and polar environments can all have less than 125 mm annual precipitation; but deserts are hot and have a high ratio of potential evapotranspiration/precipitation, whereas the polar environment is cold and has a lower ratio. Polar, pluvial tundra, pluvial woodland, and wet rainforest biomes may all have PET/precipitation <0.5, but vary in their mean annual rainfall.

Many local factors can produce marked perturbations in the local climate. For example, mountain ranges can create a rainshadow, or orographic desert, to their leeward side. The peaks of mountains may be isolated pockets of alpine habitat in an otherwise moderate environment. Coastal upwellings of cold seawater on the western margins of some continents produce a cold, coastal desert with frequent fogs.

Air temperature can vary more dramatically than water temperature, ranging from less than $-70°$ C in extreme polar to greater than $+50°$ C in desert environments. This is because air has a much lower heat capacity than water. Only 0.001 joule of energy is needed to warm one ml of air by 1° C, whereas 4.2 J is needed to warm 1 ml of water by 1° C. Air temperature also decreases with altitude, the lapse rate being about 6 to 10° C km^{-1} elevation in altitude. Daily cycles in air temperature are a common aspect of terrestrial environments. The air temperature in certain microclimates can vary even more dramatically than the meterological air temperature (measured by the standard Stephenson screen located 1.5 m above ground). For example, in equatorial deserts air temperatures may rise to 80° C or more near the ground because the intense incident solar radiation is absorbed at the ground surface and heats the nearby air (Cloudsley-Thompson 1975).

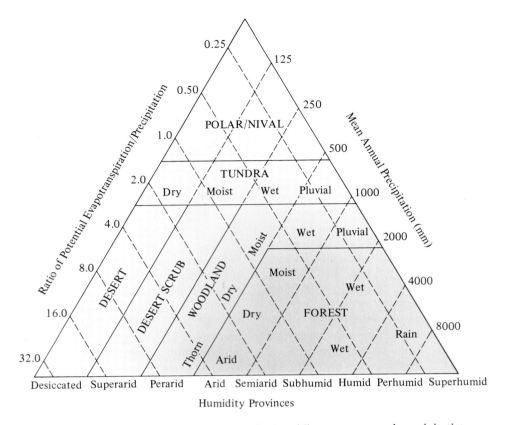

FIGURE 2–1 Classification of major biomes by humidity, mean annual precipitation, and ratio of potential evapotranspiration to precipitation. *(Modified from Holdridge 1947.)*

The water vapor content of air, the relative humidity (% of saturation water vapor content at that temperature), and the absolute humidity (mg water liter air^{-1}) vary dramatically, from very low values in cold air (1 mg liter^{-1}) to high values in humid, warm tropical air (30 mg l^{-1}). The rainfall and atmospheric humidity vary geographically. Hot equatorial regions, with upwardly-moving humid air masses, tend to have a high rainfall because the rising air is cooled by the lapse rate and water condenses to fall as rain. Again, localized microclimates can differ markedly from the ambient climate. Even in deserts, humid microclimates can exist near vegetation and underground.

The gaseous composition of subterranean air is often lower in O_2 (hypoxic) and higher in CO_2 (hypercapnic) than normal ambient air because of micro-organism and animal metabolism. In particular, fossorial birds and mammals may substantially deplete the local O_2 to 10 to 15%, and 5 to 10% CO_2 may accumulate because of their high metabolic rates (Withers 1976).

The Internal Environment

The internal environment, found within the bodies of animals, is substantially different from the external environment. Water is almost invariably the main constituent. It is the universal solvent and is the predominant molecule found in the body (60 to 90% of total body mass is water in most animals). Body water contains a variety of solute molecules dissolved in solution (inorganic and organic electrolytes, such as Na^+, K^+, Cl^-, amino acids, proteins; organic nonelectrolytes, such as urea, glucose), nondissolved solids (such as lipid membranes and cellular inclusions of glycogen and lipid), and the organic and inorganic constituents of skeletons. An almost infinite variety of organic and inorganic molecules form the body of animals, but four elements make up over 99% of the total by weight (Figure 2–2): hydrogen (63%), oxygen (25.5%), carbon (9.5%), and nitrogen (1.4%). Minor elements (in terms of % content) comprising 0.01 to 1% include calcium (0.31%), phosphorus (0.22), chlo-

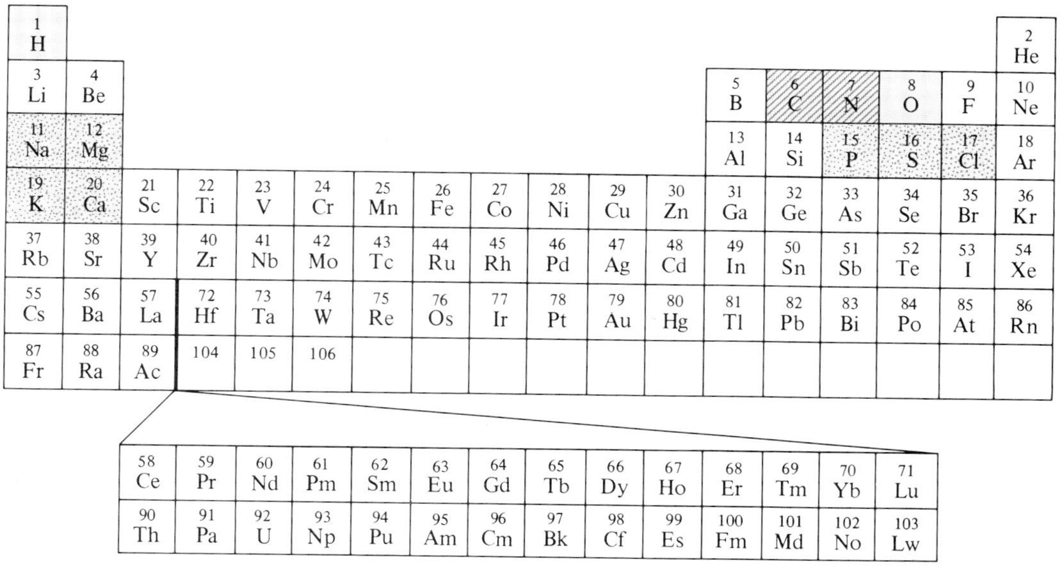

FIGURE 2–2 Periodic table of elements showing the most common by weight in biological systems: >10%, solid color; 1 to 10%; crosshatched; 0.01 to 1%, stippled.

rine (0.03), potassium (0.06), sulfur (0.05), sodium (0.03), and magnesium (0.01). A large number of elements, although of specific and great physiological importance, are present in only trace amounts of <0.01%, e.g., iron, fluorine, zinc, and iodine. Carbohydrates are structural components and energy stores. Lipids form membranes and are deposited as energy stores. Proteins form structural components and are catalytic enzymes. A more complete description of these macromolecules is provided in Chapter 3.

The physiological adaptations of animals for regulating their internal environments in extreme environments, such as deserts and polar regions, high altitude and great depth under water, and highly concentrated and dilute aquatic environments, will be examined in considerable detail in the following chapters.

The Extracellular Environment

The extracellular environment for a unicellular organism is, of course, the external medium. For a multicellular metazoan animal, the extracellular environment is not the external medium but the extracellular fluid between cells. Two potential extracellular fluid compartments of animals are the blastocoel and the coelom, a secondarily-derived body cavity. The blastocoel is the space within the hollow ball of cells formed during cleavage of the fertilized egg. The coelom is a space formed by splitting, or pocketing, of the mesoderm. There is a general evolutionary trend for development of an extensive extracellular fluid compartment, or compartments. Let us briefly examine the development of the extracellular fluid spaces and then consider the basic composition of the extracellular fluid.

The primitive radiate phyla, sponges, cnidarians, and ctenophores have a thin and poorly developed extracellular space located between the outer body layer of cells (epidermis) and the inner body layer of cells (endodermis). The mesohyl of sponges is a gelatinous proteinaceous matrix containing amoeboid cells and skeletal spicules. The mesoglea of cnidarians and ctenophores varies from a thin, acellular membrane to a thick, jelly-like matrix with amoeboid cells. The ionic and osmotic composition of the mesoglea of marine species is virtually identical to the external medium (see Chapter 16).

The structurally more complex flatworms and nemertean worms lack a specialized body cavity; the blastocoel is filled with a loose tissue parenchyma, and there is no coelom; they are **acoelomate** animals (Figure 2–3). Their intercellular fluid has a similar composition as the external environment for primitive marine species, but not for advanced marine species. Freshwater and terrestrial species must regulate the composition of their extracellular body fluids.

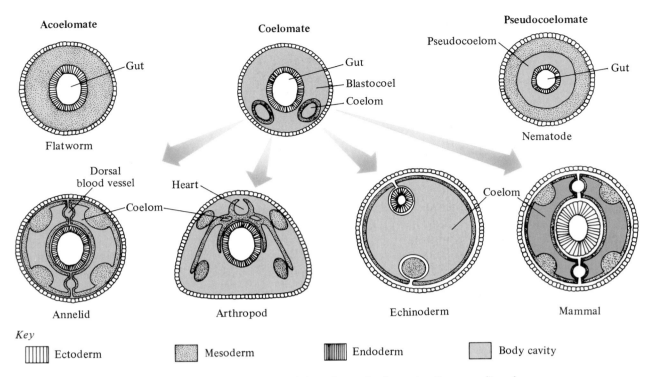

Key

Ectoderm		Mesoderm		Endoderm		Body cavity

FIGURE 2–3   Representation of the development of the primary body cavity (hemocoel) and the coelom, and the relative development of the two body cavities in the adult animal for an acoelomate (e.g., flatworm), for a pseudocoelomate (e.g., nematode) and for the following coelomates: an annelid (e.g., earthworm), an arthropod (e.g., insect), an echinoderm (e.g., sea urchin) and a vertebrate (e.g., human).   *(Modified from Ramsay 1964.)*

The most complex arrangement of the extracellular spaces is found in the more advanced animals that have body cavities. These **pseudocoelomate** and **coelomate** animals also have better homeostatic regulatory mechanisms for extracellular fluid composition. The body cavity of pseudocoelomate animals (e.g., nematodes, rotifers, gastrotrichs) is derived from the blastocoel and has a different structure than the body cavity of coelomates (Figure 2–3). It is small in most pseudocoelomates, but is large and functions as a hydrostatic skeleton in some (e.g., nematodes), in similar fashion as the coelom of many coelomates (e.g., annelids).

In some coelomate animals, the secondary body cavity remains small, persisting as a pericardial or genital organ cavity, or is even secondarily lost altogether. The blastocoel remains the major body cavity and is called the hemocoel since it contains blood. The circulatory system is open; the fluid filling the vascular system flows through the channels, then percolates through the rest of the hemocoel. Consequently, the composition of the intravascular blood is the same as the fluid in the hemocoel. The open circulatory system is formed as a meso-dermal tube and merely separates one portion of the hemocoel (inside the tube) from the remaining hemocoel space.

In contrast, the body cavity of other coelomates expands in volume until it severely restricts the vascular (blastocoel) space. The closed circulatory system (i.e., arteries, capillaries, veins, heart chambers) contains blood. The fluid outside the circulatory system is the interstitial fluid, or lymph. Lymph has a similar composition to blood except that the blood cells are absent, and large molecules such as proteins are found at a lower concentration. The body cavity of many coelomates (e.g., annelids and echinoderms) is extensive and comprises a major fraction of the extracellular space, but the coelom is reduced in volume in other coelomates (e.g., vertebrates).

The extracellular fluids of animals vary greatly in ionic concentration and total solute concentration, i.e., osmotic concentration. Animals whose body fluid ion concentrations are the same as those of the external medium are said to **ionoconform**. Very few animals ionoconform. Animals whose body fluids have different ionic concentrations from the ambient medium **ionoregulate**. Most animals are good iono-regulators. Osmotic concentration is the total concentration of osmolytes in solution; any dissolved solute is an osmolyte. Animals whose body fluids are at the same osmotic concentration of the medium are said to **osmoconform**. For example, the shrimp *Callianassa* osmoconforms over a wide range of environmental osmotic concentrations (Figure 2–4). Animals whose osmotic concentration is different from that of the medium **osmoregulate**. For example, the shrimp *Upogebia* osmoregulates over a range of environmental osmotic concentrations, although it osmoconforms at high ambient concentrations.

Three major patterns of osmo- and ionoregulation can be discerned in aquatic animals. Some marine animals have the same osmotic concentration and similar ionic concentrations as seawater, i.e., they osmoconform and ionoconform. The jellyfish *Aurelia* and the primitive vertebrate hagfish *Myxine* are examples. Many animals have the same osmotic concentration as seawater but markedly lower ionic concentrations, i.e., they osmoconform but ionoregulate. The balance of their body fluid osmotic concentration is made up by organic molecules, either compatible solutes such as amino acids (e.g., the crab *Carcinus*) or counteracting solutes such as urea, TMAO, and betaine (e.g., the dogfish *Squalus*). Many marine vertebrates have a lower osmotic concentration than seawater, i.e., they osmoregulate and ionoregulate. The salmon *Salmo* in oceanic habitats is an example. These animals do not require high concentrations of compatible or counteracting solutes. Freshwater animals must osmoregulate and ionoregulate to avoid an intolerable dilution of their body fluids and cells. The salmon *Salmo* in fresh water and the freshwater lamellibranch mollusk *Anodonta* are examples. Terrestrial animals, like freshwater animals, must osmoregulate and ionoregulate.

The Intracellular Environment

The total concentration of solutes within animal cells must be equal to the solute concentration of the extracellular fluids. Otherwise, the cell will shrink or expand by movement of water or solutes until the osmotic concentrations are equal (see Chapters 6 and 16). This is because an osmotic difference across an animal cell membrane could only be maintained by a hydrostatic pressure difference, but the thin, flimsy lipid membranes of animal cells are unable to mechanically resist a hydrostatic pressure difference. Plant cells, in contrast, have rigid cellulose cell walls that can resist positive internal pressures. Consequently, plant cells can have higher internal osmotic concentrations and hydrostatic pressures than the external medium.

The intracellular environment is invariably very different from the extracellular environment despite their equal osmotic concentrations. Energy must be continually expended to maintain this disequilibrium in solute concentrations. This necessity for energy expenditure suggests that various solutes (e.g., Na^+, K^+) differ in their "adaptiveness" or "usefulness" as intracellular solutes. Otherwise, cells would not be so selective about their intracellular solutes.

The constituents of the internal environment vary markedly for different animals, depending on the taxonomic position, the nature of the external environment, and the pattern of ionic and osmotic regulation or conformation. Nevertheless, a number of generalizations can be made. Water is the most abundant molecule within cells. There usually is a high concentration of amino acids and protein and varying, but often quite significant, levels of other

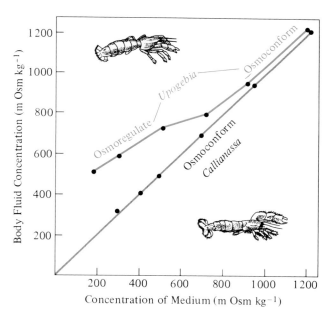

FIGURE 2–4 Body fluid osmotic concentration (of the hemolymph) as a function of ambient osmotic concentration, for the mud shrimp *Upogebia pugettensis* and *Callianassa californiensis*. *Upogebia* is a limited osmoregulator at ambient osmotic concentrations less than seawater (about 1000 mOsm) and is an osmoconformer at higher osmotic concentrations. *Callianassa* is an osmoconformer throughout its entire range of ambient osmotic concentration tolerance. Both species are euryhaline. *(Modified from Thompson and Pritchard 1969)*

nitrogenous compounds, such as urea, trimethylamine oxide, betaine, and sarcosine, or carbohydrates such as glycerol, sorbitol, and mannitol. The total intracellular ionic content is fairly low and constant. Where there are large increases of intracellular solute concentration, it is typically the concentration of organic molecules that varies. The ionic environment of animal cells is always dominated by K^+, rather than Na^+, and generally has a low Cl^- concentration. The intracellular pH is always close to neutrality. This basic pattern of high organic content, low ionic content, high K^+ relative to Na^+, low Cl^-, and neutral pH is consistent for protozoans and all animals, from freshwater to the most advanced terrestrial animals, such as mammals and insects. It therefore seems safe to conclude that the internal environment of the unicellular organisms that evolved about a billion years ago would have been similar in basic composition to that of extant protozoans and animals.

Homeostasis and Regulation

The maintenance of an intracellular environment that differs from the extracellular space and the external environment is, as we have seen, a characteristic of most animals. The compositions of the intracellular and extracellular fluids are often maintained constant, or at least relatively constant, even in the face of fluctuating external conditions. Such a constancy of conditions is called **homeostasis**, or "la fixité du milieu intérieur" as described by Claude Bernard. The term, homeostasis, was originally coined to describe the stability of conditions of mammalian body fluids. Referring to the extracellular fluid environment of humans, Cannon (1929) stated that

> This "internal environment" as Claude Bernard called it, has developed as organisms have developed; and with it there have evolved remarkable physiologic devices which operate to keep it constant. . . . So long as this personal, individual sack of salty water, in which each one of us lives and moves and has his being, is protected from change, we are freed from serious peril. I have suggested that the stable state of the fluid matrix be given the name of homeostasis.

Examples of homeostasis include constancy of body temperature in mammals and birds, constancy of blood gases (pO_2 and pCO_2) in body fluids of many animals, blood pressure in animals with circulatory systems, and body water content in most animals.

Such examples of homeostasis are more apparent, and operate more precisely, in higher animals (such as mammals and birds, but also in certain insects and mollusks), compared with lower invertebrates that are characterized by less constant internal conditions and homeostasis of fewer physical and chemical variables.

Homeostasis, or constancy, of physiological conditions should not be confused with the concept of regulation. Certain aspects of the internal environment can be constant, even in the total absence of a regulatory mechanism, and regulation of a variable does not necessarily imply constancy. For example, the body temperature of an icefish, which lives in the ice-cold antarctic seawaters, is $-1.9°$ C and might not vary over a year by more than $1°$ C, a variation less than that of human body temperature over a 24 hour period! The constancy of the body temperature of an icefish is a consequence of its living in water that has an extremely stable temperature of $-1.9°$ C, since the fish's body temperature is virtually identical to the water temperature. We should not conclude that the icefish has a better thermoregulatory system than a mammal! Similarly, the ionic composition of the extracellular fluids of echinoderms is quite constant, but this does not imply great powers of regulation. In fact, the ionic composition of the extracellular fluids is nearly identical to that of seawater. Again, constancy or homeostasis is an attribute of the external environment, not of the animal. Changing the temperature of the water in which the icefish lives, or changing the ionic composition of seawater surrounding an echinoderm, will invariably alter the internal environment of the icefish (temperature) and the echinoderm (ionic composition).

Regulation does not necessarily imply absolute homeostasis. In fact, few regulatory systems are "perfect," and so regulated variables invariably are not perfectly constant, i.e., there is not complete homeostasis. Nevertheless, these variables are maintained more constant by the regulatory systems than if there was no regulation. For example, the osmotic concentration of the body fluids of the shrimp *Upogebia* is regulated, but not perfectly, at a higher concentration than the medium, at least at environmental osmotic concentrations <1000 milliosmolal, by physiological mechanisms; the body fluid osmotic concentration depends on the ambient medium although it is not equal to it (Figure 2–4). Physiological mechanisms maintain the body fluids. The mechanisms for osmoregulation are discussed further in Chapter 16. Measurement of the body fluid pH of most lower animals indicates a marked effect of temperature, of about -0.014 U $°$ C^{-1}, i.e., a 10° C increase in temperature lowers the pH by about 0.14 units (which is equivalent to a 38% increase in hydrogen ion concentration). Consequently, we might suspect the lack of precise pH regulation but these animals actually are precisely regulating the ratio of H^+ to OH^- (relative alkalinity) regardless of temperature. Their pH changes parallel that of pure water at differing temperature, for which there is also about a -0.14 U ° C^{-1}. The mechanism for the regulation of H^+/OH^- is provided by the metabolic production and the respiratory excretion of CO_2 (see Chapter 15).

Thus, there are two basic patterns of how physiological variables alter with changes in the external environment: **conformation** when the internal variable is always equal to the environmental variable and **regulation** when the internal variable is kept different from the external value. If the physiological variable in question is body temperature, then the respective terms would be thermoconformation and thermoregulation. Similarly, for osmotic concentration, the terms would be osmoconformation and osmoregulation, for ionic concentration the terms would be ionoconformation and ionoregulation, etc. Animals such as *Upogebia* and *Callianassa,* which survive over a wide range of external salinities, are said to be euryhaline, in contrast to stenohaline animals, which survive over a more restrictive, narrow range of external salinities. The prefixes **eury-** and **steno-** are also applied to other physiological variables, e.g., temperature: eurythermal and stenothermal.

Other prefixes used to describe the constancy or change in an internal variable are **poikilo-** and **homeo-**. For example, an animal whose body temperature is variable is poikilothermic; the body temperature may be equal to ambient temperature, or it may alter in some other fashion as ambient temperature alters, or it may even alter while ambient temperature remains constant. The important point is that the body temperature of a poikilotherm is variable rather than constant. An animal whose body temperature remains constant is, in contrast, a homeotherm. Similar pairs of terms would be poikilo-osmotic and homeo-osmotic and poikilo-ionic and homeoionic. It is important to appreciate that the prefix poikilo- and the term conformation do not convey the same meaning. For example, the shrimp *Upogebia* is poikilo-osmotic at external salinities <1000 mOsm but is not osmoconforming; at higher salinities, it is both poikilo-osmotic and osmoconforming. The prefix homeo- and the term regulation are also not synonymous. For example, an icefish is a good homeotherm, but does not thermoregulate.

In summary, the prefixes poikilo- and homeo- are useful for describing whether a physiological variable is constant or variable, but they do not indicate whether there are regulatory mechanisms for that physiological variable. The terms conform and regulate are preferable when describing the absence or presence of a specific regulatory mechanism. The prefixes eury- and steno- describe whether a wide or narrow range of values is tolerated.

A final term has recently been introduced to describe a situation in which normal functioning is maintained in the absence of homeostasis. **Enantiostasis** is the maintenance of function when the effect of a change in one physiological variable is counteracted by a change in another physiological variable (Mangum and Towle 1977). For example, blue crabs transferred from seawater to brackish water experience a decrease in body fluid osmotic and ionic concentrations and an increase in the pH because of higher hemolymph ammonia levels. The lower salt concentration of the hemolymph decreases the affinity of its respiratory pigment hemocyanin for oxygen but the higher pH increases the affinity of hemocyanin and counteracts the effect of lowered salt concentration. Oxygen transport by the respiratory pigment is therefore not as compromised by transfer to brackish water as expected. Homeostasis of function, or enantiostasis, has occurred in the absence of homeostasis of two important physiological variables, salt concentration and pH. The significance of this example of enantiostasis is to emphasize the complexity of physiological functioning and the important interactions that can occur between different physiological variables. The recent introduction of this new term illustrates the limitations of terminologies in describing complex, interrelated physiological processes.

Tolerance and Resistance

Animals either conform or regulate in the face of moderate environmental change. Eventually they become unable to tolerate further extreme change and die; regulators are unable to regulate and their internal environment alters to an extent incompatible with normal cellular functioning; conformers experience sufficient changes in their internal environment that normal cellular functioning is disrupted.

The range of any specific environmental variable that an animal can survive is called its **range of tolerance**. Above and below the critical values at the extremes of the ranges of tolerance are **ranges of resistance**, in which the animal is not quickly killed but will eventually die. An example of ranges

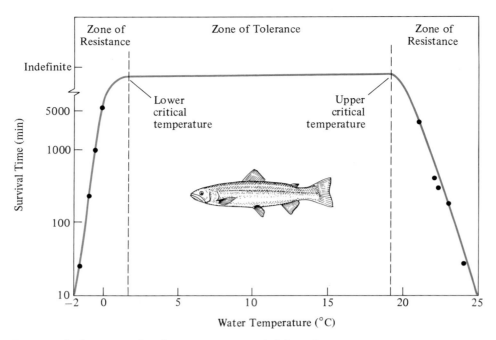

FIGURE 2–5   Ranges of tolerance and resistance to high and low temperatures for young chum salmon *(Oncorhynchus keta)* acclimated to a water temperature of 5° C. An experimentally determined survival time longer than 5000 minutes was interpreted to indicate in-definite tolerance to that temperature. The lower and upper critical temperatures are approximately 0° C and 20° C respectively, with this acclimation regimen. *(Data from Brett and Alderice 1958; Brett 1952.)*

of tolerance and resistance is illustrated by the temperature tolerance of fish (Figure 2–5). The survival time is indefinite for temperatures within the range of tolerance, but declines to zero at the extremes of the lower and upper ranges of resistance.

Experimental determination of survival, above the critical maximal value or below the critical minimal value, is fairly straightforward, although there is individual variability in tolerance. The relationship between survival time and level of exposure is usually sigmoidal (S-shaped) over a range of physiological values. The classical means for determining the average tolerance to death is to estimate the level that is lethal for 50% of the experimental subjects. This value for 50% survival (and 50% mortality) is called the LT_{50} (lethal tolerance for 50%). Similar sigmoidal relationships are observed for response as a function of drug dosage (LD_{50}), and lethal concentrations of toxin (LC_{50}), as a function of dosage. Typically, a graph of the accumulated mortality yields a sigmoidal or curvilinear relationship with the physiological variable, e.g., mortality of carp as a function of ammonia

concentration (Figure 2–6A). The LD_{50} can be determined from such a graph by extrapolation of the data through the 50% line. The sigmoidal or curvilinear relationship can be converted to a straight line, to facilitate determination of the LD_{50} value, by a probit transformation (Figure 2–6B). Physiologists generally prefer straight-line relationships over curvilinear relationships because of their greater ease of statistical analysis by least-squares linear regression, which allows the determination of the intercept (± its standard error), slope (± its standard error), correlation coefficient (*r*), and 95% confidence limits.

Acclimatization and Acclimation

The range of tolerance for a species, or even individuals within a specific population, is not rigid and unalterable. In fact, animals are generally able to readjust their range of tolerance over a period of time in response to alterations in their external environment, i.e., animals adjust to their environment. **Acclimatization** is the term used to describe this readjustment phenomenon when it is observed

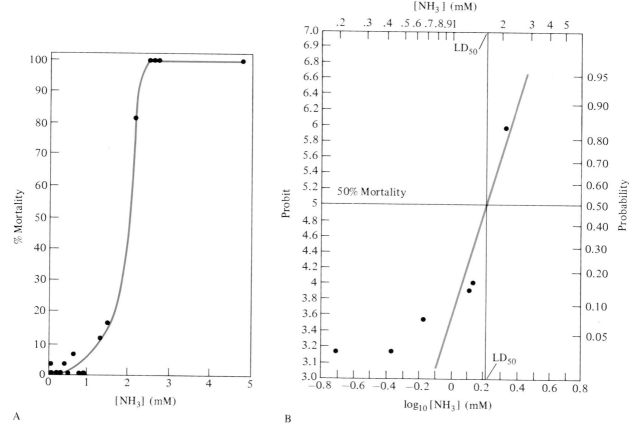

FIGURE 2–6(A)   Dosage % mortality relationship for the toxic effects of ammonia on carp. Note the curvilinear relationship between % mortality and dosage on arithmetic scales. **(B)** The linear relationship using logarithmic scales and probit plot for the dosage % mortality curve for carp (data from part A). The latter relationship allows determination of the regression line (solid line). The LD_{50} is 1.68 mM NH_3. *(Data from Hasan and Macintosh 1986.)*

in the natural environment. For example, an alteration in the upper critical temperature of a species of fish in response to seasonal changes in water temperature is thermal acclimatization (see Chapter 5). The physiological adaptation of an animal's range of tolerance to an altered environment, when investigated under the controlled conditions of a research laboratory, is called **acclimation**, rather than acclimatization.

Acclimation and acclimatization generally occur in the "appropriate" direction. For example, the upper critical temperature of an aquatic fish is higher during summer when water temperatures are higher, and the lower critical temperature is lower during winter when the water temperatures are lower. Similar adaptive alterations are observed in the critical thermal maxima and minima of fish acclimated in the laboratory to varying water tempera-

tures (Figure 2–7). It may be more appropriate for other physiological functions, such as rate functions (metabolic rate, heart rate, respiratory rate, etc.), to alter in different ways from parameters such as upper and lower critical temperatures. For example, it might be appropriate for metabolic rate to remain unaltered by a change in environmental temperature (see thermal acclimation and metabolic acclimation in Chapter 5).

Regulatory Mechanisms

We have already seen that homeostasis does not necessarily require a regulatory mechanism; the constant body temperature of an Antarctic icefish illustrated this point. Any such example of an internal variable being in equilibrium with an external environmental variable could result in **equilib-**

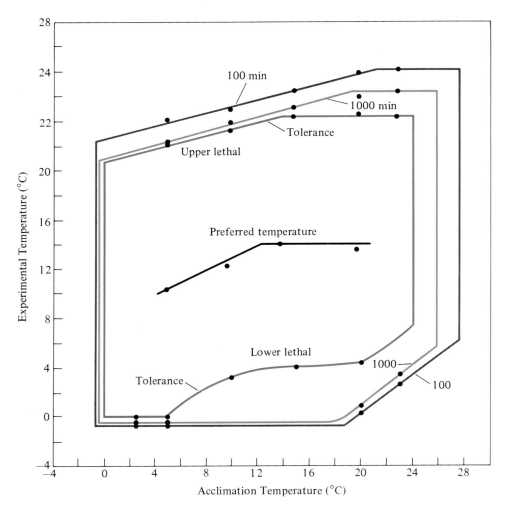

FIGURE 2–7   The thermal tolerance polygon for young chum salmon *(Oncorhynchus keta)* acclimated to water temperatures from 0 to 25° C shows the upper and lower lethal temperatures for indefinite tolerance (inner color polygon) and the temperatures for 1000 and 100 min tolerance (outer polygons). For example, a salmon acclimated to 12° C has an upper lethal temperature of 22° C and a lower lethal temperature of 4° C for indefinite exposure, but will survive 100 minutes at 22.2° C and −0.3° C. The preferred body temperature selected by the salmon in a thermal gradient is also indicated.   *(Modified from Fry and Hochachka 1970; after Brett 1952; Brett and Alderdice 1958.)*

rium homeostasis, without the necessity of any regulatory mechanism (Figure 2–8A). However, few physiological variables in only some animals are ever in equilibrium with the external environment.

Homeostasis can occur in the absence of a regulatory mechanism even if the animal is not in equilibrium with the external environment as long as a steady-state condition is maintained. Consider a molecule "X" that is liberated into the body fluids at a constant rate (e.g., X might be a metabolic end product). X is eliminated from the body fluids at a rate proportional to its concentration in the body fluids; the concentration of X will increase until the

rate of elimination equals the rate of release. Then, the concentration of X will remain constant at a steady-state value. The actual concentration of X depends on the rate of release and the concentration dependence of its rate of elimination. This **steady-state homeostasis** is an open system, since there is no apparent regulatory mechanism linking the concentration of X to either its production or elimination (Figure 2–8B).

An example of a steady-state open system is the average blood concentration of an important reproductive hormone, luteinizing hormone (LH), in female mammals. LH is released into the blood

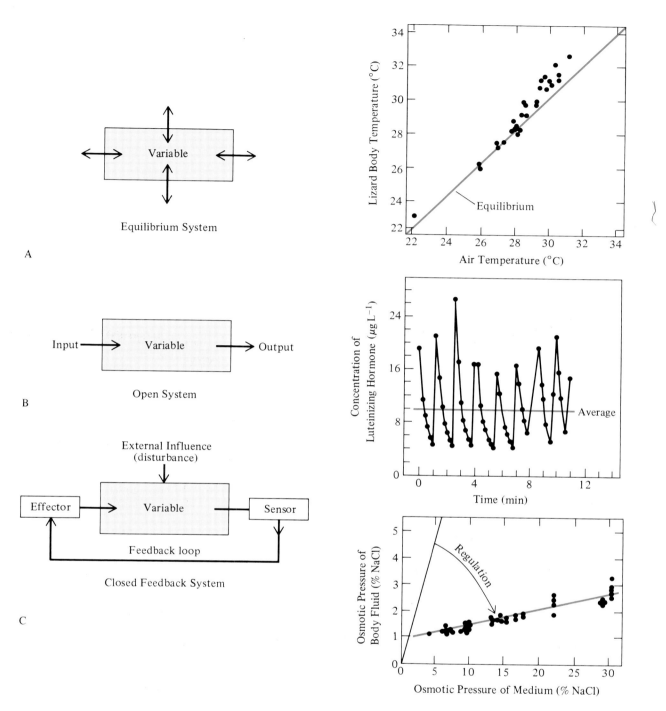

FIGURE 2–8 Three models resulting in relative homeostasis. (**A**) For a static equilibrium system in a constant environment, the value of an internal variable is constant if it is in equilibrium with an external value that is constant. For example, the body temperature of a forest-dwelling lizard is essentially the same (dots) as the air temperature (line). (**B**) In an open loop dynamic equilibrium system, the value of an internal variable is determined by the rate of input and the rate of elimination, which is concentration dependent. For example, the average blood LH concentration in ewes is deter-mined by the rate of pulsatile release. (**C**) A feedback control loop results in a dynamic equilibrium between the effect of an external disturbance with the value of the variable determined by an internal setpoint and monitored by a sensor system that provides information to an effector system. The level of the regulated vari-able is kept relatively constant despite dramatic changes in the disturbance. For example, the body fluid composition of brine shrimp is maintained relatively constant (dots) in external media of widely differing concentration (black line).

from the anterior pituitary gland in pulses; it is then eliminated from the blood by a variety of mechanisms (tissue uptake, metabolism, urinary excretion). The concentration of LH in the blood increases during periods of release, then declines in the intervening periods of nonsecretion owing to its clearance from the blood, but the average concentration of LH in the blood is quite consistent over time, with the mean value depending on the frequency of episodic release. For example, ewes with a higher frequency of LH release have a higher mean concentration of LH than ewes with a low frequency of release (Karsch 1980). There are also regulatory mechanisms for the control of blood LH levels (see Chapter 11, Endocrinology).

Another mechanism for homeostasis involves a specific physiological mechanism for the regulation. A particular physiological variable is regulated at, or near, a specified value of the variable, i.e., a setpoint. The actual value of the physiological variable is monitored by a sensory system. Information from the sensory system is used to control the value of the variable through some effector system. There is a feedback loop from the value of the variable, via sensory and effector mechanisms, that

controls the value of the variable at the setpoint despite external disturbances to the variable. The four essential elements of such a regulatory system—setpoint, sensor, feedback, and effector—are unique to this form of regulated, feedback homeostasis. This **regulatory feedback homeostasis** is a closed system, in contrast to the open equilibrium and steady-state systems. An example of feedback regulation is the control of hemolymph osmotic concentration by the brine shrimp *Artemia* (Figure 2–8C). The regulated variable is hemolymph osmotic concentration and the effector mechanism is branchial salt pumps.

Let us now examine the basic principles involved in feedback regulation of a closed system by example. A hot water bath maintained at a prescribed temperature by a heating element is a simple, and familiar, engineering regulatory system (Figure 2–9A). The components corresponding to the four elements of a feedback regulatory system are a setpoint (the position of a switch on a thermometer scale), the sensor (a bimetallic spiral strip thermometer), feedback (the open or closed position of the switch), and the effector (a current source and heater element). The setting of the thermostat control of

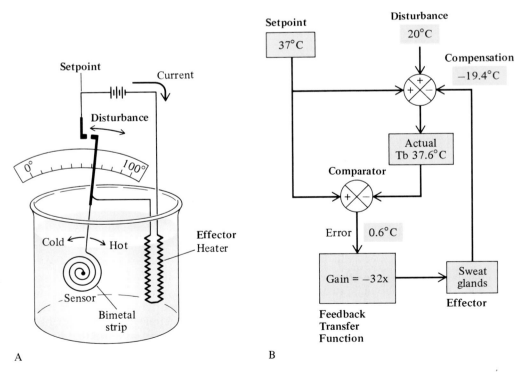

FIGURE 2–9(A)   The principles of negative feedback regulatory systems are well illustrated by a simple engineering example of feedback regulation of water temperature by a water bath heater. **(B)** Schematic outline of the elements of a negative feedback regulatory system.

the bath water determines the theoretical value at which the water will be maintained, e.g., $T_{setpoint} = 50°$ C. The sensor switch keeps power supplied to the heater element as long as the temperature of the water is less than the $T_{setpoint}$ of 50° C, because the tension of the bimetallic strip keeps the switch contacts closed. Heating stops when the temperature recorded by the thermometer exceeds the setpoint and the switch contacts open. This is a relatively simple "On–Off" regulatory system, capable of only moderate precision of water temperature regulation (e.g., ± 2° C might be typical for such a water bath).

Some physiological control systems operate as "On–Off" regulators, but systems with proportional control are capable of far greater precision of regulation (Figure 2–9B). Proportional control means that the magnitude of the effector response (e.g., rate of heating) is proportional to the error (deviation of the actual temperature from the setpoint), thus providing rapid heating when the water temperature is much lower than the setpoint (e.g., 20° C) but lower heating when the temperature is closer to the $T_{setpoint}$ (e.g., 45° C).

The setpoint value of a proportional physiological regulatory system is exactly analogous to that of the "On–Off" bath water example, i.e., it is the value at which the regulatory system is attempting to maintain the physiological variable. For example, the setpoint for the human temperature regulatory system is about 37° C; for the arterial blood pressure system it is about 13.2 kPa (100 torr). The actual value of the physiological variable (temperature, blood pressure, etc.) is determined by a combination of the setpoint value, the extent of disturbances on the system, and the degree of compensation by the regulatory system. The convention in a model of a control system is to assign the setpoint value a positive sign, the disturbance a positive sign indicating that it would increase the variable value, and the compensation a negative sign, since it will counteract the disturbance. A substantial increase in air temperature or a radiative heat load, for example, might be sufficient to elevate a human's body temperature by 20° C in the absence of a thermoregulatory system; the disturbance is + 20. In this particular example of body temperature regulation, the compensation might be − 19.4° C (see below). The actual value of the body temperature is then

$$
\begin{aligned}
\text{actual body temperature} &= \text{setpoint} \\
&\quad + \text{disturbance} \\
&\quad - \text{compensation} \\
&= 37 + 20 - 19.4 \\
&= 37.6° \text{C} \quad\quad (2.2)
\end{aligned}
$$

The actual body temperature is compared with the setpoint by a comparator in order to determine the difference (error) between the two terms, e.g., error $= 37.6 − 37 = 0.6°$ C. The error is then processed according to some specific relationship between error and degree of compensation. This relationship, called a **transfer function**, is specific for the particular regulatory system involved. It can entail amplification (or multiplication by a constant factor); addition, subtraction, or multiplication by numerous inputs; multiple algebraic operators; integration; etc. The net effect of the transfer function is to convert the error signal into a compensation signal. In the human thermoregulatory system, for example, the transfer function is multiplication by the constant value of 32, i.e., compensation is 32 × error.

The gain of the regulatory control system is defined as the compensation relative to the remaining error.

$$\text{gain} = \text{compensation/remaining error} \quad (2.3)$$

Gain reflects the conversion of the error into a compensation; a gain >1 indicates amplification and <1 reflects attenuation. The negative sign indicates that the gain is inverting, so that the compensation is in the opposite direction to the error. For the thermoregulatory system, gain $= − 19.4/0.6 = − 32$. This is a fairly high gain for a control system, indicating that body temperature is quite precisely regulated. The gain of the baroreceptor system for regulation of arterial blood pressure is substantially lower, at about − 7, reflecting relatively poor regulation. However, other regulatory mechanisms also operate in concert with the baroreceptors for precise, long-term regulation of mean arterial blood pressure. The gains for other physiological regulatory systems vary dramatically from close to 0, to − 20, to − 50, or even to − infinity.

A general consequence of feedback regulation is that the actual value of the physiological variable is different from the setpoint value, except when there is no disturbance. The magnitude of the difference between the actual and setpoint values at steady-state depends primarily on the gain. The greater the gain, the lower the deviation of actual value from setpoint value.

A few regulatory control systems are perfect. The actual value of a variable is regulated equal to the setpoint value, i.e., there is negative infinite gain. One example of a perfect regulatory system is the regulation of arterial blood pressure by the human kidney (Figure 2–10). The amount of water excreted by the kidneys is a function of the mean arterial blood pressure; an elevation in pressure

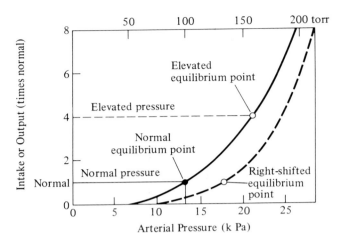

FIGURE 2–10 There is infinite-gain regulation of mean arterial blood pressure in humans by the renal control of fluid loss. The normal equilibrium point is eventually regained after any disturbance that either increases arterial blood pressure (and results in elevated fluid excretion) or decreases blood pressure (resulting in a renal fluid loss less than fluid intake). The mean arterial blood pressure can be elevated by either a chronic increase in fluid intake and loss (thin dashed line) or a right shift in the curve caused by elevated blood angiotensin levels (thick dashed curve).

renin – angiotensin – aldosterone system

elicits a greater renal fluid loss. Part of the excreted fluid comes from the circulatory system, and any decrease in circulatory fluid volume causes blood pressure to decline. Thus, if arterial pressure is increased by a disturbance, then the rate of renal water excretion is elevated, the circulatory fluid volume is reduced, and blood pressure declines to the setpoint value. The opposite scenario serves to elevate blood pressure if some disturbance causes arterial pressure to drop. This is the normal renal regulatory mechanism for long-term blood pressure regulation. Chronic elevation of blood pressure can be caused by two mechanisms: increased fluid and salt intake, and alteration of the position of the relationship between renal fluid excretion rate and mean arterial pressure. An increase in water intake necessitates an increased renal loss, which is accomplished by/causes an elevation of mean arterial blood pressure. Alternatively, a shift in the position of the relationship—a right-shift caused, for example, by elevated levels of a hormone (angiotensin)—will result in an elevated mean arterial pressure at the normal rate of fluid intake and loss.

Negative and Positive Feedback. The direction of compensation must be opposite to the direction of the disturbance for a feedback system to regulate

the value of a physiological variable; the feedback must be negative. For example, the compensation must decrease body temperature if the disturbance elevates body temperature, and vice versa. There are innumerable examples of **negative feedback regulation** in physiological systems: blood glucose regulation by insulin, regulation of mean arterial pressure by baroreceptors and renal fluid loss, plasma calcium level regulated by the hormone parathormone (and also calcitonin), body temperature regulation in mammals and birds, and chemical regulation of respiration.

Positive feedback regulation occurs when the compensation augments the disturbance. Positive feedback clearly has detrimental consequences for homeostasis; it destabilizes the regulated variable. For example, a considerable blood loss decreases the blood volume to such an extent that there is insufficient blood flow to the heart muscle. This inadequate blood flow causes anoxia of the heart muscle, and so the heart fails to pump at a normal rate and pressure, lowering cardiac output. This in turn decreases blood pressure, causing a further drop in blood supply to the heart. Thus, a positive feedback, or vicious cycle, is initiated and can rapidly lead to circulatory collapse and death.

Positive feedback cycles are often used for beneficial, though nonhomeostatic, functions by animals. Positive feedback can initiate an extremely rapid change. For example, the rapid generation of an action potential by a nerve or muscle cell is due to positive feedback between membrane electrical potential and the permeability of the membrane to sodium ions. The membrane electrical potential rapidly increases during an action potential from about -100 mV to about $+40$ mV. The details of the neurophysiological processes involved are presented in Chapter 6, but a general discussion of the concept of positive feedback mechanisms is of interest here. If some external stimulus depolarizes the resting membrane potential (e.g., changes it from -100 to -90 mV), then this increases the sodium ion permeability by opening sodium channels through the membrane (Figure 2–11). The increased sodium permeability causes an influx of Na^+ into the cell, further depolarizing the membrane and, in a regenerative fashion, increasing the sodium permeability even more. A positive feedback cycle is thus initiated, and it causes a rapid increase in membrane potential. The timing of the cycle is also important; the early phases of change in sodium permeability are more rapid than the final phases. The positive feedback cycle ends only when all of the sodium channels are open and sodium permeability is maximal. How does the sodium permeability

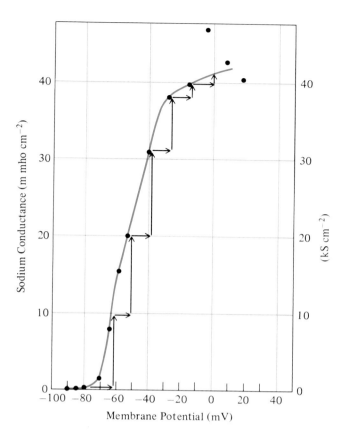

FIGURE 2–11 There is a positive feedback relationship between the resting membrane potential of nerve cells (E_m) and the sodium conductance (g_{Na}^+). A depolarization of E_m towards 0 mV results in an increased g_{Na}^+, which in turn causes a further depolarization of E_m. The positive feedback effect is shown as a series of step changes in E_m and g_{Na}^+ (light arrows), but in reality the positive feedback results in a smooth depolarization of E_m and increased g_{Na}^+ (solid line). Units for Na$^+$ conductance are $k\Omega^{-1}$ (or mmho) cm^{-2} and kSiemens cm^{-2} (1 S = 1 mho). *(Data from Hodgkin and Huxley 1952.)*

ever return to normal? Positive feedback would prevent any decrease in sodium permeability, so some other factor is involved. The sodium channels have an inherent property of automatically closing after they have opened, regardless of the membrane potential, and this ensures a return to initial resting conditions.

A second example of positive feedback involves one aspect of the endocrine regulation of reproduction in female mammals. Estrogen secreted from the ovaries causes a rapid surge in secretion of LH (luteinizing hormone) from the anterior pituitary immediately prior to ovulation. The elements for

positive feedback are high estrogen levels promote LH secretion by the anterior pituitary, then elevated LH levels promote estrogen secretion by the ovaries. Other examples of positive feedback include the emptying of body cavities—swallowing, defecation, birth—and blood clotting. These are all non-homeostatic functions.

Multiple Control Systems. The level of important physiological variables is often controlled by not one, but multiple, regulatory effectors. Multiple control systems increase the precision of regulation and provide greater flexibility of regulation through the higher complexity of the sensory, feedback, and effector systems.

The behavioral thermoregulatory system of many lizards provides a clear example of a dual regulatory system, even at the level of simply observing the behavior of the lizards. Most lizards thermoregulate by basking in direct sunlight or on warm substrates (see Chapter 5). Lizards will move into a favorable basking location when their body temperature is below some lower setpoint in order to warm up to their "preferred" body temperature. However, the lizard will move to a cooler location if its body temperature rises above an upper setpoint. Lizards can be readily observed in the field and laboratory to "shuttle" with some regularity between cool and warm locations in order to keep their body temperatures within the narrow "preferred" body temperature range. The thermoregulatory mechanisms of such lizards have been hypothesized to have a dual thermostat (Figure 2–12). A low setpoint thermostat determines movement to a warm environment if body temperature < low setpoint; the movement causes a heat gain that elevates body temperature into the preferred range. A high setpoint thermostat initiates movements to a cooler environment, if body temperature > high setpoint, and lowers body temperature into the preferred range. There is also input of peripheral temperature receptors into the thermoregulatory control system via the reticular formation in the brain, which modifies the operation of the dual setpoint hypothalamic system. The dual thermostat system provides a fairly precise regulation of body temperature regulation. The precision of thermoregulation depends upon how close the upper and lower setpoints are.

A second example of a dual thermostat thermoregulatory system, but of entirely different mechanisms, is observed in mammals. Mammals can regulate their body temperatures through physiological, in addition to behavioral, means. Internal heat production is increased by elevated cellular metabolism to elevate body temperature if it drops

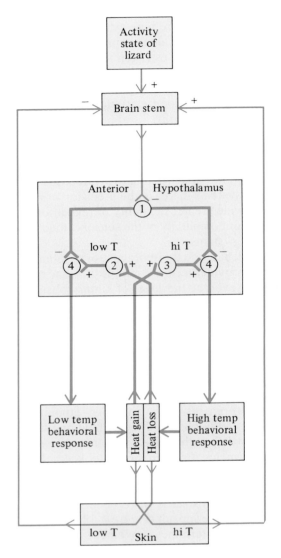

FIGURE 2–12 Hypothetical neural network illustrating the dual setpoint model of body temperature regulation by lizards. The anterior hypothalamus contains temperature-insensitive neurons (1), cold-sensitive neurons (2), warm-sensitive neurons (3), and motor neurons (4) responsible for low and high temperature shuttling activity. Shuttling into a warm environment causes heat gain, which elevates the anterior hypothalamic temperature and stimulates the high temperature receptors, eventually eliciting heat-avoidance behavior (heavy lines). The peripheral temperature-sensitive neurons of the skin feed information back to the reticular formation of the brain stem (light lines), thence to the anterior hypothalamus, that reduces the firing activity of the temperature-insensitive neurons. Shuttling into a warm environment elicits the opposite responses. The general activity state of the lizard could also influence the activity of the reticular formation; a high activity level would increase the activity of the reticular formation. *(Modified from Berk and Heath 1975.)*

below a lower setpoint. Heat is dissipated through evaporative (and other) means if body temperature exceeds an upper setpoint. Both of the thermostats are located in the hypothalamus, deep within the brain, and consequently the regulated variable is hypothalamic temperature rather than body temperature measured elsewhere (e.g., oral, esophogeal, rectal). The low setpoint for metabolic heat production is approximately 36.8° C (if skin temperature is 30° C; see below); heat production is increased at body temperatures below this setpoint, and remains stable at body temperatures above the setpoint (Figure 2–13). The high setpoint that initiates evaporative heat loss, by sweating, is about 37.3° C (if skin temperature is 30° C); body temperatures above this setpoint result in progressively higher rates of evaporative heat loss. These hypothalamic thermoreceptors are clearly not the only temperature sensors in the body; for example, the integument is well endowed with both cold and hot thermoreceptors. These peripheral thermoreceptors also contribute to the operation of the thermoregulatory control system by altering the lower and upper setpoints for metabolic heat production and evaporative heat loss, and also by altering the gains for the two transfer functions (the slopes of MHP and EHL versus hypothalamic temperature in Figure 2–13).

Oscillation and Damping. We have seen an example of how a physiological variable, the blood concentration of luteinizing hormone (LH), can **oscillate** owing to episodic pulsing of LH release with subsequent clearance (Figure 2–8B); this oscillation occurs in an open system. Similar oscillation, or "hunting," can occur in negative feedback, closed systems if some of the elements have inertia (slowness of response), or if there are delays in the feedback loop. Let us return to our example of a hot water bath. If the water bath had a small, unstirred volume of water and a massive heating element with a high rate of heat output, then it is easy to imagine that the rapid rate of heating of the element would quickly add so much heat to the water that, because of the large amount of heat remaining in the heater element, the temperature would not only rise rapidly to the setpoint of 50° C, but would overshoot the setpoint even after the thermostat opened the contacts and turned off the heater. The temperature would then slowly decline by passive heat loss to the environment until it fell below the setpoint of 50° C. The heater element would then be turned on, rapidly overheat the water, and the extreme temperature cycle would repeat itself. This tendency to oscillate, and the magnitude

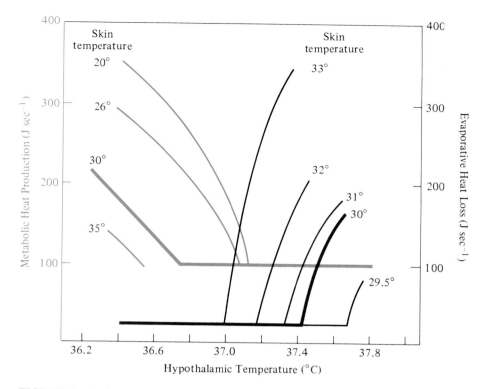

FIGURE 2–13 Body temperature regulation in humans involves elevation of metabolic heat production during cold stress (decreased anterior hypothalamic temperature) and evaporative heat loss during heat stress (elevated anterior hypothalamic temperature). The low setpoint (T_{low}) is about 36.8° C for control of metabolic heat production, and the high setpoint (T_{high}) is about 37.3° C for control of evaporative heat loss (heavy lines are relationships for a skin temperature of 30° C). Peripheral skin temperature receptors also influence the low and high setpoints and the slopes of the relationships of metabolic heat production and evaporative heat loss to hypothalamic temperature (light lines). *(Modified from Benzinger 1964.)*

of the oscillation, would be diminished by decreasing the inertia of the regulatory effector system, i.e., reducing the thermal capacity of the heater element or increasing the water volume. A delay between turning the heater ON and OFF and closing the temperature sensor contacts (e.g., poor water mixing) could also cause oscillation. Elimination of such time delays, for example, by keeping the water well-stirred, would minimize the tendency for oscillation. Finally, oscillation can be exacerbated by an extremely high gain.

Oscillation in physiological control systems is well illustrated by the knee jerk reflex and other stretch reflexes. When the knee jerk reflex is initiated by striking the patellar tendon, the quadriceps muscle is stretched. It then contracts in response to its stretching, and this causes the lower leg to jerk forward. The myogram (recording of muscle length) of this dynamic stretch reflex (Figure 2–14A) shows a rapidly disappearing oscillatory cycle of quadriceps contraction and relaxation. The tendency to oscillate becomes more pronounced if the dynamic stretch reflex is facilitated by stimulation from the brain. Sensitization of the reflex can result in such a high gain that a prolonged oscillation of the ankle, called clonus, may occur. Normally, the tendency for oscillation is minimized by damping and clonus doesn't occur. However, clonus can occur under pathological conditions and can be demonstrated experimentally in decerebrate animals (animals in which the higher centers of the brain are severed from the brainstem). Oscillation can also be demonstrated in other physiological systems under experimental, or abnormal, conditions. The respiratory rhythm is generated by the respiratory center in the brain stem and normally is extremely regular. The basic respiratory rhythm is modified by a chemoreceptor (chemical sensor) regulatory

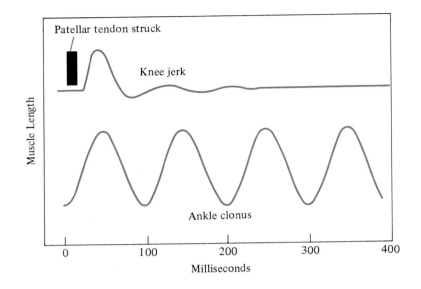

A

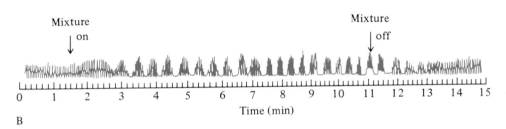

B

FIGURE 2–14(A) Recordings of muscle length (myograms) showing the normal minor oscillation in quadriceps length after the patellar tendon is struck (upper) and the continual oscillation of gastrocnemius length during ankle clonus. *(From Guyton 1986.)* **(B)** Normal respiratory rhythm of J. S. Haldane and periodic, oscillatory respiratory pattern when breathing an air mixture of 11.05% O_2 and 0.7% CO_2. *(From Haldane 1935. Respiration. Clarendon Press, Oxford. Copyright Oxford University Press; reproduced by permission of the Oxford University Press.)*

mechanism, which monitors the pO_2 and pCO_2 of arterial blood, and the pH of brain tissue near the respiratory center (see Chapter 13). Breathing air of a low pO_2 (but not a correspondingly high pCO_2) can cause a prolonged instability of the respiratory rhythm. The recording of the respiratory rhythm shows such a pronounced oscillation when breathing air containing 11.05% O_2 and 0.7% CO_2 (Figure 2–14B). Respiratory oscillations like this are called Cheyne-Stokes breathing, and consist of episodes of rapid breathing, punctuated by pauses. They can occur when breathing normal air at high altitudes, especially at night during sleep (low pO_2, low pCO_2); with cardiac insufficiency, since this increases the amount of time required to transport blood from the lungs to the brain (i.e., increased delay); or after brain damage that increases the gain of the respiratory regulatory mechanism.

The tendency of a negative feedback system to oscillate can be reduced by **damping** the correction signal in proportion to the magnitude of the correction signal. Damping may sometimes be accomplished by hydraulic or viscous, mechanical means. Consider a linear spring as a mechanical model for damping. The relationship between a force F_k applied to one end, and the overall extension of the spring D, is given by Hooke's law as

$$F_k = k (D_2 - D_1) = k \, \Delta D \qquad (2.4)$$

where D_1 and D_2 are the displacements of each end of the spring, ΔD is the net stretching of the spring, and k is the spring constant (units of force per length of spring). Application of a force will rapidly stretch the spring to its equilibrium net displacement, ΔD, i.e., $\delta D/\delta t$ is high. The rate of stretching can be reduced by incorporating a dashpot (an oil-filled

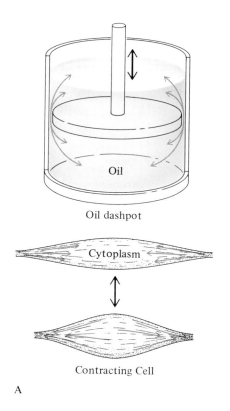

Oil dashpot

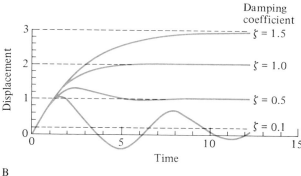

Contracting Cell

A

B

FIGURE 2–15(A)   Viscous damping in a mechanical dashpot due to viscous flow of oil past the piston, and in a muscle cell due to cytoplasmic flow. (B) Effects of the damping coefficient (ζ) on the error of a negative feedback mechanism. A ζ of 1.0 results in rapid attainment of the new equilibrium error value after a disturbance; higher damping ($\zeta = 1.5$) causes slow movement to the new equilibrium; less damping results in slight oscillation prior to equilibrium ($\zeta = 0.5$) or pronounced oscillation ($\zeta = 0.1$). The damping coefficient is a function of inertia, time delays, and gain of the regulatory system. *(Modified from Bayliss 1966.)*

cylinder with a piston) across the ends of the spring (Figure 2–15A). Now, to stretch the spring, there must be a corresponding movement of the piston through the oil, but the high viscosity of the oil slows down the movement of the piston and therefore the rate of stretching of the spring, i.e., $\delta D/\delta t$ is reduced. Damping doesn't prevent the eventual stretching of the spring by displacement ΔD. The rate of stretching is damped by the dashpot exerting a force F_{damp}

$$F_{damp} = -B(\delta D/\delta t) \qquad (2.5)$$

where $-B$ is the damping constant, and the negative sign indicates that the damping force resists the stretching force. A dashpot is a simple mechanical example of a damping system. It is analogous however to the damping of muscle cell contraction by the necessary viscous movement of the muscle cell contents (Figure 2-15A). The damping coefficient (ζ) is a measure of the degree of damping in control systems. It is determined in a complex fashion by inertia, time delays, and gain (Bayliss 1966). A damping coefficient of 1 results in rapid adjustment after a disturbance to the new equilibrium error; coefficients >1 cause slow adjustment to the new equilibrium; coefficients <1 cause rapid adjustment with oscillation (Figure 2–15B).

Damping of biological control systems is generally accomplished by neural rather than mechanical means. The damping effect of the muscle spindle apparatus during muscle contraction is a good example. The muscle spindle is essentially a highly modified muscle cell with sensory input to the central nervous system; it monitors the degree of stretch of the muscle (Chapter 9). The muscle spindle also has motor innervation from the central nervous system (the gamma efferent fibers) that cause the muscle spindle fiber to shorten (thus mimicking stretch of the spindle by contraction of the muscle). A complex feedback of information between a muscle and the CNS via the muscle spindle normally results in smooth muscle contractions rather than jerky motions. However, the muscle contraction is jerky if the muscle spindles are destroyed.

Summary

The general physical properties of aerial, aquatic, and terrestrial environments are described, since these either influence the environment that animal cells experience or require regulatory systems to maintain relative constancy of the internal environment.

The extracellular fluids, which provide the internal environment that animal cells experience, create an internal environment that is different from the external environment. Many aspects of the internal environment are quite constant, in contrast to the external environment, i.e., there is homeostasis of the internal environment.

The mechanisms that are responsible for the maintenance of the integrity of the internal environment can include passive or dynamic equilibrium with the external environment in open systems, but generally involve feedback regulation in closed control systems. Feedback systems react to a disturbance in the value of a regulated variable. They have a fixed setpoint value, near which the variable is regulated; a comparator determines the error between the setpoint and the actual value of the variable, and the error is converted into a compensation according to some transfer function. Generally, the compensation is antagonistic to the disturbance, i.e., the system has a negative feedback system.

In some control systems the compensation augments the disturbance and destabilizes the variable, i.e., the system has a positive feedback and is nonhomeostatic. Positive feedback may be useful in producing a rapid or sustained change in a physiological parameter (e.g., action potential, emptying of a body cavity, blood clotting) or mav be detrimental and compromise the regulation of a physiological parameter (e.g., circulatory collapse).

Many control systems have multiple, rather than single, negative feedback loops; thermoregulation by reptiles and mammals is an example of a multiple control system.

The particular properties of a negative feedback system can damp changes in a variable or can induce oscillation.

Recommended Reading

Bayliss, L. E. 1966. *Living control systems*. London: English Universities Press.

Cannon, W. B. 1929. Organization for physiological homeostasis. *Physiol. Rev.* 9:399–431.

Mangum, C. P. and D. W. Towle. 1977. Physiological adaptation to unstable environments. *Am. Sci.* 65: 67–75.

Chapter 3

Cellular Energetics

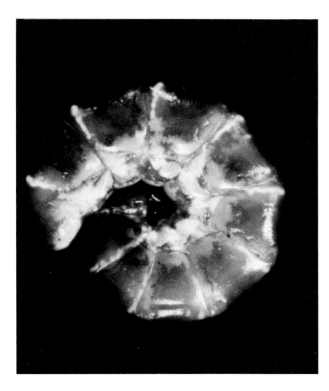

The bioluminescent larva of the beetle *Phengodes* emits light from subdermal oenocyte cells. *(Photo courtesy of Bassot/Jacana/Photo Researchers, Inc.)*

E nergy is central to virtually all physiological processes, and the utilization of chemical energy is one of the fundamental characteristics of living animals. Any definition of life would have to include the use of exogenous energy sources for the maintenance of cellular order. The exogenous source of energy for animals is the chemical energy of ingested food. It is essential to have a basic knowledge of energy, energy transformations, and the biochemical means of converting chemical energy into biochemical useful forms of energy because animals, like machines, must conform to the basic laws of physics and chemistry. An understanding of the principles of energy transformation is necessary to understanding the physiology of animals.

TABLE 3–1

Common units for energy and their joule equivalence.	
1 electron volt	$1.6 \ 10^{-19}$ joule
1 electron's energy[1]	$8.2 \ 10^{-14}$
1 H atom's energy[1]	$1.49 \ 10^{-10}$
1 erg (dyne-cm)	$1 \ 10^{-7}$
1 newton-meter	1
1 foot-pound	1.36
1 calorie	4.18
1 ml O_2	20.1
1 British thermal unit (BTU)	$1.05 \ 10^{+3}$
1 watt-hour	$3.6 \ 10^{+3}$
1 horsepower-hour	$2.69 \ 10^{+6}$
1 gram of matter[1]	$9 \ 10^{+16}$

[1] $E = mc^2$

Thermodynamics

Thermodynamics is the study of the quantitative relationships between heat and other forms of energy, particularly work and energy transformations. The principles of energy transformations are summarized to provide a framework for understanding biochemical and animal energetics, although a thorough understanding of thermodynamics is not essential for the following discussion of cellular metabolism (this chapter) and animal energetics (Chapter 4). Thermodynamics provides the theoretical basis for understanding the principles of energy transformation and the transport processes, including diffusion, osmosis, the flow of liquids through tubes, the formation of electrical potentials across membranes, heat exchange, directions and rates of chemical reactions, and many other processes central to animal physiology. The remainder of this section summarizes these principles and the applications of thermodynamics that will be used in successive chapters to describe the physiological functioning of animals.

Thermodynamics, at least as it is applied to animal physiology, need not be exceedingly mathematical and complex. In fact, many of the basic principles of thermodynamics are apparent in our daily lives. There are many forms of energy; common examples include kinetic, potential, gravitational, electrical, chemical, heat, and light energy.

Energy must be conserved. It cannot be created or destroyed, although it can be transformed from one form to another. Matter and energy are interconvertible in the realm of nuclear physics but not in the realm of physiology.

The SI unit for energy is the **joule**; one joule is the amount of energy released by dropping a weight of one newton (1 kg m s^{-2} = 1/9.8 kg force) through a distance of 1 meter, i.e., 1 J = 1 newton-meter. The calorie, a non-SI unit for energy, is the amount of energy required to warm 1 gram of water from 14.5 to 15.5° C. There is an equivalence between various forms of energy, and 1 calorie (defined above in terms of heat) is equivalent to 4.18 joules (defined above in terms of force × distance). A variety of energy units, and their joule equivalents, ranging from the fission energy of a single atom, to a horsepower, to the fission energy of 1 gram of matter, are listed in Table 3–1. Having established the units for energy, let us now examine the principles of energy transformation.

Work from Heat

Temperature gradients can be used to perform useful work even though heat is a "low quality" form of energy. For example, a Carnot cycle heat engine uses energy from a thermal gradient to perform useful mechanical work. Temperature gradients do exist between animals and their environment and even between different parts of the same animal, but animals cannot perform significant amounts of work by using temperature gradients, i.e., animals are not heat engines. The reason for this has a solid basis in thermodynamics. The theoretical maximum efficiency (E_{max}) of a cyclic heat engine is

$$E_{max} = (T_1 - T_0)/T_1 \qquad (3.1)$$

where T_1 is the highest temperature during the cycle and T_0 is the lowest temperature. Other thermodynamic considerations make the actual efficiency of heat engines even lower than the predicted E_{max}. However, the point is that even an ideal animal heat engine would require a substantial temperature gradient to operate at a reasonable efficiency. For example, if we accept 20% as an adequate efficiency and a body temperature of 20° C (293° K), then the temperature gradient must be between 59° and 73° C (depending on whether the body temperature is the highest or lowest temperature), i.e., the ambient temperature would have to be either −39° or +93° C!

Biological systems must rely on chemical energy, rather than thermal energy, to perform useful work.

First Law of Thermodynamics

The **First Law of Thermodynamics** is the law of conservation of energy. It states that energy cannot be created or destroyed, but can be transformed from one form of energy to another.

A closed system (any volume in space that we wish to consider, with no exchange of energy or matter across its boundaries) contains a certain quantity of energy (E). The energy content of a closed system must be constant (although it is impossible to determine the actual energy content) and there cannot be a change in its energy content.

$$\Delta E = 0 \quad \text{(closed system)} \quad (3.2a)$$

For an open system, there may be an exchange of energy, either absorbed as heat (Q) or as work (W).

$$\Delta E = Q - W \quad \text{(open system)} \quad (3.2b)$$

Any change in the energy of an open system (e.g., an animal) must be accompanied by an equal and opposite energy change of the environment.

The first law of thermodynamics indicates the quantitative relationship between an energy change (ΔE), work done (W), and heat absorbed (Q), but gives no indication of the likely direction of energy transformations between work and heat.

Second Law of Thermodynamics

The **Second Law of Thermodynamics** states that there is an inevitable degradation of useful energy (energy able to do useful work) into heat. Heat energy can be converted into other forms of energy and can perform useful work, but not at 100% efficiency. Note, however, that other forms of energy can be converted to heat at 100% efficiency. Heat energy is the random molecular motion of atoms; it is an energy of randomness and disorder.

Entropy (S; J °K^{-1}) is a measure of the thermal randomness of molecules, i.e., it quantifies molecular disorder. The increase in entropy (ΔS) for a reversible reaction at constant temperature is the ratio of heat added (Q_{rev}) to ambient temperature (T; °K). It is independent of how the reaction occurs and equals the following.

$$\Delta S = S_{final} - S_{initial} = Q_{rev}/T \quad (3.3)$$
$$\text{or} \quad Q_{rev} = T\Delta S$$

If heat is lost rather than gained, then entropy decreases.

There is no net change in entropy for a reversible reaction, since the entropy change of the reaction is accompanied by an equal and opposite entropy change in the environment. Unfortunately, most energy transformations are irreversible, not reversible, and the net entropy change for a system and its environment is not zero, but is positive.

The second law of thermodynamics indicates that there is an inevitable increase in entropy, or random molecular motion, within any system, and "high quality" energy is inevitably degraded to "low quality" heat.

Free Energy

Free energy is a measure of the amount of energy that a system can provide. Biologists and biochemists generally use Gibbs free energy (G) rather than Helmholtz's free energy (F), although they are essentially the same for biological and biochemical processes. A release of free energy by the transformation of a system from one state to another can perform useful work. It is analogous to the potential energy released when an object falls through a gravitational field; the decrease in potential energy can be used to perform work.

The **change in Gibbs free energy** (ΔG) is essentially the maximum amount of useful energy made available by a reaction. It equals the total amount of heat released (ΔH) minus any energy expended to decrease the entropy.

$$\Delta G = \Delta H - T.\Delta S \quad (3.4)$$

The magnitude of ΔG indicates the likelihood for spontaneity of the reaction. A reaction with a highly negative ΔG (due to either a large release of heat or a large increase in disorder) tends to be spontaneous. In a biological system, a negative change in Gibbs free energy might be used to transport solutes across a membrane, to produce mechanical movement, to establish an electrical membrane potential, or even to release light energy.

The "standard" Gibbs free energy change is defined as the $\Delta G°$ at 25° C, pH = 0.0, pressure = 101 kPa (1 atm), and molar concentrations of 1 for all reactants (except water). The $\Delta G°$ at a more realistic biological condition of pH 7.0 is designated as $\Delta G°'$.

The amount of energy that is liberated in a system as heat (ΔH) by a reaction (at constant pressure) is the **enthalpy**. The enthalpy is equal to the Gibbs free energy change plus entropic energy change (Equation 3.4). Enthalpy can be measured as heat liberated by combustion of a material when all of the free energy is released as heat and none is used for useful work. The enthalpy change for oxidation of biological and biochemical materials is often measured using a bomb calorimeter (see Chapter 4). Although P is not constant in a bomb calorimeter, the error is sufficiently small ($<1\%$) that it is ignored.

Components of Free Energy. We have seen that the change in Gibbs free energy is the maximum amount of useful available energy from a reaction, but how is it used by biological systems? It can perform mechanical work (e.g., muscle cell contraction), it can move molecules against a concentration or electrical gradient, or it can release light

$$\Delta G = f.\Delta s + \mu.\Delta n + E.\Delta q + h.v, \quad (3.5)$$

where $f.\Delta s$ is mechanical work (force times displacement), $\mu.\Delta n$ is the chemical energy for movement of molecules against a chemical gradient, $E.\Delta q$ is the energy for moving ions against an electrical gradient, and $h.v$ is the energy of light emission. Helmholtz's free energy change (ΔF) is similar, but includes an additional term for the energy of pressure-volume work, $P.\Delta V$.

Different kinds of work are graphically summarized in Figure 3–1. For example, the mechanical work done by a force acting over a certain distance is the area under the curve towards the displacement axis. All of the terms in Equation 3.5 for ΔG are probably familiar, except chemical potential (μ), which requires further explanation.

Chemical Potential. Chemical potential (μ) is a property of the concentration of molecules, just as electrical potential is related to the concentration of charges. It is related to concentration (C) in a logarithmic, rather than a linear, manner (Nobel 1983).

$$\mu = \mu^\circ + RT \ln C \quad (3.6)$$

There is no fixed zero point for the chemical potential scale ($\ln C$ is undefined if $C = 0$) so μ° is used to specify the standard chemical potential. The units of μ are energy mole^{-1} per mole transported, e.g., joules mole^{-2} since $\Delta G = \mu.\Delta n$.

Generally, the chemical properties of a molecule, whether solute or solvent, are adequately described by its concentration, and chemical potential is proportional to $\ln C$. However, molecular interactions,

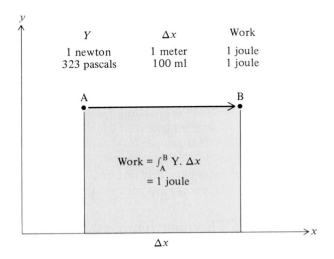

FIGURE 3–1 Representation of work done ($W = Y.\Delta x$) in moving from position A to position B; work is the area subtended by the path from A to B towards the x-axis. Some actual values for Y and x are shown, so that the work done would be 1 joule.

especially at high solute concentrations, can alter the chemical behavior of molecular species, and concentration then only approximately describes its chemical properties. It is necessary to "correct" the actual concentration of a solute (C) to an "effective concentration," or activity (**a**). This is accomplished by defining an **activity coefficient** (γ) such that $\mathbf{a}_{solute} = \gamma C_{solute}$. Activity is calculated for a solvent from the activity coefficient and mole fraction (N) as $\mathbf{a}_{solvent} = \gamma N_{solvent}$. The activity coefficient is 1 for an ideal solute and ideal solvent so activity equals concentration, or mole fraction. The activity coefficient is slightly less than 1 for most solutes, indicating that the thermodynamically effective concentration is only slightly less than the actual concentration. However, some solutes such as charged ions can have γ considerably less than 1, dependent on the actual concentration. For Na^+Cl^- at a concentration of 100 mM, γ is about 0.76; it decreases further at higher C. Activity is a thermodynamically preferable measure to concentration, but concentration is generally an adequate measure of the chemical potential for dilute biological solutions and we shall generally use C rather than **a** to calculate chemical potential.

The chemical potential of a molecular species also depends on factors other than concentration. There is an additional contribution to chemical potential by pressure, equal to $\overline{V}.P$, where $\overline{V}$ is the partial molar volume of the species. The partial molar volume is usually similar to the volume of one mole of that species. An electrical potential also contributes to the chemical potential, equal to $q.E$, where q is the amount of charge difference between the regions (in coulombs) at electrical potential E (the electrical term is also equal to $z.F.E$, where z is the charge on the ion and F is Faraday's constant). A gravitational field imparts chemical energy of $m.g.h$, where m is the mass per mole, g is the gravitational constant, and h is the height. Taking all of these terms into account, the chemical potential is as follows.

$$\mu = \mu^\circ + RT \ln C + \overline{V}.P \\ + z.F.E + m.g.h \qquad (3.7a)$$

The concept of chemical potential is extremely important because it indicates the free energy associated with molecules due to their concentration, pressure, temperature, electrical potential, or gravitational potential.

The difference in chemical potential for two regions is $\mu_2 - \mu_1$, which is equal to

$$\mu_2 - \mu_1 = RT \ln (C_2/C_1) \\ + \overline{V} \Delta P + zF\Delta E + mg\Delta h, \quad (3.7b)$$

since $\mu_2^\circ = \mu_1^\circ$; ΔP is the pressure difference, ΔE the electrical potential difference, and Δh the height difference between regions 2 and 1. The chemical potential difference indicates whether species in two regions are in equilibrium ($\mu_2 = \mu_1$) or are in disequilibrium ($\mu_2 \neq \mu_1$); it also indicates the interrelationships between different components of chemical potential.

An ion in equilibrium across a membrane has the same chemical potential on each side (at the same temperature, pressure, or gravitational potential). The difference in electrical potential can be calculated as

$$E_2 - E_1 = RT/zF \ln C_2 - RT/zF \ln C_1 \\ = RT/zF \ln (C_2/C_1). \qquad (3.8)$$

This important relationship is the **Nernst equation**. The value of RT/F is about 0.025 V at 20° C, or 58 mV if we use $\log_{10}$ rather than $\log_e$.

Water is always present in biological systems as either a solid, a liquid (usually), or a vapor. The chemical potential of liquid water is related to its concentration, C_w (or activity, a_w). The presence of solutes in water decreases the concentration of water, hence its chemical potential. At the same time, the presence of solutes increases the osmotic concentration, i.e., the total concentration of solutes. There are two common units used for measuring osmotic concentration (Π): the osmolar concentration is the moles of solutes per liter of solution, whereas the osmolal concentration is the moles of

solutes per kg of solvent. There is an inverse logarithmic relationship between the water concentration (C_w) and osmotic concentration for a solution; the presence of solutes decreases the concentration (activity) of water and increases the osmotic concentration.

$$RT \ln C_w = -V_w \Pi \qquad (3.9a)$$

The chemical potential of water vapor is related to the water vapor pressure (P_{wv}) and the saturation water vapor pressure (P_{sat}) as

$$RT \ln C_{wv} = RT \ln (P_{wv}/P_{sat}) \\ = RT \ln (RH/100) \qquad (3.9b)$$

where RH is the relative humidity ($RH = 100 \, P_{wv}/P_{sat}$).

Water potential (Ψ) is defined as the water chemical potential divided by the partial molar volume of water, i.e., $(\mu_w - \mu_w^\circ)/\overline{V}_w$. For a solution, the water potential is equal to (from Equations 3.7a and 3.9a),

$$\Psi_w = P - \Pi + \rho_w gh \qquad (3.10a)$$

where ρ_w is the density of water. The water potential of a solution is thus comprised of three components: a pressure potential (Ψ_p), an osmotic potential (Ψ_Π), and a hydrostatic potential (Ψ_h). The units of Ψ are pressure (e.g., kPa or atmospheres); 2269 kPa (22.4 atm) are equivalent to 1 osmolal osmotic concentration. The water potential for water vapor is

$$\Psi_{wv} = RT/\overline{V}_w \ln (RH/100) + \rho_w gh \quad (3.10b)$$

where ρ_w is the density of liquid water. For water vapor, there are two water potential components: a vapor pressure potential and a gravitational potential.

Diffusion Flux

Having described the various components that contribute to chemical potential, we can now examine the consequences of a difference in chemical potential between two regions. A gradient in chemical potential $\Delta\mu/\Delta x$ (i.e., a difference in chemical potential $\Delta\mu$ over a distance Δx) is a driving force for the passive movement of molecules from the region of high chemical potential to low chemical potential. Such a movement of molecules is called a **diffusional flux**. The larger the chemical potential gradient, the greater the flux; the smaller the gradient, the less the flux.

In general, diffusional flux is directly proportional to the chemical potential gradient. The unidirectional flux (moles m^{-2} sec^{-1}) from region 1 (J_1) depends on the concentration of the species there (C_1), the chemical potential gradient ($\Delta\mu/\Delta x$), and

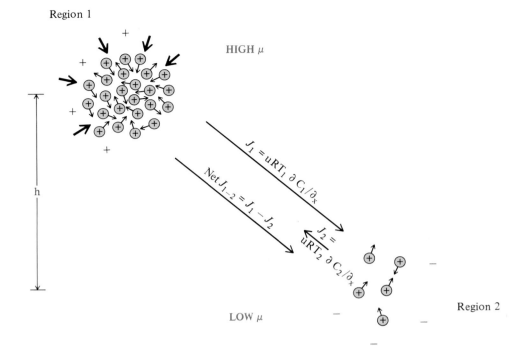

FIGURE 3–2 Representation of factors contributing to the chemical potential (μ) in two regions (1 and 2): concentration, temperature (molecular velocity indicated by arrows on molecules), electrical charge ($+$, $-$), high pressure ($\rightarrow$), and gravity (h). The unidirectional fluxes from region 1 (J_1) and region 2 (J_2) are combined to obtain the net flux from region 1 to 2 ($J_{1\text{-}2}$).

the ability of the molecules to move (their mobility, u). This flux, which is dependent on the local concentration as well as the pressure and gravitational head, is schematically depicted in Figure 3–2. The absolute flux (moles sec^{-1}) from each region is calculated as

$$J_1 = uAC_1(-\Delta\mu/\Delta x)$$
$$J_2 = uAC_2(-\Delta\mu/\Delta x), \qquad (3.11)$$

where A is the area (m^2) for diffusional exchange.

A gradient in any (or more than one, or all) of the different terms contributing to chemical potential (Equation 3.7b) will produce a diffusional flux. For example, a concentration gradient ($\Delta C/\Delta x$) will produce a diffusional flux. A pressure gradient ($\Delta P/\Delta x$) will produce a hydraulic flux. A voltage gradient ($\Delta E/\Delta x$) will produce an electrical flux of ions. A gravitational gradient (g) will produce an acceleration. A temperature gradient ($\Delta T/\Delta x$) will cause a thermal flux of molecules. The exchange of heat by a thermal gradient is described by the same equations as diffusional exchange of mass in response to a concentration gradient (but not the equation for

exchange of mass due to a temperature gradient). All of these fluxes are of physiological significance, and will be further examined in following chapters. However, the diffusional flux determined by $\Delta C/\Delta x$ is of such central significance to physiology that we will further examine it here.

The net flux between regions 1 and 2 (J_{net}) is the difference in the two unidirectional fluxes, J_1 and J_2.

$$J_{\text{net}} = J_2 - J_1 = uA(C_2 - C_1)(-\Delta\mu/\Delta x) \quad (3.12a)$$

The net flux equation can be simplified to the following.

$$J_{\text{net}} = -uRTA\Delta C/\Delta x \qquad (3.12b)$$

Consolidating the constants uRT yields **Fick's first law of diffusion** for 1-dimensional flux.

$$J_{\text{net}} = -DA\Delta C/\Delta x \qquad (3.12c)$$

The net rate of diffusion J_{net} is proportional to the diffusion coefficient (D), the area for diffusional exchange (A), the concentration difference (ΔC), and the diffusion distance (Δx). $\Delta C/\Delta x$ is the **concentration gradient**. The consolidated constant D (uRT;

TABLE 3–2

Molecule	MWt	Air	Water
H_2O	18	$2.42 \; 10^{-5}$	—
Na^+	23	—	$1.5 \; 10^{-9}$
O_2	32	$1.95 \; 10^{-5}$	$2.5 \; 10^{-9}$
K^+	39	—	$1.9 \; 10^{-9}$
CO_2	44	$1.51 \; 10^{-5}$	$1.7 \; 10^{-9}$
Glucose	180	—	$0.67 \; 10^{-9}$
Sucrose	342	—	$0.52 \; 10^{-9}$

Diffusion coefficients (D; m^2 sec^{-1}) for a variety of physiologically important molecules in air and water (at 20° C)

m^2 sec^{-1}) is the **diffusion coefficient**; it is not a physical constant, but varies with absolute temperature and molecular weight ($D \propto 1/\sqrt{MWt}$ for small molecules). Some biologically important values for D are summarized in Table 3–2. Diffusion in two or three dimensions is described by more complex equations because the concentration gradient is more complicated (see Supplement 3–1, page 74).

Fick's law is of central significance to many physiological processes, not only mass transport as we have just seen, but also heat transfer (J_h; joule sec^{-1}).

$$J_{heat} = -DA\Delta T/\Delta x \qquad (3.12d)$$

where D has units of joule cm^{-1} sec^{-1} K^{-1} and $\Delta T/\Delta x$ is the temperature gradient. Electric charge transfer (J_E; coulomb sec^{-1}) is similarly determined as

$$J_E = -DA\Delta E/\Delta x \qquad (3.12e)$$

where D has units of coulomb cm^{-1} sec^{-1} $volt^{-1}$ and $\Delta E/\Delta x$ is the voltage gradient. Momentum transfer (J_{mom}; newton) is

$$J_{mom} = -\eta \, A\Delta M/\Delta x \qquad (3.12f)$$

where η is the viscosity and has units of (kg m^{-1} sec^{-1}), and $\Delta M/\Delta x$ is the gradient in momentum transfer (newton kg^{-1} sec^{-1}).

Fick's second law of diffusion describes diffusion over time, i.e., nonsteady-state conditions. Material will diffuse from an initially high concentration over time, and the concentration decreases at the source and increases at more distant points. The change in concentration is greater for longer diffusion times. This is illustrated in Figure 3–3 for diffusion from an initial concentration of material in a plane. The continuity equation relates the change in flux density at any particular location (x) to the change in local concentration over time (t)

$$\delta J/\delta x = -\delta C/\delta t, \qquad (3.13a)$$

i.e., the partial derivative of J with respect to x is equal to the negative partial derivative of concentration with respect to time. For the simple case of 1-dimensional diffusion away from a plane, with an initial amount of M moles per unit area of plane at the x-origin ($x = 0$), the concentration (C) at any time (t) at distance x from the original plane is

$$C = \frac{M}{2(\pi Dt)^{1/2}} e^{-x^2/4Dt}. \qquad (3.13b)$$

One index of the rate of diffusion from the original plane ($x = 0$) is the distance from the plane (x_e) at which the concentration is $1/e$ (37%) of the highest value. From Equation 3.13b, this is when

$$x_e^2 = 4Dt. \qquad (3.13c)$$

This equation indicates a fundamental property of diffusion; the time required for the diffusional movement of a molecule increases with the square of the required distance.

For example, how long would it take for an oxygen molecule to diffuse across the lung epithelium ($d = 0.5 \; \mu$) compared to diffusion from the

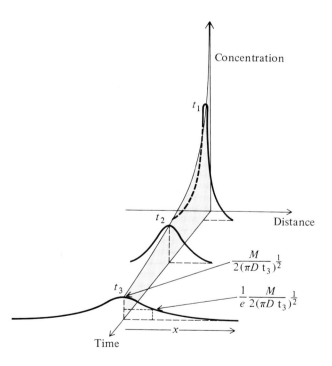

FIGURE 3–3 Concentration (C) as a function of distance from original planar distribution (x) and time (t); M is the original concentration of material, and D the diffusion coefficient. *(Modified from Nobel 1983.)*

lung to the big toe of a human (d = 1.5 m)? If D is 1.8×10^{-9} m^2 sec^{-1}, then 37% of the oxygen molecules would diffuse across the lung epithelium in t = (0.5 10^{-6})^2/(4)(1.8 10^{-9}) = 3 10^{-5} seconds. In marked contrast, 37% of the O$_2$ molecules would diffuse from the lung to the big toe in t = (1.5)^2/(4)(1.8 10^{-9}) = 3 10^8 sec = 9.9 years.

It is clear that diffusional exchange allows rapid mass transport over short distances (i.e., at the cellular and subcellular scale) but is grossly inadequate for mass transport over larger distances, such as cm or m.

Chemical Reactions

The principles of thermodynamics are directly applicable to the study of chemical and biochemical reactions. Three aspects of chemical and biochemical reactions merit discussion here: free energy changes, reaction rate, and activation energy.

Free Energy Change

The change in standard Gibbs free energy ($\Delta G°$) for an equilibrium chemical reaction of the general type

$$A + B \rightleftharpoons C + D \qquad (3.14)$$

is equal to the free energy of the products ($G_c°$ + $G_d°$) less that of the reactants ($G_a°$ + $G_b°$). The change in Gibbs free energy (ΔG) depends on $\Delta G°$ and the concentration of the reactants and products

$$\Delta G = \Delta G° + RT \ln [C][D]/[A][B]$$
$$= \Delta G° + RT \ln K_{eq} \qquad (3.15a)$$

where K_{eq} is the **equilibrium constant**. At equilibrium, ΔG = 0, so

$$\Delta G° = -RT \ln [C][D]/[A][B]$$
$$= -RT \ln K_{eq} \qquad (3.15b)$$

The standard Gibbs free energy change for biochemical reactions under biologically-relevant conditions ($\Delta G°'$) typically varies from −40 to +40 kJ mol^{-1} (Table 3–3).

It is important to appreciate that ΔG values can differ substantially from $\Delta G°$, depending on the concentrations of the reactants and products and on the temperature. Consider, for example, the equilibrium between glucose-1-phosphate and glucose-6-phosphate at 38° C (West 1961). The $\Delta G°$ is −7500 J mol^{-1} ($-RT \ln K_{eq}$ = −8.314 × 311 × ln (19)). If the concentrations are 0.01 M for G-1-PO$_4$ and 0.0001 M for G-6-PO$_4$, then ΔG = −19410

TABLE 3–3

Gibbs free energy change ($\Delta G°'$) for some chemical and biochemical reactions, including hydrolysis of high-energy phosphate molecules (pH 7.0, 25° C, 101 kPa, 1 M concentration of reactants except H$^+$ and water).

Reaction	$\Delta G°'$(kJ mole^{-1})
Fructose-1,6-diPO$_4$ $\rightleftharpoons$ dihydroxyacetone PO$_4$ + glyceraldehyde-3-PO$_4$	+23.6
Ice $\rightleftharpoons$ water (10° C)	+0.21
Ice $\rightleftharpoons$ water (0° C)	0
Ice $\rightleftharpoons$ water (−10° C)	−0.21
0.01M H$^+$ $\rightleftharpoons$ 0.001 M H$^+$ by dilution[1]	−1.4
Internal polypeptide hydrolysis	−2.1
ADP + creatinePO$_4$ $\rightleftharpoons$ ATP + creatine	−6.2
Glycerol-1-phosphate hydrolysis	−9.6
Glucose-6-phosphate hydrolysis	−13.8
AMP hydrolysis to adenosine	−14.2
Lactic acid $\rightleftharpoons$ lactate + H$^+$[2]	−18.0
CO$_2$ + H$_2$O $\rightleftharpoons$ H$^+$ + HCO$_3^-$	−20.9
Acetylcholine hydrolysis	−25.1
Pyruvate + NADH/H$^+$ $\rightleftharpoons$ lactate + NAD$^+$	−25.1
Sucrose hydrolysis	−27.6
ATP hydrolysis to ADP	−30.5
ADP hydrolysis to AMP	−30.5
ATP hydrolysis to AMP	−32.2
Arginine phosphate hydrolysis	−32.2
Pyrophosphate (PP$_i$) hydrolysis	−33.5
Creatine phosphate hydrolysis	−43.1
Phosphoenolpyruvate hydrolysis	−61.9

[1] $\Delta G°'$ = RT ln(C_2/C_1)
[2] $\Delta G°'$ = 2.3 RT (pK-pH); pH = 7

J mol^{-1}. However, if the concentrations are 0.0001 M for G-1-PO$_4$ and 0.01 for G-6-PO$_4$, then ΔG = +4410 J mol^{-1}.

Reaction Kinetics

It is obvious that molecules must come into contact before they can chemically react. Consequently, the rate of a chemical reaction must depend to some extent on the frequency of molecular collisions. As a first approximation, the rate of a chemical reaction is proportional to the number of collisions per unit time, which in turn depends on the concentration of the reacting molecules.

In a simple (first order) reaction of the kind

$$A \underset{k_2}{\overset{k_1}{\rightleftharpoons}} B \qquad (3.16a)$$

the rate of the forward reaction (A → B) is $k_1.[A]$, where k_1 is the **rate constant**. The rate of the reverse reaction is $k_2.[B]$. The net rate of the forward reaction is equal to the rate of decrease in A, $-\Delta[A]/\Delta t$, and depends on the relative rates at which A is converted to B ($k_1.[A]$) and B to A ($k_2.[B]$). At equilibrium ($-[A]/t = 0$),

$$k_1[A] = k_2[B] \quad \text{or} \quad k_1/k_2 = [B]/[A] = K_{eq} \tag{3.16b}$$

The value of the specific rate constant (k) depends on the temperature. It is independent of time or concentrations of substrates or products, but it can be altered by the presence of a catalyst (see below). For a first-order reaction, the unit for k is sec^{-1}.

A second-order reaction is more complex, since two different molecules (A and B) must collide to form a product (P)

$$A + B \underset{k_2}{\overset{k_1}{\rightleftharpoons}} P \tag{3.17a}$$

The rate of the forward reaction is $k_1[A][B]$, and that for the reverse reaction is $k_2[P]$. At equilibrium,

$$k_1[A][B] = k_2[P] \quad \text{or} \quad k_1/k_2 = [P]/[A][B] = K_{eq} \tag{3.17b}$$

The units for a second-order rate constant are liter mole^{-1} sec^{-1}.

In general, for the complex reaction of substrates A, B, C, . . . to form products P, Q, R, . . ., we have the following.

$$aA + bB + cC + \ldots \underset{k_2}{\overset{k_1}{\rightleftharpoons}} pP + qQ + rR + \ldots \tag{3.18a}$$

The equilibrium constant (K_{eq}) for this reaction can be calculated as follows.

$$K_{eq} = [P]^p[Q]^q[R]^r \ldots /[A]^a[B]^b[C]^c \ldots \tag{3.18b}$$

Reactions can be induced or accelerated by the presence of a **catalyst**, which is itself unaltered by the reaction. The treatment of a catalytic reaction when the catalyst itself actually combines with a reactant is complex. In the special but biologically relevant case of enzyme-catalyzed reactions, the substrate (S) combines with an enzyme (E) to form a product (P) via an intermediate enzyme-substrate complex (ES);

$$E + S \rightleftharpoons ES \rightleftharpoons E + P \tag{3.19a}$$

The rate of the forward reaction (V_f) is $k_1[S][E]$ or $k_1'[S]$ (k_1' is not independent of [E]). The relationship between reaction velocity and [S] for enzyme-

catalyzed biochemical reactions is hyperbolic (Figure 3–4A) if only the rate of the forward reaction is considered (i.e., we assume the rate of the reverse reaction is negligible). The Michaelis-Menten equation describes the hyperbolic curve as follows.

$$V_f = V_{max}.[S]/(K_m + [S]) \tag{3.19b}$$

This relationship indicates a maximal reaction velocity at high (infinite) substrate concentrations, i.e., V_{max}. The **Michaelis-Menten constant** (K_m) is the substrate concentration at $1/2\ V_{max}$. In practice, it is difficult to estimate K_m and V_{max} from a graph of V_f and [S] (e.g., Figure 3–4A), but this hyperbolic relationship can be analyzed in various other ways (Cornish-Bowden 1979). A rearrangement of the Michaelis-Menten equation shows that a graph of $1/V$ against $1/[S]$ yields a straight line that intercepts the $1/[S]$ axis at $-1/K_m$ and intercepts the $1/V_f$ axis at $1/V_{max}$.

$$1/V_f = K_m/V_{max}.(S) + 1/V_{max} \tag{3.19c}$$

This graph is a Lineweaver-Burk plot (Figure 3–4B). Another graphical method for calculation of K_m and V_{max} is the Eadie-Hofstee plot (Figure 3–4C).

$$V_f = V_{max} - \frac{K_m V_f}{[S]} \tag{3.19d}$$

In the Cornish-Bowden plot, V_{max} and K_m are treated as variables, and [S] and V_f are constants.

$$V_{max} = V_f + \frac{V_f K_m}{[S]} \tag{3.19e}$$

The K_m and V_{max} are calculated from the average point of intersection of the various lines for each pair of V_f and [S] values (Figure 3–4D).

In reality, the concentration of reaction products is rarely insignificant and the reverse reaction is not necessarily negligible. The kinetics of the forward and reverse reactions are not independent of each other but reflect the equilibrium constant K_{eq}, which equals $V_{max,f} K_{m,r}/V_{max,r} K_{m,f}$; V_{max} is the maximal velocity and K_m the Michaelis-Menten constants for the forward (f) and reverse (r) reactions. The velocity of the back reaction can be minimized by a large K_{eq} (and a correspondingly large $-\Delta G$) and by $K_{m,r} \gg K_{m,f}$.

The above analysis of chemical reactions based on free energy change and rate of molecular collisions provides a useful basis for determining whether a chemical reaction is energetically favorable and for determining the effects of concentration on reaction rate. Unfortunately, it doesn't necessar-

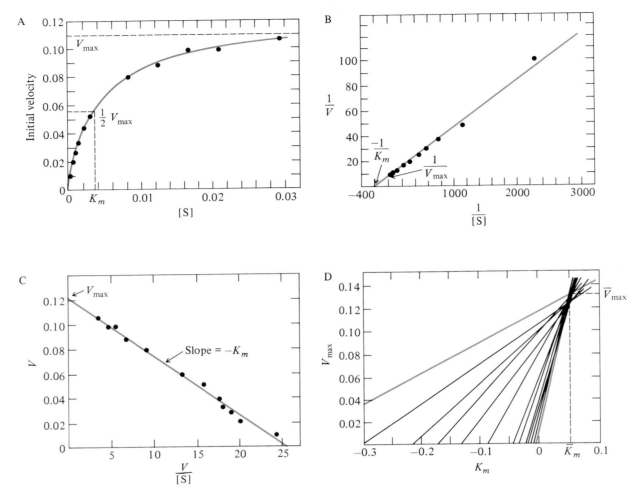

FIGURE 3–4 **(A)** Hyperbolic relationship between initial reaction velocity (V) and substrate concentration ($[S]$) for an enzyme-catalyzed reaction. **(B)** Lineweaver-Burk plot showing the relationship between $1/V$ and $1/[S]$. Note that the $1/V$ intercept is equal to $1/V_{max}$ (since the substrate concentration is infinite) and the $1/[S]$ intercept is $1/K_m$, where K_m is the Michaelis-Menten constant. **(C)** Eadie-Hofstee plot for calculation of K_m and V_{max} from the relationship between V and $V/[S]$. **(D)** Direct linear plot for determination of K_m and V_{max} assuming V and $[S]$ are variables. The median point of intersection of the lines is at the mean K_m ($\overline{K}_m$) and mean V_{max} ($\overline{V}_{max}$).

ily indicate whether a reaction will occur, or at what rate!

Activation Energy

Some energetically favorable reactions (with $\Delta G^{\circ\prime} < 0$) never occur spontaneously. For example, glucose does not spontaneously react with oxygen to form CO_2 and water despite the highly favorable ΔG (-2870 kJ mole^{-1}). This is because the overall reaction requires the initial addition of energy (often

to form an activated transitional complex) before the products are formed and energy is released. Reactions with unfavorable ΔG can also occur if there is an external source of energy, such as heat.

Reactions that proceed through some transitional state, such as an enzyme-substrate complex, are often limited by the rate at which the transitional complex is formed rather than the rate at which the transitional complex forms the products. Each reactant molecule must have a certain amount of energy, the **activation energy**, before it can form the

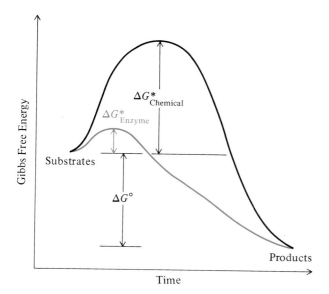

FIGURE 3–5 Representation of the free energy changes for a noncatalyzed chemical reaction ($\Delta G^*_{chemical}$) from the free energy of the reactants (S) to that of the products (P), compared with the lower free energy of activation for an enzyme-catalyzed chemical reaction ΔG^*_{enzyme} from the reactants to products.

activated enzyme-substrate complex ES* (Figure 3–5);

$$E + S \underset{}{\overset{\text{activation}}{\rightleftharpoons}} ES^* \underset{}{\overset{\text{reaction}}{\rightleftharpoons}} E + \text{products} \quad (3.20a)$$

The equilibrium constant for the initial portion of this reaction is

$$K^*_{eq} = [ES^*]/[E][S]$$
$$= e^{-\Delta G^*/RT}. \quad (3.20b)$$

The $K_{eq} = [ES^*]$ at standard conditions when [E] and [S] = 1.

The free energy of activation ΔG^* (or E_α) is $-RT \ln K^*_{eq}$, just as $\Delta G = -RT \ln K_{eq}$ (but note that ΔG^* is not related to ΔG). Thus, K^*_{eq} is equal to $e^{-\Delta G^*/RT}$. The rate constant, k, equals kTK^*_{eq}/h, where k is Boltzmann's constant and h is Planck's constant. Consequently,

$$k = (kT/h)e^{-\Delta G^*/RT}. \quad (3.20c)$$

Thus, the rate constant depends on T (which occurs in the equation twice) and inversely on ΔG^*. Any factor that decreases ΔG^* will increase the rate constant.

Catalysts and **enzymes** increase reaction rates by decreasing ΔG^* or E_a (Figure 3–5). The reduction in E_a is often dramatic. The nonenzyme-catalyzed

reaction of glucose (0.0003 M) with ATP (0.002 M) to form glucose-6-phosphate occurs at a rate of $<10^{-13}$ mol min^{-1} ($k < 1.7 \ 10^{-7}$ liter mole^{-1} min^{-1}) but the presence of an enzyme (hexokinase) can increase the rate to $1.3 \ 10^{-3}$ mol min^{-1} and k to $2.2 \ 10^3$, an increase in rate of more than 10000000000!

The above equation (3.20c) was formulated in the late nineteenth century by Svente Arrhenius in a slightly different form

$$k = k'e^{-A/RT} \quad (3.20d)$$

where k' is a constant (equal to (kT/h)) and A is a constant (equal to the activation energy). The relationship between the reaction rate at two different temperatures is obtained by taking the natural logarithm for k (from Equation 3.20d) for each of two temperatures.

$$\ln k_1/k_2 = -E_a(T_1 T_2)/R(T_1 - T_2) \quad (3.21a)$$

A graph of ln(k) as a function of $1/T$ yields a straight line with a slope of $-E_a$. An example of such an **Arrhenius plot** is shown for a biochemical reaction in Figure 3–6A.

The Q_{10} is also used to describe the effects of temperature on reaction rate. The Q_{10} is the ratio of the reaction rates when the temperature difference is 10° (see Chapter 5) i.e.,

$$Q_{10} = k_{(T+10°)}/k_{(T°)} \quad (3.21b)$$

Arrhenius's empirically derived relationship between reaction rate and temperature can be applied not only to biochemical reactions (when the value of Arrhenius's A constant is E_a) but also to complex biological reactions, such as heart rate, metabolic rate, and respiration rate (e.g., Figure 3–6B). Here, the term E_a is replaced by μ, the critical thermal increment, or "apparent activation energy." The clear analogy between μ and E_a led early physiologists to speculate that complex biological phenomena, such as the heart beat, could be analyzed in terms of a series of chemical reactions, the slowest of which was termed the "master," or "controlling," reaction. Many biological phenomena showed a discontinuity in the Arrhenius plot (see Figure 3–6A and 3–6B, gill rate). This discontinuity was interpreted as a shift in "master reactions" from one with a low μ_1 at high temperatures to one with a high μ_2 at low temperatures. The identification of these "master reactions" was never accomplished, and it seems fruitless to analyze complex biological phenomena as a sum of a series of discrete biochemical reactions. Many other physical and chemical processes, such as diffusive and convective trans-

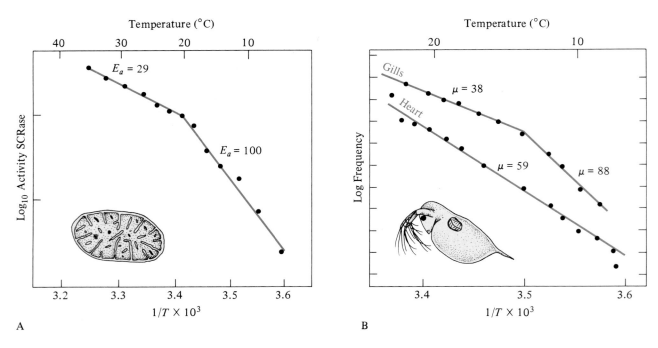

FIGURE 3–6 (A) Arrhenius plot for the effect of temperature on the activity of heart mitochondrial succinate:cytochrome C reductase activity for a marsupial mouse *Sminthopsis crassicaudata*. The break between the high and low temperature curves at 16° C is interpreted as a change in fluidity of the mitochondrial membrane system and the mobility of membrane-bound enzymes. The numbers next to the lines are the activation energies (slope $= -E_a/RT$; E_a, kJ mole^{-1}). The minimal T_b of the marsupial mouse during torpor is about 16° C, and so the E_a is always low (29 kJ mole^{-1}). (*Modified from Geiser and McMurchie 1986.*) (B) Arrhenius plot for the effect of temperature on heart rate and gill beating rate for a water flea *Daphnia*. There is a transition temperature for gill beating but not heart rate. The value of the critical thermal increment μ (slope $= -\mu/RT$; μ, kJ mole^{-1}) is indicated next to the lines. (*Modified from Stier and Wolf 1932.*)

port, viscosity, and membrane structure, are also involved.

Oxidative Metabolism

Cellular metabolic processes are responsible for the synthesis of all of the macromolecules in the body by anabolic metabolism, as well as the breakdown of macromolecules for production of usable energy by catabolic metabolism. Examples of anabolism that we will examine in more detail include synthesis from ammonia of nitrogenous wastes (urea and uric acid) and tissue production (Chapter 4). The free energy (ΔG) released by catabolism is used for a wide variety of cellular processes, including active transport of molecules across membranes; muscle contraction; ciliary and flagellar movement; and the production of heat, electricity and even light. We shall concentrate here on catabolic metabolism.

Cellular reactions typically need about 20 to 40 kJ of energy per mole of reactants. Such energy needs are considerably less than the total energy yield from the complete oxidation of a typical metabolic substrate. For example, oxidation of glucose to carbon dioxide and water has a $\Delta G^{\circ\prime}$ of -2870 kJ mole^{-1}. Cellular metabolism of macromolecules such as glucose must convert a significant fraction of the ΔG to smaller amounts of chemical energy (20 to 40 kJ mole^{-1}) that can be used by cells. High-energy phosphate compounds (phosphagens) such as ATP are chemical energy stores of 20 to 40 kJ mole^{-1}. These high-energy phosphate compounds are essentially the "energy currency" that the cells "spend" to do useful work.

High-Energy Phosphate Compounds

Animal cells contain a variety of high-energy phosphate compounds, or **phosphagens**. The $\Delta G^{\circ\prime}$ for hydrolysis of the various phosphagens common in

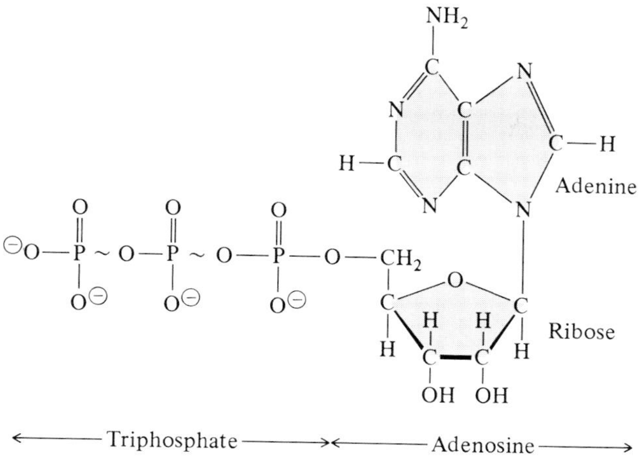

Adenosine Triphosphate (ATP)

FIGURE 3–7 Structure of adenosine triphosphate (ATP), the "energy currency" of the cell.

animal cells ranges from about -10 to -50 kJ mole^{-1}.

Adenosine triphosphate (ATP) is the most common high-energy phosphate molecule. It is a complex molecule containing a purine base (adenine), a five-carbon sugar (ribose), and three phosphate groups linked to the ribose by high-energy ester and acid-anhydride bonds (Figure 3–7). The adenine + ribose is called adenosine.

The free energy released by hydrolysis of the terminal phosphate of ATP to form adenosine diphosphate (ADP) and inorganic phosphate (P_i) is 30.5 kJ mole^{-1} under standard biological conditions.

$$ATP \rightleftharpoons ADP + P_i \quad \Delta G^{\circ \prime} = -30.5 \text{ kJ mole}^{-1} \tag{3.22}$$

However, it is important to appreciate that ATP is not generally present in animal cells at standard conditions, and so the actual amount of energy released by hydrolysis of one mole of ATP can differ substantially from $\Delta G^{\circ \prime}$. A high ratio of (ATP)/(ADP), as commonly occurs in cells, results in ΔG values more negative than $\Delta G^{\circ \prime}$, often -40 to -60 kJ mole^{-1}. This makes even more energy available for useful work than under standard conditions. The hydrolysis of ATP is readily reversible, provided the necessary free energy is available. Consequently, there is a cyclic formation of ATP from ADP (by cellular metabolism) and subsequent breakdown of ATP for energy-requiring processes.

The energy released by hydrolysis of ATP can be used in energy-requiring reactions. For example,

the phosphorylation of glucose to form glucose-6-phosphate requires ATP, since the $\Delta G^{\circ \prime}$ is unfavorable.

$$\text{Glucose} + P_i \rightleftharpoons \text{glucose-6-phosphate}$$
$$\Delta G^{\circ \prime} = +13.8 \text{ kJ mole}^{-1} \tag{3.23a}$$

Coupling the reaction with hydrolysis of ATP produces an overall $\Delta G^{\circ \prime}$ (about -16 kJ mole^{-1}), which not only makes the reaction favorable but essentially makes it irreversible.

$$\text{glucose} + ATP \rightleftharpoons \text{glucose-6-phosphate}$$
$$+ ADP + P_i \tag{3.23b}$$
$$\Delta G^{\circ \prime} = -16.7 \text{ kJ mole}^{-1}$$

This coupled reaction is, in fact, the first step in the metabolism of glucose (glycolysis).

ATP is not the only high-energy phosphate molecule that is used for the storage of chemical energy. ADP and AMP also store significant amounts of chemical energy in cells. Creatine phosphate and arginine phosphate are commonly used by animal cells to store chemical energy. For example, vertebrate muscle cells contain high concentrations of creatine phosphate (about 17 mM) compared to ATP (about 6 mM). Hydrolysis of creatine phosphate (CP) makes the regeneration of ATP from ADP energetically favorable.

$$\begin{array}{lc} & \Delta G^{\circ \prime} \\ ADP + P_i \rightleftharpoons ATP & 30.5 \\ CP \rightleftharpoons C + P_i & -43.1 \\ \hline ADP + CP \rightleftharpoons ATP + C & -12.6 \end{array} \tag{3.24a}$$

Similarly, invertebrates such as cephalopods store energy as arginine phosphate, which can be used to regenerate ATP

$$\text{arginine phosphate} + ADP \rightleftharpoons \text{arginine} + ATP \tag{3.24b}$$

with an overall $\Delta G^{\circ \prime}$ of -6.3 kJ mole^{-1}.

ATP is the source of energy for most cellular reactions, but other phosphagens are used to store energy. But why isn't energy stored as ATP? First, only a small amount of adenosine is present in cells and it is constantly being cycled from ADP to ATP by metabolic processes; ATP storage would require much higher levels of adenosine. Second, accumulation of high concentrations of ATP would greatly alter the ATP/ADP and ATP/AMP ratios in the cell and unfavorably increase the free energy requirements for ATP synthesis. Third, storage of energy as another phosphagen, e.g., creatine phosphate, allows energy transfer during metabolic demand to form ATP at a much higher rate than would be possible by aerobic or anaerobic metabolism.

Metabolic Substrates

Carbohydrates. Carbohydrates are an important, commonly occurring group of organic compounds, based on the empirical formula $(CH_2O)_n$ where n is an integer greater than 2. Carbohydrates are of central significance to animals. Essentially all organic foods are ultimately derived from the photosynthetic synthesis of carbohydrates by plants. Carbohydrates have structural roles (e.g., chitin, membrane constituents) and are a source of metabolic energy in animals (e.g., liver glycogen stores).

There are three basic kinds of carbohydrates: monosaccharides, oligosaccharides, and polysaccharides. Monosaccharides contain three to nine carbon atoms. Oligosaccharides typically contain two to ten monosaccharides. Polysaccharides are highly polymerized carbohydrates, containing more than ten monosaccharides, and frequently many thousands. Structures of some important monosaccharides, oligosaccharides, and polysaccharides are shown in Figure 3–8.

Lipids. Lipids are a complex, heterogeneous assemblage of oily, waxy, or greasy compounds that are extractable from animal tissues by nonpolar solvents, such as chloroform and benzene (Figure 3–9).

Free fatty acids are generally monocarboxylic acids, with an even numbered and unbranched carbon chain. Saturated fatty acids contain only single bonds, e.g., acetic, propionic, butyric, palmitic and stearic acids. The carbon chain of unsaturated fatty acids has double bonds, often between C_9 and C_{10} (e.g., oleic acid) or multiple double bonds (e.g., linoleic acid). Some fatty acids have a cyclic arrangement of the carbons, or OH groups (other than that of the carboxyl COOH). Fatty acids are rarely found free but are generally combined with other molecules. Fatty acids are the fundamental components of many structural lipids (e.g., biological membranes and fat stores); they can be important metabolic substrates.

Glycerides are esters of fatty acids and the carbohydrate glycerol (CH_2OH—$CHOH$—CH_2OH). Mono-, di-, and triglycerides are glycerol combined with one, two, or three fatty acids (generally myristic, palmitic, stearic, palmitoleic, oleic, and linoleic acids). Triglycerides (neutral fats) are important lipid stores (e.g., adipose tissue).

Phospholipids are derivatives of glycerol phosphate, based on the general structure of a diglyceride-phosphate (phosphatidic acid). The PO_4^{3-} group makes these lipids water soluble, an important property for structural lipids in cell membranes.

Phosphatides have the phosphate combined via an ester linkage to choline (phosphatidyl cholines, or lecithins), ethanolamine (phosphatidyl ethanolamines), or serine (phosphatidyl serines). Phosphoinositols are similar, but contain an inositol group rather than choline, ethanolamine, or serine. Phosphoglycerols have two phosphate groups, one at each terminal C of glycerol.

Sphingolipids, which include ceramides, sphingomyelins, and glycosphingolipids, contain a long chain aliphatic base of either sphingosine or dihydrosphingosine.

Terpenes generally contain multiples of five-carbon fragments resembling isoprene (CH_2=CCH_3—CH=CH_2). Important animal terpenes and related compounds include squalene (a precursor for cholesterol synthesis); carotenes (pigments): vitamins A, E, and K; and CoQ.

Waxes are esters of fatty acids and alcohols; any naturally occurring fatty acid ester of any alcohol (except glycerol) is called a "wax." Spermaceti, the "head oil" of the sperm whale, contains the wax cetyl palmitate, formed from cetyl alcohol (CH_3—$(CH_2)_{14}$—CH_2OH) and palmitic acid. Beeswax contains myricyl palmitate.

Steroids are important as structural lipids (e.g., cholesterol is an important lipid constituent of cell membranes), hormones (e.g., androgens and estrogens are C_{18} and C_{19} steroids; progesterone and adrenal cortex hormones are C_{21} steroids; ecdysone, a molting hormone is a C_{28} steroid), vitamins (e.g., D and its active derivatives), and detergents (e.g., bile salts) are C_{24} steroids.

Proteins. Proteins are large macromolecules, formed from 20 common amino acids. They are extremely important in cellular structures (e.g., connective tissue, collagen; skin, keratin) and for physiological function (e.g., muscle contraction, enzymes).

The general structure of an amino acid is

$$H_2N-\underset{R}{\overset{H}{C}}-COOH$$

where R varies from H (e.g., glycine) to complex cyclic carbon and nitrogen groups. The structures of neutral, aromatic, imino, sulfur-containing, basic, dicarboxylic, and other amino acids are presented in Figure 3–10 on pages 46–47.

Aerobic Metabolism

A primary role of cellular metabolism is the provision of energy as ATP to provide the energy required for other cellular processes. Carbohydrate is an

(Text continues on page 44)

Monosaccharides

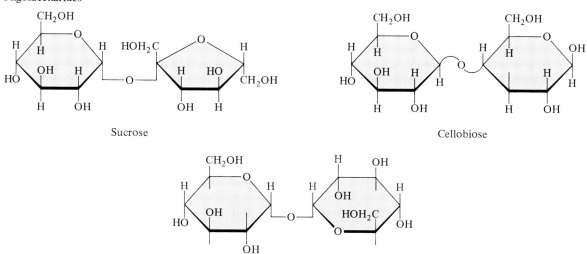

Oligosaccharides

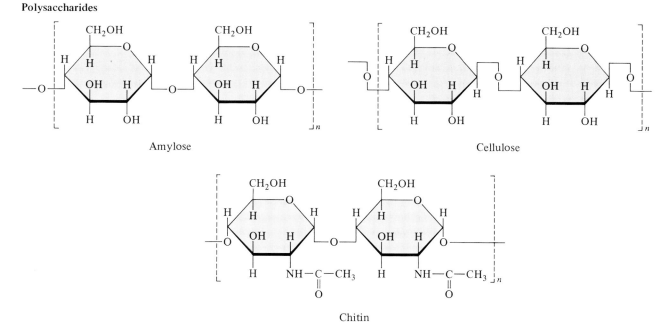

FIGURE 3–8 Structures of various carbohydrates of the general chemical formula $(CH_2O)_n$.

important energy source for ATP synthesis, although lipid stores and body proteins are also used to synthesize ATP. Metabolism of carbohydrates, lipids, and proteins often involves oxidation by oxygen; this is **aerobic metabolism**.

Carbohydrate Metabolism. Glucose and other carbohydrates are a common metabolic substrate for the cellular metabolic pathways that provide energy in the form of ATP. The aerobic metabolic pathways for glucose metabolism are also used for lipid and protein metabolism. Consequently, we shall first study the cellular metabolic pathways for glucose metabolism.

Glycolysis. The first steps in glucose metabolism involve its conversion to pyruvate via a series of biochemical reactions, the Embden-Meyerhof pathway (Figure 3–11, page 48). The final conversion of pyruvate to lactate completes the process of **glycolysis**. These reactions occur in the cell cytosol (i.e., are extramitochondrial). No oxygen is used during glycolysis and no CO_2 is liberated.

Glucose is first converted to glucose-6-phosphate, at the expense of 1 ATP/glucose. Glucosyl subunits of polysaccharides, such as glycogen, can be converted to glucose-6-phosphate by the addition of P_i; ATP is not required. Embden-Meyerhof reactions with large positive $\Delta G^{\circ\prime}$ are essentially irreversible and require ATP to be hydrolyzed to ADP (e.g., glucose to glucose-6-phosphate; fructose-6-phosphate to fructose-1,6-diphosphate). Steps with large, negative $\Delta G^{\circ\prime}$ allow ATP generation from ADP (e.g., 1,3-diphosphoglycerate to 3-phospho-

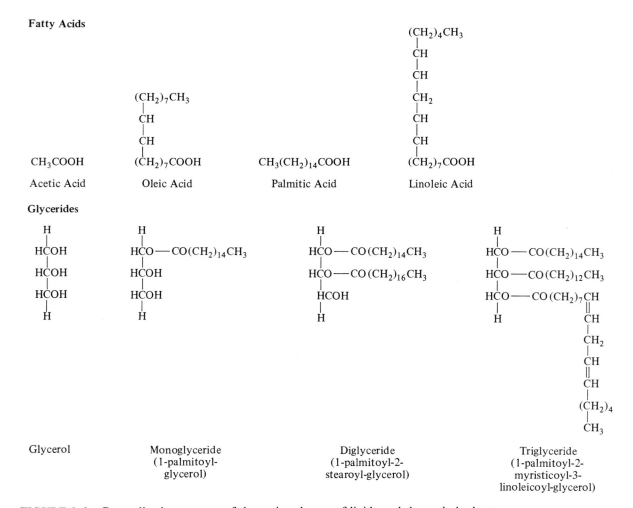

FIGURE 3–9 Generalized structures of the major classes of lipids and the carbohydrate glycerol.

(Text continues on page 47)

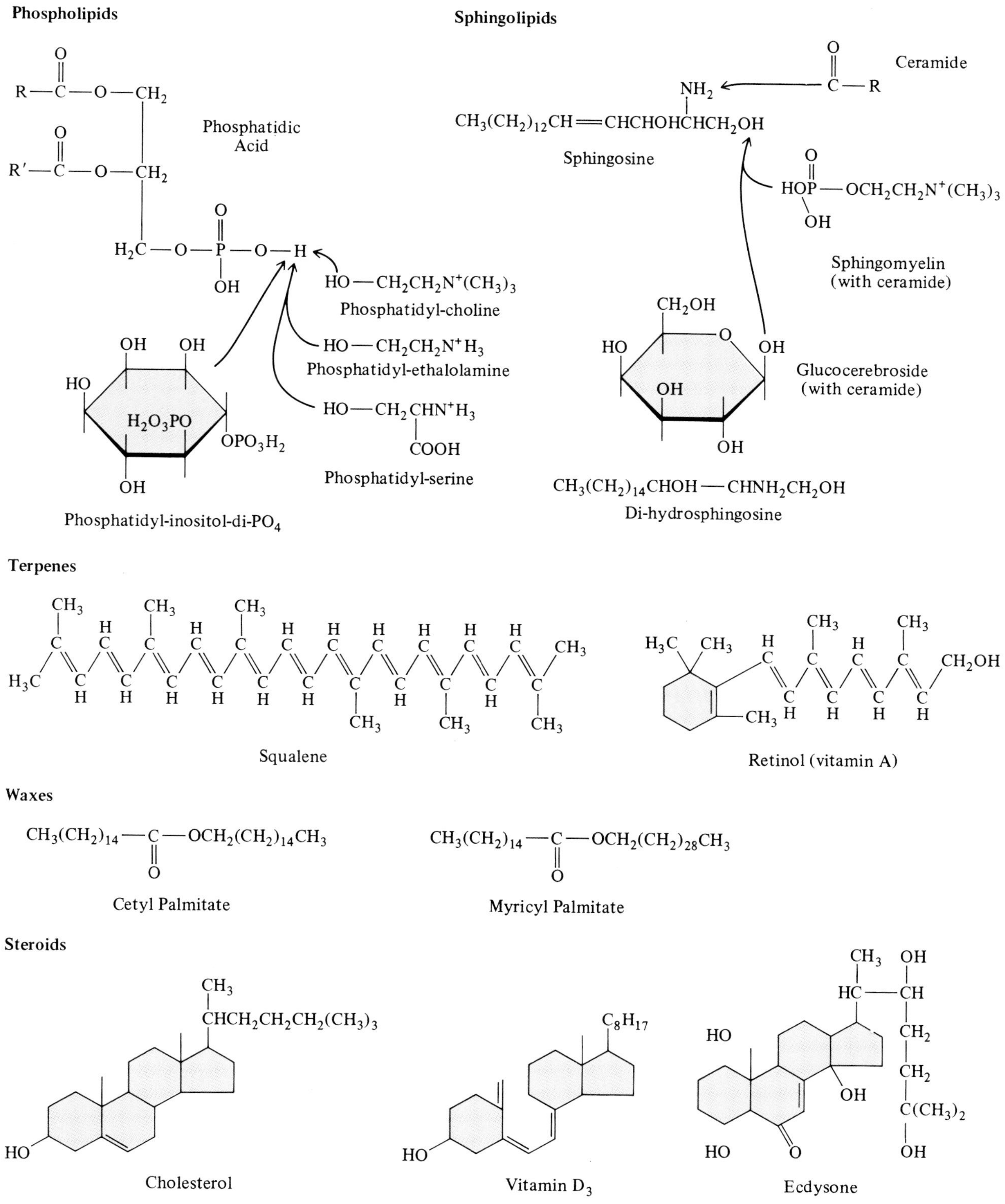
Phospholipids
Phosphatidic Acid
HO—CH2CH2N+(CH3)3
Phosphatidyl-choline
HO—CH2CH2N+H3
Phosphatidyl-ethalolamine
HO—CH2CHN+H3
COOH
Phosphatidyl-serine
H2O3PO
OPO3H2
Phosphatidyl-inositol-di-PO4
Sphingolipids
Ceramide
CH3(CH2)12CH=CHCHOHCHCH2OH
Sphingosine
HOP—OCH2CH2N+(CH3)3
Sphingomyelin
(with ceramide)
Glucocerebroside
(with ceramide)
CH3(CH2)14CHOH—CHNH2CH2OH
Di-hydrosphingosine
Terpenes
Squalene
Retinol (vitamin A)
Waxes
CH3(CH2)14—C—OCH2(CH2)14CH3
Cetyl Palmitate
CH3(CH2)14—C—OCH2(CH2)28CH3
Myricyl Palmitate
Steroids
CHCH2CH2CH2(CH3)3
Cholesterol
C8H17
Vitamin D3
C(CH3)2
Ecdysone

FIGURE 3–9 *Continued*

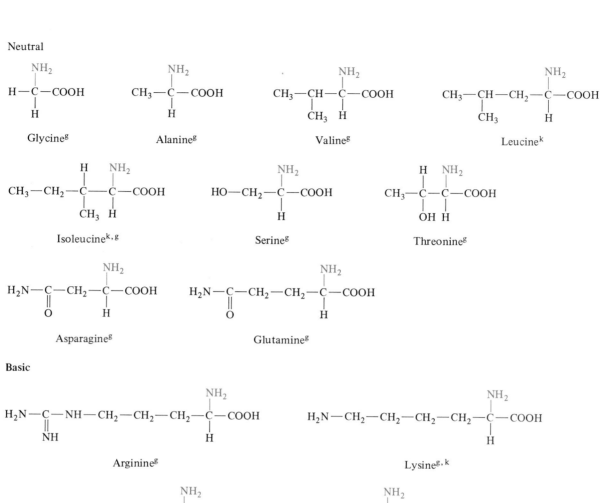

Neutral

Glycine[g]   Alanine[g]   Valine[g]   Leucine[k]

Isoleucine[k, g]   Serine[g]   Threonine[g]

Asparagine[g]   Glutamine[g]

Basic

Arginine[g]   Lysine[g, k]

Hydroxylysine[g, k]   Histi.

Aromatic

Phenylalanine[k, g]   Tyrosine[k, g]   Tryptophan[g]

FIGURE 3–10   Structure of amino acids showing the amino group ($-NH_2$) and the structures of the R-group for 26 amino acids and two important related biochemicals, ornithine and thyroxine. The metabolic fate of the amino acid is glucogenic (g), ketogenic (k), or both (k, g).

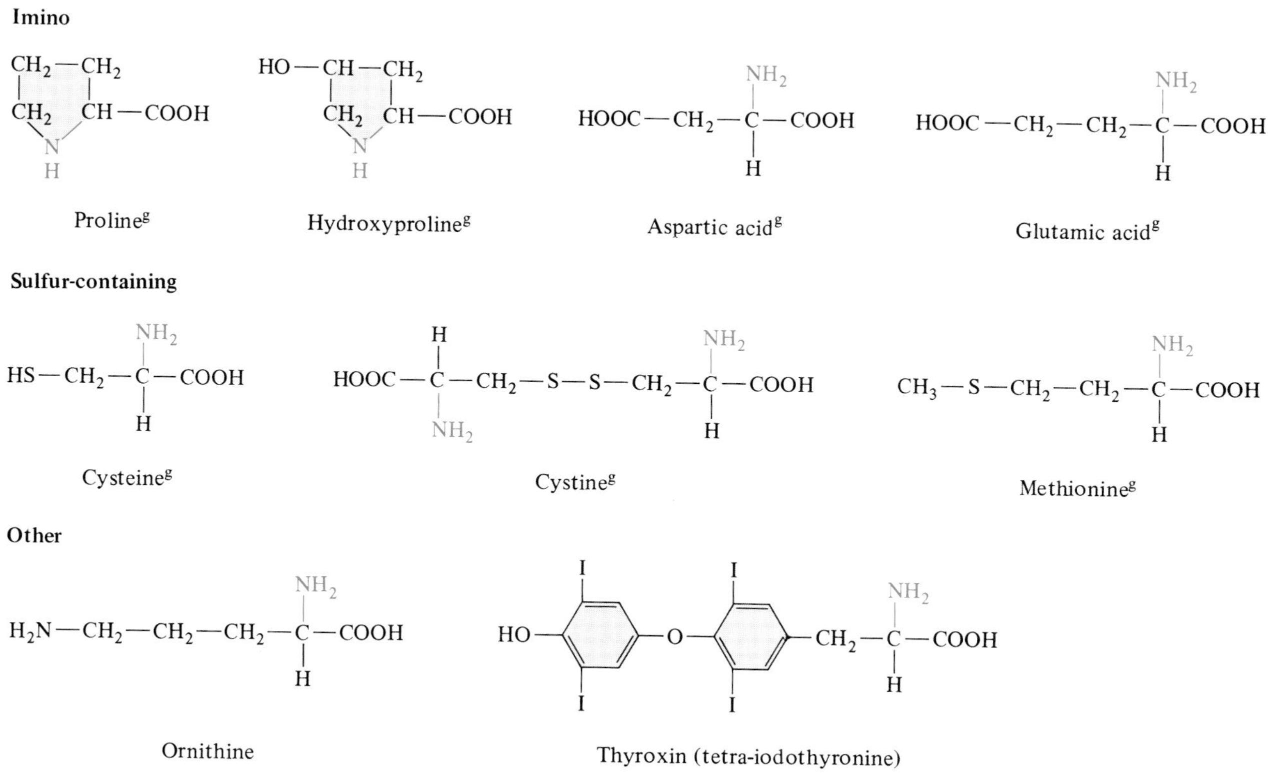

Imino

Proline[g] Hydroxyproline[g] Aspartic acid[g] Glutamic acid[g]

Sulfur-containing

Cysteine[g] Cystine[g] Methionine[g]

Other

Ornithine Thyroxin (tetra-iodothyronine)

FIGURE 3–10 *Continued*

glycerate; phosphoenolpyruvate to pyruvate). Most of the other steps in glycolysis have small negative or positive $\Delta G^{\circ\prime}$ values and are reversible.

The sum of the $\Delta G^{\circ\prime}$ values for the Embden-Meyerhof pathway (including formation of two moles ATP from ADP per mole glucose) is -72.8 kJ mole^{-1} glucose^{-1} (Table 3–4). The $\Delta G^{\circ\prime}$ is -134 kJ mole^{-1} if the two moles of ATP are hydrolyzed to ADP (at 31 kJ mole ATP^{-1}). Thus, 72.8 kJ mole^{-1} of the total 134 kJ mole^{-1} energy released by metabolism of one mole glucose to two moles of pyruvate is given off as heat, and 61.2 kJ mole^{-1} is associated with ATP synthesis from ADP. The efficiency is $100 \times (61.2/134)$, or 46% for glucose metabolism via the Embden-Meyerhof pathway. Metabolism of one glucosyl subunit to two pyruvates has an overall

TABLE 3–4

Summary of Gibbs free energy changes for glucose and glucosyl subunit metabolism by glycolysis. Values for $\Delta G^{\circ\prime}$ are kJ mole^{-1}. The $\Delta G^{\circ\prime}$ for ATP/GTP hydrolysis is assumed to be -30.5 kJ mole^{-1}.

		$\Delta G^{\circ\prime}$
Glucose + 2 P$_i$ + 2 ADP + 2 NAD$^+$ $\longrightarrow$	2 pyruvate + 2 ATP + 2 NADH/H$^+$	-72.8
Glucose $\longrightarrow$	2 pyruvate	-134
Glucose + 2 P$_i$ + 2 ADP $\longrightarrow$	2 lactate + 2 ATP	-123
Glucose $\longrightarrow$	2 lactate	-184
Glucosyl + 3 P$_i$ + 3 ADP + 2 NADH/H$^+$ $\rightarrow$	2 pyruvate + 3 ATP + 2 NADH/H$^+$	-61
Glucosyl $\longrightarrow$	2 pyruvate	-152

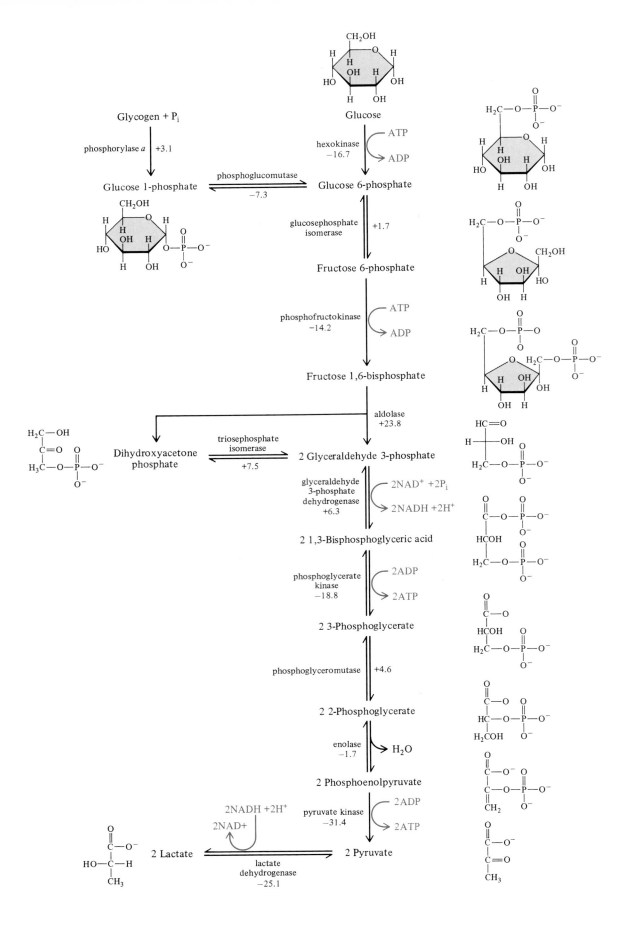

$\Delta G°'$ of -152 kJ mole^{-1} if the ATP is hydrolyzed to ADP, or -61.0 with ATP as an end product; the efficiency is about 60%.

If the pyruvate is converted to lactate, then the total $\Delta G°'$ is -184 kJ mole glucose^{-1} (or -123 kJ mole^{-1} with ATP formation); the efficiency is about 34%.

Two hydrogen atoms are transferred from glyceraldehyde-3-phosphate to an electron acceptor, nicotinamide adenine dinucleotide (NAD$^+$), during glycolysis. NAD$^+$ contains nicotinamide and adenine (Figure 3–12). It accepts one H$^+$ and two e$^-$ from two H to form NADH; the remaining H$^+$ enters the cell proton pool. We shall use the nomenclature NADH/H$^+$ to remind us of this additional H$^+$, since only one H is chemically associated with NAD$^+$. NAD$^+$ is present in cells at low concentrations, and so the regeneration of NAD$^+$ from NADH/H$^+$ is necessary for continued glycolysis. As we shall see below, the NAD$^+$ can be regenerated from NADH/H$^+$ with concomitant ATP formation by the electron transfer chain, or can be regenerated by a variety of special anaerobic metabolic pathways. In glycolysis, for example, pyruvate is converted to lactate and NAD$^+$ is regenerated. The conversion of pyruvate to lactate is readily reversible, depending on the ratio of pyruvate to lactate. The formation of lactate from pyruvate has a very favorable $\Delta G°'$ (-25.1 kJ mole^{-1}).

The total free energy change for complete metabolism of glucose to carbon dioxide and water is -2854 kJ mole^{-1}. Of this, glycolysis has made available only 61 kJ mole^{-1} as chemical energy (2 ATP), i.e., only 2.1% of the total energy change has been converted to usable energy and only 4.7% of the total energy has been released. Consequently, it is not surprising that, because of the low energy yield, animal metabolism of glucose seldom ends at the final steps of glycolysis.

Citric Acid Cycle. An alternative metabolic fate of pyruvate, which provides further ATP formation, is its conversion to acetyl-CoA rather than lactate. The acetyl-CoA is then converted to carbon dioxide and water by the **citric acid cycle** (or tricarboxylic acid cycle or Krebs cycle). Let us first examine these metabolic pathways, then examine their energetic yield.

Pyruvate is condensed with coenzyme A (CoA) by the enzyme pyruvate dehydrogenase to form acetyl-CoA and carbon dioxide. NAD$^+$ is converted to NADH/H$^+$ in the condensation reaction (Figure 3–13). The large negative $\Delta G°'$ of -33.5 kJ mole^{-1} makes the reaction essentially irreversible. The acetyl-CoA enters a cycle of biochemical reactions, starting by combining with oxaloacetate to form citrate, and ending with the formation of oxaloacetate from malate. Each of the two carbons entering

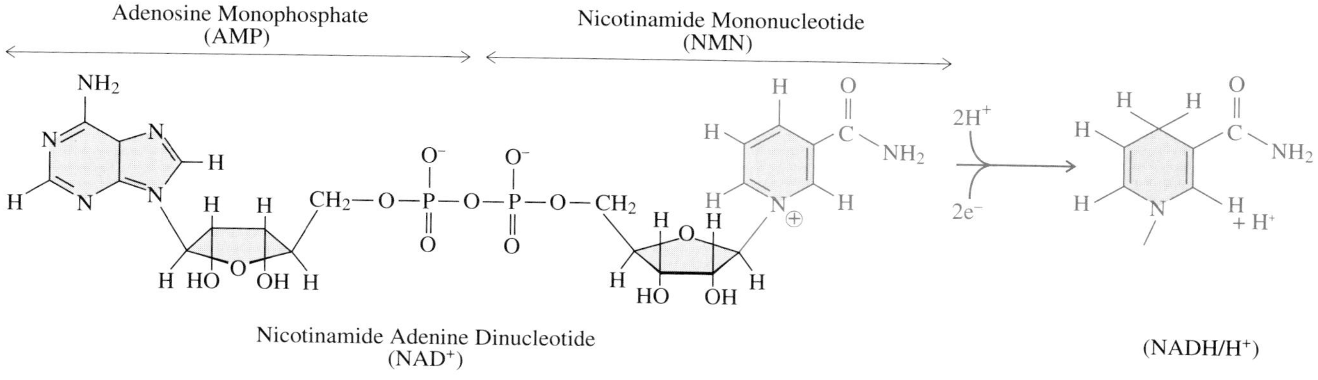

Adenosine Monophosphate (AMP)

Nicotinamide Mononucleotide (NMN)

Nicotinamide Adenine Dinucleotide (NAD$^+$)

(NADH/H$^+$)

FIGURE 3–12 Structure of NAD$^+$ and the combination of the NMN portion with 2 H$^+$ and 2 e$^-$ to form NADH/H$^+$.

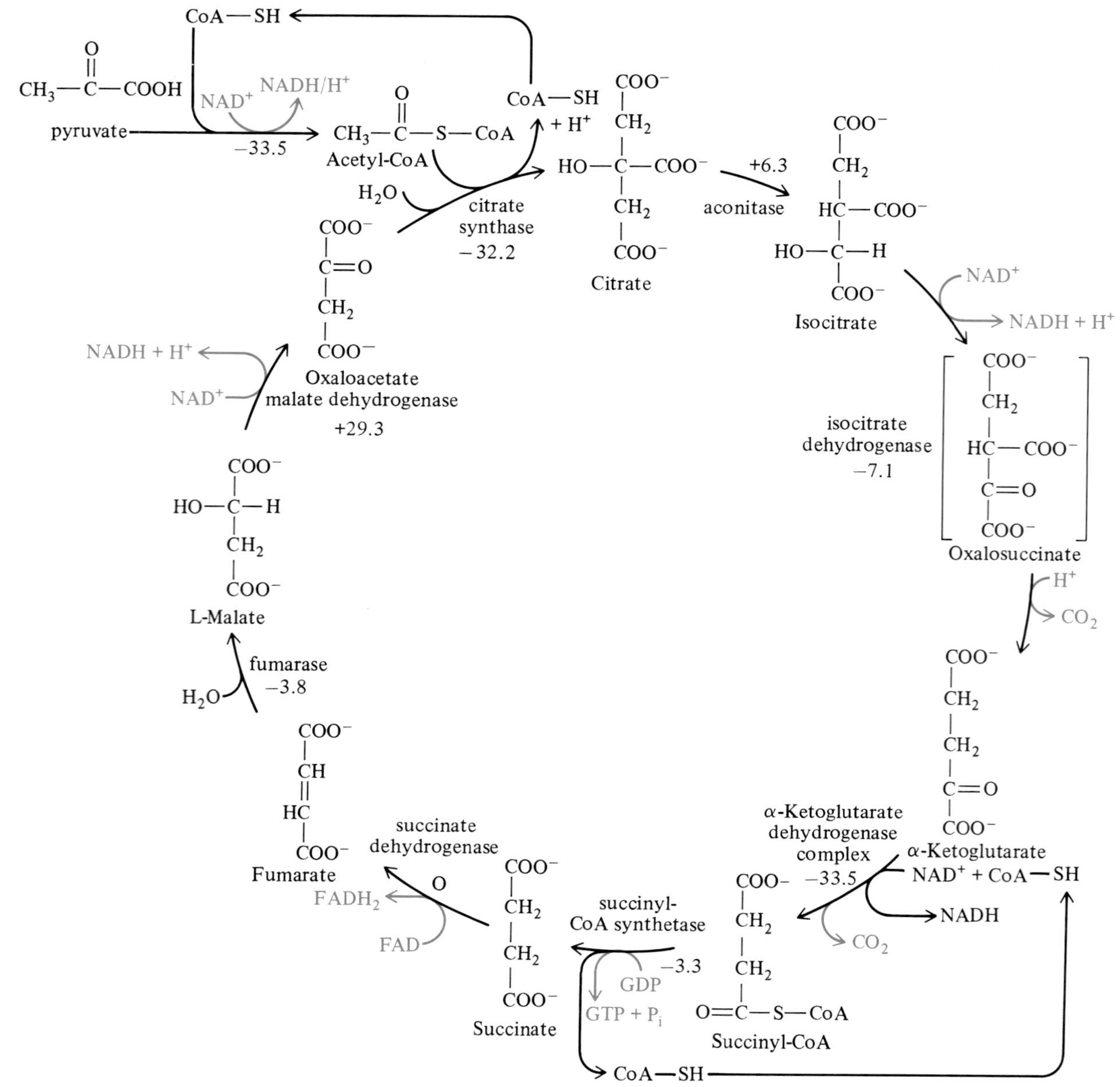

FIGURE 3–13   Schematic pathway for oxidation of acetyl-CoA by the citric acid cycle showing individual reactions, Gibbs free energy changes ($\Delta G^{\circ\prime}$; kJ mole^{-1}), chemical structures, and overall energetics.

the citric acid cycle as acetyl-CoA is released as carbon dioxide. There is only one step in the cycle in which a high-energy phosphate compound is formed. The conversion of succinyl-CoA to succinate and CoA is coupled to the formation from GDP of GTP (guanosine-triphosphate), and the high-energy phosphate bond of GTP can be readily transferred to form ATP from ADP. There are three steps during the citric acid cycle in which an NAD$^+$

accepts two H to form NADH/H$^+$, and one step in which another proton acceptor, flavin adenine dinucleotide (FAD), is converted to FADH$_2$.

The overall reaction from pyruvate through a complete turn of the citric acid cycle is

$$\text{pyruvate} + 4\text{NAD}^+ + \text{FAD} + \text{GDP} + \text{P}_i$$
$$\rightarrow 2\text{CO}_2 + 4\text{NADH/H}^+ + \text{FADH}_2 + \text{GTP}$$

$$(3.25)$$

with a $\Delta G°'$ of -78 kJ mole^{-1}. Two pyruvate molecules are formed per glucose, so 2 ATP are generated by the citric acid cycle per glucose, and additional hydrogen atoms and e$^-$ are transferred to form 8 NADH/H$^+$ and 2 FADH$_2$. The final steps in the metabolic pathway are the regeneration of NAD$^+$ and FAD by the electron transfer system and further ATP synthesis.

Electron Transfer System. Glycolysis and the citric acid cycle convert only a small fraction of the total possible energy yield from oxidation of glucose to ATP, but ten NADH/H$^+$ and two FADH$_2$ are also formed per glucose. Regeneration of the NADH/H$^+$ to NAD$^+$ and FADH$_2$ to FAD is necessary if the metabolic pathways are to operate continuously. Furthermore, the fates of NADH/H$^+$ and FADH$_2$ in the mitochondrion account for most of the ATP synthesized by oxidation of glucose (about 32 ATP/glucose). Mitochondria are the "powerhouses" of animal cells and provide 95% of the ATP obtained by aerobic metabolism of glucose (glycolysis provides the other 5%). The mitochondrion is a double membrane cell organelle, with the inner membrane folded to form cristae (Figure 3–14A).

Cells of most, but not all, eukaryotic organisms (which include animals, plants, and fungi) contain mitochondria. However, more than 1000 species of protozoans, and some fungi, lack mitochondria (Cavalier-Smith 1987). Mitochondria are thought to have arisen from a symbiotic relationship between a host cell and small aerobic bacteria (Margulis 1975). The host cells evolved into eukaryote cells, and the aerobic bacteria became mitochondria. The host cell may well have been a swimming, biciliate metamonad protozoan similar to *Giardia* (an important cause of dysentery in humans). The prokaryote endosymbiont that became the mitochondria was probably a purple bacteria of the α-subdivision (which includes rhizobacteria, agrobacteria, and rickettsias). Significantly, all of these α-division prokaryotes have close relationships with eukaryote cells (Yang et al. 1985). Other eukaryote cell organelles are also thought to have arisen by similar endosymbiosis, e.g., chloroplasts (plastids), flagella, and perhaps peroxisomes. The manner of ATP synthesis by aerobic bacteria, which have a similar electron transport system as the animal mitochondrion, supports the theory for an endosymbiotic bacterial origin of mitochondria.

The **chemiosmotic theory** describes the mechanism for coupling of NAD$^+$ regeneration from NADH/H$^+$ and FAD from FADH$_2$ with ATP formation in the mitochondrion (Mitchell 1979; Elthon and Stewart 1983; Hinckle and McCarty 1978). A similar process occurs in aerobic bacteria and plant chloroplasts. The coupling is accomplished by the establishment of a H$^+$ concentration gradient (hence a chemical potential and osmotic gradient, or chemiosmotic gradient) across the inner mitochondrial

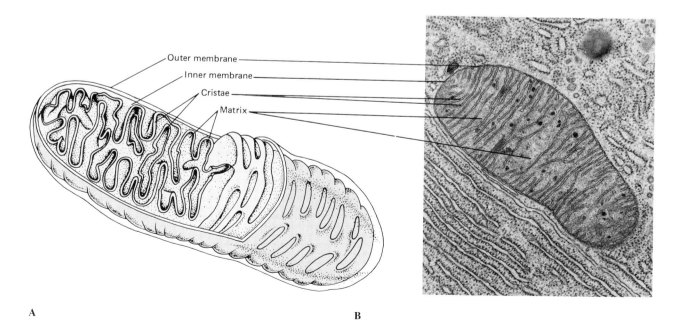

A   **B**

FIGURE 3–14 (A) Three-dimensional representation and **(B)** electron micrograph of a mitochondrion. *(Photograph by Keith Porter.)*

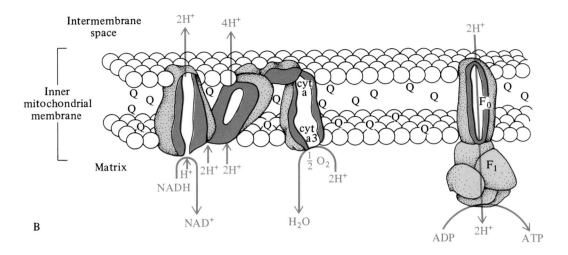

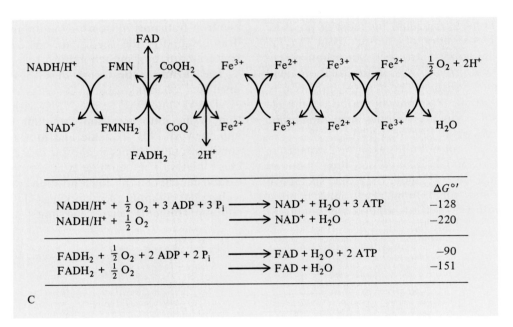

FIGURE 3–14 (B)   Schematic outline of the electron transport system showing the electron-shuttling role of FMN, coenzyme Q, and the cytochromes; structure of the inner mitochondrial membrane showing the role of the electron transfer system in pumping a net six H^+ from the mitochondrial matrix into the intermembrane space, thereby establishing a 12 H^+ gradient and the overall energetics of NADH/H^+ and $FADH_2$ oxidation in mitochondria. ATP is synthesized by the passive flux of H^+ through the Fo/Fi protein complex. **(C)** Summary of electron transfer along the electron transfer system and the role of O_2 as the terminal e^- acceptor.

membrane between the matrix and the intermembrane space. The "pumps" that move H^+ from the matrix to the intermembrane space are associated with a complex of cytochromes (a, a_3, b, c, c_1), coenzymes (ubiquinone, or CoQ), iron sulfide containing proteins, copper ions, and a flavoprotein (flavin mononucleotide, FMN; Figure 3–14B).

In the mitochondrial matrix, NADH/H^+ is oxidized to NAD^+, and the two protons (H^+) and two electrons (e^-) are transferred to the FMN molecule (Figure 3–14B,C). The FMN molecule extends through the inner mitochondrial membrane, and the two H^+ are transported into the intermembrane space. The two electrons are transferred separately (i.e., one at a time) along a series of iron-sulfur proteins. In this way, the FMN is regenerated. Ubiquinone (CoQ) is a very important component in the electron transfer steps because it is responsible for further transport of H^+ from the mitochondrial matrix to the intermembrane space. Two CoQ molecules accept one e^- from the iron sulfur proteins, one e^- from a cytochrome b, and pick up two H^+ from the matrix to form two $CoQH_2$ (dihydroquinone). These $CoQH_2$ then migrate through the inner membrane towards the intermembrane space where they each donate an e^- to a cytochrome c_1 and release a H^+ into the intermembrane space and form CoQH (a semiquinone). The two CoQH then release their last H^+ into the intermembrane space and transfer the e^- to cytochrome b (from where one e^- originally came), to regenerate CoQ. The e^-'s are then transferred from cytochrome c_1 to cytochrome c, then to cytochromes a and a_3. The cytochrome a_3 and its e^- combine with molecular oxygen ($1/2\ O_2$) to form $1/2\ O_2^-$. Addition of another e^- and two H^+ to O_2^- forms water (H_2O). The above described sequence of e^- transfer and H^+ transport from the matrix to the intermembrane space moves six H^+ from the mitochondrial matrix into the intermembrane space (four by CoQ and two from NADH/H^+). The net change in the H^+ gradient is therefore 12 H^+ (2 × 6).

No ATP is synthesized by the process of electron transfer, but much of the chemical energy of the NADH/H^+ and $FADH_2$ has been converted into the chemical/osmotic energy of a H^+ concentration gradient across the inner mitochondrial membrane. This chemical energy is now used by a special ATPase enzyme, ATP synthetase (or F_0-F_1 protein complex) to synthesize ATP. The F_0 part of the protein complex forms a hydrophobic channel for H^+ to flow from the intermembrane space to the matrix (the direction of flow from the intermembrane space to the matrix is determined by the direction of the H^+ gradient). The F_1 part of the complex has five subunits, the largest two having ATPase activity. The process by which movement of H^+ through the F_0 protein results in ATP synthesis is unclear. Movement of two H^+ might remove an oxygen molecule from P_i to form H_2O, with the O-less P_i combining with ADP to form ATP. Alternatively, the movement of H^+ through the F_0 might cause a conformational change in the F_1 complex, which overcomes the unfavorable ΔG for ATP synthesis from ADP and P_i. Interestingly, the coupling of H^+ flux through the F_0 with ATP synthesis is reversible; a flux of H^+ from the matrix to the intermembrane space converts ATP to ADP! Regardless of the actual mechanism, it is thought

that 1 ATP may be synthesized by transfer of as few as two H^+ through the F_0-F_1 complex.

In a typical mitochondrion, there is about a 10 × difference in H^+ concentration between the intermembrane space (high [H^+]) and the matrix (low [H^+]); the pH difference is about 1 unit (pH = $-\log$ [H^+]). There consequently is a difference in transmembrane potential (E_m) of about 60 mV due to H^+ ($E_m = 60\Delta pH$; this is a modified form of the Nernst equation). There is an additional E_m of about 160 mV due to transmembrane concentration gradients for other ions. Thus, the total electrochemical gradient causing H^+ movement from the intermembrane space to the matrix (the so-called proton-motive force) is about 220 mV. The potential useful energy released by moving one mole of H^+ down an electrical gradient of 220 mV is about 0.22 × 6.02 10^{23} electron volts or 21.2 kJ. Six moles of H^+ are transported across the inner mitochondrial membrane per mole of NADH/H^+ oxidized, so the potential energy yield is 6 × 21.2 = 127 kJ mole^{-1} of NADH/H^+, or 42.4 kJ per 2 H^+. If two H^+ must pass through the F_1 to form one ATP, then the energy yield is about 42.4 kJ/2 moles H^+. It would appear to be thermodynamically possible to synthesize 1 ATP per 2 H^+, or 3 ATP per NADH/H^+, since the $\Delta G^{\circ\prime}$ for ATP synthesis under physiological conditions is about 42 to 50 kJ mole^{-1}.

Transfer of protons and electrons from $FADH_2$ to the intermembrane FMN does not result in the movement of two H^+ from the $FADH_2$ into the intermembrane space; only the four H^+ (from the two $CoQH_2$) are moved across the inner mitochondrial membrane. Consequently, probably only 2 ATP are synthesized per $FADH_2$.

The basic concept of ATP synthesis by a proton motive force, and the general validity of the chemiosmotic theory, are well accepted and understood. However, the specific details of electron transfer in the mitochondrial membrane and, in particular, the exact stoichiometry of ATP synthesis (i.e., 1 ATP per 2 H^+ or 1 ATP per 3 H^+?) are controversial.

Within the cell there is a complex partitioning of the locations for various parts of the glycolytic and citric acid pathways. Glycolysis proceeds as far as pyruvate formation (with some NADH/H^+ production) in the cell cytoplasm. Further oxidation of pyruvate to acetyl-CoA and carbon dioxide by the pyruvate dehydrogenase complex, then the citric acid cycle, proceeds in the mitochondrial matrix. Oxidation of the NADH/H^+ and $FADH_2$ from glycolysis and the citric acid cycle proceed in the mitochondrial matrix. However, the ATP that is formed in the mitochondria is not utilized there, but in the cytoplasm.

TABLE 3–5

Summary of Gibbs free energy changes for glucose and glucosyl metabolism. Only the following major reactions are shown: glycolysis, citric acid cycle, electron transport, and the major metabolites. Of the overall Gibbs free energy change (-2854 kJ mole^{-1}), 1693 is initially released as heat and 1161 is stored as ATP/GTP. Values for $\Delta G^{\circ\prime}$ are kJ mole^{-1}. The $\Delta G^{\circ\prime}$ for ATP/GTP hydrolysis is assumed to be -30.5 kJ mole^{-1}.

	$\Delta G^{\circ\prime}$
Glucose + 2 NAD$^+$ + 2 ADP $\longrightarrow$ 2 pyruvate + 2 NADH/H$^+$ + 2 ATP	-72.8
2 pyruvate + 2 CoA + 2 NAD$^+$ $\longrightarrow$ 2 acetylCoA + 2 CO$_2$ + 2NADH/H$^+$	-66.9
2 acetylCoA + 6 NAD$^+$ + 2 FAD + 2 GDP $\longrightarrow$ 4 CO$_2$ + 2 CoA + 6 NADH/H$^+$ + 2 FADH$_2$ + 2 GTP	-88.7
10 (NADH/H$^+$ + $\frac{1}{2}$ O$_2$ + 3 ADP $\rightarrow$ NAD$^+$ + 3 ATP + H$_2$O)	-1284
2 (FADH$_2$ + $\frac{1}{2}$ O$_2$ + 2 ADP $\longrightarrow$ FAD + 2 ATP + H$_2$O)	-181
Glucose + 2 GDP + 36 ADP + 6 O$_2$ $\longrightarrow$ 6 CO$_2$ + 6 H$_2$O + 2 GTP + 36 ATP	-1693
ATP/GTP Hydrolysis	
2 GTP + 36 ATP $\longrightarrow$ 2 GDP + 36 ADP + 38 PP$_i$	-1161
Overall Reaction	
Glucose + 6 O$_2$ $\longrightarrow$ 6 CO$_2$ + 6 H$_2$O	-2854

In some cells, such as the brown fat of mammals, ATP synthesis is uncoupled from the chemiosmotic processes. Special transport proteins of the inner mitochondrial membrane allow H$^+$ to move back into the mitochondrial matrix from the intermembrane space, without concomitant ATP synthesis. The chemical energy of the H$^+$ gradient is released directly as heat, which is used for thermoregulation. Conversion of the H$^+$ gradient energy into energy via an intermediate chemical form is unnecessary and undesirable (see below).

Overall Energetics.   The combination of the Embden-Meyerhof pathway, the citric acid cycle, and the electron transfer system comprises the aerobic metabolic pathway. The overall free energy change for the complete oxidation of glucose to carbon dioxide and water is -2854 kJ mole^{-1} (Table 3–5). Of this, 1693 is released as heat, and the remaining 1161 is converted to chemical energy (essentially 38 ATP); the efficiency is 41%. The remaining 59% of the energy is released as heat and is therefore not "usable" energy, although heat production can be useful to endothermic animals for thermoregulation (see Chapter 5).

Lipid Metabolism.   Triglycerides are an important energy store and source of chemical energy for ATP synthesis. Their metabolic fate will be considered here.

Triglycerides yield glycerol and three fatty acids on hydrolysis. Glycerol, being a carbohydrate, is metabolized by the Embden-Meyerhof pathway. It is converted to glycero-phosphate by glycerokinase, then by glycerolphosphate dehydrogenase to dihy-droxyacetone-phosphate, an intermediate in the Embden-Meyerhof pathway.

Fatty acids are mainly metabolized by the β-oxidation pathway, after they are converted to a fatty acyl-SCoA (Figure 3–15). This requires ATP and thiokinase enzymes. The length of the R-group determines which thiokinse is required: acetate thiokinase (R = CH$_3$), medium chain fatty acid thiokinase (C$_4$ to C$_{12}$), or long chain fatty acid thiokinase (>C$_{12}$). A C$_2$ fragment is removed from the fatty acyl-SCoA to form acetyl-CoA, which then enters the citric acid cycle. ATP can be synthesized via the citric acid cycle from the acetyl CoA as well as from the NADH/H$^+$ and FADH$_2$ produced in formation and fragmentation of fatty acyl-SCoA. The overall energy yield for oxidation of palmitic acid, for example, is 9791 kJ mole^{-1}, of which 5851 is released as heat and 3940 is used for ATP synthesis. The efficiency of palmitate metabolism is about 40%.

An alternate fate of acetyl-CoA, for example in the liver, is the formation of ketone bodies: acetoacetate, β-hydroxybutyrate, and acetone. Two acetyl-CoA molecules are condensed by thiolase to form acetoacetyl CoA, which is then converted to the ketone bodies. The ketone bodies circulate to peripheral tissues (especially muscle) where they are reconverted to acetyl-CoA and metabolized to produce ATP.

Protein Metabolism.   Protein metabolism initially involves breakdown of the protein into its individual amino acids, which are then metabolized by conversion of most of their constituent carbons to either pyruvate or acetyl-CoA. However, amino acid (and

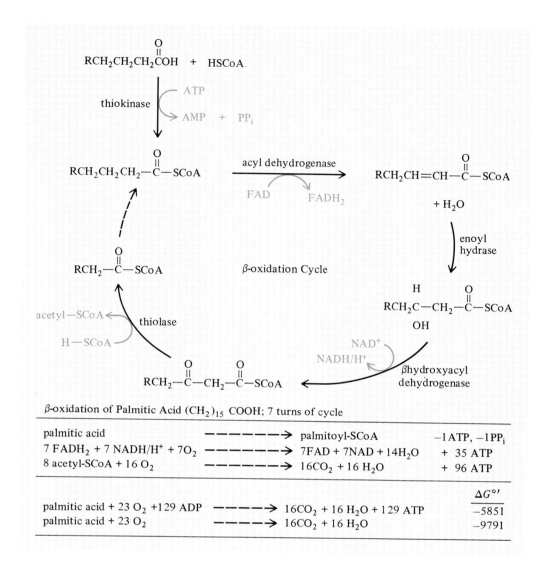

β-oxidation of Palmitic Acid $(CH_2)_{15}$ COOH; 7 turns of cycle

palmitic acid	$- - - - - \rightarrow$	palmitoyl-SCoA	$-1\,ATP, -1\,PP_i$
$7\ FADH_2 + 7\ NADH/H^+ + 7O_2$	$- - - - - \rightarrow$	$7FAD + 7NAD + 14H_2O$	$+\ 35\ ATP$
8 acetyl-SCoA $+ 16\ O_2$	$- - - - - \rightarrow$	$16CO_2 + 16\ H_2O$	$+\ 96\ ATP$

			$\Delta G^{\circ\prime}$
palmitic acid $+ 23\ O_2 + 129\ ADP$	$- - - - \rightarrow$	$16CO_2 + 16\ H_2O + 129\ ATP$	-5851
palmitic acid $+ 23\ O_2$	$- - - - \rightarrow$	$16CO_2 + 16\ H_2O$	-9791

FIGURE 3–15 β-oxidation pathway for fatty acid oxidation to acetyl-SCoA and a C_2 shorter fatty acid.

protein) metabolism (unlike carbohydrate and triglyceride metabolism) is complicated by the presence of elements other than H, C, and O, i.e., nitrogen (N) and sulfur (S).

The initial step in amino acid metabolism is removal of the amino (NH_2) group, either by transamination or deamination. Further metabolism leads either to pyruvate then glucose synthesis (the glycogenic pathway), or to acetyl-CoA and ketone bodies (the ketogenic pathway). Figure 3–10 indicates whether the fate of amino acids is glucogenic, ketogenic, or both.

Transamination. The amino group of an amino acid can be transferred to α-ketoglutarate, forming glutamate and converting the amino acid to an α-ketoacid.

amino acid $+$ α-ketoglutarate $\longrightarrow$

R	COOH
HCNH$_2$	CH$_2$
COOH	CH$_2$
	CO
	COOH

α-ketoacid $+$ glutamate

R	COOH
CO	CH$_2$
COOH	CH$_2$
	HCNH$_2$
	COOH

(3.27a)

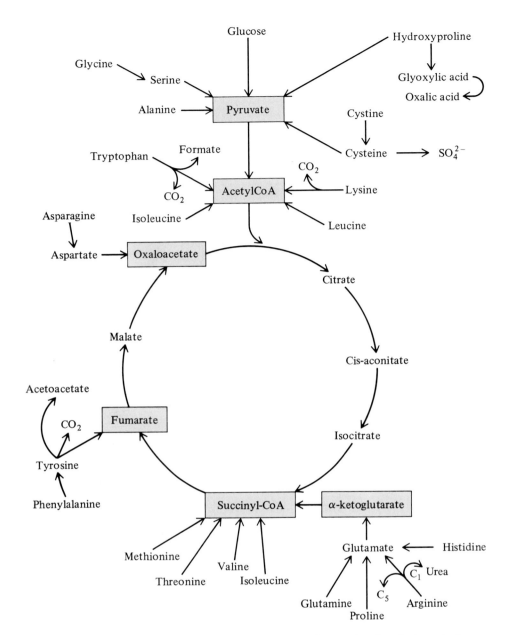

FIGURE 3–16   Metabolic fates of amino acids in relation to the glycolytic and citric acid cycle metabolites.

This deamination reaction allows further metabolism of the ketoacid, but does not indicate the ultimate excretory fate of the NH_2 group. Generally, the glutamate is reconverted to α-ketoglutarate and ammonia (NH_3) by the process of deamination.

Deamination.   The overall reaction for deamination of an amino acid is

$$R\text{-}NH_2 + NADP^+ + H_2O$$
$$\rightarrow R\text{-}OH + NADPH/H^+ + NH_3 \quad (3.27b)$$

In the specific case of glutamate deamination (the glutamate perhaps formed by transamination; see above), the ketoacid is α-ketoglutarate.

The metabolic fates of the carbon skeletons of the amino acids are generally conversion to pyruvate, acetyl-CoA, or citric acid cycle intermediates (summarized in Figure 3–16). Pyruvate can be converted to glucose by a modified reversal of glycolysis called gluconeogenesis or it can enter the citric acid cycle for ATP synthesis. The citric acid cycle intermediates can be converted to glucose, fatty acids, and carbon dioxide. Acetyl-CoA can enter into the citric acid cycle, but there can be no net incorporation of its two carbons into glucose because (a) the conversion of pyruvate to acetyl-CoA and CO_2 is essentially irreversible and (b) once acetyl-CoA enters the cycle, both of the carbons are released as CO_2.

The metabolic fates of amino acid N (both amino nitrogen and other nitrogens in the side group, R) include transamination to form glutamate from α-ketoglutaric acid and deamination to form ammonia, or urea in the case of arginine only (Figure 3–17). Further aspects of nitrogenous waste excretion are examined in Chapter 17.

Energy Yield.   The energy yield of amino acid and protein metabolism depends on the specific amino acid(s) involved. The possible intermediate metabolites and end products for three representative amino acids, and meat protein, are summarized in Table 3–6. The ATP production from the various intermediate metabolites differs markedly for the three amino acids, as does the energy released by combustion in a bomb calorimeter (heat of combustion). The efficiency of metabolism for amino acids and protein tends to be lower than for carbohydrate and lipid metabolism because of the formation of

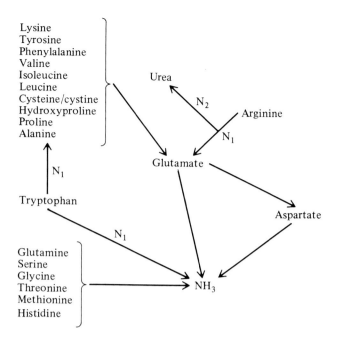

FIGURE 3–17   Fate of amino acid nitrogen in production of ammonia and urea (excluding the potential conversion of ammonia to urea by the urea cycle). The number of nitrogens transferred by each step is indicated as N_1, N_2.

TABLE 3–6

Comparison of the usable energy yields (ATP mole^{-1}) of three representative amino acids, and meat protein, with the total heat energy released by combustion (H_c°; kJ mole^{-1}). The possible metabolic intermediates, final products, ATP yield (moles ATP/mole substrate except for meat protein), and efficiency of catabolism are also indicated.

	Intermediates	**Products**	**ATP**	**H_c°**	**Efficiency**
ALANINE	Pyruvate	CO_2	18		
	NADH/H$^+$		3		
	NH$_3$	$\frac{1}{2}$ urea	$\underline{-2}$		
			19	1619	36%
TRYPTOPHAN	Alanine	Formate	0		
	Acetyl-CoA	CO_2, $\frac{1}{2}$ urea	19		
	NH$_3$	CO_2	30		
		$\frac{1}{2}$ urea	$\underline{-2}$		
			47	5627	25%
CYSTEINE		SO_4^{2-}	1		
	Pyruvate	CO_2	18		
	NH$_3$	$\frac{1}{2}$ urea	$\underline{-2}$		
			17	1652	32%
MEAT PROTEIN	Amino acids	4.11 CO_2			
		0.70 urea			
		0.034 SO_4^{2-}			
			22.2	2536	27%

nitrogenous waste products, such as ammonia, urea, or uric acid; typical efficiency values are 25 to 35% with urea synthesis (Table 3–6).

Anaerobic Metabolism

The complete oxidation of glucose to carbon dioxide and water requires oxygen, for the regeneration of NAD^+ and FAD and synthesis of ATP. However, oxygen is sometimes in limited supply (hypoxia) or unavailable (anoxia) for some animals. Consequently, these animals have alternative pathways for **anaerobic metabolism** to form ATP in the absence of oxygen.

Many animals rely on aerobic metabolism for at least their resting ATP requirements; they are strict, or obligate, aerobes. However, even strict aerobes can use anaerobic metabolic pathways for supplemental ATP formation under some circumstances, e.g., intense activity. For example, almost all vertebrates rely on aerobic metabolism to sustain their resting metabolic rate, but can produce ATP during activity by anaerobic metabolism.

Some animals rely on anaerobic metabolism for extended periods of time; they are facultative anaerobes. For example, many intertidal invertebrates experience cyclic submersion and emersion and are facultative anaerobes.

Some animals are strict, or obligate, anaerobes. For example, many parasitic invertebrates are obligate anaerobes and cannot survive in the presence of significant amounts of oxygen.

Anaerobic Metabolic Strategies

Glycolysis is essentially the only important anaerobic metabolic pathway for higher vertebrates. Some invertebrates use glycolysis for anaerobic metabolism, but many have alternate anaerobic pathways for ATP production. Some of these pathways produce more ATP/glucose, and some can be maintained as steady-state processes.

Glycolysis. Glycolysis is one anaerobic metabolic pathway that is commonly used by vertebrates. Two ATP are formed per glucose at an efficiency of about 50% of the free energy change. The coenzyme NAD^+ accepts two hydrogen atoms during glycolysis to form $NADH/H^+$, but is rapidly depleted because it is present in very low concentrations. Glycolysis would quickly stop unless the NAD^+ were regenerated. In glycolysis, this regeneration involves the conversion of pyruvate to lactate. Pyruvate acts as a hydrogen acceptor.

$$\overset{\text{lactate}}{\overset{\text{dehydrogenase}}{\text{pyruvate} + NADH/H^+ \quad \rightleftharpoons \quad \text{lactate} + NAD^+}}$$

$$\begin{array}{ll} CH_3 & CH_3 \\ C{=}O & HCOH \\ COO^- & COO^- \end{array} \quad (3.28)$$

Consequently, glycolysis can continue unabated, at least in terms of NAD^+ availability.

The equilibrium between pyruvate and lactate favors lactate ($\Delta G° = -25$ kJ $mole^{-1}$), so high concentrations of lactate can accumulate before the reverse reaction (lactate $\rightarrow$ pyruvate) becomes significant. However, lactate production is not a steady-state anaerobic pathway, and lactate can accumulate to such high levels that pyruvate also accumulates and inhibits further glycolysis.

Linkage with the Citric Acid Cycle. Glycolysis can be linked with the metabolic pathways of the citric acid cycle to form a variety of end products and additional ATP (Figure 3–18).

Glycolysis proceeds normally to phosphoenolpyruvate, but formation of pyruvate is minimized by low levels of pyruvate kinase (PK). A low level of lactate dehydrogenase (LDH) minimizes lactate formation. The main metabolic pathway is formation of oxaloacetate by CO_2 fixation; high levels of phosphoenolpyruvate carboxy kinase (PEPCK) favor this pathway. Oxaloacetate is reduced to malate by malate dehydrogenase. Malate then enters the mitochondria, where about half is converted to pyruvate and CO_2 and the rest to fumarate. Fumarate conversion to succinate (by fumarate reductase) results in ATP generation. The pyruvate and succinate are converted (via CoA) to acetate and propionate which, in turn, are converted to 2-methylvalerate and 2-methylbutyrate; ATP may also be formed in these steps.

The ATP production is high for these pathways. The formation of succinate yields at most 4 ATP per glucose, with 2 ATP formed at the level of phosphoglycerate kinase, 2 ATP at PEPCK, and 2 ATP by electron transfer at fumarase reductase (with 2 ATP used at glucokinase and phosphofructokinase). The further conversion of succinate to propionate produces an additional 2 ATP per glucose, for a total of 6 ATP/glucose. Further synthesis of 2-methylbutyrate and 2-methylvalerate might yield an additional ATP.

Pyruvate Condensation with Amino Acids. The general reaction for the combination of pyruvate (or any $R{-}C{=}O$ compound) with an amine (NH_2) allows the regeneration of NAD^+ from $NADH/H^+$ to sustain anaerobic metabolism;

$$R\!-\!\underset{\underset{R'}{|}}{C}\!=\!O + H_2N\!-\!R'' + NADH/H^+$$

$$\rightleftharpoons R\!-\!\underset{\underset{R'}{|}}{CH}\!-\!NH\!-\!R'' + H_2O + NAD^+ \tag{3.29}$$

Clearly, any amino acid could readily combine with pyruvate and at the same time regenerate NAD$^+$ via this reaction. In fact, a variety of amino acids are used by various animals (principally mollusks) to achieve this (Table 3–7).

Aspartate Metabolism. Aspartate can be an important anaerobic metabolic substrate (Figure 3–19). The aspartate is deaminated by α-ketoglutarate to form oxaloacetate, which is then converted to propionate (as described above via linkage to the citric acid cycle). To sustain this pathway, the deaminating mechanism must be maintained, i.e., α-ketoglutarate must be re-formed from glutamate. This is accomplished by coupling transamination of

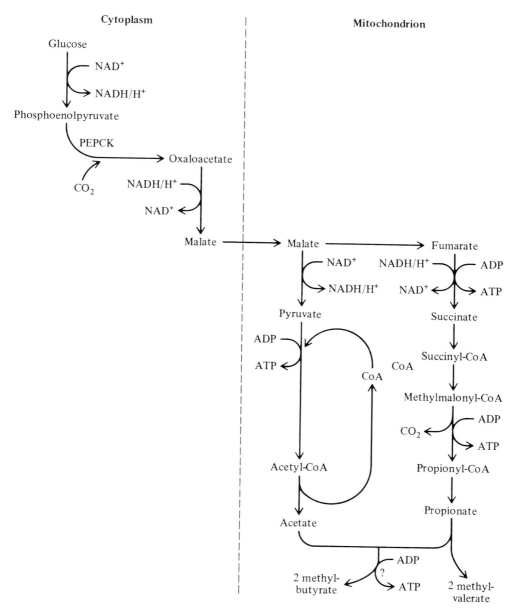

FIGURE 3–18 Linkage of glycolysis with citric acid cycle pathways is an anaerobic metabolic pathway that provides additional ATP formation. These pathways are found, for example, in many platyhelminth worms.

TABLE 3–7

Amino Acid	Enzyme	End Product
Amino acids (H_2N-R) are used by bivalve and cephalopod mollusks for removal of pyruvate ($CH_3COCOOH$) and regeneration of NAD^+ from $NADH/H^+$ during anaerobiosis, using the general reaction illustrated below.		

Amino Acid	Enzyme	End Product
H_2N-R + $NADH/H^+$ + CH_3 $C=O$ $COOH$	enzyme $\rightleftharpoons$	CH_3 HC-NH-R + H_2O + NAD^+ $COOH$
Alanine[1]	Alanine dehydrogenase	Alanopine
Glycine[1]	Strombine dehydrogenase	Strombine
Lysine[1]	Lysopine dehydrogenase	Lysopine
Arginine[2]	Octopine dehydrogenase	Octopine

[1] Bivalves.

[2] Cephalopods.

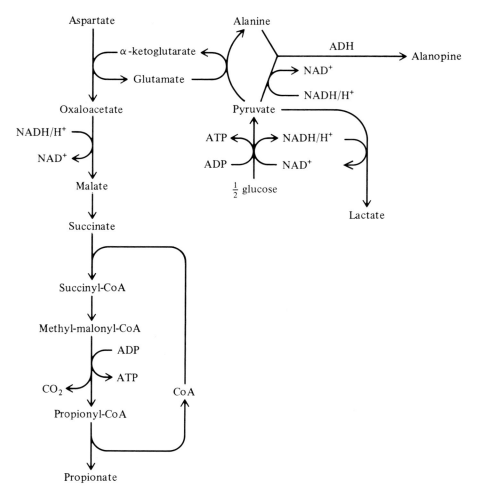

FIGURE 3–19   Anaerobic metabolic pathway of bivalve mollusks showing coupling of aspartate and glucose metabolism to produce propionate and alanopine.

pyruvate to alanine. Low concentrations of LDH and PEPCK channel pyruvate from glycolysis towards alanine, rather than oxaloacetate (which is formed from aspartate) or lactate. Glycolysis proceeds without pyruvate or lactate accumulation, but alanine would eventually accumulate and inhibit both aspartate and glucose metabolism.

Arginine Phosphate Stores. Arginine phosphate is an important phosphagen in some animals. It can be an important source of energy over short periods of anaerobic metabolism, although the amount of arginine phosphate present in muscle cells is insufficient to maintain ATP supplies for more than brief periods (even though it can be as high as 70 μmole g^{-1}).

Hydrolysis of arginine phosphate stores to arginine is energetically coupled to the regeneration of ATP from ADP. However, a high concentration of arginine would have detrimental effects on cells, since arginine is a very basic amino acid. It would disrupt intracellular acid-base balance and might inhibit the catalytic properties of enzymes. However, arginine is an amino acid and participates in the general reaction outlined above (Equation 3.29) for the elimination of pyruvate and regeneration of NAD$^+$. The end product is octopine, and the catalytic enzyme is octopine dehydrogenase (Table 3–7). Thus, utilization of high-energy arginine phosphate stores and anaerobic metabolism can be conveniently coupled (Figure 3–20). A small amount of lactate also accumulates as some LDH is present in cephalopod muscle.

Acetate, Acetaldehyde, and Ethanol. Another anaerobic metabolic strategy is to form ethanol as the

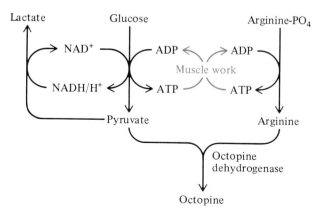

FIGURE 3–20 High-energy phosphate stores (arginine-PO$_4$) and glycolysis are coupled as an anaerobic metabolic pathway, with octopine as a major metabolic end product.

end product. Pyruvate is converted to acetyl-CoA, releasing CO_2. Subsequent synthesis of acetate (by acetate thiokinase) regenerates CoA and forms ATP.

$$\text{pyruvate} + \text{CoA} \xrightarrow{\text{CO}_2} \text{acetyl-CoA}$$
$$\xrightarrow[\text{ADP ATP}]{} \text{acetate} + \text{CoA}$$

$$\begin{matrix} \text{CH}_3 \\ \text{C=O} \\ \text{O}^- \end{matrix} \quad (3.30a)$$

Acid-base balance is not disturbed as much by acetate formation as by lactate production. Acetic acid has a higher pK (4.8) than lactic acid (3.7); the pK is the pH at which 50% of an acid, HA, is present as the anion A$^-$, and 50% as acid HA. Furthermore, acetate can be further metabolized to acetaldehyde, then ethanol.

$$\text{acetate} \xrightarrow[\text{NADH/H}^+ \text{ NAD}^+]{\text{acetaldehyde dehydrogenase}} \text{acetaldehyde} + \text{H}_2\text{O}$$

$$\begin{matrix} \text{CH}_3 \\ \text{C=O} \\ \text{O}^- \end{matrix} \qquad \begin{matrix} \text{CH}_3 \\ \text{C=O} \\ \text{H} \end{matrix}$$

$$\xrightarrow[\text{NADH/H}^+ \text{ NAD}^+]{\text{ethanol dehydrogenase}} \text{ethanol}$$

$$\begin{matrix} \text{CH}_3 \\ \text{CHOH} \\ \text{H} \end{matrix}$$
$$(3.30b)$$

Each step regenerates NAD$^+$ from NADH/H$^+$.

Invertebrates

Invertebrate animals use a variety of the above described strategies for anaerobic metabolism. In general, there is an evolutionary trend towards a greater activity level in the higher invertebrates, hence a greater reliance on aerobic metabolism. Associated with this trend is the tendency for a lesser reliance on sustainable anaerobic pathways, and greater utilization of phosphagen stores.

Free-living platyhelminthes (turbellarian worms such as planarians) have little need for a well-developed anaerobic capacity, but parasitic platyhelminths (flukes, tapeworms) often encounter hypoxic or anoxic environments (e.g., the gut) and have well-developed anaerobic metabolic pathways (Saz 1981). Parasitic helminths can be obligate aerobes (e.g., *Litomosoides*, *Nippostrongylus*), can support resting metabolism anaerobically but require aerobic metabolism for motility (e.g., microfilarial stages of *Litomosoides*), or do not require oxygen (e.g., *Ascaris*, *Schistosoma*). The anaerobic pathways of parasitic helminthes are citric acid pathways linked with glycolysis, and there are various end products. For example, the common metabolic end products for *Ascaris* are succinate, propionate, acetate, 2-methylvalerate, and 2-methylbutyrate.

Many mollusks, especially intertidal bivalves (mussels, oysters, clams), routinely experience prolonged hypoxia. Some bivalves, such as oysters, can live indefinitely in the complete absence of oxygen! These bivalves obviously have a well-developed anaerobic capacity. The amino acid aspartate is a major metabolic substrate for anaerobiosis, with succinate and alanine as end products, e.g., the oyster ventricle is a well-studied anaerobic molluskan tissue (Collicutt and Hochachka 1977). Aspartate and glucose metabolism can continue in many bivalve mollusks because alanine combines with pyruvate to form alanopine; this also regenerates NAD^+ from $NADH/H^+$. The enzyme that catalyzes this reaction is alanopine dehydrogenase. Littoral, sessile bivalves rely on carbohydrate rather than aspartate as their anaerobic energy substrate. Glycolysis is linked to the citric acid cycle via oxaloacetate and malate, by conversion of phosphoenolpyruvate to oxaloacetate. The latter reaction is made possible by the important enzyme, PEPCK. The normal (aerobic) energy pathway is replaced by different anaerobic pathways during short-term and long-term hypoxia (de Zwaan 1983).

Squid and octopus are considerably more active than bivalve mollusks, and their greater metabolic demands suggest a greater reliance on aerobic metabolism. Nevertheless, they have a considerable anaerobic capacity because periods of relative hypoxia still occur, e.g., during strenuous swimming. The primary phosphagen of cephalopod muscle is arginine phosphate (rather than creatine phosphate); hydrolysis of arginine phosphate is coupled with ATP formation. The arginine is removed by its combination with pyruvate to form octopine. There are general similarities in aerobic metabolism of cephalopods and vertebrates despite differences in their specific anaerobic reactions. Glucose (or glycogen) is the only significant metabolic substrates, not amino acids. Only 2 ATP are produced per glucose. Anaerobic metabolism is not a steady-state process; there is an accumulation of end products (octopine and lactate) that eventually inhibit further glycolysis.

Vertebrates

The only significant anaerobic metabolic pathway of most vertebrates is lactate formation. However, some fish have other important anaerobic metabolic pathways for periods of inactivity coupled with hypoxia.

Fish, like other vertebrates, are generally active and aerobic animals. Their resting metabolism is maintained by aerobic pathways (with a few exceptions; see below), but a significant anaerobic capacity is advantageous in two circumstances: during bursts of strenuous activity and during hypoxia. However, very different anaerobic strategies are employed by fish in these two different situations. Bursts of intense activity are generally of short duration with an extremely high rate of ATP demand. Periods of hypoxia (such as being frozen under ice during winter) are typically of much longer duration with a low ATP demand rate.

Fish utilize the lactate anaerobic pathway during periods of strenuous activity. There is a positive relationship between the level of activity and anaerobic capacity for fish muscle, as shown by their concentrations of glycolytic enzymes and metabolite levels. The white muscle of active fish, such as tuna, has considerably higher levels of some glycolytic enzymes, including LDH, and citric acid cycle enzymes than does white muscle of an inactive fish (Table 3–8).

There is also adaptive variation for enzyme and metabolite levels in different muscles of individual fish. Red muscle of tuna has lower levels of some anaerobic glycolytic enzymes (PK and LDH) and higher levels of some aerobic citric acid cycle

TABLE 3–8

Levels for anaerobic (glycolytic) and aerobic (citric acid cycle) enzymes of white and red muscle for an inactive and active species of fish. Units for enzyme activity are μmoles per minute per gram wet tissue weight. *(Values for Hochachka et al. 1978; Guppy, Hulbert, and Hochachka 1979.)*

	Inactive Fish Arapaima *Arapaima*		Active Fish Skipjack tuna *Euthynnus*	
	White	Red	White	Red
Anaerobic				
Pyruvate kinase	103	134	1295	195
Lactate dehydrogenase	260	263	5492	514
Aerobic				
Citrate synthetase	1.7	3.3	2.1	20.6
Malate dehydrogenase	140	221	718	723
Glutamate dehydrogenase	1.3	3.1	3.0	5.9
Glutamate-oxaloacetate transaminase	11.2	54.4	43	102

enzymes (CS, GDH, and GOT) than does white muscle (Table 3–8). These enzymatic differences are reflected by changes in the metabolite levels of tuna red and white muscle, at rest and after burst activity (Table 3–9). There is a substantial depletion of muscle energy stores (glycogen, creatine phosphate, and ATP) and an accumulation of anaerobic end product (lactate), but only small changes in the concentration of glycolytic and citric acid cycle intermediary metabolites. White muscle has a greater depletion of energy stores and greater accumulation of lactate than red muscle. Red muscle also has a higher myoglobin content and is more vascular than white muscle; it is used for sustained aerobic metabolism. White muscle has a high anaerobic capacity and is used for bursts of anaerobic metabolism.

Fish that survive long periods of environmental hypoxia rely on alternate anaerobic pathways. Apparently, lactate anaerobiosis is not suitable for long-term bouts of hypoxia. For example, lactate accumulation would eventually inhibit glycolysis. The capacity for anaerobic metabolism would determine the likelihood and duration for survival of extreme hypoxia. Carp, goldfish, and some other

cyprinid fish can, under certain circumstances, survive hypoxia for 100 days or more. Two observations of fish kept under hypoxic conditions indicate that lactate accumulation is not the major source of ATP synthesis, and there are alternate metabolic pathways for long-term tolerance of anoxia. First, blood lactate levels of goldfish after five days of anoxia are only 20 mM; this level of glycolysis is apparently insufficient to support resting metabolism (Hochachka 1980). Second, anoxic fish excrete carbon dioxide, but CO_2 is not evolved by the glycolytic pathway to lactate. Acetaldehyde dehydrogenase and ethanol dehydrogenase are present in goldfish tissues at sufficient concentrations to form ethanol during anoxia. The ethanol is then excreted into the environment, avoiding the accumulation of a metabolic end product.

Other possible anaerobic end products for fish include succinate, lipids (elongation of fatty acids by condensation with acetyl-CoA), and ammonia. The fatty acids could be derived from carbohydrate (oxidation of glucose or glycogen to acetyl-CoA) and the NH_3 from amino acid deamination, in coupled reactions.

There may be some partitioning by different organs within the fish with respect to specific pathways occurring in specific organs (Hochachka 1980). For example, peripheral tissues receiving a low blood supply (such as white muscle) might predominantly produce lactate as their anaerobic end product. The lactate could diffuse to more highly perfused and aerobic tissues (such as liver, red muscle, heart) and be further oxidized to other anaerobic end products, such as ethanol or fatty acids. There may be coupling of anaerobiosis with amino acid metabolism, particularly in the liver.

Anaerobic metabolism in tetrapods (i.e., amphibians, reptiles, birds, and mammals) is almost completely limited to glycogen utilization and lactate accumulation. Only small amounts of other anaerobic end products, such as glycolytic or citric acid cycle intermediates, pyruvate, succinate, alanine, and ethanol have been reported (Hochachka et al. 1975; Felig and Wahren 1971). For example, preliminary studies of diving vertebrates (a sea turtle, seal, sea lion, and porpoise) indicate a minor role of amino acid metabolism compared with glycolytic metabolism (Figure 3–21). Although the quantitative contribution of this form of anaerobic metabolism has not been evaluated, it is likely to be low (1 to 2% of the total anaerobic energy production).

The accumulation of lactate by amphibians during activity well illustrates the important role of muscle glycogen as the primary source of anaerobic energy (Table 3–10). The production of lactate not only

TABLE 3–9

Changes in metabolite concentrations of red and white muscle from skipjack tuna (*Euthynnus pelamis*) after burst activity. Units for concentrations are μmole g wet weight^{-1}. *(Data from Guppy, Hulbert, and Hochachka 1979.)*

Metabolites	Red	White
High-Energy PO₄ Stores		
Creatine PO₄	−1.73	−12.90
ATP	−0.87	−2.60
Substrate Stores		
Glycogen	−1.70	−22.8
Glycolytic Intermediates		
Glucose	−0.13	+2.01
Glucose-6-PO₄	+0.41	+2.30
Fructose-6-PO₄	−0.21	−0.45
Fructose-1,6-diPO₄	−0.06	−0.34
Di(OH)acetonePO₄	−0.05	−0.02
Glyceraldehyde-3-PO₄	−0.04	−0.01
Citric Acid Cycle Intermediates		
Citrate	+0.08	−0.05
α-Ketoglutarate	−0.02	−0.07
Malate	+0.13	+0.12
Anaerobic End Products		
Lactate	+5.9	+70.95

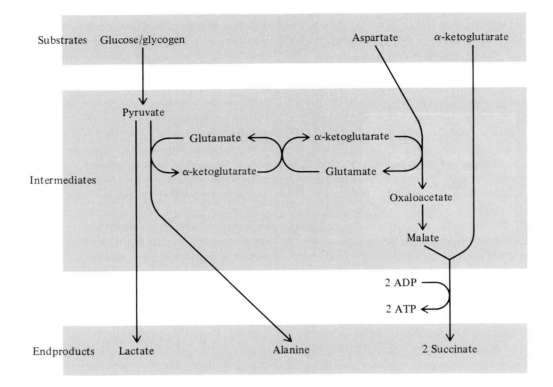

Changes in Metabolite Levels:

	µm before dive	µm after dive	µM change
Aspartate	96	73	− 23
α-ketoglutarate	200	110	− 90
Succinate	40	280	+ 240
Alanine	300	650	+ 350
Lactate	90000	160000	+ 70000

FIGURE 3–21 Schematic for metabolism of substrates to anaerobic end products (lactate, alanine, succinate) for higher vertebrates, and levels of blood metabolites (µm) in seal blood before and after a dive, illustrating the predominance of lactate as the anaerobic end product. *(Values from Hochachka et al. 1975; Hochachka 1980. Photo by George Whitely/Photo Researchers, Inc.)*

TABLE 3–10

Comparison of whole body and tissue lactate levels, glycogen stores, and blood pH for a resting and fatigued clawed frog (*Xenopus laevis*). *(Data from Putnam 1979a; Putnam 1979b.)*

	Rest	Fatigued
Liver glycogen (g%)	10.4	9.3
Gastrocnemius glycogen (g%)	1.8	0.7
Whole body lactate (mg%)	11	213
Blood lactate (mg%)	42	177
Liver lactate (mg%)	29	144
Gastrocnemius lactate (mg%)	98	289
Blood pH	7.62	6.89

eventually leads to end product inhibition of glycolysis, but also has profound effects on acid-base balance. The dissociation of 20 mM lactic acid would be accompanied by formation of 20 mM H^+, and pH would decline by 5 units if the excess H^+ were not buffered! Fortunately, the body has well-developed buffer systems (see Chapter 15), and so the decline in pH due to lactate accumulation is considerably less than 1 pH unit. Nevertheless, pH does decline and this could be an important contributing factor to onset of muscle fatigue.

Anaerobic metabolism is a graded rather than an all-or-none process. There are significant levels of lactate in the blood even at rest, indicating limited lactate formation in muscle and its diffusion into the blood stream. The blood lactate level of humans during graded activity remains at a low, resting level (1 to 2 mM) until about 70% of the maximal work load. After this **anaerobic threshold** is reached, the lactate level increases linearly with work load. The elevated blood lactate levels at high work loads are indisputably the result of muscle anaerobiosis and loss of lactate from muscle, but the "anaerobic threshold" does not necessarily represent the work load at which the muscle first becomes anaerobic. At rest, blood lactate is converted in the liver to glucose for recirculation to the muscle (the Cori cycle). A decrease in blood flow to the liver at moderate work loads could diminish the removal of resting lactate levels from the blood and result in blood lactate accumulation. The anaerobic threshold might also reflect a change in the types of muscle cells that are recruited or a change in substrate from fat to carbohydrate (Davis 1985; Brooks 1985). Consequently, the anaerobic threshold is perhaps more appropriately called a "hyperlactemia threshold," reflecting the observation of elevated blood

lactate levels without implying a mechanism for lactate buildup.

The cycle of lactate production in one tissue (e.g., muscle) and its reconversion to glucose in another (e.g., liver) can be of great significance during long-term anaerobic conditions, such as diving (Hochachka 1980). In the Weddell seal, for example, lactate produced in the brain and muscle during diving enters the circulation and is reconverted to glucose in the liver and kidney; the glucose then recirculates to the brain and muscle.

Anaerobic metabolism is important to higher vertebrates not only during physical activity. Some amphibians and reptiles are very tolerant of long-term anoxia, and their lactate levels can be much higher during chronic anoxia than during intense and exhaustive activity. This extreme hyperlactemia is accompanied by a massive acid-base imbalance, which has numerous physiological consequences, including dissolution of bone and elevation of plasma calcium. For example, blood lactate levels exceed 200 mM in freshwater turtles *Chrysemys* after 180 days of submergence in nitrogen-equilibrated water, and total plasma calcium increases markedly from normal levels of about 4 mM, to 120 mM or more (Ultsch and Jackson 1982; Jackson and Ultsch 1982). Up to 2/3 of the total calcium may be bound to lactate, so that as little as 1/3 of the calcium is present in the free, ionized form (Ca^{2+}). Nevertheless, such ionized Ca^{2+} levels would be expected to markedly modify nerve cell excitability and synaptic transmission (see Chapter 6) and muscle contractility (see Chapter 9). For example, elevated Ca^{2+} levels increase the cardiac muscle contractility of the turtle heart *in vitro* (Yee and Jackson 1984).

Energetics of Anaerobic Metabolism

No anaerobic pathways produce as much ATP from a substrate as aerobic metabolism. Anaerobic metabolic pathways have potential ATP yields of 2 to 6 ATP per C_6 unit (Table 3–11). Anaerobic metabolism of glucose to lactate produces 2 moles lactate/mole glucose and 2 moles ATP/mole glucose, i.e., 1 mg lactate $\equiv$ 11 μmole ATP. For glucosyl subunits, the yield is 2 lactate/glucosyl and 3 ATP/glucosyl, i.e., 1 mg lactate $\equiv$ 17 μmole ATP. Aerobic metabolism can potentially yield 38 ATP mole^{-1} of glucose, or 140 to 150 mole ATP mg^{-1} of CO_2. Aerobic metabolism of glucose consumes 6 moles O_2/mole glucose and yields 38 ATP, i.e., 1 ml O_2 (STPD) $\equiv$ 283 μmole ATP. Consequently, anaerobic energy production is generally used to either support a low metabolic rate for a long period or a high metabolic rate for a short period.

TABLE 3–11

		μm ATP mg^{-1}
Stoichiometry of ATP synthesis by various anaerobic metabolic pathways for glucose (Glu) and glycogen (Glc) compared with aerobic metabolism; the ATP yield is expressed as μmole mg^{-1} of the end product. *(Modified from de Zwaan 1983.)*		
Anaerobic Metabolism		
Glu + 2 ADP ⟶	2 lactate + 2 ADP	11.2
Glc + 3 ADP ⟶	2 lactate + 3 ATP	16.7
Glc + 3 ADP ⟶	2 lysopine + 3 ATP	7.3
Glc + 2 Arg + 3 ADP ⟶	2 octopine + 3 ATP	6.4
Glc + 2 Gly + 3 ADP ⟶	2 strombine + 3 ATP	11.1
Glc + 2 Ala + 3 ADP ⟶	2 alanopine + 3 ATP	10.1
Glu + 4 ADP ⟶	2 acetate + 4 ATP	33.9
Glc + 2 Asp + 4.71 ADP ⟶	1.71 succinate + 1.14 CO_2 + 2 Ala + 4.71 ATP	23.7
Glc + 0.86 CO_2 + 4.71 ADP →	1.71 succinate + 4.71 ATP	23.7
Glc + 6.43 ADP ⟶	1.71 propionate + 0.86 CO_2 + 6.43 ATP	51.5
Aerobic Metabolism		
Glu + 6 O_2 + 38 ADP ⟶	6 CO_2 + H_2O + 38 ATP	144
Glc + 6 O_2 + 39 ADP ⟶	6 CO_2 + 5 H_2O + 39 ATP	148

Metabolic Fates of Anaerobic End Products

So far we have considered the pathways for anaerobic metabolism and the nature of the end products, but what is the fate of these end products after anaerobic metabolism stops?

Some bivalve mollusks accumulate anaerobic end products in their muscle tissues during anaerobiosis, and these are later reconverted *in situ* to the original substrates when O_2 is available. In other bivalves, however, the anaerobic end products (especially propionate and acetate) are distributed by the hemolymph to other body organs, and they may be used by these other tissues as substrates for aerobic metabolism or for reconversion to the original substrate. Gastropod mollusks, which use the lactate, opine (octopine, alanopine, and strombine), and succinate pathways during anaerobiosis, may also distribute these end products via the hemolymph for subsequent utilization by other tissues (such as liver).

The octopine produced anaerobically in muscle by cephalopods during activity can be used by other tissues as their aerobic metabolic substrate (Storey and Storey 1983). At least some octopine is metabolized *in situ*, but some fraction (perhaps high) is released into the blood after the cessation of muscle anaerobiosis. The octopine is taken up by tissues that either aerobically metabolize the arginine and pyruvate portions (e.g., kidney) or resynthesize

glucose and recycle it to muscle (e.g., hepatopancreas). These processes are analogous to the fates of lactate in the Cori cycle of vertebrates.

Vertebrates can accumulate large quantities of lactate during intense activity, and this can result in muscle fatigue either through end product inhibition of glycolysis or pH imbalance. There clearly is adaptive significance to the rapid restoration of normal acid-base status and cellular metabolite levels and replenishment of muscle glycogen stores after activity. However, the removal of lactate after the cessation of activity poses certain problems, particularly for ectothermic animals with low resting metabolic rates. The amounts of lactate that accumulate during activity would sustain resting metabolism for many hours if all of the lactate were slowly oxidized to CO_2 and H_2O for maximum energy economy. For example, the lactate accumulated during a short activity period by the salamander *Amphiuma* would sustain resting metabolism for well over a day (Table 3–12)! A further disadvantage of this strategy for lactate elimination would be the eventual depletion of body glycogen stores. An alternative strategy would be to reconvert all of the lactate to glycogen and glucose, but this requires 6 ATP per glucose and there would be a net loss of 4 ATP in the biochemically futile cycle.

$$\text{glucose} \underset{\substack{\text{recovery}\\\text{6 ATP}}}{\overset{\substack{\text{2 ATP}\\\text{activity}}}{\rightleftharpoons}} \text{2 lactate} \qquad (3.31)$$

TABLE 3–12

Comparison of resting oxygen consumption rate, lactate levels accumulated after activity, and time calculated for elimination of lactate via resting aerobic metabolism for an amphibian (salamander, *Amphiuma*) and a mammal (rat, *Rattus*). Each has a body mass of about 100 g.

	Salamander	Rat
Aerobic Metabolic Rate[1]		
ml O_2 hr^{-1}	3.9	120
mmole ATP hr^{-1}	1.09	33.6
Anaerobic Metabolism[2]		
Lactate accumulated (mg)	153	90
ATP from lactate (mmole)	2.6	1.5
Aerobic Recovery[3]		
ATP from aerobic lactate metabolism (mmole)	30.6	18
Time to eliminate lactate	28 hr	32 min

[1] Assuming 1 ml O_2 (STPD) yields 0.28 mmole ATP.

[2] Assuming 1 mg lactate formed from glycogen yields 0.017 mmole ATP.

[3] Assuming 1 mg lactate yields 0.20 mmole ATP when metabolized to $CO_2 + H_2O$.

The ATP required for glucose resynthesis could be derived from fat or protein metabolism. An advantage of this strategy would be the conservation of body carbohydrate stores for subsequent anaerobic metabolism; remember that glucose and glycogen are the only significant anaerobic metabolic substrates for tetrapod vertebrates.

Another strategy would be the oxidation of some of the lactate to CO_2 and H_2O, to provide the ATP necessary for reconversion of the remaining lactate to glucose or glycogen. The optimal stoichiometry for reconversion of lactate to glucose, by ATP produced from some lactate oxidation, is to oxidize about 0.15 mole of lactate per 0.85 mole reconverted to glucose. This ratio of moles lactate reconverted/mole lactate oxidized is the Meyerhof quotient, e.g., 0.85/0.15 = 5.7.

In endothermic mammals and birds, most lactate is probably oxidized because the high resting metabolic rate can rapidly deplete the lactate (Table 3–12) and restore normal conditions; this has been demonstrated for the rat. In ectotherms, the low resting metabolic rate would only slowly deplete the lactate and the conservation of body carbohydrate stores for further anaerobic activity might be a high priority. There is less oxidation of lactate and

more reconversion to glucose/glycogen in toads and lizards.

Bioluminescence

Bioluminescence is the emission by animals of visible light derived from the chemical release of photons. For example, the "flashlight" fish *Photoblepharon* has a large bioluminescent organ under each eye (Figure 3–22). Incandescence, in contrast to bioluminescence, is the emission of visible light at high temperatures. All animals emit nonvisible infrared radiation by incandescence (see Chapter 5), but only some animals are able to emit visible light by the chemical processes of bioluminescence.

The phylogenetically widespread yet scattered distribution of bioluminescence among animals and the diversity of bioluminescent mechanisms indicate the probable independent evolution of light emission many times among these diverse groups of organisms. Nevertheless, there are three basic patterns of bioluminescence among animals: those of bacterial origin, those that are extracellular, and those that are intracellular.

Some squid and teleost fish rely on symbiotic bacteria for light production. For example, the rat fish *Malacocephalus* has anterior and posterior light organs containing bacteria (Figure 3–23A). The

FIGURE 3–22 The bioluminescent "flashlight" fish, *Photoblepharon*, has a light-emitting organ under each eye. The organ emits light continuously by bacterial action, but the organ can be covered and uncovered by a shutter mechanism. *(Photo by David C. Powell, Monterey Bay Aquarium)*

bacteria generally luminesce continuously, and so the light-emitting organs of these animals continuously glow. However, some animals have a mechanism to cover the light-emitting organ. The "flashlight fish" can cover and uncover the large light-emitting organ under each eye with a shutter.

Many marine invertebrates, myriapods, oligochete worms, and teleost fish exude luminescent secretions, and the bioluminescent reaction occurs extracellularly. For example, the earthworm *Eisenia submontana* produces a luminous slime from oral, anal, and dorsal pores in response to irritating stimuli; the source of the secretion is coelomic fluid.

Bioluminescence of some animals is intracellular and may be associated with very specialized structures, including reflecting surfaces and lenses. This pattern of bioluminescence is most developed in teleost fish, cephalopods, and terrestrial insects. The toad fish *Porichthys* has dorsal, lateral, and ventral rows of light-emitting photophores (Figure 3–23B).

Bioluminescence appears to have evolved as a fortuitous by-product of the exogenous metabolic pathways of organisms. The general principle for most forms of bioluminescence is the oxidation (usually, but not always, involving oxygen) of a complex-structured, high-energy organic molecule ("**luciferin**"), catalyzed by a specific enzyme ("**luciferase**") to release photons of light (see Supplement 3–2, page 76). Luciferin and luciferase are general terms used to describe a wide variety of high-energy organic molecules and enzymes.

A well-studied example of extracellular bioluminescence is that of a crustacean, the ostracod *Cypridina*. The luciferin and luciferase are produced by two different kinds of secretory cells in a large light-producing organ located near the mouth; both are simultaneously ejected by muscular contractions. The luciferin is oxidized by O_2 to produce oxyluciferin and carbon dioxide; light is also released (λ_{max} = 460 nm). The *Renilla* coelenterate-type luciferin has a similar reaction sequence (see Supplement 3–2). Accumulation of end products is not a problem because they are secreted into the external medium. All luminescent crustaceans and many fish use this *Cypridina*-type mechanism, although the luminescence is intracellular in some animals (e.g., the fish *Porichthys*).

The energy released as light by bioluminescent reactions is considerable

$$E = h\nu = hc/\lambda = 119560/\lambda \text{ kJ mole}^{-1} \quad (3.32)$$

where E is the light energy (kJ mole photons^{-1}), h is Planck's constant, ν is the frequency, λ is the wave length (nm), and c is the velocity of light.

There is a considerable variation in the λ and E emitted by bioluminescence (Figure 3–24). E varies from about 254 kJ mole photons^{-1} for blue light (λ = 470 nm) to 234 kJ mole^{-1} for green light (λ = 510). The rarer red bioluminescence of the "railway worm" (λ = 700 nm) has the lowest energy (of about 171 kJ mole^{-1}).

The chemical energy for even the lowest energy bioluminescence cannot come from ATP on a mole ATP/mole photons basis ($\Delta G^{\circ\prime}$ for ATP hydrolysis is 30 to 60 kJ mole^{-1}). In bacteria, only 0.15 to 0.20 moles of photons are produced per mole of substrate ($FMNH_2$). The chemical energy yield of $FMNH_2$ is about 150 kJ mole^{-1} and that of the emitted light is about 40 to 50 kJ mole^{-1}, and so this bioluminesce system is energetically feasible without additional energy input. *Cypridina*-type and coelenterate-type bioluminescence require O_2 for the direct reaction with luciferin (not for the cellular production of ATP). *Renilla* has a quantum yield of about 0.055 moles photons per mole substrate. This is equivalent to light production of about 0.61 J per ml O_2 used (*cf* 20.1 J ml O_2^{-1} for aerobic metabolism). The *Cypridina* luciferin has a quantum yield of about 0.3, which is equivalent to about 3.5 J ml O_2^{-1}. For the firefly, each of the reacting luciferins may release a photon (214 kJ mole^{-1}), with the hydrolysis of only 1 ATP (30 to 40 kJ mole^{-1}), to form oxyluciferin. The synthesis of luciferin and reconversion of oxyluciferin to luciferin must therefore require considerable energy expenditure.

The metabolic cost of *Cypridina*-type luminescence has been estimated for photophores of the teleost fish, *Porichthys* (Mellefet and Baguet 1984). Quiescient photophores have a VO_2 of about 0.07 nmole O_2 min^{-1} photophore^{-1}, which is about 2% of its resting O_2 requirement. Luminescence increases the O_2 requirements of the photophores to about 0.25 nmol O_2 photophore^{-1} (about a five times increase). About 1 nmol O_2 photophore^{-1} provides for the production of about 76 megaquanta of photons (76 10^6 photons) at a wavelength of about 470 nm; this is equivalent to about 1.4 J ml O_2^{-1}. The deep sea fish *Argyropelecus* appears to maintain a prolonged luminescence of photophores, without an increase in O_2 requirement (Mellefet and Baguet 1985). The bioluminescence of *Argyropelecus* may involve a coelenterate-like, preactivated Ca^{2+}-sensitive photoprotein intermediate, rather than the direct oxidation of luciferin by O_2, so that the metabolic cost of "preactivation" is expended over long periods between luminescence activity.

The many different functions proposed for bioluminescence (Buck, 1978) can be divided into three general categories: (1) roles of light emission by an

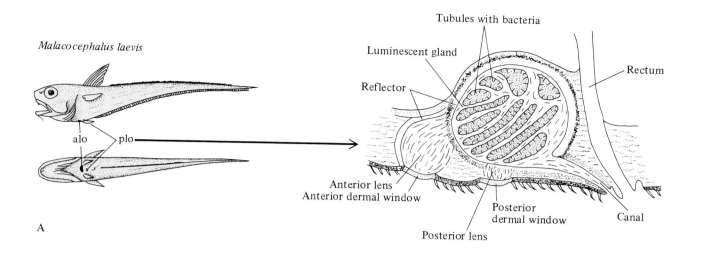

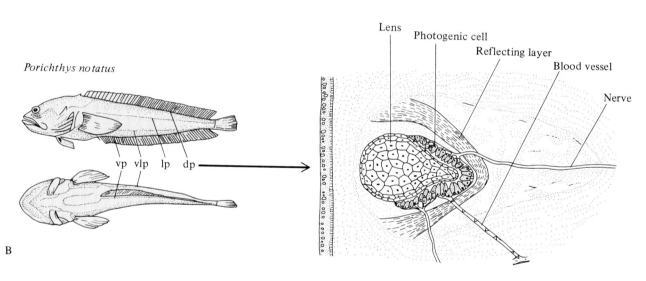

FIGURE 3–23 **(A)** Bacterial light organ of the rat-tail fish, *Malacocephalus laevis*; alo, anterior light organ; plo, posterior light organ. **(B)** Intracellular bioluminescent organ of the toad fish, *Porichthys notatus*; vp, ventral photophore row; vlp, ventro-lateral photophore row; lp, lateral photophore row; dp, dorsal photophore row. *(From Herring and Morin 1978.)*

individual; (2) intraspecific communication; and (3) interspecific interactions. Emitted bioluminescence can be used for food gathering by attracting prey to light (e.g., deep-sea angler fish have a luminous "lure," and various fish have luminous photophores on their lips or within their mouth). Some bioluminescent animals might benefit during food gathering by improved vision owing to their emitted light. A variety of animals use emitted light for defense, either by startling and confusing predators (e.g., the flashlight fish *Photoblepharon* has a "blink-and-run" behavior to confuse predators), as camouflage, as

concealment by counter-shading the ventral surface to match the surroundings, as a decoy to confuse predators, or as a repellent to light-avoiding predators. Many examples of bioluminescence are involved with intraspecific communication for either courtship (e.g., many female fireflies attract males by a species-specific flashing pattern) or congregation of many individuals into large groups (e.g., some Oriental fireflies aggregate into courtship congregations with synchronized light-flashing; many schools or shoals of fish, squid, and crustaceans have nonsynchronized bioluminescence). The pred-

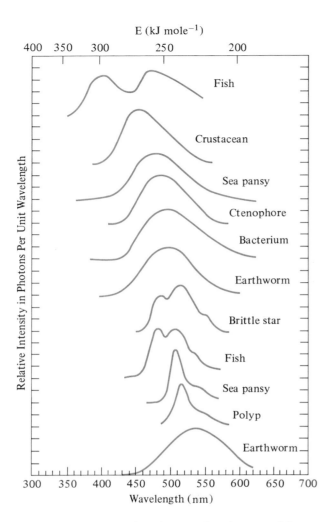

FIGURE 3–24 Wavelength (λ; nm) and energy (kJ mole photons⁻¹) emitted by bioluminescent organisms. Searsid fish exudate, crustacean *Cypridina* reaction *in vitro*, sea pansy *Renilla* reaction *in vitro*, ctenophore *Mnemiopsin* emission *in vivo*, photobacterium emission *in vivo*, earthworm *Diplocardia* emission *in vivo*, brittle star emission *in vivo*, fish *Porichthys* emission *in vivo*, sea pansy *Renilla* emission *in vivo*, hydropolyp *Clytia* emission *in vivo*, earthworm *Diplotrema* emission *in vivo*. *(From Wampler 1978.)*

FIGURE 3–25 A female firefly *Photuris versicolor* consumes a male firefly *Photinus tanytoxus*, which she has attracted with false signals that mimic that of the female *Photinus tanytoxus*. *(Photograph by J. E. Lloyd.)*

atory use of bioluminescence by some female fireflies is an example of interspecific communication; females of several species of *Photuris* fireflies lure male *Photinus* fireflies by mimicking the courtship flash pattern of the female *Photinus* (Figure 3-25).

Thermogenesis

Endothermic animals use chemical energy to generate heat for thermoregulation. A high rate of heat production enables some animals (mammals, birds, a few reptiles, many insects, and some large elasmobranch and teleost fish) to elevate body temperature above ambient temperature and often to precisely regulate body temperature independent of ambient temperature (see Chapter 5). Insects and fish produce heat when flying, walking, or swimming by contracting locomotory muscles. Nonmuscular sources of heat are also very important in mammals and birds for the regulation of body temperature since they are continuously endothermic (although some abandon endothermy for short periods of daily torpor or hibernation).

There are at least two nonmuscular mechanisms for the specialized production of metabolic heat: futile cycles of ATP synthesis and degradation, e.g., Na^+-K^+-ATPase and leaky cell membranes, and "proton leaks" in the inner mitochondrial membrane, e.g., brown adipose tissue (BAT).

Leaky Membranes

Endothermic mammals and birds have about 4 to 5 × the metabolic rate of similar-sized ectotherms (e.g., lizards) at the same body temperature (see Chapter 4). This difference in metabolic rate is also evident *in vitro* for tissue homogenates and slices. A large portion of the energy expended by the tissues of endotherms appears to be used by the Na^+-K^+-ATPase to actively pump K^+ into and Na^+ out of the cells. However, the transcellular membrane K^+ and Na^+ gradient is similar for endotherms and ectotherms (despite the higher energy expenditure of endotherm cells for ion pumping) and so the membranes of endotherm cells appear to be considerably more "leaky" to K^+ and Na^+ (Else and Hulbert 1987). Thus, a futile cycle of passive influx of Na^+ and K^+ efflux, and active transport of Na^+ out and K^+ in, provides a major source of metabolic heat. This futile cycle, and perhaps other similar ATP-requiring futile cycles, provide metabolic heat for thermoregulation. In contrast, brown adipose tissue provides heat without requiring the futile synthesis and degradation of ATP.

Brown Adipose Tissue

Brown adipose tissue (BAT) is a highly specialized form of fat, found only in certain mammals. Some hibernators, some cold-adapted mammals, and some newborn mammals (including humans) have brown fat. BAT is the only animal tissue with the sole function of heat production. Brown fat is generally deposited in the thorax, and its vascular drainage is directed to the vital organs (heart, lungs, etc). The BAT content of mammals can vary with age and between species, but it is generally present in those mammals, and at those stages of development, when the capacity for metabolic heat production is of critical importance, e.g., hibernating mammals that need to warm up and newborn mammals that have a high surface/volume, hence high heat loss.

BAT cells are quite distinctive from normal white lipid deposits; they have numerous small lipid droplets and are packed with circular mitochondria, characterized by many cristae (Figure 3–26A). The normal coupling of electron transport and the proton gradient across the inner mitochondrial membrane, resulting in ATP synthesis, can be short-circuited in BAT mitochondria. Heat can therefore be produced by electron transport without concomitant ATP synthesis.

Activation of BAT is primarily hormonal, through the β-adrenergic pathway (Cannon and Nedergaard 1985). Norepinephrine (noradrenalin) binds to β-adrenergic receptors on the cell membrane and causes cyclic-AMP synthesis (Figure 3–26B). The cyclic-AMP activates protein kinase, which in turn activates a hormone-sensitive lipase (HSL) that hydrolyzes intracellular triglyceride stores to free fatty acids. The cyclic-AMP also activates lipoprotein lipase, which hydrolyzes circulating triglycerides to fatty acids. Thus, fatty acids are made available to the cells for β-oxidation to acetyl-CoA, which then enters the citric acid cycle and produces $NADH/H^+$ and $FADH_2$.

But how is electron transport uncoupled from ATP synthesis? Free fatty acids appear to control the activity of a special protein, **thermogenin**, which is found only in the inner mitochondrial membrane of BAT mitochondria. Thermogenin can bind purine nucleotides (e.g., ATP) but in particular binds GDP; it also modifies the ionic permeability of the inner mitochondrial membrane by providing channels for Cl^- and H^+ exchange. A low concentration of GDP increases the permeability of the thermogenin protein to H^+ and Cl^-, but the GDP concentration of the cells is probably always sufficiently high to maintain minimal H^+ and Cl^- permeabilities. However, the presence of even submicromolar concentrations of fatty acids affects thermogenin and increases the permeability of the inner mitochondrial membrane to H^+ and Cl^-, effectively short-circuiting the proton gradient established by the electron transport chain. This allows β-oxidation and maximal rates of heat production without concomitant ATP synthesis.

BAT is an important component of the elevated capacity of cold-acclimated mammals to produce heat by nonshivering thermogenesis. There is both an increased thermogenin content of BAT mitochondria and increased mitochondrial density with cold acclimation (Sundin et al. 1987).

Defense

Thermal energy can be used as a defense mechanism as well as for thermoregulation. For example, the defensive noxious quinone spray of carabid "bombardier" beetles (Figure 3–27A) is made more effective by its high temperature, 60 to 100° C! The explosive discharge of the hot spray is clearly audible, hence the name bombardier beetle.

Most carabid beetles discharge a cold defensive secretion, but bombardier beetles belonging to three separate subfamilies—the Carabinae (*Brachinus*), Paussinae (*Goniotropis*), and Metriinae (*Metrius*)—have a hot discharge. It is not clear if the bombardier strategy evolved independently in these three separate subfamilies, since the biochemistry and thermodynamics are similar for bombardier beetles of each subfamily (Aneshansley et al. 1983).

The secretory portions of the bombardier beetle's glands (Figure 3–27B) synthesize hydroquinone (and methylhydroquinone); hydrogen peroxide; and a hydrocarbon, pentadecane (Schildknecht and Holoubek 1961). The secretion is stored in a reservoir portion of the double-chambered gland. The second chamber of the gland is the "reaction" chamber, which contains a mixture of catalase and peroxidase enzymes. To discharge a hot, noxious spray, the beetle contracts a small muscle to open a valve between the reservoir and reaction chambers. This

A

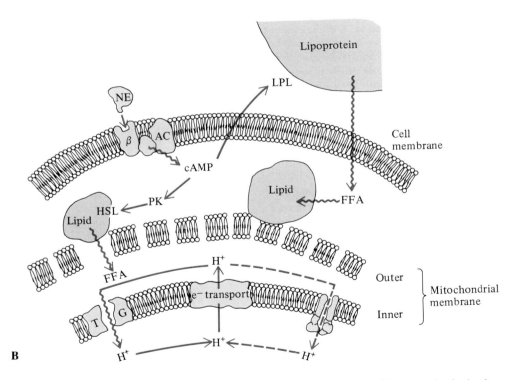

B

FIGURE 3–26   (A) Brown adipose tissue (BAT) has an abundance of large, spherical mitochondria with cristae extending across the width of the mitochondrion. Electron micrograph is of BAT from a recently aroused bat.   (*From Fawcett 1986.*) **(B)** Schematic details of the β-adrenergic pathway for thermogenesis by brown adipose tissue (BAT) showing norepinephrine (NE) stimulation of surface receptors, activation of lipoprotein lipase (LPL) and hormone-sensitive lipase (HSL), formation of free fatty acids (FFA), and activation of thermogenin (TG) to increase the permeability of the inner mitochondrial membrane to H^+ and Cl^-.   (*Modified from Cannon and Nedergaard 1985.*)

A

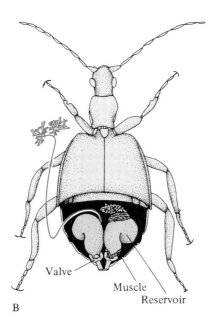

Valve

Muscle

Reservoir

B

FIGURE 3–27 (A) The Bombardier beetle can accurately discharge a hot, noxious defensive spray of p-benzoquinones in pulses of 500 sec⁻¹. *(From Dean et al. 1990.)* **(B)** The Bombardier beetle has paired defensive glands, each with a secretory gland (uncoiled on the left) and reservoir. A muscle controls the release of hydroquinones and hydrogen peroxide into a reaction chamber where the enzymatic catalytic production of quinones, oxygen, and water releases heat. Sufficient heat is released to vaporize some of the reactant solution. *(Modified from Eisner 1970.)*

allows some of the reactant fluid (25% hydrogen peroxide, 10% hydroquinones) to enter the reaction chamber where peroxidases promote oxidation of the hydroquinones to quinones, and catalases decompose the hydrogen peroxide to water and oxygen.

The reactions that occur in the reaction chamber are very exothermic. For example, the reaction of hydroquinone with peroxide to form quinone and water has a $\Delta G°$ of -203 kJ mole⁻¹. The actual heat production of the reaction solution is about 790 J g⁻¹ solution. This is sufficient heat to warm the solution to 100° C (about 334 J g⁻¹) and to vaporize about 1/5 of the solution (about 455 J g⁻¹).

Bombardier beetles are not only able to emit a hot, noxious spray, but they have a very accurate aim! *Brachinus* rotates the tip of its abdomen (where the glands open) to effect an accurate aim! *Goniotropis* has a "high technology" targeting system, consisting of a pair of curved and grooved abdominal flanges. The ejected spray follows the curvature of the flanges and is directed forward towards the target (Eisner and Aneshansley 1982).

The release of cornicle wax by aphids is another even more bizarre defense system, utilizing an exothermic reaction (Edwards 1966). The aphid secretes from the cornicles a globule of liquid wax, which crystallizes on contact and deters or immobilizes predators. The waxes have a sufficiently high melting point (38° C for *Aphis* to 48° C for *Macrosiphum*) that they appear to be secreted in a supercooled state. The supercooled wax is released on contact with a predator, such as the hymenopteran wasp *Aphidius*, by a reflex contraction of the cornicle muscle that opens the terminal cornicle valve. A positive abdominal hydrostatic pressure is thought to then expel the liquid wax. A foreign object such as the wasp acts as a nucleating point to immediately crystallize the wax. The thermodynamics of this spontaneous change of state from liquid to solid wax is similar to that for supercooled water crystallizing to ice, when it is "nucleated" by a small ice crystal. Such a charge of state for a supercooled solution is an exothermic reaction and is irreversible.

Summary

Animals use chemical energy to support energy-requiring cellular processes (e.g., ion pumps, biosynthesis) and to do work on the external environment (e.g., muscle, ciliary movement).

The Gibbs free energy change (ΔG) for a chemical reaction depends on the change in heat energy, or enthalpy (ΔH), temperature (T), and degree of disorder, or entropy (ΔS). A negative ΔG (i.e., energy released by the reaction) can be used in biological systems to do mechanical work ($f.\Delta s$), to transport molecules against a chemical gradient ($\mu.\Delta n$), to transport ions against an electrical gradient ($E.\Delta q$), or to emit light ($h\mu$);

$$\Delta G = H - T.\Delta S = f.\Delta s + \mu.\Delta n + E.\Delta q + h\nu$$

Diffusion is passive mass transport from a high concentration to a low concentration. Fick's first law of diffusion states that the diffusion flux (J) is dependent on the diffusion coefficient (D), the area for diffusional exchange (A), the concentration difference (ΔC), and the path length for diffusion (Δx). For 1-dimensional planar diffusion, we have the following.

$$J = -DA\Delta C/\Delta x$$

The standard Gibbs free energy change ($\Delta G°$) for complete oxidation of typical metabolic substrates is considerably negative, e.g., for glucose, -2854 kJ mole^{-1}; palmitate, -9791 kJ mole^{-1}; and alanine, -1619 kJ mole^{-1}. The $\Delta G°$ for individual biochemical reactions is considerably less, about $+40$ to -40 kJ mole^{-1}. Biochemical reactions with negative $\Delta G°$, about -20 to -40 kJ mole^{-1}, provide the energy for active cellular reactions. Adenosine-triphosphate (ATP) hydrolysis is the most common energy-providing cellular reaction ($\Delta G° = -30.5$ kJ mole^{-1}).

Oxidation of glucose to CO_2 and water yields about 2854 kJ mole^{-1} of energy; about 38 mole of ATP are formed from ADP per mole of glucose. Two ATP are formed by the Embden-Meyerhof pathway (glucose $\rightarrow$ 2 pyruvate). A further 2 GTP are formed by the citric acid cycle and 34 ATP are formed via the electron transport system from NADH/H$^+$ and FADH$_2$. About 1693 of the total 2854 kJ mole^{-1} of energy is released as heat by cellular metabolism of glucose, and 1161 is converted to the chemical energy of ATP; the efficiency of cellular glucose metabolism is about 41%. Cellular metabolism of lipids (e.g., fatty acids) and proteins (and amino acids) has an efficiency of about 60% and 30%, respectively.

The cellular metabolic pathways of many animals are adapted for ATP synthesis in the absence of oxygen, due to either limitations of the O_2 delivery system during intense activity or to environmental hypoxia or anoxia. Glycolysis is a common anaerobic pathway. Glucose is metabolized to pyruvate, then the end product lactate; 2 ATP are synthesized per glucose. Most vertebrates and many invertebrates utilize mainly glycolysis for anaerobic metabolism. Many invertebrates however, have other anaerobic pathways for increased ATP synthesis from glucose, utilization of additional substrates (e.g., amino acids), and to form other nonaccumulated end products (e.g., propionate, succinate, acetate, methyl-butyrate, and ethanol).

In some biochemical reactions, part of the free energy change is released as photons of light, rather than as heat. The chemical reactions generally involve an activated, or high-energy, substrate (luciferin) and an enzyme (luciferase) that oxidize the substrate; light is released by the oxidation reaction. Bioluminescence has evolved independently in a number of diverse protozoans and animals. Some animals utilize symbiotic luminescent bacteria for their light production. Some exude mixtures of luciferin and luciferase for extracellular light production. Some animals have intracellular biochemical pathways for light production.

The biochemical production of heat has no useful role in ectothermic animals, but endothermic animals use metabolic heat to regulate body temperature. These endotherms can augment metabolic heat production by increasing the activity of ATP-requiring cellular processes, such as muscle contraction (e.g., shivering) or ion pumping (e.g., cellular Na-K ATPase). Some mammals have a special lipid store, brown adipose tissue (BAT), the sole function of which is metabolic heat production. In BAT, heat is produced by short-circuiting the synthesis of ATP in mitochondria, rather than by using ATP hydrolysis to release heat. A protein, thermogenin, is stimulated by fatty acids to increase the permeability of the inner mitochondrial membrane to H$^+$, thereby short-circuiting the proton gradient across the membrane.

Biochemical heat production is also important as a defense mechanism for the bombardier beetle, which ejects a hot, noxious spray. The heat produced in the defensive secretion by the reaction of hydrogen peroxide with hydroquinones is sufficient to vaporize about 1/5 of the secretion.

Supplement 3–1

Diffusion in Various Geometries

· ·

The general form of Fick's law of diffusion is

$$J_j = -D_jA\frac{\delta C}{\delta x} - D_jA\frac{\delta C}{\delta y} - D_jA\frac{\delta C}{\delta z},$$

where J_j is the flux of species j; D_j is the diffusion coefficient for species j; and $\delta C/\delta x$, $\delta C/\delta y$, and $\delta C/\delta z$ are the partial derivatives of the concentration with respect to the x, y, and z coordinates. The negative sign indicates that the flux is from the high to the low concentration. The exact geometrical form of the diffusional exchange surface determines the complexity of the diffusion equation. For example, in simple 1-dimensional planar geometry, the general diffusion equation reduces to

$$J_j = -D_j A \Delta C / \Delta x.$$

Biological examples of such simple diffusion geometries would be gas exchange across the lung alveolar surface in mammals or diffusional exchange of ions across the body surface of aquatic animals.

The equations for diffusional flux across a variety of physiologically relevant spatial organizations of exchange surfaces are given below, with some biologically relevant examples of such exchange geometries. Note that all equations are of the same general form

$$J_j = -FD\Delta C,$$

where F is the diffusion factor and ΔC the concentration difference. The same equations apply to heat transfer as well as mass flux by diffusion.

Configuration	Flux	Biological Examples
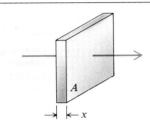	$J = -\dfrac{A}{x}D\Delta C$	Alveolar surface exchange in higher vertebrates (not birds). x = thickness of alveolar membrane; A = area.
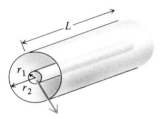	$J = -\dfrac{2\pi L}{ln(r_2/r_1)}D\Delta C$	Cutaneous gas exchange in long, thin animals; r_2 is the boundary layer thickness.
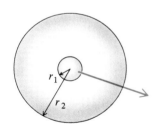	$J = -\dfrac{4\pi}{(1/r_1) - (1/r_2)}D\Delta C$	Cutaneous gas exchange in spherical animals; r_2 is the boundary layer thickness.
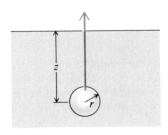	$J = -\dfrac{4\pi r}{1 - r/2z}D\Delta C$	Gas exchange from subterranean animal nest chamber ($z > r$).
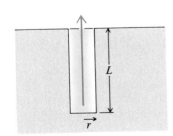	$J = -\dfrac{2\pi L}{ln(2z/r)}D\Delta C$	Insect tracheal system; air capillaries of bird lung ($L \gg r$).

Configuration	Flux	Biological Examples
	$$J = -\frac{2\pi L}{ln(2L/r)}D\Delta C$$	Blood vessel under surface of skin ($L \gg r$; $z \gg 3r$).
	$$J = -\frac{2\pi L}{\cosh_{-1}\frac{(w^2 - r_1^2 - r_2^2)}{(2\,r_1\,r_2)}}D\Delta C$$	Countercurrent exchange between parallel artery and veins ($L \gg r_1, r_2$; $l \gg w$)
	$$J = -\frac{2\pi L}{ln(4z/\pi r)}D\Delta C$$	Capillary in fish gill lamella ($z > r$).

Supplement 3–2

Mechanisms of Bioluminescence

Bioluminescence occurs when some of the chemical energy of biochemical reactions is released not as heat but as photons of light. This conversion of chemical energy to light is typically due to highly strained chemical structures of luciferins, often with peroxide bonds (—O—O—). The light energy is emitted when the chemical structure returns from an excited, high-energy state to a lower energy, stable ground state. Different bioluminescent organisms have various types of luciferins that are used in different biochemical pathways to release light.

There are a variety of types of luciferin, the bioluminescent substrate for light production, although there are some general similarities in structure for some luciferins from some very different animals. For example, note the basic similarity of the reactive portions of the *Renilla, Aequorea,* and *Cypridina* luciferins (shaded regions).

The luciferin of bioluminescent bacteria is a complex formed from a flavoprotein ($FMNH_2$) and a long-chain aldehyde (R—CHO; R is longer than C_6). The FMN-RCHO complex is oxidized to FMN and a carboxylic acid in the presence of a luciferase and oxygen, producing H_2O and releasing photons of visible light (A). Most of the photons have a wavelength of about 480 nm (λ_{max}) with energy ($h\nu$) of 249 kJ mole photons^{-1}. The overall stoichiometry of the reaction is

$$FMNH_2 + O_2 + RCHO$$
$$\xrightarrow{\text{luciferase}} 0.15\,h\nu + FMN + H_2O + RCOOH.$$

The bioluminescence system of an anthozoan coelenterate, the sea pansy *Renilla*, requires oxygen but not ATP. The oxidation of reduced oxyluciferin releases light. In some coelenterates, including *Renilla*, the emitted light is green ($\lambda_{max} = 510$ nm) *in vivo* but is blue ($\lambda_{max} = 470$) *in vitro*, indicating a transformation of the chemical energy from the oxyluciferin complex by a green fluorescent protein (GFP).

Calcium-activated photoproteins emit light in the presence of calcium ions. Examples are aequorin from the hydrozoan coelenterate *Aequorea* and mnemiopsin from the ctenophore *Mnemiopsis*. The photoprotein consists of protein and a coelenterate-type luciferin. Oxygen is incorporated into the structure of the photoprotein during its synthesis, and so free O_2 is not required for the luminescence process (unlike other coelenterate-type luminescent reactions). A model for Ca^{2+}-activated bioluminescence of aequorin and mnemiopsin is essentially identical to *Renilla* bioluminescence except for the preincorporation of O_2 and the excitatory role of Ca^{2+}. The fluorescent properties alter only slightly in the absence of Ca^{2+} to a bright blue (*in vitro*) or green (*in vivo*) in the

Renilla (sea pen)

Cypridina (Ostracod)

Latia (Limpet)

Aequorea/Mnemiopsin (hydrozoan/comb jelly)

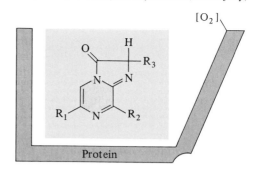

Photinus (firefly)

Structure of luciferins.

presence of Ca^{2+}. This mechanism has enabled physiologists to measure the amount of Ca^{2+} present inside cells by injecting aequorin into the cell and then measuring the light emission.

The bioluminescence systems of terrestrial earthworms such as *Diplocardia* are poorly known. Hydrogen peroxide reacts with the luciferin to release light, and it is likely that O_2 is converted to H_2O_2 by an oxidase enzyme when coelomic cells containing the luciferin are lysed. This occurs prior to the exudation of coelomic fluid from the mouth, anus, and dorsal pores. Hydrogen peroxide also activates the bioluminescent system of the freshwater limpet, *Latia*. The *Latia* luciferin combines with purple protein, then reacts with H_2O_2 to release light ($\lambda_{max} = 535$ nm).

A very different series of reactions requiring ATP is seen for insect luminescence, typified by that of the firefly beetle *Photinus* (family Lampyridae). The reduced form of the luciferin combines with ATP in the presence of a luciferase to form a luciferyl adenylate complex. This complex then decomposes to produce the oxidized luciferin carbon dioxide and light ($\lambda_{max} = 560$ nm). The luciferin reacts with ATP to form a luciferyl-adenylate intermediate, which then reacts with O_2 to form oxyluciferin and CO_2, and releases light. (See Cormier 1978; McElroy and DeLuca 1981; Hastings 1978; Lehninger 1982.)

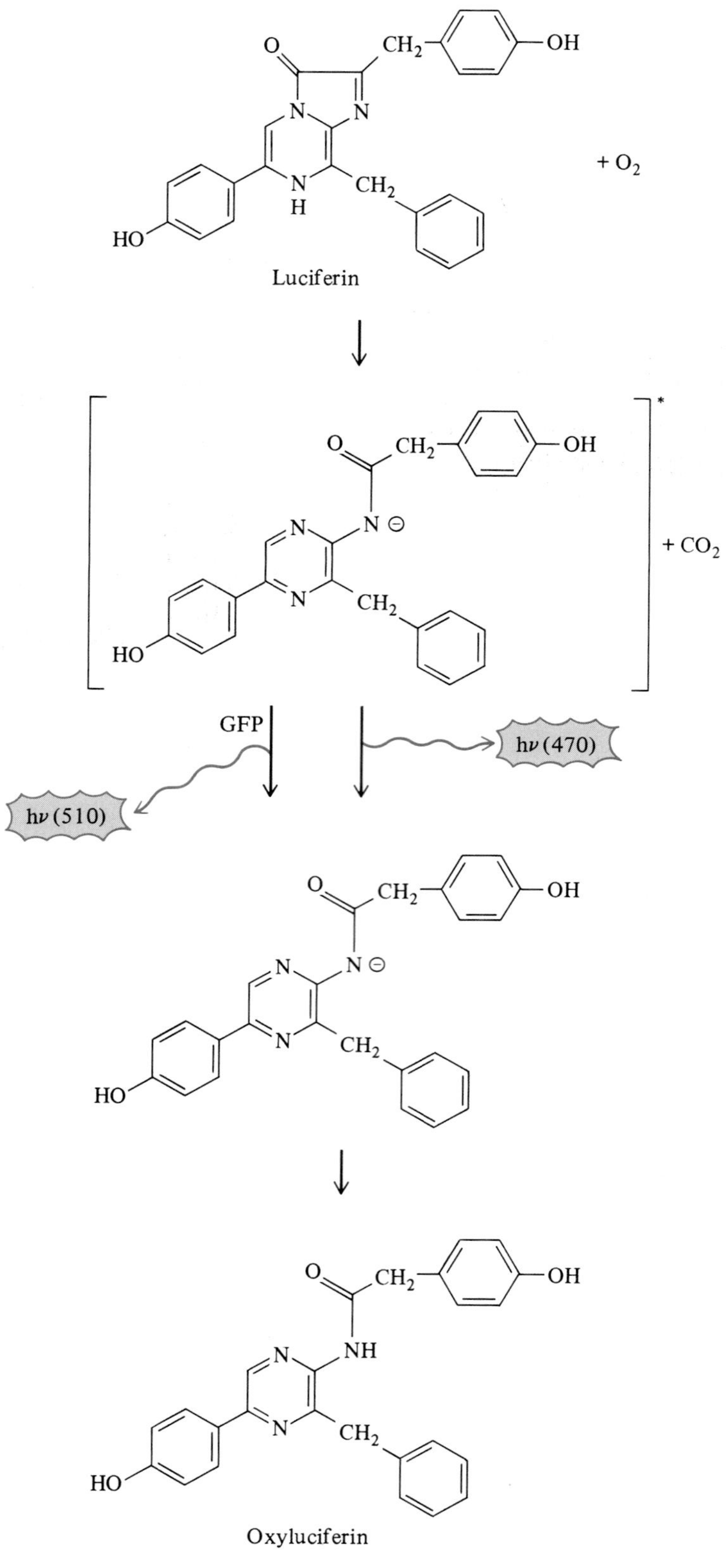

Luciferin

+ O$_2$

+ CO$_2$

GFP

hν (470)

hν (510)

Oxyluciferin

Bioluminescence in the coelenterate *Renilla* emits either blue light (470 nm) directly from the blue fluorescent photoprotein or green light (510 nm) by the green fluorescent protein associated with the luciferin complex.

Luciferin (reduced)

Bioluminescent system of coelenterate-type calcium-activated photoproteins, such as aequorin and mnemiopsin. Light is continually emitted from the high-energy activated complex as long as calcium is present.

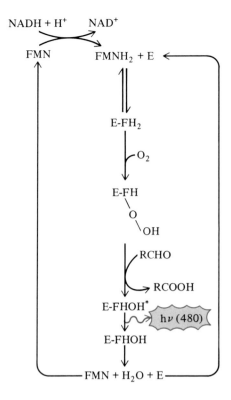

Reaction sequence for bacterial bioluminescent release of light (hν).

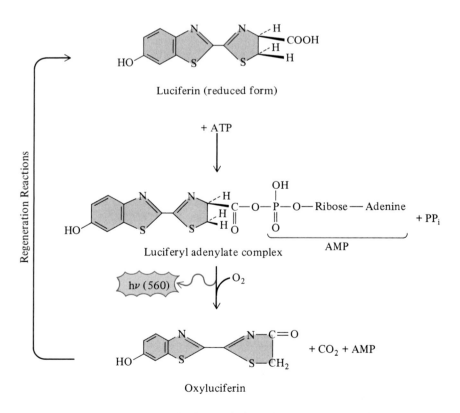

Bioluminescent reactions for firefly luciferin and ATP to release light.

Recommended Reading

Armstrong, F. B. 1983. *Biochemistry*. New York: Oxford University Press.

Braefield, A. E., and M. J. Llewellyn. 1982. *Animal energetics*. Glasgow: Blackie.

Harvey, E. N. 1952. *Bioluminescence*. New York: Academic Press.

Herring, P. J. 1978. *Bioluminescence in action*. New York: Academic Press.

Hochachka, P. W. 1980. *Living without oxygen*. Cambridge: Harvard University Press.

Hochachka, P. W., and G. N. Somero. 1984. *Biochemical adaptation*. Princeton: Princeton Univ. Press.

Klotz, I. M. 1967. *Energy changes in biochemical reactions*. New York: Academic Press.

Mahler, H. R., and E. H. Cordes. 1971. *Biological chemistry*. New York: Harper & Row.

McGilvery, R. W., and G. W. Goldstein. 1983. *Biochemistry: A functional approach*. Philadelphia: W. B. Saunders Co.

Nobel, P. S. 1983. *Biophysical plant physiology and ecology*. New York: W. H. Freeman & Co.

Patton, A. R. 1965. *Biochemical energetics*. Philadelphia: W. B. Saunders Co.

Peusner, L. 1974. *Concepts in bioenergetics*. Englewood Cliffs: Prentice-Hall.

West, E. S. 1961. *Textbook of biophysical chemistry*. New York: Macmillan.

White, A., P. Handler, and E. L. Smith. 1968. *Principles of biochemistry*. New York: McGraw-Hill.

Chapter 4

Animal Energetics

· ·

Hovering by a large animal, as does this rufous hummingbird, is one of the most energy-demanding activities of animals. *(T. Ulrich/H. Armstrong Roberts, Inc.)*

The initial investigations of animal metabolism closely followed advances in the physical sciences related to heat and energy and considerably preceded our concepts of modern chemistry and biochemistry. For example, Lavoisier used an ice bath calorimeter to measure the metabolic rate of animals and named oxygen as a constituent of air in the 1770s, well before oxygen was identified as an element and before the details of atomic and molecular structure and biochemistry were known.

Sanctorio Sanctorio, an Italian professor of medicine, conducted a simple yet pioneering experiment in about 1590. Using a large scale to weigh himself, Sanctorio found that the weight of the food and drink that he consumed was greater than the weight lost as feces and urine, and that his body weight did not increase to account for the difference in weight between intake and loss. Sanctorio accounted for the difference in weight as "insensible perspiration": "If eight pounds of meat and drink are taken in a day, the quantity that usually goes off by insensible perspiration in that time is five pounds" (Clendening 1933). Sanctorio's explanation is partially correct, but it was not until a number of centuries later that much of the weight loss could be correctly ascribed to metabolism.

In the centuries after Sanctorio, a number of experimenters (Boyle, Mayow, Lavoisier, and others) demonstrated a fundamental similarity between animal metabolism and inanimate combustion: both require the same constituent of air to live or burn. The analogy between animal metabolism and inanimate combustion is aptly epitomized by Kleiber's phrase describing animal metabolism as the "Fire of Life." Mayow showed that a candle within an airtight bell jar would extinguish at about the same time that a mouse in the jar would die. Priestley, in 1774, isolated a gas that supported both fire and animal life; he called it "dephlogisticated air." Lavoisier, by 1777, had also isolated "dephlogisticated air," described its properties, and named it **oxygen**. He further described how animal metabolism consumes oxygen, liberates "fixed air" (**carbon dioxide**), and releases heat. These studies provided the framework for the modern study of animal energetics.

Calorimetry

Examination of the methods used for measuring heat exchange and animal metabolism provides a historical perspective to the technical development of animal energetics and provides the conceptual framework required for an understanding of animal energetics.

Lavoisier determined the metabolic rate of animals, such as guinea pigs, mice, and sparrows; he directly determined their rate of heat (joule) production by measuring their rate of heat loss to their environment. This type of measurement of animal metabolism is termed **direct calorimetry** because it directly measures heat production. Lavoisier used a double-jacket ice bath calorimeter (Figure 4–1). The inner ice jacket maintained the animal at a constant 0° C, and heat produced by the animal melted the ice. The rate at which water dripped from the inner jacket was converted to heat production according to the latent heat of fusion for ice (334 J g^{-1} at 0° C). The outer ice jacket prevented conduction of heat from the external environment into the inner jacket, which would also have melted the ice. Direct calorimetry is still commonly used to measure animal metabolism, although in a considerably more sophisticated form than Laviosier's ice jacket calorimeter.

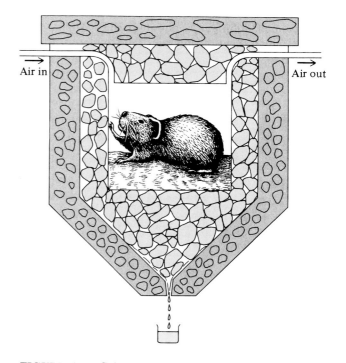

FIGURE 4–1 Schematic of ice bath calorimeter used by Lavoisier to measure the metabolic heat production of mammals and birds by direct calorimetry. The rate of the melting of the ice from the inner adiabatic jacket was converted to the rate of heat production (latent heat of fusion = 334 kJ g^{-1}). *(From Kleiber 1961.)*

In principle, there are many other methods for measuring the metabolic rate of animals. Consider, for example, the metabolism of glucose to carbon dioxide and water (see Chapter 3). Lavoisier's direct calorimetry experiments measured the heat produced by metabolism, i.e., the 2874 kJ mole^{-1} heat of combustion for glucose. Similarly, the rate of oxygen consumption (VO_2) or production of carbon dioxide (VCO_2) would provide a measure of metabolic rate. Measures of metabolism such as VO_2 and VCO_2 are not directly equivalent to a rate of heat production, but can be related to heat production through the stoichiometry of the chemical reaction. Consequently, such techniques for the measurement of metabolic rate are termed **indirect calorimetry**.

Another of Lavoisier's experiments illustrates the concept of indirect calorimetry (Figure 4–2). A large bell jar is floated on mercury to make it airtight. The initial gas volume (V_1) is determined by measuring the height of the top of the bell jar. A guinea pig is then introduced into the bell jar through the mercury seal, where it sits on a piece of cork (which floats on the mercury); the indicated volume is now V_2. The guinea pig is removed via the mercury seal when its metabolism has sufficiently depleted the air of oxygen (its respiration becomes labored). The gas volume is then redetermined (V_3) and is found to be slightly lower than the initial volume. Finally, sodium hydroxide is introduced into the bell jar to absorb carbon dioxide, and the volume of air decreases to V_4. The overall decrease in gas volume ($V_1 - V_4$) is equal to the amount of oxygen consumed by the guinea pig; the amount of carbon dioxide produced by the guinea pig is $V_3 - V_4$. The rates of oxygen consumption and carbon dioxide production, both of which are estimates of metabolic rate, are then calculated from these gas volumes and the length of time that the guinea pig was inside the bell jar. Many currently used manometric methods for measurement of metabolic rate use this same general principle, but in a more sophisticated form than Lavoisier's bell jar and mercury bath. Direct measurement of oxygen or carbon dioxide concentration by gas analyzers, rather than volume change, also are commonly used for the measurement of metabolic rate.

Metabolic water production is, theoretically, as good a measure of metabolism as is oxygen consumption or carbon dioxide production. However, it generally is not possible to accurately measure metabolic water production since it is a small amount

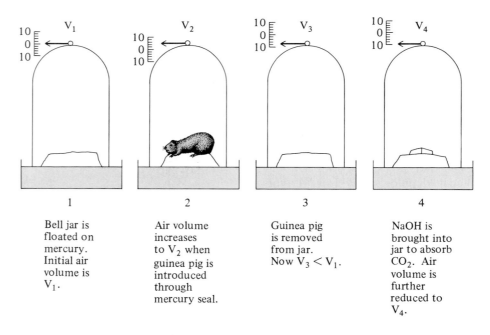

1	2	3	4
Bell jar is floated on mercury. Initial air volume is V_1.	Air volume increases to V_2 when guinea pig is introduced through mercury seal.	Guinea pig is removed from jar. Now $V_3 < V_1$.	NaOH is brought into jar to absorb CO_2. Air volume is further reduced to V_4.

FIGURE 4–2   Schematic of technique used by Lavoisier for measurement of metabolic rate for a guinea pig by indirect calorimetry. Measurement of the volume of air in the bell jar (using the top scale) before, after a guinea pig is placed in the chamber, and after a CO_2 absorbent is placed in the chamber yields the amount of CO_2 produced ($V_3 - V_4$) and O_2 consumed ($V_1 - V_4$). *(From Kleiber 1961.)*

of water diluted into a much larger total body water content of the animal, although isotope trace techniques can measure metabolic water production under favorable circumstances. It also is not generally feasible to measure metabolic rate as the rate of utilization of a particular metabolic substrate, e.g., glucose. An animal would most likely use a complex combination of various carbohydrates, lipids, and amino acids and not just glucose as its metabolic substrate. It also is not possible to measure the metabolite (e.g., glucose) content of an animal before and after a prescribed time period.

A final indirect calorimetric method for measurement of metabolic rate relies on the concept of conservation of energy (the first law of thermodynamics) and the determination of energy content by bomb calorimetry (see Chapter 3). The total amount of energy released by a chemical reaction must be the same, irrespective of the number or nature of the intermediate steps. This statement is Hess's law of sums. For example, combustion of glucose in a bomb calorimeter to CO_2 and H_2O releases the same amount of energy as conversion of glucose to the same end products by any other pathway, e.g., conversion through a carbon monoxide intermediate or by the multiple steps of cellular metabolism.

$$C_6H_{12}O_6 + 6O_2 \longrightarrow 6CO_2 + 6H_2O + 2874 \text{ kJ mole}^{-1}$$

$$C_6H_{12}O_6 + 3O_2 \longrightarrow 6CO + 3O_2 \longrightarrow 6CO_2 + 6H_2O + 2874 \text{ kJ mole}^{-1} \quad (4.1)$$

$$C_6H_{12}O_6 + 6O_2 \longrightarrow \longrightarrow \longrightarrow \longrightarrow 6CO_2 + 6H_2O + 2874 \text{ kJ mole}^{-1}$$

All chemical energy consumed by an animal as food (or drink) must be conserved, i.e., it must be excreted as feces or urine, stored as tissue production, lost as metabolic heat, or used to do external work. This is, of course, a modern version of Sanctorio's experiment but utilizing energy balance rather than weight balance. The complete balance equation for chemical energy intake (E_{in}) is as follows.

$$E_{in} = E_{food} + E_{drink}$$
$$= E_{out} = E_{feces} + E_{urine} + E_{growth} \quad (4.2a)$$
$$+ E_{metabolism} + E_{external work}$$

This general equation can be simplified if there is no energy intake via drinking, no growth, negligible urine energy, and negligible external work.

$$E_{metabolism} = E_{food} - E_{feces} \quad (4.2b)$$

The necessary measurements of energy content for food and feces can be readily accomplished with

a bomb calorimeter. A sample of the material is placed at a high pressure of pure oxygen and ignited (with a fuse wire heated by an electrical circuit). The heat energy released by complete combustion of the sample is measured by the temperature rise of the calorimeter. Careful calibration of the instrument with a standard material of known energy content (e.g., benzoic acid; 26.4 kJ g^{-1}) allows accurate measurement of energy content for various materials. The heat liberated during combustion of a substance in a bomb calorimeter is only an estimate, not an exact measure, of the enthalpy change (ΔH) because a bomb calorimeter is a constant volume, variable pressure system. A true enthalpy change is measured at constant pressure (see Chapter 3) but a bomb calorimetric measure is sufficiently close ($<1\%$ error) to be adequate for the study of animal energetics.

The energy content (measured by bomb calorimetry) is markedly different for carbohydrate (17 kJ g^{-1}), lipid (38 kJ g^{-1}), and protein (23 kJ g^{-1}; Table 4–1). The energy content of materials is generally expressed per gram ash-free dry weight because of the variable, and often high water content, and the small but significant fraction of dry weight that is inorganic (and does not contribute to the joule content). Foodstuffs are rarely pure, but generally are mixtures of carbohydrate, lipid, and protein. The energy content of biological materials

TABLE 4–1

Joule energy content (kJ g ash-free dry weight^{-1}) for various food items and animals, measured by bomb calorimetry.

	kJ g^{-1}
Carbohydrate	17
Millet seed	17.1
Gekko adult	19.4
Aquatic algae	19.7
Soy beans	21.8
Protein	23
Copepods	23.1
Mollusks (shell-free)	23.1
Earthworms	23.8
Mite	24.3
Guppy	24.4
Cnidaria	24.7
Ciliates	24.8
Beetle	26.4
Gekko egg	26.5
Platyhelminthes	26.6
Sponges	27.2
Spittlebugs	29.1
Lipid	38

varies from about 17 kJ g^{-1} to 30 kJ g^{-1}. These energy equivalence values can be used to calculate the metabolic rates of animals by the indirect calorimetric method of energy balance ($E_{metabolism} = E_{food} - E_{feces}$).

Glucose metabolism releases 2874 kJ mole^{-1} or 15.9 kJ g^{-1}; the equivalence between energy and indirect calorimetric measures is 21.4 kJ liter O$_2^{-1}$ and 21.4 kJ liter CO$_2^{-1}$ (see Supplement 4–1, page 118). The metabolism of C$_6$-subunits of glycogen yields slightly different energy equivalence values because there is one less H$_2$O per C$_6$-subunit. Lipid metabolism has very different indirect calorimetric ratios; the energy yield from palmitate metabolism is 39.2 kJ g^{-1}. The energetics of protein metabolism is complex because of its N content (see Supplement 4–2, page 118). The metabolism energy yield is 20.1 kJ g^{-1} if urea is the nitrogenous waste, but is slightly higher at 23.8 kJ g^{-1} if measured by bomb calorimetry because the nitrogenous product is not urea.

The ratio of volume of CO$_2$ produced per oxygen consumed is the **respiratory quotient (RQ)**. It is 1 for carbohydrate metabolism (6 moles CO$_2$/6 moles O$_2$ per mole), about 0.7 for lipids, and about 0.84 for protein. RQ is a useful index to the general nature of the metabolic substrate because it is lower for lipid and protein metabolism than for carbohydrate. Animals rarely use pure carbohydrate or lipid as their energy source and often do not utilize protein. The relationship between RQ and the joule equivalence of O$_2$ and CO$_2$ for varying proportions of carbohydrate and lipid can be calculated by combining the stoichiometric equations (Table 4–2). At lower RQ, fewer kJ are released per liter of O$_2$ and more kJ are released per liter of CO$_2$.

TABLE 4–2

Joule equivalence and RQ for various mixtures of carbohydrate (glucose) and lipid (glycerol tripalmitate).

Proportions of		Joule Equivalence		
Carbohydrate	**Lipid**	***kJ L O$_2^{-1}$***	***kJ L CO$_2^{-1}$***	**RQ**
1	0	21.4	21.4	1.00
0.8	0.2	20.9	23.7	0.88
0.6	0.4	20.5	25.6	0.80
0.4	0.6	20.1	26.4	0.76
0.2	0.8	19.9	27.2	0.73
0	1	19.5	27.8	0.70

[1] STPD.

Energy Budgets

Energy budgets are an account, or balance sheet, of all sources of useful energy gain, storage, and loss. As we shall see, sources of energy include food and drink; we will represent these sources by **consumption** as C. Energy is lost via **feces** (F) and **urine** (U). In addition, energy can be stored by tissue growth, i.e., **production** (P). Production is generally positive but can be negative if the animal metabolizes body tissues. Energy can also be used by an animal to perform external **work** (W_{out}), such as lifting a weight or overcoming frictional drag. External work is generally small or negligible. Work can also be done on an animal by the external environment, but this is not biologically useful and will not be further considered. The last, but by no means least, avenue for energy loss is **metabolism** (M); metabolic energy is dissipated to the environment as heat. Metabolism is often called "internal respiration," but the term "external respiration" is also used to describe gas exchange (Chapters 12, 13); to avoid possible confusion we will not use either of these terms.

The overall equation for balance of energy is

$$C = P + U + F + M + W_{out} \qquad (4.3)$$

Measurement of all components of the above equation enables the complete documentation of the energy budget.

The energy budget of a polychete worm (*Neanthes*) illustrates the concept of balance for an energy budget (Figure 4–3). The rate of energy consumption was determined by measurement of the rate of food consumption (a small crustacean). Production (P) and fecal energy losses (F) were determined by bomb calorimetry. Urinary energy losses (U) were calculated from the nitrogen excreted as ammonia, and the energy content of the ammonia. Metabolic energy (M) was calculated from the oxygen consumption of the fish, assuming an appropriate RQ and joule equivalent for oxygen. The measured values yield the satisfying result that C is closely approximated by $P + M + F + U$; the small differences can be explained by experimental variability in measurement, rather than by invoking useful work being done on the worm by its environment. In this example, all components of the energy balance equation were measured, and so the verification that $C = P + M + F + U$ was obtained. More often, all components except one are measured, and that component is calculated by assuming energy balance; e.g., metabolism could be measured as $M = C - (F + U)$.

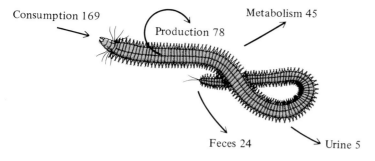

FIGURE 4–3 Total energy budget for a polychete worm *Neanthes*, illustrating the avenues for energy gain (consumption), storage (production), and loss (metabolism, feces, urine). Values are J day^{-1}. *(Data from Kay and Braefield 1973.)*

The energy consumed by an animal is apportioned to one of a number of different fates (Table 4–3). The difference between energy consumption and fecal energy loss is the **apparent digestible energy** or apparent assimilated energy (*A*), i.e.,

$$A = C - F \tag{4.4}$$

The percentage of the energy consumption that is assimilated is the apparent assimilation efficiency or the apparent digestibility or digestibility coefficient,

i.e., $100\ A/C = 100\ (C - F)/C$. The true assimilated energy is the energy absorbed across the gut. This is different from $C - F$ if materials are added to the feces from the body, for example by secretion of material and by sloughing of the intestinal lining. The true assimilation efficiency, or true digestibility, is $100 \times$ true assimilated energy/C.

Some of the assimilated energy is subsequently excreted in the urine, primarily as nitrogenous waste products. The remaining assimilated energy that is

TABLE 4–3

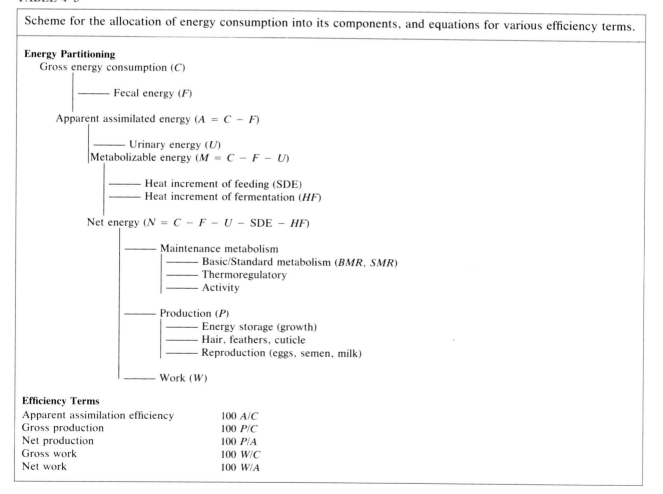

Scheme for the allocation of energy consumption into its components, and equations for various efficiency terms.

Energy Partitioning
    Gross energy consumption (*C*)
        —— Fecal energy (*F*)
    Apparent assimilated energy ($A = C - F$)
        —— Urinary energy (*U*)
    Metabolizable energy ($M = C - F - U$)
        —— Heat increment of feeding (SDE)
        —— Heat increment of fermentation (*HF*)
    Net energy ($N = C - F - U - \text{SDE} - HF$)
        —— Maintenance metabolism
            —— Basic/Standard metabolism (*BMR, SMR*)
            —— Thermoregulatory
            —— Activity
        —— Production (*P*)
            —— Energy storage (growth)
            —— Hair, feathers, cuticle
            —— Reproduction (eggs, semen, milk)
        —— Work (*W*)

Efficiency Terms

Apparent assimilation efficiency	$100\ A/C$
Gross production	$100\ P/C$
Net production	$100\ P/A$
Gross work	$100\ W/C$
Net work	$100\ W/A$

available for metabolic purposes is the **metabolizable energy**, and equals $C - F - U$; the metabolizable energy efficiency is $100 (C - F - U)/C$. Some of the metabolizable energy is utilized by the digestive processes themselves and is released as heat; this is the specific dynamic effect (SDE). The energy of SDE must be subtracted from the metabolizable energy to calculate the net energy that is available for metabolism and production. Energy of fermentation must also be subtracted from the metabolizable energy for ruminant, pseudoruminant, and post-gastric fermenting animals.

Production and metabolism are the major fates of assimilated energy; the energy lost in urine (U) is often negligible. Urine and feces are sometimes excreted as a combined mixture in, for example, lizards, birds, and insects, and so it is difficult to separate U from F. This, and other reasons, have led to the mistaken definition of "A" as $C - F - U$ rather than $C - F$, and this incorrect definition of A is unfortunately entrenched in some scientific literature.

It is important, of course, that all of the estimates of C, F, U, P, and M are measured under comparable conditions, preferably simultaneously. Otherwise, the total energy budget is unlikely to balance. For example, M cannot be measured for standard laboratory conditions and compared with other components of the energy balance equation measured under different conditions. Such a hybrid energy budget will not necessarily balance, as has been demonstrated for some insects (Wightman 1981).

Determination of energy budgets in absolute units is of value in a quantitative sense, but expression in percentage terms allows ready comparison for animals with very different absolute values of C, P, M, U, and F. For example, the apparent assimilation efficiency is the percentage of the energy consumed that is made available for production, metabolism, and urine formation. This value is primarily determined by the chemical nature of the food and the physiological digestive processes of the animal (see Chapter 18). Values for $100 \ A/C$ are relatively low for herbivores (40 to 60%) because their food contains cellulose, which is difficult to digest. Carnivores tend to have higher values because their food is rich in protein. Insectivores, like carnivores, have high values, especially if they do not ingest the chitinous exoskeleton, which is difficult to digest. Animals with extremely specialized liquid diets, such as nectar, phloem, xylem, or blood, or suckling mammals drinking milk generally have very high assimilation efficiencies.

The assimilated energy is apportioned to a number of physiological processes, including production, metabolism, and urine. Ratios of production to assimilation (P/A) tend to be less variable than A/C because P/A is determined mainly by the biochemistry of macromolecular synthesis (which will be fairly similar for all animals), whereas A/C depends to a great extent on the specific diet and digestive physiology. However, there is considerable variation in P/A for different taxa. For example, endothermic mammals and birds generally have lower P/A ratios than ectothermic animals because their higher metabolism (M) uses a larger fraction of the assimilated energy, leaving a smaller fraction for P.

The energy budget of a spittlebug (Table 4–4) shows that nymphs and adults have a similar daily energy consumption by sucking phloem (C) of about 16 to 20 J day^{-1}, but nymphs have a lower apparent assimilation efficiency than adults. The rapidly growing nymphs have a production of about 3 J day^{-1}, whereas the adults have a neglible production. Reproducing adults must expend some assimilated energy on the production of reproductive tissues and secretions, mainly eggs. The energy expended by male spittlebugs on sperm is negligible. However, to ensure mating success, males of some species produce large, energetically significant materials. For example, some male crickets produce a nutritious capsule surrounding the spermatophore (the spermatophylax), which provides a meal for the female. The spermatophylax can be a significant fraction of the body mass, and is a considerable investment of energy in the female and, ultimately, in the offspring.

Energy budgets can be determined not just for individual animals but also for populations of animals (e.g. Table 4–4). The energy budget estimated for a population of polychete worms (*Neanthes*) inhabiting an estuarine mud flat indicates an annual energy consumption of 314 kJ per m^2 (Kay and Braefield 1973). Most of this consumption is used for production (189 kJ m^{-2} year^{-1}; 60% of C), a smaller percentage for metabolism (70 kJ m^{-2} year^{-1}; 22%), and the remainder is feces (47 kJ m^{-2} year^{-1}; 15%) and urine (8 kJ m^{-2} year^{-1}; 3%). This energy budget balances exactly, but not because all components were accurately measured; the energy budget was calculated from values for individual polychetes.

It is conceptually straightforward to extend energy budgets from individuals or populations to entire ecosystems. The classical approach to ecosystem energetics is to divide the organisms of an

TABLE 4–4

	Consumption C	Metabolism M	Production P	Excretion $F + U$	Assimilation Efficiency $100\,A/C$	Production Efficiency $100\,P/A$
Individual and populational energy budget for the meadow spittlebug. *(From Weigert 1964.)*						
Individuals (J day⁻¹)						
Nymph	18.9	3.2	3.1	9.7	44%	15%
Adult	16.1	4.2	Negligible	1.7	71%	Negligible
Population (J m⁻² year⁻¹)						
Nymph	1084	159	193[1] + 8[2]	724	33%	56%
Adult	3130	2075	17[3]	1038	67%	1%

[1] Growth.
[2] Exuvae.
[3] Eggs.

ecosystem into trophic levels, with each level representing the organisms occupying similar "rungs" of the **food chain** (Elton 1927). The first trophic level is plants, the second is herbivores, the third is the primary carnivores consuming the herbivores, etc. There is a decreasing magnitude in numbers of organisms, biomass, and productivity for successive trophic levels. This simple approach yields trophic pyramids with the successive trophic levels forming smaller and smaller blocks on the large trophic base of plants (Figure 4–4A).

A more complex examination of ecological energetics analyzes each trophic level in the same manner as the energy budgets of individual animals, or populations, i.e., quantifies consumption, assimilation, metabolism, production, and excretion (Figure 4–4B). Each trophic level has consumption and fecal losses (radiative energy losses for the plants!) that determine, by difference, the assimilated energy. The assimilated energy is apportioned into metabolism, urinary and fecal energy losses, and production. Part of the production is diverted to detritivores and the remainder becomes the consumption of the next trophic level. A major advantage of this analysis of ecosystem energetics (apart from its greater detail, hence reality) is that it includes detritivores (e.g., earthworms, mites, springtails, fungi, and bacteria), which consume dead plant and animal material. Detritivores are very important in most ecosystems, eventually consuming up to 90% of the energy produced by plants.

We can extend our analogy of animal energetics to ecosystem energetics by including the efficiencies of energy transfers within and especially between trophic levels. The ratio $100\,A/C$ for the plant trophic level is the gross primary production efficiency; $100\,A/C$ is the apparent assimilation efficiency for the higher trophic levels, and the ratios $100\,P/C$ and $100\,P/A$ are the gross and net productivity efficiencies. Ratios of similar, or even greater, importance to ecosystems are the proportions of consumption, assimilation, or production of successive trophic levels, e.g., $100\,C_2/C_1$, $100\,A_2/A_1$, $100\,P_2/P_1$, $100\,C_3/C_2$, $100\,A_4/A_3$, etc. A general rule of thumb is that these ratios are approximately 0.10, e.g., 10% of the energy consumed by one trophic level is consumed by the next trophic level; 10% of the production by one trophic level is channeled to production in the next trophic level. This is an extremely broad generalization and should be used with great caution. Another useful and informative ecological energetic ratio is $100\,C_n/P_{n-1}$ (e.g., $100\,C_3/P_2$, $100\,C_4/P_3$). This is the percentage of energy produced by a trophic level that is consumed by the next trophic level; the difference ($100 - C_n/P_{n-1}$) is the percentage of energy diverted to the detritivores. Estimated values for this ratio, the exploitation efficiency, range from less than 5% (e.g., for many herbivorous animals, such as voles, grasshoppers, rabbits, and deer) to over 75% (e.g., algae-eating snails and granivores, such as desert rodents and ants).

However, even such schemes as that shown in Figure 4–4B are still highly simplified. The complexity of interactions between organisms of different trophic levels, of organisms within trophic levels,

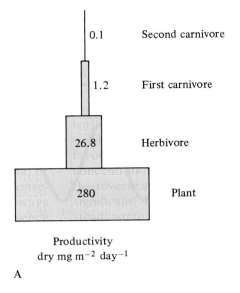

Productivity
dry mg m⁻² day⁻¹

A

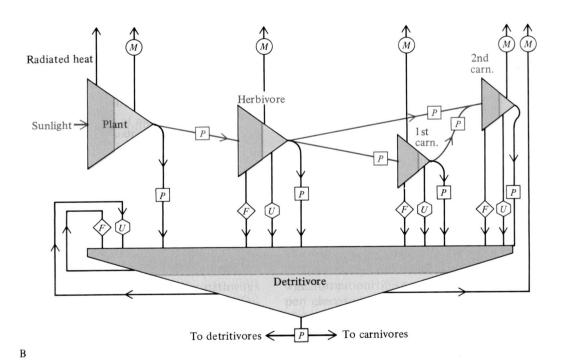

B

FIGURE 4–4 (A)   Trophic pyramid for productivity (mg dry mass m⁻²
day⁻¹) of an experimental pond. **(B)** Schematic for energy flow through an
ecosystem, from the primary producers, plants, to a second level carnivore.
Colored region indicates the gross energy input; grey region the assimilated
energy input (gross input − fecal energy loss). *M* = metabolism, *F* = feces,
U = urine, and *P* = production.   *(Modified from Braefield and Llewellyn 1982.)*

and of organisms that are actually in a number of trophic levels (e.g., omnivores) makes the modeling of ecosystems an extremely complex proposition.

Production

Production is that part of the assimilated energy that is directed to tissue growth. It can be manifested as a weight gain (or loss, if production is negative), as reproductive energy investment (e.g., ova, sperm, and often associated secretions or tissues), or it can be eliminated for some useful purpose (e.g., moulting of an exoskeleton or skin, spider's webs, bee's wax, deer antlers, mucus secreted for locomotion by gastropods). We have already examined in general terms the concepts of, and values for, P/A and P/C ratios for various animals. The P/C ratio is of ecological interest because it indicates that fraction of energy consumed by a trophic level that is available to the next. The ratio P/A is of energetic interest because it indicates the apportionment of assimilated energy to production. The ratio 100 P/A is often called the **net production efficiency** and 100 P/C is **gross production efficiency**. A measure of the efficiency of the physiological processes responsible for production is the energy content of tissues relative to the energy content of the tissues plus that fraction of the metabolism required for their production, 100 $P/(P + M_p)$. This is a physiological production efficiency. Sometimes, 100 $P/(P + M_p)$ is called net production efficiency and 100 P/metabolizable energy is called the gross production efficiency.

Production is the synthesis of tissue macromolecules, such as proteins, from smaller precursor molecules. The calculated energy cost for biosynthesis is actually quite low and the physiological production efficiency is quite high for biosynthesis. For example, the average energy required for biosynthesis of tissue from monomers is only about 0.24 kJ g^{-1}, compared to the energy content of the monomers of about 22.86 kJ g^{-1}. Only a small fraction (about 1%) of the energy content of the monomers is used for macromolecular synthesis (Calow 1977). The efficiency of macromolecular synthesis is still about 95%, even if we take into account the inefficiencies in ATP synthesis and utilization. However, taking into account other energy requirements for biosynthesis, metabolism, and excretion can substantially reduce the efficiency of production.

Growth of animal embryos is an example of extremely rapid tissue production that often occurs in a closed environment, i.e., there is no external energy input other than the yolk present in the egg (except O_2) and no elimination from the egg of waste products (except CO_2). Energetic studies of developing embryos can readily determine the initial energy content of eggs, the production (embryo growth), and the metabolized energy (initial energy − production − energy of remaining tissues, including embryo urinary and fecal wastes). The ratio of production/metabolized energy is an estimate of net production. Consider a chicken egg. The energy of combustion (in a bomb calorimeter) of an average hen's egg is about 364 kJ. Of this, 159 kJ is assimilated into the chick (production) and 109 kJ remains as unused material (yolk, membranes, embryonic wastes, etc). The remainder (96 kJ) is the energy lost through metabolism. The net production efficiency is about 100 × (159/(159 + 96)), or 62%. Similar estimates of net production efficiency have been made for embryos of gastropods (62 to 67%), silkworms (63%), frogs (51%), sea urchins (59%), and herring (70%). For the herring embryos, the metabolic cost of growth is about 11 µg O_2 for a 14-day embryo weighing 72 mg, and the total metabolism is about 48 µg O_2, i.e., about 25% of the metabolism is devoted to production and 75% to maintenance (Kiorboe and Mohlenberg 1987). The physiological production efficiency is about 91%, which approaches the theoretical value of about 95%.

The energetic efficiency of egg, milk, and tissue production by female animals is of considerable agricultural importance. For leghorn hybrid hens, about 27% of the energy consumed is lost in feces, 4% in urine, 17% as eggs, and 54% as metabolism (Hoffman and Schliemann 1973). Thus, $A = C - F = 73\%$ and $P = 17\%$ of C. Net production efficiency is 23% (100 × 17/73) and gross production efficiency is 17%. Taking into account the high proportion of assimilated energy expended on metabolism (chickens are endotherms) yields a physiological production efficiency of 70 to 80%.

Mammalian milk is an important energy and nutrient source for the suckling young, and it tends to be rich in lipids and carbohydrates. The net production efficiency for dairy cows is about 30% (closer to 50% for champion cows). A net production efficiency of about 50% appears to be the upper limit, indicating that as much assimilated energy is devoted to milk production as metabolism and growth. The physiological production efficiency for milk is about 60 to 65%. Physiological production efficiencies for weight gain by mammals and chickens range from 50 to 90%, varying with species, diet, and age. In general, ectotherms have a con-

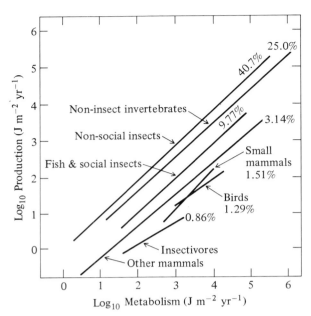

FIGURE 4–5 Relationship between production (J m^{-2} year^{-1}) and metabolic rate (J m^{-2} year^{-1}) for populations of a variety of animals: the number next to each regression line is the mean value for net production efficiency (100 *P/A*). *(Modified from Humphreys 1979.)*

siderably higher net production efficiency than endotherms.

A statistical summary of a large amount of data for production and metabolism for populations of various animals indicates that the ratio 100 *P/M* is independent of body mass, but varies dramatically from <1% for insectivorous mammals to >40% for nonsocial insects (Figure 4–5). There is a linear relationship between $\log_{10} P$ and $\log_{10} M$, with slopes not significantly different from 1.0 for animals divided into seven groupings (based on *P* and *M*), i.e., there is no effect of mass on 100 *P/M* or *P/(P + M)*. However, for a given production level (e.g., 50 J m^{-2} year^{-1}) there is a dramatically lower *M* for nonsocial insects (about 72 J m^{-2} y^{-1}) compared with insectivorous mammals (about 5050 J m^{-2} y^{-1}), with these figures corresponding to net production efficiencies of about 41% and 1% respectively.

Aerobic Metabolism

Metabolic rate is one of the most commonly measured physiological variables; there is an immense body of information concerning metabolic rates for many different animals under a wide variety of conditions. Metabolic rate measurements invariably fall in a range between a minimal value, called the standard metabolic rate or basal metabolic rate, and an upper value, often called summit metabolic rate or maximal metabolic rate. A number of different levels can be distinguished within this range.

Standard metabolic rate (SMR) is the value measured when an ectothermic animal is quiet, inactive, not digesting a meal, and not experiencing any stress (either physical, thermal, psychological, etc). **Basal metabolic rate (BMR)** is the equivalent minimal metabolic rate measured for an endothermic animal. The distinction between the terms standard and basal is necessary because the metabolic rate of ectotherms is temperature dependent, and so the SMR is the minimal metabolic rate at a particular temperature. The basal metabolic rate of an endotherm is measured within a range of ambient temperatures that is thermally neutral (the thermoneutral zone) to avoid temperature stress (see Chapter 5).

Resting metabolic rate (RMR) is somewhat similar to standard, or basal, except that the animals are simply perceived to be "resting," whereas the conditions for standard or basal metabolic rate are more stringently defined. RMR may be substantially higher than standard/basal, e.g., up to two times standard or basal. For example, the resting metabolic rate of a standing human is about 440 kJ hr^{-1}, whereas basal metabolic rate is about 270 kJ hr^{-1}. The metabolic rate even lying still but awake is about 322 kJ hr^{-1} and "at rest" is about 420 kJ hr^{-1}. Resting metabolic rate is measured with uncontrolled, but minimal, activity. **Average daily metabolic rate (ADMR)** is the metabolic rate averaged for the routine activities of a 24 hour period.

Activity metabolic rate is that measured during some form of activity, ranging from slight, to moderate, to intense, to the highest metabolic rate, summit or **maximal metabolic rate (MMR)**. This can be determined for animals during intense physical exercise, locomotion at high speeds, or with extreme cold stress for endotherms. However, it is difficult to show that the highest metabolic rate measured is actually the highest achievable metabolic rate for an animal. Furthermore, no activity requires the simultaneous activity of all muscles, and so the maximal metabolic rate for one activity (e.g., locomotion) is not necessarily the same as for other activities (e.g., thermoregulation). Motivation is another factor that makes the estimation of maximal metabolic rate difficult.

The metabolic rate of animals is determined not only by their physiological state and level of activity, but also by a multitude of other factors, including developmental stage, body mass, food or oxygen availability, the nature of their diet, photoperiod, hormonal balance, salinity for aquatic animals, and taxonomy. Some of these important determinants of metabolic rate are discussed next.

Body Size

One of the most intriguing, and yet unresolved, problems in comparative animal physiology is the observed relationship between metabolic rate and body mass. Metabolic rate must be greater for animals of larger mass. An elephant is bigger than a mouse and has a proportionately higher metabolic rate. But the fundamental question concerns the rule of proportionality.

The relationship between metabolic rate (e.g., VO_2) and body mass is of the general form of a power curve,

$$Y = a\,Mass^b \qquad (4.5a)$$

where **a** is the intercept (the metabolic rate when mass = 1) and **b** is the mass exponent. This power curve can be transformed to a linear relationship by taking the $\log_{10}$ of both metabolic rate and body mass values (see Supplement 4–3, page 119). The slope of the linear relationship is equal to the mass exponent of Equation 4.5a.

$$(\log_{10} Y) = (\log_{10} a) + b(\log_{10} Mass) \quad (4.5b)$$

A classical analysis for mammals and birds (Kleiber 1932) showed that the relationship between metabolic rate and body mass was not linear. The metabolic rate of large mammals and birds was considerably lower than expected from a direct proportionality, and that of small mammals and birds was greater than expected. The difference between the metabolic rate predicted from a linear relationship and actual metabolic rate is immense over many orders of magnitude, e.g., a mouse compared to an elephant. The relationship was linear for $\log_{10}$ metabolic rate as a function of $\log_{10}$ mass (Figure 4–6). The equation describing this relationship is

$$VO_2 = 3.9\,g^{0.738}\,(ml\,O_2\,hr^{-1})$$
$$= 1.9\,g^{0.738}\,(kJ\,day^{-1}) \qquad (4.6)$$

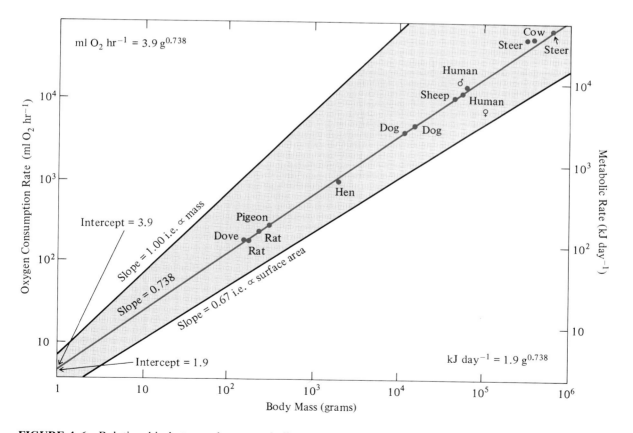

FIGURE 4–6 Relationship between $\log_{10}$ metabolic rate and $\log_{10}$ body mass for mammals and birds. *(Modified from Kleiber 1932.)*

where g is the body mass in grams. The mass exponent, or slope, is 0.738 and the intercept is 3.9 ml O_2 hr^{-1}, or 1.9 kJ day^{-1} (i.e., this is the metabolic rate for a 1-gram mammal or bird). It has become common practice as a consequence of this study to summarize metabolic rate data not only in absolute units (e.g., ml O_2 hr^{-1}) and mass-specific units (e.g., ml O_2 g^{-1} hr^{-1}), but also in mass-independent units (e.g., ml O_2 $g^{-0.738}$ hr^{-1} in the above example of mammals and birds). The mass-independent units give a metabolic rate that is independent of body mass and is therefore convenient for comparing the metabolic rates of animals of differing body mass. However, $mass^{-0.738}$ is not necessarily the most appropriate correction factor to standardize metabolic rate for all animals (see below).

The allometric relationship between metabolic rate and body mass is probably the best documented but least understood topic in comparative animal physiology. There are many studies of the allometry of metabolic rate for numerous taxa of animals, and for single species (Table 4–5).

This comprehensive summary of interspecific and intraspecific allometric relationships for various animals is presented because of the considerable interest of comparative physiologists in this topic. Compilation of such a table of data is fraught with difficulties, necessary extrapolations, and calculations from the original data because of the great diversity in experimental approaches, conditions, the nature of the animals studied, and the units used for metabolic rate. Body temperature is also an important variable in determining the standard metabolic rate of ectotherms (see below) and so the **a** values for ectotherms have been converted to a temperature of 20° C if necessary. The **a** values for endotherms are for their normal body temperatures (generally 35° to 41° C). Mass was standardized to grams, but it is important to appreciate that the extrapolation of allometric relationships from pg mass (e.g., unicells) or ktonne (e.g., large mammals) to an intercept value of 1 gram mass may result in an almost meaningless **a** value. Only **a** values for wet mass are included in Table 4–5. The **a** value depends on the unit for mass, e.g., wet mass, dry mass, ash-free dry mass, soft body mass, shell-free dry mass, or even in terms of nitrogen content; for example, **a** is about five times higher for dry mass than wet mass. Perusal of this large summary of allometric data reveals considerable taxonomic diversity in allometric relationships, with respect to both **a** and **b** values, even for taxonomically similar animals. For example, placental mammals have higher **a** values than marsupials, which in turn are higher than monotreme **a** values. Even within placental mammals, there are considerable differ-

ences in **a** values for various taxa. Birds exhibit similar **a** taxonomic groupings based on metabolic rate, i.e., passerine birds (highest), nonpasserine carinate birds (intermediate), and ratite birds (lowest). Lower vertebrates and invertebrates also exhibit great diversity in **a** values.

A summary and analysis of these numerous data for allometric slopes (Figure 4–7) and intercepts

(Text continues on page 99)

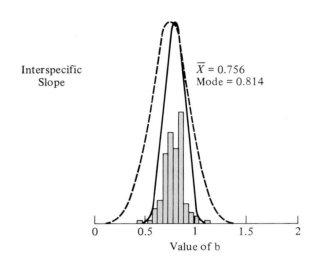

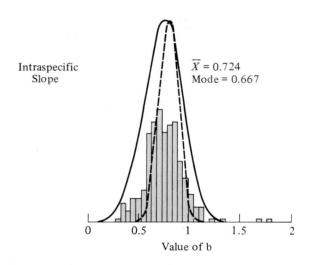

FIGURE 4–7 Summary of slope coefficients **b** for the allometric relationship between basal metabolic rate (BMR; J h^{-1}) and body mass (M; grams). Allometric equations are of the form BMR = aM^b. The solid envelope is the normal distribution fitted to the data, and the broken envelope is the distribution for the other data set. Both distributions have similar means, but the mode for intraspecific analyses is 0.67 and the mode for interspecific analyses is 0.81. Sample size is 107 for interspecific allometric analyses and 220 for intraspecific analyses.

TABLE 4–5

Interspecific and intraspecific allometry of metabolic rate (J hr^{-1}) for various animals. Intercept values for ectotherms are corrected to a temperature of 20° C if necessary; the Q_{10} was obtained from the original source or assumed to be 2.5. In some instances, intercept values were estimated or calculated from the original data or figure (or the slope was assumed to be 0.75). Body mass (M) is in grams.

INTERSPECIFIC			INTRASPECIFIC		
MAMMALS					
Placentals					
All	63.6	$M^{0.76}$	*Peromyscus*	12.8	$M^{0.72}$
Edentates	63.9	$M^{0.66}$	*Mus*	58	$M^{0.91}$
Ground squirrels	65.1	$M^{0.66}$	*Rattus*	146	$M^{0.64}$
Megachiroptera	67.4	$M^{0.73}$	*Ovis*	147	$M^{0.72}$
Chiroptera	73.1	$M^{0.71}$	*Canis*	207	$M^{0.65}$
Heteromyids	74.3	$M^{0.72}$	*Felis*	220	$M^{0.58}$
Small mustelids	74.3	$M^{0.72}$	*Homo*	382	$M^{0.60}$
All	89.6	$M^{0.70}$	*Bos*	456	$M^{0.56}$
Crocidurine shrews	136	$M^{0.75}$			
Soricine shrews	272	$M^{0.75}$			
Large mustelids	374	$M^{0.55}$			
Marsupials					
All	47.6	$M^{0.75}$			
Dasyurids	49.2	$M^{0.74}$			
All	50.1	$M^{0.75}$			
Monotremes			*Zaglossus*	17.4	$M^{0.75}$
			Tachyglossus	19.8	$M^{0.75}$
			Ornithorhynchus	44.7	$M^{0.75}$
BIRDS					
Passerines					
All	132	$M^{0.73}$			
All	151	$M^{0.72}$			
Nonpasserines					
All	80.5	$M^{0.73}$			
All	80.5	$M^{0.72}$			
Honey eaters	134	$M^{0.68}$			
Small species	138	$M^{0.72}$			
Procellariiformes	153	$M^{0.66}$			
Ratites					
All	50.4	$M^{0.73}$			
REPTILES					
Lizards					
All	1.5	$M^{0.80}$	*Ctenosaura*	1.4	$M^{0.86}$
Varanids	1.9	$M^{0.82}$	*Chalcides*	1.7	$M^{0.65}$
Xantusids	2.2	$M^{0.56}$	*Amblyrhynchus*	2.9	$M^{0.78}$
Lacertids	2.6	$M^{0.76}$	*Lygosoma*	6.2	$M^{0.63}$
Snakes					
All	2.4	$M^{0.77}$	*Lampropeltis*	3.9	$M^{0.65}$
Boids	0.8	$M^{0.81}$			
Colubrids	0.8	$M^{0.98}$			
Turtles					
All	1.3	$M^{0.86}$	*Geochelone*	1.9	$M^{0.82}$
			Chelonia	2.5	$M^{0.83}$
Crocodilians			*Caiman*	1.1	$M^{0.93}$
			Crocodylus	6.0	$M^{0.65}$

continues

TABLE 4-5

continued

INTERSPECIFIC			INTRASPECIFIC		
AMPHIBIANS					
Apoda			*Typhlonectes*	1.5	$M^{0.78}$
Salamanders					
Neotropical	0.7	$M^{0.86}$			
Temperate	1.1	$M^{0.82}$			
Lungless	1.2	$M^{0.78}$			
Lunged	3.5	$M^{0.86}$			
Lungless	3.8	$M^{0.72}$			
Anura					
All	4.2	$M^{0.68}$	*Rana*	1.7	$M^{1.06}$
			Xenopus	1.8	$M^{1.08}$
			Bufo	3.7	$M^{0.75}$
FISH					
All	2.5	$M^{0.70}$	*Cyprinus*	0.1	$M^{0.98}$
All	4.8	$M^{0.88}$	*Ictalurus*	0.1	$M^{1.00}$
			Catostomus	0.1	$M^{0.99}$
			Notothenia	0.4	$M^{0.96}$
			Carassius	0.6	$M^{0.91}$
			Notothenia	1.6	$M^{0.79}$
			Gobius	1.6	$M^{0.89}$
			Kuhlia	2.6	$M^{0.79}$
			Cirrhinus	2.8	$M^{0.80}$
			Onchorhynchus	9.2	$M^{0.78}$
			Scophthalmus	13.4	$M^{0.71}$
			Gadus	16.5	$M^{0.83}$
ARTHROPODS					
Crustaceans	3.4	$M^{0.81}$			
Horseshoe crab			*Limulus*	1.7	$M^{0.81}$
Crabs	0.9	$M^{0.80}$	*Uca*	0.3	$M^{0.62}$
			Hemigrapsis	1.2	$M^{0.56}$
			Ocypoda	2.2	$M^{0.56}$
			Hemigrapsis	2.6	$M^{0.32}$
			Carcinus water	2.8	$M^{0.52}$
			Carcinus air	3.0	$M^{0.61}$
			Pachygrapsis	3.0	$M^{0.61}$
Amphipods					
Temperate	2.9	$M^{0.85}$	*Vibilia*	0.6	$M^{0.58}$
Seawater	3.0	$M^{0.7}$	*Talitrus*	0.6	$M^{0.65}$
All	3.8	$M^{0.57}$	*Lygia*	1.4	$M^{0.73}$
			Ampelisca	4.1	$M^{0.69}$
			Mesothidia	1.8	$M^{0.85}$
Fresh water	5.1	$M^{0.7}$	*Gammarus*	7.0	$M^{0.70}$
Cold water	5.3	$M^{0.86}$			
Antarctic	6.4	$M^{0.81}$			
Euphasiids	6.7	$M^{0.50}$	*Artemia* male	1.0	$M^{0.62}$
			Artemia female	2.1	$M^{0.72}$
			Palaemonetes	3.4	$M^{0.76}$
Spiders					
All	1.0	$M^{0.65}$	*Sericopelma*	0.7	$M^{0.80}$
Wolf spiders	1.7	$M^{0.67}$			

TABLE 4-5

continued

INTERSPECIFIC			INTRASPECIFIC		
ARTHROPODS					
Insects					
Collembola	1.3	$M^{0.74}$	*Chironomus*	1.4	$M^{0.47}$
Beetles	3.5	$M^{0.86}$	*Blateria*	2.9	$M^{0.80}$
Saturniids	5.4	$M^{0.67}$	*Pterinoxylus*	2.9	$M^{0.80}$
Sphingids	6.4	$M^{0.81}$	*Blaberus*	4.6	$M^{0.98}$
Dragonflies	7.4	$M^{0.91}$	*Apis*	7.0	$M^{0.75}$
Coleoptera	10.7	$M^{0.77}$			
All	11.3	$M^{0.67}$			
All	13.7	$M^{0.62}$			
Millipedes			*Arthrosphaena*	0.2	$M^{0.32}$
			Spirostreptus	0.5	$M^{0.48}$
			Rhinocricus	1.0	$M^{0.86}$
			Plusioporus	2.5	$M^{0.36}$
			Plusioporus	2.7	$M^{0.47}$
MOLLUSKS					
Mollusks					
Operculates	0.3	$M^{0.67}$	*Mercenaria*	1.2	$M^{0.66}$
Pulmonates	0.8	$M^{0.67}$	*Patella*	1.7	$M^{0.73}$
All	4.2	$M^{0.75}$	*Otala*	2.4	$M^{0.57}$
			Patella	2.5	$M^{0.70}$
			Lymnaea	2.6	$M^{1.00}$
			Lymnaeate	2.9	$M^{0.76}$
			Lymnaea	5.1	$M^{0.94}$
ANNELIDS					
Polychetes	1.9	$M^{0.74}$	*Clymenella*	0.9	$M^{0.48}$
Polychetes	2.6	$M^{0.81}$			
Oligochaetes	1.9	$M^{0.61}$			
Hirudinea	2.1	$M^{0.82}$			
NEMATODES					
	0.6	$M^{0.72}$	*Caenorhabditis*	0.6	$M^{0.70}$
			Cysticercus	0.4	$M^{0.90}$
CESTODES			*Schistocephalus*	0.5	$M^{0.37}$
			Taenia	1.0	$M^{0.76}$
			Oikopleura	0.05	$M^{0.63}$
TUNICATES			*Pyura*	0.60	$M^{0.95}$
ECHINODERMS					
Ophiuroids			*Ophionereis*	1.1	$M^{0.71}$
			Ophionereis	1.3	$M^{0.50}$
			Ophionereis	1.4	$M^{0.42}$
Holothuroids			*Scherodactylus*	2.9	$M^{0.85}$
			Cucumaria	4.6	$M^{0.82}$
Asteroids			*Strongylocentrodus*		
0.2 $M^{0.80}$			*Strongylocentrodus*		
0.7 $M^{0.69}$			*Strongylocentrodus*		
1.1 $M^{0.65}$			*Mellita*	2.2	$M^{0.51}$

continues

TABLE 4-5

continued		
INTERSPECIFIC		**INTRASPECIFIC**
COELENTERATES		
Anthozoa	0.69 $M^{0.86}$	
UNICELLS		
All	0.001 $M^{0.42}$	
All	0.025 $M^{0.55}$	
All	0.03 $M^{0.66}$	
All	0.35 $M^{0.75}$	
All	0.56 $M^{0.68}$	
All	1.58 $M^{0.83}$	

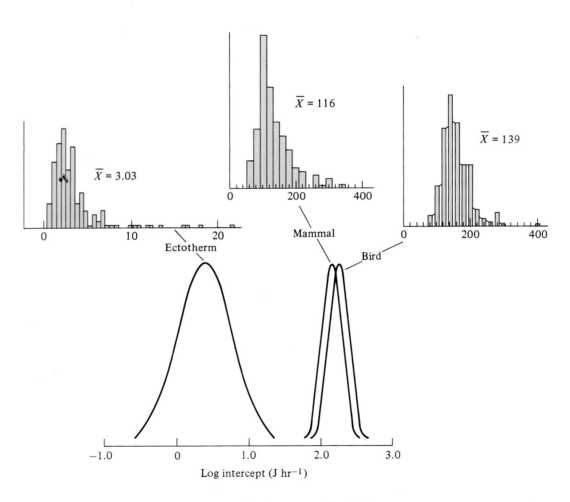

FIGURE 4–8   The frequency histograms for intercept value (**a**) for the relationship between $\log_{10}$ metabolic rate as a function of $\log_{10}$ body mass are positively skewed for ectotherms, birds, and mammals. The **a** values for ectotherms are considerably lower than the values for mammals and birds, as is evident from the graph of the normal curves fitted to the frequency histograms for $\log_{10}$ **a** of ectotherms, mammals, and birds.

(Figure 4–8) reveal some consistent trends. The slope varies from <0.5 to >1.0, but on average is about 0.76 for interspecific relationships (i.e., for various species of animals), and about 0.72 for intraspecific relationships (i.e., a single species). Why the slope is about 0.75 is one of the most perplexing questions in biology. The intercept values vary from <1 to >400 J hr^{-1}, with a mean of 3 for ectotherms (at 20° C), 116 for mammals, and 139 for birds. Why there is such a variation in intercept values is a little easier to explain than the value of the slope.

Unicells, Ectotherms, and Endotherms. There is a profound consistency in the relationship between metabolic rate and body mass for animals of varying taxonomic position.

There are three major "grades" of animals, based on the allometry of their metabolic rates. This is apparent from the relationship between log metabolic rate and log mass observed for virtually all kinds of animals, as is apparent from a reanalysis (Phillipson 1981) of the classical studies of Hemmingsen (1950) and Zuethen (1953) for unicellular ectotherms, multicellular ectotherms, and endotherms (Figure 4–9). The data for ectotherms are corrected to a body temperature of 10° C (the approximate mean annual temperature for these animals), and for endotherms the data are corrected to a body temperature of 39° C.

A number of salient features are illustrated by Figure 4–9. First, there is considerable overlap in body mass for unicellular and multicellular ectotherms. Second, the relationships for the three grades of organisms are fairly discrete. Third, the slopes of the three relationships are not equal to one for any grade of organism. The log$_{10}$-transformed relationship between VO$_2$ (μl O$_2$ hr^{-1}) and mass (g) for unicellular ectotherms (bacteria, fungi, flagellates, ciliates, rhizopods) has a slope of 0.66 (with 95% confidence limits of ±0.09) and an intercept of 0.59 μl O$_2$ hr^{-1} (0.012 J hr^{-1}). The relationship for multicellular ectothermic animals (see legend for Figure 4–9) has a higher slope (0.88 ±0.00002) and intercept value of 14.8 μl O$_2$ hr^{-1} (0.297 J hr^{-1}). The relationship for endotherms has a slope of 0.69 (±0.0017) and an intercept of 2630 μl O$_2$ hr^{-1} (52.9 J hr^{-1}). Finally, the major metabolic step between unicellular and multicellular ectotherms is about three times, and the metabolic step between multicellular ectotherms (at 10° C) and endotherms (corrected to 10° C) is even larger, at about eight times.

Why are there three metabolic grades of animals? The metabolic step from unicellular ectotherm to multicellular ectotherm has been explained by their differing surface areas (Phillipson 1981). Consider a theoretical single-celled, cuboidal animal with cell dimensions 1μ × 1μ × 1 μ; the surface area is 6 μ^2 (6 × 1 × 1). If a multicellular animal of the same mass had 1000 cubic cells, then the total cellular surface area would be 60 μ^2 (1000 × (6 × 1/10 × 1/10)), i.e., 10 × the area of the unicell. The metabolic rate of the multicellular animal might be expected to be 10 × that of the unicell if adequate

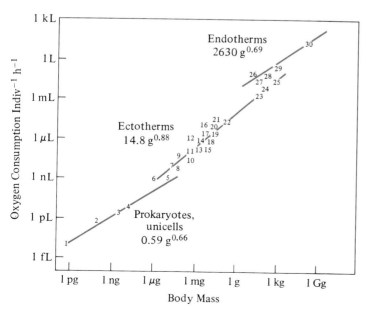

FIGURE 4–9 Log$_{10}$ transformed relationship between rate of oxygen consumption (from 1 femtoliter to 1 kiloliter hr^{-1}) and body mass (from 1 picogram to 1 gigagram) for prokaryote and unicellular eukaryote organisms, multicellular ectothermic animals, and endothermic animals. Numbers indicate the pivotal point for the regression line for the following taxa: 1 bacteria; 2 fungi, 3 flagellates, 4 ciliates, 5 rhizopods, 6 nematodes, 7 microcrustaceans, 8 acari, 9 collembolans, 10 isopteran larvae, 11 enchytraeids, 12 coleopteran larvae, 13 isopteran adults, 14 formicid workers, 15 lumbricid cocoons, 16 phalangiids, 17 diplopods, 18 araneans, 19 isopods, 20 mollusks, 21 coleopteran adults, 22 lumbricid adults, 23 macrocrustaceans, 24 fish, 25 reptiles, 26 small mammals, 27 chiropterans, 28 birds, 29 primitive mammals, and 30 large mammals. Data for unicells are corrected to 10° C, and data for endothermic animals are corrected to 39° C. *(From Phillipson 1981.)*

A

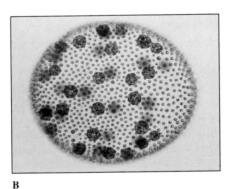

B

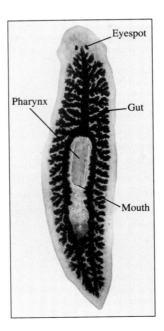

C

TABLE 4–6

Comparison between a dragon lizard and a mouse for the following: body temperature; organ weights; standard/basal metabolic rate of whole animal; *in vitro* metabolic rate of liver; component of *in vitro* metabolic rate of liver for Na$^+$ transport (*i.e.* blocked by ouabain); and mitochondrial volume, surface area and enzyme activity. The ratio of the mammalian to lizard value (M/L) is also given.
(Modified from Else and Hulbert 1981.)

	LIZARD *Amphibolurus nuchalis*	**MOUSE** *Mus musculus*	**M/L**
Body weight (g)	34.3	32.1	1
Body temperature (°C)	37.0	36.8	1
Liver mass (% total)	10.7	20.7	1.9
Kidney mass (% total)	1.9	5.9	3.1
Heart mass (% total)	1.1	2.8	2.5
Brain mass (% total)	1.7	5.5	3.3
VO$_2$ (ml O$_2$ g^{-1} h^{-1})			
In Vivo—whole animal	0.20	1.62	8.1
In Vitro[1]—liver	0.90	4.59	5.1
In Vitro[1] Na$^+$ transport	0.22	1.97	9.0
% Mitochondrial volume	12.4	16.0	1.3
Mitochondrial Surface Area			
Cristae (m^2 g^{-1})	15.5	22.9	1.5
Inner (m^2 g^{-1})	0.79	1.34	1.7
Total area (m^2)	3.4	10.2	3.0
Cytochrome oxidase (nmol O$_2$ mg^{-1} min^{-1})	11.2	30.0	2.7

[1]Per gram dry weight.

O$_2$ could be delivered to the entire cellular surface area. An alternative argument is to assume that the metabolic rate of a cell in a multicellular animal is equal to that of a unicell of the same size as that cell. The multicellular animal would then have a higher metabolic rate than the same-sized unicell. If the multicellular animal again has 1000 cells, then each cell would have a metabolic rate not 0.001 × that of the unicell but about 0.0098 ×, if the slope of the allometric relationship is 0.67. The total metabolic rate of the multicellular animal would

FIGURE 4–10 The protozoan *Amoeba proteus* (**A**), the colonial protozoan *Volvox* (**B**), and a planarian flatworm (**C**) illustrate the possible evolutionary sequence from a unicellular organism to a multicellular animal, via a colonial intermediate form. *(Photographs A and C, Courtesy of Carolina Biological Supply Co., Inc.; B Courtesy of Richard Starr.)*

therefore be 9.8 × that of the same-sized unicell. Such arguments may provide a plausible explanation for the metabolic grade between unicells and multicellular animals, particularly if the intermediate evolutionary stage was a colony of small unicells (Figure 4–10).

The metabolic grade between ectotherms and endotherms cannot be so easily explained by surface area effects, since the total cellular surface area is presumably similar for ectothermic and endothermic multicellular animals. The difference might be attributed to differences in cellular metabolic machinery. For example, the scaling of $VO_{2,rest}$ and $VO_{2,max}$ in mammals is closely related to the scaling of their visceral and total mitochondrial surface areas respectively (Figure 4–11). There is a similar relationship for reptiles. The mitochondria of endotherms are similar in morphology and biochemistry to those of ectotherms, although mammalian mitochondria have slightly greater membrane areas (both cristae and inner membrane). Endotherm tissues also have

slightly greater mitochondrial volumes, due to greater numbers of mitochondria and/or larger mitochondria. Consequently, endothermic tissues have greater mitochondrial enzyme activity than ectotherm tissues, hence a potentially higher metabolic rate. The significant difference between *in vivo* and *in vitro* metabolic rates for tissues of house mice and similar-sized lizards (at the same body temperature) are explained by the summed effects of a number of small differences, including minor differences in mitochondrial morphology, numbers of mitochondria, mitochondrial enzyme activity (cytochrome oxidase), and the greater mass of internal organs: liver, kidneys, heart, and brain (Table 4–6). The visceral mass of endotherms provides a major fraction of the resting metabolic requirements, disproportionate to their mass (72% and 8% respectively in humans).

The metabolic differences between ectotherms and endotherms are apparent for whole animals and for *in vitro* homogenates or tissue slices. The scaling

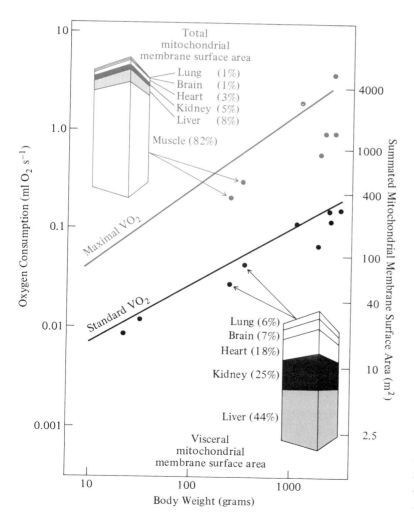

FIGURE 4–11 Relationship between $\log_{10}$-transformed resting oxygen consumption rate and body mass for a mammal (solid line), and between visceral organ mitochondrial surface area (liver + brain + kidney + heart) and body mass (●). Shown also is the relationship between maximal oxygen consumption rate (colored line) and total mitochondrial surface area (visceral organs + skeletal muscle; ●) with body mass. The mitochondrial surface for visceral organs and skeletal muscle is shown. *(Modified from Else and Hulbert 1981.)*

relationships observed *in vivo* for whole animals are sometimes, but not always, observed *in vitro* with tissue slices or homogenates. For example, mammalian liver slices show the same systematic increase in metabolic rate with body weight (the exponent is about 0.75; Kleiber 1945) but fish organs and tissue homogenates do not (Vernberg 1954).

Why Isn't Metabolism Proportional to Mass? Why is the slope about 0.75 for the relationship between log metabolic rate and log body mass for virtually all animal taxa? This is a fundamental question in biology, but unfortunately we lack a convincing answer to this question. There are, however, a number of hypotheses to explain why the slope is about 0.75.

Geometric Similarity. Geometric similarity predicts a slope of 2/3, or 0.67, for the metabolic rate–mass

relationship. Consider a sphere of radius r; its surface area is $4\pi r^2$ and volume is $4/3\ \pi r^3$. If the density of the sphere (ρ) is independent of its size, then its mass is $4/3\ \pi\rho r^3$, i.e., its weight is proportional to r^3 and r is proportional to mass$^{1/3}$. The surface area, which is proportional to r^2, is therefore proportional to mass$^{2/3}$. This relationship between mass and surface area is found not only for spheres, but for any geometrically similar (i.e., same-shaped) animal. For example, a small cockroach is geometrically similar to a large cockroach that might weigh 100 or even 1000 times more, and their surface area/mass is proportional to mass$^{2/3}$. For many species of mammal, the slope is about 0.67, although the inter-specific slope is about 0.75 (Figure 4–12).

Metabolic rate should be proportional to surface area, or mass$^{2/3}$, if an important metabolic process is dependent on surface area. Many important meta-

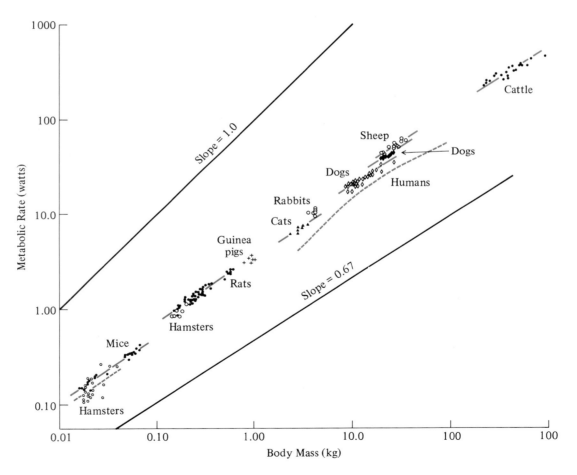

FIGURE 4–12 Log$_{10}$-transformed relationship between metabolic rate and body mass for a variety of mammals, showing the lower generally intraspecific scaling relationship (slopes from 0.51 to 0.91) compared to the interspecific scaling relationship (slope of 0.776). *(From Heusner 1982; Hill and Rahimtulla 1965.)*

bolic processes, such as O_2 and CO_2 exchange and digestive absorption, are dependent either on the surface area of the animal or other internal surface areas, such as the lung surface or digestive tract surface. Thus, we might expect metabolic rate to be proportional to mass$^{2/3}$ for geometrically similar animals; for example, based on a relationship between respiratory surface area and mass (Hughes 1984). The heat loss of endothermic animals is a function of their surface area, not their volume. Since the metabolic rate of endotherms is essentially proportional to the rate of heat loss, then metabolic rate might be expected to be proportional to mass$^{2/3}$.

The metabolic rate of unicellular organisms (which tend to be geometrically similar) is proportional to mass$^{0.66}$ (see Figure 4–9, page 99), which is essentially the slope expected if metabolism is proportional to surface area, for geometrically similar animals. For multicellular ectotherms, which are generally not geometrically similar, metabolic rate is not proportional to mass$^{0.67}$, but the slope is 0.88 ($\pm 95\%$ confidence limits of 0.00002). However, the slope for endotherms is 0.69, which is again similar to the 2/3 value expected from geometric similarity.

Structural Support. The structural support properties of materials influence body shape, muscle size, and power output (McMahon 1973). The ability of a cylinder to resist bending or buckling depends on its length, radius, and elastic properties. The critical length at which a cylinder will buckle (l_{crit}) is proportional to $r^{2/3}$. Consequently, r^2 is proportional to l_{crit}^3. The weight of the cylinder of critical length is equal to $\pi r^2 \rho l_{crit}$ since weight = volume × density (ρ). The cylinder weight is thus proportional to $r^2.r^{2/3}$, or $r^{8/3}$, and r is proportional to weight$^{3/8}$. The maximal force exerted by a muscle is proportional to its cross-sectional area (see Chapter 9) and the maximal energy expenditure of a muscle is therefore proportional to r^2 or weight$^{3/4}$. There appears to be a fairly consistent proportionality between resting and maximal metabolic rate and so resting metabolism is also proportional to mass$^{0.75}$.

This hypothesis predicts the allometric relationship for the metabolic rate of animals that might experience buckling deformation, but it is unlikely to be universally applicable to all animals, from aquatic unicells to terrestrial endotherms.

Additive Scaling. The metabolic scaling relationship may be the additive result of two different influences (Swan 1972; Yates 1981), a surface-area specific effect ($VO_2 \propto$ mass$^{0.67}$) and a mass-specific effect ($VO_2 \propto$ mass$^{1.0}$).

$$VO_{2,total} = k_1 \, mass^{0.67} + k_2 \, mass^{1.0}$$
$$\propto k_3 \, mass^{0.75} \qquad (4.7)$$

Complex explanations based on heat exchange have been presented to account for the slope of 0.75 for mammals (Swan 1972; Gray 1981). The minimal, and essential, energy requirement for cells (M_{eer}) is assumed to be proportional to mass, i.e., $M_{eer} = e$Mass$^{1.0}$ where e is a constant. An additional energy production is required by endotherms for regulation of body temperature, since M_{eer} is too low; the metabolic rate for obligate heat production (M_{ohp}) is proportional to surface area because heat loss is proportional to surface area, i.e., $M_{ohp} = h$Mass$^{2/3}$ where h is a constant. The total metabolic rate is $M_{tot} = M_{eer} + M_{ohp}$. Larger endotherms are not geometrically similar to small endotherms, and they have a higher proportion of supporting tissues with lower metabolic requirements than other tissues, i.e., the values of e and h are mass-dependent. Appropriate values for e and h as a function of mass closely predict the observed relationship between M_{tot} and body mass for mammals, i.e., $M_{tot} = a$Mass$^{0.75}$.

Unfortunately for the above argument, ectothermic animals also have **b** values similar to 0.75 but do not have a functional relationship between body temperature, heat exchange, and surface area. Either there are different explanations for the allometry of metabolic rate for endotherms and ectotherms resulting in similar predicted **b** values, or there is a single explanation and the above thermal argument for mammals is irrelevant. An obsession with "Ockham's razor" entices comparative physiologists to seek for the single, unifying theory of metabolic allometry. Ockham's doctrine of nomination suggests that explanations should not be unnecessarily complex.

Four-dimensional Scaling. Another explanation for a slope of 0.75 is afforded by considering area in four dimensions rather than the three dimensions we are used to (Blum 1977). In n-dimensional space, the surface area/volume is proportional to radius$^{(n-1)/n}$. Surface area/volume $\propto r^{2/3}$ for three dimensions (as we saw above) and $\propto r^{3/4}$ for four dimensions, i.e., metabolic rate $\propto$ mass$^{3/4}$ if metabolic rate is proportional to four-dimensional surface area.

This is a relatively straightforward derivation of the 3/4 slope, but what is the fourth dimension? Time is one suggestion, especially as the life span of animals is proportional to mass$^{0.25}$ (Calder 1984). Thus, the total energy expended by an animal over its entire life span is proportional to metabolic rate × life span, or mass$^{1.0}$. But, is time as important a determinant of metabolic rate as the linear dimensions of an animal? Another suggestion for the fourth dimension is a ratio of quantities required for mechanical stability, such as the ratio of energy cost of ion pumps per unit area/energy cost of assembling/maintaining metabolic machinery per unit volume.

Fractal Scaling.   Fractal dimensional analysis has also been applied to metabolic scaling. In essence, the fractal dimension is a scaling exponent that alters, depending on the scale of the measurement (see Chapter 13). For metabolic scaling, $VO_2 = aM^{b-f}$, where **b** is the normal scaling exponent and **f** is a fractal scaling exponent, which may change with the value of M (Sernetz, Gelleri, and Hoffman 1985). Fractal scaling may provide yet another procedure for fitting a curve to the metabolic mass data, but unfortunately it does not provide a fundamental insight into the reason for the scaling relationship.

The best summary for the allometry of metabolic rate that seems possible at present is that metabolic rate increases with body mass at less than direct proportionality (i.e., **b** < 1.0), sometimes in accord with surface area proportionality (i.e., **b** = 0.67), but generally with an intermediate proportionality (i.e., **b** = 0.7 to 0.8). Different explanations for **b** may be required for interspecific and intraspecific analyses and for different taxonomic groups of animals (Economos 1982). It seems highly unlikely at present that there is any "universal **b** value" for all animal groups.

Mass-Specific Metabolism.   The slope for the scaling of mass-specific metabolic rate with body mass (**b′**) is different from that for absolute metabolic rate (**b**); **b′** = 1 − **b**. The intercept value **a** is the same. For example, Kleiber's mouse-to-elephant equation for mass-specific metabolism of mammals is as follows.

$$VO_2 = 3.2\,g^{-0.24}\,ml\,O_2\,g^{-1}\,hr^{-1}$$
$$= 64\,g^{-0.24}\,J\,g^{-1}\,hr^{-1} \qquad (4.8)$$

The allometry of mass-specific metabolism illustrates the energetic constraints of small size for endotherms. The mass-specific metabolic rate of a small mammal is substantially greater than that for a large mammal. For example, a 2 g shrew has a mass-specific metabolism of 141 J g^{-1} hr^{-1}, whereas a 4 tonne elephant has a mass-specific metabolic rate of about 1.66 J g^{-1} hr^{-1}. The shrew expends in one hour about 1% of its total body energy, whereas an elephant expends 1% of its total body energy in about 5 days. Obviously, the rate of food consump-

FIGURE 4–13   The adult bumblebee bat weighs only 2 grams. Its size is about the minimum lower limit for an endothermic mammal or bird.   *(Photograph courtesy of Bat Conservation International, Merlin D. Tuttle, BCI.)*

tion has to match the energy expenditure, and so the shrew has to eat almost continuously to support its high mass-specific metabolism, whereas the elephant could starve for many days or weeks without severely depleting its body energy stores.

A body mass of 1 to 2 grams would appear to be an energetic lower limit to size for endotherms. The smallest adult mammals (bumblebee bat and Etruscan shrew) and bird (bee-hummingbird) weigh about 2 g (Figure 4–13). There are many extremely small endothermic invertebrates (e.g., moths weighing only a few mg), but these insects are intermittent rather than continuous endotherms. The largest size of terrestrial animals (African elephant, about 5000 kg) is probably determined by strength and mechanical constraints, rather than by metabolic effects of size. Marine animals can attain much larger sizes than terrestrial animals. For example, the largest mammal is the blue whale, which weighs about 150000 kg.

Temporal and Geographic Effects

Most animals have a pronounced circadian (daily) cycle in activity. Nocturnal animals are active at night and sleep during the day, whereas diurnal animals are active during the day and sleep at night. Crepuscular animals are active near dawn and dusk. The minimal metabolic rate measured for mammals and birds during their active phase of the circadian cycle (called the α phase) is typically 25 to 30%

TABLE 4-7

Variation of basal metabolic rate (BMR) for various mammals and birds in their activity phase (α) and inactivity phase (ρ) of the photoperiodic cycle. Regression equations are of the form BMR = aM^b, where M is mass in grams. *(From Aschoff 1981; Kenagy 1982; Aschoff and Pohl 1970.)*

	α	ρ	α/ρ
Mammals			
Primates	285 $M^{0.51}$	158 $M^{0.56}$	1.80 $M^{-0.05}$
Nonprimates	402 $M^{0.53}$	268 $M^{0.49}$	1.50 $M^{0.04}$
Small mammals	143 $M^{0.61}$	112 $M^{0.61}$	1.28 $M^{0.00}$
Birds			
Passerines	190 $M^{0.70}$	133 $M^{0.73}$	1.23 $M^{-0.03}$
Nonpasserines	103 $M^{0.73}$	81 $M^{0.73}$	1.24 $M^{0.00}$

higher than the minimal values during the inactive (ρ) phase (Table 4-7). There are some exceptions to this general trend among small mammals. Continuously fossorial mammals (e.g., gophers) and herbivorous species that consume low-quality food (e.g., voles) do not have day–night activity cycles, and exhibit smaller (5 to 10%) differences in day–night metabolic rates (Kenagy and Vleck 1982). Very small mammals (e.g., shrews), which must continuously feed day and night to support their high mass-specific metabolic rate, also have small (about 7%) differences between day and night minimal metabolic rates.

There are undoubtedly seasonal and geographic variations in basal metabolic rate for individual species and for groups of similar species (e.g., passerine birds), but it is not clear how general these variations are. Altitude, season, and latitude can influence the basal metabolic rate of some birds by about 20%, with higher rates in colder conditions, i.e., high altitude, winter, high latitude (Weathers 1979). Similarly, some desert mammals and birds have low basal rates of metabolism, perhaps as a consequence of their thermal environment.

Food and Oxygen Availability

The standard/basal metabolic rate of an animal reflects its minimal energy requirement for the steady-state maintenance of necessary metabolic processes. The rate of a biochemical reaction is dependent on the concentration of substrates and end products, but we might not expect in an analogous manner that the availability of oxygen or metabolic substrates would influence the standard or basal metabolic rate of animals. However, the availability of oxygen or metabolic substrates does influence the standard/basal metabolic rate for some animals under nonsteady-state conditions. For example, some animals often have a reduced standard/basal metabolic rate during partial or complete O_2 deprivation (hypoxia or anoxia) or food deprivation (food restriction or starvation).

There are two basic patterns for the effect of O_2 availability on standard/basal metabolic rate. Many animals, called **metabolic regulators**, maintain their normal standard/basal metabolic rate as pO_2 is reduced, down to some critical value (P_{crit}) below which the standard/basal metabolic rate declines markedly (often in direct proportion to the pO_2). The P_{crit} depends on many variables, including temperature, acclimation state, body mass, and standard/basal metabolic rate (see also Chapter 12). Many protozoans, annelids, mollusks, crustaceans, and vertebrates are metabolic regulators; a typical example is the worm *Lumbricus* (Figure 4-14). The critical pO_2 for metabolic regulators is elevated at metabolic rates above resting values; hovering honeybees have a much higher metabolic rate than resting honeybees, and their critical pO_2 is much higher than at rest. In other animals, called metabolic conformers, the standard/basal metabolic rate is directly proportional to ambient pO_2. Many invertebrates and some aquatic vertebrates are **metabolic conformers**; a typical example is the pycnogonid crustacean *Decolopoda*. Some animals have responses intermediate to metabolic regulators and metabolic conformers.

Starvation depresses basal metabolic rate in a wide variety of animals (Table 4-8). Absolute metabolic rate (i.e., ml O_2 hr^{-1}) often declines dramatically with starvation, but a decline in body mass complicates the analysis of starvation effects. A decline in mass should slightly increase the mass-specific metabolic rate (i.e., ml O_2 g^{-1} hr^{-1}), based on the allometry of metabolic rate. However, mass-specific metabolic rate tends to decline during starvation, although not by so much as absolute metabolic rate.

Short-term starvation can markedly reduce many avenues of energy expenditure, such as activity and food processing, and may reduce the standard/basal metabolic rate. In the white rat, for example, 24 hr starvation reduces total daily metabolic rate by 10%; there is a 12% increase in nonfeeding activity, an 87% decline in metabolic energy expended on feeding, and an 8% reduction in basal metabolic rate (Morrison 1968). Prolonged starvation results in a greater depression of basal metabolic rate in rats (Westerterp 1978). The badger has a reduced metabolism when starved, due to diminished activity, reduced food processing, and lowered maintenance

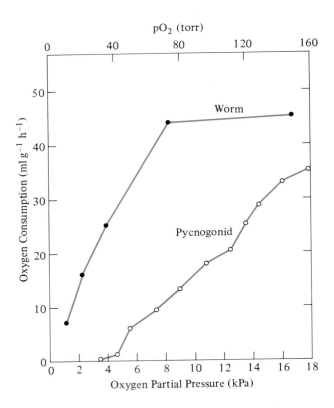

FIGURE 4–14 Relationship between oxygen consumption rate and ambient pO_2 for metabolic conformers and metabolic regulators. The pycnogonid (*Decolopoda*) is a typical metabolic conformer whose resting oxygen consumption rate depends on ambient pO_2 from 0 to 21.3 kPa; there is no critical pO_2. The earthworm (*Lumbricus*) is a typical metabolic regulator whose resting oxygen consumption rate is independent of ambient pO_2 at high values (10.6 to 21.3 kPa) but not at lower pO_2's; the critical pO_2 is about 10.6 kPa. *(Earthworm data from Johnson 1942; pycnogonid data from Davenport et al. 1987.)*

costs (Figure 4–15). A fraction of the decrease in basal metabolic rate during starvation can be attributed to a lower body temperature and its effect on metabolic rate; this is significant to the basal energy savings of starving rats and badgers.

Animals that consume food of low nutritive value or widely spaced and scarce food items may be regarded, in a sense, as experiencing chronic "partial starvation." For example, many spiders are "sit-and-wait" predators and therefore prone to extended periods between meals; spiders have a lower metabolic rate than is expected (Anderson 1970). There is a general relationship for mammals between diet and the allometry of metabolic rate. Frugivorous bats have higher basal metabolic rates than insectivorous bats; blood- or nectar-feeding bats have intermediate metabolic rates (McNab 1983). The insectivorous bats are probably more influenced by seasonal abundance of their food than fruit-eating bats.

Temperature

Body temperature has a marked effect on metabolic rate, just as temperature affects chemical and biochemical reaction rates. In general, there is an exponential relationship between metabolic rate and temperature, with a 2 to 3 × increase in VO_2 per 10° C increase in body temperature, i.e. the Q_{10} is 2 to 3.

Ambient temperature has a profound effect on the standard metabolic rate of ectotherms because their body temperature is often the same as ambient temperature. Typical examples of the relationship between metabolic rate and body temperature are illustrated in Figure 4–16A for a variety of ectotherms. The VO_2 does not rise as rapidly, and may even decline, at high ambient temperatures,

TABLE 4–8

Effects of starvation on basal/standard metabolic rate for a variety of animals. Metabolic rate is given in both absolute (kJ day⁻¹) and mass-specific (kJ g⁻¹ day⁻¹) units.						
Species	**Absolute Metabolism**			**Mass-Specific Metabolism**		
	Normal	*Starved*	*N/S*	*Normal*	*Starved*	*N/S*
Spider	0.002	0.001	0.35	0.21	0.08	0.41
Mussel	0.03	0.019	0.63	0.021	0.024	1.14
Spider	0.057	0.021	0.37	0.061	0.026	0.43
Sparrow	58	32	0.56	2.14	1.5	0.7
Sand rat	70	50	0.65	0.33	0.27	0.81
Rat	162	67	0.41	0.56	0.35	0.62
Badger	244	136	0.56	0.03	0.02	0.73
Goose	984	577	0.59	0.53	0.37	0.70
Human	6347	4414	0.79	0.1	0.09	0.85

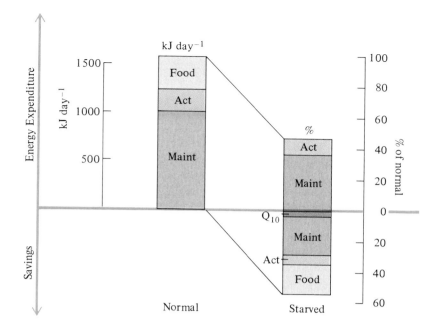

FIGURE 4–15 Effects of starvation on metabolic rate (kJ day^{-1}) and the partitioning of metabolic rate between food acquisition (Food), activity (Act), and maintenance (Maint) for a badger (starved for 30 days). The percentage decline in metabolic rate is primarily due to reduced maintenance metabolism, part of which reflects the reduced body temperature (Q_{10}). *(Data from Harlow 1981.)*

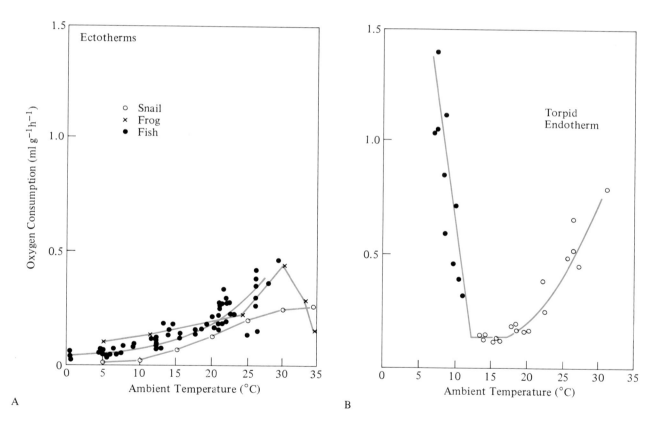

FIGURE 4–16 (A) Exponential relationship between metabolic rate and ambient temperature (body temperature) for ectotherms (snails, frog, and fish). The metabolic rate plateaus or even declines at high ambient temperatures, indicating thermal inhibition of metabolism. **(B)** There is a complex relationship between metabolic rate and ambient temperature for an endotherm, such as the marsupial mouse *Antechinomys*, when torpid. There is an exponential relationship between metabolic rate and ambient temperature, except at ambient temperatures less than about 10° C, when the marsupial mouse thermoregulates to prevent body temperature from dropping below 10° C. *(Snail data from Santos, Penteado, and Mendes 1987; frog data from Jusiak and Poczopko 1972; fish data from Haugaard and Irving 1943; marsupial mouse data from Geiser, 1986.)*

indicating thermal inhibition of metabolism by enzyme inactivation.

The metabolic rate of endotherms is also dependent on body temperature, but this effect is generally obscured by their precise regulation of body temperature. However, many endothermic mammals and birds hibernate, estivate, or become torpid (see Chapter 5) and then the exponential effect of body temperature on metabolic rate becomes apparent (Figure 4–16B). The metabolic rate of a hibernating mammal, corrected to a body temperature of 10° C, is 3.09 $g^{0.69}$ J hr^{-1}, which is about 50 × lower than the metabolic rate of mammals at 37° C (157 $g^{0.62}$ J

hr^{-1}; Malan 1986). The metabolic rate of a large hibernating endotherm is similar to that expected for an ectotherm, but small hibernating mammals have a higher metabolic rate than expected for an ectotherm.

The taxonomic differences in the basal metabolic rates of animals are partly due to differences in body temperature (Table 4–9). For example, the differences in basal VO_2 for the platypus, marsupials, primitive insectivorous mammals, "general" mammals, and nonpasserine birds are reduced when corrected to a body temperature of 38° C, i.e., the corrected metabolic rate is 3 to 4 watts $kg^{-0.75}$. However, the substantial differences in basal metabolic rates of other endotherms, e.g., shrews and passerine birds (which have high VO_2's) and the monotremes and ratite birds (which have a low VO_2) are not eliminated by correction to a body temperature of 38° C. Reptiles and amphibians have substantially lower standard VO_2 values at, or when extrapolated to, body temperatures of 38° C (0.3 to 1.1 watts $kg^{-0.75}$).

Specific Dynamic Effect

Lavoisier found that the metabolic rate of a human subject was increased after a meal to about 150% of their fasting metabolic rate. He explained this increase as the "work of digestion." Many different terms have been subsequently used to describe the observed increase in metabolic rate as a consequence of a meal. "Work of digestion" is unsuitable because it implies an incorrect digestive origin for the metabolic increase (as we shall see below). The term **specific dynamic effect (SDE)** was used by Rubner; the similar term specific dynamic action (SDA) is a poor translation of his original German term "spezifisch-dynamische Wirkung." Numerous other terms also have been used, e.g., heat of nutrient metabolism, postprandial thermogenesis, calorigenic effect, and dietary-induced thermogenesis.

The specific dynamic effect is of varying magnitude for a wide variety of animals. Rubner's classical studies indicated that the fasting metabolic rate of a dog (about 3105 kJ day^{-1}) was increased by 1272 kJ (41%) after ingestion of 2 kg meat. The magnitude of the SDE accounted for 15.8% of the metabolic energy content of the ingested meat. For man, the SDE is often a 30% increase in metabolic rate after a meal, which is sustained for a few hours. Fish (plaice) have a twofold increase in metabolic rate for 1 to 3 days after feeding (Figure 4–17). In many instances, the magnitude of a SDE is greatest for a high protein meal, and is often proportional to the

TABLE 4–9

Basal and standard metabolic rates of ectothemic and endothermic vertebrates, measured at their body temperature and corrected to a body temperature of 38° C. All values are expressed as watts $kg^{-0.75}$. A Q_{10} of 2.5 was assumed for temperature correction unless a different Q_{10} was cited in the original reference.

	Normal T_b (°C)	Normal Metabolic Rate	Metabolic Rate at (38° C)
Mammals			
Monotreme *Zaglossus*	32	0.86	1.53
Monotreme *Tachyglossus*	32	0.98	1.81
Edentates	33	1.69	2.66
Marsupials	35	2.37	3.00
General	38	3.34	3.34
Primitive insectivores	35	2.76	3.63
Golden mole	35	2.86	3.76
Monotreme *Ornithorhynchus*	32	2.21	3.80
Crocidurine shrews	36	6.7	8.0
Soricine shrews	38	13.4	13.4
Birds			
Ratites	37	2.10	2.31
Nonpasserines	40	4.05	3.37
Passerines	41	7.40	5.62
Reptiles			
Turtles	20	0.15	0.58
Lizards	30	0.40	1.02
Snakes	30	0.48	1.02
Crocodiles	23	0.29	1.06
Amphibians			
Apoda	25	0.15	0.25
Salamanders	20	0.06	0.33
Lungless salamanders	20	0.07	0.36
Anurans	20	0.21	1.07

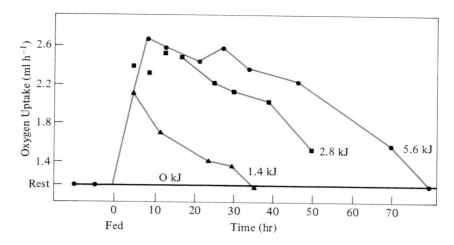

FIGURE 4–17 Specific dynamic effect of three levels of energy consumption by the plaice. Numbers indicate the energy content of the meal. *(From Jobling and Davis 1980.)*

amount of protein ingested. Many invertebrates have an SDE. In general, the SDE is proportional to the amount of food ingested, as seen for the plaice. An SDE of 494 J per mmol N ingested (about 35 kJ g N^{-1}) has been reported for frogs. Many factors contribute to the wide range of observed SDE values, including whether a fasting or maintenance ratio is used as the baseline metabolic state, how much of the additional nutrient is metabolized or stored, the ambient temperature, etc.

The history of the study of the "work of digestion" has been long and complex, and there still is not a universally accepted explanation for its mechanism. This probably is because there is not a single explanation for this complex phenomenon. Original notions regarding the "work of digestion" were that SDE represented mastication, muscle movements for food transport through the gut, the cost of secretion of digestive fluids and enzymes, and the metabolic cost of active nutrient absorption from the gut. However, the lack of a metabolic effect of feeding bones to dogs and the increase in metabolic rate after injection of amino acids into the blood stream argue against this "work of digestion" theory. The "plethora" theory advocates a generalized metabolic stimulation after a meal by the presence of high concentrations of metabolites; similar theories advocate a metabolic stimulation by specific nutrients. The metabolic basis for SDE has also been attributed to a surge of metabolic synthesis, particularly protein synthesis, after a meal and to the metabolic cost of amino acid deamination. Ruminant and pseudoruminant mammals may also have a

significant component of SDE due to the "microbial fermentative costs of digestion" (see Chapter 18).

The metabolic cost of SDE can be separated into a mechanical component that is due to physical processing and movement of food through the digestive tract, anabolic and catabolic biochemical processes, and metabolic pathways for nitrogen excretion. The mechanical component of SDE has been estimated for fish to be from 10 to 30% of the total SDE, depending on the meal size (Tandler and Beamish 1979). The energy released by hydrolysis of glucosidic bonds (about 14 to 18 kJ $mole^{-1}$), peptide bonds (10 to 24), and ester bonds (10) is so low (about 1/2%) compared to the chemical energy content of foodstuffs that SDE does not reflect the costs of hydrolysis or resynthesis of macromolecular bonds. The possible contribution of the biochemical costs of nitrogen excretion to SDE merit detailed examination.

Early studies indicated that the SDE for dogs was about 31% of resting metabolism for meat, 13% for lipid, and 6% for carbohydrate (Brody 1945). The SDE was also reported to be proportional to the amount of nitrogenous waste excreted. For example, the urinary nitrogen excretion of dogs closely parallels the SDE (Figure 4–18) and indicates a conversion factor of about 46 J g N^{-1} between SDE and urinary N excretion. The SDE component of the increased metabolism of mussels after feeding corresponds to the observed rate of nitrogen excretion; the conversion factor is about 94 kJ g N^{-1} (Bayne and Scullard 1977). The metabolism of alanine provides a typical example of the energy

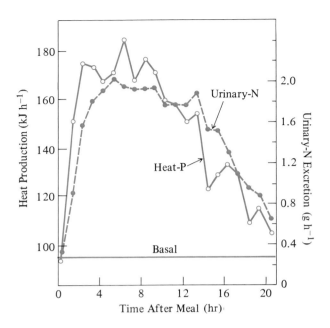

FIGURE 4–18 Relationship between increment in metabolic rate (kJ hr^{-1}) and urinary nitrogen excretion (g N hr^{-1}) for a fasting dog after being fed 1200 g meat. *(Modified from Brody 1945.)*

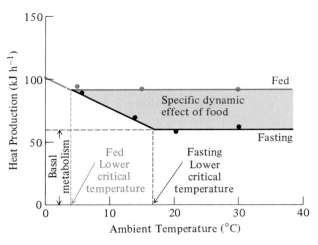

FIGURE 4–19 Schematic diagram of the increment in metabolic heat production for fasting dogs and for dogs fed 320 g meat, showing the substitution by heat of specific dynamic effect for thermoregulatory heat production. There appears to be no specific dynamic effect at low ambient temperatures. *(Data from Rubner 1902.)*

requirements for amino acid metabolism. The overall reaction

$$alanine + 1/2\,O_2 \rightarrow NH_3 + CH_3COOH \quad (4.9)$$

yields about 17 kJ heat g N^{-1}. The conversion of NH$_3$ to urea releases about 17 kJ g N^{-1}, and the renal excretion of urea releases perhaps a further 4 to 8 kJ g N^{-1}. The total energy released by deamination and urea excretion is therefore about 40 kJ heat g N^{-1}. About 126 kJ of potentially useful energy is yielded by oxidation of the deaminized alanine fragment. Thus, heat equivalent to about 28% of the energy content of alanine is liberated by deamination and urea excretion.

The metabolic expenditure due to SDE may be of no value to the metabolic processes of animals, but the energy of SDE may be of physiological consequence to endotherms. The magnitude of SDE appears to depend on the ambient temperature; the fasting rate of heat production increases by 50% for dogs after feeding at an ambient temperature of 30° C but does not increase at 7° C. This is because the heat production of SDE substitutes for the thermoregulatory increase in metabolic heat production of a cold-stressed endotherm. The magnitude of SDE appears to decline at ambient temperature

below the critical thermal minimum temperature and is absent when SDE equals or is less than the thermogenic heat requirement (Figure 4–19; see Chapter 5).

In summary, SDE reflects the energetic requirements of many processes that occur as a consequence of food digestion, including mechanical processing, energy exchange through catabolic and anabolic biochemical pathways, and amino acid deamination and nitrogen excretion.

Activity

Activity is probably the most important determinant of metabolic rate for an animal. There is a continuum of activity levels from none to maximal (or summit) and the level of activity determines the elevation of metabolic rate above the basal value.

The metabolic cost of graded activity has been most thoroughly documented for humans; a typical range of metabolic rates from minimal values to maximal values is summarized in Table 4–10. Similarly detailed compilations are not possible for animals, but one commonly measured metabolic rate is the rate during maximal activity or cold stress. The maximal metabolic rate sustained by aerobic metabolism is generally about 5 to 10 × the resting, basal, or standard metabolic rate. The

TABLE 4–10

Human basal metabolic rate and metabolic rate with various forms of graded activity. Values are J min⁻¹. *(Data from Passmore and Durnin 1955.)*	
Basal	4.2
Lying at ease	6.3
Sitting at ease	6.7
Standing at ease	7.1
Walking: 1 km hr⁻¹	8.4
Driving car	11.7
Walking: 4 km hr⁻¹	14.2
Walking: 6 km hr⁻¹	20.9
Cricket batting	25.1
Walking: + 15% incline/3 km hr⁻¹	26.4
Tennis	29.7
Walking: 8 km hr⁻¹	33.5
Rapid marching	40.6
Squash	42.7
Climbing vertical ladder	48.1
Walking in loose snow: 20 kg load	84.5
Ax work: 51 blows min⁻¹	100.9
Carrying 60 kg upstairs	128.4

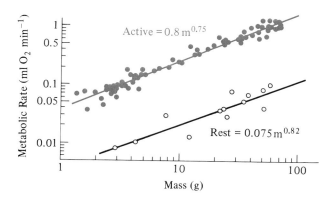

FIGURE 4–20   Relationship between metabolic rate (MR; ml min⁻¹) for the toad *Bufo boreas* at rest and during activity.   *(From Hillman and Withers 1979.)*

ratio $VO_{2,max}/VO_{2,basal}$ is the **factorial aerobic scope**, whereas $VO_{2,max} - VO_{2,basal}$ is the **absolute aerobic scope**.

The maximal metabolic rate of small mammals (<4 kg) is proportional to mass$^{0.72}$ with an intercept value (at 1 gram) of 602 J hr⁻¹; this is 8.3 × the intercept for basal metabolic rate of 72.6 J hr⁻¹ (Lechner 1978). Larger mammals, dogs, humans, and horses appear to have higher maximal rates ($VO_{2,max} = 789$ mass$^{0.79}$) and high factorial scopes of 11.5 × basal (Pasquis, Lacaisse, and Dejours 1970). Wild and domesticated mammals appear to have slightly different scaling relationships for $VO_{2,max}$ (Taylor et al. 1981). Dasyurid marsupials have a factorial scope of 8.7 × basal despite their lower basal VO_2 of 2.45 ml O_2 hr⁻¹ mass⁻⁰·⁸² and lower $VO_{2,max}$ of 21.2 ml O_2 hr⁻¹ mass⁻⁰·⁹⁵ (MacMillen and Nelson 1969; Baudinette, Nagle, and Scott 1976).

Factorial metabolic scopes during activity for other animals are generally of similar magnitude, about 5 to 10. Anuran amphibians demonstrate similar metabolic scopes for activity, even with a single species, over a wide range of body masses; *Bufo boreas* ranging in mass from 1 to 69 g have almost parallel allometric relationships for resting VO_2, with an almost constant factorial scope of 10 × (Figure 4–20).

Insects and crustaceans have aerobic scopes for walking and running that are similar to vertebrates.

For example, beetles have a factorial scope ranging from 8.0 times (0.1 g) to 33.3 times (10 g) due to a differing allometric relationship for resting VO_2 (ml O_2 hr⁻¹ = 0.23 mass$^{0.86}$) and active VO_2 (37.6 mass$^{1.17}$; Bartholomew and Casey 1977). This size dependence of factorial scope is due to the larger beetles having higher body temperatures than the smaller beetles. Cockroaches, tarantula spiders, and land crabs have aerobic scopes of two to ten times resting during locomotion (Herreid 1981; Full 1987).

Flying insects have much larger factorial metabolic scopes than other animals because of the high metabolic cost of flight, particularly hovering (Figure 4–21). Hovering honeybees have a factorial

FIGURE 4–21   A hovering honeybee has an extremely high metabolic cost.   *(Photograph courtesy of C. Ellington and A. R. Thomas.)*

scope of up to 40 × rest ($VO_{2,hov}$ = 120 ml O_2 g^{-1} hr^{-1}; $VO_{2,rest}$ = 3 ml O_2 g^{-1} hr^{-1}). An elevation of body temperature during hovering is not a major contributing factor to this high factorial scope (Withers 1981). Hovering sphingid moths have a 172 × increase in VO_2 over rest, but part of this immense factorial scope is due to a substantial elevation of thoracic temperature from 23° to 42° C (Bartholomew and Casey 1978). The factorial scope is still high, about 29 × rest when $VO_{2,rest}$ is corrected to 42° C assuming a Q_{10} of 2.5. Similar scopes for saturniid moths are 127 × total factorial scope, and 29 times when resting VO_2 is corrected to the same thoracic temperature as hovering moths. These factorial scopes are not necessarily the maximal values, since hovering VO_2 is probably not the maximal rate.

Free-Living Metabolism

One aspect of animal metabolism that has proven very difficult to measure is the average daily metabolic rate of animals under natural conditions, i.e., **field metabolism**. However, the measurement of isotopic turnover of doubly labelled water has recently allowed the determination of field metabolic rate for many animals, primarily terrestrial vertebrates.

The field metabolic rate of terrestrial vertebrates is typically two to three times the basal metabolic rate (of endotherms) or standard metabolic rate (of ectotherms at the appropriate body temperature). Field metabolic rate (FMR) often scales with body mass in a similar fashion as BMR or SMR. For example, the **b** value for FMR (0.749) is similar to **b** for BMR of nonpasserine birds (Nagy 1987). However, the allometric relationships are not always similar; for example, the **b** value for FMR is greater than that for BMR in placental mammals and passerine birds, and **b** is less for FMR than BMR in marsupial mammals. Because of variation in **b** the intercept **a** values are not necessarily two to three times higher for FMR than BMR. The ratio of **a** for FMR/BMR varies from 1.2 (nonpasserine birds), 1.8 (passerine birds), and 1.9 (placentals) to 10.3 (marsupials). For iguanid lizards, field and standard values for **b** are virtually the same, and FMR = 9.3 $M^{0.799}$ and SMR = 1.5$M^{0.80}$ (at 20° C). The high ratio of **a** for FMR/SMR of 6.2 is due in part to field body temperature exceeding 20° C. The expected difference in **a** for ectotherms and endotherms is also apparent for FMR. For example, the ratio of **a** for eutherian mammals to iguanids is 15.1, and for passerine birds to iguanids the ratio is 39.8.

Anaerobic Metabolism

The biochemical details of anaerobic metabolism were discussed in Chapter 3, but the energetic significance of anaerobic metabolism to animal energetics will be discussed here.

Invertebrates

The resting metabolic rate of many bivalve mollusks is only partly supported by aerobic metabolism (Table 4–11). The ratio of aerobic heat production (Q_{ox}) to total heat production (Q_{tot}) varies from 0.5 to 0.7 for many pelycopods but is about 1 for *Mya* and many other marine invertebrates.

Metabolic responses to environmental anoxia are well documented for intertidal mollusks (de Zwaan 1983; Schick, de Zwaan, and de Bont 1983; Livingston and de Zwaan 1983; de Zwaan and van den Thillart 1985; de Zwaan and Putzer 1985). For

TABLE 4–11

Comparison of aerobic (Q_{ox}) and total heat production (Q_{tot}) for various marine invertebrates during rest, with oxygen available. *(Data from Hammen 1979; Hammen 1980; Grainger 1968; Becker and Lamprecht 1977; Gnaiger 1980.)*

	HEAT PRODUCTION (J g^{-1} h^{-1})		
	Aerobic (Q_{ox})[1]	Total (Q_{tot})	Ratio Q_{ox}/Q_{tot}
Pelecypods			
Mytilus	1.18	1.59	0.74
Crassostrea	7.26	14.03	0.54
Mya	1.85	1.90	0.97
Mercenaria	0.66	1.02	0.65
Modiolus	1.39	2.10	0.66
Petricola	3.13	5.15	0.61
Gastropods			
Littorina	1.75	1.65	1.08
Nucella	2.77	2.74	1.01
Biomphalaria	3.16	2.66	1.19
Oligochetes			
Lumbricus	3.20	3.10	1.03
Crustaceans			
Libinia	1.18	1.24	0.95
Asellus	2.45	2.23	1.10
Echinoderms			
Asterias	1.28	1.25	1.02

[1] Assuming 1 ml O_2 is equivalent to 20.1 kJ.

example, the mussel *Mytilus edulis* routinely experiences anoxia during tidal emmersion (Figure 4–22), and relies heavily on anaerobic metabolism. The aerobic metabolic rate of the intertidal mussel *Mytilus* when submerged is about 50 μl O_2 g^{-1} hr^{-1}; this is equivalent to about 14.7 μm ATP g^{-1} hr^{-1}. This aerobic metabolic rate is only about 74% of the total metabolism in water (Table 4–11), and so there must be a significant anaerobic component (26% of total) to metabolism even at rest with O_2 available. During emmersion and anoxia, the metabolic rate declines dramatically to about 1/20 of the submerged normoxic rate, and so anaerobic metabolism can support the reduced energy requirement. After 48 hours of anoxia, various anaerobic metabolic pathways have produced about 36 μM ATP g^{-1} and a variety of end products (0.9 μm lactate, 9.0 alanine, 12.2 succinate, 13.5 propionate). About 0.6 μm of ATP stores and 1.2 μm of phosphoarginine stores are also used. The total ATP turnover for the 48 hours is about 37.8 μm ATP g^{-1} (the submerged rate would have been about 706 μm ATP g^{-1}).

A *Mytilus* with a typical glycogen store of about 550 μm glucosyl units g^{-1} could survive at the submerged metabolic rate for about three days if lactate were the metabolic end product (2 ATP/glucosyl unit). Fermentation yielding succinate would double the survival time to about six days (4 ATP/glycosyl unit) or nine days for propionate formation. In contrast, the marked metabolic depression during emmersion prolongs anoxia tolerance by twentyfold, i.e., from 3 to 60 days for lactate production, 120 days for succinate, and 150 days for propionate production! Unfortunately, this analysis of aerobic and anaerobic metabolic rates is slightly complicated by the fact that not all of the

resting metabolic rate is aerobic, and that the various metabolic end products accumulated after 48 hours of anoxia (3.27 J mussel^{-1} 2 days^{-1}) accounts for only about 37% of the measured heat production (8.82). Consequently, the decline in heat production over 48 hours of anoxia is only about 1/10, rather than 1/20 as calculated from the accumulated end products. There are similar discrepancies between heat production and aerobic and anaerobic metabolic rates in other invertebrates. The ratio of anoxic to normoxic metabolism is about 0.48 for the oligochete worm *Lumbriculus*, but energy production calculated from the accumulation of anaerobic end products/energy from oxygen consumption is much lower, at about 0.15 (Gnaiger 1980).

Other intertidal invertebrates show a metabolic depression and reliance on anaerobic metabolism during anoxia similar to that of *Mytilus*. The ATP turnover rate may decline to 1/10 of the normoxic value in annelids and crustaceans and to even lower levels in bivalve mollusks (de Zwaan and van den Thillart 1985), although several hours may be required for the metabolic depression to be fully expressed. The bivalve *Modiolus* has an aerobic metabolism of 7.64 kJ hr^{-1} and an anaerobic metabolism rate of only 0.58 kJ O_2 hr^{-1}, i.e., 1/13.3 of aerobic. The cockle *Cardium* can reduce its metabolic rate from 1.8 μm ATP g^{-1} hr^{-1} (control, aerobic) to 0.23 (24 to 48 hour anoxia, anaerobic). A reliance on ATP and arginine phosphate stores becomes less important as the period of anoxia increases, even though the metabolic rate becomes further depressed. The percentage of ATP derived from arginine phosphate declines from 56% over four hours of anoxia to only 15% for 24 to 48 hours of anoxia.

Activity is typically of a much shorter duration (seconds or minutes) than anoxia (hours or days). The metabolic rate during activity is 8 to 30 times the resting metabolic rate, whereas it declines to 1/10 to 1/20 times during anoxia in a mollusk (*Placopecten*), an annelid worm (*Arenicola*), and a sipunculid (*Sipunculus*; Table 4–12). Anaerobic metabolism during activity can use the same fundamental metabolic pathways and energy stores as occurs during environmental anoxia, but there typically is a greater reliance on ATP and phosphagens (e.g., arginine phosphate) stores during activity and less reliance on anaerobic metabolic pathways.

Vertebrates

Vertebrates are generally less tolerant of environmental anoxia than invertebrates, although some have an exceptional tolerance of hypoxia and even

FIGURE 4–22 The bivalve mollusk *Mytilus edulis* routinely experiences anoxia during tidal emmersion. *(Photo courtesy of Eugene N. Kozloff.)*

TABLE 4–12

Comparison of rates of ATP production at rest, during activity, and during prolonged environmental anoxia for three invertebrates, showing the factorial metabolic scope.			

	Mollusk *Placopecten*	Annelid *Arenicola*	Sipunculid *Sipunculus*
Rest (μm ATP g^{-1} min^{-1})	0.43	0.276	0.084
Exercise (μm ATP g^{-1} min^{-1} (X resting)	12.65 29X	2.93 8X	0.68 8X
Anoxia (μm ATP g^{-1} min^{-1} (X resting)	0.115 0.26X	0.040 0.15X	0.028 0.33X

complete anoxia, particularly if it is accompanied by a low ambient temperature to depress metabolic rate. Some fish (goldfish, carp) can survive many hours, days or even months under anoxic conditions, particularly at temperatures of 0° to 4° C (van den Thillart 1982). The cyprinid fish *Rasbora* can survive over 100 days sealed in a jar, at 29° to 33° C (Mathur 1967). Some amphibians are also tolerant of hours or even days of anoxia and certain turtles are remarkably tolerant of anoxia; *Chrysemys picta* can survive over 126 days in anoxic water at 3° C (Figure 4–23). A considerable metabolic depression (to about 1/5 the normal metabolic rate) contributes

to this tolerance of anoxia (Ultsch and Jackson 1982), although the metabolic depression may not be as pronounced as is observed for invertebrates (especially bivalve mollusks). The heat production of a diving turtle is depressed from 1.75 J g^{-1} hr^{-1} (normal) to 0.25 after four hours of diving (Jackson 1968), and the metabolic depression after 120 days of anoxia at 3° C may be much more profound. The primary anaerobic metabolic end product that accumulates during anoxia is lactate, although other end products (succinate, alanine) accumulate in small amounts. Goldfish accumulate a novel (for vertebrates and invertebrates) metabolic end product—ethanol—and may utilize protein to a significant extent as an anaerobic metabolic substrate (van den Thillart, Henegouwen, and Kesbeke 1983).

Mammals, reptiles, amphibians, and the mollusk *Limaria* rely on anaerobic metabolism to a considerable extent during short-term intense activity (five sec to five min or more); 20 to 80% of the total ATP production comes from anaerobic pathways (Table 4–13). All of the materials necessary for anaerobic metabolism are located in the cell cytoplasm, where the ATP is required. No materials such as O_2 need to be transported into the cells. Thus, anaerobic metabolism can be extremely rapid. However, only a finite amount of energy can be derived from anaerobic pathways before the accumulation of anaerobic end-products inhibits further anaerobic

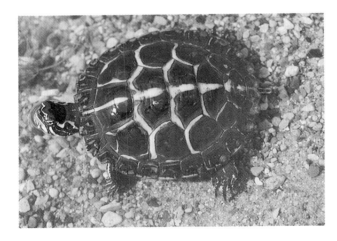

FIGURE 4–23 The turtle *Chrysemys picta* relies extensively on anaerobic metabolism during long periods of submersion. *(Photograph courtesy of James H. Harding, Michigan State University Museum.)*

TABLE 4–13

Aerobic and anaerobic ATP production (μmoles g^{-1} min^{-1}) and % contribution of anaerobic metabolism to total ATP production during short bursts of activity.

	Duration	ATP Production Aerobic	Anaerobic	Total	% Anaerobic
Mammal *Microtus*	30 sec	23.2	12.6	35.8	35%
Snake *Crotalus*	5 min	2.1	3.2	5.4	60%
Lizard *Aniella*	2 min	2.0	7.1	9.2	78%
Amphibian *Rana*	3 min	2.9	0.8	3.7	21%
File Shell *Limaria*	5 min	4.3	1.7	6.0	28%

metabolism. It is therefore not surprising that the significance of anaerobic metabolism in higher vertebrates is usually limited to short bursts of intense activity.

A limited amount of energy is always present in cells as ATP and creatine-phosphate (or other phosphagens) but considerably greater amounts of energy are stored as glycogen or lipid. There also is a considerable store of protein, but this is not generally available for metabolic energy production. Anaerobic metabolism enables the rapid utilization of at least some of the glycogen energy stores, but not the lipid stores. There is about five times as much energy present as creatine-phosphate than ATP, about 1400 times as much energy present as glycogen, and 2500 times as much present as lipid.

The significance of duration of activity to the relative contribution of anaerobic metabolism to ATP production is best illustrated by studies of human exercise. The typical maximal aerobic metabolic rate is about 5000 ml O_2 min^{-1} (101 kJ min^{-1}). The maximal anaerobic metabolic expenditure is about 188 kJ (equivalent to about 9000 ml O_2). The anaerobic expenditure is a quantity of energy, not a rate. The maximum anaerobic metabolic rate depends on the duration of the activity. The rate is 376 kJ min^{-1} if the 188 kJ is produced over 30 sec (i.e., about four times the maximal aerobic metabolic rate). The rate is about 94 kJ min^{-1} if the 188 kJ is produced over two minutes, i.e., similar to maximal aerobic metabolic rate. A series of similar calculations yield the general relationship between aerobic and anaerobic contributions to energy production as a function of duration of activity, shown in Figure 4–24.

Anaerobic metabolism is often considered to be important only during periods of anoxia or intense activity. However, for vertebrates, lactate is released by some tissues even at rest, at a rate of about 100 mg min^{-1} in humans. The lactate is removed at the same rate by other tissues, such as the liver. It is either aerobically oxidized or reconverted to glucose for redistribution to other tissues (see Chapter 3). Anaerobic metabolism is a significant source of energy for muscles, not only during intense activity when the work load exceeds the aerobic capacity, but at submaximal work loads. The blood lactate concentration is elevated over the resting level at a work load of 50 to 80% maximal (Figure 4–25). The anaerobic threshold is about 1700 ml O_2 min^{-1} for an adult male human (about 60% of $VO_{2,max}$). Other physiological variables alter at the same work load; for example, the slope of the relationship between respiratory ventilation (V_E)

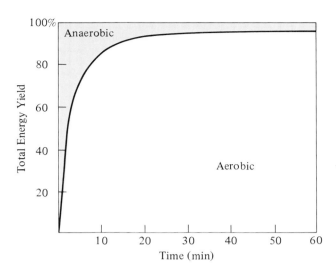

FIGURE 4–24 Anaerobic and aerobic contributions to total energy yield for a human during maximal activity of durations up to 60 min. Anaerobic metabolism provides more than 50% of the total energy yield at less than about 2 min duration, anaerobic and aerobic contributions are equal at about two min duration, and aerobic metabolism provides more than 50% of the total energy yield for activity durations greater than about 2 min.

and work load increases at the anaerobic threshold; the ratio of V_E/VO_2, V_E/VCO_2, RQ, and the end-tidal pO_2 (PEO_2) also alter at the anaerobic threshold.

Other vertebrates also have an aerobic threshold, although at differing percentages of the maximum work load. For example, the monitor lizard *Varanus* has an anaerobic threshold at about 1.5 km hr^{-1}, at which the VO_2 is about 80% of $VO_{2,max}$ (Figure 4–26).

Scaling of Cellular Metabolism

We might expect the anaerobic metabolic capacity of cells to be similar for different-sized animals because enzyme and substrate concentrations are similar for cells of large and small animals, ectotherms, and endotherms (Coulson, Hernandez, and Herbert 1977). Consequently, the anaerobic metabolic rate of animals might scale directly with body mass because it is an entirely endogenous metabolic process. There is no requirement for transport of metabolites into or out of cells, hence there are no "surface area" scaling requirements. In contrast, aerobic metabolic rate is often assumed to be limited by the rate at which substrates (including O_2) are

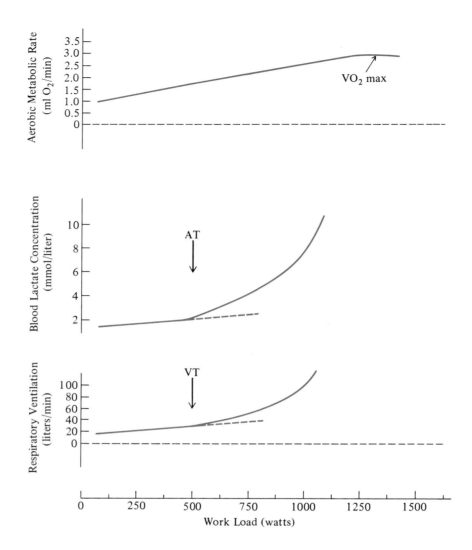

FIGURE 4–25   Relationship between work load and aerobic metabolic rate, blood lactate concentration, and respiratory ventilation for human athletes. The anaerobic threshold (AT) occurs at about 60% of the maximal aerobic metabolic rate, at the same work load as the ventilatory threshold (VT). *(Modified from Lamb 1984.)*

delivered to cells, rather than by the cellular metabolic capacity. The orders of magnitude difference in absolute metabolic rates of large and small animals are not matched by similar differences in enzyme or substrate concentrations.

There is considerably less information available concerning the allometry of anaerobic metabolism than aerobic metabolism, but it is consistent with either a direct proportionality of anaerobic metabolic capacity with mass, i.e., $b = 1$ (mass-specific anaerobic metabolic rate is mass independent) or a higher absolute anaerobic metabolic rate for larger

animals, i.e., $b > 1$. The anaerobic metabolic rate of bivalve mollusks and amphibians is proportional to mass ($b = 1$). For water snakes and crocodiles, $b > 1$. Considerably more data are required before generalizations can be made, but it appears that anaerobic metabolism scales in a fundamentally different manner than aerobic metabolic rate.

There are significant enzymatic correlates of the scaling relationships for aerobic and anaerobic metabolic rate (Somero and Childress 1980; Emmett and Hochachka 1981; Somero and Childress 1985). Citrate synthase (CS) is a key aerobic enzyme

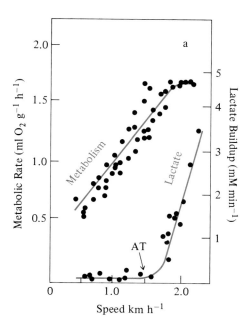

FIGURE 4–26 Anaerobic threshold (increase in rate of lactate production; AT) for the varanid lizard *Varanus exanthematicus* is about 95% of maximal oxygen consumption rate. *(From Seeherman, Dmi'el, and Gleeson 1983.)*

of the citric acid cycle (Chapter 3). CS activity (measured as total units of activity) scales with body mass; in various species of fish, CS is proportional to $M^{0.77}$ to $M^{0.95}$, and in mammals $M^{0.89}$. Similar scaling exponents are apparent for other important aerobic enzymes and are observed for diverse tissues, such as brain and muscle. In contrast, the key anaerobic (glycolytic) enzymes, such as lactate dehydrogenase (LDH) and pyruvate kinase (PK), have mass-specific scaling exponents close to 1 or >1. For example, the exponent is from 1.08 to 1.66 for fish and 1.15 for mammals. Significantly, this is observed only for tissues that rely upon anaerobic metabolism, such as muscle, and not for aerobic tissues, such as brain. Thus, the enzymatic capacities of tissues are well correlated with the allometric relationships observed for aerobic and anaerobic metabolism.

Summary

The metabolic rate of animals can be measured by direct calorimetry as metabolic heat production or by indirect calorimetry as O_2 consumption or CO_2 production. Metabolic rate can also be determined as the difference between the energy ingested and the energy excreted as urine and feces.

The energy content of biological materials can be determined by bomb calorimetry. Carbohydrate, lipid, and protein differ markedly in their stoichiometry of metabolism and energetic equivalents, i.e., kJ g^{-1}, kJ liter O_2^{-1}, kJ liter CO_2^{-1}, RQ. The stoichiometry for protein metabolism is complicated by the final form of the nitrogen, e.g., ammonia, urea, uric acid. Energetically, ammonia is the most expedient nitrogenous waste product for animals.

The energy budget of an animal accounts for all avenues of chemical energy gain and loss. Energy consumption (*C*) equals the sum of all fates for energy loss: urine (*U*), feces (*F*), tissue production (*P*), metabolism (*M*), and external work (W_{out}); $C = U + F + P + M + W_{out}$. The apparent assimilated energy is ($C - F$) and the apparent assimilation efficiency is $100 (C - F)/C$. The metabolizable energy is $C - F - U$. Gross production efficiency is $100 \, P/C$ and net production efficiency is $100 \, P/(C - F)$. Gross work efficiency is $100 \, W_{out}/C$ and net work efficiency is $100 \, W_{out}/(C - F)$.

Metabolic rate (*MR*) of animals is not directly proportional to body mass (grams), but is related to mass by the power equation $MR = \mathbf{a}m^{\mathbf{b}}$ where **a** is the intercept and **b** is the mass exponent, or slope for the log-transformed relationship. The value of **a** varies for different animal taxa, although animals can be divided into three broad categories of **a** values: unicells have the lowest **a**, multicellular ectotherms have higher **a**, and multicellular endotherms have the highest **a**. Values of **b** are generally about 0.7 to 0.8.

Metabolic rate can be influenced by photoperiod, availability of oxygen and food, body and air temperature, digestion and processing of food, and level of activity. The metabolic rate of an animal ranges from the minimal value when totally resting and fasted (standard metabolism for ectotherms; basal metabolism for endotherms) to the maximal or summit metabolism.

Anaerobic metabolism is a significant fraction of resting normoxic metabolic rate for only a few animals (some bivalve mollusks), although it can provide substantial amounts of ATP, especially during hypoxia/anoxia or during intense activity. Many invertebrates have a substantial capacity for anaerobic metabolism during anoxia and activity, e.g., bivalve mollusks, annelid worms, sipunculids. Most vertebrates also have a substantial anaerobic energy production during intense activity, and some have a considerable tolerance of hypoxia/anoxia (e.g., some fish, amphibians, and reptiles).

Supplement 4–1

Stoichiometry for Carbohydrate, Lipid, and Protein Metabolism

· ·

The complete oxidation of one mole of glucose consumes six moles of oxygen, releases six moles of carbon dioxide, and liberates 2874 kJ energy.

$$C_6H_{12}O_6 + 6O_2 \rightarrow 6CO_2 + 6H_2O + 2874 \text{ kJ mole}^{-1}$$

The energy equivalence of glucose is 2874 kJ mole^{-1} or 15.9 kJ g^{-1}. Other useful ratios calculated from this stoichiometry of glucose metabolism are 21.4 kJ liter O_2^{-1}, 21.4 kJ liter CO_2^{-1}, and RQ = 1. The metabolism of C_6-subunits of glycogen yields slightly different values for kJ O_2^{-1} and kJ CO_2^{-1}, because there is one less H_2O per C_6-subunit.

$$C_6H_{10}O_5 + 6O_2 \rightarrow 6CO_2 + 5H_2O + 2874 \text{ kJ mole}^{-1}$$

	Glucose	Glucosyl Unit	Lipid	Protein
kJ g^{-1}	15.9	17.7	39.2	20.1
kJ mole^{-1}	2870	2870	10042	64400
L O_2 g^{-1}	0.75	0.83	2.01	1.07
L CO_2 g^{-1}	0.75	0.83	1.40	0.91
kJ L O_2^{-1}	21.4	21.4	19.5	18.8
kJ L CO_2^{-1}	21.4	21.4	27.9	24.0
RQ	1.0	1.00	0.70	0.84
g H_2O kJ^{-1}	0.038	0.031	0.029	0.021
M urea mole^{-1}	0	0	0	19.5
kJ g N^{-1}	0	0	0	126

The stoichiometry for lipid metabolism is quite different, and the indirect calorimetric ratios (kJ g^{-1}, kJ O_2^{-1}, kJ CO_2^{-1}, RQ) differ substantially from those for glucose and glycogen metabolism. For example, combustion of palmitate, a long-chain fatty acid, is as follows.

$$C_{16}H_{32}O_2 + 23O_2 \rightarrow 16CO_2 + 16H_2O + 10042 \text{ kJ mole}^{-1}$$

The energy equivalents for palmitate are 39.2 kJ g^{-1}, 19.5 kJ liter O_2^{-1}, 27.9 kJ liter CO_2^{-1}, and RQ = 0.70. Not all lipids have the same stoichiometry, however. For ex-ample, metabolism of a short-chain triglyceride, $C_3H_5(CH_3CH_2COO)_3$ has an RQ of 0.81.

The stoichiometry of protein metabolism is complex because it contains N and S as well as C, H, and O. There is about 52g C, 7g H, 23g O, 17g N, and 1g S in 100 g of "meat protein". The final form of the N and S complicates the stoichiometry. An approximate stoichiometry for metabolism of 1 mole (3205) g of meat protein is

$$C_{139}H_{224}O_{46}N_{39}S + 142.6 O_2 \rightarrow 119.5 CO_2 + 68.7 H_2O$$
$$+ 19.5 CO(NH_2)_2 + H_2SO_4$$
$$+ 64400 \text{ kJ mole}^{-1}$$

if the N is converted to urea and the S to SO_4^{2-}. The metabolic energy yield is 20.1 kJ g^{-1}, but the energy yield is higher when measured by bomb calorimetry (about 23.8 kJ g^{-1}) because the nitrogenous product is not urea. The RQ for metabolism of "meat protein" is 0.84. The RQ for a typical amino acid, alanine, is 0.81.

The stoichiometry for protein metabolism depends on whether ammonia (NH_3), urea ($CO(NH_2)_2$), or uric acid ($C_5H_4O_3N_4$) is the nitrogenous waste product. Consider the metabolism of 100 g protein (2364 kJ energy).

$$C_{4.42}H_7O_{1.44}N_{1.14}$$
$$+ 4.6 O_2$$

↓ammonia	↓urea	↓uric acid
1.14 NH_3	0.57 $CO[NH_2]_2$	0.285 $C_5H_4O_3N_4$
+ 4.42 CO_2	+ 3.85 CO_2	+ 3.00 CO_2
+ 1.79 H_2O	+ 2.36 H_2O	+ 2.93 H_2O
+ 1967 kJ	+ 2002 kJ	+ 1815 kJ

None of these pathways yield 2364 kJ of energy. The energy yield is less because of the energy content of the nitrogenous waste: ammonia, 347.9 kJ/1.14 mole; urea, 362 kJ/0.57 mole; uric acid, 549 kJ/0.285 mole. The joule equivalent for protein is fairly similar for each pathway; 19.1 (ammonia), 17.6 (urea), and 19.4 (uric acid) kJ liter O_2^{-1}. The RQ values are quite different, being 0.96, 0.84, and 0.72, respectively.

Supplement 4–2

Energetics of Nitrogenous Waste Products

· ·

The three main nitrogenous wastes from protein metabolism, ammonia, urea, and uric acid differ considerably in their energy contents, but the comparison is complicated by the variety of ways to express energy content.

The energy content of ammonia, urea, and uric acid, measured by bomb calorimetry, is different from that calculated from their biosynthetic pathways from ammonia.

		Ammonia NH_3	Urea $CO(NH_2)_2$	Uric Acid $C_5H_4O_3N_4$
	Mol Wt	*17*	*60*	*168*
Energy	kJ g^{-1}	20.5	10.6	11.5
Content	kJ mole^{-1}	347.9	634.3	1926
	kJ g N^{-1}	24.9	22.7	34.5
Energy of	kJ g^{-1}	0	3.1	1.7
Biosynthesis[1]	kJ mole^{-1}	0	122	244
from NH_3	kJ g N^{-1}	0	8.7	4.9

[1] Biosynthesis from ammonia, assuming 1 mole ATP is equivalent to 30.5 kJ and urea synthesis requires 4 ATP per urea and uric acid synthesis requires 8 ATP per molecule.

Ammonia has the highest energy content per gram but the lowest per mole (measured by bomb calorimetry). Ammonia has an intermediate energy content per gram of nitrogen between urea (lowest value) and uric acid (highest value). Uric acid would therefore appear to be the most energetically wasteful and urea the most energetically conserving waste product if the primary objective of N excretion is to minimize the energy loss. However, much of the chemical energy associated with nitrogen waste products is unavailable to animals because they cannot further oxidize ammonia despite its 20.5 kJ g^{-1} energy content.

Another important consideration concerning the energetics of nitrogen waste products is the biological cost for synthesis of the waste product. This is not taken into account by the bomb calorimetric energy content. Ammonia is the initial waste product of amino acid metabolism (except for arginine). Energy is therefore required for conversion of ammonia to urea or uric acid, regardless of their respective joule energy content. The overall conversion of ammonia to urea by the urea cycle (see Chapter 17) requires 4 moles of ATP per mole of urea, i.e., about 122 kJ mole^{-1}.

$$2\,NH_3 + CO_2 \xrightarrow[\underset{4\,ATP \quad 4\,ADP}{}]{urea\ cycle} CO(NH_2)_2 + H_2O$$

The overall conversion of ammonia to uric acid requires 8 moles of ATP per mole of uric acid, i.e., about 244 kJ mole^{-1}.

$$\begin{aligned}2\,NH_4^+ + 2\,formate \\ + HCO_3^- + glycine \\ + aspartate + 2\,O_2\end{aligned} \xrightarrow[\underset{8\,ATP \quad 8\,ADP}{}]{uricogenesis} \begin{aligned}uric\ acid + fumarate \\ + 2\,H^+ + 2\,O_2^{2-}\end{aligned}$$

These estimated costs for biochemical conversion of ammonia to urea and uric acid are quite substantial compared to their joule contents, and they provide a clear energetic incentive for ammonia excretion rather than urea or uric acid excretion.

Energy content is, however, only one aspect concerning the relative advantages and disadvantages of the various nitrogenous waste products. Water conservation and osmoregulation are probably much more important than energetics in determining the most favorable nitrogenous waste product for particularly terrestrial animals (see Chapters 16 and 17; see Pilgrim 1954).

Supplement 4–3

Allometric Analysis of Metabolic Rate

Allometry is the study of the way that a dependent variable Y (e.g., metabolic rate) varies in relation to an independent variable X (e.g., body mass); it is the study of scale effects. For a simple linear relationship between X and Y,

$$Y = a + bX$$

where **b** is the slope $(Y_2 - Y_1)/(X_2 - X_1)$ and **a** is the Y-intercept (the Y value when $X = 0$). Analysis of linear data is often accomplished by the method of least-squares linear regression analysis. The informative statistics summarizing the goodness of fit for a linear regression relationship include the slope of the relationship (**b**); the Y-intercept (**a**); the correlation coefficient (r), which varies from -1 (perfect inverse correlation) to 0 (no significant correlation) to $+1$ (perfect positive correlation); the square of the correlation coefficient (r^2), which is the fraction of the variation in Y values that is explained by the variation in the X values; and the standard errors or 95% confidence limits for the slope (**b**) and intercept (**a**).

The relationship between metabolic rate and body mass is generally not linear, but is curvilinear. For example, the metabolic rate (kJ day^{-1}) of a variety of mammals and birds is definitely curvilinear, when graphed as a function of body mass (kg) using normal (nonlogarithmic) axes. Three different panels are required to graph the relationship over a mass range from 0.01 kg to 5000 kg. One solid line (**b** = 1) indicates the expected metabolic rate if it were directly proportional to body mass (calculated from the metabolic rate of a canary). The other line (**b** = 0.67) is the expected metabolic rate if it were proportional to mass$^{0.67}$.

These curvilinear data can be fitted by various types of curve (e.g., quadratic, polynomial, hyperbolic, exponential) but a power curve provides an excellent and convenient fit.

$$Y = \mathbf{a}X^{\mathbf{b}}$$

For the metabolic data,

$$\text{kJ day}^{-1} = 289 \text{ kg}^{0.734}$$

It is not possible to derive statistics for a curvilinear relationship as conveniently as for a linear regression analysis. However, curvilinear relationships can generally be made into a linear relationship by appropriately transforming the X and Y variables.

A power curve can be transformed into a straight line by a logarithmic transformation of both X and Y, to give the following.

$$\begin{aligned}(\log Y) &= (\log \mathbf{a}) + \mathbf{b}\,(\log X)\\ &= \mathbf{a}' + \mathbf{b}\,(\log X)\end{aligned}$$

Either $\log_{10}$ or $\log_e$ can be used; this changes the value of $\mathbf{a}$ but not $\mathbf{b}$. For example, the curvilinear relationship between metabolic rate and body mass is linearized by transforming and plotting $\log_{10}$ (metabolic rate) as a function of $\log_{10}$ (body mass). For the metabolic data,

$$\log_{10} \text{kJ day}^{-1} = 2.46 + 0.734\,(\log_{10} \text{kg})$$

A $\log_{10}$-$\log_{10}$ transformation of data is commonly used to summarize and analyze many physiological relationships.

Transformation to a linear relationship allows the calculation of the linear regression coefficients, e.g., $\mathbf{a}'$ and $\mathbf{b}$. The relationship can then be "detransformed" to the original data form

$$Y = \mathbf{a}X^{\mathbf{b}}$$

where $\mathbf{a} = 10^{\mathbf{a}'}$. The detransformed equation shows the curvilinear relationship between metabolic rate and body mass (e.g., with arithmetic axes) but the statistical analysis of the transformed data cannot be validly applied to the detransformed data.

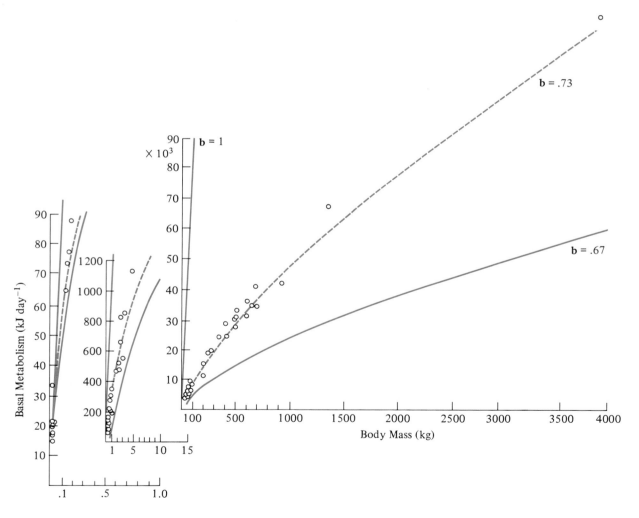

Linear (**b** = 1) and power (**b** = 0.67 and **b** = 0.75) relationships between metabolic rate and body mass for birds and mammals.

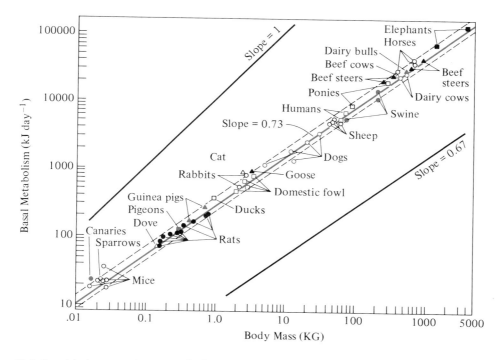

Relationship between $\log_{10}$ metabolic rate and $\log_{10}$ body mass for birds and mammals. *(Modified from Brody 1945.)*

Recommended Reading

Blaxter, K. 1989. *Energy metabolism in animals and man.* Cambridge: Cambridge University Press.

Braefield, A. E., and M. J. Llewellyn. 1982. *Animal energetics.* Glasgow and London: Blackie.

Brody, S. 1945. *Bioenergetics and growth.* New York: Reinhold.

Calder, W. A. 1984. *Size, function, and life history.* Cambridge: Harvard University Press.

Elton, C. 1927. *Animal ecology.* New York: Macmillan.

Hochachka, P. W., and M. Guppy. 1987. *Metabolic arrest and the control of biological time.* Cambridge: Harvard University Press.

Kleiber, M. 1961. *The fire of life: An introduction to animal energetics.* New York: Wiley & Sons.

Phillipson, J. 1966. *Ecological energetics.* London: Edward Arnold.

Rickleffs, R. E. 1973. *Ecology.* Newton: Chiron Press.

Schmidt-Nielsen, K. 1984. *Scaling: Why is animal size important?* Cambridge: Cambridge University Press.

Sibley, R. M., and P. Calow. 1986. *Physiological ecology of animals: An evolutionary approach.* Oxford: Blackwell Scientific.

Townsend, C. R. and P. Calow. 1981. *Physiological ecology: An evolutionary approach to resource use.* Oxford: Blackwell Scientific.

Wiegert, R. G. 1976. *Ecological energetics.* Benchmark Papers in Ecology/4. Stroudsburg: Dowden, Hutchinson & Ross.

Chapter 5

Temperature

· ·

The pile, or 'fur', on the thorax of the moth is effective insulation.
(Photograph courtesy of P. Withers and T. Stewart, Zoology Department, The University of Western Australia.)

The **body temperature** (T_b) of active animals ranges from about $-2°$ C to $+50°$ C, although some can survive at lower or higher temperatures in an inactive (dormant) state. A few animals can survive extremely low body temperatures (e.g., $-200°$ C) and some can survive moderately elevated temperatures (60 to 70° C).

The body temperature of many animals is similar to the **ambient**, or air, temperature (T_a). Other animals, however, raise their body temperature above T_a and may precisely regulate T_b by either behavioral or physiological means.

Body temperature has a profound effect on the physiology of animals. Let us first consider the physical effects of temperature on physiology, and then let us examine the mechanisms for heat exchange between an animal and its environment and the regulation of heat exchange.

Physical Effects of Temperature

Temperature Scales

Temperature is a measure of the average thermally induced molecular motion; molecules vibrate faster at higher temperatures. Our sensory system provides us with a qualitative indication of temperature. An object may feel "hot" or "cold" to our touch, but this thermal sensory information is subjective, qualitative, and unreliable! Consider the sensation when we remove a metal tray of ice cubes from a freezer; the metal tray feels "colder" than the ice cubes, although both are at the same temperature. This is because the metal tray conducts heat better than ice, and our hand loses heat faster to the metal than to ice. Our sensory perception of temperature is thus biased by the rate of heat transfer.

Thermometers are mechanical, electrical, or optical devices that measure temperature. They measure some physical material property that is a function of temperature, rather than temperature *per se*, e.g., a change in volume of a liquid (mercury or alcohol thermometer), a change in length of a solid (bimetallic strip), a change in voltage across a junction of two dissimilar metals (thermocouple), or a change in emitted radiation (pyranometer).

There are a number of temperature scales. The SI unit for temperature is the degree Kelvin, or K (named after Lord Kelvin). The Celsius (or centigrade) scale defines 0° C as the freezing point of pure water and 100° C as the boiling point of water (at standard pressure); this scale is named

after Anders Celsius. The K scale and C scale are related as follows.

$$°C = °K - 273.15$$
$$°K = °C + 273.15 \qquad (5.1a)$$

Absolute zero (0° K $= -273.15$ °C) is the lowest possible temperature; it is the temperature at which thermal motion ceases. Another non-SI temperature scale in common usage is the Fahrenheit scale, °F (named after Gabriel Fahrenheit); the freezing point of water is 32° C and the boiling point is 212° F. Hence,

$$°F = \tfrac{9}{5} °C + 32 \qquad (5.1b)$$

Reaction Rates

Temperature is an important and all-pervasive physical property of the environment; it measures the motion and thermal kinetic energy of molecules. The average kinetic energy of molecules is proportional to the absolute temperature. For an ideal gas, the mean kinetic energy (E; J mole^{-1}) is proportional to the temperature (K)

$$E = \tfrac{1}{2}mv^2 = 1.5kT \qquad (5.2)$$

where m is the molecular weight, v is the mean molecular velocity (m sec^{-1}), and k is the Boltzmann constant. However, not all molecules at a given temperature have exactly the same kinetic energy. The distribution of energies for molecules in gases or liquids at a particular temperature has a maximum at a specific velocity, and is skewed towards higher velocities and energies (Figure 5–1A).

Temperature can have a profound effect on physical, chemical, and biochemical reaction rates as we have already seen (see Chapter 3, page 30). At a given temperature, a certain fraction of molecules have kinetic energy greater than the required **activation energy** (E_a) and react when they collide. The fraction of molecules with velocity/energy greater than any particular value can be calculated from the Maxwellian distribution formula (Figure 5–1A). Alternatively, the specific rate constant, k, can be determined from the Arrhenius equation (see Chapter 3, page 39) and related to temperature as

$$\frac{k_2}{k_1} = e^{\frac{-E_a}{R}\left(\frac{1}{T_1} - \frac{1}{T_2}\right)} \qquad (5.3a)$$

where k_2 and k_1 are the reaction rates at temperatures T_2 and T_1, and E_a is the activation energy. The **critical thermal increment**, μ, rather than activation energy, is used to describe the effects of temperature on complex physiological processes such as heart rate, respiratory rate, or metabolic rate.

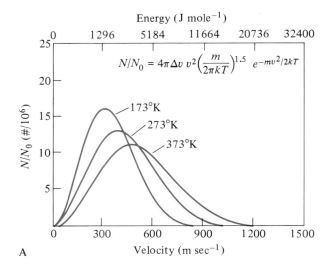

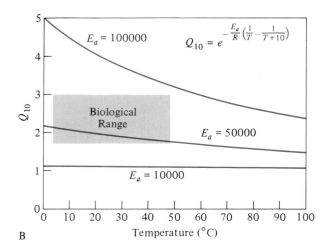

FIGURE 5–1 (A) Velocity and energy distribution of air molecules at three temperatures. The number of molecules per 10^6 molecules (N/N_0) was calculated using the Maxwellian energy distribution formula

$$N/N_0 = 4\pi\Delta v\, v^2(m/2\pi kT)^{1.5}e^{-(mv^2/2kT)}$$

where N_0 is the number of molecules (i.e., 10^6), Δv is the velocity range, v is the velocity (m sec^{-1}), m is the molecular weight ($7.78\ 10^{-26}$ kg molecule^{-1} for air), k is Boltzmann's constant, and T is temperature (°K).
(B) Q_{10} values calculated as a function of activation energy (E_a; J mole^{-1}) and temperature (°C). The shaded region indicates the approximate biological range.

The ratio k_2/k_1 for a 10° K difference in temperature (from 303° K to 313° K) is often about 2, because E_a is about 50 kJ mole^{-1}. The value for a 10° difference in temperature is the Q_{10}; it can be calculated if the temperature difference is not exactly 10° C as the following.

$$Q_{10} = \frac{k(T+10)}{k(T)} = (k_2/k_1)^{10/(T_2-T_1)} \quad (5.3b)$$

The Q_{10} should decrease at increasing temperatures (since $1/T - 1/(T+10)$ is smaller at higher T) and increase with E_a or μ (Figure 5–1B).

TABLE 5–1

Q_{10} and activation energy (E_a)/critical thermal increment (μ) for physical, biochemical, and physiological rates. E_a or $\mu = 0.1\ RT\ (T+10)\ \ln Q_{10}$

	Temperature (°C)	Q_{10}	E_a/μ (kJ mole^{-1})
Physical Reactions			
Diffusion[1]	20	1.03	—
Biochemical Reactions			
Pyruvate kinase V_{max}			
Rat	>25	1.75	41.9
Rat	<25	3.23	83.7
Cytochrome reductase			
Possum	>20	1.52	33.0
Possum	<20	2.53	64.0
Albumin coagulation	69	635	646
Hemoglobin coagulation	60	13.8	249
Leukocyte heat death	38	28.8	279
Protozoan heat death	36	900–1000	
Physiological Reactions			
Water flea			
Gill movement	5	3.73	87.6
	13	1.72	38.1
Crayfish heart rate	5	2.4	60.3
	10	1.9	44.2
	15	1.6	33.5
	20	1.4	24.8
	25	0.8	−17.0
	30	0.5	−54.7
Potato beetle VO$_2$	10	2.4	60.4
	15	2.5	65.4
	20	2.1	54.8
	25	1.3	20.0
Torpid mammal VO$_2$	0	2.85	67.3
	10	3.71	90.4
	20	4.11	104.3

[1] $D = RT/6\pi N\eta r$; R = gas constant, T = °K, $N = 6.02\ 10^{23}$, η = viscosity, r = radius of diffusing molecule.

The Q_{10} for physical processes, such as diffusion, is about 1. That for biochemical reactions and many physiological rates (e.g., metabolic rate, respiratory rate, heart rate) is typically 2 to 3 (Table 5–1), but often depends on T in a more complex manner than indicated by Equation 5.3b. For example, the Q_{10} for the heart rate of crustaceans is greatest (>3) below 10° C and is about 2 from 10° to 20° C; Q_{10} declines to 1 to 2 at temperatures above 20° C. There is a similar rapid decline in Q_{10} at temperatures above 20° C for the VO_2 of a potato beetle. The gill ventilation rate of a water flea *Daphnia* shows a marked transition in Q_{10} from about 3.7 to about 1.7 at 13° C. The value of μ declines markedly from typical biological values (50 to 60 kJ mole $^{-1}$) at high temperatures (e.g., *Daphnia* and potato beetle).

The Thermal Environment

The thermal environment of animals is generally complex. Consider, for example, a large lizard in a typical habitat (Figure 5–2). Radiative heat exchange is dominated by direct solar radiation but there is also scattered and reflected solar radiative heat gain and radiative heat gain by long-wave radiation from objects in the environment; the animal loses heat by long-wave radiation from its surface. There is conductive heat transfer between the animal and the ground through the feet and the body if it is pressed to the ground. Convective heat transfer occurs by forced and/or free convection. Evaporative heat loss occurs across the skin and from the respiratory tract. The relative magnitudes of these various avenues for heat gain and loss are determined by the particular biotic and abiotic conditions.

The effective thermal environment is best determined by empirical measurement; this can be accomplished using models of animals placed in the natural environment. The effective environmental temperature depends on air temperature (T_a) and the conductive, convective, and radiative thermal regimens. For example, the environmental temperature is effectively $>T_a$ if there is a significant radiative heat load, and is $<T_a$ if there is a considerable radiative heat loss. The **operative temperature**

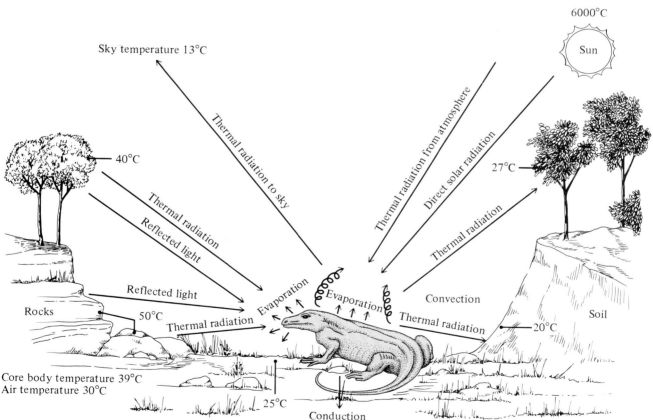

FIGURE 5–2 Representation of conductive, convective, radiative, and evaporative heat exchange for a terrestrial reptile in a typical terrestrial environment. *(From Heatwole and Taylor 1987.)*

(T_e) provides a measure of the effective environmental temperature. It is the temperature of an isothermal (same temperature throughout) black body in an identical conductive, convective, and radiative environment as the animal occupies (Bakken 1980), i.e.,

$$M - E = C_e (T_b - T_e) = C_{es} (T_b - T_{es}) \quad (5.4)$$

where M is metabolic heat production, E is evaporative heat loss, and C_e is the operative thermal conductance (cf. the nonevaporative total heat transfer coefficient, h; see below). The **standard operative temperature** (T_{es}) is the operative temperature under standardized convective conditions (usually defined as free convection), where C_{es} is the standard operative thermal conductance.

T_{es} and T_e account for the radiative and convective characteristics of the thermal environment. For example, the VO$_2$ of white-crowned sparrows (*Zonotrichia*) is decreased at low T_a and is increased at high T_a if there is a significant radiative heat load (Figure 5–3A). There is a linear relationship between ($M - E$) and T_e (Figure 5–3B), as predicted by Equation 5.4 when evaporative heat loss and the

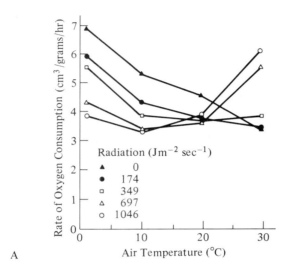

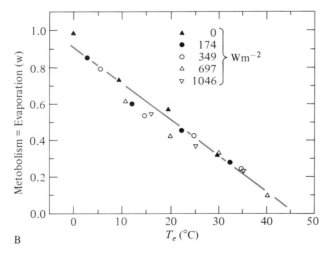

A

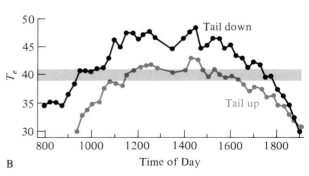

B

FIGURE 5–3 (A) Oxygen consumption rate of the white-crowned sparrow (*Zonotrichia leucophrys*) at varying air temperatures in the absence of near incident infrared radiation (solid triangles) and varying levels of incident IR and visible radiation (other symbols). **(B)** Net nonevaporative heat flux (watts) for white-crowned sparrows as a function of operative temperature (T_e). *(From Calder and King 1974; after DeJong 1971; from Bakken 1980.)*

FIGURE 5–4 (A) The Cape ground squirrel uses its tail as a 'parasol' to shade its body from direct solar radiation. **(B)** The 'parasol' thermoregulatory behavior of the ground squirrel considerably reduces the operative temperature (T_e; °C) for a model of the ground squirrel. *(Photograph by P. Withers; data from Bennett et al, 1984.)*

convective/radiative regimes are taken into account (convection was free in this example so $T_e = T_{es}$). For the antelope ground squirrel in the Sonoran desert, the T_{es} can rise to 70° C on a hot sunny day; the T_e is less because there is forced convective heat loss, but still reaches about 65° C; air temperature is lower, reaching 55° C (Chappell and Bartholomew 1981).

T_e provides a better index of the actual thermal environmental than does T_a, and T_{es} indicates the potential thermal environment in the absence of forced convection. The Cape ground squirrel uses its tail as a parasol to shade its body from direct solar radiation (Figure 5–4A). This has important thermoregulatory consequences, since shading the body reduces the T_e by about 5° C (Figure 5–4B).

Patterns of Temperature Regulation

Most animals are unable to control their body temperature, and T_b passively conforms to their thermal environment (T_e); these animals **thermoconform**. There must be a thermal gradient between the body and ambient environment for all animals to dissipate the metabolic heat production, but this is biologically insignificant (typically <1° C) for most thermoconformers.

Some animals, in contrast, regulate their body temperature often against a substantial thermal gradient between their body and the environment; they **thermoregulate**. Thermoregulation can be accomplished in essentially two different ways: by either having a high heat gain or a large metabolic heat production. An insulating layer of fur (mammals), feathers (birds), fat (birds and mammals), or chitin hairs (insects) facilitates thermoregulation. Many terrestrial animals thermoregulate by basking in the sun. Birds and mammals, a few reptiles and fish, a number of insects, and even a few plants are able to generate sufficient metabolic heat to thermoregulate.

We must carefully define the terminology for patterns of temperature variation before we examine patterns of body temperature regulation or conformation for different animals. Many of the terms that have been used to describe patterns of body temperature are relative terms and lack a mechanistic basis (Cowles 1940, 1962). "**Cold blooded**" and "**warm blooded**" are poor terms, although, unfortunately, they still are used. We intuitively know what these terms mean; birds and mammals are warm blooded and snails, crabs, and frogs are cold blooded. However, what is the temperature that separates warm from cold blooded? Is a desert snail with a T_b of 40° C cold blooded or warm

blooded? Is a hibernating mammal with a T_b of 5° C cold blooded or warm blooded?

The terms **poikilotherm** and **homiotherm** are also vague and ambiguous. A poikilotherm has a variable body temperature (from the greek "poikilos," which means changeable) and a homiotherm has a constant T_b ("homoios" is similar). Again, we intuitively know that snails, crabs, and frogs are poikilotherms and birds and mammals are homiotherms, but a fish that lives deep in the ocean at a constant water temperature is a homiotherm, although not in the same physiological sense that a mammal is a homiotherm. A hibernating mammal can have a markedly variable T_b, but it is not a poikilotherm in the same sense that a fish is a poikilotherm.

The terms **ectotherm** and **endotherm** are perhaps the most useful for describing the thermal capabilities of animals because they have a mechanistic basis. An ectotherm is an animal whose thermal balance is predominated by external sources of heat, and its metabolic heat production is insignificant. Ectotherms that absorb heat by basking in sunlight are called **heliotherms**, whereas ectotherms that absorb heat from the substrate are called **thigmotherms**. Essentially all animals, except birds, mammals, and a number of insects (and a very few reptiles and fishes), are ectothermic. Endotherms are animals whose thermal balance is predominated by their endogenous metabolic heat production. Adult birds, mammals, and a number of insects (and a very few reptiles and fish) are endothermic. Some endotherms temporarily become ectothermic when cold stressed or deprived of food or water; they are called **heterotherms**. Endothermy is not synonymous with thermoregulation and ectothermy is not synonymous with thermoconformation. In general, endotherms thermoregulate and thermoconformers are ectothermic, but many ectotherms thermoregulate (see Table 5–2).

The distinction between ectotherm and endotherm is not absolute and clear cut. All ectotherms have some metabolic heat production but their $T_b - T_a$ differential is generally negligible, whereas endotherms can sustain much larger $T_b - T_a$ differentials, e.g., 40° C or more. However, some large and active ectotherms have a sufficiently high metabolic production and low heat loss that their metabolic $T_b - T_a$ differential is significant, although still generally small, i.e., 3 to 10° C. Furthermore, endothermy evolved from ectothermy and so there must have been a gradual transition from ectotherm to endotherm. It is difficult to define the specific point at which these transitional animals would cease being ectothermic and become endothermic. Nevertheless, the concepts of ectotherm and endo-

TABLE 5–2

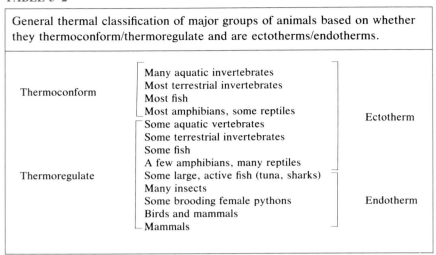

General thermal classification of major groups of animals based on whether they thermoconform/thermoregulate and are ectotherms/endotherms.

| Thermoconform | Many aquatic invertebrates
Most terrestrial invertebrates
Most fish
Most amphibians, some reptiles | Ectotherm |
| Thermoregulate | Some aquatic vertebrates
Some terrestrial invertebrates
Some fish
A few amphibians, many reptiles
Some large, active fish (tuna, sharks)
Many insects
Some brooding female pythons
Birds and mammals
Mammals | Endotherm |

therm provide a mechanistic and relatively objective definition for distinguishing between two fundamentally different thermal strategies.

The terms **eurythermal** and **stenothermal** define the range of thermal tolerance. An animal that tolerates or is active over only a narrow range of T_b is a stenotherm. For example, many mammals and birds are stenothermic; they cannot tolerate a change in T_b by more than a few degrees. Some lizards, such as the desert iguana, are stenothermic thermoregulators; they regulate T_b within a narrow range (e.g., 37° to 39° C) while active. The alligator lizard, in contrast, is eurythermal; it is active over a wide range of T_b.

Let us examine the mechanisms for heat exchange, the various thermal strategies of ectothermy/endothermy and thermoconformation/thermoregulation, the physiological consequences of these patterns, and the biochemical and physiological adaptations of animals to their thermal environment.

Heat Exchange

The heat exchange between two inanimate objects is proportional to the difference in their temperatures. Furthermore, animals continually produce metabolic heat (Chapter 4), which must be dissipated to the environment. Animals also evaporate water and this can significantly influence their thermal balance.

Animals cannot avoid heat exchange between their body tissues and the environment, but they can manipulate the various avenues of heat exchange to

their thermal and physiological advantage. It is therefore important to understand the basic mechanisms for the three main avenues of heat exchange: **conduction, convection, and radiation.** We must also consider a change in state of water (i.e., **evaporation/condensation** and **freezing/melting**) as an additional avenue for heat exchange.

Conduction

The direct transfer of heat between two solid materials in physical contact is **conduction.** Heat flows from a region of high temperature to a region of lower temperature. The heat transfer occurs on an atomic scale as the exchange of kinetic energy between adjacent molecules. The rate of exchange (flux) by conduction between two objects depends on their area of physical contact, the difference in their temperatures, and their thermal conductive properties.

Fourier's law of heat transfer (for 1-dimensional heat exchange) summarizes the determinants of conductive heat flow

$$Q_{\text{cond}} = -kA(T_2 - T_1)/x = C\Delta T \quad (5.5a)$$

where Q_{cond} is the rate of heat exchange (J sec^{-1}), k is the thermal conductivity (J sec^{-1} °C^{-1} cm^{-1}), A is the contact area for conduction (cm^2), ΔT is the temperature difference (°C), and x is the distance between the two temperatures (cm). The negative sign indicates that heat flow occurs from the highest temperature to the lowest temperature. The **temperature gradient** is the difference in temperature per unit distance for conduction, i.e., $(T_2 - T_1)/x$. The thermal conductance (C; J sec^{-1} °C^{-1} or various other units) is often used instead of ($-kA/x$) because A and x are often difficult to measure. It is also called a conductive heat transfer coefficient ch$_{cond}$.

Fourier's law for heat transfer by conduction is equivalent in form to Fick's first law for 1-dimensional exchange of material by diffusion (Chapter 3); the thermal conductivity (k) is equivalent to the diffusion coefficient (D) and the temperature difference (ΔT) to the concentration difference (ΔC). Heat conduction can be considered to be thermal energy diffusion from a "high concentration" of heat to a "lower concentration" of heat. Heat conduction can also be considered to be propagation of heat waves (like photons of light) rather than diffusion, and heat waves have been experimentally verified in low temperature liquids and crystals (Maddox 1989). Fortunately, Fourier's law remains valid for most practical purposes.

Thermal conductivity (k) is a material property; it varies dramatically for different materials from 2.4 10^{-2} for still air to >100 J sec^{-1} °C^{-1} m^{-1} for metals (Table 5–3). The **insulation** is the reciprocal of the total heat flux per m^2 of surface area per °C temperature difference. The SI unit for insulation is °C m^2 sec J^{-1}. The industrial unit for insulation is the clo ("clothing unit"); 1 clo = 0.155 °C m^2 sec J^{-1}.

The **resistance** of an insulative layer (r; sec cm^{-1}) to heat flux indicates both the thermal conductivity property of the layer and its thickness

$$Q_{cond} = \rho C_p A \Delta T / r \qquad (5.5b)$$

where ρC_p is the volumetric specific heat capacity of the medium (air; 1200 J m^{-3} °C^{-1}; Walsberg, Campbell, and King 1978). There is an inverse relationship between thermal resistance and thermal conductivity.

$$r = \rho C_p l / k \qquad (5.5c)$$

Values for r per meter (i.e., sec m^{-2}; Table 5–3) are essentially the reciprocal of thermal conductivity and are analogous to the inverse of a diffusion coefficient (D; m^2 sec^{-1}). Most animal and synthetic insulation has an insulative value/thermal conductivity similar to that of air. Indeed, it is the still air trapped within insulative materials that provides their low conductivity. The insulative value of a material increases with resistance and thickness. The insulative value of fur (or feathers) and thermal resistance increases with the thickness of the fur layer (Figure 5–5). Shrew fur (<0.5 cm thick) has a lower insulation than polar bear fur (6 cm thick). However, the insulation per thickness is lower for

thicker fur; shrew fur has a higher insulation per thickness than polar bear fur. Fur or feathers immersed in water have a lower insulative value than in air because (1) water has a higher specific heat and thermal conductivity, (2) they are compressed by the water and have a lower thickness, and (3) they cannot be pilo-erected (fur) or ptilo-erected (feathers). For example, the insulation of eider duck feathers is about 1.62 °C m^{-2} sec J^{-1} in air. The insulation of peripheral tissues contributes a further 0.20 °C m^{-2} sec J^{-1}, and so the total insulation is 1.82 (Jenssen, Ekker, and Bech 1989). The insulation is reduced in water by about 1/2 (to 0.83) for feathers but increased for tissues (0.35); the total insulation is lower in water (1.18) than in air (1.82).

TABLE 5–3

Thermal conductivity (k), resistance (r) to heat loss per meter, and insulation (Clo) per meter, for air, a variety of animal insulation materials (fur, feathers), and other materials.

Material	Thermal Conductivity (J sec^{-1} m^{-1} °C^{-1})	Resistance/m[a] r m^{-1} (sec m^{-2})	Insulation/m[b] Clo m^{-1} (°C m sec J^{-1})
Vaccuum	0	∞	∞
Still air	2.4 10^{-2}	5.0 10^4	269
Red fox fur	3.6 10^{-2}	3.3 10^4	179
Lynx fur	3.8 10^{-2}	3.2 10^4	170
Husky dog fur	4.1 10^{-2}	2.9 10^4	157
80/20 Goose down	5.3 10^{-2}	2.3 10^4	122
Polyolefin	5.7 10^{-2}	2.1 10^4	113
Sheep wool	6.3 10^{-2}	1.9 10^4	102
Dacron II	6.5 10^{-2}	1.8 10^4	99
Pigeon feathers (flat)	6.5 10^{-2}	1.8 10^4	99
Wood	1.3 10^{-1}	9.2 10^3	50
Galloway cattle	1.3 10^{-1}	9.2 10^3	50
Helium	1.4 10^{-1}	8.6 10^3	46
Pigeon feathers (erect)	1.6 10^{-1}	7.5 10^3	40
Fat	1.7 10^{-1}	7.1 10^4	38
Rubber	1.7 10^{-1}	7.1 10^3	38
Dry soil	3.3 10^{-1}	3.6 10^3	20
Human tissue	4.6 10^{-1}	2.6 10^3	14
Water	5.9 10^{-1}	2.0 10^3	11
Glass	1.0	1.2 10^3	6.5
Ice	2.2	5.5 10^2	2.9
Steel	46	26	1.4 10^{-1}
Aluminum	240	5	2.7 10^{-2}
Silver	430	2.8	1.5 10^{-2}

[a] r m^{-1} = 1200/k.

[b] Clo m^{-1} = 6.45/k.

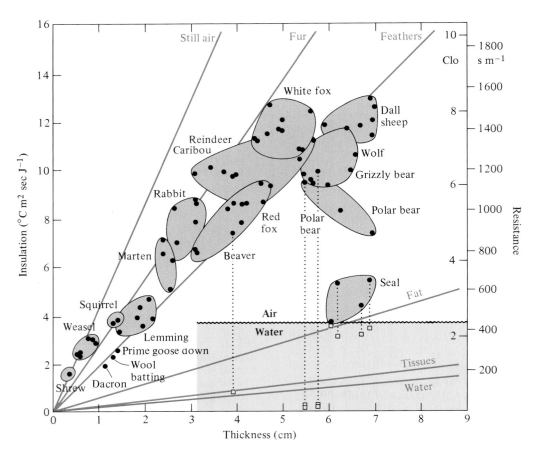

FIGURE 5–5   The insulative value (°C m² sec J⁻¹) of animal fur varies with its thickness. Fur has a markedly reduced insulative value in water (dotted lines; squares). Seal skin has a low insulative value in air compared to fur but retains most of its insulative value in water. Solid lines indicate the expected values for still air, average values for fur and feathers, fat, tissues, and still water; the slopes of these lines equal 1/thermal conductivity. Solid points indicate some insulative values for artificial insulation (prime 80/20 goose down, wool fiber batting, and Dacron). The alternative insulation scales of resistance (*r*; sec m⁻¹) and Clo (1 Clo = 0.155 °C m² sec J⁻¹) are also indicated.   *(Modified from Scholander et al. 1950; artificial insulation values from Kaufman, Bothe, and Meyer 1982.)*

Convection

The transfer of heat by the movement of a fluid (liquid or gas) is **convection**. Convective heat exchange is essentially conductive heat exchange across the fluid **boundary layer** (δ), which is the layer of air surrounding the object that varies in temperature from that of the object to that of the free-stream air (see Supplement 5–1, page 187)

$$Q_{conv} = -kA(T_b - T_a)/\delta$$
$$= h_{conv}(T_b - T_a) \qquad (5.6)$$

where h_{conv} is a convective heat transfer coefficient. The value of h_{conv} depends on many complex factors

and should be empirically determined. However, it can be estimated at varying wind velocity from tabulated values for objects of various shapes and sizes.

Convection is **forced** if the fluid movement is produced by an external force (e.g., gravity or a pump). **Free** convection occurs in the absence of forced fluid movement. Fluid near a hot object will be warmer and have a lower density than free-stream fluid; hence, it will rise up from the object. For example, there is a mass flow of air from a naked human in still air of up to 600 liters min⁻¹; the hot plume of air may extend for 1.5 m above the head, and have a velocity of 0.5 m sec⁻¹ (Clark and Toy 1975). Air near a cold object will cool, have a higher density, and flow down from the object.

The importance of free convection depends on the temperature difference between the object and fluid ($T_b - T_a$), the dimension of the object (x), the fluid's coefficient of thermal expansion ($a = 1/273$ for an ideal gas), and the kinematic viscosity (v).

Radiation

All objects with a surface temperature greater than 0° K emit light of specific wavelengths that depends on the surface temperature. The predominant wavelength (λ_{max}; nm) is inversely proportional to the surface temperature (Wien's law).

$$\lambda_{max} = 2.898 \; 10^6/T \qquad (5.7)$$

The spectral emission has a λ_{max} of 500 nm for the sun (surface temperature = 5800° K), 1999 nm for an incandescent lamp (2900° K), and 9722 nm for an object at room temperature (293° K). Animals emit long wavelength infrared radiation (IR) because their surface temperature is low. A thermogram measures the emitted IR radiation and represents it as visible wavelengths. The thermograms of a mouse and a lizard, for example, show the variation in surface temperature (Figure 5–6). The energy of the emitted light (kJ mole photons^{-1}) can be calculated from wavelength (nm) as 119660/λ.

The total rate at which an object emits radiant energy is proportional to the fourth power of its surface temperature T_s (Stefan's law)

$$Q_{rad} = \sigma \epsilon A T_s^4 \qquad (5.8a)$$

where σ is the Stefan-Boltzmann constant, ϵ is the emissivity (dimensionless), and A is the radiative surface area (m²). An ideal radiator (a "black body") has $\epsilon = 1$; most biological materials have an ϵ of 0.90 to 0.95. Objects at temperatures in the biological range emit relatively little energy (300 to 500 J sec^{-1} m^{-2}) compared to very hot objects such as the sun (6.1 10^7 J sec^{-1} m^{-2}). However, 300 to 500 J sec^{-1} m^{-2} is a lot of energy, compared to the metabolic rate of animals. For example, a mammal with a fur surface area of 1 m² would emit about 480 J sec^{-1} (if surface temperature was 35° C); the basal metabolic rate would be about 20 J sec^{-1} (if it weighed 10 kg) and the free convective heat loss would be about 20 J sec^{-1} (if $T_b - T_a$ was 10° C and the fur was 2 cm thick).

Animals not only emit radiation but also receive radiation from their environment, i.e., they have a net radiative heat exchange. The net radiative heat exchange between an animal and its environment can be quite complex. Consider first the simple example of an object surrounded by a large surface, and separated from it only by a gas (which has no effect on radiative heat transfer). The **net radiative heat exchange** is

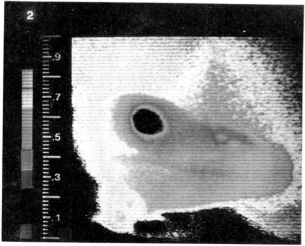

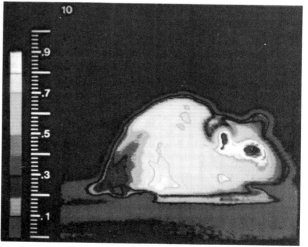

LIZARD   MOUSE

FIGURE 5–6 Thermogram of a mouse (right) and the head of a monitor lizard (*Varanus salvator*; left) showing the variation in surface temperature. The color corresponds to the elevation of surface temperature above background (about 25° C); the scale for the lizard corresponds to a temperature difference of 2° C, whereas the scale for the mouse is 10° C. *(From den Bosch 1983.)*

$$Q_{\text{rad,net}} = \sigma\epsilon A(T_s^4 - T_{\text{sur}}^4) \qquad (5.8b)$$

where T_{sur} is the temperature of the surrounding surface. If T_s is approximately equal to T_{sur}, then there is little net radiative heat exchange because the object gains an equivalent amount of radiative heat from its environment.

The equation for $Q_{\text{rad,net}}$ is of a different form from those for conduction and convection, where Q was proportional to $(T_b - T_a)$. However, it can be converted to a similar form by defining a radiative heat transfer coefficient (h_{rad}) so that

$$Q_{\text{rad,net}} = h_{\text{rad}}(T_s - T_{\text{sur}}) \qquad (5.8c)$$

However, we must appreciate that h_{rad} depends markedly on both T_s and T_{sur} because

$$h_{\text{rad}} = \sigma\epsilon(T_s + T_{\text{sur}})(T_s^2 + T_{\text{sur}}^2) \qquad (5.8d)$$

In contrast, h_{cond} is independent of temperature, and h_{conv} is only weakly dependent on temperature.

Evaporation/Condensation

Evaporation of water dissipates considerable heat. The latent heat of evaporation is 2500 J g^{-1} at 0° C, 2400 at 40° C, and 2260 at 100° C. This is much more energy than is required to melt ice (latent heat of fusion is 334 J g^{-1}) or to heat water from 0 to 100° C (about 418 J g^{-1}).

The evaporation of water depends not on the temperature difference between the animal and its environment, but on (1) the difference in water vapor density between the animal and the ambient air, and (2) the resistance to water loss from the surface

$$Q_{\text{evap}} = \frac{DA(\chi_b - \chi_a)}{x} = \frac{A(\chi_b - \chi_a)}{r}, \qquad (5.9)$$

where D is the diffusion coefficient for water vapor, A is the area, χ is the absolute water vapor density (g m^{-3}), x is the pathlength for water loss, and r is the resistance to water loss. We generally assume that the air at the animals' surface is saturated (100% RH) at the surface temperature. Consequently, χ_b depends on surface temperature, and is proportional to body temperature.

Condensation occurs if $\chi_b < \chi_a$, and heat is gained rather than lost.

Heat Balance

The thermal balance of an animal is determined by the net exchange of heat by all of the avenues for heat exchange that we have discussed, as well as by metabolic heat production. Animals must, in general, be in heat balance, i.e., heat gain = heat loss. Otherwise, body temperature would decrease (heat gain < heat loss) or increase (heat gain > heat loss). However, an animal can be in transient positive or negative heat balance and changes in body temperature can buffer the thermal imbalance.

The general equation describing heat balance is

$$
\begin{aligned}
M = {} & h_{\text{cond}}(T_b - T_a) \\
& + h_{\text{conv}}(T_s - T_a) \\
& + h_{\text{rad}}(T_s - T_{\text{sur}}) \\
& \pm E \\
& \pm S
\end{aligned} \qquad (5.10a)
$$

where M is metabolic heat production, E is evaporative/condensative heat transfer, and S is storage or loss of heat by a change in body tissue temperature. The various heat transfer coefficients (h_{cond}, h_{conv}, h_{rad}) vary markedly depending on the specific circumstances, but approximate ranges of values indicate that conduction, free convection, and radiation have similar h values, forced convection in gases has a higher h, forced convection in liquids has an even higher h, and forced convection with evaporation can have extremely high h values (Table 5–4).

The general equation for heat balance can be greatly simplified for steady-state ($S = 0$) if T_s is approximately equal to T_{sur} and we ignore condensation

$$M - E = h(T_b - T_a) \qquad (5.10b)$$

where h is the total nonevaporative heat transfer coefficient.

TABLE 5–4

Typical values for heat transfer coefficients by conduction, free or forced convection, convection with evaporation, and radiation.	
	h (J sec^{-1} m^{-2} °K^{-1})
Conduction	
Air	0.5–5000
Water	50–5000
Convection	
Free	2–25
Forced—air, dry	25–250
Air, with evaporation	2500–100000
Water	50–20000
Radiation (excluding direct solar)[1]	4–8

[1] Solar radiation heat gain can be about 200—1000 J sec^{-1} m^{-2} depending on atmospheric conditions and latitude/altitude (the solar constant is 1353 J sec^{-1} m^{-2}).

Ectotherms

The metabolic heat production of ectotherms is generally negligible and their heat balance equation (see Equation 5.10a) reduces to

$$0 = h_{\text{cond}}(T_b - T_a) + h_{\text{conv}}(T_b - T_a)$$
$$+ h_{\text{rad}}(T_b - T_{\text{sur}}) \pm E \pm S \quad (5.11a)$$

One solution to this equation is $T_b = T_a$ if E and S are negligible and $T_{\text{sur}} = T_a$, i.e., the animal thermoconforms. However, there are other solutions. For example, a high radiative heat gain ($T_{\text{sur}} > T_a$) can significantly elevate T_b above T_a until the heat loss by conduction, convection, and evaporation balances the radiative heat gain.

Let us first examine the thermal relations of aquatic ectotherms that generally do not experience a significant radiative heat load and, therefore, thermoconform, and then let us examine terrestrial ectotherms that often experience a significant radiative heat load and thermoregulate.

Aquatic Ectotherms

The thermal relationships of aquatic ectotherms are constrained not only by their insignificant metabolic heat production, but also by the high thermal conductivity and specific heat of water, by the relative insignificance of thermal radiative heat gain, and by their inability to dissipate heat by evaporation. Water has a much higher thermal conductivity than air, and the specific heat of water and animal and plant tissues is also considerably greater than that of air and other gases, especially on the basis of J liter^{-1} °C^{-1} (Table 5–5).

TABLE 5–5

Specific heat (J g^{-1} °C^{-1} and J liter^{-1} °C^{-1}) for selected materials. *(Modified from Serway 1983; Incropera and Dewitt 1981.)*		

	Density (kg m^{-3})	Specific Heat J g^{-1} °C^{-1}	J liter^{-1} °C^{-1}
Helium	0.18	5.19	0.93
Water vapor	0.59	1.97	1.46
Air	1.29	1.006	1.2
Oxygen	1.43	0.92	1.22
Carbon dioxide	1.90	0.83	1.63
Cotton	80	1.30	104
Water	1000	4.186	4186
Tissues	1000	3.9	3900
Clay	1460	0.88	1285
Gold	19300	0.13	2509

Heat dissipation across gills is so effective for most aquatic animals that metabolic heat production has no thermal significance and T_b is similar to T_{water}. Metabolic heat production has no thermal consequences for aquatic ectotherms because the heat dissipation capacity of gill ventilation of water far exceeds the capacity for metabolic heat production. The temperature differential between the body and water (ΔT) can be calculated from the oxygen solubility coefficient for water (α_{O_2}; 0.0115 μM L^{-1} Pa^{-1} for seawater at 20° C), the ambient O_2 partial pressure (pO_2; 21200 Pa), and percentage O_2 extraction from the water (E$_{O_2}$; often about 50%) as 1.08 10^{-6} O_2 pO_2 E$_{O_2}$. The ΔT is less than 0.03° C even if the α_{O_2} extraction is 100%. In contrast, the ventilatory heat loss of an air-breathing animal due to the warming of the respiratory air flow is 0.00154 pO_2 E$_{O_2}$. The ΔT in air is about 163° C for a typical O_2 extraction of 5% and is over 3000° C for 100% extraction!

Aquatic ectotherms thus have thermal constraints that generally preclude significant elevation of T_b above T_{water}. Consequently, the heat exchange equation for aquatic ectotherms reduces to the simple form of

$$0 = h_{\text{cond}}(T_b - T_{\text{water}}) + h_{\text{conv}}(T_b - T_{\text{water}}) \quad (5.11b)$$

and so $T_b = T_{\text{water}}$ since the thermal gradient is the same for both conductive and convective heat exchange (an aquatic animal cannot gain heat by conduction but at the same time lose heat by convection). For example, the body temperature of aquatic ectotherms is generally not very different from water or substrate temperature (Figure 5–7A). However, some large fish can selectively warm certain tissues and restrict heat flow to the gills, and some large air-breathing sea turtles can have T_b substantially higher than T_{water}.

Many aquatic ectotherms are able to precisely thermoregulate at a **preferred body temperature** ($T_{b,\text{pref}}$). The thermal preferences for a number of fish and other aquatic ectotherms are summarized in Table 5–6. This regulation of $T_{b,\text{pref}}$ is accomplished by behavioral selection of a suitable water temperature, rather than thermoregulation by physiological means. Many fish maintain body temperature in a narrow range if there is a suitable gradient in water temperature. For example, bluegill sunfish select a T_b of about 26° C in a thermal gradient providing a range of T_{water} from 6° to 34° C (Figure 5–7B).

Amphibians are a particularly interesting group of ectotherms because many species are transitional between aquatic and terrestrial environments. They

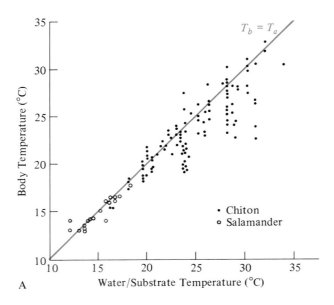

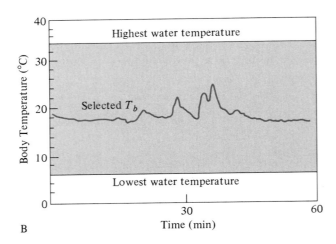

FIGURE 5–7 (A) Relationship between body temperature (T_b) and water temperature (T_w) for an aquatic salamander *Taricha torosa*, and between T_b and substrate temperature (T_{sub}) for a chiton *Clavarizona hirtosa*.
(B) Continuous recording of body temperature (dorsal muscle) of a bluegill sunfish in a thermal gradient (6° to 36° C temperature range).   *(From Brattstrom 1963; Kenney 1958; modified from Crawshaw 1975.)*

TABLE 5–6

Preferred body temperatures (°C) for some invertebrate and lower vertebrate ectotherms.	
Arthropods	
Wireworm	17
Blowfly	20–25
Flour beetle	25–30
Earwig	25–30
Louse	29–30
Leather beetle	30
Housefly larva	30–37
Chicken louse	42.5
Fish	
Largemouth bass	24
Opal-eye shorefish	26
Bluegill sunfish	26
Brown bullhead	26
Carp	22–28
Amphibians	
Salamander tadpoles	25
Grass frog tadpoles	23–30
Bullfrog tadpoles	24–30
Bullfrogs	22–28
Gray treefrog	up to 38
Phyllomedusa	up to 40
Aquatic Reptiles	
Box turtle	21–25
Painted turtle	29–35
Alligator	32–35

have T_b's from about 0° C (e.g., a salamander walking over a snow field) to over 40° C (a basking "waterproof" frog). As we would expect, aquatic amphibians have a T_b similar to T_{water}, but many select a preferred T_b if there is a suitable thermal gradient in T_{water}. For example, many tadpoles select the warmer water around the edges of a pond and avoid the cooler, deeper water (Brattstrom 1970). *Hyla regilla* tadpoles orient so that their dorsal surface is exposed to solar radiation. Aggregations of tadpoles can absorb sufficient solar radiation to increase the local T_{water} (Brattstrom 1962). Tadpoles also select a preferred range of T_b in a thermal gradient (Table 5–6).

Terrestrial amphibians have the potential to raise T_b by choosing suitable microclimates or by basking in sunlight, but their high cutaneous evaporative loss tends to decrease the efficiency of heliothermic basking. It has been suggested that basking is thermally ineffective and behavioral thermoregulation would be crude, at best, for amphibians because their evaporative water loss would dissipate essentially all of the solar heat gain (Tracy 1976). Nevertheless, many terrestrial amphibians behaviorally thermoregulate by choosing thermally-favorable microclimates and by basking in the sun. For example, juvenile green toads (*Bufo debilis*) basking in the sun have a T_b 10° to 15° C above that of animals in the shade. Juvenile bullfrogs (*Rana catesbeiana*) select a T_b of 27° to 28° C in a thermal gradient and bullfrogs in the field have a T_b of about 30° C. These patterns of thermoregulation are, however, relatively crude compared to the thermoregulatory capacity of many other terrestrial invertebrates and vertebrates (see below).

Terrestrial Ectotherms

Many terrestrial ectotherms have a considerable capacity for behavioral and physiological thermoregulation. Air has a lower thermal conductivity and heat capacity than water, so it is easier for terrestrial ectotherms to maintain a thermal gradient between T_b and T_a, and T_b will equilibrate more slowly with T_a. The rate of cooling (or warming) of an object, or animal, in a cooler (or warmer) environment is determined by the temperature differential, the surface properties (i.e., surface area and nature of insulation) and the thermal properties of the medium (i.e., conductivity and specific heat).

The rate of change in T_b is proportional to $(T_b - T_a)$ and decreases as T_b approaches T_a. The T_b changes in an exponential fashion (Figure 5–8A) towards an equilibrium temperature (that is close to, but not necessarily equal to, the ambient temperature). The general equation for conductive or convective heat loss is

$$Q = h(T_b - T_a) = C_p M\delta(T_b - T_a)/\delta t \quad (5.12a)$$

where C_p is the specific heat (J g^{-1}), M is mass (g), and $\delta(T_b - T_a)/\delta t$ is the rate of change in $(T_b - T_a)$

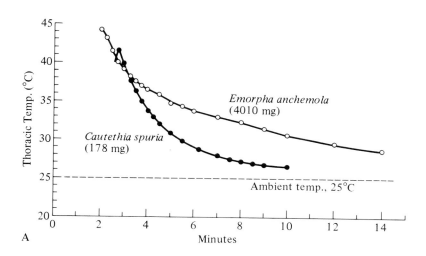

A

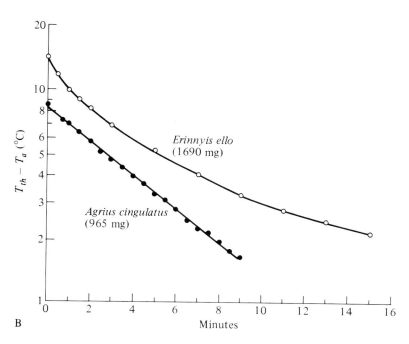

B

FIGURE 5–8 (A) Time course for thoracic cooling of the sphinx moths *Emorpha* and *Cautethia*. **(B)** Semilogarithmic plot of cooling curves for the sphinx moths *Erinnyis* and *Agrius*; the initial steep part of the cooling curve for *Erinnyis* indicates facilitated heat loss. *(From Bartholomew and Epting 1975.)*

over time (t). This equation can be rearranged and integrated to yield

$$\ln(T_b - T_a) = \frac{h}{C_p M}t + \text{con} \qquad (5.12b)$$
$$= CCt + \text{con}$$

where t is time (min) and con is a constant of integration. Consequently, the exponential change in T_b over time for a cooling or warming curve can be linearized (Figure 5–8B) by graphing $\ln(T_b - T_a)$ against time (Morrison and Tietz 1957). The slope of this line ($h/C_p M$) is called the cooling constant (CC; h^{-1}). The cooling constant is converted to a thermal conductance (C; J g^{-1} h^{-1} °C^{-1}) by accounting for the specific heat of tissues (C_p; generally about 3.4 J g^{-1} °C^{-1}); $C = 3.4\,CC$.

Many variables affect the thermal conductance of animals. The most important are body size (hence surface area for heat exchange), presence of insulation, and the nature of the medium (air or water). Body mass is important since the surface area:volume ratio of the animal determines the ratio of heat dissipation capacity:heat content. Large animals have a lower surface area:volume than small animals, and therefore cool (or warm) slower, e.g., the larger (4 g) sphinx moth *Emorpha* cools more slowly than the smaller (0.2 g) sphinx moth *Cautethia* (Figure 5–8A). The scaling of thermal conductance varies for different groups of animals (e.g., insects, lizards, mammals, birds) but there is a general trend that C is proportional to mass$^{-0.5}$, with an intercept value (i.e., C for a 1 g animal) of about 30 J g^{-1} h^{-1} °C^{-1}. For insects, the thoracic mass is often used rather than the body mass because it is primarily the thoracic temperature that is regulated.

The inverse relationship between C and body mass has profound biological significance. Small animals have considerable difficulty in maintaining a temperature differential; for example, a 20 mg bee will cool almost to ambient temperature within a few minutes. In contrast, a large animal has such a low conductance that its T_b is relatively constant. A 2000 kg dinosaur would hardly warm or cool over a 48 h period, by virtue of its gargantuan size. The thermal conductance of such a large animal is also the same in air and water.

Factors other than body size also influence thermal conductance. The intercept value is often higher for live than dead insects, perhaps reflecting the contribution of blood flow in live animals to heat dissipation from the thorax. Dead insects also have a greater mass dependence of thermal conductance than live insects. Conductance is also expected to be marginally higher for dead than for live animals because of metabolic heat production.

Insulation affects thermal conductance. For example, bees with their body hairs removed have a higher conductance than normal bees. However, insulation is often not as important as might be expected (May 1976). For example, tabanid flies have only very short hairs whereas sphinx moths have a very long, dense cover of scales, but their conductances are similar. Species of bees that are virtually hairless have conductance similar to hairy species. This is probably because the boundary layer of air around an insect's body is probably of greater significance than the insulation (at least under the experimental conditions of low convection). The thermal conductance of the insulated insects would be expected to be lower than that of the naked insects at high wind speeds when the boundary layer is less important. For example, the seta of the gypsy moth caterpillar decrease the thermal conductance more at high wind velocity than at lower velocity, or with free convection. The air boundary layer resistance of lizards is approximately independent of the body mass and is relatively less important in determining heating/cooling rates for larger lizards (Bell 1980).

Animals generally heat faster than they cool; this is particularly apparent for lizards and crocodilians. There is also a similar hysteresis in heart rate (faster during heating) and metabolic rate (higher during heating). Cardiovascular and metabolic adjustments presumably facilitate heating and retard cooling. For example, lizards and crocodiles have peripheral vasodilation during heating and vasoconstriction during cooling (Morgareidge and White 1969; Grigg and Alchin 1976). Such cardiovascular adjustment requires an augmented peripheral blood flow that is unrelated to metabolic demand. A right-to-left intracardiac shunt increases the ventricular systemic outflow for the iguana during heating (Baker and White 1970).

The allometry of thermal conductance for most insects, birds, and mammals is similar to that predicted for free convective heat loss in air, i.e., their thermal conductance is near the minimal possible value (Figure 5–9). In contrast, reptiles in water have a higher thermal conductance, close to that predicted for a cylindrical object in water with maximum convective heat loss. Small reptiles in air have an intermediate thermal conductance. Very large reptiles would have a thermal conductance similar to that of large mammals, and the free convective minimal value, whether they are in air or water.

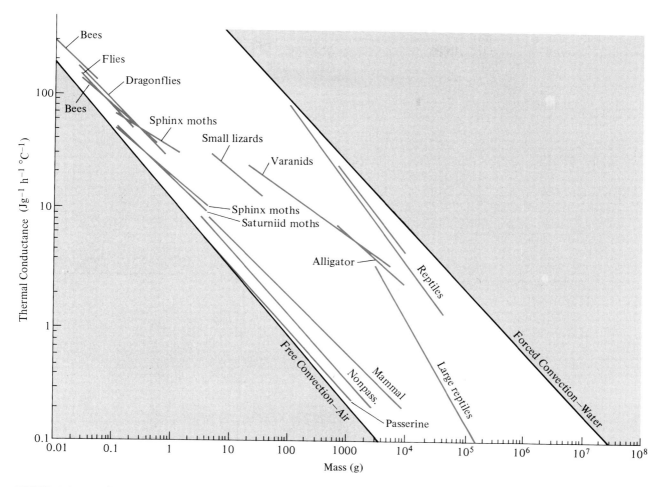

FIGURE 5–9 Allometric relationships for minimal thermal conductance of insects and vertebrates. Thin lines indicate relationships for air; heavy lines indicate relationships for water. Free convection air line indicates the minimal thermal conductance for free convective heat loss from a sphere in still air. Forced convection water line indicates the maximal thermal conductance for forced convective heat loss of a cylinder in water.

The extent to which animals can behaviorally or physiologically thermoregulate is indicated by the relationship between T_b and T_a (Figure 5–10). The slope is 1 for total dependence of T_b on T_a, i.e., thermal conformation, and 0 for thermal independence. The precision of thermoregulation is indicated by the extent of scatter of data points around the regression line.

Basking Insects. Body posture and orientation of the wings can markedly affect the body temperature of basking insects. "Perching" dragonflies bask at low air temperatures and regulate their radiative heat gain by postural adjustment (Figure 5–11A). Several species bask with the wings positioned forward and downward to reduce convective heat loss. Other dragonflies ("fliers") use endogenous heat production for thermoregulation (see Endothermic Insects, page 173). Many butterflies can elevate thoracic temperature by as much as 15° C above T_a by postural adjustment, although some species also use endogenous heat production to elevate T_{th}. Many bask with the wings held to the side of the body and the thorax perpendicular to the sun ("dorsal baskers"), with the wings held vertically over the body and perpendicular to the sun ("lateral baskers"), or with the forewings vertical and the hindwings flattened ("body baskers"). The wing area nearest the body is the most effective in facilitating heat gain by the thorax (Figure 5–11B). Syrphid

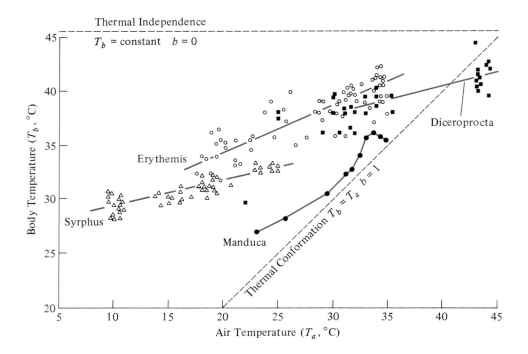

FIGURE 5–10   Thoracic temperature as a function of air temperature in desert cicadas (*Diceroprocta*), hoverflies (*Syrphus*), sphinx moth caterpillars (*Manduca*), and dragonflies (*Erythemis*).

hoverflies use a combination of behavior (perching in sun or shade) and endogenous heat production to thermoregulate. Caterpillars rely on behavioral rather than endogenous mechanisms for thermoregulation. Many tenebrionid beetles thermoregulate behaviorally (although some are endothermic).

Ground-dwelling insects are especially prone to overheating because they are in the boundary layer of the ground surface. They intercept considerable direct and reflected solar radiation, and also infrared radiation, conductive and convective heat from the ground. Consequently, many different types of ground-dwelling arthropods (tenebrionid beetles, locusts, scorpions) raise their bodies as high off the ground as possible ("stilt") to minimize heat gain from the ground, to move the body into the cooler part of the thermal boundary layer, and to increase convective heat loss in the higher velocity part of the velocity boundary layer. For example, the locust *Schistocerca* adopts a crouched posture when the ground is warmer than the air (to warm up) and adopts a semistilted (head off ground) or stilted posture (entire body off ground) when the ground is extremely hot. Locusts and some caterpillars also orient with reference to both the sun and wind, to vary both radiative heat gain and convective heat loss.

Color has significant effects on thermoregulation because about 50% of the radiant energy from the sun is in the visible spectrum. Consequently, the visible reflectance (hence color) influences radiative heat gain. A black surface reflects less radiant energy than a white surface. For example, the Namib desert tenebrionid beetle *Stenocara* has black elytra with a mean reflectance of about 23%, whereas its white sides have a higher mean reflectance of about 35% (Figure 5–12A). Black animals would be expected to absorb more radiation and therefore have a higher T_b than white animals. Tenebrionid beetles (*Onymacris*) with white elytra have a lower T_b than species with black elytra (Figure 5–12B). The black beetles seem to be more active earlier in the day because they absorb more radiation and heat faster, whereas white beetles are more active in the hotter parts of the day because they have less of a radiative heat load. The wing base of some lateral basking butterflies is melanized at the ventral part of the fore and hind wings, and increases the T_b when basking.

Basking Reptiles.   The body temperature of reptiles is quite variable, ranging from only a few degrees to over 40 degrees. Nocturnal and some diurnal lizards thermoconform, with T_b similar to T_a. For

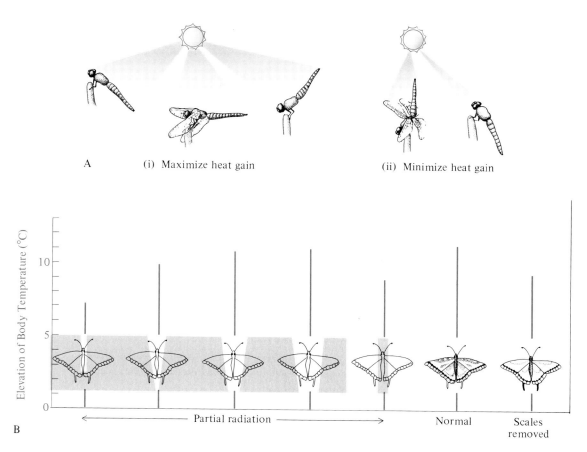

A   (i) Maximize heat gain   (ii) Minimize heat gain

B ← Partial radiation →   Normal   Scales removed

FIGURE 5–11 (A) Postures adopted by a dragonfly to maximize heat gain (i) by heliothermy or (ii) to minimize heat gain. **(B)** Effects of thermal radiation on wings of the butterfly Papilio to elevation of thoracic temperature above ambient temperature, and the effects of removing the inner wing scales. *(Modified from May 1976b; Wasserthal 1975.)*

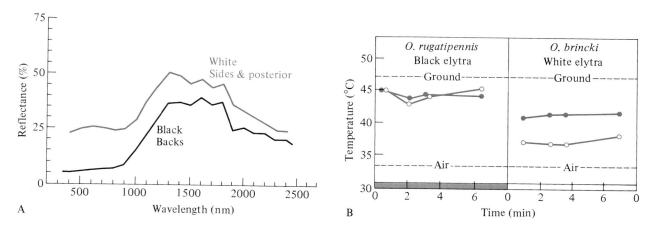

FIGURE 5–12 (A) Reflectance of the black back and whiter side of the Namib desert tenebrionid beetle *Stenocara phalangium*; the mean reflectance is 35% for the whitish side and 23% for the black back. **(B)** Temperature of the thorax (solid symbols) and abdomen (open symbols) of the Namib desert tenebrionid beetles *Onymacris rugatipennis* (black elytra) and *O. brincki* (white elytra) in natural sunlight. *(From Henwood 1975; modified from Edney 1971.)*

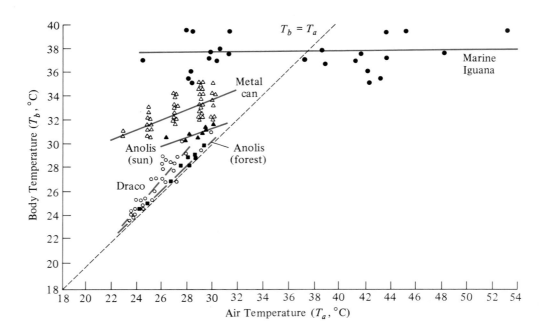

FIGURE 5–13 Relationship between body temperature and air temperature for thermo-conforming lizards (*Draco, Anolis,* in shaded forest habitat) and thermoregulating lizards (*Amblyrhynchus, Anolis,* in open habitats); also shown is the relationship for an inanimate object (water-filled metal can).

example, the agamid lizard *Draco* and the forest *Anolis* species (Figure 5–13) have T_b similar to T_a. Nocturnal lizards have no solar radiative heat source (although some bask during the day while inactive, e.g., geckos bask under tree bark). Many forest lizards cannot bask because they live in shaded habitats (e.g., some forest floor anoles) or do not bask even if patches of sunlight are available (e.g., the forest floor skink *Sphenomorphus*). The tropical anole *A. cristatellus* is a thermal opportunist; it is a thermoconformer in shaded habitats but a thermoregulator in open park habitats (Huey 1974). Some diurnal lizards do not bask because they are active late in the day when the ambient temperature is high enough for normal activity, e.g., the great plains skink *Eumeces*. Some skinks (*Eumeces*) and the alligator lizard (*Gerrhonotus*) absorb heat by ventral contact with a sun-heated substrate or dorsal contact with the undersides of sun-heated rocks. These lizards are **thigmotherms** because their main heat gain is conductive rather than radiative.

Most diurnal lizards bask in sunlight and are accomplished thermoregulators with preferred T_b about 35° to 40° C (e.g., the chuckwalla; Figure 5–14A). These **heliothermic** lizards manipulate their thermal exchange in a number of behavioral and

physiological ways. These include varying whether they are in sun or shade by movement (shuttling), their orientation to the sun, their skin color, body contour, peripheral vasoconstriction or vasodilation, heart rate, etc. The complex suite of behavioral and physiological mechanisms for heating, regulating T_b, or cooling is well illustrated by the behavioral repertoire of the horned lizard *Phrynosoma* (Figure 5–14B).

The preferred T_b, or eccritic T_b, is not necessarily the same for lizards in nature as in the laboratory. This reflects the additional complexities of survival in the natural environment, e.g., suboptimal thermal regimes (e.g., cloudy days), and time allocated to activities other than thermoregulation (e.g., feeding, defending territories, finding mates, avoiding predation). There are temporal, physiological, and ecological costs to thermoregulation (Huey and Slatkin 1976; Huey 1982). Thermoregulatory precision is compromised if any, or all, of these costs are substantial. Laboratory experiments have clearly demonstrated a dependence of both the level and precision of T_b regulation on the metabolic cost for thermoregulation. For example, the eurythermal lizard *Gerrhonotus* can precisely regulate a relatively high T_b of 27.8 ± sd 2.0° C in a low-cost environment, but has a lower T_b and less precision

A

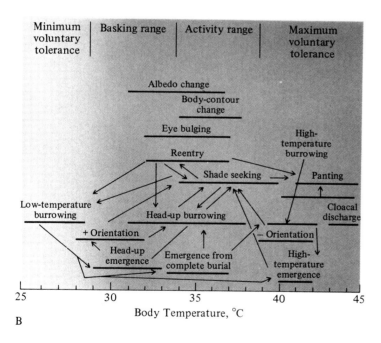

B

FIGURE 5–14 (A) A chuckwalla (*Sauromalus*) showing typical heliothermic behavior basking on a rock pile in direct solar radiation. **(B)** The horned lizard (*Phrynosoma*) uses a complex suite of behavioral and physiological responses to thermoregulate. *(Courtesy, Department of Library Sciences, American Museum of Natural History/Raymond L. Ditmars, Neg. no. 16555 and Heath (1965) with permission of the Regents of the University of California.)*

(20.1 ± sd 3.1° C) in a high-cost environment (Campbell 1985). The stenothermic lizard *Dipsosaurus* thermoregulates precisely at a high T_b of 39.1 ± sd 2.0 in a low-cost environment, but has a lower T_b and less precision of 32.9 ± sd 4.0 in a high-cost environment (Withers and Campbell 1985).

Heliothermic lizards balance one set of behaviors and physiology for heat gain against a different set for heat dissipation. Laboratory studies suggest that

there are separate thermostats for control of heat gain and heat loss, i.e., there is a dual-setpoint regulatory system (see also Chapter 2). However, the distribution of T_b is not normally distributed about the $T_{b,pref}$, but is **negatively skewed** if the regulatory precision is not the same for the low and high setpoints. For example, the desert iguana *Dipsosaurus* has a modal T_b of about 39° C and a negatively skewed T_b distribution, i.e., there are more T_b values <39° C than >39° C (Figure 5–15). The lower body temperature setpoint (LBTS) for triggering mechanisms that increase T_b is 36.4 ± sd 1.9° C and the higher body temperature setpoint (HBTS) for mechanisms that lower T_b is 41.7 ± sd 1.3. On average then, the T_b of the lizard will rise to 41.7° C and trigger the heat loss mechanisms (e.g., shuttle into the shade), and T_b will decline until it reaches 36.4° C at which point the mechanisms for heating are triggered (e.g., shuttle into sunlight) and T_b will rise. However, the LBTS and HBTS values of 36.4 and 41.7 are only average values, and the actual minimum and maximum T_b will vary in different thermal cycles. The net effect is that the mean T_b distribution of *Dipsosaurus* is negatively skewed because the LBTS is more variable than the HBTS. The T_b distribution would be normally distributed if LBTS and HBTS had the same variation, and would be positively skewed if HBTS was more variable than LBTS.

The dual-setpoint model for T_b regulation by heliothermic lizards provides a mechanism to negatively skew the T_b distribution, but what is the physiological significance of a negatively skewed T_b distribution? It is more important to a lizard that T_b not rise above $T_{b,\text{pref}}$ and exceed its critical thermal

maximum temperature rather than for T_b to decline below $T_{b,\text{pref}}$; consequently, a negatively skewed distribution seems adaptive. Another explanation is that physiological processes typically have a Q_{10} of about 2 to 3. An increase in T_b elevates metabolic rate more than the same decline depresses metabolic rate. For example, if metabolic rate is 1 at 38° C, then a 2° C increase in T_b to 40° C increases VO_2 to 1.20 (a 20% increase) whereas a 2° C decline in T_b to 36° C decreases VO_2 to 0.83 (a 17% decrease), if $Q_{10} = 2.5$. Consequently, a negatively skewed T_b distribution may result in a normal distribution of the physiological function (such as metabolic rate), whereas a normal T_b distribution would result in a positively skewed distribution of the physiological function (DeWitt and Friedman 1979). We should also note that even inanimate objects (cans of water) "basking" in sunlight may have a negatively skewed temperature distribution (Heath 1964), and so neither a thermoregulatory capacity nor an adaptive advantage (both lacking for cans of water) need be invoked to explain observed temperatures or a negatively skewed temperature distribution.

Heliothermic lizards often have a narrow range of $T_{b,\text{pref}}$, but their **thermal performance breadth** is a wider range of T_b's for which physiological

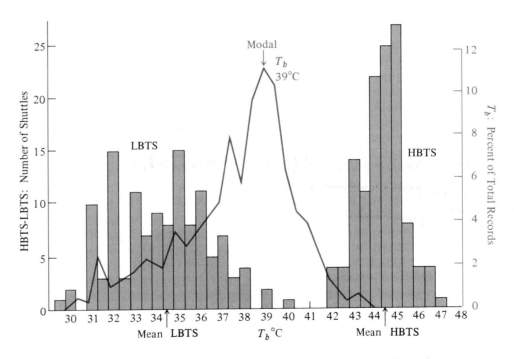

FIGURE 5–15 Frequency distribution for body temperatures of the desert iguana (*Dipsosaurus*) in a laboratory thermal gradient, and the frequency distributions for an individual lizard of its low body temperatures that initiate a heating cycle (LBTS) and the high body temperatures that initiate a cooling cycle (HBTS). *(Modified from DeWitt 1967; Berk and Heath 1975.)*

performance is "fairly high." This breadth is arbitrarily defined depending on the physiology and ecology of the species in question; the temperature range for 80% of the maximum value is a common value. The **thermal tolerance range** is the broadest range of temperatures over which the animal can survive indefinitely. Finally, the **thermal survival zone** is defined by the lethal critical thermal minimum temperature (CT_{min}) and the critical thermal maximum (CT_{max}).

There are physiological advantages to elevated T_b that explain why heliothermy and high $T_{b,pref}$ have evolved in many different animal groups. Essentially all biochemical reactions have a Q_{10} of 2 to 3 and consequently so do many physiological processes. Muscles contract faster at a higher temperature, enzymes cleave substrates faster, digestion is faster, animals can run faster and for longer periods of time, and metabolic rate is higher. There is, however, an **optimal temperature** for many of these biochemical and physiological functions, and rates plateau or decline at even higher temperatures (Figure 5–16) because of the thermal instability of protein structure and function. There generally is a close correspondence between the optimum temperature for many biochemical and physiological processes and the $T_{b,pref}$. For example, there is a marked thermal dependency of digestive efficiency for the herbivorous lizard *Dipsosaurus*: 54% at 33° C, 63%

at 37° C, and 71% at 41° C. Lizards kept at 22° C do not pass food from their stomach and die. Insectivorous lizards have a higher digestive efficiency (see also Chapter 18) and a lesser thermal dependence: 83 to 91% for 26 to 33° C (no food passed at 21° C). There is little effect of temperature on digestive efficiency for the eurythermal thigmothermic lizard *Gerrhonotus* (92 to 94% for 18° to 30° C).

Adaptations to Cold

Many ectotherms are biochemically and physiologically adapted to survive and even function normally at low ambient temperatures. However, temperatures that are cold enough to freeze animal tissues are potentially lethal and must either be avoided or require specific adaptations for survival. Many ectotherms simply avoid freezing conditions by migrating or retreating to warmer microclimates, but some occasionally or routinely experience freezing temperatures. Aquatic ectotherms can experience only mild freezing conditions, since water or seawater freezes at 0° to −2° C, and the high latent heat of fusion buffers further temperature change. Terrestrial ectotherms, in contrast, are particularly susceptible to freezing because the air temperature can fall considerably below 0° C (e.g., −20° to −50° C).

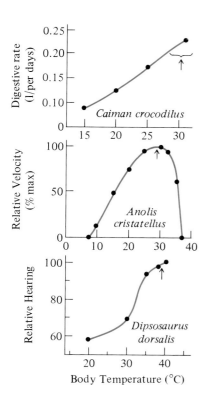
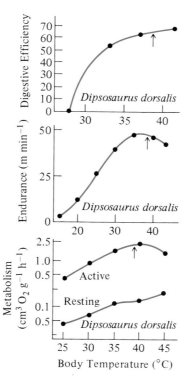

FIGURE 5–16 Relationship between various physiological functions and body temperature for reptiles, showing the general relationship between maximum physiological performance and preferred body temperature (vertical arrow). *(From various sources, as adapted by Huey 1982.)*

There are a variety of strategies that enable ectothermic animals to survive freezing conditions.

Anti-Freeze Strategy.   Various ectotherms, such as polar arthropods and icefish (Figure 5–17A), avoid freezing by lowering the freezing point of their body fluids below that of the ambient medium, i.e., **freezing point depression**, or by allowing the body fluids to **supercool** below their normal freezing point. Alternatively, some ectotherms such as insects, frogs, and baby turtles allow selective extracellular freezing of their body fluids, but avoid freezing of their intracellular fluids. Let us now examine each of these strategies in more detail.

Osmotic Depression of Freezing Point.   Fresh water freezes at 0° C. The freezing point of a solution is depressed below the normal freezing point of water by the presence of solute molecules. The magnitude of the freezing point depression (Δ_{fp}) is proportional to the osmolal concentration ($C_{osmolal}$) as

$$\Delta_{fp} = -1.86 C_{osmolal} \qquad (5.13)$$

Ectotherms can counteract a potentially freezing ambient temperature by increasing the osmotic concentration of their body fluids so that its freezing point is depressed below the ambient temperature. The normal body fluid concentration of ectotherms confers a very limited protection against freezing in fresh water because the freezing point of tissues is generally $-0.6°$ to $-0.7°$ C. Animals in seawater are more prone to freezing because their tissues are generally not hyperosmotic, and therefore will freeze at the same, or a higher, temperature than seawater ($-1.86°$ C).

Some ectotherms accumulate high concentrations of specific solutes to depress their freezing point by <1 to $>10°$ C. These solutes are typically sugars (glucose, fructose, trehalose) or sugar alcohols (polyols: glycerol, sorbitol, mannitol, erythritol, myo-inositol). Their low molecular weight maximizes the freezing point depression per mass of solute. These sugars and polyols may also have cryoprotectant effects; they protect membranes and enzymes against cold denaturation and cold shock injury, inhibit lipid phase transitions, and promote desiccation resistance. Many solutes also promote supercooling. Some animals also have antifreeze proteins that induce a hysteresis between the freezing point and the melting point of the body fluids.

Supercooling.   If water is cooled, its temperature is expected to decline in proportion to the rate of heat removal and the specific heat of water (4.2 J g^{-1} °C^{-1}) until ice begins to form at the freezing point (0° C). The rate of temperature decline is reduced during freezing because latent heat is released by the freezing water (the latent heat of fusion is about 334 J g^{-1}). When all of the water is frozen, the temperature declines in proportion to the rate of

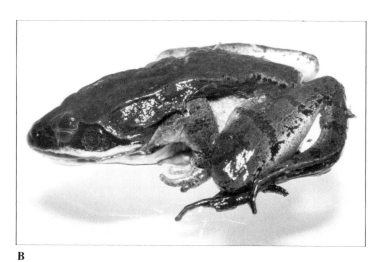

A   B

FIGURE 5–17   Examples of animals that use different physiological strategies for adapting to freezing ambient temperatures. **(A)** The supercooled icefish *Trematomus* has antifreeze proteins that reduce the freezing point of its tissues to below the freezing point of seawater. **(B)** The frozen wood frog *Rana sylvatica*; this frog survived when thawed.   *(Photographs courtesy of: (A) Dr. Arthur L. DeVries, University of Illinois at Urbana-Champaign, 1988. (B) Courtesy of Dr. F. H. Pough, Laboratory of Functional Ecology, Cornell University, © F. H. Pough.)*

heat loss and the specific heat of ice (about 2.1 J g^{-1} $°C^{-1}$). We might expect a similar pattern in temperature change for a cooling animal, except that the temperature would decline in proportion to the specific heat of animal tissues (about 3.4 J g^{-1} $°C^{-1}$) and the freezing point would be lower than 0° C, depending on the osmotic concentration of the body fluids.

Pure water and solutions do not necessarily freeze at their nominal "freezing points," but may **supercool**, i.e., remain liquid at below their normal freezing temperature. For example, pure water can be easily supercooled by a few degrees, and with care to $< -20°$ C. Similarly, animal tissues can supercool to below their expected freezing point (Figure 5–18A). The capacity to supercool seems to be a fairly general property of solutions and animal tissues. For example, many reptiles and the blood of fish, frogs, reptiles, birds, and mammals will supercool to $-5°$ to $< -10°$ C, although their melting temperatures are $-0.4°$ to $-1.0°$ C (Lowe et al. 1971). The melting temperature of ice or frozen tissues is always at its theoretical freezing temperature, determined by their osmotic concentration.

For example, the hemolymph of the supercooling insect *Bracon* has a supercooling temperature that is much lower than its theoretical freezing temperature and the melting temperature because of the accumulation of high concentrations of glycerol (Figure 5–18B).

Some ectotherms, and even an endotherm (the arctic ground squirrel), rely on their supercooling capacity to avoid freezing. For example, benthic fish in the fjords of Labrador supercool below their body fluid freezing point (about $-0.7°$ C) to a T_b near $-1.86°$ C, the freezing point of seawater (Scholander et al. 1953). Contact with exogenous ice crystals, for example at the gills, would "seed" the supercooled body fluids and initiate rapid freezing of the entire fish, but this does not normally occur because the fish are benthic and do not come into contact with ice crystals in the water, since ice floats.

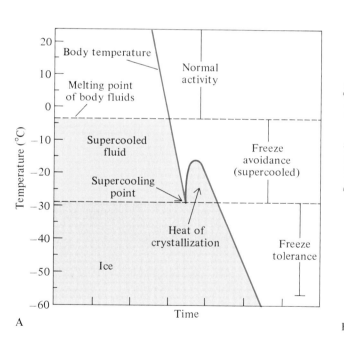

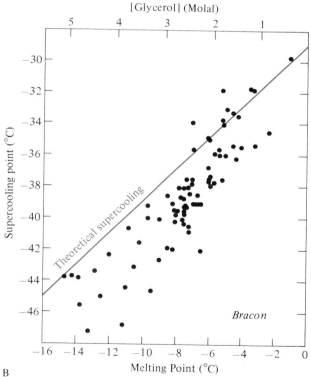

FIGURE 5–18 (A) Schematic response of an animal to cooling to subfreezing temperatures. The body temperature declines to below the freezing point, i.e., supercools. When freezing is initiated, the heat of crystallization is released and temporarily increases the temperature. After all fluid is frozen, the tissue continues to freeze. **(B)** Relationship between the supercooling point and melting temperatures for the hemolymph of a cold-hardy parasitic wasp, showing also the theoretical concentration of glycerol required to depress the supercooling point by a constant amount. *(From Lee 1989; modified from Salt 1959.)*

Overwintering insects are generally categorized as either freezing intolerant or freezing tolerant. Many freezing-tolerant species have a spectacular capacity to supercool to avoid tissue freezing, and some are also tolerant of tissue freezing (Baust 1986). They produce an often complex array of low molecular weight antifreeze osmolytes that depress the freezing point and lower the supercooling temperature and cryoprotective solutes (Table 5–7).

The few studies of cold resistance in spiders indicate that they can supercool to −20° C and below (Kirchner 1973). The capacity to supercool is proportional to the location of the overwintering sites and the likely extent of cold stress. The lowest supercooling temperatures are for spiders overwintering in vegetation (−16° to −26° C) or under tree bark (−13° to −15° C). The highest supercooling temperatures are for spiders overwintering in caves or under stones (−4° to −8° C). Whether, or what, antifreeze proteins, polyols, or other osmolytes are involved in supercooling is unclear. Glycerol and hemolymph proteins are not

TABLE 5–7

Supercooling temperatures and accumulated solutes for a variety of ectothermic invertebrates and vertebrates that are exposed to freezing conditions. The approximate concentrations of solutes in the hemolymph or plasma (h), whole body water (wbw), or whole body (wb) are also indicated. AFP, antifreeze protein; INP, ice nucleating protein; INLP, ice nucleating lipoprotein; glyc, glycerol; sorb, sorbitol; tre, trehalose; ery, erythritol; glu, glucose; ala, alanine; aa, amino acids.

Species	SCT (°C)	Solutes
Spiders		
Sac spider *Clubiona*	−15.4	glyc (4.4%[h]), AFP
Crab spider *Philodromus*	−26.2	glyc (3.3%[h]), AFP
Insects		
Fly *Xylophagus*	−6.0	sugars (5.9%[wbb]), aa (0.8%)
Beetle *Pytho*	−6.6	glyc (4.5%[wb]), sugars (1.4%)
Crane fly *Tipula*	−7	sorb (0.4 M), INP, INLP
Gall fly *Euura*	−7.1	no glyc
Sawfly *Trichiocampus*	−8.6	tre (9%[h])
Carabid beetle *Pterostichus*	−10.0	glyc (25%[h])
Darkling beetle *Meracantha*	−10.3	AFP
Goldenrod gall fly *Eurosta*	−10.3	glyc (5.9%), sorb (9.1%), tre (0.4%)
Pyrochroid beetle *Dendroides*	−12	glyc (14%[h]), sorb (2.5%), AFP
Wooly bear *Isia*	−18.2	glyc (4.44%[wb]), sorb (0.83%)
Grain moth *Nemapogon*	−26.1	tre (0.2m[h]), ala (0.07M)
White cabbage butterfly *Pieris*	−26.2	tre (0.08M[h]), ala 0.07M
Leaf-cutter bee *Megachile*	−27.7	glyc (2,2% [wbw])
Carpenter ant *Camponotus*	−28.7	glyc (5.8%[wbw])
Mite *Alaskozetes*	−30	polyols (3–5%[wb]), glyc (0.5M)
Tortricid moth *Laspeyresia*	−31.5	glyc (0.5M[h]), tre 0.1M)
Bark beetle *Ips*	−32.4	ethylene glycol (2.9M)
Rose root gall fly *Diplolepis*	−32.7	glyc (6.4%[wbw])
Pine beetle *Dendorctonus*	−34	glyc (23.4%[wbw])
Wasp *Bracon*	−41.2	glyc (2.7M[wbw])
Willow aphid *Pterocomma*	−41.9	glyc (15.5%[wbw])
Cankerworm *Alsophila*	−44.6	glyc (15.5%[wbm])
Gall midge *Rhabdophaga*	−49.1	glyc (32.4%[wbw])
Wasp *Eurytoma*	−49.2	glyc (23.4%[wbw])
Beetle *Pytho*	−54	glyc (13.2%[wb]), sugars (5.5%)
Vertebrates		
Gray tree frog *Hyla*	−2.0	glu (0.02M[h])
Chorus frog *Pseudacris*	−2.0	glu (0.06M[h])
Spring peeper *Hyla*	−2.2	glu (0.18M[h])
Wood frog *Rana*	−3.0	glu (0.41M[h])
Turtle	−3.3	glu (0.01M[h]), aa (0.047M)
Arctic ground squirrel	−2.9	none (?)

involved with supercooling in *Araneus*. The hemolymph of the crab spider (*Philodromus*) and sac spider (*Clubiona*) in winter contain a thermal hysteresis protein (i.e., antifreeze protein, AFP) and glycerol (see below).

Little is known of the mechanisms for cold resistance in terrestrial gastropods, e.g., pulmonate snails. The snail *Arianta* survives subzero temperatures by supercooling to $-3.5°$ to $-9°$ C (Stover 1973). These snails have some tolerance to tissue freezing, which is greater in winter than summer and in high altitude populations, but the tolerance is minimal and is probably of little adaptive significance in nature.

Antifreeze Proteins. Antarctic fish can't rely on supercooling because ice frequently occurs in their environment, both at the surface and at the ocean bottom. The freezing point of their body tissues can be as low as $-2.7°$ C (which provides complete protection against freezing in seawater) but the melting temperature is only about $-0.9°$ C (Table 5–8). The lowered freezing point and the hysteresis between freezing point and melting point are attributed to specific "antifreeze" proteins in the blood and other body fluids. The extent that fish antifreeze proteins depress the melting temperature depends on their molal concentration as expected, whereas the freezing point is lowered in a noncolligative manner (i.e., the freezing point depression is not proportional to the osmotic concentration). The antifreeze proteins of most antarctic fish and some northern fish are glycoproteins (AFGP), and for other polar fish are proteins (AFP). Similar AFPs are responsible for a thermal hysteresis in freezing and thawing temperature for some spiders and insects.

Antifreeze glycoproteins are polymers of a repeating tripeptide (alanyl-alanyl-threonine) linked to a galactose-N-acetylgalactosamine carbohydrate moiety; repetition of up to 50 of these subunits results in a molecular weight from 2.6 to 33 10^3 Da (Figure 5–19A). There is minor variation among small AFGPs with proline and/or arginine replacing alanine or threonine at some positions. The AFGPs of the Antarctic nototheniid *Pagothenia borchgrevinki* vary in molecular weight; 32700 (AFGP1), 28800 (AFGP2), 21500 (3), 17000 (4), 10500 (5), 7900 (6), 3500 (7), and 2600 (8).

Antifreeze proteins (AFPs) lack a carbohydrate moiety. They vary considerably in structure, but there are three general types: those containing much alanine (e.g., winter flounder, shorthorn sculpin) with an α-helical structure and molecular weight from 3.3 to 4.5 10^3 Da (Figure 5–19A); those containing much cystine (e.g., sea raven) and a β-structure with molecular weight 11 to 13 10^3 Da; and the remainder that are neither alanine or cystine rich, with a compact structure and molecular weight of 6 10^3. The three AFPs (1,2,3) of the Antarctic eel pout *Rhigophila* have a molecular weight of 6900 but differ in amino acid residues.

Antifreeze proteins depress the freezing temperature (but not the thawing temperature) by about $-0.6°$ C. This freezing point depression is 200 to 300 × that expected from their osmotic concentration (Figure 5–19B). The maximum freezing point depression is about $-120°$ C/osmolal (cf. about $-1.86°$ C/osm for NaCl). On a mass basis, all of the glycoproteins have a similar capacity for freezing point depression except the highest molecular weight AFGPs, which are less effective. However, all antifreeze proteins (AFGPs and AFPs) are slightly more effective on a mass basis than NaCl in depressing the freezing point to a concentration of about 10 g l^{-1}, but are less effective at higher concentrations.

A high concentration of polar groups facilitates a strong interaction of antifreeze proteins with ice crystals and inhibits the further addition of water molecules to the ice front, thereby "poisoning" the growth of an ice crystal. This binding to ice crystals is suggested by the fact that antifreeze proteins do not freeze out of solution (i.e., do not become

TABLE 5–8

Thermal hysteresis in freezing temperature and melting temperature for hemolymph of spiders and insects and for plasma of icefish.			
	Freezing Temperature	Melting Temperature	Thermal Hysteresis	Supercooling Temperature
Fish				
Gadus	-1.1	-0.7	0.4	
Chaenocephalus	-1.5	-0.9	0.6	
Rhigophalia	-2.0	-0.9	1.1	
Myxocephalus	-2.0	-1.1	0.9	
Notothenia	-2.1	-1.1	1.0	
Eleginus	-2.2	-1.2	1.0	
Pagothenia	-2.7	-1.1	1.6	
Spiders				
Philodromus	-5.19	-2.74	2.45	-26.2
Clubiona	-4.76	-2.87	1.89	-15.4
Insects				
Meracantha	-5.01	-1.31	3.71	-10.3
Dendroides	-7.03	-3.37	3.66	-9.5

Antifreeze Glycoprotein: Antarctic cod (*D. mawsoni*)

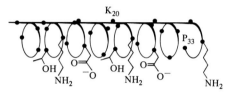

ALA—ALA—THR—ALA—ALA—THR—ALA—ALA—THR—ALA—
  | | |
 Gal—A   Gal—A   Gal—A
  | | |
 Gal   Gal   Gal

Antifreeze Protein: grubby sculpin (*M. aenaeus*)

A

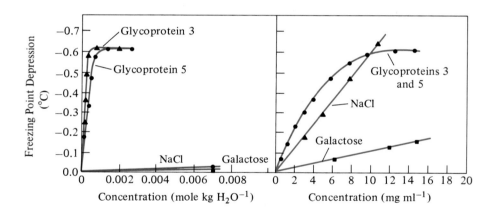

B

FIGURE 5–19 (A) Schematic structure of the antifreeze glycoprotein (AFGP) of the antarctic cod (*Dissostichus mawsoni*) and a schematic representation of part of the antifreeze protein of the grubby sculpin (*Myoxocephalus aenaeus*) showing the hydrophilic side chains extending below the hydrophobic α-helix chain. **(B)** Freezing point depression as a function of osmolal concentration (mole kg water^{-1}) and mass concentration (mg ml^{-1}) for antifreeze glycoproteins, sodium chloride, and galactose. *(From DeVries 1980b; Chakrabartty and Hew 1988; DeVries 1971.)*

concentrated in the remaining liquid water as do normal colligative solutes) but are frozen into the ice.

Antifreeze proteins may prevent the entry of ice crystals across the gills, integument, or gut of ice fish, but there is evidence that ice crystals are normally present in their body tissues. Fish taken from the ice water environment and cooled to −2.7° C will rapidly freeze. This is not because attached ice crystals of seawater precipitate tissue freezing, because the fish will still freeze even if they are initially kept at −1.2° C (which would melt seawater ice crystals but not body fluid ice crystals). If the fish are initially kept at 0° C (to melt all ice crystals, whether seawater or body tissues), then the fish can supercool to −7° C. The body ice crystals are apparently not in the blood (which will not freeze at −2.7° C) but elsewhere.

Antifreeze proteins are synthesized in the liver and are secreted into the circulation for distribution to most of the other body fluid compartments except the intracellular fluid, the endolymph (inner ear fluid), and the urine. Antifreeze proteins are present at low concentrations in pericardial, peritoneal, extradural, and cerebrospinal fluid and at very low concentrations in the ocular fluid. Presumably antifreeze proteins are not required in fluid compartments that are surrounded by antifreeze-protected fluids, e.g., intracellular and intraocular fluid (it is not clear why CSF has antifreeze proteins, since it is similarly protected by surrounding extracellular fluids).

Urine is not antifreeze protected, and is therefore normally supercooled by about 1° C. The urine lacks antifreeze proteins because (1) the kidneys are aglomerular and antifreeze proteins are not secreted into the tubule, e.g., Nototheniids; or (2) the filtration barrier is thick and Bowman's capsule is not connected to the nephron tubule, e.g., the Antarctic eel pout; or (3) charge repulsion prevents filtration of the acidic AFP across the anionic filtration barrier, e.g., northern winter flounder.

Freezing Tolerance. A number of invertebrates, and a very few vertebrates, are able to tolerate the freezing of a significant fraction of their body water (Figure 5–20). The few freeze-tolerant frogs and reptiles can tolerate less ice formation (35 to 50% frozen) than intertidal mollusks (54 to 76%) and insects (>90%).

It is important in these frozen animals that the freezing of body water is restricted to the extracellular fluids. For example, the cells of frozen intertidal mollusks are shrunken and distorted, but do not contain ice crystals (Kanwisher 1955). Ice crystal formation in the intracellular fluids inevitably results in death. Intracellular ice crystals apparently destroy the integrity of intracellular membranes and organelles. In contrast, extracellular freezing does not disrupt essential cell structures and is not lethal (at least to the freeze-tolerant species).

The ability of tissues to avoid intracellular freezing depends on the rate of cooling; rapid cooling promotes intracellular freezing (Mazur 1963). It is potentially disadvantageous, therefore, for freeze-tolerant animals to supercool, because when freezing occurs it is rapid and likely to occur in the intracellular fluids as well as the extracellular fluids. Consequently, a number of freeze-tolerant animals contain specific **nucleating agents** that promote freezing and retard supercooling. Freeze-tolerant frogs (see Figure 5–17**B**) do not have nucleating agents, but their large size (1 to 15 g) ensures that they cool

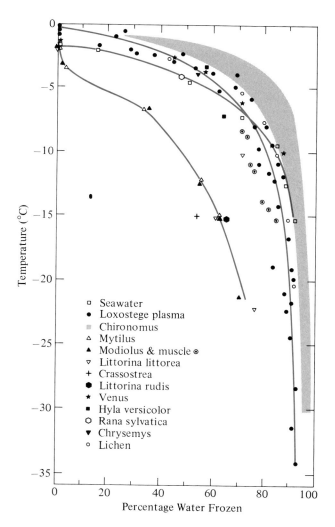

FIGURE 5–20 Percentage of water frozen as a function of temperature for seawater, plasma of an insect (*Loxostege*), insect larvae (*Chironomus*), marine mollusks (*Mytilus, Crassostrea, Modiolus, Littorina, Venus*), frogs (*Hyla, Rana*), and a lichen (*Cetraria*).

slowly, minimizing the likelihood of supercooling (Storey 1986). The extracellular water of their limbs freezes first (the limbs would cool faster because of their higher surface:volume ratio). Ice crystals then form under the skin and become interspersed with leg muscles, and a large mass of ice fills the abdominal cavity. The frozen frog has no respiratory cycle, no (or only infrequent) heart beat, and frozen blood.

Freezing of the extracellular fluid has severe osmotic consequences for the intracellular environment. As the extracellular fluid freezes, ions and other solutes tend to remain in solution in the unfrozen fluid, hence the osmotic concentration of the unfrozen water increases. Intracellular water would be osmotically withdrawn from the intracellular space by the elevated extracellular osmotic concentration. There is generally an exponential relationship between the percentage ice content and temperature (Figure 5–20), and this results in a fairly linear relationship between the osmotic concentration of the extracellular fluid and temperature (Scholander et al. 1953). Thus, freeze-tolerant animals must also be very dehydration tolerant. For example, a marine mollusk with 75% of its water frozen would have an osmotic concentration in the remaining 25% of the unfrozen water of 4 × seawater. Seasonal acclimation of freezing tolerance in intertidal mollusks involves greater tolerance of tissue dehydration and elevation of body fluid osmotic concentration, rather than reduction of the amount of water lost from the cells. For example, *Modiolus* acclimated to 23° C die when 35% of their intracellular water is osmotically withdrawn into the partially frozen extracellular space, but animals acclimated to 0° C can tolerate an even higher loss of 41% intracellular water (Murphy and Pierce 1975). In contrast, adaptation to low salinity (12‰) decreases the percentage of frozen water and the freezing tolerance of *Modiolus* (to −5.1° C) compared to −10.3° and −12.2° C at 34 and 46‰. The high freeze tolerance of these animals may involve polyols (like glycerol) acting as cryoprotectants.

Adaptations to Heat

Ectotherms exposed to high temperature can either attempt to regulate their T_b below the T_a or rely on biochemical adaptations for tolerance of high T_b. The former strategy of thermoregulation can only be accomplished by evaporative cooling, hence is unavailable to aquatic animals. The second strategy increases the upper lethal temperature (CT_{max}).

Critical Thermal Maximum. The ultimate thermal limit for living cells is the boiling point of water (100° C at normal atmospheric pressure). No animals or plants are able to survive temperatures approaching the boiling point, but some prokaryotic organisms tolerate temperatures close to the boiling point. For example, thermophilic bacteria live in the Yellowstone Hot Springs, close to the boiling point of 93.3° C (the high altitude lowers the boiling point). Eukaryotic organisms are much more sensitive to high temperature. A few animals have maximum temperatures of 50° C or greater, but most animals have a lower CT_{max} (Table 5–9).

The upper thermal limit can be measured in a variety of ways, but the lethal temperature for 50%

TABLE 5–9

Critical thermal maximum (CT_{max}) for a variety of invertebrates and vertebrates.

Invertebrates		Vertebrates	
Prokaryotes		**Fish**	
Bacteria		*Fundulus*	35
Aquatic	73		
Thermophilic	91	**Amphibians**	
Bluegreen algae	75	Salamanders	
		Rhyacotriton	29
Echinoderms		*Plethodon*	31.5
Asterias	32	*Ensatina*	34
Ophioderma	37	*Aneides*	34.3
Arbacia	37	*Ambystoma*	35.6
		Anurans	
Annelids		*Hyla*	36
Lumbricus	29	*Rana*	37
		Eupemphix	37.8
Crustaceans		*Bufo*	41
Palaemonetes	34		
Porcellio	39.0	**Reptiles**	
Uca	39.5	Alligator	38
Porcellio	41.0	Turtles	41
Armadillidum	41.4	Lizards	43.4
Uca	45.2	Xantusids	40.1
		Scincids	41.2
Mollusks		Varanids	42.0
Modiolus	37.8	Anguids	42.4
Nassa	42	Helodermatids	42.5
Clavarizona	43	Gekkonids	43.7
		Iguanids	45.0
Insects		Agamids	45.4
Lepisma	36	Teids	46.0
Thermobia	>40	Pygopodids	46.0
Sphingonotus	41	Acertids	46.9
Bembex	42	Snakes	40.4
Aiolopus	45	Elapids	40.4
Trigonopus	45	Crotalids	41.3
Ctenolepisma	48	Colubrids	41.6
Onymacris	49		
Onymacris	50		
Onymacris	51		
Dasymutilla	52		
Arachnids			
Buthotus	45		
Leiurus	47		

mortality (LT_{50}; see Chapter 2) is one common measure. Both the temperature and the time of thermal exposure are important determinants of LT_{50}. For example, heat death occurs in the starfish *Asterias* after 40 min at 32° C, but after only 9 min at 42° C; the times to heat death for the gastropod *Nassa* are 30 min at 42° C but only 9 min at 46° C. An index of thermal incapacitation, rather than death, is often used to define an upper critical temperature. For example, the temperature at which an animal loses its coordination and locomotor ability is often defined as the CT_{max}. The maximum temperature that an animal will voluntarily select is called the experimental voluntary maximum (T_{EVM}). A panting threshold temperature (T_{PT}) may be used to indicate heat stress. The temperature for onset of muscle spasms (T_{OMS}) is another possible index to define a CT_{max}, as is the temperature for cessation of breathing (T_{CB}).

Most invertebrates and vertebrates have CT_{max} values over 30° C, and some have CT_{max} values over 40° C (Table 5–9). CT_{max} is generally consistent within a taxonomic group, e.g., scincid lizards have a lower CT_{max} than lacertid lizards, but there is considerable variability within a taxonomic group according to habits and thermal environment. For example, temperate fossorial skinks have lower CT_{max} than terrestrial skinks and fossorial desert skinks. The CT_{max} is lower for forest anole lizards than open habitat species; edge species have intermediate CT_{max} (Table 5–10). Ambient relative humidity also affects the CT_{max}, which is generally lower in humid air than in dry air. CT_{max} also varies latitudinally for anuran amphibians. CT_{max} is different for larval and adult amphibians; *Rana pipiens* tadpoles are more sensitive to high temperature than are the adults. Furthermore, different tissues may have varying thermal sensitivity. The skeletal muscle of *R. pipiens* has about the same thermal tolerance as the intact adult, but sciatic nerve and heart have a higher thermal tolerance. Acclimation temperature also influences the CT_{max} (see below).

TABLE 5–10

Effects of habitat on critical thermal maximum (CT_{max}) in closely-related anole lizards (*Anolis*). (*Data from Hertz 1979.*)	
Open Habitat	
A. cooki	38.9
A. cybotes	39.3
A. cristatellus	39.4
A. auratus	40.0
A. carolinenis	40.7
Edge Habitat	
A. tropidogaster	33.4
A. frenatus	35.7
A. krugi	36.0
Forest Habitat	
A. limifrons	33.4
A. gundlachi	34.5

Evaporative Cooling. Evaporation of water dissipates heat (2456 J g^{-1} at 20° C, 2432 at 30° C, 2408 at 40° C, and 2260 at 100° C). Consequently, evaporative water loss can dissipate a considerable amount of heat and lower the temperature. The extent of evaporative cooling depends on the ambient relative humidity and temperature. For example, objects that evaporate water as if they were a free water surface have a wet bulb temperature that is lower than the dry bulb air temperature. The wet bulb temperature is that measured by a thermometer with a moist covering, with maximum forced convection. The dry bulb temperature is the normal air temperature, with no evaporative cooling effect.

The skin of many ectothermic invertebrates and vertebrates evaporates water as if it were a free water surface, and these animals act as wet bulb thermometers. For example, most frogs, toads, and slugs have T_b closer to T_{wet} than to T_{dry} (Figure 5–21). Active snails (or snails with their shell removed) have T_b similar to T_{wet}, but inactive snails withdrawn into their shell have T_b closer to T_{dry}. Earthworms have T_b similar to T_{wet}, but T_b is closer to T_{dry} if their skin becomes desiccated. Fiddler and ghost crabs can have a T_b 3° to 4° C lower than T_{dry}. The T_b of the isopod *Ligia* can be depressed by as much as 8° C, but more terrestrial species have lower rates of water loss and a lesser T_b depression (4° C for *Porcellio* and 4° to 5° C for *Oniscus*). Some bivalves, such as *Modiolus*, maintain a 1° to 2° C temperature gradient by keeping their valves partly open. These animals are passive evaporators; their T_b depression is not an active thermoregulatory response. However, some ectotherms have mechanisms to keep their skin moist during evaporation. For example, some frogs frequently discharge skin mucous glands to prevent dehydration of the skin, and a high cutaneous blood flow maintains the hydration level of the skin. Terrestrial ectotherms tend to have reduced cutaneous evaporation to prevent rapid dehydration, e.g., terrestrial isopods, crustaceans, waterproof frogs, and lizards (see Chapter 16). Their T_b is normally closer to T_{dry} than to T_{wet} (e.g., reptiles; Figure 5–21).

Relatively waterproof terrestrial animals are generally able to enhance evaporative cooling for T_b regulation at high T_a by markedly increasing their cutaneous or respiratory water loss. For example, the lizard *Dipsosaurus* totally dissipates its metabolic heat production by panting at T_a >40° C, although it is not able to significantly reduce its T_b below T_a. In contrast, the large agamid lizard *Amphibolurus* can reduce its T_b about 3.2° C below T_a (42.5° C). The chuckwalla *Sauromalus* can reduce

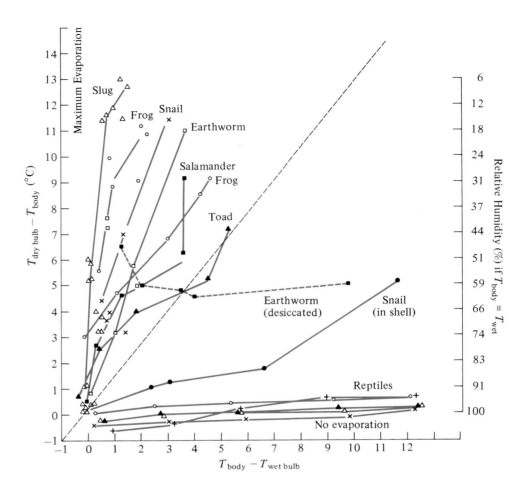

FIGURE 5–21   Body temperatures of some invertebrates and vertebrates graphed as a function of the difference between body temperature (T_b) and wet bulb temperature (T_{wet}), and the difference between body temperature and dry bulb temperature (T_{dry}). The ordinate ($T_b - T_{wet} = 0$) indicates the maximum extent of evaporative cooling and the abscissa ($T_b - T_{dry} = 0$) indicates no evaporative cooling.

its T_b about 0.9° C below T_a (at 45° C) and brain temperature (T_{br}) about 2.7° C below T_a by evaporative cooling by panting (Figure 5–22). A chuckwalla prevented from panting, or a dead lizard, has T_b and T_{br} similar to T_a.

The waterproof frogs *Chiromantis* and *Phyllomedusa* have markedly lower evaporative water loss than other frogs, consequently their T_b is similar to T_a, but is lower at high T_a (Figure 5–23). Both *Chiromantis* and *Phyllomedusa* can dramatically increase water loss when heat stressed to precisely regulate T_b below T_a. *Chiromantis* discharges cutaneous mucous glands to elevate evaporation water loss. *Phyllomedusa*, in contrast, is waterproofed by an epidermal wax layer; its EWL increases when the epidermal wax layer melts at high T_a (McClanahan,

Stinner, and Shoemaker 1978); there is also a mucous gland discharge that contributes to the increased evaporative water loss (Shoemaker et al. 1987).

Advantages of Ectothermy

There are a number of thermal and energetic advantages to ectothermy (Pough 1980). We need to distinguish two types of ectotherms: (1) eurytherms that do not thermoregulate (at least very well) and whose T_b is similar to T_a and (2) stenotherms that thermoregulate T_b precisely within a narrow range using heliothermy or thigmothermy.

The primary advantage of eurythermal ectothermy is that no metabolic energy or time is

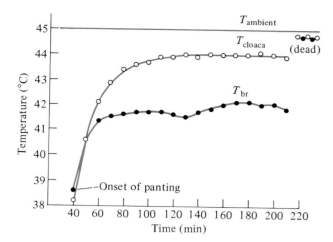

FIGURE 5–22 The body temperature (T_b) and brain temperature (T_{br}) of the chuckwalla lizard *Sauromalus obseus* are considerably lower than the ambient temperature ($T_{ambient}$) of 45° C; the T_b and T_{br} of a dead lizard are about 0.5° C below $T_{ambient}$ because there is some evaporative heat dissipation. *(From Crawford 1972.)*

experienced by a eurythermal ectotherm. Many physiological functions (locomotion, digestion, growth, excretion, membrane and action potentials) are similarly compromised over a wide range of temperatures. Thus, a "jack-of-all-temperatures" may be a master of none (Huey and Hertz 1984).

A stenothermal ectotherm precisely thermoregulates using behavioral means when environmental conditions are appropriate, and can have as high and precisely regulated T_b as endotherms. There is no direct metabolic cost to these ectotherms because most of the heat required for thermoregulation is from solar radiation, but there may be costs for shuttling movement (Huey and Slatkin 1976). Stenothermal ectotherms have a metabolic rate about 1/10 that of endotherms of the same size at the same T_b. Furthermore, the T_b of stenothermic ectotherms declines during the night and further depresses metabolic rate. Consequently, the daily energy expenditure of stenothermal ectotherms is only about 1/20 that of endotherms, and the ectotherm can convert much more of its energy intake into production (growth, reproduction) rather than metabolism (see Chapter 4). However, there are costs and disadvantages to stenothermal ectothermy. Activity is limited by daily and seasonal solar radiation cycles; digestion can be limited by low body temperatures during the inactivity phase; there is a time and energy cost for thermoregulatory activities; realized field T_b values may not be equal to either

expended on thermoregulation, but there are biochemical, physiological, and energetic disadvantages. For example, enzyme systems cannot function optimally over the wide range of body temperatures

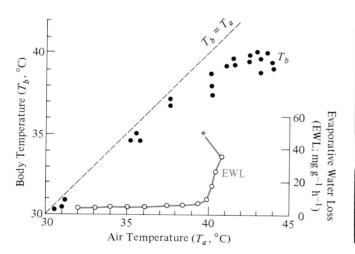

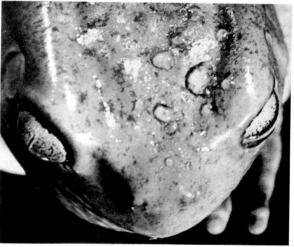

FIGURE 5–23 The body temperature (T_b; solid symbols) of the waterproof frog *Phyllomedusa sauvagei* increases linearly with air temperature to about 38° C when T_b is regulated lower than T_a. The evaporative water loss (EWL; open symbols) is low until temperature regulation begins; EWL increases markedly when the frog moves (*). The increase in EWL is a consequence of glandular secretion, appearing in the photograph as beads of water. *(From Shoemaker et al. 1987.)*

the physiologically or ecologically optimum temperatures, depending upon ambient conditions. Few ectotherms can sustain a high aerobic capacity for prolonged periods and tend to rely on anaerobic metabolism for burst activities.

Endotherms

Endotherms derive all (or most) of their heat content from metabolic heat production, rather than from the external environment. Endogenously produced metabolic heat can allow precise regulation of T_b and most endotherms are also good thermoregulators (although this isn't a necessary consequence). The extent of thermoregulation is indicated by both the intercept value and the slope for the relationship $T_b = a + bT_a$. For thermoconformers, a is close to 0° C and b is about 1, whereas a is about 20 to 40 and b is about 0 for thermoregulators (Table 5–11).

Let us now examine patterns of endothermy for various animals, then we will consider the more speculative questions of why endothermy evolved, how it evolved, and why the T_b of endothermic animals is commonly in the 35° to 40° C range.

Mammals and Birds

Mammals and birds are endothermic, with a high and precisely regulated T_b (35° to 42° C). Partly because of this, birds and mammals were originally classified (by Owen in 1866) as members of a single taxon, the Haemothermia. This classification scheme has been rejected by virtually all zoologists in this century in favor of a separate derivation of birds from theropod archosaurs (hence birds are closely related to dinosaurs and crocodilians) and mammals from synapsid reptiles (and therefore mammals are not closely related to birds). Mammals and birds must have independently evolved endothermy. The taxon Haemothermia has recently been resurrected (Gardner 1982) but is this generally rejected as an outrageous hypothesis (Kemp 1988). We shall examine endothermy of mammals and birds together, despite their separate evolution, since both groups have essentially the same pattern of thermoregulation.

Body Temperature.   Birds and mammals have independently evolved a similar pattern of endothermic physiological regulation, but there is considerable variation in their T_b (Figure 5–24). Monotremes typically have the lowest T_b of mammals (28° to 32° C); edentates also have low T_b's (33° C). Marsupials

TABLE 5–11

Coefficients for the relationship between body temperature (T_b) and ambient temperature (T_a) of the form $T_b = a + bT_a$ for animals and a plant (*Philodendron*) that thermoconform or thermoregulate, and an inanimate object (water-filled metal can) that passively responds to the thermal environment.

	a	*b*	Comments
Forest anole lizard	−3.8	1.17	Ectothermic
Tenebrionid beetle	1.3	1.01	Ectothermic
Aquatic salamander	0	1.0	Ectothermic
Asian honeybee	18.3	0.81	Endothermic
Male diamond python	9.6	0.67	Ectothermic
Bullfrog	12.0	0.60	Ectothermic
Nonbrooding diamond python	14.5	0.48	Ectothermic
Desert cicada	24.8	0.44	Ectothermic
Open habitat anole lizard	16.4	0.43	Ectothermic
Dragonfly	26.3	0.43	Ectothermic
Naked mole rat	20.6	0.41	Endothermic
Sphinx moth caterpillar	23.7	0.40	Ectothermic
Water-filled metal can	24.3	0.30	Passive
Cuculinid moth	30.8	0.28	Endothermic
Bluefin tuna	25.5	0.24	Endothermic
Planigale marsupial	27.4	0.21	Endothermic
Dragonfly	35.1	0.21	Endothermic
Philodendron	38	0.18	Endothermic
Poorwill	33.6	0.17	Endothermic
Brooding diamond python	28.0	0.12	Endothermic
Pocket mouse	35.4	0.081	Endothermic
Honey possum	35.1	0.07	Endothermic
Queen bumblebee (brooding)	35.0	0.07	Endothermic
Cactus mouse	35.8	0.059	Endothermic
Chilean tinamou	35.4	0.05	Endothermic
House finch	40.0	0.05	Endothermic
Parrot	39.9	0.05	Endothermic
Rosy finch	41.0	0.02	Endothermic
Amazonian parrot	40.9	0.01	Endothermic
Least weasel	39.5	0	Endothermic
Mallee fowl	40.3	−0.04	Endothermic

and insectivores (excluding shrews) have a low T_b of about 34° to 36° C. Members of a number of mammalian orders have mean T_b values about 37° C (e.g., bats) or 38° C (primates). Among birds, the more primitive ratites have low T_b's (38° to 39° C). There is a considerable diversity in the mean T_b for various groups of nonpasserine birds, T_b may be less than 40° C (penguins), 40° to 41° C (owls), 41° to 42° C (parrots), or 42° C (woodpeckers). Passerine birds have a T_b of about 42° C.

The T_b of endotherms is essentially determined by mass-specific metabolism and thermal conductance. Birds, especially passerines, have a lower thermal conductance than mammals (Aschoff 1981).

Mammals: $C = 20.5\,g^{-0.426}$
Nonpasserine birds: $C = 19.0\,g^{-0.583}$ (5.14)
Passerine birds: $C = 11.6\,g^{-0.576}$

where C is thermal conductance ($J\,g^{-1}\,hr^{-1}\,°C^{-1}$) and g is body mass in grams. The scaling of VO_2 and C varies for mammals, nonpasserine birds, and passerine birds. We would expect a variation in their T_b since $VO_2/C = 3.1\,M^{0.186}$ (mammals), 4.9 $M^{0.303}$ (nonpasserines), and 13 $M^{0.298}$ (passerines). Consequently, passerines should have a higher T_b than nonpasserines which should have a higher T_b than mammals. We would also expect a slight mass-dependence of T_b, although there is not any obvious relationship between T_b and mass for mammals or birds.

There are differences in allometric scaling of VO_2 and C for activity and resting phases of the circadian cycle, and so we might expect a corresponding circadian cycle in T_b. For example, the T_b of the gray jay *Perisoreus* is about 41.2° C during the night (when VO_2 is lower) and 42.4° C during the day (when VO_2 is higher; Veghte 1964).

Body Temperature Regulation. The endothermic mechanisms of mammals and birds are best examined by considering the relationship between metabolic heat production (MHP or VO_2) and T_a. The

$VO_2 - T_a$ relationship for the monk parakeet (Figure 5–25) illustrates the typical endothermic responses of mammals and birds to T_a. Metabolic heat production is high at low T_a to compensate for the high rate of heat loss. There is a range of T_a at which metabolic rate is minimal (basal metabolic rate, BMR) and constant; this is the thermoneutral zone. Metabolic rate increases at T_a below the lower critical temperature (T_{lc}) of the thermoneutral zone, and at T_a above the upper critical temperature (T_{uc}).

The relationship between VO_2 and T_a below T_{lc} generally conforms to a physical model for endothermic heat balance

$$VO_2 = C(T_b - T_a) \qquad (5.15)$$

where C is the thermal conductance (see Supplement 5–2, page 188). The relationship is not necessarily linear. The body temperature is not necessarily constant, but may decrease at lower T_a. Thermal conductance is not a physical constant, but can be varied by behavioral and physiological means. The wet thermal conductance (C_{wet}), which includes heat loss by evaporation, is not constant but increases at higher T_a as the evaporative water loss increases. The dry thermal conductance (C_{dry}) is less dependent on T_a than is C_{wet}, but it can also increase at high T_a (Figure 5–25) by physiological adjustments (e.g.,

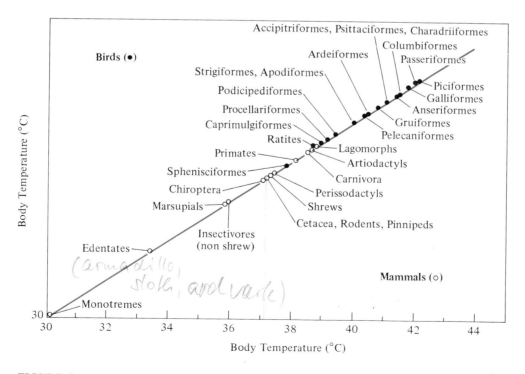

FIGURE 5–24 Average body temperatures for birds and mammals.

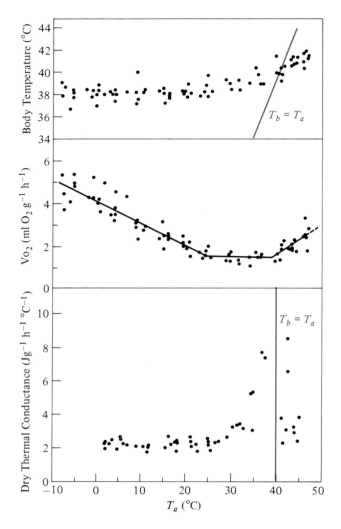

FIGURE 5–25   Metabolic rate (VO_2), body temperature (T_b), and dry thermal conductance (C_{dry}) as a function of air temperature (T_a) for the monk parakeet.   *(From Weathers and Caccamise 1975.)*

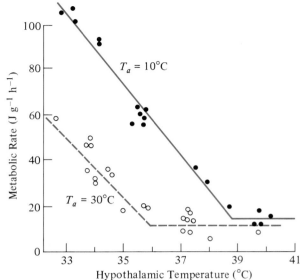

FIGURE 5–26   Metabolic heat production of a kangaroo rat, as a function of hypothalamic temperature, at two ambient temperatures.   *(From Glotzbach and Heller 1975.)*

patterns of peripheral blood flow) and behavioral adjustments (e.g., posture). Finally, the relationship between VO_2 and $T_a (<T_{lc})$ often does not extrapolate to zero VO_2 at $T_a = T_b$; this is especially apparent for birds.

Most mammals and birds do not regulate T_b exactly constant over a range of T_a (e.g., Figure 5–25). The gain for thermoregulatory control systems is often about -4 to -8, and the thermoregulatory response is proportional to an error term (depends on $T_b - T_a$). We therefore should expect a slight (but significant) positive relationship between T_b and T_a for endotherms.

Body temperature is controlled primarily by hypothalamic and spinal cord thermoreceptors, with some input from peripheral thermoreceptors (see Chapter 2). The nature and extent of thermoregulatory responses (R; e.g., metabolic heat production, evaporative heat loss) are proportional to the difference between hypothalamic temperature (T_{hypo}) and the hypothalamic setpoint ($T_{b,set}$)

$$R = \alpha(T_{hypo} - T_{b,set}) \qquad (5.16)$$

where α is the proportionality constant for each thermoregulatory response for heat gain (α is negative) or heat loss (α is positive). For example, the metabolic heat production of the kangaroo rat *Dipodomys* has a hypothalamic setpoint of about $36°$ C and α is -10.83 J g^{-1} hr^{-1} $°C^{-1}$ at a T_a of $30°$ C; at T_a of $10°$ C the $T_{b,set}$ is about $39°$ C and α is -13.7 (Figure 5–26).

Adaptations to the Cold.   Endothermic mammals and birds have essentially three strategies for surviving in the cold; they either decrease their rate of heat loss, increase their rate of heat production, or abandon thermoregulation of the normal T_b and allow it to decline (hypothermia, or torpor). Virtually all mammals and birds use the first two strategies, and some mammals and birds use the last strategy. Cold-stressed endotherms may also bask in the sun to supplement endogenous heat production.

Mechanisms to Decrease Heat Loss. One way to reduce heat loss is to decrease thermal conductance (an alternative way is to reduce T_b; see heterothermy). A decrease in thermal conductance (at the same body mass) can reduce the T_{lc} and VO_2 substantially.

Body size is an important determinant of thermal conductance; larger animals have a lower mass-specific thermal conductance by virtue of their lower surface:volume ratio, although they have a higher absolute thermal conductance. Larger mammals have a lower T_{lc} than do smaller mammals, and their VO_2 is relatively less affected by low T_a (Figure 5–27). Increased body mass is not necessarily a useful response to cold stress for an individual endotherm. Increasing mass by 50% only decreases T_{lc} by about 2° C and by about 3° C for a 100% increase. However, endotherms that live in cold climates often have a higher body mass than individuals from warmer climates. This is Bergmann's rule, one of the bioclimatic laws postulated in the 19th century (see Supplement 5–3, page 190).

Thermal conductance can be minimized at low T_a by postural adjustments. A sphere has the minimum surface area:volume ratio, hence it is the energetically most effective shape. For example, mammals may curl up, and birds may retract their head and fluff up their feathers to assume a spherical shape (Figure 5–28). Many birds draw the legs into their more-or-less spherical insulation layer and tuck their bill under a wing to minimize heat loss from the uninsulated appendages.

Endotherms can further reduce their thermal conductance by using external insulation, i.e., nesting material. For example, lemmings with a nest of cotton wool can reduce their effective thermal conductance by 40%. Weasels will use lemming nests when resting and add to the insulation of the nest with lemming fur. Bank voles at a T_a of 4° C can reduce their daily energy expenditure from 3.5 kJ g^{-1} day^{-1} without nesting material to 2.3 kJ g^{-1} day^{-1} with nesting material.

A group of individuals can collectively reduce their thermal conductance by huddling together. Bank voles at 4° C can reduce their daily energy expenditure from 3.5 to 3.0 kJ g^{-1} day^{-1} by huddling with three other voles.

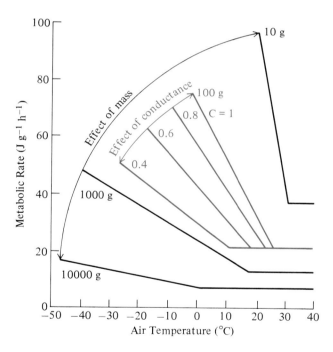

FIGURE 5–27 Predicted relationship between metabolic rate and air temperature for a 10 g, 100 g, 1000 g, and 10000 g mammal showing the decrease in basal metabolic rate, thermal conductance, and lower critical temperature (T_{lc}) with higher body mass. Shown also is the effect of decreasing thermal conductance for a 100 g mammal from 1.0 to 0.8, 0.6, and 0.4 × the normal value.

FIGURE 5–28 The typical hibernation posture of the western pygmy possum is an almost spherical ball. This minimizes the surface-to-volume ratio. *(From Geiser 1985.)*

Some mammals and birds seasonally alter their insulation to reduce conductance in winter (and at the same time change coat color to white for cryptic coloration). For example, some arctic mammals have a seasonal change in thermal conductance

$$\text{Winter: } C = 13.9 \, g^{-0.534}$$
$$\text{Summer: } C = 23.5 \, g^{-0.534} \quad (5.17)$$

(calculated from Casey, Withers, and Casey 1979). However, some arctic mammals such as least weasels do not have a summer–winter change in thermal conductance. These weasels also have a higher than expected conductance for a mammal because of their elongate shape and have a high metabolic rate.

A thick layer of subcutaneous fat will not only add insulation but will also decrease the surface:volume ratio with little increase in metabolic rate (since fat has a low metabolic rate). Thermal conductance can be reduced by restricting blood flow to the skin and by allowing skin temperature to decline because heat loss across the fur/feather insulation is proportional to $(T_{skin} - T_a)$ rather than $(T_b - T_a)$.

Evaporative water loss generally doesn't contribute significantly to thermal conductance at low T_a. Cutaneous evaporative water loss is low since T_{skin} is reduced and the ambient relative humidity is high. Respiratory evaporation is potentially high because of the elevated VO_2 but is minimized by a reduced expired air temperature (T_{exp}) and nasal countercurrent heat exchange (see Chapter 16). For example, the grey seal has a T_{exp} as low as 6° C (at $T_a = -30°$ C) and can conserve up to 70% of the heat that would be lost if $T_{exp} = T_b$ (Folkow and Blix 1987). Arctic mammals similarly conserve respiratory heat loss. For example, lemmings can reduce respiratory heat loss to <5% of total heat loss at low T_a.

Mechanisms to Increase Heat Production. The principal means for increasing heat production is an increased metabolic heat production by skeletal muscle. Skeletal muscle has a high aerobic metabolic capacity and is a considerable proportion of the body mass; hence, it is capable of considerable supplementary heat production.

Skeletal muscles usually contract to move parts of the body (see Chapter 10) but it is relatively straightforward for muscle contractions to be rendered nonlocomotory. **Shivering** has been reported for many mammals (monotremes, marsupials, and placentals) and birds. In birds, there tends to be a clear inverse relationship between shivering activity (measured as electromyograph activity) and T_a, suggesting that shivering is the primary thermogenic response to cold (Figure 5–29).

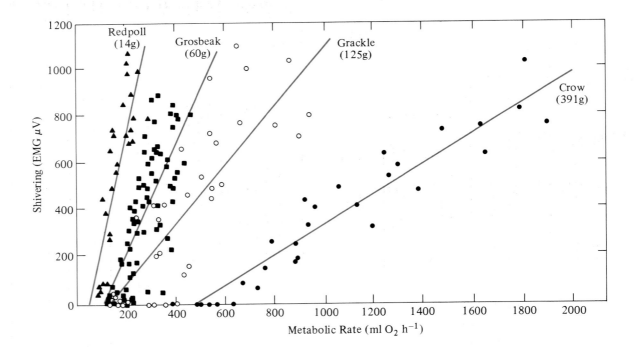

FIGURE 5–29   Linear relationship between degree of shivering (measured as electromyogram electrical activity) and metabolic rate for species of four birds.   *(From West 1965.)*

Shivering is a myotactic reflex oscillation due to muscle spindle activation by γ-efferent nerve fibers to the muscle spindle (see Chapter 8). Higher brain centers (perhaps the cerebellum) are required for the oscillatory muscle contractions of shivering; shivering does not occur below the level of spinal cord transection. A few endothermic pythons also shiver and some endothermic insects "shiver" with their wings (see below). Shivering thus appears to be a fairly generalized response of many diverse endotherms to cold. It may have evolved from a generalized activity at low temperature through the acquisition of a neural capacity to repetitively stimulate muscle contraction with no gross movement (Whittow 1973).

Nonshivering thermogenesis (NST) is a second mechanism for augmented metabolic heat production. It has been reported for many placental mammals, some marsupials, and a few birds. NST is more important in small mammals, and can increase VO_2 to about 2 to 4 × basal metabolic rate (Figure 5–30).

Brown adipose tissue (BAT, or brown fat) is a type of mammalian adipose tissue that is specialized for metabolic heat production (Chapter 3). It has been identified only in placental mammals (chiropterans, insectivores, rodents, lagomorphs, artiodactyls, carnivores, and primates; Smith and Horowitz 1969). A superficially similar adipose tissue has been reported for a marsupial (Loudon, Rothwell, and Stock 1985). It has not been identified in birds. BAT is present in some hibernating mammals, some cold-adapted mammals, and some newborn mammals. It releases heat from a futile mitochondrial electron-

TABLE 5–12

Total oxygen consumption rate (VO_2; ml O_2 g^{-1} hr^{-1}) and oxygen delivered to brown adipose tissue (BAT) and other tissues in cold-acclimated rats at varying air temperatures. Oxygen delivery to tissues is calculated from tissue blood flow and arterial O_2 content (16.3 ml/100 ml blood) and is expressed as a percentage of the total cardiac O_2 delivery (% COD). *(Data from Foster and Frydman 1979.)*

	Air Temperature (°C)			
	−19	*−6*	*+6*	*+21*
Blood flow (% COD):				
BAT	25.0	22.6	20.2	5.6
Skeletal muscle	15.5	14.2	15.8	17.3
Heart	5.2	4.0	3.4	3.1
Kidneys	11.3	12.1	13.6	15.7
Brain	1.6	1.5	1.5	1.4
Hepatosplanchnic	14.9	12.1	15.8	19.8
Total VO_2	3.52	2.96	2.26	1.28

transport cycle that produces heat without the necessity of ATP synthesis and degradation. BAT can produce remarkable amounts of heat, up to 500 J sec^{-1} kg^{-1} (cf. active skeletal muscle produces 50 to 60 J sec^{-1} kg^{-1}). Such high metabolic rates require a substantial O_2 supply; the BAT of cold-stressed rats can receive up to 1/4 of the total cardiac output (Table 5–12). Not surprisingly, BAT is an important component of nonshivering thermogenesis in many mammals, although other tissues (e.g.,

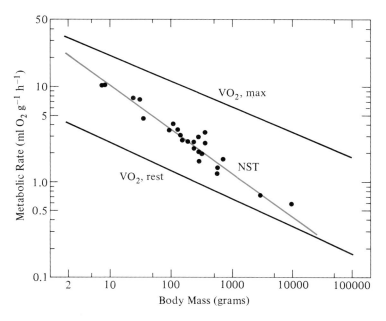

FIGURE 5–30 Nonshivering thermogenesis in rodents, bats, hedgehog, dog, and rabbit as a function of body mass; NST VO_2 (ml O_2 g^{-1} h^{-1}) = 30 g$^{-0.454}$. Upper line indicates the predicted maximal VO_2 and the lower line indicates the predicted resting VO_2. *(From Heldemaier 1971.)*

liver) can also contribute to NST at moderate T_a. The Na$^+$/K$^+$ "leak" of endotherm cells could also provide an important mechanism for NST.

Some young and adult birds appear to have NST, but not BAT. For example, king penguin chicks have a lower shivering threshold temperature (STT = $-18.5°$ C) when cold acclimated than when maintained at 25° C (STT = $-9.1°$ C: duChamp et al. 1989). The site for NST may be visceral organs (e.g., liver) or skeletal muscle. Some cold-acclimated birds do have a specialized fat tissue (it is highly vascular and multilocular), but this is probably a highly mobilizable fat store, rather than a thermogenic tissue.

Heat production by other biochemical processes may substitute for shivering and nonshivering thermogenesis. For example, the specific dynamic effect (SDE) of digestion can reduce the required shivering and nonshivering thermogenesis (see Chapter 4). The heat of fermentation can substitute for shivering and NST thermogenesis for ruminant and pseudoruminant mammals.

Heat produced by general activity (e.g., during the activity phase of the circadian cycle) might also be expected to contribute to thermoregulation if generalized activity was the evolutionary precursor activity for shivering. However, an increase in general activity would at the same time promote heat loss by altering the body posture (e.g., extension of appendages for locomotion); minimizing boundary layer thickness; and disrupting the fur or feather insulation layer, i.e., generalized activity would increase thermal conductance. It is significant in this respect that both mammals and birds have lower thermal conductance in their inactivity phase than in their activity phase. Activity would also prevent shivering thermogenesis (which doesn't have associated detrimental effects on thermal conductance). In practice, exercise appears to partly substitute for shivering and NST. For example, exercise can partly substitute for shivering/NST at $T_a <$ 10° C for white rats acclimated to 30° C, but only at lower T_a ($< -20°$ C) for rats acclimated to 6° C because of their greater NST (Jansky and Hart 1963). Exercise thermogenesis apparently only partially substitutes for shivering and NST since T_b declines at low T_a.

Finally, solar radiative heat gain can substitute for metabolic heat production. Cold-stressed endotherms can bask in sunlight to minimize their metabolic heat production requirements. For example, herring gulls have a lower VO$_2$ at low T_a when exposed to a radiative heat load, and their T_i is reduced from about 20° C to $< -5°$ C (Lustick et al. 1979; see also Figure 5–3A).

Heterothermy. There is a considerable metabolic cost to endothermy at low T_a despite the above-mentioned means for reducing heat loss. Consequently, some endotherms will allow the temperature of peripheral tissues, such as skin and appendages, to decline to below core T_b; this is peripheral, or regional, heterothermy. Some mammals and birds allow the core T_b to decline; this is temporal heterothermy, or torpor.

Appendages tend to have a high surface:volume ratio and high heat loss. Cold-adapted animals tend to have reduced appendages (Allen's rule; Supplement 5–3, page 190).

Heat loss from appendages is often exacerbated by the ineffectiveness or lack of insulation, e.g., mammals may have naked or poorly furred digits and tails, and birds have naked beaks and poorly feathered legs. Many endotherms cover these "thermal windows" by postural adjustments when cold stressed, but some do not, particularly if they are

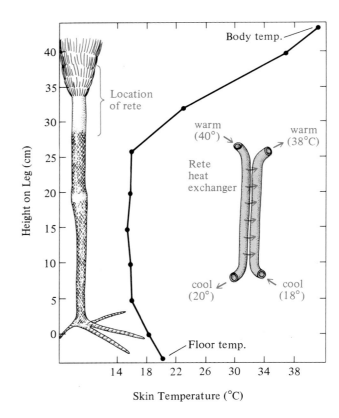

FIGURE 5–31 Skin temperature of the leg of the wood stork at an ambient temperature of 12° C and a floor temperature of 20° C. Inset shows the principle of countercurrent heat exchange in an arterial venous rete, such as that located in the stork leg. *(Modified from Kahl 1963.)*

active. For example, the glaucous-winged gull at a T_a of $-16°$ C has a core T_b of about 37.8° C but the foot skin temperature is as low as 0 to 4.9° C at the feet (Irving and Krog 1955). Similarly, the skin temperature of the wood stork's leg is close to T_a (Figure 5–31); this minimizes heat loss to the environment.

The capacity to lower the skin temperature of appendages and to minimize the loss of core body heat to hypothermic limbs is achieved by a countercurrent exchange of heat between warm arterial blood ($T_{art} = T_b$) and cold venous blood returning from the limbs ($T_a < T_{ven} < T_b$). The warm arterial blood flows in the opposite direction to the cool venous blood, and there is conductive heat transfer across the walls of the artery and veins so that the

heat is lost from the arterial blood and warms the venous blood returning to the body (Figure 5–31, inset). Such a countercurrent heat exchange system has been described for limbs of a variety of mammals and birds: sloth limbs, whale fins, human arms, beaver tail, fox legs, monkey tails, and gull and stork legs.

The countercurrent exchange of heat between arterial and venous blood is facilitated by the anatomical arrangement of the blood vessels. There are three general types of countercurrent heat exchangers. In venae comitantes, a central artery is surrounded by a number of anastomosing veins, e.g., beaver hindlimbs and the forelimb of penguins (Figure 5–32A). The vascular heat exchangers in limbs and flukes of porpoises and whales have 15

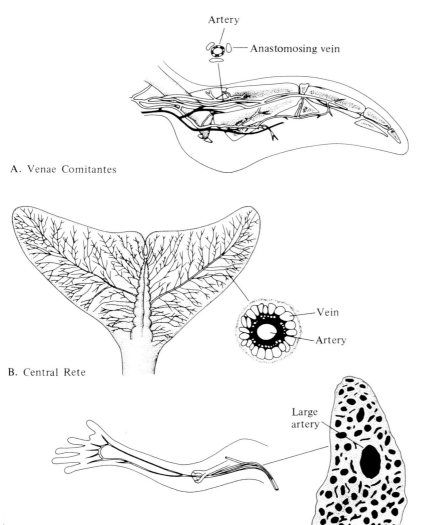

A. Venae Comitantes

B. Central Rete

C. Artery-vein Rete

FIGURE 5–32 Three types of countercurrent heat exchangers in mammal limbs. **(A)** A venae comitantes in which two, three, or more anastomosing veins surround a central artery, e.g., the flipper of the jackass penguin; **(B)** a central artery surrounded by many small veins, e.g., the fins and flukes of cetaceans; and **(C)** a rete of small interdigitating arteries and veins, e.g., the limb vascular bundle of the loris. *(From Frost, Siegfried, and Greenwood 1975; Scholander and Schevill 1955; Scholander 1957.)*

to 20 small veins surrounding a central artery (Figure 5–32B). The most specialized and effective heat exchanger is a rete of small, intertwining arteries and veins, e.g., legs of wading birds, manatee limbs, beaver tail, and loris limb (Figure 5–32C).

At subfreezing air temperatures, limb countercurrent heat exchange could allow freezing of the peripheral tissues, causing frostbite and tissue death (vertebrate tissues are generally unable to survive freezing). Peripheral freezing can be prevented by conveying sufficient body heat to the limbs via arterial blood. For example, control of blood flow to the limbs enables the arctic wolf to avoid freezing of its peripheral tissues (Henshaw, Underwood, and Casey 1972). The normal countercurrent heat exchange system must be bypassed to allow warm arterial blood to reach the peripheral tissues. For example, the beaver has two countercurrent heat exchangers, a venae comitantes of arteries and veins to the hindlimbs, and a rete of small arteries and veins in the tail (Cutright and McKean 1979). Each heat exchange system has a bypass vein that returns blood to the central venous system without it passing through the countercurrent heat exchange system. These bypasses also allow heat dissipation in, for example, warm seasons or during activity.

A diverse assemblage of birds and mammals are temporal heterotherms (Table 5–13). They abandon thermoregulation at the normal T_b when cold stressed and T_b declines. In some mammals and birds, the decline in T_b may be only 4° to 8° C and T_b remains above about 30° C, but in others the decline in T_b is more profound; the T_b may decline to almost equal T_a and may be less than 5° C.

An adaptive decline in core T_b is **hypothermia**. There is a considerable, and confusing, terminology for the various patterns of hypothermia. Hypothermia may be natural or experimentally induced. The criterion for natural hypothermia is that the animal is able to spontaneously arouse (rewarm) to its normal T_b using endogenous heat production (typically shivering or NST). **Torpor** is a pronounced natural hypothermia accompanied by a substantial depression of metabolism, respiratory rate, heart rate, and lack of motor coordination and response to external stimulation. The term dormancy is also applied to natural hypothermia, but has other usages (e.g., dormancy of plant seeds, winter dormancy of amphibians and reptiles). Hibernation is also a widely used term to describe long-term torpor in response to winter cold and food deprivation. Unfortunately, it is also used to describe winter dormancy of reptiles (a different physiological phenomenon) and the mild hypothermia of bears. To avoid confu-

sion, the term torpor is used here to describe the physiological state associated with pronounced hypothermia ($T_b < 30°$ C) as opposed to mild hypothermia ($T_b > 30°$ C). **Estivation** is a torpid state induced by lack of food and/or water during high environmental temperatures; it is physiologically indistinguishable from torpor, except for the higher

TABLE 5–13

Taxonomic distribution of temporal heterothermy in mammals and birds, including mild hypothermia (in parentheses) and torpor.

Birds		
Nonpasserines	Columbiformes	(inca dove)
	Cuculiformes	(ani, roadrunner)
	Strigiformes	(snowy owl)
	Falconiformes	(turkey vulture)
	Caprimulgiformes	poorwill
	Apodiformes	Apodidae; swifts
	Trochiliformes	Trochilidae; hummingbirds
	Coliformes	Mouse-birds
Passerines	Passeriformes	Hirundinidae; swallows
		(Paridae; chickadees, tits)
		(Ploceidae; sparrows, weavers)
		(Fringillidae; finches)
		(Nectarinidae; sunbirds)
		(Pipridae; manakins)
		(Icteridae; new-world blackbirds)
Mammals		
Monotremes	Tachyglossidae	echidna
Marsupials	Didelphidae	American marsupials
	Dasyuridae	marsupial mice
	Phalangeridae	possums
	Tarsipedidae	honeypossum
Placentals	Rodentia	Sciuridae; squirrels
		Muridae; murids
		Cricetidae; cricetids
		Heteromyidae; kangaroo rats, mice
		Dipodidae; jerboas
		Gliridae; dormice
		Zapodidae; jumping mice
	Primates	mouse lemurs
	Megachiroptera	fruit bats
	Microchiroptera	other bats
	Insectivora	insectivores
	Carnivora	(bears, badgers)

T_b during estivation. It will be briefly described below as an adaptation to heat stress.

A number of birds and mammals use moderate hypothermia as a short-term response to cold stress. For example, the tropical manakins *Manacus* and *Pipra* (small frugivorous passerines) have a normal T_b of about 37.9° C but starved birds will become hypothermic at night, with T_b dropping to 27° to 36° C. This hypothermia significantly reduces the metabolic rate to 58% of the normal value (Bartholomew, Vleck, and Bucher 1983). A number of other birds use short-term hypothermia to conserve energy; these include the turkey vulture, smooth-billed ani, inca dove, and the snowy owl. Some mammals also use moderate hypothermia in response to cold. For example, the marsupial mouse *Antechinus stuartii* may have a moderate depression of T_b (up to 5° C) when inactive in moderate and cold environments. Hibernating bears and some other carnivores, such as badgers, have only a moderate (5° to 10° C) decline in T_b when inactive in the cold. For example, the T_b of black bears declines to 31° to 35° C during winter dormancy; heart rate declines from 50 to 60 min^{-1} to 8 to 12 min^{-1} and VO$_2$ declines to 32% of normal levels. The term "carnivore lethargy" is sometimes used to distinguish this moderate hypothermia from torpor.

Deep hypothermia, or torpor, has been reported for a variety of mammal and bird families. A typical torpor cycle has three stages: **entry** into torpor, the **prolonged period** of torpor, and **arousal** from torpor (Figure 5–33). T_b declines markedly during torpor, often to within 1° to 2° C of T_a. Associated with the profound decline in T_b is a marked decline in metabolism. There is an obvious energetic savings associated with torpor.

Torpor is not an abandonment of thermoregulation and assumption of ectothermy but is a controlled physiological state. This is evident from the relationship between T_b and VO$_2$ of torpid endotherms at low T_a. The T_b is generally close to T_a (T_b must be a degree or so above T_a to dissipate metabolic heat). However, there is a minimum **critical body temperature** ($T_{b,crit}$), which a torpid endotherm will maintain during torpor, even if T_a drops below the $T_{b,crit}$. Endogenous heat production is increased to maintain T_b at $T_{b,crit}$, and the elevation in heat production is proportional to ($T_b - T_a$). For example, the hummingbird *Eulampis* has a $T_{b,crit}$ of about 15° C; T_b is similar to T_a when torpid at 15 < T_a < 30° C, but the T_b is regulated at about 15° C by elevated metabolic heat production when T_a < 15° C (Figure 5–34). The increase in VO$_2$ at T_a < 15° C is parallel to that observed for euthermic hummingbirds, since the slope of this relationship is the thermal conductance (which is similar for euthermic and torpid birds). A very similar pattern of T_b regulation during torpor is observed for mammals during torpor, e.g., the shrew *Suncus* has a $T_{b,crit}$ slightly less than 15° C. Many torpid birds and mammals have lower $T_{b,crit}$; for example, the pygmy and honey possums have a $T_{b,crit}$ of about 5° C. Torpor at T_a < $T_{b,crit}$ is thus a highly regulated physiological state equivalent to normal thermoregulation, except that the setpoint for T_b regulation is lowered to $T_{b,crit}$.

Entry into torpor is generally very rapid, particularly for small birds and mammals. Some animals show a few "test-drops" of VO$_2$ then rapidly enter torpor. It is energetically advantageous to enter torpor as rapidly as possible because this maximizes the energy savings. Entry into torpor appears to be a fairly passive cooling response to the abandonment of normal thermoregulation. Animals do not, for example, shiver during entry into torpor to regulate the rate of cooling. Hummingbirds and the poorwill enter torpor at a rate determined by their passive cooling properties. The rate of entry into torpor could be maximized by increasing thermal conductance, for example by a postural adjustment of ptilo/pilo-depression, but there is little evidence that this occurs. Consequently, the rate of entry into torpor is proportional to the normal thermal conductance, i.e., $\propto$ mass$^{-0.5}$. Small mammals and birds (2 to 10 g) therefore enter torpor much more rapidly than

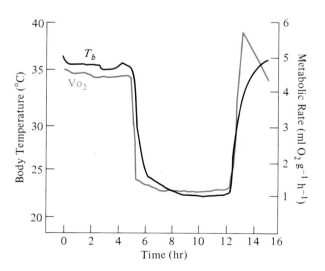

FIGURE 5–33 Metabolic rate and body temperature of a deer mouse during a typical daily torpor cycle. *(From Nestler 1990.)*

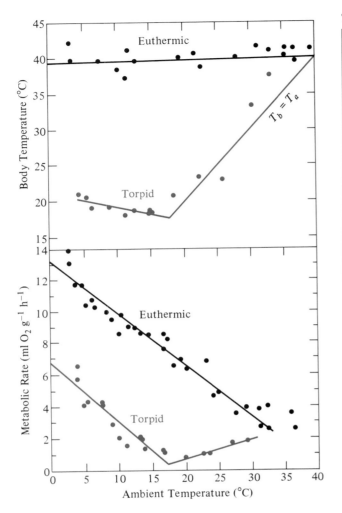

FIGURE 5–34 Relationship between metabolic rate, body temperature, and ambient temperature for euthermic and torpid hummingbirds. *(Modified from Hainsworth and Wolf 1970.)*

TABLE 5–14

Times required for entry into torpor and arousal from torpor, calculated from allometric relationships between rates of entry into torpor and arousal for a variety of birds and mammals of varying body mass. Calculations are for cooling and arousing at an ambient temperature of 15° C, with body temperature changing from 37° C to 17° C (note that some of the animals listed can only tolerate mild hypothermia).

	Body mass (grams)	Entry Time[1] (min)	Arousal Time[2] (min)
Shrew *Suncus*	2	35	13
Hummingbird *Archilochus*	4	59	17
Honey possum *Tarsipes*	10	80	24
Poorwill *Phalaenoptilus*	40	224	41
Nightjar *Eurostopodus*	86	350	55
Turkey vulture *Cathartes*[3]	230	2336 (39h)	190 (3.2h)
Echidna *Tachyglossus*	3500	1648 (27h)	226 (3.8h)
Marmot *Marmota*	4000	1766 (29h)	237 (4.0h)
Badger *Taxidea*[3]	9000	2685 (45h)	323 (5.4h)
Bear *Ursus*[3]	80000	8307 (138h)	741 (12.3h)

[1] Calculated from the thermal conductance.

[2] Calculated from arousal rate (°C min^{-1}) = $1.97 \, g^{-0.38}$.

[3] Can only tolerate mild hypothermia.·

larger mammals and birds. A 2 g shrew would cool from 37° to 17° C in about 35 min at a T_a of 15° C, whereas an 80 kg bear would require about 138 hr to cool by the same amount (Table 5–14).

The metabolic rate during torpor is profoundly depressed, to 1/20 to 1/100 of the normal euthermic value. The magnitude of the decrease in VO_2 reflects two factors. First, there is a marked decline in VO_2 due to the abandonment of thermoregulation at low T_a, i.e., the difference between cold-stressed VO_2 and $VO_{2,basal}$. Second, there is a subsequent decline in T_b, hence VO_2, due to a Q_{10} effect as T_b declines. The energy saving, due to a torpor cycle, is considerable, depending on body mass, T_a, and Q_{10}. The

energy saving is more for a small mammal or bird because small endotherms have a higher thermal conductance and T_{lc}, and their metabolic rate is increased more by low T_a (Figure 5–35). However, the energy savings from a decline in T_b is similar for a small and large endotherm (if the Q_{10} were the same). The Q_{10} for metabolic rate depends on the body mass, whether the torpor is short or long term, and on the T_a. The Q_{10} is about 2.2 for daily torpor regardless of the T_b and body mass, but tends to be over 3 for small mammals and birds during prolonged torpor. This suggests that there are additional physiological mechanisms for depressing metabolic rate during prolonged torpor, at least for small mammals and birds.

The $T_{b,crit}$ is generally much higher than the freezing point of tissues. Is the role of such high $T_{b,crit}$ to prevent tissue freezing? Endotherm tissues

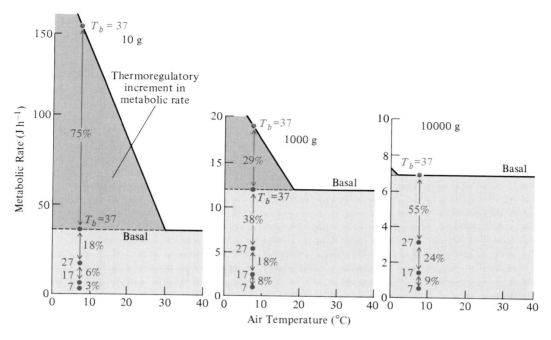

FIGURE 5–35 Predicted relationships between metabolic rate and temperature for euthermic and torpid mammals varying in mass from 10 to 10000 grams and showing the energy savings of torpor partitioned into that accruing from the abandonment of thermoregulation and that from the decline in body temperature.

can supercool to avoid freezing; for example, the torpid arctic ground squirrel can safely supercool to a core temperature of $-2.9°$ C (Barnes 1989). This suggests that the role of $T_{b,crit}$ is not in antifreeze protection. It more likely reflects the difficulty of arousal from torpor (see below).

There are a number of different patterns of torpor, with respect to its depth and duration. Some species have a shallow torpor ($T_{b,crit} > 12°$ C), whereas others have much deeper torpor ($T_{b,crit} < 12°$ C). There is, in fact, a continuum in the $T_{b,crit}$ of various torpid endotherms but $12°$ C seems to be a reasonable arbitrary breakpoint to distinguish shallow from deep torpor. Many birds and mammals become torpid on a circadian cycle, and arousal occurs within 16 to 24 hours of the onset of torpor. In other birds and mammals the torpor bout can be prolonged, lasting many days, weeks, or even months. Daily torpor is often of the shallow type (e.g., dasyurid marsupials, many cricetid rodents, shrews) but may be deep (e.g., the marsupial honey possum and pygmy possum, heteromyid rodents). Long-term torpor is generally deep (e.g., pygmy possums, squirrels) but may be shallow (e.g., some insectivores, carnivores).

Torpor is terminated by spontaneous arousal. The metabolic rate increases rapidly and T_b rises to the normal levels. The torpid endotherm must generate sufficient metabolic heat to arouse, but its maximal metabolic rate depends on T_b. For example, $VO_{2,max}$ (ml g^{-1} hr^{-1}) of a pocket mouse *Perognathus* is $0.38 T_b - 2.8$, and for a honey possum *Tarsipes* it is $0.45 T_b - 2.1$. Consequently, there is a minimum T_b below which arousal is not possible; this would be similar to, or lower than, the $T_{b,crit}$.

The arousal rate should depend on the initial T_b and body mass, since mass-specific metabolic rate is proportional to mass$^{-0.25}$. At a T_b of $20°$ to $25°$ C the arousal rate (°C min^{-1}) of torpid birds and marsupial and placental mammals is about 1.8 g$^{-0.3}$ (Figure 5–36). Large endotherms have longer arousal times than small endotherms (Table 5–14). Very large mammals (e.g., bears) have a very long arousal time. This, and their long entry time, indicate a minimum torpor duration of about 6 days for a bear. This would save perhaps 50% of the energy that would have been expended at normal T_b. The energy savings would be 80% if $T_b = T_a$ for the entire torpor period. Insects warm up at a rate proportional to their thoracic mass$^{0.084}$ rather than

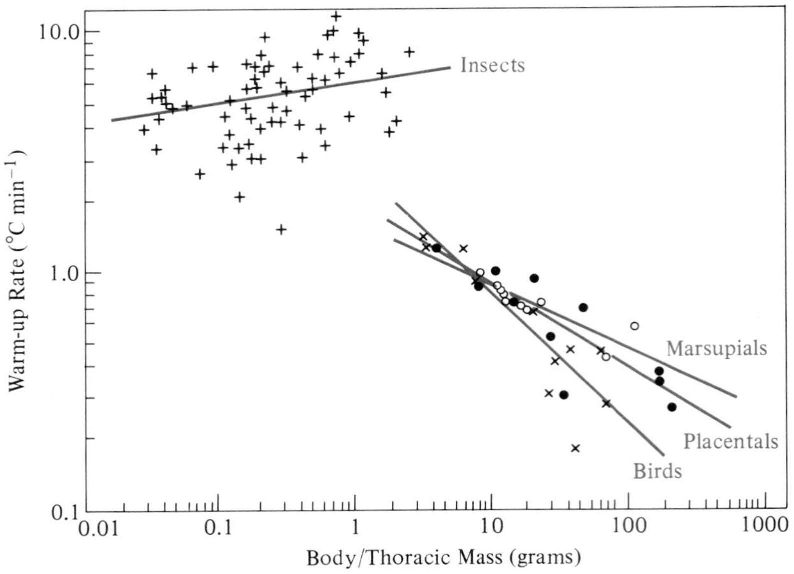

FIGURE 5–36 Rate of warming (*WR*; °C min^{-1}) for heterothermic insects during warm up and for torpid mammals (placentals and marsupials) and birds during arousal from torpor, as a function of body mass (grams).
Insects: $WR = 5.9 \, g^{0.084}$
Birds: $WR = 2.7 \, g^{-0.545}$
Marsupials: $WR = 1.5 \, g^{-0.269}$
Placentals: $WR = 1.9 \, g^{-0.351}$

being inversely proportional to mass, but their warm-up rates are similar to those predicted from the mammal/bird relationships extrapolated to the small mass of insects.

Adaptations to Heat. Most mammals and birds can readily tolerate moderate heat stress for long periods of time, but the T_a that constitutes moderate heat stress depends on their body mass. A large, well-insulated mammal would experience heat stress at a much lower air temperature (e.g., 25° C) than a small mammal or bird (e.g., >35° C). Thermal conductance is elevated at the upper end of the thermoneutral zone by nonmetabolic means such as postural change, increased peripheral blood flow, pilo or ptilo-depression, and a moderate increase in evaporative water loss. These mechanisms are no longer sufficient at $T_a > T_{uc}$ and more metabolically costly mechanisms are used, e.g., panting, gular flutter, sweating, or salivation. Body temperature may also increase to maintain a $T_b - T_a$ differential for passive heat loss (or minimize the differential for passive heat gain), but this also increases the metabolic heat production through a Q_{10} effect.

Mammals and birds are able to tolerate very high T_a (40° to 60° C) at least for short periods of time. Conduction, convection, and evaporation are avenues for heat dissipation only if $T_b > T_a$, but evaporation is the only mechanism for heat dissipation if $T_b < T_a$. Heat storage by a temporary increase in T_b (**hyperthermia**) is also an effective, though nonsteady-state mechanism employed by many mammals and birds when heat stressed.

Evaporative Heat Loss. The evaporative heat loss of endotherms increases exponentially with T_a, and evaporative heat loss can exceed metabolic heat production at a T_a of about 40° C for many birds (Figure 5–37). For birds, the maximum *EWL* (mg min^{-1}) and the corresponding maximum evaporative heat loss (J min^{-1}) are as follows (Calder and King 1974).

$$EWL_{max} = 1.03 \, g^{0.80}$$
$$EHL_{max} = 2.47 \, g^{0.80} \qquad (5.18)$$

The cutaneous evaporative water loss (CEWL) of birds and mammals is generally low because of the high resistance of their keratinized epidermis to water loss (see Chapter 16). Many mammals (but not rodents or birds) can markedly increase their CEWL by **sweating**. For example, humans can increase their CEWL from about 100 mg m^{-2} min^{-1} to about 23000 mg m^{-2} min^{-1} by sweating. Many other mammals also substantially increase their CEWL by sweating. Even some birds can moderately elevate their CEWL, presumably by increasing skin temperature and cutaneous blood flow.

Respiratory evaporative water loss (REWL) can be a major avenue for evaporative water loss, especially for nonsweating mammals and birds. Mammals can elevate REWL by panting (typically at a high, resonant frequency). Birds pant or **gular flutter**. Gular flutter is the movement of the moist gular (throat) region by the hyoid apparatus; it can occur in synchrony with panting (e.g., pigeons, ducks, geese, chickens) or can be independent of panting (e.g., cormorants, pelicans). Respiratory

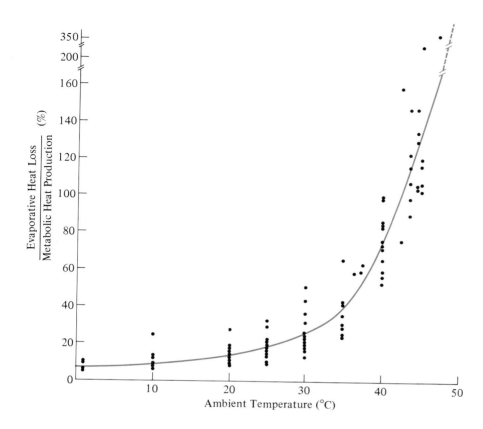

FIGURE 5–37 Evaporative heat loss per metabolic heat production at varying ambient temperatures for birds. *(From Calder and King 1974.)*

evaporative heat loss can be increased by reducing nasal countercurrent water and heat exchange (see above and Chapter 16). For example, dogs inspire through the nose and expire through the mouth when shallow panting. Many panting/gular fluttering birds do so with the mouth open. Breathing through the mouth rather than the nose also decreases the resistance to air flow.

Panting often occurs at a considerably higher frequency than resting respiration, with a rapid change from resting to panting frequency. For example, dogs abruptly increase respiratory frequency from 32 at rest to 320 min^{-1} when panting (f_{pant}/f_{rest} = 10). The panting frequency is similar to the resonant frequency of the lungs (317 min^{-1}). The pigeon and ostrich show a similar abrupt increase in respiratory frequency from rest to panting (29 to 612 min^{-1} and 4 to 40 min^{-1}, respectively). The pigeon's panting frequency is similar to the resonant frequency of its lungs (564 min^{-1}).

An increased respiratory ventilation during panting could disturb acid-base balance, causing a respiratory alkalosis (decreased blood pCO_2 and elevated pH; see Chapter 15). Some mammals and birds do experience a respiratory alkalosis, but there is little change in acid-base balance for many (see Figure 15–33, page 163). Respiratory alkalosis can be minimized, or avoided, if the increased ventilation is limited to the respiratory dead space rather than the respiratory exchange surface (alveoli in mammals, parabronchi in birds). For example, greater flamingos reduce their tidal volume when panting to 15% of the normal value, and their acid-base balance is not disturbed despite the 23 $\times$ increase in respiratory rate and 3 $\times$ increase in ventilation.

There are other possible sources of water for evaporative cooling. Some mammals (e.g., rodents and marsupials) salivate profusely when heat stressed. Some birds (e.g., the wood stork, turkey vulture, black vulture) urinate on their legs (urohidrosis) when heat stressed.

Heat Storage. Virtually all mammals and birds become hyperthermic at high T_a. This maintains a $T_b - T_a$ gradient for passive heat loss at high T_a, and minimizes the gradient for heat gain when $T_b < T_a$, but it adds to the metabolic heat load.

Hyperthermia can confer an important non-steady-state thermal advantage. Its potential significance to heat balance can be estimated as follows. An increase in T_b of 1° C absorbs about 3.5 J

g^{-1}. This is equivalent to a fractional dissipation of the hourly metabolic heat production (h^{-1}) of the following.

$$\text{Mammals: } 0.055\ g^{0.24}$$
$$\text{Nonpasserine birds: } 0.038\ g^{0.28} \qquad (5.19)$$
$$\text{Passerine birds: } 0.023\ g^{0.28}$$

Thus, a mammal can store more of its metabolic heat production by hyperthermia than a nonpasserine bird, which can store more of its metabolic heat production than a passerine bird. A small mammal (or bird) stores relatively less of its metabolic heat production by a 1° C hyperthermia than would a large mammal (or bird). A 10 g mammal can store 0.10 of its MHP by a 1° C hyperthermia over one hour, compared to 0.50 for a 10 kg mammal and 1.28 for a 500 kg mammal. Consequently, small mammals and birds use hyperthermia only for short-term tolerance of high T_a but larger animals can use hyperthermia for longer-term tolerance. For example, the antelope ground squirrel (100 g) is hyperthermic, with T_b up to 43° C for short periods then it returns to its burrow to cool; this cyclic hyperthermia can be repeated a number of times during the day (Figure 5–38A). In contrast, the camel (260 kg) takes a day to heat from a nighttime T_b of <35° C to a late afternoon T_b of >40° C. This daily hyperthermia cycle of the camel confers a number of advantages. First, it reduces the amount of water that must be evaporated to prevent, or minimize, changes in T_b. Second, a higher T_b facilitates heat loss to the environment and thereby minimizes the environmental heat load. Camels dehydrated by water deprivation have a more pronounced daily cycle in T_b compared to hydrated camels. This results in (1) a greater amount of heat stored in the body tissues during the day, (2) decreased evaporative water loss, and (3) reduced heat gain from the environment (Figure 5–38B).

Heat storage by hyperthermia is also important during exercise when metabolic heat production can exceed the capacity for heat dissipation. For example, gazelles running at 6 km h^{-1} store 8% of their metabolic heat by hyperthermia, but at 20 km h^{-1} the heat storage rises to 77% of metabolic heat production.

Brain Temperature Regulation. Hyperthermia may be an effective strategy for tolerating high temperatures and minimizing evaporative water loss, but it sometimes results in very high T_b's (>42° C) that could compromise the function of some body tissues. For example, the human brain tolerates temperatures up to about 40.5° C, but human core temperature may exceed 42° C (e.g., during a mara-

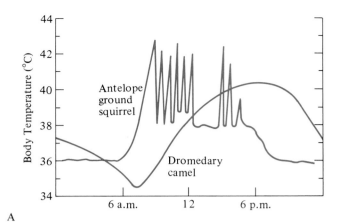

A

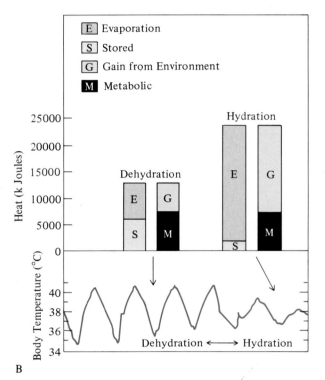

B

FIGURE 5–38 (A) Schematic representation of the use of short-term hyperthermia by the antelope ground squirrel *Ammospermophilus* and daily hyperthermia by the dromedary camel *Camelus*. **(B)** Effect of dehydration and hydration on daily fluctuations in body temperature of the dromedary camel *Camelus*, and a heat budget showing the magnitude of heat stored by hyperthermia, dissipated by evaporation, heat gained from the environment, and metabolic heat production. *(From Bartholomew 1964; Schmidt-Nielsen et al. 1957.)*

thon run). Gazelles and some other mammals experience even higher T_b values.

Many mammals and birds precisely regulate their brain temperature (T_{br}) at a lower value than body

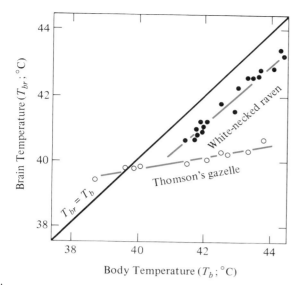

A

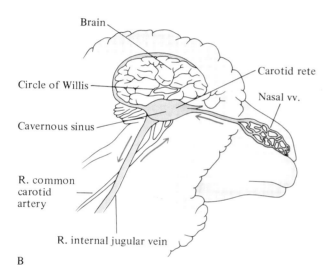

B

FIGURE 5–39 (A) Brain temperature of the white-necked raven and Thomson's gazelle in relation to body temperature. **(B)** Carotid rete in the brain of a sheep allows countercurrent heat exchange between cool blood draining from the nasal mucosa and warm arterial blood passing to the brain. *(From Kilgore, Bernstein, and Hudson 1976; Taylor and Lyman 1972; Cabanac 1986.)*

temperature during hyperthermia (Figure 5–39A). This capacity is not limited to mammals and birds, but is also observed in reptiles and ectothermic flying insects. The mechanism for T_{br} regulation by mammals and birds involves evaporative cooling of blood in the nasal mucosa and subsequent heat exchange between the cool venous blood draining the nasal mucosa and warm arterial blood passing

to the brain. In many mammals, the internal carotid artery forms a carotid rete of small arteries where it passes a large venous sinus, the sinus cavernosus (Figure 5–39B). Humans lack a carotid rete and do not pant when heat stressed. Nevertheless, there is probably a significant cooling of venous blood draining from the face (by sweating), and countercurrent heat exchange may occur between cool venous blood at the sinus cavernosus and internal jugular vein and warm arterial blood in the internal carotid artery. Countercurrent heat exchange in the bird brain occurs in the opthalmic rete, a network of small arteries and veins near the eye. The opthalmic rete may also exchange oxygen and carbon dioxide between the venous blood draining the nasal mucosa and arterial blood traveling to the brain (Bernstein, Duman, and Pinshow 1984).

Estivation. A variety of small fossorial mammals **estivate** during hot, dry periods. They are able to minimize thermal stress and evaporative water loss by withdrawing to the relatively cool and humid microclimate of their burrows, but prolonged periods of inactivity result in both food and water deprivation. In some species, this stress is avoided by shallow torpor cycles that occur at relatively high burrow temperatures (20° to 30° C). This shallow, summertime torpor, called estivation, is physiologically similar to winter torpor.

The cactus mouse *Peromyscus eremicus* becomes torpid during winter in response to cold stress and food restriction, and during the summer it estivates in response to either food or water restriction or negative water balance (MacMillen 1965). Estivation by the cactus mouse is characterized by relatively high burrow temperatures (about 20° C) and a high minimal T_a from which they can arouse (about 16° C). Other rodents such as ground squirrels (*Citellus*) and kangaroo mice (*Microdipodops*) estivate during the summer; kangaroo mice will estivate at a T_a as high as 28° C and as low as 5° C.

Ontogeny of Thermoregulation. Newborn and young mammals and birds generally have a poorer thermoregulatory capacity than adults. This is in part due to their smaller size, hence higher surface:volume ratio. Clearly, the young of smaller species will have more difficulty thermoregulating than the young of larger species. The generally poor thermoregulatory capacity of newborn and young mammals and birds is bolstered by a number of behavioral responses of the young (huddling) and the adults (nest building, brooding, shading).

Precocial young are relatively large at birth, have a well-developed insulation (fur or down feathers), can move easily, and rapidly thermoregulate at adult capacity. For example, Western gull chicks (*Larus occidentalis*) are fully covered by down feathers and can move about and behaviorally thermoregulate (e.g., seek shade) within minutes of hatching. Newborn guinea pigs (that weigh 60 to 100 g) have a good fur coat, their eyes are open, and they can readily move about; they can also effectively thermoregulate over a wide range of T_a.

Altricial young tend to be small, naked, and totally dependent on their parents for survival. They are generally unable to thermoregulate by physiological or behavioral means, and rely on their parents for elevation and regulation of T_b. The young of small mammals and birds (adults < 20 g) inevitably are altricial because their very small size would confer a prohibitively high metabolic rate if they were endothermic. For example, many rodents and passerine birds have altricial young.

The development (ontogeny) of thermoregulation by altricial young is of particular interest. For example, the chicks of the masked booby *Sula dactylatra* are naked and essentially ectothermic, but their T_b is regulated at about 38° C by brooding and other behavior of the parents. They essentially have no capacity to thermoregulate until their mass exceeds about 200 g. The vesper sparrow *Pooecetes gramineus* has much smaller young (about 2 g) that are also altricial. For the first four days they are ectothermic but rapidly develop endothermy by about seven days. The small dasyurid marsupial *Dasyuroides byrnei* has extremely altricial young, as do all marsupials. The young remain in the mother's pouch for about 30 days. They are essentially ectothermic until 55 days, when they are left in the nest. Endothermy develops slowly over the next 30 or so days; the young are not fully endothermic until about 90 days old. Many placental mammals also have altricial young. The rabbit is 50 to 70 g at birth and has a sparse covering of fur. By ten days, it is better insulated and weighs 200 g and is a better endotherm than when newborn. This is primarily because of its better insulation rather than its endogenous metabolic capacity (which is actually lower per gram than that of the newborn).

There are relative advantages and disadvantages to both precocial and altricial development. The altricial strategy allows a shorter gestation period and smaller birth size for mammals, and smaller egg size for birds. Altricial species consequently tend to have larger litter/clutch sizes. The maintenance metabolism of altricial young is low because they are essentially ectothermic, so more of the ingested energy is channeled into production (growth). For example, vesper sparrow young grow at 40% mass day^{-1} while "ectothermic" (first four days). The parents provide the energy for thermoregulation, at little additional cost to their energy budget.

Reptiles

We might expect some reptiles to be endothermic because birds and mammals evolved independently from reptiles, so endothermy could well have also evolved in other reptilian groups. Furthermore, there must have been an evolutionary continuum from ectothermic reptiles to endothermic mammals and birds, so we might expect there to be some extant reptiles derived from either of these transitional endothermic lineages. However, there are few bona fide endothermic reptiles. The female Indian python *Python molurus* can regulate its T_b at about 5° to 7° C above T_a (Figure 5–40) when brooding its clutch of eggs; the T_b is elevated above T_a by shivering thermogenesis (Hutchison et al. 1966; van Mierop and Barnard 1978). The brooding diamond python *P. spilotes* is able to maintain a high T_b of about 31° C when brooding (Slip and Shine 1988).

There is not a hard-and-fast criterion distinguishing endothermy from ectothermy, and this is readily apparent for large reptiles. The extent to which reptiles can elevate T_b above T_a depends on two factors: their rate of endogenous heat production and their thermal conductance. For resting reptiles, metabolic heat production is 1.5 g$^{0.8}$ J hr^{-1} at $T_b = 20°$ C (see Chapter 4). Thus, heat production increases with large size. The thermal conductance of large reptiles is 2603 g$^{0.148}$ J g^{-1} hr^{-1} °C^{-1} for large reptiles (> 10 kg). We can approximately calculate $(T_b - T_a)$ as VO$_2$/C, or

$$T_b - T_a = 0.00058 \text{ g}^{0.652} \qquad (5.20)$$

This is a minimum estimate of $T_b - T_a$ because the metabolic rate can be elevated above resting by, for example, activity. How does the calculated $(T_b - T_a)$ compare with actual observations of T_b in reptiles? There is a clear trend for $(T_b - T_a)$ to be higher in large reptiles, even aquatic ones (Table 5–15).

Dinosaurs, the largest reptiles, are estimated to have weighed up to 6000 kg. There is no doubt that large reptiles (e.g., weighing over 100 kg) would be homeothermic because of their massive thermal inertia (Spotila et al. 1973), but there is considerable conjecture and debate concerning whether any dinosaurs were endothermic in the sense that mammals and birds are endotherms. That is, did any dinosaurs

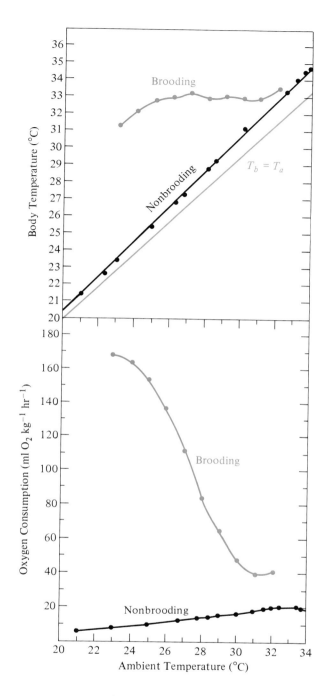

FIGURE 5–40 Relationship for body temperature and metabolic rate with ambient temperature for the non-brooding and brooding female python. *(Modified from Van Mierop and Barnard 1978.)*

TABLE 5–15

Differential between body temperature (T_b) and ambient temperature (T_a) observed for various large reptiles, and the predicted $T_b - T_a$ differential for resting reptiles in air at 20° C; for aquatic animals the predicted $T_b - T_a$ is less than, or similar to, the predicted value, depending on the body mass. Dinosaurs are shown in **bold**.

	Mass (kg)	T_a (°C)	T_b–T_a (°C)	Predicted T_b–T_a (°C)
Monitor lizard	7	25	0.2–0.4	0.2
Monitor lizard	12	25	0.2–0.5	0.3
Moschorhinid	20			0.4
Monitor lizard	35	25	0.2–0.6	0.5
Pristerignathid	50			0.7
Hawksbill/ Ridley turtles	120	28	1–3	1.2
Green turtle	127	20–30	3	1.2
Dimetrodon	150			1.4
Galapagos tortoise	170	20–30	4.1	1.5
Leatherback turtle	417	7.5	3–18	2.6
Tyrannosaur	2000			7.4
Allosaur	3000			9.6
Ceratopsid	4300			12.2
Hadrosaur	5600			14.5

physiological evidence is difficult to glean from the scanty fossil record. Advocates for the theory that dinosaurs were endothermic argue that some had an erect gait, had a Haversian bone histology, had predator-prey ratios typical of carnivorous mammals rather than carnivorous ectotherms, had a large brain, and lacked a pineal eye (as do most mammals and birds; Bakker 1971, 1972; Benton 1979). The numerous antagonists of the endothermic dinosaur theory have refuted, or at least brought into serious doubt, the validity of most of the evidence in favor of endothermy, but they generally concede that these large reptiles were at least homeothermic (McGowan 1979; Thomas and Olson 1980).

Fish

Some large and active fish can produce and retain sufficient metabolic heat to elevate tissue temperatures considerably above T_{water}. For example, bluefin tuna are large, actively swimming fish with a high metabolic rate; their muscle temperature can be 10°

have high a resting VO_2 and use endogenous heat production to precisely regulate a high T_b? Whether any dinosaurs were endothermic is not entirely a subject of speculation, although anatomical and

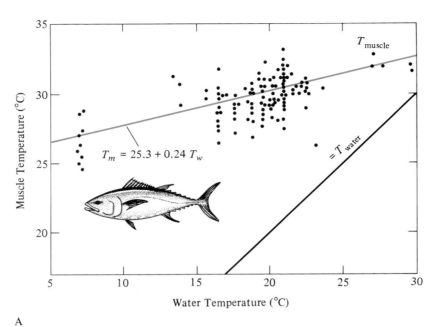

A

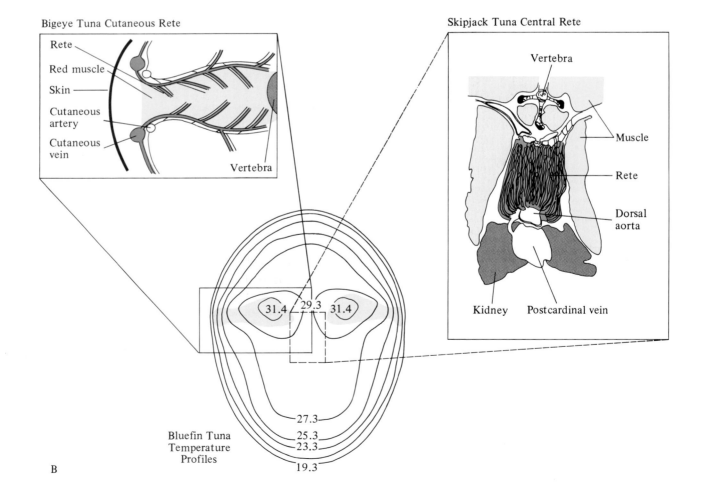

B

FIGURE 5–41 (A) Relationship between muscle temperature and water temperature for bluefin tuna (*Thunnus thynnus*). **(B)** Temperature distribution in a bluefin tuna and arrangement of the cutaneous (bigeyed tuna) and central (skipjack tuna) countercurrent heat exchange retia. *(Modified from Stevens and Neill 1978; Carey et al. 1971; Stevens, Lam, and Kendall 1974; Carey and Teal 1966.)*

C or more above T_{water} (Figure 5–41A). Other tuna are also able to regulate T_{muscle} a considerable amount above T_{water}. The visceral temperature (e.g., liver) of some tuna is also elevated above T_a as are the brain and eyes (although they are not as warm as muscle). The $(T_b - T_{water})$ is highest at low T_{water}, suggesting regulation of heat retention in the muscle.

Tuna do not regulate their metabolic heat production to regulate T_{muscle} because the metabolic cost of swimming is not temperature dependent, and so their metabolic rate is essentially independent of T_{water}. This is in contrast to endothermic mammals and birds that regulate their metabolic rate to maintain T_b constant. Rather, tuna retain metabolic heat in their swimming muscle by countercurrent heat exchange in a variety of circulatory retia (Figure 5–41B). The very large tuna (e.g., bluefin) tend to have cutaneous retia, whereas smaller tuna (e.g., yellowfin) tend to have central retia (Stevens and Neill 1978). The cutaneous retia consist of a cutaneous artery and vein; the arterioles form a dense and continuous sheet that enters the muscle mass and interdigitates with a network of venules draining blood from the muscle to the cutaneous vein. The brain, eye, liver, and gut vasculature often have complex retia for countercurrent heat exchange and retention of heat in the viscera. The central retia of the smaller tuna is located beneath the vertebral column. Cool arterial blood from the gills passes from the dorsal aorta through a rete of small arteries then to segmental arteries; warm venous blood from the swimming muscle drains into segmental veins and then passes through the venous rete vessels to exchange heat with the arterial blood before entering the postcardinal vein.

A variety of other fish are also spatial endotherms, i.e., they maintain a high temperature in specific tissues. The mako, great white, and porbeagle sharks have a high visceral temperature, e.g., the stomach temperature of a mako shark may be up to 8° C warmer than T_{water}. These lamnid sharks have a peculiar routing of arterial blood through a paired vascular rete anterior to the liver. An enlarged pericardial artery forms a rete of small arteries in the lumen of a large venous space. The swordfish has a countercurrent heat exchange rete to regulate brain temperature about 4.7° C warmer and eyes 3.4° C warmer than T_{water} (Figure 5–42). There is also a mass of brown tissue associated with one of the extrinsic eye muscles; this tissue has a high density of mitochondria and cytochromes (hence the brown color) and appears to function in thermogenesis, i.e., it is similar in function and structure to brown fat of mammals.

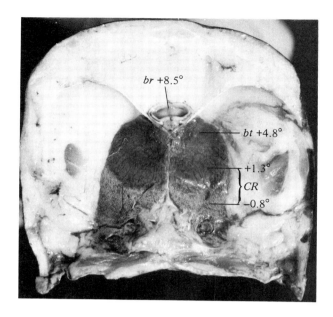

FIGURE 5–42 The brain of a swordfish (*br*) is surrounded ventrally by brown thermogenic tissue associated with the extrinsic eye muscles (*bt*). A carotid rete (*CR*) retains metabolic heat in the brain tissue for brain temperature regulation. The approximate temperatures in different parts of the head are indicated. (*Courtesy of Dr. Frank Carey, Woods Hole Oceanographic Institute 1982.*)

Insects

Many insects are endothermic and capable of precise regulation of body temperature over a wide range of T_a. Most of these are spatial endotherms that regulate a constant thoracic temperature (T_{th}) but not abdominal temperature (T_{ab}). Thermoregulation by flying insects is accomplished by the regulation of heat loss from the thorax, rather than by the regulation of metabolic heat production. This is because the metabolic cost of flight is essentially independent of T_a. Thus, endothermic insects regulate heat loss not heat production (like endothermic fish).

Perhaps the most striking example of an endothermic insect is winter-flying moths, which are active at about 0° C (Heinrich 1987). The small noctuid moth *Eupsilia* has a high T_{th} of about 30° C when flying at subzero T_a (Figure 5–43A). This is remarkable because it only weighs 100 to 200 mg. The capacity to thermoregulate at low T_a is a consequence of its effective thermal insulation on the thorax and head, and the thermal isolation of the

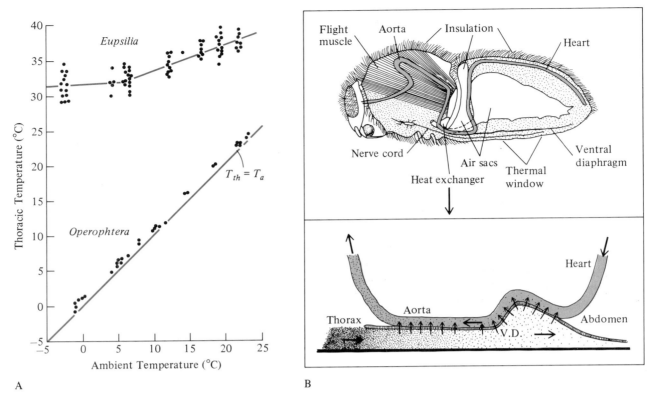

FIGURE 5–43 (A) Relationship between thoracic temperature and air temperature for two winter-flying moths, the noctuid *Eupsilia* and the geometrid *Operophtera*. **(B)** Diagrammatic cross section of a bumblebee (*Bombus vosnesenskii*) showing the thermal isolation of the thorax from the abdomen by air sacs, and the countercurrent heat exchange between blood flowing to the thorax and returning to the abdomen. *(From Heinrich and Mommsen 1985; Heinrich 1987; Heinrich 1976.)*

thorax from the abdomen by air sacs. The only significant connections between the thorax and abdomen are the esophagus, ventral nerve cord, and ventral aorta. There is countercurrent heat exchange between cool hemolymph entering the thorax from the abdomen and warm hemolymph returning to the abdomen. There is also a thoracic heat exchanger consisting of a vertical hairpin loop of the aorta in the thoracic muscle. The sophisticated thermoregulatory system of *Eupsilia* is contrasted with the winter-flying moth *Operophtera*, which thermoconforms. This small (<10 mg) geometrid moth has a T_{th} only a few degrees above T_a even at 0° C. It is able to fly at such low temperatures because of its low energy cost of flight; its wing loading per mass is only 3.2 mg cm^{-2}, compared with 43 mg cm^{-2} for *Eupsilia*. *Operophtera* does not appear to have any enzymatic/metabolic specializations of the thoracic muscle to facilitate functioning at low T_a.

The bumblebee has a petiolar countercurrent heat exchange that is similar to that of the winter-flying moth *Eupsilia* (Figure 5–43B). This minimizes heat loss from the thorax to the abdomen at low T_a but can be bypassed at high T_a to dissipate thoracic heat and prevent an excessive increase in T_{th}.

Many moths warm up prior to flight; they elevate T_{th} by shivering contractions of thoracic flight muscle until it is high enough to sustain flight. Bumblebee flight muscle has an equivalent capacity to mammalian nonshivering thermogenesis. The flight muscle contains two enzymes, phosphofructokinase (PFK) and fructose bisphosphatase (FBPase). PFK catalyzes the phosphorylation of fructose-6-PO$_4$ to fructose-1,6-diPO$_4$, and FBPase catalyzes the reverse reaction (Hochachka and Somero 1984). The net effect of this futile cycle is hydrolysis of ATP to ADP + P$_i$. This futile cycle can be initiated at low muscle temperatures (<30° C) to release heat and warm the thorax. When the muscles are sufficiently warm, the intracellular Ca^{2+} concentration is elevated; this inhibits FBPase and eliminates the futile thermogenic cycle.

The sphinx moth *Manduca* precisely regulates T_{th} at about 38° to 40° C while flying by using the abdomen as a heat sink at high T_a. Heat exchange between the thorax and abdomen is controlled by the nervous system. The abdominal heart responds to high T_{th} by increasing its pulsations, circulating hemolymph into and out of the thorax. Ligating the blood vessel compromises T_{th} regulation at high T_a.

A variety of other insects regulate a high body temperature by metabolic heat production during flight or while walking. For example, dung beetles can maintain a high T_{th} of 38° to 42° C while flying and rolling dung balls; tropical beetles can have T_{th} 4° to 16° C higher than T_a (the largest beetles have the highest $T_{th} - T_a$) during terrestrial activity; scarabid and cerambycid beetles can endogenously raise T_{th} by 5° to 7° C above T_a. The elephant beetle *Megasoma* is a large beetle (10 to 35 g), which can endothermically maintain T_{th} above T_a independent of locomotor activity, by a cyclic elevation of VO_2 rather than a sustained elevation of locomotory VO_2. Worker honeybees and incubating queen bumblebees also maintain homeothermy by nonlocomotory activity. Honeybees have the physiological capacity to elevate metabolic rate at low T_a and thereby maintain a high temperature of the hive or of a bee cluster. The relationship between VO_2 and T_a is essentially the same as that observed for an endothermic mammal or bird, and metabolic rate increases with cluster mass in the same fashion, and at an intermediate level, as the VO_2 of mammals and birds (Southwick and Heldmaier 1987).

Plants

Endothermy is not restricted to animals. Some plants produce sufficient metabolic heat to raise their floral temperature significantly above T_a. For example, the inflorescence of *Philodendron* can raise its temperature to 38° to 46° C at T_a of 4° to 39° C, by metabolic heat production of male sterile flowers (Nagy et al. 1972). The voodoo lily elevates its flower temperature by up to 22° C above T_a to volatilize its putrescent odor and attract insect pollinators. This is followed by a second phase of elevated (but lower) temperature when pollen is shed to warm the insect pollinators (Raskin et al. 1987). Some plants that flower in the snow also have a marked thermoregulatory capacity.

Evolution of Endothermy

We have seen essentially two different patterns of endothermy. Some large reptiles and fish, and many small insects, are endothermic as a consequence of activity, i.e., swimming, flying, or walking. Their metabolic rate is not regulated at varying T_a to maintain T_b constant. Homeothermy, if it is achieved by these animals, is by the physiological regulation of heat loss. In contrast, birds and mammals, brooding pythons, and some insects are endothermic by virtue of their physiological regulation of heat production (by shivering and nonshivering thermogenesis). The physiological regulation of heat loss is much less important.

The endothermic strategy of active insects and fish apparently evolved because locomotion produced sufficient metabolic heat that thermoregulatory strategies could be subsequently evolved. Elevated locomotory metabolism could well have preceded the evolution of thermoregulation. How did the latter endothermic strategy of birds and mammals evolve? It has also been suggested to have been the consequence of sustained activity, but there are several, although inconclusive, arguments against this. Birds and mammals rely on nonshivering thermogenesis, not shivering, for thermoregulation, except when cold stressed. Did nonshivering thermogenesis independently evolve as a precursor to endothermic homeothermy in birds and mammals, or did it independently evolve after shivering thermogenesis as an alternative mechanism? We have already seen that activity does not effectively substitute for thermoregulatory heat production in mammals, and so the hypothesis for the evolution of the endothermic strategy of birds and mammals from activity-derived metabolic heat is not compelling.

There are other hypotheses for the evolution of endothermy by mammals and birds. For example, mammalian homeothermy may have evolved in two steps (Crompton, Taylor, and Jagger 1978). First, small mammals (30 to 40 g) invaded the nocturnal niche. They regulated T_b at only about 25° to 30° C and their ($T_b - T_a$) gradient was probably only 10° C or less, so they did not require a marked capacity for thermogenesis. Tenrecs, which are nocturnal and have a low VO_2 and low T_b, may be an example of this thermal strategy. The second evolutionary step was a consequence of these nocturnal animals invading a diurnal niche; their T_b was raised to about 38° to 40° C to avoid the need to evaporatively dissipate water for thermoregulation of a low T_b (25° to 30° C) when subjected to a radiant heat load.

But, when and how did the mammalian metabolic machinery and insulation evolve? The metabolic rate of primitive mammals is the same as that of advanced mammals at equivalent T_b, i.e., the difference between "primitive" and "advanced" mammals is their T_b not their basic metabolic machinery (see Chapter 4). Why do living primitive mammals (including tenrecs) have essentially the same metabolic capacity and insulation as advanced mammals, if they represent the initial stage in the evolution of endothermy?

A second and very different scenario for the evolution of endothermy in mammals is the conversion of inertial homeothermy to endothermic homeothermy (McNab 1978). This theory begins with the reasonable assertions that the large reptilian ancestors of mammals were inertial homiotherms and the first endothermic mammals were small (shrew-sized). Homeothermy could only be maintained during this progressive reduction in size from reptile to mammal by the substitution of endothermic homeothermy for inertial homeothermy. This theory has the attraction that both insulation and increased basal metabolism could evolve gradually as body mass slowly declined from large reptiles to smaller mammals.

Advantages of a Constant Body Temperature. There are obvious advantages to endothermy and a constant T_b. An increased rate of enzymatic catalysis is a fundamental selective advantage to a higher body temperature. Enzymes can be adapted to function at low temperatures, but catalytic rates are nevertheless higher at elevated temperatures. A high T_b means that force and velocity of muscle contraction, and activity metabolic rate, are greater than at low T_b. A constant T_b allows enzymes to always be at their optimal temperature for catalysis. A high and stable T_b means that activity can be sustained irrespective of T_a, and so cold environments (nocturnal, high altitude, and latidude) can be exploited better by endotherms than ectotherms.

There are also costs to endothermic homeothermy. The principal disadvantage is the high energy expenditure for thermoregulation during periods of inactivity and low T_a. Endotherms must expend a greater fraction of their energy turnover on respiration rather than production (see Chapter 4).

There is an obvious evolutionary trend for the T_b of endothermic homeotherms to be regulated at successively higher values. This trend is apparent among mammals (e.g., monotremes and edentates have T_b of 30° to 32° C; cf. primates and lagomorphs with T_b of 38° to 39° C) and birds (e.g., ratites, 38° C;

cf. passerines, 42° C). Ectothermic thermoregulators also tend to have similar, high T_b levels while thermoregulating, e.g., many lizards, 38° to 39° C; walking beetles and flying insects, 35° to 40° C. Why have these diverse thermoregulating animals evolved preferred T_b values in this general range of 35° to 40° C?

One disadvantage of a low preferred T_b is that it is more likely for T_a to approach or exceed T_b, thereby requiring evaporative cooling for T_b regulation. A higher T_b will minimize the likelihood of thermal stress. A higher T_b also increases the catalytic rate of reactions, up to a point. Enzymes can be adapted to higher temperatures (even up to 80° to 90° C) and so the ultimate limit to enzymatic/protein function is certainly not 40° to 45° C, even in higher eukaryote animals. What determines the maximum tolerable T_b? Why haven't animals evolved preferred T_b considerably higher than 35° to 40° C? Will they evolve even higher $T_{b,pref}$ values in the future millennia?

There are a number of disadvantages to a T_b that is too high. If $T_b \gg T_a$ a marked endogenous heat production is required to regulate T_b. If T_b did decline close to T_a (e.g., daily torpor) then this reduction in T_b would dramatically compromise the structure and function of enzymes and membranes. Maximal metabolism during torpor might be so reduced that arousal to a high T_b would be impossible, or at least energetically costly and slow. A very high T_b would require a correspondingly high energy acquisition. The physiological advantages of having a T_b of 50° C might not be sufficient compared to having a T_b of 40° C to justify the additional energy demands.

Other types of arguements have been offered to explain why T_b's often fall in the range 35° to 40° C. At about 37° C, any change in temperature will alter the free enthalpy of activation (ΔH^*) and free entropy of activation (ΔS^*) in an offsetting fashion, so that the free energy of activation (ΔG^*) is approximately compensated to a constant value (Hochachka and Somero 1984). The temperature at which ΔH^* and ΔG^* exactly offset each other is the compensation temperature (see below). Compensation temperatures for various enzymes are often about 35° to 55° C. Thus, the preferred T_b for endothermy may be adapted to minimize the overall effects of temperature on enthalphy and entropy for activation.

A less compelling arguement for the T_b range 35° to 40° C is provided by thermodynamic properties of water (Calloway 1976). A temperature of 37° C is consistent with some thermodynamic properties of water, e.g., the specific heat of water is minimal at about 35° C; 38.5° C is the halfway temperature between the temperature of minimal thermal expansivity (4° C) and maximum (100° C); 40° C is the halfway temperature for kinetic reaction rates between 0° and 100° C. However, the physiological significance to an animal of such minimal or halfway temperatures is not clear. Even less compelling are observations such as the difference between the freezing and boiling points of water, divided by e (2.718) is 36.8° C, and the freezing point of water (in °K) divided by e^2 is 37.0° C.

Fever

Mammals, a wide variety of other vertebrates, and many invertebrates have a **fever** response (Table 5–16). Fever is an important and apparently general response of animals in which the thermoregulatory setpoint temperature is elevated by endogenous and exogenous pyrogens. For example, mammals generally have a rapid increase in T_b after the administration of bacterial toxins. Fever increases the hypothalamic setpoint and also the setpoint for onset of cutaneous vasodilation (for heat dissipation) and shivering (for heat production; e.g., rabbit). The hypothalamic thermostat is thought to be reset by a small protein, interkeukin, that is released from white blood cells in response to a variety of pathogens, such as bacteria and viruses.

Fever presumably has beneficial effects, especially as it is such a phylogenetically diverse phenomenon (Kluger 1979). The increase in T_b may enhance the activity of the immune system (see Chapter 15), e.g., the mobility and activity of white blood cells, stimulation and effect of interferon production, and activation of T-lymphocytes. Lizards (*Dipsosaurus*) injected with bacterial pyrogens have a higher survival at higher T_b (42° C) compared with lizards at lower T_b (e.g., 40°, 38°, 36°, and 34° C), suggesting that the higher T_b is advantageous.

Cryogens have the opposite effect as pyrogens, i.e., they lower the thermoregulatory setpoint. Mammals, including man, produce endogenous cryogens that induce a mild and transient hypothermia if injected into other mammals. For example, injection of human cryogens (present in the urine) can decrease the T_b of rabbits by 0.5° C.

Acclimation and Acclimatization

Temperature affects the rates of most physical, biochemical, and physiological functions, generally with a Q_{10} of 2 to 3. However, the biochemistry and

TABLE 5–16

Effects of fever on body temperature (T_b) of invertebrates and vertebrates.

	Normal Preferred T_b	Fever T_b	$\triangle T$
Dog *Canis*	38.2	39.4	1.2
Monkey	38.9	40.1	1.2
Rabbit	39.5	40.8	1.3
Beetle *Onymacris*	33	34.5	1.5
Desert iguana *Dipsosaurus* HBTS	41.0	42.7	1.7
Chimpanzee *Pan*	38.3	40.0	1.7
Pigeon *Columba*	39.7	41.5	1.8
Crayfish *Cambarus*	22.1	23.9	1.8
Bluegill sunfish *Lepomis*	30.1	32.2	2.1
Desert iguana *Dipsosaurus* LBTS	37.4	39.6	2.2
Large-mouth bass *Micropterus*	29.6	31.9	2.3
Frog *Hyla*	25.5	27.9	2.4
Tadpole *Rana*	28.5	31.2	2.7
Cockroach *Gromphadorhina*	32.3	35.9	3.6
Man *Homo*	37.4	41.3	3.9
Shrimp *Penaeus*	31	35.5	4.5
Lobster *Homarus*	16	20.7	4.7
Frog *Rana*	25.0	30.3	5.3
Horseshoe crab *Limulus*	27.0	33.0	6.0
Leech *Nephelopsis*	20.5	30.0	9.5
Scorpion *Androctonus*	24.8	28.8–39.8	≤15
Scorpion *Buthus*	25.1	25.1–43.1	≤18

physiology of animals are not necessarily at the mercy of the thermal environment and their T_b because biochemical and physiological rates can be adjusted to compensate for variations in temperature. Such compensation for temperature is called **acclimatization** if it occurs in nature, and **acclimation** if it is induced in the laboratory. For example, the $T_{b,pref}$ of many fish is lower during the colder parts of the year than during the warmer parts of the year (acclimatization). Fish kept in the laboratory at various water temperatures show a similar decrease in $T_{b,pref}$ in the colder water (acclimation). Many biochemical and physiological processes show thermal acclimation/acclimatization, e.g., enzyme reaction rates, heart rate, metabolic rate, respiratory rate, preferred T_b, CT_{min}, and CT_{max}. There is also a suite of morphological, enzymatic, and physiological adaptations for thermal acclimation in plants (Chabot 1979).

In general, acclimation and acclimatization maintain similar rates at varying temperatures, i.e., the rate is the same for a cold-acclimated animal at its T_a and a warm-acclimated animal at its higher T_a. However, there are a number of different patterns in acclimation/acclimatization (Precht 1958; Prosser 1958). The types of acclimation/acclimatization recognized by Precht are summarized in Figure 5–44A. Type 2, or perfect acclimation, results in a rate after thermal acclimation that is exactly the same as the initial rate. We might expect this to be the most prevalent, and ideal, type of acclimation but there are many circumstances in which perfect acclima-

tion is not optimal. Type 3, or partial, acclimation occurs if an acclimation response partially returns the rate to the preacclimation value. There are many examples of incomplete acclimation, perhaps reflecting the biochemical difficulty in completely compensating for a change in temperature, or the physiological unnecessity for perfect acclimation. Type 4 is no acclimation; there is a lack of any acclimation response to a change in temperature. Type 5 acclimation is inverse acclimation. Such an acclimation pattern might be adaptive, for example, during winter dormancy when the metabolic rate is depressed by the lowered temperature and is further depressed by the inverse acclimation. For type 1, or over-acclimation, the rate after acclimation is higher than the initial rate after acclimation to a

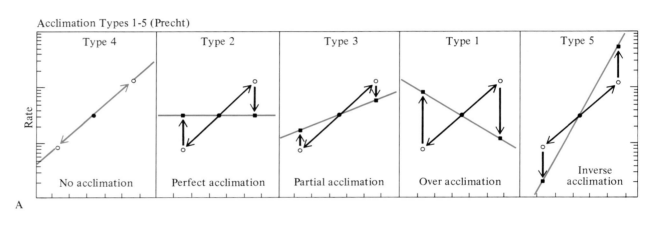

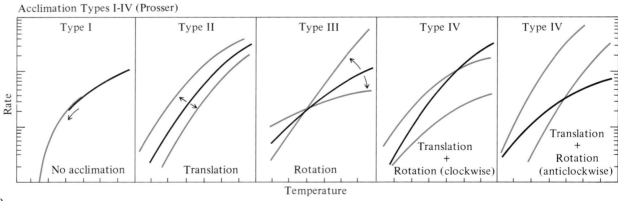

FIGURE 5–44 (A) Patterns of acclimation as defined by Precht. The solid circles indicate the initial rate/temperature and the open circles indicate the initial, acute change in rate/temperature (thin arrows); the solid squares and colored lines indicate the rate/temperature after thermal acclimation (thick arrows). **(B)** Patterns of acclimation as defined by Prosser. The solid lines indicate the initial rate-temperature curve; the colored lines indicate the rate-temperature curve after thermal acclimation. *(Modified from Precht 1958; Prosser 1958.)*

lower temperature (or lower than the initial rate after acclimation to a higher temperature). Such an acclimation response might be useful, for example, in limiting the effect of high temperature ($>T_{b,\text{pref}}$) on reaction rates.

The acclimation scheme of Prosser (Figure 5–44B) considers the effect of acclimation over a range of temperatures, whereas the Precht classification scheme considered the change between only two temperatures. The relationship between rate over a variety of temperatures may be unaltered by acclimation (Type I, no effect ≡ Precht Type 4), may move up or down (Type II, translation ≡ Precht Type 1, 3, 4, or 5), may rotate about a constant rate at one temperature (Type III), or may translate and rotate (Type IV).

Many invertebrates show thermal acclimation (Cloudsley-Thompson 1970). The CT_{max} of the earthworm *Pheretima* increases by 0.3° C per 1° C rise in acclimation temperature. The slug *Arion circumscripta* shows metabolic acclimation. Many arthropods also show various types of thermal acclimation. The CT_{min} and CT_{max} of two isopods are influenced by acclimation temperature (Table 5–17). Thermal acclimation may occur quite rapidly (<24 hr). Cockroaches (*Blatella*) transferred from a warm to a cold environment show almost complete thermal acclimation within a few to about 24 hours for transfer from 25° to 15° C, but require longer to acclimate from 35° to 25° and 35° to 15° C (Figure 5–45A). Thermal acclimatization may also occur on a latitudinal or altitudinal gradient. For example, there are seasonal changes in the type of the third chromosome in the Californian fruit fly *Drosophila pseudoobscura* (the SI, AR, and CH types). At 30° C, ST-type pupae have a high survival and ST-adults have a greater longevity than CH flies. A laboratory population of flies showed a shift in frequency to 70% ST when transferred from 17° to 25° C. Northern populations (i.e., cooler climates) of the European fruit fly *D. funebris* are more resistant to lower temperatures, and southern populations (i.e., warmer climate) are more resistant to higher temperatures. Eggs of the tortricid moth *Acrolite* from eastern Norway accumulate more glycerol (for freezing tolerance) than do eggs of western Norway moths that experience a milder climate. Orbatid mites from West Africa have a higher CT_{max} (37° C) than mites from North America (30° C).

Fish generally show a substantial metabolic acclimatization, e.g., comparing species from cold and warm climates, and also for individuals acclimatized (or acclimated) to various temperatures. For example, temperate fish have a similar metabolic rate as arctic and antarctic fish despite marked differences in their ambient temperature (Figure 5–45B). However, tropical fish tend to have higher metabolic rates than temperate fish, showing more of a Q_{10} effect than an acclimatory compensation. There is also marked acclimation in both CT_{min} and CT_{max} of fish. For example, the CT_{max} of the salmon *Oncorhynchus keta* varies from about 22° to 24° C at acclimation temperatures of 0° to 40° C; CT_{min} varies from 0° to 7° C. A temperature polygon showing similar changes in CT_{min} and CT_{max} for *O. nerka* was described in Chapter 2.

Amphibians and reptiles generally show thermal acclimation of metabolic rate, CT_{min} and CT_{max}. For example, there is thermal acclimation of CT_{max} in temperate and tropical anuran amphibians, although CT_{max} is higher in the tropical species at equivalent acclimation temperatures. There is also a general trend for CT_{min} to decrease for amphibians at higher (colder) latitudes.

Rates of acclimation are quite variable but generally follow a hyperbolic curve with complete acclimation in two to four days. The thermal response ratio (TRR) is the change in CT_{max} per change in acclimation temperature ($\Delta CT_{\text{max}}/\Delta T_a$). The TRR and the time for 50% acclimation (1/2 AT) are commonly used to describe the time course and magnitude of acclimation. For amphibians, the TRR varies from about 0.07 to 0.44 and 1/2 AT varies from 0.12 to 2.8 days (Table 5–18). Both CT_{max} and CT_{min} of reptiles shows thermal acclimation. In turtles, there is a daily variation in CT_{max}. The $T_{b,\text{pref}}$ of lizards can also vary with time of day or season, with age, and with hormonal and physiological state.

Endotherms also show thermal acclimation of many physiological and biochemical variables, although in response to variation in T_a rather than T_b. For example, acclimation responses include

TABLE 5–17

Critical thermal minimum *(CT_{min})* and critical thermal maximum *(CT_{max})* temperatures for the terrestrial isopods *(Porcellio laevis, Armadillidum vulgare)*, as a function of acclimation temperature. *(Data from Edney 1964.)*

Acclimation Temperature	Porcellio laevis		Armadillidum vulgare	
	CT_{min}	CT_{max}	CT_{min}	CT_{max}
10° C	−2.4° C	37.4° C	−2.7° C	38.3° C
30° C	5.5° C	41.6° C	3.0° C	41.6° C

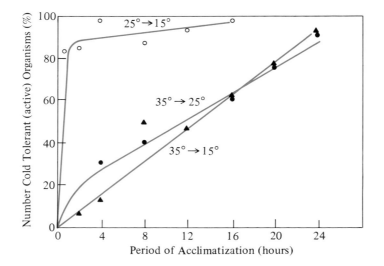

A

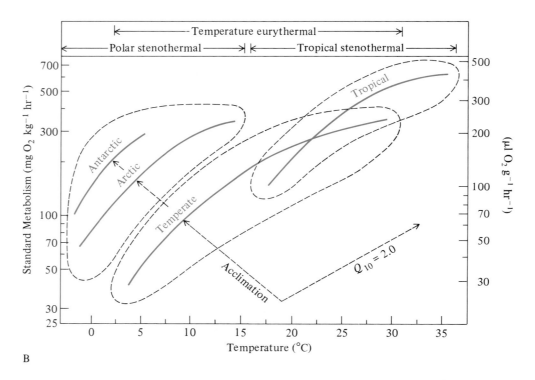

B

FIGURE 5–45 (A) Time course for acclimation of cold tolerance in the cockroach *Blatella* from high ambient temperature (25° or 35° C) to a lower ambient temperature (25° or 15° C). **(B)** Thermal acclimation of metabolic rate for polar, temperate, and tropical fish. *(From Colhoun 1960; modified from Brett and Groves 1979.)*

nonshivering thermogenesis, fur and feather thickness and color, nerve conduction velocity, ability to become torpid and $T_{b,\text{crit}}$ during torpor, the melting point of lipids in hypothermic limb extremities, sweating rate, and lowering of basal metabolic rate.

Biochemical Adaptations to Temperature

Thermal acclimation is an important physiological phenomenom that allows cold-adapted animals to have "normal" biochemical and physiological rates

TABLE 5–18

The magnitude and rate of thermal acclimation in critical thermal maximum *(CT*max*)* for salamanders. The magnitude of thermal acclimation in CT_{max} is indicated as the thermal response ratio (change in CT_{max} per change in acclimation temperature $\Delta CT_{max}/\Delta T_{acc}$); the rate of acclimation is indicated as the half time for attainment of the new CT_{max} ($\frac{1}{2}$ AT; days). *(Data from Claussen 1977.)*

	Acclimation Temperature ° C	Thermal Response Ratio $\Delta CT_{max}/\Delta T_{acc}$	Half Time $\frac{1}{2}$ AT
C. alleghaniensis	5→25/25→5	0.20/0.23	2.80/2.15
N. maculosus	5→25/25→5	0.21/0.16	1.54/1.18
N. viridescens	5→25/25→5	0.16/0.17	0.17/2.24
C. multidentatus	5→25/25→5	0.17/0.19	0.39/1.00

at low temperature, compared to warm-adapted animals. There are a number of adaptive changes in both enzyme structure and function, and lipid membrane structure and physical properties, that occur during thermal acclimation.

Enzymes. There are a number of potential strategies to manipulate enzyme catalytic rate and achieve thermal acclimation. These include altering the following: (1) enzyme concentration; (2) substrate concentration; (3) catalytic efficiency of enzymes; and (4) the intracellular environment, e.g., ionic concentration and pH. Enzyme concentration and catalytic efficiency are generally the more important adjustments during thermal acclimation (Hochachka and Somero 1984).

Many ectotherms and endotherms show acclimatory changes in enzyme concentration. The metabolic rate of cold-acclimated ectotherms can be equivalent to that of warm-acclimated ectotherms if there is an increased concentration of the key, rate-limiting enzymes. Not all enzyme concentrations have to be increased, just those for enzymes of reactions that limit the overall reaction rate. The key enzymes for aerobic metabolism may show a marked temperature compensation (Hazel and Prosser 1974). The mitochondrial protein content of eel liver increases with cold acclimation, from <4 mg g^{-1} at 25° C to 6 mg g^{-1} at 7° C; eels also show partial thermal acclimation in VO_2 to lowered temperature (Wodkte 1973). However, the increase in mitochondrial protein does not result in a propor-

tional maintenance of VO_2 because the specific activity of the protein is reduced at the lower temperature, and so the total VO_2 of the liver mitochondria is lower at 7° C than 25° C. The cytochrome oxidase activity of goldfish skeletal muscle of 45 μmol sec^{-1} mg protein^{-1} at an acclimation temperature of 5° C is higher than the value of 20 at an acclimation temperature at 25° C. An adaptive change in cytochrome oxidase concentration might be inferred from these results, but it is important to appreciate that a change in catalytic rate is not direct evidence for a change in enzyme concentration *per se*. Nevertheless, there are demonstrated changes in cytochrome oxidase concentration for the green sunfish during thermal acclimation (Sidell 1977), and so we can conclude that changes in enzyme concentration are sometimes one of the mechanisms for thermal acclimation. However, the general utility of achieving thermal acclimation by adjustment of enzyme concentrations is questionable. Synthesizing high concentrations of enzyme to compensate for its thermally induced catalytic inefficiency is not necessarily an optimal solution. For example, glycolytic enzymes may show little, or even inverse, acclimation in concentration.

Substrate concentration can significantly influence reaction rates and their temperature dependence. Low substrate concentrations reduce the Q_{10} effect, often substantially below the Q_{10} for V_{max}. For example, the Q_{10} of LDH varies from <1.5 (at <0.1 mM pyruvate) to >1.5 (at >0.8 mM pyruvate) for fish and lizards. The Q_{10} for pyruvate kinase of a crab varies from <2.0 (at <2.0 mM PEP) to >3 (at >0.5 mM PEP). However, substrate concentration tends to be remarkably similar among different species and adjustment in K_m is a more important mechanism for maintaining catalytic rates.

The catalytic efficiency of enzymes is generally temperature dependent. However, homologous enzymes from different individuals, or species, can counteract the effect of temperature on catalytic rate by variation in their free energy of activation (ΔG^*). For example, cold-adapted enzymes may have a lower ΔG^* than warm-adapted enzymes to minimize the change in catalytic efficiency. Reaction velocity is not so dependent on temperature if $\Delta G^*/T$ is fairly constant, since

$$V = \frac{kT}{h}e^{-\Delta G^*/RT} \qquad (5.21)$$

where k is the Boltzmann constant and h is Planck's constant. For example, the $\Delta G^*/T$ increases slightly for cold-adapted enzymes (0.216 for LDH of the ice fish *Pagothenia* compared with 0.194 for the rabbit) so that V decreases for the cold-adapted enzyme, but not by as much as it would if ΔG^* did not decrease. For Mg^{2+}-Ca^{2+} myofibrillar ATPase, the ΔG^* declines more dramatically with lowered temperature and the $\Delta G^*/T$ decreases for the low temperature enzyme; the V is slightly higher for the cold-adapted species whereas it would have been much lower if the ΔG^* had not been lower for the cold-adapted enzyme. Thus, modification in ΔG^* can have profound effects on thermal sensitivity of enzymes.

How is ΔG^* varied for different enzymes? The structures of substrate molecules and cofactors are invariant, and the chemistry of the reaction at the active site is also likely to be invariable for homologous enzymes. It is enzyme structure that is modified (Hochachka and Somero 1984). The flexibility of an enzyme's structure reflects the degree of covalent and noncovalent bonding between the constituent amino acids. Enzyme catalytic efficiency is related to its structural stability, due to weak bonding. Weak bonds have a low free energy of formation (Van der Waals forces, -4.2 kJ mole^{-1};

hydrogen bonds, -20.9; ionic bonds -20.9; hydrophobic interactions, $+8.4$) and are susceptible to thermal perturbation. Covalent bonds, in contrast, have a high free energy change (e.g., C—C, -350 kJ mole^{-1}; S—S, -210) and confer a considerable structural stability at high temperatures. Increased thermal stability by more weak bonding decreases the flexibility of enzymes, hence decreases their catalytic and regulatory capacity. This is indicated by the strong correlation between activation enthalpy (ΔH^*) and activation entropy (ΔS^*) for homologous enzymes (Figure 5–46). This graph is called a compensation plot because it shows that an increase in ΔH^* is compensated by an increase in ΔS^* ($\Delta G^* = \Delta H^* - T\Delta S^*$). The compensation temperature is the slope of the compensation plot; it indicates the temperature at which the enthalphy change for homologous enzymes is exactly balanced by entropy change. Compensation temperatures are generally 25° to 60° C.

Temperature not only affects the reaction velocity but also affects the Michaelis-Menten coefficient

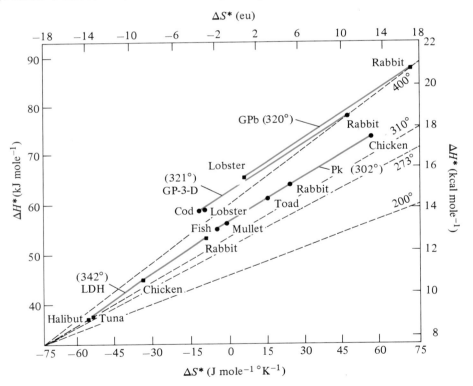

FIGURE 5–46 Compensation plots (ΔH^* as a function of ΔS^*) for homologous forms of the enzymes pyruvate kinase (Pk), glycogen phosphorylase b (GPb), lactate dehydrogenase (LDH), and glyceraldehyde-3-phosphate dehydrogenase (GP-3-D) from various vertebrates. The slopes of the lines (compensation temperature, °K) are indicated for each enzyme. The dotted lines indicate theoretical relationships for the indicated compensation temperatures. *(Modified from Somero and Low 1976.)*

(K_m), i.e., the affinity of the enzyme for its substrate. There may be an optimal temperature at which K_m is minimal and catalytic efficiency is maximal, e.g., acetylcholinesterase from trout brain (Figure 5–47A) or K_m continues to decline at lower tempera-

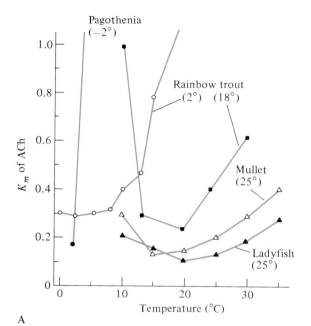

A

tures, e.g., LDH of bluefin tuna (Figure 5–47B). In the former example, the increase in K_m at temperatures below the optimum value (about 20° C) will result in high Q_{10} values since a decrease in temperature will not only decrease reaction rate by the normal Q_{10} effect but also decreases the affinity of the enzyme for its substrate; this is called negative thermal modulation. The effect of elevated temperature on reaction rate is minimized by an increase in K_m; with elevated temperature (apparent for both trout brain AChE at $T > 20°$ C, and bluefin tuna LDH). This is positive thermal modulation and keeps the Q_{10} low.

Cold-adapted enzymes tend to have a similar K_m as warm-adapted enzymes at their respective temperatures; this requires a significant shift in the K_m temperature curve. For example, the congeneric barracuda (*Sphyraena argentea, S. lucasana, S. ensis*) occur in temperate, subtropical, and tropical areas of the western coast of North America, respectively. The kinetic properties (K_m, k_{cat}) of the muscle LDH of these fish vary when measured at the same temperature (e.g., 25° C) but are similar at the temperature appropriate to their natural environmental temperature (Table 5–19). The k_{cat} is the turnover number per active site, or moles of substrate converted per mole enzyme per unit time.

Adaptive changes in the thermal sensitivity of homologous enzymes from different populations, or different species, may reflect allelic variation in the structure of the enzyme; such allelic variants are called **allozymes**. A convincing example of allozymic thermoadaptation is the heart-type LDH of the fish *Fundulus* (Place and Powers 1979). The LDH$_a$ gene from southern populations is replaced progressively by the LDH$_b$ gene in more northern populations. The ratio of k_{cat}/K_m is an *in vivo* measure of catalytic efficiency. This ratio is maximal at 20° C for the cold-adapted LDH$_b$ and at 30° C for the warm-

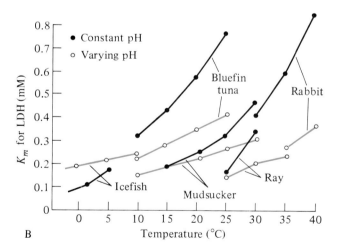

B

FIGURE 5–47 (A) Effect of temperature on the K_m of acetylcholine for acetylcholinesterase (AChE) of fishes from varying thermal environments; the trout has two isozymes adapted to 2° C and 20° C. **(B)** Effect of temperature on K_m for pyruvate of LDH from several vertebrates adapted to varying temperatures, at constant pH (pH = 7.4; solid circles) and at a pH buffered by imidazole to conform to the thermal effect on the neutral point for water (open circles). *(From Hochachka and Somero 1984; Yancey and Somero 1978.)*

TABLE 5–19

Kinetic parameters at 25° C for LDH of three barracuda *(Sphyraena spp)* from different thermal environments, and the kinetic parameters at the normal temperature midrange *(T_M)* for each species. *(Data from Graves and Somero 1982.)*

	S. argentea	*S. lucasana*	*S. ensis*
T_M	18° C	23° C	26° C
K_m at 25° C	0.34 mM	0.26 mM	0.20 mM
K_m at T_M	0.24 mM	0.24 mM	0.23 mM
k_{cat} at 25° C	893 sec^{-1}	730 sec^{-1}	658 sec^{-1}
k_{cat} at T_M	667 sec^{-1}	682 sec^{-1}	700 sec^{-1}

adapted LDH_a. The skeletal muscle LDH of congeneric barracuda provide a similar example of allozymic variation in enzyme catalytic properties.

Some examples of thermal adaptation for enzymes of individual animals can reflect variation in the thermal properties of different forms of the enzymes. For example, rainbow trout (*Salmo gairdneri*) acclimated to varying temperatures have different forms of acetylcholinesterase (e.g., minimum K_m at 2° and 18° C; Figure 5–47A). The 2° C AChE has a low K_m at low T_a whereas the 18° C AChE has a low K_m at high T_a. These enzyme variants of individual animals are called **isozymes** because they represent variation in the multiple copies of the genetic code for the enzyme (e.g., LDH). The multiple isozyme strategy is not very common; most species do not have "cold" and "warm" isozymes, perhaps because of the additional genetic load of multiple copies of enzymes (the rainbow trout is tetraploid, rather than diploid, and therefore has essentially twice as much genetic material coding for enzymes). However, many species have various isozymes in different tissues or organs, e.g., muscle and heart LDH.

A final potential mechanism for compensation of thermal modulation of enzyme catalytic efficiency is modification of the intracellular environment in which the enzymes function. For example, there are changes in ionic concentration of intra- and extracellular fluids that accompany thermal acclimation (Behrisch 1973). The K_m of LDH from the yellowfin sole (*Limanda*) has a minimum value at 4° C (measured with no K^+ present). The yellowfin sole lives in water ranging in temperature from −1.86° C in winter to 4° to 5° C in summer, and so it might seem that its LDH functions at submaximal catalytic efficiency for most of the year. However, the K_m measured in 150 mM K^+ (a more physiologically relevant condition) has a minimum at about −2° to 0° C; the normal ionic K^+ concentration is required for optimal enzyme function.

Temperature has a marked effect on the pH of neutral water, and the pH of intracellular and extracellular body fluids (see Chapter 12). This pH temperature dependence provides significant stabilization of enzyme kinetics (e.g., K_m). For example, the K_m of muscle LDH has a marked temperature dependence when pH is kept constant at 7.4 by a phosphate buffer (Figure 5–46B). For many animals, such a constant pH is not physiologically relevant since pH increases at lower temperatures. There is a lesser dependence of K_m on temperature in an imidazole buffer, which mimics the *in vivo* temperature pH relationship for many animals.

The thermal stability of proteins is often implicated as a cause of thermal death. There is a correlation between the denaturation temperature for proteins (melting temperature, T_m) and the CT_{max}, but enzyme catalytic efficiency is likely to decline markedly and cause death well before proteins actually denature. Enzyme structure is generally quite flexible to maintain catalytic efficiency at low temperature, and so proteins may become too flexible at high temperature. The proteins of thermophilic bacteria are remarkably heat tolerant; some can tolerate 90° C *in vitro*. Their high stability may be due to increased ionic bond stabilization, reduced surface hydrophobicity and enhanced internal hydrophobicity, or increased covalent stabilization. The bacterium *Thermus* also contains thermoprotective polyamines that protect its proteins from thermal denaturation. There would appear to be only a limited capacity for animal proteins to decrease their thermal sensitivity because their catalytic functions would be reduced at lower temperatures.

Lipid Membranes. Biological membranes are essentially a bilayer of amphiphilic lipids (one end is polar and the other is nonpolar) such as phospholipids and sphingolipids (see Chapters 3 and 6). Phospholipids, for example, consist of two fatty acids (the nonpolar end) bound to a glycerol, and a phosphate with a polar headgroup such as choline or ethanolamine (the polar end). Associated with the lipid bilayer are a variety of structural and enzymatic proteins that may be an integral part of the membrane or located more peripherally on the surface.

The physical properties of lipids, and especially lipid bilayer membranes, are markedly influenced by temperature. Biological membranes exist in a "liquid-crystalline" state that is functionally intermediate between a rigid, solid lipid (e.g., lipids at low temperature) and a highly fluid lipid state (e.g., lipids at high temperature). The phospholipid extracts of a bacterial membrane (*E. coli*) clearly illustrate a phase transition between a low viscosity at high temperature to a high viscosity at low temperature Figure 5–48). The maintenance of a normal **homeoviscous** lipid state in biological membranes is essential to the functioning of the membrane.

One of the major variables contributing to the homeoviscous state is the fatty acid composition of the lipid bilayer. Fatty acids vary in both chain length and degree of double bonding of the carbon backbone. Shorter chain fatty acids are more liquid than longer chain fatty acids. Unsaturated fatty acids (with double bonds) are more liquid than

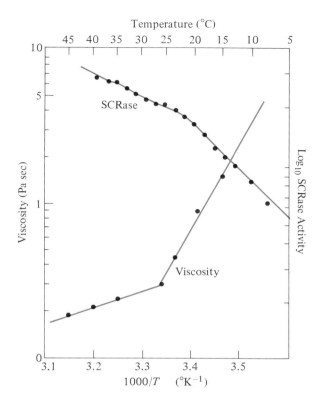

FIGURE 5–48 Effect of temperature (Arrhenius plot) on the viscosity of phospholipids extracted from a bacterium showing a phase transition at about 27° C, and on the activity of mitochondrial cytochrome C reductase (SCRase) from the guinea pig. *(Modified from Sinensky 1974; Geiser and McMurchie 1984.)*

TABLE 5–20

Effects of ambient temperature on the ratio of saturated to unsaturated fatty acids for three phospholipids (choline, ethanolamine, serine/inositol) isolated from brain lipid membranes of a range of vertebrates. *(Data from Cossins and Prosser 1978.)*

| | | Phospholipids | | |
	Temperature	*Choline*	*Ethanolamine*	*Serine/Inositol*
Arctic sculpin	0°C	0.593	0.947	0.811
Goldfish	5°C	0.659	0.340	0.459
	25°C	0.817	0.506	0.633
Desert pupfish	34°C	0.990	0.568	0.616
Rat	37°C	1.218	0.651	0.664

saturated fatty acids because their "kinked" shape reduces stabilization of adjacent molecules. The lipid fluidity is correspondingly adapted to the normal environmental temperature.

The membrane lipids of animals and plants from varying ambient temperatures show changes in fatty acid composition that are adaptive for maintenance of the homeoviscous state (Table 5–20). The variation in fatty acid composition of membrane lipids is accomplished by regulating the chain length and polar head group composition or degree of saturation (see desaturase enzymes below). Rainbow trout acclimated to 20° C then cooled to 5° C have a decline in phosphatidyl choline level and an increase in phosphatidyl ethanolamine; there is a significant change in the PC/PE ratio within three days. The fatty acid composition of the nerve cord of an insect has a decreased level of three saturated fatty acids (myristic, pentadecanoic, and palmitic) at low temperatures, and increased content of three unsatu-

rated fatty acids at high temperature (linoleic, eicosadienoic, and arachidonic); the fat body lipid does not show a corresponding temperature-dependent change in fatty acid content. There are extremely rapid changes (<12 hours) in head group composition for ectotherms that experience marked daily temperature fluctuations (Carey and Hazel, 1989).

Desaturase enzymes regulate the degree of unsaturation of fatty acids. The activity of desaturase enzymes can be rapidly modified in response to temperature changes, either by a direct thermal effect on their activity (high temperature inactivates the desaturase enzyme) or by a change in the membrane location of the desaturase (the active site is exposed in low fluidity membranes but hidden within the lipid bilayer in high fluidity membranes). For example, carp initially acclimated to 30° C have an enhanced desaturase activity and concentration after transfer to 10° C, and this results in the restructuring of the rough endoplasmic reticulum lipids within two days of cooling to 10° C (Wodtke, Teichert, and Konig 1986).

The homeoviscous state of a membrane has great significance to membrane-bound enzyme catalysis. Membrane fluidity would influence the catalytic activity of enzymes that must undergo conformational changes during catalysis, e.g., membrane-bound transport proteins. The fluidity of the lipid bilayer surrounding the enzyme clearly would influence the ability to undergo conformational change. Membrane fluidity may also influence the location and exposure of the active site. This concept agrees well with the observation that the activation energy (E_a) increases for many membrane-bound enzymes below a critical temperature; this is readily apparent from Arrhenius plots (Figure 5–48). At temperatures below the breakpoint, the enthalpy of activation (ΔH^*) is increased dramatically; this is compensated to some extent by the increased activation entropy (ΔS^*) and so the free energy of activation ($\Delta G^* = E_a$) is not so much affected as is the ΔH^*.

Summary

Temperature is a measure of the average thermal motion of molecules. It dramatically affects reaction rate, as indicated by the activation energy (E_a) and Q_{10} values of biochemical and physiological rates. Heat exchange can occur by conduction, convection, radiation, and a change in state of water (evaporation/condensation or freezing/thawing).

Animals either conform to their thermal environment or thermoregulate. The T_b of ectothermic animals is determined by heat exchange from their environment; their metabolic heat production is negligible. Many ectotherms are thermoconformers. Aquatic ectotherms may thermoregulate by selecting water of the appropriate temperature. Terrestrial ectotherms can be accomplished thermoregulators by either basking in sunlight or gaining heat from the substrate by conduction. Endothermic animals thermoregulate by virtue of their endogenous metabolic heat production.

Conductive heat exchange depends on the thermal conductivity, area of contact, distance for heat transfer, and the temperature difference between the objects. The thermal conductivity of animal insulation (fur, feathers, and chitin hairs) is similar to that for still air; their resistance to heat exchange, and insulative value, depend on the thickness of the insulating layer. Subcutaneous fat has a low insulative value in air but is a better insulator in water than fur or feathers.

Convective heat exchange is essentially conduction across the boundary layer, in proportion to the temperature difference and convective heat transfer coefficient. Forced convection transfers heat more rapidly than free convection.

Net radiative heat exchange depends on the surface temperature of the animal, the average temperature of the surroundings, area, and emissivity. The radiative heat transfer coefficient is markedly dependent on temperature.

Evaporation dissipates about 2500 J g^{-1}; condensation releases the same amount of heat. Freezing releases about 334 J g^{-1}; whereas melting absorbs the same amount of heat.

The thermal environment of animals is complex. The operative temperature is the effective environmental temperature determined by conductive, convective, and radiative heat exchange. The standard operative temperature is the operative temperature with standardized convective conditions (usually free convection). These are better measures of the environmental temperature than is air temperature.

Ectotherms avoid freezing by behavioral avoidance or physiological adaptation. Strategies for physiological adaptation include freezing point depression by the accumulation of specific osmolytes (e.g., sugars, polyols), supercooling to temperatures below the freezing point without ice formation (often facilitated by specific accumulated osmolytes), inhibition of freezing by antifreeze proteins (a noncolligative effect), and tolerance of extracellular freezing. Ectotherms adapt to high temperatures by enhanced evaporative cooling or biochemical acclimation.

Endothermic mammals and birds regulate body temperature by control of their endogenous heat production. Body temperature is generally regulated at 35° to 42° C, depending on the taxonomy of the mammal or bird. Metabolic heat production is highest at low T_a, is minimal (basal) in the thermoneutral zone, and is elevated at high T_a. Endotherms have three strategies for survival in the cold: (1) they decrease heat loss, (2) they increase heat production, or (3) they decrease body temperature. The hypothermia may involve only the peripheral appendages (the core T_b is maintained; regional heterothermy), or there may be a reduced core T_b (temporal heterothermy). The core T_b may be decreased slightly, as in moderate hypothermia, or markedly depressed, as in torpor. Torpor is the abandonment of normal T_b thermoregulation, and T_b declines to near T_a. At low T_a the T_b is regulated at a minimal $T_{b,crit}$ value. Endotherms respond to high T_a by enhancing evaporative heat loss. The T_b may be increased (hyperthermia) to facilitate heat dissipation. Brain temperature is often regulated below core T_b to avoid nervous system disfunction; there is countercurrent heat exchange between warm arterial blood and cool venous blood returning from evaporative surfaces of the head (nasal cavity, skin, eyes).

The only living endothermic reptiles are some brooding female pythons, which shiver to regulate body and egg temperature relatively independent of T_a. Large reptiles ($>$100 kg) are homiothermic by virtue of their high mass and thermal inertia, and low thermal conductance. Their T_b can considerably exceed T_a because of passive constraints to thermal dissipation. Large dinosaurs may have been endothermic and regulated their T_b by physiological means, including control of metabolic heat production.

Large, active fish, such as tuna, sharks, and swordfish, are regional endotherms. Metabolic heat production of, for example, skeletal muscle is retained within tissues by vascular countercurrent heat exchange. Muscle, brain, eye, or visceral temperature can thus be maintained considerably above the ambient water temperature. The rate of metabolic heat production is not varied to regulate T_b. Rather, metabolic heat production is constant and the rate of heat loss is controlled.

Many flying, running, or walking insects are endothermic and regulate thoracic temperature. Generally, the rate of metabolic heat production is determined by the intensity and type of locomotion and is not controlled to regulate thoracic temperature. Thoracic insulation and countercurrent heat exchange between thorax and abdomen facilitate preflight warm up and thoracic temperature regulation during flight at low T_a.

Endothermy has evolved in a variety of animals. One possible scheme for the evolution of endothermy is the initial elevation of T_b by metabolic heat production from locomotion followed by the acquisition of insulation or vascular heat exchangers for further elevation and regulation of T_b. Alternatively, mammalian endothermy may have evolved during a change in niche from nocturnality to diurnality, or the progressive reduction in body size from large, inertially-homiothermic reptiles to smaller mammals.

The advantages of endothermy include optimal biochemical adaptation to the constant and high T_b, high muscle force, velocity and power expenditure, and independence of activity from ambient thermal conditions. The principal disadvantage of endothermy is the high metabolic cost for thermoregulation.

The T_b of endotherms is generally 35° to 42° C. This range presumably reflects the catalytic advantages of higher temperature for biochemical and physiological processes and the disadvantages of too high a T_b of excessive metabolic expenditure for thermoregulation and excessively high demands for energy consumption.

Endogenous and exogenous pyrogens increase the T_b of a variety of ectotherms and endotherms. For example, heliothermic lizards will select higher temperatures in a thermal gradient, and mammals will shiver to elevate T_b above the normal preferred temperature. The presumed selective advantage for the hyperthermia is an enhanced immune response and/or diminished viability of the infecting pathogen.

Biochemical and physiological reactions are thermally dependent but can acclimate (in the laboratory) or acclimatize (in nature) with prolonged exposure to differing temperatures. There are a number of patterns of thermal acclimation. Generally, the acclimation response is adaptive; the rate after acclimation to a new temperature is more similar to the initial rate than was the rate immediately after the temperature change. Many physiological processes show thermal acclimation, e.g., metabolic rate, respiratory rate, heart rate, preferred T_b, critical thermal maximum and minimum temperatures. There are a variety of biochemical mechanisms for thermal acclimation: changes in enzyme concentration, alteration of substrate concentration, change in catalytic efficiency, change in the intracellular environment, and modification of the lipid membrane structure and function.

Supplement 5–1

Convective Heat Transfer

. . . .

Convection is the transfer of heat by movement of a fluid (either a liquid or a gas). Consider a flat plate of temperature T_{fp} that is immersed in a moving fluid with a free-stream temperature T_∞ (the ∞ doesn't mean the temperature is infinite, but is measured at an infinite distance from the plate). The fluid has a free-stream velocity (i.e., at infinite distance from the plate) of V_∞. Motion of the fluid will establish a boundary layer on the surface of the flat plate. The fluid is stationary at the immediate surface (zero velocity; this is the no-slip condition) and the velocity profile extends away from the plate until it equals the free-stream velocity. The boundary layer is defined as that region with velocity >0 and <0.99 V_∞. The thickness of the boundary layer (δ) increases with distance from the leading edge of the plate (x); it is thinnest at the leading edge and thickest at the trailing edge. The boundary layer thickness depends on many variables, including the free-stream velocity (V_∞), distance from the leading edge of the plate (x), and the fluid density (ρ) and viscosity (η).

The boundary layer is initially streamlined (laminar) but becomes turbulent at a critical distance (L_c) from the leading edge. The critical distance depends on the local Reynolds' number, a dimensionless coefficient defined as

$$R_e = V_\infty x_c/(\rho/\eta) = V_\infty x_c/\nu$$

where ν is the kinematic viscosity ($1.5 \ 10^{-5}$ for air and $1.0 \ 10^{-6}$ m^2 sec^{-1} for water at 20° C). The Reynolds'

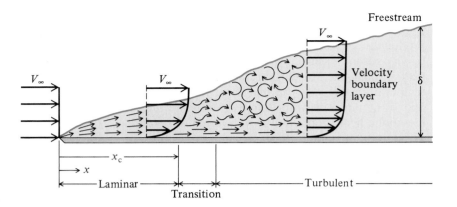

Velocity boundary layer.

number is essentially the ratio of inertial forces ($V_\infty x_c$) to viscous forces (ρ/η), and it indicates whether the flow is laminar (streamline flow, at low R_e) or turbulent (eddy flow, at high R_e). The critical R_e is generally about $5 \cdot 10^5$. The laminar boundary layer becomes turbulent at about 4.5 m from the leading edge and is about 0.04 m, for a flat plate in air ($V_\infty = 1$ m sec^{-1}).

There are not only velocity boundary layers around objects. For example, there is a thermal boundary layer around an object if there is a temperature difference between it and the fluid. There is a boundary layer of O_2-depleted water around aquatic animals. There is a relative humidity boundary layer around a moist-skinned animal in dry air. The same general considerations determine the thickness of these boundary layers as for velocity boundary layers, but it is important to appreciate that the thicknesses of the different types of boundary layers are not necessarily the same, or even of the same order of magnitude. (*See Incropera and Dewitt 1981.*)

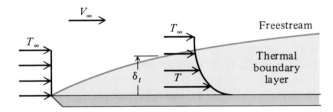

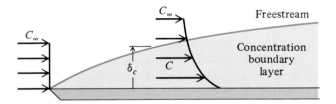

Thermal and concentration boundary layers.

Supplement 5–2

Newtonian Model for Thermoregulation in Endothermic Mammals and Birds

• •

The thermoregulatory strategy of endothermic mammals and birds is readily apparent from consideration of the effects of air temperature on the thermal balance of inanimate objects. The heat loss of an object is proportional to ($T_{obj} - T_a$), and therefore increases in a linear fashion at lowered T_a. The heat input needed to keep T_{obj} constant increases in a corresponding linear fashion with decreased T_a, i.e., heat input is proportional to ($T_{obj} - T_a$). A graph of heat loss as a function of ($T_{obj} - T_a$) for such a simple, Newtonian system would extrapolate to zero heat loss at $T_{obj} = T_a$. The slope of the relationship between heat input and T_a is thermal conductance (C);

Heat Input = $C(T_{obj} - T_a)$

This simple Newtonian model for heat balance of an inanimate object (with no evaporation of water) applies in principle to endothermic mammals and birds.

Let us consider a hypothetical endothermic mammal or bird that maintains a constant T_b over a wide range of T_a but has an elevated T_b when stressed at high T_a (facultative hyperthermia). The T_b is kept constant by the control of metabolic heat production. There is a similar hypothetical relationship between metabolic heat production (MHP) and body temperature (T_b) as a function of air temperature (T_a), for an endothermic mammal or bird, as for the inanimate object

$$MHP = C_{wet}(T_b - T_a)$$

where C_{wet} is the wet thermal conductance (since animals invariably have some evaporative heat loss that contributes to heat dissipation). The MHP doesn't decline to 0 at $T_a = T_b$, but plateaus at a minimum value, the basal metabolic rate (BMR). The BMR is constant over a range of T_a, from the lower critical temperature (T_{lc}) to the upper critical temperature (T_{uc}). The relationship between MHP and $T_a < T_{lc}$ has a slope equal to −(thermal conductance) and extrapolates to T_b at MHP = 0. The figure inset

shows the same relationship, but MHP is graphed as a function of $T_b - T_a$.

The wet thermal conductance is the thermal conductance uncorrected for evaporative heat loss; its units are ml O_2 g^{-1} hr^{-1} $°C^{-1}$ or J g^{-1} hr^{-1} $°C^{-1}$. The value of C_{wet} alters through the thermoneutral zone (being lowest at T_{lc} and highest at T_{uc}). This conductance change involves nonenergy requiring physiological changes, such as redistribution of blood flow to the skin, increased respiratory water loss, and behavioral responses (posture adjustment, pilo- or ptilo-depression of the fur/feathers). The VO_2 increases above T_{uc} because (1) T_b tends to increase, hence VO_2 is increased by a Q_{10} effect, and (2) there is a significant metabolic cost for many physiological responses to heat stress (panting, sweating, etc). The dry thermal conductance (C_{dry}) is the conductance corrected for evaporative heat loss (EHL; J g^{-1} h^{-1}).

$$(MHP - EHL) = C_{dry}(T_b - T_a)$$

Evaporative heat loss is calculated from the evaporative water loss (e.g., g H_2O g^{-1} hr^{-1}) and the latent heat of fusion (e.g., 2400 J g^{-1}).

Some endothermic mammals and birds conform to this hypothetical physical model for heat exchange but many

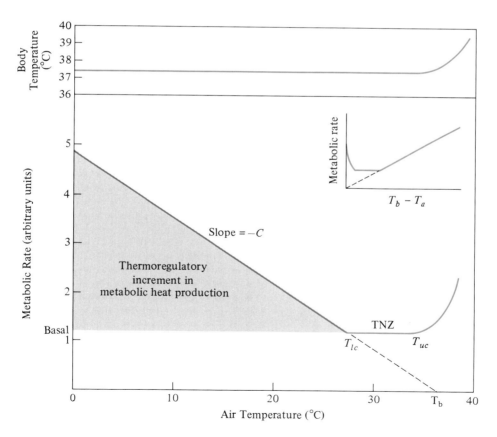

Effects of ambient temperature on metabolic rate and body temperature for a Newtonian model.

deviate somewhat from the model. For example, T_b often declines slightly at low T_a reflecting the gain of the thermoregulatory system. The slope of the relationship between MHP and $T_a < T_{lc}$ is not necessarily constant,

and does not necessarily extrapolate to zero MHP at T_b. The dry thermal conductance can be markedly increased at high T_a, to facilitate passive heat dissipation.

Supplement 5–3

Bioclimatic Rules

· ·

Size, shape, and insulation are major determinants of heat exchange for endotherms. Consequently, it is logical to think that adaptation to cold climates might affect these aspects of an endotherm's morphology. A series of bioclimatic rules or laws have been proposed to explain adaptive climatic variation in body morphology. These rules are of interest because they usually reflect a mechanism for the adaptive modification of heat exchange, even though many may not be generally applicable but reflect specific adaptations in only certain animal taxa.

The bioclimatic laws were generally based on theory or circumstantial evidence, such as observed climatic trends in the morphology of certain species of endotherms, or of different geographic populations for a single species. Direct developmental evidence for bioclimatic rules has sometimes been obtained by raising endotherms (e.g., littermates) in differing T_a environments, and showing a direct ontogenetic effect of climate on body morphology.

In 1839, Sarrus and Rameaux postulated their surface rule: larger animals have a lower surface-to-volume ratio than do smaller animals. Consequently, larger endotherms would have a lower mass-specific heat loss and this would presumably be adaptive in a cold climate. Bergmann's rule makes the similar assertion that it is energetically less expensive for a large endotherm to survive in a cold climate because its mass-specific heat loss is lower, i.e.,

endotherms should be bigger in cold climates. There is evidence for Bergmann's rule in some taxa. For example, wood rats (*Neotoma*) are larger in colder climates, and also have a lower thermal conductance and a lower CT_{min}. The mean body mass of male humans increases by about 0.5 kg for every 1° C decline in mean annual temperature. However, similar climatic trends in body size are not observed for many species of endotherm. There is also no conclusive developmental evidence for Bergmann's rule in endotherms. Perhaps Bergmann's rule is not universal because many other cold-adaptations of endotherms can override body size effects. Another body mass rule, Cope's rule, relates body size to evolutionary history; the evolutionary trend within many taxa is towards larger size.

Allen's rule suggests that the heat loss is reduced for cold-adapted endotherms by a reduction in the size of their appendages, e.g., ears, digits, limbs. The lower surface-to-volume ratio of a smaller appendage reduces its heat loss. Circumstantial evidence for Allen's rule is again obtained from climatic trends in body morphology of related endothermic species. For example, arctic foxes have small ears compared to desert-adapted kit foxes; the desert fennec and bat-eared fox have extremely large ears, and temperate foxes have intermediate-sized ears. Arctic rabbits have smaller ears than desert jackrabbits.

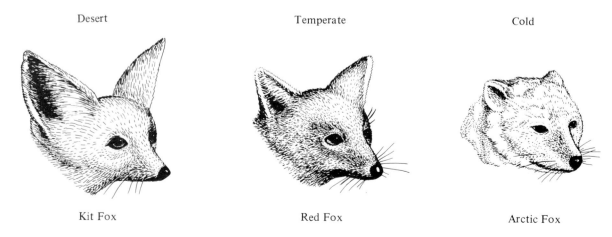

Desert   Temperate   Cold

Kit Fox   Red Fox   Arctic Fox

Ear sizes for foxes from different climates.

Even cold-adapted races of humans tend to have shorter limbs and stockier bodies than desert-adapted races. Thomson's nose rule suggests that the human nose becomes relatively narrower at lower mean annual temperatures; the climatic significance of nose shape may be related to susceptibility to respiratory infections, or nasal countercurrent and water exchange. However, there may be reasons other than thermoregulation for appendage size; for example, bat-eared foxes have acute hearing for locating underground prey. There is developmental evidence for Allen's rule. Mice raised at low T_a have shorter tails than mice raised at high T_a. Pigs raised at low T_a have shorter tails, smaller ears, and stockier bodies than pigs raised at high T_a. A simple mechanism for these developmental changes is the effect of peripheral hypothermia on blood flow. Peripheral vasoconstriction would decrease nutrient delivery to the appendages and retard growth. For example, the fingernails of humans grow more slowly in cold climates.

Wilson's rule relates the thickness of insulation to climate. A thicker insulative layer is clearly adaptive to endotherms in cold climates. Arctic species have thicker coats than tropical species. There are also seasonal changes in the coat thickness for many arctic mammals; the coat is thinner in summer and thicker in winter. There is again developmental evidence for Wilson's rule. Pigs raised at low T_a have more hair than pigs raised at high T_a.

Gloger's rule suggests that animals have a lighter coat color in cold, wet climates and darker coats in warm, dry climates. Coat color clearly has many different roles, but it can affect thermal exchange. For example, the white fur of arctic mammals reflects solar radiation deep into the coat and facilitates thermoregulation. The white fur of polar bears may act as a light guide to facilitate deep penetration of light. (*See Ley 1971; McNab 1971; Hafez 1968; Damon 1975; Calder 1984; Kleiber 1975.*)

Recommended Reading

Alexandrov, V. Y. 1977. *Cells, molecules and temperature*. Berlin: Springer-Verlag.

Bakken, G. S. 1976. An improved method for determining thermal conductance and equilibrium body temperature with cooling curve experiments. *J. Thermal Biol.* 1:169–75.

Bakker, R. T. 1971. Dinosaur physiology and the origins of mammals. *Evolution* 25:636–58.

Bakker, R. T. 1972. Anatomical and ecological evidence of endothermy in dinosaurs. *Nature* 238:81–5.

Bartholomew, G. A. 1981. Physiological thermoregulation. In *Biology of the reptilia*. Vol. 12, *Physiological ecology*, edited by C. Gans and F. H. Pough, 167–212. New York: Academic Press.

Benton, M. J. 1979. Ectothermy and the success of dinosaurs. *Evolution* 33:983–97.

Calder, W. A., and J. R. King. 1974. Thermal and caloric relations of birds. In *Avian biology*, edited by D. S. Farner and J. R. King, Vol. IV, 259–413. New York: Academic Press.

Carey, F. G. 1982. A brain heater in swordfish. *Science* 216:1327–29.

Casey, T. M. 1981. Behavioral mechanisms of thermoregulation. In *Insect thermoregulation*, edited by B. Heinrich, 80–114. New York: John Wiley & Sons.

Cossins, A. R., and K. Bowler. 1987. *Temperature biology of animals*. London: Chapman & Hall.

Hardy, R. N. 1978. *Temperature in animal life*. Baltimore: Univ. Park Press.

Hazel, J. R., and C. L. Prosser. 1974. Molecular mechanisms of temperature compensation in poikilotherms. *Physiol. Rev.* 54:620–77.

Heinrich, B. 1981. *Insect thermoregulation*. New York: John Wiley & Sons.

Hew, C. L., G. K. Scott, and P. L. Davies. 1986. Molecular biology of antifreeze. In *Living in the cold: Physiological and biochemical adaptations*, edited by H. C. Heller, X. J. Musacchia, and L. C. H. Wang, 117–23. New York: Elsevier.

Hochachka, P. W., and G. N. Somero. 1984. *Biochemical adaptation*. Princeton: Princeton University Press.

Huey, R. B. 1982. Temperature, physiology, and the ecology of reptiles. In *Biology of the reptilia*. Vol. 12, *Physiological ecology*, edited by F. H. Pough, 26–90. London: Academic Press.

McGowan, C. 1979. Selection pressure for high body temperatures: Implications for dinosaurs. *Paleobiology* 5:285–95.

Monteith, J. L. 1973. *Principles of environmental physics*. London: Edward Arnold.

Pough, F. H. 1980. The advantages of ectothermy for tetrapods. *Am. Nat.* 115:92–112.

Precht, H., J. Christophersen, H. Hensel, and W. Larcher. 1973. *Temperature and life*. Berlin: Springer-Verlag.

Stevens, E. D., and W. H. Neill. 1978. Body temperature relations of tuna, especially skipjack. In *Fish Physiology*. Vol. VII, *Locomotion*, edited by W. S. Hoar and D. J. Randall, 315–424. New York: Academic Press.

Thomas, R. D. K., and E. C. Olson. 1980. *A cold look at the warm-blooded dinosaurs*, AAAS Selected Symp 28. Washington: AAAS.

Chapter 6

Membrane Physiology

· ·

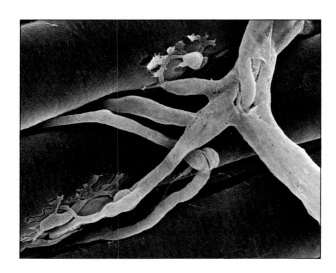

Scanning electron micrograph of a motor nerve and two end-plates on adjacent muscle fibers. *(Photo courtesy of D. W. Fawcett/Desaki & Venara/ Photo Researchers, Inc.)*

The **plasma membrane** is the interface between the internal cytoplasmic environment and the extracellular environment. This cell membrane must limit the exchange of many solutes, often against immense concentration gradients that favor exchange between the cytoplasm and external medium. At the same time, it must allow ready transport of many nutrients and waste products into and out of the cell; consequently, there are often specific membrane transport systems for passive or active exchange.

Cells have a complex internal system of interconnected membrane spaces, membrane-bound vesicles, and organelles (Figure 6–1). Mitochondria and lysosomes are examples of membrane-bound organelles. In fact, most of the total membrane content of a cell is intracellular, not the surface plasma membrane. For example, a single liver cell of a rat (Weiner et al. 1968) has an approximate surface area of 2000 μ^2 and volume of 5100 μ^3. The 1100 or so mitochondria make up 20% of the cytoplasmic volume (995 μ^3) and have membrane areas of 7470 μ^2 (outer membrane) and 39600 μ^2 (inner membrane). Endoplasmic reticulum also has an extensive surface area of 17000 μ^2 (smooth ER) and 30400 μ^2 (rough ER). The nucleus is about 4% of the cellular volume and has a surface area of about 200 μ^2.

The functions of the membranous structures in a typical animal cell include protein synthesis, transport, formation of storage vesicles, release of

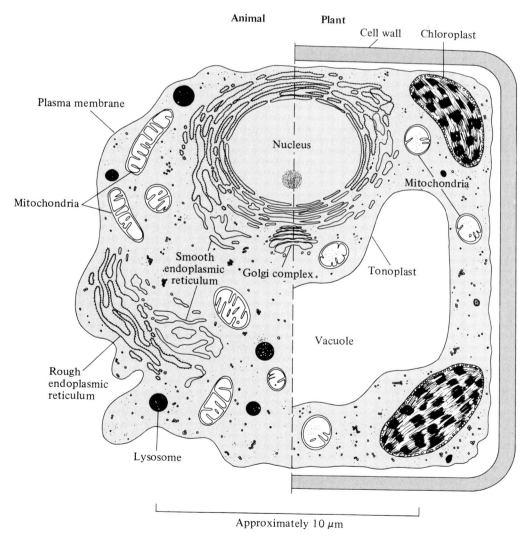

Animal **Plant**

Cell wall   Chloroplast

Plasma membrane

Nucleus

Mitochondria

Mitochondria

Smooth endoplasmic reticulum

Golgi complex

Tonoplast

Rough endoplasmic reticulum

Vacuole

Lysosome

Approximately 10 μm

FIGURE 6–1 Diagrammatic illustration of the membranous structures and organelles of an animal cell (left) and a plant cell (right). *(From Finean, Coleman, and Michell.)*

TABLE 6–1

Membrane structures of animal cells and their primary functions. *(Modified from Fawcett, 1986; Lockwood 1978).*

Plasma Membrane	Barrier between intracellular and extracellular fluids, controls passive and active transport of many solutes, and has electrically excitable properties in many cells.
Nuclear Envelope	Double membrane barrier separating the nucleus from the remainder of the cytoplasm; is perforated by large pores to allow diffusional exchange.
Mitochondria	Membrane-bound organelles that synthesize ATP from $NADH^+$ and $FADH_2$; they have their own DNA and are self-replicating.
Rough Endoplasmic Reticulum	Membrane-lined reticulum of spaces containing ribosomes; the site of protein synthesis.
Smooth Endoplasmic Reticulum	Membrane-lined reticulum continuous with rough ER; the site of steroid metabolism; acts as transport route for movement of products of rough ER to Golgi apparatus.
Golgi Complex	Membrane-lined spaces continuous with rough and smooth ER; concentrates, modifies, and packages secretory products into membrane-bound vesicles for secretion.
Lysosomes	Membrane-bound vesicles containing hydrolytic enzymes at acid pH for intracellular breakdown of materials engulfed by phagocytosis or pinocytosis for removal of damaged cell organelles and destruction of engulfed bacteria.
Peroxisomes	Small spherical membrane-bound vesicles containing enzymes that synthesize hydrogen peroxide; may be involved in uric acid metabolism.
Phagosomes	Membrane-bound vesicles containing particulate material engulfed from the outside of the cell by phagocytosis.
Pinocytotic Vesicles	Small membrane-bound vesicles containing materials absorbed into the plasma membrane to form a vesicle.

materials by exocytosis, uptake of materials by phagocytosis, and mitochondrial ATP synthesis (Table 6–1). The nuclear membrane, endoplasmic reticulum, and Golgi complex are examples of interconnected membranous spaces with different functions. Plant cells are similar to animal cells except for their stiff cell wall, the often large vacuole bounded by the tonoplast membrane, and the photosynthetic organelles (chloroplasts).

Membrane Structure

The existence and important role of the plasma membrane was recognized in the late nineteenth century, and the basic **lipid bilayer** structure of membranes was deduced early in the twentieth century. The amount of lipid that could be extracted from, for example, a red blood cell could be reconstituted as a lipid monolayer with an area that was about twice that of the original red cell. The conclusion of such experiments was that the plasma membrane was a bilayer of lipid molecules; this is clearly evident in electron micrographs of cell membranes (Figure 6–2A). The structural model of the plasma membrane structure, proposed by Danielli and Davson (1935), was a bilayer of lipid molecules, with the hydrophobic portions ("water-avoiding," i.e., not water soluble) oriented towards the inside of the lipid bilayer and the hydrophilic portions ("water-loving," i.e., water soluble) forming the outer surfaces in contact with either the cytoplasmic or extracellular fluids. They further speculated that there was a protein coating at the aqueous surfaces, some of which penetrated through the lipid bilayer, since this was consistent with the observations of facilitated transport of various solutes.

Subsequent studies of plasma and also various intracellular membranes have variously modified Danielli and Davson's basic lipid-bilayer model into more complex models (Figure 6–2B). The "fluid-mosaic" model (Singer and Nicholson, 1972) represents the membrane as a disorderly, fluid-like lipid bilayer that contains complex proteins with hydrophobic regions located deep within the lipid hydrocarbon zone and the polar regions at the surface. There are ordered aggregates of both protein and lipid called domains covering the membrane surface. The outer surface of the plasma membrane contains many glycoproteins, proteins with short (about 10 monosaccharide residues) polysaccharides. These carbohydrates "bristle" from the cell surface and may assist in keeping apart adjacent cells since they

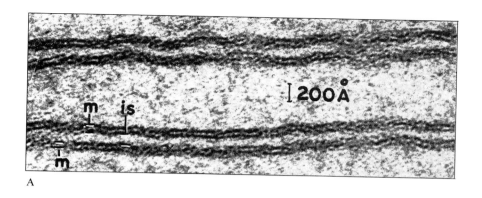

A

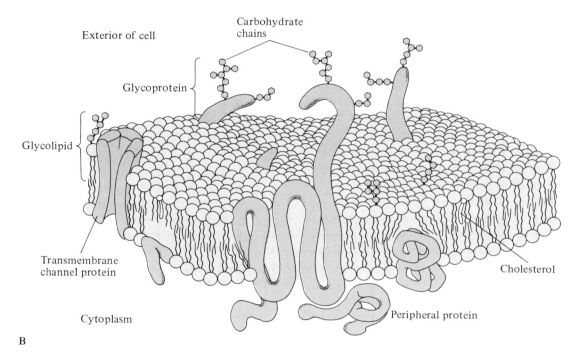

B

FIGURE 6–2 **(A)** An electron micrograph of a cell membrane (four separate membranes are shown, labelled M) shows the 10 nm thick, double layer of dark lines representing the hydrophilic lipid heads and the intervening space of hydrophobic tails. **(B)** A schematic representation of the fluid-mosaic model of a cell membrane showing the lipid constituents and their arrangement as the bilayer portion of the membrane and the protein constituents of the membrane.

are negatively charged. The inner surface of the membrane has other loosely associated proteins and microfilaments and microtubules that provide lateral movement of the membrane domains (Nicholson 1976).

About 25 to 50% of the membrane dry weight is lipid, mostly compound lipids such as glycerophosphatides (phosphatidyl choline, phosphatidyl ethanolamine, phosphatidyl serine, phosphatidyl inositol), sphingolipids (which resemble glycerophosphatides but contain a nitrogenous base, sphingosine, instead of glycerol), and cholesterol (Lockwood 1978). The polar ends of these lipids are arranged on the aqueous sides of the lipid bilayer. The cholesterol, which is found mainly in the outer layer of the bilayer, most likely interacts structurally with the phospholipids, such as phosphatidyl ethanolamine (Figure 6–2B). The local concentration of particular lipids can create zones of greater or lesser mobility. There is also considerable variation in specific lipid constituents between membranes of different types of cells and membranes of different parts of cells. For example, the myelin sheath of some nerve cells has a higher sphingolipid content and lower glycerophosphatide content than the mitochondrial membrane and is rich in saturated, long-chain fatty acids.

Changes in the specific lipid constituents of membranes are an important aspect of thermal acclimation by membranes (see Chapter 5). The chain length and extent of saturation of the lipids determine the melting point, or phase transition temperature, of the membrane. Short chains and low saturation decrease the phase transition temperature. Cholesterol has the interesting property of decreasing the molecular mobility of membrane constituents in the liquid phase, but increasing their mobility in the solid phase. Consequently, the presence of high levels of cholesterol in a membrane "blurs" the temperature transition between the liquid ("melted") phase and solid ("frozen") phase.

Membrane-associated proteins are often enzymes but many have a structural, rather than a catalytic, function. Many proteins bridge, or span, the lipid bilayer (Figure 6–2B). The acetylcholine (ACh) receptor is a good example of a structural protein in a membrane. It usually is a dimer of two pentameric subunits, joined by a disulfide bridge and associated proteins on the intracellular side of the membrane. The pentameric subunit consists of four glycopeptides. The ACh receptor has two binding sites for ACh and the transmembrane bridging portion forms a gated ionic pore through the lipid bilayer; the gate opens to allow cation flux when two acetylcholine molecules are bound to the receptor portion (see below). Other examples of bridging proteins that function as specific transport channels are the proton-pumping bacteriorhodopsin proteins found in the purple membrane of certain halophytic bacteria, plasma membrane connexons for intercellular transport, ATP synthetase, and cytochrome oxidase. Many other examples of specific transport systems will be described below. For example, the Na^+-K^+ ATPase enzyme is a characteristic constituent of all plasma membranes and is of vital importance in maintaining the intracellular-extracellular Na^+ and K^+ concentration gradients. Another important enzyme links carbohydrates with protein to form glycoproteins, which often form a 5 to 7 nm thick bristling coat over the cell surface. Blood group factors are one class of glycoproteins; other glycoproteins are released from the plasma membrane and become components of serum proteins, cartilage, mucus, and the vitreous humor of the vertebrate eye.

Membrane Permeation

The plasma membrane influences the passage of all solutes and water that enter or leave the cell. It must be selectively permeable, allowing ready passage of certain molecules, such as O_2, CO_2, glucose, NH_3, and urea, yet impeding the loss of important intracellular solutes, such as K^+, ATP, amino acids, and proteins.

An artificial pure lipid bilayer is the simplest model of plasma membrane transport. Solutes that are lipid soluble will be able to dissolve into, hence pass through, the lipid bilayer; they will have a high permeability. Solutes that are not lipid soluble will not traverse the lipid bilayer. Oxygen, for example, is quite lipid soluble and passes relatively freely through a hexadecanol membrane. However, O_2 is about 100 × more permeable to a plasma membrane than to simple lipid bilayers. This is probably because the close packing of lipids in the plasma membrane is partly disrupted by other large, "awkwardly shaped" molecules, thus providing pores, or gaps. The random thermal motion of the lipids would also create temporary pores large enough for transport of small molecules, such as O_2, CO_2, and water. It is therefore important to examine properties of membrane permeation with a realistic model of, or an actual, plasma membrane, and appreciate that there are a variety of mechanisms for membrane permeation (Figure 6–3).

Diffusion

The passive permeability of a lipid membrane varies for different solutes; it depends on the **permeability coefficient** (*P*). The permeability of a solute should be proportional to its lipid solubility, or **partition coefficient** (*K*). The partition coefficient is the ratio of the solubility of a solute in a solvent (e.g., olive oil or ether) to the solubility in water. There is a strong correlation between the permeability coefficient and the partition coefficient, and a strong influence of molecular weight on *P*. Lipid-soluble solutes are passively permeable to cell membranes because they dissolve into the membrane lipid bilayer. Their passage through cell membranes is essentially a diffusional process, described by Fick's first law of diffusion. Some small water-soluble solutes may also diffuse through small pores in the cell membrane, but most water-soluble solutes are too large for this.

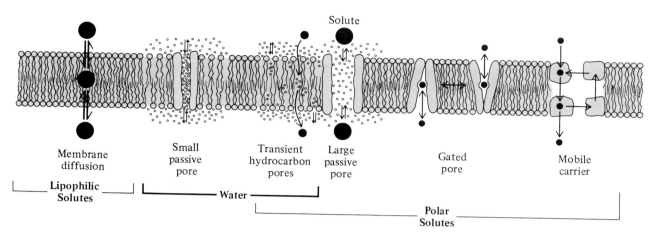

FIGURE 6–3 Possible mechanisms for permeation of cell membranes by water, lipophilic solutes, and polar solutes (e.g., ions). Lipophilic solutes can diffuse directly through the membrane because they are lipid soluble. Water may permeate through the spaces between hydrophobic lipid molecules, specific water pores, or other pores; polar solutes may permeate through the lipid bilayer but are more likely to permeate through specific pores, gated pores, or "ferryboat" mobile carriers.

The passive permeation of lipid-soluble solutes through cell membranes depends on their partition coefficient. It is relatively straightforward to rearrange Fick's first law of diffusion to a form that includes the partition coefficient and the diffusion coefficient of the membrane for the solute (D_m; cm² sec⁻¹);

$$J_m = D_m K A_m (C_{m2} - C_{m1})/\Delta x_m$$
$$= P_m A_m (C_{m2} - C_{m1}) \qquad (6.1a)$$

where J_m is the net transmembrane flux (mol sec⁻¹), A_m is the membrane surface area (cm²), C_m is the solute concentration on each side of the membrane (mol cm⁻³), and Δx_m is the thickness of the membrane (cm). The membrane diffusion coefficient, partition coefficient, and membrane thickness can be consolidated into a single coefficient, the permeability (P_m; cm sec⁻¹), i.e.,

$$P_m = D_m K/\Delta x_m \qquad (6.1b)$$

The membrane permeability should be directly proportional to the partition coefficient and the membrane diffusion coefficient, and inversely proportional to the membrane thickness. The diffusion coefficient also depends on other variables, including the molecular weight of the solute; in general, $D \propto \sqrt{1/\mathrm{MWt}}$. Thus, P and P/K should also depend on the molecular weight. Such an inverse dependence of P/K on MWt is observed for many membranes.

The permeability of solutes is not well correlated with their partition coefficient if there are specific transport mechanisms, either passive or active. Small nonlipophilic molecules, such as water and urea, may also diffuse through the plasma membrane pores. In theory, pores of diameter ≈ 0.8 nm form when the hydrocarbon tails of membrane lipids or a cholesterol molecule are temporarily displaced by random thermal motion; this would allow passage of water (<0.2 nm), urea (0.2 nm), hydrated Na⁺ (0.56 nm), and K⁺ (0.38 nm), but not Ca²⁺ (0.96 nm) or Mg²⁺ (1 nm). Because of specific transport mechanisms, many other polar solutes have a permeability higher than predicted by their K and molecular weight (see below).

There can be a flow of water through a membrane's transient hydrocarbon or protein pores due to a hydrostatic pressure difference or osmotic concentration difference; this hydraulic water flow is called **bulk flow**. There is a fundamental difference at the molecular level between the permeability coefficient of water for diffusional and hydraulic flow. The hydraulic permeability (P_f, or filtration permeability) is generally higher than the diffusional permeability (P_d; Sha'afi 1981). Bulk water movement thus "enhances" the rate of diffusion of water across a membrane; it also enhances the rate of exchange of other small solutes, such as urea, which are drawn along by the water flow. This enhancement of solute flux by bulk water flow is

solvent drag. The incorporation of cholesterol into membranes generally decreases their diffusional permeability to water; remember that cholesterol also generally decreases molecular mobility of the liquid membrane phase.

Many polar solutes are transported by specific membrane pores, and so their permeability is not dependent on their partition coefficient or molecular weight. Some permeation might occur through the same thermally formed hydrocarbon pores that allow water exchange, but the observation that water and small polar solute flux can be experimentally disassociated suggests that "water pores" are too small for the larger polar solutes and that there must be specific larger diameter pores. Urea has a low partition coefficient ($K_{ether} < 0.003$) but is highly permeable to most, although not all, biological membranes. There appear to be specific urea transport pores in at least some membranes. For example, the urea permeability of toad urinary bladder and the mammalian nephron can be increased by antidiuretic hormone (as is water permeability), but there appear to be different pores for water (smaller diameter and more of them) and urea (larger diameter and fewer of them).

Ions are another category of small polar molecules that can be passively transported across membranes through membrane pores or specific ion channels, e.g., Na^+, K^+, Cl^-, and Ca^{2+}. For ions, the combined electrical and chemical driving forces, the **electrochemical gradient**, drives diffusion. This is significant because most membranes generally have a resting membrane potential of about -90 mV, which favors ion flux even in the absence of a concentration gradient. Specific mediated-transport mechanisms for ions will be further discussed below.

Mediated Transport

The permeation mechanisms discussed so far are driven by diffusion; the flux is proportional to the permeability and the concentration or electrochemical gradient. However, membrane permeation of many solutes is not explained by simple diffusion; the permeation flux may be much faster than predicted, may be a nonlinear function of the concentration difference, may be inhibited by chemical analogs or reagents that alter protein structure, may become saturated, and may occur against a concentration gradient. Transport of many solutes is mediated by specific membrane proteins, in a fashion resembling enzyme-substrate interactions; this is **mediated transport**.

In **facilitated diffusion**, the flux is mediated transport but is passive and in the direction down the electrochemical gradient, i.e., "downhill" from a high concentration to a low concentration. In contrast, **active transport** is mediated transport but is an "uphill" flux against an electrochemical gradient, i.e., from a low to a high concentration.

The basic concept of mediated transport is simple. A protein carrier molecule binds the solute molecule, then a conformational alteration of the protein-solute complex translocates the solute across the membrane; the solute then separates from the carrier on the opposite side of the membrane (West 1983). The essential features of such a mediated-transport mechanism are:

1. there is a transport protein with a solute-specific binding site,
2. the binding site alternates in location from one side of the membrane to the other, and
3. the kinetic events of transition from one state to the other have specific rate constants.

The characteristic properties of such a system are:

1. the rate of flux is greater than predicted by a solute's partition coefficient, size, or molecular weight;
2. transport becomes saturated at high solute concentrations;
3. chemically or sterically similar solutes can competitively inhibit the flux; and
4. reagents that interfere with the protein carrier can noncompetitively inhibit solute flux.

Some facilitated exchange systems have more complex properties, including exchange diffusion and countertransport (see below).

Mediated transport can involve a mobile intramembrane carrier, i.e., is **mobile carrier-mediated**, or a transmembrane pore that has a closing mechanism, i.e., a **gated channel-mediated** pore (Figure 6–3). In practice, the end result is the same for carrier-mediated and channel-mediated transport, and it is experimentally difficult to distinguish between the two mechanisms. However, there are well-established examples of both types of mediated transport and some solutes may be transported by a combination of both mechanisms.

Channel-mediated transport resembles the kinetics of a simple Michaelis-Menten enzymatic reaction (Chapter 3)

$$C + S_i \rightleftharpoons CS \rightleftharpoons C + S_o \qquad (6.2a)$$

where C is the channel protein, S_i is the substrate on the inside of the membrane, and S_o is the substrate translocated to the outside. The "enzyme reaction" is the translocation of the substrate from one side to the other, rather than a chemical reaction. The flux (J) is related in a hyperbolic fashion to substrate concentration ([S]).

$$J = \frac{V_{max}[S]}{K_t + [S]} \qquad (6.2b)$$

where V_{max} is the maximal velocity, K_t is analogous to the Michaelis-Menten constant, and [S] is the substrate concentration.

Mobile carrier-mediated transport is more complex; the carrier is located alternately at each side of the membrane and may bind or release the solute. There are eight possible "reactions" (each with a specific rate constant, a, b, c, . . . , g, h) for the mobile carrier and the solute (Figure 6–4A). If there is no substrate on the inside ($S_i = 0$), then the rate of flux (J) is related to the total carrier concentration

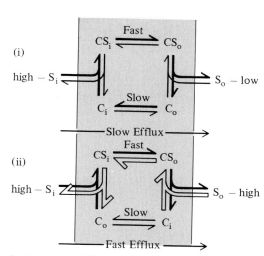

B. Exchange Diffusion

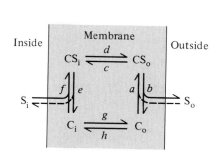

A. Mobile Carrier Transport System

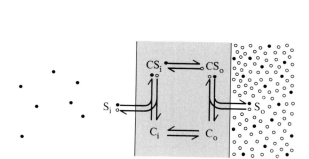

C. Counter Transport of ●

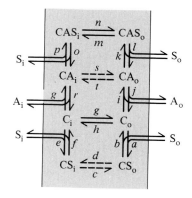

D. Cotransport

FIGURE 6–4 **(A)** General scheme for a mobile carrier transport system in the cell membrane; C represents the carrier molecule and S the substrate molecule, and i and o the two sides of the membrane. There are eight rate constants for the various reversible reactions (a–h) that can occur during transport of the substrate. **(B)** Exchange diffusion can transport a substrate faster from one side (e.g., inside, i) of a membrane to the other side (e.g., outside, o) if there is a higher concentration on the outside because there is a higher concentration of the transport intermediate CS_o, which is rapidly transported across the membrane to become CS_i. Solid arrows indicate the preferential efflux route; open arrows indicate the influx route. **(C)** Countertransport can produce a transient, passive, "uphill" transport of an isotopic or chemical analog (●) of a solute (o) from side i to side o. **(D)** Cotransport by a mobile carrier membrane transport system; the mobile carrier binds two substrates, S and A. There are 20 different rate constants (a–t) but some (s, t; c, d) are unimportant (or else the system would not be cotransport).

(C$_t$), the outside concentration (S$_o$), and the eight reaction constants as follows.

$$J = \frac{C_t[S_o]aceg/(a(ce + cg + dg + eg))}{[S_o] + (h + g)(bd + be + ce)/(a(ce + cg + dg + eg))}$$

(6.3)

This equation resembles the Michaelis-Menten hyperbolic kinetic equation (see Equation 6.2b).

The rate constant for the mobile carrier moving from one side of the membrane to the other may be equal to the rate constant for carrier-solute movement in the same direction (i.e., $d = g$, $c = h$) but this need not be so. In fact, the rate constants can differ by a factor of 10 to 100 $\times$, e.g., the loaded carrier may reorient faster than the unloaded carrier. The rate of efflux ($J_{i \to o}$) may be low even if the internal concentration (S$_i$) is high and the external concentration (S$_o$) is low because the rate of reorientation of the unloaded carrier from the outside to the inside is slow (Figure 6–4Bi). The rate of efflux will increase if the concentration of solute on the outside is high because of the faster reorientation of the carrier-solute complex towards the inside (Figure 6–4Bii). The influx $J_{o \to i}$ also increases at high [S$_o$]. The unidirectional flux of a solute can thus depend on the unidirectional flux in the opposite direction if there is an asymmetry in the rate constants; this is **exchange diffusion**.

Countertransport is passive transport against a concentration gradient; it can occur if there is an asymmetry in concentrations of a solute and a similar solute (or radioisotope) that is also transported (Figure 6–4C). Let us assume that there is a very high concentration of solute (S) on the outside and none on the inside, and there is a low concentration of the analog (A) on the inside and a higher concentration on the outside (i.e., $0 = S_i < A_i < A_o \ll S_o$). The carrier system will preferentially transport S inwards (since $S_o \gg A_o$ and will competitively inhibit inward transport of A) and preferentially transport A outwards (since $A_i > S_i = 0$). There is an efflux of A even though $A_o > A_i$. Remember that the carrier cannot distinguish A from S, although we can. This passive uphill transport of A is a transient phenomenon that will cease when the concentrations of A and S come to equilibrium across the membrane. The occurrence of countertransport can be used as a criterion to distinguish between pore exchange and mobile carrier exchange, since it would not occur for a gated-pore transport mechanism.

Cotransport of two solutes (S and S′) is considerably more complex than mediated transport of one solute. There are 20 possible reactions and rate constants for cotransport (Figure 6–4D), although four (*c*, *d*, *s*, *t*) must be insignificant by definition, since for cotransport both S and S′ are required for transport. There are two possible routes from C to CSS′

$$C + S \to CS + S' \to CSS'$$

and

$$C + S' \to CS' + S \to CS'S$$

The cotransport is said to be ordered if there is a required sequence for the formation of CSS′ and is random if either pathway to CSS′ occurs. The Na$^+$-K$^+$ ATPase enzyme/ion pump is one example of a cotransport system; another is the Na$^+$-linked active transport of amino acids and some monosaccharides across the intestinal mucosa. These examples will be further discussed below.

Facilitated Diffusion. Facilitated diffusion is a mediated, "downhill" transport process that is better described by a hyperbolic flux concentration relationship (Equation 6.2b) than by Fick's first law of diffusion. A typical example of the flux concentration relationship showing saturation kinetics is observed for the facilitated uptake of galactose by the human erythrocyte (Figure 6–5A).

Facilitated transport mechanisms include both mobile carrier and gated pore models. This variation in transport mechanism is well exemplified by ionophores, compounds of bacterial origin that function as ion-selective pores when introduced into a membrane. These ionophores are heterogeneous in terms of their specific molecular structure, but all are characterized by a folded structure that has a hydrophobic exterior and an interior lined by oxygen atoms that can bind cations. Valinomycin is a mobile carrier ionophore that is highly specific for K$^+$ rather than Na$^+$ (by a preference of over 10,000 to 1); this high specificity is remarkable considering the similar sizes of hydrated K$^+$ and Na$^+$ ions (see below). Valinomycin clearly is a mobile carrier, requiring normal membrane fluidity (Figure 6–5B). In contrast, gramicidin is a channel ionophore. Its protein chain appears to form a spiral (about 3 nm long and 0.4 nm diameter) that spans the hydrophobic portion of the membrane and forms a water-filled, ion-conducting pore that allows the permeation of water and many small ions. Gramicidin is not as discriminating for solutes as is valinomycin.

The glucose carrier of the human erythrocyte membrane is another example of a facilitated diffusion system but is somewhat intermediate between a mobile carrier and a gated pore model. The carrier

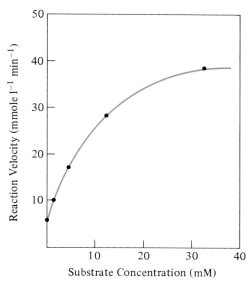

FIGURE 6–5 **(A)** Example of carrier-mediated transport showing the nonlinear relationship between transport rate (J; mmole l^{-1} cells^{-1} min^{-1}) and intracellular substrate concentration ([S]; mM), and saturation kinetics for the human red blood cell. **(B)** Schematic representation of a mobile ionophore (valinomycin), a channel (pore-forming) ionophore (gramicidin), and a somewhat intermediate membrane transport system, the gated pore glucose carrier. *(Data from Ginsburg and Stein, 1975; modified from West 1983; Harrison and Lunt 1980.)*

A

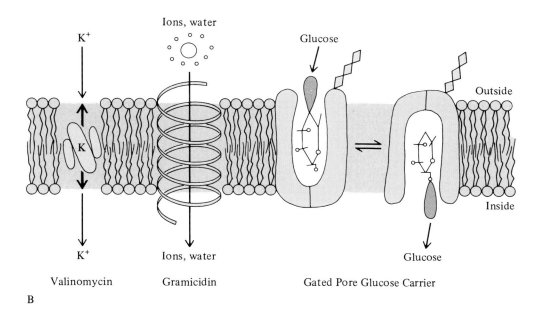

B

is asymmetric in that a carbohydrate moiety is located on the outside of the membrane; it therefore is not a freely mobile carrier. However, the active site for glucose binding appears to be alternately exposed to the outside and inside membrane surfaces, and in this sense it is like a mobile carrier (Figure 6–5B).

Active Transport. Active transport mechanisms move solutes against an electrochemical gradient (i.e., "uphill") and require the expenditure of metabolic energy. It is a primary active transport system

if the free energy change for the uphill solute transport by the carrier is provided directly by the metabolic energy stores (e.g., ATP). It is a secondary active transport system if the free energy change is provided by other energy sources, such as ion electrochemical gradients.

Probably the best understood primary active transport mechanism is the Na$^+$-K$^+$ ATPase (Na$^+$ pump) of animal plasma membranes. The Na$^+$-K$^+$ pump generally transports 3 Na$^+$ out and 2 K$^+$ in, per ATP hydrolyzed. The net transport of one positive charge out per ATP hydrolyzed means that the pump establishes an electrical gradient, i.e., it is **electrogenic**. Sometimes there is a 1:1 ratio of Na^+ to K^+ transport (e.g., 2 Na^+:2 K^+ or 3 Na^+:3 K^+ per ATP) and the pump is **electroneutral**. However, the 1:1 rather than 3:2 ratio of Na^+:K^+ transport may be a consequence of the membrane being leaky to Na^+, allowing some of the extruded Na^+ to reenter the cell and giving the impression of no net charge transfer by the pump. The Na^+-K^+ ATPase is a tetramer, $\alpha_2\beta_2$, of α-subunits (molecular weight about 10^5) and β-subunits (molecular weight about 5 10^4). The larger subunits span the plasma membrane from the inside to outside surfaces; the binding sites for ATP and Na^+ are located on the inside, and for K^+ on the outside. Ouabain, a cardiac glycoside that is a specific reversible inhibitor of the Na^+-K^+ ATPase, binds to the outside of the plasma membrane. There appear to be two forms of the Na^+-K^+ ATPase enzyme, E_1 and E_2, that differ significantly in binding properties. The chemical events that occur in the Na^+-K^+ transport cycle are fairly well understood (Figure 6–6) although the details of translocation process across the membrane are not. The two slowest steps are the interconversion of E_1 and E_2, i.e., $E_1 \rightarrow E_2$ and $E_2 \rightarrow E_1$; these are suspected to be the actual translocation processes.

Secondary active transport uses other forms of energy, such as electrochemical gradients, to provide free energy for active transport. The halophytic bacterium *Halobacterium halobium* has a unique light-driven proton pump, bacteriorhodopsin, which transports H^+ ions out of the bacterium; the energy is derived from illumination. This proton pump can supplement or even replace the respiratory chain proton pumps during anoxia. In animal cells, the electrochemical gradient for secondary active transport is most commonly the Na^+ gradient (it can be a H^+ gradient in plant and bacterial cells; Nobel 1983).

The Na^+-linked uptake of amino acids and some monosaccharides (glucose, galactose) by the epithelial cells of the intestine, or integuement, of certain marine invertebrates is an example of secondary active transport (Figure 6–7). The Na^+ pump is located on the serosal membrane of the epithelium (blood side) and actively transports Na^+ from inside the epithelial cell into the body fluids. The low intracellular (Na^+) and negative intracellular charge provide a substantial electrochemical gradient for Na^+ influx into the epithelial cells. A "facilitated diffusion" type of transport mechanism binds an amino acid (or monosaccharide) and an Na^+ at the same time; this is a **symport**. Alanine-Na^+ cotransport is a typical example. Alanine will slowly diffuse into an epithelial cell in the absence of Na^+ until the intracellular and luminal concentrations are equal. The presence of Na^+ increases the rate

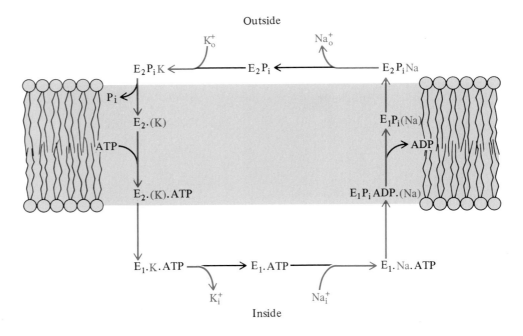

FIGURE 6–6 Proposed movement of the Na^+-K^+ ATPase (E_1 and E_2) to exchange Na^+ and K^+ across the cell membrane and the role of ATP hydrolysis. *(Modified from Karlish, Yates, and Glynn 1978.)*

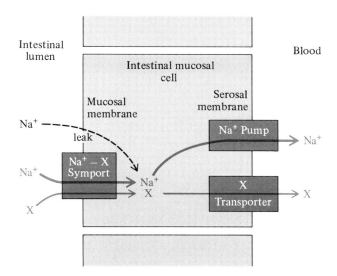

FIGURE 6–7 Cotransport of solute "X" (e.g., glucose, amino acids) and sodium by the intestinal mucosal cell across the lumen membrane by a symport carrier molecule, then separate transport across the mucosal membrane by the Na^+ pump and a transport protein for "X." *(Modified from Harrison and Lunt 1980.)*

of uptake, and the internal alanine concentration approaches ten times that of the luminal concentration, i.e., there is rapid "uphill" transport of alanine driven by the Na^+ gradient. The Lineweaver-Burk plot for alanine transport, with and without Na^+, indicates that transport is independent of the $[Na^+]$ at infinite [alanine], but the absence of Na^+ decreases the affinity of the transport system (K_t, which is analogous to the Michaelis-Menten constant for enzymatic reactions, K_m). The absence of Na^+ is effectively acting as a competitive inhibitor of transport.

Water can diffuse across membranes via small pores and there can also be a bulk flow of water due to either a hydrostatic pressure difference or an osmotic concentration difference (see Chapter 16). Water flow can be osmotically linked to active transport of solutes, especially Na^+, through establishment of local osmotic gradients (see Chapter 16). No specific carrier for active transport of water has yet been identified, although active transport of water has occasionally been hypothesized.

Dynamics of Semipermeable Membranes

Biological membranes vary greatly in their permeability to different solutes, i.e., they are complex **semipermeable membranes**. The semipermeability of biological membranes has far-reaching consequences (Donnan 1927). Consider two water com-

partments separated by a semipermeable membrane (Figure 6–8A). If a solute, for example KCl, is added to one compartment (side 1), then $[K^+]_1 > [K^+]_2 = 0$ and $[Cl^-]_1 > [Cl^-]_2 = 0$. If the membrane is freely permeable to both K^+ and Cl^-, then both ions will diffuse from side 1 to side 2 until $[K^+]_1 = [K^+]_2 = [Cl^-]_1 = [Cl^-]_2$. The addition of potassium-proteinate to side 1 and KCl to side 2 so that $[K^+]_1 = [K^+]_2$ produces a complex situation if the membrane is permeable to K^+ and Cl^- but not Pr^- (Figure 6–8B). The Pr^- cannot diffuse from side 1 to 2 despite its concentration difference, and K^+ would not be expected to diffuse because the concentrations are the same. The Cl^- concentration is higher on side 2 and it is not in equilibrium; it will diffuse from side 2 to side 1. However, Cl^- will not continue to diffuse until $[Cl^-]_1 = [Cl^-]_2$ because movement of Cl^- from side 2 to side 1 establishes a change imbalance, making side 2 negative and side 1 positive. The negative charge on side 1 will repel Cl^- from side 2 to side 1, and will also influence the distribution of K^+ ions even though the $[K^+]$ was initially in concentration equilibrium across the membrane. K^+ will be electrically attracted towards side 1 ($-$) and repelled from side 2 ($+$). The Pr^- distribution is unaffected because it is impermeable. When the K^+ and Cl^- ions come to equilibrium, there will be (1) a concentration gradient for K^+, Cl^- (and Pr^-) across the membrane; (2) a net positive charge on side 2 and a negative charge on side 1; and (3) a higher osmotic concentration on side 1 than side 2. This equilibrium distribution of permeable and impermeable ions across a semipermeable membrane is called a **Donnan equilibrium**.

The magnitude of the electrical gradient across a semipermeable membrane due to a Donnan equilibrium can be calculated using the Nernst equation

$$E_m = \frac{RT}{zF} \ln \frac{C_2}{C_1} = \frac{2.303RT}{zF} \log_{10} \frac{C_2}{C_1} \quad (6.4)$$

where E_m is the membrane potential (volts), R is the gas constant, T is the temperature (°K), z is the charge on the ion (e.g., $+1$ for Na^+ and K^+, -1 for Cl^-, $+2$ for Ca^{2+}), F is Faraday's constant, and C_1 and C_2 are the ion concentrations on each side of the membrane. The value of RT/F is about 0.025 volt at T = 20° C, or 58 mV if we use $\log_{10}$ rather than natural log (ln). But, in the above example, do we use $[K^+]_1$ and $[K^+]_2$ or $[Cl^-]_1$ and $[Cl^-]_2$ to calculate E_m? We can use either, because both ions are distributed at equilibrium across the membrane. Therefore,

$$E_m = \frac{RT}{(+1)F} \ln \frac{[K^+]_2}{[K^+]_1} = \frac{RT}{(-1)F} \ln \frac{[Cl^-]_2}{[Cl^-]_1} \quad (6.5a)$$

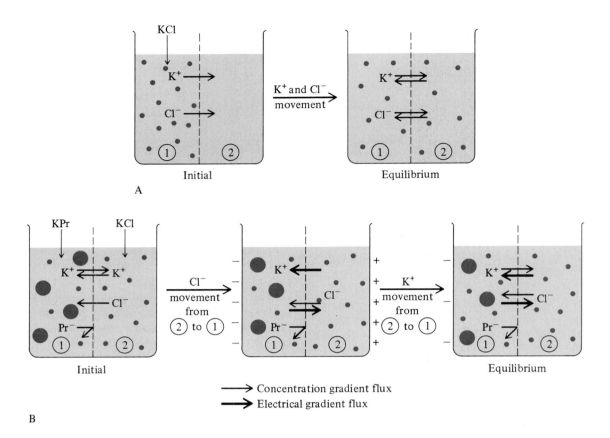

FIGURE 6–8 **(A)** Distribution of KCl solution added to one side (1) of a permeable membrane; there are equal concentrations of K^+ and Cl^- on each side of the membrane at equilibrium. **(B)** Donnan distribution of potassium proteinate solution and equal concentration of KCl added across a semipermeable membrane (permeable to K^+ and Cl^- but not Pr^-). At equilibrium, there is a concentration gradient (thin arrows) for K^+ and Cl^- balanced by an electrical gradient (thick arrows).

Rearranging this equation yields the characteristic reciprocal distribution of the cation and anion for a Donnan equilibrium.

$$\frac{[K^+]_2}{[K^+]_1} = \frac{[Cl^-]_1}{[Cl^-]_2} \quad \text{or} \qquad (6.5b)$$
$$[K^+]_2[Cl^-]_2 = [K^+]_1[Cl^-]_1$$

Electroneutrality must also be maintained on each side of the membrane (although there is a small charge imbalance due to the membrane potential; see below) and so

$$[Pr^-]_1 + [Cl^-]_1 = [K^+]_1 \quad \text{and} \qquad (6.5c)$$
$$[Cl^-]_2 = [K^+]_2$$

The magnitude of the Donnan potential can be estimated from Equations 6.5a, 6.5b, and 6.5c. Let us assume that the intracellular fluid has a protein concentration of 40 mEq l^{-1}, a K^+ concentration of 140 mEq l^{-1}, and a Cl^- concentration of 100 mEq l^{-1} (these values are realistic for most cells, but we will unrealistically ignore Na^+). If the outside $[Pr^-]$ is 0, then $[Cl^-]_o = [K^+]_o = 118$ mEq l^{-1}. The E_m due to the Donnan equilibrium can now be calculated as $RT/F \ln [K^+]_i/[K^+]_o = -4.2$ mV.

Resting Membrane Potentials

Electrical potentials, of the order of a few millivolts (mV) to over 100 mV, commonly occur across cell membranes. The resting membrane potential (E_m) remains constant for many types of cells but alters dramatically for excitable (irritable) cells, such as sensory, nerve, and muscle. Essentially three mechanisms contribute to these potentials: (1) the electrogenic ion pump, (2) the Donnan equilibrium, and (3) diffusion potentials.

The electrogenic Na^+-K^+ pump exchanges 3 Na^+ for 2 K^+ per ATP hydrolyzed. There is thus a net transfer of 1 + charge, and this contributes to the normal membrane potential. However, it is not the immediate cause of the normal membrane potential, as is apparent when the Na^+-K^+ pump is poisoned by specific blockers, such as ouabain. There is little or no effect of ouabain on E_m (although the E_m ultimately declines to 0 as the Na^+ and K^+ concentration gradients dissipate; see below).

A Donnan equilibrium contributes a small membrane potential because the intracellular fluid has a higher protein (Pr^-) concentration than the extracellular fluid. The Donnan potential is generally not responsible for the resting E_m.

The third and most important mechanism contributing to the E_m is the existence of marked diffusion potentials for various ions across the membrane; these occur because there are large ionic concentration differences for ions that are permeable across the membrane.

Diffusion Potentials

There are three important aspects to the electrical contribution of ions to membrane potentials: (1) ion mobility, (2) selective ion permeability, and (3) ion concentration gradients.

Ion Mobility. All ions are not equally mobile in solution. Ions experience frictional forces when in motion, and these frictional forces retard their movement. The magnitude of the frictional force depends on the size of the ion, and so mobility (u) is a strong function of size. The absolute ion mobility is defined as the average velocity (cm sec^{-1}) in an electrical field of 1 V cm^{-1}. Perhaps contrary to expectation, the ions of smaller atomic radius have lower mobilities (Table 6–2). This is because ions are generally covered with a surface "hydration layer" of water molecules, attracted by the charge density of the ion. Larger ions have a lower charge density (because they have a greater surface area) and therefore have a thinner hydration layer and a smaller hydrated radius. It is possible to estimate the hydrated radius and number of water-of-hydration molecules in the hydration layer, e.g., Li^+, 0.06 nm hydrated radius, 6-12 H_2O; Na^+, 0.095 nm radius, 4.5 H_2O; K^+, 0.133 nm radius, 3 H_2O. The water of hydration is in a dynamic exchange with other water molecules, and so the hydrated ion should not be thought of as a fixed volume sphere with a central ion and peripheral hydration layer. The diffusion coefficient is directly proportional to mobility.

$$D = RTu/F$$

Ionic Permeability. The electrical properties of cell membranes depend on the selective (and often changing) permeabilities of the membrane to different ions. There are many specific ion channels, or pores, that make the cell membrane highly but selectively permeable to ions. For example, there are Na^+ channels, K^+ channels, Ca^{2+} channels, Cl^- channels, etc. These channels are usually very selective. The Na^+ channel is quite specific for Na^+; it is considerably less permeable to similar ions such as K^+, Rb^+, and Cs^+ (Table 6–3). The K^+ channel is highly specific for K^+ and relatively impermeable to Na^+ and Li^+. The selective permeability of channels is not just due to molecular weight and size, but also to hydrated radius and

TABLE 6–2

Mobility (u; μ sec^{-1} per Volt cm^{-1}) and diffusion coefficient (D; cm sec^{-1}; $= RTu/F$) for various important ions are dependent on atomic mass and hydrated ionic radius (r_{hyd}, nm). Values are for 25° C. *(Data from Hille 1984; Davis, Gailey, and Whitten 1984.)*

	Atomic Mass	Hydrated Radius r_{hyd}	Mobility u	Diffusion Coefficient D
Ca^{2+}	40.08	0.099	3.08 10^{-4}	0.79 10^{-5}
Li^+	6.94	0.060	4.01 10^{-4}	1.03 10^{-5}
SO_4^{2-}	96.06		4.15 10^{-4}	1.06 10^{-5}
Na^+	22.99	0.095	5.19 10^{-4}	1.33 10^{-5}
K^+	39.10	0.133	7.62 10^{-4}	1.96 10^{-5}
Cl^-	35.45	0.181	7.92 10^{-4}	2.03 10^{-5}
I^-	126.90	0.216	7.96 10^{-4}	2.04 10^{-5}
Br^-	79.90	0.195	8.09 10^{-4}	2.08 10^{-5}

TABLE 6–3

Relative permeabilities of sodium and potassium channels in an excitable cell membrane of a frog axon to various ions.

	Na^+ Channel P/P_{Na}	K^+ Channel P/P_k
Li^+	0.930	<0.018
Na^+	1.000	<0.010
Th^+	0.330	2.30
K^+	0.086	1.000
NH_4^+	0.160	0.13
Rb^+	<0.012	0.91
Cs^+	<0.013	<0.077

charge. Some ion channels are relatively nonselective, e.g., are general cation or anion channels.

What is the structure of such an ion pore, and how can it be so selective for particular ions? The pore has a specific transmembrane protein structure that has a selectivity filter at some point along its length (Figure 6–9A). The selectivity filter discrimi-

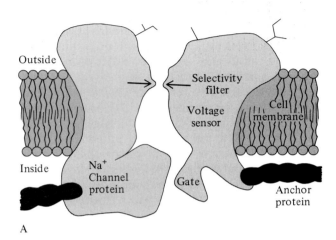

A

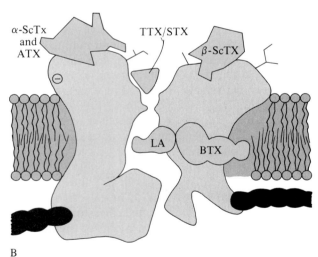

B

FIGURE 6–9 (A) Diagrammatic representation of an ionic channel with typical properties of selectivity (due to a size or charge-dependent selectivity filter) and a gating mechanism that is voltage dependent; a sensor portion of the protein channel acts as the sensor that monitors the transmembrane potential and alters the gating mechanism accordingly. (B) Tetrodotoxin (TTX) and saxitoxin (STX) are highly specific blockers of sodium channels; both bind to the external opening of the channel. Other toxins, such as batrachatoxin (BTX), scorpion toxins (α- and ß-ScTX), and anemone toxins (ATX) bind to other parts of the Na⁺ channel and increase Na⁺ conductance. LA, local anesthetic. *(Modified from Albuquerque, Daly, and Warnick 1988.)*

nates between ions mainly on hydrated size and charge. The differing hydrated radii of various ions are one important aspect to ion selectivity. Pores may also have a gating mechanism that can, by conformational change, open or close the pore to ion flow. A sensor, which may be voltage sensitive, controls the gating mechanism.

The identification of specific ion channels was aided greatly by the discovery of highly specific channel blocking agents. For example, tetrodotoxin (TTX), which is obtained from puffer fish and other fish of the Order Tetraodontiformes, and saxitoxin (STX) from the dinoflagellates *Gonyaulax*, specifically block the Na⁺ channel when applied to the outside of the membrane (Figure 6–9B). Other agents also affect the Na⁺ channel permeability. Various scorpion and anemone toxins (ScTX and ATX) prevent Na⁺ channel inactivation. Batrachatoxin (BTX) from poison-arrow frogs increases Na⁺ channel permeability.

Ionic Concentration Differences. In general, the intracellular fluid has a lower Na and Cl⁻ concentration and a higher K⁺ concentration than the extracellular fluid. Ion gradients are summarized for frog and insect muscle in Table 6–4. The intracellular K⁺ concentrations are about 10 to 60 × higher than the extracellular concentrations, whereas intracellu-

TABLE 6–4

Intracellular and extracellular ion concentrations for frog and insect muscle.			
	[ ]$_{inside}$	[ ]$_{outside}$	[ ]$_i$ / [ ]$_o$
Frog Muscle[1]			
	Muscle Cell	*Plasma*	
K⁺	124	2.3	55
Na⁺	10	109	0.092
Cl⁻	1.5	77.5	0.019
Ca²⁺	4.9[3]	2.1[3]	—
Insect Muscle[2]			
	Muscle Cell	*Hemolymph*	
K⁺	86	11	7.6
Na⁺	22	117	0.19
Ca²⁺	3.4[3]	3.6[3]	—

[1] *(From Conway 1957.)*
[2] *(From Natochin and Parnova 1987.)*
[3] The total intracellular calcium concentration is given; the cytoplasmic calcium ion activity is considerably lower due to sequestration of calcium in sarcoplasmic reticulum and it is not meaningful to calculate the [ ]$_i$ / [ ]$_o$ from these values.

lar Na⁺ and Cl⁻ concentrations are only 0.02 to
0.30 × the extracellular values. There are similar
differences for concentrations of other ions across
the cell membrane.

Equilibrium Potentials

There are marked ionic concentration differences
across cell membranes. The ultimate cause of these
ion concentration gradients is the Na⁺-K⁺ ATPase,
but it is irrelevant in this respect whether the pump
is electrogenic or neutral. There are concentration
gradients for many ions, K⁺, Na⁺, Cl⁻, Ca²⁺,
Mg²⁺, etc., but let us initially consider K⁺ because
it is normally quite permeable (so is Cl⁻ but why
we needn't consider Cl⁻ here will be explained
below).

Consider a selectively permeable membrane (permeable to K⁺ but not Cl⁻) separating a 1.0 M KCl
solution (side 1) from a 0.1 M KCl solution (side 2;
Figure 6–10A). K⁺ ions tend to diffuse down their
concentration gradient from the 1 M side to the 0.1
M side, thereby establishing a positive electrical

potential on the 0.1 M side and a negative potential
on the 1 M side. The K⁺ ions will not continue to
diffuse until the K⁺ concentrations are equal on
either side of the membrane because the establishment of an electrical potential causes K⁺ movement
from side 1 to side 2. At equilibrium, the magnitude
of the membrane potential due to K⁺ movement
can be calculated from the Nernst equation

$$E_m = \frac{RT}{zF} \ln \frac{C_2}{C_1}$$
$$= 0.058 \log_{10} \frac{0.1}{1.0} = -58 \text{ mV} \quad (6.6)$$

at 20° C. Which side of the membrane is negative
and which is positive is indicated from the calculated
sign of E_m if we remember that the potential of side
1 relative to side 2 is calculated if the concentration
ratio is C_2/C_1 as above; side 1 had the high [K⁺]
and its potential was negative with respect to side
2. If we had used the concentration ratio of C_1/C_2
or considered anion flux ($z = -1$), then we would
calculate the potential of side 2 relative to side 1,
as +58 mV. If in doubt about the polarity of the

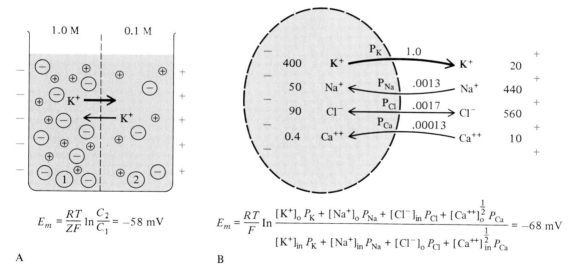

$$E_m = \frac{RT}{ZF} \ln \frac{C_2}{C_1} = -58 \text{ mV}$$

A

$$E_m = \frac{RT}{F} \ln \frac{[K^+]_o P_K + [Na^+]_o P_{Na} + [Cl^-]_{in} P_{Cl} + [Ca^{++}]_o^{\frac{1}{2}} P_{Ca}}{[K^+]_{in} P_K + [Na^+]_{in} P_{Na} + [Cl^-]_o P_{Cl} + [Ca^{++}]_{in}^{\frac{1}{2}} P_{Ca}} = -68 \text{ mV}$$

B

FIGURE 6–10 (A) Establishment of a membrane potential (E_m) due to passive distribution of
potassium ions across a semipermeable membrane (permeable to K⁺ but not Cl⁻). The concentration-driven flux of K⁺ (thin arrow) from the side with a concentration of 1.0 M KCl towards the side with a concentration of 0.1 M KCl is countered at equilibrium by an electrical-
driven flux (thick arrow). The magnitude of the E_m can be calculated using the Nernst equation (at 20° C)

$$E_m = \frac{RT}{zF} \ln \frac{[K^+]_2}{[K^+]_1} = 58.2 \log_{10} \frac{[K^+]_2}{[K^+]_1} \text{mV}$$

(B) Establishment of a membrane potential across a cell membrane by passive distribution of
ion species with differing permeabilities. The membrane potential can be calculated from the
concentration differences and permeabilities for all ion species using the Goldman-Hodgkin-
Katz equation; the greatest contribution to resting E_m is by K⁺ (shown in bold) because of its
high permeability. Concentration and permeability values are for the giant squid axon.

calculated membrane potential, just consider the direction of movement and charge for the permeable ion to determine which side is positive and which is negative.

What would happen if the membrane was permeable to both K^+ and Cl^-? Both ions would diffuse from side 1 to side 2, and at equilibrium their concentrations would be equal on either side of the membrane and there would be no membrane potential. However, it is of interest to consider in passing what happens during the equilibration process as well as at the equilibrium state. If the membrane was equally permeable to both K^+ and Cl^-, then we might expect no electrical potential to develop across the membrane. However, Cl^- ions have a higher mobility than K^+ ions ($7.92 \cdot 10^{-8}$ and $7.62 \cdot 10^{-8}$ cm^2 volt^{-1} sec^{-1} respectively), and so Cl^- ions diffuse more rapidly from side 1 to side 2 than do K^+ ions. This temporarily establishes a small electrical potential, negative on side 2 and positive on side 1. The magnitude of the transient membrane potential is calculated as

$$E_m = \frac{RT}{F} \frac{u^+ - u^-}{u^+ + u^-} \ln \frac{C_2}{C_1} \qquad (6.7a)$$

where u is the mobility of the $+$ and $-$ charged ions; for our example from above, side 1 is $+1.1$ mV with respect to side 2.

$$E_m = 58 \frac{(7.62 - 7.92)}{(7.62 + 7.92)} \log_{10} \frac{0.1}{1.0} = +1.1 \text{ mV} \quad (6.7b)$$

What if we had used NaCl rather than KCl? Na^+ has a lower mobility than K^+ and Cl^-, of $5.19 \cdot 10^{-8}$, and so the transient E_m is 12.1 mV, not 1.1 mV. The NaCl diffusion potential is much greater than the KCl diffusion potential. This is partly the reason for filling microelectrodes with a KCl solution rather than NaCl or some other solution; the small KCl diffusion potential that exists across the microelectrode tip due to differing ionic permeabilities doesn't affect the measurement of the membrane potential as much as would the higher NaCl diffusion potential.

The membrane potential calculated from the internal and external K^+ concentrations and Nernst equation (Equation 6.6) is called the **potassium equilibrium potential** (E_K), because it is the E_m if only the distribution of K^+ ions across the membrane contributes to the membrane potential. The value of E_K is readily calculated for the mammalian muscle cell as -98 mV (Table 6–5). The potassium equilibrium potential is approximately equal to the resting E_m because the potassium permeability is much greater than that of the other ions at rest, and as a first approximation K^+ is the only ion contributing to the E_m.

TABLE 6–5

Intracellular ($[\ ]_i$) and extracellular ($[\ ]_o$) concentrations and equilibrium potentials for major ions of mammalian skeletal muscle. *(Data from Hille 1984.)*

	$[\ ]_o$	$[\ ]_i$	$[\ ]_o / [\ ]_i$	E_{eq}[1]
Na^+	145	12	12	$+67$ mV
K^+	4	155	0.026	-98 mV
Cl^-	123	4.2	30	-90 mV
Ca^{2+}	1.5	$<10^{-7}$	>15000	$+128$ mV

[1] $E_{eq} = \dfrac{RT}{zF} \ln \dfrac{[\ ]_o}{[\ ]_i} = 61.54 \log_{10} \dfrac{[\ ]_o}{[\ ]_i}$ mV at 37° C.

If Na^+ had been the more important ion, then the E_m would be closely approximated by the **sodium equilibrium potential** ($E_{Na} = +67$ mV). The resting membrane potential is not close to the sodium equilibrium potential, but the membrane potential of excitable cells does approach the E_{Na} during an action potential (see below).

We can similarly calculate the **chloride equilibrium potential** (E_{Cl}) as -90 mV. The chloride equilibrium potential, like the potassium equilibrium potential, is close to the resting membrane potential. This is because chloride is passively distributed across the membrane. The intracellular-extracellular ratio of Cl^- is not due to active transport but to passive distribution in response to the normal E_m; the intracellular potential is negative, so the intracellular chloride concentration is low. This is why chloride can be omitted from the Goldman-Hodgkin-Katz equation without serious error. In some cells, P_{Cl} is fairly low whereas it is high in other cells.

An equilibrium potential can be calculated for any other ion present in biological systems. For example, the calcium equilibrium potential (E_{Ca}) is $+128$ mV.

Goldman-Hodgkin-Katz Potential

Animal cells are not such simple systems as K^+Cl^- solutions; many other ions are present, especially Na^+. Fortunately, it is relatively easy to extend the Nernst equation to a more complex form that can describe the electrical potential across biological membranes. All we need to appreciate is that (1) any $+$ charged ion is equivalent to any other $+$ charged ion, electrically speaking, i.e., Na^+ is electrically equivalent to K^+, and (2) the contribution of any ion species to the membrane potential depends not only on its charge, but also on its permeability, i.e., freely permeable ions will readily diffuse down their concentration gradient and contribute to E_m, whereas impermeable ions will not diffuse across the membrane and cannot contribute to the E_m. The net contribution of any ion species (X) by movement in one direction across a membrane is therefore equal to the product of its concentration on that side and its permeability, i.e., $[X].P_X$. The Goldman-Hodgkin-Katz equation calculates the E_m from all species of ion present by taking into account for each ion its concentration on each side of the membrane and its permeability; the general form is as follows.

$$E_m = \frac{RT}{F} \ln \frac{[K^+]_o.P_K + [Na^+]_o.P_{Na} + [Cl^-]_i.P_{Cl} + [Ca^{2+}]_o^{\frac{1}{2}}.P_{Ca} + \ldots}{[K^+]_i.P_K + [Na^+]_i.P_{Na} + [Cl^-]_o.P_{Cl} + [Ca^{2+}]_i^{\frac{1}{2}}.P_{Ca} +} \quad (6.8a)$$

Figure 6–10B shows this concept diagrammatically for four of the major ions: K^+, Na^+, Cl^-, and Ca^{2+}. The extracellular, or outside, concentration of ions is placed in the numerator by convention so that the calculated potential is that of the inside relative to the outside. Note that for anions the outside concentration is in the denominator. The charge term z must be incorporated into the log term. Hence, the $[Ca^{2+}]$ term is raised to the power $1/z$, i.e., $1/2$; the power terms for Na^+ and K^+ are omitted for clarity since $1/z$ is 1. For Cl^-, $z = -1$ and so the $[Cl^-]_i$ is in the numerator and $[Cl^-]_o$ is in the denominator. Fortunately, this equation can be greatly simplified to

$$E_m = \frac{RT}{F} \ln \frac{[K^+]_o.P_K + [Na^+]_o.P_{Na}}{[K^+]_i.P_K + [Na^+]_i.P_{Na}} \quad (6.8b)$$

This is possible because many ions (e.g., Ca^{2+}) are not very permeable, and Cl^- can generally be ignored (even though it may be fairly permeable) because it is passively distributed across the membrane and its permeability remains constant (see below). Furthermore, K^+ is normally much more permeable than Na^+, and so a further simplification is possible.

$$E_m = \frac{RT}{F} \ln \frac{[K^+]_o}{[K^+]_i} \quad (6.8c)$$

How well does the measured E_m conform to the value calculated from the above equations? Very well, in fact! Table 6–6 summarizes ion concentrations and permeabilities for squid giant axons, and shows that the measured E_m of -68 mV is closely matched by the calculated E_m from Equation 6.8a (-74 mV) and Equation 6.8b (-74 mV). The most simplified form of the Goldman-Hodgkin-Katz

TABLE 6–6

Intracellular and extracellular ionic concentrations (mEq l^{-1}), and relative membrane permeabilities for the squid giant axon. These data are used to calculate the resting membrane potential using the Goldman-Hodgkin-Katz equation using all four ions, the simplified equation for only Na^+ and K^+, and only K^+ (i.e., the potassium equilibrium potential). The equilibrium potentials for each ion are calculated using the Nernst equation. Values are calculated at 15° C assuming $\frac{RT}{F} \log_{10} = 57.17$.
(Data from Curtis and Cole 1942; Hodgkin 1958; Meves and Vogel 1973).

	[ ]$_i$	[ ]$_o$	**Relative P**
Na^+	50	440	0.0013
K^+	400	20	1.0
Cl^-	90	560	0.0017
Ca^{2+}	0.4	10	0.00013

Goldman-Hodgkin-Katz Equation
$$E_m = \frac{RT}{F} \ln \frac{[K^+]_o P_K + [Na^+]_o P_{Na} + [Cl^-]_i P_{Cl} + [Ca^{2+}]_o^{\frac{1}{2}} P_{Ca}}{[K^+]_i P_K + [Na^+]_i P_{Na} + [Cl^-]_o P_{Cl} + [Ca^{2+}]_i^{\frac{1}{2}} P_{Ca}}$$
$$= -73.6 \text{ mV}$$

Short version of Goldman-Hodgkin-Katz Equation
$$E_m = \frac{RT}{F} \ln \frac{[K^+]_o P_K + [Na^+]_o P_{Na}}{[K^+]_i P_K + [Na^+]_o P_{Na}} = -73.7 \text{ mV}$$

Equilibrium Potentials
$$E_K = \frac{RT}{F} \ln \frac{[K^+]_O}{[K^+]_i} = -74.4 \text{ mV}$$
$$E_{Na} = \frac{RT}{F} \ln \frac{[Na^+]_O}{[Na^+]_i} = +54.0 \text{ mV}$$
$$E_{Cl} = \frac{RT}{F} \ln \frac{[Cl^-]_i}{[Cl^-]_O} = -45.4 \text{ mV}$$
$$E_{Ca} = \frac{RT}{2F} \ln \frac{[Ca^{2+}]_O}{[Ca^{2+}]_i} = +40.0 \text{ mV}$$
Actual $E_m = -68$ mV

equation (Equation 6.8c; -74 mV) also closely estimates the E_m.

Further experimental evidence for the applicability of the Goldman-Hodgkin-Katz equation is the close match between measured and calculated E_m when the external $[K^+]$ and $[Na^+]$ are altered from their normal values (Figure 6–11). The internal $[K^+]_i$ is assumed to remain constant at 140 mEq l^{-1}, but the external $[K^+]_o$ and $[Na^+]_o$ concentrations can be readily manipulated by changing the bathing medium; the E_m can be calculated using either Equation 6.8b or Equation 6.8c, by assuming $P_K = 1.0$ and $P_{Na} = 0.01$ (the denominator term $[Na^+]_i.P_{Na}$ is totally ignored because both $[Na^+]_i$ and P_{Na} are small). There is a close correspondence between measured E_m and calculated E_m using Equation

FIGURE 6–11 The predominant role of K^+ distribution across a muscle cell membrane in determining the membrane potential is apparent from experimental manipulation of the extracellular K^+ concentration ($[K^+]_o$), from the normal value of about 5 mM to as low as 0.5 mM and over 100 mM. The measured E_m is approximately calculated using the Nernst equation for K^+ distribution (color line); the deviation of actual E_m from estimated E_m at low $[K^+]_o$ is due to the relatively greater importance of the Na^+ distribution even though the sodium permeability is lower than the potassium permeability. The measured E_m is closely estimated by the Goldman-Hodgkin-Katz equation accounting for K^+ and Na^+ concentrations, assuming a ratio of permeabilities of 1:0.01 (K^+:Na^+) and an internal $[Na^+]$ of 0 (black line). *(Modified from Hodgkin and Horowicz 1959.)*

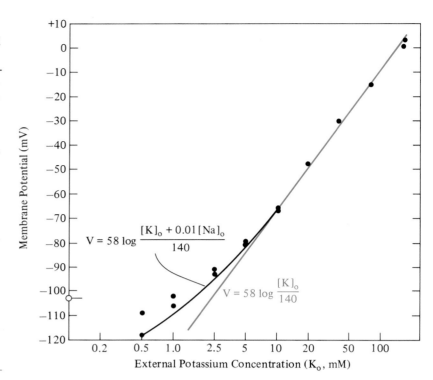

6.8B, and a good fit using Equation 6.8C, except at the lowest $[K^+]_o$ where the contribution of $[Na^+]$ becomes significant.

Conductance, Current, and Capacitance

The complex biological membrane can be represented as a simple electrical circuit that has equivalent electrical properties as the membrane. The membrane is essentially a low-voltage battery that establishes an electrical potential; the ion-specific pores of the membrane are pathways for electrical flow equivalent to resistors (or conductors; conductance = 1/resistance); the lipid bilayer portion of the membrane is a capacitor, keeping electrical charges separated on either side of the membrane. Such an electrical analog of the cell membrane is shown in Figure 6–12, superimposed over a representation of the structure of a membrane. We have already discussed the origin of the membrane potential (the "battery").

The membrane conductance (g; mho cm^{-2}) is the electrical analog of membrane permeability. The units for conductance are the inverse of resistance, since $g = 1/R$; if the membrane resistance to electrical flow is measured in ohms cm^{-2}, then g has units of ohm^{-1} cm^2, or mho cm^2. The terms "permeability" and "conductance" are not identical in meaning although they are related. Conductance depends on the permeability and driving force for movement. A low permeability means that conductance is low, but a high permeability does not necessarily mean that the conductance is high. Ions will only move across a membrane if they are present and if there is a driving force, regardless of their permeability. For example, a membrane might have a high permeability to lithium (Li^+) through Na^+ channels, but the Li^+ conductance would be low because Li^+ ions are not present in biological fluids.

Ohm's law summarizes the relationship between voltage difference (E), resistance (R), conductance (g), and current (I) for an electrical resistor.

$$I = E/R \qquad \text{and} \qquad I = Eg \qquad (6.9a)$$

The relationship is not as simple for biological membranes because current flow across a membrane, due to any particular ion species, depends on both the ionic conductance (g) and the electrochemical gradient (E). There is no current flow when the E_m is equal to the equilibrium potential of that that ion, but there is current flow if $E_m = 0$ mV and $E_{eq} \neq 0$. The electrical potential that drives the current flow of an ion is therefore not E_m but $E_m - E_{eq}$. The relationship between current flow, conductance, and E_m is different for each ion species, depending on their equilibrium potential.

$$
\begin{aligned}
I_K &= g_K(E_m - E_K) \\
I_{Na} &= g_{Na}(E_m - E_{Na}) \\
I_{Cl} &= g_{Cl}(E_m - E_{Cl})
\end{aligned} \qquad (6.9b)
$$

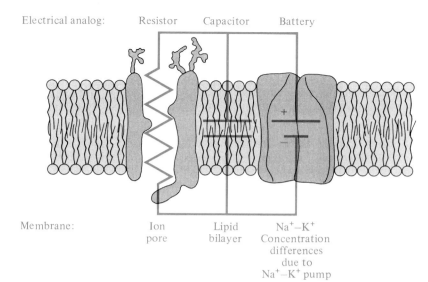

FIGURE 6–12 The basic electrical analog model of a cell membrane consists of an electrochemical gradient equivalent to an electrical potential (or battery) due to the Na^+/K^+ concentration differences across the membrane, a capacitor (due to the membrane lipid bilayer), and a resistor (ion-specific pores allow ion flow and are the membrane "resistors").

Current flow is low if conductance (and permeability) is low and/or the electrical gradient is low.

Capacitance (C; Farads) is a measure of how much charge (Q; Coulombs) can be maintained across an insulating gap at a certain voltage (E; Volts). A 1 farad capacitor can separate 1 coulomb of charge (1 coulomb = 6.24×10^{18} charges) with a 1 volt difference. The capacitance of two parallel plates depends on their separation distance (d), area of the plates (A), and the dielectric constant of the medium separating the plates (k)

$$C = Q/E = k\epsilon_o A/d \qquad (6.10)$$

where ϵ_o is the permittivity of free space (8.85×10^{-12} C V^{-1} m^{-1}). For example, a 1 cm^2 capacitor separated by 10 nm of air ($k = 1$) has a capacitance of about 0.09 μF.

The lipid bilayer of biological membranes is an effective capacitor because it separates charged ions across a very thin gap (about 10 nm) and has a high-dielectric constant ($k = 10$). The capacitance of cell membranes is generally about 1 μF cm^{-2}. Cell membranes sustain a high-voltage gradient, e.g., 100 mV/10 nm is equivalent to 10^7 V m^{-1}. This exceeds the dielectric strength (the maximum voltage gradient without electrical breakdown) of air (3×10^6 V m^{-1}).

The capacitance of membranes is important because it contributes an additional transient current whenever there is a change in membrane potential. Whenever a capacitor charges or discharges, there is a transient current flow until the voltage reaches the new equilibrium value. The capacitor-discharge current depends on the rate of voltage change ($\Delta E/\Delta t$)

$$I = C\Delta E/\Delta t \qquad (6.11a)$$

The voltage across a discharging capacitor declines in an exponential manner over time

$$E = E_0 e^{-t/RC} \qquad (6.11b)$$

where E is the instantaneous voltage, E_0 is the initial voltage, t is time, C is the capacitance, and R is the resistance discharging the capacitor. The exponential discharge of such a resistance-capacitance (RC) circuit is shown in Figure 6–13A. The values of R and C determine the rate of discharge; a high time constant ($\tau = RC$) confers a low rate of charge or discharge. The voltage decays to $1/e$ (0.367) of the initial E_0 after 1 time constant (RC seconds); the voltage further decreases to $1/e$ of the value after each successive time constant.

Membranes, although more complex in structure than an electrical capacitor, behave similarly; the value of $R_m C_m$ for a cell membrane is the **membrane time constant**, τ_m. For biological membranes, τ_m varies from about 10 μsec to 1 sec, corresponding to an R_m of 10 to 10^6 Ω cm^2. An example of the measurement of the membrane parameters, τ_m, C_m, and R_m, for a protozoan (*Paramecium*) is shown in Figure 6–13B. Injection of a current (I_m) into the protozoan causes an exponential change in E_m, with a time constant of about 60 msec. Since C is approximately 1 μF cm^{-2} then R must be about 60000 Ω cm^2. A current of about 0.23 nA results in a voltage change of about 23 mV, hence the membrane resistance is about 10^8 Ω. The membrane area of *Paramecium* is therefore about 6×10^{-4} cm^2.

The net current flow across a membrane is the sum of all ionic currents and any charging or discharging current flow if there is a change in membrane potential.

$$I_{total} = \Sigma I_{ion} + C_{membrane} \Delta E/\Delta t \qquad (6.12)$$

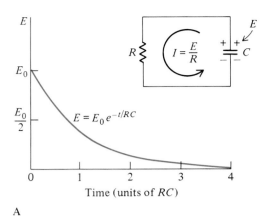

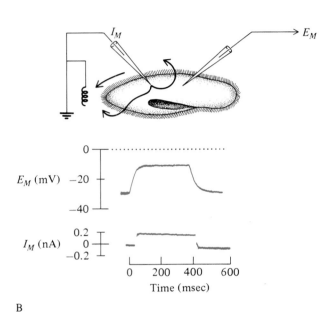

FIGURE 6–13 **(A)** An *RC* circuit (above) is a simple electrical analog showing the exponential nature of capacitor (*C*) discharging through a resistance (*R*). The magnitude of the current flow (*I*) depends on the voltage across the capacitor (*E*). The direction of the "conventional" current is indicated (i.e., direction of flow of + charges). The capacitor discharges in an exponential fashion (below) as the initial electrical potential across the *RC* circuit (E_o) declines to zero, with a time course dependent on the values for *R* and *C*; the value of *E* is 36.7% of E_o after one time constant (*RC*). **(B)** The cell membrane of the protozoan *Paramecium* acts as an *RC* circuit, showing exponential charging and discharging when an exogenous current is injected across the cell membrane. The membrane time constant τ_m is estimated to be about 60 msec, and the membrane resistance about 10^8 Ω since an injected current of 0.23 namp causes a change in E_m of 23 mV. *(From Hille 1984; Kung and Eckert 1972.)*

If the membrane voltage is held constant (i.e., "clamped" at a specified value by an external electrical circuit), then $\Delta E/\Delta t = 0$ and any measured current is an ionic flux current rather than a capacitance current, i.e., voltage clamping a membrane enables the direct measurement of ionic current. A voltage clamp only works if the time constant for the clamping circuit is considerably shorter than the membrane time constant.

Only a small number of ions are involved with cell electrical phenomena. Consider a cell membrane permeable to K^+ but not Cl^-, with $R_m = 1000$ Ω cm^{-2}; $C_m = 1$ μF cm^{-2}; $\tau_m = 1$ msec. If we add two KCl solutions with a concentration ratio of 1:52 across the membrane, then the membrane potential increases exponentially over time from 0 to 100 mV ($E_m = RT/F \ln(1/52) = 100$ mV) with a time constant of 1 msec. The system reaches equilibrium after only a few msec and the amount of charge now distributed across the membrane is equal to 10^{-7} coulomb cm^{-2} ($Q = E.C_m = 0.1 \cdot 1$ μF $cm^{-2} = 10^{-7}$ coulomb cm^{-2}). How many K^+ ions moved? The number is Q/F, or 10^{-12} moles K^+ cm^{-2}. This is such a tiny amount of K^+ that it would hardly affect the original concentration of K^+ on either side of the membrane. Thus, significant electrical potentials can be generated very rapidly by the movement of only a few ions through a few membrane pores. The actual effect on ionic concentration of such an ionic flux through a membrane to establish a membrane potential depends on the ratio of membrane surface to cytoplasmic volume. For a 1000 μ diameter giant squid axon, a K^+ flux of 10^{-12} moles cm^{-2} alters the K^+ concentrations by only about $1/10^5$. For a tiny cell dendrite, e.g., 0.1 μ diameter, the surface-to-volume ratio is about 10^4 times higher and the K^+ flux might alter the K^+ concentrations significantly, by 10% or so.

Excitable Cell Membranes

All cell membranes have an electrical potential because there is an ionic concentration gradient across membranes. The membrane potential is stable in some cells but often transiently increases, forming an **action potential** in excitable cells (primarily sensory, nerve, and muscle cells).

Action Potentials

The cell membrane is essentially an electrical circuit with a capacitance and parallel resistances for the flow of each ion, e.g., K^+, Na^+, and Cl^-. The most important ions are, as we shall see, Na^+ and K^+, and their resistances are variable. In addition, each ionic resistance is in series with the equilibrium potential of each ion. For example, there is no current flow due to potassium ions if $E_m = E_K$; the E_m for no K current flow is not 0 mV!

The resting membrane potential is approximately equal to the potassium equilibrium potential. Let us now consider what happens if K^+ is not the most permeable ion. If Cl^- becomes the most permeable ion, then the rather uninteresting effect is to shift the E_m towards the E_{Cl}, which is very similar to the resting E_m. However, what if Na^+ became the most permeable ion? The sodium equilibrium potential is very different (about $+65$ mV) from the resting E_m, E_K, and E_{Cl}. Thus, increasing Na^+ conductance (g_{Na}) by 1000 × would dramatically shift the E_m from -90 mV to $+65$ mV (Figure 6–14a). This change in E_m from resting E_m towards 0 mV and positive mV values (i.e., towards E_{Na}) is **depolarization**. If the g_{Na} returns to normal, then the E_m returns to resting E_m; this is **repolarization**. The membrane potential becomes even more negative than resting E_m (moves towards E_K) if the K^+ conductance is increased; this is **hyperpolarization**. This rather simplistic analysis of depolarization and hyperpolarization in response to increases in g_{Na} and g_K conductance is actually quite similar to the events during an action potential.

An action potential is a rapid, transient (about 1 to 2 msec) depolarization of the cell membrane potential, followed by a rapid repolarization and transient hyperpolarization (Figure 6–14B). The action potential transfers information from one cell (e.g., a sensory cell or neuron) to another (e.g., a neuron or effector cell). The mechanisms for information exchange will be detailed more thoroughly in Chapter 7. Here, it is sufficient to note that all action potentials are essentially identical and information is transferred by the frequency of action potential firing, rather than by the specific shape or amplitude. This information transfer is more like digital information in a computer (i.e., "On"–"Off", or "0"–"1" information) rather than analog voltage information (i.e., any voltage within a continuous range, such as from 0 to 5 volts).

Action potentials are initiated by the depolarization of a resting membrane. There can be many causes of an initial depolarization. For sensory cells, the depolarizing stimulus may be photons of light, heat, sound, movement, etc. For nerve cells, it may be an action potential in a sensory cell or another nerve cell. For muscle or secretory cells, it is usually an action potential in a nerve cell. If the depolarization reaches or exceeds a critical E_m, called the **threshold**, then an action potential occurs

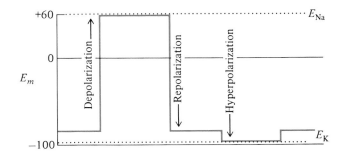

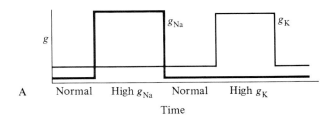

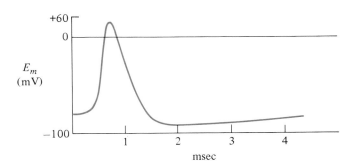

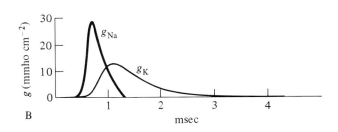

FIGURE 6–14 **(A)** A simple model of changes in membrane conductance (g) that alter the membrane potential. A transient increase in Na^+ permeability depolarizes the E_m towards the sodium equilibrium potential (E_{Na}); a subsequent transient increase in potassium permeability hyperpolarizes the E_m towards E_K. The actual changes in E_m and g_{Na}/g_K are much more complex (see part B). **(B)** Changes in membrane conductance (mmho cm^{-2}) to sodium (g_{Na}) and potassium (g_K during a nerve action potential, and the change in cell membrane potential (E_m; mV) due to ionic conductance changes.

(see below). If the depolarization does not reach threshold (is subthreshold), then an action potential will not be initiated; the E_m will return to the normal resting value. An action potential is "all-or-none"; it either occurs or it doesn't. There is (normally) no such thing as a "small" action potential or a "big" action potential.

The mechanism for an action potential is a change in ionic conductance, in particular Na$^+$ conductance (g_{Na}). The membrane potential E_m affects the sodium permeability of the cell membrane; the sodium permeability is **voltage dependent** (Figure 6–15). There is also a voltage dependence of the potassium permeability; the magnitude of the change is similar for K$^+$ but the change occurs considerably more slowly. The voltage-dependent increase in Na$^+$ conductance is responsible for initiating an action potential. An increase in Na$^+$ conductance causes the E_m to move towards E_{Na}; this change in E_m will itself further increase g_{Na}. If depolarization reaches threshold, there is a rapid progressive opening of all Na$^+$ channels such that Na$^+$ conductance increases to about 1000 × resting; this exceeds the resting potassium conductance by about 100 ×. This positive feedback between depolarization and increased g_{Na} results in a rapid and maximal increase

in E_m and g_{Na}. The E_m approaches the sodium equilibrium potential since $g_{Na} \gg g_K$, but rapidly repolarizes because the Na$^+$ channels automatically close and g_{Na} declines. The E_m at any point during an action potential can be approximately calculated from the instantaneous g_{Na} and g_K values using the short version of the Goldman-Hodgkin-Katz equation (Equation 6.8b). A rapid rise then fall in g_{Na} would cause a fairly symmetrical action potential, but in practice the action potential is followed by a short period of hyperpolarization due to a delayed increase of potassium permeability. Supplement 6–1 (page 248) summarizes the model of an action potential proposed by Hodgkin and Huxley.

There is a similar voltage dependence of K$^+$ conductance on E_m and also a feedback cycle for K$^+$ conductance. However, the K$^+$ cycle is a negative feedback or stabilizing cycle. Depolarization increases g_K causing K$^+$ efflux from the cell, and the K$^+$ efflux repolarizes the E_m. One role of the increase in K$^+$ conductance and hyperpolarization after the action potential spike is to return the E_m to normal and to stabilize it from further changes. This, and the property of the Na$^+$ channels to temporarily remain inactivated after activation, ex-

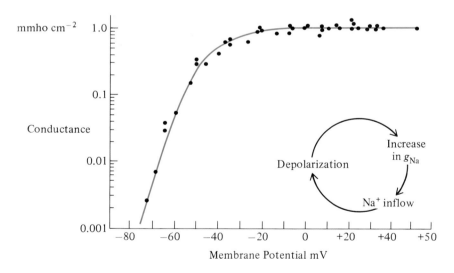

FIGURE 6–15   The ionic conductance of the giant squid axon for Na$^+$ (and K$^+$) is highly voltage dependent; g_{Na} increases from about 0.002 for the resting membrane to 1.0 mmho cm^{-2} for a membrane depolarization of 120 mV, and g_K increases from 0.02 to 1.0 mho cm^{-2}. The rate of change in conductance (mmho cm^{-2} sec^{-1}) is about 15 times faster for g_{NA} than g_K; the change in g_K is therefore limited during an action potential by the rapid repolarization (after about 0.5 to 1.0 msec). The voltage dependence of the g_{Na} leads to a positive-feedback relationship between membrane potential and g_{Na} (whereas the voltage dependence of g_K is a negative-feedback, stabilizing effect). *(Modified from Hodgkin and Huxley 1952.)*

plains the inability of a membrane to support a second action potential during, or soon after, a previous action potential. This period when a membrane cannot be stimulated to support an action potential is called the **absolute refractory period**.

The extracellular fluids of most animals contain high levels of Na^+ and low levels of K^+. However, many of the "higher" insects, especially phytophagous insects that consume copious amounts of plant material, have remarkably different extracellular ion concentrations than other insects; the extracellular K^+ and Mg^{2+} concentration is high, and the Na^+ level is low (Florkin and Jeuniaux 1974; Treherne and Maddrell 1967; Mullins 1985). This perhaps reflects the low Na^+ and high K^+ levels and high ingestion rates of food. These unusual extracellular ion concentrations would have dramatic implications for the electrical properties of excitable cells, both resting membrane potential and action potential spikes. For example, the stick insect *Carausius* has a higher extracellular $[K^+]$ than $[Na^+]$ (Table 6–7). The resting E_m estimated as the potassium equilibrium potential for hemolymph-intracellular fluid is -70 mV, whereas the actual value is -25 mV. The

calculated peak E (i.e., E_{Na}) for an action potential is -37 mV; this is more negative than the resting E_m! Either the hemolymph and/or intracellular Na^+ and K^+ concentrations used to calculate these E_m's are incorrect, or there is a fundamentally different mechanism for action potential initiation and propagation in phytophagous insects, or the extracellular fluid around the nerve cells has a different composition from hemolymph.

The nerve cord of the stick insect is surrounded by a sheath of fat-body cells, which forms an extraneural space between the axons and the hemolymph (Figure 6–16). The fluid in this extraneural space has a higher Na^+ and K^+ than hemolymph because of ionoregulation by the fat-body cell membranes. The E_K calculated using the extraneural and intracellular K^+ concentrations is -38 mV, which is considerably closer to the observed E_m than the value of -70 mV obtained using the hemolymph concentration. The peak E_m during an action potential (E_{Na}) is calculated to be $+23$ mV, which is

TABLE 6–7

Estimated intracellular and extracellular ionic concentrations (mM) for the stick insect at 18° C, and potassium and sodium equilibrium potentials calculated using the ratio of hemolymph:intracellular concentration and nerve cord fluid:intracellular fluid. The resting E_m is more closely calculated from the nerve cord fluid:intracellular K^+ concentrations and action potential spike from the nerve cord fluid:hemolymph Na^+ concentrations. *(Data from Treherne 1965; Treherne and Maddrell 1967.)*

	Hemolymph **(H)**	**Extracellular (in sheath around nerve cord) (E)**	**Intracellular (neuron) (I)**
Na^+	20.1	212.4	86.3
K^+	33.7	124.5	555.8
Ca^{2+}	6.4	2.2	61.8
Mg^{2+}	61.8	117.4	10.7

	H/I [1]	E/I [2]	
E_K	-70.4	-37.6	Resting $E_m = -25$ mV
E_{Na}	-36.6	$+22.6$	Peak E_m $+59$ mV

[1] $E_m = 57.8 \log_{10}$ Hemolymph/Intracellular
[2] $E_m = 57.8 \log_{10}$ Extracellular/Intracellular

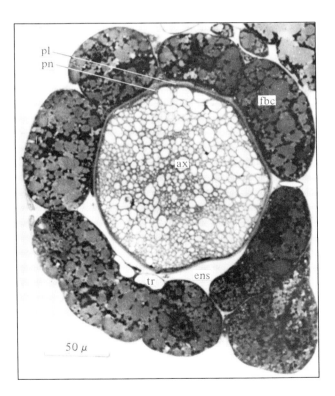

FIGURE 6–16 A sheath of fat-body cells (*fbc*) surrounds the interganglionic nerve connective between thoracic ganglia of the stick insect, leaving only a narrow connection to the extraneural space (*ens*); *ax*, axons; *pl*, acellular nerve cord perilemma; *pn*, cellular perineurial layer surrounding the nerve cord; *tr*, trachea. *(From Treherne and Maddrell 1967.)*

closer to the observed $+59$ mV than is -37 mV, the E_{Na} calculated from hemolymph concentrations. Thus, consideration of the composition of the extraneural fluid better explains the electrical properties of the muscle cells of this phytophagous insect.

Role of Ion Channels. There are many different types of ion channels (Table 6–8). Na^+ channels are relatively less diverse than K^+ and Ca^{2+} channels

and have only one major function—to initiate rapid depolarization of the transmembrane potential of excitable cells. Their ion selectivity is relatively constant ($Na^+ = Li^+ > K^+$), and the pharmacology of various receptor sites of the Na^+ channel is constant (there is an external TTX/STX receptor, an external polypeptide receptor, a hydrophobic receptor, and an internal receptor for local anesthetics). Potassium channels are much more diverse,

TABLE 6–8

Properties of various ion channels for excitable cell membranes that exhibit voltage-dependent permeability change.

Channel	Properties	Role	Blocker
Sodium I_{Na}	Rapidly activated by depolarization with voltage-dependent activation and inactivation.	Current responsible for action potential, e.g., nerve, muscle membranes.	TTX[1]
Potassium Delayed rectifier $I_{K(V)}$	Delayed activation by depolarization.	Rapidly repolarizes membrane after an action potential.	Ca^{2+} TEA[2]
Fast transient $I_{K(A)}$	Activated by depolarization, inactivated by prolonged depolarization; inactivation removed by hyperpolarization.	Repetitive pacemakers.	Low TEA[2]
Calcium Activated $I_{K(Ca)}$	Voltage-dependent and activated by intracellular Ca^{2+}; higher conductance than other K channels.	Terminates Ca^{2+} entry into cells; causes long hyperpolarizing pauses, e.g., bursting pacemakers.	Apamin TEA[2] Cs^+
Inward rectifying $I_{K(X_1)}$ $I_{K(K_2)}$	Inactivated by depolarization; activated by hyperpolarization.	Decreases K conductance during prolonged depolarization for energy economy and to maximize I_{Na}, e.g., cardiac muscle.	Cs^+ TEA[2] Rb^+
Calcium I_{Ca}	Activated by strong depolarization; low current; slow and incomplete inactivation with prolonged depolarization; intracellular Ca^{2+}-dependent inactivation.	Regulation of intracellular Ca^{2+} for control of a variety of cellular functions, i.e., secretion, muscle contraction, ionic gating.	Verapamil Nifedipine D-600
Chloride I_{Cl}	Some are weakly voltage dependent ("background" channels) or strongly voltage dependent.	"Background" channels stabilize resting E_m.	Anthracene-9-carboxylic acid Thiocyanate

[1] TTX = tetrodotoxin.
[2] TEA = tetraethylammonium.

functioning in many different roles: some open rapidly, some slowly, most open if the E_m becomes less negative, but some open if it becomes more negative. They can help stabilize the resting membrane potential, make the E_m more negative, or produce a plateau potential (see below). There are four main types of K channels: (1) delayed rectifiers, (2) Ca^{2+}-dependent K channels, (3) A channels, and (4) inward rectifier channels. These are defined mainly by their gating characteristics, and their functions will be described below. Chloride channels generally lack voltage dependence, and simply stabilize the resting cell membrane potential, although some giant algae apparently have Cl^- channels that initiate action potentials. Many synaptic membranes also have Cl^- channels. Ca^{2+} channels are found in all excitable cells; they close more slowly than Na^+ channels, and therefore they can produce long-term depolarization. They also have the essential role of transducing membrane depolarization into nonelectrical events, such as glandular secretion, vesicle exocytosis, and muscle contraction.

Sodium channels. The single most important event during an action potential is the initial rapid increase in Na^+ conductance. The activation gating mechanism for the Na^+ channel is voltage dependent, i.e., the g_{Na} depends on E_m. An increase in g_{Na} further depolarizes the membrane, increasing g_{Na}. Thus, a positive-feedback cycle is initiated once the depolarization reaches some critical E_m and g_{Na}; there follows a rapid and complete activation of all Na^+ channels. This critical E_m is the threshold value.

An equally important property of the Na^+ channels is inactivation. Once an Na^+ channel is activated, it is automatically inactivated soon afterward by a mechanism that is essentially independent of the activating mechanism.

Potassium channels. There are a number of different K^+ channels that confer a variety of different properties to cell membranes (Table 6–8). The different K^+ channels stabilize the E_m, ensure that the length of an action potential is kept short, terminate periods of rapid action potential initiation, vary the rate of repetitive action potential occurrence, or lower the excitability of the cell membrane.

Probably the two most important properties of action potentials are a high velocity of spread (this will be described in detail below) and rapid inactivation to assure a short action potential duration. The latter property requires a rapid inactivation of the Na^+ channels and a rapid increase in potassium permeability; this latter role is performed by the rapidly activating, **delayed rectifying K^+**

channel, which provides the $I_{K(V)}$ current. Despite their daunting name, the function of these K^+ channels is fairly straightforward; their current flow is voltage dependent, i.e., rectified. Any electrical circuit where resistance and conductance vary with voltage is a rectifying circuit; the Na^+ channel is also rectified. It is easier for K^+ ions to move out through the cell membrane than in, and so the new E_m is more rapidly reached after a depolarizing stimulus than after a repolarizing stimulus.

Neural information is generally encoded by the frequency of action potentials, which is quantitatively related to the stimulus intensity (see Chapter 7). However, axon membranes generally will not produce a chain of action potentials whose rate is proportional to the magnitude of a stimulus current, i.e., they will not encode voltage information into action potential frequency information. The information encoding membrane, located elsewhere in the neuron, has **fast transient K^+ channels**, or A channels (providing the $I_{K(A)}$ current). The A channels can encode a sustained depolarizing stimulus into a sustained rate of action potentials. They are inactivated at the end of an action potential, but are "primed" by the hyperpolarization to open when the membrane repolarizes to the normal E_m. The $I_{K(A)}$ current opposes the depolarizing stimulus current and temporarily keeps E_m at the normal resting value. The A channels then automatically close, and the depolarizing stimulus now depolarizes the membrane and initiates an action potential. This cycle is repeated for as long as the stimulus current is maintained.

Bursting pacemakers of mollusks fire with a complex pattern of alternating bursts of firing and quiescent periods (Figure 6–17). These bursts of firing are due to the influx of Ca^{2+} from the external medium. The Ca^{2+} influx exceeds the rate at which it can be pumped out of the neuron, and so the intracellular Ca^{2+} concentration rises until eventually it activates Ca^{2+}-dependent K^+ channels. The resultant $I_{K(Ca)}$ hyperpolarizes the neuron, inhibits further action potentials, and allows the Ca^{2+} to be pumped out of the neuron. The cycle of burst activity begins again as the intracellular Ca^{2+} concentration falls.

The **inward rectifying K^+ channel** has been identified in many types of cells, including vertebrate cardiac muscle fibers, frog skeletal muscle, starfish and tunicate eggs, and the electric organ of electric eels. There are at least two types of inward rectifying K^+ channels in cardiac muscle that differ in their E_m range for activation. K_2 channels close if E_m is -90 to -60 mV and inactivate slowly (the halftime is about 3 sec). X_1 channels close if E_m is -50 to

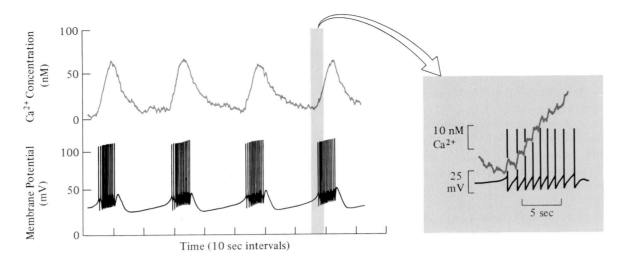

FIGURE 6–17 Bursting pacemaker of the sea hare *Aplysia* exhibits periods of repetitive action potentials interspersed with quiescent periods (lower trace). There is an increase in intracellular Ca^{2+} concentration with each action potential (upper trace), until the intracellular Ca^{2+} concentration becomes sufficiently high to open calcium-dependent potassium channels; the $I_{K(Ca)}$ hyperpolarizes the membrane and inhibits further action potentials and Ca^{2+} influx. *(From Gorman and Thomas 1978.)*

+ 10 mV, and inactivate faster. Vertebrate cardiac muscle cells are depolarized for about 1/2 of the time; the Purkinje muscle fiber has a typical plateau action potential, with each depolarization lasting from 100 to 600 msec. The cardiac action potential is initiated by a depolarization, then positive feedback increases the sodium conductance above threshold (about − 60 mV). This causes a brief "fast Na^+ current" of about 600 μA cm^{-2}, which is replaced by a slower Na^+ and Ca^{2+} current. Depolarization inactivates the K^+ channels and decreases g_K, so the membrane E_m quickly repolarizes to about − 10 to − 30 mV. A progressive activation of X_1 K^+ channels (since E_m is in their activation range) provides sufficient current flow ($I_{K(X_1)}$) to eventually repolarize the E_m to about − 90 mV. This E_m activates the K_2 K^+ channels and the repolarization is stabilized. However, the K_2 channels slowly inactivate, and this slowly depolarizes E_m to threshold, initiating another action potential. A continual influx of Na^+ and efflux of K^+ during the long depolarized plateau would require considerable energy expenditure by the Na^+-K^+ pump to reestablish the normal ion concentration gradients. This is avoided by the low density of most ionic channels, which restricts the current flow to 1 to 10 μA cm^{-2} (although the Na^+ channels pass up to 1 to 2 mA for the brief periods during which they are open).

Inward rectifying K^+ channels are also present in other cells. Some egg membranes have them to minimize K^+ efflux during the long depolarization after fertilization that prevents additional sperm from entering the fertilized egg. The electric organ of the electric eel has inward rectifying K^+ channels that allow the maximum development of Na^+ current to shock the eel's prey. In frog skeletal muscle, these K^+ channels may prevent excessive hyperpolarization of the cell membrane by very active electrogenic Na^+-K^+ pumps.

Calcium channels. Ca^{2+} channels were accidentally discovered in crab leg muscle cells, which have Ca^{2+}-driven action potentials rather than the typical Na^+-driven action potential (Fatt and Katz 1951). The absence of Na^+ in the external ringers actually increased the magnitude of the action potential in the crab leg muscle cell! An action potential can be driven by changes in Ca^{2+} rather than Na^+ permeability and flux, since the calcium equilibrium potential is positive, like the sodium equilibrium potential. For example, in the crustacean skeletal muscle either Ca^{2+} or the similar ions Ba^{2+} or Sr^{2+} are required for an action potential to occur normally, but Na^+ is not; the action potential is a "Ca^{2+} spike" and not a "Na^+ spike" (Fatt and Ginsborg 1958).

Ca^{2+} channels are present in virtually every type of excitable cell. They support a Ca^{2+} action potential in some cells (arthropod, molluskan, tunicate, and nematode muscle). In other cells, they

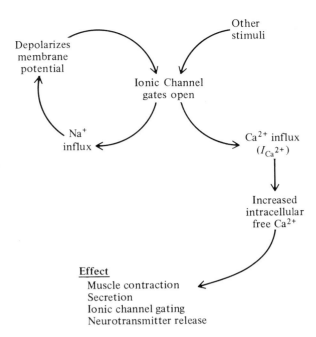

FIGURE 6–18 Membrane calcium channels are activated by strong depolarization (e.g., an action potential) and the low-current influx of Ca^{2+} (I_{Ca}) dramatically increases the normally low intracellular concentration ($< 10^{-7}$ M) by perhaps 20 ×. The increase in intracellular Ca^{2+} concentration is the signal initiating a number of important biological processes in different cells, from muscle contraction to secretion.

allow influx of Ca^{2+} into the cell to induce some nonelectrical event (e.g., muscle contraction, glandular secretion, vesicle exocytosis; Figure 6–18).

Ca^{2+} channels are activated by depolarization. They usually require a greater depolarization to open than do Na^+ channels. Their inactivation is slow and incomplete if the stimulus depolarization is maintained. The intracellular Ca^{2+} concentration is normally very low (e.g., 10^{-7} M) and so the Ca^{2+} influx can easily increase the intracellular concentration by 20 × or more during a single depolarization. The level of the intracellular Ca^{2+} is buffered to low levels by special Ca^{2+}-binding proteins (e.g., calmodulin, troponin) that are very sensitive to Ca^{2+} concentration changes of the order of 10^{-7} to 10^{-6} M. In muscle cells, the increased Ca^{2+} concentration due to influx from either the extracellular fluid or intracellular membranous vesicles (the sarcoplasmic reticulum) stimulates shortening of the muscle sarcomere units. In nonmuscle cells, the Ca^{2+}-calmodulin complex has effects on the intracellular cytoskeleton, hence intracellular motility (e.g., mitotic movement, ciliary and flagellar

movement). The release of neurotransmitter at chemical synapses is a consequence of Ca^{2+} influx through Ca^{2+} channels. Ca^{2+} influx affects many hormone systems, including the adrenergic system. For example, the effect of epinephrine on cardiac muscle to enhance the force of contraction is due in part to an increased Ca^{2+} current. Thus, gating of the Ca^{2+} channels by various hormones and other molecules can modify cellular function.

Chloride channels. Chloride is the most abundant anion of body fluids and there are some physiological roles for Cl^- channels. However, Cl^- is generally distributed across animal cell membranes in equilibrium with the resting E_m, i.e., $E_{Cl} = E_m$. Chloride channels therefore stabilize the E_m against depolarization (like K^+ channels). In many cells, such as vertebrate twitch muscle, the g_{Cl} is up to 10 × higher than g_K, and much higher than g_{Na}, and so the Cl^- channels play a "background" role in stabilizing the resting E_m. Some animal cells have voltage-dependent Cl^- channels (i.e., are rectified) but their role is unclear. The slow action potentials of the giant algae *Nitella* and *Chara* are thought to be due to the regenerative opening of Cl^- channels in response to a depolarizing current (i.e., analogous to the Na^+-driven action potential of animal cells) and repolarization due to a delayed increase in g_K. The ionic distributions across these algal cell membranes must be very different from those across animal cells in order for Cl^- flux to initiate an action potential (Nobel 1983); E_{Cl} is about $+99$ mV and the resting E_m is about -138 mV. It is possible that the increase in g_{Cl} is not actually voltage dependent but due to influx of Ca^{2+} due to voltage-dependent Ca^{2+} channels (Lunevsky et al. 1983).

Role of the Na^+-K^+ Pump. The Na^+-K^+ pump of excitable cell membranes (and also all other cells) exchanges Na^+ for K^+ across the cell membrane. The typical Na^+-K^+ pump exchanges 3 Na^+ for 2 K^+ per ATP hydrolyzed. It is electrogenic and contributes to the resting cell membrane potential. For example, injection of Na^+ into a snail neuron causes a rapid hyperpolarization due to the action of the Na^+-K^+ pump in pumping Na^+ out of the cell in a 3:2 exchange for K^+, thus reinforcing the normal -90 mV resting E_m. This hyperpolarization is blocked by ouabain (a cardiac glycoside that selectively blocks the Na^+-K^+ pump) and by removal of extracellular K^+, indicating that hyperpolarization is due to Na^+-K^+ exchange. The membrane potential of neurons of the gastropod mollusk *Anisodoris* can be separated by experiments at differing temperature into two components: one determined by ionic permeability and ionic concentration differences, and one due to the electrogenic pump (Marmor and Gorman 1970). In frog nerves, the E_m can become even more negative than E_K when there is a high rate of Na^+-K^+ exchange, presumably because of the electrogenic nature of the pump.

The Na^+-K^+ pump is not normally important in producing or maintaining the resting E_m, at least directly. For example, metabolic inhibitors, which block the Na^+-K^+ pump, have little or no immediate effect on E_m, and blocked cells can sustain many action potentials. However, blocked cells exhibit a slow, progressive decline in E_m towards 0 mV because of the slow dissipation of the Na^+, K^+, and other ion gradients. Thus, the Na^+-K^+ pump is essential as the ultimate mechanism for establishment of the ion concentration gradients on which the resting E_m and action potentials are dependent but is of little direct consequence to E_m and action potentials.

Properties of Action Potentials

There are a number of general properties of action potentials, such as threshold, shape, and frequency, that merit further discussion. The properties of membranes, enzymes, and protein channels are also influenced by the physical environment; two important physical parameters of the environment that can dramatically influence the electrical characteristics of membranes are temperature and pressure.

Temperature has a marked effect on excitable cell membranes. One minor effect is the influence of temperature on the mobility of ions and the establishment of a membrane potential by diffusion gradients as calculated by the Goldman-Hodgkin-Katz equation. This is a weak effect; increasing the temperature by 10° C increases the E_m by only 3 to 5%, i.e., the Q_{10} is about 1.035. Temperature affects the activity of important membrane enzymes. For example, the Na^+-K^+ pump of neurons of the mollusk *Anisodoris* is temperature dependent and so its electrogenic contribution to the resting membrane potential increases at elevated temperature.

Ion channel conductance is temperature dependent; Q_{10} values are generally about 1.3, but can range from 1.0 to 2.5. For example, an Arrhenius plot of conductance (g; pSiemens) as a function of $1/T$ yields a linear inverse relationship at T > 6° C with a Q_{10} of about 1.2, but a sharp transition at about 6° C to a more negative slope with Q_{10} about 7.2 (Figure 6–19). The activation energy for ion movement through the channels increases dramatically from about 25.6 kJ mole^{-1} at T > 6° C to 194.6 kJ mole^{-1} at T < 6° C. There is a similar temperature

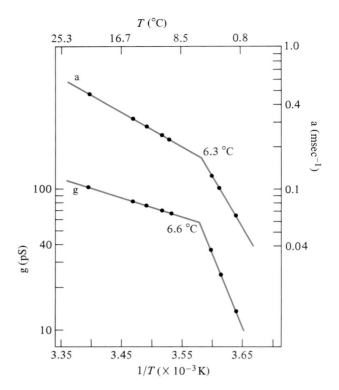

FIGURE 6–19 Effect of temperature on conductance and ion channel closing. *(Modified from Anderson, Cull-Candy, and Miledi 1977.)*

dependence of the rate of ion channel closing (*a*, msec^{-1}), also with a transition temperature of about 6° C. The activation energy increases from 47.7 kJ mole^{-1} (Q_{10} = 1.8) to 139.3 kJ mole^{-1} (Q_{10} = 6.3). These temperature-dependent properties of ion channels are possibly due to a change in the membrane fluidity, i.e., there is a solid-liquid phase transition at 6° C or a conformational change in the receptor-ion channel protein. Various ion channels can respond differently to a temperature change, and so the ratio of conductances of two ions can alter at differing temperatures.

Hydrostatic pressure has a similar effect on ionic mobility as lowered temperature. There is, in fact, a formal relationship between temperature and pressure effects

$$K_p/K_0 = e^{-P\Delta V^*/RT} \qquad (6.13)$$

where K_p is the reaction rate constant at pressure P, K_0 is the rate at standard pressure (0.1 MPa), and ΔV^* is the volume change of reactants as they are activated. Hydrostatic pressure often does not affect the resting E_m or threshold, although prolonged high pressure depolarizes the resting E_m of a crustacean axon by 10 to 15 mV, possibly due to effects on the electrogenic pump.

High pressure prolongs the duration of an action potential and decreases the rate of depolarization and repolarization. Vertebrate cardiac muscle has a decreased excitability at high pressure, a lower conduction velocity of action potentials, and increased action potential duration. Hydrostatic pressure can also affect the membrane current during an action potential. For example, a pressure of 20 MPa slightly diminishes the delayed K^+ current and especially the Na^+ current across a squid giant axon during an action potential. This is probably because the gating properties of the ion channels are altered, rather than because of a change in the number of open ion channels during an action potential.

High pressure generally has a lesser effect on excitable membranes of deep-sea fish than shallow-water fish. For example, high pressure decreases the amplitude of action potentials in axons of shallow-water fish but not deep-sea fish. The action potential amplitude for the shallow-water cod (*Gadus*) is reduced at 31 MPa to about 30% of the normal value at 0.1 MPa, whereas there is no effect at 42 MPa for the deep-sea *Bathysaurus* and *Coryphaenoides*. *Mora*, a fish of intermediate depth, has an intermediate depression of action potential amplitude. The duration, threshold, and absolute refractory period increase for axons of all species regardless of normal depth. The conduction velocity decreases for all species regardless of depth, although the effect is less for the deep-sea fish.

Threshold. A depolarization must reach a certain, critical level before the feedback cycle between depolarization and increase in g_{Na} becomes positive. This critical depolarization level is the **threshold** (Figure 6–20A). A subthreshold depolarization fails to reach threshold and does not elicit an action potential. A suprathreshold depolarization exceeds threshold and will elicit an action potential. In general, a faster rate of depolarization will cause a more rapid attainment of threshold and action potential initiation.

All-or-None Response. Normally, all action potentials of a given neuron are exactly equivalent in shape, i.e., have the same duration and amplitude. The effect of membrane depolarization is "all-or-none"; there is an action potential if E_m reaches threshold ("all"), but no action potential if threshold is not reached ("none").

There are some other exceptions to the all-or-none form of an action potential. One is the reduction in amplitude of action potentials that occurs in the relative refractory period, soon after a previous action potential (see below). A second example is the experimental manipulation of the external environment, e.g., replacing the extracellular Na^+ with choline diminishes the amplitude of the action potential because E_{Na} is reduced; application of small doses of tetrodotoxin also diminishes the action potential amplitude because some Na^+ channels are blocked.

Latency. The **latency** is the time period between the onset of the stimulus current and the peak of the ensuing action potential. Latency decreases with increasing current strength because the depolarization to threshold occurs faster.

Strength-Duration Relationship. A stimulating current must be of at least a minimal value to depolarize a cell membrane to threshold, but both the strength and the duration of the stimulus determine whether threshold is reached (Figure 6–20B). A short duration of a high-intensity current might have the same threshold action as a longer duration of a lower-intensity current. The inverse relationship between stimulus strength and duration is of the general form

$$E_{th} = \frac{a}{t} + b \qquad (6.14)$$

where E_{th} is the threshold voltage, t is the stimulus duration (msec), and a is a constant. The other constant, b, is the **rheobase**; it is the lowest current that will initiate an action potential. The **chronaxie** is the minimum required stimulus duration when the current is 2 × rheobase. Rheobase and chronaxie define the shape of the strength-duration relationship.

Accommodation. A constant depolarizing or hyperpolarizing current can modify the kinetics of Na^+ and K^+ channels and alter the threshold. A maintained depolarization causes the threshold to rise towards 0 mV. A slowly increasing depolarizing current may not initiate an action potential even though it may rise to an intensity much greater than threshold for a rapidly rising stimulating current because accommodation occurs faster than the current intensity rises. A maintained hyperpolarization causes the threshold to decrease further from 0 mV. The threshold may even fall below the normal resting E_m, so that an action potential is initiated when the hyperpolarizing current is removed because the return to resting E_m passes the lowered threshold; this is anodal block excitation.

Refractory Period. It is not possible to elicit a second action potential for a brief period after an action potential (Figure 6–20C). This short period is called the absolute refractory period. It occurs because all of the Na⁺ channels are inactivated, and cannot be activated; they are refractory to depolarization. As some Na⁺ channels become capable of activation, another depolarizing stimulus may initiate a second action potential, but its amplitude will be less than the normal amplitude because

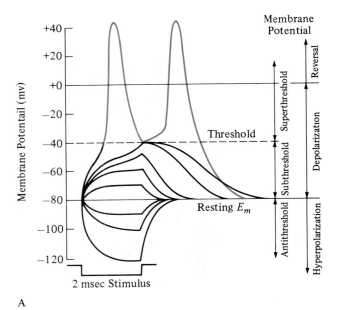

A

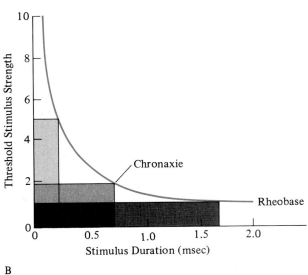

B

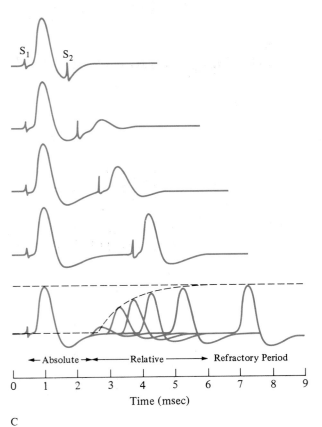

C

FIGURE 6–20 **(A)** A local depolarization only initiates an action potential if it reaches or exceeds the threshold value ($E_{threshold}$). Hyperpolarizing changes are antithreshold; no action potential is initiated. Depolarizing changes can be subthreshold (no action potential elicited) or superthreshold (action potential elicited). Depolarizations that increase the E_m above zero (reversal) are typically superthreshold and elicit an action potential. **(B)** Effect of strength and duration of a stimulating current on the threshold stimulation. Three combinations of strength and duration that elicit threshold are shown. The rheobase (minimum voltage at long duration) and chronaxie (pulse duration at a voltage of 2 × rheobase) are indicated. **(C)** The absolute refractory of an isolated whole nerve is the period after an initial stimulus (S_1) during which a second stimulus (S_2) fails to elicit a second action potential; the relative refractory period is that during which the second stimulus fails to elicit an action potential of normal amplitude. *(From Katz 1966.)*

not all of the Na⁺ channels can be activated; some are still inactivated and refractory. This second action potential, which occurs in the relative refractory period, has a lower than normal amplitude. A progressively more normal amplitude action potential occurs as the duration between successive stimuli is increased because the second stimulus occurs when more and more Na⁺ channels are capable of being activated. Eventually, the delay is sufficient that the second stimulus elicits a normal action potential.

Axonal Propagation

Excitable cell membranes not only sustain an action potential, but also allow its spread, or **propagation**. The spread of an action potential along a nerve cell axon well illustrates the mechanism of propagation. The axon is an elongate process of a neuron extending a variable distance from the cell body. A fairly typical multipolar neuron with a long axon is shown in Figure 6–21. An action potential is initiated at the axon hillock, where the axon extends from the cell soma. The action potential propagates along the length of the axon until it reaches the terminal synapse.

An axon can be represented by a sequence of equivalent membrane electrical circuits (Figure 6–22A), just as we have previously used an $R_m C_m$ equivalent circuit to represent the cell membrane except that we must also consider the external resistance to electrical flow (through the extracellular fluid, R_o) and the internal resistance to electrical flow (through the cytoplasm of the axon, R_i).

There is both a passive spread of electrical depolarization along the equivalent circuit of resistors and capacitors, as well as a regenerative spread of an action potential. The passive electrical spread is similar to the conduction of electricity through wires, resistors, and capacitors; it is **electrotonic spread**. The extent of passive electrotonic spread is determined by the values of the membrane resistances and capacitances. These values determine the cable properties of the axon, i.e., how the axon would act if it were a passive electrical cable. Some

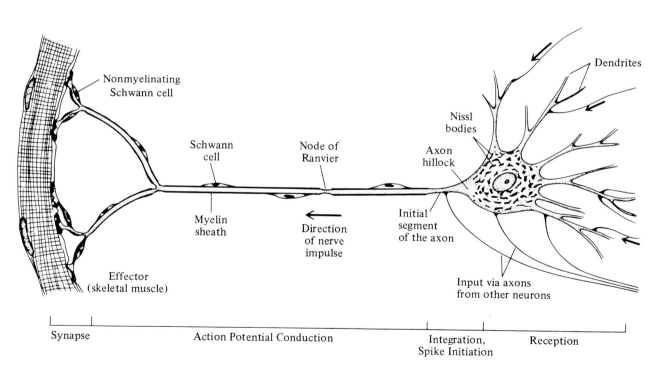

FIGURE 6–21 Schematic representation of a multipolar neuron showing the many dendritic afferents where sensory information is received, the axon hillock where this information is integrated to initiate action potential spikes, the axon conducting portion, and the terminal portion where neurotransmitter is released at the effector cell. *(Drawing from Fawcett, D. W. (1986) after Bunge from Bailey's Textbook of Histology 16ᵗʰ ed.)*

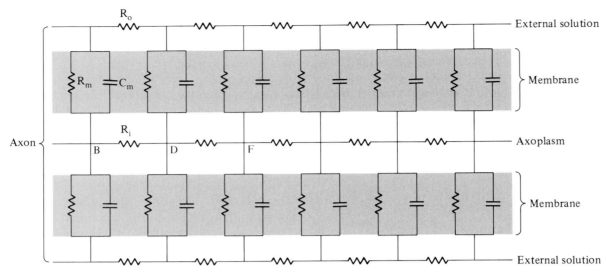

A

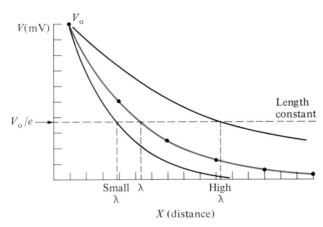

B

FIGURE 6–22 **(A)** Electrical circuit analog for the axon of neuronal cells. **(B)** Electronic spread of an applied electrical potential in the electrical circuit analog of an axon. The voltage (E) declines to 36.7% of the applied voltage (E_o) at a distance equal to the length constant (λ). *(Modified from Aidley 1989.)*

typical cable property values for R_i, R_o, R_m, and C_m are summarized in Table 6–9. The current flow due to a voltage E_o applied between R_i and R_o through a cable network transiently charges the membrane capacitors, then is a sustained, smaller current through the successive $R_m C_m$ circuits for as long as E_o is applied. The voltage (E_x) measured at a point x along the axon decreases exponentially as a function of distance from where E_o is applied (Figure 6–22B).

$$E_x = E_o e^{-x/\lambda} \qquad (6.15a)$$

The constant λ, the length constant, determines how quickly the voltage E_x decreases with distance from E_o. If $x = \lambda$, then $E_x = 0.368\ E_o$, i.e., the length constant is the distance at which E_x is decreased (or attenuated) to 37% of E_o. The value

of the length constant is determined by the values of R_i, R_o, and R_m

$$\lambda = \sqrt{R_m/(R_i + R_o)} = \sqrt{R_m/R_l} \qquad (6.15b)$$

where R_l is the summed longitudinal resistance of the axon "cable" ($R_i + R_o$). If R_m is high, then the E_o will electrotonically spread over a long distance because little current will flow through the membrane. If R_i and R_o are high, then E_o will not spread as far along the cable.

An axon shows exactly the same electrotonic spread as the equivalent circuit, although this is only apparent if the depolarization is subthreshold or the initiation of an action potential is blocked, for example by local cooling (Figure 6–23). The length constant for the axon illustrated in Figure 6–23 is about 0.6 mm.

TABLE 6–9

Cell	Dia μ	Length Constant λ mm	Time Constant τ msec	Capacitance μF cm⁻²	RESISTANCE		
					Membrane Ω cm²	Cytoplasm Ω cm	Extracellular Ω cm
Marine worm nerve	560	5.4	0.9	0.75	1200	57	—
Squid nerve	500	5	0.7	1	700	30	22
Lobster nerve	75	2.5	2	1	2000	60	22
Crab nerve	30	2.5	5	1	5000	60	22
Crab muscle	334	1.6	14.5	37	465	157	—

Cable properties of cell membranes at a temperature of about 20° C.   *(Data from Katz 1966; Hays, Lang, and Gainer 1968; Rall 1977.)*

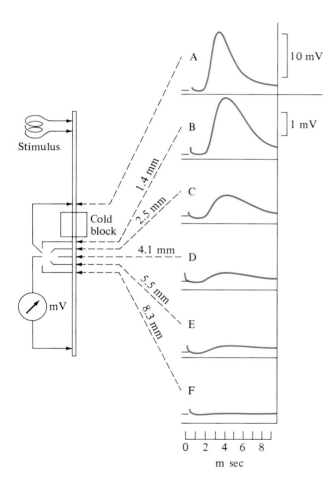

FIGURE 6–23 Electronic spread is readily observed for an axon with action potential propagation blocked by cold. There is an exponential decrease in voltage amplitude with distance from the block. The length constant is about 0.6 mm (distance for attenuation of voltage to 36.7% of E_o).   *(Modified from Hodgkin 1937.)*

The velocity of electrotonic spread is extremely rapid, as is electrical current flow in conducting wires and electrolyte solutions. This is why Johannes Muller declared in the late 1830s that the velocity of action potential propagation would never be accurately measured. However, within a couple of decades one of his students, von Helmholtz, had measured the velocity of an action potential as 3 10^3 cm sec⁻¹. Action potential propagation is clearly much slower than electrotonic spread. It is also a regenerative process; the voltage amplitude of an action potential at any point, x, is identical, i.e., a -90 to $+50$ mV signal, in contrast to the exponential voltage decline with x observed for electrotonic spread.

An action potential occurring at any particular point on an axon induces a local current flow, an inward Na^+ current followed by an outward K^+ current (Figure 6–24). There is also an electrotonic spread of current forward (and also backward) relative to the direction of propagation. This depolarizes the axon membrane to threshold, and a new action potential is initiated at a point ahead of the present action potential. Another action potential is induced further in front by electrotonic current spread. There actually is a smooth forward progression of the action potential, rather than a jumping motion as depicted for convenience in Figure 6–24.

Action potentials only travel in one direction along an axon from the neuron soma towards the axonal tip. This one-way propagation is not an inherent property of the axon membrane or action potential propagation mechanism. This can be readily demonstrated experimentally; an axon stimulated in the middle will propagate an action potential in each direction. An action potential traveling in the "correct" direction, from neuron soma to axon terminus, is described as **orthodromic**, whereas an

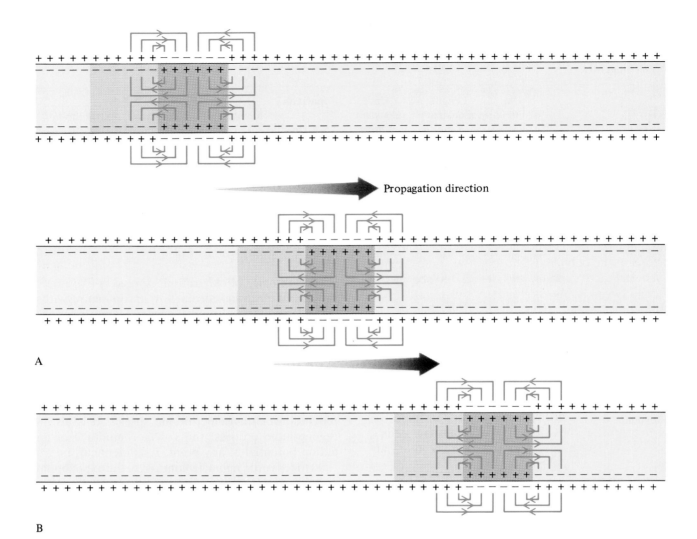

FIGURE 6–24 Local current flow during the propagation of an action potential along an axon. Arrows indicate direction of conventional current flow (i.e., flow of positive charges). The region of action potential depolarization is shown by dark color; the refractory portion is indicated by intermediate color. The one-way direction of propagation is maintained by the refractory period of the membrane after an action potential has occurred.

action potential traveling in the abnormal direction is **antidromic**. An action potential doesn't suddenly reverse direction as it spreads along an axon because the membrane where the action potential has just occurred is refractory; the Na^+ channels are inactivated and there is an outward K^+ current that hyperpolarizes the membrane, counteracting any electrotonic depolarization. In front of the action potential there is no repolarizing K^+ current and the Na^+ channels are not refractory.

The velocity of propagation of a regenerative action potential is much less than that of electrotonic spread because the membrane capacitance discharges at a finite rate and because complex protein configurational changes must occur. These processes are extremely slow compared to electrotonic spread, and so the action potential velocity is much less. The velocity of an action potential depends in part on the rate at which the membrane ahead of the action potential depolarizes to threshold. This in turn depends on the cable properties of the axon and the actual threshold level. The longitudinal resistance of the axoplasm (R_i) depends on both the intrinsic, or specific, resistance of the axoplasm (about 60 Ω cm for most axons) and the cross-sectional area of the axon (πa^2, where a is the axon radius). A 25 μ radius axon would have an internal resistance of about 3×10^6 Ω cm^{-1}, whereas a 100 μ

radius axon would have a lower resistance of about $2 \ 10^5 \ \Omega \ cm^{-1}$. The lower internal resistance might be expected to increase conduction velocity. The theoretical relationship between propagation velocity (V) and fiber diameter (D), if all other cable constants remain the same, is as follows.

$$V \propto \sqrt{D} \qquad (6.16)$$

One way to increase the conduction velocity is to increase axon diameter, although there clearly is a practical upper limit to the velocity that can be attained in this manner without the axon becoming impossibly large. Nevertheless, many invertebrates and lower vertebrates have giant axons that have high conduction velocities (Table 6–10). The giant axons of the polychete *Myxicola* are up to 1 mm in diameter and conduct at 6 to 20 m sec^{-1}! The function of giant fibers is predominantly rapid behavioral reflexes for protection from predators; they trigger massive behavioral responses, such as body flexion (fish) or abdominal flexion (crayfish) rather than controlled, finely graded movement.

Giant axons have independently evolved many times, but there are three basic patterns (Dorsett 1980). First, paired giant axons run longitudinally along the nerve cord with the cell bodies located in the central nervous system, e.g., Mauthner cells of fish, giant fibers of the polychete *Protula*, and

TABLE 6–10

Diameter (μ) and conduction velocity for giant axons of various invertebrates and vertebrates. Hydrozoa (*Nanomia, Aglantha*), polychete worms (*Nereis, Myxicola*), cockroach (*Periplaneta*), crayfish (*Cambarus*), lobster (*Homarus*), squid (*Loligo*), oligochete (*Lumbricus*), cyclostome (*Enterosphenus*), teleost (*Cyprinus*). (*Data from Dorsett 1980*).

	Dia (μ)	Velocity (m sec^{-1})
Nereis Paramedial	9	2.5
Aglantha ring	35	2.6
Nanomia stolon	30	3
Aglantha motor giant	40	4
Nereis Median	18	4.5
Nereis Lateral	35	5
Enterosphenus Muller cell	50	5
Lumbricus lateral	60	11.3
Periplaneta giant fiber	40	12
Cambarus laterals	150	15
Homarus median	125	18
Myxicola	1000	20
Lumbricus median	90	25
Loligo third order giant fiber	450	30
Cyprinus Mauthner	65	55

cockroach giant axons. Second, segmental giant neurons form junctions with adjacent neurons to act as a single functional unit that can conduct in either direction, e.g., median and lateral giant fibers of the earthworm and lateral giant fibers of crayfish. The junctions between the giant fibers of the earthworm slow down the conduction velocity to about 8 m sec^{-1} from 25 m sec^{-1}, and there is a slight reduction of the action potential amplitude across the gap junction. Third, a large number of axons from different neurons fuse to form a single giant fiber, e.g., the median giant fiber of the polychete *Nereis* and the giant axon of squid. Giant axons are not always a suitable solution for maximizing conduction velocity. For example, it would be very cumbersome for the optic nerve of vertebrates to contain giant fibers, and the spinal cord would assume huge dimensions if all of its nerve fibers were giant axons.

Many vertebrate axons and some invertebrate axons are covered with a series of insulating sleeves and intervening, uninsulated nodes. Special **Schwann sheath cells** surround the axons to form the insulation. The arrangement of the Schwann cells ranges from one Schwann cell per axon, to many axons per Schwann cell, to many Schwann cells per axon (Figure 6–25A). The Schwann cell membrane of many axons becomes repeatedly wound round the axon, forming a thick layer of cell membrane (Figure 6–25B). The thick lipid sheath of myelin is interrupted by gaps between the adjacent Schwann sheath cells, called the **nodes of Ranvier**. This arrangement of Schwann sheath cells, myelin, and nodes of Ranvier of vertebrate axons (Figure 6–25C) is closely paralleled by the myelination of some invertebrate axons, such as those of the prawn *Palaemonetes* (Figure 6–25D). The presence of a myelin sheath considerably alters the electrical properties of the axon (Rall 1977). There is an additional capacitance and resistance of the myelin sheath ($0.004 \ \mu F \ cm^{-2}$; $0.01 \ 10^6 \ \Omega \ cm^2$) compared to the internodal membrane ($5 \ \mu F \ cm^{-2}$; $15 \ \Omega \ cm^2$).

Myelination of axons is an alternative means of increasing conduction velocity. The conduction velocity of myelinated axons is very high because of the important role of electrotonic spread. Consider the effect of covering an axon with a series of insulating sleeves to increase R_m and λ, but leaving "nodes" where the axon membrane is exposed to the extracellular fluid (Figure 6–26A). The inward Na^+ current due to an action potential occurring at a node cannot exit from the axon except at the adjacent nodes because of the insulation; the current therefore rapidly spreads electrotonically from one node to the next. There is then a time delay as the membrane capacitance discharges and the Na^+

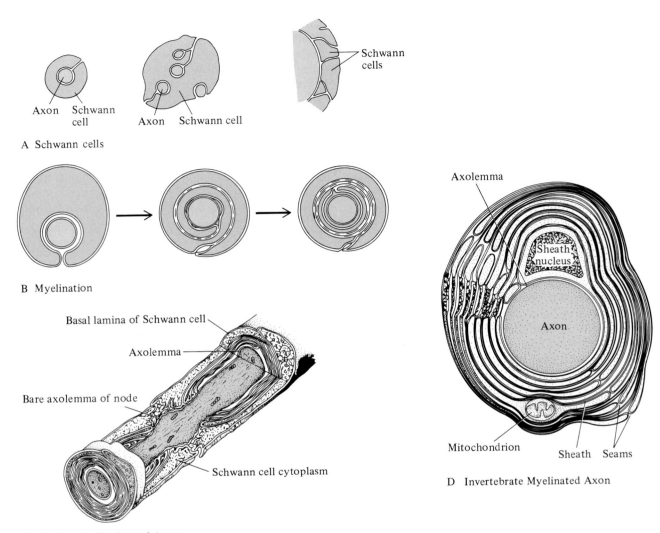

A Schwann cells

B Myelination

Basal lamina of Schwann cell

Axolemma

Bare axolemma of node

Schwann cell cytoplasm

C Vertebrate Myelinated Axon

Axolemma

Sheath nucleus

Axon

Mitochondrion

Sheath Seams

D Invertebrate Myelinated Axon

FIGURE 6–25 **(A)** Various arrangements of Schwann sheath cells around a single or multiple axons, and multiple Schwann cells around a giant axon. **(B)** The myelin sheath of myelinated axons is formed by an encircling extension of the inner Schwann cell membrane. **(C)** The myelin sheath of a vertebrate myelinated axon is surrounded by multiple membrane layers of the Schwann sheath cells, with Nodes of Ranvier between adjacent Schwann cells. **(D)** Cross section of a myelinated axon from an invertebrate (a prawn) is superficially similar to the myelin sheath of a vertebrate axon. *(From Aidley 1989; Fawcett 1986; Huesner and Doggenweiler 1966.)*

channels open, as occurs in naked axons, but then the inward Na^+ current rapidly spreads electrotonically to the next node. The net effect is an overall higher conduction velocity than for a naked axon of the same diameter because the action potential jumps from node to node at high velocity. This jumping propagation is **saltation** or **saltatory conduction**. A second advantage to saltatory conduction is the energy economy resulting from having only

a small portion of the axon membrane actually supporting an action potential. The I_{Na} and I_K are smaller and the Na^+-K^+ pump doesn't expend as much ATP to restore the normal ionic concentration differences, as if the whole axon membrane sustained an action potential.

The action potential conduction velocity (V) of a myelinated axon depends on its diameter (D) (Figure 6–26B). There is a fairly linear relationship between

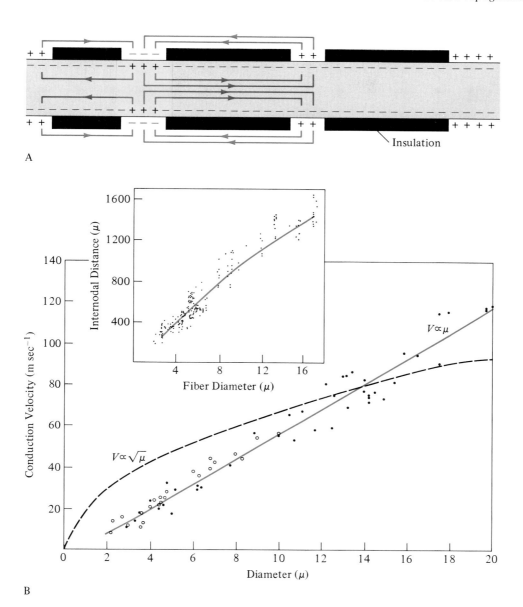

FIGURE 6–26 **(A)** Local current flow in an axon covered with discontinuous insulation. Current travels electrotonically along the axon to the node between insulating layers, rather than through the axon membrane where it is covered with the insulator. **(B)** Essentially there is a linear relationship between the fiber diameter (μ) and conduction velocity (V) for vertebrate (cat) myelinated axons rather than a square root relationship (dashed line). This linear relationship is due to the linear relationship between internode distance and fiber diameter (inset). *(Modified from Hursh 1939.)*

conduction velocity and diameter because the internode distance is correlated with the axon diameter. There is a minimum theoretical diameter, about 1 μ, below which myelinated axons have a slower propagation than nonmyelinated axons (Rushton 1951). This corresponds well with the observed minimum diameter of unmyelinated axons found in peripheral nerves of mammals, but smaller myelinated axons (about 0.2 μ diameter) are found in the mammalian central nervous system.

The peripheral nerves of animals contain many different axons of varying diameter and presence/absence of myelination. Consequently, the axons of these compounds vary in conduction velocity. This is readily apparent from recording the **compound action potential** of a compound nerve, such as the frog sciatic. The action potentials, recorded at a distance from the point of stimulation, are separated into discrete classes of action potentials, depending on the characteristic conduction velocity (i.e., axon diameter and myelination). The large, myelinated Aα axons have the highest conduction velocity (4.2×10^3 cm sec^{-1}) and their action potentials are recorded first (Figure 6–27). The unmyelinated C axons have the lowest conduction velocity (4-5×10^1 cm sec^{-1}) and their action potentials are recorded considerably after the Aα axon action potentials. The Aβ, Aγ, Aδ, and B axons have intermediate diameters and conduction velocities.

Sloths, especially the three-toed sloth (*Bradypus tridactylus*), are renowned for their slow motion movements; but is their "slothfulness" physiological or motivational, i.e., would sloths move more rapidly if appropriately motivated? Their axonal conduction velocity (6 to 35 m sec^{-1}), action potential duration (2 to 3 msec for skeletal muscle membrane), and synaptic delay (about 2 to 3 msec for skeletal muscle membrane) are a little slower than are found in other mammals, but not enough to explain their slothfulness (Enger and Bullock 1965). A slow muscle contraction velocity is probably the reason for the slowness of sloths (see Chapter 9).

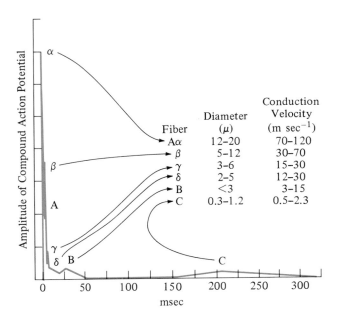

FIGURE 6–27   Hypothetical compound action potential recorded from a peripheral compound nerve shows the peaks for different-sized nerve fibers. *(Modified from Ganong 1969.)*

Many neurons that lack axons, or have short axons, are nonspiking, or local circuit, neurons. They do not sustain action potentials; rather, they rely on electrotonic spread because only short distances are involved and the electrical signal remains sufficiently great to elicit the appropriate response. These small cells also tend to have a high membrane resistance to minimize the attenuation of the electrotonic spread. Local neurons are found in the vertebrate retina and barnacle eye, the vertebrate and insect central nervous systems, and the stomatogastric ganglion of crustaceans. Electrotonic spread also occurs in the invaginated muscle cell membrane, called the t-tubule system.

Synaptic Transmission

Action potentials are transmitted from one cell to another. There are essentially two different mechanisms for transfer of electrical information from one cell (the presynaptic cell) to the next (the postsynaptic cell): (1) electrical synapses and (2) chemical synapses (Figure 6–28). In electrical synapses, the action potential jumps electrotonically from the presynaptic cell membrane to the postsynaptic cell membrane. Electrical synapses are more simple in principle, but are less common, than chemical synapses. The more common chemical synapse involves release of a special chemical, a neurotransmitter, from the presynaptic cell. The neurotransmitter has an electrical effect at the postsynaptic membrane, often depolarizing the membrane and initiating an action potential.

Electrical Synapses

Electrical synapses have a very specific anatomical organization and specialized membrane properties to electronically transmit an action potential from the presynaptic membrane to the postsynaptic membrane without so much attenuation that the postsynaptic cell fails to be depolarized to threshold (Katz 1966). The presynaptic membrane at an electrical synapse is closely apposed to the postsynaptic membrane to form a **gap junction** (Figure 6–29A). These gap junctions are composed of numerous **connexons** that allow direct movement of ions and small molecules from the presynaptic cell into the postsynaptic cell (Figure 6–29B). The connexons allow passage of small molecules (up to a molecular weight of about 800). The permeability of the connexon can be regulated, perhaps by the intracellular Ca^{2+} concentration. The electrical connection usually allows current flow in either direction, but this

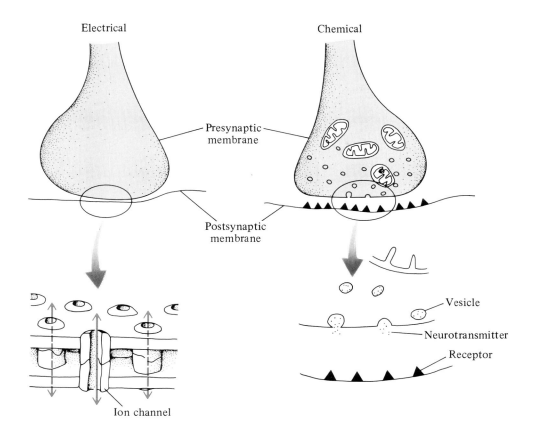

FIGURE 6–28 Comparison of an electrical synapse with gap junctions for electrical continuity between adjacent cells (left) and a chemical synapse where neurotransmitter is released from vesicles in the presynaptic terminal and binds to a receptor on the post-synaptic membrane (right). *(Gap junction from Loewenstein 1976.)*

is not always so. Preferential current flow in one direction is **rectification**.

Electrical junctions between neurons rely on local currents from the presynaptic membrane to the postsynaptic membrane (see Supplement 6–2, page 251). Sufficient current must flow from the first cell to depolarize the second cell to threshold. It is unlikely that an action potential in a small diameter axon could generate sufficient current flow to depolarize to threshold a large postsynaptic neuron with a high membrane area and low input resistance. Thus, an axon could probably not initiate an action potential across, for example, a neuromuscular junction (from a small neuron end terminal to a very large muscle cell). This limitation of electrical transmission is most likely one reason for the relative rarity of electrical synapses compared with the widespread occurrence of chemical synapses. Electrical synapses also do not allow complex signal integration.

The direct electrical coupling of neurons is often observed when there is a requirement for the close synchronization of effector organs (Bennett 1966). Examples are the cells of the lobster heart, the electric organ of mormyrid fish, the sound production muscle of toadfish, and the escape response of some invertebrates (e.g., cockroachs and crayfish) and vertebrates (e.g., fish).

The "tail-flick" escape response of the crayfish (Figure 6–30) involves giant axons in the nerve cord; they are giant to maximize conduction velocity. These axons stimulate large motor axons that inner-vate the abdominal musculature. There is an almost immediate depolarization of the giant motor fiber when the lateral giant fiber is stimulated. There is virtually no time delay, or latency, in transmission of electrical depolarization from the presynaptic to postsynaptic neuron. A small "kink" in the postsynaptic depolarization indicates the point at which the depolarization of the postsynaptic mem-

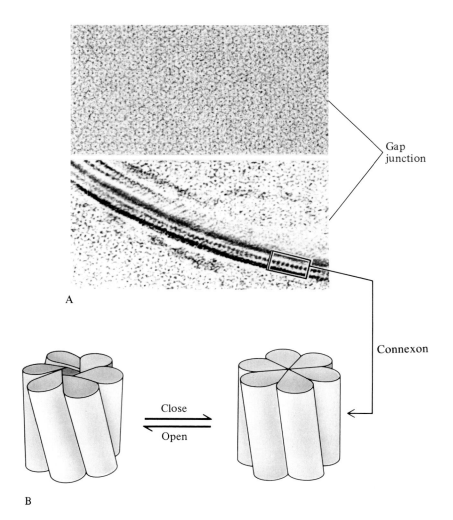

Gap
junction

Connexon

Close

Open

A

B

FIGURE 6–29 **(A)** Electron micrographs of the gap junction region viewed from a perpendicular direction (upper) and a parallel direction (lower). The connections appear as hexagonal units from above and as bridges connecting the two cell membranes from the side. **(B)** A schematic model of the six subunits of a connection suggest a mechanism for opening and closing of the connexon pore by a sliding and rotation of the subunits. *(From Unwin and Zampighi 1980.)*

brane reaches threshold. The normal direction of synaptic transmission is from the presynaptic membrane to the postsynaptic membrane (i.e., orthodromic). If the giant motor axon is stimulated, then little depolarization spreads to the lateral giant axons, i.e., there is little antidromic spread of the depolarization. This is a rectified electrical synapse.

Some electrical synapses are inhibitory. Many fish, when startled, rapidly flex their bodies by a massive synchronous contraction of lateral muscles on one side, followed by a tail flip. This startle response involves two large interneurons, the Mauthner fibers, with cell bodies in the brain and spinal axons that innervate the lateral musculature.

It obviously is important that only muscles on one side of the body contract during the startle reflex, and so there must be rapid inhibition of one Mauthner cell when the other is active. There appears to be a hyperpolarizing and inhibitory potential near the axon hillock of a Mauthner cell when the other is firing (Furukawa and Furshpan 1963).

Chemical Synapses

Chemical synapses connect sensory cells to neurons, neurons to other neurons, and neurons to effector cells. They have a more complex structure than the simple gap junctions of electrical synapses.

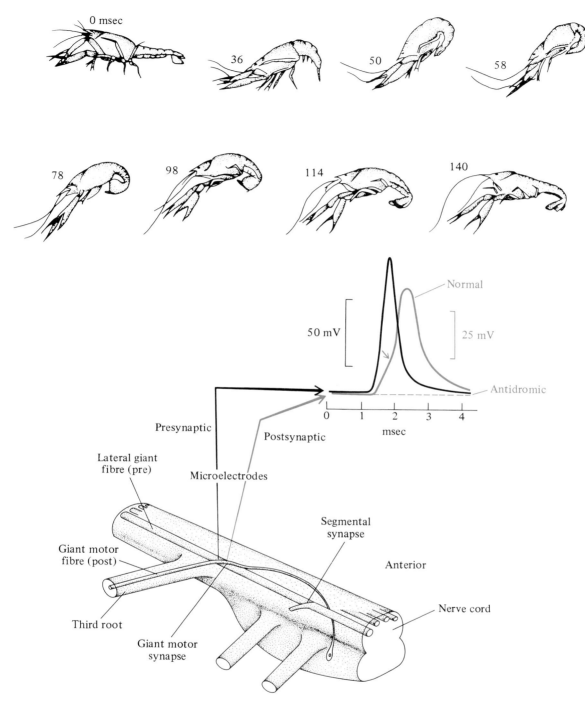

FIGURE 6–30 The "tail-flick" startle response of the crayfish (top panel) is controlled by lateral giant axons that innervate the tail flexor muscles via large motor fibers. The diagrammatic representation of part of the abdominal nerve cord of the crayfish shows the course of one motor axon from the cell body in the ventral nerve cord to the third ganglionic root on the other side of the nerve cord. The electrical synapse of the motor axon with the giant lateral axon is shown (not shown are additional electrical synapses between the motor axon and two medial giant axons). The recording microelectrodes are shown in the giant lateral axon and the motor axon. The normal stimulation of the giant fiber results in an extremely rapid depolarization of the motor axon; the postsynaptic potential initiates an action potential at the point indicated with an arrow. Antidromic stimulation of the motor axon results in a negligible (about 0.3 mV) depolarization of the lateral giant axon indicating that this electrical synapse is capable of considerable rectification.   *(From Wine and Krasne 1982; Furshpan and Potter 1959.)*

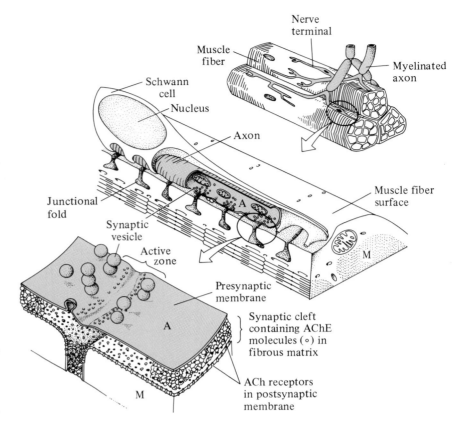

FIGURE 6–31 Representation of the end plate of a typical neuromuscular junction. Branches of the innervating axon form multiple end plates on one or more muscle cells; each axonal branch has an incomplete Schwann cell sheath. The synaptic vesicles are located in the axon terminal and are concentrated at synaptic junctional folds of the postsynaptic membrane, where neurotransmitter (acetylcholine) receptors are abundant. The extracellular matrix of the synaptic cleft contains acetylcholinesterase. *(From Hille 1984.)*

The neuromuscular junction is a typical example of a chemical synapse (Figure 6–31) between a presynaptic neuron membrane and a postsynaptic muscle cell membrane. The synaptic portion of the muscle cell membrane is the **end plate**. The axon terminal characteristically contains many mitochondria and synaptic vesicles, small membrane-bound vesicles about 40 nm diameter containing 1 to 5 10^4 molecules of neurotransmitter. There is a well-defined space, the synaptic cleft, between the presynaptic and postsynaptic membranes. This cleft is filled with mucopolysaccharide that attaches to the pre- and postsynaptic membranes.

The basic sequence of events during chemical synaptic transmission (Figure 6–32) is:

(A) the presynaptic action potential depolarizes the presynaptic membrane;

(B) the increased Ca^{2+} permeability of the depolarized presynaptic membrane allows Ca^{2+} influx into the axon terminal;

(C) the elevated intracellular Ca^{2+} concentration causes the release of neurotransmitter from synaptic vesicles into the synaptic cleft;

(D) neurotransmitter molecules diffuse across the synaptic cleft to the postsynaptic membrane, and they reversibly bind to specific receptors on the postsynaptic membrane;

(E) the receptor-neurotransmitter complex increases the permeability of the postsynaptic membrane to ions (e.g., Na^+, Ca^{2+}, Cl^-) and depolarizes or hyperpolarizes the end plate; and

(F) the end plate potential spreads electrotonically and initiates an action potential that propagates along the postsynaptic membrane, and the neurotransmitter is removed from the synaptic cleft by uptake into the presynaptic terminal and by enzymatic hydrolysis in the synaptic cleft.

Let us now examine some of these steps in greater detail.

Transmitter Release. Ca^{2+} has a central role in neurotransmitter release. There is a strong correlation between the postsynaptic depolarization and presynaptic intracellular Ca^{2+} concentration. A low extracellular Ca^{2+} concentration, presence of competing ions such as Mg^{2+} and La^{2+}, or depolarization of the E_m to E_{Ca} reduce neurotransmitter release,

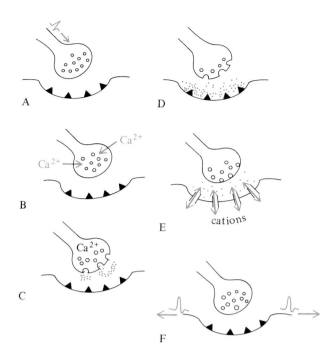

A

B

C

D

E

cations

F

FIGURE 6–32 Representation of the sequence of events that occurs at a chemical synapse from an action potential reaching the axon terminal to initiation of a subsequent action potential on the postsynaptic membrane of a typical motor end plate. **(A)** Action potential reaches the presynaptic axon terminal. **(B)** Depolarization of axon terminal allows Ca^{2+} influx. **(C)** Increased intracellular Ca^{2+} concentration causes synaptic vesicles to release neurotransmitter into the synaptic cleft, by exocytosis. **(D)** Neurotransmitter diffuses to postsynaptic membrane. **(E)** Neurotransmitter combines with specific receptors on the postsynaptic membrane and opens fairly nonspecific cation channels. **(F)** Local current flow through cation channels depolarizes the postsynaptic membrane and the postsynaptic potential spreads electrotonically to the adjacent postsynaptic cell membrane and depolarizes it to threshold; an action potential is initiated and propagates over the postsynaptic cell membrane. The neurotransmitter is removed from the synaptic cleft and the cation channels close.

whereas microinjection of Ca^{2+} into the presynaptic terminal elicits additional neurotransmitter release.

The intracellular Ca^{2+} is required for the synaptic vesicle to fuse with the presynaptic membrane and release its contents (about 1 to 5 10^4 neurotransmitter molecules and about 1/5 as many ATP molecules) by exocytosis. The Ca^{2+} probably forms some intermediate, Ca"X" (or perhaps Ca_4"X"). There appear to be 2 to 4 Ca^{2+} per active site because the

rate of neurotransmitter release is proportional to $(Ca^{2+})^n$, where n varies from 2 to 4 (Dodge and Rahamimoff 1967). The Ca^{2+} may activate a protein complex in the presynaptic terminal resembling actin and myosin of muscle cells. Brain presynaptic terminals contain two such proteins: neurin (associated with the presynaptic membrane) and stenin (associated with the synaptic vesicles) that may be involved in exocytotic neurotransmitter release.

There is considerable evidence for the quantal release of vesicles containing neurotransmitter, i.e., the release of the contents of 1 synaptic vesicle, or 2 vesicles, or 3, or 4, etc. (see Supplement 6–3, page 252). That the entire contents of a vesicle is released is not surprising, nor is the observation that either 1, 2, 3, 4, etc. vesicles release their contents at a time. This property of quantum release enabled the estimation of the number of neurotransmitter molecules per vesicle to be about 1 to 6 10^4.

Receptor Binding. The neuromuscular junction is a typical example of a chemical synapse. The postsynaptic membrane in the immediate vicinity of the synapse, the **motor end plate**, has proteinaceous receptors that specifically bind the neurotransmitter, acetylcholine (ACh). Each acetylcholine receptor has an ionic channel (Figure 6–33). The conductance of the motor end plate would increase linearly with (ACh) if only one ACh was required to bind to the receptor and open each channel. However, the conductance is proportional to $(ACh)^2$, suggesting that two ACh molecules are needed to open each ionic channel. The activation sequence for the receptor (R) by ACh would be something like the following.

$$R \underset{}{\overset{ACh}{\rightleftharpoons}} R\text{-}ACh \underset{}{\overset{ACh}{\rightleftharpoons}} R\text{-}ACh_2 \rightleftharpoons R\text{-}ACh_2^* \quad (6.17)$$

Only the activated $R\text{-}ACh_2^*$ complex opens the ionic channel. Once activated, the receptor becomes desensitized for a variable period of time during which the ionic channel remains closed even in the presence of ACh. The desensitized state may last briefly (fast desensitized state) or for seconds or more (slow desensitized state). The ACh-ionic channel can be specifically and irreversibly blocked by α-bungarotoxin (BuTX), which is one component of the venom of the krait, a highly venomous cobra snake. d-Tubocurarine, the active ingredient of curare, inhibits the action of ACh. It competitively binds to the ACh receptor but doesn't open the ionic channel. (Curare is a poison used by some South American Indians; it is a crude mixture of various plant extracts.)

The end plate ACh-ionic channel, when activated and open, is relatively unselective to cations. Any

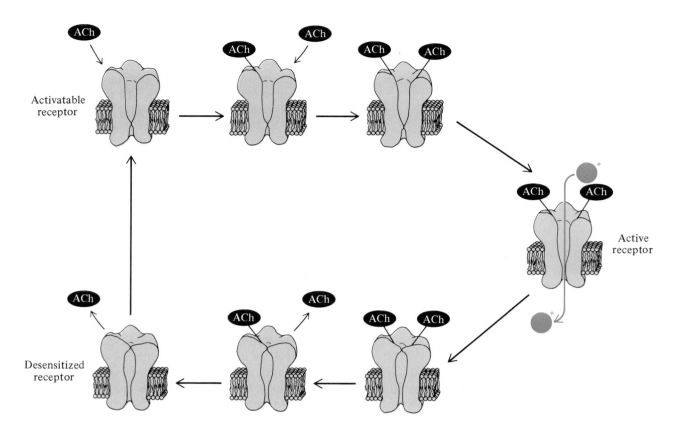

FIGURE 6–33   There is a complex series of reactions between the neurotransmitter acetyl-choline and its receptor. This schematic diagram of these reactions shows the main cycle of activation, with two ACh molecules binding to a receptor and activating it, opening a cation channel through the membrane and allowing Na⁺ influx. The ACh-receptor complex becomes inactivated and the ion channel closes, the ACh molecules detach from the receptor, and the receptor is temporarily desensitized to further binding of ACh and opening of the cation channel.   *(From Montal, Anholt, and Labarca 1986.)*

monovalent and divalent cations are permeable if they can fit through a 0.65×0.65 nm pore. This includes not only the alkali metals (Li^+, Na^+, K^+, Rb^+, Cs^+, Fr^+) and alkaline earth metals (Be^{2+}, Mg^{2+}, Ca^{2+}, Sr^{2+}, Ba^{2+}, Ra^{2+}) but also various organic cations such as choline (Table 6–11). The channel rejects anions. It is this relatively unselective permeability to cations that results in the reversal potential of the end plate being different from the equilibrium potential of either K^+, Na^+, or Ca^{2+} (see below).

The synaptic cleft of the neuromuscular junction contains high concentrations of an enzyme that hydrolyzes ACh, **acetylcholinesterase** (ChE). The ChE rapidly removes ACh by hydrolysis; it can hydrolyze 10^{-14} mole ACh ($6 \ 10^9$ molecules) in 5 msec.

Inhibitory synapses have anion channels, of relatively unselective permeability, e.g., glycine and GABA channels (Table 6–11).

End Plate Potentials.   The **end plate potential** (EPP) is the electrical depolarization of the postsynaptic end plate. This membrane cannot sustain a regenerative action potential; it can only electrotonically propagate a depolarization. The remainder of the postsynaptic membrane can, however, propagate a regenerative action potential.

The separate recording of end plate potentials and action potentials at the postsynaptic membrane of a skeletal muscle cell can be accomplished by recording at different distances from the end plate, and by administering curare to reversibly block the action potentials (Figure 6–34A). At a considerable distance from the end plate, only the regenerative action potential is observed with low curare levels. Nearer the end plate, however, the regenerative action potential is recorded at low enough levels of curare to still allow depolarization to threshold, but at higher curare levels only subthreshold depolarizations due to electrotonic spread of the end plate

TABLE 6-11

Relative permeability of excitatory end plate ion channels to different ions in frog skeletal muscle in response to acetylcholine activation, and two inhibitory neuron ion channels of the mouse spinal cord that respond to the inhibitory neurotransmitters glycine and GABA.

Channels Neurotransmitter Permeability Ratio	Excitatory ACh P/P_{Na}		Inhibitory Glycine P/P_{Cl}	Inhibitory GABA P/P_{Cl}
Tl^+	2.51	SCN^-	7.0	7.3
NH_4^+	1.79	I^-	1.8	2.8
K^+	1.11	NO_3^-	1.9	2.1
Na^+	1.0	Br^-	1.4	1.5
Li^+	0.87	Cl^-	1.0	1.0
Trimethylamine	0.36	$Formate^-$	0.33	0.50
Ca^{2+}	0.22	HCO_3^-	0.11	0.18
$Acetate^-$	<0.01	$Acetate^-$	0.035	0.08
$Choline^+$	<0.15	F^-	0.025	0.02
Cl^-	<0.01	K^+	<0.05	<0.05

potential are observed. There is an exponential decrease in the amplitude of the end plate potential with distance from the end plate and an increasing time delay to peak depolarization, as would be expected for electrotonic spread (Figure 6–34B).

Small fluctuations in the end plate potential can be measured even in the absence of any stimulation of the presynaptic membrane. These spontaneous miniature end plate potentials have the typical shape of normal end plate potentials evoked by stimulation of the presynaptic neuron, but are of much smaller magnitude (typically 1 to 1.5 mV) than the normal end plate potentials (about 50 mV). These miniEPPs reflect the spontaneous release of one, or a few, synaptic vesicles. Neurotransmitter is released in a quantum fashion, i.e., one vesicle releases neurotransmitter, or two vesicles, or three vesicles, etc. (see Supplement 6–3, page 252). Exogenous administration of about 6000 molecules of ACh depolarizes the motor end plate by about 1 mV. An average miniEPP thus reflects the release of about 10000 molecules of ACh, and this must therefore be

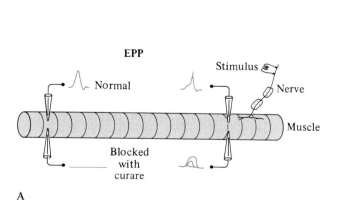

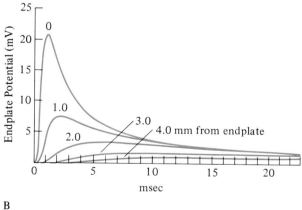

A   B

FIGURE 6–34 (A) Procedure shown in diagrammatic fashion for recording end plate potentials and action potentials of the postsynaptic membrane of a neuromuscular junction. Application of different levels of curare reversibly inactivates the postsynaptic membrane cation channels from opening, decreasing the amplitude of the end plate potential below threshold and inhibiting action potential initiation. (B) End plate potentials recorded at different distances from the end plate show electrotonic spread. The end plate potential amplitude diminishes with distance from the end plate, and the time delay increases. *(Modified from Katz 1966; Fatt and Katz 1951.)*

the number of neurotransmitter molecules in each synaptic vesicle (Kuffler and Yoshikami 1975).

The end plate does not sustain an action potential, and so the E_m will not rise to the action potential peak of $+40$ to 50 mV. The peak voltage of an end plate potential depends on the ion permeabilities of the ACh-receptor ion channel. This channel is relatively unspecific for cations (Table 6–11). There is a K^+ flux, and the I_K would hyperpolarize the E_m. Likewise, there are Na^+ and Ca^{2+} fluxes that depolarize the E_m. The E_m does not approach the equilibrium potential of any specific ion (e.g., E_K, E_{Na}, E_{Ca}) but approaches a potential determined by the relative permeabilities of all cations. This potential, called the **reversal potential** (E_{rev}) can be calculated from the Goldman-Hodgkin-Katz equation. For example, E_{rev} for the ACh-receptor ion channel can be estimated for mammalian skeletal muscle from the relative permeabilities of the channel to K^+, Na^+, and Ca^{2+} (Table 6–11) and the transmembrane ion concentrations (Table 6–5, page 208), considering only K^+, Na^+, and Ca^{2+} as follows.

$$E_{rev}$$
$$= \frac{RT}{F} \ln \frac{[Na^+]_o P_{Na} + [K^+]_o P_K + [Ca^{2+}]_o^{\frac{1}{2}} P_{Ca}}{[Na^+]_i P_{Na} + [K^+]_i P_K + [Ca^{2+}]_i^{\frac{1}{2}} P_{Ca}}$$
$$= 58 \log_{10} \frac{(145)(1.0) + (4.0)(1.11) + (1.5)^{\frac{1}{2}}(0.22)}{(12)(1.0) + (155)(1.11) + (10^{-7})^{\frac{1}{2}}(0.22)}$$
$$= -5.2 \, mV \qquad (6.18)$$

The normal E_m is depolarized toward E_{rev} when ACh is released at the motor end plate, although it doesn't necessarily reach E_{rev} if there is insufficient current flow. The E_m returns to normal when the ACh is removed by hydrolysis (Figure 6–35). If the E_m of the endplate is experimentally clamped to a more negative value, e.g., -50 mV, then the addition of ACh will cause a depolarization that will move E_m closer to E_{rev}. However, if the E_m is clamped to a value more positive than E_{rev}, e.g., $+20$ mV, then addition of ACh will hyperpolarize the E_m towards the E_{rev}. The direction of change in E_m is now opposite, or reversed, to the normal direction of change; this is why it is called the reversal potential.

The reversal potential is not the same for all postsynaptic ionic channels. E_{rev} depends on the nature of the transmitter and the specific properties of the ionic channels. Values for E_{rev} vary for different neurotransmitters/ionic channels from a hyperpolarizing -105 mV to a highly depolarizing $+6$ mV (Table 6–12). This variation in E_{rev} reflects the specific permeabilities of the ionic channel for various ions, either cations or anions. For example, the effects of different ratios of P_{Na} to P_K are shown in Figure 6–35.

Whether the effect of neurotransmitter release at a chemical synapse is excitatory or inhibitory is not an inherent property of the particular neurotransmitter. For example, acetylcholine can be excitatory at some synapses (e.g., the neuromuscular motor end plate) but inhibitory at other synapses (e.g., parasympathetic synapses in the heart).

Synaptic Delay. Depolarization of a presynaptic membrane is followed virtually instantaneously by a depolarization of the postsynaptic membrane at an electrical synapse; any small time delay could be ascribed to the cable constant properties of the membranes involved. In contrast, there is a considerable delay at a chemical synapse between

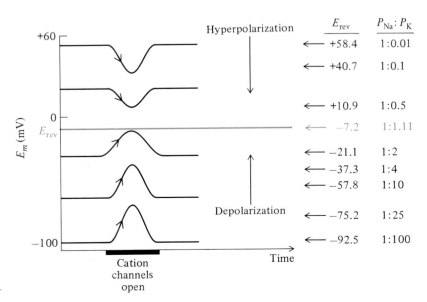

FIGURE 6–35 Reversal potential (E_{rev}) of a membrane clamped at varying membrane potentials is the potential towards which depolarization ($E_m < E_{rev}$) or hyperpolarization ($E_m > E_{rev}$) occurs when the cation channels are opened. The E_{rev} (-7.2 mV) shown in this example is for frog skeletal muscle end plate with a 1.0:1.11 ratio of $P_{Na}:P_K$. Different values of E_{rev} are shown for varying ratios of $P_{Na}:P_K$ on the right. E values are calculated at 25° C using the short version of the Goldman-Hodgkin-Katz equation

$$E_{rev} = 59.16 \log_{10} \frac{[K^+]_o P_{K:Na} + [Na^+]_o}{[K^+]_i P_{K:Na} + [Na^+]_i}$$

TABLE 6–12

Summary of various end plate channels with differing receptors and electrical responses to neurotransmitters because of differences in ionic conductance changes and reversal potentials.   *(Adapted from Hille 1984.)*

End Plate	Receptor	Response	Conductance Change	E_{rev}
Crayfish leg	Glutamate	EPSP	Cations	+ 6 mV
Frog skeletal muscle	ACh[1]	EPP	Cations	− 5 mV
Aplysia buccal ganglion cell	ACh	Rapid IPSP	Anions	− 60 mV
Crayfish leg	GABA	IPSP	Anions	− 72 mV
Frog sympathetic ganglion cell	ACh[2]	Slow EPP	K^+	− 86 mV
Mudpuppy parasympathetic ganglion cell	ACh[2]	Slow IPSP	K^+	− 105 mV

[1] Nicotinic.
[2] Muscarinic.

electrical activity arriving at the presynaptic terminal and the regenerative action potential propagating away from the postsynaptic membrane. Considering the complex sequence of events involved in chemical synaptic transmission (A–F listed on page 234), it is not surprising that there is a time delay, especially as some events involve physical movement of synaptic vesicles and protein conformational changes. Any disadvantage of a time delay is apparently compensated for by the one-way transmission across synapses and the capacity for complex integration of numerous inputs.

The pre- and postsynaptic depolarizations of a frog-toe neuromuscular junction show considerable variability in the magnitude of the EPP (as expected reflecting the quantum release of synaptic vesicles) and also a considerable variability in the synaptic time delay, from 0.5 msec to 2.0 msec or even more. Why is there such a long synaptic delay, and why is it so variable? The time for diffusion of neurotransmitter from the synaptic cleft to the receptors can be estimated to be about 0.05 msec, i.e., it is insignificant. Postsynaptic depolarization begins within about 0.15 msec of the addition of neurotransmitter to the postsynaptic membrane, and so this is not the major time delay. The major portion of the synaptic time delay is most likely the opening of Ca^{2+} channels and the release of neurotransmitter from the presynaptic terminal. Probably the movement of the vesicles to the membrane and exocytosis is the major, variable time delay in chemical synaptic transmission.

Neurotransmitters.   Identification of a substance as a neurotransmitter is often difficult. The criteria for probable identification of a substance as a neurotransmitter include: (1) it must be released

from the presynaptic terminal during action potential transmission, (2) it must elicit the normal postsynaptic depolarization, and (3) the effect of the substance must be blocked by the same agents that block synaptic transmission. There are many different neurotransmitters, which fall into four general categories: (1) acetylcholine, (2) biogenic amines, (3) amino acids, and (4) peptides.

Acetylcholine is the neurotransmitter at the vertebrate neuromuscular junction; we have already examined in detail this neuromuscular junction as a typical example of a chemical synapse. ACh is also the neurotransmitter at the preganglionic synapse of the sympathetic and parasympathetic nervous system and the neurotransmitter at the postganglionic synapse of the parasympathetic nervous system (and also rarely at the postganglionic synapse of the sympathetic nervous system; see Chapter 8). The structure of ACh and the dynamics of its release, hydrolysis, and resynthesis are shown in Figure 6–36A. Not all ACh receptors are identical. For example, the vertebrate postganglionic parasympathetic ACh receptor (and the less common cholinergic sympathetic receptor) is stimulated by muscarine (derived from the mushroom *Amanita muscaria*); it is a **muscarinic** ACh receptor. In contrast, the preganglionic parasympathetic and sympathetic synapses, and the neuromuscular end plate are unaffected by muscarine but are stimulated by nicotine; these are **nicotinic** ACh receptors. The muscarinic and nicotinic receptors respond to different structural aspects of the same ACh molecule. Muscarine and nicotine compete with ACh for the receptor site, but muscarinic and nicotinic receptors have different inhibitors and potentiators.

The catecholamines, epinephrine (adrenaline), norepinephrine (noradrenaline), and dopamine con-

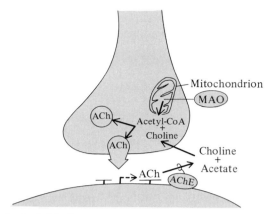

A   Acetylcholine synapse

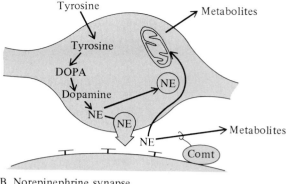

B   Norepinephrine synapse

FIGURE 6–36 **(A)** Dynamics of acetylcholine (ACh) release at a neuromuscular junction showing release, receptor binding, hydrolysis by acetylcholine esterase (ChE), and recycling of choline and acetate into the presynaptic nerve terminal. **(B)** Dynamics of norepinephrine synthesis, release, and recycling at a postganglionic axon swelling. The enzymes for norepinephrine hydrolysis (MAO, monoamine oxidase; COMT, catechol-o-methyl transferase are located intracellularly, not in the synaptic cleft). **(C)** Dynamics of a 5-hydroxytryptaminergic (5-HT) synapse showing synthesis, release, and recycling.

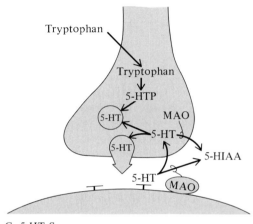

C   5-HT Synapse

tain an amine and a catechol group (a benzene ring with two adjacent hydroxyl groups). They are closely associated by their common synthetic pathway from phenylalanine. Norepinephrine is the neurotransmitter of the postganglionic sympathetic synapse for most vertebrates (epinephrine is the transmitter in some, such as amphibians). Epinephrine and norepinephrine are also circulating hormones, released from the adrenal medulla. The pharmacology of catecholamine receptors is even more complex than that of ACh receptors. There are α- and β-type norepinephrine and epinephrine receptors and even subclasses of β receptor, e.g., β_1 and β_2 receptors. The dynamic cycle of norepinephrine release, hydrolysis, and resynthesis is summarized in Figure 6–36B. The catechols are enzymatically inactivated by monoamine oxidase (MAO; located within the mitochondria) and catechol-o-methyltransferase (COMT; in the cytoplasm). Both MAO and COMT are intracellular enzymes and therefore are not the primary means of inactivating catecholamines after release into the

synaptic cleft. The catechols are removed from the synaptic cleft mainly by reincorporation into presynaptic vesicles or by diffusion into the general circulation.

Serotonin (5-hydroxytryptamine or 5-HT) is also an amine neurotransmitter, but it does not contain a catechol group. It is synthesized from the amino acid tryptophan (Figure 6–36C). The nervous systems of many mollusks contain 5-HT, and it is probably the neurotransmitter of the molluskan catch muscle synapse. It is also found in high levels in the central nervous system of vertebrates. There are at least three subtypes of 5-HT receptors: 5-HT_1, 5-HT_2, and 5-HT_3. These receptors for amine neurotransmitters are generally thought to transduce their effects via GTP binding proteins, but the 5-HT_3 receptor is also the actual ion channel (Derkach, Suprenant, and North 1989). Histamine is a neurotransmitter in arthropod photoreceptors (Hardie 1989).

A number of amino acids are known, or are suspected, to be neurotransmitters. Glutamate, aspartate, glycine and cysteic acid, and GABA (γ-aminobutyric acid) may be neurotransmitters at various synapses in different animals.

A wide variety of neuropeptides have been putatively identified as neurotransmitters, although they can also be neuromodulators or hormones. The first neuropeptide identified as a neurotransmitter, Substance P, has acetylcholine-like effects but is not blocked by ACh antagonists. Other noteworthy neuropeptides that may act as neurotransmitters are antidiuretic hormone (ADH), oxytocin, angiotensin II, LH-releasing hormone, and cholecystokinin. Endorphins and enkephalins are neuropeptides that bind at opioid receptors on the surface membranes of some neurons. Met-enkephalin, leu-enkephalin, β-endorphin, and other related opioid peptides are generally thought to be the natural messengers that bind to opioid receptors in animal tissues to which morphine (a plant opioid) also binds (Kosterlitz 1985). These opioid receptors probably only coincidentally bind exogenous narcotic opiates (opium, morphine, heroin) although endogenous morphine is present in animal tissues (especially the brain) at low levels. There are three opioid receptors: μ, γ, and κ receptors. Morphine binds very selectively at low K_m to μ-receptors (selectivity index = 0.98; $K_m = 0.56$ nM^{-1}), whereas β-endorphin and met-enkephalin have lower selectivity (0.52 and 0.09, respectively). Naloxone is a competitive inhibitor of opioid receptors, and thus interferes with the actions of exogenous and endogenous endorphins and enkephalins.

The classical concept of chemical synaptic transmission is the release of a single neurotransmitter that binds to a specific receptor on the postsynaptic membrane (i.e., Figure 6–37A). This simple concept has recently been elaborated in a number of ways. There may be a number of different receptors on the postsynaptic membrane that have different responses on synaptic transmission, e.g., α and β receptors (Figure 6–37B). In addition, there may be receptors for the neurotransmitter on the presynaptic membrane that influence subsequent synaptic transmission (Figure 6–37C). Many synapses have multiple neurotransmitters, i.e., there may be two or more neurotransmitters coexisting in the presynaptic terminal in the same vesicles or different vesicles (Figure 6–37D).

Many neuropeptides are found at the same synapse with some of the "more classical" neurotransmitters (e.g., norepinephrine, ACh, GABA; Table 6–13) or with other neuropeptides at synapses that lack a "classical" neurotransmitter (e.g., the hypothalamic neurosecretory cells of the paraventricular and supraoptic nuclei of the vertebrate central

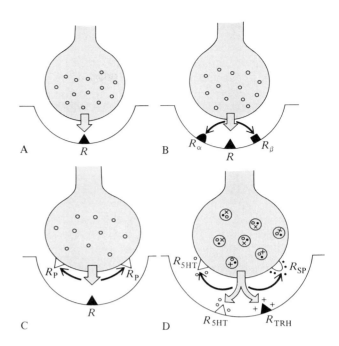

FIGURE 6–37 Representation of current concepts for chemical synaptic transmission. **(A)** Classical view of neurotransmitter release and binding to postsynaptic receptor (R). **(B)** There may be additional types of postsynaptic receptors for the same neurotransmitter that have different effects on the postsynaptic membrane (R$_\alpha$ and R$_\beta$). **(C)** There may be presynaptic receptors for the neurotransmitter that influence subsequent release of neurotransmitter (R$_p$). **(D)** There may be multiple neurotransmitters released at a single synapse with both postsynaptic and presynaptic receptors. Shown is a 5-HT neuron synapse in the ventral spinal cord that also contains thyroxin-releasing hormone (TRH) and Substance P. The smaller vesicles (about 50 nm dia) only contain 5-HT, whereas larger vesicles (about 100 nm dia) contain all three messengers. 5-HT acts on both postsynaptic excitatory receptors and presynaptic inhibitory receptors. TRH may act on a postsynaptic excitatory receptor (similar to the 5-HT receptor). Substance P may block the presynaptic inhibitory receptor of 5-HT. *(Modified from Hokfelt et al. 1986.)*

nervous system). Some neuron terminals contain more than one classical neurotransmitter (e.g., GABA and 5-HT, or GABA and dopamine can occur in the same presynaptic terminal). In some instances, ATP has been considered to be a neurotransmitter since it is often present in presynaptic vesicles (e.g., with ACh or norepinephrine). The functional role of multiple messengers at chemical synapses is not clear at present, but there are many possibilities. Multiple messenger systems may

TABLE 6–13

Coexistence of classical neurotransmitters and peptides in the central nervous system (based on immunohistochemical evidence). *(Modified from Hokfelt et al. 1986.)*

Classical	Peptide Neurotransmitter
Dopamine	CCK[1], neurotensin
Norepinephrine	Enkephalin, NPY[2], vasopressin
Epinephrine	CCK, Substance P, neurotensin, NPY
5-HT	CCK, enkephalin, Substance P, TRH[3]
ACh	Enkephalin, Substance P
GABA	CCK, enkephalin, NPY
Glycine	Neurotensin

[1] CCK = cholecystokinin.
[2] NPY = neuropeptide Y.
[3] TRH = thyrotropin releasing hormone.

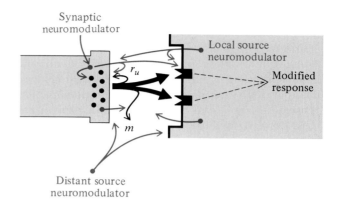

FIGURE 6–38   The presence of a neuromodulator (•) produces a modified synaptic response to transmission at a chemical synapse. The neuromodulator can be released from the synapse, from a local source (e.g., the postsynaptic cell), or from a distant sourse. The possible mechanisms for action of neuromodulator may be on the synaptic vesicles, on vesicular release of neurotransmitter (•), on the interaction of the neurotransmitter with its receptor (◣), on reuptake (r_u) of the neurotransmitter, or on the metabolic removal (m) of the neurotransmitter. *(Modified from Gorski 1983.)*

increase the capacity for complexity of information transfer. Multiple receptor systems on the postsynaptic membrane can have different effects on transmission properties if they have different reversal potentials; receptors, or multiple receptors on the presynaptic membrane, could also have different effects on subsequent synaptic transmission. The neuropeptides may not be involved with the transmission of action potentials *per se*, but may have other roles, e.g., growth (trophic) effects, induction of long-term changes in synaptic function (neuromodulation; see below), etc.

A variety of neurotransmitters and neuropeptides act as **neuromodulators** because they alter, or modify, the functioning of synapses (Figure 6–38). For example, serotonin, GABA, norepinephrine, and octopamine are neurotransmitters that are thought to be such modulating agents; several opiates and enkephalins are also thought to be modulating agents. These neuromodulating agents that alter synaptic transmission may be released at that synapse or may be released from a different synapse, i.e., they have a heterosynaptic effect. Their action may be to change the Ca^{2+} influx into the presynaptic terminal, thereby influencing synaptic vesicle release of neurotransmitter.

Synaptic Agonists and Antagonists.   Modification of any of the sequential processes of synaptic transmission can block or potentiate transmission. Chemicals agents, or drugs, that have the same effect as a neurotransmitter are **agonists** (or mimetics), whereas chemicals that reduce or prevent synaptic transmission are **antagonists** (or lytics).

Agonists often mimic the neurotransmitter at the receptor site. For example, choline, carbachol, succinylcholine, nicotine, and muscarine are all agonists of acetylcholine; part of their chemical structure mimics that of ACh, hence they mimic ACh at the receptor site (Figure 6–39). Agonistic effects can occur by mechanisms other than mimicking the neurotransmitters at the receptor. For example, the venom of the black widow spider causes massive release of ACh from the presynaptic vesicles, thereby mimicking normal ACh synaptic transmission (this is followed by a block in transmission because synaptic vesicles are depleted by the action of the venom). A number of drugs (eserine, neostigmine) and organophosphates (e.g., diisopropylphosphofluoridate) inhibit acetylcholine esterase, thereby potentiating the effect of normally released ACh.

Antagonists may competitively or noncompetitively inhibit the neurotransmitter at the receptor site. Curare, a crude mixture of various plant alkaloids, including d-tubocurarine, competitively blocks ACh at the receptor. α-bungarotoxin, a polypeptide from the venom of the krait *Bungarus multicinctus*, noncompetitively binds to the ACh receptor. Hemicholinium prevents synthesis of ACh. Botulinum toxin (produced by the bacterium

Agonists of Acetylcholine

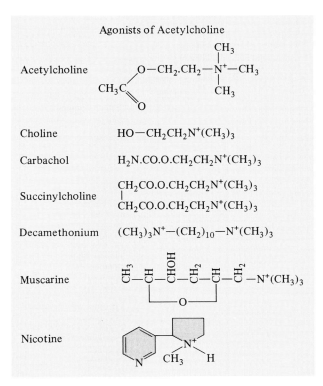

Acetylcholine

Choline $HO-CH_2CH_2N^+(CH_3)_3$

Carbachol $H_2N.CO.O.CH_2CH_2N^+(CH_3)_3$

Succinylcholine
$CH_2CO.O.CH_2CH_2N^+(CH_3)_3$
$|$
$CH_2CO.O.CH_2CH_2N^+(CH_3)_3$

Decamethonium $(CH_3)_3N^+-(CH_2)_{10}-N^+(CH_3)_3$

Muscarine

Nicotine

FIGURE 6–39 Chemical structure of acetylcholine compared with that of a variety of agonists.

Clostridium botulinum) prevents the release of ACh from presynaptic vesicles.

Cone shells (*Conus*) have a potent venom that they inject into their prey (fish) to rapidly immobilize them. The venom contains a variety of toxic peptides, called conotoxins (Olivera et al. 1987). These are small polypeptides (13 to 29 amino acids), are strongly basic, and are highly cross-linked. They act at various steps in neuromuscular transmission. The ω-conotoxin prevents the voltage-dependent influx of Ca^{2+} into the presynaptic terminal, α-conotoxins inhibit the ACh receptor, and μ-conotoxins directly inhibit the postsynaptic (muscle cell) action potential.

Neural Integration

Synapses integrate incoming action potentials, often in exceedingly complex fashions. A single neuron can have hundreds, or even thousands, of presynaptic terminals. This immense amount of incoming information is integrated at the axon hillock (where the axon extends from the cell soma) to generate the

efferent information (action potentials) that travel along its axon. Effector cells generally do not integrate information by having numerous presynaptic terminals, but simply respond to the presence or absence of afferent information from a single neuron. For example, most types of muscle cell have a single, or a few, excitatory motor end plates. However, some effector cells (e.g., invertebrate striated muscle) are innervated by two or even more types of neurons, which are excitatory and inhibitory.

The α motoneuron of the vertebrate spinal cord has been well studied with respect to neural integration. The cell body, which is located in the ventral horn of the spinal cord, receives thousands of excitatory and inhibitory synapses, the net integration of which determines its rate of action potential firing. The axon extends from the spinal cord to innervate a peripheral skeletal muscle cell. The action potentials that arise in the neuron are initiated at its axon hillock. This membrane region has a lower threshold and therefore a greater sensitivity to depolarization than the rest of the cell.

An excitatory synaptic event increases the likelihood of an action potential being initiated at the axon hillock, i.e., the reversal potential is more positive than the threshold. However, the depolarization of a single synapse generally does not even approach the reversal potential because of the low number of postsynaptic ionic channels opened. The stimulation of a single presynaptic terminal will release a few synaptic vesicles of neurotransmitter and depolarize the axon hillock membrane by only a mV or so towards the reversal potential; this is an **excitatory postsynaptic potential** (EPSP). This is in marked contrast to the neuromuscular junction previously described where presynaptic stimulation typically released 200 or so synaptic vesicles and depolarized the end plate membrane by about 40 to 60 mV. The small depolarization of the postsynaptic membrane by a single presynaptic terminal is analagous to the mini-end plate potentials of the neuromuscular junction end plate. Postsynaptic currents spread electrotonically from the dendrites in accord with the cable properties of the cell membrane. Depolarizing currents from different synapses are attenuated to varying extents when they reach the axon hillock. For example, a long, slender dendrite will have greater attenuation of its EPSP than would a short, thick dendrite or a synapse nearer the axon hillock. Some incoming action potentials are thus more important than others.

A large number of presynaptic terminals must be depolarized before the axon hillock of a neuron is sufficiently depolarized to initiate an action potential. The activity of one presynaptic terminal is added to the depolarizations of other presynaptic terminals; this is **spatial summation**, a summation of different stimuli occurring at different places (Figure 6–40A). Alternatively, the sequential activity of two presynaptic terminals will add, or superimpose, the depolarization of the second to that of the first; this is **temporal summation** (Figure 6–40B).

An **inhibitory postsynaptic potential** (IPSP) has a reversal potential that is more negative than threshold. The IPSP will hyperpolarize the E_m if the reversal potential is more negative than E_m. A postsynaptic potential with E_{rev} equal to resting E_m will cause no change in E_m. If the reversal potential lies between threshold and E_m, then the membrane depolarizes but the IPSP will reduce the depolarizing effect of a simultaneous EPSP and inhibit an action potential that might have been initiated had the IPSP not occurred (Figure 6–40C). The neurotransmitter GABA causes postsynaptic inhibition because its receptor-gated, postsynaptic membrane channels allow Cl⁻ flow. Consequently, GABA-activated

receptors stabilize E_m close to the resting value and are inhibitory because E_{Cl} is similar to resting E_m. Inhibitory postsynaptic potentials are also summed (actually, subtracted) with EPSPs at the axon hillock.

Some GABA Cl⁻ channels are also permeable to HCO_3^- (see Table 6–11, page 237). The large size of HCO_3^- ions suggests that these HCO_3^--permeable Cl⁻ channels are about 0.52 nm diameter. There is a substantial electrochemical force driving HCO_3^- out of the cell (Kaila and Voipio 1987). The intracellular and extracellular HCO_3^- concentrations are about 15 to 20 mM in crayfish and mammalian neurons, and so the resting E_m of about −90 mV is a large outward driving force for HCO_3^-. The E_{rev} for HCO_3^- is about 0 mV; the E_{rev} for the GABA-activated Cl⁻/HCO_3^- channels is about 10 to 15 mV more positive than E_{Cl}. Some glycine-activated Cl⁻ channels are also permeable to HCO_3^-. An outward flux of HCO_3^- could increase the intracellular pH and modify the sensitivity of ion channels and even alter intracellular ion concentrations.

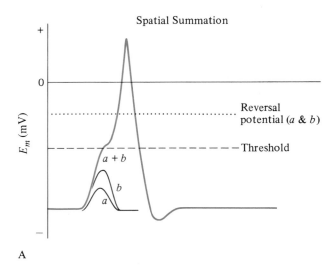

A

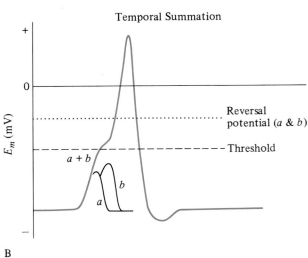

B

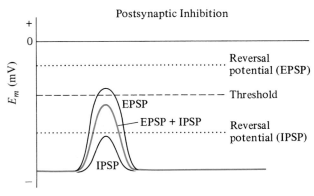

C

FIGURE 6–40 Representation of integration of various excitatory and inhibitory postsynaptic potentials at an axon hillock. (**A**) Spatial summation of two excitatory postsynaptic potentials (ESPs), *a* and *b*, to produce a threshold-attaining EPSP that initiates an action potential. (**B**) Temporal summation of two EPSPs to produce a threshold EPSP that initiates an action potential. (**C**) An inhibitory postsynaptic potential (IPSP) reduces the effect of a threshold EPSP to produce a subthreshold EPSP that does not initiate an action potential.

The summation and inhibition processes described so far occur at the postsynaptic membrane; they are **postsynaptic excitation** and **postsynaptic inhibition**. In contrast, **presynaptic inhibition** occurs when an inhibitory nerve terminal synapses with an excitatory nerve terminal, and activity of the inhibitory nerve terminal diminishes the amount of transmitter released at the excitatory synapse. There are many examples of presynaptic inhibition, but a particularly interesting one is the crustacean striated muscle fiber, which has not only an excitatory innervating axon but also an inhibitory innervating axon; this is a relatively uncommon example of multiple innervation of an effector cell (Figure 6–41A). The inhibitory axons responsible for postsynaptic inhibition also send collateral branches that terminate on the excitatory innervating axons; this occurs at both the muscle and within the central nervous system. The presynaptic inhibitory synapse appears to increase the permeability of the excitatory presynaptic membrane to K^+ and/or Cl^-, thereby decreasing the magnitude of the action potential spike, of Ca^{2+} influx, and of neurotransmitter release. There is therefore a reduced postsynaptic potential. About 16 quantum units of postsynaptic current can be measured for a crustacean striated

muscle fiber in response to stimulation of the excitatory neuron; this is reduced to almost zero by presynaptic inhibition (Figure 6–41B). Presynaptic inhibition occurs in the vertebrate and invertebrate central nervous systems, as well as the crustacean striated muscle fiber.

The effect of a postsynaptic depolarization depends not only on the cable properties but also on the recent history of synaptic activity. There can be temporary functional alterations in synaptic transmission dependent upon its prior activity. Facilitation, depression, and post-tetanic potentiation are examples of such use-dependent changes in synaptic function.

Facilitation occurs when the effect of a presynaptic stimulus is enhanced by another presynaptic stimulus, i.e., the effect of the second stimulus is greater than the effect of the first, and the sum of the two is greater than twice the first stimulus (Figure 6–42). Facilitation occurs because some calcium ions that entered the nerve terminal during the first stimulus are still present during the second; the intracellular $[Ca^{2+}]$ is therefore greater for the second stimulus and so more neurotransmitter is released (neurotransmitter release is proportional to $[Ca^{2+}]^n$ where n is 2 to 4).

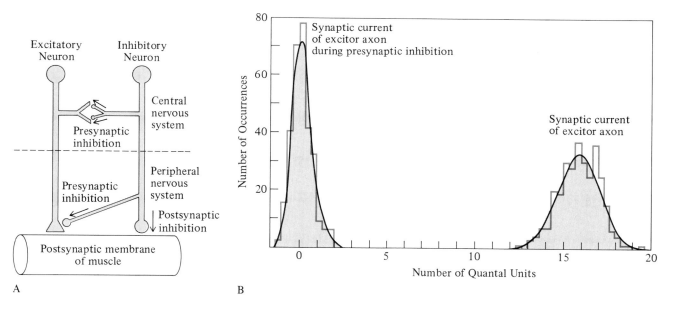

FIGURE 6–41 **(A)** Organization of the neuromuscular excitator and inhibitor neurons of the crayfish claw. The excitator muscle that opens the claw is presynaptically inhibited at both the muscle fiber synapse and within the central nervous system. **(B)** Recording from synapses of the crab leg muscle neuromuscular junction indicates that stimulation of the excitatory nerve causes release of about 16 quanta of neurotransmitter (with a high probability of release; $P = 0.99$). The probability of quantum release is markedly decreased by presynaptic inhibition ($P = 0.16$) and far fewer quanta are released. *(From Tse and Atwood 1986.)*

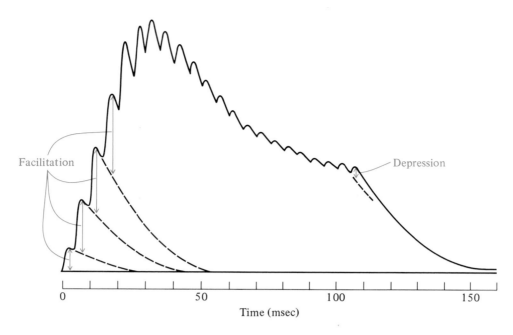

FIGURE 6–42 Repetitive stimulation initially causes a facilitation of the response (vertical bar with arrowheads) but eventually causes depression of the response to each individual stimulus. *(From Katz 1966.)*

Synaptic depression is the decrease in magnitude of the postsynaptic potential that occurs with repeated stimulation of the synapse because of depletion of synaptic vesicles in the presynaptic terminal. Fewer synaptic vesicles fuse with the presynaptic membrane, less neurotransmitter is released, and the postsynaptic potential is correspondingly reduced. Depression does not occur during magnesium block (elevated extracellular $[Mg^{2+}]$) because the Mg^{2+} diminishes the rate of neurotransmitter

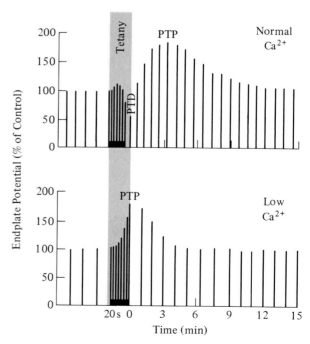

FIGURE 6–43 Tetanic stimulation of the frog neuromuscular junction normally results in post-tetanic depression, followed by post-tetanic potentiation at slower stimulation rate. The post-tetanic depression (PTD) is due to synaptic vesicle depletion, and post-tetanic potentiation (PTP) is due to residual elevation of intracellular Ca^{2+} concentration and reformation of synaptic vesicles. With low extracellular Ca^{2+} ($1/12$ normal), there is no post-tetanic depression (since synaptic vesicles are not depleted) but post-tetanic potentiation occurs sooner because of the residual intracellular Ca^{2+}. *(Modified from Rosethal 1969.)*

release and synaptic vesicles are not depleted by repetitive stimulation.

Post-tetanic potentiation occurs after repetitive high-frequency stimulation of a presynaptic terminal (Figure 6–43). The postsynaptic potential declines with repetitive stimulation due to synaptic vesicle depletion. After the tetanic stimulation is removed, there is a delayed increase in the postsynaptic potential which greatly exceeds that of the control postsynaptic potential (by about 2 ×). This post-tetanic potentiation of the postsynaptic potential is probably due to residual Ca^{2+} ions in the presynaptic terminal that augment release of the neurotransmitter once it is again present in synaptic vesicles. If this experiment is repeated with a low external $[Ca^{2+}]$, there is a smaller postsynaptic potential because the intracellular $[Ca^{2+}]$ is lower, there is no synaptic depression because there has not been any depletion of synaptic vesicles, and there is an immediate (not delayed) post-tetanic potentiation because of the residual Ca^{2+} in the presynaptic terminal.

Summary

Cells have an outer cell membrane (plasmalemma) and a complex array of inner membrane structures. Membranes are responsible for control of entry of solutes into and out of cells, and the compartmentalism of the cell into regions of specialized structure and function. Membranes are lipid bilayers and associated proteins that are often enzymes, transport mechanisms, or ion channels. Solute permeation of membranes can be passive diffusional exchange, due to lipid solubility or exchange through pores; passive transport by a facilitated diffusion carrier; or active transport by an energy-requiring carrier mechanism.

There generally is an electrical potential across biological membranes (resting membrane potential, E_m). A Donnan effect and electrogenic pumps contribute to the E_m, but the most important source of biopotentials is ionic equilibrium potentials. The ionic equilibrium potential (E_{eq}) for a particular ion species depends on its concentration difference across the membrane; its magnitude is calculated by the Nernst equation. The membrane potential is determined by the relative mobility, permeability, and concentration differences for each ion species between the inside and outside of the cell membrane; its magnitude is calculated by the Hodgkin-Huxley-Katz equation. The normal, resting E_m is approximately equal to the K^+ equilibrium potential, about −90 mV (inside relative to outside).

Action potentials are rapid, transient changes in E_m (about 1 msec duration) from the K^+ equilibrium potential (−90 mV) to about the Na^+ equilibrium potential (+60 mV). The action potential is initiated by depolarization of the E_m to a critical threshold value, and a consequent positive feedback increase in membrane Na^+ permeability. The Na^+ permeability is voltage dependent and increases with depolarization; the increase in Na^+ permeability then depolarizes the membrane further, initiating a positive-feedback cycle of continued depolarization. The action potential is terminated by an automatic decrease in membrane Na^+ permeability and by a transient increase in K^+ permeability that hyperpolarizes the membrane.

Action potentials are induced in excitable membranes by a depolarization to threshold. They are (generally) all-or-none transient increases in E_m, i.e., either an action potential occurs in response to depolarization or it doesn't. Information is conveyed by the rate of action potential firing, not by their shape or amplitude. There is an inverse relationship between the depolarization intensity and duration required to initiate an action potential, and there is a minimum depolarization voltage (rheobase) required to elicit an action potential. There is a latency period between the depolarizing stimulus and the action potential. Excitable membranes have an absolute refractory period during which they are unresponsive to successive depolarizing stimuli and cannot sustain an action potential. This is followed by the relative refractory period in which an action potential of reduced amplitude can be initiated.

Excitable membranes, such as neuron axons, can propagate an action potential at a velocity dependent on their cable properties, i.e., membrane capacitance and resistance. The conduction velocity for unmyelinated axons is proportional to the square root of the diameter. Many invertebrates and vertebrates have giant axons that provide rapid conduction of important reflexive information (e.g., predator escape responses). Myelination of axons by Schwann sheath cells also increases the conduction velocity by making action potentials saltate between adjacent nodes (lacking the myelin sheath). The conduction velocity of myelinated axons is proportional to internodal distance and axon diameter.

Electrical activity is transmitted between excitable cells at synapses. Electrical synapses allow rapid coupling of neurons for highly synchronized responses. Chemical synapses have a more complex structure and functioning. A specific chemical messenger, the neurotransmitter, is released from vesicles in the presynaptic terminus. The neurotransmitter diffuses to the postsynaptic membrane and binds to specific postsynaptic receptors. Activation of the receptors modifies the permeability of relatively nonspecific anion and cation channels, which causes depolarization or hyperpolarization of the postsynaptic membrane. Depolarization of the postsynaptic membrane to threshold will elicit an action potential in the postsynaptic cell. Chemical synapses allow complex integration of information of the synapse and unidirectional information transfer.

Supplement 6–1

Hodgkin-Huxley Model of Action Potentials

. . . .

Changes in ionic conductance and E_m during an action potential were elegantly measured and analyzed by Hodgkin and Huxley in the 1950s using voltage clamp experiments. Their results illustrate the basic principles of action potential generation and the relationships between E_m, ionic conductance, and ionic currents.

In a representative experiment, the resting membrane potential of a squid giant axon was depolarized and clamped to about -60 mV; the membrane potential was then rapidly depolarized and held at 0 mV. There was a transient inward current of about 2 mA followed by a maintained, outward current of about 2 mA. A different current signal was recorded if the experiment was repeated with the axon in a fluid bath with a low Na^+ and high K^+ concentration rather than normal squid Ringers because there could be no current due to Na^+ movement. The difference between these two current signals, normal and low Na^+, represents the Na^+ current (I_{Na}) that flows during the action potential. If all of the external Na^+ were replaced with choline [an impermeant cation, $HOCH_2CH_2N^+(CH_3)_3$], then there is actually a small outward Na^+ current when the Na^+ permeability increases because of Na^+ efflux. The potassium current could also be readily determined during an action potential by similar experimental manipulations. The Na^+ and K^+ conductances during an action potential were calculated from the currents, I_{Na} and I_K, as

$$I_{Na} = g_{Na}(E_m - E_{Na})$$
$$I_K = g_K(E_m - E_K)$$

The data for g_{Na}, g_K, I_{Na}, and I_K at a variety of voltage-clamped values of E_m were used to formulate a mathematical model of the membrane potential, current flow, and permeability changes that occur during an action potential. The changes in sodium conductance were analyzed as being due to two discrete events: Na^+ channel activation and inactivation. The activation event is due to opening of three channel gates (m) and inactivation by a single gate (h). The potassium conductance was analyzed as being activated by the opening of four n gates. The conductance of the Na^+ and K^+ channels is

$$g_{Na} = g_{Na,max}m^3h$$
$$g_K = g_{K,max}n^4$$

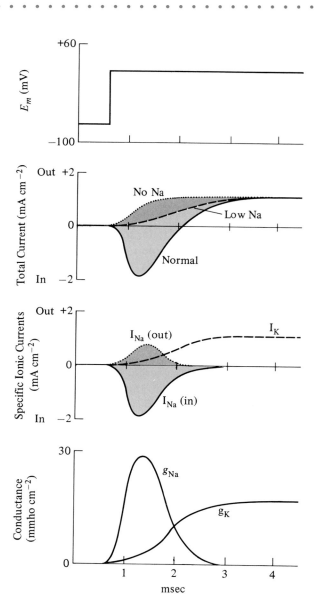

Currents during an action potential.

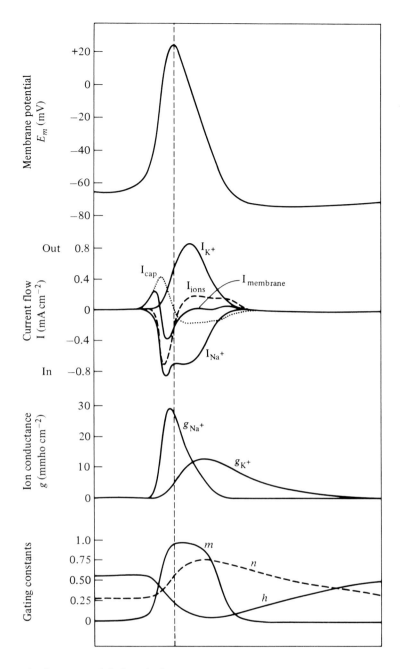

Action potential electrical events.

where $g_{Na,max}$ is the maximum Na$^+$ conductance and $g_{K,max}$ is the maximal K$^+$ conductance. The mathematical model enabled Hodgkin and Huxley to calculate g_{Na}, g_K, and E_m over time, from m, h, and n.

The variables m, h, and n can be imagined to be membrane-bound particles that form part of the Na$^+$ and K$^+$ channels, and change from closed to open and open to closed in an exponential fashion. If 3 m particles must each open to form a single Na$^+$ channel, then the transition from closed to open occurs in a sigmoidal (m^3) fashion since all subunits need to open for Na$^+$ flux; only the single h gate needs to close to prevent Na$^+$ movement through the Na$^+$ channel.

A major limitation of this model is that it describes the electrical events during an action potential, but not the mechanisms. Are there really m, h, and n gates in the cell membrane that collectively control the Na$^+$ and K$^+$ channels? The details of gating mechanisms for Na$^+$ and K$^+$ channels at the molecular level are still poorly understood, but models based on the empirical observations and original assumptions of Hodgkin and Huxley are still useful diagrammatic simplifications that at least enable us to visualize possible mechanisms. (*See Hodgkin and Huxley 1952a; Hodgkin and Huxley 1952b; Benzanilla, Rojas, and Taylor 1970; Aidley 1989; Hille 1984.*)

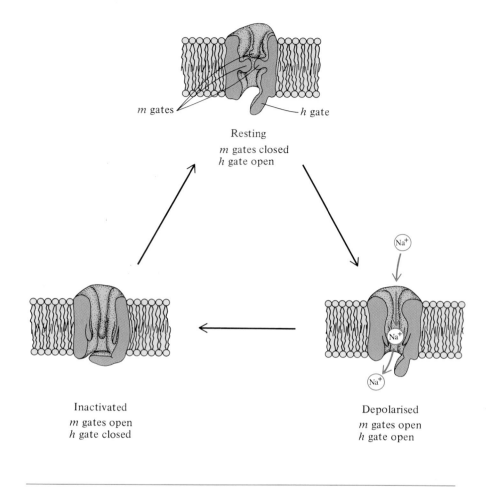

m gates ——— h gate

Resting
m gates closed
h gate open

Inactivated
m gates open
h gate closed

Depolarised
m gates open
h gate open

Supplement 6–2

Electrical Synapses

· ·

Signal transmission across an electrical synapse is difficult because of the high resistance of the intermembrane space to current flow. The typical action potential amplitude at the presynaptic membrane is about 140 mV (changing from -90 to $+50$ mV) and the depolarization of the postsynaptic membrane must reach about -70 mV to initiate an action potential. Thus, the 140 mV amplitude of the action potential can be reduced to about 20 mV and still result in propagation of an action potential, i.e., the attenuation can be 20/140 or 14%.

Let us consider electrical transmission across a typical 5μ radius axon with an area-specific membrane resistance (R_m) of 2000 Ω cm^2, membrane capacitance (C_m) of 1 μF cm^{-2}, and specific cytoplasmic resistance (R_i) of 200 Ω cm. An action potential traveling along this axon would be attenuated by current leakage through the cell mem-

brane and eventually dissipate if there was not a regenerative membrane process. The overall resistance of the axon to current loss, the input impedance (Z), is about 2 10^7 Ω. Impedance is essentially the equivalent in an AC circuit of resistance in a DC circuit; it is a complex function of both the membrane and cytoplasmic resistances, axon radius (μ), and the frequency of the action potential signal (f; about 250 sec^{-1});

$$Z = \sqrt{\frac{R_m R_i}{2\pi^2\mu^2 \sqrt{1 + (4\pi^2 f^2 R_m^2 C_m^2)}}}$$

There is little attenuation of the action potential along the axon, but there is extreme attenuation across the axon membrane. If we place a single membrane septum across the axon, then the high impedance will obviously reduce longitudinal spread along the axon, i.e., there will be

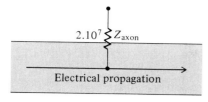

Attenuation = 0

Axon (5μ)

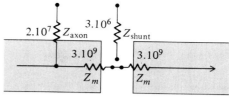

$$\text{Attenuation} = 100 \frac{(2.10^7)(3.10^6)}{(3.10^9)(3.10^9)} = 0.0007\%$$

12 μ Extracellular Gap

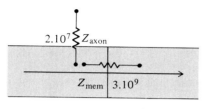

$$\text{Attenuation} = 100 \frac{2.10^7}{3.10^9} = 0.7\%$$

Axon + single membrane

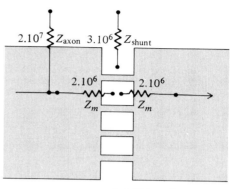

$$\text{Attenuation} = 100 \frac{(2.10^7)(3.10^6)}{(2.10^6)(2.10^6)} = 15\%$$

Gap + cross bridges

Extracellular gap and cross-bridge electrical synapses.

greater attenuation. The attenuation can be calculated from the ratio of radial to longitudinal impedance, i.e., attenuation = $(2\ 10^7)/(3\ 10^9)$ = 0.007 = 0.7%. This extent of attenuation would never allow propagation of an action potential across the membrane septum. Electrotonic transmission is made even more difficult by adding a second membrane septum and having a 10 nm gap of extracellular fluid between the two septa, i.e., a realistic electrical synapse. There is now a radial loss of current via the extracellular fluid, which we shall label as a shunt impedance (Z_{shunt}). The overall attenuation at such a synapse reduces the presynaptic depolarization to about 0.0007% at the postsynaptic membrane, which is too small for transmission of an action potential.

An electrical synapse must have a specific anatomical organization and special membrane properties to be able to pass action potentials. Increasing the axon diameter

facilitates electrical transmission since Z_m is decreased in proportion to the increase in cross-section area. Increasing the radius from 5 to 150 μ increases the attenuation to 7%, which is closer to the required value of 14%. Thus, giant axons are better able to have electrical synapses. A few electrical cross-bridges between adjacent axons facilitate electrical transmission by decreasing the resistance at the junction. Only about 1 pore per μ² (1/10,000 of the cross-section area of the gap) will decrease the membrane resistance to a value (e.g., 2 10^6 Ω) such that signal attenuation is greater than 14%. (*See Katz 1966.*)

Supplement 6–3

Quantal Release of Neurotransmitter Vesicles

. . . .

The quantal nature of neurotransmitter release from the presynaptic terminus of chemical synapses has been demonstrated by a statistical analysis of the amplitudes of miniature end plate potentials (miniEPPs). Not all synaptic vesicles contain exactly the same number of

neurotransmitter molecules, and so not all miniEPPs have exactly the same amplitude; rather, there is a normal distribution of miniEPP amplitudes. The amplitude of EPPs elicited by presynaptic neural stimulation of a frog skeletal muscle end plate showed a distribution that was

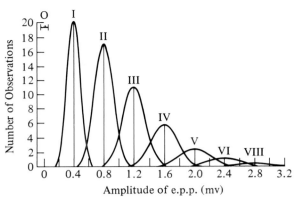

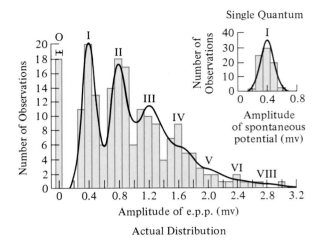

Theoretical Quantal Distribution

Actual Distribution

Theoretical and actual quantal distributions.   *(From Boyd and Martin 1956.)*

consistent with the predicted distribution based on (1) a normal distribution of miniEPP amplitudes, and (2) by assuming quantal release of neurotransmitter, i.e., 1 vesicle, or 2 vesicles, or 3, etc. In this experiment, the neuromuscular junction was partially blocked by a high external Mg^{2+} concentration to decrease the number of vesicles released per stimulation. In fact, the majority of stimuli did not elicit an EPP equivalent to a normal miniEPP (i.e., were "failures"; $n = 0$). (*See Boyd and Martin 1956; del Castillo and Katz 1954.*)

Recommended Reading

Aidley, D. J. 1989. *The physiology of excitable cells.* Cambridge: Cambridge University Press.

Bennett, M. V. L. 1966. Physiology of electrotonic junctions. *Ann. N. Y. Acad. Sci.* 137:509–39.

Bryant, H. J., and J. E. Blankenship. 1979. Action potentials in single axons: Effects of hyperbaric air and hydrostatic pressure. *J. Appl. Physiol.* 47:561–67.

Cooper, J. R., F. E. Bloom, and R. Roth. 1986. *The biochemistry of neuropharmacology.* New York: Oxford University Press.

Diamond, J. M., and E. M. Wright. 1969. Biological membranes: The physical basis of ion and non-electrolyte selectivity. *Ann. Rev. Physiol.* 31:581–646.

Dorsett, D. A. 1980. Design and function of giant fibre systems. *TINS* 3:205–8.

Finean, J. B., R. Coleman, and R. H. Mitchell. 1978. *Membranes and their cellular functions.* Oxford: Blackwell.

Hall, Z., J. G. Hildebrand, and E. Kravitz. 1974. *The chemistry of synaptic transmission.* Newton: Chiron.

Harper, A. A., et al. 1987. The pressure tolerance of deep sea fish axons: Results of Challenger cruise 6B/85. *Comp. Biochem. Physiol.* 88A:647–53.

Henderson, J. V., and D. L. Gilbert. 1975. Slowing of ionic current in the voltage-clamped squid axon by helium pressure. *Nature* 258:351–52.

Hille, B. 1982. Membrane excitability: Action potential and ionic channels. In *Physics and biophysics*, edited by T. C. Ruch and H. D. Patton, vol. IV, 68–100. Philadelphia: Saunders.

Hille, B. 1984. *Ionic channels of excitable membranes.* Sunderland: Sinauer Assoc.

Hodgkin, A. L. 1958. Ionic movements and electrical activity in giant nerve fibers. *Proc. Royal Soc. Lond. B.* 148:1–37.

Hodgkin, A. L. 1964. *The conduction of the nervous impulse.* Springfield: Thomas.

Hokfelt T., B. Everitt, B. Meister, T. Melander, M. Schalling, O. Johansen, J. M. Lundberg, A-L. Hulting, S. Werner, C. Cuello, H. Hemmings, C. Ouimet, I. Walaas, P. Greengard, and M. Goldstein. 1986. Neurons with multiple messengers with special reference to neuroendocrine systems. *Rec. Prog. Hormone Res.* 42:1–70.

Kandell, E. R., and J. H. Schwartz. 1985. *Principles of neural science.* New York: Elsevier.

Katz, B. 1966. *Nerve, muscle, and synapse.* New York: McGraw-Hill.

Kravitz, E. A., and J. E. Treherne. 1980. *Neurotransmission, neurotransmitters and neuromodulators.* Cambridge: Cambridge University Press.

Lockwood, A. P. M. 1978. *The membranes of animal cells.* London: Edward Arnold.

MacDonald, A. G., and A. R. Cossins. 1985. The theory of homeoviscous adaptation of membranes applied to deep-sea animals. In "Physiological adaptations of marine animals," edited by M. S. Laverack. *Symp. Soc. Exp. Biol.* 39:301–22.

Moore, J. W., and T. Narahashi. 1967. Tetrodotoxin's highly selective blockage of an ionic channel. *Fed. Proc.* 26:1655–63.

Rall, W. 1977. Core conductor theory and cable properties of neurons. In *Nervous system: Handbook of physiology,* edited by J. M. Brookhart and V. B. Mountcastle, sect. 1, vol. 1, pt. 1. Bethesda: Am. Physiol. Soc.

Sha'afi, R. I. 1981. Permeability for water and other polar molecules. In *Membrane transport,* edited by S. L. Bonting and J. J. H. H. M. de Pont, 29–60. Amsterdam: Elsevier/North Holland Biomedical Press.

Singer, S. J., and G. Nicholson. 1972. The fluid mosaic model of the structure of cell membranes. *Science* 175:720–31.

West, I. C. 1983. *The biochemistry of membrane transport.* London: Chapman & Hall.

Chapter 7

Sensory Physiology

The compound, ommatidial eye of a housefly provides high acuity, binocular vision. *(A. Devaney, Inc., N.Y.)*

Sensory cells detect information that allows animals to respond to their external and internal environments. The sensory information is first detected by a specialized cell membrane then transduced into an electrical receptor potential. The receptor potential is then encoded into a discharge of action potentials that is transmitted via peripheral nerves to the central nervous system. The central nervous system interprets information and elicits appropriate effector responses, such as muscle contraction, secretion, etc. Some sensory organs have a complex organization of neurons for peripheral processing of information; the retina of the vertebrate eye is a good example of a sensory system with complex peripheral processing.

Structural and Functional Classification

There are a number of ways to classify sensory systems. One is the **modality**, or nature, of the sensory signal to which the receptors respond. There are six basic modalities of information: chemicals, temperature, mechanical movement, electrical fields, magnetic fields, and light (Table 7–1). The threshold sensitivity of various sensory receptors is remarkably low, e.g., single molecules for chemoreceptors, single photons for photoreceptors. The minimum amounts of energy detectable by receptors are about 10 to 20 J receptor^{-1}. Pain is a further, poorly defined sensory modality; it identifies tissue damage. It can be induced by stimulation of specific pain receptors (nociceptors), and can be elicited by excessive stimulation of any of the other modalities, e.g., intense pressure, heat, or light. Pain is transmitted via specific afferent neural pathways to the central nervous system (see Chapter 8).

The sensory modality of a receptor is a property of the receptor itself. The sensory membrane, sensory cell, and specialized supporting cells are specifically designed to respond best to a particular sensory modality. The afferent nerve fiber conveys information from a sensory receptor to the central nervous system as a series of action potentials, but the transmitted information does not itself indicate the sensory modality. Rather, the modality is determined by the point of termination in the central nervous system of the afferent nerve fiber. This interpretation of a specific modality by the point of termination of individual nerve axons is the **labeled line principle**. For example, a photoreceptor in the vertebrate retina normally responds only to light, and so neurons in the visual cortex of the central

nervous system interpret action potentials transmitted from a photoreceptor as the detection of light. Photoreceptors can also respond to intense stimuli of other sensory modalities (e.g., pressure, mechanical movement) but the visual cortex interprets all incoming electrical activity as the reception of light regardless of the actual stimulus modality. This and other examples of sensory inputs to the central nervous system with specific labeled line organization will be discussed in Chapter 8.

Each sensory modality is characterized by certain sensory qualities. **Intensity** is a fundamental quality for all sensory modalities. Other specific sensory qualities for different modalities include: sweet, salt, sour, and bitter for gustatory chemoreceptors; cold and hot for thermoreceptors; direction and rate of deformation for mechanoreceptors; voltage gradient for electroreceptors; and color for photoreceptors.

Exteroceptors are sensory receptors that respond to the external environment. They can be subdivided into distance exteroceptors that detect information from a distance and contact exteroceptors that detect information by contact with the body surface. For example, olfaction (smell), audition (hearing), and vision are distance exteroceptive senses because they detect chemicals, sound, and light emitted from a source at a distance (although the

TABLE 7–1

Sensory modalities of animals. Also shown are the lowest limit of detection (threshold level) and the approximate minimal energy level for the lowest threshold stimulus.

Modality	Threshold	Energy Level
Chemoreception		
Chemicals	Single molecule	10^{-20} J receptor^{-1}
Humidity	1–2% RH	
Temperature		
Noninfrared	0.01° C	
Infrared radiation	0.003° C	0.1 mW cm^{-2}
Mechanoreception		
Bending amplitude	10^{-9} cm	10^{-18} J receptor^{-1}
Sound amplitude	10^{-12} cm	2 10^{-21} J receptor^{-1}
Sound pressure	10 Pa	2 10^{-21} J receptor^{-1}
Electroreception	10^{-8} V cm^{-1}	
Magnetoreception	0.05 10^{-4} weber m^{-2}	
Photoreception	Single photon	4 10^{-19} J receptor^{-1}
Nociception (Pain)	—	—

chemicals, air vibrations, and photons of light must obviously contact the sensory receptor cells to be detected). In contrast, gustation (taste) and tactile (touch) sensation require direct physical contact of the emitting object with the sensory receptor. However, the distinction between distance and contact exteroceptors can be somewhat arbitrary. For example, differences between olfaction and taste are difficult to distinguish in a fish; humans can "hear" vibrations transmitted to our ears by touch as well as by air transmission.

Interoceptors are sensory receptors that monitor the internal environment, e.g., baroreceptors monitor arterial and venous blood pressure; chemoreceptors monitor blood pO_2, pCO_2, and pH; hypothalamic temperature receptors monitor body temperature; hypothalamic iono- and osmoreceptors monitor blood Na^+ and osmotic concentration. Interocep-

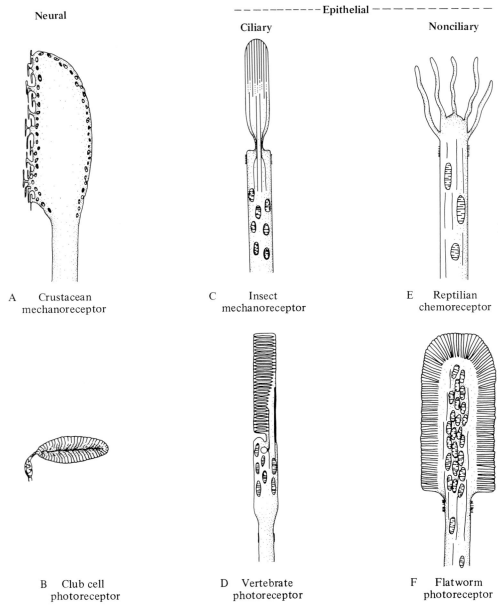

FIGURE 7–1   Examples of nonepithelial and epithelial receptor cells. **(A)** crustacean mechanoreceptor, **(B)** club cell (bipolar neuron), **(C)** insect mechanoreceptor, **(D)** vertebrate photoreceptor, **(E)** reptile chemoreceptor, and **(F)** flatworm photoreceptor. *(Modified from Thurm 1983; Hyman 1951.)*

tors often function as the sensory element in negative-feedback loops and are often only apparent at the subconscious level. Other interoceptors have important subconscious and conscious sensory roles. For example, limb proprioceptors monitor limb position and equilibrium receptors monitor body orientation and movement.

Sensory receptors can be classified by whether they are neuron endings or specialized sensory cells. Many sensory receptors are the peripheral endings of sensory neurons (Figure 7–1 A, B); they are part of the peripheral nervous system (see Chapter 8). These nerve endings may be "naked," i.e., they do not have a myelin sheath and are not associated with other types of cells. Most thermoreceptors and pain receptors are naked nerve endings. Many neuron endings have accessory structures that direct or modify the sensory stimulus. For example, the Pacinian corpuscle (a vertebrate pressure receptor) and cuticular hairs of arthropods have accessory cells that are involved in sensory detection.

Secondary sense cells are specialized, non-neuronal receptor cells that communicate sensory information to neurons. For example, vertebrate taste buds, photoreceptors (rods and cones), and the eccentric cell of the eye of the horseshoe crab *Limulus* are secondary sense cells, not neurons. Specialized receptor cells sometimes have no accessory cells (e.g., some surface chemoreceptors), but they often have complex accessory structures (e.g., the iris and lens of the eye, the cochlea of the vertebrate ear).

The specialized receptor membrane of epithelial receptor cells can be part of a modified cilium; a cilium is a cell surface organelle usually involved with cell movement (see Chapter 9). Ciliary derived receptors include mechano-, chemo-, and photoreceptors. The invertebrate mechanoreceptor and vertebrate rod photoreceptor are typical examples (Figure 7–1C, D). The basic ciliary structures may be so specialized as to be unrecognizable except for remnants of the ciliary microtubules and basal body. Alternatively, other parts of the plasma membrane, not associated with cilia, become specialized for sensory reception; examples are the reptile chemoreceptor and the flatworm photoreceptor (Figure 7–1E, F).

Other classification schemes for vertebrate sensory receptors include the anatomical site of the receptors and the diameter of their afferent axons. **Special senses** are those detected by specialized sensory organs that communicate the information to the central nervous system via cranial nerves (e.g., vision, audition). **Superficial senses** are those modalities sensed in the skin and supporting tissues;

deep senses are located in the central tissues, such as the viscera. Vertebrate sensory physiologists have classified afferent sensory systems by their axon diameter into types I, II, III, and IV. Type I fibers (10 to 20 μ diameter) convey information on position and movement of skeletal muscle (annulospiral muscle spindle and Golgi tendon organ). Type II axons (5 to 10 μ) convey information from the flower-spray endings of muscle spindles and from discrete cutaneous touch receptors. Type III axons (1 to 5 μ) carry temperature and pricking pain sensation. Group IV fibers (0.5 to 2 μ, unmyelinated) carry pain, temperature, and crude pressure sensation.

Sensory Coding

Sensory receptor cells detect a sensory modality and **transduce** it into information for transmission to the central nervous system (Figure 7–2). There

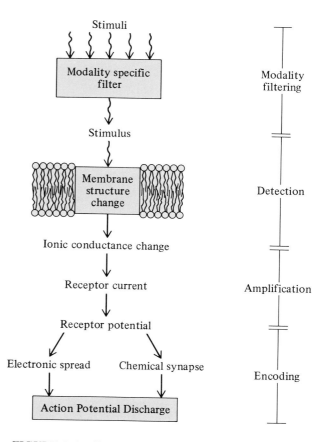

FIGURE 7–2 General functions of a sensory receptor: stimulus filtering, detection, amplification, and encoding of sensory information as action potentials.

are four basic steps in the process: filtering to isolate the modality of that receptor, detection of the filtered information, amplification of the detected signal, and encoding of the information into action potentials. Let us first examine the general concept of sensory transduction of stimuli into electrical activity; many specific examples of sensory transduction will be described in detail below.

Transduction

Sensory receptors generally respond to only a specific stimulus modality, although they are continuously exposed to many kinds of stimuli, e.g., temperature, pressure, light, chemicals, movement, and electrical gradients. Receptor selectivity can be accomplished by the specific molecular structure of the membrane, the structure of the receptor cell, or the specific organization of the receptor cells with other accessory cells. For example, a mechanoreceptor cell has a specialized cilium, whose membrane is sensitive to stretch; a photoreceptor membrane contains specific pigment molecules whose structure is altered by absorbed light.

Stimulation of the receptor membrane by the appropriate sensory modality transiently changes the membrane permeability and ionic conductance. The stimulus is transduced into a membrane permeability change in a variety of ways for different sensory modalities (Figure 7–3). Mechanoreceptors may physically deform ion channels and alter their ionic permeability, e.g., stretch may enlarge ion channels. Thermoreceptors may have temperature-dependent oscillations in ion channel permeability or temperature-dependent Na^+-K^+ ATPase activity. Chemoreceptors (e.g., taste and smell) have receptor proteins that regulate the permeability of ion channels through G proteins, adenyl cyclase, and cyclic AMP. Photoreceptors have a similar role for a G protein (transducin), adenyl cyclase, and cAMP in transduction of absorption of a photon of light by the protein receptor, rhodopsin, into an altered Na^+ permeability. Electroreceptors have voltage-dependent ion channels.

Stimulation of a sensory receptor temporarily opens or closes ion channels in the receptor membrane. The change in conductance transiently depolarizes or hyperpolarizes the receptor membrane; this is the **receptor potential**. Whether the receptor potential is a depolarization or hyperpolarization depends on whether the sensory stimulus opens or closes ion channels on the charge of the ions moving through the membrane, on the normal resting membrane potential, and on the equilibrium potential of the permeant ions. For example, absorption of light increases the Na^+ conductance of the invertebrate

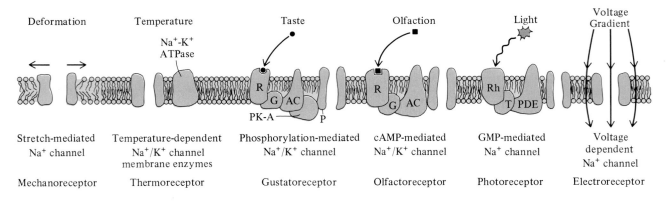

FIGURE 7–3   Generalized schema for the transduction of a sensory stimulus into a change in sensory membrane permeability to ions. Mechanoreceptors respond to physical deformation of membrane ion channels; thermoreceptors have temperature-dependent effects on ion channel permeability or membrane-bound enzymes; stimulation of membrane-bound taste receptors initiates a biochemical cascade (G protein, adenyl cyclase, protein kinase) that regulates the permeability of ion channels; stimulation of membrane-bound olfactory receptors initiates a biochemical cascade (G protein, adenyl cyclase) that regulates the permeability of ion channels; the rhodopsin of photoreceptors absorbs light and initiates a biochemical cascade (transducin, phosphodiesterase) that regulates ionic permeability; and electric field strength affects the permeability of voltage-dependent ion channels of electroreceptors.

photoreceptor and depolarizes the membrane but decreases the Na^+ conductance and hyperpolarizes the vertebrate photoreceptor.

Receptor cells typically amplify a weak stimulus into a stronger signal. Photoreceptors are an excellent example of high amplification. Some photoreceptors can detect a few photons or even a single photon. A single photon of visible light has an exceedingly small energy, about 3 to 5 10^{-19} joule ($E = h\nu$; h is Planck's constant and ν is the frequency of light). Reception of a single photon might open 10^3 to 10^4 ion channels and cause a transient (10 msec) current flow of about 40 picoAmp (10^6 ions). The current energy is about 4 10^{-15} J, so the energy amplification is about 10^4 (4 10^{-15}/4 10^{-19}). Olfaction is a similarly sensitive sensory modality. The binding energy of odorous molecules to receptors is about 10^{-20} J per receptor and the energy of the resultant receptor current is about 10^{-17} J; amplification is about 10^3 (Kaissling 1983).

All aspects of sensory receptor response so far discussed (membrane ion channel modification, membrane conductance change, receptor current flow, and receptor potential) are graded processes that are proportional to the stimulus intensity. Receptor potentials can be graded depolarizations or hyperpolarizations, but are not regenerative, all-or-none potentials. They are therefore more like end plate potentials than action potentials (see Chapter 6). This is not surprising because sensory receptors must not only detect the presence of stimulation, but they must also convey intensity information.

Conversion of receptor potentials into a pattern of neuronal action potentials can occur in one of two ways, depending on whether the receptor is a neuron terminal or a secondary sensory cell. Let us first consider neuronal receptors, using as an example a vertebrate pressure receptor, the Pacinian corpuscle. The sensory receptor of the Pacinian corpuscle is an unmyelinated nerve ending that is surrounded by concentric lamellae of accessory cells (Figure 7–4A). Pressure causes an electrical depolarization of the free nerve terminal, and this receptor potential is electrotonically propagated to the myelinated portion of the nerve where regenerative action potentials are initiated (Figure 7–4B). The electrical depolarization that is electrotonically propagated to the myelinated nerve fiber is the stimulus that initiates (or generates) an action potential; it is a **generator potential**. The difference between the receptor potential of the free nerve ending and the generator potential of the myelinated portion of the axon might seem unimportant here, but the distinction becomes more obvious and important for secondary sensory cells; the receptor potential

is the depolarization of the secondary sensory cell whereas the generator potential is the depolarization of the neuron.

The Pacinian corpuscle has numerous accessory cells that envelop the nerve terminal like concentric onion skin layers. These lamellae seem to act as a series of springs and dashpot dampers that attenuate pressures applied to the surface of the corpuscle. Much of the stimulus energy is stored in the lamellar "springs" and is highly attenuated by the time it reaches the central nerve ending. The Pacinian corpuscle has an initial response to stimulation, i.e., an "ON" response, but no further response until the stimulus is removed, i.e., an "OFF" response (Figure 7–4C). Such an ON–OFF responsiveness means that the Pacinian corpuscle does not relay a constant stream of potentially unchanging information that the central nervous system has to monitor and interpret; it only sends information when there is a change in stimulation. Removal of the lamellae alters the ON–OFF response to a continual ON response (with slow adaptation; see below) and no OFF response.

Transduction of a sensory stimulus into a generator potential is more complex for secondary sensory cells because the receptor potential of the sensory cell must be transferred to the nerve cell to form a generator potential. This transfer of the receptor potential is generally accomplished by a chemical synapse. For example, photoreceptors, taste buds, and acoustico-lateralis receptors of vertebrates use chemical synaptic transmission.

Relationship Between Stimulus Intensity and Response

Stimulation of a sensory receptor elicits a receptor and generator potential that is encoded into an action potential discharge; but what is the relationship between the stimulus intensity and the discharge of action potentials in the neuron carrying the information to the central nervous system? Is it linear or curvilinear? Is it the same for all receptors? Is it constant or can it be modified?

The relationship between stimulus intensity and amplitude of the receptor and generator potentials is graded, although not necessarily linear. For example, the relationship between the stimulus intensity (amplitude of compression) and the amplitude of the generator potential is graded but curvilinear for the Pacinian corpuscle (Figure 7–5). This kind of curvilinear relationship between stimulus intensity and receptor or generator potential is observed for many receptor cells.

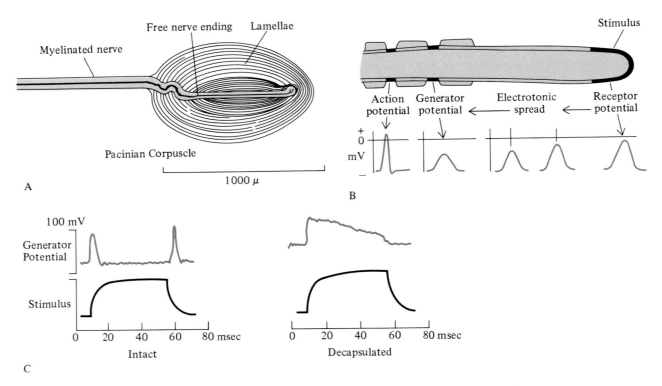

FIGURE 7–4 **(A)** The Pacinian corpuscle is a large pressure-sensitive mechanoreceptor. The free nerve ending is encapsulated by lamellae of accessory cells. **(B)** The dendrite nerve ending of a Pacinian corpuscle depolarizes in response to mechanical stimulation (receptor potential) and the electrotonic spread of the receptor potential to the first myelin internode forms a generator potential that initiates action potentials. **(C)** An intact Pacinian corpuscle has an ON–OFF response in generator potential (upper trace) to mechanical stimulation (lower trace), whereas a decapsulated dendritic ending without its accessory cells responds with an adapting ON response and no OFF response. *(Modified from Quilliam and Sato 1955; Lowenstein and Mendelson 1965.)*

There often is a linear relationship between the receptor or generator potential and action potential discharge. Consequently, there often is a curvilinear relationship between the rate of action potentials and stimulus intensity, and there is a fairly linear relationship between action potential discharge and $\log_{10}$ intensity. This linear semilogarithmic relationship between stimulus intensity and response is the **Weber-Fechner relationship**. For example, the relationship between $\log_{10}$ stimulus intensity and action potential frequency for the Pacinian corpuscle is fairly linear, except at very low and very high intensities.

In 1846, Weber investigated the ability of human subjects to distinguish by weight between pairs of objects. He found that subjects could distinguish between objects so long as they differed by at least 2 1/2% in weight, i.e., the threshold difference was

a relative one (one weight as a proportion of the other) rather than an absolute one (a difference of 1 gram, or 10 gram, etc). The minimum weight difference of 2 1/2% is the increment threshold. In 1862, Fechner formalized the results of such studies by the relationship

$$S = a\log I + b \qquad (7.1)$$

where S is the sensation, I is the stimulus intensity, and a and b are constants. This relationship, the Weber-Fechner "law," indicates that a sensory response is proportional to the logarithm of the stimulus intensity. This relationship is actually not an invariable law, but it is generally applicable to many examples of sensory reception.

The general applicability of the Weber-Fechner relationship between intensity and sensory response

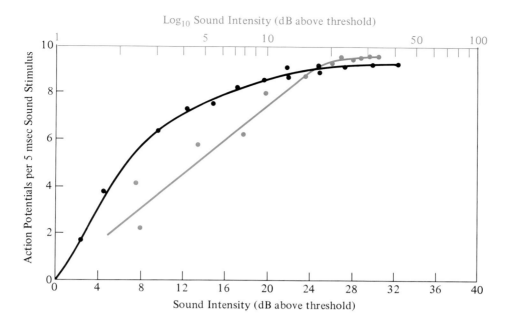

FIGURE 7–5 There is a sigmoidal relationship between the stimulus intensity and action potential frequency for an invertebrate acoustic detector, but the relationship between $\log_{10}$ stimulus intensity and action potential is more linear, reflecting the Weber-Fechner relationship.

has considerable physiological significance. It reflects a good compromise between high sensitivity of detection and wide range of detection. Sensory receptors become "saturated" by a high-stimulus intensity; the maximum possible action potential frequency is about 500 to 1000 sec^{-1} because the duration of an action potential is 1 to 2 msec. Saturated sensory receptors are unable to discriminate between high-intensity stimuli. If there were a linear relationship between response and intensity, then we could have either a fine discrimination between different intensities but a low intensity for saturation (Figure 7–6, curve fd) or have a low discriminating power but a high saturation intensity (curve ld). The semilogarithmic relationship (curve log) gives a fairly constant relative sensitivity regardless of intensity and a high saturation intensity. The useful intensity range of the semilogarithmic relationship is much greater than that of the fine-discriminating linear relationship.

Adaptation

The action potential frequency of most sensory receptors declines if the stimulus intensity is maintained constant. This decline in response is **adaptation**. Some receptors adapt rapidly and some adapt

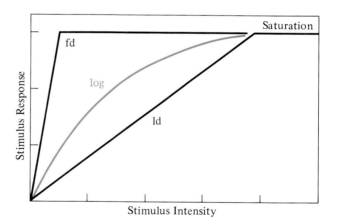

FIGURE 7–6 Hypothetical comparison of the sensory response to varying intensity stimuli for a fine-discriminating receptor (fd) that reaches maximal response at low intensity, a low-discrimination receptor (ld) that saturates at high stimulus intensity but is relatively insensitive, and a logarithmic relationship receptor (log) that provides fine discrimination at low-stimulus intensity but has a high-stimulus intensity at saturation.

slowly. For example, the stretch receptor of the vertebrate muscle spindle has a slow decline in action potential discharge if a steady stimulus is maintained (Figure 7–7A). When the stimulus is removed, the action potential discharge declines transiently to less than its normal spontaneous value, indicating a momentary inhibition of action potential initiation.

A **tonic receptor** has a slow, or no, decline in response to a maintained stimulus whereas a **phasic receptor** rapidly adapts and has a faster decline in action potential discharge. The stretch receptor system of the crayfish abdominal muscles (Figure 7–7B) is an excellent example of tonic and phasic receptors. Two pairs of muscle fibers (one pair per side) stretch between adjacent intersegmental

membranes; each muscle fiber has a stretch sense organ. One fiber of each pair has a phasic stretch receptor; the other has a tonic receptor. Muscle stretch excites both receptors, but the generator potential of the phasic receptor rapidly declines to below threshold and the action potential discharge declines. In contrast, the generator potential of the tonic receptor remains above threshold for the duration of the stretch stimulus, and the action potential discharge continues with only a slight decline in rate. Inhibitory axons to the muscle modify these stretch receptor responses.

Tonic and phasic receptors have different functional roles. Phasic receptors convey information concerning changes in stimulus intensity because they rapidly adapt. This is particularly important,

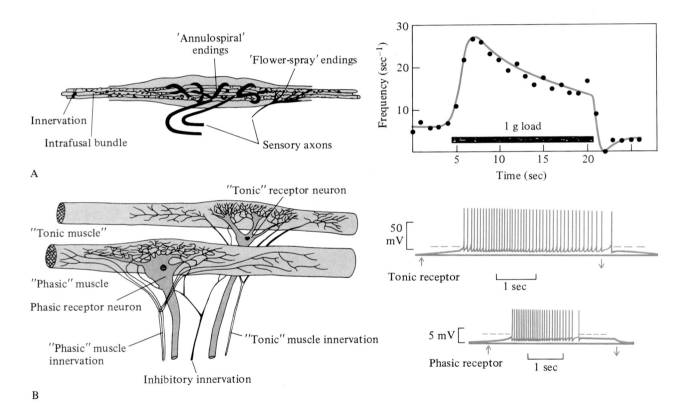

FIGURE 7–7   (A) The frog muscle spindle consists of specialized intrafusal muscle fibers that are noncontractile at their center; it has annulospiral and flower-spray sensory nerve endings. Stimulation of the spindle by a sustained 1 gram load causes an increase in action potential discharge of the afferent sensory neuron that is slow adapting. Removal of the load initially causes a decrease in action potential discharge to below the normal spontaneous activity. **(B)** The abdominal muscle stretch receptor system of the crayfish is a pair of phasic and tonic receptors each associated with its "phasic" or "tonic" muscle fiber. The phasic receptor rapidly adapts to a sustained stretch, whereas the tonic receptor has a sustained response to the stretch and only slowly adapts. The muscle fibers are also innervated by excitatory motor axons and an inhibitory axon.   *(Modified from Gray 1957; Adrian and Zotterman 1926; Burkhardt 1958; Eyzaguire and Kuffler 1955.)*

for example, to mechanoreceptors that respond to touch or pressure. We are initially aware of the touch of clothes against our skin when we first dress, but the cutaneous sensory receptors rapidly adapt and we lose the sensation of touch. Tonic receptors convey information concerning a maintained stimulus intensity; they slowly adapt, or do not adapt at all. For example, the proprioceptors that monitor limb position do not rapidly adapt. It is essential that many enteroceptors do not adapt at all. For example, physiological homeostatic mechanisms require a continued, accurate monitoring of the absolute value of a physiological variable (e.g., body temperature, blood pressure, blood sodium concentration, blood osmolality). Long-term homeostasis would be impossible if these sensory receptors adapted even slowly.

There are a number of mechanisms for sensory adaptation. First, depletion or inactivation of the receptor structures can cause adaptation. For example, the bleaching of the photoreceptor pigment rhodopsin by light reduces the sensitivity of photoreceptors by up to 10^5 times. Second, the electrical properties of the receptor membrane may be time dependent. The accumulation of calcium ions within receptor cells during prolonged stimulation can inactivate receptors and their ion channels, or activate K^+ channels and hyperpolarize the membrane. Third, the accessory structures associated with many receptor cells may have time-dependent properties, e.g., the lamellae of the Pacinian corpuscle greatly modify the pressure stimulus at the sensory nerve ending.

Sensory Threshold

The minimum stimulus intensity that can be detected by a receptor is its **sensory threshold**. This depends on many factors, including receptor properties and signal processing capabilities of the central nervous system.

Very sensitive sensory receptors can detect small stimuli. However, their sensory information is only useful if they can discriminate between sensory stimulation and random background noise. The ratio of the sensory signal to the background noise must be greater than 1 and preferably much greater than 1. For intense stimuli (e.g., loud noise or bright light), the signal to noise ratio is high. At low stimulus intensities (e.g., faint noises or single photons of light) the signal to noise ratio is less favorable for signal detection. The auditory hair cell can detect sound energy levels of 1 kJ mole^{-1}, which is similar to the background energy of random thermal motion. However, it is not useful for the auditory sensory receptors to be so sensitive that they detect background noise, i.e., "hear" random molecular motion. The maximum sensitivity of the auditory cells is therefore determined by background Brownian motion.

The ability to discriminate low intensity stimuli from background noise is essentially the result of event probabilities (Aidley 1989). Receptors can optimize signal detection by maximizing their sensitivity, at least to the point where they can detect background noise, or they can maximize their signal to noise ratio for sensory detection by averaging a number of signals (signal averaging) or by increasing the number of receptors and the number of neurons conveying the sensory information to the central nervous system (multiple channels).

High Sensitivity. There is a stimulus intensity below which the probability of detection is 0, and there is an intensity above which the probability of detection is 1. In between, there is some curvilinear relationship for stimulus intensity and probability of detection (Figure 7–8A). The sensory threshold is the stimulus intensity for a given probability of detection; this probability at the sensory threshold can be arbitrarily defined as 0.5. Shifting the probability of the detection-stimulus intensity curve to the left lowers the sensory threshold and shifts the most sensitive part of the probability-intensity curve (the steepest part) to lower intensity values (Figure 7–8B). However, sensitive receptor systems often have action potentials even in the absence of stimulation, i.e., they have **spontaneous activity**.

Spontaneous activity is not necessarily undesirable noise. It can be useful for sensory systems that monitor positive and negative responses. For example, the action potential discharge of an electroreceptor might be increased above a spontaneous level by an electric field of one polarity, and decreased below the spontaneous level by an electric field of the opposite polarity.

Signal Averaging. The detection and averaging of many separate or successive stimuli, i.e., **signal averaging,** allows reliable discrimination of low intensity signals from background noise. For example, vertebrate hair cells discriminate sounds of low signal to noise ratio by signal averaging many successive auditory stimuli. This is possible because noises generally have a much longer duration than the frequencies of their individual components. Consequently, a single noise has many discrete auditory signals that can be averaged.

Multiple Channels. A **multiple channel** sensory system has a number of receptors that function in

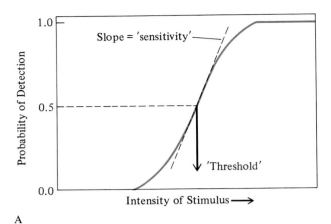

A

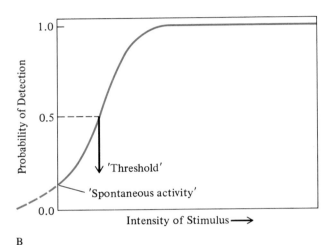

B

FIGURE 7–8 **(A)** The probability of stimulus detection increases with stimulus intensity in a curvilinear fashion between no detection ($P = 0$) and certain detection ($P = 1$). The "threshold" for detection can be arbitrarily defined as that stimulus intensity for a given probability of detection (*e.g.*, $P = 0.5$). **(B)** Decreasing the threshold for signal detection at $P = 0.5$ results in spontaneous activity.   *(From Aidley 1989.)*

parallel. The amount of information that can be carried by a multiple channel system is greater than for a single channel, and so the sensory threshold can be lower.

Increasing the number of channels has two effects. First, it increases the probability of detection and decreases the sensory threshold. Second, it increases the "sharpness" of threshold detection. An illustration of the enhanced sensitivity of multiple channel systems is the detection of sucrose in a solution by a fly's tarsal chemoreceptors. Flies taste with their feet, and flies using only one leg ("one channel") have a higher threshold for detection of

sucrose than flies using two legs ("two channels"). There is a five-fold decrease in the 50% response threshold for flies using two legs compared to flies using one leg, for 0.01 M sucrose.

Multiple channel sensory systems can also provide **spatial resolution** and **range fractionation** if each unit conveys information from a different point in space. The large number of photodetectors in some eyes allows spatial resolution of small objects. The equilibrium-detecting organ of crustaceans, joint proprioceptors of vertebrates, and frequency resolution by ears are examples of range fractionation. Range fractionation by a number of sensory receptors allows the higher and more linear resolution of a sensory modality over a much wider range between threshold and saturation intensities, than does a single sensory receptor over its individual threshold to saturation range. The multiple sensory system has a wider linear dynamic range than a single receptor.

The numerous, closely packed photoreceptors of both vertebrate and invertebrate eyes are a multiple channel system that provides considerable spatial resolution of the visual image. Each photoreceptor "sees" a different point in the visual field, so having more photoreceptors increases the spatial resolution. Interactions between adjacent receptor units can further enhance their resolving power. For example, the compound eye of the horseshoe crab *Limulus* has a large number of closely packed photoreceptor units (ommatidia). Adjacent ommatidia are mutually inhibitory through their inhibitory lateral plexus, a system of collateral branches of axons from the sensory cells that have inhibitory synapses on adjacent photoreceptor cells (Figure 7–9). The

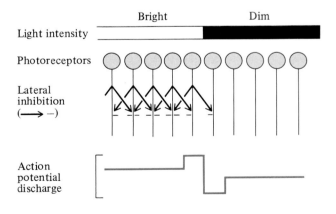

FIGURE 7–9 Representation of lateral inhibition showing the accentuated contrast at the border of the brightly and dimly illuminated regions of the eye. *(Modified from Aidley 1989.)*

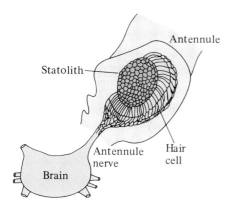

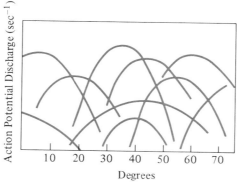

FIGURE 7–10 The statocyst organ in the antellule base of the lobster (left) has numerous hair cells that respond to orientation of the statocyst at varying angles. *(From Horridge 1968; Cohen 1960.)*

sensory response of an ommatidium is decreased by illumination of an adjacent ommatidium. One effect of this **lateral inhibition** is to emphasize the sensory response at the border of illuminated and nonilluminated areas. Strongly illuminated ommatidia exert strong lateral inhibition on adjacent ommatidia; weakly illuminated ommatidia exert weaker lateral inhibition. Consequently, a strongly illuminated ommatidium adjacent to a poorly illuminated ommatidium has the greatest response, and a poorly illuminated ommatidium adjacent to a strongly illuminated ommatidium has the weakest response. This results in a high contrast of the border of the illuminated and nonilluminated areas.

Most decapod crustaceans have a pair of **statocysts** at the base of their first pair of antennae that provide information of body orientation with respect to gravity, i.e., static equilibrium sense (Figure 7–10). The statocysts are invaginated sacs, containing a **statolith**. The statolith is formed from secretions of the statocyst wall or by accretion of fine sand grains bound together by secreted material. The floor of the statocyst has rows of sensory hairs innervated by a branch of the antennule nerve. The precise orientation of the decapod determines the position of the statolith within the statocyst and, hence, which sensory hairs are stimulated. Different sensory hairs will be stimulated by varying orientations, e.g., some hairs are maximally stimulated at 0° orientation, others at 20°, 40°, etc. Thus, the overall range of the sensory stimulus (through 360° of orientation) is divided, or fractionated, into specific regions, each maximally detected by one, or a few, sensory hair receptors. The vertebrate joint proprioceptor has extremely similar range fractionation.

Frequency Filtering. Many sensory receptors, particularly those that respond to frequency, can filter out unwanted frequency ranges.

Some filtering systems eliminate frequencies below a specific value, the cutoff frequency. These are **high-pass filters**. They pass wavelengths lower than a specific cutoff wavelength. For example, the oil droplet in some photoreceptors of birds passes high frequency (short wavelength) light, i.e., blue, but eliminates low frequency (long wavelength) light, i.e., red (Figure 7–11A). The filtering properties are reflected by the particular cutoff frequency/wavelength, and the midfrequency/wavelength that passes 50% of that frequency/wavelength light.

A **low-pass filter** passes frequencies lower than a cutoff frequency and eliminates higher frequencies. It passes wavelengths greater than a cutoff wavelength and eliminates shorter wavelengths.

A **band-pass filter** passes frequencies between a lower cutoff and an upper cutoff frequency. For example, each of the various sensory neurons of the vertebrate ear is specifically "tuned" to a particular frequency and passes only a narrow band of frequencies on either side of this frequency (Figure 7–11B).

The filtering capacity of a sensory receptor is an inherent structural property of the receptor cell or its associated cells. For example, the low-pass optical characteristics of the avian cone photoreceptors are due to the optical properties of the oil droplet in the cone. The band-pass properties of vertebrate auditory receptors are a property of their mechanical attachment to adjacent cells and the properties of the supporting basilar membrane.

Central Control of Sensory Reception

The sensory function of many receptors can be modified by the central nervous system. Efferent neurons to the sensory receptors or associated structures convey motor information that modifies their sensory properties.

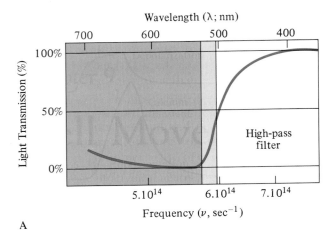

A

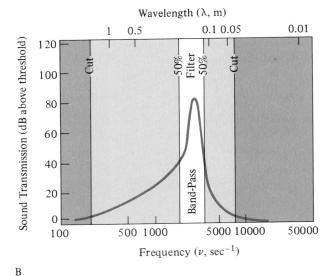

B

FIGURE 7–11 **(A)** Representation of the cutoff wavelength and frequency, and wavelength and frequency for 50% transmission for a high-pass filter, the oil droplet of an avian cone photoreceptor. **(B)** The auditory response of a "tuned" acoustic neuron from the ear of a guinea pig illustrates the band-pass filtering properties of the acoustic sensory organ; each auditory neuron has a specific band-pass frequency, and lower and higher frequencies are eliminated. *(Modified from Partridge 1989; Evans 1972.)*

There are many examples of central nervous modification of sensory reception: the amount of light reaching the retina of the vertebrate eye is controlled by the pupil diameter, and the hearing threshold of the vertebrate middle ear is controlled by the stapedius and tensor tympani muscles.

The vertebrate muscle spindle (see Figure 7–7A, page 262) and crustacean muscle stretch receptors (see Figure 7–7B) illustrate the two different modes of central action on sensory reception, i.e., effects on the receptors themselves (invertebrate muscle stretch receptor) or effects on accessory structures (vertebrate muscle spindle). The inhibitory fibers to the crustacean stretch receptors synapse on the sensory neuron itself; their effect is similar to presynaptic inhibition. Efferent fibers to the auditory hair cells of the vertebrate ear synapse both on the sensory hair cell and the neuron. Efferent neurons synapse on the taste and olfactory sensory cells of vertebrates. In contrast, the inhibitory fibers to the vertebrate muscle spindle do not synapse with the receptor neuron but with specialized intrafusal muscle fibers whose contraction stimulates the stretch receptors. The efferent fibers are directed to accessory cells, rather than to the receptor or neuron cells.

Epithelial Modification of the Receptor Current Circuit

Many epithelia have nonsensory portions that can modify the electrical environment of the epithelial sensory cells (Thurm 1983). The insect mechanoreceptor and the vertebrate cochlea are two examples. The cochlea has a nonsensory epithelium, the stria vascularis, that pumps K^+ ions into the cochlear duct to a concentration of about 150 mM. This is an unusually high K^+ concentration for an extracellular fluid and it establishes a transepithelial electrical potential of about 80 mV (Figure 7–12A). The high external K^+ concentration and +80 mV electrical potential essentially neutralizes the normal tendency for K^+ to diffuse out of the apical surface of the sensory hair cells, and so there is little resting K^+ current at the apical surface. When the sensory membrane is stimulated and Na^+ channels open, the high external potential increases the inward Na^+ current flow. Thus, the electrical environment of hair cells increases their sensitivity to stimulation. The role of the stria vascularis may also provide a non-neural mechanism for modification of the sensory cell function, e.g., hormones could stimulate the K^+ pumping of the epithelium, raising the sensory sensitivity.

The insect mechanoreceptor has a similar means of modifying the ionic environment of the receptor cell (Figure 7–12B). The tormogen cell secretes K^+ into the mechanoreceptor lymph space, to a concentration of 120 to 180 mM. There is a substantial transcellular electrical potential between the lymph space and general body tissues of 50 to 100 mV, and the apical sensory membrane of the

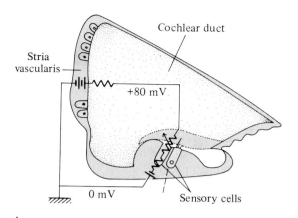

A

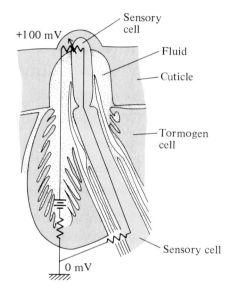

B

FIGURE 7–12   (A) The stria vascularis epithelium of the mammalian cochlea modifies the receptor-current circuit by establishing an 80 mV potential across the epithelium; this is short-circuited through the sensory hair cells of the Organ of Corti. **(B)** The tormogen cell of the insect mechanoreceptor establishes a 50 to 100 mV transepithelial potential that is short-circuited through the sensory cell; sensory stimulation of the dendrite of the receptor cell (outer ciliary segment) modifies the current flow through the receptor-current circuit.   *(Modified from Davis 1965; Thurm 1983.)*

mechanoreceptor cell has a correspondingly reduced membrane potential. Mechanical stimulation increases the ionic conductance of the apical membrane, allowing Na^+ flow into the sensory cell; this reduces the transepithelial potential and depolarizes the basal membrane region of the sensory cell,

initiating action potentials. Factors that alter the magnitude of the transepithelial potential will modify the mechanosensitivity of the sensory cells. For example, the hormone serotonin increases the transepithelial potential in the cockroach.

Chemoreception

Chemoreception is the sensory detection of specific chemicals. Many specific receptor molecules are proteins, and this is generally true for taste and smell chemoreceptors. Membrane-bound chemoreceptor proteins have highly specific shapes that respond to specific chemicals. The binding of the chemical to the receptor protein opens ion channels in the membrane, causing depolarization and initiation of action potentials.

Chemical synaptic transmission is an example of chemoreception that we have already discussed in detail (Chapter 6). Sensory monitoring of O_2, CO_2, and H^+ by cardiovascular chemoreceptors will be discussed in Chapters 12 and 13. Hormones and their receptors, enzymes and their substrates, and solute pumps and the specific solutes that they transport also involve chemical recognition. This discussion of chemoreception is limited to taste and smell, and humidity.

Gustation and Olfaction

Gustation (taste) and **olfaction** (smell) are perhaps the most universal of sensory modalities to animals. They are generally important for the most basic of animal activities, such as location of food and mates, and avoidance of predators. Both taste and smell rely on the dissolving of odor/taste molecules, either in saliva at the taste bud pore or in mucus covering the olfactory epithelium or sensillum.

Sensitivity to sweet taste is widespread in both invertebrates and vertebrates, e.g., insects, fish, frogs, and mammals (Pfaffman 1975). Bitter taste also appears to be a phylogenetically widespread, hence ancient, sensation (Garcia and Hankins 1975). The rejection of quinine, a bitter tasting alkaloid toxin, is observed in protozoans, sponges, annelids, crustaceans, mollusks, echinoderms, and all classes of chordates. Most, if not all, animals respond to sour taste (acids), i.e., they have H^+ receptors (Biedler 1975). Sour taste is a general sensation; sour receptors do not exhibit the marked specificity of, for example, sweet receptors.

Many roles of chemoreception involve either interspecific or intraspecific communication (Mackie and Grant 1974). Interspecific roles include attraction to food; recognition of predators; stimulation of defensive secretions; and location by commensal, symbiotic, or parasitic animals of their hosts. For example, the miracidium of the parasitic trematode *Schistosoma* is attracted towards its snail host (*Astralorbis*) by chemicals such as short-chain fatty acids and some amino acids. The polychete *Arctonoe* is chemically attracted to its commensal starfish host *Evasterias*. A directed movement in response to a chemical source is called **chemotaxis**.

Pheromones are specific chemicals used for various forms of intraspecific communication, including finding mates; they provide a "chemical language." For example, eggs of some coelenterates release a sperm attractant. Many female moths release a pheromone to attract males; the moth *Pyrrharctia* is unusual in that it releases its pheromone not by volatilization but as an aerosol spray (Krasnoff and Roelofs 1988). Sexual differentiation can be controlled by chemicals. Larvae of the echiurian worm *Bonnella* develop into small, parasitic males if they settle onto a female but develop into females if they settle on some other substrate. The chemical messenger for sex determination of *Bonella* is called bonellinin.

From our human perspective, the sensory modalities of taste and smell are anatomically quite discrete; taste receptors are located in taste buds mainly on the tongue (but also in the mouth lining and pharynx), whereas smell receptors are located in the nasal mucosa. Taste is detected by the physical contact of food or drink with the taste buds. Smell is the detection of distant objects by the diffusive or convective transport of molecules to the olfactory epithelium, i.e., taste is contact chemoreception and smell is distance chemoreception. Taste and smell receptors also have quite different structures and innervation. However, the functional distinction between taste and smell is less clear-cut. Much of our perception of "taste" is actually olfaction (hence the bland taste of food when our sense of smell is impaired by a cold or nose infection). For aquatic animals, the distinction between taste and smell becomes even less clear-cut, e.g., does a catfish "taste" or "smell" with its cutaneous chemoreceptors?

Sensory Receptors. Chemosensory cells have probably evolved (independently) a number of times from neurons and ciliated and nonciliated cells. It is therefore not surprising that chemoreceptors vary greatly in structure (Laverack 1974). Chemoreceptor cells are often small primary sense neurons that have axons extending to the central nervous system. The mammalian olfactory sense cells are bipolar neurons with from one to hundreds of sensory cilia

(the number is species dependent) extending into the surface mucus layer. Ciliated cells have evolved into a variety of types of sensory receptors, including chemoreceptors. Most "ciliary" chemoreceptors retain the basic "9 + 2" arrangement of ciliary microtubule filaments, although it is reduced to "9 + 0" in some. Some chemoreceptors have evolved from microvilli. The mammalian taste receptor is a secondary sense cell, with a cluster of coarse taste hairs (microvilli) that project into the taste pore. Cilia and microvilli may be preadapted to evolve sensory functions because they have a high surface area. Animals may have more than one type of chemosensory receptor although different kinds are found in different organs, e.g., the vertebrate taste buds have microvillar sensory receptors whereas the olfactory epithelium has ciliary receptors. Chemosensory cells are arranged in many animals as a single cell or a few clustered cells, but chemosensory units may be present in enormous numbers located close together and conveying information in parallel to the central nervous system.

Olfactory and gustatory receptors undoubtedly have highly specific receptor molecules that reversibly bind specific odor molecules and regulate the permeability of ion channels (Kaissling 1983). Most receptor molecules are probably proteins, although not all chemical receptors are proteins. For example, the salt receptor of vertebrate taste buds may be a phospholipid. The proteinaceous chemoreceptors are similar in function to other well-understood protein receptors, such as the acetylcholine and insulin receptors. Proteins that bind sugar and bitter substances have been isolated from mammalian tongues. Fly taste receptors contain the enzyme glucosidase, which may function as the taste receptor. The bacterium *Escherischia coli* has many different receptor proteins located on its cell wall that can detect low concentrations (10^{-8} to 10^{-6} mol l^{-1}) of specific chemicals that elicit positive chemotaxis (e.g., fructose, glucose, serine) or negative chemotaxis (e.g., butyric acid, leucine, indole, H^+).

Binding of an odor molecule to its specific receptor is transduced into an electrical signal. Receptor binding causes a conformational change that modifies the permeability of the membrane ion channels; the altered membrane conductance depolarizes (or sometimes hyperpolarizes) the membrane. Some insect olfactory receptors are sufficiently sensitive that a single odor molecule can initiate an action potential. The action potential is preceded by a transient, low-voltage elementary potential change (200 to 300 μV) that could reflect the opening of a single ion channel for about 40 msec. The energy of such an elementary potential change is about 10^{-17} J, which represents an amplification of about 1000 times if the binding energy of the receptor and odor molecule is about 10^{-20} J molecule^{-1}.

The process of chemoreceptive transduction involves protein phosphorylation in bacteria (Ames et al. 1988; Hess et al. 1988). The membrane-bound protein has a receptor site, four sites for methylation (addition of a CH$_3$ group), and a signaling site (Figure 7–13A). The bacterium *E. coli* has a family of such methyl-accepting chemotaxis proteins (MCPs). The MCPs elicit chemotaxis by modifying the rate of phosphorylation of a chemotaxis protein (CheA), which controls the phosphorylation of other proteins

that control the rotation of the bacterial flagellum (CheY and CheZ) and demethylation of the receptor, causing sensory adaptation (CheB).

The transduction of taste and smell in animals involves a stimulatory GTP-binding protein and cyclic AMP as a secondary messenger (Lancet et al. 1988). The olfactant (or gustatory) molecule dissolves in a surface mucus layer and interacts with a protein receptor on the chemoreceptor membrane (Figure 7–13B). There may be specific odor/taste-binding proteins in the mucus that facilitate interaction of the olfactant with the receptor protein. Binding of the olfactant to the receptor activates an adjacent GTP-binding protein, which activates

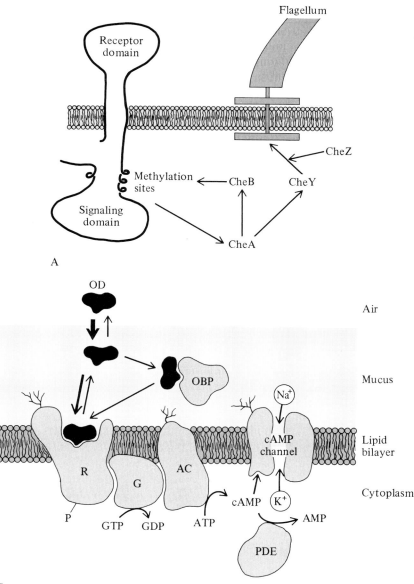

FIGURE 7–13 **(A)** Model for the chemoreceptor of the bacterium *E. coli* showing an extracellular receptor domain, intracellular methylation sites (⊖), and a signaling domain. Chemotransduction occurs by phosphorylation of the chemotaxic protein CheA then CheY and CheZ, which affect the bacterial flagellum. CheB methylesterase regulates the receptor activity. **(B)** Representation of transduction by olfactory receptors. Odorants (OD) dissolve into the mucus lining of the olfactory epithelium and diffuse, or are transported by an olfactory binding protein (OBP), to the membrane-bound olfactory receptor (R). Binding of an olfactant to the receptor protein activates a G protein, which in turn activates adenyl cyclase to form cyclic AMP (cAMP). The cAMP controls the Na$^+$ permeability of membrane ion channels. *(Modified from Ames et al. 1988; Hess et al. 1988; Lancet et al. 1988.)*

adenyl cyclase. The elevated cAMP level modifies the ionic permeability of the sensory membrane. Taste, photoreception, and the action of the β-adrenergic receptor similarly involve G proteins and adenyl cyclase.

Invertebrates.   Protozoans have a chemical sense; they respond by avoidance behavior to acid, alkali, and salt stimuli. Predatory ciliates are attracted to their prey by specific chemicals. All animals have at least a general chemical sense. Contact or distance chemoreception is important for location of food or hosts in coelenterates, platyhelminth worms, annelids, mollusks, arthropods, echinoderms, and vertebrates. The feeding response of a hydra (a sessile coelenterate) has provided much valuable information concerning chemoreception (Lenhoff and Lindstedt 1974). Food extracts initiate the feeding response of *Hydra littoralis*; the active principle of the food extract is reduced glutathione (a tripeptide, glutamate-alanine-glycine). Glutathione is also attractive to the Portuguese man-of-war *Physalia*. Other coelenterates respond to proline, valine, glutamine, or leucine.

Gustation and olfaction are extremely important senses for arthropods, especially insects. Much sensory physiological research has been undertaken with insect chemoreceptors because their anatomical arrangement facilitates sensory recording (see Supplement 7–1, page 323). Contact and distance chemoreceptors are generally similar in structure, but let us first examine contact chemoreceptors.

Insect contact chemoreceptors may be located on any part of the body, but they are generally concentrated on the mouthparts and the tarsi of the legs. The receptor is a cuticle-covered, peg-shaped projection of the cuticle with a pore at the tip (Figure 7–14). Each receptor, or sensillum, generally contains four to five neurons. The sensory dendrites extend up the cone of the peg; they are surrounded by the scolopale (a cuticular sheath continuous with the exocuticle of the peg). The scolopale may be fused to one side of the core of the peg. The neurons are ensheathed by two specialized epidermal cells: the trichogen cell, which secretes the cuticle peg and scolopale, and the tormogen cell, which forms the wall of the fluid-filled cavity surrounding the sensory receptor neuron. The dendrite tips are

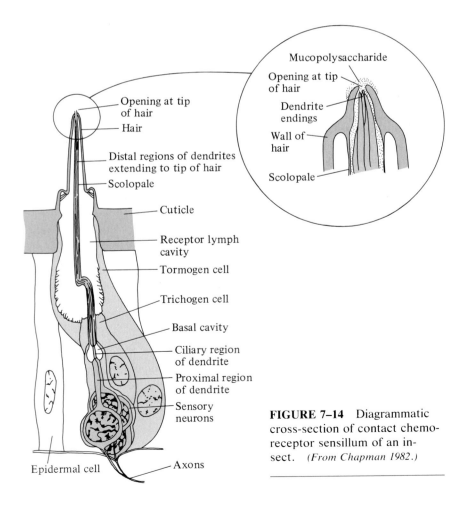

FIGURE 7–14 Diagrammatic cross-section of contact chemoreceptor sensillum of an insect. *(From Chapman 1982.)*

protected by a viscous mucopolysaccharide, which is probably secreted at the base of the fluid-filled cavity and flows to the tip.

Much information is available concerning the responses of insect taste chemoreceptors to different chemicals and threshold detection levels because of the relative simplicity of behavioral taste experiments, e.g., proboscis extension. The flies *Phormia* and *Calliphora* have the lowest thresholds for certain disaccharides (sucrose, maltose) and monosaccharides (glucose, fructose); other sugars may have no effect or may even inhibit feeding. Inorganic cations inhibit proboscis extension; their potency is proportional to their ionic mobility ($H^+ > NH_4^+ > K^+ > Na^+$). Anions also inhibit proboscis extension ($OH^- > I^- > Cl^- > PO_4^{3-}$). The stimulating effect of fatty acids is proportional to their chain length.

Some contact chemosensory neurons are specific for certain types of stimulatory molecules. The medial maxillary sensillum of *Pieris*, the cabbage butterfly, has four sensory sensilla: one (the sinigrin receptor) responds specifically to glycosides from mustard oil, two respond to inorganic ions, and one responds to deterrent chemicals (Schoonhoven 1972). One neuron of the tarsal receptor of *Phormia* responds to salt, one to sugars, and one to water.

Chemoreceptors can respond to chemicals other than their primary stimulants. For example, a single locust sensillum can respond to salts; C_5, C_6, and C_{12} sugars; amino acids; inorganic and organic acids; terpenes; and alkaloids. Each sensory cell appears to have its own spectrum of chemical responses. Consequently, a single sensillum with four or five neurons may be capable of discriminating many different tastes, not just a few. Taste discrimination can also occur at the level of the central nervous system. We have already examined one example of multiple channels for insect taste receptors in which simultaneous stimulation of two sets of receptors lowers the sensory threshold. In similar fashion, stimulation of one set of receptors by an inhibitory chemical can change the threshold of other receptors.

Insect olfactory receptors are found mainly on the antennae. There are four basic types of olfactory sensilla that resemble, in general, the gustatory sensillum just described but differ in specific cuticular structure and number of sensory neurons. The sensillum trichodeum is a long, thin, tapering hair-like sensillum; there are 1 to 3 sensory neurons whose dendrites extend to pores at the tip of the sensillum. The sensillum basiconicum is shorter, thicker, and has 1 to 50 neurons whose dendrites divide into fine branches. The sensillum coeloconicum, or pit-peg, has the sensillum opening into a chamber formed by the cuticle. The sensillum

placodeum is a plate-like or scale-shaped cuticular process with many (15 to 30) neurons.

Olfactory receptors, like gustatory receptors, can be extremely specific or can be more general and respond to many chemical stimuli. Others respond to a considerable variety of chemicals. Generalist receptors that differ markedly in their pattern of response to a variety of chemicals enable a multineuronal sensillum to discriminate a wide range of odors. Some sensory neurons are extremely specialized responding primarily to only one specific molecule, although they may respond submaximally to other similar odors. For example, the male silkworm moth *Bombyx mori* is remarkably sensitive to the female pheromone bombykol (hexadecadien-10-trans 12-cis-ol; $CH_3(CH_2)_2CHCHCHCH(CH_2)_8CH_2OH$. The male's antennae are morphologically elaborate to increase the number of sensilla trichodea (Figure 7–15A), many of which are specific for bombykol. Female silkworm moths lack the specific bombykol receptors (or else their bombykol receptors are so adapted to their own bombykol that they are unresponsive to experimental stimulation by exogenous bombykol). The specificity of the male's bombykol receptors is sufficient to discriminate it from similar chemicals, including stereoisomers (Figure 7–15B).

Vertebrates. Taste and smell are important senses for both aquatic and terrestrial vertebrates. The olfactory organ of many fish has a very elaborate surface with high densities of olfactory receptors, e.g., $5 \cdot 10^4$ to 10^5 mm^{-2}. Olfaction can be important for detection of food or predators, intraspecific communication, and the remarkable homing ability of many newts and fish (e.g., salmon).

The vertebrate sensations of taste and smell are functionally similar to those of invertebrates, although there are structural differences (especially for gustation). The olfactory receptor is a sensory bipolar neuron located in the olfactory mucosa with sensory cilia and an axon (Figure 7–16A); these have a structural resemblance to those of invertebrates. Not all vertebrate olfactory sensors are ciliary, however; a microvillar receptor of a reptile was illustrated in Figure 7–1E (see page 256). Vertebrate taste receptors (Figure 7–16B), in contrast, are specialized secondary sense cells that convey their sensory information to a neuron, which itself is not chemosensitive (except to neurotransmitters).

Vertebrate taste receptors typically occur in the mouth and pharynx but do occur elsewhere; fish have them on their gills, skin, and fins. Fish have a general chemical sensitivity to acid, alkali, and salt, as well as specific tastes, often of very low threshold,

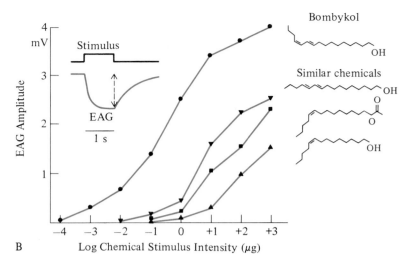

FIGURE 7–15 **(A)** The male silk moth *Bombyx mori* has elaborate antennae for chemoreception of the female pheromone bombykol. **(B)** Electroantennogram (EAG) response of the antenna of the male moth *Bombyx mori* to bombykol (upper trace) and to similar chemicals. *(From Kaissling 1987; Kaissling 1983. Photograph courtesy of R.A. Steinbrecht.)*

e.g., the minnow has thresholds of 2×10^{-5} M for sucrose and 4×10^{-5} M for NaCl. Homing salmon have extremely low thresholds for chemoreception to "taste" their way to their spawning grounds. Mammals generally have four types of taste receptor: sweet, salt, sour, and bitter; some also have a water taste. Reptiles vary in the occurrence of taste buds, and many appear to rely more on olfaction (and vomeronasal chemoreception; see below) than on taste. Birds, like reptiles, generally have a poorly developed sense of gustation, although nectar-feeding birds can discriminate sweet tastes.

The teleost fish *Heterotis* has a unique organization of its taste receptors and corresponding sensory portion of the brain (Braford 1986). *Heterotis*, and many other fish, have paired epibranchial organs (outgrowths of the pharynx) that aid in food concentration and swallowing. The epibranchial organ of *Heterotis* is a blind-ended, flat tube that is highly coiled; the number of spiral turns increases with age and size of the fish. The taste buds of the epibranchial organ are innervated by the vagus nerve, which projects to a specialized portion of the vagal lobe of the brain, the epibranchial portion. This part of the central nervous system is also a spiraled structure, which receives direct connections from the corresponding position of the epibranchial spiral. Like the epibranchial organ itself, the epibranchial portion of the vagal lobe has indeterminate growth. This organization of a taste organ

and its CNS connections is unique. Not even the cochlea, a highly coiled auditory structure of mammals, has such a corresponding coiled CNS projection. The mammalian cochlea also has a fixed, or determinate, growth and forms its final spiral shape before it develops its connections with the CNS.

A vomeronasal organ (Jacobson's organ) is present in many amphibians, reptiles, and mammals, but it is most highly developed in snakes and some lizards (Parsons 1970); it is absent in adult birds and crocodilians and is vestigial in many reptiles. Jacobson's organ arises as a medial outpocket of the developing nasal cavity and olfactory epithelium, hence it is functionally allied with the olfactory system. The organ develops into a spherical structure with the ventral side invaginated into the sphere as a mushroom-shaped body, leaving a narrow lumen (Figure 7–17). A narrow duct connects the interior of Jacobson's organ to the oral cavity, but how odoriferous molecules reach the sensory epithelium in Jacobson's organ is unclear. In many lizards, fluid draining from the eye may bring odoriferous molecules into contact with the sensory epithelium of Jacobson's organ. Snakes and some lizards (e.g., varanids) may directly transfer odoriferous molecules from their tongue to Jacobson's organ.

Vertebrate olfactory receptors have a considerable specificity for certain stimuli and are considerably more varied than taste receptors (which had

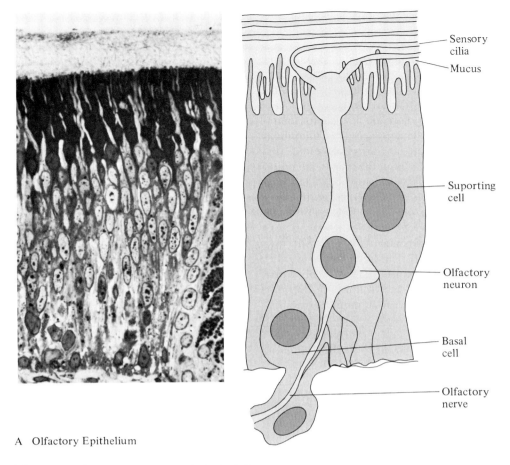

A Olfactory Epithelium

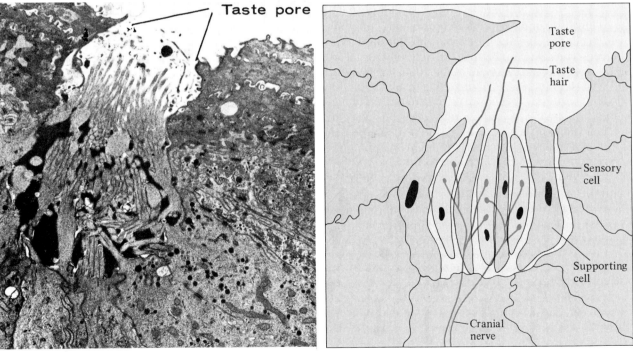

B Taste Bud

FIGURE 7–16 **(A)** Photomicrograph (left) and diagrammatic representation (right) of the mammalian olfactory epithelium. **(B)** Electronmicrograph (left) and diagrammatic representation (right) of a mammalian taste bud synapse with neurons. *(Photographs courtesy of M. Coppe and T.J. Reese.)*

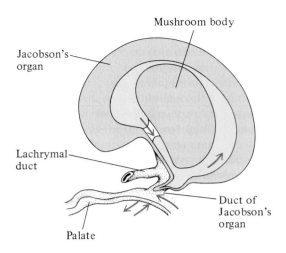

Mushroom body

Jacobson's organ

Lachrymal duct

Duct of Jacobson's organ

Palate

FIGURE 7–17 Representation of Jacobson's organ in a monitor lizard *Varanus bengalensis.* *(From Parsons 1970.)*

four basic taste sensations). Humans, for example, can distinguish about ten thousand different odors, although there is a smaller number of basic or **primary odors**. One classification scheme has seven primary odors: musky, floral, pepperminty, camphoraceous, ethereal, pungent, and putrid. The ability to discriminate many more specific odors probably does not imply an equivalent number, i.e., thousands of different receptor neurons, each specialized for only one specific odor. It is more likely that each sensory cell has a slightly different response to a number of similar odors, i.e., they have different olfactory spectrums. Nevertheless, there appear to be a large number (100 to 1000) of different olfactory receptor proteins, so the ability to discriminate 10,000 odors is due to a combination of many receptor proteins and central nervous system interpretation of olfactory spectrums.

Humidity

A number of terrestrial animals have cutaneous receptors that detect the water content of air, i.e., **hygroreceptors**. Some insects, for example, can detect changes in ambient relative humidity of as little as 2% RH. This enables them to seek out specific hygric environments or to modify their physiology and behavior with respect to the ambient humidity, e.g., to control the spiracular opening at varying relative humidity.

A variety of hygrosensory structures have been identified on the antennae, palps, underside of the body, and near the spiracles of insects. The

receptors may be a specialized basiconic peg sensilla, branched hairs, or specialized neurons incorporated into other sensory sensilla. The olfactory sensillum of the mosquito *Aedes* contains a hygrosensory as well as chemosensory neurons. The triad sensory sensillum of the cockroach antenna contains three sensory cells: a "dry" air hygroreceptor (low humidity increases its firing rate), a "moist" air hygroreceptor (high humidity increases its firing rate), and a cold thermoreceptor neuron (Figure 7–18).

How humidity is transduced by a hygrosensory receptor into action potentials is unclear. The receptor membrane may respond directly to water molecules, or it may respond to local evaporation that decreases the local temperature (i.e., like a wet-dry bulb thermometer system for measuring relative humidity) or increases the local osmotic or ionic concentration. Alternatively, hygroreceptors may contain a hygroscopic material whose water content is proportional to the humidity and whose structure alters with humidity to stimulate mechanoreceptors.

Temperature

Temperature is one of the most important physical aspects of the external environment because it determines the rate of many physical and physiological processes (see Chapter 5). Consequently, temperature is an important sensory modality to most animals; it is sensed by **thermoreceptors**. Some thermoreceptors can discriminate temperature changes as small as 0.01° C.

Temperature has a direct effect on the membrane potential of excitable cells (see Nernst equation, Chapter 6) but its effect on excitable cells can be more complex. For example, an elevated temperature increases the firing rate of some insect chemoreceptors but decreases the firing rate of others. This variable thermosensitivity of chemoreceptors is apparently due to changes in temperature of the neuron cell body where action potentials are initiated, rather than the temperature of the sensory nerve endings where the receptor potential occurs (Dethier 1963). Such temperature sensitivity of other sensory modalities may provide useful thermal information to an animal, but animals also have specific thermoreceptors that provide unambiguous temperature information.

Temperature-sensitive neurons can be classified into two types: (1) those for which a lowered temperature increases the action potential discharge, i.e., cold thermoreceptors, and (2) those for which an elevated temperature increases the action potential discharge, i.e., hot thermoreceptors. Some

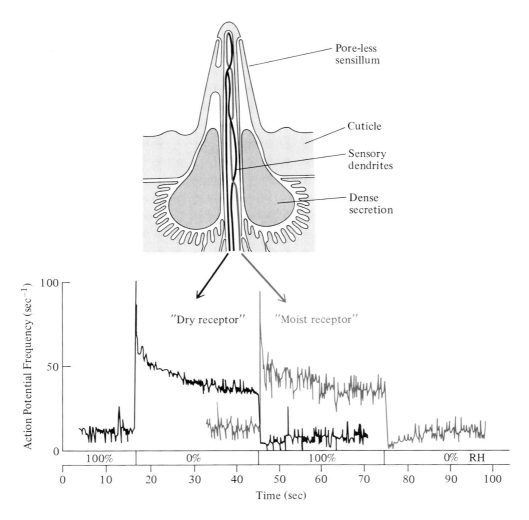

FIGURE 7–18 The "cold-moist-dry" triad sensory sensillum of the cockroach contains three bipolar sensory neurons; one neuron of the hygroreceptor responds to high humidity ("moist" receptor) and one to low humidity ("dry" receptor). The receptor cavity of the poreless sensillum is filled with a dense secretion. *(Modified from Yokohari and Tateda 1976; Schaller 1978.)*

thermoreceptors also have a marked phasic response to both increases and decreases in temperature. For example, a cold receptor found on a cat's tongue has a rapid increase in firing frequency when cooled, a rapid decrease in frequency when heated, and tonic sensitivity to long-term changes in temperature. The extent of the phasic and tonic responses is proportional to the magnitude of the temperature change.

Vertebrates have thermoreceptors in their skin and deep within their body (e.g., hypothalamus, spinal cord). Many of these thermoreceptors are naked nerve endings without specialized accessory structures. For example, cold thermoreceptors on the cat's tongue have naked nerve endings that penetrate a few mm into the epidermal epithelium (Figure 7–19A). Warm thermoreceptors have not been specifically identified in structure, although their physiological characteristics are well described. Mammalian cold receptors have a maximal firing frequency at about 25° C, with decreasing frequency at higher and lower temperatures; hot receptors are similar except that the maximum firing frequency is at about 45° C (Figure 7–19B). In addition, pain receptors (nociceptors) respond to very low (cold pain) or high (hot pain) temperatures. Thus, thermal sensation is the result of the pattern of activity of a number of different thermoreceptors and nociceptors. Some thermosensitive neurons have rapid phasic responses to changes in temperature as well as tonic responses.

Many thermoreceptors adapt to a sustained change in temperature but some do not adapt. For example, the hypothalamic thermoreceptor of mammals, which is responsible for regulation of body temperature, does not adapt; an adapting thermoreceptor could not provide precise, long-term regulation of body temperature.

The mechanism for thermal detection by thermoreceptors probably involves metabolic modifications of neuron function by changes in temperature. A typical Q_{10} of 2 to 3 (see Chapter 5) for an intracellular metabolic process could dramatically alter the functioning of an excitable cell. For example, the receptor potential of the cat's cold receptor continually oscillates, initiating spontaneous action potentials. The oscillation of E_m is due to temperature-dependent periodic changes in Na^+ and K^+ permeability of the cell membrane (Braun, Bade, and Hensel 1980). An increase in P_{Na} depolarizes the cell membrane, causing an influx of Ca^{2+}, which opens Ca^{2+}-dependent K^+ channels and repolarizes the cell membrane. Thermal sensitivity of the electrogenic pump can also modify the action potential frequency of thermoreceptor cells.

Infrared Reception

The detection of infrared thermal radiation by a number of snakes and some invertebrates is appropriately discussed here rather than as an example of vision (i.e., electromagnetic radiation detection) since the sensory mechanism is thermal detection.

Many snakes have infrared receptors (Barrett 1970; Gamow and Harris 1973; Harris and Gamow 1971; Breipohl 1984). The infrared sense of pit vipers (Crotalidae, e.g., rattlesnakes) and some pythons (Boidae) enables them to detect endothermic prey, such as small birds and rodents. Covering the eyes of a rattlesnake does not prevent prey location; a

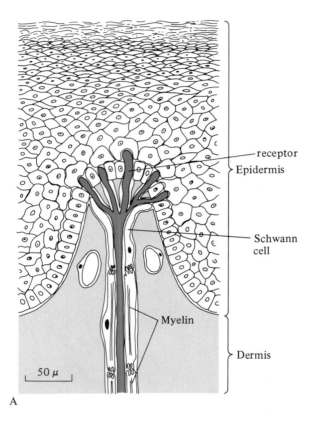

A

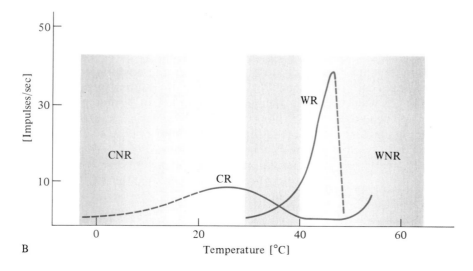

B

FIGURE 7–19   (A) Representation of a cold thermoreceptor, consisting of a dendritic arborization at the base of the epidermis. **(B)** Cold and warm receptors of mammals have maximal responses at low (e.g., 25° C) and high (e.g., 45° C) temperatures, respectively. Cold pain and warm pain nociceptors respond to more extreme freezing or burning temperatures. *(From Breipohl 1984; Breipohl (op. cit.) after Necker 1981.)*

blinded rattlesnake will strike at any thermally-appropriate "prey," such as a moving lightbulb or soldering iron.

The paired facial organs of pit vipers are located between the eyes and nostrils (Figure 7–20A). Some boid snakes have labial pits (specialized labial scales) along the upper and lower jaws. The crotaline pit organ (Figure 7–20B) has an outer chamber separated from an inner chamber by a thin membrane (10 μ thick) that contains many free thermoreceptive nerve endings, each of which spreads palmately over an area of about 1500 μ². The labial infrared receptors of boids have a similar structure, with an outer and inner chamber and a heat-sensitive membrane. Infrared radiation strikes and causes local warming of the thin membrane; the inner air-filled chamber provides insulation to retard rapid conduction of heat to the underlying tissues. The local increase in temperature of the membrane is transduced by the thermosensitive free nerve endings into action potentials and transmitted to the brain via the trigeminal nerve. This thermal detection system is sensitive to infrared radiation (IR; 0.7 to 15 μ) but not visible light (0.4 to 0.7 μ; Goris and Nomoto 1967). Temperature changes as small as 0.003° C can be detected; this is equivalent to an energy threshold of 0.1 mW cm⁻². An increase in temperature of 0.4° C increases the action potential discharge from 18 sec⁻¹ to 68 sec⁻¹. The response of the pit organ sensory neurons to heat is a rapid phasic increase in activity, followed by a maintained tonic increase in activity.

The bilateral arrangement of pit organs provides binocular "IR vision" since the receptive fields of the pit organs overlap. There are a number of similarities in the neural organization of binocularity for pit organs and eyes. The pit organ neurons pass to the optic tectum, which also receives visual information from the eyes. Some neurons of the optic tectum respond to infrared irradiation of the pit organ on the same side of the head, whereas others respond to irradiation of the pit organ on the

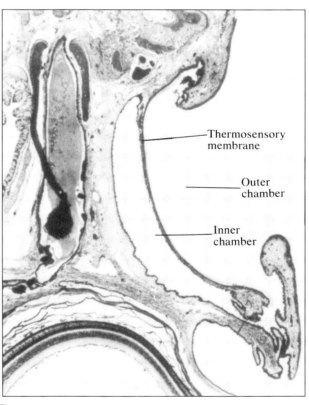

A   B

FIGURE 7–20   (A) The pit organ of the black-tailed rattlesnake *Crotalus molossus* is clearly seen between the eye and nostril. (B) Cross section of the crotaline pit organ of *Agkistrodon mocassen*.   *(Photo: Zoological Society of San Diego 1985 by Ron Garrison; photomicrograph: Cordier 1964.)*

other side of the head (this is similar to the pattern of processing for visual information). The IR information is mapped into the optic tectum to provide overlaying IR and visual maps; some neurons respond to both infrared radiation of pit organs and visual stimulation of the eyes. The integration of the IR and visual information would presumably provide more precise localization of prey than either sensory modality alone.

Some insects detect infrared radiation. The buprestid beetle *Melanophila* has sensory pits contiguous with the coxal cavities of the mesothoracic legs (Evans 1964). Brief pulses (1/300 to 1/2 sec) of infrared radiation (λ = 0.8 to 6 μ) on these pit sense organs elicit a behavioral response (antennal twitching). The antennae also respond to light ($\lambda <$ 1.25 μ) but this effect is attributed to the stimulation of antennal thermoreceptors rather than specific IR receptors. The size of the antennal organs of the corn earworm moth *Heliothis* is consistent with IR detection in the 1 to 5, 6, and 9 to 11 μ wavelength bands; this would enable detection of the IR emitted by the thorax of other moths that may be 1–10° C warmer than the ambient temperature (Callahan 1965). The bloodsucking bug *Triatoma* can estimate the temperature and distance of warm objects by their radiant energy (Lazzari and Nunez 1989), although warm air gradients, convective currents, and chemoreception (CO_2) are also important cues.

Other animals may also be able to detect infrared radiation. For example, the accessory retina of the slug *Agriolimax* may respond to IR radiation (Newell and Newell 1968).

Mechanoreception

Mechanoreceptors measure force and displacement. They vary greatly in structural complexity, from simple naked neurons that respond to movement (e.g., dermal pressure receptors), neurons connected to a single bristle or hair (e.g., invertebrate and mammalian sensory hairs), nerve endings with relatively simple accessory structures (e.g., Pacinian corpuscles), or extremely elaborate sensory structures (e.g., invertebrate and vertebrate ears). Many mechanoreceptors have a cilium (or ciliary-derived) or cilia-like structures. The vertebrate hair cell has a highly specialized cilium (kinocilium) and stereocilia, which are actually microvilli.

Mechanoreceptors respond to physical displacement, but this sensory modality can be used to detect many different types of information, e.g., stretching and bending of connective tissue layers, movement of hairs and bristles, muscle stretch, fluid movement, static and dynamic equilibrium, and

audition. Mechanoreceptors have also become highly modified for electric field reception, and possibly for magnetic field detection.

Mechanotransduction

Mechanosensitivity is common in animal cells but the transduction process is not clear. The simplest scheme for mechanosensory transduction is stretching of the receptor cell membrane by physical movement, thereby opening ion channels and increasing ionic conductance (Goldman 1965). For example, a membrane Na^+ channel might normally be slightly smaller than a Na^+ ion, but stretching the membrane expands the channel and allows passage of Na^+ ions. The sensory membranes of epithelial mechanoreceptor cells (both vertebrate and invertebrate) have cation-selective ionic channels that are stretch sensitive (Thurm 1983). These channels are relatively unselective for alkali cations, although larger organic ions are less permeable; the membrane channels have a diameter of about 0.7 nm when open. The short latency between mechanical stimulation and sensor current flow (15 to 100 μsec) suggests a direct coupling of the mechanical "sensor" and membrane ion channels.

Mechanosensitivity may be due to mechanical stimulation of specific sensory receptors on the membrane, rather than a general membrane response to stretch. The minimal energy requirement for mechanostimulation is about 10^{-18} joule (e.g., human ear and cockroach vibration-sensitive organ); this is similar to the minimum energy threshold for a photoreceptor (4 10^{-19} J). This similarity in threshold suggests that the mechanical energy might also be concentrated, or localized, at specific receptor locations on the membrane, as is a photon of light. Many mechanoreceptors have specialized receptor structures (finger shaped membrane extensions, microvilli, cilia, or highly modified cilia) that may have specialized receptor molecules. For example, the sensory dendrite of the filiform sensillum of a cricket (*Acheta*) contains microtubular structures between the sensory membrane and an interior amorphous connecting material; the microtubular structures are connected to the membrane by cone-shaped bridges, which are probably integral membrane proteins (Figure 7–21). The membrane cones may be the specific sensory receptor proteins whose physical distortion controls ion channel permeability.

Invertebrate Mechanoreceptors

The body surface of invertebrates has a variety of mechanoreceptors for touch, sound, and movement detection. For example, ctenophores have several

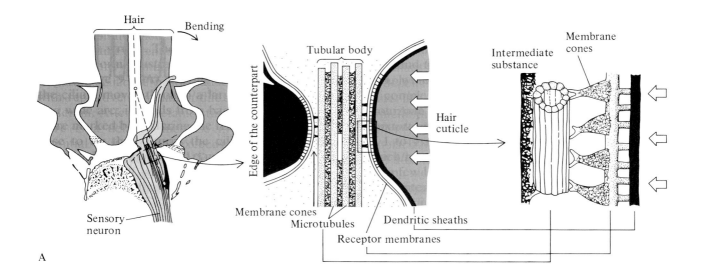

A

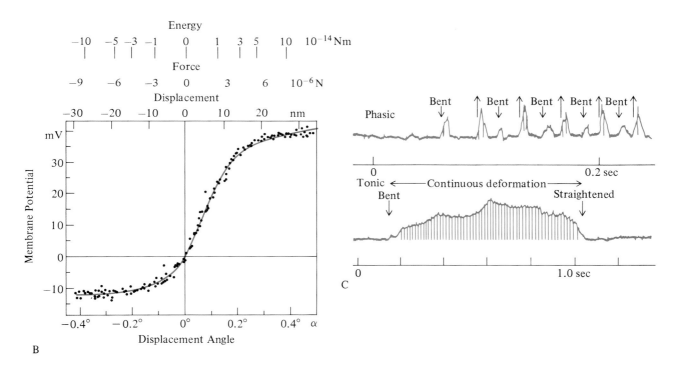

B

C

FIGURE 7–21 **(A)** Organization of the joint region of the filiform hair sensillum of the cricket *Acheta* showing the base of the hair, the direction of bending of the hair and the resultant compression force on the tubular body between the counterpart cushion (solid) and the edge of the hair cuticle. Dashed lines indicate the axis of the hair and its bending deflection. The compression is transmitted by microtubule filaments of the sensory dendrite via membrane cones, whose mechanical deformation may open ion channels in the receptor membrane. **(B)** The maximum amplitude of a receptor membrane potential is positive for the normal displacement direction ($+\alpha$) and negative for displacement in the opposite direction ($-\alpha$). The force at the hair tip (F, 10^{-8} Newton), the displacement of the hair tip (S, nm), and the energy required to bend the hair (W, 10^{-6} N m) are also shown. **(C)** Responses to deformation of a phasic mechanoreceptor (wing hair of the fly *Sarcophaga*) and a tonic mechanoreceptor (clasper of the male fly *Phormia*).   *(From Thurm 1983a; Thurm 1983b; Wolbarsht 1960.)*

types of ciliated epithelial sensory cells that synapse with underlying neurons. Chaetognath worms detect water movement with ciliated neuroepithelial receptor cells, i.e., rheoreceptors. Turbellarian worms detect water movement with four groups of large rheoreceptor cells underneath the epidermis; 20 to 100 long dendrite bristles project through the epithelial cells and beyond the ciliated surface of their epithelium. They also have tactile receptor cells scattered over the body surface and chemosensory cells with single, short dendrites concentrated at the anterior of the animal.

The sensory tactile hairs of arthropods are complex structures that articulate with the stiff chitinous exoskeleton. For example, the trichobothria (1 mm long filiform hairs) on the posterior appendages (cerci) of the cricket are sensitive **trichoid sensilla** that detect air currents and low frequency sound. The bipolar mechanoreceptor neuron has its cell body located near the base of the cuticular hair, and the dendrite is inside the hair. The dendrite is surrounded by a cuticular sheath, the scolopale. The very thin cuticular support of these hairs is elliptical in shape, and this confines movement to essentially one plane. Movement in this plane compresses the tubular body of the dendrite between the hair cuticle and an opposing elastic cushion and produces a generator potential that has a peak polarity and amplitude determined by the direction of hair deflection (Figure 7–21B). The mechanoreceptor has a greater response to bending in one direction than in the opposite direction; it is asymmetric.

Some insect mechanosensitive hairs are phasic, responding to changes in position, whereas others are tonic, continuously responding to a deformation (Figure 7–21C). The phasic response is most useful when a rapid directional response is required, as in flight, whereas a tonic response allows monitoring of a constant behavior, such as clasping. Invertebrate hair cells are often grouped to form hair beds, which can function as a wind direction sensor (e.g., for flying locusts), a proprioceptor when stimulated by an adjoining body surface, and for orientation to gravity (e.g., the hair beds of honeybees).

The insect campaniform sensillum has a dome-shaped section of thin cuticle; the inner surface has a scolopale that encloses a sensory neuron dendrite. Campaniform receptors are often arranged in groups of 10 to 50 sensilla and respond to shearing stresses in the stiff exoskeleton cuticle. They function as proprioceptors. Examples are the motion detectors of wings, the gyroscopic halteres of dipteran flies, and leg motion detectors. The socket membrane of the large tactile leg hairs of cockroaches contacts a campaniform sensillum; bending the hair probably

folds the socket membrane and presses on the campaniform sensillum.

The **scolopidium** is a third type of mechanoreceptor. It forms the functional unit of the insect **chordotonal organs** (e.g., subgenual, Johnston's, and tympanal organs). The scolopidium is a highly specialized bipolar sensory neuron with complex accessory structures, the scolopale and cap cells. The complex structure of the scolopidium is well illustrated by that of the locust ear (Figure 7–22A). The sensory neuron soma and axon are surrounded by a Schwann sheath cell (each Schwann cell enfolds several neurons). The neuron dendrite is surrounded by a fibrous sheath cell and a scolopale cell. A sensory cilium extends from the dendrite and terminates in the scolopale cap, which is connected via attachment cells to the cuticle.

The tympanal organ is a chordotonal organ specialized for hearing. It detects vibration of a thin cuticular membrane (the tympanum), which is usually separated from underlying tissues by an air space. There can be from 1 to over 1000 individual scolopidia in the tympanal organ. The location of the tympanum (ear drum) varies; it can be on the first abdominal segment adjacent to the spiracle (locusts), ventrolateral on the first abdominal segment (cicadas), on the ventral side of the medial vein of the forewing (lace-wing), on the posterior metathorax (noctuid moths), or on the proximal tibia (tettigoniid bush crickets).

The auditory system of a tettigoniid bush cricket well illustrates the structure and functioning of a tympanal organ. Their tympanal organ (there is also a proprioceptive subgenual organ and an intermediate chordotonal organ of unclear function) is a tent-shaped structure called the crista acoustica, which extends along the tympanum between the anterior wall of the tibia and a trachea (Figure 7–22B). The chordotonal sensilla decrease in size along the crista acoustica, and they have very different frequency responses to sound. There is a close correspondence between the anatomical location of scolopidia and their frequency response; this is a **tonotopic** organization. The larger, more proximal scolopidia respond to lower frequencies of sound, whereas the smaller, more distal scolopidia respond to higher frequencies. The frequency threshold characteristics of eight neurons of the crista acoustica of a bush cricket illustrate this tonotopic organization of receptors tuned to different frequencies (Figure 7–22C).

The receptor neurons from the crista acoustica run into the prothoracic ganglion and terminate in a well-defined area, the anterior ring tract (aRT). There is also a tonotopic organization of neurons in the aRT with different regions receiving information

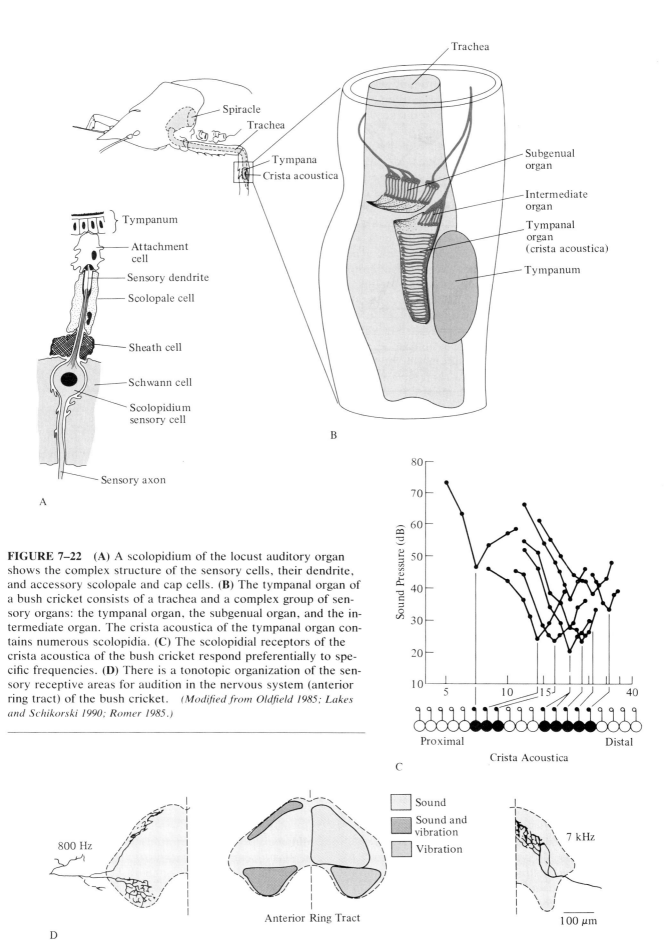

FIGURE 7–22 **(A)** A scolopidium of the locust auditory organ shows the complex structure of the sensory cells, their dendrite, and accessory scolopale and cap cells. **(B)** The tympanal organ of a bush cricket consists of a trachea and a complex group of sensory organs: the tympanal organ, the subgenual organ, and the intermediate organ. The crista acoustica of the tympanal organ contains numerous scolopidia. **(C)** The scolopidial receptors of the crista acoustica of the bush cricket respond preferentially to specific frequencies. **(D)** There is a tonotopic organization of the sensory receptive areas for audition in the nervous system (anterior ring tract) of the bush cricket. *(Modified from Oldfield 1985; Lakes and Schikorski 1990; Romer 1985.)*

281

for high-frequency sound, low-frequency vibration, or both. Projection areas of neuron fibers in the anterior ring tract form a high-frequency sound area, a low-frequency vibration area, and common projection areas (Figure 7–22D). The frequency-tuned neurons innervate specific areas of the aRT.

Many chorodotonal organs are proprioceptive, detecting joint movements. The subgenual organ, which is located on the proximal part of the tibia of many insects, is a vibration detector. Its sensory scolopidia lie free in the hemolymph, with their attachment cells joined to accessory cells whose long endings attach to the exoskeleton cuticle. Subgenual organs are sensitive to extremely small vibrational displacements. For example, the cockroach *Periplaneta* responds to displacements of 10^{-9} to 10^{-7} cm at frequencies up to 8000 Hz with an optimal sensitivity of 1500 Hz.

Johnston's organ is a chordotonal organ located in the second segment of the antenna; it senses antennal movement. All adult insects (except Collembola and Diplura) and many larvae have Johnston's organs, but it is best developed in mosquitos (Culicidae) and midges (Chironomidae). Johnston's organ senses wind speed and is important for control of flight velocity, detection of sound vibration, gravity, or orientation. Male mosquitos use it to detect and localize females. The sense of directionality can be determined by a single antenna. The diving beetle *Notonecta* uses its Johnston's organ to orient with respect to gravity when underwater, by monitoring the position of its air bubble with its antennae.

Statocysts are gravity receptors that enable an animal to determine up and down and the angle of orientation. They are present in many invertebrates (coelenterates, nemertean and annelid worms, mollusks and crustaceans, echinoderms, but rarely in insects). There are essentially two different types of statocyst. The lithostyle type has a number of sensory epithelial cells, each containing a statolith. The statolith is a dense calcium carbonate accretion. Altering the orientation of the statocyst with respect to gravity causes the epidermal cells to be bent at the apex, thus stimulating their sensory dendrites. Alternatively, the statocyst has a single statolith that is surrounded by a layer of sensory hair cells (see Figure 7–10, page 265). The epithelial invagination may retain an opening with the exterior (open type) or be totally sealed (closed type). The statolith is a dense secretion of calcium carbonate or a dense accretion of sand grains. The position of the statolith within the statocyst depends on the orientation of the animal; different hair cells are stimulated by contact with the statolith in varying positions and

each sensory cell responds maximally at some optimal angle of the animal with respect to gravity. We have previously examined the sensory physiology of statocysts as an example of range fractionation.

Many aquatic insects have water pressure/depth receptors. For example, *Aphelocheirus* has a group of large hairs that trap an air bubble. The hairs are bent by compression of the air bubble, and the extent of their bending is detected by mechanosensory hairs. The heteropteran *Nepa* has a pressure sensory system that monitors the relative depth of different parts of the abdomen. Three pairs of pressure sensors (on abdominal segments 3, 4, and 5) have airspaces interconnected by spiracular openings and tracheae. If the insect is horizontal, then each pressure sensor is equally stimulated. If the head is higher than the tail, the more anterior airspace is at a lower pressure than the posterior airspace; the membrane over the anterior airspace is pushed outwards and that of the posterior airspace is pushed inwards. The opposite happens if the head is lower than the tail. This provides mechanoreceptive information of body position.

Most invertebrate mechanosensory neurons are bipolar, i.e., they have two cell processes—a sensory dendrite and an afferent axon. However, the muscle stretch receptors of invertebrates are multipolar sensory neurons, i.e., they have a number of cell dendritic processes. For example, the stretch receptor of the caterpillar *Antheracea* is a modified muscle cell (with its innervating axon) and an associated connective tract secreted by a special tract cell (see Figure 7–7B, page 262). The multipolar sensory neuron has two to four dendrites that run along the connective tract and are bound to it. The sensory cell is unstimulated if there is no tension on the receptor, but it shows a marked phasic response to stretch and a prolonged tonic response to a maintained stretch.

Vertebrate Mechanoreceptors

The sensory roles of vertebrate mechanoreceptors are as diverse as those of invertebrate mechanoreceptors. Most vertebrate mechanoreceptors are hair cells, which have a very different structure to invertebrate bipolar mechanoreceptors. Multipolar sensory cells are also uncommon in vertebrates; the muscle spindle is an example of a vertebrate multipolar mechanoreceptor.

Muscle Spindle.   The muscle spindle is specialized for detection and control of muscle stretch; its structure varies in different vertebrate classes.

Amphibians are the most primitive vertebrates to commonly possess muscle spindles, although a few fish may possess primitive muscle spindles (Maeda, Miyoshi, and Toh 1983). The frog muscle spindle (see Figure 7–7A, page 262) has 3 to 12 small intrafusal muscle fibers that are attached at their ends to the normal extrafusal muscle fibers. There are two types of intrafusal fibers: nuclear bag fibers, which have nuclei aggregated in a central dilated region, and nuclear chain fibers, which have nuclei distributed more evenly along the fiber. The central portion of these intrafusal fibers is noncontractile (lacks actin and myosin filaments) but is innervated by two types of sensory nerve endings: the annulospiral endings (nuclear bag and chain fibers) and the flow-spray endings (nuclear chain fibers only). The muscle spindle is stretched or shortened if the entire muscle is stretched or shortened. Both types of nerve ending have a tonic increase in action potential discharge when the intrafusal fibers are stretched, but only annulospiral endings have a phasic response. Thus, the muscle

spindle provides sensory information concerning limb movement and position.

The muscle spindle also regulates muscle contraction (see also Chapter 2). The ends of the intrafusal fibers are innervated by γ motor neurons and are contractile. Different γ axons innervate the nuclear bag (γ-dynamic) and nuclear chain (γ-static) intrafusal fibers. Stimulation of these axons contracts the ends of the intrafusal muscle fibers and stimulates the central sensory portions of the intrafusal fibers as if the entire muscle had been stretched.

Acoustico-lateralis System.   The vertebrate hair cell is a highly specialized secondary sensory cell, i.e., it is non-neural. It transfers sensory information (receptor potential) to a neuron by chemical synaptic transmission; the generator potential of the neuron initiates action potentials (Figure 7–23A). An efferent neuron synapses with the hair cell to modify the response of the sensory cell. The surface of the hair cell usually has a single long cilium, the **kinocilium** (with the characteristic $9 + 2$ microtubular structure

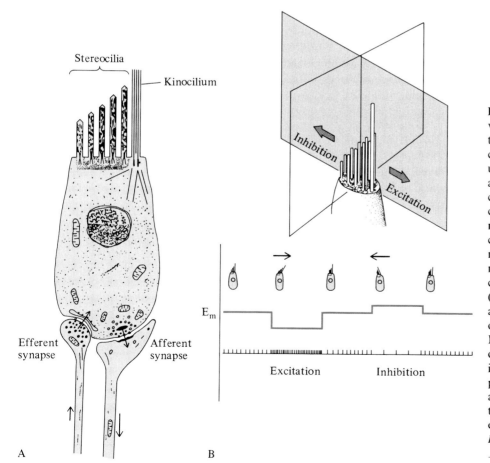

FIGURE 7–23 **(A)** Typical vertebrate hair cell with multiple stereocilia and one true cilium, the kinocilium. Stimulation of the hair cell causes a receptor potential that is communicated by a granular chemical synapse to the innervating neuron. The hair cell also receives an efferent neuron from the central nervous system for modification of hair cell sensitivity. **(B)** The hair cell is directionally sensitive to displacement of the stereocilia/kinocilium. Movement in the excitatory direction (shortest stereocilium to kinocilium) causes depolarization of the hair cell and an increased action potential frequency in the afferent neuron. *(Modified from Flock 1965.)*

of cilia) and a number of shorter **stereocilia**, which are actually microvilli containing actin filaments. The stereocilia vary in length, the longest being adjacent to the kinocilium. Some hair cells lack a kinocilium (e.g., in the mammalian cochlea).

Bending of the stereocilia induces a receptor potential in the hair cell; bending towards the longest stereocilium depolarizes the hair cell and bending towards the shortest stereocilium hyperpolarizes the hair cell (Figure 7–23B). The receptor potential is relayed to the innervating neuron by a chemical synapse, and the generator potential of the neuron increases the action potential discharge (hair cell depolarization) or decreases it (hair cell hyperpolarization). Displacement towards the longest stereocilium causes a greater depolarization of the hair cell and a greater action potential discharge than the hyperpolarization and decreases in discharge caused by displacement in the other direction, i.e., the hair cell is asymmetric (as is the invertebrate hair cell).

Groups of hair cells (from a few to over 100) are often arranged with accessory cells to form a **neuromast**. The free-standing neuromast cell has its stereocilia and kinocilium embedded in an elongate, plate-shaped gelatinous mass, the **cupula**. The cupula is preferentially deformed in one direction by water movement (Figure 7–24A). A number of more complex sensory structures have evolved from free neuromasts. The lateral line system of fish has neuromasts located within canals (Figure 7–24B). The vertebrate semicircular canal is a bulb-shaped expansion of the canal with an occluding neuromast (Figure 7–24C). The otolith organ of fish has neuromasts that are stimulated by the displacement by gravity of a dense otolith located within the gelatinous capsule (Figure 7–24D). The hair cells in the ear of higher vertebrates responds to vibration of their basal membrane (Figure 7–24E).

Free neuromasts, with the cupula projecting into the ambient medium, are present in the **lateral line organs** of cyclostomes, some of the more advanced fish, and aquatic amphibians. For example, the lateral line system of the amphibian *Xenopus* has free neuromasts located on the surface of the skin. The sensory neurons have a spontaneous discharge, but local water currents elicit bursts of activity. The lateral line of many fish is not free neuromasts but a complex subepithelial canal system that contains neuromasts (Figure 7–25). These canals communicate with the ambient medium through pores. The neuromasts located in the canals of the lateral line system provide highly directional information on local water movements at the animals' surface. Specialized organs for equilibrium and gravity detection, audition, electroreception, and magnetoreception have evolved from the lateral line system.

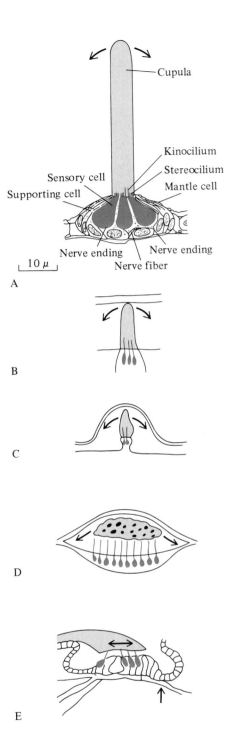

FIGURE 7–24 Representation of the different arrangements of hair cells in the acoustico-lateralis sensory system of vertebrates. **(A)** The freestanding neuromast of the teleost fish *Oryzias* detects water currents. **(B)** The lateral line canal organ of fish detects water currents. **(C)** The semicircular canal of the inner ear provides equilibrium sensation. **(D)** The otolith organ of a teleost fish detects gravity. **(E)** The Organ of Corti detects sound vibration of the cochlear membrane. *(Modified from Florey 1966; Hudspeth 1985.)*

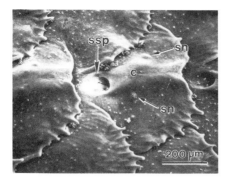

A

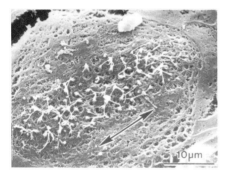

B

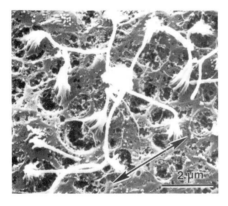

C

FIGURE 7–25 The lateral line scales of the cichlid fish *Cichlasoma nigrofasciatum* have a canal that contains a group of neuromasts sensitive to water flow. Each hair cell has numerous stereocilia and a kinocilium. **(A)** Lateral line scale with lateral line canal (c), suprascalar pore (ssp) that allows water entry into the canal, and two superficial neuromasts (sn). **(B)** A canal neuromast, which is sensitive to water movement in the direction of the arrow, is located inside the lateral line canal. **(C)** The hair cells of canal neuromasts each have a kinocilium and numerous stereocilia. The placement of the kinocilium relative to the stereocilia on the surface of the hair cell determines its axis of physiological sensitivity (arrow). *(From Webb 1989.)*

The internal ear of tetrapod vertebrates has evolved from a lateral line system. The embryonic ear sac differentiates into two sacs: a (more primitive) ventral sacculus and a dorsal utriculus (Figure 7–26A). A series of semicircular-shaped canals (one in myxinoid cyclostomes, two in petromyzontid cyclostomes, and three in other vertebrates) extend from the utriculus. One of the openings of each semicircular canal into the utriculus has an enlarged bulb-like ampulla containing a neuromast organ, the **crista ampullaris**. The sacculus has a small ventral lobe, the lagena, which becomes more elongate in reptiles and birds and coiled in mammals. The sacculus and lagena have neuromast sensory structures **(maculae)**, and a **basilar papilla** that develops into the basilar organ (Organ of Corti) of the reptile, bird, and mammal ear. The tetrapod ear thus contains neuromast organs specialized for detection of gravity (the maculae of the sacculus and utriculus), body movement (the cristae ampullarae of each semicircular canal), and audition (Organ of Corti).

The cristae ampullarae of the semicircular canals function in a similar fashion as canal neuromasts of the fish lateral line system. Body movement induces relative movement of the fluid within the semicircular canals and bends the gelatinous capsule. The fluid actually remains stationary because of its inertia, and the canal wall and crista move past the fluid.

The neuromast of the fish otolith organ, and the macula in the sacculus and utriculus of higher vertebrates, function in the same manner as the invertebrate lithostyle. Dense otoliths, or otoconia, in the surface gelatinous cupula, respond to orientation by bending the gelatinous mass, thereby bending the stereocilia and kinocilia of the hair cells.

The neuromast organ of the saccule lagena becomes highly specialized for audition in reptiles, birds, and mammals. The basilar papilla, or Organ of Corti, transduces vibrational fluid movement into electrical depolarizations of the sensory hair cells that are encoded to transmit acoustic parameters of frequency and intensity to the brain.

The human ear is a good example of the complex auditory structure of mammals (Figure 7–27A). The external ear (pinna and external auditory meatus) collects sound vibrations and directs them to the middle ear. The outer ear provides some acoustic directionality and filtering. Many mammals can move the pinna to aid sound localization. The external auditory meatus preferentially transmits sound of wavelength equal to 4 times its length (i.e., about 4 kHz for humans).

The middle ear consists of a tympanic membrane (ear drum), the middle ear bones (ossicles), and the bony wall of the inner ear. The middle ear transmits

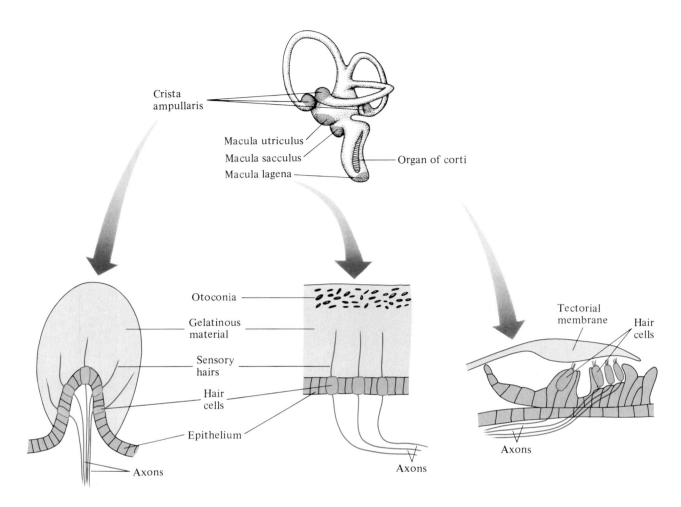

FIGURE 7–26   The inner ear of vertebrates, for example, a bird, has semicircular canals with cristae ampullarae for dynamic equilibrium sensation, a macula utriculus, macula sacculus, and macula lagena for static equilibrium sensation, and an Organ of Corti for audition.

sound-induced vibrations of the tympanic membrane to the oval window portion of the bony wall of the inner ear (Figure 7–27B). The middle ear ossicles (malleus, incus, and stapes) are a system of levers that transmit vibration from the large tympanic membrane (about 45 mm²) to the small oval window (3 mm²).

The malleus contacts the tympanum and the stapes contacts the oval window. An important role of the middle ear is to match the high impedance of air particle vibration in the outer ear to the low impedance of inner ear fluid vibration. Air particles oscillate with a high amplitude and low force whereas fluid in the inner ear vibrates with a low amplitude and high force. The vibrational energy is amplified by about 15:1 and the vibrational amplitude is decreased by about 1.3:1 (tympanic membrane:oval window displacement). These ratios of displacement and relative areas (in man) provide

optimal impedance matching in the middle frequency range (about 1 kHz).

The inner ear is an endolymphatic fluid space comprised of the sacculus, utriculus, semicircular canals and cochlea, and a perilymphatic fluid space located between these structures and the bony support of the ear (Figure 7–27B). The mammalian **cochlea** is a coiled structure, comprising an endolymph canal and two perilymphatic canals; the endolymphatic canal is separated from one perilymphatic canal (the scala vestibuli) by Reissner's membrane and from the other (scala tympani) by the basilar membrane. The Organ of Corti is located on the basilar membrane; it has two rows of inner hair cells and three rows of outer hair cells that run the length of the organ. The inner hair cells provide about 95% of the afferent neurons carrying auditory information to the brain. The stereocilia of the inner hair cells are covered by a gelatinous structure, the

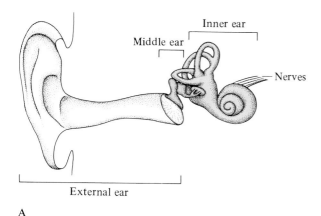

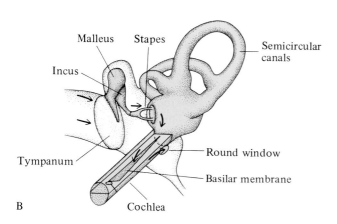

FIGURE 7–27 **(A)** External, middle, and inner ear of the human (*Homo sapiens*). **(B)** Diagram of the middle ear (tympanic membrane and ear ossicles) and inner ear (semicircular canals and cochlea). For clarity, the cochlea is shown uncoiled. The direction of vibrational sound transmissional movement from the auditory canal to the round window is indicated by solid arrows. (*Modified from Vick 1984; from Davis 1965.*)

tectorial membrane. The outer hair cells provide only about 5% of the afferent innervation to the brain. Their cilia are embedded in the tectorial membrane. The outer hair cells may be effector rather than sensory cells; efferent innervation from the brain may stimulate contractile proteins and increase the stiffness of the hair cells and cilia, thereby controlling the flexibility of the Organ of Corti.

Details of mechanoelectrical transduction by the cochlear hair cells are not well understood. The shearing displacement of stereocilia exerts a tensile force on the adjacent stereocilium membrane, opening ion channels. Each stereocilium appears to have a fine filamentous attachment to its adjacent longer stereocilium, and one model of mechanoelectrical transduction suggests a trapdoor-spring analogy for the opening of the ion channels by tension on these fine filaments (Figure 7–28). The opening of only a few specific ion channels (there are only about 280 mechanosensitive ion channels per hair cell in the frog sacculus) depolarizes the hair cell membrane. These ion channels appear to have a diameter of about 0.7 nm since they are permeable to alkali cations (Na^+, K^+, Rb^+, Cs^+), divalent cations (Ca^{2+}, Ba^{2+}, Sr^{2+}), and some organic cations (choline$^+$, tetramethylammonium$^+$). Shearing of the hair cell sensory tip towards the longest stereocilium depolarizes the hair cell by about 20 mV; shear in the opposite direction hyperpolarizes the cell by about 5 mV; shear at right angles has no effect, i.e., the hair cell is asymmetric.

The transduction and encoding mechanisms of the Organ of Corti are complex (Hudspeth 1985). Movement of the oval window by the stapes is transmitted to the perilymph in the scala vestibuli, which in turns displaces the basilar membrane and hair cells. The scale tympani completes the circuit,

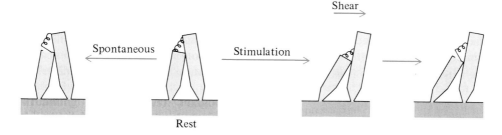

FIGURE 7–28 A simple model for the mechanoelectrical transduction of hair cell bundle displacement suggests that the unstimulated stereocilia normally have closed ion channels because of low tension on the opening "spring" (center) although they can spontaneously open (left). Auditory stimulation by a shearing motion of the basilar membrane changes the orientation of the stereocilia and stretches the "spring," facilitating ion channel opening (right). (*Modified from Hudspeth 1985.*)

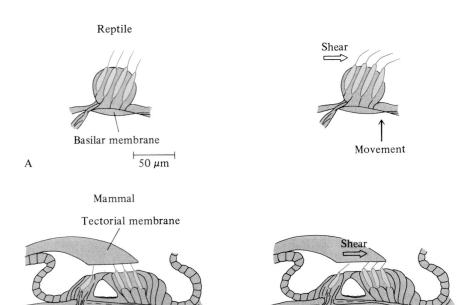

A

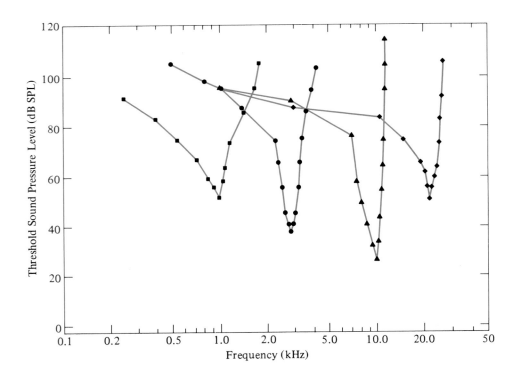

B

FIGURE 7–29   **(A)** The hair cells of the reptilian basilar membrane (shown for a lizard *Gerrhonotus multicarinatus*) are freestanding; vibrational movement of the basilar membrane (solid arrow) causes a shearing motion of the hair cell bundles (open arrow). **(B)** Vibrational movement of the mammalian basilar membrane (solid arrow) causes a shear between the hair cell bundles and the tectorial membrane (open arrow). *(Modified from Hudspeth 1985.)*

FIGURE 7–30   Responses of various sensory neurons tuned to different frequencies in the guinea pig, measured in the auditory nerve.   *(Modified from Evans 1972.)*

transmitting vibrational movement of the perilymph to the round window of the middle ear. In reptiles, the hair cells are arranged on the basilar membrane with the stereocilia projecting into the endolymphatic fluid i.e., the hair bundles are freestanding. Vertical displacement of the basilar membrane causes a lateral movement of the hair cell sensory ending; the hydrodynamic drag of the stereocilia causes them to bend and to stimulate the hair cell (Figure 7–29A). The hair cells of the mammalian basilar membrane are covered by the gelatinous tectorial membrane, and vibrational movement of the basilar membrane causes a shearing movement between the hair cells and the tectorial membrane (Figure 7–29B).

There is a progressive increase in the length of the hair cells from the base of the basilar organ to the tip. These differing length hair cells respond differently to acoustic stimulation; the shortest hair bundles are preferentially deflected by high frequency sound (about 4 kHz), whereas the longer hair bundles are preferentially stimulated by lower frequencies (about 1 kHz). The physical properties of the hair cells bundles thus "tune" the hair cells to different frequencies. That each hair cell is tuned to a specific frequency is shown by recordings of action potential response to specific sound frequencies for individual cochlear nerves (Figure 7–30).

Auditory receptors can detect incredibly small vibrational displacements and low energy levels (Figure 7-31). Air particle displacement at the auditory threshold is about 10^{-9} to 10^{-6} cm and the vibrational amplitude of the basilar membrane is about 10^{-11} to 10^{-10} cm (depending on frequency). A hair cell can therefore respond to a displacement much less than the diameter of a hydrogen atom (about 10^{-8} cm)! This corresponds to a remarkably low stimulus energy, similar to the energy of random Brownian motion (Harris 1968).

Sound intensity is the difference in pressure between the peak compression and expansion of the sound wave. The human ear can detect (at its optimum frequency) a pressure difference of about 20 μPa (0.0002 dyne m^{-2}); this is the **threshold of hearing**. The energy of a sound wave is proportional to the square of the pressure difference

$$E = \frac{P^2}{2\rho V} \quad (7.2a)$$

where E is the energy (W m^{-2}), P is pressure (Pa), ρ is air density (1.2 kg m^{-3}), and V is the velocity of sound (343 m sec^{-1}). The threshold of hearing is equivalent to about 5 10^{-13} W m^{-2}. The loudest sound that the human ear can tolerate, the threshold of pain, is 30 Pa, or about 1 W m^{-2}.

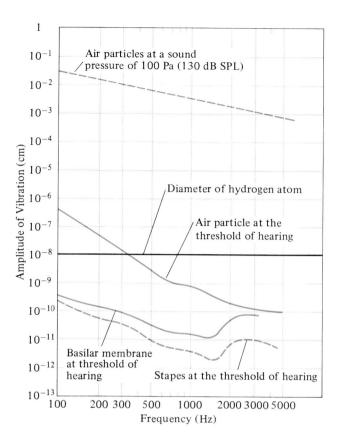

FIGURE 7–31 Vibrational amplitudes for air particles—the stapes (middle ear ossicle) and the basilar membrane at the threshold of hearing and air particles for a loud sound. Most of these vibrational amplitudes are less than the diameter of a hydrogen atom! *(Modified from Davis 1965.)*

The sound energy level is usually measured using a logarithmic ($\log_{10}$) bel scale; the sound energy is expressed relative to a standard energy level. For example, 1 bel has 10 times more sound energy than the standard sound energy level. The **decibel** (1 dB) is 1/10 of a bel or a 1.26 times higher energy level ($1.26^{10} = 10$). The sound pressure level (SPL, dB) is calculated as

$$SPL = 10 \log_{10} (P/P_0)^2 = 10 \log_{10} (E/E_0) \quad (7.2b)$$

where P is the pressure of the sound (Pa), P_0 is the standard pressure level, E is the energy, and E_0 is the standard energy level. The standard pressure level is 20 μPa (although sometimes 1 dyne cm^{-2} is used as the "standard pressure level"). The threshold of sensation for the human ear is 0 dB SPL at its optimal frequency.

The random Brownian motion of air vibrates the tympanic membrane with an energy of about -18 dB, i.e., it is considerably less than the auditory threshold (Harris 1968). Random noise due to Brownian motion of air would appear to be insignificant. However, the sound detector is the hair cells, not the tympanic membrane, and the vibrational energy of the tympanic membrane vibration is dissipated to thousands of individual hair cells. What then is the ratio of threshold signal to Brownian noise for an individual hair cell? Estimates vary widely, depending on whether hair cells are assumed to be mechanically independent of adjacent hair cells (weakly coupled) or firmly attached to neighbors (strongly coupled). Brownian noise for a weakly coupled hair cell may be about 33 dB higher than the threshold for signal detection, i.e., the hair cell would easily detect random Brownian motion as noise. However, it would not be useful for our ear to continually detect random background noise. Hair cells are connected to adjacent hair cells, and a tight coupling could reduce their random noise level to about 22 dB below the auditory threshold. Thus, the sensitivity of the ear is potentially high enough to hear Brownian motion, but the hair cells are mechanically coupled and do not "hear" random molecular motion.

Echolocation. A few vertebrates can detect the presence of distant objects by hearing echoes of sound that they emit and that bounces off the objects. This **echolocation**, or biosonar, is similar in principle to radar but uses sound rather than radio waves. It is a spectacular biological example of a "transmitter–receiver" sensory system, whereby an animal detects energy which it has emitted and is reflected back, or modified, by distant objects in its environment. Weakly electric fish, which detect objects that modify the shape of their electric field, and bioluminescent fish, which see by their own emitted light, are other examples of transmitter–receiver sensory systems.

Animals that echolocate are either active at night and forage for food over a wide range, or live in caves, or are aquatic. For example, a number of nocturnal mammals, some marine mammals, and a few cave-dwelling birds use echolocation (Table 7–2). Echolocation is used for either prey capture (e.g., insectivorous bats, dolphins) or obstacle detection (e.g., shrews, cave-dwelling oilbirds).

Bats are the best-studied echolocating animals. Microchiropteran bats emit sound signals from their larynx, whereas megachiropteran fruit bats click with their tongues. The sounds can be emitted from the mouth or nose. The horseshoe bats (and some other bats) have a complex nose leaf for emission

TABLE 7–2

Frequencies of sound used by various echolocating animals.

Animal	Dominant Frequency
Bats	
Microchiroptera	
European mouse-eared bat	
Myotis	25–100 kHz (FM)[1]
Greater horseshoe bat	35–45, 75–90 kHz (CF [2]/
Rhinolophus	FM)
False vampire *Megaderma*	30–125 kHz (USP)[3]
Painted bat *Kerivoula*	233–243 kHz (UHF)[4]
Megachiroptera	
Fruit bat *Rousettus*	13, 26, 38 kHz
Cetaceans	
Dolphin *Tursiops*	15–130 kHz
Amazon river dolphin *Inia*	60–65 kHz
Killer whale *Orcinus*	14 kHz
Other Mammals	
Shrew *Sorex*	18–60 kHz
Shrew *Crocidura*	70–110 kHz
Tenrec *Centetes*	5–17 kHz
Birds	
Oilbird *Steatornis*	1.5–2.5, 5–7 kHz
Cave swiftlet *Collocalia*	4–7.5 kHz

[1] FM = frequency modulated.
[2] CF = constant frequency.
[3] USP = ultrashort duration sound pulse.
[4] UHF = ultrahigh frequency.

of the echolocating sound pulse (Figure 7–32A). Bats emit short pulses of high frequency sound (20 to 140 kHz) every 100 msec or so until their small prey, or an obstacle, is detected. Then, the duration of the sound pulses is decreased and the interval between pulses is reduced to about 10 msec. The increased signal repetition rate allows higher precision of echolocation. For example, the horseshoe bat *Rhinolophus* emits a long duration, pure frequency (82 kHz) sound pulse when hunting (Figure 7–32B). When a prey insect is detected, shorter pulses of 82 kHz pulses are emitted until the prey is captured. A series of short, frequency modulated pulses are emitted after capture.

Bats and other echolocating animals must obviously be able to hear the frequencies of the sounds they emit. A bat emitting a 100 kHz sound pulse would not be able to hear the echo if its ear had the frequency response characteristics of the human ear (we cannot hear above about 20 kHz). However, the auditory threshold of the bat ear is highly tuned

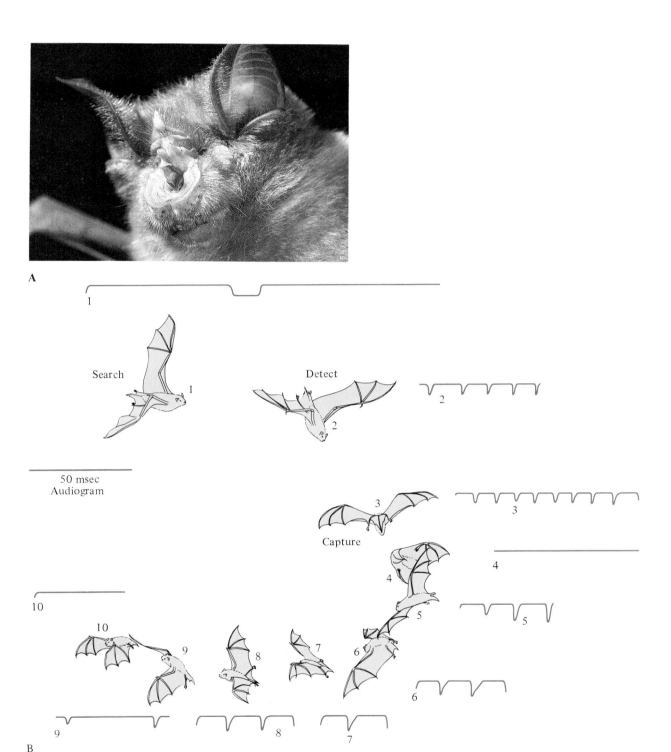

FIGURE 7–32 (A) The horseshoe bat *Rhinolophus ferrumequinum* has a complex nose leaf—the bats fly with their mouth closed, and echolocating sound is emitted through the nostrils. The echo is collected by the large pinnae and directed to the middle ear *(Stephen Dalton/Photo Researchers, Inc.).* (B) Flight sequence for the horseshoe bat catching a moth. The audiograms of a hunting bat in the search phase (1) are long, pure tones (82 kHz) with no FM component; the sounds are shorter and more frequent, but also pure 82 kHz, in the detect phase (2) and capture phase (3, 4); the long pure tones are for detection of the prey's wing beats. No sound is emitted when the prey is caught (4). Sounds after the capture (5 to 9) include initial and final FM components. This capture sequence lasted about 0.9 sec. *(From Neuweiler 1983.)*

to the high frequencies of sound that it emits for echolocation. The bat also has a broad range of auditory sensitivity in addition to its sharply tuned high frequency sensitivity.

The intensity of sound decreases with the square of the distance traveled from the sound source (if we consider the sound wave to propagate as a spherical wave front). The intensity of sound emitted by a bat can be 80 to 120 dB SPL, and it decreases by about 20 dB for every 9 meters. The echo of a 90 dB SPL sound pulse (at 100 kHz) emitted by a bat and reflected from an object 5 m distant would be attenuated to about 4 dB SPL at the bat's ear; reflection from an object at 20 m would be hardly detectable.

Echolocation requires emission of high-energy sound pulses (e.g., 120 dB) but detection of low-energy pulses (e.g., a few dB). How does the bat's sensitive ear cope with the high intensity of its emitted sound? Two muscles of the middle ear contract just before the sound is emitted and dampen oscillations of the tympanic membrane (the tensor tympani muscle) and the stapes (tensor stapedius muscle) so that the bat is not deafened by its own noise. The muscles relax at the end of the sound emission, allowing detection of the low-intensity echoes.

The sound pulse used by the 800 or so species of bats is extremely complex and variable with respect to the main (dominant) frequency, frequency variation during a sound pulse, and structure of the sound pulse. Bats that hunt in the open sky tend to emit short duration (1 to 5 msec) sound pulses with frequency decreasing by over an octave during the pulse. These frequency modulated (FM) bat calls may also emit different harmonics (frequency components at 2, 3, 4, etc. times the dominant frequency). Bats that hunt close to the ground have ultrashort sound pulses to avoid overlap of the emitted call with the echo from the ground. Several bats, such as the horseshoe bat *Rhinolophus*, have a composite sound pulse of an initial, short FM pulse (1 to 2 msec of increasing frequency), a long pulse of constant frequency (15 to 200 msec), and a final short 1 to 2 msec FM pulse; these bats are called constant frequency/frequency modulated (CF/FM) bats.

What factors are responsible for this variability in frequency and structure of echolocating sound pulses? Air is not the optimal medium for echolocating because the atmosphere strongly absorbs ultrasonic sound waves (Neuweiler 1983). The absorption, or attenuation, of sound depends on the composition, temperature, and relative humidity of the air and the frequency of the sound. Attenuation

is greater at higher frequencies, and this limits the usefulness of high frequencies for echolocation. Nevertheless, some bats have ultrahigh frequency sound pulses (100 to 250 kHz; Table 7–2) compared to most bats (< 100 kHz). These ultrahigh frequency bats tend to emit low intensity sounds, for short-distance echolocation.

Why don't all bats use lower frequencies for echolocation to benefit from the lesser attenuation? High frequency sound provides better directionality for echo detection, whereas low frequency sound produces diffuse echoes. Furthermore, the minimal detectable size of an object is inversely proportional to the wavelength (λ) of the sound pulse, hence proportional to the frequency. For example, the minimum size detected by the little brown bat (*M. lucifugus*) is about 0.03 cm, at a frequency of 100 kHz ($\lambda = 0.343$ cm), i.e., the bat can detect objects about 1/10 λ (Griffin 1958).

The insect prey of bats can try to evade capture if they can hear the high frequency echolocating pulses of their predators. Many noctuid moths, for example, are able to detect and attempt to avoid foraging bats because their ears (chordotonal organs) are tuned to the high frequencies used by echolocating bats (Roeder 1966). Bats that echolocate at a frequency that their prey cannot detect would be at a definite advantage. For example, the ultrahigh frequency bats "cheat" their prey by using an undetectably high frequency, although the cost of this ultrahigh frequency strategy is a drastic attenuation of their echolocating signal.

Shrews echolocate objects in their environment using high frequency sound (about 50 kHz; Table 7–2). Oilbirds and cave swiftlets use low frequency pulses (about 7 kHz) to echolocate obstacles as they fly through dark caves to their roosting sites; the echolocating clicks are clearly audible to humans. Birds generally have a similar auditory frequency discrimination as humans, although they are able to detect more rapid frequency changes. Some birds (e.g., pigeons) can detect **infrasound** (very low frequency sound), which may aid them in long-distance migration and navigation.

The principles of echolocation are similar in air and water, but there is less sound attenuation in water. Many cetaceans (whales, dolphins) and some seals (Weddell seal) echolocate prey or other objects in their environment, e.g., Weddell seals may echolocate breathing holes in surface ice. Dolphins (Figure 7–33A), for example, emit a series of clicks (about 10 sec^{-1} for 3 to 4 secs) of high frequency (30 to 150 kHz); a typical echolocation click is shown in Figure 7–33B. Echolocating marine mammals use either specialized nasal structures or the larynx (but

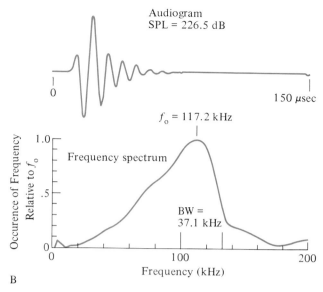

FIGURE 7–33 **(A)** The head of the bottlenosed dolphin *Tursiops truncatus*. *(A. W. Ambler, National Audubon Society/Photo Researchers, Inc.)* **(B)** Typical echolocatory click made by a bottlenosed dolphin is a short-duration (70 sec), high-intensity (227 dB SPL) pulse of about 9 cycles of a high-frequency sound. The dominant frequency (f_o) is 117 kHz with a band width (BW) of 37 kHz. *(Modified from Au 1980.)*

not the normal mammalian vocal cords) to produce their echolocating clicks (Green, Ridgway, and Evans 1980). The arytenoid cartilages of the larynx are modified as vibratory, sound-producing structures. The sound is emitted in a highly directional manner, essentially in front of the animal as a cone of sound with limited sideways and vertical spread (Au 1980). Auditory reception of the echo is substantially different in aquatic mammals than terrestrial mammals because the external auditory meatus is not the preferential pathway for sound transmission to the middle and inner ears. Rather, dolphins use bone vibration (especially by the lower jaw) to transmit sound from the water to the inner ear (McCormick et al. 1980). The exceptionally dense bone surrounding the middle ear, the bulla, is anatomically isolated from the rest of the cranial bone (although attached by connective tissue, cartilage, and fat). Sound causes vibrational movement of the dense bony bulla relative to the rest of the skull. The middle ear ossicles transmit the vibrations of the bulla to the inner ear, which is located within a cranial bone (the petrous portion of the temporal bone).

Electroreception

Some predatory fish, such as sharks and rays, detect the small electric currents that are produced by the tissues (e.g., heart, gill ventilation muscles) of their prey. Some weakly-electric fish produce an endogenous electric field to electrolocate objects in their immediate vicinity and for intraspecific communication. The strongly-electric fish, such as electric eels and electric rays, use powerful electrical discharges to immobilize prey (Chapter 9).

The predatory and weakly-electric fish have ampullary electroreceptors distributed over their body, in close association with the lateral line system from which the electroreceptors have evolved (Figure 7–34A). The electroreceptive system of the elasmobranchs consists of **ampullae of Lorenzini**. The ampullae tend to be grouped into a few capsules on the dorsal and ventral sides, with ducts radiating away to the external openings that are distributed over the body surface. Many teleost fish also have surface electroreceptors for the detection of exogenous electric fields or endogenously produced electric fields. For example, the transparent catfish

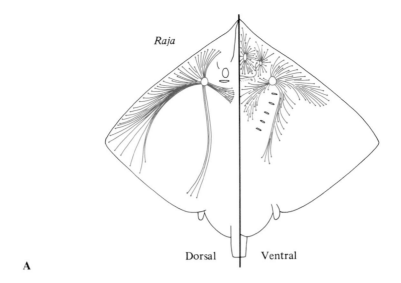

Raja

Dorsal   Ventral

A

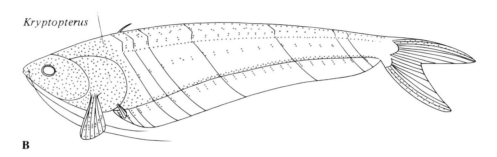

Kryptopterus

B

FIGURE 7–34   **(A)** Distribution of electosensory cells for the ray *Raja clavata;* ampullae of Lorenzini are located in capsules (open circles), from which the electroconductive tubules radiate to external openings (solid dots). **(B)** The transparent catfish *Kryptopterus bicirrhus* has tonic ampullary electroreceptors (solid dots) distributed over the body surface, especially the head and lateral lines.   *(From Murray 1960; Wachtel and Szamier 1969.)*

Kryptopterus has electroreceptors distributed over its body surface, particularly on the head, lateral line, and the caudal fin (Figure 7–34B). Probably all aquatic amphibians, and a few mammals, such as the platypus and echidna (Scheich et al. 1986; Gregory et al. 1989), have ampullary electroreceptors.

Predatory sharks, rays, and some teleosts detect the weak electric field of prey. A dogfish *Scyliorhinus*, for example, can detect a flounder buried under sand at a distance of about 15 cm by electrodetection (Kalmijn 1971). The flounder emits a current of about 4 μamp. The dogfish can locate an artificial prey item with a similar magnitude electric current. The very sensitive electroreceptors can respond to electric gradients of as little as 10^{-8} V cm^{-1}; this is similar to an electrical gradient along a wire as long as the diameter of the earth (40000 km) with the ends connected to a 12 V battery (about 3 10^{-9} V cm^{-1})!

Electroreceptors are modified hair cells of lateral line neuromasts. **Ampullary receptors** have flask-shaped canals, filled with a gelatinous material; the sensory cells are located at the base of the ampulla such that only a small part of the cell surface is in contact with the gelatinous matrix (Figure 7–35A). Each ampullary electroreceptor of elasmobranchs is innervated by 4 to 15 nerve fibers from the lateral line nerves. Ampullary organs are often tonic receptors, responding to low fre-

A

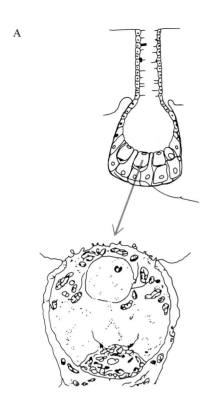

B

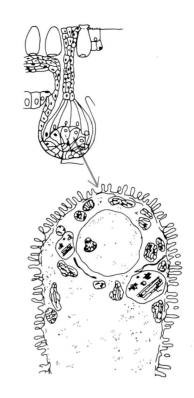

Tonic ampullary electroreceptor

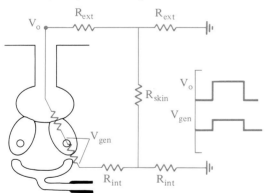

Phasic tuberous electroreceptor

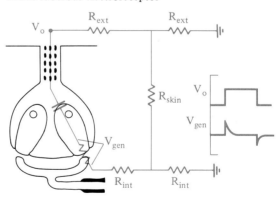

FIGURE 7–35 Electrosensory receptor units for (**A**) an ampullary organ and (**B**) a tuberous organ of a gymnotid weakly-electric fish. Shown enlarged in the center are the electroreceptor cells of the ampullary and tuberous organs. Shown below is a schematic of the tonic ampullary and phasic tuberous electroreceptors, with superimposed equivalent electrical circuits. An external voltage (V_o) at the electroreceptor pore induces a current flow through the electroreceptor cell, a generator voltage gradient across the cell (V_{gen}), and a current flow through the skin and external medium to ground. The generator voltage V_{gen} in the tonic ampullary receptor is of the same waveform as the external voltage V_o. In the phasic tuberous receptor cell, the V_{gen} only occurs in response to a change in V_o because of the capacitance role of the tuberous electroreceptor cell. R_{ext}, external resistance; R_{skin}, skin resistance; R_{int}, internal resistance. *(From Szabo 1965; Szamier and Watchel 1969; Bennett 1967.)*

quency AC or DC electric stimuli. Predatory fish, such as sharks and catfish, possess only this type of electroreceptor.

The ampullary receptors of freshwater fish have very short canals, just long enough to penetrate the skin, whereas marine fish have ampullary receptors with much longer canals. This is because of the different resistivities of fresh water, seawater, and body fluids. The body fluids of marine fish have a higher resistance to current flow than seawater, and so lines of electric field diverge away from a marine fish. Long ampullary canals maximize the voltage difference between the different electrosensory receptors, and allow the detection of weak electric fields. In contrast, freshwater fish have a lower resistance than their medium, and electric fields tend to converge on them. A short ampullary canal maximizes the current reaching the receptor cells. The skin of weakly-electric freshwater fish also has a higher resistance to current flow (3 to 50 kΩ cm^{-2}) than the skin of nonelectric freshwater fish (<1 kΩ cm^{-2} for goldfish); this also maximizes current flow to the electroreceptors.

Weakly-electric fish, such as the gymnotids and mormyrids, have tonic ampullary electroreceptors distributed over the body surface and a large number of a second type of **tuberous electroreceptor** (Figure 7–35B). The tuberous receptors are phasic, responding to high frequency AC currents and are insensitive to low frequency AC and DC currents. They are used for electrolocation by endogenously produced high frequency AC discharges of the weakly-electric organ. The canal of tuberous receptors is filled with loosely packed epithelial cells. The sensory cells are located at the base of the canals, most of their surface exposed and elaborated with microvilli. The innervating neurons at the base of the receptor cells receive the sensory information from the receptor cells via a chemical synapse (rarely by electrical synapse). The tuberous electroreceptors often have a high resting discharge of action potentials (about 100 sec^{-1}) and respond in a phasic fashion to an increase or decrease in the local electric field with an increase or decrease in action potential discharge. The phasic response of the tuberous electroreceptor is due to a membrane capacitance effect of the outer face of the receptor cells that is exposed in the organ cavity. As a result, the generator voltage at the inner membrane of the receptor cell shows a transient increase to an

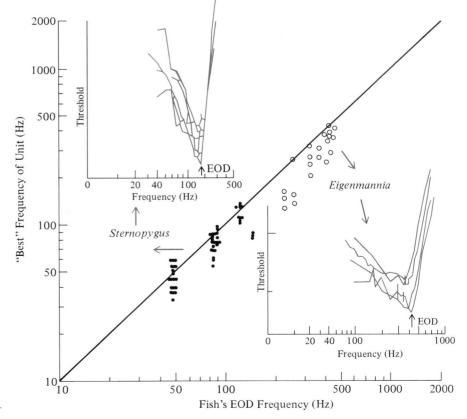

FIGURE 7–36 The electroreceptor organs of weakly-electric fish are "tuned" to the normal discharge frequency of their electric organs, as is apparent for *Sternopygus* and *Eigenmannia*. Shown as insets are the individual receptor responses to varying frequencies of electrical signals. *(Modified from Hopkins 1976.)*

externally applied voltage, returns to baseline during a prolonged stimulus, and shows a transient decrease when the stimulus ceases. This is in marked contrast to the tonic ampullary receptor, in which the generator potential is maintained at the inner membrane of the receptor cell for as long as the stimulus is maintained.

Weakly-electric fish generally have an electric organ (usually modified muscle, but sometimes nerve) at the base of the tail or laterally along the side of the body. This generates a high frequency AC discharge (50 to 2000 Hz) depending on the species of fish (see Chapter 9). The fish's electroreceptors respond best to the frequency of the electric organ discharge (Hopkins 1976). For example, *Sternopygus*'s electroreceptors respond most to 100 to 150 Hz signals, and the frequency of the electric discharge is about 120 Hz; *Eigenmannia*'s electroreceptors respond most to about 400 Hz and its electric organ discharge is about 400 Hz (Figure 7–36). The electroreceptor regions along the sides of the body respond to the endogenously produced electric field. Conducting or insulating objects that move into the electric field alter its shape and the pattern of stimulation of the fish's electroreceptors. The fish is thus able to detect the presence of electro-opaque objects.

Weakly-electric fish of the same species have similar frequencies of electric organ discharge and conspecifics might interfere with electrolocation. However, the jamming avoidance response shifts the discharge frequency away from the frequency used by an interfering conspecific (Bullock, Hamstra, and Scheich 1972). The electroreceptive system of weakly-electric fish may also enable communication between individuals, e.g., sex recognition (Westby 1981). For example, male *Sternopygus* have a frequency of discharge about 1 octave lower than the females; some other weakly-electric fish also have sexually dimorphic patterns of electric organ discharge.

Magnetoreception

The earth has a magnetic field that is approximately aligned with its geographic axis; the north pole of a magnet points towards the magnetic and geographic south pole, and the magnet's south pole points to the magnetic and geographic north pole (Figure 7–37). The magnetic field has frequently reversed in the past (e.g., it has reversed several times in the last million years). The Earth's magnetic field is ever present, but weak. Its strength at the surface is about $0.5 \ 10^{-4}$ weber m^{-2} (or 0.5 gauss). A magnetosensitivity would allow animals to determine the north–south orientation and perhaps even latitude, but it is not a common sensory modality for animals.

The electroreceptors of many fish are sufficiently sensitive to detect the earth's magnetic field. A current flow is induced in an electrical conductor

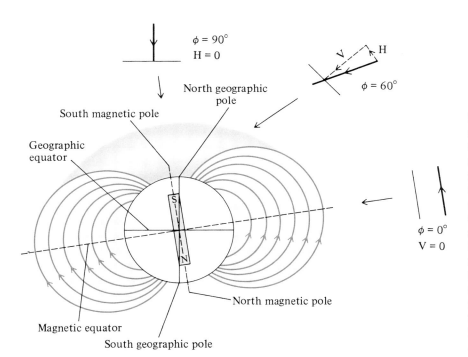

FIGURE 7–37 Orientation of magnetic field lines with respect to magnetic and geographic north–south axes. Note that the magnetic South Pole is actually at the geographic North Pole. The angle of the magnetic lines to the earth's surface (ϕ) and the vector components of the magnetic field lines into horizontal (H) and vertical (V) components are shown for three latitudes, the magnetic equator ($\phi = 0°$, V $= 0°$), a mid-northern latitude ($\phi = 60°$), and the magnetic North Pole ($\phi = 90°$, H $= 0$).

moving through a magnetic field. A marine fish swimming through seawater acts as an electrical conductor moving through the earth's magnetic field, and the induced current could be detected by its electroreceptors. The seawater provides the return circuit for maintenance of current flow; otherwise, the fish would act as a capacitor rather than a conductor, and a current flow could not be sustained. Magnetic field detection has been experimentally demonstrated for the ampullae of Lorenzini of sharks and rays (Kalmijn 1978). It is unlikely that freshwater animals could detect magnetic fields in similar fashion because their medium has a higher resistance to current flow. Terrestrial animals also could not detect magnetic fields because air will not allow any return current flow.

Certain bacteria (*Spirillium*) have a well-developed magnetic sense, which they use to avoid moving out of their anaerobic mud environment into less hypoxic water (Blackmore 1975; Blackmore et al. 1979; Blackmore et al. 1980). These bacteria contain chains of small magnetic particles that passively align the bacteria northwards with respect to the magnetic field (dead bacteria similarly align northwards). The bacteria do not use magnetic field orientation in order to swim north, but rather to swim down. The magnetic field varies in orientation to the earth's surface depending on latitude. At the magnetic poles, the lines of magnetic field are perpendicular ($\phi = 90°$) to the horizontal; there is a vertical component (V) to the magnetic field but no horizontal component (H). At the magnetic equator, $\phi = 0°$ and there is no vertical component, only a horizontal component. Elsewhere, the magnetic field can be resolved into a vertical and a horizontal component, e.g., $0 < \phi < 90°$ (Figure 7–37). The orientation of the earth's magnetic field in the Northern Hemisphere where these bacteria were studied is at about 60° to the horizontal, and so swimming along the lines of magnetic field also means swimming down into the substrate as well as swimming north. Why don't the bacteria simply use gravity? They are too small to experience a significant gravitational force, even if their density is very different from that of their environment. Magnetic bacteria from the Southern Hemisphere similarly orient to the earth's magnetic field, but swim south.

Many animals, including birds and bees, can orient to the earth's magnetic field. Magnetodetection is one intriguing aspect of bird navigation. Migratory or homing birds are able to navigate using celestial (star) and sun compass information and other directional cues, including infrasound (ultralow-frequency sound) and olfaction. Experiments with birds in artificial magnetic fields of varying vertical and horizontal components also indicate that they are able to orient to the lines of magnetic field, but (unlike the bacteria) are not able to determine the direction (polarity) of the field, i.e., they can determine the north–south axis but cannot tell which direction is north and which is south. For example, north is the appropriate migratory direction for a European robin in the northern hemisphere in spring. The robin is able to orient to the magnetic field lines, but to migrate in the appropriate direction (north) it must orient to the magnetic field lines such that the lines pass through the body from behind and above, to forward and below (Figure 7–38). In autumn, the bird must reorient so that the field lines run from front/above to behind/below to migrate south. If the polarity of

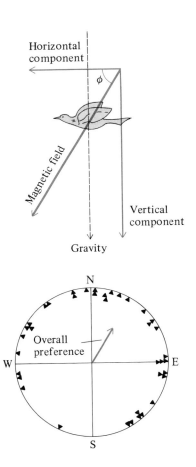

FIGURE 7–38 The migratory direction of the European robin is normally NE; it orients to the dip angle of the magnetic field lines, not the N–S polarity. *(Modified from Wiltschkow and Wiltschkow 1972; Markl 1983.)*

the magnetic field were reversed for the spring robin, then it would still fly north. If either V or H, but not both, is reversed, then the bird will fly south instead of north. If the bird were moved to the Southern Hemisphere, then H is the same but V is reversed, so the bird would fly south; note that this is the "appropriate" direction to fly, away from the equator. Birds at the equator orient at random.

Honeybees detect magnetic fields. Bees recently arrived at their hive communicate the direction of a food source to other bees by their "waggle" dance. The angle between the axis of the dance and the gravitational axis (α') indicates the angle between the sun and the food source (α) (Figure 7–39). The flight direction is towards the sun if the waggle dance is up and away from the sun if the dance is down. There is a small random error between the angle of the waggle dance (α') and the orientation angle (α), as would be expected, but there is also a systematic variation that changes over the daily activity cycle. This systematic error depends on the earth's magnetic field; it is eliminated if the earth's magnetic field is neutralized by an external magnetic field. This response of the honeybees towards a

magnetic field is, like that of birds, independent of polarity; it is only an orientation to the lines of magnetic field, not to polarity.

The sensory mechanism for magnetic field detection is unknown, except for elasmobranchs (ampullae of Lorenzini) and magnetic bacteria (magnetite crystal chains). Both honeybees and pigeons contain magnetite, which could provide a torque force in a magnetic field and stimulate mechanoreceptors, but there is no experimental evidence for such a sensory role of magnetic particles (except for the passive orientation of magnetic bacteria). There are other possible, although perhaps even less likely, mechanisms for magnetic field detection. Electromagnetic induction (moving an electrical conductor through a magnetic field) is unlikely because it should be sensitive to magnetic field polarity (which birds and bees aren't). Paramagnetic molecules, such as O_2, orient with respect to a magnetic field; it is possible that such paramagnetic materials would also be stimulated by light energy, but how animals could use such material properties is not known. The separation of flowing charged particles by a magnetic field (the Hall effect) and superconducting materials

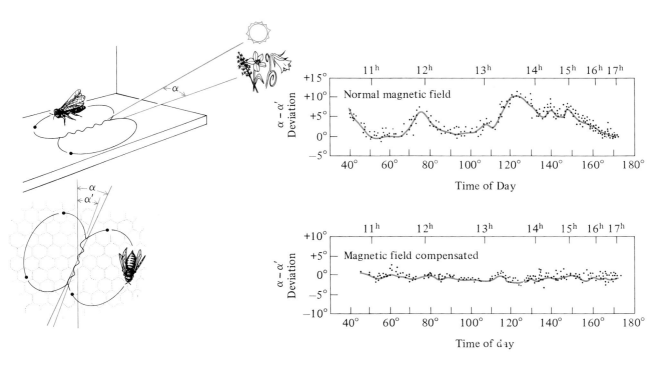

FIGURE 7–39 The angle of the honeybee's "waggle dance" (α') signifies the direction of a food source from the bee hive (α). There is a systematic daily deviation in the difference between α' and α (upper panel), which is reduced to a random deviation if the earth's magnetic field is artificially compensated by an exogenous magnetic field (lower panel). *(Modified from Lindauer and Martin 1968; Markl 1983.)*

have also been suggested to provide mechanisms for magnetodetection.

Photoreception

The detection of light is a widespread sensory modality, ranging from diffuse photosensitivity by nonspecialized tissues (including protoplasm and neurons) to highly sensitive photosensory cells that are arranged in specialized organs and are capable of detecting single photons of light, discriminating color, and forming sharp, focused images.

Light

Visible light is electromagnetic radiation within the range of wavelengths of 400 nm (violet) to 700 nm (red); this definition of visible light is based on the sensitivity of the human visual system. Some animals are also able to detect ultraviolet or infrared radiation. Light is only a small segment of the continuous range of electromagnetic radiation (EMR) from very high frequency gamma rays to low frequency, long wave radiation (Figure 7–40A).

The velocity of light is so high ($c = 2.998 \ 10^8$ m sec^{-1}) that light transmission from an object to a photoreceptor is virtually instantaneous. The energy of a photon of light is determined by frequency (and wavelength)

$$E = h\nu = hc/\lambda \qquad (7.3a)$$

where E is the energy of a photon (joules), h is Planck's constant, ν is frequency (sec^{-1}), and λ is wavelength (m). Rearranging this equation yields the following.

$$E \ (\text{kJ mole}^{-1}) = 119660/\lambda \ (\text{nm}) \qquad (7.3b)$$

The energy of visible light varies from about 170 kJ mole^{-1} (red) to 300 kJ mole^{-1} (violet).

Light behaves both as particles (photons, or quanta) and waves (of specific ν and λ). The wave

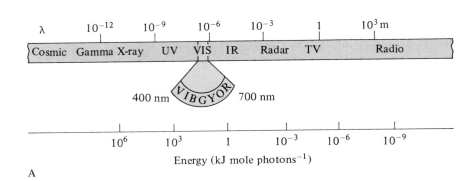

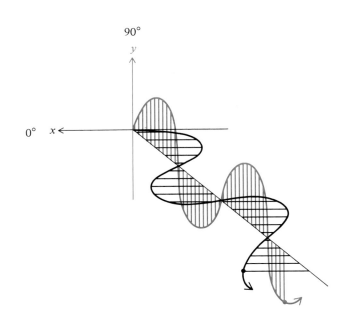

FIGURE 7–40   **(A)** Spectrum of electromagnetic radiation from short wavelength cosmic rays ($\lambda < 10^{-12}$ m) to long radio waves ($\lambda > 10^3$ m). Visible light is a small part of this spectrum, from 400 10^{-9} (violet) to 700 10^{-9} m (red). **(B)** One of the wave properties of light is the plane of polarization (the axis of oscillation of the wave). Shown here are two waves whose plane of polarization differs by 90°. Normally, light is comprised of photons whose plane of oscillation varies at random from 0° to 180°; light may be polarized by passage through crystals or polarizing filters or by reflection or scattering from objects.

property of a photon of light means that it has an axis of oscillating movement, perpendicular to its direction of propagation; this axis is normally random for photons within the full range of 0° to 180°; two such waves at 0° and 90° are shown in Figure 7–40B. The light is **polarized** if all of the photons have their axis of oscillation in the same plane. Normal light, which isn't polarized, can be polarized by passing it through a filter, which has a specific axis of transmission; by reflection from a surface; by passage through a double-refracting material, such as a calcite crystal; or by scattering from small particles or gas molecules. The latter is the cause of polarization of light by the earth's atmosphere when the light comes from directly overhead. Shorter wavelengths (blue light) are scattered more than longer wavelengths (red), and so the sky appears blue.

Photoreceptors

Light affects many nonsensory cells and even cytoplasm. A sensitivity of the body surface to light, i.e., a **dermal light sense** has been reported in almost all animal phyla. Dermal photoreceptors are difficult to identify, but they are thought to be photosensitive nerve endings within, or just beneath, the translucent skin. Haem and carotenoid pigments may be the photoreceptive pigments of these nerves, but the evidence is not compelling. A diffuse dermal light sense would indicate whether it was night or day, and enable the determination of day length. By detecting shadows, it could indicate to a benthic animal the presence of a predator or food. A dermal light sense may be the only sensation of light for many animals, or it may be interpreted in concert with light detection by specialized photoreceptors. The diffuseness of the dermal light sense and the lack of obvious, let alone elaborate, photoreceptors and photopigments infer that the dermal light sense is evolutionary primitive.

Specialized, **photoreceptor organs** (eye spots and eyes) provide more discrete photic information than a dermal light sense. The localization of photoreceptors in a small spot or sheet, with a light shade of accessory pigment cells, provides directional information. The eye spot of the protozoan *Euglena* is a light-sensitive cytoplasmic swelling with a curtain of orange-red pigment. The photoreceptors of some multicellular animals are merely sheets of receptor cells, often backed by a pigmented epithelium (Figure 7–41). These simple sheets of photoreceptors are found in many animal groups but are generally primitive in function. Cup-shaped or vesicular eyes, with narrowed light apertures, provide

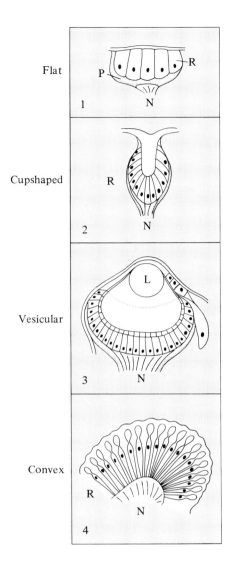

FIGURE 7– 41 Four basic types of animal eyes: flat eyes, cup-shaped eyes, vesicular eyes, and convex (compound) eyes. All examples are for annelid worms. L = lens, P = pigment cells, R = retina, N = optic nerve. Animals are 1 *Nais*; 2, *Ranzania*; 3, *Vanadis*; 4, *Sabella*. *(Modified from Novikoff 1953.)*

considerable directionality of light detection and the possibility of image formation.

The simple cup-shaped eye of the turbellarian worm *Dugesia* has an outer cup of pigment cells and an inner lining of a few sensory nerve endings with rod-shaped structures; the axons exit the cup via the same opening as light enters. This is an **inverse eye**. The eyes of a land planarian have more photoreceptors, with complex sensory endings called retinal clubs (see Figure 7–1B, page 256); many of the photoreceptor cell axons leave the eyecup between the pigment cells rather than through the cup opening. This is a **converse eye**.

Vesicular eyes have a transparent cornea and a lens to focus light on a photoreceptor layer, the **retina**. The eye of the slug *Agriolimax* is a simple vesicular eye with a retina of photoreceptor cells and pigmented cells; the light-sensitive portions of the receptor cells are located above the pigment layer (Figure 7–42A). The accessory retina of the slug eye would not intercept light passing through the lens, but it may detect infrared radiation (Newell and Newell 1968). The alciopid polychete worms also have well-developed eyes (Figure 7–42B). Highly specialized vesicular eyes are also present in other invertebrates (e.g., cephalopods) and vertebrates.

Many vertebrates have a third eye, the pineal or **median eye**; some have only the vestige of this median eye, the pineal gland (Eakin 1973). It is best developed in lampreys and some reptiles (e.g., *Sphenodon* and some other lizards). The pineal eye is part of a complex of pineal organs—the pineal, parapineal, and paraphysis—that develop from the diencephalon region of the brain (see Chapter 8).

The pineal and parapineal organs once may have been paired-eye structures, but the pineal is a single, third eye located in a socket on the forehead of living lower vertebrates and fossil vertebrates. The most specialized pineal eyes (of lizards) have a miniature cornea, lens, and retina, but these (and especially the other less specialized median eyes) probably only detect the presence/absence of light, and do not form clear visual images. The median eye could function for detecting predators or food above the animal, or for sensing circadian light/dark cycles and day length. The pineal of most fish and amphibians, many reptiles, birds, and mammals, is reduced to a glandular structure inside the brain case with no remnants of cornea, lens, or retina. It may retain an endocrine function controlled by photoreception via innervation from the eyes or by light penetrating into the brain.

Many annelids, mollusks, and arthropods have **convex eyes**. These are a radical departure from the cup-shaped or vesicular eyes of other animals. The **ommatidia** (individual photoreceptor units) radiate outwards from a central base; each ommatidium has its own lens, light guide, pigment "light shade," and

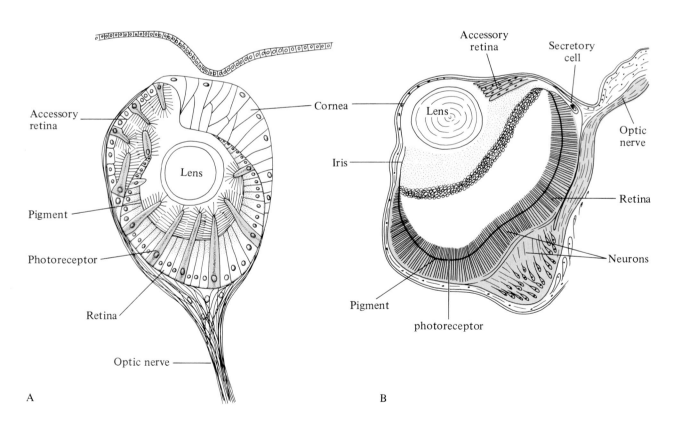

A

B

FIGURE 7–42 **(A)** Simple vesicular eye of a slug. **(B)** Well-developed vesicular eye of a pelagic alciopid polychete. *(Modified from Newell and Newell 1968; Hermans and Eakin 1974.)*

6 to 9 photoreceptor cells (Figure 7–43A, B). The ommatidia often function as independent units. In fact, the eye of a worker ant (*Pomera*) has only one ommatidium. In contrast, the eyes of dragonflies have over 10000 ommatidia.

The optical part of the ommatidium is a cuticular biconvex lens and (often) a crystalline cone. Both structures direct light to the photoreceptor cells. Some types of ommatidia have no crystalline cone (acone) or a gelatinous fluid cone (eucone). The light receptive, **retinular cells** of the ommatidium have a highly folded light-sensitive membrane, the **rhabdomere**. The rhabdomere consists of closely packed microvilli, oriented at right angles to the long axis of the retinular cell and the direction of light passage. The rhabdomeres of adjacent retinular cells may fuse to form a rod-shaped photosensitive unit, the **rhabdome**. The retinular cells of an ommatidium may function as a single photoreceptor unit

or as independent units (see below). The convex eyes of arthropods, although of a radically different design, can have a similar visual sensory capability as the best vesicular invertebrate (e.g., cephalopod) and vertebrate (e.g., bird, mammal) eyes.

Image Formation

Vesicular eyes and convex eyes can form sharp images. Some simple eyes work like a pinhole camera. In a pinhole camera, light enters through a small hole and forms a sharp but low intensity image on the back of the camera; there is no lens, iris, or focusing mechanism, and the image is low intensity because little light enters the narrow pinhole aperture. The eye of the nautilus (a cephalopod mollusk) works as a simple "pinhole camera" eye (Muntz and Raj 1984). The large size of the pupil (1 mm) relative to the depth of the eye (8.7 mm) and photographs

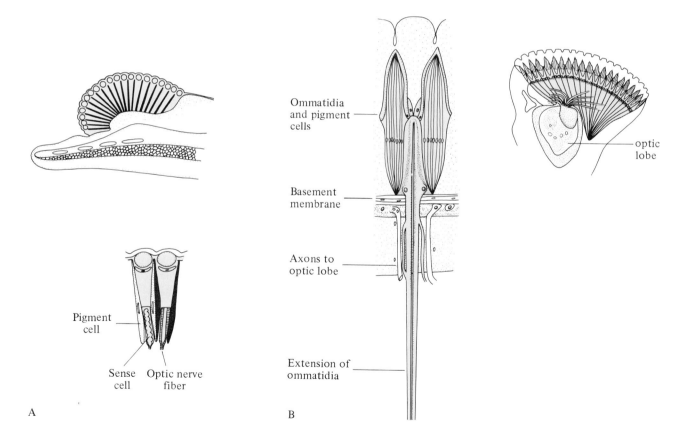

FIGURE 7–43 **(A)** Compound eye of the sabellid worm *Branchiomma* has photoreceptor units arranged quasiradially to provide a remarkably wide field of view. Shown also are details of two individual ommatidia. **(B)** Compound eye of the dipteran insect *Simulium* has elongate photoreceptor cells projecting below the basement membrane and axons passing to the optic lobe. *(Modified from Bullock and Horridge 1965.)*

made using a model of the nautilus eye, suggest a poor quality of its visual image. Calculation of the extent of blurring suggests that the photoreceptors cannot distinguish objects smaller than about 6.5° of arc. The behavioral (optomotor) response of the nautilus also indicates a poor visual acuity (5.5° to 11.25°). However, the spacing of photoreceptors in the retina would indicate a much greater visual acuity, equivalent to that of the octopus (17/60° or 17′).

A lens and an iris considerably improve the image-forming capacity compared to a pinhole camera eye (see Supplement 7–2, page 325). Many invertebrate and vertebrate eyes form images with a high visual acuity. For example, the eyes of vertebrates and advanced cephalopods have a lens to focus light, an iris to regulate the amount of light reaching the photoreceptors, an array of photoreceptors forming a retina in the plane of focus, and a mechanism to control the focus. The image formed on the retina is inverted, i.e., upside down. The convex insect eye contains many ommatidial units and does not use a common lens to focus the light; it does not form an inverted image but rather an upright image.

The image-forming eyes of cephalopods and vertebrates (Figure 7–44A, B) have a well-developed **iris** to adjust the amount of light entering the eye. The iris controls the diameter of the pupil (d), and the amount of light entering the eye is proportional to the area of the pupil, i.e., $\pi d^2/4$. The pupil diameter of the human eye can increase from about 2 mm in bright light to about 8 mm in the dark; the 4 times increase in d causes a 16 times increase (4^2) in the amount of light entering the eye.

The **retina** is a specialized layer of photoreceptors and neurons; it is usually in the plane of focus. The cephalopod retina is a relatively simple epithelium of two main cell types: the photoreceptors and supporting (pigmented) cells. The visual cells are long and cylindrical (2 μ diameter × several hundred μ long) oriented so that the nucleated proximal segment lies on a basement membrane at the outer surface of the retina (inner means towards the center of the eye, outer away from the center of the eye). The rod portion of the photoreceptor contains a photosensory rhabdome; it is located at the inner surface of the retina. Light entering the eye and impinging on the retina first reaches the photoreceptive portion of the retina; this is a **converse eye**. Pigmented supporting cells, and also pigment within the photoreceptor cells (especially the basal nucleated portion), absorb light not intercepted by the rhabdomes. Axons arising from the base of the photoreceptor cells proceed along the outer wall of the eye directly to the optic lobe of the brain.

The vertebrate retina has a very different structure (Figure 7–44B). The outer layer of the retina is a pigmented epithelium. The inner neural layer contains six kinds of nerve cells; from outer to inner they are: (1) photoreceptive rods and cones, (2) horizontal cells, (3) bipolar cells, (4) amacrine cells, (5) interplexiform cells, and (6) ganglion cells. The rods and cones are arranged so that their photoreceptive segments lie next to the outer pigment epithelium. Their nuclear and synaptic regions lie above the outer segments, in the direction from which the light travels. This is an **inverse eye** because the light passes through most of the retina before it reaches the photoreceptors. Horizontal and bipolar cells synapse with the rods and cones in the outer plexiform layer. Bipolar, amacrine, and ganglion cells synapse with each other in the inner plexiform layer. The ganglion cells give rise to the axons that form the optic nerve and convey visual information to the brain. There is a vertical pathway for information transfer through the retina from the rods and cones, via the bipolar cells, to the ganglion cells. The interplexiform neurons receive synaptic input from amacrine cells, and synapse with amacrine cells, horizontal cells, and bipolar cells; their input is only from the inner plexiform layer, but their output is to both the inner and outer plexiform layers. There is also horizontal transfer of information via the amacrine and horizontal cells.

The human retina has about 100 10^6 rods, 6 10^6 cones, and about 10^6 ganglion cells and optic nerve axons. There must therefore be considerable convergence of information from the rods and cones to the less numerous ganglion cells (a ratio of about 100:1 and 6:1). Rods have considerable convergence onto bipolar and ganglion cells, whereas cones have some convergence, but their ratio with bipolar and ganglion cells is closer to 1:1. The vertebrate retina is not just a sensory structure, but is essentially an extension of the central nervous system that also has an interpretative role. It is capable of very sophisticated signal processing of the visual information from the rods and cones (see Chapter 8).

Cup-shaped and vesicular eyes have a single lens that focuses the image on the retina. The range of distances at which objects are simultaneously in focus is the **depth of field**; this depends on the diameter of the iris and the depth of the eye. The focusing mechanisms of an eye must be adjusted if it is to maintain sharp focus for near and distant objects. **Accommodation** is the ability to adjust the focusing mechanisms of the eye so that objects at different distances can be focused clearly (although not at the same time). Accommodation can be accomplished by altering the focal length of the lens, or by either moving the photoreceptor layer to the new focal plane or moving the lens so that the plane of focus remains in the receptor plane, if the lens has a fixed focal length. Convex eyes are always in focus, and do not require an accommodating capability.

Moving the retina to accommodate is generally impracticable, although the eyes of alciopid annelids accommodate in this way. For example, the eye of the polychete *Vanadis* (see Figure 7–42B, page 302) accommodates by changing the fluid volume in one of the optic chambers. Fluid is secreted or reabsorbed to slowly alter the distance between the lens and the photoreceptor layer.

Moving the lens to accommodate is a common mechanism used by many invertebrate and some lower vertebrate eyes. The cephalopod eye accom-

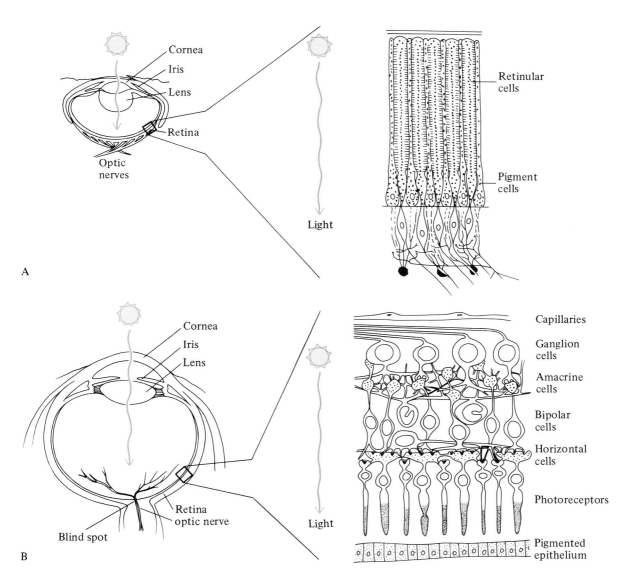

FIGURE 7–44 **(A)** Light passing to the retina of a cephalopod (*Octopus*) eye impinges first upon the photosensitive portions of the retinular cells; the photoreceptor cell nuclei, synapse, and neurons are located below the photoreceptor cells. This is a converse retina. **(B)** Light passing to the retina of a vertebrate (*Homo*) eye travels through the blood vessels, neurons, and synaptic and nuclear segments of the photoreceptor cells before reaching the photosensitive portions of the rods and cones. The retina has a blind spot where blood vessels and axons enter the eye. This is an inverse eye.   *(Modified from Bullock and Horridge 1965; Dowling and Boycott 1966.)*

modates by squeezing the eye to move the lens forward or contracting ciliary muscles to move the lens back towards the retina.

Lizards, birds, and mammals have an alternative, and probably quicker and more effective, means of accommodation. The soft and pliant lens is contained within a transparent capsule that is attached by suspensory ligaments to two sets of ciliary muscles (some circular and some meridional); contraction of these muscles regulates the tension on the suspensory ligaments and the shape of the lens capsule. The lens is more spherical in shape when it is not under tension from the suspensory ligaments, and focuses on near objects. Relaxation of the ciliary muscles increases the tension on the suspensory ligaments and stretches the lens into a less convex shape for focusing on distant objects.

The eye lens of aquatic animals must be more spherical than that of terrestrial animals because light moving from the ambient medium through the cornea is not refracted as strongly in water as in air (Land 1987). Thus, the lens of marine annelids (see Figure 7–42B, page 302), cephalopods (Figure 7–44A), and other aquatic animals are more spherical than those of terrestrial animals, such as the human (Figure 7–44B). However, spherical lenses do not form clear images but distort the image; such **spherical aberration** can be minimized by a gradient in density from the outside of the lens to the center. An optically inhomogeneous spherical lens has evolved independently in aquatic vertebrates, cephalopods, gastropods, alciopid annelids, and a copepod crustacean. Alternatively, a nonspherical lens, with a reduced radius of curvature for the outer surface, can overcome the spherical aberration but this severely restricts the field of view. Nevertheless, a few aquatic animals, such as the copepod *Pontella*, have such asymmetric lenses.

Amphibious animals have a special visual problem; their cornea and lens must be highly convex to focus light on the retina for underwater vision, but in air the visual image would be focused in front of the retina i.e., the eye would be **myopic**. An eye can function well in air and water if its cornea is flat (has no focusing power) or if it has an especially strong accommodation power. Eyes with a flat cornea must have a spherical lens, but they work equally well in air and water. The mudskipper *Periophthalmus* has a slightly flattened cornea and a very convex lens. The clinid fish *Mnierpes* has a flat cornea, but this distorts the peripheral image (Graham and Rosenblatt 1970). The flying fish *Cypselurus* has a cornea with three flat. triangular surfaces; these form a wide, but disjointed image. Some amphibious vertebrates do not have a flat

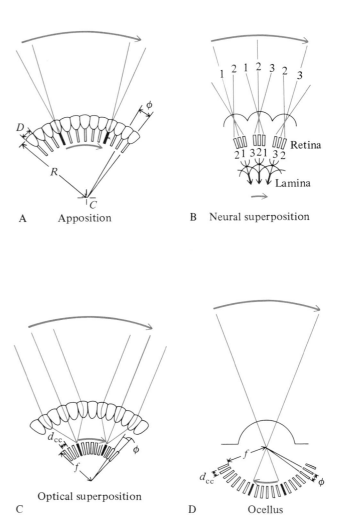

FIGURE 7–45 There are four basic types of insect eye. **(A)** Apposition; each ommatidium of this type of compound eye functions as a separate unit; a noninverted image is formed and the angular resolution of the eye $\phi = D/R$, where D is the ommatidial facet diameter and R is the radius of curvature. **(B)** Neural superposition; rhabdomeres of retinular cells in the compound eye receive light from the same position in the visual field (e.g., 1, 2, 3); no clear optical image is formed, but the pattern of stimulation of individual retinular cells is interpreted as an "image" by the neural organization of the lamina. **(C)** Optical superposition; light from adjacent ommatidia is brought to a focus on photoreceptor cells located deep within the eye; the visual acuity $\phi = d_{cc/f}$, where d_{cc} is the distance between adjacent photoreceptor units and f is the focal length of the ommatidia. **(D)** The ocellus is a simple flat or cup-shaped eye, with a lens that forms an inverted image; the visual acuity is calculated as for the optical superposition eye; $\phi = d_{cc/f}$. A focused larval ocellus is illustrated; adult ocellar eyes are often defocused, i.e., the focused image does not coincide with the plane of the photoreceptor cells. *(From Land 1985.)*

cornea but rely on powerful accommodation to adjust vision in air and water. A few diving birds have a very powerful accommodation. The merganser (*Mergus*) accomplishes this by the powerful action of the ciliary muscles, which squeeze the lens partially through the rigid iris, greatly increasing its radius of curvature. The "four-eyed" fish *Anableps* can simultaneously focus a terrestrial and an aquatic image by having a pyriform (pear-shaped) lens and two separate pupils—the upper pupil forms the aerial image and the lower pupil forms the aquatic image. The amphibious blenny *Dialommus* also has a double pupil, but the openings are side-by-side rather than vertical.

The optics of image formation by the ommatidial eye is straightforward because there is not a common lens through which all light passes and the image is not inverted. Each ommatidium of a simple compound eye has its own lens and detects light in a certain small arc of vision (ϕ). All objects are simultaneously in focus regardless of their distance.

There are four basic types of insect eye. The simplest type is the **apposition eye** (Figure 7–45A). The rhabdome of each ommatidium is located immediately underneath the ommatidial lens and crystalline cone. The retinular cells of each ommatidium act as a single unit and respond to light only from a small arc (ϕ) of the complete field of view (visual field). Diurnal insects typically have apposition eyes (Table 7–3). **Superposition eyes** have a functional overlap of adjacent ommatidia. The retinular cells of an ommatidium, instead of acting as a single functional unit, may be physically and optically isolated from each other (the rhabdome is "open"). The axons of the retinular cells from adjacent ommatidia that "see" the same point in space are reorganized outside the photoreceptor layer to produce not a visual image but a neural image; this is a **neural superposition eye** (Figure 7–45B). The ommatidial retinular cells of **optical superposition eyes** (Figure 7–45C) form a single rhabdome (as in apposition eyes), but each rhabdome receives light from a number of adjacent ommatidial cones. This requires that the photoreceptor cells lie a considerable distance below the ommatidial lenses. The final type of insect eye is the **vesicular ocellus** (Figure 7–45D).

The compound convex eyes of insects do not have an iris or a pupil. However, some insect eyes appear to a close observer to have a dark spot, or **pseudopupil**, which moves over the surface of the eye as the observer moves. The pseudopupil is the

TABLE 7–3

Types of eyes in adult insects. For brevity, only some orders are included. Table indicates whether ocelli are present in the adults. *(Modified from Land 1985.)*

None	Reduced to Single Lens	Apposition	Superposition Neural	Superposition Optical	Ocelli
Thysanura		Thysanura			no
Diplura					no
Protura					no
		Collembola			no
		Ephemeroptera			3
		Odonata			3
		Plecoptera			3
		Orthoptera			Yes[1]
		Phasmida			Yes[1]
		Isoptera			2[1]
	Hymenoptera	Hymenoptera			3
	Mallophaga				no
	Siphunculata				no
		Hemiptera	Hemiptera?		0–3
		Neuroptera?		Neuroptera	0 or 3
		Lepidoptera (butterflies)	Lepidoptera (moths)		2
			Diptera		3
				Coleoptera (nocturnal)	0–2
		Coleoptera (diurnal)			no
Strepsistera ♀	*Strepsistera* ♂				

[1] Present in winged forms only.

dark image cast by the observer on the ommatidial eye.

Many insect eyes can regulate the intensity of light reaching their photoreceptors. Radial movement of pigment in the superposition eye controls how many cuticular facets contribute light to a single receptor cell (Figure 7–46A). An outward movement of pigment to lie between the crystalline cones allows light from many facets to reach a single retinular cell; this is functionally equivalent to pupil dilation in a vesicular eye. An inward movement of pigment prevents light reaching a retinular cell from many facets; this is functionally equivalent to pupil constriction. The light-adapted eye of the moth *Ephestia* reduces the effective "pupil size" to a single facet, but the dark-adapted eye has a "pupil size" of about 180 facets; this is about 10 times the intensity range provided by the iris of the human

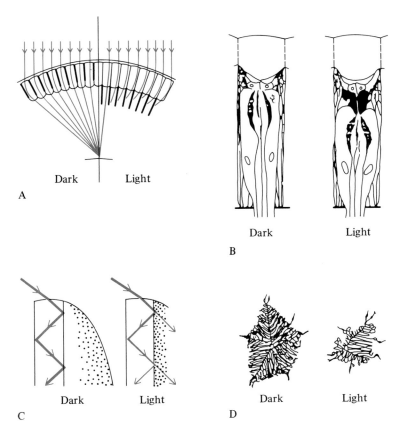

FIGURE 7–46   Light-restricting mechanisms ("pupils") of insect eyes. **(A)** The inward radial movement of pigment in the light-adapted (LA) eye prevents most light reaching the photoreceptor cells from adjacent ommatidia of the optical superposition eye, whereas light from adjacent ommatidia is normally focused on the receptor cell in the dark-adapted (DA) eye. **(B)** Pigment cells at the outer tip of the rhabdom can constrict to reduce the amount of light entering the light-adapted ommatidium or dilate to allow entry of more light into the dark-adapted ommatidium. There may also be gross movements of the pigment cells and retinular cells. **(C)** Light can be "bled" from the wave-guiding rhabdome of the dipteran eye by apposition of material (mitochondria, pigment cells) with a high refractive index to the edge of the rhabdom in the light-adapted eye. **(D)** The amount of photosensitive membrane can be reduced in the light-adapted eye by membrane and photopigment breakdown and increased in the dark-adapted eye by membrane/photopigment synthesis.   *(Modified from Land 1985.)*

eye. Many insects with apposition eyes have a pigment "iris" at the distal end of the rhabdome; contraction of the pigment cells can create a sphincter that reduces the effective diameter of the rhabdome (Figure 7–46B). The pigment cells of the eye of the giant waterbug *Lethocerus* form a 5 μ diameter light path under bright illumination and a 20 μ diameter light path in dim light, i.e., a 16 times sensitivity change. This is not effective for insect eyes with small diameter rhabdoms (e.g., the honeybee with a 1.5 μ diameter rhabdom), and opening the pigment aperture also increases the rhabdom's field of view (from about 3.5° in the light-adapted ommatidium of *Lethocerus* to 90° in the dark-adapted eye).

Another type of "pupil" in the eyes of many diurnal insects reduces light reaching the rhabdome by bringing a high-refractive index material into contact with the rhabdome (Figure 7–46C). This decreases the effectiveness of the rhabdome as a light guide and allows more light to escape from the sides of the rhabdome. The high-refractive index material may be mitochondria (in locust eyes) or pigment granules (butterfly eyes). Finally, the size of the rhabdome itself may be changed in response to light intensity (Figure 7–46D). For example, there is about a 5 times increase in rhabdome cross-sectional area between day and night for the locust eye.

The dorsal ocellus of adult insects can be quite complex in structure; it has a lens and often an extensive retina with up to 10000 receptors, may have an iris, and may have a tapetum to reflect light back into the photoreceptive retina (see below). However, these seemingly "good" eyes are apparently out of focus; the focused image is calculated to lie behind the photoreceptor layer (Land 1987). For example, the receptors of the dorsal ocellus of *Calliphora* are located at 40 to 100 μ behind the lens, but the image is focused about 120 μ behind the lens. The conclusion is that the dorsal ocelli are deliberately defocused, but it is not clear why. Larval insects have ocelli, but not other eyes. Their ocelli, being the only photoreceptors, are often focused, or at least tend to be better focused, than adult ocelli. The ocellus of the larval catepillar *Isia* has a corneal and a crystalline lens, but the plane of focus is behind the photoreceptor layer and a fuzzy image is formed (Dethier 1963). However, the fuzzy image forms somewhere along the photoreceptive portion of the receptor cells regardless of how near or far the object is from the ocellus. This could be regarded as a primitive form of accommodation but is not associated with a sharp image.

An eye can have high sensitivity or high resolution but not both, because of optical constraints on photoreceptor morphology (see Supplement 7–3, page 327). Resolution is increased by having a long focal length lens (f), but sensitivity is inversely proportional to f^2. Resolution can be increased by decreasing the diameter of the photoreceptors, but sensitivity is proportional to diameter^2. The human and honeybee photoreceptors have high resolution compared to the moth eye but lower sensitivity.

Structure and Evolution of Photoreceptors

The photoreceptor cells for dermal light sense have not been identified, although they may be free nerve endings. In contrast, the structures of specialized photoreceptor cells are well known. These photoreceptors have a highly folded membrane, with a high surface area to maximize light interception.

There are two general trends in the evolution of photoreceptor cells of animals that correspond to two ways of increasing membrane surface area. Ciliary photoreceptors have the ciliary membrane highly folded to form lamellae or disks; the cilium may be so specialized that it is unrecognizable except for remnants such as the basal body (Figure 7–47A). Rhabdomeric-type photoreceptors have microvillar elaborations of the cell membrane. A variety of ciliary and rhabdomeric photoreceptors are illustrated in Figure 7–47B. A few animals (e.g., some rotifers and *Branchiostoma*) have a third type of photoreceptor cell, a ganglionic, or diverticular cell, that is not ciliary or rhabdomeric (see Figure 7–1B, page 256).

Protostome animals were originally thought to have rhabdomeric photoreceptors and deuterostome animals to have ciliary photoreceptors (Eakin 1982). However, a number of exceptions and inconsistencies make a strict phylogenetic distribution of these photoreceptor types untenable. Alternate theories are that all photoreceptors, including rhabdomeric ones, are of ciliary origin, and so there is a monophyletic origin of photoreceptors in the animal kingdom (Vanfleteren 1982), or that ciliary, rhabdomeric, and diverticular photoreceptors have independently evolved from a primitive, diffuse photosensitive cell that was monociliate with microvilli (Salvini-Plawen 1982).

Photochemistry

The highly folded membranes of photoreceptor cells contain chromoproteins (protein pigments). The most common is **rhodopsin** (visual purple). Rhodop-

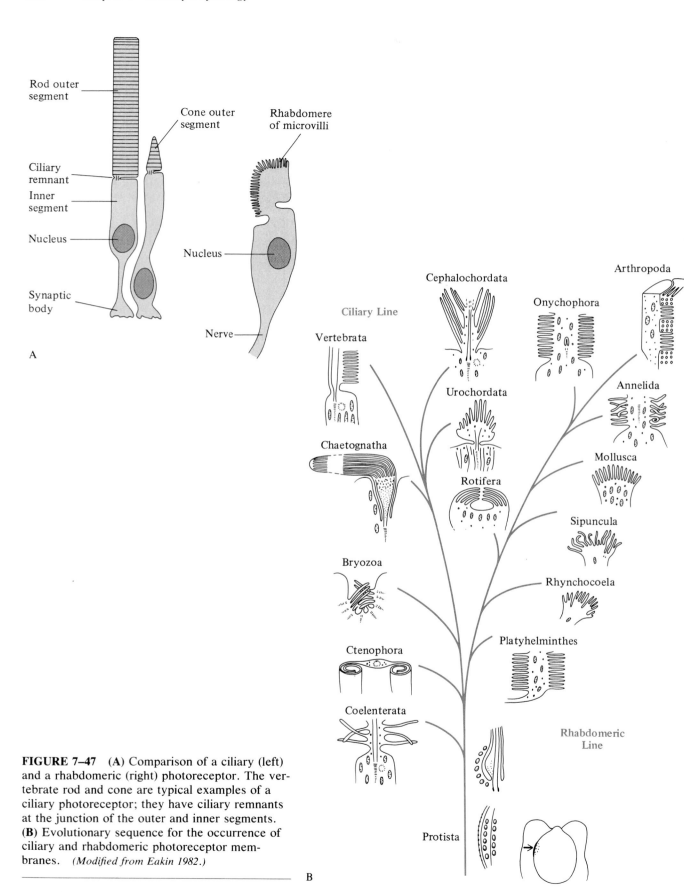

FIGURE 7–47 **(A)** Comparison of a ciliary (left) and a rhabdomeric (right) photoreceptor. The vertebrate rod and cone are typical examples of a ciliary photoreceptor; they have ciliary remnants at the junction of the outer and inner segments. **(B)** Evolutionary sequence for the occurrence of ciliary and rhabdomeric photoreceptor membranes.   *(Modified from Eakin 1982.)*

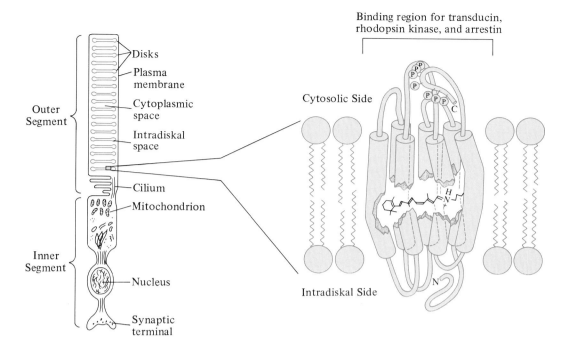

Binding region for transducin, rhodopsin kinase, and arrestin

Outer Segment
- Disks
- Plasma membrane
- Cytoplasmic space
- Intradiskal space
- Cilium

Inner Segment
- Mitochondrion
- Nucleus
- Synaptic terminal

Cytosolic Side

Intradiskal Side

FIGURE 7–48   Representation of rhodopsin in the membrane disk of the mammalian rod. *(From Stryer 1986. Ann. Rev. Neurosci. 9:87–119. Reproduced, with permission, from the Annual Review of Neuroscience Vol. 9, © 1986 by Annual Reviews Inc.)*

sin contains a pigment (11-cis-retinal), a polypeptide (opsin), and two polysaccharide chains. Rhodopsin has a specific organization within the photoreceptor membrane (Figure 7–48). The opsin portion of rhodopsin from various animals has a variable number and sequence of amino acids. **Iodopsins** are the photopigment polypeptides of cone cells; they differ from rhodopsin in their amino acid composition.

Rhodopsin doesn't absorb all wavelengths of visible light equally well, but it has a maximum absorbance at about 500 nm. It has an additional absorbance peak at about 280 nm (UV) due to opsin (Figure 7–49A). The specific wavelength for maximum absorption (λ_{max}) varies from about 350 to 550 nm in different species.

Animal eyes that discriminate different colors contain a number of photoreceptors with different pigments (Figure 7–49B). For example, λ_{max} is 498 nm for rhodopsin of rods of the human eye. There are three types of iodopsins in cones, each with a different maximum absorbance wavelength (λ_{max} = 440, 535, and 575 nm). Other animals, such as the sphingid moth *Deilephila* and the deep-sea fish *Bathylagus* also have multiple rhodopsins with differing λ_{max}. The sphingid moth has a UV-sensitive rhodopsin (λ_{max} = 350 nm) in addition to violet (440

nm) and green (525 nm) sensitive photopigments (Figure 7–49B). The rhodopsins of *Bathylagus* have λ_{max} of 466 and 500 nm. Other deep-sea fish have multiple rhodopsins with λ_{max} up to 540 to 550 nm, which may function to detect either their own bioluminescent red emissions from their suborbital photophores or to detect red crustaceans.

The absorption of photons by a vertebrate rhodopsin molecule initiates a complex sequence of chemical reactions (Figure 7–50A). The rhodopsin is converted to bathyrhodopsin within 2 picoseconds of absorbing a photon, and there is a change in λ_{max} (from 498 to 543 nm for bovine rhodopsin). A further series of reactions (with intermediates lumi-, meta I-, meta II-, meta III-rhodopsin) forms all-trans retinal and opsin. The overall reaction converts rhodopsin to retinal + opsin. The conversion of rhodopsin to all-trans retinal and opsin is often called **bleaching** because free retinal and opsin do not absorb visible light. There are also changes in the structural conformation of the opsin. Rhodopsin is regenerated by the enzymatic isomerization of all-trans retinal to 11-cis retinal; enzymatic regeneration of rhodopsin has a halftime of 5 to 30 min. Rhodopsin can also be regenerated by absorption of further photons by any of the rhodopsin pho-

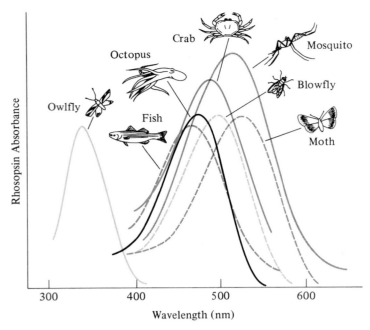

A

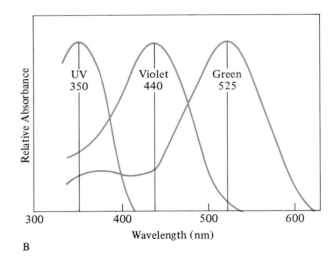

FIGURE 7–49 **(A)** Examples of absorption spectra for rhodopsin from a variety of animals. **(B)** The sphingid moth *Deilephila elpenor* has three color photoreceptor pigments sensitive to UV, violet, and green light. *(Modified from Hoglund, Hamdorf, and Rosner 1973.)*

B

toproducts preceding the splitting of all-trans retinal and opsin. This is **photoregeneration**, but its physiological significance is uncertain. One of the cascading reactions of rhodopsin conversion to the all-trans retinal + opsin changes the membrane properties of the photoreceptor cell. Details of this membrane change are unclear, but it occurs before the retinal is split from the opsin.

The sequence of photochemical reactions is somewhat different for rhodopsin of invertebrates and *Halobacterium* (Figure 7–50B). Invertebrate rhodopsins generally do not bleach and are not converted to all-trans retinal and opsin. Absorption of a photon transforms rhodopsin into all-trans lumirhodopsin and then all-trans metarhodopsin. All-trans metarhodopsin is converted by light to 11-cis metarhodopsin and then rhodopsin. There is an equilibrium under constant illumination between the concentrations of rhodopsin and metarhodopsin, since each is converted to the other by a photon. The proportion of rhodopsin:all-trans metarhodopsin depends on the specific wavelength of illumina-

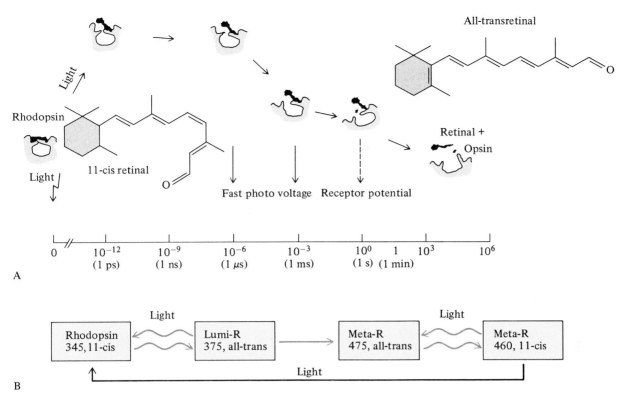

FIGURE 7–50 **(A)** Representation for a vertebrate rhodopsin of the reaction steps following absorption of light. The initial photochemical reaction is followed by a cascade of dark reactions that rapidly produce a fast photovoltage (early receptor potential; at about 1 msec) and a slower receptor potential (at about 1 sec). Possible molecular conformations are shown, with the maximum absorption wavelength indicated for each intermediate. The structures of 11-cis-retinal and all-trans-retinal are also shown. **(B)** Reaction scheme for light activation of invertebrate rhodopsin to lumirhodopsin then all-trans metarhodopsin by photons (λ_{max} = 345 nm). Conversion of all-trans metarhodopsin to 11-cis metarhodopsin then rhodopsin is due to the further absorption of a photon (λ_{max} = 475 nm). There may be, at least in some invertebrates, a slow enzymatic regeneration of rhodopsin that provides an alternative pathway to photoregeneration. *(Modified from Stieve 1983; Hamdorf, Paulsen, and Schwemer 1973.)*

tion, since each photopigment has a different λ_{max} (345 nm and 460 nm, respectively). For example, there is normally about 90% rhodopsin and 10% all-trans metarhodopsin for the owlfly *Ascalaphus* because skylight has a maximum in the blue rather than the UV and the ommatidial pigment cells transmit blue but screen UV. It is the conversion of rhodopsin to all-trans metarhodopsin that initiates the receptor potential. For invertebrates, photoregeneration is the principal means for regeneration of rhodopsin (White 1985). There is evidence for "dark regeneration" of rhodopsin in some invertebrates, presumably by enzymatic reactions analagous to those in vertebrates, or even membrane and rhodopsin synthesis.

Electrophysiology

The resting membrane potential of photoreceptor cells is determined by ion concentration gradients and ion-specific membrane permeabilities; the E_m is primarily a diffusion potential (see Chapter 6). The ion-specific permeability of photoreceptor membranes is due to specialized ion channels and may vary markedly in different regions of the photoreceptor cell. This is important in the transduction of photons into action potentials (see below). Let us first investigate the electrophysiology of invertebrate photoreceptors.

Invertebrate Photoreceptors. The **electroretinogram** (ERG) is the overall electrical response of the eye to illumination. During illumination, the surface of the cornea becomes negative relative to an indifferent electrode due to radial flow of photocurrents. The ERG voltage amplitude can be quite large (e.g., 10 mV) because of the large number of current-generating photoreceptors; underlying neural cells also contribute to the ERG. The form of the ERG can vary dramatically. For example, the ERG of the dragonfly *Agriocnemius* has a simple monophasic ON-hyperpolarization and OFF-repolarization; intracellular recording from a retinular cell (whose rhabdomere is fused with those of adjacent retinular cells) indicates a marked depolarization (Figure 7–51). The eye of a fly (*Lucilia*) has a more complex ERG with a phasic positive ON and a negative OFF response, although the retinular cell depolarization resembles that of the *Agriocnemius* eye.

Stimulation by light causes depolarization of the resting E_m, after a small transient delay of several milliseconds. The peak amplitude of the depolarization is proportional to the stimulus intensity (Figure 7–52A). In addition, a number of small "bumps" in E_m are apparent at low-stimulus intensity. These "bumps" represent depolarization due to the variation in the number of photons impinging on the photoreceptor at any particular time, i.e., there may be 0, 1, 2, etc. photons striking the membrane at any instant. The depolarization amplitude is proportional to the number of photons striking the membrane; a single photon elicits a "$n = 1$" quantum "bump," two photons striking the membrane elicit an (approximately) twice as large "$n = 2$ bump," etc. Bumps also occur in the dark; these "dark bumps" presumably reflect spontaneous thermal (or other) excitation of the photoreceptor membrane in the total absence of photons of light. The photoreceptors of *Locusta* are particularly quiet in the dark; they have only 10 or so "dark bumps" per hour.

The electrical depolarization of photoreceptors by light is due to a change in the membrane ion conductance. The resting E_m of about -40 to -60 mV is similar to the K$^+$ equilibrium potential because the membrane is permeable to K$^+$ ions in the dark. The depolarization by illumination is due, in most invertebrate photoreceptors, to an influx of Na$^+$ ions (this is readily demonstrated by changing the extracellular Na$^+$ concentration or substituting other ions for Na$^+$). The light-activated ion channels are Na$^+$ channels, or at least the channels have a

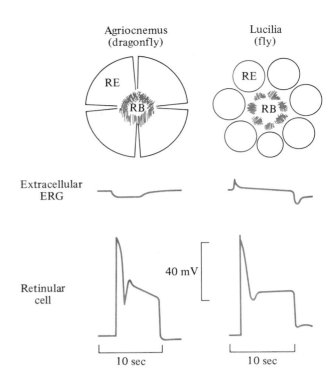

FIGURE 7–51 Comparison of extracellular electroretinogram (ERG) and intracellular depolarization for the eye of a dragonfly *Agriocnemius* and a dipteran fly *Lucilia* to a short (<100 msec) light flash. The retinular cells (RE) of the dragonfly eye have fused rhabdomeres (RB), whereas the retinular cells of the fly are separate. *(Modified from Naka 1961.)*

2:1 ratio of Na$^+$:K$^+$ permeability. The roles of the dark K$^+$ channels and light-activated Na$^+$ channels have been clearly demonstrated for *Limulus* photoreceptors using voltage clamp experiments (Figure 7–52B). In the dark, the current flow elicited by a change in E_m is normally outward, with a reversal potential (E_{dark}) of about -50 mV, i.e., outward K$^+$ flow. There is a nonlinear voltage dependence of K$^+$ current flow on E_m. With illumination, the current flow is normally inward, with a reversal potential (E_{light}) of about $+10$ mV, i.e., inward Na$^+$ flow; there is again a nonlinear voltage dependence of the Na$^+$ current.

Vertebrate Photoreceptors. The vertebrate ERG is a complex summation of the electrical activity of many photoreceptors and neurons. The initial response to illumination is an *a* **wave** due to receptor current of the rods and cones, then a *b* **wave** due to activity of the retinal neurons synaptically stimulated by the photoreceptors, and finally a slow *c*

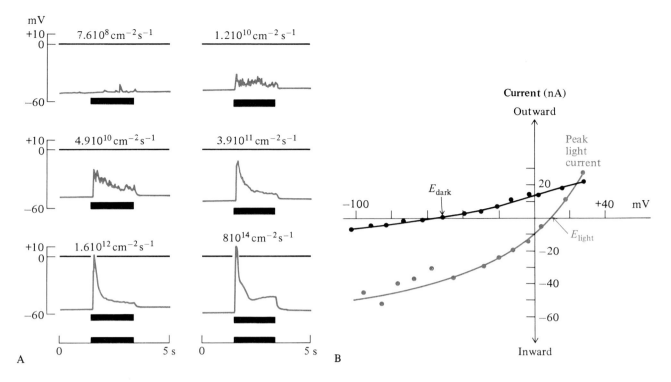

FIGURE 7–52 **(A)** The effect of increasing the intensity of a 2 sec light flash (solid bar) on the membrane potential of the ventral nerve photoreceptor of the horseshoe crab *Limulus*. The intensity of each flash (*I*) is given as the number of photons $cm^{-2} sec^{-1}$ at 550 nm wavelength. At low light intensity, individual "bumps" are clearly apparent. **(B)** The current-voltage relationship measured for the ventral nerve photoreceptor of *Limulus* in the dark (black line) and under constant illumination (colored line) shows that the dark-adapted eye has a negative reversal potential (E_{dark}) and the light-adapted eye has a positive reversal potential (E_{light}). *(From Stieve 1983; modified from Millechia and Mauro 1969.)*

---

wave due to extra-retinal pigment cells that contact the outer segment of the photoreceptors (Figure 7–53A). In addition, there is an exceedingly fast, low amplitude **early receptor potential** (ERP) that is due to charge movement within the photoreceptor membrane and is therefore a step in the phototransduction process rather than a part of the receptor potential.

Vertebrate photoreceptor cells are strikingly different in electrophysiology to invertebrate photoreceptor cells. The outer segment of, for example, a rod is the photoreceptive portion packed with membrane disks and rhodopsin. In the absence of illumination, the outer membrane of the rod has a membrane potential of about −20 mV because it has a high Na^+ permeability. The membrane of the inner segment, in contrast, is more permeable to K^+, like most other resting membranes. There is,

in the dark, a continual inward Na^+ current flow at the outer segment that tends to depolarize the E_m and an outward K^+ current flow at the inner segment that tends to hyperpolarize the E_m; the resultant E_m is about −20 mV. There is an intracellular current flow from the outer segment to the inner segment and an extracellular current flow that completes the circuit. The synaptic end of the rod functions in a similar fashion as the inner segment. The inward Na^+ and outward K^+ current flow requires the continual operation of the Na^+-K^+ pump and expenditure of ATP. When illuminated, the Na^+ conductance of the outer segment membrane decreases, thereby reducing the depolarizing Na^+ inward current and hyperpolarizing the E_m. This is the opposite electrical response to illumination that occurs in invertebrate photoreceptor cells. In the dark or at low light levels, there are small fluctuations in the

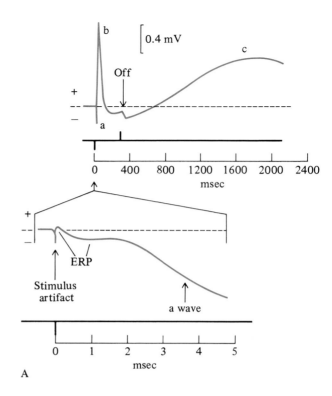

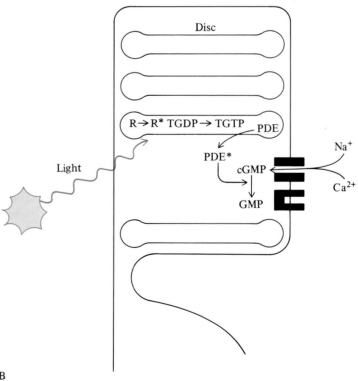

FIGURE 7–53 **(A)** Electroretinogram (ERG) of a cat eye, measured for a light-adapted eye with a recording electrode in the vitreous humor. The ERG is comprised of a rapid negative *a* wave, a positive *b* wave, and a slow (after stimulus OFF) *c* wave. The early receptor potential (ERP) occurs at 1 to 2 msec after stimulation and is the direct electrical response of the photoreceptor pigment/membrane to light stimulation. **(B)** Schematic representation of phototransduction in the vertebrate rod. Rhodopsin in the disk membrane absorbs photons and activates transducin (a G protein) and then phosphodiesterase (PDE), which converts 3,5 cyclic GMP to 5 GMP. The lowered intracellular concentration of cGMP closes Na^+ channels in the outer rod membrane. *(Modified from Brown 1968; Stryer 1986; Attwell 1985.)*

resting E_m of the vertebrate photoreceptor cells. These "bumps" (or rather "antibumps") resemble those observed for invertebrate photoreceptors, but are hyperpolarizing rather than depolarizing. They are caused by random, thermal noise and by the absorption of single photons; the conductance change during each antibump is 20 to 30 picoSiemens and reduces the total outer segment membrane ion conductance by about 5%; this conductance change is probably equivalent to the closing of about 100 Na^+ channels.

In the rods, the absorption of light usually occurs in the disk membrane, not the outer plasma membrane of the outer segment, although it is the Na^+ conductance of the outer plasma membrane that is decreased. The membrane response to absorption of a photon must therefore be transmitted over a short distance (a few microns from the disk lamellae) to the site of the Na^+ channels. The mechanism of phototransduction, i.e., how absorption of a photon by a rhodopsin molecule opens Na^+ channels has only recently been elucidated (Figure 7–53B). In rods, rhodopsin in the disk membrane absorbs photons and this activates a GTP-binding protein (transferrin, a G protein) that in turn activates a phosphodiesterase enzyme (PDE). The PDE hydrolyzes 3,5 cyclic GMP to 5-GMP. The lowered 3,5 cGMP concentration closes ion channels of the rod outer cell membrane thereby reducing the Na^+ current and hyperpolarizing the rod (Stryer 1986; Gilman 1987).

Thermal noise may preclude photoreceptors from detecting single photons. If so, we would expect animals with a low body temperature to have a better (lower) threshold of light detection than animals with a high body temperature. There is, in fact, such a relationship between the absolute visual threshold and the rate of thermal isomerization of rhodopsin for amphibians and humans (Aho et al. 1988).

Color Vision

Representatives of all vertebrate classes, and many invertebrates, discriminate colors. A visual pigment, such as rhodopsin, has a specific spectral absorbance; it maximally absorbs light of a particular wavelength and absorbs lesser amounts of different λ. A single photopigment cannot discriminate wavelength (i.e., detect color) because of the compounding effect of light intensity on photoreceptor response. For example, the absorbance by a photoreceptor of photons at λ_{max} for a given light intensity can be equal to the absorbance of higher intensity light at $\lambda \neq \lambda_{max}$ (Figure 7–54A). The receptor

cannot discriminate between the lower intensity of λ_{max} light and the higher intensity of $\lambda \neq \lambda_{max}$ light. At least two different photopigments, differing in spectral absorbance, are required to discriminate between different wavelengths of light (Figure 7–54B).

How many different photopigments are required to discriminate color? The human eye can distinguish about 1500 "color hues," from $\lambda = 400$ nm (blue) to 700 nm (red), i.e., differences of 0.2 to 0.3 nm can be discriminated. However, we do not have 1500 different types of color photopigments, but only three with λ_{max} of 445 (blue cone), 535 (green cone), and 570 (yellow cone); there is also the rhodopsin of rods, with λ_{max} of 498 nm. The three color photopigments enable extremely precise color discrimination.

Color vision is based on the differential spectral sensitivity of multiple photoreceptor pigments. The three cone pigments of humans form the basis of **trichromatic color vision** (Figure 7–54C). Pure blue, green, or red light will maximally stimulate the blue, green, or red cones. Other wavelengths of light will have submaximal stimulatory effects on one or more of the cones. For example, orange light (580 nm) markedly stimulates red cones (e.g., 99% of maximal) and submaximally stimulates the green cones (e.g., 42% of maximum); it does not affect the blue cones (e.g., 0%). This ratio of cone response to orange light, 0:42:99 (blue:green:red), defines the sensation of orange light. Other wavelengths of light would cause differing ratios of response. A red and green light shining on the retina is interpreted as a shade of yellow, since the ratio 0:100:100 corresponds not to red (0:0:100) or green (0:100:0) but to yellow. Thus, we interpret color from the ratio of response of blue, green, and red cones. Three photoreceptor pigments can interpret a wide range of colors, given the appropriate central nervous system processing of the photoreceptor information (see Chapter 8).

The specific λ_{max} for photopigments varies; some animals have only two photopigments and presumably poorer color discrimination, whereas others appear to have four or more photopigments. There are at least three to four spectral types of cone photoreceptors in reptiles and birds. Some fish have five specific types of photoreceptors, some flies have five types, and some butterflies also have pentachromatic vision. The mantis shrimp *Pseudosquilla* has a retina with many types of photoreceptors (10 or more) due to a variety of photopigments and light filters (Cronin and Marshall 1989). Different photopigments are present in different photoreceptor cells. It is difficult to imagine how a

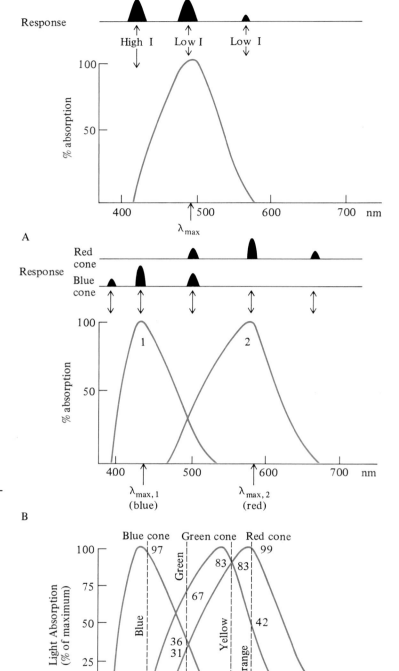

FIGURE 7–54 **(A)** A single photoreceptor pigment (e.g., human rod rhodopsin) cannot discriminate a low-intensity light flash of $\lambda = \lambda_{max}$ from a higher-intensity flash of $\lambda < \lambda_{max}$. **(B)** Two photoreceptor pigments (e.g., human blue and red cones) can discriminate between equal intensity pulses of light from about 400 to 700 nm. **(C)** Trichromatic color vision in the human retina depends on the relative ratio of stimulation by light of three cone photoreceptors (blue, green, and red cones). Indicated on the figure for blue, green, yellow, and orange light is the approximate percentage of maximal stimulation by that wavelength of each type of cone, e.g., green light has a ratio of 36:67:31 stimulation of blue:green:red cones.

photoreceptor cell containing more than one type of photopigment could itself discriminate between one pigment absorbing a photon at its λ_{max} and another pigment absorbing a photon at its λ_{max}. Nevertheless, a few invertebrate photoreceptors contain more than one photopigment.

The cones of some vertebrates (some amphibians, reptiles, and birds) contain small oil droplets of varying color (Figure 7–55A). The oil drops act as high-pass filters, allowing light of shorter λ than $\lambda_{50\%}$ to pass and block light with λ greater than $\lambda_{50\%}$ (Bowman 1980); $\lambda_{50\%}$ is the wavelength at which

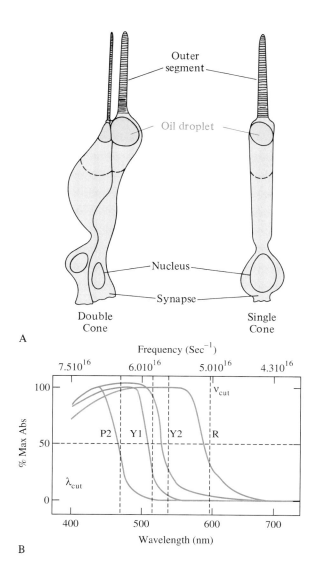

A

B

FIGURE 7–55 **(A)** Diagram of avian single
and double cones with oil droplets. The double cone
has a second accessory cone. **(B)** Cutoff frequency (v_{cut})
and cutoff wavelength (λ_{cut}) and 50% transmission frequency (v_{mid}) and 50% transmission wavelength (λ_{mid})
for various cone oil droplets from the mute swan (P2 =
pale 2; Y1 = yellow 1; Y2 = yellow 2; R = red).
(Modified from Bowmaker 1980; Partridge 1989.)

the wavelength sensitivity of cones, thereby increasing their spectral specificity.

The eyes of the grasshopper *Phlaeoba* contain
only a single rhodopsin (λ_{max} = 525 nm) but the
animal can distinguish green from yellow; they are
attracted to green (White 1985). This grasshopper's
eye is divided into clear and brown portions; the
ommatidia of the brown portion contain an accessory pigment (λ_{max} = 500 nm). The receptors in the
clear region respond maximally to λ = 525 nm but
those in the brown portion respond best to λ = 545
nm, presumably because of the selective filtering by
the accessory pigment.

The deep-water fish, *Malacosteus*, has a visual
pigment that appears to detect its own bioluminescence (Crescitelli 1989). Retinal extracts have two
photopigments, with a λ_{max} of 556 and 514 nm.

Polarized Light

Light emitted from the sun is not polarized but
passage through various media (certain crystals,
glass, plastics) and the earth's atmosphere can
polarize sunlight so that the wave motion is confined
to a specific plane, i.e., is **plane polarized**. The
polarizing effect of the atmosphere is most pronounced at right angles to the path of the sun's rays,
i.e., looking vertically up at sunrise or sunset yields
distinct patterns in both angle of polarization and
percentage of light that is polarized. A number of
crustaceans, insects, cephalopods, and teleost fish
can orient with respect to this polarized light. The
crayfish *Procambarus* can detect the plane of light
polarization. The microvilli of the adjacent retinular
cells interdigitate at 90° to adjacent microvilli in
a highly organized pattern (Figure 7–56A). The
absorption of polarized light is greatest when the
plane of wave motion coincides with the long axis
of the microvilli (Figure 7–56B). The UV-sensitive
retinular cells of ants and bees can detect the
plane of polarization. These retinular cells and their
rhabdomeres are spirally twisted through about 40°
(by about 1° per micron length); this makes them
responsive to the plane of polarization. Various
desert ants are able to navigate with incredible
accuracy by using the plane of polarized light.

Photopic and Scotopic Vision

Most animals are generally adapted to being active
by day (diurnal) or by night (nocturnal). Diurnal
animals live in a very different optical environment
from nocturnal animals. Eyes are morphologically
and physiologically adapted in different ways to the
high or low light levels that characterize their
environment. **Photopic** (daylight) eyes operate at

50% of the incident light is transmitted. Clear oil
droplets pass most wavelengths, including ultraviolet (to which birds are sensitive); A, C, and red
droplets have $\lambda_{50\%}$ of 473, 570, and 610 nm, respectively; B droplets have a variable $\lambda_{50\%}$, from 522 to
554 nm. Diurnal birds have at least four visual
pigments, and so the different oil droplets are
not required for color vision. Rather, the droplets
probably minimize chromatic aberration and narrow

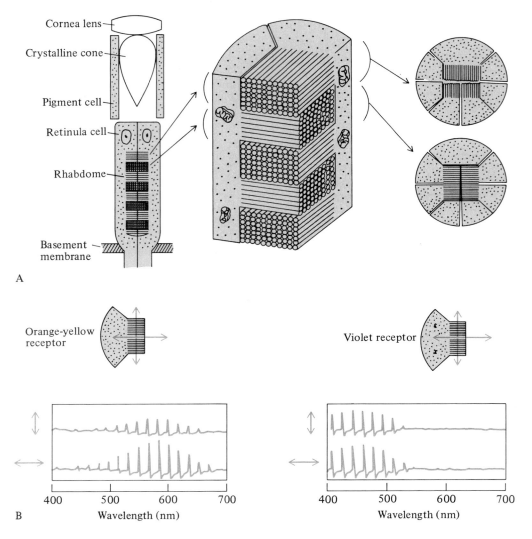

FIGURE 7–56 **(A)** Representation of an ommatidium of the eye of the crayfish *Procambarus clarkii* showing the detailed arrangement of microvilli forming the rhabdome. There are alternating layers of rhabdomeres, arranged perpendicularly, that form the rhabdome. For example, rhabdomeres of retinular cells 1, 4, and 5 (upper right) alternate with rhabdomeres of retinular cells 2, 3, 6, and 7 (lower right). **(B)** Intracellular electrical recording from two types of photoreceptor cells (orange-yellow and violet) of the crayfish indicate that they are more sensitive to light polarized parallel to the microvilli than perpendicular to the microvilli. (The light is traveling in the direction perpendicular to the plane of the diagram.) *(From Eguchi 1965; modified from Waterman and Fernandez 1970.)*

high light intensity; threshold detection is not limiting and visual acuity and color vision can be maximized. **Scotopic** (dark) eyes must minimize the threshold of light detection at the expense of acuity and color vision.

The photopic vertebrate eye typically has a high density of cones. Cones are thicker than rods, with an elliptical surface shape; these morphological properties appear to converge light from a particular direction on their visual pigments (this is sometimes aided by the inclusion of a tiny oil droplet in the cone). Rods are better adapted for capturing diffuse light. Cones also have a more direct spatial transfer of information to the central nervous system than the more diffuse spatial summation of rods. Thus, cones are better suited for high visual acuity than are rods, which are better suited for detection of low light levels.

The distribution of cones in the vertebrate retina is often highly concentrated to one (or sometimes two) region, the **fovea**. The fovea of lower vertebrates is a retinal thickening with a high density of cones. The fovea of vertebrates with high visual acuity has a thin retinal layer; the network of nerves normally overlying the photoreceptors are moved to the side of the fovea so that light impinges directly on the photoreceptors. There is an accumulation of a yellow pigment in the fovea of some primates, hence it is called the **macula lutea** (yellow spot); the yellow pigment corrects for chromatic aberration. The foveae are often associated with binocular vision for predation (e.g., the temporal foveae of hawks) or manipulative skills (e.g., the fovea centralis of primates). Many vertebrates have monocular foveae (e.g., some birds and fish) or visual streaks that correspond to the projection of the horizon on the retina.

Diurnal birds have the highest visual acuity of vertebrates; cone density may approach 10^6 mm^{-2} (cf. 1.45 10^5 mm^{-2} in the human fovea) and there is a temporal as well as a central fovea (Meyer 1986). The avian retina has no blood vessels for nutrient supply. Rather, nutrients are supplied by capillaries in the choroid layer of the eye and by a peculiar intraocular structure, the **pecten**. The pecten is an accordion-like, pleated structure of capillaries and pigment cells arising from the site of exit of the optic nerve from the eye, i.e., the blind spot. The size and number of pleats in the pecten is related not to eye size, but to the reliance on vision of the bird. Active diurnal birds have a large pecten; nocturnal species (and primitive birds) have a smaller, simple pecten. The major function of the pecten is to provide nutrients to the retinal cells since the retina is completely avascular. The pecten not only secretes nutrients but stirs the ocular fluid, facilitating nutrient transport (Pettigrew, Wallman, and Wildsoet 1990). Birds have rapid (saccadic) eye movements, every 0.5 to 40 sec, consisting of up to 13 oscillations at 15 to 30 sec^{-1}. The pecten acts as an agitator during the saccadic movements, propelling its perfusate toward the central retina. Other functions have also been proposed for the pecten, including non-nutritive secretion, light absorption to prevent internal light reflection, casting of a shadow onto the retina to enhance movement perception, and regulation of intraocular fluid pressure.

The insect apposition eye is best suited as a photopic eye and is found in mainly diurnal insects. Each ommatidium is optically isolated from its neighbors, has a fused rhabdome, and often has receptors specialized for color discrimination (having up to four pigments, e.g., some ants and butterflies), UV detection, and polarized light detection. The neural supposition eye of dipteran flies (which

are also diurnal) is also a high-acuity eye and often has a specialized region, the "invertebrate fovea." These foveal areas have enlarged facets (i.e., can collect more light per ommatidium) and a decreased radius of curvature (i.e., greater visual acuity). This foveation is often restricted to males and may be a moderate enlargement of the eye, as in some syrphid flies, or extreme enlargement of the eye, such as the "turbanate" eyes of *Chloean*, an ephemeropteran. The male syrphid hover-fly *Syritta* has enlarged facets (up to 40 μ between adjacent centers) compared to normal facets (16 to 20 μ); the larger-facet ommatidia are also arranged so that the interommatidial angle is as low as 0.5° compared to the normal value of about 1.5° (Collett and Land 1975). The male hover-flies therefore have a high visual acuity within the forward visual field of 0° to 5° on each side of the midline. This is important for visual tracking and flight pursuit of female flies for mating. Males pursue another fly until it lands, then the male accelerates rapidly towards the motionless fly, turns to be in the copulatory position just prior to landing, and mates with the motionless fly regardless of whether it is female or male!

Crepuscular animals (active at dawn and dusk) and nocturnal animals typically have eyes adapted to low light levels, i.e., they have maximal sensitivity. As a consequence, their eyes have low visual acuity and poor or no color vision. Vertebrate scotopic eyes often have elongate rod photoreceptors, many of which converge their sensory information onto a single ganglion cell. The retina of bats, for example, may have up to 1000 rods converging onto a single ganglion cell. This results in an exceptionally high sensitivity, but poor acuity.

The "eyeless" shrimp *Rimicaris exoculata* is a deep-sea shrimp, found around the hot (350°) hydrothermal "black smoker" vents of the mid-Atlantic ridge. It lacks the usual eyes of shrimp, but has an unusual thoracic "eye" that is well adapted for vision in dim light (Van Dover et al. 1989). Its main role might be to detect the dim, black-body radiation emitted by the 350° vent water, rather than the dim light emitted by bioluminescent organisms (Pelli and Chamberlain 1989).

Vertebrate scotopic eyes may also have a **tapetum**, or reflecting layer, behind the retina to reflect back any photons of light that are not initially absorbed by the photoreceptors. The tapetum can be formed from guanine or riboflavine crystals, white collagenous fibers, or minute spheres of triglyceride. The eyes of many insects, both diurnal and nocturnal, also have a tapetum to increase photoreceptor sensitivity. The butterfly tapetum reflects light that passes through the rhabdome, thus effectively doubling the rhabdome length. The

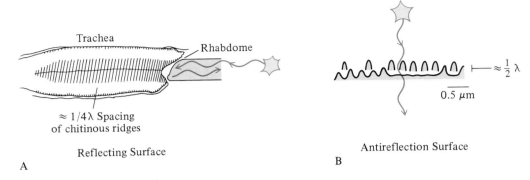

FIGURE 7–57 The optical reflecting surface of the tapetum of a butterfly eye (**A**) has a ¼ λ spacing of optically different surfaces (chitin and air) whereas the antireflecting surface of a moth eye (**B**) has a ½ λ thick layer of small, conical projections. *(Modified from Miller 1979; Miller, Mollner, and Bernhard 1966; Land 1985.)*

tapetum of diurnal moths surrounds each rhabdome and presumably entraps light within the individual ommatidia. The corneal layers of many diurnal insects are mirrored and produce colored facets or eye stripes, perhaps for display or camouflage. The tapetum of many eyes, and various other structures, such as fish scales, the swim-bladder wall, and some bird feathers and butterfly wings, are excellent mirrors. Their high reflectance is based on the 1/4 wavelength spacing of alternating lamellae of high and low refractive index (Figure 7–57A). A small proportion of incident light is reflected at each optical surface (for an air-chitin interface, 4 to 5% is reflected). The proportion of reflected light is maximized (up to 17% for an air-chitin interface) if the optical layer is 1/4 λ thick. Only seven alternating layers of high and low refractive index (each 1/4 λ thick) are needed to reflect more than 99% of the incident light.

Some biological surfaces are modified to minimize the reflectance of light. Reflection can be reduced by coating the surface with a layer 1/2 λ thick of a material whose refractive index is intermediate between that of the two materials. The moth eye has a coating that reduces reflection from about 4% (e.g., for a fly or bee eye) to about 0.08% (Miller 1979). This almost nonreflecting property is not due to a 1/2 λ thick special coating but to a dense array of small conical projections on the surface of the cornea, each about 200 nm high and 200 nm apart (Figure 7–57B). These projections act like a 1/2 λ coating, but their reflective property is not so dependent on the specific wavelength of light. Their function may be to reduce internal reflection of light within the eye, rather than to maximize the entry of light into the eye.

Summary

Sensory cells transduce specific kinds of sensory information into patterns of action potential discharge that are transmitted to the central nervous system for analysis and interpretation. The receptor cells generally respond to only one kind of sensory modality, e.g., chemicals, mechanical deformation, electric field, magnetic field, light, or temperature.

The transduction process at the sensory membrane may be a physical deformation of the membrane, a temperature-sensitive biochemical process, a voltage-dependent ion channel, or a receptor-mediated cascade of biochemical reactions. Transduction depolarizes (or hyperpolarizes) the sensory membrane; this receptor potential induces a generator potential in a neural membrane that encodes the stimulus response into an action potential discharge.

There generally is a logarithmic relationship between stimulus intensity and sensory response. This maximizes the sensitivity of the sensory receptor at low stimulus intensity and provides a wide range of response to varying stimulus intensity. Many sensory receptors adapt, or accommodate, to a sustained stimulus. Phasic receptors rapidly adapt, whereas tonic receptors adapt slowly or not at all.

Receptor sensitivity at low stimulus intensity can be enhanced by signal averaging many separate receptor responses or many successive stimuli. The central nervous system often is able to modify the receptive sensitivity via an efferent innervation to the receptor cells. Sensory sensitivity can also be modified by epithelial cell regulation of the extracellular environment of the sensory receptor cells.

Chemoreception (taste and smell) is a widespread sensory modality. Olfactant and taste molecules bind with specific membrane receptor proteins that control the ionic permeability of the sensory membrane via a G protein, adenyl cyclase, and the second messenger, cyclic AMP (and also a protein kinase for taste). Some chemosensory cells are sufficiently sensitive to detect single odorant molecules. Hygroreceptors respond to relative humidity; there are "dry" and "moist" hygroreceptors.

Thermoreception is probably transduced by thermally-dependent biochemical processes, e.g., the Na^+-K^+ ATPase or ion channels are temperature sensitive. There are hot and cold thermoreceptors and also hot and cold pain nociceptors. Infrared detection by some snakes and invertebrates is accomplished by sensitive temperature detection of infrared radiation, rather than by photoreception.

Mechanoreceptors transduce physical stimulation (bending, stretching) into sensory information. The transduction process may involve physical distortion of ion channels or deformation of specialized receptor proteins that modulate ion channel permeability. Invertebrate mechanoreceptors are diverse in structure and function, e.g., proprioception (body position), rheoception (water current detection), audition (sound reception), vibration detection, pressure detection, gravity, and equilibrium. Vertebrate mechanoreceptors are hair cells, which have similarly diverse functions. The tetrapod ear is a highly specialized mechanosensory organ that detects static and dynamic equilibrium and provides precise frequency and intensity analysis of sound. Some terrestrial and aquatic mammals and birds are able to echolocate by hearing the echoes from distant objects of high-frequency sound pulses which they emit.

Vertebrate electroreceptors are specialized hair cells that transduce exogenous current flow into sensory information. Electroreceptors can detect extremely small electrical gradients, e.g., 0.01 μV cm^{-1}. Fish use electroreceptors for prey location, electrolocation of objects in their endogenously produced electric field, and for interspecific communication.

Some animals can orient to magnetic fields but the mechanisms for magnetoreception are unclear. Some marine fish may detect magnetic fields by electrodetection of electrical currents induced by their swimming motion.

Photoreception of visible light (400 to 700 nm) is a widespread sensory modality. Photoreceptors may be photosensitive nerve endings (general dermal light sense) or specialized rhabdomeric or ciliary photosensory cells. Photosensitive cells contain a photopigment, rhodopsin, or various similar photopigments of differing spectral photosensitivity. Photoreceptor cells are often arranged into visual organs, e.g., eyespots, vesicular eyes, and convex eyes. Vesicular eyes are invaginated structures with a transparent cornea and a lens to focus light on the photoreceptive retina. Formation of a clear image of near and distant objects requires the control of the focusing power of the lens or the distance between the lens and the retina, i.e., accommodation. Convex eyes consist of numerous independent photoreceptor units (ommatidia) arranged in a cluster; each ommatidium responds to a small part of the visual field. These eyes do not require a lens or a means of accommodation for near and far vision. Photoreceptors contain rhodopsin, which absorbs photons of light and activates a G protein, transferrin. Transferrin activates phosphodiesterase, which controls the intracellular concentration of cyclic GMP; this in turn regulates the permeability of ion channels. Color vision requires multiple photoreceptive pigments with differing spectral sensitivities. The trichromatic visual system of humans has three color photoreceptors: blue, green, and red cones. These photoreceptors and complex neural processing enable the discrimination of about 1500 hues of color. Some vertebrates and invertebrates are sensitive to UV light, or polarized light.

Supplement 7–1

Electrophysiological Recording from Sensory Receptors

Electrophysiological recording of the response of chemoreceptors has employed two basic methods for measuring the electrical response to stimulation. With the electrode-tip method, a micropipette containing a test solution is placed over the end pore of a taste receptor. The solution itself is used as the recording electrode (hence the solution must be conductive and therefore contain electrolytes; a reference electrode is placed in the tissue underlying the receptor. The side-wall cracking technique uses a test solution in a microelectrode to stimulate the receptor, but

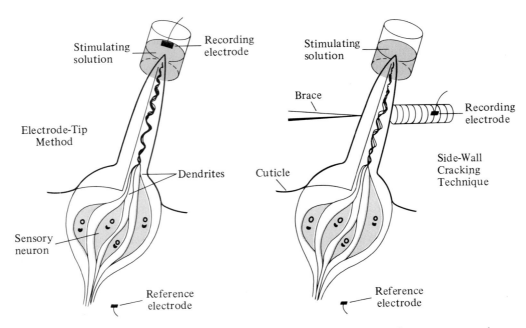

Two techniques for stimulating and recording the electrical responses of taste receptors in insects involve the use of the stimulating solution as the recording electrode (left) of fracturing the wall of the receptor hair to record electrical activity of sensory neuron dendrites (right). *(Modified from Wolbartsht 1965.)*

a separate recording electrode is pushed against the side of the receptor and contacts the receptor membrane through fractures in the exocuticle. These procedures record a general depolarization of the tip/wall of the receptor with respect to the reference electrode and action potentials that are initiated in the sensory neuron; these are superimposed on the general depolarization. This complex electrical signal is a consequence of the manner of recording; the electrode is not intracellular but extracellular, and so action potentials are recorded as biphasic signals superimposed on any DC electrical changes.

The response of antennal chemosensory cells of a male silk moth *Bombyx* to repeated stimulation by puffs of air containing the pheromone bombykol (E-6, Z-11 hexadecadienyl acetate) shows the generalized DC depolarization with superimposed action potentials; this is an electroantennogram (EAG). The upper trace is the 0 volt baseline; the center trace is the DC signal of receptor potential with superimposed AC action potentials of efferent neurons, recorded by the electrode tip method; the bottom trace is the stimulus marker for puffs of air.

The electro-olfactogram (EOG) of the vertebrate olfactory mucosa can be readily measured using a microelectrode placed in the surface mucus layer of the epithelium with an indifferent electrode in the tissue underlying the epithelium. The EOG is typically a long duration (e.g., 1 to 10 sec) depolarization. Action potential responses of individual sensory receptors can be measured using microelectrodes placed within the epithelium near the sensory cell bodies and efferent axons. Simultaneous recording of the EOG and individual sensory receptors can demonstrate many important aspects of sensory function, e.g., the relationship between stimulus intensity and magnitude of EOG depolarization, action potential

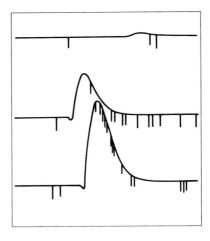

Electro-olfactogram recording from the olfactory epithelium of a frog shows a depolarization and action potentials in response to olfactory stimulation. *(From Gesteland et al. 1963.)*

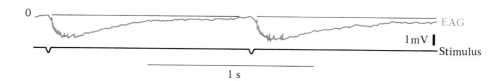

Electroantennogram (EAG) recording of the moth *Bombyx mori* shows an electrical depolarizing response to puffs of the pheromone, bombykol. *(From Kaissling 1986.)*

frequency, adaptation, and specialization of receptor cells for certain odors.

Other examples of extracellular recording of sensory receptor activity include the electrogustatogram (EGG) for taste chemoreceptors and electroretinogram (ERG) for photoreceptors. (*See Wolbarsht 1965; Kaissling 1986; Dodd and Squirrel 1980; Gesteland et al. 1963.*)

Supplement 7–2

Lenses, Image Formation, and Visual Resolution

A convex lens converges parallel rays of light (e.g., from a distant object) onto a single point; the distance from the midline of the lens to the point of focus is the focal length. Light rays from a point source close to a convex lens are also focused to a spot behind the lens, but the focal length is longer than for parallel rays of light. This difference in focal length for near and distant objects is inconvenient for an eye or camera, since it requires the retina (or film) to move relative to the lens when focusing between distant and near objects.

A lens can focus light because it has a different refractive index than the medium. The refractive index of a material is the ratio of the velocity of light in the material to the velocity of light through a vacuum. Light

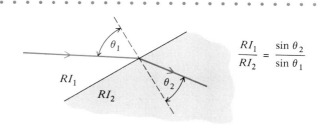

$$\frac{RI_1}{RI_2} = \frac{\sin \theta_2}{\sin \theta_1}$$

continues to travel in a straight line when it passes between two media of differing refractive index, only if it passes perpendicularly to the interface. Light is refracted, or bent, from its original direction if it strikes the

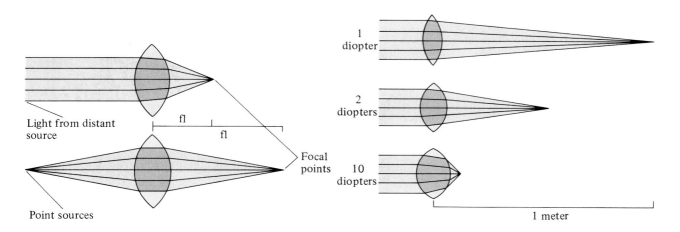

The focusing power of lenses. Fl = focal length.

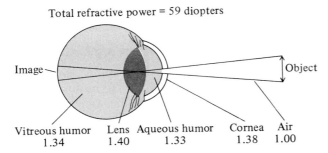

Total refractive power = 59 diopters

Vitreous humor 1.34	Lens 1.40	Aqueous humor 1.33	Cornea 1.38	Air 1.00

The different parts of the human eye that contribute to its focusing power. Values are refractive index.

interface at an angle other than 90°. The angle of refraction depends on the ratio of the refractive indices for the two media and the angle between the light path and the interface boundary (see figure). A thick convex lens has greater focusing power than a thin convex lens. The focusing power of a lens is measured in diopters, the reciprocal of the focal length (in meters). A 1 diopter lens focuses parallel light at 1 meter behind the lens; a 2 diopter lens focuses light at 0.5 meter behind the lens; 10 diopters focuses at 0.1 m behind the lens. The human eye has a total diopter value of about 59 (focal length = 1.7 cm). This diopter value is the sum of the refractive powers of each interface between media of differing refractive index that light passes through (i.e., the air-cornea, cornea-aqueous humor, aqueous humor-lens, and lens-vitreous humor interfaces all function as "lenses"). Most of the refractive power of the human eye is provided by the cornea, since the greatest difference in refractive index is for the air-cornea interface. The lens, despite its high curvature, has a diopter value of only about 15, one quarter of the total value (note that the diopter value of the lens would be about 90 if it were in air rather than surrounded by the aqueous and vitreous humors). The important role of the lens is its ability to modify the focal length of the eye, i.e. change its diopter value, allowing the eye to focus on both near and far objects; this is accommodation.

A simple eye, such as an insect ocellus or vesicular eye of cephalopods and vertebrates, forms an inverted image on the photoreceptive retina. A clear image of objects is formed if the eye can focus images on the retina and has a large number of photoreceptors. The apposition eye of insects and some other invertebrates has numerous independently functioning ommatidia; the eye essentially forms a noninverted, mosaic image that is a composite of the responses of the individual ommatidia. The apposition eye does not have a lens, and so near and distant vision are equally clear.

Visual acuity (spatial resolution) is an important property of eyes. The ability of an eye to resolve fine detail depends on the angle (ϕ) between the optical axes of adjacent photoreceptors (whether rods and cones in the vertebrate vesicular eye or ommatidia of compound eyes). The eye can resolve alternate white/dark lines at as low as 2 ϕ radians between adjacent dark (or light) lines, where ϕ is the angle (in radians) between axes of adjacent photoreceptors (360° = 2π radians). The value of ϕ for a simple eye (cup or vesicular eye) is calculated as

$$\phi = d_{cc}/f \qquad \text{(rads)}$$
$$= 57.3 \, d_{cc}/f \qquad \text{(degrees)}$$

where d_{cc} is the distance between centers of adjacent photoreceptors and f is the focal length of the eye. For example, the rods and cones of the human eye are spaced at about 2.5 μ apart and the focal length is 16.7 mm, so ϕ is about 1/120 degrees and 2ϕ is 1/60 degrees (or 1 minute, 1'). The human eye has a high visual acuity. The simpler ocellar eye of an insect larva (*Perga*) has a d_{cc} of 20 μ, f of 220 μ, and ϕ is 5.2 degrees; it has a much lower

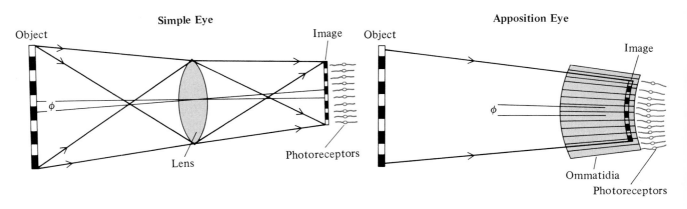

A simple eye uses a lens to focus and form an inverted image, but an apposition eye forms an upright image without a lens.

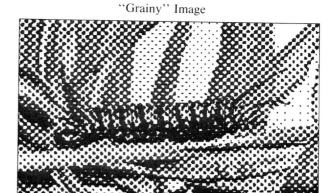

The "sharpness" or "graininess" of the image formed by an eye depends on the size of the receptive field of each photoreceptor. *(Philip Withers and Stan Hopwood, University of Western Australia.)*

---

visual acuity than the human eye. An optical superposition eye is equivalent to a simple eye with the focal length equal to the distance out from the center of curvature of the eye to the level of the image (f is approximately 1/2 eye radius). For a noctuid moth, d_{cc} is about 8 μ and f is 170 μ, so ϕ is about 2.7 degrees.

For a convex apposition eye, ϕ is calculated as

$$\phi = D/R \qquad \text{(rads)}$$
$$= 57.3 \, D/R \qquad \text{(degrees)}$$

where R is the radius of the eye and D is the diameter of the ommatidial facet (this assumes that the eye is spherical). For the honeybee eye, D is about 25 μ and R is about 1 mm, so ϕ is approximately 1.4 degrees. This is considerably less than the visual acuity of the human eye. The visual image for such an eye would be "grainy," consisting of a mosaic of coarse picture elements (pixels).

There is a physical limit to the angular resolving power of a lens (ϕ_{min}). This is dependent on the wavelength (λ; m) of light to be resolved and the diameter of the lens (D).

$$\phi_{min} = 1.22 \, \lambda/D \qquad \text{(rads)}$$
$$= 69.9 \, \lambda/D \qquad \text{(degrees)}$$

This limit is due to diffraction of light at the surface of even an optically-perfect lens. For the human eye, ϕ_{min} is about 0.0035 degrees (0.2') resolving green-blue light;

this is less than the angular resolving power of the photoreceptors (about 1'). For the simple ocellar eye of *Perga*, ϕ_{min} is greater (0.194 degrees) because the diameter of the ommatidial lens is less than the diameter of the human eye. The much poorer physical resolution limit of *Perga* is still less than the anatomical resolving power of the eye; 2ϕ is 10.4° Advanced insects have eyes with smaller facets and higher visual acuity than *Perga*. The honeybee eye is close to its physical limit since ϕ_{min} is about 1.4 degrees ($D = 25 \, \mu$) and $2\phi = 2.8$ degrees. The eye of the sandwasp *Bembix* operates at the resolution determined by the optical properties of its lenses; 2ϕ is 0.66°; ϕ_{min} is higher (0.87°) than 2ϕ for $\lambda = 500$ nm, although it is reduced to 0.61 for $\lambda = 350$ nm (i.e., in the UV).

Material	Refractive Index (RI)
Air	1.000293
Glass	1.5
Water	1.33
Diamond	2.42
Human Eye	
Cornea	1.38
Aqueous humor	1.33
Lens	1.40
Vitreous humor	1.34

Supplement 7–3

Visual Sensitivity

. . . .

The eyes of especially nocturnal animals have a high visual sensitivity. The sensitivity (S) of a photoreceptor is determined by the number of photons reaching it per second, relative to the luminance, or ambient light intensity

$$S = F_{ph}/L = (\pi/4)^2 (D/F)^2 d_r^2 (1 - e^{-kl})$$

where F_{ph} is the photon flux reaching the photoreceptor (photons sec^{-1}) and L is the ambient luminance (photons m^{-2} sr^{-1} sec^{-1}). Ambient luminance varies markedly,

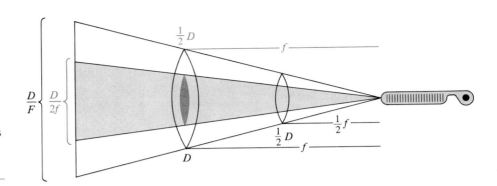

The light-collecting capacity of a lens is determined by its ratio of diameter to focal length (D/f).

from 0 in absolute darkness to 10^{10} in overcast starlight, 10^{14} in moonlight, and 10^{20} for a white surface in bright sunlight. The value of S can be calculated from the F number of the lens (D/F), the diameter of the photoreceptor (d_r), and the light-absorbing properties of the photoreceptor. The amount of light impinging on a photoreceptor that is absorbed is $1 - e^{-kl}$, where l is the length of the photosensitive portion of the cell (e.g., the rhabdome length) in the direction of light movement, and k is the extinction coefficient of the photosensitive portion of the cell, i.e., the fraction of light that remains after passing through a unit length of the absorbing material. The optically-dense rods of vertebrates have k about 0.03 μ^{-1}; for the lobster rhabdome, k is about 0.0067 μ^{-1}. A photoreceptor region 100 μ long absorbs 49% of the incident light if $k = 0.0067$ $(1 - e^{-0.0067(100)} = 0.488 = 49\%)$; a 200 μ photoreceptor region absorbs 74% $(1 - e^{-0.0067(200)} = 0.738)$; a 1 mm rod absorbs 99.9% $(1 - e^{-0.0067(1000)} = 0.999)$.

The sensitivity of the human eye is about 0.046 μ^2 or 0.046 10^{-12} m^2 ($D = 2$ mm pupil; $r = 2.5$ μ, $l = 60$ μ). Thus, in bright sunlight the human eye receives about 4.6 10^6 photons sec^{-1} $(0.046\ 10^{-12} \times 10^{20})$. The honeybee eye has a higher sensitivity; $S = 0.20$ μ^2; the superposition eye of a noctuid moth has an even greater sensitivity, $S = 114$ μ^2, primarily because of the much larger diameter of the effective pupil (400 μ) and the large diameter of the photoreceptor (8 μ).

Recommended Reading

Aidley, D. J. 1989. *The physiology of excitable cells.* Cambridge: Cambridge University Press.

Barrett, R. 1970. The pit organs of snakes. In *Biology of the Reptilia.* Vol. 2, *Morphology B,* edited by C. Gans and T. S. Parsons, 277–300. New York: Academic Press.

Bennett M. V. L. 1971. Electric organs. In *Fish physiology,* edited by W. S. Hoar and D. J. Randall, vol. 5:347–491. New York: Academic Press.

Bennett, M. V. L. 1971. Electroreception. In *Fish physiology,* edited by W. S. Hoar and D. J. Randall, vol. 5:493–574. New York: Academic Press.

Bowmaker, J. K. 1980. Colour vision in birds and the role of oil droplets. *TINS* 3:196–99.

Dethier, V. G. 1963. *The physiology of insect senses.* London: Methuen.

Eakin, R. M. 1973. *The third eye.* Berkeley: University California Press.

Gamow, R. I., and J. F. Harris. 1973. The infrared receptors of snakes. *Sci. Am.* 228(5):94–101.

Griffin, D. R. 1958. Listening in the dark: the acoustic orientation of bats and men. New Haven: Yale University Press.

Lancet, D. et al. 1988. Molecular transduction in smell and taste. *Cold Spring Harbor Symp. Quant. Biol.* 53:343–48.

Land, M. F. 1985. The eye: optics. In *Comprehensive insect physiology, biochemistry and pharmacology,* edited by G. A. Kerkut and L. I. Gilbert. Vol. 6, *Nervous system: Sensory,* 225–76. Oxford: Pergamon Press.

Land, M. F. 1987. Vision in air and water. In *Comparative physiology: life in water and on land,* edited by P. Dejours et al. Fidia res. Ser. 9. Berlin: Springer-Verlag: Liviana Press, Padora.

Mazokhin-Porshnyakov, G. A. 1969. *Insect vision.* New York: Plenum.

Neuweiler, G. 1983. Echolocation. In *Biophysics,* edited by W. Hoppe et al., 683–97. Berlin: Springer-Verlag.

Nillson, D.-E. 1989. Vision optics and evolution. *Bioscience* 39:298–307.

Chapter 8

Nervous Systems

· ·

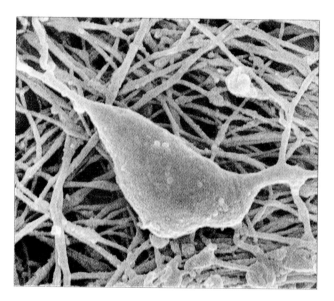

This multipolar neuron of the myenteric plexus is a typical excitable cell of nervous systems. *(Scanning electron micrograph courtesy of D. W. Fawcett/Komuro, Science Source.)*

The **nervous system** consists of groups of nerve cells (neurons) that provide an interface between the sensory reception of information and the motor response systems. Sensory cells generally do not have motor effects, such as secretion or contraction. It is also unusual for a sensory neuron to directly communicate its information to an effector cell.

The nervous system of primitive animals consists of simple sensory-motor circuits. There usually are at least three cells in a sensory-motor circuit. One is the neuron receptor, that is specialized for filtering and recording information; another is an effector cell such as a muscle or secretory cell; and another is a motor neuron (motoneuron) that conveys the sensory information from the receptor neuron to the effector cell (Figure 8–1). The monosynaptic reflex arc of vertebrates is an example of a simple three-cell sensory-motor circuit. It has one synapse within the nervous system between the sensory neuron and the motor neuron and a second peripheral synapse between the motor neuron and the effector cell.

Nervous systems generally have a greater functional complexity than a three-cell, sensory-motor circuit due to additional **interneurons** that are interposed between the sensory neuron and the motor cell. The nervous system often has a very large number of interneurons that can provide an exceedingly complex interpretation of sensory information, generate complex behavior patterns, and exert complex motor control.

Evolution

The general organization of the nervous system as a sensory-interneuron-effector cell circuit is apparent at varying levels of complexity in animals. The organization of sensory receptors, neurons, and effector (myoepithelial) cells in the epidermis of coelenterates well illustrates this general organization. Sensory receptors relay their information to a network of conducting neurons, which synapse

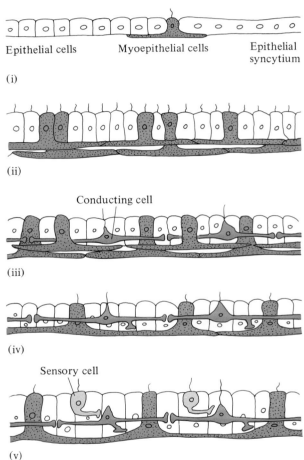

FIGURE 8–2 Possible scenarios for the origin of nerve cells from a conducting epithelium: (i) conducting epithelium with a contractile myoepithelial cell (stippled)—the conducting epithelium might be individual cells (left) or a syncytium (right); (ii) myoepithelial cells are connected by basal extensions; (iii) specialized, conducting epithelial cells (not stippled) form a high-velocity conduction bridge with other epithelial cells; (iv) specialized conducting cells synapse with myoepithelial cells; and (v) sensory cells (light shading) synapse with conducting cells and then myoepithelial cells. *(Modified from Horridge 1968.)*

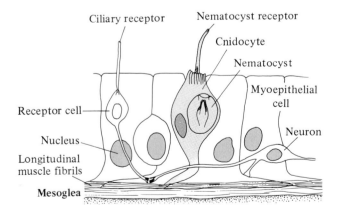

FIGURE 8–1 The body wall of a hydroid coelenterate has sensory receptor cells that are connected to neurons and then myoepithelial cells. Cnidocytes are independent effector cells; they have both sensory and motor (nematocyst stinging) roles.

with the contractile myoepithelial cells. But, which elements of the circuit evolved first, and from what precursor cells? One possible sequence for the evolution of these elements of the nervous system from an epithelium is shown in Figure 8–2. The specialized sensory neurons, motor neurons, and effector cells would have been derived from undifferentiated epithelial cells that performed similar functions but in a simpler fashion. For example, the contractile myoepithelial cell of coelenterates may be the precursor of muscle cells in higher animals.

Some effector cells are independent of sensory and neuronal cells; they do not require a sensory or neural innervation to function. For example, the cnidocyte (stinging) cells of coelenterates (Figure 8–1) have a mechano/chemosensory receptor and a stinging thread that is expelled from the nematocyst. Epidermal glandular cells, the myogenic hearts of many invertebrates, the dermal pores of sponges, and the smooth muscle in the iris of some lower vertebrates are also examples of independent effector cells. Effector cells without separate sensory or neural control can only have simple, stereotyped responses to specific kinds of stimuli. The existence and variety of independent effector cells in various animals suggests that specialized effector cells probably evolved before circuits of receptor cells, interneurons, and motor neurons.

The nervous system of primitive animals is often arranged as a diffuse network of interneurons, or **nerve net**, interposed between a series of epithelial sensory cells and epidermal effector cells. In more advanced animals, there is a reduction in the reflex motor response to peripheral sensory reception and a greater centralization of sensory interpretation and of motor control. Sensory nerves relay their information to clusters of specialized neurons; these neuron clusters are called **ganglia** if the neuron bodies are located outside the central nervous system and **nuclei** if the neuron bodies lie within the central nervous system. The ganglionic and nuclear neurons communicate with other neurons via cords of nervous tissue and also innervate the effector cells via motor neurons.

The general nerve net of primitive animals becomes concentrated into six to eight longitudinal **nerve cords** in higher animals. These nerve cords are interconnected by transverse **commissures**. The nervous system of some primitive turbellarian worms, e.g., *Bothrioplana*, (Figure 8–3), shows

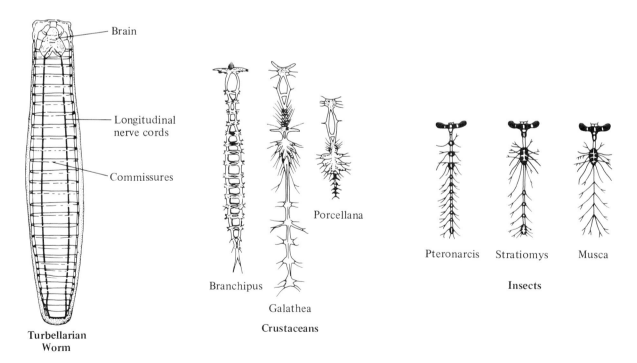

FIGURE 8–3 The nervous system of an allocoel turbellarian worm *Bothrioplana* has an anterior cephalization (brain) and an orthogonal arrangement of pairs of longitudinal nerve cords linked by transverse commissures. The process of cephalization of ventral nerve cord ganglia to form a more complex anterior (or posterior) brain is illustrated by crustaceans and insects. (*Modified from Bullock and Horridge 1965.*)

such an orthogonal organization. The number of longitudinal nerve cords is reduced in higher animals to two or even one cord, and the number of commissural connectives is accordingly reduced. Animals with radial, rather than bilateral, symmetry have a reduced longitudinal organization of their nerve cords (e.g., echinoderms, adult tunicates).

A second specialization of the nervous system in advanced animals is the development of a **brain**. The central nervous system develops an anterior concentration of neural tissue that is connected with the longitudinal nerve cords. This process of cephalization has occurred in all animals with bilateral symmetry. The progressive reduction in the interpretive role of the peripheral nervous system (e.g., nerve cords) and concentration of neural function into an anterior brain has occurred independently in a number of vertebrate and invertebrate groups (e.g., crustaceans, insects, arachnids, gastropods, cephalopods) by the anteriorward migration of nerve cord ganglia (Figure 8–3).

Nervous Tissue

Two important types of cells in the nervous system are **neurons** and supporting cells (**glia**, or neuroglia). The neurons are the excitable cells of the nervous system. Glia are a complex assortment of cells with varying functions, which may include protection and nourishment of neurons, ionic regulation, and electrical activity. Connective tissue cells are also present in the nervous system and blood vessels are sometimes present (e.g., arthropods and vertebrates).

Neurons

Neurons are excitable cells specialized for the repetitive transfer of membrane potentials between cells (often receptor or effector cells) and also between other neurons. The neuron's cell body (soma) generally has two types of cell processes: **dendrites** receive nerve impulses from sensory cells or other neurons; an **axon** conveys signals to other cells. This topographical polarization of the neuron, with "inward" and "outward" directed processes, was established by the Spanish histologist Ramon ý Cajal, who also realized that nerve cells were independent cellular units rather than a network of cells interconnected by cytoplasmic bridges. Specializations for conduction of action potentials and transfer of information to other cells (e.g., electrical or chemical synapses) are the major structures defining a cell as a neuron. However, dendritic arborizations and axons are not invariate cytological characters of neurons.

Neurons can be classified in various ways. In one morphological scheme, three types of neurons are defined by the number and pattern of cell body processes (Figure 8–4). **Unipolar** neurons have a fairly spherical soma with a single process that bifurcates into a dendritic arborization and an axon with terminal synapse. The majority of invertebrate interneurons and motoneurons are unipolar; the unipolar neurons of vertebrates are primarily sensory, but they are also present in cranial and spinal ganglia. **Bipolar** neurons have a dendritic process and an axonal process; sensory neurons are commonly bipolar. **Multipolar** neurons have a number of dendritic processes and a single axon. Many interneurons and motoneurons are multipolar, but only a few types of sensory neurons are multipolar (e.g., muscle stretch receptors).

A more functional classification divides neurons into sensory, interneuron, and motor types. Interneurons can be **projection** neurons, which communicate anteriorly or posteriorly to the other neurons; **commissural** neurons, which send an axon to a corresponding location in the nervous system but on the opposite side; or **intrinsic** neurons, which are confined to the same level and side of the central nervous system. Many sensory, motor, and projection axons cross, or **decussate**, to the opposite side of the body (i.e., are contralateral), whereas others remain on the same side of the body (i.e., are ipsilateral). For example, most of the retinal ganglion cell axons of vertebrates decussate to the opposite side of the brain in the optic chiasma, although some are ipsilateral.

Nerve cells may also be classified by their size and ratio of nucleus-to-cytoplasm diameters. Small neurons may be as little as 3 μ in diameter with a nucleus:cytoplasm ratio close to 1, e.g., globuli cells of invertebrates and granule cells of vertebrates. At the opposite extreme, giant neurons may approach 1000 μ dia with a nucleus:cytoplasm ratio of about 0.5, e.g., giant neurons of the marine gastropod *Aplysia*. There is a continuum of intermediate-sized neurons. The giant axons of many invertebrates are not necessarily associated with giant cell bodies. For example, the 300 μ dia giant axon of the polychete *Protula* arises from single cells of about 50 μ dia; the 800 μ dia giant squid axons arise by fusion of axons of hundreds of small neurons, each about 40 μ in dia.

Neurons may occur as single solitary cells although they are commonly organized into tissues, e.g., in epithelial layers, nerve plexuses, peripheral and central ganglia, nerve cords, and commissures.

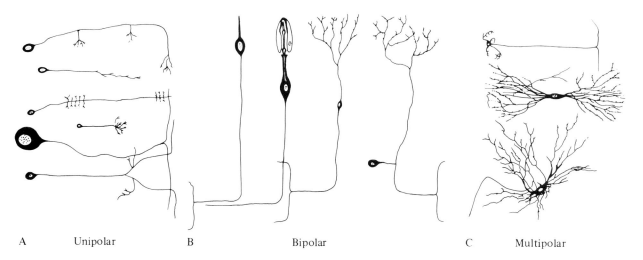

A   Unipolar   B   Bipolar   C   Multipolar

FIGURE 8–4 Three basic types of neurons. (**A**) Unipolar neurons have only a single cytoplasmic extension; they are commonly interneurons (upper four) or motoneurons (lower two) in higher invertebrates. (**B**) Bipolar neurons have a distal dendritic process conveying information to the cell soma and an axon carrying information away from the soma; many sensory neurons (shown here) are bipolar. (**C**) Multipolar neurons have a number of afferent arborizations and a single efferent axon; the upper two are interneurons and the lower one is a motoneuron. *(From Bullock and Horridge 1965.)*

The nervous system of lower animals (e.g., coelenterates) is a nerve net of isolated neurons in synaptic contact with other neurons. Neurons may be part of a simple epithelium with only the motor axons extending from the epithelial layer to underlying effectors. Higher animals have neurons grouped into a more organized central nervous system and peripheral ganglia. Ganglia may be purely sensory (e.g., the vertebrate dorsal root ganglion), purely motor (e.g., the vertebrate autonomic ganglion), or mixed (e.g., the podial ganglion of annelids). Nerves are bundles of axons connecting peripheral sensory and motor cells with the central nervous system.

The central nervous system of invertebrates is divided into an inner layer, the **core**, and an outer layer, the **rind**. The neuron somas are located in the rind, and the neuron fibers and synapses are in the core. The organization of the vertebrate central nervous system is fundamentally different. The **gray matter**, which contains neuron cell bodies, dendrites, axons, and synapses, is located (at least in primitive vertebrates) inside the nervous system. The **white matter** is axon tracts that, at least primitively, are on the outside of the nervous system. The plexus of axons and dendrites, where the synapses are located, is the **neuropile**; it is found in the core of the invertebrate and in the gray matter of vertebrate nervous systems.

The earliest theories of neuronal function suggested that nerves were hollow tubes that transported a substance, "psychic pneuma." We now understand that information flows along a nerve axon by electrical propagation of an action potential, not the movement of some material. However, there is a complex axoplasmic movement of material along axons both from the soma to the synapse (anterograde flow) and from the synapse to the soma (retrograde flow; Figure 8–5). Axonal flow transports neurotransmitter to the synapse and recycles synaptic terminal materials to the soma for reuse. The slow anterograde flow (0.1 to 1.0 cm day^{-1}) of axoplasm is readily demonstrated by the accumulation of axoplasm on the soma side of a ligature that constricts an axon. The rapid anterograde flow (5 to 40 cm day^{-1}) is especially particulate, rather than soluble, material, e.g., protein, glycoprotein, phospholipid, and enzymes. There is also a slow retrograde flow (8 cm day^{-1}) of material. Membrane constituents, toxins, and viruses can be transported to the soma in this manner. Some materials, such as acetylcholine, are transported by both the rapid and slow anterograde and by the retrograde flow systems. The mechanism for axonal flow is unclear but most likely involves actin filaments and microtubules (see Chapter 9) that run the length of axons.

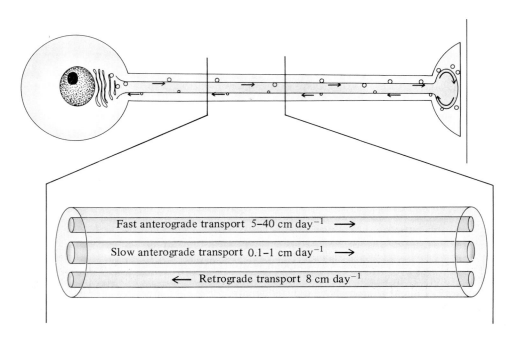

FIGURE 8–5 Axoplasmic flow between neuron soma and axon terminal transfers neurotransmitter to the terminal (anterograde flow) and recycles membranous organelles from the terminal (retrograde flow). There is a slow and a fast anterograde transport. *(Modified from Ottoson 1983.)*

Glia

The **glia** are non-neuronal cells found in the central nervous system in close association with the neurons of the invertebrate nervous system and the neurons and capillaries in the vertebrate nervous system. Neurons of the central nervous system are also closely associated with other non-neural cells that form a protective outer sheath of dense connective tissue (just as peripheral axons have accessory Schwann cells). These perilemma cells are not classified as glia.

Vertebrates have three types of glia. **Astrocytes** have numerous cytoplasmic processes that surround

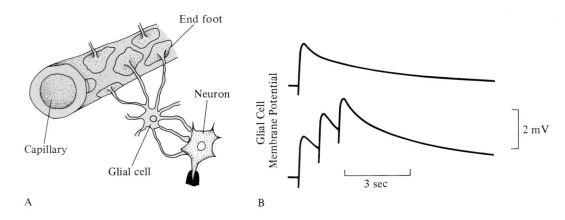

FIGURE 8–6 **(A)** Organization of neuroglial cells (astrocytes) in the vertebrate brain; the end feet of glial cells cover the capillaries, leaving small intercellular clefts; other extensions of the glia contact the neurons. **(B)** Changes in membrane potential of neuroglial cells in the optic nerve of *Necturus* in response to a single (upper) or multiple (lower) nerve action potentials. *(Modified from Kuffler 1967; Orkand, Nicholls, and Kuffler 1966.)*

both the capillaries and the neurons (Figure 8–6A). Extensive intercellular clefts between the astrocyte processes (10 to 20 nm dia) allow rapid diffusion of small solutes. Astrocytes may control the distribution of large solutes between capillaries and neurons. **Oligodendrocytes** are found mainly in the white matter where they form the myelin sheath of axons. **Microglia** are phagocytic, motile cells that engulf and destroy cellular debris and microbes within the central nervous system. Invertebrates also have a number of distinct types of neuroglia.

Neuroglial cells, unlike neurons, are not electrically excitable. However, they do respond electrically to local nerve cell activity. The resting cell membrane potential of glial cells is generally more negative than that of the neurons, e.g., E_m is about -75 and -50 mV for the neuroglia and neurons of the leech ventral nerve cord (Nishi and Koketsu 1960); glial and neuron E_m's are -90 and -70 mV, respectively, for the amphibian *Necturus* (Orkand, Nicholls, and Kuffler 1966). The E_m of glial cells alters in response to local neural activity, although the changes are much slower and of a smaller magnitude than the neural action potential (Figure 8–6B). Glial depolarization is not an electrical response to neural activity *per se* but to a local increase in the extracellular K^+ concentration because of the transient release of K^+ from neurons.

Integrative Neurophysiology

The nervous system of even primitive animals is capable of complex integration of multiple channels of information. This can be a property of the way in which a number of neurons are interconnected to form neural circuits, but even single neurons have a complex integrative capability. Let us first examine the integrative capabilities of single neurons and then examine how neural circuits can accomplish even more complex integrative actions.

Interneurons

Interneurons typically have synaptic inputs and axonal connections with many other neurons; the synapses on their dendrites or cell soma may be either excitatory or inhibitory (Figure 8–7A). One, or a few, excitatory synapses may elicit an action potential, or many excitatory synapses may be required to elicit an action potential. The activity of a single inhibitory neuron might prevent the neuron from initiating an action potential, or many inhibitory synapses may be required to block an action potential.

How do neurons integrate the effects of their many synaptic inputs of information? Each synapse, whether excitatory or inhibitory, produces a change in the postsynaptic membrane permeability, causing a local current flow and a **postsynaptic potential** (PSP). This local inward (or outward) current flow must ultimately leave (or enter) the neuron at some other point. This is one of Kirchhoff's rules; the sum of currents entering a point must equal the sum of currents leaving that point. All of the individual synaptic currents are summed at the base of the axon where it leaves the cell soma, the **axon hillock**. If the net current flow at the axon hillock is sufficient to depolarize the E_m to threshold, then an action potential occurs. If the depolarization is inadequate to reach threshold, then the E_m repolarizes to normal and no action potential occurs. An inhibitory synapse, if it hyperpolarizes the E_m, would make it more unlikely that an action potential would be initiated.

An **excitatory postsynaptic potential** (EPSP) is a graded depolarization and an **inhibitory postsynaptic potential** (IPSP) is a graded hyperpolarization. The EPSP resembles the end plate potential of the neuromuscular junction (see Chapter 6). The EPSP causes a fairly rapid increase in E_m, followed by a slow, exponential decay; its total duration is about 10 msec (Figure 8–7B). If the EPSP exceeds threshold, then an action potential is initiated; the higher the amplitude of the EPSP, the sooner the E_m reaches threshold and the sooner the action potential is initiated. The amplitude of the EPSP depends on the initial E_m; the depolarization of an EPSP at the normal E_m becomes a hyperpolarization if the E_m is experimentally clamped to, for example, $+60$ mV. The EPSP has a reversal potential (E_{rev}; see Chapter 6) of about 0 mV. For example, the EPSP of the frog sympathetic ganglion neuron has an E_{rev} of about -10 mV (Nishi and Koketsu 1960). The inhibitory postsynaptic potential (IPSP) is an initial rapid hyperpolarization followed by a slower exponential return to resting E_m. Its reversal potential is more negative than the normal E_m. For example, it is about -80 mV for the cat motoneuron, which has a resting E_m of about -75 mV (Coombs, Eccles, and Fatt 1955).

It is the overall, summed effect of EPSPs and IPSPs at the axon hillock that determines whether the E_m is depolarized to threshold and an action potential occurs. The axon hillock is the major site of initiation of action potentials because it has the lowest threshold, hence it is preferentially excited by EPSP depolarization. An action potential initiated at the axon hillock then propagates along the axon.

Two or more simultaneous excitatory EPSPs can summate to depolarize E_m to threshold and elicit an

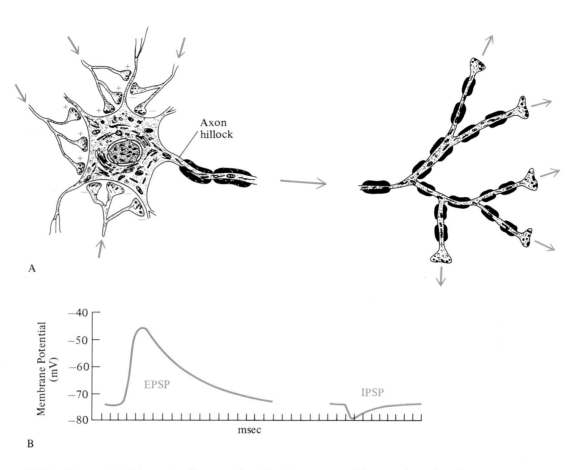

A

B

FIGURE 8–7 (A) Schematic diagram of multipolar neuron with a number of excitatory (+) and inhibitory (−) synaptic inputs and axonal output. (B) Excitatory postsynaptic potential (EPSP) recorded in the cat biceps-semitendinosus motoneuron in response to electrical stimulation of the afferent fibers from the muscle. Inhibitory postsynaptic potential (IPSP) recorded from the same motoneuron in response to stimulation of the afferent fibers from an antagonistic muscle (the quadriceps).

action potential even if each EPSP is individually subthreshold; this is **spatial summation** (Figure 8–8). Alternatively, EPSPs can reach threshold and initiate an action potential if an initial subthreshold EPSP is rapidly followed by a second subthreshold EPSP; the second EPSP sums with the first to elicit an action potential. This is **temporal summation**.

EPSPs and IPSPs also "summate," except that the IPSP reduces the likelihood of an action potential being initiated in response to an EPSP, depending on the spatial and temporal overlap of the IPSP and EPSP (Figure 8–8). This is well illustrated by the antagonist effects of extensor and flexor muscles during the monosynaptic stretch reflex of spinal motoneurons. Stimulation of muscle sensory stretch fibers (e.g., the flexor biceps-semitendinosus muscles) monosynaptically stimulates the muscle to contract, and at the same time inhibits contraction

of its antagonist muscle (the extensor quadriceps) by an inhibitory interneuron. EPSPs of the biceps-semitendinosus motoneurons due to stimulation of its sensory neurons "summate" with IPSPs caused by stimulation of sensory neurons of the antagonistic quadriceps muscle; whether an action potential is elicited in the motoneuron depends on the specific temporal relationship between the EPSP and IPSP.

The amount of neurotransmitter that is released from presynaptic vesicles at a synapse depends on the amplitude of the presynaptic action potential. If the E_m at the presynaptic terminal is slightly depolarized before the action potential arrives, then the action potential amplitude is diminished and less neurotransmitter is released. The consequence of this is a reduced EPSP amplitude. **Presynaptic inhibition** is the result of this partial depolarization of the presynaptic membrane.

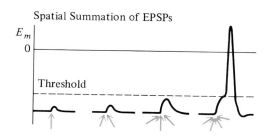

Spatial Summation of EPSPs

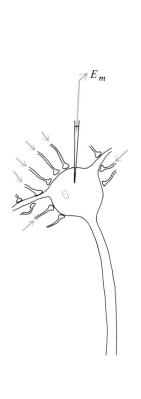

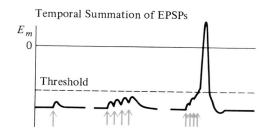

Temporal Summation of EPSPs

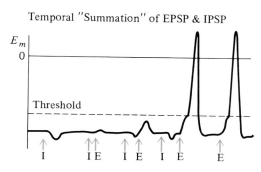

Temporal "Summation" of EPSP & IPSP

FIGURE 8–8 Temporal and spatial summation at the axon hillock of a neuron. With spatial summation, the EPSP increases in amplitude with an increased number of stimulated synapses until threshold (dashed line) is reached and an action potential is initiated. With temporal EPSP summation, multiple rapid synaptic stimulation increases the EPSP amplitude to threshold. The precise timing of an IPSP and EPSP determines whether their interaction results in attainment of threshold and initiation of an action potential or whether the resultant postsynaptic potential is subthreshold.

Neural Circuits

The neurons of a nervous system are typically organized in specific patterns, called **neural circuits**, which receive sensory information or generate spontaneous information, integrate information, and control effector cells. Neural circuits of different animals vary dramatically in their complexity, from the simple and diffuse nerve net of coelenterates to the advanced and exceedingly complex nervous systems of higher vertebrates and invertebrates.

The simplest neuronal circuit consists of a sensory receptor that relays its information directly to an effector cell (Figure 8–9A). There is only a single synapse in this simple neuronal circuit at the effector cell. There are no interneurons and no sensory neuron-motoneuron synapse. Such a neuronal circuit provides a fairly direct "wiring" of an effector to a sensory cell and clearly has little capacity for integration of sensory information before it becomes a motor command. Simple neuronal circuits are relatively uncommon in animals. A more complex, and more common, neuronal circuit has a motor neuron interposed between the sensory cell and the

effector cell (Figure 8–9B); this is a **monosynaptic reflex arc**, or direct reflex arc. A reflex arc has an automatic response to sensory input; the motor effect is a reflex action. The presence of the motoneuron provides a capacity for some integration of various sensory inputs before the motor command is determined. For example, the motoneurons of the mammalian monosynaptic reflex arc receive both excitatory and inhibitory inputs.

Most neurons, and particularly those of higher animals, are arranged in more complex fashions than reflex arcs, and they provide varying degrees of integration and interpretation of sensory information. Sensory and also other neurons frequently transmit their information to a number of other neurons by multiple branches of their axon; this is **divergence** (Figure 8–9C). A multiple neuronal chain circuit is another example of divergence (Figure 8–9D); each neuron has only a few diverging axon branchs, but the information may ultimately be diverged to many neurons. Conversely, information from a number of different neurons may be directed to a single neuron; this is **convergence** (Figure 8–9E). The multiple inputs of information are analyzed

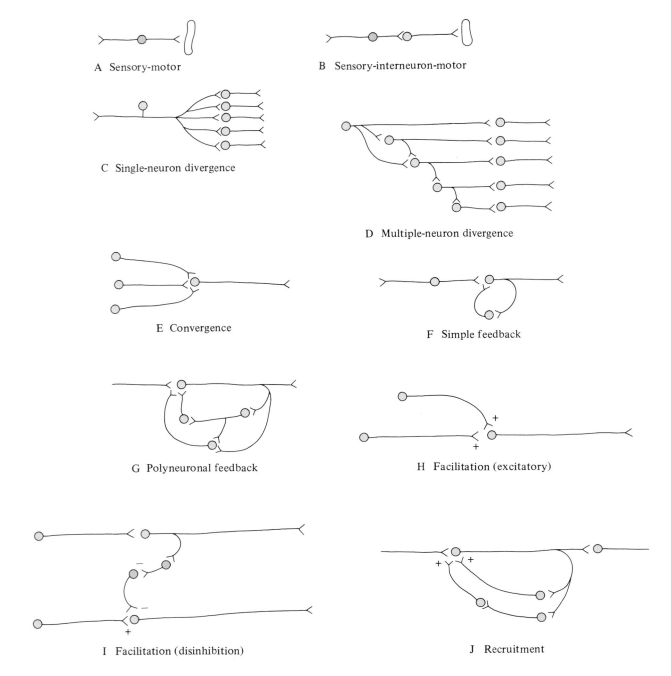

FIGURE 8–9 Examples of neuronal circuits. **(A)** The simple sensory-motor circuit consists of a sensory neuron synapsing directly onto a motor cell; such simple circuits are extremely rare. **(B)** The sensory-motor neuron-effector circuit is common, for example, in monosynaptic reflex arcs. **(C)** Divergence occurs when many axon branches from a single neuron synapse with multiple neurons. **(D)** Another form of divergence where most neurons have only two axonal branches but chain to form a considerable divergence from the single neuron. **(E)** Convergence occurs when a number of axonal projections synapse with a single neuron. **(F)** A simple feedback loop has a collateral axonal branch that returns action potentials to the neuron soma via an interneuron. **(G)** A polyneuronal feedback loop has a number of interneurons. **(H)** Facilitation of synaptic transmission by an excitatory synapse. **(I)** Facilitation of synaptic transmission by removal of tonic inhibition. **(J)** Recruitment occurs by excitatory feedback via interneurons to the neuron receiving an excitatory input.

by the recipient neuron, which produces an integrated response.

The neural circuits so far described are **open circuits** because there is a linear flow of information from the afferent neurons through the interneuron/motoneurons to the effector cells. **Feedback circuits** provide an additional level of complexity. In a simple feedback system, efferent information is fed via collateral branches of the efferent axon back to the soma of the same neuron via an interneuron (Figure 8–9F). No neuron has collateral branches that synapse directly on its own soma. Initial activity of the neuron can result in a prolonged activity if the interneuron is excitatory; instead of a single action potential in the efferent axon there are many action potentials forming an "after discharge." Closed feedback circuits may be even more complex if there are a number of interneurons in the feedback loop (Figure 8–9G).

An example of an inhibitory feedback loop is the Renshaw cell of the vertebrate spinal cord. It is an interneuron, which receives collaterals from skeletal muscle motoneurons, and its axon terminates on the motoneuron soma. The Renshaw cell has a prolonged action potential discharge in response to stimulation by the motoneuron collateral, which inhibits further motoneuron activity. The role of the Renshaw cell apparently is to limit the duration of the activity of the motoneurons and hence of the muscle contraction. Strychnine interferes with the inhibitory synapses of Renshaw cells and causes convulsions and spastic paralysis. However, even the seemingly simple feedback system of Renshaw cells is actually quite complex in its actions. Renshaw cells can also be stimulated, and inhibited, by skin and muscle sensory receptors and by the higher centers of the central nervous system. They usually inhibit motoneurons but can, under some conditions, facilitate activity in adjacent motoneurons.

The Mauthner giant interneurons of some teleost fish and salamanders are responsible for their startle response; simultaneous activity of both Mauthner interneurons is prevented by inhibitory feedback (see Supplement 8–1, page 387).

Facilitation occurs if a neuron is induced to initiate an action potential in response to synaptic stimulation if it would not normally have initiated an action potential. Facilitation can be produced by either the excitatory effect of one neuron on another or by the removal of inhibition. Excitatory facilitation occurs when two excitatory neurons synapse with a third (Figure 8–9H). **Disinhibition** occurs if a neuron is normally inhibited by an interneuron. If a neuron stimulates an interneuron, which in turn inhibits the inhibitory neuron, then there is temporary disinhibition (Figure 8–9I).

Neuronal **recruitment** is a form of facilitation involving closed, multineuronal circuits; a neuronal response is elicited only after repetitive stimulation (Figure 8–9J). For example, a single afferent impulse reaches an interneuron, but the interneuron's response is insufficient to elicit motor action potentials in a motoneuron. However, the initial impulse also activates a multineuronal closed feedback loop, which restimulates the interneuron after a short time delay. The restimulation might also be subthreshold but in conjunction with further afferent impulses would elicit motor activity.

Nerve Networks

The nervous systems of animals are networks, or aggregations, of sensory, internuncial, and motor neurons. There is an enormous diversity in the structure and functioning of nervous systems in different animals. The more primitive animals, such as coelenterates and flatworms, have correspondingly simple nervous systems that are basically nerve networks diffusely connecting the sensory structures with the motor structures. These simple nerve net systems provide the basis for examining the more specialized nervous systems of higher invertebrates and vertebrates where the nervous tissue is condensed into nerve cords and a centrally located ganglia of neurons, the brain.

Primitive Nervous Systems

Many protozoans have cilia, which beat in a variety of ways and generally have a locomotory or feeding function (Chapter 9). Many cilia beat in only a single plane and either beat or do not, i.e., there is no control of direction or frequency. The cilia of some protozoans and metazoans can reverse their direction of beating (e.g., the ciliated cells in the pharynx of sea anemones) or may beat in any orientation. The normal beating rhythm of cilia is under some form of "nervous control." Each cilium can apparently beat spontaneously, and that with the highest frequency becomes the "pacemaker," driving the beating of adjacent cilia at the same frequency through mechanical coupling of the adjacent cilia by the viscosity of water (water is a highly viscous fluid at the scale of a cilium, i.e., at low Reynold's number; see Chapter 10).

The coordination of ciliary activity in some protozoans is not achieved by mechanical coupling of adjacent cilia. For example, the oral cilia of *Stentor* beat in a coordinated metachronal wave that is not affected by the viscosity of the medium, temperature, or various agents (including Mg^{2+} and drugs).

Such coordination of ciliary activity is probably due to subsurface cellular structures, perhaps cytoplasmic fibrils. Fibrillar transmission may involve fibrillar tension forces, rather than nerve-like electrical depolarization.

The direction of ciliary beating of the parasitic ciliate *Opalina* is determined by its membrane potential. The sensory detection of external stimuli presumably regulates E_m to provide the appropriate direction of ciliary beating, but the details of this supposed regulation are not known. The ciliate protozoan *Paramecium* exhibits an avoidance reaction if it bumps into an object while swimming; it reverses the beating direction of its cilia, backs away from the object, then resumes forward locomotion to avoid the obstacle. The escape reaction is an increased forward swimming in response to mechanical stimulation of the posterior end of the cell. Mechanical stimulation of the anterior part of the cell opens Ca^{2+} channels and the cell membrane depolarizes, causing ciliary reversal; stimulation of the posterior end opens K^+ channels and hyperpolarizes the E_m, increasing the rate of beating (Eckert 1972).

Sponges, although multicellular, represent a cellular grade of organization where the individual is a loose aggregation of cells rather than cells organized as specialized tissues. The outer epidermis of some sponges contains porocytes—pore cells that surround the incurrent water pores. The inner epithelium contains choanocytes—flagellar cells that are partially responsible for water movement through the sponge. There generally is a specialized excurrent opening (osculum). The mesenchyme, a layer between the inner and outer epithelia, contains skeletal spicules and several types of cells. Sponges are generally considered to lack nervous and sensory cells, although they do have effector cells (pore and oscular sphincter cells). There is, however, some evidence for neuron-like cells in sponges that may be functional neurons or the progenitors of true neurons in higher animals (Lentz 1968). Sponges also have some capacity for intercellular communication. The hexactinellid sponge *Rhabdocalyptus* can arrest its water current in response to electrical or mechanical stimulation (Lawn, Mackie, and Silver 1981). Stimulation of any part of the sponge produces a 30 to 100 sec arrest of water flow after a latency period of about 20 to 50 sec. The velocity of propagation of the stimulation is slow, about 0.2 to 0.3 cm sec^{-1}, and no intracellular or extracellular electrical depolarization is apparent. Any portion of the sponge tissue can be stimulated, and there is a generalized spread of conduction over the entire sponge, i.e., conduction is not polarized.

Nerve Nets

Cnidarians (hydra, medusae, and corals) and ctenophores (comb jellies) are radially symmetrical, with a complex body wall structure. The outer epidermis of cnidarians contains sensory cells, stinging nematocyst cells, contractile myoepithelial cells, and often specialized neurons (see Figure 8–1, page 330). The inner body wall (gastrodermis) contains mainly secretory and digestive cells and is separated from the epidermis by an amorphous mesoglea layer. The epithelium of parts of siphonophore coelenterates is devoid of sensory and neural cells but will conduct an electrical stimulation in all directions at a velocity of about 35 to 50 cm sec^{-1} (Figure 8–10).

Most cnidarians have a well-developed **nerve net**, a meshwork arrangement of bipolar or multipolar neurons that rapidly conducts sensory information along a diffuse or specific pathway and elicits motor actions, such as swimming contractions or tentacular movements. There can be a number of fairly independent nerve nets. The **fast specific nerve net** of large bipolar (and occasionally tripolar) neurons has a preferential direction of electrical propagation at a relatively high-conduction velocity of 1 to 2 m sec^{-1}. The fast nerve net contains many pacemaker units in the marginal ganglia that regulate the rhythmic pulsation of the entire bell. The **slow diffuse nerve net** is a second, well-developed and semi-independent nerve net of small multipolar neurons. The slow nerve net regulates other kinds of activity, especially slow and local movements of tentacles and mouth parts. There may also be other nerve nets, associated with the oral disk, tentacles, pharynx, gastrodermis, and pedal disk.

The first physiological evidence for separate nerve nets was provided by Romanes in 1877 using pieces of medusae cut into complex shapes. A complex pattern of cuts provides evidence that conduction of activity is via a diffuse nerve net, rather than discrete tracts of nerve fibers. Gentle mechanical stimulation of the epidermis initiates a slow wave of tentacular contraction that passes away from the point of stimulation. If a marginal body is present in the piece of medusa, it is stimulated by the slow wave of contraction and initiates a faster wave (at about 2 times the slow wave velocity) of bell contraction. There is a difference in the functions of the fast and slow nerve nets, but they are not functionally isolated from each other.

Active medusae have marginal ganglia of neurons or two parallel, circularly oriented nerve nets. The upper nerve ring receives information from the tentacles, nearby sensory cells, statocysts, and

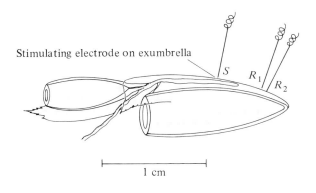

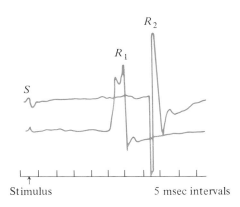

Stimulating electrode on exumbrella

1 cm

S R_1 R_2

R_2

R_1

S

Stimulus 5 msec intervals

FIGURE 8–10 Conduction of electrical activity by the epithelium of a siphonophore coelenterate *Chelophyes* from the stimulating electrode (*S*) to the recording electrode (*R*); multiple recording (*R₁*, *R₂*) allows the determination of the velocity of conduction. *(Modified from Mackie 1965.)*

ocelli, and most likely provides rapid circular conduction of activity between tentacles. The lower nerve ring is primarily motoneurons to the subumbrella and velum. The nerve rings, or marginal ganglia, often have three or four types of specialized sensory cells: one or more types of olfactory pits, statocysts, and ocelli. The coelenterate *Geryonia* has a complex system of nerve nets; there is a radial nerve net of upper and lower nerve rings, a tentacle

nerve net, and a manubrium nerve net (Figure 8–11). Numerous radial statocysts and pacemakers control the swimming rhythm. There are also two separate radial, no-net nerve pathways for unidirectional spread of activity for control of swimming and contractions of the manubrium.

Comb jellies (ctenophores) have eight prominent rows of ciliated comb plates. Each comb plate has an underlying nerve plexus that coordinates the

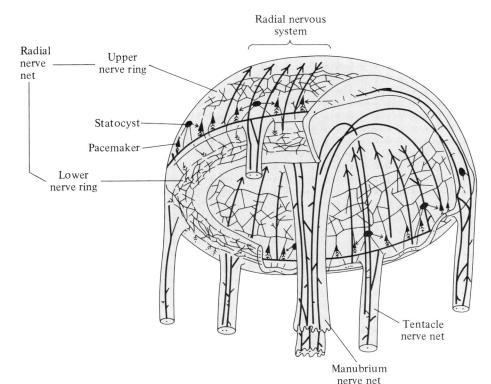

Radial nervous system

Radial nerve net

Upper nerve ring

Statocyst

Pacemaker

Lower nerve ring

Tentacle nerve net

Manubrium nerve net

FIGURE 8–11 Diagram of the multiple nerve-net nervous system of a coelenterate *Geryonia*. The statocysts and pacemakers communicate with both the tentacle-manubrium nerve net (heavy lines) and the nerve net of the velum (thin lines). *(Modified from Bullock and Horridge 1965.)*

beating of each plate, usually from the aboral to the oral end (Bullock and Horridge 1965). The apical region presumably contains pacemaker cells that initiate the ciliary rhythm. At low beating rates, each comb plate has a rhythm independent of the other comb plates, but all plates become synchronized at high frequencies through the general nerve plexus covering the entire body surface. There also is, in some species (e.g., *Beroë*), a circular nerve plexus and ring of sensory cells around the mouth.

The beroid ctenophores are predatory on other ctenophores. Around the inside of their oral lips they have a row of macrocilia, which are large, finger-shaped structures composed of hundreds of ciliary shafts joined in a hexagonal array (Tamm 1988). Local mechanical stimulation and sensory stimulation (e.g., by the juice of pulverized ctenophores) induces beating of the macrocilia. Electrical stimulation of the oral nerve net stimulates macrociliary beating, although the individual ciliated cells are not responsive to direct electrical stimulation. The brilliant luminescence of many ctenophores, such as *Mnemiopsis*, is neurally mediated; spread of luminescence occurs at about 5 to 25 cm sec^{-1} and is accompanied by action potentials.

The neurons and nerve nets of cnidarians and ctenophores illustrate many of the basic properties of the central nervous systems of higher animals. The neurons show integrative mechanisms: spatial and temporal summation, facilitation, inhibition, and high-velocity conduction. The nerve nets have pacemaker regions, specialized ganglia, and associated sensory structures; there also are simple reflex arcs associated with movement and feeding. Whether complex neural structures such as these should be regarded as a central nervous system depends more on our particular definition of a central nervous system than the demonstrated specialized role of these neural structures. However, the neural organization of cnidarians and ctenophores has not reached the stage of nerve cords and brains. This is probably due to their radial symmetry; nerve cords and brains have evolved in bilaterally-symmetrical animals.

Nerve Cords and Brains

Animals of most phyla higher than Cnidaria have a bilateral rather than radial symmetry, i.e., they have an anterior and posterior end and a left and right side. Bilateral symmetry is accompanied by the process of **cephalization**, the specialization of the anterior end into a head with an aggregation of neurons to form a **brain**. The brain is connected to a variable number of well-developed nerve cords that may be interconnected by commissures.

Platyhelminth worms (flatworms) well illustrate the development of a bilaterally-symmetrical, cephalized nervous system, although they exhibit a wide diversity in neural organization. They have two nerve plexuses, an often poorly developed subepidermal plexus and a well-defined submuscular plexus; from one to four pairs of longitudinal nerve cords; and a more-or-less distinct anterior neural specialization—a brain (Figure 8–12A). The primitive acoel flatworms have about five pairs of longitudinal cords of nerve tissue with irregular commissures. More advanced flatworms have four pairs of well-developed nerve cords with numerous commissures (Figure 8–12B). The polyclad flatworms have a well-developed pair of ventral nerve cords and a less distinct pair of dorsal nerve cords. In marine triclad flatworms, the ventral nerve cords are important, but the lateral (and sometimes dorsal) nerve cords are reduced, or absent; there also are additional connectives between the dorsal and ventral nerve cords. Terricolous triclad flatworms have an additional well-developed plexus, the ventral nerve plate.

The brain appears first in the animal kingdom in turbellarian flatworms. A brain may be lacking; may be merely a node of nerve fibers; or may be a large, well-developed ganglion of neurons and fibers. There essentially are two different theories for the evolution of a brain. According to one theory, a sensory statocyst acts as the primary organizer of a cerebral ganglion. The statocyst is originally epidermal, but it and its surrounding neurons move underneath the epidermis to form an endom (brain), which ultimately loses all contact with the epidermis. In contrast, the orthogonal brain theory suggests that the brain originates as a cephalic development of a longitudinal nerve cord (or cords). These theories are not mutually exclusive. Some acoel flatworms have both a primordial endom and an orthogonal brain. The brain of catenulid turbellarians has a well-developed periphery of cell bodies and a central neuropile of unmyelinated dendrites and axons (Moraczewski, Czubaj, and Bakowska 1977). Some have a brain with a statocyst (e.g., the large posterior portion of the brain of *Catenula* contains a statocyst) or without a statocyst (e.g., *Stenostomum*).

The brain of many platyhelminth worms is not essential for normal, or even near normal, function. For example, removal of the brain has little effect in many planarians (triclad, terricolous turbellarians), except to decrease locomotor rate and impair chemosensory detection of food. The planarian can still reflexly swallow food, is still negatively phototaxic, and can still locomote fairly normally (although slower). The brain does not appear to be required

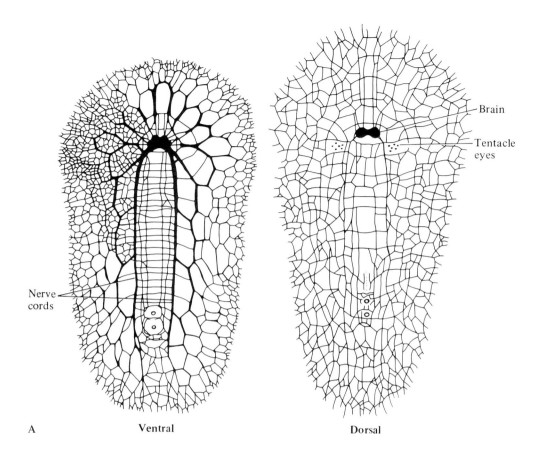

Nerve
cords

Brain

Tentacle
eyes

A   **Ventral**   **Dorsal**

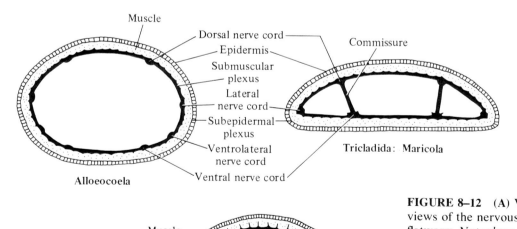

Muscle

Dorsal nerve cord

Epidermis

Submuscular
plexus

Lateral
nerve cord

Subepidermal
plexus

Ventrolateral
nerve cord

Ventral nerve cord

Commissure

Alloeocoela

Tricladida: Maricola

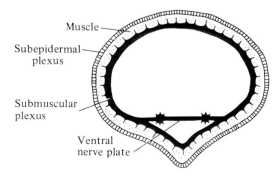

Muscle

Subepidermal
plexus

Submuscular
plexus

Ventral
nerve plate

Tricladida: Terricola

FIGURE 8–12   (**A**) Ventral and dorsal
views of the nervous system of a polyclad
flatworm *Notoplana* showing the nerve
cords and "brain." (**B**) Schematic cross-
section of the nervous system in various
types of flatworm. The Alloeocoela have
four pairs of longitudinal nerve cords connected by the submuscular nerve plexus. In
other flatworms, the number of pairs of
nerve cords is reduced and there may be
additional commissures connecting the
nerve cords. *(From Bullock and Horridge
1965.)*

B

for specific actions, although it maintains a nervous level of excitability, or "tone." Nevertheless, planarians are common subjects for behavioral experiments involving locomotory responses to various stimuli (e.g., phototaxis, geotaxis, chemotaxis) and even learning. The latter phenomenon, in particular, would seem to be a complex neural role of a central nervous system, but planarians can be "trained" (conditioned; see below). For example, a planarian can be conditioned to remain motionless during illumination but not if its brain has been removed (Hovey 1929). In contrast, the planarian *Dugesia* does not require its brain to "remember" a conditioned response; control planarians as well as the

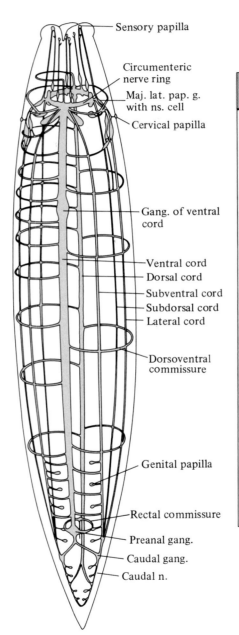

Ganglia, or Peripheral structure	Nerve Cell Count
Central Nervous System	
Subdorsal papillary ganglia	7 × 2 = 14
Subventral papillary ganglia	7 × 2 = 14
Lateral papillary ganglia	4 × 2 = 8
Amphidial ganglia	11 × 2 = 22
Internal lateral ganglia	11 × 2 = 22
External lateral ganglia	13 × 2 = 26
Circumenteric nerve ring	4
Dorsal ganglion	2
Subdorsal ganglia	2 × 2 = 4
Ventral ganglion	33
Retrovesicular ganglion	13
	Total 162
Tail Structures	
Preanal ganglion	11
Preanal sensory cells (Bursal papillae) ♂	appr 15 × 2 = 30
Lumbar ganglia	appr. 6
Postanal sensory cells (incl. phasmidial ganglion) ♂	7 × 2 = 14
Latero-caudal nerves	3 × 2 = 6
Enteric Systems	
Esophageal "sympathetic"	appr. 17
Rectal "sympathetic"	2 × 2 + 4 = 8
Grand total	appr. 254

Figure labels: Sensory papilla; Circumenteric nerve ring; Maj. lat. pap. g. with ns. cell; Cervical papilla; Gang. of ventral cord; Ventral cord; Dorsal cord; Subventral cord; Subdorsal cord; Lateral cord; Dorsoventral commissure; Genital papilla; Rectal commissure; Preanal gang.; Caudal gang.; Caudal n.

FIGURE 8–13 The nervous system of the parasitic nematode *Ascaris* resembles that of flatworms but has a greater anterior cephalization and also posterior specialization of the nervous system. The neurons (including sensory receptors) have been completely mapped; the approximate neuron counts for different parts of the central nervous system, tail, and enteric nervous systems are indicated. *(Modified from Bullock and Horridge 1965.)*

regenerated anterior and posterior parts of a planarian cut in halves are able to relearn a conditional response with equal facility (McConnell, Jacobson, and Kimble 1959).

Nemertine worms, nematodes, and annelid worms have a nervous system that is distinctly more organized than that of platyhelminths. Their brain is well developed, with a variable number of nerve cords. Nevertheless, the nervous system is still sufficiently simple and with a small enough number of neurons that all of the neurons and their synaptic connections can be mapped. For example, the

nematode *Ascaris* has about 254 neurons divided between 11 ganglia of the central nervous system, 5 ganglia of the tail region, and the esophageal and rectal enteric ganglia (Figure 8–13).

Annelid worms have a bilobed brain located beneath the dorsal epithelium and a double ventral nerve cord that is fused to varying degrees. There is a single ganglionic swelling of the nerve cord per body segment that forms three or four pairs of lateral nerves (Figure 8–14A). The ganglion consists of a peripheral rind of interneuron and motoneuron cell bodies (about 500 per ganglion) and a central

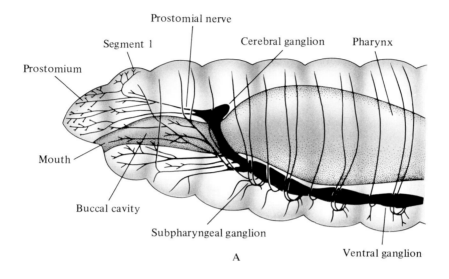

A

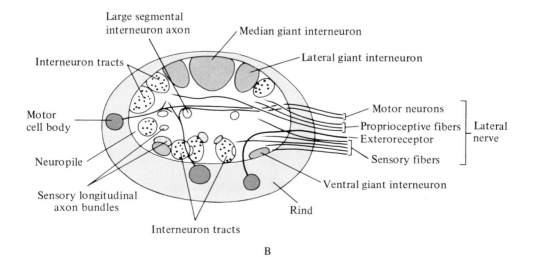

B

FIGURE 8–14 **(A)** Lateral view of the anterior nervous system of an oligochete annelid worm *Lumbricus* showing the anterior cephalization of the central nervous system to form a cerebral ganglion. **(B)** Diagrammatic cross-section of a ventral nerve cord ganglion showing the organization of an outer rind composed of cell bodies of interneurons and motoneurons and an inner neuropile containing giant axons, interneuron axons, and sensory axons.

neuropile of dendrites and axons. The neuropile contains giant axons, interneuron fiber bundles, and sensory fiber bundles (Figure 8–14B). The relatively simple nervous system of leeches is particularly amenable to neurophysiological study; many of its neurons have been mapped and their specific sensory or motor roles determined. The neural circuits responsible for swimming motions have also been mapped (see Supplement 8–2, page 388).

The giant axons of annelids rapidly conduct information for reflex muscle contraction during defensive avoidance. There is a considerable diversity in the arrangement of giant fiber systems of polychete worms (Nicol 1948). Nereid polychetes characteristically have five giant, longitudinal fibers: two large laterals, one medium-sized medial, and two small paramedial fibers (Figure 8–15). The lateral pair of giant axons of *Nereis* are divided segmentally by transverse septa; presumably there is a separate cell soma per segment. The smaller paramedial giant fibers extend anteriorly through two adjacent segments. Some giant axons are intersegmental motoneurons that arise in cell bodies of each segment and enter the peripheral nerves after decussating, e.g., *Euthalenessa*. Most polychetes also have giant intersegmental axons running longitudinally between segments. These giant fibers are interneurons, not motoneurons. *Euthalenessa* has a number of such anteroposterior giant interneurons; most arise from single cell bodies, but one arises from the fusion of axons from two cells. *Eunice* has a single medial giant axon and paired lateral giant axons running the length of the nerve cord. *Myxicola* has a single longitudinal giant axon (the largest in the animal kingdom, at about 1.5 mm dia) that arises by fusion of two anterior decussating axons; there are pairs of segmentally arranged motoneurons also with cytoplasmic fusion with the giant longitudinal fiber.

The general plan of the arthropod nervous system resembles that of annelids. There is a dorsal, anterior brain that has circumesophageal connectives and a ventral nerve cord with segmental thoracic and abdominal ganglia, commissural connectives, and peripheral nerves. The ventral nerve cord has segmentally arranged ganglia, joined longitudinally by connectives and transversely by one (or more primitively two) commissures. There is a marked tendency in higher crustaceans and insects for anterior fusion (and also for posterior fusion) of the segmental ganglia (see Figure 8–3B, page 331).

The detailed structure of the arthropod brain differs from that of annelids, in part because of the different roles of sensory organs in the two groups. The arthropod brain consists of three main regions

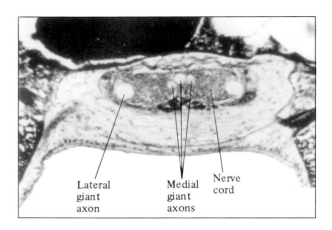

Lateral giant axon   Medial giant axons   Nerve cord

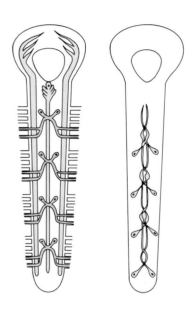

FIGURE 8–15   The ventral nerve cord of the polychete annelid *Nereis virens* has five giant axons. Two large lateral axons and a smaller medial axon extend the length of the nerve cord; the large lateral axons contact a pair of axons in each segment that enter peripheral nerves. There is also a pair of small medial gaint axons that extend anteriorly a few segments.   *(From Nicol 1948.)*

(Figure 8–16A). The **protocerebrum** has several neuropile areas (paired optic lobes, median central body, and protocerebral bridge) for sensory integration, especially for the eyes, and for control of motor activity in response to that sensory integration (Figure 8–16B). Complex forms of behavior may also be initiated by the protocerebrum. The **deuterocerebrum** has neuropile areas for reception of antennal sensory information. The **tritocerebrum** has neuropile areas for the second pair of antennae (in

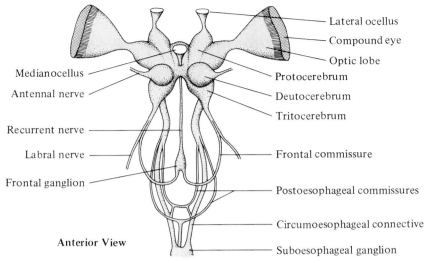

Anterior View

A

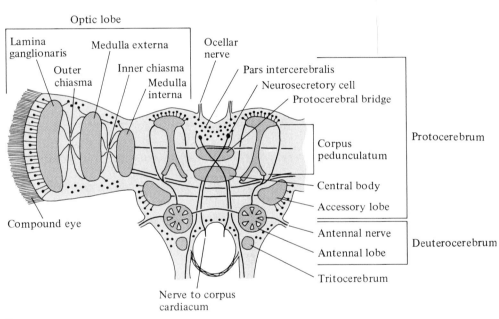

B

FIGURE 8–16 (**A**) The brain of the locust *Locusta* shows considerable cephalization. It has three main lobes (proto-, deutero-, and tritocerebrum) and a large optic lobe for each compound eye. The bilobed protocerebrum is the most complex part, being continuous with the important optic lobes. (**B**) The main neuropile areas (stippled) of the insect brain and interconnecting nerve tracts; some neuron cell bodies are also indicated.
(From Chapman 1982.)

crustaceans) and the anterior alimentary canal and mouth parts. The first ganglion, the subesophageal ganglion, is actually a fusion of the "segmental" ganglia corresponding to the different mouthparts; it is responsible for chewing movements and also tonically excites the more posterior ganglia.

Giant fiber systems are common in many crustaceans and some insects. For example, the tail-flick escape response of crayfish and the rapid escape movements of dragonfly nymphs, cockroaches, and dipterans are mediated by giant interneuron axons. The cockroach jumps forward when a puff of air stimulates the anal appendages, the cerci. The sensory fibers from the cerci synapse with the intersegmental giant axons, which rapidly conduct activity to the metathoracic ganglia where they synapse with the motoneurons responsible for jumping forward (Figure 8–17). This rapid reflex can be modified by descending activity from the brain. The giant axons of other insects have a similar role in minimizing the delay between sensory-interneuron-motor neuron transmission.

The arthropod brain generally does not initiate important aspects of peripheral nervous system function, although it may modify neural activity. For example, gut motility, walking, flying, and respiratory ventilation are primarily controlled by pattern generators in the segmental ganglia, rather than by the brain.

A **central pattern generator** (CPG) is a group of neurons organized in the central nervous system (brain or nerve cord ganglia) that produces an often complex series of neural activity patterns responsible for motor actions. Central pattern generators can be modified, or modulated, by other neural inputs or hormones to provide flexibility of motor

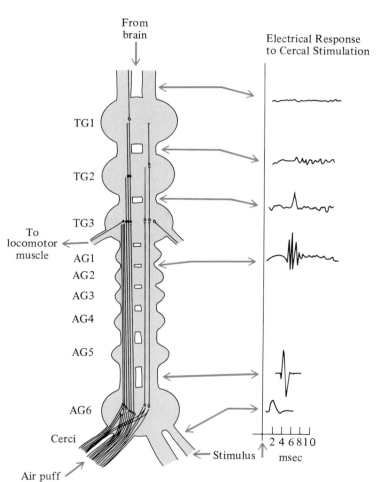

FIGURE 8–17   The rapid escape response of the cockroach is triggered by sensory stimulation of the anal cerci (indicated by arrows) by a puff of wind or low-frequency vibration. The cercal nerves synapse in the sixth abdominal ganglion (AG6) with giant axons that extend to the third thoracic ganglion (TG3) where they stimulate the locomotory muscles. Stimulation of the cerci is rapidly transmitted to the third thoracic ganglion as indicated by electrical activity in the nerve cord (right).   *(Roeder 1948; Lynwood M. Chance, National Audubon Society/Photo Researchers, Inc.)*

output under changing internal or external conditions (Dickinson, Mecsas and Marder, 1990; Meyrand, Simmens and Moulins, 1991).

Two well-studied, model CPGs are the pyloric pattern generator and the gastric mill networks of the stomatogastric ganglion of decapod crustaceans (Dickinson, Nagy, and Moulins 1988). The pyloric pattern generator controls the filtering movement of the pyloric stomach. Its activity can be altered by modulatory neurons, such as the anterior pyloric interneuron, the proctolin neuron, and the pyloric suppression neuron. The gastric mill network controls the chewing movements of the lateral and medial teeth of the gastric mill. Synaptic interactions between a network of neurons are responsible for the rhythmic pattern of gastric mill movement, and extrinsic rhythmic inputs also modulate the gastric mill rhythmn (Figure 8–18). The gastric mill has two systems of motor neurons, one system controlling the power and recovery strokes of the lateral teeth (neurons LG, MG, and LPG) and the other controlling the power and recovery strokes of the single median tooth (GM, DG, and AM). These two systems are linked by an interneuron (Int 1) and also by direct electrical connections of the motoneurons. The stomatogastric network of these neurons may produce its own intrinsic rhythm, but rhythmic inputs from commissural ganglia (neurons CG, P) are also important in forming the normal oscillatory cycle of gastric mill activity. In addition, the anterior pyloric neuron (APM) in the esophageal ganglion can activate the rhythmic oscillatory system of the gastric mill. These examples of central pattern generators illustrate convergence, whereby several modulatory inputs converge on a single motor network and also divergence where a single modulatory neuron (APM) can influence more than one CPG.

Walking and flying in insects is controlled to a large extent by sensory-motoneuron feedback loops for each individual leg via its associated nerve cord ganglion. For example, in the walking cockroach there are nonoverlapping bursts of activity in the coxal (leg) levator and depressor motoneurons (Bowerman 1977); the basic pattern persists even if all afferent neurons in the legs are eliminated, hence the basic rhythm is locally generated. However, there are important inputs from tonic proprioceptors monitoring limb position and movement, and campaniform sensilla (cuticular stretch proprioceptors), to the rhythm generating system, as well as a command fiber system that descends from higher levels in the brain. For example, each of the legs of the stick insect *Cuniculina* has its own pattern generator for walking movements. Sensory inputs

from the leg include leg position (femoral chordotonal organ) and loading of the leg (campaniform sensilla). The six different pattern generator systems, one for each leg, must be coordinated through central nervous system processes but how this is accomplished is not understood. Wing movements are generally initiated by loss of tarsal contact with the substrate when the insect initially jumps into the air. The absence of tarsal contact is sufficient to maintain wing movements for some insects (e.g., *Drosophila*) but most insects require additional sensory stimulation by wind movement over the hair beds on the face (locusts) or the antennal segments (dipterans) to maintain wing movements. Interneurons of the ventral nerve cord ganglia may provide an inherent oscillatory rhythmn for wing movement but this cycle is modified by sensory input from wing position sensors.

Control of abdominal ventilation and spiracular opening by the locust provides another example of local ganglionic control of motor function. Accumulation of CO_2 and, to a lesser extent, depletion of O_2 directly stimulates "respiratory center" neurons in the ganglia of the ventral nerve cord. Each abdominal ganglion can initiate and maintain an autonomous rhythm, but in some insects (*Schistocerca* and *Periplaneta*) a bursting neuron in the metathoracic ganglion (G3) acts as a coordinating pacemaker through an interneuron in each ventral connective that extends to the terminal abdominal ganglion. This interneuron activates the expiratory muscles via an intrasegmental interneuron and weakly inhibits the inspiratory muscles. The intersegmental interneuron is continually active except when inhibited by the bursting neuron. Therefore, the expiratory muscles are also continually active and the inspiratory muscles inactive except when the bursting neuron is active and the intersegmental interneuron is inhibited. The activity of the respiratory muscles is coordinated with the opening and closing of the spiracles.

The molluskan nervous system shows a tendency for cephalization, the fusion of the nerve cord ganglia into a few, large central neuron masses (Figure 8–19). The basic plan of the nervous system of a gastropod mollusk, for example, is a circumesophageal nerve ring with a ventral pair of pedal nerve cords innervating the foot and a dorsal pair of visceral nerve cords innervating the viscera and mantle. The main ganglia are the cerebral (or supraesophageal) ganglia innervating the head and neck; the buccal ganglia (or subesophageal ganglia) innervating the pharynx, esophagus, stomach, and salivary glands; the pedal ganglia innervating the

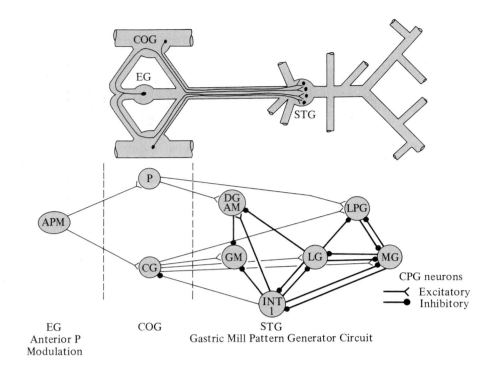

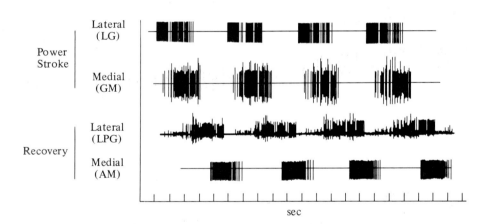

FIGURE 8–18 Neuronal circuit model for the central pattern generator controlling the gastric mill of the crustacean *Palinurus* is comprised of neurons in the stomatogastric ganglion (STG) with extrinsic inputs from the commissural ganglion (COG) and esophageal ganglion (EG). The basic STG neuronal circuit is a series of six interconnected neurons (with excitatory and inhibitory synapses); these include motoneurons controlling the power stroke of the medial tooth (GM) and lateral teeth (LG and MG), the recovery stroke of the medial tooth (DG and AM) and lateral teeth (LPG), and an interneuron (Int 1). There is a coordinated rhythm of power stroke and return stroke for the medial and lateral teeth. The basic gastric mill rhythm generated by these neurons is modulated by P and CG neurons in the commissural ganglion and the APM neuron in the esophageal ganglion. Abbreviations for individual neurons: AM = anterior median neuron; DG = dorsal gastric neuron; GM = gastric mill neuron; LG = lateral gastric neuron; MG = median gastric neuron; LPG = lateral posterior gastric neuron; Int 1 = interneuron 1; CG = commissural neuron; P = P neuron; and APM = anterior pyloric modulator neuron. *(Modified from Dickinson, Nagy, and Moulins 1988.)*

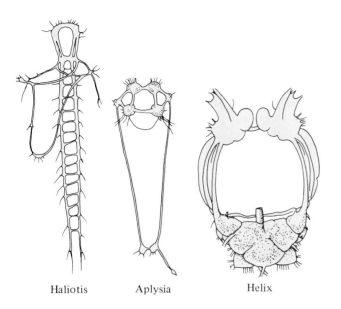

Haliotis   Aplysia   Helix

Helix

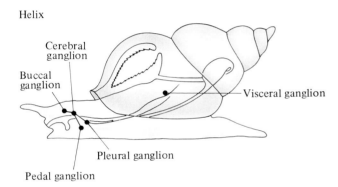

Cerebral
ganglion

Buccal
ganglion

Visceral ganglion

Pleural ganglion

Pedal ganglion

FIGURE 8–19 The nervous system becomes more cephalized in higher mollusks. The generalized gastropod molluscan nervous system is illustrated by that of the land snail *Helix.* *(Modified from Bullock and Horridge 1965; Barnes 1986.)*

muscles and skin of the foot; the parietal ganglia innervating the lateral body wall and mantle; and the visceral ganglia (or ganglion) innervating the viscera, posterior gut, and posterior body. Giant neurons are found in many gastropod mollusks (Gillette, 1991).

The nervous system of cephalopod mollusks (nautilus, squid, octopus) is basically similar to that of the other mollusks (with homologous areas to the cerebral, buccal, pedal, parietal, and visceral ganglia), but it is superficially arranged quite differently. There are additional branchial, optic, olfactory, and peduncle ganglia in the brain and additional

peripheral ganglia (e.g., the stellate ganglion). Cephalopods are the only invertebrates whose circumesophageal ganglionic masses, or brain, essentially include the entire central nervous system. The well-developed and highly organized central nervous system is correlated with their being among the most active invertebrate animals; they are predatory, with a highly developed visual sense and varied behavioral repertoire (including recognition, learning, color change, and bioluminescence). Giant axons provide a rapid escape response mechanism (see Supplement 8–3, page 390).

Chordate Nervous System

The organization of the vertebrate nervous system is fundamentally different from that of invertebrates. The primitive chordates (urochordates) and the presumably distantly related hemichordates have a nervous system more like that of primitive invertebrates and do not well illustrate an intermediate stage in the evolution of the vertebrate nervous system.

Hemichordates (enteropneust worms and pterobranchs) have an intradermal nerve plexus with localized thickenings of nervous tissue and two nerve cords in the trunk. The collar portion of the dorsal nerve cord is hollow (perhaps a forerunner of the hollow dorsal nerve cord of vertebrates). Giant unipolar neurons with giant axons innervate the longitudinal muscles of the trunk and initiate body and possibly proboscis shortening. The sensory nervous system is poorly differentiated, consisting of individual epithelial sensory cells. Some species have a modified epithelial structure called the preoral ciliary organ, which has a concentration of epidermal sensory cells. The bioluminescent photophores are under complex control by nerve cells. The urochordate nervous system is generally invertebrate-like and primitive (e.g., in tunicates) but the hollow dorsal nervous system of the tunicate larva and appendicularians is reminiscent of the primitive vertebrate nervous system.

Vertebrates have a well-organized hollow dorsal nervous system. The central nervous system is an anterior brain and a single spinal nerve cord. The nervous system develops as a tubular structure with an inner mass of neurons (the gray matter) and an outer covering of axons (the white matter). This is the opposite organization to the outer rind and inner core of invertebrate nervous systems. The peripheral nervous system of vertebrates consists of numerous paired nerves extending from the spinal cord with some peripheral ganglia of neurons.

Peripheral Nervous System. The peripheral nervous system consists of nerves that carry afferent sensory information to the central nervous system and efferent motor commands from the central nervous system to the peripheral effectors. The peripheral nerves connect to both the brain (**cranial nerves**) and the spinal cord (**spinal nerves**).

The spinal nerves are both sensory (afferent) and motor (efferent). The sensory fibers can be divided for convenience into **somatic sensory** fibers that carry proprioceptive and exteroceptive sensation and **visceral sensory** fibers that carry interoceptive sensation. The neuron bodies of the sensory nerves lie outside but near the spinal cord in the dorsal root ganglion; there is one pair of dorsal root ganglia per vertebra. The motor fibers can be similarly divided into **somatic motor** and **visceral motor** fibers. The efferent somatic motor fibers innervate the striated muscles subject to voluntary control, and the visceral motor fibers innervate the involuntary effector systems (gut, blood vessels, glands, etc.).

The spinal cord has separate dorsal and ventral nerves that connect at the dorsal and ventral roots. In primitive chordates (e.g., amphioxus and lampreys), the dorsal nerve and dorsal root carry sensory (somatic and visceral) and visceral motor axons (Figure 8–20). In higher vertebrates, there is a reorganization of this primitive pattern such that the visceral motor fibers exit the spinal cord at the ventral root with the somatic motor fibers, rather than at the dorsal root. Consequently, only afferent (sensory) neurons enter the spinal cord through the dorsal roots, and only efferent (motor) fibers leave through the ventral roots; this is the Bell-Magendie rule.

The cranial nerves serve a variety of sensory and motor functions—they are afferent, efferent, or mixed nerves (Table 8–1). Some of the cranial nerves correspond to the two types of spinal nerves of primitive vertebrates (dorsal root and ventral root; see below), but others are special somatic sensory nerves (from the nose, eye, and acoustico-lateralis systems).

The somatic motor fibers of the **somatic nervous system** arise from neuron bodies located in the brain and ventral portion of the spinal cord, and action potentials are transmitted along a single axon to the voluntary efferent organ (e.g., a striated muscle cell). In contrast, the visceral motor fibers of the **autonomic nervous system** innervate the involuntary visceral organs by a two-neuron chain; there is a peripheral synapse between the spinal cord and the effector organ, and a preganglionic neuron and a postganglionic neuron.

In amphioxus, each segmentally arranged dorsal root has efferent visceral motor fibers that travel to the organs where the peripheral synapse occurs. Elasmobranchs have a more complicated autonomic nervous system because some of the visceral motor fibers synapse in ganglia close to the spinal cord, whereas other fibers synapse at the effector organ (Figure 8–21). The length of the postganglionic fibers varies; it is long for neurons synapsing in the ganglia close to the spinal cord but short for the cranial and those spinal nerves synapsing in ganglia at the target organs. The cranial fibers innervate anterior structures (e.g., eye, pharynx) as well as posterior structures (gut) via the vagus nerve. There is no differentiation in function and no overlap in innervation pattern of long or short postganglionic fibers

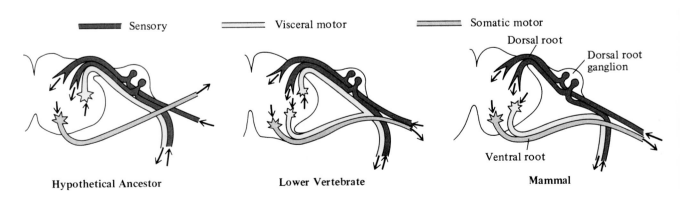

FIGURE 8–20 Schematic arrangement of the spinal nerves in a probable primitive ancestor (left), a lower vertebrate (center), and a mammal (right) showing the reorganization of the visceral motor fibers from the dorsal root nerve (visceral and somatic sensory fibers) to the ventral root nerve (somatic motor fibers). *(Modified from Romer and Parsons 1986.)*

TABLE 8–1

__	Cranial nerves of the vertebrate brain, based on the human brain.	
0	Terminalis[1]	Sensory and motor; from near olfactory bulb to diencephalon.
I	Olfactory	Sensory; from olfactory epithelium.
II	Optic	Sensory; from eye.
III	Oculomotor	Motor; to four of the extrinsic muscles of eye.
IV	Trochlear	Motor; to superior oblique eye muscle.
V	Trigeminal	Sensory and motor; to head.
VI	Abducens	Sensory and motor; from nucleus in pons to posterior rectus eye muscle.
VII	Facial	Sensory and motor; to face.
VIII	Acoustic[2]	Sensory; from inner ear.
IX	Glossopharyngeal	Mainly sensory; to tongue and pharynx.
X	Vagus	Sensory and motor; to head and viscera.
XI	Accessory[3]	Motor; accessory to the vagus.
XII	Hypoglossal	Motor; to muscles of tongue.

[1] A small and poorly understood cranial nerve (see Demski and Schwanzel-Fukuda, 1987).
[2] Also called auditory or stato-acoustic nerve.
[3] Also called spinal accessory nerve.

to any particular organ. In teleost fish, the long postganglionic axons of the trunk autonomic ganglia also innervate those organs innervated by the more cranial autonomic nerves and the cranial vagus nerve; this foreshadows the autonomic nervous system of higher vertebrates. The urinary bladder of teleosts is innervated by long preganglionic fibers from the spinal cord that synapse at the bladder, not by the postganglionic fibers of a trunk ganglion.

The mammalian autonomic nervous system illustrates the more complex organization of the autonomic nervous system in higher vertebrates. It is divided into two functional branches: the **parasympathetic** branch and the **sympathetic** branch (Figure 8–22). The preganglionic fibers of the parasympathetic system exit the central nervous system via cranial nerves and sacral spinal nerves and synapse at the peripheral organ. The preganglionic fibers of the sympathetic system leave the spinal cord via thoracic and lumbar spinal nerves and either synapse close to the spinal cord in the sympathetic chain ganglia or synapse in more peripheral sympathetic ganglia. These fibers leave the ventral root of the spinal cord and enter the sympathetic chain ganglia via the white ramus (the axons are myelinated and the tissue appears white). Many synapse in the sympathetic chain ganglion and the postganglionic fibers leave the sympathetic ganglion by the gray ramus (these axons are unmyelinated) and pass to the peripheral organs. Some preganglionic fibers pass via connectives to more anterior or posterior autonomic chain ganglia or to peripheral sympathetic ganglia, where they synapse.

The sympathetic and parasympathetic systems have antagonistic actions (Table 8–2). In general, the sympathetic nervous system prepares the body for response to stressful or dangerous situations; it initiates the fight-or-flight reaction, e.g., elevation of heart rate and increased force of contraction,

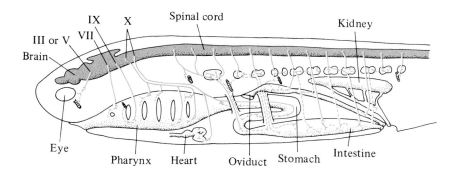

FIGURE 8–21 Representation of the autonomic nervous system of a shark showing a combination of long preganglionic fibers (parasympathetic type) synapsing with the postganglionic neuron near the target organ and short preganglionic fibers (sympathetic type) synapsing in sympathetic ganglia near the spinal cord. Preganglionic fibers are shown as solid lines, postganglionic fibers as dashed lines. Note that there is no pattern of antagonistic innervation of organs by parasympathetic and sympathetic systems.

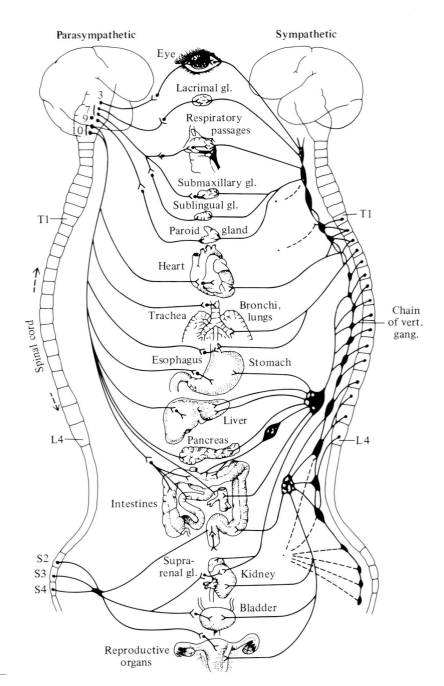

Parasympathetic

Sympathetic

Eye

Lacrimal gl.

Respiratory passages

3
7
9
10

Submaxillary gl.

Sublingual gl.

T1

Paroid gland

T1

Heart

Bronchi, lungs

Trachea

Spinal cord

Chain of vert. gang.

Esophagus

Stomach

Liver

Pancreas

L4

L4

Intestines

S2
S3
S4

Supra-renal gl.

Kidney

Bladder

Reproductive organs

FIGURE 8–22 The mammalian (human) autonomic nervous system has antagonistic parasympathetic and sympathetic branches. *(Modified from Millard, King, and Showers 1956.)*

peripheral vasoconstriction, and sweating. In contrast, the parasympathetic nervous system controls general bodily functions, such as digestion. Sympathetic stimulation is not always excitatory, and parasympathetic stimulation is not always inhibitory, but their actions are generally antagonistic.

The sympathetic and parasympathetic nervous systems have different neurotransmitters at their postganglionic synapse with the effector organ. The neurotransmitter of the postganglionic parasympathetic synapse is acetylcholine, so the postsynaptic

receptors of these synapses are **cholinergic**. The synapse is also stimulated by muscarine, hence these postsynaptic receptors are also **muscarinic**. The neurotransmitter of the sympathetic postganglionic synapse is usually norepinephrine (noradrenalin), so these are **adrenergic**. The sympathetic neurotransmitter is epinephrine (adrenalin) rather than norepinephrine in some vertebrates. Some sympathetic synapses are cholinergic (e.g., fibers to sweat glands, ptiloerector muscles, and some blood vessels). The neurotransmitter at the preganglionic

TABLE 8–2

Effects of the sympathetic and parasympathetic branches of the autonomic nervous system on some of the organs of the body.

Organ	Sympathetic Stimulation	Parasympathetic Stimulation
Eye:pupil	Dilation	Constriction
Ciliary Muscle	Slight relaxation	Constriction
Heart:muscle	Increased rate Increased force	Decreased rate Decreased force
Coronary Vessels	Dilation (β) or constriction (α)	Dilation
Gut:lumen	Decreased peristalsis	Increased peristalsis
Sphincters	Increased tone	Decreased tone
Glands:nasal	Slight secretion	Copious secretion
Salivary	Slight secretion	Copious secretion
Gastric	Slight secretion	Copious secretion
Pancreatic	Slight secretion	Slight secretion
Sweat	Copious sweating	None
Apocrine	Copious secretion	None
Adrenal Medulla	Stimulates secretion	None
Basal Metabolism	Increased	None

synapse of both the parasympathetic and sympathetic branches is acetylcholine. This cholinergic synapse is stimulated by nicotine, but not muscarine; its postsynaptic receptors are **nicotinic**.

There are two different types of acetylcholine receptor: the nicotinic receptor binds nicotine and the muscarinic receptor binds muscarine. There are also two basic types of epinephrine/norepinephrine receptor: α and β receptors. In mammals, α receptors are stimulated by norepinephrine and epinephrine, whereas β receptors are stimulated mainly by norepinephrine. Some effects of α receptors are excitatory (e.g., vasoconstriction), whereas some are inhibitory (e.g., intestinal relaxation). Similarly, some effects of β receptors are excitatory (cardioacceleration) and some are inhibitory (vasodilation).

Spinal Cord. The spinal cord is the part of the central nervous system that is surrounded and protected by the vertebral column. It is a slightly flattened hollow nerve cord with a series of dorsal and ventral roots, divided (in humans) into cervical (8 pairs), thoracic (12), lumbar (5), sacral (5), and coccygeal (1) nerves.

In cross section, the spinal cord is organized as a butterfly shaped mass of **gray matter** (neuron cell bodies and unmyelinated fibers) surrounded by **white matter** (myelinated and unmyelinated nerve tracts running longitudinally along the spinal cord; Figure 8–23). The neuron cell bodies are organized into three discrete areas of the gray matter: the dorsal horn, the intermediate zone, and the ventral horn; each area contains a number of nuclei. The substantia gelatinosa of the dorsal horn receives input from somatic afferents and descending fibers from the brain, and sends fibers to the gray matter and into the tracts of the white matter. The interomediolateral nucleus of the intermediate zone contains the preganglionic neurons of the autonomic nervous system. The ventral horn contains the motoneuron bodies for the somatic motor system. There also are interneuron and commissural neuron bodies in the gray matter.

The white matter contains many sensory tracts ascending to the brain and motoneurons descending from the brain (Figure 8–23). The main afferent sensory tracts (ascending tracts) are the anterior and lateral spinothalamic, dorsal column (relaying in the nucleus gracilis and nucleus cuneatus), and anterior and lateral spinocerebellar tracts. These and other fiber tracts carry sensory information to the brain, as well as to other parts of the spinal cord. For example, the lateral spinothalamic tract has fibers that enter the dorsal root; their cell bodies are located in the dorsal root ganglion. These fibers synapse in the gray matter and decussate through the gray matter and enter the white matter of the lateral spinothalamic tract (Figure 8–24). Efferent (descending) motor pathways from the cerebral cortex, midbrain, and cerebellum synapse with the motoneurons in the ventral horn of the gray matter. The main motor tracts are the corticospinal, rubrospinal, tectospinal tracts and the vestibulospinal and reticulospinal tract. For example, fibers from the anterior and lateral corticospinal tracts enter the gray matter and synapse with motoneurons in the ventral gray matter; their axons leave the spinal cord via the ventral root.

The spinal cord is not merely a pathway for sensory and motor connections between the brain and the peripheral nervous system. It is also responsible for sensory-motor coordination, particularly reflex arcs. The **extensor reflex** (the familiar knee jerk or myotactic reflex) is a two-neuron (sensory and motor) monosynaptic reflex arc (Figure 8–25A). Sudden stretch of the quadriceps femoris muscle (caused by tapping the patellar tendon) initiates a **monosynaptic** (only one synapse in the central nervous system), **ipsilateral** (the response is only on the same side of the body), and **intrasegmental** (only one spinal cord segment is involved) reflex contraction of the quadriceps femoris muscle, causing sudden extension of the lower leg. The Renshaw cell, an inhibitory interneuron that receives collaterals from

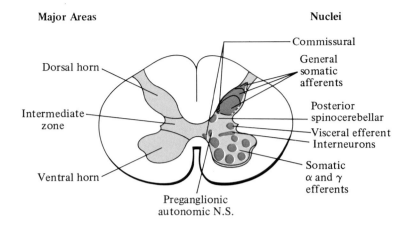

Major Areas

Nuclei

Commissural

General somatic afferents

Posterior spinocerebellar

Visceral efferent

Interneurons

Somatic α and γ efferents

Dorsal horn

Intermediate zone

Ventral horn

Preganglionic autonomic N.S.

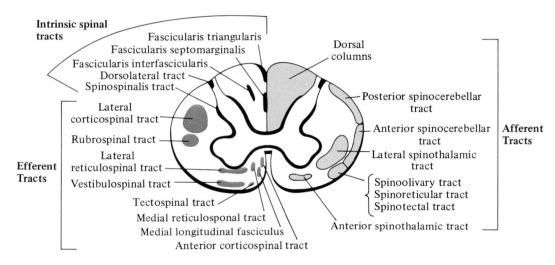

Intrinsic spinal tracts

Fascicularis triangularis
Fascicularis septomarginalis
Fascicularis interfascicularis
Dorsolateral tract
Spinospinalis tract

Dorsal columns

Posterior spinocerebellar tract

Lateral corticospinal tract

Rubrospinal tract

Lateral reticulospinal tract

Vestibulospinal tract

Anterior spinocerebellar tract

Lateral spinothalamic tract

Spinoolivary tract
Spinoreticular tract
Spinotectal tract

Afferent Tracts

Efferent Tracts

Tectospinal tract
Medial reticulosponal tract
Medial longitudinal fasciculus
Anterior corticospinal tract

Anterior spinothalamic tract

FIGURE 8–23 Schematic location of the major areas, nuclei, and nerve tracts in the spinal cord. The neuron fibers in the white matter of the spinal cord are organized into afferent sensory tracts ascending to more anterior parts of the central nervous system, efferent motor tracts descending from more anterior parts of the central nervous system, and ascending or descending fibers intrinsic to the spinal cord. *(Modified from Noback 1967.)*

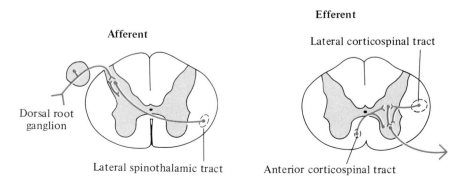

Efferent

Afferent

Lateral corticospinal tract

Dorsal root ganglion

Lateral spinothalamic tract

Anterior corticospinal tract

FIGURE 8–24 Examples of afferent and efferent spinal pathways showing the location of neuron cell bodies and axon tracts. *(Modified from Noback 1967.)*

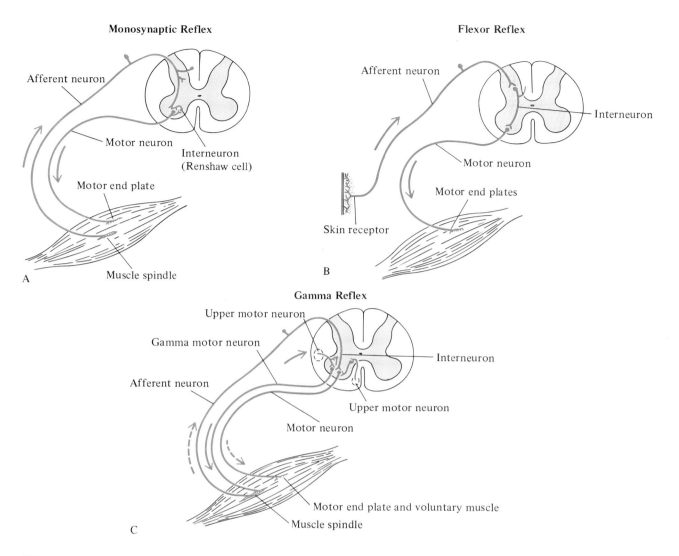

Monosynaptic Reflex

Afferent neuron

Motor neuron

Interneuron
(Renshaw cell)

Motor end plate

Muscle spindle

A

Flexor Reflex

Afferent neuron

Interneuron

Motor neuron

Motor end plates

Skin receptor

B

Gamma Reflex

Upper motor neuron

Gamma motor neuron

Interneuron

Afferent neuron

Upper motor neuron

Motor neuron

Motor end plate and voluntary muscle

Muscle spindle

C

FIGURE 8–25 (**A**) The extensor (knee jerk) reflex is a two-neuron, monosynaptic, ipsilateral, intrasegmental reflex consisting of a sensory input from the annulospiral ending of the muscle spindle and a motoneuron. Recurrent feedback via the inhibitory interneuron (Renshaw cell) is also shown. (**B**) The flexor reflex is a three-neuron, disynaptic, ipsilateral, intersegmental reflex consisting of a sensory input from a cutaneous receptor, interneuron, and a motoneuron. The interneurons have branches to other segments of the spinal cord. (**C**) The gamma reflex loop has an afferent muscle spindle neuron and an efferent motor gamma (γ) fiber which innervates the muscle spindle. Stimulation of the muscle spindle by the γ motor neuron initiates afferent sensory information from the muscle spindle and monosynaptically stimulates the muscle motoneuron. *(From Noback 1967.)*

the motoneuron and prevents prolonged motoneuron activity, is also involved in this extensor reflex circuit. This "fast knee jerk" reflex is a rapid, phasic response to simultaneous stimulation of many muscle spindles in the quadriceps. The "slow knee jerk" is a tonic response to asynchronous discharge of muscle spindles, and it is important in postural adjustment.

The **flexor reflex** loop, involved, for example, in the reflex withdrawal of the hand from a painful stimulus, is a three-neuron, polysynaptic reflex (Figure 8–25B) that is ipsilateral and intersegmental. For example, stimulation of a cutaneous receptor can reflexly elicit withdrawal of the limb. The consequent stimulation of the muscle spindle endings (flower-spray receptors) can facilitate the reflex.

The **gamma reflex** (Figure 8–25C) is a complex reflex loop involved with continuous stretch of a muscle. Stretch receptors in the muscle (muscle spindles) and also the tendon (Golgi tendon organ) control the state of contraction of the muscle fibers. Efferent fibers descending from the brain also are involved with initiation of muscle contraction.

Some reflex loops, such as the crossed-extensor reflex, have contralateral as well as ipsilateral actions. The crossed-extensor reflex uses commissural interneurons in the spinal cord to relay neural information (e.g., a painful stimulus) to the opposite side of the body. For example, a painful stimulus to the left foot can elicit extension of the contralateral leg (i.e., right), to support the body as the flexed left foot is withdrawn from the noxious stimulus.

Brain. The vertebrate brain is a complex, anterior development of the hollow dorsal nerve cord. It may have primitively coordinated local reflexes to the head and pharynx (as in amphioxus); and gathered interpreted sensory information for motor responses. However, there has been a marked tendency in higher vertebrates for the brain to assume control of almost all sensory and motor functions (except, for example, simple spinal reflexes).

The chordate brain is divided into three main parts: the **prosencephalon** (forebrain), **mesencephalon** (midbrain), and **rhombencephalon** (hindbrain; Figure 8–26A). Each of these three parts of the brain in primitive vertebrates has a dorsal area of gray matter associated with the three main senses: olfaction (forebrain), vision (midbrain), and the acoustico-lateralis system (hindbrain). Further subdivision of these parts and their specialized structures in higher vertebrates are summarized in Table 8–3. The basic architecture of the vertebrate brain does not in itself provide an understanding of the functioning of the brain, but it does illustrate the profound evolutionary reorganization and development that has occurred from the more primitive vertebrates (e.g., lamprey, amphibian) to the more advanced vertebrates (e.g., mammals; Figure 8–26B, C, D, E, F).

Hindbrain. The rhombencephalon is divided into the more posterior myelencephalon (most of the medulla) and the more anterior metencephalon (cerebellum, pons, and part of the medulla).

The **medulla**, especially in lower vertebrates, is similar in structure to the spinal cord except that the central canal is enlarged to form a dilated, fluid-filled space, the IVth ventricle. The roof of the IVth ventricle is a thin vascular membrane, the posterior choroid plexus; this and similar anterior choroid plexuses in the third and lateral ventricles secrete

TABLE 8–3

Major divisions of the structure of the vertebrate brain.

Prosencephalon	.... Telencephalon ...... Cerebral hemispheres, including olfactory bulbs and cortex, basal nuclei, cerebral cortex.
	.... Diencephalon ...... Epithalamus, thalamus, hypothalamus.
Mesencephalon	.... Tectum, including optic lobes and tegmentum.
Rhombencephalon	.... Metencephalon ...... Cerebellum, pons, part of medulla oblongata.
	.... Myelencephalon ...... most of medulla oblongata.

the cerebrospinal fluid that fills the ventricles of the brain and the subarachnoid space surrounding the brain. The columns of gray matter of the medulla are continuous with the corresponding gray matter of the spinal cord but tend to become isolated into discrete nuclei. These nuclei are the primary receptor areas for the acoustico-lateralis system. In mammals, there is a separate vestibular nucleus (for static and dynamic equilibrium) and cochlear nucleus (for hearing). Medullary nuclei are also centers for control of specific motor functions, such as heart rate, respiration, and salivation. The pons is a connective tract that is associated with the cerebellum.

The medullary **respiratory center** generates the rhythmic cycles of inspiration and expiration; it is a good example of motor pattern generation and information processing by the brain stem. The respiratory center is actually several widely dispersed nuclei in the medulla and pons. The dorsal respiratory neurons, located dorsally along the entire length of the medulla, control inspiration. The ventral respiratory neurons, also running the entire length but located more ventrolaterally, control both inspiration and expiration. The basic respiratory rhythm is generated by the inspiratory dorsal neurons, which have periods of inactivity (about 3 sec) followed by weak activity that steadily increases in strength, for about 2 secs, then abruptly ceases, until the next "ramp cycle" of activity is initiated. The ventrolateral neurons are almost totally inactive during the normal respiratory cycle but are activated during forced breathing, and they contribute to both the inspiratory part of the cycle and also active expiration (expiration is normally a passive recoil of the lungs, not requiring the contraction of expiratory muscles). The basic respiratory rhythm is continuously modified by the action of the **pneumotaxic**

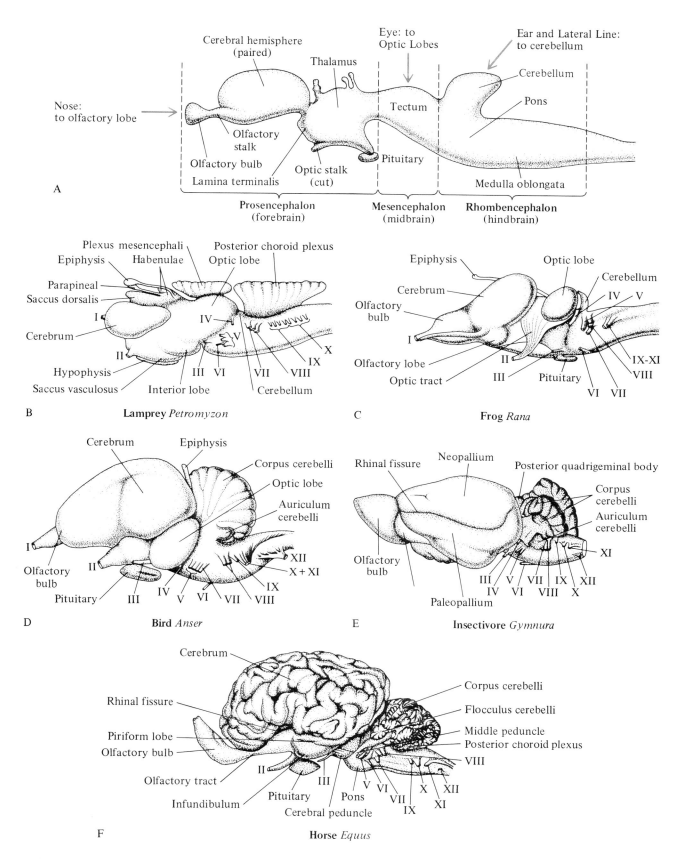

FIGURE 8–26 **(A)** Schematic lateral view of the vertebrate brain showing the three major divisions (prosencephalon, mesencephalon, and rhombencephalon), generalized sensory inputs, and some specific areas. **(B)** Brain of a primitive vertebrate, the lamprey *Petromyzon*. **(C)** Brain of an amphibian *Rana*. **(D)** Brain of a bird *Anser*. **(E)** Brain of a primitive mammal, the insectivore *Gymnura*. **(F)** Brain of a mammal, the horse *Equus*. *(From Romer and Parson 1986.)*

center, which is located in the upper pons. This center controls the duration of the "inspiratory ramp cycle" of the dorsal inspiratory neurons, hence determines respiratory rate. There is also an **apneustic center** in the lower pons that can prolong inspiration (apneusis, lasting up to 20 sec), although the normal role of the apneustic center is unclear. The basic respiratory rhythm is also subject to modification by other inputs. Stretch receptors in the lung bronchi and bronchioles prevent over inspiration by reflex inhibition of the inspiratory muscles (Hering-Breuer reflex). The chemical sensors (O_2, CO_2, and H^+) of the peripheral aortic bodies and the central respiratory center modify the respiratory rhythm. Motor cortex and muscle proprioceptors stimulate respiration during exercise. The cerebral cortex can voluntarily modify the basic respiratory rhythm.

The **cerebellum** is a large anterior and dorsal structure of the hindbrain. It coordinates and regulates the more automatic motor activities, such as maintenance of posture and locomotion. It does not initiate motor activity but responds to activity from other parts of the brain. A number of sensory (e.g., visual, auditory, tactile) inputs are integrated by cerebellar neural circuits and feedback loops. Specific portions of the outer layer of the cerebellum (which contains cell bodies not nerve fibers) receive sensory input from the spinocerebellar, pontocerebellar, and vestibulocerebellar tracts.

The cerebellar cortex is the first part of the brain, which we have so far discussed, to show a functional **lamination**. The neurons of the cerebellar cortex are organized as a thin sheet, rather than a compacted spherical mass (Figure 8–27A). The architectural arrangement of strata, or layers, of neurons allows processing of information at a number of levels (the layers) while preserving a spatial, two-dimensional organization (a map). The tactile inputs to the cerebellum form a miniature "map" of the body surface on the surface of the cerebellum (Figure 8–27B). Extensive folding of the cerebellar cortex forms folia, which packs a maximum laminar area of neurons into a minimum space. The surface of the cerebral cortex is also highly folded for the same reason. The human cerebellum has 3/4 the surface area of the cerebrum, despite its smaller size, because of its more extensive folding. A laminar organization is a fundamental characteristic of many brains, both vertebrate and invertebrate.

Sensory inputs to the cerebellar cortex from the climbing fibers and mossy fibers are integrated by a variety of neurons (basket cells, stellate cells, Golgi cells) and the output of the Purkinje cells inhibits the deeper cerebellar nuclei (Figure 8–27C). The motor output of the cerebellum, which is only inhibitory, is generated by the deep cerebellar nuclei.

Midbrain. The mesencephalon consists of the **tectum**, thickened gray matter above the central canal, and the **tegmentum**, thinner lateral areas of gray matter. In mammals, major fiber tracts from the cerebral hemispheres form the cerebral peduncles of the midbrain. The midbrain is important in lower vertebrates for the processing of visual information (tectum) and motor output (tegmentum).

The midbrain integrates different modes of information as overlaying spatial "maps." For example, some parts of the tectum correlate visual, auditory, and somatic sensory information; other parts overlay equilibrium, lateral line, and electroreceptor information. The important role of the tectum for visual processing is mainly usurped in mammals by their forebrain, although it retains a visual and auditory role for control of simple reflex movements. In mammals, the corpora quadrigemina (four rounded structures, or colliculi, of the tectum) control reflex movement of the eyes and head in response to visual and other stimuli (the superior colliculi) and reflex control of the head and trunk in response to auditory stimuli (the inferior colliculi).

The tegmentum is essentially an anterior extension of the motor areas of the medulla. It and other similar parts of the brain form a diffuse system of interlacing neurons, the **reticular formation**. This system receives a variety of sensory inputs (spinal cord, cranial nerves, cerebrum, hypothalamus) and has motor output, essentially connecting the anterior brain to the motor centers of the medulla and spinal cord and the α and γ motoneurons via the reticulospinal tract. These descending motor functions are especially important in the lower vertebrates since other motor tracts from the anterior brain are poorly developed.

Forebrain. The prosencephalon is divided into the **diencephalon** and the **telencephalon**. The dorsal portion of the diencephalon is the epithalamus, the lateral portion the thalamus, and the ventral portion the hypothalamus. The epithalamus contains a small nucleus that transmits olfactory information to the brain stem, the pineal and parapineal bodies, and the anterior choroid plexus.

The dorsal thalamus of lower vertebrates is essentially an anterior extension of the tectum, but in mammals the importance of the tectum is reduced and most sensory information is transferred to the cerebral cortex via parts of the dorsal thalamus, the lateral geniculate body (LGB, for vision), medial geniculate body (MGB, for hearing), and ventral nucleus (for somatic sensation). Consequently, the

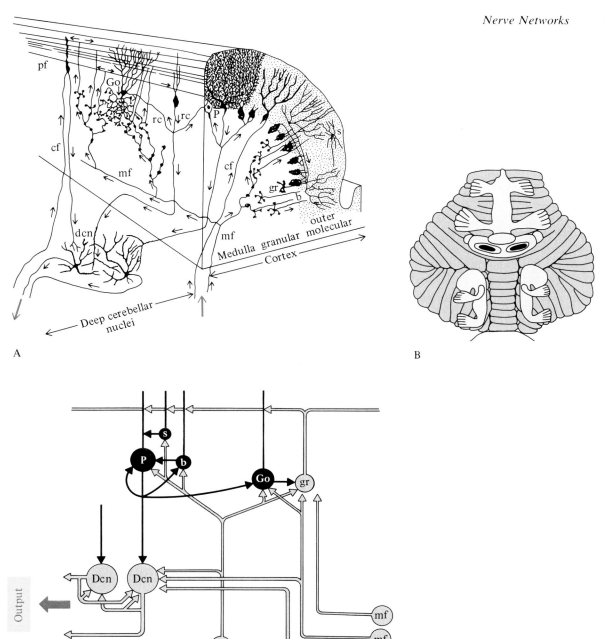

A

B

C

FIGURE 8–27 (**A**) Schematic of the cerebellar cortex showing the interneuronal connections within the three layers of the cortex (inner medullary, granular, and outer molecular layers) and the deep cerebellar nuclei. (**B**) The areas of cerebellar cortex that receive tactile information from the spinocerebellar tract (in the cat) form a somatotopic map. Proprioceptive sensation forms a corresponding, overlapping map, as do projections from the cerebral motor cortex via the pons. (**C**) A simplified circuit of the cerebellar connections shows that the output of the cortex to the deep cerebellar nuclei (Dcn) is entirely inhibitory. Input to the cortex from the mossy fibers (mf) and climbing fibers (cf) is processed by granule neurons (gr), Golgi neurons (Go), basket neurons (b), and stellate neurons (s). The output of the cortex is via the Purkinje neurons (P) to the deep cerebellar neurons (dcn). Excitatory synapses are shown in light shading, inhibitory synapses in dark shading. *(Modified from Crosby, Humphrey, and Lauer 1962; Snider 1950; Thatch 1980.)*

dorsal thalamus is the major integrative station between the cerebral cortex and subcortical structures in mammals.

The floor of the diencephalon contains the optic chiasma, where the optic nerves cross and enter the brain, and the hypothalamus, which is the highest integrative center of the autonomic nervous system. The hypothalamus is only about 1% of the brain mass (in humans) but it is one of the brain's most important motor output paths. It controls most vegetative and endocrine functions of the body, as well as emotions. A number of discrete hypothalamic areas, or nuclei, control specific functions (Figure 8–28), although these areas are poorly delimited. The preoptic area is involved in temperature regulation. The lateral hypothalamus contains a "thirst center" and the supraoptic nucleus secretes antidiuretic hormone, which is released from the posterior pituitary. The paraventricular nucleus secretes ADH and oxytocin, a female reproductive

hormone involved in birth and lactation. There is a "hunger center" in the lateral hypothalamus and a "satiety center" in the ventromedial nucleus. The mammillary bodies activate many feeding reflexes.

The **limbic system** consists of parts of the central nervous system that are structurally and functionally related to the hypothalamus. It includes the most primitive cortical areas (paleocortex, orbitofrontal area, cingulate gyrus, parahippocampal gyrus) and subcortical areas (epithalamus, anterior hypothalamus, basal ganglia, hippocampus, amygdala). The limbic system is involved with many of the roles of the hypothalamus, as well as other functions, e.g., amygdala, olfaction; the hippocampus, emotions and learning; the limbic cortex, association area for behavior.

The telencephalon consists of the cerebral hemispheres, basal nuclei (corpus striatum), and the olfactory bulb. In primitive vertebrates, the cerebral hemispheres are mainly involved with olfaction and

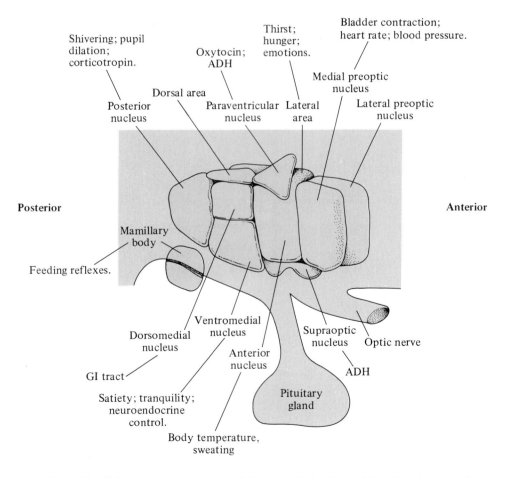

FIGURE 8–28   Schematic arrangement of the hypothalamic nuclei and their general functions.

related motor responses. The telencephalon shows the most dramatic evolutionary modification and development of all parts of the brain (see Supplement 8–4, page 392).

In higher vertebrates, the gray matter of the cerebral hemispheres, except for the basal ganglia, is moved to the surface, forming the cerebral cortex. In mammals, the cortex forms the majority of the cerebral hemispheres. There are billions of neurons in the mammalian cortex and they interact with virtually every other region of the central nervous system! The cortex is a thin (2.5 to 4.0 mm), multilayered sheet of neurons that is convoluted to form a considerable surface area (about 2500 cm^2 in humans). It is arranged as six cellular layers, and the synapses transfer information in essentially a vertical direction (Figure 8–29). Groups of neurons are arranged as vertical columns with afferent information coming in at the bottom of the column, ascending vertically, and leaving at the top. The neural processing ability of the cortex depends on the number of interconnected neurons, and it appears that these cells are optimally arranged in

two-dimensional sheets at the surface rather than as compacted three-dimensional conglomerations of neurons. Hence, both the cerebellar and cerebral cortex are a highly folded sheet of neurons at the surface of the brain, not internal nuclei as in the primitive arrangement of the vertebrate nervous system.

The various regions of the cortex have very specific functions; the three general areas are the sensory cortex, the association cortex, and the motor cortex (Figure 8–30). The **primary sensory cortex** receives sensory information, e.g., thalamo-cortical projection fibers, association fibers from the cortical areas, and commissural fibers from the other side of the brain. Sensory information is relayed through the spinal cord to the cerebral cortex by a number of pathways. The spinothalamic pathway conveys temperature and pain sensation via the lateral spinothalamic tract. The dorsal column pathway conveys touch and proprioceptive sensation. The motor areas give rise to motor fibers that travel to specific parts of the body (pyramidal motor pathways). The corticospinal tracts descend through

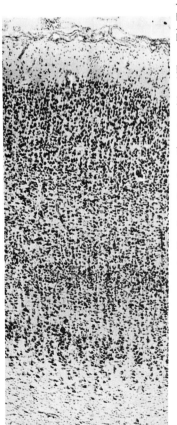

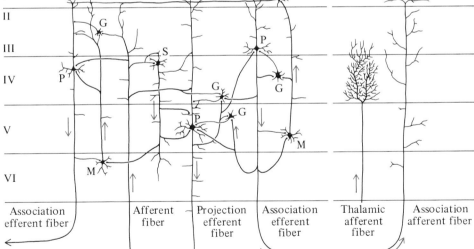

FIGURE 8–29 Organization of neurons in the six layers of the cerebral cortex showing afferent, intracortical, and efferent circuits. The neurons are as follows: S = stellate, M = Martinotti, G = granule, and P = pyramidal *(From Noback 1967; photograph courtesy of P. Bailey.)*

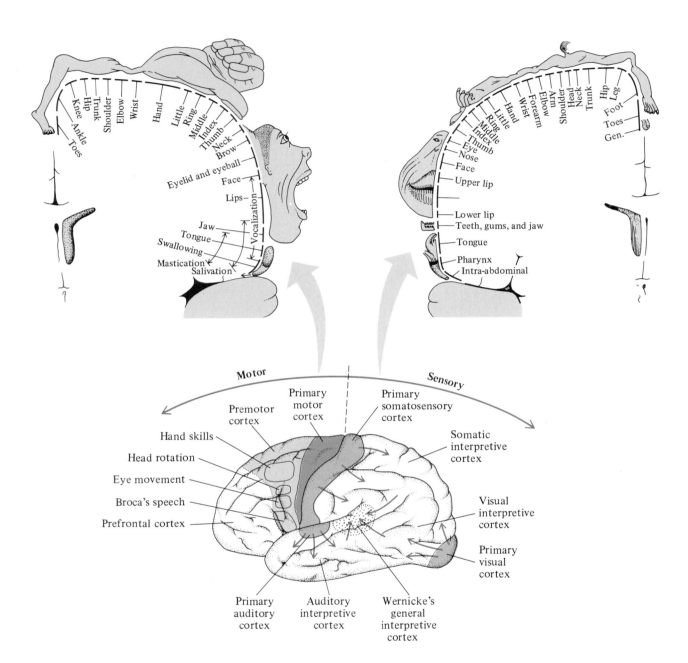

FIGURE 8–30   The human cerebral cortex is organized into areas with specific functions. The primary sensory areas include somatic sensation (e.g., tactile information), vision, and audition. The somatosensory cortex receives sensory information that is organized on the cortex as a map, or homunculus. The primary sensory areas relay information to their respective sensory interpretive areas, then to a general sensory interpretive area (Wernicke's cortex). Much of the motor output of the cortex is from the primary motor cortex, which has a highly organized map. The primary motor cortex receives information from the premotor cortex and other parts of the brain. The premotor cortex has specific areas for control of functions such as speech (Broca's area), eye movement, head rotation, and hand skills. It receives information from the sensory associative areas as well as other areas of the brain. The prefrontal cortex has more diffuse effects, affecting "depth of feeling" and "elaboration of thought."

the medullary pyramids (hence their alternate name, pyramidal tracts) and then through the lateral-corticospinal tract to the spinal nerves.

Sensory fibers from the dorsal columns convey discriminative touch, weight perception, vibrational sense, and position sense to the main somatosensory cortex, **somatosensory area I**. This cortical area has a precise topographic map of the different parts of the body. The map resembles the form of the body, except that some areas are over-represented because they have a greater density of sensory receptors (e.g., face, hands) and others are under-represented because they have a lower density of sensory receptors (e.g., back of head, lower limbs). There are other specific cortical sensory areas. The **somatosensory area II** is a separate and less topographically precise area than somatosensory area I. The **visual cortex** of the occipital lobe has a topographic organization corresponding to different parts of the retina (this will be described in more detail below). The **auditory cortex** has a number of tonotopic maps with different regions corresponding to different frequency ranges. The **olfactory cortex** is part of the "primitive" paleocortex; it is the only part of the cortex that receives sensory information that is not relayed through the thalamus.

There is a **sensory association area**, or secondary sensory area, around each primary sensory cortex area. These cortical areas provide a more complex interpretation of sensations than do the primary sensory areas. For example, damage to the somatosensory association area impairs spatial perception by different parts of the body and greatly decreases hand skills. Damage to the visual association cortex reduces the ability to interpret visual information even though the primary visual sensory processes are normal. Dyslexia, an inability to interpret the meaning of words, is an example of impaired visual association. Damage to the auditory association area impairs the ability to understand spoken words, although they are still heard.

The motor areas of the cortex are also spatially organized, like the primary sensory areas (Figure 8–30). The **primary motor cortex** controls discrete muscle control; it is located immediately anterior to the primary somatic sensory area I and has a precise topographic organization. The human motor map is quite different from that of "lower" mammals in the high degree of representation of the hands, mouth, and facial regions; this is associated with the human capacity for hand skills and speech. There also is a supplemental motor area that elicits coordinated contractions of different muscle groups, rather than discrete muscle contractions. The **premotor cortex** is located anterior to the primary motor cortex. This association area elicits often complex, coordinated contractions of groups of muscles. It is connected to the sensory association areas, the primary motor cortex, the thalamic areas relaying information to the sensory cortex, and the basal ganglia and cerebellum. Broca's area controls word formation and vocalization of complex sounds, and an associated cortex area coordinates vocalization with respiration. Other motor association areas control eye movement, head rotation, and hand skills.

Overall organization. The human brain contains about 10^{12} neurons; the cortex alone has about $2 \ 10^7$ neurons cm^{-3} and 10^{12} synapses cm^{-3}. The neurons of the brain are arranged into nuclei of often quite specific function, but these are interconnected in very complicated patterns. It has been estimated that each neuron is interconnected to every other neuron in the brain by an average of five neurons. Even the relatively "simple" brains of primitive vertebrates are enormously complex when the numbers of neurons and their connections are considered. Nevertheless, some basic sensory and motor pathways are evident in vertebrate brains, although there have been dramatic modifications throughout the evolutionary progression of the vertebrate brain.

In lower vertebrates, the medulla is the primary area for reception of sensory information from equilibrium and lateral line sense organs; the olfactory bulbs receive olfactory information and the tectum receives visual information. The cerebellum, in association with these and other sensory inputs, controls posture and body movement; its motor output is mostly through the midbrain and reticular formation. These basic patterns of sensory-interpretive-motor connections are clearly evident, for example, in the brain of a reptile (Figure 8–31A).

In birds, the cerebral hemispheres are important association areas and the corpus striatum and hyperstriatum are prominent centers for control of higher functions. Sensory information is relayed via the dorsal thalamus, and important motor pathways develop from the cortex to the brain stem and spinal cord (Figure 8–31B). In mammals, there is a similar development but the cerebral cortex, rather than the corpus striatum, becomes the dominant center for sensory, interpretive, and motor function (Figure 8–31C).

Visual Processing

The vertebrate visual system is an excellent example of the sophisticated capacity of the nervous system to process sensory information. The complexity of visual interpretation includes pattern discrimination, movement, color, and depth perception. The

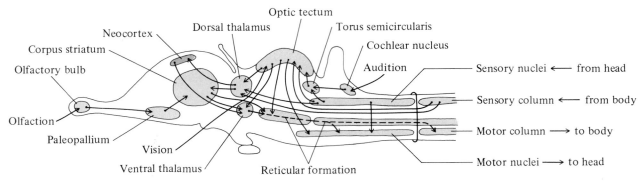

A **Primitive Reptile**

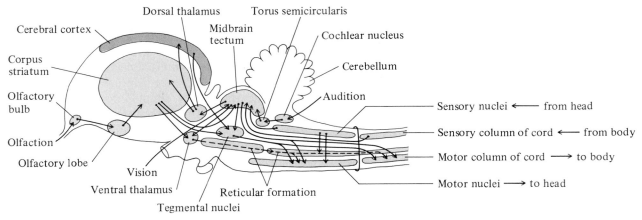

B **Bird**

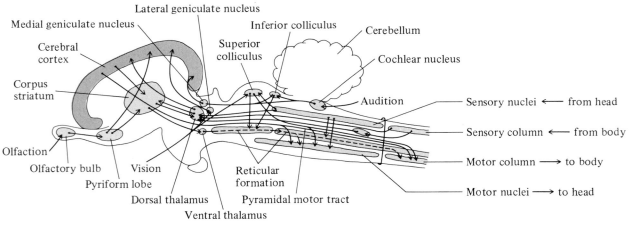

C **Mammal**

FIGURE 8–31 **(A)** A highly simplified schematic of the major brain centers and their interconnections in a reptile. The major sensory inuts (solid shading) are olfaction to the olfactory bulb and paleopallium vision to the tectum, and equilibrium and hearing to the medulla. The cerebral neocortex (color) is of little significance and the corpus striatum is of some importance, but the midbrain tectal region plays the dominant role for higher functions. The reticular formation is an important motor pathway from the midbrain. **(B)** Similar schematic diagram of the bird brain showing the greater role of the corpus striatum in brain function. **(C)** Similar schematic for the mammalian brain showing the predominance of the cerebral cortex and drastically lessened role of the midbrain tectum to a minor reflex center.   *(Modified from Romer and Parsons 1986.)*

exceptional processing capacity of the visual pathways is not due to any special properties of the visual interneurons but to the specific organization of the retinal receptors and visual interneurons at various levels of the central nervous system. The retina, which is actually a part of the central nervous system, at least in terms of its embryonic origins, is anatomically isolated from the central nervous system and is therefore more amenable to study than are most other areas of the central nervous system.

Retinal Processing

In lower vertebrates, visual information is received by the retinal photoreceptors, initially processed by the nonphotoreceptive retinal neurons, and transmitted to the optic lobe (tectum; Figure 8–32A).

The retinal processing of a visual stimulus has been elegantly analyzed for the salamander *Necturus*, whose retinal ganglion neurons are unusually large (Werblin and Dowling 1969). Many of the retinal neurons in the early stages of visual processing do not have action potentials but only graded depolarizations or hyperpolarizations; only amacrine and ganglionic neurons conduct action potentials (Figure 8–32B). Why don't the other retinal neurons sustain action potentials? Theoretically, signal processing is more efficiently accomplished by graded, not "all-or-none," impulses since graded transfer avoids the necessity for transfer of the receptor potential from an amplitude-modulated receptor potential (graded) into a frequency-modulated signal (action potential discharge) at the synapse, after which the postsynaptic potential is again an amplitude-modulated signal. The retinal photore-

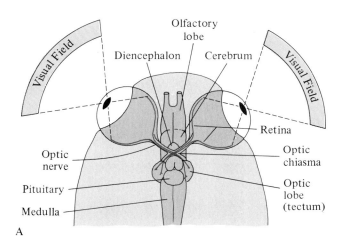

A

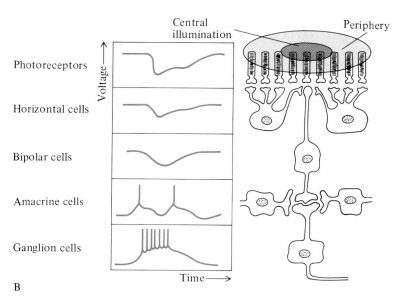

B

FIGURE 8–32 (**A**) Ventral view of the frog brain showing the visual pathway from the retina via the optic nerve to the tectum; ganglion neuron axons decussate at the optic chiasma. (**B**) Visual processing in the frog retina is accomplished by the photoreceptors and horizontal, bipolar, amacrine, and ganglion neurons. The receptive field of a ganglion neuron consists of a number of photoreceptors arranged as a central zone with an annular surround. Receptor, horizontal, and bipolar neurons respond to stimulation with graded hyperpolarization or depolarization, and amacrine and ganglion neurons respond with depolarization and action potentials. *(Modified from Ottoson 1983.)*

ceptors and neurons are thus very efficient in processing information because they use graded, not "all-or-none," electrical signals. However, the transfer of graded electrical information is only effective over short distances, as in the retina. Long-distance transfer of information to the tectum requires "all-or-none," frequency-modulated action potentials. Therefore, the retinal neurons at the final stages of visual information processing (amacrine and ganglion neurons) have action potentials.

Retinal photoreceptors respond only to a light stimulus falling in their immediate vicinity, i.e., each photoreceptor has a narrow receptive field. The other nonphotoreceptive sensory-processing neurons of the retina also have their own receptive fields, but these are larger than the receptive fields of individual photoreceptors because many receptors converge onto fewer nonphotoreceptive neurons. Horizontal neurons have a larger receptive field than the photoreceptors, but like the photoreceptors have only an OFF (unstimulated) or ON (stimulated) response. Many amacrine neurons also have only an OFF or ON receptive field. The response of other amacrine neurons, bipolar neurons, and ganglion neurons depends on which part of the receptive field is stimulated because their receptive fields are organized as a central zone surrounded by an annular zone (Figure 8–32B). There are two types of receptive field neurons depending on whether illumination by light of the central zone causes depolarization or hyperpolarization. Illumination of the annular surround has the opposite effect as illumination of the central zone.

The retinal ganglion neurons have remarkably sophisticated responses to different types of visual stimulation (Horridge 1968). The axons in the optic nerve do not transmit a point-to-point representation of the pattern of light on the retina but do transmit the results of retinal analysis of variation in light intensity, movement of edges of images, curvature of edges, and contrast. Six types of bipolar neurons have been described in frogs, each responding to a different type of stimulus: (1) sustained edge detectors, (2) convex edge detectors, (3) changing contrast detectors, (4) dimming detectors, (5) dark detectors, and (6) blue-sensitive detectors. **Sustained edge detectors** respond to a contrasting edge in a narrow visual field (of 1° to 3° arc) with a sustained firing if the edge is stationary and with bursts of firing if the edge moves. The respiratory movements of a frog cause eye movement, which would make a stationary edge in the visual field appear to move with each respiration, but habituation of the nervous system to the regular respiratory rhythm filters out these respiratory movements. **Convex edge detectors** respond to movement of small, dark objects with a

sharp edge within a 2° to 5° visual field. **Changing contrast detectors** respond to stimulation by an edge (dark or light) moving across a wide visual field (7° to 12°); there is an optimal direction and speed of movement, and there is no response to a stationary edge. **Dimming detectors** have wide fields of up to 15° and respond for many minutes to a decreased light intensity. **Dark detectors** have very large receptive fields and respond with decreased activity to reduced light intensity. The **blue-sensitive** ganglion neurons send axons to the thalamus; axons of other types of ganglion neurons go to the tectum. The retina of birds also has six types of ganglion neurons, although they differ somewhat from the frog types, being even more complex and subtle in their interpretation of visual information.

Central Nervous System Processing

The retinal ganglion cells of nonmammalian vertebrates, e.g., frogs, send information primarily to the tectum of the midbrain. These tectal neurons also have large receptive fields with quite complex information-processing properties. There are two main types of tectal neurons in the frog. **Newness neurons** respond to nonrepetitive, jerky movements of objects; some respond best to particular image sizes. **Sameness neurons** respond to an image almost anywhere within the entire visual field of the eye but have a null region where the stimulus is not effective. The neurons will suddenly respond with a burst of activity when they "notice" a moving object, then continue to respond to its movement unless it moves into the null region or is stationary for a long period of time. Much of the visual information received by the retinal photoreceptors of lower vertebrates is interpreted at the level of the retinal ganglion neurons, and detail of the visual image is not transferred to the central nervous system (tectum). The basic information is therefore "lost" and cannot be subsequently used by the central nervous system.

The visual pathway of mammals de-emphasizes the role of the tectum and emphasizes the role of the lateral geniculate body (LGB) of the thalamus and the visual cortex. There is also a reduced role for retinal processing of visual information, but there still is not a point-to-point transmission of information from retinal photoreceptors to the lateral geniculate body or visual cortex. Retinal ganglion neurons have mainly OFF-center and ON-center receptive fields with opposite-responding annular surrounds, although there are other types specialized for movement or orientation detection.

The retinal ganglion neurons do not relay their visual information directly to the cortex but send it to an intermediate relay center, the lateral geniculate body (Figure 8–33A). The LGB neurons have complex receptive fields similar to the retinal ganglion receptive fields, i.e., ON- or OFF-center with opposite surround. For example, an ON-center, OFF-surround LGB neuron responds to illumination of its central field with an increased action potential discharge, to peripheral field illumination with a decreased action potential discharge, and to both with an intermediate response (Figure 8–33B).

There is a complex processing of visual information by the visual cortex neurons. The visual cortex is also the first level at which there is integration of visual information from the left and right eyes. There are four general types of neurons in the visual cortex: (1) circularly symmetrical neurons, (2) simple neurons, (3) complex neurons, and (4) hypercomplex neurons.

The **circularly symmetrical neurons** have receptive fields that are similar to those of the ganglion and LGB neurons. The **simple neurons** also are similar, except that their receptive field is generally elongate, not circular; there is a central excitatory

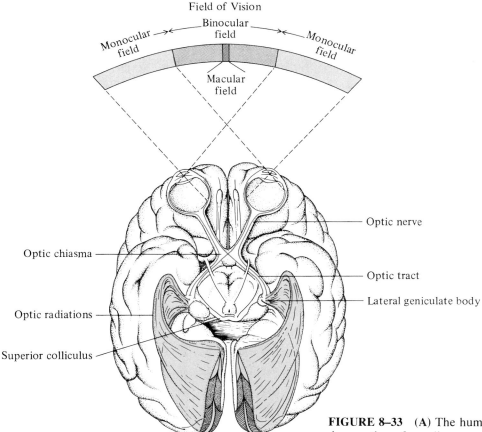

A

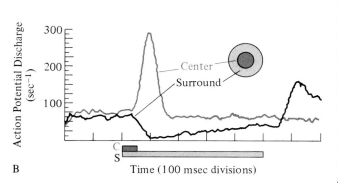

B

FIGURE 8–33 (A) The human visual pathway involves decussation of ganglion neuron axons from the nasal segment of the visual field of each eye, but axons from the temporal field do not decussate. Most axons terminate in the lateral geniculate body (LGB) but some terminate in the superior colliculus (part of the tectum). (B) The lateral geniculate neurons have either ON-center OFF-surround or OFF-center ON-surround visual fields. Stimulation of an ON center (0.60 diameter) produces a marked increase in action potential discharge for a cat LGB neuron with a central excitatory surround with an annular inhibitory surround. Stimulation of the inhibitory surround (1 to 2.70 inner-outer diameters, 500 msec) produces an inhibition of spontaneous action potential discharge. *(From Noback 1967; modified from Poggio et al. 1969.)*

region with an inhibitory surround, or vice versa, or an opposite left and right side (Figure 8–34). These simple neurons, like ganglion and LGB neurons, detect differences in illumination but respond best to lines, slits, or edges because of their elliptical shape; they have been called feature detectors. They respond optimally to edges of specific orientation and may respond best to edges moving in specific directions (Figure 8–35). Some show end inhibition, making them responsive to edges of specific length.

Complex neurons have large receptive fields without specific ON-OFF areas (Hubel and Wiesel 1962).

They respond to edges, or moving edges, of specific orientation, regardless of where they fall within the receptive field. Some respond equally well to a stimulus, e.g., an illuminated slit wherever it falls within the receptive field. Others do not respond

Receptive Fields

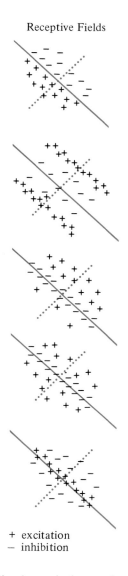

+ excitation
− inhibition

FIGURE 8–34 Simple cortical receptive fields are elliptical in shape, with a variety of ON and OFF arrangements. + = excitation, − = inhibition. *(Modified from Hubel and Wiesel 1962.)*

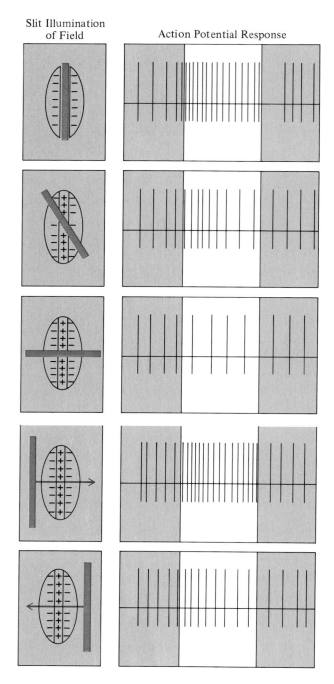

FIGURE 8–35 Simple cortical neurons respond maximally to specific orientation of bars and slits or edges, often dependent on the direction of movement. *(Modified from Ottoson 1983.)*

equally well regardless of the specific location of the stimulus within the receptive field and show opposite effects to mirror image stimuli. **Hypercomplex neurons** provide even more sophisticated processing capabilities (Hubel and Wiesel 1965). Their large receptive fields have activating and inhibiting regions that interact in complex ways. They respond best to moving edge stimuli of specific orientation and have a selective sensitivity for the length of the edge; some respond specifically to angles or corners, hence are called corner detectors. Some neurons have even more complex responses and are called **higher-order hypercomplex neurons**.

Animals with color vision have a number of different photoreceptor types with differing spectral characteristics, e.g., mammals have three types of cones (blue, red, green) with differing λ_{max} (see Chapter 7). There appear to be two basically different means for interpretation of color vision based on the functional properties of retinal ganglion neurons. One type of retinal ganglion neuron receives input from red and green cones and from blue cones at high-stimulus intensity. The effects of the different wavelengths are additive, hence these **broad-band ganglion neurons** convey information on intensity or brightness. The other type of retinal ganglion neuron receives two antagonistic color inputs, either red and green or yellow and blue. There are therefore four types of these **color-opponent ganglion neurons** (and corresponding LGB neurons) depending on which of the pair of colors is excitatory $(+)$ or inhibitory $(-)$, i.e., R^+G^-, R^-G^+, Y^+B^-, and Y^-B^+. The responses of two color-opponent ganglion neurons (R^+G^- and R^-G^+) to red and green stimulation are shown in Figure 8–36A. Note that the "color coding" of these neurons does not correspond to the retinal cone types; there is no yellow cone (at least in mammals) and the spectral sensitivities of the four ganglion color-opponent neurons do not correspond to the spectral sensitivities of the photopigments (Figure 8–36B). Few of the visual cortex neurons appear to be color coded (unlike the color-opponent and LGB color neurons), but many are broadbanded with an overall spectral sensitivity that is similar to the behavioral sensitivity of the animal to color.

Binocular Processing

In mammals, each eye has its own monocular visual field, but their visual fields overlap to form a central region of binocular vision (see Figure 8–33A, page 369). Much of the retina of each eye receives the same visual information as does the retina of the other eye because there is a substantial overlap in the visual fields. However, the overlapping visual fields are not exactly the same because the eyes are about 5 cm apart (see below).

The visual cortex on each side of the brain would receive the complete image from the contralateral eye if all of the axons in the optic nerve decussated in the optic chiasma, but there is not a complete decussation; axons from the temporal part of the retina (that see the nasal part of the visual field) do not decussate. Thus, the LGB on each side of the brain receives the ipsilateral nasal visual field and the contralateral temporal visual field. There is complete segregation of decussated and nondecussated axons in the LGB, but these separate tracts are combined in the visual cortex to form a topographic map of the visual field.

Most cortical neurons are binocular (i.e., receive input from left and right eyes) because of the extensive overlap of left and right visual fields. These overlapping visual fields from each eye provide for depth perception by binocular parallax, or **stereopsis**. The visual images of the two retinae are not identical because the eyes are about 5 cm apart. If both eyes are focused on the same object, then its image falls on the fovea of each retina. Distant objects and close objects will form images at different locations on each retina, relative to the foveal position. Only those objects that fall on a circle passing through the object on which the eyes are focused, and the optical axis of each eye, will form images that fall on the same part of the retina of each eye; this circular path is called the **horopter**. Most objects do not lie on the horopter, hence their images are slightly offset in the two retinae. The binocular offset of the two images (parallax) is interpreted by the visual cortex to provide depth perception, since the extent of parallax depends on the distance of the object from the eye (up to about 60 m, beyond which the binocular parallax is too small to provide depth perception).

Animals without binocular vision, and binocular animals blinded in one eye, are still capable of depth perception, but not by stereopsis. The size of the retinal image of an object of known dimensions provides **dimensional depth perception**; the closer the object is to the eye, the larger is its image. However, a familiar-looking object of unfamiliar size will "trick" the brain into perceiving it as being closer if it is larger than normal or more distant if it is smaller than normal. Moving the eye from side to side also provides depth perception by **moving parallax**. An object close to the eye is perceived to move considerably, whereas an object farther from the eye moves less.

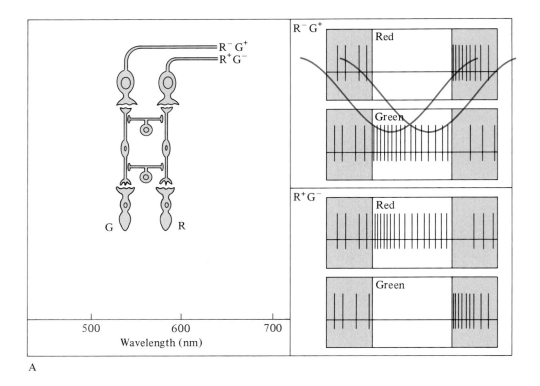

A

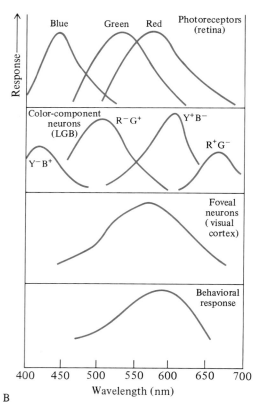

B

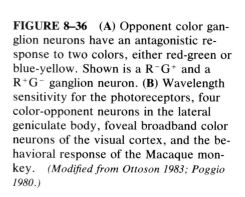

FIGURE 8–36 **(A)** Opponent color ganglion neurons have an antagonistic response to two colors, either red-green or blue-yellow. Shown is a R⁻G⁺ and a R⁺G⁻ ganglion neuron. **(B)** Wavelength sensitivity for the photoreceptors, four color-opponent neurons in the lateral geniculate body, foveal broadband color neurons of the visual cortex, and the behavioral response of the Macaque monkey. *(Modified from Ottoson 1983; Poggio 1980.)*

Memory and Learning

Memory is the mechanism whereby a wide variety of types of sensory information is retained. The change in the nervous system that corresponds to a "memory" is an **engram**. It can be a short-term or a long-term phenomenon.

Learning is a modification of behavior pattern in response to a previous experience; it requires a memory. The effects of the first stimulus must be "remembered" by the nervous system to allow association with subsequent stimuli. Learning, like memory, can be short term or long term. In its simplest form, learning is the process of association between two stimuli. The response to a second stimulus is modified by the previous experience of the stimulus. As we shall see below, there are many levels of complexity to learning, ranging from very short-term modification of synaptic functioning (e.g., post-tetanic potentiation) to habituation of response to repetitive stimuli, reflex sensitization, and long-term learning.

Memory and learning are exceedingly complex neurophysiological phenomena; this is evidenced by our poor understanding of the mechanisms for even the simplest examples in primitive animals. The incomprehensible complexity of advanced nervous systems at present defies a definitive understanding of the neurophysiological processes involved with learning and memory. In contrast, much is known of the abilities of different animals to learn and remember from a psychological rather than a physiological perspective. Let us first examine some examples of memory and learning in different animals before addressing their possible neurophysiological basis.

Learning can, for convenience, be classified into a number of types. A primary dichotomy is whether the learning processes occurs in response to a stimulus in the absence of an association of this stimulus with another kind of stimulus, i.e., **nonassociative learning**, or whether it involves an association with another stimulus, i.e., **associative learning**.

Post-tetanic potentiation is an example of nonassociative learning. It is perhaps the simplest example of memory, albeit very transient and short term. The postsynaptic responsiveness to a presynaptic action potential may be enhanced after repetitive presynaptic stimulation (see Chapter 7). Post-tetanic potentiation has been observed in neuromuscular synapses as well as neuron-neuron synapses, e.g., the EPSPs of motoneurons of a reflex arc are increased by tetanic stimulation. The accumulation of Ca^{2+} ions within the presynaptic terminal is responsible for the release of more neurotransmitter and an increased EPSP amplitude. This transient increase in Ca^{2+} concentration is a short-term "memory" of previous stimulation, and the synapse has at least temporarily "learned" to facilitate transmission of presynaptic information.

Habituation and **sensitization** are examples of nonassociative learning where repetition of a single type of stimulus results in either a diminution of the animals' response (habituation) or an increase in response (sensitization). Habituation occurs commonly in primitive animals with simple nervous systems, such as hydra, anemones, and even protozoans. For example, the protozoan *Stentor* will move in response to a mechanical stimulus, by bending and then contracting, to release itself from the substrate and reattach at a different point. *Stentor* rapidly habituates to such mechanical stimulation (Wood 1970). The probability of movement in response to stimulation declines with repetition of the stimulus; the average response of 414 *Stentor* repetitively stimulated once per minute by a mechanical prod shows a rapid decline in the probability of response, from about 0.94 for the first stimulus to 0.2 for the 60th (Figure 8–37A). The reduction in response to stimulation is associated with a decline in electrical response of the cell membrane. The initial stimulus elicits a small amplitude prepotential (about 1 mV) and a larger amplitude biphasic spike potential (about 10 to 20 mV), but the amplitude of both potentials declines with repetition of the mechanical stimulus (Figure 8–37B). The protozoan recovers its responsiveness to mechanical stimulation rapidly after the repetitive stimulation is stopped. There is a rapid increase in probability of response within 30 min, and by six hours the protozoan has about the same probability of response as it did for the first 10 stimuli.

There are many similar examples of habituation in both invertebrates and vertebrates. Gastropod and pelecypod mollusks show habituation of retraction of their gill or mantle in response to mechanical stimulation. In the sea hare *Aplysia*, the mantle, siphon, and gill contract vigorously and withdraw into the mantle cavity if the mantle or siphon is touched. This reflex can be behaviorly modified by nonassociative and associative learning. If the mantle or siphon is repetitively touched, the gill withdrawal reflex habituates after 10 to 15 stimuli (Figure 8–38). However, a noxious stimulus to the tail enhances (sensitizes) the gill withdrawal reflex to the next touch. Repetitive stimulation of large neurons of *Aplysia* elicits EPSP habituation, probably due to presynaptic effects, such as lesser

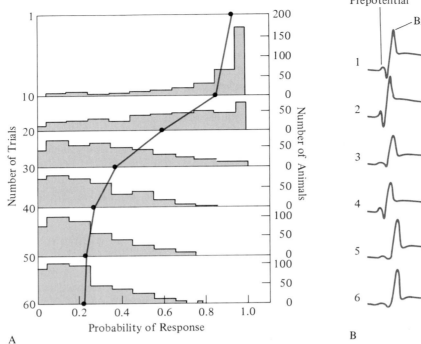

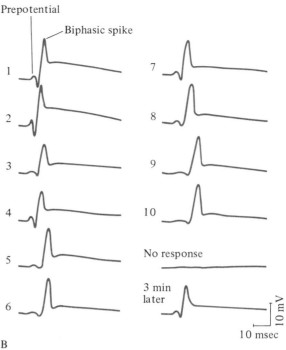

FIGURE 8–37 (A) The protozoan *Stentor* rapidly habituates to repeated mechanical stimulation. The probability of response declines markedly with repetition of the stimulus at one min intervals. The probability of a response by 414 individuals to stimuli (histograms), summarized in groups of 10 stimuli, shows a dramatic shift from an initially high proportion of individuals responding to all 10 consecutive stimuli to a low proportion within 60 stimuli. The mean probability of response declines in a similar fashion. (B) The membrane potential of *Stentor* shows a small prepotential and a larger magnitude biphasic spike potential in response to mechanical stimulation (1). The magnitude of the electrical response declines with repeated stimulation (2 to 10) until no response is observed. After three min of rest, there is a partial recovery of the electrical response. *(From Wood 1970; Wood 1971.)*

quantities of neurotransmitter release (Kupfermann et al. 1970).

The squid stellate ganglion shows similar habituation. Visual neurons in the locust brain habituate to movement in the contralateral eye. Habituation is not due to sensory adaptation, although habituation of the flexor escape reflex of the crayfish is due to interneuron habituation rather than to giant fiber or motoneuron habituation. Habituation is readily reversed by the prolonged absence of stimulation, i.e., the memory for habituation is short. Considerable recovery can occur within three min for *Stentor* and nine min for *Aplysia*. Rapid dishabituation can be elicited by changing the location of the stimulus, or by some other generalized stimulus. Dishabituation may be caused by post-tetanic potentiation or by heterosynaptic facilitation.

Associative learning associates one stimulus with a second type of stimulus. **Classical conditioning** takes advantage of a normal stimulus-response reflex. For example, an unconditioned stimulus, such as a puff of air on the cornea, will elicit an unconditioned response, such as blinking. Similar unconditioned stimulus/response reflexes include painful stimulation/limb withdrawal and sight of food/salivation. Presentation before the unconditioned stimulus of a different type of stimulus, one which would normally not elicit the unconditioned response (e.g., a noise or light flash), can cause an animal to associate the new, conditioned stimulus with the unconditioned stimulus to such an extent that presentation of the conditioned stimulus alone will elicit the unconditioned response. The classical example of such conditioning is an experiment of Pavlov. He trained dogs to associate noises or flashes of colored light (the conditioned stimulus) with food presentation (the unconditioned stimulus), such that the dogs were conditioned to salivate (the unconditioned

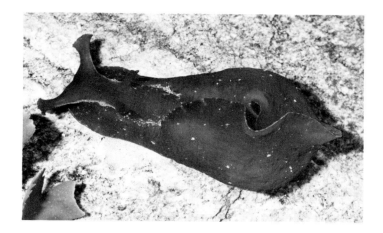

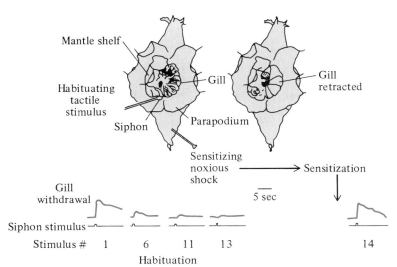

FIGURE 8–38 Dorsal view of the marine gastropod *Aplysia*. Ilustration on left shows the mantle shelf retracted to reveal the gill. Short-term sensitization of the gill withdrawal reflex occurs in response to a noxious stimulation of the neck or tail (mechanical or electrical stimulus) after habituation of the gill-withdrawal reflex to continued mechanical stimulation of the siphon. After 13 stimuli (at 1.5 min intervals) to the siphon, there is no gill withdrawal response (i.e., habituation) but a sensitizing noxious stimulus to the tail restores the gill withdrawal reflex to the next siphon stimulus.   *(Modified from Kandel and Schwartz 1982.)*

response) on the ringing of a bell or the flashing of a light even when no food was offered. This conditioning is lost if the conditioned stimulus is repeatedly presented without the unconditioned stimulus. The *Aplysia* gill withdrawal reflex is also amenable to associative learning, e.g., classical conditioning (Kandel et al. 1983). The conditioned stimulus is touching the mantle; the unconditioned stimulus is shock to the tail. Learning is optimal if the conditioned stimulus precedes the unconditioned stimulus by about 0.5 sec.

A second type of associative learning is **operant conditioning**. Presentation of a reward for performing a specific action increases the frequency of that action. The classic example of this type of learning is a rat in a "Skinner box"; the rat has to learn to press a lever in the box to gain a reward of a food pellet. Maze learning is another common example; an animal learns to find its way through a labyrinth maze to obtain some reward. Alterna-

tively, use of a punishment, such as electrical shock, results in avoidance conditioning.

There are many examples of learning in animals and even in protozoans (McConnell 1966). Many microinvertebrates can be habituated to mechanical shock or can learn avoidance of electrical shock. Many annelids show maze or alley learning, respond to classical conditioning, or show habituation to a variety of stimuli. Cephalopods, which are among the most advanced invertebrates in terms of their central nervous system and vision, are capable of fairly precise visual and tactile learning. The learning process is often accomplished by association with tactile examination of objects or presentation of visual patterns of a reward (food) or a punishment (electrical shock).

Octopus readily learn by sight, and this has enabled the detailed examination of the visual capacity of the octopus eye. They can discriminate many shapes and also size, irrespective of distance, but they do not use binocular vision to accomplish this, unlike squid and decapod crustaceans. The octopus can discriminate vertical from horizontal rectangles (i.e., | from —) but not rectangles at 45° angles (i.e. not / from \). This unusual ability to discriminate vertical and horizontal, but not oblique, rectangles depends on the orientation of the eye's slit-pupil. The animals interpret horizontal and vertical relative to the orientation of the pupil. The pupil is normally kept horizontal regardless of the orientation of the head, but the pupil is not kept horizontal if the octopus's statocysts (balance organs) are removed. Blind octopuses readily learn to discriminate objects, such as plastic cylinders, by touch (Wells and Wells 1957). Cylinders with varying patterns of grooves can be discriminated by the proportion of the surface that is grooved (Figure 8–39). Octopuses, which have learned such discrimination tasks, retain the memory for many days or weeks. Many verte-

brates, especially mammals, are capable of equivalent and even more complex learning tasks. Humans, at least as far as we can tell, are capable of the most complex learning tasks.

Little is known or understood at present of the neurophysiological mechanisms for learning and memory. Even in relatively simple nervous systems the processes involved with learning and memory are exceedingly difficult to discern. A simple preparation for the demonstration of learning is the thoracic ganglion of an insect (Horridge 1962). The insect is suspended above a saline solution such that it receives an electrical shock every time the leg is lowered to touch the saline. Insects, and even headless insects with the thoracic ganglia isolated from the remainder of the nervous system, rapidly learn to lift (flex) their leg to avoid electrical shocks. Insect locomotion, as we have seen previously in this chapter, is primarily programed by the thoracic

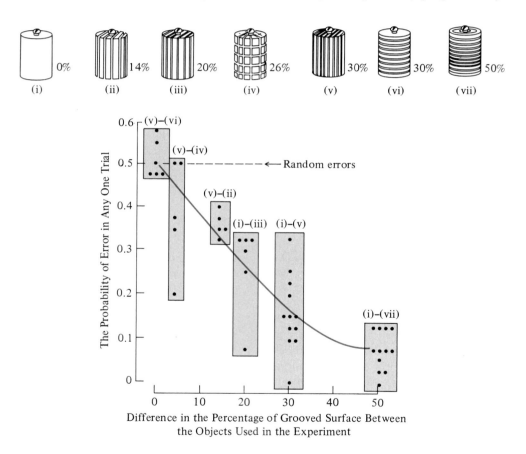

FIGURE 8–39 Tactile discrimination by the octopus can be shown by the ability of blinded animals to learn to discriminate between objects on the basis of surface texture. Cylinders with a similar percentage of grooved surface cannot be distinguished by touch regardless of the orientation of the grooves (e.g., cylinders iv, v, and vi) but cylinders of a differing percentage of grooved surface can be readily distinguished (e.g., i, ii, iii, and iv. *(From Wells 1962.)*

ganglia, not the head. The memory remains for several hours as shown by retesting. There is also some transfer of the memory between legs. There is a spontaneous discharge of action potentials from the thoracic ganglion via the motoneuron to the metathoracic anterior coxal adductor muscle, one of the important flexor muscles. An elevated action potential discharge raises the leg, and a lowered discharge lowers the leg. The thoracic ganglion can be trained to produce an action potential frequency within any prescribed range by shocking the ganglion if the frequency falls outside the prescribed range (Hoyle 1965).

Another simple preparation for the analysis of learning is the sensitization of the gill withdrawal reflex of *Aplysia* (Kandel and Schwartz 1982). Most of the neurons mediating the gill withdrawal reflex, including sensory neurons, excitatory interneurons, facilitating interneurons, and motoneurons, have been identified (Figure 8–40A). During habituation, the excitatory synapse to the motoneurons is depressed by lessened release of neurotransmitter. Sensitization is caused by other synapses, i.e., it is heterosynaptic.

Facilitation in *Aplysia* of the sensory neurons is mediated by specific and identified modulator neurons, and the cascade of molecular events is known (Schwartz et al. 1983). The neurotransmitter serotonin acts on the sensory neuron terminal and elevates the intracellular concentration of cyclic AMP, which in turn enhances the release of neurotransmitter (Figure 8–40B). The serotonin and cAMP modulate the permeability of a novel type of K^+ channel, the "serotonin-dependent K^+ channel." The K^+ channels are normally open and contribute to the rapid repolarization of an action potential, but serotonin and cAMP reduce the K^+ permeability and extend the duration of the action potential by 10 to 20%. This allows greater influx of Ca^{2+} into the presynaptic terminal and greater release of neurotransmitter, hence a greater response. The cAMP is an intermediate messenger in this cascade of molecular events leading to sensitization, and it may constitute the "memory" of the sensory neuron. There is an elevation of intracellular cAMP concentration for more than one hour after a single noxious stimulus, which parallels the period of memory for presynaptic facilitation. Longer-term memory is induced by multiple noxious stimuli: four consecutive noxious stimuli invoke a memory lasting about one day, 16 consecutive stimuli a memory of several days, 16 spaced stimuli (four per day for four days) a memory of several weeks. The long-term memory may involve new specific changes in gene expression, i.e., synthesis of a specific protein that may, for example, act as an alternate regulator of adenyl-cyclase and cAMP synthesis.

The neuronal circuitry and cellular mechanisms for classical conditioning in *Aplysia* are similar to those for nonassociative learning. There is a greater increase of intracellular Ca^{2+} concentration and subsequent inactivation of K^+ channels than with sensitization, perhaps because adenyl-cyclase activity is enhanced by an elevated intracellular Ca^{2+} concentration. An intracellular Ca^{2+}-binding protein, like calmodulin, may bind in the presence of Ca^{2+} to the adenyl-cyclase, enhancing the synthesis of cAMP.

Are these subcellular mechanisms for learning in *Aplysia* applicable to other animals? Learning is such a variable phenomenon in different animals that there may be a multitude of different mechanisms. However, there is evidence in similar animals, such as the marine gastropod *Hermissenda*, and very different animals, such as mammals, for similar cellular changes during learning. In *Hermissenda*, sensory receptor K^+ currents are reduced in response to an as yet unidentified neurotransmitter (serotonin or norepinephrine?) by the action of two different protein kinase systems: a Ca^{2+}/calmodulin type II kinase and a protein kinase C decrease the permeability of K^+ channels (Farley and Auerbach 1986). Neurons in the hippocampus of the mammalian brain have an increased cell excitability and reduced post-action potential after hyperpolarization by administration of the neurotransmitters norepinephrine and acetylcholine (Hopkins and Johnston 1984). Both neurotransmitters inhibit Ca^{2+}-dependent K^+ conductance through two protein kinase systems: a cAMP-dependent kinase and a phospholipid-dependent kinase. There is a long-lasting change in synaptic transmission (long-term potentiation, LTP) which may be the result of presynaptic changes (e.g., increased amount of transmitter release) or postsynaptic changes (e.g., enhanced sensitivity of the postsynaptic membrane to neurotransmitter; Bekkers and Stevens, 1990). LTP can also occur at electrical synapses (junctions), and is not limited to chemical synapses.

Alteration of synaptic properties, whether by presynaptic or postsynaptic facilitation, is clearly involved in some examples of short-term memory. However, humans, and presumably many other long-lived animals, have both short-term and long-term memory. Long-term memory may persevere for over 100 years in some very long-lived animals. Do short- and long-term memory share the same mechanism, or is short-term memory fundamentally different from long-term memory? Is the long-term

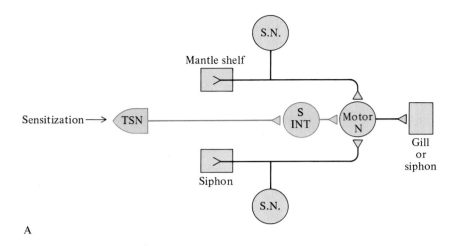

A

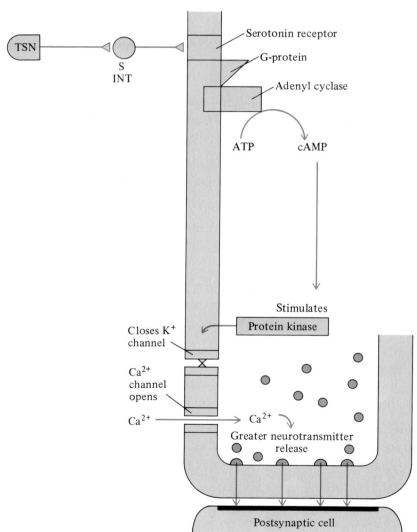

FIGURE 8–40 **(A)** Simplified circuit diagram for the gill withdrawal reflex in *Aplysia*. Stimulation of the siphon or mantle activates sensory neurons (SN) that synapse with motoneurons to the mantle and gill (Motor N). Stimulation of the tail sensory neuron (Tail SN) activates a facilitatory interneuron (S Int), which sensitizes the motoneurons. **(B)** Model of molecular cascade of events responsible for sensitization of the gill withdrawal reflex by noxious stimulation of the tail of the sea hare. A serotoninergic interneuron induces an increase in intracellular cAMP concentration, which activates a protein kinase and closes serotonin-specific K^+ channels. This allows longer depolarization of the synapse during an action potential and greater Ca^{2+} increases the amount of neurotransmitter released. *(From Kandel and Schwartz 1982; Kandel et al. 1983.)*

B

memory trace, or engram, a more physical change in neurons or their synapses than in short-term memory? Short-term memory may involve transient synaptic changes or reverberating neuronal circuits; long-term memory may involve establishment/loss of specific synapses or synthesis of specific proteins or RNA. Much is known concerning the psychological aspects of memory in, for example, humans and other mammals but there is little evidence for its neurophysiological mechanism.

Human memory has been categorized into several discrete stages (Table 8–4). **Sensory memory** is an extremely transient stage of collecting and holding raw sensory information, for perhaps 100 to 200 msec, before it is either discarded (by spontaneous decay or erased by new sensory information) or is transferred to a more stable form. If a verbal label is assigned to the sensory information, it becomes assigned to another limited capacity, short-term **primary memory**, lasting several seconds. This memory span is about nine binary figures (e.g., 101101011), or eight decimal figures (e.g., 37564551), or seven letters of the alphabet (e.g., NPTAWQE), or five monosyllabic words (e.g., no, up, ten, yes, of). The capacity of this primary memory span is 7 ± 2 symbols, but the information content is quite different for the various symbols. The lengths of the "alphabets" for the different symbol types are two (binary), 10 (numbers), 26 (letters), and 1000 (monosyllabic words). This information is either lost by decay or replacement with new primary memory or it is transferred to a more permanent memory system. The **secondary memory** system is

larger and more permanent than primary memory; a secondary memory may persist for several minutes or years. Forgetting of this type of memory is not spontaneous, but requires unlearning. Retrieval of secondary memory is relatively slow. A final category is **tertiary memory**, which is a very large, permanent, rapid-retrieval system. It contains highly overlearned memories, e.g., names, one's social security number, how to read and write, etc.

There is little anatomical, physiological, or biochemical evidence for specific memory storage mechanisms. In mammals, the cerebral cortex is undoubtedly the primary region involved with learning and memory. The human cerebral cortex is immensely complex. The entire brain contains about 10^{12} neurons and the cortex about $2 \cdot 10^7$ per cm^3; each cortical neuron has an average of 50000 synapses! Memory was once thought to be stored in the association areas. Extreme advocates of memory localization theories suggested that individual memories were contained within particular cortical neurons. A less extreme theory suggested that reflexes, conditioned reflexes, and "memories" involved specific facilitated neuronal pathways between the sensory and motor cortex. In cephalopods, tactile and visual memory may consist of complex neural circuits in various parts of the brain; damage to any of the parts proportionally reduces the memory capacity (Young, 1991).

There is some anatomical evidence for a general relationship between memory and cerebral cortex structure. Rats raised in an enriched or an impoverished environment exhibit significant differences

TABLE 8–4

Types of memory processes in the human. *(Modified from Werner 1980.)*				
	Sensory	**Primary**	**Secondary**	**Tertiary**
Duration	Fractions of a second	Several seconds	Several minutes to years	May be permanent
Capacity	Limited by receptor transmission	7 ± 2 symbols	Very large	Very large
Storage entry	Automatic with reception	Verbal recoding	Rehearsal	Overlearning
Information	Sensory	Verbal	All types	All types
Accessibility	Limited only by speed of readout	Very rapid	Relatively slow	Very rapid
Forgetting	Spontaneous decay and erasure	New information replaces old	Interference: retroactive and proactive	None?

in brain structure (Rosenzweig 1970), presumably reflecting their different learning histories and memory stores. There also are biochemical differences (e.g., acetylcholine activity) and histological differences (numbers of neuroglia cells). However, experiments involving cuts to isolate specific cortex areas, or lesions that destroy specific areas, indicate virtually no specific localities for learned behaviors (Lashley 1950). Rather, loss of a learned habit (e.g., maze learning) is proportional to the area of cortex ablated, not its specific location.

No specific location for a memory engram has been found; neither has the structure of the engram been elucidated. Short-term memory may involve transient synaptic changes or reverberating circuits, but long-term memory is more likely to involve specific structural changes, e.g., synthesis of new protein or RNA. There is, in fact, considerable but controversial evidence that inhibition of protein synthesis impairs or prevents learning and that transfer of RNA between animals may confer some aspect of "memory" and learning ability from the donor to the recipient (Byrne 1970; Pribram and Broadbent 1970). Planarians can retain a conditioned response after transection and regeneration (McConnell, Jacobson, and Kimble 1959). The planarians were conditioned to avoid a light; they were then transected and each half was allowed to regenerate. Both the head and tail ends showed significant retention of the conditioning compared to control planaria. The brain is required for learning to take place but is not the repository of learned information. How does the tail of the planarian retain anything? Is the conditioned response built into the new cerebral ganglia? The questions raised by this study are still unanswered.

Biological Clocks

Biological rhythms are widespread among animals, plants, protozoans, and microorganisms. Rhythmic activities are apparent at every level of organization of animals, from the subcellular level, e.g., action potential discharge of pacemaker neurons, to the physiology and behavior of animals, e.g., rhythms in metabolic rate, to the coordination of activities for populations of animals, e.g., breeding synchrony. The **periodicity** of rhythms varies markedly for different activities and in different organisms. For example, the rhythmic action potential discharge of a neuron may have a periodicity of one to two milliseconds; the neural mechanisms responsible for such pacemaker neurons have been

previously described (Chapter 6). Biological rhythms range from a periodicity of about one day, to an intertidal periodicity (24.8 hr), to one lunar cycle (29.5 days), or to one year.

Rhythms are usually coordinated with the normal environmental cycle. The natural day–night cycle of light intensity is an important **zeitgeber**, or environmental cue, that coordinates biological rhythms with the environmental rhythmn. Light is also involved in rhythmicities with other periodicities, e.g., the lunar cycle and the annual cycle. Other environmental cues may also contribute to the coordination of biological rhythms with environmental rhythms, e.g., temperature or subtle geophysical variables, such as electric or magnetic fields.

Biological rhythms do not simply track the environmental cycle. Many rhythms are endogenous and persist even in the absence of the environmental cue. These endogenous rhythms must therefore be controlled by a **biological clock**. The mechanisms of the biological clock are poorly understood, but in many instances it is clear that it is located in the nervous system and its effector control is mediated by the nervous system.

Circadian Rhythms

There are countless examples of daily rhythms. For example, the chaffinch *Fringilla*, a small passerine bird, has a daily rhythmn of metabolic rate, activity,

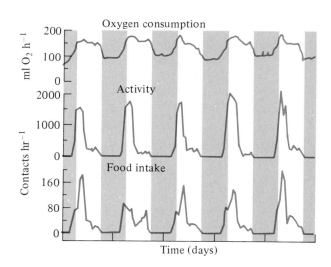

FIGURE 8–41 Daily rhythm of oxygen consumption rate, activity, and food intake for a chaffinch (*Fringilla coelebs*) with an articifial light-dark cycle of LD 12:12. *(From Aschoff and Pohl 1970.)*

and feeding. Metabolic rate, activity, and feeding are highest during the day, i.e., the light (L) cycle, and lowest during the night, i.e., the dark (D) cycle (Figure 8–41). The artificial light cycle used in this experiment was 12 hours of light and 12 hours of dark, i.e., LD 12:12. The chaffinch clearly anticipates the light coming on because its metabolic rate and activity increase before the light actually comes on. This suggests that the chaffinch must have an endogenous sense of time. Its biological clock enables it to anticipate the time at which environmental cues change. Many other birds and mammals have a similarly pronounced daily rhythm of metabolic rate (see also Chapter 4).

Many daily rhythms persist even if the light-dark cycle is altered to constant conditions, either constant light (LL) or constant dark (DD), i.e., the rhythm is allowed to **free-run**. The chaffinch, for example, continues to have a rhythm in metabolic rate, activity, and feeding under constant light (Figure 8–42A). However, the periodicity of the rhythm is no longer exactly 24 hours, but is about 22.9 hours (at least for birds at a constant light intensity of 14 lux). This change in rhythm periodicity is apparent in Figure 8–42A, but is more clearly demonstrated by visually comparing the timing of the midpoint of the activity period on consecutive days (Figure 8–42B). It is a general property of

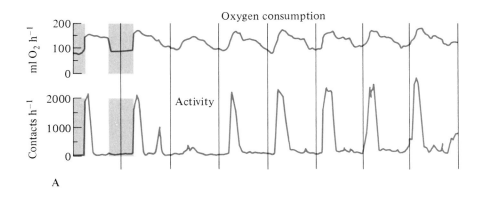

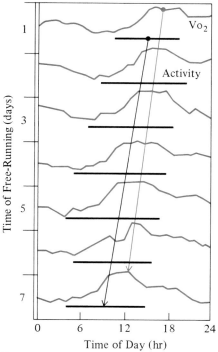

FIGURE 8–42 (**A**) Endogenous circadian rhythm in oxygen consumption rate and activity for a chaffinch in a constant light (LL) cycle of 14 lux intensity. (**B**) Circadian rhythm in oxygen consumption (colored line) and activity (horizontal bar) for the chaffinch in constant light (LL; 14 lux). *(From Aschoff and Pohl 1970.)*

biological rhythms that the periodicity of the free-running rhythm is not exactly 24 hours but is a little shorter or a little longer than 24 hours. Consequently, these biological rhythms are called **circadian rhythms** (*circa*, about; *diem*, day). Diurnal vertebrates tend to have a free-running periodicity slightly shorter than 24 hours, whereas nocturnal vertebrates tend to have a free-running periodicity slightly longer than 24 hours; this is Aschoff's rule.

There is an important relationship between the timing of the zeitgeber for a biological rhythm and where it occurs in the cycle. The timing of the zeitgeber can shift the phase of the rhythm, and the magnitude of the phase shift depends on the timing of the zeitgeber within the period of the rhythm. For example, pupae of the fruit fly *Drosophila pseudoobscura* eclose (the adults emerge from the pupal stage) at a time of day that is determined by a circadian rhythm. Eclosion occurs just before "dawn" if the light component of the photoperiod is shorter than 7 hr, e.g., LD 6:18, or just after "dawn" if the light component is longer than 7 hr, e.g., LD 8:16. The rhythm persists in constant dark (DD) but not in constant light (LL). If the pupae are kept in constant dark, then a short pulse of light (e.g., 15-min duration) can act as a zeitgeber and shift the phase of the rhythm that dictates the time of eclosion. If the light pulse occurs during the first

three or so hours of the "day" portion of the circadian cycle, then the phase of the circadian rhythm is advanced ($\Delta\phi > 0$); remember that the pupae are in constant dark, so "day" refers to the period of the day that would normally be light, although it actually is not. The phase is delayed ($\Delta\phi < 0$) if the zeitgeber occurs during the initial part of the "dark" phase. The phase response curve of pupal eclosion for *D. pseudoobscura* (Figure 8–43) shows this fundamental feature of all biological rhythms, a phase dependence of the cycle on the timing of the zeitgeber.

A circadian clock not only allows the timing of important activities and even the anticipation of the environmental cues (e.g., fruit flies eclosing before dawn; the chaffinch becoming active before dawn), but can allow animals to continuously "know" the time of day or night. This is of considerable importance, for example, to animals that navigate using the sun (sun compass) or stars (celestial compass). The position of the sun (or stars) can be used by animals to orient to food sources and their nests (ants, honeybees) or for migration (birds) if the animals can compensate for the sun/star apparent movement by knowing the time of day. For example, north is to the left of the sun in the morning and to the right of the sun in the afternoon (in the northern hemisphere). Those animals that navigate continu-

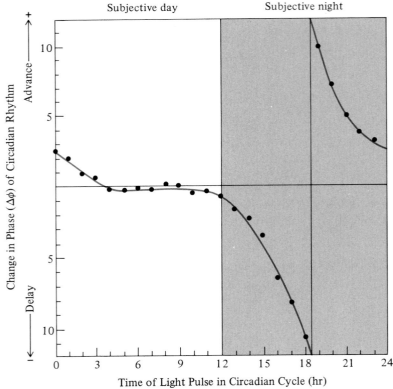

FIGURE 8–43 The phase response for pupal eclosion by the fruit fly *Drosophila pseudoobscura* shows that a short (15 min) pulse of light advances or delays the timing of pupal eclosion by pupae maintained in constant darkness (DD). The relative timing of the light pulse within the circadian cycle determines whether the phase is advanced or delayed. The endogenous cycle for eclosion is divided into subjective day and night. *(From Saunders 1977.)*

ously "consult" their biological clock for time compensation.

The structure and mechanisms of biological clocks are not well understood, and in many instances the location of the clock is not even known. However, there are a number of animals in which the location of the biological clock is precisely known. In cockroaches (*Leucophaea*), the clock is in the optic lobe; it is connected to the compound eyes for acquisition of the environmental zeitgeber, the LD cycle, and via the rest of the brain and circumesophageal ganglia to the thorax and legs to dictate a circadian cycle of locomotor activity (Nishiitsustuji-Uwo and Pittendrigh 1968a,b). In the giant silk moths *Antheraea* and *Hyalophora*, the clock also resides in the brain, but its input is direct photoreception by the brain (not the eyes) and the motor output is via the endocrine systems (Truman and Riddiford 1970). The sea hare *Aplysia* has a circadian rhythm of neural discharges in the abdominal ganglion and another "clock" in its eye that has a periodicity of 27.5 hr.

In birds and mammals, the biological clock resides in a small suprachiasmatic nucleus of the hypothalamus. Transplantation of pieces of the suprachiasmatic nucleus transfer the circadian rhythmicity of the donor to the recipient (Ralph et al. 1990). Nervous connections of the suprachiasmatic nucleus with the eyes provide the light-dark zeitgeber for entrainment of the circadian cycle with the environmental light-dark cycle in mammals. The suprachiasmatic-eye nervous tract is independent of the normal visual pathways. In birds, both retinal and extraretinal (brain) photoreceptors mediate the entrainment of the circadian rhythm with the exogenous light-dark cycle. The pineal gland is also involved with circadian rhythmicity; its removal abolishes a free-running circadian rhythmn. In mammals, there is a neural connection of the pineal to the suprachiasmatic nucleus, which is necessary for the minor circadian role of the pineal (e.g., in reproduction). Birds may not require this pathway; their pineal appears to be a self-sustained circadian oscillator with an important circadian role (Takahashi and Menaker 1979).

The mammalian fetus has a functional biological clock, despite the fact that the suprachiasmatic-eye tract is not functional before birth. Their clocks are normally synchronized with that of the mother; the zeitgeber is not the normal environmental LD cycle but some signal from the mother (Reppert and Schwartz 1983). The maternal suprachiasmatic nucleus is necessary for the entrainment of the biological rhythm of the offspring to the mother's rhythm; its removal from the mother destroys her circadian rhythm and the fetal clock no longer "knows" what time it is (removal of other maternal structures, such as pineal, adrenals, pituitary, thyroid, parathyroid, and ovaries have no effect on rhythms).

The mechanism of biological clocks remains obscure, but there are two general hypotheses for the general principle of biological clocks: the hourglass and the Bunning models. The **hourglass model**, or interval-timer model, suggests that some biological process is initiated and the process takes a specific amount of time to reach completion. This model is analogous to the running of sand through an hourglass. Some examples of biological clocks support such a mechanism. For example, the duration of the dark period determines whether eggs of some aphids (*Megoura*) develop into sexual, egg-laying females (>9.5 hr darkness) or parthenogenetic females (<9.5 hr darkness). The length of the light period is not involved; it can be experimentally extended to up to 30 hr, but this has no effect on the nature of egg development. Thus, nighttime measurement appears to involve an interval timer that is initiated at dusk and runs a specified time course (9.5 hr); interruption of the timer before it completes its cycle results in parthenogenetic egg development, whereas completion of the timer cycle results in egg-laying female development.

The **Bunning model** suggests that the day-night lengths, i.e., the light-dark periodicities, measure time. Thus, time measurement involves circadian oscillators but there are a number of possible schemes for the operation of such oscillator clocks. For example, the parasitic wasp *Nasonia* apparently has a "dawn" and a "dusk" oscillator, and the phase relationship between the two oscillators indicates time, i.e., there is internal coincidence of two oscillators. Alternatively, the flesh fly *Sarcophaga* appears to have a photoperiod oscillator whose phase is set by light. The endogenous circadian cycle has a photosensitive period that initiates pupal eclosion if the next light pulse occurs within the photosensitive period; this is external coincidence. This is the same principle as the control of eclosion of *Drosophila* pupae already described. The photoperiodic control of gonads in Japanese quail is another example of external coincidence. A circadian sensitivity to light is initiated by light (Figure 8–44). If the next light period falls within the photosensitive period, then gonadal growth is stimulated; if the next light period falls in the photoinsensitive period, then gonadal growth is not initiated.

But what is the cellular/subcellular mechanism for a biological clock? Many different models have been proposed; these models can be grouped into four general categories (Edmunds 1988). Strict **mo-**

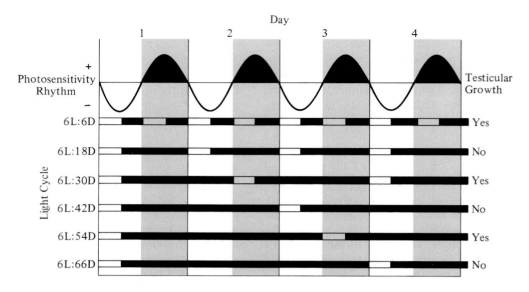

FIGURE 8–44 Testicular development in the Japanese quail *Coturnix* is stimulated if a light pulse occurs within a specific photosensitive portion of the circadian cycle, but not in the photoinsensitive portion of the cycle. *(Modified from Lofts, Follett, and Mutron 1970.)*

lecular models suggest that the property of molecules generates a persistent 24 hr cycle. **Feedback-loop models** hypothesize oscillations in cellular energy production or other biochemical pathways; these involve regulation of energy stores and allosteric constants and turnover numbers for key enzymes. **Transcriptional models** suggest that translation of polycistronic pieces of DNA and rate-limiting diffusion steps could provide a clock mechanism. Various **membrane models** assume a transport process across a membrane is responsible for the circadian rhythmicity. These various models are not mutually exclusive, and each has evidence for and against it. At present, there is no conclusive evidence for the molecular mechanism of biological clocks.

Lunar Rhythms

The lunar cycle has a periodicity of 29.5 days. The phase of the moon determines the light intensity at night and might therefore be expected to have some effect on the activity rhythms of nocturnal animals. For example, nocturnal small mammals may be less active on nights with a full moon. Many marine organisms have lunar reproductive cycles, e.g., the moonlight swarming of palolo worms *Eunice*.

The lunar cycle also controls the ocean tides. There are one or two high and two low tides per day depending on locality, but the basic tidal cycle occurs with a 24.8 hr periodicity. Tides vary in amplitude depending on the relative positions of the earth, moon, and sun. Many marine and intertidal animals have biological rhythms that are coincident with the tidal cycle. For example, many intertidal bivalve mollusks and crabs are inactive during low tide to avoid desiccation and to survive the anoxia associated with immersion. In contrast, many intertidal animals are active during low tide, foraging for food, e.g., fiddler crabs and littoral birds. Many of these animals continue to show activity rhythms that follow an approximate lunar cycle if removed from their natural environment and maintained in the laboratory with no environmental cues for the tidal rhythm.

Tidal rhythms are not synchronized with the light-dark cycle. Many tidal cycles are observed to persist in constant LL or DD or in various LD cycles. Activity rhythms can occur totally independently of the light-dark cycle. For example, fiddler crabs are active during low tide regardless of the photoperiod. However, the light-dark cycle can influence the tidal activity cycle. For example, shore birds have a strong diurnal control of their essentially lunar-based activity cycles because they require light to see as they forage. Light can also influence the lunar cycle. Altering the light-dark cycle phase shifts not only circadian cycles but also tidal cycles, i.e., the tidal cycle appears to be functionally associated with the circadian cycle.

Circannual Rhythms

The yearly cycle in day and night length is a final environmental rhythm that we shall consider. Many animals exhibit precise annual cycles in activity, breeding, etc. (Gwinner 1986). For example, many species are brought into reproductive condition by the lengthening days and shortening nights of spring; these are "long-day" species. "Short-day" species breed in the autumn when the day length is shortening and the nights are lengthening. Many mammals hibernate on a circannual cycle. The carpet beetle is a long-lived insect that overwinters as a dormant larva for its first two winters and pupates after its third overwintering; there is a circannual rhythm in pupation. These annual cycles often persist in constant laboratory conditions, again suggesting an endogenous **circannual rhythm**. Free-running ground squirrels *Spermophilus* maintain an approximately circannual rhythm of hibernation with a periodicity of about 365 days, but not exactly 365 days. Some individuals have a longer periodicity, and some have a shorter periodicity (Figure 8–45). Carpet beetles have a free-running periodicity of about 10 to 11 months.

The determination of time of year, in some instances, apparently involves the circadian clock. The change in timing of dawn and dusk with the 24 hr circadian cycle provides the environmental cue indicating the time of year. Temperature is generally less important than light in indicating time of year. Some annual cycles appear to be remarkably constant in periodicity, and synchronized with the actual environmental cycle, i.e., are not cued by photoperiod, temperature, etc. The precision of these cycles suggests a role of some other, subtle environmental cue that has not been eliminated by the experimental protocols.

Summary

The nervous system is the interface between sensory reception of information and motor responses. There generally is at least one interneuron interposed between a sensory neuron and an effector cell, e.g., simple reflex arcs, but usually the nervous system is a complex association of many interneurons. The nervous system of primitive animals is arranged as a diffuse nerve net, but in higher animals it has an anterior specialization, the brain, and the nerves are arranged as interconnected, longitudinal nerve cords.

The nervous system is capable of the complex integration of many sensory inputs and the coordination of many effector systems. Individual neurons integrate their afferent sensory information by spatial and temporal summation and presynaptic inhibition. The net response of the neuron is determined at a single encoding membrane for action potential initiation, e.g., the axon hillock.

The organization of neurons also provides complex integration of senses and effectors. The complex arrangement of neuronal circuits allows divergence of information to many neurons or convergence of much information onto a single neuron. Collateral branches of axons provide excitatory or inhibitory feedback loops via interneurons. Facilitation, disinhibition, and recruitment are other functions of neural circuits.

Sponges have a generalized epidermal system for the conduction of depolarizations. The nervous system of cnidarians and ctenophores is a nerve net or multiple nerve nets of varying location,

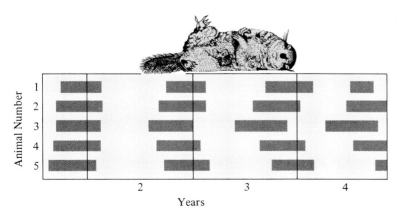

FIGURE 8–45 Golden-mantled ground squirrels *Spermophilus lateralis*, isolated from birth and kept in constant dark (DD) at 3° C, hibernate (dark bars) with an endogenous circannual rhythm that has a periodicity of about 365 days. The periodicity varies for different individuals, e.g., about 300 days for individual number 3. *(From Palmer 1976.)*

conduction velocity, and function. The more advanced, bilaterally-symmetrical animals have a longitudinal arrangement of the nervous system with nerve cord and segmental ganglia. There is a general trend among higher animals for consolidation of nerve cord ganglia into an anterior brain and into fused thoracic or abdominal ganglia. Cephalopods and vertebrates have a complete consolidation of the nervous system into a central brain.

The vertebrate nervous system consists of a peripheral nervous system (sensory and motor nerves) and a central nervous system (brain and spinal cord). The peripheral nervous system communicates with the central nervous system via cranial and spinal nerves. It includes somatic and visceral sensory nerves, somatic motor nerves, and the visceral nerves of the autonomic nervous system (sympathetic and parasympathetic branches). The spinal cord is organized as a discrete region of grey matter, which consists of neuron cell bodies, and a region of white matter, which is axon tracts. The spinal cord contains sensory tracts ascending to the brain, motor tracts descending to the peripheral nervous system, and local spinal reflex tracts.

The brain can be divided into the hindbrain (including the medulla and cerebellum), the midbrain (tectum, tegmentum), and the forebrain (thalamus, hypothalamus, cerebral ganglia, and cerebral cortex). The medulla is the primary receptive area for the acoustico-lateralis sensory information, and it contains centers for the regulation of basic body functions (e.g., respiration and heart rate). The cerebellum is responsible for the integration of subconscious sensory information and the control of subconscious body movement and posture. The midbrain is the primary sensory area for vision (at least in lower vertebrates) and has effector control through the spinal cord and the hypothalamus. The hypothalamus, and the associated structures of the limbic system, are important centers for the control of most vegetative and endocrine systems and emotions. The forebrain is primitively a receptive area for olfaction and the control of associated effector functions. In higher vertebrates, the cerebral ganglia and/or cerebral cortex are highly elaborated and control most higher sensory, integrative, and motor roles of the central nervous system.

The visual system provides many examples of the complex processing of sensory information. Retinal ganglion neurons process and integrate visual information in complex ways, e.g., some ganglion neurons (in frogs) are sustained edge, convex, changing contrast, diminishing, dark, or blue-sensitive detectors. Neurons at the next level of visual processing (tectum in lower vertebrates, thalamus and visual cortex in higher vertebrates) also have complex responses to visual information, e.g., "sameness" or "newness" detectors (frog tectum); "ON" or "OFF" center visual fields (mammalian lateral geniculate body); and circularly-symmetrical, simple, complex, or hypercomplex neurons (mammalian visual cortex). Color vision and binocular vision also involve complex visual processing pathways.

Memory and learning are poorly understood processes of the nervous system. Memory is the retention of information by the central nervous system, but the nature of the memory information is not clear. Short-term memory may involve post-tetanic potentiation and reverberating neural circuits, and longer-term memory may involve synaptic facilitation of specific neural pathways or synthesis of proteins or RNA. Habituation and sensitization are examples of nonassociative learning of responses to stimuli, i.e., they are not associated with simultaneous stimuli. The nonassociative sensitization by the gastropod mollusk *Aplysia* of gill or mantle retraction in response to stimulation involves serotonin receptors, elevated intracellular cAMP levels, and modulation of the permeability of "serotonin-sensitive" K^+ channels. Associative learning involves an association between an experimental stimulus (the conditioned stimulus) and another stimulus (the unconditioned stimulus) that normally elicits a reflex or the association of a reward or punishment with an action. The classical conditioning of gill or mantle retraction by *Aplysia* involves intracellular pathways similar to those for nonassociative learning.

Biological rhythms are widespread among animals. Their periodicity can be very short (e.g., 1 to 2 msec for pacemaker neurons), 24 hr (circadian rhythms), 24.8 hr (tidal rhythms), 29.5 days (lunar rhythms), or 365 days (circannual rhythms). Most long-period biological rhythms are synchronized with some geophysical cycle, such as tides or photoperiod, but will persist in the absence of the geophysical cues. This suggests that these animals have an endogenous biological clock. The periodicity of the endogenous rhythm is generally similar to, but not exactly equal to, the exogenous periodicity. Some biological clocks have been shown to reside in the nervous system. In vertebrates, the biological clock is located in the hypothalamic suprachiasmatic nucleus; it receives photic input from the eyes. Some biological clocks appear to function as interval timers, or hourglasses, whereas other clocks act as single or multiple oscillators.

Supplement 8–1

Inhibition by Mauthner Interneurons

· ·

Many fish and urodele amphibians react to a sudden stimulus with a rapid "startle response," which is a very rapid flexion of the body followed by a tail-flip to propel the animal forward. It is important that only the muscles on one side of the fish or amphibian are stimulated to contract during the startle response.

The startle response is controlled by a pair of giant Mauthner interneurons. Each Mauthner cell body, which is located in the brainstem, has two large dendrites and a giant axon that arises at a prominent axon hillock. The giant axons decussate then extend down the spinal cord to innervate motoneurons of the segmental muscles. The region around the axon hillock, the axon cap, has numerous glial cells and inhibitory synapses from interneurons. The main sensory input to the Mauthner interneurons is from the auditory and lateral line systems. The ipsilateral branches of the VIIIth (acoustic) cranial nerve terminate on the lateral dendrite of the Mauthner interneuron and have a stimulatory effect. Contralateral branches of the VIIIth nerve, and axons from the contralateral vestibular nucleus terminate on the Mauthner interneuron soma; these have an inhibitory effect.

The role of the Mauthner neurons is to elicit the startle response, and so the activity of the other Mauthner

The startle response of the kelp bass is a rapid flexion of the tail that changes the orientation of the fish (shown) then propels it away from the initiating stimulus. *(Modified from Eaton, Bombarderi, and Meyer 1977.)*

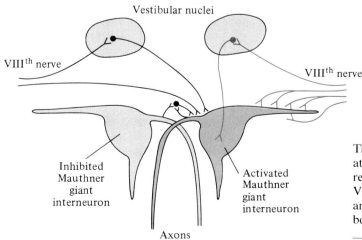

The paired giant Mauthner interneurons have excitatory input from their ipsilateral VIIIth nerve, and receive inhibitory inputs from the contralateral VIIIth nerve, the contralateral vestibular nucleus, and an interneuron that receives collaterals from both Mauthner interneurons.

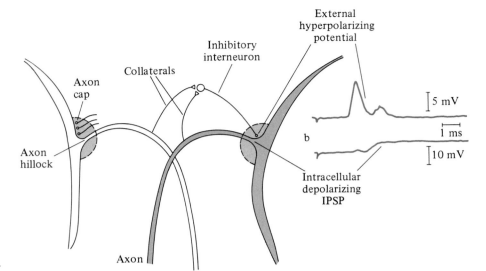

The synapse of the inhibitory interneuron at the axon cap causes an external hyperpolarizing potential, which produces an inhibitory postsynaptic potential at the axon hillock. *(Modified from Aidley 1989.)*

interneuron must be simultaneously inhibited. Stimulation of the contralateral Mauthner interneuron elicits an external hyperpolarizing potential (EHP) in the axon cap region of the Mauthner interneuron. Collateral branches of the contralateral Mauthner giant axon stimulate inhibitory interneurons that elicit the EHP at the axon cap region of the Mauthner interneuron. There also are collateral branches of the giant axon that stimulate these interneurons, thereby preventing further activity in that giant axon. The high conduction velocity of the giant Mauthner axons (50 to 100 m sec⁻¹) ensures a rapid and fairly simultaneous contraction of the segmental muscles on one side of the body to elicit the startle response. (*See Eaton, Bombardieri, and Meyer 1977; Retzlaff 1954; Aidley 1989.*)

Supplement 8–2

The Nervous System of the Leech

The leeches (Hirudinea) are fairly homogeneous, yet specialized, annelids. Their relatively uncomplicated nervous system consists of an anterior (cerebral) ganglion and posterior (caudal) ganglion connected by a chain of 21 segmental ganglia.

Many of the neurons of the leech nervous system have been mapped, and their functions studied with microelectrodes, which can record action potentials, EPSPs, and IPSPs from individual neurons. Each ganglion contains seven pairs of sensory neurons: three for touch (T), two for pressure (P), and two for nociception (pain, N). The touch receptor neurons are extremely sensitive; they respond to an 0.005 cm indentation of the skin and even movement of fluid over the skin. These cells rapidly adapt (within a fraction of a second) to a continuous stimulus. The pressure receptor neurons respond to only marked skin deformation and have a tonic response to stimulation maintained for 10 to 20 sec. The nociceptor neurons require even stronger stimulation to elicit a response; they do not adapt rapidly. Each sensory neuron only responds to stimulation over a discrete area of the body surface; each specific receptive field on the body surface overlaps slightly with that of the adjacent receptive

field of a different, but same modality (e.g., P), sensory neuron.

Motoneurons can be similarly identified, and their functions mapped. For example, the large longitudinal motoneuron (L) produces reflex shortening of the body segment; its axon decussates and innervates a group of longitudinal muscles. The paired L cells are electrically coupled so that their effects are synchronous. The muscle cells stimulated by the L cells are not used for other locomotory functions (e.g., swimming, bending).

The T, P, and N sensory neurons innervate the L motoneuron, creating three reflex loops. The synaptic connections are either chemical (N), electrical (T), or both (P). There also are longitudinal connections between the sensory cells and the motor cells of adjacent ganglia, e.g., a more anterior N neuron also innervates the L motoneuron of the next posterior ganglion. The N neuron of the posterior ganglion also innervates the L motoneuron of the anterior ganglion, although the physiological effect is not the same as the connection in the reverse direction.

Swimming movement is generated by an ensemble of bilaterally symmetrical pairs of ganglionic motoneurons. Four pairs of dorsal excitatory motoneurons (neurons 3,

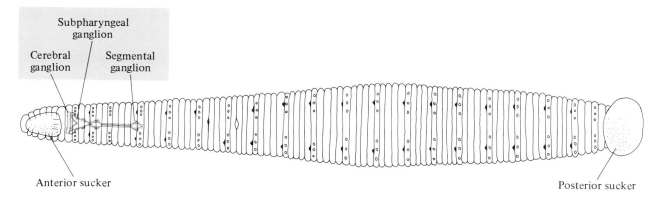

The nervous system of the medicinal leech *Hirudo medicinalis*.

5, 7, 107) innervate the dorsal body wall segmental longitudinal muscles, and three pairs (4, 8, 108) innervate the ventral muscles. There are two pairs of inhibitory neurons (1, 102) to the dorsal muscles, and two pairs (2, 119) to the ventral muscles. Finally, there is a pair of dorsoventral excitatory neurons (109) that innervate the right and left dorsoventral muscles. The dorsoventral excitator neuron is continuously depolarized during swimming, thereby flattening and extending the body. The dorsal and ventral excitator motoneurons have an oscilla-

tory E_m during swimming, as do the dorsal and ventral inhibitors. A network circuit of these motoneurons, and of oscillatory interneurons (27, 28, 33, 123), is responsible for the swimming pattern and the coordination of adjacent body segments. Basically, the oscillatory circuit is the intersegmental connection of five neurons—28 and 123 of the more anterior ganglion and 28, 123, and 27 of the more posterior ganglion. An oscillatory rhythm of activity and inactivity is maintained by recurrent cyclic inhibition between this odd number of neurons (five), each of which

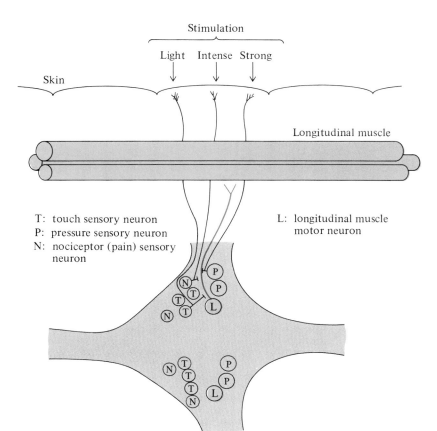

T: touch sensory neuron
P: pressure sensory neuron
N: nociceptor (pain) sensory neuron

L: longitudinal muscle motor neuron

Stimulation of touch (T), nociceptor (N), and pressure (P) receptors activate the motoneuron (L) that innervates the longitudinal muscles. The locations of the specific sensory and motor neurons have been mapped. *(Modified from Nicholls and van Essen 1974.)*

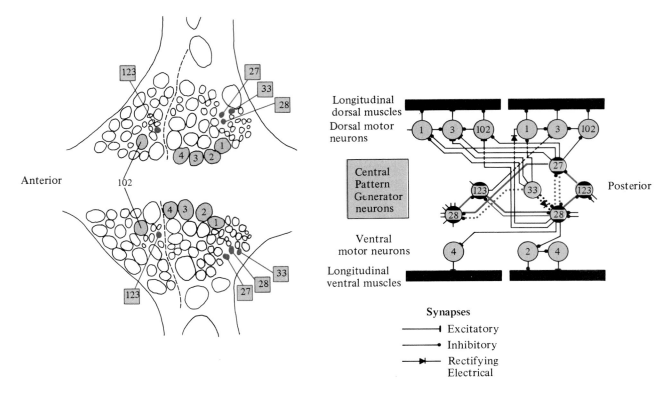

The central pattern generator neurons (27, 28, 33, 128) in the leech's nerve cord determine the pattern of activation of motor neurons (1, 3, 102 and 2, 4) and locomotor muscles involved in swimming.   *(Modified from Stent et al. 1978.)*

is tonically excited but makes an inhibitory synapse with the next neuron in the circuit and receives an inhibitory synapse from the previous neuron in the circuit. The addition of neuron 33, the mutually inhibitory connection between neurons 27 and 28, and the intersegmental connection of neurons 28 and 33, modify the oscillation into a four-step cycle. (*See Nicholls and Van Essen 1974; Stent et al. 1978.*)

Supplement 8–3

The Giant Fiber System of Cephalopods

The giant fibers of squid form a complex system of three types of giant axons, innervating various muscle groups. The first-order giant fibers originate from large multipolar neurons in the palliovisceral ganglion of the brain; these large neurons receive sensory information from the statocysts, eyes, skin, and cerebral ganglia. These giant axons decussate and have an unusual cytoplasmic connection between the two axons where they cross. Each first-order giant fiber synapses with short collaterals of the axons of seven second-order giant axons. These giant axons also have other presynaptic inputs. Most of the second-order giant axons are motoneurons that innervate the head and funnel retractor muscles. Two second-order giant axons pass to the stellate ganglion where one chemically syn-

apses with the axons of the third-order giant neurons. Large accessory axons from the brain also synapse with the third-order giant axons in the stellate ganglion. The third-order giant axons are each formed by the fusion of axons of many (300 to 1500) small unipolar neurons. Each third-order axon innervates about 1/10 of the mantle muscle on its side of the body. There is some variation in diameter of the third-order axons; the larger ones, with the higher conduction velocity, innervate the more posterior parts of the mantle. This ensures an almost simultaneous contraction of the entire mantle.

The complex giant fiber system provides a simple, rapid triggering mechanism for quick escape from a threatening stimulus. The chemical synaptic delay of the

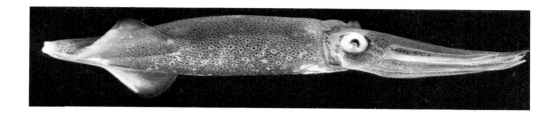

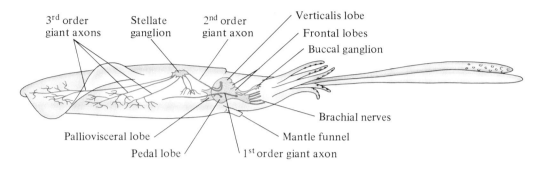

The giant axon system of the squid *Loligo.* *(Tom McHugh, Steinhart Aquarium/Photo Researchers, Inc.)*

second to third-order giant fibers is short, about 1/2 msec. Normally, a presynaptic action potential in the second-order axon produces a postsynaptic potential in the third-order axon with about a 0.4 msec delay. The conduction velocity of the giant axons is high, from 5 to 25 m sec^{-1}. The short synaptic delay and high-conduction velocity of the giant axons probably reduce the response time be-

tween stimulus and escape by about 1/2. A single action potential of a giant axon elicits a maximal contraction of the mantle muscle that it innervates. Mantle contraction rapidly expels water from the mantle cavity, jetting the squid forward or backward, depending on the orientation of the mantle funnel. *(See Bullock and Horridge 1965; Katz and Miledi 1966.)*

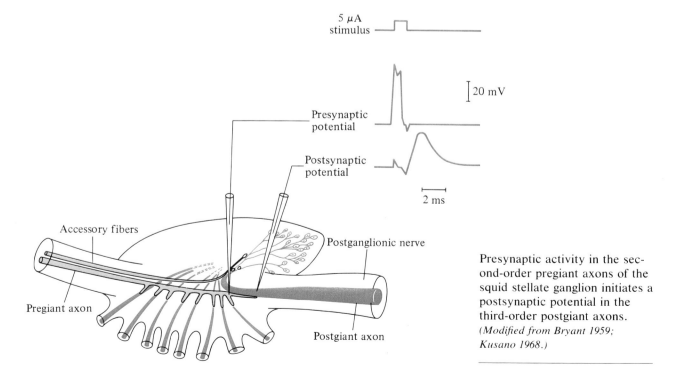

Presynaptic activity in the second-order pregiant axons of the squid stellate ganglion initiates a postsynaptic potential in the third-order postgiant axons. *(Modified from Bryant 1959; Kusano 1968.)*

Supplement 8–4

Development of the Cerebral Cortex in Vertebrates

• •

The telencephalon of primitive vertebrates has an olfactory bulb, with projections from the olfactory sense cells, and a cerebral hemisphere connected to the olfactory bulb by an olfactory tract. The cerebral hemisphere is primarily olfactory cortex (paleopallium), which processes olfactory information and relays it to other brain centers. In more advanced vertebrates (sharks, lungfish, amphibians), the hemispheres can be divided into three areas, each still mainly olfactory in function: the basal nuclei, the paleopallium, and the archipallium (which is antecedent to the hippocampus). In tetrapods, the basal nuclei move centrally and the archipallium and paleopallium move outwards to form the cerebral cortex. In most bony fish, the grey matter of the forebrain develops downward and inward to form structures that bulge into the ventricle; the roof of the hemispheres is a thin, non-neural membrane forming the dorsal boundary of the ventricle. The cerebral "hemispheres" are inverted in structure compared to the tetrapod hemispheres.

Some advanced reptiles show traces of a new type of cortex, the neocortex, or neopallium, which develops between the paleopallium and archipallium. This is an

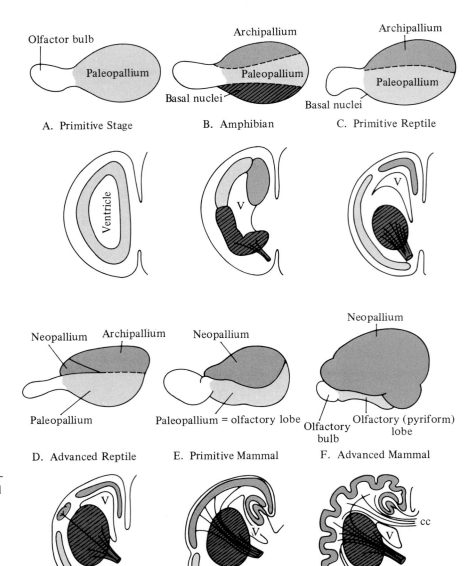

The evolutionary sequence in progressive expansion of the paleopallium into archipallium and basal basal nuclei, then the re-reorganization occurring with expansion of neocortex, is shown in lateral view (above) and cross section (below). *From Romer and Parsons 1986.)*

association center that receives sensory information and sends motor information to the motor columns. In birds, the basal ganglia are enlarged to form the corpus striatum and a hyperstriatum (with powers of memory and learning, like the mammalian neocortex). The mammalian brain has a marked expansion of the neocortex, with restriction of the archipallium to form the hippocampus, and of the paleopallium to form the pyriform lobe. The basal ganglia, or corpus striatum, persist as a center for more automatic reactions. (*See Romer and Parsons 1986.*)

Recommended Reading

Applewhite, P. B., and H. J. Morowitz. 1967. Memory and the microinvertebrates. In *Chemistry of learning*, edited by W. C. Corning and S. C. Ratner, 329–40. New York: Plenum Press.

Baringa, M. 1990. The tide of memory, turning. Science 248:1603–1605.

Bullock, T. H., and G. A. Horridge. 1965. *Structure and function in the nervous systems of invertebrates.* Vols. 1 and 2. San Francisco: W. H. Freeman & Co.

Bullock, T. H., R. Orkand, and A. Grinnell. 1977. *Introduction to nervous systems.* San Francisco: W. H. Freeman & Co.

Byrne, W. L. 1970. *Molecular approaches to learning and memory.* New York: Academic Press.

Cloudsley-Thompson, J. L. 1980. *Biological clocks: Their functions in nature.* London: Weidenfeld & Nicholson.

Edmunds, L. N. 1988. *Cellular and molecular bases of biological clocks: Models and mechanisms for circadian timekeeping.* Berlin: Springer-Verlag.

Horridge, G. A. 1968. *Interneurons.* London: Freeman & Co.

Hovey, H. B. 1929. Associative hysteresis in marine flatworms. *Physiol. Zool.* 2:322–33.

Kandel, E. R., and J. H. Schwartz. 1982. Molecular biology of learning. *Science* 218:433–43.

Lashley, K. S. 1950. In search of the engram. *Symp. Soc. Exp. Biol.* 4:454–82.

Lentz, T. L. 1968. *Primitive nervous systems.* New Haven: Yale University Press.

McConnell, J. V. 1966. Learning in invertebrates. *Ann. Rev. Physiol.* 28:107–36.

Mill, P. J. 1982. *Comparative neurobiology.* London: Edward Arnold.

Palmer, J. D. 1976. *An introduction to biological rhythms.* New York: Academic Press.

Pribram, K. H., and D. E. Broadbent. 1970. *Biology of memory.* New York: Academic Press.

Ratner, S. C. 1967. Annelids and learning: a critical review. In *Chemistry of learning*, edited by W. C. Corning and S. C. Ratner, 391–406. New York: Plenum Press.

Saunders, D. S. 1982. *Insect clocks.* Oxford: Pergamon Press.

Stein, D. G., and J. J. Rosen. 1974. *Learning and memory.* New York: Macmillan Publishing Co.

Suda, M., O. Hayaishi, and H. Nakagawa. 1979. *Biological rhythms and their central mechanism.* Amsterdam: Elsevier/North-Holland Biomedical Press.

Usherwood, P. N. R., and D. R. Newth. 1975. *"Simple" nervous systems.* London: Edward Arnold.

Wells, M. J. 1962. *Brain and behaviour in cephalopods.* Stanford: Stanford University Press.

Chapter 9

Cell Movement

- -

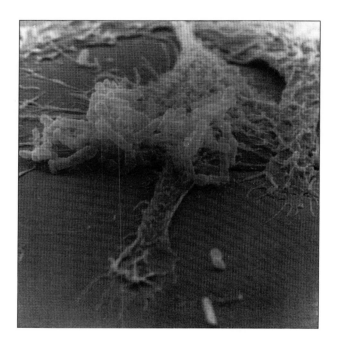

A human macrophage ingesting a microbe, ***Pseudomonas.*** *(Courtesy D. M. Phillips/The Population Council, Science Source.)*

Chapter Outline

The Cytoskeleton
  Microtubules
  Microfilaments
  Intermediate Filaments
  Principles of Cellular Movement
Microtubular Movement
  Cytoplasmic Streaming
  Cell Division
  Cilia and Flagella
Microfilament Movement
  Amoeboid Movement
  Muscle
Summary

A nimal cells have an inherent stable structure, supported by a subcellular skeleton, but they also have the capacity for movement—either of their intracellular constituents or of the cells themselves. There are three general types of cell movement: ciliary (and flagellar) movement, amoeboid movement, and muscle contraction.

The capacity of cells for movement has been described and detailed for many years. The mechanisms of muscle cell contraction have been understood, at least in general detail, for many years because of their highly organized intracellular structure, but the mechanisms for ciliary and flagellar movement have only recently been elucidated. The mechanisms of amoeboid movement are only now being detailed.

The Cytoskeleton

Cells, after histological preparation, appear to have an extensive internal skeleton, a **cytoskeleton**, composed of highly structured microtrabeculae that link the plasmalemma, organelles, and filamentous

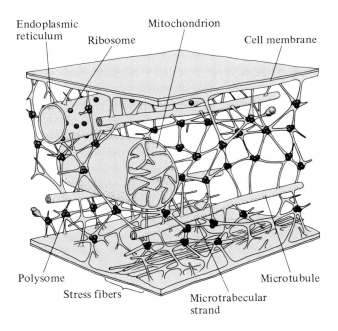

Endoplasmic reticulum
Ribosome
Mitochondrion
Cell membrane
Polysome
Stress fibers
Microtubule
Microtrabecular strand

FIGURE 9–1 Model of the cytoplasmic skeleton, showing fiber bundles that are continuous with actin filament bundles, microtubules, intermediate filaments, and endoplasmic reticulum.

components into a structural organization, the cytoplast (Figure 9–1). The structure and role of the apparent cytoskeleton is unclear for living cells because it is not clear whether the interconnecting network of cytoskeletal trabeculae exists inside living cells. The composition of the microtrabeculae is also unclear, but they probably are continuous with microtubule, microfilament, and intermediate filament bundles and may contain all of these filament types. The term cytoskeleton implies a static, rigid structure but it is probably a highly dynamic, changing structure.

Microtubules

Microtubules are long, slender cylinders about 20 to 30 nm in diameter; the wall is 6 to 7.5 nm thick (Figure 9–2A). A microtubule is composed of 13 parallel protofilaments, and each protofilament is a polymer of tubulin subunits. Each tubulin subunit (molecular weight = 1.2×10^5) is a dimer of an α tubulin and a β tubulin (MWt 5.5 to 6×10^4). The tubulin dimer subunits of adjacent protofilaments are slightly offset so that they appear to form a helical spiral around the microtubule. The α and β tubulins are closely related proteins, whose structure is extremely conservative throughout the animal kingdom, e.g., chick tubulins are almost identical to sea urchin tubulins.

The microtubule is polarized because the spiral can be left-handed or right-handed. The microtubule is further polarized because α and β tubulin subunits are consistently aligned along the microtubule, and so the helical structure has an end directed towards the α tubulins and the other end towards the β tubulins.

Why do microtubules have 13 protofilaments rather than an even number such as 12 or 14 or some other number? Microtubules are very dynamic polymeric structures, being constantly polymerized and depolymerized, and a prime number of protofilaments (e.g., 5, 7, 11, 13, 17, 19, 21) probably confers some inherent instability to a microtubule structure because the intertubulin bridges must be strained to attain symmetry (Fujiwara and Tilney 1975). This instability may be useful for a dynamic microtubular structure. The intrinsic asymmetry of an odd number of protofilaments may also allow better passive elastic recoil of microtubules to their original structure after deformation than would a more symmetric microtubule. A microtubule with 5, 7, 9, or 11 protofilaments might be too strained to be stable and with 17, 19, or 21 might not be

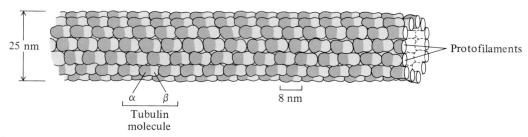

25 nm

Protofilaments

α   β

8 nm

Tubulin
molecule

A

Self-assembly

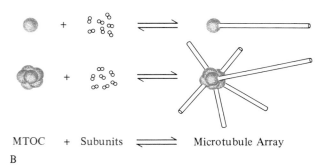

Subunits

[Nucleus]

Microtubule

Site-initiated Assembly

MTOC   +   Subunits   ⇌   Microtubule Array

B

Interphase Cell

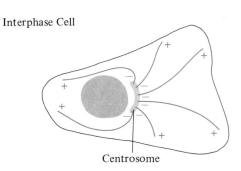

Centrosome

Dividing Cell

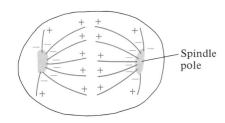

Spindle
pole

Ciliated Cell

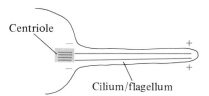

Centriole

Cilium/flagellum

Nerve Cell

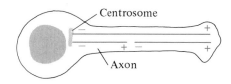

Centrosome

Axon

C

FIGURE 9–2   **(A)** Model structure of a microtubule comprised of protofilaments; each protofilament subunit contains an α and β tubulin. **(B)** Self-assembly of microtubules from tubulin subunits through nucleation and elongation by subsequent addition of tubulin subunits is by self-assembly, whereas site-initiated assembly of microtubules is controlled by a nondiffusable nucleating element (stippled region) that can form a constellation of assembly regions, called a microtubule organizing center (MTOC). **(C)** Schematic arrangement of the organization of plus (fast growing) and minus ends of microtubules in interphase and dividing cells and in cilia and nerve cells.   *(From Alberts et al. 1989; Borisy and Gould 1977.)*

strained enough to be unstable; perhaps 13 provides a suitable degree of instability.

Microtubules can be self-assembled from the tubulin subunits; a **microtubule-nucleating center** acts as a nucleus for polymerization (Figure 9–2B). Alternatively, microtubule formation may be initiated by a specific structure or constellation of structures, which induce microtubule synthesis in a specific 2- or 3-dimensional organization; the **microtubule-organizing center** (MTOC) is an example of such an organizing system. Microtubules have an inherent functional polarity, determined by which end of the tubule more readily grows by tubular assembly; the fast growing end is the **plus** end and the other is the **minus** end. In general, the plus end extends away from the cell center and away from an MTOC, e.g., centrosome (Figure 9–2C). This functional polarity is important because it usually determines the direction of movement along a microtubule by the ATPase "motors."

Microtubules form a variety of structures in animal cells; some are part of the intracellular cytoskeleton, and some are arranged as bundles in the cytoplasm or in cilia and flagella. These are relatively stable microtubules. Labile microtubules appear during cell division (mitosis or meiosis) and then disappear. Neurons contain similar microtubular structures, neurotubules, that differ slightly from normal microtubules in their dimensions and nature of the tubulin subunits. Many microtubules have associated ATPase enzymes that are the "motors" responsible for movement.

Microfilaments

Microfilaments, like microtubules, are polymers of protein subunits. However, microfilaments, such as actin and myosin, are thin strands of often helically intertwined filaments rather than tubules.

Actin is typically 5 to 7 nm in diameter and has two helically intertwined protofilaments comprised of 100 or so globular subunits (Figure 9–3A). It has been isolated from virtually every kind of animal cell, but it is present in skeletal and cardiac muscle cells at high concentration and in a highly organized structure (the sarcomere). Actin is also present in a less regularly organized fashion in smooth muscle cells and in nonmuscular cells. For example, up to 1/3 of the protein of amoeba protozoans and amoeboid cells of multicellular animals (e.g., fibroblasts, blood platelets) is actin. **G-actin** is the globular actin subunit; it is similar in chemical composition to tubulin. G-actin subunits are polymerized to form filamentous **F-actin**, which is two linear protofilaments that are helically intertwined (Figure 9–3B) rather than a microtubular arrangement of protofilaments. F-actin thus resembles two strings of beads (the G-actin) twisted together. The distance between G-actin subunits is about 5.5 nm, and the repeating distance for the twisted helix is about 72 nm. Cytoplasmic actin appears to be in as dynamic a state as we saw for microtubules (see Supplement 9–1, pages 443–444). The structure of actin is quite conservative throughout the animal and plant kingdom. There are small differences in amino acid composition of actins from different animals, but the few changes are quite conservative. Thus, the actin of a slime mold, for example, can functionally replace that of an invertebrate or mammalian muscle cell.

Troponin and **tropomyosin** are other proteins associated with F-actin, and they regulate the interaction of actin with myosin; their role will be discussed in detail below. Tropomyosin is also found in nonmuscle cells, but only when the actin is "immobile," i.e., cross-linked; it may be involved in maintaining the shape of the cell. Troponin is a Ca^{2+}-binding protein that is present only in muscle cells. Calmodulin is a Ca^{2+}-binding protein, similar to the C-subunit of tropinin that is found in all animal cells. It activates many proteins, such as Ca^{2+}-dependent enzymes when complexed with Ca^{2+}.

Another microfilament, **myosin**, is the enzymatic "motor" for actin movement. Myosin, unlike actin, has ATPase activity (although actin and other regulatory proteins can modify the enzymatic activity of the myosin). The myosin molecule is a long (15 to 19 nm), rod-shaped molecule of two protein chains (Figure 9–3C). Each protein is divided into a light meromyosin (LMM) and a heavy meromyosin (HMM) portion. The **myosin head**, a globular part of the heavy meromyosin end (HMM-S1), extends laterally at the end of the molecule; the remainder of the heavy meromyosin is called HMM-S2. The myosin head has the ATPase activity; it is equivalent to the dynein and kinesin "motor" system of microtubules. Heavy meromyosin fragments, enzymatically separated from LMM, bind to actin in a unidirectional manner, forming a "pointed end" and a "barbed end."

Myosin-1 is monomeric myosin that causes actin movement past membranes and active sliding of adjacent actins. Myosin molecules are often polymerized to form a thick, bipolar filament (**myosin-2**) with myosin heads extending in a helical fashion and facing opposite directions at each end. Myosin, the "thick filament" of muscle cells, is one example of such a polymerized myosin (actin is the "thin filament"). In muscle contraction, and some other types of intracellular movement, actin microfilaments do not slide past adjacent actin filaments but slide past polymeric myosin.

Intermediate Filaments

Intermediate filaments are α-helical coiled proteins, about 10 nm in diameter (Figure 9–4). They are

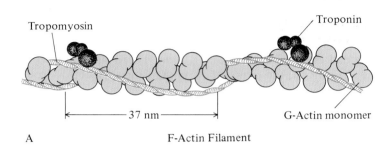

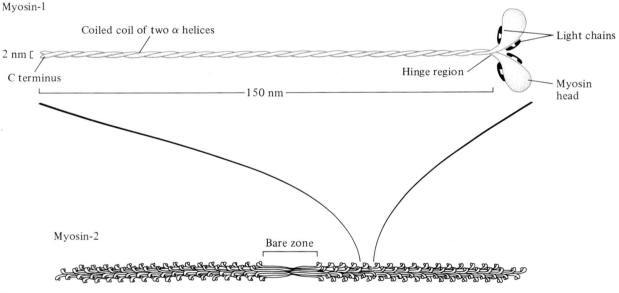

FIGURE 9–3 (**A**) Structure of a microfilament; an F-actin filament is comprised of two strands each of which is comprised of globular G-actin subunits, twisted with a pitch of about 36.5 nm. The F-actin microfilaments of striated muscle are a polymeric arrangement of globular G-actin subunits with accessory proteins, tropomyosin and troponin. (**B**) The myosin-2 microfilament is a polymer of myosin-1 subunits, which is a twisted helix of two proteins, each with a light and heavy meromyosin portion and a myosin head.   (*From Alberts et al. 1989; Fawcett 1986.*)

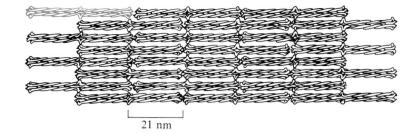

FIGURE 9–4 Structure of an intermediate filament comprised of eight protofilaments, each made up of protein subunits. *(From Alberts et al. 1989.)*

21 nm

components of the cytoskeleton in many cells. All intermediate filaments have similar morphological, physical, and chemical properties, but there is a wide diversity in their polypeptide composition. The five main types of intermediate filaments are keratins in epithelial cells, neurofilament polypeptides in neurons, desmin and skeletin in muscle, vimentin in mesenchyme cells, and glial fibrillar acidic protein (GFA) in glial cells. Tektin of cilia and flagella is also an intermediate filament. Keratins are a large subgroup of intermediate filaments. These relatively insoluble proteins are abundant in epithelial cells of many vertebrates, and their presence confers a mechanical strength and decreases the evaporative water loss (see Chapter 16).

Many intermediate filaments have associated proteins. For example, filaggrin is a specific cross-bridging protein of intermediate filaments in skin; desmoplakin may function similarly in membrane desmosomes. Neurofilaments are composed of four subunit strands around a narrow central space (3 nm dia) with lateral arms extending away from the axis of the filament.

Principles of Cellular Movement

Animal cells are not just rigid structures with a fixed cytoskeleton but are dynamic and move in a variety of ways. Brownian motion of cytoplasmic molecules and small particles is a random movement due to thermal motion; it does not require energy expenditure, is not an ordered movement, and is not responsible for cell movement, so it will not be discussed further here (but it is important for diffusion; see Chapter 2).

There are many types of directed cell movement. There is a continual movement within cells of cytoplasm and intracellular organelles, of secretory vesicles and phagocytic and pinocytotic vesicles, and of pigment granules. The movement of subcellular particles is often rapid and discontinuous (salta-

tory). Cytoplasmic streaming is a deformation and movement of cytoplasm. Axonal flow is a rapid movement of fluid, chemicals, and vesicles along axons. There is a massive coordinated movement of chromosomes and other intracellular structures during cell division. Many cells locomote by the amoeboid movement of pseudopodia. Other types of cells, such as muscle, can contract and shorten. Finally, many cells have specialized projections, cilia and flagella, that, by a beating action, can move the cell through the medium or move the medium over the surface of the cell.

Let us first consider an amusing but insightful analogy to possible mechanisms for cellular movement (Figure 9–5). Imagine that we want to move a train locomotive and carriage along a track (the carriage represents a cell organelle and the track is the desired intracellular path). There are a number of different ways to move the train carriage. The first (and actual mechanism for a real train) is for the engine of the locomotive to pull the carriage (scheme 1), i.e., the locomotive and carriage are mobile with the power provided independent of any other systems. However, this is only one of the ways to move the carriage, and it apparently is not analagous to any actual mechanism for cellular motility. Alternatively, the locomotive and carriage can be passively moved by a combination of three systems: the track on which they run (cf. microtubules), railway men that supply the physical power (cf. microtubule and microfilament enzymatic ATPase "motor"), and ropes (cf. microfilaments). Other ways to move the train involve track assembly/disassembly (schemes 2, 3) or track treadmilling (scheme 4), transport of the track (scheme 5), mechanotransduction by the railway men on the track (scheme 6), track-associated mechanotransduction (scheme 7), rope-based contraction with mechanotransduction (scheme 8), rope-based contraction with track-associated mechanotransduction (scheme 9), movement of the tracks and attached railway men by ropes and independent mechano-transduction (scheme 10), or a screw mechanism (scheme 11).

Many of the diverse types of cell movement share a similar mechanism, although the mechanisms for some types of movement (e.g., muscle contraction) are much better understood than for other types (e.g., mitotic division). There essentially are two different mechanisms for movement: one based on microtubules and one on microfilaments. Both mechanisms have ATPase "motors" that either move adjacent microtubules or adjacent microfilaments or slide membrane particles along a microtubule or a microfilament (Table 9–1).

Microtubular Movement

Many types of intracellular movement are based on microtubular structures and ATPase proteins. Some types of microtubular movement appear to rely on, or at least require, the dynamic assembly of microtubules from, and/or disassembly into, tubulin subunits. The "motor" responsible for this movement is the assembly/disassembly process. Other examples of intracellular movement do not involve a change in the length of the microtubules but a change in the relative position of adjacent microtubules or microtubules and organelles; the motor is an ATPase enzyme.

The "motor" that slides adjacent ciliary or flagellar microtubules in opposite directions is an unusually large ATPase enzyme, **axonemal dynein** (see Table 9–1). **Cytoplasmic dynein** is a similar "motor" ATPase, but it mediates microtubule membrane movement rather than the sliding movement of adjacent microtubules. Dyneins generally transport towards the minus end of microtubules. Another motor ATPase, **kinesin**, moves various cellular organelles (e.g., mitochondria, chromosomes, vesicles) along cytoplasmic microtubules towards the plus end. However, some kinesins are "minus-end motors" (Malik and Vale, 1990). A single kinesin can

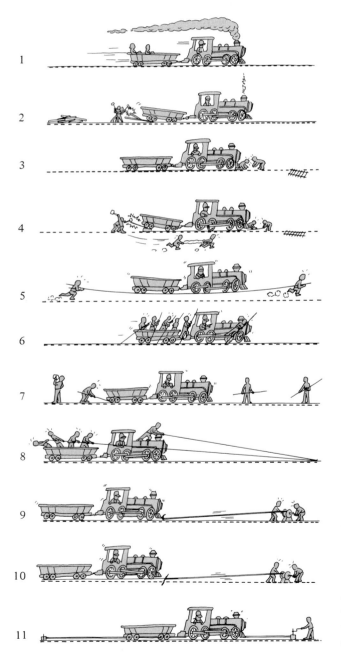

FIGURE 9–5 Representation of various models for intracellular movement for a self-powered cytoplasmic particle (1) or an externally powered cytoplasmic particle (2–11); the steam engine represents the self-powered unit and the carriage represents the cytoplasmic particle, the track is a cytoskeletal pathway, the ropes are microfilaments, and the men are the mechanical transducers. Scheme 1 represents independent particle mobility (this is the only model presently considered not to be applicable to cytoplasmic particle movement); scheme 2, microtubule assembly; scheme 3, microtubule assembly/disassembly; scheme 4, treadmilling; scheme 5, microtubule transport; scheme 6, particle-associated mechanochemical transducers; scheme 7, microtubule-associated mechanochemical transducers; scheme 8, microfilament-associated contraction with particle-associated transducers; scheme 9, microfilament-associated contraction with microtubule-associated transducers; scheme 10, microfilament-associated contractility with transducers moving the microtubules; and scheme 11, screw mechanics. *(From Allen 1981.)*

TABLE 9–1

Summary of microtubule and microfilament types of cellular movement involving ATPase "motors." *(Based on Scholey 1990.)*	
"Motor"	**Function**
Microtube Based	
Microtubule assembly/disassembly	Particle movement
Axonemal dynein	Axonemal microtubule sliding in cilia and flagella; movement is towards "minus" end
Cytoplasmic dynein	Particle movement along microtubules towards "minus" end
Kinesin	Particle movement along microtubules towards "plus" end
Dynamin	Sliding of adjacent microtubules
Microfilament Based	
Myosin-1 (monomer)	Moves membrane particles along actin filaments, and actin-actin filament sliding
Myosin-2 (polymer filament)	Muscle contraction, cortical tension, cytokinesis

move a microtubule several micrometers (Howard, Hudspeth, and Vale 1989). Kinesin and dynein "motors" move microtubules by a ratcheting action of their two "foot" structures (Figure 9–6A). **Dynamin** is a cytoplasmic ATPase protein that causes active sliding of adjacent microtubules (Shpetner and Vallee 1989). The sliding action is apparently unlike the dynein-kinesin movement (Figure 9–6B). Dynamin also has GTP binding sites, suggesting that it and other mechanochemical ATPases (dynein,

kinesin) may be derived from GTP binding proteins (Hollenbeck, 1990).

Cytoplasmic Streaming

The cytoplasm of most animal cells is constantly moving; this **cytoplasmic streaming** is often most apparent and dramatic for neurons, which tend to be large cells (Porter 1976). Plant cells and especially giant algae such as *Nitella* have an even more rapid

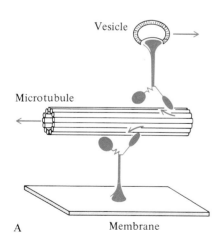

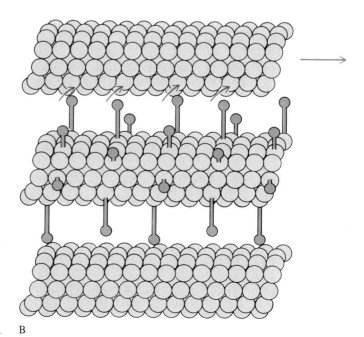

FIGURE 9–6   **(A)** Representation of kinesin movement of a microtubule along a membrane and a particle along a microtubule. **(B)** Representation of sliding movement of microtubules by dynamin. *(From Scholey 1990; Shpetner and Vallee, 1989.)*

cytoplasmic streaming than neurons. Cytoplasmic extensions of heliozoan protozoans have a rapid microtubule-associated streaming (at 2 to 3 μ sec^{-1}). Each extension, or **axopod**, has a central core of microtubules, and cytoplasmic streaming occurs between this microtubule core and the cell membrane. The microtubular structures appear to provide channels for direct cytoplasmic movement but not the mechanism for movement.

The movement of many specific intracellular structures, such as pigment granules and chromosomes, depends on microtubules. For example, many animals can dramatically change color as a consequence of the aggregation or dispersion of intracellular pigment granules, primarily melanin (see Chapter 11). The movement of melanin granules within these melanophore cells is readily observable under the microscope. The melanin granules are aggregated into a tiny speck at the center of the melanophore when the animal is pale, and the granules are dispersed widely throughout the cytoplasm when the animal is dark (Figure 9–7A). The aggregation of pigment occurs relatively rapidly, e.g., in 2 to 3 seconds at a steady velocity of 15 to 20μ sec^{-1} for melanocytes of the red squirrel fish. Dispersion takes about twice as long and involves an irregular, saltatory motion of the melanin granules. The pathways for melanin granule dispersion and aggregation are strictly radial.

Pigment dispersion involves microtubule assembly. The melanocytes contain radially arranged microtubules that emanate from central sickle-shaped dense bodies to the outer cell membrane (Figure 9–7B). These radially arranged microtubules disassemble during aggregation and reassemble during dispersion. The cell also changes shape during pigment aggregation from a disk to a sphere. The microtubules apparently exert a "pushing" force (i.e., they resist compression) during the dispersed state and stretch the cell into a discoidal shape. Pigment aggregation is caused by microtubule disassembly, and the elastic recoil of the cell membrane provides the force for smooth inward movement of the pigment granules and the assumption of a spherical shape by the cell. The rapidity of melanin aggregation and dispersion indicates the marked dynamic nature of the microtubules involved.

Neuron axons transport vesicles and cytoplasmic proteins (see also Chapter 8). There are two well-developed transport systems: a fast (100 to 400 mm day^{-1}) anterograde system and a slow (1 to 4 mm day^{-1}) anterograde and retrograde system. Neurotubules would appear to be an essential component of the rapid vesicular transport system, but there is not a direct translocation of the microtubular

cytoskeleton of the axon, i.e., there isn't microtubule synthesis at the soma and disassembly at the terminus (Okabe and Hirokawa 1990). Rather, a "motor" appears to provide the movement of specific structures along the axon. Kinesin binds to active sites on the microtubules and "walks" along the microtubule lattice towards the plus end of microtubules (which is usually distal to the neuron soma). The axoplasm also contains another motile protein, retrograde factor, that "walks" towards the minus end of the microtubules. The slow transport of cytoskeletal proteins involves microfilaments and neurofilaments, not microtubules.

Cell Division

Cell division is another example of microtubular-based intracellular movement. All cells of the body arise by **mitosis** (cell division) from precursor cells. In a mitotic division, the chromosomes replicate and a copy of each chromosome moves to opposite sides of the cell. The cell then constricts and separates into two daughter cells, each of which has a complete set of the chromosomes. In adult cells, there typically are 2n chromosomes, where n is an integer, since there are two copies of each chromosome (e.g., 2n = 46 for an adult human). Meiosis, or reduction division, is a cell division event similar to mitosis, except that the first cell division is followed by a second reduction division in which the pairs of chromosomes separate prior to a division of each daughter cell. Four daughter cells are formed, each of which has n chromosomes. Ova and sperm are formed in this manner. Subsequent fusion of an egg and sperm forms a 2n zygote that develops into a 2n adult.

Some adult cells, once formed, do not subsequently divide but are long-lived and are not replaced if they die, e.g., neurons. Other cells continue to divide by mitosis at a slow rate, to replace dying cells, e.g., liver hepatocytes. Other cells divide constantly and often very rapidly, e.g., epithelial cells of the skin and gastrointestinal tract. A typical division cycle for one of the latter types of cell (Figure 9–8A) includes a period of DNA replication preparatory to mitotic division (S-phase), a short G2-phase for other mitotic preparations, a rapid mitotic division phase (M-phase), and a long phase of RNA and protein synthesis associated with growth of the daughter cells (G-phase).

Cell division requires the replication of each chromosome and movement of each replicate (chromatid) to opposite poles of the cell prior to cytoplasmic division, as well as constriction of the cell into the two daughter cells. Chromosome movement

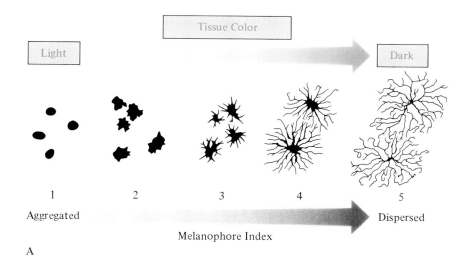

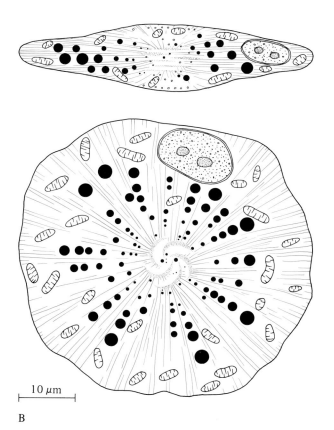

FIGURE 9–7 **(A)** Melanophore aggregation and dispersion measured by Hogben's dispersion index. **(B)** Lateral and dorsal views of the erythrophore of the red squirrel fish illustrate the location of melanin pigment and the radial microtubules when pigment is dispersed. *(From Hogben and Slone 1931; Porter 1976.)*

involves microtubules and will be described here; cytoplasmic constriction involves microfilaments (such as actin), not microtubules. The **centrosome** (a pair of centrioles and surrounding pericentriolar region) is located near the nucleus and acts as an origin for the cytoplasmic microtubular skeleton. Each centriole is a cylindrical organelle 120 to 150 nm in diameter and up to 200 nm long. It consists of nine symmetrically arranged fibers, each made up of three partially fused, 25 nm dia subfibrillar microtubules of tubulin (Figure 9–8B). The inner circular microtubules of each triplet are connected to the outer microtubule of an adjacent triplet and to a star-shaped center of the centriole by arms. Around each centriole is an amorphous pericentriolar region; this, plus the centriole, forms the

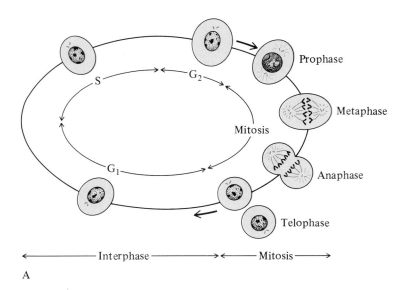

A

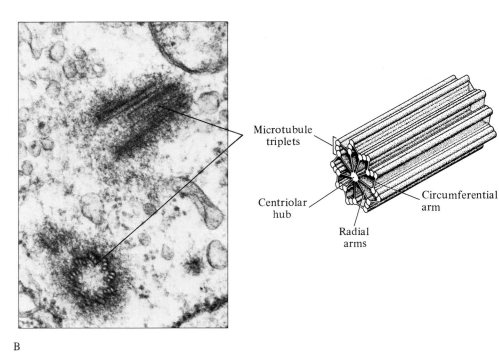

B

FIGURE 9–8 **(A)** Schematic summary of the cell division cycle (shown here for a proliferating bone cell with a cycle length of 25 hours). **(B)** The centriole has nine triplets of microtubules forming a cylinder; the two centrioles lie perpendicular to each other. **(C)** Schematic model for the kinetochore of a chromosome where microtubules are assembled/disassembled depending on whether an external force pulls on the microtubule (causing assembly) or an external force is absent (disassembly).

C

centrosome. The structure of the centriole resembles that of the basal body of cilia and flagella, and it has been suggested that the mitotic spindle apparatus may be an evolutionary development of a cilium.

During interphase, the nuclear material begins to condense into separate dense aggregations. In prophase, the nuclear membrane disintegrates. The centrioles replicate and the centrosome splits into two (each with a pair of centrioles). The centrosomes separate towards opposite poles of the dividing cell and produce a spindle-shaped array of microtubules. The mitotic spindle is a continuous microtubular array in many protozoans and fungi, but in most animal cells the microtubules do not form a continuous spindle from one centrosome to the other. Each chromosome has, at its point of constriction, a pair of microtubular-organizing centers, the **kinetochores** (Figure 9–8C), which cap the plus end of the microtubules from the centrosomes (minus end). The kinetochore of each chromatid is connected to one centrosome by 30 to 40 microtubules.

The chromosomal kinetochore controls a dynamic assembly/disassembly of the microtubules (Mitchison 1986). The kinetochore remains attached to the microtubule but allows assembly (when chromosome movement is away from the pole towards the equatorial plate, i.e., metaphase) and disassembly (when chromosome movement is towards the pole, i.e., anaphase). During metaphase, microtubule assembly occurs continually at the kinetochore, even though the length of the microtubule from pole to kinetochore remains constant until the chromosome is at the equator. This is because the tubulin components are assembled to form a microtubule at the kinetochore and move in a "treadmill" fashion to the centrosome at the pole, where they are disassembled. The kinetochore may assemble microtubules when tension is exerted on the microtubule but may disassemble the microtubule in the absence of tension. The kinetochore also slides over the microtubules. There are two different microtubule-based motors on the kinetochore, which move microtubules in opposite directions. The "minus-end motor" is dynein. The activities of these motors are affected by phosphorylation (Hyman and Mitchison, 1991).

The chromosomes align along the equatorial plane of the cell between the polar centrosomes during metaphase. The aligning process involves an increase in the total number of microtubules, as well as the making and breaking of microtubules, and much jerky movement of the chromosomes. The chromosomes appear to be pulled into the equatorial plane by equal tensions of the microtubules from each centrosome to the kinetochores.

The force exerted on each kinetochore appears to be proportional to the length of the microtubule from the centrosome to the kinetochore.

In anaphase, the chromosomes split into the two chromatids, each with a kinetochore. The chromatids are drawn to their respective centrosome; this is usually brought about by both a shortening of the microtubules connecting the chromosomes to the centrosome and a further separation of the two centrosomes. The signal for anaphase may be an increased intracellular concentration of Ca^{2+}. The cell then constricts and divides into two daughter cells during telophase. The mechanism for chromosome movement to the poles during anaphase is complex (McIntosh 1984; Nicklaus 1988). Initially, the centrosome-kinetochore microtubules shorten and separate the chromatids; tubulin is disassembled at the kinetochore. The centrosomes also move apart by elongation (assembly) of centrosome-centrosome microtubules and this further separates the chromatids. The astral microtubules (the centrosome microtubules radiating from the centrosome away from the mitotic spindle) may also pull the centrosomes apart by disassembly.

In telophase, a nuclear envelope re-forms around each group of chromosomes, now located at opposite poles of the cell. The cytoplasm furrows then divides the cell into two daughter cells by a process called **cytokinesis**. Cell cleavage is caused by the contraction of a ring of actin microfilaments that is bound to the cell membrane. Myosin is involved with contraction of the actin ring, but the nature of the actin-myosin interaction is not clear.

Cilia and Flagella

Cilia and **flagella** are motile organelles extending from the cell surface (Figure 9–9). They are identical in basic structure but differ in their pattern of movement. Cilia and flagella are generally from 5 to 100 μ long, although some insect spermatozoa may have a flagellum up to 1 to 2 mm long. Flagella were apparently the original organelles from which cilia evolved. Bacterial flagella are superficially similar to animal flagella, but they are fundamentally different in structure and mechanism for propulsion (see below).

Pattern of Movement. Flagella have symmetrical undulatory movements that propagate along their length. Some flagella beat with the wave of propagation restricted to one plane; others beat in a helical fashion. For example, the dinoflagellate protozoan *Ceratium* has a longitudinal flagellum with a planar beating motion and a transverse flagellum that has a helical beating motion of its lateral fin. Many

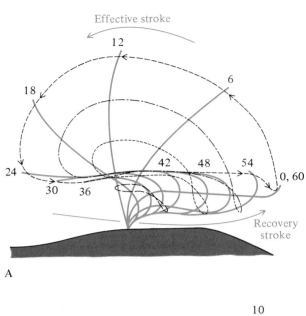

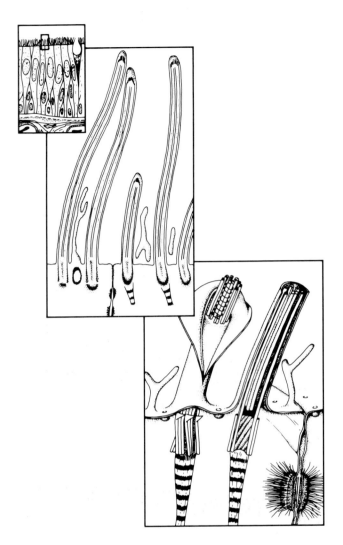

FIGURE 9–9   Schematic structure of cilia. Shown are the ciliated epithelial cells of the mammalian trachea (upper left) and a detail of the membrane-bound ciliary processes. Each cilium has nine peripheral and two central pairs of microtubules. The peripheral pairs connect to the basal body, which has nine microtubular triplets, and a basal foot, which has periodic striations.   (*Modified from Krstic 1979.*)

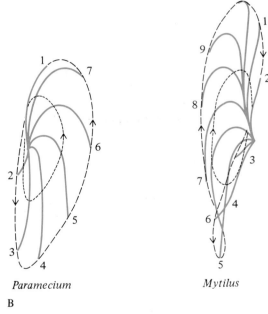

FIGURE 9–10   (**A**) Side view of the movements of a typical planar gill cilium of a sabellid worm. Values are msec intervals. (**B**) The cilia of *Paramecium* (left) and *Mytilus* (right) have a nearly vertical effective stroke (stages 1 to 3 and 1 to 4, respectively) and an eccentric recovery stroke (stages 4 to 7 and 5 to 10, respectively).   (*Modified from Sleigh 1974a.*)

sperm and protozoans have a flagellum that moves in only a single plane with waves traveling from the base to the tip. The trypanosome flagellum can propagate a wave in either direction. The flagellum of cattle sperm has a helical beat; this causes the sperm to rotate about its long axis as it swims. Some sperm also rotate when swimming because they have a prominent spiral keel structure, not because of a helical beating of the flagellum. The wave of flagellar beating is approximately sinusoidal in shape, ranging from about one wave to nine waves along the length of the flagellum. The wavelength, velocity, and frequency vary considerably for different flagella.

Cilia, unlike flagella, generally have a single wave, with a separate effective (power) stroke and a recovery stroke. The beating of a gill cilium of a sabellid worm illustrates the basic planar beat of cilia (Figure 9–10A). During the effective stroke, the cilium moves through a large volume of water in a wide arc; it bends from base to tip. There is a more marked bend during the recovery stroke from base to tip that returns the cilium to its upright position by bending and "unbending" through a small volume of water. The effective stroke is of shorter duration than the recovery stroke.

Both the effective and recovery stroke occur in a single plane for many cilia, but some beat in an eccentric rather than a planar manner. The effective stroke is essentially in a straight line, but during the recovery stroke the tip of the cilium moves in either a counterclockwise or a clockwise pattern depending on the species (Figure 9–10B). Such eccentric ciliary beating occurs in a wide variety of protozoans and animals, e.g., *Paramecium, Opalina,* and *Corella* (counterclockwise) and *Mytilus* and nudibranch veliger larvae (clockwise).

Measurements of ciliary-induced fluid flow indicate a maximum velocity at about the level of the cilium tip, of about 1/5 to 1/3 the velocity of the cilium tip (Figure 9–11). Ciliary propulsion of a mucus layer is similar; the mucus appears to be concentrated at the level of the ciliary tip, and there is a more watery fluid layer between the mucus and the surface of the ciliated cell.

Cilia are generally arranged in large numbers on a cell surface. For example, the ciliate protozoans *Paramecium* and *Opalina* swim with their numerous cilia, and the protozoan *Stylochonia* "walks" using cirri, which are groups of cilia. Cilia often beat in complex metachronal rhythms. The flagellate protozoan *Myxotricha,* a gut symbiont of termites *Mastotermes,* is covered by regularly spaced groups of 1 to 4 symbiotic flagellate spirochetes, each of which has a helically-rotating flagellum. There also are a few larger spirochetes, and the protozoan has four anterior flagella. The spirochetes beat in a synchronized wave from the anterior to the posterior, and this propels the protozoan anteriorly.

Metachronism is the coordinated temporo-spatial movement by a group of cilia or flagella; it is observed as a series of waves traveling over a ciliated (or flagellated) surface. A two-dimensional arrangement of cilia will show lines of cilia of the same stage of the beating cycle (Figure 9–12A); the wavelength (λ) is the perpendicular distance between adjacent lines of cilia at the same stage of the cycle. The direction of metachronal wave propagation can vary relative to the direction of the effective stroke of the cilia, even within organisms (Machemer 1974).

The cilia of the protozoan *Paramecium* form a highly versatile propulsion system, capable of a number of complex patterns of metachrony. Metachronal waves ($\lambda = 10$ to 14 μ) proceed around the protozoan from the animal's posterior left to the

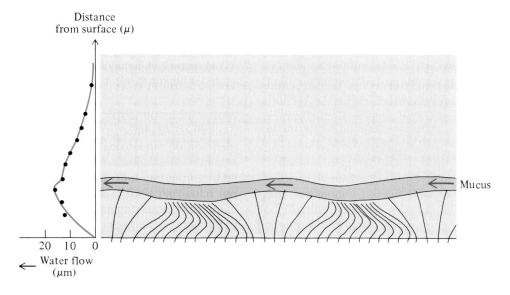

FIGURE 9–11 Propulsion of mucus by a ciliated surface. There is little flow near the surface and mucus flow is concentrated around and beyond the ciliary tips (dashed line). *(Modified from Sleigh 1974b.)*

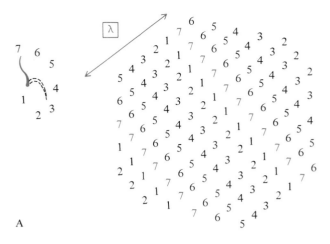

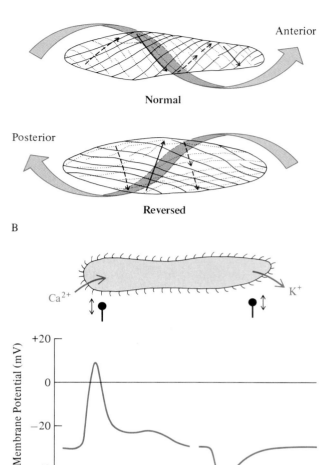

FIGURE 9–12   **(A)** Metachronal coordination of a field of cilia. The stage of ciliary beat (left) for a field of cilia is organized along two axes: the line of ciliary synchronization and the line of metachronal movement along which the wavelength of metachronism (λ) is minimal. **(B)** The pattern of ciliary metachronism of the ciliate protozoan *Paramecium* can alter, to propel the organism forwards or backwards. **(C)** Mechanical stimulation of the anterior region of a *Paramecium* impaled on a microprobe causes membrane depolarization due to Ca^{2+} influx, whereas posterior stimulation causes hyperpolarization due to K^+ influx.   *(Modified from Machemer 1974; Naitoh and Eckert 1974.)*

anterior right side when swimming forward and is reversed when swimming backward (Figure 9–12B). During the avoidance reaction, *Paramecium* back away from an object, rotate around the posterior end, and then swim away in a different direction. The transitory swerving phase is the result of competing actions of simultaneous anteriorly and posteriorly directed ciliary beating. These complex changes in ciliary beating pattern of *Paramecium* require regulation of the direction and rate of beating. This is accomplished by the membrane potential. A *Paramecium* placed in an electric field reorients so that it faces the cathode; this movement is **galvanotaxis**. Stimulation of the anterior end of *Paramecium* slowly depolarizes its cell membrane, and this is followed by a rapid action potential-like change in membrane potential due to an influx of Ca^{2+} (Figure 9–12C); this initiates the avoidance response. Stimulation of the posterior end causes hyperpolarization due to K^+ efflux and accelerates forward motion.

Metachronism also commonly occurs in metazoans where ciliary fields typically consist of many adjacent epithelial cells, whose function is to transport water (e.g., the gills of many mollusks) or mucus (e.g., the vertebrate trachea). The presence of cell boundaries does not cause irregularity or discontinuity in the ciliary metachronism, although the presence of nonciliated cells (e.g., mucus-secreting goblet cells in the respiratory epithelium of vertebrates) does disrupt metachronal continuity. Ctenophores have eight meridional comb plates; each comb plate is a rectangular giant compound cilium, containing hundreds of thousands of individual cilia. The comb plates have metachronal waves of beating that pass from the apical organ at the top of the animal and down a ciliated groove, then along the comb rows to the mouth (Figure 9–13); the cilia have an upward effective stroke. The comb plates, unlike most ciliary structures, remain stationary unless activated by an external stimulus. Movement of the more anterior comb plate draws water in

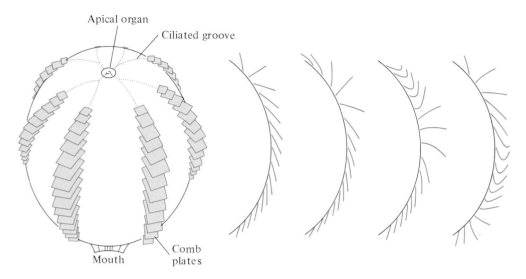

FIGURE 9–13 The comb plates of the ctenophore *Pleurobranchia* beat in waves that pass from the apical organ towards the mouth. *(From Sleigh 1974.)*

behind it and passively lifts the adjacent plate, which is activated by its movement to beat.

Ciliary control in most lower metazoans is probably similar to that in protozoans. Mechanical, electrical, or chemical stimulation can reverse ciliary beating in echinoid larvae in association with changes in epithelial electrical potential. Neuronal reflex inhibition of ciliary beating, and often reversal of beating, has also been reported in other metazoans. Low-frequency stimulation (2 to 10 sec^{-1}) of the branchial nerve accelerates ciliary beating, and high-frequency stimulation (25 to 50 sec^{-1}) depresses beating of the gill cilia of bivalve mollusks. Dopamine administration inhibits ciliary beating, so there probably is an inhibitory dopaminergic innervation. The cilia of primitive chordates (*Branchiostoma*, a cephalochordate), hemichordates (*Saccoglossus*), and urochordates (*Ascidia*) are probably under neural control, but there is a tendency among higher vertebrates for decreased neural control of ciliary activity. The oropharyngeal and esophogeal cilia of amphibians are under excitatory neural control, but there appears to be no significant neural control of cilia in the oral cavity, respiratory tract, or esophagus.

The Axoneme. The bending of a cilium or flagellum is due to a sliding movement of a central array of microtubules that forms the **axoneme**. The axoneme generally consists of two central microtubules and nine equidistantly spaced doublets of microtubules that are radially arranged around the two central ones; this is the 9 + 2 arrangement (Figure 9–

14). There is some variation in the basic 9 + 2 arrangement, but the variants (which will be described below) function in essentially the same manner as the basic 9 + 2 cilium or flagellum.

Each of the 9 + 2 doublet microtubules consists of one circular microtubule, subfiber A, fused with a crescent-shaped microtubule, subfiber B. Subfiber A has 12 protofilaments and subfiber B has 10. These are continuous with the 9 doublet microtubules of the basal body of the cilium or flagellum. A long intermediate fiber, tektin, appears to stabilize the junction of the A and B subfibers; it is a 2 to 3 nm dia filament with a pitch of 16 nm (28 amino acid residues) that matches the 8.2 nm length of the tubulin dimer. Each A subfiber has an inner and outer dynein arm and a spoke-like radial link to a sheath surrounding the two central microtubules. Dynein is a large protein of 9 to 12 polypeptides; its base binds to an A tubule and its three globular heads move to the minus end of adjacent microtubules. It has ATPase activity that is Mg^{2+}-dependent (like muscle myosin ATPase) and is fairly specific for ATP, rather than GPT, UTP, ITP, or CTP (unlike muscle myosin ATPase). There are also fine interdoublet links of nexin between the 9 radial microtubule doublets connecting the inner dynein arm with its adjacent B subfiber. These nexin links may be part of the dynein cross-bridge system that causes sliding of adjacent doublets or may be a static structural member that supports the axoneme. Regardless of the exact role of these interdoublet links, they must break and re-form during ciliary/flagellar movement. There are three radial linkage

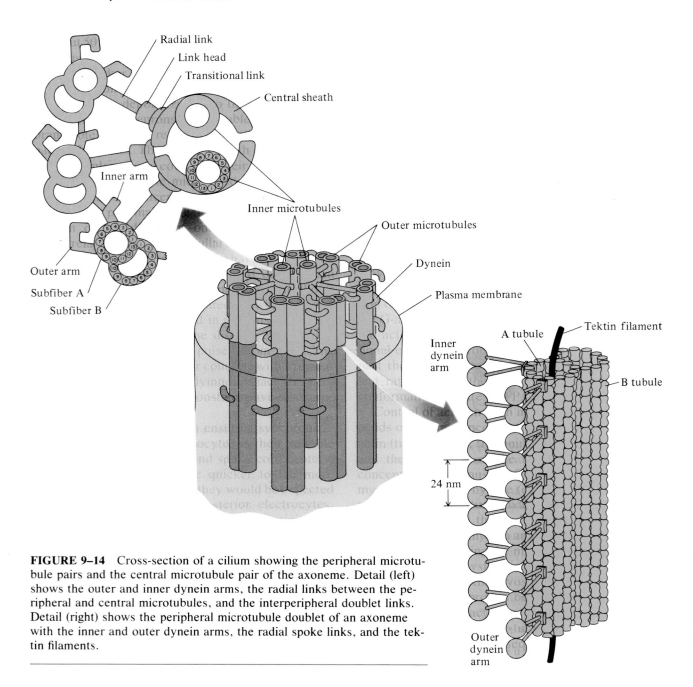

FIGURE 9–14 Cross-section of a cilium showing the peripheral microtubule pairs and the central microtubule pair of the axoneme. Detail (left) shows the outer and inner dynein arms, the radial links between the peripheral and central microtubules, and the interperipheral doublet links. Detail (right) shows the peripheral microtubule doublet of an axoneme with the inner and outer dynein arms, the radial spoke links, and the tektin filaments.

proteins: the radial link; the link head; and the transitional protein, which connects the link head to the central sheath. The structure of the axoneme is not radially symmetrical but is enantiomorphic; a transverse section of the axoneme viewed from the base towards the tip shows the doublet dynein arms directed clockwise whereas the arms are counterclockwise when viewed from the tip towards the base.

The 9 + 2 microtubular arrangement of cilia and flagella has remained remarkably constant through-

out the evolutionary development of animals, although there are some variations in structure. Many insects have a sperm flagellum with an additional 9 microtubules, i.e., a 9 + 9 + 2 arrangement. The sperm flagellum of mayflies lacks central microtubules (9 + 9 + 0). Some protozoans and sperm of flatworms and scorpions have a 9 + 0 arrangement. There are numerous other patterns, e.g., 9 + 9 + 1 (mosquito sperm), 9 + 1 (some flatworms), 9 + 3 (some spiders), and 9 + 7 (some mayflies). *Sciaria* and *Rhynchosciara* sperm flagella have a large number of peripheral doublet and singlet microtubules, e.g., 60 and about 360 respectively; the centrioles of *Sciara* also contain 60 to 90 singlet tubules. The spermatozoa of a thrips insect has 18 doublet and 4 singlet microtubules arranged in a completely disorganized fashion. Coccoid sperm have 20 to 250 singlet microtubules arranged in circles, single spirals, double spirals, or more complex patterns. Treehopper sperm have a branching flagellum; pscocids have spirally arranged 9 + 9 + 2 microtubules. Despite wide variations in structure, most of these sperm flagella are motile and presumably use the same basic mechanism for movement as the typical 9 + 2 flagella and cilia.

Sliding Movement. Ciliary movement was originally thought to be due to a rapid shortening of axoneme tubules (the contractile hypothesis), but it actually involves sliding interactions of the axoneme microtubules (the **sliding filament model**). The sliding filament model was derived in part by analogy with the sliding filament model for muscle contraction (see below).

The sliding filament model is supported by the action of isolated dynein (Figure 9–15A). An axoneme, with the ciliary membrane removed by detergent and the circumferential and radial links removed by mild trypsin digestion, will show sliding movements when ATP is added (Summers and Gibbons 1971). The axoneme elongates as adjacent microtubules slide past one another until the doublets become totally separated. This sliding movement would bend an intact axoneme with interdoublet connections. The A subfibers (which have the dynein arms) move towards the base of the adjacent B subfiber, i.e., the dynein arm "walks" to the minus end of the B subfiber. The dynein arm presumably swings towards the tip of the adjacent B subfiber before attaching and sliding the B subfiber past the A subfiber and then detaches and returns to its original orientation. The specific nature of the dynein-B subfiber cross-bridges and mechanism for movement are unclear (in contrast to the actin-myosin cross-bridge structure and movement in muscle). Dynein isolated from *Tetrahymena* flagella has three globular heads, each attached by a strand to the base of the A subfiber. Each head has one ATPase site, and all heads interact with B microtubules in an ATP-dependent reaction.

Structural evidence also supports the sliding filament model (Satir 1968). The axoneme at the cilium tip has a short length with only A subfibers and no B subfibers or dynein. During ciliary bending, there is a change in the morphological configuration of the A and B subfibers at the axoneme tip (Figure

9–15B). In cross-section at the appropriate level, there are microtubule doublets (i.e., A and B subfibers) on the side towards the bending direction but only A subfibers on the other side.

Ciliary and flagellar bending probably involves sliding of adjacent A and B subfibers on one side of the axoneme and no movement of the subfibers on the opposite side; if so, the sliding movement would be unidirectional (as in actin-myosin interaction of muscle). The subfibers on one side of the axoneme could actively slide in the opposite direction but this requires a bidirectional sliding mechanism.

The propagation of bending along a cilium or flagellum does not involve an electrical depolarization or repolarization of the axoneme cell membrane since demembranated axonemes beat normally. Passive bending might stimulate inactive regions to become active and thereby propagate the bending longitudinally along the cilium. Mathematical models based on this concept can simulate a fairly normal flagellar beat (Brokaw 1972).

Energy Expenditure. The mechanics and energetics of bending are well documented for the abfrontal cilium of the mussel *Mytilus*, a large compound cilium containing about 36 individual cilia (Baba and Hiramoto 1970). During bending, each part of the cilium exerts a force on the medium. The magnitude of this viscous force depends on (1) the velocity of movement, (2) the shape of the segment of the cilium in question, and (3) the viscosity of the medium. The sum of the viscous forces along the length of the entire cilium is transmitted through the cilium to its point of attachment to the cell. This force at the point of attachment can be resolved into two vectors, one tangential to the cell surface and one normal to the cell surface (Figure 9–16). Forces equal in magnitude but opposite in direction are exerted by the cilium onto the medium. The cilium of a fixed cell will transport the medium over the surface of the cell, e.g., the ciliated respiratory epithelium of the vertebrate respiratory tract. The amount of fluid moved per beat is independent of the velocity and frequency of ciliary movement. The cilium of a movable cell will move both the medium and the cell, e.g., a swimming ciliate protozoan.

The rate of energy expenditure by the compound abfrontal cilium of *Mytilus* to overcome viscous drag has been calculated to be greatest during the end of the effective stroke, at about 10^{-12} J sec^{-1}. The total work done during one beat of the compound cilium is calculated to be 3.6 10^{-14} J. The total energy expenditure of the compound cilium has been measured as 1.6 10^{-13} J beat^{-1} (1 to 2 ATP

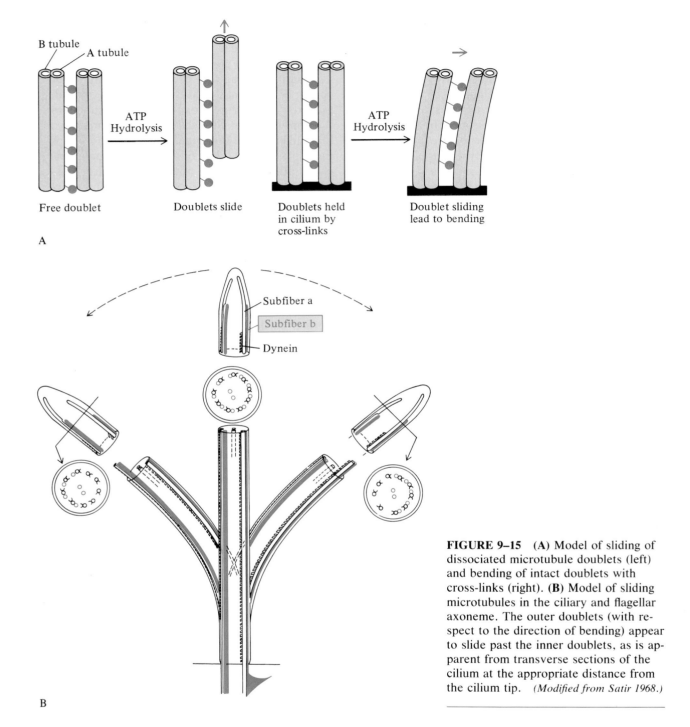

B tubule

A tubule

ATP Hydrolysis

Free doublet

Doublets slide

Doublets held in cilium by cross-links

ATP Hydrolysis

Doublet sliding lead to bending

A

Subfiber a

Subfiber b

Dynein

B

FIGURE 9–15 **(A)** Model of sliding of dissociated microtubule doublets (left) and bending of intact doublets with cross-links (right). **(B)** Model of sliding microtubules in the ciliary and flagellar axoneme. The outer doublets (with respect to the direction of bending) appear to slide past the inner doublets, as is apparent from transverse sections of the cilium at the appropriate distance from the cilium tip. *(Modified from Satir 1968.)*

per dynein arm per beat). This energy expenditure accounts for the viscous work done if the mechanical efficiency of the abfrontal cilium is about 22.5% (but this does not account for work done on elastic deformation of the cilium during the beat, and so the efficiency may be higher than 22.5%).

Ciliated surfaces can perform useful work, for example moving the medium or even solid objects over the surface. The ciliated gills of the oyster *Ostrea* maintain a stream of water over the gills for feeding and respiration; the energy expended in maintaining the fluid flow is about $1.4 \ 10^{-6}$ J sec^{-1}. The ciliated epithelium of the esophagus of a frog can move weights placed on its surface. The velocity of movement is inversely proportional to the force required, and the energy expenditure depends on

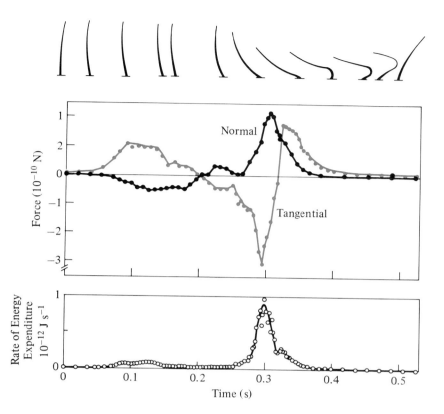

FIGURE 9–16 The force exerted on a cell by a beating cilium (middle panel; beat is from left to right) can be resolved into a component tangential to the cell surface and a component normal to the surface; equal and opposite forces are exerted on the medium and generate a fluid current. The stages of ciliary movement are shown in the top panel. The rate of energy expenditure by the cilium is shown in the bottom panel. *(Modified from Hiramoto 1974.)*

the product of force and velocity (Figure 9–17). The greatest rate of energy expenditure is about $5 \cdot 10^{-6}$ J min^{-1} for a weight of 53 mN (5.4 g). The mechanical work done by a beating cilium is independent of its velocity. A cilium does not expend more energy per stroke if it beats faster, although its total energy expenditure per second is increased at a higher beating frequency.

Bacterial Flagella. The bacterial flagellum merits a brief description here because it has a function similar to animal flagella but a very different struc-

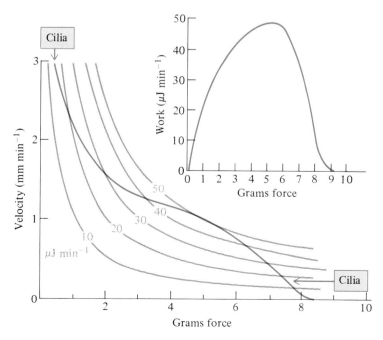

FIGURE 9–17 Force–velocity diagram for the transport of weights by cilia on the frog esophagus; the colored curve shows that the velocity of movement is inversely related to the weight of the object placed on the epithelium. The grey curves show lines of equal work expenditure (μJ min^{-1}). The inset shows the relationship between work done (μJ min^{-1}) and force for the ciliated epithelium. *(Calculated from data of Maxwell cited by Rivera 1962.)*

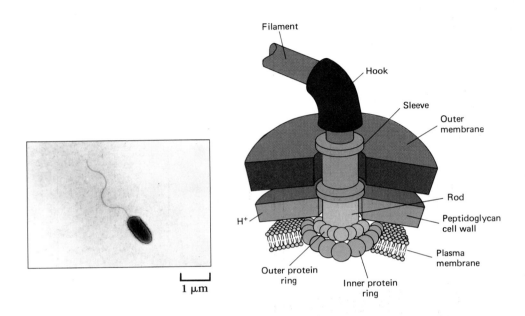

FIGURE 9–18   Photograph of a bacterium, *Pseudomonas* showing its flagellum, and a diagrammatic scheme for the rotating base of the bacterial flagellum that is a unique rotary "motor."   *(From Villee et al. 1991.)*

ture. Some protozoans also rely on flagellar beating of symbiotic bacteria for locomotion (e.g., *Myxotricha* has symbiotic flagellate spirochaetes).

A casual observer could easily confuse the structure and beating motion of the bacterial flagellum with that of an animal flagellum; both appear to have helical waves of bending movement. However, the bacterial flagellum is actually a rigid, spiral-shaped structure that rotates (Berg and Anderson 1977). The rotating motion of the flagellum is difficult

to discern using a microscope, but bacteria agglutinated to a surface by their flagellum clearly rotate.

The bacterial flagellum is an extracellular structure; it is not surrounded by the cell membrane. The flagellum itself is a long, thin helically twisted rod, about 13 to 14 nm diameter (Figure 9–18). It is connected to the cell by a short hook-shaped structure and a complex basal body. The basal body has two ring structures (L and P), associated with the outer membrane and peptidoglycan layer of the bacterial cell wall, and two inner rings (S and M), associated with the cell membrane. A shaft runs through these rings and connects to the extracellular hook. The basal body is probably the "motor" that rotates the flagellum. The exact mechanism for flagellar rotation is not clear, but the energy is derived from an electrochemical proton gradient across the bacterial cell membrane, not from ATP hydrolysis. Proton flux through a protein pore may cause a mechanical movement that is transduced into rotation of the basal body rings or the shaft passing through the rings.

Microfilament Movement

The general principle of microfilament movement involves a sliding of adjacent filaments or microfilament and membrane structures, similar to that postulated for microtubule-based movement. A major difference between microfilament and microtubule movement is that microfilaments can only shorten; they cannot push (extend) in the manner that microtubules can.

There are many examples of cell motility due to microfilament (actin-myosin) interaction, but the best understood is the contraction of striated muscle cells because they have a highly-regular organization of microfilaments. Amoeboid cell motility, axonal movement of materials, and cell cleavage involve actin and myosin, but their mechanism is less clear. Actin and myosin have also been implicated in other aspects of cell mobility, such as the mitotic spindle, but their structural and functional roles are speculative at present.

Amoeboid Movement

Many protozoans (e.g., *Amoeba, Chaos*) and isolated cells of multicellular animals (e.g., fibroblasts, leukocytes, epithelial cells) are highly motile. Some examples of **amoeboid movement** involve obvious cytoplasmic streaming into finger-like extensions of the cell, called pseudopods or delicate cytoplasmic lamellae on the leading edge, although some small amoebas glide slowly over the substrate without cytoplasmic streaming (*Hyalodiscus*). Phagocytic leukocytes (white blood cells) have a clear ring of peripheral cytoplasm (it lacks subcellular organelles and membranes) (Figure 9–19). This clear zone contains all of the necessary structures for cell movement. It can detach itself from the rest of the cell if the leukocyte is heated to above its normal temperature and "walk away," leaving the rest of the cell rendered immobile!

Numerous models have been proposed to explain amoeboid motility but a likely general explanation is offered by an early theory that fluid cytoplasm (a "sol") has an inherent ability to stiffen (form a "gel"). "Gelled" cytoplasm is very rigid, able to withstand up to 1 kg cm^{-2} compression. Recent studies have suggested a mechanism for sol-gel transition that involves actin, myosin, and other related microfilament proteins that are present in amoeboid cells. The actin filaments form a rigid cytoskeleton.

One model of epidermal cell movement (Bereiter-Hahn and Strohmeier 1986) combines features of a number of previous models for cell movement. A positive hydrostatic pressure is generated in the cell by contraction of fibrillar acto-myosin meshworks near the dorsal plasmalemma anterior to the nucleus and forms cytoplasmic extensions at weakenings of the plasmalemma. The edge of the lamella is a structurally weak point because it has a high radius of curvature and so "sol" cytoplasm is pushed into the extending leading edge of the lamella, and the meshwork is "gelled" by an influx of Ca^{2+} ions. A sliding movement of acto-myosin may provide the motile mechanism, but myosin does not seem to be required for movement in many cells.

Another model of movement by phagocytic leukocytes does not require a myosin-actin interaction (Stossel 1990). Solation of the actin cytoskeleton is stimulated at the cell surface, for example by the binding of an antibody to a receptor, and this elicits formation from phosphatidylinositol of a mobile C$_6$ messenger that liberates Ca^{2+} from intracellular storage vesicles. The Ca^{2+} ions activate gelsolin, which dismembers the actin cytoskeletal network (Figure 9–19Bi). The elevated local concentration of actin molecules causes an osmotic influx of water, which extends the cell membrane (Figure 9–19Bii). Finally, the cellular actin cytoskeleton is re-formed in the extended region. Actin/gelsolin molecules bind to polyphosphoinositols in the cell membrane, and this breaks up the actin-gelsolin complex, promoting resynthesis of the actin cytoskeletal network (Figure 9–19Biii). A similarly dynamic actin filament meshwork may be responsible for the rapid movement of amoeboid cells (Theriot and Mitchison, 1991).

Muscle

Muscle cells are specialized to actively shorten (contract). They are only able to actively shorten, and they elongate as a passive response to an external force provided, for example, by an antagonistic muscle or a hydrostatic force.

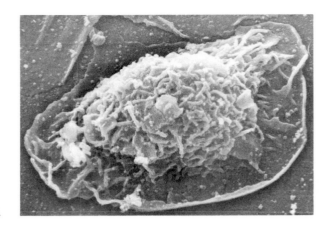

A

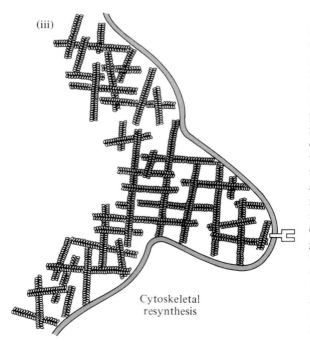

(i)

Ca^{2+}

Ca^{2+}

Ca^{2+}

Ca^{2+}

← Antibody

Cytoskeletal
breakdown
initiated

(ii)

Osmosis

Osmotic
water movement

(iii)

Cytoskeletal
resynthesis

B

FIGURE 9–19 (A) A phagocytic leucocyte moves using a
peripheral zone of clear cytoplasm. **(B)** Schematic sequence
of events during amoeboid extension of a pseudopod by a
phagocytic leucocyte: (i) binding of an antibody stimulates
calcium release from vesicles,
the activation of gelsolin, and
the breakdown of the cellular
actin cytoskeleton; (ii) osmotic
influx of water causes extension of the cell membrane;
(iii) dissociation of actin-
gelsolin causes reformation of
the intracellular actin cytoskeleton. *(Photograph courtesy of
Terry A. Robertson, Department
of Pathology, The University of
Western Australia; modified from
Stossel 1990.)*

TABLE 9–2

Summary of major types of muscle

Skeletal muscle (striated)
  Tonic
  Phasic
    Slow (slow contracting, slow fatiguing)
    Fast: glycolytic (fast contracting, fast fatiguing); oxidative (fast contracting, fatigue resistant)
  Asynchronous insect flight muscle

Cardiac muscle (striated)

Smooth muscle
  Vertebrate
    Unitary (visceral)
    Multiunit (ciliary body, iris, pilomotor)
  Invertebrate
    Classical smooth muscle
    Helical
    Oblique
    Paramyosin catch muscle

There are a variety of types of muscle cells (Table 9–2). Muscle cells can be divided into two general types based on their microscopical appearance: **striated muscle** and **smooth muscle**. Striated muscle can be further divided into skeletal and cardiac muscle. This categorization is based not on differences in basic contractile mechanism but on the degree of orderly arrangement of the actin and myosin contractile proteins. These microfilaments are highly ordered in striated muscle, and this confers a striking pattern of alternate dark and light striations. There are many other differences in the structure and physiology of skeletal, cardiac, and smooth muscle, including the nature of their innervative, the form of their action potentials, and their relation to the generation of a contractile force. The two types of vertebrate striated muscle, skeletal and cardiac muscle, differ more in their innervation pattern and mechanical and electrical properties than in basic ultrastructure.

Skeletal muscle is usually under voluntary control. Individual skeletal muscle cells, or fibers, are large (10 to 100 dia), long, multinucleate cells grouped into bundles (fascicles) that are visible to the naked eye. Individual fibers generally do not run the entire length of the muscle, although they may in some muscles that do not taper at the ends (e.g., frog sartorius muscle). Skeletal muscle can be further classified into types by innervation pattern, structure, and function. The skeletal muscle of vertebrates is generally organized into groups, termed a **motor unit** (Figure 9–20). Each muscle cell in a motor unit receives one or a few synaptic connections from a single motoneuron, whose soma is located in the spinal cord. An action potential is propagated over the surface of each muscle cell in response to an axonal action potential, and an "all-or-none" twitch is elicited.

Cardiac muscle cells differ somewhat in morphology from skeletal muscle cells (Figure 9–20). They are smaller, bifurcating cells that are uni- or binucleate. These cells are joined end to end with other cardiac muscle cells by specialized intercalated disks to form a 3-dimensional network. Cardiac muscle is under involuntary control. The cardiac action potential has an elongate plateau and the muscle contraction is of similar, long duration. Some cardiac muscle cells are specialized to conduct electrical depolarization, rather than to contract. Purkinje fibers and AV (atrioventricular) node fibers are both morphologically and physiologically specialized for conduction. Cardiac muscle has a longer contractile duration than skeletal muscle. Most cardiac muscle cells do not have any innervation but have an inherent rhythmicity of contraction. Each cardiac cell is electrically connected to adjacent cells, and they form an electrical syncytium so the entire cardiac muscle contracts as a unit. Some cardiac muscle cells are specialized as pacemaker cells; their inherent rhythmicity of electrical depolarization is normally faster than that of the other cardiac muscle cells, and so their rhythm determines the rate of contraction of the entire cardiac muscle system.

Nearly all invertebrates have striated muscle, although some have mainly nonstriated muscle (e.g., mollusks, annelids). The vertebrate pattern of striated skeletal muscle being voluntary and striated cardiac and smooth muscle being involuntary is not consistent among the invertebrates. For example, the alimentary canal muscles of arthropods are usually striated, and many locomotory muscles of annelids and cephalopods are smooth muscle.

Invertebrate striated muscle has relatively few, but large, fibers that are innervated by a small number of motoneurons. Muscle fibers are up to 1 to 2 mm diameter in barnacles and 4 mm in king crabs. An entire muscle may be innervated by as few as 2 motoneurons. The axons do not form discrete terminal synapses with end plates but release neurotransmitter at many contact points between the axon and muscle fibers. The arrangement of actin and myosin is essentially the same for invertebrate and vertebrate striated muscle, as are the changes in banding pattern that occurs during muscle contraction; the mechanism for contraction is also identical.

Smooth muscle cells lack the obvious transverse striations of striated muscle but they contain both

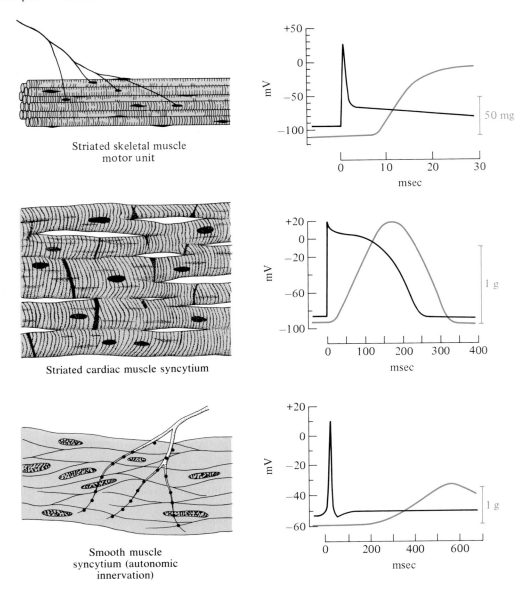

Striated skeletal muscle
motor unit

Striated cardiac muscle syncytium

Smooth muscle
syncytium (autonomic
innervation)

FIGURE 9–20   Major types of muscle: striated skeletal fibers, striated cardiac fibers, and smooth fibers showing details of action potential and time course of the muscle twitch.

actin and myosin. Typical smooth muscle cells, such as vertebrate visceral muscle, are small and spindle shaped with a single central nucleus (Figure 9–20). The cells may occur singly but they typically are associated with other smooth muscle cells with the thick central portion of one joined to the thin ends of adjacent smooth muscle cells. Vertebrate smooth muscle can be divided into functional subtypes: visceral (single unit, or unitary) smooth muscle and multiunit smooth muscle. **Visceral smooth muscle** cells generally are arranged as sheets, or bundles, of many cells that are electrically inter-

connected by gap junctions to form a single electrical syncytium (similar to cardiac muscle cells). These syncytial cells contract in unison, hence the arrangement is called single-unit smooth muscle. Electrical activity spreads as action potentials. These syncytia of smooth muscle cells have a diffuse innervation by varicosity synapses, synaptic terminals spread along autonomic nerves. They also respond to circulating hormones (e.g., epinephrine and norepinephrine), physical stretch, and temperature. In contrast, **multiunit smooth muscle** cells are organized as discrete fibers. Each cell is innervated by a single nerve ending and operates independently of the others. Their sarcolemma generally does not support an action potential but relies on electrotonic propagation. The cell generally responds only to its neuronal innervation and not to other factors, such as circulating hormones, stretch, or temperature.

Striated Muscle. The most obvious characteristic of a striated muscle fiber is its regular transverse striations (Figure 9–21). The cytoplasmic matrix, or **sarcoplasm**, contains numerous long, cylindrical myofibrils (1 to 2 μ dia), which also are striated. Most of the sarcoplasm is myofibrils, although there also are organelles, e.g., mitochondria, nucleus, and deposits of glycogen and lipid. The myofibrils have alternating light and dark bands; because of their properties under polarized light, the light-staining bands are called I (isotropic) bands and the dark-staining bands are A (anisotropic) bands. A dark transverse line, the Z line, bisects each clear I band into two halves. The sarcomere is the portion of the myofibril between adjacent Z lines.

The sarcomere. The **sarcomere** is both the structural and functional contractile unit of striated muscle. It is a highly ordered array of thin actin filaments and thick myosin filaments (Figure 9–21). The actin molecules (about 5 nm dia, 1 μ long) are firmly attached to the Z line at each end of the sarcomere. Each actin filament appears to be connected to the Z line by four Z filaments that run through the Z line and connect to actin filaments of the adjacent sarcomere. The protein α-actinin is also present in the Z line, and binds the actins together. Adjacent myofibrils are interconnected at their Z lines by desmin and vimentin filaments. The thick myosin filaments (15 nm dia, 1.5 μ long) extend in parallel between the ends of the sarcomere, at about 45 nm apart. The myosin filaments are thicker in the middle and taper towards their ends. The myosin heads project transversely from the ends; they can form cross-bridge links to the actin filaments. Myosin filaments are held in their regular arrangement by slender cross connections at their centers (these form the M band).

The area of overlap of the actin and myosin in the center of the sarcomere forms the A (anisotropic) band. The H band, located in the middle of the A band, is the central region of the myosin where there is no overlap with actin. In the center of the H band is the M band, the central segment of the myosin that lacks cross-bridges.

Striated muscle cells have an important membrane system, the **sarcotubules** and **sarcoplasmic reticulum** (Figure 9–21). These membrane structures are important in the transmission of electrical depolarization from the cell surface to the interior and with Ca^{2+} regulation of muscle contraction. Sarcotubules are long, thin, blind-ended membrane tubules that invaginate from the cell surface and branch and extend within the cell to form a ring of narrow (about 0.1 nm dia) tubules running around the perimeter of each sarcomere; these transverse sarcotubules are called **t-tubules**. The sarcoplasmic reticulum (SR) is an intracellular membrane system between adjacent t-tubules. It forms an irregular, hollow collar-shaped sheath around the sarcomeres. The SR of muscle cells is a special type of endoplasmic reticulum that lacks ribosomes. The terminal cisterna at each end of the SR is in intimate contact with a t-tubule. The SR of one sarcomere, the t-tubule, and the SR of the adjacent sarcomere are closely associated to form the **triad**. The SR normally sequesters Ca^{2+} and releases it to induce muscle contraction.

In amphibian skeletal muscle, each Z line has an adjacent t-tubule, hence each sarcomere has one triad. In mammalian skeletal muscle, the t-tubules are adjacent to the I through A band junction, hence each sarcomere has two triads. Mammalian cardiac muscle has one triad per sarcomere, at the Z line. The t-tubule and sarcoplasmic reticulum systems are less well developed in the cardiac muscle of lower vertebrates. For example, the amphibian cardiac muscle cell has no t-tubule system and only a poorly developed SR.

Sliding filament model. An early theory of muscle contraction proposed that specific muscle proteins shortened by some structural rearrangement. This theory was abandoned in favor of the **sliding filament model**, in part due to the microscopical evidence of sarcomere contraction and the dual myofilament nature (i.e., actin + myosin) of myofibrillar structure (Huxley and Niedergierke 1954; Huxley and Hanson 1954). The sliding filament model is the correct mechanism for muscle contraction, and the actin-myosin sliding filament model has provided inspiration for the examination and interpretation of movement systems in other cells (e.g., smooth muscle cells) and microtubular movement (e.g., cilia).

Sarcomere contraction is due to the sliding movement of adjacent thin actin filaments and thick myosin filaments. Neither the actin nor myosin filaments change length during sarcomere shortening but the Z lines move closer together (Figure 9–22A). The A band always has the same length, but the relative lengths of the I and H bands vary from short (contracted muscle cell) to long (extended

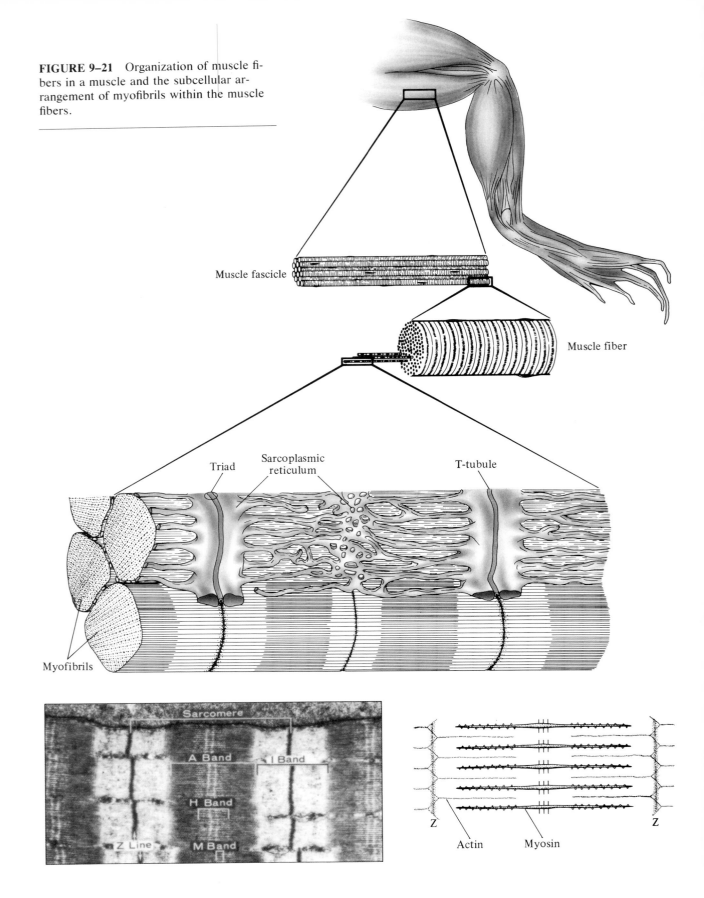

FIGURE 9–21 Organization of muscle fibers in a muscle and the subcellular arrangement of myofibrils within the muscle fibers.

Muscle fascicle

Muscle fiber

Triad

Sarcoplasmic reticulum

T-tubule

Myofibrils

Sarcomere

A Band

I Band

H Band

Z Line

M Band

Actin

Myosin

Z

Z

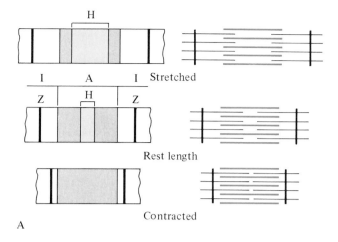

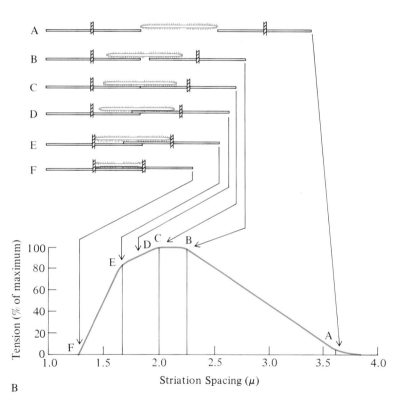

FIGURE 9–22 **(A)** Schematic representation of sarcomeric shortening showing the change in proportions of the I and A bands and the H zone as the Z lines move closer together. **(B)** Relationship between length of sarcomeres (indicated by striation spacing) and tension generated (as percentage of maximum tension). *(Modified from Gordon, Huxley, and Julian 1966.)*

muscle cell). The appearance of the sarcomere, and these changes that occur with contraction, are consistent with what we now understand of the molecular structure and function of actin and myosin. In fact, the histological appearance of sarcomeres provided much of the initial evidence for the sliding filament model of muscle contraction.

The sliding filament model predicted from microscopical evidence suggested that the molecular force generators (the "motors") were spaced along the thick filaments and moved the thin filaments by cyclical attachment of the thick to the thin filaments. The existence of such cross-bridges between myosin and actin was subsequently demonstrated by electron microscopy. Further studies have demonstrated cyclic positional changes of the cross-bridges that correspond to the relaxed and contracted (rigor) states. Part of the myosin head forms the cross-bridge with the actin binding site.

The sliding filament model predicts that (1) the number of cross-bridges varies with the length of the sarcomere; (2) the force generated by a sarcomere should be proportional to the number of cross-bridges, i.e., the length; (3) the myosin heads in the M band do not contact an actin filament and parts of the actin filament (in the I band) are not in contact with myosin heads; and (4) no force can be generated when the sarcomere is so shortened that the ends of the myosin filaments abut against each Z line. Measurement of the relationship between force of contraction and sarcomere length has verified these predictions of the sliding filament model (Figure 9–22B).

Cross-bridge mechanics. The cross-bridges that form between the myosin head and actin convert ATP's chemical energy into mechanical work and provide the mechanism for filament sliding. The myosin head (HMM-S1) has ATPase activity. Pure myosin actually has a low ATPase activity, hydrolyzing about six substrate molecules (a Mg^{2+}-ATP complex) per minute. The rate constant k_1 varies with ATP concentration; it is about $10^{-6}\,M^{-1}\,sec^{-1}$. k_2 is independent of ATP concentration, at about $100\ sec^{-1}$. The overall $\Delta G°$ is about -36.4 kJ $mole^{-1}$, i.e., is very favorable for ATP hydrolysis.

$$\text{HMM-S1} + \text{ATP} \xrightarrow[\substack{\text{(fast)}\\ \Delta G° = -66}]{k_1} \text{S1.ATP*} \xrightarrow[\substack{\text{(fast)}\\ -16.4}]{k_2} \text{S1*.ADP.Pi}$$

$$\xrightarrow[\substack{\text{slow}\\ +46\ \text{kJ mole}^{-1}}]{k_3} \text{S1} + \text{ADP} + \text{Pi} \quad (9.1)$$

A mixture of pure F-actin and myosin (an actomyosin complex) has about $100 \times$ the ATPase activity of pure myosin, i.e., actin activates the myosin ATPase. However, F-actin, which is associated with troponin and tropomyosin, does not stimulate the myosin ATPase. The inhibition by troponin and tropomyosin of myosin activation by actin is removed if Ca^{2+} is present at 10^{-7} to 10^{-6} M. The actions of actin, troponin/tropomyosin, and Ca^{2+} on myosin ATPase activity are not enzymatic effects, i.e., they are not cofactors. Rather, they are conformational regulators of the ability of myosin heads to bind to the actin binding site and to form cross-bridges.

The reaction of F-actin with myosin-ATPase is complex, but there is a favored cycle that includes ATP hydrolysis and attachment/detachment of the myosin head (Figure 9–23). A myosin head attaches to the actin filament, undergoes a conformational change (rotation) and slides the actin filament past the myosin, detaches from the actin filament, then rerotates to its original orientation so it can attach to another actin binding site. It is not clear whether both heads of each myosin molecule act independently of each other or whether they act in concert. One possible sequence of events for the myosin head rotation involves sequential attachment of four different binding sites of the myosin head (M_1, M_2, M_3, M_4); the myosin head rotates during the attachment and detachment of successive binding sites. Other models for myosin-actin sliding have the same overall effect.

Myosin head rotation is transduced to a longitudinal displacement of actin and myosin through the cross-bridge link between the myosin head and the

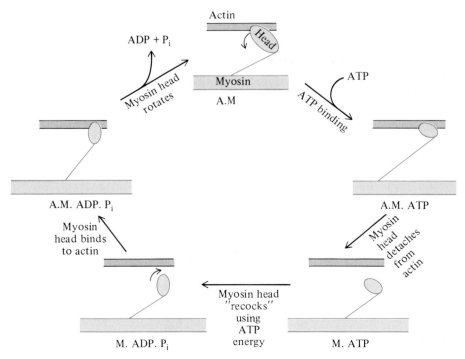

FIGURE 9–23 Dynamics of actin (A) and myosin head (M) resulting in net movement of actin and myosin filaments by cross-bridge formation, rotational movement of the myosin head, and cross-bridge detachment. The addition of ATP begins the cycle.

actin filament. The power stroke, i.e., myosin head rotation, does not require ATP hydrolysis although this is the step in the sliding filament cycle that performs useful work. The chemical energy that is converted to mechanical work was stored in the myosin head by prior ATP hydrolysis. ATP hydrolysis actually occurs during the step in the cycle when the myosin head detaches from the actin binding site and repositions in readiness for the next attachment to an actin binding site, i.e., ATP is hydrolyzed when the myosin "recocks" its position.

Tilting of the 19 nm-long myosin head would be expected to produce about 12 nm of actin-myosin filament sliding. However, recent estimates suggest that the actual filament sliding distance per ATP hydrolyzed is about 40 nm, or even more (Higuchi and Goldman, 1991). The 28 nm additional movement may be due to filament sliding while the cross-bridges bear a negative force, i.e., the myosin head is passively dragged 28 nm during filament sliding and active head rotation contributes a further 12 nm sliding. Alternatively, there may be multiple myosin head power strokes per ATP hydrolyzed.

Acto-myosin regulation. Calcium and magnesium ions are required for muscle contraction, and it was initially thought that both were enzymatic cofactors for ATPase activity. The Mg^{2+} does have this cofactor role; it forms an ATP-Mg^{2+} substrate for the myosin ATPase.

The regulatory role of Ca^{2+} is more complex than that of Mg^{2+}. Myosin head ATPase activity is Ca^{2+} dependent with little activity at a Ca^{2+} concentration of $<10^{-7}$ and maximal activity at concentrations $>10^{-5}$. For example, both mammalian and squid muscle show a superficially similar dependence of ATPase activity on Ca^{2+} concentration (Figure 9–24). Muscle cells have a total intracellular Ca^{2+} content equivalent to an average concentration of 1 to 5 10^{-3} M, but most of the intracellular Ca^{2+} is normally sequestered within vesicles and the normal cytoplasmic Ca^{2+} concentration is $<10^{-7}$ M. Consequently, there normally is no Ca^{2+} activation of actin-myosin and cross-bridge formation.

There are two different mechanisms for Ca^{2+} regulation of actin-myosin interaction. One mechanism depends on actin's capacity to bind to myosin heads, i.e., the myosin is **actin regulated**. The other mechanism is myosin-dependent control of its ATPase activity, i.e., the myosin is **myosin regulated**. Vertebrate striated muscle and the locomotor muscles of many invertebrates have only actin-dependent regulation, whereas myosin-dependent regulation is common in muscles of many other invertebrates (Figure 9–25). A number of the higher

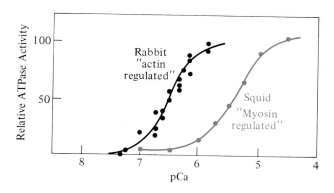

FIGURE 9–24 Mg-ATPase activity is dependent on the Ca^{2+} concentration ($pCa^{2+} = -\log_{10} [Ca^{2+}]$). For rabbit the actin is Ca^{2+} regulated, whereas for squid there is myosin regulation (and also some actin regulation). *(Modified from Bendall 1968; Konno, Arai, and Watanabe 1980.)*

invertebrate groups (e.g., annelids, mollusks, many arthropods) have myosin-dependent regulation. Single regulatory systems tend to be found in more primitive animals and double regulatory systems in higher invertebrates (Lehman and Szent-Gyorgyi 1975). Both the myosin and actin regulatory systems occur in primitive phyla, and so neither appears to be the "ancestral" regulatory system, although myosin regulation involves only one regulatory protein and actin regulation involves a number of regulatory proteins. Vertebrate smooth muscle has a different regulatory system, with both actin and myosin regulation by caldesmon (see below). Let us first examine actin-mediated regulation by Ca^{2+} in vertebrate striated muscle, since it is the sole regulatory mechanism.

The actin-mediated response to Ca^{2+} of vertebrate striated muscle requires the presence of tropomyosin and troponin in the F-actin filament (Figure 9–26A). **Tropomyosin** molecules are positioned end-to-end in the grooves of the double stranded F-actin with each tropomyosin traversing seven G-actin subunits. This position of the tropomyosin in the actin groove precludes the correct steric interaction of the myosin head with the G-actin binding site for myosin and prevents cross-bridge formation. Attached to one end of each tropomyosin is **troponin**, a complex globular protein. Troponin has three subunits: troponin C, which binds Ca^{2+}; troponin T, which binds to the tropomyosin; and troponin I, which inhibits the actin binding sites. Troponin C has four Ca^{2+} binding sites; two high-affinity sites bind Ca^{2+} or Mg^{2+} and two lower affinity sites bind only Ca^{2+}. The activating effect of Ca^{2+} on troponin

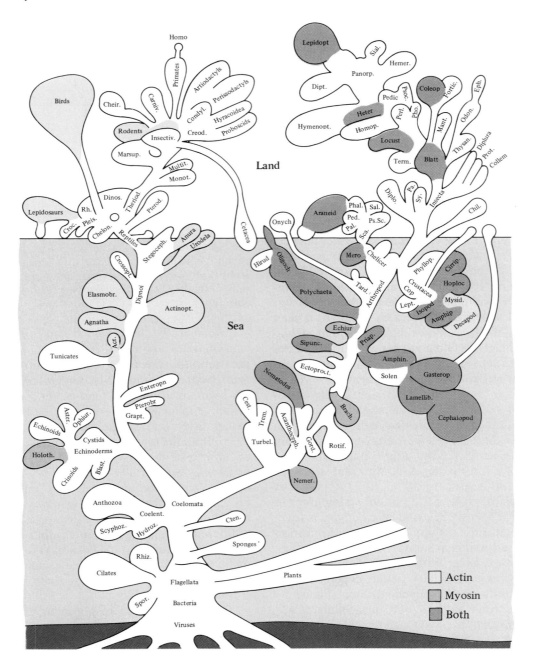

FIGURE 9–25   Phylogenetic distribution of myosin-linked (medium color), actin-linked (light color), or both types (dark color) of regulation of myosin Mg-ATPase in animals. Only regulation of vertebrate striated muscle and various locomotor muscles is included; vertebrate smooth muscle has a different regulatory system.   *(Adapted from Lehman and Szent-Gyorgyi 1975; Lehman 1983.)*

C causes a change in the structure of troponin I. This involves movement of the tropomyosin so that the myosin heads can bind to the G-actin binding sites and form cross-bridges (Figure 9–26B).

Myosin control of muscle contraction is observed in some invertebrates (e.g., echinoderms, nemer-

tines, brachiopods, mollusks, echiuroids). The myosin ATPase is directly activated by Ca^{2+}; it is not activated by pure actin. The Ca^{2+} binds to a small regulatory protein (MWt 1.7×10^4) that normally is tightly bound to the myosin. Removal of the Ca^{2+}-binding protein from myosin prevents ATPase activ-

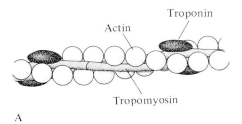

A

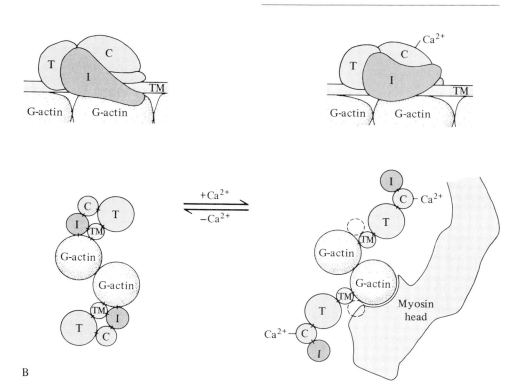

B

FIGURE 9–26 **(A)** F-actin is a twisted bifilament of globular G-actin subunits with accessory proteins tropomyosin and troponin complex. **(B)** Representation of the structural rearrangement of troponin I in response to the binding of Ca^{2+} by troponin C, which results in the uncovering of the myosin head binding site on the G-actin subunit. *(Modified from Ohtsuki 1980; Gergely and Lewis 1980.)*

ity, even in the presence of excess Ca^{2+}, but replacement of the Ca^{2+}-binding protein restores the ATPase activity at high Ca^{2+}. There is one Ca^{2+}-binding site per myosin molecule (i.e., per two myosin heads) and it binds one or two Ca^{2+}. The myosin ATPase activity is minimal at the normal, low intracellular Ca^{2+} concentration ($<10^{-7}$ M) and is elevated at higher intracellular Ca^{2+} concentrations ($>10^{-7}$ M). This effect is superficially similar to the role of Ca^{2+} in an actin-activated system.

Myosin activity is regulated in the nematode *Caenorhabditis* by a very large muscle protein, twitchin (Benian et al 1989). Twitchin is a protein kinase enzyme that probably is associated with the myosin (it is located in the sarcomere A bands), whose function it regulates. *Caenorhabditis* muscle apparently also has both actin and myosin regulation by Ca^{2+}.

Excitation-contraction coupling. So far we have examined the molecular mechanisms for muscle con-

traction and the role of intracellular Ca^{2+} in initiating sliding filament movement, but how does synaptic depolarization of the muscle end plate elevate the intracellular Ca^{2+} concentration and initiate a muscle contraction? Stimulation of the presynaptic axon depolarizes the end plate, the intracellular Ca^{2+} concentration transiently increases, and the muscle cell develops tension (Figure 9–27). The sequence of events linking end plate depolarization and muscle contraction is **excitation-contraction coupling**.

The sarcolemmal membrane potential determines the force of contraction. Normally, an action potential is propagated over the sarcolemma, but some invertebrate muscle cell membranes do not support regenerative action potentials and only graded depolarizations spread over the sarcolemma. Localized depolarization of openings of the t-tubules at the sarcolemma surface elicits a localized sarcomeric contraction, but similar depolarizations at other sarcolemmal sites have no effect. The surface membrane depolarization moves by electrotonic spread

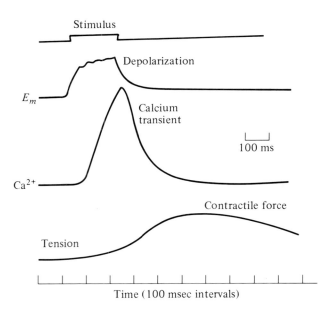

Stimulus

Depolarization

E_m

Calcium
transient

|—| 100 ms

Ca^{2+}

Contractile force

Tension

Time (100 msec intervals)

FIGURE 9–27   Relationship between intracellular Ca^{2+} concentration and muscle fiber tension in response to an electrical stimulus and membrane depolarization.   *(Modified from Keynes and Aidley 1981.)*

down the t-tubules and releases Ca^{2+} from the adjacent sarcoplasmic reticulum. The elevated intracellular Ca^{2+} then elicits a localized sarcomeric contraction.

The t-tubules are in intimate contact with the sarcoplasmic reticulum membrane at the triads (Figure 9–28). There are specific t-tubular membrane proteins, arranged in tetrads, and also integral sarcoplasmic reticulum proteins. The sarcoplasmic reticulum tetrads are arranged in two parallel rows along the t-tubule, but only half contact t-tubular tetrads. Ca^{2+} efflux from the sarcoplasmic reticulum is mediated by a specific Ca^{2+} channel, the **"Ca^{2+}-release channel,"** that is distinct from the sarcoplasmic reticulum Ca^{2+} pump (Ca^{2+}-Mg^{2+}-ATPase), which sequesters Ca^{2+} within the sarcoplasmic reticulum.

It is not clear how depolarization of the t-tubule membrane transduces Ca^{2+} efflux from the sarcoplasmic reticulum. The signal transduction is thought to involve either a chemical messenger of unknown nature or mechanical coupling. Several recent (and not mutually exclusive) hypotheses for the transduction process include electrical depolarization of the sarcoplasmic reticular membrane; change in pH; Ca^{2+}-induced Ca^{2+} release; charge displacement and consequent mechanical movement of the feet bridging the t-tubule/sarcoplasmic membrane space; a chemical messenger, inositol

1,4,5 triphosphate (IP_3); and change in oxidation state of sulfydryl groups on the Ca^{2+}-release channel (Trimm, Salama, and Abramson 1986). The t-tubule protein tetrads may be voltage sensors, and depolarization of the t-tubule may induce a conformational change, thereby gating the abutting sarcoplasmic reticular foot to allow Ca^{2+} efflux from the sarcoplasmic reticulum. The sarcoplasmic reticular feet that do not contact t-tubular tetrads may have a different gating mechanism, perhaps Ca^{2+}, i.e., they may be Ca^{2+}-induced Ca^{2+} channels. IP_3 is a phospholipid metabolite that couples a variety of extracellular signals with a transient increase in intracellular Ca^{2+}, by stimulating Ca^{2+} release from intracellular stores or influx across the cell membrane. This occurs in brain neurons and hepatocytes, as well as probably muscle.

The details of excitation-contraction coupling are slightly different in cardiac muscle cells because there is a less extensive system of t-tubules and sarcoplasmic reticulum. In mammalian and avian heart muscle, some of the elevated intracellular Ca^{2+} concentration during an action potential is due to the influx of Ca^{2+} from the extracellular fluid through sarcolemmal Ca^{2+} channels during the plateau phase of the cardiac action potential. The influx of extracellular Ca^{2+} triggers further release of Ca^{2+} from the sarcoplasmic reticulum, i.e., there is Ca^{2+}-induced Ca^{2+} release. This is particularly important for mammalian atrial and Purkinje fibers and for avian cardiac fibers, since these cells lack functional t-tubules. Frog cardiac muscle has a sparse sarcoplasmic reticular system; most of the increase in intracellular Ca^{2+} in amphibian cardiac muscle is due to influx from the extracellular fluid.

Smooth Muscle Contraction.   Smooth muscle cells contain numerous thin actin filaments that are generally oriented parallel to the long axis of the cell. Attached to actin filaments are dense bodies, which are scattered throughout the cytoplasm and at intervals along the inner surface of the sarcolemma. The thick myosin filaments of smooth muscle cells have all of their heads projecting in the same direction along their length (unlike the myosin of striated muscle cells).

Actin is the predominant intracellular protein filament of smooth muscle. The molar ratio of actin to myosin varies from 12:1 to 50:1 for smooth muscle (Figure 9–29A) compared to 4:1 in striated muscle. The actin is not homogeneously distributed throughout the cell, but is partitioned between two domains: the **acto-myosin domain** and the **filamin domain** (Sparrow 1988). The acto-myosin domain contains myosin, actin (plus tropomyosin), and caldesmon;

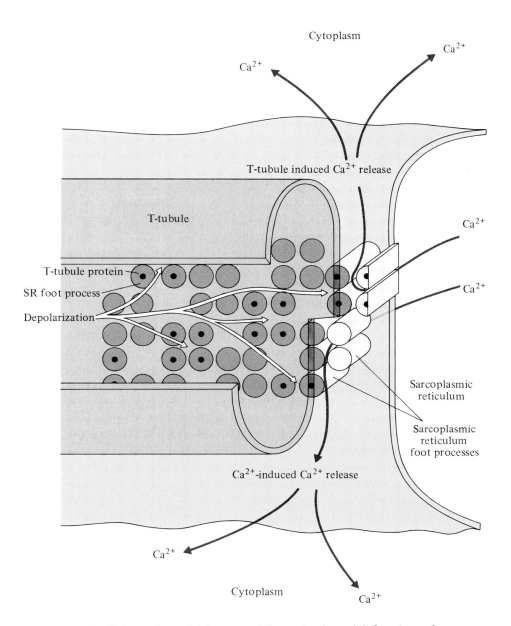

FIGURE 9–28 Schematic model for a possible mechanism of Ca^{2+} release from sarcoplasmic reticulum. Electrotonic depolarization spreads down t-tubules and activates voltage-dependent, t-tubular membrane proteins (solid). This includes a conformational change in underlying proteins of the sarcoplasmic reticulum membrane; units of four SR proteins (a SR "foot") allows Ca^{2+} to diffuse into the cytoplasm. The alternating SR feet without t-tubular, voltage-sensitive proteins may also undergo a conformational change, in response to the t-tubule-induced Ca^{2+} release, and allow further Ca^{2+} to diffuse from the SR into the cytoplasm, i.e., this is Ca^{2+}-induced Ca^{2+} release. *(Based in part on Rios and Pizaro 1988.)*

it is responsible for force development during contraction. The filamin domain contains filamin, actin (plus tropomyosin), desmin, and dense bodies (α-actinin); it is responsible for tonic maintenance of tension.

The sequence of events from sarcolemma depolarization to actin-myosin cross-bridge formation is generally similar for smooth muscle and striated muscle, although there are some differences. The intracellular Ca^{2+} concentration is the regulator

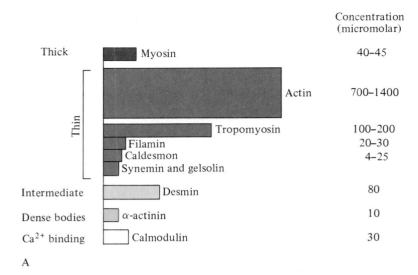

A

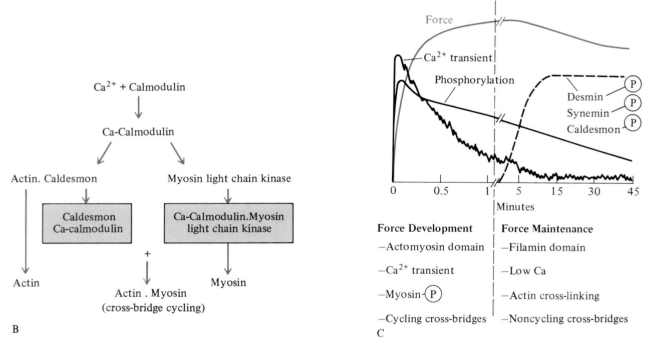

B

C

FIGURE 9–29 **(A)** Relative proportions of various contraction-related proteins in smooth muscle. **(B)** There is a dual calcium regulation of smooth muscle contraction by the effects of Ca^{2+}-calmodulin on the myosin (right) and actin (left). **(C)** Events during smooth muscle contraction initiated by a transient increase in intracellular Ca^{2+} concentration are an initial period of force development due to the acto-myosin domain (related to the Ca^{2+} transient, phosphorylation, and cross-bridge cycling) followed by a period of force maintenance due to the filamin domain (related to actin cross-linking and no cross-bridge cycling). *(From Sparrow 1988.)*

of actin-myosin cross-bridge formation in smooth muscle, as in striated muscle; it is increased from about 10^{-7} to 10^{-6} after stimulation. The rudimentary sarcoplasmic reticulum of smooth muscle is one source of the increased Ca^{2+} concentration. Intracellular Ca^{2+} is also elevated by the entry of extracellular Ca^{2+} into the smooth muscle cell through the sarcolemma.

The SR-sequestered Ca^{2+} is released by inositol 1,4,5 triphosphate (IP_3). The IP_3 initiates Ca^{2+} release from SR vesicles of mammalian smooth muscle but apparently not striated muscle vesicles (Ehrlich and Watras 1988). The IP_3-gated Ca^{2+} channel of smooth muscle vesicles is very different from the Ca^{2+}-activated Ca^{2+} channel of striated muscle.

The intracellular Ca^{2+}-binding protein of smooth muscle, **calmodulin,** has four Ca^{2+}-binding sites. The Ca^{2+}-calmodulin complex binds to a specific region of the myosin light chain kinase and exposes the catalytic binding site responsible for binding of myosin with ATP (Figure 9–29B right). The myosin binds to actin, cross-bridges form, the myosin heads rotate and detach, and ATP is hydrolyzed. This Ca^{2+}-regulated mechanism for smooth muscle contraction corresponds closely to the myosin-activation scheme for some striated muscles.

Smooth muscle also has another Ca^{2+}-regulatory scheme, at least *in vitro* (Figure 9–29B left). Smooth muscle actin is an F-actin double filament containing tropomyosin but has the protein caldesmon rather than troponin. Caldesmon normally binds strongly to the actin-tropomyosin complex and inhibits cross-bridge formation. The ratio of caldesmon:tropomyosin:G-actin subunits is 1:4:28. An increased intracellular Ca^{2+} forms a Ca^{2+}-calmodulin complex, which binds to caldesmon and weakens its binding to actin. This allows formation of myosin-actin cross-bridges.

Smooth muscle, unlike striated muscle, maintains tension even after the intracellular Ca^{2+} returns to the resting level (Figure 9–29C). The tension is maintained at a low energy cost, i.e., the rate of ATP hydrolysis is low. There may be some form of slow cross-bridge cycling or even formation of latched cross-bridges. Alternatively, the filamin domain, rather than the acto-myosin domain, may be responsible for the maintenance of tension. Proteins such as gelsolin may decrease the cytoskeletal rigidity during the Ca^{2+} transient of smooth muscle contraction, and the cytoskeleton is then rigidified after the Ca^{2+} transient. Phosphorylated proteins (desmin, synemin, and caldesmon) may maintain the resting tension; diacyl-glycerol (which is formed concurrently with IP_3 during the Ca^{2+} transient) stimulates their phosphorylation.

Many invertebrate smooth muscle cells resemble those of vertebrates, at least in general appearance, but **helical smooth muscle** and **paramyosin smooth muscle** differ in structure from typical smooth and striated muscle cells. The locomotory muscles of annelids and cephalopods have smooth muscle with helically arranged myofibrils. Some other mollusks and animals in other phyla (e.g., echinoderms, tunicates) also have helical smooth muscle but these have a "double-oblique" striation pattern due to the appearance of both aspects of the helix in the same focal plane. The adductor muscle of lamellibranch mollusks contains paramyosin smooth muscle cells. The adductor muscle closes the shell and maintains tension against a springy hinge; it can maintain tension for many hours at a very low metabolic cost. For example, an oyster adductor muscle can maintain 0.56 kg cm^{-2} for 20 to 30 days!

Properties of Muscle Contraction. The mechanical properties of muscle cells, as well as muscles, vary dramatically depending on the type of muscle and the metabolic capacity of the muscle cell. Vertebrate skeletal muscle has been studied in greatest detail with respect to its mechanical properties. We shall first concentrate on its general mechanical properties and then compare these properties with those of other types of skeletal muscle, cardiac muscle, and smooth muscle.

Threshold stimulation. Excitable cells, such as muscle and nerve, can be stimulated by an electrical impulse. The efficacy of the electrical impulse in eliciting a contraction is determined by both its magnitude and duration. A high-voltage impulse may be ineffective in eliciting a muscle contraction if it has a very short duration (i.e., subminimal duration), whereas a longer duration impulse may be equally ineffective if the voltage is too low (i.e., subthreshold voltage). The threshold voltage for very long duration impulses that elicits a response is the **rheobase**; the shortest duration that elicits a contraction of a rheobase-level stimulus is the **utilization time** (Figure 9–30). In practice, the utilization time is difficult to determine accurately, and so the minimum duration for a $2 \times$ rheobase voltage (**chronaxie**) is used to indicate the "excitability" of the cell; a low chronaxie indicates a high excitability. Different types of muscle vary in rheobase and chronaxie; the five examples illustrated in Figure 9–30 are superimposed on the same curve by normalizing the voltage scale to rheobase and the duration scale to the chronaxie.

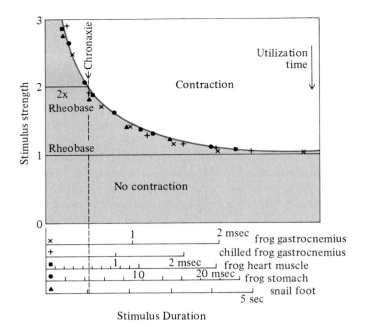

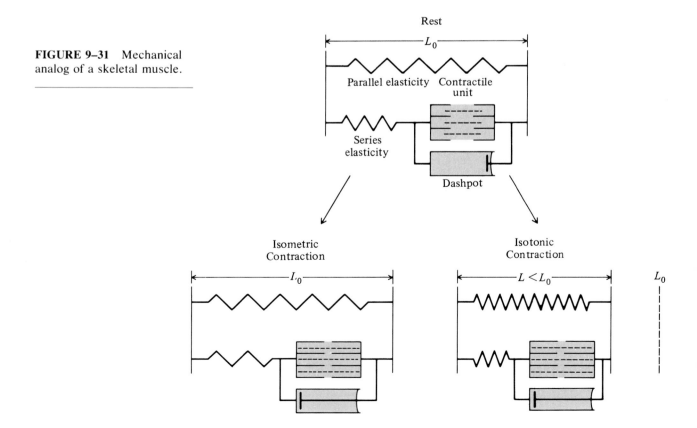

FIGURE 9–30 Relationship between duration of an electrical stimulus and the stimulus strength required to elicit a threshold response. Five strength-duration curves are superimposed by adjusting the time scale and using a relative strength scale (1 = threshold at infinite duration = rheobase). Chronaxie is the duration for 2 × rheobase strength.

Mechanical model. There are two different experimental types of muscle contraction: isometric and isotonic (Figure 9–31). In an isometric contraction, the muscle generates tension but the length of the muscle remains constant. In an isotonic contraction, the muscle shortens as it maintains a constant force.

The muscles of an animal seldom have a pure isometric or isotonic contraction but contract in an intermediate fashion. Nevertheless, it is still convenient to study the properties of isolated muscles during isometric or isotonic contractions because this allows different aspects of muscle con-

FIGURE 9–31 Mechanical analog of a skeletal muscle.

traction to be examined. Experimental results of force generated during isometric muscle contraction (*F*) are usually expressed as a ratio of the maximum force that can be generated by the muscle measured during continual maximal stimulation (F_0), i.e., F/F_0, or for isotonic experiments as a ratio of length (*L*) to the resting length (L_0) i.e., L/L_0.

The muscle cell contents must move as the cell contracts because the length of the cell decreases and its diameter increases. The viscous resistance to the shortening of muscle cells decreases the maximum force that they can generate. This viscous damping element of the cell is illustrated in Figure 9–31 as its mechanical analog, an oil pot damper, or dashpot. Another important mechanical property of muscle is its elasticity, i.e., parts of the muscle act as springs. A muscle contains two different types of "spring"; the **series elastic element** is a spring in series with the contractile sarcomere units and the **parallel elastic element** is a spring in parallel. The series elastic element consists of springy biological materials that are stretched as an immediate consequence of sarcomere shortening. The sarcomere Z band, the sarcolemma connections to myofibrils, and the tendons and other connective tissues that attach muscle cells to the bones act as series

elastic elements. The parallel elastic element is muscle connective tissue and the sarcolemma; it is not influenced if the length of the muscle remains constant (i.e., in an isometric contraction). The properties of the elastic elements and viscous dashpot significantly determine the mechanical properties of a muscle.

An ideal spring has a linear relationship between tension and length; this is expressed by Hook's law

$$F = kL \tag{9.2a}$$

where *F* is the force exerted on the spring; *L* is the displacement; and *k* is a constant, the spring factor. A resting muscle is generally quite elastic, reflecting the flexibility of its parallel and series elastic elements as well as some stretch of the sarcomeres. Elastic tension is generated in an exponential manner as the muscle is passively stretched (Figure 9–32); the relationship between resting force F_r and stretch (L/L_0) is

$$F_r = ke^{c(L/L_0)} \tag{9.2b}$$

where *c* is an additional constant. Thus, resting muscles do not obey Hook's law. For example, the resting frog sartorius muscle (at 0° C) can be stretched to 1.1 to 1.2 × L_0 by small forces, but

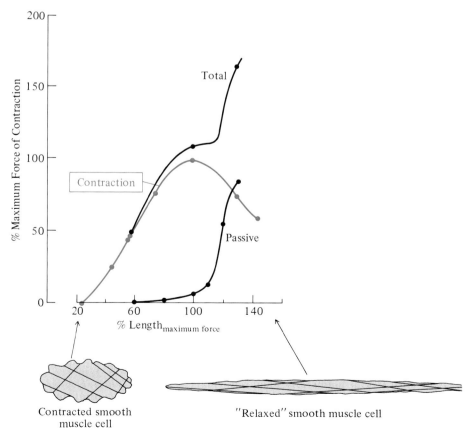

FIGURE 9–32 Passive tension, total tension (passive + contractile), and contractile tension of a smooth muscle cell as a function of initial length. *(Modified from Fay 1976.)*

requires a considerable force to be stretched to >1.4 L_0. A contracting muscle can generate a considerably greater force than at rest, up to a moderate degree of stretch, but the contraction force declines above a critical initial length, generally about 1.25 L_0. The difference between the active force and the resting force curves is the actual force generated by the contractile elements. This active force-generated curve has a strikingly similar shape to that derived for sarcomeres of contraction force as a function of sarcomere length (see Figure 9–22B, page 421). This similarity is to be expected because both reflect the basic contractile capacity of the sarcomeres.

There is a considerable variation in the passive length-tension curves for different muscles, depending on the particular sarcomeric organization (striated muscles) or acto-myosin dense bodies (smooth muscle) and the spring factor of the parallel and series elasticities. For example, insect flight muscles are very stiff compared to vertebrate striated muscle; the passive force increases dramatically with only short extension to lengths over the maximum *in vivo* length. This stiffness allows for considerable storage of elastic energy during the high-frequency wing beat cycle. Smooth muscle has a low spring factor and requires little force for extension to over 1.5 × the maximum *in vivo* length. The marked passive extensibility of smooth muscle cells is related to its absence of an organized sarcomeric structure and greater capacity for shape change. The maximal contraction force generally occurs at about the maximal *in vivo* length and is about 2 × that for maximal passive stretch (which occurs at much greater lengths).

Contraction time course. A single electrical stimulation of a muscle elicits a single contraction, or twitch (Figure 9–33). There is a brief (5 to 10 msec) **latency period** between the electrical stimulation and the first increase in tension. This corresponds to the time required for electrochemical coupling, Ca^{2+} binding to troponin, tropomyosin movement to allow myosin heads to cross-bridge with the G-actin binding sites, and stretch of the series elastic element. The maximum isometric tension occurs at about 150 msec after the stimulus, and the relaxation is complete after about 900 msec. The isometric contraction period is considerably shorter (150 msec) than the relaxation period (about 750 msec). For isotonic contractions, the time for maximum shortening is considerably longer than the time to maximum tension for an isometric contraction, and it is increased by elevated load.

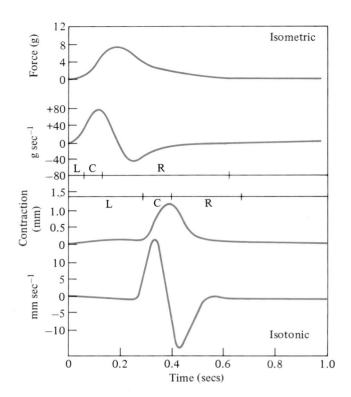

FIGURE 9–33 Single isometric and isotonic twitch for the gastrocnemius muscle of a marine toad *(Bufo marinus)* showing latency (L), contraction (C), and relaxation (R).

The contraction time varies markedly for different muscles. One functional classification of muscles is based on whether they are "fast" or "slow." The fastest mammalian muscle is the eye oculomotor muscle (5 to 6 msec contraction time). Some vertebrate muscles are extremely fast (e.g., the muscles of the puffer fish that are used for sound production; 1 to 1.5 msec). The soleus is an intermediate muscle (70 msec). The sloth claw retractor muscle is a slow muscle (150 to 300 msec). Many smooth muscles are very slow (e.g., gut, 100 to 300000 msec), although some are quite fast (e.g., trachea, 17 msec). Invertebrate muscles have a similar range in contraction time, from a few msec for fast striated muscles (e.g., cockroach coxal muscle) to 30000 msec (coelenterate smooth muscle). The tension generated by a muscle is generally related in an inverse fashion to the speed of contraction; fast muscles tend to have low tension development.

The rapid events at the beginning of a muscle contraction are quite complex. There is a very slight decrease in muscle tension after the stimulating action potential and towards the end of the latency

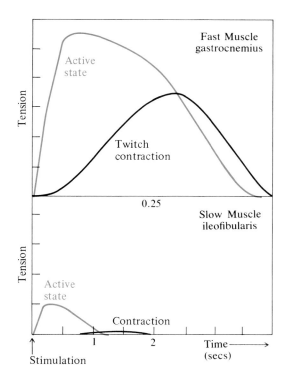

FIGURE 9–34 Active state, the actual contraction of sarcomere units, occurs rapidly after muscle stimulation; the mechanical development of tension occurs rapidly after muscle stimulation; the mechanical development of tension occurs more slowly, and achieves a lower maximal tension than does active state. The slower development, and lower maximal tension, is more pronounced for slow muscles (such as the frog ileofibularis and crayfish claw opener), than for fast muscle (such as the frog gastrocnemius and crayfish claw closing muscle).

period; this is **latency relaxation**. The tension rapidly returns to normal after the latency relaxation, then increases as the muscle begins to contract.

The period of force generation by a sarcomere is the **active state**. The intracellular Ca^{2+} is elevated, there is cross-bridge cycling, and ATP is hydrolyzed by the myosin ATPase during active state. In contrast, the cycle of tension generation by a single muscle twitch is considerably delayed compared to the active state of the sarcomeres, and the tension is considerably attenuated. The series elastic element of muscle is responsible for the delay and damping. Shortening of the sarcomeres first stretches the series elastic springs of the muscle. A high elasticity will greatly dampen the magnitude of a single twitch relative to maximal tetanic tension (which indicates the actual contraction force of the contractile units). For example, the ratio of single twitch tension to active state tension is about 0.2 for the frog sartorius muscle and 0.5 for the less elastic locust flight muscle; both have a considerable time delay to maximum twitch tension. The delay and attenuation are considerably greater in a slow muscle compared to a fast muscle. For example, the frog gastrocnemius muscle (a fast muscle) is still in active state at the end of its latency period and twitch tension is about 0.5 of active state tension (Figure 9–34). In contrast, a slow muscle (crayfish claw opener) has completed active state well before the end of the latency period and the twitch tension is much less than active state tension.

Muscle twitches are an "all-or-none" phenomenon. Successive twitches generate the same tension, as long as they are spaced at more than a minimal interval (Figure 9–35). Summation occurs if a second twitch is initiated while the muscle is still generating tension from the first twitch. The force of contraction is greater if the duration between successive stimuli is decreased. Both mechanical and contractile mechanisms contribute to summation. The first stimulus prestretches the elastic components of a muscle and allows a more rapid mechanical response to the second stimulus. An elevated intracellular Ca^{2+} due to the first stimulus will result in a higher Ca^{2+} in response to the second, hence a greater force of contraction.

A series of successive stimuli will cause continual summation of individually distinguishable twitches until a constant force of contraction is achieved

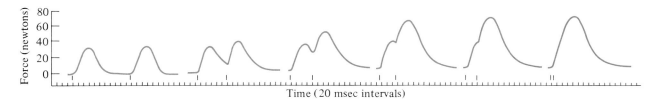

FIGURE 9–35 Mechanical summation of two muscle twitches by the cat gastrocnemius as a function of interval between double shocks. *(Modified from Cooper and Eccles 1930.)*

(although individual twitches can still be perceived); this is incomplete **tetany** (Figure 9–36). The individual contractions are no longer discernible and the muscle shows complete tetany if the frequency of stimulation is further increased to the fusion (critical) frequency. A fast muscle maintains discrete twitches at higher frequencies than does a slow muscle, i.e., the fast muscle has a higher fusion frequency.

Maximum force.   The maximum force exerted by muscles varies widely, reflecting differences in their size rather than intrinsic properties. The maximum force expressed per cross-sectional area (F_{max}; N cm^{-2}) is relatively constant at about 10 to 30 N cm^{-2} for many vertebrate and invertebrate striated muscles (Table 9–3). Some muscles (e.g., mollusk catch muscles) can produce greater peak forces, over 100 N cm^{-2}. The remarkable tension that can be generated by these muscles is thought to be related to the presence of paramyosin, although paramyosin is not only found in molluskan catch muscle. It also occurs in byssal and pedal retractor

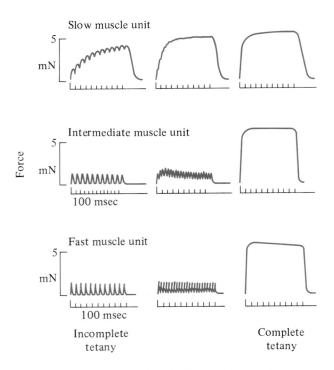

FIGURE 9–36 Isometric twitches and incomplete/complete tetany for a slow muscle unit and an intermediate muscle unit of a rat soleus muscle, and a fast muscle unit from a rat extensor digitorum longus. *(Modified from Close 1967.)*

muscles of bivalve mollusks, retractor muscles of sipunculids and snails, the body wall of gordian worms, and the notochord of amphioxus. Paramyosin is also found in the striated and smooth muscle of nematodes and various arthropods including horseshoe crabs, spiders, scorpions, and insects.

Bivalve mollusks use one or more adductor muscles to close their shells against a springy elastic hinge, and these adductor muscles can generate remarkably high peak tensions of 100 to 120 N cm^{-2}. The adductor muscles generally have a translucent (or colored or sometimes striated) phasic muscle and a white, opaque tonic **catch muscle** (Hanson and Lowy 1960). The catch muscles contain a large amount of a special myofilament protein, paramyosin (Chantler 1983). The paramyosin molecules are rod-shaped, double α-helix coils of two polypeptides. They form an organized array that acts as a core which is covered by myosin molecules (perhaps organized as a single layer); the ratio of paramyosin to myosin can vary from 2:1 to 10:1 (by mass). The myosin-paramyosin thick filaments are much longer and thicker than the typical myosin filament, being 20 to 40 μ long and 60 to 80 nm diameter (cf. myosin of vertebrate striated muscle is 1.5 μ long and 15 nm dia).

The "catch" state is characterized by muscle rigidity unlike the flaccidity of the normal relaxed muscle. It was initially thought that catch did not require metabolic energy expenditure, but there actually is a low elevation of metabolism during catch. For example, the resting metabolic rate of the anterior byssus retractor muscle of *Mytilus* is 75 μm O$_2$ g^{-1} min^{-1}, and in catch there is a 26 μm g^{-1} min^{-1} elevation in metabolic rate. In contrast, phasic contractions (50 N cm^{-2}) elevate metabolic rate by up to 83 μm g^{-1} min^{-1}.

Although the catch mechanism is not fully understood, the high force generated at low metabolic cost during catch appears to involve the stabilization of acto-myosin cross-bridges through some structural change of the paramyosin core of the myosin filaments, perhaps induced by paramyosin phosphorylation (Watanabe and Hartshorne 1990). There most likely is a slow cycling of the myosin-actin cross-bridges; normal muscle contraction involves a fast cross-bridge cycling. Stimulation of catch muscle by cholinergic nerves increases the intracellular Ca^{2+} and initiates a typical muscle contraction (Figure 9–37). There is a rapid cycling of myosin-actin cross-bridges. Subsequent reduction of the intracellular Ca^{2+} concentration leads to catch in the absence of serotonin and to relaxation in the presence of serotonin. In the catch state, the muscle tension is maintained at a low metabolic rate. It is

TABLE 9–3

			Maximum Force ($N\ cm^{-2}$)
Maximum force developed by muscles (force per cross-sectional area). 1 N cm^{-2} = 0.10 kg cm^{-2}. *(Modified from Prosser 1973.)*			
Animal	**Muscle**	**Type**	**Maximum Force (N cm^{-2})**
Rat	Heart	Cardiac	0.2
Lobster	Fast remotor	Striated	0.2
Toadfish	Sonic	Striated	1.0
Rabbit	Uterus	Smooth	1.3
Rat	Soleus	Striated	2.2
Cat	Papillary	Cardiac	7.8
Dog	Tracheal	Smooth	7.8
Cockroach	Coxal	Striated	7.8
Rabbit	Taenia coli	Smooth	8.7
Cat	Tenuissimus	Striated	13.7
Guinea pig	Taenia coli	Smooth	14.7
Sloth	Gastrocnemius	Striated	15.9
Rat	Gastrocnemius	Striated	17.6
Frog	Sartorius	Striated	19.6
Sloth	Diaphragm	Striated	20.6
Lobster	Slow remotor	Striated	27.4
Rat	Extensor digitorum	Striated	29.4
Oyster	Adductor	Catch	5.9 (tonic)
Oyster	Adductor	Catch	117.6 (peak)

not clear how catch is maintained; perhaps the large size and the high number of cross-bridges of the myosin/paramyosin complex confers the capacity to "catch." Stimulation of serotoninergic nerves releases serotonin and elevates the intracellular cAMP level. The high cAMP activates a protein kinase, which may phosphorylate paramyosin and myosin to induce relaxation.

Velocity, power, and energy. The average velocity (V) of contraction during an isotonic muscle twitch depends on the contraction time (Δt) and the distance shortened (ΔS); $V = \Delta S/\Delta t$. For an isometric contraction, ΔS is 0 so there is no velocity of shortening; however, the rate of change in tension ($\Delta F/\Delta t$) is an analogous measure of muscle contractile activity.

The velocity of shortening increases during an isotonic contraction to a maximum value about halfway through the contraction time, then declines to zero at the end of the contraction time. The velocity is negative during the relaxation time (i.e., the muscle lengthens rather than shortens) and is greatest about halfway through relaxation time. The maximum rate of shortening is generally higher than the maximum rate of lengthening because the contraction time is shorter than the relaxation time. The velocity of shortening depends on the load (F) against which the muscle has to work, i.e., the weight that the muscle has to lift (Figure 9–38). The velocity of shortening is less for greater forces. There is a hyperbolic relationship between force and velocity (see Supplement 9–2, pages 445–447). The power (P) output of a muscle can be readily calculated from the $V - F$ diagram as force × velocity. It is zero at no velocity (isometric) and maximum velocity (isotonic, no load).

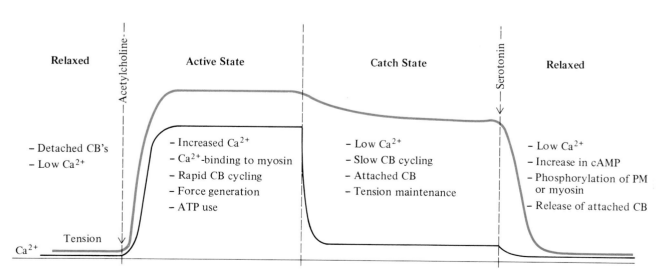

FIGURE 9–37 Possible model for contractile tension generation, "catch," and relaxation of catch muscle. *(Modified from Watanabe and Hartshorne 1990.)*

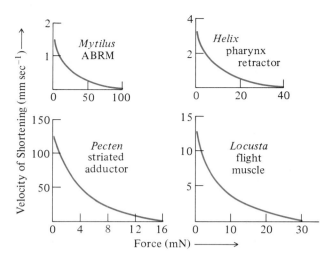

FIGURE 9–38 Hyperbolic force-velocity curves for a variety of vertebrate and invertebrate muscles: *Mytilus* anterior byssal retractor muscle, *Helix* pharynx retractor muscle, *Pecten* striated adductor muscle, and *Schistocerca* flight muscle. *(Modified from Hanson and Lowy 1960.)*

The energy expended by a muscle during a contraction can be measured by the heat produced or the depletion of high energy stores (e.g., ATP or creatine phosphate). Changes in the chemical energy balance of muscle cells (i.e., phosphagen levels, particularly creatine phosphate) can be related to total energy expenditure (work done + heat liberated). A value of 46.4 kJ mole^{-1} creatine phosphate is obtained whether the muscles contract isometrically (twitch or tetany) or isotonically, performing positive or negative work.

There is a complicated pattern of heat release by a muscle during and after a twitch. There is an initial, rapid release of heat soon after the initiation of a tetanic, isometric contraction even before tension is generated (Figure 9–39); this is the **activation heat** (H_a). The muscle soon reaches a steady-state rate of heat production, the **stable heat** (H_s). A third component of heat production, the **labile heat** (H_l), decays after about 3 to 4 msec. After the tetanic contraction ceases and the muscle relaxes, there is a slow but prolonged release of **recovery heat**, which can be approximately equal in magnitude to the total contraction heat. The activation heat may be due to Ca^{2+} release, Ca^{2+} binding to troponin, and rearrangement of the tropomyosin, as well as the internal work done in shortening the sarcomeres and taking up any slack in the series elasticity (there is some sarcomere shortening even if a muscle contraction is isometric). The stable

(and perhaps labile) heat is the energy required to maintain steady-state tension. The recovery heat reflects the bioenergetic cost for removal of lactate that accumulated during the contraction.

The energy expenditure of a muscle is not equal to the mechanical work done, partly because of biochemical inefficiency (aerobic metabolism is only about 41% efficient; see Chapter 3) and partly for mechanical reasons. The **contractile mechanical efficiency** is the ratio of mechanical work done to energy expended for the contractile process. It is difficult to measure the contractile mechanical efficiency because many other energy-requiring processes occur simultaneously. The contractile mechanical efficiency is 0% for an isometric contraction and for an isotonic contraction against zero load and increases to a maximum at intermediate loads. Mammalian fast muscle (e.g., biceps brachii) has a maximum contractile mechanical efficiency of about 55% at a fast shortening velocity of about 5 muscle lengths sec^{-1} (Alexander and Goldspink 1977). A slow, phasic muscle, such as the soleus, has a maximal contractile efficiency of about 75% at a shortening velocity of about 1.5 muscle lengths sec^{-1}. Maximal contractile efficiencies of 75 to 80% have been measured for tortoise muscles and values up to 45% for frog and toad muscles. Thus, energy transduction by cross-bridges into mechanical work can be a highly efficient process. In contrast, the **overall mechanical efficiency** of muscle, the ratio of mechanical work done to total energy expenditure,

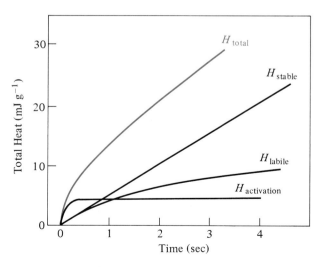

FIGURE 9–39 Total heat production by a toad muscle during an isometric contraction showing the various components: activation heat, stable heat, and labile heat. *(Modified from Carlson and Wilkie 1974.)*

is considerably lower because many processes other than cross-bridge cycling require energy expenditure. The maximal overall mechanical efficiency is generally 10 to 35%; e.g., 28% for a human pedaling a bicycle, 30% for *in vitro* muscle, 35% for tortoise muscle, and 10% for insect flight muscle.

Determinants of muscle properties. As we have seen, there are three general types of muscle: skeletal, cardiac, and smooth. However, there are many specific biochemical and mechanical differences for muscles even within one of these three general types. For example, vertebrate striated muscle fibers vary in their major structural/mechanical properties and are divided into tonic and phasic fibers. **Tonic fibers** have very slow conduction; they generally do not propagate a sarcolemmal action potential and require multiple stimulations to contract and can respond with a graded (not all-or-none) contraction. They normally are involved with maintaining posture rather than locomotory movement. **Phasic fibers** are either slow or fast fibers. Fast phasic fibers can be either oxidative or glycolytic. Slow phasic fibers (red muscle) contract relatively slowly (but faster than tonic fibers) and are slow to fatigue. Fast phasic glycolytic fibers (white muscle) are fast contracting and fatigue rapidly. Fast phasic oxidative fibers are fast contracting and fatigue more slowly than the fast phasic glycolytic fibers. There are numerous differences between these skeletal muscle fiber types in innervation pattern, electrical properties of the sarcolemma, enzymatic properties, metabolic capacity, and blood flow pattern (Table 9–4).

Differences in structure and organization of the contractile machinery can contribute to differences in mechanical properties of muscle cells, but the basis for many of the functional differences is the innervation pattern; the biochemical properties of muscle cells are determined by the innervation. The properties of fast muscle are induced by the nature of its innervating axon; "fast" axons innervate fast muscles, and "slow" axons innervate slow muscles. Denervation of a fast muscle and reinnervation by a slow axon will induce slow muscle properties.

Skeletal muscle fibers are organized as motor units, and all fibers of one motor unit are of the same type. A muscle will generally contain motor units of all types (Figure 9–40), although the proportions vary. For example, the locomotory muscles of mammals with rapid, sustained activity (e.g., wolves, dogs, ungulates) have a high percentage of fast oxidative fibers; the lion, in contrast, has fast glycolytic fibers. Fast-cruising fish (mackerel, tuna) have a high percentage of slow oxidative fibers, whereas stealthy, rapidly-striking fish (pike) have a high percentage of slow phasic fibers. The recruitment of different type units in a muscle is probably hierarchical, i.e., slow fibers are activated first, then fast oxidative fibers, then fast glycolytic fibers.

Invertebrate striated muscle differs from vertebrate striated muscle in fiber size and innervation pattern. It has relatively few, but large, fibers that are innervated by a small number of motoneurons;

TABLE 9–4

Types of phasic vertebrate striated muscle fiber and some of their important metabolic and biochemical properties. *(Modified from Goldspink 1977.)*			
	Fast Phasic Glycolytic	**Fast Phasic Oxidative**	**Slow Phasic Oxidative**
Structure			
Mitochondrial content	Low	High	Intermediate
Z line	Narrow	Wide	Intermediate
Neuromuscular junction	Large, complex	Small, simple	Intermediate
Enzymes			
Oxidative activity	Low	High	Intermediate/high
Glycolytic activity	High	Low	Intermediate
Myofibrillar ATPase	High	High	Low
Glycogen content	Intermediate	High	Low
Myoglobin content	Low	High	High
Mechanical			
Contraction velocity	Fast	Fast/intermediate	Slow
Fatigue time	Very short	Fairly long	Long
Efficiency	Fairly high	?	High

FIGURE 9–40 A human muscle illustrates the presence of different types of muscle fibers in a muscle. The different staining appearance of three fiber types is accomplished by special histochemical staining of ATPase. *(Courtesy of P. C. Withers and K. Cole.)*

one slow innervation; or two fast; or one inhibitory and one fast; or two slow; or two slow and one fast. One motoneuron may stimulate two different muscles, e.g., the opener and stretcher muscles of the crayfish claw share a single excitatory motoneuron; which muscle responds to nerve stimulation can depend on the frequency of stimulation. The closer, bender, and extensor muscles of the crab claw have one slow, one fast, and one inhibitory axon.

Electric Organs.   Electric organs are muscle (or nerve) cells that are specialized for producing external electric fields. About eight different groups of fish have evolved electric organs (Table 9–5).

Strongly-electric fish can produce an external electric field of sufficient current and voltage to stun prey or deter predators, e.g., electric and torpedo rays can produce up to 60 V and 1 kW power, the electric catfish more than 300 V, the electric eel over 500 V. Strongly-electric fish tend to have simple, discrete monophasic impulses of high-power output (Figure 9–41A). **Weakly-electric** fish use their electric organ as an electrosensory system, for communication, and to locate objects. They have a lower voltage, current, and power discharge, which is either a "pulse type" with a rapid (5 to 120 msec) series of short pulses (0.5 to 2 msec duration) or a "wave type" of a continuous low frequency (Figure 9–41B), a constant high frequency (Figure 9–41C), or a variable frequency (Figure 9–41D).

The **electrocytes** are the cells of the electric organ that generate the current/voltage discharge. The electrocytes are modified muscle cells in all electric

an entire muscle may be innervated by as few as two motoneurons. The motoneurons are either excitatory or inhibitory (they are only excitatory in vertebrate muscle). There is considerable variation between muscles, even for different muscle cells in the same muscle in the innervation pattern. For example, there may be one inhibitory, one fast and

TABLE 9–5

Types of electric fish, their distribution, and the nature of their electric discharge. *(Modified from Bennett 1971.)*

Common Name	Family	Distribution	Electric Organ Discharge
Skates	Rajidae	Marine	Weak pulse
Mormyrids	Mormyridae	Freshwater	Weak pulse
Gymnarchus	Gymnarchidae	Freshwater	Weak wave
Gymnotid eels	Gymnotidae, Sternopygidae, Rhamphichthyidae, Apteronotidae[1]	Freshwater	Weak pulse and wave
Stargazers	Uranoscopidae	Marine	Strong (5 V) pulse
Electric rays	Torpedinidae	Marine	Strong (60 V) pulse
Electric catfish	Malapteruridae	Freshwater	Strong (300 V) pulse
Electric eel	Electrophoridae	Freshwater	Strong (>500 V) pulse

[1] Neurogenic electric organ.

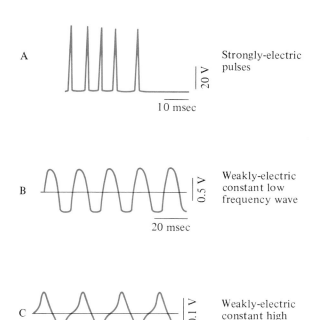

A   Strongly-electric pulses

20 V

10 msec

B   Weakly-electric constant low frequency wave

0.5 V

20 msec

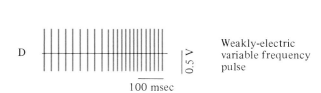

C   Weakly-electric constant high frequency wave

0.1 V

1 msec

D   Weakly-electric variable frequency pulse

0.5 V

100 msec

FIGURE 9–41 Pattern of electrical discharge by the electric organ of the following: **(A)** Pulses of the strongly-electric catfish *Malapterurus*. **(B)** Low-frequency, wave-type continuous discharge of a weakly-electric gymnotid, *Sternopygus*. **(C)** Constant, high-frequency wave discharge of the gymnotid, *Sternarchus*. **(D)** Variable frequency pulses of the gymnotid *Gymnotus*. *(Modified from Bennett 1969.)*

fish except apteronotids (sternarchids), which have electric organs formed by neurons. Myogenic electrocytes are often flattened, disk-shaped cells although some cells are cup shaped or have other complex shapes. Electrocytes produce electric fields using the same principles of bioelectricity already described for nerve and muscle cells, i.e., ion concentration gradients and selective ion permeability. A resting electrocyte has no external current flow (I_{ext}) or internal current flow (I_{int}) because the resting membrane potential of opposite sides of the electrocyte are equal but opposite in polarity (Figure 9–42). There is a constant, but low-current flow across each cell membrane (due mainly to K^+). The nerve innervating the electrocyte synapse depolarizes the adjacent cell membrane but not the opposite cell membrane. This creates an external electrical gradient and external current flow (I_{ext}) as well as an intracellular current flow (I_{int}). Some electrocytes have a more complex generation of their electric field, including a role of the noninnervated membrane and cell stalks. The high voltage and current discharge of strongly-electric organs is not due to an extraordinary membrane potential change, but to a combination of (1) the anatomical arrangement and synchronous activity of hundreds or even thousands of electrocytes to maximize the external field, (2) a low-electrical membrane resistance of electrocytes, and (3) accessory structures that channel external current flow.

The electric organs of the three marine groups of strongly-electric fish, electric rays (rajids), torpedo rays (torpedinids), and stargazers (*Astroscopus*; Figure 9–43) use the simple mechanism described above for electric organ discharge. The electric organ is densely innervated by numerous branches of the oculomotor nerve. The electrocytes are large, flattened cells horizontally arranged in about 150 to 1000 layers of cells that are stacked to form long columns. The dorsal, innervated membrane of the electrocyte depolarizes during organ discharge, but

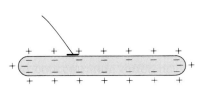

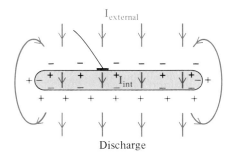

$I_{external}$

I_{int}

Rest

Discharge

FIGURE 9–42 Representation of a simple type of electrocyte from an electric discharge organ showing the pattern of membrane potential at rest (left) and during discharge (right). During discharge, there is an intracellular current flow (I_{int}) and an extracellular current flow (I_{ext}).

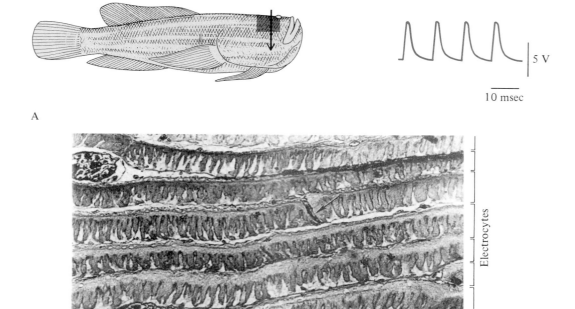

Astroscopus   Electric organ

A

Electrocytes

0.1 mm

B

FIGURE 9–43 **(A)** Location of the electric organ and an electromicrograph of the stacked electrocytes of the electric organ for the strongly-electric stargazer *Astroscopus*. **(B)** The smooth dorsal membrane (top) is innervated; two innervating nerve bundles can be seen towards the left. Arrow indicates a probable junction between adjacent electrocytes. *(Modified from Bennett 1969.)*

the ventral membrane is not innervated and does not depolarize.

The strongly-electric silurid catfish *Malapterurus* has two electric organs, each containing millions of electrocytes. Each electrocyte is a flat disk with a central stalk that is innervated by a single giant neuron and axon. The stalk side of the electrocyte faces posteriorly. The electric organ discharge is a brief, 1 to 2 msec, high-voltage (up to 350 V) pulse. The depolarization of the posterior stalk face of the electrocytes is responsible for an initial small head positive potential, but the nonstalk (anterior) side of the electrocyte rapidly depolarizes and creates a strong negative anterior discharge.

The freshwater fish of tropical South America include the electric eel, the best-known strongly-electric fish, and a wide variety of weakly-electric fish. The electric eel *Electrophorus* emits monophasic pulses of either low amplitude (10 V) for electrosensory detection or high amplitude (100 to 500 V, depending on the length of the eel) for offensive and defensive actions. *Gymnotus* is a weakly-electric freshwater fish with a pulse discharge of about 0.5 V amplitude and 1 msec duration at about 50 sec^{-1}.

The apteronotid (sternarchid) fishes, a group of about nine genera including *Apteronotus*, are also freshwater weakly-electric fish. They have the highest frequencies of diphasic wave discharge of all electric fish (700 to 1700 sec^{-1}). Their electrocytes are spinal neurons rather than modified muscle cells (see Supplement 9–3, pages 446–447). The South American weakly-electric rhamphichthyid fish are poorly known. The electric organ of *Gymnorhamphicthys* resembles that of sternopygids. It emits pulses at a fairly constant rate of 10 to 15 sec^{-1}.

The weakly-electric African fish *Gymnarchus* is closely related to the more numerous mormyrid fish but superficially resembles the South American gymnotids. Its electric organ discharge is a wave about 1 msec duration and 1 V amplitude; the discharge is head positive during pulses and head negative between pulses. The mormyrids are weakly-electric fish of tropical Africa. Their electrocytes discharge with irregular pulses (1 to 2 V, 0.5 to 1 msec; Figure 9–44A). The organ consists of four columns of electrocytes, each containing 100 to 200 cells. The electrocytes often have complex posterior stalks that fuse into a lower number of stalks before

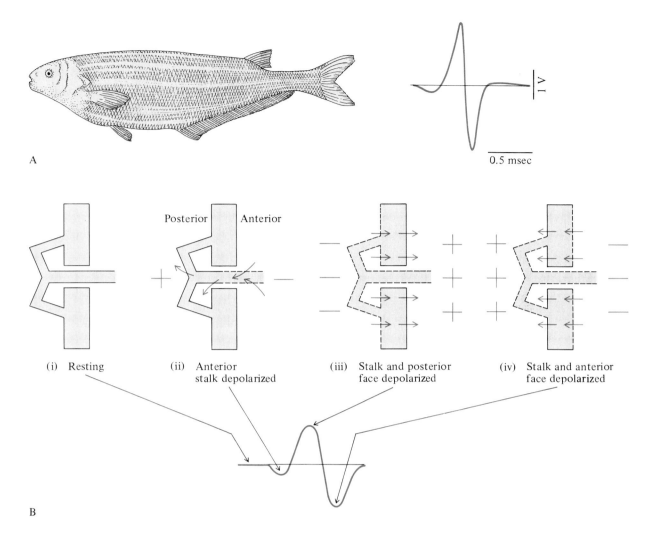

FIGURE 9–44 (**A**) The weakly-electric mormyrid fish *Hyperopisus* has a triphasic electric organ discharge. (**B**) The electric organ discharge has a triphasic discharge pattern due to a complex pattern of depolarization by the complicated, penetrating stalk and the anterior and posterior membranes of the electrocytes. *(Modified from Bennett 1969.)*

being innervated. In some species, the stalks pass anteriorly through holes in the electrocytes before being innervated. The form of the electric organ discharge depends on the extent of stalk penetration, varying from biphasic (no penetrating stalks) to triphasic (many penetrating stalks). In some species with many penetrating stalks, the head positive phase of electric discharge is caused by activity passing through the stalk from the anterior to the posterior face of the electrocyte (Figure 9–44B). Head positivity follows when the impulse from the stalk exits the posterior face; head negativity is later produced by excitation of the anterior face. Stalked electrocytes have independently evolved in South American and African weakly-electric fish.

The capacity of strongly-electric fish to emit high-voltage potentials depends on their ability to synchronously discharge many hundreds or even thousands of electrocytes. The control of electrocytes by the central nervous system involves two neurons in the first spinal segment, each of which controls all of the electrocytes on one side of the body. The two neurons are tightly electrically coupled to synchronize their outputs. The weakly-electric gymnotid and mormyrid fish have more complex central control because electric discharge is more continuous and requires a pacemaker to determine the characteristic frequency. In gymnotids, there are 30 to 200 pacemaker cells in the midline of the medulla that activate a relay center of about 50 neurons, which in turn controls hundreds or thousands of spinal neurons to the electric organs. The pacemaker cells have an inherent rhythm of depolarization to threshold, firing, hyperpolarization, and then depolarization to threshold. In contrast, the relay neurons have stable resting membrane potentials and respond only to input from the pacemaker cells. The mormyrid fish have an even more complex neural control of their electric organ. Bilateral nuclei in the midbrain contain the pacemaker cells; one pacemaker is dominant, and the subordinate pacemaker nucleus is entrained to the rhythm of the dominant one. These pacemaker nuclei relay to a midline medullary nucleus thence to the spinal nerves. Mormyrids also have a connection between the sensory processing part of the brain and the pacemaker command nucleus that allows the sensory system to be prepared for when afferent information will be provided in response to pulses from the electric organ. The discharges can vary in frequency (audibilized pulses from a resting mormyrid sound like a geiger counter, with irregular electric organ discharges). Gymnotids lack this neural connection and have a constant wave discharge of the electric organ.

A further complication in ensuring synchronization of the numerous electrocytes is their variable distance from the medulla and spinal cord centers. Conduction time should be quicker to the more anterior electrocytes, hence they would be expected to discharge before more posterior electrocytes. This potential desynchronization can be minimized by either having a more tortuous pathway of axons to the anterior electrocytes or a lower conduction velocity to anterior electrocytes.

Summary

The cell cytoskeleton and mechanisms for cell movement depend on three general types of structural protein filaments: microtubules, microfilaments, and intermediate filaments. Microtubules are tubular fibers consisting of 13 protofilaments of α and β tubulin subunits. Microfilaments (actin and myosin) are helically twisted strands of subunits. Intermediate filaments (keratin, neurofilaments, desmin, vimentin) are complex arrangements of α-helical coiled protein subunits.

Microtubular movement can be due to tubular synthesis or breakdown, or enzymatic "motors" that move membrane particles along microtubules, or are responsible for the sliding of adjacent microtubules. Axoneme dynein is a "motor" ATPase responsible for the bending of cilia and flagella. These cell organelles beat with undulating waves (flagella) or a power stroke and recovery stroke (cilia). The sliding filament model of ciliary and flagella beating is supported by microscopical and biochemical evidence.

Microfilament movement by actin and myosin is best understood for the sarcomeres of muscle cells, but many other types of cell movement involve cytoplasmic actin and myosin, e.g., cleavage during cell division, amoeboid movement. F-actin is a filamentous polymer of globular G-actin subunits. Myosin-1 is a rod-shaped molecule of two helically twisted protein subunits, each with a globular head region that has ATPase activity and binds to actin. Myosin molecules are often aggregated to form thick myosin microfilaments (myosin-2). Microfilament movement in striated muscle is a sliding of adjacent actin and myosin molecules by the repeated formation of cross-bridges between the myosin heads and the G-actin binding sites. The myosin head binds to the actin binding site and undergoes a conformational change and rotates, thereby sliding the actin past the myosin. Detachment of the myosin head and ratcheting of the myosin head to its original conformation requires ATP.

Control of actin-myosin interaction generally depends on the presence of other proteins attached to actin (tropomyosin, troponin, calmodulin, twitchin) and the regulation of the free intracellular Ca^{2+} concentration. In muscle, Ca^{2+} activates the actin-myosin interaction by one (or both) of two general mechanisms. In actin-activated systems, the Ca^{2+} binds to troponin; this induces a conformational change in the tropomyosin and uncovers the G-actin binding sites, allowing the myosin heads to bind. In myosin-activated systems, the Ca^{2+} regulates the ATPase activity of the myosin by the mediation of a small regulatory protein bound to the myosin.

Excitation-contraction coupling of striated muscle primarily involves the t-tubules and sarcoplasmic reticulum (SR). The SR sequesters Ca^{2+} and normally maintains a low cytoplasmic Ca^{2+} concentration ($<10^{-7}$ M). Electrotonic depolarization of the t-tubules induces Ca^{2+} release from the SR. There are several hypotheses for the coupling of t-tubular depolarization and Ca^{2+} release, including Ca^{2+}-induced opening of SR Ca^{2+} channels. In smooth muscle, opening of the SR Ca^{2+} channels is due to inositol 1,4,5 triphosphate (IP$_3$), rather than being Ca^{2+} induced.

A muscle maintains a constant length during an isometric contraction but shortens against a constant force in an isotonic contraction. The mechanical properties of muscle cells reflect their viscous resistance and also the elastic properties of their cellular and subcellular materials. The series elastic element is the sum of all elastic elements in series with the contractile units of the muscle cell. The parallel elastic element is the total effect of all elastic materials in parallel with the contractile units. Different muscles vary greatly in mechanical properties; some are very stiff and highly inelastic, e.g., insect flight muscle. Others are very stretchy, with a marked capacity for shape change, e.g., smooth muscle.

Single muscle twitches can summate if successive stimuli occur more frequently than a minimal inter-stimulus duration. Repetitive summation results in incomplete tetany or complete tetany if the stimuli occur so frequently that the muscle does not noticeably relax between stimuli. The maximum tension generated by tetanized muscles is generally 10 to 30 N cm^{-2}. Some muscles, such as bivalve mollusk catch muscle, can generate much higher peak tensions (up to 100 N cm^{-2}) and maintain tension at a low metabolic cost. There appears to be either a reduced cycle time for actin-myosin cross-bridges, or a complete latching of cross-bridges during "catch."

The velocity of shortening by a muscle is inversely related to the load. The power expenditure, calculated as force × velocity, is zero at no load (maximal velocity) and maximal load (no velocity), and is maximal at an intermediate velocity and load. There is a similar-shaped relationship between mechanical work done and velocity. The maximum contractile mechanical efficiency of muscle is high, at about 70 to 80% of the energy expended for contraction, but the overall mechanical efficiency for a muscle contraction is 20 to 30% of the total energy expenditure by a muscle.

Electric fish generate substantial electrical discharges for either intraspecific communication or electrolocation (weakly-electric fish) or for stunning prey and repelling predators (strongly-electric fish). The electric organ discharge is either a variable rate of pulses (strongly-electric and some weakly-electric fish) or a constant frequency wave discharge (some weakly-electric fish). The electrocyte cells of electric organs are generally modified muscle cells (but are modified neurons in apteronotid fish). The electric organ discharge is the summed depolarization of many (even thousands) of electrocytes arranged in series. Each electrocyte has an action potential-like discharge (about 90 mV), and the summed discharge can be a few volts or many volts. The discharge can be monophasic, biphasic, or triphasic, depending on the arrangement of electrocytes in the electric organ and their pattern of innervation and depolarization. Many electric fish have complex central nervous system involvement with the generation of electric organ discharges (pacemaker command centers) and the coordination of sensory reception with electric organ discharge (sensory gating).

Supplement 9–1

Dynamics of Actin

· ·

Actin is a ubiquitous intercellular microfilament that is involved in cell movement. Cytoplasmic actin appears to be in a dynamic state. Polymerization of G-actin into F-actin filaments occurs spontaneously at nucleation sites. Polymerization can be inhibited by proteins, called profilin. Self polymerization of G-actin forms filamentous strands of F-actin, each end of which may be in dynamic equilibrium between G- and F-actin. Different ends of F-actin may have differing rates of polymerization and depolymerization. For example, the so-called "barbed end" of F-actin may have a higher rate of depolymerization than polymerization; in this circumstance, the G-actin subunits "treadmill" from the pointed end, with a higher rate of polymerization than depolymerization, to the "barbed end," with a higher rate of depolymerization than polymerization.

F-actin can be organized into bundles or 3-dimensional networks by the process of gelation. A variety of proteins can cause gelation; some require Ca^{2+} ions (e.g., filamin, spectrin, fascin, fimbrin, nonmuscle α-actinin), whereas others do not (e.g., muscle α-actinin, actinogelin, vinculin, villin). Filamins are strand-shaped, actin-binding proteins of smooth muscle and other cells (e.g., blood cells: macrophages, neutrophils, platelets) that induce polymerization and cross-linking of actin. Spectrin is a similar protein of red blood cells; α-actinins are rod-shaped actin cross-linking proteins. In striated muscle, α-actinin is located in the Z bands of sarcomeres where the actins are interconnected. In nonmuscle cells, α-actinin is involved in binding actin to membrane surfaces possibly by cross-linking and spacing the actin filaments. Vinculin may also be involved with binding actin to membranes. Gelated actin can be depolymerized by "depolymerizing protein" or severed into smaller microfilaments by solation proteins (e.g., gelsolin, α-actinin, fragmin, villin). Actin fragments can be stabilized by capping proteins that bind to the end and prevent further polymerizing ("end-capping proteins"). Some properties of these various proteins involved with the dynamics of actin structure are summarized in the following table.

Actin-binding proteins (excluding those from muscle cells). *(See Weeds 1982.)*

Type	Ca²⁺ Sensitivity	Type	Ca²⁺ Sensitivity
Gelation Proteins		**Severing/Capping Proteins**	
Filamin	No	α-actinin	No
Spectrin	No	Gelsolin	Yes
		Villin	Yes
Gelation and Bundling Proteins		Fragmin	Yes
α-actinin	Yes/No	Capping Protein	No
Actinogelin	Yes	Depolymerizing Protein	No
Vinculin	Yes/No		
		G-Actin Stabilizing Proteins	
Bundling Proteins		Profilin	No
Fascin	No		
Fimbrin	No		

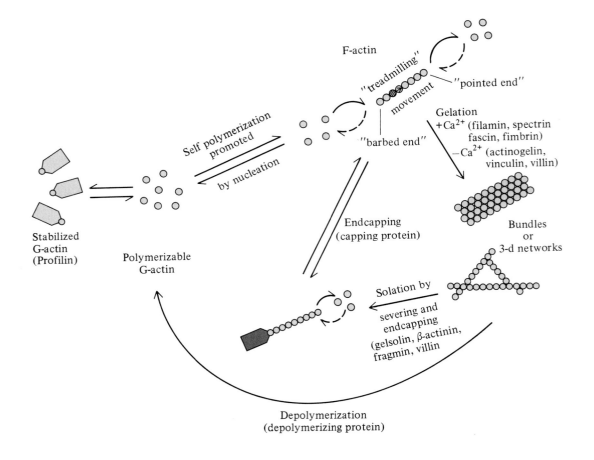

Dynamics of actin polymerization and depolymerization.

Supplement 9–2

Muscle Force, Velocity, Power, and Work Done

· ·

An isotonically contracting muscle shortens against a load. The curvilinear relationship between load (F) and velocity (V) is closely fitted by a hyperbolic equation (developed by A. V. Hill),

$$(F + a)(V + b) = (F_0 + a)b$$

where F_0 is the maximum force developed by an isometric contraction, and a and b are constants. Hill's equation is a good fit for small loads but not for high loads because V is not constant throughout the muscle twitch. The constants a (with units of force, e.g., Newtons) and b (with units of velocity, i.e., m sec^{-1}) can be obtained by linearizing the relationship.

$$F = \frac{b(F_0 - F)}{V} - a$$

Graphing $(F_0 - F)/V$ as a function of F yields a straight line whose slope is b and whose intercept is $-a$. The values of a and b vary for different muscles; a is generally about 0.15 to $0.25F_0$ for vertebrate muscles.

A muscle will lengthen if the load applied to it exceeds its maximum capacity (F_0). Muscles that are mechanically stretched during a contraction will develop a tension greater than the isometric tension (see the extension of the force–velocity curve for frog sartorius muscle) and work is done on the muscle during its extension. The Hill equation is not a good fit for the lengthening curve.

The power (P) output of a muscle is equal to the force generated (F) $\times$ velocity (V); it can be readily calculated from the $V - F$ diagram. The load for maximum power output (F_{mp}) can be calculated from Hill's hyperbolic equation as follows.

$$F_{mp} = (a^2 + aF_0)^{1/2} - a$$

The value of F_{mp} is $0.29\ F_0$ if $a = 0.2F_0$.

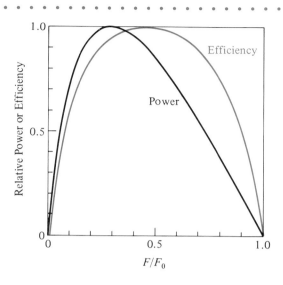

Relative power and efficiency of muscle contraction as a function of relative force of contraction (F/F_0).

The mechanical work (W) done by a muscle (or the work done on a muscle) is equal to the integral of power output over time.

$$W = \int_{t_0}^{t} \text{Power} . dt = P\Delta t$$

Alternatively, the work done can be calculated as the integral of force through a displacement, or force $\times$ velocity $\times$ time.

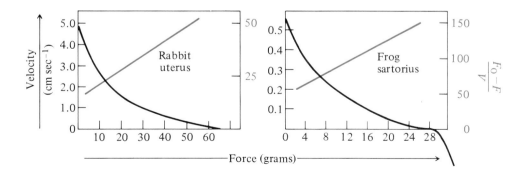

There is a hyperbolic relationship between the velocity *(V)* and force of muscle contraction *(F)* with a maximum contractile force *(F_0)* at zero velocity. There is a linear relationship between the relative force $(F_0 - F)/V$ and force. *(Modified from Csapo 1970.)*

$$W = \int_{s_0}^{s} \text{Force} . ds = F\Delta s$$

Hill's equation for the relationship between force and velocity can be rearranged to calculate the work done.

$$W = bFt(F_0 - F)/(F + a)$$

The maximum work can be shown to be done when

$$F = \sqrt{a(F_0 + a)} - a$$

Hence, mechanical work is maximal when $F \approx 0.31\, F_0$, since $a \approx 0.25\, F_0$.

Muscle efficiency, defined as 100 mechanical work/total energy released, increases from 0% at $F/F_0 = 0$ and 1, to a maximum at intermediate F. This efficiency curve has a very broad top, from about $F = 0.24$ to $0.7\, F_0$ for >90% of maximal efficiency. The maximum efficiency is about 25 to 30% for many muscles. (*See White 1977.*)

Supplement 9–3

Neurogenic Weakly-Electric Fish

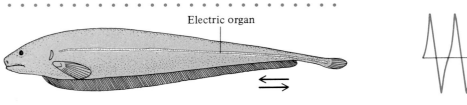

Location of the neurogenic electric organ and its discharge pattern for the weakly-electric fish *Apteronotus*. (*Modified from Bennett 1969.*)

The apteronotid (sternarchid) weakly-electric freshwater fishes of South America have neurogenic, not myogenic, electric organs and a continuous, wave-type electric organ discharge, e.g., *Apteronotus*. The discharge is a diphasic wave (initially head positive) with an exceptionally high frequency (700 to 1700 sec^{-1}, depending on the temperature and the species) for weakly-electric fish.

The electrocytes are highly modified spinal nerves. Spinal nerves enter the electric organ, run anteriorly for a few segments, then turn and run posteriorly to about

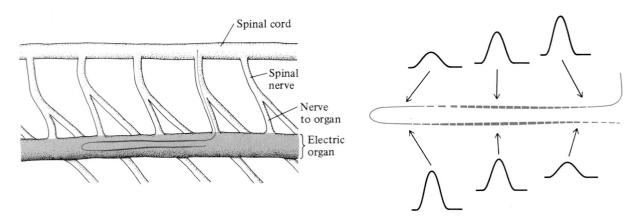

The electrocyte neurons pass from the spinal cord to the electric organ via spinal nerves and form a U-shaped terminus that produces the electric organ discharge (only one such electrocyte is shown). (*Modified from Bennett 1969.*)

the level at which they entered the electric organ. The myelinated spinal nerves are of normal size where they enter the electric organ (10 to 20 μ dia) but dilate to as much as 100 μ at the center of the anterior and posterior portions within the electric organ, then taper gradually until they terminate. Intracellular electrical recordings from the modified nerve fiber indicate a large amplitude action potential where the axon enters the electric organ but a smaller amplitude and delayed action potential as it travels anteriorly along the axon. The central dilated portion of the axon is presumably inexcitable and the depolarization spreads electrotonically. This part of the discharge cycle produces the head positive potential. The posteriorly running portion of the axon also conducts an impulse anteriorly; the magnitude is greater for the more anterior portions. Its impulse is delayed compared to that of the anteriorly running portion of the axon and is presumably caused by excitation from the anterior running segment of the axon. It generates the second part of the electric discharge, the head negative potential. Current runs posteriorly from the large anterior depolarization to the smaller posterior depolarization. (*Modified from Bennett 1971.*)

Recommended Reading

Alberts, B., et al. 1989. *Molecular biology of the cell.* New York: Garland Publishers.

Allen, R. D., et al. 1980. Cytoplasmic transport: Moving ultrastructural elements common to many cell types revealed by video-enhanced microscopy. *Cold Spring Harbor Symp. Quant. Biol.* 46:85–7.

Baba, S. A., and Y. Hiramoto. 1970. A quantitative analysis of ciliary movement by means of high-speed micro-cinematography. *J. exp. Biol.* 52:645–90.

Bagshaw, C. R. 1982. *Muscle contraction.* London: Chapman & Hall.

Baserga, R. 1985. *The biology of cell reproduction.* Cambridge: Harvard University Press.

Bennett, M. V. L. 1971. Electric organs. In *Fish physiology.* Vol. 5, *Sensory systems and electric organs,* edited by W. S. Hoar and D. J. Randall, 347–491. New York: Academic Press.

Bereiter-Hahn, J., and R. Strohmeier. 1986. Biophysical aspects of motive force generation in tissue culture cells and protozoa. In *Nature and functions of cytoskeletal proteins in motility and transport,* edited by K. E. Wohlfarth-Botterman, 1–16. Stuttgart: Gustav Fischer Verlag.

Berg, H. C., and R. A. Anderson. 1977. Bacteria swim by rotating their flagellar filaments. *Nature* 245:380–82.

Carlson, F. D., and D. R. Wilkie. 1974. *Muscle physiology.* Englewood Cliffs: Prentice-Hall.

Goldman, R. D., et al. 1986. Intermediate filament networks: Organization and possible functions of a diverse group of cytoskeletal elements. *J. Cell. Sci. Suppl.* 5:69–97.

McIntosh, J. R. 1984. Mechanics of mitosis. *Trends in Biochem. Sci.* 9:195–98.

Mitchison, T. J. 1986. The role of microtubule polarity in the movement of kinesin and kinetochores. *J. Cell Sci. Suppl.* 5:121–28.

Naitoh, Y., and R. Eckert. 1974. The control of ciliary activity in protozoa. In *Cilia and flagella,* edited by M. A. Sleigh, 305–52. London: Academic Press.

Nicklaus, R. B. 1988. The forces that move chromosomes in mitosis. *Ann. Rev. Biophys. Biophys. Chem.* 17:431–50.

Porter, K. R. 1976. Introduction: Motility in cells. In *Cell motility.* Book A, Vol. 3, *Motility, muscle and non-muscle cells,* edited by R. Goldman, T. Pollard, and J. Rosenbaum, 1–28. Cold Spring Harbor Laboratory: Cold Spring Harbor Conf. Cell Prolif.

Rios, E., and G. Pizarro. 1988. Voltage sensors and calcium channels of excitation-contraction coupling. *NIPS* 3:223–27.

Scholey, J. M. 1990. Multiple microtubule motors. *Nature* 343:118–20.

Sleigh, M. A. 1974. *Cilia and flagella.* New York: Academic Press.

Stossel, T. P. 1990. How cells crawl. *Amer. Zool.* 78:408–23.

Vallee, R. 1990. Dynein and the kinetochore. *Nature* 345:206–207.

Watanabe, S., and D. J. Hartshorne. 1990. Paramyosin and the catch mechanism. *Comp. Biochem. Physiol.* 96B:639–46.

Weeds, A. 1982. Actin-binding proteins—regulators of cell architecture and motility. *Nature* 296:811–16.

Wilkie, D. R. 1956. The mechanical properties of muscle. *Brit. Med. Bull.* 12:177–82.

Chapter 10

Support and Locomotion

· ·

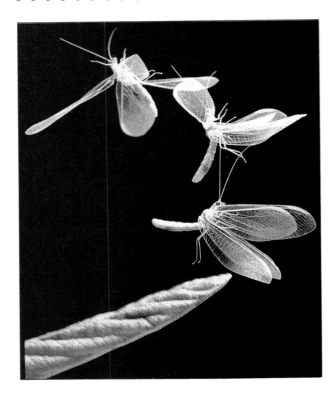

Flight acrobatics of a green lace wing, during take-off. *(S. Dalton/ Photo Researchers, Inc.)*

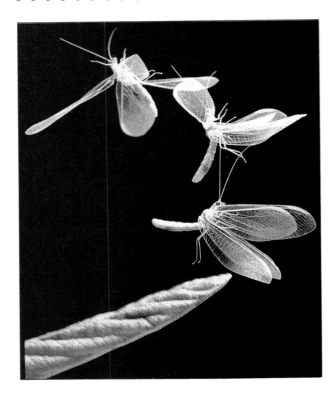

448

The **shape** of animals is extremely important for many reasons. For example, the shape of an animal reflects morphological adaptations for locomotion and feeding. To maintain a constant body shape, the body tissues must be supported against external forces that would deform them and also against internal muscular forces. Terrestrial and aerial animals must also resist the force of gravity. Aquatic animals must resist water currents, which can be very powerful, but gravity is not a major force in their environment. A **skeleton** provides structural support for the maintenance of an animal's body shape. Most animal skeletons are not just static support structures but are dynamic structures that move in response to muscle contractions. The musculoskeletal system of animals is responsible for both support of body shape and locomotor movement. **Locomotion**, or animal movement, is the way that animals use their muscles and skeletons to move.

There are three basic designs for skeletal systems. First, the skeleton may be a pressurized fluid compartment that acts as a rigid structure because of its high internal pressure; this is a hydrostatic skeleton, or **hydrostat** (Figure 10–1A). Alternatively, the skeleton may be a framework of solid elements (bone, chitin, calcareous, or silicous mate-

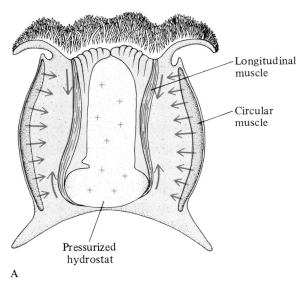

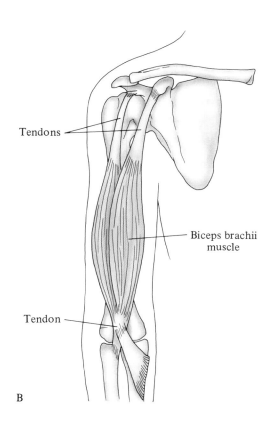

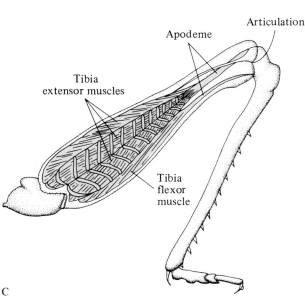

FIGURE 10–1 **(A)** The hydrostatic skeleton of a sea anemone consists of the gastrovascular cavity (filled with seawater), which is pressurized by contraction of circular and longitudinal muscles. **(B)** The endoskeleton and muscular arrangement in vertebrates typically has a muscle attached by tendons to two skeletal elements that articulate at a joint. **(C)** The exoskeleton and muscular arrangement in an arthropod consists of muscles attached to the interior of the exoskeletal elements by a chitinous apodeme. *(Modified from Tortora 1983; Chapman 1982.)*

rials). The framework skeleton can be an internal **endoskeleton** (Figure 10–1B) or an outer **exoskeleton** (Figure 10–1C).

Support

Aquatic animals do not have to support their body against a gravitational force because they generally are almost neutrally buoyant, but their tissues must resist deformation by water currents and other external forces. Hydrostatic skeletons are often adequate to accomplish this, but many aquatic animals have a solid skeleton for additional support or protection. Terrestrial animals must primarily resist the force of gravity and most have a solid skeleton, although some have hydrostats.

The mechanical and contractile properties of muscle and the mechanical properties of the various tissues that form the skeleton and connect it with the muscles determine the strength and functions of the musculoskeletal system. We have already described the contractile and mechanical properties of muscle (Chapter 9). Now we need to examine the mechanical properties of skeletal and other biological materials and see how the skeleton is a structure that provides mechanical support.

Material Properties

A skeletal structure experiences tension, compression, bending, twisting, and shearing forces. The mechanical properties of skeletal materials determine the ability of skeletal structures to resist these forces. Density, elastic modulus, elasticity, viscoelasticity, and strength are some of the most important material properties.

The **density** of biological structural materials (ρ = kg m^{-3} or grams cm^{-3}) determines the weight of a structure that has a certain volume. The **specific gravity** is the ratio of the material density to the density of water (1000 kg m^{-3} at 4° C); it is nearly the same as the density. Body fluids generally have a density of about 1000; flexible biological materials have a density of about 1300 to 1500; and rigid, mineralized materials have a density of about 3000 (Table 10–1). Steel, concrete, and glass are high-density engineering materials that also are rigid. The weight of an animal's skeleton is of bioenergetic significance since many animals have to carry their skeleton around. Consequently, the skeletal material of mobile animals should be as light as possible. A hydrostatic skeleton that consists primarily of a body fluid space and some flexible wall materials is much lighter than a solid, rigid skeleton of the same volume. However, solid skeletons may be smaller

TABLE 10–1

Some important mechanical properties of biological materials and tissues. Density is the ratio of mass to volume. The modulus of elasticity is the ratio of stress/strain for the elastic portion of the stress-strain curve. The ultimate strength is the stress at which the material falls. Values are for tension, unless specified for compression (comp).

	Density[1] kg m^{-3}	Elastic Modulus N m^{-2}	Ultimate Strength N m^{-2}
Engineering Materials			
Steel	7800	2.1 10^{11}	1.5 10^{9}
Concrete	2300	1.7 10^{10}	4 10^{6}
Glass	2500	7 10^{10}	1 10^{8}
Rubber	1100	7 10^{6}	7 10^{6}
Biological Materials			
Rubbers			
Elastin	1300	5.9 10^{5}	—
Resilin	1300	2.0 10^{6}	2.9 10^{6}
Proteins			
Collagen	1400	1.2 10^{9}	5.5 10^{8}
Keratin	1300	2.9 10^{8}	2.0 10^{8}
Silk	1300	9.8 10^{9}	5.9 10^{8}
Polysaccharides			
Cellulose	1600	4.9 10^{10}	1.1 10^{9}
Chitin	1600	4.4 10^{10}	5.7 10^{8}
Minerals			
Apatite	3200	1.8 10^{11}	—
Calcite	2700	1.4 10^{11}	—
Biological Tissues			
Verbetrates			
Cartilage	1100	1.3 10^{7}	5.9 10^{5}
Tendon	1300	1.9 10^{8}	9.8 10^{7}
Bone	2000	1.2 10^{10}	1.2 10^{8}
Bone^{comp}	2000	6.5 10^{9}	1.1 10^{8}
Arthropods			
Locust cuticle	1200	9.4 10^{9}	9.4 10^{7}
Crab exoskeleton	1900	1.3 10^{10}	3.4 10^{7}
Mollusk			
Bivalve shell	2700	4.4 10^{10}	3.8 10^{7}
Gastropod shell	2700	6.8 10^{10}	3.7 10^{7}
Echinoderm			
Spine	2000	6.7 10^{9}	8.2 10^{7}
Coral			
Branched^{comp}	—	2.2 10^{10}	4.8 10^{10}
Massive^{comp}	—	1.1 10^{10}	2.2 10^{10}

[1] 1000 kg m^{-3} = 1 gram cm^{-3}.

than hydrostat skeletons and so are not necessarily heavier than a larger-volume hydrostat.

For a material experiencing a tensile (stretching) force (Figure 10–2A), the **tensile stress** (σ) is the force per unit cross-section and the **tensile strain** (ϵ) is the amount of deformation or relative change in size

$$\sigma = F/A$$
$$\epsilon = \Delta l/l \qquad (10.1)$$

where F is the force, A is the cross-sectional area, l is the initial length, and Δl is the increase in length.

Stress has units of N m^{-2} (or Pa) and strain is dimensionless.

There generally is a linear relationship between stress and strain, at least for nonbiological materials and for many biological materials at small ϵ. For example, there is a linear relationship between stress and strain for a bone loaded in tension up to its yield value (at about 150 MPa and 0.005 strain), when it begins to rapidly deform (Figure 10–2B). The slope of the linear part of the relationship (Δstress/Δstrain) is the **modulus of elasticity** (E), or Young's modulus.

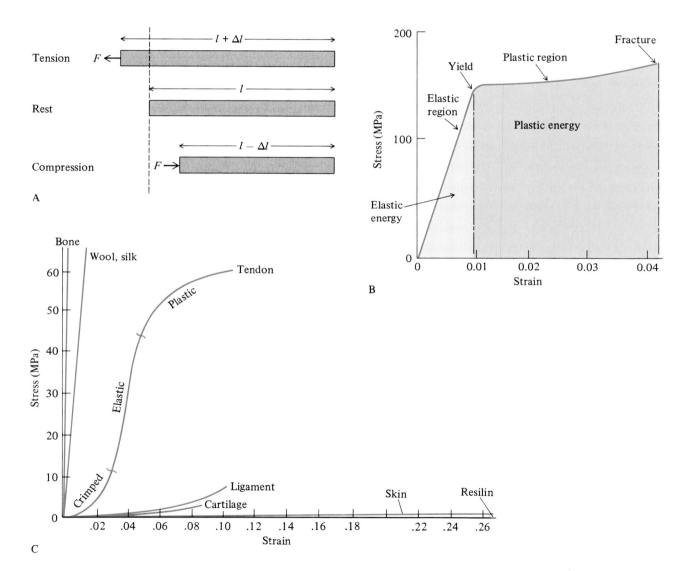

FIGURE 10–2 **(A)** Representation of tension and compression for a rod-shaped skeletal element. **(B)** Relationship between stress and strain for a bone showing the elastic and plastic regions of the stress-strain relationship and the yield and fracture points. **(C)** Relationship between stress and strain for biological tissues showing the considerable differences in their elastic modulus and the complex shape of the relationship for tendon. *(Modified from Currey 1984.)*

The linear part of the stress-strain relationship for a material is called the **elastic** region (Figure 10–2B). The area under the elastic region represents the elastic energy that is absorbed by the stressed bone. A material that releases all of the stored elastic energy when it relaxes is, in a biomechanical sense, elastic, i.e., the stress-strain relationship is the same for loading and unloading. Bone and many other biological materials are elastic. Forces that exceed the yield stress and yield strain considerably deform the bone; this is the **plastic** region. The area under the plastic region represents the absorbed plastic energy. After plastic deformation, the bone has a different unloading stress-strain curve than for loading, and not all of the energy absorbed during loading is released. Further stress will eventually break the bone at its breaking stress and breaking strain.

Biological tissues vary markedly in their relationship between stress and strain and elastic modulus (Figure 10–2C). Rubbery materials, such as elastin, abductin, and resilin, have a low-elastic modulus ($<10^7$ N m^{-2}) and are very stretchy compared to other proteins, such as collagen, silk, and keratin, which in turn have a higher elastic modulus ($>10^8$) and are more stretchy than structural carbohydrates (cellulose, chitin) and minerals (apatite, calcite; see Table 10–1, page 450).

For many biological materials, the initial stress-strain relationship is not linear, but is J shaped (e.g., tendon, ligament, skin; Figure 10–2C). This means that very small stresses initially deform the material considerably. For tendon, this initial nonlinear region might correspond to stretching kinks out of the collagen fibers of the tendon or stretch of the tendon matrix, not collagen; the subsequent linear stress-strain portion might correspond to stretch of the straight collagen fibers. There may not be any *in vivo* significance of the J-shaped foot of the stress/strain curve for collagen because the J-shaped region is only present for slow rates of strain. High rates of strain, such as occur *in vivo*, tend to make the stress-strain curve linear. Skin also has a markedly J-shaped stress-strain curve (although this is not apparent in Figure 10–2C because of the scales used). The nonlinear initial part of the curve reflects the initial stretching of elastin fibers (with low elastic modulus) before the collagen fibers are aligned with the stress. The steeper linear part of the relationship reflects stretch of the collagen fibers (with a higher elastic modulus).

The stress-strain relationship for biological materials generally is complicated by the viscous damping of strain; elastic modulus varies with the rate of strain. **Viscoelasticity** is a time dependence of the elastic modulus, which is called creep (or stress relaxation). The viscoelastic property of biological materials is a consequence of their containing large, complex polymeric molecules (e.g., proteins, polysaccharides). We have already examined a viscoelastic model for a muscle cell (see Figure 9–30, page 430). The effect of viscoelastic behavior is to reduce the rate of strain in response to a stress, although the ultimate strain is the same for a viscoelastic material and a normally elastic material. A viscoelastic material can be stiff when it is deformed rapidly and for a short duration, but less stiff when it is deformed slowly. The mesoglea of sea anemones provides a good example of viscoelasticity. The mesoglea is quite rigid at high-strain rates, and this enables it to withstand short-term environmental stresses, e.g., wave surge. In contrast, small, long-term stresses can produce over 300% strain, and this allows the sea anemone to inflate itself by low-pressure ciliary pumping.

The stress at which a material breaks, the **ultimate strength**, is much greater for some materials (e.g., apatite) than for others (e.g., elastin). The area under the stress-strain curve is roughly proportional to the amount of energy that the material absorbs before it breaks. In general, more "stretchy" materials have a lower ultimate tensile strength, i.e., they are broken by smaller stresses than are less "stretchy" materials.

Skeletons

Skeletons support the body against external forces and allow the forces developed by muscles to move specific parts of the body. Hydrostatic skeletons utilize the incompressibility of biological fluids (essentially water) or tissues (e.g., muscle cells), whereas solid skeletons consist of materials with a high-elastic modulus (e.g., bone and chitin).

A hydrostatic skeleton is a fluid-filled space surrounded by a flexible wall. The fluid is pressurized by the action of circular and/or longitudinal muscles. The hydrostat of sea anemones is the gastrovascular cavity filled with seawater. The hydrostat of many animals is a body cavity filled with body fluids. The body wall, which is the flexible boundary of the hydrostat, has multiple layers of fibrous connective tissue, usually collagen. The fiber wrapping stops the pressurized hydrostat from becoming spherical and prevents bulges (aneurysms) from forming in the wall (Wainwright 1982). The fibers of the hydrostat wall are typically arranged as a double helical array (Figure 10–3A). This allows the hydrostat to change length and bend without kinking. The angle between the longitudinal axis and the helically-wound fibers, the **fiber angle**, is 54°44′ for a cylinder of maximal volume.

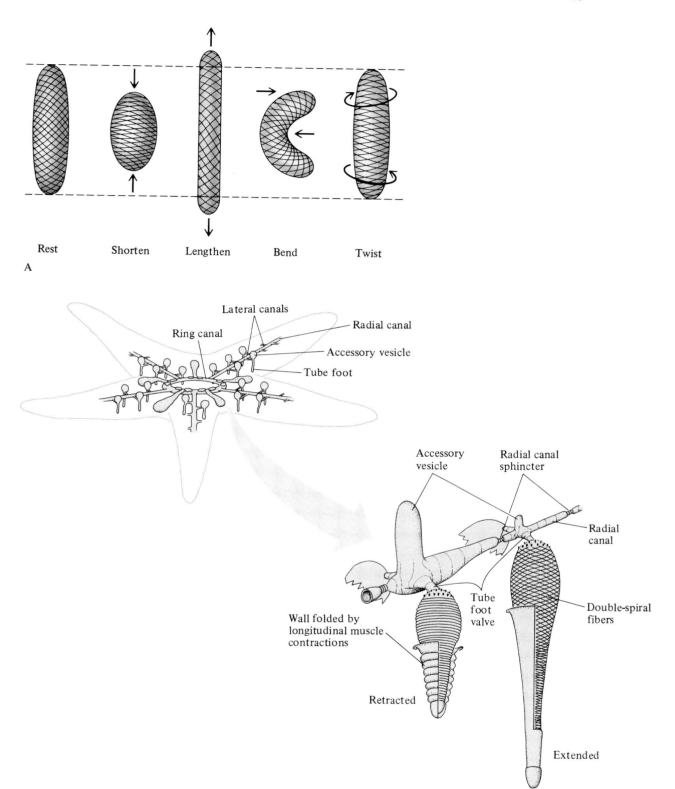

FIGURE 10–3 **(A)** A hydrostat that has a double-helical fiber wrapping can shorten, lengthen, twist, and bend without kinking. **(B)** The tube feet of the ophiuroid *Amphiura* are hydrostats. They retract (left) when longitudinal tube foot muscles contract and force fluid out of the tube foot into the elastic accessory vesicle and elongate (right) when fluid is expelled from the accessory vesicle into the tube foot by elastic recoil. *(From Wainwright 1982; Woodley 1980.)*

The tube feet of echinoderms are very simple hydrostats; they are essentially fibrous balloons that are expanded by fluid pressure of the water vascular system (Figure 10–3B). The tube foot has a fibrous wall with a double helical array of collagen fibers and an outer layer of longitudinal muscle. Contraction of the longitudinal muscle pressurizes the fluid in the tube foot. Fluid is forced into the corresponding segment of the radial canal if the tube foot valve is open. The segment is expanded and stores energy in its elastic walls. This energy reinflates the tube foot when its muscles relax and the valve is open.

Some animal hydrostats have a high-fiber angle. For example, the cuticular collagen of the nematode worm *Ascaris* has a fiber angle of about 75°; the collagen almost forms circular belts around the nematode. Contraction of longitudinal muscles on one side of the body shortens that side. This elevates the internal fluid pressure of the hydrostat, increases the diameter of the hydrostat, and slightly tenses the collagen fibers. When the muscle relaxes, the elastic energy stored in the collagen fibers returns the nematode to its resting shape. The nematode does not need circular muscles to regain its original shape; indeed, it does not have any circular muscles! Sharks and whales also utilize a hydrostatic skeleton during swimming, and fibers in their body walls have a high-fiber angle and function in a similar fashion. Some animals have a low-fiber angle. Squid mantle has a fiber angle of 23°, i.e., the fibers are almost longitudinal. Contraction of circular muscles decreases the mantle diameter and elongates the fiber array. When the muscles relax, the elastic recoil of the fiber array returns the mantle to its normal shape.

The pressurized space of some hydrostats is not a free fluid but consists of muscle cells. These **muscular hydrostats** can dramatically alter their shape by the contraction of longitudinal and circular muscles. Many soft appendages function as a muscular hydrostat. For example, the trunk of an elephant, the tongue of vertebrates, and the tentacles of squid are muscular hydrostats (Smith and Kier 1989).

Bryozoans, small colonial animals that live in rigid calcareous boxes, have both a hydrostatic and a rigid skeleton. The filter-feeding organ, the lophophore, is extended by contraction of muscles attached between the side of the calcareous box and a thin membrane forming the dorsal surface of the animal. Contraction of the muscles compresses the body fluid space causing the lophophore to be everted. Some bryozoans have a more solid dorsal covering and draw water into a compensation chamber when the lophophore is everted (Figure 10–4).

Solid skeletons consist of rigid materials with a high mechanical strength, e.g., apatite or chitin. The rigid material may be on the outside of the body, such as the chitinous exoskeleton of arthropods, or may be located internally, such as the bony endoskeleton of vertebrates. There are numerous advantages and disadvantages to both types of rigid skeletal designs but exoskeletons definitely resist bending better.

Bending and twisting are the most common ways that stiff skeletons break. We can best analyze the role of stiff skeletal elements by considering the mechanics of bending for a long, thin element, i.e., a beam (see Supplement 10–1, page 487). A rigid cylindrical beam that is supported at one end and is loaded at the other end by a force will sag at the

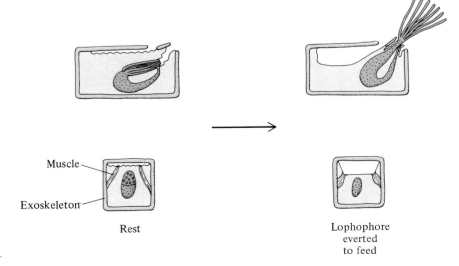

FIGURE 10–4 The skeleton of some bryozoans combines a pressurized hydrostat with a rigid, box-like external skeleton and a compensation chamber that allows the dorsal surface to be protected by exoskeleton. Contraction of the dorsolateral muscles pulls on the dorsal membrane, which compresses the body fluids and everts the lophophore, and also fills the compensation with seawater. *(From Currey 1970.)*

Muscle

Exoskeleton

Rest

Lophophore everted to feed

loaded end. The force stretches the top of the beam and compresses the bottom. The beam breaks if the force is excessive. It first fails where the stress and strain are greatest, i.e., at the very top or very bottom of the beam. Material near the center of the beam contributes little to its strength because it is not compressed or stretched very much. Material farther from the center of the beam contributes more mechanical strength. Consequently, the optimal design for many skeletal elements is a hollow tube rather than a solid rod. The actual structure of bones is a complex optimization to maximally resist the various forces that it experiences. A larger diameter cylinder has greater rigidity than a cylinder of smaller diameter. For example, a leg with an exoskeleton of radius $2r$ would be about $6 \times$ more rigid than one with a solid endoskeleton of radius r. The hollow leg is about $3 \times$ stronger than the solid leg.

Exoskeletons have mechanical advantages compared to endoskeletons, but other factors mitigate the disadvantages of endoskeletons. The wall of a hollow cylinder becomes thinner as its radius increases, if we assume a constant cross-sectional skeletal area, and this makes the cylinder more susceptible to compressive buckling. The tendency to buckle is also increased by any small scratches or cracks that cause local weakening of the exoskeleton. A thin exoskeleton is also more prone to puncture by sudden impact, whereas the outer soft tissues of animals with endoskeletons are better able to absorb impact stresses. Exoskeletons also limit the extent of body growth by enclosing the outside of the animal in a rigid case. Animals with exoskeletons must periodically molt the exoskeleton

to grow in size, and this temporarily reduces all of the advantages accruing from the exoskeleton, such as resistance to predator attack and mechanical strength. Most vertebrates are larger than most invertebrates, and so the rigidity and strength disadvantages of an endoskeleton appear to be outweighed by the other advantages over exoskeletons.

Shapes and Structure of Bones. There is an almost infinite variety in the shapes of bones because they are used in many different structural roles by different vertebrates. However, there are four general shapes of bones: (1) long thin bones, e.g., the limb bones; (2) compact-shaped bones that are about as long as they are wide, e.g., the carpal (wrist) and tarsal (ankle) bones; (3) flat thin bones, e.g., the scapula and skull bones; and (4) vertebrae.

Bone is a suitable material for skeletal elements involved with support and movement because it has high-tensile and compressive strengths. However, it is a heavily mineralized tissue (density about 2000) and so it is important to minimize the amount of bone (hence its weight) in structural elements without compromising the mechanical strength required for their particular role. The general mechanical principles that determine the shape and structure of a bone are well illustrated by the mechanical properties of the femur, a long limb bone of the vertebrate leg.

The femur is a long, thin bone that is expanded at each end for articulation at the hip and knee joints (Figure 10–5). Its shape reflects the nature of the forces that it experiences, and the surface features (e.g., spines, ridges, tubercles) are for attachment to muscles. A bone cannot be made lighter simply

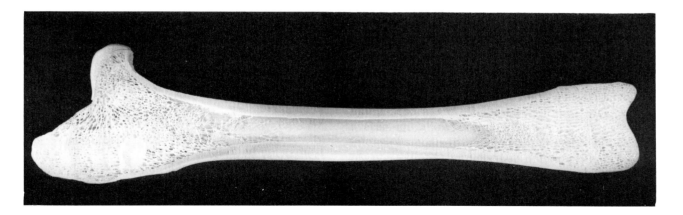

FIGURE 10–5 The femur of a western gray kangaroo illustrates the typical structure of this hollow bone; there is an outer layer of compact bone and an inner cancellous bone layer. The orientation of the cancellous bone lamellae reflects the directions of forces that the femur experiences. *(Photograph by P. C. Withers and J. O'Shea.)*

by making it smaller because this compromises its structural roles. However, the internal structure of a bone can be modified to maintain its structural integrity but minimize its weight. Beam theory (see Supplement 10–1, page 487) shows that a hollow bone has a greater mechanical strength than a solid bone of the same mass. Not surprisingly, the femur, like most bones, is hollow (Figure 10–5). The cavity is filled with bone marrow, which is either hemato-poietic tissue (forms blood cells) or is fat, or contains an air-filled extension of the air sacs (birds). Whether fatty bone marrow has a biomechanical role is not clear.

Long bones, such as the femur, will buckle under compression or bending if the wall is too thin relative to the diameter of the bone. It is complex to estimate the optimal ratio of diameter (D) to wall thickness (t), since this depends on many factors, including the nature of the loading force. One estimate for the optimal D/t for long bones, filled with marrow and experiencing an impact load, is about 4.6. It is about 6.1 for a marrow-filled bone with a static load for maximal strength and about 8 for maximal stiffness.

A hollow, air-filled bone will buckle at D/t greater than about 35. In practice, D/t is about 4.4 for many marrow-filled bones of nonflying mammals, about 7 for birds, and about 13 for air-filled bird bones (Figure 10–6). Some bones of pterodactyls (extinct flying reptiles) had extremely high D/t ratios; these were probably air-filled bones and would have been very susceptible to buckling.

A second weight economy is gained from the structure of the bone itself. **Compact bone** is very dense, almost solid bone (Figure 10–7). The only spaces in the bone are for bone cells (osteocytes) and their cell processes, for capillaries and nerves, and for sites of bone remodeling. It has a maximum tensile and compressive strength but it is also very dense. Consequently, compact bone is found where maximum compressive or tensile forces must be resisted. **Cancellous bone** (or spongy bone) has numerous large spaces in the bone. For example, one type of cancellous bone has randomly oriented struts (about 0.1 mm diameter × 1 mm long) that interconnect to form a 3-dimensional meshwork. This is generally found deep within bones, away

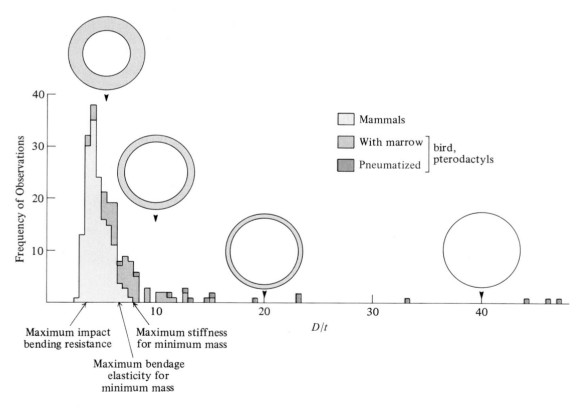

FIGURE 10–6   Most vertebrate long bones have a ratio of diameter (D) to wall thickness (t) of about 4 to 5 but bird and pterodactyl wing bones have higher D/t ratios.   *(Modified from Currey 1984.)*

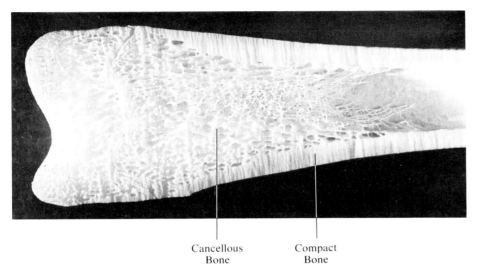

Cancellous
Bone

Compact
Bone

FIGURE 10–7 Gross structure of compact and cancellous bone (femur of a western gray kangaroo). *(Photograph by P. C. Withers and J. O'Shea.)*

from the directional surface loading. Another type of cancellous bone consists of thin sheets of bone occasionally connected to adjacent sheets by fine struts. This is generally found near bone surfaces where the pattern of loading is fairly directional and consistent. The femur has compact bone at the outside surface and cancellous bone in the center, particularly at each end. The struts of the cancellous bone are clearly oriented in particular directions, which correspond to the directions of mechanical forces that the femur experiences.

The ends of long bones, such as the femur, are quite expanded and of complex structure where they articulate with adjacent skeletal elements. The articulating surfaces must sustain high forces, often much higher than the overall force exerted by the muscle, depending on its mechanical advantage. The load-bearing surface of joints is usually cartilage, not bone, to provide a smooth and somewhat pliant surface. Cartilage has a relatively low tensile strength (about 20 MPa). The high forces experienced, and the low tensile strength of cartilage, mean that there must be a high load-bearing surface to widely distribute the forces. Consequently, the femur is greatly expanded at its ends to form the articulating surfaces. The bone underneath the cartilage surface is cancellous to minimize bone mass and also to minimize the tendency for high loads to excessively squeeze the cartilage surface. The underlying cancellous bony support deforms a little and distributes the load widely through the bone. A thin supporting layer of compact bone would deform too much, and a thick supporting layer of compact bone would deform too little and squeeze the cartilage with impact forces.

Joints

Stiff skeletons have **joints**, or articulations, between the skeletal elements across which muscles act to produce movement. Joints that transmit large forces have a limited plane of rotation to prevent sideways bending. Joints that allow movement in many planes of motion can only support low forces without being damaged.

There are two basic types of joints: sliding and elastic. **Sliding joints** are subject to wear and require lubricating; vertebrate joints tend to be sliding joints. For example, the joint between two articulating mammal bones is contained within a synovial cavity that contains a lubricating synovial fluid and the articulating surfaces of the bones are covered by cartilage to provide a smooth sliding surface (Figure 10–8A). **Elastic joints** consist of soft, pliable materials that allow part of an otherwise rigid skeletal element to bend. The deformation of elastic joints absorbs some of the energy required for movement. Arthropods tend to have elastic joints. For example, many insect joints have soft, pliable regions of cuticle that either allow deformation in any direction or limit motion to a specific plane by having double ball-in-socket joints (Figure 10–8B).

Most muscles extend over at least one articulation joint between two adjacent skeletal elements and their contraction moves the distal ends of the skeletal elements closer together. Usually one skeletal element remains fairly stationary (it is the **origin**) and the other element (the **insertion**) moves more. The insertion element is a **lever**, which moves about the fixed articulation point, the **fulcrum**. The muscle contraction effort moves a weight, or overcomes a

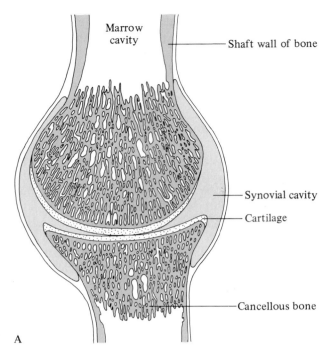

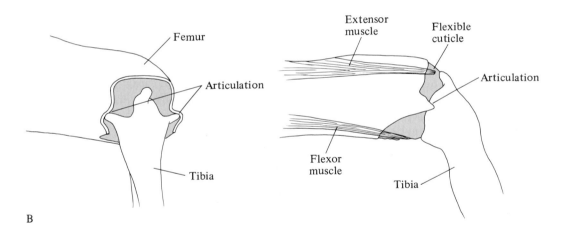

FIGURE 10–8 **(A)** A typical articulation between two bones has a
large articulating surface area covered by hyaline cartilage. It is enclosed within a synovial cavity that
is filled with a lubricating synovial
fluid. **(B)** A typical arthropod joint
has two small articulating surfaces
that define the plane of movement
and surrounding areas of flexible
cuticle (grey).   *(From Currey 1970;
modified from Chapman 1982.)*

load, by rotating about the fulcrum (Figure 10–9A).
Lever systems are able to apply greater, or lesser,
forces than the muscles themselves can generate,
depending on the particular arrangement of the
muscle effort and load relative to the fulcrum. Lever
systems can be classified by the relative positions
of the muscle contraction effort, the load, and the
fulcrum. A first class lever system has the fulcrum
located between the muscle effort and the load (like
a seesaw). A second class lever has the load located
between the fulcrum and the muscle effort (like a
wheelbarrow; Figure 10–9B). A third class lever has
the muscle effort located between the fulcrum and
the load (like a crane with a jib; Figure 10–9C).

The principle of **leverage** is that a lever system
with the muscle effort directed farther from the

fulcrum (distance ΔM) than is the load (distance
ΔL) can exert a greater force than can the muscle.
Alternatively, if the muscle effort is applied closer
to the fulcrum than is the load, then the force on
the resistance is less than that generated by the
muscle. The relationship between the muscle effort
and load and their respective distances from the
fulcrum is

$$M.\Delta M = L.\Delta L \qquad (10.2a)$$

The **mechanical advantage** (MA) of a fulcrum system
is the ratio of the load to the muscle effort, i.e.,

$$MA = L/M = \Delta M/\Delta L \qquad (10.2b)$$

The **range of motion** (ROM) is the distance that a
load is moved by a muscle contracting a distance

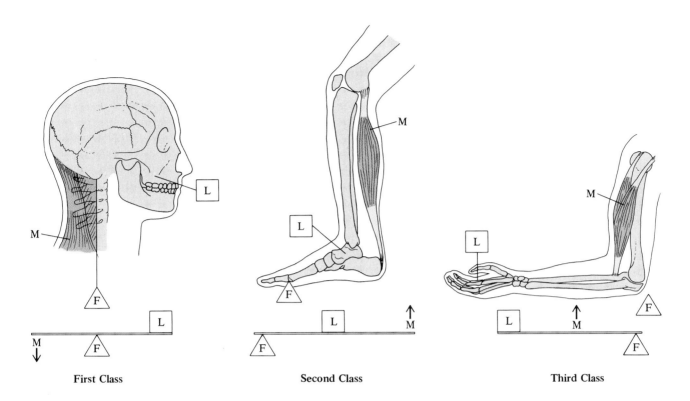

First Class   Second Class   Third Class

FIGURE 10–9 (**A**) Example of a first class lever. (**B**) A second class lever. (**C**) A third class lever. Each lever type is defined by the relative positions of the fulcrum (F), muscle contraction (M), and load (L). *(Modified from Tortora 1983.)*

Δx; it is proportional to the distance of the load from the fulcrum.

$$ROM = \Delta x \Delta L / \Delta M = \Delta x / MA \qquad (10.2c)$$

The range of motion is inversely proportional to the mechanical advantage. The velocity of movement is also proportional to the range of motion and inversely proportional to the mechanical advantage. Consequently, maximal strength (proportional to MA) and maximal range of motion/velocity (proportional to 1/MA) are inversely related; they are incompatible in the same lever system.

A musculoskeletal system designed for a high-power output (e.g., a digging limb, such as the forearm of an armadillo) has a larger ΔM than ΔL, a high mechanical advantage, and a low range of motion and velocity (Figure 10–10). Conversely, a musculoskeletal system designed for rapid motion (e.g., a limb of a cursorial animal, such as a horse) has a low mechanical advantage but a high range of motion and velocity. The mechanical advantage of a musculoskeletal system is not constant but alters with the relative location of the skeletal elements. For example, the muscle that extends the femorotib-

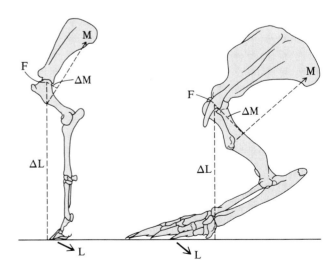

FIGURE 10–10 The skeletal arrangement of the foreleg of two mammals, the cursorial horse (*Equus;* left) and the burrowing armadillo (*Dasypus;* right), shows the difference in mechanical advantage (L/M = $\Delta M / \Delta L$) of the teres major muscle. The muscular effort (M) is applied between the fulcrum point (F), and the load applied to the ground (L). *(Modified from Smith and Savage 1956.)*

ial joint ("knee") of the locust's hind leg has a mechanical advantage of 1/60 if the knee is flexed, 1/35 if the knee is partly extended, and 1/60 when it is fully extended.

The mechanical advantage of a musculoskeletal system can also be modified by the arrangement of the muscle fibers within the muscle. The simplest way that muscle fibers can be arranged is as essentially parallel fibers extending from one end of the muscle to the other; this is a **fusiform** muscle. The ends of the muscle attach either directly to the skeletal element or tendon, or to a chitinous apodeme (in arthropods). When the muscle contracts it shortens and becomes thicker. The fibers of a **pennate** muscle are attached to a tendon at an oblique angle to the direction of the muscle. When a pennate muscle contracts, the short fibers each contract and become more perpendicular to the central tendon/apodeme. A pennate organization increases the maximum force of contraction of a muscle compared to a fusiform muscle of the same mass, i.e., has the same effect as an increased mechanical advantage (see Supplement 10–2, pages 489–90). Pennate muscles have a further advantage over fusiform muscles; they maintain a constant cross-sectional area during muscle shortening. In contrast, fusiform muscles become thicker when they contract. Consequently, it is especially advantageous for arthropod muscles to be pennate because they are enclosed within a rigid, tubular exoskeleton.

Locomotion

Animals have evolved a diverse array of locomotory methods depending to a large extent on whether they are aquatic (e.g., swimming, rowing, floating), terrestrial (e.g., burrowing, crawling, walking, running, hopping, jumping), or aerial (e.g., gliding, flapping flight). We will examine terrestrial locomotion first because it generally involves hydrostats or simple lever-joint systems. Aquatic locomotion uses hydrodynamic forces, mainly drag but sometimes lift. Aerial locomotion exploits the same fluid dynamic principles (aerodynamics) but lift generally predominates over drag.

Terrestrial Locomotion

Terrestrial animals use a variety of techniques for moving over, or through, their solid, terrestrial substrate, e.g., they crawl, burrow, walk, run, hop, or jump. The relative velocity of these types of terrestrial locomotion varies from slow (e.g., crawling) to fast (running, hopping, jumping). Animals

with soft bodies and hydrostats generally crawl or burrow, e.g., worms. Animals with solid skeletons often walk or run, e.g., many vertebrates and arthropods. Their solid framework skeleton is used as a lever system to transmit muscle forces to the substrate.

Crawling. Many soft-bodied animals, and also some animals with rigid skeletons, use their fluid-filled body cavities as a hydrostatic skeleton when they crawl. Some caterpillars use their anterior legs and posterior prolegs to crawl by a "two-anchor" technique (Figure 10–11A). The prolegs are fixed to the ground and the body is extended anteriorly by contraction of circular muscles on the hydrostat. The anterior legs then contact the ground and the prolegs are raised; the posterior end of the body is then drawn forward by looping the body. Leeches crawl in a similar fashion but use anterior and posterior suckers for attachment to the substrate.

Two-anchor burrowing and peristaltic crawling similarly rely on a hydrostatic skeleton to alter the shape of the soft body. Bivalve mollusks well illustrate two-anchor burrowing (Figure 10–11B). The shells gape open when the shell closing muscle is relaxed because of the elastic shell hinge. The muscular foot extends from the opening and probes into the substrate, then is expanded when the shell adductor muscle contracts and compresses the body spaces, forcing fluid into the foot and expanding it to form an anchor. In the final stage of burrowing, the foot is shortened and this draws the shell towards the anchored foot. Some bivalves can burrow extremely rapidly in this manner, burying themselves in about 4 sec, although most bivalves are considerably slower, taking up to 60 sec to burrow into the substrate.

Earthworms crawl with a peristaltic action of the body wall muscles that lengthens then shortens the body segments. Each segment of the worm is essentially a separate, fluid-filled hydrostatic skeleton. Contraction of its circular muscles extends the segment, and contraction of its longitudinal muscles shortens it. The fluid volume remains essentially the same during these locomotory efforts (the holes in septa between adjacent segments that accommodate the nerve cord are kept shut by special muscles). Segments that are constricted (thicker) support most of the body mass and the segmental setae also help to anchor these segments to the substrate. When a segment elongates, it pushes the more anterior portion of the body forward (Figure 10–11C). The sequence of constriction/extension forms a wave of thickening that moves posteriorly, and this moves the worm forward; this is retrograde

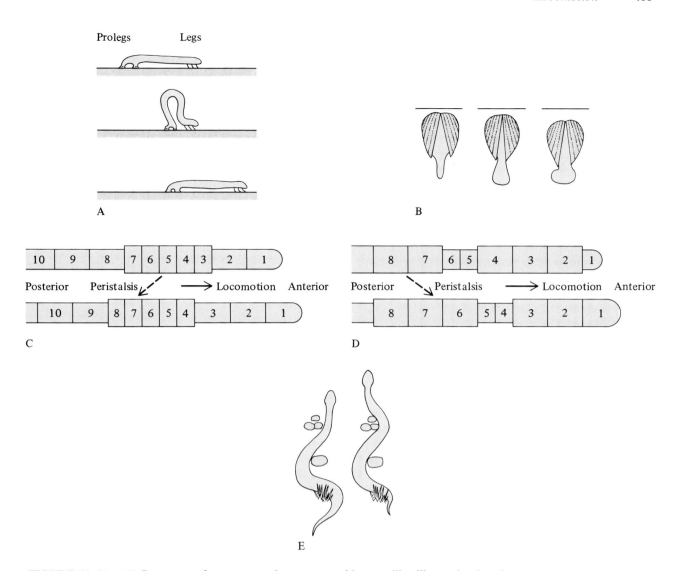

FIGURE 10–11 **(A)** Sequence of movements by a geometrid caterpillar illustrating looping movement by the two-anchor principle. **(B)** Burrowing movements of a bivalve mollusk using the two-anchor principle. **(C)** A crawling earthworm uses retrograde peristalsis; there is no change in segment volume during elongation or shortening. **(D)** Crawling movements of some worms use direct peristalsis; there is change in segment volume during elongation and shortening. **(E)** Serpentine crawling by a snake. *(From Alexander 1982.)*

peristalsis. Even unsegmented, soft-bodied animals, such as nemertine worms, can move by this means as long as they can constrict/elongate and dilate/shorten different parts of the body at the same time. An alternative pattern of peristaltic movement is direct peristalsis, where the peristaltic wave moves anteriorly, in the direction of locomotion (Figure 10–11D). The individual segments are alternatively increased and decreased in length and volume by moving the body fluid between adjacent segments. Some segmented polychetes (with incomplete septa) and unsegmented holothurians move in this manner. Many chitons and gastropod mollusks move by crawling on their large "foot." Waves of muscular activity move the foot in a manner similar to peristaltic movement by earthworms. The pedal waves are retrograde in chitons but anterograde in some gastropods (e.g., *Helix*).

Serpentine crawling uses waves of bending of the body to push past objects (Figure 10–11E). Snakes, for example, use lateral bends of the body to push past objects, such as stones, grass tussocks, or other substrate irregularities. Nematodes similarly move past soil particles. Snakes less commonly use concertina crawling or side winding to move. Concertina crawling is analogous to retrograde peristaltic crawling by earthworms and can be used to crawl up trees or through tubes. Part of the body is wedged in position by lateral folds, and the more anterior part of the body is thrust forward. Side winding is a more complex locomotory pattern, with only parts of the body in contact with the ground; the other parts are raised into the air as loops.

Walking and Running.   Walking and running are the typical locomotory pattern for terrestrial animals with rigid skeletons and appendages. This is faster than crawling. The body mass is supported on only a few points of contact with the substrate, rather than having much or all of the body resting on the ground. The body mass is typically supported at any time by fewer than the total number of legs since the legs are alternately raised and lowered. It is therefore necessary to maintain body equilibrium during walking and running. A walking salamander well illustrates these principles (Figure 10–12).

The particular pattern of limb movement during locomotion, the **gait**, can be very different for different animals, and at different speeds. The gait is usually described by two variables for each foot: the **relative phase** between the cycle of each foot and the **duty factor**, the fraction of time that the foot is on the ground. For example, humans are bipedal animals; we walk on two legs. Walking at slow speeds involves regular cycles of leg movements. If we use the left foot as the reference (its relative phase is 0) then the right foot has a relative phase of 0.5, i.e., the right foot has the opposite phase as the left foot (Table 10–2). The duty factor is 0.6 for each foot, i.e., each foot is on the ground for 60% of the walking cycle. There are times during the walking cycle when both feet are on the ground. In running, the relative phase remains the same but the duty factor is less than 0.5, i.e., there are times when both feet are off the ground. For animals that hop, the relative phase is the same for both feet and the duty factor can be less than 0.5. Gaits can be expressed similarly for quadrupedal animals. A slowly walking horse has a duty factor that is >0.5 and each pair of legs works in opposite phase, with the hind legs offset by 0.25 in phase from the forelegs. When trotting, the duty factor is reduced to <0.5 and the pairs of limbs are in opposite phase, with the left-front/right-rear and right-front/left-rear feet in phase. Galloping also has a low duty factor but has a complex phase pattern for the feet.

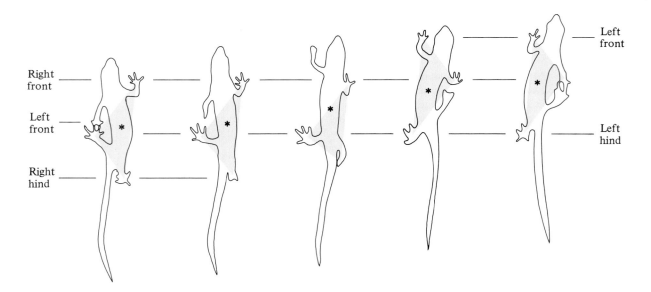

FIGURE 10–12   The locomotory strides of a salamander involve repeated undulation of the body from side to side and alternate movement of the feet. The bends in the body do not travel along the body, as in serpentine locomotion, but are stationary waves. The center of gravity (∗) is kept within a triangle formed by the feet that remain on the ground.   *(Modified from Roos 1964.)*

TABLE 10–2

Gait pattern for various animals with differing numbers of legs at slow speeds. The duty factor (**df**) is the time that each foot is on the ground; it is approximately the same for all feet, and only an averge value is given. The relative phase of the limbs (**rp**) is expressed relative to the left front (0). L = left, R = right; number indicates the pair of legs numbered from the front. *(Modified from Alexander 1977.)*

			rp	
Biped (human)	**df**	**Leg Pair**	*L*	*R*
	0.6	1	0	0.5

			rp	
Quadruped (horse)	**df**	**Leg Pair**	*L*	*R*
	>0.5	1	0	0.5
		2	0.75	0.25

			rp	
Hexiped (cockroach)	**df**	**Leg Pair**	*L*	*R*
	0.8	1	0	0.5
		2	0.6	0.1
		3	0.2	0.7

			rp	
Multiped (centipede)	**df**	**Leg Pair**	*L*	*R*
	0.5	1	0	0.5
		2	0.33	0.83
		3	0.67	0.17
		4	0	0.5
		5	0.33	0.83
		6	0.67	0.17
		7	0	0.50
		etc		

Animals with more than four legs (e.g., a cockroach with six, a centipede with many) have even more complex gait patterns that also vary dramatically with speed.

An animal that walks very slowly must always keep its body position in equilibrium throughout its stride or else it would fall over. Bipedal animals must have at least one foot on the ground and must keep their center of gravity over that foot. For other animals to maintain equilibrium, they must have at least three feet on the ground at all times and the center of mass must be within a triangle of support provided by these three feet (see Figure 10–12, page 462). A slow-moving tortoise could in theory maintain a constant equilibrium while walking, but in practice it doesn't because the muscle contrac-

tions are too slow for an optimal gait pattern. At times, the tortoise has only two feet on the ground and is momentarily in disequilibrium.

Fast-moving animals do not maintain body equilibrium but are in disequilibrium for often a significant fraction of the time (with even no feet on the ground during part of the stride). Low duty factors are often associated with fast speed, since this allows greater forward motion when the foot is not in contact with the ground. However, fast-moving insects do not have much lower duty factors than at low speed (values of about 0.5 or more). They also always have at least three feet on the ground; the six feet act in two groups of three, moving alternately. The high duty factor of insects means that their stride length is necessarily short, and so their only way to achieve high speed is to have a high frequency of limb movement (e.g., 20 to 24 sec^{-1}). Thus, "running" insects actually walk at a high frequency. Running vertebrates, in contrast, have low duty factors and achieve high speeds by having a long stride. There is a striking change in the pattern of limb movement (gait) at different speeds, e.g., in humans, walking and running; in quadrupeds, walking, trotting, galloping; etc.

Hopping and Jumping.   Many bipedal animals hop, e.g., kangaroos, many birds, some rodents (jerboas, kangaroo rats, hopping mice). The movement of both legs is in phase and the duty factor is quite low (Figure 10–13A). As we have seen, a low duty factor is adaptive for fast locomotion. The force exerted by the kangaroo on the ground peaks at about 300 N in a vertical direction and at about 50 N in a horizontal direction (Figure 10–13B). Calculations from such force measurements indicate that the kangaroo does work on the ground when landing at about 190 watts, and work of about 190 watts is done on it when it leaps into the air (Alexander 1982). The metabolic expenditure of a hopping kangaroo is calculated to be about 918 watts, but the measured metabolic rate is only about 380 watts or 40% of the calculated estimate. This discrepancy occurs because energy is stored in muscle tendons during the landing phase of hopping and is released during the leaping phase. As the kangaroo lands, the major leg muscles (gastrocnemius and plantaris) and their elastic tendon (the Achilles tendon) are stretched by the impact (Figure 10–13C), i.e., work is done on the muscle. When the kangaroo jumps, contraction of the muscles accelerates the body vertically and forward; the muscles do work. Recovery of elastic energy that had been stored by stretching the Achilles tendon also contributes to the energy for the jump. Elastic

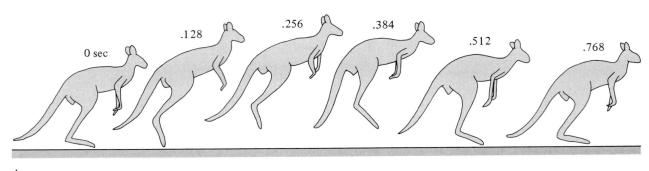

A

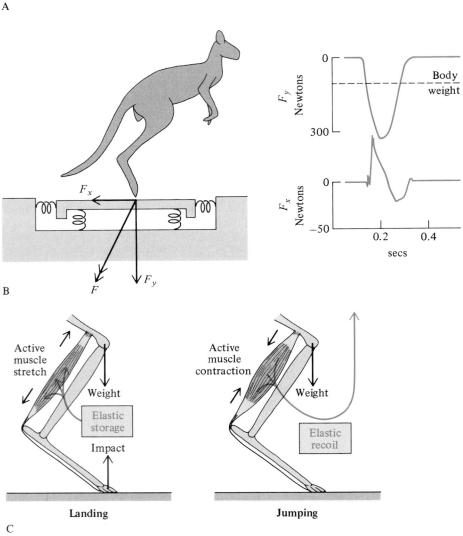

B

C

FIGURE 10–13 **(A)** Kinematics of a kangaroo hopping; numbers are time in msec from take-off. **(B)** Recordings of the forces exerted on the ground by a hopping kangaroo recorded using a force platform; the kangaroo exerts a forward then backward horizontal force on the ground and a vertical force. **(C)** Schematic of the arrangement of the limb bones, muscles (gastrocne-mius and plantaris), and Achilles tendon for a hopping kangaroo when landing and jumping. When landing, the impact force and weight of the kangaroo is absorbed by active stretching of the muscle and elastic stretch of the Achilles tendon. When jumping, the weight is accelerated by a recoil force due to active muscle contraction and elastic recoil of the Achilles tendon. *(Drawn from Muybridge 1957; modified from Alexander 1977.)*

recoil appears to provide about 60% of the energy required to jump.

Many animals, small and large, can jump to remarkable heights. Jumping is a more extreme disequilibrium movement than fast running, with even lower duty factors. The height that an animal can jump is related to its body mass; a small rabbit flea (<1 mg) can jump about 35 mm; a human flea (1 mg) about 130 mm; a locust (5 g) about 400 cm; and a bushbaby (*Galago*; about 1 kg) about 2.3 m. The maximum vertical height to which an animal can jump is determined by its kinetic energy (1/2 mv_0^2) when it leaves the ground, since this is converted to potential energy (mgh) at the top of the jump when the vertical velocity is 0. Ignoring air resistance for the moment,

$$\tfrac{1}{2}mv_0^2 = mgh \quad \text{and} \quad h = v_0^2/2g \quad (10.3)$$

where m is body mass (grams), v_0 is initial velocity (m sec^{-1}), g is the gravitational constant (9.8 m sec^{-2}), and h is the jump height (m). A bushbaby (Figure 10–14) has an initial jump velocity of about 6.7 m sec^{-1} and a maximum jump height of about 2.26 m (ignoring air resistance). About 36% of its mass is muscle, with a maximal energy output of about 200 K kg^{-1}, so the maximum energy available for leaping would be 72 J kg^{-1}. In fact, the initial kinetic energy per unit mass (1/2 v_0^2) is only about 22 J kg^{-1}. Many factors contribute to the initial kinetic energy being less than that theoretically available for jumping; not all muscles are used for jumping, and not all muscles are working at their optimal length.

All animals with the same initial takeoff velocity v_0 should attain the same height, h. However, this is not true in practice because the air resistance is not negligible, especially for small animals. Aerodynamic drag is an additional decelerating force to gravity and decreases the jump height. It is a complex force, depending on the velocity (v); the frontal area of the animal (A_f); the air density (ρ); and a drag coefficient (C_d), which depends on the shape and size of the animal (drag and drag coefficients will be described in more detail when we discuss hydrodynamics and aerodynamics). Small animals have a greater A_f/m ratio than do larger animals, and so they cannot leap as close to their theoretical maximum height as can larger animals.

Jumping muscles must contract rapidly. Animals that jump by extending their legs (of length l) accelerate their bodies through a distance that equals the average velocity during acceleration multiplied by time; this distance must be less than l.

$$l \geq \tfrac{1}{2}v_0 t \quad (10.4a)$$

where 1/2 v_0 is the average velocity. Hence,

$$t \leq 2l/v_0 \quad (10.4b)$$

For the bushbaby, t is about 30 msec (2 × 0.1/6.7), which is a realistic muscle contraction time. For a

FIGURE 10–14 The bushbaby *Galago* can leap up to 2.3 m vertically. *(A. W. Ambler, National Audobon Society/Photo Researchers, Inc.)*

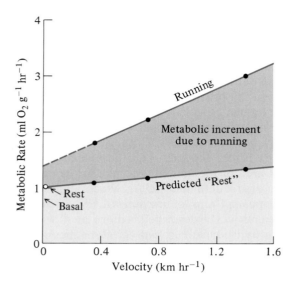

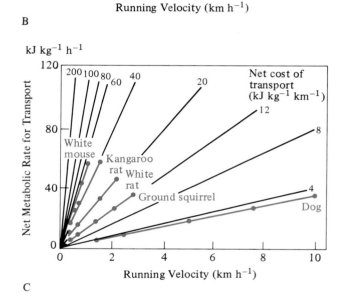

FIGURE 10–15 **(A)** Relationship for a white rat between metabolic cost of running and velocity. The linear relationship extrapolated to zero velocity intercepts at above the resting metabolic rate. Resting metabolic rate is predicted to increase slightly with running speed because of the elevated body temperature. **(B)** Metabolic cost when running for a variety of mammals of differing body mass. **(C)** Net cost of locomotion for a variety of mammals of differing body mass. *(Modified from Taylor, Schmidt-Nielsen, and Raab 1970.)*

small flea, *t* is about 1 msec (2 × 0.004/1), which is an unreasonably short contraction time for even a very fast muscle (except insect fibrillar flight muscle). Fleas and other small animals jump using stored elastic energy and a catapult mechanism, rather than by using power supplied directly from muscle contractions. The energy of a muscle contraction is temporarily stored by an elastic mechanism, then rapidly released.

Metabolic Cost.   The metabolic cost of locomotion depends on the velocity. For example, the metabolic rate of the white rat increases linearly with velocity (Figure 10–15A). Its metabolic rate during locomotion does not extrapolate at zero speed to its resting metabolic rate, probably because there is a

metabolic increment above resting for the maintenance of locomotor posture and stance. The metabolic cost of locomotion at any velocity can be calculated as $VO_{2,\text{locomotion}} - VO_{2,\text{rest}}$. The $VO_{2,\text{rest}}$ increases slightly when extrapolated to various running speeds because body temperature is elevated by running.

Small animals cannot run as fast as large animals, and they expend a relatively greater amount of energy on locomotion (Figure 10–15B). For example, a 21 g white mouse has a higher mass-specific resting VO_2 than an 18 kg dog and a greater increase in metabolic rate when running even at a much lower velocity than a dog at a much higher velocity.

The total **cost of transport** (COT_{tot}) is calculated at any velocity (*v*) as follows.

$$\text{COT}_{\text{tot}} = \text{VO}_{2,\text{locomotion}}/v \qquad (10.5a)$$

It has mass-independent units of ml O_2 km^{-1} if VO_2 is absolute, i.e., ml O_2 hr^{-1} and velocity is in km hr^{-1}; 1 ml O_2 km^{-1} = 20.1 mJ m^{-1}. The mass-specific COT_{tot} has units of ml O_2 kg^{-1} km^{-1} or mJ kg^{-1} m^{-1}. The COT_{tot} changes with velocity from infinity at $v = 0$ to lower values at high v (see Figure 10–15B).

The **net cost of transport** (COT_{net}) is calculated from the increment in VO_2 above resting due to locomotion

$$\text{COT}_{\text{net}} = (\text{VO}_{2,\text{locomotion}} - \text{VO}_{2,\text{rest}})/v \quad (10.5b)$$

i.e., COT_{net} includes only the metabolic increment above resting.

The COT_{net} is equal to the slope of the relationship between VO_2 and v if this relationship is linear (as it is for the mammals shown in Figure 10–15C but isn't for the kangaroo; see below). The COT_{net} is a better comparator for locomotory cost at varying velocity because it is not velocity dependent for many mammals and is not dependent on the resting metabolic rate. The scaling effects of body mass on the COT_{net} are remarkably uniform for a variety of taxa that walk or run (Figure 10–16).

The relationship between VO_2 and velocity is often, but not always, linear. Different walking/running gaits are most efficient for limited velocity ranges, and so there are discontinuities in the metabolic rate at velocities where gait patterns change. This means that the net metabolic cost of locomotion is not constant, but alters with velocity. This is also true for swimming and flying (see below). Burrowing, in contrast, has a higher cost of transport, reflecting the additional metabolic cost of moving the substrate out of the way. It also depends on the nature of the substrate, e.g., the cost of locomotion is lower for mollusks in soft marine sediment than for mammals burrowing in hard soil.

Locomotion is more economical (in cost per unit mass) for large than small animals. The net cost

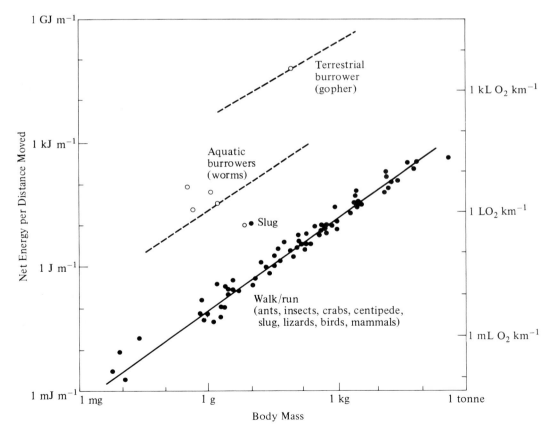

FIGURE 10–16 Metabolic cost of terrestrial locomotion and burrowing as a function of body mass. The cost of locomotion is expressed as the total energy expended per distance traveled, i.e., joules per meter or ml O_2 per km. *(Modified from Alexander 1982.)*

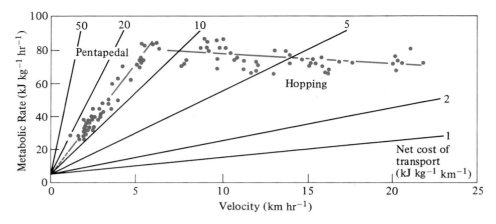

FIGURE 10–17 Relationship between metabolic rate and velocity for a kangaroo when using pentapedal (forelimbs, hindlimbs, and tail) locomotion at low speed and bipedal hopping at high speed. *(Modified from Dawson and Taylor 1973.)*

of locomotion for hopping animals is markedly dependent on their velocity. For example, the metabolic rate of a kangaroo increases markedly at low v, then actually declines at higher v (Figure 10–17).

Aquatic Locomotion

The principles of aquatic locomotion are quite different from those for terrestrial locomotion. The density of a fluid medium provides a substantial buoyant force that almost eliminates the gravitational force on the body. The movement of objects through a fluid medium generates drag and sometimes lift forces that can facilitate movement or provide a locomotor mechanism.

Buoyancy. Animals immersed in fresh- or seawater experience a considerable **buoyancy** force that counteracts most of their body weight. Freshwater has a density of 1000 kg m^{-3}; seawater has a higher density (1026 kg m^{-3}) because of its dissolved salts. Air, with a lower density of 1.18 kg m^{-3}, provides far less buoyancy.

Aquatic animals are slightly more dense than fresh water (Table 10–3) because they contain denser materials, e.g., solutes, proteins, structural carbohydrates, bone (see Table 10–1, page 450). They also contain some materials that are less dense than water (e.g., lipids) but their overall density is usually greater than that of water. Marine animals are generally a little more dense than seawater primarily because of their skeleton (e.g., bone or shell), but even tissues such as muscle are slightly denser than seawater. Consequently, aquatic animals need to counteract their negative buoyancy;

many attain neutral or even positive buoyancy by a variety of means. Most use low-density materials (lipids, specific ions) or gas-filled bladders (swimbladders, gas floats). Some generate lift using hydrofoils to achieve neutral buoyancy while they are swimming.

Body tissues that have a lower density than fresh- and seawater provide positive buoyancy in

TABLE 10–3

Density for air, freshwater, seawater, ionic solutions, tissues, and animals. *(Values are from Alexander 1982.)*

	Density (kg m^{-3})
Air	1.18
Freshwater	1000
Seawater	1026
Ammonium chloride[1]	1007
Sodium chloride[1]	1018
Sodium sulphate[1]	1040
Trimethylamine[1]	989
Muscle	1060
Bone	2000
Lipid—fat	930
—squalene/wax esters	860
Nautilus shell	2700
Dogfish	1075
Squid	1070
Tuna	1090
Copepod	1080

[1] For solution isoosmotic with seawater.

proportion to their volume in the body. Most lipids have a density of about 930 kg m^{-3} but some (squalene and wax esters) are even less dense (about 860 kg m^{-3}). Some ionic solutions also have a lower density than seawater (but not freshwater). For example, a 1 osmolar NH_4Cl solution has a density of about 1007 kg m^{-3}. Some solutions even have a lower density than freshwater e.g. trimethylamine.

A relative large proportion of an animal's body must be of a low-density material to achieve neutral buoyancy. For example, an animal of volume V_a and typical density ρ_a (about 1.075) can reduce its density to equal that of the water (ρ_w) if it adds an additional volume (V_{ld}) of low-density materials of density ρ_{ld}. The efficacy of a low-density material in making an animal more buoyant depends on how different its density is from that of the ambient medium. The relative volume of low-density material required to attain neutral buoyancy can be calculated as follows.

$$V_{ld}/V_a = (\rho_a - \rho_w)/(\rho_w - \rho_{ld}) \qquad (10.6)$$

An animal of density 1075 kg m^{-3} would have to add an additional 51% body volume of fat ($\rho_{ld} = 930$) to achieve neutral buoyancy in seawater and 107% to be neutrally buoyant in freshwater. A smaller volume is needed for lower-density materials (e.g., squalene, wax esters; $\rho_{ld} = 860$). Many fish and invertebrates use fats, including squalene and wax esters, to maintain neutral buoyancy. For example, some sharks have large, fatty livers (up to 30% of body volume) that contain much squalene. Some fish (coelacanth and lantern fish) have very oily tissues and are nearly neutrally buoyant. Many invertebrates use oil droplets to regulate their buoyancy.

Some ionic solutions have a lower density than body fluids of an equivalent concentration. For example, solutions of NH_4Cl (MWt 51.5) are less dense than equivalent solutions of $NaCl$ (MWt 58.5) because of the lower molecular weight. However, the density of solutions does not only depend on the molecular weight of the solutes. Solutions of some solutes have a lower density than the solvent they are dissolved in!

The volume of a solution is equal to the sum of the volumes of the solvent (solvent volume) and the solute (partial molar volume, $\overline{V}_0$; cm^3 mole^{-1}). The solution volume equals the solvent volume if the solute has a zero partial molal volume. The volume of the solution is greater than that of the solvent if the solute has a positive partial molal volume. The solution volume is less for a solute with a negative partial molal volume.

Whether an ion in solution provides positive or negative buoyancy thus depends on both its molecular weight and its partial molar volume. Solution density is less than solvent density for a solute with a partial molar volume greater than its molecular weight. Sodium and calcium ions have negative partial molar volumes. Both ammonium and chloride ions have a positive partial molar volume but their V_0 is less than their molecular weight. For ammonium, molecular weight (18) is almost the same as partial molar volume (17.5 cm^3 mole^{-1}) and so the density of 0.5 M NH_4^+ is about 1.000. For 0.5 M Cl$^-$ (MWt 35.3), V_0 is only +16.7, and so the density is 1.009. The predicted density of 0.5 M NH_4^+ Cl$^-$ is therefore 1.005. Consequently, body fluids with a high concentration of NH_4Cl will contribute a slight negative buoyancy force compared to pure water and a slight positive buoyancy compared to body fluids and seawater. It is impossible to attain neutral buoyancy in fresh water with even pure NH_4Cl solutions. For trimethylamine, $(CH_3)_3NH^+$, molecular weight (60) is less than the partial molar volume (+71.54) and so the density of a 1 M solution is only 0.989.

The use of an ammoniacal, or even a trimethylamine solution, to attain neutral buoyancy in seawater requires an extremely large body fluid volume. The deep sea squid *Helicocranchia* has an enormously enlarged body cavity containing a highly ammoniacal body fluid of density 1010 kg m^{-3}. Decreasing the content of high molecular weight/low partial molar volume solutes (*e.g.* Na$^+$, Ca^{2+}, Mg^{2+}, SO$_4^{2-}$) also reduces the density. For example, jellyfish have low body fluid sulphate levels, which slightly reduces their negative buoyancy.

The deep sea pelagic copepod *Notostomus* (Figure 10–18) has a positive buoyancy unlike most pelagic crustaceans. It is positively buoyant because its enlarged dorsal carapace contains a considerable volume of a low-density body fluid (Sanders and Childress 1988). About 40% of the animal's water is located within this dorsally enlarged carapace. The low-density carapace fluid differs markedly from normal body fluids in solute composition (Table 10–4). The unusual solute composition provides considerable positive buoyancy, about 17.7 mg ml^{-1}. The low density is partly accomplished by the replacement of normal body fluid ions by lower-density ions. NH_4^+ (MWt 18) replaces some Na$^+$ (MWt 23), and there are reduced concentrations of Ca^{2+}, Mg^{2+}, and SO$_4^{2-}$. This contributes about 5.4 mg ml^{-1} negative buoyancy. A second and more important mechanism for lowering the density of the carapace fluid is the replacement of normal body fluid ions with ions that have a positive partial molar volume. The carapace fluid of *Notostomus* contains large amounts of trimethylamine (Me$_3$NH$^+$; MWt 60) and also NH_4^+. These ions replace nearly 90%

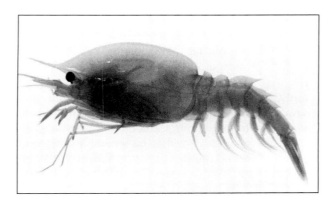

FIGURE 10–18 The marine copepod *Notostomus* is neutrally buoyant because its enlarged carapace contains a low-density body fluid. *(From Saunders and Childress 1988.)*

of the Na^+ that would be present in the carapace fluid if it had the same ionic composition as plasma. They have a positive partial molal volume (e.g., V_0 = $+71.5$ for Me_3NH^+ and $+17.5$ for NH_4^+) and therefore provide a negative buoyancy of 10.9 mg ml^{-1}. The ions they replace have negative partial molar volumes (e.g., V_0 = -2.90 for Na^+, -19.26

for Ca^{2+}, and -21.66 for Mg^{2+}). The density of the carapace fluid can be calculated from the solute concentrations, their molecular weights, and their partial molar volumes. The solutes in 1 liter of carapace water would weigh 33.199 g, and increase its volume by 22.659 ml, i.e., its density is $(1000 + 33.199)/(1000 + 22.659) = 1.0103$ g ml^{-1}. In contrast, for seawater the solutes in 1 liter of water would weigh 35.081 g and add 6.804 ml volume, i.e., its density is $(1000 + 35.081)/(1000 + 6.804)$ = 1.0281 g ml^{-1}. The high concentrations of urea and trimethylamine oxide in elasmobranchs may also contribute significant negative buoyancy.

A gas, such as air, is a far superior material compared with fat or salt solutions to use for achieving neutral buoyancy because it has a very low density. Only about 5 to 8% gas volume needs to be added to an animal to make it neutrally buoyant in seawater or freshwater. Consequently, many animals, and particularly those found near the water surface, have gas-filled chambers. Most teleost fish have a gas-filled swim-bladder that contains a variable concentration of N_2, O_2, and CO_2. Some cephalopods have a rigid, gas-filled shell (the cuttlebone or shell).

There is a major disadvantage to the use of a gas-filled buoyancy chamber. Air is very compressible,

TABLE 10–4

The unusual solute composition of carapace fluid of the copepod *Notostomus* has a relatively low density (1.010) and provides positive buoyancy relative to seawater (density = 1.028). *(Modified from Sanders and Childress 1988).*

Solute	Molecular weight (g mol^{-1})	Partial Molar Volume (cm^3 mol^{-1})	Carapace Fluid[1] Solute Concentration (g l^{-1})	Carapace Fluid[1] Solute Volume (ml l^{-1})
Na^+	23.0	-2.90	1.428	-0.180
NH_4^+	18.0	$+17.49$	5.328	$+5.176$
K^+	39.1	$+7.59$	0.422	$+0.082$
Me_3NH^+	60.0	$+71.54$	7.656	$+9.284$
Mg^{2+}	24.3	-21.66	0.085	-0.076
Ca^{2+}	40.1	-19.26	0.000	0.000
Cl^-	35.3	$+16.65$	18.165	$+8.520$
SO_4^{2-}	96.1	$+11.70$	0.115	$+0.014$
Total ions			33.199	$+22.659$
Water			1000	1000
Solution			1033.199	1022.659

Solution density (g cm^{-3}) 1.0103 (1033.199/1022.659)
Seawater density (g cm^{-3}) 1.0281

[1] Solute concentration and volume per liter (1000 g) of water.

unlike water and body tissues, and so the increased hydrostatic pressure associated with greater depth tends to collapse the gas chamber; hydrostatic pressure increases by 101 kPa (1 atm) for every 10 m depth. Hydrostatic compression reduces the volume and efficacy of a flexible gas chamber in providing neutral buoyancy and could crush a rigid-walled gas chamber. Fish and cephalopods have dealt with this limitation in very different ways. Fish swim-bladders have flexible walls (except the coelacanth has an ossified swim-bladder that is filled with lipid), and a constant gas volume is maintained by balancing gas secretion into the swim-bladder with gas loss; the gas pressure in the swim-bladder is always equal to the external hydrostatic pressure. Cephalopods, in contrast, have a rigid-walled gas chamber with a gas pressure less than the external hydrostatic pressure.

The teleost swim-bladder. The **swim-bladder** of teleost fish is a gas-filled chamber. It is located in about the center of the body (Figure 10–19A) so that its positive buoyancy acts at about the center of gravity of the fish to minimize any tendency to pitch or roll the body. The swim-bladder contains a **gas-secreting gland** and often a **gas-reabsorbing gland**.

Swim-bladders and lungs evolved from a gut diverticulum. Physostome teleosts retain a connection between their swim-bladder and their esophagus (Figure 10–19B). The more advanced physoclist teleosts lack this connection; these teleosts have a gas (oval) reabsorbing gland. The pressure in the swim-bladder must be greater than 101 kPa, except when at the surface, e.g., at 10 m depth, the pressure is 202 kPa; at 1000 m, the pressure is 10201 kPa. The partial pressures of gases in the swim-bladder are therefore considerably higher than atmospheric and water gas partial pressures (depending not only on the hydrostatic pressure but also on the specific gas composition). There is a considerable partial pressure gradient for diffusional loss of gases through the swim-bladder wall into the body fluids and external water and for convective loss via blood leaving the swim-bladder gland. There is an inevitable loss of gas from the swim-bladder, although the losses can be minimized by reducing the gas permeability of the swim-bladder wall (e.g., with a coating of flat guanine crystals) and by a countercurrent arrangement of the blood supply to the swim-bladder.

Physostome fish can add gas to their swim-bladder by gulping air at the surface or by secreting gas into the swim-bladder against a considerable partial pressure gradient. Physoclist fish rely on gas secretion. The hydrostatic pressure decreases and

the swim-bladder expands if the fish swims up through the water column. Diffusional losses of gas are too slow for rapid compensation of such increases in volume. Physostome fish can eliminate the excess gas by "burping" air out of their swim-bladder. The swim-bladder of physoclist fish has a gas-reabsorbing gland to reabsorb excess gas into the blood. Secretion and reabsorption of gases are relatively slow processes; re-equilibration after the swim-bladder volume alters can take 4 to 6 hours. The swim-bladder of a deep-sea physoclist teleost that is rapidly brought to the surface will be enormously expanded by the rapid decrease in hydrostatic pressure. The massive expansion of the swim-bladder in the abdominal cavity may eviscerate the fish.

The functioning of swim-bladders relies on a countercurrent anatomical arrangement of the blood supply to the swim-bladder, the **rete mirabile**, and a special O_2 transport property of blood, the Root effect. The arterial vascular supply to the gas gland splits into a large number of small, parallel blood vessels, and the venous drainage similarly consists of numerous small, parallel vessels that are interspersed with the arterial vessels. This specialized anatomical arrangement of countercurrent blood vessels is the rete mirabile (Figure 10–19C). The rete capillaries are very long, often 2 cm or more, and the length is positively correlated with the gas pressures required in the swim-bladder (i.e., deep-sea fish have a longer rete).

The rete mirabile provides considerable opportunity for a countercurrent exchange of materials to which their walls are sufficiently permeable (e.g., O_2, other gases, small organic molecules, heat). For example, the arterial O_2 tension may be 20 kPa and may increase to 10000 kPa at the inner surface of the gas gland. As blood exits the gas gland along the venous vessels, its O_2 tension is greater than that in adjacent arterial vessels, and so O_2 will exchange from the venous to the arterial vessels (Figure 10–20Ai). This O_2 exchange reduces the O_2 tension of blood leaving the gas gland, from say 10000 kPa to perhaps 40 kPa at the terminus of the rete. The countercurrent rete obviously confers a considerable reduction in the blood convective loss of O_2 (and also other gases) by blood flow from the swim-bladder.

The rete of the gas gland not only minimizes passive gas losses but also provides the mechanism for gas secretion (otherwise, why would the vascular supply be there!). Blood entering the arterial vessels of the rete has a certain partial pressure of O_2 (e.g., 20 kPa) and blood O_2 content (e.g., 10 ml O_2 per 100 ml blood or 10 volumes percentage). Blood in

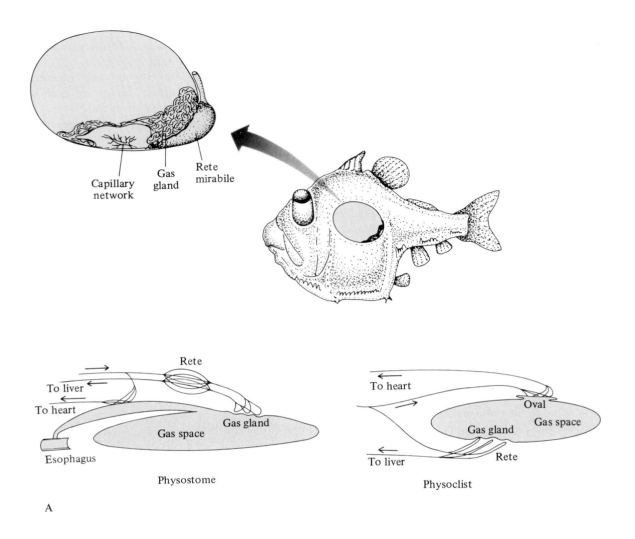

Capillary
network

Gas
gland

Rete
mirabile

Rete

To liver

To heart

Esophagus

Gas gland

Gas space

Physostome

To heart

Oval

Gas space

Gas gland

To liver

Rete

Physoclist

A

Swim-bladder
lumen

Gas gland

Rete mirabile

B

FIGURE 10–19 **(A)** Anatomical arrangement and gross structure for the swim-bladder of a deep-sea fish, *Argyropelecus aculeatus.* **(B)** Representation of swim-bladders from a physostome fish (the eel *Anguilla vulgaris)* and a physoclist fish (the perch *Perca fluviatilis).* **(C)** Detailed representation of a swim-bladder showing the rete mirabile.   *(From Denton 1961; Wittenberg 1958.)*

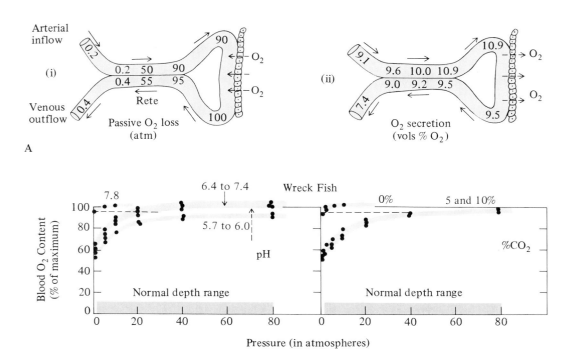

FIGURE 10–20 (A) Representation of the swim-bladder showing passive O_2 loss (left; values are O_2 partial pressure in atm) and secretion of O_2 into the swim-bladder (right; values are O_2 content of the blood in ml per 100 ml blood). (B) The Root effect for hemoglobin of the wreck fish decreases the O_2 capacity with decreased pH (left) or elevated CO_2 level (right).   *(Modified from Denton 1961; from Scholander and Van Dam 1954.)*

the gas gland has a very high pO_2 (e.g., 10000 kPa) and a higher O_2 content. Blood leaving the venous portion of the rete can have a pO_2 lower than blood in the gland but it must have a higher pO_2 than arterial blood because O_2 cannot be actively transported from the venous into the arterial blood. However, the venous blood must have a lower O_2 content than arterial blood if O_2 is to be secreted into the swim-bladder (Figure 10–20Aii). Venous blood must have a lower O_2 content than arterial blood despite having a higher pO_2. We will assume that the gas gland does not have a significant O_2 consumption; metabolic consumption of O_2 by the gas gland would reduce the O_2 content of the venous blood relative to the arterial blood, but this won't contribute to gas secretion.

The relationship between pO_2 and percent saturation of hemoglobin (the hemoglobin-oxygen equilibrium curve) is influenced by many variables, including temperature, pH, and pCO_2 (see Chapter 15). The O_2-carrying capacity of blood from some animals is also influenced by pH. A decrease in pH can reduce the O_2-carrying capacity of hemoglobin,

even when the hemoglobin is 100% saturated; this is the positive **Root effect** (Figure 10–20B). The Root effect is of great functional significance to swim-bladders. During gas secretion, the gas gland liberates lactic acid as its metabolic end product of anaerobic metabolism (hence O_2 consumption of the gas gland is negligible). The lactate and H^+ ions reduce the pH of the blood. The Root effect reduces the O_2 carrying-capacity of the blood and forces some O_2 off hemoglobin and into solution; this increases the pO_2 of the blood and transports O_2 into the swim-bladder. The venous blood thus has a higher pO_2 than arterial blood but has a lower O_2 content because of the Root effect.

Lactate is exchanged from the efferent venous vessels into the afferent arterial vessels. This maintains a high concentration of lactate in the gas gland to unload O_2, but we might expect the magnitude of the Root effect to be reduced along the length of the venous vessel as lactate is exchanged into the arterial vessel. However, the loading of O_2 onto hemoglobin as pH increases (the Root-On effect) is slow, with a halftime of about 10 to 20 sec, whereas the unloading of O_2 from hemoglobin (the Root-Off effect) is fast, with a halftime of about 50 msec.

The swim-bladder rete system is a **countercurrent multiplier**; there is countercurrent exchange between afferent and efferent blood vessels, and there is addition of lactate along the length of the rete. There is a small exchange of O_2 (due to a small O_2 concentration gradient) in each short segment of the rete, but this is multiplied into a large change over the entire length of the rete capillaries. This produces a substantial overall O_2 transport from venous to arterial blood against a marked O_2 partial pressure gradient.

Cephalopod shells.   The cephalopod buoyancy organ is a rigid gas-filled chamber. The nautilus has a complete shell; the outer part houses the animal and the inner part is its buoyancy organ. Cuttlefish, such as *Sepia*, have a reduced, internal shell that is highly modified into a cuttlebone. Squid (*Loligo*) have the shell reduced to a long, flattened chitinous plate, the pen, and octopods have no remnant of the shell at all. The cephalopod shell/cuttlebone provides neutral buoyancy in the same fashion as a teleost swim-bladder, but it works by a fundamentally different principle.

The cuttlebone of *Sepia* is about 9.3% of its body volume and has a density of about 620 kg m^{-3}; the remainder of the cuttlefish has a density of about 1067 kg m^{-3}. The overall density is about 1026 kg m^{-3}, i.e., the cuttlefish is neutrally buoyant in seawater. The cuttlebone has about 100 plates of calcified lamellae that form a series of chambers; the chamber walls are held apart by sturdy pillars and are further divided by thin membranes (Figure 10–21). A strong, calcified membrane covers the entire structure except for a vascular epithelium on the posteroventral surface. This epithelium is an osmotic pump and provides the mechanism for regulation of buoyancy. The chamber volume (about 38 ml) contains about 14 ml of fluid and 24 ml of gas. The gas is enclosed within the rigid structure and has a subatmospheric pressure, e.g., about 70 kPa. The gas pressure is always less than the ambient hydrostatic pressure regardless of the depth. Consequently, the shell must be sufficiently strong to withstand considerable hydrostatic pressures. Cuttlebone will not collapse at less than 2400 kPa, which is adequate because cuttlefish do not swim deeper than about 150 m (1600 kPa pressure). The shell of *Nautilus* can withstand up to 6500 kPa; this is sufficient to withstand the maximum pressure that it would experience (about 5000 kPa at 500 m depth).

Cephalopods do not secrete gas into their shell but osmotically withdraw body fluid to leave a partial vacuum into which gases diffuse from the body fluids. Osmotic withdrawal of ions from fluid in the chamber causes an osmotic efflux of water from the chamber into the body fluids. This produces a negative pressure inside the chamber and causes the gases dissolved in the body fluids to diffuse into the gas space of the chamber. The chamber gas would tend to come into equilibrium with body fluid gas composition, e.g., pO$_2$ about 20 kPa, pN$_2$ about 80 kPa, but the metabolic consumption of O_2 reduces the pressure to below 100 kPa. The external hydrostatic pressure tends to force body fluids into the air space, but this is avoided by the large hydrostatic pressure gradient that can be sustained by a small osmotic gradient. Even at a depth of 72 m and a hydrostatic pressure of 820 kPa, a pressure difference of about 750 kPa can be sustained by an osmotic gradient of only 335 mOsm (1 Osm is equivalent to about 2240 kPa). The normal body fluid concentration of cephalopods is about iso-osmotic with seawater, i.e., 1000 mOsm (see Chapter 17) and it is relatively straightforward to reduce the osmotic concentration of the cuttlebone fluid by 335 mOsm to 665 mOsm.

Swimming.   Aquatic animals are immersed in a fluid medium that provides near-neutral buoyancy and allows the generation of a thrust force for swimming by pushing against the fluid. The mechanics for generating thrust depend in a complex fashion on the properties of the medium (density, viscosity, etc), the size of the organism, and the type of swimming mechanism, e.g., rowing, undulatory movements, hydrofoils, jet propulsion.

Size and scale are extremely important in determining the significance of thrust generation. Water "seems" to be a very viscous fluid to small animals; viscous forces predominate over weight, buoyancy, and inertial forces. Water does not "seem" to be very viscous to large animals; inertial rather than viscous forces predominate. Viscous forces are proportional to the size of an animal (measured by some characteristic dimension, e.g., length, l), the velocity of motion (v), and the viscosity of water (η).

$$\text{viscous force} \propto lv/\eta \qquad (10.7a)$$

Inertial forces are proportional to the square of the characteristic length (l^2), the square of the velocity (v^2), and the density of water (ρ).

$$\text{inertial force} \propto l^2v^2\rho \qquad (10.7b)$$

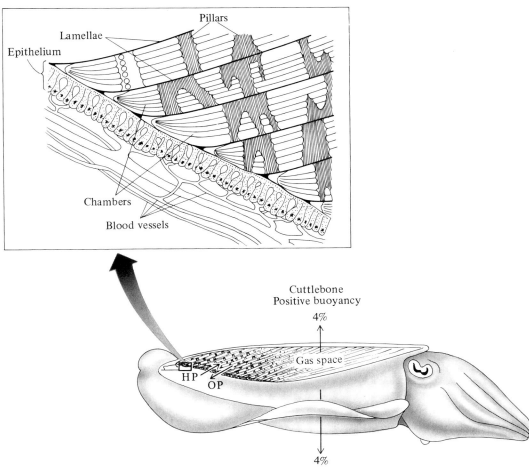

FIGURE 10–21 The cuttlefish *Sepia officinalis* uses its cuttlebone to attain neutral buoyancy. The cuttlebone is divided into rigidly supported chambers by lamellae and pillars. An underlying epithelium regulates the osmotic composition of body fluid that partially fills the chambers. The volume of gas in the cuttlebone depends on the equilibrium between the hydrostatic pressure (HP) that forces body fluids into the cuttlebone and the osmotic pressure (OP) that withdraws fluid from the cuttlebone. The buoyancy provided by the cuttlebone is about 4% of body weight (in air). *(From Denton 1961; Denton and Gilpin-Brown 1961.)*

The **Reynold's number** (R_e) is the ratio of the inertial forces to the viscous forces. It is calculated as

$$R_e = \text{inertial force/viscous force}$$
$$= lv\rho/\eta \qquad (10.7c)$$
$$= lv/\nu$$

where ν is the kinematic viscosity ($\eta/\rho = 1 \; 10^{-6} \; m^2 \; sec^{-1}$ for water at 20° C). The R_e varies dramatically from 10^{-6} for single cells to $>10^8$ for large whales (Figure 10–22). At low R_e, viscous forces predominate in locomotory mechanisms, whereas at high R_e inertial forces determine the efficacy of the locomotory mechanism. The typical cruising velocity is also highly dependent on R_e, being less than 10^{-6} m sec^{-1} at low R_e and greater than 10 m sec^{-1} at high R_e.

An object moving through a fluid, or with a fluid moving past it, experiences a **drag** force. Drag is a complex phenomenon; it is a force proportional to velocity at low R_e and to velocity^2 at high R_e. The

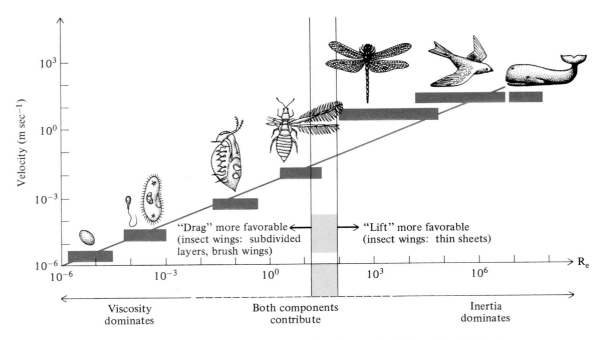

FIGURE 10–22   Scaling of velocity and Reynold's number from unicellular animals (R_e of 10^{-6}) to large swimming animals (R_e of 10^8).   *(From Nachtigall 1983.)*

direction of the drag force is, by definition, parallel to the direction of the local fluid flow; this is not necessarily the direction of movement by the animal. The total drag (D_{tot}) of an object is due to **friction drag** at its surface (D_f) and **pressure drag** (D_p). The friction drag component is proportional to the total surface area of the body. Pressure drag is caused by the inertia of the fluid flowing past the object, forming a low-pressure turbulent wake; consequently, pressure drag depends on shape. **Induced drag** (D_{in}) is caused by the generation of lift by hydrofoils; it will be further described below.

Many animals swim using drag. The drag force on part of the body is resolved into a thrust, propelling the animal forward, and a sideways force. The overall sideways force is generally insignificant when averaged over an entire swimming cycle. Rowing animals, such as water beetles (Dysticidae, Gyrinidae) and bugs (Notonectidae, Corixidae), use synchronized rowing strokes of their rear legs to generate drag, which results in a propulsive thrust force. The leg motions of a whirling beetle (*Gyrinus*; Figure 10–23) generate a low drag during the recovery part of the cycle (stages 1–4) and the spread in readiness for the power stroke (4–6), and a high drag during the power stroke (7–14). This high drag is resolved into a forward thrust parallel to the long axis of the body and the swimming direction.

Many animals generate drag and thrust by undulatory movements of the body that "push" against the medium (Figure 10–24A) rather than by rowing with appendages. Spermatozoans and flagellate protozoans, as well as larger animals (eels, other fish, snakes, leeches, and some worms), swim using undulatory body movements. Flagellates and spermatozoans swim at low R_e and so drag is proportional to velocity. The motion of the flagellum pushes against the water, and the drag of the flagellum generates an equal and opposite-directed locomotor force that can be resolved into a transverse thrust and a longitudinal thrust (Figure 10–24B). The sideways thrust summed over the length of the flagellum, and over time, becomes negligible, so the spermatozoan (or flagellate) does not move to the side. However, there is a net forward thrust that propels the organism forward.

The principle of undulatory movement is similar for larger animals (worms, fish, etc.) except that the R_e is higher and drag is proportional to v^2. A typical fish with a large caudal fin (e.g., a trout) swims forward at a velocity v by undulations of the body that travel posteriorly along the body at a different velocity, c; lateral movements of the tail alternately force water to the left then the right (Figure 10–25A). The hydrodynamic force of the fish's tail is resolved into a drag and lift force or alternatively

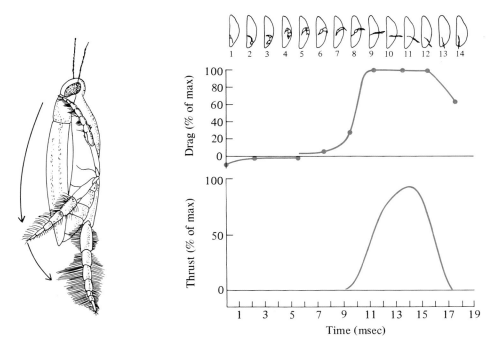

FIGURE 10–23 Diving beetles swim by rowing; drag from the rowing movements of the rear right leg produces hydrodynamic drag and forward thrust during the forward phase of swimming. *(From Nachtigall 1960; Nachtigall 1983.)*

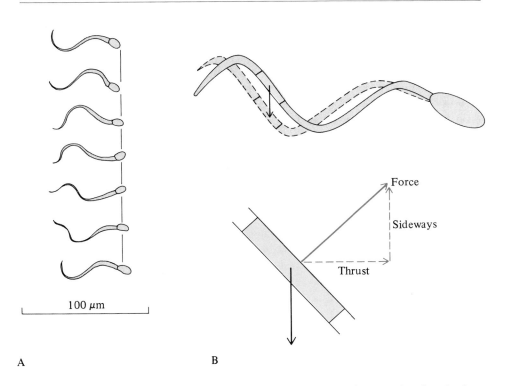

A   B

FIGURE 10–24 **(A)** Undulatory movements of a sperm. **(B)** Diagram showing the hydrodynamic force acting on a segment of the spermatozoan flagellum during swimming and the resolution of the force into a forward-directed thrust and a sideways force. *(Modified from Alexander 1982.)*

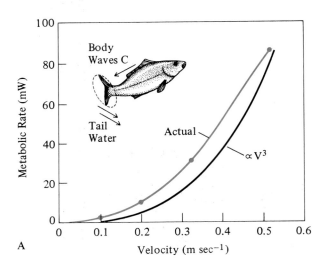

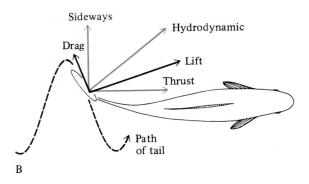

FIGURE 10–25 (A) Undulatory movements of a teleost fish's tail. The actual metabolic rate for swimming closely matches the calculated rate, proportional to velocity³. **(B)** The streamlined body form of a skipjack tuna minimizes drag. Its undulating trail acts as a hydrofoil and generates lift and drag forces. The resultant hydrodynamic force is resolved into a forward-directed thrust, and a sideways force. *(Modified from Alexander 1977.)*

into a useful forward thrust and a nonuseful sideways thrust (that averages over time to 0).

Some fish, such as tunny and tuna, have a hydrodynamically sophisticated propulsive mechanism that emphasizes the role of lift forces to produce thrust, rather than drag forces (Figure 10–25B). The tail is stiff and shaped like a wing, i.e., it is a **hydrofoil**; it moves about 6 × its chord length in each left-to-right and right-to-left movement. The hydrofoil tail generates lift and drag. Lift (L) is force exerted on an object by a fluid, perpendicular to the fluid motion, whereas drag is exerted parallel to the fluid movement (see Aerial Locomotion, page 479, for a more detailed discussion of lift). A hydrofoil

must move a distance of about 2 × its chord before it creates a normal fluid circulation pattern around it and achieves stable lift generation; this is the Wagner effect. Hence, it is significant that the hydrofoil tail of tunny and tuna moves by about 6 chords per sideways stroke.

Many other actively swimming, large animals also swim using hydrofoils because they operate at high R_e and so lift is substantial. Marine turtles and penguins use their flippers/wings as hydrofoils to generate lift. The up and down movements of their flippers generate vertical forces that cancel over a flipper/wing beat cycle but generate a net forward thrust. Whales also move their tail flukes up and down when they swim, rather than from side to side as do fish.

Some fish and cephalopods use lift from hydrofoils to achieve neutral buoyancy (Figure 10–26). For example, the pectoral fins of sharks and tunny and the tail fins of squid generate a lift that counteracts their negative buoyancy. Additional lift must also be generated by other body surfaces to maintain body equilibrium since the lift of the hydrofoils is not at the center of mass. The heterocercal tail of sharks, the caudal peduncle of tunny, and the jet propulsion thrust of squid provide these additional lift forces. Hydrofoil-generated lift is not a suitable mechanism for less-active animals to attain neutral buoyancy.

Jet propulsion is another mechanism for aquatic locomotion (Wells 1990). Thrust can be generated by squirting water from a mantle cavity (e.g., cephalopods, such as squid *Loligo*), rectum (e.g., dragonfly larvae), gastrovascular cavity (e.g., jellyfish), or rapidly closing shells (e.g., bivalve scallops). Squid draw water into their mantle cavity through a wide slit in the mantle wall and squirt it out of a tubular funnel. Gentle rhythmic contractions of the mantle cavity maintain a one-way flow of water over the gills. Strong contractions of the mantle rapidly eject seawater and jet the squid forwards, or backwards, depending on the direction of the funnel opening.

Metabolic Cost. The power required to swim is equal to drag × velocity. The metabolic cost of swimming is the metabolic power required to sustain this mechanical power output. The metabolic cost of swimming obviously varies for different animals, for different swimming mechanisms, at differing R_e, and at differing swimming velocities. At low R_e, drag is proportional to v and so power is proportional to velocity². For high R_e, drag is proportional to v^2 and so power should be proportional to v^3 (see Figure 10–25A, page 478).

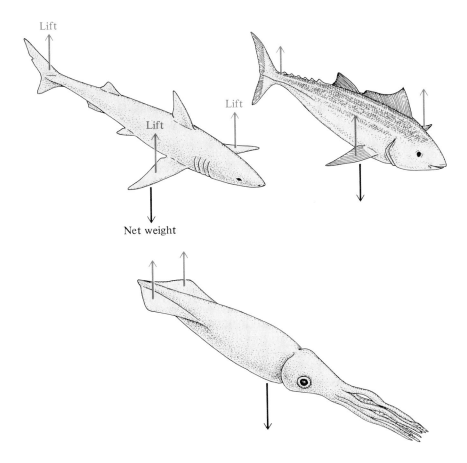

Net weight

FIGURE 10–26 The shark, tunny, and squid are denser than seawater but swimming generates a lift force that reduces their negative buoyancy. *(Modified from Alexander 1982.)*

The net cost of transport depends on swimming velocity because locomotor power is not linearly proportional to velocity (as it was for many walking and running animals). Nevertheless, it is possible to compare the minimum cost of transport for many swimming animals at a velocity appropriate to their size. The relationship between minimum net cost of transport and body mass is remarkably consistent for swimmers from sperm to large fish; the absolute cost increases with mass (Figure 10–27). Swimming at the water surface requires considerably more energy because of the energy dissipation as the wake (e.g., a paddling duck or a swimming human).

Aerial Locomotion

Animals that glide or fly are constrained by the same general fluid dynamic principles as swimming animals. Drag and lift are the fluid dynamic forces that provide the propulsive force for locomotion (see Supplement 10–3, pages 490–92). However, there are two major differences between animals that move in air and in water. First, air has a much lower density (1.18 kg m^{-3}) than water (1000 kg m^{-3}), and so it provides essentially no buoyancy

compared to water. Therefore, animals that fly horizontally must generate sufficient lift to counterbalance their body weight. Even gliding animals, which do not maintain their horizontal height (unless gliding in an updraft), must generate considerable lift to avoid descending too quickly. Second, the Reynold's number is lower for air than water at similar velocity (v) and characteristic length (l) because air has a higher kinematic viscosity ($\nu = 16 \times 10^{-6}$ m^2 sec^{-1}) than water (10^{-6} m^2 sec^{-1}).

The upwards force that counteracts the body weight of a gliding or flying animal is produced by the downward acceleration of air. In theory, this could be accomplished by using either drag (e.g., "rowing") or lift, but all fliers except the smallest rely on aerodynamic lift rather than drag because it is energetically more economical.

The wings of extremely small insects are only 0.05 to 0.2 mm long. They consist of a central support with a marginal row of hairs (Figure 10–28A). Their R_e is so low that the viscous properties of air predominate over inertial forces, and so the wings do not have to be solid structures. Air is too "thick" to flow between the marginal hairs. These small wings would produce little lift, and drag is the

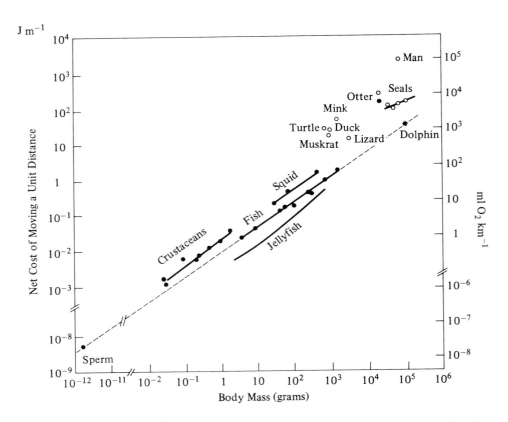

FIGURE 10-27 Minimum cost of transport as a function of body mass for swimming animals expressed as the metabolic expenditure to move a unit distance (J per m or ml O₂ per km). •, swimming underwater; ○, swimming at the surface; *, calculated assuming laminar flow.

main propulsive force (Weis-Fogh 1976). The lift coefficient is <0.3 and the drag coefficient >0.3; the lift/drag ratio is less than 1. The wings of larger flying animals are more conventional aerofoils (Figure 10–28B–E), and lift is the predominant aerodynamic force. The lift coefficient exceeds 1.0 and the drag coefficient is less than 0.1; the lift/drag ratio exceeds 10.

Gliding. Many animals glide. Some also use flapping flight (e.g., insects, birds, bats), but many do not. There are gliding mammals (e.g., possums, rodents, flying lemurs), reptiles (e.g., snakes, geckos, and other lizards), and even some gliding amphibians and fish.

A flat or cambered aerofoil that is oriented at an angle to a moving fluid will experience on aerodynamic force (A) perpendicular to its surface (Figure 10–29). For a gliding animal, such as a bird, that is moving forward and downward, the relative air path is upward and backward. The aerodynamic force A is directed forward and upward relative to

its body. It can be resolved into a lift and a drag force. The lifting force (L) is, by definition, perpendicular to the direction of fluid movement and drag (D) is parallel to the fluid motion. Generally, L >> D for an aerodynamically shaped aerofoil at high R_e. The overall aerodynamic force can also be resolved into a vertical component (V) opposed to gravity and a horizontal thrust component (T).

A gliding animal keeps its wing (or gliding membranes) rigid. It can maintain a constant forward velocity by inclining the relative air path to the horizontal so that the forward thrust balances the horizontal retarding force. The resultant vertical force also counterbalances the weight, and so the glider decreases in altitude at constant vertical velocity at a glide angle (θ = arctan [drag/lift]). The glider can maintain a constant altitude if the relative path of the air is horizontal (view Figure 10–29, page 481, rotated 45° clockwise!). The lift force now balances the weight, but the drag is not counterbalanced by any thrust and therefore the glider decelerates. This obviously is not aerodynamically stable.

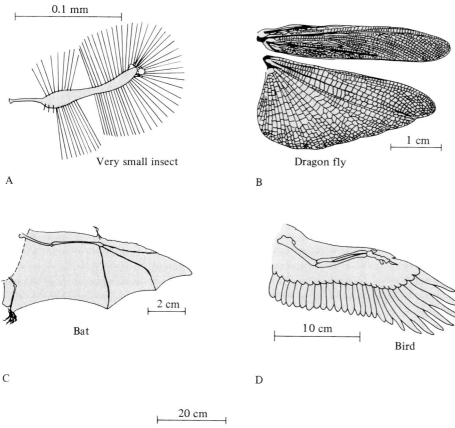

0.1 mm

Very small insect

A

Dragon fly

1 cm

B

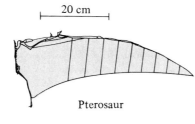

2 cm

Bat

C

10 cm

Bird

D

20 cm

Pterosaur

E

FIGURE 10–28 Scale of different animal wings. **(A)** A very small insect. **(B)** A small insect. **(C)** A medium-sized bat. **(D)** A large bird. **(E)** A very large pterosaur.

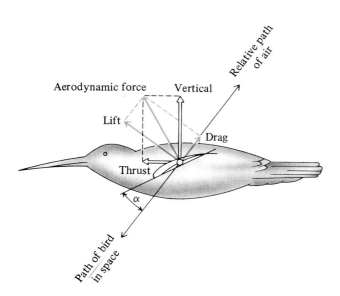

Relative path of air

Aerodynamic force   Vertical

Lift

Drag

Thrust

α

Path of bird in space

FIGURE 10–29 Schematic of the aerodynamic force induced on an aerofoil. This force is resolved into lift perpendicular to the direction of air flow (which is the result of the forward velocity of the animal and any flapping movements of the wing), and drag parallel to air flow. The aerodynamic force can also be resolved into a vertical component that counteracts the body weight and a forward thrust that counterbalances the backward retarding force. α is the angle of attack of the wing. *(Modified from Nachtigall 1983.)*

A glider can only maintain both forward velocity and altitude if the air has a vertical component, e.g., gliding in an updraft above a slope or in a thermal.

There is no direct metabolic cost to gliding. The generation of the aerodynamic force on a wing does not require metabolic energy expenditure; it is a fluid dynamic phenomenom. The energy is derived from the decrease in altitude (potential energy) of the glider. However, a gliding animal expends some muscular energy in keeping its wings rigid. The metabolic cost of gliding for birds is only about 2 × the resting metabolic rate (Baudinette and Schmidt-Nielsen 1974).

Flapping Flight. Flying insects, birds, and bats can maintain a constant forward velocity and altitude by using muscular energy expenditure to generate aerodynamic lift; they flap their wings to fly. Moving the wing down in concert with the forward motion of the animal produces an inclined path of air movement relative to the wing (see Figure 10–29, but read "path of wing in space" rather than "path of bird in space"). Gliding produces this same relative air path but by a decrease in altitude. It is much more difficult to analyze the aerodynamics of flapping flight than gliding with a fixed wing because the up and down motion of the wings must be vectorially added to the forward velocity of the wing relative to the air, and there are unsteady-state aerodynamic effects (especially at the top of up-stroke and bottom of downstroke). The up and down wing velocity also varies from a maximal value at the wing tip to a minimum value at the wing base. For a large flying bird, such as a stork, the vertical wing movements are relatively minor compared to the forward velocity and most of its wing generates forward thrust during both upstroke and downstroke (Figure 10–30). Smaller animals, such as the locust, only generate forward thrust during the downstroke and the wings generate negative thrust during the upstroke.

The energy expenditure of flapping flight overcomes the aerodynamic drag. The total aerodynamic power required for flapping flight (P_{total}) depends on the total aerodynamic drag, D_{total}, and velocity v.

$$P_{total} = D_{total}v \qquad (10.8)$$

A flying animal experiences three types of drag force: drag on its body (parasite drag), drag on its wings (profile drag), and drag due to lift generation (induced drag).

Parasite drag (D_{par}) is the combined friction and pressure drag of air moving over the body (pressure drag predominates at high R_e). It is calculated from the frontal area (A_f) of the body, velocity, and drag coefficient (see Supplement 10–3, pages 490–92).

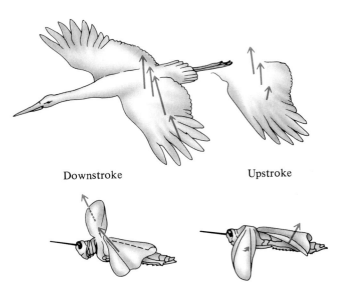

Downstroke   Upstroke

FIGURE 10–30 Aerodynamics of fast forward flight by a stork results in almost constant lift and thrust throughout the wing beat cycle, whereas the upstroke of the wings of a locust generates significant drag and some lift. *(From Weis-Fogh 1976.)*

Parasite drag increases with the square of velocity, and so parasite power increases with the cube of velocity (Figure 10–31A).

The generation of lift entails an additional drag component, the **induced drag** (D_{in}). Induced drag is proportional to the square of the lift, and is inversely proportional to the square of the velocity and square of the wing span. Thus, induced power is greater for larger body mass animals, at lower velocity, and for short wingspans.

Profile drag (D_{pro}) is a complex combination of friction and pressure drag of the wing (pressure drag predominates at high R_e). Profile drag is less well understood than the other drag terms. Profile power is sometimes assumed to be constant regardless of v or to vary with v in proportion to the sum of P_{par} and P_{in}.

The **total aerodynamic power requirement** for flight (P_{aero}) is the sum of $P_{par} + P_{in} + P_{pro}$; an example of the calculated P_{aero} curve and the individual components is given in Figure 10–31A. There is a specific velocity for minimum power (V_{min}; about 5 m sec^{-1} for the pigeon). The velocity for moving the maximum range per unit power expended (V_{mr}) is the velocity at the tangent from the origin to the P_{aero} curve. V_{mr} is higher than V_{min} (about 13 m sec^{-1} cf. 5 m sec^{-1} for the pigeon). Unfortunately, aerodynamic theory does not neces-

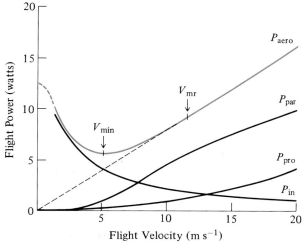

A

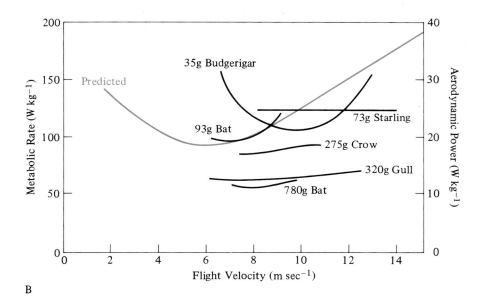

B

FIGURE 10–31 (A) Calculated power requirements for flight by the pigeon (mass 0.333 kg) as a function of velocity. The total power requirement (P_{aero}) is the sum of parasite drag on the body (P_{par}), profile drag of the wings (P_{pro}), and induced drag (P_{in}). (B) Measured metabolic cost of flight for birds and a bat compared with the general shape of the curve calculated from aerodynamic theory. *(From Rayner 1979; Alexander 1982.)*

sarily describe well the actual relationship between the power expended by a flying bird and flight velocity; the actual metabolic rate curves tend to be flatter than the P_{aero} curves for those few birds that have been studied (Figure 10–31B). Many factors contribute to the total cost of flying other than the aerodynamic costs (see below).

Metabolic Cost. The minimum aerodynamic energy cost calculated for flight ($P_{min,aero}$) increases with body mass; for birds,

$$P_{min,aero} = 15.4 \, M^{1.10} \quad (10.9a)$$

where P has units of watts and mass of kg (Rayner 1979). However, the net aerodynamic cost of transport for birds, the aerodynamic work required to transport a unit weight through a unit distance (COT_{aero}), declines with mass

$$\begin{aligned} COT_{aero} &= P_{min,aero}/(Mass \cdot g \cdot v) \\ &= 0.212 \, M^{-0.07} \end{aligned} \quad (10.9b)$$

where COT_{aero} is dimensionless.

The power expended by a flying animal, the power input P_i, is not equal to the calculated $P_{aero}(P_{par} + P_{in} + P_{pro})$ for a number of reasons (Figure 10–32). First, a flying animal expends some muscle energy overcoming aerodynamic drag (P_{aero}) but also expends considerable energy on physiological functions, some of which are related to flight (e.g., elevated heart and respiratory rate), some of which are a consequence of flight activity (e.g., increased VO_2 due to an elevated body temperature

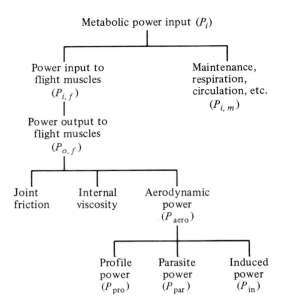

FIGURE 10–32 Partitioning of total power input of a flying animal (P_i) into flight muscles and maintenance, and the partitioning of mechanical power output of the flight muscles. *(From Tucker 1975.)*

during flight), and some of which are unrelated to flight (e.g., liver and kidney metabolism). These nonaerodynamic metabolic expenditures can be combined as a maintenance power input ($P_{i,m}$). Second, the energy input to the flight muscle ($P_{i,f}$) is more than the energy output for mechanical work ($P_{o,f}$) because muscle has a relatively low efficiency of mechanical work output, about 20 to 30%, and energy is also dissipated by joint friction and internal viscosity. Thus, only a fraction of the energy input to the muscle is available for aerodynamic work (P_{aero}); the remainder of the energy used by a muscle is liberated as heat. Therefore, power input by the bird (P_i) is about four to five times the aerodynamic power output.

The total and net costs of transport for flight are markedly dependent on flight velocity, being infinite for hovering ($v = 0$), minimal for v_{mr}, and high at $v > v_{mr}$. The minimal net cost of transport for invertebrate and vertebrate fliers declines with body mass. The cost of hovering, however, increases with mass, and only relatively small fliers (<20 grams or so) can hover (except in an updraft) for significant periods of time.

Comparison of Locomotory Costs

There are many different ways to express the cost of locomotion. The total cost of transport (COT_{tot}) is the measured metabolic rate divided by the locomotor velocity, e.g., ml O_2 km^{-1}, J kg^{-1} m^{-1}. Alternatively, the net cost of locomotion (COT_{net}) is calculated from the metabolic increment above resting.

Each approach to measuring or calculating a cost of transport has its relative advantages and disadvantages. Probably the best estimate that allows us to compare the cost of transport for a variety of different types of locomotion, by different types of animals, and of varying body mass, is the total metabolic cost of transport. Further, by using mass-specific units, i.e., ml O_2 kg^{-1} km^{-1}, we can readily gauge the relative cost of transporting a unit mass (1 kg) a unit distance (1 km). We can convert this unit system into a dimensionless one by converting ml O_2 to joules (1 ml O_2 is equivalent to about 20.1 J) and converting mass to weight (1 kg mass = 9.8 N); 1 ml O_2 kg^{-1} km^{-1} is equivalent to a total metabolic cost of transport of 0.0021 (dimensionless).

A second complication in comparing the cost for different types of locomotion by different animals of differing body mass is the dependence of locomotor cost on velocity. This is less of a problem for the net cost of transport for many walkers and runners because this is independent of velocity, but it is not velocity independent for all forms of terrestrial locomotion (e.g., bipedal hopping) and is strongly dependent on velocity for swimming and flying. Consequently, we need to estimate the minimum total metabolic cost of transport for these various examples so that we can compare the most economical cost of transport for different animals. For swimmers, cost is proportional to v^3, so the cost of transport is often calculated at the normal cruising velocity.

The minimum total metabolic cost of transport calculated in the above manner is inversely related to body mass for terrestrial locomotion (walking, running), for swimming, and for flying (Figure 10–33). There are marked differences, however, in the metabolic cost of transport for the different kinds of locomotion at the same body mass (i.e., there are intercept differences for the regression equations). Swimming is the most economical form of locomotion, and burrowing is the least economical.

Why is swimming so much more economical than other forms of transport? First, swimmers are neutrally buoyant, or nearly so, and no locomotory energy is required to support body mass. Second, swimmers generally move at low velocities; the cost of transport increases with v^3 for swimming, and so the cost of transport by swimming would be much higher at velocities equivalent to flight velocities.

Why is flight more economical than walking and running? Much of the energetic cost of flying

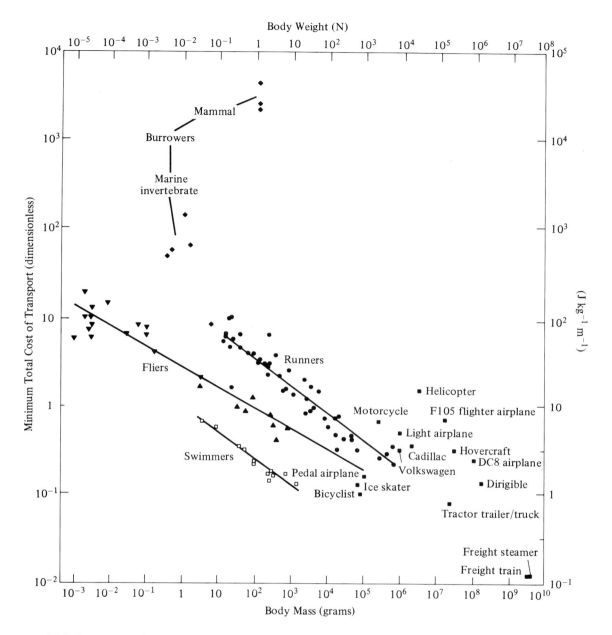

FIGURE 10–33 Comparison of the minimum cost of transport for swimmers, fliers, runners, and some man-made machines. The cost of transport is a dimensionless unit. It is calculated as total metabolic expenditure divided by body weight (mass times the gravitational constant) and velocity, at the most economical velocity. *(Modified from Tucker 1975.)*

overcomes aerodynamic drag (of the body, wings, and induced drag). Sufficient lift must be generated to balance body mass but lift does not require energy expenditure *per se*; it is a consequence of air flow over the wings. Aerofoils typically have favorable lift/drag ratios, and so considerable lift is generated at minimal drag cost. In contrast, walking and running animals have to support their body mass against gravity, otherwise they would fall over.

Muscle tension is required to maintain the posture and erect stance of walking and running animals and to accelerate the center of mass during locomotion; this involves considerable energy expenditure. Walkers and runners expend very little energy in overcoming aerodynamic drag, in marked contrast to fliers, because their locomotory velocities are generally quite low. Furthermore, the metabolic cost of transport for flying is very high, but this is at a higher velocity than for running so the cost of transport is low.

Why does it cost so much energy to burrow? Marine invertebrates that burrow in soft, water-saturated substrate must not only move across the substrate (i.e., walk) but also have to push the substrate material out of the way; this obviously requires more energy than just walking. Mammalian burrowers have a considerably higher cost of transport. They burrow through harder, drier soil and more energy is expended on shearing the soil to make a tunnel to walk through. They must also move the excavated soil out of the way; this requires additional locomotion up and down the burrow.

Summary

Animals have a skeleton to resist deformation of their body shape against external and internal forces and for support during locomotion. Terrestrial animals must resist the force of gravity and other external and internal forces. Aquatic animals do not experience the full force of gravity because water provides a buoyant force, but they do experience other external forces as well as internal forces. The skeleton can be a pressurized fluid compartment (hydrostat) or a rigid skeletal framework.

Biological materials differ dramatically in their mechanical properties, such as density, modulus of elasticity, strength, elasticity/plasticity, and visco-elasticity. Bone and chitin have a high modulus of elasticity and a high strength; they are suitable materials for rigid skeletons. Collagen fibers have a high tensile strength and are present in structures that experience tensile forces, e.g., the walls of hydrostats, tendons. Elastin, abductin, and resilin are rubbery polymeric proteins that have a very low modulus of elasticity; they are present in tissues that deform considerably (e.g., elastin in skin) or deform and elastically store energy (e.g., resilin in the insect thorax and abductin in the shell valve of scallops).

Hydrostat skeletons are a pressurized fluid space. Some are balloon-like cavities filled with either body fluid or water, whereas muscular hydrostats are masses of longitudinal and circular muscle cells. Solid skeletons have a high tensile and compressive strength of materials. They can be an exoskeleton that covers the outside of the body, e.g., the arthropod cuticle, or an endoskeleton that is located within the tissues, e.g., the vertebrate bony skeleton. Many of the solid skeletal elements act as beams and resist bending forces. These are often hollow, rather than solid, structures because a hollow beam has a greater stiffness and strength than a solid beam of the same mass and length. The rigid skeletal elements form a framework with moveable joints between many of the separate elements.

Many rigid skeletal elements have either articulating or elastic joints. Separate articulating skeletal elements either slide or roll past each other, and so their contact surfaces must be lubricated. Elastic joints rely on the deformability of a specialized region of the skeletal material that allows the joint to bend; articulations may also be present to limit the plane of motion of the joint. The mechanical properties of joints depend on the relative distances of the point of insertion of the muscle, and the position of the load, from the fulcrum. A joint has a high mechanical advantage if the muscle insertion is at a greater distance from the fulcrum than the application of the load. A joint has a low mechanical advantage if the muscle insertion is closer to the fulcrum than the load.

Animals have evolved a number of diverse mechanisms for movement. Terrestrial animals generally crawl, walk, run, hop, or leap. Aquatic animals often swim, although some crawl or walk over the substrate. Aerial animals either glide, or use flapping flight. Terrestrial animals with hydrostats can use a two-anchor principle to push or pull their body forward between the separate anchor points. Some use peristaltic body contractions to crawl. Those with framework skeletons use their appendages as supports and levers to walk or run. Their pattern of limb movement (gait) is quite variable, depending on the number of legs (e.g., 2, 4, 6, many) and the velocity of locomotion. In general, the metabolic expenditure for walking/running increases linearly with speed, and the metabolic cost of locomotion (energy cost to move a unit distance) increases with increasing mass. The energy cost of locomotion is low for some bipedal hopping animals, such as kangaroos, especially at high velocities because of elastic storage of energy by tendons.

The body tissues of aquatic animals generally have a negative buoyancy, but many can regulate their body density and achieve neutral buoyancy. There are three basic mechanisms for buoyancy control. Many deep-sea fish and sharks accumulate low-density lipids. Deep-sea squid, copepods, and coelenterates replace body fluid solutes with lower density solutes. Many teleost fish and some cephalopods have an internal gas-filled float. The gas pressure inside the soft-walled teleost swim-bladder is equal to the exterior hydrostatic pressure. Gas is secreted into the swim-bladder against considerable partial pressure gradients by a gas-secreting gland. Loss of gas from the swim-bladder is minimized by a low permeability of its wall and by a countercurrent arrangement of the vascular supply to the gas gland, the rete mirabile. The decreased O_2 capacity of blood at low pH (the Root effect) provides the mechanism for transporting gas into the swim-bladder against a pO_2 gradient. The shell of the nautilus and cuttlefish is a rigid-walled float. It contains both fluid and a gas that is always at a lower pressure than the ambient. The shell has sufficient mechanical strength to resist the external hydrostatic pressure. The gas volume is regulated by osmotic withdrawal of solutes from the fluid inside the chamber that balances the hydrostatic pressure pushing fluid against the gas space.

Swimming is accomplished by a variety of mechanisms. Undulatory swimming by waves of movement along the body and tail, and rowing with oar-like appendages, produce a drag force that is resolved into a forward thrust. Some large, active swimmers use sideways or vertical movements of hydrofoil appendages (flippers, tail fins, or flukes) to generate lift that is resolved into a forward thrust force. Some animals squirt jets of water to move. The metabolic cost of swimming increases in approximate proportion to the velocity[3].

Aerial animals glide or flap their wings to fly. The aerodynamic force on the wing is resolved into a lift and drag force that resists the gravitational force (weight) and provides a horizontal force for forward motion. In gliding, the aerodynamic force is resolved into a vertical force that counters gravity and a forward thrust that counters the drag of the animal, but a gliding animal cannot maintain both vertical height and forward velocity in still air. Either height is lost to maintain a constant forward velocity, or forward velocity is lost to maintain a constant height. Flapping flight converts muscular movements of the wings into an additional aerodynamic force; both vertical height and horizontal speed are maintained. Some flying animals can hover. The aerodynamic costs of flapping flight overcome the parasite drag of the body, the profile drag of the wings, and the induced drag resulting from lift generation. The total aerodynamic power for flight is high at zero velocity (i.e., hovering), is minimal at some intermediate velocity, and increases at higher velocities.

The metabolic expenditure for locomotion varies dramatically for different modes of locomotion and for animals of differing body mass. The most economical form of transport is swimming, but the typical swimming velocity is quite low. Flying is not quite as economical as swimming, but flight is accomplished at much higher velocities than swimming. Walking and running are not as economical as swimming or flying. Burrowing is very uneconomical because it involves not only forward movement but also requires the breaking apart of the substrate to form a path and often the transport of the removed substrate out of the way. Large animals have a lower cost of transport than smaller animals, i.e., it is more economical for a large animal to move 1 kg of body mass over 1 km.

Supplement 10–1

Analysis of Bending Forces

• •

Bending is one of the most common ways that stiff skeletons are deformed, sometimes to the point that they break. Consider a rigid cylindrical beam (length l and radius r) that is long and thin. If the beam is supported at one end and loaded by a force (F) at the other end, then it will sag at the loaded end. The deflection of the beam (Δx) can be calculated as

$$\Delta x = \frac{Fl^3}{3EI}$$

where the radius of deflection is R, E is the elastic modulus, and I is the second moment of inertia. The bending force stretches the top of the beam to length $l + \Delta l$ and compresses the bottom to a length $l - \Delta l$. The stress required to deform the top of the beam from l to $l + \Delta l$ depends on the elastic modulus and the ratio of the beam radius (r) to the bending radius (R), i.e., stress = Er/R since the strain is r/R ($\Delta l/l$). There is a corresponding stress compressing the bottom of the beam, also equal to Er/R. The neutral plane in the center of the beam experiences no stress. The stresses at intermediate distances from the neutral plane are less than the stress at the top and bottom of the beam.

The bending moment (M) is the sum of all of the forces that tense the top of a bending beam and compress the bottom. This moment acts about the neutral plane of the beam, which is neither compressed nor stretched. The bending moment depends on the elastic modulus and the second moment of inertia (I) and is inversely dependent on the radius of bending of the beam (R).

$$M = EI/R$$

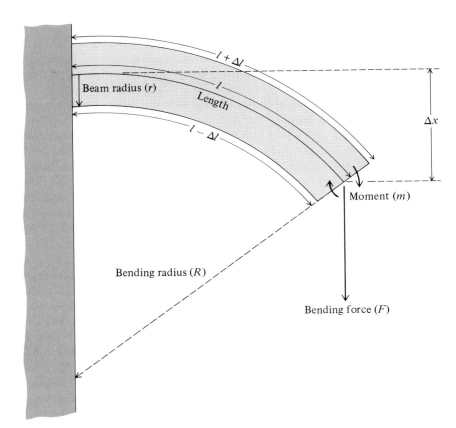

Beam radius (r)

Length

$l + \Delta l$

l

$l - \Delta l$

Δx

Moment (m)

Bending radius (R)

Bending force (F)

Schematic analysis of the forces involved with stretching the top of a beam from length l to $l + \Delta l$ and compressing the bottom to length $l - \Delta l$. These tensile and compressive forces are a moment of forces acting about the neutral, mid plane of the beam, which is neither stretched or compressed; it remains at length l.

The second moment of inertia is the integral of the cross-sectional area of the beam multiplied by the square of the distance of the area from the center line of the beam. If we divide the cross-section of the beam into a large number of small areas (each of area A and distance r^2 from the center), then I is the sum of r^2A. The second moment of inertia indicates the strength of the beam in resisting bending; a high I will result in a high R (little bending) for a constant E and M.

The beam breaks if the load is excessive. It first fails where the stress and strain are greatest, i.e., at the very top or very bottom of the beam. Whether it fails at the bottom, where the material is compressed, or at the top, where the material is stretched, depends on the relative tensile and compressive strengths of the material. The maximum stress (σ_{max}) depends on the bending moment (M) and the second moment of inertia (I). It occurs at the very top and very bottom of the beam (at distance r from the neutral axis).

$$\sigma_{max} = Er/R = Mr/I$$

The concept of the second moment of inertia is important to our understanding of the optimal design of skeletal elements. Material near the neutral plane at the center of a beam contributes little to the strength of the beam because it is not compressed or stretched very much. A certain cross-sectional segment of the beam contributes more mechanical strength if it is further from

the neutral axis. The center of a solid cylinder contributes little to its mechanical strength and the outer parts contribute most. The weight of skeletal elements is important to animals because of the costs of construction

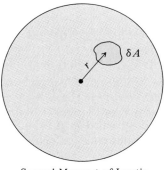

Second Moment of Inertia
= $\Sigma r^2 \delta A$

The second moment of inertia of a small part of the cross-section of a beam is equal to the square of its distance (r^2) from the center multiplied by its area (A).
The second moment of inertia for the entire beam is the sum of $r^2 A$ for all of the skeletal elements, i.e., $\Sigma r^2 \delta A$.

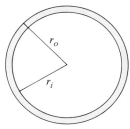

Outer radius	$r_o = 1$		$r_o = 2$	$(2\ \times)$
Inner radius	$r_i = 0.5$		$r_i = 1.8$	$(3.6 \times)$
Rigidity	$I = 0.736$		$I = 4.27$	$(5.8 \times)$
Strength	$I/r_o = 0.736$		$I/r_o = 2.14$	$(2.9 \times)$

The rigidity and strength of a hollow exoskeletal element depends on the inner and outer radii (r_i and r_o) and thickness (t). The amount of skeletal material is assumed to remain constant (i.e., the total cross-sectional area of the element is constant).

and maintenance of the skeleton and because of the energetic cost of moving the skeleton about. Consequently, the optimal design for many skeletal elements is a hollow tube rather than a solid rod. The actual structure of bones is a complex optimization to maximally resist the various forces that it experiences.

The second moment of inertia for a solid cylinder is

$$I = \frac{\pi r^4}{4}$$

The second moment of inertia for a hollow cylinder of outer radius r_o and inner radius r_i is as follows.

$$I = \frac{\pi(r_o^4 - r_i^4)}{4}$$

For hollow cylinders of varying wall thickness but of constant length and cross-sectional area A, i.e., containing the same amount of skeletal material,

$$I = \tfrac{1}{2}A(\pi r_o - (A/2\pi))$$

The rigidity of a hollow cylinder is proportional to I, so a larger diameter cylinder has greater rigidity. For example, using the same amount of material, a leg with an exoskeleton would be about 6 $\times$ more rigid than a leg with endoskeleton of 1/2 the external skeletal radius, r_o. Maximal strength is proportional to I/r, because it is inversely proportional to the maximum stress σ_{max}. A hollow leg with 2 $\times$ the diameter has about 3 $\times$ the strength of a solid leg. Note that rigidity increases with r faster than strength increases.

Supplement 10–2

Mechanics of Fusiform and Pennate Muscles

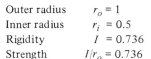

A fusiform muscle has its cells oriented along the length of the muscle so that the contraction of each muscle cell contributes to the overall shortening of the muscle. A pennate muscle has a more complex arrangement of its

muscle cells than the simple parallel arrangement of a fusiform muscle. The pennate muscle cells are arranged at an oblique angle to the direction of the muscle; the cells converge from a central tendon or apodeme. This

(i) (ii) (iii) (iv) (v)

Different arrangements of muscle fibers in (i) a simple fusiform muscle, (ii) a double-headed fusiform muscle, (iii) a unipennate muscle, (iv) a bipennate muscle, and (v) a multipennate muscle. *(From Goldspink 1977.)*

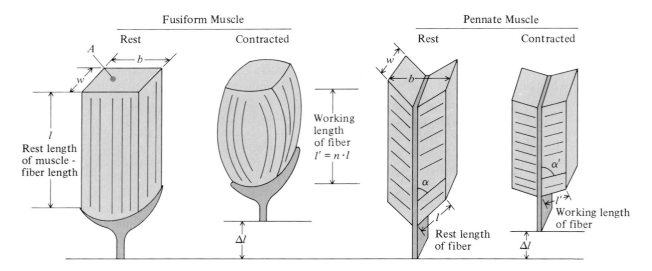

Comparison of changes in muscle dimensions with contraction for a fusiform and pennate muscle.   *(From Nachtigall 1983.)*

means that the cells are shorter, have a larger total cross-sectional area, and that there are more of them than in a fusiform muscle of the same mass. When a pennate muscle contracts, the short fibers each contract and shorten the muscle, becoming more perpendicular to the central tendon/apodeme. The direction of pull by pennate muscle cells is not in line with the direction of the muscle.

A fusiform muscle of breadth b and width w consists of parallel muscle cells; each with a cross-section A. The number of cells is bw/A. During shortening, the resting length l shortens by a length Δl. The contraction ratio, n, is the fractional shortening of the muscle $n = (l - \Delta l)/l$. The relative contraction ($\Delta l/l$) equals $(1 - n)/l$. The contractile force of the muscle is $(bw/A)F_m$, where F_m is the contractile force of a single cell. The total contractile force of the muscle along the length of the tendon is the same as the total muscle contractile force. Calculation of contraction distances and forces is more complex for a pennate muscle than for a fusiform muscle because the pinnate muscle fibers are at an angle (α) to the central tendon/apodeme and the angle alters with the extent of contraction.

Calculation indicates that a pennate muscle can generate a greater force than a fusiform muscle.

$$F_{penn}/F_{fusi} = \frac{2l \sin \alpha \sqrt{n^2 - \sin^2 \alpha}}{nb}$$

A pennate muscle has the same effect of increasing the force of motion as does a greater mechanical advantage. For example, the hind leg of the locust *Schistocerca gregaria* has a pennate muscle with α of 15°, length 17 mm, breadth 2 mm, and change in length of 1 mm. Thus, n is 0.89 ($1 - \Delta l/l = 16/17$). From the above equation, we have the following.

$$\begin{aligned} F_{penn}/F_{fusi} &= \frac{2(1.7 \ 10^{-2})(\sin 15°)\sqrt{(0.89^2 - \sin^2 15°)}}{(0.89)(2 \ 10^{-3})} \\ &= 4.2 \end{aligned}$$

It develops a force that is 4.2 times greater than that of a similar fusiform muscle. The claw closer muscle of a crab is another example of a pennate muscle that can generate a much higher contraction force than a simple fusiform muscle. (*See Alexander 1968; Nachtigall 1983.*)

Supplement 10–3

Aerodynamic Forces

. . . .

An aerofoil is able to generate a lifting force because it deflects a stream of air downwards; the downward change in momentum of air induces an equal-and-opposite momentum change for the aerofoil. Most wings are cambered in cross section, rather than being flat plates, because camber reduces the drag of the wing. The net aerodynamic

force acting on the wing can be resolved into a drag force that is parallel to the direction of the air flow and a lifting force that is perpendicular to the air flow.

The lift force of an aerofoil (or hydrofoil) depends on the dynamic pressure of the air ($1/2 \ \rho v^2$), the plan area (A_p), and a dimensionless lift coefficient (C_l).

$$L = \tfrac{1}{2}\rho v^2 A_p C_l$$

The plan area is the area of the aerofoil projected in dorsal or ventral plan view. The drag force of an aerofoil depends on the dynamic air pressure, the plan area, and a dimensionless drag coefficient (C_d).

$$D = \tfrac{1}{2}\rho v^2 A_p C_d$$

The drag coefficient of nonlifting bodies is often calculated with respect to the total surface area (wetted area, A_w) or the frontal projected area (A_f) rather than the plan area.

The dimensionless lift and drag coefficients are insensitive to size and velocity but they do depend on the Reynold's number. The maximum lift coefficient ($C_{l,max}$) is about 1.0 at R_e about 10^3 and 1.5 at R_e about 10^6. Even nonstreamlined bodies have fairly high lift coefficients, but they are not effective aerofoils because they have a very high pressure drag coefficient.

The C_d is determined by the shape of the object and the Reynold's number; C_d can be as high as 1.3 for blunt objects but is considerably lower (≤ 0.2) for streamlined objects (depending on the R_e). Objects of some shapes have a weak dependence of C_d on R_e (they are "R_e-insensitive" shapes), whereas others have a critical R_e above which the C_d declines dramatically (they are "R_e-sensitive" shapes). A sphere is an example of an R_e-sensitive shape ($R_{e,crit}$ is about 10^5); a model of a seagull is also "R_e-sensitive," with an $R_{e,crit}$ of about $4 \cdot 10^5$.

The magnitude of lift and drag, and the values of the lift and drag coefficients, vary with the angle between the wing and the air flow, the angle of attack α. A graph of C_l and C_d as a function of α (and also C_l/C_d) shows that a typical bird wing has a minimum drag coefficient at low, positive α and a maximum C_l of about 1.1 at higher α, about 30° to 40°. The lift/drag ratio is maximal at moderate C_l values and declines markedly, even at α for $C_{l,max}$.

The aerodynamic polar diagram, which graphs C_l as a

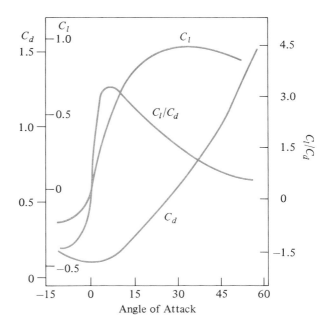

Aeorodynamic polar diagrams of lift coefficient (C_l) and drag coefficient (C_d) as a function of angle of attack (α) for a hawk wing; also shown is the C_l/C_d ratio. *(From Withers 1981b.)*

function of C_d, clearly shows that the minimum C_d value occurs at low C_l and that there is a marked increase in C_d at $C_{l,max}$. Different aerofoils have slightly different aerodynamic properties, e.g., $C_{d,min}$, $C_{l,max}$, $(C_l/C_d)_{max}$, etc., as is readily seen from a comparison of the polar diagrams for an NACA aerofoil, a hawk wing, a locust rear wing, and a fruit fly wing. $C_{l,max}$ decreases and $C_{d,min}$ increases at lower R_e.

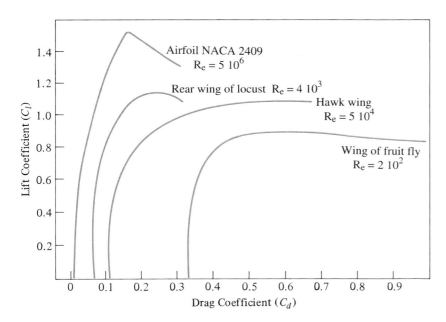

Aerodynamic polar diagram of lift coefficient as a function of drag coefficient for an airfoil, a hawk wing, the rear wing of a locust, and a fruit fly wing showing the effects of Reynold's number on maximum lift coefficient and minimum drag coefficient. *(Modified from Nachtigall 1983.)*

The actual lift and drag forces generated by a wing depend not only on the wing's lift and drag coefficients but also on its size (plan area, A_p) and the local air velocity (v). The lift that a wing must generate (body mass) is approximately proportional to l^3 (body volume), whereas wing area is approximately proportional to l^2, where l is a characteristic dimension of the body. Consequently, wing area decreases relative to body mass with increasing size, i.e., the wing loading (wing area/body mass) increases with mass.

The energy expenditure of a flying animal is used largely to overcome the aerodynamic drag of its wings. There are two components to the aerodynamic drag of a wing. Profile drag is due to the drop in pressure behind a wing as a result of the disruption of air flow and friction between air flowing over the wing and the wing surface.

Induced drag (D_{in}) is a second component of drag that is related to the lift generated by the wing

$$D_{in} = \frac{L^2}{\frac{1}{2}\rho v^2 s^2} = \frac{L^2}{\frac{1}{2}\rho v^2 \text{AR}}$$

where s is span of the aerofoil, A is the characteristic area (the product of span and chord, sc), and AR is the aspect ratio (the ratio of span/chord, s/c). Thus, a wing with a high aspect ratio generates less induced drag for a given lift than does a wing of lower aspect ratio but the same area. Induced drag is inversely proportional to v^2. For hovering flight, the induced drag is not infinite, but is calculated as

$$D_{in} = W^{3/2}/2\rho A$$

where W is body weight and A is planar wing area.

Recommended Reading

Alexander, R. McN. 1970. *Animal mechanics*. Seattle: University of Washington Press.

Alexander, R. McN. 1982. *Locomotion of animals*. Glasgow: Blackie.

Alexander, R. McN. 1988. *Elastic mechanisms in animal movement*. Cambridge: Cambridge University Press.

Alexander, R. McN., and G. Goldspink. 1977. *Mechanics and energetics of animal locomotion*. London: Chapman & Hall.

Currey, J. 1970. *Animal skeletons*. New York: St. Martin's Press.

Currey, J. D. 1984. *The mechanical adaptations of bones*. Princeton: Princeton University Press.

Denton, E. J. 1961. The buoyancy of fish and cephalopods. *Progr. Biophysics Biophysical Chemistry* 11:178–233.

French, M. J. 1988. *Invention and evolution: Design in nature and engineering*. Cambridge: Cambridge University Press.

Herreid, C. F., and C. R. Fourtner. 1981. *Locomotion and energetics in arthropods*. New York: Plenum Press.

Hildebrand, M. 1980. The adaptive significance of tetrapod gait selection. *Amer. Zool.* 20:255–67.

Hoppe, W., et al. 1983. *Biophysics*. Berlin: Springer-Verlag.

McMahon, T. A. 1984. *Muscles, reflexes, and locomotion*. Princeton: Princeton University Press.

Rainey, R. C. 1976. Insect flight. *Symp. R. Ent. Soc.* 7. London: Blackwell Scientific.

Sanders, N. K., and J. J. Childress. 1988. Ion replacement as a buoyancy mechanism in a pelagic deep sea crustacean. *J. exp. Biol.* 138:333–43.

Schmidt-Nielsen, K. 1972. Locomotion: Energy cost of swimming, flying, and running. *Science* 177:222–28.

Smith, K. K., and W. M. Kier. 1989. Trunks, tongues, and tentacles: Moving with skeletons of muscle. *Amer. Sci.* 77:328–35.

Taylor, C. R., et al. 1970. Scaling of energetic cost of running to body size in mammals. *Am. J. Physiol.* 219:1104–7.

Tucker, V. A. 1975. The energetic cost of moving about. *Amer. Sci.* 63:413–19.

Vogel, S. 1988. *Life's devices: The physical world of animals and plants*. Princeton: Princeton University Press.

Wainwright, S. A. 1988. *Axis and circumference: The cylindrical shape of plants and animals*. Cambridge: Harvard University Press.

Wainwright, S. A. et al. 1976. *Mechanical design in organisms*. New York: John Wiley & Sons.

Wu TY-T., C. J. Brokaw, and C. Brennen. 1974. *Swimming and flying in nature*. New York: Plenum Press.

Chapter 11

Endocrinology

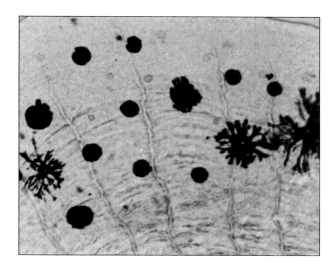

Chromatophores in the scale of a fish, the gluttonous goby *Chasmichthys gulosus.* *(Photo courtesy of R. Fujii, Department of Biomolecular Science, Toho University.)*

493

The endocrine system consists of various ductless glandular tissues. Although endocrine glands lack an obvious anatomical role, the physiological role of one endocrine gland, the testes, in the development of male sexual characteristics and reproduction has been recognized since before the twentieth century when the effects of castration had been repeatedly documented in both domestic mammals and man. Experiments by Berthold in the 1840s indicated that the testes of roosters released something into the body fluids that maintained secondary sex characteristics (such as the comb) and male behavior. About 10 years later, Claude Bernard coined the term "internal secretion," although he used it to describe the secretion of glucose into the blood, rather than to describe release of a hormone. Bayliss and Starling demonstrated in the early twentieth century that a chemical messenger (secretin) was released into the blood from the duodenum and conveyed to the pancreas where it initiated the rapid release of pancreatic juices into the pancreatic duct. Starling used the term **hormone** (from the Greek word "hormon," which means "to excite, to set in motion") to describe the blood-borne chemical messenger. Soon after, Pende introduced the term **endocrinology** (from the Greek words "endon," which means "within," and "krinein," which means "to separate") to describe the study of internal secretions from ductless glands, such as the testes. Initial studies of invertebrates were less promising; the absence of the effects of castration in insects seemed to exclude vertebrate-like endocrine functions (at least for the sex glands) and appeared to indicate that insects did not have hormonal systems. However, this hypothesis was soon abandoned and a considerable number of endocrine systems have now been described in invertebrates.

The **endocrine system** is a diverse assemblage of neurons and other cells that secrete specific chemical messengers or hormones. Hormones are secreted into and circulated throughout the body fluids and elicit responses in specific **target cells**. For example, the thyroid gland of vertebrates secretes a hormone, thyroxin, that affects the metabolism of most cells in the body. Exocrine glands also consist of secretory cells, but in contrast to endocrine glands they have a well-defined duct system to

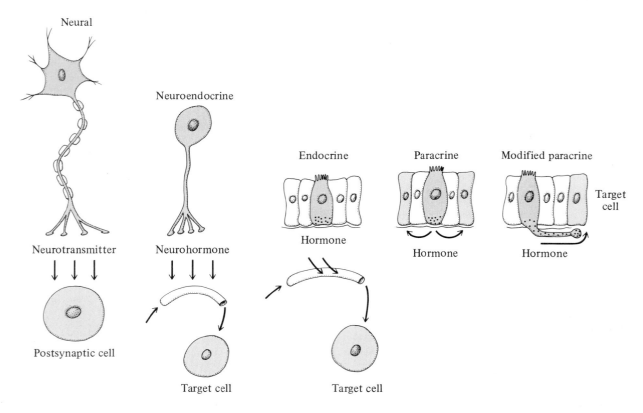

FIGURE 11–1 Schematic comparison of secretion systems for neurotransmitters, neurohormones, and hormones. *(Modified from Dockray 1981.)*

convey their secretions to the site of release. The salivary glands are an example of a typical exocrine gland. Saliva is secreted by the glandular cells and conveyed to the oral cavity via salivary ducts.

The endocrine system is an important effector system. Most animals have a variety of endocrine glands that secrete a number of hormones into the blood. These hormones often control many diverse body functions, including cellular metabolism, growth, iono- and osmoregulation, reproduction, color, cardiovascular function, and digestion. Hormones are present in the body fluids only at low concentrations because they are diluted into a large body fluid volume. Secretion of hormones only slowly increases their circulating concentration, and the hormone concentration is only slowly decreased by tissue metabolism or excretion. Consequently, endocrine regulation is generally slow and sustained.

The endocrine and nervous systems are similar (Figure 11–1) in that the messenger in each system is a chemical (hormone or neurotransmitter) secreted by a cell (endocrine cell or neuron), and there are specific target cells for the chemical messenger (peripheral target cell or postsynaptic cell). Neurosecretory neurons are somewhat intermediate, releasing their "neurotransmitter" into the blood to act on a target cell. **Paracrine** endocrine cells secrete a hormone that diffuses only a short distance to target cells. Some paracrine cells are modified so that cell extensions convey their hormone closer to the target cell.

The endocrine system is as important an effector system as the nervous system. Nevertheless, there are many important differences between the two systems. In the endocrine system, hormones are present in the body fluids at low concentrations and endocrine regulation is slow but sustained. In contrast, neurons elicit rapid and short-term motor responses via action potential transmission to peripheral motor structures, e.g., muscles. The endocrine control system differs thus from the nervous system in its mechanism, rapidity of response, duration of response, and nonelectrical nature of the message.

Endocrine Systems

There are two general types of endocrine systems. The **neuroendocrine system** consists of neurons in the CNS that are specialized for synthesis, storage, and secretion of large amounts of a neurohormone, their "neurotransmitter," into the blood (Figure 11–1). The axon terminals of these neurosecretory cells often form specialized neurohemal areas, or

organs, where the neurosecretion diffuses into the blood. The **classical endocrine system** consists of non-neural (often epithelial) cells that release hormones into the blood. These endocrine glands lack a duct system; the body fluids transfer their secretions to target sites. Classical endocrine glands are only present in higher invertebrates and vertebrates, suggesting that neurosecretory systems evolved before classical endocrine systems.

The neurohormones of neurosecretory cells and the hormones of classical endocrine cells are released into the body fluids and are distributed throughout the body by the circulatory system. They come into contact with virtually all of the cells in the body but affect only those specific target cells that are under their control. This specificity of reaction by target cells to a widely distributed hormone is due to specific receptor molecules located either on the surface or inside the target cells. Binding of a hormone with its specific receptor initiates a sequence of intracellular events that elicits the required effector response of the cell, e.g., a change in membrane permeability, glandular secretion, or protein synthesis.

The neural reflex arc is a relatively straightforward example of transduction by stimulation of a sensory cell into an effector response of a motor cell via an interneuron (Figure 11–2i). The neurosecretory and classical endocrine systems have comparable "reflex arcs," or effector pathways. Neuroendocrine pathways of invertebrates are often first order loops; the neurosecretion has a direct action on the target organ (Figure 11–2ii). Two typical examples are the inhibition of reproductive development in annelids by release of a neurohormone into the hemolymph and control of the distal retinal pigments in crustacean eyes. First order neuroendocrine loops are not common in vertebrates, but release of arginine vasopressin and oxytocin from the neurohypophysis are two examples. Neurosecretions are often only intermediary messengers and do not control the target organ. Rather, they regulate hormone secretion by a classical endocrine gland. This is a second order endocrine loop (Figure 11–2iii). An example is the control of molting by insects via neurosecretory control of release of a classical hormone, ecdysone, from the corpora allata. Second order loops are common in both invertebrates and vertebrates. The third order loop involves two classical endocrine organs interposed between the CNS neurosecretion and the effector cell (Figure 11–2iv). Such complex multiendocrine control systems are most evident in vertebrates. The gonadal release of steroid hormones in response to anterior pituitary hormones, which in turn are

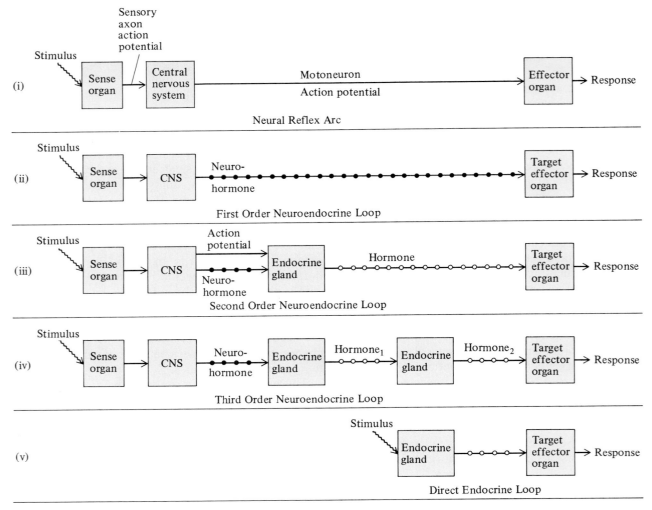

FIGURE 11–2 Generalized schemes comparing a neural reflex arc **(i)** with endocrine control loops **(ii–v)**: **(ii)**, a first order neuroendocrine loop where the neurohormone directly affects the target effector organ; **(iii)**, a second order neuroendocrine loop where the neurohormone (or sometimes a direct neural innervation) regulates an intermediary classical endocrine gland, whose secretion acts on the target effector; **(iv)**, a third order neuroendocrine loop where the neurohormone regulates an endocrine gland, whose hormone regulates a second endocrine gland, the hormone of which affects the target effector; and **(v)**, a direct endocrine loop where the receptosensory endocrine cell is both the sensory receptor and the secretory cell—there is no involvement of the nervous system in this endocrine loop. Not shown are the various possible feedback pathways whereby the effector response (or intermediate stages in the loops) can modulate the stimulus. *(Modified from Frye 1967.)*

regulated by a neurosecretion of the hypothalamus, is an example. Finally, some classical glands are not regulated by the CNS (at least not directly by neurohormones; Figure 11–2v).

Neurosecretion

The generalized neuroendocrine system has a neurosecretory center of neuron bodies that synthesize neurohormone and axonal pathways that transfer the neurohormone to a release site, typically a specialized neurohemal organ (Figure 11–3). The neurohormone is typically a peptide.

Neurosecretory neurons appear to be present in all vertebrates and invertebrates. They had been identified in the nervous systems of many invertebrates and vertebrates for a considerable period before the concept of neurosecretion of hormones was appreciated (Gabe 1966; Orchard and Loughton 1985). Neurosecretory cells have the typical anat-

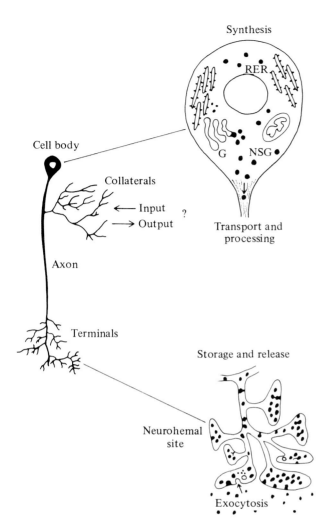

Synthesis

RER

Cell body

Collaterals

← Input

→ Output

?

G   NSG

Transport and processing

Axon

Terminals

Storage and release

Neurohemal site

Exocytosis

FIGURE 11–3 Schematic of a typical neurosecretory cell from an insect central nervous system shows rough endoplasmic reticulum (RER), where neurosecretory products are synthesized, and the Golgi complex (G), which packages the products into neurosecretory granules (NSG). These granules are transported along the axon to swollen storage terminals and neurohemal areas where the neurosecretion is released by exocytosis. Other branches of the axon serve for sensory input or as alternative output pathways. *(From Orchard and Loughton 1985.)*

omy of neurons. Their cell body may be monopolar or multipolar. Many neurosecretory cells are larger than conventional neurons, but large size is not a general criterion for distinguishing neurosecretory cells. They have dendritic arborizations and axons that may travel considerable distances to the site of neurohormone release. Neurosecretory cells are generally identified by their histological appearance and histochemical staining properties, rather than by their cell structure. Their essential feature is an

abundance of neurosecretory granules (100 to 300 nm dia), or particles, droplets, or even "colloidal pools" of material. Specific histochemical stains, such as chrome-hematoxylin-phloxine or paraldehyde fuchsin and azan, stain neurosecretory cells and their axonal connections to the release site. The stains most likely differentiate carrier proteins for the hormones, rather than the peptide hormones themselves.

Neurosecretory granules originate in the rough endoplasmic reticulum where neurohormone is synthesized and packaged by the Golgi complex into membrane-bound granules. The granules most likely contain a carrier protein as well as the neurohormone. The granules appear to be "processed" as they are transported to the axon terminus at the neurohemal site. The axon terminals are often outside the CNS (outside of the blood-brain barrier) and are typically swollen with membrane-bound granules. The contents of the granules are released by exocytosis.

The neurosecretion may be released at unspecialized axon terminals, but there often is a specific anatomical organization of a neurohemal area/organ to facilitate diffusion of neurohormone into the vascular system. Invertebrates have a variety of types of neurohemal organs. Polychete worms have a relatively primitive cerebrovascular neurohemal structure at the base of the cerebral ganglion of the brain. Axons of the neurosecretory cells located in the posterior portion of the cerebral ganglion are thought to transport a "juvenile hormone" to the dorsal blood vessel at the cerebrovascular neurohemal organ. In mollusks, neurohormones are not released at just a few specific neurohemal organs, but there are a large number of sites for widespread release (Joosse and Geraerts 1983). This is in marked contrast to the few localized neurohemal areas or organs of most other invertebrates and vertebrates and may reflect the lack of a blood-brain barrier in mollusks (Maddrell and Nordman 1979). Gastropod and cephalopod mollusks have neurohemal areas at the periphery of the central and peripheral nervous system (Figure 11–4A). The neurohormones diffuse through the perineurium covering the nervous system before reaching the blood. Some axons also enter peripheral tissues and release their neurohormones directly at the target organs, such as renal tubules.

The various hypothalamic neurosecretory cells of vertebrates have three general types of neurohemal organs. Neurosecretory neurons of the supraoptic and paraventricular nuclei have long axons that extend to and terminate near fenestrated capillaries in the posterior pituitary (Figure 11–4B). Dilations

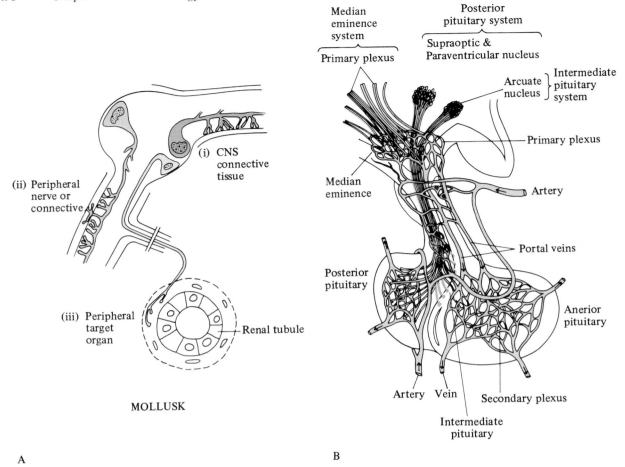

FIGURE 11–4 **(A)** Neurosecretory cells of pulmonate gastropods have a variety of types of neurohemal areas. The axons can ramify at the connective sheath surrounding the nervous system, nerves, connectives, and commissures or in peripheral target organs. **(B)** The mammalian hypothalamo-hypophyseal axis has three neurosecretory systems. The paraventricular and supraoptic nuclei are a typical neuroendocrine system that distributes neurohormones into a capillary bed of the posterior pituitary and then the general circulation. Axons of other hypothalamic neurons terminate on a primary capillary plexus where they release their neurohormones. These are distributed by a special portal circulation to a second capillary plexus surrounding their target organ cells in the anterior pituitary; the neurohormones reach their target cells before being distributed in the general body circulation. Axons of neurosecretory cells in the arcuate nucleus release their hormone directly at their target cells in the intermediate pituitary.   *(From Joosse and Geraerts 1983; modified from Guillemin 1980.)*

of the axon terminals act as neurohormone storage sites (Herring bodies) within the posterior pituitary. The neurohormones are released into the general vascular space and then transported to their target cells (kidney, smooth muscle). Axons of other hypothalamic neurosecretory cells terminate near capillaries of the median eminence, and their neurosecretions (releasing factors) are carried directly to their target cells in the anterior pituitary by a special portal circulation. This ensures that the concentrations of the releasing factors are much higher at the target organs than they would be after distribution throughout the entire body fluid space. Some advanced vertebrates have a third type of neurohormone system. Neurosecretory axon terminals directly contact the intermediate pituitary endocrine cells whose actions they regulate; this is a direct "synaptic-type" role of neurosecretory cells.

Neurosecretory cells have bioelectrical properties similar to conventional neurons. Their resting E_m is generally -40 to -60 mV and they have typical action potentials. The rate of neurohormone release appears to be proportional to the action potential discharge. For example, the action potential discharge of the paraventricular and supraoptic neurons of the hypothalamus is proportional to the elevation of plasma osmotic concentration above normal, as is the circulating concentration of arginine vasopressin. The obvious inference is that the rate of arginine vasopressin secretion is determined by the frequency of action potentials.

Classical Endocrine Glands

The classical endocrine system consists of various ductless glandular tissues. The diversity of classical endocrine systems seems to be greatest in vertebrates, whereas invertebrates rely more extensively on first order neuroendocrine regulation of peripheral tissues. Vertebrate neurosecretions (from the hypothalamus) control many peripheral physiological functions, as in invertebrates, but generally through second and third order neuroendocrine loops involving classical endocrine glands.

Endocrinological control of integument molting in two invertebrates (crustaceans, insects) and a vertebrate (amphibians) provides a comparative example of classical endocrine systems as well as their neuroendocrine control (Figure 11–5). Arthropods must slough their exoskeleton to allow for continued body growth. The sloughing of the old cuticle is called **molting**, or ecdysis. Molting by crustaceans is influenced by a number of external factors, including light intensity, photoperiod, temperature, salinity, food availability, and interactions with conspecifics. Molting is under the control of two hormones in adult decapod crustaceans, such as crabs, shrimp, and lobster. A molt-inhibiting hormone (MIH) is a peptide produced by neurosecretory neurons of the optic lobe (the X-organ) and released at a neurohemal organ in the eye stalk (the sinus gland). A molt-inducing hormone (MH) is secreted by a classical endocrine gland (the Y-organ), located in the antennary segment. The Y-organ secretes molting hormone when the inhibitory influence of MIH is removed.

Insects also molt periodically, at least up to the adult stage, but their molting is complicated by the concomitant process of development (metamorphosis; see below). Nevertheless, the endocrine control system for molting in insects is quite similar to that of crustaceans. Neurosecretory cells in the pars intercerebralis of the brain secrete a peptide hormone (prothoracicotropic hormone, PTTH, or ecdysiotropin). Release of PTTH from the neurohemal organ (the corpora cardiaca) stimulates the prothoracic gland to secrete a molting hormone, ecdysone.

Many vertebrates also periodically shed their outer epidermis, although for different reasons than do arthropods. The cells of the vertebrate epidermis are regularly replaced by cell division of the inner

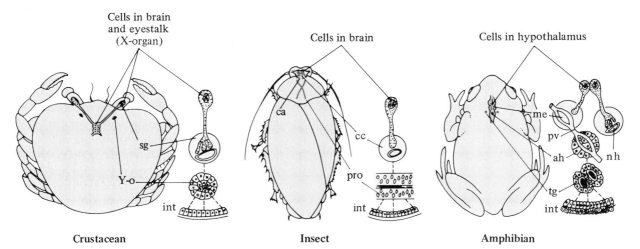

FIGURE 11–5 Diagrammatic comparison of neuroendocrine regulation of molting in arthropods (crustacean and insect) and vertebrates (amphibian). The central nervous system activates neurosecretory cells that release their neurohormone via a neurohemal organ (crustacean sinus gland, sg; insect corpora cardiacum, cc; amphibian median eminence, me) to control the activity of a peripheral endocrine gland (crustacean Y-organ, Y-o; insect prothoracic gland, pro; amphibian adenohypophysis, ah; neurohypophysis, nh; and thyroid gland, tg). The hormones of these respective glands induce molting of the integument (int). *(Modified from Scharrer 1959.)*

layer of the epithelium, hence the outer layers are either continually sloughed (as in mammals) or periodically sloughed at intervals of days (amphibians) or weeks to months (reptiles). The regular molting cycle of fish, amphibians, and reptiles is thought to be an autonomous rhythm of the epidermis, under permissive regulation of various endocrine glands, including the adenohypophysis, thyroid, and gonads. In the amphibian, hypothalamic neurosecretory cells release a thyrotropin-stimulating factor that stimulates the secretion of thyroid-stimulating hormone (TSH) from the adenohypophysis. TSH stimulates the thyroid gland to secrete thyroxin, which initiates molting. In the toad *Bufo*, corticotropin and interrenal gland steroids, and AVT affect molting, rather than TSH and thyroid hormones. There are clear analogies between the endocrine control of molting in vertebrates and invertebrates, but neither the endocrine glands nor the hormones are homologous.

Classical endocrine glands release their hormone secretions into the body fluids; they do not have specific neurohemal sites. The endocrine glands of vertebrates have a rich blood supply and the secretory cells are often arranged in cords of cells (often no more than two layers thick) that are separated from adjacent cords by capillaries or by enlarged and irregular blood spaces called sinusoids (Figure 11–6A). Each cell has a contact region with the blood space for direct secretion into the blood. One notable exception to this organization is the thyroid gland, which has a follicular organization. The follicle is a hollow sphere of a single layer of cells; the center of the follicle stores a colloidal material, thyroglobulin, and thyroid hormones diffuse from the colloid into a dense basketwork of capillaries that surround the follicle.

The endocrine glands of invertebrates, such as the crustacean Y-gland and insect prothoracic gland, do not have a vascular supply but are bathed in hemolymph. The secretory cells of invertebrate endocrine glands do not have a morphology that maximizes exposure of cell surfaces for secretion (unlike vertebrate endocrine cells). The glands are generally strands or globular, conical, or sheetlike clusters of cells, often closely associated with trachea (Figure 11–6B). Hormones diffuse from the inner cells to the surface of the gland to be circulated

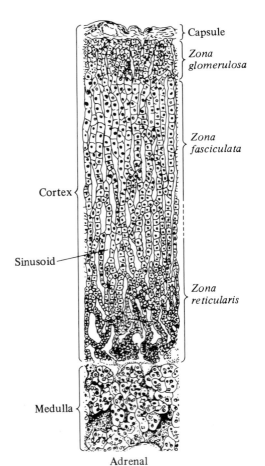

FIGURE 11–6 **(A)** The general organization of vertebrate endocrine glands is cell cords interspersed with capillaries or sinusoids, as illustrated by the cortex and medulla of the mammalian adrenal gland. **(B)** Arthropod endocrine glands, such as the prothoracic gland, are relatively compact masses of cells, rather than cell strands, and are not interspersed with capillaries. *(From Turner and Bagnara 1976; Scharrer 1948.)*

A

B

Capsule

Zona glomerulosa

Zona fasciculata

Cortex

Sinusoid

Zona reticularis

Medulla

Adrenal

Trachea Nerve

Prothoracic Gland

by the hemolymph. In insects, the connective tissue sheath of the viscera and endocrine glands may form preferential channels to direct flow of secretions towards the hemolymph or specific target cells. Some dipteran insects have a retrocerebral complex, or ring gland, consisting of fused corpora allata, corpora cardiaca, and prothoracic gland. Channels in the basement membrane of the ring gland and the connective tissue may direct secretions to the heart for rapid dispersion by the blood (Whitten 1964). Secretions of pericardial cells may be conveyed to the wall of the heart along fine channels in the connective tissue attaching the pericardial cells to the heart; these secretions control the heart beat.

Mechanisms of Hormone Action

The **specificity** of the message carried to the target cell by a hormone is an inherent property of its 3-dimensional structure and ability to bind correctly to a specific receptor protein of the target cell. There is no intrinsic limitation to the chemical nature of a hormone except that its structure must be sufficiently complex that it can stimulate a specific receptor to elicit its effect. It is therefore not surprising that there is a considerable variety in the structure of hormones.

Many classical hormones (and also neurohormones) are **peptides** (e.g., distal retinal pigment hormone; Figure 11–7A). The physical and biochem-

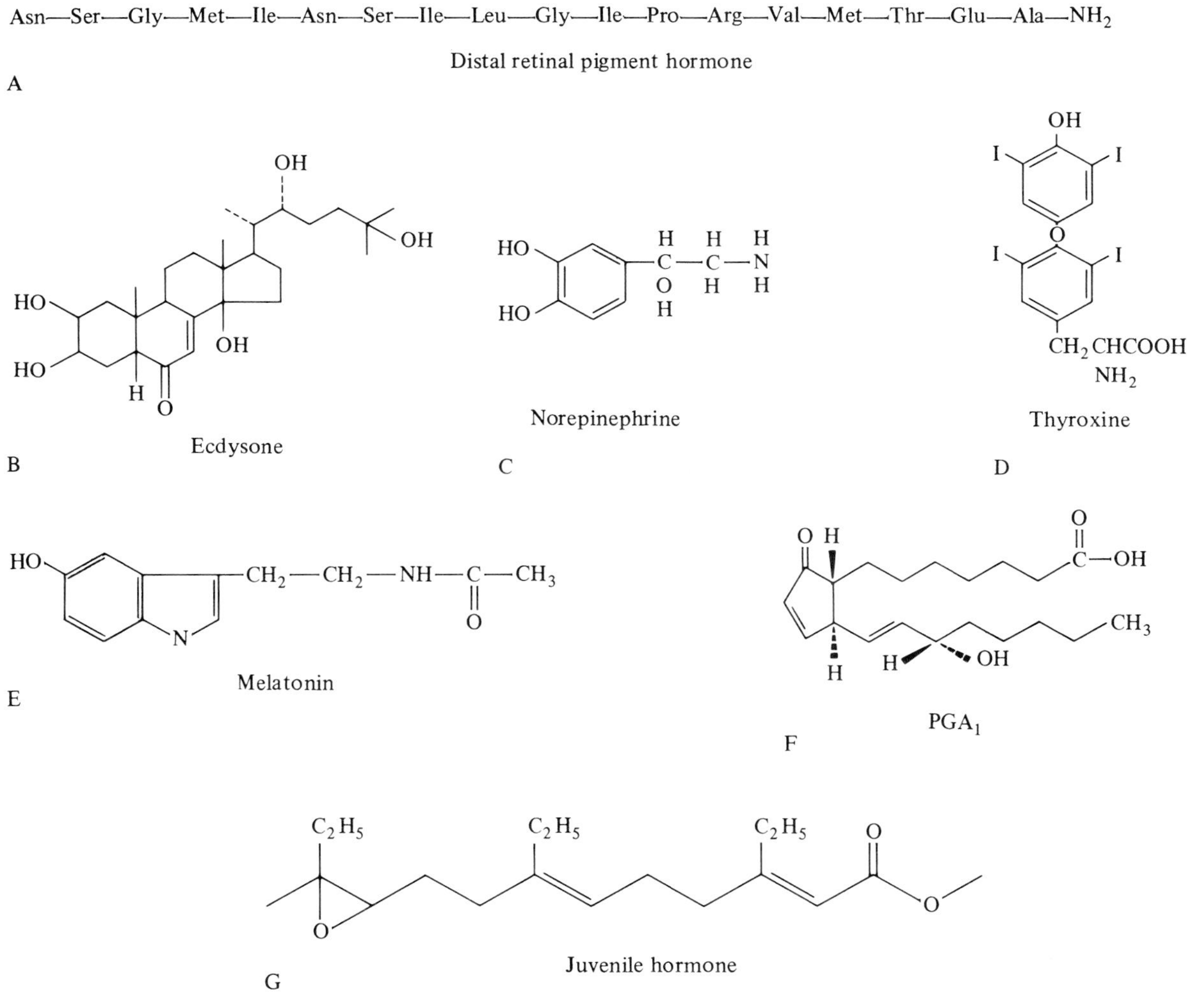

Asn—Ser—Gly—Met—Ile—Asn—Ser—Ile—Leu—Gly—Ile—Pro—Arg—Val—Met—Thr—Glu—Ala—NH$_2$

Distal retinal pigment hormone

A

FIGURE 11–7 **(A)** Amino acid sequence of a peptide hormone, distal retinal pigment hormone (DRPH). **(B)** Structure of a steroid hormone, ecdysone. **(C)** The hormone norepinephrine is a catecholamine. **(D)** Thyroxin is a thyroid hormone synthesized from tyrosine. **(E)** The pineal hormone melatonin is synthesized from tryptophan. **(F)** The prostaglandin PGA$_1$ is synthesized from prostanoic acid. **(G)** This is one of a variety of juvenile hormones.

ical determinants of molecular shape, even for small peptides, can produce very specific binding properties. Larger peptides and proteins and covalently linked multiple proteins can assume any of an almost infinite variety of shapes through variation in amino acid sequence, secondary structure, tertiary structure, and quaternary structure.

The "families" of peptide hormones (Table 11–1) share homologous sequences of amino acids but differ in function. For example, secretin and glucagon form one such family that also includes gastric inhibitory peptide and vasoactive intestinal peptide. Secretin and glucagon have the same amino acids at 14 of 27 positions, but their 3-dimensional structures are sufficiently different (despite the 52% amino acid homology) that they have completely separate and noninteracting functions. Many peptide hormone families have each evolved from a single precursor amino acid sequence. Similarly structured proteins may retain similar or identical functions in different species, or similar proteins may assume very different functions within an individual. Some families of proteins share a common precursor protein from which different members of the family are formed by cleavage of different amino acid sequences. For example, the family of pro-opiocortin peptides from the adenohypophysis (including ACTH, α-MSH, β-MSH, β-LPH, β-endorphin, CLIP) are all derived from the same precursor protein, pro-opiocortin.

A single amino acid substitution would greatly alter the structure of a small peptide, and so their amino acid sequences would be expected to be highly conservative. There is considerably more scope for variation in the amino acid sequence for larger peptide hormones, especially as many amino acid substitutions would be neutral and not affect the structure of the active site(s). For example, amino acid sequences for salmon, eel, and human calcitonins have considerable variability although some amino acids are the same in all calcitonins. However, there can be some variation in active site structure. Receptor molecules will bind peptides with similar but not identical active sites, although possibly not with the same affinity. Receptor molecules also show species-to-species variability in their structure, hence their specificity for different hormones. For example, growth hormone (GH, 188 amino acid residues) of humans has some effect in fish, but fish GH has no effect in humans. Mammalian GH receptors apparently do not recognize fish GH, although the fish GH receptors show some binding of human GH. Usually, a hormone receptor responds optimally to its normal hormone, but this is not always true. For example, salmon calcitonin is about 50 × more potent in affect-

TABLE 11–1

Main peptide and protein hormone families. *(From Gorbman et al. 1987.)*	
PRL	Prolactin
GH	Growth hormone
LTH	Chorionic somatomammotropin (placental lactogen)
LH	Luteinizing hormone
FSH	Follicle stimulating hormone
TSH	Thyroid stimulating hormone
HCG	Human chorionic gonadotropin
ACTH	Adrenocorticotropic hormone
αMSH	α-melanocyte stimulating hormone
βMSH	β-melanocyte stimulating hormone
βLPH	β-lipotropin
	α-endorphin
	β-endorphin
	Enkephalins
AVP	Arginine vasopressin
LVP	Lysine vasopressin
AVT	Arginine vasotocin
	Phenypressin
	Oxytocin
	Mesotocin
	Valitocin
	Isotocin
	Glumitocin
	Aspartocin
PG	Glucagon
	Secretin
GIP	Gastric inhibitory peptide
VIP	Vasoactive intestinal peptide
	Gastrin
CCK	Cholecystokinin
ECSP	Substance P
	Neurotensin
	Caerulein
	Bombesin
I	Insulin
	Somatomedin A
	Somatomedin C
IGF	Insulin-like growth factors I, II
	Relaxin
PP	Pancreatic polypeptide
PYY	Peptide YY
M	Motilin

ing calcium metabolism in humans as is human calcitonin!

A second major group of classical hormones are those based on the lipid, cholesterol, e.g., ecdysone

(Figure 11–7B). These steroid hormones are commonly sex hormones in both vertebrates and invertebrates but also have other functions. Some of the major vertebrate steroid hormones include androgens (C_{19} steroids such as testosterone), estrogens (C_{18} steroids such as estradiol and estrone), progestogens (C_{21} steroids including progesterone), and corticosteroids (C_{21} hydroxylated steroids such as cortisol, cortisone, and aldosterone). Ecdysone and 20-hydroxyecdysone are examples of invertebrate steroids.

A third loose group of hormones are those produced through chemical modification of amino acids. The vertebrate adrenal medullary hormone norepinephrine (noradrenalin) is one example (Figure 11–7C). Thyroxin (Figure 11–7D) and tri-iodothyronine are synthesized in the vertebrate thyroid gland from tyrosine. Melatonin (Figure 11–7E) and related indoles are synthesized from tryptophan via serotonin (5-hydroxytryptamine, 5-HT). Serotonin may also be a neurohormone secreted by the pericardial organ of crustaceans to elevate heart rate.

Finally, there are a variety of dissimilar hormones synthesized from various precursor molecules. For example, prostaglandins A, B, E, and F (Figure 7–11F) are derived from a C_{20} fatty acid, prostanoic acid. Juvenile hormone (Figure 11–7G) of insects has a dihomosesquiterpinoid skeleton that is synthesized from acetylCoA and propionylCoA.

The **specificity** of hormone action is a consequence of the target cells having specific, high-molecular weight protein receptors for that hormone. These receptors are found either at the cell surface, i.e., are membrane-bound proteins, or are intracellular, i.e., are cytoplasmic or nucleoplasmic proteins (Figure 11–8). In general, the water soluble hormones (peptides and amino acid derivatives) bind to membrane receptors at the cell surface. Cytoplasmic receptors would not be effective for these hormones because they do not readily penetrate the cell membrane. In contrast, the lipid soluble steroid and thyroid hormones are permeable to cell membranes and have intracellular receptors. All cells in the body presumably have similar cytoplasmic levels of steroid and thyroid hormones, but only those with the specific receptor molecules respond to the hormones.

A complex series of biochemical events is initiated by hormone-receptor binding and a conformational change of receptor structure. The sequence is substantially different for the surface membrane receptor systems (e.g., peptides) and for intracellular receptor systems (e.g., steroids), although the final effect may be similar (change in membrane permeability, secretion, protein synthesis, etc.)

Hormone Binding

Binding of a hormone to its specific receptor initiates a conformational change of the receptor protein. A high **affinity** of the receptor for its hormone ensures a specificity of response and sufficient response at low-circulatory concentrations typical of hormones. The activation of a receptor by its hormone is analogous to enzyme activation by a substrate or cofactor, neurotransmitter activation of postsynaptic membrane receptors (e.g., the ACh receptor), and an antibody binding to an antigen.

The precise measurement of hormone concentrations at physiological (i.e., very low) concentrations by radioimmunoassay (RIA) has been one of the most important recent methodological advances for the study of endocrinology. This technique well illustrates the role for hormones of binding specificity to receptor proteins, in this case, antibodies (see Supplement 11–1, pages 560–63). Hormone levels in the circulating body fluids are generally very low, in the nanomole per liter (10^{-9} M) or picomole per liter (10^{-12} M) range. Consequently, their receptors generally have a high affinity for the hormone. The dissociation constant (K_d) is the ratio of the concentration product of the receptor (R) and hormone (H) to the concentration of the receptor-hormone complex (R.H), i.e.,

$$K_d = [R][H]/[R.H] \qquad (11.1)$$

A high K_d indicates a low affinity (low [R.H]) and a low K_d reflects a high affinity (high [RH]). K_d's typically are 10^{-8} to 10^{-9} M. A further variable of considerable significance is the number of receptors per cell. Generally there are about 1 to 100 femtomoles of receptor per 100 μg of DNA (10^{-19} to 10^{-17} moles μg^{-1}).

Membrane-Bound Receptor Systems. Peptide and most amino acid-derived hormones do not readily cross the cell membrane because they are hydrophilic and often are large molecules. The receptors for these hormones are high-molecular weight (about 10^5) integral cell membrane proteins. The receptors have a high specificity and affinity for particular hormones.

Hormone-receptor binding initiates a sequence of biochemical events that culminate in the required cellular response (Figure 11–8). Binding of the hormone to its receptor activates a transducer protein through a reaction of guanosine triphosphate (GTP); hence, it is called a **G protein**. The G proteins are activated so long as they are bound to GTP; hydrolysis of the GTP to GDP inactivates the G protein. An intracellular, membrane-bound enzyme, **adenyl cyclase**, is activated by hormone-receptor

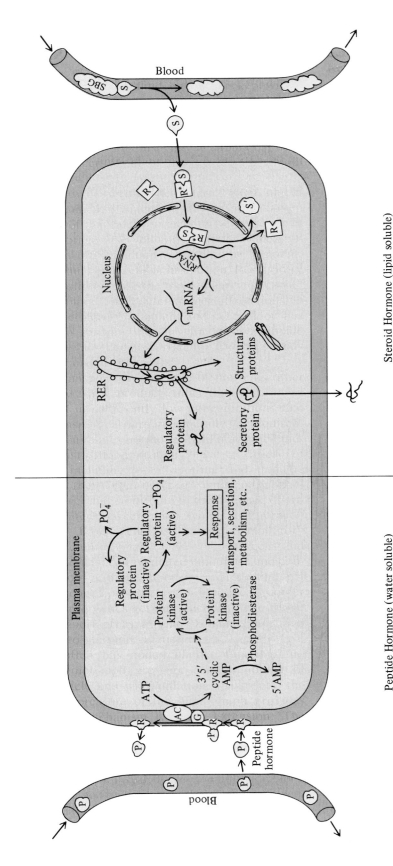

FIGURE 11-8 Generalized scheme for the modes of action of peptide hormones (left) and steroid hormones (right). Peptide hormones (P) diffuse from the blood and bind to their receptor (R) on the outer surface of the plasmalemma. Hormone-receptor binding activates adenyl cyclase (AC) via a G protein (G) to synthesize 3,5 cyclic AMP. The cAMP activates protein kinase enzymes, which in turn phosphorylate and activate the regulatory proteins that initiate physiological responses in the target cell; cAMP is removed by a phosphodiesterase enzyme. Steroid hormones (S) generally are bound to plasma steroid-binding globulins (SBP). The steroid hormone diffuses into the cytoplasm where it binds to a specific receptor (R) and forms an "activated" steroid-receptor complex (SR*) that diffuses into the nucleus. The SR* complex initiates mRNA synthesis by RNA polymerase, and the mRNA diffuses to the ribosomes located on rough endoplasmic reticulum (RER) and initiates protein synthesis. The synthesized proteins are either regulatory and elicit physiological effects in the cell or are structural and contribute to an intracellular structure (e.g., microtubules) or are secreted by the cell. The SR* complex subsequently dissociates and the steroid hormone is metabolized to an inactive form (S'). *(Modified from Gorbman et al. 1983.)*

binding through an intermediary G protein, and converts ATP into cyclic 3'5'AMP, **cAMP**. The cAMP activates a protein kinase enzyme that phosphorylates and activates an inactive regulary protein. The activated regulatory protein then initiates the cellular response to hormone stimulation, which may be stimulation of cell secretion, membrane transport, a change in cellular metabolism, a change in membrane permeability, etc. cAMP is the "second messenger" in this biochemical sequence of events. The hormone is the "first messenger" and it is very specific for the target cell and response in question. The second messenger is a general intracellular messenger common to almost all peptide-hormone systems. Sometimes, cyclic GMP, inositol-PO_4, prostaglandins, or Ca^{2+} are the intracellular second messenger rather than cAMP (see below).

There are a number of types of G protein (Gilman 1987). The stimulatory G_s protein activates adenyl cyclase, whereas the inhibitory G_i protein inhibits adenyl cyclase. Another G protein, G_o, is present in the brain and heart tissue; its specific function is not yet known, but its existence suggests a broader role for G proteins than merely stimulation/inhibi-

tion of adenyl cyclase. One possible role of G_o may be the direct control of permeability of membrane ion channels, mediated by a protein kinase or a phosphoprotein phosphatase. For example, G_o may mediate a neurotransmitter-induced inhibition of voltage-dependent Ca^{2+} channels in dorsal root ganglion cells. Transducin (G_t protein) regulates a light-activated GTPase activity in the retina (see Chapter 7).

G proteins contain three subunits: an α and β subunit and a smaller γ subunit that is tightly associated with the β subunit. The α subunits of different G proteins are generally similar in amino acid sequence, although there are some sequence differences concentrated in three variable regions. The $\beta\gamma$ subunits are less variable, and are functionally interchangeable between different G proteins. The interaction of receptor (R) and G protein is activated by the appropriate hormone (H) through the subunits (Figure 11–9). The activated H-R complex causes the phosphorylation of G-GDP in the presence of GTP to form a H-R-$G_{\alpha\beta\gamma}$-GTP complex that dissociates into H + R + $G_{\beta\gamma}$ + G_α-GTP. The G_α-GTP has a relatively long half-life (many seconds), which considerably amplifies the H-R signal. A similar cycle exists for inhibitory effects involving R, G_i, and $G_{i\alpha}$-GTP. Two well-described effects of G-GTP involve adenyl cyclase (i.e., G_s in endocrine target cells) and phosphodiesterase (G_t in photoreceptor cells).

The $G_{\beta\gamma}$ subunits can also modulate intracellular responses. For example, the $\beta\gamma$ subunit can stimulate membrane-bound phospholipase A_2 (PLA_2) to produce arachidonic-acid metabolites (e.g., leukotrienes) that open K^+ channels in retinal, heart, and neuronal cells. In the yeast *Saccharomyces cerevisiae*, a peptide pheromone from cells of one mating type initiates a programmed response in other cell types for conjugation. The pheromone receptors interact with G proteins, whose $G_{\beta\gamma}$ subunits induce the response.

Adenyl cyclase is activated by $G_{s\alpha}$-GTP and inhibited by $G_{i\alpha}$-GTP. It cleaves pyrophosphate from ATP to form cyclic 3'5'AMP. The cAMP is subsequently inactivated to adenosine-5-PO_4 by phosphodiesterase. cAMP activates a phosphorylating enzyme, cAMP-dependent protein kinase (cAMP-PrK or A-kinase), by binding to an inhibitory regulatory subunit of the protein kinase. The activated cAMP-PrK then phosphorylates (using ATP) and activates phosphoproteins. Some phosphoproteins are enzymes but others may be membrane channels, regulatory proteins, or structural proteins.

There are second messenger systems other than cAMP. Cyclic GMP, produced by guanylate cyclase

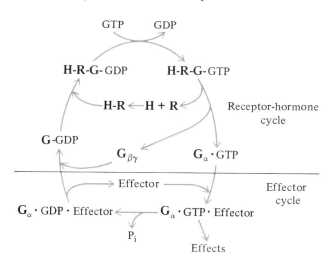

FIGURE 11–9 The role of the G proteins (G) is to transduce the hormone (H) and receptor (R) interaction (i.e., formation of H-R) to an effector system (E). The receptor-hormone activation cycle combines G.GDP and H-R form H-R-G-GDP then H-R-G-GTP, from which H and R dissociate and the G-GTP separates into its subunits G_α.GTP and $G_{\beta\gamma}$. The G_α.GTP and some effector molecule (E) form a G_α.GTP.E complex that initiates the next step in the appropriate sequence of events leading to the cellular effector response. This complex is eventually dephosphorylated to G_α.GDP.E, which releases E and recombines with the other G protein subunits to form G.GDP. *(Modified from Gilman 1987.)*

in a manner analogous to cAMP formation, has effects opposite to cAMP. For example, a membrane-bound guanylate cyclase functions as both the receptor for atrial natriuretic peptides (ANPs) and the enzyme for cGTP synthesis (Chinkers et al. 1989). The synthesis of cGMP is stimulated at high Ca^{2+}; cAMP synthesis is inhibited by high Ca^{2+}. Membrane lipids can also be second messengers. Hydrolysis of phosphatidylinositol 4,5 biphosphate (PIP_2) by phospholipase C forms two second messengers, diacylglycerol (DG) and inositol triphosphate (IP_3). DG remains in the membrane and interacts with a membrane-bound protein kinase C (C-kinase). IP_3 diffuses into the cytoplasm and elicits the release of Ca^{2+} from intracellular stores, such as endoplasmic reticulum (it has also been implicated in sarcoplasmic reticular release of Ca^{2+} in skeletal muscle cells; see Chapter 9). Ca^{2+} also stimulates C-kinase activity. C-kinase phosphorylates phosphoproteins, as does cAMP-PrK, and thereby elicits a variety of cellular responses.

A good example of hormone action involving G protein and adenyl cyclase is the regulation of glycogen breakdown to glucose by mammalian skeletal muscle cells in response to hormones, e.g., epinephrine and insulin (Figure 11–10). The regulation of glycogen breakdown by liver hepatocytes in response to glucagon is almost identical. Epinephrine diffuses from the blood to its specific β-adrenergic receptor on the outside of the muscle cell membrane. This activates adenyl cyclase via the stimulating G_s protein. cAMP then activates cAMP-PrK, which phosphorylates phosphoproteins. The main phosphoprotein in skeletal muscle cells and hepatocytes is a regulatory protein, phosphorylase kinase. This enzyme is the modulator of both hormonal and neural control of glycogen breakdown (Cohen 1982). The enzyme has a tetrameric subunit structure $(\alpha\beta\gamma\delta)_4$; a serine residue on each of the α and β subunits is phosphorylated by cAMP-PrK. The α subunit is regulatory. The β subunit is calmodulin, a Ca^{2+}-binding protein. The catalytic γ subunit is activated either by phosphorylation of the α and β subunits or by Ca^{2+} binding. Activated phosphorylase kinase then phosphorylates and activates a normally inactive enzyme, glycogen phosphorylase (or phosphorylase b), to an active enzyme (phosphorylase a). This activated enzyme promotes glycogen breakdown (glycogenolysis) to glucose-1-PO_4 then glucose-6-PO_4. The glucose-6-PO_4 then enters the citric acid cycle to provide ATP for muscle contraction. In hepatocytes, the glucose-6-PO_4 is converted to glucose, which diffuses out of the cell and into the blood for distribution to other tissues. cAMP-PrK also phosphorylates the enzyme glycogen synthetase, but this inhibits its enzyme activity; glycogen synthetase promotes glycogen synthesis from glucose (glycogenesis).

The control of glycogenolysis in skeletal muscle well illustrates the complex cascade of biochemical events in response to hormone-receptor binding, the role of G_s and G_i proteins, and stimulatory or inhibitory effects of protein phosphorylation. It also illustrates the interaction of the typical hormone (first messenger)-cAMP (second messenger) system with another important and ubiquitous intracellular messenger, Ca^{2+}. The intracellular Ca^{2+} concentration is usually very low ($<10^{-7}$ M) compared to the extracellular concentration (about 10^{-3} M). Membrane depolarization or cAMP-PrK phosphorylation can induce influx of Ca^{2+} into the cytoplasm from the extracellular fluids or intracellular stores. Much of this Ca^{2+} is bound to nonregulatory sites; this buffers the changes in intracellular Ca^{2+} concentration. Some Ca^{2+} binds to Ca^{2+}-binding proteins, e.g., calmodulin; their conformational change elicits an effector response. Calmodulin binds four Ca^{2+} ions; the affinity of two Ca^{2+}-binding sites in the N-terminal region is about 10 × higher than the two sites in the C-terminal region, but all sites have a sufficiently high affinity that they bind Ca^{2+} in micromolar concentrations. The δ subunit of phosphorylase kinase is identical to calmodulin and binds Ca^{2+}. A second calmodulin, the δ' subunit, binds to phosphorylase kinase in the presence of Ca^{2+} and interacts with the α and β subunits (the sites for phosphorylation) and strongly activates the kinase even when it is dephosphorylated.

The regulation of adrenal cortex cells further illustrates the multiple regulation of a cell response (synthesis and release of a steroid hormone, aldosterone) by a variety of hormone-receptor systems and various second messenger systems (Quinn and Williams 1988). Several pituitary hormones (ACTH, αMSH, βMSH, β-endorphins) elicit aldosterone secretion, but ACTH is the most potent. ACTH stimulation of aldosterone secretion involves the G_s-adenyl cyclase system. Dopamine, angiotensin II, and somatostatin also inhibit aldosterone secretion through the G_i-adenyl cyclase system. Arginine vasopressin stimulates aldosterone secretion through an unidentified G protein and the PIP_2-IP_3/DG system. Extracellular K^+ elicits aldosterone secretion, presumably by an influx of Ca^{2+} via voltage-dependent Ca^{2+} channels in the plasma membrane. Atrial natriuretic peptide (ANP) inhibits aldosterone secretion through specific membrane receptors and increases cGMP and decreases cAMP levels. The ACTH/receptor/G protein/cAMP system also promotes mRNA synthesis from nuclear DNA (Simpson and Waterman 1988). The mRNA codes for the synthesis of steroid hydroxylase inducing

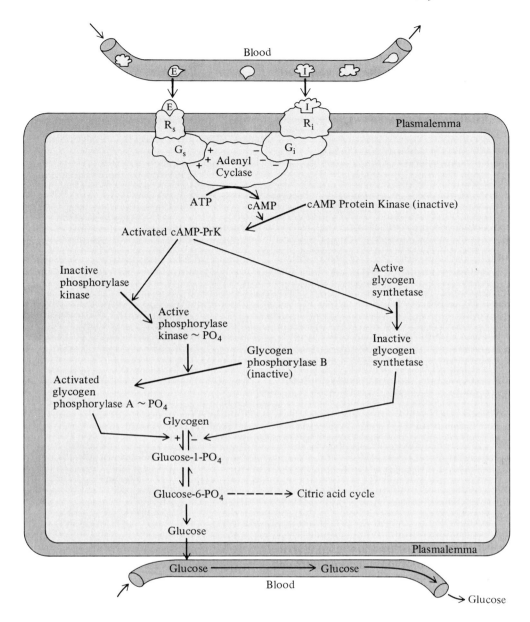

FIGURE 11–10 Representation of the modes of action of stimulatory hormones (e.g., epinephrine, E) and inhibitory hormones (e.g., insulin) on adenyl cyclase in skeletal muscle cells and hepatocytes as mediated by the excitatory and inhibitory G proteins (G_s and G_i, respectively). Cyclic AMP produced by adenyl cyclase activates cAMP protein kinase, which phosphorylates and activates phosphorylase kinase, which in turn phosphorylates and activates glycogen phosphorylase. The effect of the sequence of events is to promote glycogenolysis and to increase intracellular glucose-6-PO_4 availability for oxidation by the citric acid cycle (and transport of glucose to the blood stream in hepatocytes). Glycogen synthetase is also inhibited by activated cAMP-PrK, which also increases the availability of glucose-6-PO_4.

protein (SHIP), which initiates mRNA synthesis coding for P_{450} enzymes; these enzymes regulate the synthesis of aldosterone from cholesterol.

Cytosolic Receptor Systems. Steroid hormones (and also some protein hormones) have their effects by directly activating or suppressing the expression of specific genes. The protein products of these genes are responsible for the effector response of the cell.

Steroid hormones are small lipophilic molecules of about 300 molecular weight. They readily diffuse through cell membranes and so their intracellular (cytosolic) concentration is similar to their extracellular concentration. Steroid hormones, like peptide hormones, have high-affinity, high-specificity proteinaceous receptors but steroid receptors are found inside target cells. The receptor cells contain rela-

tively small numbers of receptors (e.g., 10 to 10000 per cell).

The mechanism for steroid hormone action (and for some other lipid soluble hormones such as thyroxin) is described by a "two-step" receptor model, based primarily on biochemical evidence

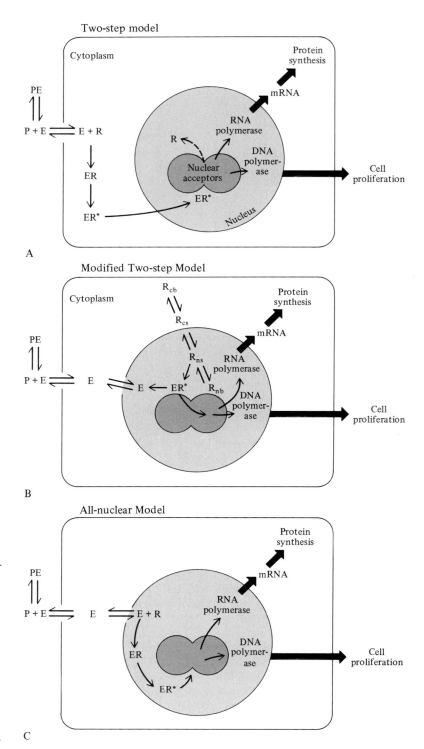

FIGURE 11–11 (A) Classical two-step model for mode of action of steroid hormone using estrogen (E) as an example. The hormone dissociates from plasma-binding proteins (P) and diffuses into the cytoplasm where it binds to a receptor. The activated receptor-hormone complex is translocated to the nucleus where it interacts with nuclear acceptors and initiates mRNA transcription. (B) Modified model of receptor dynamics showing an equilibrium between cytoplasmic receptors that are bound (R_{cb}) and freely soluble (R_{cs}), and nuclear bound (R_{nb}) and soluble (R_{ns}) receptors. There is an equilibrium across the nuclear membrane between R_{cs} and R_{ns}. The hormone (E) could interact with any of these receptor populations. (C) Revised model with only nuclear steroid receptor. The hormone diffuses directly into the nucleus where it binds to the nucleus-confined receptor. *(Modified from Welshons and Jordan 1987; Sheridan and Martin 1987; Barrack 1987.)*

(Jensen et al. 1968; Gorski et al. 1968; O'Malley and Schrader 1976). High-affinity receptor molecules in the cytoplasm combine with the cytoplasmic hormone molecules, and the hormone-receptor complex is somehow "transformed" or "activated" and acquires a high affinity for DNA (Figure 11–11A). Consequent accumulation of the "activated" receptor-hormone complex in the nucleus modifies DNA transcription and translation, hence protein synthesis. Proteins synthesized in response to steroid hormones may activate intracellular regulatory proteins, may be cellular structural components, or may be secreted from the cell.

The "two-step" model for the action of steroid hormones is now the subject of controversy because recent immunohistochemical and biochemical evidence suggests that the steroid receptors are found preferentially, or only, in the nucleus independent of whether they are bound to their hormone or are free (Clark 1987). A "modified two-step" model (Figure 11–11B) suggests a dynamic equilibrium between soluble (s) and bound (b) steroid receptors (R) in both the cytoplasm (c) and nucleus (n).

$$R_{cs} \rightleftharpoons R_{cb} \qquad R_{ns} \rightleftharpoons R_{nb} \qquad (11.2)$$

The fraction of nuclear receptors increases in the presence of steroid hormone because the steroid-receptor complex has a higher affinity for chromatin than does the unbound receptor and there is a net movement of receptor from the cytoplasm into the nucleus. Alternatively, the "all nuclear" model suggests that the unoccupied receptors are located exclusively in the nucleus and not in the cytoplasm (Figure 11–11C). The steroid binds to a receptor within the nucleus in a functional association with the acceptor site and DNA. The nature of the acceptor site is not clear. It may be a region of DNA associated with the nuclear matrix or a protein associated with the matrix to which the receptor is attached. The receptor may interact with matrix-associated DNA and a matrix-associated acceptor protein. The acceptor may be a matrix-bound protein, and DNA is associated with the matrix only when the steroid-receptor complex is associated with the acceptor.

Invertebrate Endocrine Systems

The endocrine systems of invertebrates are predominantly neurosecretory rather than classical. A second trend in the endocrine systems of invertebrates is an increasing complexity in the higher invertebrate phyla in terms of the numbers of neuro- and classical hormones and the number of physiological functions that they regulate. In general, lower invertebrates (e.g., cnidarians, platyhelminths) have a limited number of neurohormones involved with the regulation of morphogenetic processes, i.e., development, growth, regeneration, and gonadal maturation. More complex endocrine control of functions, such as egg development and laying, osmoregulation, heart rate, hemolymph metabolite levels, and color change, are observed in higher invertebrate phyla (e.g., annelids, mollusks, arthropods). These animals have more developed circulatory systems (see Chapter 14) and therefore have more rapid and effective circulation of hormones.

There is a considerable diversity in the role of endocrine systems in invertebrates. Furthermore, it generally is not clear if endocrine systems with similar roles in different invertebrates are homologous or are independently evolved systems. Consequently, we shall separately examine invertebrate endocrinology for the lower invertebrate phyla, for the annelids, mollusks, and for two of the major arthropod groups, the crustaceans and insects.

Lower Invertebrates

Sponges are generally considered to not have neurons and consequently do not have neurosecretory cells; they also do not have classical endocrine glands.

Hydrozoans (coelenterates) have neurosecretory cells whose activity is associated with normal growth, asexual reproduction, and regeneration. In *Hydra*, if the head is removed, dense neurosecretory granules migrate to axon terminals and are released into the intercellular spaces during regeneration of the tentacles and mouth. The neurosecretion apparently regulates growth and differentiation during regeneration. Transected hydras regenerate additional heads if incubated in a homogenate containing neurosecretory granules.

Platyhelminth worms have neurosecretory cells. The cerebral neurosecretory cells of turbellarians show changes in activity during caudal regeneration (as in *Hydra*) but are not essential for regeneration. The caudal end of a transected turbellarian can regenerate its head without the benefit of cerebral neurosecretion! A neurosecretory role has been suggested for ocular regeneration but, again, the evidence is equivocal since extracts from parts of the body other than the brain also promote eye regeneration. Alteration of the osmotic environment (particularly hypo-osmotic stress) also influences neurosecretory activity, and so a neurohormone may be involved with osmoregulation. There is good evidence for neuroendocrine control of reproduction; reproductive planarians have two to three times as many active neurosecretory cells as reproductively inactive planaria. Neurosecretory cells in the scolex of cestode tapeworms may control shedding of the first proglottid (reproductive) segment or the initiation of strobilization (formation of the array of proglottid segments).

Nemertine (ribbon) worms have a greater cephalization than platyhelminth worms and a larger brain composed of a dorsal and ventral pair of ganglia connected by a nerve ring. The nemertines are the first group, in our phylogenetic survey, to have classical endocrine cells, in the paired cerebral organs. These are ectodermal pockets connected to the exterior by a cephalic canal that opens through a cephalic slit. The cerebral organs have a ciliated canal and glandular cells in loose association with, or even fused to, the cerebral ganglia. The cerebral ganglia/cerebral organ appears to be an endocrine system with inhibitory control of gonadal development and regulation of water balance.

Nematode (round) worms are covered by a complex cuticle that is molted four times. They have variable numbers of neurosecretory cells with somas located in the anterior ganglion and axons extending along the commissures and longitudinal nerves. There are no specific neurohemal areas, although neurosecretory granules appear concentrated in axons near the excretory pore in *Haemonchus*. The neurosecretion apparently controls ecdysis of the

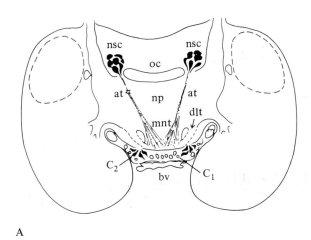

A

FIGURE 11–12 (A) Representation of the neuroendocrine and infracerebral endocrine systems of the polychete *Eulalia* (frontal section). The neurosecretory cell bodies (nsc) near the optic commissure (oc) send an axon tract (at) to the neurosecretory neuropile. The median neurosecretory tract (mnt) of these axons and additional dorsolateral neurosecretory tracts (dlt) abut the connective tissue capsule near the infracerebral gland and blood vessel (bv). The infracerebral gland contains two cell types, C_1 and C_2. (B) Detailed representation of the cerebrovascular complex of *Eulalia*. The neurosecretory axon tract has two types of axon terminals: the type 1 granular endings (ge) presumably secrete peptide hormones, and the type 2 secretory end feet (sef) may secrete nonpeptide hormones. The infracerebral gland contains two cell types: C_1 cells are more common and appear to be endocrine cells; C_2 cells are restricted to a small area of the gland and appear to have a proximal "process" (PP) that passes through the capsule into the brain. *(From Whittle and Golding 1974.)*

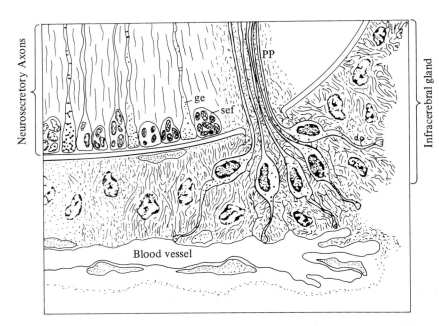

B

old cuticle through a first order neuroendocrine loop. The hormone is released after the new cuticle is produced and stimulates the excretory gland to secrete an enzyme, leucine aminopeptidase, into the space between the old and new cuticle. The accumulation of fluid in this space causes the old cuticle to split and be shed. No classical endocrine glands have been reported for nematodes.

Annelids

Annelid worms have a well-developed and cephalized nervous system, a well-developed circulatory system, and a large coelom. It is not surprising, therefore, that they have a correspondingly well-developed endocrine control of physiological functions (Highnam and Hill 1977; Olive and Clark 1978). Neurosecretory cells have been described in all classes of annelids (Polychaeta, Oligochaeta, and Hirudinea). The various endocrine systems of annelids are generally involved with morphogenetic processes: development, growth, regeneration, and gonadal maturation.

In polychetes, neurosecretory neurons are found in the posterior ventral and anterior dorsal bodies of the cerebral ganglia, supraesophogeal ganglion, and ventral chain ganglia. Many of the cerebral axons form tracts that run ventrally to terminate at the neural lamella in a cerebrovascular neurohemal area (Figure 11–12A). The neurohemal area has a relatively primitive organization for release of neurohormones into the blood and/or coelomic fluid (Figure 11–12B). There are two general types of axonal endings. The Type 1 endings contain many neurosecretory granules and presumably secrete a peptide neurohormone; the type 2 endings (secretory end feet) contain few granules but numerous mitochondria, and may secrete a nonpeptide substance. Closely associated with the neurohemal area is an epithelial glandular structure, the infracerebral gland, which is also of probable endocrine function (Baskin 1976; Whittle and Golding 1974). The infracerebral gland contains epithelial (C_1) cells and C_2 cells of similar appearance to the neurosecretory cells but with short "axons" that presumably secrete hormones. Some axons cross the neural lamella; these may innervate the infracerebral gland or pass from the infracerebral glandular cells into the brain.

Immature polychete worms regularly form additional body segments during early growth. Amputation of posterior segments leads to regeneration of new segments. This regeneration is under neurohormonal control and involves secretions of specific neurons (nuclei 20 and 21) in the supraesophageal ganglion. A general growth hormone is continually secreted during immature growth, resulting in the regular addition of new body segments. Loss of body segments accelerates segment proliferation.

Polychete worms generally, and nereids in particular, reproduce by a synchronized spawning migration in which the mature male and female worms release their gametes into the sea for external fertilization. There is a marked external transformation (epitoky) from the immature, benthic worm (the atoke) to a more active swimming pelagic adult (the epitoke). The epitoke has a larger body, enlarged eyes, modified parapodia, and reduced cephalic appendages (Figure 11–13); there are also internal changes. Epitoky is controlled by a cerebral ganglionic neurohormone. A high level of this neurohormone inhibits growth and development of the worm, but a low level must be maintained during epitoky to complete metamorphosis. This is an inhibitory control system of sexual development, and the neurohormone could therefore be called a "juvenile hormone." The gonads of some polychetes mature and form large numbers of spermatozoans/oocytes during epitoky. These gametes are stored in the coelom until their release. A cerebroganglionic

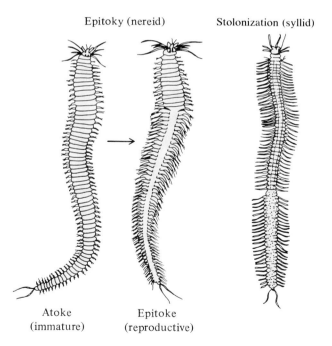

FIGURE 11–13 Metamorphosis in polychetes involves extensive external and internal modifications of the immature atoke to form the mature epitoke (left). Some polychetes, such as syllids, reproduce by stolonization; the posterior end of the body forms a "bud" that develops into a separate animal. *(From Charniaux-Cotton and Kleinholtz 1964.)*

neurohormone retards maturation of sex cells. In some polychetes, there is a stimulatory effect of cerebral ganglion neurosecretions on oocyte growth, meiosis, and gonadal and somatic changes associated with the reproductive period. Some polychetes, such as syllids, reproduce through a budding process, stolonization; the posterior body segments break off from the parent and form a separate individual that grows a new head (Figure 11–13). The proventricle/pharynx region produces a hormone that inhibits stolonization. The small eunicid worm *Ophryotrocha puerilis* is protandric; the female form is maintained indefinitely by isolated individuals, but when reared together the smaller individual becomes a male. This sexual reversal is under the control of a cerebral neurosecretion, whose production is controlled by a pheromone (Pfannenstiel 1975).

Oligochetes have neurosecretory cells in their cerebral and subesophageal ganglia; there is no specialized neurohemal site because the brain has an external and internal blood supply (unlike the polychete brain). Reproductive development of oligochete worms is controlled by neurohormones. Differentiation of gonadal tract, vitellogenesis (yolk formation), and development of the secondary sexual characteristics (e.g., clitellum) occur in concert with increased neurosecretory activity of certain neurons. Other neurosecretory cells produce a neurohormone for regeneration of amputated posterior segments. There may be a separate neurohormone that inhibits diapause and stimulates sexual reproduction. Osmoregulatory hormones have been reported in oligochetes. A hyperglycemic hormone that maintains a high blood glucose level has been reported for the oligochete *Lumbricus*.

Leeches have neurosecretory neurons in their cerebral and subesophageal ganglia. Their axons extend into segmental nerves and form a primitive neurohemal area on the posterior surface of the dorsal commissure in close association with a transverse blood vessel. Activity of specific neurosecretory neurons is associated with the onset of reproductive activity and so their secretions are believed to exert a gonadotropic influence.

Mollusks

Molluskan endocrine systems are best studied in the gastropod snails, prosobranchs, opisthobranchs, and pulmonates (Joosse and Geraerts 1983). Neurosecretory hormones are important in first order responses and also second order and some third order systems. Much molluskan endocrinological research has been devoted to understanding reproductive control. This is especially complex because many mollusks are hermaphroditic, i.e., individuals are simultaneously male and female. Control of male and female reproductive systems must therefore occur simultaneously.

Neurosecretory cells are present in all ganglia of prosobranch snails, except the buccal and pedal ganglia. Their neurohormones are primarily involved with regulation of reproduction. Most species have fixed sexes (are gonochoristic) but a number have protandric sex reversal with an initial male phase, an intervening hermaphroditic phase, and a final female phase. In protandrous species, the immature gonad is bisexual. At sexual maturity, the male sex cells differentiate first under the influence of a cerebral ganglion androgenic factor. The tentacles appear to produce a hormone that suppresses male activity, resulting in seasonal periods of cessation of spermatogenesis and sex reversal. In the female phase, activation of the ovary and oogenesis is also under the control of a cerebral ganglion factor. Sex reversal not only requires development of the testes instead of an ovary but also requires replacement of the male accessory sex organs (sperm duct, seminal vesicles, external sperm groove, penis) by female secondary sex organs (oviduct, gonopericardial canal, receptaculum seminis, uterus, vagina). These secondary sex changes are under direct control of neurohormones. For example, a neurohormone from the pedal ganglion causes differentiation of the penis in the male stage. The transition to the female phase is controlled by a neurohormone from the mediodorsal region of the pedal ganglion. A probable neuropeptide from the parietal ganglion, called egg capsule-laying substance, induces egg laying in the gonochoristic prosobranch *Busycon*.

Opisthobranchs are hermaphroditic marine gastropods (e.g., *Aplysia*). Their reproductive endocrinology is almost unknown except for the role of an egg-laying hormone (ELH), which is the most thoroughly studied molluskan hormone. ELH is secreted by neurosecretory cells located near the pleuroabdominal connectives of the parietovisceral ganglion. Injection of ELH induces the complete behavioral and physiological repertoire for ovulation of oocytes, their transport, fertilization and packaging in the egg string, and egg string extrusion and fixation to the substrate.

Gastropod pulmonate snails have a centralized nervous system, with cerebral, pedal, pleural, parietal, and visceral ganglia arranged as a circumesophageal ring (Figure 11–14). Neurosecretory cells are common in the pulmonate nervous system. Histochemically similar types of neurosecretory

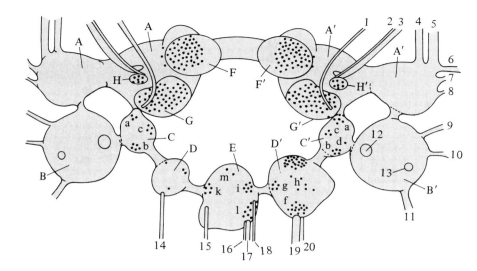

FIGURE 11–14 Location of putative neurosecretory cells in the brain of the snail *Lymnaea*. A and A', cerebral ganglia; B and B', pedal ganglia; C and C', pleural ganglia; D and D', parietal ganglia; E visceral ganglion; F and F', mediodorsal bodies; G and G', laterodorsal bodies; H and H', lateral lobes. a through m are groups of putative neurosecretory cells. 1 through 20 are nerves. *(After Lever et al. 1961; Tombes 1970.)*

cells are located in the mediodorsal and laterodorsal cell bodies (DC) and there is a different type in the caudodorsal cell (CDC). Axons from the DC pass to the median lip nerve and the bulb-shaped axon terminals form a neurohemal organ under the perineurium. The CDC axons pass to the peripheral surface of the intercerebral commissure.

All pulmonate snails are hermaphrodites and the mature gonad contains only one type of stem cell. Hermaphroditism is successive in the Stylommatophora, with the male stage preceding the female stage. The gonadal cells will autodifferentiate into female cells, but they differentiate into male cells under the influence of a factor from the cerebral ganglia and the optic tentacles (this factor also suppresses the differentiation of female sex cells). The initial onset of male reproductive activity is initiated by release of this neurohormone, then subsequent decreased secretion induces development of the female sex cells. Vitellogenesis is initiated by a hormone from the dorsal body (DBH) under neural control from the cerebral ganglia. Cellular differentiation and growth of the male and female accessory sex organs is due to respective male and female gonadal hormones. Secretion by the female accessory sex organs may be under direct control of the DBH. Figure 11–15A summarizes these endocrinological control systems in stylommatophorans.

Hermaphroditism occurs more simultaneously in basommatophoran pulmonates than in the stylommatophorans, and endocrine control of reproduction is substantially different. The lateral lobes (LL) apparently initiate simultaneous development of both sexual cells through the caudodorsal cells (CDC) and dorsal body (females) and neurosecre-

tory cells (males). The CDC hormone (CDCH) controls ovulation and oviposition. Dorsal body hormone functions as in stylommatophorans, i.e., induces female sex cell development and accessory sex organ development and functioning. These endocrine relationships are summarized in Figure 11–15B. There is no gonadal control of accessory sex organs in basommatophoran pulmonates, in contrast to stylommatophoran pulmonates.

For cephalopods, reproduction is a terminal event; probably all cephalopods die after they mate. The control of reproduction is remarkably simple, involving only one hormone from the optic glands, which are of neural origin. The optic gland produces a gonadotropic hormone when released from neural inhibition by the subpedunculate lobe of the brain. The hormone secreted by the optic gland may be a peptide (the gland has a neural origin) or a steroid. The effects of the optic gland hormone include spermatophore production and development of the male reproductive ducts, and production of proteinaceous yolk in oocytes and development of accessory sex organs in females. It also promotes protein breakdown and mobilization of amino acids for use by reproductive organs. The gonads of cephalopods synthesize steroids, but their role may be limited to control of secondary metabolic activities. Thus, the highly developed nervous system of cephalopods, in concert with the terminal reproductive strategy, has led to a marked simplification of the endocrine control of reproduction.

Mollusks have a number of nonreproductive endocrine systems. Body and shell growth are under endocrine control by growth hormones that affect all tissues, including the shell. Their primary effect on shell growth is calcium transport through the

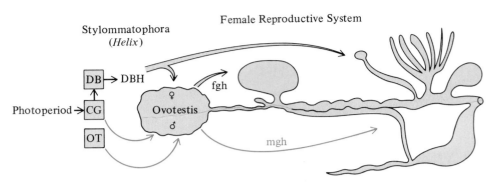

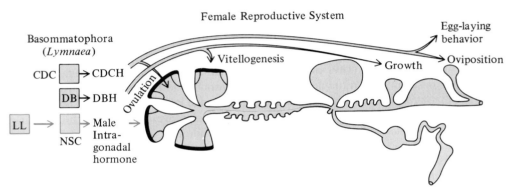

FIGURE 11–15 **(A)** Structure of the reproductive system of a stylommatophoran pulmonate snail and summary of the endocrine control of reproduction. The ovotestis is connected to the male and female accessory sex organs by a hermaphroditic duct. The female and male reproductive tracts diverge and may, or may not, re-fuse to form a common genital opening. The female accessory sex organs are shown in gray. The male accessory sex organs are shown in color. Dorsal body hormone (DBH) from the dorsal body (DB), which is under neural control of the cerebral ganglia (CG), promotes vitellogenesis and functioning of the female secondary sex organs. A secretion of the optic tentacles (OT) inhibits female sex cell development and promotes male sex cell differentiation. A female gonadal hormone (fgh) controls development of the female accessory sex organs and a male gonadal hormone (mgh) controls development of the male accessory sex organs. **(B)** Structure of the reproductive system of a basommatophoran pulmonate snail and a summary of the endocrine control of reproduction. The ovotestis is connected to the male and female accessory sex organs by a hermaphroditic duct. The female and male reproductive tracts diverge at the fertilization pocket and may, or may not, re-fuse to form a common genital opening. The female accessory sex organs are indicated in gray. The male accessory sex organs are in color. Dorsal body hormone (DBH) from the dorsal body (DB) promotes female sex cell development, vitellogenesis, and the development of female accessory sex organs. Caudodorsal hormone (CDCH) from the caudodorsal cells (CDC) promotes ovulation, oviposition, and egg-laying behavior. A secretion of the lateral lobes (LL) promotes male sex cell maturation, most likely through stimulation of a neurohormone from neurosecretory cells (NSC). *(Modified from Joosse and Geraerts 1983.)*

mantle edge into the extrapallial fluid; this is often accompanied by elevated levels of calcium-binding proteins.

Energy metabolism is also under hormonal control. The visceral ganglia of *Mytilis* secrete a neurohormone that promotes storage of glycogen, lipid, and protein in glycogen and adipogranular cells. The cerebral ganglia and other cells secrete an insulin-like substance (ILS) that promotes glycogen storage (by stimulating glycogen synthetase).

Most mollusks have cardioactive neuropeptides. Bivalves have a tetrapeptide amide (Phe-Met-Arg-Phe-NH$_2$, FMRFamide) that stimulates the heart. Two snail cardioexcitatory peptides (SCP) of *Helix* are secreted by subesophageal ganglion cells and released from a neurohemal area located at the atrioventricular junction of the heart. Neurosecretory cells of cephalopods release a cardioexcitatory neurohormone at extensive neurohemal areas in the anterior vena cava and pharyngo-ophthalmic vein. The molluskan neurohemal areas that release cardioactive peptides also contain vertebrate peptides, or at least peptides that react with antibodies to the vertebrate peptides; α-MSH; arginine vasopressin; neurophysin; met-enkephalin and leu-enkephalin in cephalopods; vasotocin in gastropods; and met-enkephalin, leu-enkephalin, and met-enkephalin-ArgPhe in bivalves. The role of these possibly cardioactive peptides is not clear.

Hormones are involved in hemolymph ionoregulation and control of body water content in gastropod snails. Dark green cells (DGC) of *Lymnaea* secrete a neurohormone with a diuretic action (i.e., increases urine flow); they are active in freshwater and inactive in saline media. Most DGC axons project to the ventral and anterior parts of the head, foot, neck, and mantle, suggesting that the skin is a target organ. The DGC hormone is similar to thyrotropin releasing hormone (TRH) of vertebrates. Other neurohormones may control ion reabsorption from the renal ultrafiltrate. *Helosoma* also appears to have an osmoregulatory (diuretic) neurohormone. The pleuropedal ganglia of slugs contain an arginine-vasotocin-like peptide that enhances the water permeability of the body wall. In the prosobranch *Aplysia*, neurosecretion has an antidiuretic effect, causing a rapid uptake of water; hypo-osmotic stimulation decreases the release of this neurohormone.

Crustaceans

The crustacean endocrine system, like that of other invertebrates, has neurosecretory neurons and some classical endocrine glands (Fingerman 1987; Figure 11–16). Neurohemal release sites are best developed in the malacostracans (crabs, lobster, shrimp, isopods, amphipods).

The cerebral ganglion of crustaceans varies considerably in morphology and location of neurosecretory cells. The crayfish, for example, has neurosecretory cells in the optic lobe; cerebral ganglion; and the subesophageal, thoracic, and abdominal ganglia. The main neurosecretory body of the optic lobe, in the eyestalk, is the X-organ (or organ of Hanstrom) located on the medulla terminalis; this is the medullary terminalis X-organ, or MT-XO. There are also neurosecretory cells in the medulla interior (MI-XO) and medulla externa (ME-XO). The poorly understood sensory pore X-organ (organ of Bellonci, SP-XO) may be neurosecretory, sensory, or both.

Malacostracan crustaceans have three neurohemal organs. The sinus gland of the eye stalk receives neurosecretory axons from cells in the optic lobe and also the brain. The postcommissure organ at the posterior fusion of the circumesophageal connective receives axons from the brain. The ribbon-like pericardial organ, which is stretched across openings of gill veins into the pericardium, receives axons from neurosecretory cell bodies located in the ventral ganglion and also from other cell bodies. The eyestalk of brachiopods may have a sinus gland but most other crustaceans appear to lack this neurohemal organ. For example, barnacles have neurosecretory cells but lack a sinus gland; a neurohemal area is located at the junction of the tritocerebrum and ventral nerve cord.

Classical endocrine glands also have important roles in crustaceans. The small, paired Y-organ, located in the maxillary or antennary segment, has a single cell type similar to steroid-secreting cells of vertebrates. The androgenic gland of male malacostracans is attached to the ejaculatory regions of the vas deferens. The ovaries of amphipods and isopods appear to secrete hormones that control differentiation of the female secondary sex characters (see below). The paired mandibular organ, which is generally located near but more anterior to the Y-organ, has two cell types that resemble lipid-secreting cells. Methyl farnesoate, a precursor for juvenile hormones, is one secretion of the mandibular gland.

The endocrinological roles of the neurosecretory and classical endocrine glands of crustaceans include control of molting, regeneration, heart rate, ionic and osmotic regulation, blood metabolite levels, neuronal activity, and pigment dispersion.

Molting.   The rigid exoskeleton of crustaceans (and other arthropods) limits growth and must therefore be periodically shed by **ecdysis**, or exuviation (Skin-

CLASSICAL ENDOCRINE SYSTEMS

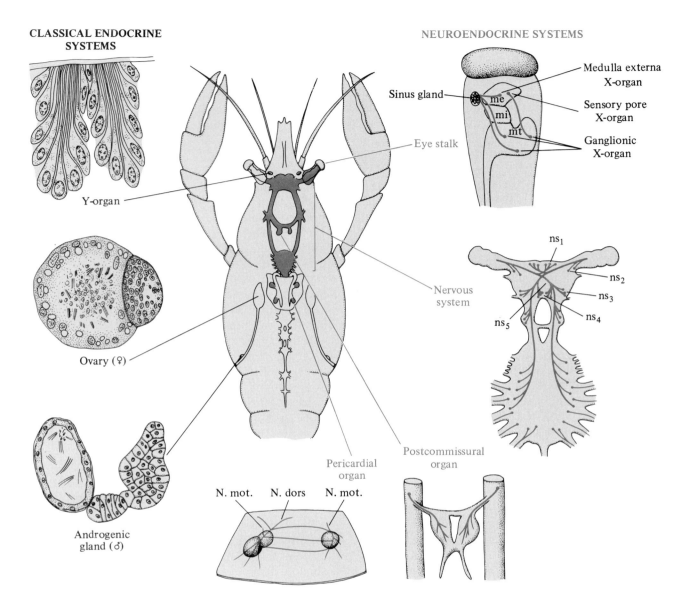

FIGURE 11–16   The generalized endocrine system of a crustacean. The principal endocrine areas are as follows. (1) The eye stalk contains a number of X-organs (sensory pore; medulla externa, me; medulla interna, mi; medulla terminalis, mt) and the neurohemal organ, the sinus gland (sg). Axons from neurosecretory cells in the brain and other parts of the nervous system pass to the sinus gland via a neurosecretory tract (bt). (2) The postcommissural organs receive axons from four neurosecretory cells in each side of the circumesophageal connectives; only one is shown on the left side. (3) The pericardial organs are located over the openings of branchiocardiac veins into the aorta; numerous nerves run from the nervous system to the organs (colored lines). The dorsal nerve of the heart (n dors) and nerves to muscles (n mot) are also shown. (4) The androgenic gland often is a vermiform mass of secretory cells attached to the distal portion of the vas deferens. The ovary secretes female sex steroids in many species. (5) The generalized neurosecretory systems (except the eyestalk; see above) consist of cerebral neurosecretory cells (ns 1 through ns 5) and thoracic ganglia neurosecretory cells. Axons of neurosecretory cells in all parts of the brain pass to the sinus gland, but only axons from ganglionic neurosecretory cells pass out the pedal nerves. (6) The Y-organ is an endocrine gland with no innervation; it is an ovoid disk of hypertrophied epidermis which secretes molting hormone.   *(From Carlisle and Knowles 1959; modified from Gorbman and Bern 1964; Alexandrowicz 1953; Fryer 1960; Matsumoto 1958; Burghause 1975.)*

ner 1985). The entire molt cycle involves not only ecdysis but also major metabolic events in the integument before and after ecdysis (Figure 11–17).

In proecdysis, the start of the molt cycle, the epidermis begins to separate from the old exoskeleton by dissolution of the membranous layer of the exoskeleton, and epidermal cells enlarge and begin to secrete the new exoskeleton. Many nutrients and minerals are resorbed from the old exoskeleton, to be reincorporated into the new exoskeleton. The hemolymph is an important repository for amino acids resorbed from the old exoskeleton (and also from muscle) and a red carotenoid pigment astaxanthin (which crustaceans cannot synthesize). Calcium is temporarily deposited as one or two gastroliths or as deposits in sternal plates (in some isopods), posterior cecae (in some amphipods), or concretions in midgut glands or epidermis (the lobster *Panilurus*). Many crustaceans also eat the shed exoskeleton after ecdysis to recycle nutrients and minerals.

The muscles of the major limb segments (propodite) atrophy during proecdysis so the limb can be withdrawn through the narrow basal segment at ecdysis. Regeneration of lost limbs also begins during proecdysis.

At ecdysis, the old exoskeleton is shed and the animal emerges with a pale, plastic exoskeleton that is stretched by the uptake of water and an expanded hemocoel. During metecdysis, the endocuticle layer of the exoskeleton is formed, calcium is redeposited in the new exoskeleton, and limb muscle hypertrophies. During the prolonged intermolt period (anecdysis II), energy reserves of glycogen and lipid are stored in the midgut glands and muscle in preparation for the next molt.

Molting is under neurohormonal control (see Figure 11–5B, page 499). It is influenced by external factors such as light intensity and photoperiod, temperature, salinity, ambient humidity, food availability, and the presence of conspecific animals.

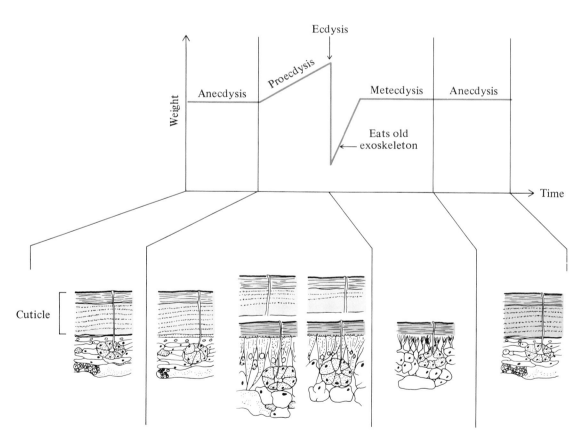

FIGURE 11–17 Molting cycle of the Bermuda land crab *Gecarcinus* showing change in weight during anecdysis, proecdysis, acdysis, and metecdysis, and associated changes in epidermal structure. *(Modified from Skinner 1985.)*

Two antagonistic hormones control the molt cycle: molt-inhibiting hormone (MIH) and molt-inducing hormone (MH).

MIH is apparently a neurohormone that is released from the sinus gland, since a precocious molt can generally be induced by removal of the eye stalks. MIH appears to be a peptide of MWt 1 to 5×10^3, perhaps structurally related to vertebrate octapeptides since lysine vasopressin, arginine vasopressin, vasotocin, and oxytocin mimic MIH. Release of MIH is controlled by the neurotransmitter 5-HT. MIH acts through a second order endocrine loop and cAMP to inhibit the secretion of molting hormone (MH) from the Y-organ.

The active MH is β-ecdysone (20-hydroxyecdysone). The Y-organ cells produce α-ecdysone, which is converted peripherally to β-ecdysone. Cells of the Y-organ hypertrophy, their cytoplasmic volume increases and the nuclear volume may increase, and their Golgi apparatus becomes more highly developed during molting. The hemolymph ecdysteroid level rises during proecdysis and peaks just before ecdysis, then declines during metecdysis and anecdysis. About 95% of the ecdysteroids circulate in the free, unbound form. This is in contrast to the important role of steroid-binding proteins in vertebrates (see below). The Y-organ of crustaceans with a terminal ecdysis degenerates after the last molt.

A third hormone, molt-accelerating hormone, appears to accelerate molting by stimulating ecdysteroid production by the Y-organ. An exuviation factor from the Y-organ may also be required to initiate ecdysis, in similar fashion to the initiation of ecdysis by eclosion hormone in insects (see below).

Limb Autotomy.   Crustaceans can autotomize (self-amputate) their limbs in response to injury or by volition. Autotomy occurs at a specific location where regeneration can occur. The regenerating limb of isopods is internal and encased by a protective sheath that is lost at ecdysis. In crabs, epidermal tissue rapidly differentiates into a papilla (a miniature limb, with epidermis, muscle, and nerves) that is external and enclosed by a protective sheath. The papilla remains quiescent until the next proecdysial period when the limb regenerate grows rapidly. There is a reciprocal relationship between regeneration and molting; regeneration occurs only during proecdysis. Autotomy of a critical number of limbs induces an early molt and can actually be a more potent stimulus for molting than eyestalk removal. The stimulus may be neuronal inhibition of MIH secretion depending on the severing of a critical number of limb nerves, although the loss of appendages is a more effective stimulus than simply severing the limb nerves.

Reproduction.   The endocrine control of reproduction involves both neurohormones and classical hormones (Adiyodi 1985). The sinus gland of decapod crustaceans releases a gonad-inhibiting peptide hormone, and a gonad-stimulating hormone is released from the brain and thoracic ganglia.

The male androgenic gland controls differentiation of the male reproductive system as well as the development and functioning of male secondary sex characters. The hormone is possibly steroidal (perhaps farnesyl-lactone) or a peptide. A number of endocrine functions have been ascribed to the female ovaries but not to male testes. An ovarian hormone from the secondary follicle cells in isopods and amphipods controls the appearance of temporary female reproductive structures, the ovigerous setae on the oostegites of the brood pouch. There is evidence in some crustaceans for ecdysteroid synthesis by ovarian tissue, or at least the accumulation of ecdysteroids in ovarian tissue, and a role of ecdysteroids in the control of vitellogenesis. The primary follicle cells in ovaries of amphipods (but not isopods) produce a hormone responsible for the normal development of the brood pouch, a permanent external female sex character (formed from the oostegites). In isopods, the permanent female sex characters develop in the absence of ovaries (the androgenic gland hormone inhibits their development in males). The secondary follicle cells of the ovary in amphipods and isopods may also secrete a vitellogenin-stimulating hormone (VSOH).

Most adult crustaceans continue to molt and there generally is an antagonistic relationship between molt and reproduction. In some, the ovarian cycle and oviposition precedes high ecdysteroid levels and ecdysis. In brachyurans, somatic growth and the molt cycle occur after successive ovarian cycles. In amphipods and isopods, ovarian and somatic growth are synergistic rather than antagonistic. In barnacles, molting is frequent compared to the long ovarian cycle.

Color.   Many crustaceans have the capacity for both physiological and morphological color change (Rao 1985). Color change enables them to cryptically adapt to their background color or shade. The shrimp *Crangon vulgaris* can strikingly color adapt to white, black, gray, yellow, red, or orange backgrounds, but many species can only change their shade from light to dark. Illumination intensity can induce color adaptation independent of, or in concert with, background color. Temperature can influence chromatophores. White chromatophores tend to disperse at high temperatures, blanching the animal to a pale color. This may have thermoregulatory significance in some species; pale *Uca* crabs have a body temperature 2° C lower than dark *Uca* after five min exposure to sunlight. There are also behavior-related color changes. The claw waving display of some fiddler crabs involves a color change. The amphipod *Hyperia* changes color in response to tactile stimulation. The shrimp *Palaemon* has excitement-related color changes. Shrimp sometimes copy the color of other individuals, rather than their background color.

Color and color change are consequences of pigments found in specialized cells called **chromatophores**. These pigment cells are located in the epidermis and also in deeper body tissues, such as gonad, gut, and nerve cord. Chromatophores have a central cell body with a number of radiating, often branched, dendritic processes. Some chromatophores have a permanent cellular shape regardless of the state of pigment dispersion or aggregation, but other chromatophores are semiamoeboid. In general, chromatophores are monochromatic and contain only one type of pigment (Figure 11–18A), e.g., melanophores have a black/brown pigment, leukophores have a white pigment, erythrophores a red one, and xanthophores a yellow one. **Polychromatic chromatophores** contain more than one pigment. For example, the erythrophores of some decapods have a red pigment and a blue pigment that is a protein bound form of the red pigment. Some polychromatic chromatophores may contain as many as four pigments (e.g., red, yellow, white, and blue). Different-colored chromatophores that are organized in a tight association can appear to be polychromatic; these are **chromatosomes**. They

FIGURE 11–18   (A) Types of crustacean chromatophores. Monochromatic chromatophores (melanophores, erythrophores, xanthophores, leucophores) contain a single type of pigment. Polychromatic chromatophores have two or more kinds of pigments (e.g., some erythrophores contain red and blue pigments). Chromatosomes are clusters of interdigitated monochromatic chromatophores. (B) Melanophores of scales from a fish, the topmouth gudgeon, illustrate the varying degrees of chromatophore dispersal. Photographs from Nagai et al (1986).

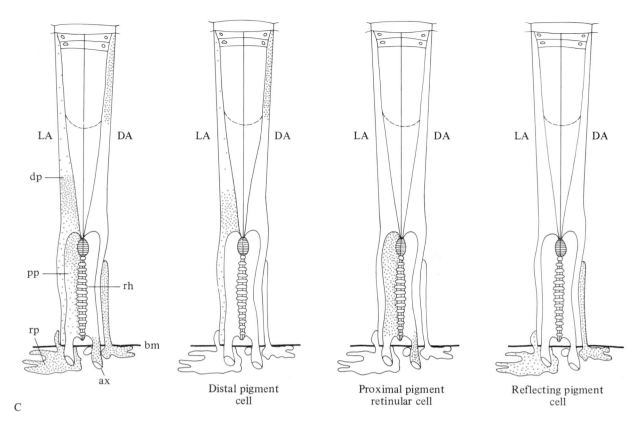

C

FIGURE 11-18 Continued **(C)** Effect of light adaptation (LA, left) and dark adaptation (DA, right) on the distribution of pigments in the ommatidium of a prawn (left) and the individual responses to light/dark adaptation for the distal pigment cell, proximal pigment cell, and reflecting pigment cell (right panels).   *(Modified from Rao 1985; Nagai, Oshima, and Fujii 1986.)*

may be combinations of two colors of chromatophores (e.g., b-r, r-w, r-y, y-w), three colors (e.g., b-r-w, b-r-y, r-w-y), or four colors (e.g., b-r-w-y).

Morphological color change is a slow process due to a quantitative change in the number, type, and distribution of chromatophores or the type and amount of exoskeletal pigments. Some examples of morphological color change involve deposition of pigment or dietary carotenoids in the exoskeleton and do not involve chromatophores. Chromatophore-dependent morphological color change is a consequence of altered numbers of chromatophores, as well as cellular changes in the quantity or type of pigment. The control of morphological color change is poorly understood but probably is similar to that of physiological color change.

Physiological color change is due to chromatophore pigment movement; pigment dispersion or aggregation enhances or obscures the coloring effect of the chromatophore (Figure 11-18B). Some physiological color changes are slow and rhythmic and

some are rapid, occurring within minutes or hours. Injection of hemolymph from dark *Crangon* into light individuals causes darkening, even if the shrimp are kept on a white background. Eyestalk removal also promotes chromatophore darkening, indicating a neurosecretory control. The regulation of physiological color change is well understood, particularly for decapods.

Chromatophores are controlled by two antagonistic hemolymph-born hormones. For example, in *Crangon*, one causes pigment dispersion ("*Crangon* darkening hormone," CDH) and one causes aggregation ("*Crangon* body lightening hormone," BLH). An antagonistic control system is also involved in the regulation of various chromatophores, e.g., red pigment dispersing hormone (RPDH) and red pigment concentrating hormone (RPCH), white pigment dispersing hormone (WPDH) and white pigment concentrating hormone (WPCH), and black pigment dispersing hormone (BPDH) and black pigment concentrating hormone (BPCH). In some crustaceans, there is a general pigment dispersing hormone that affects all chromatophores. In fiddler crabs *Uca*, 5-HT indirectly induces pigment dispersion by promoting release of the specific RPDH. Norepinephrine similarly induces black pigment dispersion, and dopamine stimulates both black and red pigment dispersion. The various crustacean chromatophorin hormones, like other peptide neurohormones, exert their effect on the target chromatophores via cell surface receptors. Both Ca^{2+} and cAMP modulate pigment motility through separate effects on the cytoskeleton.

Light-dark adaptation of the crustacean compound eye involves morphological and physiological color changes in retinular cells and reversible movements of screening pigment granules (Figure 11–18C). There also are photochemical and neural mechanisms for dark-light adaptation. The retinular cells contain a black or brown intraphotoreceptor pigment, called proximal pigment. The proximal pigment approaches the rhabdomere closely in the light to decrease light transmission to the rhabdomere and migrates away from the rhabdomere, sometimes into the retinular cell axons beneath the basement membrane to maximize light transmission to the rhabdomere in the dark.

Malacostracans also have additional screening pigments: a distal pigment in distal pigment cells and a light pigment in reflecting pigment cells. In apposition eyes, there is some pigment movement in extraretinular cells. Superposition eyes (e.g., of the prawn; Figure 11–18C) show more pronounced extraretinular pigment movement. With light adaptation, distal pigment moves proximally to form a sleeve around the crystalline cone; during light adaptation, the reflecting pigment is mainly below the basement membrane. With dark adaptation, the distal pigment aggregates distally and the reflecting pigment surrounds the base of the ommatidium. These pigment movements increase the effective aperture and field of view of the light-adapted ommatidia. The retinal pigments also act as wavelength selective filters. The spectral sensitivity peak of light-adapted pigments is shifted to longer wavelengths. Some malacostracan ommatidia contain additional pigment cells; proximal pigment cells, green cells, proximal absorbing pigment cells, basal absorbing pigment cells, light distal pigment cells, and basal red pigment cells.

A direct action of light is thought largely to control proximal pigment movement within the retinular cells, but distal pigment migration is mediated by humoral factors. The eyestalk contains distal retinal pigment hormones (DRPH) that elicit light adaptation and dark adaptation.

Other Endocrine Systems.   The pericardial organs contain at least two cardioexcitatory substances that increase both rate and force of contraction of the heart beat.

The sinus gland releases a neuro-depressing peptide that modulates the circadian activity of the central nervous system; it might initiate the low-activity day phase in the circadian cycle of nocturnal crustaceans.

Crustacean hyperglycemic hormone (CHH), a peptide released from the sinus gland, increases the hemolymph glucose concentration. A hypoglycemic factor has been reported for the crab *Paratelphusa*.

Osmotic and ionic regulation is under endocrine control. The rapid water uptake of molting crustaceans is presumably under the control of ecdysteroids. For intermolt individuals, exposure to hypo-osmotic media induces the release from the eyestalk of a diuretic factor that prevents weight gain and eliminates water. An antagonistic antidiuretic hormone that increases water uptake has been reported for the crab *Thalamita*. Brine shrimp *Artemia salina* in diluted seawater secrete a neurohormone that may promote salt retention.

Insects

Neurosecretory cells are the main endocrine system of insects and other terrestrial mandibulate arthropods. The endocrine roles of these neurosecretory cells are best understood for insects, and so we shall concentrate on this group. Many physiological systems in insects have endocrine control; these include molting and metamorphosis, reproduction, diapause, hemolymph metabolite levels, osmoregulation, and color changes.

There are three main groups of neurosecretory cells in the insect nervous system (Figure 11–19). One near the median furrow of the protocerebrum sends axons via the NCCI nerve to the paired corpora cardiaca, an important neurohemal organ just posterior to the supraesophageal ganglion. The **corpora cardiaca** also contain neurons whose axons (as well as axons from the supraesophageal ganglion) pass to the more posterior corpora allata; in some insects there also are secretory cells in the corpora cardiaca that have a likely endocrine function. A second lateral group of protocerebral neurosecretory cells send axons to the corpora cardiaca via the NCCII nerve. Some phasmid insects have an additional pair of protocerebral neurosecretory cells with axons passing to the corpora cardiaca via an NCCIII nerve. Virtually all insects have neurosecretory cells in the subesophageal ganglion, with axons passing to the corpora allata. There also are neurose-

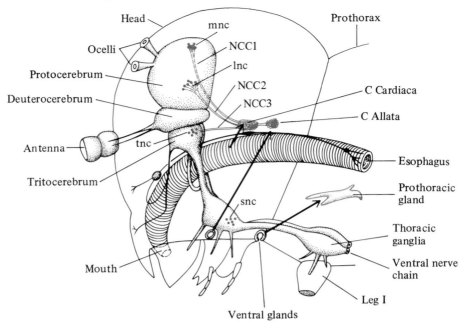

FIGURE 11–19 Generalized central nervous system and endocrine systems of an insect. The nervous system contains three major groups of neurosecretory cells: the median (mnc), lateral (lnc), and subesophageal (snc), which are connected with the corpora cardiaca via NCCI and NCCII nerves; phasmid insects have additional neurosecretory cells in the tritocerebrum (tnc) connected to the corpora cardiaca via NCCIII nerves. The corpora cardiaca are a neurohemal organ for the release of neurosecretions; they arise from stomodeal ectoderm (shown by black arrows). The corpora allata are classical endocrine glands that arise from ectodermal invaginations near the maxillae (black arrows). The prothoracic gland is another important classical endocrine gland; it arises (black arrow) from ventral glands (which are present in primitive insects). *(Modified from Jenkin 1962; after Knowles 1963.)*

cretory cells in the thoracic and abdominal ganglia. Their axons pass along interganglionic connectives and peripheral nerves to neurohemal organs on peripheral nerves or pass directly to effector organs.

The paired **corpora allata** (or corpus allatum in those insects with a fused structure) are classical endocrine glands, although they do have a neural innervation. The gland is usually a solid mass of cells, but in lepidopterans they are loose clusters of cells and in phasmids are a hollow gland. In higher dipterans, the corpus allatum becomes incorporated into the ring gland.

The second important classical endocrine gland in insects is the paired ecdysial, or prothoracic, gland. The gland arises in the embryo from ectodermal cells at the base of the mandible and in the adult has a variable structure and location. In general, the prothoracic gland is compact and located in the head of primitive insects but is more diffuse and posterior (in the thorax) in advanced

insects. In higher dipterans, the prothoracic gland is incorporated into the ring gland.

Molting and Metamorphosis. Insects, like other arthropods, develop in stages and molt at the end of each instar to allow further body growth. Molting does not cease in the adults of a few primitive insects (Protura, Collembola); this pattern is described as **ametabolous**. In most insects, a final molt occurs when the adult characteristics develop and the juvenile characteristics are lost; this change from juvenile to adult is **metamorphosis**. If the juvenile is similar to the adult and metamorphosis occurs in graded steps, then development is **hemimetabolous** (Figure 11–20). If the juvenile and adult have a very different morphology, then an intervening pupal stage occurs before metamorphosis; this is **holometabolous** development.

The controls for molting and metamorphosis are clearly interrelated and are best discussed together rather than as separate endocrinological control

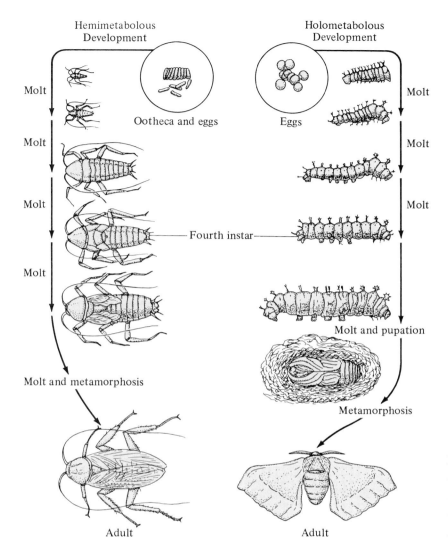

Hemimetabolous Development

Holometabolous Development

Molt

Molt

Molt

Molt

Ootheca and eggs

Eggs

Molt

Molt

—Fourth instar—

Molt

Molt

Molt

Molt and pupation

Molt and metamorphosis

Metamorphosis

Adult

Adult

FIGURE 11–20 Comparison of growth and development in a hemimetabolous insect (cockroach) and a holometabolous insect (silkworm). *(From Turner 1966.)*

systems. Median neurosecretory cells of the protocerebrum secrete a peptide prothoracicotropic hormone (PTTH, also called ecdysiotropin), which is released by the corpora cardiaca. There actually are a number of PTTH peptides (Bollenbacher and Granger 1985). PTTH stimulates the prothoracic gland to secrete a molt-inducing hormone, ecdysone.

Ecdysone initiates the epidermal changes necessary for molting and also causes differentiation of the body tissues to adult structures. However, the differentiation towards adult structures can be inhibited by a juvenile hormone that is secreted by the corpora allata. High hemolymph levels of JH result in development of an immature animal; a low JH level will result in intermediate stages in hemimetabolous insects, or a pupa in holometabolous insects (Figure 11–21). Control of JH levels is linked to the insect developmental sequence, culminating in an adult. For example, allatectomy

of a third instar *Bombyx* results in premature pupation and metamorphosis and forms a small adult; a supernumerary "sixth" instar larva (induced by implantation of corpora allata into a fifth instar larva) pupates and develops into a giant adult.

The control of corpora allata secretion is not well understood. Its activity must be inhibited prior to the last larval molt for adult development, but the corpora allata resume activity in the adult to regulate reproductive development (see below). The gland may be under neural or neurosecretory control (via the corpora cardiaca nerves, the nervi allati). JH synthesis can be maintained or inhibited, depending on the species, by signals reaching the corpora allata via the NCA-I tract (from the corpora cardiaca) but apparently not by signals from the subesophageal ganglion via NCA-II. The level of JH has a negative feedback effect on its secretion by the corpora allata; the feedback is probably not a direct effect

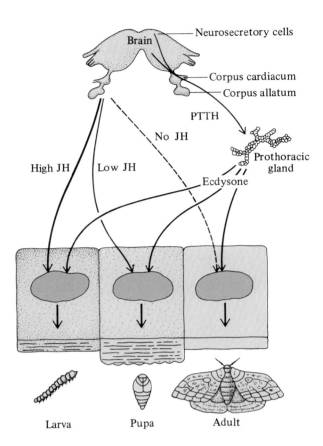

Brain

Neurosecretory cells

Corpus cardiacum

Corpus allatum

PTTH

No JH

Prothoracic gland

High JH

Low JH

Ecdysone

Larva

Pupa

Adult

FIGURE 11–21 Representation of endocrine events controlling larval, pupal, and adult molting and development. A brain neurohormone (prothoracicotropin, PTTH) stimulates the prothoracic gland to secrete a molt-inducing hormone (ecdysone), which stimulates the epidermal molt cycle. A second hormone (juvenile hormone, JH) from the corpora allata determines the developmental transition at molting; high JH levels maintain larval molt, low levels of JH induce pupation, and no JH induces the adult molt. *(Modified from Schneiderman and Gilbert 1964.)*

of JH on the corpora allata but involves brain neurosecretory cells. In adults, the ovaries exert both stimulatory and inhibitory effects on the corpora allata.

A number of hormonal controls are involved with molting and metamorphosis (Truman 1985). Towards the end of the premolt period, the two cuticle layers are separated by a molting fluid which, when activated, digests the outer (old) endocuticle and transports the resorbed materials across the old cuticle. Those parts of the new cuticle that sustain major stress forces during ecdysis are strengthened by pre-ecdysial tanning. Just prior to eclosion, the levels of ecdysteroids decline and eclosion hormone (EH) is released from the corpora cardiaca to trigger eclosion; EH is a peptide. Release of EH and tissue

sensitization to EH is initiated by the decline in ecdysteroids (which also stimulates other terminal developmental events, such as endocuticle digestion and atrophy of intersegmental muscles). The role of EH is best studied in lepidopterans but is probably similar throughout the insects.

There generally are three phases to ecdysis: a preparatory phase (selection of ecdysis site, movement to loosen the old cuticle from the new cuticle, air swallowing), the ecdysial phase (stereotyped behavioral movements to rupture and escape the exuvium), and the postecdysial period (cuticular expansion, hardening, and darkening). EH initiates the behavioral aspects of eclosion as well as physiological adjustments. It also may trigger secretion of "cement substances" from the Verson's glands; deposition of the cement substances on the cuticle surface waterproofs the new cuticle after eclosion. A tanning hormone, bursicon, is released from the segmental perivisceral organs associated with the abdominal ganglia about one to two hours after the release of EH. Two cardioactive peptides are released from the perivisceral organ at the time of wing expansion to facilitate blood flow into the wings for their expansion by sustained anterior waves of heart contraction and increased activity of the wing accessory pulsatile organs.

Higher dipteran flies have a specialized pupal developmental stage; the pupa remains inside the shed cuticle of the third (last) larval instar (Zdarek 1985). The pupal case, which is called a puparium, is a protective, cocoon-like structure. Pupal and adult development are completed within the puparium, and pupal ecdysis and adult eclosion occur at the same time when the adult emerges from the puparium. In *Sarcophagus*, incipient pupariation of the third larval instar is first observed in the "red-spiracled larvae" as an orange-red tanning of the cuticle of the posterior spiracles, then other spiracles. Activity then diminishes and the puparium forms. Pupariation is induced by ecdysone. Anterior retraction factor (ARF) is a neurosecretory peptide that causes contraction of the body into the puparial shape. Puparium tanning factor (PTF, a peptide) causes tanning of the puparium. It may be related to, but is not identical with, bursicon. Puparium-immobilizing factor (PIF) is also a protein that causes immobilization of postfeeding but preredspiracle stage larvae. Puparium-stimulating factor (PSF) stimulates the somatoneuromuscular system; its properties and effects are poorly understood.

Diapause. Many insects have a behavioral and physiological state of inactivity and lowered metabolism, called **diapause**, that enables them to withstand periods of environmental stress, such as low or high temperature and drought (Denlinger 1985). Diapause generally is a developmental option, rather than an obligate developmental stage. It can occur in various insects at any stage, but the diapause stage (egg, larva, pupa, or adult) is quite specific for a particular species.

The onset of diapause is often controlled by photoperiod and is very abrupt. For example, a day length of 13½ hr induces diapause in flesh flies. Temperature is also an important environmental cue for onset of diapause and often modifies the photoperiod cue.

Embryonic diapause often occurs prior to the maturation of the embryonic neuro- and classical endocrine systems. This, and the lack of a photoreceptive system, suggests that diapause is not controlled by an endogenous endocrine response. In older embryos, ecdysone and JH might induce diapause. Maternal determination of diapause in the silk moth *Bombyx* is the best known example of endocrinological control of embryonic diapause. Adult female silkmoths lay diapausing eggs if they have been exposed to long day lengths and high temperatures. A diapause hormone (DH) is released from the subesophageal ganglion of such females and stimulates their ovarioles to produce diapausing eggs. Termination of diapause requires one to three weeks of chilling at 5° C.

Larval diapause is best known in lepidopterans but also occurs in many other insects. The last larval instar commonly enters diapause, but other instars may also diapause. Maternal determination of larval diapause has been reported in a few insects (e.g., *Calliphora*) but usually diapause is induced by environmental cues acting on the larva. The environmental cues act via the brain and inhibit prothoracic gland secretion of ecdysone. In some insects, JH also plays a role in diapause induction. The hemolymph JH level both initiates and maintains diapause in the corn borer *Diatraea*. In the corn borer *Ostrinia*, JH only initiates diapause.

Pupal diapause is common in Lepidoptera and higher Diptera. There are diverse controls for pupal diapause, but maternal induction is rare. Generally, the prothoracic gland is inactive during diapause, and ecdysteroid levels are consequently low. Diapause can be terminated in both lepidopterans and dipterans by exogenous ecdysteroids. In some species, the brain releases PTTH to initiate adult development. In other species, the prothoracic gland is activated in the absence of PTTH. Environmental factors, such as temperature, may have a direct stimulatory effect on the prothoracic gland, which is the regulatory site determining the duration of diapause.

Adult diapause is common in many insect groups, especially coleopterans. Reproduction is invariably suppressed during adult diapause. Juvenile hormone is likely involved with the control of diapause since it is the primary regulator of reproduction in adult insects. The corpora allata are inhibited by the brain during diapause and levels of JH are low. The regulation of the corpora allata is hormonal in some species but is neural in others.

Reproduction. The male reproductive system of insects consists of a pair of testes and lateral ducts (or vas deferens), which open into a common ejaculatory duct (Figure 11–22). A seminal vesicle is often present as a dilation of the vas deferens.

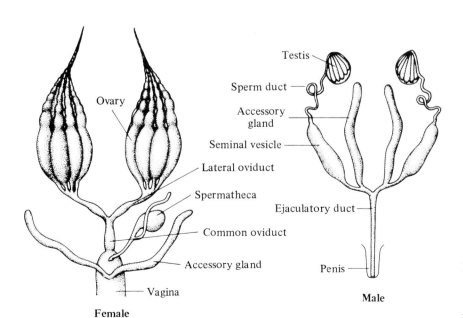

Ovary

Testis

Sperm duct

Accessory gland

Seminal vesicle

Lateral oviduct

Spermatheca

Ejaculatory duct

Common oviduct

Accessory gland

Penis

Vagina

Female

Male

FIGURE 11–22 Generalized scheme of the reproductive system of female and male insects.

The female reproductive system has two ovaries connected to a genital chamber, or vagina, by lateral oviducts. Diverticula of the vagina or oviducts form a sperm receptacle (spermatheca), paired accessory glands, and a receptacle for the penis of the male (bursa copulatorix). Each ovary has a variable (1 to >2000) number of ovarioles, each a linear array of developing oocytes surrounded by follicular cells. The oocytes are surrounded by a one-, two-, or three-cell thick layer of cells, forming a follicle. The accessory (colleterial) glands produce a substance that attaches eggs to the substrate or forms an egg case (ootheca) in cockroaches. The right and left glands have a different morphology and structure. The left colleterial gland secretes structural proteins (oothecins), protocatechuric acid glycoside, calcium oxalate, and phenol oxidase. The right gland secretes betaglucosidase. When the components of the two glands mix, the betaglucosidase hydrolyzes protocatechuric acid glucoside to glucose and protocatechuric acid. The phenol oxidase converts protocatechuric acid (and O_2) to a reactive quinone that binds to oothecins and links them by phenolic bridges; the protein is tanned to form a cuticle-like protective case.

The endocrine control of reproduction in insects involves a number of neurosecretory and classical hormones (Koeppe et al. 1985). Juvenile hormone from the corpora allata is one important reproductive hormone that has a number of different roles. JH has a "gonadotropic" effect on the ovary, being required for oocyte maturation. JH has some direct effects on protein, lipid, and RNA/DNA synthesis in the ovaries. In some insects, the ovaries secrete ecdysteroids under the control of JH. The left colleterial gland is also under JH regulation. JH stimulates oothecin synthesis and regulates the accumulation of protocatechuric acid betaglucoside, calcium oxalate.

A major target organ of JH is the fat body, a mass of mesodermal cells that is freely suspended in the hemolymph but is enclosed within a membranous sheath. It is an important regulator of hemolymph metabolite levels under hormonal control. It also produces in females the vitellogenin proteins, glycoproteins (1 to 14% carbohydrate and 6 to 12% lipid) that are incorporated into the oocyte, to form the main egg yolk proteins, vitellins. Vitellogenins (V_g) are released from the fat body into the hemolymph and are absorbed through spaces between the follicular cells and across the oocyte plasmalemma. V_g's are absorbed by the oocyte in preference to other hemolymph proteins and to exogenous V_g's from other insects.

Vitellogenin production by the fat body and V_g uptake by oocytes is regulated by three hormones:

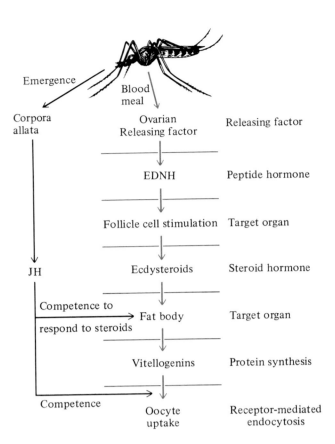

FIGURE 11–23 Regulation of egg development in the mosquito *Aedes* by two environmental signals (emergence and blood meal) and three endocrine systems (juvenile hormone, egg development neurosecretory hormone, and ecdysone).

EDNH, ecdysteroids, and JH. A similar protein synthesis/transport/oocyte uptake system with similar endocrine control is seen in oviparous vertebrates; the liver produces vitellogenins that are converted by the oocytes into phosphovitin and lipovitin.

The endocrine control of egg development in mosquitos (*Aedes*) involves two environmental stimuli, emergence and a blood meal, and three hormones, JH, EDNH, and ecdysterone (Figure 11–23). After emergence, JH is synthesized by the corpora allata and promotes feeding and mating behavior, follicle development to a resting stage, and competence of the gonad to respond to EDNH and of the fat body to respond to 20-hydroxyecdysterone. A blood meal releases EDNH, which stimulates the ovary to secrete ecdysone. The ecdysone is converted to 20-hydroxyecdysone and stimulates vitellogenin synthesis by the fat body. JH promotes vitellogenin uptake by oocytes.

The male reproductive system is also influenced by JH. The effects of JH on spermatogenesis are conflicting; mature sperm can be produced by some species in the absence of JH, and spermatogenesis is accelerated in some species by low JH levels. In other species, JH accelerates spermatogenesis. JH may also promote secretion by the male accessory glands. The fat body of adult male insects does not synthesize vitellogenins (although some larval male insects can be induced to synthesize vitellogenins by JH).

There is a considerable amount of recent evidence for a role of gonadal hormones in insects (Hagedorn 1985). The apical tissues of the testes secrete an androgenic hormone in *Lampyris*; this controls the primary and secondary male sex characteristics. The ovaries of houseflies and mosquitos release an oostatic hormone when they contain mature eggs that prevents the maturation of further oocytes. Ecdysteroids also affect reproduction. The presence of ecdysteroids in adult insects was at first puzzling because the source of ecdysteroids in immature insects is the prothoracic gland, which degenerates in adults. The adult ecdysteroids are synthesized by the ovaries and testes (and other tissues in a few insects). Ecdysteroids control vitellogenin synthesis and spermatogenesis and probably stimulate previtellogenic growth, ovulation, and sclerotization of eggs.

Color. The morphological color of some insects is under hormonal control. Many insects match their color to that of their background; this is **direct homochromy**. For example, the acridid *Oedipoda* varies in color from yellow, to red, to gray, to black depending on its background color; it cannot, unlike many other acrids, be green. Not surprisingly, these *Oedipoda* are not naturally found in green meadows. The ability to color match the background depends on visual input; severing the optic nerves eliminates the homochromic response. Many insects can approximately match the background brightness as well as color. There is a strong correlation between the background color and the quantity of pigments (ommochromes) in the cuticle. The homochromic response is mediated by a neurosecretion from the median neurosecretory cells in the pars intercerebralis that is released from the corpora cardiaca. Morphological color change in other insects is also under hormonal control. For example, the tritocerebrum of the stick insect secretes a darkening factor that is released from the corpora cardiaca.

Some insects have physiological color control. *Acrida pellucida* and the solitary phase of *Locusta migratoria* are bright green at high relative humidity; *Mantis religiosa* larvae are green at low light intensity. *Carausius* adapt quickly to light or dark backgrounds (at high relative humidity of 70 to 100%). Their dark adaptation is due to movement of pigment (ommochrome) granules and also partial migration of carotenoid pigments (Figure 11–24). With a light background, ommochrome granules are grouped deep in the proximal part of the epidermal cells; pterin granules near the surface impart a light shade to the cuticle. The corpora allata and JH are often involved with physiological color adaptation in insects.

Other Endocrine Systems. Hemolymph metabolite levels (lipid, carbohydrate, and amino acids) are

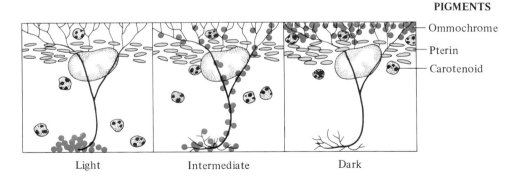

PIGMENTS

— Ommochrome
— Pterin
— Carotenoid

Light   Intermediate   Dark

FIGURE 11–24 Physiological adaptation to background color in the stick insect *Carausius* (at high humidity to facilitate color change). On a light background, the ommochromes are in the deep proximal part of the epidermal cell near the carotenoid pigments, and the more dorsal pterin granules impart a light coloration. In the dark, the ommochromes migrate along microtubule systems to lie above the pterin granules and impart a dark color. The carotenoids also migrate, but to a lesser extent than the ommochromes. *(From Fuzeau-Braesch 1985.)*

under hormonal control (Steele 1985). Insect hemolymph contains high levels of diglycerides, mostly bound to a protein lipophorin. Two hyperlipemic agents, or adipokinetic hormones (AKH), are released from the glandular lobes of the corpora cardiaca under neural (not neurosecretory) control. A high level of lipids inhibits AKH release and a low level of trehalose stimulates AKH release. The mode of action of AKH is thought to be release of diglycerides from the fat body (Figure 11–25). An antagonistic hypolipemic agent is produced by medial neurosecretory cells; it may be similar to insulin (a vertebrate hypoglycemic hormone). Its mode of action may be to accelerate diglyceride uptake and triglyceride synthesis or to decrease diglyceride mobilization from the fat body.

Utilization of fat body glycogen is also under hormonal control of the corpora cardiaca; hypertrehalosemic factor (trehalagon) stimulates the conversion of glycogen to a disaccharide, trehalose (Figure 11–25). A hypotrehalosemic factor, with insulin-like properties, is synthesized by median neurosecretory cells; it may stimulate trehalose uptake by tissues (as insulin promotes glucose uptake in vertebrates). A hormone from the medial neurosecretory cells (MNCH) suppresses glycogen synthesis.

Endocrine control of hemolymph amino acid levels has received little attention, except for tyro-sine, which is important in cuticular tanning since it is readily converted to a quinone-like tanning agent, n-acetyldopamine. Tyrosine is stored in the hemolymph as tyrosine-o-phosphate, which is more soluble than tyrosine and prevents metabolic removal of tyrosine from the hemolymph. Two hormones, puparium tanning factor (PTF) and bursicon, regulate the use of tyrosine by epidermal cells for tanning. Protein synthesis is inhibited by adipokinetic hormone in *Locusta* and there is an antagonistic protein synthesis stimulating hormone; both originate in the corpora cardiaca.

There are also endocrine control mechanisms for osmoregulation. Blood-sucking insects must rapidly excrete a large volume of water ingested as the blood meal. For example, *Rhodnius* eliminates excess water after a blood meal at a rate equivalent to its body mass every 30 min. The potent diuretic hormone of *Rhodnius* is released by neurosecretory cells in a pulse after each infrequent blood meal. Stretch of the abdominal wall stimulates stretch receptors and induces the release of the diuretic hormone close to the Malpighian tubules, their target organ. The hormone diffuses from the neurohemal area across the nerve sheath directly into the hemolymph and to the nearby Malpighian tubules. In contrast, the stick insect *Carausius*, which feeds almost continuously on moist plant material, has a

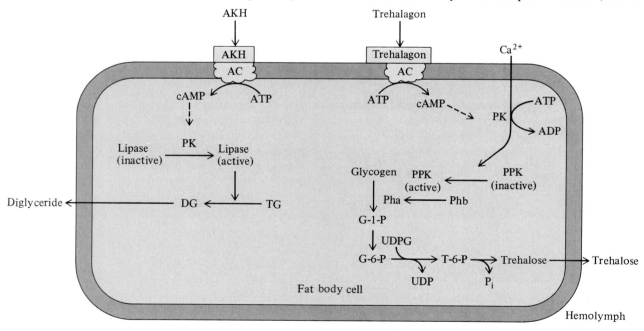

FIGURE 11–25 Possible scheme for mode of action of adipokinetic hormone (AKH) in stimulating diglyceride production, and trehalagon in stimulating trehalose release from fat body cells. Abbreviations are as follows: AC, adenyl cyclase; TG, triglyceride; DG, diglyceride; PK, protein kinase; PPK, Phosphorylase kinase; Ph, phosphorylase; G-1-P, glucose-1-phosphate; G-6-P, glucose-6-phosphate; UDPG, uridine diphosphophosphate; UDP, uridine diphosphate; T-6-P, trehalose-6-phosphate. *(Modified from Steele 1985.)*

weak diuretic hormone that is continuously released from the corpora cardiaca at a considerable distance from the target organ. Antidiuretic hormones reduce fluid secretion by the Malpighian tubules and promote water reabsorption by the rectum. A chloride transport stimulating neurohormone is released by the corpora cardiaca and promotes fluid reabsorption by the rectum. Water balance may also be influenced by hormonal control of cuticular water permeability. A hormone from the brain appears to decrease the cuticular water loss of the cockroach *Periplaneta* (Noble-Nesbitt and Al-Shuker 1988). The brain synthesizes, stores, and releases a water loss promoting factor (BHP) and a water loss restricting factor (BHR). The physiological state of the animal determines which hormone is released; hydrated cockroaches release BHP but not BHR and have a high cuticular water loss. Desiccation arrests BHP release and promotes BHR release, decreasing the rate of cuticular water loss.

Polyphenism is the occurrence of two or more distinct phenotypes that are induced by extrinsic factors (Hardie and Lees 1985). For example, aphids often have five or more distinct adult forms to enable them to adapt and survive in ephemeral habitats. Environmental factors such as population density, nutrition, photoperiod, and temperature influence wing polymorphism, either winged (alate) or wingless (apterous). Apterousness is a larval resemblance and is often considered to be a neotenic condition. Photoperiod in particular influences a sexual/parthenogenetic reproductive polymorphism. The production of one morph, the virginopara (parthenogenetic females giving birth to parthenogenetic offspring) is induced by a neurosecretion, virginopara-inducing hormone, either by a direct effect or via an allatotropic effect. JH is also involved in many examples of polyphenism in nonsocial and social insects.

Echinoderms

Echinoderms, being deuterostome animals, are somewhat more closely allied to chordates than are the protostome invertebrates. However, the endocrine systems of echinoderms provide few insights into the evolution of chordate endocrine systems because the echinoderm hormones and endocrine glands are very different from those of chordates.

The radial nerves of starfish produce a gonad-stimulating substance (GSS) that causes the ovaries/testes to secrete a maturation-inducing substance (MIS). MIS induces oocyte maturation and spawning of gametes. GSS is a polypeptide; MIS is 1-methyadenine. Maturing oocytes produce a maturation-promoting factor (MPF) that promotes meiotic maturation and breakdown of the oocyte germinal vesicle. MPF is quite similar in nature and function to a maturation-promoting factor in frog oocytes that is formed under the influence of progesterone. The chemical nature of MPF is not known. The ovaries, testes, and pyloric cecae contain enzymes for synthesis of C_{26}, C_{27}, C_{28}, and C_{29} steroids. Estrogens appear to promote protein synthesis in the pyloric cecae and their transport to the ovary for incorporation into developing oocytes. This role of ovarian steroids is similar in echinoids and vertebrates, perhaps reflecting the deuterostome lineage.

Chordate Endocrine Systems

The vertebrate endocrine system is, in general, quite different from that of invertebrates. First, neurosecretory cells in higher vertebrates are reduced in importance from being widespread in higher invertebrates and primitive vertebrates (lampreys). Second, many peripheral physiological functions of vertebrates are under classical endocrine control. Many of these endocrine glands are controlled by a "master endocrine gland," the anterior pituitary (adenohypophysis). The nervous system does, however, retain an important regulating influence of peripheral endocrine glands by its control of adenohypophyseal secretions. Third, the classical endocrine glands have often evolved from scattered groups of cells, and there is a marked tendency for the formation of compact, discrete endocrine glands and even the consolidation of different endocrine organs into a single structure (e.g., thyroid/parathyroid, adrenal cortex/medulla).

There is a general phyletic conservatism for vertebrate endocrine glands and their hormones, although the roles of some hormones have altered dramatically during the course of vertebrate evolution. The general adage of the comparative endocrinologist that "it is not the hormones that evolve, but the uses to which they are put" is generally true for vertebrates. We could also add that new endocrine organs do not often evolve, but "old" hormones are put to "new" uses. However, some new endocrine glands have evolved during the phylogenetic development of vertebrates (e.g., ultimobranchial glands first appear in elasmobranchs, and parathyroids in teleosts), and some "old" endocrine glands have disappeared (e.g., urophysis and corpuscles of Stannius).

The endocrine systems of protochordates contribute little to our understanding of the evolution of the vertebrate endocrine systems. Endocrinologists have searched for the origins of the vertebrate endocrine systems in cephalochordates with little success (Barrington 1964; Olsson 1969). Cephalochordates do not have neurosecretory cells. Osmoregulation and perhaps spawning are under classical hormonal control. The endostyle, a mucus-secreting organ of cephalochordates and urochordates, is of particular interest because it is the precursor of the thyroid gland of vertebrates. The endostyle contains iodine and thyroid hormones (and their precursors). Hatschek's pit is homologous with the vertebrate pituitary. The cerebral ganglia of urochordates contains neurosecretory cells that may influence oocyte maturation, but their hormones are not similar to vertebrate neurohyphyseal octapeptides. The neural and asymmetric glands are epithelial structures of endocrine significance; the neural gland is considered to be homologous with the vertebrate pituitary.

We must envisage a "hypothetical archetypal vertebrate" to illustrate the development of the basic vertebrate endocrine system (Hoar 1965). There are three major areas of evolutionary change in the endocrine systems of vertebrates: the nervous system (especially the hypothalamus), the anterior gut, and the nephrogenic tissues (Figure 11–26).

Neurosecretory cells are widespread in the lamprey brain, but other lower vertebrates have only four groups of neurosecretory cells and tetrapods have three groups. One group, the neurohypophysis (posterior pituitary), secretes octapeptide neurohormones. These peptide neurosecretions are released into the blood at a neurohemal organ and elicit effector responses in distant target cells. A second group of anterior hypothalamic neurosecretory cells control the endocrine secretions of the nearby epithelial endocrine organ, the adenohypophysis. A third group of neurosecretory cells form the pineal gland/organ. The fourth group of caudal spinal neurosecretory cells are the Dahlgren cells of elasmobranchs and the urophysis of teleost fish.

Many classical vertebrate endocrine glands are derived from the gut. The thyroid is homologous with the endostyle of cephalochordates; the thyroid of adult lampreys develops at metamorphosis from the mucus-secreting endostyle of the ammocoete larva. Diffuse cells in the primitive gastrointestinal tract provide endocrine control of digestion; pockets of glandular gut tissue develop into the pancreas. Gill pouches 3 and 4 form the ultimobranchial and parathyroid glands. The adenohypophysis develops as a pouch of gut epithelium in the roof of the embryonic mouth (Hatschek's pouch).

Kidney (nephrogenic) tissues develop into the interrenal gland of lower vertebrates and the adrenal cortex of mammals. Chromaffin tissue, another endocrine system closely allied with the kidneys, secretes catecholamines; it forms the adrenal medulla in mammals. Chromaffin tissue is derived from neural crest cells; consequently it is homologous with a sympathetic nerve ganglion and secretes the hormones epinephrine (adrenaline) and norepinephrine (noradrenaline).

The generalized endocrine system of vertebrates thus consists of the neurohypophysis and its octapeptide hormones, the adenohypophysis, which is under neurosecretory control by the hypothalamus and controls many classical endocrine glands, such as the thyroid gland and various peripheral endocrine glands (Figure 11–27). Let us now examine these endocrine systems in more detail.

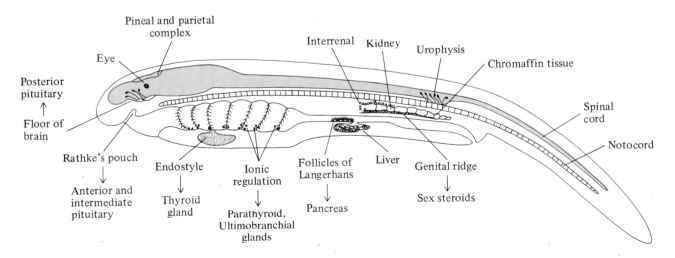

FIGURE 11–26 The major vertebrate endocrine system of a hypothetical ancestral vertebrate. *(Modified from Hoar 1965.)*

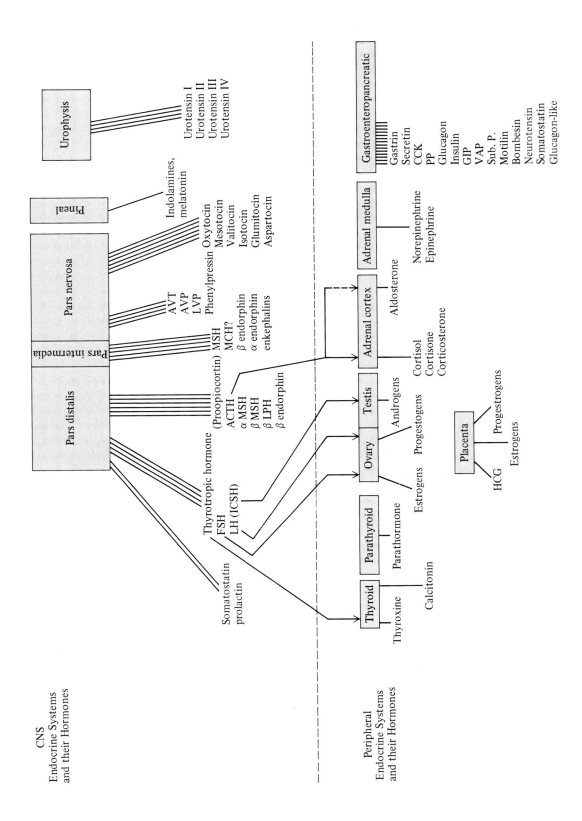

FIGURE 11–27 Generalized scheme for neuroendocrine and classical endocrine systems in vertebrates (based primarily on mammals, except for the urophysis, which is present in teleost fish).

531

Hypothalamo-Hypophyseal Axis

The **hypothalamus** is responsible for integration of sensory information from other brain centers and controls visceral functions in part by neurosecretory endocrine relationships with the **pituitary** (hypophysis). The hypothalamus and the pituitary gland are connected by neural and vascular links (see Figure 11–4B, page 498). The pituitary has a neural portion, the **neurohypophysis** (or pars nervosa, pars posterior, posterior pituitary), and a classical endocrine portion, the **adenohypophysis** (or pars distalis, ante-

rior pituitary). The neurohypophysis consists of three regions: the distal neural lobe, a neural stem, and the median eminence. The latter two regions are collectively called the infundibulum.

The adenohypophysis originates as an outpocketing of the epithelium in the roof of the mouth (Rathke's pouch). The cells closest to the neural lobe become the pars intermedia, the pars tuberalis is lateral outgrowths of the outpocketed tissue, the pars distalis is the anterior wall of the pouch. The pars distalis is generally the largest and most important part of the adenohypophysis. There is

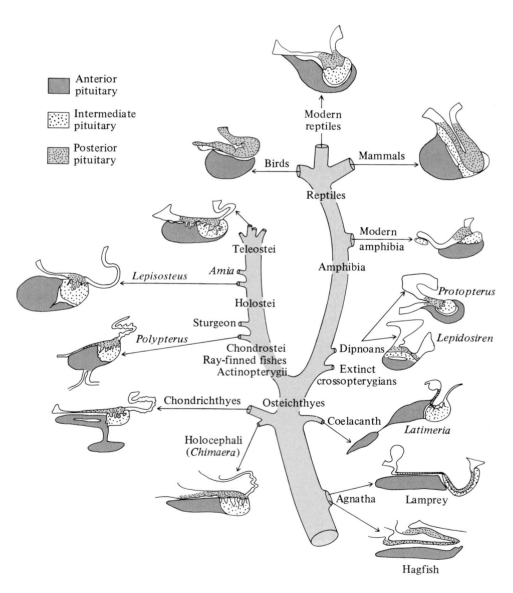

FIGURE 11–28 Vertebrate phylogenetic tree with diagrammatic representations of the "typical" pituitary morphology for the major groups. *(From Gorbman et al. 1983.)*

considerable variety in the structure and organization of the adeno- and neurohyphophyses (Figure 11–28).

Neurohypophysis. The neurohypophyseal hormones are octapeptides. The first to be identified were arginine vasopressin and oxytocin but there are a number of structurally similar octapeptides containing nine amino acids (Figure 11–29). The four basic octapeptides (arginine vasopressin, AVP; arginine vasotocin, AVT; lysine vasopressin, LVP; phenypressin) have antidiuretic effects. The six neutral octapeptides (oxytocin, mesotocin, valitocin, isotocin, glumitocin, aspartocin) have oxytocin-like effects.

Any particular species of vertebrate generally has only a few neurohypophyseal octapeptides, including a basic and a neutral octapeptide, although cyclostomes have only arginine vasotocin (Figure 11–30). AVT is the basic octapeptide in all vertebrates except mammals. Most adult mammals have AVP; some pigs have LVP, AVP, or both; some mice have LVP; macropod marsupials have LVP and phenypressin, whereas phalangerid marsupials have AVP. The neutral peptides show an even more phylogenetically diverse distribution.

Why is there such a complex phylogenetic distribution of neurohypophyseal octapeptides in vertebrates? Variation allows greater separation and complexity in the roles of the basic and neutral octapeptides. For example, the basic octapeptides of mammals are antidiuretic hormones, whereas the neutral octapeptide oxytocin stimulates uterine smooth muscle and mammary milk ejection. AVP

has a much greater specificity than AVT for antidiuretic effects rather than oxytocin-like effects (Table 11–2) and so the presence of AVP rather than AVT better separates the roles of the basic and neutral octapeptides in mammals.

Neurohypophyseal octapeptides have carrier proteins, **neurophysins**, which are formed from a precursor peptide that also contains the octapeptide neuro-

TABLE 11–2

Biological specificity of basic and neutral neuro-octapeptides for antidiuretic, uterine contracting, and milk ejection actions in mammals; the most common mammalian octapeptides, AVP and oxytocin, are emphasized in bold. The basic octapeptides are generally specific for antidiuretic effects and the neutral octapeptides for uterine contracting and milk ejection effects. Values are activity in International Units per μmole. *(Modified from Archer 1974.)*

Octapeptide	Antidiuretic	Uterine Contracting	Milk Ejection
Basic			
AVP	**465**	**17**	**69**
AVT	260	120	220
LVP	260	5	63
Neutral			
Oxytocin	**5**	**450**	**450**
Mesotocin	1	291	330
Valitocin	0.8	199	308

Basic Peptides

                    1 2 3 4 5 6 7 8 9

1 Arginine vasotocin (AVT)   Cys-Tyr-Ileu-Gln-Asn-Cys-Pro-Arg- Gly-NH_2

2 Arginine vasopressin (AVP)   Cys-Tyr-Phe-Gln-Asn-Cys-Pro-Arg- Gly-NH_2

3 Lysine vasopressin (LVP)   Cys-Tyr-Phe-Gln-Asn-Cys-Pro-Lys- Gly-NH_2

4 Phenypressin   Cys-Phe-Phe-Gln-Asn-Cys-Pro-Arg- Gly-NH_2

Neutral Peptides

1 Oxytocin   Cys-Tyr-Ileu-Gln-Asn-Cys-Pro-Leu-Gly-NH_2

2 Mesotocin   Cys-Tyr-Ileu-Gln-Asn-Cys-Pro-Ileu-Gly-NH_2

3 Valitocin   Cys-Tyr-Ileu-Gln-Asn-Cys-Pro-Val- Gly-NH_2

4 Isotocin   Cys-Tyr-Ileu-Ser-Asn-Cys-Pro-Ileu-Gly-NH_2

5 Glumitocin   Cys-Tyr-Ileu-Ser-Asn-Cys-Pro-Gln-Gly-NH_2

6 Aspartocin   Cys-Tyr-Ileu-Asn-Asn-Cys-Pro-Val-Gly-NH_2

FIGURE 11–29 Structures of neurohypophyseal octapeptide hormones.

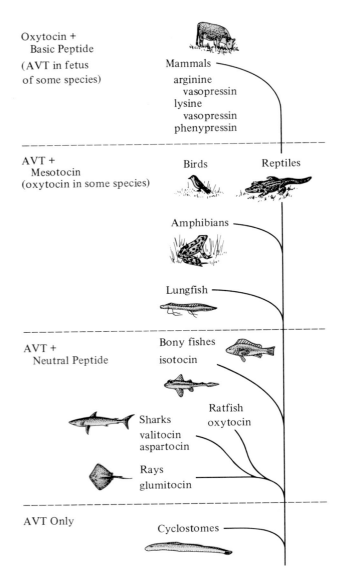

Oxytocin +
Basic Peptide
(AVT in fetus
of some species)

Mammals
arginine
vasopressin
lysine
vasopressin
phenypressin

AVT +
Mesotocin
(oxytocin in some species)

Birds   Reptiles

Amphibians

Lungfish

AVT +
Neutral Peptide

Bony fishes
isotocin

Ratfish
oxytocin

Sharks
valitocin
aspartocin

Rays
glumitocin

AVT Only

Cyclostomes

FIGURE 11–30 Phylogenetic distribution of neurohypophyseal basic and neutral octapeptides. *(From Gorbman et al. 1983.)*

and collecting duct of nephrons to water and promotes reabsorption by osmosis (see Chapter 17). In birds and reptiles, AVT is also an antidiuretic, promoting nephron reabsorption of water (although not by the same mechanism as in mammals), and in reptiles it also decreases glomerular filtration rate. AVT has little effect in aquatic amphibians, such as the clawed toad *Xenopus*, but stimulates water uptake across the skin in more terrestrial anuran amphibians. This water balance response, or Brunn effect, is a consequence of AVT's effect on the skin, kidney, and bladder. AVT dramatically increases cutaneous water uptake, especially across the highly vascular pelvic patch. The control of blood flow to the pelvic patch also regulates the water balance response (Yokota and Hillman 1984). AVT also controls the water permeability of the bladder, which is an important water store for amphibians. AVT has a dual effect on amphibian nephrons; it decreases glomerular filtration rate and increases tubular reabsorption, thereby reducing urine flow rate.

In amphibians, reptiles, and birds, AVT is a potent inducer of oviduct or uterine contractions and may be involved with egg-laying and parturition. It is not clear if neurohypophyseal octapeptides have similar reproductive roles in fish. Oxytocin stimulates uterine (myometrium) contractine in mammals immediately prior to parturition. There is a dramatic increase in the number of oxytocin receptors and in oxytocin bound to receptors in myometrium muscle cells before parturition. Oxytocin also promotes milk ejection from the mammary glands. The milk-secreting glandular alveoli of the mammary gland are covered by stellate myoepithelial cells and the ducts by spindle-shaped myoepithelial cells. Oxytocin causes these myoepithelial cells to contract.

Adenohypophysis. The pars distalis is the main part of the adenohypophysis. The pars intermedia, when present, is closely associated with the neurohypophysis despite its common origin with the other adenohypophyseal cells, and it has a separate hormonal role from the pars distalis.

Pars distalis. The major adenohypophyseal secretions are thyroid stimulating hormone (TSH); two gonadotropic hormones, follicle stimulating hormone (FSH) and luteinizing hormone (LH); growth hormone (GH); prolactin; and adrenocorticotropic hormone (ACTH). These hormones can be divided into a number of families. Glycoproteins (TSH, FSH, LH) consist of two α and β peptide units linked by noncovalent bonds and a carbohydrate

hormones. In mammals, there are MESL- and VLDV-neurophysins. For example, the ox has a "preprovasopressin" containing AVP, MESL-neurophysin, and a glycopeptide copeptin, and a "preprooxytocin" containing oxytocin, VLDV-neurophysin, and histidine. Neurophysins are released with the octapeptides but they have no known physiological role.

A major role of AVP in mammals is to promote water reabsorption by the kidney (hence AVP is also called antidiuretic hormone, ADH). AVP increases the water permeability of the distal tubule

moiety. The α subunits are similar, but the β subunits vary to confer the biological specificity of the different hormones. Tropic hormones (GH and prolactin) are a small, polypeptide chain (GH is also called somatotropin hormone, SH). Melanocorticotropins are derived from a common glycoprotein, pro-opiocortin, that is cleaved in pars distalis cells to ACTH and β-lipotropin (and in pars intermedia cells to MSH, CLIP, endorphins, and MSH).

Secretion by adenohypophyseal cells is regulated by **hypophysiotropic hormones** from the hypothalamus. Thyrotropin-releasing hormone (TRH) stimulates the release of TSH. Somatostatin (or somatotropin release inhibiting factor, SRIF) inhibits the release of GH, whereas growth hormone-releasing hormone (GHRH) elicits GH release. Luteinizing hormone-releasing hormone (LHRH) is a decapeptide that promotes release of especially LH but also FSH. It is not clear whether there is a separate FSH releasing hormone (i.e., FSHRH) or whether LHRH is the only gonadotropic releasing hormone (and therefore could be called gonadotropin releasing hormone, GnRH). Corticotropin-releasing hormone (CRH) stimulates ACTH secretion. It is unclear if prolactin release inhibiting factor (PIF) is a catecholamine, or SRIF, or a separate peptide. Melanocyte hormone inhibiting factors (MIF) and stimulating factors (MSF) have been implicated in the control of MSH release.

The release of hypophysiotropic hormones is best illustrated by a pituitary target hormone model with three types of feedback loop: long loop, short loop, and ultrashort loop (Figure 11–31). In long-loop feedback, an increase in the blood concentration of the target cell hormone above a critical threshold level inhibits the release of a stimulatory hypophysiotropic factor. This decreases the release of adenohypophyseal tropic hormones and secretion by the target cell hormone, lowering the blood hormone concentration. These long loops are usually negative feedback, but in rare instances are positive feedback, resulting in a rapid surge in hormone secretion. Some pituitary hormones have short-loop negative feedback, the hypophysiotropic hormone directly influences the secretory activity of the adenocyte cells. Some hypothalamic releasing factors may also

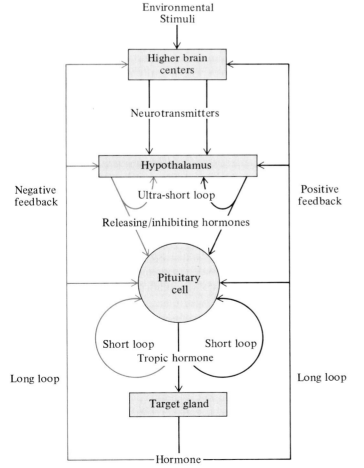

FIGURE 11–31 Generalized schematic of negative and positive feedback pathways that control pituitary (pars distalis) hormones. *(From Chester-Jones, Ingleton, and Phillips 1987.)*

control their own release by ultrashort negative feedback loops.

Thyrotropin-releasing hormone (TSH) has its primary action on the thyroid (glycoproteins generally have only one or a few target cell types). TSH promotes iodine uptake by the thyroid, thyroglobulin synthesis, and release of **thyroxin** (T_4) and **tri-iodothyronine** (T_3) into the blood. TSH induces an increase in both thyroid follicle cell size (hypertrophy) and number (hyperplasia). In lower vertebrates, LH and FSH also have some TSH activity, reflecting the structural similarity of this glycoprotein hormone family. Lower vertebrate TSH has no activity in mammals, suggesting a considerable evolutionary modification in both their TSH and receptor structures. TSH secretion is primarily under control of TRH (and TIF and SRIF) from the hypothalamus, with modification by corticosteroids, estrogen, neurotensin, opioids, and higher brain centers.

The gonadotropic hormones LH and FSH (like TSH) have actions restricted to one, or a few, target cells. FSH is primarily involved with gametogenesis, and LH with steroidogenesis by the gonads. For example, the Sertoli cells of the testes have FSH receptors; LH stimulates Leydig cells to secrete testosterone. In ovaries, FSH is primarily required for oogenesis, and LH for luteal steroidogenesis (see below). The role of the gonadotropic hormone system can be quite different in nonmammalian vertebrates. There may be only one gonadotropic hormone, with other hormones (e.g., prolactin and steroids) being important regulators of reproductive function. Gonadotropin release is primarily controlled by LHRH (GnRH), but many other factors also modulate LH/FSH release.

Prolactin has many roles in vertebrates. This illustrates the tendency for hormones to acquire new functions, rather than the evolution of new hormones. The many actions of prolactin, which often occur in concert with steroid hormones, can be summarized as three types of effect: (1) growth and development, (2) osmoregulation, and (3) reproduction (Table 11–3). Prolactin has been reported

TABLE 11–3

Actions of prolactin involve growth, water, electrolyte balance, and reproduction. *(Modified from Chester-Jones, Engleton, and Phillips 1987.)*		
Growth	**Water/Electrolyte**	**Reproduction**
Teleosts		
Melanocytes	Mucus secretion	Skin mucus secretion
Seminal vesicles	Reduced gill Na^+ flux	Seminal vesicle secretion
Renal tubules	Reduced gill water flux	Parental behavior
	Increased urine flow	Gonadotropic effects
	Decreased salt excretion	
Amphibians		
Water drive	Water drive	Water drive
Tail and gill	Skin water/ion flux	Antispermatogenic
Limb regeneration	Bladder water/ion flux	Stimulates cloacal gland
Melanocytes	Plasma Na^+ level	Brain
Reptiles		
Tail regeneration	Plasma Na^+ level	Antigonadotropic effects
Skin shedding		
Birds		
Pigeon crop sac	Salt gland secretion	Crop milk secretion
Brood patch		Antigonadal effects
Feather growth		Premigratory restlessness
		Reproductive tract
Mammals		
Mammary gland	Lactation	Lactation
Sebaceous growth	Renal Na^+ retention	Luteotropic effects
Hair growth		Vaginal mucification
Renotropic effects		Male accessory sex glands

to enhance growth, especially in young vertebrates. It has an osmoregulatory role in all vertebrate groups, by effects on epithelia of gut, skin, urinary bladder, and kidney. It is vital for survival of euryhaline fish in freshwater, affects water and Na$^+$ exchange in amphibians and reptiles, stimulates the nasal salt gland in birds, and promotes renal Na$^+$ retention in mammals. Prolactin has a variety of reproductive effects, often stimulating secretion, e.g., skin mucus in fish, crop "milk" in some birds, milk in mammals. Most effects promote gonadal function, but some are antigonadotropic in some vertebrates. Many factors are involved in the feedback regulation of prolactin synthesis. Prolactin secretory cells are, unusually, under inhibitory control by SRIF and dopamine from the hypothalamus as well as blood osmotic concentration (in fish) and its own feedback effect. TRH and the purported PRF stimulate prolactin release, in addition to low blood osmotic concentration (in fish), estrogens, and suckling in mammals.

Growth hormone (GH) is a growth-promoting hormone in all vertebrates. It has a variety of metabolic effects involving lipids, carbohydrates, and proteins. In mammals, GH elicits release of somatomedin (SM) from the liver. SM promotes cell replication especially in the epiphyseal plates of long bones. Release of GH is under antagonistic control (SRIF, GHRH), negative feedback via somatomedin from the liver, and thyroxin.

Adrenocorticotropic hormone (ACTH) is produced by proteolytic cleavage of a larger precursor peptide, pro-opiocortin (Jackson et al. 1981). Cleavage at the residue 172 to 173 position forms a lipotropic hormone (β-LPH) and further cleavage at 104 to 105 produces γ-MSH and ACTH (Figure 11–32). Consequently, ACTH secretion is also accompanied by β-LPH and γ-MSH release; their possible physiological effects are unknown. Small amounts of β-endorphin, α-MSH, and corticotropin-like intermediate lobe peptide (CLIP) are also released with ACTH, but these are generally produced in more significant amounts by the intermediate lobe (see below). ACTH appears to maintain a basal level of steroid hormone secretion in all vertebrates, mainly cortisol in fish, aldosterone in amphibians, and corticosterone in reptiles and birds. In mammals, ACTH primarily maintains the secretory activity of the glucocorticoid secreting cells (cortisol, cortisone, corticosterone) rather than mineralocorticoid secreting cells (aldosterone). Secretion of ACTH is regulated by a long-loop negative feedback via corticosteroids (aldosterone in amphibians), hypothalamic CRF, and pituitary ACTH. There is also an influence of the higher brain centers on hypothalamic CRF secretion, mediated by GABA and norepinephrine (inhibitory) and serotonin (stimulatory).

Pars intermedia. The adult **pars intermedia** is usually a layer of cells at the surface of the neural lobe, although it is embryologically a part of the adenohypophysis. There is considerable variation among vertebrates in the location, structure, and importance of the pars intermedia. It is large in some species (e.g., rat), a one- or two-cell thick layer in some (e.g., geckos, marsupials), and absent in some (e.g., whales, adult humans). The intermediate lobe cells of the human involute and disappear at about birth.

Pars intermedia cells, like the corticotroph cells of the pars distalis, synthesize pro-opiocortin. This is proteolytically cleaved to β-LPH, ACTH, and γ-MSH and then to a number of other peptides,

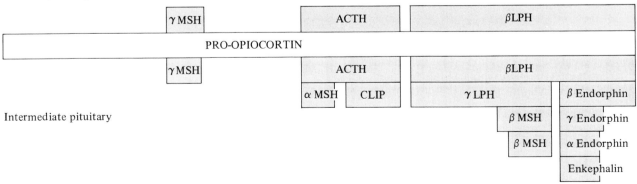

FIGURE 11–32 Major peptides derived from the pro-opiocortin protein in anterior pituitary cells (upper) and intermediate pituitary (lower). Note that the pars distalis cells also produce small amounts of αMSH, CLIP, endorphin, and other peptides. *(From Lerner 1981.)*

including α-MSH and β-MSH (Figure 11–32), the major hormones of the intermediate pituitary. α-MSH and β-MSH are melanotropic hormones in lower vertebrates (cyclostomes, fishes, amphibians) and control melanin synthesis and dispersion. Absence of light, or a dark background, increases MSH secretion. The posterior pituitary hormone, melanophore-concentrating hormone (MCH) of lower vertebrates, has antagonistic effects to MSH on chromatophores (Sherbrooke, Hadley, and del Castrucci 1988). In higher vertebrates, MSH may be involved with Na$^+$ regulation (reptiles, mammals) and suckling (mammals). This effect may be related to the close morphological relationship between the pars intermedia and the neurohypophysis with its antidiuretic/oxytocin effects. Fetal growth of mammals is influenced by α-MSH.

Control of MSH release by the pars intermedia cells has less important hypothalamic hormonal control than do pars distalis hormones because the pars intermedia has a poorly developed vascular supply. Rather, there is direct and indirect innerva-

tion by peptidergic and aminergic hypothalamic neurons of the pars intermedia. MIF, norepinephrine, and dopamine inhibit MSH secretion, whereas MRF promotes MSH secretion.

The identification of opiate (plant alkaloid) receptors in the mammalian brain stimulated the identification of two endogenous pentapeptides (met-enkephalin and leu-enkephalin) with opiate-like activity (Hughes et al. 1975). Met-enkephalin is the 61-65 amino acid sequence of β-LPH. Endorphins are also endogenous opioids; β-endorphin is the 31 C terminal amino acid of β-LPH. Three families of neurons produce endogenous opioids by proteolytic cleavage of a precursor polypeptide (Imura et al. 1985). Pro-opiocortin neurons in the arcuate nucleus/tuberal region synthesize pro-opiocortin and release β-endorphin. Proenkephalin A neurons, which are widely distributed in the brain (including the hypothalamus), form met-enkephalin, leu-enkephalin, and peptides E and F from a preproenkephalin A protein. The proenkephalin B neurons, also found in several areas of the brain including

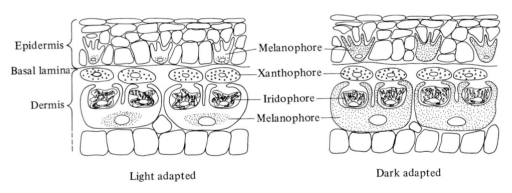

Epidermis

Basal lamina

Dermis

Melanophore

Xanthophore

Iridophore

Melanophore

Light adapted

Dark adapted

A

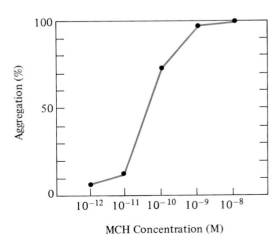

B

FIGURE 11–33   (A) Epidermal and dermal chromatophores of frog skin when light adapted and dark adapted. **(B)** Melanin-concentrating hormone (MCH) causes aggregation of chromatophores in scales of the glass fish, *Kryptopterus*. *(From Gorbman et al. 1983; Nagai, Oshima, and Fujii 1986.)*

the hypothalamus, form leu-enkephalin, dynorphin, leumorphin, and α-neo-endorphin from preproenkephalin B.

There appear to be at least three different types of opioid receptors: δ, μ, and κ; β-endorphin binds selectively to μ, but also to δ receptors; enkephalins bind to δ receptors. Dynorphin and α-neo-endorphin bind to κ receptors. These endogenous opioids have a number of physiological functions, including regulation of secretion of the hypothalamic neuroendocrine cells (e.g., GnRH release) and other endocrine systems, analgesic defense against noxious stimulation, modulation of the vegetative nervous system, and behavior.

In lower vertebrates, morphological color change in response to background color is due to increased numbers of melanophores as well as greater melanin production. In addition, many epidermal melanocytes deposit melanin extracellularly, as melanosomes that are taken up by nonmelanocyte epidermal cells. Physiological color change is a rapid, endocrine-mediated color change of the epidermis. Frog skin, for example, contains many epidermal melanophores as well as dermal chromatophore units (Figure 11–33A). The dermal chromatophore units are stacks of three pigment cells: an upper yellow xanthophore, a central reflecting iridophore, and a lower, basket-like black melanophore. The melanophores are under hormonal control. In light-adapted animals, the absence of MSH causes melanin aggregation in melanophores and expansion of the iridophores. The dermal melanin granules concentrate underneath the iridophores and only green light is reflected from the iridophores (other wavelengths are filtered out by xanthophores); consequently, the skin appears green. In dark-adapted animals, MSH causes melanin dispersion in melanophores and contraction of the iridophores. Dispersion of dermal melanophores covers the iridophore cells and blocks their reflection of light, making the skin appear dark.

Tadpoles can exhibit an unexpected blanching response when kept in the dark, as may some fish. This is because of a developmental change in the ability to inhibit MSH secretion and the responsiveness of the dermal melanocytes to melatonin (a hormone released from the pineal gland; see below). There are two stages of tadpoles based on their coloration capabilities. First-stage tadpoles cannot adapt to their background color but can blanch in the dark. They are unable to inhibit MSH secretion, hence appear dark irrespective of the background color. However, in the dark, release of melatonin causes contraction of the dermal melanophores (but not the epidermal melanophores) and the iridophores are contracted, i.e., the tadpole blanches.

Second-stage tadpoles can reduce MSH secretion and adapt to their background color and can also blanch in the dark. In the dark, melatonin causes dermal melanophore contraction and reduced MSH levels cause partial contraction of the epidermal melanophores and partial expansion of the iridophores. Adult frogs can adapt to their background color (MSH response) but do not blanch in the dark because their dermal melanophores are now unresponsive to melatonin.

Catecholamines, serotonin, melatonin, and an adrenergic (α and β) or cholinergic innervation cause melanin aggregation or dispersion. Teleosts also have an antagonistic hormone to MSH, melanin-concentrating hormone (MCH). MCH is a small hypothalamic heptadecapeptide (i.e., has 17 amino acids) released from the posterior pituitary. It causes pigment concentration in melanophores, xanthophores, and erythrophores in many teleost fish at very low concentrations (Figure 11–33B). It also causes pigment dispersion in leukophores.

The epidermal melanocytes of birds and mammals are unlike those of lower vertebrates. The epidermal melanin units consist of a melanocyte and several dozen keratinocytes. The melanocytes synthesize and release melanin-containing melanosomes that keratinocytes take up by phagocytosis. In a representation of melanin synthesis by an epidermal melanophore in the human epidermis (Figure 11–34), premelanin is synthesized and pack-

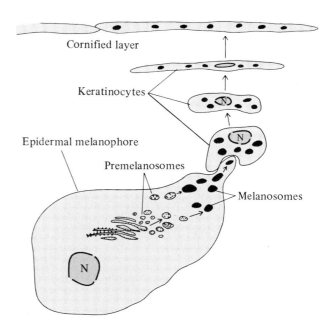

FIGURE 11–34 Epidermal melanophores of mammals synthesize and extrude melanosomes, which are phagocytosed by keratinocytes. *(From Gorbman et al. 1983.)*

aged into premelanosomes, which develop into melanosomes when melanin synthesis is completed. Melanosomes are extruded from the melanophore and phagocytosed by keratinocytes for incorporation in the keratin layer. This morphological color change (i.e., tanning) is induced by ultraviolet radiation which stimulates melanocyte production. In general, skin, hair, and feather color is determined either genetically or by direct effects of ultraviolet radiation on the epidermal melanin units, rather than by MSH-mediated effects.

Pineal Gland

The **pineal organ** is in the roof of the midbrain; it is called a pineal gland if it has a secretory function. The pineal complex of lower vertebrates includes the pineal organ and a parapineal organ (which forms the parietal, or "third" eye, of some lower vertebrates). The pineal organ may be the right remnant of a pair of "eyes," the parietal being the left.

The presence and structure of the pineal organ varies dramatically among vertebrates, suggesting that it has varying photoreceptive/neurosecretory functions (Figure 11–35). Some vertebrates appear to lack a pineal organ, e.g., hagfish, crocodiles, armadillos, sloths. The pineal organ of lampreys and most fish, amphibians, and lizards has sensory cells (pinealocytes) that are similar to retinal photoreceptors. The ganglion cells and neural connections of the pineal organ to the brain are reduced then disappear in the evolutionary progression from lower to higher vertebrates, and the pinealocyte structure changes from a photoreceptor, in contact with the lumen of the pineal eye, to a secretory cell with no photoreceptive role.

The pineal glands of birds and mammals are the best studied, at least with respect to endocrine function. The first indication of a physiologically active secretion by the mammalian pineal was the observation that extracts of cow pineal glands induced blanching of frog tadpoles; this blanching factor is an indolic compound, **melatonin** (5-methoxy-N-acetyltryptamine). Melanin is synthesized from tryptophan then serotonin by the enzymes serotonin-N-acetyltransferase (NAT) and hydroxy-indole-o-methyltransferase (HIOMT). Mammalian pineal glands also contain octapeptides (AVT, AVP) and hypophysiotropic hormones (TRH, LHRH, PIH, PRH) as well as renin and vasoactive peptide; their physiological roles in the pineal are not known.

Pineal melatonin is implicated in the control of color ("dark blanching response") in cyclostomes, teleost fish, and frog tadpoles. In birds, the pineal responds directly to light transmitted through the cranium into the brain and influences photoperiodic control of reproduction. Secretions of the pineal of mammals have been implicated as an antigonadal agent and as a modulator of thermoregulation. Nonpineal tissues (e.g., retina and gut) can also synthesize and release melatonin into the blood. Animals without a pineal gland (either naturally, e.g., crocodiles, armadillos, or after experimental pinealectomy) still have significant levels of plasma melatonin.

Urophysis

Most elasmobranchs have neurosecretory **Dahlgren cells** in their caudal spinal cord. These giant neurons send axons to the ventral surface of the spinal cord to make contact with blood vessels at diffuse neurohemal areas (Figure 11–36). Many teleost fish have a group of neurosecretory cells located in the spinal cord near the caudal vertebrae (Bern and Nandi 1964). Axons of these cells pass out of the spinal column to a well-developed neurohemal organ, the **urophysis**, located in the ventral midline, laterally, or both. Cyclostomes, dipnoan lungfish, and tetrapods do not have caudal neurosecretory cells. The teleost urophysis has a morphological and histological appearance similar to the neurohypophysis, suggesting a similar neuroendocrine role.

The urophysis secretes a number of peptide neurohormones, **urotensins**, that appear to control osmoregulation and smooth muscle contraction of the urogenital tract (e.g., spawning). A variety of urotensin peptides (I, II, III, IV) have been identified in the urophysis. These have various pharmacologic effects, including increased blood pressure (UTI), smooth muscle contraction (UTII), gill Na^+ uptake (UTIII), and vasotocin-like effects (UTIV).

Thyroid Gland

The **thyroid gland** and some other endocrine glands (ultimobranchial and parathyroid glands) arise embryologically from pharyngeal pouches; the thyroid forms from the fourth pouch. The thyroid gland is variable in shape but has a similar histological structure in all vertebrates. It is a highly vascular assemblage of **follicles**. Each follicle is a hollow ball formed from a single layer of epithelial cells that envelops a space filled with a colloid material. In cyclostomes, the follicles are loosely scattered about the pharynx, but in other vertebrates the follicles are organized as a discrete organ (Figure 11–37).

The follicle epithelial cells absorb iodine from the blood and synthesize a protein, **thryoglobulin**, and

Phylogenetic Cladogram for Vertebrates

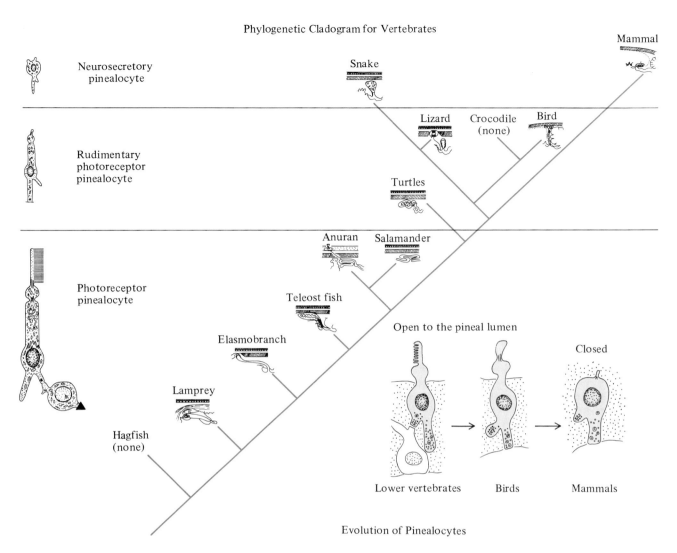

Evolution of Pinealocytes

FIGURE 11–35 Phylogenetic cladogram for vertebrates showing the development of the morphology of the pineal organ, the structure of the pinealocyte (whether photoreceptive or neurosecretory), and a scheme for the evolution of secretory pinealocytes from open-type photoreceptor cells to open-type secretory cells to closed-type secretory cells. *(Modified from Gorbman et al. 1983; Fujita, Kobayashi, and Iwanaga 1980.)*

the thyroid hormones **thyroxine** (tetraiodothyronine, T_4) and **triiodothyronine** (T_3) from tyrosine (Figure 11–37). Thyroglobulin is a large protein that, because of its size, cannot normally diffuse from the center of the follicle into the blood. It elicits an antigenic response if it is present in the blood (due to disease or mechanical damage). An iodide pump actively accumulates iodine inside the follicle cells, and a peroxidase enzyme reacts the iodine with tyrosine residues of the thyroglobulin protein. This reaction probably occurs on the colloidal side of the apical cell membrane of the follicle cells. It combines

some thyroglobulin tyrosine residues with one iodine (monoiodotyrosine, MIT) or two iodines (diiodotyrosine, DIT). Coupling between MIT and DIT, either on the same or different thyroglobulins, produces T_3 and T_4. The thyroglobulin-bound thyroid hormones are retained in the follicle until they are split from the thyroglobulin by complete proteolytic hydrolysis of the thyroglobulin. The principal thyroprotease enzyme is a cathepsin (its pH optimum is acidic, about pH 4.0). The T_3 and T_4 released by proteolysis diffuse into the blood; any MIT and DIT is deiodinated by a specific enzyme (desiodase) and

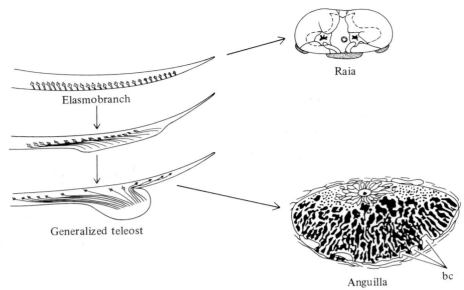

FIGURE 11–36 Possible evolutionary sequence for development of the teleost urophysis from the Dahlgren cells in the elasmobranch caudal spinal cord. The neurohemal area of an elasmobranch (*Raia*; stippled areas are the vascular bed) and a teleost (*Anguilla*; dark areas are neurosecretory material; bc, blood capillary). (*Modified from Fridberg and Bern 1968; Enami 1959.*)

the iodine returned to the thyroid iodine pool. TSH influences a number of aspects of T_3 and T_4 synthesis and release, e.g., the iodine pump, thyroprotease levels, and desiodase level.

Most T_3 and T_4 in the blood is bound to proteins. In humans, only 0.5% of T_4 and 0.3% of T_3 is free in plasma. The rest is bound to three thyroxine binding proteins: thyroxine-binding globulin (TBG), thyroxine-binding prealbumin (TBPA), and serum albumin (ALB). TBG is the least abundant but carries about 75% of the protein bound hormones because it has the highest affinity for T_4 and T_3. Its affinity for T_3 is about 1/10 that for T_4, and so T_3 is more readily lost to peripheral tissues than is T_4. Furthermore, the T_4 receptor has about the same affinity as TBG, but the T_3 receptor has a higher affinity. T_3 is thus more mobile and active than T_4. Other mammals, reptiles, and birds lack TBG, and ALB is their main thyroid hormone carrier.

Thyroxine affects more organs than perhaps any other hormone! It stimulates metamorphosis of tadpoles into adult frogs, and the metamorphosis of some fish (e.g., salmon) is partly controlled by thyroxine. It has many maturational effects. Morphological effects include enhanced developmental growth rate, bone growth, tooth growth and eruption, horn and antler growth, fin ray development in fish, epidermal cell proliferation, promotion of molting, epidermal pigment deposition, development of the reproductive system, limb regeneration, development of the central nervous system, and development of the gut. Thyroid hormones have many physiological and metabolic effects, including nervous system function (spontaneous electrical activity, sensitivity threshold to stimuli; reflex time;

motor response; mental acuity). T_4 stimulates food movement through the gut and uptake of glucose, galactose, oleic acid, and vitamin A; diuresis; increased metabolic rate in endotherms (but only a few ectotherms); and nitrogen metabolism (either positive or negative nitrogen balance depending on T_4 level).

The Parathyroid Gland, the Ultimobranchial Gland, and Corpuscles of Stannius

Hormones from the parathyroid gland, the ultimobranchial gland, and the Corpuscles of Stannius of teleost fish are involved with Ca^{2+} homeostasis. This is achieved by their actions on various target organs (bone, kidney, gut) involved with Ca^{2+} metabolism.

Tetrapod vertebrates have a **parathyroid gland**. Urodele amphibians have a pair of parathyroids located near the carotid and systemic arches. Anuran amphibians have two pairs of parathyroids of quite different histological appearance. Reptiles have one or two pairs of parathyroids; birds have one pair near the thyroid and ultimobranchial glands; mammals have two pairs, embedded in the thyroid gland tissue. The parathyroid gland secretes a parathyroid hormone, **parathormone** (PTH), which increases blood Ca^{2+} levels. PTH is a straight chain polypeptide of 84 amino acid residues; amino acids 1 through 34 provide its biological activity. PTH is formed by cleavage of a preproparathyroid hormone in the endoplasmic reticulum during transport to the Golgi complex. In cyclostomes and teleosts, the

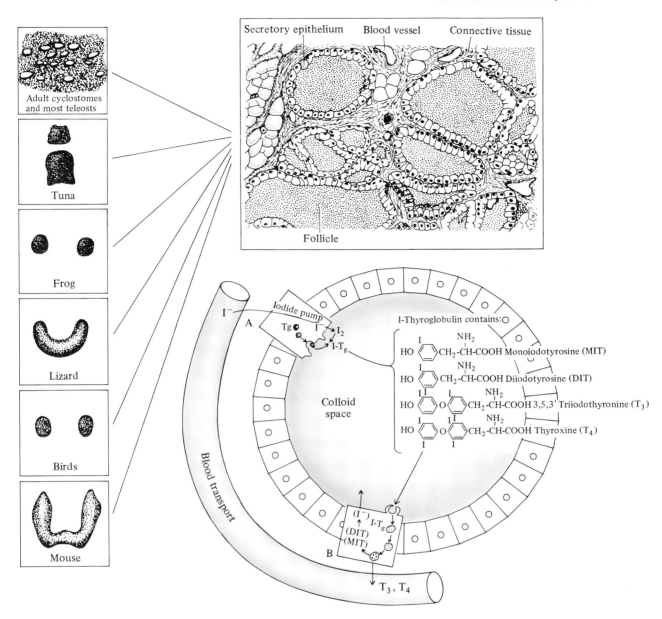

FIGURE 11–37 Morphology of the thyroid glands of various vertebrates and structure and function of the follicles in the mammalian thyroid gland. *(Modified from Gorbman et al. 1983; Turner and Bagnara 1976; Fawcett 1986.)*

role of PTH is filled by other hormones (ACTH, prolactin).

The plasma Ca^{2+} level of vertebrates is normally regulated precisely. For example, the plasma Ca^{2+} concentration is 2.4 mM in humans—about 1.2 mM is free, ionized Ca^{2+}; about 0.2 mM is un-ionized, and combined with other ions (e.g., phosphate, citrate); and about 1.0 mM is un-ionized and bound to plasma proteins so that it is not permeable to the capillary wall. Hypocalcemia (low blood Ca^{2+}) causes uncontrolled muscle tetany due to neuron

hyperexcitability and hypercalcemia (high blood Ca^{2+}) depresses the nervous system and initiates Ca deposition. Symptoms of hypocalcemia occur in humans if the total Ca^{2+} concentration is less than 1.75 mM and of hypercalcemia if it exceeds 3.1 mM.

Ca^{2+} regulation is linked to PO_4^{3-} regulation since both ions are the major constituents of bone. Bone contains hydroxyapatite crystals ($Ca_{10-x}^{2+}H_3O_{2x}^+$ $(PO_4^{3-})_6(OH^-)_2$) deposited on a collagenous organic matrix. Phosphate is present in plasma as free monobasic ions (0.26 mM HPO_4^{2-}) and dibasic ions

(1.05 mM $H_2PO_4^-$), as sodium biphosphate, and as protein-bound phosphate. The total phosphorus content of human plasma is about 40 mg L^{-1}. Phosphate is a minor pH buffer system of the extracellular fluid (and a major pH buffer in urine).

The plasma level of PTH is determined by direct feedback of plasma Ca^{2+} level on the parathyroid cells; low plasma Ca^{2+} elevates PTH secretion and vice versa. PTH elevates blood Ca^{2+} through its actions on three target organs: bone, kidney, and gut (Figure 11–38). The plasma Ca^{2+} and PO_4^{3-} concentrations are high enough to deposit calcium phosphate in bone, i.e., the plasma is supersaturated. However, precipitation inhibitors, such as pyrophosphate, prevent Ca^{2+} and PO_4^{3-} precipitation in the blood and other tissues.

The exchange of Ca^{2+} between plasma and bone is complex. Bone resorption and deposition are mediated by bone cells. Osteoblasts synthesize collagen and deposit bone; they also form a cellular barrier separating the "bone matrix fluid" from the extracellular fluid. Osteocytes regulate the dynamic balance between bone deposition and resorption. Osteoclasts resorb bone. There is also an exchange of Ca^{2+} ions at the bone surface that is independent of bone resorption/deposition. The bone matrix fluid has a lower [Ca^{2+}] than the extracellular fluid, so Ca^{2+} constantly diffuses from the extracellular fluid into the bone matrix fluid via channels between the surface osteoblast barrier. Osteoblast Ca^{2+} pumps return Ca^{2+} to the extracellular fluid. This dynamic interchange of Ca^{2+} between body fluids and the exchangeable Ca^{2+} pool of the bone matrix fluid enables the regulation and modification of plasma Ca^{2+} independent of osteocyte and osteoclast activity.

PTH also affects the kidney and gut. It stimulates Ca^{2+} resorption and PO_4^{3-} excretion by nephrons (in mammals at least). A decrease in body fluid PO_4^{3-} level elevates the Ca^{2+} level by altering the equilibrium between bone hydroxyapatite and body fluid Ca^{2+}/PO_4^{3-}. High levels of PTH enhance Ca^{2+} absorption from the gut, but the physiological effect may be indirect, mediated through **vitamin D** (cholecalciferol is vitamin D_3). Vitamin D was first recognized as a vitamin required for gut absorption of Ca^{2+} and mobilization of bone calcium. It is actually a prohormone for the calcium regulating hormone, 1,25-dihydroxycholecalciferol (1,25-di(OH)D_3). Vitamin D_3 is synthesized in the skin from 7-dehydroxy-cholesterol under the influence of ultraviolet radiation and is also obtained in the diet. It is hydroxylated to 25-(OH)D_3 in the liver, then to 1,25-di(OH)D_3 in the kidney. The latter hydroxylation is stimulated by PTH, estrogen, and low plasma phosphate concentration. The bioactive 1,25-di(OH)D_3 has major effects on the gut and bone. It stimulates active Ca^{2+} and PO_4^{3-} absorption by the gut mucosal cells (especially duodenal cells). It probably stimulates the transcription and translation of a Ca^{2+}-binding protein that transports Ca^{2+} across the brush border into the mucosal cells. 1,25-di(OH)$_2D_3$ also promotes bone resorption and is thought to be required for bone remodeling.

In amphibians, PTH has a similar role in Ca^{2+} regulation as in higher vertebrates but prolactin may, in addition, have a hypercalcemic role. Fish do not have parathyroid glands and do not secrete PTH; they have other hormones for Ca^{2+} homeostasis. Prolactin has a slow hypercalcemic effect, promoting Ca^{2+} uptake by the gills, but ACTH may be a more physiologically relevant hypercalcemic hormone. They also have calcitonin and hypocalcin.

All vertebrates except cyclostomes have a pair of **ultimobranchial bodies**. In mammals, these are incorporated into the thyroid gland as C-cells. The ultimobranchial gland/C-cells secrete **calcitonin**, a remarkably variable 32 amino acid polypeptide. In mammals, calcitonin has a hypocalcemic effect; it enhances Ca^{2+} excretion by the nephrons and inhibits bone resorption but does not appear to affect the gut. There is a negative feedback control of calcitonin secretion by plasma Ca^{2+} level, but calci-

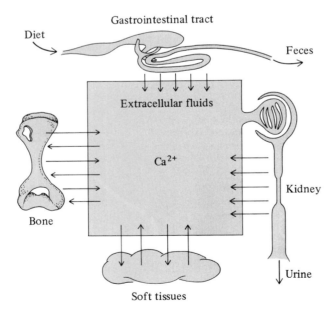

Diet

Gastrointestinal tract

Feces

Extracellular fluids

Ca^{2+}

Bone

Kidney

Urine

Soft tissues

FIGURE 11–38 Dynamics of extracellular fluid (including plasma) Ca^{2+} exchange between bone, soft tissues, intestinal contents, and the kidney filtrate.

tonin has only slight effects on Ca^{2+} regulation and its physiological significance is unclear. Calcitonin generally has no effect on plasma Ca^{2+} regulation in nonmammalian vertebrates, although the calcitonins of lower vertebrates are potent hypocalcemic agents in mammals. For example, salmon calcitonin actually has a much greater hypocalcemia potency in mammals than endogenous mammalian calcitonin! Calcitonin of nonmammalian vertebrates may control electrolyte or water balance rather than Ca^{2+}. Calcitonin is apparently not an evolutionary relic because it first appears in elasmobranchs and presumably has a significant physiological role.

Bony fish have **corpuscles of Stannius**, small organs attached to, or embedded within, the kidney tissue. These secrete a hypocalcemic hormone, **hypocalcin**, a protein that enhances Ca^{2+} uptake by the gills.

Gastroenteropancreatic Cells

Many peptide endocrine cells of the gastrointestinal tract and endocrine pancreas share common morphological and functional characteristics. These **gas-troenteropancreatic cells** (GEP) are best studied in mammals (Table 11–4) but have been identified in all major vertebrate groups, in protochordates, in some invertebrates, and even in some protozoans. GEP cells produce peptide hormones and possibly amines. They may be found scattered throughout a tissue (e.g., secretin cells in the duodenal mucosa), as discrete aggregates of cells (e.g., pancreatic islets), or associated with other GEP cells (e.g., somatostatin cells associated with gastrin cells). The GEP cells are probably derived from endoderm of the gut, but may be APUD cells derived from the neural crest (see below).

GEP cells are recepto-secretory cells that synthesize peptides. The various GEP peptides are delivered to their target cells by classical endocrine, neuroendocrine, and neurotransmitter modes, as well as by paracrine secretion (Figure 11–39A). For example, somatostatin (a tetrapeptide from the gut and pancreas) inhibits the release of pancreatic hormones and gastrointestinal hormones in this fashion.

The peptide-secreting characteristic of GEP cells associates them with neurons, and GEP cells have

TABLE 11–4

Classification and function of established gastroenteropancreatic cells and candidate cells. Other possible secretory cells are not listed. *(Modified from Chester-Jones, Engleton, and Phillips 1987.)*

Hormone		Function
Pancreas		
I	Insulin	Decreases blood glucose level
PG	Glucagon	Increases blood glucose level
SOM	Somatostatin	Inhibits secretion of insulin, glucagon, and growth hormone
PP	Pancreatic polypeptide	Inhibits pancreatic enzyme secretion and gallbladder contraction
Stomach		
SOM	Somatostatin	Inhibits secretion and contraction
AG	Antral gastrin	Stimulates gastric HCl secretion
Small Intestine		
S	Secretin	Stimulates pancreatic HCO_3^- secretion
SOM	Somatostatin	Inhibits secretion of insulin, glucagon, and growth hormone
GIP	Gastric inhibitory peptide	Inhibits gastrin secretion, stimulates insulin secretion
CCK	Cholecystokinin	Stimulates pancreatic enzyme secretion and gallbladder contraction
N	Neurotensin	Inhibits gastric motor activity
EG	Enteroglucagon	Stimulates growth of intestinal mucosa
IG	Intestinal gastrin	
M,EMM	Motilin	Stimulates upper GI tract contractions
ECSP	Substance P	Stimulates smooth muscle contraction
Large Intestine		
EG	Enteroglucagon	Stimulates growth of intestinal mucosa
SOM	Somatostatin	Inhibits secretion of insulin, glucagon, and growth hormone
PP	Pancreatic polypeptide	

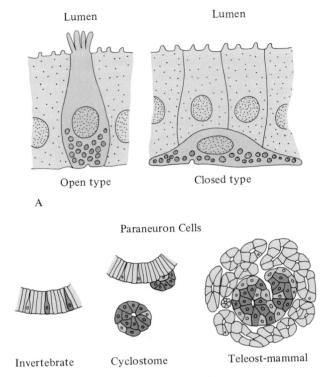

Lumen   Lumen

Open type   Closed type

A

Paraneuron Cells

Invertebrate   Cyclostome   Teleost-mammal

Evolution of Pancreatic Islets

B

FIGURE 11–39 (A) Gut paraneuron-type endocrine cells of open and closed type. (B) Scheme for evolution of pancreatic Islets of Langerhans from open-type paraneuron cells.

been termed **paraneurons** (Fujita, Kobayashi, and Iwanaga 1980). Other endocrine cells that have also been classified as paraneurons include pituitary cells, pinealocytes, and some coelenterate dermal and gastrodermal secretory cells. Open-type paraneurons have their apical membrane in contact with the lumen of a hollow organ, whereas closed paraneurons have no direct contact with the lumen (Figure 11–39A). The pancreatic islet cells of mammals are closed aggregates of secretory cells that have been derived from open type cells (e.g., in invertebrates) through an intermediate closed stage (e.g., in cyclostomes; Figure 11–39B). Similarly, pinealocytes show a clear phylogenetic change from open-type photoreceptors to closed-type secretory cells (see Figure 11–35, page 541).

GEP cells are probably best classified by the molecular structure of their peptide hormone. The best characterized are the gastrin and glucagon families (see Table 11–1, page 502). Pancreatic polypeptide (PP) and peptide YY (PYY) are another

GEP family. The insulin family includes GEP insulin (insulin-like growth factors GFI and GFII, relaxin and nerve growth factor are similar peptides but are not part of the GEP system). The GEP hormones motilin and somatostatin are not clearly associated with any of the other families. Other peptides of enteric neurons have been classified by whether they occur, or have counterparts, in amphibian skin, e.g., bombesin, substance P, and neurotensin. Met- and leu-enkephalins are also produced by GEP cells in the duodenum and pancreas, but from different precursor molecules from pro-opiocortin (cf. adenohypophyseal cells).

The general role of gut hormones is to regulate the breakdown of ingested food to readily absorbable nutrients, gallbladder contraction, pancreatic exocrine secretion, and gut motility. In mammals, the initial (cephalic) phase of gastric secretion is triggered by cholinergic vagal nerves in response to sensory input (sight, taste, smell, anticipation of food; Figure 11–40A). The presence of food in the stomach elicits the second (gastric) phase, by physical distension of the stomach lining and by the presence of specific food breakdown products, called secretagogues. During the third intestinal phase, food is present in the duodenum. These stimuli elicit gastric secretion by the stomach parietal cells. In mammals, enzyme release from pancreatic zymogen cells is stimulated by cholinergic vagal nerves and by hormones from the small intestine (Figure 11–40B).

Gastrin, a stomach hormone, is released in the gastric phase of digestion. It regulates acid secretion from the parietal cells of the stomach lining in higher vertebrates. Gastrin secretion can be further stimulated by gastrin-releasing peptide (GRP) or inhibited by secretin, cholecystokinin (CCK), and gastric inhibitory peptide (GIP) in the intestinal phase. Gastrin primarily promotes HCl secretion by parietal cells in the stomach lining but also promotes pepsinogen secretion by chief cells. Pepsinogen is the precursor of a proteolytic enzyme, pepsin (see Chapter 18). Lower vertebrates (fish, amphibians) do not have gastrin. Rather, cholecystokinin has a "gastrin-like" effect of stimulating pepsinogen and HCl secretion. We would expect CCK to be the more primitive hormone since the intestine (which secretes CCK) is a more primitive structure than the stomach (which secretes gastrin).

The **pancreas** of cartilaginous and bony fish is a diffuse, thin organ; it becomes consolidated into a more compact organ in tetrapod vertebrates. The exocrine part of the pancreas secretes a variety of enzymes that hydrolyze food and bicarbonate that neutralizes the acid stomach chyme when it enters

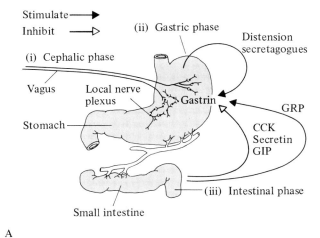

A

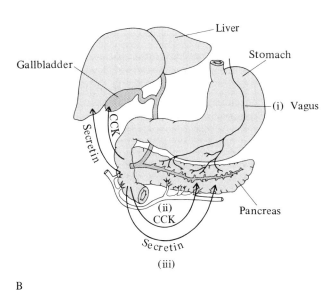

B

FIGURE 11–40 **(A)** Neural and endocrine regulation of gastric function in mammals: (i) cephalic phase is vagal stimulation of gastric secretion, (ii) gastric phase is stimulation of gastric secretion by distension of stomach and presence of secretagogues, (iii) intestinal phase is neural and hormonal of gastric secretion by the small intestine. **(B)** Neural and endocrine regulation of pancreatic, liver, and gallbladder function in mammals: (i) vagal stimulation of pancreatic acini and liver cells, (ii) cholecystokinin (CCK) stimulation of pancreatic acini and gallbladder smooth muscle, (iii) secretin stimulation of pancreatic duct cells and liver cells.

the small intestine. **Cholecystokinin** stimulates pancreatic zymogen cells to secrete digestive enzymes. The secretion of CCK by the small intestine is induced by the presence of various food breakdown products, such as proteins, peptides, amino acids (especially leucine and phenylalanine) and fatty

acids. **Secretin**, another small intestine hormone, promotes secretion of electrolytes and water. Secretin release is induced by the acidity of the stomach chyme as it enters the duodenum. Cyclostomes, and also cephalochordates and some other invertebrates, have zymogen cells scattered throughout the mucosal epithelium of their gut. These cells are the precursor of the pancreas in higher vertebrates. The presence of CCK-like and secretin-like proteins suggests hormonal control of these cells.

Another major role of the gastrointestinal tract is food movement along the gut by gastrointestinal motility and contractility. The gut mechanically mixes its contents to physically break down food and distribute enzymes throughout the food and to transport breakdown products to the intestinal mucosa for absorption. There is both neural and hormonal control of gut motility. Many neuropeptides (substance P, VIP, neurotensin, somatostatin, bombesin, enkephalin) affect gut contractility in nonmammalian vertebrates. These peptides, as well as PHI, CCK, and NPY may serve as local hormones, or neurotransmitters/neuromodulators of gut contractility in mammals. Motilin stimulates gastric and small intestine motility.

The emulsification of ingested fats into small droplets facilitates lipid digestion. **Bile**, an important emulsifying agent, is continually secreted by the liver cells. It is stored and concentrated in the gallbladder, then released into the small intestine. Vagal stimulation causes a weak contraction of the gallbladder, but CCK induces a strong contraction and emptying of the gallbladder and relaxation of the sphincter of Oddi that constricts the opening of the bile duct into the small intestine. Secretin has a mild stimulatory effect on bile secretion by the liver cells.

The endocrine pancreas consists of aggregates of several types of cells scattered throughout the exocrine pancreas as islets of Langerhans. The four cell types in the mammalian islets secrete glucagon (A cells), insulin (B cells), somatostatin (D cells), and pancreatic polypeptide (PP cells). Many other peptides have also been found in the endocrine pancreas but their roles are uncertain. Secretion by the islet cells is controlled by circulating metabolites and hormones, by neural stimulation, and by inhibitory paracrine secretions of adjacent islet cells. For example, D cells secrete somatostatin and inhibit the release of insulin, glucagon, and PP from adjacent B, A, and PP cells. Precursors of the islet hormones (and many other GEP hormones) are present in invertebrates and protochordates. For example, amphioxus has open-type GEP cells (including A, B, and PP cells) scattered throughout its gut. Insulin and somatostatin cells are postulated to have evolved first, followed by glucagon cells and then PP cells.

Insulin is a hypoglycemic hormone. In mammals, an increase in blood glucose (hyperglycemia) above the normal level of 5.5 mM stimulates B cells to secrete insulin. Glucose reacts with a putative glucose receptor on the cell membrane, initiating action potentials and increasing the membrane permeability to Ca^{2+}. The increased intracellular Ca^{2+} level activates a microtubule/microfilament system that translocates insulin-containing β-granules to the cell surface. Hyperglycemia also increases insulin biosynthesis. A precursor peptide, preproinsulin, is first synthesized; the "pre" sequence is degraded cotranslationally and the proinsulin is transported to the Golgi complex where the C-peptide is removed, and the insulin and C-peptide are packaged in the β-granules. Insulin stimulates the transcellular transport of sugars and amino acids; it also stimulates the intracellular phosphorylation of glucose by glucokinase thereby "trapping" the phosphorylated glucose inside the cells. Insulin also increases the activity/synthesis of enzymes for glycogenesis, lipogenesis, and protein synthesis. The effects of insulin are generally to promote anabolic (synthetic) pathways for carbohydrates to form glycogen, fat to form triglycerides, and amino acids to form proteins. The effects of insulin are like those of growth hormone.

Glucagon is a hyperglycemic hormone; its release is stimulated by low blood glucose (hypoglycemia). Glucagon has two main effects: the breakdown of liver glycogen and increased gluconeogenesis in the liver. Actions of glucagon are not antagonistic to all of the actions of insulin, although the end result (elevated blood glucose) is antagonistic to the effect of insulin (lowered blood glucose). Glycogenolysis by liver cells is promoted by the normal cascade of cAMP-second messenger events. Glucagon activates adenyl cyclase; the cAMP then activates protein kinase and then phosphorylase b kinase, which converts phosphorylase b to a, which degrades glycogen to glucose-1-PO_4. Glucagon also stimulates gluconeogenesis.

Chromaffin Tissues

Chromaffin cells are derived from neural crest cells and, in effect, are modified neurons of the sympathetic autonomic nervous system. Chromaffin cells lack postganglionic axons and secrete their "neurotransmitter" into the blood as hormones. In cyclostomes, chromaffin cells are associated with veins in the peritoneal cavity and the heart, but in higher vertebrates they are closely associated with the kidneys. In elasmobranchs, there are segmental clusters of chromaffin cells along the medial border of the kidneys (Figure 11–41A). In teleosts, the "head kidneys" have clusters of chromaffin cells, as do kidneys of urodele amphibians. In reptiles, birds, and mammals, the chromaffin cells are closely associated with the corticosteroidogenic tissues (adrenal cortex of mammals) to form discrete adrenal glands, generally at the anterior ends of the kidneys.

Chromaffin cells secrete the **catecholamines** epinephrine (adrenaline) and norepinephrine (noradrenaline), which are synthesized from phenylalanine and contain a catechol ring (Figure 11–41B). The catecholamines are stored in membrane-bound granules, some containing epinephrine and others containing norepinephrine. The catecholamine content of chromaffin tissues varies from 66 to 100% norepinephrine in elasmobranchs to a generally lower proportion of norepinephrine in higher vertebrates: amphibians 40 to 60%, reptiles 60%, birds 0 to 80%, primates 0 to 20%, and lagomorphs 0 to 12%. The catecholamines, once released into the blood, are inactivated by catechol-o-methyl transferase (COMT) in the liver and kidneys and monoamine oxidase (MAO) in the sympathetic nerves. These enzymes result in a short half-life for catecholamines in the blood. Plasma norepinephrine and epinephrine levels increase with a variety of physiological stresses, such as physical and mental activity, hypoxia, hemorrhage, dehydration, and fasting.

Catecholamine levels in the blood are determined not only by adrenal secretion but also by release from autonomic nerve terminals. In mammals, norepinephrine is the neurotransmitter of the autonomic nervous system, and the resting plasma norepinephrine level is higher than the epinephrine level. Most of the plasma norepinephrine is due to sympathetic release; about 40% of the plasma norepinephrine is released from the pulmonary tissues, 17% from the kidneys, 8% from the hepatomesenteric tissues, 3% from the heart, and only a little from the adrenal medulla (Esler et al. 1984). Adrenalectomy has little effect on plasma norepinephrine level, but the epinephrine level declines to almost zero. In amphibians, the sympathetic neurotransmitter is epinephrine, and the plasma epinephrine level exceeds the norepinephrine level.

Many tissues are target organs for plasma catecholamines but their response depends on the type of receptor present and the nature of the intracellular metabolic response initiated by the catecholamine-receptor complex. There are three general catecholamine receptor types: α receptors, β receptors, and dopamine receptors. The β receptors are subdivided into two types ($β_1$, $β_2$), depending on their pharmacologic responses. Each type of receptor responds

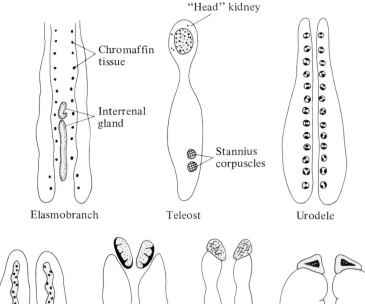
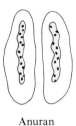

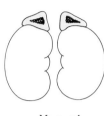

A

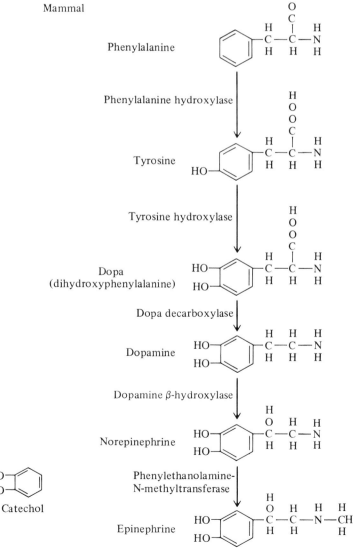

FIGURE 11–41 **(A)** Schematic comparison of the arrangement of interrenal gland (mammalian adrenal cortex; color) and chromaffin tissue (mammalian adrenal medulla; solid) in various vertebrates. **(B)** Synthesis pathway for the catecholamines—dopamine, norepinephrine, and epinephrine—from phenyalanine. Shown also is the structure of catechol.

differently to norepinephrine and epinephrine. Norepinephrine is a potent α and β_1 stimulator but is less effective for β_2 receptors. Epinephrine has a weak α response and a strong β_1 and β_2 effect. Thus, tissues with different receptors (or combinations of receptors) have different responses to elevated plasma catecholamine levels. The number of catecholamine receptors on target cell membranes is sensitive to the circulating concentration of catecholamines. Chronic high plasma levels of catecholamines decreases the number of receptors. This "down regulation," or desensitization, may protect the target cells from chronic stimulation. Some other hormone-receptor systems have similar down regulation, e.g., insulin and endogenous opioids. The nature of the metabolic response of a cell to the catecholamine-receptor complex determines the type of cellular response. For example, some α functions are excitatory (intestinal sphincter contraction), whereas others are inhibitory (intestinal relaxation); β functions may also be excitatory or inhibitory.

The adrenal medullary hormones regulate responses appropriate to "fight-or-flight" emergency conditions. These are short-term effects, such as hyperglycemia, elevated cardiac output, redistribution of blood flow, and sweating, which are appropriate not only to exercise (fight, flight) but also other stresses, such as hypoxia, hypoglycemia, and cold.

Steroid Hormones

Steroid hormones are synthesized from cholesterol (Figure 11–42). The initial step is removal of the C_6 side-chain attached to carbon number 20, then a complex series of reactions synthesize a wide variety of steroid hormones. The C_{19} androgens (male sex steroids) are usually formed from C_{21} steroid intermediates. The C_{18} estrogens (female sex steroids) are generally produced from androgens by several enzymatic steps.

The specificity of steroid hormone receptors for steroids is remarkable, given the often only minor structural differences between steroids that have very different functions. For example, the only difference between two steroids with very different functions, progesterone (a female sex steroid) and deoxycortisone (DOC, an adrenal mineralocorticoid), is the presence of an OH group on C_{21} of DOC. Addition of a methyl group to C_2 and a fluoro group to C_9 of cortisol increases its glucocorticoid activity by about $40\times$. To have progesterone activity, a pregnane steroid must have a 4-en-3-one configuration in the A-ring, an oxygen on C_{20}, and a methyl group on the terminal C_{21}.

Steroid hormones reversibly bind to several plasma protein carriers, usually albumins or globulins. The low-water solubility of steroids means that binding to proteins facilitates their transport; binding also increases the size of the hormone pool (reservoir) in the blood. Albumins have a low-binding affinity for steroids, whereas globulins have a high affinity. There are two categories of steroid-binding globulins: corticosteroid-binding globulins (CBG or transcortin) and sex hormone-binding globulins (SHBG). These have different dissociation constants (K_d), hence different affinities for the glucocorticosteroids and sex steroids. Target tissue receptors have a lower K_d (higher affinity) than plasma-binding proteins.

Steroid hormones are secreted by two groups of endocrine organs: the interrenal gland (adrenal cortex) and the gonads (ovary, testes).

Interrenal Gland/Adrenal Cortex. Lower vertebrates have separate **interrenal glands** and chromaffin tissue. The interrenal glands contain cells arranged as interlaced cords or loops. The **adrenal cortex** of higher vertebrates is the outer part of the compound adrenal gland that contains interrenal tissue. In reptiles and birds, there is a morphological zonation of the cords of cells that extend from the periphery to the interior in a parallel arrangement but there is not a functional zonation. The mammalian adrenal cortex has a highly organized 3-tier pattern of cells (see Figure 11–6A, page 500). The outer zona glomerulosa is whorls of cell cords, the central zona fasciculata has columnar cords of cells, and the inner zona reticularis has a network arrangement of cell cords. These morphological adrenal cortical zones correspond with functional zones; the zona glomerulosa secretes mineralocorticoids and the zona fasciculata and reticularis secrete glucocorticoids. Other zones are present in the adrenal of fetal mice (X zone) and marsupials.

The interrenal gland/adrenal cortex synthesizes steroid hormones with two general types of action: metabolic effects (the glucocorticoid steroids) and sodium regulation (mineralocorticoids). The glucocorticoids (e.g., cortisol, cortisone, corticosterone) have a $=O$ or $—OH$ group on carbon 11; the mineralocorticoids (e.g., aldosterone, 11-deoxycorticosterone) are deoxy at that position. We can only generalize about the occurrence and importance of these various steroids in different vertebrates because of the limited characterization of the role of these steroids in lower vertebrates. In general, most vertebrates have a number of glucocorticoids and at least one mineralocorticoid. Corticosteroid levels are very low in cyclostomes, and their physiological actions are not clear. Elasmobranchs have

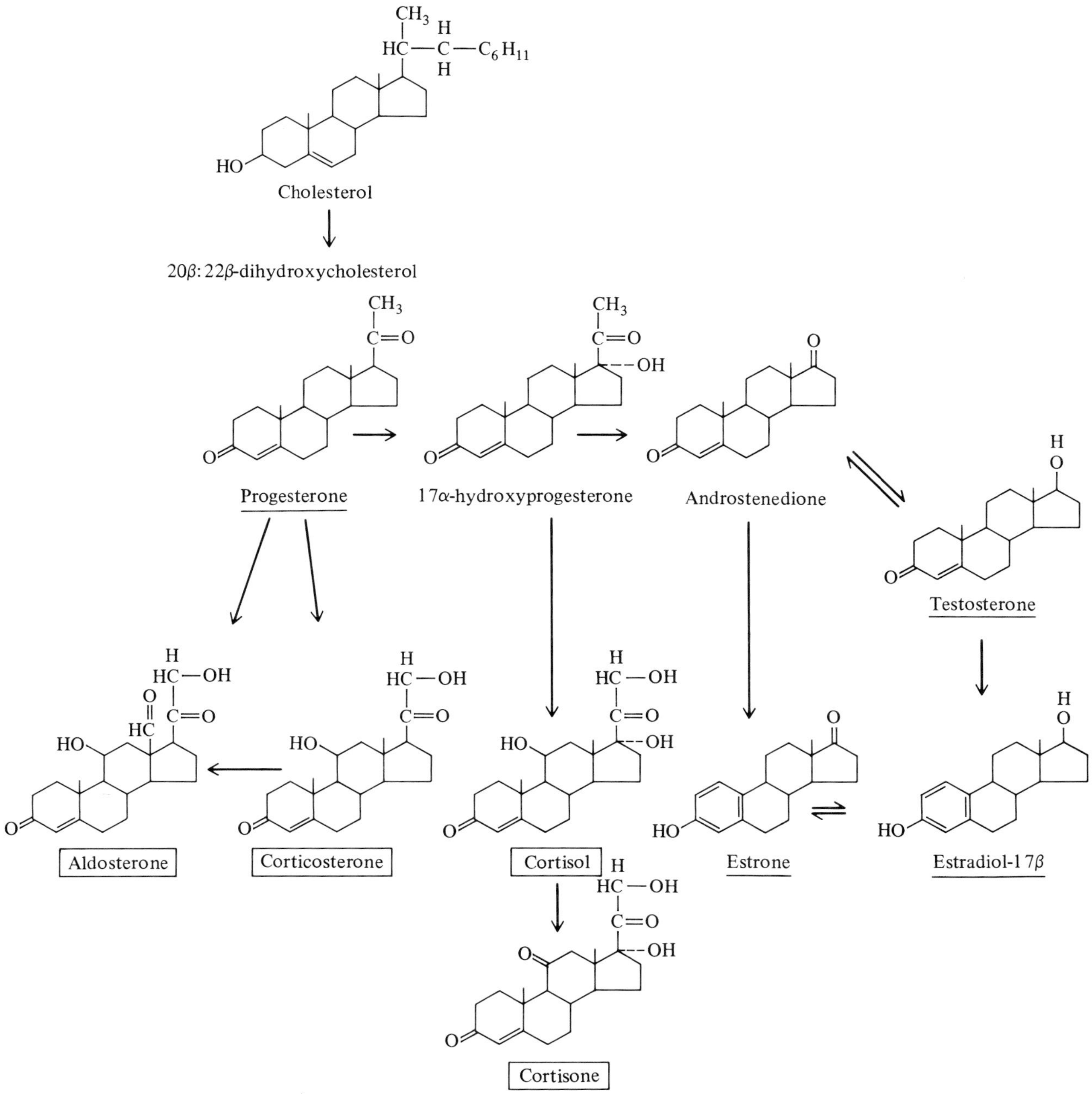

FIGURE 11–42 Synthetic pathways for the main adrenal and gonadal steroids; the main secretory products are indicated in boxes (adrenal steroids) or underlined (gonadal steroids). *(From Gorbman et al. 1983.)*

an unusual steroid, 1-hydroxycorticosterone, that is not found in other vertebrates. The major corticosteroid of teleosts is cortisol, and they also have low levels of aldosterone. In higher vertebrates, the important glucocorticoid is corticosterone and aldosterone is the major mineralocorticoid.

Glucocorticoids primarily have metabolic and developmental effects, e.g., gluconeogenesis, lipoly-

sis, protein degradation, retinal differentiation, mammary gland differentiation, and induction of lung surfactants.

Mineralocorticoids (primarily aldosterone) promote Na^+ retention and K^+ excretion by acting on a variety of organs, including the nephron and intestine and the sweat and salivary glands. In the kidney, aldosterone promotes Na^+ reabsorption especially by the distal convoluted tubule and probably from the proximal convoluted tubule.

Corticosteroids have a wide range of functions in lower vertebrates, including metabolic, iono-, and osmoregulation and migratory behavior. Lampreys in freshwater reduce Na^+ loss in response to aldosterone administration (this is not, however, one of their endogenous circulating steroids). In elasmobranchs, 1α-deoxycorticosterone stimulates Na^+ excretion by the rectal gland. The effects of corticosteroids varies in different teleost fish, but the eel *Anguilla* well illustrates the role of corticosteroids in freshwater and marine environments. Eels in freshwater tend to lose ions and gain water across their gills, skin, and gut (see Chapter 16). Cortisol stimulates Na^+ uptake by the gills, gut, and urinary bladder, and prolactin synergistically reduces the water permeability of the skin, gills, gut, and bladder. Eels adapted to seawater drink the seawater to replace the water lost by osmosis and urine. The ingested ions are absorbed across the gut (which is also permeable to water due to low-prolactin levels) and are excreted by gill Na^+ pumps. Cortisol stimulates the outward Na^+ transport by gills of eels adapted to seawater (but promotes inward transport in freshwater eels). Aldosterone promotes Na^+ uptake across the skin of amphibians as well as Na^+ reabsorption from the urinary bladder. In reptiles and birds, aldosterone promotes Na^+ retention. The kidney, urinary bladder, and cloaca are probably the important sites of action. In addition, many reptiles and birds have salt glands for extrarenal excretion of NaCl (see Chapter 17). The salt glands secrete a hyperosmotic NaCl solution in response to parasympathetic stimulation as well as by corticosteroids. Marine birds have larger salt glands than freshwater birds, and also have larger adrenal glands.

Adrenocorticosteroid secretion is controlled by two different systems in mammals; ACTH regulates glucocorticoid secretion and has a lesser influence on aldosterone secretion. Aldosterone secretion is primarily regulated by a diffuse endocrine system, the **renin-angiotensin system** (RAS), and plasma K^+ level. The RAS consists of (1) juxtaglomerular cells located in the afferent (and sometimes the efferent) glomerular arteriole supply to the nephron glomerulus (see Chapter 17); (2) the specialized macula densa region of the distal convoluted tubule where it contacts the juxtaglomerular apparatus; (3) a blood-borne glycoprotein angiotensinogen; and (4) converting enzymes in the liver, lung, and vascular bed. The juxtaglomerular cells contain an enzyme (renin) that is released into the blood when the macula densa is stimulated by low Na^+ levels in the distal tubule fluid (due to a low blood Na^+ level or low blood pressure in the glomerular arteriole). The renin converts the plasma angiotensinogen into the decapeptide **angiotensin I**, which is then cleaved by converting enzymes to smaller peptides (e.g., angiotensin II, angiotensin III). Angiotensin II is a potent vasopressor. Angiotensin II and III also stimulate the adrenal cortex to secrete aldosterone, which promotes Na^+ reabsorption and facilitates maintenance of blood Na^+ levels. The RAS is absent in cyclostomes and elasmobranchs, but JG cells are present in other chondrichthyean fish and all other vertebrates. The macula densa is only present in some amphibians and higher tetrapods, suggesting that it evolved after juxtaglomerular cells. The RAS may therefore have evolved initially as a blood pressure regulatory system, and control of adrenal steroidogenesis and Na^+ regulation was developed secondarily (Nishimura 1980).

Gonadal Steroids.   Reproduction by vertebrates is almost exclusively sexual. A few lizards reproduce parthenogenetically; only females are present in their populations. Hermaphroditism is relatively rare, unlike its common occurrence in invertebrates. Vertebrates frequently exhibit secondary sex characteristics; the males are different in morphology from females, e.g., in body size, plumage coloration, antlers and horns, etc.

In the primitive vertebrate reproductive pattern, eggs are layed and development is external (the **oviparous pattern**). Sometimes the egg is retained internally for embryonic development and there is a transfer of nutrients from the mother to the developing offspring; the young are born "alive" (**viviparous pattern**). A more-or-less intermediate situation is for development to be internal but with no substantial transfer of nourishment from the mother to the developing embryo (**ovoviviparous pattern**).

The gonads (testes in males, ovaries in females) are derived from a ridge of mesoderm in the body coelom. In addition, there are (except in cyclostomes) gonadal ducts and accessory reproductive structures associated with the gonads.

The **ovary** usually is a paired structure, but in cyclostomes the gonads fuse to form a single median structure. In some elasmobranchs, the left ovary remains relatively undeveloped, and in almost all birds and the platypus the left ovary is the only one to mature. The ovary contains a large number of oocytes, each often surrounded by a cluster of other cells that form a follicle. The follicle cells assist in nutrient supply to the oocyte and secrete estrogenic steroids. The oocyte and follicle increase in size as they mature. The oocyte remains small in viviparous vertebrates, although the follicle enlarges and forms a fluid-filled space around the oocyte. The follicle size increases dramatically in species with large-yolked eggs. During the reproductive cycle, the follicles "ripen" at the ovarian surface and burst, thereby releasing the egg into the coelomic cavity; this process is **ovulation** (Figure 11–43A). In almost all lower vertebrates, the follicle cells rapidly degenerate and are resorbed. In mammals and some elasmobranchs, the follicle cells temporarily persist as the corpus luteum ("yellow body") that secretes progesterone. Generally, only a small number (2 to 10) of oocytes mature in a single breeding cycle, but in amphibians often hundreds or thousands of ripe eggs are produced. Some fish release even more

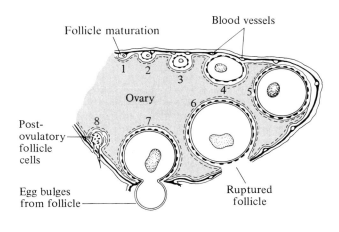

A

B

FIGURE 11–43 (A) Representation of a frog ovary showing stages of follicular growth (1 through 5), rupture of follicle (6), emergence of egg (7), and postovulatory follicle (8). (B) Schematic of the mammalian testis showing seminiferous tubules, epididymis, and vas deferens. The fine structure of the seminiferous tubule is also shown.

ripe eggs, e.g., the codfish can lay 4 10^6 eggs per reproductive season (but some oysters and *Aplysia* can produce 400 10^6 eggs per year!).

The paired **testes** (medial in cyclostomes) contain either fairly spherical, hollow sperm ampullae in lower vertebrates or elongate seminiferous tubules in some teleost fish and amniotes. The epithelium of sperm ampullae produces countless numbers of sperm in each breeding cycle; seminiferous tubules often continuously produce sperm. Supporting cells (Sertoli cells) and steroid synthesizing endocrine cells (Leydig cells) are also present in the seminiferous tubule epithelium (Figure 11–43B).

In amphioxus, eggs and sperm are shed from the gonads directly into the external environment. In cyclostomes, eggs and sperm are shed into the coelomic cavity and exit via posterior genital pores. In other vertebrates, the sperm are conducted to the exterior via closed tubes (the vas deferens), but the eggs are shed into the coelom then drawn through a funnel-shaped structure (infundibulum) into the oviduct (Mullerian duct) for transmission through the remainder of the reproductive tract to the exterior (Figure 11–44). The female and male reproductive tracts develop in a complex association with the kidneys. In lower vertebrates, the anterior end of the oviduct is a ciliated infundibulum and ovulated eggs are drawn into the relatively unspecialized oviduct. An ovisac may be present for egg storage (or retention of the developing egg until "hatching" in ovoviviparous fish, amphibians, and coelacanth). In teleosts, the entire ovary is enveloped, and there is a duct to the exterior; this avoids choking the body cavity with the large numbers of eggs shed by many fish. In amniotes, most of the oviduct is a broad, muscular tube, the uterus. In egg-laying amniotes, the oviduct secretes albumen (egg white protein) and the uterus secretes the shell.

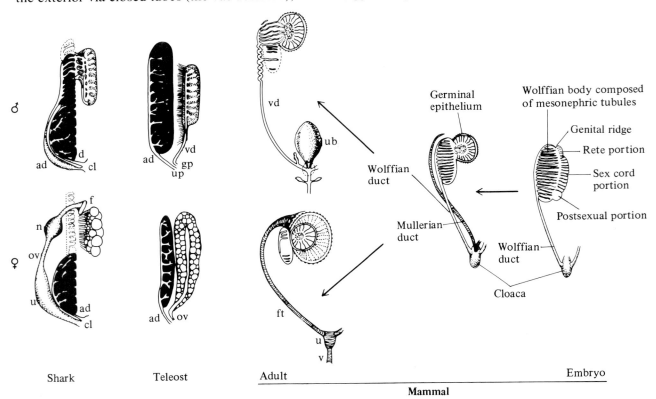

FIGURE 11–44 Male and female urogenital systems of a shark and teleost fish showing the relationship between testis (stippled) or ovary (open circles), kidney (solid), and the urogenital ducts. Abbreviations are as follows: archinephric duct (ad), special duct draining the shark kidney (d), special duct draining the teleost testis (vd), female reproductive duct (funnel, f; shell gland, n; oviduct, ov; uterus, u), cloaca (cl), genital pore (gp), and urinary pore (up). The mammalian male and female reproductive systems show the development of the male vas deferens (vd) from the Wolffian duct (archinephric duct) and the female fallopian tube (ft), uterus (u), and vagina (v) from the Mullerian duct. The kidney has an entirely separate drainage, the ureter, to the urinary bladder (ub).

In marsupial and placental mammals, the paired reproductive tract fuses to form a bipartite uterus or a single uterus (in higher primates).

The male reproductive tract (except in cyclostomes) is a complete tube that is specialized for transmission of sperm and ejaculation; it is actually a modified urinary duct. The developing testis has close ties with the embryonic mesonephric kidney which produces urine that is drained to the exterior by the archinephric (or Wolffian) duct (Figure 11–44). In primitive kidneys (e.g., cyclostomes) only urine is drained via the archinephric ducts; the sperm are shed into the coelomic cavity and exit via the gonadal pores. This "unsatisfactory" method of sperm shedding has led to a "struggle" between the male genital and urinary systems for use of the archinephric ducts. In teleosts the archinephric ducts retain their urinary transport role, but in other vertebrates the archinephric duct is usurped partially or completely by the testes for sperm transport. The kidney develops a separate ureter system for urine drainage.

The steroidal secretions of the gonads are male androgens and female estrogens (follicular steroids) and progestogens (luteal steroids). Control of gonadal steroid secretion is almost entirely by the pars distalis and the hypothalamus. The typical ovarian steroid hormones are estradiol-17β (a folliculoid) and progesterone (a luteoid). The corpus luteum also secretes relaxin, a polypeptide hormone responsible for softening the pubic symphysis ligament prior to birth. The Leydig (interstitial) cells secrete testosterone and a variety of other androgenic steroids (as well as estrogens). The Sertoli cells of the testis also synthesize small amounts of androgens; they cannot synthesize testosterone from cholesterol, but require the neighboring Leydig cells to synthesize the pregnenolone precursor. The main role of the Sertoli cells is to provide an appropriate nutritive environment for the germinal cells and for developing sperm. They also play a special role in sperm formation (spermiation) by removing cytoplasm from the spermatids and perhaps by even physically shaping the head and tail of the sperm.

Testosterone maintains spermatogenesis in most vertebrates once it is initiated by a "priming" action of FSH. The role of FSH is to (1) sensitize gonadal germ cells to synthesize androgen receptors, (2) stimulate Sertoli cells to secrete androgen-binding protein (ABP) into the lumen of the seminiferous tubules, and (3) stimulate the synthesis of LH receptors in Leydig cells. The ABP is required for transport and retention of testosterone in the seminiferous tubules; the testosterone is required for final development of the sperm. Testosterone promotes the functioning of various male reproductive glands (seminal vesicles and cloacal glands in fish; seminal vesicles, prostate, bulbourethral, and coagulating glands in mammals). Testosterone also promotes the development of many accessory male structures, e.g., modified fins or pigment patterns of male fish; the dorsal crest of male *Triton* salamanders; nuptial pads of male anuran amphibians; dorsal crest or gular skin folds of male reptiles; beak color, comb development in birds; antlers, horns, and hair growth in mammals.

The ovarian hormones have developmental and regulatory influences on the female reproductive duct system, as well as developmental effects on the maturing oocytes themselves. In vertebrates that produce large yolk eggs, estrogens stimulate the liver to synthesize vitellogenin lipoproteins, which are transmitted to the ovaries for incorporation into the yolking eggs. For example, in the clawed frog *Xenopus*, follicular estrogens are secreted under the influence of hypothalamic GnRH and FSH; these estrogens induce vitellogenesis by liver cells (Figure 11–45). The liver must be made competent to respond to estrogens, by thyroxin, and vitellogenin uptake by the oocytes is dependent on gonadotropins from the anterior pituitary (compare with mosquito oogenesis, Figure 11–23).

The effects of ovarian hormones are complicated by the often cyclic nature of the female reproductive cycle; the mammalian female reproductive cycle will be described below. Estrogens and progesterone promote the development of the oviducts, uterus, and vagina in mammals and also have a variety of other functions, e.g., number and activity of oviduct ciliated cells, contraction of oviduct smooth muscle, and development and secretion of the glandular and vascularized lining of the uterus (endometrium). Progesterone generally promotes changes related to oocyte maturation. Mammary gland development is under sex steroid control; estrogens promote duct development and progesterone stimulates development of the secretory alveoli. Hair growth and dermal collagen are reduced by estrogens. In some monkeys, estrogens promote reddening of the perineal skin.

The female mammalian reproductive cycle has a variable pattern. Females of some species have only one reproductive (estrous) cycle per year. These monestrous species breed at the time of the year most favorable for reproduction and rearing of the young. Many mammals are polyestrous; the estrous cycle is repeated until a successful fertilization interrupts the cycles. Generally, the estrous cycles occur irrespective of season (unless fertilization occurs), but sometimes the estrous cycles are limited

AMPHIBIAN (Xenopus)

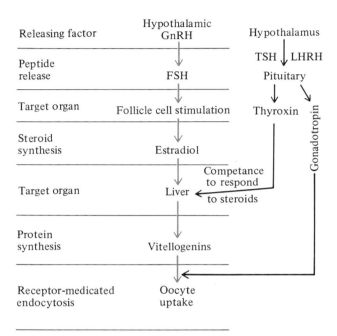

FIGURE 11–45 In an amphibian, the clawed toad *Xenopus*, hypothalmic gonadotropin releasing hormone (GnRH) stimulates follicle development and vitellogenesis by the liver, but thyroxin is required to enable the liver to respond to follicular estrogens, and pituitary gonadotropins are required for vitellogenin uptake by the oocyte.

to certain times of the year. In sheep, short photoperiods limit the ovarian cycles to autumn and winter; lambs are therefore born in spring or early summer. In some mammals, the estrous cycle is arrested just prior to ovulation until mating occurs. Copulation provides an afferent neural stimulus that releases hypothalamic LHRH, and thence adenohypophyseal gonadotropins. The rabbit is an example of a mammal with induced ovulation.

The mammalian female reproductive cycle is well illustrated by that of a human (Figure 11–46). Beginning at the end of menstruation, there is a progressive increase in the plasma estradiol level due to follicular secretion of estrogens but little change in progesterone level. The glandular and vascular endometrial lining thickens during this

phase. There appears to be a temporary positive feedback between LH release and high-estradiol levels during this period and after about 14 days there is a peak in estradiol level and a sudden surge in LH secretion by the adenohypophysis. The estradiol levels decline after the LH surge (and a smaller FSH surge). Progesterone levels rise after the LH surge due to luteal secretion. The endometrial glandular tissue becomes active during this phase. After another 14 days, the corpus luteum begins to degenerate and the estradiol and progesterone levels decline; the blood supply to the endometrium is restricted and the endometrial lining is shed (menstruation).

These cyclic hormonal changes are somewhat different in other mammals. In the rat, there are two progesterone peaks: one before ovulation (due to follicular progesterone) and a lesser, but more prolonged, peak after ovulation due to luteal secretion. Preovulatory progesterone controls estrous behavior. In sheep, progesterone levels are usually high but must decrease before estrous behavior or the LH surge can occur. The FSH cycle is more complex, and is often difficult to understand. In the rat, there are two FSH peaks: one immediately postovulatory (and after the LH peak) and one during the luteal phase. The estrous cycle of the rat may be so short that the earliest phases of the follicular cycle must begin during the luteal phase of the previous cycle. In sheep, the FSH peak seems to occur between LH peaks. The disparity in LH and FSH cycles is puzzling because there is only one known hypophysiotropic releasing hormone, LHRH, that regulates the release of both gonadotropins (LH and FSH).

The endometrial lining of the uterus provides a nourishing and protective environment for the implantation and development of a fertilized ovum. Shedding of the endometrial lining 14 days after ovulation and fertilization would obviously terminate the pregnancy, and so the endometrial lining must be maintained after fertilization occurs. This is accomplished by maintaining the estrogen and progesterone secretory activity of the corpus luteum. By eight to nine days after fertilization (i.e., still before menstruation would have occurred), the trophoblast cells of the developing ovum begin to secrete human chorionic gonadotropin (HCG), and HCG release increases to a maximum at about eight weeks after fertilization. HCG is a glycoprotein that is very similar in structure and function to LH. It delays the degeneration of the corpus luteum and hence maintains estrogen and progesterone levels during the initial period of pregnancy. By about 10 weeks of pregnancy, the placenta secretes estrogens

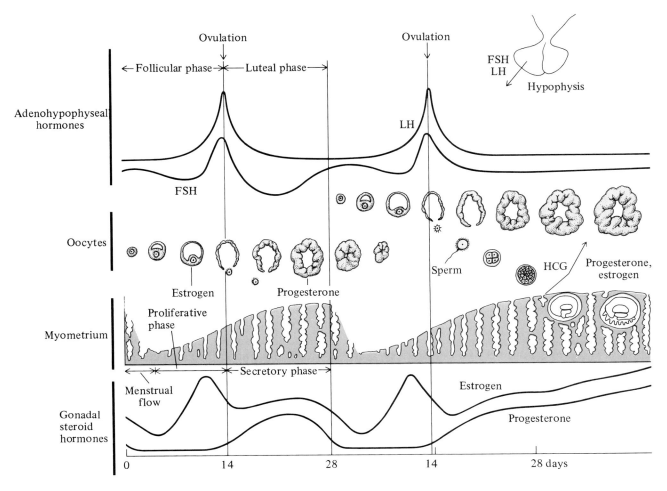

FIGURE 11–46 Estrous cycle of a polyestrous mammal (e.g., human) showing the cycle of follicle development, endometrial thickening and shedding, pars distalis secretion of gonadotrophins (FSH, LH), and ovarian hormones (estrogens, progesterone). *(From McNaught and Callander 1975.)*

and progesterone, and the corpus luteum degenerates as HCG levels decline. The estrogens secreted by the placenta are not synthesized *de novo* from cholesterol in the placenta, but from an androgenic steroid (dehydroepiandrosterone) formed in the maternal and fetal adrenal cortex.

Evolution of Endocrine Systems

For many years, the nervous and classical endocrine systems were considered to be distinct, with separate roles. The nervous system controlled rapid, short-term responses and the endocrine system controlled slower, long-term affects. The initial concept of a hormone was that it was a molecule released into the circulation by classical endocrine cells, was transported in the circulatory fluids at a low level, and elicited specific physiological responses in distant target cells. Neurons, in contrast, communicated directly with target cells by axons and direct synaptic contact, where a neurotransmitter was released to diffuse the short distance across the synaptic cleft to receptors on the postsynaptic membrane of the "target" cell. Thus, a perceived difference between hormonal and neural transmission was the scale of the transport of the chemical messenger from the originating cell to the target cell; it is a long distance in hormonal systems and a short distance in neural systems. There is however, in reality, a continuum in this scale of transport; for example, some hormones released by paracrine secretory cells diffuse only short distances and have local effects.

A synthetic theory for the origins and interrelationships of a variety of neural and non-neural endocrine cells is provided by the concept of an **APUD system** (Pearse 1969). APUD (Amine Precursor Uptake and Decarboxylation) cells absorb 5-hydroxytryptophan (5-HTP) from the blood and decarboxylate it to 5-hydroxytryptamine (5-HT, serotonin). The APUD cell series includes (1) neuroendocrine cells of the hypothalamus and the corticotroph-, somatotroph-, and prolactin-secreting cells of the adenohypophysis; (2) peripheral endocrine cells of the GEP system (A, B, D cells of pancreas, gastrin cells of stomach, motilin cells of intestine) and thyroid (calcitonin) cells; and (3) some peripheral cells in gut, lung, and urinary tract. The APUD series includes most of the paraneuron cells but not all of them. There would seem to be compelling evidence in favor of the general concepts of an APUD system, and it is possible that all peptide secreting cells have evolved from APUD cells, which may be derived from neural crest cells. However, there is considerable controversy concerning such a broad synthetic theory.

There are many basic similarities between the endocrine systems of invertebrates and vertebrates, and there are also many examples of convergent evolution of control systems. Neuroendocrine systems are important in all animals, but tend to be more numerous and fundamental to direct control of physiological function in invertebrates, whereas neurosecretions are generally restricted in vertebrates to the neurohypophyseal octapeptides and hypothalamic releasing factors. There is a general evolutionary progression from first order neuroendocrine loops in both invertebrates and vertebrates towards more complex second and third order neuroendocrine loops and independently functioning classical endocrine glands. Another trend, also seen in both invertebrates and vertebrates, is the development of a feedback role for gonadal steroid hormones on reproductive function. Compare, for example, the gonadal steroid control of gamete maturation in some mollusks, insects, and vertebrates and vitellogenin secretion and mobilization from fat body (insects) or liver (vertebrates) for oocyte incorporation.

Another apparent trend in the evolution of endocrine systems is the frequent occurrence of similar or identical hormones in often widely divergent animals. For example, vertebrate insulin-like molecules are present in insect brains and molluskan guts as well as in other invertebrates; even slime molds and protozoans contain insulin-like molecules (LeRoith and Roth 1984). Other "typically vertebrate" hormones are also present in invertebrates

and/or unicellular organisms: somatostatin, TSH, HCG, neurotensin, ACTH, β-endorphin, relaxin, and calcitonin. The common occurrence of these molecules in a wide variety of organisms suggests a possible early role for these molecules as intercellular messengers and their subsequent utilization by metazoans as either neurotransmitters or hormones or both. Evidence for their intercellular communication role in unicellular organisms is sparse but suggestive. Myxobacteria use soluble messengers (probably lipids) to communicate with adjacent cells and to modify their feeding behavior. A slime mold uses cAMP as an intercellular messenger to regulate food intake. Sexual reproduction in *Streptococcus* and *Saccharomyces* is facilitated by peptide sex pheromones.

There are a number of generalizations that can be made concerning peptide hormone evolution (Niall 1982). Many different hormones are produced by many different cells. We no longer have the reassuringly simple picture of endocrine systems in which each endocrine gland cell produces a specific hormone. Rather, many endocrine glands produce a variety of hormones. Neurons in the brain contain pituitary hormones, the hypothalamus contains somatostatin, neurons may contain CCK and gastrin, and lung tumors produce AVP. The mammalian placenta produces many different hormones. Amphibian skin produces a multitude of peptides. Every cell in the body contains the genetic information to code for synthesis of every hormone, so the widespread distribution of many hormones in various cells is comprehensible, although the physiological significance is not necessarily clear.

Families of peptide hormones (e.g., glycoproteins, prolactins, insulin-like peptides) probably arose by gene duplication, a common event in the genome that produces redundant copies of hormone coding sequences. The duplicate gene product can mutate and acquire different receptors and elicit different physiological responses in the same, or different, target cells. The evolution of peptide hormones is not so much an evolution of the hormones but the uses to which they are put. For example, prolactin, AVP, and calcitonin are hormones whose functions have altered throughout vertebrate evolution.

Steroid hormones may also have had a diverse evolutionary history (Sandor and Mehdi 1979). There is considerable conservatism in a wide variety invertebrates and vertebrates their biosynthesis of steroids and, to some extent, in the physiological roles of the steroid hormones. Bacteria, algae, and plants also contain steroids. A steroid may well have been one of the primordial molecules (along with 20 or so amino acids, 2 purines, 2 pyrimidines, 2 sugars, 1 fatty acid) present in the first biomolecular assemblage from which self-replicating, energy-utilizing living cells evolved. The structural role of steroids in prokaryote organisms is well established, but it also is possible that steroids have functional roles in, for example, gene expression. The steroids could interact with the prokaryote genome to regulate gene expression, as occurs in metazoans.

Thus, we have at least a basis to speculate that peptides, steroids, and cAMP were present and had potential to act as intercellular messengers in prokaryotes. This could have provided the basis for the evolution of general steroid and peptide/cAMP endocrine systems in metazoans.

Summary

The neuroendocrine system consists of central nervous system cells that secrete neurohormones which enter the body fluids and are distributed to target cells by the circulatory system. The classical endocrine system consists of glandular (non-neural) cells that secrete hormones into the body fluids for distribution to target cells. The neuroendocrine and classical endocrine systems are important effector systems. Endocrine regulation is slower and longer term than nervous regulation.

Neurosecretory cells are typical neurons but contain neurosecretory granules, droplets, or colloidal pools. The neurosecretion is often liberated into the body fluids at a specialized neurohemal organ. Neuroendocrine regulation is more common in invertebrates than in vertebrates. Neurohormones are peptides.

Classical endocrine glands are found especially in higher invertebrates and vertebrates. They do not have specialized neurohemal organs for release of the hormone, but the cells of the glands are often organized as thin sheets or strands to facilitate hormone liberation into the body fluids. Classical hormones are peptides, derivatives of amino acids, steroid lipids, prostaglandins, or other chemicals.

The target cells of neuro- and classical hormones have specific receptors located on the cell surface (especially peptide hormones) or in the cytoplasm (especially steroid hormones). Hormone-receptor binding initiates a sequence of biochemical reactions.

Peptides and most amino acid-derived hormones bind to surface receptors. Binding often stimulates a G protein, which either activates or inhibits adenyl cyclase. Adenyl cyclase forms cyclic AMP from ATP. Cyclic AMP is the internal second messenger that initiates a variety of intracellular reactions, such as protein phosphorylation, and produces the effector response. There are other second messenger systems in addition to cAMP. Steroid hormones bind to intracellular receptors. The hormone-receptor complex modifies DNA transcription and translation in the nucleus; they directly activate or suppress gene expression.

Invertebrate endocrine systems are predominated by neurohormones, although especially the more advanced invertebrates also have complex endocrine control by classical endocrine systems. Invertebrate endocrine systems typically regulate regeneration, molting, metabolism, color change, reproduction, and iono- and osmoregulation. The endocrine systems of nonvertebrate deuterostomes (e.g., echinoderms and cephalochordates) provide few insights into the evolution of the vertebrate endocrine systems.

Vertebrate endocrine systems are, in general, quite different than invertebrate endocrine systems. The role of peripheral classical endocrine glands is emphasized over neuroendocrine control, although most classical endocrine glands are regulated by neurohormones from the pituitary.

The hypothalamo-hypophyseal system (hypothalamus-pituitary) is an important effector control mechanism. The posterior pituitary is the neurohemal release site for basic octapeptides (e.g., AVP) and neutral octapeptides (e.g., oxytocin). The intermediate pituitary releases MSH under neural control from the arcuate nucleus. The pineal organ, which is derived from the nervous system, is photoreceptive in lower vertebrates but neurosecretory in higher vertebrates. It secretes melatonin, which affects chromatophores, and may have antigonadal and thermoregulatory effects in birds and mammals. The Dahlgren cells of elasmobranchs and the urophysis of teleost fish form a spinal neuroendocrine system. A variety of urotensins have been identified, with a variety of functions, including blood pressure regulation, Na^+ uptake, and smooth muscle contraction.

The principal classical endocrine glands of vertebrates are the anterior pituitary, thyroid, parathyroid and ultimobranchial glands, gastroenteropancreatic glands (stomach, small intestine, pancreas), chromaffin tissues, interrenal tissues, and the gonads. Many of the vertebrate classical endocrine systems are derived from the gut (e.g., anterior pituitary, thyroid, pancreas, and ultimobranchial glands). Chromaffin tissue is derived from the neural crest cells and is homologous with the autonomic ganglia. The interrenal tissue is derived from the kidney. The anterior pituitary cells secrete a variety of hormones (e.g., TSH, ACTH, LH and FSH, GH, endorphins) under control by releasing factors from the hypothalamus, which either regulate other endocrine glands (e.g., TSH, LH, FSH) or control effector cells (e.g., MSH, GH). The thyroid secretes thyroxin and T_3, which have various widespread metabolic effects. The parathyroids secrete parathormone, which elevates the blood Ca^{2+} level. The ultimobranchial glands and corpuscles of Stannius also secrete hormones that affect the blood Ca^{2+} level. The gastroenteropancreatic cells secrete hormones involved with secretion of H^+, proteolytic and other enzymes, and regulation of gut motility (e.g., gastrin, cholecystokinin, secretin). Chromaffin tissues (adrenal medulla in higher vertebrates) secrete catecholamines (adrenaline, noradrenaline, dopamine) that initiate the fight-flight reflex reactions (e.g., elevated heart rate, reduced gut motility, and cellular metabolic changes, such as conversion of glycogen to glucose). The inter-renal tissues (adrenal cortex in higher vertebrates) secrete steroids that affect carbohydrate metabolism (glucocorticoids) and electrolyte balance (mineralocorticoids). The testis and ovary secrete many steroid hormones, such as testosterone, estrogens, progesterone, that are involved with control of reproduction.

The variety of endocrine systems of invertebrates and vertebrates indicates the multiple evolution of neurohormonal and classical hormonal systems. Many endocrine systems are clearly derived from neurons, and there is a continuum in "endocrine" cell function from typical axons that secrete a neurotransmitter at a synapse, to cells that have short axons and secrete hormones with local effects, to cells with a neurohemal organ and a hormone that diffuses long distances to remote target organs. The concept of paraneurons encompasses this range of structural and functional diversity. Many diverse endocrine cells that secrete peptide hormones are thought to be derived from APUD cells.

There is a widespread occurrence in animals of many hormones that were once thought to be specific hormones in particular animal groups. For example, insulin was thought to be a typical vertebrate hormone, but insulin-like hormones are found in many invertebrates and even some protozoans and slime molds. Many other "vertebrate" hormones, such as TSH, ACTH, neurotensin, and calcitonin, are also found in many invertebrates. There apparently is a very diverse evolutionary history of many of these peptide hormones and also their general intracellular "second messenger," cAMP. Many lipid hormones appear to have a similarly diverse evolutionary history.

Supplement 11–1

Radioimmunoassay: A Technique Using Hormone-Binding Specificity

Radioimmunoassay (RIA) is an important endocrinological technique that allows the measurement of low levels of specific hormones. It relies on the specificity of binding of a hormone to an antibody, rather than the normal receptor protein. Variations in the technique allow the measurement of receptor proteins.

RIA requires two reagents: an antibody to the hormone and a hormone that is labeled, usually with a radioisotope such as ^{125}I. The labeled hormone will compete with unlabeled hormone for antibody binding sites. A standard curve is prepared by adding known amounts of labeled hormone and antibody to a series of test tubes. Varying amounts of unlabeled hormone are also added to the series of tubes (e.g., 1 to 100 ng). The labeled and unlabeled hormone molecules bind to the antibodies at equilibrium in a proportion determined by their relative concentrations, i.e., a tube containing a large amount of unlabeled hormone will have a small amount of antibody-bound labeled hormone and a tube with a small amount of unlabeled hormone will have a high amount of antibody-bound labeled hormone. Separation of the bound hormone (e.g., by precipitation of the antibody proteins) after a sufficient incubation period (typically several hours), enables the determination of the amount of bound labeled hormone, and a standard curve can be constructed to relate the amount of bound labeled hormone to the concentration of the unlabeled hormone. The standard RIA curve relates the percent of labeled hormone radioactivity that is recovered bound to antibody (in the pellet) to the concentration of unlabeled hormone present in the standard incubation solution. The hormone concentration in an unknown sample can be determined from the percentage of radioactivity recovered in the pellet using the standard curve. Repeating the RIA procedure with biological samples of unknown hormone content (instead of adding known amounts of unlabeled hormone) allows the determination of the hormone concentration in the sample.

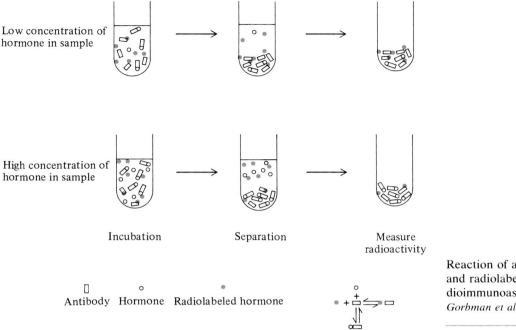

Low concentration of hormone in sample

High concentration of hormone in sample

Incubation   Separation   Measure radioactivity

▯ Antibody   ○ Hormone   • Radiolabeled hormone

Reaction of antibody, hormone, and radiolabeled hormone for radioimmunoassay (RIA). *(From Gorbman et al. 1983.)*

A similar procedure allows the estimation of the numbers of hormone receptors in target tissues. The receptors are isolated by cellular disruption and centrifugation. They are present in low concentration but have a high hormone-binding affinity. Unfortunately, other constituents of the cell fraction also nonspecifically bind hormones. These nonspecific binding materials are present in high concentration but have a low affinity for the hormone.

A series of test tubes containing the cell fraction are incubated with varying amounts of labeled, and a high concentration of unlabeled, hormone; the dynamic equilibrium between bound and unbound hormone is then determined by centrifuging to separate the bound hormone (in the pellet) from the unbound hormone (in the supernatant). The large amount of unlabeled hormone ensures that all receptor binding sites are occupied, and so the labeled hormone is essentially only able to bind nonspecifically to the numerous, low affinity sites. Repeating the assay, but with a low concentration of unlabeled hormone, allows some labeled hormone to bind to the specific receptor as well as to the nonspecific sites.

A nonspecific binding curve is obtained by graphing the amount of bound labeled hormone as a function of

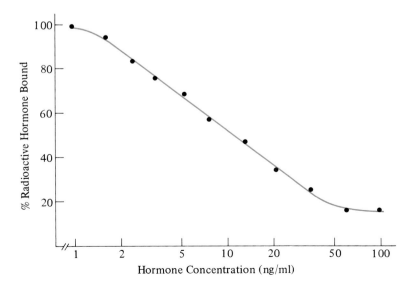

Example of a standard curve for RIA showing the relationship between percent of radiolabeled hormone that is bound and the hormone concentration. *(From Gorbman et al. 1983.)*

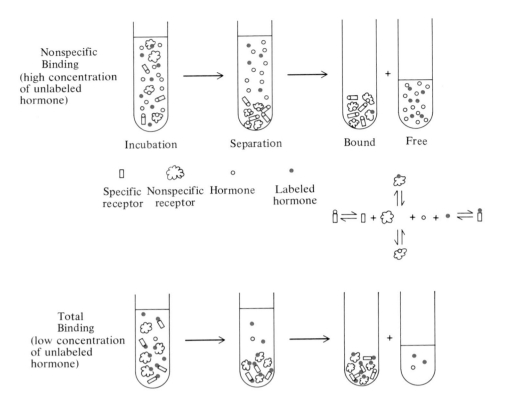

Principle for estimation of numbers of hormone receptors in tissues.

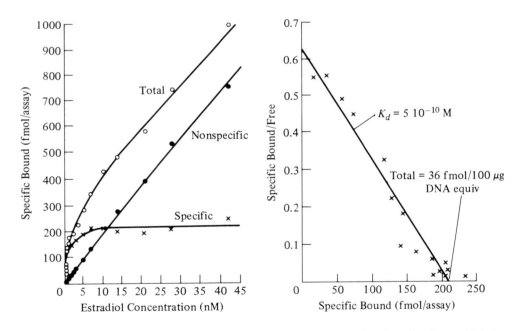

Total nonspecific and specific binding curves (left) used to graph a Scatchard plot (right) for the calculation of the numbers of hormone (estradiol) receptors in a tissue; K_d is the dissociation constant for the hormone and receptor.   *(Modified from Barrack and Coffey 1982.)*

the amount of free labeled hormone. A total binding curve is obtained by graphing the amount of bound labeled hormone as a function of the free labeled hormone concentration. The difference between the total and nonspecific binding curves represents the saturation binding curve for the receptors. There is little specific binding at low hormone concentrations, and specific binding increases at higher hormone concentrations until the receptors become saturated. This curve is similar in shape to the relationship between enzyme reaction velocity and substrate concentration. Further graphical analysis of the ratio of bound/free labeled hormone to the hormone bound to receptors (a Scatchard plot) allows calculation of the dissociation constant K_d for the receptor-hormone complex and the number of receptors present. The slope of the Scatchard plot is $-K_d$; a low K_d reflects a high receptor-hormone affinity. The X-intercept indicates the amount of hormone bound to receptors when they are saturated; multiplying the X-intercept by 6.02×10^{23} yields the number of receptor sites (assuming 1 hormone molecule per receptor site). (*See Gorbman et al. 1983; Barrack and Coffey 1982.*)

Recommended Reading

Barrington, E. J. W. 1964. *Hormones and evolution.* London: English Universities Press.

Bentley, P. J. 1971. *Endocrines and osmoregulation: A comparative account of the regulation of water and salt in vertebrates.* New York: Springer-Verlag.

Bourne, H. R. 1989. G-protein subunits. Who carries what message? *Nature* 337:504–5.

Chester-Jones I., P. M. Ingleton, and J. G. Phillips. 1987. *Fundamentals of comparative vertebrate endocrinology.* New York: Plenum Press.

Clark, C. R. 1987. *Steroid hormone receptors: Their intracellular localization.* Ellis Horwood Series in Biomedicine. England: Ellis Horwood, and Weinheim: VCH Verlagsgesellschaft.

Cohen, P. 1982. The role of protein phosphorylation in neural and hormonal control of cellular activity. *Nature* 296:613–20.

Evered, G., and G. Lawrensen. 1981. *Peptides of the pars intermedia.* CIBA Symp. 81. London: Pitman Medical.

Fingerman, M. 1987. The endocrine mechanisms of crustaceans. *J. Crust. Biol.* 7:1–24.

Frye, B. E. 1967. *Hormonal control in vertebrates.* New York: Macmillan.

Gabe, M. 1966. *Neurosecretion.* Int. Ser. Monog. Biol. 28. Oxford: Pergamon Press.

Gilman, A. G. 1987. G proteins: Transducers of receptor-generated signals. *Ann. Rev. Biochem.* 56:615–49.

Gorbman, A., and H. A. Bern. 1964. *A textbook of comparative endocrinology.* New York: John Wiley & Sons.

Gorbman, A., et al. 1983. *Comparative endocrinology.* New York: John Wiley & Sons.

Highnam, K. C., and L. Hill. 1977. *The comparative endocrinology of the invertebrates.* London: Edward Arnold.

Imura H., et al. 1985. Endogenous opioids and related peptides: From molecular biology to clinical medicine. *J. Endocrinol.* 107:147–57.

Ishii, S., T. Hirano, and M. Wade. 1980. *Hormones, Adaptation, and Evolution.* Japanese Scientific Society Press, Tokyo, and Springer-Verlag, Berlin.

Maddrell, S. H. P., and J. J. Nordman. 1979. *Neurosecretion.* Glasgow: Blackie & Son.

Niall, H. D. 1982. The evolution of peptide hormones. *Ann. Rev. Physiol.* 44:615–24.

Olsson, R. 1969. Endocrinology of the Agnatha and Protochordata and problems of evolution of vertebrate endocrine systems. *Gen. Comp. Endocrinol. Suppl.* 2:485–99.

O'Malley, B. W., and W. T. Schrader. 1976. The receptors of steroid hormones. *Sci. Am.* 234(2):32–43.

Pang, P. K. T., and A. Epple. 1980. *Evolution of vertebrate endocrine systems.* Lubbock: Texas Tech. Press.

Pearse, A. G. A. 1976. Neurotransmission and the APUD concept. In *Chromaffin, enterochromaffin, and related cells,* edited by R. E. Coupland and T. Fujita, 147–54. Amsterdam: Elsevier.

Sandor, T., and A. Z. Mehdi. 1979. Steroids and evolution. In *Hormones and evolution,* ed. by E. J. W. Barrington, 1–72. New York: Academic Press.

Tombes, A. S. 1970. *An introduction to invertebrate endocrinology.* New York: Academic Press.

Chapter 12

Aquatic Respiration

A nudibranch, such as *Glaucus atlanticus*, uses its cutaneous surface for gas exchange. *(Photo Researchers, Inc.)*

Respiration is the exchange of gases (mainly O_2 and CO_2) between the ambient medium and the body fluids. The processes of cellular metabolism are also often called "internal respiration" and gas exchange is often called "external respiration," but we will restrict the definition of respiration to gas exchange, to avoid potential confusion.

Respiratory Gases

Most animals rely on aerobic metabolism to sustain their resting energy requirements for the simple reason that almost 20 times as much ATP is synthesized by aerobic metabolism compared to the common anaerobic pathways leading to lactic acid (see Chapter 3). Furthermore, substrates other than carbohydrate (i.e., fats and protein) can be readily metabolized aerobically but not anaerobically. Oxygen is also freely available in the atmosphere, comprising about 21% of air by volume. Aerobic metabolism produces CO_2 as a waste product in about equal volumes to the O_2 consumed (the respiratory quotient **RQ** $\approx$ 1). Oxygen uptake is therefore accompanied by an equivalent CO_2 excretion. The amount of CO_2 in the environment is generally low, and this and the high solubility of CO_2 in body fluids greatly facilitates its excretion.

Oxygen uptake from the environment is a relatively straightforward process for small, aquatic, unicellular animals. Uptake of oxygen across the body surface from the environment is adequate to supply their metabolic demands. Diffusion is also adequate for distributing oxygen throughout the cytoplasm. Consequently, small animals do not require any specialized respiratory structures for oxygen uptake or carbon dioxide loss or an internal circulatory system. The evolution of larger and more complex animals placed impossible demands on their body surface for maintaining an adequate respiratory supply of oxygen to their tissues.

Three major evolutionary trends within the animal phyla have required extensive modification of the respiratory exchange surface. First, there has been a general trend towards higher cellular metabolic rates (compare unicells, ectotherms, and endotherms in Figure 4–9, page 99). Second, there has been a marked trend in larger multicellular animals for specialization of certain areas of the surface for specific functions, e.g., food gathering, digestion, reproduction, locomotion, vision, and information transfer. This reduces the surface available for gas exchange and often requires specific mechanisms for maintaining an adequate oxygen supply to those cells. A larger body size decreases the body surface area available for respiratory exchange relative to body mass (providing the general body shape is maintained). Consider, for example, a spherical animal of radius r; its surface area is $4\pi r^2$, whereas its volume is $4/3 \ \pi r^3$. An increase in radius by a factor of 2 increases the surface area by 4 times and the volume by 8 times. Consequently, the surface area/volume ratio is decreased by 1/2. This surface area/volume effect is mitigated to some extent by the almost universal dependence among animals of metabolic rate on body mass as $VO_2 \propto Mass^{0.75}$ (see Chapter 4). Nevertheless, the relative decrease in metabolic rate for the larger animal does not completely compensate for its lower surface area/volume ratio. Consequently, progressively larger animals are less able to rely on diffusion across their body surface and through their body tissues. A third major evolutionary trend was the shift from breathing water to breathing air that accompanied the invasion of land. As we shall see, this has necessitated major changes in the basic design of the respiratory system, from gills to lungs or tracheae.

The evolution of larger and more metabolically active animals is invariably associated with progressively more complex respiratory structures specialized for gas exchange, both oxygen uptake and carbon dioxide excretion. A respiratory system is only the first step in oxygen transport for large animals, however. Of equal, or even greater, significance is the ability to circulate the oxygen within the body by movement of specialized fluids (blood, hemolymph) within a vascular system (the circulation) through the action of a pump, or pumps (heart). The circulatory system of animals is described in Chapter 14 and the composition and role of blood is discussed in Chapter 15.

Composition of Air

The composition of normal, dry atmospheric air (Table 12–1) is 20.95% oxygen, 0.03% carbon dioxide, and 78.08% nitrogen. All of the other physiologically inert gases (argon, krypton, neon, etc.) are generally included with nitrogen as a total of 79.02% of the air. We will therefore consider normal dry air to consist of only three gases: the physiologically active gases (oxygen and carbon dioxide) and physiologically inert nitrogen.

The composition of the atmosphere has been fairly constant over historical time, although dramatic changes have occurred over the previous 4 billion years of the earth's history. The primitive atmosphere of the earth, judging by the present

TABLE 12–1

Gas	% Content	Mole Fraction
Composition of normal (dry) atmospheric air at sea level. *(From Verniani 1966; Williamson 1973; Walker 1977.)*		
Oxygen	20.948	0.2095
Carbon dioxide	0.0315	0.0003
Nitrogen	78.084	0.7808
Argon	0.934	0.0093
Neon	0.002	0.00002
Helium	0.005	0.000005
Methane	0.0002	0.000002
Krypton	0.0001	0.000001
Total	100.00	1.00

composition of volcanic gases (83.3% carbon dioxide, 0.9% sulfur dioxide, 13.7% hydrogen, 0.8% nitrogen; Walker 1977) was considerably different from the present atmosphere. Even in historic times there have been small (but not insignificant) changes in the composition of the air. Burning of fossil fuels, deforestation, and changes in agricultural patterns have caused a slight decline in the oxygen content and an increase in the carbon dioxide and methane content of air. These changes are of no direct respiratory significance to animals but they ultimately will have major effects on the earth's climate and biota by a **greenhouse effect** if they continue unabated.

The normal atmospheric pressure at sea level is about 101 kPa (= 1 atmosphere = 760 torr = 1032 bars). The SI unit for pressure is the Pascal (Pa), but it is not yet commonly used. Fortunately, 1 kPa is about 1% of atmospheric pressure, so this is fairly easy to remember. One torr (the unit is named after Evangelista Torricelli) is the pressure of 1 millimeter of mercury, or 1.36 cm of water. An atmospheric pressure of 101 kPa is called the **standard pressure** at which 1 mole of gas occupies 22.4 liters (at 0° C, the **standard temperature**). The concept of standard pressure is very useful because it enables the comparison of amounts of gas (i.e., moles) even if their volume is measured at different pressures. Gas volumes are therefore generally expressed as the volume at standard temperature and pressure (STP). However, other conventions are sometimes used, e.g., volume at ambient temperature and pressure (ATP), volume at body temperature and pressure (BTP).

The total pressure exerted by a gas mixture is equal to the sum of pressures exerted by each constituent gas (**Dalton's law**). The **partial pressure** exerted by a gas (indicated by the abbreviation p) is determined by its fractional content (F) or mole fraction and the total barometric pressure (P_b);

$$pO_2 = FO_2.P_b$$
$$pCO_2 = FO_2.P_b \qquad (12.1)$$
$$pN_2 = FN_2.P_b$$

For normal, dry air, pO_2 = 21.2 kPa (159.2 torr), pCO_2 = 0.03 kPa (0.23 torr), and pN_2 = 79.8 kPa (600.6 torr) at P_b = 101 kPa (760 torr).

The normal composition of air can alter markedly in terms of its fractional content and atmospheric pressure, hence partial pressures (Table 12–2). For example, different fractional contents of oxygen and carbon dioxide occur underground or underwater. High altitude reduces the total atmospheric pressure and atmospheric pressure increases markedly under water, so the partial pressures of gases in an air space can change although the percentage composition is unaltered. Addition of water vapor decreases the partial pressures of the other gases, sometimes significantly.

The metabolism of soil microbes and small soil animals, and also soil chemical reactions, can reduce the fractional oxygen content and elevate the carbon dioxide content of soil air. The O_2 content can be further reduced in the vicinity of large animals that burrow, or bury themselves, in the soil. Under extreme conditions, the oxygen content in animal burrows can be depleted to less than 10% and the carbon dioxide can accumulate to exceed 10%. The extent to which burrow gas differs from ambient air depends on the complex interaction of physiological variables (e.g., metabolic rate and RQ), the dimensions of the underground burrows and nest chambers, and the soil properties (Withers 1976).

Changes in barometric pressure due to elevation above sea level, or descent below sea level, can dramatically alter the partial pressures of gases, although the fractional composition is unaltered. The atmospheric pressure at sea level reflects the weight of the air column above. The atmospheric pressure is 1/2 of the sea level value, i.e., 50.5 kPa (380 torr) at an elevation of about 5500 meters. The partial pressures of all gases are decreased proportionately; pO_2 = 10.6, pCO_2 = 0.0016, pN_2 = 39.9 kPa. The relationship between altitude and pressure is not linear but is exponential; the pressure decreases by 1/2 for every 5500 m increase in altitude (Figure 12–1). An ambient air sample taken underwater in a collapsible container experiences the additional weight of the water column above it. The extra weight of a mere 10 meters of water is

TABLE 12–2

Gas composition of air from unusual environments, such as underground environments, at high altitude and deep under water, and the O_2 content of various bodies of water, such as swamps and the ocean O_2 minimum layer. The O_2 and CO_2 partial pressures (pO_2, pCO_2) and their fractional content (FO_2, FCO_2) vary; barometric pressure (P$_b$) varies with altitude and depth underwater. Values for air are kPa and for water are parts per million (by weight; ppm); 1ppm O_2 = 2 to 2.7 kPa (at 15° to 30° C).

Air	pO$_2$	FO$_2$	pCO$_2$	FCO$_2$	P$_b$
Normal	21.1	0.2095	0.03	0.0003	760
5500 m altitude	10.6	0.2095	0.01	0.0003	380
8848 m altitude (Everest)	6.9	0.2095	0.01	0.0003	250
10 m underwater	41.1	0.2095	0.06	0.0003	1520
100 m underwater	231.5	0.2095	0.33	0.0003	8360
1000 m underwater	2135.8	0.2095	3.06	0.0003	76760
Mammal Burrows					
Gopher	15.9	0.155	3.85	0.038	760
Ground squirrel	10.9	0.137	6.25	0.062	760
Mole rat	14.1	0.140	4.78	0.048	760
Marsupial pouch	15.8	0.157	5.32	0.053	760

Water	ppm O$_2$	ppm CO$_2$
Normal	8–10	0.02
1000 m depth (in equilibrium with normal air)	8–10	0.02
Water hyacinth swamp	2–7	0–9
Bahr el Gabel swamp	0.7–6.2	2.8–18
Guinea grass swamp	0.22–1.2	8.4–9.2
East Atlantic O$_2$ minimum	1.5–3.0	0.02

equivalent to the whole atmosphere, i.e., 10 m water = 101 kPa. Consequently, the pressure at 10 m depth is 202 kPa; pO_2 = 42.3, pCO_2 = 0.06, pN_2 = 159.6 kPa. At greater depths, extreme pressures are experienced, e.g., at 100 m, P = 1111 kPa = 8360 torr = 11 atm. However, the partial pressures of gases dissolved in water do not increase in the same manner with depth. In theory, the partial pressures of dissolved gases are the same throughout the water column regardless of depth and pressure. In practice, there is some minor variation in the partial pressures of dissolved gases with depth due to animal metabolism and chemical reactions.

Atmospheric air always contains some water vapor; even very cold polar air has a significant, although low, water vapor content. The water exerts a partial pressure, the **water vapor pressure**, just as other gases in air have a partial pressure. The saturation partial pressure of water vapor (pH$_2$O$_{sat}$) for air in equilibrium with water (or ice) at a pressure

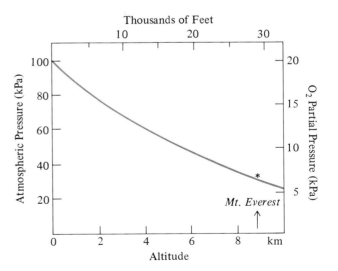

FIGURE 12–1 Relationship of barometric pressure (P$_b$) and O_2 partial pressure to altitude. *(Modified from Bouverot 1985.)*

TABLE 12–3

Saturation water vapor pressure (pH_2O_{sat}; kPa) and absolute humidity (χ_{sat}; mg liter^{-1}) of air in equilibrium with water or ice at 21.1 kPa ambient pressure at varying ambient temperatures (T; °K).

$$\text{saturation WVP (kPa)} = 0.1\ T^a\ 10^{c+(b/T)},$$
$$\text{saturation } \chi\ (\text{mg cm}^{-3}) = k\ T^{a-1} 10^{c+(b/T)},$$

where the constants over water/ice are a, $-4928/-0.3229$; b, $-2937/-2705$; c, $23.552/11.482$; k, 0.21668. *(All values are calculated from Parrish and Putnam 1977.)*

T_{air}	pH_2O_{sat}	χ_{sat}	T_{air}	pH_2O_{sat}	χ_{sat}
-10	0.259	2.14	35	5.61	39.6
-5	0.400	3.25	40	7.36	51.1
0	0.609	4.85	45	9.56	65.4
5	0.869	6.80	50	12.3	82.9
10	1.22	9.41	60	19.9	129.7
15	1.70	12.8	70	31.1	198.
20	2.33	17.3	80	47.0	290.
25	3.16	23.1	90	69.5	417.
30	4.24	30.4	100	101	585.

of 101 kPa is dependent on the ambient temperature (Table 12–3). The **relative humidity** (RH) of any air sample of a particular water vapor pressure (pH_2O) is equal to $100 \times (pH_2O/pH_2O_{sat})$. Absolutely dry air, of course, has a relative humidity of 0%. Air in equilibrium with water (or ice) has a relative humidity of 100%. The **absolute humidity** of air (χ) is the molar concentration of water in the air; for example, 100% RH air at 20° C has an absolute humidity of about 1 millimole H_2O per liter air, or 17.3 mg liter^{-1}. The relationship between absolute humidity (χ), absolute saturation humidity (χ_{sat}), and relative humidity at a given temperature is straightforward:

$$\chi = (RH/100)\chi_{sat} \qquad (12.2)$$

The absolute humidity can be calculated approximately from the ideal gas law, $PV = nRT$, since $(n/V) = P/RT$; R is the gas constant (8.314 m^3 Pa mol^{-1} K^{-1}) and T is the absolute temperature (°K). Unfortunately, water vapor is not an ideal gas and accurate calculations of saturation water vapor pressure and absolute humidity involve the use of more complex formulas (Parrish and Putnam, 1977). Some values of RH and χ calculated from these equations are presented in Table 12–3.

The presence of water vapor in air usually has only a small effect on the partial pressures of the other gases. Calculation of the PO_2, pCO_2, and pN_2 for moist air is accomplished using Dalton's law

$$P_b = pO_2 + pCO_2 + pN_2 + pH_2O \qquad (12.3)$$

(plus the partial pressures of any other gases present). The pO_2 is equal to $0.2095(P_b - pH_2O)$; the pCO_2 and pN_2 are calculated similarly. For example, normal air saturated with water vapor has a pO_2 of 20.7 kPa, pCO_2 of 0.023 kPa, pN_2 of 78.0 kPa, and pH_2O of 2.3 kPa at 20° C. Values for water-saturated air at human body temperature (37° C) are fairly similar ($pO_2 = 19.9$ kPa, $pCO_2 = 0.03$, and $pN_2 = 74.9$) but pH_2O is considerably higher (6.25 kPa). A very significant fraction of the air can be water vapor at higher temperatures or lower pressures. At 80° C, the pH_2O is 470 kPa and almost 50% of the air is water vapor; at 100° C, the water vapor pressure is 101 kPa and all of the air should be water vapor. High altitude can also drastically reduce the partial pressures of O_2 and N_2 of dry air and especially of saturated air. At an altitude of about 20000 m, the atmospheric pressure of 0.62 kPa is equal to the water vapor pressure in the human lung; only water vapor can be present!

Gases Dissolved in Water

The gas composition of water is slightly more complex to analyze than the composition of air. Water in equilibrium with air will have the same gas partial pressures as the saturated air but may have considerably different molar concentrations of the gases. The molar concentration of a gas in water is described by **Henry's law**

$$[A] = pA\ \alpha_A \qquad (12.4)$$

TABLE 12–4

| Solubility coefficients for the physiologically important gases as a function of temperature and ionic concentration. Units are $\mu mol \ liter^{-1} \ kPa^{-1}$. 1 $\mu mol \ liter^{-1} \ kPa^{-1}$ = 0.0226 $ml \ liter^{-1} atm^{-1}$. |

Solubility of Gases in Distilled Water				
°C	*Oxygen*	*Carbon Dioxide*	*Nitrogen*	*Helium*
0	21.7	767.5	—	—
10	16.9	531.2	—	—
20	13.7	386.8	6.82	—
30	11.6	294.9	—	—
37	10.6	250.5	5.61	3.75
40	10.2	234.8	—	—

Effect of Salinity on Oxygen Solubility						
°C	*Salinity*	*0‰*	*10‰*	*20‰*	*30‰*	*40‰*
0		21.7	20.2	18.9	17.7	16.6
10		16.9	15.8	14.8	13.9	13.1
20		13.7	12.9	12.2	11.5	10.8
30		11.6	11.0	10.4	9.86	9.33
40		10.2	9.71	9.26	8.73	8.35

where [A] is the molar concentration of A (moles $liter^{-1}$), pA is the partial pressure of A (kPa), and α_A is the solubility coefficient for A (moles $liter^{-1}$ kPa^{-1}). The value of the solubility coefficient depends on the gas in question (e.g., O_2, CO_2, N_2, etc.), the nature of the solvent (e.g., water, lipids), and many other variables, such as temperature and ionic strength. Values for α have been empirically determined for the physiologically important gases under physiological conditions (Table 12–4). Solubility is decreased by an elevation in temperature or ionic strength.

O_2 and N_2 are present in water at much lower molar concentrations that in air, but CO_2 is present at about the same concentration in air and water (Table 12–5). The volumes of O_2 and CO_2 in water are not equal at the same partial pressures because their solubilities differ. At high temperatures, there is less CO_2 in water than air, but more CO_2 is present in water than air at low temperatures. The volumes of O_2 and CO_2 in air are equal at the same partial pressure, i.e., $[O_2]/pO_2 = [CO_2]/pCO_2 = 1/RT$ (Figure 12–2). The gas content per kPa is called the **capacitance coefficient** (β); it is equivalent to the solubility coefficient for liquids. All gases have the same capacitance coefficient in air.

The different solubilities of O_2 and CO_2 in water result in markedly disparate changes in dissolved pO_2 and pCO_2 as a result of animal metabolism. If the RQ is 1, then the same molar volume of CO_2 is added to the water as is O_2 removed, but the increase in pCO_2 of the water is small compared to the accompanying decrease in pO_2. The slope of the relationship between ΔpCO_2 and ΔpO_2 is the ratio of their solubility coefficients, i.e., about 28 at 20° C. In air, there is an equal increase in pCO_2 for a decrease in pO_2 (if RQ = 1) and the slope of the relationship is 1 (Figure 12–3).

TABLE 12–5

| Comparison of the solubility (α) of gases in water and capacitance in air (β; $ml \ liter^{-1} \ kPa^{-1}$), and the concentrations ($ml \ liter^{-1}$) in air and water at 20° C. |

	Solubility/Capacitance		**Concentration**	
	Water	*Air*[1]	*Water*	*Air*
O_2	0.331	9.88	6.98	209.5
CO_2	9.30	9.88	0.31	0.30
N_2	0.164	9.88	13.5	790.2

[1] $\beta = 1/RT = 0.120/T \ mol \ Pa^{-1} \ m^{-3}$.

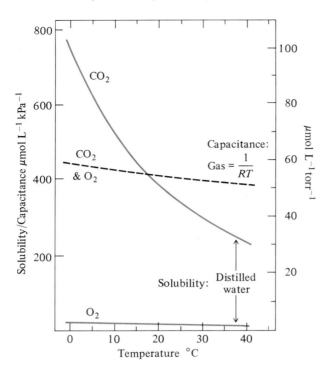

FIGURE 12–2 Relationship between ambient temperature and O_2 and CO_2 solubility/capacitance coefficient of air (gas content of air and water per kPa vapor pressure; μMoles liter^{-1} kPa^{-1}). For water, the O_2 and CO_2 content per kPa^{-1} is called the solubility (or capacitance) coefficient and the relationship between content and partial pressure is described by Henry's law. For air, the content of O_2 and CO_2 per kPa is called the capacitance and is described by the ideal gas law; it is inversely proportional to the absolute temperature. *(From Dejours 1976.)*

Diffusion

All molecules at a temperature greater than absolute zero are in constant random thermal motion and their velocity is proportional to the absolute temperature (see Chapter 2). This random motion was first described by Robert Brown in 1817 and is termed **Brownian motion**. It is more apparent in gases than liquids or solids because of the higher velocities of molecules in gases. Brownian motion is the mechanism for the ubiquitous process of diffusion. **Diffusion** is simply the movement of molecules from one region to another due to their random thermal motion. It is a process reflecting the probability of occurrence of random events.

Fick's first law of diffusion describes the net diffusional flux of a respiratory gas (i.e., O_2 or CO_2) from one region to a second in one dimension, for example from one side to the other side of a large flat membrane

$$J/A = -D\Delta C/\Delta x = -D(C_1 - C_2)/x \quad (12.5a)$$

where J/A is the O_2 or CO_2 flux per unit area of exchange; ΔC is the concentration difference; Δx is the diffusion pathlength difference; C_1 is the concentration in region 1; C_2 is the concentration in region 2; x is the distance over which diffusion occurs; and D is the **diffusion coefficient**, a physical constant. The negative sign indicates that net diffusion is from the higher concentration to the lower concentration. Commonly used units for this equation are J/A, moles cm^{-2} sec^{-1}; C, moles cm^{-3}; x, cm; and D, cm^2 sec^{-1}. If the area through which diffusion is taking place is A (cm^2), then the total flux J (moles sec^{-1}) is

$$J = -DA(C_2 - C_1)/x \quad (12.5b)$$

The value of D (cm^2 sec^{-1}) in Fick's law for respiratory gas transport depends on the nature of the diffusing gas, the solvent, and the temperature (Table 12–6). D is considerably higher for air than water, increases in approximate proportion to absolute temperature, and is approximately inversely proportional to the square root of the solute molecular weight. The diffusion coefficients for many biological materials are similar to that of water (e.g., muscle, connective tissue, lungs) but for some tissues it is substantially reduced (e.g., chitin, mucus, egg capsules and egg cases).

Krogh's diffusion coefficient (K) is another common form of D. It allows calculation of the flux from a partial pressure gradient rather than a molar concentration gradient; it has units of nanomole cm^{-1} kPa^{-1} sec^{-1}. These two types of diffusion coefficient (D and K) are readily interconvertible via the solubility or capacitance coefficient, β, since K = βD. CO_2 has a much higher rate of diffusion than oxygen in water for an equivalent partial pressure gradient because it is much more soluble in water than is oxygen, i.e., KCO$_2$ >> KO$_2$ since αCO$_2$/αO$_2$ = 251/10.6 (at 37° C). D is inversely proportional to the square root of the molecular weight, so $DCO_2/DO_2 = \sqrt{32/44}$. Hence, KCO$_2$/KO$_2$ = (23.6)·(0.85) = 20.1. Note that β differs for gases dissolved in liquids but is the same for all gases in air (441 at 20° C and 388 nmol ml^{-1} kPa^{-1} at 37° C).

So far, we have dealt with a simple example of respiratory gas diffusion in one dimension. More complex spatial arrangements are encountered in animals. The general form of the diffusion equations is similar to the 1-dimensional equation; the flux per length (l) of concentric cylinders of inner radius r_1 and outer radius r_2 is

$$J_{cyl}/l = 2\pi D(C_1 - C_2)/\ln(r_2/r_1). \quad (12.6a)$$

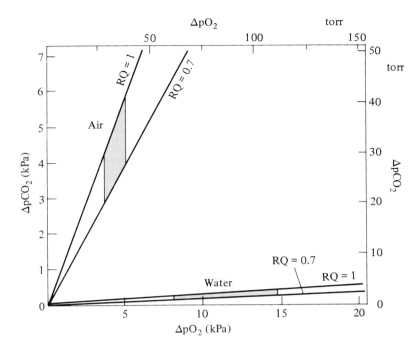

FIGURE 12–3 Relationship between the pCO_2 gradient and the pO_2 gradient for aquatic respiration (i.e., gradient = partial pressure difference between incurrent and excurrent water) and aerial respiration (i.e., gradient = partial pressure difference between inhalant and exhalant air). The normal inhalant pO_2 is 21.1 kPa (20.6 kPa for incurrent water); the normal inhalant pCO_2 is 0.03 kPa for air and water. The relationships were calculated from the ratio of O_2 solubility/CO_2 solubility (0.035) for water and the ratio of O_2/CO_2 capacitance (1.0) for air. Ambient temperature is assumed to be 20° C; respiratory quotient (RQ) = 0.7 to 1.0. The dark shaded region between lines indicates typical values.

The flux for concentric spheres is

$$J_{sph} = 4\pi D(C_1 - C_2)/((1/r_1) - (1/r_2)) \quad (12.6b)$$

(See also Supplement 3–1, page 74.)

Diffusion Limitations. We now can evaluate the effectiveness of diffusion for gas exchange in animals, having established the principles governing diffusional exchange. August Krogh demonstrated many years ago (Krogh 1941) that diffusion is generally adequate for oxygen exchange only in very small animals (<1 mm diameter). Let us examine this analysis with some modification.

The metabolic consumption of O_2 by an animal depletes O_2 from a layer of surrounding water, and a pO_2 gradient is established around the animal. This O_2-depleted layer extends at equilibrium for an infinite distance from the animal's surface. This layer of O_2-depleted water around the animal is called a **boundary layer** (see Supplement 12–1, page 603). The pO_2 gradient in this water boundary layer can be a substantial fraction of the maximum possible O_2 gradient between the ambient medium and the center of the animal (21 kPa). We can estimate that a small, spherical animal in an infinite volume of unstirred water can only attain a radius of 0.055 cm even if the boundary layer pO_2 gradient is 21 kPa (Figure 12–4A). The external boundary layer clearly exacerbates the problem of relying on diffusion for gas exchange. However, the thickness of the boundary layer can be reduced by movements of the animal through the medium or of the water past the animal. Decreasing the thickness of the boundary layer from ∞ to 10 × that of the animals radius, or 1 × or 0.1 ×, greatly increases the size that the animal can achieve. The diffusion coefficient for O_2 in air is so much higher than in water

TABLE 12–6

Diffusion coefficients (D; cm^2 sec^{-1}) for various gases in water and a variety of biological tissues. Krogh's diffusion constant (K; nmol cm^{-1} kPa^{-1} sec^{-1}) allows calculation of the flux from a partial pressure gradient rather than a concentration gradient. To convert D (cm^2 sec^{-1}) to K (nmol sec^{-1} cm^{-1} kPa^{-1}), multiply by the solubility of the gas (nmol ml^{-1} kPa^{-1}), e.g., O_2 in air, 411; O_2 in water, 13.7; CO_2 in air, 411; CO_2 in water, 387 at 20° C.

	Oxygen	Carbon Dioxide	Water
Air (0° C)	0.178	0.139	0.239
(20° C)	0.20	—	—
Water (20° C)	20 10^{-6}	18 10^{-6}	—
(37° C)	33 10^{-6}	—	—
Human lung tissue (37° C)	23 10^{-6}	—	—
Muscle (20° C)	14 10^{-6}	—	—
Salamander skin (25° C)	14 10^{-6}	—	—
Connective tissue (20° C)	12 10^{-6}	—	—
Frog egg jelly (20° C)	10.2 10^{-6}	—	—
Bream egg capsule (20° C)	6.3 10^{-6}	—	—
Dogfish egg case (15° C)	3.0 10^{-6}	—	—
Eel skin (14° C)	2.4 10^{-6}	—	—
Salmon egg capsule (5–15° C)	1.8 10^{-6}	—	—
Chitin (20° C)	0.7 10^{-6}	—	—

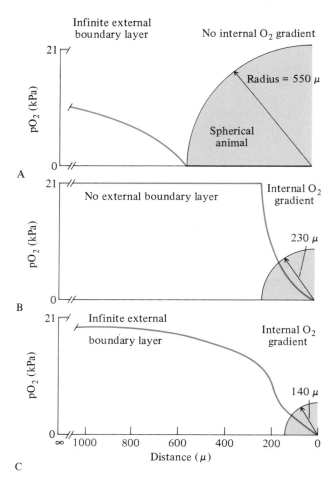

A

B

C

FIGURE 12–4 Hypothetical model of a spherical animal in an aquatic medium. The maximum radius of the animal is indicated for each example. (A) pO_2 gradient in the external medium if there was infinite mixing of O_2 within the organism and the internal $pO_2 = 0$ kPa. (B) pO_2 gradient within the animal if the medium is perfectly mixed, i.e., the pO_2 at the surface of the animal is 21.2 kPa. (C) pO_2 gradient in the external medium, if unstirred, and within the animal.

that our hypothetical animal placed in still air is equivalent to the animal in well-stirred water, i.e., the effect of the external boundary layer is minimal in air.

The problem of O_2 diffusion is also exacerbated by the establishment of an internal pO_2 gradient. The tissues of the animal consume O_2 at a rate determined by its body mass (see Chapter 4), and the local tissue pO_2 declines as the tissues consume O_2 until the overall pO_2 gradient is just sufficient to maintain the necessary rate of O_2 diffusion into the animal. If the water at the animal's surface had a pO_2 of 21 kPa (i.e., normal atmospheric pO_2), then

the animal could be about 0.023 cm in radius before the pO_2 at its center would be reduced to 0 atm by diffusion (Figure 12–4B). The PO_2 declines in an exponential fashion, rather than a linear fashion, within the animal. An internal circulation of fluid will decrease the pO_2 gradient within the animal, just as mixing the external medium reduces the boundary layer thickness. For most spherical animals, the combination of an external boundary layer and an internal diffusion gradient limits body radius to about 0.014 cm (Figure 12–4C). Animals that are long and thin, or very flat, can be very much bigger than a spherical animal weight (by weight) because their internal diffusion distances are reduced.

This shape adjustment, i.e., being very thin, is not practical for many large animals. Consequently, they have evolved mechanisms for maintaining a flow of water over their respiratory surface to minimize the external boundary layer, they have evolved circulatory systems to maximize internal mixing of body fluids, and they have evolved specialized respiratory surfaces with high-surface areas because the body surface area is inadequate for diffusional gas exchange and it is impractical anyway for large animals to devote their entire body surface to gas exchange. The respiratory surface is generally a thin, delicate epithelium (thin to maximize the O_2 diffusion from water to the blood) with blood vessels lying very near the surface. It is therefore very susceptible to physical damage and is generally well-protected against physical damage.

Bulk Flow

Large animals maintain a flow of water over their respiratory surface to minimize the boundary layer thickness and have circulatory systems to distribute the O_2 to the body tissues. Such transport of O_2 by the physical movement of water and/or blood is **bulk flow**. The transport of O_2 is due to movement of the medium, i.e., convection, rather than diffusion. The rate of convective transport (Q; e.g., ml O_2 min^{-1}) from a region of high O_2 concentration to one of low O_2 concentration depends on the rate of mass flow of water (V_w; e.g., ml water min^{-1}) and the concentration difference between the two regions.

$$Q = V_w(C_2 - C_1) \qquad (12.7)$$

A high convective transport is achieved by either a high water flow or a high O_2 concentration difference or both.

The rate of convective transport is much greater than for diffusional transport, but an expenditure of energy is required to maintain the flow of the fluid.

Diffusion, being a manifestation of random thermal motion, is "free" and doesn't require a source of energy other than the ambient temperature.

Respiratory Systems

We have seen that larger animals do not rely on the body surface for diffusional O_2 exchange but have evolved specialized respiratory surfaces, ventilatory mechanisms to pump water over the respiratory surface, and circulatory systems to distribute O_2 to the tissues. There are three general types of respiratory systems: gills, lungs, and tracheae.

Generalized Respiratory Systems

Gills and lungs are two common respiratory structures that increase the respiratory surface area.

A **gill** is an evaginated extension of the body surface that can be highly folded to increase the surface area (Figure 12–5A). Although the gill is an external structure, it is often protected by a specialized cover and is not externally visible (e.g., in many crustaceans and in fish). An internal circulatory system distributes blood through the gill and body. The external circulation of water over the gill can be accomplished by pumping movements, by the action of surface ciliated cells, or by moving the gills through the water.

A **lung**, in contrast to a gill, is an invaginated, internalized surface. Lungs are often highly folded to maximize their surface area for gas exchange (Figure 12–5B). An internal circulatory system distributes blood to the lungs and body. A ventilation mechanism is often used to move air convectively into and out of the lung. Some air-breathing animals rely on diffusion for lung gas exchange, since the diffusion coefficients for gases are much higher in air than water.

Tracheae are a third type of respiratory system, used by many air-breathing invertebrates, particularly arthropods. A tracheal system consists of a series of air-filled tubes (the tracheae) that extend into the body tissues from an opening(s) (spiracle) at the body surface (Figure 12–5C). The tracheae branch internally to form numerous small tubes (tracheoles) whose blind ends lie close to all cells of the body. An internal circulatory system is not required to distribute O_2 to the tissues (but is still usually present to distribute other nutrients, waste products, heat, etc.; see Chapter 14).

There are three generalized models (and an additional fourth specialized model) of gas exchange that correspond to these three basic types of respiratory

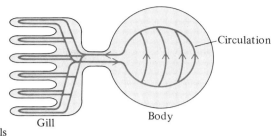

A. Gills

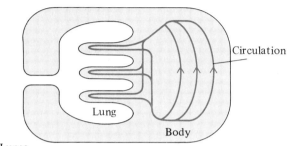

B. Lungs

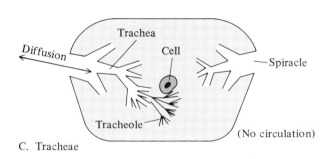

C. Tracheae

FIGURE 12–5 Generalized structures of respiratory systems. **(A)** Evaginated gills; **(B)** invaginated lungs; **(C)** tracheae.

surfaces (Piiper and Scheid 1982). For animals using their general body surface (e.g., small animals, many amphibians), there is an infinite "pool" of water (or air) that is the source of O_2; there is no inspiratory or expiratory respiratory flow of the medium (Figure 12–6A). The cutaneous blood vessels absorb O_2 by diffusion across the skin and the blood pO_2 increases from the venous value to an arterial value, which is necessarily lower than that of the ambient medium. Gills often have a flow of water over their surface that is in the opposite direction to blood flow; this is a **countercurrent flow** (Figure 12–6B). The pO_2 of blood increases along the length of the gill capillary and the arterial blood pO_2 may closely approximate the ambient water pO_2. A co-current flow (i.e., parallel flow in the same direction) is less effective in oxygenating gill

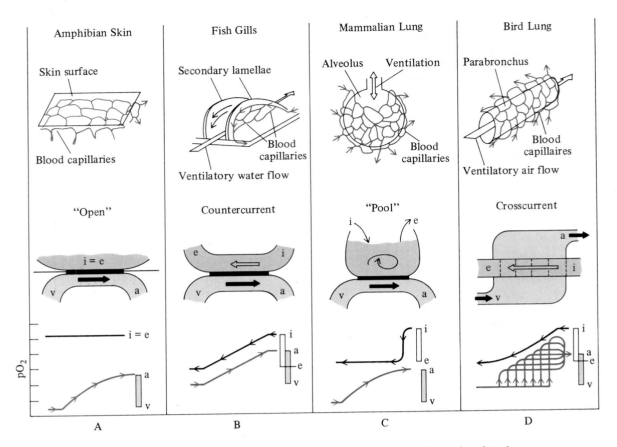

FIGURE 12–6 Models of gas exchange for various types of respiratory surfaces showing the generalized organization of the respiratory surface, the blood supply, the arrangement of the ambient fluid with respect to the vascular fluid, and the pO_2 gradients for the ambient fluid and the vascular fluid. (**A**) A cutaneous respiratory surface, such as frog skin. (**B**) A gill surface, such as that of fish. (**C**) A lung surface, such as a pulmonate mantle cavity or a mammalian alveolus. (**D**) The highly specialized respiratory surface of a bird lung. Abbreviations are as follows: i = incurrent/inhalant, e = excurrent/exhalant, a = arterial, and v = venous. *(From Piiper and Scheid 1982.)*

blood. The relative merits of countercurrent flow are further discussed in Supplement 12–2 (page 605). Gas within lungs is essentially a circulated "pool"; the O_2 of lung gas is less than the ambient pO_2 (Figure 12–6C). Bird lungs are a very specialized gas exchange system; there is a one-way flow of air through the lungs but blood flow is essentially crosscurrent (perpendicular to the air flow) rather than countercurrent or co-current. Consequently, the blood in lung capillaries at different parts on the air tubes attain different pO_2s. The arterial blood is a pooled mixture and has an average O_2 content (Figure 12–6D; see Chapter 13 for more detail of bird respiration).

Let us now examine the respiratory systems of aquatic animals. The respiratory systems of air-breathing animals will be further examined in Chapter 13.

Invertebrates

We have already discussed the diffusional possibilities and limitations for very small, spherical animals (<1 mm diameter). Some larger animals can also rely on diffusion by limiting the maximum diffusion distance from the surface to any point within their body by being elongate cylinders (e.g., trematode flatworms) or very flat (e.g., acoelous flatworms). However, most large animals have either gills or lungs; some have both.

Sponges. Sponges (Phylum Porifera) vary greatly in size, from about a millimeter to over a meter in length. The architecture of sponges is relatively simple. They have an outer layer of protective pinacoderm, a middle gelatinous protein layer called the mesohyl, an inner layer of choanocyte cells, and

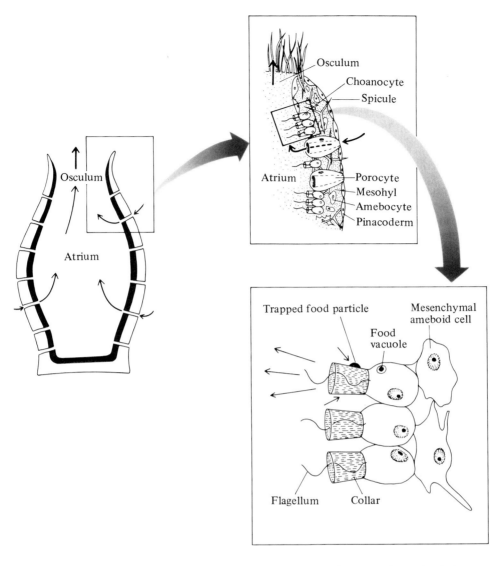

FIGURE 12–7 Representation of an asconoid sponge showing the primary body layers (choanocytes, dark area; pinacoderm and mesohyl, light area) and pattern of water circulation; detailed arrangement of the choanocytes, mesohyl, and pinacoderm; and the pattern of water flow in the incurrent pores, through the porocytes to the interior of the sponge, and out the osculum. Schematic of choanocytes with flagellum and microvillar collar. *(From Barnes 1987.)*

a skeleton of calcareous or silicous spicules. The sponge is essentially constructed around a series of water canals. The general structure of sponges is most conveniently illustrated by the primitive asconoid type (Figure 12–7).

Water flows into a sponge through side openings and out of the top opening, the osculum. The channels are formed by donut-shaped porocyte cells. Choanocytes, or neck cells, of the inner sponge layer generate the water movement and collect food. Their long flagellum draws water through the sieve-like collar of the cell and out of the collar neck (Figure 12–7). The water current provides both food and oxygen, removes waste products, and distributes sperm and eggs. Food particles are trapped as water passes through the microvillar mesh of the collar. All of the cells in the sponge obtain their oxygen from the circulated water.

Sponges also have an alternative mechanism for maintaining a water flow, in essence parasitizing the ocean currents as an energy source (Vogel 1974). Consider a sponge attached to a flat surface in a moving fluid with a velocity boundary layer (see also Supplement 12–1, page 603). The base and sides of the sponge are in a region of lower flow velocity than the osculum. **Bernoulli's principle** of conservation of energy (see Chapter 14) is highly relevant to the sponge. The higher velocity fluid flow at the osculum has a lower pressure, and so water is drawn out the osculum and in the sides of the sponge even if the choanocytes are inactive. Interestingly, a similar principle induces an air flow through the burrows of some rodents (Vogel, Ellington, and Kilgore 1973).

Worms. Flatworms (Phylum Platyhelminthes) are dorso-ventrally flattened worms whose outer surface is often the only respiratory surface. Turbellar-

ian flatworms range in size from minute animals lacking a digestive system to large (5 cm long) polyclad worms. Their general body surface functions for gas exchange, but the highly branched digestive tracts of triclad and polyclad worms are also significant gas exchange surfaces.

Many annelid worms (Phylum Annelida), being small and cylindrical in shape, can rely on diffusion across their cutaneous surface for gas exchange. This is true of most oligochetes (terrestrial worms), leeches, and many polychetes (marine worms). A few oligochete worms have true gills; these are long filamentous extensions of the body wall. Leeches (Hirudinea) generally rely on cutaneous diffusion. They can promote gas exchange by undulating their

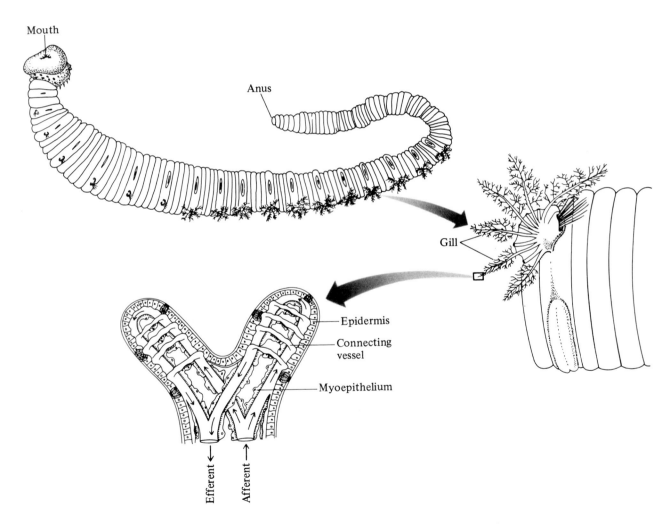

FIGURE 12–8 Respiratory structures of a burrowing annelid worm, *Arenicola*. The primary branches (PB) of the filamentous gills divide into secondary branches then tertiary branches and then quaternary branches that provide most of the respiratory surface area. The diagrammatic representation of the gill shows the pattern of blood flow through the distal gill branches. *(From Barnes 1987; Kozloff 1990; Joiun and Toulmond 1989.)*

bodies while fixed to the substrate with the posterior sucker. The fish-parasite leeches (piscidolids) have lateral extensions of the body wall that function as gills.

Marine polychete worms commonly have gills, and the considerable diversity in gill structure and location indicate their independent evolution in many groups. The gills are most often modifications of the parapodia (lateral appendages) and have a wide variety of shapes, e.g., cones, flattened lobes, branched filaments, cirriform, pectinate, or spiral branched. A constant flow of water is often maintained over the parapod gills by ciliary action. Burrowing polychetes, such as *Arenicola*, use undulatory movements of their body to maintain a flow of water through the burrow and over their filamentous gills (Figure 12–8). There is a well-defined vascular circuit of hemolymph through the gills. Cirratulid worms have long thread-like gills extending from various parts of the body wall, and terrebellid worms have branching gills extending from the anterior body segments.

Mollusks.   A generalized mollusk has a number of pairs of bipectinate gills located within its mantle cavity (Figure 12–9). Each gill has numerous **gill lamellae** that are triangular-shaped plates attached to a supporting axis. The frontal surface of the lamellae, which faces the incoming water current, contains the efferent blood vessel; the afferent

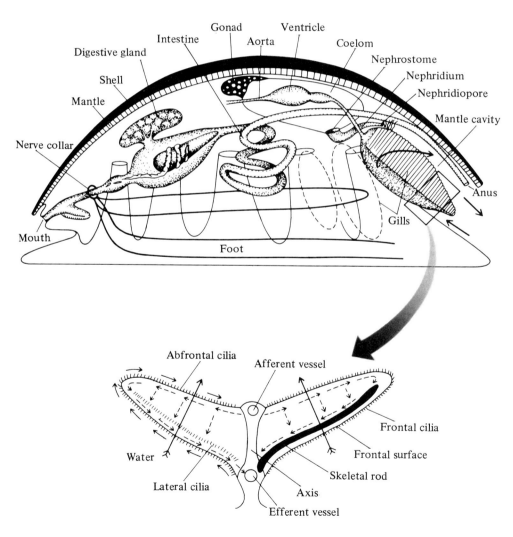

FIGURE 12–9   Generalized body plan of a primitive mollusk showing the location of the posterior gills. The expanded view of a single bipectinate gill filament shows the cilia and pattern of water movement and position of the afferent and efferent blood vessels.   *(From Barnes 1987.)*

vessel is located on the abfrontal surface. Blood flows through the gill lamellae from the afferent to the efferent blood vessels via thin, indistinct channels. Blood flow is countercurrent to the water flow. This arrangement of opposite directions for water and blood flow is highly efficient for gas exchange; it will be discussed in detail later (see Supplement 12–2, page 605, and discussion of fish gills, page 594). The gill filaments have thin chitinous rods behind the frontal edge to prevent deformation by the water flow. The water current is produced

by a band of lateral cilia, located just behind the frontal edge of the gill lamellae. The gills are primitively bipectinate (have respiratory surfaces on both sides of the supporting structure) but are unipectinate (have respiratory surfaces on only one side of the supporting structure) in many mollusks.

Primitive mollusks have five to six pairs of unipectinate gills (*Neopilina*) to as many as 26 pairs of bipectinate gills (chitons) extending from the ventral body wall into the mantle cavity. In chitons, water enters the trough-shaped mantle cavity at the

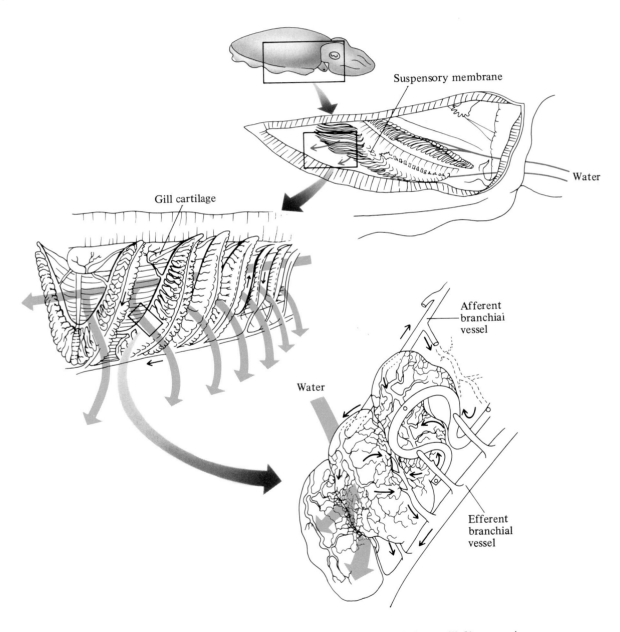

FIGURE 12–10 General structure of the gills and detailed structure of three gill filaments in a cephalopod mollusk, the cuttlefish *Sepia*. *(Modified from Wells 1983.)*

anterior of the body into a ventro-lateral inhalant chamber, passes through the gills into the dorso-medial exhalant chamber, and exits posteriorly through one or two exhalant siphons. The water flow is maintained by a band of lateral cilia on the gill filaments.

Solenogasters (Aplacophora) are unusual worm-shaped mollusks; burrowing forms have a pair of gills located in an invaginated cavity at the posterior end of the body that might represent a highly modified mantle cavity. The burrowing marine mollusks called tusk or tooth shells (Scaphopoda) lack gills and rely on diffusion across the mantle surface.

Cephalopods are active, free-swimming mollusks; their basic gill structure (Figure 12–10) resembles that of other mollusks but is modified according to their swimming habits. There is one pair (squid, octopuses) or two pairs (nautilus) of gills in the mantle cavity. The gills lack cilia, and the water current (and locomotion) is provided by contraction of the highly muscular mantle. The gill filaments have supporting chitinous rods on the abfrontal margin (not the frontal margin as in other mollusks) and better-developed channels between the afferent and efferent blood vessels to increase the efficiency of gill gas exchange. Cephalopods that swim using webbed arms (e.g., octopuses) have vestigial gills and rely on their body surface for gas exchange.

The gills of gastropods and bivalves separate the mantle cavity into inhalant and exhalant chambers. The bipectinate gills of primitive bivalves (protobranchs) resemble those of chitons and gastropods, but the gills of advanced bivalve mollusks show marked structural adaptations for filter feeding in addition to their normal respiratory function. The advanced bivalves have a lengthening and folding of the gill filaments and the attachment of adjacent filaments, to form a W shaped sheet-like gill, perforated by water channels (Figure 12–11A). Structural support of the complex lamellibranch gill is provided by ciliary tufts and tissue bridges connecting the V-shaped sections of each gill filament, between adjacent filaments, and between the filaments and the mantle or foot. In the **filibranch gill**, the individ-

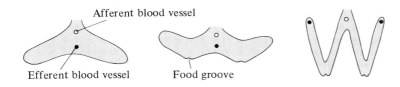

A

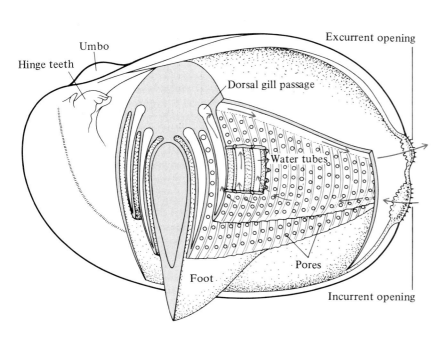

B

FIGURE 12–11 **(A)** Evolution of the W-shaped gill filament in eulamellibranch mollusks. **(B)** Location of the eulamellibranch gill within the animal, e.g., a clam, showing the pattern of water flow in the incurrent siphon through the gill pores and water tubes into the exhalant chamber and out the exhalant siphon. The eulamellibranch gill has the gill filaments fused to form an almost continuous gill sheet perforated by pores for water flow. *(From Barnes 1987; modified from Buchsbaum 1948.)*

ual gill filaments remain more or less separate, but in the most specialized **eulamellibranch gill** the adjacent filaments are fused to such a degree that they appear to be a continuous sheet with only occasional pores (ostia) for water entry (Figure 12–11B). Water passes from the inhalant chamber through the ostia into the interlamellar space, which now consists of vertically oriented water tubes; these exit into the suprabranchial chamber and exhalant chamber.

Primitive prosobranch gastropod mollusks have two bipectinate gills. Water enters the mantle cavity anteriorly, passes through the gill, and exits upward and outward. More advanced prosobranchs have reduced gills (only a left unibranchiate gill) or no gills, relying instead on a vascular mantle cavity (a lung) for gas exchange. Opisthobranch and pulmonate gastropods are probably derived from prosobranchs that had a single, left unibranchiate gill, but have a reduced gill, or no gill. Some have secondarily evolved **anal gills** (e.g., dorid nudibranchs) or **cerrata**, club-shaped or grape-like extensions of the dorsal body wall (e.g., eolid nudibranchs). Pulmonate snails lack gills; instead, they have a vascularized mantle cavity. Marine and freshwater pulmonates (e.g., limpets) have secondarily derived gills in the mantle cavity or simply rely on diffusion of gases between the body and the fluid-filled mantle cavity. Terrestrial pulmonates, in which the mantle cavity functions as a lung, will be described later (Chapter 13).

Crustaceans. Many small or filter-feeding crustaceans lack gills and rely on diffusion across their body surface for gas exchange (e.g., ostracods, copepods, barnacles). The gills of crustaceans, when present, are typically specialized portions of the thoracic appendages or less commonly the abdominal appendages. Crustaceans have appendages associated with the head (e.g., antennae, maxillae, mandibles), the thorax (**periopods**), the abdomen (**pleopods**), and the telson (**uropod**; Figure 12–12A). The generalized crustacean appendage has a basal protopodite of two segments, the basopodite and coxopodite (Figure 12–12B). There is an inner (endopodite) and outer (exopodite) branch attached to the basopodite. Various extensions of the coxopodite (called epipodites), exopodite (called exites), or endopodites (called endites) may also be present. The epipodite of the coxa in branchiopods (fairy, tadpole, and clam shrimps; water fleas) is a thin, lamellar gill (Figure 12–12C).

The Malacostraca (shrimp, crabs, lobster, crayfish) are the largest group of crustaceans and exhibit a considerable diversity in gill structure and location. The gills generally are modifications of the thoracic appendages. The thoracic gills vary in structure from flattened lamellar epipodites (in leptostracans) to four pairs of elaborate gills per appendage (primitive decapods). The opossum shrimp (Mysidacea) have branched gills extending from the thoracic coxa (Figure 12–12D) or rely on the inner surface of the carapace for gas exchange. Stomatopod crustaceans have filamentous gills on their pleopods (Figure 12–12E) and isopods also use their pleopods for gas exchange. The abdominal pleopods of isopods are variously modified to form a cover (operculum) and the lamellar gill surfaces (the area of which may be increased by filamentous projections). Gas exchange in terrestrial isopods will be discussed further in Chapter 13.

Advanced decapods have reduced numbers of gills per appendage and also a reduced number of appendages that bear gills (Figure 12–13A). The gills are of three basic forms. Filamentous (**trichobranchiate**) gills have unbranched filaments extending from the gill axis (Figure 12–13B). In **dendrobranchiate** gills, the gill axis bears two main branches of further branching filaments. **Phyllobranchiate** gills have flattened lamellae extending from the gill axis. The gill axis contains an afferent and efferent blood vessel and the filament contains channels or fine sinus networks for blood flow between the afferent and efferent vessel. The gill structure of these crustaceans is very similar to that of many fish (see below). The gill area of crabs generally is correlated with habitat (gills are reduced in number and area, and the interlamellar spacing is greater in amphibious and terrestrial crabs), oxygen demand (a high level of activity and ambient temperature require a greater gill area), and osmoregulatory requirements (some of the gills have ion pumps). The gill area of a variety of crustaceans increases in proportion to body mass$^{0.70}$ but is considerably less than that of similar-sized teleost fish (see Figure 12–23); gill area is generally about 5 to 10 cm^2 g^{-1}. The diffusion path-length for O_2 from water to blood is also an important determinant of the rate of diffusion and is generally quite low, e.g., 10 to 12 μ in *Carcinus*. However, about 1/2 of this distance is chitin, which has a low diffusion coefficient (see Table 12–6, page 571).

The gills of malacostracans are generally located and protected within the **carapace**. Water flow through the brachial cavity and over the gills is produced by the beating action of an oar-like extension of the second maxilla appendage, the gill bailer or **scaphognathite**. There is considerable variation

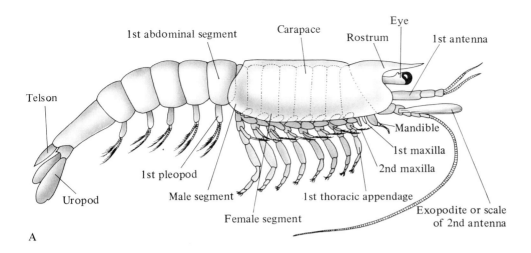

A

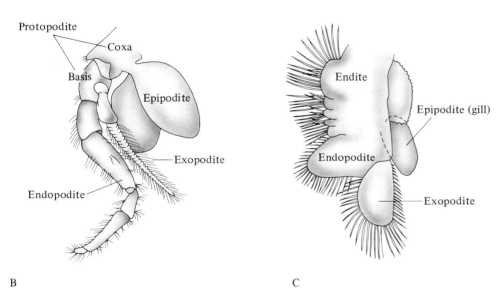

B

C

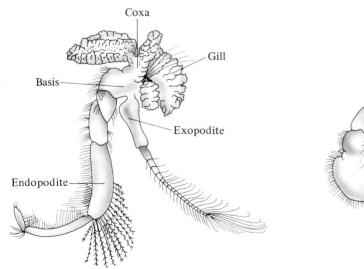

D

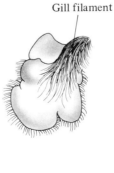

E

FIGURE 12–12 **(A)** Generalized structure of a crustacean. **(B)** Generalized structure of a crustacean appendage. **(C)** Lamellar gill on the thoracic appendages of a branchiopod crustacean. **(D)** Arborescent gill of a mysidacean crustacean's appendage. **(E)** Filamentous gill of a stomatopod crustacean. *(From Barnes 1987.)*

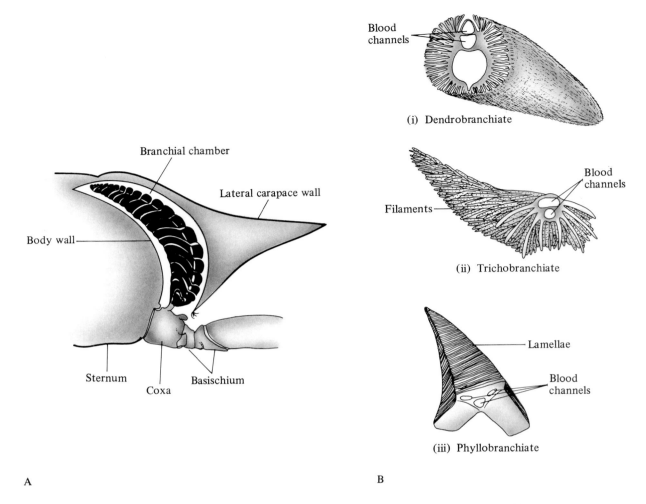

FIGURE 12–13 **(A)** Cross-section through the gill chamber of a decapod crustacean (a crab) to show the gill extending from the coxa of the limb appendage. **(B)** Types of gill filaments in decapod crustaceans: (i) dendrobranchiate gills (branched filaments); (ii) trichobranchiate gills (unbranched filaments); (iii) phyllobranchiate gills (plate-like filaments). *(From Barnes 1987 and Mill 1972.)*

in the pattern of water flow through the branchial chamber depending on how tightly the carapace fits against the sides of the body. In shrimp, the loose fit allows water to be drawn into the branchial chamber all along the posterior and ventral edges, but the points of water entry are restricted in brachyurans (e.g., crabs) to a small anterior inhalant opening (Figure 12–14). The blade surface of the gill bailer undergoes complex movements that produce a biphasic, oscillatory pressure change within the branchial chamber; this pressure is generally negative, from -0.03 to -0.44 kPa (0.3 to 4 cm H_2O). Even in terrestrial crabs that breath air, such as hermit crabs *Coenobita*, there is a similar oscillatory pressure, but it is <0.01 kPa in amplitude.

The rate of water flow through the branchial chamber (V_w; ml min^{-1}) is determined by the frequency of scaphognathite movements (f_{sc}; min^{-1}) and the stroke volume (V_{sv}; ml). The volume of water pumped per gill-bailer movement (the stroke volume) is fairly independent of the frequency, i.e., the gill bailer acts as a fixed volume pump. The stroke volume depends markedly on body mass; for six species of decapods, there is a linear relationship between stroke volume and body mass (M, kg); V_{sv} = 2.94 M − 0.076 (McMahon and Wilkens, 1983). Variation in V_w is accomplished more by variation in f_{sc} than variation in V_{sv} (e.g., *Cancer magister*, Table 12–7). The relationship between gill water flow and mass is $\log V_w = 2.529 \log M - 0.094$. The

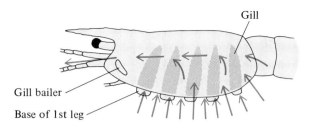

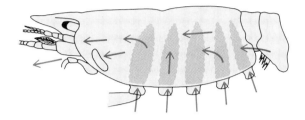

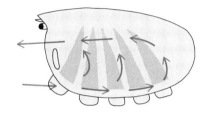

FIGURE 12–14 Pattern of water flow through the branchial chamber of various decapod crustaceans, a shrimp, a crayfish, and a crab. *(From Barnes 1987.)*

flow, associated with a positive intrabranchial pressure, probably serves to clear the gill lamellar spaces and inlet filters of foreign material. Similar reversal of flows are often observed in gills of other animals, including bivalves and fish. The resistance to water flow is greater in the reverse direction than in the normal direction of flow.

Insects. Insects generally have a tracheal system for aerial respiration (see Chapter 13), but many aquatic species or aquatic life stages (larvae, pupae) have tracheal systems modified for aquatic gas exchange (Mill 1972). Many aquatic insect larvae have **tracheal gills**, extensions of the body surface that contain tracheae, for O_2 extraction from their aquatic medium, rather than gaining O_2 through the spiracles as in conventional tracheal systems. Stonefly larvae commonly have lateral abdominal gills, and caddis fly larvae have filamentous gills. A few coleopterans and some dipterans have filamentous gills. Mayfly larvae have simple tracheal gills often consisting of pairs of simple lamellar gills on the abdominal segments (Figure 12–15). Sometimes, these have tufts of gills or elaborate fringes. Some also have gill filaments on the head or thoracic appendages. Most damselfly larvae have caudal gills, and sometimes they have abdominal gills.

Spiracular gills occur mainly in some dipteran and coleopteran pupae but are also present in some larvae. The spiracle, or surrounding cuticle, or both, form a projection of highly variable form in different species. The pupae of the fly *Taphrophila* has an anterior spiracular gill with 8 branches (Figure 12–16). Extensions of the tracheae enter these gill branches and small tubes (**aeropyles**) extend from the tracheal branches to the gill surface, where they form long, partly open surface tubes (plastron lines). These function as the air-water interface for gas exchange (see Chapter 13, plastron respiration).

resistance to water flow through the gills (R; kPa min ml^{-1}) is calculated as $\Delta P/V_w$, where ΔP is the pressure gradient for water flow. In the crab *Cancer*, the gill resistance remains fairly constant, at about 0.055 kPa min ml^{-1}, regardless of the variation in gill bailer frequency. Occasional reversal of water

TABLE 12–7

Relationship between scaphognathite (gill bailer) frequency (f_{sc}; min^{-1}), stroke volume (V_{sv}; ml), branchial pressure (Pa), and gill ventilation rate (V_w: ml min^{-1}) for the decapod crab *Cancer magister* with normal forward water flow and reversed flow. Resistance (kPa min ml^{-1}) to water flow is calculated as branchial pressure/V_w. *(From McDonald, McMahon, and Wood 1977; McMahon and Wilkens 1983.)*

	Forward			Reverse	
f_{sc}	50	100	150	100	150
V_{sv}	5.2	5.0	4.9	4.2	4.3
Branchial pressure	−1.4	−2.8	−4.1	+2.9	+5.9
V_w	260	500	740	420	640
Resistance	0.054	0.058	0.057	0.071	0.094

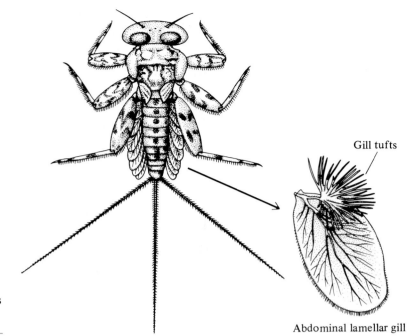

Gill tufts

Abdominal lamellar gill

FIGURE 12–15   The abdominal gills of the mayfly *Ecdyurus* are flat lamellae often with a highly branched, filamentous extension.   *(From Eaton 1885.)*

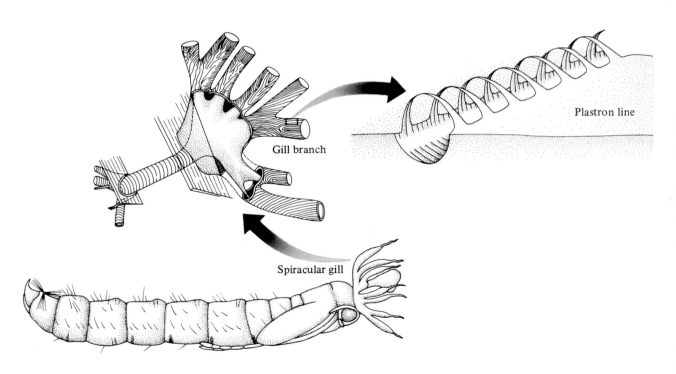

Gill branch

Plastron line

Spiracular gill

FIGURE 12–16   Spiracular gill of the fly *Taphrophila* showing the tracheal branches extending into the cuticular gill surface, the aeropyle extensions of the tracheal branches to the gill surface, and the complex plastron lines of partly open surface tubes.   *(From Hinton 1957.)*

Dragonfly larvae have complex **rectal gills** associated with six longitudinal folds of the rectal wall. The shape of the gill varies (it is a useful taxonomic character) but each contains a tracheole loop with both ends connecting to an efferent tracheole at the base. There is a tidal respiratory water current into and out of the rectum via the anus.

A few aquatic insects have **blood gills**, which are small, blood-filled sacs formed by evagination of thin sections of the integument or rectal lining. These blood gills are filled with hemolymph but are devoid of tracheae. They are probably not respiratory organs but are responsible for active uptake of Cl^- from the water.

Echinoderms. Echinoderms are a diverse group of marine, largely bottom-dwelling, and fairly familiar animals: sea stars (Asteroidea), basket and brittle stars (Ophiuroidea), sand dollars and sea urchins (Echinoidea), sea cucumbers (Holothuroidea), and sea lilies (Crinoidea). The respiratory system varies greatly within each class but often involves ciliated evaginations of the coelomic cavity that extend into the seawater.

In sea stars, for example, the dorsal surface is covered with **papulae**—small, thin projections of the coelomic wall. The outer surface is ciliated to maintain a constant flow of seawater and the inner surface is ciliated to circulate coelomic fluid. The **podia** (tube feet) are also important sites of gas exchange. These extensions of the complex water vascular system are derived from the coelom and have a ciliated epithelium. The podia also function for locomotion and prey capture.

Crinoids rely on their tube feet and the considerable area of their branching arms for gas exchange; the branching arms are used for filter feeding. Echinoids also rely on their podia for gas exchange; many of the dorsal podia are specialized for circulation of water vascular fluid, often being partitioned for a two-way blood flow and even a countercurrent exchange between blood and seawater. Many echinoids also have five **peristomial gills**, circum-oral evaginations of the body wall with a ciliated epithelium on the inner and outer surfaces (resembling sea star papulae). Coelomic fluid can be pumped into and out of these gills to promote gas exchange (Fenner 1973). Heart urchins and sand dollars lack peristomial gills. Instead, they have modified podia in the dorsal petaloid region that act as countercurrent exchangers between seawater and water vascular fluid circulated through the short, flat, sheet-like podia (Figure 12–17). The seawater current is produced by external cilia. Ophiuroids have 10 invaginations of the oral disk forming sacs or bursae. Each **bursa** opens to the exterior via a slit, which is often ciliated. Water is drawn into the slit at the peripheral end of the slit and is pumped out at the oral end, either by the action of cilia, or by a pumping action of special disk muscles, or by raising and lowering the oral wall.

Burrowing and pelagic sea cucumbers rely on their body surface for gas exchange, but other sea cucumbers have taken the internalized bursae a step further than have ophiuroids. These animals have two internal **respiratory trees**, one on each side of the digestive tract (Figure 12–18). Each respiratory tree has a main trunk and many side branches, each terminating in a vesicle. The main trunks open separately or through a common duct into the cloaca. Water is pumped into the respiratory tree by multiple contractions of the cloaca (e.g., 6 to 10 contractions in *Holothuria*), each contraction lasting a minute or more. Water is expelled by a single contraction of the respiratory tree. These respiratory trees of sea cucumbers have become a convenient daytime abode for tropical pearlfish. The fish leaves its host at night to feed, then forces its way back into the respiratory tree to shelter during the day. These respiratory trees of sea cucumbers, and the bursae of ophiuroids, are clearly "lungs" rather than gills, being invaginated structures.

Primitive Chordates. The subphylum Urochordata (tunicates) and Cephalochordata (amphioxus) more closely resemble nonchordate invertebrates than vertebrates in respiratory system structure. Gas exchange occurs over the entire body surface and the extensive pharyngeal gill slits; specialized gills are lacking.

Vertebrates

Aquatic vertebrates rely on one, or a combination of, the following surfaces for gas exchange: the cutaneous body surface, external filamentous gills, and internal lamellar gills.

A variety of fish and amphibians, and even some reptiles, rely to a variable extent on cutaneous respiration. For example, eels, some catfish, and bullheads, can obtain sufficient O_2 across their skin to maintain a normal resting metabolic rate. Some fish larvae lack gills and use cutaneous gas exchange.

The salamander *Siren* (which also has small external gills and lungs) can rely on cutaneous diffusion even at body masses approaching 3000 g due to their thin skin, the close proximity of many capillaries to the surface, and their well-developed

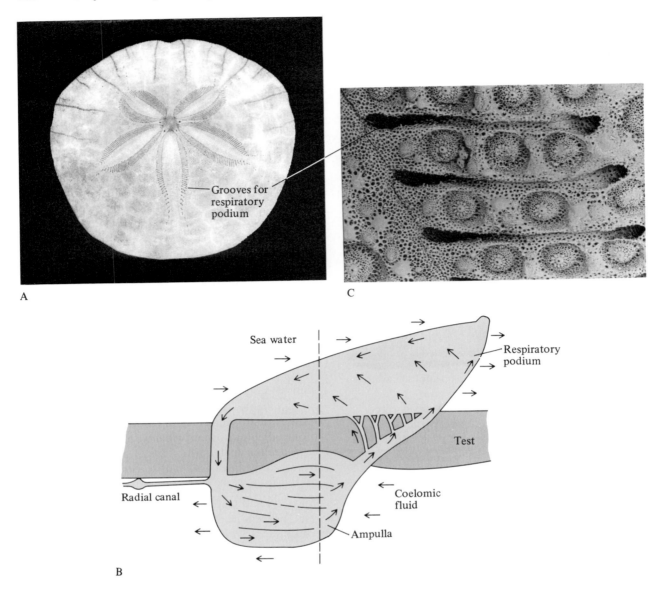

FIGURE 12–17 **(A)** Structure of the skeleton of the petaloid of a sand dollar (*Dendraster*). **(B)** A diagrammatic section of the respiratory podium (showing seawater and blood flow). **(C)** Grooves in test for respiratory podia. *(A) From Kozloff; (B) from Fenner 1973; (C) from Withers and Stewart, Department of Zoology and Electron Microscope Center of the University of Western Australia.)*

circulatory system. Their small gills are of little significance to gas exchange in comparison with their body surface area. Body movements, even if infrequent, would minimize the development of a boundary layer of low O_2 water. Nevertheless, the unfavorable ratio of surface area to body mass in large *Siren* clearly illustrates the limitations of cutaneous respiration, especially at low ambient pO_2 (Figure 12–19). Oxygen consumption rate is

independent of high ambient pO_2s, but there is a linear relationship between VO_2 at ambient pO_2s less than a critical value (P_c). *Siren* greater than about 1000 g cannot obtain sufficient O_2 via their skin to support their resting VO_2. Only *Siren* smaller than 1000 g mass can maintain their normal VO_2 in hypoxic water. For *Siren*, resting VO_2 is proportional to body mass$^{0.66}$ (unlike most animals with $VO_2 \propto Mass^{0.75}$; see Chapter 4); this relationship is

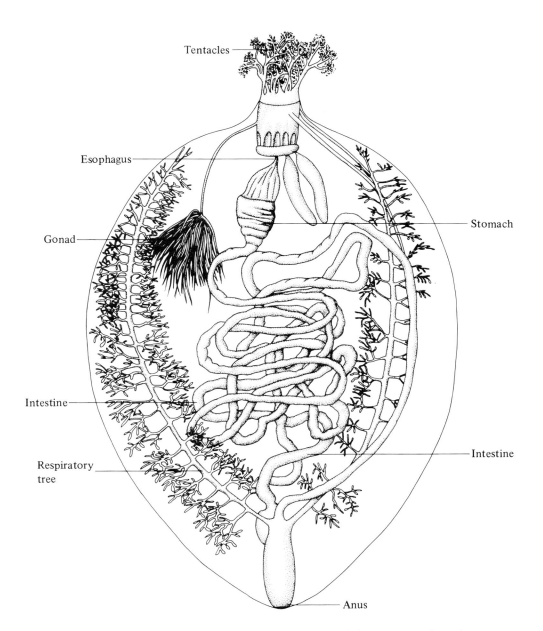

FIGURE 12–18 The respiratory tree (invaginated "lung") of the sea cucumber *Thyone* consists of two respiratory trees that open separately into the cloaca.

parallel to that between surface area and body mass (surface area $\propto$ mass$^{0.66}$; Ultsch 1974). Large *Siren* compensate for their low surface/volume ratio through pulmonary respiration and use of their gills.

A number of amphibians have extensive foldings of the skin that increase their cutaneous surface area. The Lake Titicaca frog *Telmatobius* has extensive skin folds that increase its cutaneous area for respiratory exchange (Figure 12–20; Hutchison,

Haines, and Engbretson 1976). At normal pO$_2$, the frogs can rely entirely on cutaneous gas exchange, but at low pO$_2$ they also use gas exchange across their reduced lungs.

The hellbender salamander *Cryptobranchus* also has extensive skin folds and uses peculiar rocking and swaying movements of its body to promote respiratory exchange, presumably by maintaining a layer of well-mixed water near the skin. The fre-

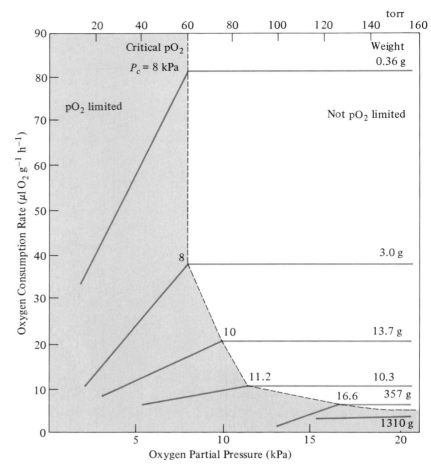

FIGURE 12–19 Effect of body mass on the relationship between resting oxygen consumption rate and ambient pO$_2$ in the salamander *Siren*, restricted to cutaneous respiration by submergence. *(Data from Ultsch 1974.)*

FIGURE 12–20 The Lake Titicaca frog *Telmatobius* has markedly folded skin to augment cutaneous gas exchange. *(Photograph courtesy of Prof. Hutchison, Department of Zoology, University of Oklahoma.)*

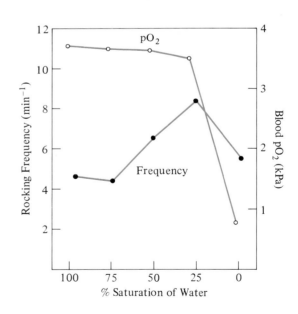

FIGURE 12–21 Relationship between frequency of body rocking movements, blood pO$_2$, and percent saturation of the ambient water for the hellbender *Cryptobranchus*. *(From Harlan and Wilkinson 1981.)*

quency of rocking movements increase at lowered ambient pO_2, and blood pO_2 is maintained almost constant (Figure 12–21). At very low ambient pO_2, the frequency of rocking movements and the blood pO_2 decline dramatically. The relative contributions of skin, gills, and lungs will be further discussed later (and see Chapter 13).

A few vertebrates have external **filamentous gills** and these are often present only during periods of increased activity and O_2 demand. Elasmobranch embryos have filamentous gills that extend from their internal gill chamber into the surrounding albuminous fluid; these filaments probably function for gas and nutrient absorption. Various catfish, sturgeons, paddlefish, and the climbing perch *Anabas* have filaments of vascular epithelium in their branchial chambers for gas exchange when breathing air. Adult male lungfish (*Lepidosiren*) supplement gill and lung O_2 uptake with filamentous gills that develop on their pelvic fins. This only occurs when the male fish is attending the nest. Many

aquatic salamanders have external gills. The larvae of a number of fish also rely upon external filamentous gills but the adults invariably have internal, lamellar gills. Male African "hairy" frogs (*Astylosternus*) develop numerous filamentous extensions of the skin during the breeding season (Figure 12–22). The highly vascular "hairs" are undoubtedly important respiratory surfaces that supplement the limited gas exchange across the reduced lungs (Noble 1925). Numerous other amphibians have highly vascular skin of similar respiratory role. Larval amphibians (tadpoles) have external gills, as do some adults that fail to metamorphose. Adult *Necturus* have well-developed external gills, which they move through the water to promote gas exchange; the frequency of movement increases at low ambient pO_2.

Most fish have some cutaneous gas exchange but rely on internal **lamellar gills** for most gas exchange. In carp, the cutaneous O_2 uptake varies from about 6% of the total at low ambient pO_2, to about 12%

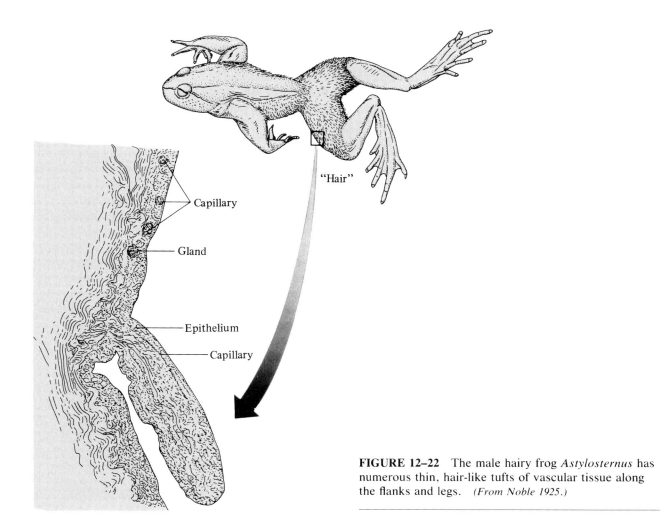

"Hair"

Capillary

Gland

Epithelium

Capillary

FIGURE 12–22 The male hairy frog *Astylosternus* has numerous thin, hair-like tufts of vascular tissue along the flanks and legs. *(From Noble 1925.)*

at normal pO_2, to over 20% at higher than normal pO_2 (Takeda 1989). In dogfish, the cutaneous O_2 and CO_2 exchange is <5% of total exchange and nearly matches the O_2 consumption and CO_2 production of the skin itself (Toulmond, Dejours, and Truchot 1982). Many freshwater and seawater teleosts also have a cutaneous gas exchange that matches the skin's metabolic activity, but some seawater fish and amphibians have a much higher cutaneous O_2 uptake than is consumed by the skin (Nonnette and Kirsch 1978). For example, flounder, sole, and eels have a considerable cutaneous contribution to gas exchange. The bullfrog tadpole *Rana*, when totally aquatic, gains about 60% of its O_2 across the skin and loses about 60% of its CO_2

across the skin (Burggren and West 1982). Older tadpoles also rely on their lungs for additional gas exchange capacity.

The cyclostomes (lamprey and hagfish) have from 6 to 14 pairs of gill pouches (marsupibranchs) that extend from the pharynx (Figure 12–23A). The gill pouches, which receive water from the outer gill openings in lampreys and from the pharynx in hagfish, open to the exterior via separate gill slits in lampreys (e.g., *Petromyzon*) but the excurrent gill ducts fuse to form a single excurrent gill slit in hagfish (e.g., *Myxine*). Epithelial ridges that cover the entire internal surface of the gill pouch are the actual respiratory surface. In hagfish, the direction of blood flow in the wall of the gill pouch is

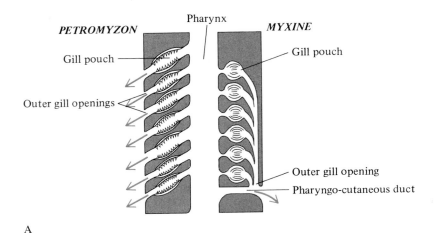

A

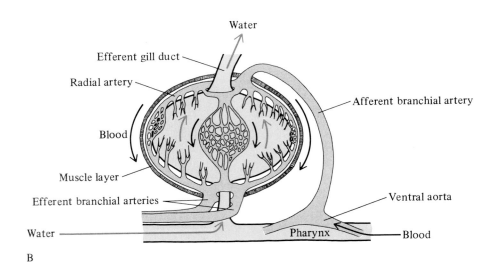

B

FIGURE 12–23 **(A)** Schematic arrangement of the gill pouches, pharynx, and external gill slits(s) in lampreys (e.g., *Petromyzon*) and hagfish (e.g., *Myxine*). **(B)** Countercurrent arrangement of water flow through the gill pouch and blood flow over the gill pouch epithelium in the hagfish *Myxine*. *(From Waterman et al. 1971.)*

countercurrent to the water flow (Figure 12–23B). Their inhalant water flow is potentially impeded during feeding when the head is embedded in the body of its prey, but the wide external naris of hagfish is connected to the roof of the pharynx and allows a continuous one-way flow of water over the gills. When lampreys feed, a retrograde flow of water into the excurrent gill ducts provides gill pouch ventilation. The muscular walls of the gill pouches constrict the gill pouch and its surrounding cartilaginous skeleton, forcing water out of the pouches. When the muscles relax, the elastic recoil of the cartilaginous support expands the gill pouch and draws fresh water in via the excurrent gill slits.

Elasmobranch and teleost fish have a considerably more elaborate gill structure than do cyclostomes. The general anatomical arrangement of gill lamellae is similar in elasmobranchs and bony fish, although the structural organization of the gill arches and gill filaments differ in detail (Figure 12–24). Elasmobranchs have an anterior spiracle opening that connects the buccal cavity to the exterior, and they typically have five pairs of gill slits (there rarely are six or seven pairs). The **gill arches**, the tissue lying between successive gill clefts that provides skeletal support for the gills, contain the afferent and efferent vasculature and support the long interbranchial septum. The interbranchial septum is

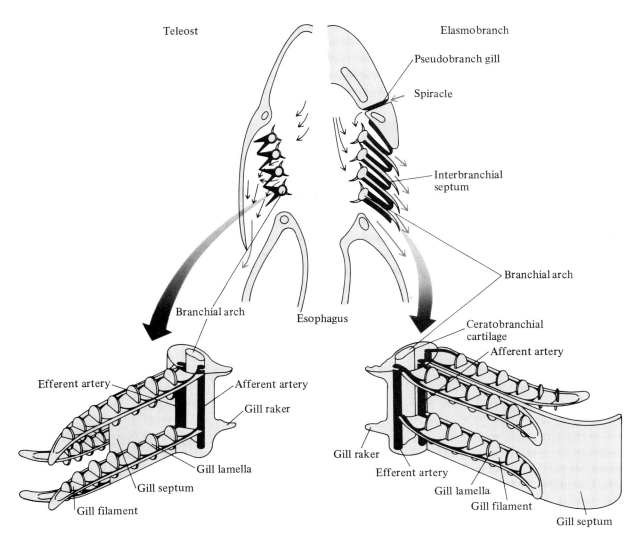

FIGURE 12–24 Arrangement of the branchial (gill) arches, interbranchial septa, and gill lamellae in an elasmobranch (right) and a teleost (left) and the detailed structures of the gill filaments. *(Modified from Waterman et al. 1971; King and Custance 1982.)*

clearly seen externally as a tissue flap. On each side of the interbranchial septum is a series of **gill filaments**, at the top and bottom of which are small, vertically oriented semicircular plates, the **secondary gill lamellae**.

The secondary gill lamellae (Figure 12–25A) are the principal site of gas exchange. The thin epithelial layers of each side of the lamella are separated by pillar cells, which have extensions that contact adjacent pillar cells and surround the capillaries (Hughes and Grimstone 1965; Figure 12–25B).

In bony fish, the pairs of gill arches (there are generally four but the number varies somewhat) provide the support and vascular supply to the gills. The gill filaments extend from each arch in the direction of water flow (Figure 12–24). The interbranchial septum is reduced to a short supporting rod, and both sides of the gill filaments are free along most of their length. The dorsal and ventral surface of each filament has vertically oriented semicircular secondary gill lamellae, which are the gas exchange surface. The complete gill structure is protected under an operculum, and thus lies between the buccal chamber and an opercular chamber. The total **gill surface area** depends on the number of gill arches, the number of gill filaments, the number of gill lamellae, and the surface area of each lamella. There is a considerable variation in total gill surface area of fish that is generally correlated with their metabolic demand. Active fish tend to have more gill filaments and more secondary lamellae per filament than do sluggish fish and air-breathing fish (Table 12–8).

Gill surface area is also highly correlated with body mass; the relationship between $\log_{10}$ surface area and $\log_{10}$ mass generally has a slope of about 0.7 to 0.8, but the intercept value varies for different taxa (Figure 12–26). A similar scaling relationship

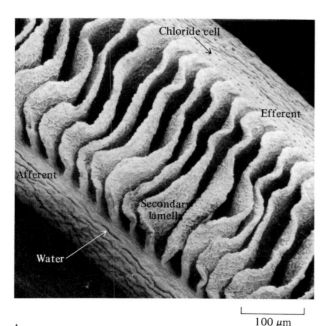

A

100 μm

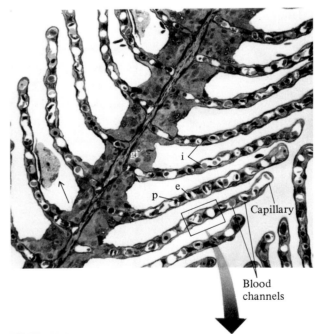

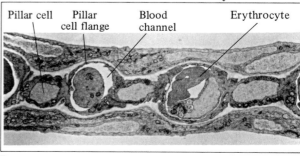

B

FIGURE 12–25 (A) The gill filament of the fish *Solea* shows the parallel plate-like stacking arrangement of the secondary lamellae on the gill filament. The chloride cells are clearly present on the efferent water (afferent blood) side of the gill lamellae. (B) Fine structure of the secondary lamellae of the gill of the codfish *Gadus*. The secondary lamellae (l) generally alternate along the gill filament (gf). Each lamella has a marginal capillary (bounded by endothelium) and medial blood channels (bounded by pillar cells, P) in which erythrocytes (e) can be seen. The arrow shows a peritrich ciliate protozoan. *(Photographs by Dunel, from Laurent 1982 and from Hughes and Grimstone 1965.)*

TABLE 12–8

Number of gill filaments, number of secondary gill lamellae (per one side), gill surface area, and water-to-blood diffusion distance in active and sluggish fish and air-breathing fish. *(From Hughes and Morgan 1973.)*

Species	Mass (g)	Total Number of Filaments	Number of Lamellae mm^{-1}	Area $(cm^2\ g^{-1})$	Diffusion Distance (μ)
Active Species					
Trachurus	26	1665	39	7.8	2.2
Lucioperca	70	1811	15	18	—
Salmo	394	1606	19	2.0	6.4
Katsuwonus	3258	6066	32	13.5	0.6
Thunnus	26600	6480	24	8.9	—
Sluggish Species					
Callionymus	39	478	16	2.1	—
Ictalurus	239	—	10	1.2	—
Opsanus	251	660	11	1.9	5
Tinca	268	1764	22	1.8	2.5
Air-Breathing Fish					
Saccobranchus	42	658	23	0.7	3.6
Anabas	54	567	21	0.6	10

is observed for the surface area of the gills of some crustaceans, and their gill surface area is generally similar to that for fish.

The **diffusion path-length** between water and lamellar blood is also significant in determining the rate of oxygen exchange, since flux is inversely proportional to diffusion path-length. Consequently, path-length is quite low in both active fish and sluggish fish (Table 12–8). Water flow over fish gills develops a laminar boundary layer at the lamellar surface (the Reynolds number, R_e, is about 100 to 300). The thickness of the boundary layer is, at most, 1/2 the distance between lamellae (i.e., about 10 μ) and is probably much lower than this. The resistance of the water boundary to O_2 exchange is probably very low compared to the diffusional resistance of the gill tissues, and the pO_2 across the water boundary layer is probably only a few tenths of a kPa (Randall and Daxboeck 1984). However, a thick layer of mucus over the gills (as occurs at low pH) may compromise gas exchange. Mucus has almost the same O_2 diffusion coefficient as water (see Table 12–6, page 571) but a 5 μ thick mucus layer markedly reduces the gas exchange across gill lamellae by adding a nonconvective boundary layer that is a significant barrier compared to the water boundary layer (Ultsch and Gros 1979).

Elasmobranchs draw water into their buccal cavity through the mouth and spiracle by negative pressure, due to muscular expansion of the buccal cavity. Muscular contraction of the buccal cavity walls and closure of the mouth/spiracle then force water over the gill filaments and out of the gill slits. Active, fast-swimming sharks have a greatly reduced spiracle and rely on water being forced into their open mouth and over their gills by their swimming movement. This pattern of forced ventilation is called **ram ventilation**.

In teleost fish, a complex coordination of expansion and contraction of the buccal and opercular chambers maintains an almost continuous one-way flow of water over the gills (Figure 12–27). Water is drawn into the buccal chamber when the buccal floor is lowered and then is forced over the gills by buccal constriction. The mouth opens and closes to create a one-way flow. The opercular pump draws water over the gills from the buccal cavity when the opercular chamber is expanded by abduction (opening) of the opercular flap. Water is expelled from the opercular chamber to the exterior when the opercular flap is open. The pressure cycles of the buccal and opercular pumps are synchronized such that water is forced over the gills by a favorable buccal-opercular pressure gradient throughout most of the respiratory cycle. The sturgeon *Acipenser* has an unusual respiratory adaptation for bottom feeding. Intake of water is by retrograde flow from the opercular opening, similar to that in lampreys.

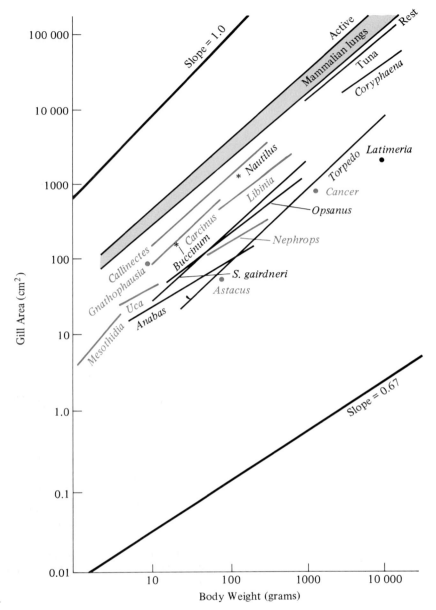

FIGURE 12–26 Relationship between gill area and body weight in a variety of fish, crustaceans (color), and mollusks (*). *(Modified from Hughes 1978; Hughes 1969; Bergmiller and Bielawski 1970; Belman and Childress 1976; Babula, Bielawski, and Sobieski 1978; Hughes 1982.)*

The retrograde flow is as effective as the normal flow in total water flow (ml min^{-1}) and O_2 extraction efficiency.

Many active fish utilize ram ventilation as an alternate means of gill ventilation. These active fish swim with their mouth open, and water is forced over the gills without buccal or opercular pumping. The energy for water flow over the gills is derived from the swimming muscles, rather than the buccal and opercular muscles. Many fish utilize buccal-opercular pumping when stationary or swimming slowly and use ram ventilation at higher swimming speeds; there is a transition from buccal-opercular to ram ventilation at intermediate speeds (Figure

12–28). Many active teleost fish (e.g., tuna) cannot maintain sufficient gill ventilation by buccal-opercular pumping and suffocate if forced to stop swimming.

The rate of water flow over the gills (V_w) is related to the body size, the metabolic demand, and the extent of O_2 extraction from the water for gill-breathing aquatic invertebrates and vertebrates (Table 12–9). A high metabolic rate requires a correspondingly high gill ventilation rate. For fish, the ratio of V_w/VO_2 is about 100 to 300; for invertebrates, the ratio is about 500 to 1000 ml H_2O ml O_2^{-1}.

Blood flow through the gill lamellae is generally in a countercurrent direction to the water flow

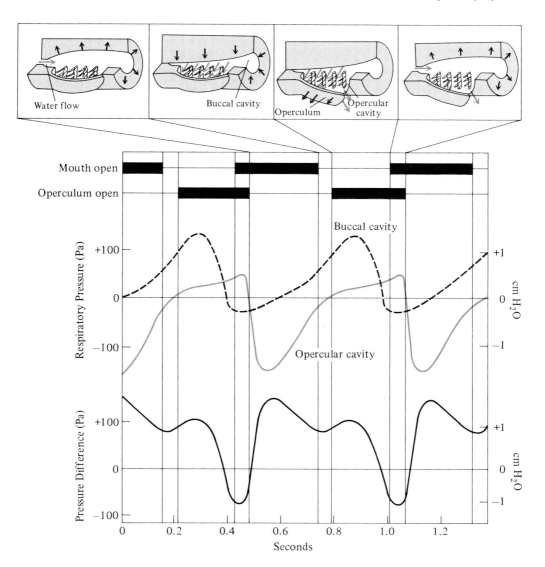

FIGURE 12–27 Buccal-opercular pumping cycles and pressure cycles in relation to opening and closing of the buccal and opercular valves. There is a net positive pressure for gill ventilation for most of the duration of the pumping cycle. *(Modified from King and Custance 1982; Hughes and Shelton 1958.)*

because this provides a high efficiency of O_2 exchange from water to blood (see Supplement 12–2, page 605), if the rates of gill water flow (V_w) and blood flow (V_b) are suitably matched (i.e., there is an optimal ventilation/perfusion ratio). The optimal ventilation/perfusion ratio is about 10 to 20 because water has a lower O_2-carrying capacity than blood (see Chapter 15). A countercurrent flow potentially enables the blood to attain almost as high a pO_2 as the incurrent water. A co-current flow (water and blood move in the same direction) results in the pO_2 of water and blood equilibrating to some common

value, thereby possibly limiting the maximal extraction of O_2 from the water.

The **O_2 extraction efficiency** (E) of gills is the percentage of the O_2 removed from the incurrent water flow, relative to the incurrent O_2 content. It is calculated as

$$E = 100(P_iO_2 - P_eO_2)/P_iO_2 \qquad (12.8)$$

where P_iO_2 is the incurrent pO_2 of the water and P_eO_2 is the excurrent pO_2. Values for fish gills are typically 20 to 60% and a similar range of values is found for invertebrate gills (Table 12–9).

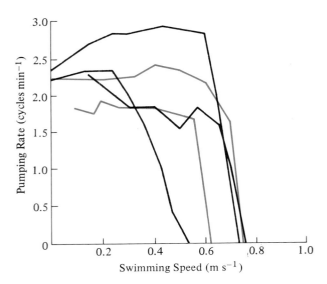

FIGURE 12–28 Transition from opercular/buccal pumping to ram ventilation at high swimming speeds in the mackerel, as determined for five individuals. Opercular pumping ceases at speeds of 0.5 to 0.8 m sec⁻¹ and the fish then rely entirely upon ram ventilation. *(From Roberts 1975.)*

The rate of water flow through the buccal cavity, gills, and opercular cavity is related to the metabolic requirements of the fish, but in a mechanical sense the V_w is determined by the resistance to water flow and the pressure gradient across the buccal cavity, gills, and opercular cavity. The gill resistance is much greater than the resistance of either the buccal or opercular chambers because of the high surface area of the lamellae and the narrow interlamellar spaces. Consequently, the V_w is primarily determined by the gill resistance. The pressure gradient across gills is typically 0.1 to 0.15 kPa but varies with ventilatory demand (see below). The resistance of the gills to water flow can be calculated from the dimensions of the gill lamellae using the Poiseuille flow equation for rectangular tubes (Hughes 1966; Hughes and Morgan 1973)

$$V_w = 2.1\Delta P d^3 b/\eta l = \Delta P/R \qquad (12.9a)$$

where V_w is the gill water flow (ml sec⁻¹), ΔP is the pressure gradient (Pa), d is the width, b is the height, l is the length (cm) of the tube, and η is the viscosity (Poise). Consequently, the resistance is equal to the following

$$R = 0.48\eta l/d^3 b \qquad (12.9b)$$

Active fish are expected to have a higher gill resistance than sluggish fish because of the closer spacing of their gill lamellae. Unfortunately, the V_w values calculated for fish are about 10 × greater than the measured V_w values, and the R values can be lower for active fish, such as tuna, compared to sluggish fish, such as the dragonet (Table 12–10). The V_w of a tuna is about 1770 ml water kg⁻¹ min⁻¹; the gill resistance is 0.101 Pa min kg ml⁻¹. Corresponding values for the less active dragonet are 300 ml H₂O kg⁻¹ min⁻¹; R is about 0.533 Pa min kg ml⁻¹. Gill resistances of crustaceans are generally similar for fish, e.g., *Cancer*, 0.484 Pa min kg ml⁻¹.

TABLE 12–9

Comparison of respiratory variables for a variety of aquatic animals, both invertebrate and vertebrate; body mass (g); oxygen consumption rate (VO₂; ml O₂ min⁻¹); gill water flow (V_w; ml H₂O min⁻¹); ratio of gill water flow to oxygen consumption rate (V_w/VO₂); pO₂ of excurrent water (P_eO_2; kPa); O₂ extraction efficiency (Eff) at varying temperature (Temp; °C).

Species	Mass	VO₂	V_w	V_w/VO₂	P_eO_2	Eff	Temp
Rana; tadpole	4.5	0.002	0.68	333	7.98	59	20
	5.3	0.0024	1.38	570	6.38	64	20
Callionymus; dragonet	100	0.108	30	278	10.64	55	12
Scyliorhinus; dogfish	200	0.134	24	179	—	40–70	12
Callinectes; crab	200	0.206	111	539	9.71	53	22
Salmo; trout	210	0.193	31	159	11.17	46	8
Entosphenus; lamprey	400	0.232	216	933	15.96	19	15
Homarus; lobster	460	0.225	224	994	15.43	23	12
Cancer; crab	900	0.504	288	514	13.57	34	8
Acipenser; sturgeon	900	0.86	379	439	13.97	30	15
Katsuwonas; tuna	2260	13.27	4000	302	14.50	56	24
Octopus; octopus	10000	2.8	2050	732	12.50	27	11

TABLE 12–10

Comparison of body weight (kg), gill ventilation (V_w; ml kg^{-1} min^{-1}), branchial pressure gradient (Pa), gill resistance (pressure gradient/ventilation; Pa min kg ml^{-1}), mechanical cost of gill ventilation (ml O$_2$ kg^{-1} min^{-1}), resting metabolic rate (ml O$_2$ kg^{-1} min^{-1}), and metabolic cost of ventilation as a percentage of resting metabolic rate. *(From McDonald et al. 1977; Hughes and Umezawa 1968; Stevens 1972.)*

	Crustacean *Cancer*	Dragonet *Callionymus*	Tuna *Euthynnus*
Mass	0.90	0.10	2.26
Gill water flow	289	300	1770
Branchial pressure gradient	140	160	178
Gill resistance	0.484	0.533	0.101
Mechanical cost of ventilation[1]	0.0018	0.0024	0.156
Resting metabolic rate	0.57	1.08	8.07
% mechanical cost of ventilation	0.36%	0.22%	0.19%
% metabolic cost of ventilation[2]	3.6%	2.2%	1.9%

[1] Assuming 1 ml O$_2$ = 20.1 joules.

[2] Assuming 10% muscle efficiency.

Not all of the water flowing through the gills is available for gas exchange. Some water flows out between the ends of the gill lamellae from adjacent gill arches (Figure 12–29). This fraction of the V_w is called the **anatomical dead space**. Not all of the water flowing between the lamellae is necessarily involved in gas exchange. A wide spacing of lamellae, exceeding 2 × the water boundary layer thickness, allows the central stream of water to pass through the gills without exchanging any O$_2$; this is called a **physiological dead space**. Closer spacing of the lamellae reduces the physiological dead space and increases the total respiratory surface area but also increases the resistance to flow. Another source of dead space is a nonoptimal ventilation/perfusion ratio (V_w/V_b). A too high V_w/V_b is equivalent to an anatomical dead space; there is too much water flow and O$_2$ is not extracted from most of it. A too low V_g/V_b is equivalent to a physiological dead space; the water flow is too low to adequately oxygenate the blood.

It is technically difficult to measure experimentally the metabolic cost of breathing for fish. A minimal estimate of the energy required to move water over the gills can be calculated as

$$\text{Work} = 10^{-6} V_w \Delta P$$
$$= 10^{-6} \Delta P^2/R \qquad (12.10)$$

where the units for work are Joules min^{-1}. The metabolic cost (in ml O$_2$ min^{-1}) can be calculated from the ventilation rate and pressure gradient since 20.1 joule $\equiv$ 1 ml O$_2$. Estimates of the mechanical cost of respiration for fish and crustaceans are typically less than 1% of the resting VO$_2$ (Table 12–10). However, these estimates of the mechanical work are only about 1/5, or even less, of the actual metabolic cost to the animal because the mechanical efficiency of muscle is only about 20%, or even less. The mechanical efficiency of ventilation may actually be quite low, from 1 to 10% (Jones and Schwartzfeld 1974). Consequently, the cost of ventilation calculated from mechanical estimates is typically 10 to 100 × higher than the mechanical costs, i.e., about 2 to 20% of resting VO$_2$. Unfortunately, attempts to measure directly the cost of breathing by indirect calorimetry yield extremely variable results, from about 10 to 40% of the resting metabolic rate, perhaps because of disturbance to the animals by the experimental protocol. The actual cost of respiration is probably fairly low, at about 5 to 15% of resting metabolism.

The increasing ventilatory demand of activity requires an additional energy expenditure for respiration. It would be most economical to increase V_w by decreasing R rather than increasing the pressure gradient driving ventilation, since $V_w = \Delta P/R$ and work $= V_w \Delta P = (\Delta P)^2/R$. However, there is a limit to how much R can be reduced without diminishing gas exchange, and so there generally is both a decrease in R and an increase in ΔP for active fish (Stevens 1972). The cost of respiration increases in both absolute terms (e.g., ml O$_2$ or J min^{-1}) and as a percent of the metabolic rate. In trout, for example, the cost of respiration (calculated from mechanical

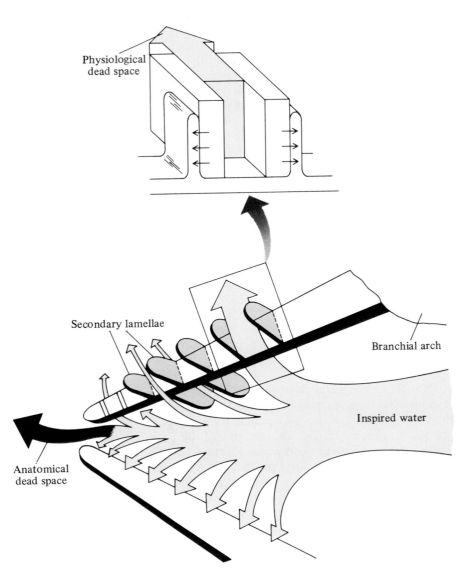

FIGURE 12–29 Anatomical and physiological dead space in fish gills. The water flow that does not encounter the gill secondary lamellae represents the anatomical dead space. The physiological dead space is that water flow between adjacent lamellae that does not undergo any gas exchange. *(Modified from Waterman et al. 1971.)*

work) increases from 0.5% at rest to 15% of VO_2 during activity when VO_2 is 5 × greater than rest (Alexander 1967).

The metabolic demands of fish generally vary considerably depending primarily on their level of activity, from the standard metabolic rate to the maximal metabolic rate, which may be 10 to 20 × higher (see Chapter 4). This increase in metabolic demand must be matched by increases in oxygen uptake across the gills. Since O_2 exchange across the gills ultimately is a diffusion process, an increase in flux must be accomplished within the confines of Fick's law of diffusion.

$$VO_2 = DA(C_2 - C_1)/x \qquad (12.11)$$

The diffusion coefficient D is a physical coefficient and cannot be altered. The diffusion area A varies

dramatically between species but little variation is possible for an individual animal in response to elevated metabolic demand, although the effective gill area might be increased during activity by minimizing anatomical and physiological dead spaces and maintaining an optimal ventilation/perfusion ratio. The diffusion path-length is an anatomical constant that cannot be altered in response to activity. The gill surface area and diffusion path-length may alter in response to chronic increased O_2 demand. For example, the lengths of external gill filaments are increased in some amphibians in response to hypoxia (Krogh 1941).

Oxygen uptake across the gills can only be increased by an increased concentration gradient or rate of water flow. The water pO_2 at the surface must be maintained as high as possible by increasing

V_w, thereby increasing the velocity of flow and reducing the boundary layer thickness and avoiding depletion of O_2 in the water. An increased gill ventilation is therefore the primary respiratory response to increased metabolic demand. Animals whose ventilation depends on flagellar or ciliary action (e.g., sponges, annelids, some mollusks) must increase their beating activity. Animals with bailing or pump systems (e.g., some mollusks, crustaceans, fish, amphibians) must increase ventilation through elevated frequency, stroke volume, or both. Crustaceans generally increase rate, and stroke volume remains relatively constant (see Table 12–7, page 583). Many fish increase both rate and stroke volume. The important role of increased blood flow through the gills will be further discussed (Chapter 15), but a high blood flow is clearly important in keeping the lamellar pO_2 low, thereby increasing O_2 uptake by diffusion.

Regulation of Respiration

Matching of respiratory ventilation with the metabolic demand requires a regulatory system. Respiratory systems generally have a group of **pacemaker neurons** that have spontaneous activity and produce the basic respiratory cycle. Respiratory regulatory systems, whether they are located in the central nervous system or are local to the ventilatory mechanism, require a sensory detector and a motor effector system. Generally, there also is some degree of intermediate neural processing. A sensory system monitoring the decline in blood or branchial pO_2 rather than pCO_2 would be most appropriate for aquatic animals because the pCO_2 increases only

slightly (due to its high solubility). The respiration of aquatic animals is, in general, more responsive to pO_2 than pCO_2; elevated pCO_2 levels often have no effect on respiration or may even inhibit respiration by direct inhibitory action on nerve or muscle cells.

The beating action of cilia and flagella is to a large extent inherent and automatic. For example, small isolated pieces of a lamellibranch gill continue to have ciliary activity for long periods. Nevertheless, there can be both neural and hormonal regulation of ciliary activity, which might contribute to respiratory regulation (see Chapter 9).

Animals whose ventilatory mechanism relies on skeletal muscles generally have a sophisticated neural center for respiratory control, especially with regard to a pacemaker that is responsible for the basic respiratory rhythm. For example, a group of neurons in the ventral nerve cord of the burrowing lugworm *Arenicola* comprise the pacemaker responsible for the respiratory rhythm. Oscillatory neurons in the subesophageal ganglion of decapods are responsible for the beating rhythm of their gill bailer. Insects have pacemakers in certain ventral nerve cord ganglia. Vertebrates have pacemakers in the medulla of their brainstem. The respiratory center of teleost fish is located in the medulla (as in all other vertebrates) but there apparently are no other extramedullary ganglia that modulate the activity of the medullary center (thus unlike higher vertebrates). The rhythmic activity of the respiratory pacemaker is well illustrated by the dogfish *Squalus*; respiratory neuron activity precedes the inward movements of the first gill flaps (Figure 12–30).

The rhythmic respiratory cycle, established by the respiratory center, must be modulated according

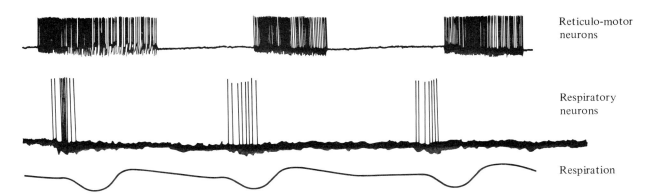

Reticulo-motor neurons

Respiratory neurons

Respiration

FIGURE 12–30 Neural regulation of gill ventilation in the elasmobranch *Squalus*. Upper trace shows activity in the reticulomotor neurons, middle trace shows rhythmic electrical activity in the respiratory neurons, and lower trace shows the respiratory cycle. *(Modified from Satchell 1968.)*

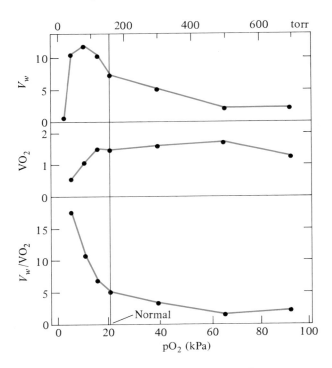

FIGURE 12–31 Relationship between respiratory ventilation (V_w; ml H_2O g^{-1} hr^{-1}) of the lugworm *Arenicola* as a function of the ambient pO_2. Also shown is the effect of ambient pO_2 on metabolic rate (VO_2; ml O_2 g^{-1} hr^{-1}), since this also influences the respiratory ventilation requirements. *(From Toulmond and Tchernigovtzeff 1984.)*

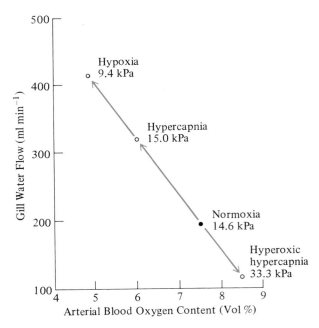

FIGURE 12–32 Relationship between gill ventilation and the arterial blood oxygen content in the trout *Salmo gairdneri*, with variation in the pO_2 and pCO_2 of the water. The numbers indicate the arterial pO_2 (kPa). *(Modified from Smith and Jones 1982; Randall and Daxboek 1984.)*

to O_2 availability and demands. For example, depth and/or rate of ventilation must increase when O_2 demand is high or the ambient water is hypoxic. The most direct means of monitoring the efficiency of respiratory exchange is by pO_2 (less commonly pCO_2) of the excurrent branchial water or the blood or tissues. Hypoxia stimulates gill ventilation of the lugworm, down to a critical pO_2 below which ventilation declines; high pO_2 inhibits ventilation (Figure 12–31). Since VO_2 also declines with hypoxia, the ratio of V_w/VO_2 is a better index than V_w of the relative stress of hypoxia on gill ventilation. The V_w/VO_2 increases markedly even at pO_2s slightly less than normal. Hyperoxia slightly reduces the V_w/VO_2 ratio.

In aquatic crustaceans, a decrease in ambient pO_2 stimulates ventilation, whereas hyperoxia ($pO_2 > 19.9$ kPa) inhibits ventilation. A similar effect of pO_2 on V_w is observed in many fish (Figure 12–32); elevation of the pCO_2 also has a slight stimulatory effect. The blood O_2 content of the trout declines with hypoxia or hypercapnia despite the increase in V_w and it increases with hyperoxia despite the decreased V_w.

The nature and location of the pO_2 (and pCO_2) receptors is not clear for many animals. In fish, the pO_2 receptors may be in the buccal cavity, gills or opercular cavity, arterial blood vessels, venous vessels, or brain. In rainbow trout, the O_2 receptors for ventilatory regulation appear to be in the brain and aorta, but O_2 receptors that influence heart rate are located in the branchial cavity.

Many stimuli other than pO_2 and pCO_2 influence the respiratory rhythm, e.g., stretch of the branchial cavity, osmotic stress, mechanical stress, chemical stimuli, and alarm. Temperature has two important influences on respiration. An increase in temperature lowers the oxygen content of the water and at the same time increases the metabolic rate because of the higher body temperature; both effects increase the respiratory demand. This important role of temperature is clearly observed in carp exposed to differing water temperatures; V_w increases when the water temperature is elevated, even if deep body temperature is constant. The change in ventilation is less when the O_2 content of the water is kept constant (by reducing the pO_2 at lower temperatures; Crawshaw 1976). The temperature of the anterior brainstem also plays an important role in respiratory regulation. Increasing the brainstem temperature (but not the temperature of the rest of the fish

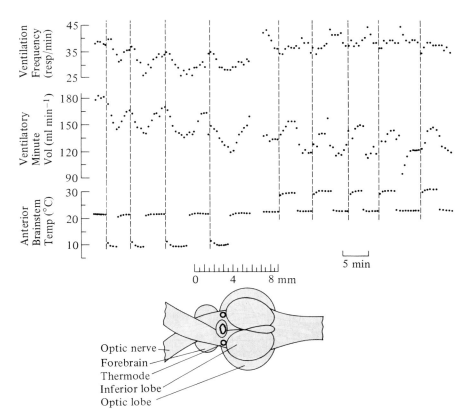

FIGURE 12–33 Effects of temperature change of the anterior brainstem in the scorpion fish *Scorpaena guttata* on gill ventilation. *(From Crawshaw, Hammel, and Garey 1973.)*

or the water) dramatically elevates the ventilation (Figure 12–33).

Physiological Implications of Aquatic Respiration

The physical properties of water, as the medium for aquatic animals, have considerable significance to the design of aquatic respiratory systems and implications for the physiology of aquatic animals (Dejours 1976). A comparison of the relevant physical properties of water and air (Table 12–11) suggests the following important consequences of water as the respiratory medium.

It is anatomically possible, mechanically practical, and physiologically efficient to maintain a one-way flow of water over aquatic respiratory surfaces because of the high density and viscosity of water. Most gill systems have one-way water flow with separate inhalant and exhalant pathways. The mechanical work required to decelerate then reaccelerate water flow in a tidal system (like sea cucumbers and feeding lampreys) would be much greater than for air flow.

A one-way flow of water can be coupled to a countercurrent flow of blood through the gill lamel-

lae. This greatly enhances the efficiency of gas exchange to typically 30 to 60% O_2 extraction (e.g., crustacean, fish, and amphibian gills).

Respiratory regulation generally is a response to altered pO_2 rather than pCO_2 because of the much higher solubility of CO_2 compared to O_2. The low blood pCO_2 of aquatic animals also has important implications for acid base balance (see Chapter 15).

Aquatic respiration has a high metabolic cost because water is a dense, viscous fluid. Estimates of the mechanical cost of respiration are generally <1% of the resting respiration rate but up to 50%

TABLE 12–11

Comparison of some physical properties of water and air. *(From Dejours 1981.)*

	Water	Air
Density (kg liter^{-1})	0.999	0.00123
Viscosity (centiPoise)	1.14	0.018
O_2 diffusion coefficient (cm^2 sec^{-1})	0.000025	0.198
O_2 content (ml liter^{-1})	1–6	210
Specific heat (J liter^{-1} °C^{-1})	4184	1.234
Thermal conductivity (mJ cm^{-1} sec^{-1} °C^{-1})	6025	252

for the metabolic cost of respiration (estimated from measurement of VO_2). The metabolic cost of respiration is probably from 5 to 10% of the resting metabolic rate for resting fish. The relative cost of respiration increases during activity.

Gills must have a relatively high rate of water flow over their surfaces to minimize the boundary layer thickness. The optimal ventilation/perfusion ratio of water flow to blood flow is about 10 to 20 because of the low O_2 content of water compared to air.

The low resting VO_2 and maximal VO_2 of aquatic animals (see also Chapter 4) are principally consequences of the low O_2 content of water and the high cost of ventilation. An aquatic animal must move about 1 liter of water, weighing 1000 g, to obtain 1 ml O_2, whereas air-breathing animals need only transport about 25 ml air (weighing 25 mg) per ml O_2 (if each animal extracts 20% of the oxygen present in the air or water).

Water has a high heat capacity compared to air. The metabolic heat production of aquatic and air-breathing animals is equivalent (about 20.1 J per ml O_2). The metabolic heat production from aerobic metabolism of 1 ml O_2 is potentially dissipated to 1000 g of water for aquatic animals, producing an elevation in temperature of about 0.005° C. Consequently, it is not surprising that aquatic animals have difficulty thermoregulating at a body temperature different from water temperature. For an air-breathing animal, the metabolic heat, if dissipated only into the ventilated air, would increase its temperature by 800° C! Terrestrial animals often thermoregulate with comparative ease.

Summary

Most animals use aerobic metabolism for ATP synthesis, and consequently they must obtain O_2 from their external environment and must excrete metabolic CO_2. The normal ambient air contains 20.95% O_2 and 0.03% CO_2; the balance is mainly nitrogen. The partial pressures of normal atmospheric air (pressure = 101 kPa) are O_2, 21.2 kPa; CO_2, 0.03 kPa; and N_2, 79.8 kPa.

The normal composition of atmospheric air is modified by altitude (same percent composition but lower partial pressures) and for an air space at depth under water (same percent composition but higher partial pressures). Some special terrestrial environments have air with a different percent composition to normal air, e.g., soil air, animal burrows, and marsupial pouches.

O_2 uptake and CO_2 excretion occur by diffusion across a respiratory surface. Diffusion across the general body surface is sufficient for many, particularly small, animals. The thickness of the external O_2 boundary layer and internal pO_2 gradients limit the size of animals that rely on diffusion across the body surface to less than about 0.1 cm diameter, unless they are elongate cylinders or very flat. Large animals have an unfavorable body surface/volume ratio and have specialized respiratory surfaces for gas exchange, e.g., gills, lungs, or tracheae. Most aquatic animals have gills.

Sponges and most small, worm-like animals rely on gas exchange across their body surface or digestive surface for O_2 and CO_2 exchange. Most annelids also rely on cutaneous gas exchange, although some oligochetes and many polychetes have gills; some leeches have lateral extensions of the body wall that act as a respiratory surface. Mollusks have a variable number of gills. Each gill has a large number of triangular gill lamellae, often with a countercurrent arrangement of water flow across the gill and blood flow through the lamella. The gills of bivalve mollusks are highly specialized for filter feeding. Many prosobranch mollusks and pulmonate mollusks lack a gill but use the vascular lining of the mantle cavity for gas exchange.

Small crustaceans use their body surface for gas exchange, but larger species use various modified appendages, often of the thoracic appendages, as gills. These gills may be flat, plate-like structures or elaborate structures with numerous plate-like or filamentous gill lamellae. The advanced decapod crustaceans have a reduced number of highly elaborate gills that have either filamentous, dendritic, or phyllobranch lamellae. The gills are protected by the carapace, and a constant flow of water is maintained over the gills by the beating action of scaphognathites (gill bailers). Variation in water flow is accomplished mainly by adjusting the frequency of scaphognathite beating, rather than by adjusting their stroke volume.

Most insects are terrestrial and use a tracheal respiratory system. However, many aquatic insects (especially larvae and pupae) have tracheal gills, spiracular gills, or rectal gills for aquatic respiration.

The respiratory surfaces of echinoderms are generally evaginations of the body wall that extend into the seawater, e.g., papulae of sea stars, tube feet, peristomial gills, or invaginations, e.g., bursa. Some sea cucumbers have an internal respiratory tree that functions as a tidal, water-filled lung.

The primitive chordates, such as cephalochordates and urochordates, rely on their body surface and filter-feeding pharynx for gas exchange. Aquatic vertebrates use either the body surface, filamentous gills, or lamellar gills for gas exchange. A number of fish and amphibians have significant gas exchange across their skin and do not require gills, at least for resting gas exchange. Some amphibians have extensive folding of the skin to increase the respiratory surface area. Some vertebrates have external, filamentous gills in the branchial cavity or on their fins (e.g., some elasmobranch embryos, some fish, male lungfish) and the male hairy frog has cutaneous filamentous gills on the lateral body and legs.

Cyclostome fish have internal gill pouches that extend from the pharynx; water flows through the gill pouches in a countercurrent fashion to the blood flow through the vascular pouch wall.

Elasmobranch and teleost fish have lamellar gills consisting of a variable number of gill arches that support numerous gill filaments, each of which is covered with numerous plate-like lamellae. There is a countercurrent flow of water over the gills and of blood through the lamellae. The flow of water through the gills is maintained by either a combination of buccal and opercular pumping or by ram perfusion of the gills due to the forward swimming movements of the fish. The rate of gill water flow is related to the body size, metabolic requirements, and O_2 extraction efficiency of the gills. For fish the ratio of gill water flow (ml min^{-1}) to oxygen consumption (ml min^{-1}) is about 100 to 300; it is about 500 to 1000 for crustaceans. The resistance of water flow through the gills determines the flow rate and pressure difference required to generate the flow; fish and crustacean gills have a similar resistance. The mechanical cost of respiration is quite low (<1% of resting metabolic rate) but the metabolic cost of respiration is a considerable fraction of the resting metabolism (probably 5 to 15%) because of the mechanical inefficiency of respiration.

Regulation of respiratory exchange matches the respiratory exchange capacity with the metabolic demand. These regulatory systems may be located in the central nervous system or may be peripheral control systems. A sensory receptor monitors the pO_2 of either the excurrent branchial water flow or the arterial blood; the pCO_2 of aquatic animals is generally very low because of the high solubility of CO_2, and so respiratory regulation is generally not controlled by CO_2. Many invertebrates and most vertebrates have a central nervous generation of the respiratory cycle and sophisticated processing of sensory pO_2 information for regulation of the respiratory rhythm. Hypoxia stimulates ventilation and hyperoxia inhibits respiration.

The physical properties of water determine many aspects of the design and functioning of aquatic gas exchange systems. Water is dense and viscous, with a low O_2 availability. The water flow over a gill is generally unidirectional to avoid the energy cost of reversing the water flow. This one-way flow of water makes it mechanically practical to have a countercurrent flow of blood, and this can greatly increase the physiological efficiency of gas exchange. The flow of water over the respiratory surface should be about 20 times more than the blood flow through the respiratory surface. There is a high metabolic cost to aquatic respiration because water is a dense, viscous fluid. The low O_2 content of water (relative to air) and high cost of ventilation generally limits aquatic animals to having a low-metabolic rate. The high heat capacity of water generally precludes thermoregulation by aquatic animals because the metabolic heat is rapidly dissipated into the water ventilated through the gills.

Supplement 12–1

Boundary Layers

. . . .

Oxygen diffuses across a respiratory surface from the adjacent water into the body fluids. This causes a local depletion of O_2 in the water immediately adjacent to the respiratory surface. As O_2 is removed from the water immediately adjacent to the respiratory surface, O_2 diffuses from the surrounding water. Eventually, a layer of O_2-depleted water extends from the respiratory surface into the adjacent water. This layer of water is called a **boundary layer**. If the water is not stirred, and diffusion is the only means of O_2 (and water) movement, then the boundary layer of O_2-depleted water becomes very thick. The boundary layer, once steady-state exchange is established, is infinitely thick. However, the conventional definition of the thickness of a boundary layer is the

distance from the respiratory surface to 99% of the overall concentration gradient.

However, water is seldom unstirred and animals are seldom motionless. What happens to the boundary layer thickness in stirred water? It is perhaps conceptually easiest to first consider boundary layers defined by the fluid **velocity** around an object rather than pO_2 boundary layers. Imagine a flat plate submerged in a moving fluid, with velocity V_∞ (the ∞ does not imply an infinite velocity but the velocity at an infinite distance from the object in undisturbed fluid). Fluid immediately in contact with the object must be stationary ($V = 0$) because the molecules next to the plate are stationary; this is the "no-slip" condition. There must clearly be some velocity gradient

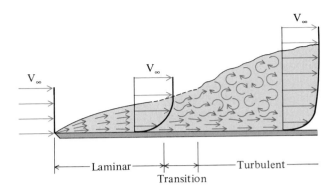

Representation of laminar and turbulent regions of the boundary layer.

in the fluid extending away from the object until V equals V_x; this is shown at two points along the plate. Note that the velocity gradient is greatest near the surface and is lower further away from the plate. Again, the boundary layer is defined as the distance from the plate to where V is 99% of V_x. The thickness of the boundary layer increases along the surface of the plate from the leading (front) edge. The thickness of the boundary layer (δ) depends on many variables, including the ambient fluid velocity (V_x), the distance from the leading edge (x), and the fluid properties (density, ρ, and viscosity, η);

$$\delta = \frac{4.9x}{\sqrt{R_e}}$$

The Reynolds number, R_e, is a dimensionless quantity calculated as $R_e = V_x x/(\rho/\eta)$, where ρ is the density, η is the viscosity, and ρ/η is the kinematic viscosity (v; 1.5 10^{-5} m^2 sec^{-1} for air and 1.0 10^{-6} for water at 20° C; see Chapter 10). R_e is essentially the ratio of inertial forces (V_x) to viscous forces (ρ/η). Its value reflects whether flow is laminar and streamlined (low R_e) or turbulent (high R_e). For a flat plate in air, the boundary layer is initially laminar at the leading edge but becomes turbulent at the local R_e of about 3 10^5, i.e., 4.5 m from the leading edge where d is 0.04 m if V_x is 1 m sec^{-1}. We will further

encounter R_e when we examine circulation, and we have already seen its significance in animal locomotion.

Velocity boundary layers, defined as the region where V varies from 0 to 0.99 V_x, are fairly familiar and visible in everyday life. For example, the wind velocity is greater far above the ground than near the ground. Rain drops on a car hood are within the thick boundary layer at low speeds, but the boundary layer thickness diminishes at higher speeds and the drops are eventually exposed to higher air speeds and blown off the hood. The rain drops are first affected at the front of the hood where the boundary layer is thinnest.

We have examined above the formation of a velocity boundary layer. There can be other types of boundary layers. For example, a thermal boundary layer forms around an object if there is a temperature difference between the object and the fluid. An O_2-concentration boundary layer forms around an animal in water. A relative humidity boundary layer forms around a moist-skinned animal in dry air. The same basic considerations determine the thickness of the boundary layer as were described above for a velocity boundary layer, but it is important to appreciate that the thicknesses of different types of boundary layers are not necessarily the same or even of similar magnitude. (*See Incropera and Dewitt 1981.*)

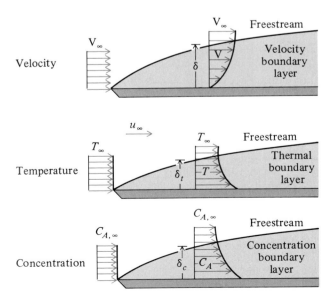

Comparison of boundary layers for velocity, temperature, and concentration gradients.

Supplement 12–2

Countercurrent Exchange

. . . .

There is an exchange of material (e.g., O_2 or CO_2) or heat between two streams of flowing fluid (e.g., water and blood in a gill) if they are separated by a permeable membrane (e.g., the gill lamellar epithelium and capillary endothelium) and if there is a diffusion gradient (e.g., a pO_2, pCO_2, or temperature difference). It would seem better for exchange if the fluid streams were flowing in opposite directions, i.e., **countercurrent** rather than in the same direction, i.e., **co-current**. This general principle of countercurrent exchange between two fluid flows is well established in biology and engineering. There are many biological examples of countercurrent material and heat exchange. For example, many annelids, mollusks, crustaceans, echinoderms, and fish gills have a countercurrent flow of water and blood. Fish larvae that lack gills may have a countercurrent flow of water over their vascularized pectoral fins. Many endothermic animals have countercurrent exchange of heat between blood vessels.

The gradient driving diffusional exchange of O_2 (or CO_2 or heat) in a gill depends on the difference in the O_2 concentration of the incurrent water (p_iO_2) and the incurrent venous blood (pVO_2). For co-current exchange, this gradient is maximal when the blood and water first contact across the O_2-permeable lamellar tissue barrier and the magnitude of the O_2 gradient declines along the vessels. Water leaves the gill with a lower pO_2 than when it entered (p_eO_2) and the arterial blood leaves with a higher

pO_2 (pAO_2) than the venous blood. The extent of O_2 exchange, the decline in water pO_2, and the increase in blood pO_2 depends on the flow rates of water and blood, the O_2-carrying capacity of water and blood (blood has a considerably higher O_2 capacity than water), and the area and permeability of the tissue barrier. If there is substantial O_2 exchange, then the water and blood pO_2 approach equality at the end of the gill lamella, i.e., $pAO_2 \approx p_eO_2$. Countercurrent flow would seem to be more favorable for O_2 exchange than co-current flow. The pO_2 gradient between water and blood remains fairly constant along the lamella, and the pAO_2 can (with appropriate conditions) approach the incurrent water pO_2, i.e., $pAO_2 \approx p_iO_2$. Countercurrent exchange thus appears to be more advantageous than co-current flow because of the potentially high extraction of O_2 from the incurrent water stream. The O_2 extraction efficiency of countercurrent gills generally is quite high. For some invertebrate and fish gills, O_2 extraction efficiency (calculated as $100(p_iO_2 - p_eO_2)/p_iO_2$) is often >50%, and can approach 80%.

The obvious advantages of countercurrent exchange to O_2 extraction efficiency (in the figure below) are lessened, however, if we consider the absolute rate of O_2 exchange. A simple model of countercurrent exchange predicts that a low volume of flow of water and blood in a countercurrent direction yields a high O_2 extraction efficiency (53% in the following example), but there is only about 33% of the theoretical maximal O_2 exchange

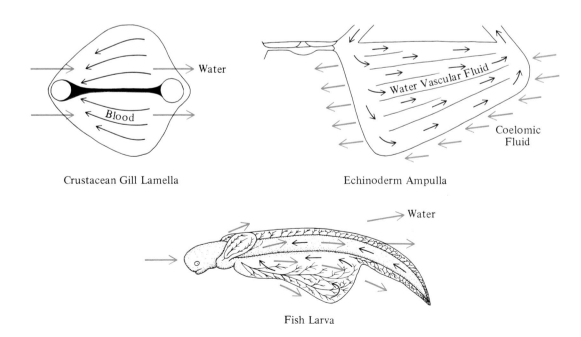

Crustacean Gill Lamella   Echinoderm Ampulla

Fish Larva

Three examples of countercurrent water and blood flow for respiratory gas exchange.

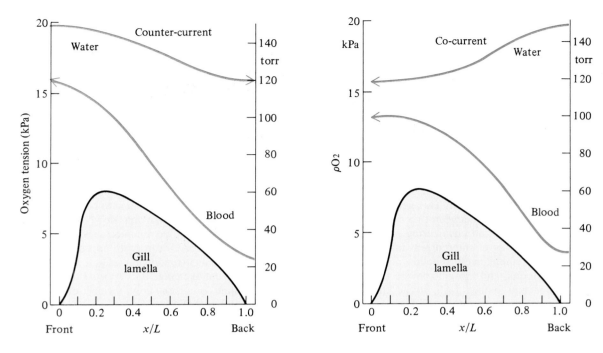

Comparison of the pO$_2$ for blood and water during lamellar gas exchange for countercurrent and co-current flow of blood and water.

(M/K = 0.333, where M is the O$_2$ exchange and K is the maximum theoretical O$_2$ exchange if the water and blood flows were infinite). For co-current exchange at a similar, low-flow rate, the O$_2$ extraction is lower (about 39%) and the O$_2$ exchange rate is also lower (M/K = 0.245). There is a clear efficiency and absolute exchange advantage to countercurrent exchange. If the volume flow is increased, then the efficiency of the countercurrent exchange system is reduced (to about 13% in the example above) but the rate of exchange is increased (M/K = 0.833). A high-volume flow in a co-current pattern also has a low efficiency (about 13%) and a rate of exchange (M/K =

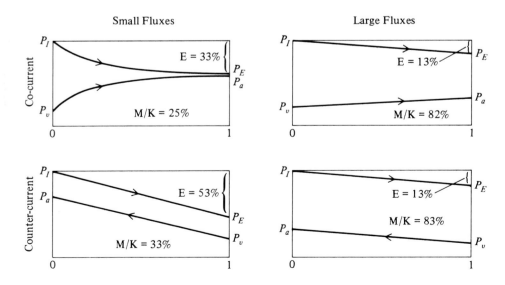

Comparison of the efficiency (E) and fraction of gas exchange relative to the theoretical maximum (M/K) for countercurrent and co-current exchange.

0.824) almost as high as for countercurrent exchange. The apparent advantage of countercurrent exchange seems less at high-flow rates. Calculations for fish gills suggest that M/K $\approx$ 0.50 to 0.60 and O_2 extraction $\approx$ 10 to 60%. The ratio of countercurrent to co-current exchange ($M_{counter}/M_{co}$) $\approx$ 1.05 to 1.15.

	O_2 Extraction	M/K	$M_{counter}/M_{co}$
Dogfish			
Rest	64%	0.45	1.17
Active	38%	0.63	1.07
Trout			
Rest	10%	0.63	1.03
Active	10%	0.66	1.03

The seemingly small M/K advantage of countercurrent exchange over co-current exchange gains greater importance if we consider realistic *in vivo* conditions. If we assume that a fish must maintain a constant VO_2 whether it has counter- or co-current exchange, and that it maintains a constant arterial pO_2, then it can be shown that the gill ventilation flow rate for co-current flow is higher than for countercurrent flow. About 2.25 times as much energy is required to maintain this higher co-current flow than to maintain countercurrent flow, i.e., there is a substantial energetic advantage to countercurrent flow, even if the exchange seems only minimally more effective. (*See Scheid and Piiper 1976; Piiper and Scheid 1982; Layton 1987.*)

Recommended Reading

Bouverot, P. 1975. Adaptation to altitude—hypoxia in vertebrates. *Zoophysiology* 16. Berlin: Springer-Verlag.

Dejours, P. 1981. *Principles of comparative respiratory physiology.* Amsterdam: Elsevier/North-Holland Biomedical Press.

Hoar, W. S., and D. J. Randall. 1984. Gills. In Anatomy, gas transfer, and acid-base regulation. *Fish Physiology*, vol. X, part A. New York: Academic Press.

Houlihan, D. F., J. C. Rankin, and T. J. Shuttleworth. 1982. *Gills.* Cambridge: Cambridge University Press.

Hughes, G. M. 1974. *Comparative physiology of vertebrate respiration.* London: Heinemann.

Jones, J. D. 1972. *Comparative physiology of respiration.* London: Edward Arnold.

Krogh, A. 1941. *The comparative physiology of respiratory mechanisms.* Philadelphia: University of Pennsylvania Press.

Piiper, J., and P. Scheid. 1982. Models for a comparative functional analysis of gas exchange organs in vertebrates. *J. Appl. Physiol.* 53:1321–29.

Chapter 13

Aerial Respiration

• •

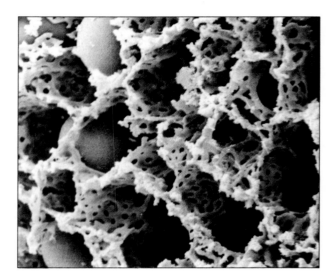

Scanning electron micrograph of a vascular cast of a rat lung shows the structure of alveoli. *(Photograph courtesy of J. O'Shea, Department of Zoology, The University of Western Australia.)*

The extremely important differences in the physical properties of air and water (see Table 12–11, page 601) significantly alter the principles of gas exchange in air compared to water. **Aerial respiration** confers a greatly facilitated rate of gas exchange at a lower metabolic cost compared to aquatic respiration for the various reasons discussed in Chapter 12. However, an important consequence of aerial respiration is the potential loss of water by evaporation from the respiratory surfaces. This new avenue of water loss for terrestrial animals generally has detrimental consequences but can be an advantageous source of evaporative heat loss under conditions of heat stress (see Chapter 5).

The **transition** from water to air breathing has occurred independently in a number of animal groups. Three animal taxa that have successfully made the transition from water to land and are found in arid environments are the higher vertebrates (reptiles, birds, mammals), some gastropod mollusks, and many arthropods (spiders, scorpions, insects). A number of other animals, both vertebrate (fish, amphibians) and invertebrate (annelids, crustaceans), have less successfully invaded the land, and terrestrial species are restricted to moist environments. Let us now examine separately the transition from water to land in vertebrates and then in invertebrates.

Vertebrates

The first vertebrates undoubtedly had gills (or a pharyngeal basket) for aquatic respiration. The most primitive extant vertebrates have either marsupibranch gills (e.g., cyclostomes) or lamellar gills (e.g., elasmobranchs, teleosts). The vertebrate transition to land has sometimes involved the modification of these lamellar gills, but more commonly it has involved the evolution of a new respiratory structure more suited to aerial respiration.

The Vertebrate Transition from Water to Land

A number of different kinds of fish have made the transition from water to air breathing, the most significant being the lungfish (Dipnoi). These are the surviving representatives of a group of lunged vertebrates that were ancestral to the so-called higher vertebrates, i.e., amphibians, reptiles, birds, and mammals. Let us first, however, examine the diversity of air-breathing mechanisms in other fish.

Fish. Many fish, mostly teleosts, have evolved mechanisms for air breathing (Table 13–1). Air breathing would be advantageous if the water in which the fish lived was temporarily reduced to small ponds or dried up completely (e.g., some salamander fish), if the water had a low pO_2 because of vegetation decaying at high temperature (e.g., many tropical freshwater fish), or for fish that occasionally migrate over land (e.g., eels, walking catfish).

Gills are not well suited for aerial respiration because they collapse under their own weight and their lamellae tend to stick together by surface tension. This obviously reduces their surface area for gas exchange and compromises O_2 uptake and CO_2 excretion. Not surprisingly, gas exchange in air-breathing fish generally does not occur across the gills but across another vascular surface; many air-breathing fish actually have a reduced gill surface area. Possible respiratory surfaces for air-breathing fish include their **cutaneous surface**; buccal, opercular, stomach, and intestinal **mucosa**; **pharyngeal air sacs**; modified gills, including complex suprabranchial **labyrinth organs; swim-bladders**; and **lungs** (see Table 13–1). Vascularized body surfaces that serve a respiratory role are the skin (e.g., eels; *Mniepes*); buccal, branchial, or opercular mucosa (e.g., *Electrophorus*, *Clarias*, *Periophthalmus*); stomach or intestine (e.g., *Plecostomus*, *Misgurnus*, *Hoplosternum*). The gills themselves are modified in many air-breathing fish for aerial exchange. Some fish have filamentous gills in addition to their lamellar gills. These filamentous gills are rosettes or branching vascularized epithelial projections in the branchial chamber. For example, the air-breathing catfish *Clarias* has fan-like structures that are formed from modified lamellae and also complex arborescent extensions of two gill arches for aerial respiration (Figure 13–1A). The suprabranchial labyrinthine organs of some air-breathing fish have elaborate modifications of the gill arches (e.g., *Osphronemus*; Figure 13–1B).

Lungs appear to have evolved as ventral outgrowths of the pharynx (e.g., *Polypterus*) and "migrated" dorsally to form the more advanced lung or swim-bladders. This evolutionary scenario is supported by the fact that the most "primitive" fish (e.g., dipnoi; *Polypterus*) have lungs, whereas swim-bladders are best developed in advanced teleosts (Figure 13–2). The swim-bladder is an elongate, distensible, air-filled sac(s) arising from the anterior digestive tract; its primary function in most fish is the regulation of buoyancy. The swim-bladder is a respiratory, as well as a buoyancy, organ in some fish. The swim-bladder, if used for gas exchange,

(*text continues on page 612*)

TABLE 13–1

Taxonomic status, habitat, and the nature of the respiratory surface/organ for a variety of air-breathing fish. The order is in boldface. *(Adapted in part from Dehadrai and Tripathi 1976; Munshi 1976; taxonomic classification after Nelson 1984.)*

	Habitat	Respiratory Surface/Organ
Gymnotiformes		
Electrophorus	SAm rivers, swamps	Buccopharyngeal chamber
Hypopomus	SAm swamps	Vascular opercular chamber
Polypteriformes		
Polypterus	Afr, fresh water	Air sacs
Calamoichthys	Afr, fresh water	Air sacs
Synbranchiformes		
Synbranchus	SAm, swamps	Suprabranchial air sacs
Amphipnous	Asia, ponds, swamps	Suprabranchial air sacs
Siluriformes		
Saccobranchus	Asia, ponds, swamps	Suprabranchial air sacs
Clarius	Afr., Asia, ponds, swamps	Arborescent gill organ
Plecostomus	SAm, swamps	Stomach
Ancistrus	SAm, swamps	Stomach
Doras	SAm, rivers, swamps	Intestine
Erythrinus	SAm, swamps	Swim-bladder
Hoplyerythrinus	SAm, swamps	Swim-bladder
Piabucina	SAm, swamps	Swim-bladder
Pangasius	Asia, fresh water	Swim-bladder
Cyrpiniformes		
Misgurnus	Eur, Asia, rivers, pools	Intestine
Lepidocephalichthys	Asia, fresh water	Intestine
Hoplosternum	SAm, rivers, pools	Intestine
Brochis	SAm, benthic still water	Intestine
Perciformes		
Ophicephalus	Afr, Asia, tropical ponds	Suprabranchial air sac
Pseudapocryptes	Asia, estuaries	Vascularized opercular chamber
Periophthalmus	Asia, estuaries	Skin, opercular chamber
Gillichthys	NAm, estuaries	Bucco-opercular chamber
Macropodus	Asia, tropical ponds	Suprabranchial labyrinth
Colisa	Asia, fresh water	Suprabranchial labyrinth
Betta	Asia, fresh water	Suprabranchial labyrinth
Trichogaster	Asia, fresh water	Suprabranchial labyrinth
Osphronemus	Asia, fresh water	Suprabranchial labyrinth
Anabas	Asia, Afr, swamps	Suprabranchial labyrinth
Mnierpes	SAm, rocky shores	Skin
Anguilliformes		
Anguilla	Eur, Afr, Asia, NAm, rivers	Skin
Amiiformes		
Amia	NAm, fresh water	Swim-bladder
Lepisosteiformes		
Lepisosteus	NAm, fresh water	Swim-bladder
Salmoniiformes		
Umbra	Eur, NAm, stagnant water	Swim-bladder
Osteoglossiformes		
Arapaima	NAm, fresh water	Swim-bladder
Heterotis	Afr, swamps	Swim-bladder
Pantodon	West Afr, fresh water	Swim-bladder
Notopterus	Asia, fresh water	Swim-bladder
Gymnarcus	Afr, swamps, rivers	Swim-bladder
Gonorynchiformes		
Neoceratodus	Aust, rivers	Lung
Lepidosireniformes		
Lepidosiren	SAm, rivers	Lung
Protopterus	Afr, rivers	Lung

[1] SAm = South America, NAm = North America, Eur = Europe, Aust = Australia, and Afr = Africa.

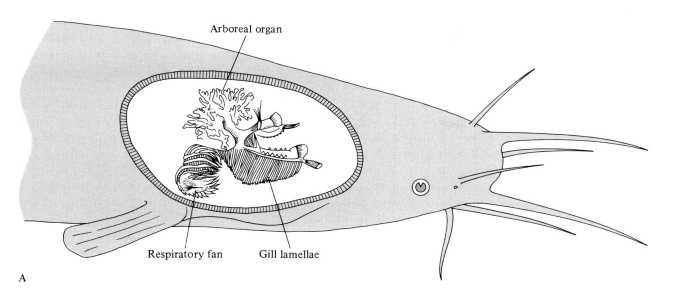

Arboreal organ

Respiratory fan   Gill lamellae

A

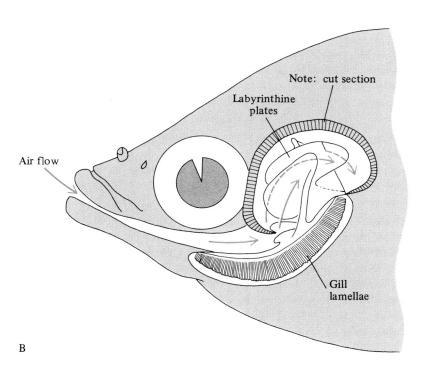

Note: cut section

Labyrinthine
plates

Air flow

Gill
lamellae

B

FIGURE 13–1 **(A)** Lateral view of the fourth gill arch of *Clarias batrachus* showing the arborescent respiratory organ and a secondary respiratory fan derived
from the gill lamellae. **(B)** Labyrinthine accessory respiratory organ of *Osphronemus goramy* showing the pattern of air flow over the labyrinthine plates and the
location of the lamellar gill.   *(Modified from Munshi 1976; Peters 1976.)*

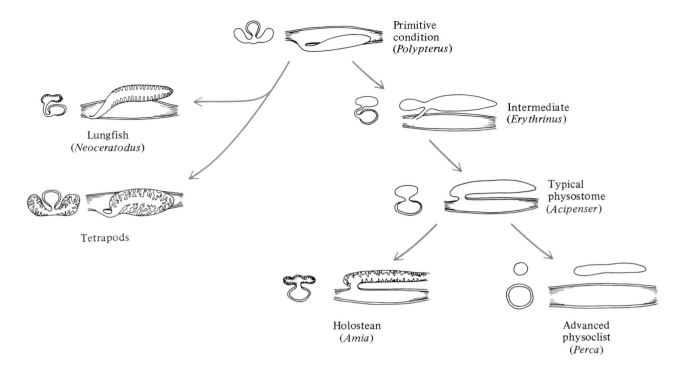

FIGURE 13–2 Possible evolutionary sequence for derivation of the lungs of a variety of air-breathing tetrapod vertebrates and the swim-bladders of fish from the primitive ventral outgrowths of the pharynx. The primitive condition is represented by *Polypterus*, the lungfish by *Neoceratodus*, a possible intermediate form with a lateral connection to the pharynx by *Erythrinus*, the typical physostomous swim-bladder by *Acipenser*, the holostean fish "lung" by *Amia*, and the more advanced physoclist swim-bladder (with no connection to the pharynx) is represented by *Perca*. *(Modified from Romer and Parsons 1986.)*

generally has a vascularized wall and a well-developed ventilation mechanism. For example, the swim-bladder of *Arapaima* has a highly vascular dorsal wall and the ventral wall acts as a movable septum responsible for inspiration (Farrell and Randall 1978). Some air-breathing fish have lung-like pharyngeal air sacs (e.g., *Saccobranchus*).

A number of fish, in normoxic water, rely on their lungs for significant gas exchange (Table 13–2). For example, 96% of the O_2 uptake by the lungfish *Lepidosiren* is across its lungs. However, much less of the CO_2 excretion, about 43%, occurs across the lungs. The respiratory quotient (RQ) for pulmonary gas exchange is low, at 0.45, whereas the RQ for gill gas exchange is high, at 6.7; the overall RQ is 0.73. *Amia*, in contrast to *Protopterus*, relies less on aerial exchange (35% for O_2) and has aerial and aquatic RQs closer to the total RQ of 1. *Neoceratodus*, another lungfish, does not normally have any aerial gas exchange.

There is a clear distinction between the roles of gills and lungs in O_2 and CO_2 transport in air-breathing fish. CO_2, because of its higher solubility

and apparent diffusion coefficient (K), is more readily exchanged across the gills or skin than is O_2. Consequently, less CO_2 is exchanged via the lungs than is O_2, and the pulmonary RQ $\ll 1$ for fish with a high percentage of aerial gas exchange; in contrast, the cutaneous or gill RQ $\gg 1$ (Figure 13–3). If most O_2 exchange is aerial, then both the aerial and aquatic RQs are close to 1. The body fluid pCO_2 is generally low when gills or skin are the avenue for loss, rather than the lungs. This has profound significance with respect to regulation of respiration (see below) and acid base balance (Chapter 14).

Many fish rely primarily on their gills for gas exchange as long as the water pO_2 is high but use air breathing as a facultative option during hypoxia. For example, *Piabucina* is a facultative air-breathing fish; in hypoxic water, it becomes progressively more reliant on air breathing, generally being 100% air breathing at low pO_2 and high pCO_2 (Figure 13–4). A number of air-breathing fish, especially the larger species, are obligate air breathers even under normoxic conditions. *Arapaima*, for example, relies

TABLE 13–2

Partitioning of air and water O_2 exchange (%A/W) and respiratory quotient (RQ) in air, water, and total for air-breathing fish in normoxic water. The original units for VO_2 and VCO_2 are used to illustrate the diversity of units (for conversion, 1 ml O_2 (STPD) = 44.6 μmole = 1.43 mg).

	Mass (g)	**Total VO_2**	**Units**	**% A/W (O_2)**	**RQ Air/Water/Total**
Lepidosiren	500	0.37	ml kg^{-1} m^{-1}	96/4	0.45/6.7/0.73
Protopterus	3250	0.19	ml kg^{-1} m^{-1}	89/11	0.25/4.7/0.75
Arapaima	2000	103	mg kg^{-1} hr^{-1}	78/22	0.45/2.26/0.60
Lepisosteus	600	0.89	ml kg^{-1} m^{-1}	73/27	0.09/2.7/0.80
Anabas	40	113	ml kg^{-1} hr^{-1}	54/46	0.20/2.29/1.17
Trichogaster	8	5.2	μm g^{-1} hr^{-1}	40/60	0.30/1.12/0.80
Amia	1200	1.5	ml kg^{-1} m^{-1}	35/65	0.60/1.40/1.00
Neoceratodus	6000	0.25	ml kg^{-1} m^{-1}	0/100	− /0.72/0.72

on its lungs for about 60% of its O_2 exchange even in normoxic water, and this reliance increases to 100% in hypoxic water.

Air-breathing fish generally rely on their buccal pump for movement of air, whether into the buccal, opercular, or other air spaces. Some fish, however, have separate muscle pumps for aspirating air into the swim-bladder. The ventral surface of the swim-bladder of *Arapaima* has a diaphragm-like septum connected to the lateral body walls. Outward pull on the septum most likely aspirates air into the

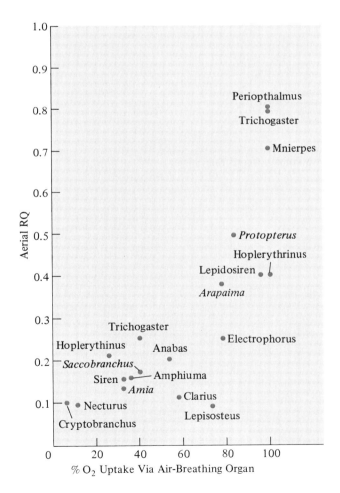

FIGURE 13–3 Aerial respiratory quotient (RQ) as a function of the percentage of gas exchange across the air-breathing organ (lungs, swim-bladder, opercular chamber, etc.) for bimodal (water and air breathing) animals, including lungfish, fish, and amphibians. *(Modified from Rahn and Howell 1976; Randall et al. 1981.)*

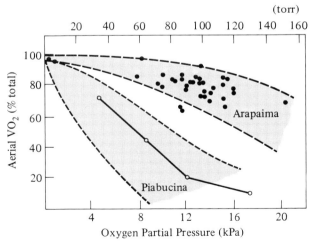

FIGURE 13–4 Transition from water breathing to air breathing as a function of the pO_2 of the water in two species of fish, one relying mainly on gill respiration under normal conditions (*Piabucina*) and one relying on mainly aerial respiration regardless of the water pO_2 (*Arapaima*). *(Data from Graham, Kramer, and Pineda 1977; Stevens and Holeton 1978.)*

swim-bladder; movements of the viscera might also contribute to aspiration (Farrell and Randall 1978).

Most air-breathing fish exhale before they inhale fresh air to minimize the mixing of the fresh inspired air with the O_2-depleted, CO_2-rich air from the lung. However, the jeju (*Hoplerythrinus*) and *Piabucina* inspire first, then exhale. The jeju apparently minimizes the mixing of fresh air and dead space air by having its swim-bladder subdivided into an anterior section and a posterior section; the posterior section is further subdivided into an anterior respiratory portion and a nonrespiratory posterior sac (Figure 13–5). Air is forced during inspiration from the buccal cavity mainly into the anterior chamber, which is closed off from the posterior chamber; O_2-depleted air is then expired from the posterior chamber. Finally, fresh air from the anterior chamber passes into the posterior chamber for gas exchange to occur across the anterior portion. The relatively large tidal volume (11 to 35 ml kg^{-1}) is a

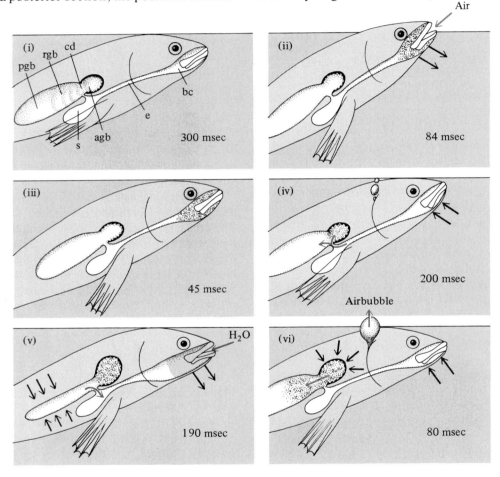

FIGURE 13–5 Ventilation cycle in the air-breathing fish *Hoplerythrinus unitaeniatus.* The approximate time required for each part of the cycle is shown. (1) Fish approaching surface to take a breath of air. Abbreviations are buccal cavity, bc; esophagus, e; stomach, s; anterior gas bladder, agb; respiratory gas bladder, rgb; posterior nonrespiratory gas bladder, pgb; communicating duct between anterior and posterior gas bladders, cd. (2) Fish opens mouth and fills buccal cavity with inspired air (stippling). (3) Fish closes mouth. (4) Fish raises buccal floor and forces inspired air primarily into anterior gas bladder; some air enters the posterior gas bladder and some may be expelled from the operculum as small air bubbles. (5) The posterior gas bladder is compressed and the buccal floor lowered—posterior gas bladder air enters the pharynx and the buccal cavity fills with water. (6) The buccal floor is retracted and water enters the pharynx, forcing the air out of the operculum as a large bubble; the anterior gas bladder is compressed, forcing the inspired air into the respiratory portion of the posterior gas bladder.   *(Modified from Kramer 1978.)*

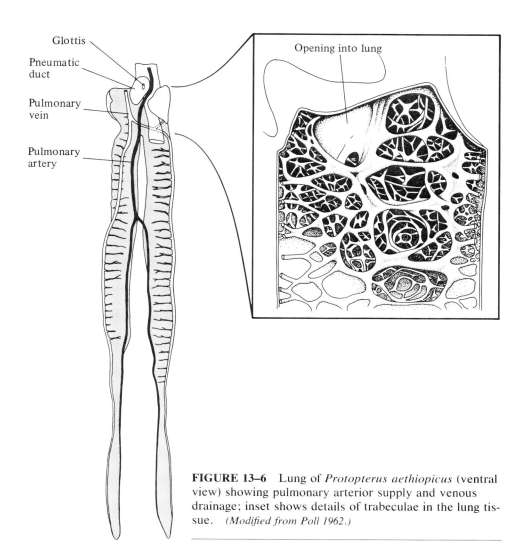

Glottis

Pneumatic
duct

Pulmonary
vein

Pulmonary
artery

Opening into lung

FIGURE 13–6 Lung of *Protopterus aethiopicus* (ventral view) showing pulmonary arterior supply and venous drainage; inset shows details of trabeculae in the lung tissue. *(Modified from Poll 1962.)*

high fraction (about 50%) of the total air-bladder volume (30 to 80 ml kg^{-1}), and this ensures a fairly high pO_2 of the bladder air. This respiratory cycle of *Hoplyerythrinus* more than superficially resembles that of anuran amphibians (see below).

Lungfish. The dipnoan **lung**, like that of other vertebrates, develops as a ventral outgrowth of the lower pharynx. In the Australian lungfish *Neoceratodus*, the single (unpaired) lung sac lies dorsal to the gastrointestinal tract and is a highly vascularized and moderately trabeculated structure. The African and South American lungfish (*Protopterus* and *Lepidosiren*) have a considerably more advanced lung, a paired structure with much more internal compartmentalization (Figure 13–6).

Neoceratodus relies almost entirely on its gills for both O_2 and CO_2 exchange when submerged under water, even with access to air. It is unable to

rely on air breathing for a significant fraction of its O_2 exchange and becomes frantic when removed from water. In contrast, *Protopterus* and *Lepidosiren* rely primarily on their lungs for O_2 exchange, although about 50% of their CO_2 exchange occurs across their gills (see Table 13–2), just as we saw in many other air-breathing fish. They are able to maintain an almost normal O_2 uptake rate when out of water and can maintain a high saturation percentage of arterial blood.

Amphibians. The living amphibians, frogs and toads (Anura), salamanders (Salienta), and caecilians (Apoda) exhibit considerable diversity in their respiratory systems, reflecting their amphibious heritage and habits.

Aquatic and terrestrial amphibians rely to a variable extent on cutaneous, buccal, gill, and lung gas exchange. For example, for gas exchange the

TABLE 13–3

	VO₂	VCO₂

Partitioning of pulmonary, branchial, and cutaneous gas exchange in a salamander (*Necturus*; mean body mass = 150 g) at an ambient temperature of 25° C. *(Data from Guimond and Hutchison 1972.)*

	VO₂	**VCO₂**
Pulmonary	10%	12%
Branchial	60%	61%
Cutaneous	30%	27%
Total[1]	26.1	23.7

[1] Units are $\mu l\ O_2\ g^{-1}\ hr^{-1}$.

aquatic salamander *Necturus* relies primarily on its external gills and cutaneous surface (to about equal degrees) and minimally on its lungs (Table 13–3); we have already examined aspects of water breathing in *Necturus*. Even terrestrial amphibians can rely on cutaneous gas exchange. Plethodontid salamanders, for example, lack lungs and have only cutaneous and buccal gas exchange. The small toad *Bufo americanus* primarily uses its lungs for O_2 exchange (80 $\mu l\ O_2\ g^{-1}\ hr^{-1}$), although cutaneous uptake is also significant (44); the skin, however, is the primary route for CO_2 elimination (95 $\mu l\ g^{-1}$ hr^{-1}) rather than the lungs (22). As in air-breathing fish, the pulmonary RQ is low (0.28) and cutaneous RQ is high (2.16). There is a general trend in the more terrestrial amphibians for higher metabolic rates, perhaps reflecting the greater O_2 availability in air and decreased cost of aerial respiration compared to water breathing.

The higher metabolic demands of activity (up to a 10 to 20 × increase over resting VO_2; see Chapter 4) require an equivalent increase in respiratory gas exchange. The increased O_2 demand is matched by elevated pulmonary gas exchange in air-breathing amphibians, such as the salamanders *Desmognathus* and *Amphiuma*; their cutaneous O_2 uptake is relatively unaltered during activity. Plethodontids that must rely on cutaneous and buccal gas exchange have lower capacities to increase VO_2 during activity than do lunged amphibians.

The structure of amphibian lungs varies from a simple, noncompartmentalized and poorly vascularized sac to a considerably compartmentalized and well-vascularized lung (Figure 13–7). In a simple sac lung, the respiratory surface area for gas exchange more closely approximates the surface area of the pulmonary capillaries rather than the surface area of the lung. In the more compartmentalized lungs, the surface area of the lung is considerably greater than that of a simple sac.

Amphibians, like most air-breathing fish, use a **positive pressure buccal pump** to force air into their

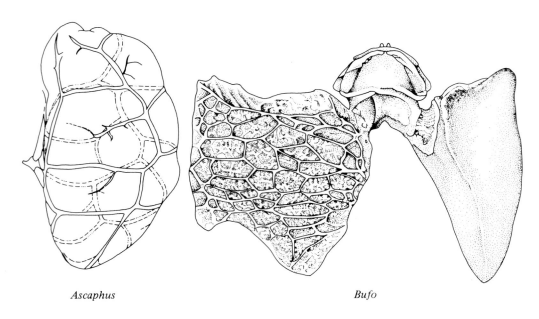

Ascaphus   *Bufo*

FIGURE 13–7 Simple sac lung of the amphibian *Ascaphus* and the moderately compartmentalized lung of *Bufo marinus*, the marine toad. *(From Noble 1931.)*

lungs; they do not aspirate air into their lungs by a negative intrapulmonary pressure. Studies of the frogs *Rana catesbeiana* and *R. pipiens* illustrate the buccal pump mechanism (Gans, De Jongh, and Farber 1969; Vitalis and Shelton 1990). During buccal ventilation (when the glottis is closed, sealing the lung from the buccal cavity), air is drawn in through the nares by lowering the floor of the buccal cavity and expelled by raising the buccal floor (Figure 13–8). These buccal movements ventilate the buccal cavity with fresh air in preparation for lung ventilation. The first phase of lung ventilation is inspiration; a large volume of air is drawn into the buccal cavity, as occurs during buccal ventilation. Then, during the second phase, lung emptying and expiration 1, the glottis opens so that air exits the lung, due to its passive collapse, and the expelled lung gas mixes with the buccal gas. The nostrils rapidly close, so little gas leaves the buccal cavity. In the third phase, lung filling, the buccal floor is raised and the resulting positive pressure in the buccal cavity fills the lung with gas; the nostrils

open at the end of this phase. In the final phase, expiration 2, the glottis closes (which keeps the lung inflated at a positive pressure) and the gas volume of the buccal cavity declines as air exits via the nares. Whether there is an overall increase in lung volume (inflation) or an overall decrease in lung volume (deflation) depends on the precise timing of opening and closing of the nares, in phase 3. The buccal ventilation frequency is often three to five times the pulmonary filling frequency. The lungs are generally filled by a number of successive lung inflation cycles, each as described above, whereas lung deflation is usually accomplished by usually one single, passive expiration.

Reptiles. The lungs of primitive reptiles resemble those of amphibians, but the lungs of advanced reptiles are more highly compartmentalized and their lung tissue has a spongy appearance. The caudal portion of the lung is often less compartmentalized and vascular than the anterior portion. This caudal portion is poorly ventilated with fresh air

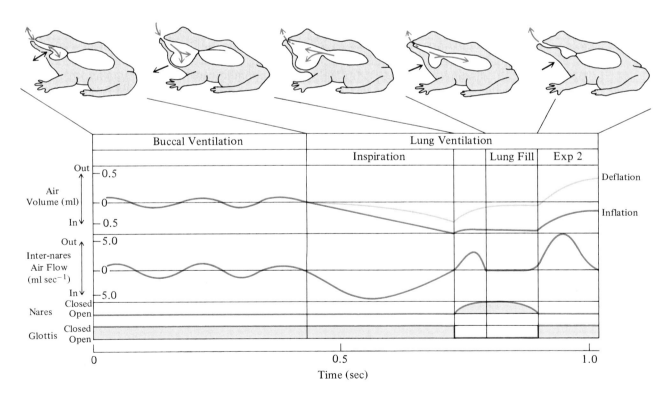

FIGURE 13–8 Pattern of air flow during successive stages of buccal and lung ventilation in a frog. Air initially flows into the buccal cavity, then is forced by the positive pressure buccal pump into the lungs after the lung air is expired. *(Modified from Gans, DeJongh, and Farber 1969; Vitalis and Shelton 1990.)*

and contains mainly residual air; its function is more as a bellows reservoir for ventilation of the anterior portion (e.g., chameleons and snakes; Figure 13–9A). Some reptiles have air sac extensions of the respiratory system; their function is unknown, but they anticipate the important air sac system of birds (see below).

A major respiratory modification in reptiles, compared to amphibians, is in their ventilatory mechanism. Reptiles (and birds and mammals) aspirate air into their lungs by a **negative intrapulmonary pressure**, rather than by forcing air from the buccal chamber into the lungs under positive pressure. Reptiles have well-developed ribs that generally articulate with the vertebral column (except in chelonians; see below). The action of rib muscles causes the rib cage to expand outwards and forward, thereby expanding the lungs. The inspiratory phase is characterized by a negative (subatmospheric) intrapulmonary and intra-abdominal pressure. The glottis closes at the end of inspiration and the lungs and rib cage passively recoil, resulting in a slight positive intrapulmonary pressure (0.02 to 0.04 kPa). Expiration is initiated after a variable period of breath holding by opening the glottis; there is a

passive expulsion of air by elastic recoil of the lung tissue and rib cage. Lizards and snakes do not have a true diaphragm membrane separating the thoracic and abdominal cavities, although the pressures in these two cavities are not necessarily equal during the respiratory cycle.

The relative roles of the anterior vascularized portion of the lung (partitioned into radial chambers or **faveoli**) and the saccular, nonvascularized posterior portion (air sac) have been clearly demonstrated in the Palestine viper *Vipera xanthina* (Gratz, Ar, and Geiser 1981). There are larger changes in O_2 and CO_2 concentrations (0.5 to 2.0%) in the faveolar region during a breath than in the air sac region (Figure 13–9B). The bellows action of the air sac results in complete replacement after inspiration of air in the anterior faveolar region with fresh air. The posterior air sac is filled with residual faveolar lung air, hence its lower pO_2 and higher pCO_2. Air is pumped back and forth between the faveolar and air sac portions of the lung during extended periods of breath holding, making all of the lung air available for gas exchange.

The rigid, or semirigid, carapace and fixed ribs of chelonians (turtles, tortoises) require a somewhat

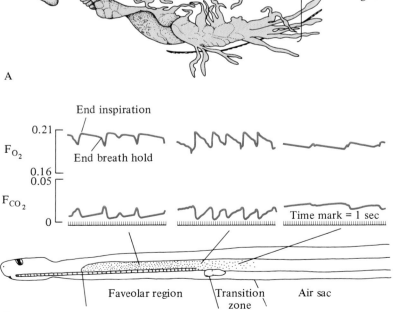

FIGURE 13–9 **(A)** Gross structure of the lung of the lizard *Chamaeleon zeylanicus* showing the anterior alveolar region and the posterior saccular region. **(B)** Intrapulmonary O_2 and CO_2 in the anterior faveolar and posterior saccular regions of the lung in the Palestine viper *Vipera xanthina*. The O_2 and CO_2 fluctuate with the respiratory cycle in the faveolar region but the low pO_2 and high CO_2 of the posterior saccular lung show little fluctuation with the respiratory cycle since it is primarily filled with dead space air from the anterior faveolar region. *(From George and Shah 1965; Gratz, Ar, and Geiser 1981.)*

modified ventilation mechanism involving a **diaphragm** (Gans and Hughes, 1967; Gaunt and Gans, 1969). The chelonian diaphragm is a nonmuscular septum between the thoracic cavity and the abdominal cavity. Extension of the pectoral and pelvic limbs out of the carapace expands the thoracic cavity and aspirates air into the lungs under negative pressure (Figure 13–10A). Exhalation (which usually preceeds inhalation and a long period of breath holding, or apnea) is caused by limb retraction. Opening and closing of the glottis is important for regulation of the appropriate inspiration/expiration air flow since any limb movements (e.g., walking) will change the intrapulmonary pressure. Respiratory ventilation is somewhat modified in aquatic turtles, such as the snapping turtle *Chelydra*, because their reduced plastron does not provide sufficient mechanical support on land for the visceral mass; the lungs are expanded by the pull of the viscera and intrapulmonary pressures are negative. Inspiration is therefore primarily passive and expiration involves retraction of the pelvic girdle (Figure 13–10B). When the snapping turtle is submerged and breathing by snorkeling, the hydrostatic pressure of the water on the body (0.10 kPa per cm water depth) causes passive expiration; inspiration is an active process due to extension of the pelvic girdle.

Crocodiles have well-developed, articulated ribs and intercostal muscles, but there apparently is little expansion of the rib cage during their respiratory cycle. Crocodiles, like lizards and snakes, lack a diaphragm. Contraction of various abdominal muscles pulls the liver forward against the lungs, causing expiration, and contraction of the diaphragmaticus muscle (which attaches the liver to the pelvic girdle) retracts the liver and expands the

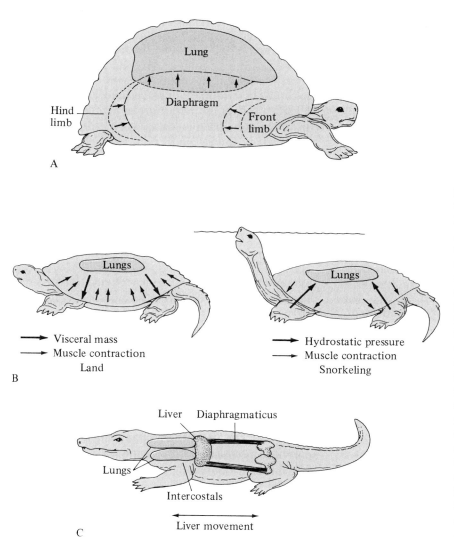

FIGURE 13–10 (A) Respiratory movements in a land tortoise by limb and visceral movements. **(B)** A snapping turtle on land experiences a passive inspiratory force due to the weight of the visceral mass expanding the lungs; when snorkeling under water, there is a passive expiratory force due to the hydrostatic pressure collapsing the lungs. Inspiration under water, and expiration on land, are caused by muscle contractions (small arrows). **(C)** The diaphragmaticus muscle of the crocodilian *Caiman* draws the liver posteriorly and expands the lungs for inspiration. Expiration is due to an anterior movement of the liver caused by contraction of various abdominal muscles, which compresses the lungs. *(From Gans and Hughes 1967; Gaunt and Gans 1969; Gans and Clark 1976.)*

lungs for inspiration (Figure 13–10C). Submerged crocodiles, like submerged turtles, utilize the hydrostatic pressure for passive expiration.

Brief mention should be made of the role of cutaneous gas exchange in some aquatic reptiles. The general cutaneous surface can be a significant avenue for gas exchange and in particular for CO_2 excretion. For example, up to 94% of the VCO_2 and 33% of the VO_2 of sea snakes occurs across their skin when they are submerged; the remainder is from pulmonary exchange (Graham 1974). Some aquatic turtles (*Amyda, Aspidonotus*) ventilate their bucco-pharyngeal mucosa with water when submerged, and the amazon turtle *Podocnemys* can obtain up to 90% of its VO_2 from rhythmic irrigation of the cloacal bursae under certain circumstances (Steen 1971; Jones 1972). Submerged soft-shelled turtles (*Trionyx*) use bucco-pharyngeal aquatic respiration and, to a lesser extent, cutaneous respiration across the skin (Zhao-Xian, Ning-Zhan, and Wen-Tang 1989). In fact, even terrestrial reptiles (and other vertebrates also) can exchange significant quantities of especially CO_2 across the skin (Feder and Burggren 1986).

Mammals. The mammalian lung is much more compartmentalized than reptile lungs; its gross appearance is more like dense foam rubber than the spongy appearance of lungfish and reptile lungs. Inspired air is drawn in through the trachea and into each lung via a bronchus. Within the lungs, the bronchi divide into a number of secondary bronchi that further divide and ramify into many smaller tertiary bronchi, bronchioles. The final extensions of the terminal bronchioles are small sac-like **alveoli** (Figure 13–11). The alveoli are the primary site of gas exchange in the lungs. In each lung of a human, for example, there are about 150,000,000 alveoli, each about 150 to 300 μ in diameter. The total surface area of the alveoli is about 80 m². The morphology of the "respiratory tree" of mammals is essentially a successive series of dichotomously branching tubes; there is a fluid-dynamic basis to the complex pattern of branching (see Supplement 13–1, page 662).

Ventilation is provided by the combination of contraction of a **muscular diaphragm** and movement of the rib cage. The muscular diaphragm forms a hemispherical septum between the thoracic and abdominal cavities. Contraction of the diaphragm causes it to flatten, pulling the lungs into a more expanded shape; the consequent negative intrapulmonary pressure draws air into the lungs (i.e., inspiration). Expiration is generally passive due to elastic recoil of the lung tissues and thorax and relaxation of the diaphragm. Two sets of intercostal

muscles between the ribs (external and internal intercostals) are also involved in the respiratory cycle. Contraction of the external intercostal muscles causes the rib cage to expand outwards and anteriorly, increasing the volume of the lungs and causing inspiration. Again, expiration is passive due to elastic recoil. Expiration becomes active when respiration is greatly increased, for example during activity; the internal intercostal muscles and a variety of other upper body muscles collapse the rib cage, facilitating expiration.

In some mammals, locomotion assists respiratory ventilation. For example, movement of the viscera of a hopping kangaroo has a piston-like effect on the lung; the lungs are compressed (expiration) every time the kangaroo lands, and the lungs are expanded (inspiration) on takeoff (Baudinette et al. 1987). Respiration is also linked to gait in a 1:1, 1:2, or other ratio in other animals including humans and some quadrupeds and with the wing-beat cycle in birds (1:1, 1:3, 1:5 pattern).

The alveoli of the mammalian lung are not homogeneous in their anatomical location with respect to the location of the heart and the trachea and in their physiological functioning. Alveoli in different parts of the lung differ in blood flow and air flow; they have different **ventilation/perfusion** characteristics. In the human lung, for example, the alveoli in the top of the lung have a lower gas ventilation and a considerably lower blood flow than alveoli in the bottom of the lung. The ventilation/perfusion ratio for these alveoli is high (about 3 × the optimal value), whereas alveoli at the bottom have a ventilation/perfusion ratio of about 0.6 × the optimal value. These imbalances in ventilation/perfusion ratio decrease the overall effectiveness of gas exchange in the lung.

We will deal with other important aspects of pulmonary ventilation in more detail after we have described the very different respiratory system of birds.

Birds. The bird lung is the most specialized respiratory system in vertebrates, and in morphological and physiological complexity it is probably unrivaled by any animal! First, the avian lung, which is the site of gas exchange as in other air-breathing vertebrates, has a relatively constant volume during the respiratory cycle. The bellows mechanism for ventilation of the lung is provided by a complex series of thin-walled **air sacs** connected to the trachea and lung in a complex fashion. The air sacs surround the viscera and even extend into the large long bones (e.g., humerus, femur). The main air sacs are the cervical, interclavicular, anterior and posterior thoracic, and abdominal (Figure 13–12A). The air sac system of

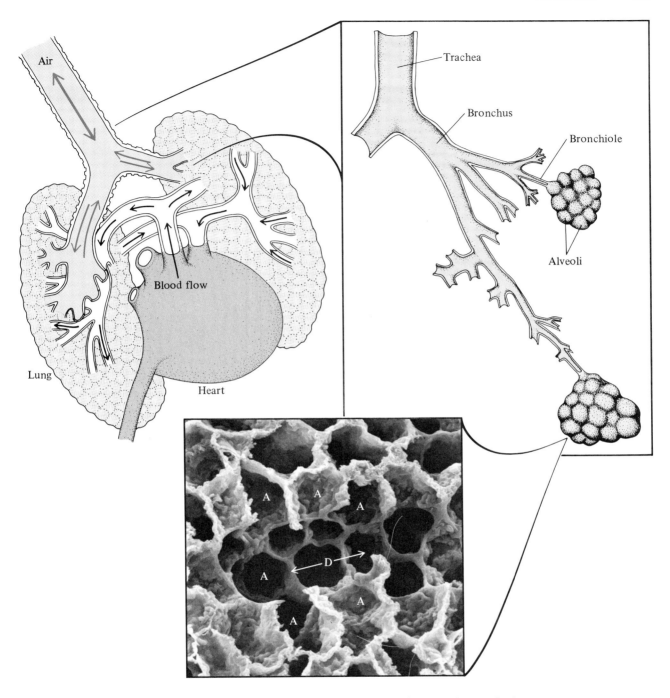

FIGURE 13–11 The organization of the major airways and blood supply to the lungs of a human is a continual branching of the trachea into smaller diameter bronchi and bronchioles that terminate in groups of alveoli. The expanded view of one group of alveoli shows this branching network of bronchi and bronchioles supplying air. A, alveolus; D, airway duct. *(Modified from Waterman et al. 1971; photograph courtesy of Weibel 1984.)*

birds is somewhat reminiscent of that in some reptiles, but it is far more extensive. The trachea divides into left and right bronchi, as in mammals, but here the similarity ends! Each primary bronchus passes through its respective lung, giving rise to four ventral secondary bronchi (the **ventrobronchi**) and then a variable number (7 to 10) of dorsal secondary bronchi (**dorsobronchi**; Figure 13–12A). Each primary bronchus connects to the abdominal air sac. The dorsobronchi and ventrobronchi are

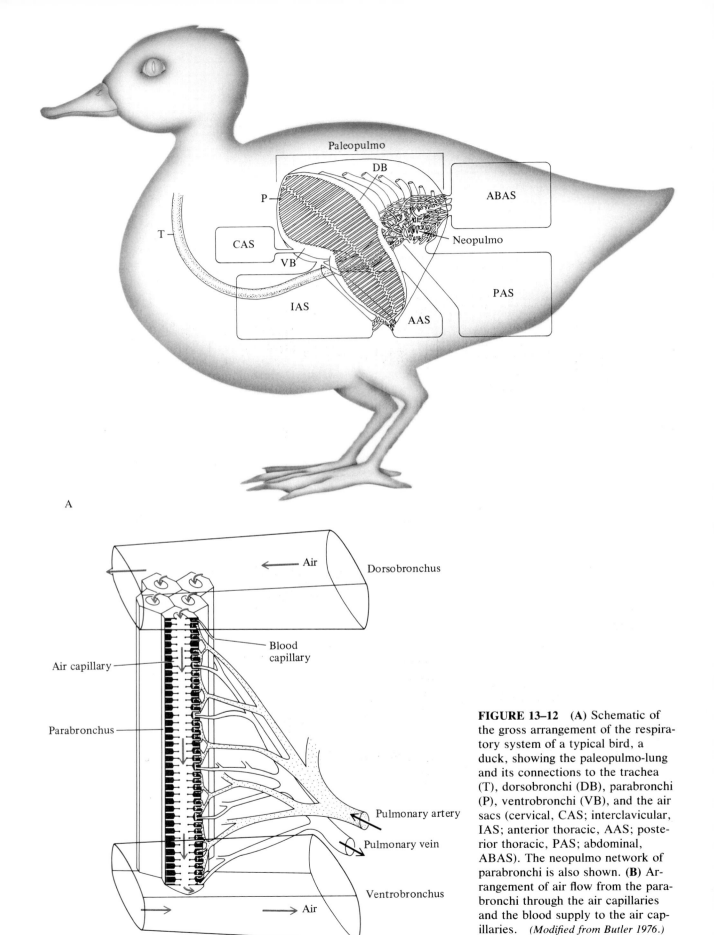

FIGURE 13–12 **(A)** Schematic of the gross arrangement of the respiratory system of a typical bird, a duck, showing the paleopulmo-lung and its connections to the trachea (T), dorsobronchi (DB), parabronchi (P), ventrobronchi (VB), and the air sacs (cervical, CAS; interclavicular, IAS; anterior thoracic, AAS; posterior thoracic, PAS; abdominal, ABAS). The neopulmo network of parabronchi is also shown. **(B)** Arrangement of air flow from the parabronchi through the air capillaries and the blood supply to the air capillaries. *(Modified from Butler 1976.)*

joined by numerous small diameter tubes (**parabronchi**) from which branch the functional gas exchange surfaces, the **air capillaries** (Figure 13–12B).

The parabronchi are a series of parallel, cylindrical tubes through which there is a unidirectional air flow (see below); each parabronchus has a diameter of about 500 μ (Figure 13–13). The air capillaries of the avian lung are small (about 100 μ diameter), blind-ended tubes that extend radially from the parabronchi. There is diffusional gas exchange between the blind-ended air capillaries and the air flowing through the parabronchi. Each air capillary is surrounded by a blood capillary network.

The parabronchi and air capillaries, as just described, form the **paleopulmo-lung,** or "old lung." In primitive birds (such as emus), this is the only lung present. In most other birds a second **neopulmo-lung** originates as an additional complex of parabronchi leading from the lateral side of the bronchus and dorsobronchi to the posterior air sacs. This neopulmo-lung is best developed in galliform and passeriform birds, where it comprises up to 25% of the total lung volume. The complex structure of the avian respiratory system should serve as a forewarning of the complex patterns of air flow through the air sacs and the lung.

For simplicity, and without sacrificing too much accuracy, we shall consider the complex air sac system to be composed of only two functional air sacs, an anterior sac and a posterior air sac, connected to the lungs (Figure 13–14A). The air sacs are not sites for gas exchange, as was elegantly demonstrated by Paul Bert. He filled the air sacs of a bird with carbon monoxide and inferred that there was insignificant exchange of CO (hence O_2) across the air sacs because the bird did not develop symptoms of carbon monoxide poisoning. Rather, the air sacs provide air ventilation of the lung. Expansion of the air sacs during inspiration draws air in through the trachea and primary bronchi. Some of the air passes to the posterior air sac and some passes through the dorsobronchi, parabronchi, and ventrobronchi to the anterior air sac. Collapse of the air sacs during expiration forces air from the posterior air sac through the dorsobronchi, parabronchi, and ventrobronchi and out of the trachea. Air from the anterior air sac is also expired through the trachea.

Air flow through the parabronchi of the lungs is unidirectional and is maintained throughout both inspiration and expiration. This proposed pattern of air flow through the respiratory system requires that

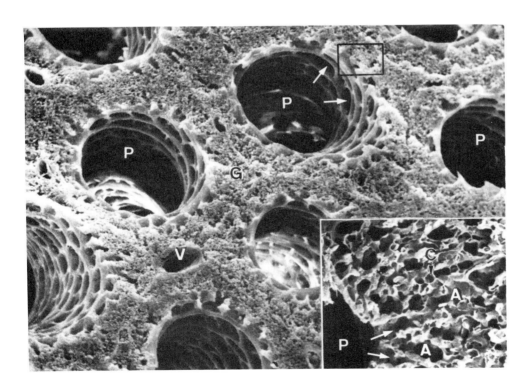

FIGURE 13–13 Cross section of a bird lung, showing the organization of the tubular parabronchi (P) and air capillaries (A) of the gas exchange tissue (G). Large blood vessels (V) and capillaries (C) are also shown. *(Photograph courtesy H. R. Dunker and Weibel 1984.)*

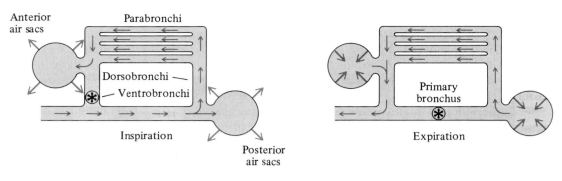

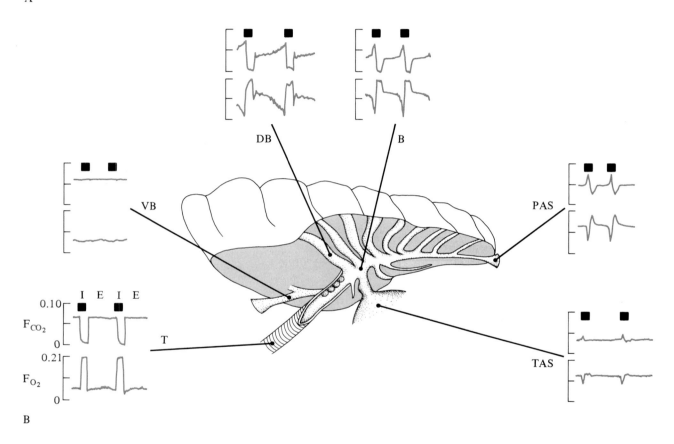

FIGURE 13–14 **(A)** Schematic outline of the pattern of air flow through the avian respiratory system during inspiration and expiration. Regions of proposed aerodynamic valving, preventing bidirectional air flow, are illustrated as ✳. **(B)** O_2 and CO_2 gas tensions in the avian respiratory tract measured over two breathing cycles. The locations of the measurement locations are T, trachea; B, main bronchus; PAS, opening to posterior air sac; TAS, opening to thoracic air sac; VB, ventrobronchus; DB, dorsobronchus. *(Modified from Jones, Effmann, and Schmidt-Nielsen 1981; Powell et al. 1981.)*

(1) air is inspired through the primary bronchus into the posterior air sac but is not expired by the same route, and (2) air is expired from the anterior air sac through the ventrobronchi but is not inspired into the anterior air sac from these same ventrobronchi.

Thus, some valving mechanism is required in these two locations (indicated by an asterisk in Figure 13–14A) to ensure that the postulated pattern of air flow actually occurs. Unfortunately, there are no anatomical valves in these locations, so it has been suggested that aerodynamic valves ensure the correct pattern of air flow (Jones, Effmann, and Schmidt-Nielsen 1981; Kuethe, 1988).

There is now considerable empirical evidence for this postulated pattern of air flow in the avian respiratory system. Direct measurement of pO_2 and pCO_2 in the primary bronchus, a dorsobronchus, and a ventrobronchus (Powell et al. 1981), indicates that during inspiration no significant fraction of the primary bronchus air ($pCO_2 = 0$) passes into the ventrobronchus or in the opposite direction (Figure 13–14B). Similarly, most of the expiratory air flow from the posterior air sacs passes through the paleopulmo-lung and little passes through the primary bronchus. Close examination of Figure 13–14B indicates the expected changes in pO_2 and pCO_2 in the trachea during the respiratory cycle; tracheal pO_2 and pCO_2 equal ambient values during inspiration and pO_2 is lowered to about 13.3 kPa (100 torr) and pCO_2 is elevated to about 7.0 kPa (53 torr) when expired air exits through the trachea (Figure 13–14B, see T). A somewhat different pO_2/pCO_2 cycle is observed in the primary bronchus, past the openings of the ventrobronchi. There initially is an elevated pCO_2 and depressed pO_2 during inspiration because of the residual (expired) air from the tracheal dead space being inhaled before the fresh inspired air (Figure 13–14B, see B). A similar but damped cycle in pO_2 and pCO_2 is observed at the openings of the thoracic and posterior air sacs (Figure 13–14B, see TAS, PAS). There is no respiratory cycle in pO_2 or pCO_2 in the ventrobronchus (Figure 13–14B, see VB) or anterior air sac because they always contain exhalant air from the parabronchi. The pO_2/pCO_2 pattern in the dorsobronchus is fairly complex (Figure 13–14B, see DB); pO_2 and pCO_2 approach ambient values during inspiration, reflecting the passage of fresh air from the trachea into the dorsobronchus. During expiration, the pO_2 is lower and the pCO_2 is higher than ambient, reflecting the gas composition of air from the posterior air sac. The pO_2 becomes progressively lower and the pCO_2 higher during expiration. (The high-frequency oscillations in PO_2 and pCO_2 in Figure 13–14B correspond to the heart beats).

Air in the posterior air sacs clearly has a lower pO_2 and a higher pCO_2 than ambient air, although the basic model of air flow suggests that this air should be equivalent to ambient air. This discrepancy is most likely due to the significant volume of tracheal dead space air that precedes the fresh inspired air into the posterior air sacs. There is also a stratification of air in the posterior air sacs from lowest pO_2 deepest in the air sacs to fresh ambient air nearest the air sac opening (Torre-Bueno, Geiser,

and Scheid 1980). Incomplete mixing of air within the posterior air sacs means that the residual air composition of the air sacs would approach the composition of tracheal dead space air.

The one-way flow of air through the parabronchi is not associated with a countercurrent flow of blood; the actual sites for gas exchange are the air capillaries that extend radially from the parabronchi. These blind-ended tubes exchange O_2 and CO_2 with the parabronchi by diffusion. Furthermore, the air capillaries are invested by many short blood capillaries; the long length of the parabronchi would preclude a countercurrent arrangement of capillaries along its length. The actual pattern of air-to-capillary exchange is a crosscurrent arrangement (see Figure 12–6, page 574). The capillaries investing the parabronchi closest to the dorsobronchi equilibrate with air of higher pO_2 and lower pCO_2 than the capillaries closest to the ventrobronchi. Consequently, the blood in the former capillaries attains a higher pO_2 and the latter a lower pO_2; the pooled capillary blood returning to the heart has an intermediate pO_2.

Principles of Pulmonary Ventilation

The same basic principles of ventilation mechanics and allometry apply to most types of vertebrate lungs, from the simple sac lung of some air-breathing fish and amphibians to the highly compartmentalized lungs of mammals. Let us now examine some of these basic principles governing the shape and stability of lungs, the gas composition of the intrapulmonary air, the factors influencing the rate of gas exchange across the lung surface, the allometry of respiratory variables, and the metabolic cost of pulmonary respiration.

Stabilizing Lungs and Alveoli. The lung surface, whether it is a simple sac or the millions of alveoli of the mammalian lung, is analogous to a soap bubble in that its water-lined surface tends to collapse because of surface tension, just as a soap bubble collapses when punctured.

Within the interior of a liquid, such as water, there are cohesive forces between the solvent molecules that result in an enormous pressure of many thousands of atmospheres! Fortunately, this pressure is not exerted on objects immersed in the liquid because there also are mutual cohesive forces between the object and the liquid. At an air-water interface, the surface water molecules experience more cohesive forces that draw them into the interior of the liquid so that the surface seems to contract spontaneously when disturbed. This

surface tension is an extremely strong force due to the high cohesive forces between the water molecules. Small objects that are denser than water can float on a water surface, and water can be retained by a fine sieve because of surface tension. The surface tension of some common liquids and biological surface films is listed in Table 13–4. Note the dramatically decreased surface tension of water containing a detergent (soap) and the extremely low surface tensions of lung surface films.

In a soap bubble, the surface tension of the soap film tends to decrease the surface area of the bubble, i.e., collapse it. However, the air pressure inside the bubble will increase until it exactly counterbalances the surface tension force. Soap bubbles thus reach an equilibrium when the internal pressure balances the surface tension force. This relationship is quantified by the law of Laplace;

$$T = \frac{1}{4}\Delta P R \quad \text{or} \quad \Delta P = 4T/R \quad (13.1a)$$

where T is the surface tension (N m^{-1}), ΔP is the pressure gradient from inside to outside of the bubble (N m^{-2}, or Pa), and R is the radius of curvature (m). For nonspherical bubbles, such as sausage-shaped lungs,

$$\Delta P = 4T/((1/R_1) - (1/R_2)) \quad (13.1b)$$

where R_1 and R_2 are the two radii of curvature. A soap bubble slowly collapses because the high pressure inside causes air to diffuse through the surface film. This decreases the volume and increases the pressure, promoting further diffusional loss.

TABLE 13–4

Surface tension of water at different temperatures and for a variety of surface films containing surfactant materials. Note that the surface tension varies with the area of the surfactant film, and so values in this table are approximate.	
	Surface Tension (mN m^{-1})
Water at 0° C	76.6
Water at 20° C	72.8
Water at 40° C	69.6
Soap solution	25
10% butyl alcohol	26
Water on cow dung	50
Water on decomposing fish	40
Goldfish swim-bladder surface film	10–20
Amia lung surface film	3.9
Lepidosiren lung surface film	1
Mammal lung surface film	1

Surface tension is very significant for alveoli of small radii because a high ΔP is required to support the surface water film. However, the pressure is only 1/2 that for an equivalent-sized soap bubble ($\Delta P = 2T/R$), since a soap bubble has surface tension acting on each side (inside and outside) of the soap film; an alveolus has surface tension acting only on the inside of the alveolar membrane. The ΔP for a water bubble equivalent to a shrew's alveolus (about 30 μ diameter) is about 9.3 kPa (70 torr); the pressure for water bubbles equivalent to human alveoli (100 μ) is about 2.8 kPa (21 torr). Such high intra-alveolar pressures cannot be sustained, so vertebrate lungs contain surfactants, agents that markedly reduce the surface tension.

Surfactants are phospholipids, such as dipalmityl lecithin, combined with protein. In theory, surfactants could form a monomolecular layer over the lung surface, but in practice there is about two to three times as much present compared to a theoretical monomolecular layer. These surfactants markedly reduce the pressure necessary to keep a small alveolus expanded. In the human lung, the actual pressure required to keep an alveolus expanded is about 0.5 kPa (4 torr) rather than 2.8 kPa (21 torr). Another important consequence of reduced surface tension in the lungs is the reduced tendency of water to infiltrate the alveoli from the alveolar tissue and capillaries due to surface tension trying to collapse the surface water film and pulling away from the alveolar wall.

A positive internal pressure is required to counterbalance surface tension and to stabilize alveoli; for example, human alveoli require about 0.5 kPa. However, the average intra-alveolar pressure is equal to ambient pressure (at least in humans; amphibians and reptiles can have slight positive pressures). How then are the alveoli stabilized? There must be a negative pressure of about −0.5 kPa outside the alveoli to keep them expanded. There is, in fact, a negative fluid pressure of about −1.33 kPa between the lining of the thoracic cavity (the visceral pleura) and the covering of the lungs (parietal pleura; Figure 13–15). This negative pressure is derived from the balance of forces between the interstitial fluid and the capillaries (see Chapter 14, Starling's equilibrium). The overall pressure, then, is a negative 0.8 kPa; this keeps the lungs firmly attached to the thoracic wall.

Mechanics of Ventilation. In the tidal lungs of mammals, and in some other tetrapods (but not birds, amphibians, and some reptiles), the volume of air within the respiratory system (V_{lung}) is mixed with the inspired **tidal volume** (V_t) on each inspiration. A high V_t/V_{lung} ratio will result in more com-

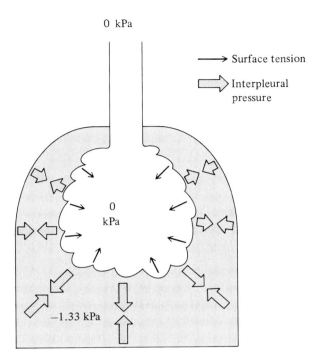

0 kPa

→ Surface tension

⇨ Interpleural pressure

0 kPa

−1.33 kPa

FIGURE 13–15 Schematic representation of the balance of forces in the lung that keep the alveoli expanded against the forces of surface tension and keep the lungs closely adpressed to the walls of the thoracic cavity. The surface tension of 0.53 kPa (4 torr) is counterbalanced by the interpleural pressure of −1.33 kPa (−10 torr), resulting in a net pressure of −0.80 kPa (−6 torr) that keeps the pleura attached.

plete replacement of lung air per breath and will maximize the pulmonary pO_2 after inspiration. The disadvantage is that the intrapulmonary pO_2 and pCO_2 fluctuate dramatically during a respiratory cycle. In the air-breathing fish *Hoplerythrinus*, for example, V_t is about 50% of V_{lung} and the gas composition of the lung air alters dramatically during the respiratory cycle. The **residual volume** of the lungs at the end of expiration is about 50% in *Hoplerythrinus*. In contrast, mammals have a high-residual volume and so the alveolar gas composition, and blood pO_2 and pCO_2, do not fluctuate markedly during the respiratory cycle. For example, the tidal volume of a human (about 500 ml) is only about 20% of the lung volume (2800 ml) whereas residual volume is 80%. Consequently, the composition of lung air does not fluctuate dramatically throughout the respiratory cycle.

The inherent inefficiency of gas exchange due to the residual volume can be minimized by separating the residual air from the fresh, inspired air. This allows the gas exchange structures to be concentrated at the fresh air rather than the residual air

space. We have already seen in some fish and reptiles (e.g., *Hoplerythrinus*, chameleon, and viper) that the posterior portion of the lung is relatively avascular and is mainly filled with the residual air so that the more vascular anterior portion is filled with mainly fresh air. In birds, the inspired air enters the posterior air sacs where there is a considerable stratification of residual and fresh air in these air sacs and relatively fresh rather than mixed gas then enters the lungs.

In tidal lungs, not all of the inspired air reaches the gas exchange surface. For example, the fresh air that fills the trachea at the end of inspiration does not participate in gas exchange. The inspired air that does not undergo gas exchange, as a consequence of the anatomical arrangement of the respiratory system, is called the **anatomical dead space**. In the adult human lung, for example, only 350 ml of the tidal volume reaches the alveolar exchange surface; the remaining 150 ml is located in the anatomical dead space. This is an important limitation to the effectiveness of a tidal ventilatory system. There is an additional limitation to the efficacy of gas exchange within the alveoli. If the blood supply to an alveolus is inadequate, then gas exchange is impeded for physiological, not anatomical, reasons. This air in the alveoli undergoing insufficient gas exchange is called **physiological dead space.** The important role of matching lung perfusion to ventilation will be further discussed in Chapter 14. The total anatomical and physiological dead space (V_{ds}) reduces the efficiency of ventilation.

The composition of lung gas is determined by the balance between the rate of fresh air entry to the alveoli and the rate of oxygen removal and CO_2 addition. The amount of air inspired per minute (**inspiratory minute volume, V_I**) equals the respiratory rate (f) multiplied by tidal volume (V_t). The amount of air reaching the alveoli is less; the alveolar ventilation rate (V_A) equals the respiratory rate (f) multiplied by ($V_t − V_{ds}$). Oxygen is removed from the alveoli at a rate equal to VO_2 and CO_2 is added at VCO_2. Consequently, the alveolar pO_2 (p_AO_2) and pCO_2 (p_ACO_2) can be calculated for body temperature and saturated conditions (BTPS) as

$$p_AO_2 = p_iO_2 - \frac{(101)(273 + T_b)VO_2}{(273)V_A} \quad (13.2a)$$

$$p_ACO_2 = p_iCO_2 + \frac{(101)(273 + T_b)VCO_2}{(273)V_A}, \quad (13.2b)$$

where p_iO_2 is the BTPS inspired pO_2; p_iCO_2 is the BTPS inspired pCO_2; T_b is the body temperature (°C); and V_A is the BTPS alveolar ventilation rate.

TABLE 13–5

	Air	Dead Space[1]	Alveolar[1]	Expired[1]
Partial pressures (kPa) of atmospheric air, dead space gas, alveolar gas, and expired gas for a human at sea level.				
Oxygen	21.1	19.8	13.8	15.9
Carbon Dioxide	0.039	0.037	5.32	3.59
Water Vapor	0.53	6.25	6.25	6.25
Nitrogen	79.3	74.8	75.6	75.2
Total	101	101	101	101

[1] Saturated with water vapor at 37° C.

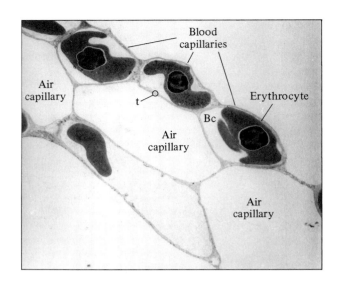

FIGURE 13–16   There is a transalveolar diffusion barrier separating the alveolar air from the red blood cells in an avian (singing cisticola) lung. The average diffusion path (*t*) length is about 0.12 μ for the bird lung. *(From Maina 1984.)*

Expired gas is a mixture of dead space gas and alveolar gas and consequently has intermediate pO_2 and pCO_2 (Table 13–5).

The efficiency of oxygen removal by the lungs (*E*, %), calculated in the same fashion as for gills, is as follows.

$$E = 100(p_iO_2 - p_eO_2)/p_iO_2 \qquad (13.3)$$

E is about 20 to 25% for mammals. For example, the value for humans is 23%. Extraction efficiency can also be calculated as 100 $VO_2/(0.21\ V_I)$, where $0.21\ V_I$ is the minute oxygen ventilation; VO_2 and V_I should have the same units, i.e., both are STPD or BTPS. Extraction efficiency calculated for mammals from the allometric relationships for VO_2 and V_I (see below) is essentially independent of body mass at 20.2%. For birds, the extraction efficiency is higher at 41.3%. The higher oxygen extraction efficiency for birds reflects their more efficient one-way flow of air through the parabronchi, the fact that the pO_2 is higher for air entering the parabronchi (18.6 kPa) than for mammalian alveoli (13.8 kPa), and the lower proportion of tracheal volume to total lung volume in birds (about 2.5%) than in mammals (about 13.9%; see below).

Transalveolar Gas Exchange.   It was originally believed that oxygen was transported across the alveolar membrane by an active secretion process, but August and Marie Krogh showed that oxygen exchange was a diffusional process (Krogh and Krogh 1910). Thus, Fick's law describes transalveolar gas exchange

$$VO_2 = DO_2\ A(p_AO_2 - p_aO_2)/t$$
$$VCO_2 = DCO_2\ A(p_ACO_2 - p_aCO_2)/t \qquad (13.4)$$

where DO_2 and DCO_2 are the diffusion coefficients; *A* is the alveolar surface area; *t* is the transalveolar thickness; and p_aO_2 and p_aCO_2 are the average blood

pO_2 and pCO_2 values. The surface area of some amphibian lungs, but especially bird and mammalian lungs, is high because of the considerable compartmentalization of the lung surface. The path length for diffusion (*t*) includes the thickness of the alveolar epithelium and underlying basement membrane and the capillary endothelium (for example, the bird lung; Figure 13–16). The diffusion distance is about 0.36 to 2.5 μ for the human lung and less for other mammalian lungs (Table 13–6). The diffusion distances are considerably less in bird lungs and fish gills have considerably greater diffusion distances. Additional diffusion resistances would be added by the air boundary layer at the alveolar surface and the distance of the red blood cells from the capillary surface. The latter distance is not negligible, especially in bird lungs, being about 0.09 to 0.17 μ (cf. the transalveolar diffusion thicknesses of 0.09 μ or more; Maina 1984).

Fick's law can be rearranged to separate the physiological and anatomical/physical variables as follows.

$$VO_2/(p_AO_2 - p_aO_2) = DO_2.A/t = PDC_{O_2}$$
$$VCO_2/(p_ACO_2 - p_aCO_2) = DCO_2.A/t = PDC_{CO_2}$$
$$(13.5)$$

The term on each side of the equation is the **pulmonary diffusing capacity** (PDC). Since *A* and *t* are the same for O_2 and CO_2 exchange, then

TABLE 13–6

Diffusion distances between ambient medium and capillary blood in a variety of air-breathing animals.	
	Diffusion Distance μm
Air-breathing Fish	
Haplochromia	0.31–2.0
Saccobranchus	
Gill	3.6
Air sac	1.6
Skin	98.0
Anabas	
Gill	10.0
Suprabranchial chamber	0.21
Labyrinthine organ	0.21
Toad	1.3–3.0
Birds	
Pigeon	0.1–1.4
Swallow	0.09
Shrike	0.17
Mammals	
Rat	0.13–0.26
Human	0.36–2.5
Shrew	0.27

TABLE 13–7

Pulmonary diffusing capacity for oxygen (PDC_{O_2}) for a variety of vertebrates; $PDC_{O_2} = VO_2/(p_AO_2 - p_aO_2) = DO_2\, A/t$; the average pO_2 gradient between alveolar air and pulmonary capillary blood is calculated as VO_2/PDC.	
	PDC_{O_2} ml min^{-1} kPa^{-1} kg^{-1}
Air-breathing fish	
Saccobranchus	
Gills	0.024
Air sac	0.029
Skin	0.003
Anabas	
Gills	0.007
Suprabranchial organ	0.054
Labyrinthine organ	0.229
Amphibians	
Bullfrog (*Rana*)	0.027
Reptiles	
Lizard	
Varanus	0.072
Tupinambis	0.049
Turtle *Pseudemys*	0.066
Tortoise *Testudo*	0.114
Birds	
Chicken *Gallus*	0.580
Sparrow *Passer*	70
Starling *Sternus*	48
Mammals	
Mouse (*Mus*) active	7.5
Shrew (*Suncus*) active	11.2
Human rest	0.3
Human active	3.59

$PDC_{O_2}/PDC_{CO_2} = D_{O_2}/D_{CO_2} = 1/20$. There are considerable taxonomic differences in PDC for air-breathing vertebrates, from <0.005 ml O_2 min^{-1} kPa^{-1} kg^{-1} for skin to >10 for some mammal and bird lungs (Table 13–7). These differences reflect variation in A and t. However, PDC and VO_2 are highly correlated, as would be expected, and so the VO_2/PDC (= $p_AO_2 - p_aO_2$) is fairly constant, at about 2.7 kPa for most vertebrates.

It is important to appreciate that the pO_2 and pCO_2 vary along the length of the pulmonary capillaries, from the pulmonary arterial value ($p_{pa}O_2$) to the pulmonary vein value ($p_{pv}O_2$). It is impossible to sample blood and measure the pO_2 and pCO_2 along the length of a pulmonary capillary, but computer simulations for human and amphibian lungs indicate that the pO_2 and pCO_2 gradients are not linear along the capillary at rest but become more linear during activity (Millhorn and Pulley 1968; Withers and Hillman 1988).

Allometry of Respiratory Variables. The **allometry**, or scaling, of physiological variables as a function of body mass provides a very useful tool for comparing the effects of body mass on different physiological variables and in comparing physiological variables from different taxonomic groups (as we have

seen in Chapter 4). Physiological variables can be related to body mass in the form of variable = a Mass^b. Respiratory variables generally scale with mass in one of four ways: $\propto$ Mass$^{1.0}$, $\propto$ Mass$^{0.75}$, $\propto$ Mass$^{0.0}$, $\propto$ Mass$^{-0.25}$. These four general scaling relationships are illustrated for mammals in Figure 13–17. The scaling relationships for lung volume, tidal volume, respiratory surface area, minute volume, and respiratory rate are summarized for a variety of vertebrate taxa in Table 13–8. The slopes (b) of the scaling relationships are generally quite similar for the same physiological variable in different taxa, with variations in intercept (a) generally reflecting taxonomic variation in the differences in VO_2.

Cost of Pulmonary Respiration. The mechanical work required for one respiratory cycle of, for

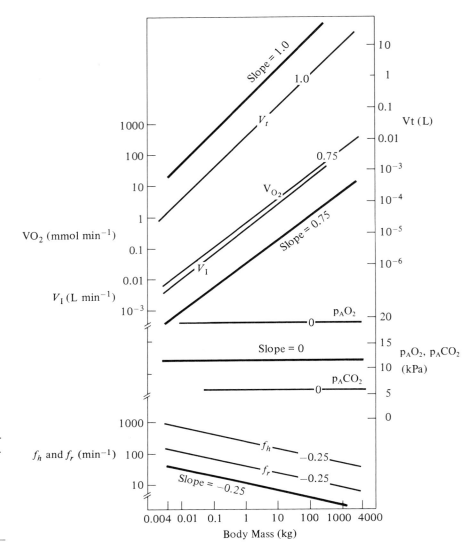

FIGURE 13–17 Allometric relationships for respiratory variables in mammals showing the four principle scaling relationships: $\propto$ Mass$^{1.0}$, $\propto$ Mass$^{0.75}$, $\propto$ Mass^{0}, $\propto$ Mass$^{-0.25}$. *(Modified from Dejours 1981.)*

example, a mammal can be calculated from the lung pressure-volume diagram (Figure 13–18). The work done in moving from one point on the P-V diagram to a second point is the area under the curve traced by the path between the two points towards the volume axis (since mechanical work = $P.\Delta V$ not $V.\Delta P$). On inspiration, the increase in volume due to the decrease in interpleural pressure (between the parietal and visceral pleura) requires expenditure of work; this is the **compliance work.** Further, the movement of air through the respiratory passages dissipates energy; the airway **resistance work** equals air flow rate multiplied by the pressure gradient; it can be graphically represented on the P-V graph. The change in shape of the lung tissues and alveolar surface film during inspiration also requires energy expenditure for **viscous work;** this can also be represented graphically. During expiration, the com-

pliance work released is equal to the compliance work done during inspiration. In contrast, more airway resistance and viscous work are expended during expiration, and so some of the compliance work released is partly required to overcome these energy requirements. Remember that expiration is normally passive; the source of energy for expiration is the compliance work. The net mechanical work done during one respiratory cycle is represented by the area on the P-V diagram between the inspiratory and expiratory curves. About 80% of this work is done to overcome airway resistance and about 20% for tissue viscosity work; no net compliance work is done.

The metabolic cost of respiration can be determined by experiments with cooperative human subjects from their lung pressure-volume curves. The mechanical work of a single respiratory cycle is

TABLE 13–8

Allometric relationships for respiratory variables. Relationship is of the form variable $= a \, Mass^b$, where mass is in grams.		
Variable	***a***	***b***
Lung Volume (ml)		
slope ≈ 1		
Mammals	0.035	1.06
Birds	0.034	0.97
Lizards	1.237	0.75
Amphibians	0.255	1.05
Tidal Volume (ml)		
slope ≈ 1		
Mammals	0.0075	1.0
Birds	0.0076	1.08
Lizards	0.020	0.80
Respiratory Area (cm²)		
slope ≈ 0.75		
Mammals (alveolar surface area)	119	0.75
Amphibians (capillary surface area)	0.04	0.98
Air-breathing Fish		
Anabas (labyrinthine organ)	0.807	0.80
Amphipnous (air sac)	0.125	0.80
Saccobranchus (air sac)	1.46	0.66
Channus (suprabranchial organ)	1.59	0.70
Minute Volume (cm³ min⁻¹)		
slope ≈ 0.75		
Mammals	1.57	0.75
Birds	1.19	0.77
Reptiles	0.41	0.76
Respiratory Rate (min⁻¹)		
slope ≈ -0.25		
Mammals	209	-0.25
Birds	146	-0.31
Reptiles	20.6	-0.04

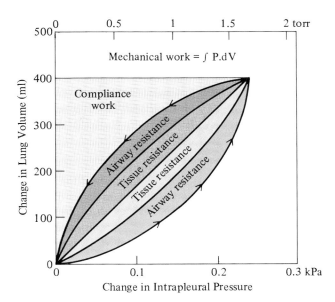

FIGURE 13–18 Pressure-volume diagram for a human lung during one breathing cycle showing the compliance work (triangular area from start of inspiration to end of inspiration, subtended towards the volume axis; grey shading), airway resistance work, and tissue resistance work (color). The area enclosed by the inspiratory and expiratory loops represents the net mechanical work for one breathing cycle. *(Modified from Guyton 1986.)*

about 0.29 J; the work per minute is about 2.9 J min⁻¹ if the respiratory rate is 10 min⁻¹. The calculated mechanical cost for breathing is about 1.5 ml O_2 min⁻¹, which is a metabolic cost of 0.5% of resting VO_2 assuming a muscle efficiency of 10% (Kao 1972). The metabolic cost of respiration is calculated to rise to about 0.8% of VO_2 if VO_2 rises to 2600 ml min⁻¹ and to about 3.8% at a VO_2 of 3500 ml min⁻¹. Although some of the compliance work released during expiration is used to overcome airway resistance and tissue viscosity, the remainder of the work is not useful and cannot be recovered. Consequently, the actual metabolic expenditure for a breathing cycle is greater than that calculated from the mechanical work difference between the inspiratory and expiratory cycles.

The metabolic cost of respiration has been estimated for the grass frog *Rana pipiens*, from the mechanical work of breathing, to be about 5% of the resting VO_2, based on the pressure-volume curves of a respiratory cycle and a calculated muscle efficiency of 8% (West and Jones 1975). This is similar to the estimated metabolic cost of breathing for some fish, at about 5 to 10% of resting metabolism (see Chapter 12).

Invertebrates

The most successful terrestrial animals, at least in terms of numbers of individuals and species, are the insects (see Chapter 1). The chelicerate arthropods (spiders, scorpions) are also quite numerous and diverse. Some crustaceans have also made the transition to air breathing. Other invertebrate groups have also made the transition from water to land, but generally with less success than the arthropods.

Invertebrate Transition from Water to Land

A number of invertebrates have made a more, or less, successful transition from water to land (see Table 13–9), including annelids, pulmonate snails, onychophorans, and a wide variety of arthropods (spiders and scorpions, ticks and mites, insects). There is a general trend for loss of the original gas exchange organs (gills) and evolution of lungs, book lungs, or tracheal/pseudo-tracheal respiratory systems. Tracheal respiratory systems and pseudo-tracheal systems have evolved independently in a number of invertebrate groups, e.g., isopod crustaceans, insects and myriapods, onycophorans, and arachnids.

Annelids. Gas exchange in amphibious and terrestrial oligochete worms occurs across their vascularized body surface. Cutaneous exchange appears

TABLE 13–9

Summary of the invertebrates that have successfully invaded the land.		
Class	**Description**	**Gas Exchange Organ**
Annelida (Oligochetes) Haplotaxidae, Alluroididae, Enchytraeidae	Amphibious and terrestrial worms	Cutaneous surface
Lumbricidae, Megascolecidae, Glossoscolecidae	Terrestrial and burrowing worms	Cutaneous surface
Mollusca (Prosobranchiates) Helicinidae, Cyclophoridae, Pomatiasidae	Operculate land snails	Vascularized mantle cavity
Ampullariidae	Amphibious snails	Gills and lung
(Pulmonates) Systellommatophora, Basommatophora	Land snails	Pulmonate lung
Veronicellidae	Tropical slugs	Cutaneous surface
Athoracophoridae	Tropical slugs	Tubular mantle
Onchidiidae	Intertidal slugs	Pulmonary sac
Onycophora	Tropical, temperate, moist habitats	Pseudotracheae
Arthropoda Isopoda	Pillbugs, sowbugs	Pleopods, pseudotracheae
Amphipods	Beach/leaf hoppers	Pleopods
Decapoda: Macrura	Crayfish	Gills, branchial surface
Anomura	Hermit and coconut crabs	Gills, branchial surface
Brachyura	Fresh water, land, ghost, fiddler, sand-bubbler, soldier crabs	Gills, branchial surface, gas windows
Merostomata	Horeshoe crabs, extinct eurypterids	Book gills and pseudo-tracheae
Arachnida	Scorpions, spiders, ticks	Book lungs and trachea
Chilopoda	Centipedes	Trachea
Diplopoda	Millipedes	Trachea
Insecta	Insects	Book lungs, sieve trachea, trachea

to be adequate to support the low metabolic demands of annelids, even in the largest species, which may exceed 3 m in length. Their well-developed, closed vascular system provides effective internal transport of O_2 once it has diffused across the skin into the cutaneous blood vessels (see Chapter 15).

A swamp earthworm from East Africa (*Alma*) uses its tail as a primitive "lung" when its burrow is inundated with water. The posterior end of the body is extended into the air and rolls up to produce a vascular, tube-shaped respiratory organ (Beadle 1957).

Gastropod Mollusks. Terrestriality, or at least air breathing, has evolved at least twice in gastropod mollusks, in the prosobranch and pulmonate snails. Many land snails are prosobranchs, but the majority are pulmonates. The land prosobranch snails possess an operculum, which can seal the shell aperture. These gastropods do not have a lung but rely for gas exchange on their mantle surface.

Some of the more primitive pulmonate snails are aquatic (freshwater) but come to the surface to breathe air. The opening of the lung (the pneumostome) is closed during submergence. Some deep water species have reverted to water breathing and a secondary gill supplants the mantle surface for gas exchange. Most of the higher pulmonates are truly terrestrial, some being found in even harsh desert environments. Almost all rely on their lungs for gas exchange. The lung is formed by fusion of the mantle cavity to the back of the animal, except for the pneumostome, which is a small opening on the right side (Figure 13–19). The lung may be ventilated to some extent by arching and then flattening the body, but most gas exchange between ambient air and lung air occurs by diffusion through the pneumostome, which is open most of the time. The pneumostome is most clearly observed in those gastropods with reduced shells, such as slugs.

Various tropical slugs have a modified respiratory system; some rely on cutaneous gas exchange and do not have a mantle cavity. Others have the mantle cavity modified into a series of tubes that invaginate the surrounding tissues or utilize a posterior air sac for air breathing (Barnes 1986).

Onychophorans. Onychophorans, or velvet worms (*Peripatus*), have resemblances to both annelids and arthropods, but their tracheal respiratory system is more arthropod-like than annelid-like. Their body surface is covered with many small openings, called **spiracles**, that open into a short atrium then ramify into the surrounding tissues as minute tubes, or **tracheae** (Figure 13–20). The spiracles are permanently open, and gas exchange occurs between the ambient air and tissues by diffusion. This kind of tracheal respiratory system of onychophorans is more highly developed in the insects.

Crustaceans. There are three groups of crustaceans that less successfully (amphipods, decapods) or more successfully (isopods) invaded the land. The terrestrial amphipods primarily inhabit beach-litter (e.g., beach hoppers) or leaf-litter (e.g., leaf hoppers). They have retained thoracic lamellar gills and probably also have some cutaneous gas exchange. Their terrestrial adaptations are primarily behavioral; they are restricted to moist habitats and are active at night to avoid desiccation.

The various amphibious and terrestrial decapods include some of the macruran crayfish (Astacidae, Parastacidae, Austrostacidae, Thallassinidae); anomuran crabs (e.g., hermit and coconut crabs); various brachyuran crabs; freshwater crabs (Potamidae, Pseudothelpusidae, Trychodactylidae); grapsid crabs (Grapsidae); land crabs (Gecarcinidae); ghost, fiddler, and sand-bubbler crabs (Ocypodidae); and soldier crabs (Mictyridae). Many amphibious and terrestrial decapods are actually unable to sustain

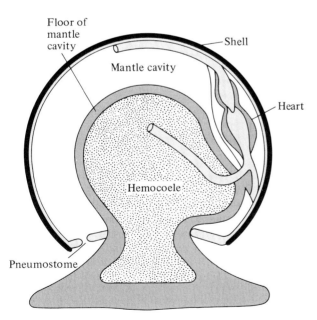

FIGURE 13–19 Diagrammatic cross section of a pulmonate land snail (*Helix*) showing the general arrangement of the lung (mantle cavity) on the right and left sides of the body and the heart and large blood vessels. (*Modified from Sommerville 1973.*)

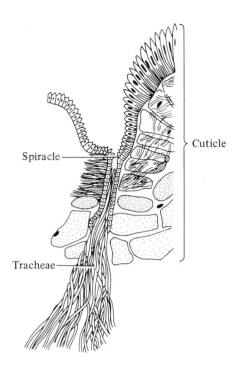

FIGURE 13–20   Tracheal system of an onycophoran, *Peripatus trinitatis*, showing one of the spiracles located in a slight cuticular depression and the bundles of tracheae extending into the tissues from the spiracle.   *(From Barrington 1967; after Grasse 1949.)*

adequate gas exchange and drown when submerged (Bliss 1968). One soldier crab, which inhabits intertidal mudflats, is also an obligate air breather; it builds a small mud "igloo" in order to store an air bubble for respiration when the tide is in.

The lamellae of crustacean gills, like those of fish gills, tend to collapse when exposed to air. In some crabs, the lamellae are strengthened and held apart in air by chitin thickenings; this can allow adequate gas exchange, but the gill surface is generally reduced in terrestrial crabs and the **branchial wall** assumes a gas exchange role, i.e., is a lung. To maximize gas exchange, the branchial chamber surface is expanded (and sometimes folded) and has a rich vascular lining. For example, the volume of the branchial chamber of terrestrial crabs increases approximately with mass$^{1.0}$ (Diaz and Rodriguez 1977), a similar allometric relationship as lung volume in vertebrates.

The potential surface area of the branchial wall is increased in pseudothelpusid crabs from South America by numerous perforations of the dorsal branchial wall. The perforations are about 0.05 to 0.25 mm in diameter, 1 mm long, and there are

about 10 to 40 per mm^2. The surface area of the pore openings is about 0.6 mm^2 per mm^2 carapace, and the total surface area lining the pores is about 13 mm^2 per mm^2 carapace. The perforations are lined by vascular tissue, and so gas exchange via the pores could be very significant. The coconut crab (*Birgus*) has much reduced gills (their surgical removal has little effect, at least on O_2 exchange) and the branchial chamber surface is highly expanded and vascularized, with epithelial tufts ("lung tissue") extending into the branchial chamber. The lung of the soldier crab *Mictyris* is formed by the inner lining of the branchial cavity and an epibranchial membrane (Figure 13–21); the cuticle has a highly vascular lining with a very short (0.09 to 0.48 m) blood-air distance (Farrelly and Greenaway 1987). The lung lining of an Australian arid-zone crab *Holthuisana* does not have vascular folds (e.g., as in *Ocypode*), perforations (e.g., as in thelphusids), or labyrinthine channels (e.g., as in *Birgus*) but has an extremely thin lining (0.25 to 0.3 μ blood-air distance; Taylor and Greenaway 1979). The highly vascular lung of *Pseudothelphusa* has complex airways, a very thin lining (0.4 μ blood-air distance), and has air ventilation by an internal "bellows" ventilation system somewhat reminiscent of the avian air-sac system (Innes, Taylor, and El Haj 1987). The efficacy of this respiratory system is indicated by the fact that gas exchange is limited by the capacity to perfuse the lung with hemolymph, rather than the diffusing capacity of the respiratory surface area.

The gills and branchial chamber of amphibious crabs are kept moist by periodic immersion in water; water remains in the branchial chamber during terrestrial excursions. This water is aerated by movements of the gill bailers that draw air into the branchial chamber through specialized posterior openings and out the anterior openings (normally the inhalant opening). In *Sesarma*, the gill bailer forces water out of the exhalant opening over the dorsal body surface via surface channels; the aerated water then returns to the branchial chamber (Barnes 1986). The gills of terrestrial crabs must be kept moist, otherwise the rate of gas exchange decreases dramatically, but little or no water is carried in the branchial chamber. In *Gecarcinus*, the pericardial sacs appear to conduct moisture (e.g., dew or rain drops) into the branchial chamber to maintain a high humidity. Hermit crabs keep their gills moist with extracorporeal water stored in the shell (McMahon and Burggren 1979); other terrestrial crabs rely on glandular secretions (Harms 1932).

Most terrestrial crabs rely on the same ventilatory system as aquatic crabs, the oscillatory motions of the gill bailers. The gill bailers generate a lower air

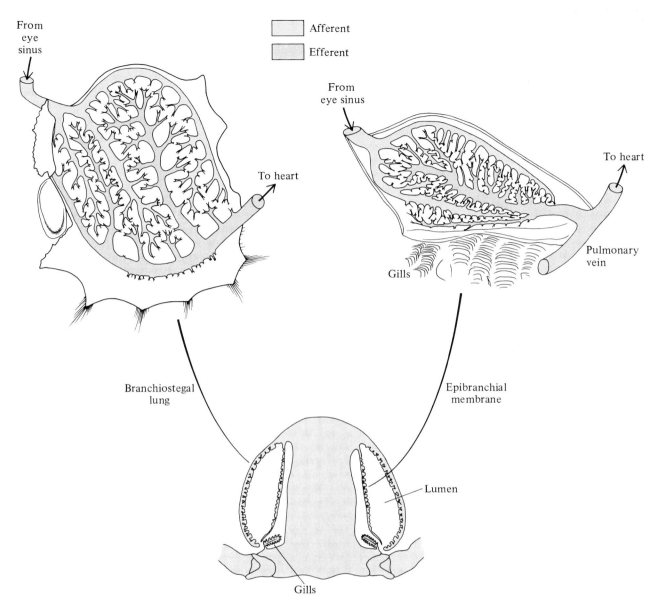

FIGURE 13–21 Vertical cross section of the branchial chamber of a soldier crab showing the reduced gills and vascular brachiostegal lung and epibranchial membrane of the branchial chamber and the main vasculature of the latter "lung" structures. *(From Farrelly and Greenaway 1987.)*

pressure gradient for branchial ventilation (<0.01 kPa) than in aquatic crabs. The ventilatory characteristics of aquatic and terrestrial crabs are compared in Table 13–10. Gills of the normally aquatic blue crab (*Callinectes*) are not adequate for aerial gas exchange, and the metabolic rate declines during air exposure; the extraction efficiency of oxygen is considerably lower in air (2%) than in water (40%). *Cardisoma*, a gecarcinid crab, has a slightly reduced VO_2 in water; this species does not drown when submerged. Note the same large difference between extraction of O_2 in air and water. *Gecarcinus* and *Gecarcoidea* (brachyuran crabs) and *Birgus* (an anomuran crab) are obligate air breathers and show similar low O_2 extractions in air. There is a general tendency for air-breathing crabs to have higher resting metabolic rates than aquatic crabs (like amphibians), and this appears to be true for isolated tissues as well as whole-animal metabolism. This trend might reflect the higher content of O_2 in air

TABLE 13-10

Metabolic rate (VO_2; ml O_2 min^{-1}), gill ventilation rate (V_w; ml water min^{-1}), inhalant and exhalant pO_2 (p_iO_2, p_eO_2; kPa), and gill extraction efficiency (E; %) for some aquatic and terrestrial crabs. *Callinectes* is an aquatic crab; the others are amphibious or terrestrial. *(After O'Mahoney 1977; Cameron 1975; Cameron and Mecklenberg 1973.)*

	VO_2	V_w	p_iO_2	p_eO_2	E
Callinectes					
Water	0.21	95	150	92	39
Air	0.072	71	50	147	1.8
Cardisoma					
Water	0.095	47	150	105	30
Air	0.109	18	150	144	3.9
Gecarcinus					
Air	0.094	21	—	—	2.7
Water	0.050	12	156	152	2.3
Gecarcoides					
Air	0.196	31	—	—	3.6
Birgus					
Air	0.707	41	53	145	5.2

and the lower cost of respiration, despite the lower O_2 extraction efficiencies and ventilation rates in air.

The sand-bubbler crabs (*Scopimera, Dotilla*) are small semiterrestrial crabs that feed during low tide and shelter during high tide in burrows that contain a trapped air bubble. These semiterrestrial crabs do not have lungs but have unusual, large membranous disks on the meral leg segments (Figure 13–22). These were thought to be tympanal organs for hearing but actually are respiratory surfaces, called **gas windows** (Maitland 1986). Deoxygenated hemolymph is carried to the windows in large vessels that branch to form a thin layer of small vessels lying adjacent to the thin (0.6 μ thick) gas window. Oxygenated hemolymph enters an interdigitating series of larger vessels and then drains into collecting vessels that convey it to the body.

Isopods (pillbugs, woodlice, snowbugs) are the most terrestrial crustaceans, having invaded land from the littoral zone. Many are restricted to moist habitats, and these species rely on their abdominal pleopod gills for gas exchange. The more terrestrial species (and more tolerant of desiccating conditions) have a lung-like cavity or **pseudotracheae** on some of their pleopods (Figure 13–23). The pleopods are more important for gas exchange in terrestrial isopods than littoral isopods, as is readily observed when pleopod gas exchange is blocked. The more terrestrial isopods are also able to maintain an almost normal VO_2 when exposed to dry air, unlike the littoral isopods. Associated with this ability is a much reduced evaporative water loss in those species with pseudotracheae.

Chelicerates

The horseshoe crab is a living representative of primitive, aquatic chelicerates. Its gills are modified ventral abdominal appendages; the underside of each appendage is highly folded into many (about

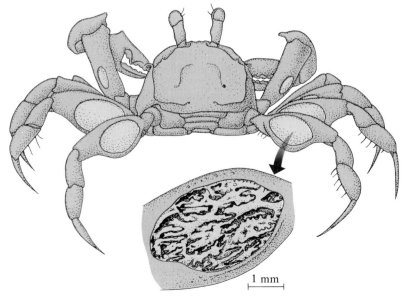

FIGURE 13–22 The sand-bubbler crab *Scopimera inflata* has membraneous gas windows in each leg meral segment (left) that consist of a thin cuticular window (about 0.6 μ thick) subtended by a network of large interdigitating blood channels connected by small channels. *(Modified from Maitland 1986.)*

1 mm

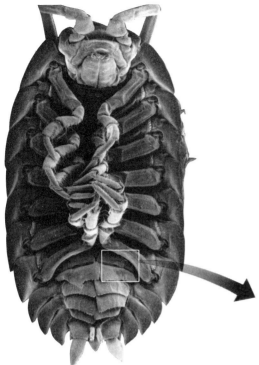

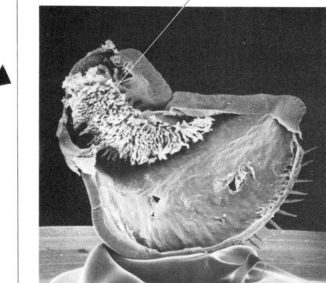

Pleopod   Pseudotracheae

FIGURE 13–23   Ventral view of a terrestrial isopod, *Porcellio scaber*, and the internal pseudotracheal respiratory system.   *(Photographs courtesy of Withers and Stewart, Department of Zoology, The University of Western Australia; and Hoese 1982.)*

150) lamellae, like pages of a book, hence the name **book gills.** The extinct Eurypterida resembled horseshoe crabs in body plan, but some were probably amphibious or semiterrestrial. The gills of these eurypterids were protected from desiccation by plate-like gill appendages, similar to those of horseshoe crabs (which can also come onto land for short periods). The surfaces of the eurypterid gills were covered with evaginated cone structures for aquatic respiration and by other strange invaginated masses of spongy chitin, which may have functioned as pseudotracheae for aerial gas exchange (Figure 13–24A).

The other chelicerates—spiders, scorpions, ticks, and mites—are essentially all air breathing (at least the adults). Gas exchange generally occurs via one of two respiratory designs: book lungs (the more

primitive design) or tracheae (more advanced). Some species possess both book lungs and tracheae. Scorpions have up to four pairs of book lungs and whip-scorpions and amblypogids two pairs; they all lack a tracheal system. False-scorpions, sun-spiders, ticks, and mites have no book lungs but have a tracheal system. Spiders vary in their respiratory structure; some have two pairs, one pair, or no book lungs, some have sieve tracheae, and some have tubular tracheae.

Book lungs, which greatly resemble book gills as their name suggests, are paired structures located on the ventral surface of the abdomen that consist of a system of invaginated lamellae. An invagination of the cuticle forms a spiracular opening and atrial chamber (Figure 13–24B). Book lungs may have evolved from book gills, often being located in the

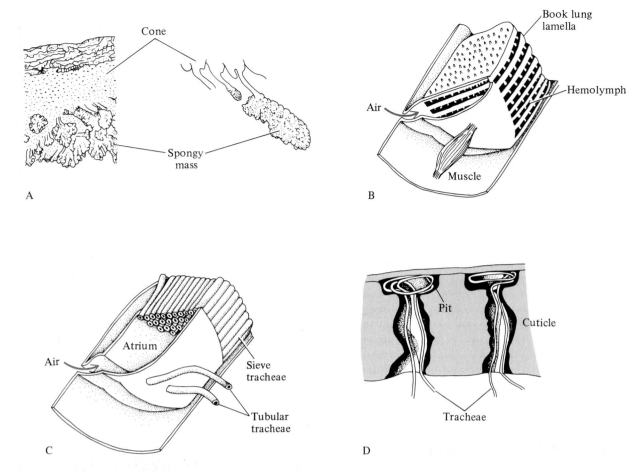

FIGURE 13–24 (A) Gill structure of a fossil eurypterid arthropod, *Parahughmilleria*, as seen from the ventral gill surface. The small evaginated cones increase the gill surface area for aquatic respiration; the invaginated spongy masses may have functioned as pseudotracheae for aerial respiration. (B) Structure of the book lung in a spider; air enters through the spiracle into the atrium by diffusion and ventilation due to muscle contraction. Air diffuses from the atrium into the lamellar spaces; hemolymph circulates through the blood lamellar spaces that alternate with air lamellar spaces. The lamellar surfaces are held apart by small peg-like surface projections. (C) Sieve tracheae structure of some spiders and pseudoscorpions showing the bundles of tube tracheae that branch from the air-filled atrium. (D) Cross-sectional view of the cuticle of the water mite *Arrenurua* showing the cuticular pit with a spiralled tracheal loop. *(Modified from Stormer 1977; Levi 1967; Mitchel 1972.)*

same position. Embryological evidence in spiders also suggests the possible evolution of book lungs from book gills. The book lamellae project from one side of the atrium forming air-filled pockets on one side of the lamellae and hemolymph-filled pockets on the other side. The lamellar plates are held apart by peg-like cuticular projections. Some air ventilation is provided by the action of a muscle inserting on the atrium surface opposite to the lamellae. Book lungs are considered to be the primitive gas exchange system in air-breathing

arachnids; variable numbers of book lungs are the only gas exchange system in a variety of arachnids.

Primitive spiders retain book lungs as their only gas exchange system, but more advanced spiders have evolved from their posterior book lungs a system of **tracheae**, or there is a complete transformation of both pairs of book lungs into tracheae. Some spiders have **sieve tracheae**, bundles of tracheal tubes that branch from a tubular cavity (Figure 13–24C). Sieve tracheae are possibly an intermediate stage in the evolution of tube tracheae. **Tube** tracheae are relatively simple and generally un-branched tubes, with spiracle openings on the abdomen or legs. The evolution of book lungs into tracheae has apparently occurred independently a number of times in advanced spiders (Levi 1967). In some spiders, the tracheae are limited to the abdomen, and gas exchange to the remainder of the body is accomplished via the anterior book lungs and hemolymph transport. In those spiders without book lungs, the tracheae invest all body tissues, including the head, antennae, and legs. Other advanced arachnids also lack book lungs and have various numbers of spiracles and tracheae.

The spiracles of the tracheal system in hard-bodied ticks (Ixodidae; the more mesic ticks) are covered by a raised plate that is perforated by numerous air pores, called **aeropyles** (Woolley 1972). The atrium of the spiracle opening is a complex system of interconnected chambers into which the tracheae open. The soft-bodied ticks (Argasidae; the more xeric ticks) lack perforate spiracular plates; their spiracular opening is a slit-like opening that communicates with a simple hollow chambered atrium, thence the tracheae. Ticks do not have a spiracular closure mechanism.

Terrestrial water mites lack tracheae, but the aquatic species generally have tracheae branching from large air chambers, with spiracles on the anterior of the body. Some water mites lack spiracles, and O_2 uptake occurs across the cuticle into the tracheae by diffusion. The tracheae are blind-ended tubes with portions lying directly under the cuticle; one or both ends of the tracheae extend into the body tissues. Water mites with thick cuticle have special pits in the cuticle that contain a spiralled extension of a trachea, the other end of which invaginates the body tissues (Figure 13–24D).

Millipedes, Centipedes, and Insects

Millipedes, centipedes, and insects (the uniramous arthropods) share a common mode of respiration, **spiracles** and a **tracheal system**. The related, minute, soft-bodied pauropods lack a respiratory and circulatory system. The small symphylans, which are also related to millipedes, centipedes, and insects, have a single pair of spiracles that open on the head. These spiracles have a system of tracheae that supplies only the first three trunk segments.

The millipedes (Diplopoda) and many centipedes (Chilopoda) have many pairs of spiracles (2 pairs per diplosegment in millipedes, 1 pair per segment in centipedes). The unclosable spiracles open into an atrium and then into the tracheal system. The long-legged centipedes (scutigerimorphs) have evolved a quite different tracheal system, somewhat resembling lungs. The single, mid-dorsal spiracle opens into an atrium that gives rise to two bundles of tracheal tubes that bathe the hemolymph of the pericardial cavity.

Insects are the most abundant and successful terrestrial arthropods; they also have the most extensive and best-studied tracheal systems (Figure 13–25A). The number of pairs of spiracles is highly variable; adult insects have up to 10 pairs (2 thoracic pairs above the second and third pairs of legs; 8 abdominal pairs) but some have no spiracles, although the tracheal system is still present. Some embryonic insects have 12 pairs (3 thoracic and 9 abdominal). The basic organization and functioning

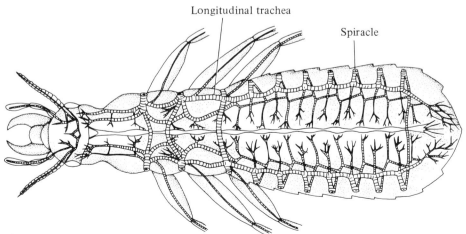

Longitudinal trachea

Spiracle

FIGURE 13–25 **(A)** Generalized tracheal system of an insect.

A

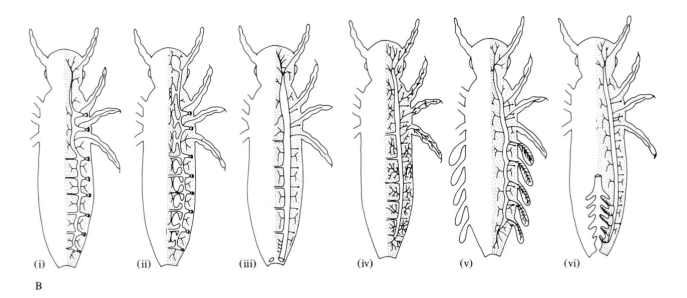

(i)   (ii)   (iii)   (iv)   (v)   (vi)

B

FIGURE 13–25 (*Continued*)   **(B)** Schematic of some different types of respiratory systems in insects: (i) typical anastomosing tracheae and numerous pairs of spiracles; (ii) tracheal system with air sacs for ventilation; (iii) only a single terminal pair of functional spiracles; (iv) no functional spiracles, diffusion occurs across the cuticle; (v) as in (iv), with abdominal tracheal gills; (vi) as in (iv), with rectal tracheal gills.   (*From Barnes 1986; Wigglesworth 1984.*)

of the tracheal system is highly modified in many insects for specialized functions. Some variations in the basic arrangement of tracheal systems without air sacs (Figure 13–25Bi) and with air sacs (Figure 13–25Bii) include one pair of functioning spiracles (often the most posterior pair; Figure 13–25Biii); no functional spiracles, with gas exchange occurring across the cuticle (Figure 13–25Biv); external abdominal gills invested with tracheae (Figure 13–25Bv); or internal rectal gills (Figure 13–25Bvi).

The spiracles of primitive insects (e.g., silverfish, springtails) are simple openings of the tracheae, but the spiracles of more advanced insects open into an atrium from which the tracheae extend; there may be a filtering apparatus and a closable valve. The spiracles of most insects can be closed by a flap-like valve or a constricting muscle. The closure mechanism in an orthopteran insect (*Dissosteira*) is two external valves connected by a ventral lobe; elasticity keeps the spiracle open, and an occlusor muscle closes it. In contrast, the spiracle of an ant (a hymenopteran) is closed internally by muscular constriction of an occlusable chamber; the chamber is opened by a second muscle (Figure 13–26).

Closure of the spiracles is an important water conservation mechanism (see below, and Chapter 16). The spiracles are normally closed, and open for short periods whenever necessary for gas exchange. The pairs of spiracles can also open and close in complex patterns, such that some are inspiratory and others are expiratory. The important role of the spiracles in limiting water loss is easily demonstrated by adding CO_2 to the air. The classic experiment of Bursell (1957) with tsetse flies also well illustrates this. Tsetse flies in low relative humidity air (low rh) are able to reduce their water loss by closing their spiracles for long periods, from 50% of the time at 80% rh, to 90% of the time at 0% rh (Figure 13–27). There is a dramatic increase in water loss if the spiracles are forced open by adding CO_2 to the air. The spiracles remain open for longer periods if O_2 demand is elevated by high temperature, activity, during digestion, and during egg production.

The insect tracheal system generally has a complex system of tracheae branching from the spiracles and atria, but the pattern of tracheae investing a tissue varies. For example, the tracheae of flight muscles of the locust *Schistocerca* can run along the center of the muscle mass and have radial branches to the muscle cells (centro-radial), can run along the edge of the muscle and have radial branches (latero-radial), or can form a lateral air sac with parallel branches (latero-linear; Figure 13–28).

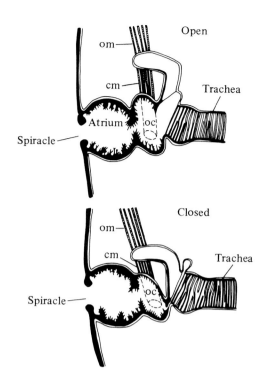

FIGURE 13–26 Abdominal spiracle of a hymenopteran insect, an ant, showing the open and closed positions. The internal occluding chamber (oc) can be constricted by the closing muscle (cm) and dilated by the opening muscle (om). *(From Richards and Davies 1977.)*

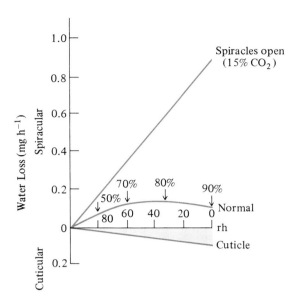

FIGURE 13–27 Relationship between ambient relative humidity and spiracular (above rh axis) and cuticular (below rh axis) evaporative water loss for a normal tsetse fly and a tsetse fly induced to keep its spiracles open by exposure to 15% CO_2. An estimate of the percent closure of the spiracles for the normal fly is also indicated. *(Modified from Bursell 1957.)*

The smallest tracheae, about 2 to 5 μ diameter, terminate in a tuft of small **tracheoles,** each less than 1 μ in diameter. The tracheoles closely invest the individual cells, often invaginating into the cells themselves (especially active cells such as flight muscle). The large tracheae are supported by annular or spiral thickenings of the cuticle. The tracheoles are also supported by microscopic spiral or annular thickenings. The tracheoles are the only significant area of gas exchange because of their large surface area and thin cuticle wall. Insects shed the lining of the large tracheae whenever they molt. The tracheole linings, however, are not shed but are reconnected to the new tracheae lining by a "cement substance."

The terminal ends of the tracheoles are often filled with interstitial fluid. There is a balance of forces between the colloid osmotic pressure of the interstitial fluid and the capillary forces of the lipophilic tracheole lining. Increased ionic strength or acidification of the interstitial fluid withdraws fluid from the tracheoles, facilitating gas exchange during activity or varying pO_2 (Figure 13–29).

The tracheal system of insects is essentially a simple diffusion system and its structure is well suited for this. For example, August Krogh (1920) carefully measured the various parts of the tracheal

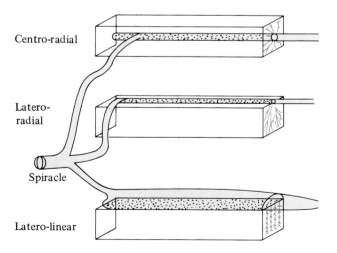

FIGURE 13–28 General types of tracheal supply to the flight muscles of *Schistocerca gregaria*. *(Modified from Weis-Fogh 1964a.)*

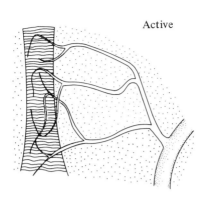

Rest

Active

FIGURE 13–29 Regulation of the interstitial fluid content in the terminal ends of the tracheoles during rest and activity. *(From Wigglesworth 1984.)*

system of a large goat moth larva (*Cossus*) and found that all portions had the same total cross-sectional area, about 6.7 mm², from the 9 pairs of spiracles to the tracheoles. The effective length of the tracheal system was about 6 mm. Thus, the tracheal system of *Cossus* is essentially a tube of 6.7 mm² in area and 6 mm long. If all the spiracles of *Cossus* were open, the metabolic rate of 18 μl O_2 min^{-1} could be supported by a pO_2 gradient of only 1.5 kPa from the spiracles to the ends of the tracheoles. There is a similar constancy of cross-sectional area in *Rhodnius*. This pattern is not true of all insect tracheal systems, however. In silk worms, the area is not constant but initially increases to over 8 $\times$ the spiracular area and then is reduced to less than the spiracular area at the tracheoles.

Insects are generally precluded from becoming very large by diffusion limitations of their tracheal system. The rate of diffusion varies inversely with path-length; consequently, gas exchange in large insects would eventually become limited by the long diffusion path-length from the spiracles to the cells. The largest living insects are tropical beetles, about 15 cm in length. Fossils of tropical dragonflies from the Carboniferous period had wing spans up to 2 feet but their long slender bodies were only about 3 cm in diameter. High tropical temperatures would marginally increase the diffusion coefficient and facilitate diffusional gas exchange but would also increase the metabolic demand for O_2.

Many insects ameliorate the diffusional limitation of their tracheal system by ventilating the larger diameter tracheae, i.e., there is bulk flow of air through at least parts of the respiratory system. The tracheal system of many insects has spiral and annular thickenings that make them highly resistant to collapse or expansion, and so there is no air ventilation through the system. The large tracheae of some insects, particularly the more active ones, are modified to form collapsible regions that are ellipsoidal or ribbon-like in cross section or have large dilatations that form **air sacs.** For example, the worker honeybee has an extensive air sac system. Alternate expansion and collapse of these regions provides a ventilatory air flow that effectively supplements diffusional exchange in these larger sections of the tracheal system. The air sacs can be compressed by muscle contraction (e.g., contractions of the flight muscles of the locust *Schistocerca* cause tracheal ventilation; 7 μl per cycle) or even by hydrostatic pressure changes of the blood. The massive flying insects such as *Goliathus* rely on strong abdominal pumping for ventilation. Air is forced through the large tracheae of the second and third spiracles of *Petrognathus* during flight by a ducting effect (**ram ventilation;** see Bernoulli effect, Chapter 14.) Large numbers of secondary tracheae provide diffusional exchange with the flight muscles. A considerable fraction of the total tracheal air volume can be expelled by ventilation movements, about 1/3 in *Melolontha* and 2/3 in *Dytiscus* (Wigglesworth 1984).

A combination of spiracle opening and closing and ventilation movements can produce a unidirectional flow through the larger tracheae (but not the tracheoles!). For example, air is inspired through the anterior four pairs of spiracles and expired via the six posterior pairs in the locust *Schistocerca*. In mantids, a unidirectional air flow is maintained through the prothoracic tracheae; air is drawn in the first pair of spiracles, passes anteriorly through the head, then posteriorly to the abdomen for expiration (Miller 1973).

Insect flight muscle is one of the most metabolically active tissues. For example, the metabolic rate of the flight muscle of the desert locust *Schistocerca gregaria* can be as high as 2.8 ml O_2 g^{-1} min^{-1} and 7.3 ml g muscle^{-1} min^{-1} for honeybees. The arrangement of the O_2 supply to these tissues is an excellent example of a mixed ventilation/diffusion tracheal system pushed to its physical limits. The actual organization of the tracheal supply to the locust's flight muscles is exceedingly complex (Weis-Fogh 1964a, 1964b). The spiracles connect through tracheae to large lateral and medial air sacs, close to the flight depressor muscles. These air sacs directly invest the muscles with numerous small tracheae in a latero-linear pattern. As much as 7% of the muscle surface may be covered by invaginating tracheae. Other muscles (such as the wing elevators) are not close to air sacs; their tracheal supply invaginates the muscle then subdivides into smaller tracheae (a centro-radial pattern), or runs along the muscle surface with small branches extending into the muscle (latero-radial supply). The intramuscular tracheae branch into smaller diameter tracheae that thoroughly invest the entire muscle; few muscle cells are more than 200 μ from these branches. Most of the spaces between the muscle cells contain hemolymph or tracheoles. In *Schistocerca*, the tracheoles invaginate the individual mus-

cle cells and then extend for a considerable distance inside the muscle fiber axis. The tracheoles do not actually penetrate the muscle cell membrane but are covered by a tubular cell membrane sheath.

Many insects, especially arid-adapted species (e.g., *Tenebrio;* silkworm pupae; larval and adult lepidopterans), have a pattern of **discontinuous respiration.** Oxygen is absorbed at a fairly constant rate but carbon dioxide is released in short bursts. The metabolic production of CO_2 is not a cyclic event but is continuous, at the cellular level, and all CO_2 exchange occurs via the spiracles. The mechanism for discontinuous CO_2 release is a simple consequence of the relative solubilities of O_2 and CO_2 and the pattern of spiracular opening. The spiracles are tightly closed for long periods of time (between bursts of CO_2 release), except for brief periods of slight opening (spiracular flutter; Figure 13–30). During the closed spiracle period of the cycle, O_2 is consumed and CO_2 produced at often a slightly lower rate (RQ = VCO_2/VO_2 = 0.7 to 1.0). CO_2 is

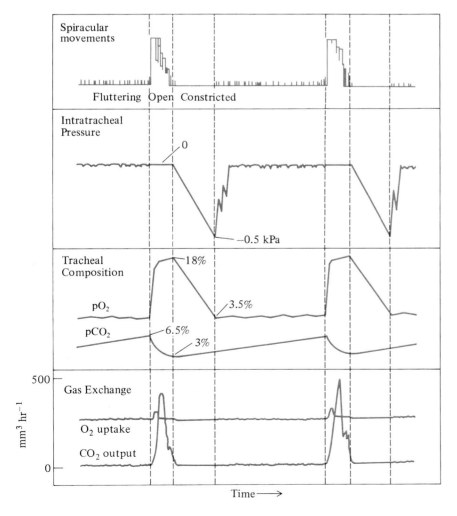

FIGURE 13–30 Discontinuous respiration in the silkworm pupa showing the spiracular movements (top), intratracheal pressure (second from top), pattern of pO_2 and pCO_2 change in the tracheae (second from bottom), and pattern of O_2 uptake and CO_2 output determined by external respirometry (bottom). *(From Levy and Schneiderman 1966.)*

more soluble than O_2 in body fluids, so the air pressure in the tracheae declines to subatmospheric values. During this period, the pO_2 declines markedly in the tracheae and the pCO_2 increases slightly. When a critical pO_2 is reached, the spiracles flutter and the subatmospheric tracheal pressure causes bulk flow of air into the tracheae. The inward bulk flow prevents any outward diffusion of CO_2. The spiracular flutter and inward bulk flow thus maintain a fairly constant tracheal pO_2 but the pCO_2 continues to increase slowly. Eventually, the pCO_2 reaches a critical value that stimulates the spiracles to open for a longer period of time than during flutter. The tracheal pressure equalizes with ambient pressure and CO_2 now diffuses out of the tracheae, returning the pCO_2 to a low value. The spiracles close and the cycle begins again. This discontinuous pattern of gas exchange provides O_2 on a fairly continuous basis but limits CO_2 and, more importantly, H_2O loss due to the short periods of spiracular opening. The overall water loss is much lower than if the spiracles had remained open permanently or for longer periods of time.

Physical Gills and Plastrons

Many insects, including nymphal, larval, and adult stages, are aquatic and breathe water (see also Chapter 12). Larval caddis flies and nymphal stone flies, mayflies, and dragonflies have abdominal or rectal gills for gas exchange, and other aquatic larvae (e.g., *Chironomus*) rely on cutaneous diffusion. However, a number of adult insects, mostly beetles and bugs, dive under water with an air bubble positioned so that it contacts one or more pairs of spiracles. Some other arachnids, such as spiders and a few centipedes, also use air bubbles for gas exchange when submerged. Some air-breathing intertidal crabs also use an air pocket to breathe when the tide is in. These air bubbles clearly act as an **O_2 store.** For example, the O_2 content of the air bubble in the subelytral space of the beetle *Dytiscus* declines from 19.5% O_2 to 1% or less within 3 to 4 minutes of diving (Wigglesworth 1984). In many instances, however, a storage role underestimates the significance of the air bubble; the bubble in fact functions as a **physical gill.**

The simple but elegant experiments of Ege (1915) with the backswimmer *Notonecta* clearly demonstrated the role of an air bubble as a physical gill. A *Notonecta* with its ventral abdominal air bubble can survive submersion in water equilibrated with normal air for about six hours. Survival time is only about five minutes if the water has been previously equilibrated with nitrogen. If given a bubble of pure O_2 in O_2-equilibrated water, the survival time is 35

minutes, only 1/10 as long as with an air bubble! The bubble is clearly an O_2 store, but why does a bubble of normal air last so much longer than a bubble of pure O_2? During diving, oxygen diffuses from the air bubble into the tracheal system and to the cells; an almost equivalent volume of CO_2 is produced and diffuses into the air bubble. However, the higher solubility of CO_2 compared to O_2 allows virtually all of the CO_2 to rapidly diffuse into the water. Consequently, the volume of the air bubble decreases over time as O_2 is removed. The lowered pO_2 of the air bubble causes O_2 to diffuse into the bubble from the surrounding water. Nitrogen also diffuses out of the air bubble due to the higher pN_2 of the air (after O_2 is removed). Consequently, the air bubble is an O_2 store, and also promotes O_2 extraction from the water.

The tendency for a gas to diffuse between air and water (its invasion rate) depends on the product of its solubility and diffusion coefficients. The invasion rate for O_2 ($IO_2 = \alpha O_2 . DO_2$) is much less, about $7.05 \ 10^{-9}$ cm^2 sec^{-1} kPa^{-1}, than for CO_2 ($ICO_2 = 1.56 \ 10^{-7}$); nitrogen has an even lower invasion rate than O_2 ($IN_2 = 3.22 \ 10^{-9}$). Consequently, N_2 diffuses out of a bubble more slowly than O_2 diffuses in. The net effect, as can be shown with a relatively straightforward mathematical analysis (Ege 1915; Rahn and Paganelli 1968), is that the O_2 supplied throughout the lifetime of an air bubble is about 8.3 times the amount of O_2 originally present. The **gill factor** is the ratio of total O_2 supplied by the air bubble to the O_2 initially present. It is equal to the ratio of O_2 and N_2 invasion rates and the fractional N_2 and O_2 contents in the bubble as follows.

$$
\begin{aligned}
GF &= (IO_2 / IN_2) / (fO_2 / fN_2) \\
&= (7.05/3.22)/(.21/.79) \quad (13.6) \\
&= 8.28
\end{aligned}
$$

An increase in hydrostatic pressure under water (about 101 kPa per 10 m) diminishes the utility of a physical gill since the bubble pO_2 and pN_2 are increased by the hydrostatic pressure but the water pO_2 and pN_2 are not. This promotes greater diffusional loss of both O_2 and N_2 from the bubble into the water.

An African water beetle (*Potamodytes*) that lives in fast-flowing streams uses Bernoulli's principle and an air bubble to its infinite advantage (Stride 1955). The submerged beetle traps an air bubble against the ventral body surface using its legs (Figure 13–31). The air bubble decreases in volume over time if the beetle is kept in still water and provides a sufficient O_2 store for about three to six hours. However, in fast-flowing water the air bubble does not diminish in volume over time and is used for 10 to 12 hours. The water velocity around the air

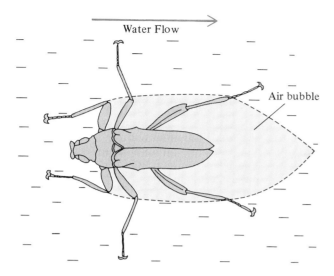

FIGURE 13–31 "Permanent" physical air bubble gill of the African beetle *Potamodytes tuberosus* in a swift-flowing stream. Bubble pressure at a velocity of 200 cm sec^{-1} is about -0.005 kPa. *(From Stride 1955.)*

bubble results in a subatmospheric pressure of up to 0.04 kPa at a velocity of 200 cm sec^{-1}. This low pressure appears to be adequate to stabilize the volume of the bubble by O_2 and N_2 diffusion into the bubble from the water. The air bubble is thus a **permanent gill.**

A more structurally stable permanent gill is provided by the **plastron** of many aquatic insects, such as the diving beetles *Haemonia* and *Elmis* and the bug *Aphelocheirus*. The plastron of these insects is a thin layer of cuticular hairs or scales, often bent at the tips or feathered (Figure 13–32A). The plastron of the hemipteran *Aphelocheirus*, for example, has about 2.5 10^8 hairs per cm^2, each about 10 μ long with a 90° bend at the tip. The nine pairs of spiracles open into longitudinal grooves, or rosettes of radiating canals, all lined with the fine cuticular hairs. In the elmid beetle, spiracular grooves communicate with the plastron and the subelytral space; the plastron covers most of the lateral and latero-ventral surface. Its shorter plastron hairs hold a thin air film (the microplastron) but a thicker air film is initially present on submergence. This larger air film is rapidly depleted by the metabolism of the insect. The plastron is a perforate cuticular surface in some insects (Figure 13–32B) or a perforate covering over cuticular grooves (see Figure 12–16B, page 584). The perforate spiracular plate of some ticks may also function as a plastron air-water interface.

The plastron holds a thin film of air that gives the insect a silvery iridescent appearance. The air film of the plastron is permanent; it does not collapse but maintains a constant volume and acts as a permanent physical gill. The ability of the plastron to resist wetting depends on the surface tension of water; it is proportional to the strength and spacing of the cuticular hairs or the diameters of the surface perforations. The high surface tension of water prevents the air film from collapsing away from the outer plastron surface, even at hydrostatic pressures over 200 kPa. The plastron of *Aphelocheirus* resists collapse at pressures up to 350 to 500 kPa, whereas the microplastron of *Haemonia* resists only 50 to 200 kPa. Reduction of the surface tension of water (by adding butyl alcohol) wets and destroys the plastron at low hydrostatic pressures. Reducing the surface tension of water from 70 mN m^{-1} to 26 (with 10% butyl alcohol) wets the cuticle of *Aphelocheirus*, and 7 to 9% butyl alcohol wets that of *Haemonia*.

One measure of the efficacy of a plastron gill is the O_2 gradient required to support the VO_2 for a given plastron area and thickness, i.e.,

$$\Delta pO_2 = VO_2/A.IO_2 \qquad (13.7)$$

where A is the area and IO_2 the oxygen invasion coefficient. Values of ΔpO_2 are lowest (<1 kPa)

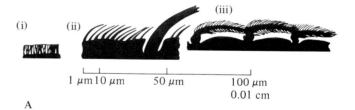

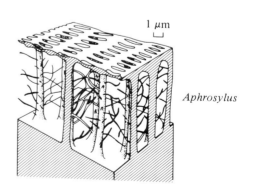

FIGURE 13–32 Various forms of plastrons, permanent physical gills, in insects. **(A)** Cuticular hairs of (i) *Aphelocheirus*, (ii) *Elmis*, and (iii) *Phytobius*. **(B)** Perforate cuticular surface forming a plastron over the surface of *Aphrosylus* cuticle. *(From Crisp and Thorpe 1948; Hinton 1967.)*

for insects with the most developed plastrons and greatest (>10 kPa) for insects with compressible gills. The eggs of many insects, which also have plastrons (see below), require a high ΔpO_2. For a vertebrate comparison, a chicken egg has a ΔpO_2 from about 1 to 9 kPa, depending on its stage of development and VO_2.

Plastrons only favor O_2 delivery if the water is well oxygenated. At low water pO_2, a plastron favors diffusion of O_2 from the tracheae into the water! Thus, most adult insects with plastrons are found in well-oxygenated water, e.g., fast-running water, the intertidal zone, or the edges of lakes. The beetle *Aphelocheirus* requires an O_2 concentration of 18.6% to avoid asphyxia. Insect eggs with plastrons are also limited to similar environments with well-oxygenated water.

Insects are not the only arthropods that exploit physical gills. For example, some intertidal centipedes and diving spiders use air bubbles as compressible physical gills. The perforated spiracular plates and complex atria of hard-bodied ticks predispose plastron breathing when submerged (Woolley 1972). Intertidal mites use a plastron (Hinton 1971; Krantz 1974).

Regulation of Respiration

We have already seen that various chemoreceptors (primarily for O_2 but also for CO_2) modify the respiratory rhythm for aquatic animals. Their primary respiratory stimulant was O_2 (i.e., hypoxia) rather than CO_2 (i.e., hypercapnia). A profound change in the basic chemosensory role has occurred in air-breathing vertebrates and invertebrates because of the markedly different capacitances of air and water for O_2 and CO_2.

The solubility coefficients for O_2 and CO_2 in water differ by about 20 ×. Aerobic metabolism produces about the same molar volume of CO_2 as the molar volume of O_2 consumed (RQ = VCO_2/VO_2 = 0.7 to 1.0). Consequently, the change in pO_2 due to metabolic activity in water is much greater than the increase in pCO_2 (see Figure 12–3, page 571). The relationship between the change in pO_2 and pCO_2 for inhaled (*i*) and exhaled (*e*) water is

$$\alpha O_2(p_iO_2 - p_eO_2) = \alpha CO_2(p_eCO_2 - p_iCO_2)$$
$$(13.8a)$$

if RQ = 1. In air, the molar production of CO_2 per molar O_2 consumption is also reflective of the RQ, but the partial pressure change in CO_2 is equal to that of O_2 because air has the same capacitance for O_2 and CO_2. Thus,

$$p_iO_2 - p_eO_2 = p_eCO_2 - p_iCO_2 \quad (13.8b)$$

if RQ = 1. These relationships for air and water are shown in Figure 12-3, with approximate ranges in aquatic and terrestrial animals. The higher pCO_2 values for terrestrial animals have profound significance for acid base regulation and also for regulation of respiration.

There is a consistent trend in air-breathing animals for CO_2 to become the primary respiratory stimulant. There are various possible reasons for this. First, blood pO_2 is not necessarily an accurate predictor of blood O_2 content, especially at high pO_2, because hemoglobin becomes saturated with O_2 (see Chapter 15). The pO_2 can decrease substantially below the normal arterial pO_2 without a substantial reduction in the O_2 content of the blood. Blood pCO_2 is a more accurate indicator of CO_2 and O_2 transport because of the steeper relationship between pCO_2 and blood CO_2 content at typical arterial values. Second, CO_2 is intimately involved with H^+ regulation; accurate monitoring of blood H^+ concomitantly confers accurate monitoring of blood CO_2. Third, chemoreceptors typically respond to the ratio of a signal to a setpoint, rather than the arithmetic difference (see Chapters 2 and 7). If alveolar pO_2 declines from its setpoint of 13.3 kPa to 12.0 kPa, the ratio is 12.0/13.3 or 0.9; the corresponding increase in pCO_2 from its setpoint of 5.3 to 6.7 kPa is 6.7/5.3 or 1.25. A CO_2 receptor might therefore be more responsive to an equivalent partial pressure change than a pO_2 receptor.

Regardless of the reasons, there is a consistent trend towards CO_2 being the primary respiratory stimulant in air-breathing fish, amphibians, birds, and mammals.

Vertebrates

The basic respiratory cycle in air-breathing vertebrates, as in aquatic vertebrates, is due to rhythmic activity of neurons located in the medulla of the brainstem, the **respiratory center** (Figure 13–33). This respiratory center receives sensory information from many other central and peripheral neurons but is nevertheless responsible for the primary respiratory rhythm. The basic respiratory rhythm is intact even after transection (severing) of the brainstem anterior to the medulla. This is true not only for lower vertebrates, but also for birds and mammals. The anatomical location and functioning of the medullary neurons are best studied in mammals. There are two functionally separate groups of neurons: the inspiratory neurons and the expiratory neurons. The inspiratory neurons are more medial than the expiratory neurons. The inspiratory neu-

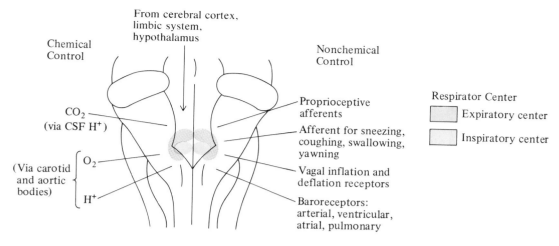

FIGURE 13–33 Location of the medullary inspiratory and expiratory centers in a mammal and a summary of the various afferent information that modifies the basic respiratory rhythm established by the respiratory center. *(Adapted from Ganong 1969.)*

rons innervate the inspiratory muscles (diaphragm and external intercostals), and the expiratory neurons innervate the expiratory muscles (mainly the internal intercostals) although these are not usually active during normal respiration. These two groups of neurons are cross-inhibitory, i.e., activity of the inspiratory neurons inhibits activity of the expiratory neurons and vice versa. This results in the basic respiratory rhythm.

The basic medullary respiratory rhythm is modified by the action of various other CNS centers and peripheral inputs. Two important centers located in the pons region of the brainstem (more anterior than the medulla) are the **apneustic** and **pneumotaxic centers**. The apneustic center stimulates the inspiratory neurons, prolonging the inspiratory part of the respiratory cycle. The pneumotaxic center, located slightly more anterior than the apneustic center, stimulates the expiratory neurons (and also inhibits the influence of the apneustic center). The combined activities of the medullary respiratory center and the apneustic and pneumotaxic centers result in the normal resting respiratory rhythm.

Many other CNS centers influence the activity of the respiratory center. For example, the hypothalamic regions active in temperature regulation can promote heat loss through additional respiratory evaporation. Humans clearly have some voluntary control of respiration via the cerebral cortex, although the normal respiratory rhythm is actually involuntary. Conscious anticipation of impending exercise can stimulate respiration even before the metabolic rate increases. Various pulmonary stretch

receptors modify the activity of the medullary respiratory center; one group of receptors inhibit inspiration to prevent over distension of the lungs; another group appear to prevent over collapse of the lungs. Stimulation of muscle joint proprioceptors elevates respiration. Blood pressure receptors (baroreceptors) diminish respiratory activity if stimulated by high blood pressure. Stimulation of pain receptors (e.g., stepping on a thumbtack) and mechanoreceptors (e.g., a foreign object lodged in the trachea or bronchi) causes us to gasp or cough. These various inputs to the respiratory center are summarized in Figure 13–33 for a mammal. Chemoreceptor input (mainly pO_2, pCO_2, and pH) from peripheral or central chemoreceptors is probably the most important, constant modifier of the basic respiratory rhythm.

Fish. Fish with bimodal breathing patterns (i.e., they breathe both air and water) are often exposed to water of low pO_2 and high pCO_2, although the air would have a normal pO_2 and pCO_2. A low aquatic pO_2 or high pCO_2 might be expected to provide the respiratory stimulus for increased air breathing and might inhibit gill ventilation so as to minimize CO_2 uptake across the gills. In one air-breathing fish, *Piabucina*, an increased pCO_2 of the water depresses gill ventilation. Air breathing is increased by these same stimuli, so there is a marked shift from water breathing to air breathing (Figure 13–34; Graham, Kramer, and Pineda 1977). The same effect of water hypoxia and hypercapnia is seen in another air-breathing fish, *Trichogaster;*

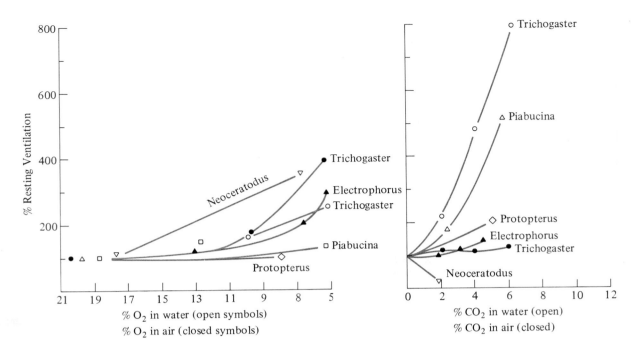

FIGURE 13–34 Role of oxygen and carbon dioxide in respiratory regulation in air-breathing fish. The effect of O_2 or CO_2 on respiration is expressed as the percent elevation of respiratory minute volume over the resting value.

aerial hypoxia also stimulates air breathing but aerial hypercapnia has little effect (Burggren 1979). The air O_2 content is the primary respiratory stimulus for the electric eel, an obligate air breather, although air CO_2 also stimulates respiration (Johansen et al. 1968).

Of the three lungfish, *Neoceratodus* relies the most on gill ventilation and least on air breathing. Gill ventilation increases with aquatic hypoxia; aquatic hypercapnia stimulates air breathing and inhibits gill ventilation. Aerial hypercapnia has no effect on ventilation, so it appears that the CO_2 receptors are located in the branchial cavity. Adult *Protopterus* rely more on air breathing than *Neoceratodus*. Water hypoxia does not stimulate either gill or lung ventilation but aerial hypoxia stimulates lung ventilation. The pO_2 receptors are located in the first three afferent gill arteries. *Protopterus* also has branchial pCO_2 receptors that inhibit gill ventilation.

Amphibians. Amphibians, like mammals, have aortic chemoreceptors that respond to both hypoxia and hypercapnia but apparently they do not have CO_2 receptors in the lungs (unlike reptiles and birds). However, stretch receptors in the lungs are

CO_2 sensitive, thus providing some direct sensory input dependent on lung pCO_2.

Many amphibians are bimodal breathers, at least during their larval stages, and control of ventilation is similar to that for some air-breathing fish. Bullfrog tadpoles, for example, exhibit a marked stimulation of lung breathing and a moderate increase in gill ventilation in response to aquatic hypoxia (Burggren and West 1982). In the adult aquatic salamander *Pseudobranchus*, lowered water pO_2 is the main stimulus for increased aerial respiration and pCO_2 has little effect. The opposite is true for a more terrestrial species, *Taricha*; CO_2 is the primary stimulus for aerial respiration. Both O_2 and CO_2 are equally important for a salamander like *Diemictylus*, that is intermediate in its habits (Wakeman and Ultsch 1975).

Reptiles. Reptiles may not have aortic and carotid chemoreceptors but they do have CO_2 receptors in their respiratory tract (Fedde, Kuhlmann, and Scheid 1977). It is difficult to generalize about the primary respiratory stimulant in reptiles. In some reptiles, O_2 is the main regulator of respiration but CO_2 is more important in other reptiles (Figure 13–35). There is little regulation of ventilation by pO_2

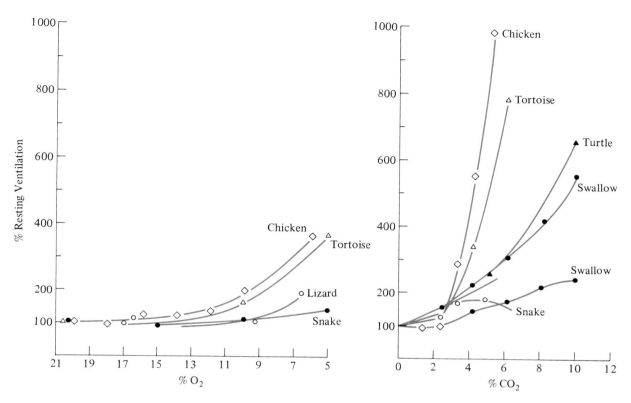

FIGURE 13–35 Role of oxygen and carbon dioxide in respiratory regulation in reptiles and birds. The effect of O_2 or CO_2 on respiration is expressed as the percent elevation of respiratory minute volume over the resting value.

in a lizard (*Lacerta*), and both ventilation and VO_2 decline with hypoxia; there is a slight increase in the ventilation/VO_2 ratio with hypoxia (Nielsen 1962). CO_2 has a moderate to major stimulatory effect in many lizards, turtles, and snakes, although excessively high CO_2 levels tend to depress ventilation and metabolism.

Birds. Ventilation in birds is markedly influenced by hypercapnia, more so than by hypoxia (Figure 13–35). There are at least two different groups of chemoreceptors—carotid and aortic bodies—that resemble those of mammals (see below) and CO_2 receptors in the respiratory tract (e.g., parabronchi) that respond only to pCO_2. These pCO_2 receptors have no, or a low, rate of discharge at high pCO_2 (>6.7 kPa) and the discharge frequency increases markedly at low pCO_2 (<2.7 to 5.3 kPa).

Mammals. Regulation of respiration is best studied in mammals. There are two main groups of chemoreceptors: (1) the peripheral aortic and carotid bodies

and (2) the central chemoreceptors located dorsal to the medullary respiratory center, near the floor of the IVth ventricle. The peripheral aortic and carotid bodies are perhaps derived from the efferent gill arch chemoreceptors of the ancestral water-breathing vertebrates. These chemoreceptors respond to pO_2, pCO_2, and pH. The central chemoreceptors respond only to the [H^+] of the cerebrospinal fluid in the IVth ventricle. However, the CSF pH is determined by the arterial pCO_2, not the arterial pH, because CO_2 is freely permeable to the "blood-brain" barrier but H^+ is not. The relationship between CO_2 and [H^+] is discussed further with respect to acid base regulation (see Chapter 15).

Ventilation is precisely regulated in mammals via the concerted effects of the peripheral and central chemoreceptors, with pCO_2 being the primary respiratory stimulant. This is of clear adaptive significance to normal, terrestrial mammals since it provides for considerable homeostasis of arterial pO_2 and pCO_2. Many mammals, such as fossorial and diving species, routinely experience hypoxia and hypercapnia. Their ventilatory response to inspired

---

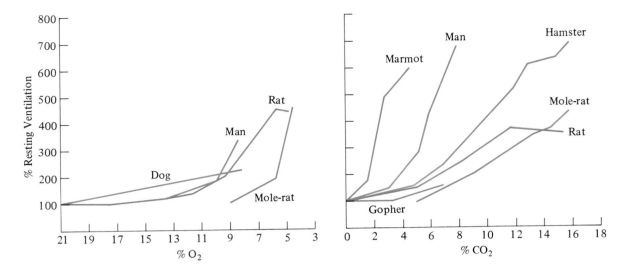

FIGURE 13–36 Role of oxygen and carbon dioxide in respiratory regulation in mammals. The effect of O_2 or CO_2 on respiration is expressed as the percent elevation of respiratory minute volume over the resting value.

CO_2 is often depressed (Figure 13–36), probably to reduce the ventilatory drive and the energetic cost of breathing in a hypercapnic environment. The "cost" to the animal of abandoning precise respiratory regulation is a more dramatic change in arterial pO_2 and pCO_2 during hypoxic or hypercapnic exposure.

Invertebrates

Regulation of ventilation by terrestrial crabs is very different from aquatic crabs; CO_2 is the primary respiratory stimulus, rather than O_2, as in the water breathers. This should not be surprising in view of the identical shift in chemosensory regulation in air-breathing vertebrates. In the aquatic crab *Callinectes*, ventilation is only regulated by pO_2, whether it is breathing water or air (Figure 13–37). Ventilation by the terrestrial crabs *Gecarcinus* and *Cardisoma*, in marked contrast, responds primarily to pCO_2, although there is a significant role of pO_2.

Many insects have sophisticated respiratory regulation; this is made possible by their capacity to close the spiracles. In fact, there are four important aspects to regulation of respiratory exchange in insects: (1) the neural origins of the respiratory rhythm and spiracular opening and closing, (2) chemosensory regulation by pO_2 and pCO_2 of spiracular opening and closing, (3) neural regulation of ventilation, and (4) control of the amount of interstitial fluid in the tracheole endings. Regulation of respiration is important not only for O_2 delivery and

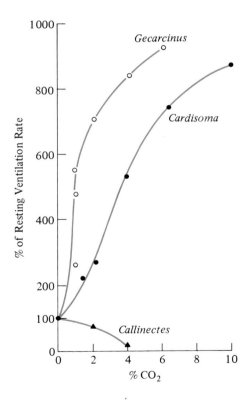

FIGURE 13–37 Ventilatory response of crabs exposed to air with a normal pO_2 and variable pCO_2; response is expressed as a percent of the resting ventilation rate. V (rest) is *Callinectes*, 120 ml min^{-1}; *Gecarcinus*, 11.8 ml min^{-1}; *Cardisoma*, 32 ml min^{-1}. Curves are typical responses for an individual of each species. *(Data are from Batterton and Cameron 1978; Cameron 1975.)*

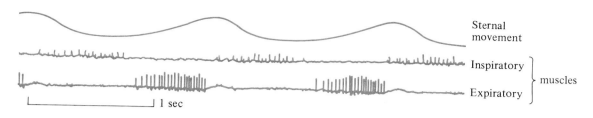

FIGURE 13–38 Synchronization of the inspiratory and expiratory nerve activity, of the nerve activity in various segments, and of the sternal movements in the dragonfly *Aeshna* during three respiratory cycles. *(Modified from Mill 1974.)*

CO_2 excretion but it also is important in limiting the concomitant loss of water. The control of spiracular closing limits the respiratory water loss in tsetse flies at low relative humidity (see Figure 13–27, page 641). Discontinuous respiration is another example of respiratory regulation that limits water loss.

The important role of the ventral nerve cord ganglia in establishing the respiratory rhythm has been clearly demonstrated in the dragonfly *Aeshna* (Mill 1973). Rhythmic bursts of action potentials in the nerves innervating the inspiratory and expiratory muscles of the spiracles result in a corresponding cycle of muscle activity (Figure 13–38). There also is a posterior-to-anterior rhythm of respiration in the different segments. In some insects, the isolated segment will continue to exhibit a rhythmic respiration due to the neural rhythmicity of the isolated ganglia. The respiratory centers will respond to elevated CO_2 levels and will stimulate spiracle opening and ventilation. For example, 0.2 to 3% CO_2 stimulates the prothoracic center and 12 to 15% CO_2 stimulates the primary respiratory center of *Carausius;* ventilation is stimulated by 10% CO_2 in cockroaches (Wigglesworth 1984). The tonic neural activity to the spiracles is often decreased by hypoxia. In *Aeshna*, 10% O_2 reduces neural discharge and 2% O_2 can totally eliminate neural activity. This also predisposes the direct response of CO_2 on the spiracular muscles (see below).

The activity of the spiracle muscles can be modified by a direct action of CO_2 and O_2. In *Schistocerca*, for example, the spiracle closing muscle is relaxed by the direct action of CO_2 and the spiracle will open due to its inherent elasticity. The initial effect of exposure to air of a high CO_2 in the cockroach *Blaberus* is the opening of the spiracles by a direct action of CO_2 on the spiracle muscles; the CO_2 then diffuses into the central nervous system and reduces the tonic neural discharge that tends to close the spiracles. The final effect of CO_2 is to stimulate ventilation.

The ventilatory mechanism of many insects is automatically stimulated during activity. For example, the action of the thoracic flight muscles regulates the ventilation cycle in many insects; the ram ventilation of the large tracheae in *Petrognathus* is automatically induced by the air flow over the thorax. However, the complex one-way patterns of air flow in some insects (e.g., *Schistocerca*, mantids) require considerable CNS coordination.

Effects of Diving and Altitude

The depth of diving has no effect on gas exchange in water-breathing animals (unless of course they descend into a region of O_2-depleted water), but diving under water can have profound physiological significance to an air-breathing animal. First, they must hold their breath for potentially long periods of time. Second, hydrostatic pressure has a profound effect on gas exchange.

The hydrostatic pressure of the water column above an air space (such as a vertebrate lung, a fish swim-bladder, or an insect plastron) exerts a pressure of about 101 kPa per 10 m of water depth. Deep diving vertebrates experience lung collapse because of the considerable hydrostatic pressures. Human scuba divers are an exception; their lungs are expanded normally while diving because the air pressure in their respiratory tract is kept equal to the ambient hydrostatic pressure by the scuba tank pressure regulator. Insects with physical gills (plastrons or air bubbles) are able to obtain O_2 from the water but are still breathing from an air space. These physical gills become less effective with depth, and the plastron can be collapsed by hydrostatic pressure.

The lung air of a deep-diving vertebrate can become a liability, rather than a useful O_2 store. The amount of a gas dissolved in body fluids depends on its partial pressure and solubility coefficient (Henry's law). For example, the O_2 and N_2 partial

TABLE 13–11

Partial pressures (kPa) for oxygen and nitrogen (dry, CO_2-free values for ambient air), alveolar carbon dioxide, and plasma-dissolved O_2 (ml O_2 per liter plasma; assuming plasma solubility of O_2 is 0.209 ml liter^{-1} kPa^{-1}) and fat N_2 content (ml N_2 per liter body fat; assuming fat solubility of N_2 is 0.67 ml liter^{-1} kPa^{-1} dissolved N_2) for a human scuba diver in equilibrium with the ambient hydrostatic pressure as a function of depth of diving.

Depth (m)	0	50	100	500
Ambient Pressure	101	202	1111	5151
pO_2	21.1	42.4	233.5	1082.6
pN_2	79.8	159.7	878.3	4072.5
Alveolar pCO_2	5.32	5.32	5.32	5.32
Plasma O_2	4.4	8.8	48.3	223.9
Fat N_2	53	106	582	2700

pressures of air, and the O_2 and N_2 contents of body fluids and fat in equilibrium with air, increase dramatically with diving depth (Table 13–11). The lung pO_2 and pN_2 will also increase in a similar manner. The lung pCO_2, however, does not increase in proportion to the air pressure with depth; the pCO_2 depends only on the rate of CO_2 production by the tissues and the alveolar ventilation rate. High pO_2 and pN_2 can have detrimental effects on a diving animal. High pO_2 can cause convulsions and be toxic; this is **oxygen toxicity.** High pN_2 has a **narcotic effect,** causing "raptures of the deep" in scuba divers. Furthermore, the additional dissolved O_2 and N_2 in the body tissues can have serious effects at the end of a dive. The return to normal atmo-

spheric (101 kPa) pressure can have the same effect on dissolved gases as removing the cap from a soft drink bottle; gas bubbles form in the supersaturated fluid. Body fluids can be three to four times supersaturated (equivalent to diving at 20 to 30 m) before gas bubbles will form, but surfacing immediately from greater depths will result in bubble formation. In **Caisson's disease** small gas bubbles in the joints cause severe pain (the "bends") and can have lethal effects if they form in the lungs or brain and block blood flow. Human scuba divers are especially prone to Caisson's disease because they dive with high pulmonary pressures and normal lung volumes; there is great potential for massive O_2 and N_2 invasion of the body tissues. Even repetitive free diving by breath-holding to only moderate depths (e.g., pearl diving) can result in mild bends.

Deep-diving marine mammals avoid excessive O_2 and N_2 invasion of the tissues during diving because their alveoli collapse during diving; the lung air is compressed into the dead space and the O_2 and N_2 invasion rates actually decline to zero at great depths because no air remains in the alveoli for gas exchange (Figure 13–39). If the dead space were to collapse, rather than the alveoli, the O_2 and N_2 invasion rates would continue to increase with depth. This is what happens in a frog experimentally subjected to high hydrostatic pressures (frogs don't normally dive to such great depths!).

Sea snakes and turtles are probably the deepest-diving reptiles. Sea snakes, which dive to depths of at least 40 m (500 kPa pressure), would be susceptible to the bends (but probably not O_2 toxicity or N_2 narcosis), but they could avoid the bends if their rate of N_2 gain from the lung is balanced by N_2 loss across the skin to the water (Seymour 1974). Venous blood returning from the body to the heart can

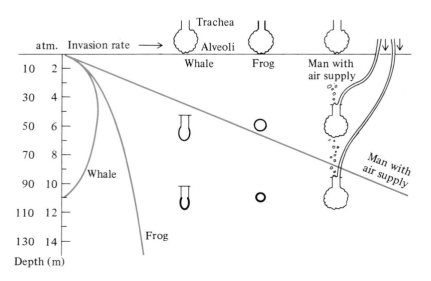

FIGURE 13–39 Relationship between diving depth and nitrogen invasion rate in a marine mammal, such as a whale, which shows alveolar collapse with depth, a frog, which has tracheal collapse with depth, and a human scuba diver, whose lungs do not collapse at depth. *(From Scholander 1940.)*

bypass the lungs and flow back to the body because there is incomplete separation of arterial and venous blood in the snake heart (see Chapter 15). More blood must bypass the lungs as the sea snake dives deeper for N_2 uptake from the lungs to be balanced by cutaneous N_2 elimination.

High altitude also has significant effects on the physiology of aquatic and terrestrial animals for a variety of reasons (Bouverot 1985). An increase in altitude diminishes the ambient barometric pressure by about 1/2 every 5500 m, and consequently the ambient pO_2 declines. The fractional O_2 content remains constant at 20.95% throughout the physiologically relevant lower reaches of the atmosphere. For example, the barometric pressure at the top of Mt. Everest (8848 m) is about 33 kPa (250 torr) and the ambient pO_2 is 6.9 kPa (53 torr). Ambient

temperature also decreases with altitude, by about 6° C per 1000 m. Dehydrational stress increases with altitude because of the lowered ambient temperature and water vapor content of the ambient air and because the diffusion coefficient for water (and other gases) increases with altitude ($D \propto 1$/barometric pressure). Nevertheless, a wide variety of animals survive at considerable altitudes (Table 13–12).

Hyperventilation is the appropriate respiratory response to altitude for O_2 exchange. The alveolar pO_2 cannot be maintained at the normal sea level value but hyperventilation keeps the alveolar pO_2 as high as possible. Unfortunately, the alveolar pCO_2 declines markedly with hyperventilation since pA,CO_2 depends only on the rate of CO_2 production and alveolar ventilation rate, not ambient pressure. This decline in alveolar and arterial blood pCO_2 will inhibit respiration. Thus, hypocapnia exerts a "braking" effect on respiration at high altitude. Fortunately, this braking effect can be compensated by metabolic acidosis (see Chapter 15); bicarbonate is actively transported out of the CSF to increase the [H^+] (the central chemoreceptor stimulant) and bicarbonate is excreted from the body via the kidneys.

The effects of altitude on some important respiratory parameters are summarized in Table 13–13. As ambient pressure and pO_2 decline, so does alveolar pO_2 and pCO_2 (the latter due to hyperventilation). The arterial percent saturation also declines; a person lapses into unconsciousness at about 50% arterial saturation. Thus, 7100 m (23000 feet) is the normal "ceiling" for an average person; higher altitudes result in loss of consciousness. However, training or breathing pure O_2 can elevate the ceiling. For example, two highly trained mountain climbers,

TABLE 13–12

Summary of the altitudinal limits for a variety of animals. *(Adapted from Bouverot 1985.)*		
	Altitude (km)	**Geographic Locality**
Amphibians		
Salamandra	3	European Alps
Telmatobius	3.8	Lake Titicaca, Peru
Eleutherodactylus	4.5	Andes, Peru
Bufo	5	Himalaya
Fish		
Trout	2.8	European Alps
Trout	3.8	Lake Titicaca, Peru
Nemachilus	4.7	Asia
Reptiles		
Sceloporine lizard	3.4	SW USA
African skink	4	Kilimanjaro, Tanzania
Iguanid lizards	4.9	Andes, Peru
Lizards	5.5	Himalaya
Birds		
Domestic fowl	4	
Various families	4–6.5	Rocky Mountains, USA and Andes, Peru
Lammergeier	5–6	Himalaya
Barheaded goose	>8.8	Himalaya
Mammals		
Deer mice	4	Rocky Mountains, USA
Human habitations	>4.5	Andes, Peru and Himalaya
Llama, guanaco, alpaca	4.8–5.4	Andes, Peru
Chinchilla	5	Andes, Peru
Yak	5.8	Asia
Taruca (deer)	6	Andes, Peru
Humans without pure O_2	8.8	Mt. Everest, Himalaya

TABLE 13–13

Effects of high altitude on atmospheric pressure (P_b; kPa), ambient pO_2 (kPa), and alveolar pO_2 and pCO_2 ($p_A O_2$, and $p_A CO_2$; kPa) for a human.				
Altitude	**P_b**	**Ambient pO_2**	**$p_A O_2$**	**$p_A CO_2$**
0	101	21.1	13.8	5.3
3100	70.6	14.6	8.9	4.8
4340	61.9	12.8	6.0	—
6200	46	9.7	5.3	3.2
7100	——	normal "ceiling"		
8848	33	6.9	4.0	1.5
9200	30	6.3	2.8	—
12300	19	3.9	1.1	—
14460	——	"ceiling" with pure O_2		——
15400	12	2.4	0.1	—
20000	6	1.3	0	0

Messner and Habeler, have climbed Mt Everest (8848 m) without supplemental O_2 (West 1986). Breathing pure O_2 raises the ceiling to about 14400 m because of the 5 × higher ambient pO_2. Aviators at higher altitudes than this must be pressurized, even if breathing pure O_2. At an altitude of about 20000 m, the ambient pressure is 6.25 kPa (47 torr). This is equal to the saturation water vapor pressure in the human lung. Consequently, the lung would only contain water vapor, no O_2, N_2, or CO_2! At even higher altitudes, water would boil from the surface of the lungs.

Water Loss from Gas Exchange Surfaces

The principal role of a respiratory surface is O_2 and CO_2 exchange. However, for terrestrial animals another important role is to minimize the respiratory water loss that occurs concomitant with O_2 and CO_2 exchange.

A markedly reduced respiratory evaporative water loss is a common characteristic of all tracheal respiratory systems and appears to have been a major selective advantage for the evolution of tracheal systems in, for example, insects. Why is water loss so reduced by tracheal or pseudotracheal respiration, and why is a tracheal system apparently better than lungs in reducing evaporative water loss?

Evaporative water loss essentially is a diffusional process, just as O_2 uptake and CO_2 excretion occur across the respiratory surface by diffusion. However, there can be convective transfer of ambient air to the respiratory surface. The water content and water vapor pressure of air at the respiratory surface are equal to the saturated water vapor content/vapor pressure at body temperature, e.g., 17 mg liter^{-1} and 2.39 kPa at 20° C. The concentration gradient between the respiratory surface and dry air

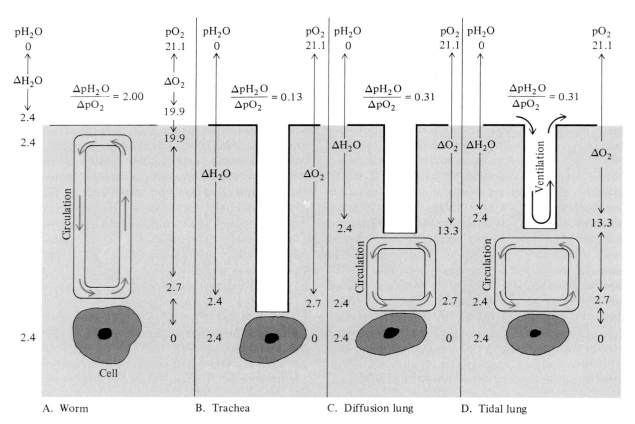

FIGURE 13–40 Relationship between diffusive oxygen uptake and water loss across the following: (**A**) a cutaneous respiratory surface (e.g., an oligochaete worm); (**B**) an insect tracheal system; (**C**) a diffusive lung (e.g., a pulmonate snail); and (**D**) a convective lung (e.g., a mammal).

(for example) would also be 17 mg liter^{-1}. The pO_2 gradient for O_2 uptake might be about 1.33 kPa if the pO_2 of air at the respiratory surface is about 18.62 kPa, a not unreasonable estimate for an oligochete worm, for example. The ratio of water loss (VH_2O) to O_2 uptake (VO_2) based on diffusional exchange is as follows.

$$VH_2O/VO_2 = (DH_2O/DO_2).(\Delta pH_2O/\Delta pO_2)$$
$$= 1.3 \, \Delta pH_2O/\Delta pO_2 \qquad (13.9)$$

For a moist-skinned animal, such as an oligochete worm (Figure 13–40), VH_2O/VO_2 is about 1.3 (2.39 to 0)/(21.13 to 19.95) = 2.6 nmol H_2O per nmol O_2 (2.1 mg H_2O per ml O_2).

If the respiratory surface is invaginated to form tracheae that extend to the individual cells, then the ΔpH_2O is the same (2.39 kPa) but the ΔpO_2 is much greater, about 18.44 (if the pO_2 at the end of the tracheae is 2.66 kPa). The ratio of VH_2O/VO_2 is now 1.3(2.39/18.44) = 0.13 nmol H_2O per nmol O_2 or 0.11 mg H_2O per ml O_2. For a lung with an intrapulmonary pO_2 of 13.3 kPa (whether diffusional or ventilated), the ratio is about 0.31 nmol/nmol or 0.25 mg/ml O_2. Thus, respiratory evaporative water loss is minimized by reducing the pO_2 at the air-fluid interface to as low a value as possible. Tracheae and pseudotracheae are clearly far superior to lungs in this respect.

Gas Exchange in Terrestrial Eggs

Many animals, including some aquatic ones (e.g., some fish, amphibians), lay their eggs on land. Some terrestrial eggs, for example, those of amphibians, are enclosed by a jelly-like capsule through which O_2 and CO_2 readily diffuse. Restricting water loss, not sustaining gas exchange, may be the limiting factor in their survival and development. In contrast, **cleidoic eggs** are enclosed by a special shell that allows gas exchange but prevents exchange of waste products. Many reptiles, birds, and insects have cleidoic eggs. These eggs, like the jelly-covered eggs, still require a fairly moist environment for survival and development, although their eggshell does greatly reduce the rate of water loss. The eggshell also restricts gas exchange.

The eggs of reptiles and birds may be leathery and flexible (e.g., many lizards and turtles) or calcified and rigid (e.g., some lizards, crocodilians, and birds). Water absorption or evaporative water loss during development changes the volume of flexible-shelled eggs. Calcified, rigid eggs develop an air cell within the egg as water is lost during development. In bird eggs, a large air space is

formed during development between the inner and outer eggshell membranes; this air cell becomes important in gas exchange for the developing embryo (Figure 13–41). The chorio-allantoic membrane of the egg, located beneath the inner egg shell membrane, becomes highly vascular for gas exchange, not only from the air cell but from the entire inner egg membrane surface.

The structure of the eggshell is important in determining its gas exchange properties. For example, a hen's egg has soft inner eggshell membranes and a thick, calcified layer. This calcified layer, which confers mechanical strength and protection, is perforated by numerous pores of varying shape. The pores allow diffusive gas exchange between the developing embryo and the external environment. Each eggshell pore provides gas exchange to a specific internal area, the eggshell respiratory unit. The morphology of egg pores can be complex; even the seemingly "simple" tubular pores are often funnel shaped or branched, particularly at the external surface. The inner constricted part of the eggshell pore is generally the primary resistance to gas exchange. The pore resistance can be considered to be a simple, tubular resistance, and so it is therefore realistic to approximate the structure of eggshell pores as a simple tube, of length T_p and radius r_p.

The outer surface of some bird eggs, for example, the chicken, has a protective coat that reduces the likelihood of bacterial invasion; this is perforated by cracks for gas exchange (Figure 13–42A). Other bird eggs have a different surface structure. For example, the ostrich eggshell has numerous discrete, complex-shaped pore openings (Figure 13–42B).

The shape and number of the eggshell pores determine the nature and amount of gas exchange. The total cross-sectional area of these pores is quite low, even though the pores are exceedingly numerous. For example, the turkey egg has a total area of about 90 cm^2; the 9600 pores, each of about 8.5 μ diameter and approximate area of 0.000023 cm^2, confer a total pore area of about 0.22 cm^2, or 0.24% of the total shell area. A chicken egg has about 10000 pores, each about 0.017 mm in diameter; the total pore surface area of about 2.3 mm^2 is less than 1/2000 of the total area of the egg, or 70 cm^2 (Wangenstein, Wilson, and Rahn 1971). The average interpore distance in the chicken egg is about 1.1 mm so that each pore provides gas exchange for about 1 mm^2.

Gas exchange in a reptile or bird egg between the ambient air and the air cell should conform well with predictions based on Fick's law of diffusion since respiratory exchange would seem to occur mainly by diffusion along thin pores of fixed diame-

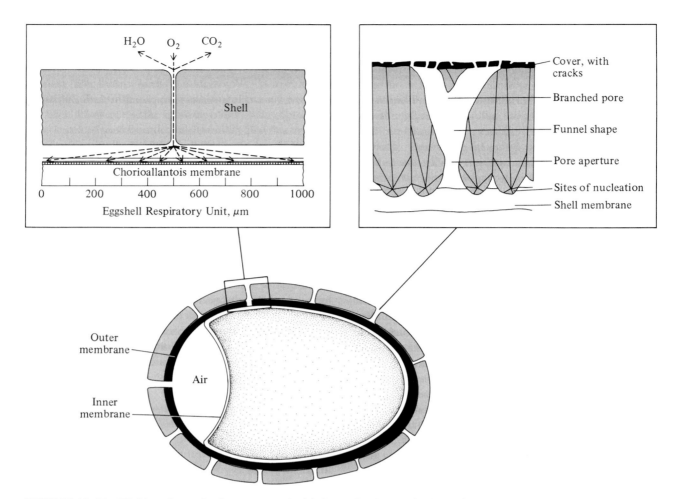

FIGURE 13–41 Highly schematized structure of a bird egg. Each pore in the chicken egg provides diffusive gas exchange to a respiratory unit area of about 1 mm². The eggshell has an outer protective coat with cracks for gas exchange, a thick calcified layer with pores for gas exchange, and an outer and inner shell membrane. The air cell space forms between the inner and outer membranes. *(Modified from Visschedijk and Rahn 1983; Sibley and Simkiss 1987.)*

ter and length, i.e., D, A, and T_p are constant. There is, in fact, the generally expected relationship between VO_2 and the pO_2 and pCO_2 gradients between ambient air and the air cell during embryonic development that would be predicted by diffusive gas exchange, i.e., for O_2

$$VO_2 = DO_2 N_p A_p (C_2 - C_1)/T_p$$
$$= DO_2 N_p A_p (C/p)(p_2 - p_1)/T_p \quad (13.10a)$$

where DO_2 is the diffusion coefficient for O_2, N_p is the number of pores, A_p is the area of each pore, $(C_2 - C_1)$ is the concentration gradient for diffusion, T_p is the diffusion path-length (or pore thickness), C/p is the concentration of O_2 in air per kPa O_2

partial pressure, and $(p_2 - p_1)$ is the O_2 partial pressure gradient. Rearranging and substituting values for the chicken egg, we have the following.

$$VO_2/(p_2 - p_1)$$
$$= (0.198)(10^4)(\pi \, 10^{-6})(4.36 \, 10^{-7})/(300 \, 10^{-4}) \quad (13.10b)$$
$$= 90.2 \text{ nmol sec}^{-1} \text{ kPa}^{-1}$$

The calculated value of 90.2 nmol sec⁻¹ kPa⁻¹ (173 ml day⁻¹ kPa⁻¹) corresponds fairly closely to the measured value of $VO_2/(p_2 - p_1)$ for a hen's egg, 105 ml day⁻¹ kPa⁻¹ (Visschedijk and Rahn 1983).

Fick's law essentially describes 1-dimensional diffusion along a long, thin pore (Figure 13–43A). In 1881, Stefan described a different relationship for

A. Surface of chicken egg

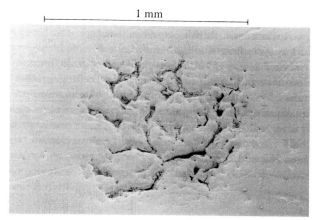

B. Surface of ostrich egg

FIGURE 13–42 (A) Outer surface of a chicken eggshell showing the surface cracks in the outer protective layer that allow gas exchange. (B) Outer surface of an ostrich egg showing the complex arrangement of the surface opening of an eggshell pore. *(Photographs courtesy of Withers and Stewart, Department of Zoology, The University of Western Australia.)*

diffusion through small apertures in relatively thin barriers; the rate of diffusion was proportional to the diameter of the aperture (or 2 × radius) and the concentration difference, but not the thickness (which is small relative to the pore diameter), i.e.,

$$VO_2 = 2DO_2 r_p N_p (C_2 - C_1) \qquad (13.11)$$

where r_p is the pore radius. Stefan's law essentially describes the radial diffusion in three dimensions through the boundary layer above the opening of a large diameter pore. This relationship generally is used to describe the rate of water loss and CO_2 uptake across the stomates of plant leaves, rather than Fick's law, because the shape and density of stomates more approximates those conditions.

Which law best describes gas exchange across the bird egg depends on the dimensions and numbers of pores (Simkiss 1986; Toien et al. 1987; Sibley and Simkiss 1987). The ratio of pore length to pore radius (T_p/r_p) is a convenient index as to whether Fick's or Stefan's law is most appropriate. The resistance to gas exchange ($R = (C_2 - C_1)/VO_2$; sec cm^{-3}) is calculated as follows.

$$R_{Fick} = T_p/(DO_2 N_p A_p) \qquad (13.12)$$
$$R_{Stefan} = 1/(2DO_2 N_p r_p) = \pi r_p/(2DO_2 N_p A_p)$$

By combining Fick's and Stefan's laws, we can readily see that the additional boundary layer resistance due to Stefan's law becomes significant at T_p/r_p of about 3 or less (Figure 13–43B).

The ratio of T_p/r_p allows us to estimate the significance of pore morphology to the nature of the diffusive exchange. For a chicken or turkey egg, T_p/r_p is about 35, whereas for a plant leaf stomate it is about 2. Thus, Fick's law is a good approximation for a turkey egg and the boundary layer resistance is low compared to the resistance of the pores, whereas for stomates Stefan's law is a necessary consideration and the boundary layer resistance is significant.

Not all gas exchange across the bird egg is diffusive (Paganelli and Rahn 1987). A chicken egg has a slight positive internal pressure (about 0.01 to 0.03 kPa) that promotes convective loss of gases. The water vapor pressure inside the egg (e.g., 6.7 kPa) is typically greater than the ambient water vapor pressure and this therefore means that the internal pN_2 is less than the external pN_2. Nitrogen should therefore constantly diffuse into the egg, but N_2 is inert and cannot continue to accumulate indefinitely. Its inward flux and the evaporation of water establish a positive hydrostatic pressure that produces an outward convective flux. This outward convective flux of N_2 balances the inward diffusive N_2 flux.

Insect eggs, like reptile and bird eggs, rely on gas diffusion through aeropyle pores in the surface of the egg shell (Hinton 1969). Some insect eggs have few pores (aeropyles) with an aggregate surface area of only 1 to 2% of the total egg surface, whereas

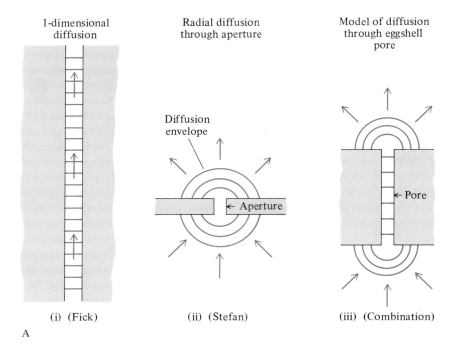

1-dimensional diffusion

Radial diffusion through aperture

Model of diffusion through eggshell pore

Diffusion envelope

← Aperture

← Pore

(i) (Fick)   (ii) (Stefan)   (iii) (Combination)

A

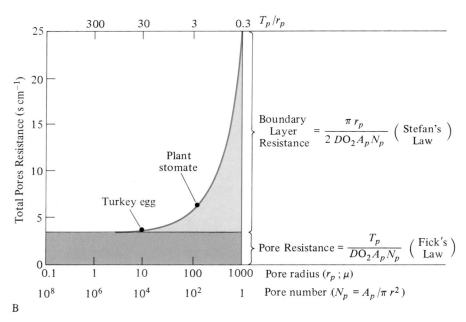

$$\text{Boundary Layer Resistance} = \frac{\pi \, r_p}{2 \, DO_2 A_p N_p} \left(\substack{\text{Stefan's} \\ \text{Law}} \right)$$

$$\text{Pore Resistance} = \frac{T_p}{DO_2 A_p N_p} \left(\substack{\text{Fick's} \\ \text{Law}} \right)$$

B

FIGURE 13–43 **(A)** Representation of diffusion through (i) a uniform tubular pore (described by Fick's law), (ii) a thin aperture (described by Stefan's law), and (iii) a short pore (described by a combination of Fick's and Stefan's laws). **(B)** Model for the diffusive resistance of eggshell pores (total pore area, A_p, of 0.314 cm^2; pore length, T_p, of 0.03 cm) as a function of the number of pores (N_p, from 10^8 to 1) and their radius (r_p, from 0.1 to 1000 μ), based on Fick's and Stefan's laws. DO_2 is the diffusion coefficient for O_2. The values for a turkey egg and a plant leaf stomate are indicated. *(Modified from Simkiss 1986; Rahn, Paganelli, and Ar 1987.)*

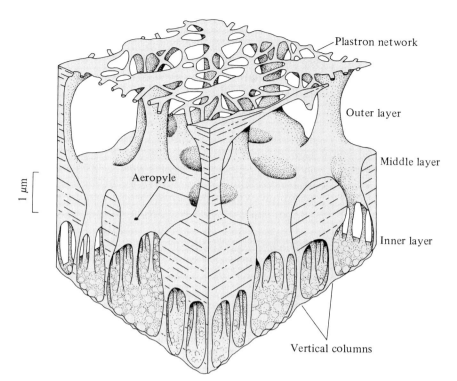

1 μm

Plastron network

Aeropyle

Outer layer

Middle layer

Inner layer

Vertical columns

FIGURE 13–44 Eggshell plastron of the blowfly showing the perforate surface and the columnar supporting structures with large airspaces (aeropyles). *(From Hinton 1963.)*

other eggs have rosettes of large aeropyles that constitute up to 12% of the total area. Many insect eggs have a very complex outer surface (Figure 13–44), or a horn-shaped extension that also acts as a plastron. These plastrons trap an air layer and function in an identical manner to the plastrons of adult aquatic beetles. Egg plastrons are moderately resistant to hydrostatic pressure and reduced surface tension. A reduced surface tension is not necessarily an unnatural occurrence for some insect eggs. For example, water in pools on cow dung or decomposing flesh can have a surface tension as low as 40 to 50 mN m^{-1} (see Table 13–4, page 626).

Physiological Implications of Air Breathing

The different physical properties of air and water result in many differences in the physiology of air-breathing animals. We have seen a number of profound alterations in structure and function of the respiratory systems of air-breathing animals.

Air has a higher O_2 content, lower density, and lower viscosity than water (see Table 12–11, page 601). There is a general trend in air-breathing animals for resting metabolic rate to be higher than for aquatic animals, most likely because of the advantages of air as the respiratory medium. The maximal metabolic rates of terrestrial, air-breathing animals also tend to be higher than those of aquatic animals.

Air-breathing animals may have a lower metabolic cost of ventilation than aquatic animals (about 1 to 5% of resting VO_2).

Countercurrent flows of air and blood are not observed in air-breathing animals despite the theoretical possibility in the bird lung where there is one-way air flow in the parabronchi. The actual arrangement of blood capillaries is crosscurrent to the parabronchial air flow. Some insects also have a one-way flow of air through the large tracheae, but there is no countercurrent exchange in the blind-ended tracheoles.

Air has a low heat capacity and thermal diffusivity, and so it is feasible for a terrestrial animal to utilize its metabolic heat production, or incident solar radiation, for thermoregulation (see Chapter 5).

Water loss across the respiratory surfaces is a virtually inescapable consequence of air breathing. Respiratory evaporative water is highest when the skin or cuticle is the respiratory surface and is greatly minimized for tracheal respiratory systems because the ratio of pO_2 gradient to pH_2O gradient is maximal. Respiratory water loss is intermediate for animals with lungs.

Air breathing has profound implications for CO_2 regulation and acid-base balance. There is a marked tendency in air-breathing animals for CO_2 to become the major respiratory stimulant, rather than O_2. Body CO_2 levels are much higher in air-breathing animals than in water-breathing animals because of the much greater CO_2 capacitance of air compared to water. The respiratory regulation of acid-base balance through CO_2/H^+ becomes important in air breathers.

Summary

The transition from water to air breathing has independently occurred a number of times in both vertebrate and invertebrate animals. Tetrapods are the most successful vertebrate group to have made the transition, and the arthropods are the most successful terrestrial invertebrates. The transition from water to air breathing has required fundamental changes to the structure and functioning of the respiratory organs. Gills are replaced in air-breathing animals by lungs, tracheal or pseudotracheal respiratory systems, or cutaneous modifications for gas transfer.

A number of fish are air breathing. Lamellar fish gills are generally ineffective for gas exchange because they collapse under their own weight and by surface tension, and so their role in gas exchange is supplemented or replaced by cutaneous or other vascular surfaces, such as the lining of the buccal, branchial, or opercular chambers; the stomach or intestine; accessory branchial structures (labyrinthine organs; vascular outgrowths of the branchial wall); or the swim-bladder.

Lungfish represent the intermediate condition of primitive lungs and a variable reliance on both aquatic (gill) and aerial (lung) respiration. The amphibians, and especially the more terrestrial anurans, represent a further specialization for aerial respiration, with often large saccular and partly compartmentalized lungs. However, amphibians use a positive-pressure buccal pump to inflate the lungs (like most air-breathing fish with lungs).

Reptiles, birds, and mammals are essentially completely reliant on aerial, lung gas exchange. A few aquatic reptiles have significant cutaneous exchange of O_2 and CO_2, and many terrestrial vertebrates (even mammals and birds) have a not insignificant cutaneous exchange of at least CO_2.

Reptiles, birds, and mammals have a negative-pressure inspiratory cycle for filling their lungs. The reptilian lung is more complex in structure than amphibian lungs, being more highly compartmentalized (the appearance is spongy) and often divided into a primarily respiratory exchange segment (faveoli) and a primarily ventilatory segment. Respiratory mechanics are somewhat different in the hard-shelled turtles and tortoises, which rely mainly on muscle contraction of limbs and viscera for pulmonary ventilation and the hydrostatic effect of the water when submerged. Crocodiles primarily use abdominal muscle contraction to move the viscera and thereby ventilate their lungs.

The mammalian lung is highly compartmentalized into millions of alveoli; its appearance is like dense foam rubber. Ventilation is accomplished by a negative-pressure inspiratory filling and a passive relaxation of the thoracic cage for expiration.

The avian respiratory system is the most complex in all tetrapods because of the additional involvement of an extensive air sac system for gas ventilation of the lung and the modified structure of the lung tissue for gas exchange. The complex connections between the trachea, the various air sacs, and the rigid lung, coupled with ventilatory movements of the thoracic rib cage, result in a unidirectional flow of air through the lung rather than the tidal lung air flow characteristic of all other tetrapod lungs. The air entering the lung via bronchi passes unidirectionally through successively more numerous and smaller bronchi until it reaches the parabronchi. Numerous blind-ended air capillaries extend radially from the parabronchi, and these form the respiratory exchange surface. There is no bulk movement of air through the air capillaries; rather, gas exchange is diffusional.

Surface tension is a significant force for lungs, particularly those with small diameter alveoli (mammals). The pressure required to keep an alveolus inflated against the collapsing surface-tension force is inversely proportional to its radius (law of Laplace). This pressure is substantial for small radii (2 to 10 kPa) at the normal surface tension of water, but is greatly reduced for alveoli by the presence of a surfactant (to <1 kPa). The balance of hydrostatic and colloid forces across the pulmonary membranes is sufficient to keep the lungs inflated against the reduced surface-tension force.

The composition of the lung gas is determined by a combination of the rate of lung ventilation and the fraction of the lung air renewed per breath and the rate of O_2 removal and CO_2 addition to the lung air. A high tidal volume/lung volume (and low-residual volume) maximizes the exchange of ambient and lung air per breath, but results in a substantial change in lung pO_2 and pCO_2 during the respiratory cycle. A lower tidal volume/lung volume (and high-residual volume) results in greater homeostasis of lung pO_2 and pCO_2 but reduces the efficacy of ambient lung gas exchange. The rate of transalveolar gas exchange depends on the pulmonary diffusing capacity (lung surface area and thickness of the alveolar membrane) and the transalveolar O_2 and CO_2 gradients.

Many invertebrate taxa have representatives that have made, more-or-less successfully, the transition from water to land. Many terrestrial annelids rely on cutaneous gas exchange; the swamp earthworm *Alma* uses its tail to form a "lung" when its burrows are filled with water.

Pulmonate land snails have a vascular mantle cavity for gas exchange; this lung is ventilated mainly by diffusion, although arching the body may provide some bulk flow of air into/out of the lung. Onycophorans have an insect-like ventilatory system consisting of spiracular openings at the body surface that extend into bundles of tubes (tracheoles) that extend into the body tissues.

Many semiterrestrial crustaceans retain lamellar gills for gas exchange. A water reservoir is held in the branchial cavity and the lamellae are strengthened by cuticular thickenings to resist collapse in air. However, the gills are reduced in most semiterrestrial crabs, which have a vascular lining of the branchial cavity. Sand-bubbler crabs have thin gas windows on their leg segments to provide gas exchange. The more terrestrial isopods have lung-like invaginations of the pleopods with numerous pseudotracheal tubes.

The chelicerate arthropods (spiders, scorpions) have either book lungs, which are gill-like invaginated respiratory structures, sieve tracheae, or tracheae for gas exchange. Insects have well-developed tracheal systems for gas exchange. The spiracles of insects are often closeable, allowing the regulation of both gas exchange and reduction of evaporative water loss from the tracheal system. The tracheal system of most insects is essentially a series of branching tubes along which O_2 and CO_2 diffuses between the ambient air and the cells, which the smallest tracheal branches (tracheoles) invest. In some, especially larger and active insects, the tracheal gas exchange is facilitated by bi- or unidirectional ventilation of the larger tracheae (but not tracheoles).

Many aquatic insects have gills on the thoracic and/or abdominal segments or have rectal gills for gas exchange. Some aquatic insects (and also spiders and other arthropods) use air bubbles for gas exchange when submerged. Many insects, including adults, larvae, and eggs, have a physical gill or a plastron that traps an air layer at the body surface and acts as a permanent "gill," allowing continuous O_2 uptake from the water and CO_2 excretion into the water.

In terrestrial animals, CO_2 is the primary respiratory stimulant rather than CO_2. This trend is apparent in both terrestrial vertebrates and invertebrates.

The normal respiratory exchange of terrestrial animals is highly modified by both depth under water and elevation above sea level. The hydrostatic pressure increases by about 101 kPa per 10 m depth of submergence, and this substantially increases the partial pressures of all gases in an air space. This in turn promotes dissolution of these gases into the body fluids (according to Henry's law). The physiological consequences are both a direct effect of the dissolved gases and partial pressure (e.g., O_2 toxicity, nitrogen narcosis) and the consequences of rapid depressurization; the supersaturated gases form small bubbles that can block blood flow (Caisson's disease; the "bends"). High altitude has important physiological consequences through the reduced total atmospheric pressure and reduced partial pressures of, especially, O_2. Hyperventilation can partially compensate for the reduced pO_2 but this results in a low pCO_2.

The water loss from a moist cutaneous respiratory surface is high. The respiratory water loss is reduced for diffusion and tidal lungs but is least for tracheal respiratory systems.

The eggshell of many reptile, bird, and insect eggs reduces the evaporative water loss but also restricts gas exchange. Eggshell pores (reptile, bird eggs) and aeropyls (insect eggs) provide an avenue for gas exchange, which is essentially diffusive and dependent on the physical properties of the shell pores (number, diameter) and the metabolic rate of the embryo.

Supplement 13–1

Branching Pattern of the Mammalian Respiratory System

. . . .

A detailed morphological examination of the branching pattern of the "respiratory tree" reveals some important physiological aspects of gas exchange. The respiratory tree has a fairly dichotomous branching pattern, i.e., each tube branches into two successive tubes. The various levels of branching can be assigned a generation number (z); the trachea is generation 0 (no branching yet); the left and right bronchi are generation 1 (the first level of branching). The terminal alveolar sacs are about generation 23, although the region of gas exchange actually ranges from about generations 17 to 23.

Two basic aspects of fluid mechanics have important implications for the dichotomous morphology of the respiratory tree. First, the angle of the branches from the main tubes should be low to minimize the resistance of air flow through the junction. Thus, the angles of the bifurcations in large diameter tubes (>0.4 cm) is about 64°. This constraint is lessened for smaller tube diameters,

and the branching angle increases to 100° in tubes <1 mm diameter. A high branching angle is desirable because it allows greater compactness of the lungs.

A second design constraint is the optimal diameter of successive generations of branches; minimum convective energy dissipation occurs if the diameters of successive branches ($D(z)$) are

$$D(z) = 2^{-z/3}D_0 = (\sqrt[3]{0.5})^z D_0$$

where z is the branch generation and D_0 is the diameter of the trachea.

This theoretical exponential relationship between diameters of successive branches for minimization of convective energy expenditure in a dichotomously branching network is indicated below by the straight line. The larger portions of the airways from a human lung (solid circles) closely conform to the expected pattern, but the respiratory bronchioles and alveolar ducts do not conform to the expected relationship (open boxes); this has been suggested to be the case because exchange occurs more by diffusion rather than by convection. Gas exchange through the terminal branches is diffusive as well as

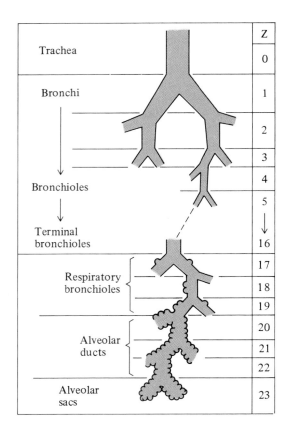

Schematic of the mammalian respiratory tree showing the branching pattern and generation number (z) from the trachea to the alveoli. *(Modified from Weibel 1984.)*

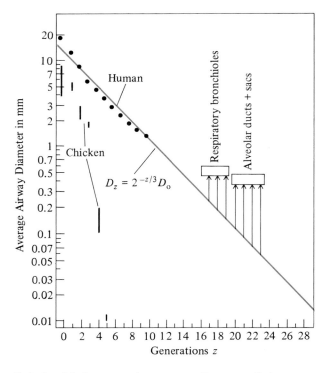

Relationship between the average diameter of airways in the human lung and the chicken lung as a function of branching generation. *(Modified from Weibel 1984.)*

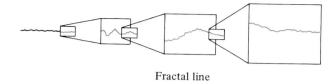

Fractal line

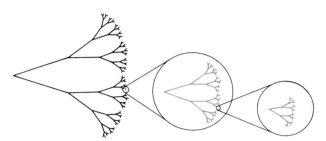

Fractal branches

The length of a fractal line, or the pattern of fractal branching, depends on the scale at which the pattern is examined.

and air capillaries, of course, are not dichotomous branches and would not be expected to conform to the optimal morphological relationship observed for dichotomously branching convective tubes.

Another explanation for the scaling relationship of airway diameter is derived from nonlinear dynamics and **fractal** analysis. Geometrical structures typically have a very simple shape, which is apparent independent of the scale of observation. For example, a perfect circle looks like a circle whether viewed from close up or at a distance. However, many natural structures have a very complex shape that depends greatly on the scale of the observation. For example, the length of the coastline of Australia is about 36750 km, according to a standard atlas. However, this length obviously depends on the scale of the measurement unit. Traveling around the entire coastline with a 1 km-long ruler might yield a similar estimate of length but using a meter-long ruler would provide a more accurate estimate, and clearly a much higher estimate (and take a lot longer to accomplish!). Using a mm-long ruler would further increase the accuracy and length of the estimated coastline. The length of coastline depends on the scale used to measure it. Similarly, a "fractal" line has a length that depends on the scale at which it is observed. Branching structures also have a structure whose detail of branching depends on the scale of the observation. The term "fractal" refers to a fractional dimension. Geometric shapes have integer dimensions, e.g., a line has 1 dimension, a rectangle 2, a cube 3, etc. Many fractal lines (e.g., a coastline) have a fractal dimension greater than 1 (the coastline of Australia is longer than the length of a simple geometrical outline that would have a dimension of 1) but less than 2 (the area of a 2-dimensional rectangle that could be drawn around Australia). Fractals, whether lines,

convective, and maintaining a high diameter maximizes diffusive exchange. The branching pattern in the respiratory tree of an adult chicken, which is markedly dissimilar in general structure from the mammalian respiratory system, does not conform well to the predicted relationship in branching diameter decrement. The trachea, air sac system, bronchi, secondary bronchi, parabronchi,

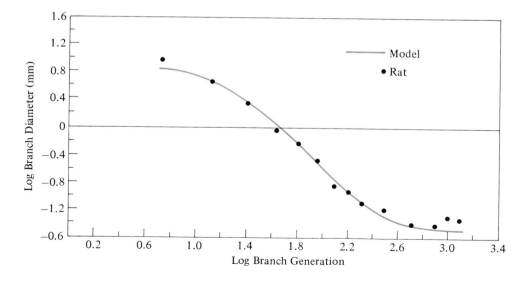

The relationship between average airway diameter and the branching generation for a rat lung is well described by a renormalized relationship. *(Modified from West, Bhargava, and Goldberger 1986.)*

branching patterns, other shapes, physiological events (e.g., a heart beat), or even time, are heterogeneous, self-similar, and lack a defined scale of length. They are heterogeneous because more structural detail is revealed the more closely they are examined. They are self-similar because the large-scale and small-scale patterns are generally similar in complexity. Their length cannot be determined independent of an arbitrarily selected scale, e.g., the size of the ruler or the magnification of the microscope objective.

Many biological structures, including the mammalian lung, have a fractal structure. Other fractal structures include vascular systems, neural networks, the biliary system of the liver, pancreatic ducts, and urinary ducts. Physiological events, such as a heart beat, also have fractal properties (although not structure).

The relationship between airway diameter and generation number for a mammalian (rat) lung is better fitted by a log-log relationship than by the above theoretical exponential relationship, but this still is not a perfect fit; there is a harmonic deviation of the data from the predicted line. A more complex, renormalized relationship closely predicts the observed scaling of airway diameter

$$D(z) = \frac{A_0 + A_1 \cos(2\,\pi \ln z/\ln \lambda)}{z^{\mu}}$$

where (for the rat lung) A_0 is 0.42 cm, A_1 is -0.14 cm, μ is 0.93, and λ is 11. (*See Weibel 1984; Horsfield and Cumming 1967; Goldberger and West 1987; West, Bhargava, and Goldberger 1986; West and Goldberger 1987.*)

Recommended Reading

Barnes, R. D. 1986. *Invertebrate zoology*. Philadelphia: Saunders.

Bliss, D. E. 1968. Transition from water to land in decapod crustaceans. *Am. Zool.* 8:355–92.

Bouverot, P. 1985. Adaptation to altitude-hypoxia in vertebrates. *Zoophysiology* 16. Berlin: Springer-Verlag.

Dejours, P. 1981. *Principles of comparative respiratory physiology*. Amsterdam: Elsevier.

Edney, E. B. 1960. Transition from water to land in isopod crustaceans. *Am. Zool.* 8:309–26.

Guyton, A. C. 1986. *Textbook of medical physiology*. Philadelphia: Saunders.

Hughes, G. M. 1976. *Respiration of amphibious vertebrates*. London: Academic Press.

Jones, J. D. 1972. *Comparative physiology of respiration*. London: Edward Arnold.

Kao, F. F. 1972. *An introduction to respiratory physiology*. Amsterdam: Excerpta Medica.

Little, C. 1990. The terrestrial invasion. An eophysiological approach to the origins of land animals. Cambridge: Cambridge University Press.

Mill, P. J. 1973. Respiration: Aquatic insects. In *The physiology of insecta*, edited by M. Rockstein, vol. 6, 346–402. New York: Academic Press.

Miller, P. L. 1973. Respiration—aerial gas transport. In *The physiology of insecta*, edited by M. Rockstein, vol. 6, 346–402. New York: Academic Press.

Randall, D. J., et al. 1981. *The evolution of air breathing in vertebrates*. Cambridge: Cambridge University Press.

Weibel, E. R. 1984. *The pathway for oxygen: Structure and function in the mammalian respiratory system*. Cambridge: Harvard University Press.

Wigglesworth, V. B. 1984. *Insect physiology*. London: Chapman and Hall.

Chapter 14

Circulation

• • • • • • • • • • • • • • • • • • • •

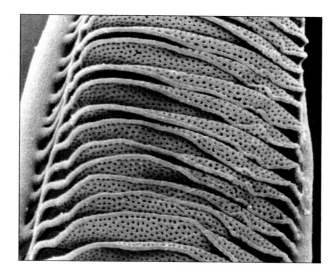

Scanning electron micrograph of a vascular cast of a fish gill illustrates the lamellae and their vascular bed. *(Photograph courtesy of J. O'Shea, Department of Zoology, The University of Western Australia.)*

Most higher metazoans rely on the **circulation** of a body fluid, the blood, for O_2 and CO_2 transport between their specialized respiratory surface(s) and their individual cells. Insects, with their tracheal respiratory system, are the obvious exception. However, O_2 and CO_2 transport are not the only, or necessarily even the major, functions of the circulatory system. Many other nutrients, metabolites, and waste products are also distributed by the circulatory system. Hormone distribution to target cells is accomplished by the circulation of body fluids. Heat distribution is an important function of the circulation in many ectothermic and endothermic animals. Blood pressure provides a hydraulic force for limb extension by spiders and for eclosion and wing expansion by insects. Many lizards clean their eyes by expanding on orbital venous sinus. The circulatory system of many higher metazoans distributes the specialized blood cells involved in immune defense and provides the fluid and solutes for nephridial or renal filtration and excretion.

Most animals thus have some kind of system that circulates a body fluid. However, there is considerable diversity among animals with respect to the functional design of their circulatory system, i.e., the structure and location of the circulatory system, the nature of the circulating body fluid (e.g., vascular or coelomic), and the means of propelling the fluid (e.g., tubular heart, chambered heart, extrinsic muscle pump, contraction of general body musculature). The nature and functions of the circulating fluid, the blood, are the topics of Chapter 15.

Design of Circulatory Systems

A circulatory system has three essential components: a **circulating fluid**; a **vascular system** through which the fluid can circulate; and some mechanism for maintaining fluid flow, such as a **heart**.

Lower metazoans lack a body cavity and generally do not have an internal circulatory system (e.g., Porifera, Cnidaria, Platyhelminthes; Figure 14–1A, B), although the acoelomate nemertean worms do have an internal, closed circulatory system. Coelenterates have an often complex pattern of gastrovascular circulation of seawater through fine channels that extend from the gut, but they do not need an internal circulatory system because their body wall is very thin (see page 677).

The body cavity of higher pseudocoelomate and coelomate animals contains a coelomic fluid. Pseudocoelomates lack a vascular circulatory system, but their body movements circulate the pseudocoelomic fluid for internal transport (e.g., nematodes, Figure 14–1C). A number of coelomate animals also lack a vascular circulatory system and rely on mixing of coelomic fluid for internal transport (e.g., some leeches).

The major body cavity (hemocoel) of some coelomate animals is derived from the embryonic blastocoel, and the true coelom remains small (or is absent). The vascular circulatory system, if present, is a mesodermal tube that only separates part of the hemocoel (i.e., inside the tube) from the rest of the hemocoel (i.e., outside the tube; e.g., a crustacean, Figure 14–1D). These **open circulatory systems** have an incomplete system of blood vessels; the circulating body fluid, which is called hemolymph in an open circulatory system, flows through the vessels and also freely percolates through the intercellular spaces. A heart(s) may be present to propel hemolymph through the vessels. In other coelomate animals, the coelom forms the main body cavity and the blastocoel is reduced to the space within the closed circulatory system. The fluid within such a **closed circulatory system** is blood; there is interstitial fluid in the intercellular spaces and lymph within lymphatic vessels. These closed circulatory systems have a complete system of vessels (Figure 14–1E, F), and the blood is completely separate and has a different composition from the interstitial fluid.

The circulation of body fluids requires a heart (or hearts) to generate a pressure gradient and to propel the body fluid. Blood flow through the circulatory system is almost always driven by a positive pressure forcing fluid away from the region of high pressure, rather than a negative pressure pump sucking fluid towards it. There are two reasons for this. First, muscle can only actively contract (stretch is passive) and the general arrangement of muscle in blood vessel walls would increase the internal pressure rather than decrease it. Second, a typical, nonrigid circulatory system would collapse with an internal negative pressure and blood would not flow through the vessels even with a highly negative driving pressure. However, there are some exceptions to this generalization that negative pressures do not drive blood flow (e.g., insect hearts and hearts with rigid pericardia).

The heart is not necessarily part of the circulatory system. For example, contraction of skeletal muscles can propel fluid through veins with valves

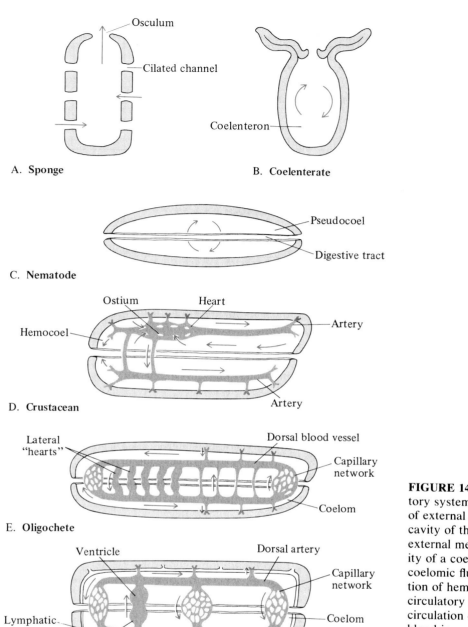

A. Sponge

B. Coelenterate

C. Nematode

D. Crustacean

E. Oligochete

F. Vertebrate

FIGURE 14–1   Diagram of the circulatory systems of animals. **(A)** Circulation of external medium in a gastrovascular cavity of the sponge; **(B)** Circulation of external medium in a gastrovascular cavity of a coelenterate; **(C)** circulation of coelomic fluid in a nematode; **(D)** circulation of hemolymph in an open vascular circulatory system of a crustacean; **(E)** circulation of both coelomic fluid and blood in separate systems of an oligochete; **(F)** circulation of blood in a closed vascular system of a vertebrate. *(From Prosser 1973.)*

(Figure 14–2A). Circulation of coelomic fluid also relies on the action of extrinsic muscles, such as locomotor or gut muscles (e.g., nematodes, leeches). The unique caudal heart of the hagfish consists of a cartilaginous rod, two chambers, and extrinsic muscles (Figure 14–2B); rhythmic contractions of the muscles alternately propel blood through the two valved chambers. Tubular pulsatile hearts propel blood by peristaltic contractions of their muscular wall (Figure 14–2C). Chambered hearts have a thickened muscular wall that expels blood from an internal chamber that typically has valves to ensure a one-way fluid flow (Figure 14–2D). The entire muscular wall contracts in synchrony, unlike

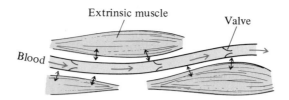

A. Extrinsic Muscle Pump

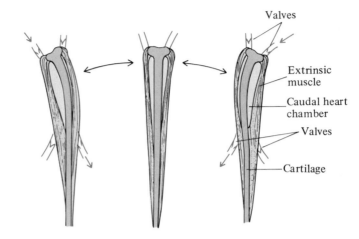

B. Hagfish Caudal Heart

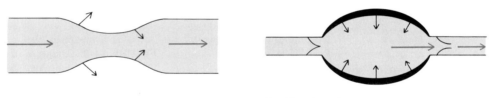

C. Pulsatile Heart

D. Chambered Heart

FIGURE 14–2   Types of hearts in animals. **(A)** External muscle pump expands and compresses a blood vessel (black arrows) and forces blood to flow along the vessel (colored arrows); valves maintain a unidirectional flow, e.g., nematode, scaphopod mollusk, and some leeches with no heart to propel blood flow, and the skeletal muscle pump of vertebrates. **(B)** Caudal heart of hagfish uses external muscles to oscillate a cartilaginous rod, which propels blood flow through valved veins. **(C)** Pulsatile tubular heart forces blood along a vessel (which may have valves to ensure one-way flow) by peristaltic contractions, e.g., insect heart. **(D)** A chambered heart propels blood by a coordinated contraction of the muscular wall; valves ensure a one-way flow of blood, e.g., most mollusk and vertebrate hearts.

the peristaltic waves that travel along tubular hearts. There are often two or more adjacent chambers: one type of muscular chamber (the **ventricle**) is specialized for pressure generation whereas the other less-muscular type of chamber (the **atrium**) collects the venous return and "primes" the ventricle with blood by low-pressure contraction. Chambered hearts are sometimes enclosed by a rigid pericardium; contraction of these hearts expels blood and at the same time draws blood into the pericardial space in preparation for the next ventricular filling cycle (e.g., *Daphnia* heart) or draws blood into the atrium (e.g., shark heart).

Rheology

The flow of blood depends on the same physical factors that determine the flow of any liquid, i.e., the pressure driving flow, the fluid viscosity, and the physical dimensions of the vessels.

Viscosity

Fluids vary in their propensity to flow, i.e., some are thick and some are thin. Water flows more easily than honey, which flows more readily than tar. **Viscosity** is the physical property of a fluid that reflects its tendency to flow. A fluid experiences a shearing motion when it flows over a surface or through a tube (Figure 14–3). The force moving the fluid (*F*; Newtons) is exerted over a certain surface area (*A*; m²). The ratio of force to area is the **shear stress**, τ; the shear stress is the relative force that moves the fluid.

$$\tau = F/A \qquad (14.1a)$$

The **shear strain** (γ; no dimensions) is the horizontal displacement (Δ*x*; m) relative to the thickness of the fluid layer (*l*; m); shear strain is a measure of the relative displacement of the fluid in response to the shear stress.

$$\gamma = \Delta x/l \qquad (14.1b)$$

The **shear strain rate** ($\dot{\gamma}$; sec^{-1}) is the rate of shear strain, i.e., shear strain per unit time (Δ*t*); the shear strain rate is thus an estimate of the relative velocity of fluid movement.

$$\dot{\gamma} = \Delta x/(l.\Delta t) \qquad (14.1c)$$

The viscosity of a fluid (η; Pa s) is the ratio of shear stress to shear strain rate;

$$\eta = \tau/\dot{\gamma} = (F\Delta t)/(Ax) \qquad (14.1d)$$

Viscosity is thus the relative force required to move a fluid compared to the relative velocity of the fluid. A thick fluid requires a large force to move it at a certain velocity, hence it has a high viscosity. A thin fluid requires a smaller force to move it at the same velocity. The units for viscosity in the SI system are Pascals seconds (Pa s) or N s m^{-2}. The viscosity of fluids is generally about 1 to 10 10^{-3} Pa s (Table 14–1). The more commonly used CGS system unit for viscosity is the centipoise; 1 poise = 1 dyne sec cm^{-2}; 1 centipoise = 0.01 dyne sec cm^{-2}. The conversion between CGS and SI units is 1 Poise = 0.1 Pa s. The viscosity of water is about 1 10^{-3} Pa s (1 cP) at 20° C; it is slightly dependent on temperature.

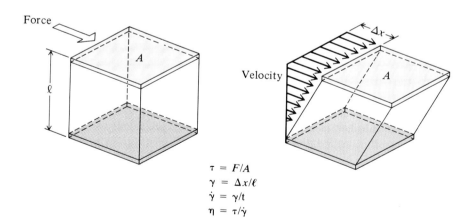

$$\tau = F/A$$
$$\gamma = \Delta x/\ell$$
$$\dot{\gamma} = \gamma/t$$
$$\eta = \tau/\dot{\gamma}$$

FIGURE 14–3 Application of a force to the top surface of a volume of fluid causes a shearing distortion of the fluid; there is a maximum displacement at the top surface and no displacement at the base, and the velocity of displacement is maximal at the top and zero at the base. The shear stress (τ) is the force per unit area (*F/A*) and the shear strain (γ) is the displacement relative to the thickness of the fluid (Δ*x/l*). The shear strain rate ($\dot{\gamma}$) is the shear strain per unit time. Viscosity (η) is $\tau/\dot{\gamma}$.

TABLE 14–1

Viscosity of water, blood, and some other fluids.
$1 \text{ Pa s} = 1 \text{ N s m}^{-2} = 10 \text{ Poise} = 10^3 \text{ centiPoise}.$

Fluid	Viscosity (Pa s)
Water 20° C	$1.0 \; 10^{-3}$
Water 100° C	$0.3 \; 10^{-3}$
Glycerine 20° C	$830 \; 10^{-3}$
10 wt motor oil 30° C	$250 \; 10^{-3}$
Albumin (30 g liter^{-1})	$1.1 \; 10^{-3}$
Dextran (MWt 26000; 30 g l^{-1})	$1.7 \; 10^{-3}$
Dextran (MWt 80000; 30 g l^{-1})	$2.3 \; 10^{-3}$
Plasma—*Amphiuma* (24 g l^{-1})	$0.9 \; 10^{-3}$
—frog (36 g l^{-1})	$1.2 \; 10^{-3}$
—man[1] (74 g l^{-1})	$1.4 \; 10^{-3}$
—man[2] (110–115 g l^{-1})	$13.1 \; 10^{-3}$
—elephant (82 g l^{-1})	$1.5 \; 10^{-3}$
Blood—human[3] 20% Hct	$2.5 \; 10^{-3}$
40% Hct	$3.8 \; 10^{-3}$
60% Hct	$6.5 \; 10^{-3}$
80% Hct	$14.4 \; 10^{-3}$

[1] Normal, 37° C.

[2] Macroglobulinemia, 24° C.

[3] 37° C.

The flow, or **rheological**, properties of blood are best described for vertebrate blood, and mammalian blood in particular (Merrill 1969). The viscosity of blood is influenced by its composition, the specific conditions of flow ($\dot{\gamma}$), and the properties and dimensions of the tube through which it flows. Blood contains **blood cells**, particularly erythrocytes (red blood cells, RBCs, in vertebrate blood) and fluid, the **plasma**. The percentage of the blood that consists of blood cells is the **hematocrit** (Hct). The viscosity of plasma is a little higher than that of water because of the plasma proteins, particularly fibrinogen; for example, human plasma has a viscosity of $1.4 \; 10^{-3}$ Pa s at 37° C (a shear strain rate of 212 sec^{-1}).

The rheological properties of blood depend on the protein concentration of the plasma; this is especially significant for blood that has its respiratory pigment free in solution. The viscosity of a protein (or other polymer) solution depends on its concentration. The complex relationship between η_{blood} and protein concentration (C) is generally modeled as

$$\eta_{\text{blood}} = 1 + [\eta]_{\text{int}}C + [\eta]_{\text{int}}^2 C^2 \qquad (14.2)$$

where $[\eta]_{\text{int}}$ is the **intrinsic viscosity** of the protein. The intrinsic viscosity is a property of the nature, molecular weight, and structure of the polymer (e.g., protein or polysaccharide, linear or branched

polymer). For example, polypeptides have a substantially lower $[\eta]_{\text{int}}$ than dextrans, which are sticky, gelatinous polymers of glucose.

The viscosity of whole blood is higher than that of plasma, depending on the hematocrit (Table 14–1; Figure 14–4A). The deformability of RBCs minimizes the blood viscosity, but fibrinogen "sticks" to red cell membranes and promotes red cell interactions, thereby increasing blood viscosity. The viscosity of blood depends on hematocrit in an exponential fashion

$$\eta_{\text{blood}} = \eta_{\text{plasma}} e^{k \, \text{Hct}} \qquad (14.3)$$

where η_{plasma} is plasma viscosity and k is a constant. Values of η_{plasma} tend to be lower for mammalian blood (at 37° C) than for blood of ectotherms (at room temperature). These values are rabbit, 0.93; elephant seal, 1.3; bullfrog, 1.38; marine toad, 1.81. Values for k are fairly similar; rabbit, 0.03; elephant seal, 0.03; bullfrog, 0.031; marine toad, 0.033.

Blood viscosity depends on the flow rate (or more precisely, $\dot{\gamma}$), particularly at low-flow rates. This nonideal, or **non-Newtonian**, behavior of blood at low $\dot{\gamma}$ is graphically illustrated by the relationship between $\tau^{1/2}$ and $\dot{\gamma}^{1/2}$ (Casson plot; Figure 14–4B). The relationship is linear at high $\dot{\gamma}^{1/2}$ (and extrapolates to the origin) but there is a transition at lower $\dot{\gamma}^{1/2}$ to a parallel line that does not pass through the origin (the slope of a line from the origin to any point on the relationship is $\eta^{1/2}$). The intercept at $\dot{\gamma}^{1/2} = 0$ is called the **yield shear stress**; it indicates that stationary blood will not begin to flow until a threshold stress is exceeded.

The viscosity of blood depends on the diameter of the tube through which it flows. This **Fahraeus-Lindqvist effect** is another non-Newtonian property of blood. This effect is especially apparent for tubes of diameter less than 0.8 mm, for which blood viscosity is substantially lower than it is in larger tubes. This effect is of particular significance in circulatory systems since many blood vessels are less than 0.8 mm diameter. Consequently, the vascular resistance to blood flow is actually less than expected when the Fahraeus-Lindqvist effect is taken into account (Snyder 1973).

Experiments with lysed mammalian blood indicate that hemoglobin solutions have a lower viscosity than normal blood at the equivalent hemoglobin concentration (Cokelet and Meiselman 1968; Schmidt-Nielsen and Taylor 1968). This is not surprising considering the importance of plasma proteins, particularly fibrinogen, in making RBCs "sticky." It suggests that hemoglobin is not packed in RBCs to reduce blood viscosity, since RBCs appear to have exactly the opposite effect. Alterna-

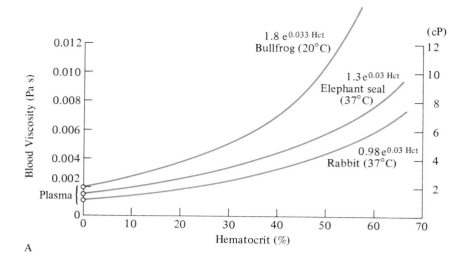

A

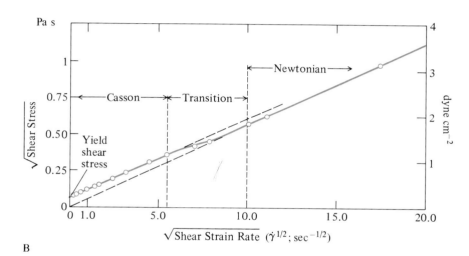

B

FIGURE 14–4 **(A)** There is an exponential relationship between blood viscosity and hematocrit for vertebrate blood (bullfrog, rabbit, elephant seal). **(B)** The Casson plot illustrates the non-Newtonian behavior of blood at low shear strain rate, the Newtonian behavior of blood at high shear strain rate, and the transition at intermediate shear strain rate. Viscosity is the ratio of shear stress to shear strain rate; it is graphically obtained as the square of the slope of the line from the origin to any point on the curve for blood. The Casson region of the graph extrapolates to the yield shear stress at zero shear strain rate; the Newtonian region of the curve extrapolates to the origin. *(Modified from Withers et al. 1991; Hedrick, Duffield, and Cornell 1986; Merrill 1969.)*

tive explanations for why vertebrate hemoglobin is packaged within RBCs are the reduction in the colloid osmotic pressure by packaging hemoglobin molecules within RBCs and prevention of hemoglobin excretion by the kidneys. However, the conclusion that packaging hemoglobin within RBCs has detrimental effects in terms of viscosity is not necessarily valid. Taking the Fahraeus-Lindqvist effect into account suggests that the vascular resistance to a hemoglobin solution is actually greater than the resistance to blood containing RBCs in small vessels (Snyder 1973).

Temperature influences the viscosity of blood. The effect is similar for water and blood at high shear strain rates (i.e., when it behaves as a Newtonian fluid), but temperature has a lesser effect on blood viscosity at lower shear strain rates (i.e., in the non-Newtonian range).

Poiseuille-Hagen Flow Formula

The Poiseuille-Hagen formula provides a simple model for blood flow through the circulatory vessels (as long as length $\gg$ diameter)

$$V_b = \Delta P \pi r^4 / 8\eta l \qquad (14.4)$$

where, in SI units, V_b is the blood flow (m^3 min^{-1}); ΔP is the blood pressure difference (N m^{-2}); r is the vessel radius (m); η is the blood viscosity (Pa s); and l is the vessel length (m). Blood flow is directly proportional to the pressure difference between the ends of the vessel and is markedly dependent on vessel radius, since $V_b \propto r^4$. For example, flow increases by 16 $\times$ if r is increased by 2$\times$. Blood flow is inversely proportional to the fluid viscosity and inversely proportional to vessel length.

Laminar and Turbulent Flow

Blood flow can be either **laminar** or **turbulent** (see Supplement 14–1, page 724). The nature of the flow depends on the fluid velocity (v), the vessel diameter (d), and the kinematic viscosity of the fluid (v; $v = \eta/\rho$, where ρ is the density). The Reynolds number (R_e) is a dimensionless coefficient that indicates whether flow is laminar or turbulent.

$$R_e = vd/v = vd\rho/\eta \qquad (14.5)$$

Flow is laminar if the R_e is less than a critical value ($R_{e,\text{crit}} \approx 200\text{--}400$) and turbulent if $R_e > 2000$. At intermediate R_e, flow can be laminar or turbulent depending on specific conditions. For example, flow may be laminar in straight vessels but may become turbulent at junctions or bends, at intermediate R_e.

Blood flow is laminar in most animals since their blood vessels are small. For example, critical R_e values calculated for various parts of the circulatory system of the dog indicate that blood flow is generally laminar, although flow is potentially turbulent in the largest vessels (the aorta and vena cava; Table 14–2). Flow can be turbulent for animals with large blood vessels and high blood flow rates. For example, the Reynolds number is about 3490 ($>R_{e,\text{crit}}$) and flow is turbulent for the human aorta (diameter = 1.8 cm; cross section = 2.4 cm²; flow velocity = 33.3 cm sec⁻¹). However, R_e is about 0.017 ($\ll R_{e,\text{crit}}$) and the flow is laminar in human capillaries (diameter = 0.00037 cm; total cross-sectional area = 108 cm²; velocity = 0.77 cm

sec⁻¹). The pulsatile nature of blood flow in the arterial system also promotes turbulence of blood flow.

Resistance

Blood flow (V_b) and the pressure gradient (ΔP) are readily measured for circulatory systems, but the other terms in the Poiseuille-Hagen flow formula are more difficult to determine (i.e., mean r, l, and η). Consequently, these latter terms and the constants (8, π) are often grouped as a single term, the **resistance** (R), where

$$R = \frac{8\eta l}{\pi r^4} \qquad (14.6a)$$

Substitution of this relationship for R into the Poiseuille-Hagen flow formula yields

$$V_b = \Delta P/R \qquad (14.6b)$$

The SI unit for resistance is N sec m⁻⁵. In the CGS system, the unit is dyne sec cm⁻⁵. Mammalian physiologists have also defined a peripheral resistance unit, or PRU, such that PRU = 1 if ΔP is 1 torr (1 mm Hg) and V_b is 1 ml sec⁻¹. In humans, V_b is about 100 ml sec⁻¹ and P is about 100 torr for the systemic circuit, hence the systemic resistance is about 1 PRU. In contrast, the normal pulmonary resistance is about 0.14 PRU (14 torr/100 ml sec⁻¹). **Conductance** is the reciprocal of resistance; it is the flow through a vessel for a given pressure difference. The most significant determinant of R and conduc-

TABLE 14–2

Vascular dimensions and Reynolds number (R_e) for a dog. Cardiac output is assumed to be 22 ml sec⁻¹ and $v = \eta/\rho = 1.16 \ 10^{-2}$ cm² sec⁻¹. *(Calculated from Burton 1966)*

Vessel	Diameter (cm)	Number	Length (cm)	Total Cross-section Area (cm²)	Total Volume (ml)	Total Surface Area (cm²)	Velocity (cm sec⁻¹)	R_e^{1}
Aorta	1.0	1	40	0.8	30	1.3 10²	28	2394
Large arteries	0.3	40	20	3.0	60	7.5 10²	7.8	200
Main arteries	0.1	600	10	5.0	50	1.9 10³	4.7	39.9
Terminal arteries	0.06	1800	1	5.0	25	3.4 10²	4.3	22.2
Arterioles	0.002	4 10⁷	0.2	125	25	5.0 10⁴	0.18	0.030
Capillaries	0.0008	1.2 10⁹	0.1	600	60	3.0 10⁵	0.036	0.0025
Venules	0.003	8 10⁷	0.2	570	110	1.5 10⁵	0.039	0.010
								8.9
Terminal veins	0.15	1800	1	30	30	8.5 10²	0.69	
Main veins	0.24	600	10	27	270	4.5 10³	0.81	16.6
Large veins	0.6	40	20	11	220	1.5 10³	1.9	99.8
Vena cava	1.25	1	40	1.2	50	1.6 10²	18	1915

¹$R_e = D \ v/v = 85.5 \ D(\text{cm}) \ v(\text{cm sec}^{-1})$.

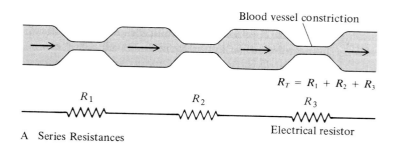

A   Series Resistances

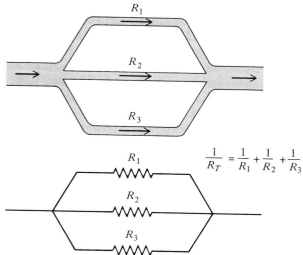

B   Parallel Resistances

FIGURE 14–5 **(A)** The total resistance (R_T) of a series of resistances is the sum of the individual resistances ($R_1 + R_2 + R_3$). Shown is a schematic of three series resistance elements in a vessel and the electrical analog circuit to vascular resistances. **(B)** The total resistance (R_T) of resistances in parallel is calculated from the sum of the reciprocal resistances for the individual resistances ($1/R_1 + 1/R_2 + 1/R_3$). Shown is a schematic of three parallel resistance elements in a vessel with the electrical analog circuit.

tance is vessel radius, since R is proportional to r^{-4} and conductance is proportional to r^4 (see Equation 14.6a).

Circulatory systems generally are a series of branching vessels, each vessel having its own resistance (determined by its r and l). For vessels in series, the total resistance (R_T) is the sum of the individual resistances (Figure 14–5A); e.g., for three vessels with resistances R_1, R_2, R_3

$$R_T = R_1 + R_2 + R_3 \qquad (14.7a)$$

For resistances in parallel, the total resistance is equal to the reciprocal of the sum of the individual reciprocal resistances (Figure 14–5B).

$$1/R_T = 1/R_1 + 1/R_2 + 1/R_3 \qquad (14.7b)$$

Both of these equations are analogous to the calculation of electrical resistances in series and parallel.

Distensibility and Compliance

Blood vessels are generally not rigid-walled tubes, but are elastic. The radius of vessels, therefore, depends on the internal blood pressure (and also the external fluid pressure), i.e., r is dependent on ΔP. The elasticity of blood vessels and their capacity to alter radius, dependent on the physiological conditions, means that the intravascular volume can alter. **Distensibility** (D; kPa^{-1}) is the relative increase in internal volume of a blood vessel per unit increase in pressure

$$D = \Delta V/(V \Delta P) \qquad (14.8a)$$

where ΔV (ml) is the change in volume (V; ml) due to a change in internal pressure (ΔP; kPa). Veins generally are 6 to 10 $\times$ more distensible than arteries.

The elasticity of blood vessels is of great significance to flow because even small changes in r have a dramatic effect on V_b, which is proportional to r^4. Consequently, there generally is not a linear relationship between V_b and P for blood vessels, but V_b increases more than in a linear proportion to P (Figure 14–6). The degree of vasoconstriction/vasodilation of the blood vessel, mediated by the sympathetic nervous system, also influences the relationship between V_b and ΔP.

Blood vessels, and in particular small arteries, tend to collapse at low ΔP, and so V_b is 0 even at some finite ΔP; this ΔP at which the vessel collapses

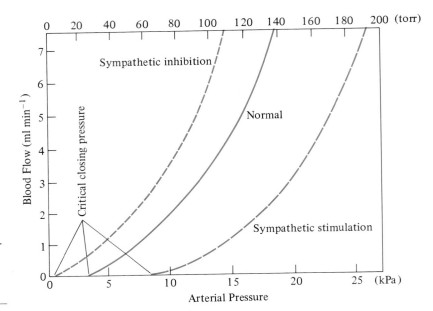

FIGURE 14–6 Relationship between blood flow and arterial pressure showing the effect of sympathetic stimulation and inhibition on critical closing pressure.

is called the **critical closing pressure**. This is due to two factors. First, vertebrate blood contains red blood cells of finite dimensions, and blood flow is impeded or even stopped if the vessel diameter is similar to the red cell diameter. Second, the law of Laplace (see Chapter 13) indicates that the wall tension (T) due to a pressure difference (ΔP) across a vessel wall is proportional to the vessel radius (r), i.e., $T = r\Delta P$. As ΔP declines, so does the wall tension, and the inherent wall elasticity causes r to decrease. At a critical P, a "vicious cycle" develops between ΔP and r, and the vessel completely collapses.

Compliance (C; ml kPa^{-1}) is the absolute change in volume of a vessel due to an alteration in pressure.

$$C = \Delta V/\Delta P \qquad (14.8b)$$

Thus, $C = DV$. Compliance has a different significance to the functioning of the circulatory system than distensibility. A low volume but very distensible vessel has a lower compliance than a larger volume vessel with the same distensibility; the larger volume vessel is able to absorb a greater change in blood volume for the same ΔP. The compliance of veins is about 24 × greater than that of arteries because they have about 3 × the volume and 8 × the distensibility. Compliance, like distensibility, is affected by the sympathetic regulation of blood vessels.

The arterial vessels (e.g., aorta) generally are very elastic and have a high distensibility and compliance. An important function of these vessels

is to damp the marked pressure cycle of the ventricular pump (i.e., 0 to 3 kPa cycle in some invertebrates and 0 to 16 kPa in some vertebrates). As early as 1773, Stephen Hales suggested that the elasticity of the arterial wall dampens the periodic cardiac pumping into a relatively steady blood pressure and flow in the more peripheral vessels. This dampening role has been classically viewed as a "**windkessel**" effect (Milnor 1982). A windkessel is a fluid compression chamber used to smooth the pumping strokes into an almost continuous stream (used in early fire engine pumps to smooth water flow and in organ pipes to smooth air flow). Blood ejected from the heart into the aorta increases the pressure and distends the elastic wall of the aorta because of the considerable downstream resistance to blood flow out of the aorta (Figure 14–7). When the ventricle ceases to eject blood, blood continues to leave the aorta as its elastic walls recoil. The vertebrate aorta and the aortas of many invertebrates have suitable elastic properties to act as windkessels (Vreugdenhil and Redmond 1987; see below).

The windkessel model can be described by simple equations involving compliance (C) and resistance (R). The pressure in the windkessel slowly declines in an exponential fashion from the maximal systolic pressure (P_{max}); the pressure at any time after ejection ceases (P_t) is calculated as follows.

$$P_t = P_{max}e^{-t/RC} \qquad (14.9)$$

The windkessel model has heuristic value but has been subsequently found to be too simplistic for

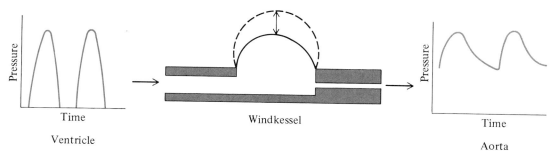

FIGURE 14–7 The windkessel effect dampens fluctuations of ventricular pressure by elastic accommodation of the pressure and volume pulses in the aorta; a high downstream resistance facilitates the pressure pulse damping in the aorta.

detailed hemodynamic analysis, and it has been replaced by more complex fluid dynamic and biomechanical models.

Pressure, Velocity, and Gravity

The energy of a moving fluid (with laminar flow) depends on the fluid velocity, the fluid pressure, and its position in the gravitational field. Hence, the energy of fluid flow (N m or kg m^2 s^{-2}) is the sum of kinetic energy, pressure energy, and gravitational (potential) energy. According to **Bernoulli's theorem**, the total energy of a moving fluid is

$$E = 1/2 \, Mv^2 + Mgh + PV \qquad (14.10a)$$

where $1/2 \, Mv^2$ is the kinetic energy (M = kg; v = m s^{-1}), Mgh is the potential energy (g = m s^{-2}; h = m), and PV is pressure energy (P = kg m^{-1} s^{-2}; V = m^3). It is instructive to examine the relative importance of these various energy terms. For 1 ml of blood in the human aorta; $1/2 \, Mv^2$ = $1/2 \, (0.001)(0.333)^2$ = 55.4×10^{-6} N m; PV = $(1.33 \times 10^4)(10^{-6})$ = 13300×10^{-6} N m; Mgh = $(0.001)(9.8)(1.3)$ = 12740×10^{-6} N m (relative to ground level). Pressure and potential energy are similar, and much greater than kinetic energy, i.e., $PV \approx Mgh \gg 1/2 \, Mv^2$.

Bernoulli's theorem suggests that the energy content of a fluid remains constant if the fluid has no viscosity and the flow is frictionless ($\Delta P = 0$) and horizontal ($\Delta Mgh = 0$), i.e.,

$$1/2 \, Mv^2 + PV = \text{constant} \qquad (14.10b)$$

This means that the PV term must decrease at a vessel constriction, since velocity increases at the constriction (mass flow is constant and velocity =

mass flow/cross-sectional area). This is the Venturi effect. Conversely, the P increases at a vessel dilation. In reality, fluids have a viscosity and flow is not frictionless, and so there is a progressive decline in fluid energy due to frictional dissipation of energy as heat as the fluid flows along a vessel.

Aquatic animals are supported by a water column of almost the same density as their body fluids, and so they experience no significant gravitational effect. In contrast, the interchange of pressure energy and potential energy can be of considerable importance to large terrestrial animals. Pressure energy is clearly interconvertible with potential energy, e.g., blood can be made to flow uphill by a high pressure, and a column of blood exerts a pressure at its base. The proportionality between pressure and potential energy is 1 m ≡ 0.1 atm ≡ 10 kPa. The arterial pressure of mammals is equivalent to about 1 m gravitational pressure. For a human lying horizontal (Figure 14–8A), there is no significant gravitational effect between the head, heart, and toes, but for an erect person there is about 1.3 m between the feet and the heart and 0.4 m between the heart and the top of the head. There is therefore a gravitational effect of the same order of magnitude as arterial blood pressure, and the blood pressure in the feet (heart blood pressure + gravitational pressure) is about 2 × the heart blood pressure and is about 0.7 × in the head. It has been suggested that blood returning from the head to the heart aids blood flowing from the heart to the head by a siphon effect, i.e., the venous return "pulls" arterial blood up into the head. Unfortunately (at least for this theory), blood vessels are not rigid tubes, and any siphon effect is negated by the collapse of the venous blood vessels (Seymour and Johansen 1987).

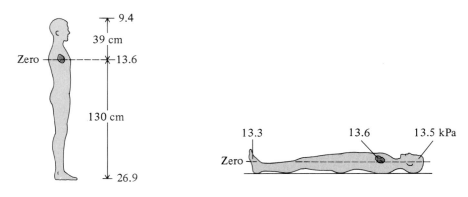

A   Gravitational Effect on Blood Pressure

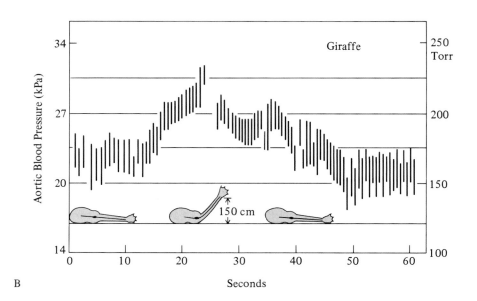

B

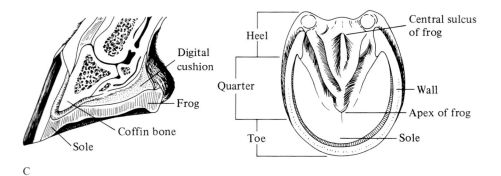

C

FIGURE 14-8 (A) Gravitational effects on hydrostatic pressure of a fluid column (e.g., the human vascular system) result in a decreased pressure at levels above the heart and increased pressures below the heart for a person while standing, but not when lying prone. The gravitational pressure effects are much greater than the small frictional effects. (B) Effect of raising the head 150 cm on the aortic pressure (measured in the aortic root near the heart) of a giraffe. (C) Structure of the foot of the horse facilitates return of venous blood from the foot to the heart. The cartilaginous "frog" is forced upward and outward by the weight of the horse, and this force is transferred to the collateral cartilages of the third phalanx thence to large veins in the foot, forcing venous blood towards the heart. Most artiodactyls and perissodactyls have similar venous pumping systems. *(From Ruch and Patton 1979; Goetz et al. 1960; Adams 1976; Ferguson 1985.)*

The significance of gravity is even greater in larger animals, e.g., a 3.5 m tall giraffe. The head–heart distance is about 160 cm for a giraffe (cf. 34 cm in humans) and the gravitational pressure difference would be 22 kPa lower in the head. If the aortic pressure at the level of the heart of the giraffe was the same as that of humans, it would support a blood column 120 cm high, i.e., 40 cm short of the giraffe's head. The arterial pressure in the head would be -6.4 kPa (this simple calculation ignores the additional pressure required to overcome vascular resistance to flow)! However, the cerebral arterial pressure of the giraffe is about $+12$ kPa and so the left ventricular pressure must be about 34 kPa; the left ventricle is hypertrophied to generate such high systolic pressures. In other parts of the body, especially the lower limbs, the blood hydrostatic pressure would be exceedingly high, e.g., $34 + 22 = 56$ kPa in the feet. Such extreme pressures would drastically alter the capillary balance of forces if they extended to the capillary bed (see Capillaries, page 696) and so there must be extensive reduction in arterial blood pressure by high-resistance arteries/arterioles. Lowering the head, for example, to drink, drastically lowers the left ventricular pressure required to pump blood to the brain. Consequently, there are marked cardiovascular adjustments in the giraffe during raising and lowering of the head; otherwise the cerebral vascular pressure would change excessively (Figure 14–8B).

Mammals with long thin extremities and only a few muscles below their knees (e.g., gazelles, horses) cannot rely on a skeletal muscle pump for venous return from the feet. Many hoofed mammals therefore have a foot pump that assists venous return. The most elastic part of the foot is a cartilaginous structure at the base of the foot, the "frog" (Figure 14–8C). This frog is compressed when the foot strikes the ground, and acts as a wedge to expand the digital cushion and compress the venous plexus of the foot. The compression of the plexus by the cartilage forces blood towards the heart. The frog also acts as a hydraulic cushion, reducing concussion of the foot.

Invertebrate Circulatory Systems

There is a considerable diversity among invertebrates in their patterns of circulation. Some invertebrates completely lack a circulatory system, whereas others have a complex closed and high-pressure circulatory system. Let us survey this remarkable diversity.

Coelenterates

The basic body plan of coelenterates, such as anemones and jellyfish, is a thin body wall of epidermal-gastrodermal cells with an intervening mesogleal layer. The epidermal and gastrodermal cells are in direct contact with the external medium (either outside the animal or inside the gut) and rely on diffusive gas exchange. Some coelenterates attain a considerable size (anemones many cm in diameter and jellyfish up to 2 m in diameter). Large anemones have many internal folds of gastrodermis, called mesenteries. These longitudinal, radiating mesenteries ensure that all gastrodermal cells are in close contact with the gut fluid for gas exchange. Large jellyfish have ciliated canals radiating from the central stomach through the mesoglea to the circular canal near the bell margin. Water flow along this gastrovascular canal system ensures a constant flow of water for gas exchange.

Nemertean Worms

Nemerteans (proboscis worms) are elongate and often flattened acoelomate worms with a closed circulatory system (Hyman 1951). Simple nemertean circulatory systems consist of two lateral blood vessels connected anteriorly by a cephalic lacuna and posteriorly by an anal lacuna. Some nemerteans have a more complex arrangement with additional longitudinal and transverse blood vessels (Figure 14–9). The walls of larger vessels have four layers: an inner epithelium of bulging cells, a thick gelatinous connective tissue layer, a circular muscle layer, and an outer acellular layer. The presence of an epithelium is unusual for invertebrate circulatory systems. Contractions of the blood vessels, as well as contractions of the body wall musculature, create an irregular forward and backward blood flow. The blood vessels are closely associated with the excretory protonephridia.

Aschelminths

The aschelminth phyla (nematodes, rotifers, gastrotrichs, etc.) are pseudocoelomate; their body cavity is derived from the embryonic blastocoel and is not a true coelom. Some aschelminths have a large, functional pseudocoel (e.g., nematodes, rotifers, gastrotrichs). Others have their pseudocoelom reduced to a thin, slit-like space. No aschelminths have a circulatory system and only those with a large coelom (i.e., nematodes) would benefit from an internal circulation of the coelomic fluid caused

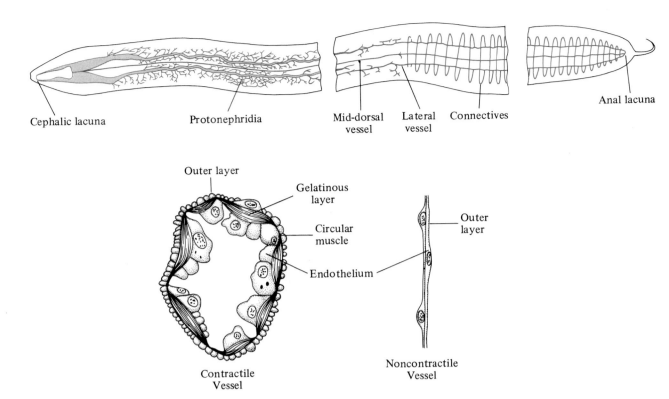

Cephalic lacuna   Protonephridia   Mid-dorsal vessel   Lateral vessel   Connectives   Anal lacuna

Outer layer
Gelatinous layer
Circular muscle
Endothelium
Outer layer
Contractile Vessel
Noncontractile Vessel

FIGURE 14–9 The circulatory system of the nemertean worm *Cerebratulus* consists of a mid-dorsal and lateral vessel with dorsal and ventral connectives. The contractile vessels have a thick wall with a circular muscle layer whereas the noncontractile vessels have a thin wall. *(From Hyman 1951. With permission of McGraw-Hill Inc., New York.)*

by body movements and contraction of body musculature (see Figure 14–1C, page 667).

Annelids

Oligochete worms have a well-developed and essentially closed circulatory system (Figure 14–10). Branches of the dorsal vessel supply blood to capillaries in the integument (for gas exchange), viscera, nephridia, and the gut. The dorsal vessel is contractile, and a variable number of anterior vessels connect the dorsal and ventral vessel and function as tubular hearts, e.g., *Lumbricus* has five pairs and *Tubifex* has a single pair of circumintestinal hearts. The hearts and also the dorsal vessel often contain valves to ensure a one-way flow. The dorsal vessel is paired in some megascolid and glossoscolid earthworms.

The giant earthworm *Glossoscolex giganteus* may weigh 500 to 600 g, be 120 cm long, and have a diameter of 2 to 3 cm; it relies on cutaneous gas exchange (Johansen and Martin 1965). Peristaltic activity of the dorsal blood vessel (6 to 8 contractions min^{-1}) develops a blood pressure of about 1.4 to 2.4 kPa. The five pairs of lateral hearts usually beat in synchrony at about 20 min^{-1}, and develop a pressure of over 10 kPa in the ventral vessel during activity. Blood pressures are affected by activity; pressure is elevated during body elongation and reduced during body shortening.

The polychete circulatory system is, like that of oligochetes, well developed and essentially closed. Most vessels, however, are relatively thin walled and are not always lined by endothelium. When present, the endothelium has its basement membrane facing the lumen rather than the endothelial cells. Blood flows anteriorly in a dorsal vessel and posteriorly in a ventral vessel. The ventral vessel forms segmental pairs of parapodial vessels supplying the appendages, the body wall, the nephridia, and several intestinal vessels that supply the gut. The dorsal vessel receives corresponding pairs of dorsal parapodial and intestinal vessels and at the anterior end is connected to the ventral vessel by

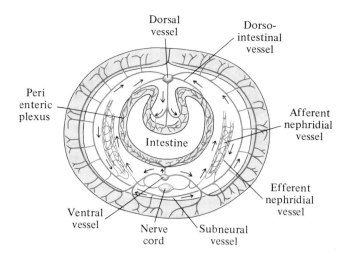

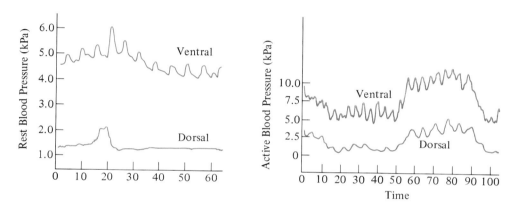

FIGURE 14–10 The closed circulatory system of an oligochete annelid (*Lumbricus*) has a dorsal and ventral vessel and capillary beds in the gut, viscera, and skin; arrows indicate the direction of blood flow. Blood pressure in the dorsal and ventral vessels of a giant earthworm increases with activity. *(Modified from Barnes 1987; Johansen and Martin 1965.)*

one or several vessels or by a network of vessels passing around the gut. Peristaltic waves of especially the dorsal vessel sustain blood flow. Some polychetes also have accessory "hearts" in various other parts of the vascular system. Gills, when present, generally have afferent and efferent vascular loops with a unidirectional blood flow. The branchial lobe of the parapodia of nereids has parallel capillaries joining dorsal and ventral lateral vessels. The radioles of fan worms, which are used for filter feeding and gas exchange, have a single blood vessel with an ebb-and-flow of blood. Polychetes also rely, to varying degrees, on circulation of their coelomic fluid for internal transport.

Leeches use their coelomic sinuses as a supplement to the normal vascular system (e.g.,

rhyncobdellid and fish leeches) or as a replacement for the vascular system (e.g., many gnathobdellid and pharyngobdellid leeches). The rhyncobdellid leeches have a dorsal blood vessel located within the dorsal sinus and a ventral blood vessel in the ventral sinus; these vessels are connected at their extremities by paired loops. The blood vessels, especially the anterior dorsal, are contractile. Dorsal, lateral, and ventral sinuses are modified to form circulatory vessels in other leeches.

The dorsal and ventral (sinus) vessels are noncontractile in the medicinal leech *Hirudo medicinalis* (Hildebrandt 1988). The dorsal vessel collects blood from the anterior lateral vessels and the organs and body wall and discharges it into the intestinal capillaries and the latero-dorsal vessels (Figure

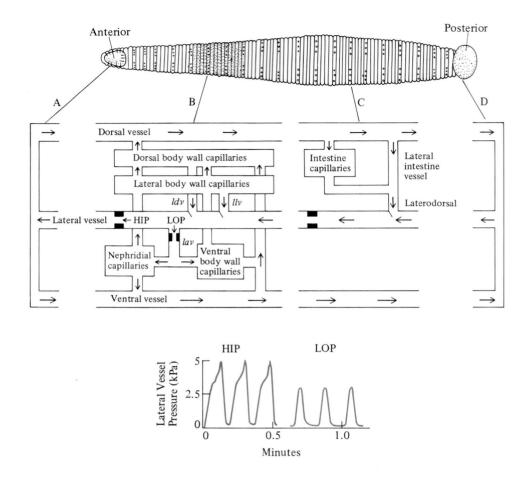

FIGURE 14–11 Schematic arrangement of the circulatory system of a blood-sucking leech. **(A)** 6 anterior fused segments; **(B)** 21 somatic segments; **(C)** 5 segments containing the intestine; **(D)** 7 posterior fused segments. Arrows indicate the direction of blood flow, HIP, direction of blood flow in high-pressure phase; *ldv*, laterodorsal vessel with valve; *llv*, laterolateral vessel with valve; *lav*, lateral abdominal vessel; LOP, direction of blood flow in low-pressure phase. *(Modified from Hildebrandt 1988.)*

14–11). The ventral vessel collects blood from the anterior lateral vessel and nephridia and discharges it into the lateral vessel. Blood flows posteriorly in the dorsal and ventral vessels. The two lateral vessels, which run the entire length, are tube hearts; they are contractile and propel the blood anteriorly. These lateral vessels of the anterior segments have peristaltic contractions, but the blood vessels of other segments contract simultaneously. Pressures recorded in the anterior lateral vessels are about 6.7/0.3 kPa during the peristaltic mode of contraction (called the high-pressure phase, or HIP) and about 3.3/0.3 kPa in the nonperistaltic low-pressure mode (LOP). The pressure pulses of the anterior vessels, in the HIP mode, exhibit an initial increase in pressure due to contraction of the more posterior segment (the "presystolic peak") and a subsequent

"systolic peak" due to contraction of that part of the vessel.

Mollusks

Mollusks generally have an open circulatory system with blood movement at least partially dependent on a chambered heart. Blood is pumped by the heart through arteries into the hemocoel and makes its way back into the heart through open sinuses. However, the circulatory system of mollusks shows considerable diversity in its structure and functioning (Hyman 1967), and only a general summary can be presented here (Figure 14–12). Scaphopod mollusks have a reduced circulatory system with only a rudimentary heart; blood is propelled through a system of sinuses by muscular contraction. At the

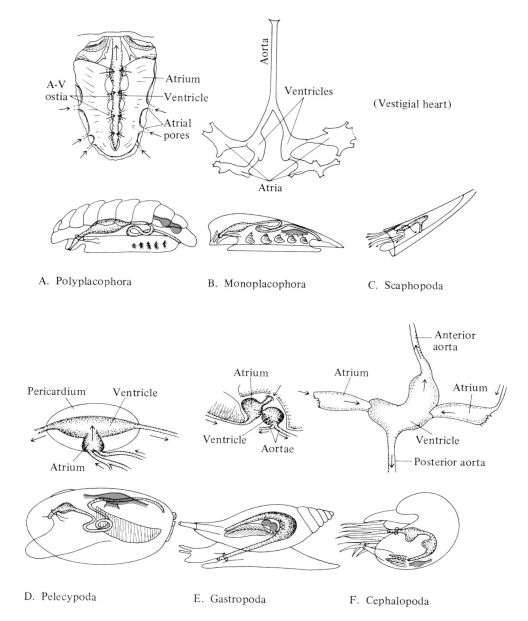

FIGURE 14–12 Summary of the general structure of the circulatory system in the six molluscan classes and a detailed structure of their heart; shown is the intestinal tract (stippled), heart (color), gill, shell, and cerebral ganglia. *(Modified from Hickman 1973; Hyman 1967; Hill and Welsh 1966.)*

other extreme, cephalopods have a well-developed, closed circulatory system with a systemic heart and a pair of branchial hearts. Let us now examine in more detail the circulatory systems of the major groups of mollusks: Polyplacophora, Monoplacophora, Scaphoda, Pelecypoda, Gastropoda, and Cephalopoda.

The circulatory system of Polyplacophora (chitons) is very open. The heart is located in the dorsal,

posterior part of the animal beneath the two most posterior valves (Figure 14–12A). The heart, located within a pericardial cavity, consists of a pair of lateral atria and a median ventricle. The atria typically open into the ventricle via two pairs of slit-like atrioventricular ostia. The ventricle continues anteriorly as a single aorta, which is an often diffuse space rather than a discrete vessel. Channels extend from the aorta into the gonad, and "intersegmental" channels supply valvular muscles. The aorta terminates at about the most anterior valve where it opens into the head sinus. Venous blood collects in the visceral sinus then venous channels of the foot, where it is directed posteriorly to the afferent branchial sinuses. After passing through the gills, blood collects in the efferent branchial sinus and most enters the atrium via atrial pores.

The circulatory system of monoplacophorans (single-shelled mollusks, e.g., *Neopilina*) is also very open. The heart is located in the dorsal posterior of the animal; there is one pair of ventricles and two pairs of lateral atria (Figure 14–12B). Each ventricle forms an anterior aorta; the two aortae fuse and extend anteriorly to the edge of the foot and open into a blood sinus. Blood drains via extensive visceral sinuses into a pallial venous sinus where it traverses the gills and then enters the atria.

Scaphopodans (elephant tusk shells) have a vestigial heart (Figure 14–12C). Rhythmic protrusion and withdrawal of the foot propel the blood through irregular coelomic spaces.

Pelecypodan (bivalve) mollusks have a well-developed circulatory system that is much less open than that of the previously described mollusks. The ventricle and two atria pump blood into an anterior aorta which forms arteries to the mantle, viscera (digestive tract and gland), foot, and anterior regions (Figure 14–12D). There sometimes is also a pair of posterior aortae (e.g., *Anodonta*). The arteries form networks of sinuses within the tissues, then the sinuses coalesce into larger sinuses that direct blood to the kidney and gills. In some species (e.g., *Anodonta*), most blood passes through the gills before it is returned directly to the atria. In other species (e.g., *Mytilus*), much of the blood returns directly to the heart from the kidneys.

The circulatory systems of gastropod mollusks (Figure 14–12E) are quite diverse. Primitive prosobranchs have a ventricle and paired atria; the left atrium is often greatly reduced and functionless (as is the left gill). Opisthobranch gastropods have a similar circulatory system (except for *Rhodope* and *Alderia*, which lack a heart, and *Kadaia*, in which the heart is not connected to any vessels!). The heart is typically located in the mantle roof near the gill and nephridia; it has a single ventricle and atrium. In *Aplysia*, the ventricle forms an anterior aorta that supplies the digestive and reproductive tracts, mantle, and foot and a posterior aorta that supplies the midgut and gonads. Venous blood collects in the tissue sinuses, then proceeds partly to the gill and partly to the kidney. Blood from the gill and kidney enters the atrium. The circulatory system of pulmonate gastropod snails, such as *Helix*

and *Lymnaea*, is less open than in other gastropods. The heart has a single atrium and ventricle enclosed by a pericardium. The ventricle forms an aorta, which divides into a small posterior aorta that supplies mainly the midgut gland and a main anterior aorta, which forms a number of well-defined arteries that pass into capillaries and then transitional channels and larger venous sinuses. All venous blood drains into the venous circle, a circular channel along the edge of the pulmonary sac, and then enters the pulmonary sac capillary bed of the lungs. Oxygenated blood passes from the lung through a pulmonary vein to the atrium.

The circulatory system of most cephalopods is closed, but *Nautilus* has an extensive system of blood sinuses. Venous blood is oxygenated in the gills and returns to the systemic heart (which consists of two lateral atria and a median ventricle), from which an anterior and posterior aorta carry oxygenated blood to the body (Figure 14–12F). Blood returns from the head via the vena cava, which divides into two lateral venae cavae near the kidney. These run through the kidney to accessory branchial hearts, located at the base of each gill, which pump blood through the gill via the afferent ctenidial vein. The muscular branchial hearts facilitate blood flow through the vasculature of the gill. Blood returns from the gills via an efferent ctenidial vein to the heart.

The return of venous blood into the molluskan heart is unlikely to be due to a significant venous hydrostatic pressure because the arterial pressure is low and the vascular resistance is high (Jones 1983). Rather, a constant volume or volume-compensating mechanism promotes venous return to the heart for cardiac refilling. Contraction of the ventricle would cause the atrium to expand simultaneously (if the pericardial volume remained constant and the pericardial membrane did not collapse) and to draw venous blood into the atrium. Conversely, atrial contraction would cause simultaneous ventricular expansion and would facilitate ventricular filling. The pericardial pressure of *Helix* is always lower than the intra-atrial and intra-ventricular pressure (but is also always positive, keeping the pericardium rigid despite the flexibility of the pericardial membrane). The pericardial pressure may be low because pericardial fluid is passed to the kidney by a ciliated renopericardial duct. The atrial wall appears to be the site for ultrafiltration of blood to form primary urine, which is subsequently modified by the kidney for excretion. The heart of the prosobranch gastropod *Littorina* (Figure 14–13A) well illustrates the complexities of cardiac function and the relationship between the heart and the

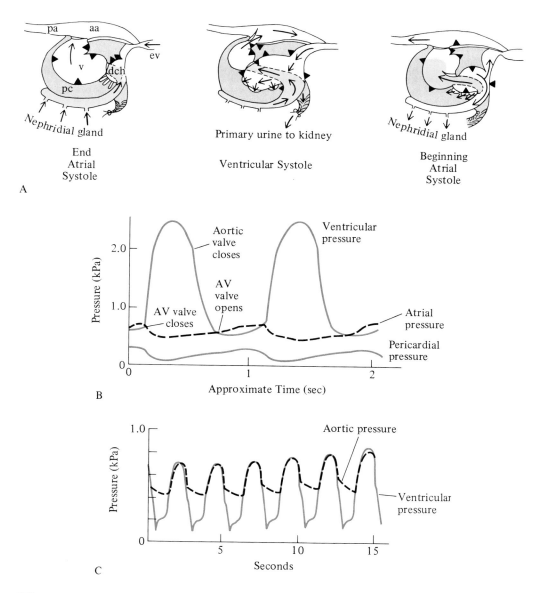

FIGURE 14–13 (A) Representation of blood flow in the heart of the prosobranch mollusk *Littorina* during the end of atrial systole, ventricular systole, and the beginning of atrial systole (dorsal view of heart). Abbreviations are as follows: aa, anterior aorta; dch, dorsal channel; ev, efferent gill vein; pa, posterior aorta; pc, pericardium; v, ventricle. (B) Changes in atrial, ventricular, and pericardial pressure during the cardiac cycle of the gastropod snail *Helix*. (C) Ventricular and aortic pressures during the cardiac cycle of the pink abalone. *(Modified from Andrews and Taylor 1988; Jones 1983; Bourne and Redmond 1977.)*

excretory system (Andrews and Taylor 1988). The atrium is, in effect, two pumps. One "pump" fills the ventricle and the other "pump" ultrafilters blood to produce primary urine for further processing by the kidney. A dorsal channel provides for a complex blood flow of atrial blood to the kidney and its return from the kidney into the ventricle.

Blood pressure is generally low in noncephalopod mollusks; about 0.1 to 0.6 kPa in bivalves and slightly higher in gastropods (0.3 to 0.8 in prosobranchs, 1.0 to 3.5 in pulmonates; Jones 1983). The cardiac cycle of *Helix* (Figure 14–13B) is typical. The cycle begins with an atrial contraction raising the internal pressure from about 0.47 to about 0.65

kPa; blood is ejected into the ventricle when the intra-atrial pressure exceeds the intraventricular pressure (0.56 kPa). The ventricular contraction follows atrial contraction. The atrioventricular valve is closed when the intra-ventricular pressure exceeds the intra-atrial pressure; the intra-ventricular pressure continues to rise to a peak of about 2.5 kPa. The high pressure opens the aortic valve and blood flows into the aorta until the ventricle relaxes and the intraventricular pressure falls to about 2.0 kPa and the aortic valve closes. The peak systolic ventricular pressure occurs when the atrial pressure is minimal, and this is when venous blood enters the atrium. The pericardial pressure is always less than the atrial and ventricular pressures, being minimal during ventricular systole. The heart of *Helix* has a mechanical efficiency of 18 to 20%, which is similar to that of mammalian hearts (Herold 1975).

The ventricular pressure cycle is considerably damped in the aorta by the elasticity of its wall. During ventricular ejection, the elastic aorta expands and absorbs some of the kinetic energy and volume of the ejected blood. In the prosobranch *Haliotis*, the maximal ventricular and aortic pressure are both about 0.8 kPa (Figure 14–13C). During ventricular diastole, after the aortic valve has closed, the aorta elastically recoils and imparts kinetic energy and flow to blood stored by its expansion; this maintains a higher diastolic pressure in the aorta (e.g., 0.4 kPa) than the minimal ventricular pressure (e.g., about 0.1 kPa). The damping effect of the aorta thus reduces the systolic/diastolic pressure pulse from 0.8/0.1 (ventricle) to 0.8/0.4 (aorta); this is another example of the windkessel effect described above.

Blood pressure decreases in the peripheral circulation, and there is only a low pressure gradient to drive blood flow through the lung (Figure 14–14). In *Helix*, the circulus venosus pressure is about 0.8 kPa and the diastolic atrial pressure is about 0.5 kPa; the pressure gradient is therefore about 0.3 kPa. The resistance of the pulmonary circuit is about $1.8 \ 10^{10}$ Pa sec m^{-3} since the blood flow is about 1 ml min^{-1} ($1.7 \ 10^{-8}$ m^3 sec^{-1}); this resistance is about 15% of the total peripheral resistance ($1.3 \ 10^{11}$ Pa s m^{-3}). In the prosobranch *Haliotis*, the gill resistance is about $3.5–6.0 \ 10^9$ Pa s m^{-3}; this is about 40% of the total peripheral resistance of $1.1–1.8 \ 10^{10}$ (Bourne and Redmond 1977); the pressure gradient across the gill is about 0.21 kPa and gill blood flow is 2.1–3.6 ml min^{-1}. These peripheral resistances of gastropods are much greater than, for example, those of mammals. We would therefore expect blood flow to be slow in mollusks compared to mammals because their peripheral resistance is high and the blood pressure is low. The cardiac output of gastropods is generally 30 to 60 ml kg^{-1}

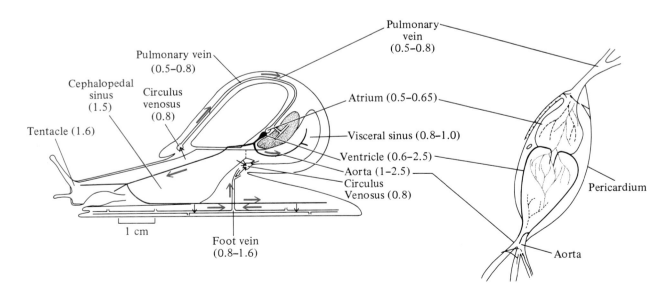

FIGURE 14–14 Representation of the heart and circulatory system of the gastropod snail *Helix* showing detail of the structure of the heart. Blood pressures are indicated (kPa). *(Modified from Dale 1973; Hill and Welsh 1966; Jones 1983.)*

min^{-1}. The circulation time, calculated as blood volume/cardiac output, is generally 5 to 30 minutes, but varies with temperature and activity state.

Cephalopods, in contrast to other mollusks, have a closed circulatory system and a high blood pressure (Figure 14–15). Resting arterial blood pressure of *Octopus dofleini* varies from a diastolic minimum of 5 kPa to a systolic maximum of 7 kPa, i.e., 7/5. The resting arterial pressure of *O. vulgaris* is about 3.5/2.5 to 5.5/3.0. Blood pressure increases with activity, e.g., to 10/4 in *O. vulgaris*. Venous blood pressure is much lower than arterial pressure; there is a pulsatile venous pressure from about 0.2 to 0.5 kPa in the anterior vena cava and 0.5 to 3.5 kPa in the arm vein. Many of the venous vessels, such as the lateral venae cavae, are pulsatile to augment venous return to the heart.

The cephalopod circulatory system has accessory, branchial hearts that assist blood flow through the gills. The afferent ctenidial pressure (i.e., blood pressure immediately after the branchial heart but before the gill) is generally lower and more variable than the arterial pressure, but there is a clear synchrony between the contractions of the branchial hearts and the systemic heart (Figure 14–15). The branchial hearts appear to initiate the heart beat cycle and to pace the systemic heart. The lateral venae cavae also pulsate, and some of the afferent

ctenidial pressure can be attributed to their activity. The branchial hearts are not only accessory hearts that facilitate blood flow through the gills but also produce the hydrostatic pressure for formation of an ultrafiltrate across the branchial heart appendages into the renopericardial canal; the ultrafiltrate is formed against a colloid osmotic pressure of about 3.4 kPa. *Nautilus* lacks branchial hearts but still has an appreciable ctenidial pressure, due apparently to contractions of the renal appendages and pericardial glands (which are synchronized with respiratory movements rather than the systemic heart).

Arthropods

The circulatory system of arthropods (e.g., crustaceans, insects, spiders, scorpions) is generally open, with hemolymph flowing freely through the hemocoelic cavity.

The circulatory system is well developed and extensive in large crustaceans but is limited or absent in smaller crustaceans. The crustacean arterial system has an anterior aorta and sometimes posterior, lateral, and ventral arteries. The venous system is a series of open sinuses; veins are not present. A muscular heart, body movements, or gut movements provide the pressure for blood circulation. The heart (when present) is a single-

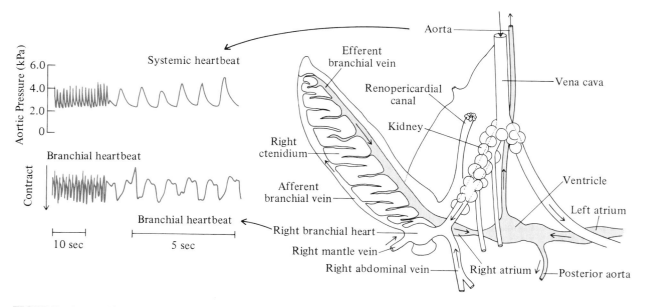

FIGURE 14–15 Circulatory system of a cephalopod mollusk showing the right gill (ctenidium), right branchial (gill) heart, and systemic heart and the relationship between systemic heart contractions (upper trace, aortic pressure) and branchial heart contractions (lower trace). *(Modified from Hickman 1967; Wells 1980.)*

chambered organ that lies dorsally above the gut, enclosed within a pericardium. It varies from a spherical muscular vesicle to a long tube; it has a variable number of openings (**ostia**) for entry of blood.

The circulatory system of branchiopod crustaceans varies from a long tubular heart with 14 to 18 pairs of ostia in anostracans (e.g., *Artemia*, the brine shrimp; Figure 14–16A) to a globular heart with one pair of ostia and a short or no aorta in cladocerans (e.g., water fleas; Figure 14–16B). The circulatory system of ostracods is similar to that of cladocerans, with no aorta in some species and a long aorta in others. Some copepods have a short

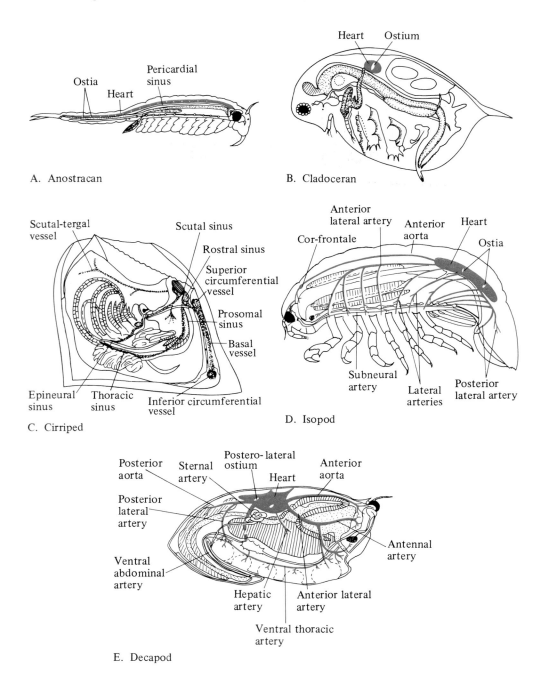

FIGURE 14–16 Circulatory systems of various crustaceans illustrating the various degrees of development of the arterial and venous systems. **(A)** Anostracan; **(B)** Cladoceran; **(C)** Cirriped; **(D)** Isopod; and **(E)** Decapod.   *(From McLaughlin 1983.)*

heart with a lateral pair of ostia and a ventral ostium and a short aorta. Others have an abdominal blood pump or sinus or a thoracic "central organ." Some lernaeopodoidid copepods have a system of closed blood vessels filled with a red fluid (free of corpuscles). There is no heart, but propulsion of fluid is due to peristaltic gut activity and elasticity of the vessel walls (Maynard 1960).

Barnacles (cirripeds) are structurally very different from other crustaceans and so is their circulatory system (Figure 14–16C). No cirripeds have a heart, although some may have a sinus that acts as a blood pump. Rather, rhythmic contractions of thoracic muscles circulate body fluids. Pedunculate cirripeds have a complex arrangement of vessels and sinuses. Blood passes through the rostral vessel via the peduncular vessel (which may represent the aorta of other crustaceans) to various regions (e.g., ovarian, mantle, scutal vessels) and then returns to the prosomal sinus. Large channels, with valves, return blood from the prosomal sinus to the rostral vessel. In sessile barnacles, blood flows from a rostral sinus into a large blood vessel (perhaps equivalent to the peduncular vessel or aorta) that divides into two pairs of circumferential vessels that form thoracic and peripheral circulations.

The circulation of malacostracans is generally well developed with a tubular heart and an arterial system. It is well represented by the isopod circulatory system (Figure 14–16D). The tubular heart has two paired segmental ostia and arteries. A large anterior aorta and a pair of anterior lateral arteries supply the anterior region, a large ventral subneural artery supplies the ventral tissues; there is a small posterior aorta. Blood return to the heart first passes through a large ventral sinus into the pleopod gills; it then passes dorsally through venous sinuses to the pericardial sinus. There is a general trend in advanced malacostracans, such as isopods, for reduction in the length of the heart and the number of ostia. Isopods and many other malacostracans have a **cor frontale**. This is an accessory blood pump, formed by an enlargement of the anterior artery at the anterior border of the stomach. It facilitates blood flow through the extensive vascular networks of the cerebral ganglion and the eyes.

In decapods (Figure 14–16E), blood enters the heart from the pericardial sinus through pairs of valved ostia; there is considerable variability in the location of these ostia (e.g., one dorsal pair, two lateral pairs in brachyurans; one dorsal pair, one lateral pair, and one ventral pair in crayfish and lobsters). The arterial system has an anterior and posterior aorta, anterior lateral arteries, hepatic artery, and sternal artery.

The peak ventricular pressure of decapod hearts is generally 1 to 2 kPa. The diastolic pressure is 0.1 to 0.2 kPa above the pressure in the ventral sinus, and this facilitates venous return. Arterial systolic and diastolic pressures remain high due to the windkessel effect of the aorta (cf. mollusks). *Homarus* has a cardiac output of about 10 to 30 ml min^{-1} and a circulatory time of 2 to 8 min. Circulatory time has been estimated as 10 to 20 sec in *Daphnia*, *Palaeomon*, and *Carcinus*, and 40 to 60 sec in large decapods.

The insect circulatory system distributes metabolites, hormones, and other solutes but not O_2 and CO_2, which are transported by the tracheal respiratory system (see Chapter 13). It also distributes heat for thoracic temperature regulation (see Chapter 5). Localized increases in blood pressure are responsible for wing removal in termites and eversion of various organs, including unrolling of the proboscis in lepidopterans, fecal pellet egestion, body swelling during molting, and wing unfolding after metamorphosis.

The insect circulatory system (Figure 14–17A) resembles the general plan of crustaceans with long tubular hearts. The dorsal, tubular heart of insects is often a simple vessel extending along the first nine abdominal segments. It can have segmental, bulbous swellings or cephalic tubular extensions or lateral bends; vertical kinks; or a series of small, tight coils. The heart is usually closed posteriorly, but sometimes it has posterior vessels that supply the anal and genital aperture muscles. The anterior of the heart opens into an aorta, which carries blood to the head. Incurrent ostia allow blood to flow into the heart. These ostia may be simple and slit like, funnel shaped and inverted into the heart lumen or internal flaps and interventricular valves. A variable number (2 to 12 pairs) of fan-shaped alary muscles connect the heart to the lateral body wall (Figure 14–17B). Contraction of these muscles expands the heart and draws blood in through the ostia. Contractions of the heart propel blood anteriorly towards the head from which it percolates through the body sinuses to be returned to the heart.

Some insect hearts have excurrent ostia that allow blood to flow out of the heart. These ostia are often simple, noncontractile openings through which blood passively leaves the heart. *Periplaneta americana* has segmental vessels that open from its heart through functional excurrent valves (Figure 14–17C). These excurrent ostia have a separate innervation that keeps the valves open for about six heart beats then closed for six beats.

Accessory pulsatile hearts (or organs) are present in the head, thorax, legs, and wings to assist blood

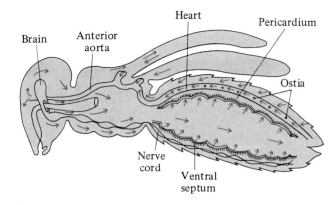

A

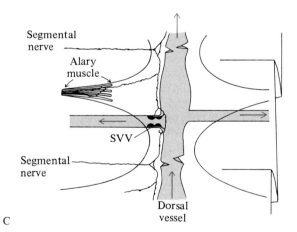

B

FIGURE 14-17 **(A)** Generalized structure of the insect circulatory system. **(B)** The heart and alary muscles of the cockroach *Blaberus*. **(C)** Dorsal vessel of the cockroach *Periplaneta* showing the segmental vessel valve (SVV) in the segmental vessel of the fourth abdominal heart chamber. *(Modified from Martin and Johansen 1965; Nutting 1958; Miller 1985.)*

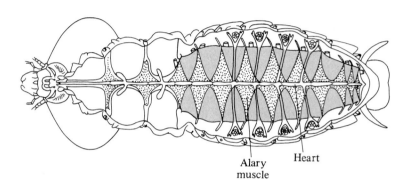

C

flow. These are either membranous or muscular hearts. The cephalic pulsatile organ is associated with the antennae and sometimes the optic lobe of the brain. The antennal pulsatile organ of the cockchafer *Melolonta* has an ampulla that is compressed by circular ampullary muscles and expanded by an elastic band of tissue (Figure 14–18). Thoracic pulsatile organs are often small, arched, striated muscles attached to the scutellum. One to five pulsatile organs may pump blood through the wing

veins. Many insects have leg pulsatile organs. This organ is a delicate fan-shaped muscle arching across the femoro-tibial joint in *Triatoma*.

The tubular heart of scorpions extends anteriorly and posteriorly as aortae and extensive lateral arteries distribute blood to the segmental structures and viscera (Figure 14–19). Blood accumulates in a ventral sinus surrounding the lung books and is circulated through the lung books by muscle contraction. Oxygenated blood then returns to the heart

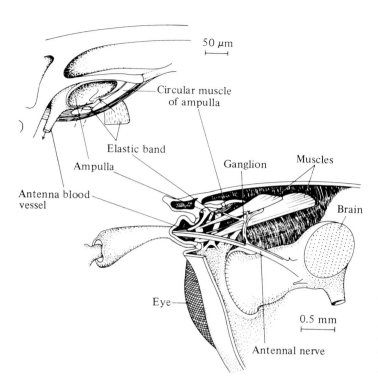

FIGURE 14–18 Antennal pulsatile organ of the cockchafer beetle showing the location of the ampulla of the accessory pulsatile organ in the heart (upper) and the detailed structure of the ampulla, the circular ampullary muscles, which compress the ampulla, and the elastic band, which expands the ampulla. *(From Pass 1980.)*

through special channels. The spider circulatory system is similar to that of scorpions but the lung books tend to be reduced (or absent) and are replaced by increasingly complex tracheal systems. Corresponding to this increased reliance on tracheal respiration is a reduction in the number of heart chambers, ostia, and arteries. The hearts of exclusively tracheate spiders have only two ostia.

Arthropods have an open circulation and are enclosed by a rigid exoskeleton. Consequently, their

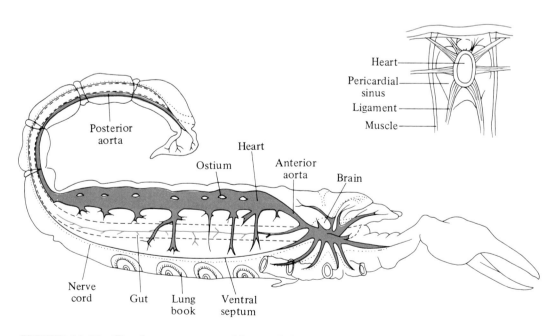

FIGURE 14–19 Circulatory system and heart of the scorpion *Buthus*. *(From Martin and Johansen 1965.)*

hemolymph pressure is generally similar to the general internal pressure of the body, which can be markedly influenced by forces acting on the exoskeleton. The hemolymph pressure of insects is generally about 0 kPa (Miller 1985; Slama 1976). Mean hemolymph pressure averages a little more than 0 kPa in the larva of *Sarcophaga* and less than 0 in adult *Pyrrhocoris* (Figure 14–20A). There are small variations in pressure in differing parts of the circulatory system. In *Locusta*, for example, the pressure in the abdominal heart is 0.94/0 kPa and in the dorsal vessel is 0.86/0.32 kPa; the hemocoel pressure varies from 1.13 kPa in the ventral thorax to − 1.52 kPa in the dorsal ampulla for wing circulation (Bayer 1968). Regulation of blood pressure can be achieved by control of the heart, by the total internal pressure, and by the blood/water volume of the insect. In *Tenebrio*, neural control by the mesothoracic ganglion of the abdominal segmental muscles regulates the hemocoel pressure. Contraction of the abdominal muscles causes marked increases in hemolymph pressure of *Tenebrio* pupa. The hemolymph pressure of *Pyrrhocoris* depends on its body water volume; drinking increases pressure and excretion decreases pressure. There are substantial variations in mean hemolymph pressure with developmental stage, locomotion, feeding, and ecdysis. Hemolymph pressures rise to high values during ecdysis (4 to 6 kPa in some species); the hemolymph pressure slowly declines from over 4 kPa in *Dermestes* during the larval-pupal ecdysis (Figure 14–20B). Hemolymph pressure is about 3 kPa during puparium formation by the fly *Sarcophaga*.

Echinoderms

Echinoderms have four different internal body fluid compartments: the **water vascular system**, the **perivisceral coelomic fluid system**, the **perihemal system**, and the **hemal system**, none of which is particularly effective for fluid transport (Lawrence 1987). The low metabolic rate of echinoderms appears to explain this lack of internal circulatory capacity. Nevertheless, the four fluid compartment systems will be briefly described here.

The water vascular system is a series of vessels, lined by flagellate cells; it is best studied in starfish (Binyon 1972). Radial vessels of each arm are interconnected by a circumoral water ring (Figure 14–21A). A series of lateral canals pass from the

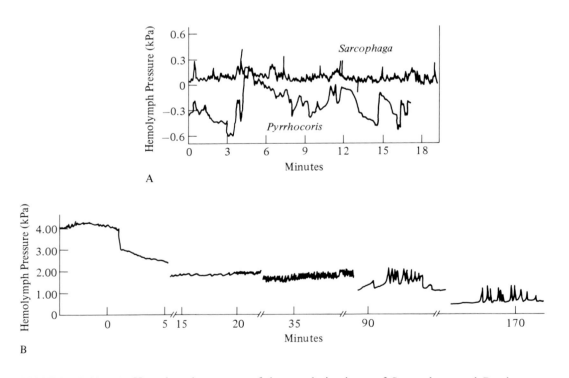

FIGURE 14–20 **(A)** Hemolymph pressure of the wandering larva of *Sarcophaga* and *Pyrrhocoris*. **(B)** Decline in hemolymph pressure during the larval-pupal ecdysis of *Dermestes*. *(Modified from Slama 1976.)*

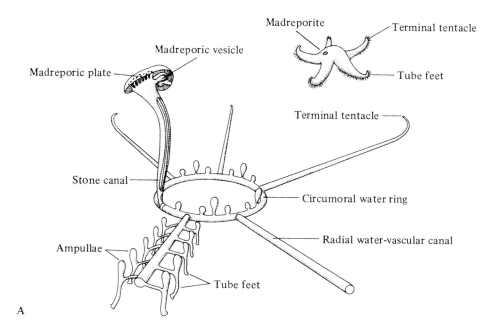

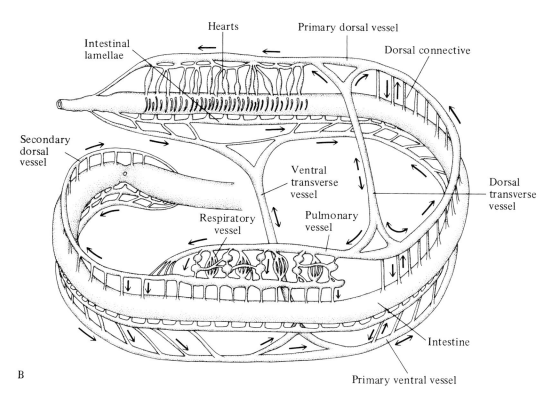

FIGURE 14–21 **(A)** Schematic representation of the water vascular system of an asteroid echinoderm; tube feet are shown for only one arm. **(B)** Blood vascular system of a holothuroid echinoderm. *(From Binyon 1964; Herreid, LaRussa, and DeFesi 1976.)*

radial canal to the tube foot-ampulla system in each leg. A flap-like valve prevents the backflow of fluid from the tube foot-ampulla system into the radial canal. The circumoral ring is connected by a stone canal to the external seawater via a porous plate, the madreporite. In general, there is no, or only a slow, flow of fluid along the stone canal. The tube foot-ampulla system provides hydraulic extension of the tube foot by contraction of ampullary muscles and retraction of the tube foot by contraction of longitudinal foot muscles. A positive hydrostatic pressure in the tube foot-ampulla system prevents the flap valve from opening and allows fluid to flow into the circumoral canal. Fluid is continually lost from the tube feet by ultrafiltration because of the elevated hydrostatic pressure (2 to 4 kPa) and the high permeability to water, e.g., about $2.55 \ 10^{-9}$ kg H_2O m^{-2} sec^{-1} Pa^{-1} (Binyon 1964). This hydraulic permeability is less than that of the mammalian glomerulus ($5.6 \ 10^{-8}$) but is greater than that of the kidney tubule of the salamander *Necturus* ($1.5 \ 10^{-9}$). Thus, a 50 g starfish with 17.6 cm^2 of tube foot area should lose about 0.05 ml water per hour (at 3 kPa pressure). We might expect that this fluid loss would be replenished via the madreporite and the stone canal, but it appears to come instead from the axial sinuses or by osmotic influx of water across the walls of the water vascular system in response to active K$^+$ transport. The water vascular system, however it functions, is not an effective fluid circulating system, especially as the radial canals are blind ended.

The perihemal system consists of coelomic spaces surrounding the oral nerve ring and radial nerves, the gonadal ducts and gonads, and the axial gland. It is lined by a flagellated epithelium that produces weak fluid currents, but there is little compelling evidence for significant transport of nutrients (or O_2) via this system.

The hemal system is a well-developed, but diffuse, series of interconnecting spaces. Its role in circulatory transport is unclear because it generally is not a fluid-filled space but contains amoeboid phagocytic cells and connective tissue ground substance. Consequently, there is little or no flow of fluid in the hemal system. However, the holothuroid *Isostichopus* has a well-developed hemal circulation (Figure 14–21B). Peristaltic waves of the dorsal vessel move hemal fluid into a series of vascular lamellae in the digestive tract aided by numerous hearts (between the dorsal vessel and interstitial lamellae). A posterior vessel forms a respiratory plexus supplying the left respiratory tree and a terminal vascular tuft attached to the respiratory tree.

The perivisceral coelomic fluid is the most likely candidate for a functional circulatory system. O_2 diffuses across the body wall into the perivisceral coelomic fluid (nutrients also are actively absorbed from seawater by the body wall and may be transported into the perivisceral fluid). The perivisceral coelomic fluid may also transport nutrients between internal organs. Flagellar currents and changes in body shape, gut contractions, or emptying and filling of the respiratory tree may produce perivisceral fluid circulation.

Hemichordates

Hemichordates (enteropneust worms and pterobranchs) have an open circulatory system. A middorsal vessel carries blood anteriorly and a midventral vessel carries blood posteriorly. A pulsating vessel ("heart") in the proboscis propels blood, as do contractions of the dorsal and ventral vessels. The "heart" is a contractile epidermal sac lying above the central blood sinus; its contractions force blood from the central sinus but it is not a true heart since no blood enters the structure.

Chordate Circulatory Systems

Urochordates have a unique open circulatory system (Figure 14–22). The heart is a short, U-shaped vessel located at the base of the digestive tract; each end opens (one dorsally and one ventrally) into large channels. The dorsal channel connects to the subendostylar channel and the pharyngeal basket plexus of blood vessels or lacunae. The ventral channel connects to an abdominal sinus. Circulation has a triangular pattern of flow from the heart to the pharynx to the abdominal sinus to the heart. The tubular heart, which has a pacemaker at each end, beats rhythmically but regularly reverses its direction of blood flow. The heart beats for a few minutes (about 100 contractions), milking blood through the heart in one direction, then the beat slows and stops. The heart then begins to beat again but blood flows in the opposite direction for a few minutes until it again stops and then begins to beat again in the original direction.

The circulatory system of the cephalochordate *Branchiostoma* (Figure 14–23) bears some resemblance to the general circulatory pattern of vertebrates. The major differences are that *Branchiostoma* lacks a single, centralized heart and has no capillary system. A main ventral vessel below the gut collects blood and passes it forward to the liver

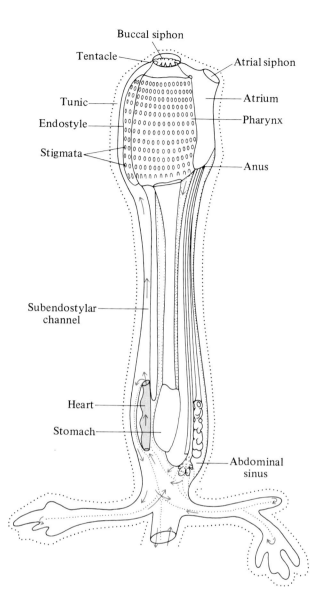

Buccal siphon

Tentacle

Atrial siphon

Tunic

Atrium

Endostyle

Pharynx

Stigmata

Anus

Subendostylar channel

Heart

Stomach

Abdominal sinus

FIGURE 14–22 Schematic representation of the circulation of the ascidian urochordate *Clavelina*. *(From Barnes 1987.)*

caudal and cardinal veins, which drain into the gut plexus and sinus venosus (by the ducts of Cuvier), respectively. Only the dorsal aorta has an endothelial lining. There is no central heart, but blood is propelled by contractions of the large vessels and the accessory branchial hearts.

Vertebrates

The closed circulatory system of vertebrates is derived from a common ancestral system, somewhat resembling that of the cephalochordates but with the addition of a central heart, hepatic (liver) circulation, and capillaries. The vertebrate heart developed from a ventral median vessel in the branchial (gill) region. The circulatory system has an arterial, a capillary, a venous, and a lymphatic section. Let us first examine this general functional organization of the vertebrate circulatory system in more detail.

Functional Organization. Vertebrates have a single, muscular heart, which propels blood through the high-pressure arterial system to the capillary beds of the tissues. Some vertebrates, such as hagfish, also have accessory hearts. The capillaries are the site of gas, nutrient, and waste exchange between the blood and interstitial fluid. The lymphatic system provides an additional return to the circulatory system of fluid and especially for protein that leaks from the circulation into the interstitial fluid.

The aorta progressively branches into many arteries, arterioles, and capillaries; the capillaries coalesce into fewer venules, veins, and then the venae cavae (Figure 14–24). There is a dramatic increase in the number of vessels with a corresponding decrease in the diameter of vessels and their length. For example, the aorta of the dog (see Table 14–2, page 672) is about 1 cm diameter and the capillaries about 0.0008 cm; the length decreases from 40 cm for the aorta to 0.1 cm each for the capillaries; the volume of blood in the vessels, their total cross-sectional area, and their surface area dramatically increase for the smaller vessels compared to the largest vessel.

The vertebrate circulatory system is closed and so the blood flow rate (V_b; ml sec^{-1}) must be the same through all levels of the circulatory system, i.e., the blood flow through the aorta must equal the total blood flow through all capillaries. However, the velocity of flow (v; cm sec^{-1}) must vary dramatically in vessels of different cross-sectional area (A; cm^2) since

$$V_b = vA \qquad \text{and} \qquad v = V_b/A \qquad (14.11)$$

via a hepatic portal vein. Blood draining from the liver via the hepatic vein empties into a sinus venosus that also receives blood from the posterior of the body. A ventral aorta runs forward to the branchial arches where a small branchial heart at the base of each arch pumps blood through the gill bars for gas exchange. Oxygenated blood collects in the paired lateral aortae, which fuse behind the pharynx to form a single dorsal aorta. The aorta supplies blood to various tissue lacunae; there are no capillaries. Blood collects from the lacunae into

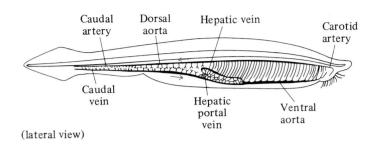

(lateral view)

FIGURE 14–23 Lateral and dorsal views of the circulatory system of the cephalochordate *Branchiostoma.* *(From Grove et al. 1969.)*

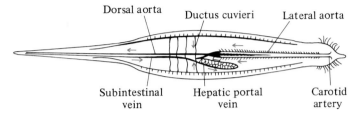

(dorsal view)

Capillary velocity is consequently much less than aortic velocity. This low velocity and the high-surface area of capillaries maximize their exchange of gases and other materials.

The fantastic magnitude of this geometric change in different levels of the circulatory system merits further illustration. In man, the aorta has a lumen diameter of about 2.5 cm and an area of about 4.5 cm^2; the blood flow velocity is about 18.5 cm sec^{-1} (since $V_b = 83$ cm^3 sec^{-1}). A capillary has a lumen diameter of about 0.0006 cm (cross-sectional area of 3 10^{-7} cm^2) but the total cross-sectional area of the approximately 1.6 10^{10} capillaries is about 4500 cm^2. The velocity of capillary blood flow is about 0.019 cm sec^{-1}, i.e., 1/1000 that of the aorta. Blood remains in the capillary for about 1 to 3 sec (they are about 0.03 to 0.1 cm long). It would take 1 ml of blood about 2 years to traverse a single capillary, although 5000 × this amount is pumped by the heart per minute!

The structures of the circulatory vessels vary dramatically, reflecting their different functions (Figure 14–25). The large arteries have more elastic tissue in their walls, especially in the aorta, because of their windkessel role. Smaller arteries have less elastin and more smooth muscle. Arterioles have mostly smooth muscle in their walls; this reflects their role in controlling blood flow through the microcirculation. Capillary walls are essentially only a basement membrane and a thin, even perforate,

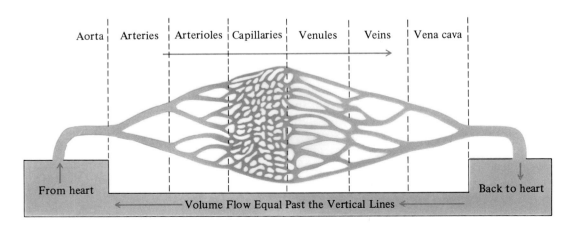

FIGURE 14–24 Representation of the systemic vascular bed in vertebrates.

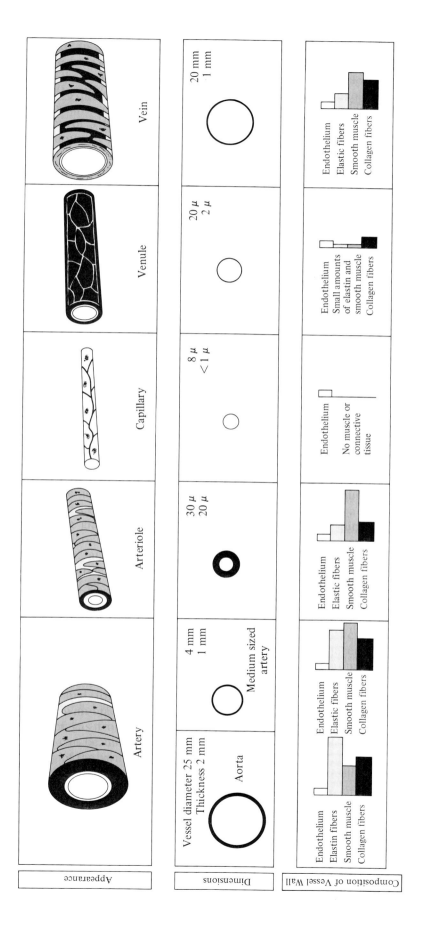

FIGURE 14-25 Appearance, dimensions, and composition of the vessel wall for vertebrate circulatory vessels. *(Modified from Waterman et al. 1971.)*

endothelial cell layer. Venules have some elastin, smooth muscle, and collagen. Veins have more smooth muscle and collagen; they are an important region of blood storage, and the smooth muscle allows regulation of venous volume.

Arterial System. The aorta and large arteries are passive conduits for blood flow to the periphery, but the aorta especially has an important role in dampening the ventricular pressure cycle. The aorta wall has a high elasticity and during cardiac systole it stretches and stores the kinetic energy of the ejected blood. During diastole, the aorta contracts, continuing to propel blood and maintain a high arterial pressure. In humans, the windkessel role of the aorta dampens the intraventricular pressure cycle (about 0 to 12 kPa) to an arterial cycle of about 8 to 12 kPa. Such a windkessel role is apparent for the aorta, or bulbus, of all vertebrates, from hagfish to mammals (Figure 14–26).

The small arteries and **arterioles** have less elastin in their walls and more smooth muscle. The smooth muscle allows regulation of the vessel diameter by the autonomic nervous system. The metabolic demands of various tissues alter, and so the capacity for selective redistribution of blood flow is an extremely important role especially for the arterioles (see below).

Capillaries. The **capillaries** are the site of gas, nutrient, and waste exchange. The geometry of capillaries maximizes the opportunity for exchange; they have a high surface area, low flow velocity, and a relatively long transit time.

The capillary wall is a single layer of very flat endothelial cells connected by an intercellular cement (Figure 14–27A). A large capillary seen in cross section has two or three endothelial cells, but a small capillary is a single cell. The capillary wall is about 0.2 to 0.4 μ thick; the cell is so thin that its nucleus bulges into the lumen. The endothelium of **continuous capillaries** (found in muscles, nervous tissue, connective tissue) forms a continuous, uninterrupted layer around the circumference except

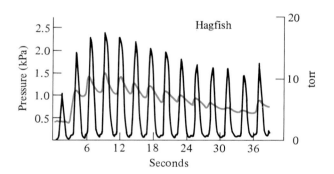

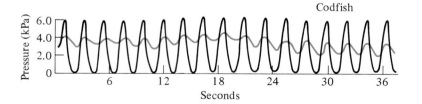

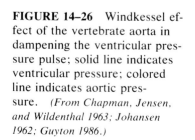

FIGURE 14–26 Windkessel effect of the vertebrate aorta in dampening the ventricular pressure pulse; solid line indicates ventricular pressure; colored line indicates aortic pressure. *(From Chapman, Jensen, and Wildenthal 1963; Johansen 1962; Guyton 1986.)*

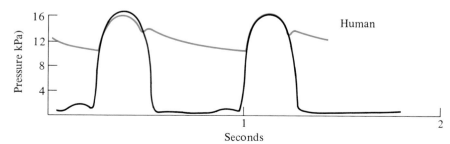

at the endothelial cell junctions. In contrast, the endothelium of **fenestrated capillaries** (found in pancreas, intestine, endocrine organs, and the kidney) has a variable thickness with extremely attenuated regions and circular pores of 80 to 100 nm diameter covered only by a very thin diaphragm membrane. This diaphragm membrane is absent in the glomerular capillaries.

There are three general routes for exchange of materials across the capillary endothelium. Lipid-soluble molecules (e.g., O_2, CO_2) can readily diffuse through the lipid bilayer of endothelial cell membranes, and their rate of diffusional exchange is much greater than that of lipid-insoluble materials (e.g., water, ions, glucose). Water also diffuses rapidly across the endothelium, directly through small pores in the lipid bilayer (as well as through the interendothelial clefts and fenestrae, if present). Other lipid-insoluble molecules, being larger than water molecules, are unable to penetrate the endo-

thelial wall through these small membrane pores. Their route(s) for transendothelial exchange has received considerable attention. The observed rates of exchange suggest two sizes of pores: small pores of about 9 nm in dia and large pores of about 70 nm in dia. The structural equivalents of these pores would appear to be the intercellular clefts (small pores) and intracellular vesicles (large pores). The intracellular vesicles fuse with one plasmalemma of the endothelial cell and take up materials and then migrate into the cytoplasm; their subsequent fusion with the opposite plasmalemma releases the materials on the opposite side of the endothelium (Figure 14–27B left). This process of micropinocytosis-like transport has been termed **transcytosis** to distinguish it from normal pinocytosis. The micropinocytotic vesicles may not form spontaneously at the plasmalemma but may be special vesicles formed by the Golgi complex that retain their structural integrity while repeatedly shuttling between the two plas-

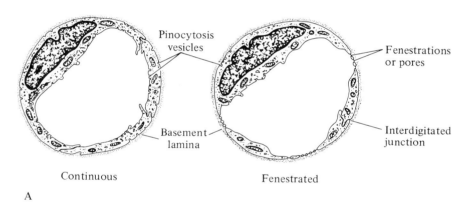

A

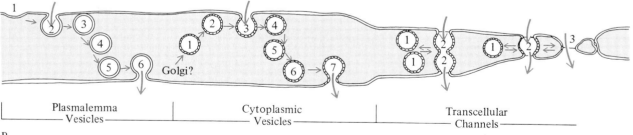

B

FIGURE 14–27 (A) Schematic representation of continuous (left) and fenestrated (right) capillary walls. (B) Schematic representation of models for transport of hydrophilic solutes across the capillary endothelium. Left: formation of plasmalemma vesicles and subsequent detachment, transit, and fusion with opposite plasmalemma. Center: transport mediated by cytoplasmic vesicles (from the Golgi complex?) fusing with one plasmalemma and then subsequently fusing with the opposite plasmalemma. Right: transcellular movement through channels or fenestrae formed from plasmalemma or cytoplasmic vesicles. Numbers indicate the sequence of movements. *(Modified from Fawcett 1986.)*

TABLE 14–3

	MWt (Da)	Radius of Equivalent Sphere[1] (nm)	Relative Permeability
Water	18	—	1.00
NaCl	59	14	0.96
Urea	60	16	0.80
Glucose	180	36	0.60
Sucrose	342	44	0.40
Inulin	5000	152	0.20
Myoglobin	17600	190	0.03
Hemoglobin	68000	310	0.01
Albumin	69000	—	<0.0001

Relative permeability of the muscle capillary to different-sized hydrophilic (lipid-insoluble) molecules. *(From Pappenheimer 1953; Guyton 1986.)*

[1] Calculated from the diffusion rate of the solute.

exchange is 100 × greater in these capillaries compared to continuous capillaries.

Molecular size is a primary determinant of transepithelial permeability for nonlipid-soluble materials (Table 14–3), although other factors such as charge also influence permeability. Cerebral capillaries are continuous, with a paucity of transport vesicles and tightly occluding intercellular clefts. The cerebral capillaries are therefore less permeable to many molecules than are other capillaries and form a **blood-brain barrier**. There are also blood-ocular and blood-thymus barriers.

Diffusive exchange across capillaries is rapid. There may also be convective exchange of water and solutes across the pores of the endothelial wall. The hydrostatic and osmotic pressures inside and outside the capillary determine whether there is a bulk flow of water and solutes through capillary pores. A bulk flow of water would also cause a bulk

malemmas of the endothelial cells (Figure 14–27B center). If the vesicles formed spontaneously at the plasmalemmas, there would be a continual intermixing of the two plasmalemma membranes, but this apparently does not occur. Transient pores may be formed by the fusion of adjacent vesicles or even by a single vesicle in a highly attenuated wall (Figure 14–27B right). Capillary fenestrae are also large pores; the fenestrae of glomerular capillaries in the kidney lack the pore diaphragm and water

TABLE 14–4

Average and range of values for plasma colloid osmotic pressure (kPa) in a variety of animals.

	Mean (Range)
Sipunculids	0
Cnidarians	0.005
Echinoderms (coelomic fluid)	0.009 (0–0.02)
Urochordates	0.05
Mollusks (noncephalopods)	0.26 (0.001–1.7)
Elasmobranchs	0.31 (0.2–0.5)
Annelids (coelomic fluid)	0.31 (0–0.2)
Mollusks (cephalopods)	0.36 (0.07–0.5)
Annelids (blood)	0.45 (0.09–1.02)
Crustaceans	0.95 (0.1–9.9)
Amphibians	1.01 (0.5–1.6)
Reptiles	1.16 (0.5–1.6)
Cyclostomes	1.30 (1.2–1.4)
Birds	1.30 (1.1–1.5)
Teleosts	1.32 (0.4–2.7)
Mammals	2.88 (2.1–3.7)
Insects	8.35 (3.1–13.6)

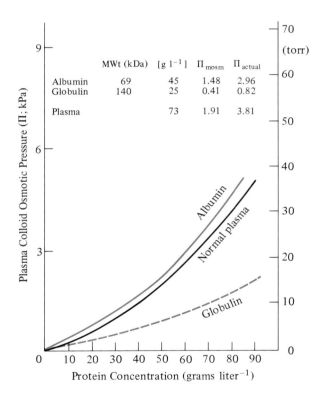

	MWt (kDa)	[g l^{-1}]	Π_{mosm}	Π_{actual}
Albumin	69	45	1.48	2.96
Globulin	140	25	0.41	0.82
Plasma		73	1.91	3.81

FIGURE 14–28 Relationship between plasma colloid osmotic pressure (kPa) and plasma protein concentration (gms liter^{-1}) for normal plasma and two important plasma proteins, albumin and globulin. Inset table gives the molecular weight, concentration, colloid osmotic pressure calculated from the protein milliosmolar concentration (Π_{mosm}), and the actual osmotic pressure (Π_{actual}).

flow of those solutes that are small enough to pass through the pores.

Solutes that are impermeant to the capillary pores can have an osmotic concentration difference across the capillary wall, whereas permeant solutes can freely diffuse across the capillary wall. Consequently, impermeant solutes that have different concentrations inside and outside of the capillary will cause an osmotically driven movement of water across capillaries. The primary impermeant solute of blood is plasma protein, which establishes a **colloid osmotic pressure** across the capillary wall. The plasma colloid osmotic pressure varies dramatically for various animals, being generally low for invertebrates and lower vertebrates, and increasing in higher vertebrates (Table 14–4). For humans, the concentration of plasma proteins is 73 g liter^{-1} and the colloid osmotic pressure is about 3.8 kPa.

The colloid osmotic pressure is a function of both the protein concentration and molecular weight. For example, the different types of human plasma proteins contribute a varying fraction of the total colloid osmotic pressure depending on their concentration and molecular weight; albumin, 3.0 kPa; globulins, 0.82 kPa; fibrinogen, 0.04 kPa (Figure 14–28). The actual colloid osmotic pressure is greater than that calculated from the milliosmolar protein concentration, and there is a nonlinear relationship between protein concentration and colloid osmotic pressure because part of the colloid osmotic pressure is due to a redistribution of ions in response to the protein concentration imbalance and the net negative charge of protein (Donnan effect). The magnitude of the Donnan effect increases with greater protein concentration, and so the colloid osmotic pressure is about 2 × the value predicted from the osmolar concentration of these plasma proteins.

The average hydrostatic pressure in the mammalian capillary is about 2.3 kPa, and the interstitial hydrostatic pressure is about −0.73 kPa (Figure 14–29). The net hydrostatic pressure gradient of 3.03 kPa forces fluid out of the capillary. This hydrostatic pressure gradient is countered by a colloid osmotic pressure gradient. The intracapillary colloid osmotic pressure of 3.81 kPa draws water into the capillary, and the interstitial colloid osmotic pressure of 0.82 kPa draws fluid out of the capillary; the net colloid osmotic pressure of 2.99 kPa draws fluid into the capillary. There is an imbalance of 0.04 kPa between the net outward force of 3.85 kPa (3.03 + 0.82) and the net inward force of 3.81 kPa

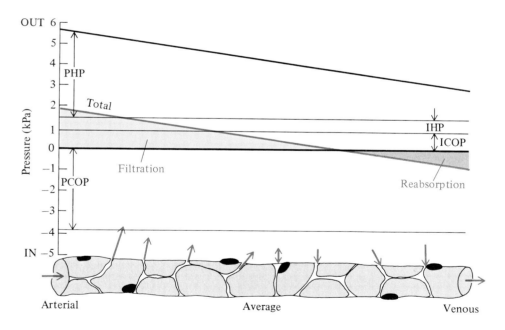

FIGURE 14–29 The balance of hydrostatic and colloid osmotic forces across a capillary causes filtration at the arterial end of the capillary and reabsorption at the venous end; there is an overall filtration of fluid from the capillary ("average"). Abbreviations are as follows: PHP, plasma hydrostatic pressure; IHP, interstitial hydrostatic pressure; ICOP, interstitial colloid osmotic pressure; and PCOP, plasma colloid osmotic pressure. Total is the net result of all hydrostatic and colloid osmotic forces.

that results in an overall outward bulk flow of water. In a human, this results in a net filtration of about 2 ml fluid min^{-1} from the capillaries (excluding the glomerular capillaries). This net fluid loss is returned to the circulation by the lymphatic circulation. The filtration coefficient is 2 ml min^{-1}/0.04 kPa = 49 ml min^{-1} kPa^{-1} for the entire body (except the kidneys). However, the filtration coefficient varies 100 × for different tissues, being low in brain and skeletal muscle, high for subcutaneous tissues, higher for the intestine, and highest for the liver and kidney glomeruli (which have fenestrated capillaries). The filtration coefficient for the kidney glomeruli is about 92 ml min^{-1} kPa^{-1}; the glomerular filtration rate is about 125 ml min^{-1} and the net pressure difference is about 1.36 kPa (see Chapter 17).

The nonglomerular capillaries are close to fluid balance in terms of bulk flow, at least considering the average capillary balance of forces. However, the capillary hydrostatic pressure is higher at the arterial end (about 4.1 kPa) than the venous end (1.4 kPa), hence there is a very different balance of forces along the length of the capillary. There is a high filtration pressure of 1.81 kPa at the arterial end (4.08 + 0.72 + 0.82 − 3.81) causing bulk flow of water out of the capillary, and a significant filtration pressure at the venous end of 0.91 kPa (1.36 + 0.72 + 0.82 − 3.81) causing bulk flow of water into the capillary.

In amphibians, there is a greater imbalance of transcapillary forces with a net outward pressure of about 0.56 kPa in hydrated toads (*Bufo*) and 0.78 kPa in bullfrogs (*Rana*; Hillman, Zygmunt, and Baustian 1987). The transcapillary balance of forces is reversed for *Bufo* after 22% mass loss by dehydration to a reabsorption force of 1.33 kPa at 30% mass loss. In *Rana*, the balance of forces continues to promote filtration regardless of dehydrational stress, although the magnitude of the filtration force declines to 0.39 kPa at 20% mass loss.

Vertebrates (and other animals with closed circulatory systems) maintain capillary fluid balance (or near balance) if the transcapillary hydrostatic pressure and colloid osmotic pressure are similar. We would therefore expect a general correlation between the level of the arterial blood pressure and plasma colloid osmotic pressure. There is, in fact, a reasonable correlation between mean arterial pressure and plasma colloid osmotic pressure in vertebrates (MAP/PCOP = 3 to 6), except for birds with a value of 9 to 15, and an invertebrate with a closed circulatory system (cephalopods; with a value of 13; Table 14–5). Birds and cephalopods must either have a high arterial resistance so that their capillary

TABLE 14–5

Arterial systolic/diastolic blood pressure, plasma colloid osmotic pressure, and mean arterial pressure/colloid osmotic pressure for a variety of animals with closed circulatory systems.

	Systolic/ Diastolic Arterial Blood Pressure (kPa)	Plasma Colloid Osmotic Pressure (kPa)	MAP/ PCOP
Mammals			
Humans	16.3/10.9	3.81	3.6
Sheep	18.4/15.2	2.99	5.6
Dog	15.2/7.6	2.72	4.2
Birds			
Chicken	20.3/5.8	1.50	8.7
Dove	18.4/14.3	1.10	14.8
Reptiles			
Turtle	5.7/4.4	0.87	5.8
Amphibians			
Frog	4.1/2.7	0.69	4.9
Toad	4.4/2.6	1.28	2.5
Fish			
Codfish	3.9/2.5	1.13	2.8
Mollusk			
Cephalopod	6.0/3.0	0.36	16.7

pressure is not so high as might be expected from the arterial pressures, or must have capillaries with a low filtration coefficient, or must have a high filtration of fluid from their capillaries. The "capillaries" of cephalopods, which are not such anatomically recognizable vessels as vertebrate capillaries, have a very complete cover of endothelial cells over the basement membrane to retain hemocyanin (the respiratory pigment; see Chapter 15) in the circulation; there is also a very low density of capillaries in the tissues (Browning 1982). The "capillary" filtration coefficient is probably therefore very low, compensating for the high MAP/COP.

Venous System.   The small vessels into which the capillaries coalesce, the **venules**, are similar to arterioles except that they have less smooth muscle in their walls. They regulate the lower pressure end of the capillaries, hence can modify the balance of hydrostatic/colloid forces and fluid balance of the capillaries.

Veins return blood to the heart. They are also an important reservoir for blood, having a high compliance. The veins are thin walled with a large cross-sectional area and low resistance to flow. Consequently, only a small pressure gradient is required to return venous blood from the periphery to the heart. Most veins have valves to maintain a one-way flow of blood towards the heart, and contractions of surrounding muscles (the "skeletal muscle pump") facilitate the movement of blood along veins. This muscle pump action is especially important for large terrestrial animals in countering the significant hydrostatic effects of large size.

Lymphatic System.   The vertebrate lymphatic system parallels the veins in structure, function, and often in topography, but it has no connection with the arterial system. Fluid, called **lymph**, enters the blind-ended lymph capillaries from the interstitial spaces by diffusion and sometimes by a slightly negative suction pressure. The wall of the lymphatic capillaries consists of a single endothelial cell layer with a discontinuous basement membrane; there are large pores between adjacent lymphatic capillary cells. The lymphatic vessels, being a low-pressure circulation, have very thin walls and little connective tissue or musculature. Valves maintain one-way flow of fluid along the lymphatic system. The lymphatic vessels generally enter the main anterior veins (anterior cardinal veins or jugular veins near the heart where the venous pressure is the lowest), but in some tetrapods they may enter the middle veins (posterior cardinals or vena cava, especially in urodele amphibians) or posterior veins (pelvic veins, in fish and birds).

Agnathan and chondrichthyean fish have a system of thin-walled sinusoids that drain into veins. These may represent a primitive lymphatic system. Other fish and all tetrapods have a well-developed lymphatic system. Amphibians have an especially well-developed lymphatic system, with one to many pairs of high-flow rate lymph hearts (Figure 14–30A).

The **lymph hearts** are small, two-chambered muscular structures that return fluid from the well-developed subdermal lymphatic spaces into the venous circulation. They are present in some fish (eels, *Silurus*), all amphibians and reptiles, and bird embryos and some adult birds (e.g., ostrich). Most adult birds and mammals lack lymph hearts; instead, they rely on the contractions of nearby skeletal muscles to squeeze the lymph vessels and propel the lymph towards the venous circulation. The tail lymph heart of an eel well illustrates the basic structure and function of a lymph heart (Figure 14–30B).

The lymphatic circulation collects the fluid that

A

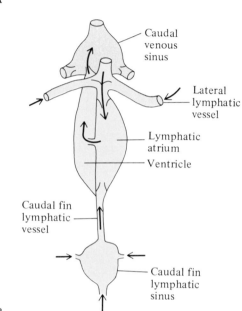

B

FIGURE 14–30   **(A)** Radioxerograph of the dorsum of the tree frog *Litorea* 10 sec after injection of a radio-opaque dye. Shown are the subcutaneous lymphatic vessels and the posterior lymph hearts (arrows). **(B)** Diagram of the lymph heart of an eel. Beating of the lymph heart by contractions of its muscular wall propels lymph from the caudal lymphatic sinus into the caudal venous sinus.   *(From Carter 1979; modified from Kampmeier 1969.)*

is filtered out of capillaries and returns it to the circulation. This is obviously important because it prevents the accumulation of fluid in the tissues (edema). An extremely important role of the lymphatic system is to return to the circulation the protein that leaks from the capillaries into the tissues. The slow leakage of protein from the capillaries would, if uncorrected, elevate the interstitial colloid osmotic pressure and dramatically upset the balance of forces across the capillary, resulting in severe edema. Thus, the lymphatic system returns plasma proteins to the circulation and maintains the high plasma colloid osmotic pressure and the lower interstitial colloid osmotic pressure.

Other roles of the lymphatic system include the transport of chylomicrons (digested and absorbed lipids) from the intestinal lining into the circulation (see Chapter 18). The lymphatic system, especially in mammals, is involved with cellular immunity (see Chapter 15). The lymphatic system of amphibians may be essential in returning water, absorbed across the skin by osmosis, into the circulation for storage in the urinary bladder (Carter 1979).

Circulatory Patterns

The organization of the vertebrate circulatory system and the structure of the heart varies, depending on taxonomy and the mode of respiration (Figure 14–31A–H).

The general arrangement of the primitive vertebrate circulation is a single circuit with a heart that pumps deoxygenated blood to the gills. Oxygenated blood passes from the gills via the dorsal aorta to the body tissues. Deoxygenated blood returns from the body tissues to the heart via the venous system. There is a portal circulation of blood from the gut to the liver. A portal circulation is characterized by blood passing through two capillary beds (e.g., gut and liver) in a single systemic circuit. Other portal systems are the renal portal system carrying blood from the posterior of the body to the kidneys and the hypothalamo-hypophyseal portal system carrying blood from the hypothalamus to the median eminence of the pituitary.

The arrangement of heart, dorsal aorta, and branchial blood vessels has evolved from six gill arches in a primitive (hypothetical) ancestral fish to a reduced number of gills and arches in elasmobranchs, teleosts, and lungfish. Air-breathing tetrapods have acquired pulmonary vessels and lost the gills and most gill arches.

In hagfish, the arterial circuit is from the gills via the dorsal aorta to the head, viscera, and posterior body wall. The head and subcutaneous sinuses make the circulatory system partly open. Venous return to the branchial heart is assisted by accessory hearts (**cardinal**, **portal**, and **caudal hearts**). The liver receives blood from the gut and head sinus via the portal heart.

Cartilaginous and body fish lack the extensive blood sinuses of the hagfish and have no accessory hearts. There is a portal system for blood flow from the gut to the liver and from the posterior body to the kidneys. The arterial system also supplies the liver. Lungfish have a similar cardiovascular system, except for the lung circuit. The oxygenated blood flows from the lung to a partially separated left atrium and partially divided ventricle and flows preferentially into the dorsal aorta. In amphibians, the lungs and skin have respiratory roles and there is a partial separation of oxygenated blood (from the lungs) and deoxygenated blood (from the body) in the heart, which has completely separated left and right atria but a single ventricle.

Reptiles generally do not have significant cutaneous respiration but rely on their lungs. There is a complex and functional separation of oxygenated and deoxygenated blood flow through the heart, although there is not an anatomical separation. Crocodile hearts have completely divided right and left atria and ventricles, but an extracardiac shunt (the foramen of Panizza) allows oxygenated blood from the left atrium to flow to the body via the right and left systemic arches. Birds and mammals have a complete anatomical and functional separation of oxygenated and deoxygenated blood flow through the heart.

A clear functional sequence is apparent when we systematically examine the circulatory systems of lower and higher vertebrates.

Fish. The circulatory system of cyclostomes is unusual, being partly open with large blood sinuses. The branchial heart (the "typical" systemic vertebrate heart) has a **sinus venosus** that receives deoxygenated blood anteriorly from the jugular vein and posteriorly from the hepatic, intestinal, ovarian, and posterior cardinal veins (Figure 14–32A). Blood passes from the sinus venosus into the atrium and then the ventricle via the atrioventricular canal. Blood is ejected from the ventricle and flows along the ventral aorta to the gills, and then the oxygenated blood passes to the dorsal aorta for anterior and posterior distribution. The atrium develops a pressure of about 0.2 kPa. The ventricular pressure is generally 0.1 to 0.8 kPa but maximal pressures are about 4 kPa.

Several accessory hearts assist venous return. The portal heart pumps blood from the cardinal veins and gut to the liver. The contraction rates of the systemic and portal hearts are generally similar, at about 20 to 30 min^{-1}, but are not synchronized. The portal heart generates about 0.2 kPa pressure. The cardinal hearts pump blood along the cardinal veins. The caudal hearts (see Figure 14–2B, page 668) pump blood along the caudal veins. They are often inactive, but when they are active they generally beat at a much higher frequency than the

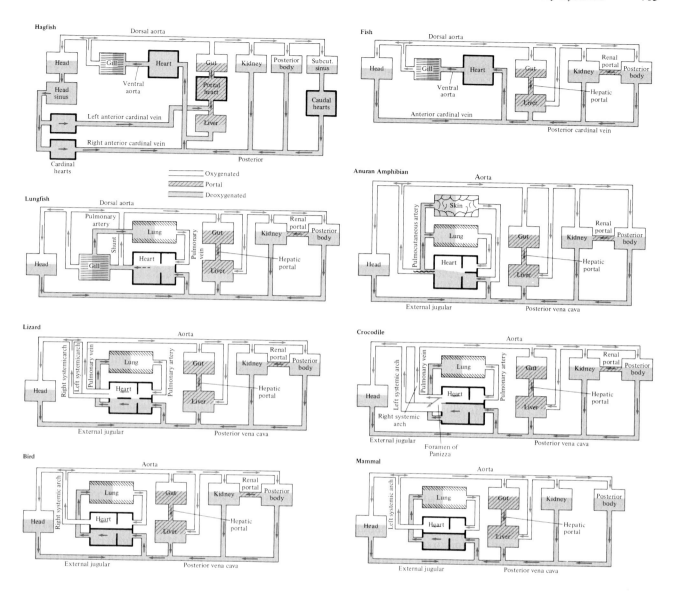

FIGURE 14–31 Representation of the circulatory systems of vertebrates. (**A**) hagfish; (**B**) fish; (**C**) lungfish; (**D**) amphibian; (**E**) lizard; (**F**) crocodile; (**G**) bird; and (**H**) mammal. Arterial blood is shown as open, venous as solid, and portal blood is cross-hatched.

systemic and portal hearts, e.g., 40 to 80 min^{-1}, and often have an alternating contraction pattern (Figure 14–32B).

The systemic heart of cyclostomes has no innervation and does not respond to epinephrine, but it does respond to varying venous return in the same manner as other vertebrate hearts, i.e., its force of contraction increases if the venous return increases (the Frank-Starling principle). The absence of a neural control of the systemic heart of cyclostomes is exceptional for vertebrates and perhaps reflects the presence and neural control of the other hearts

(portal, caudal, and cardinal) that control venous return. The portal heart has a more consistent contraction rate than the systemic heart and caudal hearts; it also responds to increased filling with an increased force of contraction. The variability in contraction rate of the caudal hearts and their permanent cessation of beating if the spinal cord is destroyed suggest a neural reflex regulation of their activity, perhaps activated by high pressure in the lateral sinuses.

The circulatory system of cartilaginous and bony fish is closed, like that of all other vertebrates except

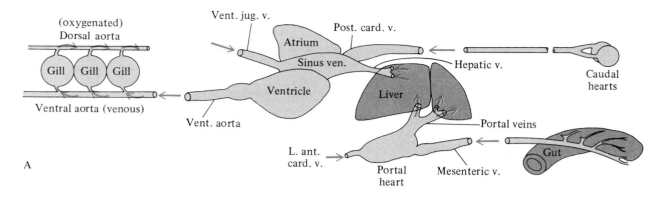

A

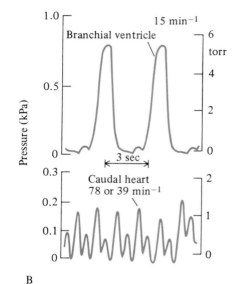

FIGURE 14–32 **(A)** Diagrammatic scheme of the circulatory system of the Pacific hagfish *Eptatretus* showing the location of the caudal hearts, portal heart, and systemic heart. **(B)** Branchial ventricular pressure and caudal heart pressure of the hagfish; the caudal hearts are beating in an alternating pattern. The small increase in branchial systemic pressure prior to ventricular contraction is due to atrial contraction. *(From Chapman, Jensen, and Wildenthal 1963.)*

B

cyclostomes. Their heart (e.g., elasmobranch, Figure 14–33A) has four chambers: the sinus venosus, atrium, ventricle, and bulbus cordis (cartilaginous fish) or bulbus arteriosus (bony fish). Do not confuse this four-chambered, single-circuit heart structure with the very different four-chambered, dual-circuit heart of birds and mammals (see below). The chambers are enclosed by a pericardial sac, which is rigid in chondrichthyeans and chondrosteans. The sinus venosus is a thin-walled chamber that receives blood returning from the ducts of Cuvier (and hepatic veins in elasmobranchs). There are no valves on the inflow side of the sinus venosus, but there is little regurgitation of blood back into the ducts of Cuvier/hepatic veins when the sinus venosus contracts because of the higher peripheral resistance to retrograde flow compared to anterograde flow into the atrium. The more muscular atrium has valves on its inflow and outflow channels; it has a large volume capacity and functions as an important priming pump to fill the ventricle. The ventricle is a thick-walled, muscular chamber; it is the primary propulsive pump. The spongy myocardium lining of the fish ventricle forms thousands of small, blood-filled compartments; this trabeculated arrangement may minimize the wall tension since each small compartment has a smaller diameter than the entire ventricle (see the law of Laplace).

Blood is ejected from the ventricle of elasmobranchs into the bulbus cordis (also termed the conus arteriosus). This is a muscular extension of the ventricular cardiac muscle which (1) provides muscular support for a series of valves that prevent backflow, (2) dampens the ventricular pressure cycle (windkessel effect), and (3) acts as an auxiliary pump. This cordis is replaced in hearts of higher fish by a bulbous swelling of the aorta, the bulbus cordis or bulbus arteriosus, that has smooth muscle and elastin in its wall; it also has a windkessel effect, damping the aortic blood pressure (Figure 14–33B).

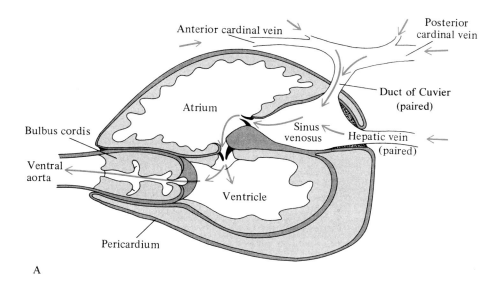

A

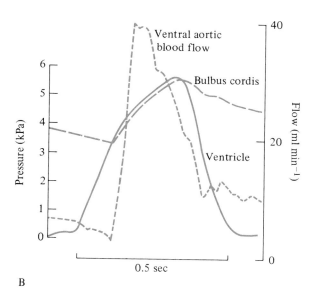

B

FIGURE 14–33 **(A)** Representation of the elasmobranch heart. The atrium and ventricle are surrounded by a rigid pericardium. Venous blood returns to the heart through anterior and posterior cardinal veins and hepatic vein. Blood pools in the sinus venosus and then is pumped by the atrium, ventricle, and bulbus cordis into the aorta. **(B)** Shown are pressures in the ventricle and bulbus cordis of the heart of the lingcod as well as aortic blood flow. *(Modified from Waterman et al. 1971; Stevens et al. 1972.)*

The rigid pericardium of elasmobranchs facilitates venous return to the heart. For example, when the ventricle contracts it simultaneously ejects blood from the ventricle and acts as a suction pump to draw venous blood into the atrium (cf. the constant volume mechanism of the molluskan heart). This dual ejection/suction role also occurs to a lesser extent in teleost and dipnoan fish hearts, as well as in some amphibian hearts.

Lungfish. Lungfish (dipnoans) and also some other fish have developed a lung for air breathing as a supplement to branchial respiration (see Chapter 13). This bimodal respiratory system requires a modification of the circulatory system for efficient circulation of deoxygenated blood to the lung and oxygenated blood from the lung (Figure 14–34A). A branch of the most posterior branchial artery supplies blood to the lungs. Oxygenated blood returns from the lungs to the left side of a partially divided atrium, whereas deoxygenated blood from the body returns to the right side. The partial atrial septum limits mixing of the deoxygenated and oxygenated blood. This partial separation of blood streams is maintained in the ventricle, which also has an incomplete septum extending from the apex of the heart (Figure 14–34B). Functional separation of the deoxygenated and oxygenated blood is maintained in the bulbus arteriosus by two spiral folds that fuse anteriorly to form two completely separated

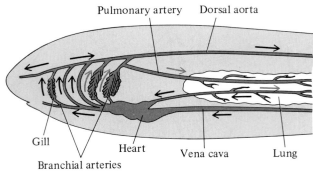

A

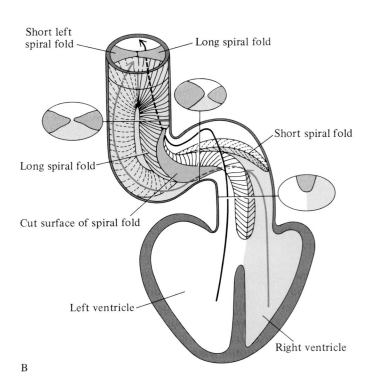

B

FIGURE 14–34 **(A)** Schematic of the central circulation of the African lungfish *Protopterus*. The most anterior aortic arch has a gill, the second and third arches do not have gills and function as direct shunts for oxygenated blood from the lungs (via the heart) to the dorsal aorta, the fourth and fifth arches have gills, and the fifth arch provides blood flow to the lungs. **(B)** Representation of the ventricle and bulbus cordis of the lungfish heart showing the anatomical separation of deoxygenated blood (in color) and oxygenated blood (no color) by the long and short spiral folds. *(From Waterman et al. 1971.)*

channels. The ventral channel conveys oxygenated blood to the anterior aortic arches and directly to the dorsal aorta, whereas the dorsal channel distributes deoxygenated blood to the posterior gill-bearing branchial arches and lung. This pattern of blood circulation can be physiologically adjusted to suit different environmental conditions, e.g., deoxygenated water or aerial exposure.

Air-breathing vertebrates have a partially or completely separate pulmonary circuit for blood flow through the lungs. The pulmonary blood pressure of lungfish is similar to its systemic blood pressure, but in tetrapods the pulmonary pressure declines, whereas the systemic pressures increase (Figure 14–35).

Amphibians. The amphibian circulatory system has a completely divided atrium but is little advanced over the dipnoan circulatory pattern with respect to anatomical or functional separation of oxygenated (pulmonary) and deoxygenated (systemic) blood flow. This partly reflects the diversity of respiratory exchange organs in amphibians (e.g., external gills, lungs, bucco-pharyngeal and cutaneous surfaces). In larval and neotenic amphibians, the aortic arches 3, 4, and 5 usually branch to gill vessels, but a connection (the ductus arteriosus or ductus Botalli) remains between the 6th aortic arch and the dorsal aorta (to bypass the lung). In adult urodeles that lack gills, all aortic arches have direct connections to the dorsal aorta; arch 3 represents the carotid

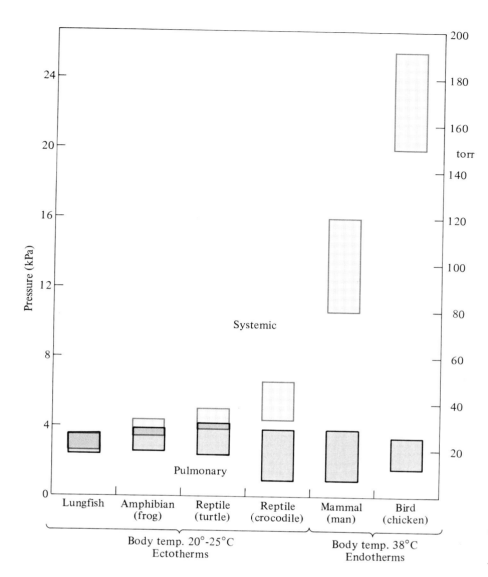

Systemic and Pulmonary pressures

Systemic

Pulmonary

Lungfish | Amphibian (frog) | Reptile (turtle) | Reptile (crocodile) | Mammal (man) | Bird (chicken)

Body temp. 20°-25°C
Ectotherms

Body temp. 38°C
Endotherms

FIGURE 14–35 Phylogenetic development of systemic and pulmonary pressures in vertebrates. *(Modified from Johansen, Lenfant, and Hanson 1970; Sturkie 1986.)*

circulation, arches 4 and 5 the aorta, and arch 6 primarily functions as a pulmonary artery and a cutaneous artery.

The pulmonary veins drain into the left atrium and systemic blood drains via a sinus venosus into the right atrium (Figure 14–36A). The atrium of the amphibian heart is typically completely divided into left and right sides. The ventricle is not divided by a septum, but is functionally subdivided by a dense trabeculation of its spongy myocardium. The short conus arteriosus (bulbus cordis) has a spiral fold that separates deoxygenated and oxygenated outflow from the ventricle. There is thus a functional, though labile, separation of oxygenated and deoxygenated blood flows from the left/right atria to the pulmonary arteries/systemic arches respectively with minimal mixing (Figure 14–36B). Oxygenated

blood returning from the skin is not, as might be expected, routed to the left atrium but flows into the systemic blood and then into the right atrium. The cutaneous circulation is apparently an unimportant means of gas exchange during elevated metabolism when most gas exchange occurs across the lungs (Withers and Hillman 1988). During inactivity, when the metabolic rate is low and cutaneous gas exchange is significant compared to pulmonary gas exchange, the mixing of oxygenated cutaneous drainage with deoxygenated systemic blood is apparently not a critical limitation.

A few salamanders, which are aquatic and use their lungs infrequently, have a fenestrated interatrial septum and no spiral valve. Plethodontid salamanders lack lungs, and their heart has a greatly fenestrated interatrial septum and no spiral valve;

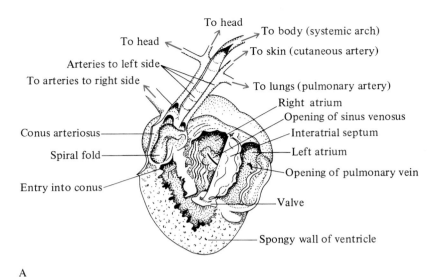

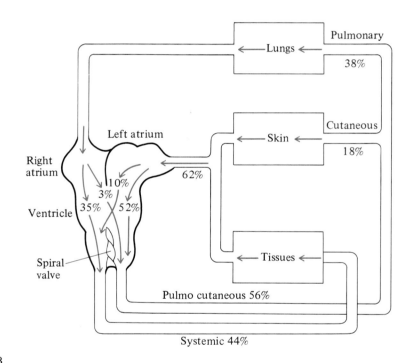

FIGURE 14–36 **(A)** Representation of the amphibian (*Rana*) heart. **(B)** Distribution of total blood flow (20.5 ml min⁻¹) through the circulatory system of the bullfrog and functional separation of oxygenated and deoxygenated blood in the heart. *(From Rogers 1986; modified from Tazawa, Mochizuki, and Piiper 1979.)*

the left atrium is also reduced in size. Caecilians also have a similar heart with regressive modifications for at least partially bypassing the pulmonary circuit.

Reptiles.   The circulatory system of reptiles is quite different from that of amphibians, reflecting their shift to predominantly pulmonary respiration. The crocodilian heart is unique for reptiles in having a completely divided ventricle and will be discussed separately below.

The heart of lizards, snakes, and turtles (and crocodiles) lack a conus arteriosus; the systemic arches and pulmonary arteries arise directly from the ventricle. The atria are completely separate, as in amphibians, but the ventricle has a particularly complex structure with three interconnected chambers (White 1968). The right atrium opens into the **cavum venosum** and the left atrium into the **cavum arteriosum** (Figure 14–37A). The cavum arteriosum also has a direct connection to the cavum venosum (without which the cavum arteriosum would be

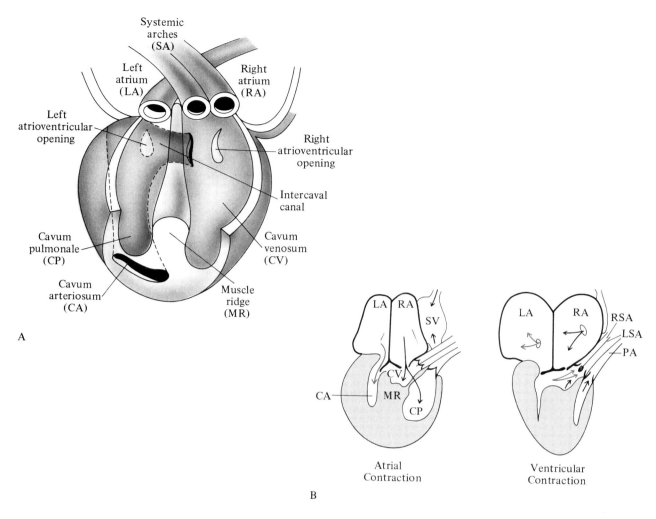

A

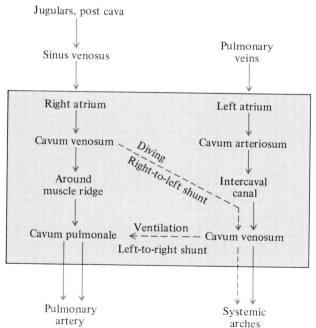

C

FIGURE 14–37 **(A)** Representation of the squamate reptile heart (e.g., a lizard) as seen from the ventral side. **(B)** Representation of the blood flow through the heart of a varanid lizard during atrial contraction (left) and ventricular contraction (right). Colored arrows indicate the direction of oxygenated blood flow, and black arrows indicate the direction of deoxygenated blood flow. **(C)** Normal pattern of blood flow (solid arrows) through the squamate reptile heart and pattern of blood flow for a left-to-right intracardiac shunt (e.g., during apnea, such as diving) and a right-to-left shunt (e.g., during pulmonary ventilation). *(Modified from Waterman et al. 1971; Romer and Parsons 1986; White 1968.)*

blind ended). The patency of the intercaval opening between the cavum arteriosum and cavum venosum is influenced by the atrioventricular valves (see below). The cavum venosum is partially separated from a third ventricular compartment, the **cavum pulmonale**, by a thick muscular ridge.

The functional separation of deoxygenated and oxygenated blood during passage through this reptilian heart is accomplished through a complex change in the relationship between these three chambers (Figure 14–37B). During atrial contraction, deoxygenated blood is ejected from the right atrium into the cavum venosum and over the muscular ridge into the cavum pulmonale. Oxygenated blood is ejected from the left atrium into the cavum arteriosum and is retained there because the atrioventricular valves occlude the connection to the cavum venosum. During ventricular systole, deoxygenated blood is first ejected into the pulmonary arteries from the cavum pulmonale and cavum venosum because the pulmonary circuit has a lower resistance than the systemic circuit. During ejection, the cavum venosum is separated from the cavum pulmonale when the muscular ridge contacts the ventricular wall; this prevents subsequent backflow of blood from the cavum pulmonale to the cavum venosum. As the ventricle continues to contract, the atrioventricular valves close the atrioventricular opening and thereby open the intercaval canal. This allows oxygenated blood from the cavum arteriosum to enter the cavum venosum and then the systemic arches. There is a high degree of functional separation of oxygenated and deoxygenated blood flow through this reptilian heart. Figure 14–37C summarizes this pattern of blood flow during the contraction cycle.

Why hasn't the reptile heart (other than the crocodile heart) evolved a complete functional separation of oxygenated and deoxygenated blood flow? First, the complex pattern of flow through the three ventricular compartments precludes a simple anatomical separation; both oxygenated and deoxygenated blood flow through the cavum venosum at different parts of the cardiac cycle. Second, there is a functional advantage to maintaining an anatomically incomplete separation; there is the potential for shunting blood from one side of the heart to the other. Normally, such shunts are minimized but may be useful under specific circumstances. For example, the lung O_2 stores may become depleted during prolonged apnea, e.g., diving, and so it is not necessary to maintain blood flow to the lungs, or during basking, when heat transfer may become a major role of blood flow, i.e., a **right-to-left shunt** may be advantageous (Figure 14–37C). A **left-to-right shunt** may occur during lung ventilation to preferentially return partially oxygenated blood to the lungs for further oxygenation. For example, pulmonary blood flow stops completely during prolonged, quiescent diving in the file snake *Acrocordis* but increases markedly during breathing, when the cardiac output largely bypasses the systemic circulation (Lillywhite and Donald 1989).

The crocodilian heart has a complete anatomical separation of the atria and ventricles into a right (deoxygenated) and left (oxygenated) side (Figure 14–38A). Deoxygenated blood flows into the right atrium, the right ventricle, and then the pulmonary arteries to the lungs. Oxygenated blood flows into the left atrium, left ventricle, and then the right systemic arch to the body. However, the left systemic arch arises from the right ventricle and should therefore carry deoxygenated blood to the body! This is prevented by an extracardiac communication between the left and right systemic arches, the **foramen of Panizza**, located at the base of the arches where they arise from the ventricles. Oxygenated blood from the right systemic arch flows into the left systemic arch via the foramen. Deoxygenated blood from the right ventricle does not normally enter the left systemic arch because the higher pressure of the left ventricular outflow compared to the right ventricular outflow closes a valve between the right ventriculo-left systemic arch.

Why do crocodiles retain the connection between the right ventricle and left systemic arch? It provides the flexibility for occasions when the pulmonary circuit may be bypassed, e.g., during prolonged diving or anoxia. Constriction of the pulmonary vasculature circuit increases the right ventricular pressure and allows the right ventriculo-left systemic arch valve to open, thereby bypassing the lung (Figure 14–38B).

Birds and Mammals. The adult mammalian and avian circulatory systems have a completely separated double circuit of oxygenated and deoxygenated blood. The heart has anatomically divided atria and ventricles with no possibility for intra- or extracardiac shunting.

In the mammalian heart, systemic blood enters the right atrium and then passes through a membranous right atrioventricular valve (tricuspid valve, since it has 3 cusps) into the right ventricle (Figure 14–39A). Blood is ejected through the pulmonary semilunar valves via the left and right pulmonary arteries for oxygenation in the lung. The oxygenated blood returns via the two right and two left pulmonary veins into the left atrium and then into the left ventricle via the membranous mitral atrioventricular

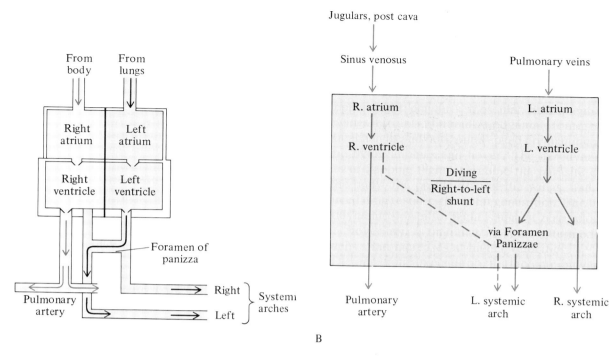

FIGURE 14–38 (A) Representation of the pattern of blood flow through the crocodile heart showing the normal shunting of oxygenated blood from the right to the left systemic arch via the foramen of Panizza. (B) Normal pattern of blood flow through the crocodilian heart (solid arrows) and probable right-to-left shunt during diving. *(Modified from Waterman et al. 1971; White 1968.)*

valve (or bicuspid as it has two cusps). The thick-walled muscular left ventricle ejects blood into the aorta (through the aortic semilunar valves) for systemic distribution. The membranous atrioventricular valves are prevented from being forced "inside-out" during ventricular systole by having their free edges attached via chordae tendineae to papillary muscles, which are projections of the ventricular wall. The heart has its own coronary circulation to supply oxygenated blood to the myocardium because only the inner layer of the myocardium can obtain nutrients directly from the blood flowing through the atria and ventricles.

The cardiac cycle of the mammalian heart consists of an atrial contraction (systole) and then refilling (diastole), coordinated with a ventricular systole and diastole. The pressure cycle of the left side of the heart is summarized in Figure 14–39B; the right side has a similar but lower-amplitude pressure cycle. Blood normally flows continually from the venae cavae and pulmonary veins into the atria, and much blood flows into the ventricles even before atrial systole. Atrial contraction only accomplishes the final 20 to 30% of ventricular

filling. Clearly, the ventricles can function adequately (at least at rest) without atrial contraction. The atrial pressure wave has three phases. The a-wave is due to atrial contraction, with a peak pressure of 0.5 to 0.8 kPa (right) and 1.0 to 1.1 kPa (left). The c-wave is caused by ventricular contraction; there may be a little backflow of blood into the atria as the atrioventricular valves close and the ventricular contraction exerts tension on the atria. The v-wave is a slow accumulation of blood in the atria while the atrioventricular valves are closed.

Ventricular filling occurs in three phases. The first phase is rapid filling when the atrioventricular valves open and blood rapidly enters the ventricle. This is followed by a second phase of a short period of relatively little blood flow (diastasis). The final, third phase of filling is due to atrial contraction. At the beginning of ventricular systole, the intraventricular pressure increases rapidly, causing the atrioventricular valves to close (Figure 14–39C). Pressure continues to increase until it exceeds the aortic pressure (about 11 kPa), at which point the semilunar valve opens and blood is ejected into the aorta.

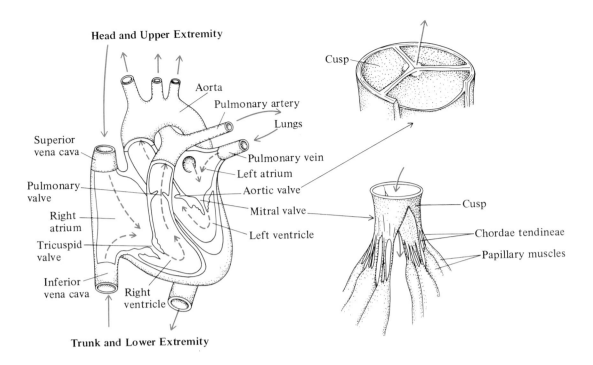

A

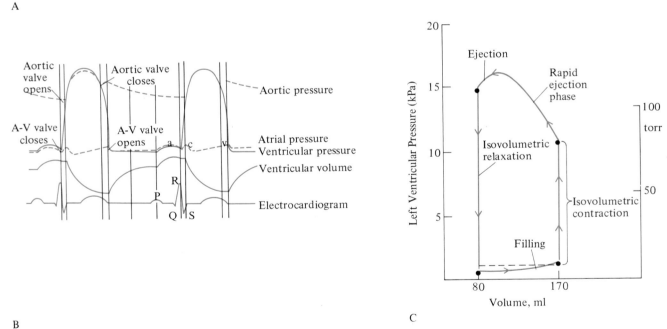

B

C

FIGURE 14–39 (A) Structure and pattern of circulation through the four-chambered mammalian heart. (B) Cardiac cycle of the human heart showing the atrial, ventricular, and aortic pressures, ventricular volume, and electrocardiogram. (C) Left ventricular pressure-volume curve for the human heart showing the isovolemic contraction and relaxation phases and the filling and emptying phases.

About 75% of the stroke volume (the blood ejected by the ventricle) is ejected in the first 1/3 of the ejection period (rapid ejection period) and then the final 25% is ejected throughout the remaining slow ejection period. At the end of ventricular systole, the intraventricular pressure declines and the semilunar valve snaps shut when the pressure drops below the aortic pressure (the incisura point). A period of isovolemic (constant volume) relaxation follows, as the intraventricular pressure declines, until the atrioventricular valves open and the ventricle is filled and the cardiac cycle is repeated. The right ventricle pumps deoxygenated blood to the lungs. The right ventricle and pulmonary vascular pressures are much lower than the left ventricular and systemic pressures.

The embryonic circulation of mammals is modified so blood bypasses the nonfunctional lungs and liver. Gas, nutrient, and waste exchange occurs across the **placenta**, where fetal and maternal vascular beds come into close proximity to facilitate exchange. During the embryologic development of the mammalian heart, the atria are initially separated by a primary partition (similar to that of the amphibian heart) which, as soon as it becomes complete, develops an opening to allow blood to flow from the right atrium into the left atrium. Later, a second partition develops to the right of the primary partition, but this remains incomplete with an opening (the **foramen ovale**) in its lower part. About 1/3 of the blood entering the right atrium passes through the foramen ovale into the left atrium and bypasses the pulmonary circuit. The blood that enters the right ventricle passes into the pulmonary artery but most enters the aorta through a connection, the ductus arteriosus, between the pulmonary artery and aorta. Only a little blood actually passes to the lungs, providing nutrients for pulmonary growth and development. Fetal blood flow to the placenta is provided by a branch of each internal iliac artery that forms two umbilical arteries. Blood returns from the placenta via the umbilical vein to the liver. Most of this oxygenated blood enters the vena cava through one branch of the umbilical vein, the ductus venosus, and mixes with deoxygenated blood from the rest of the fetal circulation and is recirculated by the heart. Some of the oxygenated blood enters the liver (which is not functional for digestion, but does synthesize red blood cells) via a second branch of the umbilical vein that joins the hepatic portal vein and then returns to the vena cava via the hepatic vein. After birth, these circulatory modifications are lost; the interatrial foramen ovale is closed and sealed and the ductus arteriosus, the umbilical arteries and the umbilical vein, and the ductus venosus constrict and degenerate.

The avian heart has independently evolved a similar four-chambered double circulation like the mammalian heart. The right atrium receives deoxygenated blood from three great veins (two anterior venae cavae and a posterior vena cava) that enter the right atrium through a clearly recognizable sinus venosus region (Figure 14–40A). Blood passes from the right atrium into the right ventricle through a muscular (not membranous) valve (Figure 14–40B; a similar muscular valve is present in the hearts of crocodilians and monotremes). The blood is ejected into the pulmonary trunk through a semilunar valve and then into the right and left pulmonary arteries. Oxygenated blood returns from the lungs via the pulmonary veins and enters the left atrium through a membranous bicuspid (or mitral) valve that actually has two large cusps and a third smaller cusp. The mitral valve is prevented from being turned inside out by attachment of its free edges to the papillary muscles of the left ventricle by chordae tendineae. Blood from the left ventricle is ejected through aortic semilunar valves into the aorta. The heart has a coronary circulation (coronary artery, coronary sinus, and coronary veins).

As we would expect, the avian embryo also has complex additional circulatory pathways that bypass the nonfunctional lung and provide gas exchange with the environment. The interatrial septum has an aperture that functions in the same manner as the foramen ovale of the fetal mammalian heart, allowing blood to pass from the right atrium to the left atrium. There is also a ductus arteriosus between the pulmonary artery and aorta that bypasses the lungs. Initially, the avian circulation has two ducti arteriosi but one degenerates early in fetal development and the other remains patent until after hatching. The gas exchange surface equivalent to the placenta is initially the area vasculosa, which develops around the embryo. This consists of blood vessels connected to the venous circulatory system by a pair of vitelline veins and to the arterial system via a pair of vitelline arteries that branch off the dorsal aorta. Later, the allantoic circulation becomes the predominant respiratory supply. After hatching, the interatrial aperture closes by growth of the myocardium (there is no equivalent to the secondary septum of the mammalian heart), and the ductus arteriosus constricts and then degenerates. The allantois dries up and the artery and vein constrict and then degenerate.

Allometry. Heart rate varies inversely with body mass in mammals ($\propto$ Mass$^{-0.25}$) and also birds ($\propto$ Mass$^{-0.23}$; Table 14–6). Both heart mass and stroke volume are proportional to body mass (about mass$^{1.0}$) in both mammals and birds. Consequently, cardiac output is proportional to Mass$^{0.81}$ in mammals and Mass$^{0.78}$ in birds. These allometric exponents are very similar to those for the scaling of metabolic rate and respiratory variables in birds and mammals (Table 4–5, page 95; Table 13–8, page 631). The intercept values for scaling of cardiac output in mammals (187 ml blood min^{-1}) and birds (285) are in about the same proportion as for

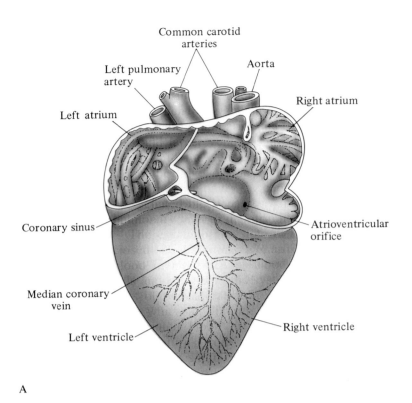

A

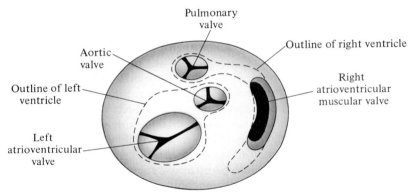

B

FIGURE 14–40 **(A)** Dorsal view of the avian (chicken) heart with the atria opened to show the internal detail. **(B)** Schematic drawing of the chicken heart at the level of the atrioventricular valves and the pulmonary and aortic valves. *(From Simons 1960.)*

metabolic rate, indicating approximately similar values for O_2 extraction from the mammal and bird blood flow. Such a constant relationship between metabolic rate and cardiac output is expected because O_2 transport by the blood is equal to the product of cardiac output and (arterial-venous) blood O_2 difference

$$VO_2 = V_b(CaO_2 - CvO_2) \qquad (14.12)$$

where VO_2 is the O_2 consumption (ml O_2 min^{-1}), V_b is the cardiac output (ml min^{-1}), and CaO_2 and CvO_2 are the arterial and venous blood O_2 contents, respectively (ml O_2 ml blood^{-1}). The arterial-venous blood O_2 difference is approximately independent of mass and is similar for mammals and birds.

Arterial blood pressure tends to be higher in birds than in mammals (Table 14–6), but is essentially independent of body mass. Birds require a higher systemic pressure to maintain their higher cardiac output since the peripheral resistance of birds and mammals is similar. The systemic peripheral resistance is inversely dependent on body mass ($\propto$ Mass$^{-.74 \text{ to } -0.78}$). The pulmonary blood pressures of birds and mammals are substantially lower than the systemic blood pressures (see Figure 14–35, page 707) and so the pulmonary resistances are substantially lower than the systemic resistances. In humans, pulmonary resistance (2.5 10^7 Pa s m^{-3}) is about 0.15 $\times$ the systemic resistance (1.6 10^8), and in the chicken the pulmonary resistance (5 10^8) is about 0.14 $\times$ the systemic resistance (3.6 10^9).

TABLE 14-6

Comparison of the allometric relationships for cardiovascular variables in birds and mammals. Allometric constants are for the equation $Y = aX^b$, where Y is the variable and X is body mass (kg). *(From Stahl 1967; Calder 1981.)*

	BIRD		MAMMAL	
	a	*b*	*a*	*b*
Blood volume (ml)	—	—	65.6	1.02
Heart mass (g)	8.6	0.94	5.8	0.98
Blood pressure (kPa)	21	0.036	12	0.032
Heart rate (min^{-1})	156	-0.23	241	-0.25
Stroke volume (ml)	1.83	1.01	0.78	1.06
Cardiac output (ml min^{-1})	285	0.78	187	0.81
O$_2$ consumption (ml min^{-1})	15.8	0.72	10.0	0.76
CaO$_2$-CvO$_2$ (ml O$_2$ ml blood^{-1})	0.055	-0.06	0.053	-0.05
Peripheral resistance (Pa s m^{-3})	4.42 10^9	-0.74	3.85 10^9	-0.78

Regulation of the Cardiovascular System

The cardiovascular system must be regulated to match blood circulation with the metabolic requirements. Even animals with a tracheal system must have some matching of the circulation with metabolism to distribute other nutrients and waste products, hormones, etc. The principal aspects of the circulatory system that are closely regulated are the heart (e.g., heart rate and force of contraction), blood pressure (by control of the heart and by regulation of the peripheral resistance), and the pattern of distribution of blood flow through the circulatory system (by selective control of peripheral resistance for different vascular beds).

The Heart

Myogenic and **neurogenic** hearts differ in terms of the origin of their heart beat. Myogenic hearts have a non-neural origin of contraction (although they may also have neural regulation of heart function). A specialized group of muscle cells often acts as a pacemaker, driving the contractile rhythm of the entire heart. Tunicates, vertebrates, lamellibranch and gastropod mollusks, and some insects have myogenic hearts. Neurogenic hearts, in contrast, rely on an extrinsic innervation for neural initiation of the heart contraction cycle. Amphibian and fish lymph hearts, crustacean hearts, some insect hearts, and spider hearts are neurogenic.

Probably all molluskan hearts are myogenic. The heart does not have a discrete pacemaker region of muscle cells but a contraction can be initiated in almost any part of the heart. In *Dolabella*, the atrioventricular region may act as a dominant pacemaker but other areas of the heart can also act as pacemakers. The aortic region of the gastropod ventricle may be a pacemaker. It is not clear how the atrium is caused to contract before the ventricle. It is also not clear how paired atria are synchronized. All molluskan hearts are innervated but the anatomical detail varies considerably. The innervation in bivalves is from the cerebrovisceral connective to each atrium. In gastropods, cardiac nerves enter the heart at the aortic and venous extremities where the heart contacts the pericardium or through the aortic connection. Most of the heart appears to be innervated, but some regions (e.g., atrioventricular valve, atrial end of the aorta) have a denser innervation. Both inhibitory and stimulatory fibers innervate the heart. In *Aplysia*, there are two inhibitory and two excitatory nerves in the abdominal ganglion and three vasoconstrictor nerves. Acetylcholine has cardioinhibitory effects in some species (e.g., *Tapes*, *Aplysia*, most gastropods) and excitatory effects in others (e.g., mytilids). Serotonin (5-HT) generally is an excitatory neurotransmitter. Dopamine can have either excitatory or inhibitory effects; other catecholamines generally have little effect, although norepinephrine is as active as dopamine in *Helix* and *Limax*. Cardioactive peptides (e.g., FMRFamide) are neurosecretions that are excitatory in most species.

In cephalopod hearts, the wave of excitation begins in the region between the atria and spreads out in an organized pattern (Wells 1983). No neurons are present in this pacemaker region, so the origin of the cephalopod cardiac cycle is clearly myogenic. Acetylcholine decreases cardioactivity and 5-HT and catecholamines have excitatory effects. A variety of cardiopeptide neurosecretions modulate the heart function. There also is direct neural control of the heart; the cardiac ganglion apparently is the normal pacemaker for the heart. It is especially important in coordinating the contractions of the two branchial hearts and the systemic heart.

The crustacean heart generally has an intrinsic neural system concentrated in the medial cardiac ganglion on the dorsal wall of the heart (but not in branchiopods, copepods, and branchiurans). Axons from these neurons innervate the cardiac and ostial muscles. There also are extrinsic nerves (one pair of inhibitors and one pair of excitators) innervating the cardiac ganglion and perhaps the myocardium, the cardiac and arterial valves, and the alary muscles. The origin of the heart beat is clearly neurogenic; isolated portions of the heart will only contract if one or more nerves are present. The cardiac ganglion neurons normally are the pacemaker. Stretch of the cardiac muscle increases heart rate and amplitude. Temperature affects heart rate, with Q_{10} values from 3 to 5 at 5° to 10° C, about 2 at 10° to 25° C, and 1 to 2 at >25° C. Heart rate (min^{-1}) is inversely proportional to body mass (grams); rate = 160 g$^{-0.12}$ for a variety of crustaceans.

Some insect hearts are neurogenic, whereas others are myogenic. In some myogenic hearts, the entire dorsal vessel is capable of initiating a contraction, but only certain areas can initiate contractions in others, e.g., the vessel in segments 5 to 8 of *Chaoborus* can initiate a contraction but not in other segments. Many hearts lack an innervation and continue to beat when semi-isolated so long as they are mechanically supported by suspensory ligaments or alary muscles. Most insect hearts have an innervation, but it is often difficult to determine whether the heart is neurogenic or myogenic. The heart may continue to beat if all neural connections are cut, but still be neurogenic rather than myogenic if there are neurons closely attached to the heart (cf. crustacean hearts). The neural influence may be a response to a reflex initiated in the heart itself, e.g., the cardiac ganglion of *Periplaneta* is stimulated by distension of the heart to enhance the myogenic activity. The insect heart often shows reversal of the direction of the heart beat, but hemolymph probably does not flow in the reverse direction because of the action of the ostial valves.

The vertebrate heart is myogenic but usually has an innervation which modulates the heart beat. The hagfish heart lacks an innervation, and the lamprey heart has only a cholinergic innervation. The cardiac pacemaker is a specialized group of cardiac muscle cells located in the sinus venosus of fish and amphibian hearts and the sinoatrial (SA) node in the right atrium of tetrapod hearts. Most cardiac muscle cells have their own inherent rate of rhythmic contraction, but the pacemaker cells have the highest frequency and, hence, entrain all other cardiac muscle cells to their rate. The inherent rates of rhythmic activity in the human heart are AV node fibers, 40 to 60 min^{-1}; Purkinje fibers, 15 to 40 min^{-1}; and SA node pacemaker, 70 to 80 min^{-1}. Sometimes the pacemaker role shifts from the SA node to another group of hyperexcitable muscle fibers; this ectopic focus causes an abnormal sequence of electrical conduction through the heart.

Electrical depolarization spreads from the pacemaker throughout the syncytium of cardiac muscle cells; in the mammalian heart the pacemaker is in the right atrium (Figure 14–41). Its electrical activity spreads throughout the right and left atrial muscle syncytium and is propagated at about 1 m sec^{-1} through specialized conduction pathways to the left atrium and to the atrioventricular (AV) node. These specialized conduction pathways consist of muscle cells that resemble the Purkinje fibers (see below). The electrical depolarization does not spread directly from atrial to ventricular muscle because of the insulating action of an atrioventricular connective tissue septum. The AV node is a second group of specialized muscle cells near the AV septum; it

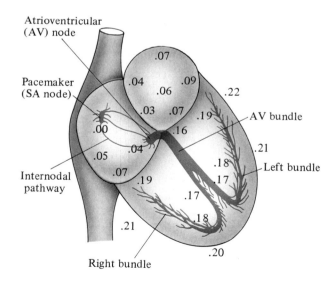

FIGURE 14–41 Pathways in the human heart for the spread of electrical depolarization from the sinoatrial (SA) node to the atrioventricular (AV) node via specialized conduction pathways, to the ventricle via the Purkinje fibers (AV, left and right bundles), and the timing of electrical transmission through the conducting systems of the human heart (values are fractions of a second after SA node depolarization). *(From Guyton 1986.)*

consists of a variety of specialized muscle cells. The small junctional fibers have a very low conduction velocity (about 0.02 m sec^{-1}) and the other nodal cells also have a low conduction velocity (about 0.1 m sec^{-1}). The low conduction velocity of these cells is due to both morphological and physiological adaptations. The AV node cells are small and therefore have a low conduction velocity; they also have few functional fast Na$^+$ channels, and action potentials are initiated by the slower Na$^+$-Ca^{2+} channels. This considerably delays the propagation of the cardiac electrical depolarization to other nodal fibers, the transitional fibers, thence the Purkinje fibers of the atrioventricular bundle. The effect of the AV node is to delay by about 0.1 sec the spread of electrical activity from the atria to the ventricles so that the ventricles contract after the atria. The Purkinje fibers extend from the transitional fibers of the AV node across the atrioventricular septum into the right ventricle where one branch spreads to the left ventricle. The Purkinje fibers have a high conduction velocity (about 1.5–4 m sec^{-1}) and almost immediately transmit the electrical depolarization to the ventricular muscle. The final distribution of electrical depolarization spreads from the Purkinje fibers throughout the ventricular muscle at about 0.3–0.5 m sec^{-1}.

Most vertebrate hearts have an antagonistic pattern of innervation, the cholinergic and parasympathetic vagi and the noradrenergic and sympathetic nerves. Parasympathetic stimulation primarily decreases heart rate (to as low as 20 to 30 min^{-1} in humans) and sympathetic stimulation elevates heart rate (up to 250 min^{-1}). Parasympathetic stimulation also decreases the force of contraction of the cardiac muscle (especially ventricle), whereas sympathetic stimulation increases the force of contraction.

Cardiac Output

Both heart rate and stroke volume of molluskan hearts (and also many other hearts) is influenced by the intracardiac pressure. Stretch of cardiac muscle by preloading the heart with a high pressure increases its force of contraction, hence stroke volume (at least up to some maximum), and also increases the rate of contraction (Figure 14–42A). The heart is able to intrinsically adjust its pumping to prevent the pooling of venous blood in the heart. It can only function within mechanical limits imposed by backpressure, however; increasing the afterload pressure (i.e., aortic pressure) decreases the stroke volume.

Cardiac pumping is also under intrinsic, mechanical control in vertebrates; the heart tends to pump

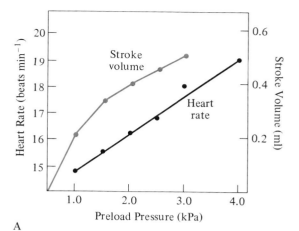

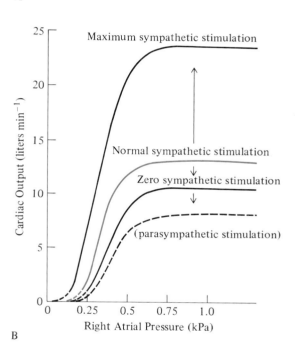

FIGURE 14–42 **(A)** The *in vivo* output of the ventricle of the octopus can be affected by the preload pressure, i.e., venous return and distension of the ventricle; both heart rate and stroke volume increase as a function of preload pressure (with a constant afterload pressure of 3 kPa). **(B)** Effects of sympathetic stimulation and inhibition and of parasympathetic stimulation on cardiac output of the human heart. *(Modified from Smith 1981; Guyton 1986.)*

all of the blood that is returned to it to prevent venous pooling. The Frank-Starling law of the heart states that the greater the filling of the heart, the greater is the amount of blood pumped into the aorta. Stretch of the cardiac muscle increases both the force of contraction and the rate of contraction. The effect of filling on force of contraction is a direct effect on the cardiac muscle fibers, but the rate of contraction is regulated through a neural reflex, the Bainbridge reflex, mediated by the central nervous system. This intrinsic regulation of cardiac output by the heart is apparent from the pronounced effect of right atrial pressure on cardiac output (Figure 14–42B). Cardiac output is increased by sympathetic stimulation and decreased by parasympathetic stimulation.

Activity requires an immediate (acute) increase in cardiac output to match the increased metabolic rate. In higher vertebrates, much of the increase in cardiac output is due to elevated heart rate rather than stroke volume, whereas stroke volume increases are more pronounced in lower vertebrates (Table 14–7). In none of the vertebrates listed in

TABLE 14–7

Comparison of cardiovascular parameters at rest and during activity in a variety of vertebrates and in an invertebrate. The factorial increase is given for each parameter and also its relative contribution to the increment in metabolic rate during activity. The percent contribution is calculated as $100 \times$ (factorial scope -1)/(sum of factorial scopes -3). Units are heart rate, min^{-1}; stroke volume, ml; AV difference, ml O_2 per 100 ml blood; VO_2, ml O_2 min^{-1}.

		Rest	Active	Factorial Scope	% Contribution to VO_2 Increase
Mammal					
Possum (1.48 kg)	Heart rate	143	321	2.2 ×	51%
	Stroke volume	2.43	2.29	0.9 ×	−2%
	AV difference	4.5	10.1	2.2 ×	51%
	VO_2	19.5	100	5.1 ×	
Bird					
Pigeon (0.44 kg)	Heart rate	115	670	5.8 ×	87%
	Stroke volume	1.70	1.59	0.9 ×	−1%
	AV difference	4.6	8.3	1.8 ×	14%
	VO_2	8.9	88	9.9 ×	
Reptile					
Lizard (1.03 kg)	Heart rate	50	108	2.2 ×	41%
	Stroke volume	2.3	3.1	1.3 ×	12%
	AV difference	2.6	6.1	2.3 ×	47%
	VO_2	3.3	21.6	6.6 ×	
Amphibian					
Toad (0.25 kg)	Heart rate	26	47	1.8 ×	16%
	Stroke volume	0.34	0.32	0.9 ×	−1%
	AV difference	2.1	10.2	4.9 ×	84%
	VO_2	0.18	1.53	8.5 ×	
Fish					
Trout (1.00 kg)	Heart rate	38	51	1.4 ×	11%
	Stroke volume	0.46	1.03	2.2 ×	39%
	AV difference	3.2	8.3	2.6 ×	50%
	VO_2	0.56	4.35	7.8 ×	
Crustacean					
Crab (0.79 kg)	Heart rate	59	69	1.2 ×	15%
	Stroke volume	1.56	2.46	1.6 ×	53%
	AV difference	6.0	8.1	1.4 ×	32%
	VO_2	0.55	1.36	2.4 ×	

Table 14–7 is the increase in cardiac output sufficient to match the increased metabolic rate, and there is a substantial increase in the arterial-venous difference in O_2 content. There also is a considerable redistribution of blood flow during activity, with increased blood flow to active tissues (e.g., muscles) and reduced blood flow to organs such as the gut and kidneys.

Blood Pressure

The systemic blood pressure in most animals is determined by a multitude of cardiac factors, including (1) magnitude of the ventricular pressure cycle, (2) cardiac output (heart rate × stroke volume), (3) compliance of the peripheral circulation, (4) vascular resistance of the peripheral circulation, and (5) blood volume. The basic relationship between pressure (i.e., blood pressure, BP), flow (i.e., cardiac output, CO), and resistance (i.e., peripheral resistance, PR), from the Poiseuille-Hagen flow formula, is as follows.

$$BP = CO\ PR \qquad (14.13)$$

A variety of mechanisms regulate arterial blood pressure in vertebrates. These can generally be divided into short-term (acute) and long-term (chronic) mechanisms. Regulation of blood pressure can be accomplished through factors affecting either cardiac output or peripheral resistance, but the most significant short-term regulation of blood pressure is through heart rate and vascular resistance. Control of stroke volume and the ventricular pressure cycle also have lesser roles in regulation of blood pressure. The factors affecting acute regulation of cardiac output (heart rate and force of contraction) have already been described; they include the innervation and mechanical aspects, such as preload/afterload pressures. Control of blood volume is an important long-term regulator of blood pressure but is not involved in short-term regulation.

The arterial baroreceptor system of vertebrates is an important acute neural reflex for regulation of blood pressure. Baroreceptors are nerve endings that are stimulated by stretch of the walls of arteries. Baroreceptors are present in many of the larger arteries but are especially plentiful in the aortic arch and each internal carotid artery just above the carotid bifurcation. These stretch receptors relay sensory information to the medulla of the brain where compensatory actions are initiated. The rate of discharge of the baroreceptors increases with pressure (wall stretch) up to a maximum at about 27 kPa. The baroreceptors are most responsive at about the normal arterial blood pressure, about 14 kPa. Baroreceptor activity vasodilates the periph-

eral circulation and decreases heart rate and force of contraction. The baroreceptors are important for short-term regulation of arterial pressure, as is apparent from an increased variability in arterial pressure after denervation of the baroreceptors. However, denervation of the baroreceptors does not alter the average, long-term blood pressure. Other mechanisms are responsible for chronic regulation of blood pressure.

The atria and pulmonary vessels have low-pressure baroreceptors that function in a similar manner to the high-pressure arterial baroreceptors. Stimulation of the atrial baroreceptors is relayed to the medulla via the vagus nerve and causes a reflex increase in heart rate and force of contraction. This Bainbridge reflex acts in concert with the direct effect of stretch on the atrial muscle to prevent pooling of blood in the atria. Atrial stretch also causes reflex dilation of the afferent arterioles in the renal glomeruli and inhibits vasopressin secretion. These effects promote renal ultrafiltration and urine formation, which decrease blood volume and blood pressure (see below).

Two intrinsic properties of the circulatory system also facilitate regulation of blood pressure. First, an alteration in arterial pressure will affect capillary fluid balance, e.g., an increased blood pressure promotes fluid filtration. Second, many blood vessels, especially veins in the important blood storage areas, such as liver, spleen, and lungs, adapt to changes in blood pressure by dilation (with increased pressure) or constriction (with decreased pressure). This automatic stress-relaxation contributes to blood pressure regulation.

Blood flow to the brain also has a direct role in blood pressure regulation by influencing the medullary vasomotor center. A reduction in brain blood flow (e.g., due to decreased arterial pressure) allows the accumulation of CO_2, which stimulates the medullary vasomotor center, promoting sympathetic stimulation of the heart and peripheral vasoconstriction (see below). This CNS ischemic response is one of the most potent activators of the sympathetic vasoconstrictor system. It is, however, only an emergency response, being activated at very low arterial pressures (e.g., <8 kPa). Chemoreceptors in the carotid and aortic bodies and the central nervous system also regulate blood pressure, at least when it dramatically decreases (e.g., <1 kPa) by stimulation of the medullary vasomotor center.

Mechanisms for acute regulation of blood pressure generally act for a few seconds to a few minutes or hours; these mechanisms rapidly adapt. For example, baroreceptor regulation of arterial pressure rapidly adapts and is not an effective long-term mechanism for regulation of blood pressure. The gain of the baroreceptor control system is about −7 (see Chapter 2); this is a relatively low gain, indicating an only moderate capacity of the baroreceptors to compensate for changes in blood pressure.

The mechanism for chronic, long-term regulation of blood pressure involves a relationship between blood volume, arterial blood pressure, and renal excretion (Figure 14–43A). It differs from the acute mechanisms in three ways: (1) it takes hours or days to act, rather than seconds or minutes; (2) it does not adapt; and (3) it has an infinite negative gain, i.e., it is a "perfect" regulatory mechanism. Simply put, a rise in blood volume and arterial pressure will promote a higher than normal rate of renal ultrafiltration and urine flow rate according to the renal function curve (Figure 14–43B). This pressure diuresis decreases the body water volume and the blood volume; this reduces the mean circulatory filling pressure and venous return and so cardiac output is decreased. This in turn decreases the arterial blood pressure. Conversely, a decrease in arterial blood pressure reduces ultrafiltration and

urinary excretion of water, allowing body fluids to accumulate (so long as there is still a water intake, e.g., drinking) and raises the blood pressure to the normal value. The long-term, mean arterial pressure can only be altered by changing one of the two basic determinants of the mechanism, i.e., water intake or the location of the renal function curve. A chronic, even slight, increase in fluid intake will increase the mean arterial blood pressure, especially after the acute mechanisms have adapted. A right-shift in the renal function curve (e.g., by elevated circulating levels of angiotensin, or glomerular damage) will increase the mean arterial blood pressure until the renal excretion rate matches the intake rate.

This straightforward and automatic regulation of blood pressure and blood volume by a direct relationship between arterial pressure and urine flow rate is also a phylogenetically ancient regulatory system. It operates, for example, in hagfish. It should also function efficiently in invertebrates whose excretory systems depend on ultrafiltration of circulating body fluids (e.g., mollusks).

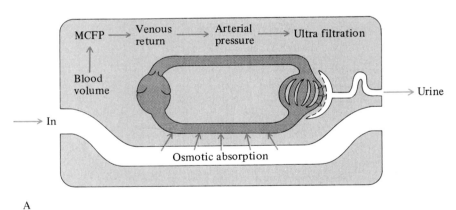

A

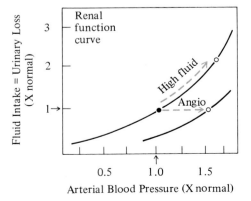

B

FIGURE 14–43 (A) Generalized schematic of how drinking and urinary loss influence fluid balance, blood volume, and long-term regulation of blood pressure. (B) The renal function curve shows how the mean arterial pressure can be elevated by an increased fluid intake or by angiotensin, which causes a right-shift in the position of the curve.

Selective Distribution of Blood Flow

The circulatory system of vertebrates has three general mechanisms for regulating the distribution of blood flow to various organs: (1) local metabolic control, (2) sympathetic control of the vascular bed, and (3) endocrine control of the vascular bed. The innervation of the microcirculation is concentrated at the arterioles (and to a lesser extent the venules), indicating the importance of their role in sympathetic control of blood flow. The precapillary smooth muscle sphincters have a sparse, or no, innervation suggesting a local rather than a neural control.

The blood flow to most tissues is proportional to their metabolic needs rather than to their size (Table 14–8). For most tissues, the ratio of percent of cardiac output to percent of VO_2 is close to 1 (or about 20 ml blood per ml O_2). The close matching of blood flow to metabolic demand suggests a local regulation of the local microcirculation. The kidneys, however, have a much higher ratio of blood flow to metabolism because they are excretory organs. The carotid body (a peripheral chemoreceptor) also has a very high blood flow relative to its metabolism, again reflecting its highly specialized function. Endocrine glands and O_2 chemoreceptors also have a very high blood flow rate relative to their size and metabolism.

A rate of blood flow that is inadequate to supply the metabolic demands of a tissue would result in the depletion of nutrients (e.g., O_2, glucose, etc.) and the accumulation of metabolites (CO_2, lactate,

adenosine, etc.). The vasodilator theory for control of local blood flow suggests that it is the accumulation of metabolites (especially adenosine) that causes local vasodilation of the microcirculation. The nutrient demand theory suggests that it is the depletion of nutrients (especially O_2) that acts as the vasodilator stimulus. These mechanisms for local control of blood flow function independently of the central nervous system and do not involve the adjustment of blood flow through the larger vessels.

The sympathetic and parasympathetic nervous systems can regulate cardiac output and peripheral blood flow through central nervous system control. A **vasomotor center**, which is located in the medulla oblongata and pons, controls the activity of sympathetic vasoconstrictor fibers that innervate virtually all blood vessels. The vasomotor center has a vasoconstrictor area, whose axons secrete norepinephrine and excite the sympathetic vasoconstrictor neurons, and a vasodilator area, whose neurons also secrete norepinephrine but inhibit the vasoconstrictor area. Normally, a net vasoconstrictor effect maintains a partial constriction of the peripheral blood vessels; this is the **vasomotor tone**. The vasomotor center also regulates the heart rate via sympathetic innervation (increases heart rate) and vagal innervation (decreases heart rate). Skeletal muscle is also innervated by sympathetic vasodilator fibers; these axons can have norepinephrine or acetylcholine as their neurotransmitter. The neural control of peripheral vasoconstriction/vasodilation is important in arterial pressure reflexes (e.g., baroreceptor reflex), blood volume regulation (e.g., control of AVP secretion, renal blood flow, and urine flow), and also thermoregulation (e.g., peripheral vasoconstriction or vasodilation).

Endocrine control of blood flow is due primarily to secretions of the adrenal medulla (norepinephrine, epinephrine) and angiotensin, which is one of the most potent vasoconstrictor agents known. Arginine vasopressin has vasoconstrictor effects at very high concentrations, such as during severe hemorrhage.

A dramatic circulatory adjustment of air-breathing vertebrates (seals, ducks, turtles, etc.) is the **diving response**. Laboratory studies indicate that the commencement of diving causes a reflex decrease in heart rate. Fish exhibit a similar diving response when they are removed from water; however, the air-breathing mudskipper *Periophthalmus* has a typical terrestrial diving response, i.e., its heart rate declines with submersion. An important component of the diving response is a dramatic reflex decrease in heart rate called diving bradycardia (Figure

TABLE 14–8

The blood flow (as a percentage of total cardiac output, CO; 5000 ml min^{-1}) expressed in proportion to mass (as a percentage of total mass; 70 kg) and metabolic rate (as a percentage of total metabolism, VO_2; 230 ml O_2 min^{-1}) for various tissues of a human. *(Modified from Guyton 1986; Ross 1982.)*

	% CO	% Mass	% VO_2	% CO_2/ % Mass	% CO/ %VO_2
Brain	15	2.0	18	7.5	0.83
Heart	4	0.5	10	8.0	0.40
Skin	6	6.6	8	1.0	0.75
Skeletal muscle	16	40.0	20	0.4	0.80
Kidneys	20	0.5	8	40.0	2.50
Viscera	25	6.0	25	4.2	1.00
Thyroid	1	0.045		22	
Adrenals	0.5	0.012		42	
Rest	14	45	11	0.31	1.27

14–44A). Cardiac output declines markedly with the bradycardia, but this is not accompanied by a similarly marked decline in blood pressure because there is also a pronounced peripheral vasoconstriction during diving, including the vasoconstriction of some of the major arteries (e.g., renal artery). Blood flow to the essential organs (e.g., brain, eyes) is virtually unaltered during diving but blood flow to other organs (e.g., viscera and heart) declines markedly (Figure 14–44B). Most muscle receives little blood flow and lactate accumulates in the muscle during the dive (and does not reach the circulation). Blood O_2 levels decline during the dive and CO_2 levels increase. At the end of a dive,

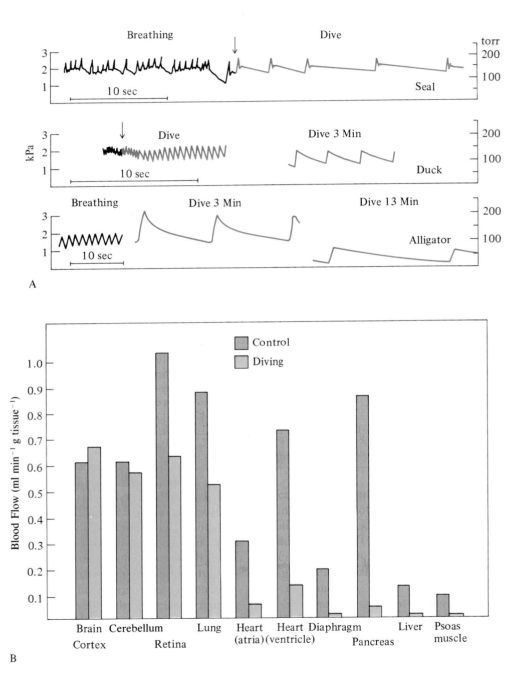

FIGURE 14–44 **(A)** Blood pressure before and during diving for a seal, a duck, and an alligator. **(B)** Selective redistribution of blood flow during diving in a seal.   *(From Scholander 1964; Zapol et al. 1979.)*

heart rate returns to above normal, the peripheral vasoconstriction is relaxed, and lactate is flushed into the circulation from the muscles. There is a period of elevated O_2 consumption and CO_2 release that roughly corresponds to the magnitude of the O_2 and CO_2 "debt" that occurred during the dive.

Unrestrained, free-diving animals often show a less extreme diving response; the bradycardia is less pronounced and there is no post-dive hyperlactemia (Guppy et al. 1986). The more pronounced diving response observed in the laboratory situation, often with restrained or traumatized animals, probably demonstrates the maximal diving response capability when stressed. This does have physiological relevance, however, since it demonstrates the capacity of the animals when stressed in nature, e.g., when pursued by predators, trapped under ice, during strenuous pursuit of prey, etc. More routine dives would require only a partial diving response.

Summary

Most animals have a circulatory system that distributes gases, nutrients, waste products, and other materials throughout the body. The circulatory space is either a blood-filled vascular space or the coelomic fluid space. Generally, one or more hearts propel the blood through the circulatory system.

The flow of blood is determined by the pressure gradient, the blood viscosity, and the dimensions of the vessels (radius, length). Flow is laminar at low Reynolds numbers but there is a transition to turbulent flow at high Reynolds numbers. Blood flow generally is laminar in the circulatory system of animals but is turbulent in the larger arteries of large animals.

The energy of a moving fluid is determined by the kinetic energy, gravitational energy, and pressure energy. Generally, the pressure and gravitational energies of blood are much greater than its kinetic energy. Gravitational energy is insignificant for aquatic animals but can have marked effects on vascular pressures, especially for large, terrestrial animals. Every meter of hydrostatic head increases the hydrostatic pressure by 10 kPa.

Most invertebrates have some mechanism to circulate body fluids but the design and role of their circulatory system vary markedly. Some coelenterates have a poorly developed "vascular" system that circulates seawater through their gastrovascular cavity. Pseudocoelomate animals generally lack a circulatory system but those with a large pseudocoelomic space have internal circulation of the coelomic fluid; nemertean worms have a closed circulatory system. Annelids have a well-developed circulatory system that is either closed (polychetes and oligochetes) or open (leeches). The circulatory system of mollusks varies from a generally open system of heart, major arteries, tissue sinuses, and vascular channels to the gills and then to the heart, to the closed system of cephalopods, which has both systemic and branchial hearts. Arthropods also have diverse circulatory arrangements. All have an open circulation, with varying development of an arterial system and tissue sinuses for percolation of hemolymph to the body tissues and gills (if present) and back to the heart. Accessory hearts may facilitate blood flow through the appendages, brain, and wings. Echinoderms have a poorly defined and ineffective circulatory capacity, although four different fluid systems may contribute to internal circulation.

Cephalochordates lack a systemic (central) heart; rather, branchial hearts at the base of each gill arch pump blood through the gills. Oxygenated blood is distributed to the tissues via the aorta and tissue lacunae. They also have a tissue sinus system rather than a capillary system.

The vertebrate circulatory system is derived from a cephalochordate-like system. It is closed (except in hagfish); it has a systemic heart with an atrium (or atria) and a ventricle (or ventricles); it may have accessory hearts (e.g., hagfish); and it has well-organized arterial, capillary, and venous systems. The high-pressure arterial system dampens the ventricular pressure fluctuations and regulates the flow of blood to the peripheral tissue vascular beds. The capillaries are the site of gas, nutrient, and waste exchange. The venous system collects blood from the capillary beds and delivers it to the gills/lungs and then to the heart.

There is a clear evolutionary development of the vertebrate circulatory system and systemic heart from the primitive cephalochordate-like circulatory system. The systemic heart initially has only a single circuit of deoxygenated blood from the tissues; blood pools in the atrium, which acts as a low-pressure priming pump for the muscular ventricle. Blood is pumped from the ventricle to the gill arches for oxygenation and then distributed to the tissues.

In lungfish, the oxygenated blood from the lungs passes to an incompletely divided left atrium, is kept functionally separated from the deoxygenated blood in the partially divided ventricle, and passes through the short and long spiral folds to the dorsal aorta. Deoxygenated blood returns to the right atrium and is preferentially directed to the gills and lung.

The amphibian heart has completely divided left and right atria and considerable functional separation of oxygenated (from the lung) and deoxygenated blood (from the tissues, but also oxygenated blood from the skin) in the ventricle.

Reptiles (except crocodiles) also have separate left and right atria and have an effective functional separation of oxygenated blood and deoxygenated blood in their complex three-chambered ventricle. Crocodilian hearts have a completely divided right and left atrium and right and left ventricle; an extracardiac shunt via the foramen of Panizza allows both systemic arches to carry oxygenated blood to the body. The reptilian circulatory system allows shunting of blood between the right and left sides of the heart; a left-to-right shunt allows blood to bypass the lungs (e.g., during diving apnea) and a right-to-left shunt allows preferential delivery of blood to the lungs (e.g., during pulmonary ventilation).

The mammalian and avian circulatory systems have a complete anatomical separation of the oxygenated blood (left atrium to left ventricle to body) and deoxygenated blood (right atrium to right ventricle to lungs). This complete separation is circumvented during embryonic/fetal development by an interatrial shunt (the foramen ovale) and an extracardiac shunt (the ductus arteriosus).

The cardiovascular system must be regulated so that blood flow to tissues matches their metabolic demand. The principal means of cardiovascular regulation are control of cardiac output by regulation of heart rate and force of contraction, control of blood pressure, and selective arterial distribution of blood to tissue beds. The control of heart rate can be intrinsic to the heart itself, can be entirely neural in origin, or can be a combination of intrinsic and neural control. There is also endocrine control of the heart. Blood pressure is generally determined by a combination of the cardiac output, the resistance and compliance of the peripheral circulatory vessels, and the blood volume. Many short-term, intermediate-term, and long-term factors contribute to blood pressure regulation. The distribution of blood flow to the peripheral vascular beds can be determined by either local control by the metabolic activity of the tissue bed or by neural control of the arterial supply to the tissue bed. Local control could be mediated, for example, by the depletion of nutrients such as O_2 or the accumulation of wastes such as CO_2. The dramatic redistribution of blood flow in birds and mammals during diving involves a pronounced vasoconstrictor control of the central nervous system on the arterial supply to the tissue beds.

Supplement 14–1

Laminar, Turbulent, and Pulsatile Flow

· ·

Blood flow in circulatory systems can be either steady, i.e., constant velocity, or pulsatile, i.e., stop-and-go flow. Blood flow can also be either laminar (streamlined) or turbulent (chaotic eddy flow), depending on the blood velocity and vessel dimensions. The laminar flow of a fluid can be envisioned as a sliding movement of adjacent layers of fluid in the direction of the fluid flow with no transverse movement of fluid between adjacent layers; it occurs for steady flow at low Reynolds numbers. The relationship between laminar blood flow (V_b) and pressure gradient (ΔP) is approximated by the Poiseuille-Hagen flow formula

$$V_b = \frac{\Delta P \pi r^4}{8 \eta l}$$

where V_b is the blood flow (m³ min⁻¹), ΔP is the blood pressure difference (N m⁻²), r is the vessel radius (m), η is the blood viscosity (Pa s), and l is the vessel length (m). In the CGS rather than SI system, the units are V_b,

ml sec⁻¹; P, dynes cm⁻²; r, cm; η, poise (dyne sec cm⁻²); l, cm. Blood flow is directly proportional to the pressure difference between the ends of the vessel, as we might expect. It is strongly dependent on vessel radius, since $V_b \propto r^4$, i.e., flow increases by 16 × if r is increased by 2 ×. Blood flow is inversely proportional to the fluid viscosity and vessel length, again as we might expect.

The velocity of blood varies for laminar flow in a parabolic fashion from 0 at the vessel wall to a maximum in the center of the vessel

$$v_x = \frac{\Delta P(r^2 - r_x^2)}{4\eta l} = \frac{2V_b(r^2 - r_x^2)}{\pi r^4}$$

where r is the vessel radius and r_x is the radius at the point in question ($0 \leq r_x \leq r$). The maximal velocity (v_{max}) is calculated as follows.

$$v_{max} = \frac{\Delta P r^2}{4\eta l}$$

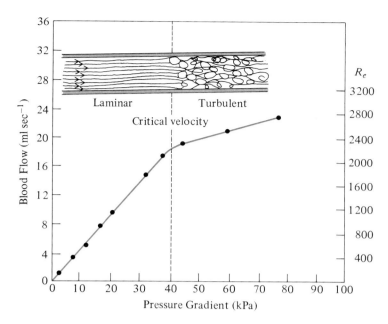

There is a linear relationship between blood flow rate and pressure gradient for bovine blood at 15° C for laminar flow; the blood flow becomes turbulent at Reynolds numbers higher than the critical value of about 2200, and the blood flow rate is less than would occur if flow were laminar. $R_e = v.D/\bar{v}$, where v is velocity (m sec^{-1}), D is vessel diameter (0.00252 m), and $\bar{v}$ is the kinematic velocity (5.7 10^{-6} m^2 s^{-1}). *(Data from Coulter and Pappenheimer 1949.)*

The average velocity (v_{avg}) is calculated as follows.

$$v_{avg} = \frac{\Delta P r^2}{2\eta l}$$

It is important to appreciate that there are a number of assumptions that must be met for the Poiseuille-Hagen flow formula to be applicable: (1) the vessel is a rigid, cylindrical tube with $l \gg r$; (2) the fluid is "ideal" (Newtonian), i.e., viscosity is not dependent on the shear rate; (3) the flow is steady and laminar (not pulsatile and/or turbulent); (4) the fluid velocity is 0 at the vessel wall. In general, blood vessels have $l \gg r$, and the natural taper and wall elasticity tend to stabilize laminar flow; this is especially true for arterial vessels (e.g., aorta) whose walls are very elastic (for a windkessel effect). Blood is not an "ideal" fluid, and its viscosity is shear dependent (see Chapter 12). Flow is steady and laminar in most blood vessels but not in some. Condition (4) is assumed to always be correct, i.e., there is zero blood

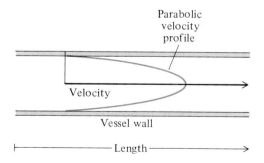

Parabolic velocity profile

Velocity

Vessel wall

Length

The velocity profile for laminar blood flow is parabolic with a maximum in the center of the vessel and zero velocity at the walls.

flow at the vessel wall. Despite the biological difficulties in accepting all of the assumptions of the Poiseuille-Hagen flow formula, it is generally an excellent approximation for the description of blood flow through the circulatory system.

At high Reynolds numbers, there is a transition from laminar flow to turbulent flow there is a movement of fluid transverse to the direction of flow due to the presence of swirling, fluid eddies; the fluid flows in a chaotic rather than an orderly manner. The Poiseuille-Hagen flow formula is not valid for turbulent flow, and $V_{b,\text{turbulent}} < V_{b,\text{laminar}}$ for the same values of ΔP, r, η, and l. The velocity profile for turbulent flow is not parabolic but is blunt, with most of the profile near v_{max}.

For pulsatile flow, the instantaneous blood flow velocity varies in a complex fashion over time during each heart beat cycle. Blood flow in the aorta of large mammals, for example, is pulsatile and has a time-dependent blunt velocity profile. The Poiseuille-Hagen flow equation is also not valid for pulsatile flow but it is generally assumed that the mean pulsatile blood flow ($V_{b,\text{pulsatile}}$) is related to the mean pressure difference (ΔP) and mean values for other relevant variables in the Poiseuille-Hagen flow formula. The oscillations in flow with each cardiac cycle can be modeled in various ways; e.g., as a Fourier series of harmonics. The sum of blood flows can be calculated for an infinite number of harmonics ($V_{b,n}$; $n = 1$ to infinity) from a Fourier series

$$V_{b,n} = \frac{\Delta P_n \pi r^4}{8\eta l} \Phi_n \sin(n\omega t + \phi_n + \beta_n)$$

where $V_{b,n}$ is the mean blood flow for various harmonics ($n = 1, 2, 3, 4$, etc.) of the pulsatile flow, ω is the frequency (i.e., heart rate), Φ_n is the phase of the harmonic pressure gradient (ΔP_n), and ϕ_n and β_n are functions of r, ω, η, and the biomechanical properties of the vessel wall.

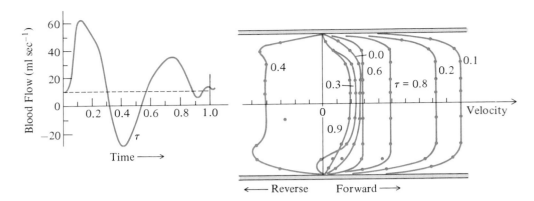

Blood flow in the descending aorta of the pig is pulsatile and turbulent; the flow velocity alters with time in a complex fashion through a single cardiac cycle. Flow is maximal at about 0.1 sec with a second smaller peak at about 0.6 sec; flow reverses at 0.4 sec. The velocity profile for flow is blunt with zero wall velocity.   *(From Ling et al. 1968.)*

Recommended Reading

Barnes, R. D. 1987. *Invertebrate zoology*. Philadelphia: Saunders College Publishing.

Binyon, J. 1972. *Physiology of echinoderms*. Oxford: Pergamon Press.

Chapman, C. B., D. Jensen, and K. Wildenthal. 1963. On circulatory control mechanisms in the Pacific hagfish. *Circ. Res.* 12:427–40.

Edwards, C. A., and J. R. Lofty. 1977. *Biology of earthworms*. London: Chapman & Hall.

Fawcett, D. W. 1986. *A textbook of histology*. Philadelphia: Saunders.

Hildebrandt, J. P. 1988. Circulation in the leech, *Hirudo medicinalis L. J. exp. Biol.* 134:235–46.

Hill, R. B., and J. H. Welsch. 1966. Heart, circulation, and blood cells. In *Physiology of the Mollusca*, edited by K. M. Wilbur and C. M. Yonge, vol. 2, 125–74. New York: Academic Press.

Hyman, L. H. 1951. *The invertebrates: Playtyhelminthes and Rhyncocoela*, vol. 2. New York: McGraw-Hill.

Hyman, L. H. 1967. The invertebrates. In *Mollusca I*. vol. 6. New York: McGraw-Hill.

Johansen, K., and W. W. Burggren. 1985. *Cardiovascular shunts*. Copenhagen: Munksgaard.

Jones, H. D. 1983. The circulatory systems of gastropods and bivalves. In *The Mollusca*, vol. 5 of *Physiology*, part 2, edited by A. S. M. Saleuddin and K. M. Wilbur, 189–238. New York: Academic Press.

Jones, J. C. 1964. The circulatory system of insects. In *The physiology of Insecta*, edited by M. Rockstein, vol. 3, 1–107. New York: Academic Press.

Jones, J. C. 1977. *The circulatory system of insects*. Springfield: C.C. Thomas Publishing Company.

Lawrence, J. 1987. *A functional biology of echinoderms*. London: Croon Helm.

Martin, A. W., and K. Johansen. 1965. Adaptations of the circulation in invertebrate animals. In *Handbook of physiology*, sec. 2 of *Circulation*, vol. 3, edited by W. F. Hamilton and P. Dow, 2545–81. Washington: American Physiological Society.

Maynard, D. M. 1960. Circulation and heart function. In *The physiology of Crustacea*, edited by T. H. Waterman, vol. 1, 161–226. New York: Academic Press.

McLaughlin, P. A. 1983. Internal anatomy. In *The biology of Crustacea*, vol. 5 of *Internal anatomy and physiological regulation*, edited by L. H. Mantel, 1–52. New York: Academic Press.

Meglitsch, P. A. 1972. *Invertebrate zoology*. New York: Oxford Press.

Miller, T. A. 1985. Structure and physiology of the circulatory system. In *Comprehensive insect physiology, biochemistry, and pharmacology*, vol. 3 of *Integument, respiration, and circulation*, edited by G. A. Kerkut and L. I. Gilbert, 289–353. Oxford: Pergamon Press.

Milnor, W. R. 1982. *Hemodynamics*. Baltimore: Williams & Wilkins.

Mountcastle, V. B. 1974. *Medical physiology*, vol. 2. St Louis: CV Mosby Company.

Ramsay, J. A. 1964. *Physiological approach to the lower animals*. Cambridge: Cambridge University Press.

Simons, J. R. 1960. The blood vascular system. In *Biology and comparative physiology of birds*, edited by A. J. Marshall, 345–62. New York: Academic Press.

Wells, M. J. 1983. Circulation in cephalopods. In *The Mollusca*, vol. 5 of *Physiology*, part 2, edited by A. S. M. Saleuddin and K. M. Wilbur, 239–90. New York: Academic Press.

White, F. N. 1968. Functional anatomy of the heart of reptiles. *Amer. Zool.* 8:211–19.

Chapter 15

Blood

· ·

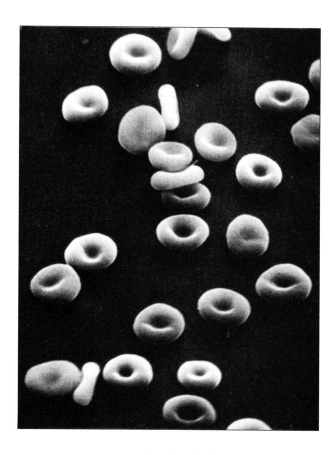

Scanning electron micrograph shows the biconcave, donut form of human red blood corpuscles. *(Photo/ Researchers, Inc.)*

Blood is a general term used to describe a fluid that is circulated within the body to transport gases and other nutrient or waste products. In vertebrates, the circulating fluid, or blood, is kept within a closed circulatory system, whereas **hemolymph** is the circulating fluid of animals with an open circulatory system, e.g., insects. The coelomic fluid of some invertebrates also has a circulatory role either in concert with, or instead of, the blood/hemolymph.

Blood provides a controlled internal environment that is the milieu in which cells function. It is often the source of fluid and products to be excreted. It also is a storage reservoir for a variety of metabolites. Blood often contains a respiratory pigment, present within blood cells or in solution. Respiratory pigments both increase the oxygen-carrying capacity of the blood and confer a special relationship between the blood O_2 content and pO_2. Blood has such a high carrying capacity for carbon dioxide that there is no specific carrier for CO_2 analogous to the O_2 transport pigments.

Blood transports a multitude of nutrients other than O_2 and waste products other than CO_2; these are present in sufficiently low concentrations that specific transport carriers are generally not required. The blood circulates hormones (and hormone-binding proteins). Blood proteins have a variety of functions, including blood clotting and immune responses; they often have a significant colloid osmotic effect and contribute to the dynamic distribution of water between blood, lymph, and intracellular fluids. Blood cells also have nonrespiratory functions, such as phagocytosis of dead cells and foreign materials, cell-mediated immunity, and control of blood clotting. Blood is an important pH buffer system. Constituents of blood can confer antifreeze properties (cryoprotection). The blood may be important as a heat exchange fluid and contribute to thermoregulation.

Blood often has noncirculatory functions. In insects, hemolymph pressure assists molting of the old cuticle and wing inflation. In spiders, the hydrostatic pressure of the hemolymph provides a hydraulic mechanism for limb extension. The coelomic fluids of annelids can be a hydrostatic skeleton. Pressure in the venous sinus of the eye helps lizards clean their eye surfaces. A few animals (some insects; the lizard *Phrynosoma*) exude blood when aggravated by predators. This reflexive bleeding presumably deters or distracts a predator, or the blood may contain toxic or distasteful materials.

The **blood volume** is the volume occupied by the blood; it is usually expressed as a percentage of the body mass (shell-free mass for those animals with heavy calcified shells, e.g., mollusks). Blood volume varies dramatically, depending on the nature of the circulatory system (open, closed, coelomic) and the extent of reliance on the circulatory system for gas and/or nutrient exchange. It is about 6% in cephalopod mollusks, 3–16% in vertebrates, and 30% or more in arthropods. For animals with an open circulatory system, the blood volume is essentially the same as the extracellular fluid volume but blood volume is only a fraction of the extracellular volume for animals with a closed circulation. Blood volume includes the volume of hemocytes/blood cells, i.e., it is both an extracellular and an intracellular volume. The percentage of the blood that consists of cells is the **hematocrit**; it is usually determined by centrifugation of blood to sediment the cells. **Plasma volume** is the volume of extracellular fluid in the circulatory system i.e., blood volume − blood cell volume, or blood volume × (1 − hematocrit/100).

Composition

Blood is primarily water with low concentrations of ions and various organic solutes (nutrients, waste products, proteins) and blood cells. Blood generally has a closely regulated solute and osmotic composition that is very different from the external environment of aquatic animals and from the intracellular fluid. The protein content of blood is important for osmotic reasons as well as for the functional roles of different blood proteins. Nonelectrolyte solutes are distributed at equal concentrations across the capillary interface (e.g., urea, glucose). Some animals use the coelomic fluid as a supplementary or sole circulatory system; coelomic fluid can differ in composition from that of the circulatory fluids and interstitial fluids, particularly with respect to proteins and cells.

Proteins

Many animals, such as jellyfish, echinoderms, and sipunculids, have little protein in their blood or hemolymph (typically 0.2 to 2.0 g L^{-1}). Decapod crustaceans have a higher concentration (10 to 90 g L^{-1}), most of which is due to the respiratory pigment hemocyanin, although some other blood proteins (e.g., for clotting) are present. Vertebrates (30 to 80 g L^{-1}) and cephalopod mollusks (up to 110 g L^{-1}) have high plasma protein concentrations.

A high blood protein concentration has important osmotic effects. The high molecular weight of proteins restricts their mobility across the semipermeable capillaries and coelomic wall. The consequent osmotic pressure due to proteins is called the **colloid osmotic pressure** (COP; unit is kPa). The effect of protein on the distribution of other charged solutes (the Donnan effect) is also important. The magnitude of the COP is proportional to the osmotic concentration, hence is inversely proportional to the protein's molecular weight. For example, the high molecular weights of decapod hemocyanins minimize their colloid osmotic pressure despite their high concentration; there is no COP difference between blood and interstitial fluid because the circulatory system of decapods is open. Similarly, the high hemolymph protein concentration of the gastropod *Planorbis* (50 g L^{-1}) has only a small COP (0.03 kPa) because of the high molecular weight. *Chironomus* midge larvae have up to 80 g L^{-1} of hemoglobin in solution; its low molecular weight (16 10^3 Da) results in a very high COP, about 13.6 kPa. The chironomid's circulatory system is open and so there is no COP imbalance between blood and interstitial fluid, but the high-hemolymph COP would significantly affect the hemolymph-intracellular distribution of water. Cephalopods and vertebrates have a high concentration of plasma proteins and a closed circulatory system so the COP potentially affects the distribution of water between the blood and interstitial fluids. The plasma proteins of vertebrates have a fairly low molecular weight and exert a significant colloid osmotic pressure (1 to 4 kPa). The high molecular weight of cephalopod hemocyanins (about 2.8 10^6 Da) minimizes their colloid osmotic effect to about 0.4 kPa.

Cells

The blood of most animals contain circulating cells, called blood cells or **hemocytes**. Some blood cells contain hemoglobin (**erythrocytes**, or red blood cells); these are often present in high numbers to facilitate oxygen transport. Some blood cells do not contain respiratory pigments but have other functions, such as phagocytosis and blood clotting. The different types of blood cells vary dramatically in concentration in different animals.

Annelid blood contains hemocytes, which are phagocytic or store glycogen. Leech blood has about 5 10^4 cells per mm^3. A few annelids, e.g., *Magelona*, have erythrocytes but most have their hemoglobin free in solution. The coelomic fluid commonly contains erythrocytes as well as a variety of other coelomocytes (amebocytes, eleocytes, lamprocytes, linocytes). These coelomocytes have various roles, including phagocytosis, glycogen storage, encapsulation, wound plugging, immune responses, and excretion. Sipunculids and echiuroids have a variety of coelomic cells with similar functions.

Coelomocytes of lophophore animals include hemerythrin-containing hemocytes and phagocytic leukocytes.

Molluskan blood has two general types of hemocytes—amebocytes and granulocytes—that have most of the above-mentioned functions as well as nacrezation (pearl formation) in some bivalves. There are from 10^5 to 7 10^6 hemocytes per mm^3 hemolymph. The blood is unusual in that it does not have a well-developed clotting system. The blood of cephalopods contains about 1 to 2% of a type of hemocyte that plugs damaged vessels until fibrocytes repair the wound. Their blood lacks specific agglutinating proteins but does contain a substance (hemocyanin?) that coats foreign matter and promotes phagocytosis.

Onycophorans and arthropods have many types of hemocytes. Onycophorans and centipedes have prohemocytes, which are progenitor cells for hemocytes, granulocytes, plasmatocytes, and spherule cells. These hemocytes are involved with phagocytosis, encapsulation, and clotting. Millipedes also have lipid storage hemocytes, oenocytes and adipohemocytes. Crustacean blood contains amoeboid hemocytes (10^3 to 5 10^4 mm^{-3}) of at least two types: a small agranular cell and a larger granular cell (although these may be different stages of the same cell type). Semigranular cells may be a third type of hemocyte. The blood cells phagocytose foreign particles, agglutinate to form blood clots, may release a thrombin-like coagulant, transport lipids, store nutrients, and form connective tissue cells. Insect hemolymph contains large numbers (10^2 to 10^5 mm^{-3}) of various hemocyte types (Figure 15–1A). Prohemocytes form the other types: plasmatocytes, granulocytes, podocytes, coagulocytes, adipohemocytes, spherule cells, oenocytes, and vermiform cells. They also have most of the above-mentioned functions. Chelicerates also have a variety of blood cells, including cyanocytes that synthesize and release hemocyanin; there are about 10^3 to 4 10^3 hemocytes mm^{-3}.

Echinoderms have a variety of hemocytes and coelomocytes that have some resemblances in structure and function to chordate blood cells. The hemocytes may contain hemoglobin and function in oxygen transport, but the very low metabolic rates of echinoderms make a significant O_2 transport role unlikely. The coelomocytes (progenitor cells, amebocytes, crystal cells, pigment cells) are involved in phagocytosis, cell clumping, encapsulation, wound repair, clotting, and perhaps digestion and storage of nutrients.

Cephalochordate blood lacks cells but urochordate blood has about 2 10^3 to 8 10^3 cells mm^{-3}. Tunicate blood cells can be divided into hemoblasts,

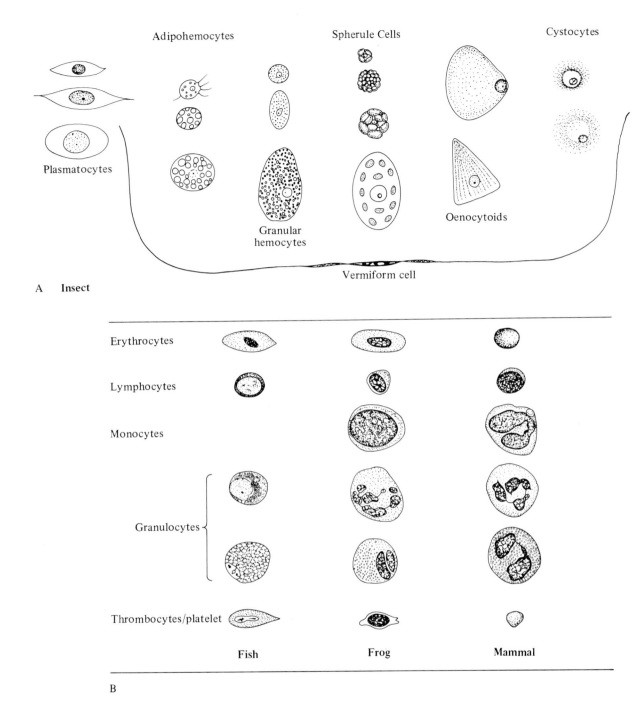

Adipohemocytes

Spherule Cells

Cystocytes

Plasmatocytes

Granular
hemocytes

Oenocytoids

Vermiform cell

A   Insect

Erythrocytes

Lymphocytes

Monocytes

Granulocytes

Thrombocytes/platelet

Fish   **Frog**   **Mammal**

B

FIGURE 15–1   **(A)** Blood cells of insects. **(B)** Blood cells of vertebrates: fish, *Labrax*; frog, *Rana*; mammal, *Homo.*   *(From Jones 1977; Romer and Parsons 1986.)*

lymphocytes, leukocytes, vacuolated cells (including morula cells), pigment cells, and nephrocytes. Hemoblasts are progenitor cells for other blood cells as well as new buds and gonadal cells.

Lymphocytes (which are universally present and very important blood cells in vertebrates) are involved with antigen recognition, inflammation, phagocytosis, encapsulation, and graft rejection.

Nephrocytes accumulate wastes as they circulate and then tend to become fixed in specific regions of the body (mantle, gonads, digestive loop). The morula cells of ascidiids and perophorids accumulate high concentrations of vanadium (up to 3700 ppm; cf. seawater concentration of 0.3 to 3 ppm) and the cells of pyurids accumulate iron. The vanadium and iron act as reducing agents for the deposition of tunicin microfibrils in the tunic when the morula cells degenerate. The pigment cells may provide protection against ultraviolet light or may act as specific wavelengths filters.

All vertebrates have blood cells (Figure 15–1B). These can be divided into a variety of leukocytes (white blood cells; WBCs) and erythrocytes (red blood cells; RBCs). White blood cells are present in lower numbers than are RBCs, generally being 1 to 2% of the blood by volume. WBCs are divided into agranular, lymphoid leukocytes (lymphocytes and monocytes) and granular, polymorphonuclear leukocytes (eosinophils, basophils, and neutrophils). Thrombocytes, which are associated with blood clotting, are also present. In mammals, their role is fulfilled by platelets, which are small, enucleate fragments of megakaryocyte cells.

The size, shape, and number of erythrocytes differ dramatically in different vertebrates (Figure 15–2). The RBCs of most vertebrates are generally nucleated, but mammalian RBCs are enucleate. Some fish and amphibians also have enucleate RBCs. The salamander *Amphiuma* has the largest RBC known: 65 × 30 × 13.5 μ. Avian RBCs are oval shaped, nucleated, and larger than mammalian RBCs (generally 10 to 15 × 5 to 8 μ). The ostrich has the largest avian RBCs: 16 × 10 μ. Mammalian red blood corpuscles (they are enucleate, hence are not cells) are generally biconcave disks. The camel and llama have elliptical, rather than oval, RBCs, and mouse deer have small, spherical RBCs.

Lower vertebrates tend to have fewer, larger RBCs than higher vertebrates; they have lower hematocrits (percent packed cell volume) and hemoglobin content, lower RBC counts, larger mean cell volume and mean cell hemoglobin content, but similar mean cell hemoglobin concentration in comparison with birds and mammals (Table 15–1).

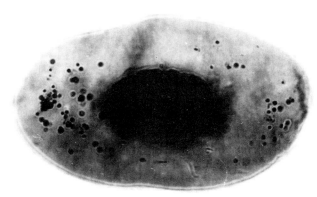

Salamander

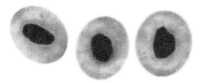

Snake

Ostrich

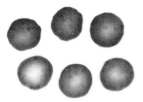

Red Kangaroo

Camel ⊢———⊣ 10μ

FIGURE 15–2 Comparison of the form of red blood cells from a variety of vertebrates; a salamander *Amphiuma* (nucleated cell); a snake (nucleated); the ostrich (nucleated); a red kangaroo (enucleate biconcave disk); and a camel (enucleate ellipsoid). *(Photograph courtesy of P.C. Withers, T.S. Stewart, and S. Hopwood.)*

TABLE 15–1

Hematological parameters for some representative and unusual vertebrate bloods; species are arranged in order of descending mean cell volume. Parameters are mean cell volume (MCV = 10 Hct/RBC; 10^{-15} L), red blood cell count (RBC; 10^{12} cells L^{-1}), hematocrit (HCT; %), hemoglobin content (Hb; g L^{-1}), mean cell hemoglobin content (MCH = Hb/RBC; picograms), mean cell hemoglobin concentration (MCHC = 100 Hb/Hct; g L^{-1}) and blood oxygen capacity (BOC; ml O_2 L^{-1}; calculated as 1.34 × Hb).

Species	MCV fL	RBC 10^{12} L^{-1}	Hct %	Hb g L^{-1}	MCH pg	MCHC g L^{-1}	BOC ml O_2 L^{-1}
Amphiuma[1]	10800	0.027	29	76	2830	262	102
Bullfrog	845	0.296	25	70	232	285	94
Turtle	394	0.520	21	69	133	335	92
Camel[2]	319	11.0	35	158	14	494	211
African Elephant[3]	141	3.03	43	153	51	356	205
Turkey	136	2.72	37	123	45	336	165
Beluga[4]	134	3.34	46	193	57	427	259
Man	90	4.99	44	146	29	335	196
House Mouse	60	8.3	50	159	19	320	213
Pocket Mouse	45	11.8	54	193	17	367	259
Shrew[5]	31	11.5	36	162	15	477	217
Goat	18	16.1	29	104	7	356	139
Mouse Deer[6]	6	55.9	31	117	2	380	154

[1] Largest vertebrate erythrocyte.

[2] Oval erythrocytes.

[3] Largest terrestrial mammal.

[4] Diving mammal.

[5] Very small mammal.

[6] Small, spherical erythrocytes.

Respiratory Pigments

There are four different types of respiratory pigment: **hemoglobin, chlorocruorin, hemerythrin,** and **hemocyanin.** These differ markedly in physical properties, structure, O_2-binding site, O_2-binding capacity, molecular weight, and location but all reversibly bind O_2 and function in O_2 transport, storage, or transfer (Table 15–2).

Respiratory pigments are so named because of their often pronounced color, which is due to a very specific absorbance at particular wavelengths and often changes with the degree of oxygenation. Hemoglobin is generally bright red when oxygenated and red-blue when deoxygenated. It maximally absorbs at about 540 and 575 nm when oxygenated and 550 when deoxygenated (Figure 15–3). Chlorocruorin is green, often having a very vivid metallic color. It has a similar absorbance spectrum to hemoglobin due to its virtually identical structure. Hemerythrin is violet-pink when oxygenated and colorless when deoxygenated; it has absorbance maxima at 330 and 500 nm when oxygenated and about 400 when deoxygenated. Hemocyanin is blue when oxygenated and colorless when deoxygenated; it has an absorbance peak at about 575 nm when oxygenated.

Respiratory pigments vary in the structure of their protein and heme group (if present), and the type of the metal ion and its O_2-binding capacity (either 1 ion/O_2 or 2 ions/O_2). Oxygen molecules bind reversibly to the metal ion (either Fe^{2+} or Cu^{2+}) associated with the protein chain of the respiratory pigment (RP).

$$RP + O_2 \underset{\text{Low pO}_2}{\overset{\text{High pO}_2}{\rightleftharpoons}} RP\text{-}O_2 \qquad (15.1)$$

The extent of O_2 binding depends on the O_2 partial pressure. At high pO_2, the pigments bind O_2 maximally (are saturated) and at low pO_2 they bind less O_2 (are unsaturated).

Respiratory pigments are proteins that are specialized for the transport of O_2 which is one of the major roles of the circulatory system in many animals. Water and body fluids are a poor transport medium, especially for O_2, because of their low solubility coefficients. At 10° C, water and body fluids contain about 360 μmoles O_2 L^{-1} (0.7 vols

TABLE 15–2

General properties of respiratory pigments: color, composition, O_2-binding/metal ion site, O_2-binding capacity (ml O_2 g pigment^{-1}), molecular weight (Da), and location in erythrocytes or in solution.

Pigment	Color	Structure	O_2-binding site	O_2 capacity	MWT	Location
Hemoglobin	Red—blue	Protein + Heme + Fe^{2+}	1 O_2/Fe^{2+}	1.29	16 10^3 to 10^6	Erythrocytes or solution
Chlorocruorin	Green	Protein + Heme + Fe^{2+}	1 O_2/Fe^{2+}	0.76	3 10^6	Solution
Hemerythrin	Violet—colorless	Protein + Fe^{2+}	1 O_2/2 Fe^{2+}	1.67	16 10^3 to 122 10^3	Erythrocytes or solution
Hemocyanin	Blue—colorless	Protein + Cu^{2+}	1 O_2/2 Cu^{2+}	0.38	25 10^3 to 7 10^6	Solution

percent) and 16 μm CO_2 L^{-1} (0.04 vols percent) at their normal ambient partial pressures (O_2 and CO_2 contents of blood are often expressed as ml per 100 ml blood or vols percent). Even if all of the dissolved O_2 could be extracted from blood by the tissues (which it can't), a considerable flow would be

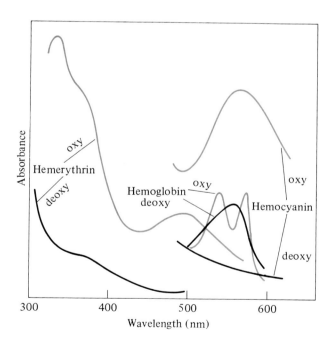

FIGURE 15–3 Absorbance spectra for oxygenated and deoxygenated states of the respiratory pigments, hemoglobin (of the giant earthworm *Megascolides*), hemerythrin (of the sipunculid worm *Phascolosoma*), and hemocyanin (of the cephalopod *Octopus*). *(Modified from Weber and Baldwin 1985; Klotz, Klotz, and Fiess 1957; Bonaventura and Bonaventura 1983.)*

required to supply even the modest O_2 demands of a resting ectothermic animal. Consequently, few animals rely on O_2 in solution for circulatory transport. Only a very few vertebrates can rely on dissolved O_2. Antarctic ice fish, which live at very low water temperatures and have modest metabolic rates, lack a circulating hemoglobin. Occasionally, some frogs (*Xenopus*) also lack circulating hemoglobin.

The O_2-binding capacity is lowest for hemocyanins (0.3 to 0.5 ml g^{-1}) and highest for hemerythrins (1.6–1.8 ml g^{-1}); hemoglobins (1.2 to 1.4) and chlorocruorins (0.6 to 0.9) are intermediate. Respiratory pigments can dramatically increase the O_2 capacity of blood depending on their O_2-binding capacity and concentration. The O_2 content of blood is generally 2 volumes percent or more, and values are >10 vols percent for blood of some invertebrates and endothermic vertebrates (Table 15–3) cf. 0.5 to 0.7 vols percent for plasma.

The four types of respiratory pigment are distributed in a complex fashion through the animal kingdom (Figure 15–4). Myoglobin, which resembles the subunit of hemoglobin, is found in fungi (yeast), plant/bacterial symbionts (root nodules), protozoans, and animals. It and hemoglobin are the only respiratory pigments of deuterostomes. Hemocyanin is found in many diverse protostomes. Hemerythrin is found in only a few protostomes. Annelida is the only phylum with representatives that have all types of respiratory pigment.

Hemoglobin

Hemoglobin and **myoglobin** are the most common respiratory pigments in animals. Hemoglobin is the only type of respiratory pigment in vertebrate blood

TABLE 15–3

Oxygen carrying capacity (vols percent) of blood compared to water (at pO_2 = 21.8 kPa for H_2O and saturation for bloods). Values are ml O_2 per 100 ml blood.

	ml O_2 100 ml^{-1}
Water (20° C)	0.65
Nematodes	
Hemoglobin	1.2–3
Annelids	
Hemoglobin	0.07–20
Chlorocruorin	7.2–10.2
Hemerythrin	3.6–6.2
Echiura	
Hemoglobin	4.5
Sipunculids	
Hemerythrin	2.1–3.5
Priapulids	
Hemerythrin	1.4
Mollusks	
Hemoglobin	??
Hemocyanin (cephalopods)	3.1–3.5
Hemocyanin (others)	0.9–3.3
Arthropods	
Chironomus hemoglobin	5.4–11.6
Crustacean hemoglobin	2.3–3.2
Crustacean hemocyanin	0.5–3.7
Chordates	
Cyclostomes	1–1.2
Chondrichthyes	4.4–4.5
Osteichthyes	4.9–19.7
Amphibians	6.3–10.4
Reptiles	6.6–12.5
Birds	10–22
Mammals	14–32

and it is found in many invertebrates. Myoglobin is a form of hemoglobin located within muscle cells.

Circulating hemoglobin is commonly present in the blood of annelids with closed circulatory systems; it is dissolved in plasma for many species and located in hemocytes for a few. Annelids without a functional circulatory system (e.g., *Glycera*) often have coelomocytes containing hemoglobin. Hemoglobin is absent in some groups of annelids (e.g., Aphroditidae, Syllidae, Chaetopteridae). Some echinoderms (echiuroids, holothuroids) have corpuscular hemoglobin of doubtful circulatory significance. Phoronids have hemoglobin in erythrocytes.

Pogonophorans have hemoglobin dissolved in their hemolymph. Among crustaceans, only the entomostracans (e.g., *Daphnia*, *Artemia*) have hemoglobin, which is free in their plasma. Only a few insects have circulating hemoglobin, which is free in solution. A few mollusks have hemoglobin in corpuscles or in solution.

Tissue-located myoglobins and hemoglobins are widely distributed throughout the animal kingdom. Myoglobin is present in vertebrate-striated muscle, gastropod radular muscle, and polychete body wall muscle. Hemoglobins are also found in nerve cells (nemertines, nematodes, mollusks, annelids, crustaceans, insects), tracheal cells (insects), and eggs and ovaries (crustaceans, insects).

Hemoglobins consist of a variable number of subunit molecules. Each subunit contains a **heme** group and a polypeptide. The heme consists of a protoporphyrin molecule containing four pyrrole groups and a ferrous ion (Fe^{2+}) at the center of the porphyrin (Figure 15–5). The Fe^{2+} can reversibly bind one O_2 molecule. The protein is organized as segments of right-handed α-helix joined by short connecting pieces of reverse turns and nonordered segments. For example, the α-chain of human hemoglobin has 7 helical segments and the β-chain has 8 segments; myoglobin has 8 segments. There are often differences in the protein chains of polymeric vertebrate hemoglobins. For example, adult mammalian hemoglobin has two α- and two β-chains, i.e., $\alpha_2\beta_2$ tetramers; the α-chain has 141 amino acid residues and the β-chain has 146. Human fetal hemoglobin has γ-chains instead of β-chains, i.e., $\alpha_2\gamma_2$. Myoglobin is a monomeric "hemoglobin" with approximately 150 amino acid residues.

The monomeric subunit of vertebrate hemoglobins (and myoglobin) has a molecular weight of about 16000 Da. Cyclostome hemoglobins circulate as monomers (e.g., *Lampetra*, 19 10³ Da; *Myxine*, 23 10³ Da) or dimers or trimers (i.e., Hb$_2$ or Hb$_3$), depending on the degree of oxygenation. The hemoglobins of other vertebrates are usually tetramers (i.e., Hb$_4$) with a molecular weight of about 64 10³ (4 × 16 10³). Some vertebrate hemoglobins are octamers (Hb$_8$).

Invertebrate hemoglobins, which are sometimes called **erythrocruorins**, may have many more than four heme subunits. They can be divided into four general types: (1) single domain; (2) two domain; (3) multi-domain; and (4) multisubunit (Figure 15–6A; Vinogradov 1985). For example, the hemoglobin of planorbid gastropods (MWt 1800 10³ Da) can be dissociated into 10 subunits (MWt 180 10³ Da), each of which can be proteolytically reduced to 10 to 12 O_2-binding domains (MWt 16 10³ Da; Figure 15–

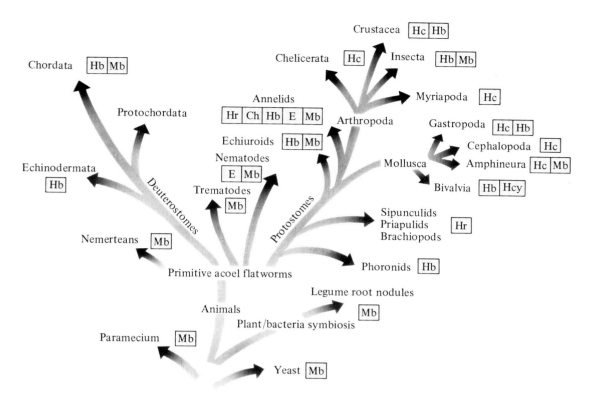

FIGURE 15–4 Schematic summary of the phylogenetic distribution of respiratory pigments in animals. Abbreviations are as follows: Hb, hemoglobin; Mb, myoglobin; E, erythrocruorin (invertebrate hemoglobin); Ch, chlorocruorin; Hr, hemerythrin; and Hc, hemocyanin. *(Modified from van Holde and Miller 1982.)*

6B), each of which is a single heme bound to a single protein.

Single-domain hemoglobins are monomeric with a single heme group bound to a single polypeptide protein, like myoglobin. Such hemoglobins are present in protozoans (*Paramecium*, *Tetrahymena*), a platyhelminth fluke (*Dicrocoelium*), in nerve tissue of the annelid *Aphrodite* and mollusk *Aplysia*, and in insects (*Buenoa* and *Chironomus*).

Single-domain multisubunit hemoglobins are aggregates of domain subunits, with a MWt of $2.2 \ 10^6$ to $3.8 \ 10^6$ Da; the minimum subunit MWt is about $23 \ 10^3$ Da. These hemoglobins are common in all three classes of annelid worms.

Two-domain hemoglobins are single polypeptides that have two heme groups. These two-domain hemoglobins are probably formed by duplication of the ancestral globin gene. They are generally aggregated to form multiple-subunit hemoglobins, but the smallest unit (about $35 \ 10^3$ Da) has two hemes and the molecular weight per heme group is about 15 to $17 \ 10^3$. The hemoglobins of the branchiopod crustaceans *Cyzicus* ($280 \ 10^3$ Da; per heme = $15 \ 10^3$ Da), *Caenestheria* ($302 \ 10^3$ Da; $15 \ 10^3$ Da per heme), and *Lepidurus* ($798 \ 10^3$ Da, $17 \ 10^3$ Da) are two-domain, multisubunit hemoglobins. The arcid clam *Barbatia* has the largest known intracellular hemoglobin (MWt $430 \ 10^3$ Da; subunit MWt is 32 to $37 \ 10^3$ Da). The chlorocruorins of various annelids have a similar structure to the two-domain multiunit annelid hemoglobins (see below).

Multidomain hemoglobins have a polypeptide that contains more than two heme units (e.g., 8 to 20); these polypeptide subunits are aggregated to form multiple unit hemoglobins. Multidomain hemoglobins presumably arose by multiple gene duplication of the precursor globin gene. The planorbid hemoglobin subunit has 10 to 12 heme domains, and there are 10 subunits per functioning hemoglobin. The hemoglobin of anostracans (*Artemia*) has a MWt of $250 \ 10^3$ Da and consists of two subunits (α and β) that each contain 8 heme groups; the dimeric hemoglobin can be α_2, β_2, or $\alpha\beta$. The extracellular hemoglobins of the clam *Candita* are $300 \ 10^3$

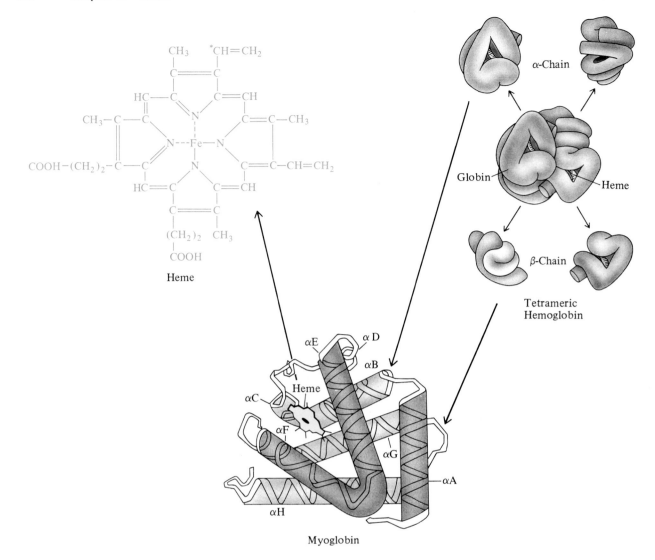

Heme

Myoglobin

FIGURE 15–5 The structure of myoglobin illustrates the general structure of hemoglobin-type respiratory pigments (myoglobins, hemoglobins, erythrocruorins, chlorocruorins). Myoglobin has eight segments of α-helix (A-H) with connecting pieces of reverse turns and unordered segments. The polypeptide chain folds around the heme prosthetic group, which consists of the porphyrin ring and a ferrous (Fe^{2+}) ion. The prosthetic group of chlorocruorin is identical except that a formy group (C-CHO) is substituted for the indicated vinyl group ($*$). The tetrameric structure of hemoglobin contains two α chains and two β chains, each chain resembling myoglobin.

Da polypeptides with up to 20 hemes per polypeptide.

Intracellular hemoglobins generally have a lower molecular weight than extracellular, dissolved hemoglobins. They are often monomers, dimers, or tetramers of a 15 to 17 10^3 Da monomer, i.e., their MWt is 16, 32, or 48 10^3 Da. There are only a few examples of animals with intracellular hemoglobins

with MWt greater than 64 10^3 Da, i.e., more than four subunits (Grinich and Terwilliger 1980); *Barbatia* is one exception, with a hemoglobin molecular weight of 430 10^3.

Hemoglobins that are free in solution typically have multiple heme groups and a high molecular weight ($>10^6$ Da). The large size of these hemoglobins minimizes their osmotic colloid effect and

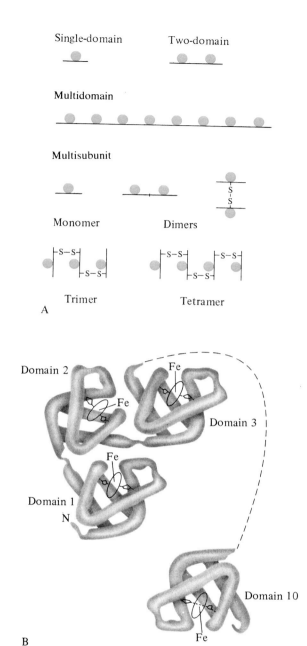

A

B

FIGURE 15–6 **(A)** Generalized classificatory scheme for hemoglobins (and erythrocruorins) of various invertebrates based on a single heme/polypeptide, myoglobin-like subunit structure (single-domain), a two heme/polypeptide subunit (two-domain), or a multiple heme/polypeptide unit (multidomain). Multisubunit hemoglobins can consist of dimers, trimers, tetramers, etc., of single-domain subunits, or of multiples of two- or multidomain subunits (not shown). **(B)** Representation of the structure of a multidomain hemoglobin of planorbid gastropod mollusks. Each of the 10 domains consists of a myoglobin-like subunit of polypeptide chain, porphyrin, and Fe^{2+} ion. *(From Vinogradov 1985; Bonaventura and Bonaventura 1983.)*

prevents their excretion. There is at least one exception to the generalization, however. The hemolymph of *Chironomus* midge larvae (small insects) has a high concentration of low molecular weight hemoglobin (16 and 22 10^3 Da) dissolved in the hemolymph; the colloid osmotic pressure would be extremely high, about 13.6 kPa (100 torr). Notably, these midge larvae lack excretory organs and have an open circulatory system (so there is no significance of colloid pressure to transcapillary fluid balance; see Chapter 14).

The coelomic hemoglobins have a low MWt and are almost always intracellular to avoid excretion via the coelomoducts. However, the annelid *Nephtys* has a low MWt hemoglobin (a trimer of 50 10^3 Da and tetramer of 65 10^3 Da) in its coelomic fluid; notably, this annelid lacks coelomoducts and has closed nephridia, which prevent excretion of the hemoglobin. The vascular dissolved hemoglobins of annelids are of high MWt, which would minimize colloid pressure effects (although the concentrations are generally low anyway) and prevent leakage from their closed circulatory system.

Chlorocruorin

Chlorocruorins are found in four polychete families: serpulids, sabellids, chlorhaemids, and ampharetids. These pigments are green in dilute solution and red in concentrated solution; they have an absorbance spectrum like that of hemoglobin, but shifted 20 to 25 nm towards the red.

Chlorocruorins are very similar in general structure to hemoglobins, particularly the two-domain, multisubunit annelid hemoglobins. The porphyrin group is identical to that of hemoglobins except a formyl group (—CHO) is substituted for a vinyl group (—CHCH$_2$) on the porphyrin ring. As in hemoglobin, each Fe^{2+} can bind one O_2. Chlorocruorins, like high molecular weight annelid hemoglobins, are multisubunit structures with a molecular weight about 3 10^6 (Terwilliger, Terwilliger, and Schabtad 1976). The structure is a two-tiered arrangement of two 6-membered rings, i.e., 12 subunits of about 250 10^3 Da each. The minimum subunit molecular weight calculated per iron content is about 22 to 34 10^3, but there apparently are lower molecular weight subunits, about 13 to 16 10^3. There is apparently not a 1:1 correspondence between heme groups and polypeptides, i.e., not all of the smallest protein subunits contain a heme group.

The close relationship between chlorocruorin and hemoglobin is illustrated by the distribution of both pigments in *Spirorbis* and *Serpula*. Some *Spirorbis* have chlorocruorin, some have hemoglobin, some have neither, and one species of *Serpula* has both chlorocruorin and hemoglobin in the ratio 3:2 (each pigment has a minimum MWt of 24.7 10^3 Da).

Hemerythrin

Hemerythrin is a violet-pink pigment when oxygenated and colorless when deoxygenated. Oxyhemerythrin has an absorbance peak at 330 nm and 500 nm (see Figure 15–3, page 733). Two Fe^{2+} ions bind a single O_2. This respiratory pigment has a limited distribution in a few phyla—Sipuncula, Priapula, and Brachiopoda—and in one genus of Annelida (Hendrickson, Smith, and Sheriff 1985; Klotz, Klippenstein, and Hendrickson 1976). Sipunculids have three forms of hemerythrin: one in blood, one in coelomic fluid, and a myohemerythrin in retractor muscle. The circulating hemerythrin is always located within blood corpuscles.

The structure of hemerythrins is based on a polypeptide (13 to 14 10^3 Da; either 113 or 118 amino acid residues) arranged as four antiparallel α-helices with an N-terminal "tail" and a C-terminal "stub" (Figure 15–7). Two Fe^{2+} ions are located about 1/3 of the way from the C-terminus. There is no porphyrin ring or heme group in hemerythrins. Sipunculid hemerythrins can be monomeric (e.g., myohemerythrin of *Dendrostomum*), dimeric, trimeric or tetrameric (e.g., *Phascolosoma*), or octameric (*Phascolopsis* coelomic hemerythrin). The coelomic hemerythrins of the brachiopod *Lingula* are octameric. That of a priapulid (*Priapulis*) is a trimer in equilibrium with its monomeric subunit (MWt about 14 to 15 10^3 Da) and is a hexamer at high concentrations; it also has a nonaggregating 10^4 Da hemerythrin.

Hemocyanin

Hemocyanins are colorless when deoxygenated and blue when oxygenated. Oxygenated hemocyanins have a broad absorption peak at 570 nm and a strong absorption band at 340 to 350 nm (see Figure 15–3, page 733). Hemocyanins are found free in the hemolymph of various members of four classes of arthropods: Crustacea, Merostomata, Chilopoda, and Arachnida. Among crustaceans, the malacostracans mainly have hemocyanin, although some amphipods, isopods, and naked barnacles also have it. Many mollusks (chitons, opisthobranchs, a few bivalves, pulmonate gastropods, and cephalopods) have hemocyanin, but monoplacophorans, aplacophorans, scaphopods, and most bivalves do not (van Holde and Miller 1982; Ellerton, Ellerton, and Robinson 1983).

Hemocyanins are very large proteins with Cu^{2+} rather than Fe^{2+} as the prosthetic group; there is no equivalent to the heme group of hemoglobin and chlorocruorin. Two Cu^{2+} ions bind one O_2 (as in hemerythrin, which also lacks a heme but has Fe^{2+} rather than Cu^{2+}). The subunit structures are similar for molluskan and arthropod hemocyanins but the molecular weight per O_2-binding site is about 50 10^3 Da in mollusks and 75 10^3 in arthropods. The hierarchical organization of the subunits to form polymeric hemocyanins is quite different in mollusks and arthropods.

Arthropod hemocyanins typically have MWts > 100 10^3 Da, but deoxygenated hemocyanin tends

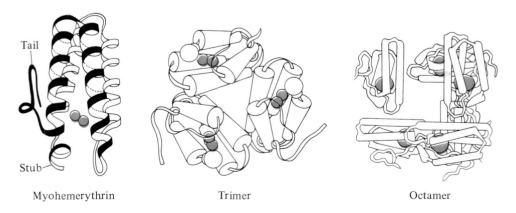

Myohemerythrin   Trimer   Octamer

FIGURE 15–7 Hemerythrins can have a monomeric structure (e.g., myohemerythrin) consisting of a polypeptide and two Fe^{2+} ions (note the absence of a porphyrin group) or a polymeric structure (e.g., trimer or octamer; for clarity, only five subunits are shown for the octamer). *(From Hendrickson, Smith, and Sheriff 1985.)*

to dissociate at high pH into subunits of about 65 to 75 10^3 Da. For example, lobster hemocyanin has a MWt of about 825 10^3 Da and dissociates into 69 10^3 Da subunits; the subunits consist of either two proteins (each with one Cu^{2+}) or three proteins (only two of which have a Cu^{2+}). The hemocyanins are formed by successive dimerization of a unit comprised of six kidney-shaped subunits (each 70 to 75 10^3 Da; Figure 15–8; Lamy et al. 1985). This hexamer, the 1 × 6 mer structural unit, is the constituent of both crustacean and chelicerate hemocyanins. Crustacean hemocyanins are relatively simple, being occasionally 1 × 6 mer, often 2 × 6 mer, and sometimes 4 × 6 mer. Chelicerate hemocyanins are either 2 × 6 mer in modern spiders, 4 × 6 mer in primitive spiders and scorpions, or 8 × 6 mer in the horseshoe crab.

Molluskan hemocyanins have a very different structure and are more heterogeneous than the arthropod hemocyanins (Figure 15–9; Bonaventura and Bonaventura 1985). There is one Cu^{2+} per 50 10^3 Da and each polypeptide chain is about 400 10^3 Da, i.e., there are about 8 Cu^{2+} per subunit and

there are usually 10 to 20 subunits per unit. The units are large and of complex structure; they are cylinders about 30 nm diameter and 38 nm high. In gastropods, for example, the subunits are 300 to 400 10^3 Da (about 1/20 that of the unit molecule that has a MWt of about 8 10^6). The hemocyanins of cephalopods are also cylindrical molecules about the same diameter but 1/2 the height as those of gastropods.

The evolutionary development of hemocyanins is not clear. They most likely evolved from tyrosinase enzymes, which they closely resemble. Tyrosinase is a Cu^{2+} protein with a similar active site as hemocyanin, and it can even reversibly bind O_2 under certain circumstances. Tyrosinase also is widely distributed in animals and plants. It is likely that hemocyanin evolved independently in mollusks and arthropods in view of the marked differences in their structure.

The modes of synthesis of hemocyanins also indicate a separate molluskan and arthropod origin. In gastropods, "pore cells" scattered throughout the connective tissue appear to synthesize and release

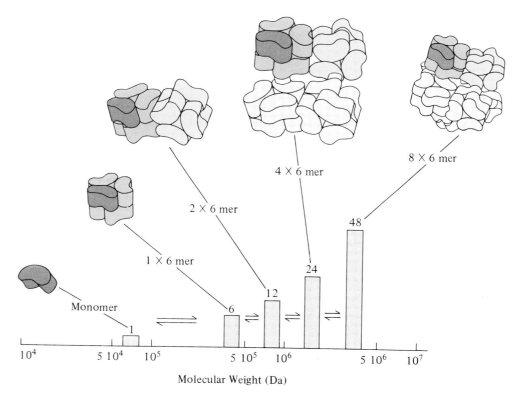

FIGURE 15–8 Arthropod hemocyanins are based on a subunit structure (MWt about 5 10^4) organized as a hexameric unit (the 1 × 6 mer). More complex hemocyanin structures are based on aggregations of the 1 × 6 mer to form 2 × 6 mer (i.e., 12 monomer subunits), 4 × 6 mer (24 subunits), or 8 × 6 mer (48 subunits). *(Modified from Lamy et al. 1985; Ellerton, Ellerton, and Robinson 1983.)*

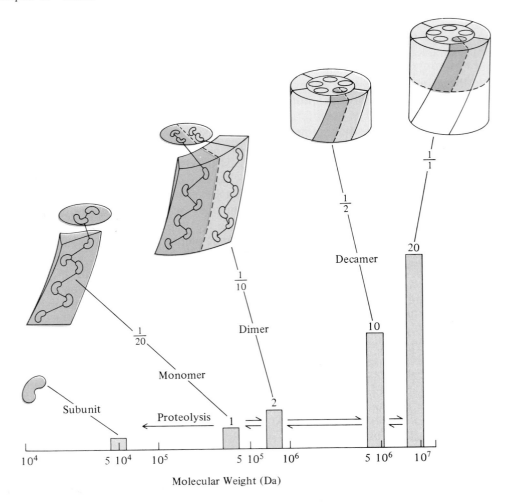

FIGURE 15–9 Molluskan hemocyanin structures are based upon a multidomain sub-unit comprised of eight subunits (each subunit has two Cu^{2+} ions and binds a single O_2). The multidomain subunits (termed 1/20 units) are organized in pairs to form a 1/10 unit, which are organized in groups of 5 to form a cylindrical 1/2 unit. Some hemocyanins have two 1/2 units stacked to form a 1/1 unit. *(Modified from Bonaventura and Bonaventura 1983; Ellerton, Ellerton, and Robinson 1983.)*

hemocyanin by exocytosis. Hemocyanin of cephalopods is synthesized in branchial gland cells and released by exocytosis. In arthropods, hemocytes, called cyanoblasts and cyanocytes, synthesize hemocyanin and release it by cell lysis.

Oxygen-Binding Properties

Let us start our examination of oxygen binding to respiratory pigments by concentrating on hemoglobin. The degree of oxygenation of hemoglobin depends on the oxygen partial pressure; a high pO_2 promotes O_2 binding and a low pO_2 reduces O_2 binding. The hemoglobin becomes 100% saturated at some high pO_2 when every O_2-binding site is occupied. The extent of saturation is calculated as $100 \times [HbO_2]/[Hb + HbO_2]$. The relationship between percent saturation and pO_2 for respiratory pigments is called the **O_2 equilibrium curve** (or O_2 loading curve or O_2 dissociation curve).

We might expect the relationship between oxygen binding by a monomeric hemoglobin (e.g., myoglobin) and pO_2 to be nonlinear. At low pO_2 and low percent saturation, most of the hemoglobin is deoxygenated and there is a high probability that an O_2 molecule will strike a deoxygenated hemoglobin. At higher pO_2 and percent saturation, the chance of an O_2 molecule striking a deoxygenated hemoglobin is lower, and so increasingly more O_2 molecules are required to increase the percent saturation by a given value (e.g., '1%). Thus, we might expect a hyperbolic rather than a linear relationship between percent saturation and pO_2 (see Supplement 15–1, page 773).

Many hemoglobins are not monomeric but are dimeric, tetrameric, octameric, etc. Consequently, the range of reactions between the hemoglobin and O_2 is more complex. For example, for tetrameric hemoglobin,

$$
\begin{aligned}
Hb_4 + O_2 &\underset{k_1'}{\overset{k_1}{\rightleftharpoons}} Hb_4O_2 \\
Hb_4O_2 + O_2 &\underset{k_2'}{\overset{k_2}{\rightleftharpoons}} Hb_4O_4 \\
Hb_4O_4 + O_2 &\underset{k_3'}{\overset{k_3}{\rightleftharpoons}} Hb_4O_6 \\
Hb_4O_6 + O_2 &\underset{k_4'}{\overset{k_4}{\rightleftharpoons}} Hb_4O_8
\end{aligned}
\qquad (15.2)
$$

where k and k' are rate constants. There generally is a sigmoidal (S-shaped) relationship between the percent saturation of polymeric hemoglobin and the pO_2. The amount of deviation of the S-shaped curve from the hyperbolic curve depends on the value of k and a cooperativity coefficient, n (see Supplement 15–1).

The **affinity** of hemoglobin for O_2 is indicated by the position of the O_2 dissociation curve. A high-affinity hemoglobin has a dissociation curve further to the left (lower P_{50}) than a low-affinity hemoglobin located more to the right (higher P_{50}). The P_{50} value is the pO_2 for 50% saturation; it is determined by both the k and n values (see Supplement 15–1).

The O_2-binding properties of the different respiratory pigments vary dramatically. There are marked differences in P_{50} for pigments of even closely related animals and even for animals of the same species. This is not surprising in view of the diversity in function of noncirculating or circulating respiratory pigments and intracellular or extracellular pigments that function for either transport or storage (see below). The P_{50} of a respiratory pigment is apparently a very "plastic" property and is easily adjusted in different species to maximize the O_2 transport capacity or storage role, subjected to particular environmental circumstances. It is also plastic within a species, depending on pCO_2, pH, and temperature. Nevertheless, there are some taxonomic trends in P_{50}. Invertebrate hemoglobins generally have a lower P_{50} (0.01 to 1 kPa) than vertebrate hemoglobins (3 to 5 kPa). The value of n varies from 1 to 6, being about 1 in insects, 1 to 3 in crustaceans and mollusks, 1 to 6 in annelids, 1 to 2 in lower vertebrates, and 2 to 3 in higher vertebrates. Intracellular myoglobins typically have

low P_{50} values (0.1 to 0.2 kPa) and low n (about 1 to 1.6; most myoglobins are monomeric but some are dimeric). Other noncirculating intracellular hemoglobins (e.g., in nerve and tracheal cells) also have low P_{50} values and n about 1 to 1.4. In annelids, coelomic hemoglobins have low P_{50} (0.2 to 1.0 kPa) and low n (1 to 1.8), and the vascular pigments have a similar P_{50} but generally higher n (up to 6). This difference in n reflects the lower molecular weights and heme numbers of the coelomic hemoglobins (16 to 60 10^3 Da) compared to vascular hemoglobins (2400 to 3400 10^3 Da). Annelid chlorocruorins have high P_{50} values (about 5 kPa) and high n (about 4, ranging from 1 to 5). Hemerythrins have low P_{50} (about 0.7 kPa) and n values close to 1. Molluskan and crustacean hemocyanins generally have intermediate P_{50} values (1 to 2 kPa) and n values of 1 to 4 (mollusks) or 2 to 8 (crustaceans).

A variety of factors other than subunit and unit structure modify the O_2 position of the dissociation curve and the P_{50} value. These factors include CO_2 partial pressure, H^+ concentration, ionic concentration, concentration of various organic phosphates, temperature, respiratory pigment concentration, and variation in amino acid sequence of the respiratory pigment protein. Some of these factors, especially pCO_2, $[H^+]$, and inorganic phosphates, have great relevance to physiological adaptation to varying environmental conditions.

The pCO_2 and H^+ of blood have a dramatic effect on the position of the O_2 equilibrium curve. For example, increased pCO_2 and $[H^+]$ (or decreased pH) shifts the equilibrium curve of many hemoglobins and hemocyanins to the right, i.e., higher P_{50}, lower n (Figure 15–10A). The magnitude of this **Bohr effect** is expressed as the change in P_{50} per unit change in pH, i.e., the Bohr effect factor (ϕ) is equal to $\Delta \log P_{50}/\Delta pH$. Some representative values for ϕ are given in Table 15–4. The Bohr effect increases the unloading of O_2 from capillary blood to the tissues, since the O_2 content of venous blood is lower when the Bohr effect is taken into account.

The Bohr effect is said to be normal, or negative, if a decrease in pH causes a right shift ($\phi < 0$) and reversed, or positive, if it causes a left shift ($\phi > 0$). Most respiratory pigments have a negative Bohr effect, but a number have no Bohr effect and some have a positive Bohr effect. The Bohr effect of some hemoglobins depends on the pH. For example, adult bullfrog hemoglobin has a normal Bohr effect at pH > 6.5, no Bohr effect at pH = 6.5, and a reversed Bohr effect at pH < 6.5 (Figure 15–10B). Other respiratory pigments, including hemoglobin of rats and humans and hemocyanin of spiders and crustaceans, have a similar pH dependence. Some hemo-

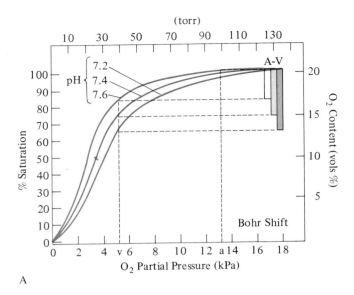

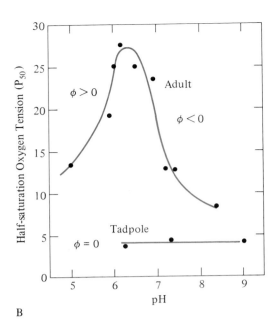

A

B

FIGURE 15–10 **(A)** The Bohr effect is a shift to the right of the O_2-equilibrium curve due to increased pCO_2 and $[H^+]$ (or decreased pH). The O_2-equilibrium curve for human blood illustrates the significance of the Bohr shift to O_2 transport at the normal arterial (a) and venous (v) pO_2s. The arterial-venous O_2 difference (A-V) is about 3 vols percent at pH 7.6 but is about 7 vols percent if a Bohr shift to pH 7.2 is taken into account. The actual significance of the *in vivo* arterial-venous pH shift (from about 7.4 to 7.3) is less. **(B)** The 1/2 saturation pO_2 (P_{50}) for hemoglobin of adult bullfrogs (*Rana catesbeiana*) is sensitive to pH; there is a positive Bohr shift ($\phi > 0$) at pH < 6.5 and a negative Bohr shift ($\phi < 0$) at pH > 6.5. In contrast, the hemoglobin of the bullfrog tadpole is insensitive to pH (no Bohr shift, $\phi = 0$). *(From Riggs 1951.)*

cyanins have reverse Bohr effects at physiological pH and normal Bohr effects at alkaline pH (*Limulus, Busycon, Helix*). The hemoglobin of bullfrog tadpoles has no Bohr effect regardless of pH. The hemoglobins of *Urechis* and *Eupolymnia* (polychetes) and *Polistotrema* (hagfish) also have no Bohr effect. Hemerythrins of sipunculids and hemocyanins of some amphineuran mollusks have no Bohr effect. Myoglobins also have no Bohr effect.

An increase in pCO_2 decreases the O_2-carrying capacity of some hemoglobins; this is the **Root effect** (Figure 15–11). It was first described for hemocyanins of squid and horseshoe crabs (Redfield and Mason 1928; Redfield, Coolidge, and Hurt 1926) and fish hemoglobins (Root 1931). Other vertebrate hemoglobins (e.g., of amphibians and reptiles) and some other hemocyanins (e.g., of crustaceans) also have a Root effect. The Root effect, like the Bohr effect, facilitates O_2 unloading to the tissues. In fish, the Root effect also provides a mechanism for secretion of gas into the swim-bladder (see below). An unusual Root effect has been reported for the marine gastropod *Buccinus undatum* (Brix 1983);

the O_2-carrying capacity of blood increases substantially at higher pCO_2/lower pH (i.e., has a "reverse" Root effect). The ability to increase O_2 capacity is not due to an exceptionally high O_2-binding capacity at high pCO_2 but to an unusually low O_2-binding capacity at low pCO_2. The hemocyanin of *B. undatum* normally carries only about 30% of the expected amount of O_2. Thus, the benefit of increased O_2 capacity at high pCO_2 occurs at the expense of a low O_2 capacity at low pCO_2.

Ionic concentration can alter the O_2-binding properties of respiratory pigments. Increased ionic concentration increases the P_{50} of mammalian hemoglobin but decreases the P_{50} for annelid hemoglobins. Molluskan and crustacean hemocyanins are also generally sensitive to ionic concentration. Ions increase the O_2 affinity of crustacean hemocyanins, which have a normal Bohr effect, but decrease the affinity for *Limulus* and gastropod hemocyanins, which have a reversed Bohr effect. The physiological significance of ionic effects on P_{50} is unclear since blood ionic composition is unlikely to be varied to regulate the respiratory function of blood.

TABLE 15–4

Bohr effect factor ($\phi = \Delta\log P_{50}/\Delta pH$) for respiratory pigments of various animals in the physiological pH range. Hb, hemoglobin; Mb, myoglobin; Hr, hemerythrin; Hcy, hemocyanin; and Ch, chlorocruorin.		
Annelids	Coelomic Hb	−0.4 to 0
	Extracellular Hb	−1.3 to 0
	Mb	−0.03 to 0
	Ch	−1.72 to −0.45
	Hr	0
Echiurids	Hb	−0.11
	Mb	0
Sipunculids	Hr	0
Brachiopods	Hr	−0.4
Priapulids	Hr	0
Mollusks	Hb	− to +
	Hcy	−1.4 to +1.3
Crustaceans	Hb	−0.23
	Hcy	−1.7 to +0.11
Arachnids	Hcy	−1.2 to +0.34
Vertebrates:		
Fish	Hb	−0.54 to −0.31
Amphibians	Hb	−0.29 to 0
Reptiles	Hb	−0.52 to −0.13
Birds	Hb	−0.5 to −0.4
Mammals	Hb	−0.96 to −0.32
	Mb	0

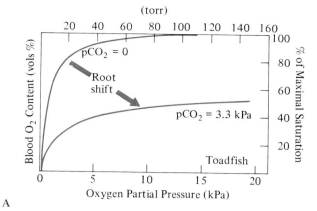

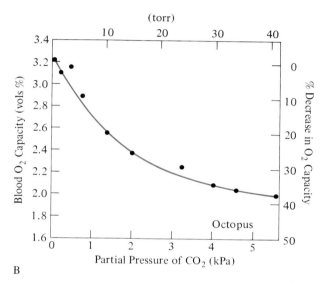

Consequently, the dependence of P_{50} on ionic strength is considered to be a perturbing effect rather than a regulatory effect.

Organic phosphates are another important modifier of respiratory pigment function (Bartlett 1980; Isaacks and Harkness 1980). For intracellular pigments, in particular, there is a considerable capacity to alter the local environment of respiratory pigments through shifts in metabolic pathways. For example, 2-3 diphosphoglycerate (DPG) is an intermediate organophosphate of the citric acid cycle (see Chapter 3); high concentrations of DPG shift the O_2 dissociation curve of amphibian and mammalian hemoglobin to the right, i.e., increase P_{50}. Human red blood corpuscles normally contain about 0.5 mM DPG or approximately 1 DPG molecule per hemoglobin molecule. Other organic phosphates, such as inositol pentaphosphate (IP$_5$; fish, birds), inositol tetraphosphate (IP$_4$; ostrich), ATP (fish, amphibians, turtles, squamate reptiles), and GTP (fish) are also important modulators of hemoglobin function.

There is a considerable variation among mammals in the DPG concentration and its modulatory effect (Bunn 1980). The red blood corpuscles of humans

FIGURE 15–11 The Root effect is a right-shift in the O_2 equilibrium curve due to increased pCO_2 or $[H^+]$ with a concomitant decrease in the O_2-binding capacity of the respiratory pigment. **(A)** The Root shift for toadfish hemoglobin. **(B)** The effect of pCO_2 on the O_2 capacity of octopus hemocyanin. *(Modified from Root, Irving, and Black 1939; Lenfant and Johansen 1965.)*

have a higher DPG level when adapted to high altitude. There is also a significant role of DPG in the rat, rabbit, dog, guinea pig, and horse but little effect in the sheep, cow, cat, and goat. In general, species with low P_{50} use DPG to decrease the affinity and to optimize O_2 transport to the tissues, whereas species with high P_{50} (felids and some artiodactyls) do not use DPG as a modulator of hemoglobin function.

In birds, there is an increase in P_{50} at about the time of hatching due to changes in intraerythrocyte ATP and IP$_5$. The P_{50} of embryonic turtle hemoglobin is modulated by DPG and IP_5 but adult turtle hemoglobin is insensitive to both organophosphates. ATP and GTP appear to be the most important organic phosphate modulators for fish hemoglobin. In contrast, the properties of more primitive hemoglobins of annelids and mollusks, and hemocyanins, do not appear to be modulated by organic phosphates, although other metabolites can have significant modulatory effects. For example, the anaerobic metabolic end product, lactate, increases the O_2 affinity of crab hemocyanins, and so lactate and possibly other metabolic end products may be of physiological significance in increasing the O_2 transport capacity of the hemolymph (Truchot 1980).

Temperature is a potentially important physiological modulator of respiratory pigment function. Oxygen binding to respiratory pigments is an exothermic process and temperature almost universally increases the P_{50} (decreases O_2 affinity). The effect of temperature on the O_2 dissociation curve for the tracheal hemoglobin of the insect *Anisops* is typical (Figure 15–12). The influence of temperature (T; °K) on P_{50} is characterized by the apparent heat of oxygenation (enthalpy change, ΔH), calculated as

$$\Delta H = -2.303R \frac{\Delta \log P_{50}}{\Delta(1/T)} \qquad (15.3)$$

(Weber 1978). Typical values for ΔH of invertebrate and vertebrate hemoglobins are -42 to -59 kJ $mole^{-1}$. The effect of temperature on P_{50} automatically decreases the affinity at higher temperatures, thereby promoting unloading of O_2 to the tissues but potentially inhibiting O_2 loading at the respiratory surface. At low temperatures, O_2 unloading to the tissues can be compromised by the low P_{50}. Even

for endotherms, the effect of temperature on P_{50} can have physiological significance. For example, the temperature of muscle in an exercising human can exceed 40.7° C; this, coupled with the drop in pH to 7.27, increases the venous P_{50} to 5.2 kPa from the normal venous P_{50} value of 3.5 kPa and this would facilitate O_2 delivery to tissues.

Hemoglobins of some ectotherms have a reduced ΔH for oxygenation that presumably is adaptive (Wood 1980). Tuna hemoglobin has a very low ΔH of -7.5 kJ $mole^{-1}$; chum salmon have two hemoglobin types, one with a normal temperature effect on P_{50} and one with a reduced temperature effect. The low thermal dependence of P_{50} may maintain normal O_2 loading at the gill and prevent excessive O_2 unloading in the warmer muscle tissues (at least during activity; see Chapter 5). The ΔH for a stenothermal lungfish (*Protopterus*) is typical, but that for the eurythermal *Neocerotadus* is low. The low ΔH for hemoglobin of the lizard *Iguana* may prevent an excessive increase in P_{50} at its preferred body temperature of 37° C. The low ΔH for hemoglobin of *Arenicola*, *Tubifex*, and *Pista* (-21 to -29 kJ $mole^{-1}$) and chlorocruorin of *Spirographis* (-19) would minimize the effect of high water temperatures on P_{50}.

Hemerythrins of sipunculids have ΔH values higher than those of hemoglobins (-42 to -75 kJ $mole^{-1}$), but whether this has an adaptive significance is not clear. Hemocyanins have a similar temperature dependence of P_{50}, e.g., *Carcinus*, -30; *Busycon*, -71. *Carcinus* inhabiting cold water have hemocyanin with a lower affinity, presumably to counteract the reduced P_{50} at low temperatures. In *Busycon*, the P_{50} at 10° C is so low (<0.27 kPa)

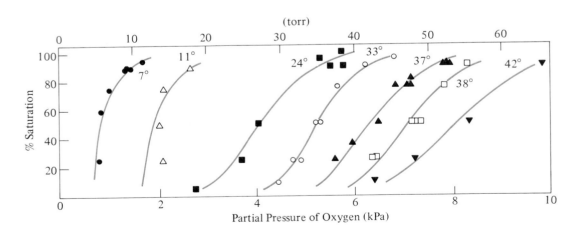

FIGURE 15–12   Effect of temperature on the O_2 equilibrium curve for hemoglobin of an insect *Anisops*. *(From Miller 1966. With permission of the Company of Biologists, Ltd.)*

that tissue oxygenation is effectively precluded and the animal becomes inactive, stops feeding and moving, and its hemolymph Cu^{2+} level falls and the O_2-carrying capacity declines. The ΔH may alter seasonally in some ectotherms.

Gas Transport

The main role of respiratory pigments is O_2 transport and the reversible O_2-binding properties of respiratory pigments are adapted to this role. O_2 is loaded onto respiratory pigments at the respiratory surface and unloaded in the tissues. Some intracellular (noncirculating) pigments facilitate the diffusion of O_2 into the cells from the body fluids and some act as intracellular O_2 stores. Sometimes respiratory pigments transport gases other than O_2. CO_2 transport is also influenced by hemoglobin.

Oxygen

Respiratory pigments in blood (or coelomic fluids) often transport O_2 from the respiratory surface to the tissues, but can also transfer O_2 from blood (or coelomic fluid) into cells or act as intracellular O_2 stores, e.g., vertebrate and invertebrate myoglobins and neuroglobins. It would be impossible to present an exhaustive summary of the diverse O_2 transport systems of various animals, but a few examples will illustrate some basic patterns of respiratory pigment function.

Oxygen Transport by Blood. Many pigments function at high pO_2, i.e., have a high P_{50}. Vertebrate hemoglobins, chlorocruorins of tubicolous polychetes, and cephalopod hemocyanins would be typical examples of these high pO_2 transport pigments. Some respiratory pigments have lower P_{50} but still are involved with O_2 transport, e.g., decapod crustacean hemocyanins. Some pigments are adapted for transport of O_2 from hypoxic media and consequently have very low P_{50}, e.g., some tubicolous worms and *Chironomus* midges.

The O_2 transport role of hemoglobin, and the significance of the shape of the sigmoidal equilibrium curve, are well illustrated by the human O_2 equilibrium curve and the normal arterial and venous pO_2 values (Figure 15–13). The hemoglobin has a high percent saturation (>95%) in arterial blood (pO_2 = 13.6 kPa) and a lower percent saturation (about 75%) in venous blood (pO_2 = 5.4 kPa); the arterial blood has a higher O_2 content (19.8 vols percent) than venous blood (15.2 vols percent). The difference between the arterial and venous O_2 contents

of 4.6 vols percent reflects the amount of O_2 "unloaded" to the tissues from the blood. If the hemoglobin had a hyperbolic loading curve ($n = 1$) and the same percent saturation at the arterial pO_2, then the venous blood would be about 94% saturated and the arterial-venous O_2 difference would only be about 1 vol percent; O_2 transport would not be very effective. If the dissociation curve was hyperbolic and had the same P_{50}, then the arterial-venous O_2 difference would be about the same (5 vols percent) but the arterial blood would only be about 78% saturated. For the normal O_2 equilibrium curve, the Bohr effect, due to a 0.82 kPa increase in venous pCO_2, only contributes about 1 to 2% additional unsaturation, i.e., <10% of the O_2 unloaded is due to the Bohr effect. The cardiac output is approximately 5000 ml blood min^{-1}, hence the rate of O_2 delivery to the tissues (which must equal the aerobic metabolic rate) is $5000 \times 4.6/100 = 230$ ml $O_2\ min^{-1}$.

The blood of squid provides an example of a similar O_2 transport role of the blood for an invertebrate (Figure 15–14). Arterial blood contains 4.27 vols percent O_2 at 16.3 kPa and venous blood 0.37 vols percent at 6.5 kPa; the pCO_2 difference between arterial and venous blood is about 0.54 kPa. The A-V difference of 3.9 vols percent is a 3.75 difference due to hemocyanin-bound O_2 and 0.15 vols percent dissolved O_2. About 37% of the O_2 unloaded is attributed to the large Bohr effect; the pCO_2 difference between arterial and venous blood is relatively small, but the Bohr effect factor is relatively large for squid blood ($\phi = -1.8$). The A-V difference and cardiac output of about 11 ml $kg^{-1}\ min^{-1}$ indicate a circulatory O_2 transport (and aerobic metabolic rate) of about 0.43 ml $O_2\ kg^{-1}\ min^{-1}$. During activity, the metabolic rate must rise to 20 or 30 ml $O_2\ kg^{-1}\ min^{-1}$, but the blood O_2 transport system provides little scope for an increase in A-V O_2 difference because the resting venous blood is only 8% saturated and so there is little "venous reserve" for activity (compare this with a venous reserve of 75% saturation in man). The O_2 transport system of squid would also be readily compromised by an (unlikely) increase in ambient pCO_2 because the pronounced Bohr shift would drastically decrease the arterial percent saturation and O_2 content.

The O_2 equilibrium curve influences both O_2 uptake across the lungs and O_2 unloading to the tissues. For example, increasing the P_{50} (lowering O_2 affinity) would increase the pO_2 for tissue diffusion (if arterial and venous O_2 contents remained the same) but would require a higher arterial pO_2 (and lung pO_2); decreasing the P_{50} would facilitate

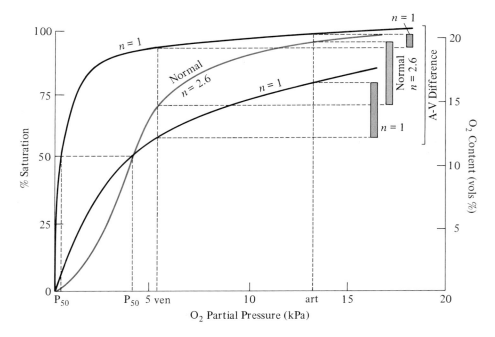

FIGURE 15–13   Normal sigmoidal O_2 dissociation curve for human blood (colored curve; P_{50} = 3.8 kPa; n = 2.6) and calculated hyperbolic curves (black curves) for blood with the same P_{50} and with the same arterial percent saturation as the sigmoidal curve, but n = 1. The arterial-venous difference in percent saturation is shown for each curve. The hyperbolic curve with P_{50} of 0.4 kPa is clearly unsuitable for O_2 transport at the normal human arterial and venous pO_2 values; the A-V difference is only about 1 vols percent. The hyperbolic curve with P_{50} of 3.8 kPa has almost the same A-V difference (3.9 vols percent) as the normal curve but has a lower venous reserve and requires a substantial decrease in venous pO_2 to unload further O_2. The normal O_2 equilibrium curve has a considerable A-V difference (4.3 vols percent) and the venous pO_2 falls on a very steep part of the curve facilitating utilization of the venous reserve with minimal decrease in venous pO_2.

FIGURE 15–14   Oxygen transport by squid blood (*Loligo*) at normal ambient pCO_2 (about 0.2 kPa) delivers about 2.2 vols percent O_2 in the absence of a Bohr shift (left grey bar) and about 3.9 vols percent O_2 when the large Bohr shift is taken into account (left color bar); there is virtually no venous reserve for enhanced O_2 transport. Hypercapnia induces a Root and Bohr shift and decreases the A-V O_2 difference (right grey bar). Abbreviations are art, arterial pO_2; ven, venous pO_2. *(Modified from Jones 1972.)*

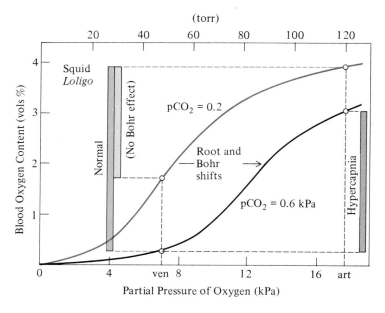

O_2 loading in the lungs but would retard O_2 delivery to the tissues.

The dependence of metabolic rate, P_{50}, and the Bohr effect factor on body mass well illustrates these conflicting effects of P_{50} on O_2 loading and unloading (Figure 15–15). A 10 g mammal (e.g., a shrew) has a VO_2 of about 20 ml O_2 hr^{-1} or 2 ml O_2 g^{-1} hr^{-1}, whereas a 10000000 g elephant has a VO_2 of about 52550 ml O_2 hr^{-1} or 0.05 ml O_2 g^{-1} hr^{-1}. Thus, 1 g of shrew tissue has about 40 × the metabolic rate of 1 g of elephant tissue. To maintain O_2 delivery to the tissues by diffusion, the shrew must have either a higher intracellular-venous pO_2 or a shorter diffusion distance, or both. There is most likely some reduction in diffusion distance between capillaries and cells in shrew tissues but a 40 × shorter diffusion distance is unlikely; it is probably a 2 to 3 × difference. The P_{50} of shrew blood (5.7 kPa) is higher than for elephant blood (3.1 kPa); this raises the venous pO_2 and facilitates diffusive O_2 delivery to the tissues. However, it also raises the arterial pO_2 that is required to keep arterial blood near saturated with O_2. The greater Bohr effect factor for shrew blood also contributes to the high venous pO_2.

The positive Bohr effect of some respiratory pigments may be an example of maladaptation, but the remarkable plasticity of respiratory pigment properties suggests that a positive Bohr effect would have some adaptive role and not be a maladaptive accident of evolutionary history. A positive Bohr effect would facilitate O_2 loading of arterial blood during environmental hypoxia and hypercapnia, but it requires a lower venous pO_2 for O_2 unloading to the tissues (Figure 15–16). The marine gastropod *Busycon* has a substantial A-V O_2 difference if the water is saturated with O_2, but the A-V difference drastically declines if the pO_2 is lowered by emersion (in the absence of any Bohr effect). However, a positive Bohr shift would maintain a substantial A-V O_2 difference if the aerial hypoxia was accompanied by hypercapnia. This adaptive role of a positive Bohr shift assumes that tissue O_2 unloading is not compromised by the low venous pO_2 values.

The hemolymph pO_2s vary dramatically for crustaceans, from <1.4 kPa to >13.6 kPa; much of this variation can be attributed to the physiological state of the animal (McMahon and Wilkins 1983). Crabs, such as *Cancer*, have two types of ventilatory pattern. When "quiescent," only one gill chamber

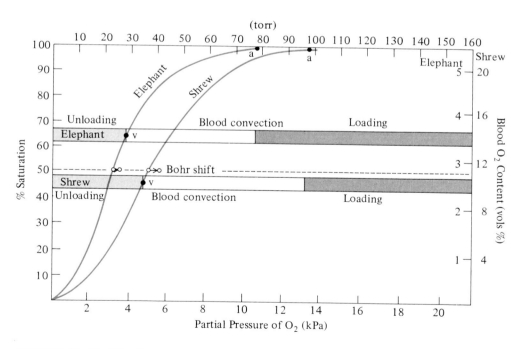

FIGURE 15–15 The blood of small mammals (e.g., shrew) has a right-shifted O_2 equilibrium curve compared to that of larger mammals (e.g., elephant). This provides a larger venous-cell pO_2 gradient for diffusive O_2 unloading from blood to the tissues but provides a smaller ambient arterial pO_2 gradient to load O_2 at the lungs. The arterial-venous pO_2 gradient for convective blood O_2 transport is similar for small and large mammals. *(Equilibrium curves from Bartels, Schmelzle, and Ulrich 1969; Dhindsa, Sedgwick, and Metcalfe 1972.)*

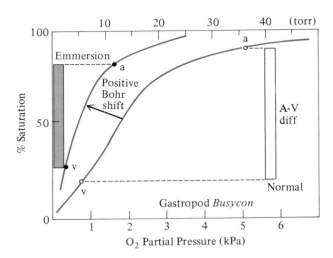

FIGURE 15–16 The positive Bohr shift for blood of the marine gastropod *Busycon* may function to maintain a relatively high arterial-venous O_2 difference during hypoxia and hypercapnia accompanying emmersion (left, closed symbols), although the venous (v) and arterial (a) pO_2 tensions must decline markedly. The normal O_2 dissociation curve (right, open symbols) provides a considerable A-V difference at the normal arterial and venous pO_2s, but the reduced arterial and venous pO_2s during hypercapnia would provide a smaller A-V difference if there were no reverse Bohr shift. *(Modified from Jones 1972.)*

is ventilated and "arterial" hemolymph pO_2 is fairly low (about 4.4 kPa) due to admixture of oxygenated blood from the ventilated gill (pO_2 about 8.5 kPa) with deoxygenated blood from the nonventilated gill (pO_2 about 1.5 kPa); the prebranchial pO_2 is also quite low, at about 0.95 kPa. Consequently, the "arterial" pO_2 is just sufficient to maintain an adequate O_2 content and the "venous" pO_2 leaves only a small venous reserve (Figure 15–17). Hemocyanin-bound O_2 accounts for 50 to 90% of the total O_2 transport; the remainder is due to dissolved O_2. When a crab is "alert," both gill chambers are ventilated and the "arterial" pO_2 is dramatically elevated to 10 kPa and the "venous" pO_2 is elevated to 2.0 kPa (because there is only a small increase in VO_2 despite the 2 × increase in ventilation). The "arterial" blood has a pO_2 well above the saturation level, and the "venous" blood has such a high pO_2 that there is now a substantial venous reserve. During the "alert" state, hemocyanin is responsible for only 30 to 40% of the circulatory O_2 transport. During activity, VO_2, blood flow, and gill ventilation increase. The "venous" pO_2 and O_2 content decline markedly, leaving no venous reserve. The "arterial"

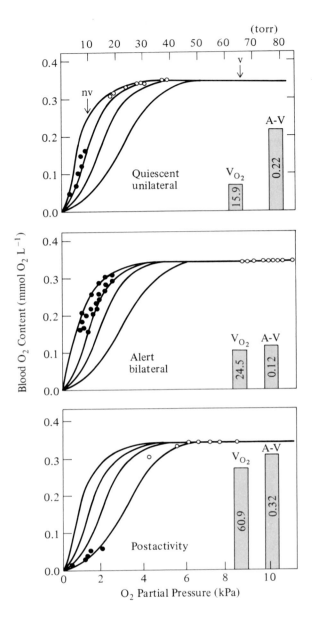

FIGURE 15–17 Oxygen tension measured in the prebranchial hemolymph (closed circles) and postbranchial hemolymph (open circles) of the crab *Cancer magister* at rest (quiescent) with unilateral gill ventilation (upper panel), at rest (alert) with bilateral gill ventilation (center), and immediately following strenuous activity (lower). The oxygen equilibrium curves at varying pH (from left to right, pH 8.17, 7.88, 7.65, 7.43) are superimposed on the pO_2 values. The rate of O_2 consumption ($\mu mol\ kg^{-1}\ L^{-1}$) and A-V difference are also shown. In the upper panel, v indicates the mean pO_2 of postbranchial hemolymph from the ventilated gills and nv the mean pO_2 for unventilated gills. *(Modified from McMahon and Wilkins 1983.)*

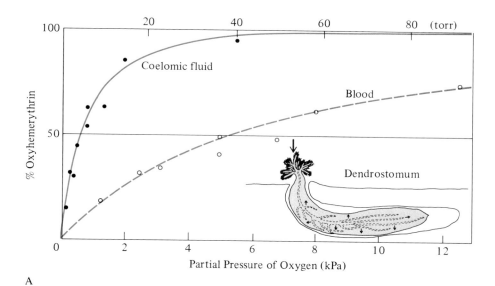

A

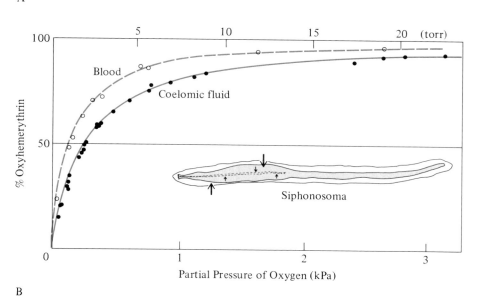

B

FIGURE 15–18 **(A)** O_2 equilibrium curves for hemerythrin in the blood (open circles, broken line) and coelomic fluid (solid circles, solid line) of the surface filter-feeding sipunculid *Dendrostomum zostericolum.* **(B)** O_2 equilibrium curves for hemerythrin in the blood (open circles, broken line) and coelomic fluid (solid circles, solid line) of the burrowing sipunculid *Siphonosoma ingens.* Thick arrows indicate O_2 uptake from medium; thin arrows indicate blood-coelomic O_2 transfer. *(From Manwell 1960.)*

pO_2 is just sufficient to maintain saturation O_2 content. A concomitant decrease in hemolymph pH due to metabolism and respiratory acidosis also enhances tissue oxygenation due to a Bohr shift.

Sipunculid worms have intracellular hemerythrins in their coelomic fluid and their vascular fluid; they also have an intramuscular hemerythrin in their radular muscle (Manwell 1960). The P_{50}s of the vascular and coelomic hemerythrins vary from about 10.54 to 2.72 kPa; their cooperativity coefficient is low (*n* is 1.1 to 1.3) and they have no Bohr shift. Respiratory function varies dramatically with the ecology of different sipunculids. *Dendrostomum* burrow in sand and frequently expand their tentacles

at the surface. There is a pronounced circulation of vascular fluid through the thin-walled tentacles, which provide for ciliary feeding and respiration. The coelomic hemerythrin has a lower P_{50} than the vascular hemerythrin, and so coelomic fluid extracts O_2 from the vascular fluid and transports it to the tissues or is an O_2 store (Figure 15–18A). In contrast, *Siphonosoma* are always burrowed and use their thick-walled tentacles to aid ingestion of sand; their vascular system is restricted to the vicinity of the retractor muscles and nerve cord. The P_{50} of coelomic hemerythrin is higher than that for the vascular pigment, indicating that O_2 uptake occurs across the body wall into the coelomic fluid and is then transferred to the tissues and to the vascular fluid (Figure 15–18B). The P_{50}s of coelomic and vascular hemerythrins of *Siphonosoma* are lower than for *Dendrostomum*, probably reflecting the lower ambient pO_2s for the burrowing *Siphonosoma* compared to the interface located *Dendrostomum*.

Optimal oxygen transport. The rate of O_2 transport by the circulatory system depends on the rate of blood flow and the O_2 contents of arterial blood and venous blood. The O_2 content of blood is determined by the hemoglobin content and hematocrit; a high hemoglobin (or hematocrit) increases the O_2 transport capacity. However, a high blood hemoglobin (or hematocrit) also increases the blood viscosity and decreases the rate of blood flow. There is thus a complex effect of blood hemoglobin content on O_2 transport capacity.

The maximum potential for O_2 delivery occurs if the arterial blood is completely oxygenated and the venous blood is completely deoxygenated; this O_2 transport capacity (OT; ml O_2 min^{-1}) is calculated for vertebrate blood as

$$OT = V_{bl}k_1\text{Hct} \qquad (15.4a)$$

where V_{bl} is the flow of blood (ml min^{-1}), k_1 is the O_2-binding capacity of blood relative to the hematocrit (ml O_2 ml blood^{-1} Hct^{-1}), and Hct is the hematocrit (%). However, the rate of blood flow is determined by both the mechanical pumping capacity of the heart(s) and the rheological properties of blood, particularly the blood viscosity (η_{blood})

$$V_{bl} = k_2/\eta_{\text{blood}} \qquad (15.4b)$$

where k_2 is a constant reflecting the driving pressure and resistance (dimensions) of the vascular system (cf. Equation 14.4, where $k_2 = \Delta P\pi r^4/8l$); k_2 has units of ml min^{-1} cP^{-1}. For vertebrate red cells, there is an exponential relationship between blood viscosity and hematocrit

$$\eta_{\text{blood}} = \eta_{\text{plasma}}e^{k_3\text{Hct}} \qquad (15.4c)$$

where η_{plasma} is plasma viscosity and k_3 is a constant. Therefore,

$$OT = k_1k_2\text{Hct}e^{-k_3\text{Hct}}/\eta_{\text{plasma}} \qquad (15.4d)$$

The ratio of maximal oxygen transport capacity per unit blood flow, relative to plasma viscosity, ($OT\eta_{\text{plasma}}/k_2$) is:

$$OT\eta_{\text{plasma}}/k_2 = k_1\text{Hct}e^{-k_3\text{Hct}} \qquad (15.5)$$

The $OT\eta_{\text{plasma}}/k_2$ is a measure of the relative O_2 transport capacity of blood. A graph of $OT\eta_{\text{plasma}}/k_2$ as a function of hematocrit clearly demonstrates that there is a hematocrit at which the O_2 transport

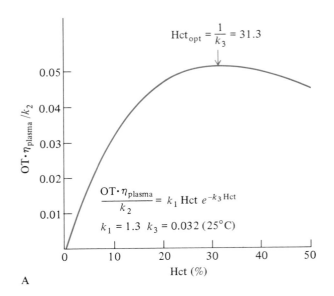

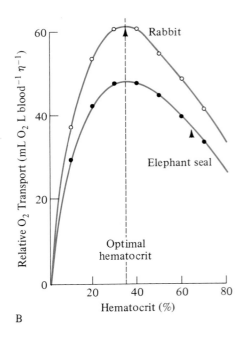

FIGURE 15–19 (A) Theoretical relationship between blood oxygen transport capacity $OT\eta_{\text{plasma}}/k_2$ and hematocrit (Hct) showing the optimal hematocrit. (B) The O_2 transport capacity of blood is maximized at an optimal hematocrit for the rabbit and elephant seal. The *in vivo* hematocrit (arrow) is similar to the optimal hematocrit for the rabbit but the *in vivo* hematocrit of elephant seal blood is considerably greater than the optimal hematocrit, presumably to maximize O_2 storage capacity of the blood. *(From Hedrick, Duffield, and Cornell 1986. Reproduced by permission of the National Research Council of Canada.)*

capacity is maximal, i.e., there is an optimal hematocrit, Hct_{opt}. The value of Hct_{opt} is calculated by differentiating the equation relating $OT\eta_{plasma}/k_2$ to Hct, and setting this to 0; $Hct_{opt} = 1/k_3$ (Figure 15–19A).

The complex relationship between maximal O_2 transport capacity and blood hematocrit is best illustrated by a graphical example; Figure 15–19B shows the calculated O_2 transport capacity for rabbit and elephant seal blood. There is an optimal hematocrit at which the O_2 transport capacity is maximal. At lower hematocrit, the blood transports less oxygen because of its low O_2 content despite the low viscosity; at higher hemoglobin content, the blood transports less O_2 despite its high O_2 content because of the high blood viscosity. In general, the blood hemoglobin content and hematocrit of animals are close to their optimal values (e.g., rabbit) but a greater than optimal hemoglobin content maximizes O_2 storage for diving mammals (e.g., elephant seal).

There is also an optimal concentration if the pigment is free in solution. The oxygen transport capacity for a pigment solution is calculated as

$$OT = V_{bl} k_1 C \qquad (15.6)$$

where k_5 is a constant and C is the pigment concentration (g ml^{-1}). The relative viscosity of a pigment solution (η_{blood}) is determined by the intrinsic viscosity ($[\eta]_{int}$) and the pigment concentration as

$$\eta_{blood}/\eta_{H_2O} = 1 + [\eta]_{int} C + [\eta]^2_{int} C^2 \qquad (15.7)$$

The intrinsic viscosity depends on the molecular weight (M) being $k_6 M^a$, where k_6 and a are constants. Hence, the O_2 transport capacity per unit blood flow and relative to the viscosity of water is

$$OT\eta_{H_2O}/k_2 = k_1 C/(1 + [\eta]_{int} C + [\eta]^2_{int} C^2) \quad (15.8)$$

Values for $[\eta]_{int}$ vary for proteins and other polymers. It is 3.6 for hemoglobin; for polypeptides in general it is equal to $0.0238\ M^{0.5}$.

The value of $OT\eta_{H_2O}/k_2$ is an estimate of the oxygen transport capacity of a respiratory pigment solution. A graph of $OT\eta_{H_2O}/k_2$ clearly indicates that there is an optimal pigment concentration at which the oxygen transport capacity is maximal, (C_{opt} Figure 15–20A). The value of C_{opt} is equal to $1/[\eta]_{int}$.

The optimal protein concentration for O_2 transport is inversely dependent on the pigment molecular weight (Figure 15–20B); both the optimal transport capacity and the optimal concentration of pigment are greater for lower molecular weight pigments. We would therefore expect to have a low molecular weight pigment in solution if hemolymph viscosity was important. Why then are the hemocyanins of mollusks and arthropods very large polymers

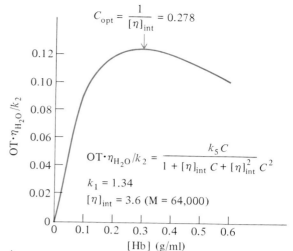

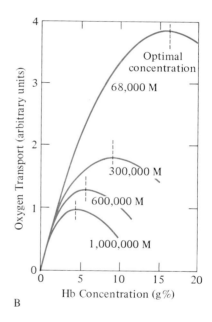

FIGURE 15–20 **(A)** Theoretical relationship for oxygen transport capacity of solutions of varying concentrations of a respiratory pigment ([Hb]) showing the optimal concentration C_{opt} (see text for more detail). **(B)** The oxygen transport capacity $OT\eta_{H_2O}/K_2$ of hemoglobin solutions depends on the molecular weight and concentration. The lower molecular weight hemoglobins have a higher maximal O_2 transport capacity, which occurs at a higher hemoglobin concentration. *(From Snyder 1977.)*

of smaller subunits? A high molecular weight respiratory pigment will not be excreted by an ultrafiltration excretory system, whereas a low molecular weight pigment might be filtered and excreted.

Another consideration is that the colloid osmotic pressure would be far greater for a high concentration of low molecular weight pigment and this could dramatically upset the capillary fluid balance of a closed circulatory system and also the intracellular-hemolymph fluid balance. It seems that the high viscosity of high molecular weight pigments in solution is not a major problem.

There is presumably some functional significance as to whether respiratory pigments are located within erythrocytes or appear free in solution. Hemocyanins, chlorocruorins, and some hemoglobins are synthesized within cells but are released to circulate in solution. Many hemoglobins (especially those of vertebrates) and hemerythrins circulate in blood cells. We can only speculate as to the reasons why some pigments are dissolved in solution and others are located within blood cells, but such speculation provides an instructive review of the basic principles of the physics of circulatory flow, respiratory pigment structure and function, and optimal hematocrit theory.

The intraerythrocyte hemoglobins of vertebrates and hemerythrins are low molecular weight pigments that could be in solution without compromising blood viscosity. In fact, the viscosity of lysed mammalian blood is less than that of normal blood (Cokelet and Meiselman 1968; Schmidt-Nielsen and Taylor 1968), so there appears to be a viscosity cost ot having red blood corpuscles. However, having hemoglobin located within erythrocytes is advantageous for three reasons. First, a low molecular weight pigment would be excreted by the kidneys and therefore must be packaged inside erythrocytes. Second, small proteins would exert a considerable colloid osmotic pressure and upset plasma-interstitial fluid balance; their restriction within cells (which are impermeable to cations) makes them osmotically inactive. Third, the Fahraeus-Lindqvist effect reduces the viscosity of blood in small vessels and the effective *in vivo* blood viscosity is lower for blood corpuscles than for a hemoglobin solution (Snyder 1973).

Oxygen Transfer.   Respiratory pigments can transfer O_2 from one tissue (with a high P_{50} pigment) to another (with a lower P_{50} pigment). For example, O_2 is transferred between vascular and coelomic hemerythrins in *Dendrostomum* and *Siphonophora* (see above). Various annelids also have vascular and coelomic pigments, probably with similar transfer functions. O_2 is transferred by the mammalian placenta (e.g., goat) from the maternal blood (P_{50} about 5.2 kPa) to the fetal blood (P_{50} about 3.1 kPa); the maternal P_{50} is normally about 4.2 kPa, but is right-shifted during pregnancy. In sheep, there is a larger

P_{50} difference between maternal and fetal blood, presumably because of the greater diffusion distance between the maternal and fetal blood across the sheep placenta compared with the goat placenta. The steep portion of the fetal O_2 dissociation curve is progressively utilized to maintain fetal O_2 transport during pregnancy as fetal O_2 demand outstrips placental O_2 diffusion capacity (Figure 15–21).

There are numerous examples of O_2 transfer between circulating pigments and intracellular muscle or nerve pigments, e.g., hemoglobin and myoglobin in vertebrates, hemocyanin and myoglobin in chitons, hemerythrin and myohemerythrin in sipunculids, and chlorocruorin and myoglobin in the sabellid worm *Potamilla*. The role of the intracellular pigment may be twofold. It may function as a short-term O_2 store during transitory anoxia. For example, vertebrate muscle when active experiences a transitory cessation of blood flow during each muscle contraction and the intracellular myo-

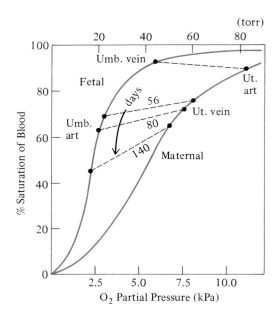

FIGURE 15–21 The fetal-maternal shift in O_2 equilibrium curve of the sheep facilitates O_2 loading by the fetal blood at the placenta (umbilical vein) from the maternal blood (uterine artery). The left-shifted fetal curve allows a higher fetal arterial blood percent saturation than for maternal arterial blood and a lower fetal venous blood percent saturation than the maternal blood, although the fetal arterial and venous pO_2 are lower than for maternal blood. As the term of pregnancy increases, there is a progressive decline in the uterine vein/umbilical artery pO_2 because the placenta does not develop concomitantly with the increased fetal O_2 demand. *(Modified from Barron and Meschia 1954.)*

globin would function as a short-term (a few seconds) O_2 store. An intracellular respiratory pigment can also facilitate diffusion of O_2 from the interstitial fluid into the cell. *In vitro* experiments with membranes containing hemoglobin, myoglobin, and hemerythrin solutions have demonstrated that the presence of the pigment facilitates the diffusion of O_2 across the membrane, in comparison with the diffusion of N_2 (Kreuzer 1970; Wittenberg 1963; Scholander 1960). Facilitation of O_2 diffusion by the respiratory pigment (plasma does not facilitate O_2 transfer) depends on the concentration of the pigment and its mobility. Mobility is inversely proportional to molecular weight; myoglobins are often monomers of low molecular weight. Hemocyanins, which generally have a much higher molecular weight than hemoglobins and hemerythrins, do not facilitate diffusion. This most likely explains why there are no intramuscular, intraneural, or intratracheal hemocyanins.

Facilitation of O_2 transfer occurs best when the pO_2 difference across the membrane occurs at the steepest part of the O_2 dissociation curve. Facilitated diffusion may have little physiological significance for hemoglobin (or hemerythrin) located inside erythrocytes, since the pO_2s that they experience are high and not optimal for facilitation. However, other intracellular pigments are generally at low pO_2 and facilitated diffusion is likely to be significant. For example, myoglobin facilitates O_2 diffusion into pigeon breast muscle (Wittenberg, Wittenberg, and Caldwell 1975) and may facilitate O_2 uptake by parasitic, hemoglobin-containing nematodes whose VO_2 is greater than can be explained by simple diffusion (Rogers 1949). The tissue distribution (e.g., body wall) and sufficiently high P_{50} of some parasite hemoglobins (>0.05 kPa) also suggest a facilitated diffusion role. However, the P_{50} of some hemoglobins is so low that they cannot be deoxygenated even by completely anoxic conditions or a vacuum! The hemoglobin in the perenteric fluid of *Ascaris lumbricoides* (P_{50} about 0.001 kPa) apparently has no role in O_2 transport, facilitated diffusion, or storage but is a metabolic source of hematin. Hemoglobin of the larvae of *Gasterophilus*, which live in the stomach of horses, apparently transports O_2 even though the P_{50} is only about 0.003 kPa, based on its wide distribution throughout the tissues and its high concentration in clusters of cells surrounding tracheole branches.

The potential significance of facilitated diffusion by intracellular respiratory pigments may clarify the puzzling distribution of hemoglobin within animals (and also protozoans and plants; Schmidt-Nielsen 1983). Hemoglobins occur very disjunctly among animal phyla, often being present in only a few

members of taxonomic groups, or in parasitic but not free-living members, or as intracellular pigments in animals with no circulating pigments (e.g., insects) or other respiratory pigments (e.g., hemocyanins). Porphyrins are common intracellular molecules, being readily synthesized from glycine and succinyl-CoA. It would seem "relatively easy" to evolve an intracellular diffusion-facilitating hemoglobin by a combination of the ubiquitous porphyrin, an Fe^{2+}, and a protein (globin). Circulating erythrocytes may have evolved from fixed, hemoglobin-containing precursor cells. The evolution of hemocyanins probably did not follow such a sequence because they do not facilitate O_2 diffusion and are not found in muscle or nerve cells.

Oxygen Storage. The respiratory pigments of some animals clearly have an O_2 storage role. This is particularly true for animals that experience periodic anoxia, e.g., diving vertebrates. Diving mammals and birds rely totally on pulmonary respiration for O_2 transport and experience transient and complete periods of respiratory apnea while submerged (Anderson 1966; Lenfant, Johansen, and Torrence 1970). During diving, blood is the most important potential O_2 store in the body because of its high O_2 capacity (1.3 ml O_2 per gram of hemoglobin), the very high hemoglobin content of blood (>15 g% hemoglobin), and high blood volume (>10% body mass; Table 15–5). Muscle myoglobin is the second most important O_2 store because of the often high intramuscular concentration (>5%) and the high skeletal muscle content of the body (20 to 40% body mass). Lung gas is only the third most important O_2 store despite its high O_2 content because lung volume is often reduced during diving to avoid excessive buoyancy and occurrence of the "bends." O_2 dissolved in body fluids is a relatively unimportant O_2 store. As expected, diving mammals and birds have higher O_2 stores than nondivers (*cf* man and divers in Table 15–5). The blood hemoglobin content of some marine mammals is actually higher than the optimal value for circulatory O_2 transport because of the importance of O_2 storage by blood (see Figure 15–19B).

Vascular respiratory pigments probably have an important O_2 storage role in many burrowing, aquatic invertebrates, such as benthic polychetes, sipunculids, and chironomid midge larvae. The storage time (time that pigment-bound O_2 would sustain resting VO_2) varies from about 15 min to 90 min and is therefore too short to sustain aerobic metabolism during periods of intertidal anoxia (e.g., six hours), but could be important for short-term anoxia, e.g., between periods of gill irrigation or during predator avoidance or feeding. Hemoglobin has a complex

TABLE 15–5

Oxygen stores of diving mammals and birds. Values are ml O_2 per kg body mass.

	Lungs	Arterial Blood	Venous Blood	Muscle	Dissolved	Total
Man	12.2	2.8	10.1	0.9	1.9	27.9
Mallard	12.3	4.7	10.2	1.8	2	31.0
Sea lion	16.5	5.0	7.2	8.1	2	39.4
Tufted duck	19.8	6.3	13.6	1.8	2	43.5
Walrus	17.4	7.7	11.5	10.0	2	48.6
Fur seal	21.8	6.7	9.9	11.7	2	52.1
Harbor seal	13.6	12.2	18.8	18.6	2	65.2
Sea otter	51.2	6.1	9.4	8.8	2	77.5
Ribbon seal	12.6	14.3	22.6	27.2	2	78.7

transport and storage role in *Chironomus* larvae. These midge larvae (e.g., *C. plumosa*) inhabit mud burrows in still, often stagnant, water. When the ambient water has a high pO_2 (about 20 kPa), the larvae spend about 50% of their time in respiratory activity (intermittent bursts of irrigation activity separated by pauses of up to 30 min), 35% filter feeding (an energetically expensive cycle of net spinning, violent burrow irrigation, and net eating), and 15% immobile. Respiratory activity increases if the ambient pO_2 declines below 6 kPa and feeding ceases if pO_2 declines below 2 kPa.

The hemoglobin of *Chironomus* has a hyperbolic O_2 equilibrium curve (n is about 1.1–1.2) with a very low P_{50} (0.014 to 0.08 kPa) and a relatively large Bohr shift ($\phi = -0.5$ to -1.0; Figure 15–22). The blood hemoglobin has no role in O_2 transport at high pO_2; the arterial and venous pO_2 difference provides no A-V O_2 content difference for the saturated hemoglobin, but a substantial A-V difference for dissolved O_2. At low pO_2, there is a substantial hemoglobin A-V O_2 difference provided by a small arterial to venous pO_2 difference, but there is a negligible A-V O_2 difference for dissolved O_2. At intermediate pO_2, there is a contribution by both hemoglobin and dissolved O_2 to the A-V O_2 difference.

The chironomid hemoglobin would obviously function for O_2 transport only at the progressively lower pO_2s during the nonirrigating phase of the respiratory ventilation cycle. The role of hemoglobin is clearly complex, altering through the irrigation cycle and depending on the ambient pO_2. At high ambient pO_2 and during irrigation, the O_2 requirements are met by dissolved O_2 transport; the hemoglobin-bound O_2 is a store. During an irrigation pause, the pO_2 in the burrow declines as the O_2 is absorbed and consumed, and the hemoglobin-O_2 store is utilized as the arterial and venous pO_2 decline. At lower pO_2s, an arterial-venous difference would develop and the hemoglobin assumes a significant role in O_2 delivery to the tissues. The venous pO_2 declines to the steep part of the O_2 dissociation curve and then there is little further decline. This "buffering" of venous pO_2 maintains the intracellular-venous pO_2 gradient required for O_2 transport into the cells.

Oxygen Secretion by Swim-Bladders. The swim-bladder of teleost fish provides neutral buoyancy control by regulating the volume of gas in the swim-bladder at various water depths (see Chapter 10). The amount of gas in the swim-bladder must vary dramatically with depth because of variation in hydrostatic pressure (10 m = 1 atm = 101 kPa), and so swim-bladders have a gas-secreting mechanism to maintain a constant gas volume and many also have a gland for gas reabsorption.

Secretion of gases from the blood (which has a relatively low O_2, pCO_2, and pN_2) into the swim-bladder is a formidable physiological problem, especially at great depths when the swim-bladder pO_2, pCO_2, and pN_2 are exceedingly high. There is no active transport mechanism for any gas; only passive transport mechanisms, such as diffusion and convection, can transport gases. However, the Root effect of hemoglobin provides a passive mechanism for gas transport into the swim-bladder against a partial pressure gradient because the saturation O_2 content of blood depends on the pCO_2 and pH.

Consider a fish at a depth of 1000 m (pressure = 101 atm). Blood entering the rete to the gas-secreting gland of the swim-bladder is saturated and has an O_2 content of about 10 ml O_2/100 ml blood at a pO_2 of 0.14 atm (Table 15–6). Air in the swim-bladder has a pO_2 of about 21 atm (if it has the same

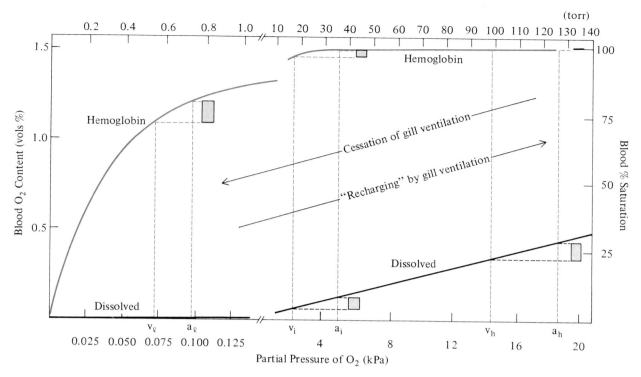

FIGURE 15–22 The O_2 equilibrium curve for hemoglobin of a chironomid midge *Chironomus plumosa* shows no role of the hemoglobin in O_2 transport at high pO_2 and a reliance on dissolved O_2 for transport but a major transport role for hemoglobin at low pO_2 during cessation of gill ventilation when dissolved O_2 does not provide significant O_2 transport. Abbreviations are: a, arterial; v, venous; ℓ, low; i, intermediate; h, high.

TABLE 15–6

Hypothetical values for the blood O_2 content (vols percent) of the afferent rete blood, blood in the fish swim-bladder gas gland with the pH reduced by lactate secretion to induce a Root effect, and for efferent rete blood. Shown are the values for normal secretion of O_2 and the effects of reduced blood O_2 capacity and decreased counter-current exchange between efferent and afferent blood. Temperature is 10° C; O_2 solubility is 2.24 mol O_2 L^{-1} atm^{-1}.

	Afferent	**Gas Gland**	**Efferent**	**Unloaded**
Normal Gas Secreting Conditions				
Hb-O_2	10	8	8	2.0
Dissolved	0.5	76	2.1	−1.6
Total	10.5	84	10.1	0.4
Decreased Blood O_2 Capacity				
Hb-O_2	8	6.4	6.4	1.6
Dissolved	0.5	76	2.1	−1.6
Total	8.5	82.4	8.5	0
Decreased Rete Exchange				
Hb-O_2	10	8	8	2.0
Dissolved	0.5	76	2.5	2.0
Total	10.5	84	10.5	0

percent composition as ambient air; see below) but the blood would still have the same saturation O_2 content (10 vols percent). Anaerobic metabolism by the gas gland produces lactate$^-$ and H$^+$, which decreases the blood pH. The Root effect, which is particularly pronounced for fish blood (see Figure 15–11A, page 743), decreases the saturation O_2 content of hemoglobin, e.g., from 10 to 8 vols percent. This means that 2 vols percent of O_2 is unloaded from the blood hemoglobin into physical solution, and the much greater amount of dissolved O_2 dramatically increases the pO_2 in the swim-bladder. When blood leaves the gas gland along the rete, it re-equilibrates with incoming blood and exchanges O_2 into the afferent blood. The pO_2 of the efferent blood decreases, although it is always greater than that of the incoming blood at the same part of the rete. When the efferent blood leaves the rete, it has a higher pO_2 and more dissolved O_2 than afferent blood (about 1.6 vols percent more) but a lower blood O_2 content; about 0.4 vols percent of O_2 has been unloaded into the swim-bladder.

Gas secretion by the swim-bladder gas gland requires that the Root effect has a sufficient magnitude to unload more O_2 than the additional dissolved O_2 content of efferent blood. Thus, (1) the afferent-efferent pO_2 difference must be minimal, i.e., there must be effective countercurrent exchange of O_2 between afferent and efferent blood, and (2) the blood must have a sufficiently high O_2 content that the difference in O_2 content of afferent and efferent blood is substantial compared to dissolved O_2. The example of gas secretion in Table 15–6 provides no O_2 secretion if the O_2 content of afferent blood is reduced from 10 to 8 vols percent, because all O_2 unloaded by the Root effect is lost as additional dissolved O_2 in the efferent blood. Similarly, there is no O_2 secretion in this example if the efferent pO_2 is sufficiently high that the added dissolved O_2 exceeds that unloaded from the hemoglobin.

There are very high amounts of O_2 dissolved in the blood along the rete and in the gas gland because of the extremely high pO_2 values. Changes in the rate of gas secretion therefore cannot be very fast because these large amounts of dissolved O_2 in the blood "buffer" the effect of secretion of O_2 from the blood into the swim-bladder or reabsorption from the swim-bladder.

The gas gland rete would not only exchange O_2 but also lactate$^-$ and H$^+$. Loss of lactate$^-$ would potentially reverse the Root effect along the efferent capillary and therefore impair O_2 transfer from the efferent to afferent blood. However, the unloading of O_2 from hemoglobin by the Root effect (i.e., the "Root-Off" reaction) is faster than the reverse ("Root-On") reaction. The 1/2 time for the Off

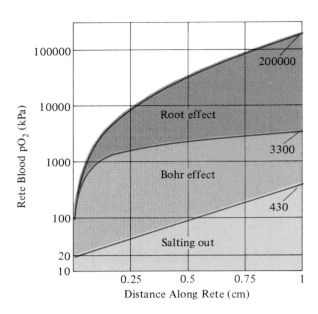

FIGURE 15–23   Theoretical relationships between pO_2 and distance along swim-bladder rete for the salting-out effect; the Bohr effect; and the combined Root, Bohr, and salting-out effects.   *(Modified from Kuhn et al. 1963.)*

reaction is about 50 msec and is 10 to 20 sec for the On reaction.

Gases other than O_2, including the inert gases such as N_2 and Ar, are also accumulated against partial pressure gradients in the swim-bladder, although O_2 is often the major constituent of swim-bladder gas. Even if the percent composition of the swim-bladder gas were the same as ambient air (e.g., $pN_2 = 80$ kPa), then all of the swim-bladder gases must still be accumulated against a partial pressure gradient. The accumulation of these gases other than O_2 is in part explained by a second mechanism for gas secretion, the salting-out effect. Addition of a solute (e.g., lactate$^-$) decreases the solubility of gases (see Chapter 12), thereby driving gases out of solution even at a constant partial pressure. The salting-out effect is, however, much less important than the Root effect (Figure 15–23). The salting-out effect of 20 mM lactate could concentrate N_2 to 25 atm, but only 5 mM lactate$^-$ could concentrate O_2 to 3000 atm by the Root effect (Kuhn et al. 1963). The Bohr effect also contributes to the elevation of swim-bladder pO_2.

Carbon Dioxide

The general principle of CO_2 transport is the same as for O_2 transport; an arterial-venous difference in the CO_2 content of blood is responsible for net convective transfer of CO_2 from the tissues to the respiratory surface. However, the mechanism of CO_2 transport is fundamentally different from O_2 transport. CO_2 is much more soluble than O_2 and has a higher diffusion coefficient (see Chapter 12), and so considerable amounts of CO_2 are present in blood even at low pCO_2. CO_2 can be readily eliminated by cutaneous diffusion without the need for a specialized respiratory surface. There is significant cutaneous CO_2 loss even in terrestrial vertebrates that rely on lungs for O_2 uptake.

Most of the CO_2 present in blood (or water) is actually not dissolved CO_2 but is chemically combined with water to form **bicarbonate** ions; there is only a small amount of the intermediate, carbonic acid because it rapidly dissociates into bicarbonate (Figure 15–24).

$$CO_2 + H_2O \rightleftharpoons H_2CO_3 \rightleftharpoons H^+ + HCO_3^- \quad (15.9a)$$

The initial chemical reaction of CO_2 and H_2O is relatively slow, whereas the second reaction is virtually instantaneous. The rate of CO_2 combining with venous blood and of CO_2 leaving arterial blood might therefore be expected to be rate limited by the first reaction. However, the presence of an enzyme, **carbonic anhydrase**, inside red cells dramatically increases the reaction rate and facilitates the formation of HCO_3^-. Small mammals typically have higher levels of carbonic anhydrase in their red blood corpuscles compared to larger mammals,

reflecting their high mass-specific CO_2 transport. However, the activity of carbonic anhydrase is not essential for CO_2 transport (inhibitors such as acetazolamide have minimal effects on CO_2 transport and acid base balance).

A significant amount of **carbonate** ion is present at high pH (>8.5).

$$CO_2 + OH^- \rightleftharpoons HCO_3^- \rightleftharpoons H^+ + CO_3^{2-} \quad (15.9b)$$

Some CO_2 combines with the $-NH_2$ groups of proteins to form **carbamino** compounds.

$$CO_2 + -NH_2 \rightleftharpoons -NHCOO^- + H^+ \quad (15.9c)$$

This binding is more important in deoxygenated blood than in oxygenated blood.

For human arterial blood, most of the "CO_2" is present in the plasma as HCO_3^-, with about 1/3 of the HCO_3^- located in the erythrocytes (Table 15–7). There is a low concentration of carbamino-CO_2 in erythrocytes and none in plasma. The CO_2 concentration is low in plasma and even lower in erythrocytes. The partitioning of total "CO_2" in venous blood is similar, except all concentrations are slightly higher than for arterial blood. Most of the arterial-venous (A-V) difference in CO_2 concentration is plasma HCO_3^- (57%) and erythrocyte carbamino-CO_2 (27%); plasma and erythrocyte CO_2 and erythrocyte HCO_3^- each contribute less than 10% of the A-V CO_2 difference.

Let us first consider CO_2 transport in vertebrate blood because it is well understood. There is no specific carrier for CO_2 in blood, hence blood does not become saturated with CO_2. The CO_2 dissociation curve for human blood (which is typical of vertebrate and invertebrate blood) has a rapid increase in CO_2 content at low pCO_2 and a continued, but slower, increase in CO_2 content at higher pCO_2 (Figure 15–25). The total CO_2 content of blood is the sum of dissolved CO_2, HCO_3^- (and CO_3^{2-}), and carbamino-CO_2. For oxygenated human blood, the CO_2 content is about 48.2 vols percent at the arterial pCO_2 of 5.4 kPa and the CO_2 content is about 50 vols percent at the venous pCO_2 of 6.3 kPa; the A-V difference is 1.8 vols percent. However, the CO_2 dissociation curve of deoxygenated blood is left-shifted and deoxygenated blood has a higher CO_2 content than oxygenated blood at the same pCO_2 (the **Haldane effect**). As hemoglobin unloads O_2, its negative charge increases and this facilitates H^+ buffering. This in turn facilitates the combination of CO_2 with water to form HCO_3^- and H^+ and increases the CO_2 content of blood. The Haldane effect approximately doubles the A-V CO_2 content difference for constant arterial and venous pCO_2. From the dissociation curve for venous blood (75% saturated hemoglobin), the venous CO_2 content is

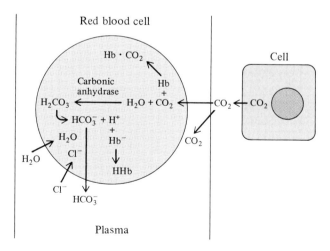

FIGURE 15–24 Transport of CO_2 in the blood is modified by the role of carbonic anhydrase in red blood corpuscles. CO_2 is rapidly hydrated to form H^+ and HCO_3^- inside red blood corpuscles; the HCO_3^- diffuses into the plasma in exchange for Cl^- (chloride, or Hamburger, shift) and the H^+ is buffered by hemoglobin. There is a net uptake of H_2O by the red corpuscle.

TABLE 15–7

	Arterial	Venous	A-V Difference Vols %	A-V Difference (% of total)
Partitioning of CO_2 content (ml L^{-1}) of arterial and venous human blood between plasma and erythrocytes as dissolved CO_2, HCO_3^-, and carbamino compounds. *(From Comroe 1965.)*				
Plasma				
CO_2	16	18	2	5%
HCO_3^-	341	363	22	57%
Carbamino	—	—	—	—
Total plasma	357	381	24	—
Erythrocyte				
CO_2	8	9	1	3%
HCO_3^-	96	99	3	8%
Carbamino	22	32	10	27%
Total erythrocyte	126	139	4	—
Whole Blood				
CO_2	24	27	3	8%
HCO_3^-	437	461	24	65%
Carbamino	22	32	10	27%
Total whole blood	482	520	38	100%

52:0 rather than 48 vols percent. The A-V CO_2 difference is thus 3.8 vols percent and the CO_2 transport is 3.8/100 × 5000 or 188 ml CO_2 min^{-1} (cardiac output is 5000 ml min^{-1}; resting VO_2 is 230 ml min^{-1}; RQ is 0.82).

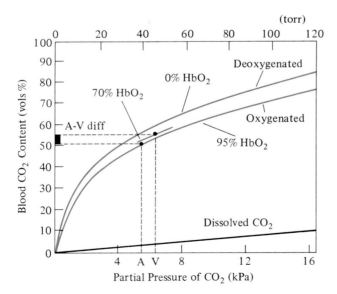

FIGURE 15–25 CO_2 dissociation curve for human blood showing the effect of hemoglobin oxygenation on CO_2 content (Haldane effect) and the arterial-venous (A-V) difference in CO_2 content.

The red cell membrane is relatively impermeable to cations (e.g., Na$^+$, K$^+$, H$^+$) and so the H$^+$ formed by hydration of CO_2 cannot easily diffuse into the blood. Hemoglobin buffers the H$^+$ ions. The buildup of HCO_3^- inside the RBC is minimized by the exchange of RBC HCO_3^- for plasma Cl$^-$; this is the chloride or **Hamburger shift**. The elevated RBC Cl$^-$ concentration causes water to osmotically enter the red cells and so their volume increases slightly; thus, venous blood has a slightly higher hematocrit than arterial blood.

CO_2 transport is similar in principle for other vertebrates. The general shape of the CO_2 dissociation curve is similar for fish (Figure 15–26); the CO_2 dissociation curve of salmon blood has a Haldane effect, whereas that of dogfish has no Haldane effect (Albers 1970). Fish red blood cells contain carbonic anhydrase, and the concentration is related to the metabolic activity of the species. Some CO_2 is also present as CO_3^{2-} at low temperatures. It is not clear whether significant amounts of carbamino compounds form in fish blood.

Invertebrate CO_2 transport is generally similar to that for vertebrates. Many invertebrates have hemocyanin, which is as effective a H$^+$ buffer as hemoglobin. The total CO_2 capacity of hemolymph is proportional to its hemocyanin content. It is not clear whether significant amounts of CO_2 are transported as carbamino compounds, although the low isoelectric points of hemocyanins suggest that

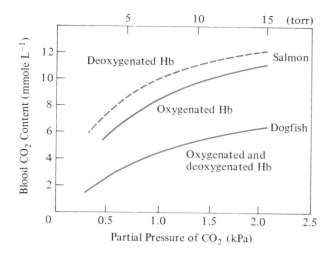

FIGURE 15–26 CO_2 dissociation curve for blood of the Atlantic salmon *Salmo salar* has a pronounced Haldane effect, whereas blood of the dogfish *Mustelus canis* has no Haldane effect. *(From Albers 1970.)*

their amino acids may be unprotonated at physiological pH and therefore capable of forming carbamino compounds.

Transport of Other Gases

Respiratory pigments can transport gases other than O_2 and CO_2. Carbon monoxide generally binds to the prosthetic group of hemoglobin and chlorocruorin with a much higher affinity than O_2. This high affinity of hemoglobin for CO effectively blocks O_2 transport, hence CO exposure has been a useful experimental tool for evaluation of the role of respiratory pigments in O_2 transport. Hemerythrin and hemocyanin have a low affinity for CO because of their different ligand-binding schemes.

The hemoglobin of certain mollusks may transport H_2S (Doeller et al. 1988). The bivalve mollusk *Solemya velum*, which inhabits marine sediments containing O_2 and H_2S, has intracellular chemoautotrophic bacteria that oxidize H_2S. *Solemya* has two hemoglobin types; in the absence of H_2S, they both transport O_2 but in the presence of H_2S one is converted to ferric hemoglobin (Fe^{3+}) and reversibly binds H_2S. This unusual hemoglobin reaction may transport H_2S to the symbiotic bacteria. Sulfur transport has also been ascribed to the colorless blood of a giant clam *Calyptogena* (Arp et al. 1984) and the hemoglobin of a tube worm *Riftia* (Arp, Childress, and Fisher 1987), but the prosthetic group of the pigment is not involved in H_2S transport.

Acid–Base Balance

Water molecules spontaneously dissociate, to a minor extent, into H^+ and OH^-.

$$H_2O \rightleftharpoons H^+ + OH^- \qquad (15.10a)$$

The apparent ion product ($K'_{eq} = [H^+][OH^-]$) is about 10^{-14}. The K'_{eq} depends on the temperature (see below) and ionic concentration. At 25° C and neutrality, pH = pOH = 7. The pH of animal body fluids is generally about 0.6 U greater than the neutral pH; intracellular pH is generally more acidic than extracellular fluids and is at about the neutral point of water.

Acids dissociate into H^+ and anions.

$$HA \rightleftharpoons H^+ + A^- \qquad (15.10b)$$

The apparent equilibrium constant ($K'_{eq} = [H^+][A^-]/[HA]$) is large for strong acids (e.g., HCl) and small for weak acids (e.g., most organic acids). Application of the law of mass action yields the following relationship.

$$pH = pK + \log \frac{[A^-]}{[HA]} \qquad (15.10c)$$

The *pK* is the pH at which $[A^-] = [HA]$ (Figure 15–27). At lower pH, there is more HA than A^- and at higher pH there is more A^- than HA. A weak acid is an effective pH buffer because the addition of a strong acid will conjugate the H^+ with A^- to form HA effectively removing H^+ and limiting the pH change. Alternatively, adding a strong base will conjugate the OH^- with HA to form A^- and H_2O.

The **buffer capacity** (β) is the amount of strong base (or strong acid) required to change the pH by 1 unit.

$$\beta = \Delta mmole/\Delta pH \qquad (15.11)$$

The unit for buffer capacity is mmol L^{-1} pH^{-1}, or "Slyke." The buffer capacity depends on both the [HA] + [A^-] and pH; it is maximal when [HA] = [A^-], or pH = p*K* (Figure 15–27). Weak acidic groups of proteins (e.g., imidazole), which have pKs of about 7 to 8, are important blood buffers. The buffer capacity of body fluids varies dramatically from almost 0 for fluids containing little protein (e.g., *Urechis* plasma and echinoderm plasma) to about 32 Slykes (Table 15–8). Hemoglobins have a buffer capacity of about 0.12 to 0.23 Slyke g^{-1} and hemocyanins about 0.14 Slyke g^{-1}. The main intracellular buffer in mammals is histidine residues of protein.

The total concentration of HCO_3^-, phosphates, and protein anions is called the **buffer base** because

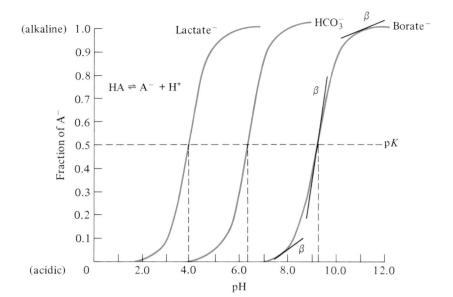

FIGURE 15–27 Titration curves for lactate ($CH_3CHOHCOO^-$), bicarbonate (HCO_3^-), and borate ($H_2BO_3^-$); the buffering capacity (β) is maximal at the pK (lactate, 3.7; bicarbonate, 6.2; borate, 9.1).

they are the main weak acids/bases responsible for buffering. The remaining ions, such as Na^+, K^+, Ca^{2+}, Cl^-, and SO_4^{2-}, are essentially completely dissociated at physiological pH and do not contribute any buffer capacity (i.e., there is no NaOH, HCl, etc. present). These completely dissociated ions are called **strong ions**. The difference in concentration between strong cations and strong anions is the **strong ion difference**; it must equal the buffer base concentration since blood is electroneutral. Thus, a change in blood buffer base must be accompanied by a change in strong ion difference and vice versa.

TABLE 15–8

Buffer capacity (β) for plasma and blood of various invertebrates and vertebrates. Units for buffer capacity are Slykes (mmol HCO_3^- L^{-1} pH^{-1}).		
		Buffer Capacity
Annelid	*Urechis* plasma	0
	blood	4.9
Sipunculid	*Sipunculus*	3.5
Crustacean	*Carcinus*	2.4–10.6
Lungfish	*Protopterus*	15.2
Elasmobranchs	*Squalus* plasma	6.5
	blood	9
Teleosts	*Opsanus*	6.7
	Scomber	14.8
Amphibians	*Necturus*	8.0
	Rana	16.4
Reptiles	Alligator	22.6
Mammals	Beaver	27
	Man plasma	6.5
	blood	30.8

Acid–base balance is inextricably linked to CO_2 production, transport, and excretion, and strong-ion regulation. At low to moderate pHs, the relationship between pH and CO_2 concentration is determined by the first dissociation constant (K_1) for CO_2 and water;

$$pH = pK_1 + \log \frac{[HCO_3^-]}{[CO_2]}$$
$$= pK_1 + \log \frac{[HCO_3^-]}{\alpha CO_2 \, pCO_2} \quad (15.12a)$$

The (CO_2) is determined by the CO_2 solubility (αCO_2) and the partial pressure (pCO_2). At high pH, a second reaction occurs and

$$pH = pK_2 + \log \frac{[HCO_3^-]}{[CO_3^{2-}]} \quad (15.12b)$$

Both pK_1 and pK_2 are temperature and salinity dependent; they decrease with increasing temperature and increasing salinity (Truchot 1976).

The complex relationship between HCO_3^- concentration, pCO_2, and pH can be summarized by the pH-bicarbonate diagram (Davenport diagram; Figure 15–28). At any CO_2 content, the HCO_3^- concentration increases exponentially with pH; higher pCO_2 values yield a higher HCO_3^- at a given pH. Acidosis occurs if excess H^+ is present and alkalosis occurs if excess OH^- is present (compared to the normal H^+ and OH^- values). The Davenport diagram readily allows the interpretation of acid base imbalance by indicating the nature of the acid base disturbance. Hypoventilation or hyperventilation will alter the blood pCO_2 level and result in respiratory acidosis (high pCO_2) or respiratory

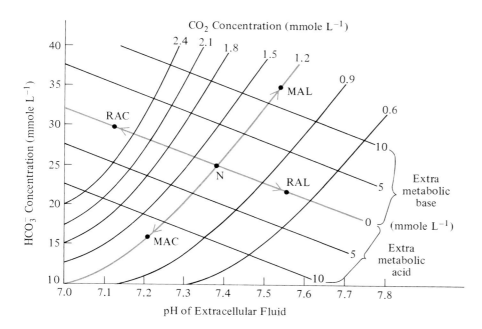

FIGURE 15–28 The pH-bicarbonate relationship (Davenport diagram) for human blood. The normal acid–base balance (position N; 1.2 mmol L^{-1} CO_2, pH 7.4; 5.44 kPa pCO_2) can be disturbed by respiratory acidosis (RAC) or alkalosis (RAL), or metabolic acidosis (MAC) or alkalosis (MAL).

alkalosis (low pCO_2). The normal acid base balance, indicated by position N in Figure 15–28, can be shifted by hyperventilation or hypoventilation to respiratory acidosis (RAC) or alkalosis (RAL). Alternatively, metabolic disorders (addition or excretion of excess acid or base by gills, kidneys, or chemical buffering by tissues) cause metabolic acidosis (MAC) or metabolic alkalosis (MAL). Acid base disturbances can be partially or even completely compensated. For example, a metabolic acidosis (MAC) can be partially compensated by a respiratory alkalosis; a metabolic alkalosis can be partially compensated by a respiratory acidosis. Similarly, respiratory alkalosis or acidosis can be partially compensated by metabolic acidosis or alkalosis.

Control of acid base balance is similar in principle for invertebrates (Figure 15–29). For example, the crab *Carcinus* develops a respiratory acidosis when exposed to air (pCO_2 increases from 0.16 to 0.68 kPa) but the blood pH and $[HCO_3^-]$ increase (metabolic compensation). Similar acid base imbalances and compensation occur in crustaceans in response to increased pCO_2 or pO_2 of the water, hypoxia, exercise, or variation through the molting cycle.

The blood pH of vertebrates and invertebrates is generally similar (at least at the same temperature; see below), but the details of acid–base balance differ markedly for aquatic and terrestrial animals. For example, compare the normal acid base status of a water-breathing crab *Carcinus* (see Figure 15–29) with that of an air-breathing human (see Figure 15–28), on the same Davenport diagram; the "normal" values are found at very different positions for

the crab and human. For water-breathing animals, the blood pCO_2 is low because of the relatively high solubility of CO_2 compared to O_2 (see Chapter 12) and so the blood HCO_3^- concentration is low (Figure

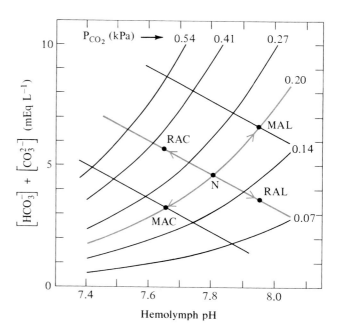

FIGURE 15–29 Davenport diagram for blood of the crab *Carcinus*. The normal acid–base balance (N; 4.5 mEq L^{-1} HCO_3^- + HCO_3^{2-}; pH 7.8; 0.20 kPa pCO_2) can be disturbed by respiratory acidosis (RAC) or alkalosis (RAL), or metabolic acidosis (MAC) or alkalosis (MAL). *(Modified from Truchot 1983.)*

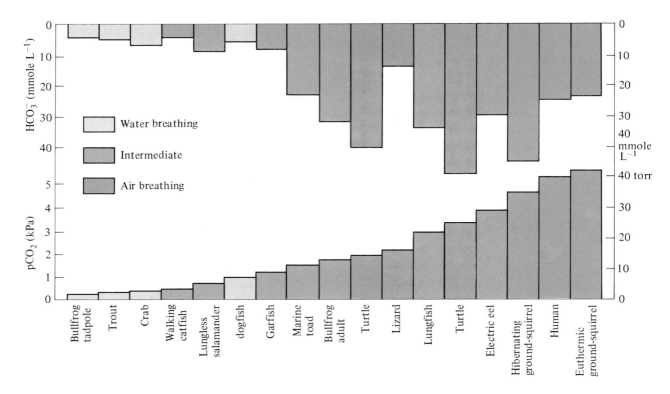

FIGURE 15–30 Blood pCO$_2$ and HCO$_3^-$ concentrations for some water-breathing and air-breathing vertebrates.

15–30). Consequently, regulation of blood pH is not readily accomplished by variation in blood pCO$_2$. Rather, counter-ion exchange of Na$^+$ for H$^+$ and NH$_4^+$, and Cl$^-$ for HCO$_3^-$ (and perhaps even OH$^-$) is used to regulate acid–base balance. The gills, rather than the kidneys, are responsible for most counter-ion exchange in fish.

For air-breathing animals, the blood pCO$_2$ is fairly high, as is the blood HCO$_3^-$ concentration. Regulation of blood pH is readily accomplished by variation in respiration, e.g., hyperventilation causes a respiratory alkalosis and hypoventilation a respiratory acidosis. In addition, acid–base balance can also be regulated by the excretion of H$^+$, HCO$_3^-$, and other ions via the kidneys or intestine.

Temperature influences the physicochemical equilibrium between H$_2$O and H$^+$/OH$^-$, e.g., the pH for neutrality varies from 7.47 at 0° C to 6.63 at 50° C (Harned and Robinson 1940), i.e., the pH of pure water decreases with an increase in temperature. The effect of temperature on pH is not linear, but the ΔpH/ΔT is about -0.017 U °C^{-1} (Figure 15–31A). Some buffer solutions have a similar pH-temperature dependence as pure water (e.g., imida-zole) but many do not (e.g., bicarbonate, phosphate). The ΔpH/ΔT is about -0.019 to -0.020 for samples of hemolymph of crustaceans in a closed system (i.e., no exposure to air), and the *in vivo* effect of temperature on hemolymph is about -0.016 U °C^{-1}.

The effect of temperature on the pH of animal body fluids is potentially more complex than for water because animals can regulate their acid–base balance by physiological means (e.g., metabolism, respiration, ion exchange, and excretion). The pH of extracellular body fluid is generally about 0.6 U higher than the neutral pH for water, but body fluids often conform to the same temperature-dependent change in pH; intracellular pH also generally conforms to the temperature dependence of water pH but has about the same pH as neutral water (Figure 15–31B). The ΔpH/ΔT for hemolymph of many, but not all, animals is also about -0.017 to -0.020.

The obvious physiological consequence of the temperature dependence of body fluid pH is the maintenance of a constant ratio of [OH$^-$]/[H$^+$], i.e., maintenance of relative alkalinity. A constant relative alkalinity maintains the net charge of pro-

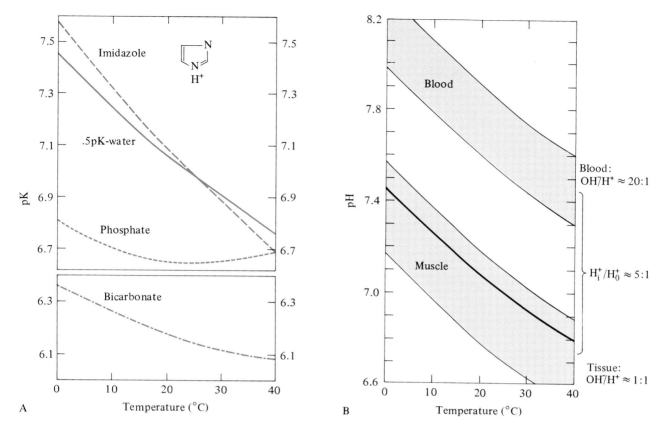

FIGURE 15–31 (A) Effect of temperature on the dissociation constant (pK) for neutral water and bicarbonate, phosphate, and imidazole buffer solutions. (B) General effect of temperature on blood pH and intracellular (muscle) pH; for blood, relative alkalinity is maintained at about $OH^-/H^+ = 20:1$ and intracellular relative alkalinity is maintained at about 1:1. *(From Rahn 1974; Rahn, Reeves, and Howell 1975. With permission of the American Review of Respiratory Disease.)*

teins, preserving molecular structure and function and constant Donnan ratios despite changes in actual pH. The ratio of $[OH^-]/[H^+]$ is about 20:1 for extracellular fluids and about 1:1 for intracellular fluids. The normal pH of intracellular fluids falls within an "ionization window" where most low-molecular weight, biologically important molecules are ionized. Ionization probably provides an important means for restricting the movement of these molecules across biological membranes (Davis 1958).

The maintenance of relative alkalinity of animal body fluids at varying temperatures is the result of precise regulation of respiratory, metabolic, and excretory processes so that relative alkalinity is preserved; it is not a physicochemical consequence of the dissociation of water. The decrease in pH of body fluids at higher temperature is a consequence of elevated pCO_2 and/or decreased blood

$[HCO_3^-]$ and $[CO_3^{2-}]$; this is apparent for such diverse animals as amphibians and crustaceans (Figure 15–32).

Some animals conform to the temperature dependence of the neutral point of water but many do not (Table 15–9). For example, many hibernating mammals maintain a constant blood pH despite their marked decline in body temperature. The hibernating mammal is "relatively" acidotic even though the pH is constant. For the little pocket mouse *Perognathus*, pH during hibernation at 10° C is 7.51 and when euthermic is 7.28; relative alkalinity is fairly constant (Withers 1977). The pCO_2 of pocket mouse blood declines with lowered body temperature but the blood $[HCO_3^-]$ is constant, resulting in relative acidosis at low body temperature. A similar relative acidosis is also seen in various other animals at lowered body temperature. In some other hibernating mammals, there is no

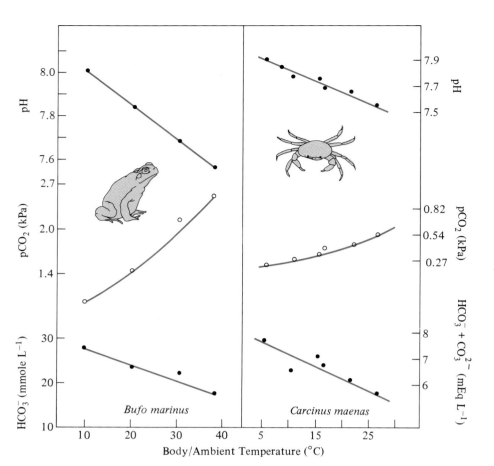

FIGURE 15–32 Blood pH, pCO_2, and HCO_3^- as a function of ambient/body temperature for an amphibian (marine toad *Bufo marinus*) and a crustacean (crab *Cancer maenas*). *(Data from Reeves 1969; Truchot 1978.)*

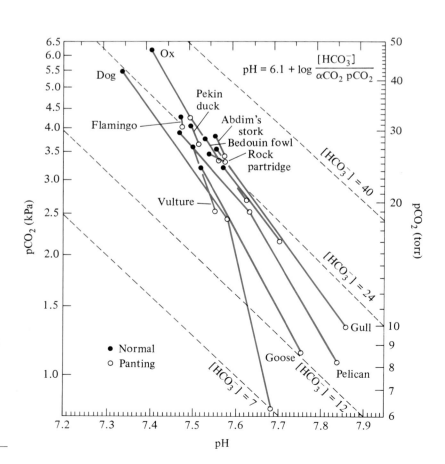

FIGURE 15–33 Relationship between pCO_2 and pH for blood of normal (closed circles) and panting (open circles) birds and mammals. Dotted lines indicate the uncompensated relationship between pCO_2 and pH for the indicated HCO_3^- concentration.

TABLE 15–9

Examples of temperature dependence of blood pH in various animals in comparison with water and some buffer solutions.

	$\Delta pH/\Delta°C$
Crab	−0.026
Carp	−0.019
Crab (*in vitro* open system)	−0.019
Turtle *in vitro*	−0.018
Alligator	−0.018
Water	**−0.017**
Frog *in vitro*	−0.017
Trout	−0.017
Iguana	−0.017
American locust (>25° C)	−0.017
Turtle	−0.016
Crab	−0.016
Toad	−0.016
HCO_3^-/imidazole buffer (25 mM + 20 mM)	−0.015
Frog	−0.013
Crab	−0.009
Bat[1]	−0.009
Pocket mouse[1]	−0.009
Desert iguana	−0.008
HCO_3^- buffer (25 mM)	−0.005
HCO_3^-/PO_4^{3-} buffer (25 mM + 20 mM)	−0.004
Sloth	−0.003
Monitor lizard	−0.002
Crab (*in vitro* closed system)	−0.002
Hedgehog[1]	−0.002
Hamster[1]	−0.002
Bat[1]	0
Ground squirrel[1]	0
American locust (<25° C)	0

[1] Euthermic and hibernating.

change in pH between euthermy and hibernation, whereas others have a $\Delta pH/\Delta°C$ of about −0.017.

Acid–base balance can become compromised by thermoregulatory panting. Hyperventilation by panting increases the evaporative heat loss for body temperature regulation, but could also "blow off" CO_2 and cause hypocapnia and respiratory alkalosis. Such an alkalotic effect of panting is observed in many mammals and birds (Figure 15–33). However, the acid base disturbance of panting can be minimized by hyperventilating the respiratory dead space, rather than the gas exchange surface, and by gular flutter in birds.

Defense

Blood can have a number of different defense roles. One is the restriction of blood loss in response to physical damage to the circulatory system (hemosta-sis). Another is to counter invasion by foreign materials, microorganisms, and parasites; this role can be generally described as immunity. Some animals use their blood as a physical defense system to deter predators.

Hemostasis

Several mechanisms can prevent the continued loss of blood from a ruptured blood vessel. The least effective is the loss of blood pressure due to excessive blood loss. Ruptured blood vessels may contract to stop blood loss. This is the main mechanism for stopping blood loss from cut blood vessels in cephalopods and many other mollusks since their blood does not clot well. In most animals, the blood itself forms a clot that temporarily plugs the vessel and blocks blood flow until the tissue damage is repaired.

The mechanism for blood clotting, or **coagulation**, is well studied for mammalian blood. The clotting process involves first the agglutination of platelets at the wound and then the formation of a fibrin clot that later contracts to constrict the blood vessel and form a solid plug (Figure 15–34). Blood clotting is a complex phenomenon, involving over 40 procoagulants and anticoagulants. The balance between these two antagonistic actions is normally anticoagulant, since blood clots must not form spontaneously in undamaged blood vessels. The effects of procoagulants predominate at the site of blood vessel damage and a clot forms.

There are essentially three steps in clotting of mammalian blood. First, a prothrombin activator substance(s) is formed by extrinsic or intrinsic means. In the extrinsic pathway, tissue damage releases thromboplastin, a complex group of substances including phospholipids and a proteolytic glycoprotein. Through a positive feedback cycle involving various intermediate substances, prothrombin activator is formed. In the intrinsic pathway, trauma to the blood activates platelets and Factor XII, then a series of factors are activated to form prothrombin activator. Prothrombin activator forms thrombin from prothrombin. Thrombin converts a large soluble plasma protein, **fibrinogen** (MWt 340 10^3), into an insoluble, fibrous protein, **fibrin**, by removing four low molecular weight peptides from fibrinogen. Fibrin polymerizes with other fibrin molecules. Fibrin-stabilizing factor strengthens the fibrin polymers by promoting covalent bonding between the fibrins. Once a blood clot forms, it initiates a positive feedback to promote further blood clotting, at least in the immediate vicinity where blood flow is stopped. The clotting process does not become a positive feedback cycle if the

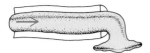

A. Severed vessel

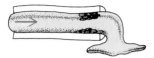

B. Platelets agglutinate

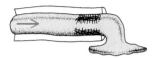

C. Fibrin appears

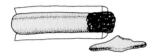

D. Fibrin clot forms

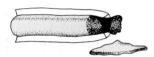

E. Clot retraction occurs

FIGURE 15–34 Clotting process in mammalian blood begins with platelet agglutination and fibrin formation; formation of a fibrin clot and then clot retraction plug the vessel and stop blood flow. *(Modified from Seegers and Sharp, 1948.)*

blood continues to flow because the concentrations of procoagulants remain lower than a critical level. Within a few minutes of the clot forming, the fibrin begins to contract because of platelets attached to the fibrin strands that continue to release fibrin-stabilizing factor and also by contraction of platelet actin-myosin molecules.

Invertebrates generally have blood clotting mechanisms, usually involving agglutination of blood corpuscles and sometimes blood proteins. In some annelids, even minor wounds may be lethal if they puncture the coelom because of the role of the coelomic fluid as a hydrostatic skeleton (Dales 1978). However, the ability to constrict the body at

septal junctions allows some annelids to survive fairly severe injuries and even amputation of anterior or posterior segments. There is no evidence for blood clotting in annelids, although some coelomic proteins may contribute to wound closure. Certain cutaneous secretions may also assist wound closure. In leeches, the coelom is occluded by body tissues and the blood vessels and sinuses can be occluded by muscle contraction at wounds. Cells migrate to the wound site within a few hours and seal off the wound. Scar tissue is formed but is not replaced by tissue regeneration. In glossophoriids and some other leeches, the male leech hypodermically impregnates the female (these leeches are hermaphroditic but protandrous) and the sperm migrate through the tissues to the ovisacs. In *Placobdella*, the tissue damage caused by sperm migration is apparently repaired, whereas other wounds are not.

Molluskan blood clots rather ineffectively by the aggregation of blood corpuscles that draw together to form a solid plug in the wound. The plug is then invaded by phagocytes and connective tissue cells, which heal the wound. This mechanism may be of value in small wounds or abrasions but is not of value when there is damage to large vessels or sinuses.

In arthropods, including the insects, blood coagulation is caused by hemocytic agglutination, plasma gelation, or both (Gupta 1985). The coagulocytes (or cystocytes) either discharge materials that promote plasma gelation or extrude long, thread-like cytoplasmic strands that form a meshwork to agglutinate hemocytes, or both. The hemocytes appear to be the source of the clotting factors. Vertebrate-like clotting factors, such as thrombin, prothrombin, or thromboplastin, have not been found in arthropods. Various mucoproteins or glycoproteins have been suggested to be the arthropod clotting protein, and a soluble fibrinogen has been reported in one crustacean (*Astacus*).

Immunology

Animals have various mechanisms for self-recognition of tissues and for defensive protection against invasion by foreign materials, whether inorganic, viral, microbial, plant, or animal (Cooper 1990). This ability to resist invasion is called **general immunity**. All animals have some innate immunity. This general defensive system requires self-recognition and uses a combination of various mechanisms, including phagocytosis by blood and tissue cells, destruction by digestive enzymes and/or acid stomach secretions, encapsulation responses, resistance of the skin or cuticle to invasion, or the presence in the blood of certain types of materials (e.g., complement proteins, antibodies) that attach to and destroy foreign materials (Table 15–10). Some animals, principally vertebrates, have an acquired immune response where antibodies and activated lymphocytes attack and destroy very specific toxins or organisms. The field of immunology, particularly acquired immunology, has advanced so rapidly in recent years that it is impossible to more than summarize some of the basic principles here. Let us now briefly examine the general innate immune responses of various animals, and then the acquired immune response of vertebrates.

Protozoans are able to recognize and reject foreign (transplanted) nuclei. Sponges have species specific reaggregation of dispersed colony cells, i.e., there must be individual cell recognition. Sponge colonies reject tissue grafts (allografts) from other colonies but not isografts from themselves (Hildemann et al 1979). Various coelenterates also have been reported to reject allografts but not isografts. Subsequent rejection of grafts occurs faster than the initial rejection, indicating a memory component to the rejection response (Hildemann et al. 1977).

Annelids have most of the mechanisms for innate immunity that were described above. Bacteria can fairly readily enter oligochetes through their dorsal pores and reach the coelomic fluids or can enter through their skin or gut. Most types of bacteria are readily phagocytosed by amebocytes. There is no enhanced response to subsequent infection. Polychetes are generally less septic than oligochetes, probably because of the more aseptic properties of seawater as well as their lack of dorsal pores. Microorganisms can invade polychetes via their nephridia (but those species with closed solenocytes have an effective barrier to invasion), by migration through their gut wall, or across their skin. Phagocytosis is the main defense against micro-organisms, but the body fluids may also contain general bactericidal substances. Annelids reject tissue grafts, indicating some specificity of the phagocytic/humoral lytic mechanisms. The early stages of rejection resemble those of tissue injury, i.e., phagocyte invasion. There is some evidence for a humoral-mediated rejection response. As in other invertebrates, annelids commonly respond to parasite invasion by encapsulation. For example, nematodes (as well as inert material) are surrounded by amebocytes and then encapsulated in a fibrous coat which may later become calcified.

Mollusks generally lack a strong adaptive response to foreign particles or bacterial invasion. This may reflect their reliance on muco-ciliary defense; cleansing sheets of mucus are produced in response to irritation. Nevertheless, mollusks are capable of diverse and complex responses to damage and invasion by foreign materials. Their blood contains a variety of hemocytes, e.g., amebocytes, granulocytes, and hyalinocytes. The main blood proteins are hemocyanin and/or hemoglobin, but there are also agglutinins that agglutinate particles and may enhance phagocytosis and encapsulation. Mollusks have both innate and acquired defense responses. Phagocytosis is an important response to foreign inert materials, protein, viruses, bacteria, metazoan parasites, and fungi. Surprisingly, mollusks lack a rejection response to implants of foreign tissue (arthropods are the only other metazoans to lack such a rejection of allogenic transplants). Gastropods often are hosts, or intermediate hosts, to metazoan parasites, e.g., nematodes and trema-

TABLE 15–10

Summary of the development of immunity in animals. *(Modified from Stites, Caldwell, and Pavia 1980.)*								
	Graft Rejection	Immunologic Specificity of Graft Rejection	Immunologic Memory	Phago-cytosis	Encapsu-lation	Non-specific Humoral Factors	Phagocytic Ameboid Coelomo-cytes	Leukocyte Differen-tiation	Anti-bodies
Protozoans	Yes	No	No	Yes	No	No	No	No	No
Poriferans	Yes	Yes	Yes	No	Yes	No	No	No	No
Cnidarians	Yes	Yes	Yes	No	Yes	No	No	No	No
Annelids	Yes	Yes	Yes	Yes	Yes	Yes	Yes	Probable	No
Mollusks	Yes	?	?	Yes	Yes	Yes	Yes	No	No
Arthropods	Yes	?	?	Yes	Yes	Yes	Yes	No	No
Echinoderms	Yes	Yes	Yes	Yes	Yes	Yes	Yes	Yes	No
Tunicates	Yes	Probable	Yes	Yes	Yes	Yes	Yes	Yes	No
Vertebrates	Yes	Yes	Yes	Yes	No	Yes	Yes	Yes	Yes

todes. Nematodes are initially surrounded by hemocytes, but the capsule often becomes thinned and loosened and does not kill the nematode. Responses to trematodes vary with the degree of resistance ranging from no cellular response to encapsulation and destruction. There is a cellular response involved with recognition of foreign materials (e.g., phagocytosis) but the humoral mechanisms are also very important.

The cuticle of arthropods is an effective barrier against invasion by micro-organisms and impregnation by foreign particles. All external surfaces, including the foregut and hindgut are protected by cuticle; even the midgut of many arthropods is protected by a peritrophic membrane. Insects have both cellular and humoral mechanisms for immune responses (Gotz and Boman 1985). Phagocytosis is an important defense mechanism. There is some selectivity to phagocytosis since some types of particles elicit phagocytosis whereas others do not. Hemocytes also encapsulate foreign objects, e.g., parasites such as trematodes, nematodes, cestodes, insects, and fungi. The multicellular envelope of the capsule consists of three zones of hemocytes: an innermost zone about 10 cells thick, a middle zone of 20 to 40 layers of flattened hemocytes, and an outer zone of 10 layers of normal appearing hemocytes. Encapsulation of living organisms, tissue grafts, and certain inanimate materials involves melanization of the capsule. Tyrosine derivatives (dopamines) are oxidized to quinones in the cuticle by phenyloxidases, which cross-link and polymerize proteins to form a resistant protein-polyphenol complex. Phenyloxidases in hemolymph can also cross-link a variety of hemolymph proteins to form a resistant melanized capsule. The capsule forms a very tight seal around the invading organism and would interrupt its nutrient, O_2, and waste exchange. Some pathogens develop resistance against this otherwise formidable defense mechanism. Sometimes the pathogens are phagocytosed but resist intracellular digestion and grow and develop within the phagocytes. Some pathogens have exceptionally rapid growth and development, e.g., the ichneumonid wasp *Phaeogenes* completes its first three larval stages within one day of the egg being layed in its host insect. Some pathogens rapidly invade the central nervous system and use it as a temporary refuge to avoid encapsulation; they later return to the body fluids and for some reason are then immune to encapsulation. The ichneumonid wasp *Nemeritis* confers immunity to its eggs against encapsulation by the host moth *Ephestia* by introducing a virus along with its eggs; the virus infects the fat body and interferes with the production of immune factors.

Insects also have humoral immunity; the factors are present in either the hemolymph or in hemocytes from which they are released. There are hemagglutinins (called lectens) that agglutinate certain membrane sugar residues; they presumably agglutinate microorganisms to facilitate phagocytosis or encapsulation. Insects also have induced immunity, i.e., they acquire immunity to a subsequent infection.

Echinoderms have the basic elements of an immune response; agglutination, phagocytosis, and encapsulation play a role in defense. Another effective strategy of starfish is to concentrate foreign materials into specific limbs that are then autotomized.

Vertebrates have the most sophisticated defense systems. These include phagocytosis; stomach acid and digestive enzymes; humoral antimicrobial agents, such as lysozymes, polypeptides, and complement proteins; skin; and an **acquired immune response** of exceptional specificity and effectiveness. There are two types of acquired immune response: one is called humoral immunity and involves **antibodies** and the other is cellular immunity whereby specific lymphocytes attack the foreign materials. Both acquired immune systems respond to specific **antigens** that are usually large (MWt $> 8 \ 10^3$ Da) proteins or polysaccharides. Lower molecular weight haptens can also act as antigens by binding to a large molecule and causing an immune response in concert with the larger molecule. Various drugs, chemicals in dust, industrial chemicals, and poison ivy toxin are haptens.

Specific blood cells, the **lymphocytes**, are responsible for the acquired immune response. Lymphoid tissues are present in vertebrates (but progenitors of lymphoid tissues are seen in echinoderms and protochordates). Lymphoid tissues and organs (lymph nodes, tonsils, gastrointestinal tract, lymphoid masses, spleen, bone marrow) essentially filter the lymph, which drains from the tissues via the lymphatic system. The lymph nodes are small, bean-shaped structures containing a cortex of densely packed lymphocytes and a medulla of strands of lymphocytes. Lymphocytes are formed from hemopoietic stem cells that are "processed" in the thymus to form **T-lymphocytes** or in other areas (bone marrow, liver?) to form **B-lymphocytes**. In birds, the B-lymphocytes are processed in the bursa of Fabricius, hence the term B-lymphocytes. The first signs of lymphoid tissue are in protochordates and echinoderms. The differentiation of lymphocytes into T and B types is first apparent in teleost fish, and a clear differentiation is seen in amphibians and higher tetrapods. There are millions of different types of T- and B-lymphocytes that are highly specific for certain antigens (see below). How can millions of different types of T- and B-lymphocytes be coded from the limited amount of DNA in the genome (in addition to all of the other genes present)? Actually, only about 1000 genes code for the millions of types of lymphocyte antibodies, but during DNA transcription the mRNAs are cut into segments and then the pieces respliced in different combinations, thus providing a vast number of different lymphocyte types.

Humoral immunity is conferred by the activity of the B-lymphocytes (Figure 15–35). Tissue macrophage cells phagocytose antigens and present them to the B-lymphocytes. The B-lymphocytes that specifically respond to those antigens immediately transform into lymphoblasts, some of which become plasma cells. These plasma cells become very active and rapidly divide, producing antibodies at a rapid rate. These antibodies are released into the blood and agglutinate their specific antigens. Some of the lymphoblasts do not become plasma cells, but produce a moderate number of daughter lymphocytes that circulate throughout the body and remain dormant in various lymphoid tissues. These memory cells rapidly respond to a subsequent exposure to their specific antigen.

Cellular immunity is a consequence of the activation of T-lymphocytes (Figure 15–35). After exposure to antigens, the T-lymphocytes proliferate (as

do B-lymphocytes) and produce large numbers of activated T-lymphocytes. Some of these are memory cells whereas others circulate through the blood and body tissues seeking out antigens. T-lymphocytes have a large number (up to 10^5) of antigen receptors on their cell surface; the receptors have a variable portion similar to that of the antibodies but are firmly attached to the cell membrane. In addition to the memory T-lymphocytes, there are at least three other types of T-cell: cytotoxic T-cells, helper T-cells, and suppressor T-cells. The cytoxic (or killer) T-cells destroy invading cells (and even the body's own cells under certain circumstances); they bind closely to the antigenic foreign cells and release cytotoxic substances such as lysozymes into the foreign cell. Each killer T-cell can destroy multiple foreign cells. They also attack the body's cells that have been invaded by viruses. The numerous helper T-cells secrete lymphokines when they are activated by antigens; these increase the activation by antigens of B-cells, killer T-cells, and suppressor T-cells. One lymphokine, interleukin-2, stimulates other T-cells, and the lymphokine macrophage migration inhibition factor activates macrophages to phagocytose foreign materials. Suppressor T-cells diminish the activity of killer and helper T-cells, presumably preventing excessive immune reactions.

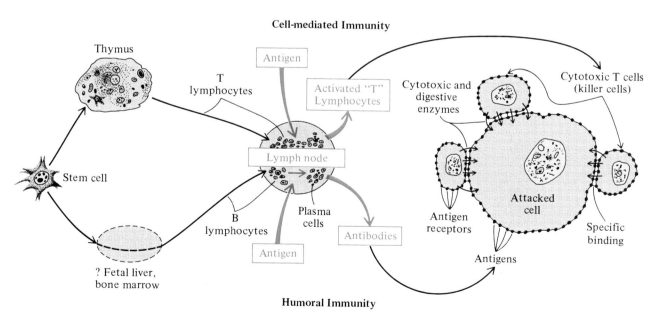

FIGURE 15–35 Representation of humoral immunity (antibody formation) and cell-mediated immunity (activated cytotoxic lymphocytes) in mammals. The thymic origin of T-lymphocytes and the possible fetal liver/bone marrow origin of B-lymphocytes is also indicated. *(From Guyton 1986).*

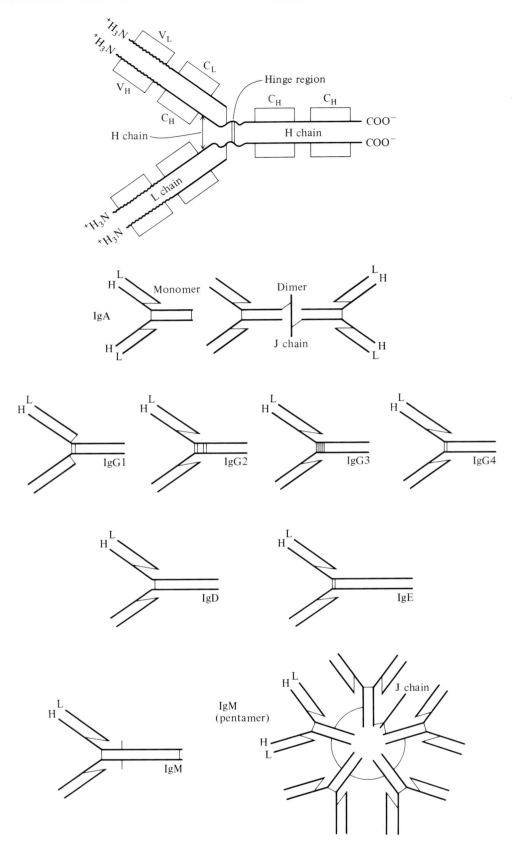

FIGURE 15–36 Representation of a human IgG antibody showing the variable (V) and constant (C) domains of the light and heavy chains and the disulphide bridges (thin lines) between heavy chains; also shown are representations of various human immunoglobulin antibodies. *(From Goodman 1980. Reproduced with permission from Appleton & Lange, Norwalk CT/San Mateo, CA.)*

Antibodies are gamma globulin plasma proteins (immunoglobulins) of molecular weight about 150 to 900 10^3 Da (about 20% of the plasma proteins are immunoglobulins). All immunoglobulins are composed of light and heavy polypeptide chains (Figure 15–36); heavy chains are associated with a light chain at one end. There are usually two pairs of heavy/light chains, but sometimes there are as many as 10 pairs in large immunoglobulins. One end of each of the heavy and light chains is a "variable" region whereas the remainder is relatively constant. It is the variable region that provides the specificity of antigen-antibody responses. The constant portion also has important functions: it determines the diffusivity and adherence properties of the antibody, it is the site of attachment for the complement complex, and it determines the permeability of cell membranes.

Mammals have five classes of immunoglobulins (Ig): IgA, IgD, IgE, IgG, and IgM. The various immunoglobulins differ in type of heavy chains and molecular weight, structure, and function. IgG is particularly important because it comprises 75% of all Ig; IgE is responsible for allergic reactions. IgM has 10 binding sites and is especially important in the initial, or primary, response to an antigen. Immunoglobulins vary in the lower vertebrates. Lampreys and hagfish have only a limited humoral response to antigens. The hagfish antibodies are not well studied, but the lamprey antibodies are unique in that they lack disulfide bonds between different chains. The elasmobranch, chondrostean, holostean, and teleost fishes have two classes (rarely only one) of immunoglobulins. The high molecular weight (HMW) immunoglobulin can be a dimer, tetramer, or pentamer and the low molecular weight (LMW) is a monomer. Lungfish, amphibians, and several reptiles have a third type of immunoglobulin, intermediate molecular weight (IMW) immunoglobulin, which resembles the mammalian IgG. Birds and reptiles also have an IgA-like immunoglobulin.

Antibodies can provide protection by either directly attacking the antigenic source or by activating complement proteins which then destroy the antigenic source. Antigens cause agglutination because they are at least bivalent (i.e., have at least two antigen binding sites) and because foreign materials typically have a number of antigenic binding sites. Antibodies precipitate antigens as insoluble complexes, neutralize the antigenic source by coating it with antibody, or lyse the cell membranes if the antibodies are particularly potent. Most protection, however, is provided by activation of the 20 or so complement proteins. When an antigen reacts with an antibody, it uncovers a binding site on the constant portion of the antibody which in turn binds to complement C1. This initiates a cascade of activations of other complement proteins and various defense reactions. Complement C3b activates phagocytosis by neutrophil white blood cells and macrophages; this activation is called opsonization. The multiple complement factor C5bC6C7 has a direct lytic effect on antigenic cell membranes. C5a causes chemotaxis attraction of neutrophils and macrophages. C3a, C4a, and C5a activate mast cells and basophils that release histamine and other factors to cause local inflammation. The complement system can sometimes be activated without the intermediation of antigen-antibody responses. For example, large polysaccharides of cell membranes react with complement B and D, activate C3, and cause the same cascade of complement activation.

The very specific antibody-antigen system of especially the higher vertebrates enables the identification and response to very specific foreign materials, including tissues of other animals, both conspecifics and other species. This is the basis for rejection of tissue grafts from other individuals of the same species (allogeneic grafts) and other species (xenogeneic grafts), but not grafts from an individual's own tissues (isografts). Mammalian mothers can even respond immunologically to their own fetuses *in utero*, although this effect is limited by the relative impermeability of the placenta to large antigens (e.g., the fetal red blood corpuscles and plasma proteins).

Temperature has a significant effect on the immune response of ectotherms. Low temperature generally depresses antibody production in vertebrates, e.g., frogs are more resistant to the bacterial infection "red leg" at higher temperatures. There is sometimes a complete inhibition of antibody production at low temperatures, e.g., alligators form antibodies against diphtheria toxin at body temperatures higher than 30° C but not at lower body temperatures.

Physical Deterrence

Many insects, and some other animals, utilize their blood as a defensive system to deter the attack of predators. In insects, the sequestration of toxins or distasteful compounds in the blood can deter predators, at least once they have tasted the blood. This may not be an effective defense strategy for the individual, especially if it is killed by the predator, but can be highly adaptive for the population as a whole, e.g., distastefulness and warning coloration in Batesian and Mullerian mimicry. Alternatively, individuals may exude blood from various parts of the body to deter predators; this is reflexive bleeding or autohemorrhage. The deterrent effect may involve distasteful or toxic compounds in the blood or mixing of the blood with defense gland secretions. Hemolymph of the cucumber beetle larvae (*Diabrotica*) is not toxic or distasteful but can still be lethal to predators. The larvae characteristically twist and spin as they exude hemolymph to entangle (and often kill) attacking fire ants. A large amount of hemolymph may be exuded for defense. *Diabrotica* larvae can exude up to 13% of their body mass. Some insects later reabsorb the exuded hemolymph to minimize hemolymph loss.

The lizard *Phrynosoma* can eject blood from its eyes for a distance up to 2 meters (Heath 1966). The blood may be mixed with a noxious secretion of the Harderian gland as it is expelled and presumably confuses or deters a potential predator. The snail *Lymnaea* ejects hemolymph via the "hemal pore," presumably as a defensive act.

Summary

Blood is the fluid contained within the circulatory system. It is mainly water and contains various ions and other solutes including protein and a variety of blood cells (or hemocytes). The blood cells have a variety of roles, including O_2 transport (e.g., erythrocytes, red blood cells). The primary role of blood is the transport of gases and other nutrients and wastes. The vascular fluids also have a variety of other roles in various animals, including a structural role as a hydrostat or in limb movement by hydraulic pressure, transfer of heat, defense by phagocytosis, encapsulation or immunity, and blood clotting.

The viscosity of blood depends primarily on the concentration of proteins and the numbers of blood cells. Blood is a non-Newtonian fluid because its viscosity depends on the rate of flow and size of blood vessels. The role of RBCs is to prevent excretion of the hemoglobin, to minimize the colloid pressure effect of hemoglobin, and to minimize the viscosity of blood in small diameter vessels.

The four types of respiratory pigment—hemoglobin, chlorocruorin, hemerythrin, and hemocyanin—are distributed in a complex pattern in animal taxa. They all contain a protein and a metal ion (Fe^{2+} or Cu^{2+}), and hemoglobin and chlorocruorin contain a heme group that surrounds the Fe^{2+} ion.

Hemoglobin is the most common respiratory pigment. It varies markedly in structure in different animals. The smallest subunit is sometimes an Fe^{2+}/heme/protein structure that can remain separate as single-domain pigments (e.g., myoglobin), or multiple subunits can be assembled into respiratory pigments (e.g., vertebrate hemoglobin). Alternatively, the subunit structure can be a protein with either two Fe^{2+}/heme structures (two-domain pigments) or more than two Fe^{2+}/hemes (multidomain pigments). The chlorocruorin pigments of some polychetes are similar to two-domain hemoglobins. Hemerythrin has two Fe^{2+} ions per protein subunit; it is found in a few different phyla (sipunculids, priapulids, brachiopods, and annelids). Hemocyanin usually has 2 proteins and 2 Cu^{2+} per subunit. The functional hemocyanin usually has multiple subunits aggregated to form large molecules. Arthropod hemocyanins typically have hexamer units (1 × 6 mers), arranged often in 2s, 3s, 4s, or 6s. Molluskan hemocyanins are typically a cylindrical assemblage of 10 or 20 subunits.

Respiratory pigments reversibly bind O_2 and increase the O_2-carrying capacity of the blood. The degree of O_2-binding depends on the pO_2; the O_2-dissociation curve is the relationship between percent saturation of the pigment and pO_2. The curve is hyperbolic for single subunit pigments but has a complex sigmoidal shape for multiple subunit pigments if there is cooperativity between subunits in O_2-binding. The P_{50} is the pO_2 at which the pigment is 50% saturated, and the cooperativity coefficient (n) reflects the extent of cooperative binding of O_2 by the multiple subunits. The P_{50} and n values differ markedly for various respiratory pigments. Numerous factors modulate the P_{50} value, including pCO_2 and H^+ (Bohr effect), temperature, ionic concentration, organic phosphates (e.g., DPG, IP_4, IP_5, ATP) and other organic solutes (e.g., lactate). The O_2-carrying capacity of some hemoglobins is reduced by an elevated pCO_2/H^+ (Root effect).

Respiratory pigments generally transport O_2 from the respiratory surface to the tissues. The arterial pO_2 and O_2 content of blood are greater than the venous values; the A-V difference indicates the amount of O_2 transported to the tissues. O_2 transport can be facilitated by the Bohr effect. Some hemoglobins have exceedingly low P_{50} values and provide emergency O_2 transport at low ambient pO_2s. Hemoglobin and hemerythrin can transfer O_2 from one tissue, or blood, to another tissue, e.g., muscle or nerve cells. Some respiratory pigments have an O_2 storage role, providing emergency O_2 during tissue hypoxia. Rarely, hemoglobins transport H_2S rather than O_2.

Blood also transports CO_2, but in a different fashion than it transports O_2. There is no specific CO_2 carrier molecule. Rather, CO_2 has a high solubility and readily dissolves in water; it also chemically combines with water to form bicarbonate and carbonate ions and with amino groups to form carbamino compounds. Most CO_2 is actually transported by the blood as bicarbonate. Blood CO_2 transport is facilitated by the lower CO_2-carrying capacity of oxygenated blood compared to deoxygenated blood (Haldane effect).

The acid–base balance of body fluids is determined by the dissociation of water molecules to form H^+ and OH^-; the action of blood buffers, particularly bicarbonate and weak organic acids; and the balance of metabolic CO_2 production and respiratory CO_2 excretion. Temperature has an important effect on the acid base status because of its effects on both the physicochemical reactions forming H^+ and metabolism and respiration. The pH of many animals alters at varying temperature in a parallel fashion to the change in the pH of neutral water; $\Delta pH/\Delta T \approx -0.017$ U $°C^{-1}$. Other animals, such as some hibernating mammals and some reptiles, maintain a relatively constant pH regardless of temperature.

Blood has an important role in defense against blood loss by physical trauma to the circulatory system. Clotting of mammalian blood involves essentially three steps: formation of prothrombin activator, formation of thrombin, and the formation of fibrin from fibrinogen. Most invertebrates also have a clotting mechanism.

Immunity is the ability to resist invasion by inorganic, viral, bacterial, plant, or animal agents. The general immune capacity involves phagocytosis by blood and tissue cells, destruction by digestive enzymes/acids, encapsulation, physical deterrence by the skin/cuticle, and antibody action. Most invertebrates and vertebrates have these immune responses. Vertebrates also have a well-developed acquired immune response involving the lymphatic system and lymphocyte white blood cells. Humoral immunity is conferred by the antibodies produced by lymphocytes. Antibodies are gamma globulin plasma proteins that respond to specific types of foreign material (antigens); they either directly attack the foreign material or activate complement proteins that destroy the antigen source. Cellular immunity is a consequence of the activation of specific lymphocytes by antigens; the activated lymphocytes proliferate and form T-lymphocytes that are involved in immune responses.

Supplement 15–1

Oxygen Equilibrium Curves

· ·

The complex equilibrium between O_2 and monomeric hemoglobin is

$$Hb + O_2 \underset{k}{\overset{k'}{\rightleftharpoons}} HbO_2$$

where k' is the forward ("on") rate constant for a second order reaction and k is the reverse ("off") rate constant for a first order reaction. The equilibrium constant K_{eq} is k'/k, or

$$K_{eq} = \frac{[HbO_2]}{[Hb][O_2]}$$

The relationship between percent saturation of hemoglobin $100\,[HbO_2]/\{[HbO_2] + [Hb]\}$ and O_2 concentration is not linear, as we can see by rearranging the above equation

$$\% \text{ saturation} = \frac{100[HbO_2]}{[Hb] + [HbO_2]} = \frac{100k pO_2}{1 + k pO_2}$$

where $k pO_2$ is substituted for $K_{eq}[O_2]$ ($[O_2]$ is proportional to pO_2). This equation describes a hyperbolic curve, and the value of k determines the relative position of the curve. To define the relative position of different hyperbolic curves (i.e., reflecting the value of k), we can determine the pO_2 at which the hemoglobin is 50% saturated; this pO_2 value is called the P_{50} (we could similarly define the P_{75} at 75% saturation, the P_{25}, P_{95}, etc.) to further describe the position of the curve, but for a hyperbolic curve only the P_{50} is necessary.

Many hemoglobins are not monomeric but are dimeric, tetrameric, octameric, etc. Consequently, the range of reactions with O_2 is more complex. For example, with tetrameric hemoglobin, there are four different reactions.

$$Hb_4 + O_2 \underset{k_1}{\overset{k_1}{\rightleftharpoons}} Hb_4O_2$$

$$Hb_4O_2 + O_2 \underset{k_2}{\overset{k_2}{\rightleftharpoons}} Hb_4O_4$$

$$Hb_4O_4 + O_2 \underset{k_3}{\overset{k_3}{\rightleftharpoons}} Hb_4O_6$$

$$Hb_4O_6 + O_2 \underset{k_4}{\overset{k_4}{\rightleftharpoons}} Hb_4O_8$$

The equilibrium constants ($K_{eq.1}$, $K_{eq.2}$, $K_{eq.3}$, $K_{eq.4}$) are calculated as before (e.g., $K_{eq.1} = k_1'/k_1$). The various equilibrium constants have different values and consequently there is not a simple hyperbolic relationship

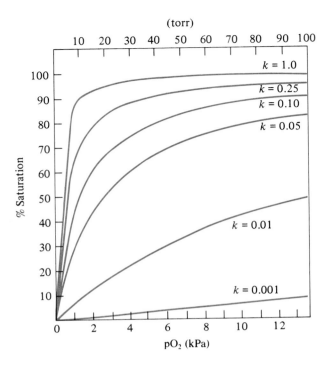

Theoretical hyperbolic O_2 dissociation curves calculated with varying values of k ($n = 1$ for all curves) to show the effect of k on P_{50}, the pO_2 for 50% saturation of the respiratory pigment. The curves were calculated using the formula % saturation = $100\ kpO_2/(1 + kpO_2)$ with units of torr.

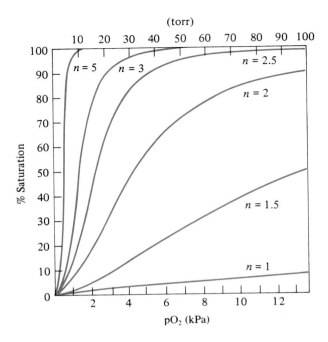

Theoretical sigmoidal O_2 dissociation curves calculated as above except that $k = 0.001$ and $n = 1, 1.5, 2, 2.5, 3,$ and 5. The curves were calculated using the formula % saturation = $100\ kpO_2^n/(C1 + kpO_2^n)$ with units of torr.

Hill plots for hemoglobin (tuna, giant earthworm), chlorocruorin (sipunculid), and hemocyanin (scorpion) are generally curvilinear with a maximum slope (n) at about the P_{50} value. Y is abbreviation for % O_2 saturation.

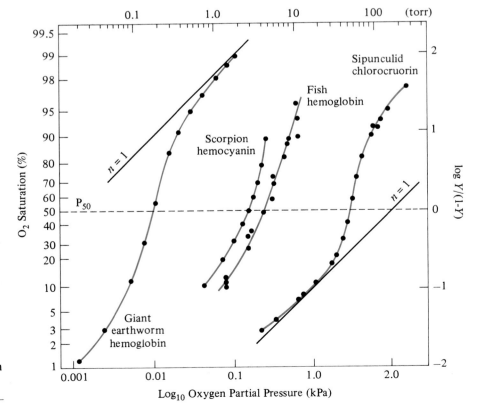

between percent saturation and pO_2. The various K_{eq} values are not necessarily equal because the hemes of the different subunits can functionally interact, such that binding of the second O_2 is facilitated by binding of the first O_2, binding of the third O_2 is facilitated by binding of the first and second, etc. Such an interaction increases the slope of the relationship between percent saturation and pO_2 at intermediate pO_2, and results in an S-shaped (sigmoidal) curve. The percent saturation is now calculated as

$$\% \text{ saturation} = \frac{100kpO_2^n}{1 + kpO_2^n}$$

The family of different S-shaped curves, obtained for various values of n, illustrates the relationship between n and the extent of sigmoidality. The constant, n, is called the cooperativity coefficient and reflects the "cooperation" of heme subunits in binding O_2. For monomeric hemoglobins, $n = 1$ because there is only one heme subunit and no cooperativity is possible. The dimeric hemoglobin of some mollusks has n about 1.4 to 1.5. For a typical mammalian hemoglobin (Hb_4), n is about 2.6. The hemoglobin of *Arenicola* has 200 hemes and n is about 4. Some hemoglobins have n values as high as 8. Note that there is a diminishing increase in n with higher numbers of subunits because of spatial limitations to the ability of hemes to physically interact. The presence of subunits does not, however, imply that cooperativity will occur. For example, n is 1 for the tetrameric hemoglobin of spiny dogfish; n is also 1 for the 200 subunit hemoglobin of *Eopolymnia*.

The hyperbolic or sigmoidal relationship between percent saturation and pO_2 can be linearized by the Hill plot, where $\log_{10}$ (percent saturation/(100 − percent saturation)) is graphed as a function of $\log_{10} pO_2$, since

$$\frac{\% \text{ saturation}}{100 - \% \text{ saturation}} = kpO_2^n$$

and so

$$\log_{10} \frac{\% \text{ saturation}}{100 - \% \text{ saturation}} = \log_{10}k + n\log_{10}pO_2$$

The Hill equation provides a reasonably good fit to actual relationships between percent saturation and pO_2 for most respiratory pigments, except at very low and very high percent saturation values. The n value is usually given at the 1/2 saturation point (i.e., n_{50}), since n is approximately 1 at very low and very high percent saturations and is maximal at 50% saturation.

A more sophisticated analysis of O_2 binding than the Hill equation is provided by the Monod-Wyman-Changeaux theory. The pigment has an initial low-affinity

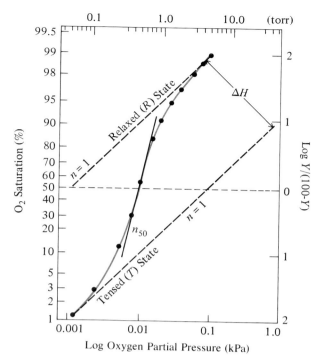

Hill plot for hemoglobin of the giant earthworm conforms to the Monod-Wyman-Changeaux model for a tense (T) state for the deoxygenated pigment, with $n = 1$, and a relaxed (R) state for the oxygenated pigment, also with $n = 1$. There is a transition between the T and R states with a maximal slope (n) at about the P_{50}. The O_2 affinity of the deoxy- and oxy-pigment can be obtained by extrapolation of the asymptotes for the T and R states to 50% saturation ($\log Y/(100 - Y) = 0$). The distance between the two asymptotes represents the free energy difference (ΔH) between the two states.

(T) state and a final high-affinity (R) state; these states correspond to the unliganded "tensed" state (hence T) and the liganded "relaxed" state (hence R). The O_2-binding kinetics depend on the state of the respiratory pigment. Cooperative binding begins with an initial "low-affinity" slope of 1, increases with the degree of O_2 binding to a maximum, and then declines to 1 at saturation. For hemoglobins, the "on" rate constant (k') is similar for the R and T states, but the "off" rate constant is faster for the R state (about 8 to 60 sec^{-1}) than the T state (about 300 to 400 sec^{-1}).

Recommended Reading

Bannister, J. V. 1977. *Structure and function of haemocyanin*. Berlin: Springer-Verlag.

Bayne, C. J. 1990. Phagocytosis and non-self recognition in invertebrates. *Bioscience* 40:723–31.

Bonaventura, J., and C. Bonaventura. 1980. Hemocyanins: Relationships in their structure, function, and assembly. *Amer. Zool.* 20:7–17.

Brix, O. 1983. Blood respiratory properties in marine gastropods. In *The Mollusks*, vol. 2, *Environmen-*

tal biochemistry and physiology, edited by P. W. Hochachka, 51–75. New York: Academic Press.

Butler, P. J., and D. R. Jones. 1982. The comparative physiology of diving in vertebrates. *Adv. Comp. Physiol. Biochem.* 8:179–364.

Dales, R. P. 1978. Defense mechanisms. In *Physiology of annelids*, edited by P. J. Mill, 479–507. New York: Academic Press.

Dunn, P. E. 1990. Humoral immunity in insects. *Bioscience* 40:738–44.

Ellerton, H. D., N. F. Ellerton, and H. A. Robinson. 1983. Hemocyanin: A current perspective. *Progr. Biophys. Mol. Biol.* 41:143–248.

Gotz, P., and H. G. Boman. 1985. Insect immunity. In *Comprehensive insect physiology biochemistry and pharmacology*, vol. 3, *Integument, respiration and circulation*, edited by G. A. Kerkut and L. Gilbert, 453–85. Oxford: Pergamon Press.

Heisler, N. 1984. Acid base regulation in fishes. In *Fish physiology*, vol. 10, *Gills: A. Anatomy, gas transfer, and acid base regulation*, edited by W. S. Hoar and D. J. Randall, 315–401. New York: Academic Press.

Hill, R. B., and J. H. Welsh. 1966. Heart, circulation and blood cells. In *Physiology of the Molluska*, edited by K. M. Wilbur and C. M. Yonge, vol. 2, 125–74. New York: Academic Press.

Jones, J. D. 1972. Comparative physiology of respiration. London: Edward Arnold.

Karp, R. D. 1990. Cell-mediated immunity in invertebrates. *Bioscience* 40:732–37.

Klotz, I. M., G. L. Klippenstein, and W. A. Hendrikson. 1976. Hemerythrin: An alternative oxygen carrier. *Science* 192:335–44.

Kreuzer, F. 1977. Facilitated diffusion of oxygen and its possible significance: A review. *Respir. Physiol.* 9:1–30.

Lamy, J., and J. L. Lamy. 1981. *Invertebrate oxygen-binding proteins*. New York: Marcel Dekker.

Lamy, J., J.-P. Truchot, and R. Gilles. 1985. *Respiratory pigments in animals: Relation, structure, and function*. Berlin: Springer-Verlag.

Malan, A. 1977. Blood acid–base state at a variable temperature: A graphical representation. *Respir. Physiol.* 31:259–75.

Marchalonis, J. J. 1977. *Comparative immunology*. Oxford: Blackwell Scientific.

McFarlane, R. G. 1970. Haemostatic mechanisms in man and other animals. London: Academic Press.

Merrill, E. W. 1969. Rheology of blood. *Physiol. Rev.* 49:863–88.

Milnor, W. R. 1982. *Hemodynamics*. Baltimore: William & Wilkins.

Rahn, H., R. B. Reeves, and B. J. Howell. 1975. Hydrogen ion regulation, temperature, and evolution. *Am. Rev. Resp. Disease* 112:165–72.

Ratcliffe, N. A., and A. F. Rowley. 1981. *Invertebrate blood cells*. New York: Academic Press.

Roughton, F. J. W. 1964. Transport of oxygen and carbon dioxide. In *Handbook of physiology*, sect. 3, *Respiration*, vol. 1, edited by W. O. Fenn and H. Rahn, 767–826. Washington: Amer. Physiol. Soc.

Schmidt-Nielsen, K., and C. R. Taylor. 1968. Red blood cells: Why or why not? *Science* 162:274–75.

Snyder, G. K. 1977. Blood corpuscles and blood hemoglobins: A possible example of coevolution. *Science* 195:412–13.

Stites, D. P., et al. 1984. *Basic and clinical immunology*. Los Altos: Lange.

Terwilliger, N. B., et al. 1987. Bivalve hemocyanins—a comparison with other molluscan hemocyanins. *Comp. Biochem. Physiol.* 89B:189–95.

Toulmond, A. 1985. Circulating respiratory pigments in marine animals. In *Physiological adaptations of marine animals*, edited by M. S. Laverack, 163–206. Cambridge: Society for Experimental Biology.

van Holde, K. E., and K. I. Miller. 1982. Haemocyanins. *Quart. Rev. Biophys.* 15:1–129.

Weber, R. E. 1978. Respiratory pigments. In *Physiology of annelids*, edited by P. J. Mill, 393–446. New York: Academic Press.

Weber, R. E., and F. B. Jensen. 1988. Functional adaptations in hemoglobins from ectothermic vertebrates. *Ann. Rev. Physiol.* 50:161–79.

Chapter 16

Water and Solute Balance

The desert cockroach *Arenivaga investigata* **is able to absorb water vapor from subsaturated air, using mouth parts that are eversible bladders.**
(Photograph courtesy of O'Donnell, Department of Biology, McMaster University.)

O ne of the most fundamental regulatory requirements for animals is to maintain the integrity of their intracellular space, especially the solute composition and cell volume. Both the osmotic and ionic composition of the body fluids are important, since the osmotic concentration determines the distribution of water between cells and the extracellular fluids and ion concentrations are important in many aspects of cell function. This chapter examines the variety of patterns and mechanisms for regulation of the water and solute environments of intracellular and extracellular fluids.

Body Fluid Composition

Water is the primary constituent of animals, being 60 to 90% of the total body mass. The body fluids contain dissolved solutes, but the specific solute composition varies dramatically between the different body fluid compartments and for different animals.

Water is distributed between the **intracellular space** and the **extracellular space**. There are considerable differences in the partitioning of water between the intra- and extracellular spaces for various animals (Figure 16–1). For example, the intracellular space is about 30% of the high body water content of the marine gastropod mollusk *Aplysia* (95% body water content), whereas it is about 80% of the lower water content for teleost fish (70% body water content). The extracellular fluid compartment is partitioned for animals with a closed circulatory system, into the intravascular volume (plasma and lymph spaces) and the extravascular volume (interstitial compartment). There are also other fluid compartments, e.g., coelomic, intraocular, and cerebrospinal spaces.

Water

Water, in liquid form, has a unique combination of physical properties that have been ascribed as reasons for its role as the **universal solvent** for life on Earth. That water is a universal solvent is not surprising, since it was the most common liquid on

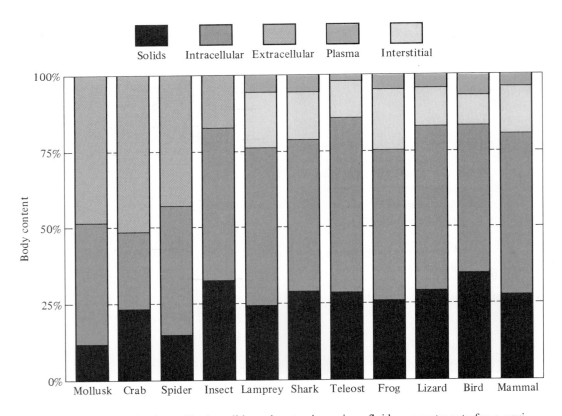

FIGURE 16–1 Distribution of body solids and water in various fluid compartments for a variety of animals.

Earth when life evolved. Whether water proves to be the universal solvent for life throughout the universe remains to be determined by extraterrestrial research!

The biologically important properties of water are a consequence of its structure. Water consists of two hydrogen atoms covalently bonded to an oxygen atom (Figure 16–2A). The O atom is more electronegative than the H atoms. Consequently, the H-O bond behaves as if it were 40% ionic and 60% covalent, and the water molecule is a **dipole**, with each H having a slight positive charge (δ^+) relative to the O ($2\delta^-$). The H—O—H angle is 104.5°.

One of the main chemical consequences of water being a dipole is its ability to form secondary bonds with adjacent molecules. In ice, the adjacent water molecules form a stable open-lattice crystalline structure with each O atom surrounded tetrahedrally by four H atoms, with its two covalently bound H atoms occupying two corners of the tetrahedron. The structure of liquid water is complex. Hydrogen bonding between the $H^{\delta+}$ and $O^{2\delta-}$ of adjacent **bulk** water molecules is extremely transient, lasting for only 10^{-10} to 10^{-11} seconds. The water molecules are essentially disorganized, although there is some transient 3-dimensional organization of water molecules, as in ice (Figure 16–2B). **Vicinal water** is structurally modified by the presence of an interface, such as a macromolecule or membrane, that tends to stabilize the clusters of water molecules within about 0.1 μ of the interface. **Bound water** is strongly attracted to ions or charged macromolecules and the water is essentially part of the ionic, or macromolecular, structure. This layer of bound water significantly increases the effective size of the ions. For

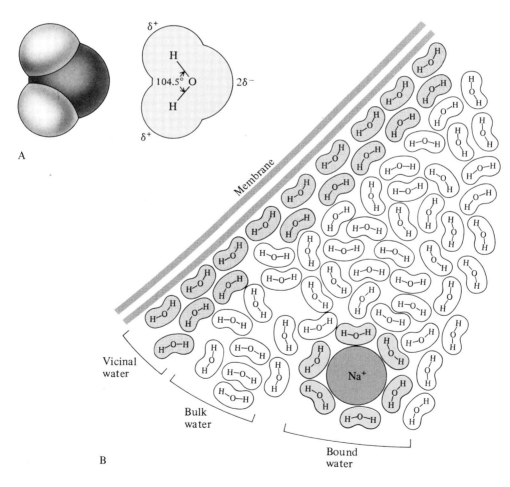

FIGURE 16–2 (A) Covalent structure of water shown with bond angle and charge distribution and as a space-filling model. (B) Schematic representation of the structure of vicinal, bulk, and bound water.

example, ionic Na^+ has a radius of 0.095 nm whereas a hydrated Na^+ has a radius of 0.512 nm; ionic K^+ has a radius of 0.133 nm compared to 0.396 nm for hydrated K^+.

At least part of the intracellular water is bound and vicinal water because of the high intracellular concentration of inorganic and organic solutes and macromolecules. The relationship of water structure to biological activity is well illustrated by the effect of water content on the metabolic rate of diapausing "eggs" of the brine-shrimp *Artemia*. The "eggs" are actually desiccated cysts delayed in development at the gastrula stage. The desiccated cysts, which have a water content less than 0.15 ml H_2O per g dry matter, are **anhydrobiotic** and essentially **ametabolic**, i.e., they have no measurable metabolic rate (Clegg 1975). The small amount of water in these cysts has been interpreted to be bound water, which does not allow mobility of solutes. At water contents from 0.15 to 0.6 ml g^{-1}, vicinal water is also present and the cysts have a restricted metabolic capacity, presumably reflecting the limited mobility of solutes in the vicinal water. Bulk water is present at water contents greater than 0.6 ml g^{-1} when the cysts become fully metabolically active. The bulk water allows effective transport of metabolites, wastes, etc. between intracellular compartments.

Solutes

Body fluids contain a variety of solutes. The primary inorganic ions of body fluids are Na^+, K^+, and Cl^-, with lower concentrations of Ca^{2+}, Mg^{2+}, SO_4^{2-}, PO_4^{3-}, and HCO_3^-. The most important organic solutes are ionic (e.g., amino acids, proteins) or nonionic (e.g., glucose, urea).

An important physico-chemical solute property of a solution is its osmotic concentration. The osmolar concentration ($C_{osmolar}$) of a solution is the total number of moles of dissolved and osmotically active solutes per liter of solution. The osmolal concentration ($C_{osmolal}$) is the number of moles of dissolved osmotically active solutes per kg of solvent. **Osmolality** is less commonly used than osmolarity, but it is preferable because the osmotic pressure (Π) is proportional to the osmolal concentration, absolute temperature (T), and gas constant (R):

$$\Pi = RT\frac{n}{v}$$
$$= RT\,C_{osmolal} \tag{16.1}$$
$$= 2479\,C_{osmolal}\ \text{kPa at 25°C}$$
$$= 24.5\,C_{osmolal}\ \text{atmospheres at 25°C}$$

This relationship is analogous to the ideal gas law, $P = RT\,n/V$ (since $PV = nRT$).

The osmolal concentration is related to the molal concentration (C_{molal}) by the dissociation coefficient, i. The value of this coefficient reflects the extent of dissociation of the solute (e.g., $i = 1$ for urea, glucose; $i = 2$ for NaCl and KCl; $i = 3$ for $CaCl_2$, H_2SO_4) and also accounts for the interaction of solutes. The value of i is generally not exactly equal the expected integer value because of nonideal properties of solutes in solution.

There are marked differences in the concentrations of most solutes for various animals and also for the intracellular and extracellular fluids (Figure 16–3). Many marine invertebrates, and some vertebrates, have essentially the same osmotic concentration as seawater; the extracellular fluid composition is also similar to seawater for some (e.g., horseshoe crab and hagfish) but not for most vertebrates (e.g., elasmobranch, frog, and fish). The body fluid osmotic concentration is considerably less than seawater for many marine vertebrates and for terrestrial animals. Freshwater animals tend to have quite low osmotic concentrations.

The major extracellular cation of animals is always Na^+ (with low K^+) and the primary intracellular cation is K^+ (with low Na^+), although there is some variation for different animals. The intracellular ionic concentration is quite variable but is always relatively low (e.g., $K^+ + Na^+$ is 10 to 200 mM), even for animals living in extremely concentrated media. A spectacular contrast to this pattern in animals is the archibacterium *Halobacterium*, which can have an intracellular K^+ of about 750 mM and Na^+ of 400 to 800 mM (Ginzburg, Sachs, and Ginzburg 1970). The relatively low ionic concentrations of animal body fluids (even those living in comparable environments to *Halobacterium*) can be attributed to the destabilizing, "salting-out" effect of high salt concentration on proteins (see page 782).

Extracellular Solutes. The extracellular fluid is a buffer between the intracellular fluid environment of cells and the external environment of the animal. It generally provides a stable environment for the cells. One of the major trends observed in the evolution of higher animals is the progressively greater regulation and homeostasis of this extracellular environment.

The osmotic concentration of the body fluids of many aquatic animals is equal to that of the medium; these animals **osmoconform** (Figure 16–4). In contrast, **osmoregulators** maintain their body fluids at a

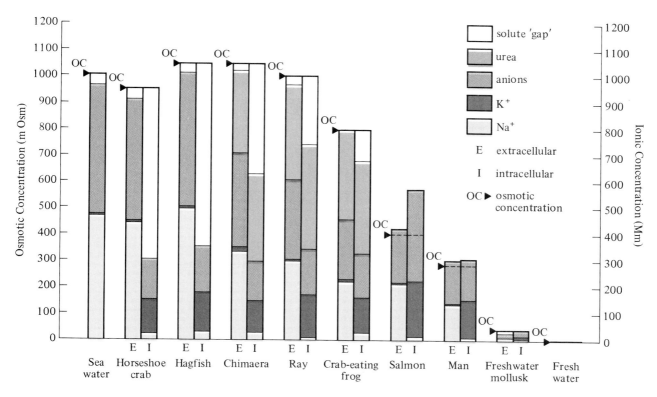

FIGURE 16–3 Extracellular and intracellular ion, urea, and osmotic concentrations of selected invertebrates and vertebrates compared with seawater and freshwater and showing the total osmotic concentration, major cation concentrations (Na$^+$, K$^+$), anions accompanying Na$^+$ and K$^+$, urea concentration, and the "osmotic gap" that is filled by various other solutes (divalent cations, other anions, amino acids, TMAO, etc.). Abbreviations are as follows: E, extracellular; I, intracellular; and OC, osmotic concentration.

different osmotic concentration from that of the medium. Animals typically regulate specific ions at very different concentrations from the medium concentration, i.e., they **ionoregulate**. A few animals approximately **ionoconform** to their external environment, but even animals with relatively primitive ionoregulatory capabilities regulate some body fluid ions. For example, the extracellular ion concentrations in coelenterates are generally similar to the concentrations in seawater, but they regulate a low [SO$_4^{2-}$] for buoyancy control.

Most marine invertebrates are osmoconformers and many are effective ionoregulators. Some marine vertebrates osmoconform (e.g., hagfish, Chondrichthyes) and others osmoregulate; virtually all fish are good ionoregulators (the hagfish is the notable exception). Freshwater animals must both osmoregulate and ionoregulate since their body fluids cannot be isoosmotic to freshwater. The lowest osmotic concentration measured for an animal is about 40 mOsm (the freshwater lamellibranch mollusk *Margaritifera*).

Some animals tolerate only a narrow range of ambient osmotic concentrations; these animals are **stenohaline**. For example, the marine crabs *Maia* and *Platycarcinus* and some *Gammarus* shrimp tolerate only a relatively narrow range of osmotic concentrations (750 to 1250 mOsm). In contrast, many animals can tolerate a wide range of external osmotic concentration, i.e., they are **euryhaline**. Examples include some polychete worms, crabs, and gammarid shrimp.

Intracellular Solutes. Inorganic ions, particularly Na$^+$ and Cl$^-$, are the main extracellular solutes for most animals (except some insects and fish), but the intracellular fluids have a lower inorganic ion concentration, consisting primarily of K$^+$. The balance of the intracellular osmolytes is organic solutes, such as amino acids, proteins, various metabolic

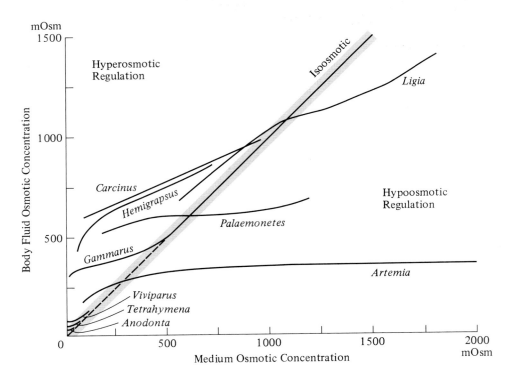

FIGURE 16–4 Relationship between ambient osmotic concentration and body fluid osmotic concentration for osmoconformers, hypoosmotic regulators, and hyperosmotic regulators.

substrates, metabolic intermediates, and waste products (Table 16–1). Intracellular fluid therefore has a markedly different solute composition from extracellular fluids although it has an identical osmotic concentration to the extracellular fluids. The intracellular and extracellular osmotic concentrations are identical because animal cell membranes are not supported by rigid cell walls (as are plant and bacterial cells) and therefore cannot resist a transmembrane hydrostatic or osmotic pressure differential.

Why is K^+ the major intracellular cation rather than Na^+? And why not Ca^{2+} or Mg^{2+}? There may be a solvent capacity problem inside cells because many solutes must be dissolved in this water space, i.e., it may be difficult to dissolve all of the intracellular solutes in the limited amount of intracellular bulk water. A considerable fraction of the intracellular water (bound and vicinal water) is sufficiently structured that it is not an effective solvent. Therefore, selection of an inorganic cation with minimal requirements for bound water would maximize the amount of bulk water available for other solvent functions. K^+ has a lower charge density than Na^+ and therefore structures water less than Na^+; K^+

also stabilizes proteins more than Na^+. The limited solvent capacity of bulk water inside cells thus favors, or even requires, ions with minimal solvent demands, i.e., K^+ rather than Na^+.

Compatible solutes are osmolytes which do not interact with the surface of a protein, and thereby stabilize its structure and function (Low 1985); these compatible osmolytes can "salt-out" proteins and remove them from the aqueous phase (Figure 16–5). Cations such as ammonium (NH_4^+) and methylated amines such as $(CH_3)_4N^+$ and $(CH_3)_2NH_2^+$ and some inorganic anions such as F^-, PO_4^{3-}, and SO_4^{2-} are compatible solutes. Some amino acids (e.g., glycine), glycerol, trehalose, trimethylamine oxide (TMAO), betaine, sarcosine, taurine, and octopine are also compatible solutes; they "look like" the compatible cations. For example, glycine looks like $(NH_4)_2SO_4$, taurine looks like ammonium acetate, TMAO and betaine look like quaternary methylamine acetate (Clark 1985). Not surprisingly, compatible solutes, rather than incompatible solutes, are typically used to increase the intracellular osmotic concentration in response to osmotic stress (see below) or to reduce the freezing point (see Chapter 5). The ability of solutes to stabilize or

TABLE 16–1

Intracellular solute composition for some invertebrates and vertebrates. All values are mmole kg⁻¹ and are corrected for extracellular fluid content. Tissues are muscle (m), nerve (n), or whole animal (a). AA, amino acids.

Species	Na$^+$	K$^+$	Cl$^-$	SO$_4^{2-}$	AA	Other
Invertebrates						
<u>Seawater</u>						
Neanthes (m)	48	234	33.3		494	
Mytilus (m)	79	152	94	8.8	289	Phosphate 39
Nephrops (m)	24.5	188	53.1	1.0	476	Betaine 66, TMAO 59, Phosphate 164
Carcinus (n)	41	422	27		444	
Limulus (m)	28.8	129	43.2	0.9	136	Lactate 22.7, TMAO 6.1, Urea 0.8, Phosphate 96
Sepia (m)	30.8	189	45	2.0	483	Betaine 108, TMAO 86
Eledone (m)	33.3	167	54.8	4.5	326	Betaine 117, TMAO 40
<u>Freshwater</u>						
Anodonta (m)	5.3	21.3	2.4		11	Phosphate 19.8
Potamon (m)	44	111	32		170	TMAO 42
Sialis (a)	1	135	<1			Mg 35
Pelomatohydra (a)	2.5	14	1.5			
Spirostomum (a)	1	7	0.3			
Vertebrates						
Myxine (m)	32.3	142	43	104	290	Betaine 72, TMAO 1.4
Raja (m)	9.6	162	8	64	612	Urea 398
Chimaera (m)	28.3	120	37	189	378	Betaine 38, Urea 335
Salmo (m)	20.5	264	3.2	46	49	
Rana (m)	15.5	126	1.2		10	Carnosine 16
Homo (m)	14	140	4	4	8	Carnosine 14

destabilize macromolecular structure and function is indicated by their position in the Hofmeister series (see Figure 16–5).

Perturbing solutes are osmolytes which bind nonspecifically with proteins and destabilize their structure by unfolding the protein to allow maximal interaction of the osmolytes with the surrounding water (Figure 16–5). Such perturbing solutes have inimical effects on protein structure and function and "salt in" (solubilize) the proteins in the aqueous phase. Urea, arginine, and the common inorganic ions Na$^+$, K$^+$, and Cl$^-$ are perturbing solutes.

Counteracting solutes, such as TMAO and betaine, are osmolytes that reduce the effects of perturbing solutes, such as urea. For example, in elasmobranchs the perturbing effects of urea are counteracted by TMAO (Yancey and Somero 1980; Yancey 1985), and methylamines (betaine and GPC) counteract urea effects in the medulla of the mammalian kidney (Yancey 1988; Yancey and Burg 1989; Yancey and Burg 1990).

The specific composition of the intracellular osmolytes presumably provides the most suitable solute environment for macromolecules with respect to their structural and catalytic properties. Cells might therefore preferentially contain osmolytes that have minimal, or stabilizing, effects on macromolecular structure and function rather than destabilizing effects (Brown and Simpson 1972).

Intracellular Volume Regulation

It is important for cells to have an iono- and osmoregulatory system that preserves the solute integrity of the intracellular environment and regulates the cell volume, especially when faced with changes in the extracellular environment.

Animal cell membranes generally are very permeable to water. In contrast, solutes are much less permeable, by about five to eight orders of magnitude! An important consequence of this semipermeability is that water will rapidly diffuse across a cell membrane if there is an osmotic concentration gradient and will change the cell volume. Regulation of cell volume therefore requires rapid control of the intracellular solute composition.

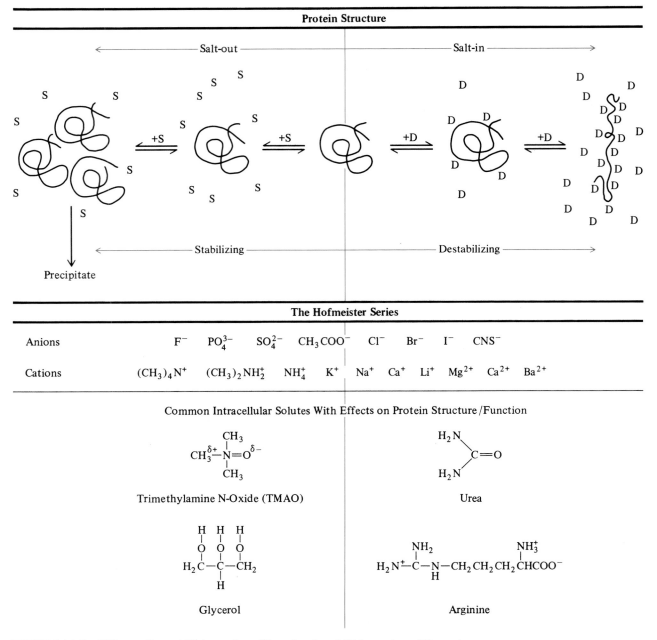

FIGURE 16–5 Effects of a stabilizing solute (S) and a destabilizing solute (D) on the structure of a protein. Stabilizing solutes are preferentially excluded from the protein surface and stabilize protein structure and promote subunit interaction and may precipitate ("salt-out") proteins. Destabilizing solutes bind nonspecifically to proteins and denature them, but promote solubility. The Hofmeister series ranks ions and organic solutes on a scale from stabilizing (salting out) to destabilizing (salting in) effects on macromolecular structure. *(Modified from Low 1985; Hochachka and Somero 1984.)*

Membrane Permeability to Water. Water readily diffuses across cell membranes. The rate of water flux, by analogy with Fick's law of diffusion, depends on the area for diffusion and the concentration difference as well as the diffusion coefficient for water and the diffusion distance (membrane thickness), i.e., permeability.

$$F_{water} = DH_2OA(C_2 - C_1)/\Delta x$$
$$= P_{diff}A(C_2 - C_1) \tag{16.2}$$

The **diffusion permeability** (P_{diff}; μ sec^{-1}) for water across membranes is quite high, about 0.1–10 μ sec^{-1}. It can be readily measured using radioisotopes of water.

The permeability of a cell membrane to water flux due to an osmotic or hydrostatic pressure gradient across the membranes is the **osmotic permeability** (P_{osm}; μ sec^{-1}). This relates the water flux to the difference in mole fraction of solute on each side of the membrane

$$F_{water} = P_{osm} A \left(\frac{n_{s,i}}{n_{w,i}} - \frac{n_{s,o}}{n_{w,o}} \right) \qquad (16.3)$$

where $n_{s,i}$ is the number of moles of solute and $n_{w,i}$ is the number of moles of water on the inside, and $n_{s,o}$ and $n_{w,o}$ are the moles of solute and water on the outside. The osmotic water permeability varies dramatically for different membranes. It is over 1000 μ sec^{-1} for very permeable, artificial colloidon membranes and is less than 0.0008 for very impermeable skin (Table 16–2). The osmotic permeability is often expressed as the day number, which is the time in days required for 1 cm^3 of water to pass through 1 cm^2 of area with a pressure difference of 1 atm.

Alternatively, membrane water permeability can be described using a hydraulic permeability coefficient (L_{hyd}; cm sec^{-1} kPa^{-1}), which relates the water flux to the osmotic pressure difference

$$F_{water} = L_{hyd} A (\Pi_i - \Pi_o) \qquad (16.4)$$

where Π_i and Π_o are the osmotic pressures on the inside and outside. The relationship between L_{hyd} and P_{osm} is $P_{osm} = L_{hyd} RT/\overline{V}_w$, where R is the gas constant, T is absolute temperature, and $\overline{V}_w$ is the partial molar volume of water ($RT/\overline{V}_w = 135 \ 10^3$ kPa or 1330 atms at 20° C).

We might expect that the diffusion permeability would be similar to the osmotic (or hydraulic) permeability, since both describe water flux across the membrane. However, the P_{osm} is generally higher than the P_{diff}. This suggests that transmembrane water flux is not a simple diffusion process. Rather, the water flux may involve bulk movement of water molecules through aqueous pores in the membrane; this would be described better by the

TABLE 16–2

Osmotic permeability and day number for various animals and animal cells. FW, freshwater; SW, seawater.

Species	P_{Osm} μ sec^{-1*}	Day Number
Colloidon membrane	1200	0.12
Echinoderm tube foot (*Asterias*)	32.2	4.5
Toad skin (*Bufo*)	23.6	6.1
Hydra (*Chlorohydra*)	12.5	11.5
Hydra (*Hydra*)	9.4	17.4
Frog bladder (*Rana*)	7.5	19.2
Frog skin (*Rana*)	5.6	25.8
Protozoan (*Zoothamnium*)	2.6	56
Crustacean (*Carcinus*)	1.4	103
Frog egg	1.2	120
Marine ciliate	1.03	140
Eel skin (*Anguilla*, FW)	0.79	183
Crustacean (*Potamon*)	0.60	240
Amoeba (*Amoeba*)	0.53	270
Zebra fish egg	0.45	320
Crustacean (*Astacus*)	0.28	515
Eel skin (*Anguilla*, SW)	0.19	759
Aquatic insect (*Sialis*)	0.045	3204
Lizard skin (*Uromastix*)	0.0061	23601
Mosquito larva (*Aedes*)	0.0056	25750
Sea snake skin (*Pelamis*)	0.0025	57680
Terrestrial snake skin	0.0008	180250

$*1 \ \mu$ sec$^{-1} = 20.76$ cm^3 cm^{-2} atm^{-1} min^{-1}
$\qquad = 2.78 \ 10^6$ mole cm^{-2} hr^{-1} mole^{-1}
$\qquad = 144.2$/day number

Poiseuille-Hagen equation for flow through tubes (flux $\propto r^4$) than Fick's law (flux $\propto r^2$). This is consistent with the observation that antidiuretic hormones (AVP, AVT) can markedly increase the P_{osm}/P_{diff} ratio; they may increase the pore radius for bulk flow, thus dramatically increasing bulk flux but not diffusional flux.

Osmosis.   Osmosis is the net movement of water as a consequence of an osmotic concentration difference across a membrane. Early studies of osmosis often used a semipermeable membrane separating water from a solution containing an impermeant solute. Water moves through the semipermeable membrane into the solution until this movement is counteracted by a hydrostatic pressure. The hydrostatic pressure that exactly counterbalances the osmotic movement of water is the **osmotic pressure** (Π). The SI unit for osmotic pressure is the Pascal; atmosphere and torr are commonly used non-SI units. A solution does not exert an osmotic pressure unless it is in contact with a semipermeable membrane, and so it is preferable to speak of the osmotic concentration of a solution rather than its osmotic pressure.

One mechanism to explain osmosis is related to the relative concentrations of solutes and water on each side of a semipermeable membrane and the presence and properties of the membrane. Molecules of the impermeant solute (like water and permeant solute molecules) experience random molecular motion and collide with the membrane pores at a rate dependent on the solute concentration, but if the solute is completely impermeant then all molecules are reflected back into the solution; the reflection coefficient (σ) for the impermeant solute is 1. Such a solute will exert an osmotic pressure as predicted from its osmotic concentration. A completely permeant solute has $\sigma = 0$; it does not exert an osmotic pressure. A solute with intermediate σ will exert an osmotic pressure proportional to its σ. Generally, a solute with a high permeability has a high σ and vice versa. Water molecules similarly impact the membrane at a rate dependent on the water concentration, but more water molecules impact the membrane pores on the water side than the solution side and so there is a net movement of water molecules into the solution. Thus, osmotic movement of water is essentially the diffusion from a region of high water concentration to a region of low water concentration across a semipermeable membrane.

A fundamentally different explanation for osmosis attributes the osmotic effect of a solute to its modification of the force of attraction between solvent molecules (Hammel and Scholander 1976). The cohesive forces between solvent molecules, such as water, are estimated to be very large— about -100 to -200 MPa (-1000 to -2000 atm). The presence of solutes enhances this solvent tension, altering the basic properties of the solvent in solution, for example increasing the osmotic pressure. An advantage of this explanation of osmosis is that all of the colligative properties of a solution (osmotic pressure, freezing point depression, boiling point elevation, and vapor pressure) are explained by the altered solvent tension. The effects of a semipermeable membrane on solute reflection do not explain the other colligative properties of solutions.

The consequences of water movement across membranes by osmosis are far reaching in biological systems, from the shrinking of plant and animal cells in concentrated media, to volume regulation by animal cells, to exchange of water across the capillaries of the animal circulation, to passive water exchange between animals and their environment, to mechanisms for water excretion and uptake, and to water vapor absorption from air.

Animals as Osmometers.   Animal cells tend to shrink when placed in a hyperosmotic solution because of osmotic loss of water, and they tend to expand when placed in a hypoosmotic solution because of osmotic water influx (their surface membrane is more permeable to water than solutes). A solution that makes a cell or animal shrink is **hypertonic**; one that makes a cell or animal expand is **hypotonic**. In this context, the terms hypertonic and hypotonic are more appropriate in describing the potential effect of immersion of a cell or animal in a solution than hyperosmotic and hypoosmotic because they describe the permeability properties of the membrane to solutes as well as the osmotic concentration of the medium.

Human red blood corpuscles generally behave as simple osmometers, shrinking in concentrated media and swelling in dilute media. Some animal cells behave as if they were a perfect osmometer, conforming to the van't Hoff equation

$$\Pi_e(V_e - b) = \Pi_i(V_i - b) \qquad (16.5)$$

where V is the cell volume (at initial conditions, i, and experimental conditions, e), b is the solid content of the cell, and Π is the osmotic pressure (Dick 1979). A graph of cell volume as a function of Π_i/Π_e should yield a straight line, with slope equal to $V_i - b$, i.e., the initial solvent volume. However, the value of $V_i - b$ determined in this manner for various cells generally does not equal the cell water content (W). The ratio of $V_i - b$ to W is called **Ponder's R** ($R = (V_i - b)/W$), which is generally <1, e.g., about 0.78 for mouse tumor cells, 0.87 for mammalian kidney slices, and 0.94 for chick

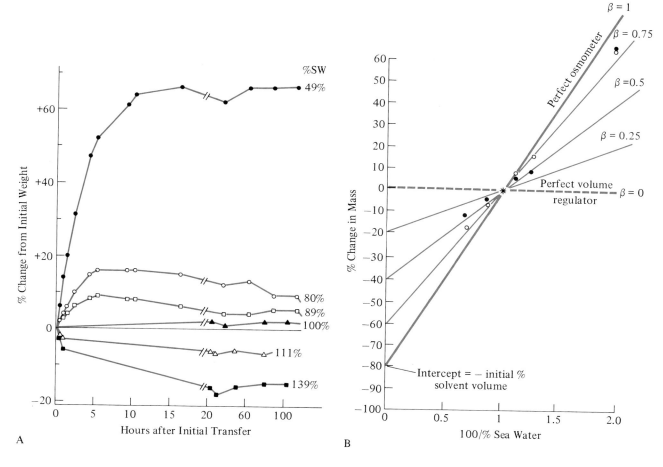

FIGURE 16–6 **(A)** Change in body mass of a sipunculid after transfer to varying concentrations of seawater. **(B)** The transient (5 to 10 hr; open circles) and long-term changes (100 hr; solid circles) in percent of mass for *Themiste* indicates a poor volume regulation compared to individuals in seawater (*). The thick solid line is the predicted relationship for an animal acting as a perfect osmometer with an initial body water content of 80% if there is no ion gain or loss, and the thin lines indicate various levels of volume regulation (β). The thick broken line indicates perfect volume regulation. *(From Oglesby 1968.)*

fibroblasts. There are a variety of explanations for this. First, solute leakage across the cell membrane (or skin for an entire animal) results in a volume change less than that predicted by Equation 16.5, i.e., the cell, or animal, is not a perfect osmometer. Second, Equation 16.5 assumes that solutes are ideal and solutions are dilute, and must be corrected for the nonideal behavior of solutes. Third, part of the intracellular water, the bound and vicinal water, is not freely available as solvent and so the solvent volume should be somewhat less than the total water content of cells. Finally, some cells have the capacity to volume regulate by controlling their intracellular solute composition.

Animals would act as a perfect osmometer (at least over a short time) if they were permeable to water but not to ions. For example, the body fluid space of a hypothetical 100 g isoosmotic marine invertebrate (with a body fluid content of 80%) transferred to 50% seawater should double to 160 ml to maintain osmotic equilibrium if its integument was permeable to water but not to ions. Any deviation from perfect osmometry indicates a significant ion permeability and/or ionic regulation. Do animals in fact, behave as osmometers?

Many animals act like osmometers when initially placed in solutions of varying osmotic concentration, i.e., they shrink or swell. For example, the sipunculid worm *Phascolosoma* gains weight when placed in media more dilute than seawater (e.g., 50% seawater), in accord with predicted changes in mass based on its water content and initial osmotic concentration. The sipunculid worm *Themiste* (which is an osmoconformer) shrinks substantially in hyperosmotic media and swells dramatically in hypoosmotic media (Figure 16–6A). However, it does not behave as a perfect osmometer; it is able to at least partially volume regulate, especially after about 100 hr of exposure (Figure 16–6B). Most animals behave even less as near-perfect osmometers than *Themiste*. The body mass of the polychete *Nereis* and the tectibranch mollusk *Aplysia* stabilizes within hours of a change in salinity, after only a moderate change in water content.

The capacity to long-term volume regulate can be expressed as the ratio of actual weight change to the predicted weight change for a perfect osmometer; this ratio, β, is 1.0 for a perfect osmometer and 0 for a perfect volume regulator (Oglesby 1981; Oglesby 1982). Values of β vary from near 1 (0.9 for *Phascolopsis* and 0.6 for *Themiste*) to near 0 (0.1 for *Nereis limnicola*); many nereids have intermediate β of 0.2 to 0.4.

There are a variety of potential mechanisms for volume regulation by animals. These include a differential permeability to water (outward osmotic loss of water is generally slower than inward osmotic gain), an adaptive reduction in the permeability to water, increased excretion of water to compensate for osmotic influx (by urine flow and hydrostatic expulsion of water across the body wall), and changes in the intracellular and extracellular osmotic composition (both ionic and nonionic solutes).

Many cells precisely regulate their volume, generally by adjusting their intracellular ionic or organic solute composition. For example, many epithelial cells regulate their volume (Spring and Siebens 1988). Some epithelial cells volume regulate when placed in hyperosmotic media by solute uptake, e.g., NaCl and volume regulate in hypoosmotic media by loss of solutes, e.g., KCl. Many tissue-cultured mammalian cells show similar volume regulation by ionic adjustment.

Volume regulation by many animal cells involves nonionic solute adjustments. For example, the euryhaline mitten crab *Eriocheir* readily acclimates from freshwater to seawater. There is some change in tissue (muscle) water content during acclimation, but the decrease is transient and cell volume is rapidly adjusted to normal (Figure 16–7). A similar regulation of cell volume by variation of amino acid concentration is observed in other crustaceans; in other invertebrates; and also in vertebrates such as hagfish, elasmobranchs, euryhaline teleosts, amphibians, and reptiles (Forster and Goldstein 1976).

The change in intracellular amino acid content during cell volume regulation is often due to changes in specific amino acids, rather than to a general change in the concentrations of all amino acids. For example, most of the 266 mOsm decline in intracellular concentration of the euryhaline crustacean *Callinectes* when transferred from seawater to

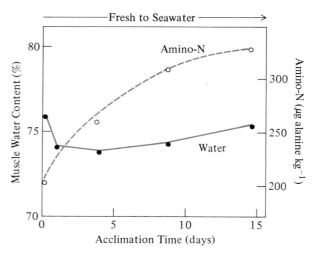

FIGURE 16–7 Changes in water content and amino-nitrogen content for muscle cells of the mitten crab *Eriocheir sinensis* during acclimation from fresh water to seawater. *(Modified from Gilles 1979.)*

50% seawater is due to a decrease in glycine (80 mOsm) and arginine and serine (40 to 50 mOsm); there are lesser changes for taurine (32), proline (25), and alanine (15 mOsm). The amino acids proline and serine increase in the larvae of the mosquito *Aedes* when acclimated to 60% seawater compared to 5% seawater, and there is also a significant increase in the concentration of the monosaccharide, trehalose. In contrast, there is only a small change in body fluid amino acid concentration for the horseshoe crab *Limulus* when acclimated to 40% seawater or seawater, but there is a large change in some unidentified nitrogenous compound.

A variety of other solutes may be used as intracellular osmolytes. The renal medullary cells of mammals experience high extracellular solute concentrations, especially during dehydrational stress. The intracellular osmotic concentration is adjusted primarily by organic solutes, not ions (Balaban and Burg 1987). For example, a variety of methylamines (betaine and GPC) and polyols (sorbitol, myo-inositol) account for intracellular osmotic balance in renal medullary cells of rodents (Yancey 1988) and rabbits (Yancey and Burg 1990).

Water and Ion Budgets

It is convenient and instructive to analyze both water and ion balance by constructing budgets for gains and losses and by examining how these budgets are kept in balance (Table 16–3). There are

TABLE 16–3

Summary of avenues for water and ion exchange.		
In	**Storage**	**Out**
Water		
Drinking	Body fluids	Feces
Food	Urine	Urine
Metabolic		Evaporation from skin
Transcutaneous diffusion		Evaporation through respiration
Water vapor		Transcutaneous diffusion
Ions		
Drinking	Body fluid ions	Feces
Food	Skeleton	Urine
Transcutaneous diffusion/active transport		Extrarenal secretion
		Transcutaneous diffusion

three main avenues for water gain: (1) **drinking**, (2) **preformed water** present in food, and (3) **metabolic water** produced as a byproduct of metabolism. Additional avenues for water gain by some animals include **integumental osmotic absorption** (e.g., amphibians absorb water across their skin) and **water vapor absorption** from the air (some insects, mites, and ticks). The main avenues for water loss are **urine** and **feces**, **integumental osmotic loss** across the body surface for hypoosmotic aquatic animals, and **evaporation** from the body surface (cutaneous water loss) and the respiratory surface (respiratory water loss) for terrestrial animals. Some animals can temporarily buffer an imbalance between gain and loss by **storage** of water or solutes.

Drinking is a ready avenue of water gain for aquatic animals; some inadvertently gain water by ingesting the medium along with their food, but many drink the medium to gain water (although this might compromise ion balance; see below). For example, drinking is an important water source for marine fish. Amphibians generally do not drink but rely on cutaneous uptake of water across the ventral "pelvic patch" of skin. Most terrestrial animals will drink freely if water is available.

The preformed water content of food varies markedly for different diets. The water content of the food is high for insectivorous and carnivorous animals. Plant material has a very variable water content, ranging from only a few percent for dry plant material to much higher values for fruit and other succulent parts of plants. Blood, plant sap,

and nectar have a very high water content and animals that feed on them require a marked renal diuresis for the rapid elimination of the excess water.

Cellular metabolism synthesizes metabolic water. Metabolic water production per energy yield is highest for metabolism of glucose and glucosyl subunits (0.031 to 0.038 ml H_2O kJ^{-1}), is intermediate for lipids (0.029 ml kJ^{-1}), and is lowest for protein (0.021 ml kJ^{-1}). Lipid has the highest metabolic water production per gram of substrate (1.13 ml g^{-1}), carbohydrates have an intermediate value (0.55–0.60 ml g^{-1}), and protein has the lowest value (0.42 ml g^{-1}).

Water gain or loss across the cutaneous surface by osmosis is unavoidable if the body fluids have a different osmotic concentration than the medium. The rate of exchange depends not only on the difference in the osmotic concentrations of the body fluids and the medium but also on the cutaneous osmotic permeability (P_{osm}), which varies for different animals by over $10^6 \times$.

Water vapor uptake from subsaturated air is thermodynamically difficult and few animals can accomplish it. However, some insects, ticks, mites, and isopods can absorb water vapor from air with RH from about 45% to 95%. Their mechanisms for water vapor absorption will be discussed below.

Evaporation occurs in terrestrial environments from the skin and respiratory surfaces. Water evaporates from the skin of many animals as if it were a free water surface, but the skin of more terrestrial animals has a much reduced evaporation rate.

Water is invariably lost via the feces and urine, but some invertebrates and vertebrates are able to markedly reduce these excretory avenues for water loss.

Animal tissues are about 60 to 80% water, and this water is a store that buffers transient imbalances in water gain and loss. The efficacy of the body fluids as a water store depends on the initial body fluid osmotic concentration, the initial body water content, and the osmotic concentration tolerance of the species (Figure 16–8). The blood osmotic concentration increases exponentially with evaporative water loss and reaches a lethal concentration sooner if the initial percentage water content is low. The body fluids are used as a substantial water store by many amphibians that can tolerate a marked increase in body fluid osmotic concentration. They also tend to have a fairly high initial body water content, about 80%, and a fairly low initial body fluid osmotic concentration (about 240 mOsm). Some animals have specialized water storage systems. Their urinary bladder stores dilute urine, which can be reabsorbed to replenish body water losses, e.g.,

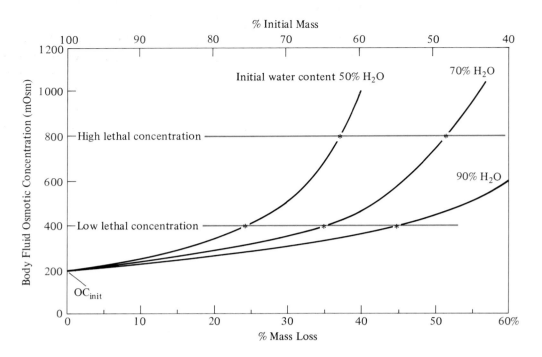

FIGURE 16–8 Theoretical relationship between the body fluid osmotic concentration and mass loss by dehydration for hypothetical animals of varying initial body water content (50%, 70%, and 90%). The increase in body fluid osmotic concentration with dehydration (OC_{deh}) is calculated from the initial osmotic concentration (OC_{init}), initial percent of body water content (BW_{init}), and percent of dehydrated mass ($Mass_{deh}$) as

$$OC_{deh} = OC_{init} \frac{BW_{init}}{Mass_{deh} - (100 - BW_{init})}$$

This assumes that water but not ions are lost from the body fluids and that there is no change in the percent of solid content during dehydration. Animals with a higher initial body water content are able to tolerate greater dehydrational mass loss and animals with a higher lethal osmotic concentration can tolerate a greater mass loss.

---

amphibians and some lizards. Lymphatic fluids can also be a water store as can fluid in the digestive tract.

Ion budgets are similar, in principle, to water budgets except that there are fewer avenues for ion gain and loss (Table 16–3). Ions are gained in drinking water and food and by cutaneous uptake for aquatic animals. Ion uptake by drinking can be substantial for marine and hypersaline animals and can greatly exceed the cutaneous diffusive load (e.g. *Aedes* and *Artemia*; see below). The ion content of food is primarily Na^+, K^+, and Cl^- for insectivores and carnivores, in about the same proportions as in their body tissues, but plant material tends to have a high K^+ content and low Na^+ content.

Fecal and especially urinary losses of ions can be substantial, although many animals eliminate a urine that is much more dilute than the body fluids and some can eliminate a urine that is much more concentrated than the body fluids.

Ions may be gained or lost by diffusion across the skin and other body surfaces (e.g., gills) if the aquatic medium is more dilute or concentrated than the body fluids. Many freshwater animals can actively absorb Na^+ (and Cl^-) from even very dilute media, e.g., amphibian skin and crustacean and fish gills.

Various extrarenal avenues eliminate ions. For example, saliva, sweat, and the secretions of other nonrenal tubular glands eliminate ions. Some reptiles and birds have nasal, orbital, lingual, or sublingual salt glands for elimination of Na^+, K^+, and Cl^-, and elasmobranchs have rectal glands for salt excretion.

Storage of ions can be accomplished by deposition of precipitates in body tissues. Some frogs store guanine (a nitrogenous waste products) in their skin and various other organs. Many insects store their nitrogenous wastes, e.g., lepidopteran larvae store purines and pteridines in their body cuticle and wing scales. Some insects and tunicates store uric acid in their body tissues for subsequent recovery of the nitrogen when required.

Aquatic Environments

Aquatic animals regulate ion and water fluxes in response to passive fluxes of ions and water across their integument. An osmotic loss of water is readily replenished by drinking the ambient medium although this may increase the ionic influx. An osmotic gain of water can be eliminated by copious urine production, although this may involve a depletion of body ions. However, integumentary ion pumps can compensate for such influxes or effluxes of ions. Consequently, ionoregulation is often the more critical problem for aquatic animals rather than osmoregulation. This is not, as we shall see below, the situation for terrestrial animals that are generally limited by the availability of water. But let us first examine the principles of ionoregulation and osmoregulation in protozoans and then in a variety of aquatic animals.

Protozoans

The intracellular environment of protozoans is separated from the external environment only by their plasmalemma; there is no equivalent to the extracellular buffering environment of multicellular animals. Protozoans are also minute organisms and therefore have a very high surface:volume ratio. Consequently, water and ions would be expected to exchange rapidly across their cell membrane and make iono- and osmoregulation difficult. Nevertheless, protozoans are found in a wide variety of aquatic environments, including freshwater, brackish water, seawater, and even hypersaline water. Some protozoans are euryhaline, surviving in both freshwater and seawater. Probably the parasitic protozoans have the most favorable and constant external environment, the body fluids of their host.

Protozoans gain water via their food and by diffusional and osmotic exchange across the plasmalemma. In general, they have a low osmotic permeability because of their thick glycocalyx coat; this limits their rate of osmotic water influx or efflux. Water is eliminated by contractile vacuoles, which expel vesicles of intracellular fluid.

Freshwater protozoans, such as some ciliates, flagellates, and amoebas, must iono- and osmoregulate because their intracellular osmotic concentration (50 to 150 mOsm) is elevated above that of the medium. Nevertheless, their osmotic and ionic concentrations are often very low (<150 mOsm). For example, *Paramecium* in dilute media has an intracellular osmotic concentration of 111 mOsm, of which 6.5 is Na^+, 28.8 is K^+, and 56.5 is amino acids. In a more concentrated medium (up to 160 mOsm sucrose), the intracellular osmotic concentration increases to over 200 mOsm so that it is always hyperosmotic to the medium, mainly as a result of elevated intracellular amino acid concentrations (Stoner and Dunham 1970). *Tetrahymena* have an osmotic concentration of about 110 mOsm in freshwater and 200 mOsm in 160 mOsm media (Figure 16–9A). *Zoothamnium* have an osmotic concentration of about 50 mOsm and *Pelomyxa* about 117 mOsm.

Most freshwater protozoans use a **contractile vacuole**, or vacuoles, to excrete water. The contractile vacuole complex is a spherical vesicle (the vacuole) with a surrounding spongiome, a system of small vesicles and tubules (see Chapter 17). The spongiome collects fluid and delivers it to the contractile vacuole, which expels it from the cell. The freshwater *Rhabdostyla* excretes about 17 μ^3 sec^{-1} but its excretion rate is reduced to 10 in 4% seawater and 0.3 in 12% seawater. Arginine vasopressin, an antidiuretic hormone in higher animals, reduces the plasmalemma water permeability and increases the volume excretion of the contractile vacuole (Couillard, Pother, and Mayers 1989).

Brackish and marine ciliates, foraminiferans, and radiolarians tend to osmoconform, thereby eliminating osmotic exchange of water, but they invariably ionoregulate. For example, the marine ciliate *Miamiensis* in seawater has an intracellular composition of 88 mM Na^+, 74 mM K^+, 3.7 mM Ca^{2+}, 29 mM Mg^{2+}, 61 mM Cl^-, and 317 mM amino acids (mainly glycine and alanine). Na^+ excretion may be the main role of the contractile vacuole for marine species. The euryhaline ciliate *Paramecium calkinsi* can survive in 10 to 2000 mOsm media. Cell volume regulation is precise, due in part to regulation of intracellular amino acid levels (especially proline and alanine; Cronkite and Pierce 1989). The marine ciliate *Zoothamnium marinum* increases in volume when moved from seawater to 40% seawater, and its contractile vacuole markedly increases water excretion from <5 to >30 μ^3 sec^{-1} (Figure 16–9B).

Invertebrates

Aquatic invertebrates occur in a wide range of media, from freshwater to markedly hypersaline water (e.g., salt lakes). Many are isoosmotic to their environment and have limited ionoregulatory requirements (e.g., many marine invertebrates), but

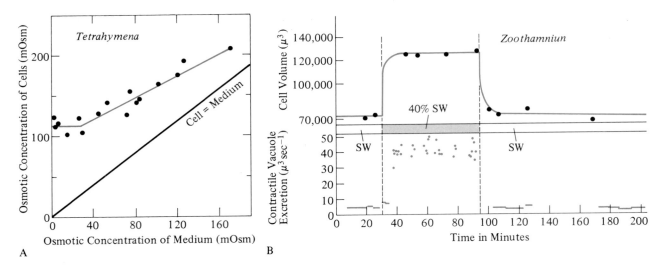

FIGURE 16–9 (A) Hyperosmotic pattern of intracellular osmoregulation in a protozoan. (B) Effect of transfer from seawater to 40% seawater on cell volume and rate of cellular fluid excretion for the protozoan *Zoothamnium*. *(From Stoner and Dunham 1970; Kitching 1934.)*

many are strong osmoregulators and ionoregulators (e.g., freshwater species).

Marine Environments. Marine invertebrates generally have about the same osmotic concentration as seawater (see Figure 16–3, page 781 and Figure 16–4, page 782). This avoids any need to osmoregulate. However, there usually is a small osmotic gain of water across the integument/gills as well as some drinking to balance the water lost as urine (e.g., *Carcinus*, Figure 16–10). Some water is also lost via the feces. Ions are mainly gained across the integument/gills but also by drinking and in the food. Ions are mainly lost via urine but some are lost across the integument.

The ionic concentrations of the body fluids of osmoconforming marine invertebrates are generally similar to those of seawater, although there are consistent differences in the concentrations of all ions and especially of K^+ and SO_4^{2-} (Table 16–4). Many of these differences are small and can be ascribed to the ionic concentration imbalance across a semipermeable membrane due to a Donnan effect, rather than to active regulation. The ionic concentrations of the extracellular body fluids are generally more similar to those for the body fluids dialyzed against seawater than to seawater. Some marine invertebrates regulate extracellular Na^+ at a higher concentration than in seawater, and almost all have a higher K^+ concentration than in seawater. The Cl^- concentration is generally higher than seawater. The SO_4^{2-} concentration is often regulated lower than seawater (e.g., the coelenterate *Aurelia*) because of its role in buoyancy regulation (see Chapter 10).

Some marine invertebrates are hypoosmotic to seawater, e.g., many grapsid crabs, such as *Pachygrapsis*, *Uca*, *Leptograpsis*, *Eriocheir*, and *Ses-*

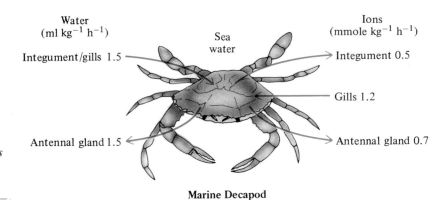

FIGURE 16–10 Water and ion budgets for a marine green crab *Carcinus maenas* (in SW). *(Values from Potts and Parry 1964.)*

Water (ml kg^{-1} h^{-1})

Integument/gills 1.5

Antennal gland 1.5

Sea water

Ions (mmole kg^{-1} h^{-1})

Integument 0.5

Gills 1.2

Antennal gland 0.7

Marine Decapod

TABLE 16–4

		Na$^+$	K$^+$	Cl$^-$	SO$_4^{2-}$	Pr	OC

Extracellular body fluid composition for some representative marine osmoconforming invertebrates. Ion concentrations are mM; protein is g L^{-1}, and osmotic concentration is mOsm.

	Na$^+$	K$^+$	Cl$^-$	SO$_4^{2-}$	Pr	OC
Seawater	*480.2*	*10.2*	*517*	*28.8*	*0*	*1033*
Cnidarian						
Aurelia	454	10.5	554	15.2	0.4	
Annelid						
Nephthys	474	18.0	560		2–4	1063
Echinoderm						
Holothuria	484	10.5	563	28.9	0.7	
Mollusk						
Mytilus	502	12.7	586	30.6		1120
Sipunculid						
Phascolosoma	498	11.2	552	26.3	0.9	
Horseshoe Crab						
Limulus	445	11.8	514	14.1		954
Crustacean						
Lithodes	477	12.5	537	25.7	38	

arma. The crab *Pachygrapsis marmoratus* in seawater has an osmotic concentration of 86% seawater and regulates Na$^+$, K$^+$, Cl$^-$, and SO$_4^{2-}$ concentrations different from seawater and dialyzed plasma. The shrimp *Palaemonetes*, *Penaeus*, and *Metapenaeus* are also weak hypoosmotic regulators in seawater. Some branchiopod crustaceans (brine shrimp; *Artemia* and *Parartemia*) are strong hypoosmotic regulators in seawater and in hypersaline water.

The octopus *O. vulgaris* is slightly hyperosmotic to seawater, at about 1300 mOsm (Wells and Wells 1989). It loses about 2.5 ml kg^{-1} hr^{-1} of water as urine (about 14% of body mass per day) and this is replaced by drinking seawater and by absorbing seawater pumped into the rectum. The seawater is absorbed by the digestive gland appendage, the "pancreas."

Marine invertebrates, despite often being isoosmotic, or nearly so, have ionic concentration gradients between body fluids of seawater of about 1.2:1 or at most 2:1. These are the result of both passive factors (differential permeability and Donnan effect) and active transport of ions. We shall see below that freshwater invertebrates ionoregulate against much greater ionic concentration gradients (100:1 to 1000:1).

The urine of marine invertebrates tends to be isoosmotic and isoionic to their hemolymph. The ion content of food contributes to the maintenance of ionic concentration differences between the extracellular fluids and seawater, although most animals can maintain the ionic concentration differences even when starved. Urinary ion losses can be substantial. For example, the lobster *Homarus* eliminates about 3.4 ml kg^{-1} hr^{-1} of urine containing 451 mM Na$^+$ (cf. hemolymph concentration of 455), 8.4 mM K$^+$ (cf. 9.3), 478 mM Cl$^-$ (cf. 473), and 14.1 mM SO$_4^{2-}$ (cf. 8.9). This is equivalent to a daily excretion of about 10% of total body water, 20% of total body Na$^+$ and Cl$^-$, 1% of total K$^+$, and 33% of total SO$_4^{2-}$. The urine also contains more Mg^{2+} (14.2 mM) than hemolymph (7.7) and less Ca^{2+} (10.1, cf. 15.7 mM). Urinary excretion tends to be lower in seawater-adapted animals (0.2 to 10 ml kg^{-1} hr^{-1}) than in brackish and freshwater animals (15 to 45 ml kg^{-1} hr^{-1}), but is nevertheless a significant avenue for water and ion loss.

Integumental active ion transport is a significant mechanism in many marine invertebrates for the maintenance of ionic concentration differences between extracellular body fluids and seawater in excess of those due to a Donnan effect.

Freshwater Environments. Freshwater invertebrates must osmo- and ionoregulate because it is impossible to be isoosmotic or isoionic with dilute media (Table 16–5). The osmotic concentrations of sponges, cnidarians, and lamellibranch mollusks (which have a high gill surface area) typically are very low ($\leq$100 mOsm). The freshwater rotifer *Asplanchna* has an osmotic concentration of 81 mOsm in lake water (18 mOsm) and 66 mOsm in distilled water. Annelids (e.g., *Lumbricus*), gastropods, and diplostracan crustaceans (e.g., *Daphnia*) have somewhat higher osmotic concentrations (200 to 300 mOsm), insect larvae even higher values (250 to 400 mOsm), and malacostracan crustaceans much higher values (400 to 600 mOsm).

The hemolymph and intracellular concentrations of Na$^+$, K$^+$, and Cl$^-$ are correspondingly low in freshwater invertebrates compared to marine invertebrates. The levels of SO$_4^{2-}$ are generally unmeasurably low (e.g., <1 mM). Ca^{2+} concentrations are relatively high in crustaceans (which have a CaCO$_3$ skeleton) and in insect larvae. Bicarbonate concentration tends to be higher in freshwater (>10 mM) than in seawater invertebrates (<10 mM). Insect larvae also have substantial levels of organic solutes, such as amino acids, in their hemolymph.

The integument of freshwater invertebrates typically has a low water and ion permeability but would still have a substantial osmotic influx of water due

TABLE 16–5

Extracellular fluid composition of freshwater invertebrates. Values are mM and mOsm.				
Animal	**Na$^+$**	**K$^+$**	**Cl$^-$**	**OC**
Sponge				
Spongilla				55
Cnidarian				
Chlorohydra				45
Rotifer				
Asplanchna	21	7		81
Mollusks				
Anodonta	15.6	0.5	11.7	44
Margaritifera	14.0	0.3	9.4	40
Pomacea	55.7	3.0	52.0	139
Annelids				
Lumbricus	75.6	4.0	42.8	299
Pheretima	42.8	6.8	54.4	152
Hirudo	136	6.0	36	201
Insect larvae				
Aedes				266
Sialis	109	4.1	199	436
Ephemera				237
Crustaceans				
Astacus	212	4.1	199	436
Eriocheir	309	5.7	279	636
Potamon	259	8.4	242	506

to the high osmotic differential between the body fluids and the medium (e.g., 100 to 600 mOsm differential) and a substantial diffusional loss of ions. For example, an aquatic insect larvae (*Aedes*) has a fairly impermeable integument, but its cuticular gain of water and loss of NaCl is about 3 and 50% of the total body content per day, respectively. The osmotic water gain by a freshwater crustacean (*Astacus*) is about 5% body weight per day.

The osmotic water gain is eliminated as urine. For example, the freshwater rotifer *Asplanchna* has a urine flow of about 0.047 nl min^{-1} in lake water and 0.060 nl min^{-1} in distilled water. The urine flow rate of other freshwater invertebrates is also high, especially compared with that of seawater invertebrates, e.g., annelid, *Lumbricus* 440; mollusk, *Anodonta* 4350; crustaceans, *Eriocheir* 22, *Gammarus* 41; insect, *Sialis* 4% day^{-1}. The high urine flow eliminates the water gained by osmotic influx, but is a source of ion loss. The urine (U) is hypoosmotic to hemolymph (H) for many of these freshwater invertebrates, e.g., *Gammarus* U/H = 0.18 although many produce near isoosmotic urine, e.g., *Eriocheir*

0.99. The U/H ratios for ions (e.g., Na$^+$ and Cl$^-$) are very low for some freshwater invertebrates, e.g., *Gammarus* 0.18. In contrast, urine is nearly isoionic in others, e.g., *Eriocheir* 0.94. The urinary ion loss is lower than the integumental loss for some, e.g., *Eriocheir* (urine is 19% of total loss) but may exceed the integumental losses for others, e.g., *Gammarus* (urine is 83% of total loss).

Integumentary and urinary losses of ions are balanced by active uptake across the integument. This active ion uptake follows the basic Michaelis-Menten type of enzyme kinetics (see also Chapter 3)

$$J = \frac{J_{max}T_m}{[S] + T_m} \qquad (16.6)$$

where J is the rate of ion flux, J_{max} is the maximal rate of flux, $[S]$ is the external concentration of the ion, and T_m is the transport constant (cf. Michaelis-Menten constant, K_m). There is considerable variability in J_{max} and T_m for active Na$^+$ uptake in various aquatic animals, generally being higher for brackish animals and lower for freshwater animals (Table 16–6). There is also an active Cl$^-$ uptake mechanism that is similar to that for Na$^+$ with similar J_{max} and T_m. Both Na$^+$ and Cl$^-$ uptake can occur from solutions containing impermeant counterions, e.g., Na$^+$ from Na$_2$SO$_4$, Cl$^-$ from KCl, and so there must be an exchange of counterions from the body fluids to maintain electroneutrality, e.g., NH$_4^+$ (or H$^+$) may be exchanged for Na$^+$ and HCO$_3^-$ (or OH$^-$) for Cl$^-$.

Brackish Environments. Brackish waters include estuarine and salt marsh habitats and large inland seas such as the Baltic, Caspian, and Aral; they vary in salinity from 5 to 30 parts per thousand (Venice system, see Chapter 2). The invertebrate fauna of such brackish environments typically includes freshwater species that can tolerate the elevated salinity of brackish waters, seawater species that can tolerate the lower salinity, and species that are found only in these brackish environments.

Many brackish water invertebrates tolerate changes in body fluid osmotic concentration, i.e., they are euryhaline osmoconformers. Some osmoregulate, at least at low or high salinities. For example, the European green crab, *Carcinus maenas*, osmo- and ionoregulates at low salinities. Its urine flow rate is inversely related to the salinity. Urinary water loss is about 8.8 ml kg^{-1} hr^{-1} in dilute media, to match the integumentary gain of about 8.8 ml kg^{-1} hr^{-1}. The urine is isoosmotic to blood, and so the urinary ion losses increase markedly in dilute media. The urinary Na$^+$ loss is

TABLE 16–6

Maximal uptake rate (J_{max}) and concentration for half-maximal uptake (T_m) for active Na^+ uptake across the integument of a variety of animals. The T_m is equivalent to a Michaelis-Menten K_m coefficient for an enzymatic reaction.

Species	Medium[1]	T_m (mM)	J_{max} (mM kg^{-1} hr^{-1})
Ascaphus (frog)	FW	0.07	0.05
Hyla (frog)	FW	0.17	0.09
Lumbricus (annelid)	FW	1.3	0.1
Astacus (crustacean)	FW	0.2	0.15
Carassius (fish)	FW	0.3	0.2
Rana (frog)	FW	0.14	0.27
Salmo (fish)	FW	0.5	0.4
Flounder (fish)	SW	400	0.42
Bufo (frog)	FW	0.23	0.48
Eriocheir (crustacean)	FW	1.0	2
Potamon (crustacean)	FW	0.1	2
Mesidotea (crustacean)	BW	12	8.5
Poecilia (fish)	SW	8	12
Gammarus duebeni (crustacean)	FW	0.4	15
Gammarus duebeni (crustacean)	FW	1.5	20
Gammarus duebeni (crustacean)	BW	2	20
Mesidotea (crustacean)	FW	1.2	20.8

[1] Abbreviations are as follows: FW = freshwater; BW = brackish water; and SW = seawater.

about 3.8 mM kg^{-1} hr^{-1} and integumental diffusive loss is about 4.4; the active integumental/gill uptake is about 8.2 mM kg^{-1} hr^{-1} (cf. Figure 16–10).

Hypersaline Environments. A number of marine invertebrates are hypoosmotic to their environment as are invertebrates in hypersaline media. For example, many (but not all) grapsid and ocypodid crabs, such as *Uca, Hemigrapsis,* and *Ocypode,* are hypoosmotic to seawater, although some *Pachygrapsis* are isoosmotic with seawater and hypoosmotic in more concentrated media. Brackish water calcanoid copepods are hypoosmotic to their medium. The prawns *Palaemonetes* and *Palaemon* are approximately isoosmotic with 70% seawater, and the prawn *Crangon* is isoosmotic with seawater but is hypoosmotic in more concentrated media.

Hypoosmotic regulators lose water to their environment by diffusion across their integument and gills. Their osmotic permeability is low to limit water loss, but they are clearly not impermeable to water. The water lost by diffusion across the integument must be replenished by drinking (and to

a minor extent by metabolic water production). The drinking rates that compensate for the integumental loss of water are generally 5–20 ml kg^{-1} hr^{-1} for these crustaceans. The gut fluid of *Uca* is hyperosmotic to the blood (and to seawater), hence water uptake must occur against an osmotic gradient by solute-linked water transport. Some of these crustaceans can excrete a urine that is slightly hyperosmotic to their blood, e.g., urine of *Pachygrapsis* is 1800 mOsm compared to 1780 mOsm hemolymph in 175% SW (1980 mOsm); *Uca* urine (1166 mOsm) is hyperosmotic to its blood (994 mOsm). The urine is hyperosmotic primarily because Mg^{2+} is excreted at a very high concentration (325 mM). Most crustaceans, however, produce isoosmotic urine, e.g., *Ocypode, Pachygrapsis;* and *Palaemonetes* probably excrete ions extrarenally across their gills (as do teleost fish; see below).

Hypersaline salt lakes have a much higher osmotic concentration than seawater. For example, the Dead Sea has a salinity of 226 parts per thousand and Na^+, Mg^{2+}, and Cl^- concentrations in excess of 4000 mEq L^{-1}. Lake Koombekine, an acid salt lake in Western Australia, has a salinity >250 ppt and an osmotic concentration of about 7000 mOsm; the Na^+ exceeds 4000 mEq L^{-1} (see Table 2–1, page 8). The invertebrates (especially insects and crustaceans) that live in hypersaline lakes must ionoregulate and osmoregulate at considerably lower concentrations than the medium.

The larvae of several insects survive in hypersaline water, e.g., the chironomid *Cricoptopus* (1.5 M NaCl); flies, such as *Coelops* (3 M NaCl); mosquito larvae *Aedes* (up to 4 M NaCl); *Ephydra* (10 M NaCl). The larvae of some *Aedes* mosquitos survive in water ranging from very dilute to several times seawater (Bradley, Strange, and Phillips 1984). For example, *A. taeniorhynchus* larvae survive in 3 × seawater, and other species in 2 to 4 × seawater.

Brine shrimp (*Artemia*, *Parartemia*) survive in even more hyperosmotic media than mosquito larvae, e.g., in 4 to 8 × seawater and even up to crystallizing salt solutions (Croghan 1958a, 1958b). The fundamental mechanism for survival by *Artemia* in such hypersaline media is active salt excretion (Figure 16–11A). Water is lost by diffusion across the cuticle and is replaced by drinking (up to 7% body mass per day); some ions are gained by diffusion across the cuticle but most of the ion load is due to drinking. Water is absorbed across the gut into the body fluids by passive osmotic flow of water in response to active ion uptake from the gut. The gut is thus the organ most responsible for water balance. The salt load due to drinking and diffusion is eliminated primarily by active excretion. These osmoregulatory mechanisms are so effective that brine shrimp can hyporegulate the body fluids at about 1 to 3% NaCl in even 30% NaCl media (Figure 16–11B).

The first 10 pairs of branchiae (appendages) in adult *Artemia* are the sites for ion excretion. This is readily demonstrated by silver or potassium permanganate staining, which shows them to be the most permeable parts of the cuticle (Croghan 1958a, 1958b). *Artemia* survive the oxidation of these branchial structures with potassium permanganate but lose their ability to osmo- and ionoregulate; these "burnt" brine shrimp become strict osmoconformers and are restricted to a narrow range of ambient osmotic concentrations (i.e., are stenohaline), in marked contrast to normal brine shrimp that are exceedingly euryhaline (see Figure 16–11B). The histological structure of the branchiae confirms their role in ion transport (see Chapter 17). The larval nauplii of *Artemia* similarly osmo- and ionoregulate in hyperosmotic media, but their small limb buds lack branchial ion pumps. Rather, their neck organ,

a large and curious structure, is responsible for ion excretion (see Chapter 17). There is a progressive development in the number of limb buds, branchiae, and branchiae with ion pumps (i.e., are stained by silver or potassium permanganate) during the ontogenetic development of *Artemia* from nauplius to adult.

Euryhalinity. Many invertebrates can survive a wide range of salinities, i.e., are euryhaline. Examples include mollusks (*Anodonta*), decapod crustaceans (*Carcinus*), brine shrimp (*Artemia*), and insects (*Aedes*). Such euryhaline capacity either reflects the regulation of extracellular fluid composition by integumental ion pumps, gut water exchange, and various excretory organs that produce urine or reflects the physiological tolerance of a wide range in body fluid osmotic and ionic concentration. Some animals achieve euryhalinity through a combination of osmoregulation at low medium strengths and by osmoconforming at higher medium concentrations.

The turbellarian *Procerodes* (= *Gunda*) *ulvae* is a small estuarine triclad flatworm that is restricted to the intertidal zone. It swells when transferred from seawater to a more dilute medium due to osmotic influx of water and efflux of ions in dilute media (<70% seawater). It dies in tap water but can survive in dilute media with low concentrations of Ca^{2+}, which appears to decrease its cutaneous permeability to water. The metabolic rate of *Procerodes* increases at <60% seawater, presumably reflecting the increased metabolic cost of osmoregulation. Much of the water absorbed by *Procerodes* is stored in vacuoles of endodermal cells, but the vacuoles are not discharged into the gut to eliminate the water. Rather, water appears to be eliminated by another route, possibly the protonephridia.

Larvae of the mosquito *Culex* osmoregulate in lower concentration media but osmoconform in media of high concentrations (>1% NaCl). The anal papillae presumably function in active Na^+ uptake from the medium and are more important in dilute media. There is a correlation between the salt content of the medium and the size of the anal papillae; in freshwater the papillae are very large but are much reduced in concentrated media (Figure 16–12).

Many *Aedes* mosquito larvae osmo- and ionoregulate in brackish water. The body fluid osmotic and ionic concentrations are regulated at about 400 mOsm and Na^+ = 100 mEq L^{-1}. The permeability of the integument of these *Aedes* larvae is lower than that of freshwater species, but they are not exceptionally impermeable to ions or water. In

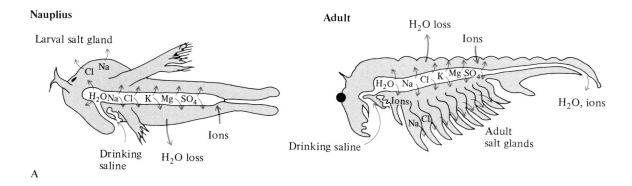

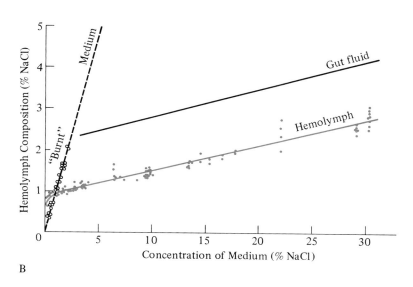

FIGURE 16–11 **(A)** Hypoosmotic regulation by the halophytic brine shrimp (*Artemia salina*) nauplius larva and adult. Integumental water loss is balanced by drinking; ion gain across the integument and by drinking is balanced by salt gland excretion (nauplius neck gland; adult gills). **(B)** Relationship between hemolymph salt concentration and salt concentration of the ambient medium for normal adult brine shrimp (color circles) that are able to hypo-osmoregulate effectively, and for brine shrimp with the salt pumps "burnt" (chemically oxidized; open circles) by potassium permanganate and unable to osmoregulate. Shown also is the salt concentration of the gut fluid.   *(From Conte 1984; modified from Croghan, 1958a; Croghan, 1958b.)*

dilute media, most of the water gain is by drinking (80%) and only 20% is across the integument. Most of the ion loss is via the integument, and this is balanced by active uptake via the anal papillae. In concentrated media, the main avenues for water loss are osmosis across the cuticle and excretion (rectal and anal papillae); drinking is the main avenue for gain. The main avenue for salt gain is drinking, and this salt load is eliminated via the rectum. The larvae in concentrated media actually drink more water than is needed to replace that

lost by integumental diffusion. For example, *A. taeniorhynchus* drinks up to 240% of its body volume per day and *A. campestris* 40% day^{-1}, which is about 12 × more than its diffusional water loss. The drinking rate may be high not only to replace the water lost by diffusion but also to provide a large uptake of dissolved nutrients. Drinking provides a much greater ion uptake than diffusion across the cuticle. *A. campestris* in hyperosmotic media gains about 0.1 μmole of Na$^+$ day^{-1} across its cuticle, but the drinking Na$^+$ load is about 1.2

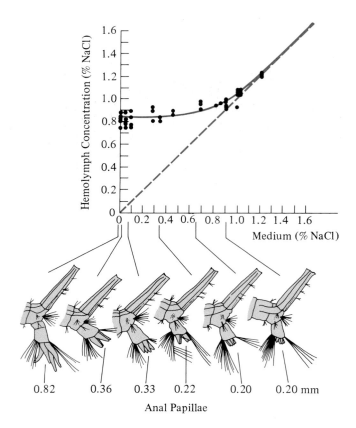

FIGURE 16–12 Relationship between the hemolymph salt concentration (percent NaCl) and the medium salt concentration for the brackish water mosquito larva *Culex* and the associated size of the anal papillae. *(From Wigglesworth 1938.)*

μmole day^{-1} (i.e., 12 $\times$ the diffusional load). This salt load is eliminated by rectal organs that excrete a urine hyperosmotic to the blood.

Vertebrates

Aquatic vertebrates inhabit a wide range of environments, and many species are euryhaline. They are a more closely related phylogenetic group than aquatic invertebrates, and it is possible to discern a phylogenetic and evolutionary progression in their iono- and osmoregulatory strategies. However, the fundamental question of whether vertebrates arose in a fresh, brackish, or marine environment is somewhat controversial.

Origins of Vertebrates. Vertebrates probably arose from a cephalochordate-like ancestor (e.g., *Branchiostoma*) or an animal resembling a tunicate larva, although there are other possible scenarios for the ancestry of vertebrates, such as echinoderm-like calcichordates. The environment in which vertebrates evolved is equally speculative. All of the potential progenitors of vertebrates were undoubtedly marine, and we therefore presume that primitive vertebrates were also marine. However, the

protovertebrates, while originally marine, may have evolved a freshwater breeding stage with a freshwater larva/immature stage. There is considerable debate about whether the major vertebrate radiations occurred in freshwater or in seawater (Griffith 1987).

The incompleteness of the early vertebrate fossil record has been variously interpreted as evidence for a marine origin of vertebrates and for a freshwater or estuarine origin. The earliest vertebrate fossils are marine, but these are fairly advanced vertebrates and their marine habits shed little light on the earlier vertebrates. Romer (1968) was an early and influential advocate for a freshwater origin of vertebrates, arguing that the sense organs, central nervous system, and swimming abilities of vertebrates evolved in animals that had to swim against a constant river or stream current. Smith (1932) also advocated a freshwater origin for vertebrates, based on the structure and functioning of the vertebrate glomerulus. Further evidence for a freshwater origin is the fact that bone forms more readily in dilute ionic environments. The most primitive extant marine vertebrate, the hagfish, has Cl^-/HCO_3^- and Na^+/H^+ ion exchange pumps that are not needed by, nor are present in, marine invertebrates (Evans

1984). Unfortunately, we shall see below that the osmoregulatory strategies of primitive vertebrates do not provide unequivocal evidence for either a seawater or a freshwater origin of vertebrates.

Freshwater Vertebrates. All freshwater vertebrates are hyperosmotic and hyperionic to their environment. They tend to lose ions by diffusion across the skin/gills and to gain water by osmosis (Figure 16–13).

The extracellular body fluids have an osmotic concentration of about 300 mOsm; Na^+ and Cl^- are the predominant ions, and they contribute a considerable fraction of the total osmotic concentration. Urea is present in very low concentrations,

i.e., <1 mM (except in the Lake Nicaragua shark and a few other freshwater elasmobranchs; see below).

Freshwater vertebrates maintain ionic and osmotic balance by active uptake of ions across their integument/gills from the medium and by elimination of water as dilute urine (as do freshwater invertebrates). Gill and cutaneous chloride cells actively transport Cl^- from the medium into the body fluids (see Chapter 17).

Marine Vertebrates. Marine vertebrates use one of three general osmoregulatory strategies: (1) they osmo- and ionoconform, (2) they osmoconform but substantially ionoregulate, or (3) they osmo- and

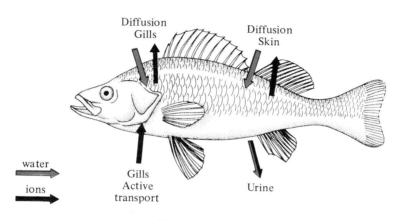

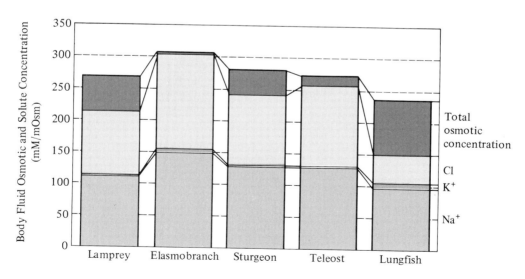

FIGURE 16–13 Schematic of water and solute balance for a freshwater teleost and the extracellular fluid composition for a variety of freshwater fish (lamprey, elasmobranch, sturgeon, teleost, and lungfish).

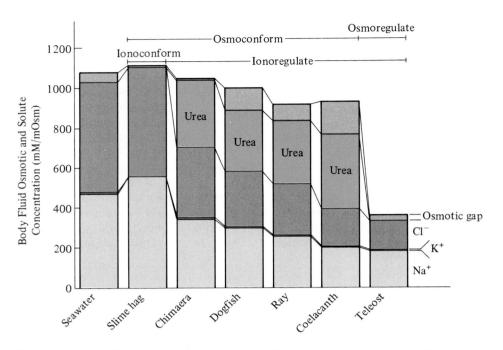

FIGURE 16–14   The extracellular fluid composition for various marine fish that osmoconform or osmoregulate, and ionoconform or ionoregulate.

ionoregulate (Figure 16–14). For example, the blood of hagfish has similar osmotic and ionic concentrations as seawater. Various elasmobranchs and the coelacanth have similar osmotic concentrations as seawater but substantially lower osmotic concentrations; the "osmotic gap" between total ion concentration and osmotic concentration is primarily filled by urea. Teleost fish and the lamprey osmoregulate and ionoregulate at substantially lower values than seawater.

Hagfish (myxinoid cyclostomes) and lampreys (petromyzontoid cyclostomes) may have independently evolved from a group of primitive bony fish, the monorhines (having a single nostril). They

certainly have very different iono- and osmoregulatory patterns. The primitive nature of hagfish is perhaps reflected by its unique (in comparison with other vertebrates) and energetically expedient pattern of osmo- and ionoconforming. There is little gill/cutaneous exchange of water since the hagfish is essentially isoosmotic, but some water is lost as urine and replaced by drinking (Figure 16–15). Ions are gained by drinking and are lost in urine (and perhaps gained by gill transport and lost in the copious slime). The plasma Na^+, K^+, and Cl^- concentrations of hagfish are similar to seawater. Intracellular ion concentrations, however, are similar to those of other vertebrates, e.g., $Na^+ = 32$,

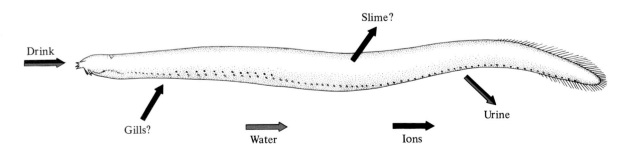

FIGURE 16–15   Schematic of water and salt exchange for a hagfish, the slime hag.

$K^+ = 142$, $Cl^- = 41.3$ mM (Robertson 1976). This iono/osmoconforming pattern of hagfish, which resembles that of many marine invertebrates, has been interpreted as evidence of a marine origin for vertebrates. Alternatively, hagfish may have secondarily reverted to an iono/osmoconforming strategy after reinvading seawater from a freshwater habitat. Some physiological attributes of hagfish are unexpected for an animal that might always have been marine, suggesting that it may have had a freshwater history. Its kidney, like that of other vertebrates, has glomerular nephrons which are thought to have evolved for the elimination of copious dilute urine (such an excretory role is required by freshwater animals but not marine animals). The low ion permeability of the hagfish integument seems unnecessary since it ionoconforms, and the branchial ion pumps have an apparently unnecessary role of ion conservation in the ionoconforming hagfish.

Many marine fish (probably including lampreys) regulate their body fluid composition at about 300 to 400 mOsm with the majority of the osmolytes being Na^+ and Cl^-, i.e., about the same body fluid composition as their freshwater counterparts. These marine fish lose water across their integument/gills (and a little water as urine) and gain ions by diffusion (Figure 16–16). The water lost by osmosis and urine is replaced by drinking seawater; metabolic water production is insignificant. Drinking seawater provides a substantial ion load in concert with the diffusional ion load. In addition, it provides a divalent cation (Ca^{2+}, Mg^{2+}) and anion (SO_4^{2-}) load. The ingested seawater is initially desalinated in the esophagus by diffusion of Na^+ and Cl^- from the lumen, as well as by active Na^+ transport (Kirsch, Humbert, and Simonneaux 1985); the esophagus is fairly impermeable to water due to its mucus lining.

In the stomach, secretory and osmotic addition of water to the lumen further dilutes the contents. Water absorption in the intestine is due to active Na^+ and Cl^- absorption. The fluid contents remain isoosmotic to plasma along the length of the intestine, and the continued absorption of Na^+, Cl^-, and water results in the accumulation of divalent ions, especially Mg^{2+} and SO_4^{2-} (about only 10% of which are absorbed). Ca^{2+} is also accumulated and $CaCO_3$ precipitates in the posterior intestine, thereby osmotically "freeing" water for further osmotic absorption. The kidneys cannot eliminate the ionic load because they are unable to produce a urine hyperosmotic to plasma, although they eliminate divalent ions. The gill and integument Cl^- pumps eliminate the ion load.

Lampreys (*Petromyzon*, *Lampetra*) breed in freshwater where their ammocoete larvae develop. Some land-locked species remain in freshwater, but adults of other species migrate into brackish water or seawater. After a period of active feeding, they return to freshwater to breed. The osmoregulatory pattern of marine lampreys is unclear because they are infrequently caught except when feeding, and they do not retain their ability to osmoregulate in seawater even soon after entering freshwater. The body fluid of a *P. marinus* from the Mediterranean Sea (1240 mOsm) had an osmotic concentration of about 350 mOsm (Burian 1910). Feeding adults of *P. marinus* acclimated to seawater (919 mOsm) have an osmotic concentration of 263 mOsm (Beamish, Strachan, and Thomas 1978). Feeding would contribute substantially to the maintenance of iono- and osmoregulation. It is not known how nonfeeding adult lampreys osmo/ionoregulate in seawater.

The third iono/osmoregulatory pattern of marine vertebrates is to osmoconform (like hagfish) but ionoregulate (like teleosts). This pattern has appar-

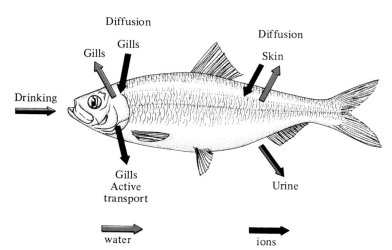

FIGURE 16–16 Schematic of water and salt exchange for a teleost fish.

ently evolved independently a number of times in two groups of fish (chondrichthyeans and the coelacanth) and in some amphibians (Griffith 1985; Griffith & Pang 1979). Plasma electrolyte levels are about 600 mOsm of the total 1000 to 1100 mOsm. The 400 to 500 mOsm "osmotic gap" is filled by various nonelectrolytes, the most important being urea (about 300 mOsm) and TMAO (about 100 mOsm). This pattern of **ureo-osmoconforming** (osmoconforming using urea as the main additional osmolyte) ensures isoosmotic water balance between body fluids and the seawater. It has a number of benefits and costs (see Supplement 16–1, page 829).

Marine elasmobranchs (sharks, rays) ureo-osmoconform. The chimaeras (rat fish) are another group of marine cartilaginous fish which, like elasmobranchs, are ureo-osmoconformers, although they tend to have slightly higher blood ion levels and lower urea/TMAO levels than elasmobranchs. Ureo-osmoconforming reduces problems of water balance, although the blood is often slightly hyperosmotic to seawater and there is a low osmotic influx of water to compensate for urinary water loss (Figure 16–17). Substantial ionic gradients between body fluids and seawater cause diffusional gain of Na^+ and Cl^- across the skin and gills. In elasmobranchs, most, but not all, of the diffusive ion load is eliminated by the rectal gland, a specialized secretory organ which can secrete a hyperionic solution, e.g., 500 to 550 mM Na^+ (cf. seawater is 440 mM). The rectal gland, however, is not essential for ionoregulation as there are other avenues for salt excretion, such as the kidneys and perhaps also the gills. There is a substantial concentration gradient for diffusional loss of urea and TMAO. This is countered by a relatively low integumentary permeability to urea, ions, and other solutes, such as TMAO. Nevertheless, most urea loss still occurs

across the gills and skin. Urea and TMAO excretion via the urine are minimized by active renal reabsorption.

Freshwater elasmobranchs, particularly the Amazon river rays (Potamotrygonidae), have abandoned ureo-osmoconforming as their osmoregulatory strategy. They have low body fluid urea levels, have low levels of urea cycle enzymes, have nonfunctional rectal glands, do not have renal reabsorption of urea, and cannot tolerate high salinities, e.g., seawater (Thorson, Cowan, and Watson 1967).

The coelacanth *Latimeria chalumnae* is a ray-finned crossopterygian fish closely related to teleosts and lungfish and only distantly related to elasmobranchs. Nevertheless, it has a ureoconforming osmoregulatory pattern similar to elasmobranchs, rather than the hyporegulating pattern of teleosts and lungfish. The coelacanth's blood ion concentrations are similar to those of teleosts (Na^+ = 197 mM, Cl^- = 187 mM) but the total osmotic concentration of 932 mOsm is only slightly less than that of seawater (1050 mOsm). The "osmotic gap" is, as in chondrichthyeans, made up primarily by urea and TMAO. The slow diffusional loss of water is replenished by drinking seawater. The drinking and diffusional ion load is probably eliminated by active secretion of the rectal gland (in the same anatomical location as that of elasmobranchs and chimaeras). The coelacanth's urine contains similar levels of both urea and TMAO as the plasma; the urea filtered in the kidney by the glomerulus is apparently not reabsorbed by the renal tubule.

The marine crab-eating frog *Rana cancrivora* inhabits coastal mangroves in southeast Asia, in contrast to other aquatic amphibians that are found in freshwater. This marine frog has independently evolved the slightly hyperosmotic, ureo-osmoconforming pattern of chondrichthyean and crossopterygian fish (Gordon and Tucker 1968). The body

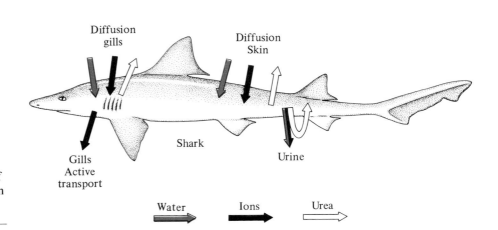

FIGURE 16–17 Schematic of water and salt exchange for an elasmobranch fish, a shark.

fluids of adults in 80% seawater contain 205 mM Na^+, 227 mM Cl^-, and 350 mM urea and the osmotic concentration is 800 mOsm; these are similar to the values for the coelacanth. *R. cancrivora* is slightly hyperosmotic to 80% seawater, thereby ensuring a constant osmotic influx of water to balance urinary water loss. Urea retention is accomplished only by a reduction in urine flow rate; there is no tubular reabsorption of urea (like the coelacanth but unlike the chondrichthyeans). The diffusive salt load is presumably eliminated by skin anion pumps. The tadpoles of *R. cancrivora* are similarly capable of living in seawater but perhaps surprisingly they rely on a totally different osmo/ionoregulatory pattern; they are hypoosmotic osmo/ionoregulators, like teleost fish (Gordon and Tucker 1965)! Their body fluids (in 80% SW) are 153 mM Na^+, 99 mM Cl^-, 5 mM K^+, and 272 mOsm. The crab-eating frog is not, however, independent of freshwater; it requires temporary freshwater ponds (that form after torrential rains) for laying their eggs, and for metamorphosis of the osmoregulating tadpoles, which do not yet synthesize urea, into ureo-osmoconforming adults that synthesize urea (Gordon and Tucker 1968; Uchiyama, Murakami, and Yoshizawa 1990).

Certain other frogs can tolerate brackish if not marine media. For example, the clawed frog *Xenopus* can be acclimated to 60% seawater; its body fluids are always slightly hyperosmotic to the medium, and the primary solute responsible for this osmoconforming pattern is urea (Romspert 1976); thus its osmo/ionoregulatory pattern is ureo-osmoconformation, as in chondrichthyeans, the coelacanth, and the adult crab-eating frog.

Euryhaline Fish. Many **euryhaline** teleosts can survive in seawater and freshwater or in fluctuating estuarine environments. **Anadromous** fish breed in freshwater but have an active adult phase in seawater (lampreys, salmon). The hypothetical vertebrate ancestor has been suggested to have been anadromous. **Catadromous** fish breed in seawater and have a freshwater adult phase (eels). For example, the catadromous eels *Anguilla vulgaris* and *A. rostrata* breed in the North Atlantic Ocean; their stenohaline marine larvae (leptocephali) drift in the ocean currents for one to three years before metamorphosing into elvers, which ascend into rivers (some remain in brackish water or in seawater). The elvers are rapacious fish, which reach maturity and then return to the Atlantic to breed.

Euryhaline fish have the same osmo/ionoregulatory pattern as freshwater fish when in freshwater and of seawater fish when in seawater, i.e., they ionoregulate by ion pumping in opposite directions

in freshwater and seawater. The intertidal blenny *Xiphister* well exhibits this reversal of iono/osmoregulatory exchanges in freshwater and in seawater (Figure 16–18). In freshwater, they have a diffusional loss of ions and gain of water; in seawater, they have a diffusional gain of ions and loss of water. The drinking rate increases in seawater to compensate for the water loss (except in the hyperosmotic elasmobranchs). The glomerular filtration rate is lower for seawater animals and urinary water loss is reduced.

Metabolic Cost of Ionoregulation

Ionic and osmotic regulation requires active transport of ions, and these active transport pumps use energy (ATP) to move ions against their electrochemical gradient. Whether ion transport is active or passive can be gauged by comparing the observed ion flux with that predicted by the diffusion (Nernst) equation for ion flux across a semipermeable membrane (Kirschner 1979). The predicted **equilibrium transepithelial potential** (E_{eq}; volts outside relative to inside) for an ion in equilibrium across a membrane is calculated as

$$E_{eq} = \frac{RT}{zF} \ln \frac{C_{bl}}{C_{medium}} \qquad (16.7)$$

where R is the gas constant, T is absolute temperature, z is the ionic charge, F is Faraday's constant, and C_{bl} and C_{medium} are the concentrations of the ion in the blood and medium respectively. An ion is probably distributed by diffusion and not active transport if its calculated equilibrium potential is the same as the actual transepithelial potential (E_{tep}). An ion's distribution is probably a consequence of active transport, not diffusion, if its equilibrium potential is not equal to the E_{tep}. This conclusion assumes that the ion flux is steady state ($J_{in} = J_{out}$) and that C_{bl} and C_m are constant.

In general, the E_{tep} and E_{eq} for Na^+ across invertebrate and vertebrate gills are similar, and so the flux of Na^+ across the gills is passive; in contrast, the E_{tep} and E_{eq} for Cl^- are not similar, indicating that Cl^- is actively transported (Table 16–7). For example, the chloride cells of the gills and skin of the fish *Platichthys*, in freshwater or in seawater, actively transport Cl^-; Na^+ transport is passive. In contrast, the transepithelial potential across frog skin is primarily due to active Na^+ uptake. The Cl^- equilibrium potential is also not equal to the E_{tep}, suggesting that Cl^- is also not in passive equilibrium across the skin. There is also an active Cl^- uptake system in frog skin with similar

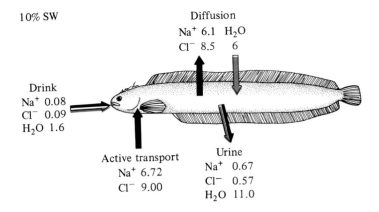

10% SW

Diffusion
Na$^+$ 6.1   H$_2$O
Cl$^-$ 8.5   6

Drink
Na$^+$ 0.08
Cl$^-$ 0.09
H$_2$O 1.6

Active transport
Na$^+$ 6.72
Cl$^-$ 9.00

Urine
Na$^+$ 0.67
Cl$^-$ 0.57
H$_2$O 11.0

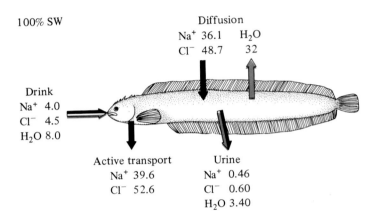

100% SW

Diffusion
Na$^+$ 36.1   H$_2$O
Cl$^-$ 48.7   32

Drink
Na$^+$ 4.0
Cl$^-$ 4.5
H$_2$O 8.0

Active transport
Na$^+$ 39.6
Cl$^-$ 52.6

Urine
Na$^+$ 0.46
Cl$^-$ 0.60
H$_2$O 3.40

FIGURE 16–18 Water and solute balance for an intertidal blenny; values are ml kg^{-1} hr^{-1} (water) and mmole kg^{-1} hr^{-1} (ions). *(Modified from Evans 1967.)*

TABLE 16–7

Transepithelial electrical potential (E_{tep}), sodium equilibrium potential (E_{Na}), and chloride equilibrium potential (E_{Cl}) for fish and a frog osmoregulating in freshwater (FW), i.e., requiring active ion uptake; and for fish osmoregulating in seawater (SW) and for brine shrimp *Artemia* osmoregulating in hypersaline water (HSW), i.e., requiring ion excretion. An E_{tep} similar to E_{Na} or E_{Cl} indicates that the ion is in passive equilibrium across the epithelium and that there is no evidence for active transport of that ion, whereas an E_{tep} substantially different from the E_{Na} or E_{Cl} (in boldface) indicates that the ion is most likely actively transported. All values are mV, inside relative to outside.

	Medium	E_{tep}	E_{Na}	E_{Cl}	Ion Transport
Anguilla	SW	+22.5	+30.5	**−36.5**	Excretion
Platichthys	SW	+33.9	+28.7	**−35**	Excretion
Salmo	SW	+10.1	+26.6	**−33.7**	Excretion
Serranus	SW	+25.2	+26.4	**−33.6**	Excretion
Artemia	HSW	+23.6	+25.7	**−32.8**	Excretion
Pholis	SW	+18	+20.5	**−27.3**	Excretion
Frog	FW	+48.8	**0**	**0**	Uptake
Pholis	FW	−6	−4.4	**+4.9**	Uptake
Platichthys	FW	−52.7	−192	**+212**	Uptake

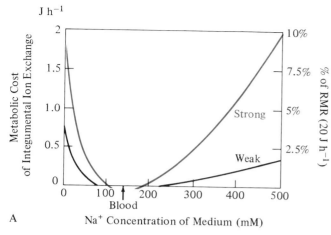

A

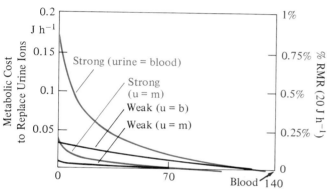

B

FIGURE 16–19 **(A)** Theoretical relationship between the metabolic cost of active integumental ion transport and Na^+ concentration of the medium for a strong ionoregulator and a weak ionoregulator. The arrow indicates the blood Na^+ when the animal is isoionic with the environment (140 mM). **(B)** Theoretical relationship for the metabolic cost of replenishing ions lost in the urine as a function of the ambient Na^+ for a strong ionoregulator and a weak regulator. If no ions are reabsorbed, the urine which is iso-osmotic with blood ($u = b$); or, ions are reabsorbed from the urine until it is iso-osmotic with the medium ($u = m$).

J_{max} and T_m as for active Na^+ uptake and carrier-mediated exchange diffusion; little of the Cl^- flux is apparently due to diffusional exchange.

The metabolic cost for ionoregulation by ion pumping can be calculated from the Nernst equation. The work required to transport ions across a membrane (W; J $mole^{-1}$), against an electrochemical gradient, can be calculated from the Nernst equation as

$$W = EzF = RT \ln C_{bl}/C_{medium} \qquad (16.8)$$

where E is the transmembrane potential (volts).

For ion exchange across the integument, the net diffusional flux (J_{int}; mole hr^{-1}) depends on the integumental permeability (K; cm hr^{-1}) and surface area (A; cm^2). The energy expenditure (E_{int}; J hr^{-1}) required to counter this integumental flux by active transport is equal to JW. If the main ions to be considered are Na^+ and Cl^- then the total energy cost is 2 × that for either of these ions, and

$$E_{int} = J_{int}W \qquad (16.9)$$
$$= 2KA(C_{bl} - C_{medium})RT \ln C_{bl}/C_{medium}$$

where C is the concentration of either Na^+ or Cl^-. The relationship between work and the blood and medium concentrations, calculated from the above equation using K and A values for the polychete *Nereis*, indicates that the metabolic cost of ionoregulation is zero if the blood and medium concentrations are the same (150 mM) and increases at concentrations lower and higher than the blood concentration (Figure 16–19A). A strong ionoregulator (i.e., maintains a constant blood Na^+) has a higher cost for ionoregulation than a weak regulator (i.e., the Na^+ is not regulated constant but alters with varying external concentrations). The calculated metabolic cost for ionoregulation (as a percent of resting metabolic rate) varies from 0% (isoionic conditions) to 10% (strong regulation in 500 mM Na^+). *Nereis* is actually intermediate between a strong and a weak regulator.

For ion balance via urine excretion, the urinary ion flux (J_{ur}; mol hr^{-1}) depends on the urine flow rate and the urine ionic concentration (C_u; mole L^{-1}). The urine flow approximately equals the rate of osmotic gain of water, which depends on the water permeability (P_{osm}; L^2 cm^{-2} $mole^{-1}$ hr^{-1}), the integument area (A), and the blood to medium concentration difference ($C_{bl} - C_{medium}$). The urine ion concentration of freshwater animals is often less

TABLE 16-8

Comparison of the calculated energetic cost for osmoregulation in a variety of aquatic invertebrates.

	Polychete *Nereis* *diversicolor*	**Crab** *Carcinus* *maenas*	**Mosquito Larva** *Aedes* *campestris*	**Brine Shrimp** *Artemia* *salina*
Environment	FW	0.4 SW	SW	Hypersaline
Ambient Na^+/Cl^-	1.4 mM	188 mM	500 mM	3000 mM
Blood Na^+/Cl^-	100 mM	300 mM	140 mM	170 mM
Ion pump energy requirement ($J\ g^{-1}\ hr^{-1}$)	0.11	0.301	0.29	1.92
Energy for urine excretion ($J\ g^{-1}\ hr^{-1}$)	0.003	0.008	1.07	—
Total osmotic energy ($J\ g^{-1}\ hr^{-1}$)	0.113	0.309	1.36	(1.92)
% metabolic rate	2.7%	11%	22%	33%
Metabolic rate ($J\ g^{-1}\ hr^{-1}$)	4.2	2.8	6.3	5.8

than that of the blood because it is energetically more economical to reabsorb ions from the urine (which is initially isoosmotic with blood) than from the medium (which is more dilute than the blood). It becomes less energetically expedient to absorb ions from the urine if it becomes more dilute than the medium.

The overall metabolic cost to replace urinary ion losses is as follows.

$$E_{ur} = RTPA(C_{bl} - C_m)(C_{bl} - C_u + C_u \ln (C_u/C_m))$$
$$(16.10)$$

For example, an animal in brackish water ($Na^+ = 140$ mM) has no metabolic cost for replacing urinary ions if the blood and urine Na^+ is also 140 mM. The energy cost increases exponentially with declining medium Na^+ and is higher for a strong regulator than a weak regulator (Figure 16-19B). Calculations (using P for the polychete *Nereis*) indicate that the energy cost of urine ion replacement increases if the urine concentration equals the blood concentration, from 0% at 140 mM Na^+ to 0.2% of the resting metabolic rate (weak regulator) or 1% (strong regulator). If ions are reabsorbed from the urine rather than the medium, and the urine concentration equals the medium concentration, then the metabolic cost for urine ion replacement is less, increasing from 0% at 100 mM Na^+ to 0.1% (weak regulator) and 0.25% (strong regulator).

The metabolic cost of ionoregulation can vary dramatically for different animals, depending on the concentration of the medium, the body fluid concentration, the permeability of the integument, and the drinking rate (Table 16-8). For example, a polychete worm *Nereis* in freshwater expends about 3% of its resting metabolic rate on osmoregulation, mainly for integumental ion uptake. The crab *Carcinus* in 40% seawater expends about 11% of its resting metabolic rate on osmoregulation, also mainly for integumental ion uptake. A mosquito larva *Aedes* in hyperosmotic water expends about 22% of its resting metabolic rate, mainly for urinary (rectal and anal papilla) excretion. The brine shrimp *Artemia* expends about 33% of its resting metabolic rate on integumental ion excretion. In fish, the metabolic cost for osmoregulation has been calculated to be from 2% to 4% of the resting metabolic rate in freshwater or in seawater. However, measurement of the actual resting metabolic rate of fish at differing salinities suggests that there is a 20 to 30% increase in metabolism associated with osmoregulation in freshwater or in seawater, compared with that in an isoosmotic environment (at which the osmoregulatory cost is zero).

Terrestrial Environments

A number of invertebrate taxa have more or less successfully invaded terrestrial habitats. The most successful terrestrial invertebrates are the arthropods, particularly the insects, spiders, scorpions, ticks, mites, centipedes, and millipedes. A few crustaceans have established a foothold in generally moist terrestrial habitats, i.e., land crabs and crayfish and some isopods. A few isopods are found in dry habitats and even deserts (i.e., pillbugs and woodlice). A number of mollusks have successfully invaded moist terrestrial habitats (e.g., pulmonate snails, slugs) and even deserts. Terrestrial oligochete worms are fossorial and restricted to relatively moist soils, although some can survive extended dry periods by estivating. Terrestrial leeches and onycophorans (*Peripatus*) are generally restricted to moist habitats. Tardigrades (water bears) are unusual, tiny animals that live in the water film of soil and plants; they are especially resistant to desiccation. The most successful terrestrial vertebrates are the reptiles, birds, and mammals; amphibians are generally restricted to moist microclimates.

The water and ion balance of terrestrial animals is quite different from that of aquatic animals because of the limited availability of water. Terrestrial animals also lose water by evaporation from their integument and are generally unable to absorb water from air. Consequently, we need to examine the factors that determine the water content of air and the exchange of water between body fluids and air in order to fully understand the water relations of terrestrial animals.

Water Vapor in Air

The water content of air can be expressed in a variety of ways. Water vapor in air behaves essentially as any other gas, i.e., it approximately obeys the ideal gas law, and the water vapor pressure is one measure of water content. The water content of air is often expressed as the relative humidity (RH; %), which is 100 × the ratio of the partial pressure of water to the saturation water vapor pressure. The absolute humidity is the water vapor density (χ; mg H_2O per liter of air). The dewpoint (DP; °C) is the temperature at which water vapor is saturated (100% RH) and water condenses as dew (or frost). The saturation water vapor pressure and water vapor density are markedly dependent on air temperature (Figure 16–20).

For aquatic animals, the driving force for water exchange is the osmotic concentration difference between the body fluids and medium (or osmotic pressure difference). The water potential (Ψ; kPa), which is related to the chemical potential of water (see Chapter 3), is equal to $-\Pi$, the osmotic pressure, if there is no hydrostatic or gravitational pressure difference. For terrestrial animals, the water potential is related to the relative humidity (air does not have an osmotic concentration or

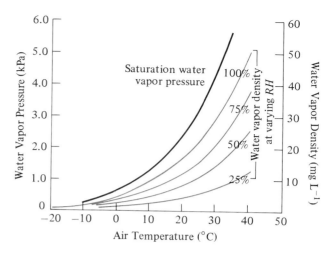

FIGURE 16–20 Relationship between saturation water vapor pressure (kPa) and water vapor density (mg L^{-1}) over a range of air temperatures (see also Table 12–3, page 568).

pressure). The water potential of air, at relative humidity RH in equilibrium with a solution of osmotic pressure Π, is the same for the solution and air

$$\Psi = RT/\overline{V}_w \ln RH/100 = -\Pi \qquad (16.11)$$

where $\overline{V}_w$ is the partial molar volume of water (18 10^{-6} m^3 $mole^{-1}$ at 20° C); $RT/\overline{V}_w$ is 135 MPa at 20° C. Table 16–9 summarizes values for Π of solutions and RH for air in equilibrium at varying Ψ.

Evaporation. The rate of evaporative water loss from an animal depends on the difference in χ for the animal and the surrounding air; by analogy with Fick's law of diffusion (Monteith and Campbell 1980)

$$EWL = -D\Delta\chi_{wv}/\Delta d \qquad (16.12)$$

where EWL is the evaporative water loss (mg cm^{-2} sec^{-1}), D is the diffusion coefficient for water vapor in air (0.242 cm^2 sec^{-1}), $\Delta\chi_{wv}$ is the difference in water vapor density (mg cm^{-3}), and Δd is the diffusion pathlength (cm). The χ for the animal is assumed to be equal to the χ for saturated air in equilibrium with body fluids; it is not the water density of body fluids. The EWL of animals is often expressed as water loss per water vapor pressure deficit (e.g., mg H_2O min^{-1} kPa^{-1}, mg min^{-1} $torr^{-1}$, or mg min^{-1} atm^{-1}) to account for the "concentration difference" driving the water movement. EWL is not proportional to the water potential difference but to the difference in water vapor density.

TABLE 16–9

Comparison at differing water potential (ψ; MPa) of the water and osmotic concentrations of a solution and equilibrium relative humidity (RH), vapor pressure (pH$_2$O) and water vapor density (χ) of air. Values are for a NaCl solution at 20° C.

$\Psi = RT/\overline{V}_w \ln(RH/100)$, where $RT/\overline{V}_w = 135$ MPa (1330 atm)

$\quad = -RT\,C_{osmol}$, where $RT = 2.44$ mPa L mol^{-1}(24.1 atm L mole^{-1})

Water Potential Ψ (MPa)	Solution Water (g kg^{-1})	Solutes (mOsm)	Air RH (%)	pH$_2$O (kPa)	χ (mg L^{-1})
0	1000	0	100.0	2.34	17.3
−0.68	996	277	99.5	2.33	17.2
−1.36	993	555	99.0	2.32	17.1
−6.92	971	2831	95.0	2.22	16.4
−14.2	942	5815	90.0	2.11	15.6
−30.1	—	—	80.0	1.87	13.8
−38.8	—	—	75.0	1.75	13.0
−93.6	—	—	50.0	1.17	8.6
−187.1	—	—	25.0	0.58	4.3
−310.9	—	—	10.0	0.23	1.7
−404.4	—	—	5.0	0.12	0.9
−621.7	—	—	1.0	0.02	0.2
−∞	—	—	0	0	0

The EWL depends inversely on the distance over which water vapor must diffuse, i.e., Δd is the boundary layer thickness for water vapor around the animal. This, in turn, is determined by the specific pattern of air flow and the size of the animal. For example, the EWL will be greater for an animal in moving air than in still air.

Cutaneous Evaporation.   Equation 16.12 can be rearranged to replace the diffusion coefficient (D) and diffusion path length (Δd) by a **resistance** term (r), making the equation for EWL analogous with Ohm's law rather than Fick's law.

$$EWL = -\Delta\chi_{wv}/r \qquad (16.13)$$

The unit for resistance is sec cm^{-1}, since EWL is mg cm^{-2} sec^{-1} and χ is mg cm^{-3}.

Some representative values for r (Table 16–10) indicate a very wide range, from minimal values of about 0 (equivalent to a free water surface) to over 1000 sec cm^{-1} for very impermeable integuments. There is a clear correlation between the degree of terrestriality and the r value for both invertebrates and vertebrates. For example, terrestrial crustaceans (pillbugs, woodlice, crabs) have low r values (<10), whereas insects, spiders, and scorpions may have r values >1000. Earthworms and active snails have low r, whereas inactive snails, withdrawn into their shell and with epiphragms covering the shell

aperture, have very high r values because of the high resistance of the shell ($r > 100,000$), the epiphragms ($r > 400$), and the shell aperture ($r > 196$). The epiphragms have a high r (about 50 each) only at low *RH*; at high *RH* the r values decline considerably (to about 0.4; Barnhart 1983). The r for terrestrial crustaceans ranges from 1.3 to 15 for terrestrial isopods and amphipods and from 4.3 to 74 for crabs. Amphibians generally have an r equivalent to that of a free water surface, about 0, but some "waterproof" frogs (*Chiromantis, Phyllomedusa, Hyperolius, Litoria*) have high cutaneous r values (>50). A few frogs, such as the casque-headed frog *Trachycephalus* have their cranial skin co-ossified (fused) with their skull bones. This slightly reduces the EWL (about trebles the resistance from 0.1 to 0.3); this may be significant for these arboreal frogs that use their head to seal the entrance to burrows and tree holes in which they live. Some amphibians, while burrowed underground and inactive (estivating), form a cocoon of accumulated layers of shed epidermis. The cocoon dramatically reduces the EWL, i.e., has a high r. Estivating lungfish also produce a cocoon of accreted mucus that similarly increases their resistance and decreases EWL. The epidermis of amniote vertebrates (reptiles, birds, mammals) is covered by a layer of dead, keratinized cells, called the stratum corneum. This layer acts as a diffusion

TABLE 16–10

Resistance values (sec cm^{-1}), for evaporative water loss from the integument of various animals, compared with a free water surface.

Species	Resistance
Free water surface	0
Lithobius (crustacean)	1.3
Bufo (toad)	1.5
Helix (snail, active)	1.6
Porcellio (isopod)	3.1
Hyla (tree frog)	3.3
Orchestia (amphipod)	3.7
Caiman (caiman)	5.5
Callinectes (aquatic crab)	6.3
Lumbricus (earthworm)	8.8
Sialis (aquatic insect)	9.7
Hemilepistus (isopod)	15
Hyperolius (reed frog)	25
Glossina (tsetse fly)	29
Ocypode (ghost crab)	32
Terrapena (box turtle)	33
Helix (dormant snail, mantle)	46
Columba (dove)	56
Cardisoma (terrestrial crab)	64
Gecarcinus (terrestrial crab)	74
Pinata (spider)	83
Litoria (tree frog)	118
Gopherus (desert tortoise)	120
Struthio (ostrich)	158
Pyxicephalus (cocooned frog)	190
Helix (snail, shell aperture)	196
Gehyra (gecko)	198
Hydrobius (aquatic insect)	207
Phyllomedusa (tree frog)	242
Iguana (lizard)	370
Chiromantis (tree frog)	404
Helix (dormant snail, epiphragm)	414
Orthoporus (millipede)	433
Pternohyla (cocooned frog)	457
Amphibolurus (lizard)	521
Scorpio (scorpion)	1317
Sauromalus (lizard)	1360
Androctonus (scorpion)	4167
Onymacris (tenebrionid beetle)	5030
Helix (snail, shell)	113418

barrier to increase *r* and reduce EWL. Aquatic reptiles generally have a low *r*, whereas terrestrial reptiles may have skin with much greater *r* values (>300). Bird and mammal skin has a moderately high *r* value (50 to 200).

The integumental resistance generally decreases at higher temperatures, particularly when animals become heat stressed and increase EWL by various physiological mechanisms (see also Chapter 5). Such increases in EWL with temperature are observed for insects, "waterproof" frogs, and some birds and mammals. There is an abrupt increase in EWL for insects at high temperatures that is generally associated with an increased permeability of the epicuticular/epidermal wax layer (Figure 16–21). In mammals, the increase in *r* at high temperatures is due to watery secretions of sweat glands. Waterproof amphibians also "sweat" when heat stressed.

Respiratory Evaporation. Water evaporates from the respiratory surfaces as well as from the external integument. The respiratory surface is highly elaborated to increase its surface area and is thin to minimize the diffusion distance. The high respiratory surface area increases the potential for evaporation and the thin, gas-permeable structure of the epidermis precludes the limitation of water loss rate by a diffusion/lipid barrier (such as a keratinized epidermal layer or a lipid layer). The respiratory surface is consequently moist and evaporates as if it were a free water surface. Respiratory water loss rate can be similar to, or even greater than, the cutaneous water loss rate.

The water exchange between the ambient air and the respiratory surface is either by diffusion (e.g., pulmonate snails, insects) or by convection (e.g., land crabs, amphibians, lizards, birds, mammals). The respiratory evaporative water loss can be con-

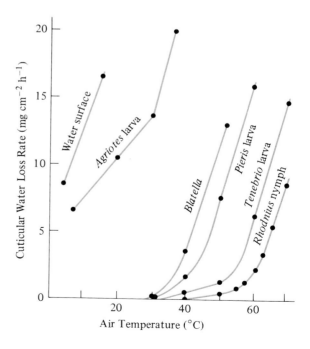

FIGURE 16–21 Rate of cuticular water loss for a variety of insects (dead) as a function of air temperature. *(Modified from Wigglesworth 1945.)*

siderably reduced by various structures of the respiratory tract and by regulation of the respiratory gas tensions. How much water is lost by evaporation per ml O_2 uptake can vary dramatically for different respiratory systems. A moist respiratory surface, such as frog skin, rapidly loses water by evaporation if exposed to dry, well-mixed air. There are substantial pO_2 and pH_2O gradients between the dry air ($pO_2 = 21.3$ kPa, $\chi = 0$ mg L^{-1}) and skin ($pO_2 = 13.3$ kPa, $\chi = 43$ mg L^{-1} at 37° C). We can calculate from the diffusion coefficient for O_2 in water that the O_2 uptake would be about 5.72 μl O_2 min^{-1} per 1 cm^2 through a 5 μ thick layer of water, and from the resistance of a free water surface to evaporation that the concomitant EWL would be about 315 μg min^{-1} (Figure 16–22A). The ratio of EWL/ml O_2 is about 55 mg H_2O ml O_2^{-1}. Terrestrial animals generally cannot sustain such a high EWL/ml O_2, and many have evolved various surface structures to reduce the ratio to more tolerable values. Even a layer of still air is an effective diffusion barrier for water (Figure 16–22B). Air has a very high diffusion coefficient for water (DH_2O) and oxygen (DO_2); the ratio for DH_2O/DO_2 is about 1.40. The EWL and O_2 values calculated for a 5 μ thick layer of still air, with the same conditions as the example of the moist epithelium, are still very high, but the ratio of EWL/ml O_2 is reduced to about 0.97. A variety of animal surface structures act as an air diffusion barrier.

Arthropod cuticle has a lower diffusion coefficient for O_2 than air and water and has a higher resistance to EWL (Figure 16–22C). This cuticular surface would also have a lower O_2 transport capacity; the EWL/ml O_2 would be about 0.98 mg ml O_2^{-1}. Estivating snails form diffusion barriers (epiphragms) across their shell aperture to reduce EWL; the epiphragm has a specialized "window," the kalkfleck, that allows O_2 and CO_2 exchange. These epiphragms have a ratio of water loss to O_2 exchange similar to that expected for an air diffusive barrier, about 0.90 mg ml O_2^{-1} (Figure 16–22D). Many frogs

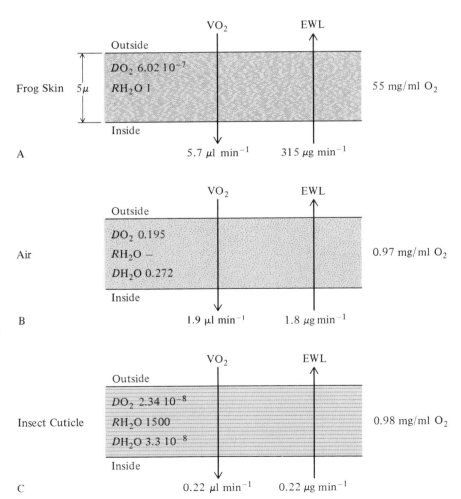

FIGURE 16–22 Schematic of the exchange of water and O_2 across various surfaces. DO_2 is the diffusion coefficient for O_2; DH_2O is the diffusion coefficient for water vapor; R is the resistance to evaporative water loss. (A) frog skin; (B) air diffusion barrier; (C) insect cuticle; (D) snail epiphragm; (E) frog cocoon; and (F) bird eggshell.

estivate during the dry season, and during this inactive, dormant phase some species form a co-coon. The cocoon is formed from multiple layers of shed epidermis that accumulate to form a thick, paper-like wrapping around the entire body surface of the frog, except for the nares. This cocoon reduces the EWL/ml O_2 to about 0.74 (Figure 16–22E). The shell of bird eggs must be permeable to respiratory gases and still limit water loss. The shell is covered with numerous small pores that act as an air-filled diffusive barrier to water loss but allow O_2 entry and CO_2 excretion. The relatively high diffusion coefficients for O_2 and H_2O through the eggshell pores result in a ratio of about 0.82 for EWL/ml O_2 (Figure 16–22F). Vertebrate convective lungs have an EWL/ml O_2 of about 0.25, whereas insect tracheal systems can have an EWL/ml O_2 of about 0.11 because the pO_2 differential between ambient air and tissues is higher (e.g., about 21.3 to 2.7 kPa) than it is for vertebrate lungs (about 21.3 to 13.3 kPa; see Figure 13–40, page 654).

This simplistic but heuristically useful analysis of EWL/ml O_2 for various respiratory surfaces emphasizes the importance of the nature of the respiratory surface and the physiological parameters (gas tensions) in determining EWL. Other factors also determine the absolute value for EWL from invertebrate and vertebrate respiratory surfaces.

Terrestrial lizards, birds, and mammals can reduce their respiratory EWL by nasal countercurrent exchange of heat and water. Inspired air is warmed to body temperature and saturated with water vapor in the nasal passages to avoid desiccation of the thin alveolar epithelium. This cools the surface of the nasal passages. Expired air is subsequently cooled by heat exchange with the nasal mucosa as it leaves the nasal passages. The extent to which the expired air is cooled depends on the anatomy of the nasal passages and the rate of air flow. There is considerable variability among birds and mammals in the cooling of expired air and the extent of nasal countercurrent heat and water exchange.

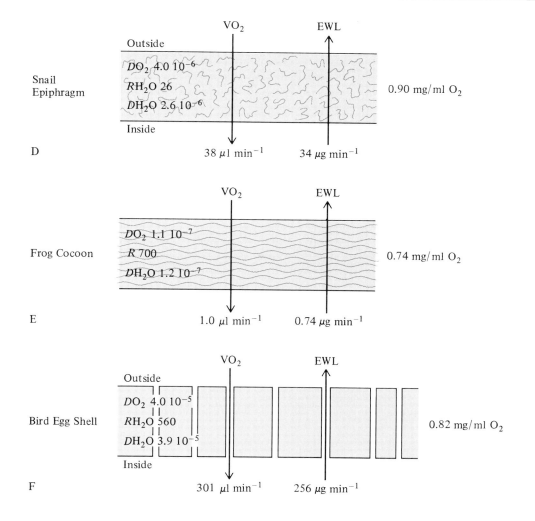

The expired air temperature can be considerably lower than the deep body temperature (Figure 16–23A). For example, the expired air of a kangaroo rat (*Dipodomys*) can actually be cooler than the inspired air; the kangaroo rat is effectively an evaporative air conditioner! Water condenses in the nasal passages as it is cooled during expiration because the expired air is initially 100% saturated with water vapor at lung temperature. This results in the countercurrent recovery of both water as well as heat from the expired air (Walker, Wells, and Merrill 1961; Schmidt-Nielsen, Hainsworth, and Murrish 1970). The dynamics of countercurrent water exchange are summarized for a human in Figure 16–23B. Water recovery is even more effective in the kangaroo rat. Air may be inspired at 30° C and 25% *RH* (with water vapor density = 13 mg L^{-1}) and lung air would be saturated at 38° C and 100% *RH* (water vapor density = 46 mg L^{-1}). The evaporative water loss would be 33 mg water per liter of air if the air was expired at body temperature; the EWL/VO$_2$ would be about 0.66 mg ml O$_2^{-1}$ (assuming an O$_2$ extraction of 20%). However, the expired air is cooled to 27° C so the water content (if 100% *RH*) is only 26 mg L^{-1}. The net respiratory water loss is thus only 13 mg L^{-1} and EWL/VO$_2$ is 0.26 mg L^{-1}.

The nasal countercurrent system also recovers the heat added to the air on inspiration to warm it to body temperature. Consequently, the nasal countercurrent exchange system is important not only for water economy in desert species but for thermal economy in arctic and antarctic species (Withers, Casey, and Casey 1979).

It is generally assumed that expired air is 100% saturated with water vapor, regardless of the extent of nasal countercurrent heat and water exchange and expired air temperature. However, there are at least two animals that can expire unsaturated air: the camel and the ostrich. Camels can exhale air at an *RH* of about 75%; this results in the recovery of about 60% of the water that is evaporated into the inspired air (Schmidt-Nielsen, Schroter, and Skolnik 1980). The ostrich can exhale air at about 36° C and 85% *RH*; this recovers about 35% of the water from lung air (Withers, Siegfried, and Louw 1981). The nasal countercurrent heat and water exchange system does not approach 100% efficiency, even if the exhaled air temperature is less than the inspired air temperature (e.g., kangaroo rat) or the exhaled *RH* is less than 100% (e.g., camel, ostrich). Animals cannot extract water from the inspired air and they always have a net respiratory evaporative water loss.

The desert iguana uses water from its nasal salt gland secretion to contribute to the humidification of the inspired air; the water content of the secretion of the nasal salt gland would otherwise have been lost without benefit to the water economy and other body fluids would have been used to humidify the inspired air.

The tracheal respiratory system of insects provides a substantial water economy through the control of spiracular opening and closing (e.g., the tsetse fly *Glossina*; see Figure 13–27, page 641) or discontinuous respiration (e.g., the silkworm pupa *Cecropia*; see Figure 13–30, page 643).

Partitioning of Evaporation.   The respiratory EWL varies from only a small fraction of the total evaporative water loss (negligible in ticks and scorpions) to about 2/3 of the total in some vertebrates (Figure 16–24).

The respiratory EWL tends to increase with activity and is greater than 60% of total EWL in flying tsetse flies and locusts and in some active insects. EWL can be augmented during heat stress by increasing cutaneous EWL (sweating in some mammals; peripheral vasodilation in some mammals and birds) or respiratory EWL (panting, gular flutter). Consequently, there is a marked disparity in the partitioning of respiratory and cutaneous water loss of these birds and mammals. Some mammals even utilize different means for evaporative cooling depending on the nature of the heat load. For example, kangaroos sweat during exercise to evaporatively cool but do not sweat during a radiative heat load; rather, they pant since this more efficiently dissipates the radiative heat load (Dawson, Robertshaw, and Taylor 1974).

Water Vapor Absorption.   Very few animals can absorb water vapor from subsaturated air, despite the virtually universal presence of water vapor in air. This is because of the thermodynamic problems associated with removing water molecules from the vapor phase, except at saturated or supersaturated conditions. Consider the equilibrium between a solution in contact with subsaturated air (Figure 16–25A); the water potential (Ψ) of the solution is equal to $-RT\ C_{osm}$. The water potential of the air is $RT/\overline{V}_w \ln RH/100$. At equilibrium, $\Psi_{air} = \Psi_{sol}$; $\Psi_{air} > \Psi_{sol}$ for the solution to absorb water vapor; $\Psi_{air} < \Psi_{sol}$ for evaporation from the solution. For an animal that can absorb water vapor, such as the feather mite *Proctophyllodes*, there is an equilibrium *RH* (RH_{equil}) at humidities greater than which the animal

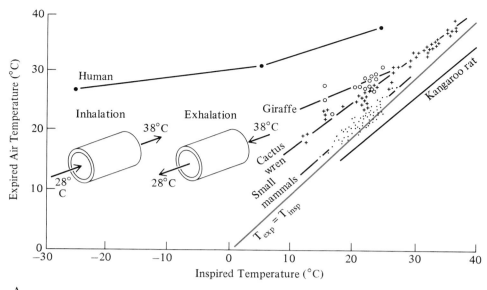

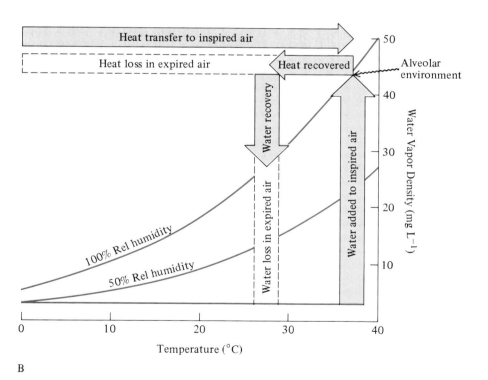

FIGURE 16–23 **(A)** Relationship between inspired and expired air temperature for a range of animals with differing efficiency of nasal countercurrent heat exchange. Inset shows the principle of nasal countercurrent heat exchange. **(B)** Effect of nasal countercurrent heat and water exchange in the human on recovery of water from the expired air (right, color), and the recovery of heat from the expired air (upper; grey).

(Data from Schmidt-Nielsen, Hainsworth, and Murrish 1970; Schmid 1976; Langman et al. 1979; modified from Walker, Wells, and Merrill 1961.)

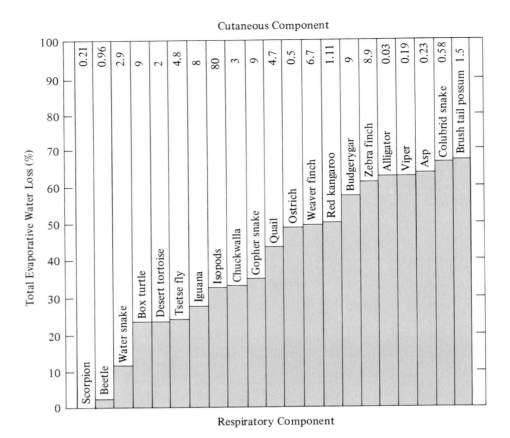

FIGURE 16–24 Partitioning of total evaporative water loss into respiratory (color) and cutaneous (open) components for a variety of animals. The total EWL (mg g^{-1} hr^{-1}) is indicated for each species.

can absorb water vapor and below which the animal loses water vapor by evaporation (Figure 16–25B).

Only a few arthropods are able to absorb water vapor from subsaturated air, e.g., some ticks, mites, and insects. The first observation of weight gain by insects at subsaturated *RH* was for the mealworm *Tenebrio*, but the explanation offered was overproduction of metabolic water not water vapor absorption (Buxton 1930). Mellanby (1932) first suggested that the mealworm larvae absorbed water vapor. Since then, a considerable number of insects (Thysanura, Mallophaga, Psocoptera, Siphonaptera, Coleoptera, Orthoptera) and acarines have been reported also to absorb water vapor from air. The equilibrium *RH* varies from about 43% to 92% for insects and from 57% to 94% for acarines.

The anatomical site, mechanism, and *RH*$_{equil}$ for water vapor absorption vary markedly for different acarines and insects; for some species the mechanism remains unclear. Three sites might be considered likely for water vapor absorption: the mouth, the anus, and the general integument. It is relatively straightforward to demonstrate which is the site for water vapor absorption by blocking the mouthparts and/or anus with paraffin wax and determining whether the capacity for water vapor absorption is eliminated.

Acarine ticks absorb water vapor via their mouth. Direct observation of dehydrated ticks indicates that they secrete a clear fluid that accumulates between the mouth parts and palps. At low *RH* this fluid dries to a white crystalline solid, but at high *RH* the crystalline solid deliquesces and is imbibed. Specific salivary gland cells, the agranular cells, are thought to secrete the fluid (which differs from normal saliva in ion composition). KCl is a major osmotic constituent of the fluid, but it is possible that specific organic solutes may also contribute to the low water potential. The *RH*$_{equil}$ varies from about 70 to 94% for ticks.

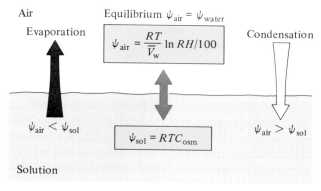

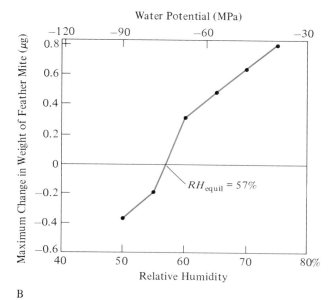

FIGURE 16–25 (A) Water vapor in air and water in solution are in equilibrium when the respective water potentials (Ψ) are equal; evaporation of water occurs if $\Psi_{air} < \Psi_{sol}$ and condensation of water vapor occurs if $\Psi_{air} > \Psi_{sol}$. (B) Equilibrium relative humidity (RH_{equil}) for the feather mite *Proctophyllodes* determined from the maximal rate of weight gain by water absorption at various ambient relative humidities. *(Data from Gaede and Knulle 1987.)*

Many mites can absorb water vapor. The critical RH_{equil} at which $\Psi_{air} = \Psi_{body}$ fluids is about 57% for the feather mite *Proctophyllodes* (Figure 16–25B). The supracoxal glands of mites appear to secrete a KCl-rich hygroscopic solution, which is exposed to ambient air on the supracoxal plate for absorption of water vapor. The diluted fluid then flows via a podocephalic canal into the prebuccal cavity where it is swallowed and pumped by the muscular pharynx into the midgut for absorption (Wharton and Furumizo 1977; Wharton 1978). The RH_{equil} varies from about 55 to 90% for mites.

All, or at least most, psocopteran insects (order Psocoptera) and some of the related biting and sucking lice (order Phthiraptera) can absorb water vapor via their mouth (Rudolph 1982; Rudolph and Knulle 1982). In *Badonnelia*, the ventral part of the preoral cavity (the salivarium) is opened during water vapor absorption and a pair of lateral sclerites is extruded and exposed to ambient air (Figure 16–26). Water vapor absorption begins immediately on exposure of the salivarium and continues until the salivarium is retracted. The lingual sclerites are smooth, except for a groove along their length that connects to a tubular filament running to the cibarial sclerite of the cibarium (part of the oral chamber). A hygroscopic fluid from dorsal labial glands appears to be spread over the lateral sclerites during water vapor absorption. This fluid is diluted by water absorption and then sucked along the central groove and tubular filament by the cibarial/epipharyngeal sclerite pump into the midgut for absorption. Rhythmic contractions of the large clypeo-epipharyngeal muscle probably pump fluid from the lateral sclerites along the tubular filament. Some, but not all, of the related biting lice can also absorb water vapor from lower RH than can psocodids. No sucking lice appear to absorb water vapor; their mouthparts are highly specialized for sucking and this may preclude a role of a lateral sclerite mechanism for water vapor absorption. The RH_{equil} for psocopterans and biting lice is about 43 to 85%.

The desert cockroach *Arenivaga* also absorbs water vapor via its mouthparts. Only larvae and neotenic, wingless female adults can absorb water vapor from $RH > 82\%$ (Edney 1966). Two eversible bladders are extruded from the anterior hypopharynx during water vapor absorption; these bladders are covered with fine, densely packed cuticular hairs (Figure 16–27). A fluid from the frontal bodies, located inside the labrum, covers the everted bladders. This fluid may be an ultrafiltrate of hemolymph and its composition therefore would not explain its capacity to absorb water vapor by an osmotic absorptive mechanism (O'Donnell 1982a, 1982b); the Na^+, K^+, and Cl^- concentrations would be two to three orders of magnitude lower than those required for osmotic absorption of water vapor. The mat of cuticular hairs appears to lower the water potential and to cause condensation when the bladders are everted; the hair cuticle may be hydrophilic, or the small, densely packed hairs may have a

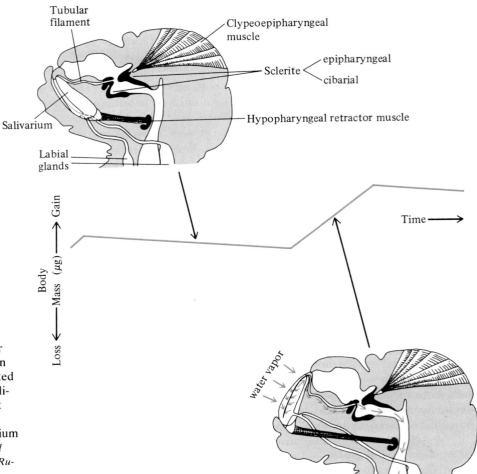

FIGURE 16–26 Water vapor absorption by the psocopteran *Badonnelia* occurs (as indicated by gain in mass) when the salivarium is exposed to ambient air (left) but water is lost by evaporation when the salivarium is retracted (right). *(Modified from Rudolph and Knulle 1982; Rudolph 1982.)*

capillary effect. When the bladders are inverted, the absorbed water is somehow removed from the bladder and ingested. Addition of the frontal body secretion may somehow cause the release of the absorbed water into the ionic solution, which is then sucked into the hypopharynx and swallowed.

A variety of insects absorb water vapor via their rectum from ambient air ventilated into the anus. For example, the rectal **cryptonephridial complex** of *Tenebrio* larvae can absorb water vapor from air of $RH > 88\%$. This rectal complex of *Tenebrio* larvae, and other beetle and lepidopteran larvae, has the distal ends of six Malpighian tubules attached to the surface of the rectum (Figure 16–28). Their outer tubule surfaces are expanded to form bubble-like dilations (boursouflures), each with a specialized Malpighian tubule cell, the leptophragmata. A perinephric membrane, which is impermeable to water, surrounds this cryptonephridial complex; the perinephric membrane is especially thin where it covers

the leptophragmata cell. The rectal epithelial cells are not particularly specialized for solute or water transport. Leptophragmata cells actively transport K^+ (and Cl^- passively follows) from the hemolymph into the Malpighian tubule distal segment; this is essentially what happens in all Malpighian tubules. However, the highly water-impermeable perinephric membrane prevents solute-linked water flux from the hemolymph, so KCl accumulates inside the Malpighian tubule to a high osmotic concentration, from 390 mOsm in hydrated insects to 4300 mOsm in dehydrated insects (cf. hemolymph osmotic concentrations of 386 to 754 mOsm respectively). An osmotic gradient of 4000 mOsm or more is sufficient for the feces to be in equilibrium with a relative humidity of 88%. This provides an effective mechanism for water vapor absorption (and also for desiccating the feces).

Other insects also absorb water vapor via their anus. The thysanuran firebrat *Thermobia* has an

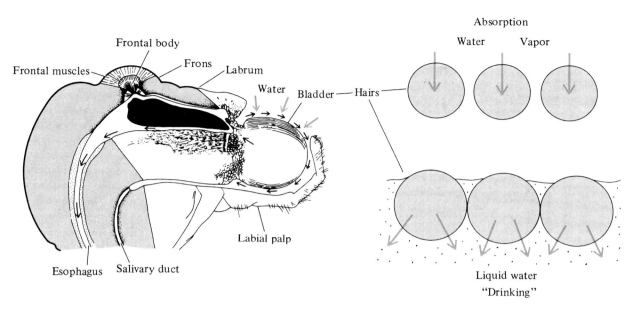

FIGURE 16–27 Structure of the head parts involved in water vapor absorption by the desert cockroach *Arenivaga*, and a possible model for water vapor absorption from air by the hygroscopic cuticular hairs of the everted bladder and the subsequent removal of water from the hairs for drinking. *(Modified from O'Donnell 1982.)*

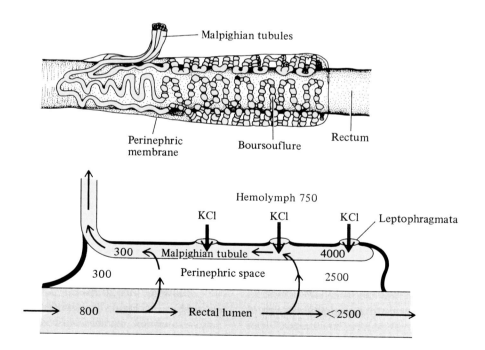

FIGURE 16–28 Structure of the cryptonephridial rectal complex of the mealworm larva *Tenebrio* and a schematic representation of how the active KCl transport into the Malpighian tubule from the hemolymph establishes a high local osmotic pressure (calculated from freezing point depression) that is coupled to the passive osmotic reabsorption of water from the rectal lumen. *(Modified from Ramsay 1964; Maddrell 1971.)*

exceptionally low RH_{equil} of 45%, but its rectum has a relatively simple epithelium and there is no cryptonephridial complex. The anal sacs of the posterior rectum appear to be responsible for water vapor absorption (Noble-Nesbitt 1977). The cuticle lining the rectal epithelium is exceedingly thin and is separated from the epithelial cell surface, forming an extensive subepicuticular space (Figure 16–29). This space is filled with a presumably hygroscopic material that is thought to passively absorb water vapor from the air in the anal sacs. Active uptake of water from this hygroscopic fluid may be driven by electro-osmosis. Apical membrane pumps secrete cations (probably K^+) into the hygroscopic subcuticular space; a transepithelial potential of $+155$ mV (lumen positive) causes passive cation flux from the subcuticular space into the epithelial cells. Water flux is linked to the ion flux, at about 7 water molecules per cation or 5.5 nanoliters per microamp of cation current flow (Kuppers and Thurm 1980).

The rat flea *Xenopsylla* can absorb water vapor from about 65% *RH* by active uptake through its rectal sacs (Bernotat-Danielowski and Knulle 1986). A rhythmical opening and closing of the anal valves is associated with peristaltic contractions from the distal to the proximal end of the enlarged, anterior rectum. The epithelium of the rectal sac is unusual, with very different cells in the dorsal and ventral portions of the gutter-like lumen, but the mechanism for water vapor absorption is not known.

Absorption of water vapor is thermodynamically difficult, but the energetic cost is actually quite low. The mechanical energy cost (E; J $mole^{-1}$) for transport of water against a water potential difference is equal to

$$E = -RT \ln P_2/P_1$$
$$= -2437 \ln P_2/P_1 \text{ (kJ } mole^{-1} \text{ at } 20° C) \quad (16.14)$$

where P_2 and P_1 are the water vapor pressures of the air and the tissues respectively. Thus, absorbing water vapor against a 50% *RH* difference (e.g.,

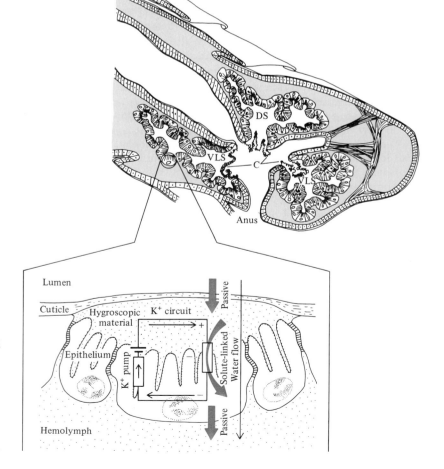

FIGURE 16–29 Structure and proposed model for water vapor absorption by the rectal epithelium of the firebrat *Thermobia* by electro-osmosis. The posterior rectal sac has a convoluted epithelium that forms three major sacs (dorsal, DS; ventro-lateral sacs, VLS) lined by cells with a highly folded apical epithelium and numerous mitochondria. The epithelial cuticle (C) is thin and separated from the epithelial surface of the rectal sacs. *(Modified from Noble-Nesbitt 1977; Kuppers and Thurm 1980.)*

ambient $RH = 50\%$, tissue $RH = 100\%$) would require about 1690 kJ mole^{-1}, or 94 J gram^{-1}. This is equivalent to a metabolic cost of about 4.6 ml O_2 g^{-1} (1 ml O_2 is equivalent to about 20.1 J). A biochemical efficiency of about 20% (see Chapter 3) increases the metabolic expenditure to about 23 ml O_2 g^{-1}.

Estimates of the metabolic expenditure for water vapor absorption suggest that it is very low compared to the daily metabolic expenditure. For ticks and *Tenebrio* larvae, it is negligible compared to the metabolic rate. *Arenivaga* is calculated to expend only a few percent of its metabolic rate to absorb water. *Thermobia* expends a similarly negligible amount of energy for water vapor absorption. There is no measurable increase in the metabolic rate of mites when absorbing water vapor. The thermodynamic difficulty in absorbing water vapor is thus not in the absolute energy required but in the mechanism for coupling energy expenditure to water vapor absorption.

Invertebrates

There are major differences among the terrestrial invertebrates in their normal body fluid osmotic concentration (Table 16–11). Semiterrestrial crustaceans, especially decapods, have a blood osmotic concentration that is similar, but slightly hypoosmotic, to seawater; this reflects their recent marine origins. The more terrestrial isopods tend to have lower osmotic concentrations of 600 to 700 mOsm. *Holthuisana* and *Sudanonautes* (freshwater/land crabs) have even lower osmotic concentrations of about 500 mOsm. In contrast, the terrestrial mollusks, insects, and arachnids have lower blood osmotic concentrations of 200 to 400 mOsm, pre-

sumably reflecting a more distant origin from marine forms, or a brackish or freshwater ancestry.

Mollusks. Terrestrial pulmonate snails and slugs have a high water loss (by evaporation and slime formation) when active. They are not well adapted to dry environments and are generally restricted when active to moist microclimates. However, many species are able to survive in very dry environments by limiting their activity to suitably moist conditions, e.g., after rain, and by being inactive (estivating) during dry periods, either burrowing underground or ascending to elevated estivation sites. During estivation, the edges of the mantle are drawn together and the shell aperature is sealed by one or more epiphragms (protective mucous membranes).

There is a clear correlation between activity level and hydration state; active pulmonates are well hydrated but inactive animals are dehydrated. The terrestrial pulmonates have an extremely variable body water content; for example, the snail *Helix* can survive 50% mass loss and the slug *Limax* 80% mass loss (Burton 1983). Clearly, hemolymph osmotic and solute concentration must alter dramatically with such large changes in body water content.

Pulmonate snails and slugs absorb water, when available, by drinking and by osmosis across their integument. Evaporation is an important avenue for water loss, at least when active, and slime is an additional source for loss of water and ions (it is approximately isoosmotic to hemolymph). Urinary water loss is quite variable, as urine flow may cease during inactivity and increase with hydration; dehydration causes increased reabsorption of water from the renal filtrate. Urine is hypoosmotic to blood due to Na$^+$ and Cl$^-$ (and also Mg^{2+} and Ca^{2+})

TABLE 16–11

Extracellular body fluid osmotic concentration (mOsm) of the main groups of semiterrestrial and terrestrial invertebrates.				
Onycophorans	**Mollusks**	**Crustaceans**	**Insects**	**Arachnids**
Peripatus 199	*Limax* 134	*Sudanonautes* 500	*Onymacris* 320	*Heterometrus* 231
	Helix 228	*Holthusiana* 524	*Alobates* 349	*Nebo* 248
		Oniscus 560	*Petrobius* 426	*Amblyomma* 330
		Porcellio 700	*Onymacris* 435	*Scorpio* 435
		Coenobita 700	*Mantis* 477	*Lycosa* 470
		Gecarcinus 940	*Tenebrio* 527	*Buthus* 471
		Ligia 1156	*Eleodes* 536	*Leiurus* 525
				Centroides 550
				Latrodectus 600
				Euscorpius 607
				Hadrurus 700

reabsorption; K^+ may be reabsorbed or secreted into the urine. The excreta of terrestrial pulmonates can be a semisolid paste, reflecting the reabsorption of water and precipitation of the remaining urate wastes.

Terrestrial prosobranch snails (helicinids, cyclophorids, and pomastiasids) can close their shell aperture with an operculum. They are as variable in water content and osmotic and solute concentration as pulmonates. Estivating operculate snails have a lower water content and higher osmotic concentration than active snails. The urine is either hypoosmotic (e.g., *Alcadia*, *Poteria*) or isoosmotic (*Pomatias*, *Tropidophora*).

Crustaceans. Most semiterrestrial crabs are restricted to beach habitats and have ready access to seawater or the water table at the bottom of their burrows. Their evaporative water loss is relatively low (*r* is 30 to 70 sec cm^{-1}). They also have a low urine flow rate when dehydrated to conserve body water. These semiterrestrial crabs require seawater for spawning and larval development.

The semiterrestrial ghost crabs *Ocypode* can extract soil interstitial water from their sandy burrows through capillary tufts of setae between the second and third pairs of walking legs and suction by the gill chamber; this essentially supplies *ad libitum* water. Gecarcinids have similar structures between the walking legs and the first abdominal segment. If the soil interstitial water is dilute, then the antennal glands are unable to excrete the excess water without incurring a substantial salt loss, since the urine is isoosmotic to blood (Wolcott and Wolcott 1985).

Semiterrestrial crabs have an extremely limited capacity to excrete other than isoosmotic urine. The fiddler crab *Uca* can excrete a slightly hypoosmotic urine; others, such as *Ocypode*, *Cardisoma*, and *Sesarma* excrete isoosmotic urine. The land crab *Gecarcinus* produces an isoosmotic urine; it does not reabsorb water or ions from its urine. Isoosmotic urine is a physiological liability for ionoregulation in situations when only freshwater is available. However, the urine can be directed to other organs (e.g., gills or gut) and made markedly hypoosmotic, i.e., 200 to 300 mOsm (cf. body fluids, 800 to 900 mOsm). The gill epithelium of *Gecarcinus* reabsorbs ions and/or water, either from water passively absorbed from the interstitial soil spaces or from antennal gland urine (Copeland 1968).

Arachnids. Most mites and ticks are parasitic and feed on fluids (e.g., vertebrate blood) that are generally more dilute than their own body fluids. Consequently, their major osmotic and ionic requirement is the rapid, diuretic elimination of excess water and ions.

Argassid ticks feed rapidly (e.g., a 100 mg *Ornithodorus moubata* can ingest 100 mg of blood in 15 to 30 min). They also rapidly excrete much of the ions and water ingested in blood, e.g., about 45% of the water and 60% of the ions are eliminated by their coxal glands in the first 60 min. The coxal fluid can be hypoosmotic or hyperosmotic to hemolymph. Ixodid ticks feed more slowly than argassid ticks, although their blood meal is of a similar magnitude. The excess water and ions are eliminated mainly by injection of saliva into the host (they lack coxal glands), although there is some elimination via feces (Figure 16–30).

The mite *Tetranychus* feeds on plant leaf tissues and has a remarkable "short-circuit" channel for direct transfer of much of the ingested water from the esophagus to the hind gut (McEnroe 1963). The filtered particulate material passes to the midgut for digestion.

The mygalomorph spider *Porrhothele* excretes about 4 μl g^{-1} day^{-1} of a slightly hypoosmotic fluid from its Malpighian tubules, midgut diverticula, and stercoral pocket when starved (Butt and Taylor 1986). During feeding, the coxal glands secrete a Na^+-rich solution into the prey item, and there is an anal diuresis, probably for ion excretion rather than water regulation.

Insects. Reduviid bugs (Hemiptera, Reduviidae) feed on vertebrate blood. They absorb an enormous blood meal, relative to their body size, extremely rapidly. For example, a *Rhodnius* can absorb 10 × its body mass in 20 min. However, the bug almost as rapidly eliminates much of the water and ions ingested. It can excrete about 50% of the blood meal volume within two to three hr of feeding; this is equivalent to 5 × its initial body mass! There is extremely rapid absorption of the blood meal in the midgut due to active Na^+ transport with passive Cl^- and water uptake. Fluid is rapidly transported into the upper Malpighian tubules, driven by the typical active transport of K^+. This primary urine is highly modified in the lower Malpighian tubule segment by rapid K^+ reabsorption so as to reduce the K^+ concentration from 70 to 5 mM and reduce the osmolality by 50%. The final excretory fluid has a high ratio of Na^+/K^+ and is hypoosmotic. During the often lengthy period between feeding, when water and nutrients need to be retained, the tubular excretory fluid is appropriately modified. Acidification in the lower Malpighian tubule (either by HCO_3^- reabsorption or H^+ secretion) precipitates uric acid, which reduces the osmotic concentration by binding various ions that were previously in

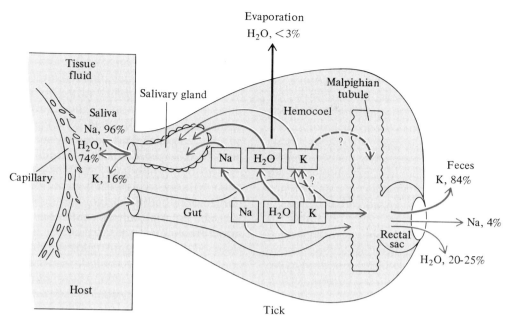

FIGURE 16–30   Schematic of the routes for ingestion and excretion of water and ions by an ixodid tick during its normal feeding cycle.   *(From Kaufman and Phillips 1973.)*

solution (e.g., Na$^+$, K$^+$, Cl$^-$) and promotes osmotic reabsorption of water.

Sap-feeding insects are also faced with the elimination of large volumes of ingested water. Sap can be obtained from either the xylem or phloem transport vessels. Xylem sap is extremely hypoosmotic to hemolymph, being 99.8 to 99.9% water and having low concentrations of ions (especially K$^+$ and Cl$^-$), amino acids, and simple sugars. Phloem has high concentrations of sugars and amino acids but low concentrations of ions. The digestive tract

of sap-feeding insects often has a "short circuit" for ingested water to pass the midgut (Cheung and Marshall 1973; Marshall and Cheung 1974). Filter chambers in the anterior of the midgut of xylem feeders have a dilated wall closely apposed to the proximal segments of the Malpighian tubules and coiled segments of the posterior midgut and hindgut and even the rectum (Figure 16–31). Ingested xylem fluid loses water by osmosis to the Malpighian tubule proximal segment and posterior midgut, where lumen fluids are hyperosmotic to the xylem.

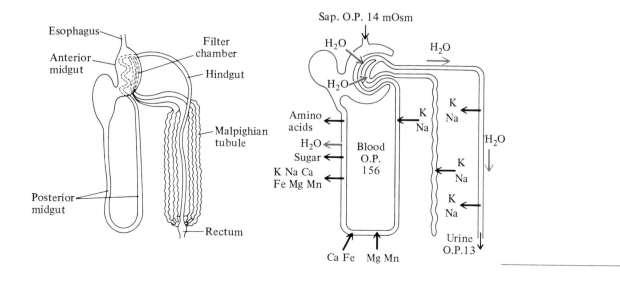

Phloem feeders have simpler filter chambers than do xylem feeders. Some aphids lack Malpighian tubules and excrete water as saliva or from their rectum as honey dew, but little is known of their osmo- and ionoregulation.

The rectal cryptonephridial complex of *Tenebrio* larvae, and other beetle and lepidopteran larvae, provides solute-linked water reabsorption. The cryptonephridial complex actively accumulates KCl inside the Malpighian tubule to a high osmotic concentration (up to 4.3 Osm in dehydrated insects), and this markedly desiccates the feces to about 14 to 17% water content.

Vertebrates

The vertebrate kidney is well adapted for the formation of copious amounts of urine, as is well exemplified by the amphibian kidney, but there is a general trend among higher vertebrates of modification of the role of the kidney for water conservation and solute excretion. Reptiles and birds have a fundamentally different strategy for osmo- and ionoregulation than do mammals. Their urinary output is highly modified by intestinal ion and water reabsorption, and salt glands provide extrarenal elimination of the reabsorbed ions. Mammals, in contrast, rely on their kidneys for water and solute excretion.

Amphibians.   Most amphibians have a high evaporation rate because their skin is essentially a free water surface. Their high EWL must be balanced by a high-water intake. Consequently, most amphibians are restricted to moist microhabitats. Amphibians do not drink, but absorb water across their skin, particularly the pelvic patch which is a highly vascular region of pelvic skin with a high rate of solute-linked water uptake. Amphibians are also able to absorb water from moist soil across their skin. Many species are remarkably tolerant of dehydration, compared with other vertebrates. There is a general trend of increased dehydrational tolerance in more terrestrial amphibians, which are more prone to experience dehydrating conditions (Figure 16–32). Some of the most terrestrial amphibians can survive up to 50% loss of body water; this excludes the additional buffering role of water stored as bladder urine.

A number of amphibians from different parts of the world reduce their cutaneous evaporative water loss, at least when inactive, by forming a protective **cocoon**. The cocoon is an accretion of many one-cell thick layers of outer epidermis that are successively shed every few days. Active frogs would normally reingest their shed skin, but the successive addition

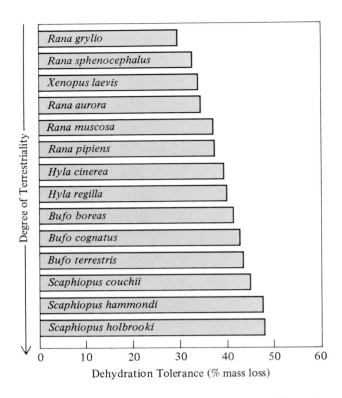

FIGURE 16–32   Relationship between terrestriality and dehydration tolerance in anuran amphibians.   *(Values from Thorson and Svhila 1943; Hillman 1980.)*

of layers of shed epidermis forms a paper-like, often thick cocoon for inactive frogs (Figure 16–33A). The cocoon, which covers the entire body surface except the nares, dramatically reduces the rate of evaporative water loss, depending on the number of layers present (Figure 16–33B).

A few amphibians have evolved an epidermal surface with a high resistance to water loss; *Chiromantis* (Loveridge 1970), *Phyllomedusa* (Shoemaker, Balding, and Ruibal 1972), *Hyperolius* (Withers et al. 1982), and *Litoria* (Withers, Hillman, and Drewes 1984) are "waterproof frogs." The waterproofing mechanism is clear only for *Phyllomedusa* (Blaylock, Ruibal and Platt-Aloia, 1976). Cutaneous lipid glands produce a waxy material that is spread over the body surface by stereotyped wiping movements to form the waterproofing layer.

Terrestrial amphibians use their dilute bladder water as a water store. Many terrestrial species have exceptionally large bladders, which can store up to 50% of their body mass as urine (Ruibal 1962). Water is reabsorbed across the bladder wall (under hormonal control by arginine vasotocin) to replenish water lost from the body fluids by evaporation. The

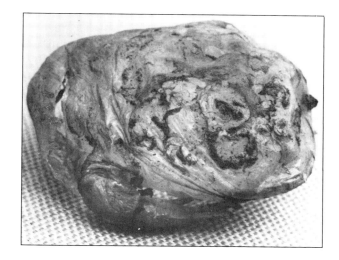

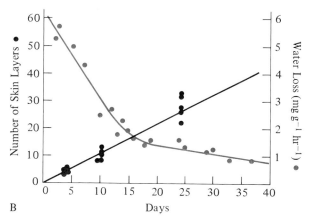

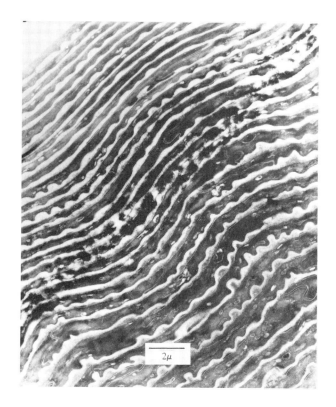

2μ

FIGURE 16–33 **(A)** The cocooned frog (*Lepidobatra-chus llanensis*) is completely enveloped by its cocoon, which consists of multiple layers of shed epidermal cells. **(B)** There is a linear relationship between the number of layers in the cocoon after initiation of cocoon formation and an inverse relationship between evaporative water loss and time.   *(From McClanahan, Ruibal, and Shoemaker 1983.)*

portant role of bladder water as a buffer for body fluids is illustrated by the effects of dehydration on body fluid composition for amphibians with, and without, a bladder urine store. Frogs without a bladder store experience an immediate increase in body fluid concentration with dehydration, but those with a bladder urine store have a delay in increase in body fluid osmotic concentration (c.f. juvenile lizards; see below).

Amphibians typically have a high urine flow rate $(10$ to 25 ml kg^{-1} $hr^{-1})$. Copious production of dilute urine is adaptive for aquatic amphibians that must eliminate their osmotic water influx, but is not adaptive for terrestrial amphibians. However, terrestrial amphibians can markedly decrease their rate

important water storage role of the dilute bladder urine is evident for the green toad *Bufo viridis* during dehydration (Table 16–12). The normal bladder urine has about 50% of the osmotic concentration of the plasma, whereas that of the dehydrated animals is over 90% of the plasma concentration. The urine to plasma ratio (U/P) for Na^+ increases from about 0.04 to 0.34. The U/P for urea, the main osmotic constituent of urine, actually decreases with dehydration from about 2.8 to 1.7. The im-

TABLE 16–12

Effects of dehydration on plasma and bladder urine for the green toad (*Bufo viridis*). *Data from Katz and Gabbay (1986).*

| | **Control** | | | **Dehydrated** | | |
	Plasma	*Urine*	*U/P*	*Plasma*	*Urine*	*U/P*
Na^+	141	6	0.04	162	55	0.34
Cl^-	110	10	0.09	161	25	0.16
Urea	32	88	2.75	272	470	1.72
mOsm	392	210	0.54	752	705	0.94

of urine formation when osmotically stressed. For example, the marine toad *Bufo marinus* dramatically decreases its urine flow rate when out of water and a waterproof frog, *Phyllomedusa sauvagei*, also has a marked (but less so than in *Bufo*) decline in urine flow rate when out of water (Shoemaker and Bickler 1979). The urine that enters the bladder is also substantially reabsorbed, further limiting the urinary water loss.

Phyllomedusa and *Chiromantis* are remarkable amphibians in that they can excrete uric acid rather than urea and this provides further water economy (see Chapter 17).

Reptiles and Birds. The avenues of water gain for terrestrial reptiles are drinking, preformed water in food, and metabolic water production (Figure 16–34). The avenues for water loss are evaporation (respiratory and cutaneous) and excretion (combined urine and feces). Avenues of salt intake are food and drinking; avenues of salt loss are salt glands and the urine/feces.

Most reptiles can survive without drinking because their preformed and metabolic water gains are sufficient to balance their low water loss. Reptiles have a low cutaneous evaporative water loss and can minimize respiratory evaporative water loss by nasal countercurrent water exchange; some lizards utilize the water from their nasal salt gland to humidify the inspired air. Urine and feces are mixed in the large intestine, where solute-linked reabsorption of water reduces the water content of the excreta (see Chapter 17).

Lizards with urinary bladders can store dilute urine, which can be used during dehydration. For example, the sand-diving lizard *Aporosaura* stores very dilute urine (25 mOsm) in its bladder. The desert tortoise *Gopherus agassizii* reabsorbs water from its bladder urine when dehydrated until the urine becomes isoosmotic with plasma. Some hatchling lizards (*Uma*, *Uta*, *Sceloporus*) have a urinary bladder that stores dilute urine (39 to 115 mOsm). In *S. jarrovi*, urine in the bladder initially is 14% of the body mass. No further urine is stored in the bladder after hatching, but the bladder water is reabsorbed at about 21 µl day^{-1} to prevent dehydration (Beuchat, Vleck, and Braun 1986). Lizards do not necessarily absorb urine water across their poorly vascularized bladder; it may be passed into the cloaca/hindgut for reabsorption.

Most birds can rely on drinking because they are able to fly or walk long distances to water, but preformed and metabolic water are additional and important avenues of intake (Figure 16–35). Some sand grouse carry water long distances to their chicks, using their specially modified breast feathers as a "sponge" to hold water. Birds minimize their evaporative water loss by nasal countercurrent water exchange, and their skin/feather layer is a significant barrier to evaporative water loss. Urinary

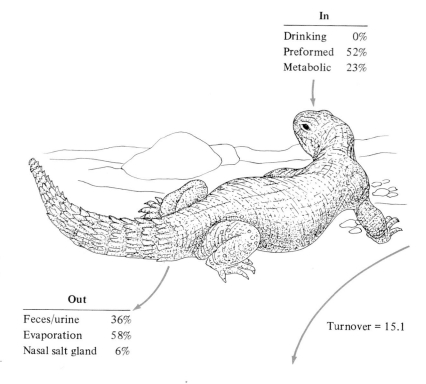

In	
Drinking	0%
Preformed	52%
Metabolic	23%

Out	
Feces/urine	36%
Evaporation	58%
Nasal salt gland	6%

Turnover = 15.1

FIGURE 16–34 Water budget for a desert lizard, *Uromastix acanthurus* (300 g). Values are percent of daily water turnover and the turnover is ml kg^{-1} hr^{-1}; note that the lizard is not in water balance. *(Data from Lemire, Grenot, and Vernet 1982.)*

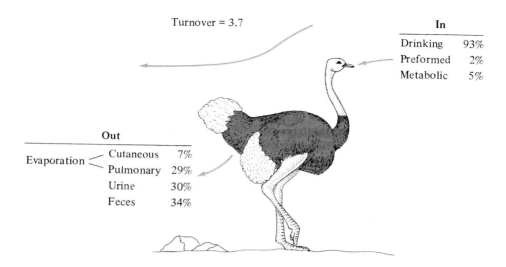

FIGURE 16–35 Water budget for a desert bird, the ostrich *Struthio camelus* (90 kg). Values are percent of daily water turnover and the turnover is ml kg^{-1} hr^{-1}. *(Data from Withers 1983.)*

and fecal water losses are minimized by solute-linked water reabsorption in the large intestine, resulting in a low fecal water content.

Two further aspects of cloacal/colonic function affect the excretory water balance of reptiles and birds. First, cloacal reabsorption of Na$^+$ and the precipitation of uric acid further minimizes urinary/fecal water loss (see Chapter 17; uricotely). Second, many reptiles and birds can excrete relatively dry feces, thereby conserving water. The fecal water content of hydrated reptiles and birds is typically 70 to 90%, but for dehydrated animals the fecal water content declines to 50 to 70% for birds and 30 to 50% for reptiles. Osmotic water reabsorption could dehydrate the feces to osmotic equilibrium between body fluid solutes and fecal solutes (residual Na$^+$, K$^+$, Cl$^-$, nonelectrolytes, uric acid, etc.), but could not make the cloacal/colonic contents hyperosmotic to the urine. In fact, the initially hyperosmotic ureteral urine of the chicken becomes less concentrated in the large intestine (see Chapter 17).

Fecal water reabsorption can be promoted by a colloid osmotic effect of the plasma proteins. The desert iguana, like most reptiles, produces a fairly dry and semisolid fecal/urinary pellet; the water content is about 45.5% for dehydrated lizards (Murrish and Schmidt-Nielsen 1970). The fecal water content is consistent with a negative hydrostatic pressure of about −2.44 kPa (−250 mm H$_2$O; Figure 16–36). This pressure represents the suction force required to draw water from the capillary spaces of

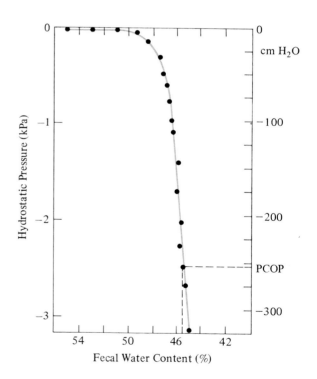

FIGURE 16–36 Effect of negative hydrostatic pressure on the water content of a fecal/urinary mixture from the lizard *Dipsosaurus* (solid circles). Note the precipitous increase in negative pressure required to dry the feces/urine mix below 50% water content. The broken lines represent the plasma colloid osmotic pressure (PCOP) of dehydrated lizards and the water content of the voided feces. *(Modified from Murrish and Schmidt-Nielsen 1970.)*

the fecal urate mass. The plasma colloid osmotic pressure of dehydrated lizards is 2.61 kPa (267 mm H_2O) and the intracloacal pressure of -2.49 kPa (-255 mm H_2O) is therefore sufficient to explain the water content of the feces. Hydrated lizards have a slightly lower plasma colloid osmotic pressure (2.10 kPa) and a slightly less negative intracloacal pressure (-2.02 kPa); presumably their feces would be more than 45.5% water. The plasma colloid osmotic pressure can account for the fecal water content of the desert iguana but would be unable to reduce the fecal water content below 45%, since exceeding high colloid pressures would be required for even a further slight reduction in water content. However, the desert iguana and other lizards can excrete feces as dry as 27 to 38% water. Furthermore, the colloid osmotic pressure difference is equivalent to an osmotic concentration difference of only a few milliosmolar (-2.44 kPa $\equiv$ 1 mOsm) and cannot account for the observed rate of water reabsorption of water (about 0.62 ml kg^{-1} hr^{-1}), unless the osmotic permeability of the intestine is very high (620 μl kg^{-1} hr^{-1} $mOsm^{-1}$). If we assume that the cloacal/colonic osmotic permeability is equivalent to that of birds, about 1–2 μl kg^{-1} hr^{-1} $mOsm^{-1}$, then the colloid osmotic pressure could only transport about 0.001–0.002 ml kg^{-1} hr^{-1}. Even a 100 $\times$ greater cloacal/colonic osmotic permeability could not account for the observed rate of water reabsorption, which is actually due to solute-linked water reabsorption (see Chapter 17).

The spectacular success of reptiles and birds in dry environments is due in part to their being uricotelic and to the role of their cloaca in solute-linked water reabsorption. However, this "strategy" for water elimination compromises ionic regulation since reptiles have no capacity for hyperosmotic urine excretion (except perhaps *Amphibolurus maculosus*) and birds have only a limited capacity to excrete a hyperosmotic urine (e.g., U/P up to 5.8). The salt gland consequently is an important nonrenal avenue for solute elimination (Chapter 17).

A final comment of comparative interest (cf. insects) is the possibility for absorption of water vapor from unsaturated air by the cloacal/colonic Na^+/H_2O reabsorptive mechanism. The total osmotic pressure of the cloacal contents would be about 815 kPa (360 mOsm) for a dehydrated desert iguana. The relative humidity of water vapor in equilibrium with 815 kPa is about 99.4% *RH*; thus, water vapor could theoretically be absorbed from air of *RH* > 99.4%. Absorption of water vapor from even 99% *RH* air would require a cloacal osmotic pressure of about 1360 kPa (600 mOsm); from 90% *RH*, about 14220 kPa (6280 mOsm). Thus, significant cloacal water vapor absorption is clearly not feasible for reptiles or birds.

Mammals.   The avenues of water gain for mammals are drinking, preformed water in the food, and metabolic water production (Figure 16–37). The avenues for water loss are evaporation (cutaneous and respiratory), urine, and feces. The relative importance of these avenues for gain and loss of water varies for different mammals.

Large mammals and mammals from mesic climates can rely on drinking. However, many mammals and especially small desert mammals cannot rely on drinking, and their water requirements must be met by preformed and metabolic water. Respiratory water loss is minimized by countercurrent water exchange, and cutaneous water loss is low because of the high resistance of mammalian skin to evaporation and the presence of the insulating fur layer.

The role of the excretory system in iono- and osmoregulation is more straightforward for mammals than for reptiles and birds because of the

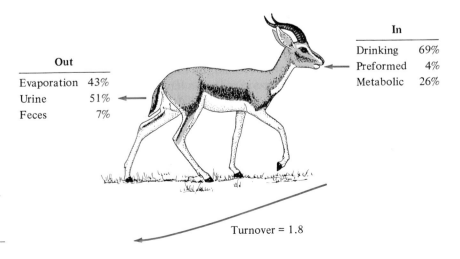

	In	
Drinking	69%
Preformed	4%
Metabolic	26%

Out		
Evaporation	43%
Urine	51%
Feces	7%

Turnover = 1.8

FIGURE 16–37   Water budget for the dorcas gazelle. Values are percent of daily turnover (in ml kg^{-1} hr^{-1}). *(Data from Ghobrial 1970.)*

absence of important extrarenal avenues for salt loss. Some NaCl and other solutes can be lost via sweat; this is associated with a thermoregulatory rather than an iono- or osmoregulatory role but there are possible ionoregulatory consequences of salt loss by sweating.

Urinary water loss is minimized by the osmotic concentrating capacity of mammals. Mammals vary dramatically in their ability to osmotically concentrate urine. Mesic mammals generally have a low capacity to concentrate urine (e.g., 1000 to 2000 mOsm), but small desert-adapted mammals can considerably exceed this capacity (5000 to 9000 mOsm). The renal concentrating capacity of these desert-adapted mammals reduces their urinary water loss to a minimal level, compared with other avenues for water loss.

The fecal water loss of mammals varies with the hydration state and can be reduced during dehydration to about 45 to 50%. Large intestine absorption of Na^+ by active transport and Cl^- by passive uptake markedly reduces the intestinal concentration of NaCl to about 1 mM and provides an osmotic gradient for water reabsorption. Plasma colloid osmotic pressure may also contribute to further dehydration of feces. Little is known concerning the role of ion reabsorption to determining the water content of feces. In a small African antelope, the dik-dik, the contents of the proximal colon are a Na^+-rich paste, slightly hyperosmotic to the plasma and containing about 83% water (Skadhauge, Clemens, and Maloiy 1980). Rapid reabsorption of Na^+ and Cl^-, and presumed solute-linked and osmotic water reabsorption along the upper intestine, reduces the water content to about 56% for feces and little Na^+ or Cl^- remains in the feces. Dehydration further reduces the fecal water content to 45%, although the mechanisms responsible for this are unclear.

Summary

All animals regulate the ionic and osmotic composition of their intracellular environment, and most regulate their extracellular environment.

Water is the primary constituent of animals, being 50 to 80% of the body mass. It is the "universal solvent" for life; its physical and chemical properties suit the required solvent properties of cells. Most of the water in cells is available as a solvent; it is bulk water. Some is vicinal water, which interacts with macromolecules and membrane surfaces and is less available for solvent activity. Some is bound water, which is tightly associated with ions and macromolecules and is not available for a solvent role.

The principal extracellular solutes are ions (mainly Na^+ and Cl^-) and low concentrations of various organic solutes (glucose, urea, amino acids, protein, etc.). The extracellular fluids of many aquatic animals have the same osmotic concentration as the external medium; they osmoconform. Particularly freshwater and hypersaline aquatic animals osmoregulate; their body fluids have a different osmotic concentration than the medium. Many animals can only tolerate a narrow range of body fluid concentration (are stenohaline), whereas others tolerate a wide range (are euryhaline).

The general intracellular solutes are ions (mainly K^+, Cl^-, SO_4^{2-}) and a variety of organic metabolites, metabolic intermediates, waste products, and enzymes. The intracellular ionic concentration is low, about 100 to 200 mM, even for animals with high intracellular osmotic concentrations (e.g., 1000 to 1100 mOsm). The "osmotic gap" of 400 to 600 mOsm is generally filled by amino acids, various other organic solutes, or urea and counteracting solutes (betaine, TMAO, etc.).

Animal cell membranes cannot resist hydrostatic pressure differences, and so the intracellular osmotic concentration is the same as the extracellular osmotic concentration. Some animal cells (and animals) passively alter their intracellular volume (shrink in hyperosmotic media, swell in hypoosmotic media). In contrast, many regulate their intracellular volume by adjusting the intracellular osmotic concentration to match the extracellular osmotic concentration. This is accomplished either by ion exchange across the membrane (e.g., Na^+, Cl^-, and K^+ in many vertebrate cells), by regulation of the concentration of intracellular organic solutes (e.g., amino acids in many osmoconforming marine invertebrates), or by the accumulation of urea, methylamines, and polyols.

There are a variety of potential avenues for water gain (drinking, in the food, metabolism, osmotic flux across the body surface, absorption of water vapor from air) and water loss (in urine, feces, by osmotic flux across the body surface, evaporation, secretions of other excretory glands). Many of these are also avenues for ion gain (drinking, food, diffusion across the body surface) and ion loss (urine, feces, across the body surface, secretions of other excretory glands). Animals are not always in water or ion balance. The body fluids and bladder water can act as buffers for imbalances between water gain and loss and between ion gain and loss. Nevertheless, animals must maintain a balance of water and ion gain/loss in the long term.

Freshwater protozoans are hyperosmotic to their environment, and they iono- and osmoregulate. They minimize passive water gain across their cell surface by having a low water permeability, and they excrete water via a contractile vacuole. Marine protozoans typically are isoosmotic to seawater. They have a lower intracellular ionic concentration than seawater and have a high intracellular concentration of organic osmolytes, particularly amino acids. Many protozoans are euryhaline.

Marine invertebrates generally osmoconform to seawater. Their extracellular ion concentrations are generally similar to those of seawater, although there are slight differences due to a Donnan effect. Some extracellular ions (e.g., SO_4^{2-}) are regulated at substantially different concentrations from seawater. The ion concentration of intracellular fluids is typically low (300 to 400 mM) and the solute "gap" is filled by various organic solutes, especially amino acids.

Freshwater invertebrates invariably iono- and osmoregulate, although some have fairly dilute body fluids. Osmotic water influx is minimized by a low cutaneous water permeability but is still often substantial. This water load is eliminated by a copious urine flow. The urine is hypoosmotic to body fluids in many species to minimize salt loss, but the urine is isoosmotic in some species. The urinary and integumental ion losses are balanced by active uptake of ions (Na^+ and Cl^-) across the integument. Estuarine invertebrates generally employ a combination of the ionic and osmotic strategies of marine and freshwater species.

Hypersaline invertebrates typically are hypoionic and hypoosmotic to their environment. Integumental water loss and salt gain are minimized by a low cutaneous permeability, but their inevitable loss of water is balanced by drinking. The diffusive and drinking salt load is eliminated by active excretion.

Aquatic vertebrates may have evolved from a freshwater cephalochordate-like ancestor. Freshwater vertebrates are iono- and osmoregulators. Their passive osmotic uptake of water is eliminated via a copious dilute urine. The urinary and diffusive salt loss of teleosts is replenished by active uptake of Cl^- and passive uptake of Na^+; amphibians have active Na^+ uptake. Marine vertebrates either osmoconform (hagfish, chondrichthyeans, coelacanth) or osmoregulate at about 300 mOsm (teleost fish). Those that osmoconform generally are strong ionoregulators and use urea and counteracting solutes (mainly TMAO) to fill the osmotic "gap." Hagfish are the exception to this ureo-osmoconforming pattern; they essentially ionoconform to seawater.

The principal water and solute problems faced by terrestrial animals are the scarcity of drinking water and the evaporative loss of water from the integument and respiratory surfaces. Some terrestrial animals survive with preformed water and metabolic water as their only avenues for water gain, but many must drink. A few arthropods are able to absorb water vapor from air. Many terrestrial animals are able to minimize their cutaneous evaporative water loss by having an integument with a lipid barrier that increases the resistance to evaporation. Estivating snails use epiphragms, or seal their shell aperture with an operculum, to reduce their evaporative water loss. Estivating lungfish and frogs form a cocoon that reduces their evaporative water loss. Reptiles, birds, and mammals reduce their respiratory water loss by nasal countercurrent exchange of water and heat; invertebrates reduce their respiratory water loss by tracheal or pseudotracheal respiratory systems, often with a discontinuous pattern of spiracular opening and closing.

A number of arthropods absorb water vapor from unsaturated air using a variety of mechanisms. The site for water vapor absorption is generally the head or rectum. Ticks and mites secrete a KCl-rich, hygroscopic saliva that absorbs water vapor and then is ingested. A few other insects use a similar osmotic mechanism for water vapor absorption. Biting and sucking lice secrete a fluid onto lingual sclerites that absorbs water vapor and is ingested. The desert cockroach secretes a fluid onto its everted hypopharyngeal bladders but the mechanism for water vapor absorption is not clear. A number of insects absorb water vapor via their rectum. Mealworm larvae use their rectal cryptonephridial system to couple KCl secretion by the Malpighian tubules with water vapor absorption from the rectum. The firebrat may use electroosmosis to absorb water vapor in concert with K^+ transport across its thin rectal cuticle. The metabolic cost of water vapor absorption is low; the difficulty in absorbing water vapor is coupling energy expenditure with absorption of water vapor.

Pulmonate mollusks are common in moist habitats and many snails occur in deserts; they are inactive and estivate during the dry season. Arthropods are the most successful terrestrial invertebrates. A few crustaceans are semiterrestrial, but they generally are restricted to beach or moist habitats. Spiders, scorpions, and insects are able to survive in the most arid terrestrial environments because of their low cutaneous and respiratory water loss, and their low excretory water loss.

Amphibians have a skin that generally evaporates water as if it were a free water surface, and they are unable to produce a concentrated urine; they primarily rely on behavioral avoidance of desiccating conditions. They also utilize their bladder water stores to buffer evaporative water loss and are remarkably tolerant of dehydration.

Reptiles and birds minimize their evaporative water loss by having an integument with a high resistance and a nasal countercurrent water exchange, and they minimize excretory water loss by refluxing urine into the large intestine, where there is solute-linked water reabsorption. The feces and nitrogenous wastes (uric acid) are eliminated as a semi-solid, relatively dry pellet (40 to 50% water content).

Mammals have a high skin resistance to reduce cutaneous evaporative water loss and have a nasal countercurrent water exchange to minimize respiratory water loss. Excretory water loss is minimized by the osmotic-concentrating capacity of the kidney. Many desert-adapted mammals can reduce the fecal water content to less than 50%. A number of desert mammals do not require drinking water; they can balance preformed and metabolic water with evaporative and excretory water losses.

Supplement 16–1

Benefits and Costs of Ureo-osmoconforming

· ·

Chondrichthyeans, the coelacanth *Latimeria*, and the crab-eating frog *Rana cancrivora* are approximately isoosmotic with their marine environment. They have slightly higher extracellular ion concentrations than teleosts, and accumulate high concentrations of urea to make their body fluids approximately isoosmotic to seawater, i.e., they **ureo-osmoconform**. In contrast, marine teleost fish hypoosmoregulate and do not accumulate significant levels of urea.

There are both benefits and costs to ureo-osmoconforming. Advantages include an energy savings and no need to drink. Energy must be expended to eliminate a diffusional ion load, and elasmobranchs (and especially chimaeras) have a lower ionic gradient than teleosts. The chondrichthyeans, coelacanth, and crab-eating frog are slightly hyperosmotic to their environment and do not have to drink; this avoids an ingested salt load. However, disadvantages include the necessity for urea tolerance, and the energy cost of urea synthesis. Ureo-osmoconformers have a constant urea loss that must be replenished by

metabolic activity. Urea is normally a nitrogenous waste product, but its synthesis from NH_3 (which teleosts excrete) requires energy (see Chapter 17).

A very approximate energy balance suggests a lower cost for iono/osmoconformation by an elasmobranch than for hypoosmoregulation by a teleost. For example, the elasmobranch *Poroderma* expends about 0.48 mM ATP kg^{-1} hr^{-1} for urea synthesis (its urea turnover is about 0.12 mM kg^{-1} hr^{-1}) and 0.07 mM ATP kg^{-1} hr^{-1} for Na^+ excretion (Na^+ turnover is about 0.185 mM kg^{-1} hr^{-1}). The total energy cost for ureo-osmoconformation is therefore about 0.55 mM ATP kg^{-1} hr^{-1}. A typical marine teleost has an Na^+ turnover of about 26 mM kg^{-1} hr^{-1} at an energy cost of about 10.4 mM ATP kg^{-1} hr^{-1}, i.e., about 20 × the cost for a ureo-osmoconformer. However, various factors other than ATP costs for iono/osmoregulation also determine whether ureo-osmoconforming is a feasible and economic strategy.

One "cost" of ureo-osmoconforming is the biochemical necessity to tolerate high urea levels. Urea is a perturbing

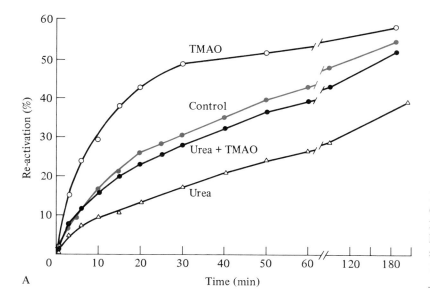

A

Effect of urea (400 mM), TMAO (200 mM), and urea + TMAO (400 + 200 mM) on the percent reactivation of lactate dehydrogenase from the white shark *Carcharodon carcharias*. (*From Yancey and Somero 1979.*)

osmolyte that destabilizes macromolecules such as proteins. Its destabilizing effects can be effectively counterbalanced by stabilizing solutes, such as betaine, sarcosine, and TMAO, or by catalytic adaptation of enzymes to urea. A ratio of TMAO:urea of about 1:2 can ameliorate the deleterious effects of urea on enzyme catalytic activity. For example, urea increases the K_m of creatine kinase (CPK) and pyruvate kinase (PK), but TMAO decreases the K_m. There is no change in the K_m at a ratio of 1:2 of TMAO:urea. There are similar counteracting effects for the V_{max} of some enzymes, e.g., the urea cycle enzymes arginosuccinate lyase (urea inhibits and TMAO activates) and LDH (urea activates and TMAO inhibits).

Chondrichthyeans, the coelacanth, and adult crab-eating frog must have a functional urea cycle. Tadpoles of the crab-eating frog, which do not urea-osmoconform but have an osmo/ionoregulatory pattern, do not have a functional urea cycle. Teleost fish, which also osmo/ionoregulate, are apparently not precluded from ureo-osmoconforming by the lack of a urea cycle. For example, the toadfish *Opsanus* and a number of other teleosts have a functional urea cycle but excrete ammonia rather than urea.

Ureo-osmoconforming is a successful strategy for various marine vertebrates, having independently evolved at least three times. It is energetically economical and biochemically possible. Why, therefore, is it not more widespread, e.g., in teleosts? First, ureo-osmoconformers generally require a low gill area (related to large size or inactivity) because their rate of urea loss is proportional to their gill/body surface area, whereas their rate of urea synthesis is proportional to their liver mass. A low surface:volume ratio is therefore favorable for ureo-osmoconformers. In contrast, hypoosmotic regulation is independent of body size; the integument area determines both the diffusional ion gain and the active integumental ion excretion capacity. Embryos of ureo-osmoconforming marine vertebrates have a high surface:volume ratio, which promotes urea loss relative to urea synthesis capacity. Consequently, internal fertilization and packaging of the embryos in a relatively impermeable packet, and birth of large young, are desirable for ureo-conforming animals. This is the reproductive strategy of elasmobranchs and the coelacanth. However, sluggish, live-bearing and large teleost fish have not become ureo-osmoconformers and small, active sharks have not become hypo-osmoregulators. Apparently, there is also physiological and biochemical "inertia" to the pattern of iono/osmoregulation. (*See Griffith and Pang 1979; Hochachka and Somero 1984; Mommsen and Walsh 1989.*)

Recommended Reading

Bradley, T. J. 1985. The excretory system: Structure and physiology. In *Comprehensive insect physiology biochemistry and pharmacology*, vol. 4, *Regulation, digestion, nutrition and excretion*, edited by G. A. Kerkut and L. I. Gilbert, 421–65. Oxford: Pergamon Press.

Bradshaw, S. D. 1986. *Ecophysiology of desert reptiles*. Sydney: Academic Press.

Burton, R. F. 1983. Ionic regulation and water balance. In *The Mollusca*, vol. 5, *Physiology*, part 2, edited by A. S. M. Saleuddin and K. M. Wilbur, 291–352. New York: Academic Press.

Conte, F. P. 1969. Salt secretion. In *Fish physiology*, vol. 1, *Excretion, ionic regulation, and metabolism*, edited by W. S. Hoar and D. J. Randall, 241–92. New York: Academic Press.

Edney, E. B. 1977. *Water balance in land arthropods*. Zoophysiology and Ecology. 9. Berlin: Springer-Verlag.

Gilles, R. 1979. *Mechanisms of osmoregulation in animals*. Chichester: John Wiley & Sons.

Gilles, R., and M. Gilles-Baillien. 1985. *Transport processes, iono- and osmoregulation*. Berlin: Springer-Verlag.

Griffith, R. W. 1987. Freshwater or marine origin of the vertebrates? *Comp. Biochem. Physiol.* 87A:523–31.

Hochachka, P. W., and G. N. Somero. 1984. *Biochemical adaptation*. Princeton: Princeton University Press.

Krogh, A. 1965. *Osmotic regulation in aquatic animals*. New York: Dover Publications.

Louw, G. N., and M. K. Seely. 1982. *Ecology of desert organisms*. London: Longman.

Maloiy, G. M. O. 1979. *Comparative physiology of osmoregulation in animals*. New York: Academic Press.

Martin, A. W., and F. M. Harrison. 1966. Excretion. In *Physiology of mollusca*, vol. 2, edited by K. M. Wilbur and C. M. Yonge, 353–86. New York: Academic Press.

Minnich, J. E. 1982. The use of water. In *Biology of the reptilia*, vol. 12, *Physiology C physiological ecology*, edited by C. Gans and F. H. Pough, 325–96. London: Academic Press.

Pequeux, A., R. Gilles, and L. Bolis. 1984. Osmoregulation in estuarine and marine animals. *Lecture notes on Coastal and estuarine studies*. 9. Berlin: Springer-Verlag.

Potts, W. T. W., and G. Parry. 1964. *Osmotic and ionic regulation in animals*. Oxford: Pergamon Press.

Shoemaker, V. H., and K. Nagy. 1977. Osmoregulation in amphibians and reptiles. *Ann. Rev. Physiol.* 39:449–71.

Skadhauge, E. 1981. *Osmoregulation in birds*. Zoophysiology, 12. Berlin: Springer-Verlag.

Yancey, P. H. 1985. Organic osmotic effectors in cartilaginous fishes. In *Transport processes, iono- and osmoregulation*, edited by R. Gilles and M. Gilles-Baillien, 424–36. Berlin: Springer-Verlag.

Chapter 17

Excretion

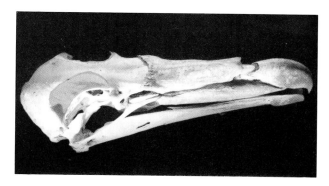

The location of supra-orbital salt glands are clearly seen in the skull of an albatross. *(Photograph courtesy of J. O'Shea and P. Withers, Department of Zoology, The University of Western Australia.)*

Animals routinely eliminate a variety of potentially deleterious waste products, including inorganic solutes (e.g., Na$^+$, K$^+$, Cl$^-$, SO$_4^{2-}$, NH$_4^+$) and organic solutes (e.g., urea, urate and other purines, benzoic acid). These wastes may be derived from the diet (e.g., ions, plant alkaloids) or may be end products of metabolic pathways.

These various waste products must be excreted to avoid their accumulation to toxic levels. Some are readily eliminated across the integument. For example, most aquatic animals can readily eliminate ammonia by diffusion across their skin or gills. Many wastes, however, are excreted in a solution that is formed by the animal specifically for the role of bulk elimination of water and/or solutes.

Excretory Organs

Almost all animals have excretory organs. These can be classified into two general types: **epithelial excretory surfaces**, such as anal papillae of insects, chloride cells of fish, and salt pumps of brine shrimp, and **tubular excretory organs**, such as nephridia, vertebrate nephrons, and insect Malpighian tubules.

Epithelial Exchange

The epithelial surface of animals is generally specialized to limit the transfer of solutes and water, but it often has specific regions that actively transport solutes, e.g., ions, glucose, amino acids, and urea. No active water pump has yet been demonstrated. Water flow is always passive, although it is often coupled with solute transport.

Solutes. Membrane mechanisms for active transport have already been described (see Chapter 6). We shall concentrate here on aspects of net solute exchange across epithelia, primarily gills.

Studies of epithelial function often use a Ussing chamber to measure electrical transport properties. The epithelium is placed between two chambers of bathing fluid and the transepithelial electrical potential (E_{tep}) is measured using agar salt bridges and a voltmeter; the current flow across the epithelium can be measured by applying an external short-circuiting current (I_{scc}) that reduces E_{tep} to zero. Ussing originally used his apparatus to study Na$^+$ transport by frog skin (Figure 17–1).

The gills of teleost fish are specialized for both respiration and iono/osmoregulation (Conte 1969). The respiratory epithelium consists of a thin epithelial layer (one to two cells thick), a basal lamina, a thin layer of connective tissue, and the sinusoids and blood capillaries (see Chapter 13). The osmoregulatory portion of the epithelium is, in contrast, two to eight cells thick and contains four types of cells: the pavement cells, the mucous goblet cells, the mitochondria-rich chloride cells, and the accessory cells (Figure 17–2A). Chloride cells are also present on the cutaneous surfaces of many fish (e.g., the opercular skin). The chloride cells either extrude Cl$^-$ (in marine fish) or absorb Cl$^-$ (in freshwater fish); a passive Na$^+$ flux is coupled to the active Cl$^-$ transport. For example, the opercular epithelium of seawater-adapted teleost fish *Sarotherodon* has a negative outward current through chloride cells but not through other epithelial cells; this indicates the outward transport of Cl$^-$ by the chloride cells (Figure 17–2B). The I_{scc} is 2 to 3 nA per cell (12 to 18 10^9 Cl$^-$ per cell per second). The relationship

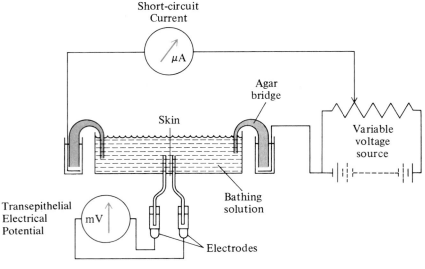

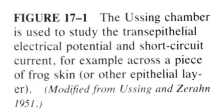

FIGURE 17–1 The Ussing chamber is used to study the transepithelial electrical potential and short-circuit current, for example across a piece of frog skin (or other epithelial layer). *(Modified from Ussing and Zerahn 1951.)*

between voltage across the opercular epithelium and the current indicates a fairly high ionic conductance ($g = I/V$; about 13.3 mSiemens cm^{-2}) over a wide voltage range (Figure 17–2C). Nonchloride epithelial cells, in marked contrast, have an extremely low ionic conductance and a high resistance.

Many crustaceans have branchial ion pumps with a similar iono/osmoregulatory role as the chloride cells of fish gills (Gilles and Pequeux

1985). Only certain gills, or parts of gills, function in salt transport. For example, silver staining suggests that the three posterior pairs of gills of the mitten crab *Eriocheir* are not respiratory but transport ions. Their ultrastructure supports this conclusion, as does the measurement of high Na$^+$ influx across these posterior gills. A tentative model for Na$^+$ and Cl$^-$ uptake across these posterior gills indicates an important role of apical Na$^+$/H$^+$ exchange (and a less important Na$^+$/NH$_4^+$ exchange)

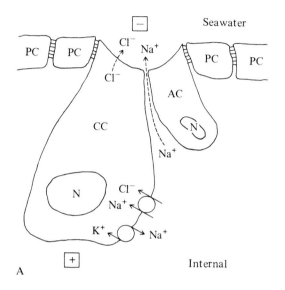

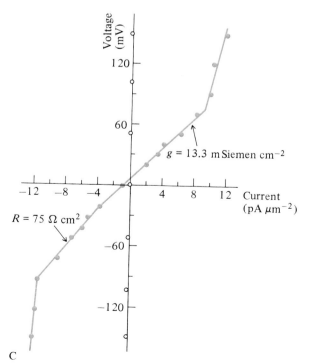

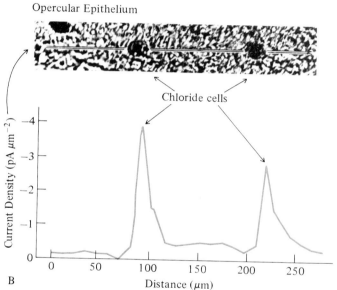

FIGURE 17–2 **(A)** Schematic model of the chloride cell (CC) of a seawater-adapted teleost fish showing its association with accessory cells (AC) and epithelial pavement cells (PC). Indicated is the pathway for C$^-$ and Na$^+$ movement. N is the cell nucleus. **(B)** A negative outward current is measured across the isolated opercular epithelium of the teleost fish *Sarotherodon* with a maximum value when the probe passes over chloride cells. The peak transepithelial current flow is adjacent to the chloride cells rather than to other epithelial cells. **(C)** Relationship between current and voltage for a chloride cell (solid circles) and a nonchloride cell (open circles) in the opercular epithelium of the teleost fish *Sarotherodon*. The chloride cell has a lower resistance (*R*) and a higher conductance (*g*) than the nonchloride cells (resistance = slope of lines; conductance = 1/slope). *(From Karnaky 1986; modified from Foskett and Scheffey 1981.)*

and a basal Na$^+$/K$^+$ exchange. The Cl$^-$ transport is independent of Na$^+$ transport and occurs via a Cl$^-$/HCO$_3^-$ exchange at the apical surface.

The brine shrimp *Artemia salina* can survive in very concentrated salt solutions. It has active Na$^+$ pumps located on the flattened, leaf-like metepipodite appendages on the phyllopodia (limbs) in adults and on the neck gland of nauplii (Figure 17–3). These cuticular surfaces stain with silver nitrate and potassium permanganate, suggesting a high cuticular permeability that allows loss of Na$^+$ and Cl$^-$ across the cuticle from the underlying salt-extruding cells. A tentative model for salt pumping and energetics of larval salt gland cells suggests active apical extrusion of Cl$^-$ coupled with passive loss of Na$^+$ via intercellular channels (Conte 1984). There may also be an Na$^+$-K$^+$ ATPase located on the apical membrane that extrudes Na$^+$ in exchange for K$^+$; such an apical location for a Na$^+$-K$^+$ ATPase is extremely unusual for animal epithelial cells. A different, more typical baso-lateral Na$^+$-K$^+$ ATPase extrudes Na$^+$ from the cell into the interstitial fluid, in exchange for K$^+$, but this does not directly contribute to salt excretion. Bicarbonate uptake at the apical surface, possibly in exchange for organic anions, amino acids, or Cl$^-$, provides CO$_2$ for fixation into oxalo-acetate and then malate and various amino acids. Intracellular glycogen and lipid

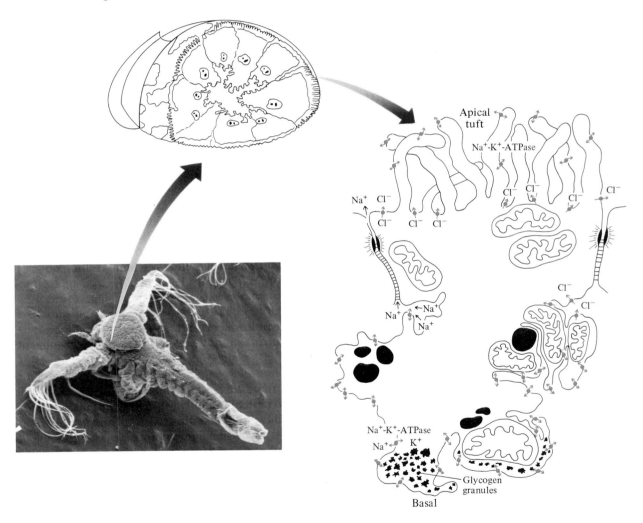

FIGURE 17–3   The first instar nauplius of the brine shrimp *Artemia salina* has an enlarged salt gland in the neck region; the schematic structure of the salt gland and a salt gland cell show the mechanisms believed to be responsible for active NaCl extrusion. A Na$^+$-K$^+$ ATP-ase is located on the basal membrane and the apical membrane to extrude Na$^+$ from the cell, and there is a basolateral Cl$^-$ pump for uptake and an apical Cl$^-$ pump for extrusion. *(Modified from Conte 1984. Photo by P. C. Withers and T. Stewart.)*

stores provide energy (aerobically or anaerobically) for operation of the Na^+-K^+ ATPase enzymes.

Water. Animals must be able to transport water across cell membranes, for example to regulate the intracellular fluid volume or to absorb water from the medium (Chapter 16). There are a variety of passive transport systems for epithelial exchange of water, including osmosis, codiffusion, double-membrane coupling (Curran model), standing-gradient flow (Diamond-Bossert model), volume pumping, formed bodies, and electro-osmosis (Figure 17–4). No active transport mechanism has yet been demonstrated for water.

Osmosis. **Osmosis** is the net movement of water as a consequence of an osmotic concentration difference across a semipermeable membrane (see Chapter 16). The membrane is permeable to water but not to solutes. Water essentially diffuses across the membrane from the side with a lower solute concentration (higher water concentration) to the higher solute concentration (lower water concentration). Animal cells (unlike bacterial and plant cells that have a rigid cell wall) must be in osmotic equilibrium with the extracellular environment because their cell membrane is not able to sustain a hydrostatic pressure gradient.

Codiffusion. Codiffusion is a rather vague concept describing diffusion of ions down a concentration gradient that is accompanied by an osmotic flow of water in the same direction (Hill 1977). Consider,

for example, a semipermeable membrane separating two isoosmotic solutions containing different concentrations of permeant and impermeant solutes. A net flux of the permeant solutes by diffusion will result in co-movement of water to maintain osmotic equilibrium. This is not a potentially important mechanism for regulation of water flux across epithelia, but may occur as a consequence of patterns of solute regulation across semipermeable epithelia.

Double-membrane Model. The double-membrane model for water transport was devised to explain the net movement of water from the lumen (mucosal) side of the rat small intestine to the blood (serosal) side (Curran 1960). The model assumes that there is a fluid space separated by two membranes (α and β) of differing permeability (Figure 17–5). Membrane α has pores sufficiently small that solutes will not passively permeate it, whereas membrane β has pores sufficiently large that solutes are freely permeable. Active transport of solute into the compartment across membrane α establishes a local osmotic gradient and water is drawn by osmosis into this compartment. Solutes and fluid are lost across membrane β (if the compartment has a fixed fluid volume).

Membrane β is needed to allow the actively transported solute to accumulate and establish a high local concentration gradient; the higher the local concentration gradient, the greater is the osmotic flux of water. For maximal water transport, the compartment bounded by membranes α and β should be as small as possible. The fluids crossing membrane α must be isoosmotic or hyperosmotic

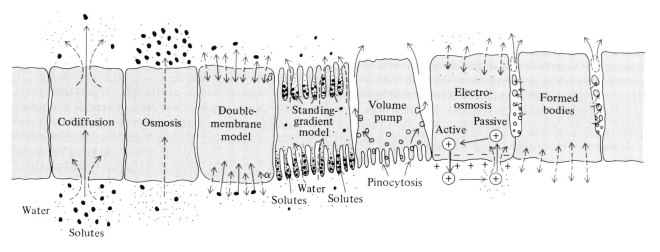

FIGURE 17–4 Representation of the various proposed mechanisms for passive transport of water across an epithelium: codiffusion, osmosis, double-membrane model, standing-gradient model, volume pump, electro-osmosis, and formed bodies.

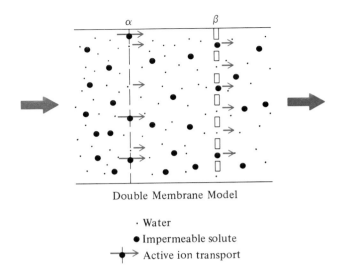

Double Membrane Model

· Water

● Impermeable solute

→ Active ion transport

FIGURE 17–5 The double-membrane hypothesis for solute-linked transport of water across a membrane. Solute is actively transported through small pores across membrane α into the intervening compartment; water follows passively by osmosis. Solute and water move across membrane β, which is freely permeable to both because the intervening compartment has a fixed volume.

to the original fluid; the fluid that crosses membrane β must be isoosmotic with the center compartment.

Standing-gradient Model. The standing-gradient model is based on studies of water flux across the epithelium of the rabbit gall bladder (Diamond and Bossert 1967). Solutes and water flow along blind-ended channels located within or adjacent to the fluid-secreting cells (Figure 17–6A). Solute is actively transported into the blind end of the channels from interstitial fluid and the contents of the channel become hyperosmotic to the cytoplasm (the blind ends of the channels have a low permeability to water). As the hyperosmotic fluid moves along the channel, there is an osmotic influx of water and fluid is pushed from the open end of the channel into the cell. The actual geometry of the channel determines whether the fluid exiting the channel is hyperosmotic or isoosmotic to the interstitial fluid. A short, wide channel allows little opportunity for osmotic equilibration of the initially hyperosmotic secretion with the interstitial fluid and the fluid exiting the channel is considerably hyperosmotic. A long, narrow channel allows more opportunity for osmotic equilibration of the channel fluid and the interstitial fluid, and so the fluid exiting the

channel may be only slightly hyperosmotic or even isoosmotic. The secreted fluid cannot be hypoosmotic to the interstitial fluid, i.e., there is no net movement of water without accompanying solute. The standing-gradient model has also been used to explain water transport by the insect Malpighian tubule; active transport of K^+ into the lumen is accompanied by passive Cl^- flux and osmotic water flux (Berridge and Oschman 1969).

A modification of the cellular structures associated with the standing-gradient model of solute and water flux can result in a net hypoosmotic water flux (Figure 17–6B). Solute is recycled into the epithelial cell, and so the fluid emptying into the lumen can be hypoosmotic to the interstitial fluid on the basal side of the cell. This model has been proposed to explain water uptake by the rectal pads of the cockroach (Wall and Oschman 1969).

Volume Pump. The volume pump model, also based on the rabbit gall bladder, proposes that a mechanical fluid pump creates a mass flow of water across the epithelium (Fredericksen and Leyssac 1969). For example, pinocytosis, or microvesicular movement caused by microtubules, would produce an iso-osmotic movement of water. However, it is difficult to explain a specificity of solute flux across the epithelium according to this model, and so this is generally considered to be an unlikely mechanism for transepithelial water movement.

Formed Bodies. Formed bodies are small, lysosomal-like extrusions of the cell membrane; they are produced by crustacean antennal glands and also by epithelia of some pulmonate snails and frogs. Some formed bodies are probably lysosomes and others also contain hydrolytic enzymes.

Formed bodies are released into spaces within or adjacent to cells, and then they swell and burst. Their swelling is probably due to an osmotic influx of water as a result of an increase in the internal osmotic concentration by hydrolysis of internal macromolecules (possibly proteins). Such swelling and lysis of formed bodies could cause a transepithelial water flux along a constricted and rigid channel if there were some directionality to the movement of swelling formed bodies and a relative restriction to water flow from the end of the channels at which the formed bodies are produced.

The fluid formed in this manner would be iso-osmotic to the cytoplasm. Fluid transport by formed bodies is potentially more rapid than solute-linked water flux because the cell membrane in the vicinity of the formed bodies can have relatively large diameter pores for water flux into the channel where

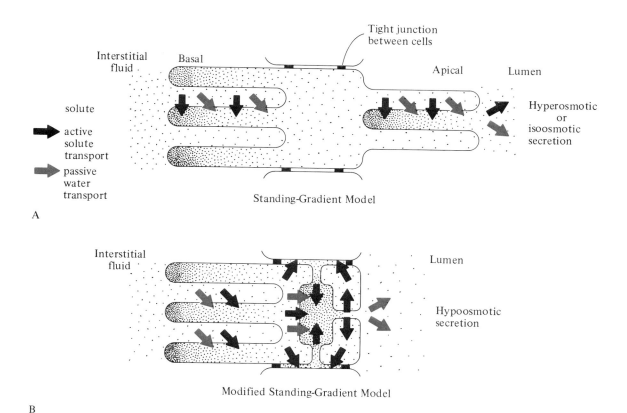

FIGURE 17–6 **(A)** The standing-gradient model for solute-linked water transport; the transported solution is either isoosmostic or hyperosmotic to the outside fluid. Solute is actively transported out of channels in the basal membrane into the cell. A local hypoosmotic environment is established in the outside of the channels, and a local hyperosmotic environment is established inside the channels. Water moves into the cell by osmosis. A similar active transport of solute causes osmotic water transport across the surface membrane. The transported fluid is isoosmotic if the channels are long and thin and hyperosmotic if the channels are short and wide; the fluid cannot be hypoosmotic, i.e., there is no net transport of water. **(B)** A highly modified standing-gradient model can explain the net transport of water against an osmotic gradient if there is solute recycling within the epithelial cell. Active transport of ions from the concentrated external solution increases the intracellular osmotic concentration and draws water osmotically across the basal membrane. Further ion transport into intercellular channels and an intracellular space draws water into the intracellular space. Ions are actively reabsorbed from this fluid as it flows to the internal fluid space (low osmotic concentration). The net effect is transport of fluid that is hypoosmotic to the external medium.

the formed bodies are swelling by osmosis. For solute-linked water flux, the cell membranes must have pores that are narrow enough to limit solute movement but wide enough to allow water flux in the direction of solute pumping (e.g., <26 nm for Na^+-linked water transport); such a narrow pore would limit the rate of water flux.

Electro-osmosis. Electro-osmosis produces a transepithelial water flux as a result of a transepithelial electrical potential (Hill 1977). The electrical potential causes the flux of specific ions through small channels, which exclude movement of counter ions. A frictional interaction between the transported ion and adjacent water molecules causes a cotransport of water with the ions. Electro-osmosis is generally not considered to be a significant mechanism for water transport across epithelia because the observed transepithelial potential is usually of the wrong polarity to transport water in the observed direction and too few H_2O molecules are transported per ion to account for the osmolality of membrane-transported fluids. However, electro-osmosis has been suggested to be responsible for water vapor absorption by the rectal epithelium of the fire-brat *Thermobia* (see Chapter 16).

Tubular Excretion

Virtually all multicellular animals have tubular excretory organs. These tubular organs primarily evolved for solute excretion rather than the elimination of nitrogenous wastes, since ammonia is the principal nitrogenous waste of primitive animals and it is readily excreted across the integument or gills because of its high solubility.

Nephridia, Coelomoducts, and Malpighian Tubules. Tubular excretory organs can be categorized into three general types: **nephridia, coelomoducts,** and **Malpighian tubules.** Nephridia are ectodermal tubes that develop into the interior of the animal from the outer surface (Goodrich 1945). Coelomoducts are mesodermal tubes that develop from the interior of the animal to the outside, thereby establishing an excretory opening with the exterior. Insect Malpighian tubules (and some other excretory tubules of various animals) are a third type of tubular excretory organ derived from the gut; they do not open directly to the exterior but empty into the hindgut.

Protonephridia are blind-ended nephridia that do not have a direct connection of the tubule lumen with the coelomic cavity (Figure 17–7A). They are considered, by Goodrich's scheme, to be evolutionarily primitive and the precursors ("proto" means first) for metanephridia (see below). Fluid is drawn into the blind-ended tube by the action of terminal cilia or flagella and then passes along the tube and is eliminated via the protonephridiopore. Protonephridia may be further classified by the nature of the cells at the blind, internal end of the tubule. Monociliate or monoflagellate cells are called solenocytes; flame cells are multiciliate, with the cell nucleus located near the base of the cell; flame bulbs are multiciliate, with the cell nucleus located laterally.

Metanephridia are tubules that develop from the ectoderm and have a direct connection of the lumen with the coelomic space. They are, according to Goodrich's scheme, evolutionarily derived from protonephridia and are present in the larger and more complex animals. They open into the coelomic space via a ciliated funnel-like opening, the nephridiostome, which draws coelomic fluid into the tubule (Figure 17–7B). Fluid then passes along the metanephridial tubule and is eliminated from the nephridiopore.

Coelomoducts are structurally similar to nephridia, but differ in their embryologic origins and development. Coelomic fluid is drawn into the ciliated coelomostome, passes through the coelomoduct, and is eliminated from the coelomopore (Figure 17–7C). The original function of coelomoducts was most likely as a route for elimination of the gametes from the coelom to the exterior, rather than as an excretory organ.

In some animals, nephridia and coelomoducts fuse to form a single nephromixium, which has both ectodermal and mesodermal origins. Fusion may be so complete as to combine both the canals and the excretory pores. The nephromixium may function for excretion and/or gonad release. A protonephromixium is formed by the fusion of a coelomoduct (coelomostome) with the canal of a protonephridium (Figure 17–7D). A metanephromixium has a coelomostome grafted onto the open nephridiostome. A mixonephridium is essentially the complete fusion of a coelomostome with the inner end of a nephridium to form a simple, large-funneled structure (Figure 17–7E).

A more recent classification of tubular excretory organs (protonephridia, metanephridia, coelomoducts) uses a functional rather than a structural/embryological basis (Ruppert and Smith 1988). Here, protonephridia are defined as blind-ended excretory tubules of animals with a single body cavity (coelom); entry of fluid into the blind-ended tubule is accomplished by the ciliary/flagellar beating (Figure 17–8A). The fluid inside the protonephridial tubule is at a negative pressure relative to the coelomic fluid, and so the tubule must be able to resist collapse. Animals with metanephridia have a coelomic space and a vascular space (Figure 17–8B). A "primary urine" (coelomic fluid) is formed by hydrostatic pressure filtration of vascular fluid into the coelomic space; a secondary urine is formed by movement of coelomic fluid into the metanephridium. Note that this functional definition of metanephridia includes embryologically defined nephridia and coelomostomes (and vertebrate nephrons). Large, complex, and generally advanced animals tend to have metanephridia, whereas smaller and generally more primitive animals have protonephridia. This is generally consistent with the view that protonephridia are ancestral to metanephridia. The importance of body size is suggested by the presence of protonephridia in larval annelids and mollusks; they are replaced by metanephridia in adults.

This functional classification of excretory tubules is generally consistent, but there are some apparent exceptions. One polychete family, the Nephtyidae,

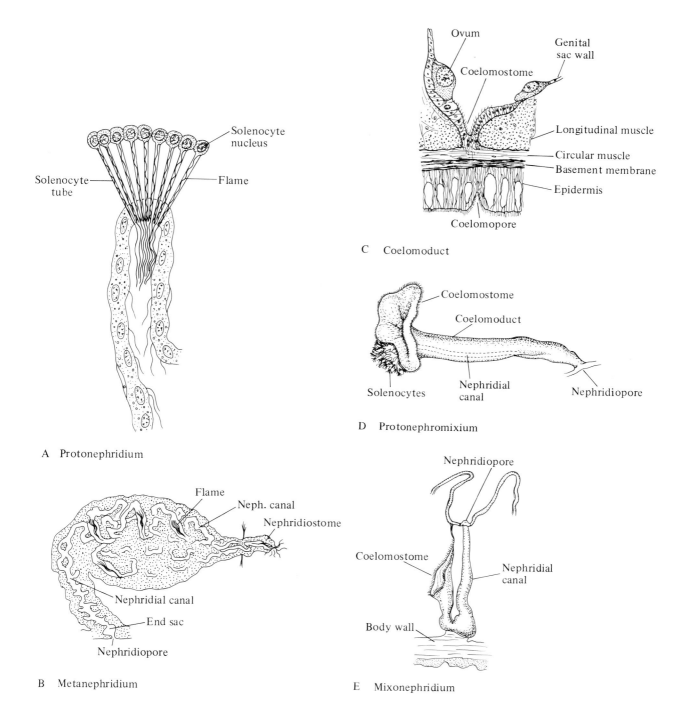

FIGURE 17–7 **(A)** The protonephridium of the polychete annelid *Phyllodoce* has a soleno-cyte cell nucleus and tube and a ciliary "flame." **(B)** The metanephridium of the oligochete an-nelid *Lenchytraeus* has cilia and "flames." **(C)** The coelomoduct of the nemertean worm *Am-phiporus* shows the coelomostome opening into the gential sac and the coelomopore. **(D)** The protonephromixium of the polychete *Phyllodoce* has a fused coelomoduct and nephridial ca-nal. **(E)** The mixonephridium of the polychete *Nerines* has a coelomostome, a nephridial ca-nal, and a nephridiopore.   *(From Goodrich 1945. With permission of the Company of Biologists.)*

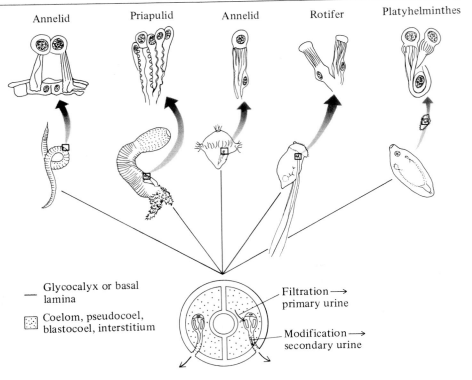

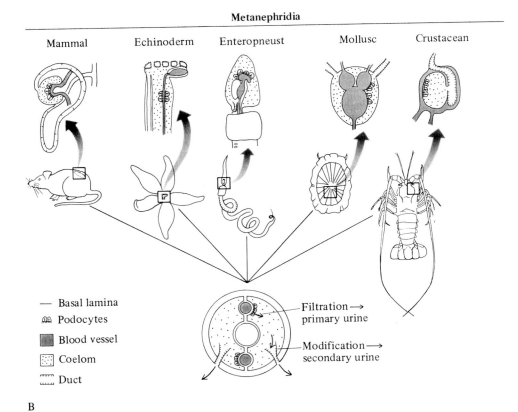

FIGURE 17–8 **(A)** Functional organization of protonephridia; primary urine is formed by filtration of the coelomic fluid. **(B)** Functional organization of metanephridia; primary urine is formed by filtration of vascular fluid to form coelomic fluid, which is then drawn into a duct to form secondary urine. *(From Ruppert and Smith 1988.)*

have a blood vascular system but have proto-nephridia rather than, as might be expected, meta-nephridia. However, these polychetes are unusual in having an extracellular coelomic hemoglobin (see Chapter 15), and perhaps protonephridia are required to prevent excretory loss of hemoglobin from the coelomic fluid. Sipunculids lack a circulatory system but have well-developed metanephridia (actually, mixonephridia). However, sipunculids may have a coelomic diverticulum which acts in an analogous fashion to the blood vascular system. Nemertean worms have a closed circulatory system but have protonephridia; however, their vascular system is unusual and may not be homologous to that in other animals.

Malphighian tubules are blind-ended excretory tubules of centipedes, millipedes, insects, and arachnids, having presumably evolved independently in these various arthropod groups. The number of Malpighian tubules varies markedly for different arthropods, from two to several hundred. The blind end of the tubule generally lies free in the hemocoele. Fluid is filtered into the blind ends of the tubules by active solute (KCl) secretion and passive osmotic influx of water, rather than a ciliary or hydrostatic filtration. Malpighian tubules do not convey their excretory fluid directly to the exterior, unlike nephridia and coelomoducts, but open into the digestive tract at the junction of the midgut and hindgut. The composition of the tubular fluid can be subsequently modified by the more proximal Malpighian tubule or the anterior hindgut, and especially the rectum.

Physiological Processes for Tubular Excretion. Urine generally forms at the internal end of the excretory tubule and flows along the tubule, being modified to a variable extent in different animals before it is eliminated through an external pore.

There are generally four physiological aspects to urine formation and its subsequent modification that require further discussion (Figure 17–9). These processes are primary urine formation by **filtration**; subsequent **reabsorption** of water, specific nutrients (glucose, amino acids), and other solutes (urea, Na^+, Cl^-); **secretion** of specific waste products into the tubule (some organic acids); and **osmotic concentration** of the urine (which occurs in the nephrons of mammals and some birds).

Filtration.   Fluid, called primary urine, can enter the excretory tubule by any of a number of means, e.g., bulk movement of fluid by ciliated funnels, hydrostatic filtration due either to ciliary activity (protonephridia) or vascular pressure (metanephridia, nephrons), formed bodies (e.g., crustacean antennal gland), or solute-linked water flux (Malpighian tubules).

Filtration is the process whereby fluid passes through pores of a filtering membrane in response to a hydrostatic or osmotic pressure difference. The blood cells are excluded from the urine by the

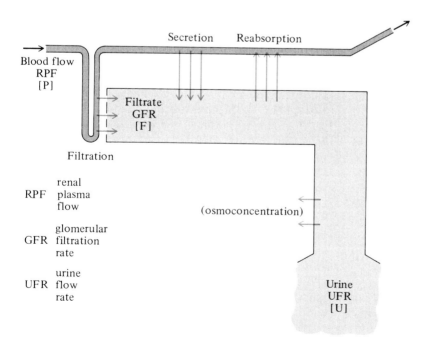

FIGURE 17–9 Schematic of the principle functions of a tubular excretory organ, such as the vertebrate nephron: filtration, secretion, and reabsorption. The avian and mammalian nephron can also osmoconcentrate urine.

pore size of the filtration membrane. Ultrafiltration occurs if the membrane pores are sufficiently small to prevent filtration of colloids and large solutes such as protein, as well as blood cells.

Protonephridial tubules form primary urine by ultrafiltration. The mechanism for primary urine formation by the protonephridium has been inferred from its structure (Figure 17–10A). A flagellum, cilium, or multiple cilia beat and create a negative pressure inside the protonephridial channel; fluid is drawn into the channel by suction through a thin ultrafiltration membrane or open pores. For example, the beating of the cilia in the flame cell of the liver fluke (a trematode) draws fluid through the basement membrane that covers a barrel-shaped structure comprised of long rods (called a cyrtocyte, or barrel-shaped cell). In the protonephridium of the cephalochordate *Branchiostoma*, the cyrtocyte contacts a blood space, and the barrel protrudes into the nephric tubule. The barrel consists of 10 rods and is not covered by a membrane.

In vascular filtration/ultrafiltration, the blood supply provides a hydrostatic pressure for fluid filtration across capillaries. The glomerulus of the mammalian nephron, for example, is an arterial ultrafiltration system (Figure 17–10B). Special cells, called **podocytes**, support the filtration membrane against the hydrostatic pressure gradient (see also the crustacean antennal gland). Filtration occurs if the sum of the forces promoting fluid passage across the filter into the tubule exceeds the sum of forces promoting loss of fluid from the tubule across the filter. These forces are hydrostatic forces and colloid osmotic forces (see also Chapter 16). The force promoting filtration out of the glomerulus is primarily hydrostatic (mean capillary blood pressure = 7.98 kPa); the colloid osmotic pressure due to protein in the filtrate is extremely low since the filter is essentially impermeable to protein (i.e., about 0 kPa). Forces promoting reabsorption include the hydrostatic pressure of the filtrate (about 2.39 kPa) and the mean plasma colloid osmotic pressure (4.26 kPa). The mean capillary colloid osmotic pressure is substantially higher than the normal plasma colloid osmotic pressure (3.72 kPa) because the plasma proteins are significantly concentrated by the filtration of much water out of glomerular blood. The filtration pressure of 1.33 kPa (10 torr) is the net balance of the forces (7.98 + 0 − 2.39 − 4.26). The brate of filtration is about 125 ml min^{-1}, so the filtration coefficient (K_f) is 94 ml min^{-1} kPa^{-1} (12.5 ml min^{-1} torr^{-1}). The balance of forces across the crustacean antennal gland is consistent with a net hydrostatic filtration (Picken 1937), although formed bodies have also been implicated in primary urine formation.

Primary urine formation by the insect Malpighian tubule occurs in a substantially different manner from ciliary or hydrostatic filtration (Figure 17–10C). Fluid secretion by most insect Malpighian tubules is the result of active transport of K$^+$ ions from the hemolymph into the cell cytoplasm, accompanied by passive Cl$^-$ flux and cotransport of water by osmosis. The K$^+$ is thought to passively enter the Malpighian tubule lumen through the basement membrane of the epithelial cell. The pathway for water transport into the Malpighian tubule lumen is not clear but is probably through the membranes of the tubule epithelial cells (i.e., a transcellular route) rather than between the epithelial cells (i.e., a paracellular route). A wide variety of other solutes are also present in the Malpighian lumen fluid, including amino acids, uric acid, mono-, di-, and polysaccharides. The pathway for the passive filtration of these solutes is thought to be between the epithelial cells (paracellular) rather than through the epithelial cell membranes. Sugars (but apparently not amino acids and other organic acids) may be actively reabsorbed by the epithelial cells, and toxic wastes may be actively secreted into the tubule (as is readily apparent for many organic dyes introduced into the body fluids).

The solute composition of the filtrate depends on the permeability of the filtering mechanism. For the mammalian nephron, solutes of molecular weight less than about 6 10^6 (about 15 μ radius) are readily filtered from the plasma into the Bowman's capsule of the renal tubule, but the passage of solutes with higher molecular weights is less (Table 17–1). For the Malpighian tubule, the filtrate:hemolymph ratio (i.e., *F/H*) is less than 1 and is inversely proportional to the solute molecular weight, consistent with passive filtration through a semipermeable membrane. The Malpighian tubule appears to be less permeable than the mammalian glomerulus, e.g., inulin is readily filtered by the mammalian nephron but not the Malpighian tubule.

Secretion. Most tubular excretory organs have some form of active secretion of waste products into the lumen. Often these are not the nitrogenous waste products (ammonia, urea, or uric acid) but are ions and complex organic acids and bases. Secretion is particularly important for those tubular organs that lack filtration mechanisms for urine formation, e.g., Malpighian tubules and aglomerular vertebrate nephrons.

The Malpighian tubules of arthropods actively secrete K$^+$ ions from hemolymph into the tubule lumen. The Malpighian tubules of some insects can produce a Na$^+$-rich fluid. For example, herbivorous

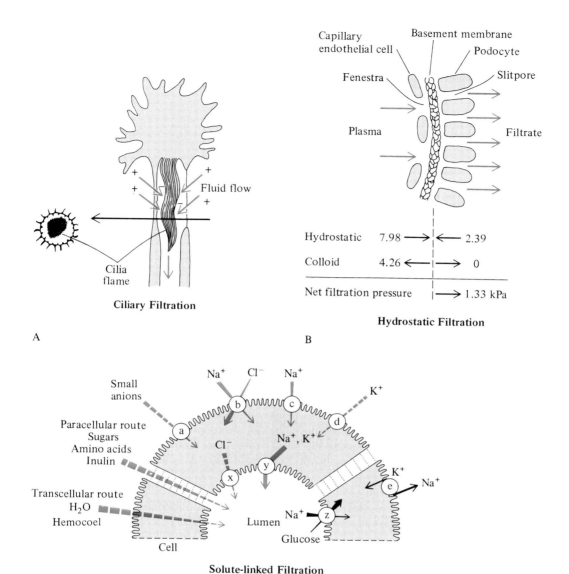

Ciliary Filtration

A

Hydrostatic Filtration

B

Solute-linked Filtration

C

FIGURE 17–10 **(A)** Probable means of ciliary filtrate formation by protonephridial excretory tubules of the liver fluke; + and − signs represent relative hydrostatic pressures and arrows indicate the probable direction of fluid flow. **(B)** Hydrostatic filtration depends on the balance of hydrostatic and colloid osmotic forces producing a net filtration pressure. Values are for the human nephron, showing a net filtration pressure of 1.33 kPa. **(C)** Schematic of the mechanism for solute-linked fluid filtration by the insect Malpighian tubule. Diffusion of potassium and small anions into the tubular cells (mechanisms a, d) and active transport of sodium, chloride, and potassium (mechanisms b, c) and sodium/potassium exchange (mechanism e) are accompanied by active transport of sodium and potassium into the tubule lumen (y), passive chloride entry into the lumen (x), and active co-linked sodium and glucose reabsorption (z). Many other solutes enter the tubule passively via a paracellular route between the epithelial cells. Water is thought to enter the tubule by passive osmosis via a transcellular route. *(Modified from Riegel 1972; Guyton 1986; Phillips 1981.)*

TABLE 17–1

Relative permeability of the mammalian nephron and the insect Malpighian tubule to solutes of varying molecular weight and size (effective radius calculated from the diffusion coefficient).   *(Data from Pitts 1974; Maddrell and Gardiner 1974.)*

	Solute MWt	Solute Radius nm	Nephron Filtrate/Plasma	Malpighian Tubule Filtrate/Hemolymph
Water	18	1.0	1.0	1.0
Urea	60	1.6	1.0	0.84
Glucose	180	3.6	1.0	
Sucrose	342	4.4	1.0	0.70
Inulin	5500	14.8	0.98	0.07
Myoglobin	17000	19.5	0.75	
Egg albumin	43500	28.5	0.22	
Hemoglobin	68000	32.5	0.03	
Serum albumin	69000	35.5	<0.01	

caterpillars (*Calpodes*) and blood-sucking insects (*Glossina, Aedes*) have active Na^+ transport into the Malpighian tubule lumen. The tsetse fly (*Glossina*) appears exceptional in that more than 90% of the cations transported into the tubule lumen are Na^+ rather than K^+. Chloride is the predominant anion in the tubular fluid and generally passively follows K^+ transport but may sometimes (e.g., in *Rhodnius*) be actively transported through the basement membrane of the epithelial cell. Phosphate and sulfate ions are actively secreted into the lumen. Toxic wastes may be actively secreted into the tubule (as is readily apparent for many organic dyes introduced into the body fluids.

Reabsorption.   Most mechanisms for formation of urine are relatively nonselective in determining the solutes that are present in the primary urine, other than on the basis of molecular weight. Nutrients (e.g., glucose and other saccharides, amino acids, fatty acids) are as likely to be present in the primary urine as waste products. It is therefore not surprising that many tubular excretory organs actively reabsorb nutrients. In the mammalian nephron, for example, glucose and amino acids are almost completely reabsorbed in the initial portion of the tubule. Malpighian tubules of some insects reabsorb K^+ or Na^+ and glucose. Water can also be passively reabsorbed by osmosis, although some excretory tubules are relatively impermeable to water and the urine produced by them is hypoosmotic to the filtrate (and blood).

Osmoconcentration.   Tubular excretory organs generally produce a hyperosmotic urine by the osmotic withdrawal of water from the urine, rather than by

pumping ions (or other solutes) into the tubule lumen. For example, mammalian and avian nephrons can produce a hyperosmotic urine by the osmotic removal of water; this requires a complex structural arrangement of the nephron to form a double countercurrent multiplication/exchange system (the loop of Henle and collecting ducts) with active transport of Cl^- out of the nephron lumen. The structure and functioning of these nephrons will be described in more detail below (see Mammals).

Flow and Clearance of Solutes and Water.   Many of the above-described aspects of tubular excretory function are readily amenable to experimental measurement. The methods used for the measurement of glomerulo-tubular function for the vertebrate nephron are generally applicable to other tubular excretory organs.

The rate of formation of primary urine (the filtrate, F) is called the **filtration rate** (FR), or **glomerular filtration rate** (GFR) for glomerular nephrons. The filtrate has an osmotic concentration (F_{osm}) and solute concentration (e.g., of solute X; F_X) that are similar to those of plasma (P_{Osm}, P_X). The glomerular filtrate is formed by filtration of the renal plasma flow (RPF; which is related to the renal blood flow, RBF, by the hematocrit) of concentrations P_{Osm} and P_X. The primary urine is subsequently modified to form the urine, with a **urine flow rate** UFR and concentrations U_{Osm} and U_X.

The ratio of urinary to plasma concentrations (U/P) indicates whether specific solutes, or total osmotic concentration, are more or less concentrated in the urine than in the plasma. The $U_{Osm}/P_{Osm} \leq 1$ for all tubular excretory organs except mammalian and avian nephrons (and an annelid metanephridium which has slightly hyperosmotic values; see below).

The glomerular filtration rate is readily calculated from the UFR and the U/P for any solute that is freely filtered, but is not subsequently reabsorbed or secreted, since the amount of this solute remains constant along the nephron. The amount of this solute that is filtered (GFR P) equals the amount of this solute eliminated in the urine (UFR U); i.e.,

$$\text{GFR} = \text{UFR } U/P \qquad (17.1a)$$

Naturally present solutes may at least approximate these requirements for the determination of GFR (i.e., freely filtered, not reabsorbed, not secreted). For example, creatinine is freely filtered and not reabsorbed, and so

$$\text{GFR} \approx \text{UFR } U_{\text{creatinine}}/P_{\text{creatinine}} \qquad (17.1b)$$

Unfortunately, there is also secretion of creatinine by some nephrons, and so it is not therefore the ideal solute for determination of GFR. Other more ideal experimental solutes, such as the polysaccharide inulin, can be introduced into the blood for the more accurate determination of GFR.

$$\text{GFR} = \text{UFR } U_{\text{inulin}}/P_{\text{inulin}} \qquad (17.1c)$$

A solute that is not only filtered, but is actively secreted into the tubule by the tubular epithelial cells, could be completely eliminated from the blood supply to the renal tubule. The total amount of this solute entering the tubule is RPF P and that eliminated is UFR U; hence,

$$\text{RPF} = \text{UFR } U/P \qquad (17.2a)$$

The organic acid para-aminohippuric acid (PAH) is essentially totally eliminated from the nephron blood; hence

$$\text{RPF} = \text{UFR } U_{\text{PAH}}/P_{\text{PAH}} \qquad (17.2b)$$

as long as the tubular secretory capacity is not exceeded by excessively high concentrations of PAH.

The value of UFR U/P represents the rate of plasma flow that has all of that particular solute removed from it, i.e., it is the amount of plasma that is cleared of that solute, per unit time. Hence, this parameter is called the **clearance** (C).

$$C_X = \text{UFR } U_X/P_X \qquad (17.3)$$

It is clearly a useful parameter for examining renal function, depending on the solute in question. The clearance of inulin (C_{inulin}) equals the GFR; the clearance of PAH (C_{PAH}) equals the RPF. If a solute is totally reabsorbed, then its clearance is 0. The clearance value for solutes thus suggests how they

are handled by the nephron, although a specific C value does not necessarily prove its glomerulo-tubular behavior. For example, $C = 0$ suggests total reabsorption (or nonfiltration); $0 < C <$ GFR suggests filtration and partial reabsorption (or minimal filtration); $C =$ GFR suggests filtration and no secretion/reabsorption (or no filtration and some secretion, or filtration and some reabsorption and some secretion); GFR $< C <$ RPF suggests filtration and some secretion (or no filtration and marked secretion); $C =$ RPF suggests filtration and total secretion, or no filtration and total secretion.

The osmotic clearance is similarly calculated, as

$$C_{\text{Osm}} = \text{UFR } U_{\text{Osm}}/P_{\text{Osm}} \qquad (17.4)$$

If $C_{\text{Osm}} <$ UFR (i.e., $U_{\text{Osm}} < P_{\text{Osm}}$) then the urine is more dilute than plasma and part of the urine represents the excretion of free water. That is, we can consider such urine to consist of a fraction that is iso-osmotic to plasma and a part that is distilled water. The free water clearance is calculated as

$$C_{\text{free water}} = \text{UFR} - C_{\text{Osm}} \qquad (17.5)$$

A final useful parameter to describe renal tubular function is the percent of reabsorption of either solutes (X), total osmolytes (Osm), or water compared to the amount filtered, i.e.,

$$R_X = 100(1 - C_X/\text{GFR}) \qquad (17.6)$$

If $C_X =$ GFR then $R = 0\%$ reabsorption; if $C_X = 0$, then $R = 100\%$ reabsorption.

Water and Solute Excretion

It is important to understand the basic organization and functioning of the various types of tubular excretory organs in major groups of animals to appreciate patterns of iono- and osmoregulation (Chapter 16) and nitrogenous waste excretion. A description of only a few types of excretory systems ignores the amazing diversity of animal excretory systems, but a systematic and phylogenetic survey would be tedious and encyclopedic. We shall compromise, and examine the variety of excretory systems in the major groups of animals.

"Lower" Invertebrates

The "lower" invertebrates are characterized by their either relying on cell contractile vacuoles for excretion (e.g., sponges, coelenterates) or by their having protonephridial excretory systems.

Protozoans excrete fluid via an organelle system, the **contractile vacuole**. This organelle is a spherical

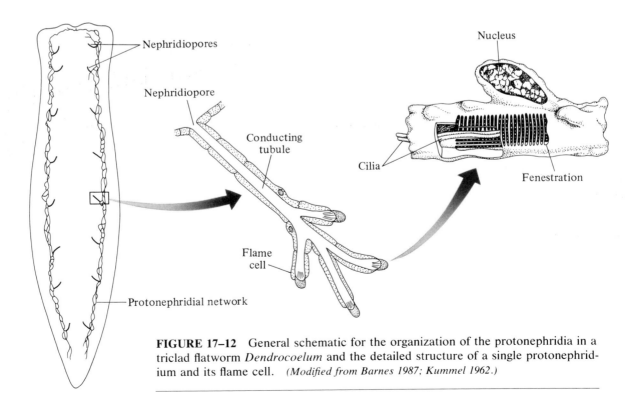

FIGURE 17–11 The rate of fluid excretion by the contractile vacuoles of the freshwater sponge is negatively related to the external osmotic concentration, being maximal in distilled water and approaching 0 in 12 mOsm water. *(Modified from Brauer 1975.)*

vesicle with a surrounding system of small tubules and vesicles, the spongiome (Patterson 1980). The spongiome collects water and delivers it to the contractile vacuole, which expels the fluid through a pore. The excretory pore is permanent in some species and is transient in others. The contractile vacuole disappears after its contraction in some amoebas and flagellates and re-forms from numerous small vesicles. In other protozoans (e.g., ciliates), the contractile vacuole collapses after discharge and is refilled with fluid from the tubules of the

spongiome. Most cells of freshwater sponges have one or more contractile vacuoles for excretion of water. Their rate of excretion is related to the external osmotic concentration (Figure 17–11).

Platyhelminth flatworms (turbellarians, flukes, cestodes) have protonephridia. Turbellarians have a variable number of typical, paired flame cell protonephridia (Figure 17–12). The flame cells (cyrtocytes) have a single cilium, or a cluster of cilia, that beat continually. The barrel-shaped wall of the cell appears to be perforated by slits (fenestrae).

FIGURE 17–12 General schematic for the organization of the protonephridia in a triclad flatworm *Dendrocoelum* and the detailed structure of a single protonephridium and its flame cell. *(Modified from Barnes 1987; Kummel 1962.)*

Fluid is drawn into the lumen of the protonephridium by the ciliary beating. The slits are actually not open, but are covered by a membrane that acts as a selective filter. Flukes generally have protonephridia with a pair of longitudinal ducts and two anterior nephridiopores (monogeneans) or a single posterior bladder and a nephridiopore (trematodes). Flame cells of cestodes drain into longitudinal ducts.

Nemertean worms have a single pair of protonephridia. The terminal cells of the protonephridia project into the wall of the lateral blood vessel, or in some cases enter the blood vessel lumen and are directly bathed by blood.

The various aschelminths typically have protonephridia. For example, the kinorhynchs have two protonephridia with solenocytes or flame bulbs (Kristensen and Hay-Schmidt 1989). The protonephridium of *Echinoderes* has three terminal cells each with two cilia and numerous microvilli. The terminal cells are folded to form a central canal containing the cilia. One side of the canal is formed by the microvilli, creating a perforate wall called the weir. The ultrafiltration pores are the weir slits between the microvilli and the basal lamina, which is an ectodermal covering of the entire protonephridium. Fluid entering through the weir passes along a central channel in the canal cell to the nephridial pore cell that contacts the cuticle of the body wall. This nephridiopore cell has a central lumen that conveys urine to an oval, porous sieve plate and then to the exterior. Priapulids have solenocyte protonephridia with a long protonephridial canal that opens into an excretory canal. Nematodes lack protonephridia, and many have no specialized excretory organs. Some nematodes have a single large cell (the Renette gland) that has a duct opening into an excretory pore. Some have a few cells that form an H-shaped tubular system that has an excretory role. Gastrotrichs have a single pair of protonephridia; these are best developed in freshwater species. Rotifers typically have two protonephridia, one on each side of the body (Figure 17–13). The two collecting ducts open into a bladder (or a dilation of the cloaca in bdellids);

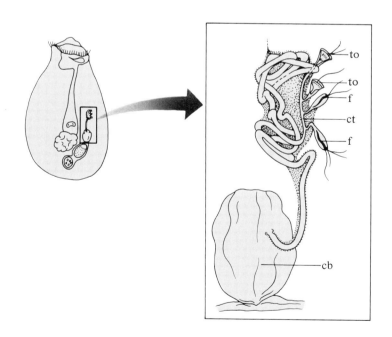 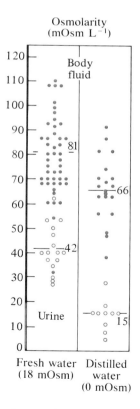

FIGURE 17–13   General structure of the protonephridial organ of a rotifer showing the terminal organ of the protonephridia (to), flame (f) and connecting tubule (ct) connecting to the contractile bladder (cb). Panel to right summarizes the body fluid (solid circles) and urine (open circles) osmolality of the rotifer in fresh water and in distilled water. *(Modified from Braun, Kummel, and Mangos 1966.)*

the considerably hypoosmotic urine is expelled by regular contractions of the bladder (or cloaca) at about 0.5 to 4 min^{-1} depending on the osmotic concentration of the medium. The freshwater rotifer *Asplanchna* forms a hypoosmotic urine (41 mOsm) compared to body fluids (81 mOsm) in lake water (18 mOsm) and its bladder contracts at about 0.56 min^{-1}; each bladder contraction expels about 0.085 nl, so the urine flow rate is about 0.047 nl min^{-1} (Braun, Kummel, and Mangos 1966). In distilled water, the urine is more hypoosmotic (15 mOsm, cf. body fluid concentration of 66 mOsm) and the bladder contracts more frequently (0.71 min^{-1}); the urine flow rate is 0.060 nl min^{-1}. The ratio of urine to body fluid concentration of inulin for animals in lake water is 1.42, indicating some reabsorption of water (about 30%). The nephridial filtration rate is about 1.42 × 0.047 = 0.067 nl min^{-1} or 0.033 nl min per protonephridium.

A variety of other lesser protostome and deuterostome adults (Loricifera, Acanthocephala, Gnathostomulida, Entoprocta) and larvae (Phoronida actinotroch, Echiura trochophore, Hemichordata early tornaria, Molluska trochophore, Annelida trochophore, Echinodermata early bipinnaria) have protonephridia. Metanephridia are present in other lesser phyla (Brachiopoda, Phoronida, Sipunculida, Echiura, Pogonophora, Echinodermata late bipinnaria larvae, Hemichordata late tornaria larvae, Onycophora) and chordates. Some phyla have other forms of excretory organs, e.g., engulfing coelomocytes (adult echinoderms, Bryozoa), a system of venous sinuses (the "glomerulus" of acorn worms; hemichordates), storage excretion (renal sacs in urochordata; but see below), intestinal glands (three "Malpighian glands" at the intestine-rectal junction in tardigrades), and metanephridial-like rectal diverticula (Echiura).

Annelids

There is considerable diversity in the excretory organs of annelid worms. Polychetes have protonephridia or metanephridia. Primitive forms have one pair of nephridia per segment, but there is a reduction in the number of pairs of nephridia to a few, or even one pair, in advanced polychetes. The terminal end of the nephridium is located in the coelom of the segment anterior to the segment with the nephridiopore. Protonephridia with solenocytes are present in some polychetes (phyllodocids, alciopids, glycerids, nephytids, etc.); the solenocytes arise in bunches from the nephridial ducts. Each solenocyte cell has a single flagellum that is enclosed

by a series of parallel rods connected with a thin lamella; this is the filtration unit. The filtrate passes down this structure and exits through a channel in the solenocyte into the nephridial canal. Other polychetes have metanephridia, with nephrostomes draining the coelom of the more anterior segment. Adult oligochete worms have metanephridia arranged segmentally as in polychetes, except in the most anterior and posterior segment. The nephridial canal is highly coiled, often with several separate groups of coils. A bladder is sometimes present near the nephridiopore. Some megascolid and glossoscolid worms have additional multiple or branched nephridia that may open externally through nephridiopores or internally into the gut (enteronephric). Leeches have 10 to 17 pairs of metanephridia located in the central body segments.

The physiology of metanephridia is best known for *Lumbricus* (an earthworm) and *Hirudo* (the medicinal leech); both produce a markedly hypoosmotic urine. The metanephridium of *Lumbricus* is a complex, multicoiled tubule (Figure 17–14) with a nephridiostome of varying morphology and a bladder before the nephridiopore. Fluid, which may be filtered from the vascular system into the coelomic space near the nephridiostome, enters the tubule through the ciliated nephridiopore. The osmotic concentration declines along the length of the tubule, to about 20% of the filtrate concentration by the distal portion of the tubule. This osmodilution is a consequence of solute reabsorption, especially active Na^+ and passive Cl^- reabsorption, without isoosmotic reabsorption of water (Zerbst-Boroffka 1977).

Leech metanephridia normally produce a markedly hypoosmotic urine, although both the urine flow and osmotic concentration increase markedly after a blood meal to eliminate the salt and water load. The mechanism for primary urine formation is complex in *Hirudo*. There is filtration of blood to form an ultrafiltrate, but the primary urine in the tubule is produced through secretion by canalicular cells of K^+ into the lumen (cf. insect Malpighian tubules). About 95% of the salts are reabsorbed, resulting in a markedly hypoosmotic urine.

Polychete proto- or metanephridia generally eliminate an isoosmotic urine, but it may be hypoosmotic for worms in dilute media. The urine produced by the single pair of metanephridia in *Glycera* is 12% hyperosmotic to the coelomic fluids; the U-shaped nephridium is associated with a local osmotic gradient in both the tubular fluid and adjacent coelomic fluid, but the mechanism for establishment of this osmotic gradient, and its significance, are not clear (Koechlin 1975).

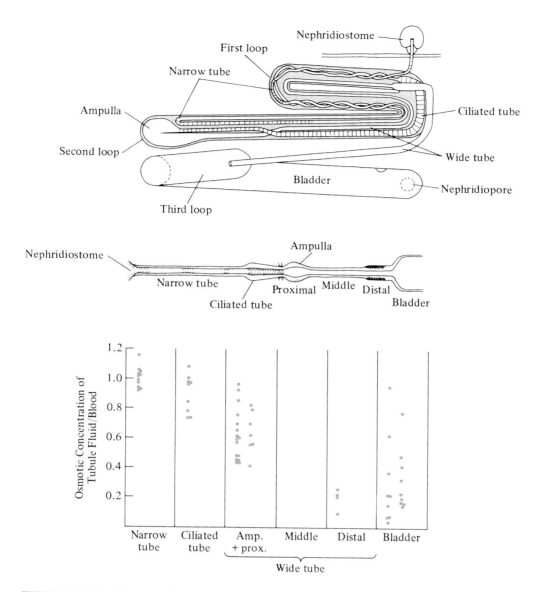

FIGURE 17–14   Diagram of the gross anatomy of the nephridium of the earthworm *Lumbricus*, and a schematic of the generalized structure and the fluid/blood osmotic concentration in various segments of the nephridium.   *(From Riegel 1972; after Graszynski 1963; Boroffka 1965.)*

Mollusks

The excretory system of mollusks includes protonephridia in larval stages and metanephridia in adults (actually, coelomoducts, according to Goodrich's scheme). The primitive mollusk *Neopilina* (Monoplacophora) has six pairs of metanephridia, with nephrostomes opening into the general body cavity (anterior pairs) and the pericardial coelom (posterior pairs 5 and 6); the nephridiopores open into the pallial groove near the gill bases. Coelomoducts open from the ovaries into nephridial pairs 3 and 4, forming mixonephridia.

Only one of the posterior pairs of nephridia are thought to persist in more advanced mollusks, forming the renopericardial–nephridial sac excretory system. In chitons, for example, there is no ciliated nephrostome, but the pericardium connects via the short, ciliated renopericardial canal to the pericardial sac, a large and diverticulated structure opening via a nephridiopore into the pallial groove; the gonads open separately via coelomoducts.

In bivalve mollusks (Figure 17–15), the primary excretory filtrate is formed by ultrafiltration from the atria into the pericardial sac via the pericardial gland and flows down the renopericardial canal into the kidney through the renopericardial canal and funnel. After renal modification, the urine exits via the excretory pore. The excretory system of prosobranch and gastropod mollusks is similar but that of opisthobranchs and pulmonates is reduced to a single kidney, and in pulmonates a secondary ureter forms to drain urine.

Cephalopod renal structures have clear homologies with those of other mollusks but are highly modified. *Nautilus* has four nephridia, and the other cephalopods have two. In the octopus, fluid enters the renal sac via the renopericardial canal from the branchial hearts and the branchial heart appendages and from the renal appendages. The latter are a series of bud-like structures associated with the vena cava and abdominal veins.

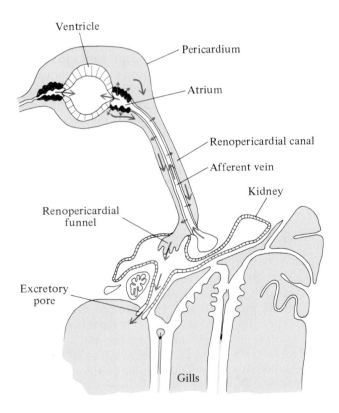

FIGURE 17–15   The excretory stucture of the bivalve mollusk *Mytilus* consists of a pericardial cavity surrounding the atria and ventricle. Pericardial fluid drains via the renopericardial canal and renopericardial funnel into the kidney and then exits via the excretory pore. *(Modified from Pirie and George 1979.)*

There is virtually no doubt that primary urine is formed by ultrafiltration in higher mollusks (lower mollusks have nephrostomes), but the anatomical site for ultrafiltration is not as clear. Experimental evidence suggests that the remnant of the coelomic nephrostomes in higher mollusks (i.e., the pericardium) is the site of filtration, and the primary urine flows via the renopericardial canal to the kidney for subsequent modification and elimination. Filtration is due to a balance of hydrostatic and colloid forces. Arterial blood pressure is generally high enough compared to blood colloid osmotic pressure that filtration occurs, e.g., 0.59 and 0.04 kPa, respectively, in *Anodonta* and 0.78 and 0.04 in *Lymnaea* (Picken 1937). Increasing the blood pressure also increases formation of pericardial fluid in *Viviparus*, as expected if filtration is due to the balance of hydrostatic and colloid forces. Formation of pericardial fluid is prevented in *Octopus* if the blood pressure is reduced to equal the colloid osmotic pressure. Urine flow can be reduced in the giant snail *Achatina* by back pressure in the ureter, which would decrease the filtration pressure differential.

In pulmonate gastropods, the filtration site appears to be not the pericardium but the kidney itself. Primary filtration may occur across the basement membrane of the kidney epithelium from the hemolymph or across capillaries that ramify through the connective tissue overlying the kidney. The filtration rate varies dramatically for different mollusks, from 0.044 ml kg^{-1} min^{-1} for *Octopus*, 0.19 for *Haliotis*, 2.0 for *Achatina*, to 3.3 for *Anodonta* (Martin, Stewart, and Harison 1965). The general trend is from low filtration rates in marine mollusks to high rates in freshwater mollusks, with terrestrial mollusks somewhat intermediate. For comparison, values for mammals are about 2 to 6 ml kg^{-1} min^{-1}. The fraction of fluid filtered relative to the tubular perfusion varies from 0.4% for *Octopus* to 35.9% for *Achatina*; values for mammals are about 30%.

The primary filtrate has similar osmotic, Na$^+$ and Cl$^-$ concentrations to blood; there is not a marked Donnan effect (i.e., the Na$^+$ and Cl$^-$ of the filtrate are not lower than the blood) suggesting that protein may be present in the filtrate. However, the Donnan effect causes only a minimal decrease in Na$^+$ (by 3%) and Cl$^-$ (by 2%) concentrations despite the low filtration of protein by nephrons of mammalian kidneys (see below). The final urine of mollusks invariably differs from blood, indicating considerable postfiltration reabsorption and/or secretion. The urine is isoosmotic for marine mollusks but varies for freshwater and terrestrial species from isoosmotic (e.g., *Achatina*, *Theodoxus*) to markedly hypoosmotic (e.g., *Viviparus*, *Helix*). Ion reabsorption occurs in tubules of the latter mollusks but not in the marine mollusks. Other solutes, such as glucose, are also reabsorbed. Some ions (i.e., SO_4^{2-}, H^+) appear to be secreted into the urine, as are many organic acids (e.g., PAH, phenol red), and NH_3 is trapped as NH_4^+ in the urine (see below, Ammonotely). Water can be reabsorbed by molluskan kidneys, and *U/H* ratios for inulin can exceed 2, indicating >50% reabsorption of water.

Arthropods

There is considerable variety among arthropods in their excretory organs. Coelomoduct renal organs are present in primitive arthropods but have been secondarily lost in higher insects. The coelomoduct organs (antennal, maxillary, labial, and coxal glands) have a terminal coelomic sac (perhaps representing the remnant of the original coelom) and an excretory tubule that is simple in noncrustaceans and lower crustaceans but is extensively folded in higher crustaceans. Gut-derived excretory structures may also be present; the best known are the Malpighian tubules of insects, but various other "digestive glands and tubules" and "gut appendages" may function for excretion.

Crustaceans have coelomoduct-type **antennal glands** and **maxillary glands** and also gut-derived rectal and cephalic glands. Maxillary glands are usually retained in some malacostracans and entomostracans and the antennal gland is retained in amphipods, mysidaceans, euphausids, and decapods (rarely are both present, e.g., in ostracods). The antennal gland of the crayfish has a terminal coelomosac, a labyrinth, a nephridial tubule (proximal, proximal-distal, and distal-distal segments) and a bladder terminating at the nephridiopore (Figure 17–16). The cells of the coelomosac (called podocytes) rest on a basement membrane with a series of interdigitating "foot" structures (pedicells); this resembles the arrangement of the glomerular filtration structure in vertebrate nephrons. The coelomosac receives an arterial blood supply and its structure suggests a filtration mechanism for formation of primary urine (the coelomosac of *Artemia*, however, does not have an arterial blood supply). Measurements of hemocoelar pressure and blood colloid osmotic pressure indicate the possibility of hydrostatic filtration, and the elimination of inulin and also dextrans and small proteins suggests a filtration mechanism rather than secretion. Cells of the coelomosac and labyrinth appear to produce formed bodies, and these may provide additional filtration. The primary filtrate of crustacean antennal glands is very similar to hemolymph but lacks high-molecular

weight hemolymph proteins and hemocytes. The *U/H* ratio for inulin is generally greater than 1, indicating some reabsorption of water. There is clearly considerable reabsorption by the renal tubule of solutes and water. Na^+ and Cl^- are about 95 to 99% reabsorbed, particularly after the proximal tubule. Virtually all glucose is reabsorbed, particularly in the bladder. Water is reabsorbed primarily along the labyrinth and proximal tubule.

The principal excretory organs of arachnids are **coxal glands** (which are related embryologically to antennal glands of crustaceans) or **Malpighian tubules** or both (Figure 17–17A). Coxal glands are thin-walled, spherical sacs that collect wastes from the hemolymph and eliminate urine through a duct opening onto the coxa. Primitive spiders have two pairs of well-developed coxal glands opening onto the first and third pairs of legs. More advanced spiders have only the anterior pair, present in varying degrees of regression. The two branched Malpighian tubules of spiders have largely replaced the coxal glands for excretion. Scorpions have a single pair of coxal glands opening onto the third coxa and two pairs of Malpighian tubules. Mites have 1 to 4 pairs of coxal glands or a pair of Malpighian tubules or both. Trombidiform mites lack Malpighian tubules and coxal glands; their hindgut is modified for excretion. Most insects have a variable number of Malpighian tubules (from 2 to about 250), but collembolans and aphids lack them. Argasid ticks have coxal glands for rapid elimination of ingested fluids, but ixodid ticks lack coxal glands.

The coxal gland has a thin-walled saccule, to which are attached coxal gland muscles, and a tubular organ with two discrete segments (Figure 17–17B). Hemolymph is filtered through the thin saccule wall, probably by a hydrostatic pressure difference due to contraction of the coxal muscles. Contraction of these muscles would establish a positive pressure outside the sacculus and a negative pressure inside the sacculus. Relaxation of the muscles probably forces fluid along the excretory tubule. Large molecules, such as casein and albumin, are not filtered, although hemoglobin is. The excretory tubule is thick-walled, has a peculiar internal chitin skeleton, and is highly tracheate; it probably reabsorbs solutes and water.

The coxal glands of the marine horseshoe crab *Limulus* excrete a fluid isoosmotic to hemolymph in normal seawater, but their secretion becomes increasingly hypoosmotic to hemolymph at lower salinities (Towles et al. 1982). The coxal gland secretion of various terrestrial arachnids can be hypoosmotic, isoosmotic, or hyperosmotic. In the tick *Argas arboreus*, most ions are slightly hyper-

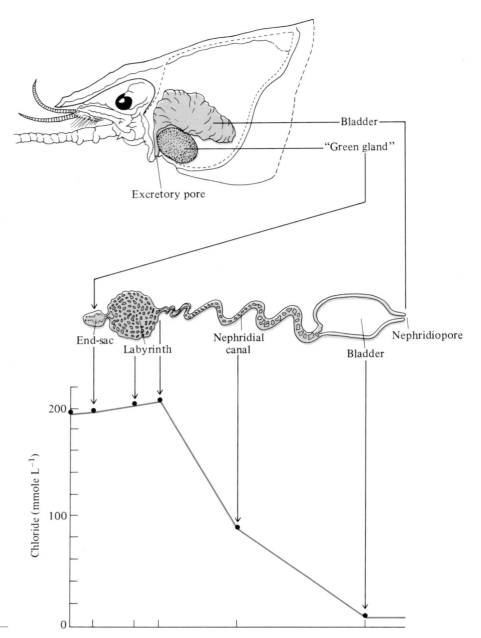

FIGURE 17–16 Location and structure of the crustacean antennal gland ("green gland") in the crayfish *Astacus,* and the chloride concentration in various segments showing the progressive ionic dilution of urine along the tubular excretory organ. *(Modified from Potts and Parry 1964; Peters 1935.)*

concentrated, whereas most amino acids are markedly hypoconcentrated, indicating selective tubular reabsorption. In ixodid ticks, which lack coxal glands, the ionic concentrations of the saliva are greater than plasma and amino acid concentrations are lower. Ticks also have Malpighian tubules that function for excretion of, for example, nitrogenous wastes. The coxal glands of orbatid mites have a sacculus and a complex tubular labyrinth. Freshwater species have a long labyrinth for maximal ion reabsorption, whereas marine and terrestrial species

have a shorter labyrinth, presumably reflecting a lesser requirement for ion reabsorption (Woodring 1973).

The coelomoduct kidneys of collembolans have an osmo- and ionoregulatory role, particularly in the xeric species (Verhoef and Prast 1989). The more xeric *Orchesella* ionoregulates during desiccation by both ion excretion and storage but is unable to osmoregulate. When hydrated, the U/H ratios are <1 for osmotic concentration (0.35), ionic concentration (0.27), and organic solutes (0.55), but during

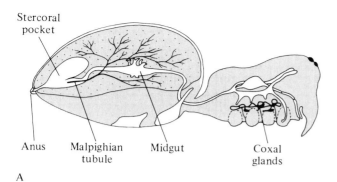

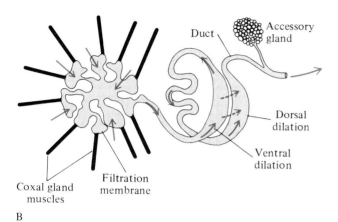

FIGURE 17–17 **(A)** Schematic of the gut and excretory systems of a mygalomorph spider. **(B)** Schematic of the coxal gland of an argassid tick *Ornithodurus*. Coxal gland muscles attached to the thin saccular membrane promote filtration (by contracting) and fluid flow (by relaxing) along the complex tubular duct, which reabsorbs solutes and water. The saccular filtration membrane actually surrounds the tubular segments *(From Butt and Taylor 1986; Berridge 1970; after Lees 1946.)*

dehydration the *U/H* ratio for osmotic concentration increases to 1 and the urine flow rate declines.

The primary excretory organs of insects are the Malpighian tubules and, as we shall see, the rectum. The blind-ended Malpighian tubules connect to the mid-hindgut junction. The primary urine is produced in the Malpighian tubules by active K^+ transport into the tubule lumen, accompanied by passive movement of Cl^-, other solutes, and water. The process of filtration and also subsequent modification of the primary urine by Malpighian tubules has been extensively studied by an ingenious *in vitro* preparation consisting of an isolated tubule immersed in paraffin (Figure 17–18). The isolated tubule continues to secrete urine that collects as a

drop at the cut end of the tubule. The fluid secreted by the tubule is always isoosmotic with the hemolymph (or bathing medium *in vitro*) and the rate of tubular secretion is inversely related to the medium osmotic concentration. The presence of solutes other than K^+, Na^+, and Cl^- is inversely proportional to their molecular weight, indicating passive movement through a paracellular filtration pathway. The *U/H* ratios of the primary urine vary from about 1 (e.g., Cl^-, urea, hexose sugars), to about 0.1, to 0.2 (Na^+, Ca^{2+}, Mg^{2+}, amino acids, hexose sugars), to very low values indicating nonfiltration or reabsorption (inulin, 0.007 to 0.05; disaccharides, 0.02 to 0.03).

The primary urine is isoosmotic to hemolymph and remains so within the Malpighian tubules, at least in most insects. Some insects (e.g., *Rhodnius*, *Calpodes*) have reabsorptive tubule segments, and the final tubular urine can be 30% hypoosmotic to hemolymph due to active reabsorption of K^+ and Cl^- without commensurate reabsorption of water.

The secretion-filtration mechanism for primary urine formation by insects is apparently less effective than hydrostatic-filtration mechanisms since the rate of primary urine formation is generally low for insect Malpighian tubules (0.1 to 13 nl mm^{-2} min^{-1}) compared with, for example, bird and mammal nephrons (10 to 1000 nl mm^{-2} min^{-1}; Phillips 1981). However, insect Malpighian tubules generally have a high surface area for secretion (2 to 10 cm^2 g^{-1}) compared with vertebrate nephrons (0.1 to 0.5 cm^2 g^{-1}) and this partially compensates for their lower area-specific rate of secretion. Consequently, the maximal rate of urine formation can be similar in insects (100 ml kg^{-1} min^{-1}) and, especially, in blood-sucking insects (200 to 600), compared with birds and mammals (10 to 500).

The Malpighian tubules may be functionally and morphologically divided into different segments. In some insects, the distal ends of the Malpighian tubules are adpressed to the rectum to form the cryptonephridial system. In *Bombyx*, there are two regions: the distal cryptonephridial segment absorbs water from the hindgut and the proximal segment excretes calcium oxalate. In *Rhodnius*, the distal segment forms the primary urine and the proximal segment precipitates uric acid. The alkali fly *Ephydra* lives in water with a high $Ca(HCO_3)_2$ and can excrete almost pure $CaCO_3$; its anterior Malpighian tubules are modified into enlarged glands, which precipitate $CaCO_3$ for storage or excretion (Herbst and Bradley 1989).

The Malpighian tubules open into the gut at the junction of the midgut and hindgut or in the rectum.

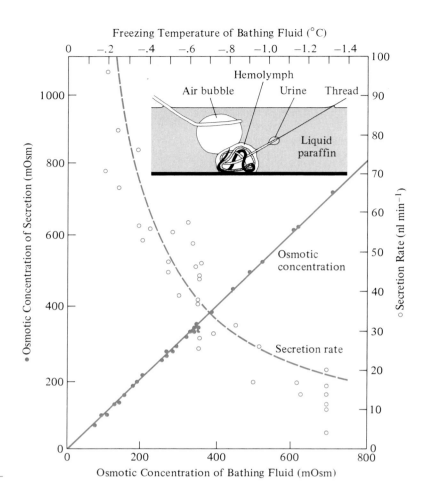

FIGURE 17–18 Effect of osmotic concentration of the bathing medium on the rate and osmotic concentration of Malpighian tubule secretion. *(Modified from Maddrell 1969; Ramsay 1954.)*

There may be further modification of the urine in the hindgut (Figure 17–19). The anterior hindgut, posterior to the midgut, is relatively narrow and is lined by cuticle so the passage of urine through it would be rapid and there would be little opportunity for modification of the urine composition. However, there is some modification of urine composition in this ileal portion of the hindgut in a variety of insects (*Schistocerca*, *Periplaneta*, *Pieris*, *Ephydrella*). For example, the urine volume may be reduced by about 10% in the midgut of stick insects, butterfly, and locust, and osmolality and K^+ concentration may be slightly reduced. An aquatic bug *Cenocorixida* and a plant-feeding *Cicada* have a long hindgut specialized for producing a very hypoosmotic urine. Water-loaded blowflies and cockroaches can also produce a moderately hypoosmotic urine in the hindgut. The rectum is the most important site for modification of the tubular urine by major bulk reabsorption of ions and other solutes (Na^+, K^+, Cl^-, proline and other amino acids, glucose, trehalose) and water. Toxins (e.g., dyes, cardiac glycosides) can be secreted into the rectal contents.

The excreta of insects can be markedly hyperosmotic to hemolymph, by 0.5 to 5.5 Osm for dehydrated terrestrial insects and by 0.7 to 2.9 Osm for saltwater species. Such an ion-reabsorbing, osmoconcentrating, and dehydrating role is analogous to the role of the mammalian and avian kidneys and the dehydrating/ion reabsorbing role of the reptilian and avian cloaca/hindgut.

The epithelium of the insect rectum is apparently the only "simple" epithelium that is capable of absorbing a hypoosmotic fluid, leaving the rectal contents more hyperosmotic (Phillips 1981). However, the structure and function of the insect rectal epithelium is not "simple." There is no definite evidence in any epithelium for active transport of water, and so hypoosmotic water uptake by the insect rectum must be explained as a solute-linked process. The absorption of a hypoosmotic fluid by the rectal epithelium of certain insects is thought to occur by local osmosis coupled to solute reabsorption (see Figure 17–6B, page 837). The mechanism for rectal reabsorption in many insects (e.g., locusts, cockroach, blowfly) involves specialized folds of

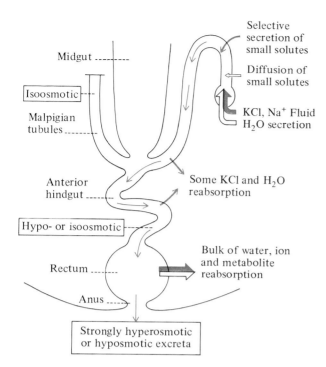

Midgut

Isoosmotic

Malpigian
tubules

Selective
secretion of
small solutes

Diffusion of
small solutes

KCl, Na⁺ Fluid
H₂O secretion

Anterior
hindgut

Some KCl and H₂O
reabsorption

Hypo- or isoosmotic

Rectum

Bulk of water, ion
and metabolite
reabsorption

Anus

Strongly hyperosmotic
or hyposmotic excreta

FIGURE 17–19 Representation of the excretory system of insects showing the formation of urine and some modification of the filtrate by Malpighian tubules, anterior hindgut modification, and extensive rectal modification of the urine. *(From Phillips 1981.)*

the rectal epithelium, the "rectal pads." A model of water uptake by rectal pads, based on the three-compartment and the standing-gradient models, has been developed from ultrastructural studies of the cockroach (*Periplaneta*) rectum (Wall 1970, 1971). The papillate rectum absorbs fluid by osmosis in response to active solute transport (e.g., Na⁺, Cl⁻, K⁺) from the rectal lumen into epithelial cells and then into intercellular spaces between the epithelial cells. The absorbed solutes and water move along the intercellular channels towards the hemocoel, and ions are reabsorbed into the epithelial cells for subsequent return to the intercellular spaces and for further osmotic absorption of water from the lumen (Gupta et al. 1980). Direct measurement of the ion concentrations in the rectal epithelium of the blowfly supports this general model. The fluid in the lateral intercellular sinuses near the lumen is hyperosmotic (by up to 700 mM) in dehydrated blowflies, but the total ion concentration declines markedly along the intercellular channel towards the hemocoel (Figure 17–20).

Brackish water mosquitos, e.g., *Aedes campestris*, can excrete urine that is hyperosmotic (1.3

M NaCl) to body fluids (0.14 M NaCl). Beadle (1939) demonstrated that the rectum formed the hyperosmotic urine, by ligating larvae at various positions along the body. Fluid entering the rectum is iso-osmotic to hemolymph (as in most insects) but is considerably hyperosmotic when it leaves the rectum. The rectum of these saltwater mosquito larvae has an entirely different mechanism for formation of a hyperosmotic urine than other insects. The rectum is divided into two distinct parts: the anterior rectal epithelium resembles that of freshwater mosquito larvae, but the posterior epithelium is highly modified for active ion secretion into the rectal lumen. Thick cells in the posterior portion in the rectum of *A. campestris* have a highly folded apical membrane and intercellular channels, which are narrow and straight. This structure is not like, for example, the rectal epithelium of the blowfly rectal pad and is unsuited for solute-linked water reabsorption. Active transport of Na⁺, K⁺, Mg²⁺, Cl⁻, and HCO₃⁻ into the rectum makes the excreta hyperosmotic. This is an effective strategy for osmoregulation in these aquatic larvae but would not be suitable for a terrestrial insect that must retain water.

Cephalochordates and Urochordates

The cephalochordate *Branchiostoma* has about 90 pairs of protonephridia in its pharyngeal region, each associated with a pair of gill slits. Each protonephridium has an upper and lower segment extending from the nephridiopore, and numerous solenocytes branch from each segment. The nephridiopores open into the atrium at the dorsal edge of the gill slit. There is also a single enlarged nephridium of Hatschek located above the dorsal blood vessel near the roof of the pharynx, which opens into the pharynx just behind the velum.

Urochordates generally excrete their nitrogenous wastes as ammonia by diffusion across their body surface. A few excrete considerable amounts of urea. They also store nitrogenous wastes as urate precipitates in the gut, mantle, gonads, atrium wall, and elsewhere. Nephrocytes circulate in the blood and accumulate urates and then become fixed in tissues and grow into large multinucleate cells with urate-containing vacuoles.

Vertebrates

The basic tubular excretory unit of the vertebrate kidney is the kidney tubule, or **nephron**. The ancestry of the vertebrate nephron is not apparent from a study of cephalochordates (e.g., *Branchiostoma*)

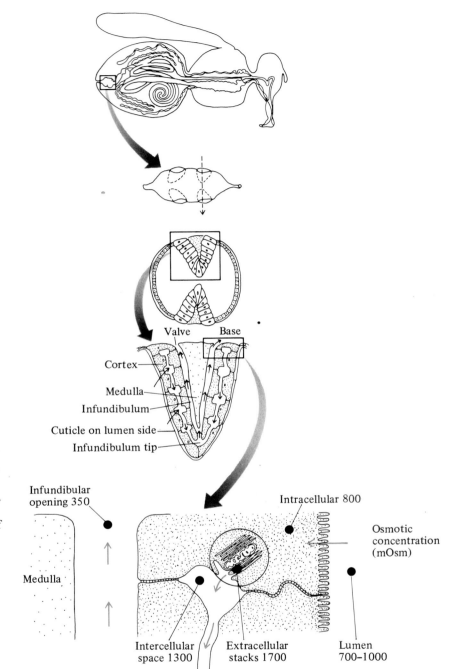

FIGURE 17–20 Representation of the organization of the rectal structures of the blowfly *Calliphora* showing the location of the rectum, the rectal papillae projecting into the rectal lumen, and the pattern of water and solute flux through the rectal pad. Also shown are the organization of the epithelial cells and their intercellular spaces and the osmotic concentrations of the lumen, the membrane stack, the intercellular space, and the infundibular space. (*Modified from Gupta et al. 1980; Maddrell 1971.*)

whose protonephridia are not homologous with vertebrate nephrons that are derived from coelomoducts.

The vertebrate nephron is derived from a coelomoduct draining the coelomic cavity (Figure 17–21). Coelomic fluid was probably formed by a glomerulus-like arrangement of capillaries near the coelomic wall but not at the coelomostome. The "glomerulus" then became associated with the tubule, as a more direct means of forming primary

urine, and the glomerular blood supply then contributed to the tubular capillary network; the coelomostome remained as an adjunct for urine formation and to drain coelomic fluid into the nephron. The role of primary urine production was then "usurped" by the glomerulus and the coelomostome was lost. In mammals, the glomerular efferent arteriole is the only blood supply to the peritubular vasculature. This evolutionary progression in nephron structure

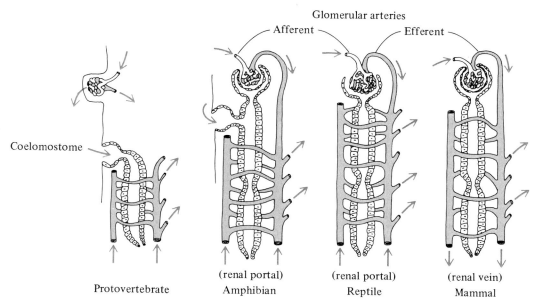

Glomerular arteries
Afferent   Efferent

Coelomostome

Protovertebrate   (renal portal)   (renal portal)   (renal vein)
                   Amphibian   Reptile   Mammal

FIGURE 17–21 Highly schematic scenario for the evolution of the vertebrate nephron from a coelomoduct. In protovertebrates, fluid is secreted into the coelom by a capillary network distant from the coelomostome opening; the renal tubule has a separate capillary supply. In vertebrates such as amphibians, there is a glomerular capillary network associated with the renal tubule and the coelomostome persists; the glomerular network also supplies blood to the renal tubule in association with a separate renal portal supply. In reptiles, birds, and mammals, the renal tubule lacks a coelomostome; in mammals, there is only a glomerular blood supply to the renal tubule (i.e., no renal portal system). *(Modified from Smith 1959.)*

is illustrated by (1) the pronephric anterior kidney of the hagfish, which has large glomeruli, a Bowman's capsule which communicates with the ureter and cardinal vein, and numerous coelomostomes; (2) the amphibian kidney, which has nephrons with both glomeruli and coelomostomes; (3) the reptilian nephron which lacks the coelomostome; and (4) the mammalian nephron, which lacks a coelomostome and does not have a renal portal blood supply to the peritubular capillaries (see below).

The vertebrate nephron typically has a renal (Malpighian) corpuscle and a highly convoluted tubule. The renal corpuscle consists of two parts: a cluster of small capillaries, the **glomerulus**, and an enveloping **Bowman's capsule**, which is essentially the blind end of the tubule invaginated to accommodate the glomerulus (Figure 17–22). The outer layer of Bowman's capsule is continuous with the endothelium of the **proximal convoluted tubule** and the intracapsular space is contiguous with the lumen of the proximal tubule. The inner layer of Bowman's capsule is highly modified **podocyte** cells that surround and support the glomerular capillaries. Each podocyte has numerous processes that extend from

the cell body and form pedicels. These pedicels interdigitate with those of adjacent podocytes to form small, slit-like spaces that cover the capillary walls. The glomerular capillaries are highly fenestrated. The capillary fenestra (pores) and the podocyte interpedicel slits form an effective filtration pore, and the basement membrane of the capillary endothelium is the filtration membrane. The renal convoluted tubule is specialized for reabsorption and secretion and for osmoconcentration of the urine in birds and mammals. The renal tubules often empty into a bladder via the ureter and urine is temporarily stored until it is voided. In some vertebrates, the bladder urine serves as a water reservoir; water is reabsorbed to replace body water lost during dehydration.

There is a marked progression in the overall organization of the kidney and the embryologic origins of the nephrons that is associated with an evolutionary progression in nephron structure. The embryological origin of the kidney is segmental trunk nephrotomes, each forming a pair of nephrons of the **pronephric kidney** (or "head kidney"). Each pronephron drains into a lateral, longitudinal duct,

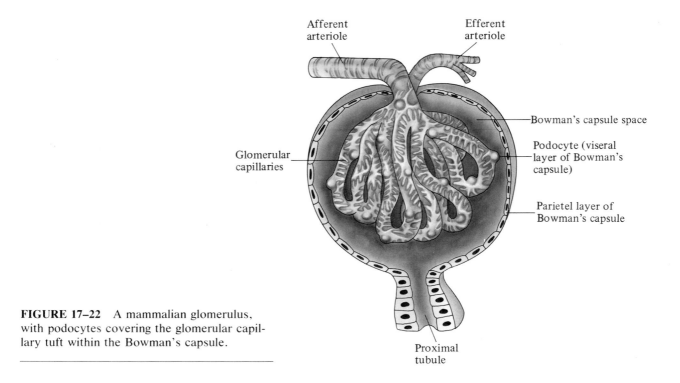

Afferent arteriole

Efferent arteriole

Bowman's capsule space

Podocyte (viseral layer of Bowman's capsule)

Parietel layer of Bowman's capsule

Glomerular capillaries

Proximal tubule

FIGURE 17–22 A mammalian glomerulus, with podocytes covering the glomerular capillary tuft within the Bowman's capsule.

the archinephric duct; these two lateral ducts often fuse before draining into the cloaca. The head kidney persists in many larval fish and amphibians. The more anterior nephrons of the pronephros degenerate, and the posterior nephrons develop to form the **opisthonephric kidney** ("back kidney"), the kidney of adult vertebrates. The pronephric tubules develop only rudimentarily and degenerate at an early stage. The next (more posterior) nephrons form the **mesonephric kidney**. The most posterior nephrons form the **metanephric kidney**. Both mesonephros and metanephros persist in anamniote vertebrates (fish, amphibians) although there is a general trend for a more major role of the metanephric tubules. In amniotes (reptiles, birds, mammals) the mesonephros functions for a considerable part of the embryonic life (and early neonatal life in some reptiles) but it and the archinephric duct are modified to form parts of the male reproductive tract and the metanephric tubules function in excretion. The metanephric tubules drain into the cloaca by an evolutionarily "new" duct, the ureter.

There is, not surprisingly, a considerable diversity in the structure and function of the adult vertebrate kidney, reflecting its complex embryologic and phylogenetic development, as well as the variation in iono- and osmoregulatory function demanded by different environments.

Fish.   The fish nephron typically has a glomerulus and a Bowman's capsule with a ciliated neck region

connecting to the rest of the nephron tubule. The cilia may assist fluid flow along the nephron, since fish typically have low filtration pressures. The neck region is also present in amphibian nephrons, but not in amniote nephrons.

In the hagfish, the neck region connects to the archinephric duct (or first proximal segment, PS I), where glucose, Ca^{2+} and macromolecules are reabsorbed, and K^+, Mg^{2+}, and SO_4^{2-} are secreted. About 0.3 ml kg^{-1} hr^{-1} of plasma (about 980 mOsm) is filtered, and the same volume of isoosmotic urine is voided. Freshwater lampreys have nephrons with large glomeruli. Their filtration rate is much higher (11 ml kg^{-1} hr^{-1}) and their urine flow is copious (6 ml kg^{-1} hr^{-1}) and hypoosmotic (20 mOsm) compared to body fluids (250 mOsm).

In freshwater teleosts, the proximal tubule is divided into two segments (Figure 17–23A). The **first proximal segment** (PS I) may reabsorb filtered protein by pinocytosis; reabsorb glucose and amino acids, Na^+, and Cl^-; and secrete divalent cations and urea. The **second proximal segment** (PS II) is the largest segment of the nephron and may have a reabsorptive and secretory role. The **intermediate segment** (IS) is a highly ciliated region (present only in nephrons of some freshwater teleosts) that may rapidly transport fluid along the nephron to minimize fluid reabsorption. The **distal segment** (DS), which is also present in nephrons of some euryhaline teleosts, elasmobranchs, and lungfish, probably actively reabsorbs Na^+ without passive water uptake,

Freshwater Teleost

Plasma Ultrafiltrate
4 ml kg^{-1} h^{-1}, 280 mOsm

Marine Teleost

Plasma Ultrafiltrate
0.5 ml kg^{-1} h^{-1}, 450 mOsm

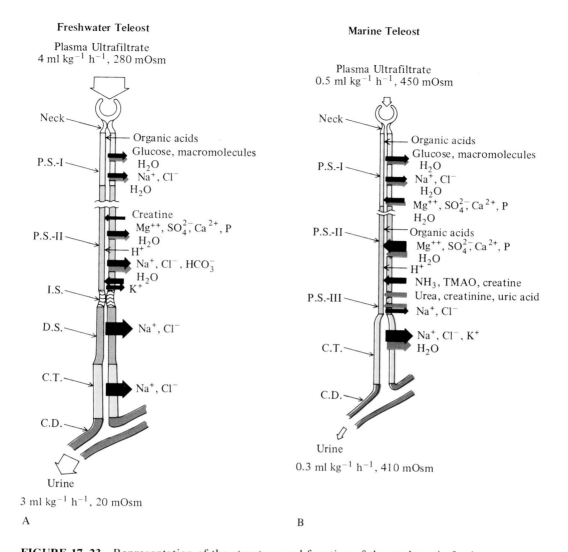

FIGURE 17–23 Representation of the structure and function of the nephron in freshwater teleosts (**A**) and marine teleosts (**B**). Indicated are the major morphological segments of the nephron, the plasma filtration rate and osmotic concentration, the tubular reabsorptive and secretive processes, and the urine flow rate and osmolality. *(From Hickman and Trump 1969.)*

i.e., it produces a hypoosmotic urine. The **collecting tubules** (CT) and **collecting ducts** (CD) reabsorb ions from the urine, also making the urine hypoosmotic. Nephrons of freshwater teleosts have a high filtration rate (4 ml kg^{-1} hr^{-1}) and urine flow rate and reabsorb ions to produce a dilute urine.

The nephrons of marine teleosts are considerably different in structure and function (Figure 7–23B). Many marine teleosts have nephrons with glomeruli that are small, markedly reduced (pauciglomerular), or even absent (aglomerular). The filtration rate is generally low (about 0.5 ml kg^{-1} hr^{-1}) as is the urine flow rate (0.3 ml kg^{-1} hr^{-1}). Aglomerular kidneys have no filtration and have a low urine flow rate (0.3 ml kg^{-1} hr^{-1}) that is formed by solute

secretion and passive water cotransport; the urine (about 400 mOsm) is nearly isoosmotic with blood (450 mOsm). Many ions, particularly the divalent cations, are actively secreted into the urine. The nephrons of marine elasmobranchs actively reabsorb many solutes, including their important osmolytes, urea, and trimethylamine oxide (TMAO).

Hagfish urine is isoosmotic to plasma with respect to Na$^+$ and Cl$^-$ and is iso-osmolar (Table 17–2). It has a higher K$^+$, Mg^{2+}, SO$_4^{2-}$, and urea concentration than plasma. The urine of elasmobranchs is about iso-ionic with plasma with respect to Na$^+$ and Cl$^-$ and is markedly hyperionic with respect to Mg^{2+} and SO$_4^{2-}$; it is slightly hyperosmotic to plasma. Urea and TMAO excretion via the

TABLE 17–2

Urine composition and urine/plasma (*U/P*) ratio for various marine fish, a hagfish *(Eptatretus)*, elasmobranch *(Squalus)*, coelacanth *(Latimeria)*, and a teleost *(Lophius)*. Values are mM or mOsm. *(Data from Griffith and Pang 1979; Brull and Nizet 1953.)*

	Hagfish	**Elasmobranch**	**Coelacanth**	**Teleost**
Na$^+$	553	240	184	11
U/P	1.0	1.0	0.9	0.1
K$^+$	11	2	9	2
U/P	1.6	0.5	1.5	0.4
Mg^{2+}	15	40	30	137
U/P	1.2	33.3	5.7	54.5
Cl$^-$	548	240	15	132
U/P	1.0	1.0	0.1	0.9
SO$_4^{2-}$	7	70	104	42
U/P	8.6	140	21.6	35
Urea	9	100	384	0.6
U/P	1.8	0.3	1.0	1.9
TMAO	—	10	94	13
U/P	—	0.14	0.77	2.5
Osmol	1051	800	932	406
U/P	1.0	0.8	1.0	0.9

TABLE 17–3

Effects of dehydration on glomerular filtration rate (GFR; ml kg^{-1} hr^{-1}) and urine flow rate (UFR; ml kg^{-1} hr^{-1}) for two amphibians, the marine toad *Bufo marinus* and the waterproof tree frog *Phyllomedusa sauvagei*. *(Data from Shoemaker and Bickler 1979.)*

	Control	**Dehydrated**
Bufo		
GFR	63.1	1.8
UFR	33.7	0.09
% reabsorption	45.1%	94.2%
Phyllomedusa		
GFR	92.3	27.1
UFR	43.9	1.9
% reabsorption	51.2%	95.4%

urine are minimized by active tubular reabsorption. For example, the dogfish *Squalus* reabsorbs about 90 to 95% of the filtered urea and 95 to 98% of the filtered TMAO. Urea retention by the nephron involves countercurrent exchange in the kidney and is also linked to Na$^+$ reabsorption; the *U/P* ratio for urea may vary from 0.07 to 0.89 (average = 0.3). Active urea reabsorption is apparently limited by the concentration difference that can be maintained across the tubular epithelium, not by a maximal tubular uptake rate. Urine is hypo-ionic to plasma with respect to Na$^+$ and Cl$^-$; hyper-ionic with respect to K$^+$, Mg^{2+}, and SO$_4^{2-}$; but has the same osmotic concentration as plasma. The coelacanth's urine contains similar levels of both urea and TMAO as the plasma; the urea filtered in the kidney by the glomerulus is apparently not reabsorbed by the renal tubule.

Amphibians. The mesonephric kidney of adult amphibians has two types of nephrons. The more ventral nephrons have a glomerulus and also a nephrostome, which collects fluid from the coelomic cavity; the more dorsal nephrons have only a glomerulus for urine formation (Figure 17–24A). The

nephron has a ciliated neck portion, a proximal tubule, a second ciliated segment (the narrow segment), and then a distal tubule that connects to the archinephric duct. In some frogs, the coelomoducts connect to the renal venous circulation rather than to the nephrons; this may facilitate the flow into the blood of water that is absorbed across the skin and transported into the coelom.

The amphibian nephron generally resembles that of freshwater teleosts in function, i.e., it produces a copious, dilute urine. Amphibians typically have a high glomerular filtration rate of 25 to 100 ml kg^{-1} hr^{-1} and a correspondingly high urine flow rate of 10 to 25 ml kg^{-1} hr^{-1} (Shoemaker and Nagy 1977). Thus, about one half of the primary filtrate is reabsorbed and one half reaches the urinary bladder. In marked contrast, about 99% of the filtered ions (e.g., Na$^+$ and Cl$^-$) are reabsorbed. The proximal convoluted tubule reabsorbs solutes (glucose, Na$^+$, Cl$^-$) and some water (*U/P* for inulin is about 1.3). The distal convoluted tubule reabsorbs salts. Consequently, the urine that drains into the bladder is considerably more dilute than the primary filtrate (Figure 17–24B).

The regulation of GFR and tubular reabsorption of Na$^+$ and water by amphibian nephrons is dramatic. Some anuran amphibians can become essentially anuric after losing a small percentage of their body water and glomerular filtration almost stops (e.g., *Bufo*; Table 17–3). Other amphibians have a dramatic, but not as complete, decline in GFR with dehydration, e.g., *Phyllomedusa*. Dehydration causes the release of arginine vasotocin, which has both glomerular and tubular effects; it controls renal aspects of the water balance response as well

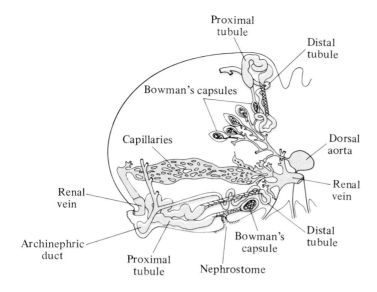

A

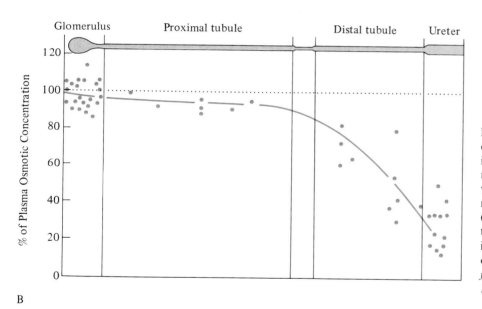

B

FIGURE 17–24 **(A)** Structure of the amphibian kidney showing a ventral coelomostome nephron and a dorsal nephron without a coelomostome; both nephrons have a glomerulus. **(B)** The total solute concentration progressively declines as it passes along the renal tubule of an amphibian. *(Modified from Noble 1931; Walker et al. 1937.)*

as other aspects of water economy (e.g., bladder reabsorption of water).

Reptiles and Birds. The nephrons of reptiles and birds are generally adapted to minimize urinary water loss and to excrete solutes (nitrogenous wastes and ions).

The reptile kidney is bilobed (Figure 17–25); the nephrons drain into ureters that empty into the large intestine. Each lobe receives an arterial blood supply from the aorta and has a venous drainage into the vena cava. In addition, each lobe receives blood

from the renal portal vein. The nephrons, like those of birds and mammals, lack coelomostomes. The reptilian glomerulus produces a primary filtrate, which passes through a convoluted tubule, an intermediary segment, a distal tubule and a connecting segment to the collecting duct and then passes into the ureter.

The avian kidney has a more complex structure; it is trilobed (cervical, middle and caudal lobes), each lobe having smaller lobules (Figure 17–26). Each lobule has a peripheral and a cortical mass of tissue and a central medullary cone that projects

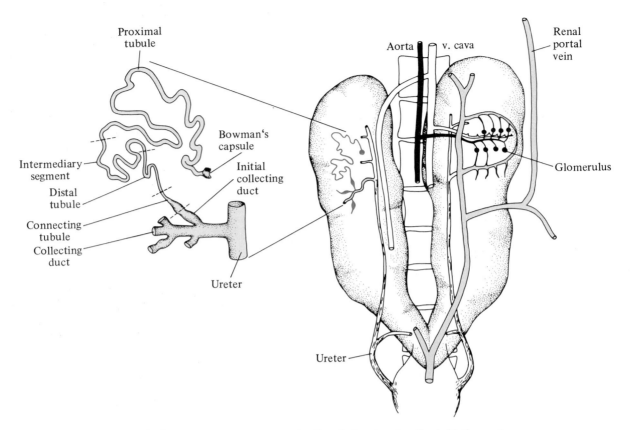

FIGURE 17–25 General structure of the kidney of a lizard showing (on the left) the ureter with one collecting duct, associated collecting tubules, and one complete nephron and (on the right) the renal and renal portal blood supply to the glomeruli and peritubular capillary bed. The detailed structure of a typical nephron is also shown. *(From Davis, Schmidt-Nielsen, and Stolte 1976.)*

into the ureteral branch draining into the ureter. The kidney cortex contains the glomeruli. Avian nephrons are functionally more specialized than those of reptiles. There are two different types of nephrons, those resembling normal reptilian nephrons (i.e., reptilian-type nephrons) and those with medullary projections of a loop of Henle (i.e., mammalian-type nephrons). The loop of Henle allows the nephron to produce a hyperosmotic urine (e.g., U/P = 1.5–6). A **juxtaglomerular apparatus** is present, with a rudimentary **macula densa**; the possible roles of the JG-apparatus in regulation of glomerulo-tubular balance are not clear (see below, for the role of the JG-apparatus in mammalian kidneys).

The kidneys of reptiles and birds have a lower glomerular filtration rate than amphibians, about 0.5 to 20 ml kg^{-1} hr^{-1} for reptiles and 1 to 4 ml kg^{-1} hr^{-1} for birds. There is some adaptive variation in

GFR by reptiles and birds, for example to dehydration, but this is not as dramatic as in amphibian nephrons. For example, the GFR of a varanid lizard is decreased 31% during dehydration, and 65% by salt loading (Bradshaw and Rice 1981). Renal control of urine production (urine flow from the kidneys, or ureteral flow) is also accomplished by increased tubular reabsorption of water and solutes. The UFR of a dehydrated varanid lizard is decreased to 27% of the control value by a 31% decline in GFR and increased tubular reabsorption of water. With salt loading, the UFR decreases to 12% of the control value due to a 65% decline in GFR and increased tubular reabsorption of water but not salts. The GFR of many birds decreases with dehydration or salt loading. For example, a decrease in GFR and UFR and increase in tubular reabsorption are observed in sparrows during dehydration (Goldstein and Braun 1988).

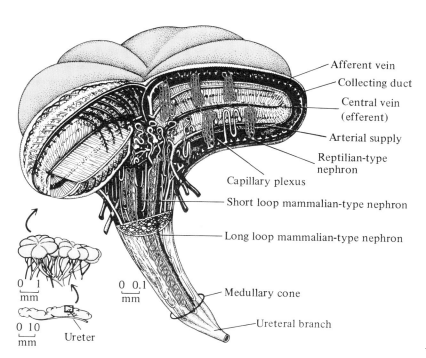

Afferent vein
Collecting duct
Central vein (efferent)
Arterial supply
Reptilian-type nephron
Capillary plexus
Short loop mammalian-type nephron
Long loop mammalian-type nephron
Medullary cone
Ureteral branch
Ureter

FIGURE 17–26 Schematized structure of the avian kidney showing its organization into kidney lobules, each containing a complex organization of two types of nephrons: the short loop ("reptilian type") and the long loop ("mammalian type"). *(From Braun and Dantzler 1972.)*

No reptile can produce a urine that is hyperosmotic to its plasma (Table 17–4). However, a lizard that inhabits salt lakes, the Lake Eyre dragon, can apparently produce a hyperosmotic excreta, although its ureteral urine is hypoosmotic to plasma (Braysher 1976). A number of birds can produce a hyperosmotic ureteral urine, from 1.5 to over 5 × plasma osmotic concentration. The house sparrow, for example, produces a slightly hypo-osmotic (325 mOsm) urine when hydrated (U/P = 0.99), whereas a dehydrated sparrow produces 826 mOsm urine (U/P = 2.24). The highest ureteral urine osmotic concentration measured for birds is 2000 mOsm for salt-loaded savannah sparrows (provided with saline drinking water); their U/P is 5.8. The significance of urine osmotic concentration to water balance is complicated in birds by the cloacal/hindgut modification of ureteral urine.

The regulation of GFR in reptiles is primarily accomplished by the number of functioning nephrons forming primary urine, rather than by the amount of primary filtrate produced per nephron; the hormone AVT is thought to be involved in regulation of nephron function. Administration of AVT reduces GFR in turtles, water snakes, crocodiles, and some lizards, presumably by vasoconstricting the afferent glomerular arteriole of certain nephrons but not others.

Tubular reabsorption of Na^+ varies considerably in some reptiles (e.g., many turtles, some lizards,

TABLE 17–4

Urine osmotic concentrations (mOsm) and urine/plasma (U/P) ratios for ureteral urine of reptiles and birds (except as indicated). Values are generally maximal ones for dehydrated animals.

	Urine	U/P
Reptiles		
Gecko	325	0.74
Crocodile	267	0.84
Iguanid	362	0.97
Horned lizard	327	0.97
Tuatara	270	0.99
Desert tortoise	337	1.0
Salt lake dragon: urine	355	0.74
excreta	671	1.5
Birds		
Emu	459	1.4
Chicken	538	1.6
Senegal dove	661	1.7
Gambels quail	669	2.0
House sparrow	826	2.2
Budgerigar	848	2.3
House finch	850	2.4
Honeyeater	925	2.4
Galah	982	2.5
Ostrich	800	2.5
Kookaburra	944	2.7
Zebra finch	1005	2.8
Savannah sparrow	2000	5.8

and snakes) but not in others (e.g., some lizards), perhaps reflecting the potential involvement of other structures (cloaca, colon, bladder, salt glands) in controlling Na^+ excretion. AVT also appears to be involved with the regulation of tubular Na^+ reabsorption since it often decreases the Na^+ clearance and free water clearance. The effect of AVT in some species is to decrease GFR, whereas in other species it promotes a tubular antidiuresis. However, the nature of the osmotic stress may influence which effect predominates. In the goanna *V. gouldii*, dehydration slightly decreases GFR and markedly decreases water clearance (i.e., predominantly a tubular effect), whereas salt loading markedly decreases GFR and also decreases free water clearance (i.e., glomerular and tubular effects). Adrenocortical hormones also play a role in regulation of tubular Na^+ balance in some reptiles and birds; corticosteroids may promote Na^+ reabsorption in at least some reptiles (Bradshaw 1986).

Water reabsorption is also an important role of renal regulation in reptiles. Reabsorption of Na^+ in the proximal tubule in accompanied by water reabsorption, such that the absorbate and lumen fluid remain isoosmotic. About 30 to 50% of the glomerular-filtered water is absorbed in the proximal tubule, coupled with Na^+ reabsorption. Na^+ and solute-linked water reabsorption in the distal tubule may vary from isoosmotic to hyperosmotic since the distal tubule is normally less permeable to water; the lumen fluid therefore remains isoosmotic or becomes hypoosmotic. The capacity for varying the osmolality of the ureteral urine varies dramatically among reptiles. Some always produce hypoosmotic urine (e.g., blue tongue skink, $U/P = 0.7$) and some produce only isoosmotic urine (e.g., desert horned lizards, $U/P = 1$), but most can vary the osmotic concentration from isoosmotic to hypoosmotic (e.g., water snake, $U/P = 0.1$ to 1.0; freshwater turtles, $U/P = 0.3$ to 1.0; tuatara, $U/P = 0.7$ to 1.0). AVT apparently controls the permeability of the distal tubule, which allows these modifications in U/P, although variation in GFR can also markedly alter the extent of Na^+ and water reabsorption (hence, U/P ratio) without changing the water permeability. In some species, cloacal or bladder modification of ureteral urine is also important; there may be hormonal regulation of cloacal/bladder reabsorption but this has not been demonstrated, except perhaps in the goanna *V. gouldii* (Braysher and Green 1970).

The glomerular filtration rate of birds is, like that of reptiles, slightly altered by the ionic/osmotic state, e.g., the GFR decreases slightly with dehydration or salt loading. For example, the GFR of the budgerigar *Melopsittacus* decreases from 266 to 194 ml kg^{-1} hr^{-1} with dehydration; the house sparrow also has a decreased GFR when dehydrated. Again, as in reptiles, GFR is regulated by AVT, which reduces GFR by up to 50%, presumably by constricting the afferent glomerular arteriole. Dehydration or salt loading also decreases the UFR and increases the ureteral urine concentration. In the budgerigar, the UFR decreases from 6.3 ml kg^{-1} hr^{-1} to 1.68 and the U/P increases from 0.73 to 2.33 for dehydrated birds. Modification of the U/P ratio is due to an AVT-induced increase in the water permeability of the collecting ducts of the "mammalian-type" nephrons. Reduced ureteral UFR and increased osmotic concentration potentially affects cloacal/colonic reabsorption.

The role of the kidney in urinary excretion is complicated in reptiles and birds by the potentially important role of the cloaca/hindgut in modifying the ureteral urine before it is finally voided. The urine of reptiles and birds passes from the ureters via two ureteral openings into the urodeum of the cloaca (Figure 17–27A). The urine may then enter cloacal bursae (accessory bladders) in some sea turtles or a urinary bladder present in all turtles, the tuatara, amphisbaenians, and some lizards (Beuchat 1986). Crocodilians and most lizards lack a urinary bladder. Birds also do not have a urinary bladder. The cloaca and cloacal bursae of freshwater reptiles actively reabsorb ions (e.g., Na^+) from the urine and also from water irrigated into these structures; the pharynx also absorbs ingested water and ions (these sites of Na^+ reabsorption may also function in respiratory gas exchange). The green turtle *Chelonia mydas* can void urine about 60 mOsm hyperosmotic to its plasma, probably by ion secretion into the bladder urine rather than by water withdrawal from the bladder against an osmotic gradient. The bladder generally has a low water permeability to minimize osmotic reabsorption of water from the hypoosmotic urine.

Urine that enters the urodeum compartment of the cloaca can be subsequently refluxed into the colon and modified by absorption of ions and water. The high concentration of inulin in both ureteral urine (as expected) and also in the colon and cecum indicates that urine is refluxed into the cloaca/colon/cecum (Figure 17–27B). An absence of water reabsorption in the hindgut seems to be indicated by the inulin concentration not increasing in the cloaca/colon over that of the plasma, but the ureteral flow into the hindgut is diluted by fluid passing from the small intestine. Consequently, the hindgut inulin concentration would be considerably lower than that of the ureteral urine if water was not absorbed.

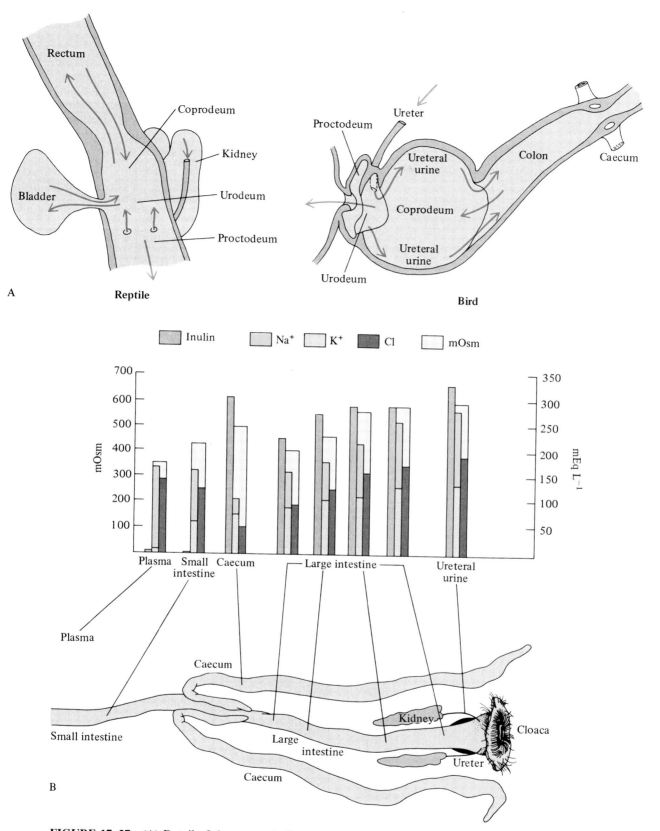

FIGURE 17–27 **(A)** Detail of the anatomical structure of the distal portion of the gut and of the ureter and urinary bladder in a lizard (*Uromastix;* left) and a bird (*Gallus;* right) showing the retrograde flow of urine into the coprodeum and colon. **(B)** Osmotic and ionic concentrations and inulin concentration in the ureteral urine, plasma, and various parts of the mid- and hindgut of the chicken. The high-inulin concentration of ureteral urine and the hindgut contents reflects reabsorption of substantial amounts of water from the renal filtrate and hindgut contents. *(Modified from Minnich 1982; after Seshadri 1957; Chosniak, Munck, and Skadhauge 1977; Skadhauge 1981.)*

The coprodeum and colon have a highly folded epithelium with irregular villi and thus are anatomically suited for absorption. Several processes modify urine in the cloaca/colon, including ion reabsorption (especially Na^+ and Cl^-), water reabsorption, and urate precipitation. Perfusion experiments have delineated the basic NaCl and water transport properties of the coprodeum and colon of various reptiles and birds. There is a reabsorptive Na^+ flux of about 100 μmole kg^{-1} hr^{-1} and a Cl^- flux of about 75 μmole kg^{-1} hr^{-1}. Water flux, which is solute linked with active ion uptake is about 0.5 μl kg^{-1} hr^{-1}. The osmotic permeability of the chicken coprodeum/colon is about 4–6 μl kg^{-1} hr^{-1} $mOsm^{-1}$, and so a 100 mOsm gradient would be required to osmotically transport the observed 0.4–0.6 μl kg^{-1} hr^{-1}.

The osmolality of the fluid reabsorbed from the coprodeum/colon can vary from 100 to 1000 mOsm, but it is always hyperosmotic to the lumen contents. For example, the reabsorbate of *Varanus gouldii* has an osmotic concentration of 140 mOsm if the perfusate is 100 mOsm. In chickens, the reabsorbate is 1160 mOsm for an isoosmotic (to plasma) perfusate of about 300 mOsm. Vertebrate epithelia are not able to transport a fluid that is hypoosmotic to

the bathing medium (although the insect rectum can; Phillips 1981). Consequently, the cloacal/colonic reabsorbate is expected to be markedly hyperosmotic (e.g., chicken) or almost isoosmotic (e.g., emu) but not hypoosmotic to the lumen fluid. Hence, cloacal reabsorption of ions and water cannot increase the osmotic concentration of the lumen fluid.

The small dragon lizard *Amphibolurus maculosus* appears to be exceptional in voiding a hyperosmotic excreta from its cloaca (Braysher 1976). When salt loaded, the ureteral urine is hypoosmotic (355 mOsm; $U/P = 0.74$) to plasma (452 mOsm) but the voided excreta (671 mOsm) is hyperosmotic ($U/P = 1.5$). Presumably, solute-linked water reabsorption and recycling of solute into the lumen osmotically concentrates the excreta. This would be essentially the same mechanism by which the rectum of various insects produces a hyperosmotic excreta.

The formation of a hyperosmotic urine by avian kidneys would promote osmotic loss of body water into the cloacal/colonic lumen in proportion to the osmotic gradient and the osmotic permeability of the epithelium. However, this potential osmotic loss of water can be balanced or exceeded by solute-linked water absorption. For example, the galah has

TABLE 17–5

Ion concentration (mM), ratio of Na^+/K^+ concentration, and flow rate for reptilian and avian salt glands. *(Data from Minnich 1979; Skadhauge 1981.)*

	Na^+ mM	K^+ mM	Na^+/K^+	Flow rate μl kg^{-1} hr^{-1}
Reptiles				
Ctenosaur lizard	67	537	0.12	2.2
Chuckwalla	121	378	0.32	2.7
Sea snake (*Hydrophis*)	509	20	25	5.5
Sea snake (*Pelamis*)	620	28	22	35.2
Varanid lizard	654	54	12	5.2
Emydid terrapin	682	32	21	2.5
Sea turtle	685	21	33	19
Sea snake (*Laticauda*)	686	57	12	4.9
Conolophus lizard	692	214	3.2	2.3
Sea snake (*Aipysurus*)	798	28	29	20.6
Desert iguana	1032	640	1.6	0.18
Marine iguana	1434	235	6.1	7.5
Birds				
Brolga	263	8	33	47
Goose	430	12	36	71
Cormorant	529	12	44	73
Swan	656	19	35	130
Pelican	698	13	54	106
Gull	727	36	20	28
Penguin	800	29	54	57
Savannah hawk	1010	16	63	100

a ureteral *U/P* of up to 2.5, but it avoids osmotic loss of water into its cloaca by having a low osmotic permeability (1.4–2.2 μl kg^{-1} hr^{-1} mOsm^{-1}) and a high rate of solute-linked water absorption of hyperosmotic reabsorbate (about 400 mOsm). The emu forms ureteral urine with a lower osmotic concentration (*U/P* about 1.4) and can absorb a greater volume of water from the cloaca/colon by solute-linked water flow; the reabsorbate is nearly isoosmotic with the cloacal/colonic contents (320 mOsm).

Cloacal/colonic water and ion reabsorption are influenced by the osmotic status of the bird. For example, dehydration alters the affinity of the transport carrier for Na$^+$ and the Na$^+$ concentration of the reabsorbate but does not affect the osmotic permeability of the cloaca, the maximal uptake rate for Na$^+$, or the transepithelial electrical potential. Aldosterone augments the rate of Na$^+$ reabsorption and may therefore be involved with the *in vivo* modification of cloacal/colonic function during dehydration. Changes in cloacal/colonic transport functions may also be passive responses to a decreased rate of ureteral flow of a more concentrated urine.

Salt Glands. A variety of reptiles and birds have **salt glands** that excrete a hyperosmotic (to blood) NaCl or KCl solution, i.e., it is an important ionoregulatory gland (Table 17–5).

Reptilian salt glands have a diverse origin, a less-specialized structure, and more general ion (Na$^+$, K$^+$, and Cl$^-$) excretion capabilities than avian salt glands, which are more complex in structure and less diverse in function. Crocodiles, marine turtles and snakes, and lizards of some families have salt glands. All iguanid lizards; many, but certainly not all, teid, scincid, lacertid, xantusid, and varanid lizards; and some agamids (e.g., *Uromastix*) have salt glands. Geckos, anguids, annielids, heloderma-tids, and possibly cordylids do not. No terrestrial turtles, snakes, or amphisbaenids have salt glands.

Reptilian salt glands are derived from any of a number of cephalic (head) glands (noncrocodilians) or from numerous lingual glands (crocodilians; Figure 17–28). The former type of salt gland may be derived from nasal glands (lizards), lachrymal glands (marine turtles, diamondback terrapin), posterior sublingual glands (sea and file snakes), and the premaxillary gland (the marine homolopsid snake *Cerberus*). The nasal salt gland of lizards is a single-lobed gland lying in the nasal cavity, lateral to the olfactory chamber. The nasal gland secretion pools anterior to a ridge in the nasal cavity and its evaporation tends to form salt encrustations; some

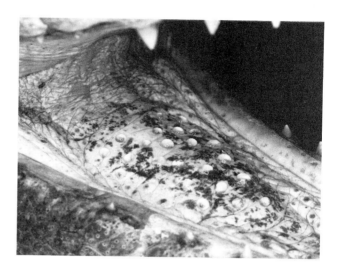

FIGURE 17–28 (A) Schematic showing the location of the lingual salt glands (arrows) of the estuarine crocodile *Crocodylus porosus*. (B) Photograph showing the secretions from the lingual salt glands. (*Courtesy of Professor Grigg, Department of Zoology, University of Queensland.*)

species sneeze to expel the nasal secretion and salt encrustation. The single-lobed gland, whose fine structure is also generally representative of that of other types of salt gland, consists of numerous tubules that branch from the main duct and then further branch three or four times to form blind-ended sacs. The numerous principle tubules, which form the majority of the gland, are the functional salt-secreting glands; the less numerous peripheral tubules probably do not function for salt secretion.

The four principal ions transported by the reptilian salt gland are Na$^+$, K$^+$, Cl$^-$, and HCO$_3^-$. Marine reptiles, as would be expected, excrete mainly Na$^+$ and Cl$^-$ at maximal rates greater than for terrestrial species (see Table 17–5), but the marine iguana excretes a substantial amount of K$^+$, reflecting the

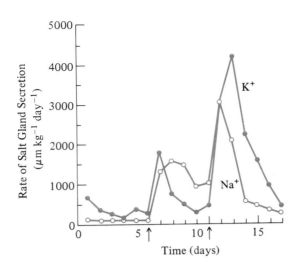

FIGURE 17–29 Effect of a salt load of 2 mmole of NaCl (arrow, day 6) and 2 mmole of KCl (arrow, day 11) on salt gland scretion of Na^+ and K^+ by the desert iguana. *(From Shoemaker, Nagy, and Bradshaw 1972.)*

high K^+ content of its herbivorous (alga) diet. Terrestrial species, in contrast, excrete relatively less Na^+, and some excrete more K^+ than Na^+, reflecting the higher K^+ content in their herbivorous diet. The desert iguana excretes considerable Na^+ and K^+ when salt loaded with either NaCl or KCl (Figure 17–29). Some terrestrial species also excrete more K^+ than Na^+ even when salt loaded with NaCl, although chronic NaCl loading may result in higher Na^+ than K^+ excretion.

The role of the avian salt gland was first described by Schmidt-Nielsen, Jorgensen, and Osaki (1958), although the presence of these supraorbital and nasal glands was previously appreciated. Functional salt glands are present in about 50 species of primarily marine birds of about 20 different orders. The avian salt gland is more complex in structure than is the reptilian salt gland but is more uniform in function. The avian nasal salt glands are orbital glands, being multilobar glandular structures with ducts opening into the nasal cavity (Figure 17–30); the salt secretion typically drips from the external narial openings. The central canal connects to numerous radiating secretory tubules that branch to form short, blind-ended sacs. As in reptiles, the salt gland excretes primarily NaCl, allowing maintenance of ionic homeostasis, particularly for marine birds. The salt glands generally secrete almost entirely NaCl, with only low concentrations of K^+ and HCO_3^-; the total osmotic concentration is gener-

ally 1000 to 1600 mOsm (i.e., 1 to 1.6 × seawater), and the ratio of Na^+/K^+ is typically 10 to 50 (see Table 17–5). A few terrestrial birds have salt glands, e.g., ostrich and roadrunner. Their nasal salt gland secretions contain less Na^+ and more K^+ than the secretions of the salt glands of marine birds, reflecting the higher dietary intake of K^+ than Na^+ in terrestrial herbivores. A number of hawks have functional salt glands that can secrete high concentrations of Na^+ and K^+ (although at low flow rates). Some juvenile birds also have functional salt glands to eliminate the ions that are left by evaporation of water for thermoregulation.

The secretory capacity of the salt gland depends on a number of factors, including the size of the gland. Some reptiles and birds have larger salt glands than do others, and chronic acclimation to a high salt intake can cause the salt gland to hypertrophy. Secretion rates vary from <1 to 200 $\mu l \ kg^{-1} \ min^{-1}$. There are at least two possible mechanisms for salt gland secretion. First, an iso-osmotic fluid is secreted and is then made hyperosmotic by subsequent modification (ion secretion and/or water removal). Second, the initial fluid may be secreted hyperosmotically across a water-impermeable membrane. Neither of these mechanisms has been proven or disproven, but the former is more likely because an ouabain-sensitive Na^+-K^+ ATPase is located on the highly folded basal and baso-lateral membrane of the secretory cells, rather than on the apical membrane. This would imply on unusual cytoplasmic concentration if the apical membrane was freely permeable to water and if it were in equilibrium with the secreted fluid. However, the rate of flow of salt gland secretion is high (about 10% of the blood flow to the gland), and if water reabsorption were the mechanism for hyperosmoconcentration, then almost all of the blood flow would initially have been secreted to form the isoosmotic primary fluid if the final secretion had a ratio of 3:1 or 4:1 with plasma. One scheme to account for nasal gland secretion is a combination of dilute secretory cells at the distal end of the blind tubules and specialized NaCl secreting cells at the proximal end that add hyperosmotic NaCl. This scheme (and also others) is consistent with the inverse relationship observed between salt gland secretion rate and concentration; the higher salt concentration is a consequence of the longer time that the fluid is in contact with the hypersecreting proximal cells.

The secretory activity of the salt gland, unlike that of the kidney, can cease. The cholinergic parasympathetic innervation of the salt gland controls

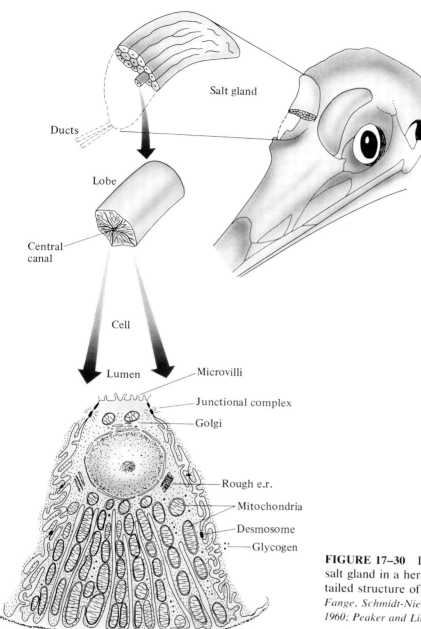

Lumen — Microvilli

— Junctional complex

— Golgi

— Rough e.r.

— Mitochondria

— Desmosome

— Glycogen

FIGURE 17–30 Location and gross structure of the salt gland in a herring gull *Larus argentatus* and the detailed structure of a salt gland cell. *(Modified from Fange, Schmidt-Nielsen, and Osaki 1960; Schmidt-Nielsen 1960; Peaker and Linzell 1975.)*

secretion, although the mechanism is unclear. There also is an adrenergic innervation, which may be antagonistic and may inhibit secretion. Secretion may be triggered by a stimulation of osmoreceptors or specific alkali metal receptors (Na^+, K^+, even Rb^+). There is also hormonal control of the salt gland. Aldosterone has its typical natriferic effect of Na^+ retention and its kaliuretic effect of K^+ excretion in reptiles (Bradshaw 1986). In birds, hormonal control of the salt gland is relatively minor (Butler, Siwanowicz, and Puskas 1989).

Mammals. Mammalian nephrons are adapted for minimizing urinary water loss and excretion of ionic and nitrogenous solutes. Mammalian nephrons have all of the standard glomerulo-tubular functions (filtration, reabsorption, and secretion) and can also markedly concentrate the urine (U/P up to 30). This latter capacity is a consequence of the presence of the loop of Henle, interposed between the proximal and distal convoluted tubules, and the specific anatomical arrangement of this nephron segment and the collecting ducts.

The mammalian kidney has a complex structural organization of its nephrons, blood supply, and the renal pelvis; there is a renal cortex and different zones of the medulla (Figure 17–31). The mammalian nephron has a glomerulus, with afferent and efferent arterioles; Bowman's capsule; PCT; a loop of Henle, consisting of a thin, descending limb and an ascending limb with thin and thick portions; a DCT; and collecting duct. The collecting ducts drain into a renal pelvis that empties into the ureter. There are two different types of mammalian nephrons: the **cortical nephrons**, which have short loops of Henle, and the **juxtamedullary nephrons**, which have long loops of Henle that extend deep into the medullary papilla. There is a well-developed juxtaglomerular apparatus. This consists of a specialized region of the DCT and afferent/efferent glomerular arterioles where they contact. The DCT epithelial cells are specialized to form a macula densa, and smooth muscle cells in the wall of the glomerular arterioles contain granules of renin.

The mammalian glomerulus is specialized for filtration and the PCT for solute and water reabsorption. The loop of Henle consists of a thin descending limb and a thin ascending limb, which apparently have little metabolic cellular activity (i.e., no active transport) and a thick ascending portion of the limb of the loop, which has markedly thickened epithelial cells that contain extensive Cl⁻ pumps and are impermeable to water. The thick ascending limb extends to the juxtaglomerular apparatus. The DCT is divided into two functional segments. The initial diluting segment is similar in function to the thick ascending limb (ion reabsorption but is impermeable to water). The late distal segment resembles the cortical collecting ducts in structure and function. These parts of the nephron are largely impermeable to urea (normally) and reabsorb Na⁺ and secrete K⁺ (largely under hormonal control by aldosterone); their water permeability is normally low in the absence of antidiuretic hormone (ADH, or arginine vasopressin, AVP) but is higher in the presence of ADH. The water permeability of the collecting duct epithelium is also under control by ADH; it also

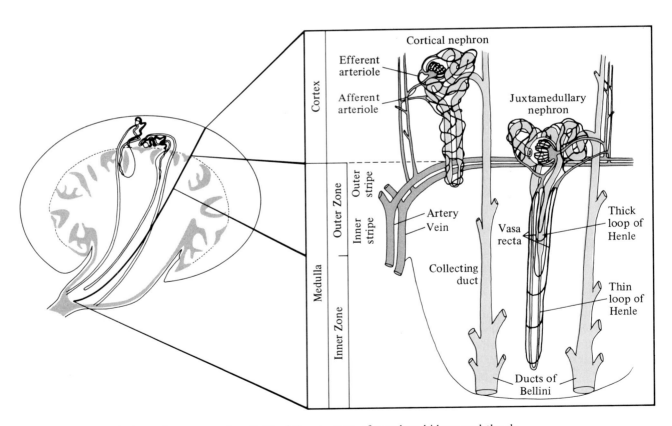

FIGURE 17–31   Schematic cross section (left) of the stucture of a rodent kidney and the detailed anatomical arrangement of the cortical and juxtamedullary nephrons (right).   *(Modified from Kaibbling et al. 1975; Pitts 1974.)*

secretes H^+ for acid-base regulation.

The principal mechanism for urine formation by the mammalian nephron is ultrafiltration, followed by reabsorption, secretion, and osmoconcentration. The glomerular filtration rate of mammals is markedly size dependent, being 10 to 600 ml kg^{-1} hr^{-1} (Yokota, Benyajati, and Dantzler 1985). Almost all of the filtered primary urine is reabsorbed, largely in the proximal convoluted tubule (PCT), but also in the distal convoluted tubule (DCT). Typical values for GFR and its composition and UFR and its composition are summarized in Table 17–6 for a macropod marsupial. The UFR is only a small fraction of GFR, indicating about 99.7% reabsorption of water. The U/P is greater than 1 for all major ions (Na$^+$, K$^+$, Cl$^-$), for total osmotic concentration, and especially for urea. Corresponding values for solute and osmotic clearance indicate that from 0.07 to 1.3 ml of plasma are cleared of that solute per minute, 0.28 ml of plasma min^{-1} are cleared of osmolytes, and 2.8 ml plasma min^{-1} are cleared of urea. The free water clearance is -0.23 ml min^{-1} since the urine concentration is greater than the plasma concentration. The filtration fractions indicate that about 99.6% of the GFR water is reabsorbed, >98% of the Na$^+$ and Cl$^-$ are reabsorbed, 88% of K$^+$ is reabsorbed, 97% of total osmolytes are reabsorbed, and 72% of the urea is reabsorbed.

The mammalian nephron has a remarkable capacity for altering the clearance and fractional reabsorption of ions, nonelectrolytes, total osmolytes, and free water. Variation in urine flow rate is accomplished primarily by regulation of tubular balance rather than by regulation of GFR, although there is some adaptive variation in GFR in response, for example, to water loading, salt loading, or dehydration. Glomerulo-tubular balance is under hormonal control. Aldosterone promotes sodium absorption and potassium secretion in the PCT, DCT, and collecting duct; water reabsorption is tightly linked to Na$^+$ reabsorption, hence the effects of aldosterone may markedly influence Na$^+$ balance but may have little effect on body fluid Na$^+$ concentration because of the concomitant reabsorption of water. Arginine vasopressin (antidiuretic hormone, ADH) also has dramatic tubular effects, including increased water reabsorption of the renal filtrate, reduced urine flow rate, and osmoconcentration of the urine. The principal effect of ADH is to increase the permeability of the collecting duct to water, allowing collecting duct fluid to osmotically equilibrate with the hyperosmotic medullary fluid. The osmotic concentration of the nephron fluid is unaltered by solute/water reabsorption through the PCT, but the osmotic concentration varies through the rest of the nephron and in the urine.

There is a dramatic variation among mammals in their ability to osmotically concentrate urine. Human urine, for example, can vary in osmotic concentration from about 60 mOsm (quite hypoosmotic) to about 1200 mOsm (considerably hyperosmotic). Other mammals can produce even more dilute, or more concentrated, urine depending on their iono-osmoregulatory demands. This is, not surprisingly, generally related to their environmen-

TABLE 17–6

Renal parameters for a macropod marsupial when hydrated and consuming a high-protein diet. Glomerular flow rate (GFR, ml min^{-1}); urine flow rate (UFR, ml min^{-1}); plasma solute and osmotic concentrations (P, mM and mOsm) and urine concentrations (U, mM); clearance (C, ml min^{-1}) for solutes and total osmolytes; and % reabsorption of water, solutes, and total osmolytes. Body mass is 2.97 kg. *(Data from Bakker and Bradshaw 1983.)*

	Plasma	Urine	U/P	Clearance[1]	% Reabs[3]
Water flow rate	10.90 (GFR)	0.038 (UFR)	0.003	-0.23[2]	99.7%[4]
Na$^+$	143	251	1.76	0.065	99.4%
K$^+$	4.4	146	33.2	1.26	87.5%
Cl$^-$	96.1	414	4.31	0.17	98.3%
Urea	7.8	568	72.8	2.80	71.9%
Inulin	—	—	287	10.90	0.0%
OC	275	2003	7.28	0.28	97.2%

[1]Clearance = UFR. U/P
[2]Free water clearance = UFR − osmolal clearance
[3]% reabsorption = 100(1 − (C/GFR))
[4]% water reabsorption = 100 (GFR − UFR)/GFR

tal availability of water. Mesic mammals, that would probably never experience a shortage of water, have maximal urine concentrations of less than 1000 mOsm, e.g., the beaver (*Castor*) and mountain beavor (*Aplodontia*). In marked contrast, many desert-adapted rodents have a maximal urine concentrating capacity that exceeds 6000 mOsm (e.g., *Desmodillus, Jaculus*) up to 8600 mOsm (*Perognathus*) and 9400 mOsm (*Notomys*). Most mammals have a urine concentrating capacity of between 3000 and 5000 mOsm.

The mammalian (and some avian) nephrons produce a hyperosmotic urine by medullary countercurrent multiplication in the loop of Henle and the collecting ducts (Figure 17–32). The adjacent medullary tissues and the peritubular blood vessels (the vasa recta) are passively involved in the countercurrent exchange process. A variety of physiological

processes establish a gradient in osmotic concentration from the cortex-medulla junction (isoosmotic to plasma) to the medullary tip (i.e., 1400 mOsm in humans and about 9000 mOsm in Australian hopping mice).

The principal mechanism for establishing this osmotic gradient is the active transport of Cl^- ions and the passive transport of Na^+ out of the thick ascending limb of the loop of Henle. Water does not passively follow since this segment of the nephron is impermeable to water. Consequently, the osmotic concentration of lumen fluid declines as it flows towards the cortex, and the osmotic concentration of the adjacent interstitial fluid increases because of the added NaCl. The local increase in medullary osmotic concentration affects the fluid passing down the descending thin limb of the loop of Henle; water is osmotically withdrawn from

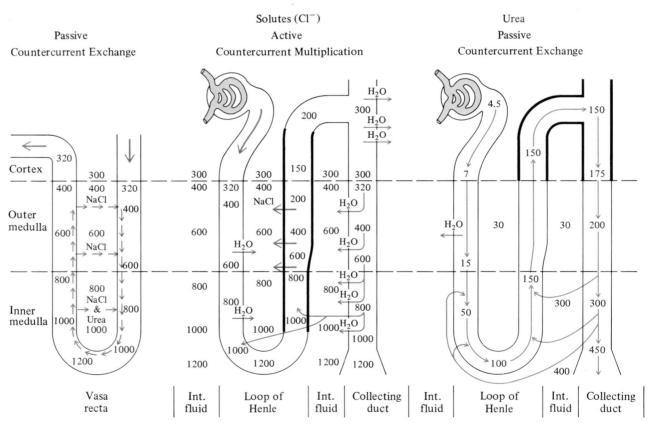

FIGURE 17–32 Mechanism for osmotic concentration of urine by countercurrent multiplication of solute concentration in the loop of Henle and adjacent interstitial fluid (int fluid) by active transport of Cl^- (center). A passive countercurrent exchange of urea contributes to the interstitial osmotic gradient (right). There is a passive countercurrent exchange of solute and urea in the peritubular blood supply (vasa recta) to limit potential dissipation of the interstitial osmotic gradient by peritubular blood flow.

the tubule into the medullary interstitium, thereby osmotically concentrating the intratubular fluid as it descends towards the medullary tip. This process of active Cl^-/passive Na^+ reabsorption from the thick ascending limb of the loop of Henle, and water withdrawal from the descending thin limb of the loop of Henle, establishes a tubular and interstitial osmotic gradient by countercurrent multiplication. Ion reabsorption in the collecting duct also contributes to the medullary osmotic gradient. The osmotic concentration of the urine is determined by the permeability of the collecting duct to water. Normally, the collecting duct is impermeable to water, and so the tubular fluid (which is isoosmotic, about 300 mOsm) at the cortical portion of the collecting duct passes through the collecting duct without major alteration of its osmotic composition. Further ion reabsorption as fluid passes through the collecting duct can produce a considerably hypoosmotic urine, about 50 to 60 mOsm. ADH increases the water permeability of the collecting duct (and also urea permeability; see below). This allows osmotic withdrawal of water from the collecting duct, increases the osmotic concentration of the fluid, and allows the final urine to be as concentrated as the medullary tip tissues.

The passive reabsorption of water from the collecting duct in the presence of ADH can also contribute to the medullary osmotic concentration gradient. The osmotic withdrawal of water markedly increases the urea concentration in the collecting duct lumen. This promotes urea diffusion from the lumen into the adjacent medullary interstitium and into the descending and ascending limbs of the loop of Henle; this occurs particularly in the inner medulla. This passive addition of urea to the medullary tissues increases the osmotic concentration gradient available for concentrating the urine. The role of urea in urine osmotic concentrating ability is very significant because of its very high concentrations in the medulla and in the urine. It might also promote further passive Na^+ and Cl^- accumulation in the inner medullary tissues from the thin limb of the loop of Henle. Addition of urea to the medullary interstitium increases the local osmotic concentration and osmotically withdraws water from the thin limb; this further concentrates Na^+ and Cl^- in the thin limb and causes diffusion of some Na^+ and Cl^- into the interstitium.

The urine concentrating ability of the mammalian kidney is closely associated with the number of "medullary" nephrons (as opposed to "cortical" nephrons) and their relative morphology, particularly the relative length of the loop of Henle and the collecting duct. This is not surprising since the loop

of Henle provides the countercurrent multiplication of ion concentration that is the principal means for establishing the medullary osmotic gradient. Mammalian kidneys with a high osmotic concentrating capacity have a typical appearance; the cortex region is small and the medullary region of the kidney is greatly expanded and the renal papilla often protrudes from the base of the kidney because of its relatively extreme length. There is a strong correlation between the maximal urine osmoconcentrating capacity and the relative medullary area (Figure 17–33; Brownfield and Wunder 1976, Beuchat 1990).

Small mammals have a higher urine concentrating capacity than larger mammals, despite the smaller absolute dimensions of their kidneys and nephrons. This may be a consequence of their higher mass-specific metabolic rates (Greenwald and Stetson 1988). For example, the Cl^--secreting cells of the thick ascending limb of the loop of Henle have a greater mitochondrial density in small mammals (e.g., a brown bat) than in large mammals (e.g., a horse). This would suggest that the thick ascending limb of the brown bat could generate a higher osmotic concentration gradient per unit length than that of the horse. Allometry indicates that the long length of the loop of Henle in large mammals ($\propto M^{0.129}$) is insufficient to offset the reduced mass-

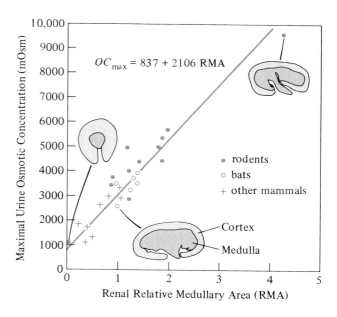

FIGURE 17–33 Relationship between maximal urine concentrating ability and relative medullary area for mammalian kidneys. Insets show a diagrammatic cross section of three kidneys, illustrating the relative sizes of the cortex and medulla.

specific metabolic rate ($\propto M^{-0.24}$) and so the urine concentrating ability declines with increasing mass (Beuchat 1990).

The peritubular capillaries of the medulla (the vasa recta) are arranged in a countercurrent fashion. This allows the blood to equilibrate osmotically with the adjacent medullary interstitium as it descends into the medulla, and then re-equilibrate as it ascends out of the medulla. This limits the potential dissipation of the medullary osmotic gradient by the local blood flow.

The **renal pelvis**, into which the urine drains from the collecting ducts, and the **bladder** are generally considered to be passive conduits and storage structures. However, both the renal pelvis and the bladder may have more complex physiological roles. Extensive projections of the renal pelvis into the medulla of the kidney may recycle urinary urea into the medullary vasa recta, thereby conserving urea. This could be of great significance for mammals in negative nitrogen balance (Kaibbling et al. 1975; Schmidt-Nielsen 1977; Schmidt-Nielsen 1988). The epithelial cells of the mammalian bladder must be specialized for storage of often markedly hyperosmotic urine (Lewis 1986). Ussing cell studies of the bladder indicate active transport of Na^+ from the lumen into the blood that is stimulated by aldosterone.

The juxta-glomerular apparatus provides a considerable automatic control of glomerular filtration rate. If the GFR is too low, the tubular solute reabsorption produces a low concentration of Na^+ and Cl^- in the ascending limb of the loop of Henle. This causes reflex afferent arteriole dilation, which increases the glomerular pressure and filtration rate. The low concentration of Na^+ and Cl^- also releases renin granules from the modified smooth muscle cells of the efferent, but mainly from the afferent, glomerular arterioles. The renin enzymatically forms angiotensin II, which constricts the efferent arteriole and increases the glomerular capillary pressure and changes the glomerular blood flow; this increases GFR. Angiotensin also increases the blood osmotic pressure and decreases the blood flow in the vasa recta, thereby promoting reabsorption of fluid into the vasa recta from the peritubular space.

Nitrogen Metabolism

Metabolism of carbohydrates and lipids forms essentially only one waste product, carbon dioxide (see Chapter 3). Amino acid metabolism, in contrast, also requires the elimination of N and S. The sulfur content of amino acids is readily eliminated as SO_4^{2-}, but the fate of the amino nitrogen requires greater attention because of the relatively high N content of amino acids. Nucleic acids (DNA and RNA), which have repeating units of (phosphate + sugar + base), are another source of excretory N; the bases are purines (adenine, guanine) and pyrimidines (thymine, cytosine). There are also a variety of other nitrogenous waste products.

The N of most amino acids is initially converted to NH_3 or NH_4^+ by deamination or transamination (Chapter 3). The subsequent fate of the NH_3 or NH_4^+ depends markedly on the type of animal (i.e., its taxonomy) and physiological state (i.e., hydration status). Potential fates include excretion as NH_3 or NH_4^+ or conversion to urea, uric acid, or guanine. The metabolic fate of the N from pyrimidine metabolism is NH_3 and for purines is one of a complex series of N-containing products. There is, of course, an interaction between the N metabolism of amino acids and nucleic acids; NH_3 formed from nucleic acid metabolism can be converted to urea, uric acid, or guanine, as might NH_3 from amino acid metabolism.

Amino N Metabolism

Amino acids contain from 0.085 g N per g (phenylalanine) to 0.322 g N g^{-1} (arginine); meat protein contains about 0.17 g N g^{-1}. Consequently, amino N excretion represents a significant fraction of the mass of amino acids and protein.

The potential metabolic fates of N from amino acids include NH_3, NH_4^+, urea, or a variety of purines (uric acid, guanine). Amino nitrogen is typically converted to NH_3 or NH_4^+ but its subsequent metabolic fate varies in different animals. Arginine metabolism directly forms urea; many animals have **urease** to degrade enzymatically the urea to ammonia.

The most common mechanism for removing amino N from amino acids is the transfer of the $-NH_2$ group to a ketoacid by a transaminase enzyme. For example, aspartate aminotransferase reversibly transfers $-NH_2$ from aspartate (to α-ketoglutarate) to form oxaloacetate (and glutamate); this is one of the most active aminotransferases. Many other aminotransferases transfer $-NH_2$ from most other amino acids to α-ketoglutarate, forming a ketoacid and glutamate. Exceptions are threonine and serine, which are not transaminated.

The $-NH_2$ group can only be removed from a few amino acids to form NH_3 or NH_4^+. Glutamate is oxidatively deaminated by glutamate dehydrogenase (GDH) to form α-ketoglutarate and NH_3; GDH is widely present in the tissues of animals and is important to N metabolism. Histidine is directly deaminated by histidase; serine and threonine are deaminated via dehydration by serine dehydratase; glutamine is hydrolytically deaminated by glutaminase to glutamine; aspartate is deaminated in the purine nucleotide cycle to form fumarate. Some deamination reactions form NH_3 and others form NH_4^+.

Nucleic Acid Metabolism

The nucleic acid bases are either double nitrogen-containing rings (purine bases; adenine, $C_5H_5N_5$; guanine, $C_5H_5ON_5$) or single nitrogen-containing rings (pyrimidine bases; thymine, $C_5H_6O_2N_2$; cytosine, $C_4H_5ON_3$; Figure 17–34A). The nitrogen content of the repeating DNA unit (phosphate + sugar + base) varies from 0.09 to 0.14 g N g^{-1} for pyrimidine bases and from 0.20 to 0.22 g N g^{-1} for purine bases.

The metabolism of purine and pyrimidine bases produces significant amounts of nitrogenous waste products. Pyrimidine metabolism is relatively straightforward, culminating in ammonia production. The pyrimidine base monophosphates (nucleotides) are first converted to nucleosides (thymidine, uridine, cytidine) and then to bases (uracil and thymine). These bases are then degraded by similar pathways to form NH_4^+ and alanine and aminoisobutyrate, which are then transaminated and the C-skeletons reduced to CO_2.

Purine metabolism is variable among different animals (Figure 17–34B). Adenine and guanine are converted (via hypoxanthine and xanthine) to uric acid, which is further degraded to allantoin (by uricase), allantoic acid (by allantoinase), urea and ureidoglycolate (by allantoicase), and then ammonia (by urease). Ureidoglycolate is reduced to urea and glyoxylate by ureidoglycolase. The actual excretory end product from this general scheme for uricolysis varies in different animals by their serial deletion of the enzymes, from urease (hence urea is excreted) to, in some instances, all of these enzymes (hence uric acid is excreted). The major animal groups excrete different end products.

Other Nitrogenous Waste Products

Animals produce a considerable variety of other nitrogenous waste products (Figure 17–35). These are often unimportant in influencing osmoregulation or nitrogen balance, but they may have other specific and important roles (e.g., TMAO, betaine). Consequently, some of these other nitrogenous waste products will be briefly described here.

The metabolic breakdown products of hemoglobin, such as bilirubin and biliverdin, can be excreted by the gut, e.g., in vertebrates. Other animal pigments, such as ommochromes (derived from tryptophan) and pteridines (from the vitamin, riboflavin) are significant nitrogen compounds. Creatine and creatinine are nitrogenous excretory products in vertebrates; creatine phosphate is an important muscle store of high-energy phosphate bonds (see Chapters 3 and 4) and creatine is a spontaneous breakdown product of creatine. Amino acids and

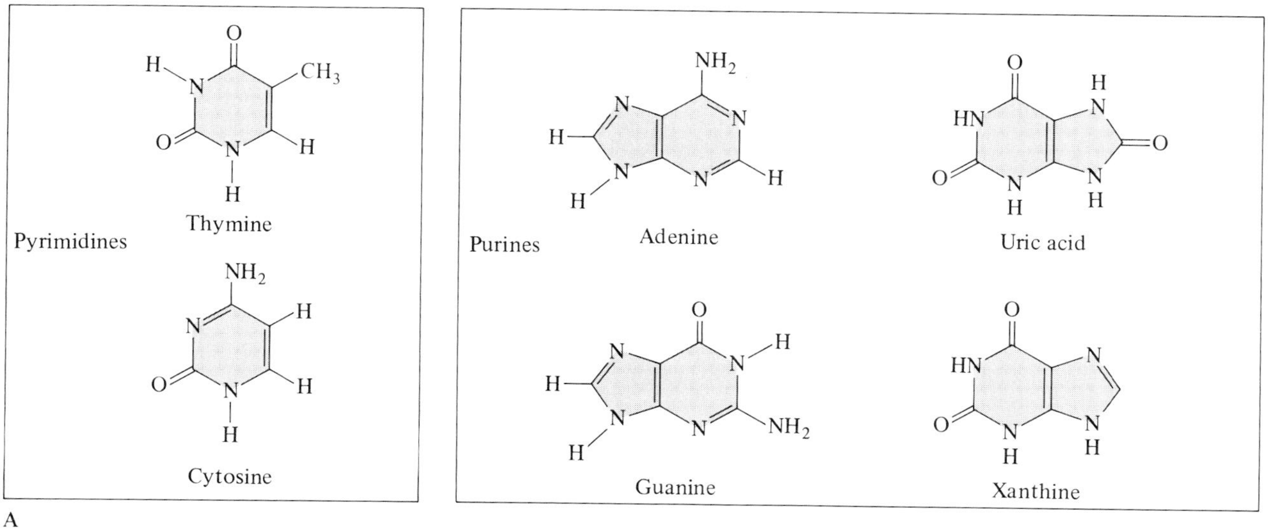

FIGURE 17–34 (A) Structure of pyrimidine bases, and purine bases and related purines.

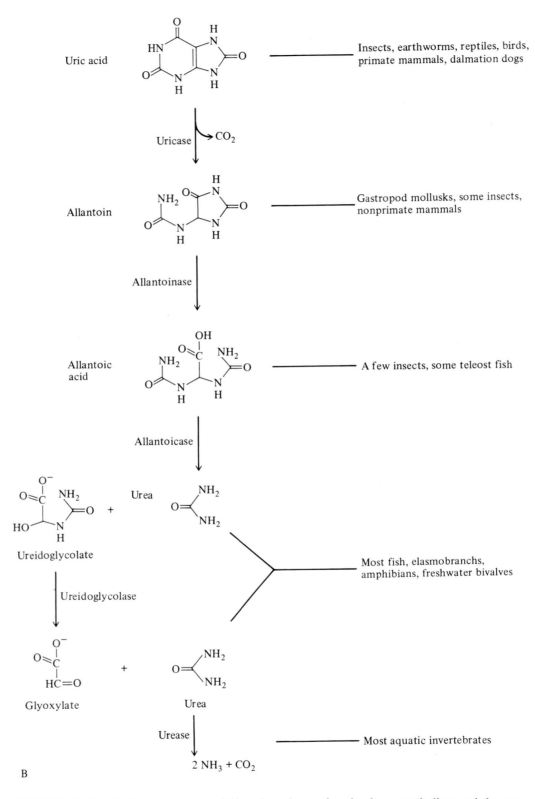

Uric acid ——————— Insects, earthworms, reptiles, birds, primate mammals, dalmation dogs

Uricase → CO_2

Allantoin ——————— Gastropod mollusks, some insects, nonprimate mammals

Allantoinase

Allantoic acid ——————— A few insects, some teleost fish

Allantoicase

Ureidoglycolate + Urea

Most fish, elasmobranchs, amphibians, freshwater bivalves

Ureidoglycolase

Glyoxylate + Urea

Urease ——————— Most aquatic invertebrates

$2 NH_3 + CO_2$

B

FIGURE 17–34   (B) Summary of possible end products of purine base catabolism and the various invertebrate and vertebrate groups that excrete these end products.

Amino acids

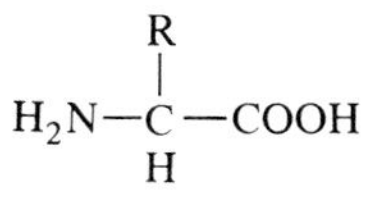

Ommochromes
(tryptophan derivative)

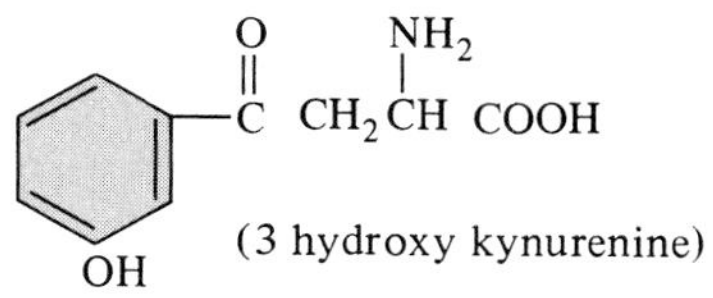

Pteridines
(riboflavin derivative)

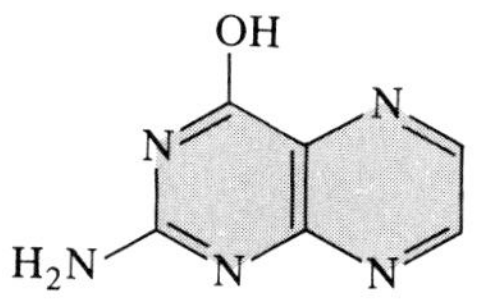

Bilirubin
(heme derivative)

Creatine and
creatinine

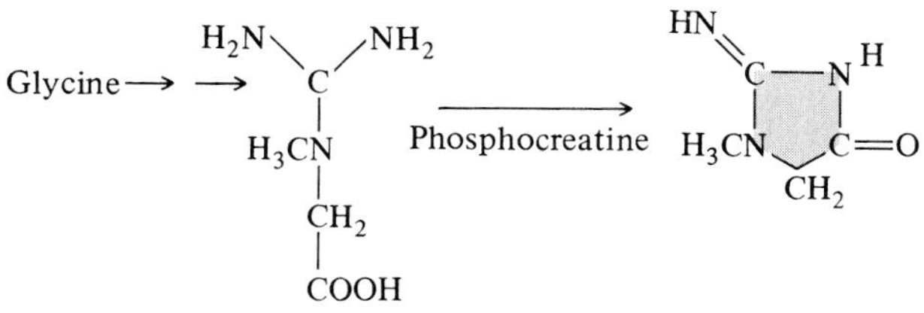

Creatine Creatinine

Choline and
derivatives

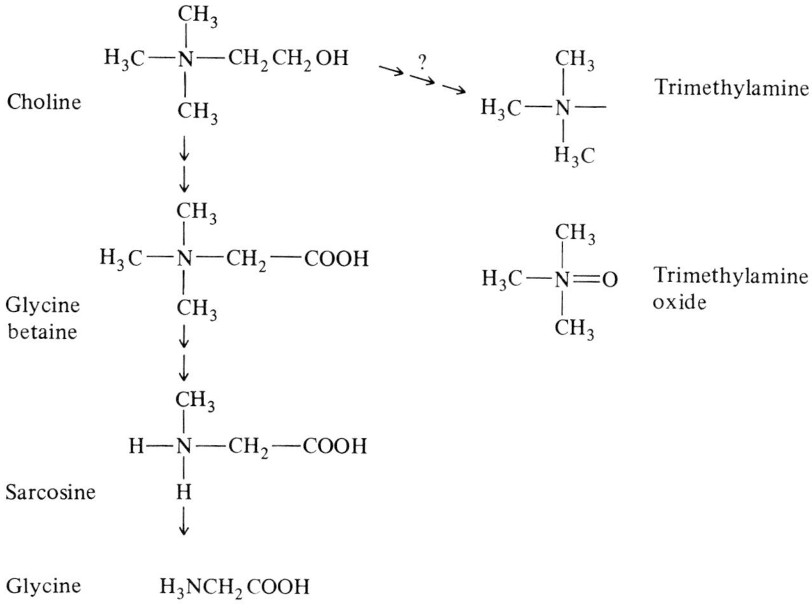

FIGURE 17–35 Schematic summary of various nitrogenous waste products in animals.

proteins are often present in excreta and thus can constitute a minor form of nitrogenous excretion.

Nitrogenous choline-like compounds are sometimes important nitrogenous products, e.g., betaine (glycine betaine), sarcosine, trimethylamine, and trimethylamine oxide. A number of invertebrates synthesize these nitrogenous compounds from choline, and some urea-retaining marine vertebrates (e.g., elasmobranchs, coelacanth) either synthesize them or obtain them in their diet. These nitrogenous compounds are often important balancing osmolytes, and they may also have counteracting solute effects, i.e., ameliorate the detrimental effects of urea on protein function (see Chapter 16).

Patterns of Nitrogen Excretion

Animals that excrete their waste N primarily as ammonia are **ammonotelic**, those that excrete mainly urea are **ureotelic**, those that excrete purines are **purinotelic** (uric acid, **uricotelic**; guanine, **guanotelic**). Let us first examine the basic biochemistry and physiology and the general question of the relative advantages and disadvantages of each N waste product before we consider the phylogenetic patterns of nitrogen excretion.

Ammonotely. Ammonia is a suitable nitrogenous waste product for aquatic animals because it has a high solubility, a high diffusion coefficient due to its low molecular weight, and biological membranes that are typically quite permeable to NH_3 (but not to NH_4^+). Thus, ammonia is readily excreted across the body surfaces into the surrounding water, i.e., across gills, skin, or even cuticle. Ammonium ions can be excreted by active transport or ion exchange. However, ammonia and ammonium ions are very toxic and so the blood levels are generally low. Total ammonia + ammonium levels for some animals are as follows: mammals, 0.03 to 0.08 mM; fish, 0.2 to 0.3 mM; aquatic frogs, 0.5 mM; insects, 0.4 to 7.1 mM; marine crabs, 0.2 to 1.4 mM; land crabs, 2 to 7 mM; terrestrial snails, 0.1 to 0.2 mM; octopus, 0.2 to 0.7 mM; and annelids, 0.5 to 4.4 mM.

Ammonia is very soluble in water (520 g L^{-1} at 25° C, i.e., 30.6 M). It reacts with water to form NH_4^+ and OH^-.

$$NH_3 + H_2O \rightleftharpoons NH_4^+ + OH^- \qquad (17.7)$$

Consequently, ammonia synthesis affects acid base balance (see below). The equilibrium constant (K_{eq}) for this reaction is about 10^{-9} to 10^{-10}, i.e., pK is about 9 to 10 (Emerson et al. 1975). Consequently, the ratio NH_3/NH_4^+ (equals 10^{pH-pK}) is markedly dependent on pH. At physiological pH (7.4), about

99% of the "NH_3" is actually NH_4^+ and only 1% is NH_3. At lower pH (e.g., mammalian urine, pH 5.6), over 99.98% is present as NH_4^+. Temperature also has an effect on pK and the NH_3/NH_4^+. Often, the measured level of "ammonia" is actually the total NH_3 and NH_4^+ (e.g., the blood levels above); the actual ammonia and ammonium levels can be calculated from the appropriate pH and pK values.

Ammonia and especially NH_4^+ are extremely toxic. In general, 0.5 to 5 mM total ammonia is lethal to both vertebrates and invertebrates. Ammonia/ammonium toxicity has been measured experimentally by administration to the external medium (aquatic animals) or directly into body fluids (terrestrial animals). The toxicity level can be measured as the dose necessary to cause 50% mortality of the experimental animals, i.e., LD_{50}, but this methodology has recently been abandoned in favor of nonlethal methods for toxicity determination. The lethal effects of ammonia/ammonium include direct effects on the nervous system, changes in membrane permeability that affect iono- and osmoregulation, inhibition of Na^+ uptake by an Na^+-NH_4^+ exchange pump, effects on O_2 transport capacity of hemocyanin, effects on carbohydrate metabolism, and effects on acid-base balance. The toxicity of ammonia is generally pH dependent because NH_4^+ is considerably more toxic than NH_3. For example, administration of alkaline ammonium salts to mammals has a more toxic effect than administration of acidic ammonium salts; the LD_{50} for ammonium chloride in mice is 7 mmole kg^{-1}, whereas that of ammonium hydroxide is considerably lower, 2.6 mmol kg^{-1}. Invertebrates have a similar pH dependence of LD_{50}. Consequently, it is important to take the pH and pK into account, and determine the proportions of NH_3 and NH_4^+ to properly account for the toxic effects of a certain total ammonia concentration.

Ammonia excretion across the gills accounts for 60 to 90% of the total N loss in fish (Forster and Goldstein 1969). Ammonia diffuses across the gill as NH_3, or NH_4^+ is exchanged for Na^+. The source of the ammonia is mainly NH_3 dissolved in blood but can also be blood amino nitrogen. Amino acid transamination in gill tissue to form glutamate, coupled with deamination by glutamate dehydrogenase, could release NH_3/NH_4^+ for excretion. Measurement of the branchial blood flow, the rate of NH_3 excretion, and the blood prebranchial-postbranchial difference in NH_3/NH_4^+ concentration suggests that ammonia extraction from the blood in some fish (e.g., carp) can account for all NH_3 excretion, whereas in other fish it only accounts for about 60% of the excretion (e.g., *Myoxocephalus*). In normal rainbow trout, nonionic diffusion (i.e., NH_3) across the gills accounts for ammonia excretion; a trans-branchial pNH_3 of only $3 \ 10^{-3}$ Pa (55 μtorr) is required to excrete the observed rate of NH_3 production given normal ambient water and plasma NH_3 concentrations (Cameron and Heisler 1983). However, active extrusion of NH_4^+ in exchange for Na^+ may be a significant route for NH_3 elimination at high environmental ammonia concentrations.

Animals can excrete gaseous NH_3 across their body surfaces by diffusion. The pNH_3 of alveolar air in mammals is the same as that of the blood, but generally only minute amounts of NH_3 are expired because the blood pNH_3 is very low (Jacquez, Poppell, and Jeltsch 1959; Robin et al. 1959). However, the guano bat has a high blood pNH_3 and excretes significant amounts of NH_3 across its lungs. Gaseous ammonia excretion is significant in some terrestrial snails and isopods. In isopods, the ammonia is not transported by the blood to the site of elimination as NH_3 or NH_4^+ since these are toxic, but as glutamate and glutamine (Wieser 1972). The NH_3 is produced at the release site by high concentrations of glutaminase. There is a close correspondence between the rate of ammonia excretion and the glutaminase concentration.

The reaction between ammonia and water forms ammonium ions and hydroxyl ions; this has major implications for acid-base balance. The ability of ammonia to act as a base

$$NH_3 + H^+ \rightleftharpoons NH_4^+ \qquad (17.8)$$

can "trap" hydrogen ions for regulation of local pH. For example, mammalian urine is generally acidic (pH 4.5 to 6.0) to excrete the acid end products of metabolism. However, even at a pH of 4.5 the urine can eliminate only about 1% of the metabolic acid load; the urine pH cannot be lower than 4.5 because this pH represents the maximum achievable pH gradient between tubular fluid and blood. To accomplish renal excretion of metabolic acids, H^+ is "trapped" in the urine by various buffer systems, one of which is ammonia (others are phosphate, urate, and citrate). The tubular epithelial cells contain glutaminase, which synthesizes NH_3 from glutamine. About 60% of the NH_3 is derived from glutamine, with the rest derived from other amino acids. The ammonia diffuses freely into the tubule where it reacts with H^+ to form NH_4^+, which cannot freely diffuse out of the tubule because of its charge.

A similar H^+ "trapping" role has been suggested for NH_3 in pulmonate snails for volatilization and excretion of gaseous ammonia. Carbonic anhydrase in the lung and mantle tissues would keep the pCO_2 low, and so blood entering these tissues would lose CO_2 to the lungs and mantle, keeping the blood alkaline. This would increase the proportion of "ammonia" that is NH_3 rather than NH_4^+ and increase the pNH_3. Consequently, NH_3 would diffuse into the lungs and mantle tissue and combine with H^+ produced by carbonic anhydrase. The high pNH_3 could also promote $CaCO_3$ deposition in the mantle fluid for shell calcification. Removal of NH_4^+ from tissues probably also promotes calcification in corals.

Ureotely. Terrestrial animals generally do not excrete ammonia because its toxicity requires excretion by dilution into large volumes of water. A common mechanism used by terrestrial animals to avoid the toxic effects of NH_3 and NH_4^+ is to convert them to a less toxic nitrogenous waste product, urea, $CO(NH_2)_2$. Urea is very soluble (1190 g L^{-1}) but is less toxic than ammonia. Urea is an important N waste product in many vertebrates; some annelids, mollusks, crustaceans, and insects excrete small amounts of urea.

Many vertebrates synthesize urea from ammonia and aspartate via the **urea cycle** (although not all vertebrates can synthesize urea). The urea cycle effectively converts 2 ammonia molecules to 1 urea molecule, requiring expenditure of 2 ATP (see Supplement 17–1, page 888). Carbamoyl phosphate is synthesized by carbamoyl phosphate synthethase (CPS) and then condensed with ornithine to form citrulline, then argino-succinate, and then arginine. The arginine is converted to urea and ornithine by arginase. The enzymes of the urea cycle are identical in all vertebrates, except that CPS requires glutamine (CPS III) and arginase is intramitochondrial in lower vertebrates, whereas higher vertebrate CPS requires NH_4^+ (CPS I) and arginase is cytosolic.

There is a complex compartmentalization of the urea cycle enzymes between the cytoplasm and the mitochondrial matrix. In lungfish, amphibians, some reptiles, and mammals (birds and some reptiles lack a complete set of urea cycle enzymes) carbamoyl phosphate synthesis (by CPS I) and the important sources of NH_4^+ for this reaction occur in the mitochondrial matrix. The rest of the urea cycle occurs outside of the mitochondrion. Consequently, citrulline must be transported out of the mitochondrion and ornithine into the matrix. Transfer of an NH_2 from cytoplasmic aspartate to citrulline forms argino-succinate and then fumarate, which enters cytosolic C-pathways (malate, pyruvate, acetyl-CoA, glucose). After formation of urea and ornithine by cytoplasmic arginase, the ornithine is retransferred into the mitochondrial matrix by the electrical gradient across the inner mitochondrial membrane at an effective metabolic cost of about 0.25 high-energy phosphate bonds.

Aquatic vertebrates are generally ammonotelic, not ureotelic, and do not have a functional urea cycle. However, elasmobranchs and the coelacanth use urea as an important osmolyte (Chapter 16). Some teleost fish, such as the toadfish *Opsanus*, the air-breathing fish *Heteropneustes*, an intertidal blenny *Blennius*, and an alkaline lake tilapia *Oreochromis*, excrete urea. All of these fish experience environmental conditions that might be detrimental to ammonia excretion (e.g., stagnant water, terrestriality, or high pH), and urea excretion is advantageous over ammonia excretion despite the higher metabolic cost (Walsh, Danulat, and Mommsen 1990). Thus, elasmobranchs, the coelacanth, and the various other fish described above have a complete, functioning urea cycle. Compared with higher vertebrates, these fish have a slightly different urea cycle (CPS III) and a slightly different compartmentalization of their urea cycle enzymes (intramitochondrial arginase) (Figure 17–36). Interestingly, the urea cycle of the coelacanth is like that of elasmobranchs rather than the more closely related tetrapod vertebrates.

Many invertebrates synthesize urea. Earthworms have all of the urea cycle enzymes, although most are cytosolic and their arginase has a lower molecular weight than the arginase of vertebrates (Campbell and Bishop 1970). Some pulmonate snails (e.g., *Helix*, *Otala*) have a complete complement of urea cycle enzymes. Their CPS and ornithine transcarbamylase enzymes are intramitochondrial with the remainder being cytosolic, as in amphibians and mammals (Speeg and Campbell 1968), but their CPS is type III. Urea synthesis in corals may remove NH_4^+ from tissues to promote calcification and also to form allantoic acid for Ca^{2+} transport as calcium allantoate and calcium allantoinate (Crossland and Barnes 1974).

A few insects excrete some urea, but its metabolic origin is unclear. The urea is probably not of dietary origin and cannot be attributed to purine degradation (because insects do not have allantoicase to convert allantoin to urea; see above). Some of the urea cycle enzymes, such as arginase, are readily demonstrated in insects, but others, such as CPS, ornithine transcarbamylase, and argino-succinate synthetase, are not as obviously present (Cochran 1985). Urea excretion may reflect dietary arginine breakdown, rather than a functional urea cycle. Alternatively, urea formation may indicate a useful biochemical pathway for proline synthesis. Arginine can be converted to ornithine (producing urea) and thence to proline, an important metabolite for flight muscle in insects.

Urea synthesis has significant acid-base implications. The metabolism of amino acids produces not only NH_3/NH_4^+ but also equivalent amounts of HCO_3^- (Atkinson and Bourke 1987). For example,

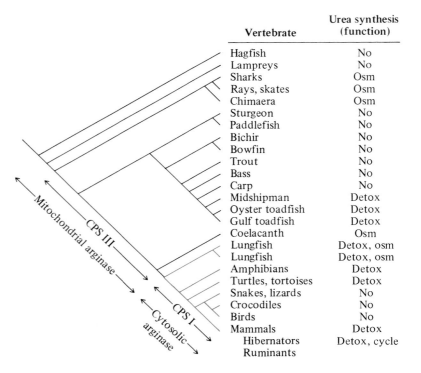

FIGURE 17–36 Schematic summary of the presence or absence of urea synthesis by the urea cycle in vertebrates, and the type of carbamoyl phosphate synthetase (CPS) and arginase. The possible functions for urea synthesis include balancing ammonia detoxification (detox), osmolytes (osm), and nitrogen cycling between the gut and liver (cycle). *(Modified from Mommsen and Walsh 1989; Mommsen, personal communication.)*

Vertebrate	Urea synthesis (function)
Hagfish	No
Lampreys	No
Sharks	Osm
Rays, skates	Osm
Chimaera	Osm
Sturgeon	No
Paddlefish	No
Bichir	No
Bowfin	No
Trout	No
Bass	No
Carp	No
Midshipman	Detox
Oyster toadfish	Detox
Gulf toadfish	Detox
Coelacanth	Osm
Lungfish	Detox, osm
Lungfish	Detox, osm
Amphibians	Detox
Turtles, tortoises	Detox
Snakes, lizards	No
Crocodiles	No
Birds	No
Mammals	Detox
Hibernators	Detox, cycle
Ruminants	

the metabolism of the amino acid alanine yields CO_2, HCO_3^-, NH_4^+, and metabolic water.

$$CH_3CHNH_3^+COO^- + 3\,O_2$$
$$\rightarrow 2\,CO_2 + HCO_3^- + NH_4^+ + H_2O \quad (17.9)$$
$$or \rightarrow 3\,CO_2 + NH_3 + 2H_2O$$

The CO_2 is readily eliminated across the lungs or gills but the fate of the HCO_3^- and NH_4^+ is complex. At physiological pH, the equilibrium between NH_3 and NH_4^+ strongly favors NH_4^+ and HCO_3^- in the above reaction. Aquatic animals may excrete sufficient NH_3 across the gills/skin/excretory organs or exchange the NH_4^+ and HCO_3^- for Na^+ and Cl^- across the integument or gills. Either of these two excretion strategies would maintain acid-base balance. Ureotelic animals convert the NH_4^+ to urea, but the remaining HCO_3^- influences the acid-base balance. For example, a human produces about 10 to 20 moles of CO_2 per day and 0.5 to 1.0 mole of HCO_3^- and NH_4^+ per day. The accumulation of 1 mole of HCO_3^- would cause lethal alkalosis. If all the HCO_3^- were excreted in the urine, it could crystallize carbonates and create another grave health danger. Fortunately, ureagenesis automatically maintains acid-base balance because it utilizes equivalent amounts of NH_4^+ and HCO_3^-. The net balance for urea synthesis is as follows.

$$2\,NH_4^+ + 2\,HCO_3^-$$
$$\rightarrow CO(NH_2)_2 + CO_2 + 3H_2O \quad (17.10)$$

Thus, ureagenesis can be considered to be an ATP-driven proton pump, transferring an H^+ from NH_4^+ to HCO_3^- and allowing CO_2 excretion by the lungs rather than HCO_3^- excretion by the kidneys. The metabolic cost of this is about 2 ATP per H^+ transfer.

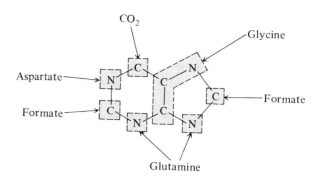

FIGURE 17–37 Metabolic precursors for uric acid synthesis. *(Modified from Hartman 1970.)*

Purinotely. Uric acid is the major purine excreted by many invertebrates as well as by many reptiles and birds. Uric acid is synthesized from glycine (C_2N), aspartate (N), glutamine (N_2), carbon dioxide (C), and the general carbon pool (e.g., formate, tetrahydrofolate, C_2; Figure 17–37). The complex biosynthetic pathway for uric acid was first determined for birds using radioisotope tracer studies (Buchanan et al. 1957). The same, or at least a similar, *de novo* synthesis pathway is thought to occur in other uricotelic animals. A nucleotide, inosine monophosphate (IMP), is first synthesized from glutamine and then the remainder of the purine ring is added (see Supplement 17–1; page 888).

Uric acid, urate salts, and other purines are exceedingly insoluble compared to ammonia and urea (Table 17–7). Such a low solubility, coupled with low toxicity, makes urates (and other purines, especially guanine) an almost ideal nitrogenous

TABLE 17–7

Solubilities of nitrogenous waste products. *(Data from Bursell 1967; McGilvery and Goldstein 1983.)*				
			Solubility	
Nitrogen Waste	**Formula**	**MWt**	**(g L^{-1})**	*mM*
Ammonia	NH_3	17	890	52.4
Ammonium bicarbonate	NH_4HCO_3	79	119	1.5
Urea	$CO(NH_2)_2$	60	1190	39.8
Allantoin	$C_4H_6O_3N_4$	158	0.6	0.015
Allantoic acid	$C_4H_8O_4N_4$	176	(slight)	(slight)
Uric acid	$C_5H_4O_3N_4$	168	0.065	0.0015
Sodium urate	$C_5H_2O_3N_4Na_2$	212	0.83	0.016
Potassium urate	$C_5H_2O_3N_4K_2$	244	(slight)	(slight)
Guanine	$C_4H_5ON_5$	151	0.039	0.0013
Xanthine	$C_5H_4O_2N_4$	152	2.6	0.068
Hypoxanthine	$C_5H_4ON_4$	136	0.7	0.021
Arginine	$C_6H_{14}O_2N_4$	174	150	3.4

waste product for maximal conservation of urinary water. Purinotely is also advantageous for animals with cleidoic eggs that must store nitrogenous wastes (insects, lizards, birds).

The urine of purinotelic animals can be a slurry, containing high amounts of precipitated urate or guanine (e.g., some mollusks, insects, arachnids, reptiles, and birds). Animals that reabsorb solutes and water from their feces in the rectum further conserve water by precipitating more urate. For example, bird urine consists of both precipitated urate and urate in the supernatant (McNabb 1974). The supernatant contains both dissolved urate (at very low concentrations) and colloidal urate, which may be present as unstable lyophobic colloids or as more stable lyophilic colloids. The solubility of urate is inversely proportional to the ion concentration; Na^+ and NH_4^+ (but not K^+) at moderate concentrations makes the urate unstable, and it tends to flocculate and precipitate. Urate coprecipitates with NH_4^+, Na^+, and K^+; most of the urinary Na^+ and K^+ are precipitated when most of the urate is precipitated (Figure 17–38). The ions appear to precipitate not only as monobasic urate salts but also as other salts (e.g., NaCl, KCl) because of physical rather than chemical reasons. Thus, urate precipitation also promotes the precipitation of other osmolytes, contributing to an even greater water economy. Urinary proteins and mucopolysaccharides, however, tend to make urate colloids hydrophilic and stable.

Uric acid may be effectively "excreted" by storage in tissues because of its low solubility, i.e., urates can be deposited permanently in body tissues but are effectively removed from the animal's active metabolic pool. For example, some adult lepidopterans permanently store purines and pteridines (see Other Nitrogenous Waste Products, above). These deposited N wastes can also function as pigments. Stored urates can even be a potential source of nitrogen for insects when their dietary nitrogen intake is low (e.g., termites). Urates deposited in the fat body, or synthesized in the fat body and released into the circulation, can be recycled by fat body or gut microbes to provide nitrogen (and also carbon) to the general metabolic pool. Stored urates have a similar role in aquatic mogulid tunicates (Saffo 1988). Urate is stored in a "renal" organ but an excretory or storage role for this organ is not obvious. Why synthesize urate, at a considerable metabolic cost, for nitrogen excretion when ammonia could be readily eliminated by these aquatic tunicates, at essentially no metabolic cost? Why store the urate in an organ without an excretory duct? The urate is probably a stored source of nitrogen; symbiotic protists (*Nephromyces*) in the "renal" organ have uricolytic activity and probably release the urate nitrogen into the symbiont and host nitrogen pools.

Phylogenetic Patterns. Each form of nitrogenous waste has relative advantages and disadvantages that are related to the general availability of water in the animal's environment. Let us briefly elaborate on these environmentally determined advantages and disadvantages (Table 17–8).

Ammonia is highly toxic, but it is a small, very soluble solute with a high diffusion coefficient. It is

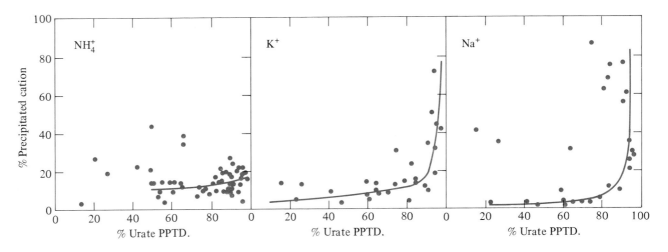

FIGURE 17–38 Percentage of urinary cations that are precipitated as a function of the percentage of precipitated urate. *(From McNabb 1974.)*

TABLE 17–8

	Ammonia	Urea	Purines
Comparison of properties of nitrogenous waste products.			
Toxicity	High	Moderate	Low
Solubility	High	High	Very low
Diffusion coefficient	High	Intermediate	Low
Metabolic cost for synthesis	None	Low	High
Means of transport	Passive (NH_3) and active (NH_4^+)	Passive and active	Passive and active
H/N	3	2	1
C/N	0	0.5	1–1.2
O/N	0	0.5	0.2–0.75
N/osmotic particle	1	2	4–5

transported across membranes by diffusion (NH_3) or by active transport (NH_4^+). There is essentially no metabolic cost involved with ammonia excretion since it is the initial form of the nitrogen waste. Ammonia has only 1 N per osmotic particle, but there is no C or O loss accompanying its excretion; however, 3 H are lost per N. These considerations suggest that ammonia is a suitable nitrogenous waste product for aquatic animals with access to large volumes of water for excretion of ammonia, either by exchange across the integument or in a copious urine flow. In contrast, urea and purines are suitable waste products for terrestrial animals because they are less toxic and their excretion requires less water. There are more N atoms per osmotic particle for these wastes than for ammonia. However, more C and O are lost per N and there is a metabolic cost associated with their synthesis.

The form of the nitrogenous waste product used by a particular animal also reflects its phylogenetic history (Baldwin 1949). Patterns of nitrogenous excretion have remained fairly conservative within taxa, although there are some unusual exceptions (e.g., there are a few uricotelic amphibians); there are also some unusual consistencies (e.g., terrestrial isopods are ammonotelic). A further significant factor related to patterns of nitrogen excretion is the mode of embryonic development, particularly for terrestrial animals (Needham 1931). There are three major strategies for embryonic development in terrestrial animals: viviparity, in which the embryo is protected and nurtured within the mother's body and is born, e.g., mammals; semiaquatic development of eggs in moist habitats, e.g., some terrestrial amphibian eggs; and cleidoic eggs, where the egg is a relatively impermeable, hence closed, system in which nitrogenous wastes must accumulate during development, e.g., insects, lizards, birds. It

is essential that the nitrogen wastes that accumulate in cleidoic eggs have a low toxicity, hence purinotely is essentially a corequisite for the evolution of cleidoic eggs.

We can summarize these factors influencing the form of the nitrogen waste product as the Baldwin-Needham hypothesis, that aquatic invertebrates tend to excrete ammonia (although some marine vertebrates tend towards ureotely) and terrestrial animals tend to be ureotelic if they don't have cleidoic eggs, or uricotelic if they have cleidoic eggs. This broad generalization for the pattern of nitrogenous excretion is generally applicable over a broad taxonomic survey (Figure 17–39).

Many aquatic invertebrates excrete ammonia as we would expect. Polychete worms are aquatic and are presumed to excrete ammonia. Their blood levels of ammonia and urea are about 0.04 to 8.0 mM and 0.04 to 6.0 mM, respectively; uric acid concentrations for blood are much lower, about 0.001 to 0.4 mM (Oglesby 1978). If urine concentrations reflected these blood concentrations, then polychetes are both ammono- and ureotelic. Oligochetes are more terrestrial, although they generally are restricted to moist habitats. Earthworms primarily excrete urea during periods of water restriction. Blood ammonia and uric acid levels are similar to those of polychetes, but blood urea levels can be much higher, 25 to 30 mM. Hirudinea (leeches) would be expected to be mainly ammonotelic because of their aquatic or moist habitat requirements.

Mollusks are quite varied in habitat (marine, freshwater, terrestrial) and consequently in their pattern of nitrogenous excretion (Bishop, Ellis, and Burcham 1983). Ammonia is the principal excretory product for bivalves and cephalopods, although other nitrogenous compounds (urea, uric acid, and amino acids) are also excreted in variable amounts

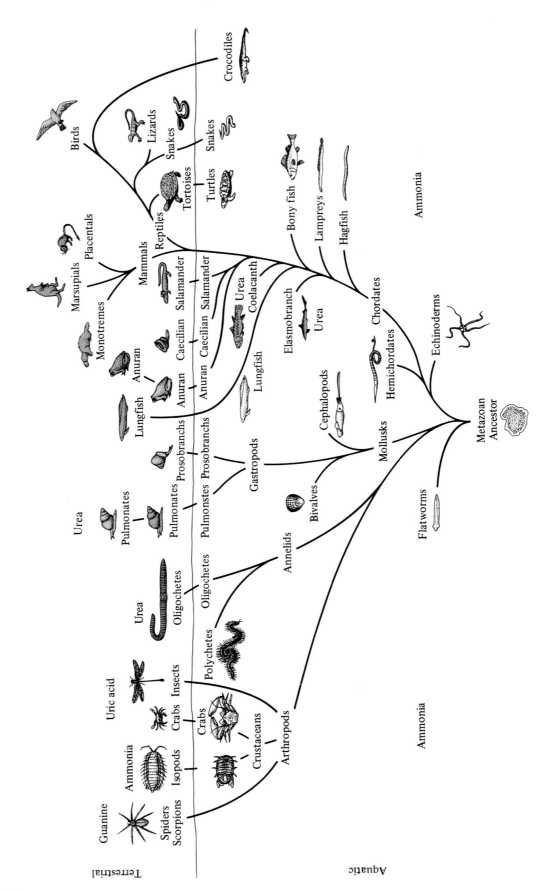

FIGURE 17–39 Highly schematic summary of the patterns of ammonotely, ureotely, and uricotely in the major groups of aquatic and terrestrial animals.

(e.g., up to 4% as urea). Prosobranch snails also tend to be ammonotelic but the semiterrestrial and terrestrial species may accumulate significant levels of uric acid (up to 1/3 total N from amino acid metabolism). The more terrestrial species tend to be uricotelic, both as adults and during egg development. There are, however, marked species-specific and phylogenetic differences that are not strictly related to terrestriality. Nitrogen excretion by pulmonate snails is a complex pattern of ammonotely, ureotely, and uricotely. Ammonia N production by aquatic *Biomphalaria* is about 4 × that of urea N when aquatic and feeding, but urea N production increases when estivating in moist terrestrial conditions. Uric acid levels are always low. Lymneids are primarily ammonotelic when aquatic but become uricotelic when estivating (e.g., 77% N is uric acid, 17% is urea, 5% is ammonia). Slugs are terrestrial but are restricted to moist environments. They are principally purinotelic; their feces contain uric acid (about 50% of the N) and xanthine (15%) as well as guanine, hypoxanthine, and adenine. However, slugs have a complete set of functional urea cycle enzymes in their tissues and excrete about 50% of their N as urea while feeding. Ammonia levels are insignificant. Terrestrial pulmonates (e.g., *Helix*, *Otala*) are primarily purinotelic (uric acid and some guanine and xanthine) and generally little or no ammonia or urea are excreted. The purines have been shown to be derived from *de novo* synthesis (i.e., rather than by nucleotide degradation). About 45% of the N excretion is ammonia in estivating snails (0.6 μatom of ammonia N g^{-1} day^{-1} and 0.7 μatom uric acid N g^{-1} day^{-1} in estivating *Otala*). All pulmonate snails can apparently form urea by *de novo* synthesis from NH_3 and by arginine metabolism, but the general absence of urea is attributed to high activities of urease. The NH_3 can be liberated through the shell.

Freshwater and marine crustaceans are primarily ammonotelic with 50 to 90% of their nitrogen excreted as ammonia. Amino acid N is a significant N waste (2 to 25% of total N). Urea excretion is generally a small, although significant fraction (1 to 20%). Uric acid and other nitrogenous wastes are generally insignificant, although TMAO is often a significant N waste. Purine metabolism forms uric acid, most of which is degraded by the complete series of purinolytic enzymes to ammonia. It is doubtful if crustaceans have a functional urea cycle. Some terrestrial crustaceans are unusual (for terrestrial animals) in retaining ammonotely. Ammonotelic isopods have already been mentioned above. Some terrestrial land crabs are consistently uricotelic (e.g., *Carcinus*), whereas others are am-

monotelic when copious water is available but are uricotelic when water is restricted (e.g., *Ocypode*, *Cardisoma*).

The onycophoran *Peripatus*, which lives in moist habitats, is uricotelic. The more terrestrial arthropods tend to be uricotelic. The centipede *Scolopendra* is uricotelic. Insects are primarily uricotelic, but some can excrete a wide range of N waste products, including ammonia, urea, allantoin, and allantoic acid (Cochran 1985). Several aquatic species excrete primarily ammonia, e.g., *Sialis* excretes about 90% of its N as ammonia. Some terrestrial species (e.g., blowflies, cockroaches) also are primarily ammonotelic via hindgut excretion and the feces, rather than by direct volatilization into the air. Blowfly larvae experience high ammonia levels in their environment and excrete ammonia-rich feces; their hindgut actively secretes NH_4^+ into their excreta. A number of insects excrete urea but generally in relatively small amounts (and it may be a metabolic byproduct rather than an N waste product). Arachnids (scorpions, whipscorpions, spiders, mites, ticks) excrete primarily guanine with small and variable amounts of uric acid. Scorpions generally excrete 40 to 90% of their total N waste as guanine and 0 to 2% as uric acid. Dietary uric acid gain (insects are a major dietary item for scorpions) may account for the low and variable uric acid content in scorpion feces. Similarly, fish-eating birds excrete some guanine derived from their diet rather than by their own metabolic pathways. The only significant purine excreted by spiders is guanine (Anderson 1966). One scorpion, *Paruroctonus*, excretes xanthine rather than guanine or uric acid (Yokota and Shoemaker 1981).

There is a similar diversity in pattern of nitrogen excretion among vertebrates. Cyclostome and teleost fish excrete primarily ammonia. Freshwater lampreys excrete primarily ammonia across their gills; urea is <1% of their total N excreted (in the urine). Freshwater teleosts excrete primarily ammonia but also some urea and TMAO across their gills. From 2.5 to 24.5% of their N eliminated via urine is in a variety of forms including (in decreasing order of importance) creatine, urea, ammonia, amino acids, uric acid, and creatinine. Marine teleosts similarly excrete mainly ammonia (but some urea and TMAO) across their gills and a variety of N compounds in the urine. The amphibious mudskipper *Periophthalmus* excretes primarily ammonia when in brackish water (0.77 mmole NH_3 kg^{-1} hr^{-1}; 0.06 mmole urea kg^{-1} hr^{-1}) but is more ureotelic in seawater (0.49 and 0.36, respectively; Gordon et al. 1965). Other amphibious fish and an alkaline lake fish also excrete urea (see above).

Marine elasmobranchs and chimaeras are ureotelic; they accumulate high concentrations of urea and TMAO as intracellular and extracellular balancing osmolytes (see Chapter 17). Urea and TMAO are actively reabsorbed by the renal tubules but some is nevertheless excreted in the urine. The primary urinary N waste of many freshwater elasmobranchs is urea, but the freshwater Amazon ray excretes ammonia rather than urea (Gerst and Thorsen 1977).

The Australian lungfish *Neoceratodus* is an obligate aquatic fish and excretes primarily ammonia. In contrast, the African lungfish *Protopterus* excretes up to 50% of its N waste as urea when aquatic and accumulates urea as its N waste when estivating out of water. The coelacanth *Latimeria*, like elasmobranchs, uses urea as a balancing osmolyte, hence is primarily ureotelic.

Patterns of N excretion in amphibians are diverse and subject to ontogenetic and environmental affects. Some amphibians are ammonotelic when in water and ureotelic on land. Aquatic tadpoles and adult amphibians are generally ammonotelic, e.g., tadpoles and adult *Xenopus* and adult *Pipa* normally excrete about 60 to 80% of their N waste as ammonia (Munro 1953; Cragg, Balinsky, and Baldwin 1961). Semiterrestrial and terrestrial adult amphibians generally excrete a larger fraction of their N waste as urea, e.g., >80% in *Bufo bufo*, although their aquatic tadpoles are primarily ammonotelic. In fact, the tadpole liver usually does not have high levels of the urea cycle enzymes and the levels do not increase until metamorphosis. However, some tadpoles are ureotelic. A few aquatic amphibians can be acclimated to high salinities (*Xenopus*) and one amphibian (the crab-eating frog *Rana cancrivora*) is semi-marine; these amphibians are primarily ureotelic, using the urea as a balancing osmolyte. Two genera of unrelated tree frogs, *Chiromantis* in Africa and *Phyllomedusa* in South America, excrete uric acid (Loveridge 1970; Shoemaker, Balding, and Ruibal 1972).

Reptiles are quite diverse with respect to their nitrogenous waste products, although all are to some extent uricotelic (Minnich 1979). Many aquatic reptiles are significantly ammonotelic. For example, the freshwater crocodile *C. niloticus* excretes 25% of its nonprotein urinary N as ammonia, 4% as urea, and 68% as uric acid. Some turtles and the tuatara can also excrete appreciable amounts of ammonia and urea. The partitioning of excreted N in turtles depends markedly on the environmental availability of water (Moyle 1949). Aquatic species excrete primarily ammonia (e.g., *Kinosternum*: 24% ammonia N, 23% urea N, 0.7% uric acid N); more terrestrial species excrete mainly urea (e.g., *Kinixys*:

6%, 61%, and 4%, respectively); dry terrestrial species excrete mainly uric acid (e.g., *Testudo graeca*: 4%, 22%, and 52%, respectively); desert species excrete primarily uric acid (e.g., *Gopherus*: 4%, 3%, and 93%, respectively). Some lizards excrete significant amounts of ammonia and urea but are primarily uricotelic, depending on the water availability. For example, *Anolis* (from a moist terrestrial environment) excretes about 25% ammonia N, 15% urea N, and 58% uric acid N; *Sauromalus* (from a desert environment) excretes about 1%, 0, and 99%, respectively. Snakes are predominantly uricotelic, although ammonia may constitute 2 to 20% of the excreted N in some species.

Birds are principally uricotelic, but significant levels of other N waste are present in their excreta, including ammonia (2 to 30% N), urea (2 to 10% N), creatine (0.2 to 8% N), and minute amounts of creatinine (Sturkie 1986).

Mammals are primarily ureotelic, although various other N wastes may be present in the urine. For example, human urine contains about 30 mM urea (1820 mg%; 60 mM N), 2.5 mM uric acid (42 mg%; 10 mM N), and 17 mM creatinine (196 mg%; 51 mM N). Some cricetid rodents, at least when water deprived, produce urine that contains a milky precipitate of white, needle-like allantoin crystals (Buffenstein, Campbell, and Jarvis 1985). For example, the cricetid *Desmodillus* excretes about 55 mg urinary N day^{-1}, of which about 19.8 is crystalline allantoin (36% of total N), 19.6 is dissolved allantoin (36% of total N), and 27.4 is urea (28%). In contrast, the murid rodent *Aethomys* excretes about 41 mg urinary N day^{-1}, of which 17.7 (33% of total N) is allantoin and 27.4 (67%) is urea. This relatively large amount of allantoin is much greater than could be accounted by purine metabolism, hence reflects a shift in N excretion pattern for amino N; these mammals are essentially allantoinotelic!

Summary

Animals must excrete ions, organic wastes, and water. There are a variety of different avenues for excretion of these solutes, generally either the body surface or tubular excretory organs.

The integument can actively transport ions, either into or out of the animal. A variety of mechanisms transport water across the integument and membranes, but none involve direct, active transport of water molecules. Osmosis is a passive mechanism for transport of water from a dilute solution to a more concentrated solution across a semipermeable barrier. Water can be cotransported with active ion exchange across membranes or transported by a double-membrane compartment system or by a standing solute-gradient system. Bulk transport of water may be accomplished by pinocytosis (the volume pump model) or formed bodies. A few animals might transport water in conjunction with ions by electroosmosis.

Most animals have tubular excretory organs that transport body fluids to the exterior, often highly modifying the fluid composition during its passage along the tubule. Nephridia are tubular excretory organs that develop from a pore in the outer ectoderm (the nephridiopore) into the interior. Protonephridia have a blind interior end that draws fluid into the tubule by ciliary action of "flame cells." Metanephridia have a ciliated, funnel-like opening (nephridiostome) that draws coelomic fluid into the nephridium. Some animals have coelomoducts that develop from the inside of the animal and empty via the coelomopore to the outside; their ciliated, funnel-like opening to the body cavity (coelomostome) draws coelomic fluid into the coelomoduct. Some animals have a nephromixium, a fused nephridium and coelomoduct. Many arthropods have Malpighian tubules, which are tubules that open into the midhindgut rather than opening to the exterior via a pore.

Urine generally is formed at the inner end of the nephridia, coelomoducts, or Malpighian tubules either by bulk water flow through ciliated nephridiostomes or coelomostomes or by a filtration process. The flow of water into the tubule for filtration can be caused by a hydrostatic pressure, difference due to ciliary activity or to the vascular pressure, or by solute-linked transport. The filtration membrane generally excludes large particles and solutes from entering the tubule (e.g., blood cells, protein) but allows smaller solutes through, in inverse proportion to their molecular size.

Many of the filtered solutes are nutrients (e.g., glucose, amino acids) or essential body ions (Na^+, K^+, Cl^-); these are often reabsorbed by the excretory tubules. Water is also often reabsorbed passively with the active solute uptake. Waste products (e.g., NH_3, urea, uric acid) are generally not reabsorbed and many are actively secreted into the excretory tubule. Excretory organs of a few animals can osmotically concentrate the fluid as it passes through the tubule. For example, the avian and mammalian nephrons can osmotically concentrate urine from 2 × to over 15 × the body fluid osmotic concentration. A few other vertebrates and invertebrates can also form a slightly hyperosmotic urine, e.g., some turtles, polychaete worms, and insects.

Protozoans and sponges lack complex excretory organ systems; contractile vacuoles excrete water and solutes. The activity of the contractile vacuole

depends on the ambient osmotic conditions; it contracts more frequently and expels a greater volume of water if the animal is in freshwater.

Many "lower" invertebrates have protonephridial excretory organs. Many "higher" invertebrate phyla (annelids, mollusks) have protonephridia or metanephridia. The proto- or metanephridia of freshwater animals often produces a copious hypoosmotic urine so as to excrete water but to conserve body fluid ions. In marine animals, the urine is often isoosmotic with body fluids and the volume is low to minimize water loss. The excretory tubules are consolidated in higher mollusks to form a discrete kidney. The tubules lack nephrostomes, and urine is formed by filtration, either from the pericardium or in the kidney itself.

Arthropods have a combination of coelomoduct excretory organs and Malpighian tubules; they may also have various other gut-derived excretory tubule systems. Crustaceans rely primarily on antennal and maxillary glands for solute excretion. Urine is formed by filtration at the terminal coelomosac, which has an arterial blood supply. There generally is considerable reabsorption of solutes and some water reabsorption along the tubule. Coxal glands are important in primitive arthropods, but their excretory role is supplanted by Malpighian tubules in higher arthropods.

The Malpighian tubules of insects and arachnids form urine by active K^+ secretion into the tubule; water and other solutes follow passively. Parts of the tubules are specialized for modification of the primary KCl-rich urine by solute and water reabsorption. The "urine" enters the gut at the midhindgut junction and is subsequently highly modified during its passage through the hindgut/rectum.

The nephron, which is the excretory tubule of vertebrates, has a Bowman's capsule that contains a capillary network, the glomerulus. Urine is formed by hydrostatic filtration of blood fluid from the glomerulus into the capsular space, which is continuous with the lumen of the nephron tubule. The structure and function of the nephron tubule varies markedly with the environment (e.g., marine, freshwater, or terrestrial). Freshwater fish have nephrons with large glomeruli that excrete excess water but conserve ions. Marine fish have nephrons with reduced or absent glomeruli to excrete ions but to minimize water loss. The nephrons of marine elasmobranchs, which use urea to osmoconform, actively reabsorb urea and TMAO from the filtrate so as to minimize their excretion.

Amphibian nephrons have a similar structure and function to freshwater fish nephrons. They produce copious amounts of dilute urine that is stored in a large bladder; the dilute urine has an important role as a water store to replace body water lost by evaporation during dehydration.

Reptilian and avian nephrons are adapted to minimize urinary water loss. The glomerular filtration rate remains relatively high and most water, nutrients, and important solutes are reabsorbed from the urine. Reptilian nephrons are not able to form a hyperosmotic urine, but some avian nephrons can osmotically concentrate urine. Reptiles and birds often do not have a bladder; urine drains from their ureters into the cloaca and is refluxed into the hindgut. The urine is considerably modified in the colon/cloaca by active solute uptake and passive water reabsorption. Many reptiles and birds have orbital (eye) or lingual (tongue) salt glands to eliminate excess ions. These salt glands generally excrete mainly NaCl in marine species and NaCl and KCl in terrestrial species.

The mammalian nephron has a considerable capacity to osmotically concentrate urine, and so the urine excretes solutes with little water loss. The loop of Henle, a hairpin bend in the nephron between proximal and distal convoluted tubules, establishes an osmotic gradient in the renal medullary tissue by active Cl^- transport out of the thick ascending limb of the loop of Henle without concomitant passive transport of water. Active transport of ions and urea diffusion out of the collecting duct further contributes to the medullary osmotic concentration gradient. The tubular fluid can osmotically equilibrate with this medullary osmotic gradient as it flows through the medulla in the collecting ducts, if the collecting ducts are permeable to water, to form a hyperosmotic urine. There is no osmotic equilibration if the collecting ducts are impermeable to water, and so the urine is dilute. The permeability of the collecting duct to water is controlled by the antidiuretic hormone (ADH).

Amino acid and nucleic acid metabolism produces nitrogenous wastes. The principal nitrogenous waste from amino acid metabolism is ammonia, NH_3 (or ammonium, NH_4^+). Ammonia is extremely toxic to animals, but aquatic species can readily excrete it across their integument or gill surface by diffusion because it is extremely soluble. A few terrestrial animals can excrete significant amounts of ammonia, either in their feces or urine, or by volatilization into the air. These animals are ammonotelic.

Semiterrestrial and terrestrial animals are generally limited by water availability and do not excrete ammonia because of its toxicity. The ammonia is synthesized into urea by the urea cycle, at a metabolic cost of about 4 ATP per urea. Urea is less toxic than ammonia but is still very soluble and readily eliminated in urine. These animals are ureotelic.

Many arthropods, reptiles, and birds convert ammonia to uric acid or other purines (guanine, xanthine). These purines are very insoluble and can be excreted at a high concentration with minimal water as a sludge or paste. These animals are purinotelic (e.g., uricotelic, guanotelic).

The metabolism of pyrimidine bases of nucleic acids forms ammonia, which can be excreted as such or converted to urea, uric acid, or guanine. Metabolism of the purine bases forms uric acid, which can then be degraded through a variety of intermediates (allantoin, allantoic acid, urea) to ammonia. The specific form of the nitrogen waste product depends on the particular species; e.g., primates, uric acid; other mammals, gastropod mollusks, allantoin; some insects and teleosts, allantoic acid; most fish, amphibians, mollusks, urea; most aquatic invertebrates, ammonia.

The particular pattern of nitrogen waste excretion depends on the taxonomy of the animal and its hydrational status. In general, aquatic animals tend to be ammonotelic with some tendency towards ureotelism in marine species. Terrestrial species tend to be either ureotelic or purinotelic. Many invertebrates and vertebrates facultatively alter their pattern of nitrogen excretion, being ammonotelic when water is readily available and ureotelic or purinotelic when water is restricted.

Supplement 17–1

Urea and Uric Acid Synthesis

· ·

Urea is formed in the urea cycle (or ornithine cycle) by a straightforward enzymatic degradation of arginine by the enzyme arginase so as to form ornithine. Continued urea synthesis requires the regeneration of arginine from ornithine, i.e., an ornithine cycle. Ornithine is reconverted via citrulline and arginosuccinate to arginine by the addition of carbamoyl phosphate and aspartate. Carbamoyl phosphate is synthesized from NH_4^+ and CO_2 by carbamoyl phosphate synthetase (CPS); the energy is derived from the conversion of 2 ATP to 2 ADP and 2 P_i. The $-CONH_2$ of carbamoyl phosphate is ultimately incorporated into urea. Citrulline is formed from carbam-

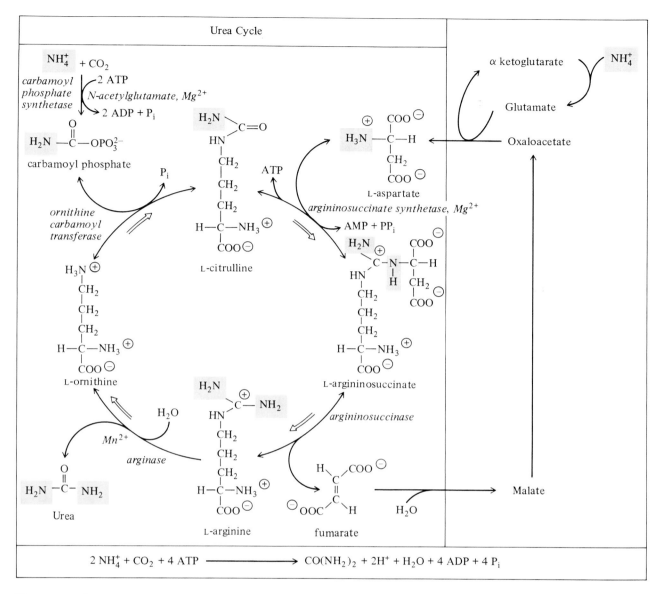

The urea cycle.

oyl phosphate and ornithine. The citrulline is condensed with aspartate to form arginosuccinate, which is split to form arginine and fumarate. Urea is then formed by the action of arginase to cleave urea from arginine, re-forming ornithine.

This cycle is not complete, however, because aspartate has been converted to fumarate; this is especially important because the aspartate is the source of one of the —NH₂ groups of the urea. The fumarate has a variety of possible metabolic fates, including conversion to pyruvate

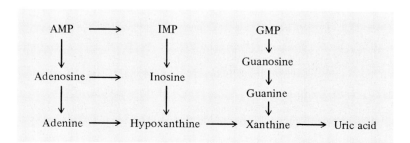

Schematic of pathways for conversion of purine nucleotides to xanthine and uric acid.

5-phospho-α-D-ribosyl pyrophosphate

inosine-5'-phosphate (IMP)

Schematic summary of the synthesis of uric acid, starting with 5-phospho-α-D-ribosylphosphate then via a number of intermediates to inosine-5'-phosphate (IMP). Only the side group is shown for the intermediate steps.

for fatty acid synthesis or acetyl-CoA formation and oxidation, conversion to phosphoenolpyruvate (PEP) for glucose synthesis, or conversion to oxaloacetate and then transamination to re-form aspartate. The latter net conversion of fumarate to oxaloacetate is part of the Krebs cycle. The oxaloacetate can then be converted to aspartate by the addition of $-NH_2$ from NH_4^+ via glutamate/α-ketoglutarate. The net urea cycle condenses two NH_4^+ with CO_2 to form urea, two H^+, and water; this requires the energy from hydrolysis of ATP.

The biosynthetic pathway for uric acid (and also other purines) proceeds via inosine monophosphate (IMP). The initial reaction is the transfer of the amide group from glutamine to form phosphoribosylamine from phosphoribosylphosphate. The remainder of the purine ring is then constructed around this nitrogen group to form IMP (a nucleotide).

IMP, adenosine-monophosphate (AMP), and guanosine-monophosphate are degraded to uric acid by a number of intermediary steps. IMP can be converted to inositol (a nucleoside) by a nucleotidase and then to the purine hypoxanthine by a nucleoside phosphorylase. Hypoxanthine is then oxidized to xanthine, which can be converted to uric acid by xanthine oxidase. Alternatively, IMP can be converted to xanthine monophosphate and then to guanosine monophosphate by addition of a further $-NH_2$ group from glutamine. The guanosine monophosphate can be converted to guanine in similar fashion as IMP conversion to uric acid. (*See McGilvery and Goldstein 1983; Cochran 1975.*)

Recommended Reading

Atkinson, D. E., and E. Bourke. 1987. Metabolic aspects of the regulation of systemic pH. *Am. J. Physiol.* 252:F947–56.

Bradshaw, S. D. 1986. *Ecophysiology of desert reptiles.* Sydney: Academic Press.

Dantzler, W. H. 1976. Renal function (with special emphasis on nitrogen excretion). In *Biology of the Reptilia,* vol. 5, edited by C. Gans and W. R. Dawson, 447–503. New York: Academic Press.

Foskett, J. K., and C. Schaffey. 1982. The chloride cell: Definitive identification as the salt-secretory cell in teleosts. *Science* 215:164–66.

Gilles, R., and M. Gilles-Baillien. 1985. *Transport processes, iono- and osmoregulation.* Berlin: Springer-Verlag.

Greger, R. 1988. *Advances in comparative and environmental physiology: NaCl transport in epithelia.* Berlin: Springer-Verlag.

Gupta, B. L., R. B. Moreton, J. L. Oschman, and B. J. Wall. 1977. *Transport of ions and water in animals.* London: Academic Press.

Hoar, W. S., and D. J. Randall. 1969. *Fish physiology,* vol. 1, *Excretion, ionic regulation, and metabolism.* New York: Academic Press.

Hochachka, P. W. 1983. *The Mollusca,* vol. 1, *Metabolic biochemistry and molecular biomechanics.* New York: Academic Press.

Imai, M., J. Taniguchi, and K. Tabei. 1987. Function of thin loops of Henle. *Kidney Int.* 31:565–79.

Kerkut, G. A., and L. I. Gilbert. 1985. *Comprehensive insect physiology biochemistry and pharmacology,* vol. 4, *Regulation, digestion, nutrition, and excretion.* Oxford: Pergamon Press.

Kinne, R. K. H. 1990. *Urinary concentrating mechanisms.* Basel: Karger.

Maloiy, G. M. O. 1979. *Comparative physiology of osmoregulation in animals.* New York: Academic Press.

McGilvery, R. W., and G. W. Goldstein. 1983. *Biochemistry: A functional approach.* Philadelphia: Saunders.

Mommsen, T. P., and P. J. Walsh. 1989. Evolution of urea synthesis in vertebrates: The piscine connection. *Science* 243:72-5.

Peaker, M., and J. L. Linzell. 1975. *Salt glands in birds and reptiles.* Cambridge: Cambridge University Press.

Phillips, J. 1981. Comparative physiology of insect renal function. *Am. J. Physiol.* 241:R241–57.

Potts, W. T. W., and G. Parry. 1964. *Osmotic and ionic regulation in animals.* Oxford: Pergamon Press.

Regnault, M. 1987. Nitrogen excretion in marine and freshwater Crustacea. *Biol. Rev.* 62:1–24.

Riegel, J. A. 1972. *Comparative physiology of renal excretion.* New York: Hafner Publ.

Ruppert, E. E., and P. R. Smith. 1988. The functional organization of filtration nephridia. *Biol. Rev.* 63:231–58.

Saleuddin, A. S. M., and K. M. Wilbur. 1983. *The Mollusca,* vol. 5, *Physiology,* part 2. New York: Academic Press.

Schmidt-Nielsen, B. 1988. Excretory mechanisms as examples of the principle "The whole is greater than the sum of its parts." *Physiol. Zool.* 61:312–21.

Shoemaker, V. H., and K. Nagy. 1977. Osmoregulation in amphibians and reptiles. *Ann. Rev. Physiol.* 39:449–71.

Skadhauge, E. 1981. *Osmoregulation in birds.* Berlin: Springer-Verlag.

Sturkie, P. D. 1986. *Avian physiology.* New York: Springer-Verlag.

Warren, K. S. 1958. The differential toxicity of ammonium salts. *J. Clinical Invest.* 37:497–501.

Chapter 18

Digestion

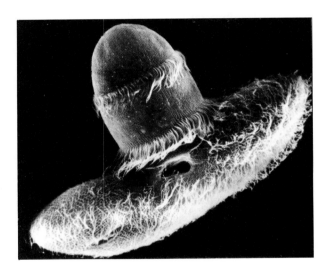

A ciliate protist *Didinium* **ingesting a** **much larger ciliate,** *Parame-* ***cium.*** *(Biophoto Associates/Science Source/Photo Researchers, Inc.)*

T he physical nature of food determines the type of food-gathering apparatus that is needed to optimize its acquisi-
• • • • • • tion. The structure of the digestive tract and the mechanisms necessary for the breakdown of the food are related to the chemical nature of food. In this chapter, we shall examine the mechanisms for acquisition of food items, the breakdown of food into its subunits (mainly carbohydrates, lipids, and proteins), how these constituents are absorbed into the body, and the nutritive roles of these materials, vitamins, and minerals.

Feeding Patterns

There is an amazing variety of food items available to animals, including micro-organisms, fungi, plants and plant products, other animals, dead or decaying plants and animals, and waste products of other plants and animals. Animals have an equally impressive array of feeding mechanisms. A detailed examination of the varied feeding strategies and structures is beyond our scope, but it is important, at least briefly, to examine the diversity of feeding patterns used by animals to obtain different types of food because of the intimate relationship between the nature of the food and the structure and function of the digestive tract.

It is convenient for further discussion to classify animal feeding mechanisms into types. This cannot be done effectively by a taxonomic scheme because an animal's feeding method depends more on the size and nature of its food than its taxonomy or level of organization. We can generally classify animal feeding mechanisms (Table 18–1) into four broad categories: (1) **suspension feeding** for collecting small food particles, (2) **manipulative mechanisms** for acquiring large food particles, (3) **sucking** fluids or soft tissues, and (4) **surface absorption** of nutrients with no specialized feeding mechanism and often no digestive tract.

Suspension Feeding

Food particles of microscopic size generally consist of bacteria, algae, spores, animal larvae, and small invertebrate animals. Suspension feeding removes these food items from the ambient medium, which is almost always water but sometimes is air, e.g., spiders "suspension feed" with their webs. Almost every animal phylum has species that suspension feed during at least one stage of their life cycle. Most suspension feeders are marine rather than freshwater because seawater contains an abundance of microscopic food particles.

There are five methods for particle capture by a web or pore (Figure 18–1). A specialized and relatively uncommon form of suspension feeding uses sieves of either fibrous webs or porous plates. The sieves remove only particles too large to pass through the pores. Sieves suffer from clogging and offer a high resistance to water flow through the pores. Suspension feeders that use sticky webs or

TABLE 18–1

A classification of feeding mechanisms in animals according to the nature of the diet.   *(Based on Yonge 1928.)*

Suspension Feeding Mechanisms for Small Particulate Diet
1. Pseudopodial   4. Tentacular
2. Flagellate   5. Mucoid
3. Ciliary   6. Setous

Manipulative Mechanisms for Ingesting Large Particulate Diet
1. Swallow inactive food   3. Seizure of prey
2. Scraping and boring   a. Swallow prey intact
                                b. Masticate prey prior to swallowing
                                c. External digestion of prey before ingestion

Pumping Mechanisms for Fluid Feeding
1. Sucking only   2. Piercing and sucking

Mechanisms for Direct Nutrient Absorption
1. Absorption across the general body surface area
2. Absorption from symbiotic micro-organisms

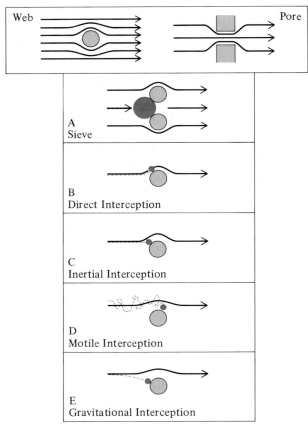

Key:

⬤ cross section of fiber   ▣ longitudinal section of pore

→ fluid streamline   • particle

----- particle path

FIGURE 18–1 Hydrodynamic principles of suspension feeding with either webs or pores. The basic pattern of water (or air) flow around a web or through a pore is responsible for particle capture by sieving (**A**), direct interception (**B**), inertial impaction (**C**), motile particle deposition (**D**), or gravitational deposition (**E**). Examples 1–5 are only shown for webs but the same principles apply to pores.   *(Modified from Rubenstein and Koehl 1977.)*

pores can collect particles even smaller than the minimum sieve size by direct interception, inertial impaction, capture of motile particles, or gravitational deposition.

The rhizopod protozoans provide an example of **pseudopodial** filter feeding; their pseudopodia are motile cytoplasmic extensions that suspension feed and also provide locomotion. Some mastigophoran protozoans and sponges have a **flagellate** feeding mechanism. For example, the flagellar beating of choanocyte cells of sponges creates a constant flow of water through the sponge into the central

chamber; suspended particles are captured by the flagellar surface and transferred to the choanocytes for digestion. The **ciliary** activity of ciliate protozoans creates water currents that bring suspended particles into contact with the "mouth" region of the cell membrane. Hemichordates suspension feed using cilia; a ciliary-induced water current directs food particles to the mouth. **Tentacular** feeding traps food particles with a mucous film. For example, holothurian echinoderms have tentacles coated with mucus. Many suspension feeders trap food particles in mucous strands, webs, or sheets secreted from the mouth or some other specialized feeding structure and ingest the food-laden mucus. For example, urochordates and cephalochordates have a large, perforate pharyngeal basket; water movement due to cilia directs food particles onto a mucous sheet that is secreted by the endostyle and is directed over the pharyngeal surface to the esophagus (Figure 18–2). Such **mucoid** feeding is generally continuous and often has little selection of food particles except on the basis of size and density. Many crustaceans, and a few insect larvae, utilize **setous** filter feeding; fringes of chitinous setae, rather than mucous webs, are used to collect suspended food particles.

Suspension feeding is observed in protozoans and in every animal phylum including the chordates. It is rare in vertebrates and is not ciliary but involves mechanical filters analogous to the chitinous setae of crustaceans. Most larval amphibians (tadpoles) filter feed. The water current that ventilates the gills carries food particles through the branchial basket where they are trapped by mucus secreted from epithelial cells. The food-laden mucus is moved from the gill filters to ciliary grooves on the margins of the pharynx and then to the esophagus. Some fish (e.g., the herring *Clupea* and the paddlefish *Polyodon*) filter feed using modified gill rakers (the gill rakers of other fish are used to keep the gills free from blockage by large particles). Filter-feeding sharks have special comb-like structures on their gills, and baleen whales have rows of keratinous plates (baleen) suspended from the roof of their mouth for filtering small crustaceans, such as krill. The highly modified beaks of filter-feeding birds, such as flamingos, act as sieves (Figure 18–3). The flamingo's lower jaw is deeply keeled (V-shaped) with the tongue running along the bottom of the V; the upper jaw is triangular in cross section and fits neatly in the top of the V. The opposing surfaces of the upper and lower jaws are covered by small fringed platelets that are mechanical filters. Water is drawn into the space between the upper and lower jaws; large particles are excluded by "excluder" marginal hooks on the upper jaw. The water is then expelled through the finer fringed platelets, sieving

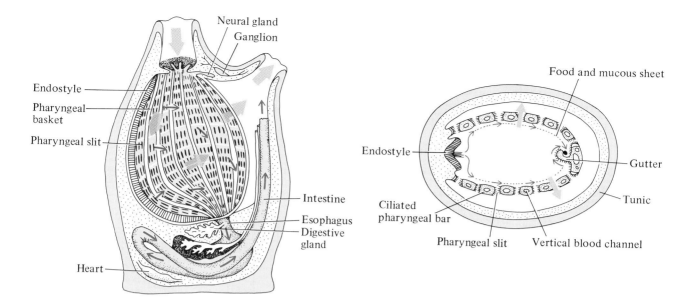

FIGURE 18–2 Schematic of the pharyngeal basket of a filter-feeding ascidian tunicate showing direction of movement of water through inlet siphon, pharyngeal basket, and outlet siphon (thick arrows) and mucous sheet from endostyle to gutter (thin arrows).

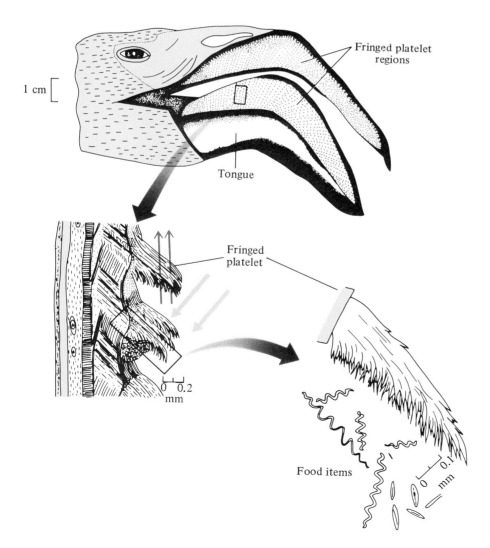

FIGURE 18–3 Sieve feeding by the lesser flamingo *Phoenicopterus minor*; a piston action of the tongue draws water into the oral cavity and pushes it out through the fringed platelets, trapping small food particles on the fringes. Typical dietary items, coiled algae *Spirulina* and diatom frustules *Navicula*, are also shown. (*Modified from Jenkin 1957.*)

out plant material (algae, diatoms, seeds) and animal material (small crustaceans, oligochetes, gastropods, nematodes, and insects). The pattern of water movement is due to an up-and-down motion of the upper jaw and a backwards-and-forwards piston action of the tongue.

There is an obvious inter-relationship between feeding and gas exchange in many suspension-feeding animals because the same gross structures perform both functions. Both filter feeding and respiration require a large surface area and a continual flow of water. An example of this evolutionary opportunism is the evolution of the ctenidial gill of primitive mollusks, which had a respiratory function, into the gill of lamellibranch mollusks, which is highly modified for suspension feeding but retains its respiratory role (see Chapter 12). The feeding basket of urochordates and cephalochordates also functions for gas exchange.

Large-Particle Feeding

A number of animals **swallow** inactive, large food particles. Some simply process soil, mud, sand, or silt for their organic contents, e.g., many annelid worms, some crustaceans. The egg-eating snake *Dasypeltis* is a remarkable example of this ability to swallow large objects. These snakes can swallow eggs that are much larger than would seem possible (Gans 1974). Not only is swallowing large eggs a seemingly impossible task, but breaking the egg shell and shell membranes by squeezing is also difficult. The egg is actually broken by a combination of squeezing and lateral undulatory motions, and ventral spinous projections of specialized vertebrae that rupture the egg shell and compress the shell into a boat-shaped pellet that is regurgitated (Figure 18–4).

Many invertebrates **scrape** food off the substrate or **bore** into large food masses. A number of mollusks have shell valves modified for boring or scraping, e.g., ship worms. Most mollusks have a **radula**, a rasp-like structure used for boring or scraping food (Figure 18–5). It is a flexible membranous belt covered with rows of horny teeth; the horns are made of magnetite, one of the hardest biological materials. The front of the radula is used to rasp the surface of substrates such as rock, or large food particles such as wood, and the eroded pieces are ingested. The radula is pulled backwards and forwards over a supporting bolster by muscles. The posterior part of the radula grows continuously so as to replace the front as it is worn away by feeding activities.

Carnivorous animals generally **seize** their prey and either ingest it whole with no form of pretreatment (seize and swallow), or masticate it to reduce the particle size for ingestion (seize and chew), or digest it externally and ingest only partially digested fluids (seize and suck). Organisms that seize and swallow often use toxins to paralyze or kill their prey and can sometimes swallow food items bigger than they are! For example, ciliate protozoans such as *Didinium* can "swallow" larger protozoans and rotifers. The gastropod mollusk *Conus* darts fish often larger than itself with a harpoon-like radula

FIGURE 18–4 Egg-eating snakes (*Dasypeltis*) can swallow relatively large eggs due to cranial modifications that increase the gape and collapse the egg by a combination of muscular squeezing and a punch-like action of specialized spinous processes of the esophageal vertebral hypapophyses that project into the lumen of the esophagus. *(Modified from Gans 1974.)*

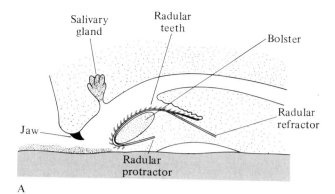

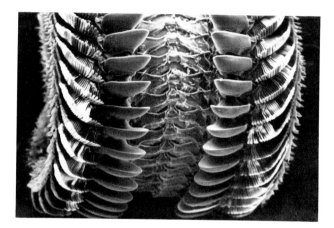

FIGURE 18–5 (A) Diagram of the buccal cavity of a gastropod mollusk showing the arrangement of the radula. (B) A scanning electronmicrograph of the radula of the gastropod *Nerita atramentosa* *(SEM photograph courtesy of Dr. R. Black and J. Kuo, Department of Zoology and Electron Microscopy Center, The University of Western Australia.)*

tooth, injects a highly neurotoxic poison, and swallows its prey whole. Snakes can swallow prey larger than their jaw gape; they have a very flexible skull and can dislocate the lower jaw, the two sides of which are held together by an elastic ligament.

Many animals masticate their food into smaller pieces before it is swallowed, e.g., rotifers, echinoderms, mollusks (some cephalopods and gastropods), arthropods (some crustaceans, insects, scorpions, and millipedes), and vertebrates. Cephalopods, like gastropods, have a radula but also have a pair of horny beak-like jaws that bite and tear food into pieces. Crustaceans, insects, and scorpions have complex mouthparts for chewing food. The mouthparts of insects exhibit remarkable specialization and diversity. The generalized insect

mouthparts are an upper (labrum) and lower (labium) lobe bounded laterally by the paired mandibles and maxillae (Figure 18–6). The mandibles have either one pair (in apterygote insects) or two pairs (in pterygotes) of articulations with the head capsule. Within the chamber formed by these appendages is a median lobe, the hypopharynx; the mouth opens into the preoral cavity anterior to the hypopharynx, and the salivary duct opens into the salivarium posterior to the hypopharynx. The mandibles of herbivorous insects have incisor cusps for snipping and molar cusps for chewing the food. Many insects have hypodermic-like mouthparts for piercing and sucking.

The teeth of vertebrates (fish, amphibians, reptiles, mammals) and the beaks of birds are often highly specialized for cutting, tearing, and chewing food. Mammalian teeth are differentiated into incisors, canines, premolars, and molars. These various forms of teeth are often highly specialized for snipping, puncturing, cutting, and grinding food (Figure 18–7). Birds (except *Archaeopteryx*) lack teeth, but their beak is specialized for food acquisition and processing. For example, granivorous birds are able to efficiently remove seed husks using their beaks, and some birds can crush very resistant seeds to obtain the nourishing kernel.

Many animal species dispense with the often elaborate mechanisms required for swallowing prey intact or mastication by partially digesting the prey externally. Examples are platyhelminth worms, echinoderms, mollusks, and arthropods, but spiders are perhaps the best illustration of this extra-corporeal feeding strategy. Spiders generally bite their prey with chelicerae, or fangs, which may also be used to hold and masticate the prey in some species. The chelicerae have associated poison glands with ducts that open near the tips of the fang; the poison immobilizes the prey. Some spiders chew their prey using teeth on the chelicerae or have a saw-like structure on other mouthparts to cut apart the prey tissues. After capture, digestive enzymes pour out of the mouthparts and coat the punctured or chewed prey. The soft tissues are liquified and then sucked into the spider's digestive tract by the action of a "pumping stomach."

Fluid Feeding

Many animals feed on fluids, or soft plant and animal tissues, by piercing and sucking. Examples are typically found in the invertebrate phyla, but one vertebrate (vampire bats) feeds specifically on blood. There are many invertebrate blood-sucking ectoparasites (e.g., the fish louse) and free-living

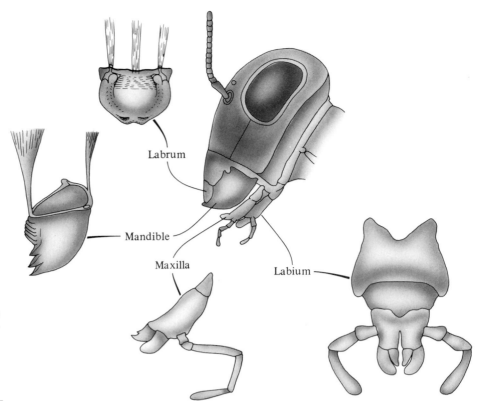

FIGURE 18–6 Insect mouth parts showing the general arrangement and structure of the labrum, mandible, maxilla, and labium. *(Modified from Chapman 1982.)*

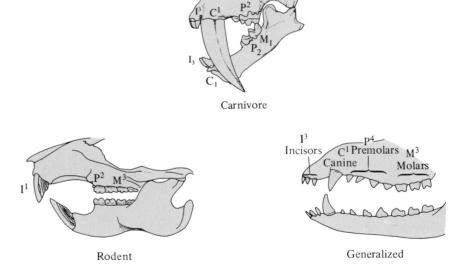

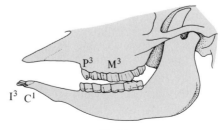

FIGURE 18–7 Generalized mammalian dentition and dental formula: incisor, I; canine, C; premolar, P; molar, M; specializations of the different teeth include chisel-shaped incisors of rodents, stabbing canines, shearing premolars and molars of carnivores (sabretoothed tiger), and grinding premolars and molars of herbivores.

blood suckers (e.g., mosquitos). The most highly developed sucking structures are seen in insects, although many nematodes, annelids, and crustaceans have a specialized pharynx, jaws, or claws for blood sucking. For example, the cicada has complex mouthparts for sucking, including a salivary pump for injecting saliva and a cibarial pump for sucking fluids. A similar elongate and complex arrangement of the mouthparts is seen in the plant-feeding bugs. The milkweed bug *Oncopeltus* has its mandibles and maxillae modified into long, piercing stylets with a separate salivary channel and food channel (Figure 18–8A). The segmented, sheath-like labium protects the stylets but is not part of the piercing structure; it folds up as the stylet penetrates the plant tissues. Blood-sucking mosquitos have a very complex beak with six piercing stylets made up of the labrum, mandible, maxillae, and hypopharynx; the labium forms a protective sheath for the piercing stylets (Figure 18–8B). The proboscis of adult lepidopterans is a long, coiled, siphoning structure formed by parts of the maxillae (Figure 18–8C). The proboscis is extended by hemolymph pressure and sucks, or siphons, liquids up the central space of the proboscis; it recoils by its own elasticity.

Most pollen- and nectar-feeding birds have long bills and tongues. The beak is often specialized for particular types of flowers (e.g., is adapted in shape, length, curvature). The tongues of some species have a brush-like tip or are hollow, or both, to collect the nectar from flower nectaries. Some nectar-feeding birds have short bills; they pierce the base of the flower and obtain nectar through the hole using their tongue. Nectar-feeding bats have a long tongue to extract nectar and have a reduced dentition. The nectar-feeding honey possum has a long, brush-tipped tongue and reduced dentition.

Surface Nutrient Absorption

A few highly specialized animals have dispensed entirely with all mechanisms for prey capture, ingestion of food particles, and digestive processes. Rather, they rely on absorption of nutrients directly across their body surface from the external medium, either the nutrient-rich ocean water, fluids in other animals' digestive tract, or body fluids of other animals (Wright and Manahan 1989). Some free-living flagellate protozoans are able to absorb all of their nutrients, often in very simple forms, across their cell surface. For example, *Chilomonas* can

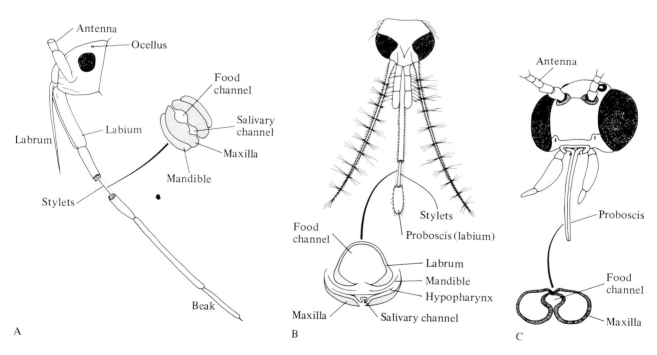

FIGURE 18–8 Structure of the piercing mouthparts of (**A**) the milkweed bug *Oncopeltus fasciatus* (**B**) the mosquito *Aedes,* and (**C**) the siphoning mouthparts of the moth *Sanninoides.* *(Adapted from Borror, De Long, and Triplehorn 1981.)*

exist in simple chemical solutions (MgSO$_4$, NH$_4$Cl, K$_2$HPO$_4$, and NaCH$_3$COO) by synthesizing all of their required nutrients from these absorbed salts. The endoparasitic protozoans absorb all of their

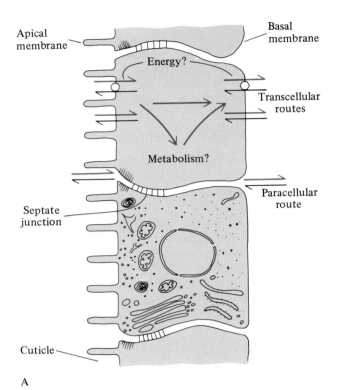

A

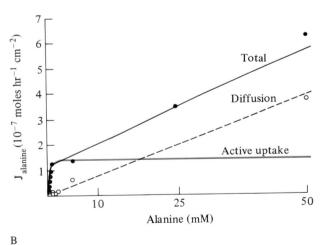

B

FIGURE 18–9 **(A)** Simplified representation of transport pathways for absorption of nutrients across the epidermal surface of the marine invertebrate *Nereis*. **(B)** There is a hyperbolic relationship between active uptake of glycine and its concentration and a linear relationship for diffusional uptake in the annelid *Glycera dibranchiata*. *(From Gomme 1982; Preston and Stevens 1982.)*

required nutrients across their body surface, as do the cestode worms, endoparasitic gastropods, and crustaceans, all of which lack mouths and guts.

A few nonparasitic multicellular animals entirely lack a digestive system and absorb nutrients across their body surface, e.g., gutless bivalves (Reid and Bernard 1980) and pogonophoran worms (Southward and Southward 1980). Many pogonophoran worms, such as *Siboglinum ekmani*, absorb nutrients from seawater across their body surface, but this source of nutrients is often supplemented by organic carbon fixed within the pogonophoran's tissues by symbiont bacteria (Felbeck, Childress, and Somero 1981; Cavanaugh 1983). These pogonophoran worms have specialized trophosomal tissue that is packed with these symbiont bacteria, which chemosynthetically fix inorganic carbon (in contrast to photosynthetic C-fixation by plants). Although these gutless pogonophorans and bivalves rely on absorption of nutrients across the body surface, many other marine organisms also gain some nutrients across their body surfaces by similar processes (Stewart 1979; Gomme 1982).

There is a striking similarity between nutrient uptake at the body surface and nutrient uptake mechanisms in the digestive tract. The cutaneous uptake of nutrients occurs from concentrations as low as 1 to 100 μM amino acid and glucose concentrations; for example, nutrient concentrations of 5 to 10 μM are found in the immediate environment of the sipunculid *S. ekmani*. Nutrient exchange between the external side of the body surface epithelium (i.e., seawater side) to the basal (body fluid) side could occur via a transcellular route or a paracellular route (Figure 18–9A). However, uptake must occur against high concentration gradients, e.g., 2000 : 1, and so the mechanism for uptake must be transcellular active transport (Buck and Schlicter 1987).

There is a hyperbolic relationship between active transport rate and substrate concentration for the carrier-mediated active uptake mechanism of amino acids and also some monosaccharides. For example, alanine uptake by the integument of an annelid *Glycera* shows such a hyperbolic relationship for active uptake at micromolar concentrations (and a linear diffusional flux that is significant at millimolar concentrations; Figure 18–9B). This hyperbolic relationship is typical of enzyme-mediated reactions, i.e.,

$$J_{act} = \frac{C_s J_{max}}{C_s + K_t} \tag{18.1}$$

where J_{act} is the rate of active transport, C_s is the concentration at the membrane of the solute being transported (amino acid or monosaccharide) at the cell surface, J_{max} is the maximal transport rate (at infinite substrate concentration), and K_t is the apparent Michaelis-Menten transport constant (analogous to K_m).

The active transport systems have properties similar to enzyme-substrate systems: specific carriers are required for transport of various sugars and amino acids, there is stereospecificity of nutrient transport in many instances, there is a high carrier affinity for their substrate (i.e., they have a low K_t, which is analogous to K_m), and carriers saturate at high nutrient concentration (i.e., have a V_{max}).

The K_t values for amino acid and monosaccharide transport by the gut epithelial cells are typically from 1 to 50 mM (Table 18–2). These K_t values are much higher than those for the transport mechanisms of those marine animals that are able to absorb nutrients directly across the body surface; these K_t values are typically 10 to 200 μM. In many instances, amino acid uptake by these marine invertebrates is also Na$^+$-dependent. Unicellular organisms (bacteria, algae) have even lower K_t values than the marine invertebrates and values about a million times lower than the vertebrate gut!

TABLE 18–2

Transport constants (K_t) for amino acid, monosaccharide, and free fatty acid uptake by micro-organisms, marine invertebrate integuments, and vertebrate intestines.

	Amino acids		Monosaccharides		Fatty Acids	
	Substrate	K_t	*Substrate*	K_t	*Substrate*	K_t
Cell Membrane						
Bacteria	Ala	98 nM	Glu	122 nM	Acet	100 nM
	Gly	280 nM	Gal	178 nM		
	Arg	236 nM				
Algae	Ala	19 μM	Glu	0.6–240 μM		
	Gly	5 μM	Lact	10 μM		
	Arg	5 μM	Gal	700 μM		
Protozoans	Phe	15 μM				
	Glut	7 μM				
Body Integument						
Sponges			Glu	100 μM		
Coelenterates	Gly	8–74 μM				
	Leu	2 μM				
Echinoderms	Ala	8–84 μM				
	Gly	74 μM				
Annelids	Ala	20–90 μM	Glu	5 M	Palmit	1 μM
	Gly	8–252 μM			Oleate	1 μM
	Arg	217 μM			Acet	102 μM
Mollusks	Ala	95 μM	Glu	36 M		
	Gly	20–126 μM				
Sipunculids	Gly	100 μM				
Pogonophorans	Ala	3–6 μM	Glu	60 μM	Acet	62 μM
	Gly	2–10 μM	Lact	45 μM		
	Phe	10 μM				
Vertebrate Intestine						
Reptiles						
Iguana			Glu	1 mM		
Chuckwalla			Glu	1 mM		
Mammals						
Rat	Ala	20 mM	Glu	10 mM		
	Lys	71 mM	Gal	38 mM		
	Pro	13 mM				
Hamster			Glu	2 mM		
			Gal	2 mM		
Rabbit	Lys	9 mM				
Mouse			Glu	3–6 mM		
Woodrat			Glu	1 mM		
Ground squirrel			Glu	1 mM		

Generalized Structure and Function of the Digestive Tract

Why a Digestive Tract?

Before examining the structures and functions of the animal digestive tract, we must first address the fundamental question of why animals have evolved a digestive tract and why it has become such a highly specialized organ system. Digestive tracts basically allow efficient extracellular digestion.

Primitive unicellular animals absorb large food particles by phagocytosis and then subsequently digest them in intracellular food vacuoles. Even some of the more primitive multicellular animals, such as sponges, rely on **intracellular digestion**. The main advantage of intracellular digestion is the ease by which the optimal intracellular environment for digestion can be provided; the optimal pH and concentration of digestive enzymes can be maintained more easily in a small food vacuole than in an alimentary canal or in the external environment. However, there are three major disadvantages to intracellular digestion: (1) it limits the size of prey, (2) there is a lack of capacity for digestive specialization, and (3) it is difficult to separate spatially different digestive processes. Only food particles small enough to be phagocytosed can be digested intracellularly. A few protozoans can ingest food items larger than themselves, but this is the exception rather than the rule. Intracellular digestion requires that each and every cell contain the digestive enzymes and specialized structures for digestion. A part of the activity of every cell therefore needs to be devoted to digestion, rather than to a more specialized function, e.g., nerve and muscle cells would have to digest food particles to obtain their nutrients. Intracellular digestion limits the capacity for the spatial separation of different digestive processes, for example initial acidic digestion followed by alkaline digestion. However, there is a spatial organization to food vacuole passage through cells that provides some capacity for spatial variation in digestive conditions.

Extracellular digestion in a gut tube allows animals to feed on larger pieces of plant or animal material. Extracellular digestion of food into its chemical subunits allows complete specialization of cellular function, although every cell must still retain the ability to absorb basic nutrients from the extracellular body fluids. The presence of a gut tube allows different digestive processes to be separated spatially as well as temporally. For example, initial acidic digestion of protein can occur in a different part of the digestive tract than subsequent alkaline or neutral digestion of carbohydrates or lipids, rather than occur in the same location (e.g., a food vacuole) at different times. Associated with this spatial organization is the advantage of a second opening of the digestive tract, the anus, which allows a one-way flow of food from the mouth (oral end) towards the anal end.

Regions of the Gut

A simple tubular digestive tract allows for complete extracellular digestion of ingested food particles and inevitably leads to the progressive specialization of portions of the tract for discrete functions and the evolutionary progression towards more complex gastrointestinal tracts in the higher vertebrates and invertebrates. Even protozoans have a specialization of intracellular pathways and processes for digestion; there is a preferential pathway for food vacuole movement from the "mouth" region towards a specialized excretory region of the cell membrane.

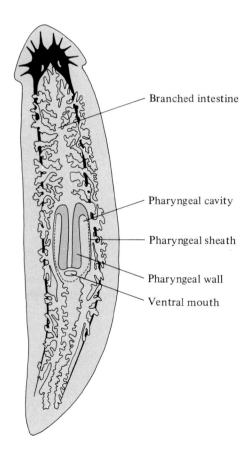

— Branched intestine

— Pharyngeal cavity

— Pharyngeal sheath

— Pharyngeal wall

— Ventral mouth

FIGURE 18–10   Structure of the branched, single-opening gut of the planarian worm *Dugesia*.

The simplest gut is an invaginated pouch with only one opening, the mouth, through which food is ingested and feces excreted, e.g., the gut of a planarian worm (Figure 18–10). There is little opportunity for specialization of parts of this gut (except by the mouthparts and pharynx for prey capture and ingestion) because all regions must perform the same functions and supply nutrients to the surrounding tissues.

The advent of a gut tube with mouth and anus and one-way flow of food provides for regional specialization. The different regions of such a gut tube can be best interpreted not only from a structural point-of-view, but also from the standpoint of function. The gut tube can be divided into the foregut, which is specialized for reception of food (prey capture, mastication, salivation) and conduction and storage of food; the midgut, which is specialized for storage and mechanical, chemical, and enzymatic digestion and nutrient absorption; and the hindgut, which is specialized for water and ion reabsorption and formation and storage of feces. The gut of an invertebrate, such as a generalized mollusk, well illustrates these regions (Figure 18–11).

The digestive processes of animals can be modeled as chemical reactor systems (Penry and Jumars 1986, 1987). There are three ideal theoretical systems: a batch reactor (BR), a plug-flow reactor (PFR), and a continuous-flow, stirred-tank reactor (CSTR). In a batch reactor, all reagents are initially loaded, mixed, then allowed to react; a digestive system with only a single opening could function as a batch reactor. In steady-flow reactors (PFR and CSTR), there is a flow of materials through the reactor; digestive systems with both mouth and anus act as steady-flow reactors. In plug-flow reactors, materials enter one end of the reaction vessel and flow in an orderly pattern as a bolus, or plug, through the reactor. In contrast, material enters one end of the reactor and is thoroughly mixed with the other contents, in continuous-flow stirred-tank reactors. No animals have a purely plug-flow or continuous-flow stirred-tank organization, but for many animals one part of the gut is essentially one type of reactor, with another part acting as the other type of reactor. For example, ruminant and pseudo-ruminant mammals are effectively a CSTR (stomach) followed by a PFR (small and large intestine). In contrast, a post-gastric fermenting mammal, such as a horse, is effectively a PFR (stomach, small intestine) followed by a CSTR (cecum).

Reception.   The reception region of the gut usually, but not always, contains or forms the structures for prey capture and mastication. In addition, **saliva** is added to the food by salivary glands.

Salivary glands are a varied assemblage of glandular tissues that empty their secretions into the gut reception region. Saliva generally contains **mucus**, a slimy mixture of water and a proteinaceous mucopolysaccharide, mucin, which lubricates food to ease its mastication and subsequent movement through the gut. Saliva can also have other functions; it can contain poisons for paralyzing or killing prey, acids for killing prey, denaturing proteins, or dissolving calcareous shells, enzymes for the chemical breakdown of food (e.g., amylase to break down the starch amylose; invertase to break down sucrose; proteases; lipases), or proteins for silk production. Saliva can contain excretory products (e.g., larval wasps *Vespa* eliminate excess water via saliva, which is removed by the worker wasps). Some ticks excrete waste products in their saliva.

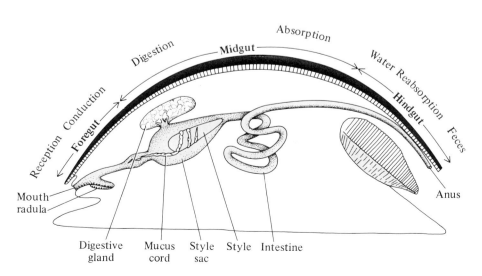

FIGURE 18–11 The generalized molluskan gut illustrates the three major divisions of the gut: foregut, midgut, and hindgut.

Saliva has a social role in other social insects that mutually exchange food (trophallaxis). Saliva is even used as a nuptial gift by male scorpion flies. Many blood-sucking insects secrete anticoagulants into their saliva to prevent coagulation of the host's blood. Blood will coagulate in the mouthparts of the tsetse fly *Glossina* if its salivary glands are removed. However, blood does not clot in the mouthparts of the mosquito *Aedes*, which does not secrete an anticoagulant, although it clots in the stomach. Many plant-fluid feeding insects secrete a viscous salivary component that hardens to form a sucking sheath and a more fluid component that contains a pectinase enzyme to break down plant tissues.

Bulk Movement.   Following the receiving portion of the gut tube is the conducting region (pharynx, esophagus) that transfers food to the digestive regions. The pharynx is a muscular tube closely associated with the oral, or buccal, cavity. It is often an eversible structure used for prey capture (e.g., ribbon worms, cone shells) or may be important in directing the passage of food to the digestive tract and air to the lungs (e.g., vertebrates).

The movement of food through the conducting region (e.g., esophagus) and all other regions of the gut tube from the pharynx to the anus requires muscle contraction or ciliary motion. This is an appropriate juncture to examine further the physiological basis for gut movement of food by ciliary and muscular action.

Animals that suspension feed using ciliary mechanisms (e.g., some polychetes and oligochetes, lamellibranch mollusks, tunicates) also have cilia throughout the entire length of their gut tube to transport food. Many worm-shaped animals rely on muscular contractions of their body wall, for example during locomotion, to indirectly force food through their gut tube, e.g., triclad turbellarian worms. Food transport by other animals depends on the contraction of muscle layers within the wall of the gut tube. A few animals rely on both ciliary beating and gut wall musculature for food movement.

Gut tubes generally have a layer of circular muscle and a layer of longitudinal muscle (Figure 18–12). The mammalian intestine provides a typical example of a gut tube with an inner circular and an outer longitudinal muscle layer. The other important layers, or tunics, in addition to smooth muscle are: an inner **mucosa** comprised of epithelial lining supported by a loose layer of connective tissue and some smooth muscle cells; a **submucosa** of vascular connective tissue that binds the mucosa to the muscle layers and provides nourishment, nervous

innervation, and lymphatic drainage; the **muscularis**, a smooth muscle layer; and the outermost **serosa** of connective tissue and epithelium that forms the peritoneum. The insect gut is organized quite differently; it has an outer circular muscle layer and an inner longitudinal muscle layer, tracheae, and a cuticle lining.

Peristalsis is the movement of contents along the gut tube by a coordinated action of longitudinal and circular muscle (smooth muscle, except in the arthropods, and the upper esophagus of vertebrates, which has some involuntary striated muscle). Transportative peristalsis by the gut is actually only one of a number of different types of peristalsis that can be more generally defined as "muscular contraction moving along a radially flexible tube in such a way that each component wave of circular, longitudinal, or oblique muscular contraction is preceded and/or followed by a period of relative relaxation of all similarly oriented muscle within a given tubular segment" (Heffernan and Wainwright 1974 page 95). In vertebrates, contraction of the inner circular muscle layer constricts the lumen of the gut tube, whereas contraction of the longitudinal muscle layer shortens the gut tube and dilates the lumen. When there is a coordinated constriction of the gut lumen behind a bolus of food and a dilation ahead of the food, then the food bolus is squeezed forward along the gut tube (Figure 18–13A). This movement is **propulsive peristalsis**, and the direction of movement is generally from the oral towards the anal end of the gut; reverse peristalsis (e.g., vomiting) can also occur. The coordinated peristaltic action squeezes the food bolus for a short distance along a segment of the gut tube and then disappears. A second type of peristalsis, called **segmentation**, also occurs in the mammalian intestine; rhythmic, local contractions of the circular muscle divide the gut tube into discrete segments, effectively mixing the gut contents (Figure 18–13B).

Peristalsis is a local property of the gut tube, with complex neural and hormonal control. Peristalsis can occur in an isolated piece of gut tube, and so its initiation must be endogenous to the gut. However, there is external modification of peristalsis by other parts of the gut tube, by neural innervation, and by circulating hormones. For example, a piece of intestine removed from a mammal and then replaced facing the opposite direction, continues to exhibit peristalsis and eventually adapts to the new direction of flow.

The inherent initiation of peristalsis by the gut tube depends on two nerve cell plexuses: Meissner's plexus is located in the submucosa near the circular smooth muscle layer, and Auerbach's plexus is

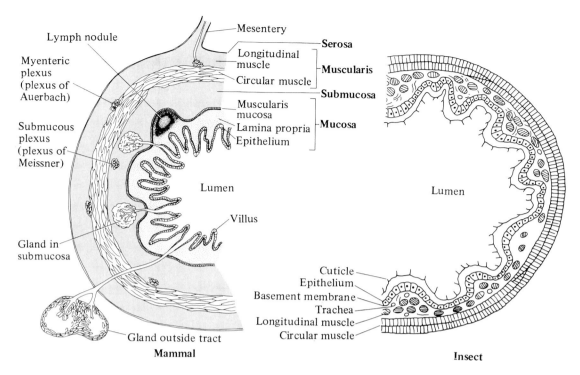

FIGURE 18–12 Generalized cross section of the mammalian gut wall (left) to show the various layers (tunics): mucosa, submucosa, muscularis, and serosa; the intramural nerve plexuses (Meissner's and Auerbach's plexus); and the location of glands in the submucosa and outside of the digestive tract. Cross section of the foregut of a grasshopper (*Dissosteira;* right) showing the cuticular intima layer covered by short spicules, the enteric epithelial layer resting upon a basement membrane, an inner layer of longitudinal smooth muscle, and an outer layer of circular smooth muscle.

located between the circular and longitudinal muscle layers. Meissner's plexus has sensory neurons that are stimulated by the presence of food in the gut; these neurons control secretion by the intestinal epithelial cells and endocrine cells and stimulate contraction of the smooth muscle for peristalsis. Auerbach's plexus has motor neurons that innervate the circular and longitudinal smooth muscle layers. Peristalsis occurs as a local reflex initiated by the sensory detection of food in the gut (by Meissner's plexus) with a motor action (due to Auerbach's plexus) of circular muscle stimulation behind the food to constrict the lumen and longitudinal muscle stimulation ahead of the food to dilate the lumen. The wave of reflex constriction squeezes the food bolus along the gut tube until the wave of peristalsis disappears.

This endogenous peristalsis is modified by the actions of the nervous and endocrine systems. There

is sympathetic and parasympathetic innervation to Auerbach's plexus (as well as blood vessels and glands in the gut tube). In general, sympathetic activity inhibits gut smooth muscle, except for various sphincters that are constricted. Parasympathetic activity generally stimulates smooth muscle contraction; this facilitates peristalsis and causes sphincters to relax, allowing food passage along the gut. These actions are consistent with the general role of sympathetic activities preparing the body for reactions to stress (the fight-or-flight syndrome) and parasympathetic activities controlling general bodily functions. The hormone norepinephrine has a similar action as sympathetic activity (norepinephrine is a neurotransmitter of the sympathetic nervous system).

The neural and hormonal effects on peristalsis are mediated through the membrane potential of the gut's smooth muscle cells. The membrane potential

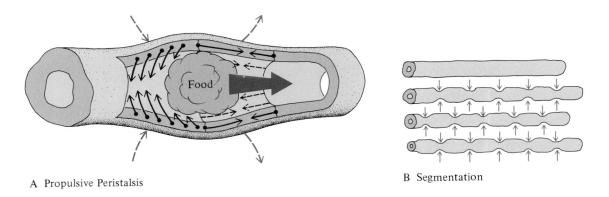

A Propulsive Peristalsis

B Segmentation

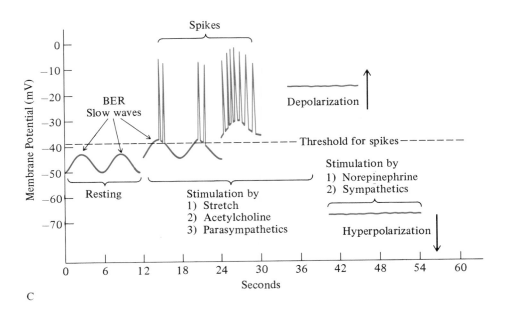

C

FIGURE 18–13 **(A)** Propulsive peristalsis. **(B)** Segmental peristalsis. **(C)** Membrane electrical activity in intestinal smooth muscle cells. The slow waves, or basic electrical rhythm (BER), are depolarized by stretch and acetylcholine and parasympathetic activity; spike potentials occur above a threshold of about −40 mV, causing muscle contraction. The slow waves are hyperpolarized by norepinephrine and sympathetic activity, inhibiting spike potential activity.

rhythmically oscillates, from about −45 to −50 mV (Figure 18–13C). This basic electrical rhythm (BER) does not cause contraction of the smooth muscle cells but an action potential occurs ("spike") and this elicits a muscle contraction if the membrane potential depolarizes during the BER to about −40 mV or more positive. Repeated action potentials and contractions will occur for as long as the BER is above this −40 mV threshold and will increase the

gut muscle tension. Thus, any factor that depolarizes the BER above the threshold increases the frequency of action potentials and muscle tension, e.g., presence of food stretching the smooth muscle, parasympathetic activity, and acetylcholine. Any factor that hyperpolarizes the BER decreases the frequency of action potentials and muscle tension, e.g., sympathetic stimulation and circulating norepinephrine.

Digestive function in many animals requires a forceful pumping action as well as peristalsis. For example, salivary pumps may be used to extrude saliva. The pharyngeal and stomach pumps of spiders suck up their partially fluidized prey; their action is partly due to extrinsic muscles from the exoskeleton to the gut. Many blood and plant juice sucking insects have a cibarial pump. This powerful sucking pump has muscles inserting from the cibarium part of the foregut to parts of the exoskeleton.

Mucus is secreted throughout the length of the gut in most animals (except insects and some other arthropods) to protect the delicate gut lining from mechanical abrasion by coarse food particles and digestion of the lining, especially by powerful proteolytic enzymes, and often to coat the feces with mucus. This is important for ciliary suspension feeding animals because their mucus-coated feces are formed into solid particles too large to be reingested by the size-selective feeding mechanism.

Insects and arachnids do not secrete mucus throughout their gut. Rather, their foregut and hindgut (which are ectodermally derived) are protected by a chitinous cuticle lining (see Figure 18–12). The midgut (which is endodermal and lacks the chitinous cuticle) is protected by the peritrophic membrane, a proteinacous sheath strengthened by chitinous fibrils. The peritrophic membrane is formed in some insects by delamination, or shedding, of the surface cells lining the midgut (e.g., orthopterans, hymenopterans) and in others it is formed at the anterior part of the midgut by specialized cells, and slowly moves along the gut until it is fragmented in the rectum and excreted. Insects that feed on fluids do not require such protection of the midgut and lack a peritrophic membrane.

The esophagus, which connects the pharynx to the stomach, is often modified to form a crop for storage of ingested food. A number of invertebrates, such as oligochete worms, gastropods, and insects, have a crop. The fluid-feeding insects often have very large crops for storage of blood, phloem, or nectar (some insects store the fluid meal in parts of the midgut, e.g., stomach). The crop can also be used to store food, mixed with salivary enzymes, i.e., it can have a digestive function. Insect pupae may store air in their crop to distend the body during metamorphosis.

Birds are the only vertebrates that have a crop; this is an expanded portion of the esophagus. Granivorous and fish-eating birds have a large crop, whereas carnivorous and insectivorous birds have a reduced, or absent, crop. Some birds use the crop to carry food to their young. Columbiform birds (pigeons and doves) produce a cheesy, lipid-rich

material from their crop; this "milk" is fed to the young until they adjust to the normal adult diet. The formation of the "milk" is controlled by the pituitary hormone, prolactin.

Digestion

The important digestive functions of the midgut are mechanical breakdown, chemical breakdown (acidic and enzymatic), and absorption.

Mechanical Breakdown. Mechanical breakdown, or triturition, physically fragments, grinds, pulverizes, and reduces food to small pieces. The high surface area of the small fragments of food promotes its subsequent chemical breakdown. Triturition is accomplished in vertebrates by the powerful muscles of the stomach wall (which possesses an oblique smooth muscle layer in addition to the normal circular and longitudinal layers of the gut). Rhythmic contractions of the stomach muscles are sufficient to reduce the stomach contents (in concert with the actions of HCl and a proteolytic enzyme pepsin) to a mushy fluid, chyme.

Stomach contractions are sometimes not able to adequately grind food, especially in herbivorous animals. Some herbivorous mammals regurgitate their stomach contents for further mastication in the oral cavity (e.g., rumination and merycism). Many birds have a gizzard, with a horny lining, that grinds food; they often swallow stones to augment the action of the gizzard. Many invertebrates also have gizzards that grind and crush. The grinding organ of rotifers, the mastax, is an oval-shaped muscular structure lined by seven interconnected plates, the trochi. In suspension feeders, two trochi with ridged surfaces form a grinding mill. All terrestrial annelids and some aquatic ones have a number of crushing gizzards (2 to 10); some freshwater pulmonate gastropods have a gizzard lined with hard cuticle, or sand grains. Most arthropods have a grinding gizzard, which often also is a complex filtering system that recycles large particles through the gizzard mill but allows small particles to pass into the rest of the gut. For example, the stomach of the crayfish *Astacus* has an anterior grinding chamber and a posterior filtering chamber. The anterior chamber has a chitinous and even calcified lining, forming a dorsal tooth and two lateral teeth, which finely mill food particles under the grinding action of powerful muscles. In the posterior filter chamber, a series of chitinous plates and bristles filter and sort food particles; large items return to the mill, indigestible items are passed to the hindgut, and small digestible items pass to the digestive cecae. The posterior part of the foregut of insects forms the proventriculus, which varies from a simple sphincter to a muscular grinding mill. The gizzard of the cockroach, for example, has six powerful teeth arranged radially around the gut tube that can be ground together by a strong compressor muscle (Figure 18–14A). A series of hairy filters posterior to the grinding mill retain food particles until they are sufficiently reduced for passage to the midgut.

The various digestive diverticula of invertebrate stomachs, when present, function for enzymatic digestion as will be discussed below. However, a specialized diverticulum of the lamellibranch mollusk stomach, the style sac, merits discussion (Figure 18–14B). Specialized cells in the style sac form a long, crystalline style made of protein; powerful carbohydrate-hydrolyzing enzymes are included in the proteinaceous style. During digestion, the crystalline style is turned within the style sac by numer-

ous cilia. The anterior end of the style is ground against a shield-shaped chitinous portion of the stomach wall, the gastric shield. This wears away the tip of the style and liberates enzymes into the stomach contents. The turning of the style also mixes the stomach contents and acts as a "windlass" to draw strands of mucus and trapped food particles into the stomach from the esophagus.

Chemical Breakdown. Chemical digestion reduces small pieces of food into its chemical constituents, e.g., monosaccharides, amino acids, and fatty acids. It can result from inorganic reactions, e.g., **high acidity**, but more generally involves the actions of **digestive enzymes**.

Digestive enzymes of various kinds are secreted by many different glandular structures throughout the gut and also by organs associated with the gut. The general categories of digestive enzymes are

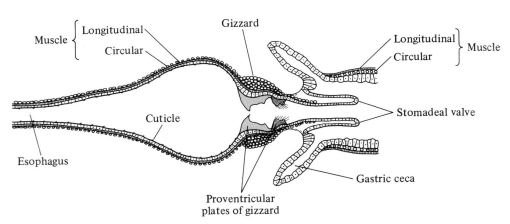

A Cockroach Gizzard

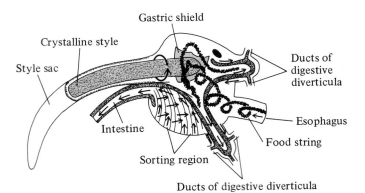

B Bivalve Mollusk Style

FIGURE 18–14 (A) Gizzard of the cockroach (*Blatta*) gut. (B) Style and gastric shield of bivalve mollusks; arrows show direction of ciliary movement of the crystalline style and food particles. *(From Snodgrass 1935; Barnes 1987.)*

TABLE 18–3

Categories of digestive enzymes. *(Modified from Florey 1966.)*

Proteases
Endopeptidases (proteinases): protein → polypeptides
   Pepsinases (1.5–2.5 pH_{opt})
   Cathepsins (1.8–6 pH_{opt})
   Tryptases (>7 pH_{opt})
Exopeptidases (peptidases): polypeptides → peptides, amino acids
   Polypeptidases
   Tripeptidases
   Dipeptidases

Carbohydrases
Polysaccharidases (polyases): high MWt carbohydrates → oligo-, di-, and monosaccharides.
   Amylases (starch)
   Cellulases (cellulose)
   Chitinases (chitin)
Oligosaccharidases (oligases): tri- and disaccharides → monosaccharides
   Glucosidases (maltose, saccharose, glucoside)
   Galactosidases (melibiose, galactoside)
   Glucosidases (cellobiose, glucoside)
   Galactosidases (lactose, galactoside)
   Fructosidases (saccharose)

Esterases
Lipases: triglycerides → fatty acids, monoglyceride, glycerol
Esterases: simple esters, complex phospholipids, cholesterol esters,
waxes → carboxylic acids, alcohols, cholesterol, fatty acids, etc.

proteases, carbohydrases, and esterases, but each category contains many specific enzymes (Table 18–3). These glandular secretions are emptied into the gut lumen through duct systems of exocrine glands. The gut tube and various associated organs also release hormones into the blood; these have effects on target organs and are important regulators of the function of the digestive processes, although they are not themselves involved in chemical digestion.

Digestive enzyme secretion by gut epithelial cells begins with transcription of DNA to messenger RNA, which is transported to the cytoplasm and translated to the amino acid sequence of the protein by ribosomes of the endoplasmic reticulum. Small vesicles containing the synthesized proteins break off from the endoplasm reticulum and coalesce in another intracellular membranous system, the Golgi complex (Figure 18–15A). The proteins are concentrated into dense secretory vesicles in the Golgi complex and released to accumulate at the apical surface of the cell, from where it is released for transport to the gut, or directly into the gut lumen, by one of three mechanisms (Figure 18–15B). In **merocrine secretion**, enzymes are released through the apical cell membrane by exocytosis with no apparent disintegration of the cell membrane or cytoplasm; the membrane of the transport vesicle fuses with the cell membrane to release its contents. In **apocrine secretion**, the apical portion of the cell is pinched off and disintegrates so as to release the secretory granules; the basal portion of the cell remains intact. In **holocrine secretion**, the entire secretory cell disintegrates, releasing the secreting granules as well as all of the other cellular contents. In glands with a duct system, the secretions of the cells can be modified during subsequent passage through the ducts. For example, the glandular portions of the mammalian salivary glands secrete an amylase enzyme, mucus, ions (mainly NaCl), and water but this fluid is modified during its passage through the ducts by active absorption of Na^+, active secretion of K^+, and passive Cl^- absorption.

All digestive enzymes are hydrolases; they break chemical bonds by the addition of water. The general reaction sequence for hydrolysis of a molecule R—R′ is

$$R—R' + H_2O \rightarrow R—OH + H—R' \quad (18.2a)$$

Specific examples are hydrolysis of peptide bonds in proteins

$$. . .—CO—NH—. . . + H_2O$$
$$\rightarrow . . .—COOH + H_2N—. . . \quad (18.2b)$$

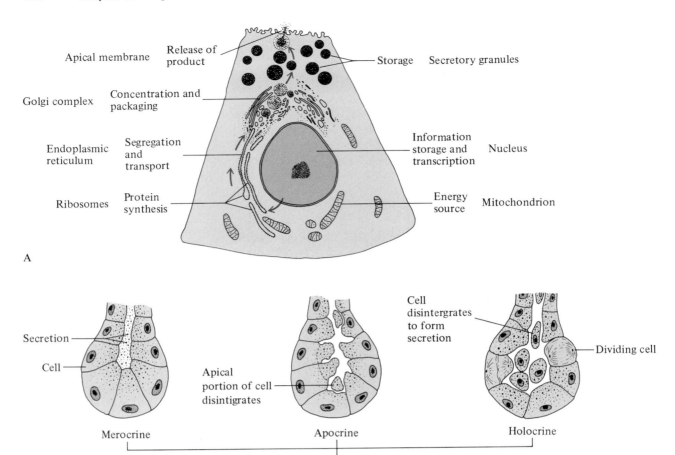

A

B

FIGURE 18–15 (A) Role of various intracellular organelles in the process of secretion. **(B)** The functional classification of secretion into merocrine secretion (release of secretory products through the apical cell membrane by exocytosis; left), apocrine secretion (release of secretory products by loss of part of apical cytoplasm; center), and holocrine secretion (release of secretory products by destruction and release of entire secretory cell; right).   *(From Tortora 1983.)*

glycosidic bonds of polysaccharides

$$...C—O—C... + H_2O$$
$$\rightarrow ...COH + HOC... \qquad (18.2c)$$

and hydrolysis of ester bonds of lipids such as triglycerides

$$..._.—COO—C—... + H_2O$$
$$\rightarrow ..._.—COOH + HOC—... \qquad (18.2d)$$

The enzymes responsible for hydrolysis reactions are generally very specific; proteolytic enzymes hydrolyze proteins but not polysaccharides or lipids; carbohydrases and lipases are also specific to their general reactions of splitting carbohydrates and lipids, respectively. However, digestive enzymes are generally not absolutely specific to one particular chemical reaction, as are many enzymes of cellular metabolism. For example, a protease enzyme may hydrolyze the peptide bonds between a variety of different amino acids; a lipase may cleave any of a number of different fatty acids from a triglyceride. Thus, digestive enzymes have **group specificity** (e.g., to protein, carbohydrate, or lipid) but not absolute specificity. Nevertheless, there are different types of protease, carbohydrase, and lipase enzymes (see Table 18–3). For example, exopeptidase enzymes cleave amino acids from the terminal ends of proteins, whereas endopeptidases cleave peptide bonds within a protein to form two shorter polypeptides.

Digestive enzymes are thus somewhat specific in function; they are also highly specific in terms of the physical and chemical conditions required for

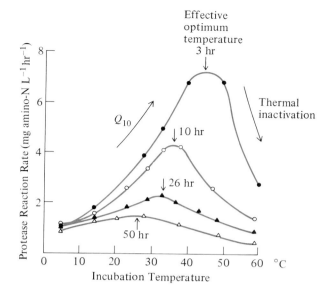

FIGURE 18–16 Temperature increases the activity of a digestive protease enzyme of a tunicate (*Tethyum*) by a Q_{10} effect, but the effective temperature optimum (*V*) decreases with increasing incubation time because of thermal inactivation of the enzyme. *(Data from Berrill 1929.)*

their optimal activity. These important conditions include temperature, pH, and inorganic ion concentrations. Temperature increases enzymatic hydrolysis rates (and also noncatalyzed reactions; see Chapters 2 and 5), up to an optimal temperature above which hydrolysis rate declines as the enzymes are inactivated and denatured. The optimal temperature for digestive enzymes is generally about the normal body temperature or normal range of body temperatures. However, the **true optimum temperature** for digestion is sometimes higher, at a temperature that inactivates the enzyme; the digestion rate is only high in short incubation experiments, since longer exposure at high temperature decreases the enzyme activity and reduces the hydrolysis rate. The **effective optimum temperature** decreases with length of incubation, reflecting the thermal inactivation of enzymes (Figure 18–16).

The vertebrate stomach, unlike that of most other animals, is very acidic, with a pH of 2 to 3. The acidophilic cells of the stomach gastric pits secrete hydrochloric acid at approximately pH 2 (an H^+ concentration of 10^{-2} M, 10^5 times as great as that of general body fluids!). This high acidity causes some chemical breakdown of food. However, most chemical digestion in vertebrates and invertebrates is the result of specific digestive enzymes. The

low pH of the vertebrate stomach has important implications to the optimal functioning of pepsin proteases.

There is an optimal relationship between hydrolysis rate and pH. The optimal pH of digestive enzymes is invariably at the normal *in vivo* pH at which the enzymes function. The pH optimum is often about 7 to 8, i.e., near neutral, but varies dramatically for enzymes from different animals and from different parts of the gut of the same animal, from about 2 to 11 (Table 18–4). For example, the vertebrate protease pepsin has an optimal pH of about 2 to 3; this closely corresponds to the low pH found in the vertebrate stomach. There are, in fact, a number of different pepsins in vertebrates; as many as 7 different pepsins have been isolated in man.

The pH optimum of pepsin varies; it is generally about 2 to 3 but is often higher (Figure 18–17). The pH optimum of pepsin also varies with different substrates, e.g., pepsin for albumin (1.5), casein (1.8), hemoglobin (2.2), and gelatin (2.3). In contrast, the proteolytic enzyme trypsin has an optimal pH of about 7 to 8, which corresponds to the slightly alkaline pH of the small intestine. Protease enzymes of invertebrates vary markedly in optimal pH. The blowfly *Musca* has a low pH protease (optimal pH about 3) that is comparable to that of vertebrate pepsin. The optimal pH is 4.9 for the blood-sucking insect *Rhodnius*, 8 for the blowfly 5 *Lucillia*, and over 10 for the clothes moth *Tineola*. Other protease enzymes, such as collagenase, have even higher optimal pH optima.

Many digestive enzymes have specific inorganic ion requirements for optimal activity. For example, human salivary and pancreatic amylases require Cl^- for activity. Other enzymes require various metallic ions for their activity.

Having examined the general properties of digestive enzymes, let us now examine proteases, carbohydrases, and lipases in more specific detail.

Proteases. These enzymes hydrolyze peptide bonds between amino acids (Figure 18–18). They can be divided into two major groups: the **endopeptidases**, which cleave peptide bonds within a protein but do not cleave the terminal amino or carboxyl amino acids, and the **exopeptidases**, which cleave the terminal amino acids with the free amino group (aminopeptidases), the amino acid with the free carboxyl group (carboxypeptidases), or dipeptides (dipeptidases).

Endopeptidases have considerable specificity with respect to the chemical groups on either side of the peptide bonds that they can cleave. For

TABLE 18–4

pH optima for various protease, carbohydrase, and lipase digestive enzymes. The actual pH optimum, or optimal range, was sometimes estimated from graphical data. The terms pepsin and trypsin are reserved for vertebrate low pH and neutral pH proteases, respectively. Vertebrate enzymes are shown in bold. Some enzymes have multiple entries, reflecting the different forms of the enzyme or the different pH optimum for different substrates.

Protease	Carbohydrase	Lipase
1.7 **Man—pepsin**		
2.0 **Toad—pepsin**		
2.2 **Man—pepsin**		
2.4 Fly		
2.6 Tick		
3.0 Bivalve mollusk		
3–4 Crayfish chitinase		
3.5 Crayfish—chitinase		
3.1 **Man—pepsin**		
3.1 Housefly		
3.7 **Man—rennin**	4.5 Bivalve β-glucosidase	
4.9 Blood-sucking bug	4.5 **α-amylase**	4.5 **Gastric lipase**
4.9 Bedbug	4.8 **Acid phosphatase**	
5.0 **Man—pepsin**	5.5 Invertase	
5.4 **Man—rennin**	5.5 Bivalve glucosidase	
5.5 Bivalve mollusk	5.5 Beetle—cellulase	
	5.6 Amylase	
	5.7 Crayfish saccharase	
	5.9 Cockroach amylase	
	5.9 Copepod amylase	
6.9 **Man—pepsin**	7.0 **β-amylase**	6.5 Bivalve lipase
7.2 Cockroach	7 Glycogenase	
7.5 **Man—carboxypeptidase**	7 **β-glucosidase**	
7.5 Onycophoran	7.5 Amylase	7.5 Onychophoran lipase
7.7 Bivalve mollusk		7.2 Bivalve lipase
7.8 **Man—trypsin**		7.5 Onycophoran esterase
7.9 Mosquito		7.8 Bivalve lipase
8.0 **Man—chymotrypsin**		8 **Lipase**
8.0 Tsetse fly		8 Cockroach lipase
8.0 Spider		
8.2 Cockroach		
8.2 Locust		
8.5 Collagenase—blowfly		
8.5 Housefly		
8.5 Blackfly		
8.5 Mealworm		8.7 Bivalve lipase
9.5 Silkworm	9.5 Silkworm amylase	
10.0 Clothes moth	9.5 **Alkaline phosphatase**	

example, pepsin and chymotrypsin hydrolyze proteins at aromatic amino acids connected to dicarboxylic acids; pepsin cleaves the protein on the free amino side of the amino acid, whereas chymotrypsin cleaves on the free carboxyl side of the aromatic amino acid. Trypsin acts on the carboxyl peptide bonds of arginine and lysine. Keratinase hydrolyzes the disulfide bonds between adjacent keratin proteins.

There are three kinds of exopeptidases. Aminopeptidases cleave the terminal amino acid in which the amino group is free, carboxypeptidases cleave the terminal amino acid in which the carboxyl group is free, and dipeptidases cleave dipeptides. There is some general specificity of these enzymes based on the specific amino acids on each side of the peptide bond to be hydrolyzed. Absolute specificity is rare but does occur, e.g., there is a dipeptidase that only cleaves the peptide bond of glycyl-glycine.

Proteases are secreted in an inactive form because they would have a devastating effect if present intracellularly in the activated state. The vertebrate

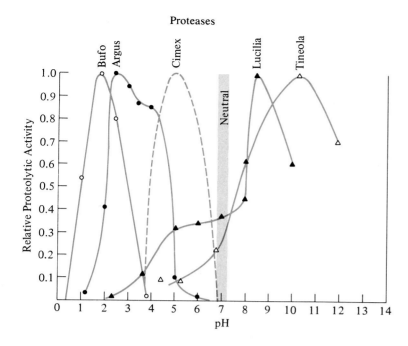

FIGURE 18-17 Effect of Ph on activity of protease digestive enzymes from a vertebrate (pepsin) and various acidic and alkaline proteases of invertebrates. *(Data from Taylor and Tyler 1986; Gooding 1975; Hobson 1931; Powning, Day, and Irzykiewicz 1951.)*

proteases (pepsin, trypsin, chymotrypsin, and carboxypeptidase) are secreted as inactive **zymogens** (pepsinogen, trypsinogen, chymotrypsinogen, and procarboxypeptidase). These inactive precursors are activated by either inorganic ions (e.g., H^+ activates pepsinogen), specific enzymes (e.g., enterokinase activates trypsinogen), other enzymes (e.g., trypsin activates chymotrypsinogen), or by the active enzyme itself (e.g., pepsin activates pepsinogen; trypsin activates trypsinogen). The latter example of self-activation is **autocatalysis**. Invertebrate endopeptidases generally are often secreted in an activated form, although their activity is further augmented by the presence of sulfydryl bonds in the gut contents. Various intracellular endopeptidases of both vertebrates and inverte-

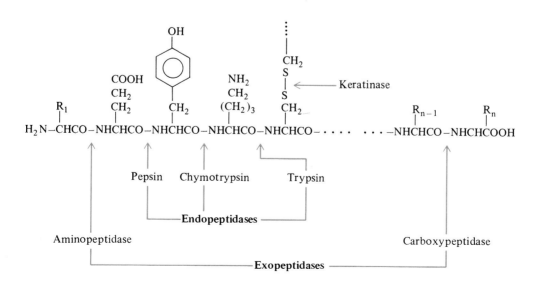

FIGURE 18-18 Cleavage points for various protease enzymes—exopeptidases (aminopeptidase and carboxypeptidase) and endopeptidases (pepsin, trypsin, chymotrypsin)—and a keratinase.

brates, called cathepsins, are produced in the active form. It is likely that the extracellular digestive endopeptidases are derived from intracellular cathepsins.

Carbohydrases.   Carbohydrate digestion, like protein digestion, involves hydrolysis of polymeric bonds and proceeds in stages until the basic units (monosaccharides) are produced. There are two classes of polysaccharide hydrolyzing enzymes: the **polysaccharidases** and the **oligosaccharidases**.

The most common polysaccharidase enzymes are amylases, which hydrolyze plant starches (amylose and amylopectin) and animal glycogen; these all are polysaccharides with α-bonds between the subunits (see Chapter 3). The α1, 4 and α1, 6 amylases cleave all but the terminal glycosidic bonds and ultimately produce (through intermediate polysaccharides of varying lengths, called dextrins and maltotriose) the disaccharide maltose and the monosaccharide glucose. Exoamylases cleave maltose residues from the ends of polysaccharides, whereas endoamylases attack bonds within the polysaccharides. Amylases generally require Cl^-.

The oligosaccharidases hydrolyze trisaccharides such as raffinose, and disaccharides such as maltose, sucrose (table sugar), lactose (milk sugar), and trehalose (a common insect sugar). Maltose, sucrose, and trehalose all contain a glucose bound by an α-bond to a second monosaccharide and are hydrolyzed by α-glucosidases, e.g., maltase, sucrase (also called invertase in invertebrates), trehalase (Figure 18–19). Glucose linked to other monosaccharides by a β-bond (e.g., lactose, cellobiose) is hydrolyzed by β-glycosidases (e.g., lactase or β-galactosidase).

Lipases and Esterases.   **Lipases** and **esterases** hydrolyze lipids in much the same manner as proteins and carbohydrates are hydrolyzed, but these enzymes are generally much less substrate specific than proteases or carbohydrases.

Lipases hydrolyze triglycerides, a common dietary lipid. Only a single lipase is required for the complete breakdown of triglycerides. Progressive hydrolysis of a triglyceride proceeds by one of two routes to form a monoglyceride and two free fatty acids; the monoglyceride may be further split to glycerol and the third fatty acid (Figure 18–20). Lipases also readily catalyze the reverse reaction, synthesis of triglycerides, and so complete hydrolysis of lipids requires that one (or both) of the products of hydrolysis (i.e., glycerol, monoglycerides, or fatty acids) are removed.

Esterases hydrolyze simple esters (e.g., ethyl butyrate) and complex lipids (phospholipids, cholesterol esters, waxes). Technically, lipase is actually an esterase enzyme because triglycerides are esters of glycerol and fatty acids, but lipase is considered a separate category of enzyme because of its specific and important role. Phosphatases are esterases that cleave phospholipids. Waxes are difficult esters to hydrolyze, although this is achieved by some animals through symbiotic micro-organisms (see below).

Absorption.   The amino acids, monosaccharides, monoglycerides, glycerol, and fatty acids that are produced by the actions of digestive enzymes, as well as ions, vitamins, and water, must be absorbed across the epithelial lining of the gut before they can enter the general body fluids. There are four essential aspects to absorption: (1) physical move-

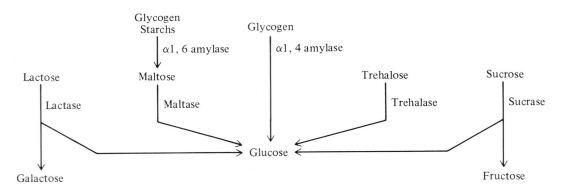

FIGURE 18–19   Hydrolysis of complex carbohydrates (starch and glycogen) to disaccharides, and hydrolysis of various important disaccharides to monosaccharides.

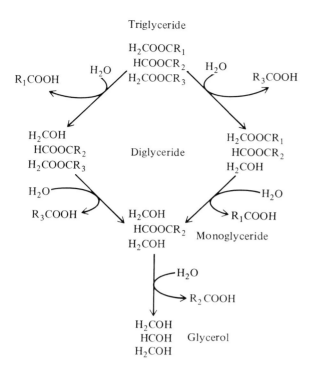

Triglyceride

FIGURE 18–20 Stages of hydrolysis of a triglyceride to likely end products of intestinal lipase digestion, 2-monoglyceride, fatty acids, and glycerol.

ment of the nutrient molecules from the gut lumen to the wall of the gut tube, (2) maximizing the area of the epithelial cells across which absorption can occur, (3) mechanisms for transport of the nutrient molecule across the cell membrane into the epithelial cell cytoplasm, and (4) transport of the nutrient molecule out of the epithelial cell into the extracellular body fluids (e.g., blood or lymph).

The important physical aspects of nutrient transport from their site of formation in the lumen to the surface of the epithelial cells reduces to two problems: convective exchange of fluid and nutrient molecules within the gut lumen and diffusional exchange across an unstirred boundary layer of fluid immediately in contact with the membrane of the epithelial cell. Convective mixing of the gut contents is promoted by the mixing movements of the gut (e.g., peristalsis) and the streaming of fluid caused by secretion and absorption by the epithelial cells. Exchange across the unstirred boundary layer of the epithelial cells is diffusional and is described by Fick's first law of diffusion. The diffusional flux (J_{diff}) is related to the difference in solute concentration difference between the lumen (C_l) and the surface

of the cell (C_s), the surface area for exchange (A), and the thickness of the unstirred boundary layer (Δx)

$$J_{diff} = -DA(C_l - C_s)/\Delta x \qquad (18.3)$$

where D is the diffusion coefficient.

Active uptake across the gut epithelium occurs in series with passive diffusion through the unstirred boundary layer. The active and diffusional fluxes (per unit area) are equal, and so from equations 18.1 and 18.3

$$J_{act} = \frac{C_s J_{max}}{C_m + K_t} = J_{diff} = \frac{-D(C_l - C_s)}{\Delta x} \qquad (18.4)$$

The effect of a thick unstirred boundary layer (large Δx) is to decrease not only J_{diff} but also the active rate of flux because C_s is decreased (Figure 18–21). A thick unstirred boundary layer is relatively more important at low C_s rather than high C_s when J_{act} approaches J_{max}.

Fick's law illustrates not only the importance of the unstirred boundary layer on exchange but also the importance of surface area to exchange. The gut tube is often modified to increase greatly its surface area for digestion and absorption. In many animals, such as the beetle *Popillia*, the gut has numerous blind-ended diverticulae (cecae) to increase the surface area for digestion and absorption (Figure 18–22A). Terrestrial oligochete worms have an inward dorsal folding of the gut tube, the typhlostyle, that increases the absorptive surface area. Increasing the length and coiling of the gut is another way to increase its surface area, but this is not very common for invertebrates. Vertebrates have a variety of structural modifications that increase the gut surface area. The spiral valve of elasmobranchs is a twisted structure, running the length of the intestine, that directs the food in a corkscrew fashion along the gut (Figure 18–22B); this effectively increases the length and absorptive surface area of the gut and slows the passage of food, allowing more time for absorption. The spiral valve of some elasmobranchs is not twisted but has extensive folds that are rolled to form a scroll-like structure that runs the length of the intestine. Other fish (e.g., chimaeras, cyclostomes, lungfish, lower actinopterygians) also have a spiral valve in their intestine, albeit often less developed than in the elasmobranchs. Some teleost fish have the remnants of the intestinal spiral valve but most teleosts and the higher vertebrates have elongated, coiled intestines, sometimes with pouches or cecae that increase the surface area. The often numerous cecae of fish are important sites of nutrient absorption (Buddington

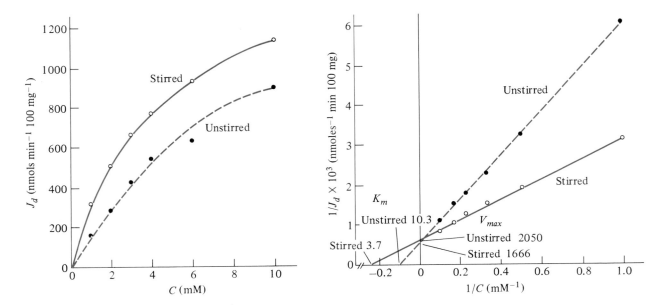

FIGURE 18–21   For active uptake of glucose by rat intestine, there is a hyperbolic relationship between flux (J_d) and glucose concentration on the intestinal side of the epithelium (C; left) and Lineweaver-Burk plot ($1/J_d$ versus $1/C$; (right) to illustrate graphically the maximum flux (J_{max}) and Michaelis-Menten transport constant (K_t). There is an effect of stirring the fluid bathing the intestinal epithelium surface on the uptake kinetics, as evidenced by the higher J_{max} and lower K_t for the stirred fluid compared to the unstirred fluid.   *(Modified from Wilson and Dietsky 1974.)*

and Diamond 1987). The mammalian small intestine is highly specialized for digestion and especially absorption. Folds of the mucosa and submucosa (**valves of Kerckring**, or valvulae conniventes) increase the surface area of the small intestine by about three-fold (Figure 18–22C). Numerous small, finger-like projections of the mucosa and submucosa, the **villi**, further increase the surface area by ten-fold. A highly folded apical cell membrane of the epithelial cells, the **microvilli** or brush border, dramatically increases in surface area by twenty-fold.

Intestinal villi (Figure 18–23) have additional roles in digestion and absorption. They are motile (there are smooth muscle cells in the villus) and their movements promote mixing and decrease the thickness of the unstirred boundary layer. Each villus contains a **lacteal**, which is a terminal branch of the lymphatic system and is an important route for absorption of lipids and water (see below). The villus also contains blood vessels that absorb nutrients for transport throughout the body. There is a countercurrent arrangement of blood flow into and out of the villus, and this allows the passive establishment of considerable gradients for ions (e.g., Na^+) and oxygen along the villus. Na^+ reab-

sorbed by the epithelial cells may diffuse into the capillaries, and countercurrent exchange into the central artery establishes an ion concentration gradient from the villus base (equal to plasma concentration) to the tip (high concentration). Countercurrent transfer of water from the central artery to the peripheral drainage would also establish an osmotic gradient along the villus. The osmotic gradient that becomes established along the villus by recycling solutes and water from the efferent venule into the afferent arteriole can be substantial, reaching 1000 mOsm kg^{-1} at the villus tip (Jodal 1977).

The surface of the microvilli is covered with a meshwork of mucopolysaccharide and glycoprotein filaments, the **glycocalyx**. This surface coat, or mat, is extremely resistant to proteolytic and mucolytic agents, hence forms a protective surface. Many of the important digestive enzymes are adsorbed into the glycocalyx meshwork, e.g., pancreatic amylase, disaccharases, and dipeptidases. Water and mucus trapped within the 0.3 μ thick glycocalyx contribute dramatically to the thickness of the unstirred epithelial cell boundary layer.

There are four basic mechanisms for uptake of nutrients from the gut lumen and glycocalyx layer across the apical membrane and into the cytoplasm

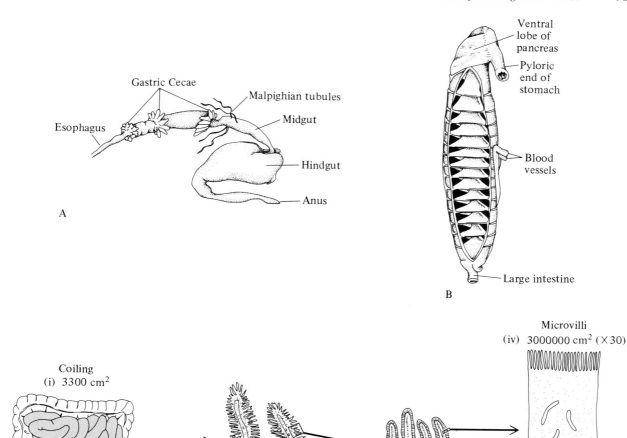

FIGURE 18–22   Modifications of digestive tracts to increase digestive and absorptive surface area. **(A)** The gut of a scarabaeid beetle larva *Popillia* has three sets of gastric cecae. **(B)** Typical "corkscrew" spiral valve of the elasmobranch *Squalus*. **(C)** Varying levels of morphological arrangements increase the surface area of the mammalian small intestine (shaded) coiling, valves of Kerckring, villi, and microvilli. The approximate surface area of the small intestine and the relative increase due to each modification are indicated.   *(From Wigglesworth 1935; Jennings 1972; Fawcett 1962.)*

of the epithelial cells: (1) passive diffusion, (2) facilitated diffusion, (3) active transport, and (4) pinocytosis (see Chapter 6). There are examples of all four mechanisms of transport operating for nutrient absorption, although active transport is the most common.

Proteins.   In general, proteins are hydrolyzed to individual amino acids before being absorbed across the gut by active transport. The uptake process is

active, being carrier mediated and saturatable. It is also linked with concomitant Na^+ transport. The energy for amino acid uptake by the gut epithelium is not expended by the amino acid carrier at the apical surface of the epithelial cell, but by the Na^+-K^+ ATPase enzyme on the baso-lateral surfaces of the cell, which maintains Na^+ and K^+ concentration gradients across the cell membrane. The low Na^+ concentration inside the cell allows Na^+ to diffuse passively into the cell from the gut lumen via a

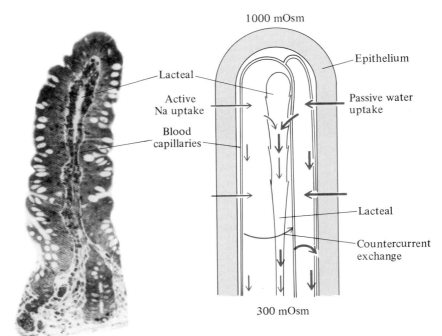

FIGURE 18–23 Structure of an intestinal villus and a diagram showing how the countercurrent exchange of ions and water in the intestinal villus establishes an osmotic gradient that promotes osmotic water absorption. *(Modified from Fawcett 1986; Jodal 1977.)*

carrier that also transports an amino acid (Figure 18–24). There are actually four different amino acid/Na$^+$ co-carriers, each specialized for a particular type of amino acid: neutral amino acids; acidic amino acids; basic amino acids; and glycine, proline, and hydroxyproline. Amino acid absorption from seawater by the epidermal cells of phylogenetically primitive marine invertebrates, such as the sea anemone (*Anemonia*), is very similar to the gut-uptake mechanisms of phylogenetically more advanced invertebrates and vertebrates.

Carriers also transport dipeptides and tripeptides into the cell cytoplasm where they are hydrolyzed into amino acids. Some large polypeptides and even intact proteins are absorbed into the epithelial cells by pinocytosis; the protein is surrounded by an evagination of the cell membrane and absorbed into the cytoplasm enclosed as a membrane-bound vesicle. The protein is then intracellularly hydrolyzed to amino acids, except in infant mammals, especially ruminants and rodents. Newborn ruminants absorb gamma globulins from their mother's colostrum (the special milk produced for the first few days after birth); this is possible because there are trypsin inhibitors in the colostrum and the neonate is achlorhydric (its stomach does not secrete HCl). In newborn rodents, the gamma globulins are absorbed by pinocytosis and released intact into the body fluids by exocytosis at the basal cell membrane. This lasts for 18 to 20 days and confers a temporary passive immunity to the neonate against antigenic materials to which the mother has been exposed. Other infant mammals, and even adults, also have a low level of intact protein absorption, but the absence of specialized uptake systems generally makes it unimportant immunologically or nutritionally.

Most amino acids are absorbed but are not metabolized inside the epithelial cells (except glutamate and aspartate); they are transported out of the epithelial cells and enter the circulation. Absorbed dipeptides and tripeptides are generally hydrolyzed intracellularly to amino acids which then enter the circulation. Absorbed proteins can enter the lymphatic system and then the circulatory system.

Carbohydrates. Carbohydrate absorption (at least of glucose and galactose) is quite similar to amino acid absorption (Figure 18–24). The process is active, carrier-mediated, requires ATP, and is linked with Na$^+$ transport. Glucose and galactose, and also other actively transported monosaccharides, compete for the same carrier molecule. Fructose and some other sugars are transported by facilitated diffusion from high concentration (normally the gut lumen) to low concentration (normally the inside of the cell). Facilitated diffusion requires a specific carrier but does not require expenditure of energy and is not linked to Na$^+$ transport. A third mechanism, hydrolase-related transport, appears to transport glucose and other monosaccharides across the apical membrane (Crane 1975). Hydrolase-related transport is Na$^+$ dependent and is active, but the membrane carrier is thought to be an actual

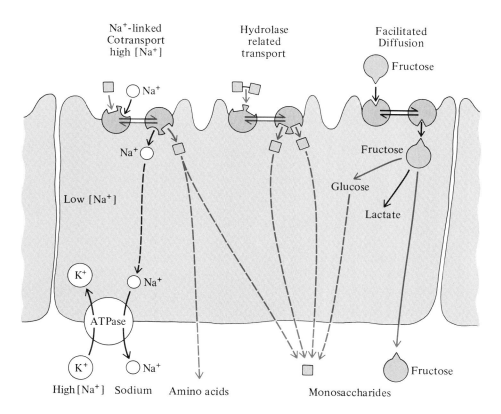

FIGURE 18–24 Representation of epithelial cell showing the active transport uptake mechanism for amino acids and some monosaccharides (Na⁺-linked cotransport utilizing the transmembrane Na⁺ gradient), one possible mechanism for hydrolase transport and disaccharides, and facilitated diffusion of fructose. *(Adapted from Crane 1975; Wilson 1962.)*

disaccharase enzyme which, in the process of hydrolyzing a dissacharide, also transports one (or both?) of the monosaccharides into the cell.

The transport of glucose and other monosaccharides out of the baso-lateral parts of the epithelial cells is not Na⁺ or energy dependent and resembles the facilitated diffusion systems of other cells (e.g., erythrocytes).

Lipids. Lipids, like carbohydrates and proteins, are cleaved by hydrolysis reactions. Otherwise, lipid digestion and absorption is fundamentally different because lipid is not water soluble and coalesces into large droplets in water. The low surface-to-volume ratio of a large lipid droplet is not conducive to effective digestion by lipases, and so **emulsification** is extremely important for lipid digestion. Emulsification stabilizes small lipid droplets in aqueous solution by coating them with molecules that are partly hydrophobic (to dissolve in the lipid droplet) and partly hydrophilic (to dissolve in water).

The liver of vertebrates produces bile, which flows into the small intestine. **Bile salts** are cholesterol conjugated with amino acids to form glycocholic acid and taurocholic acid. These are moderately effective emulsifying agents and become much more effective in concert with polar lipids such as lecithin, lysolecethin, and monoglycerides (lysolecithin and monoglycerides are the digestive products of lecithin and triglycerides, respectively). Thus, enzymatic breakdown of lipids further stabilizes the lipid emulsion. The final product of lipid emulsification is a dispersion of lipids as small (500 to 1000 mμ dia) droplets. Ingested lipids that are solid at normal body temperature are poorly emulsified and difficult to digest, but mixing of these high melting point lipids with other low melting point dietary lipids reduces the melting point of the mixture, making it easier to emulsify and digest.

Lipase enzymes attach to the emulsified lipid droplets and hydrolyze triglycerides to 2-monoglycerides, some glycerol, and free fatty acids. The monoglycerides combine with bile salts to form **micelles**, highly stable emulsified lipid droplets of only 4 to 6 mμ diameter (Figure 18–25). Bile salts are normally more concentrated in the gut lumen than the minimum threshold concentration required for micelle stabilization (the critical micellar concentration is about 1 to 5 mM). Micelles also contain other lipids, such as cholesterol and lysolecithins, and fat-soluble vitamins. Each micelle probably contains about 20 molecules. The micelles transport fatty acids and monoglycerides from the site of lipid hydrolysis in the lumen through the unstirred boundary layer to the surface of the epithelial cell membranes. Micelles are not transported into the epithelial cells, but their contents are freely soluble in the lipid bilayer cell membrane and dissolve readily through the membrane into the intracellular

fluid. The bile salts are recycled back into the gut lumen to form further micelles. Most of the bile salts are eventually absorbed by the hindgut and recycled to the liver after their role in micellar transport is completed. Lipid absorption is thus energy independent and no membrane carriers are involved.

Triglycerides are resynthesized inside the epithelial cells from the free fatty acids and monoglycerides by acyl-transferase enzymes (Figure 18–25). The fatty acids are converted to fatty-acylCoA esters; this requires energy expenditure. If fatty acids are present in micelles but monoglycerides are not (e.g., in ruminant mammals because the symbiotic micro-organisms metabolize the monoglycerides), then the glycerol must be synthesized *de novo*; this also occurs if only unesterified fatty

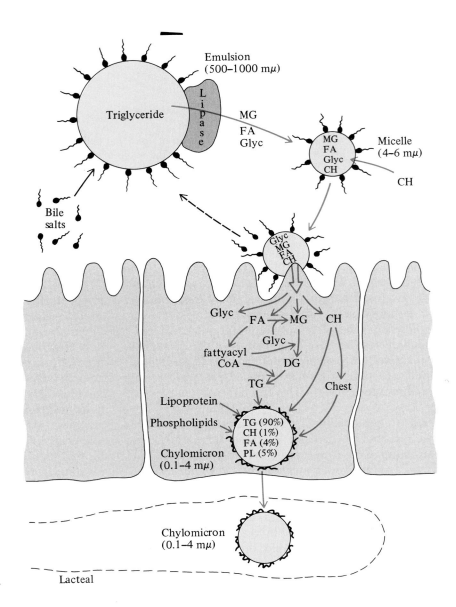

FIGURE 18–25 Digestion and absorption of lipids (triglycerides, cholesterol) involves emulsification of triglycerides; digestion by intestinal lipase; absorption into the intestinal epithelial cells; cellular modification and formation of chylomicrons; and extrusion of chylomicrons into the lymphatic vessels, the lacteals. The diameter of the droplets is indicated. Abbreviations are as follows: TG, triglyceride; DG, diglyceride; MG, monoglyceride; Glyc, glycerol; FA, fatty acid; CH, cholesterol; Chest, cholesterol esterase; PL, phospholipid.

acids are present in the diet. Lysolecithins are reconverted to lecithins in the epithelial cells. Cholesterol is esterified, mainly with oleic acid. These resynthesized complex lipids are stabilized within the epithelial cells by phospholipids (both dietary phospholipids and endogenous phospholipids) and lipoproteins, to form **chylomicrons**. Chylomicrons (0.1 to 4 mμ dia) are 80 to 95% triglyceride and covered by a phospholipid monolayer containing small amounts of protein, cholesterol, and triglycerides. The chylomicrons are then discharged from the sides and base of the epithelial cells and enter the lymphatic vessels (lacteals). Chylomicrons are too large to pass through the capillary fenestrae but do penetrate the larger spaces of the lacteal walls. They drain from the lymphatics into the circulation. Short and medium length fatty acids are water soluble and diffuse directly into the capillaries and then the general circulation.

Ions and Water. Monovalent ions are readily absorbed from the gut by active transport. Sodium absorption can occur by passive diffusion from the gut lumen into the epithelial cells if the electrochemical gradient is favorable or by active transport against an electrochemical gradient. As we have already seen, the basal membrane of the epithelial cells has an Na^+-K^+ pump that maintains a low intracellular Na^+. It is this low intracellular concentration that promotes passive influx of Na^+ from the lumen as occurs with Na^+-linked amino acid, monosaccharide, and disaccharide transport. There are also exchange systems whereby Na^+ is absorbed in exchange for K^+ or H^+ at the apical cell surface. Chloride is absorbed passively by the small intestine, in response to active Na^+ uptake; transport of Na^+ from the lumen produces a negative (lumen) to positive (intracellular) electrical gradient that favors passive Cl^- influx. In the large intestine, Cl^- from the lumen is actively exchanged with bicarbonate from the inside of the epithelial cells; a major role of the bicarbonate is to buffer the acidic metabolic products of bacteria in the large intestine.

Divalent ions, in contrast, are more difficult to absorb. For example, there is a special active uptake mechanism for iron, which is an essential mineral required for hemoglobin and enzyme function. Ferrous ions (Fe^{2+}) are absorbed more effectively than ferric ions (Fe^{3+}) because of their higher solubility. Anions, such as phosphate and oxalate that form insoluble iron salts, interfere with its absorption, whereas soluble chelating agents (e.g., fructose, vitamin C) promote iron absorption. Active uptake of iron by the small intestine occurs in two steps: mucosal absorption and then transfer to the plasma.

Excess absorption into the body is prevented by regulation of iron transfer into the plasma, not by mucosal uptake. Iron exists within the epithelial cells in two forms: a pool of Fe^{2+} and Fe^{3+} in solution and a pool of ferritin. Ferritin is iron bound to a protein, apoferritin. Transfer of iron into the blood is regulated by the ferritin pool; excess iron remains within the epithelial cell until it dies and is sloughed from the epithelium and excreted. A low iron diet results in minimal retention of iron in the epithelial cell and maximum transport to the plasma as ferritin. A balance between these two processes occurs with a normal dietary iron intake.

Calcium ions are absorbed by active transport involving a Ca^{2+}-binding protein in the microvilli. This uptake mechanism is under the control of vitamin D and parathyroid hormone (PTH). Phosphate ions are actively absorbed by the intestinal epithelium; phosphate uptake is closely related to calcium uptake because both form the major constituent of bone. Magnesium is absorbed by passive diffusion.

Water absorption by the gut is indirectly linked to solute transport, e.g., Na^+, Cl^-, amino acids, monosaccharides, etc. There is no specific transport mechanism for water, and so absorption or secretion depends on osmotic gradients to promote passive water movement. The intestines of many animals, especially the hindgut of insects, reptiles, and birds, are highly specialized for water absorption linked to active solute uptake (Chapter 17). Much of this water is absorbed into the lymphatic lacteals as well as into the capillaries.

Digestive Systems

The structures and functions of various parts of the digestive tracts of animals have now been reviewed. Some examples of complete digestive systems will now be examined so as to illustrate how the coordinated functioning of a system of different structures accomplishes the overall functions of digestion, absorption, and the elimination of waste materials. Four digestive systems will be examined to illustrate this. The incomplete digestive system of a ciliate protozoan will be described as an example of an intracellular digestive system. The complete digestive system (i.e., a tubular gut with mouth and anus) of two multicellular invertebrates, bivalve mollusks and insects, will be described briefly. Then, the digestive system of vertebrates (with an emphasis on mammals) will be described in greater detail and used to illustrate the mechanisms responsible for the control of the digestive processes.

Protozoans

Protozoans may be autotrophic (photosynthetic), saprozoic (detritus feeders), or heterotrophic (ingest food particles). There are many parallels between the feeding and digestion by ciliates (Figure 18–26) and these processes in multicellular animals. Food is directed by ciliary action into the cytostome, or "mouth," which opens into the cytopharynx. The cell membrane lining the cytopharynx enlarges to pinch off food-containing vacuoles. These detached food vacuoles then begin a directed movement through the cytoplasm. Excess water is removed from the vacuole, the contents acidified and then made alkaline, and digestive enzymes added from lysosomes. Food particles are digested in the vacuole, the nutrients absorbed into the cytoplasm, and then the waste products excreted via the cytopyge or "anus."

Many protozoans feed on small food particles or other micro-organisms. Some harvest microbes or use their metabolic products as nutrient sources, and some have internal symbiotic bacteria that are either photosynthetic or use the waste products of the protozoan as their nutrient source (Finlay and Fenchel 1989). The ciliate *Kentophoros* has a dense coat of large bacteria; it lacks a "mouth" but feeds by invaginating the cell surface to "ingest" some of the bacteria. The bacteria are autotrophic, synthesizing organic materials from O_2, H_2S, and CO_2. Some anaerobic protozoans contain methanogenic bacteria that produce methane by coupling H_2 oxidation with CO_2 reduction or splitting of acetate. The hydrogen is formed by hydrogenosomes, mitochondria-sized protozoan organelles that produce CO_2, acetate, and H_2. In some ciliates, such as *Metopus*, the methanogenic bacteria are actually coupled to the hydrogenosomes. The microbial growth provides a food source for the protozoan.

Some freshwater ciliates have a symbiotic relationship with algae, e.g., *Chlorella*, and some of the organic products of the algae are used by the

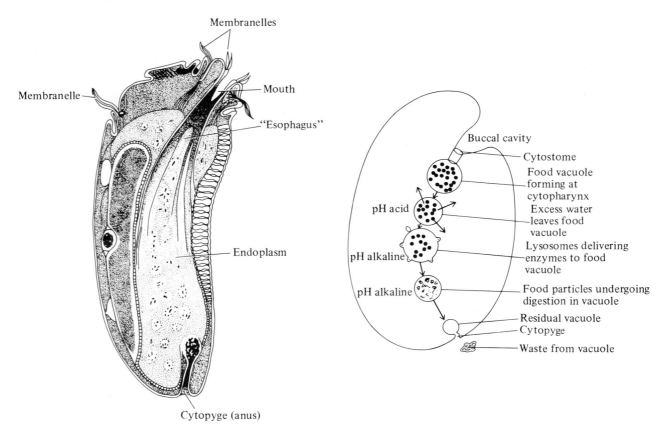

FIGURE 18–26   Representation of digestion in a ciliated protozoan (right). Food directed by ciliary membranelles towards the cytostome, or "mouth," enters the cytopharynx where food vacuoles form and detach from the cytopharynx. The detached food vacuoles undergo acidic and then alkaline digestion, and the waste laden vacuole migrates to the cytopyge, the "anus," for excretion. A typical example of such a ciliate protozoan is the entodiniomorph ciliate *Diplodinium*, which is a symbiont in the rumen of cattle (left).   *(From Barnes 1987.)*

protozoans as food. The protozoans supply nitrogen to the algae, as NH_4^+; in fact, the protozoans can control the growth rate of their algae by regulating the supply of NH_4^+. Many oligotrich protozoans "steal" chloroplasts from their algal prey; the ingested algae are broken open and their chloroplasts are retained and nutured near the cell membrane to fix CO_2 and to produce sugars, as occurred in the algae.

Bivalve Mollusks

Lamellibranchs, a common type of bivalve mollusk, suspension feed and ingest small food particles mixed with mucous strands. Their digestive strategy is unique among higher animals because both extracellular and intracellular digestion are important.

The digestive tract has a short esophagus opening into a stomach, midgut, hindgut, and rectum. The stomach contains a crystalline style and gastric shield and a ciliated, diverticulated region. The digestive diverticulae of the stomach are blind-ended sacs that have an absorptive and intracellular digestive role. The midgut, hindgut, and rectum have extracellular digestive and absorptive roles.

Food is drawn into the stomach and is mechanically reduced by the mortar-and-pestle action of the rotating crystalline style, sorted into sizes by the stomach, and subjected to extracellular enzymatic action (enzymes are released from the style and gastric shield; Figure 18–27). In addition, fragmentation spherules released by the digestive diverticulae contribute to a low pH and extracellular enzymatic digestion. As small particles are produced and

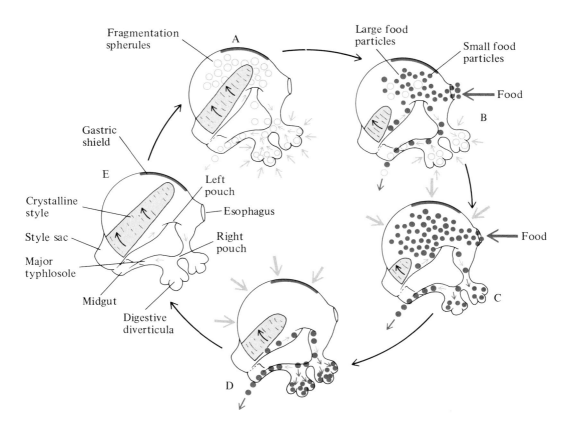

FIGURE 18–27 Processes of intracellular and extracellular digestion in the bivalve mollusk stomach and digestive diverticulae. The basic digestive cycle starts before food ingestion with dissolution of the crystalline style and presence of fragmentation spherules in the stomach (**A**). Food arriving in the stomach is mechanically reduced by the rotating style and shield and extracellular digestion begins (**B**). Particles are then passed to the digestive diverticulae for intracellular digestion (**C**). Cessation of feeding is followed by progressive passage of food particles from the stomach to the digestive diverticulae (**D**). During the quiescent interfeeding phase, the stomach empties and the crystalline style is re-formed, while intracellular digestion in the diverticulae is completed and fragmentation spherules begin to form (**E**). The movement of fragmentation spherules to the stomach starts the next cycle (**A**). *(From Morton 1983.)*

sorted by the stomach they pass into the digestive diverticulae for intracellular digestion. The crystalline style is reduced in size as feeding continues and food particles are ground and passed into the digestive diverticulae; this process continues even after the feeding bout is ended, and the mollusk closes its valves. During this quiescent phase, the stomach empties all material into the digestive diverticulae and the style is re-formed by secretion. The digestive cells of the diverticulae continue intracellular digestion and transfer the digested metabolites into the blood and release into the stomach fragmentation spherules that contain waste products. The fragmentation spherules increase the stomach acidity and begin the dissolution of the style in preparation for the next feeding bout.

The digestive cycle of lamellibranch mollusks is a coordination of three subcycles: (1) feeding, (2) extracellular digestion, and (3) intracellular digestion. The quiescent phase is preparative for extracellular digestion. The mechanical and extracellular enzymatic breakdown during feeding provides the small particles for intracellular digestion. The processes of intracellular digestion release digestive products into the blood and produce fragmentation spherules that are both excretory and reduce the pH for optimal extracellular digestion. The feeding cycle is linked to tidal immersion and emmersion of intertidal animals, and there is also a rhythmic pattern for shallow water bivalves. The level of food availability may also influence the periodicity of the digestive cycle.

Insects

The cockroach is a highly successful and representative insect. It is omnivorous and will feed on virtually any organic material. Food is first masticated by the mandibles and maxillae, passed into the buccal cavity, and then passed to the crop via the esophagus. Saliva is added to the food during mastication to lubricate it for passage through the digestive tract, and the saliva also contains an amylase enzyme (which is activated by Cl^- in the saliva). The salivary amylase continues to hydrolyze starches during food storage in the crop. Other enzymes (carbohydrases, proteases, lipases) secreted by the midgut are regurgitated into the crop; the pH is slightly acidic (pH 4 to 5) and so not only the proteolytic enzymes function under acidic conditions.

Food passes slowly from the crop through the gizzard where it is mechanically reduced and the particles are size sorted. Large particles are returned to the crop and gizzard for further processing and small particles pass to the midgut and enter the hepatic cecae where extracellular digestion is completed. Undigested food remnants are passed along the gut, combined with excretory products of the Malpighian tubules at the junction of the mid- and hindguts, and then passed into the rectum. Extensive water and ion reabsorption in the rectum produces solid fecal pellets that are extruded through the anus (see Chapter 17).

There is considerable control of the production of enzymes in insects during periods of feeding and fasting. There generally is a low level of protease activity in the gut during fasting and an increase after feeding. Such changes are especially pronounced in insects that feed only infrequently (e.g., the blood-sucking bug *Rhodnius*) compared to insects that feed frequently (e.g., caterpillars and grasshoppers). The regulation of enzyme activity can be accomplished by control of either enzyme synthesis or release. Proteases, for example, are found in gut epithelial cells at low concentrations during fasting, indicating control of synthesis. However, carbohydrases accumulate in the gut epithelial cells during fasting, indicating control of release.

There are three possible mechanisms for control of enzyme secretion and release: (1) secretagogues, (2) the nervous system, and (3) the endocrine system. The presence in the midgut of specific foods, food extracts, or breakdown products of food, collectively called **secretagogues**, can induce enzyme secretion. Secretagogue stimulation of enzyme release is probably important in blood-sucking insects. Neural regulation is unlikely to be important in regulating digestive functions because of the considerable time delay that can occur between feeding and digestion. In some insects, a secretion from the median neurosecretory cells regulates the production and/or secretion of proteases.

Vertebrates

The vertebrate digestive tract, like that of invertebrates, is highly specialized in both structure and function for the digestion of a wide array of food items (Figure 18–28). The basic components of the gut tube include the buccal cavity, pharynx, esophagus, stomach, small intestine, large intestine, rectum, and anus/cloaca. The structure of the stomach can be highly modified, depending on the diet. A few vertebrates lack a stomach (e.g., cyclostomes, lungfish) but in most it functions for storage and acidic/enzymatic digestion. It is highly modified for fermentation in ruminant and pseudoruminant mammals (see below). The intestine may have been the primary or even the sole site of digestive

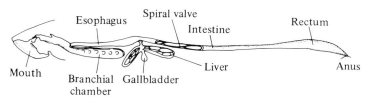

A. **Cyclostome**

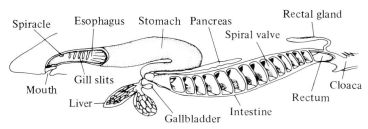

B. **Elasmobranch**

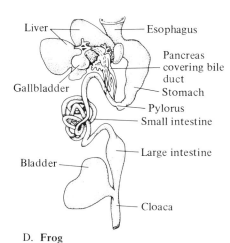

D. **Frog**

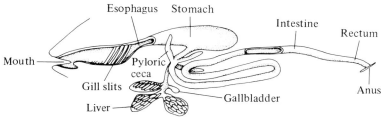

C. **Teleost**

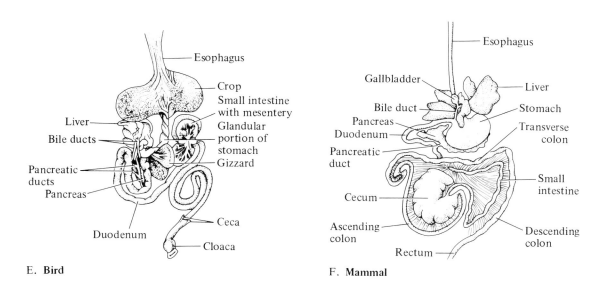

E. **Bird**

F. **Mammal**

FIGURE 18–28 Vertebrate digestive tracts showing progressive anatomical specialization for digestion (stomach and accessory digestive glands such as pancreas and liver) and absorption (surface area of small intestine). (**A**) Cyclostome, *Petromyzon*. (**B**) Elasmobranch, *Squalus*. (**C**) Teleost fish, *Perca*. (**D**) Frog, *Rana*. (**E**) Bird, *Columba*. (**F**) Mammal, *Cavia*.

enzyme production in primitive vertebrates, but its enzymatic production role has been supplemented by the stomach and pancreas (which is an outgrowth of the small intestine). Absorption is the primary function of the small intestine in higher vertebrates, although some absorption occurs in the stomach and large intestine. The short intestine of the elasmobranch has its surface area increased by the spiral valve; teleost fish have an elongate small intestine with a variable number of pouches, or pyloric cecae (up to 900!), which extend from the junction of the stomach and intestine. Higher vertebrates have a much elongated small intestine to increase the absorptive surface area. The generalized tetrapod intestine is exemplified by that of amphibians, with a long, narrow, coiled small intestine and short, broad large intestine. A small pocket, the cecum, is usually present at the junction of the small and large intestines. The cecum is greatly enlarged as a fermentation chamber in certain postgastric fermenting mammals (see below).

There are three important glandular systems associated with the vertebrate digestive tract: (1) the salivary glands; (2) the liver, gallbladder, and bile duct; and (3) the pancreas and pancreatic duct. The role of the salivary glands has already been described (see Reception).

The **liver** is a large, consolidated organ that secretes bile fluid, which may be stored in the gallbladder prior to release via the bile duct into the small intestine. Some mammals (e.g., the rat) lack a gallbladder. The bile contains bile salts for emulsification of lipids, and also some excretory products (e.g., bile pigments). The liver is also an extremely important digestive organ that stores and distributes absorbed metabolites, interconverts metabolites, and detoxifies poisons; it receives arterial blood and nutrient-rich blood from the small intestine. The hepatic diverticulum of the cephalochordate *Branchiostoma* is not homologous with the vertebrate liver; it is a site of enzyme production and food absorption but lacks the metabolic functions of the vertebrate liver.

The **pancreas**, which is often a diffuse organ (especially in fish) produces pancreatic juice that is emptied into the small intestine via the pancreatic duct. Pancreatic fluid contains a number of proteolytic enzyme precursors (trypsinogen, chymotrypsinogen, carboxypeptidase), amylase, DNAase, RNAase, cholesterol esterase, lipase, and bicarbonate to neutralize the acidic stomach chyme. Cephalochordates and cyclostomes lack a pancreas; the former has groups of cells in the anterior intestine that resemble the pancreatic cells, and the latter has small separate clusters of glandular tissue near the opening of the bile duct that presumably are

transitional to the formal pancreatic gland of most higher vertebrates.

The large intestine of mammals, which is generally more highly developed than in other vertebrates, is responsible for water reabsorption and formation of feces. Bacterial action in the hindgut can ferment cellulose and can synthesize vitamins.

Control.   The control of digestive functions in vertebrates is important, especially in birds and mammals, because of the complexity of the structures of the digestive tract (Table 18–5). There are basically three mechanisms for regulation: (1) local regulation by contents of the gut, (2) neural innervation of the gut from higher brain centers, and (3) hormonal regulation. The regulation of digestion in a typical omnivorous mammal, man, illustrates the varying role of these three control mechanisms. Neural regulation is generally more important at the proximal end of the gut (salivary glands, stomach), whereas endocrine regulation is generally more important in the midgut; local effects are important throughout the gut.

Salivary secretion is regulated by neural innervation, primarily the parasympathetic nervous system. The salivatory nuclei, located in the brain stem (medulla), receive taste and tactile sensory information, especially from the tongue, and elicit copious salivary secretion. Other areas of the brain, such as the appetite area (in the anterior hypothalamus), which responds to taste and smell, can modify the action of the salivatory nuclei. The classical conditioning experiments of Pavlov, in which dogs were conditioned to salivate in response to inappropriate (for digestion) stimuli such as bells ringing or lights flashing, demonstrate the neural reflex control of salivation. There is no important endocrine control of salivation. This is not surprising considering the rapidity with which salivation must occur when food is consumed, and even the necessity for salivary secretion to actually anticipate feeding. Neural regulation is far quicker than hormonal regulation.

Gastric secretion occurs in three phases: the cephalic phase, the gastric phase, and the intestinal phase. The cephalic phase occurs rapidly after eating and can even anticipate ingestion. It is parasympathetically controlled via the vagus nerve in response to cerebral cortex or appetite center activity. It is the least important of the three phases. The most important phase, the gastric phase, is initiated by a number of factors when food reaches the stomach. The presence of secretagogues (specific foods, food extracts, and breakdown products) as well as mechanical distension of the stomach, elicits a local neural reflex that releases a hormone, **gastrin**, from the gastric mucosa. Gastrin is distributed by

TABLE 18–5

Summary of the organs, their functions and secretions, and mechanisms of regulation of the vertebrate digestive tract.

Organ	Function	Secretions	Regulation
Oral cavity	Food acquisition Mastication	—	Voluntary and reflex neural control
Salivary glands	Salivary secretion	Mucus for lubrication Serous containing α-amylase, water	Neural reflex—salivatory nucleus, appetite center
Stomach	Storage Mechanical digestion Chemical digestion Enzymatic digestion	Mucus HCl Pepsin, amylase, tributyrase, lipase	Neural—vagal, Gastric reflex Enterogastric reflex Endocrine—gastrin; cholecystokinin, secretin
Pancreas	Exocrine gland	Serous—water; bicarbonate Enzymes—trypsin; chymotrypsin, carboxypeptidase, DNA'ase, RNA'ase, amylase, lipase, cholesterol esterase, phospholipase, trypsin inhibitor	Neural—vagal (mainly enzymes) Endocrine—secretin (enzymes); cholecystokinin (bicarbonate, water)
Liver/Gallbladder	Bile secretion Bile storage	Bile salts Metabolic processing of absorbed nutrients	Neural—weak vagal role Endocrine-cholecystokinin
Small intestine	Secretion Watery fluid Absorption	Mucus Enzymes—peptidases, disaccharases, lipase	Local neural reflex Endocrine-secretin Cholecystokinin
Large intestine	Secretion Absorption	Mucus Water	Local reflex

the blood to the rest of the gastric mucosa and promotes secretion of especially hydrochloric acid (HCl), but also pepsinogen. Gastrin also enhances peristaltic muscle contractions of the stomach and relaxes the pyloric sphincter, thereby promoting emptying of the stomach contents into the small intestine. The gastrin-mediated control of secretion is slower, but much longer lasting, than neural stimulation. The third phase of gastric secretion, the intestinal phase, has both stimulatory and inhibitory effects on the stomach. Small amounts of gastrin released from the small intestine promote gastric secretion, but this can be over-ridden by the inhibitory effects of two other hormones released from the small intestine: **cholecystokinin** and **secretin**. These hormones, as well as an enterogastric neural reflex, inhibit gastric secretion and also gastric motility, thus delaying further emptying of gastric contents so that food already present in the small intestine can be digested and absorbed.

Regulation of pancreatic secretion is both neural and endocrine, with the latter being the more important. The neural stimulation of pancreatic secretion via the parasympathetics causes release of mainly the digestive enzymes (e.g., amylase, lipase) or the inactive zymogen precursors (i.e., trypsinogen, chymotrypsinogen, procarboxypeptidase). There is little secretion of water or electrolytes and so the enzymes remain, temporarily, in the pancreatic ducts. The activation of the zymogens is inhibited by another secretory product, trypsin inhibitor. The presence of chyme in the small intestine causes release of secretin and pancreozymin (the latter is thought to be the same hormone as cholecystokinin). These hormones promote secretion of water and electrolytes, especially bicarbonate (secretin), and further pancreatic enzyme secretion (pancreozymin/cholecystokinin). Gastrin also has a cholecystokinin-like stimulatory effect on enzyme secretion.

Emptying of bile into the small intestine after a meal is due to reflex gallbladder contraction and relaxation of the sphincter where the bile duct enters the small intestine. Both vagal stimulation of the gallbladder and, more importantly, the effect of cholecystokinin, stimulate the gallbladder to contract and forces bile into the small intestine.

The copious secretions of the small intestine are primarily produced in response to a local neural reflex initiated by the presence of chyme in the small intestine. The hormones secretin and chole-cystokinin may also stimulate small intestine secretion.

Absorptive Adaptations of the Gut

The gut, like other organs of the body, adaptively responds to its environment, which is determined by the quality and quantity of the dietary intake. There are essentially three types of adaptive change exhibited by the gut (Karasov and Diamond 1983): (1) an increase in mucosal mass to facilitate absorption, e.g., an increase in gut length; (2) changes in the transport characteristics of the specific uptake mechanisms, e.g., decreased K_t and increased V_{max}; and (3) an increase in the transmucosal Na^+ gradient. (See Supplement 18–1, page 946.)

Many aspects of the structure of the digestive tract can be related to its function. For example, the **gut length** is often correlated with the nature of the diet. Carnivores tend to have a short, simple gut, whereas herbivores have a long, complex gut with large storage/fermentative regions (e.g., stomach, cecum). Birds forced to consume low-quality food show a dramatic increase, sometimes by over 50%, in the length of their gut, and there are seasonal changes in gut length of some birds that correspond to changes in diet. Carp respond to increased glucose in the diet with an increase in intestinal length and higher rates of glucose uptake. In amphibians, the surface area of the gut is increased for herbivorous species mainly by an increase in length, rather than by pyloric cecae or spiral valves. Herbivorous tadpoles, for example, have a long, thin gut tube that is twisted into a compact spiral of $2\frac{1}{2}$ to 3 double loops. Carnivorous tadpoles have a shorter digestive tract, resembling that of adults.

Another index of gut structure is the **coefficient of gut differentiation** (COD), the ratio of stomach and large intestine to small intestine (by weight, surface area, or volume). The COD is low, 0.1 to 0.4 (i.e., the small intestine is relatively large) for carnivorous and insectivorous mammals, and is high, 2 to 6 (i.e., the small intestine is relatively small) for herbivores and folivores (Chivers and Hladik 1980). Browsing birds have larger cecae than seed-eating birds. Herbivorous coleopteran beetles have larger midguts than do carnivorous beetles. The higher uptake capacity of the mammalian gut than the reptilian gut for sugars and amino acids is mainly due to an anatomical difference, the larger surface area of the mammalian gut (Karasov, Solberg, and Diamond 1985).

The **rate of feeding** is also responsive to dietary changes. In general, food is more thoroughly digested if it remains longer in the digestive tract, but the absolute amount of absorbed nutrients depends not only on digestive efficiency but also the rate of digestion. Thus, a greater rate of ingestion can compensate for the low quality and low digestibility of certain foods.

Specialized Digestive Systems

The digestive systems described so far are for omnivorous invertebrates and vertebrates, which are able to cope with most food materials. However, some types of dietary constituents are exceedingly difficult to digest and require special enzymes and digestive processes. For example, the common plant carbohydrates **cellulose** and **hemicellulose**, the polysaccharide **chitin**, the proteins **keratin** and **collagen**, and the lipid **wax** cannot be digested by many animals. Some animals are able to readily digest these materials, but they generally have a specialized gut morphology and/or unique enzymes.

Cellulose

Many plant polysaccharides and all animal polysaccharides are monosaccharides joined by α-bonds. These α-bonds can be hydrolyzed by common animal amylases (e.g., α-amylase), but these enzymes are ineffective in hydrolyzing β-bonds. Cellulose is difficult for most animals to digest because it is composed of monosaccharide units joined by β-bonds. A number of other plant materials are also difficult to digest. Hemicellulose is a complex polymer of xylose, arabinose, galactose, mannose, and other carbohydrates; its covalent bonding to lignin makes it less water soluble. Lignin is not a carbohydrate but is a phenyl-propane polymer with variable amounts of cutin (a polymeric waxy material), tannins, proteins, and silica. The association of lignin and tannins with cellulose and hemicellulose make them more resistant to fermentative breakdown by micro-organisms, and so a high lignin and tannin content is one defense of plants against herbivores. However, some herbivores have counterdefensive measures. The high proline content of saliva from some herbivores, such as deer, binds tannins and reduces their effects on inhibition of cell wall digestion (Robbins et al. 1987).

Cellulose is hydrolyzed by **cellulase** enzymes. Three different types of carbohydrase enzyme are needed to digest cellulose. Endo-β-gluconases split β-linkages within polysaccharides. Exo-β-gluconases liberate glucose or cellobiose from the end of polysaccharides. β-glucosidases hydrolyze cellobiose to glucose.

A discussion of digestion of plant material is complicated by the various terms used to describe different plant constituents reflecting, in part, their chemical extraction techniques. Soluble carbohydrate (or nitrogen-free extract, NFE) is the easily digestible carbohydrate fraction; "crude fiber" is the difficult-to-digest cell wall carbohydrate. However, "crude fiber" does not include all cellulose, hemicellulose, and lignin and the "NFE" includes some cellulose, hemicellulose, and lignin along with sugars, starch, pectin, gums, and organic acids. Neutral detergent residue (NDR) is the cell wall fraction (i.e., cellulose, hemicellulose, and lignin). Acid-detergent fiber (ADF) is the cellulose-lignin component of cell walls. Some values for these soluble cell and cell wall constituents of plant materials are given in Table 18–6.

Cellulose digestion, as indicated by the presence of the hydrolytic enzyme cellulase, occurs in many diverse animals (Yokoe and Yasumasu 1964; Elyakova 1972). In general, there are two means by which animals obtain cellulase. Some animals synthesize cellulase; examples are some crustaceans, silverfish, snails, and wood-boring beetles (Table 18–7). However, many cellulose-digesting animals do not themselves secrete the cellulase but rely on cellulase production by symbiotic micro-organisms (bacteria, protozoan ciliates, and flagellates). The micro-organisms may be ingested in the diet (as in the isopod *Philoscia*) but are usually permanent intestinal residents (as in termites, ruminant and pseudo-ruminant mammals). The symbiotic strategy for cellulose digestion has both advantages and disadvantages, as we shall see.

It is difficult when investigating the source of animal cellulase to exclude the possibility of micro-organismal secretion, and so some examples of "endogenous cellulase synthesis" may actually be due to symbiotic cellulase production. For example,

the ship worm *Teredo* was thought to secrete its own cellulase, but recent experiments have demonstrated the role of symbiotic bacteria in a structure not physically associated with the digestive tract, the gland of Deshayes (Waterbury, Calloway, and Turner 1983). Some animals, such as the silverfish *Ctenolepisma*, undoubtedly secrete their own cellulase; silverfish raised from sterilized eggs in sterile environments are able to digest cellulose (Lasker and Giese 1956).

A few omnivorous insects can digest cellulose. The cockroach *Periplaneta* can digest cellulose. A number of domestic lepismatids (silverfish) consume paper and other cellulosic materials. The fire-brat *Thermobia* can digest about 60% of ingested cellulose within 24 hr of ingestion (Zinkler and Gotze 1987). Most of the cellulose is digested in the relatively unspecialized crop (Figure 18–29A).

Many other insects are highly specialized for cellulose consumption, e.g., termites, woodroaches, wood-boring beetles, and siricid wasps (see Martin 1983). Many insect larvae consume wood. Some, such as the Lyctidae and Bostrychidae, digest the soluble cell contents but not the cellulose or hemicellulose. Larvae of the lepidopteran *Cossus* fail to grow if soluble carbohydrates have been removed from the wood; they die if the starches are also removed. Some insects, such as the Scolytidae, can digest hemicellulose. Some Anobiidae and Cerambycidae have a cellulase and therefore can utilize the cell walls as well as the cell contents. Other insects rely on either symbiotic bacteria or protozoans for their cellulase activity. The lamellicorn beetle larva *Oryctes* has a bacterial fermentation chamber in its hindgut (Figure 18–29B). The wood-feeding termites, which can normally live for long periods on pure cellulose, have a rectal pouch that contains flagellate protozoans (Figure 18–29C). The reliance of these termites on their symbiotic flagel-

TABLE 18–6

Difficult to digest contents of some common plant foods. Values are percent of dry matter. *(From Kronfeld and Van Soest 1976.)*

	"Crude Fiber"	Cellulose	Hemicellulose	Crude Lignin	NDR[1]	ADF[2]
Lettuce	13.5	14.0	1.3	2.0	17.3	16.0
Carrot	5.5	8.7	0.2	0.3	9.2	9.0
Potato	1.8	1.8	2.7	0.2	4.7	2.0
Apple	3.7	4.4	2.8	0.4	7.6	12.9
Wheat bran	12	9	34	4	47	13
Alfalfa	30	27	11	8	46	35

[1] Neutral-detergent residue = cellulose + hemicellulose + lignin.

[2] Acid-detergent fiber = cellulose + lignin.

TABLE 18-7

Examples of animals with an important role for cellulase activity in digestion and whether the cellulase activity is due to symbiotic bacteria or protozoans or is produced by the animal itself. Note that it is experimentally difficult to eliminate the possibility of symbiotic production of cellulase and some supposed examples of "endogenous" cellulase synthesis may be due to intestinal symbionts.

	Genus	Origin of Cellulase	Location of Cellulose Digestion
Protozoan			
Rhizopod	*Vampyrella*	Endogenous	Intracellular
Crustaceans			
Decapod	*Astacus*	Endogenous	Hepatopancreas
Isopods	*Limnoria*	Endogenous	Hepatopancreas
	Oniscus	Endogenous	Gut, cecae
	Philoscia	Bacteria (diet)	Gut
Mollusks			
Bivalve	*Scrobicularia*	Endogenous	Crystalline style, digestive diverticula, midgut
Gastropods	*Levantina*	Endogenous	Hepatopancreas
	Helix	Bacteria	Intestine
Pelecypod	*Teredo*	Bacteria	Gland of Deshayes
Insects			
Silverfish	*Ctenolepisma*	Endogenous	Midgut
Termites	*Termopsis, Zootermopsis*	Protozoa (flagellates)	Rectal pouch
Beetles			
Lamellicorn	*Potosia*	Bacteria (diet)	Gut
Anobiidae	*Xestobium*	Endogenous	Midgut
Cerambycid	*Cerambyx*	Endogenous	Midgut
Cerambycid	*Rhagium*	Bacteria	Gut
Cockroach	*Panesthia*	Bacteria	Crop
	Cryptocercus	Protozoa (flagellates)	Gut
Chordates			
Ruminant and ruminant-like mammals	*Bos, Ovis, Capra*	Bacteria and protozoa (ciliates)	Stomach
Cecant mammals	*Equus*	Bacteria	Cecum, colon
Coprophagous mammals	*Mus, Oryctolagus*	Bacteria	Cecum, stomach
Hoatzin bird	*Opisthocomus*	Bacteria	Crop, esophagus

lates for cellulose digestion is easily demonstrated by exposing the termites to 3.5 atm pure oxygen, or 36°C for 24 hrs; the hyperoxia or hyperthermia kills the flagellates but not the termites (Cleveland 1925). The termites, however, soon die of starvation even when provided with surplus cellulose. They can survive, however, on a diet of cellulose once they are reinfected with flagellates. The cockroach *Cryptocercus* has a similar flagellate symbiotic relationship. The advanced termites (Termitidae) are fungus farmers. Their plant diet is partially digested for them by fungi that are cultivated in underground chambers on collected plant material.

A number of vertebrates digest cellulose through the activities of symbiotic bacteria or protozoans; no vertebrate has the capacity to produce its own cellulase. For example, a variety of fish have gut cellulase activity, but this is often not correlated with their diet, digestive tract morphology, or phylogeny. No tissues have been shown to contain cellulase, and so the enzyme is probably derived from gut micro-organisms or in the diet. Two species of herbivorous fish (*Kyphosus*) have a pouch-shaped extension of the intestine near the rectum that contains large quantities of bacteria. A high level of **volatile fatty acids** (VFAs) in this pouch indicates bacterial fermentation of plant cellulose (Rimmer and Wiebe 1987). It is not known if there is any possible role of microbial fermentation for herbivorous amphibians. Many large reptiles (>300 g) are

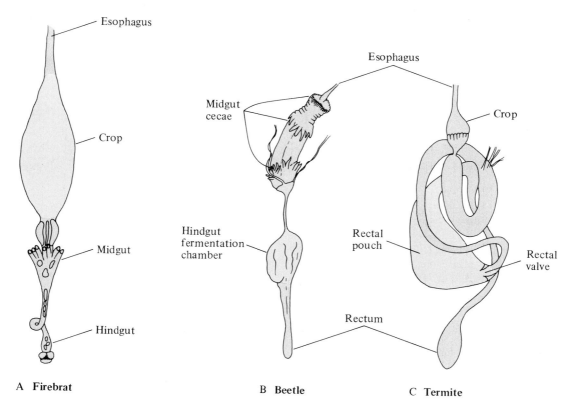

FIGURE 18–29 Specialized gut structures of invertebrates for fermentation by symbiotic bacteria or protozoans showing the main location of the symbionts (in color). **(A)** The relatively unspecialized gut of the firebrat *Thermobia* has cellulase activity in the crop. **(B)** The hindgut bacterial fermentation chamber of a lamellicorn beetle *Oryctes* is lined by cuticle with branched spines and pierced by fine canals. **(C)** The hindgut pouch of the wood-feeding termite *Eutermes* contains flagellate protozoans. *(From Zinkler and Gotze 1987; Wigglesworth 1935.)*

herbivorous, whereas the juveniles of these species and other smaller reptiles are carnivorous. VFAs are present in the cecum of these large herbivorous reptiles, indicating a potential role of microbial fermentation.

The most specialized forms of cellulose digestion occur in ruminant and ruminant-like mammals, e.g., artiodactyls of the suborder Ruminantia (bovids, cattle, sheep, goats; cervids, deer, antelope; giraffe; pronghorn; camelids). These ruminant mammals have a complex, four-chambered stomach (Figure 18–30A). The first two chambers, the rumen and the reticulum, are essentially fermentation vats. Ingested food is first masticated finely, then swallowed into the rumen where fermentation by bacteria and protozoans (ciliates and flagellates) occurs. The rumen contents can be regurgitated into the oral cavity for further mechanical processing, and then reswallowed ("chewing the cud," or rumination). The reswallowed material bypasses the rumen

and reticulum, by way of a deep fold in the anterior rumen wall, and enters the omasum where it is again physically triturated, from whence it enters the final chamber, the abomasum. The abomasum corresponds to the normal mammalian stomach, i.e., it secretes HCl and pepsin.

Not only ruminant mammals have been able to exploit the symbiotic digestion of cellulose (Table 18–7). A number of other placental mammals utilize a ruminant-like digestive process with a stomach modified for microbial fermentation, e.g., leaf-eating tree sloths, colobid and langur monkeys, and some rodents (Moir 1968, Hume 1989). The macropod marsupials (kangaroos and wallabies) also have a ruminant-like digestive system. The stomach of a typical macropod, the grey kangaroo *Macropus*, is highly specialized as a fermentation chamber (Figure 18–30B), but is a more tubular structure with a continuous flow of material through it, compared to the four-chambered ruminant stomach. Some

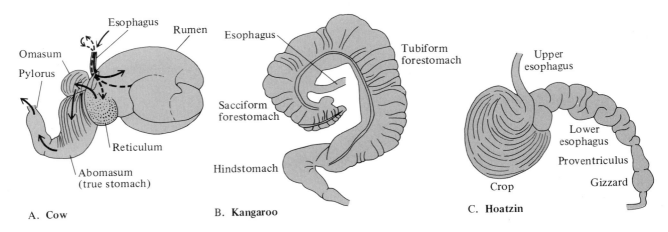

FIGURE 18–30   The specialized stomach of vertebrates is a fermentation chamber, containing symbiotic bacteria or protozoans. (**A**) The stomach of a ruminant bovid, *Bos,* has four chambers, the largest being the main fermentation chamber, the rumen; the "true stomach" is the abomasum. (**B**) The tubular stomach of the ruminant-like macropod marsupial, the grey kangaroo *Macropus,* is divided into a sacciform and tubiform forestomach, and a hindstomach. (**C**) The gut of the foregut fermenting Hoatzin bird *Opisthocomus* has an enlarged crop. *(From Romer and Parsons 1986; Hume 1982; Grajal et al. 1989.)*

material is regurgitated, rechewed, and reingested (merycism) in an manner analogous to rumination, but this is of lesser significance to macropods than to the true ruminants. The forestomach of macropods, like ruminants, contains high densities of bacteria and ciliate protozoans (and fungal sporangia, which are also found in ruminant stomachs). A neotropical leaf-eating bird, the hoatzin (*Opishocomus hoazin*) is probably the smallest endothermic "ruminant" (Grajal et al. 1989). It weighs 750 g and has a foregut fermentation system (Figure 18–30C).

The crop and esophagus of the hoatzin have bacteria in similar concentrations to the ruminant forestomach. Many other birds also have sufficient levels of micro-organisms in their gut that fermentation could be a significant energy source; for example, the willow ptarmigan might derive 4 to 11% of its energy requirements from microbial fermentation.

A comparison of the overall digestive efficiency of ruminants and macropods yields striking similarities (Table 18–8). Sheep and kangaroos (and horses

TABLE 18–8

Comparison of the digestive efficiencies of the sheep (*Ovis aries*), grey kangaroo (*Macropus giganteus*), and horse (*Equus caballus*) for various fractions of an alfalfa (lucerne) diet.[1]			
	Sheep	**Kangaroo**	**Horse**
Body weight (kg)	48.2	20.8	163
Dry matter intake (g kg^{-1} d^{-1})	23.0	26.4	21
Apparent digestibility:			
Dry matter	59%	55%	56%
Crude protein	74%	73%	71%
Acid-detergent fiber	49%	39%	41%
Digestible energy intake (kJ kg^{-1} d^{-1})	250	267	216

[1] Composition of alfalfa: 93.8% dry matter, 90.4% organic matter, 5.2% water-soluble carbohydrate, 4.9% pectin, 13.7% hemicellulose, 26.6% cellulose, 35.1% acid-detergent fiber, 8.4% lignin, 19.0% crude protein, 3.8% lipid, 0.3% silica; 18.4 kJ g dry weight^{-1}.

which are post-gastric fermenters; see below) are all able to digest 40 to 50% of the acid-detergent fiber (cellulose + lignin) content of alfalfa; crude protein digestibility is over 70%; the overall digestibility of dry matter intake is 55 to 60%.

The rumen microbial symbionts do not simply produce a cellulase that breaks down ingested cellulose to monosaccharides for absorption by the host mammal. The metabolic pathways of the bacteria reduce the monosaccharides and other substrates to a wide variety of fermentative waste products: methane; hydrogen; lactate; ethanol; succinate; and particularly the volatile fatty acids (VFAs) acetic, propionic, butyric, and valeric acids (Table 18–9). Some of the symbiotic ciliate protozoans also digest cellulose and produce VFAs. Methane (CH_4) is produced by the bacterium *Methanobacterium* from CO_2 and H_2 excreted by other micro-organisms; only a small fraction of the ingested energy is incorporated into methane. Macropods appear to produce less methane than do ruminants; the H_2 may be trapped as ethanol or formate, or eliminated as H_2.

VFAs are produced by symbiont micro-organisms as their waste products. The symbionts of ruminant and pseudoruminant mammals and the hoatzin liberate similar proportions of VFAs, primarily acetate and propionate with some butyrate. The VFAs are absorbed across the wall of the digestive tract and used in the metabolic pathways of the host as an energy source.

The energetic advantages of the ruminant digestive system are clear from the fate of ingested food (Figure 18–31). For example, cellulose is a considerable fraction (27%) of the energy ingested in alfalfa. About 50% of the cellulose is utilized by the symbiotic micro-organisms in the rumen/reticulum, and much of this energy is ultimately transferred to the host mammal. About 5% of the energy is converted to methane, which is lost to the atmosphere, and about 3% is converted to bacterial metabolic heat production (the heat of fermentation). About 17% of the ingested energy is converted to microbial production, i.e., an increase in biomass of microbes. About 33% of the ingested energy is incorporated into volatile fatty acids, waste products as far as the microorganisms are concerned, but valuable nutrients for the host mammal. The VFAs are absorbed across the rumen wall into the blood and the microbial production, as well as the so-far unused 41% of ingested energy, is passed to the rest of the digestive tract for further digestion and absorption. About 18% of this energy is absorbed mainly by the small intestine and 41% is lost in the feces. Of the total ingested energy, about 46% has been lost as waste (methane, feces), 21% has been utilized by the micro-organisms for their metabolism and population growth, and 51% has been made available to the host ruminant (VFAs, postgastric digestion); the heat of fermentation is potentially useful for temperature regulation by the ruminant if it is cold stressed.

TABLE 18–9

Substrates and the fermentation products of some rumen bacteria and protozoans and an overall (theoretical) stoichiometric equation for the fermentation of carbohydrate to the average ratios of acetic, propionic and butyric acids, and methane and carbon dioxide. *(Modified from Hungate 1968; Hungate 1966.)*

Micro-organism	Substrates	Products
Bacteria		
Ruminococcus	Cellulose, cellobiose, xylan, CO_2	Succinate, acetate, formate, lactate, ethanol, H_2
Bacteroides	Cellulose, cellobiose, glucose, CO_2	Succinate, acetate, formate
Eubacterium	Cellobiose, glucose, other C_{18}-C_6 sugars	Acetate, butyrate, formate, lactate, CO_2, H_2
Methanobacterium	Formate, CO_2, H_2	CH_4
Protozoans		
Dasytricha	Soluble sugars (sucrose, maltose, cellobiose)	CO_2, H_2, acetate, butyrate, lactate
Eudiplodinium	Starch, soluble sugars, cellulose	CO_2, H_2, acetate, butyrate, propionate, formate, lactate

Carbohydrate Fermentation Equation

58 hexose $\longrightarrow$ 62 Acetate + 22 Propionate + 16 Butyrate + 33.5 CH_4 + 60.5 CO_2
(16300 kJ) (54400 kJ) (33760 kJ) (34900 kJ) (29400 kJ)

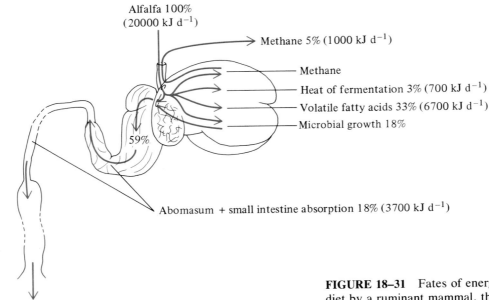

Alfalfa 100%
(20000 kJ d^{-1})

Methane 5% (1000 kJ d^{-1})

Methane

Heat of fermentation 3% (700 kJ d^{-1})

Volatile fatty acids 33% (6700 kJ d^{-1})

Microbial growth 18%

59%

Abomasum + small intestine absorption 18% (3700 kJ d^{-1})

Feces
41% (8400 kJ)

FIGURE 18–31   Fates of energy ingested as an alfalfa diet by a ruminant mammal, the sheep *Ovis*.

The major advantage of ruminant and ruminant-like fermentative digestion is the ability to assimilate at least some of the energy present in the diet as cellulose and hemicellulose. There are also other advantages. Ammonia and urea can be converted into protein by the metabolic pathways of symbiotic micro-organisms, pathways that most animals lack. Any ammonia or urea present in the food, or produced by the micro-organisms, can be converted to micro-organismic protein that can then be digested and absorbed by the mammalian host. Furthermore, any ammonia or urea synthesized by the host mammal as its nitrogenous waste product can be transported to the stomach (via saliva or the blood) and converted to microbial protein. Thus, ruminants and ruminant-like mammals can considerably reduce their urinary nitrogen waste losses when nitrogen deprived. Related to this capacity for protein synthesis from urea is the ability of the kidneys of ruminants and ruminant-like mammals to reabsorb a large fraction of the urea from the urine, rather than to excrete it as would other mammals.

A further advantage of ruminant and ruminant-like digestion is the synthesis of a number of vitamins, especially the B complex vitamins, by the symbiotic micro-organisms. The acquisition of vitamins from symbionts is important not only for the ruminant and ruminant-like mammals but for many other animals as well. Larvae of the beetle *Stegobium* are unaffected by lack of dietary vitamins, but larvae lacking their normal symbionts

have extensive vitamin and sterol requirements (Dadd 1970). Blood-sucking insects rely on their intestinal symbionts for the B complex vitamins, which normally are not present in their specialized flood. Blood-sucking insects with a free-living larval stage may rely on the larval stores of vitamins throughout their adult life and not possess symbionts for vitamin synthesis.

A disadvantage of a ruminant type of digestion arises from the sequelae of the micro-organismic breakdown of the cellulose; the micro-organisms use the cellulose and other carbohydrates as the metabolic substrates for anaerobic fermentation, the end products of which are volatile fatty acids (lipids), not carbohydrates. Ruminant/ruminant-like animals therefore do not have ready nutritional access to carbohydrates, despite the high level of polysaccharides in their diet. This is reflected by the lower levels of disaccharase enzymes in the small intestine of ruminants and macropods (Ballard, Hanson, and Kronfeld 1969) and a lower blood glucose level for ruminant and ruminant-like mammals (400 to 800 mg L^{-1}) compared to other mammals (800 to 1000 mg L^{-1}). The normally low blood glucose levels of ruminant/ruminant-like mammals are reflected by their considerable tolerance to low blood glucose levels induced by injection of insulin (hypoglycemia) and poor tolerance to hyperglycemia induced by alloxan administration. Certain tissues, such as the brain, generally require glucose as their metabolic substrate, and so it is essential that these ruminant/ruminant-like mammals have an alternate source of blood glucose. Most animals do not have a functional glyoxylate cycle, which can convert fatty acids to carbohydrates, although micro-organisms, plants, and some animals (a few nematodes, frogs, and mammals) do (Davis et al. 1990). The liver, which is a major site of glucose synthesis, is this alternate source of glucose (Figure 18–32). In a normal, omnivorous mammal such as the dog *Canis*, glucose is absorbed across the small intestine after a meal and is taken up by the liver for storage as glycogen, or conversion to lipids or amino acids; during fasting, the liver releases glucose into the blood. In ruminants (e.g., sheep) and macropods (e.g., quokka), there is a hepatic release of glucose into the blood regardless of whether the animal is feeding or fasting; glycogen stores cannot be the origin of this continual glucose release, and so metabolic synthesis of glucose (**gluconeogenesis**) must be occurring in the liver. The VFA propionate is a major gluconeogenic precursor in both ruminants and macropods.

[Glucose] in Hepatic Vein − [Glucose] in Portal Vein
(mg dL^{-1})

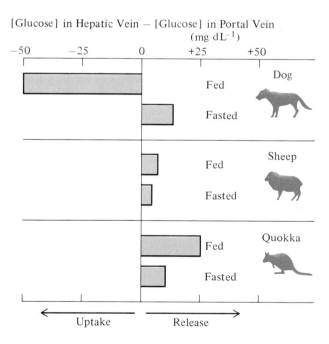

FIGURE 18–32 Hepatic absorption and release of glucose in a nonruminant mammal (the dog), a ruminant mammal (sheep), and a ruminant-like macropod (the quokka). The liver of the ruminant and ruminant-like mammals release glucose into the blood stream whether the animals are fed or fasting, whereas the dog's liver absorbs glucose from the blood after a meal and releases glucose when fasted. *(From Ballard, Hanson, and Kronfeld 1969; Hume 1982.)*

Postgastric fermentation is an alternate strategy to ruminant/ruminant-like digestion where cellulose digestion is pregastric (at least with respect to the "normal" mammalian stomach that secretes HCl and pepsin). The site for postgastric fermentation is typically the colon, or a greatly enlarged cecum or cecae, and so this is sometimes called cecant digestion. Cecant digestion occurs in many placental mammals (Perissodactyls, horses; elephant; hyrax; some rodents) and in marsupials (koala, wombat, many herbivorous or folivorous possums). For example, bacterial hydrolysis and fermentation of cellulose to VFAs occur in the cecum and large intestine of the horse, and these VFAs are absorbed across the wall of the large intestine; this is remarkably similar to ruminant digestion (Kern et al. 1974). However, the micro-organismic production of the cecant mammal is not digested and assimilated, as occurs in ruminants, because the microbial production occurs after the stomach and small intestine. Consequently, cellulose digestion is less effective in the horse compared to a ruminant, such as a cow, but horses compensate for this by having a higher rate of food intake. Postgastric fermentation also does not confer the additional advantage of ruminant digestion, nitrogen recycling, but there is some bacterial synthesis and uptake of essential vitamins.

Ruminants and horses compete in terms of dietary intake and digestive strategies. When food is scarce, cows will out-compete horses, but with abundant food the horse will compete favorably with the cow. The overall superiority of ruminant digestion over cecant digestion is reflected by the evolutionary outcome of the competition; perissodactyls adapted to a high cellulose diet in the late Paleocene before the artiodactyls, which evolved ruminant digestion in the mid-Oligocene, but most of the original perissodactyl radiation has been displaced by artiodactyls, except for the horses, which specialize on highly fibrous diets (Janis 1976).

Koalas are marsupials that have a highly specialized diet, eucalypt leaves. Koalas digest about 54% of their dry matter intake, but the apparent digestibility (about 69%) of the cell contents (soluble carbohydrates, lipid, protein, amino acids, phenolics) is considerably higher than the apparent digestibility (about 25%) of the cell walls (cellulose, hemicellulose, lignin). The lesser role of microorganismic fermentation is apparent from the low contribution of VFAs to the digestible energy intake (about 9% of 500 kJ kg$^{-0.75}$ day^{-1}). Phenolic compounds are a major component of the digestible energy intake, but there is also a significant urinary excretion of phenolics and so their contribution to metabolizable energy intake is substantially less than their contribution to digestible energy intake. Similarly, the proportion of metabolizable energy due to essential oils is less than the 16% of the digestible energy intake because of urinary excretion.

One means for avoiding the major disadvantage of postgastric fermentation is to return the microbial production to the stomach and small intestine by reingesting the feces; this is **coprophagy**. Lagomorphs and some rodents, for example, produce two types of fecal pellet; one is a soft, pale pellet of cecal material and the other is the normal, hard, dark fecal pellet. The soft, cecal pellets are reingested, thus making cecal microbial production available to the normal digestive processes. In the rabbit, the ingested fecal pellets, which are over 50% bacteria, are further fermented in the stomach to produce lactate, which is then readily absorbed into the blood (Griffiths and Davies 1963). Rabbits are also able to recycle urea into their protein pool because they can digest and absorb the bacterial protein production (Haupt 1963).

Micro-organisms are present in the digestive tracts of many animals other than cellulose-digesting invertebrates and ruminant, ruminant-like, and cecant vertebrates. Fruit-eating vertebrates often rely on intestinal bacteria for the digestion of pectins. Intestinal bacteria also have important nutritional roles, synthesizing biotin, folic acid, vitamin K, and other metabolic cofactors, many of which can be absorbed across the hindgut. The coleopteran *Stegobium* obtains B vitamins and sterols from yeasts; bacterial symbionts of the roach *Blatella* provide some amino acids as well as B vitamins; micro-organisms in homopteran and heteropteran insects are associated with nitrogen metabolism, either fixing N_2 or metabolizing the urea/uric acid wastes of the insect. Micro-organisms may even determine the sex of offspring; parthenogenetically developed eggs of the coccid *Stictococcus* become males if uninfected and females if infected!

Chitin

Chitin is the second most common biological polysaccharide, after cellulose, and is present in fungi, arthropod cuticle, mollusks, annelids, and coelenterates. Many animals are unable to digest chitin, but those whose diet typically contains considerable chitin generally have evolved an endogenous production of chitinase (e.g., some earthworms and crustaceans) or utilize bacterial chitinase (e.g., *Aplysia*, many other predaceous invertebrates). Insectivorous vertebrates (e.g., bats, some lizards, and turtles) produce chitinase in their gastric mucosa and pancreas, but closely related herbivorous mammals do not (Vonk 1937; Jeuniaux 1961). Many fish produce endogenous chitinase in their stomach and small intestine. The level of chitinase activity is correlated with the mechanical digestive capacity; those fish with a high capacity to fragment mechanically their prey have a lower level of chitinase. Some fish also have symbiotic bacteria in their digestive tract, which contribute to protein, amylose, and chitin digestion.

Wax

Larvae of the wax moth, or bee moth *Galleria mellonella*, are able to utilize about 50% of the ingested bees wax, mainly the fatty acids and hydrocarbons (Dickman 1933). Larvae without bacteria are unable to digest the myricyl-alcohol esters, a major fraction of the wax, and so symbiotic bacteria provide their normal wax digestion capacity. An African bird, the honey guide *Indicator indicator*, attracts the ratel (*Mellivora*), a large badger-like mustelid, to a bee hive; the ratel plunders the hive and the honey guide consumes the remaining honeycomb, larvae, and wax. The honey guide is able to digest wax because of its intestinal symbiotic bacteria. The wax-digesting ability can be transferred to domestic chickens by inoculation with a pure culture of the bacteria (Friedmann, Kern, and Hurst 1957). Some fish that consume copepods (which may be up to 70% wax!) have wax lipases (Sargeant and Gatten 1976).

Algal Farming

Many invertebrates grow and harvest ("farm") unicellular algae (golden-brown algae, Zooxanthellae; green algae, Zoochlorellae). Many sea anemones have zooxanthellae, zoochlorellae, or both, in the gastrodermal cells of their tentacles and oral disc; many corals have zooxanthellae within their gastrodermal cells. The turbellarian worm *Convoluta* has zooxanthellae, zoochlorellae, and diatoms within its parenchymal cells; the worm's digestive tract disappears in the adult as the symbionts multiply. Some mollusks, especially the tridacnid clams, have many zooxanthellae in their mantle tissues. The tridacnids are able to absorb nutrients directly from the seawater and have the typical bivalve filter-feeding system, but also "farm" their symbiotic zooxanthellae by blood amebocytic activity; both the zooxanthellae and their photosynthetic products are digested.

Symbiont Transfer

As we have seen, symbiotic micro-organisms are of considerable significance to digestion by many animals. Some animals lose their symbionts when they molt. For example, termites lose their symbiotic flagellates when they molt the cuticle because the lining of the hindgut is ectodermal and is lost along with the hindgut contents. However, the flagellates use the termite's molting hormone as a cue to encyst; the termite eats the molt exuvium and reinfects itself. The habit of proctodeal feeding also ensures reinfection from other termites. The cockroach *Cryptocercus* has a similar flagellate symbiotic relationship as termites, but it is not a social insect and communal reinfection after molting could be difficult. However, some of its flagellates migrate to the space between the hindgut cuticle and intima of the epithelium before molting and become the nucleus for the regenerated symbionts of the next instar.

It is important that symbiont micro-organisms are passed from one generation to the next. There are various ways for the symbionts to be passed from the parent to the offspring, including fecal contamination. The mycetocyte symbionts of many insects may be incorporated into the egg (e.g., cockroaches and the termite *Mastotermes*), or smeared on the outside of the egg when it is layed, or are present in the seminal fluid of the male.

Nutrition

Nutrients are chemicals that are important for the effective physiological functioning of an organism. Nutrients are sometimes defined to exclude chemicals that are simply an energy source (e.g., glucose), but these are the chemicals required in the greatest amounts by animals (daily food requirements for energy are often measured in grams per kg mass per day). Much of the previous discussion of digestion and animal energetics has dealt only with these "energy nutrients."

Other essential nutrients have a structural rather than energetic role, e.g., amino acids, some lipids such as fatty acids and cholesterol, some carbohydrates, and purine and pyrimidine bases. They are required in small amounts (e.g., mg kg^{-1} d^{-1}). A third category of essential nutrient are those chemicals required in only minute amounts (e.g., μg kg^{-1} d^{-1}) but which have such specific metabolic or structural factors that they are absolutely essential, e.g., vitamins, coenzymes. A final category of essential nutrients is the inorganic elements, e.g., Na$^+$, K$^+$, Cl$^+$, Fe^{2+}, Zn^{2+}, Mg^{2+}, I$^-$, etc. Water, although not usually classified as a nutrient, is also an essential part of the diet.

There is considerable diversity among animals with respect to their specific nutrient requirements for normal physiological functioning, and which of the nutrients can be synthesized by the animal's own metabolic pathways (i.e., are nonessential), which can be obtained from symbiotic micro-organisms (i.e., are essential but are not needed in the diet), and which must be obtained in the diet (i.e., are dietarily essential). The role and possible dietary requirements of amino acids, fatty acids and sterols, and vitamins will be discussed below.

Amino Acids

Animals do not have the ability to synthesize all of the 20 or so common amino acids from C, H, O, N, and S precursors. The essential amino acids must be in their diet or produced by symbiotic micro-organisms. Consequently, proteins deficient in these essential amino acids are nutritionally inadequate. The most common essential amino acids are valine, leucine, lysine, isoleucine, histidine, arginine, phenylalanine, tryptophan, threonine, and methionine. Sometimes glycine, tyrosine, proline, and cystine are also essential. Generally, alanine, serine, hydroxyproline, glutamate, aspartate, and cysteine are nonessential. Thus, for most animals, about 10 amino acids are essential and 10 are nonessential, although many nonessential amino acids are necessary for optimal growth or for specific stages of the life cycle, e.g., *Aedes* larvae require glycine to pupate. There is some interaction between the abundance of different amino acids, e.g., man requires tyrosine if the diet is low in phenylalanine, since tyrosine synthesis requires phenylalanine as a precursor; similarly, cystine and cystene are not required if there is sufficient methionine for their synthesis.

Fatty Acids and Sterols

Most lipids can be synthesized from carbohydrates and amino acids, hence the major dietary role of lipids is as an energy source. However, some fatty acids (linoleic, linolenic, and arachidonic acids) are required by a variety of animals. Sterols, the most common of which is cholesterol, can be synthesized by most animals, but many insects and a cephalopod *Sepia* are unable to synthesize it. Many insects thus have a dietary requirement for cholesterol, or similar sterols such as 7-dehydrocholesterol, 22-dehydrocholesterol, or ergosterol. The fruit fly *Drosophila pachea* feeds only on the cactus *Lophocereus schotti* and specifically requires the sterol schottenol from the cactus; cholesterol cannot replace schottenol in the diet. The sterols are required for synthesis of a wide range of steroid hormones and are incorporated into cell membranes.

Vitamins

Vitamins are required in only small amounts but are absolutely essential for normal physiological functioning. Vitamins were originally thought to be amines necessary for life, hence their name "vita" and "amine," but they are now known to include a number of other chemicals as well. The nomenclature for vitamins is based on letters of the alphabet (i.e., A, B, C, etc.); additional discovery of new vitamins sometimes required an additional subscript (i.e., B_1, B_2, B_6). The vitamins can be divided into the water-soluble vitamins (B complex and C) and the lipid-soluble vitamins (A, D, E, K). Vitamin deficiency (hypovitaminosis) or excess (hypervitaminosis) can have serious metabolic or other effects.

Water-Soluble Vitamins. The water-soluble vitamins are cofactors for specific metabolic reactions (Figure 18–33). They are generally not required for the biosynthesis of macromolecules, unlike amino acids, fatty acids, and sterols. The water-soluble vitamins, of the B complex and vitamin C, are not stored, unlike the fat-soluble vitamins, but are excreted in the urine. They must therefore be

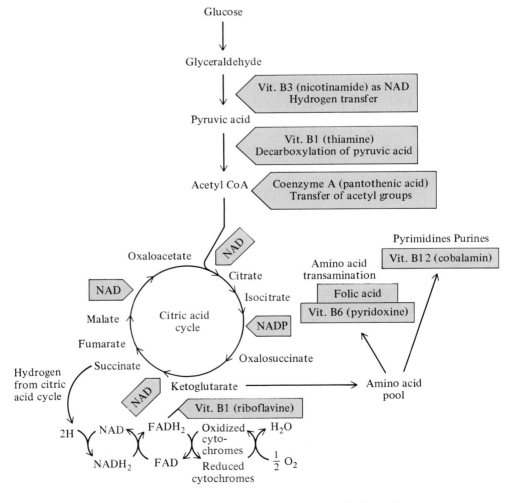

FIGURE 18–33 Roles of B-complex vitamins in cellular metabolic pathways. *(From Marshall and Hughes 1980.)*

ingested or synthesized fairly continuously, or else a vitamin deficiency will occur.

Vitamin B$_1$, **thiamine**, was the first B vitamin discovered. It is composed of a pyrimidine and a thiazole moiety. Many micro-organisms can synthesize thiamine or require either the pyrimidine or thiazole moiety. Many protozoans and all multicellular animals require thiamine, but those animals with a significant intestinal micro-organismic population (e.g., termites, ruminants) do not need to ingest thiamine but rely on its symbiotic production. Humans require about 1.5 mg per day. The thiamine is combined with ATP to form a cocarboxylase coenzyme that is required for the decarboxylation of pyruvate at the start of the citric acid cycle and also for other decarboxylation reactions. Thiamine deficiency can cause beriberi (a peripheral nervous disorder), heart failure, and edema.

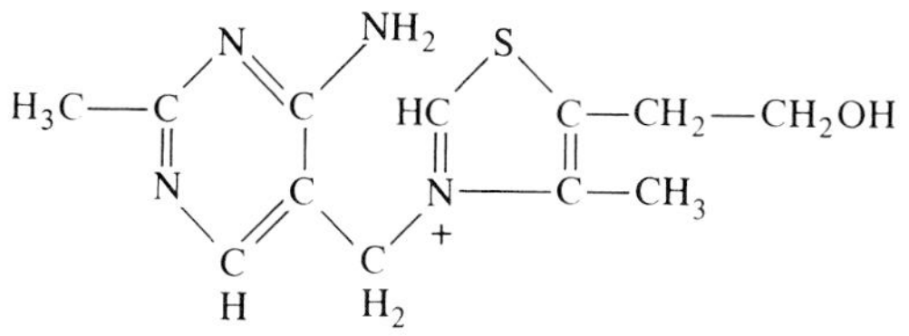

Thiamine

Vitamin B$_2$, **riboflavin**, is a derivative of alloxazine. It is synthesized by many micro-organisms and is required in the diet (or from intestinal symbionts). Humans require about 1.8 mg per day. Riboflavin is the prosthetic group in flavin mononucleotide (FMN) and flavin adenine dinucleotide (FAD), which are required as intermediates for electron transport. It is also a cofactor for xanthine oxidase (for adenine metabolism), cytochrome reductase (for electron transfer in the endoplasmic reticulum), and amino acid oxidase (for amino acid deamination). Deficiency causes lack of growth, reddened skin, dermatitis, myelin degeneration, and paralysis.

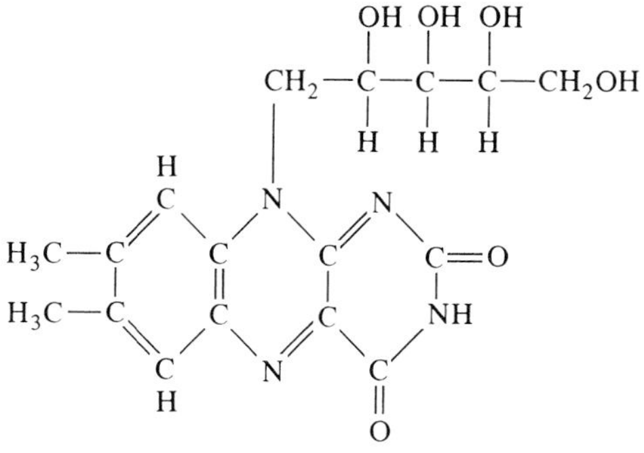

Riboflavin

Vitamin B$_3$, **niacin** or nicotinic acid, is synthesized from tryptophan by many micro-organisms and animals. Thus, niacin deficiency often occurs if the diet is deficient in tryptophan. Humans require about 20 mg per day. Nicotinic acid is combined with adenine and a pentose sugar to form the coenzyme nicotinamide adenine dinucleotide (NAD) and an additional phosphate to form NADP. These coenzymes act as hydrogen acceptors, for example, in glycolysis, the citric acid cycle, and the electron transport system. Deficiency causes pellagra ("black tongue"), loss of appetite, skin and gastrointestinal lesions, and diarrhea.

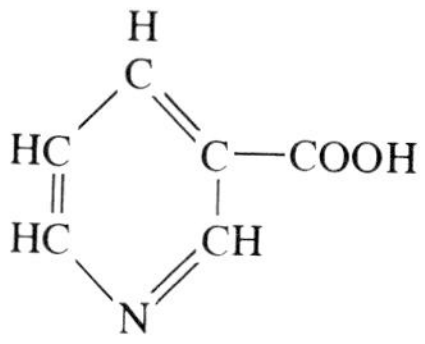

Niacin

Vitamin B$_6$, **pyridoxine**, is a pyrimidine derivative obtained in a number of forms. It is synthesized by many micro-organisms, including the gut symbionts of, for example, man. Humans require about 2 mg per day. It is an important coenzyme in amino acid transamination. Symptoms of deficiency include irritability, convulsions, muscle twitching, eye dermatitis, and kidney stones.

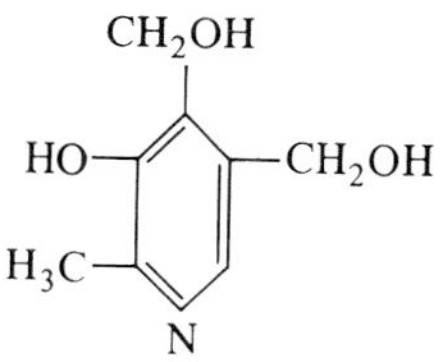

Pyridoxine

Vitamin B$_{12}$, **cyanocobalamin**, is a complex vitamin containing a porphyrin-like ring centered on cobalt, a cyanide group, and a nucleotide. It is synthesized by bacteria and is required by animals, either in the diet or from symbiotic bacteria. Humans require about 3 μg per day. Cyanocobalamin is an extremely active vitamin, required for many transamination and transmethylation reactions, for nucleoprotein and muscle metabolism, and for stimulating growth and for red blood corpuscle synthesis and maturation. It is a coenzyme for reduction of ribonucleotides to deoxyribonucleotides. Its uptake by the gastric mucosa requires intrinsic factor, which is normally secreted by the parietal cells of the gastric mucosa. Lack of intrinsic factor, and the concomitant inability to absorb cyanocobalamin, results in the failure of red blood cells to mature, causing an anemia called pernicous anemia, nervous disorders, and growth disorders.

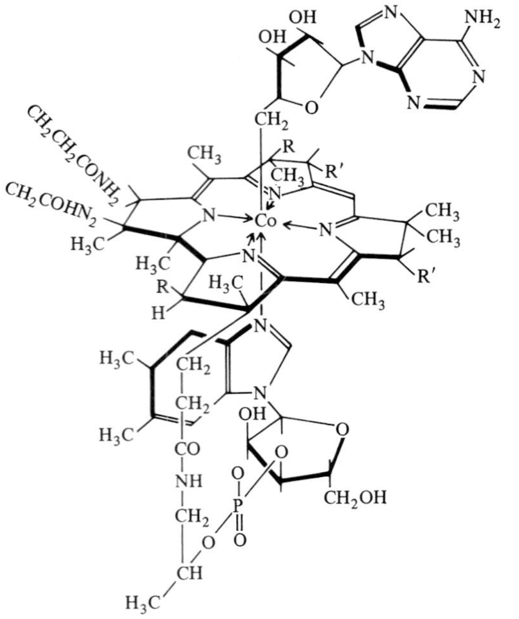

(COO—) group. Its deficiency is readily induced by raw egg white in the diet; symptoms include dermatitis, lack of appetite, and "blue slime" disease in fish.

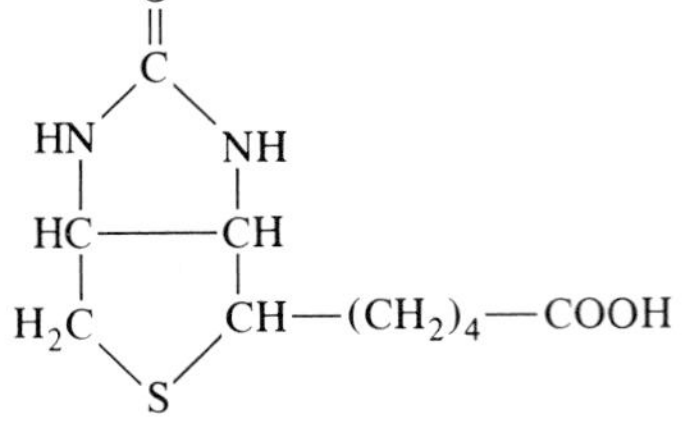

Cyanocobalamin

Biotin

Vitamin C, **ascorbic acid** ($C_6H_8O_6$), resembles glucose in structure. Humans require about 60 to 100 mg per day. It is required for the activation of folic acid and the synthesis of hydroxyproline, an important constituent of the connective tissue, collagen. It is required for normal formation of bone, cartilage, teeth, and subcutaneous connective tissue. Deficiency causes scurvy, characterized by degeneration of skin, teeth, blood vessels, and epithelial hemorrhage.

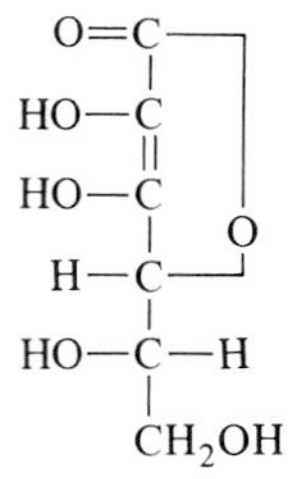

Pantothenic acid is required for synthesis of coenzyme A (CoA), in concert with adenylic acid, 2-mercaptoethylene, and 3 P_i. Humans require about 5 to 10 mg per day. CoA is required for carboxyl transfer from pyruvate into the citric acid cycle, and for the synthesis of the neurotransmitter acetylcholine from choline. Deficiency symptoms include fatigue, sleep disorders, nausea, and impaired coordination.

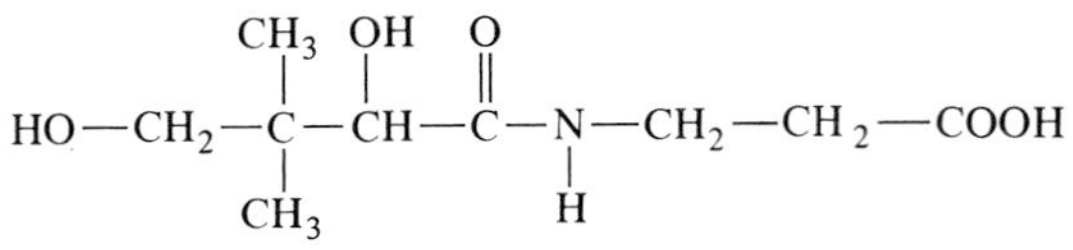

Pantothenic Acid

Ascorbic Acid

Vitamin H, **biotin**, is a sulfur-containing vitamin. Humans require about 100 to 200 µg per day. It is a prosthetic group required for numerous biosynthetic reactions involving the transfer of a carboxyl

Folic acid (pteroylglutamic acid) consists of a pterin group, glutamate, and p-amino benzoic acid. It is synthesized by micro-organisms and is required by animals. Humans require about 0.4 mg per day. Folic acid functions as a carrier of formyl and hydroxyl methyl groups (C_1 groups), which are required for the synthesis of purines and the pyrimidine thymine, hence gene replication and growth. Symptoms of deficiency include anemia, gastrointestinal disorders, and diarrhea.

Lipid-Soluble Vitamins. The lipid-soluble vitamins have roles in specific physiological functions, such as vision, calcium regulation, and blood clotting. These vitamins are stored in the body and are therefore less susceptible to depletion than the water-soluble vitamins. They have more specific roles than the general metabolic roles of the water-soluble vitamins. Vitamins A, D, and K appear to be required only by vertebrates, but vitamin E is more generally required by invertebrates also.

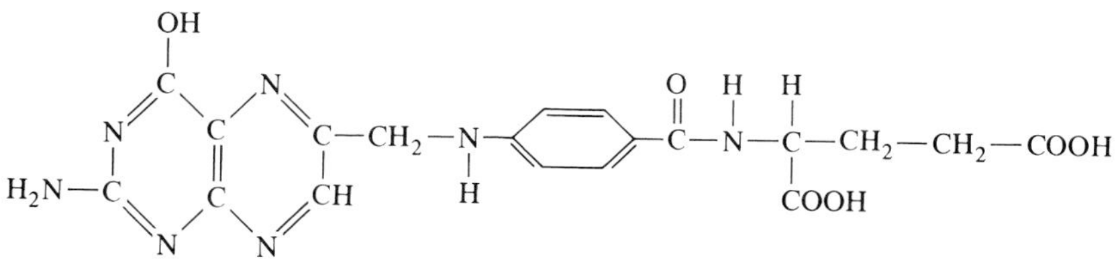

Folic Acid

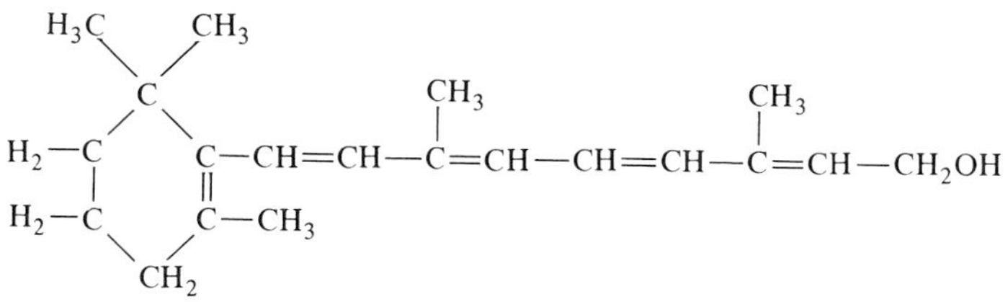

Vitamin A$_1$

Vitamin A, **retinol**, can be synthesized from dietary provitamins, the carotene pigments. Humans require about 50000 IU per day (International Units, measured by bioassay). The role of vitamin A in retinal (visual) pigments is well understood (see Chapter 7) and would appear also to have other roles, especially in epithelia. Deficiency causes keratinosis of the cornea, night blindness, permanent blindness, and scaliness of the skin.

Vitamin D is actually a class of substances, one of which is cholecalciferol (D$_3$); another is D$_2$, calciferol. In mammals, 7 dehydrocholesterol is activated to form D$_3$ by ultraviolet irradiation of the skin. D$_3$ is converted to 25 hydrocholecalciferol in the liver and then to 1,25 dihydrocholecalciferol in the kidney. The latter two metabolites are more active than D$_3$ itself. In birds, a vitamin D precursor is secreted by the preen gland and placed on the feathers where it is activated; the birds then either ingest or directly absorb the activated form. Humans require about 400 IU per day. Vitamin D is involved with Ca^{2+} absorption and metabolism, and bone formation. Deficiency causes bone deformation (e.g., rickets) and bone reabsorption.

Vitamin E, α-**tocopherol**, is one of three tocopherols that differ in number and position of methyl groups. Humans require about 15 IU per day. Vitamin E prevents oxidation of unsaturated fatty acids and is present at high levels in mitochondrial membranes. Deficiency causes effects in soft tissues, e.g., cardiac myopathy, anemia, epithelial degeneration.

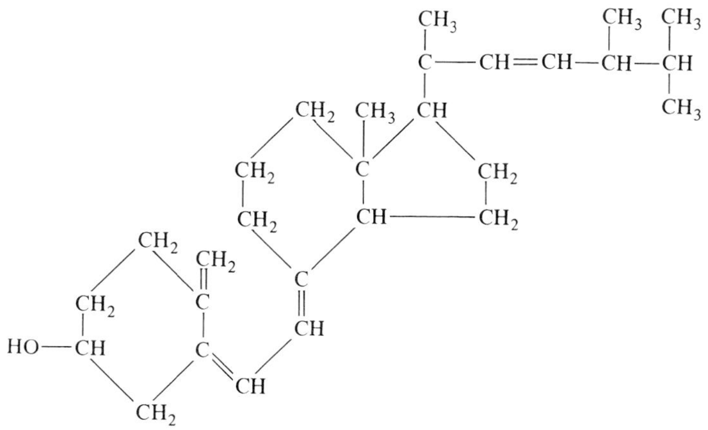

Vitamin D$_2$

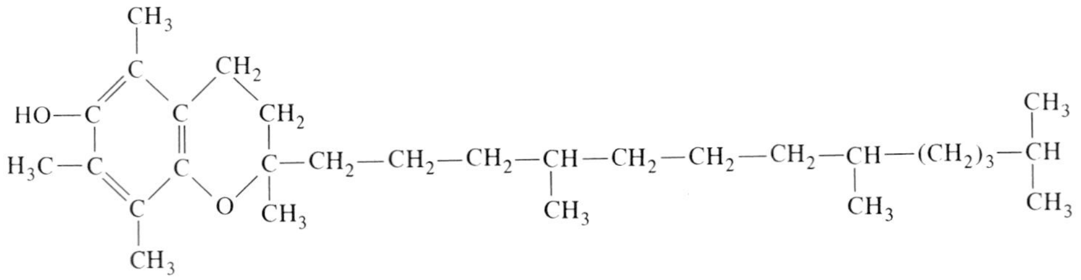

α-Tocopherol

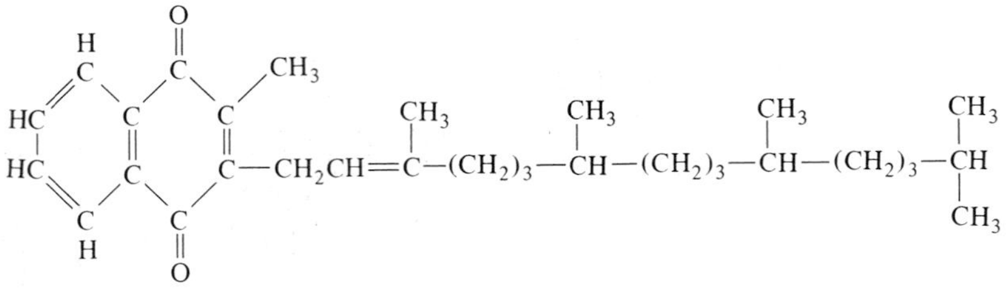

Vitamin K₁

The two forms of **vitamin K** (K$_1$, K$_2$) resemble vitamin A and vitamin E in general structure. They are widely distributed in micro-organisms and animals. It is similar to the pigment echinochrome of echinoderms. Requirements are difficult to determine because of the important role of intestinal bacteria and coprophagy in some species. Vitamin K is required for the formation in the liver of prothrombin and other blood clotting factors.

Other "Vitamins." Choline (hydroxyethyl-trimethyl-ammonium oxide) is an important source of methyl groups for the biosynthesis of various methylated compounds, e.g., acetylcholine (a neurotransmitter) and various phospholipids, lecithins, sphingomyelin, and betaine. Some animals require dietary choline.

$$(CH_3)_3N^+CH_2CH_2OH$$

Choline

Carnitine (vitamin B$_T$, or B$_7$) is a quaternary amine. It may be an essential growth factor for some animals (e.g., the mealworm *Tenebrio*, hence B$_T$) but not others (e.g., humans). It is required for fatty acid oxidation.

$$(CH_3)_3N - CH_2 - \overset{\text{OH}}{\underset{}{CH}} - CH_2 - COOH$$

Carnitine

There are a number of other nonessential, biologically active compounds. Inositol (C$_6$H$_{12}$O$_6$) is regarded as a B-complex vitamin; it is biologically active as myoinositol. Bioflavonids (vitamin P) are colored phenolic plant compounds, based on 1,4 benzopyrone. They appear to be required for maintenance of capillary integrity and permeability.

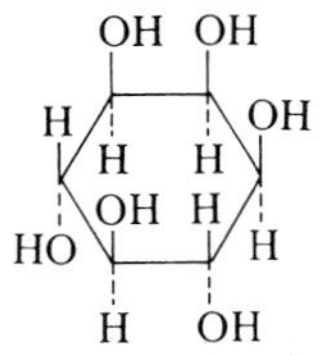

Myoinositol

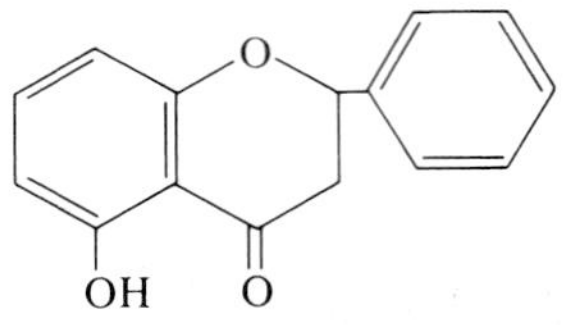

Bioflavonid

A number of other compounds have been isolated and called vitamins but they have been subsequently found to be related to other vitamins, or are not required by many animals, or are not essential for humans, e.g., "vitamins" M, B$_4$, B$_{10}$, B$_{11}$, B$_X$, T, and L.

Pseudovitamins, compounds claimed to be vitamins, often for commercial reasons, are not essential dietary nutrients, e.g., laetrile (B$_{17}$), pangamic acid (B$_{15}$), gerovital (H$_3$), and methylsulfonium salts (U).

Regulation of Nutritional Intake

The intake of energy is controlled by a physiological drive, called hunger, that initiates feeding activity. There is a complex interaction via the nervous system of sensory information from both the internal environment (e.g., levels of body energy stores, circulating concentrations of energy substrates) and the external environment (e.g., types and amounts of food available). The result of these sensory inputs is the **hunger drive**.

For example, a blowfly *Phormia* that has not recently fed will extend its proboscis to feed if its tarsal receptors contact a sugar solution, and stimulation of oral receptors then initiates sucking movements. The food passes primarily to the crop where it is stored and passes aperiodically in small slugs to the midgut for digestion and absorption. Feeding terminates when internal sensory inputs to the nervous system inhibit the external feeding stimulus, the presence of sugar solution. The internal sensory inputs are of two types: a "central" sensory monitoring of the osmotic pressure of the body fluids (it might be expected that circulating sugar level would be a more appropriate signal of satiation, but injection of sugar into the body fluids does not prevent hunger) and a sensing by stretch receptors of food traveling from the crop into the midgut. The insects overfeed (hyperphagia) to the point that they burst if the recurrent nerve that carries the latter stretch receptor information to the nervous system is cut. Mosquitos also become hyperphagic and burst if their ventral nerve cord is sectioned.

There is a similar integration of exogenous and endogenous information in vertebrates. In mammals, activity in the lateral portion of the hypothalamus causes pronounced hunger and voracious feeding, whereas activity of the ventral medial hypothalamus causes satiety ("nonhunger") and can totally inhibit feeding (aphagia). There are thus a hunger center and a satiety center in the brain. These brain centers are not involved with mechanical aspects of feeding (e.g., chewing, salivation, swallowing, etc.) that are controlled by other areas in the lower brain stem. The nutritional information relaying to the hypothalamic centers includes blood glucose level. The **glucostatic** theory of hunger suggests that an increase in blood glucose level increases the activity of the satiety center and decreases the activity of the hunger center; low blood glucose has the opposite effects. Amino acid levels and lipid levels have also been suggested to influence hunger and satiety (**aminostatic** and **lipostatic** theories). The **hepatostatic** theory of hunger control argues for a role of the liver as an important sensor and modulator of the body energy stores. Mammals also have an important gastrointestinal sensory component to determining hunger. Distension of the stomach and small intestine,

stretching of the abdominal wall, the hormone cholecystokinin, and the mechanical activity of chewing and swallowing all inhibit the hunger center.

It would not be practical for animals to experience hunger drives for each of the numerous and necessary dietary nutrients, e.g., carbohydrates, lipids, amino acids, vitamins, minerals. Especially the "micronutrients," such as vitamins, will be present in adequate amounts in the normal intake of a typical diet. For example, "cafeteria-fed" rats with a variety of foods to select from (e.g., carbohydrate, protein, fat, yeast, specific minerals) select some of each and maintain a balanced nutritional intake and grow normally (Richter and Rice 1945). Nevertheless, some animals do have hunger drives for specific nutritional requirements and feeding activities are concentrated towards those foods containing the needed nutrients. For example, the white rat will preferentially select a food containing B vitamins if its diet has been deficient in B vitamins (and if it has been prevented from coprophagy since its intestinal micro-organisms synthesize B vitamins). This example does not prove that rats can monitor the level of B vitamins in different food, however. In fact, rats learn which diets result in illness and avoid these, even if B vitamins are subsequently not limited (Rozin 1965). Sodium is another example of a specific nutrient drive. Many animals, including man, develop a salt craving if their diet is deficient in sodium.

Nutrient intake sometimes requires optimization for different nutrients. For example, energy and sodium are important nutrients for the moose *Alces*, which have a dietary choice of terrestrial or aquatic plants (Belovsky 1978). There are only limited intakes of terrestrial and aquatic plants that satisfy the constraints of required Na^+, the physical volume of intake (due to the rumen size), and energy; these closely correspond with the actual diet selected by moose.

Summary

There is a considerable variety in the nature of the food that animals consume, in the mechanisms used to capture food, and in the structure and function of the digestive tract.

Some animals feed on suspended material, using webs or pores that sieve or intercept food particles. Many protozoans, invertebrate animals, and vertebrate animals suspension feed. Some animals use sticky pseudopodia, or tentacles, or filters to suspension feed. Many animals feed on large food particles that may be sessile or active motile prey. Predatory animals may ingest the prey whole, masticate it to reduce the food particle size before swallowing, or externally digest the prey and suck in the liquid digesta. A variety of animals feed on liquid foods, e.g., sap, nectar, blood. They often have hypodermic-like mouthparts adapted for piercing and sucking. A few, generally marine or endoparasitic animals, lack a specialized digestive system and absorb nutrients across their body surface.

A digestive tract allows an animal to digest food extracellularly, rather than intracellularly. This enables them to digest food particles that are considerably larger than their cells; allows specific cells and parts of the digestive tract to become specialized for particular digestive functions, e.g., secretion, digestion, absorption; and allows a spatial specialization of the digestive for particular roles, e.g., for acidic digestion in the foregut and alkaline digestion in the midgut.

The generalized gut tube is divided into an anterior region for reception, storage, and movement of food; a section for mechanical and chemical digestion; a region for absorption of the products of digestion; and a posterior section for water and ion reabsorption and storage of feces. Food is passed along the digestive tract by peristaltic contractions of short segments of the gut. The peristaltic contractions are initiated by the presence of food in the gut but are modified by the nervous and endocrine systems.

Important roles of the midgut are mechanical digestion to reduce the food to small particles and chemical digestion to reduce the small food particles into their biochemical constituents. Some animals, particularly vertebrates, use highly acidic secretions to digest food chemically. A variety of animals have a gizzard for the physical crushing and grinding of food. Chemical digestion is accomplished by the action of specific digestive enzymes. These digestive enzymes have group specificity, i.e., are specialized for a particular substrate (e.g., carbohydrate, protein, or lipid) and have an optimal temperature and pH requirement (e.g., acidic, neutral, or alkaline pH optima; temperature optima matched to the preferred body temperature).

Proteases hydrolyze proteins. Endopeptidases internally hydrolyze proteins into smaller peptides, and exopeptidases terminally hydrolyze peptides to release an amino acid. Protease enzymes are generally quite specific with respect to the nature of the peptide bonds that they hydrolyze. They are often secreted as an inactive zymogen, to prevent digestion of the secretory cell.

Carbohydrases hydrolyze carbohydrates into smaller units and eventually into monosaccharides. Polysaccharases hydrolyze plant starches and ani-

mal glycogen into intermediate-sized dextrins, maltotriose, and disaccharides. Oligosaccharidases hydrolyze trisaccharides and disaccharases hydrolyze disaccharides into monosaccharides.

Lipases hydrolyze triglycerides into fatty acids, di- and monoglycerides, and glycerol. Esterases hydrolyze simple esters and complex lipids (e.g., phospholipids, cholesterol esters, and waxes).

The epithelial lining of the gut absorbs the products of digestion: monosaccharides, amino acids, fatty acids, as well as vitamins, water, and ions. The peristaltic mixing movements of the gut move the nutrients to the epithelium for absorption. The surface area of the gut epithelium is maximized to facilitate absorption by a variety of anatomical modifications, including lengthening of the gut tube, folding, and a brush border of the epithelial cells. Amino acids are absorbed by an active transport, colinked with Na^+ uptake. Most monosaccharides are absorbed by a similar Na^+-colinked active carrier system, although fructose is absorbed by facilitated diffusion. Lipids are freely soluble in the epithelial cell membrane and do not require active carrier uptake. However, the low aqueous solubility of lipids requires emulsification of lipid droplets into micelles for digestion and transport of digestive products to the epithelial cells and formation of stabilized intracellular lipid droplets (chylomicrons) for transport via the lymphatic and circulatory system to the body. Some ions are absorbed passively from the gut lumen if the electrochemical gradient is favorable for diffusive uptake (e.g., Na^+ and Cl^-). Many other ions are absorbed by active transport. Water absorption is passive, by osmosis linked to solute transport.

Digestion by protozoans illustrates intracellular specializations for digestion. Food is directed by ciliary action to the "mouth," which opens into the cytopharynx. Vacuoles containing food particles pass in a directed manner through the cytoplasm, being acidified and then made alkaline for digestion. Nutrients are absorbed into the cytoplasm and wastes eliminated via the cytopyge. Many protozoans have symbiotic bacteria and algae.

Bivalve mollusks utilize a complex pattern of intracellular and extracellular digestion. Food is collected by filter feeding and passed in mucous strands to the stomach to be reduced by a mortar-and-pestle action of the crystalline style and gastric shield. Wear of the crystalline style releases digestive enzymes for extracellular digestion. Small food particles are absorbed by digestive cecae for intracellular digestion.

The digestive system of the cockroach is representative of digestion in many invertebrates. Food is initially masticated by the mandibles and maxillae and then passed to the buccal cavity and then the crop. Saliva adds mucus and a carbohydrase enzyme and other enzymes are regurgitated from the midgut into the crop. Food passes from the crop into the gizzard, where it is mechanically reduced to small particles. Size sorting directs small particles to the hepatic cecae of the midgut, for completion of extracellular digestion.

The vertebrate digestive system generally consists of the buccal cavity, pharynx, esophagus, stomach, small intestine, large intestine, and rectum. Additional important digestive organs are the salivary glands, liver, and pancreas. Saliva adds mucus and a carbohydrase to the food. The stomach provides storage and acidic (HCl) and protease (pepsin) digestion. The liver secretes bile, which contains bile salts to emulsify lipids for digestion. The pancreatic secretions add many digestive enzymes (proteases, carbohydrases, lipases, DNAase, and RNAase), and HCO_3^- to neutralize the acidic stomach chyme. The small intestine is the site for final digestion of food and absorption of nutrients, ions, and vitamins. The large intestine is responsible for water and further ion absorption and the formation and storage of feces.

Regulation of digestion in vertebrates is a combination of immediate local reflex control, rapid nervous control, and relatively slow endocrine control. Control of salivation is primarily neural in response to visual and olfactory stimulation and chewing. Stomach HCl and pepsin secretion is promoted by parasympathetic stimulation, but especially by the hormone gastrin, released from the stomach mucosa by the presence of food. Release of bile into the small intestine is promoted by parasympathetic stimulation and especially by the hormone cholecystokinin, released from the small intestine. Pancreatic secretion is promoted by neural stimulation and by hormones (secretin, cholecystokinin) from the small intestine. The copious secretions of the small intestine, including fluid, enzymes, and hormones, are promoted by local neural reflexes due to the presence of food.

The structure of the gut is highly adaptive, both in different species with varying diet and in individuals in response to changes in diet. Gut length is one parameter that is very different for carnivores and herbivores and can alter with changes in diet; the coefficient of gut differentiation (stomach + large intestine/small intestine) also varies markedly in carnivores and herbivores.

Some food materials, such as the carbohydrates cellulose and chitin; some proteins, such as keratin and collagen; and the lipid waxes cannot be digested by many animals. Specific enzymatic, structural, and even symbiotic adaptations are required to digest these materials. The enzyme cellulase is required to digest cellulose. Some animals, such as silverfish, are able to synthesize cellulase. Most animals are not able to synthesize cellulase, but many nevertheless can digest cellulose because of the cellulase activity of symbiotic micro-organisms, particularly bacteria and ciliate and flagellate protozoans. Many beetles, termites, lepidopteran larvae, wood-boring mollusks, ruminant mammals, pseudoruminant mammals, and birds have symbiotic micro-organisms. In ruminant mammals, the symbiotic rumen micro-organisms ferment cellulose and release volatile fatty acids as their waste product. These, and the microbial production, are available to the host ruminant for digestion and absorption in addition to the remaining ingested nutrients. In addition, the symbionts can also recycle the nitrogenous waste (urea) of the host mammal into protein and synthesize many vitamins; these are available to the ruminant by their digestion of the microbial production. Many other mammals and a few birds have evolved a similar, pregastric fermentation system; these pseudoruminants gain similar energetic, N-recycling, and vitamin synthesis advantages from their symbionts. Some mammals have evolved a postgastric fermentation system, with symbionts generally in their cecum or colon. This is potentially less effective for the host because of the limited ability for digestion of the microbial production and absorption of volatile fatty acids. However, this disadvantage of postgastric fermentation can be obviated by ingestion of the feces (coprophagy). For example, rodents and lagomorphs form special fecal pellets containing their microbial symbiont products, for ingestion.

Digestion of chitin requires chitinase. A number of insectivorous animals have a chitinase, whereas many noninsectivorous species do not.

Probably many animals are able to digest some waxes, but a few specialize on wax digestion. For example, the wax moth can digest about 50% of ingested bees wax; symbiotic bacteria are required for the wax digestion. A bird, the honey guide, is able to digest wax because it has the appropriate micro-organismal intestinal symbionts.

Energy sources are the major nutrient required by animals, but a variety of other nutrients are essential. Many of these other nutrients are required in lesser, or extremely small, levels for structural synthesis or as cofactors, coenzymes, etc. There is considerable variation among animals in their nutrient requirements, but most animals require a number of amino acids, some lipids and carbohydrates, as well as various minerals and vitamins. Many of the water-soluble vitamins are required as metabolic cofactors of coenzymes, or for growth and development. These are not stored in the body in large amounts, and symptoms of their deficiency can readily develop. The lipid-soluble vitamins tend to have more specific structural roles and are stored in greater amounts, hence signs of their deficiency do not develop so readily.

Nutrient intake is controlled by physiological drives, e.g., hunger. Cessation of feeding is sometimes controlled by physical distension of the crop or stomach. In vertebrates, a hunger center in the central nervous system regulates the hunger drive and a center in the medial hypothalamus inhibits the hunger drive. Many nutritional factors influence the hunger drive, e.g., blood glucose levels, body lipid levels, and perhaps the liver metabolite levels. Many factors, including distension of the stomach and abdominal wall, hormones (cholecystokinin), and the mechanical activity of chewing and swallowing, can inhibit hunger. There are not specific hunger drives for every dietary requirement; these are generally satisfied by the selection of a varied diet.

Supplement 18–1

Adaptations of the Gut to Metabolic Rate and Diet

• •

The digestive and absorptive capacity of the gut must obviously match, or exceed, the metabolic demands of the animal. Otherwise, the supply of nutrients would be inadequate and the animal would die. The metabolic requirements of endothermic mammals and birds are profoundly higher than those of ectothermic lower vertebrates and invertebrates. For example, the basal mass-specific metabolic rate of a small endotherm, such as a 10 g mammal (at a body temperature at 37° C), is about 29 times greater than the standard metabolic rate of a 10 g fish (at a body temperature of 20° C). Such a disparity in metabolic rate must be matched by a corresponding disparity for digestion and absorption. Thus, there is a clear relationship for fish, reptiles, birds, and mammals between their metabolic rate (basal or standard) and nutrient absorptive capacity.

Animals diets vary markedly. The diet of carnivores and insectivores contains little carbohydrate (1 to 5%) but considerable fat and protein, whereas that of herbivores consists of more carbohydrate. The diet of nectar-feeding animals consists of mainly carbohydrate (glucose, fructose, or sucrose) and has very low levels of protein and ions. We would expect that the digestive and absorptive capacities of animals with such different diets would be adapted for the digestion and absorption of their primary dietary constituents.

What adaptations might match digestive and absorptive capacities of the gut with the metabolic demands and the nature of the ingested nutrients? Guts that require a high digestive and absorptive capacity for particular common nutrients might be expected to have (1) a high intestinal surface area, (2) more carriers for nutrient absorption, (3) higher levels of carriers for those nutrients that are more common in the diet, and (4) a high affinity of the carriers for nutrient uptake. Let us here briefly examine the roles of surface area and nature of the active transport carrier as adaptations of the gut. There are also other possible mechanisms for increasing nutrient transport, such as a high passive solute permeability, modification of cell membrane fluidity to modulate the carrier protein activity, increasing the Na^+ electrochemical gradient to regulate cotransport of nutrients, and control of retention times.

Relationship for fish, reptiles, birds, and mammals between the overall absorptive capacity of the gut and their metabolic rate. The absorptive capacity is measured as the sum of glucose and proline uptake, and the metabolic rate is that calculated from body mass at the appropriate body temperature (37° C for mammals, 41° C for birds, an average of 37° and 20° C for reptiles, and 20° C for fish; see also Table 4–5, page 95). *(For body mass and absorptive capacities, see Karasov 1987.)*

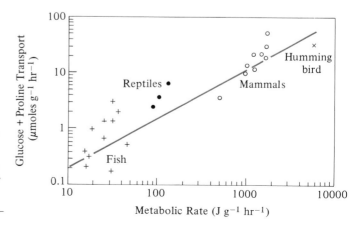

Temperature also has a profound effect on the rate of nutrient uptake, with Q_{10}'s of 1.8 to 3.4 for active uptake; the Q_{10} is about 1.2 for passive uptake.

Surface Area

The small intestine of vertebrates is essentially a tube. Its nominal surface area depends on both on the overall diameter (d) and length (l) of the tube; the nominal surface area is calculated as πdl. The surface area can be increased by cecae, blind-ended diverticulae of the small intestine. For example, many fish have pyloric cecae, which can dramatically increase the nominal surface area and absorptive capacity (16 to 70% of the total surface area and nutrient uptake). The nominal surface area of the small intestine is higher in larger animals and is about four times higher for mammals than reptiles and fish of comparable size.

The actual surface area of the small intestine also depends on the extent of internal folding of the lumen epithelium. Macroscopic folding (e.g., valves of Kerckring and villi) and microscopic folding (e.g., microvilli) dramatically increase the overall internal surface area of the small intestine. For example, in mammals and reptiles the actual surface area is increased by 3 to 20 times by villi and by 20 to 80 times by microvilli. Thus, the actual microscopical and functional surface area can considerably exceed the nominal surface area.

The anatomical gut surface area is very responsive in some animals to the nature of the diet. For example, in herbivorous carp, there is an increase in the gut length, the gut nominal surface area, and the rate of glucose absorption for fish consuming a high-glucose diet. Similar changes are not seen in trout, which are carnivorous and do not normally experience high carbohydrate levels in their diet.

Carrier Type

The nature of the diet can affect the activity of digestive enzymes and absorptive capacities for nutrients. For example, experiments by Pavlov demonstrated that the nature of the diet affects the induction of various digestive enzymes in mammals. Carnivorous fish have lower levels of amylase enzyme than do herbivorous fish. Similarly, it is well documented that the nature of the diet affects the rates of nutrient absorption. For example, in fish and reptiles, carnivorous species have lower rates of glucose uptake than herbivorous or omnivorous species.

All animals require dietary protein for body maintenance and growth, and so it is not surprising that most vertebrates have very similar levels of amino acid carriers in their intestines. Rates of proline uptake, for example, are remarkably consistent for various vertebrates, at about 200 to 400 nmoles min^{-1} cm^{-2} (nominal surface area). In contrast, there is much greater variability in the rate of glucose uptake. For carnivores, with low-carbohydrate diets, glucose uptake rates are generally less than 200 nmoles min^{-1} cm^{-2}, whereas for omnivores and herbivores the rates are generally greater than 400 nmoles min^{-1} cm^{-2}. The hummingbird has the highest rate of glucose uptake, about 1500 nmoles min^{-1} cm^{-2}. Consequently, there is a consistently higher ratio of glucose/proline uptake rates for omnivores and herbivores than carnivores. In general, carnivores have a ratio of about 0.25, whereas omnivores have values about 1.2.

It is clear that the gut is responsive to the nature of the diet with respect to nutrient uptake. However, animals are also selective with respect to what they eat, based on the nature of their digestive capabilities. For example, many birds (and also other vertebrates) feed on plant nectar, which is high in sugars. In general, hummingbird-pollinated plants produce a nectar high in sucrose, whereas passerine-pollinated plants produce a nectar that is high in glucose, fructose, or other hexoses. Sucrose

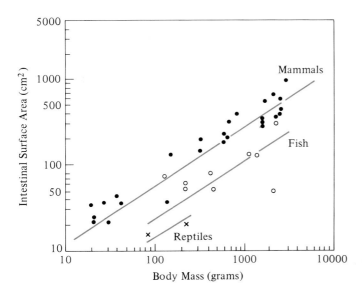

Relationship for fish, reptiles, and mammals between body mass and the nominal surface area of their small intestine (including pyloric cecae, if present). The common slope for the relationship is 0.63. The proportionality coefficients for the different taxa are fish, 1.68; reptiles, 1.06; and mammals, 3.94. *(Modified from Karasov 1987.)*

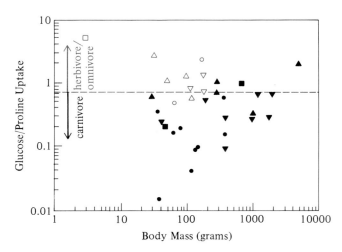

Ratio of rates of sugar (glucose) and amino acid (proline) uptake rates for herbivorous and omnivorous (open symbols) and carnivorous (solid symbols) vertebrates; ▽, fish; ○, frog/reptile; □, bird; △, mammal. *(Modified from Diamond and Buddington 1987.)*

digestion requires sucrase for its hydrolysis to monosaccharides, and so birds feeding on sucrose solutions both require this enzyme in the intestine and take longer to absorb the carbohydrate from their diet. Starlings, which lack sucrase, have a definite preference for glucose/fructose solutions and an aversion to sucrose solutions. They are unable to digest the sucrose and develop an osmotic diuresis in response to sucrose ingestion. Hummingbirds, which have sucrase, have no aversion to sucrose solutions and can digest 99% of the ingested sucrose.

Carrier Transport Rate, Affinity, and Density

It is relatively easy, at least in theory, to characterize the transport properties of intestinal transport carriers, i.e., apparent V_{max}^* and apparent K_t^*, by measuring transport rates at a variety of substrate concentrations (and by correcting for passive transport).

In general, the K_t^* varies from about 0.5 to 6 mM for the glucose carrier, and there is not any general relationship between K_t^* and the nature of the diet. For example, the K_t^* for the intestine of a hummingbird is not, as might be expected, very low (to maximize uptake of ingested nectar sugars), but is about 4.3 mM. However, it is difficult in practice to determine the actual K_t of the carrier because transport creates a boundary layer of depleted substrate around the carrier and the thickness and effect of the boundary layer depends on the rates of uptake and mixing of the fluid bathing the epithelium. Thus, the observed K_t^* value is an apparent K_t rather than the actual K_t of the carrier; $K_t^* = K_t + 0.5RV_{max}$, where R is the boundary layer resistance. The small variation in K_t^* for different vertebrates can be readily ascribed to unstirred boundary layer effects, rather than to species-specific adaptation, and the actual carrier K_t probably varies little for different vertebrates.

There is considerable variation in the absolute rate of nutrient transport by the intestines of different vertebrates, but there is less variation when the rate of nutrient transport is expressed per nominal area of the intestine, particularly when temperature is taken into account. For example, the rate of proline uptake is about 86 nmole min^{-1} cm^{-2} for fish, 271 for reptiles, and 276 for mammals (at their normal body temperatures of 20°, 37°, and 37° C, respectively). Their rates of glucose uptake are 62, 357, and 300, respectively. This variation is essentially eliminated when the rates are measured at equivalent temperatures (20° to 25°C); 62, 100, and 80, respectively. Nevertheless, there is species variation in nutrient uptake rates and even within-species variation depending on the location in the small intestine and the nature of the diet. For example, the V_{max} for the hummingbird is very high, at about 1500 nmole min^{-1} cm^{-2}. This value accounts for a total glucose uptake rate of about 58 μmole in 40 min (the mean retention time of the small intestine), which is almost exactly equal to the observed rate of glucose ingestion (57 μmole per 40 min). However, it is not usually clear whether any variation in transport rate is due to variation in the kinetic properties (e.g., V_{max}) of the carrier or just to variation in the density of carriers. Few studies have actually measured the density of carrier binding sites, but from these it is clear that a higher density of binding sites is correlated with higher rates of transport. Thus, species variation in transport rate can generally be attributed to variation in the density of carriers rather than to species-specific adaptive variation in V_{max}. (See *Karasov, Buddington, and Diamond 1985; Del Rio and Stevens 1989; Karasov, Solberg, and Diamond 1985; Karasov et al. 1986; Del Rio et al. 1988; Del Rio, Karasov, and Levey 1989; Buddington 1987; Buddington and Diamond 1987.*)

Recommended Reading

Beerstecher, E. 1964. The biochemical basis of chemical needs. In *Comparative biochemistry*, edited by M. Florkin and H. S. Mason, 120–220. New York: Academic Press.

Brindley, D. N. 1977. Absorption and transport of lipids in the small intestine. In *Intestinal permeation*, edited by M. Kramer and F. Lauterbach, 350–62. Amsterdam: Excerpta Medica.

Caspary, W. F. 1977. Mechanism and specificity of intestinal sugar transport. In *Intestinal permeation*, edited by M. Kramer and F. Lauterbach, 74–84. Amsterdam: Excerpta Medica.

Coates, M. E. 1976. The water-soluble vitamins and other accessory food factors. *Comparative Animal Nutrition* 1:136–67.

Dadd, R. H. 1970. Digestion in insects. In *Chemical zoology*, vol. 5, *Arthropoda*, part A, edited by M.

Florkin and B. T. Scheer, 117–45. New York: Academic Press.

Davenport, H. W. 1977. *Physiology of the digestive tract.* Chicago: Year Book Medical Publishers.

Dethier, V. G. 1976. *The Hungry Fly.* Cambridge: Harvard University Press.

Dougherty, R. W. 1965. *Physiology of digestion in the ruminant.* London: Butterworths.

Hoffman, R. R. 1989. Evolutionary steps of ecophysiological adaptation and diversification of ruminants: A comparative view of their digestive system. *Oecologia* 78:443–57.

Hume, I. D. 1982. *Digestive physiology and nutrition of marsupials.* Cambridge: Cambridge University Press.

Hungate, R. E. 1966. *The rumen and its microbes.* New York: Academic Press.

Jennings, J. B. 1972. *Feeding, digestion, and assimilation in animals.* London: MacMillan.

Karasov, W. H., and J. M. Diamond. 1983. Adaptive regulation of sugar and amino acid transport by vertebrate intestine. *Am. J. Physiol.* 245:G443–62.

Karasov, W. H., D. H. Solberg, and J. M. Diamond. 1985. What transport adaptations enable mammals to absorb sugars and amino acids faster than reptiles? *Am. J. Physiol.* 249:G271–83.

Machlin, L. J. 1984. *Handbook of vitamins: Nutritional, biochemical, and clinical aspects.* New York: Marcel Dekker.

Matthews, D. M. 1975. Intestinal absorption of peptides. *Physiol. Rev.* 55:537–608.

Moir, R. J. 1968. Ruminant digestion and evolution. In *Handbook of physiology,* vol. 5, sec. 6, edited by C. F. Code, 2673–94. Washington: American Physiological Society.

Morton, B. 1983. Feeding and digestion in bivalvia. In *The Mollusca,* vol. 5, *Physiology,* part 2, edited by A. S. M. Saleuddin and K. M. Wilbur, 65–147. New York: Academic Press.

Munck, B. G. 1977. Intestinal transport of amino acids. In *Intestinal permeation,* edited by M. Kramer and F. Lauterbach, 123–35. Amsterdam: Excerpta Medica.

Phillipson, A. T. 1970. *Physiology of digestion and metabolism in the ruminant.* Newcastle: Oriel Press.

Ruckebusch, Y., and P. Thivend. 1980. *Digestive physiology and metabolism in ruminants.* Lancaster: MTP Press Ltd.

Sargent, J. R., and R. R. Gatten. 1976. The distribution and metabolism of wax esters in marine invertebrates. *Biochem. Soc. (Lond.) Trans.* 4:431–33.

Sibly, R. M. 1981. Strategies of digestion and defecation. In *Physiological ecology,* edited by C. R. Townsend and P. Calow, 109–39. Oxford: Blackwell Scientific.

Steffens, W. 1989. *Principles of fish nutrition.* Chichester: Ellis Horwood.

Stevens, C. E. 1988. *Comparative physiology of the vertebrate digestive system.* Cambridge: Cambridge University Press.

Thompson, J. N. 1976. Fat-soluble vitamins. *Comparative Animal Nutrition* 1:99–135.

van Weel, P. B. 1970. Digestion in crustacea. In *Chemical zoology,* vol. 5, *Arthropoda,* part A, edited by M. Florkin and B. T. Scheer, 97–115. New York: Academic Press.

Vonk, H. J. 1964. Comparative biochemistry of digestive mechanisms. In *Comparative biochemistry,* Vol. 6, *Cells and organisms,* edited by M. Florkin and H. S. Mason, 347–401. New York: Academic Press.

Wigglesworth, V. B. 1935. *The principles of insect physiology.* London: Methuen.

Wilson, T. H. 1962. *Intestinal absorption.* Philadelphia: Saunders.

Yonge, C. M. 1928. Feeding mechanisms in the invertebrates. *Biol. Rev.* 3:21–76.

Yonge, C. M. 1937. Evolution and adaptation in the digestive system of metazoa. *Biol. Rev.* 12:87–115.

Appendix A

Classification of Living Animals

• • • • • • • • • • • • • • • • • • • •

A summary is presented of the classification of all micro-organisms, plants, and animals and the classification of animals to the taxonomic level of Phylum, Class (for major Phyla only), and Order for the major classes from insects to vertebrates. K, kingdom; Ph, phylum; Cl, class; Or, order. *(Modified from Hanson 1981; Barnes 1987; Carroll 1988; Archer 1984.)*

K. MONERA: Prokaryote cells; bacteria

K. FUNGI: Multicellular eukaryote autotrophs with a cell wall; fungi, lichens

K. PLANTAE: Multicellular eukaryote plants; red and brown algae, liverworts, mosses, ferns, conifers, flowering plants

K. PROTISTA: Solitary eukaryote cells; algae, slime molds, amoeba, ciliates, flagellates, foraminiferans

K. ANIMALIA: Multicellular eukaryote animals

Ph. PLACOZOA: *Trichoplax*

Ph. MESOZOA: Mesozoan parasites

Ph. PORIFERA: Sponges

Ph. CNIDARIA: Coelenterates
   Cl. HYDROZOA: Hydroids
   Cl. SCYPHOZOA: Jellyfish
   Cl. ANTHOZOA: Sea anemones, corals

Ph. CTENOPHORA: Comb jellies or sea walnuts

Ph. PLATYHELMINTHES: Flat worms
   Cl. TURBELLARIA: Planarians
   Cl. MONOGENEA: Monogenetic flukes
   Cl. TREMATODA: Digenetic flukes
   Cl. CESTODA: Tapeworms

Ph. NEMERTEA: Proboscis worms

Ph. GASTROTRICHA: Gastrotrichs
Ph. GNATHOSTOMULIDA: Gnathostome worms
Ph. ROTIFERA: Rotifers
Ph. KINORHYNCHA: Kinorhynchs
Ph. ACANTHOCEPHALA: Spiny-headed worms
Ph. NEMATODA: Round worms
Ph. NEMATOMORPHA: Horsehair worms
Ph. LORICIFERA: Loriciferans
Ph. PRIAPULIDA: Priapulid worms
Ph. ENTOPROCTA: Pseudocoelomate polyzoans
Ph. BRYOZOA: Coelomate polyzoans
Ph. PHORONIDA: Phoronid worms
Ph. BRACHIOPODA: Lamp shells
Ph. SIPUNCULA: Peanut worms
Ph. ECHIURA: Echiuroid worms
Ph. TARDIGRADA: Water bears
Ph. PENTASTOMIDA: Linguatulids
Ph. ONYCOPHORA: Onycophorans
Ph. POGONOPHORA: Beard worms

Ph. ANNELIDA
   Cl. POLYCHAETA: Marine worms
   Cl. OLIGOCHAETA: Earthworms
   Cl. HIRUDINEA: Leeches

Ph. MOLLUSCA
   Cl. APLACOPHORA: Solenogasters
   Cl. POLYPLACOPHORA: Chitons
   Cl. MONOPLACOPHORA: *Neopilina, Vema*
   Cl. SCAPHODA: Tusk and tooth shells
   Cl. BIVALVIA: Clams, oysters, mussels

Cl. GASTROPODA: Snails, slugs and
  nudibranchs
Cl. CEPHALOPODA: *Nautilus,* squid,
  octopus

Ph. ARTHROPODA
sub-Ph. CHELICERATA
  Cl. MEROSTOMATA: Horseshoe crabs
  Cl. PYCNOGONIDA: Sea spiders
  Cl. ARACHNIDA: Spiders, scorpions, ticks,
    mites

sub-Ph. CRUSTACEA
  Cl. CEPHALOCARIDA: Tiny, primitive
    crustaceans
  Cl. BRANCHIOPODA: Brine, fairy, tadpole,
    clam shrimp, and water fleas
  Cl. OSTRACODA: Ostracods
  Cl. COPEPODA: Copepods
  Cl. MYSTACOCARIDA: Mystacocarid
    crustaceans
  Cl. REMIPEDIA: Remipedid crustaceans
  Cl. TANTULOCARIDA: Ectoparasites
  Cl. MALACOSTRACA: Crabs, shrimp
  Cl. BRANCHIURA: Ectoparasitic crustaceans
  Cl. CIRRIPEDIA: Barnacles

sub-Ph. UNIRAMIA

  Cl. CHILOPODA: Centipedes
  Cl. DIPLOPODA: Millipedes
  Cl. PAUROPODA: Pauropods
  Cl. SYMPHYLA: Symphylans
  Cl. INSECTA:
    Or. Anopleura; sucking lice
    Or. Coleoptera; beetles
    Or. Collembola; springtails
    Or. Dermaptera; earwigs
    Or. Diptera; true flies
    Or. Embioptera; web spinners
    Or. Ephemeroptera; mayflies
    Or. Hemiptera; true bugs
    Or. Homoptera; cicadas, aphids, scale
      insects
    Or. Hymenoptera; bees, wasps, ants
    Or. Isoptera; termites
    Or. Lepidoptera; butterflies, moths
    Or. Mallophaga; biting lice
    Or. Mecoptera; scorpion flies
    Or. Neuroptera; lacewings, antlions
    Or. Odonata; dragonflies, damsel flies
    Or. Orthoptera; grasshoppers, crickets,
      roaches
    Or. Plecoptera; stone flies
    Or. Protura; proturans
    Or. Psocoptera; book and bark lice
    Or. Siphonaptera; fleas
    Or. Strepsiptera; twisted-wing insects
    Or. Thysanoptera; thrips

    Or. Thysanura; bristletails, silverfish
    Or. Tricoptera; caddis flies
    Or. Zoraptera; zorapterans

Ph. ECHINODERMATA
  Cl. CRINOIDEA: Sea lilies, feather stars
  Cl. ASTEROIDEA: Asteroids (star fish)
  Cl. OPHIUROIDEA: Brittlestars
  Cl. ECHINOIDEA: Sea and heart urchins,
    sand dollars
  Cl. HOLOTHUROIDEA: Sea cucumbers

Ph. CHAETOGNATHA: Arrow worms

Ph. HEMICHORDATA: Acorn worms,
  pterobranchs
Ph. CHORDATA
sub-Ph. UROCHORDATA
  Cl. ASCIDIACEA: Tunicates, sea squirts
  Cl. THALIACEA: Salps
  Cl. LARVACEA: Appendicularians
sub-Ph. CEPHALOCHORDATA: Lancelets,
  amphioxus
sub-Ph. VERTEBRATA
  Cl. AGNATHA:
    Or. Myxiniformes; hagfish
    Or. Petromyzontiformes; lamprey

  Cl. CHONDRICHTHYES
  sub-Cl. HOLOCEPHALI
    Or. Chimaeriformes; Chimaeras or rat fish

  sub-Cl. ELASMOBRANCHII
    Or. Heterodontiformes; Port Jackson
      sharks
    Or. Hexanchiformes; frill and cow sharks
    Or. Lamniformes; hammerheads, smooth
      dogfish, thresher, basking, mackerel, cat
      and requiem sharks
    Or. Squaliformes; dogfish, saw and angel
      sharks
    Or. Rajiformes; sawfish, guitarfish, rays,
      electric rays, skates

  Cl. OSTEICHTHYES
  sub-Cl. CHONDROSTEI
    Or. Polyteriformes; bichirs, reedfish
    Or. Acipenseriformes; sturgeon, paddlefish

  sub-Cl. HOLOSTEI
    Or. Lepisosteiformes; gars
    Or. Amiiformes; bowfin

  sub-Cl. TELEOSTEI
    Or. Anguilliformes; eels
    Or. Atheriniformes; garfish, mosquitofish,
      flying fish
    Or. Aulopiformes; *Aulopus,* greeneye,
      pearleye

Or. Batrachoidiformes; toadfish
Or. Beryciformes; squirrelfish, soldierfish, knightfish
Or. Clupeiformes; herring, sardines, anchovy
Or. Characiformes; characins
Or. Cypriniformes; minnow and carp
Or. Cyprinodontiformes; killifish, top minnows, four-eyed fish
Or. Dactylopteriformes; flying gurnards
Or. Elopiformes; tarpon, tenpounder, bonefish
Or. Gadiformes; cod, haddock, whiting, hake
Or. Gasterosteiformes; stickleback, tubesnout
Or. Gobiesociformes; clingfish, dragonet
Or. Gonorhynchiformes; milk fish
Or. Indostomiformes; *Indostomus*
Or. Lampridiformes, opah, oarfish
Or. Lophiiformes; goosefish, batfish, frogfish, anglerfish
Or. Mormyriformes; elephantfish, gymnarchids
Or. Myctophiformes; lanternfish, lizardfish
Or. Notacanthiformes; spiny eels, halosaurs
Or. Ophidiiformes; brotulids, cusk eel
Or. Osteoglossiformes; bony tongues, mooneye, knifefish
Or. Pegasiformes; sea moths, dragonfish
Or. Perciformes; snook, snapper, cichlids, crocodile ice fish, perch, remora
Or. Percopsiformes; trout perch, pirate perch, cavefish
Or. Pleuronectiformes; flounder, halibut, sole
Or. Polymixiiformes; beardfish
Or. Salmoniformes; salmon, trout, smelt, pike, galaxxiids, mud minnows, ice fish
Or. Scorpaeniformes; scorpionfish, sculpin, sea robin
Or. Siluriformes; catfish, bristlemouths, snaggletooths
Or. Stomiiformes; hatchet fish, lightfish
Or. Synbranchiformes; swamp eels, cuchia
Or. Syngnathiformes; trumpetfish, seahorse, pipefish
Or. Tetraodontiformes; pufferfish, spikefish, triggerfish
Or. Zeiformes; dories, boarfish

sub-Cl. SARCOPTERYGII
Or. Coelacanthiformes; coelacanth
Or. Ceratodiformes; Australian lungfish
Or. Lepidosireniformes; South American and African lungfish

Cl. AMPHIBIA
Or. Gymniophona; caecilians

Or. Urodela; salamanders
Or. Anura; frogs and toads

Cl. REPTILIA
Or. Crocodylia; alligators and crocodiles
Or. Chelonia; turtles and tortoises
Or. Rhyncocephalia; tuatara
Or. Squamata
sub-Or. Lacertilia; lizards
sub-Or. Amphisbaenia; blind snakes
sub-Or. Serpentes; snakes

Cl. AVES
super-Or. PALAEOGNATHAE
Or. Apterygiformes; kiwi
Or. Casuariiformes; emu, cassowary
Or. Rheiformes; rhea
Or. Struthioniformes; ostrich
Or. Tinamiformes; tinamou

super-Or. NEOGNATHAE
Or. Anseriformes; waterfowl and screamers
Or. Apodiformes; hummingbirds
Or. Caprimulgiformes; goatsuckers, nighthawks, oilbirds
Or. Charadriiformes; plovers, auks, gulls and terns
Or. Ciconiiformes; herons, storks and ibis
Or. Coliiformes; mousebirds
Or. Columbiformes; doves, pigeons
Or. Coraciiformes; kingfishers, bee-eaters and horn bills
Or. Cuculiformes; cuckoos
Or. Falconiformes; falcons, hawks, eagles, vultures, and condor
Or. Galliformes; fowl
Or. Gaviiformes; loons
Or. Gruiformes; rails, cranes and bustards
Or. Passeriformes; perching birds (e.g., sparrows, jay)
Or. Pelicaniformes; pelicans, frigate birds, tropic birds, boobies, cormorants
Or. Phoenictopteriformes; flamingos
Or. Piciformes; woodpeckers
Or. Podicipediformes; grebes
Or. Procellariiformes; albatross, shearwater petrel
Or. Psittaciformes; parrots
Or. Pteroclidiformes; sandgrouse
Or. Sphenisiformes; penguins
Or. Strigiformes; owls

Cl. MAMMALIA
sub-Cl. PROTOTHERIA
Or. Monotremata; platypus and echnida

sub-Cl. METATHERIA
Or. Dasyuridae; thylacine, numbat, dasyurid carnivorous marsupials

Or. Didelphiformes; didelphids, opossum
and microbiotherids

Or. Diprotodonta; wombat, koala,
kangaroos, wallabies and possums

Or. Notoryctemorphia; marsupial mole

Or. Paucituberculata; caenolestids

Or. Peramelina; bandicoots

sub-Cl. EUTHERIA

Or. Artiodactyla; cattle, deer, camels,
hippopotamus

Or. Carnivora; cats, dogs, bears and
hyenas

Or. Cetacea; whales and dolphin

Or. Chiroptera; bats

Or. Dermoptera; flying lemurs

Or. Edentata; anteaters, sloths, armadillos

Or. Hyracoidea; dassies

Or. Insectivora; shrews, moles and
hedgehogs

Or. Lagomorpha; rabbits, hares and pika

Or. Perissodactyla; horse, zebra,
rhinoceros

Or. Pholidota; pangolins

Or. Pinnipedia; seals, sea lions and walrus

Or. Primates; lemurs, monkeys, apes and
humans

Or. Proboscidea; elephants

Or. Rodentia; mice, rats and squirrels

Or. Sirenia; manatee and dugong

Or. Tubulidentata; aardvark

Appendix B

Units

• •

A summary is presented of the fundamental units as well as the supplementary and derived SI units adopted by the Eleventh Conference Generale des Poids et Measures (1960). Also presented are the factors to convert between SI and non-SI units. *(Based on Mechtley 1973; Nobel 1983.)*

	Unit	Dimension		Conversion Factor
Fundamental				
Current	Ampere	A		$6.24 \ 10^{18}$ charges sec^{-1}
Length	Meter	m		3.28 feet
Light	Candela	cd		1 lumen sr^{-1}
				1 candle
Mass	Kilogram	kg		2.20 pounds
Matter	Mole	mol		
Temperature	Kelvin	°K		$273.15 + °C$
				$273.15 + \frac{5}{9}(°F - 32)$
Time	Second	s		
Supplementary				
Plane angle	Radian	rad		57.30°
Solid angle	Steradian	sr		
Derived				
Capacitance	Farad	F	$A^2 s^4 kg^{-1} m^{-2}$	1 coulomb $volt^{-1}$
Charge	Coulomb	C	$s A^{-1}$	1 joule $volt^{-1}$
Energy	Joule	J	$kg \ m^2 \ s^{-2}$	0.239 calorie
				$9.49 \ 10^{-4} \ BTU$
				$6.242 \ 10^{18} \ eV$
				10^7 erg
Entropy		S	$J \ K^{-1}$	0.239 cal $°C^{-1}$
Force	Newton	N	$kg \ m \ s^{-2}$	0.102 kg mass
				10^5 dynes
Frequency	Hertz	Hz	s^{-1}	
Luminous flux	Lumen	lm	cd sr	$1.46 \ 10^{-3} \ W \ (\lambda = 540 \ nm)$
Pressure	Pascal	Pa	$kg \ m^{-1} s^{-2}$	$9.87 \ 10^{-6}$ atm
				$7.52 \ 10^{-3}$ mm Hg (torr)
				10^{-5} bar
Power	Watt	W	$kg \ m^2 \ s^{-3}$	0.239 calorie s^{-1}
				$1.34 \ 10^{-3}$ horsepower
Resistance	Ohm	Ω	$kg \ m^2 \ s^{-3} A^{-2}$	
Viscosity	Pa sec	η	$kg \ m^{-1} s^{-1}$	10 Poise
Voltage	Volt	V	$kg \ m^2 \ s^{-3} A^{-1}$	1 ampere ohm

PREFIX for Units

Tera	T	10^{12}
Giga	G	10^{9}
Mega	M	10^{6}
Kilo	k	10^{3}
Deci	d	10^{-1}
Centi	c	10^{-2}
Milli	m	10^{-3}
Micro	μ	10^{-6}
Nano	n	10^{-9}
Pico	p	10^{-12}
Fempto	f	10^{-15}
Atto	a	10^{-18}

Appendix C

Constants and Coefficients

• •

A summary is presented of the values for a variety of constants and coefficients that are commonly used in physiology. (*Based on Nobel 1983.*)

Symbol	Constant/Coefficient	Value
c	Speed of light	$2.998 \ 10^8$ m sec^{-1}
C	Specific heat of water	4187 J mol^{-1} °C^{-1}
CP_{air}	Heat capacity of air	1.21 kJ m^{-3} °C^{-1} (20° C, 101.3 kPa)
CP_{water}	Heat capacity of water	4.18 MJ m^{-3} °C^{-1} (20° C, 101.3 kPa)
DCO_2	CO_2 diffusion coefficient	$1.51 \ 10^{-5}$ m^2 s^{-1} (20° C, air)
		$1.8 \ 10^{-9}$ (20° C, water)
DO_2	O_2 diffusion coefficient	$1.95 \ 10^{-5}$ m^2 s^{-2} (20° C, air)
DH_2O	H_2O diffusion coefficient	$2.42 \ 10^{-5}$ m^2 s^{-2} (20° C, air)
e	Electron charge	$1.602 \ 10^{-19}$ coulomb
ϵ	Permittivity of a vacuum	$8.854 \ 10^{-12}$ C m^{-1} V^{-1}
F	Faraday's constant	$9.649 \ 10^4$ coulomb mole^{-1}
g	Gravitational constant	9.780 m s^{-2} (sea level, at equator)
h	Planck's constant	$6.626 \ 10^{-34}$ J s
hc		$1.986 \ 10^{-25}$ J m
H_{fusion}	Heat of fusion	51.0 kJ mol^{-1} (water, 0° C)
H_{vapor}	Heat of vaporization	44.21 kJ mol^{-1} (water, 20° C)
		43.35 kJ mol^{-1} (water, 40° C)
k	Boltzmann's constant	$1.381 \ 10^{-23}$ J molecule^{-1} K^{-1}
kT		$4.046 \ 10^{-21}$ J molecule^{-1} (20° C)
		0.0235 eV molecule^{-1} (20° C)
K_{air}	Thermal conductivity	0.0257 W m^{-1} °C^{-1} (dry air, 20° C)
		0.0255 W m^{-1} °C^{-1} (wet air, 20° C)
N	Avogadro's number	$6.0220 \ 10^{23}$ molecules mol^{-1}
pH_2O	Water vapor pressure	0.609 kPa at 0° C (saturated air)
		2.33 kPa at 20° C
		6.32 kPa at 37° C
		7.36 kPa at 40° C
R	Gas constant	8.3143 J mol^{-1} K^{-1}
RT		$2.271 \ 10^3$ J mol^{-1} (0° C)
		$2.437 \ 10^3$ J mol^{-1} (20° C)
		$2.577 \ 10^3$ J mol^{-1} (37° C)
		$2.602 \ 10^3$ J mol^{-1} (40° C)
RT/F		25.3 mV (20° C)
$RT/\overline{V}_w$		135 MPa (water; 20° C)
ρ_{air}	Density of air	1.205 kg m^{-3} (dry air, 20° C)
		1.194 kg m^{-3} (wet air, 20° C)

Symbol	Constant/Coefficient	Value
ρ_{water}	Density of water	999.8 kg m^{-3} (0° C)
		1000.0 kg m^{-3} (4° C)
		998.2 kg m^{-3} (20° C)
		992.2 kg m^{-3} (40° C)
S	Solar constant	1360 W m^{-2}
σ	Stefan-Boltzmann constant	5.67 10^{-8} W m^{-2} K^{-4}
σ_{water}	Water surface tension	0.0728 N m^{-1} (20° C)
u	Atomic mass unit	1.661 10^{-27} kg
$\overline{V}_w$	Partial molar volume	1.805 10^{-5} m^3 mol^{-1} (water, 20° C)

Glossary

α-adrenergic receptor Cell surface receptor that binds norepinephrine and epinephrine and elicits an intracellular response via a G protein, adenylcyclase, and cAMP.

β-adrenergic receptor Cell surface receptor that binds especially epinephrine and norepinephrine to a lesser extent, and elicits an intracellular response via a G protein, adenylcyclase, and cAMP.

abomasum Fourth chamber of the four-chambered stomach of a ruminant mammal; it is the "true," acid-secreting stomach.

absolute humidity (χ) Water vapor content of air, usually measured in milligrams of water per liter of air.

absolute metabolic scope The arithmetic difference between the maximal metabolic rate (MMR) and the standard metabolic rate (SMR) or basal metabolic rate (BMR), i.e., MMR − SMR or MMR − BMR.

absolute refractory period The brief period of time that must elapse after an action potential, before a second action potential can occur.

absolute zero (−273.15° C) The theoretical lowest temperature at which molecules have no thermal motion and chemical reactions do not take place.

acclimation The gradual physiological readjustment of an organism to a change in its environment, as observed in the laboratory.

acclimatization The gradual physiological readjustment of an organism to a change in its environment, as observed in nature.

accommodation (neural) A temporary change in the threshold of a sensory receptor to a stimulus during a sustained subthreshold stimulation.

accommodation (vision) Adjustment of the focal plane of the eye so that an object remains in focus on the retina, regardless of its distance from the eye.

acid A substance that releases hydrogen ions (H^+) when dissolved in water.

acidosis A condition of the blood, or other body tissue, when its H^+ concentration is higher than normal (i.e., low pH).

acoustico-lateralis system The sensory system for hearing, equilibrium, and detection of water currents by vertebrates; the hair cell is the sensory receptor.

actin A microfilament (protein) that is contractile; it consists of globular G-actin monomers aggregated to form an F-actin filament. It is found in many cells, and forms a highly organized array with myosin in the sarcomere of striated muscle cells.

action potential A transient (about 1 msec) change in membrane electrical potential of sensory and nerve cells in response to a change in membrane permeability, primarily to Na^+.

activation energy (E_a) The energy required to initiate a chemical reaction by increasing the velocity of reactant molecules.

activation heat (H_a) The initial, rapid release of heat from a muscle cell, even before tension is generated.

active site The portion of an enzyme that binds the substrate(s) and catalyzes their reaction.

active state The initial, short-term phase of muscle cell contraction when the actin-myosin cross bridging shortens the sarcomere to generate tension (which is delayed relative to the active state).

active transport The energy-requiring process of movement of a molecule against its electrochemical gradient, e.g., across a membrane from a low to a high concentration.

actomyosin domain The part of a smooth muscle cell, containing actin, myosin, troponin, and calmodulin, that is responsible for force development during contraction (cf. filamin domain).

adaptation (evolutionary) A genetically determined trait that enhances an organism's Darwinian fitness.

adaptation (neural) A decrease in sensory sensitivity during sustained stimulation.

adenine One of the purine base constituents of DNA and RNA; $C_5H_5N_5$.

adenohypophysis (anterior pituitary, anterior lobe, pars distalis) The anterior endocrine glandular portion of the vertebrate pituitary that secretes a number of hormones; derived from the ectoderm lining the roof of the mouth.

adenosine-diphosphate (ADP) An organic compound containing an adenine base, a ribose sugar, and two phosphate groups; a nucleotide. Formed from ATP by hydrolysis with the release of one high-energy phosphate bond.

adenosine-monophosphate (AMP) An organic compound containing an adenine base, a ribose sugar, and one phosphate group; a nucleotide. Formed from ADP by hydrolysis with the release of one high-energy phosphate bond.

adenosine-triphosphate (ATP) An organic compound containing an adenine base, a ribose sugar, and three phosphate groups; a nucleotide. Forms ADP when hydrolyzed, releasing one high-energy phosphate bond.

adenosine-triphosphatase (ATPase) A class of enzymes that hydrolyzes ATP to provide chemical energy for other reactions.

adenyl-cyclase A membrane-bound enzyme that is stimulated by a G protein to synthesize cyclic AMP from ATP.

adrenal cortex (inter-renal gland) The cortical adrenal tissue type that is found intermingled with the medullary adrenal tissue or forms a separate adrenal cortex; secretes various glucocorticoid and mineralocorticoid hormones.

adrenal gland (epinephric gland, suprarenal gland) Paired endocrine glands located adjacent to the kidneys and often capping them; they contain two tissue types that may be intermingled or discrete.

Adrenalin Trade name for adrenaline.

adrenaline (epinephrine) A catechol hormone released from the adrenal medulla of vertebrates and from the sympathetic nerve terminal in some vertebrates.

adrenal medulla (chromaffin tissue) The medullary adrenal tissue type that is found intermingled with the cortical adrenal tissue or forms a separate adrenal medulla; secretes the catecholamines epinephrine and norepinephrine.

adrenocorticotropic hormone (ACTH) A hormone secreted by the adenohypophysis that stimulates endocrine secretion by the adrenal cortex.

aequorin A calcium-activated photoprotein from the hydrozoan coelenterate *Aequorea* that emits blue light (λ = 469 nm) in the presence of Ca^{2+}.

aerobiosis Catabolic metabolic pathways that involve oxidation by molecular oxygen (which is an electron acceptor) to form water, e.g., complete metabolism of glucose to CO_2 and H_2O by glycolysis, the citric acid cycle, and the electron transport chain.

aeropyle A respiratory gas tube in the insect spiracular gill or egg shell plastron or the spiracular plate of some ticks.

aglomerular Referring to the type of vertebrate kidney which has nephrons that lack a glomerulus; urine is formed by active secretion of solutes and passive water transport.

agonist (drug action) Drug that mimics the action of a neurotransmitter, e.g., nicotine is an agonist of acetylcholine.

agonist (muscle) A muscle that has the same mechanical effect as another muscle.

air capillary Small tubes radiating from the parabronchi that form the gas exchange surface of the bird lung.

air sac A thin-walled, air-filled, collapsible extension of the respiratory system in some reptiles, but primarily in birds; an expanded diverticulum of the larger tracheae of the insect respiratory system; the swim-bladder of fish; the alveolus of the mammalian lung.

aldosterone A potent mineralocorticoid hormone secreted by the adrenal cortical tissue of vertebrates.

alkali a substance that forms hydroxyl ions (OH^-) when dissolved in water.

alkalosis A condition of the blood, or other body tissue, in which its H^+ concentration is lower than normal (i.e., high pH).

allantoin A nitrogenous compound ($C_4H_6N_4O_3$) formed during uricolysis; an important nitroge-

nous waste in some mammals, e.g., some cricetid rodents.

allantois A highly vascular, sac-like outgrowth of the posterior gut in embryonic reptiles, birds, and mammals; functions in gas, nutrient, and waste exchange.

allograft A tissue graft derived from a genetically different donor of the same species.

allometry (scaling) The study of the change in value, or proportion, with differing body size.

allozyme One of two or more slightly different forms of the same enzyme; different allozymes may confer an advantage under particular environmental conditions, e.g., temperature.

altricial Requiring extensive parental care after birth or hatching, e.g., newborn mammals.

alveolus (gland) The sac-like unit structure of some glandular tissues.

alveolus (respiration) The terminal air sac of the mammalian respiratory system; the main site of gas exchange in the mammalian lung.

ambient temperature and pressure (ATP) Pertaining to the volume of a gas at the environmental temperature and pressure.

ametabolous Pattern of metamorphosis in insects that have no distinct changes, other than size, at each molt.

amine precursor uptake and decarboxylation system See APUD system.

amino acid Compound containing an amino group (NH_2—) and a carboxyl group (—COOH); the constituents of proteins.

ammonia (NH_3) Nitrogenous waste product of amino acid metabolism (by deamination) and pyrimidine base metabolism; very soluble but highly toxic.

ammonotely The pattern of excretion in which ammonia is the primary nitrogenous waste.

ampere (A) The unit of current; defined as 1 coulomb of charge flow per second, which is the current flow through a 1 ohm resistance with an electrical potential difference of 1 volt.

ampulla of Lorenzini Terminal vesicle of the electroreceptive sensory system of elasmobranchs.

anabolism Chemical reactions in which simple substrates are combined to form more complex molecules resulting in the storage of energy, formation of structural molecules, or growth.

anadromous A fish that migrates from seawater to freshwater to breed, e.g., lamprey (cf. catadromous).

anaerobic metabolic end product A compound produced as the final step of anaerobic metabolism that may be excreted or stored for subsequent aerobic reprocessing, e.g., lactate.

anaerobic threshold The level of activity at which anaerobic metabolism contributes significantly to the total energy production.

anaerobiosis Catabolic metabolic pathways that do not use molecular oxygen as an electron donor but use organic molecules to regenerate NAD^+, e.g., pyruvate.

angiotensin II A bioactive octapeptide formed from angiotensin I by the enzyme renin; it is a potent vasoconstrictor and promotes aldosterone secretion.

anion A negatively charged ion that is attracted to an anode, e.g., Cl^-.

anoxia The absence of O_2.

antagonist (drug action) Drug that reduces or prevents the action of a neurotransmitter, e.g., D-tubocurarine is an antagonist of acetylcholine.

antagonist (muscle) Another muscle that has an opposing mechanical response to a muscle contraction.

antennal gland Coelomoduct-type excretory organ of crustaceans with the excretory pore located near the base of the antenna.

anterior pituitary See adenohypophysis.

antibody A blood-borne, four-chain immunoglobin protein that reacts with specific antigens; responsible for the humoral immune response.

antidiuretic hormone (ADH) See arginine vasopressin.

antifreeze protein Protein or glycoprotein that lowers the freezing point of the blood and other body fluids in a noncolligative manner; affects the melting temperature in a colligative manner.

antigen An exogenous material that initiates the formation of antibodies by the immune system.

aorta The main artery that carries blood, or hemolymph, from the heart.

apodeme An invagination of the body wall that forms a rigid support for muscle attachment in arthropods.

apparent digestive efficiency (ADE) The digestive efficiency calculated as 100 ($Food_{in}$ − $Feces_{out}$)/$Food_{in}$; values can be dry matter or energy.

apposition eye A simple type of insect eye in which the rhabdome of each ommatidium is located immediately underneath the ommatidial lens and cone.

APUD system A series of neuroendocrine and peripheral endocrine cells that absorb 5-hydroxytryptophan and convert it to serotonin.

archinephric duct (Wolffian duct) The duct draining urine from the primitive kidney (pronephros and mesonephros); its role is replaced by the ureter in metanephric kidneys.

arginine vasopressin (AVP) A neurohypophyseal octapeptide hormone of mammals that promotes water reabsorption by the collecting ducts of nephrons and decreases urine flow rate.

arginine vasotocin (AVT) A neurohypophyseal octapeptide of lower vertebrates that initiates various water-conserving responses, e.g., cutaneous water uptake, bladder water reabsorption, decreased urine flow.

Arrhenius plot Relationship between the natural logarithm of the reaction rate (ordinate) and the reciprocal of temperature (abscissa); the slope is $-E_a/R$.

arteriole A small artery located just before the capillaries.

artery A blood vessel that conveys blood from the aorta to the body.

articulation A joint between bones or exoskeletal structures.

ATP See adenosine-triphosphate and ambient temperature and pressure.

ATPase See adenosine-triphosphatase.

atrioventricular (AV) node A group of specialized cardiac muscle cells, located in the wall of the right atrium near the tricuspid valve; it transmits electrical activity from the right atrium to the ventricles, but with a slight time delay.

atrium Anterior cavity of the heart; usually collects blood returning from the body or gills; it is a priming pump for the ventricle.

autonomic nervous system A system of visceral motor neurons that regulates the more automatic bodily functions in vertebrates; comprised of the sympathetic and parasympathetic divisions.

autotomy Self-amputation of part of the body, e.g., of body segments by worms, legs by arthropods, and the tail by lizards.

average daily metabolic rate (ADMR) The average metabolic rate of an animal over a complete daily cycle of activity and inactivity.

Avogadro's law One mole of gas at standard temperature and pressure (0° C and 101 kPa) contains 6.02 10^{23} molecules.

axon (nerve fiber) The elongated process of a neuron that propagates action potentials from the cell soma to the peripheral synapse.

axoneme The microtubular structure that forms the axial filament of a flagellum or cilium.

axon hillock   The transitional region of a neuron where the axon forms.

band pass filter   A filtering effect of certain sensory cells or neurons that preferentially transmits signals within a specific frequency range.

baroreceptor   Sensory receptor that responds to changes in pressure.

basal body   A centriole-like structure located at the base of cilia and flagella; it anchors and organizes the microtubules and other components of the axoneme to the cell.

basal metabolic rate (BMR)   The minimal metabolic rate of an endothermic bird or mammal measured when inactive and undisturbed, postabsorptive, and within thermoneutrality.

base   A substance that releases hydroxyl ions (OH^-) when dissolved in water.

basilar papilla   A neuromast sensory area of the lagena of the inner ear in many amphibians and reptiles; the hair cells and their gelatinous membrane detect sound vibrations. See also organ of Corti.

Bell-Magendie rule   Only sensory axons are present in the dorsal root ganglion and only motor axons are present in the ventral root of the vertebrate spinal cord.

bends   See Caisson disease.

Bernouilli's principle   The total energy of a fluid in stream-lined, laminar flow is conserved, i.e., kinetic energy + potential energy + pressure energy = constant.

bile   A secretion of the liver; it contains bile salts for fat digestion and bile pigments for excretion.

bile pigment   A pigment, such as bilirubin (a greenish-yellow pigment), that is a breakdown product of heme; formed in the liver and transported to the intestine in the bile for excretion.

bile salt   Cholic acid conjugated with taurine or glycine in liver cells and present in bile; facilitates fat digestion in the small intestine by emulsifying it into small droplets.

biological clock   An endogenous biochemical or molecular "clock" found in the nervous system of many animals that is used to coordinate activities of the animal with an external daily, lunar, or annual cycle.

bioluminescence   Cellular metabolic production and emission of light by animals.

biorhythm   A regular rhythm of activity exhibited by an animal; it may be entrained to the external circadian/lunar or annual cycle by a zeitgeber or may free run with an inherent periodicity in the absence of a zeitgeber.

black body   An object that emits as thermal radiation all of the energy that it receives; absorptivity = emissivity = 1 for all wavelengths.

bleaching   Loss of photopigment color in response to absorption of light.

blood   A circulating fluid that transports gases, nutrients, waste products, hormones, other solutes, or heat.

blood-brain barrier   The physical barrier that prevents the transfer of specific high molecular weight solutes between the blood and the cerebrospinal fluid and brain tissues.

body temperature and pressure, saturated (BTPS)   Conditions determining the gas volume at the temperature of the body, the ambient pressure, and for gas saturated with water vapor.

Bohr effect   A change in O_2 affinity of a respiratory pigment caused by an increase in partial pressure of CO_2 or decrease in pH.

Bohr effect factor (ϕ)   The change in $\log_{10}$ of the P_{50} of a respiratory pigment, per change in pH; $\phi = \Delta \log P_{50}/\Delta pH$.

book gill   The gill of some arthropods, consisting of sheets of delicate, leaf-like lamellae that are stacked like pages of a book.

book lung   The lung of some arthropods, consisting of sheets of leaf-like lamellae in an enclosed space that opens to the exterior by a narrow slit.

boundary layer   A region in a fluid where it passes a surface and has a different physical property than the free-stream fluid (at an infinite distance from the surface), e.g., a lower velocity, a different temperature, or a different solute concentration.

bound water   Water that is structured around charged ions and other electrolytes; it is effectively prevented from having any solvent capacity.

Bowman's capsule   The double-walled invaginated spherical structure at the proximal end of the nephron, across which fluid is filtered from the glomerular capillaries into the lumen of the nephron.

Boyle's law   The product of pressure and volume of a gas is constant at a constant temperature, i.e., PV = constant.

bradycardia   A reduction in heart rate from the normal level.

brain   A mass of nervous tissue that assumes a major and coordinating role in the regulation of sensory and motor functions of the nervous system, e.g., supra- and subesophageal ganglia of invertebrates and the brain of vertebrates.

bronchus   An air tube that connects the trachea with each lung.

brown adipose tissue (BAT)   A type of fat found in some mammals that is specialized for metabolic heat production; its fat cells typically contain many spherical lipid droplets and numerous mitochondria.

brown fat   See brown adipose tissue.

Brownian motion   The random motion of small particles in solution as a result of the random thermal motion of all molecules; originally observed by Robert Brown with chalk dust.

BTPS   See body temperature and pressure, saturated.

buccal pump   Structural modification of the floor or sides of the mouth cavity that is used to pump air or water into and out of the mouth for respiration, e.g., buccal ventilation of the lungs in amphibians and/or for filter feeding as in tadpoles.

buffer A chemical mixture of an acid (HA) and salt (A$^-$) that is able to stabilize the pH against changes due to the addition of an acid or an alkali by the salt chemically combining with added H$^+$ to form acid or the acid combining with added OH$^-$ to form salt; a buffer is most effective when pH = pK.

buffer base The total concentration of all buffering anions (e.g., HCO$_3^-$, phosphate ions, protein anions) in solution; these buffering anions have a pK near the biological pH value.

buffer capacity (β) The amount of strong base (or strong acid) that is required to change the pH by 1 unit; β = Δmoles/ΔpH.

bulk flow The convective movement of water as a mass of fluid, e.g., in response to a pressure differential, rather than the movement of discrete water molecules by diffusion.

bulk water Water that is not structured by ions or membranes and has its full solvent capacity.

buoyancy The tendency to float or sink in water or air; the buoyant force equals the force on a mass of fluid of equivalent volume to the object; objects float if they have positive buoyancy and sink if they have negative buoyancy.

bursicon A neurosecretory hormone from the central nervous system of insects; it promotes tanning and hardening of the cuticle.

Caisson disease (bends) Formation during decompression of small bubbles of gas, especially nitrogen, that can block blood vessels and cause pain, dizziness, paralysis, and even death.

calcitonin (thyrocalcitonin) A protein hormone secreted by the C-cells of the thyroid in response to high blood Ca^{2+}; its effect is to decrease blood Ca^{2+}.

calmodulin A calcium-binding protein, resembling troponin, that

is found in many different types of cells.

calorie (cal) A non-SI unit for energy; it is the amount of heat required to raise the temperature of 1 gram of water from 14.5 to 15.5° C.

calorimetry The measurement of the metabolic rate of an animal as heat production.

cancellous (spongy) bone Bone with a reticular or lattice-like arrangement of bony struts; it is less dense than compact bone but retains considerable mechanical strength because the trabeculae are generally arranged parallel to the direction of stress.

capacitance (electrical) The capacity to store electrical charge separated across a thin insulating gap; a capacitance of 1 farad can store 1 coulomb at a potential difference of 1 volt.

capacitance (vascular) See compliance.

capacitance coefficient (β) The gas content of a liquid or gas per kilopascal partial pressure; for liquids, it is equal to the solubility coefficient (α); it is the same for all gases.

capillary A microscopic blood vessel typically located between the arteries and veins; it is the primary site for nutrient and waste exchange between blood and tissues.

carbamino compound The amino groups of proteins combine with CO$_2$ in the blood to form carbamino compounds, e.g., carbamino-hemoglobin.

carbonic anhydrase An enzyme that reversibly catalyzes the reaction of CO$_2$ and H$_2$O to form carbonic acid H$_2$CO$_3$.

cardiac muscle A form of vertebrate striated muscle found in the heart that has branching cells that form an electrical syncytium, and has involuntary control.

cardiac output The volume of blood pumped by the heart per unit time; for bird and mammal hearts, it is the output of one side of the heart, e.g., the systemic output.

cartilage A specialized connective tissue that forms tough, but flexible, parts of the vertebrate skeleton.

catabolism Chemical reaction in which complex molecules are hydrolyzed into simple molecules, resulting in the liberation of chemical energy and breakdown of cellular structural molecules.

catadromous Fish that migrate from freshwater to seawater to breed, e.g., eels (cf. anadromous).

catalyst An agent that can increase the rate of a chemical reaction without itself being changed by the reaction. See enzyme.

catch muscle A specialized form of muscle of invertebrates, such as bivalve mollusks, that have a "catch" mechanism that maintains tension at little metabolic cost.

catecholamine A group of biologically active compounds, derived from catechol, that have sympathetic effects, e.g., epinephrine, norepinephrine, dopamine.

cation A positively charged ion that is attracted to a cathode, e.g., Na$^+$.

cavum arteriosum A chamber in the reptilian (noncrocodilian) ventricle that receives oxygenated blood from the left atrium and passes it to the cavum venosum.

cavum pulmonale A chamber in the reptilian (noncrocodilian) heart that receives deoxygenated blood from the cavum venosum and passes it to the pulmonary arteries.

cavum venosum A chamber of the reptilian (noncrocodilian) ventricle that receives deoxygenated blood from the right atrium and passes it to the cavum pulmonale, and receives oxygenated blood from the cavum arteriosum and passes it to the systemic arches.

cecum A blind diverticulum extending from the digestive tract.

cellular immunity An immune response that is primarily a consequence of the actions of blood cells (lymphocytes and macrophages).

cellulase A group of enzymes that hydrolyze the β-bonds of cellulose;

found primarily in bacteria and fungi, but also synthesized by a few animals.

cellulose A major carbohydrate constituent of plant cell walls; a polymer $(C_6H_{10}O_5)_n$, linked by β- rather than α-type bonds.

central pattern generator (CPG) A group of neurons that forms a neural circuit that generates a rhythmic pattern of discharge.

centriole One of the pair of small, dark-staining structures of the centrosome; each replicates before mitosis and forms a pole and aster of the mitotic spindle.

centrosome The region of cytoplasm that contains the pair of centrioles; usually located near the nucleus.

cerebellum Dorsal part of the vertebrate hindbrain; involved with coordination of reflex control of posture.

cerebral hemisphere A large, paired outgrowth of the forebrain; involved primarily with olfaction in lower vertebrates; it becomes a progressively larger and more important center of sensory and motor function in higher vertebrates.

chemical potential (μ) The free energy associated with a chemical, relative to its standard chemical potential (μ°); depends on chemical activity, pressure, electrical potential and chemical charge, and gravity.

chemio-osmotic theory Hypothesis of Mitchell that a transmembrane chemical and osmotic concentration gradient in H^+ drives an H^+ flux through the membrane, which is coupled to the phosphorylation of ADP to form ATP; occurs in chloroplasts and mitochondria.

chemoreceptor A sensory receptor for chemicals; may be generalized for a class of chemicals, or highly specialized for a specific chemical.

chemotaxis Directed movement in response to a gradient in chemical concentration.

chief cell (zymogen cell) A cell of the vertebrate stomach lining that secretes the proenzyme pepsinogen.

chitin A nitrogenous carbohydrate polymer $(C_{32}H_{54}O_{21}N_4)_n$ that forms the exoskeletal material of insects and also the cell wall of fungi.

chitinase An enzyme that hydrolyzes chitin; secreted by numerous animals that have considerable chitin in their diet.

chloride cell An epithelial cell of fish gills, and other external epithelia of various animals, that actively excretes Cl^-.

chloride shift (Hamburger shift) The movement of Cl^- into an erythrocyte in the venous circulation in exchange for HCO_3^-; is associated with an osmotic influx of water that causes the erythrocyte to swell.

chlorocruorin A metallic green respiratory pigment that is very similar to hemoglobin, i.e., has a heme group and polypeptide chain; found in four polychete families.

cholecystokinin (CCK, or pancreozymin, CCK-PZ) A hormone that is released from the intestine and induces pancreatic secretion of digestive enzymes and the release of bile into the small intestine.

cholesterol An important sterol that is a precursor for steroid hormones.

chordotonal organ A mechanical, sound, or vibration sensory organ of insects that contains rod-like or bristle-like scolopidium units.

chromaffin tissue (adrenal medullary tissue) The medullary adrenal tissue type that is found intermingled with the cortical adrenal tissue or forms a separate adrenal medulla; secretes the catecholamines epinephrine and norepinephrine.

chromatocyte A pigment-containing cell.

chromatophore A pigment-containing cell that is responsible for physiological color change by aggregating the pigment (light color) or dispersing the pigment in its cell processes (dark color), e.g., melanophore (black pigment), leukophore (white pigment), erythrophore (red pigment), xanthophore (yellow pigment), or cyanophore (blue pigment).

chromatosome Two or more different-colored chromatophores that are closely associated and appear to be polychromatic.

chronaxie The latent period between electrical stimulation and muscle contraction with a stimulus that is twice the threshold value.

chylomicron A small, protein-coated lipid droplet that forms in intestinal epithelial cells and is transported by the lymphatic system into the circulation.

cilium A motile cell organelle consisting of the "9 + 2" microtubular structure (axoneme) that may be present in high numbers on the cell surface; a small flagellum.

circadian rhythm A biological rhythm with a periodicity of about 24 hours.

circannual rhythm A biological rhythm with a periodicity of about one year.

citric acid cycle An essential biochemical cycle in aerobic animals that results in ATP synthesis; it catabolizes acetyl-CoA into CO_2 and synthesizes ATP, $NADH/H^+$, and $FADH_2$.

classical conditioning A type of associative learning in which the presentation of an unconditioned stimulus (e.g., a puff of air on the cornea) becomes associated with a conditioned stimulus (e.g., noise from a buzzer), so that presentation of only the conditioned stimulus will elicit the unconditioned response (e.g., blinking).

clearance The rate at which a substance is removed from the blood and excreted in urine; calculated as the product of urine flow rate and the U/P ratio.

cleidoic egg An egg with a shell, or membrane, in which the embryo develops in isolation from its environment, except for gas exchange, e.g., insect, reptile, and bird eggs.

closed circulation A circulatory system in which the entire vascular circuit is a complete system of tubes and the vascular fluid is unable to mix with the interstitial fluid.

coagulation The clotting of a fluid, such as blood, by chemical reactions of specific coagulants, which commonly are proteins.

cochlea The coiled structure of the vertebrate inner ear that contains the sensory cells for hearing (basilar papilla, or organ of Corti).

cocoon The protective covering of many eggs, larvae, pupae, or adult animals, e.g., egg sac of oligochete worms, leeches, and spiders; pupal silk cocoon of some insects; mucous covering of estivating earthworms and lungfish; multi-layered epithelial covering of estivating frogs.

coefficient of gut differentiation (COD) An index of gut structure, equal to the ratio of stomach + large intestine to small intestine (either mass, volume, or surface area).

coelom A body cavity; the true coleom, in contrast to a pseudocoelom, forms as a cavity in the mesoderm, and all organs that lie within the true coelom are covered by a layer of peritoneum.

coelomoduct A channel leading from the coelom to the exterior; a common type of excretory organ in invertebrates and some vertebrates (cf. nephridium and nephron).

coelomostome The ciliated opening of the coelomoduct into the coelom.

collagen An important connective tissue protein, e.g., in vertebrate tendon, the organic matrix of bone, and fish scales.

collagenase A protease enzyme that hydrolyzes collagen; it has a high optimal pH (about 10 to 11).

collecting duct The terminal, and fairly straight, portion of nephrons that conveys urine to the kidney pelvis; an important site of ion reabsorption and osmotic withdrawal of water in mammal and some bird nephrons.

colligative properties The important physical characteristics of a solution that are proportional to the osmolal concentration, i.e., freezing point depression, boiling point elevation, water vapor pressure, and osmotic pressure.

colloid Fine particles (generally 1 to 500 nm dia) that are suspended in solution and remain in suspension by virtue of their large size and electric charge; particles that will not pass through the cell membrane.

colloid osmotic pressure The osmotic pressure of body fluids (or blood) due to the solutes that are not permeable to the cell membrane (or capillaries); primarily due to protein.

commissure Tract of nerve fibers that connects two parts of the nervous system on opposite sides of the body.

compact (dense) bone Solid and dense bone with no apparent spaces between the lamellae and a high mechanical strength; found on the outside of bones, covering the inner cancellous (spongy) bone.

compatible solute A solute that does not interact with the surface of proteins and therefore stabilizes its structure and function, e.g., methylated amines and NH_4^+.

complement Plasma proteins that, in concert with specific antibodies, cause the inactivation or destruction of antigens; they are the primary humoral mediator of antigen-antibody responses.

compliance The increase in vascular volume per increase in vascular pressure ($\Delta V/\Delta P$) for part of a vascular system.

compound eye The complex eye of insects and some crustaceans, consisting of many ommatidial units.

conductance (electrical, g) The reciprocal of electrical resistance, i.e., the current per voltage difference; the unit is the siemen.

conductance (thermal, C) The rate of heat flux per temperature difference; for an endotherm, it can be calculated as metabolic rate/($T_b - T_a$).

conductance (vascular) The reciprocal of vascular resistance, i.e., the blood flow per blood pressure difference.

conduction A form of heat transfer in which thermal energy is transferred from one solid object to another by direct physical contact.

cone The color photoreceptor of the vertebrate eye, so called because the outer segment of the photoreceptor is tapered; the membrane lamellae are continuous with the plasmalemma.

conformation A general term describing the situation in which a physiological variable is essentially the same value as an external variable, e.g., ionoconformation, osmoconformation, thermoconformation.

contractile vacuole One or more fluid-filled cell vacuoles that rhythmically contract to expel fluid from the cell, found especially in protozoans and sponges; have primarily an osmoregulatory role.

contralateral On opposite sides of the body.

convection A form of heat transfer in which the transfer of thermal energy is accomplished by the movement of a fluid (by free or forced convection).

converse eye A type of eye in which light that enters the eye first impinges on the photoreceptive part of the retina, e.g., the cephalopod eye (cf. inverse eye).

cooling curve The curve showing a decrease in temperature of an object over time as it cools to ambient temperature; used to calculate the thermal conductance of an object.

coprophagy The ingestion of feces, or the reingestion of bacteria-laden feces by some mammals (e.g., rodents, lagomorphs).

core (of nervous system) The inner layer of the invertebrate nervous system (cf. rind).

cor frontale An accessory blood pump of many crustaceans, formed by an enlargement of an anterior artery.

corpora allata A pair of nonneural endocrine glands of some insects located posterior to the corpora cardiaca; they secrete juvenile hormone.

corpora cardiaca   A pair of neurohemal organs located immediately posterior to the brain of insects; they liberate brain neurohormones into the hemolymph.

corpuscles of Stannius   Small organs embedded within, or attached to, the kidneys of teleost fish; they secrete a hormone, hypocalcin, which increases blood Ca^{2+} by promoting gill uptake.

corpus luteum   The "yellow body" of the mammalian ovary formed by the follicle cells after the egg is ovulated; it secretes progesterone and estrogens.

cortical nephron   A type of nephron in the mammalian kidney that has its glomerulus located in the cortex and has a short loop of Henle; it does not produce a very concentrated urine.

corticosteroid   A general name for steroid hormones secreted by cortical adrenal tissue.

cost of transport (COT)   The metabolic cost of locomotion calculated as the total metabolic rate/locomotor velocity (COT_{total}) or metabolic cost of locomotion/locomotor velocity (COT_{net}).

cotransport   A type of carrier-mediated transport in which two different molecules have separate binding sites on the carrier for transport in the same direction.

coulomb (C)   A unit of electrical charge equal to the charge moved in one second by one ampere, i.e., 6.24×10^{18} charges per second.

counteracting solute   A solute that is able to reduce the destabilizing effects of perturbing solutes on protein structure, e.g., TMAO and betaine.

countercurrent exchange   The exchange of material or temperature between two fluid streams that are traveling in opposite directions.

countercurrent multiplier   The cumulative increase in a concentration difference that is established along the length of a countercurrent exchanger by active transport of solutes (e.g., the loop of Henle of the mammalian nephron) or

other means (e.g., the Root effect in the fish swim-bladder); the exchange per unit length of the countercurrent exchanger is accumulated over its entire length to establish a much greater concentration difference than can be established by a short segment of the exchanger.

counter transport   The movement of a solute against its concentration gradient when driven by the movement of a second solute down its concentration gradient.

coxal gland   A coelomoduct-type excretory gland of some arthropods; the excretory pore is on the limb coxa.

crista ampullaris   A sensory area of the ampulla of each semicircular canal of the vertebrate inner ear; it detects fluid motion in the semicircular canals, hence is a motion detector.

critical closing pressure   The pressure at which a small blood vessel will collapse because of wall tension forces.

critical pO₂ ($pO_{2,crit}$)   The minimal pO_2 at which an animal can maintain its normal O_2 consumption rate.

critical thermal increment (μ)   A coefficient that reflects the thermal dependence of a physiological rate function, e.g., heart, respiratory, or metabolic rate; it is the equivalent in biological systems of the activation energy and is similarly calculated from an Arrhenius plot.

critical thermal maximum (CT_{max})   The maximal temperature at which an animal is able to maintain its normal physiology; it varies for different physiological functions.

critical thermal minimum (CT_{min})   The minimum temperature at which an animal is able to maintain its normal physiology; it varies for different physiological functions.

crop   A diverticulum of the gut that is specialized for storage and sometimes digestion of food.

crop milk   The crop of columbiform birds (both male and female) secretes a cheesy "milk" that is fed by regurgitation to the chicks.

cross bridge   The mechanical connection between a myosin head and its actin-binding site that is responsible for acto-myosin sliding during muscle contraction.

cryptonephridial complex   A structural modification of the Malpighian tubules and rectum of some insects that is able to osmotically withdraw water from the rectal lumen and desiccate the feces; it is responsible for water vapor absorption in some insects, e.g., *Tenebrio* larvae.

cupula   The mass of gelatinous material of neuromast organs in which the kinocilia and stereocilia are embedded.

current (electrical)   The flow of charge (electrons or ions) in response to a voltage gradient. See Ohm's law.

cuttlebone   The internal calcareous skeleton of cuttlefish (*Sepia*); it functions in buoyancy regulation.

cyclic adenosine-monophosphate (cAMP)   A cyclic nucleotide, adenosine 3,5-cyclic monophosphate, that is produced from ATP by the enzyme adenylcyclase; it is an ubiquitous intracellular messenger mediating the cellular response to many hormones and neurotransmitters.

cytopyge   A more-or-less permanent posterior pore of protozoans in which the residual contents of food vacuoles are discharged.

cytosine   One of the pyrimidine base constituents of DNA and RNA; $C_4H_5N_3O$.

cytoskeleton   The internal skeleton, comprised of microtubules, microfilaments, and intermediate filaments, that provides internal support and mechanical strength to animal cells.

cytostome   A more-or-less permanent anterior pore of protozoans in which food is ingested and formed into food vacuoles for digestion.

Dahlgren cell   A neuroendocrine cell found in the caudal spinal cord of elasmobranchs. See also urophysis.

Dalton's law   The total pressure of a gas is equal to the sum of the partial pressures for each individual gas.

dark current The steady sodium current flowing into the outer segment of the vertebrate photoreceptor in the dark; light diminishes the inward Na^+ current.

Davenport diagram A diagram showing the acid-base status, by relating the HCO_3^- concentration of body fluids to the pH.

dead space That part of the respiratory system in which air does not undergo gas exchange for morphological or physiological reasons.

deamination A chemical reaction in which an amino group ($-NH_2$) is removed from an amino acid to form either NH_3 or NH_4^+.

dendrite A slender process of a neuron on which other cells have synaptic contact.

depolarize The reduction of resting membrane potential towards 0 mV or the reversal to >0 mV.

depth-of-field The range of distances over which objects in the visual field are in focus.

deuterocerebrum The middle part of the insect brain containing the antennal lobes.

deuterostome Those animals in which the coelom originates as pockets of the mesoderm, the mouth is not derived from the blastopore, and embryonic cleavage is indeterminate; includes Chaetognatha, Chordata, Echinodermata, and Hemichordata; cf. protostome.

diapause A period of suspended development or growth during the egg, larval, nymph, or adult stage of insects or of the embryo of some mammals.

diaphragm A membrane separating the body cavity of animals; e.g., the dorsoventral septum of some annelids and arthropods; the transverse dorsal membrane of some insects; and the transverse, arched muscular and tendinous diaphragm of mammals.

diastole The phase of the atrial or ventricular cycle in which the myocardium is relaxed.

diffusion The transport of a solute by random, thermal motion of molecules down a concentration gradient; the rate of flux (J) depends on the concentration difference (ΔC), the pathlength for diffusion (x), the area for exchange (A), and the diffusion coefficient (D); $J = DA\Delta C/x$.

diffusion coefficient (D; $cm^2\ sec^{-1}$)-
The coefficient that relates the rate of diffusional flux (J) to the concentration difference (ΔC), pathlength (x), and area for diffusion exchange (A); $D = Jx/A\Delta C$.

diffusion permeability (P_{diff}) The coefficient for diffusion of water across a membrane.

digestible energy (DE) That part of the consumed energy (C) that is not excreted in the feces (F), i.e., $DE = C - F$.

dimer A molecule formed by the combination of two identical monomer subunits.

direct calorimetry The direct estimation of an animal's metabolism by measurement of the heat released from the animal.

discontinuous respiration A pattern of respiration by insects in which the spiracles are kept closed for long periods, and there is a period of spiracular flutter before the spiracles are opened for diffusive gas exchange; it reduces the respiratory evaporative water loss.

distensibility The relative increase in volume of part of a vascular system per increase in vascular pressure ($\Delta V/V\Delta P$).

divalent With an electric charge of 2, e.g., Ca^{2+}, SO_4^{2-}.

Donnan equilibrium The passive equilibrium across a semipermeable membrane of two ions, e.g., Na^+ and Cl^-, one of which is impermeable and distributed across the membrane in unequal concentrations; from the Nernst equation, $Na_1^+ \cdot Cl_1^- = Na_2^+ \cdot Cl_2^-$.

drag The vector of a fluid-dynamic force acting on an aero- or hydrofoil that is resolved parallel to the direction of fluid flow.

ductus arteriosus A small blood vessel, the remnant of the sixth aortic arch, that connects the pulmonary artery to the aorta allowing pulmonary blood flow to bypass the lungs; present in amphibians and fetal higher vertebrates, such as mammals.

ductus venosus A small blood vessel in fetal mammals that allows blood flow from the umbilical vein to bypass the liver.

duodenum First part of the small intestine of some higher vertebrates.

dynein An ATPase found in the axoneme of cilia and flagella that is responsible for the sliding movement of adjacent microtubules by its ratcheting movements; also present in the cytoplasm of some cells where it mediates microtubule membrane movement.

ecdysis (molt) The periodic shedding of the exoskeleton that allows body growth; occurs in nematodes, tardigrades, and arthropods.

ecdysone An ecdysis-inducing steroid hormone that is secreted by the prothoracic gland of insects.

echolocation The perception of distant objects by emission of sound and sensory detection of its echo.

ecto- Pertaining to the outside of the body, e.g., ectoderm, ectotherm.

ectotherm An animal whose body temperature is determined primarily by passive heat exchange with its environment.

egg development neurosecretory hormone (EDNH) An insect neurohormone that stimulates follicle cell development in the ovary and synthesis of ecdysone.

egg-laying hormone (ELH) A neurohormone secreted by the parietovisceral ganglion connectives of prosobranch mollusks, e.g., *Aplysia*; it induces the complete repertoire of oocyte ovulation, their transport, fertilization and packaging in an egg string, egg laying, and fixation to the substrate.

elastic modulus (Young's modulus)
An important material property indicating the elasticity of a material; equal to the slope of the relation-

ship between stress and strain, i.e., Δstress/Δstrain.

electroantennogram (EAG) Electrical recording of the response of antennal chemoreceptors to olfactory stimulation.

electrochemical potential The chemical potential of a solute due to chemical concentration and the electrical potential and solute charge.

electrocyte A modified muscle or nerve cell of electric fish; hundreds or thousands of electrocytes are responsible for generating the electric discharge.

electrogenic pump An active transport mechanism that establishes an electrical potential as it transports solutes, e.g., the Na-K ATPase exchanges 3 Na^+ for 2 K^+ thereby establishing an electrical potential; cf. electroneutral pump.

electrolyte A solute that has an ionic charge.

electron transfer system The mitochondrial cytochrome system that transfers electrons to an acceptor, elemental O_2, to form H_2O and establishes the proton gradient across the inner mitochondrial membrane.

electroneutral pump An active transport pump that maintains electroneutrality; cf. electrogenic pump.

electro-osmosis The movement of water across a membrane as a consequence of ion transport; a proposed mechanism for water vapor uptake by the fire-brat.

electroreceptor A sensory receptor that responds to changes in electric field.

electroretinogram (ERG) The electrical recording of the response of the retina to photic stimulation.

electrotonic spread The extremely rapid passive spread of an electrical potential along a conductor; the voltage declines exponentially with distance; cf. regenerative spread.

emulsify Stabilize fat as small droplets suspended in aqueous solution using an amphipathic molecule with both hydrophobic and hydrophilic parts, e.g., a detergent or bile salt.

enantiostasis The maintenance of a normal physiological function when the affect of a change of condition on the variable is counteracted by a simultaneous change of another condition; it is the maintenance of normal function without homeostasis.

encapsulation An immune response of insects to a foreign material or organism; the foreign material or organism is coated by hemocytes and a protective melanized capsule is formed around the foreign material/organism.

endo- Pertaining to the inside, e.g., endoderm, endotherm.

endopeptidase An enzyme that hydrolyzes the interior peptide bonds of proteins.

endorphin A variety of neurohormones with morphine-like effects; they naturally occur in the brain of vertebrates.

endoskeleton An internal skeleton, e.g., the cartilage or bony skeleton of vertebrates.

endostyle The ciliated ventral groove or grooves in the pharyngeal basket of tunicates, cephalochordates, and larvae of cyclostomes, that collects food-laden mucus and transports it to the esophagus.

endothelium A single layer of flattened cells that lines blood vessels and lymphatic vessels of vertebrates and some invertebrates.

endotherm An animal whose body temperature is substantially elevated above the ambient temperature by internal, metabolic heat production.

end plate The neuromuscular synapse of vertebrates.

end plate potential (EPP) The electrical potential recorded at the end plate in response to neurotransmitter release.

energy The capacity for doing work; exists in a variety of forms, e.g., chemical, potential, kinetic, pressure, and heat. The SI unit is the joule.

engram The change in structure or function of the nervous system that stores a unit of memory.

enkephalin A variety of neuropeptides that have morphine-like effects in the brain of vertebrates.

enterogastrone A hormone secreted by the duodenum in response to the presence of food; it suppresses gastric secretion and motility.

enthalpy The amount of energy liberated as heat by a reaction (at constant pressure).

entropy A measure of the thermal randomness of molecules, i.e., a quantification of the extent of molecular disorder.

enzyme A protein that catalyzes biological reactions.

epinephrine (adrenaline) A catechol hormone released from the adrenal medulla of vertebrates and from the sympathetic nerve terminal in some vertebrates.

epiphyseal complex A pair of evaginations of the midbrain of primitive vertebrates that forms a pair of dorsal "eyes"; the more anterior is the parietal (parapineal), the more posterior is the pineal. The eye structure is generally simple and does not form a complex image. Present in some cyclostomes, fish, amphibians, and reptiles; reduced to a midbrain secretory organ in some lower vertebrates and in birds and mammals.

equilibrium constant (K_{eq}) The ratio of the forward and reverse specific rate constants for a reversible reaction at equilibrium; equal to the product of the concentrations of the reaction products, divided by the product of the concentrations of the reactants.

equilibrium potential (E_{eq}) The electrical potential predicted by the Nernst equation from the transmembrane concentrations of an ion.

equilibrium relative humidity (RH_{equil}) The relative humidity at which the rate of water vapor uptake equals the rate of evaporative water loss.

equilibrium transepithelial potential (E_{tep}) The electrical potential predicted by the Nernst equation from the transepithelial concentrations of an ion.

erythrocruorin A high-molecular weight type of hemoglobin found in many invertebrates.

erythrocyte (red blood cell) Hemoglobin-containing blood cell.

esterase An enzyme that hydrolyzes simple esters and complex lipids.

estivation A period of inactivity in response to hot, dry environmental conditions; is observed in some earthworms, snails, lungfish and other fish, amphibians, reptiles, birds, and mammals.

estrogen A variety of female sex steroid hormones that are often synthesized by the ovary; they regulate many aspects of female reproduction, e.g., oogenesis, ovulation, and secondary sex characters.

eury- Pertaining to wide tolerance of variation in some physiological parameter, e.g., temperature (eurythermy) or salinity (euryhalinity); cf. steno-.

exchange diffusion A process in which the movement of one solute across a membrane increases the movement of another solute in the opposite direction; it probably reflects cotransport by the same carrier.

excitation-contraction coupling (EC coupling) The coupling of electrical excitation of the sarcolemma to the contraction of the sarcomere.

excitatory postsynaptic potential (EPSP) An electrical potential, measured at the postsynaptic membrane, that depolarizes the postsynaptic membrane potential towards threshold.

exopeptidase An enzyme that hydrolyzes the terminal peptide bonds of proteins.

exoskeleton An external skeleton, e.g., the arthropod cuticle.

exteroceptor A sensory receptor that responds to external stimuli; cf. interoceptor.

extracellular space The fluid space that is external to the cells (excluding the vascular volume if the circulation is closed).

facilitated diffusion Diffusional transport of a solute across a membrane and down its concentration gradient, which is promoted by a passive carrier system that increases membrane permeability.

facilitation An increase in the reactivity of a synapse in response to prior synaptic activity.

F-actin Filamentous actin, a polymer of numerous globular actin molecules.

factorial metabolic scope The ratio of maximal metabolic rate (MRR) to standard metabolic rate (SMR) or basal metabolic rate (BMR), i.e., MMR/SMR or MMR/BMR.

FAD See flavin adenine dinucleotide.

Fahraeus-Lindqvist effect A decrease in viscosity of blood as a consequence of its flow through small diameter tubes.

farad See capacitance.

faveoli The gas exchange surfaces in the spongy region of some reptilian lungs; cf. alveoli.

ferritin An iron storage protein.

fibrin A large plasma protein of vertebrates that is responsible for the formation of blood clots.

Fick's law of diffusion See diffusion.

field metabolic rate (FMR) The metabolic rate of a free-living animal, measured in natural conditions.

filamin domain The part of a smooth muscle cell containing actin, troponin, filamin, and dense bodies; it is responsible for the tonic maintenance of tension; cf. actomyosin domain.

filtration The process of the removal of suspended or colloidal material from a solution by passing it through a porous membrane that restricts the passage of the suspended or colloidal material.

filtration coefficient The amount of fluid that is filtered relative to the hydrostatic/colloid pressure difference across the filtration membrane.

filtration rate The rate at which fluid is filtered, particularly by excretory organs.

flagellum A long, whip-like motile cell organelle consisting of the "9 + 2" microtubular structure (axoneme); only one or a few are generally present on the cell surface; larger than a cilium.

flame bulb A hollow structure formed by one or several flame cells that has either a flagellum or a tuft of cilia that beat and propel fluid into an excretory tubule (protonephridium).

flame cell See flame bulb.

flavin adenine dinucleotide (FAD) A coenzyme that is a hydrogen acceptor for the citric acid cycle; the $FADH_2$ donates its hydrogens to the electron transfer chain, providing a sufficient proton gradient to synthesize about 2 ATP per $FADH_2$.

fluid-mosaic membrane model A modern model of the structure of a cell membrane in which proteins are an integral part of the membrane lipid bilayer.

flux A rate of transport of material, heat, or electricity.

FMRFamide A family of peptide hormones, best known in mollusks; they affect the excitability of nerve and muscle cells.

follicle stimulating hormone (FSH) A gonadotropic hormone secreted by the adenohypophysis; it stimulates the development of ovarian follicles in females and testicular spermatogenesis in males.

food chain (food web) The pattern of interrelationships of species as indicated by the transfer of material as food from plants to herbivores, carnivores, omnivores, and detritivores.

foramen of Panizza An extracardiac opening between the left and right systemic arches of crocodiles; it allows oxygenated blood from the left ventricle to pass to the body via the left systemic arch (which would otherwise convey deoxygenated blood to the body).

foramen ovale   An opening in the fetal bird and mammal heart that allows most blood that returns to the right atrium to enter the left atrium and bypass the pulmonary circuit.

forced convection   Convective heat loss in which the fluid flow is a consequence of external mechanical action.

Fourier's law   Relationship describing conductive heat transfer (Q_{cond}) to be proportional to the temperature difference (ΔT), area (A), and thermal conductance (C) and inversely proportional to distance for heat transfer (x), i.e., $Q_{cond} = CA\Delta T/x$; analogous to Fick's law of diffusion.

fovea centralis   A cup-like or pit depression in the retina containing densely packed cones.

fractal   A shape (line, surface, or solid) whose apparent length, area, or volume depends on the scale at which it is measured, e.g., a fractal line is jagged, not smooth, with an apparent length that keeps increasing with the amplification at which it is viewed, and it therefore does not have a definite length.

fractal dimension (D)   The dimension for a fractal line that relates its perceived length (L_δ) to the measuring unit (δ); $L_\delta = k\delta^{1-D}$, where k is a proportionality constant; surfaces and volumes can similarly have fractal dimensions.

free convection   Convective heat loss in which the fluid flow is a consequence of a temperature differential between an object and the fluid.

free energy   See chemical potential.

freezing point depression   The freezing temperature of a solution is lower than that of pure solvent, in proportion to the molal concentration of the solution; it is one of the colligative properties of a solution.

fusiform muscle   A spindle-shaped muscle cell that tapers at each end.

ΔG   See Gibb's free energy change.

$\Delta G°$ See Gibb's standard free energy change.

G-actin   The monomeric actin from which F-actin is polymerized.

gait   The pattern of limb movement during locomotion.

galvanotaxis   A directed movement with respect to an electric field.

ganglion   A group of nerve cell bodies that lies inside or outside of the central nervous system.

gap junction   A specialized junction between cells that provides electrical continuity, i.e., an electrical synapse.

gastrin   A protein hormone secreted mainly by the mucosa of the vertebrate stomach; it induces gastric secretion and contractions.

gastroenterohepatic cells (GEP cells)   A diverse array of protein hormone-secreting cells of the gastrointestinal tract and endocrine pancreas.

gas window   A peculiar large membranous disk on the meral leg segments of sand-bubbler crabs that functions for gas exchange.

Gay-Lussac's law   Either the pressure or volume of a gas is directly proportional to the absolute temperature, if the other is kept constant.

gel   A state of cytoplasm when it behaves as a stiff, highly viscous fluid because of microtubular and microfilament cytoskeletal structures.

generator potential   A receptor potential that depolarizes the region of a sensory neuron axon where action potentials are initiated.

Gibb's free energy   The free energy of chemicals at constant temperature and pressure. See chemical potential.

Gibb's free energy change (ΔG)   The change in Gibb's free energy; the biologically relevant free energy change for chemical reactions.

Gibb's standard free energy change ($\Delta G°$)   The change in Gibb's free

energy at standard conditions; at equilibrium, $\Delta G° = -RT \ln K_{eq}$.

gill   An evaginated structure that functions as an aquatic respiratory organ.

gill bailer (scaphognathite)   A modified second maxilla of many crustaceans that creates a water current (or air flow in some terrestrial species) through the branchial chamber and over the gills.

gill factor   The ratio of the total O_2 obtained from an air bubble that is used for aquatic respiration by certain insects to the initial volume of O_2 in the bubble; it is about 8.3.

gill filament (primary filament)   Long, plate-like structures extending from the gill arches; they contain the vascular supply and support the secondary gill lamellae.

gill lamella (secondary lamella)   Thin, plate-like structure arranged on the top and bottom of gill filaments; it is the gas exchange surface of the gill.

gizzard   A thick and muscular portion of the anterior digestive tract of many animals that is used for mechanical breakdown of the food.

glia   Supporting cells for neurons in the central nervous system; they are nonexcitable cells.

glomerular   Referring to the type of vertebrate kidney that has nephrons that have a glomerulus; urine is formed primarily by ultrafiltration.

glomerular filtration rate (GFR)   The rate at which glomerular fluid is filtered by the vertebrate kidney nephrons.

glomerulus   A round mass of blood vessels or nerves; in particular, the mass of capillaries enclosed by the Bowman's capsule of the vertebrate nephron.

glucagon   A protein hormone secreted by the endocrine pancreas cells; it increases the blood glucose level.

glucocorticoid   A class of steroid hormones that are secreted by the adrenal cortical tissue and affect particularly carbohydrate metabolism.

glucogenesis The synthesis of glucose from other carbohydrates.

gluconeogenesis The synthesis of glucose from noncarbohydrate precursors, e.g., amino acids or lipids.

glycogen An animal starch, or polysaccharide, comprised of many glucose subunits polymerized by α-bonds into linear and branching chains; commonly stored in liver and muscle cells.

glycogenesis The synthesis of glycogen.

glycogenolysis The breakdown of glycogen into glucose.

glycolysis (Embden-Meyerhof pathway) The metabolic pathway for catabolism of glucose to pyruvate or lactate with the production of ATP.

G protein A membrane-bound protein that transduces a signal (e.g., hormone binding to its membrane-bound receptor or light reception) into an intracellular signal (e.g., adenylcyclase activation and cAMP synthesis, or phosphodiesterase activity); the G protein is active when its α-subunit is complexed with GTP. G proteins can have excitatory (G_s) or inhibitory (G_i) effects.

gray matter That tissue of the central nervous system that consists of cell bodies, nonmyelinated axons, and glia; cf. white matter.

growth hormone (GH) A protein hormone secreted by the anterior pituitary that affects carbohydrate, protein, and fat metabolism and promotes tissue growth.

guanine One of the purine base constituents of DNA and RNA; excretory nitrogen waste product in some terrestrial invertebrates; $C_5H_5N_5O$.

guanotely The pattern of excretion in which guanine is the primary nitrogenous waste product.

gular flutter Rapid movement of the throat region that promotes evaporative water loss, especially in birds.

gustation The chemoreceptor sense of taste of molecules in solution; cf. olfaction.

habituation The progressive decrease in responsiveness to repetitive stimulation.

hair cell The mechanosensory cell of the vertebrate acoustico-lateralis sensory system; each cell has numerous stereocilia (which are microvilli) and usually a single longer kinocilium (a true cilium).

Haldane effect The total CO_2 content of blood is lower if the hemoglobin is oxygenated rather than deoxygenated (at constant pCO_2); it facilitates CO_2 loading from the tissues and CO_2 excretion from the lungs.

Hamburger shift See chloride shift.

heart A specialized muscular blood vessel, or organ, that pumps blood through the circulatory system.

heat transfer coefficient (h) A coefficient that relates heat transfer (J) by conduction, convection, or radiation to the difference between the temperature of an object and its environment; $J = h\Delta T$.

heliothermy A behavioral pattern of basking in sunlight to elevate and regulate body temperature; observed in many insects and reptiles.

hematocrit The percentage of blood that consists of erythrocytes.

heme An iron-protoporphyrin structure ($C_{34}H_{33}O_4N_4FeOH$) that is the O_2-binding site in the respiratory pigments, hemoglobin, myoglobin, and chlorocruorin.

hemerythrin A violet-pink respiratory pigment that consists of an Fe^{2+} and a polypeptide; found in only a few invertebrate phyla.

hemimetabolous Pattern of metamorphosis in insects in which the egg hatches into a nymph, and there is a series of immature forms that are generally similar in structure to the adults except for size and body proportions, e.g., hemipterans, orthopterans, cockroaches, dragonflies, and mayflies.

hemocoel A body cavity, derived from the blastocoel, through which the vascular fluid (hemolymph) circulates, e.g., in mollusks and arthropods.

hemocyanin A colorless-blue extracellular respiratory pigment containing a Cu^{2+} ion and a polypeptide chain that is often arranged in complex polymers of very high molecular weight; it is found primary in mollusks and arthropods.

hemocyte A blood cell.

hemoglobin A reddish-blue intracellular or extracellular respiratory pigment containing a Fe^{2+} ion, a heme group, and a polypeptide chain; may be monomeric, di-, tri-, tetra-, octa-, or polymeric; found in various invertebrates, and is the sole respiratory pigment of vertebrates (as well as myoglobin).

hemolymph The vascular fluid of mollusks and arthropods that circulates through the hemocoel.

Henderson-Hasselbalch equation The formula relating the pH of a buffer solution to its pCO_2 and bicarbonate concentration, and the pK for hydrolysis of CO_2; $pH = pK + \log_{10}[HCO_3^-]/[CO_2]$.

Henry's law The quantity of gas dissolved in solution is proportional to its partial pressure; it depends on the solubility coefficient (α).

hermaphrodite An individual animal that has functional male and female reproductive tracts, although not necessarily at the same time.

Hesse's law of sums The free energy change for a chemical reaction is independent of the particular pathway from reactants to products.

hibernation A period of inactivity that normally is induced by cold; observed in numerous ectotherms and also endotherms; often used to describe the specific physiological state of endotherms with a long-term (many days) cycle of lowered body temperature and depressed metabolic rate.

high-pass filter A filtering effect of certain sensory cells and neurons that preferentially transmits signals greater than a specific cutoff frequency.

holometabolous Pattern of metamorphosis in insects where there is a complex series of larval stages (instars) and usually a nonmotile, nonfeeding pupal stage between egg and adult stages, e.g., flies and hymenopterans.

homeo-(homio) Pertaining to a constant value, e.g., homiotherm.

homeostasis The occurrence of constant conditions; it may be the consequence of regulatory mechanisms or simply a consequence of constant environmental conditions.

hormone A blood-borne chemical that is synthesized and secreted by an endocrine (ductless) gland, distributed throughout the body by the vascular system, and elicits specific responses in certain target cells.

humoral immunity An immunity that is a consequence primarily of blood-borne factors in solution (antibodies and complement proteins) rather than blood cells.

hydraulic permeability (L_{hyd}) Coefficient for the membrane permeability to water calculated from the water flux (F_{water}), the osmotic concentration gradient ($\Delta\Pi$), and the membrane area (A); $F_{water} = L_{hyd}A\Delta\Pi$.

hydrolysis The splitting of a compound into two parts by the chemical addition of water, e.g., R—O—R' + H_2O → ROH + HOR'.

hydrophilic Attracted to water.

hydrophobic Repelled from water.

hydrostat A pressurized fluid space, or muscle mass, that acts as a skeleton.

hygroreceptor A sensory receptor that responds to humidity.

hyper- Pertaining to an elevation in value, e.g., hyperthermia, hypercalcemia; cf. hypo-.

hyperlactemic threshold The workload at which the blood lactate level increases.

hyperpolarize To increase the resting membrane potential away from 0 mV, i.e., make E_m more negative.

hypertonic A solution in which a cell, or animal, will shrink because of osmotic loss of water (if the cell or animal acts as a simple osmometer).

hypo- Pertaining to a lowered value, e.g., hypothermia, hypocalcemia; cf. hyper-.

hypocalcin A protein hormone secreted by the corpuscles of Stannius of teleost fish that increases blood Ca^{2+} by promoting gill uptake.

hypophysiotropic hormones Peptide hormones that are secreted by the hypothalamus of the vertebrate brain and control the endocrine activity of the adenohypophysis.

hypophysis The pituitary gland; a small endocrine and neuroendocrine structure located at the base of the hypothalamus and connected to it by the infundibular stalk.

hypothalamus Part of the vertebrate forebrain that forms the floor and part of the walls of the third ventricle; involved in regulation of the adenohypophysis, autonomic nervous system, emotions, body temperature, water balance, and appetite.

hypotonic A solution in which a cell, or animal, will swell because of osmotic loss of water (if the cell or animal behaves as a simple osmometer).

ideal gas law The theoretical relationship between the gas constant (R), the number of moles of a gas (n) and its pressure (P), the volume (V), and the temperature (T); $PV = nRT$.

immunity The capacity to resist injury, especially by foreign proteins and invading organisms.

indirect calorimetry The indirect estimation of an animal's metabolism, commonly by the measurement of the rate of O_2 consumption, CO_2 production, or difference between energy consumption and fecal excretion (in contrast to the direct measurement of heat production).

induced drag That portion of a drag force that an animal experiences that is a consequence of lift generation by an aerofoil or hydrofoil.

infrared Electromagnetic radiation of wavelength between 770 and 40000 nm; the frequency is less than the lower, red end of the visible spectrum.

infrasound Very low frequency sound that is not audible to most animals, e.g., 0.05 to 1 sec^{-1}.

inhibitory postsynaptic potential (IPSP) An electrical potential measured at the postsynaptic membrane that depolarizes or hyperpolarizes the postsynaptic membrane potential towards a reversal potential which prevents depolarization to threshold.

instar A stage during the larval or nymphal development of an insect, or more generally any stage between molt in nematodes, tardigrades, or arthropods.

insulation A thermal barrier that limits the rate of heat transfer; usually a still air layer or fat layer, e.g., hair of mammals, feathers of birds, thoracic scales of moths, subcutaneous fat of seals. The insulative value (I) is the reciprocal of thermal conductance, i.e., $1/C$.

insulin A protein hormone that is secreted by the endocrine pancreas of vertebrates; it lowers blood glucose by increasing cell permeability to glucose and affecting carbohydrate, lipid, and amino acid metabolism.

intermediate filaments Cytoplasmic protein fibers that are intermediate in size between microfilaments and microtubules and form part of the cytoplasmic skeleton, e.g., keratin.

intermediate pituitary (pars intermedia) A part of the pituitary gland that lies between the anterior pituitary and the posterior pituitary but is embryologically derived from the anterior pituitary; secretes primarily melanocyte-stimulating hormone (MSH).

interneuron   A neuron that transfers action potentials from one neuron to another neuron.

interoceptor   A sensory receptor that responds to internal stimulation; cf. exteroceptor.

inter-renal gland (adrenocortical tissue)   The cortical adrenal tissue type that is found intermingled with the medullary adrenal tissue or forms a separate adrenal cortex; secretes various glucocorticoid and mineralocorticoid hormones.

interstitial space   The body fluid compartment that is extracellular but is not vascular.

intracellular space   The body fluid compartment located within cells.

intrinsic viscosity ($[\eta_{int}]$)   A coefficient that relates the viscosity (η) of a protein solution to its concentration (C); $\eta = 1 + [\eta_{int}] \cdot C + [\eta_{int}]^2 \cdot C^2$.

inverse eye   A type of eye in which light first impinges on the nonphotoreceptive portion of the retina before it reaches the photoreceptor part, e.g., the vertebrate eye; cf. converse eye.

in vitro   Occurring within an artificial environment (literally, "in glass").

in vivo   Occurring in the natural environment (literally, "in life").

iodopsin   A visual pigment of cone photoreceptors for color vision.

ion   A charged solute particle, e.g., Na^+, Cl^-.

iono-   Pertaining to ions, e.g., ionoregulation, ionotransport.

ipsilateral   On the same side of the body.

islets of Langerhans   Small aggregations of endocrine cells scattered throughout the exocrine pancreatic tissues; they secrete insulin and glucagon.

isograft   A graft of tissue that is genetically identical, i.e., from the same individual.

isometric   A muscle contraction in which the muscle length remains constant but the force of contraction varies.

isotonic   A muscle contraction in which the muscle supports a constant load and is able to shorten.

isozyme   Different forms of an enzyme that have essentially the same catalytic properties.

Jacobson's organ   See vomeronasal organ.

Johnston's organ   A sensory chordotonal organ located in the second antennal segments of most insects; it functions for sound or vibrational detection.

joule (J)   The SI unit for energy; equivalent to the work done by a force of one newton acting over one meter (1 joule = 0.239 calorie).

juvenile hormone   A class of insect hormones (based on di-homosesquiterpene) that are secreted by the corpora allata and that promote the retention of juvenile characters during molting.

juxtaglomerular (JG) apparatus   A structure in the nephrons of higher vertebrates consisting of the macula densa and juxtaglomerular cells of the afferent and efferent arteriole; it is responsible for regulation of the glomerular filtration rate.

juxtamedullary nephron   A type of nephron in the mammalian kidney that has its glomerulus located in the cortex but near the cortex-medullary boundary and has a long loop of Henle; it produces a very concentrated urine.

kalkfleck ("lime-spot")   A gas-permeable region of the epiphragm of estivating snails, providing limited gas exchange.

K_{eq}   See equilibrium constant.

keratin   A type of intermediate filament; an especially insoluble protein found in the skin, hair, horns, hoofs, feathers, and scales of vertebrates.

kilocalorie (kcal, Cal)   1000 calories; see calorie, joule.

kinematic viscosity (v)   the ratio of viscosity to density (η/ρ); see Reynold's number.

kinesin   An ATPase, that is the "motor" responsible for the movement of cytoplasmic organelles along microtubules by a ratcheting action.

kinetic energy   The energy associated with a mass (m) moving at velocity v; $= 1/2mv^2$.

kinocilium   A long, true cilium (i.e., has a 9 + 2 or 9 + 0 axoneme structure) of vertebrate hair cells.

K_m   See Michaelis-Menten constant.

Krogh's diffusion coefficient (K; nmole cm^{-1} kPa^{-1} s^{-1})   A commonly used coefficient that indicates the rate of gas diffusion relative to the partial pressure gradient; it is related to the diffusion coefficient (D) by the capacitance coefficient (β); $K = \beta D$.

K_t   The Michaelis-Menten constant for epithelial transport of solutes.

labyrinth organ   A complex, air-breathing structure connected to the branchial cavity in some fish.

lacteal   The large, central lymph vessel of intestinal villi.

lagenar macula   Macula of the lagena of the vertebrate inner ear that is a probable sound vibration detector in some fish but has little if any role in higher vertebrates.

laminar flow   A type of fluid flow in which fluid particles move in streamlines, along the direction of flow, with no sideways movement; occurs at low Reynold's numbers.

lapse rate   The rate of decline in temperature with an increase in altitude; about 6° to 10° C per 1000 m.

larva   General term for any active developmental stage that is morphologically different from the adult; develops into an adult by metamorphosis.

latency   The period of delay between a stimulus and the response.

lateral line organ A series of small sense organs along the head and sides of fishes and some amphibians that respond to water currents, vibration, and pressure (part of the acoustico-lateralis system).

law of Laplace Describes the relationship between wall tension (T) and internal pressure (P) for fluid-filled chambers as a function of radius (r); $P = 4T/r$.

LD$_{50}$ The lethal dose for 50% mortality.

leptophragma A small, single cell that forms the thin portions of the Malpighian tubule epithelium in apposition to the thin portions of the perinephric membrane of the cryptonephridial system of certain insects; it is a K^+-transporting cell that establishes the osmotic concentration gradient responsible for water reabsorption from the rectal lumen.

leukocyte A white blood cell.

leverage Use of a lever and fulcrum to increase the mechanical advantage for moving large weights or to decrease the mechanical advantage for rapid movement.

lift The vector of a fluid-dynamic force acting on an aero- or hydrofoil that is resolved perpendicular to the direction of fluid flow.

ligament A collagenous connective tissue that attaches bones to other bones; a connective tissue band in the pseudocoel of some animals.

lipase An enzyme that hydrolyzes triglycerides to fatty acids and glycerol.

liver A large digestive gland or organ of many invertebrates and vertebrates.

loop of Henle The hairpin part of the nephron of mammals and some birds that is located between the proximal and distal convoluted tubules; it is responsible for the establishment of the medullary osmotic gradient.

low-pass filter A filtering effect of certain sensory cells and neurons that preferentially transmits signals less than a specific cutoff frequency.

luciferase An enzyme that catalyzes the conversion of a luciferin protein to oxy-luciferin, with the release of light (bioluminescence).

luciferin The protein substrate that is converted to oxy-luciferin by luciferase, with the release of light (bioluminescence).

lung An invaginated, thin-walled, or trabeculated structure for gas exchange by terrestrial animals.

luteinizing hormone (LH) A gonadotropic hormone secreted by the anterior pituitary; it induces ovulation in females and promotes the endocrine activity of testicular Leydig cells in males.

lymph The vascular fluid of the lymphatic system that is similar in composition to interstitial fluid and to plasma but with a lower protein concentration.

lymphatic system A system of blind-ended, thin-walled, and vein-like vessels that drains fluid from the interstitium and returns it to the vascular system; especially important in returning protein that leaks from the vascular system and preventing edema.

lymph heart A muscular pump that promotes flow of lymph along the large lymphatic vessels in fish, amphibians, reptiles, and some birds.

lymphocyte A type of white blood cell of vertebrates that is associated with lymph nodes and is responsible for immunity.

lysine vasopressin (LVP) A neuro-hypophyseal octapeptide hormone of a few mammals that promotes water reabsorption by the collecting ducts of nephrons and decreases urine flow rate; similar to arginine vasopressin.

macula (of inner ear) Neuromast sensory structures of the sacculus and lagena of the vertebrate inner ear.

macula densa A group of specialized sensory epithelial cells in the distal convoluted tubule of the nephron of higher vertebrates; it responds to a low Na^+ and Cl^- in the tubular fluid by stimulating the juxtaglomerular cells to secrete renin.

macula lutea (yellow spot) A spot on the vertebrate retina in which there is a high concentration of cones for maximum visual acuity.

magnetoreceptor A sensory receptor that responds to magnetic fields.

Malpighian tubule A blind-ended, sometimes branched excretory tubule that produces urine by active solute (K^+) secretion and empties into the hindgut; found in most terrestrial arthropods.

maximal metabolic rate (MMR) The highest metabolic rate of an animal, often occurring during rapid locomotion or extreme cold stress.

mechanical advantage (MA) The ratio of the force that is exerted by a lever (F_2, at a distance d_2 from the fulcrum) to the force that is exerted on the lever (F_1, at a distance d_1 from the fulcrum); MA $= F_2/F_1$ ($= d_2/d_1$ in theory, but $<d_2/d_1$ in practice).

mechanoreceptor A sensory receptor that responds to mechanical displacement.

median eye The middle, third eye of some lower vertebrates; see epiphyseal complex.

mediated transport Transport of a solute across a cell membrane that requires a specific carrier; can be active or passive transport.

medulla (medulla oblongata) That part of the brain stem that connects the spinal cord to the higher brain centers in vertebrates; it contains centers for control of heart rate, respiration, and blood pressure.

meiosis Division of a cell that produces four haploid daughter cells; occurs during gamete formation.

melanin A dark brown/black pigment usually found in melanocytes and melanophores.

melanin-concentrating hormone (MCH)   A hypothalmic heptadeca-peptide that is released from the posterior pituitary in teleost fish; it causes melanin aggregation in melanophores.

melanocyte   A cell containing melanin; common in higher vertebrates.

melanocyte stimulating hormone (MSH)   Protein hormones that control the synthesis of melanin and the dispersion of melanin in melanophores and melanocytes.

melanophore   A melanin-containing cell that controls the aggregation/dispersion of the pigment for physiological color change; common in many invertebrates and lower vertebrates.

melatonin (5-methoxy-N-acetyltryptamine)   An indolic hormone secreted by the pineal gland that promotes pigment aggregation in melanocytes in lower vertebrates and may inhibit reproductive development and affect thermoregulation in higher vertebrates.

memory   Ability to store and recall past events.

mesencephalon   The midbrain of vertebrates.

mesonephros   The "second kidney" of vertebrates that develops as segmental posterior extensions of the pronephric kidney and drains its urine into the archinephric duct; it persists as a renal organ in adult fish and amphibians and is present in embryonic reptiles, birds, and mammals.

metabolic acidosis   A condition of low body fluid pH resulting from metabolic production of H^+ (other than from CO_2 hydration) and insufficient H^+ excretion or excessive OH^- excretion.

metabolic alkalosis   A condition of high body fluid pH resulting from gastrointestinal absorption of OH^- or metabolic production of OH^- or excessive H^+ excretion.

metabolic "grade"   The grouping of animals into three general categories based on their level of metabolism: as unicells, ectothermic metazoans, and endothermic metazoans.

metabolic water   Water that is chemically synthesized by cellular metabolism, e.g., $C_6H_{12}O_6 + 6 O_2 \rightarrow 6 CO_2 + 6 H_2O$.

metabolism   The sum of all physical and chemical reactions that are used by animals for anabolic synthesis of macromolecules for cell assembly and catabolic degradation of macromolecules for energy production.

metabolizable energy (ME)   That part of the consumed energy (C) that is made available to cellular metabolism and is not excreted as feces (F) or urine (U), i.e., ME = $C - F - U$.

metachronism   Pattern of coordinated, wave-like beating of fields of cilia.

metamorphosis   An abrupt transition from one developmental stage to another, e.g., in insects and amphibians.

metanephridium   An excretory tubule that develops from the exterior towards the coelom where it has a ciliated nephridiostome that collects coelomic fluid.

metanephros   The "third kidney" that develops from the posterior, lumbar mass of nephrogenic tissue and forms the adult kidney of amniote vertebrates; urine drains to the bladder via the ureter.

metazoan   A multicellular animal.

metencephalon   The hindpart of the vertebrate brain.

micelle   A small, water-soluble particle of digested fat, present in the small intestine.

Michaelis-Menten constant (K_m)   The substrate concentration at which the reaction velocity of an enzyme-mediated reaction is 1/2 of the maximal velocity.

microfilament   Contractile proteins (actin) that are present in the cell cytoplasm and are involved with cell motility.

microtubule   Cylindrical arrangements of tubulin protein filaments that are present in cells and are involved with the cytoskeleton and cell motility.

microvilli (brush border)   Microscopic, finger-like projections of the cell surface that dramatically increase the surface area for the exchange of solutes and water.

mineralocorticoid   A class of steroid hormones that are secreted by the adrenal cortical tissue and affect electrolyte balance, e.g., aldosterone.

miniature end plate potential (mEPP)   A small depolarization (1 mV or less) of the motor end plate due to spontaneous quantal release of neurotransmitter from synaptic vesicles.

minute volume (V_I)   The volume of air inspired into the respiratory tract, per minute.

mitochondrion   A spherical or elongate cell organelle that contains the electron transfer chain and other enzymes; the site of oxidative phosphorylation and ATP synthesis.

mitosis   Division of a cell in which two diploid daughter cells are formed; consists of prophase, metaphase, anaphase, and telophase.

mixonephridium   A type of nephromixium in which the coelomoduct is fused to the inner end of a nephridium.

molality   Molal concentration of a solute in solution; calculated as moles of solute per kilogram of solvent.

molarity   molar concentration of a solute in solution; calculated as moles of solute per liter of solution.

mole   The number of molecules in a gram molecular weight of the compound, i.e., 6.023×10^{23} molecules.

molt   See ecdysis.

molting hormone (β-ecdysone, 20-hydroxyecdysone)   An active ecdysis-inducing steroid hormone that is secreted as α-ecdysone by the Y-organ of crustaceans.

molt-inhibiting hormone (MIH) A peptide neurohormone released from the sinus gland of crustaceans; it inhibits the release of molting hormone from the Y-organ.

monomer A subunit that is combined to form multiple units, e.g., dimers, trimers, tetramers, polymers.

monovalent Having a single ionic charge.

morphological color change A slow color change due to a change in the amount of pigment in pigment cells; cf. physiological color change.

motoneuron A neuron that innervates a muscle cell.

motor unit The group of muscle cells that is innervated by a single motoneuron.

mucus A viscous secretion containing primarily the protein mucin and water that is secreted to trap food particles, or to lubricate the passage of food through the gut, or for locomotion.

Müllerian duct Paired embryonic ducts that convey eggs from the coelom to the exterior in primitive vertebrates; they develop into the oviduct, uterus, and other parts of the female reproductive tract in higher vertebrates.

muscarine A chemical extracted from the mushroom *Amanita muscaria*; it is an agonist of acetylcholine at postganglionic sympathetic synapses.

myelin A fatty sheath that forms short lengths of insulated regions around some (myelinated) vertebrate axons that are regularly interrupted by nodes of Ranvier; formed from many layers of Schwann cell membrane tightly wrapped around the axon.

Myerhoff quotient The ratio of moles of lactate, formed by anaerobic metabolism, that are reconverted to glucose, to the moles that are aerobically oxidized; it is 5.7 for optimal energetic conversion of lactate back into carbohydrate.

myogenic heart A heart in which the origin of the heart beat is an inherent property of the muscle cells; cf. neurogenic heart.

myoglobin An intracellular respiratory pigment that is a monomeric form of hemoglobin.

myosin A protein filament that forms the thick filaments of muscle cells; the myosin head forms cross bridges with actin and has ATPase activity to ratchet the head and slide the actin-myosin filaments.

NAD⁺ See nicotinamide adenine dinucleotide.

negative feedback A homeostatic regulatory system in which a deviation of the value of a physiological variable from a predetermined setpoint results in a restoring action; cf. positive feedback.

nematocyst A spherical or elongate stinging capsule found in epidermal cells (cnidoblasts) of coelenterates; a stinging thread is expelled from the capsule when activated, sometimes by stimulation of an external trigger structure (cnidocil).

neopulmo lung The "new" gas-exchange portion of some bird lungs that develops from the lateral sides of the bronchus and dorsobronchi connecting to the posterior air sac.

nephridiopore The external opening of a nephridium.

nephridiostome The internal, ciliated opening of a nephridium into the coelomic cavity.

nephridium An excretory tubule that develops from the exterior of an animal and terminates at the coelom in a ciliated funnel.

nephromixium An excretory tubule consisting of a fused nephridium and coelomoduct that has mixed excretory and reproductive roles.

nephron The functional unit of the vertebrate kidney that is derived from a coelomoduct-type excretory tubule and typically has a glomerulus for formation of primary urine by ultrafiltration.

Nernst equation The relationship between membrane potential (E_m) and the concentrations of an electrolyte on each side of a membrane (C_1, C_2), its ionic charge (z), and the absolute temperature (T); $E_m = {}^{RT}\!/_{zF} \ln C_2/C_1$.

nerve A group of axons forming a bundle within a connective tissue sheath.

nerve cell See neuron.

neurogenic heart A heart in which the origin of the heart beat is an extrinsic property (of the nervous system); cf. myogenic heart.

neurohemal site The site where a neurohormone is released from the nerve ending into the circulation, e.g., crustacean sinus gland, vertebrate neurohypophysis.

neurohormone A peptide hormone synthesized by a neuron in the nervous system and released into the circulation at a neurohemal site.

neurohypophysis (posterior pituitary, posterior lobe, pars nervosa) The posterior neural portion of the vertebrate pituitary that stores and releases hypothalmic neurosecretions; derived from, and connected to, the brain.

neuromast A group of hair cells with their kinocilia/stereocilia embedded in a gelatinous cupula; see acoustico-lateralis system.

neuromodulator A neuropeptide or neurotransmitter that modifies the response of a synapse to another neurotransmitter.

neuron A nerve cell, typically consisting of a cell body (soma), dendrites, and axon.

neurophysin A protein released with neurohypophyseal octapeptides that is formed during cleavage of the octapeptide from its precursor peptide; they may have a physiological role.

neuropile A part of the nervous system where a plexus of axons and dendrites synapse.

nicotinamide adenine dinucleotide (NAD⁺) A coenzyme that is a hydrogen acceptor for glycolysis and the citric acid cycle; the NADH/H⁺ donates its hydrogens to the electron transfer chain, providing a sufficient proton gradient to synthesize about 3 ATP per NADH/H$^+$.

nicotine A poisonous plant alkaloid; it mimics the action of acetylcholine at the neuromuscular end plate and the preganglionic parasympathetic and sympathetic synapses.

nitrogen narcosis A narcotic effect due to a high partial pressure of N_2.

node of Ranvier The regularly spaced gaps in the myelin sheath of myelinated axons; the sites for saltatory movement of action potentials.

nonshivering thermogenesis Increased heat production in response to cold stress by other than muscle contractions (shivering), e.g., brown adipose tissue.

noradrenaline See norepinephrine.

norepinephrine (noradrenaline) A catechol hormone released from the adrenal medulla of vertebrates and neurotransmitter of the sympathetic nerve terminal in many vertebrates.

nucleating agent A solute that promotes freezing and inhibits supercooling.

nucleic acid Nucleotide polymers of purine and pyrimidine bases; DNA and RNA.

nucleus (cell structure) A cell organelle containing DNA.

nucleus (neural) A group of nerve cell bodies in the central nervous system.

nymph An immature stage of development in insects and arachnids.

O_2 See oxygen.

ocellus The simple eye of arthropods.

ohm (Ω) A unit of electrical resistance, equal to the resistance of a column of mercury 1 mm^2 in cross section and 106 cm long.

Ohm's law The relationship between electrical current (I), voltage (V), and resistance (R); $I = V/R$.

olfaction The chemoreceptor sense of smell of molecules transmitted in a fluid; cf. gustation.

oligosaccharidase An enzyme that hydrolyzes oligosaccharides into smaller carbohydrate units.

oligosaccharide Polymeric carbohydrate made up of a small number of monosaccharide subunits.

ommatidium The functional visual unit of the invertebrate compound eye consisting of a cuticular lens, focusing and light-guide cone, photoreceptor and pigment cells, and neuron axon.

oocyte Female sex cell derived from an oogonium that undergoes meiosis and forms an ovum.

open circulation A circulatory system in which the vascular circuit is an incomplete system of tubes, and the vascular fluid is able to mix with the interstitial fluid.

operant conditioning A type of associative learning in which the presentation of a reward increases the frequency of a specific action (or administration of a punishment reinforces a response).

operative temperature (T_e) The effective air temperature for an animal, taking into account both conductive, convective, and radiative heat exchange.

opiate A narcotic substance derived from opium poppies.

opioid A substance that has an opiate-like effect.

optimal body temperature ($T_{b,opt}$) The body temperature at which a physiological reaction is maximized.

optimal hematocrit (H_0) The hematocrit at which blood oxygen transport is maximized.

organ of Corti The elongate, tubular hearing organ of crocodiles, birds, and mammals that develops from the lagena and basilar papilla (e.g., amphibians and reptiles); it is a straight organ in crocodiles and birds and coiled in mammals.

osmo- Pertaining to osmotic concentration or pressure, e.g., osmoregulation.

osmolality The total solute concentration calculated as osmoles of solutes per kilogram of solvent.

osmolarity The total solute concentration calculated as osmoles of solutes per liter of solution.

osmole One mole of solute particles, regardless of their specific identity (e.g., ion, lipid, amino acid, protein, etc.).

osmolyte A dissolved solute that contributes to the osmotic concentration.

osmotic concentration See osmolality, osmolarity.

osmotic permeability (P_{osm}) Coefficient for the membrane permeability to water calculated from the water flux (F_{water}), the mole fraction gradient (Δn), and the membrane area (A); $F_{water} = P_{osm}A\Delta n$.

osmotic pressure (II) The hydrostatic pressure that a concentration difference of an impermeant solute can sustain across a semipermeable membrane; $\Pi = 2262$ kPa osmol^{-1} at standard temperature.

ostium Opening into the tubular heart of arthropods.

ovary The female gonad or reproductive organ.

ovoparous Reproductive pattern in which the female lays eggs from which the young hatch.

ovoviparous Reproductive pattern in which females lay large, yolky eggs that are retained in the oviducts and hatch there; occurs in some insects, sharks, lizards, and snakes.

ovulation The release of the ovum from the ovary.

ovum Mature, but unfertilized egg cell.

oxidation A compound is oxidized by either combination with elemental O_2, dehydrogenation (removal of H$^+$ and e$^-$), or by electron transfer (removal of e$^-$).

oxygen affinity (P_{50}) A measure of the intensity of O_2 binding by respiratory pigments; the pO$_2$ at which 50% of the pigment has O_2 bound to it.

oxygen debt The additional oxygen that is consumed after a period of anaerobiosis, which is required for the aerobic removal of anaerobic end products, such as lactate.

oxygen equilibrium curve The curve relating the extent of O_2 binding by respiratory pigments to the pO_2; graph of % saturation as a function of pO_2.

oxygen toxicity The toxic effect of high pO_2.

oxytocin A neurohypophyseal octapeptide with effects on female reproduction, e.g., induces uterine contractions and milk release in mammals.

P_{50} See oxygen affinity.

pacemaker An excitable cell or group of cells that has an inherent rate of discharge that drives the discharge of other cells, e.g., the SA node.

paleopulmo lung The primary gas exchange area of the bird lung, consisting of parabronchi and air capillaries in the lung.

pancreas Exocrine and endocrine organ of vertebrates with an enzymatic role in digestion (exocrine portion) and regulation of blood glucose and carbohydrate metabolism (endocrine portion).

pancreozymin See cholecystokinin.

panting Increased respiratory rate, often at a high, resonant frequency, which increases evaporative water loss; a response to heat stress in many reptiles, mammals, and birds.

parabronchus A small, air-conducting passageway in the bird lung from which the air capillaries radiate; its airflow is unidirectional.

paracrine cell An endocrine cell that secretes a hormone that diffuses only a short distance to its target cell or has cell processes that transport the hormone and release it near the target cell.

parafollicular (C) cells Calcitonin-secreting cells of the mammalian thyroid gland.

paramyosin A special myofila-ment protein found in the catch muscle of bivalve mollusks that forms a rod-like core around which myosin filaments are organized; it is thought to stabilize the acto-myosin cross bridges during catch.

paraneuron A class of peptide-secreting cells (GEP cells, pituiticytes, pinealocytes).

parasite drag That portion of a drag force that an animal experiences, which is a consequence of fluid drag on the body.

parasympathetic nervous system A division of the autonomic system with the peripheral ganglia located close to the target organs; primarily involved with the automatic maintenance of normal internal organ function (e.g., digestion).

parathormone (parathyroid hormone) A peptide hormone secreted by the parathyroid gland that increases blood Ca^{2+} by promoting bone resorption, intestinal Ca^{2+} uptake, and reduced renal Ca^{2+} excretion.

parathyroid gland Small glands (often one or two pairs) located near, or attached to, the thyroid gland; they secrete parathormone.

parietal eye The parapineal "third" eye of reptiles located in the parietal scale; generally has a simple cornea, lens, and retina. See also pineal eye, median eye.

parthenogenesis A type of unisexual reproduction in which females lay eggs that develop without fertilization; occurs in rotifers, cladocerans, some insects, and a few lizards.

partial molar volume ($\overline{V}$) The volume of a mole of a solute or solvent (e.g., 18.05 cm^3 mol^{-1} for water at 20° C); $\overline{V}$ can be negative for some solutes when dissolved in solution, i.e., addition of the solute decreases the volume of the solution.

partial pressure That fraction of the total gas pressure that is exerted by a particular constituent.

partition coefficient (K) The ratio of the solubility of a solute in different solvents (e.g., oil and water).

pascal The SI unit for pressure (101 kPa = 1 atmosphere = 760 torr).

pecten Comb-like structure in the eye of birds and many reptiles; it has a nutritive role.

pennate muscle A muscle with a feathered, rather than a fusiform, shape; muscle fibers extend from each side of the central axis.

pepsin The main protoeolytic endopeptidase enzyme secreted by the gastric mucosa of vertebrates; has a low pH optimum; similar proteases occur in some invertebrates.

pericardium The flexible or rigid connective tissue sac that surrounds the heart.

peripheral resistance unit (PRU) A unit of vascular resistance defined as the resistance that allows a blood flow of 1 ml s^{-1} if the driving pressure is 1 torr; the resistance of the human systemic circulation is about 1 PRU.

peristalsis Rhythmic waves of muscle contraction that propel the lumen contents through tubular structures, e.g., gut, ureter, reproductive tubes.

permeability The relative ability of a solute to pass through a membrane or blood vessel wall.

permeability coefficient (P) The ratio of diffusion constant to membrane thickness for diffusional exchange across a membrane.

perturbing solute A solute that binds to the surface of proteins and destabilizes its structure, e.g., urea and arginine.

pH The negative logarithm (base 10) of the H^+ concentration; $pH = -\log_{10} [H^+]$.

phagocytosis The ingestion by a cell of food particles, micro-organisms, other cells, or foreign material by endocytosis.

phasic Transient or rapid.

phenylpressin A neurohypophyseal octapeptide hormone of a few mammals that promotes water reabsorption by the collecting ducts of nephrons and decreases urine flow rate; similar to arginine vaso-pressin.

pheromone   A specific chemical that is secreted by an animal to influence the actions (development or behavior) of a conspecific individual.

phosphagen   A compound that stores chemical energy as high-energy phosphate~phosphate bonds, e.g., ATP, phosphoarginine, and phosphocreatine.

phosphoarginine   A phosphagen, consisting of arginine with a high-energy phosphate bond, that can convert ADP to ATP; present in muscle cells of some invertebrates.

phosphocreatine   A phosphagen, consisting of creatine with a high-energy phosphate bond, that can convert ADP to ATP; present in muscle cells of vertebrates.

phosphodiesterase   A cytoplasmic enzyme that hydrolyzes cAMP into AMP.

phosphoprotein   A preactivated protein that can liberate light without being oxidized by O_2; one of the mechanisms for biolumines-cence (e.g., *Aequorin, Renilla*).

phosphorylation   The addition of a PO_4^{3-} group.

photophore   A light-emitting cell or area of bioluminescent animals.

photopic   The state of an eye when light adapted.

photoreceptor   A sensory receptor for light.

physical gill   A structure of aquatic animals, particularly insects, that uses an air bubble or air film as a respiratory organ.

physiological color change   An often rapid change in color resulting from control of pigment aggregation/dispersion in chromatophores; cf. morphological color change.

pineal eye   The main "third" eye of some vertebrates, consisting of a simple cornea, lens, and retina e.g., lamprey. See also parietal eye, median eye.

pineal organ   A light-receptive structure/endocrine gland located in the roof of the third ventricle in the vertebrate brain.

pinnate muscle   See pennate muscle.

pinocytosis   The uptake of fluid at the surface of a cell by invaginations of the plasmalemma forming, small, fluid-filled vesicles.

pituitary   See hypophysis.

pK $-\log_{10} K_{eq}$.

pK$_w$ $-\log_{10} K_{eq}$ for the dissociation of water into H^+ and OH^-.

placenta   The structure in mammals through which exchange of materials occurs between the circulatory systems of the developing fetus and its mother.

plasma   The extracellular fluid component of blood.

plasmalemma   The surface membrane of a cell.

plasticity   Mechanical behavior in which there is not a linear relationship between stress and strain and the material does not return to its original configuration when the stress is removed.

plastron   A permanent physical gill of certain aquatic insects, often consisting of a thin film of air that is held to the cuticle and supported against collapse by numerous cuticular hairs with terminal bends or barbs, or various forms of tubular or sieve-like spiracular structures; used for gas exchange between the ambient water and the air of the plastron/tracheal respiratory system. See also physical gill.

poikilo-   Pertaining to a variable condition, e.g., poikilotherm.

Poiseuille's law   The relationship for laminar flow between flow rate (V) and the pressure difference driving flow (ΔP), the vessel radius (r) and length (l) and fluid viscosity (η); $V = \Delta P \pi r^4 / 8\eta l$.

polarized light   Light is polarized (plane polarized or linearly polarized) if all of the light waves have their plane of vibration aligned.

polychromatic chromatophore   A chromatophore that contains pigments of different colors.

polymer   A molecule comprised of numerous subunits.

polyphenism   The occurrence of two or more adult body types induced by extrinsic factors, e.g., some aphids have five or more distinct adult forms, depending on various environmental factors.

polysaccharidase   An enzyme that hydrolyzes polysaccharides.

Ponder's R   The ratio of solvent volume to the total water content of a cell.

portal vessel   A blood vessel that directly connects two capillary beds, e.g., hepatic, renal, and hypophyseal portal systems.

positive feedback   A destabilizing control system in which the deviation of a variable from its setpoint causes a further destabilizing response, promoting a rapid change in the variable away from its setpoint, i.e., a vicious cycle; cf. negative feedback.

posterior pituitary   See neurohypophysis.

post-tetanic depression (PTD)   Decreased responsiveness of a synapse after tetanic stimulation, probably because of the depletion of synaptic vesicles; it occurs prior to post-tetanic potentiation.

post-tetanic potentiation (PTP)   Increased responsiveness of a synapse after tetanic stimulation, probably because of residual Ca^{2+} still present in the nerve terminal; it occurs subsequent to post-tetanic depression.

potential energy   The energy associated with gravitational height.

precocial   Not requiring such extensive parental care after birth or hatching as altricial young.

preferred body temperature ($T_{b,pref}$)   The body temperature that an animal will select, given the behavioral or physiological opportunity to do so.

proboscis   Elongate mouthparts (e.g., sucking and piercing insects) or snout (e.g., tapir, elephant, elephant shrew).

production (P)   The material or

energy devoted to tissue growth or gamete production.

profile drag The portion of a drag force that an animal experiences as a consequence of fluid drag on the wings.

progesterone A steroidal female sex hormone produced by the adrenal cortical tissue and the corpus luteum and the placenta; its role is to regulate the menstrual cycle and pregnancy in mammals; it has other reproductive roles in lower vertebrates.

prolactin An adenohypophyseal peptide hormone that has various reproductive, ionoregulatory, and growth roles in vertebrates.

pronephros The "first" kidney of vertebrates, which is the most anterior and first to develop; it contains nephrons whose urine drains via the archinephric duct; it persists in some fish and amphibian larvae as the "head" kidney.

proprioceptor A sensory receptor that responds to limb position.

prosencephalon The forebrain of vertebrates.

protandry The pattern of development of male reproductive activity before female reproductive activity in hermaphroditic animals, or the appearance of males earlier in the year than females.

protease An enzyme that hydrolyzes proteins.

protein kinase An enzyme that phosphorylates a protein to form a phosphoprotein using ATP.

prothoracic gland Diffuse or compact glands in the anterior thorax of insects that secrete ecdysone.

prothoracicotropic hormone (PTTH) A peptide neurohormone of insects; synthesized by median neurosecretory cells and released by the corpora cardiaca; it stimulates the prothoracic gland to secrete ecdysone.

protocerebrum The bilobed, anterior, and most complex part of the insect brain; comprised of the lateral optic lobes and the corpora pedunculata.

protonephridium A nephridial excretory tubule that does not connect directly with the coelom but terminates with a flame bulb.

protonephromixium An excretory tubule formed by the fusion of a coelomostome with a protonephridial canal.

protostome Those animals in which the coelom originates by a splitting of the mesoderm; the mouth is derived from the blastopore and embryonic cleavage is determinate; includes acoelomate and pseudocoelomate animals, mollusks, annelids, arthropods, lophophorate animals, and other minor phyla; cf. deuterostome.

proximal convoluted tubule The initial, coiled segment of the vertebrate nephron; it is responsible for considerable solute and water reabsorption.

pseudocoelom A body cavity that is derived from the blastocoel of the embryo; organs are free within the blastocoele (there is no peritoneum).

pseudopodium A transient projection of an amoeboid cell for locomotion or feeding.

pseudopupil The apparent "pupil" of the insect compound eye, which is actually the shadow of the observer on the eye.

pseudoruminant An animal with a ruminant-like symbiont digestive system for the digestion of cellulose.

pseudotracheae A system of branching respiratory tubes resembling the tracheal respiratory system of insects.

pulmonary diffusing capacity (PDC) The capacity of the respiratory membrane for diffusive exchange of O_2 or CO_2.

pupa A nonfeeding and usually inactive developmental stage between larva and adult in insects with complete metamorphosis.

puparium A case formed by the hardening of the next-to-last larval cuticle in which the pupa develops.

pupil The circular or slit opening at the center of the iris through which light enters the eye.

purine A class of nitrogenous compounds that are components of ATP, NAD, and nucleic acids, e.g., adenine, guanine.

purinotely The pattern of excretion in which purines are the primary nitrogenous waste product.

pyrimidine A class of nitrogenous compounds that have a single C- and N-containing ring, e.g., cytosine, thymine, and uracil.

pyrogen A substance, such as a bacterial toxin, that resets the thermostat (that regulates body temperature) to a higher value, thus elevating body temperature; it produces a fever.

Q_{10} The ratio of a physiological reaction rate at one temperature to the rate at a temperature that is 10° C lower.

quantum The concept of a single "packet" or amount of material or energy, e.g., a quantum of light or a quantum of neurotransmitter (10^3 to 10^4 neurotransmitter molecules contained in a single, synaptic vesicle).

radiation All objects emit thermal energy as electromagnetic radiation; it is an important mechanism of heat transfer.

radula The file-like, flexible, rasping tongue of mollusks.

ram ventilation The ventilation of the respiratory system by fluid flow resulting from forward locomotion.

range fractionation The division of a wide range of sensory intensity into a series of narrower ranges of intensity, each of which is detected by a different sensory cell.

rate constant See specific rate constant.

Rathke's pouch Embryological pouch from the roof of the mouth that develops into the vertebrate adenohypophysis.

receptor current The ion flow across a sensory receptor membrane that is due to sensory stimulation.

receptor potential The change in membrane potential that is caused by the flow of receptor current.

rectal gland The excretory organ of elasmobranchs and the coelacanth located near the rectum; it excretes a hyperosmotic salt solution.

rectification The preferential passage of current in one direction due to a voltage-dependent resistance; the one-way flow of electrical activity across a chemical synapse from pre- to postsynaptic membrane but not vice versa.

reflex arc A neural pathway that controls the automatic and reflex motor response to a specific sensory stimulus.

regenerative spread The spread of electrical depolarization as an action potential along a cell membrane without a decrease in voltage amplitude; the result of changes in membrane permeability to primarily Na^+; cf. electrotonic spread.

regulation The active maintenance of a physiological variable at a specific value, the setpoint, by negative feedback.

relative humidity (RH) The percent value of the water vapor pressure relative to the saturation water vapor pressure.

relative refractory period The brief period of time that must elapse after an action potential when a second action potential can occur, but at a reduced amplitude compared to a normal action potential.

renin A protoeolytic enzyme released into the blood from juxtaglomerular cells of the nephron in higher vertebrates; it converts angiotensinogen to angiotensin I.

renin-angiotensin system (RAS) An endocrine regulatory system of higher vertebrates that regulates arterial pressure via the kidney (renin release) and angiotensin, a potent vasoconstrictor.

rennin A milk-coagulating enzyme present in the gastric juices of particularly young mammals.

repolarize To return the membrane potential toward the resting value.

residual volume The volume of gas that remains in the respiratory system after expiration.

resistance (electrical; R, ohms) The hindrance to current flow; 1 ohm (Ω) is the resistance that allows a current flow of 1 ampere if the electrical potential difference is 1 volt. See ohm.

resistance (heat transfer; r, sec cm^{-1}) The hindrance to heat transfer by an insulating layer.

resistance (vascular; R) The hindrance of blood flow through a vessel; the resistance depends on the fluid viscosity (η) and vessel radius (r) and length (l); $R = 8\eta l/\pi r^4$.

resistance (water loss; R, sec cm^{-1}) The hindrance to water loss by evaporation from a surface; by analogy with electrical resistance, the resistance to water loss is equal to the water vapor concentration gradient divided by the rate of water loss.

respiratory acidosis A condition of low body fluid pH, resulting from respiratory retention of CO_2, forming excess H^+.

respiratory alkalosis A condition of high body fluid pH, resulting from excessive respiratory excretion of CO_2.

respiratory pigment An O_2-binding protein that is used to transport, transfer, or store O_2 (hemoglobin, chlorocruorin, hemerythrin, hemocyanin).

respiratory quotient (RQ) The ratio of CO_2 produced to O_2 consumed by cellular metabolism.

respiratory tree The tree-like branching structure of the highly developed respiratory system of mammals (and also some holothuroid echinoderms).

resting membrane potential (E_m) The normal membrane potential of an unstimulated cell; about -90 mV for many cells.

resting metabolic rate (RMR) The metabolic rate of an animal at rest.

rete mirabile An extensive countercurrent arrangement of small arterial and venous blood vessels.

retina The inner coating of the eye containing a photoreceptor layer, as well as nerves, pigment cells, and sometimes blood vessels.

retinular cell A photoreceptive cell of the arthropod ommatidium.

reversal potential (E_{rev}) The electric potential towards which the membrane potential will move when stimulated, and the potential at which no current will flow.

Reynold's number (R_e) A dimensionless coefficient that reflects the tendency for fluid flow to be laminar, transitional, or turbulent; it is the ratio of inertial forces (length and velocity) to viscous forces (viscosity and density). $R_e = vl/v$ where v is velocity, l is a characteristic length, and v is the kinematic viscosity.

rhabdome The long, cylindrical photoreceptive structure in the base and center of the arthropod ommatidium, composed of adjacent rhabdomeres of the sensory retinular cells.

rhabdomere The photosensory portion of a retinular cell of the arthropod ommatidium consisting of a microvillus-covered portion of the cell membrane containing rhodopsin.

rheobase The minimal electrical stimulus that will stimulate a muscle or nerve.

rheology The study of fluid flow.

rhodopsin (visual purple) A purplish photopigment containing 11-cis-retinal that is bleached to all-trans-retinal by light.

rind (of nervous system) The outer layer of the invertebrate nervous system; cf. core.

rod The black/white photoreceptor of the vertebrate eye; the photoreceptive membrane lamellae are not continuous with the plasmalemma but are isolated disks; cf. cone.

Root effect The change in the saturation O_2 carrying-capacity of hemoglobin and hemocyanin, with varying pCO_2; essential to the functioning of the fish swimbladder.

rumen The first chamber of the stomach of ruminant mammals, e.g., a cow; it is a large fermentation chamber in which symbiotic microbes digest the food, including cellulose.

ruminant A mammal with a multicompartment stomach (rumen, reticulum, omasum, abomasum) for symbiotic cellulose digestion; mammals of the suborder Ruminantia, of the Order Artiodactyla (bovids, cervids, pronghorn, giraffes, and tragulid deer).

saccular macula Macula of the sacculus of the inner ear, which has neuromast cells with calcium carbonate otoliths, that provide a sense of gravity.

salinity The salt content of water.

saliva A secretion of the salivary glands.

salt gland An orbital or lingual gland of many reptiles and birds that secretes a hyperosmotic NaCl or KCl secretion; an important osmoregulatory organ, especially for marine species.

sarcolemma The plasmalemma (cell membrane) of a muscle cell.

sarcomere The contractile unit of a striated muscle cell; consists of overlapping actin and myosin myofilaments within two Z bands of a myofibril.

sarcoplasm The cytoplasm of a muscle cell containing a high-density number of myofibrils as well as the sarcoplasmic reticulum, t-tubules, and mitochondria.

sarcoplasmic reticulum A smooth, membranous system in muscle cells that normally sequesters Ca^{2+} but releases Ca^{2+} to elicit a muscle contraction.

sarcotubule A transverse membrane tubule (t-tubule) that invaginates from the sarcolemma into the sarcoplasm to electrotonically propagate an electrical depolarization of the surface membrane to the triad, where the t-tubule contacts the sarcoplasmic reticulum.

scaling See allometry.

scaphognathite See gill bailer.

Schwann sheath cell A neuroglial cell that forms a myelin sheath around an axon.

scolopidium The mechanosensory unit of insect chordotonal organs.

scotopic An eye that is dark adapted.

SDA See specific dynamic effect.

SDE See specific dynamic effect.

secretagogue A substance, generally a food constituent or breakdown product, that stimulates gastrointestinal secretion.

secretin A protein hormone secreted by the small intestine in response to the presence of stomach chyme; it stimulates the pancreatic secretion of especially bicarbonate and fluid.

semipermeable membrane A membrane that is permeable to the solvent and some solutes (e.g., small, uncharged) but not to others (e.g., large, charged).

sensillum A sensory receptive structure of the insect cuticle containing sensory neurons and chitinous hairs or other mechanosensory structures.

shivering thermogenesis The increased heat production by rapid, repeated muscle contractions (shivering) in cold-stressed endotherms.

SI unit The standardized, metric international system of units (Système Internationale, adopted by the Eleventh Conference Generale des Poids et Mesures, 1960).

siemen (S) Unit of electrical conductance (1/resistance). 1 siemen is the conductance that allows a current flow of 1 ampere if the electrical potential difference is 1 volt.

sieve tracheae The tracheal system of some spiders consisting of a series of tubular tracheae opening into an atrium.

signal average A sensory technique for detection of faint signals by averaging numerous repeated stimuli to reduce noise and enhance the signal.

sinoatrial node (SA node) A group of specialized cardiac muscle cells located in the right atrium of the heart of higher vertebrates, which is the cardiac pacemaker.

sinus gland The neurohemal organ of crustaceans, located in the eyestalk, that releases neurohormones from the antennal X-organ and brain neurosecretory cells.

sinus venosus A large, thin-walled chamber that receives blood from the major veins and transmits it to the right atrium in fish, amphibians, and reptiles; it forms the SA node in birds and mammals.

skeleton A rigid, protective, and supporting structure of animals that may be internal (endoskeleton) or external (exoskeleton).

smooth muscle cell A muscle cell that lacks the organized sarcomere structure of striated muscle, but contracts using the same basic acto-myosin sliding mechanism.

sodium-potassium ATPase An ATPase enzyme that actively transports Na^+ across the cell membrane in exchange for K^+; the exchange is often 3 Na^+:2 K^+, hence is electrogenic.

sol A state of cytoplasm when it behaves as a low-viscosity fluid because of disassembly of microtubular and microfilament cytoskeletal structures.

solubility coefficient (α; mole L^{-1} kPa^{-1}) The amount of a gas that can be dissolved in solution per partial pressure.

solute A dissolved molecule.

solvent The liquid in which solutes are dissolved, e.g., water for biological systems.

solvent drag The movement of a solute in response to the bulk movement of a solvent by the transfer of momentum from the moving solvent molecules to the solute molecules.

sound pressure level (SPL, dB) The decibel intensity of sound calculated from the ratio of sound pressure (P) to the standard sound pressure level (P_0; 20 μPa); SPL = $10 \log_{10} (P/P_0)^2$.

specific dynamic action See specific dynamic effect.

specific dynamic effect (SDE) The increment in metabolic rate as a consequence of digesting and assimilating food; it is greatest for protein.

specific rate constant (k) A constant that relates the reaction rate ($J_{reaction}$) to the product of the concentrations of reactants, e.g., C_1, C_2, C_3; $J_{reaction} = k\ C_1\ C_2\ C_3$.

spinal cord The part of the vertebrate central nervous system that is enclosed by the vertebral column.

spiracle The surface opening of the tracheal system of terrestrial arthropods; the excurrent aperture of the tadpole gill chamber; the nasal opening of cetaceans.

spiracular gill The permanent physical gill of some aquatic insects that consists of highly modified tubular, perforated, or sieve-like modifications of the spiracle, for gas exchange between the ambient water and air inside the tracheal system; see also plastron.

spiral valve A twisted, valve-like structure in the truncus arteriosus of the heart of some amphibians; a spiral partition in the gut of elasmobranchs, ganoids and lungfish that slows the passage of food and provides a high-surface area for absorption.

standard chemical potential ($\mu°$) The chemical potential at standard, reference conditions, i.e., chemical activity = 1, hydrostatic pressure equal to atmospheric pressure, zero electrical potential or no charge on the chemical, zero gravitational energy, and chemical temperature equal to ambient temperature.

standard metabolic rate (SMR) The minimal metabolic rate of an ectotherm, when resting, postabsorptive, and undisturbed at a particular temperature.

standard operative temperature (T_{es}) The effective air temperature for an animal, taking into account both conductive and radiative heat exchange, with standard (free) convective conditions.

standard pressure An ambient pressure of 1 atmosphere (101 kPa = 760 torr).

standard temperature and pressure, dry (STPD) Standard conditions at which gas volumes are often expressed; 0° C, 1 atmosphere, dry air.

statocyst A gravity-sensing organ containing sensory cells and dense statoliths (calcium carbonate granules or sand accretions).

statolith A small, dense accretion of calcium carbonate, or sand grains, found in statocysts.

Stefan's law The relationship between gas diffusion and concentration difference for an aperture in a thin barrier; the rate of exchange (V) depends on the diffusion coefficient (D), the aperture radius (r), and the concentration difference (ΔC); $V = 2Dr\Delta C$.

steno- Pertaining to tolerance of a physiological condition over only a narrow range, e.g., stenothermal, stenohaline; cf. eury-.

stereocilium Nonmotile, nonciliary, microvillar projection of a hair cell.

stereopsis Binocular vision.

steroid hormone A class of hormones synthesized from cholesterol.

STPD See standard pressure and temperature, dry.

strain Mechanical deformation, measured as the relative change in length (l), i.e., $\Delta l / l$.

stress Mechanical force that causes strain.

striated muscle cell A muscle cell in which the actin and myosin filaments are arranged in regular repeating structural units, sarcomeres; the muscle therefore appears to have bands, or striations.

stroke volume The volume of blood pumped by the heart per beat; for dual-circuit hearts (bird, mammal), it is the volume pumped by one side of the heart per beat.

strong ion Ions that are essentially completely dissociated because their pK is much lower or higher than the biological pH, e.g., Na^+, Cl^-.

strong ion difference The difference in total concentration for strong ion cations (e.g., Na^+ + K^+ + Ca^{2+} + Mg^{2+}) and strong ion anions (e.g., Cl^- + SO_4^{2-}); it must equal the buffer base.

stylet The mouthparts of piercing and sucking insects.

summation The addition of rapid, successive stimuli to generate a greater response e.g., neuron depolarization and muscle cell contraction.

supercool The ability of water and many biological solutions to remain liquid below their nominal freezing point.

superposition eye A type of insect eye in which there is functional overlap of visual fields of adjacent ommatidia; the overlap can be neural or optical.

surfactant A surface-active material that reduces the surface tension; found in many lungs and swim-bladders.

sweating A thermoregulatory response of some mammals, wherein sweat glands secrete a watery fluid onto the skin surface for evaporative cooling; some amphibians have a similar mucous discharge of cutaneous mucous glands.

swim-bladder A gas-filled air bladder of teleost fish that is used to adjust buoyancy.

symbiosis An intimate, mutually beneficial relationship between two different species of organisms.

sympathetic nervous system A division of the autonomic system with the peripheral ganglia located close to the spinal cord, in the sympathetic chain ganglia, or in a few large peripheral sympathetic ganglia; primarily involved with the "fight-or-flight" response.

synapse An area of communication between two different cells (sensory, nerve, or motor) where a membrane depolarization in the presynaptic cell influences the postsynaptic cell; the synaptic communication can be electrical or chemical.

systole The phase of the atrial or ventricular cycle in which the myocardium is contracting.

tachycardia An elevation of heart rate from the normal level.

tapetum A mirrored structure that reflects light, e.g., in mollusk, arthropod, and vertebrate eyes, fish scales, butterfly wings; consists of alternating lamellae of differing refractive index, spaced at about 1/4 λ.

target cell The specific cell that responds to hormones circulating in the body fluids.

taxis A directed orientation and movement in response to a specific stimulus.

tectorial membrane A fine, gelatinous membrane overlying the hair cells of the organ of Corti in the inner ear of vertebrates.

tendon A white fibrous dense connective tissue that attaches muscle to bone.

tensile stress A mechanical force that stresses a material in tension.

testis The male gonad or reproductive organ.

testosterone A male steroid hormone or androgen.

tetany The high-frequency repetitive contraction of a muscle, producing a much greater contraction tension than a single twitch; there may be incomplete fusion of the individual contractions, or complete fusion to produce a smooth contraction.

thermal conductance (C) The rate at which an animal loses heat, per °C temperature differential between it and the environment.

thermal neutral zone (TNZ) The range of air temperatures over which an endotherm maintains its basal metabolic rate.

thermal performance breadth A range of temperatures over which a

physiological function is maintained close to its optimal value.

thermal survival zone The range of temperatures over which an animal can survive acute exposure, i.e., between CT_{min} and CT_{max}.

thermal tolerance range The broadest range of temperatures over which an animal can survive indefinitely.

thermo- Pertaining to temperature, e.g., thermoregulate, thermoconform.

thermodynamics The study of the inter-relationships between thermal energy and other forms of energy.

thermogenin A protein that modifies the permeability of the inner membrane of mitochondria from brown adipose tissue, and short circuits the H^+ gradient, allowing maximal rates of heat production without the concommitant synthesis of ATP.

thermoreceptor A sensory receptor that responds to heat or temperature.

thigmothermy Behavioral thermoregulatory pattern of some reptiles, whereby heat is gained by conduction from the environment.

thymine One of the pyrimidine base constituents of DNA and RNA; $C_5H_6N_2O$.

thyroid gland A vertebrate endocrine gland with a follicular structure that is located near the trachea; it secretes thyroxin (T_4) and tri-iodothyronine (T_3).

thyroid-stimulating hormone (TSH) An adenohypophyseal peptide hormone that stimulates the thyroid gland.

thyroxin (T_4) A tetra-iodinated, tyrosine-derived hormone secreted by the thyroid gland; it has general effects on metabolism and tissue growth.

tidal volume The volume of air inspired into the respiratory system per breath.

time constant (τ) A measure of the time required for the decay of a physiological variable or process, to 63% of the overall change; for

electrical current, it depends on the resistance and capacitance of the circuit.

tonic Steady or slow adapting.

topic organization The spatial organization of a sensory signal that is retained within, or on, the surface of the interpretive part of the central nervous system e.g., tonotopic, somatotopic.

torpor A period of inactivity that is normally induced by cold; observed in numerous ectotherms and also endotherms; often used to describe the specific physiological state of endotherms with a circadian or short-term cycle of lowered body temperature and depressed metabolic rate; see also hibernation.

torr A non-SI unit of pressure equal to 1 mm of mercury pressure head; 1 torr = 0.133 kPa.

trachea The large respiratory passage of vertebrates leading from the pharynx to the point of division of the two bronchi; large tube of the arthropod tracheal respiratory system connecting from the spiracle to the internal tracheole branches.

tracheal gill A gill with a superficial tracheal plexus for gas exchange; found in a variety of aquatic insects, primarily larvae.

tracheole A fine terminal branch of the tracheal respiratory system of arthropods.

transamination A chemical reaction in which an amino group ($-NH_2$) is transferred from an amino acid to a keto acid to form a keto acid and another amino acid.

transduction The modulation of a sensory stimulus into another form of energy in the receptor cell and ultimately a receptor potential.

transepithelial potential (E_{tep}) The electrical potential measured across an epithelium.

trehalase A disaccharase enzyme that hydrolyzes trehalose into glucose.

trehalose A disaccharide found in insects that can be an important metabolic substrate, e.g., during flight.

trichromatic color vision A system that discriminates color by having three different types of color photoreceptors, each with a different absorption spectrum.

tri-iodothyronine (T_3) A tri-iodinated, tyrosine-derived hormone secreted by the thyroid gland; it has general effects on metabolism and tissue growth.

trimethylamine oxide (TMAO) An important osmolyte and counteracting solute in elasmobranchs and the coelacanth.

tritocerebrum Small, posterior part of the insect brain; consists of two lobes that connect via the circumesophogeal connectives to the subesophogeal ganglia.

tropomyosin A muscle cell protein that lies in the grooves of F-actin and regulates acto-myosin binding under the control of troponin.

troponin A muscle cell protein that binds Ca^{2+} and causes a conformational change in tropomyosin, thereby allowing acto-myosin cross bridges to form.

trypsin A proteolytic enzyme with pH optimum about 7 to 8; secreted by the vertebrate exocrine pancreas.

t-tubule See sarcotubule.

tubulin A small globular protein that is the monomer for the synthesis of microtubular protofilaments and microtubules.

turbulent flow A type of fluid flow at high Reynold's number in which the fluid movement is chaotic, with considerable sideways eddies of fluid movement; it dissipates more energy than laminar flow.

tympanum The tympanic membrane (eardrum) of the vertebrate middle ear; the auditory membrane of certain insects.

ultimobranchial body A small, epithelial outgrowth of the gill pouch of the vertebrate embryo that is present in all vertebrates except cyclostomes (they are incorporated into the thyroid gland in mammals as C-cells); they secrete calcitonin, a hormone that controls Ca^{2+} levels in mammals.

ultrafiltration A form of filtration in which the filtering membrane has pores that are sufficiently small to remove colloidal materials and large solutes, such as proteins.

ultraviolet Electromagnetic radiation of wavelength less than 400 nm; the frequency is greater than the upper, violet end of the visible spectrum.

uracil One of the pyrimidine base constituents of DNA and RNA; $C_4H_4N_2O_2$.

urea (CON_2H_4) A nitrogenous waste product formed from arginine and the urea cycle; an important balancing osmolyte for elasmobranchs, coelacanth, and some amphibians.

urea cycle (ornithine cycle) A biochemical cycle, starting with ornithine and carbamoylphosphate, that synthesizes urea.

urease An enzyme that hydrolyzes urea into ammonia.

ureo-osmoconform An osmoregulatory strategy of marine elasmobranchs and coelacanth, and some amphibians, that uses urea as a balancing osmolyte.

ureotely The pattern of excretion in which urea is the primary nitrogenous waste product.

ureter A muscular tube that drains urine from the metanephric kidney of amniote vertebrates and some mollusks.

uric acid A nitrogenous purine ($C_5H_4O_3N_4$) that is an important, water-conserving nitrogenous waste product in some crustaceans, many insects, a few frogs, and reptiles and birds; it has a very low solubility.

uricogenesis A complex series of metabolic reactions wherein uric acid is synthesized from a variety of precursor molecules.

uricolysis A series of chemical reactions in which uric acid (primarily derived from purine metabolism) is degraded to allantoin, then urea, then ammonia.

uricotely The pattern of excretion in which uric acid is the primary nitrogenous waste product.

urophysis A caudal spinal cord neuroendocrine gland of teleost fish, which secretes urotensins.

urotensin A group of peptide hormones secreted by the urophysis.

utricular macula Macula of the utriculus of the inner ear, which has neuromast cells with calcium carbonate otoliths, to provide a sense of gravity.

valency The charge on an ion, due to missing or extra electrons.

vasomotor center A center in the medulla that controls the vasoconstriction of blood vessels.

vasopressin See arginine vasopressin.

vasotocin See arginine vasotocin.

vein A thin-walled vessel that returns blood to the heart at low pressure; often contains valves to ensure one-way flow.

vena cava A large vein that returns blood to the heart of higher vertebrates.

ventricle A muscular chamber of the heart that receives blood from an atrium; it is the primary pump that propels blood through the circulatory system.

vesicular eye A sac-shaped eye that has a transparent cornea, a lens, and a retinal photoreceptor layer.

vicinal water Intracellular water that is structured by its close proximity to membrane surfaces and does not have its full solvent capacity.

villus A small, finger-like projection of the intestinal epithelium.

viscoelastic A material whose elastic modulus depends on the rate of strain.

viscosity (η) A physical property of a fluid that determines its tendency to flow.

visible light Light with a wavelength within the range which the human eye can detect (about 400 to 770 nm).

visual acuity   The capacity of an eye to discriminate the angle between incident light paths.

vitamin   One of many organic requirements for proper metabolism, growth, reproduction, and health of an animal; is only required in trace amounts.

vitellogenesis   The deposition of yolk in an oocyte.

vitellogenin   A yolk protein, found in the blood, that is used for oocyte yolk synthesis; synthesized by the fat body of insects and the liver of vertebrates.

viviparous   Reproductive pattern in which females produce eggs that are nourished in the oviducts or uterus and are born, e.g., mammals.

volatile fatty acid (VFA)   Short-chain fatty acids of the general structure R—COOH.

volt (V)   A unit of electrical potential; that electrical potential required for a current of 1 ampere to flow through a resistance of 1 ohm.

vomeronasal organ (Jacobson's organ)   An olfactory sensory organ of many tetrapods located in grooves or blind sacs in the roof of the mouth or posterior nasal channel.

warm up   The process of elevating body temperature by physiological or behavioral means to the normal activity body temperature.

water balance response   A suite of behavioral and physiological responses of amphibians to dehydration, including water uptake across the pelvic patch skin, withdrawal of water from the bladder urine, and reduction of urine flow rate.

water potential (Ψ)   The chemical potential of water in solution or the vapor phase.

water vapor pressure (pH_2O)   The partial pressure exerted by water vapor as a normal gaseous constituent of air.

wax   An ester of fatty acids and monohydric alcohols, e.g., beeswax $C_{30}H_{61}COC_{15}H_{31}$.

Weber-Fechner relationship   The relationship between stimulus sensation (S) and stimulus intensity (I), which is generally semilogarithmic, i.e., $S \propto \log I$.

white matter   The tissue of the central nervous system of vertebrates that consists primarily of myelinated nerve fibers.

windkessel   An elastic chamber that dampens pressure fluctuations due to the pumping cycle, e.g., the aorta.

Wolffian duct   See archinephric duct.

xanthine   A nitrogenous purine ($C_5H_4O_2N_4$) that is an important, water-conserving nitrogenous waste product in some scorpions; it has a very low solubility.

x-organ   An endocrine gland of some crustaceans; closely associated with the sinus gland.

yield shear stress   The shear that is required to initiate flow of a non-Newtonian fluid, such as blood.

Young's modulus   See elastic modulus.

zeitgeber   An environmental cue that entrains an endogenous rhythm to an environmental cycle.

zymogen   The inactive form in which some enzymes, such as proteases, are secreted, e.g., pepsinogen.

zymogen cell   See chief cell.

References

Acher, R. 1974. Chemistry of the neurohypophysial hormones: An example of molecular evolution. In *The handbook of physiology*, section 7, *Endocrinology*, vol 4, *The pituitary gland and its neuroendocrine control*, part 1, edited by R. O. Greep and E. B. Astwood, 119–30. Washington: American Physiological Society.

Adams, O. R. 1976. *Lameness in horses*. Philadelphia: Lea and Febiger.

Adiyodi, R. G. 1985. Reproduction and its control. In *The biology of Crustacea*, vol. 9, *Integument, pigments, and hormonal processes*, edited by D. E. Bliss and L. H. Mantel, 147–215. New York: Academic Press.

Adrian, E. D., and Y. Zotterman. 1926. The impulse produced by sensory nerve endings. II. The response of a single organ, *J. Physiol.* 61:151–71.

Aho, A.-C., K. Donner, C. Hyden, L. O. Larsen, and T. Reuter. 1988. Low retinal noise in animals with low body temperature allows high visual sensitivity. *Nature* 334:348–50.

Aidley, D. J. 1989. *The physiology of excitable cells*. Cambridge: Cambridge University Press.

Albers, C. 1970. Acid-base balance. In *Fish physiology*, vol. 4, *The nervous system, circulation, and respiration*, edited by W. S. Hoar and D. J. Randall, 173–208. New York: Academic Press.

Alberts, B., D. Bray, J. Lewis, M. Raff, K. Roberts, and J. D. Watson. 1989. *Molecular biology of the cell*. New York: Garland Publishers.

Albuquerque, E. X., J. W. Daly, and J. E. Warnick. 1988. Macromolecular sites for specific neurotoxins and drugs on chemosensitive synapses and electrical excitation in biological membranes. In *Ion channels*, vol. 1, edited by T. Narahashi, 95–162. New York: Plenum Press.

Alexander, R. McN. 1967. *Functional design in fishes*. London: Hutchinson.

Alexander, R. McN. 1968. *Animal mechanics*. Seattle: University of Washington Press.

Alexander, R. McN. 1977. Terrestrial locomotion. In *Mechanics and energetics of animal locomotion*, edited by R. McN. Alexander and G. Goldspink, 168–203. London: Chapman and Hall.

Alexander, R. McN. 1982. *Locomotion of animals*. Glasgow: Blackie.

Alexander, R. McN., and G. Goldspink. 1977. *Mechanics and energetics of locomotion*. London: Chapman and Hall.

Alexandrowicz, J. S. 1953. Nervous organs in the pericardial cavity of the decapod Crustacea. *J. mar. biol. Assoc. U. K.* 31:563–80.

Allen, R. D., J. L. Travis, J. H. Hayden, N. S. Allen, A. C. Breuer, and J. L. Lewis. 1980. Cytoplasmic transport: Moving ultrastructural elements common to many cell types revealed by video-enhanced microscopy. *Cold Spring Harbor Symp. Quant. Biol.* 46:85–7.

Ames, P., J. Chen, C. Wolff, and J. S. Parkinson. 1988. Structure-function studies of bacterial chemosensors. *Cold Spring Harbor Symp. Quant. Biol.* 53:59–65.

Amos, W. B., L. A. Amos, and R. W. Linck. 1986. Studies of tektin filaments from flagellar microtubules by immunoelectron microscopy. *J. Cell Sci. Suppl.* 5:55–68.

Anderson, C. R., S. G. Cull-Candy, and R. Miledi. 1977. Potential-dependent transition temperature of ionic channels induced by glutamate in locust muscle. *Nature* 268:663–65.

Anderson, H. T. 1966. Physiological adaptations of diving vertebrates. *Physiol. Rev.* 46:212–43.

Anderson, J. F. 1966. The excretion of spiders. *Comp. Biochem. Physiol.* 17:973–82.

Anderson, J. F. 1970. Metabolic rate of spiders. *Comp. Biochem. Physiol.* 33:51–72.

Andrews, E. B., and P. M. Taylor. 1988. Fine structure, mechanism of heart function and haemodynamics in the prosobranch gastropod mollusc *Littorina littoria* (L.). *J. Comp. Physiol.* 158B:247–62.

Aneshansley, D. J., T. H. Jones, D. Aslop, J. Meinwald, and T. Eisner. 1983. Thermal concomitants and biochemistry of the explosive discharge mechanism of some little known bombardier beetles. *Experientia* 39:366–68.

Archer, M. 1984. The marsupial radiation. In *Vertebrate zoogeography and evolution*, edited by M. Archer and G. Clayton, 633–808. Carlisle: Hesperian Press.

Arp, A. J., J. J. Childress, and C. R. Fisher. 1984. Metabolic and blood gas transport characteristics of the hydrothermal bivalve *Calyptogena magnifica*. *Physiol. Zool.* 57:648–62.

Arp, A. J., J. J. Childress, and R. D. Vetter. 1987. The sulphide-binding protein in the blood of the vestimentiferan tube-worm *Riftia pachyptila* is the extracellular hemoglobin. *J. exp. Biol.* 128:139–58.

Aschoff, J. 1981. Der tagesgang der korpertemperatur und des energieum satzes bei saugertieren als funktion des korpergewichtes. *Sond. Z. f. Saugetierkunde* 46:201–6.

Aschoff, J. 1981. Thermal conductance in mammals and birds: Its dependence on body size and circadian phase. *Comp. Biochem. Physiol.* 69A:611–19.

Aschoff, J., and H. Pohl. 1970. Rhythmic variations in energy metabolism. *Fed. Proc.* 29:1541–52.

Atkinson, D. E., and E. Bourke. 1987. Metabolic aspects of the regulation of systemic pH. *Am. J. Physiol.* 252:F947–56.

Atwell, D. 1985. Phototransduction changes focus. *Nature* 317:14–15.

Au, W. W. L. 1980. Echolocation signals of the Atlantic bottlenose dolphin (*Tursiops truncatus*) in open waters. In *Animal sonar systems*, edited by R.-G. Busnel and J. F. Fish, 251–82. New York: Plenum Press.

Baba, S. A., and Y. Hiramoto. 1970. A quantitative analysis of ciliary movement by means of high speed micro-cinematography. *J. exp. Biol.* 52:645–90.

Babula, A., J. Bielawski, and J. Sobieski. 1978. Oxygen consumption and gill surface area in *Mesothidea entomos* (L.) (Isopoda, Crustacea). *Comp. Biochem. Physiol.* 61A:595–97.

Baker, L. A., and F. N. White. 1970. Redistribution of cardiac output in response to heating in *Iguana iguana*. *Comp. Biochem. Physiol.* 35:253–62.

Bakken, G. S. 1976. An improved method for determining thermal conductance and equilibrium body temperature with cooling curve experiments. *J. Thermal Biol.* 1:169–75.

Bakker, H. R., and S. D. Bradshaw. 1983. Renal function in the spectacled hare-wallaby, *Lagorchestes conspicallatus*: Effects of dehydration and protein deficiency. *Aust. J. Zool.* 31:101–8.

Bakker, R. T. 1971. Dinosaur physiology and the origins of mammals. *Evolution* 25:636–58.

Bakker, R. T. 1972. Anatomical and ecological evidence of endothermy in dinosaurs. *Nature* 238:81–85.

Balaban, R. S., and M. B. Burg. 1987. Osmotically-active organic solutes in the inner renal medulla. *Kidney Int.* 31:562–64.

Baldwin, E. 1949. *An introduction to comparative biochemistry*. Cambridge: Cambridge University Press.

Ballard, F. J., R. W. Hanson, and D. S. Kronfeld. 1969. Gluconeogenesis and lipogenesis in tissue from ruminant and nonruminant animals. *Fed. Proc.* 28:218–31.

Barnes, B. M. 1989. Freeze avoidance in a mammal: Body temperatures below 0° C in an Arctic hibernator. *Science* 244:1593–95.

Barnes, R. D. 1987. *Invertebrate zoology*. Philadelphia: Saunders College Publications.

Barnhart, M. C. 1983. Gas permeability of the epiphragm of a terrestrial snail, *Otala lactea*. *Physiol. Zool.* 56:436–44.

Barrack, E. R. 1987. Localization of steroid hormone receptors in the nuclear matrix. In *Steroid hormone receptors*, edited by C. R. Clark, 86–127. Chichester: Ellis Horwood.

Barrack, E. R., and D. S. Coffey. 1982. Biological properties of the nuclear matrix: Steroid hormone binding. *Recent Prog. Hormone Research* 38:133–95.

Barrett, R. 1970. The pit organs of snakes. In *Biology of the Reptilia*, vol. 2, *Morphology B.*, edited by C. Gans and T. S. Parsons, 277–300. New York: Academic Press.

Barrington, E. J. W. 1964. *Hormones and evolution*. London: English Universities Press.

Barrington, E. J. W. 1967. *Invertebrate structure and function*. London: Nelson.

Barron, D. H., and G. Meschia. 1954. A comparative study of the exchange of the respiratory gases across the placenta. *Cold Spring Harbor Symp. Quant. Biol.* 19:93–101.

Bartels, H., R. Schmelzle, and S. Ulrich. 1969. Comparative studies of the respiratory function of mammalian blood. V. Insectivora: Shrew, mole and non-hibernating and hibernating hedgehog. *Respir. Physiol.* 7:278–86.

Bartholomew, G. A. 1964. The roles of physiology and behavior in the maintenance of homeostasis in the desert environment. *Symp. Soc. Exp. Biol. XVIII Homeostasis and Feedback Mechanisms*, 7–29.

Bartholomew, G. A. 1986. The role of natural history in contemporary biology. *Bioscience* 36:324–29.

Bartholomew, G. A., and T. M. Casey. 1977. Body temperature and oxygen consumption during rest and activity in relation to body size in some tropical beetles. *J. Thermal Biol.* 2:173–76.

Bartholomew, G. A., and T. M. Casey. 1978. Oxygen consumption of moths during rest, pre-flight warm-up, and flight in relation to body size and wing morphology. *J. exp. Biol.* 76:11–25.

Bartholomew, G. A., and R. J. Epting. 1975. Rates of post-flight cooling in sphinx moths. In *Perspectives of biophysical ecology*, edited by D. M. Gates and R. B. Schmerl, 405–15. New York: Springer-Verlag.

Bartholomew, G. A., C. M. Vleck, and T. L. Bucher. 1983. Energy metabolism and nocturnal hypothermia in two tropical passerine frugivores, *Manacus vitellinus* and *Pipra mentalis*. *Physiol. Zool.* 56:370–79.

Bartlett, G. R. 1980. Phosphate compounds in vertebrate red blood cells. *Amer. Zool.* 20:103–14.

Baskin, D. G. 1976. Neurosecretion and the endocrinology of nereid polychaetes. *Am. Zool.* 16:107–24.

Batterton, C. V., and J. N. Cameron. 1978. Characteristics of resting ventilation and response to hypoxia, hypercapnia, and emersion in the blue crab *Callinectes sapidus* (Rathbun). *J. Exp. Zool.* 203:403–18.

Baudinette, R. V., and K. Schmidt-Nielsen. 1974. Energy cost of gliding in herring gulls. *Nature* 248:83–84.

Baudinette, R. V., B. J. Gannon, W. R. Runciman, and S. Wells. 1987. Do cardiorespiratory frequencies show entrainment with hopping in the tammar wallaby? *J. exp. Biol.* 129:251–63.

Baudinette, R. V., K. A. Nagle, and R. A. D. Scott. 1976.

Locomotory energetics in dasyurid marsupials. *J. Comp. Physiol.* 109:159–68.

Baust, J. G. 1986. Insect cold hardiness: Freezing tolerance and avoidance—the *Eurosta* model. In *Living in the cold: Physiological and biochemical adaptations*, edited by H. C. Heller, X. J. Musacchia, and L. C. H. Wang, 125–30. New York: Elsevier.

Bayer, R. 1968. Untersuchungen am kreislaufsystem der wanderkeuschrecke (*Locusta migratoria migratoroides* R. et F. Orthopteroidea) mit besonderer berucksichtigung des blutdruckes. *Z. vergl. Physiol.* 58:76–136.

Bayliss, L. E. 1966. *Living control systems.* London: English Universities Press.

Bayne, B. L., and C. Scullard. 1977. An apparent specific dynamic action in *Mytilus edulis* (L.) *J. mar. biol. Assoc. U. K.* 57:371–78.

Beadle, L. C. 1939. Regulation of the haemolymph in the saline water mosquito *Aedes detritus* Edw. *J. exp. Biol.* 16:346–62.

Beadle, L. C. 1957. Respiration in the African swamp worm *Alma emini* Mich. *J. exp. Biol.* 34:1–10.

Beamish, F. H., P. D. Strachan, and E. Thomas. 1978. Osmotic and ionic performance of the anadromous sea lamprey, *Petromyzon marinus*. *Comp. Biochem. Physiol.* 60A:435–43.

Becker, W., and I. Lamprecht. 1977. Microcalorimetric investigations of the host-parasite relationship between *Biomphalaria glabrata* and *Schistosoma mansoni*. *Z. Parasitenk.* 53:297–306.

Behrisch HW. 1973. Molecular mechanisms of temperature adaptation in Arctic ectotherms and heterotherms. In *Effects of temperature on ectothermic organisms: Implications and mechanisms of compensation*, edited by W. Wieser, 123–37. Berlin: Springer-Verlag.

Bekkers, J. M., and C. F. Stevens. 1990. Presynaptic mechanism for long-term potentiation in the hippocampus. *Nature* 346:724–29.

Bell, C. J. 1980. The scaling of thermal inertia of lizards. *J. exp. Biol.* 86:79–85.

Belman, B. W., and J. J. Childress. 1976. Circulatory adaptations to the oxygen minimum layer in the bathypelagic mysid *Gnathophausia ingens*. *Biol. Bull.* 150:15–37.

Belovsky, G. E. 1978. Diet optimisation in a generalist herbivore: The moose. *Theor. Population Biology* 14:105–34.

Bendall, J. R. 1968. Muscles, molecules and movement. London: Heinemann.

Benian, G. M., J. E. Kiff, N. Neckelmann, D. G. Moerman, and R. H. Waterson. 1989. Sequence of an unusually large protein implicated in regulation of myosin activity in *C. elegans*. *Nature* 342:45–50.

Bennett, M. V. L. 1966. Physiology of electrotonic junctions. *Ann. N. Y. Acad. Sci.* 137:509–39.

Bennett, M. V. L. 1967. Mechanisms of electroreception. In *Lateral line detectors*, edited by P. Cahn, 313–98. Bloomington: Indiana University Press.

Bennett, M. V. L. 1971. Electric organs. In *Fish physiology*, vol. 5, *Sensory systems and electric organs*, edited by W. S. Hoar and D. J. Randall, 347–491. New York: Academic Press.

Benton, M. J. 1979. Ectothermy and the success of dinosaurs. *Evolution* 33:983–97.

Benzanilla, F., E. Rojas, and R. E. Taylor. 1970. Sodium and potassium conductance during a membrane action potential. *J. Physiol.* 211:729–51.

Benzinger, T. H. 1964. The thermal homeostasis of man. In Homeostasis and feedback mechanisms. *Symp. Soc. Exp. Biol.* 18:49–80.

Bereiter-Hahn, J., and R. Strohmeier. 1986. Biophysical aspects of motive force generation in tissue culture cells and protozoa. In *Nature and functions of cytoskeletal proteins in motility and transport*, edited by K. E. Wohlfarth-Botterman, 1–16. Stuttgart: Gustav Fischer Verlag.

Berg, H. C., and R. A. Anderson. 1977. Bacteria swim by rotating their flagellar filaments. *Nature* 245:380–82.

Bergmiler, E., and J. Bielawski. 1970. Role of the gills in osmotic regulation in the crayfish *Astacus leptodactylus* Esch. *Comp. Biochem. Physiol.* 37:85–91.

Berk, M. L., and J. E. Heath. 1975. An analysis of behavioral thermoregulation in the lizard, *Dipsosaurus dorsalis*. *J. Thermal Biol.* 1:15–22.

Bern, H. A., and I. R. Hagadorn. 1965. Neurosecretion. In *Structure and function in the nervous system of invertebrates*, edited by T. H. Bullock and G. A. Horridge, vol. 1, 353–429. San Francisco: Freeman.

Bern, H. A., and J. Nandi. 1964. Endocrinology of poikilothermic vertebrates. In *The hormones: Physiology, chemistry and applications*, edited by G. Pincus, K. V. Thimann, and E. B. Astwood, vol. 4, 199–298. New York: Academic Press.

Bernotat-Danielowski, S., and W. Knulle. 1986. Ultrastructure of the rectal sac, the site of water vapour absorption in larvae of the Oriental rat flea *Xenopsylla cheopis*. *Tissue and Cell* 18:437–55.

Bernstein, M. H., H. L. Duman, and B. Pinshow. 1984. Extrapulmonary gas exchange enhances brain oxygen in pigeons. *Science* 226:564–66.

Berridge, M. J. 1970. Osmoregulation in terrestrial arthropods. In *Chemical zoology*, edited by M. Florkin and B. T. Scheer, vol. 5, part A, 287–320. New York: Academic Press.

Berridge, M. J., and J. L. Oschman. 1969. A structural basis for fluid secretion by Malpighian tubules. *Tissue and Cell* 1:247–72.

Berrill, N. J. 1929. Digestion in ascidians and the influence of temperature. *J. exp. Biol.* 6:275–92.

Beuchat, C. A. 1986. Phylogenetic distribution of the urinary bladder in lizards. *Copeia* 1986:512–17.

Beuchat, C. A. 1990a. Body size, medullary thickness, and urine concentrating ability in mammals. *Am. J. Physiol.* 258:R298–308.

Beuchat, C. A. 1990b. Metabolism and the scaling of urine concentrating ability in mammals: Resolution of a paradox? *J. Theor. Biol.* 143:113–22.

Beuchat, C. A., D. Vleck, and E. J. Braun. 1986. Role of the urinary bladder in osmotic regulation of neonatal lizards. *Physiol. Zool.* 59:539–51.

Biedler, L. 1975. Phylogenetic emergence of sour taste. In *Olfaction and taste, V.*, edited by D. A. Denton and J. P. Coghlan, 71–76. New York: Academic Press.

Binyon, J. 1964. On the mode of functioning of the water vascular system of *Asterias ruber* L. *J. mar. biol. Assoc. U. K.* 44:577–88.

Binyon, J. 1972. *Physiology of echinoderms.* Oxford: Pergamon Press.

Bishop, S. H., L. L. Ellis, and J. M. Burcham. 1983. Amino acid metabolism in molluscs. In *The Mollusca*, vol. 1, *Metabolic biochemistry and molecular biomechanics*, edited by P. W. Hochachka, 243–327. New York: Academic Press.

Blaylock, L. A., R. Ruibal, and K. Platt-Aloia. 1976. Skin structure and wiping behavior of phyllomedusine frogs. *Copeia* 1976:282–95.

Bliss, D. E. 1968. Transition from water to land in decapod crustaceans. *Am. Zool.* 8:355–92.

Blum, J. J. 1977. On the geometry of four dimensions and the relationship between metabolism and body mass. *J. Theor. Biol.* 64:599–601.

Bollenbacher, W. E., and N. A. Granger. 1985. Endocrinology of the prothoracic hormone. In *Comprehensive insect physiology, biochemistry and pharmacology*, vol. 7, *Endocrinology*, edited by G. A. Kerkut and L. I. Gilbert, 109–51. Oxford: Pergamon Press.

Bonaventura, C., and J. Bonaventura. 1983. Respiratory pigments: Structure and function. In *The Mollusca*, vol. 2, *Environmental biochemistry and physiology*, edited by P. W. Hochachka, 1–50. New York: Academic Press.

Bonaventura, J., and C. Bonaventura. 1985. Physiological adaptations and subunit diversity in hemocyanins. In *Respiratory pigments in animals: Relation structure-function*, edited by J. Lamy, J.-P. Truchot, and R. Gilles, 21–34. Berlin: Springer-Verlag.

Borisy, G. G., and R. R. Gould. 1977. Microtubule-organizing centers of the mitotic spindle. In *Mitosis facts and questions*, edited by M. Little, N. Paweletz, C. Petzelt, H. Ponstingl, D. Schroeter, and H.-P. Zimmerman, 78–87. Berlin: Springer-Verlag.

Boroffka, I. 1965. Elektrolyttransport im nephridium von *Lumbricus terrestris*. *Z. vergl. Physiol.* 51:25–48.

Borror, D. J., D. M. De Long, and C. A. Triplehorn. 1981. *An introduction to the study of insects.* Philadelphia: Saunders College Publishing.

Bourne, G. B., and J. R. Redmond. 1977. Haemodynamics in the pink abalone, *Haliotis corrugatis*. I. Pressure relations and pressure gradients in intact animals. *J. Exp. Zool.* 200:9–16.

Bourne, G. B., and J. R. Redmond. 1977. Haemodynamics in the pink abalone, *Haliotis corrugatis*. II. Acute blood-flow measurements and their relationship to blood pressure. *J. Exp. Zool.* 200:17–22.

Bouverot, P. 1985. *Adaptation to altitude-hypoxia in vertebrates.* Zoophysiology 16. Berlin: Springer-Verlag.

Bowerman, R. F. 1977. The control of arthropod walking. *Comp. Biochem. Physiol.* 56A:231–47.

Bowmaker, J. K. 1980. Colour vision in birds and the role of oil droplets. *TINS* 3:196–99.

Boyd, I. A., and A. R. Martin. 1956. The end-plate potential in mammalian muscle. *J. Physiol.* 100:1–63.

Bradley, T. J., K. Strange, and J. E. Phillips. 1984. Osmotic and ionic regulation in saline-water mosquito larvae. In *Osmoregulation in estuarine and marine animals.* Lecture notes on coastal and estuarine studies 9, edited by A. Pequeux, R. Gilles, and L. Bolis, 35–50. Berlin: Springer-Verlag.

Bradshaw, S. D. 1986. Hormonal mechanisms and survival in desert reptiles. In *Endocrine regulations as adaptive mechanisms to the environment*, edited by I. Assenmacher and J. Boissin, 415–40. Paris: Ed. Centre Nat. Recherche Scientifique.

Bradshaw, S. D., and G. E. Rice. 1981. The effect of pituitary and adrenal hormones on renal and postrenal reabsorption of water and electrolytes in the lizard, *Varanus gouldii* (Gray). *Gen. Comp. Endocrinol.* 44:82–93.

Braefield, A. E., and M. J. Llewellyn. 1982. *Animal energetics.* Glasgow: Blackie.

Braford, M. R. 1986. De gustibus non est disputandem: A spiral center for taste in the brain of the teleost fish, *Heterotis niloticus*. *Science* 232:489–91.

Brattstrom, B. H. 1962. Thermal control of aggregation behavior in tadpoles. *Herpetologica* 18:38–46.

Brattstrom, B. H. 1963. A preliminary review of the thermal requirements of amphibians. *Ecology* 44:238–55.

Brattstrom, B. H. 1970. Amphibia. In *The comparative physiology of thermoregulation*, vol. 1, *Invertebrates and nonmammalian vertebrates*, edited by G. C. Whittow, 135–66. New York: Academic Press.

Brauer, E. B. 1975. Osmoregulation in the freshwater sponge, *Spongilla lacustris*. *J. Exp. Zool.* 192:181–92.

Braun, E. J., and W. H. Dantzler. 1972. Function of mammalian-type and reptilian-type nephrons in the kidney of desert quail. *Am. J. Physiol.* 222:617–29.

Braun, G., G. Kummel, and J. A. Mangos. 1966. Studies on the ultrastructure and function of a primitive excretory organ, the protonephridium of the rotifer *Asplanchna priodonta*. *Pflugers Archiv.* 289:141–54.

Braun, H. A., H. Bade, and H. Hensel. 1980. Static and dynamic discharge patterns of bursting cold fibers related to hypothetical receptor mechanisms. *Pflugers Arch.* 386:1–9.

Braysher, M. L. 1976. The excretion of hyperosmotic urine and other aspects of the electrolyte balance of the lizard *Amphibolurus maculosus*. *Comp. Biochem. Physiol.* 54A:341–45.

Braysher, M. L., and B. Green. 1970. Absorption of water and electrolytes from the cloaca of an Australian lizard, *Varanus gouldii* (Gray). *Comp. Biochem. Physiol.* 35:607–14.

Breipohl, W. 1984. Thermoreceptors. In *Biology of the integument*, edited by J. Bereiter-Hahn, A. G. Matoltsy, and K. S. Richards, 561–85. Berlin: Springer-Verlag.

Brett, J. R. 1952. Temperature tolerance in young pacific salmon, genus *Oncorhynchus*. *J. Fish. Res. Brd. Canada* 9:265–323.

Brett, J. R., and D. F. Alderice. 1958. The resistance of

cultured young chum and sockeye salmon to temperatures below 0° C. *J. Fish. Res. Brd. Canada* 15:805–13.

Brett, J. R., and T. D. D. Groves. 1979. Physiological energetics. In *Fish physiology*, vol. 8, *Bioenergetics and growth*, edited by W. S. Hoar, D. J. Randall, and J. R. Brett, 280–352. New York: Academic Press.

Brix, O. 1983. Blood respiratory properties in marine gastropods. In *The Mollusca*, vol. 2, *Environmental biochemistry and physiology*, edited by P. W. Hochachka, 51–75. New York: Academic Press.

Brody, S. 1945. *Bioenergetics and growth*. New York: Reinhold.

Brokaw, C. J. 1972. Computer simulation of flagellar movement. I. Demonstration of stable bending propogation and bend initiation by the sliding filament theory. *Biophys. J.* 12:564–86.

Brooks, G. A. 1985. Anaerobic threshold: Review of the concept and directions for future research. *Med. Sci. Sports Med.* 17:22–31.

Brown, A. R., and J. R. Simpson. 1972. Water relations of sugar-tolerant yeasts: The role of intracellular polyols. *J. Gen. Microbiol.* 72:589–91.

Brown, K. T. 1968. The electroretinogram: Its components and their origins. *Vision Res.* 8:633–77.

Brownfield, M. S., and B. A. Wunder. 1976. Relative medullary area: A new structural index for estimating urinary concentrating capacity of mammals. *Comp. Biochem. Physiol.* 55A:69–75.

Browning, J. 1982. The density and dimensions of exchange vessels in *Octopus pallidus*. *J. Zool.* 196:569–79.

Brull, L., and E. Nizet. 1953. Blood and urine constituents of *Lophios piscatorius* L. *J. mar. biol. Assoc. U. K.* 32:321–28.

Bryant, S. 1959. The function of the proximal synapses of the squid stellate ganglion. *J. Gen. Physiol.* 42:609–16.

Buchanan, J. M., J. G. Flaks, S. C. Hartman, B. Levenberg, L. N. Lukens, and L. Warren. 1957. The enzymatic synthesis of inosinic acid *de novo*. In *Chemistry and biology of purines*, edited by G. E. W. Wolstenholme and C. M. O'Connor, 233–255. Boston: Little, Brown and Company.

Buchsbaum, 1948. *Animals without backbones: An introduction to the invertebrates*. Chicago: University of Chicago Press.

Buck, J. B. 1978. Functions and evolutions of bioluminescence. In *Bioluminescence in action*, edited by P. J. Herring, 419–62. London: Academic Press.

Buck, M., and D. Schlichter. 1987. Driving forces for the uphill transport of amino acids into epidermal brush border membrane vesicles of the sea anemone, *Anemonia sulcata* (Cnidaria, Anthozoa). *Comp. Biochem. Physiol.* 88A:273–79.

Buddington, R. K. 1987. Does the natural diet influence the intestine's ability to regulate glucose absorption? *J. Comp. Physiol.* 157B:677–88.

Buddington, R. K., and J. M. Diamond. 1987. Pyloric ceca of fish: A "new" absorptive organ. *Am. J. Physiol.* 252:G54–76.

Buffenstein, R., W. E. Campbell, and J. U. M. Jarvis. 1985. Identification of crystalline allantoin of the urine of African Cricetidae (Rodentia) and its role in water economy. *J. Comp. Physiol.* 155B:493–99.

Bullock, T. H., and G. A. Horridge. 1965. Structure and function in the nervous systems of invertebrates, vols. 1. and 2. San Francisco: Freeman and Company.

Bullock, T. H., R. H. Hamstra, and H. Scheich. 1972. The jamming avoidance response of high-frequency electric fish. *J. Comp. Physiol.* 77:1–22.

Bunn, H. F. 1980. Regulation of hemoglobin function in mammals. *Amer. Zool.* 20:199–211.

Burggren, W. W. 1979. Bimodal gas exchange during variation in environmental oxygen and carbon dioxide in the air breathing fish *Trichogaster trichopterus*. *J. exp. Biol.* 82:197–213.

Burggren, W. W., and N. H. West. 1982. Changing respiratory importance of gills, lungs, and skin during metamorphosis in the bullfrog *Rana catesbeiana*. *Respir. Physiol.* 47:151–64.

Burghause, F. 1975. Das Y-organ von *Oronectes limosus* (Malacostraca, Astacura). *Z. Morphol. Tiere* 80:41–57.

Burian, R. 1910. Funktion der nierenglomeruli und ultrafiltration. *Pflugers Arch. Physiol.* 136:741–60.

Burkhardt, D. 1958. Die sinnesorgane des skeletmuskels und die nervose steuerung der muskeltatigkeit. *Ergebn. Biol.* 20:27–66.

Bursell, E. 1957. Spiracular control of water loss in the tsetse fly. *Proc. R. Ent. Soc. Lond.* (A) 32:21–29.

Bursell, E. 1967. The excretion of nitrogen in insects. *Adv. Insect Physiol.* 4:33–67.

Burton, A. C. 1966. *Physiology and biophysics of the circulation*. Chicago: Year Book Medical Publishers.

Burton, R. F. 1983. Ionic regulation and water balance. In *The Mollusca*, vol. 5, *Physiology*, part 2, edited by A. S. M. Saleuddin and K. M. Wilbur, 291–352. New York: Academic Press.

Butler, D. G., H. Siwanowicz, and D. Puskas. 1989. A reevaluation of experimental evidence for the hormonal control of avian nasal salt glands. In *Progress in avian osmoregulation*, edited by M. A. Hughes and A. Chadwick, 127–41. Leeds: Leeds Philosophical and Literary Society.

Butler, J. P., H. A. Feldman, and J. J. Fredberg. 1987. Dimensional analysis does not determine a mass exponent for metabolic scaling. *Am. J. Physiol.* 253:R195–99.

Butler, P. J. 1976. Gas exchange. In *Environmental physiology of animals*, edited by J. Bligh, J. L. Cloudsley-Thompson, and A. G. MacDonald, 163–95. New York: Wiley and Sons.

Butt, A. G., and H. H. Taylor. 1986. Salt and water balance in the spider, *Porrhothele antipodiana* (Mygalomorpha: Dipluridae): Effects of feeding upon hydrated animals. *J. exp. Biol.* 125:85–106.

Buxton, P. A. 1930. Evaporation from the meal-worm (*Tenebrio*: Coleoptera) and atmospheric humidity. *Proc. Roy. Soc. Lond.* 106B:560–77.

Byrne, W. L. 1970. *Molecular approaches to learning and memory*. New York: Academic Press.

Cabanac, M. 1986. Keeping a cool head. *NIPS* 1:41–44.

Calder, W. A. 1981. Scaling of physiological processes in homeothermic animals. *Ann. Rev. Physiol.* 43:301–22.

Calder, W. A. 1984. Size, function and life history. Cambridge: Harvard University Press.

Calder, W. A., and J. R. King. 1974. Thermal and caloric relations of birds. In *Avian biology*, vol. 4, edited by D. S. Farner and J. R. King, 259–413. New York: Academic Press.

Callahan, P. S. 1965. Intermediate and far infrared sensing of nocturnal insects. Part I. Evidences for a far infrared (FIR) electromagnetic theory of communication and sensing in moths and its relationship to the limiting biosphere of the corn earworm. *Ann. Ent. Soc. Am.* 58:727–45.

Calloway, N. O. 1976. Body temperature: Thermodynamics of homeothermism. *J. Theor. Biol.* 57:331–44.

Calow, P. 1977. Conversion efficiencies in heterotrophic organisms. *Biol. Rev.* 52:385–409.

Cameron, J. N. 1975. Aerial gas exchange in the terrestrial Brachyura *Gecarcinus lateralis* and *Cardisoma guanhumi*. *Comp. Biochem. Physiol.* 52A:129–34.

Cameron, J. N., and N. Heisler. 1983. Studies of ammonia in rainbow trout: Physico-chemical parameters, acid-base behaviour and respiratory clearance. *J. exp. Biol.* 105:107–25.

Cameron, J. N., and T. A. Mecklenberg. 1973. Aerial gas exchange in the coconut crab *Birgus latro* with some notes on *Gecarcoides lalandii*. *Respir. Physiol.* 19:245–61.

Campbell, J. D. 1985. The influence of metabolic cost upon the level and precision of behavioral thermoregulation in eurythermal lizards. *Comp. Biochem. Physiol.* 81A:597–601.

Campbell, J. W., and S. H. Bishop. 1970. Nitrogen metabolism in molluscs. In *Comparative biochemistry of nitrogen fixation*, vol. 1, edited by J. W. Campbell, 103–206. New York: Academic Press.

Cannon, B., and J. Nedergaard. 1985. Biochemical mechanisms of thermogenesis. In *Circulation, respiration, and metabolism*, edited by R. Gilles, 502–18. Berlin: Springer-Verlag.

Cannon, W. B. 1929. Organization for physiological homeostasis. *Physiol. Rev.* 9:399–431.

Carey, C., and J. R. Hazel. 1989. Diurnal variation in membrane lipid composition of Sonoran desert teleosts. *J. exp. Biol.* 147:375–91.

Carey, F. G. 1982. A brain heater in swordfish. *Science* 216:1327–29.

Carey, F. G., and J. M. Teal. 1966. Heat conservation in tuna fish muscle. *Proc. Natnl. Acad. Sci. USA* 56:1464–69.

Carey, F. G., J. M. Teal, J. W. Kanwisher, and K. D. Lawson (1971) Warm-bodied fish. *Am. Zool.* 11:137–45.

Carlisle, D. B., and F. Knowles. 1959. *Endocrine control in crustaceans*. Cambridge: Cambridge University Press.

Carlson, F. D., and D. R. Wilke. 1974. *Muscle physiology*. Englewood Cliffs: Prentice-Hall.

Carroll, R. L. 1988. *Vertebrate palaeontology and evolution*. New York: Freeman and Company.

Carter, D. B. 1979. Structure and function of the subcutaneous lymph sacs in the Anura (Amphibia). *Copeia* 1979:321–27.

Casey, T. M., P. C. Withers, and K. K. Casey. 1979. Metabolic and respiratory responses of arctic mammals to ambient temperature during the summer. *Comp. Biochem. Physiol.* 64A:331–41.

Cavalier-Smith, T. 1987. Eukaryotes with no mitochondria. *Nature* 326:332–33.

Cavanaugh, C. M. 1983. Symbiotic chemoautotrophic bacteria in marine invertebrates from sulphide-rich habitats. *Science* 302:58–61.

Chabot, B. F. 1979. Metabolic and enzymatic adaptations to low temperature. In *Comparative mechanisms of cold adaptation*, edited by L. S. Underwood, L. L. Tietzen, A. B. Callahan, and G. E. Folk, 283–301. New York: Academic Press.

Chakrabarty, A., and C. L. Hew. 1988. Primary structures of the alanine-rich antifreeze polypeptides from grubby sculpin, *Myoxocephalus aenaeus*. *Can. J. Zool.* 66:403–8.

Chantler, P. D. 1983. Biochemical and structural aspects of molluscan smooth muscle. In *The Mollusca*, vol. 4, *Physiology*, part 1, edited by A. S. M. Saleuddin and K. M. Wilbur, 77–154. New York: Academic Press.

Chapman, C. B., D. Jensen, and K. Wildenthal. 1963. On circulatory control mechanisms in the Pacific hagfish. *Circ. Res.* 12:427–40.

Chapman, R. F. 1982. *The Insects*. Cambridge: Harvard University Press.

Chappell, M. A., and G. A. Bartholomew. 1981. Standard operative temperatures and thermal energetics of the antelope ground squirrel *Ammospermophilus leucurus*. *Physiol. Zool.* 54:81–93.

Charniaux-Cotton, H., and L. H. Kleinholtz. 1964. Hormones in invertebrates other than insects. In *The hormones*, edited by G. Pincus, K. V. Thimann, and E. B. Astwood, vol. 4, 135–98. New York: Academic Press.

Chester-Jones, I., P. M. Ingleton, and J. G. Phillips. 1987. *Fundamentals of comparative vertebrate endocrinology*. New York: Plenum Press.

Cheung, W. W. L., and A. T. Marshall. 1973. Studies on water and ion transport in homopteran insects: Ultrastructure and cytochemistry of the cicadoid and cercopod midgut. *Tissue and Cell* 5:651–69.

Chinkers, M., D. L. Garbers, M.-S. Chang, D. G. Lowe, H. Chin, D. V. Goeddel, and S. Schulz. 1989. A membrane form of guanylate cyclase is an atrial natriuretic peptide receptor. *Nature* 338:78–83.

Chivers, D. J., and C. M. Hladik. 1980. Morphology of the gastrointestinal tract in primates: Comparison with other mammals in relation to diet. *J. Morph.* 166:337–86.

Chosniak, I., B. G. Munck, and E. Skadhauge. 1977. Sodium chloride transport across the chicken coprodeum: Basic characteristics and dependence on the sodium chloride intake. *J. Physiol.* 271:489–504.

Clark, C. R. 1987. *Steroid hormone receptors: Their intracellular localisation.* Ellis Horwood Series in Biomedicine. England: Ellis Horwood, and Weinheim: VCH Verlagsgesellschaft.

Clark, M. E. 1985. The osmotic role of amino acids: Discovery and function. In *Transport processes: Iono- and osmoregulation*, edited by R. Gilles and M. Gilles-Baillien, 412–23. Berlin: Springer-Verlag.

Clark, R. P., and N. Toy. 1975. Natural convection around the human head. *J. Physiol.* 244:283–93.

Claussen, D. L. 1977. Thermal acclimation in ambysto-matid salamanders. *Comp. Biochem. Physiol.* 58A: 333–40.

Clegg, J. S. 1975. Metabolic consequences and the extent and disposition of the aqueous intracellular environment. *J. Exp. Zool.* 215:303–13.

Clendening, L. 1933. *Behind the doctor.* New York: Garden City Publishers.

Cleveland, L. R. 1925. Toxicity of oxygen for protozoa *in vivo* and *in vitro*: Animals defaunated without injury. *Biol. Bull.* 48:455–68.

Close, R. J. 1967. Properties of motor units in fast and slow skeletal muscles of the rat. *J. Physiol.* 193:45–55.

Cloudsley-Thompson, J. L. 1971. Terrestrial inverte-brates. In *The comparative physiology of thermoregu-lation*, vol. 1, *Invertebrates and non-mammalian verte-brates*, edited by G. C. Whittow, 15–77. New York: Academic Press.

Cloudsley-Thompson, J. L. 1975. *Terrestrial environ-ments.* London: Croom Helmes.

Cochran, D. G. 1975. Excretion in insects. In *Insect biochemistry and function*, edited by D. J. Candy and B. A. Kilby, 177–281. London: Chapman and Hall.

Cochran, D. G. 1985. Nitrogenous excretion. In *Compre-hensive insect physiology biochemistry and pharma-cology*, vol. 4, *Regulation, digestion, nutrition and excretion*, edited by G. A. Kerkut and L. I. Gilbert, 467–506. Oxford: Pergamon Press.

Cohen, M. J. 1960. The response patterns of single receptors in the crustacean statocyst. *Proc. Roy. Soc. Lond. B.* 152:30–49.

Cohen, P. 1982. The role of protein phosphorylation in neural and hormonal control of cellular activity. *Nature* 296:613–20.

Cokelet, G. R., and H. J. Meiselman. 1968. Rheological comparison of hemoglobin solutions and erythrocyte suspensions. *Science* 162:275–77.

Colhoun, E. H. 1960. Acclimatization to cold in insects. *Entomol. Exptl. Appl.* 3:27–37.

Collett, T. S., and M. F. Land. 1975. Visual control of flight behaviour in the hoverfly, *Syritta pipiens* L. *J. Comp. Physiol.* 99:1–66.

Collicutt, J. M., and P. Hochachka. 1977. The anaerobic oyster heart: Coupling of glucose and aspartate fermen-tation. *J. Comp. Physiol.* 115:147–57.

Comroe, J. H. 1965. *Physiology of respiration.* Chicago: Year Book Medical Publishers.

Conte, F. P. 1969. Salt secretion. In *Fish physiology*, vol. 1, *Excretion, ionic regulation and metabolism*, edited

by W. S. Hoar and D. J. Randall, 241–92. New York: Academic Press.

Conte, F. P. 1984. Structure and function of the crustacean larval salt gland. *Int. Rev. Cytol.* 91:45–104.

Conway, E. J. 1957. Nature and significance of concentra-tion relationships of potassium and sodium ions in skeletal muscle. *Physiol. Rev.* 37:84–132.

Coombs, J. S., J. C. Eccles, and P. Fatt. 1955. The inhibitory suppression of reflex discharges from moto-neurones. *J. Physiol.* 130:396–413.

Cooper, E. L. 1990. Immune diversity throughout the animal kingdom. *Bioscience* 40:720–22.

Cooper, S., and J. C. Eccles. 1930. The isometric re-sponses of mammalian muscles. *J. Physiol.* 69:377–85.

Copeland, E. 1968. Fine structure of salt and water uptake in the land crab *Gecarcinus lateralis. Am. Zool.* 8:417–32.

Cordier, R. 1964. Sensory cells. In *The cell*, edited by J. Brachet and A. Mirsky, 313–86. New York: Academic Press.

Cormier, M. J. 1978. Comparative biochemistry of animal systems. In *Bioluminescence in action*, edited by P. J. Herring, 75–108. New York: Academic Press.

Cornish-Bowden, A. 1979. *Fundamentals of enzyme ki-netics.* London: Butterworths.

Cossins, A. R., and C. L. Prosser. 1978. Evolutionary adaptation of membranes to temperature. *Proc. Natnl. Acad. Sci. USA* 75:2040–43.

Couillard, P., F. Pothier, and P. Mayers. 1989. The effects of vasopressin and related peptides on osmoregulation in *Amoeba proteus. Gen. Comp. Endocrinol.* 76:106–13.

Coulson, R., T. Hernandez, and J. D. Herbert. 1977. Metabolic rate, enzyme kinetics *in vivo. Comp. Bio-chem. Physiol.* 56A:251–62.

Coulter, N. A., and J. R. Pappenheimer. 1949. Develop-ment of turbulence in flowing blood. *Am. J. Physiol.* 159:401–8.

Cowles, R. B. 1940. Additional implications of reptilian sensitivity to high temperatures. *Am. Nat.* 74:542–61.

Cowles, R. B. 1962. Semantics in biothermal studies. *Science* 135:670.

Cowles, R. B., and C. M. Bogert. 1944. Preliminary study of the thermal requirements of desert reptiles. *Am. Mus. Nat. Hist.* 83:261–96.

Cragg, M. M., J. B. Balinsky, and E. Baldwin. 1961. A comparative study of nitrogen excretion in some amphibia and reptiles. *Comp. Biochem. Physiol.* 3:227–35.

Crane, R. K. 1975. 15 years of struggle with the brush border. In *Intestinal absorption and malabsorption*, edited by T. Z. Csaky, 127–41. Newlett: Raven Press.

Crawford, E. C. 1972. Brain and body temperatures in a panting lizard. *Science* 177:431–33.

Crawshaw, L. I. 1975. Twenty-four hour records of body temperature and activity in bluegill sunfish (*Lepomis macrochirus*) and brown bullheads (*Ictalurus nebulo-sus*). *Comp. Biochem. Physiol.* 51A:11–14.

Crawshaw, L. I. 1976. Effect of rapid temperature change

on mean body temperature and gill ventilation in carp. *Am. J. Physiol.* 231:837–41.

Crawshaw, L. I., H. T. Hammel, and W. F. Garey. 1973. Brainstem temperature affects gill ventilation in the California scorpionfish. *Science* 181:579–81.

Crescitelli, F. 1989. The visual pigments of a deep-water malacosteid fish. *J. mar. biol. Assoc. U. K.* 69:43–51.

Crisp. D. J., and W. H. Thorpe. 1948. The water-protecting properties of insect hairs. *Discuss. Faraday Soc.* 3:210–20.

Croghan, P. C. 1958a. The mechanism of osmotic regulation in *Artemia salina* (L.): The physiology of the branchiae. *J. exp. Biol.* 35:234–42.

Croghan, P. C. 1958b. The osmotic and ionic regulation of *Artemia salina* (L.). *J. exp. Biol.* 35:219–33.

Crompton, A. W., C. R. Taylor, and J. A. Jagger. 1978. Evolution of homeothermy in mammals. *Nature* 272:333–36.

Cronin, T. W., and N. J. Marshall. 1989. A retina with at least ten spectral types of photoreceptors in a mantis shrimp. *Nature* 339:137–40.

Cronkite, D. L., and S. K. Pierce. 1989. Free amino acids and cell volume regulation in the euryhaline ciliate *Paramecium calkinsi. J. Exp. Zool.* 251:275–84.

Crosby, E. C., T. Humphrey, and E. W. Lauer. 1962. *Correlative anatomy of the nervous system.* New York: Macmillan Company.

Crossland, C. J., and D. J. Barnes. 1974. The role of metabolic nitrogen in coral calcification. *Marine Biol.* 28:325–32.

Csapo, A. 1970. Molecular structure and function of smooth muscle. In *The structure and function of muscle,* vol. 1, edited by G. Bourne, 229–64. New York: Academic Press.

Curran, P. F. 1960. Na, Cl, and water transport by the rat ileum *in vitro. J. Gen. Physiol.* 43:1137–48.

Currey, J. D. 1970. *Animal skeletons.* New York: St Martin's Press.

Currey, J. D. 1984. *The mechanical adaptations of bones.* Princeton: Princeton University Press.

Curtis, H. 1983. *Biology.* New York: Worth Publishers.

Curtis, H. J., and K. S. Cole. 1942. Membrane resting and action potentials from the squid giant axon. *J. Cell. Comp. Physiol.* 19:135–44.

Cutright, W. J., and T. McKean. 1979. Countercurrent blood vessel arrangement in beaver (*Castor canadensis*). *J. Morph.* 161:169–76.

Dadd, R. H. 1970. Arthropod nutrition. In *Chemical zoology,* vol. 5, *Arthropoda,* part A, edited by M. Florkin and B. T. Scheer, 35–95. New York: Academic Press.

Dale, B. 1973. Blood pressure and its hydraulic functions in *Helix pomatia. J. exp. Biol.* 59:477–90.

Dales, R. P. 1978. Defense mechanisms. In *Physiology of annelids,* edited by P. J. Mill, 479–507. New York: Academic Press.

Damon, A. 1975. *Physiological anthropology.* Oxford: Oxford University Press.

Danielli, J. F., and H. Davson. 1935. A contribution to the theory of permeability of thin films. *J. Cell. Comp. Physiol.* 5:495–508.

Davenport, J., N. Blackstock, D. A. Davies, and M. Yarrington. 1987. Observations on the physiology and integumentary structure of the Antarctic pycnogonid *Decolopoda australis. J. Zool.* 211:451–65.

Davis, B. D. 1958. On the importance of being ionized. *Arch. Biochem. Biophys.* 78:497–509.

Davis, H. 1965. A model for transducer action in the cochlea. *Cold Spring Harbor Symp. Quant. Biol.* 30:181–90.

Davis, J. A. 1985. Anaerobic threshold: Review of the concept and directions for future research. *Med. Sci. Sports Med.* 17:6–18.

Davis, L. E., B. Schmidt-Nielsen, and H. Stolte. 1976. Anatomy and ultrastructure of the excretory system of the lizard, *Sceloporus cynanogenys. J. Morph.* 149:279–326.

Davis, R. E., K. D. Gailey, and K. W. Whitten. 1984. *Principles of chemistry.* Philadelphia: Saunders College Publishing.

Davis, W. L., D. B. P. Goodman, L. A. Crawford, O. J. Cooper, and J. L. Matthews. 1990. Hibernation activates glyoxylate cycle and gluconeogenesis in black bear brown adipose tissue. *Biochim. Biophys. Acta.* 1051:276–78.

Dawson, T. J., and C. R. Taylor. 1973. Energetic cost of locomotion in kangaroos. *Nature* 246:313–14.

Dawson, T. J., D. Robertshaw, and C. R. Taylor. 1974. Sweating in the kangaroo: A cooling mechanism during exercise, but not in the heat. *Am. J. Physiol.* 227:494–98.

de Zwaan, A. 1983. Carbohydrate metabolism in bivalves. In *The Mollusca,* vol. 1, *Metabolic biochemistry and molecular biomechanics,* edited by P. W. Hochachka, 138–76. New York: Academic Press.

de Zwaan, A., and V. Putzer. 1985. Metabolic adaptations of intertidal invertebrates to environmental hypoxia (a comparison of environmental anoxia to exercise anoxia). In Physiological adaptations of marine animals, edited by M. S. Laverack. *Soc. Exp. Biol. Symposium* 39:33–62.

de Zwaan, A., and G. v. d. Thillart. 1985. Low and high power output modes of anaerobic metabolism: Invertebrate and vertebrate strategies. In *Circulation, respiration, and metabolism,* edited by R. Gilles, 166–92. Berlin, Springer-Verlag.

Dean, J., D. J. Aneshansley, H. E. Edgerton, and T. Eisner. 1990. Defensive spray of the bombardier beetle: A biological pulse jet. *Science* 248:1219–21.

Dehadrai, P. V., and S. D. Tripathi. 1976. Environment and ecology of freshwater air-breathing teleosts. In *Respiration of amphibious vertebrates,* edited by G. M. Hughes, 39–72. New York: Academic Press.

Dejours, P. 1976. Water versus air as the respiratory media. In *Respiration of amphibious vertebrates,* edited by G. M. Hughes, 1–16. London: Academic Press.

Dejours, P. 1981. Principles of comparative respiratory physiology. Amsterdam: Elsevier.

Del Castillo, J., and B. Katz. 1954. Quantal components of the end-plate potential. *J. Physiol.* 124:560–73.

Del Rio, C. M., and B. R. Stevens. 1989. Physiological constraint on feeding behavior: Intestinal membrane disaccharidases of the starling. *Science* 243:794–96.

Del Rio, C. M., B. R. Stevens, D. E. Daneke, and P. T. Andreadis. 1988. Physiological correlates of preference and aversion for sugars in three species of birds. *Physiol. Zool.* 61:222–29.

Del Rio, C. M., W. H. Karasov, and D. J. Levey. 1989. Physiological basis and ecological consequences of sugar preferences in cedar waxwings. *Auk* 106:64–71.

Demski, L. S., and M. Schwanzel-Fukuda. 1987. The terminal nerve (Nervus Terminalis): Structure, function, and evolution. *Annals New York Acad. Sci.* 519.

den Bosch, H. A. J. 1983. Snout temperatures of reptiles, with special reference to changes during feeding behaviour in *Python molurus bivittatus* (Serpentes, Boidae): A study using infrared radiation. *Amphibia-Reptilia* 4:49–61.

Denlinger, D. L. 1985. Hormonal control of diapause. In *Comprehensive insect physiology biochemistry and pharmacology*, vol. 8, *Endocrinology II*, edited by G. A. Kerkut and L. I. Gilbert, 353–412. Oxford: Pergamon Press.

Denton, E. J. 1961. The buoyancy of fish and cephalopods. *Progr. Biophysics Biophysical Chemistry* 11:178–233.

Denton, E. J., and J. B. Gilpin-Brown. 1961. The distribution of gas and liquid within the cuttlebone. *J. mar. biol. Assoc. U. K.* 41:365–81.

Derkach, V., A. Surprenant, and R. A. North. 1989. 5-HT$_3$ receptors are membrane ion channels. *Nature* 339:706–9.

Dethier, V. G. 1963. *The physiology of insect senses.* London: Methuen.

DeVries, A. L. 1971. Freezing resistance in fishes. In *Fish physiology*, vol. 6, *Environmental relations and behavior*, edited by W. S. Hoar and D. J. Randall, 157–90. New York: Academic Press.

DeVries, A. L. 1980. Biological antifreezes and survival in freezing environments. In *Animals and environmental fitness*, edited by R. Gilles, 583–607. New York: Pergamon Press.

DeVries, A. L. 1988. The role of antifreeze glycopeptides and peptides in the freezing avoidance of Antarctic fishes. *Comp. Biochem. Physiol.* 90B:611–21.

DeWitt, C. B. 1967. Precision of thermoregulation and its relation to environmental factors in the desert iguana, *Dipsosaurus dorsalis. Physiol. Zool.* 40:49–66.

DeWitt, C. B., and R. M. Friedman. 1979. Significance of skewness in ectotherm thermoregulation. *Amer. Zool.* 19:195–209.

Dhindsa, D. S., C. J. Sedgwick, and J. Metcalfe. 1972. Comparative studies of the respiratory functions of mammalian blood. VIII. Asian elephant (*Elephas maximus*) and African elephant (*Loxodonta africana africana*). *Respir. Physiol.* 14:332–42.

Diamond, J. M., and R. K. Buddington. 1987. Intestinal nutrient absorption in herbivores and carnivores. In *Life in water and on land*, edited by P. Dejours, L. Bolis, C. R. Taylor, and E. R. Weibel, 193–203. Padova: Liviana Press.

Diamond, J. M., and W. H. Bossert. 1967. Standing-gradient osmotic flow: A mechanism for coupling of water and solute transport in epithelia. *J. Gen. Physiol.* 50:2061–83.

Diaz, H., and G. Rodriguez. 1977. The branchial chamber in terrestrial crabs: A comparative study. *Biol. Bull.* 153:485–504.

Dick, D. A. T. 1979. Structure and properties of water in the cell. In *Mechanisms of osmoregulation in animals*, edited by R. Gilles, 35–45. Chichester: John Wiley and Sons.

Dickinson, P. S., C. Mecsas, and E. Marder. 1990. Neuropeptide fusion of two motor-pattern generator circuits. *Nature* 344:155–58.

Dickinson, P. S., F. Nagy, and M. Moulins. 1988. Control of central pattern generators by an identified neurone in Crustacea: Activation of the gastric mill motor pattern by a neurone known to modulate the pyloric network. *J. exp. Biol.* 136:53–87.

Dickman, A. 1933. Studies on the wax moth *Galleria mellonella*, with particular reference to the digestion of wax. *J. Cell. Comp. Physiol.* 3:223–46.

Dockray, G. J 1981. Brain-gut peptides. *Viewpoints Dig. Dis.* 13:5–8.

Dodd, G. H., and D. J. Squirrel. 1980. Structure and mechanisms in the mammalian olfactory system. *Symp. Zool. Soc. Lond.* 45:35–56.

Dodge, F. A., and R. Rahaminoff. 1967. On the relationship between calcium concentration and the amplitude of the end-plate potential. *J. Physiol.* 189:90P–92P.

Doeller, J. E., D. W. Kraus, J. M. Colacino, and J. B. Wittenberg. 1988. Gill hemoglobin may deliver sulphide to bacterial symbionts of *Solemya velum* (Bivalvia, Mollusca). *Biol. Bull.* 175:388–96.

Donnan, F. G. 1927. Concerning the applicability of thermodynamics to the phenomena of life. *J. Gen. Physiol.* 8:685–88.

Dorsett, D. A. 1980. Design and function of giant fibre systems. *TINS* 3:205–8.

Dowling, J. E., and B. B. Boycott. 1966. Organisation of the primate retina: Electron microscopy. *Proc. Roy. Soc. Lond. B* 166:80–111.

Duchamp, C., H. Barre, D. Delage, J.-L. Rouanet, F. Cohen-Adad, and Y. Minaire. 1989. Nonshivering thermogenesis and adaptation to fasting in king penguin chicks. *Am. J. Physiol.* 257:R744–51.

Eakin, R. M. 1973. *The Third Eye.* Berkeley: University of California Press.

Eakin, R. M. 1982. Continuity and diversity in photoreceptors. In *Visual cells in evolution*, edited by J. A. Westfall, 91–105. New York: Raven Press.

Eaton, A. E. 1885. A revisional monograph of recent Ephemeridae or mayflies. *Trans. Linn. Soc. Ser. 2. Zool.* 3:1–352.

Eaton, R. C., R. A. Bombarderi, and D. L. Meyer. 1977.

The Mauthner initiated startle response in teleost fish. *J. exp. Biol.* 66:65–81.

Eckert, R. 1972. Bioelectric control of ciliary activity. *Science* 176:473–81.

Economos, A. C. 1982. On the origin of biological similarity. *J. Theor. Biol.* 94:25–60.

Edmunds, L. N. 1988. *Cellular and molecular bases of biological clocks: Models and mechanisms for circadian timekeeping.* Berlin, Springer-Verlag.

Edney, E. B. 1964. Acclimation to temperature in terrestrial isopods. I. Lethal temperatures. *Physiol. Zool.* 37:364–77.

Edney, E. B. 1966. Absorption of water vapour from unsaturated air by *Arenivaga sp* (Polyphagidae, Dictyoptera). *Comp. Biochem. Physiol.* 19:387–408.

Edney, E. B. 1971. Body temperatures of tenebrionid beetles in the Namib desert of Southern Africa. *J. exp. Biol.* 55:253–72.

Edwards, J. S. 1966. Defence by smear: Supercooling in the cornicle wax of aphids. *Nature* 211:73–74.

Ege, R. 1915. On the respiratory function of the air stores carried by some aquatic insects (Corixidae, Dytiscidae, and *Notonecta*). *Z. Allg. Physiol.* 17:81–124.

Eguchi, E. 1965. Rhabdom structure and receptor potentials in single crayfish retinular cells. *J. Cell. Comp. Physiol.* 66:411–30.

Ehrlich, B. E., and J. Watras. 1988. Inositol 1,4,5-triphosphate activates a channel from smooth muscle sarcoplasmic reticulum. *Nature* 336:583–86.

Eisner, T. 1970. Chemical defense against predation in arthropods. In *Chemical ecology*, edited by E. Sondheimer and J. B. Simeon, 157–218. New York: Academic Press.

Eisner, T., and D. J. Aneshansley. 1982. Spray aiming in Bombardier beetles: Jet deflection by the Coanda effect. *Science* 215:83–85.

Ellerton, H. D., N. F. Ellerton, and H. A. Robinson. 1983. Hemocyanin: A current perspective. *Progr. Biophys. Mol. Biol.* 41:143–248.

Else, P. L., and A. J. Hulbert. 1981. Comparison of the "mammal machine" and the "reptile machine": Energy production. *Am. J. Physiol.* 204:R3–R9.

Else, P. L., and A. J. Hulbert. 1987. Evolution of mammalian endothermic metabolism: "Leaky" membranes as a source of heat. *Am. J. Physiol.* 253:R1–7.

Elthon, T. E., and C. R. Stewart. 1983. A chemiosmotic model for plant mitochondria. *Bioscience* 33:687–92.

Elton, C. 1927. *Animal ecology*. London: Sidgewick and Jackson.

Elyakova, L. A. 1972. Distribution of cellulases and chitinases in marine invertebrates. *Comp. Biochem. Physiol.* 43B:67–70.

Emerson, K., R. C. Russo, R. E. Lund, and R. V. Thurston. 1975. Aqueous ammonia equilibrium calculations: Effect of pH and temperature. *J. Fish. Res. Board Can.* 32:2379–83.

Emmett, B., and P. W. Hochachka. 1981. Scaling of oxidative and glycolytic enzymes in the shrew. *Respir. Physiol.* 45:261–67.

Enami, M. 1959. The morphology and functional significance of the caudal neurosecretory system of fishes. In *Comparative endocrinology*, edited by A. Gorbman, 697–727. New York: John Wiley and Sons.

Enger, P. S., and T. H. Bullock. 1965. Physiological basis of slothfulness in the sloth. *Hvalradets Skrifter* 48:143–60.

Esler, M., G. Jennings, P. Korner, P. Blombery, N. Sacharias, and P. Leonard. 1984. Measurement of total and organ-specific norepinephrine kinetics in humans. *Am. J. Physiol.* 247:E21–28.

Evans, D. H. 1967. Sodium, chloride and water balance of the intertidal teleost, *Xiphister atropurpureus*. *J. exp. Biol.* 47:519–34.

Evans, D. H. 1984. Gill Na^+/H^+ and Cl^-/HCO_3^- exchange systems evolved before the vertebrates entered fresh water. *J. exp. Biol.* 113:465–69.

Evans, E. F. 1972. The frequency response and the properties of single fibres in the guinea-pig cochlear nerve. *J. Physiol.* 226:263–87.

Evans, W. G. 1964. Infrared receptors in *Melanophila acuminata* DeGreer. *Nature* 202:211.

Eyzaguire, C., and S. W. Kuffler. 1955. Process of excitation in the dendrites and in the soma of single isolated nerve cells of the lobster and crayfish. *J. Gen. Physiol.* 39:87–120.

Fange, R., K. Schmidt-Nielsen, and H. Osaki. 1960. The salt gland of the herring gull. *Biol. Bull.* 115:162–71.

Farley, J., and S. Auerbach. 1986. Protein kinase C activation induces conductance changes in *Hermissenda* photoreceptors like those seen in associative learning. *Nature* 319:220–23.

Farrell, A. P., and D. J. Randall. 1978. Air-breathing mechanics in two teleosts, *Hoplerythrinus unitaeniatus* and *Arapaima gigas*. *Can. J. Zool.* 56:953–58.

Farrelly, C., and P. Greenaway. 1987. The morphology and vasculature of the lungs of the soldier crab, *Mictyuris longicarpus*. *J. Morph.* 193:285–304.

Fatt, P., and B. Katz. 1951. An analysis of the end-plate potential recorded with an intra-cellular electrode. *J. Physiol.* 115:320–70.

Fatt, P., and B. L. Ginsburg. 1958. The ionic requirements for the production of action potentials in crustacean muscle fibres. *J. Physiol.* 142:516–43.

Fawcett, D. W. 1962. Physiologically significant specialisations of the cell surface. *Circulation* 26:1105–25.

Fawcett, D. W. 1986. *A textbook of histology*. Philadelphia: Saunders Company.

Fay, F. S. 1976. Structural and functional features of isolated smooth muscle cells. In *Cell motility*, book A, *Motility, muscle and non-muscle cells*, edited by R. Goldman, T. Pollard, and J. Rosenbaum, 185–201. Cold Spring Harbor: Cold Spring Harbor Laboratory.

Fedde, M. R., W. D. Kuhlmann, and P. Scheid. 1977. Intrapulmonary receptors in the tegu lizard: I. Sensitivity to CO_2. *Respir. Physiol.* 29:35–498.

Feder, M. E., and W. W. Burggren. 1986. Skin breathing in vertebrates. *Sci. Am.* 118(11):126–42.

Felbeck, H., J. J. Childress, and G. N. Somero. 1981. Calvin-Benson cycle and sulphide oxidation enzymes in animals from sulfide-rich habitats. *Nature* 293:291–93.

Feldman, H. A., and T. A. McMahon. 1983. The 3/4 mass exponent for energy metabolism is not a statistical artifact. *Respir. Physiol.* 52:149–63.

Felig, P., and J. Wahren. 1971. Inter-relationship between amino acid and carbohydrate metabolism during exercise: The glucose-alanine cycle. In *Advances in experimental biology and medicine*, edited by B. Pernow and B. Saltin, vol. 11, 205–214.

Fenner, D. H. 1973. The respiratory adaptations of the podia and ampullae of echinoids (Echinodermata). *Biol. Bull.* 145:323–39.

Fenton, M. B., and G. P. Rall. 1981. Recognition of species of insectivorous bats by their echolocation calls. *J. Mammal.* 62:233–43.

Ferguson, J. H. 1985. *Mammalian physiology.* Columbus: Charles E. Merrill.

Finean, J. B., R. Coleman, and R. Mitchell. 1978. *Membranes and their cellular functions.* Oxford: Blackwell Scientific.

Fingerman, M. 1987. The endocrine mechanisms of crustaceans. *J. Crust. Biol.* 7:1–24.

Finlay, B., and T. Fenchel. 1989. Everlasting picnic for protozoa. *New Scientist* 123(1671):38–41.

Flock, A. 1965. Transducing mechanisms in the lateral line canal organ receptors. *Cold Spring Harbor Symp. Quant. Biol.* 30:133–44.

Florey, E. 1966. *An introduction to general and comparative animal physiology.* Philadelphia: Saunders Company.

Florkin, M., and C. Jeuniaux. 1974. Hemolymph: Composition. In *Physiology of the Insecta*, edited by M. Rockstein, vol. 5, 255–307. New York: Academic Press.

Folkow, L. P., and A. S. Blix. 1987. Nasal heat and water exchange in grey seals. *Am. J. Physiol.* 253:R883–89.

Foster, R. P., and L. Goldstein. 1969. Formation of excretory products. In *Fish physiology*, vol 1, *Excretion, ionic regulation and metabolism*, edited by W. S. Hoar and D. J. Randall, 313–50. New York: Academic Press.

Forster, R. P., and L. Goldstein. 1976. Intracellular osmoregulatory role of amino acids and urea in marine elasmobranchs. *Am. J. Physiol.* 230:925–31.

Foskett, J. K., and C. Scheffey. 1981. The chloride cell: Definitive identification as the salt-secretory cell in teleosts. *Science* 215:164–66.

Foster, D. F. D. O., and M. L. Frydman. 1979. Tissue distribution of cold-induced thermogenesis in conscious warm- or cold-acclimated rats reevaluated from changes in tissue blood flow: The dominant role of brown adipose tissue in the replacement of shivering by non-shivering thermogenesis. *Can. J. Physiol. Pharmacol.* 57:257–70.

Frederiksen, O., and P. P. Leyssac. 1969. Transcellular transport of isosmotic volumes by the rabbit gallbladder *in vitro*. *J. Physiol.* 201:201–24.

Fridberg, G., and H. A. Bern. 1968. The urophysis and the caudal neurosecretory system of fish. *Biol. Rev.* 43:175–99.

Friedmann, H., J. Kern, and J. H. Hurst. 1957. The domestic chick: A substitute for the honey-guide as a symbiont with cerolytic microorganisms. *Am. Nat.* 91:321–25.

Frost, P. G. H., W. R. Siegfried, and P. J. Greenwood. 1975. Arterio-venous heat exchange systems in the jackass penguin *Spheniscus demersus*. *J. Zool.* 175:231–41.

Fry, F. E. J., and P. W. Hochachka. 1970. Fish. In *Comparative physiology of thermoregulation*, edited by G. C. Whittow, 79–134. New York: Academic Press.

Frye, B. E. 1967. *Hormonal control in vertebrates.* New York: Macmillan.

Fryer, G. 1960. The spermatophores of *Dilops ranarum* (Crustacea, Branchiura): Their structure, formation and transfer. *Q. J. micros. Sci.* 101:149–61.

Fujita, T., S. Kobayashi, R. Yui, and T. Iwanaga. 1980. Evolution of neurons and paraneurons. In *Hormones, adaptation, and evolution*, edited by S. Ishii, T. Hirano, and M. Wada, 35–43. Tokyo: Japan Scientific Societies, and Berlin, Springer-Verlag.

Fujiwara, K., and L. G. Tilney. 1975. Substructural analysis of the microtubule and its polymeric forms. *Ann. N. Y. Acad. Sci.* 253:27–50.

Full, R. J. 1987. Locomotor energetics of the ghost crab. I. Metabolic cost and endurance. *J. exp. Biol.* 130:137–53.

Furshpan, E. J., and D. D. Potter. 1959. Transmission at the giant motor synapses of the crayfish. *J. Physiol.* 145:289–325.

Furukawa, T., and E. J. Furshpan. 1963. Two inhibitory mechanisms in the Mauthner neurons of goldfish. *J. Neurophysiol.* 26:140–76.

Fuzeau-Braesch, S. 1985. Color changes. In *Comprehensive insect physiology biochemistry and pharmacology*, vol. 9, *Behaviour*, edited by G. A. Kerkut and L. I. Gilbert, 549–89. Oxford: Pergamon Press.

Gabe, M. 1966. *Neurosecretion*. Int. Ser. Monog. Biol. 28. Oxford: Pergamon Press.

Gaede, K., and W. Knulle. 1987. Water vapour uptake from the atmosphere and critical equilibrium humidity of a feather mite. *Exp. Appl. Acarology* 3:45–52.

Gamow, R. I., and J. F. Harris. 1973. The infrared receptors of snakes. *Sci. Am.* 228(5):94–101.

Ganong, W. F. 1969. *Review of medical physiology.* Los Altos: Lange Medical Publications.

Gans, C. 1974. *Biomechanics.* Philadelphia: Lippincott.

Gans, C., and B. Clark. 1976. Studies on ventilation of *Caiman crocodilus* (Crocodilia: Reptilia). *Respir. Physiol.* 26:285–301.

Gans, C., and G. M. Hughes. 1967. The mechanism of lung ventilation in the tortoise *Testudo graeca* Linne. *J. exp. Biol.* 47:1–20.

Gans, C., H. J. DeJongh, and J. Farber. 1969. Bullfrog (*Rana catesbeiana*) ventilation: How does the frog breathe? *Science* 163:1223–25.

Garcia, J., and W. Hankins. 1975. The evolution of bitter

and the acquisition of toxiphobia. In *Olfaction and taste*, vol. 5, edited by D. A. Denton and J. P. Coghlan, 39–46. New York: Academic Press.

Gardner, B. G. 1982. Tetrapod classification. *Zool. J. Linn. Soc.* 74:207–32.

Gaunt, A. S., and C. Gans. 1969. Mechanics of respiration in the snapping turtle *Chelydra serpentina* (Linne). *J. Morphol.* 128:195–227.

Geiser, F. 1985. Hibernation in pygmy possums (Marsupialia: Burramyidae). *Comp. Biochem. Physiol.* 81A: 459–63.

Geiser, F. 1986. Thermoregulation and torpor in the kultarr, *Antechinomys laniger* (Marsupialia: Dasyuridae) *J. Comp. Physiol.* 156:751–57.

Geiser, F., and E. J. McMurchie. 1984. Differences in the thermotropic behaviour of mitochondrial membrane respiratory enzymes from homeothermic and heterothermic endotherms. *J. Comp. Physiol.* 155B:125–33.

Geiser, F., R. V. Baudinette, and E. J. McMurchie. 1986. Seasonal changes in the critical arousal temperature of the marsupial *Sminthopsis crassicaudata* correlate with the thermal transition in mitochondrial respiration. *Experientia* 42:543–47.

George, J. C., and R. V. Shah. 1965. Evolution of air sacs in Sauropsida. *J. Anim. Morphol. Physiol.* 12:255–63.

Gergely, J., and P. C. Lewis. 1980. The structure of troponin-C and thin filament regulation in striated muscle. In *Muscle contraction: Its regulatory mechanisms*, edited by S. Ebashi, K. Maruyama, and M. Endo, 191–206. Tokyo: Japan Scientific Societies Press.

Gerst, J. W., and T. B. Thorsen. 1977. Effects of saline acclimation on plasma electrolytes, urea excretion, and hepatic urea biosynthesis in a freshwater stingray, *Potamotrygon sp.* Garmay 1877. *Comp. Biochem. Physiol.* 56A:87–93.

Gesteland, R. C., J. Y. Lettvon, W. H. Pitts, and A. Rojas. 1963. Odor specificity of the frog's olfactory system. In *Olfaction and taste*, vol. 1, edited by Y. Zotterman, 19–34. Oxford: Pergamon Press.

Ghobrial, L. I. 1970. The water relations of the desert antelope *Dorcas dorcas*. *Physiol. Zool.* 43:249–56.

Gilles, R. 1979. Intracellular organic osmotic effectors. In *Mechanisms of osmoregulation in animals*, edited by R. Gilles, 111–54. Chichester: John Wiley and Sons.

Gilles, R., and A. J. R. Pequeux. 1985. Ion transport in crustacean gills: Physiological and ultrastructural approaches. In *Transport processes, iono- and osmoregulation*, edited by R. Gilles and M. Gilles-Baillien, 136–58. Berlin: Springer-Verlag.

Gillette, R. 1991. On the significance of neuronal giantism in gastropods. *Biol. Bull.* 180:234–40.

Gilman, A. G. 1987. G proteins: Transducers of receptor-generated signals. *Ann. Rev. Biochem.* 56:615–49.

Ginsburg, H., and W. D. Stein. 1975. Zero-trans and infinite-cis uptake of galactose in human erythrocytes. *Biochim. Biophys. Acta* 382:353–68.

Ginzburg, M., L. Sachs, and B. Z. Ginzburg. 1970. Ion metabolism in a *Halobacterium*. I. Influence of age of

culture on intracellular concentrations. *J. Gen. Physiol.* 55:187–207.

Glotzback, S. F., and H. C. Heller. 1975. CNS regulation of metabolic rate in the kangaroo rat *Dipodomys ingens*. *Am. J. Physiol.* 228:1880–86.

Gnaiger, E. 1980. Energetics of invertebrate anoxibiosis: Direct calorimetry in aquatic oligochaetes. *FEBS Lett.* 112:239–42.

Goetz, R. H., J. V. Warren, O. H. Gauer, J. L. Patterson, J. T. Doyle, E. N. Keen, and M. McGregor. 1960. Circulation of the giraffe. *Circ. Res.* 8:1049–58.

Goldberger, A. L., and B. J. West. 1987. Fractals in physiology and medicine. *Yale J. Biol. Med.* 60:421–35.

Goldman, D. E. 1965. The transducer action of mechano-receptor membranes. *Cold Spring Harbor Symp. Quant. Biol.* 30:59–68.

Goldman, R. D., A. E. Goldman, K. J. Green, J. C. R. Jones, S. M. Jones, and H.-Y. Yang. 1986. Intermediate filament networks: Organization and possible functions of a diverse group of cytoskeletal elements. *J. Cell. Sci. Suppl.* 5:69–97.

Goldspink, G. 1977. Design of muscles in relation to locomotion. In *Mechanics and energetics of locomotion*, edited by R. McN. Alexander and G. Goldspink, 1–22. London: Chapman and Hall.

Goldstein, D. L., and E. J. Braun. 1988. Contributions of the kidneys and intestines to water conservation, and plasma levels of antidiuretic hormone, in house sparrows (*Passer domesticus*). *J. Comp. Physiol.* 158B:353–61.

Gomme, J. 1982. Epidermal nutrient absorption in marine invertebrates: A comparative analysis. *Am. Zool.* 22:691–708.

Gooding, R. H. 1975. Digestive enzymes and their control in haematophagous arthropods. *Acta Trop.* 32:96–111.

Goodman, J. W. 1980. Immunoglobulins 1: Structure and function. In *Basic and clinical immunology*, edited by D. P. Stites, J. D. Stobo, H. H. Fundenberg, and J. V. Wells, 30–42. Los Altos: Lange Medical Publications.

Goodrich, E. S. 1945. The study of nephridia and genital ducts since 1895. *Quart. J. Micro. Sci.* 86:113–392.

Gorbman, A., and H. A. Bern. 1964. *A textbook of comparative endocrinology*. New York: John Wiley and Sons.

Gorbman, A., W. W. Dickoff, S. R. Vigna, N. B. Clark, and C. L. Ralph, 1983. *Comparative endocrinology*. New York: John Wiley and Sons.

Gordon, A. M., A. F. Huxley, and F. J. Julian. 1966. The variation in isometric tension with sarcomere length in vertebrate muscle fibres. *J. Physiol.* 184:170–92.

Gordon, M. S., and V. A. Tucker. 1965. Osmotic regulation in the tadpoles of the crab-eating frog (*Rana cancrivora*). *J. exp. Biol.* 42:437–45.

Gordon, M. S., and V. A. Tucker. 1968. Further observations on the physiology of salinity adaptation in the crab-eating frog (*Rana cancrivora*). *J. exp. Biol.* 49:185–93.

Gordon, M. S., J. Boetins, I. Boetius, D. H. Evans, R. McCarthy, and L. C. Oglesby. 1965. Salinity adaptations in the mudskipper fish, *Periophthalmus sobrinus*. *Hval. Skrifter Norske Viden-Akad. Oslo* 48:85–93.

Goris, R. C., and M. Nomoto. 1967. Infrared reception in oriental crotaline snakes. *Comp. Biochem. Physiol.* 23:879–92.

Gorman, A. L. F., and M. V. Thomas. 1978. Changes in the intracellular concentration of free calcium ions in a pace-maker neurone, measured with the metallochromic indicator dye arsenazo III. *J. Physiol.* 275:357–76.

Gorski, J., D. Toft, G. Shyamada, D. Smith, and A. Notides. 1968. Hormone receptors: Studies on the interaction of estrogen with the uterus. *Rec. Progr. Hormone Res.* 24:45–80.

Gorski, R. A. 1983. Perspectives in neuroendocrinology. In *Neuroendocrine aspects of reproduction*, edited by R. L. Norman, 395–417. New York: Academic Press.

Gotz, P., and H. G. Boman. 1985. Insect immunity. In *Comprehensive insect physiology biochemistry and pharmacology*, vol. 3, *Integument, respiration and circulation*, edited by G. A. Kerkut and L. I. Gilbert, 453–85. Oxford: Pergamon Press.

Gould, E., N. C. Negus, and A. Novick. 1964. Evidence for echolocation in shrews. *J. Exp. Zool.* 156:19–39.

Graham, J. B. 1974. Aquatic respiration in the sea snake *Pelamis platurus*. *Respir. Physiol.* 21:1–7.

Graham, J. B., and R. H. Rosenblatt. 1970. Aerial vision: Unique adaptation in an intertidal fish. *Science* 168:586–88.

Graham, J. B., D. L. Kramer, and E. Pineda. 1977. Respiration in the air breathing fish *Piabucina festae*. *J. Comp. Physiol.* 122:295–310.

Grainger, J. N. R. 1968. The relation between heat production, oxygen consumption and temperature in some poikilotherms. In *Quantitative biology of metabolism*, edited by A. Locker, 86–89. Berlin: Springer-Verlag.

Grajal, A., S. D. Strahl, R. Parra, M. G. Dominguez, and A. Neher. 1989. Foregut fermentation in the Hoatzin, a neotropical leaf-eating bird. *Science* 245:1236–38.

Grasse, P. P. 1949. *Traite de Zoologie*, vol. 6. Paris: Masson et C.

Graszynski, K. 1963. Die feinstruktur des nephridialkanals von *Lumbricus terrestris* L. Eine elektron mikroskopische untersuchung. *Zool. Beitr.* 8:189–296.

Gratz, R. K., A. Ar, and J. Geiser. 1981. Gas tension profile of the lung of the viper, *Vipera xanthina palestinae*. *Respir. Physiol.* 44:165–76.

Graves, J. E., and G. N. Somero. 1982. Electrophoretic and functional enzymic evolution in four species of eastern pacific barracudas from different thermal environments. *Evolution* 36:97–106.

Gray, B. F. 1981. On the "surface law" and basal metabolic rate. *J. Theor. Biol.* 93:757–67.

Gray, E. G. 1957. The spindle and extrafusal innervation of a frog muscle. *Proc. Roy. Soc. Lond. B.* 146:416–30.

Gray, E. G. 1961. The fine structure of the insect ear. Phil. Trans. Roy. Soc. B243: 75–94.

Green, R. F., S. H. Ridgway, and W. E. Evans. 1980. Functional and descriptive anatomy of the bottlenosed dolphin nasolaryngeal system with special reference to the musculature associated with sound production. In *Animal sonar systems*, edited by R.-G. Busnel and J. F. Fish, 199–238. New York: Plenum Press.

Greenwald, L., and D. Stetson. 1988. Urine concentration and the length of the renal papilla. *NIPS* 3:46–49.

Gregory, J. E., A. Iggo, A. M. McKintyre, and U. Proske. 1989. Responses of electroreceptors in the snout of the echidna. *J. Physiol.* 414:521–38.

Griffin, D. R. 1958. *Listening in the dark: The acoustic orientation of bats and men*. New Haven: Yale University Press.

Griffith, R. W. 1985. Habitat, phylogeny and the evolution of osmoregulatory strategies in primitive fishes. In *Evolutionary biology of primitive fishes*, edited by R. E. Foreman, A. Gorbman, J. M. Dodd, and R. Olsson, 69–80. New York: Plenum Press.

Griffith, R. W. 1987. Fresh water or marine origin of the vertebrates? *Comp. Biochem. Physiol.* 87A:523–31.

Griffith, R. W., and P. K. T. Pang. 1979. Mechanisms of osmoregulation in the coelacanth: Evolutionary implications. *Occ. Pap. Calif. Acad. Sci.* 134:79–93.

Griffiths, M., and D. Davies. 1963. The role of the soft pellets in the production of lactic acid in the rabbit stomach. *J. Nutr.* 80:171–80.

Grigg, G. C., and J. Alchin. 1976. The role of the cardiovascular system in thermoregulation of *Crocodylus johnstoni*. *Physiol. Zool.* 49:24–36.

Grinich, N. P., and R. C. Terwilliger. 1980. The quaternary structure of an unusual high molecular weight intracellular hemoglobin from the bivalve mollusc *Barbatia reeveana*. *Biochem. J.* 189:1–8.

Guillemin, R. 1980. Hypothalamic hormones: Releasing and inhibiting factors. In *Neuroendocrinology*, edited by D. T. Krieger and J. C. Hughes, 23–32. Sunderland: Sinauer Associates.

Guimond, R. W. and V. H. Hutchison. 1972. Pulmonary, branchial and cutaneous gas exchange in the mud puppy, *Necturus maculosus maculosus* (Rafinesque). *Comp. Biochem. Physiol.* 42A:367–92.

Guimond, R. W., and V. H. Hutchison. 1973. Aquatic respiration: An unusual strategy in the hellbender *Cryptobranchus alleganiensis alleganiensis* (Daudin). *Science* 182:1263–65.

Guppy, M., R. D. Hill, R. C. Schneider, J. Qvist, G. C. Liggins, W. M. Zapol, and P. W. Hochachka. 1986. Microcomputer-assisted metabolic studies of voluntary diving of Weddell seals. *Am. J. Physiol.* 250:R175–87.

Guppy, M., W. C. Hulbert, and P. W. Hochachka. 1979. Metabolic sources of heat and power in tuna muscles. *J. exp. Biol.* 82:303–20.

Gupta, A. P. 1985. Cellular elements in the hemolymph. In *Comprehensive insect physiology biochemistry and pharmacology*, vol. 3, *Integument, respiration and circulation*, edited by G. A. Kerkut and L. I. Gilbert, 401–51. Oxford: Pergamon Press.

Gupta, B. L., B. J. Wall, J. L. Oschman, and T. A. Hall. 1980. Direct microprobe evidence for local concentration gradients and recycling of electrolytes during fluid absorption in the rectal papillae of *Calliphora. J. exp. Biol.* 88:21–48.

Guyton, A. C. 1986. *Textbook of medical physiology.* Philadelphia: Saunders.

Gwinner, E. 1986. *Circannual rhythms: Endogenous annual clocks in the organization of seasonal processes.* Berlin: Springer-Verlag.

Hafez, E. S. E. 1968. *Adaptation of domestic animals.* Philadelphia: Lea and Febiger.

Hagedorn, H. H. 1985. The role of ecdysteroids in reproduction. In *Comprehensive insect physiology biochemistry and pharmacology*, vol. 8, *Endocrinology II*, edited by G. A. Kerkut and L. I. Gilbert, 205–62. Oxford: Pergamon Press.

Hainsworth, F. R., and L. L. Wolf. 1970. Regulation of oxygen consumption and body temperature during torpor in a hummingbird, *Eulampis jugularis. Science* 168:368–69.

Haldane, J. S., and J. G. Priestley. 1935. *Respiration.* Oxford: Clarendon Press.

Hamdorf, K., P. Paulsen, and J. Schwemer. 1973. Photoregeneration and sensitivity control of photoreceptors of invertebrates. In *Biochemistry and physiology of visual pigments*, edited by H. Langer, 155–66. Berlin: Springer-Verlag.

Hammel, H. T., and P. F. Scholander. 1976. *Osmosis and tensile solvent.* Berlin: Springer-Verlag.

Hammen, C. S. 1979. Metabolic rates of marine bivalve molluscs determined by calorimetry. *Comp. Biochem. Physiol.* 62A:955–59.

Hammen, C. S. 1980. Total energy metabolism of marine bivalve mollusks in anaerobic and aerobic states. *Comp. Biochem. Physiol.* 67A:617–21.

Hanson, E. A. 1981. *Understanding evolution.* Oxford: Oxford University Press.

Hanson, J., and J. Lowy. 1960. Structure and function of the contractile apparatus in the muscles of invertebrate animals. In *The structure and function of muscle*, edited by G. H. Bourne, vol. 1, *Structure*, 265–335. New York: Academic Press.

Hardie, J., and A. D. Lees. 1985. Endocrine control of polymorphism and polyphenism. In *Comprehensive insect physiology, biochemistry and pharmacology*, edited by G. A. Kerkut and L. I. Gilbert, 441–90. New York: Pergamon Press.

Hardie, R. C. 1989. A histamine-activated chloride channel involved in neurotransmission at a photoreceptor synapse. *Nature* 339:704–6.

Harlan, R. A., and R. F. Wilkinson. 1981. The effects of progressive hypoxia and rocking activity on blood oxygen tension for hellbenders, *Cryptobranchus alleganiensis. J. Herpetology* 15:383–87.

Harms, J. W. 1932. Die realisation von genen und die consecutive adaption. II. *Birgus latro* L. als landkrebs und seine beziehungen zu den coenobiten. *Zeit. f. wissensch. Zoologie* 140:12–299.

Harned, H. S., and R. A. Robinson. 1940. A note on the temperature variation of the ionisation constants of weak electrolytes. *Trans. Faraday Soc.* 36:973–78.

Harris, G. G. 1968. Brownian motion in the cochlear partition. *J. Acoust. Soc. Am.* 44:176–86.

Harris, J. F., and R. I. Gamow. 1971. Snake infrared receptors: Thermal or photochemical mechanism? *Science* 172:1252–53.

Harrison, R., and G. G. Lunt. 1980. *Biological membranes: Their structure and function.* Glasgow: Blackie.

Hart, J. S., and L. Jansky. 1963. Thermogenesis due to exercise and cold in warm- and cold-acclimated rats. *Can. J. Biochem. Physiol.* 41:629–34.

Hartline, H. K., H. G. Wagner, and F. Ratliff. 1956. Inhibition in the eye of *Limulus. J. Gen. Physiol.* 39:651–73.

Hartman, S. C. 1970. Purines and pyrimidines. In *Metabolic pathways*, edited by D. M. Greenberg, 1–68. New York: Academic Press.

Hasan, M. R., and D. J. Macintosh. 1986. Acute toxicity of ammonia to common carp fry. *Aquaculture* 54:97–107.

Hastings, J. W. 1978. Bacterial and dinoflagellate luminescent systems. In *Bioluminescence in action*, edited by P. J. Herring, 129–70. New York: Academic Press.

Haugaard, N., and L. Irving. 1943. The influence of temperature upon the oxygen consumption of the cunner (*Tautogolabrus adspersus*) in summer and in winter. *J. Cell. Comp. Physiol.* 21:19–26.

Haupt, T. R. 1963. Urea utilization by rabbits fed a low-protein ration. *Am. J. Physiol.* 205:1144–50.

Hays, E. A., M. A. Long, and H. Gainer. 1968. A reexamination of the Donnan distribution as a mechanism for membrane potentials and potassium and chloride ion distributions in crab muscle fibers. *Comp. Biochem. Physiol.* 26:761–92.

Heath, J. E. 1964. Reptile thermoregulation: Evolution of field studies. *Science* 146:784–85.

Heath, J. E. 1965. Temperature regulation and diurnal activity in horned lizards. *University Calif. Publ. Zool.* 64.

Heath, J. E. 1966. Venous shunts in the cephalic sinuses of horned lizards. *Physiol. Zool.* 39:30–35.

Heatwole, H., and J. Taylor. 1987. *Ecology of reptiles.* Chipping Norton, New South Wales: Surrey Beatty and Sons.

Hedrick, M. S., D. A. Duffield, and L. H. Cornell. 1986. Blood viscosity and optimal hematocrit in a deep-diving mammal, the northern elephant seal (*Mirounga anguirostris*). *Can. J. Zool.* 64:2081–85.

Heffernan, J. M., and S. A. Wainwright. 1974. Locomotion of the holothurian *Euapta lappa* and redefinition of peristalsis. *Biol. Bull.* 147:95–104.

Heinrich, B. 1976. Heat exchange in relation to blood flow between thorax and abdomen in bumblebees. *J. exp. Biol.* 64:561–85.

Heinrich, B. 1987. Thermoregulation by winter-flying endothermic moths. *J. exp. Biol.* 127:313–32.

Heinrich, B., and T. P. Mommsen. 1985. Flight of winter moths near 0° C. *Science* 228:177–79.

Heldemaier, G. 1971. Zitterfreie warmebildung und korpergrobe bei saugetieren. *Z. vergl. Physiol.* 73:222–48.

Hemmingsen, A. M. 1950. The relation of standard (basal) energy metabolism to total fresh weight of living organisms. *Rept. Steno Mem. Hosp. and Nord. Insulin Lab.* 4:7–58.

Henderson, I. F., and W. D. Henderson. 1975. *A Dictionary of Biological Terms*, 8th ed., edited by J. H. Kenneth. London: Longman.

Hendrickson, W. A., J. L. Smith, and S. Sheriff. 1985. Structure and function of hemerythrins. In *Respiratory pigments in animals: Relation structure-function*, edited by J. Lamy, J.-P. Truchot, and R. Gilles, 1–7. Berlin: Springer-Verlag.

Henshaw, R. E., L. D. Underwood, and T. M. Casey. 1972. Peripheral thermoregulation: Foot temperature in two Arctic canines. *Science* 175:988–90.

Henwood, K. 1975. A field-tested thermoregulation model for two diurnal Namib desert tenebrionid beetles. *Ecology* 56:1329–42.

Herbst, D. B., and T. K. Bradley. 1989. A Malpighian tubule lime gland in an insect inhabiting alkaline salt lakes. *J. exp. Biol.* 145:63–78.

Hermans, C. O., and R. M. Eakin. 1974. Fine structure of the eyes of an alciopid polychaete, *Vanadis tagensis* (Annelida). *Morph. Tiere* 79:245–67.

Herold, J. P. 1975. Myocardial efficiency in the isolated ventricle of the snail, *Helix pomatia. Comp. Biochem. Physiol.* 52A:435–40.

Herreid, C. F. 1981. Energetics of pedestrian arthropods. In *Locomotion and energetics in arthropods*, edited by C. F. Herreid and C. R. Fourtner, 491–526. New York: Plenum Press.

Herreid, C. F., V. F. LaRussa, and C. R. DeFesi. 1976. Blood vascular system of the sea cucumber, *Stichopus moebii. J. Morphol.* 150:423–52.

Herring, P. J. 1985. How to survive in the dark: Bioluminescence in the deep sea. *Symp. Soc. Exp. Biol.* 39:323–50.

Herring, P. J., and J. G. Morin. 1978. Bioluminescence in fishes. In *Bioluminescence in action*, edited by P. J. Herring, 273–329. New York: Academic Press.

Hertz, P. E. 1979. Sensitivity to high temperature in three West Indian grass anoles (Sauria, Iguanidae), with a review of heat sensitivity in the genus *Anolis. Comp. Biochem. Physiol.* 63A:217–22.

Hess, J. F., R. B. Bourret, K. Oosawa, P. Matsumura, and M. I. Simon. 1988. Protein phosphorylation and bacterial chemotaxis. *Cold Spring Harbor Symp. Quant. Biol.* 53:41–48.

Heusner, A. A. 1982. Energy metabolism and body size. I. Is the 0.75 mass exponent of Kleiber's equation a statistical artifact? *Respir. Physiol.* 48:1–12.

Heusner, J. E., and C. F. Doggenweiler. 1966. The fine structural organization of the nerve fibers, sheaths, and glial cells in the prawn *Palaemonetes vulgaris. J. Cell. Biol.* 30:381–403.

Hickman, C. P. 1967. *Biology of the invertebrates*. St. Louis: C. V. Mosby Company.

Hickman, C. P., and B. F. Trump. 1969. The kidney. In *Fish physiology*, vol. 1, *Excretion, ionic regulation,*

and metabolism, edited by W. S. Hoar and D. J. Randall, 91–240. New York: Academic Press.

Highnam, K. C., and L. Hill. 1977. *The Comparative endocrinology of the invertebrates*. London: Edward Arnold.

Higuchi, H., and Y. E. Goldman. 1991. Sliding distance between actin and myosin filaments per ATP hydrolyzed in skinned muscle fibres. *Nature* 352:352–54.

Hildebrandt, J. P. 1988. Circulation in the leech, *Hirudo medicinalis* (L.) *J. exp. Biol.* 134:235–46.

Hildemann, W. H., I. S. Johnson, and P. L. Jokiel. 1979. Immunocompetence in the lowest metazoan phylum: Transplantation immunity in sponges. *Science* 204:420–22.

Hildemann, W. H., R. L. Raison, G. Cheung, C. J. Hull, L. Akala, and J. Okamoto. 1977. Immunological specificity and memory in a scleractinian coral. *Nature* 270:219–23.

Hill, A. E. 1977. General mechanisms of salt-water coupling in epithelia. In *Transport of ions and water in animals*, edited by B. L. Gupta, R. B. Moreton, J. L. Oschman, and B. J. Wall, 183–214. London: Academic Press.

Hill, J. R., and K. A. Rahimtulla. 1965. Heat balance and the metabolic rate of new-born babies in relation to environmental temperature; and the effect of age and weight on basal metabolic rate. *J. Physiol.* 180:239–65.

Hill, R. B., and J. H. Welsh. 1966. Heart, circulation, and blood cells. In *Physiology of Mollusca*, vol. 2, edited by K. M. Wilbur and C. M. Yonge, 125–74. New York: Academic Press.

Hille, B. 1982. *Ionic channels of excitable membranes.* Sunderland, Sinauer Associates.

Hillman, S. S. 1980. Physiological correlates of differential dehydration tolerance in anuran amphibians. *Copeia* 1980:125–29.

Hillman, S. S., and P. C. Withers. 1979. An analysis of respiratory surface area as a limit to activity metabolism in anurans. *Can. J. Zool.* 57:2100–05.

Hillman, S. S., and P. C. Withers. 1987. Oxygen consumption during aerial activity in aquatic and amphibious fish. *Copeia* 1987:232–34.

Hillman, S. S., A. C. Zygmunt, and M. Baustian. 1987. Transcapillary fluid forces during dehydration in two amphibians. *Physiol. Zool.* 60:339–45.

Hinckle, P. C., and R. E. McCarty. 1978. How cells make ATP. *Sci. Am.* 238(3):104–22.

Hinton, E. A. 1957. The structure and function of the spiracular gills of the fly *Taphrophila vitripennis. Proc. Roy. Soc. Lond.* 147B:90–120.

Hinton, H. E. 1963. The respiratory system of the egg shell of the blowfly, *Calliphora erythrocephala* Meig, as seen with the electron microscope. *J. Insect Physiol.* 9:121–29.

Hinton, H. E. 1969. Respiratory systems of insect egg shells. *Ann. Rev. Entom.* 14:343–68.

Hinton, H. E. 1971. Plastron respiration of the mite, *Platyseius italicus. J. Insect Physiol.* 17:1185–99.

Hiramoto, Y. 1974. Mechanics of ciliary movement. In

Cilia and flagella, edited by M. A. Sleigh, 177–98. London: Academic Press.

Hoar, W. S. 1965. The endocrine system as a chemical link between the organism and its environment. *Trans. Roy. Soc. Canada* 3 (ser 4):175–200.

Hobson, R. P. 1931. On an enzyme from blow-fly larvae (*Lucillia sericata*) which digests collagen in alkaline solution. *Biochem. J.* 25:1458–63.

Hochachka, P. W. 1980. *Living without oxygen*. Cambridge: Harvard University Press.

Hochachka, P. W., and G. N. Somero. 1984. *Biochemical adaptation*. Princeton: Princeton University Press.

Hochachka, P. W., M. Guppy, H. E. Guderly, K. B. Storey, and W. C. Hulbert. 1978. Metabolic biochemistry of water vs air-breathing fishes: Muscle enzymes and ultrastructure. *J. Zool.* 56:736–50.

Hochachka, P. W., T. G. Owen, J. F. Allen, and G. C. Whittow. 1975. Multiple end products of anaerobiosis in diving vertebrates. *Comp. Biochem. Physiol.* 50B:17–22.

Hodgkin, A. L. 1937. Evidence for electrical transmission in nerve. Part 1. *J. Physiol.* 90:183–232.

Hodgkin, A. L. 1958. Ionic movements and electrical activity in giant nerve fibres. *Proc. Roy. Soc. Lond. B.* 148:1–37.

Hodgkin, A. L., and A. F. Huxley. 1952. A quantitative description of membrane current and its application to conduction and excitation in nerve. *J. Physiol.* 117:500–44.

Hodgkin, A. L., and P. Horowicz. 1959. The influence of potassium and chloride ions on the membrane potential of single muscle fibres. *J. Physiol.* 148:127–60.

Hoese, B. 1982. Morphologie und evolution der lungen bei den terrestrischen isopoden (Crustacea, Isopoda, Oniscoidea). *Zool. Jb. Anat.* 107:396–422.

Hoffman, L., and R. Schiemann. 1973. Die verwertung der futterenergie durch die legende henne. *Arch. Tierernahrung* 23:105–32.

Hogben, L., and D. Slone. 1931. The pituitary effector system. VI. The dual character of endocrine coordination in amphibian colour change. *Proc. Royal Soc. Lond. B.* 108:10–53.

Hoglund, G., K. Hamdorf, and G. Rosner. 1973. Trichromatic visual systems in an insect and its sensitivity control by blue light. *J. Comp. Physiol.* 86:265–79.

Hokfelt, T., B. Everitt, B. Meister, T. Melander, M. Schalling, O. Johansson, J. M. Lundberg, A.-L. Hulting, S. Werner, C. Cuello, H. Hemmings, C. Ouimet, I. Walaas, P. Greengard, and M. Goldstein. 1986. Neurons with multiple messengers with special reference to neuroendocrine systems. *Recent Prog. Hormone Res.* 42:1–70.

Holdridge, L. R. 1947. Determination of world plant formations from sample climate data. *Science* 105:367–68.

Hollenbeck, P. J. 1990. Dynamin joins the family. *Nature* 347:229.

Hopkins, C. D. 1976. Stimulus filtering and electroreception: Tuberous electroreceptors in three species of gymnotoid fish. *J. Comp. Physiol.* 111:171–207.

Hopkins, W. F., and D. Johnston. 1984. Frequency-dependent noradrenergic modulation of long-term potentiation in the hippocampus. *Science* 226:350–52.

Horridge, G. A. 1962. Learning leg position by the ventral nerve cord in headless insects. *Proc. Royal Soc. B.* 157:33–52.

Horridge, G. A. 1968. *Interneurons*. London: Freeman and Company.

Horsfield, K., and G. Cumming. 1967. Angles of branching and diameters at branches in the human bronchial tree. *Bull. Math. Biophysics* 29:245–59.

Hovey, H. B. 1929. Associative hysteresis in marine flatworms. *Physiol. Zool.* 2:322–33.

Howard, J., A. J. Hudspeth, and R. D. Vale. 1989. Movement of microtubules by single kinesin molecules. *Nature* 342:154–58.

Hoyle, G. 1965. Neurophysiological studies on "learning" in headless insects. In *The physiology of the insect central nervous system*, edited by J. E. Treherne and J. W. L. Beament, 203–32. New York: Academic Press.

Hubel, D. H., and T. N. Wiesel. 1962. Receptive fields, binocular interaction and functional architecture in the cat's visual cortex. *J. Physiol.* 160:106–54.

Hubel, D. H., and T. N. Wiesel. 1965. Receptive fields and functional architecture in two non-striate visual areas (18 and 19) of the cat. *J. Neurophysiol.* 28:229–89.

Hudspeth, A. J. 1985. The cellular basis of hearing: The biophysics of hair cells. *Science* 230:745–52.

Huey, R. B. 1974. Behavioral thermoregulation in lizards: Importance of associated costs. *Science* 184:1001–02.

Huey, R. B. 1982. Temperature, physiology, and the ecology of reptiles. In *Biology of the Reptilia*, vol. 12, *Physiology C. physiological ecology*, edited by F. H. Pough, 26–90. London: Academic Press.

Huey, R. B., and M. Slatkin. 1976. Cost and benefit of lizard thermoregulation. *Quart. Rev. Biol.* 51:363–84.

Huey, R. B., and P. E. Hertz. 1984. Is a jack-of-all-temperatures a master of none? *Evolution* 38:441–44.

Hughes, G. M. 1966. The dimensions of fish gills in relation to their function. *J. exp. Biol.* 45:177–95.

Hughes, G. M. 1978. Some features of gas transfer in fish. *Bull. Inst. Math. Appl.* 14:39–43.

Hughes, G. M. 1982. An introduction to the study of gills. In *Gills*, edited by D. F. Houlihan, J. C. Rankin, and T. J. Shuttleworth, 1–24. Cambridge: Cambridge University Press.

Hughes, G. M. 1984. Scaling of respiratory surface areas in relation to oxygen consumption of vertebrates. *Experientia* 40:519–652.

Hughes, G. M., and A. V. Grimstone. 1965. The fine structure of the secondary lamellae of the gills of *Gadus pollachius*. *Quart. J. micro. Sci.* 106:343–53.

Hughes, G. M., and G. Shelton. 1958. The mechanism of gill ventilation in three freshwater teleosts. *J. exp. Biol.* 35:812–23.

Hughes, G. M., and M. Morgan. 1973. The structure of fish gills in relation to their respiratory function. *Biol. Rev.* 48:419–75.

Hughes, G. M., and S.-I. Umezawa. 1968. On respiration in the dragonet, *Callionymus lyra* (L.) *J. exp. Biol.* 49:565–82.

Hughes, G. M., B. Knights, and C. Scammell. 1969. The distribution of pO$_2$ and hydrostatic pressure changes within the branchial chamber in relation to gill ventilation. *J. exp. Biol.* 51:203–20.

Hughes, J., T. W. Smith, H. W. Kosterlitz, L. H. Fothergill, B. A. Morgan, and H. Morris. 1975. Identification of two related pentapeptides from the brain with potent opiate agonist activity. *Nature* 258:577–80.

Hume, I. D. 1982. Digestive physiology and nutrition of marsupials. Cambridge: Cambridge University Press.

Hume, I. D. 1989. Optimal digestive strategies in mammalian herbivores. *Physiol. Zool.* 62:1145–63.

Humphreys, W. F. 1979. Production and respiration in animal populations. *J. Anim. Ecology* 48:427–53.

Hungate, R. E. 1966. The rumen and its microbes. New York: Academic Press.

Hungate, R. E. 1968. Ruminal fermentation. In *Handbook of physiology*, vol. 5, sect. 6, edited by C. F. Code, 2725–45. Washington: American Physiological Society.

Hursch, J. B. 1939. Conduction velocity and diameter of nerve fibers. *Am. J. Physiol.* 127:131–39.

Hutchison, V. H., G. H. Dowling, and A. Vinegar. 1966. Thermoregulation in brooding female Indian python, *Python molurus bivittatus*. *Science* 151:694–96.

Hutchison, V. H., H. B. Haines, and G. Engbretson. 1976. Aquatic life at high altitude: Respiratory adaptations in the Lake Titicaca frog, *Telmatobius culeus*. *Respir. Physiol.* 27:115–29.

Huxley, A. F., and R. Niedergerke. 1954. Structural changes in muscle during contraction: Interference microscopy of living muscles. *Nature* 173:971–73.

Huxley, H. E., and J. Hanson. 1954. Changes in the cross-striations of muscle during contraction and stretch and their structural interpretation. *Nature* 173:973–76.

Hyman, A. A., and T. J. Mitchison. 1991. Two different microtubule-based motor activities with opposite polarities in kinetochores. *Nature* 351:206–11.

Hyman, L. H. 1951. *The Invertebrates: Playtyhelminthes and Rhyncocoela*, vol. 2, New York: McGraw-Hill.

Hyman, L. H. 1967. *The Invertebrates*, volume 6, *Mollusca I.* New York: McGraw-Hill.

Imura, H., Y. Kato, Y. Nakai, K. Nakao, I. Tanaka, H. Jingami, T. Koh, T. Yoshimasa, T. Tsukada, M. Suda, M. Sakamoto, N. Morii, H. Takahashi, K. Tojo, and A. Sugawara. 1985. Endogenous opioids and related peptides: From molecular biology to clinical medicine. *J. Endocrinol.* 107:147–57.

Incropera, F. P., and D. P. Dewitt. 1981. *Fundamentals of heat transfer.* New York: John Wiley and Sons.

Innes, A. J., E. W. Taylor, and A. J. El Haj. 1987. Air-breathing in the Trinidad mountain crab: A quantum leap in the evolution of the invertebrate lung? *Comp. Biochem. Physiol.* 87A:1–8.

Irving, L., and J. Krog. 1955. Temperature of skin in the arctic as a regulator of heat. *J. Appl. Physiol.* 7:355–64.

Isaacks, R. E., and D. R. Harkness. 1980. Erythrocyte organic phosphates and hemoglobin function in birds, reptiles, and fishes. *Amer. Zool.* 20:115–29.

Itoh, Y., and R. Eckert. 1974. The control of ciliary activity in protozoa. In *Cilia and flagella*, edited by M. A. Sleigh, 305–352. London: Academic Press.

Jackson, D. C. 1968. Metabolic depression and oxygen depletion in the diving turtle. *J. Appl. Physiol.* 24:503–09.

Jackson, D. C., and G. R. Ultsch. 1982. Long-term submergence at 3° C of the turtle, *Chrysemys picta belli*, in normoxic and severely hypoxic water. II. Extracellular ionic responses to extreme lactic acidosis. *J. exp. Biol.* 96:29–43.

Jackson, S., J. Hope, F. Estivariz, and P. J. Lowry. 1981. Nature and control of peptide release from the pars intermedia. In *Peptides of the pars intermedia*, edited by D. Evered and G. Lawrensen, 141–62. CIBA Symp. 81. London: Pitman Medical.

Jacquez, J. A., J. W. Poppell, and R. Jeltsch. 1959. Partial pressures of ammonia in alveolar air. *Science* 129:269–70.

Janis, C. 1976. The evolutionary strategy of the Equidae and the origins of rumen and cecal digestion. *Evolution* 30:757–74.

Jenkin, P. M. 1957. The filter-feeding and food of flamingoes (Phoenicopteri). *Phil. Trans. Roy. Soc. B.* 240:401–93.

Jenkin, P. M. 1962. Animal hormones, a comparative survey, part 1, *Kinetic and metabolic hormones.* New York: Pergamon Press.

Jennings, J. B. 1972. *Feeding, digestion and assimilation in animals.* London: MacMillan.

Jensen, E. V., T. Suzuki, T. Kawashima, W. E. Stumpf, P. W. Jungblut, and E. R. Desombre. 1968. A two-step mechanism for the interaction of oestradiol with the rat uterus. *Proc. Nat. Acad. Sci.* 59:632–38.

Jenssen, B. M., M. Ekker, and C. Bech. 1989. Thermoregulation in winter—acclimatized common eiders (*Somateria mollissima*) in air and water. *Can. J. Zool.* 67:669–73.

Jeuniaux, C. 1961. Chitinase: An addition to the list of hydrolases in the digestive tract of vertebrates. *Nature* 192:135–36.

Jobling, M., and P. S. Davis. 1980. Effects of feeding on metabolic rate and the specific dynamic action in plaice, *Pleuronectes platessa*. *J. Fish Biol.* 16:629–38.

Jodal, M. 1977. The intestinal countercurrent exchanger and its influence on intestinal absorption. In *Intestinal permeation*, edited by M. Kramer and F. Lauterbach, 48–55. Amsterdam: Excerpta Medica.

Johansen, K. 1962. Cardiac output and pulsatile flow in the teleost *Gadus morhua*. *Comp. Biochem. Physiol.* 7:169–74.

Johansen, K., and A. W. Martin. 1965. Circulation in a giant earthworm, *Glossoscolex giganteus*. I. Contractile processes and pressure gradients in the large blood vessels. *J. exp. Biol.* 43:333–47.

Johansen, K., C. Lenfant, and D. Hanson. 1970. Phylogenetic development of pulmonary circulation. *Fed. Proc.* 29:1135–40.

Johansen, K., C. Lenfant, K. Schmidt-Nielsen, and J. A. Petersen. 1968. Gas exchange and control of breathing in the electric eel, *Electrophorus electricus. Z. vergl. Physiol.* 61:137–63.

Johnson, M. L. 1942. The respiratory function of the haemoglobin of the earthworm. *J. exp. Biol.* 18:266–77.

Joiun, C., and A. Toulmond. 1989. The ultrastructure of the gill of the lugworm *Arenicola marinus* (L.) (Annelida, Polychaeta). *Acta Zool.* 70:121–29.

Jones, D. R., and T. Schwartzfeld. 1974. The oxygen cost to the metabolism and efficiency of breathing in trout (*Salmo gairdneri*). *Respir. Physiol.* 21:241–54.

Jones, H. D. 1983. The circulatory systems of gastropods and bivalves. In *The Mollusca*, vol. 5, *Physiology*, part 2, edited by A. S. M. Saleuddin and K. M. Wilbur, 189–238. New York: Academic Press.

Jones, J. C. 1977. *The circulatory system of insects.* Springfield: C. C. Thomas.

Jones, J. D. 1972. *Comparative physiology of respiration.* London: Edward Arnold.

Jones, J. R., E. L. Effmann, and K. Schmidt-Nielsen. 1981. Control of air flow in bird lungs: Radiographic studies. *Respir. Physiol.* 45:121–31.

Joosse, J., and W. P. M. Geraerts. 1983. Endocrinology. In *The Mollusca*, vol. 4, *Physiology*, part 1, edited by A. S. M. Saleuddin and K. M. Wilbur, 318–406. New York: Academic Press.

Jusiak, R., and P. Poczopko. 1972. The effect of ambient temperature and season on total and tissue metabolism of the frog (*Rana esculenta* L.). *Bull. Acad. Sci. Pol. Sci. Ser. Biol.* 20:523–29.

Kahl, M. P. 1963. Thermoregulation in the wood stork, with special reference to the role of the legs. *Physiol. Zool.* 36:141–51.

Kaibbling, B., C. deRouffignac, J. M. Barrett, and W. Kriz. 1975. The structural organization of the kidney of the desert rodent *Psammomys obesus. Anat. Embryol.* 148:121–43.

Kaila, K., and J. Voipio. 1987. Postsynaptic fall in intracellular pH induced by GABA-activated bicarbonate conductance. *Nature* 330:163–65.

Kaissling, K.-E. 1983. Molecular recognition: Biophysics of chemoreception. In *Biophysics*, edited by W. Hoppe, W. Lohmann, H. Markel, and H. Ziegler, 697–709. Berlin: Springer-Verlag.

Kaissling, K.-E. 1986. Temporal characteristics of pheromone receptor cell responses in relation to orientation behaviour of moths. In *Mechanisms in insect olfaction*, edited by T. L. Payne, M. C. Birch, and C. E. J. Kennedy, 193–99. Oxford: Clarendon Press.

Kaissling, K.-E. 1987. *RH Wright lectures on insect olfaction.* Burnaby: Simon Fraser University.

Kalmijn, A. J. 1971. The electric sense of sharks and rays. *J. exp. Biol.* 55:371–83.

Kalmijn, A. J. 1978. Electric and magnetic sensory world of sharks, skates and rays. In *Sensory biology of sharks, skates and rays*, edited by E. S. Hodgson and R. T. Mathewson, 507–28. Arlington VA: ONRL Department of the Navy.

Kampmeier, O. F. 1969. *Evolution and comparative morphology of the lymphatic system.* Springfield: C. C. Thomas.

Kandel, E. R., and J. H. Schwartz. 1982. Molecular biology of learning. *Science* 218:433–43.

Kandel, E. R., T. Abrams, L. Bernier, T. J. Carew, R. D. Hawkins, and J. H. Schwartz. 1983. Classical conditioning and sensitization share aspects of the same molecular cascade in *Aplysia. Cold Spring Harbor Symp. Quant. Biol.* 43:821–30.

Kanwisher, J. W. 1955. Freezing in intertidal animals. *Biol. Bull.* 109:56–63.

Kao, F. F. 1972. *An introduction to respiratory physiology.* Amsterdam: Excerpta Medica.

Karasov, W. H. 1987. Nutrient requirements and the design and function of guts in fish, reptiles, and mammals. In *Life in water and on land*, edited by P. Dejours, L. Bolis, C. R. Taylor, and E. R. Weibel, 184–91. Padova: Liviana Press.

Karasov W. H. 1988. Nutrient transport across vertebrate intestine. In *Advances in comparative and environmental physiology*, vol. 2, edited by R. Gilles, 131–72. Berlin: Springer-Verlag.

Karasov, W. H., and J. M. Diamond. 1983. Adaptive regulation of sugar and amino acid transport by vertebrate intestine. *Am. J. Physiol.* 245:G443–62.

Karasov, W. H., D. H. Solberg, and J. M. Diamond. 1985. What transport adaptations enable mammals to absorb sugars and amino acids faster than reptiles? *Am. J. Physiol.* 249:G271–83.

Karasov W. H., D. Phan, J. M. Diamond, and F. L. Carpenter. 1986. Food passage and intestinal nutrient absorption in hummingbirds. *Auk* 103:453–64.

Karasov, W. H., R. K. Buddington, and J. M. Diamond. 1985. Adaptation of intestinal sugar and amino acid transport in vertebrate evolution. In *Transport processes, iono- and osmoregulation*, edited by R. Gilles and M. Gilles-Baillien, 227–39. Berlin: Springer-Verlag.

Karlish, S. J. D., D. W. Yates, and I. M. Glynn. 1978. Conformational transitions between Na^+-bound and K^+-bound forms of $(Na^+ + K^+)$-ATPase, studied with formycin nucleotides. *Biochem. Biophys. Acta* 525:252–64.

Karnaky, K. J. 1986. Structure and function of the chloride cell of *Fundulus heteroclitus* and other teleosts. *Amer. Zool.* 26:209–24.

Karsch, J. F. 1980. Seasonal reproduction: A saga of reversible fertility. *Physiologist* 23:29–38.

Katz, B. 1966, *Nerve, muscle, and synapse.* New York: McGraw-Hill.

Katz, B., and R. Miledi. 1966. Input/output relation of a single synapse. *Nature* 212:1242–45.

Katz, U., and S. Gabbay. 1986. Water retention and plasma and urine composition in the toad (*Bufo viridis* Laur.) under burrowing conditions. *J. Comp. Physiol.* 156B:735–40.

Kaufman, W. R., and J. R. Phillips. 1973. Ion and water balance in the ixodid tick *Dermacentor andersoni*. I. Routes of ion and water excretion. *J. exp. Biol.* 58:523–47.

Kay, D. G., and A. E. Braefield. 1973. The energy relations of the polychaete *Neanthes* (= *Nereis*) *virens* (Sars). *J. Animal Ecol.* 42:673–92.

Kemp, T. S. 1988. Haemothermia or Archosauria? The interrelationships of mammals, birds and crocodiles. *Zool. J. Linn. Soc.* 92:67–104.

Kenagy, G. J., and D. Vleck. 1982. Daily temporal organization of metabolism in small mammals: Adaptation and diversity. In *Vertebrate circadian systems*, edited by J. Aschoff, S. Daan, and G. Groos, 322–38. Berlin: Springer-Verlag.

Kenney, R. 1958. Temperature tolerance of the chiton *Clavarizona hirtosa*. *J. Roy. Soc. West. Aust.* 41:93–101.

Kern, K. L., L. L. Slyter, E. C. Leffel, J. M. Weaver, and R. R. Oltjen. 1974. Ponies vs steers: Microbial and chemical characteristics of intestinal ingesta. *J. Anim. Sci.* 38:559–64.

Keynes, R. D., and D. J. Aidley. 1981. *Nerve and muscle*. Cambridge: Cambridge University Press.

Kilgore, D. L., M. H. Bernstein, and D. M. Hudson. 1976. Brain temperatures in birds. *J. Comp. Physiol.* 110:209–15.

King, G. M., and D. R. N. Custance. 1982. *Colour atlas of vertebrate anatomy*. Oxford: Blackwell.

Kiorboe, T., and F. Mohlenberg. 1987. Partitioning of oxygen consumption between "maintenance" and "growth" in developing herring *Clupea harengus* (L.) embryos. *J. Exp. Mar. Biol. Ecol.* 111:99–108.

Kirchner, W. 1973. Ecological aspects of cold resistance in spiders (a comparative study). In *Effects of temperature on ectothermic organisms: Implications and mechanisms of compensation*, edited by W. Wieser, 271–79. Berlin: Springer-Verlag.

Kirsch, R., W. Humbert, and V. Simonneaux. 1985. The gut as an osmoregulatory organ: Comparative aspects and special references to fishes. In *Transport processes, iono- and osmoregulation*, edited by R. Gilles and M. Gilles-Baillien, 265–77. Berlin: Springer-Verlag.

Kirschner, L. B. 1979. Control mechanisms in crustaceans and fishes. In *Mechanisms of osmoregulation in animals: Maintenance of cell volume*, edited by R. Gilles, 157–222. Chichester: John Wiley and Sons.

Kitching, J. A. 1934. The physiology of contractile vacuoles. 1. Osmotic relationships. *J. exp. Biol.* 11:364–81.

Kleiber, M. 1932. Body size and animal metabolism. *Hilgardia* 6:315–53.

Kleiber, M. 1945. Body size and metabolism of liver slices *in vitro*. *Proc. Soc. Exp. Biol. Med.* 48:419–22.

Kleiber, M. 1975. *The fire of life*. New York: Krieger Publishing Company.

Klotz, I. M., G. L. Klippenstein, and W. A. Hendrickson. 1976. Hemerythrin: An alternative oxygen carrier. *Science* 192:335–44.

Klotz, I. M., T. A. Klotz, and H. A. Fiess. 1957. The nature of the active site of hemerythrin. *Arch. Biochem. Biophys.* 68:284–99.

Kluger, M. J. 1979. Phylogeny of fever. *Fed. Proc.* 38:30–34.

Knowles, F. G. W. 1963. The structure of neurosecretory systems in invertebrates. *Comp. Endocrinol.* 2:47–63.

Koechlin, N. 1975. Micropuncture studies of urine formation in a marine invertebrate *Sabella pavonia* Savigny (Polychaeta: Annelida) *Comp. Biochem. Physiol.* 52A:459–64.

Koeppe, J. K., M. Fuchs, T. T. Chen, L.-M. Hunt, G. E. Kovalick, and T. Briers. 1985. The role of juvenile hormone in reproduction. In *Comprehensive insect physiology biochemistry and pharmacology*, vol. 8, *Endocrinology II*, edited by G. A. Kerkut and L. I. Gilbert, 165–203. Oxford: Pergamon Press.

Konno, K., K.-I. Arai, and S. Watanabe. 1980. Myosin-linked calcium regulation in squid muscle. In *Muscle contraction: Its regulatory mechanisms*, edited by S. Ebashi, K. Maruyama, and M. Endo, 391–99. Tokyo: Japan Scientific Societies Press.

Kosterlitz, H. W. 1985. Has morphine a physiological role in the animal kingdom? *Nature* 317:671–72.

Kramer, D. L. 1978. Ventilation of the respiratory gas bladder in *Hoplerythrinus unitaeniatus* (Pisces, Characoidei, Erythrinidae). *Can. J. Zool.* 56:931–38.

Krantz, G. W. 1974. *Phaulodinychus mitis* (Leonardi 1899) (Acari: Uropodidae). An intertidal mite exhibiting plastron respiration. *Acaryologia* 16:11–20.

Krasnoff, S. B., and W. L. Roelofs. 1988. Sex pheromone released as an aerosol by the moth *Pyrrharctica isabella*. *Nature* 333:263–65.

Kreuzer, F. 1970. Facilitated diffusion of oxygen and its possible significance; a review. *Respir. Physiol.* 9:1–30.

Kristensen, R. M., and A. Hay-Schmidt. 1989. The protonephridia of the Arctic kinorhynch *Echinoderes aquilonius* (Cyclorhagida, Echinoderidae). *Acta Zool.* 70:13–27.

Krogh, A. 1920. Studien uber tracheenrespiration. 2 Ueber gasdiffusion in den tracheen. *Pflugers Arch.* 179:95–112.

Krogh, A. 1941. *The comparative physiology of respiratory mechanisms*. Philadelphia: University of Pennsylvania Press.

Krogh, A. 1965. Osmotic regulation in aquatic animals. New York: Dover.

Krogh, A., and M. Krogh. 1910. On the rate of diffusion of carbonic monoxide into the lungs of man. The mechanism of gas-exchange. VI. *Skand. Arch. Physiol.* 23.

Kronfeld, D. S., and P. J. Van Soest. 1976. Carbohydrate nutrition. *Comp. Anim. Nutr.* 1:23–73.

Krstic, R. V. 1979. *Ultrastructure of the mammalian cell*. Berlin: Springer-Verlag.

Kuethe, D. O. 1988. Fluid mechanical valving of air flow in bird lungs. *J. exp. Biol.* 136:1–12.

Kuffler, S. W. 1967. Neuroglial cells: Physiological properties and potassium mediated effect of neuronal activity on the glial membrane potential. *Proc. Roy. Soc. Lond. B* 168:1–21.

Kuffler, S. W., and D. Yoshikami. 1975. The number of transmitter molecules in a quantum: An estimate from iontophoretic application of acetylcholine at the neuromuscular synapse. *J. Physiol.* 251:465–82.

Kuhn, W., A. Ramel, H. J. Kuhn, and E. Marti. 1963. The filling mechanism of the swimbladder: Generation of high gas pressures through hairpin countercurrent multiplication. *Experientia* 19:497–511.

Kummel, G. 1962. Zwei neue formen von cyrtocyten vergleich der bisher bekannten cyrtocyten und erorterung des begriffes "zelltyp". *Z. Zellforsch.* 57:172–201.

Kung, C., and R. Eckert. 1972. Genetic modification of electrical properties in an excitable membrane. *Proc. Nat. Acad. Sci. USA.* 69:93–97.

Kupfermann, I., V. Castelucci, H. Pinsker, and E. Kandel. 1970. Neuronal correlates of habituation and dishabituation of the gill withdrawal reflex in *Aplysia*. *Science* 167:1743–45.

Kuppers, J., and U. Thurm. 1980. Water transport by electroosmosis. In *Insect biology in the future*, edited by M. Locke and D. Smith, 125–44. New York: Academic Press.

Kusano, K. 1968. Further study of the relationship between pre- and postsynaptic potentials in the squid giant synapse. *J. Gen. Physiol.* 52:326–45.

Lakes, R., and T. Schikorski. 1990. The neuroanatomy of tettigoniids. In *The Tettigoniidae: Biology, systematics and evolution*, edited by W. J. Bailey and D. C. F. Rentz, 166–90. Bathurst: Crawford House Press.

Lamb, D. R. 1984. *Physiology of exercise*. New York: Macmillan.

Lamy, J., J. Lamy, P.-Y. Sizaret, P. Billiald, and G. Motta. 1985. Quaternary structure of arthropod hemocyanins. In *Respiratory pigments in animals: Relation structure-function*, edited by J. Lamy, J.-P. Truchot, and R. Gilles, 73–86. Berlin: Springer-Verlag.

Lancet, D., D. Lazard, J. Heldman, M. Khen, and P. Nef. 1988. Molecular transduction in smell and taste. *Cold Spring Harbor Symp. Quant. Biol.* 53:343–48.

Land, M. F. 1985. The eye: Optics. In *Comprehensive insect physiology, biochemistry and pharmacology*, vol. 6, *Nervous system: Sensory*, edited by G. A. Kerkut and L. I. Gilbert, 225–75. Oxford: Pergamon Press.

Land, M. F. 1987. Vision in air and water. In *Comparative Physiology: Life in Water and on Land*, edited by P. Dejours, L. Bolis, C. R. Taylor, and E. R. Weibel, 289–302. Fidia Res. Ser. 9. Padora: Liviana Press, and Berlin: Springer-Verlag.

Lashley, K. S. 1950. In search of the engram. *Symp. Soc. Exp. Biol.* 4:454–82.

Lasker, R., and A. C. Giese. 1956. Cellulase digestion by the silverfish *Ctenolepisma lineata*. *J. exp. Biol.* 33:542–53.

Laurent, P. 1982. Structure of vertebrate gills. In *Gills*, edited by D. F. Houlihan, J. C. Rankin, and T. J. Shuttleworth, 25–43. Cambridge: Cambridge University Press.

Laverack, M. S. 1974. The structure and function of chemoreceptor cells. In *Chemoreception in marine organisms*, edited by P. T. Grant and A. M. Mackie, 1–48. New York: Academic Press.

Lawn, I. D., G. O. Mackie, and G. Silver. 1981. Conduction system in a sponge. *Science* 211:1169–71.

Lawrence, J. 1987. *A functional biology of echinoderms*. London: Croon Helm.

Layton, H. E. 1987. Energy advantages of countercurrent oxygen transfer in fish gills. *J. Theor. Biol.* 125:307–16.

Lazzari, C. R., and J. A. Nunez. 1989. The response to radiant heat and the estimation of the temperature of distant sources in *Triatoma infestans*. *J. Insect Physiol.* 35:525–29.

Lechner, A. 1978. The scaling of maximal oxygen consumption and pulmonary dimensions in small mammals. *Respir. Physiol.* 34:29–44.

Lee, R. E. 1989. Insect cold-hardiness: To freeze or not to freeze? *Bioscience* 39:308–13.

Lees, A. D. 1946. Chloride regulation and the function of the coxal gland in ticks. *Parasitology* 37:172–84.

Lehman, W., and A. G. Szent-Gyorgyi. 1975. Regulation of muscular contraction: Distribution of actin and myosin control in the animal kingdom. *J. Gen. Physiol.* 66:1–30.

Lehninger, A. L. 1982. *Principles of biochemistry*. New York: Worth.

Lemire, M., C. Grenot, and R. Vernet. 1982. Water and electrolyte balance of free-living Saharan lizards, *Uromastix acanthinurus* (Agamidae). *J. Comp. Physiol.* 146:81–93.

Lenfant, C., and K. Johansen. 1965. Gas transport by hemocyanin-containing blood of the octopus *Octopus dofleini*. *Am. J. Physiol.* 209:991–98.

Lenfant, C., K. Johansen, and J. D. Torrence. 1970. Gas transport and oxygen storage capacity in some pinnipeds and the sea otter. *Respir. Physiol.* 9:277–86.

Lenhoff, H. M., and K. J. Lindstedt. 1974. Chemoreception in aquatic invertebrates with special emphasis on the feeding behavior of coelenterates. In *Chemoreception in marine organisms*, edited by P. T. Grant and A. M. Mackie, 143–175. New York: Academic Press.

Lentz, T. L. 1968. *Primitive nervous systems*. New Haven: Yale University Press.

Lerner, A. B. 1981. The intermediate lobe of the pituitary gland: Introduction and background. In *Peptides of the pars intermedia*, edited by D. Evered and G. Lawrenson, 3–12. Ciba Foundation Symp. 81. London: Pitman Medical.

LeRoith, D., and J. Roth. 1984. Are messenger-like molecules in unicellular microbes the common phylogenetic ancestors of vertebrate hormones, tissue factors and neurotransmitters? In *Frontiers in physiological research*, edited by D. G. Garlick and P. I. Korner, 87–98. Canberra: Aust. Acad. Sci.

Lever, J., M. Kok, E. A. Meuleman, and J. Joosse. 1961. On the location of gomori-positive neurosecretory cells in the central ganglia of *Lymnaea stagnalis*. *Koninkl. Ned. Acad. Wetenschap. Proc. (C)* 64:640–47.

Levi, H. W. 1967. Adaptations of respiratory systems of spiders. *Evolution* 21:571–83.

Levy, R. I., and H. A. Schneiderman. 1966. Discontinuous respiration in insects. IV. Changes in intratracheal pressure during the respiratory cycle of silkworm pupae. *J. Insect Physiol.* 12:465–92.

Lewis, S. A. 1986. The mammalian urinary bladder: It's more than accomodating. *NIPS* 1:61–65.

Lillywhite, H. B., and J. A. Donald. 1989. Pulmonary blood flow regulation in an aquatic snake. *Science* 245:293–95.

Lindauer, M., and H. Martin. 1968. Die schwereorientierung der bienen unter dem einfluss des erdmagnetfelds. *Z. vergl. Physiol.* 60:219–43.

Ling, S. C., H. B. Atabek, D. L. Fry, D. J. Patel, and J. S. Janicki. 1968. Application of heated-film velocity and shear probes ot hemodynamic studies. *Circ. Res.* 23:789–801.

Livingston, R. A., and A. de Zwaan. 1983. Carbohydrate metabolism of gastropods. In *The Mollusca*, vol. 1, *Metabolic biochemistry and molecular biomechanics*, edited by P. W. Hochachka, 177–242. New York: Academic Press.

Lockwood, A. P. M. 1978. *The membranes of animal cells*. London: Edward Arnold.

Lofts, B., B. K. Follett, and R. K. Mutron. 1970. Temporal changes in the pituitary-gonadal axis. In *Hormones and the environment*, edited by G. K. Benson and J. G. Phillips, 545–72. Cambridge: Cambridge University Press.

Loudon, A., N. Rothwell, and M. Stock. 1985. Brown fat, thermogenesis and physiological birth in a marsupial. *Comp. Biochem. Physiol.* 81A:815–19.

Loveridge, J. P. 1970. Observations of nitrogenous excretion and water relations of *Chiromantis xerampelina* (Amphibia, Anura). *Arnoldia* 5:1–6.

Low, P. S. 1985. Molecular basis of the biological compatibility of nature's osmolytes. In *Transport processes, iono- and osmoregulation*, edited by R. Gilles and M. Gilles-Baillien, 469–77. Berlin: Springer-Verlag.

Lowe, C. H., P. J. Larimer, and E. A. Halpern. 1971. Supercooling in reptiles and other vertebrates. *Comp. Biochem. Physiol.* 39A:125–35.

Lowenstein, W. R. 1976. Permeable junctions. *Cold Spring Harbor Symp. Quant. Biol.* 40:49–63.

Lowenstein, W. R., and M. Mendelson. 1965. Components of receptor adaptation in a Pacinian corpuscle. *J. Physiol* 177:377–97.

Lunevsky, V. Z., O. M. Zherelova, I. Y. Vostrikov, and G. N. Berestovsky. 1983. Excitation of Characeae membranes as a result of activation of calcium and chloride channels. *J. Membrane Biol.* 72:43–58.

Lustick, S., B. Battersby, and M. Kelty. 1979. Effects of insolation on juvenile herring gull energetics and behavior. *Ecology* 60:673–78.

MacDonald, A. G., and A. R. Cossins. 1985. The theory of homeoviscous adaptation of membranes applied to deep sea animals. In *Physiological adaptations of marine animals*, edited by M. S. Laverack. *Symp. Soc. Exp. Biol.* 39:309–22.

Machemer, H. 1974. Ciliary activity and metachronism in protozoa. In *Cilia and flagella*, edited by M. A. Sleigh, 199–286. New York: Academic Press.

Mackie, A. M., and P. T. Grant. 1974. Interspecies and intraspecies chemoreception by marine invertebrates. In *Chemoreception in marine organisms*, edited by P. T. Grant and A. M. Mackie, 105–41. New York: Academic Press.

Mackie, G. O. 1965. Conduction in the nerve-free epithelia of siphonophores. *Amer. Zool.* 5:439–53.

MacMillan, R. E., and J. E. Nelson. 1969. Bioenergetics and body size in dasyurid marsupials. *Amer. J. Physiol.* 217:1246–51.

MacMillan, R. E. 1965. Aestivation in the cactus mouse, *Peromyscus eremicus*. *Comp. Biochem. Physiol.* 16:227–48.

Maddox, J. 1989. Heat conduction is a can of worms. *Nature* 338:373.

Maddrell, S. H. P. 1969. Secretion by the Malpighian tubules of *Rhodnius*. The movements of ions and water. *J. exp. Biol.* 51:71–97.

Maddrell, S. H. P. 1971. The mechanisms of insect excretory systems. *Adv. Insect Physiol.* 8:199–331.

Maddrell, S. H. P., and B. O. C. Gardiner. 1974. The passive permeability of insect Malpighian tubules to organic solutes. *J. exp. Biol.* 60:641–52.

Maddrell, S. H. P., and J. J. Nordman. 1979. *Neurosecretion*. Glasgow: Blackie and Son.

Maeda, N., S. Miyoshi, and H. Toh. 1983. First observation of a muscle spindle in fish. *Science* 302:61–62.

Maina, J. N. 1984. Morphometrics of the avian lung. 3. The structural design of the passerine lung. *Respir. Physiol.* 55:291–307.

Maitland, D. P. 1986. Crabs that breathe air with their legs—*Scopimera* and *Dotilla*. *Nature* 319:493–95.

Malan, A. 1986. pH as a control factor in hibernation. In *Living in the cold*, edited by H. C. Heller, X. J. Mussachia, and L. C. H. Wang, 61–70. New York: Elsevier.

Malik, F., and R. Vale. 1990. A new direction for kinesin. *Nature* 347:713–14.

Mallefet, J., and F. Baguet. 1984. Oxygen consumption and luminescence of *Porichthys* photophores stimulated by potassium cyanide. *J. exp. Biol.* 109:341–52.

Mallefet, J., and F. Baguet. 1985. Effects of adrenalin on the oxygen consumption and luminescence of the photophores of the mesopelagic fish *Argyropelecus hemigymnus*. *J. exp. Biol.* 118:341–49.

Mangum, C. P. 1982. The function of gills in several groups of invertebrate animals. In *Gills*, edited by D. F. Houlihan, J. C. Rankin, and T. J. Shuttleworth, 77–97. Cambridge: Cambridge University Press.

Mangum, C. P., and D. W. Towle. 1977. Physiological adaptation to unstable environments. *Am. Sci.* 65:67–75.

Manwell, C. 1960. Histological specificity of respiratory pigments. II. Oxygen transfer systems involving hemerythrins in sipunculid worms of different ecologies. *Comp. Biochem. Physiol.* 1:277–85.

Margulis, L. 1975. Symbiotic theory for the origin of

eukaryotic organelles: Criteria for proof. *Symp. Exp. Biol.* 29:21–37.

Markl, H. 1983. Geobiophysics: The effect of ambient pressure, gravity, and of the geomagnetic field on organisms. In *Biophysics*, edited by W. Hoppe, W. Lohmann, H. Markl, and H. Ziegler, 776–87. Berlin: Springer-Verlag.

Marmor, M. F., and A. L. F. Gorman. 1970. Membrane potential as the sum of ionic and metabolic components. *Science* 167:65–67.

Marshall, A. T., and W. W. K. Cheung. 1974. Studies on water and ion transport in homopteran insects: Ultrastructure and cytochemistry of the cicadoid and cercopoid Malpighian tubules and filter chamber. *Tissue and Cell* 6:153–71.

Martin, A. W., and K. Johansen. 1965. Adaptations of the circulation in invertebrate animals. In *Handbook of physiology*, sect. 2, *Circulation*, vol. 3, edited by W. F. Hamilton and P. Dow, 2545–81. Washington: American Physiological Society.

Martin, A. W., D. M. Stewart, and F. M. Harison. 1965. Urine formation in the pulmonate land snail, *Achatina fulica. J. exp. Biol.* 42:99–124.

Martin, M. M. 1983. Cellulose digestion in insects. *Comp. Biochem. Physiol.* 75A:313–24.

Mathur, G. B. 1967. Anaerobic metabolism in a cyprinoid fish, *Rasbora daniconius. Nature* 214:318–19.

Matsumoto, K. 1958. Morphological studies on the neurosecretion in crabs. *Biol. J. Okayama Univ.* 4:103–76.

May, M. L. 1976. Warming rates as a function of body size in periodic endotherms. *J. Comp. Physiol.* 111:55–70.

Maynard, D. M. 1960. Circulation and heart function. In *The physiology of Crustacea*, vol. 1, edited by T. H. Waterman, 161–226. New York: Academic Press.

Mazur, P. 1963. Kinetics of water loss from cells at subzero temperatures and the likelihood of intracellular freezing. *J. Gen. Physiol.* 47:347–69.

McClanahan, L. L., J. N. Stinner, and V. H. Shoemaker. 1978. Skin lipids, water loss, and energy metabolism in a South American tree frog (*Phyllomedusa sauvagei*). *Physiol. Zool.* 51:179–87.

McClanahan, L. L., R. Ruibal, and V. H. Shoemaker. 1983. Rate of cocoon formation and its physiological correlates in a ceratophryd frog. *Physiol. Zool.* 56:430–35.

McConnell, J. V. 1966. Learning in invertebrates. *Ann. Rev. Physiol.* 28:107–36.

McConnell, J. V., A. L. Jacobson, and D. P. Kimble. 1959. The effects of regeneration upon retention of a conditional response in the planarian. *J. Comp. Physiol. Psychol.* 52:1–5.

McCormick, J. G., E. G. Wever, S. H. Ridgway, and J. Palin. 1980. Sound reception in the porpoise as it relates to echolocation. In *Animal sonar systems*, edited by R.-G. Busnel and J. F. Fish, 199–238. New York: Plenum Press.

McDonald, D. G., B. R. McMahon, and C. M. Wood. 1977. Patterns of heart and scaphognathite activity of the crab *Cancer magister. J. Exp. Zool.* 202:33–44.

McElroy, W. D., and M. A. Deluca. 1981. Chemistry of firefly luminescence. In *Bioluminescence and chemiluminescence: A basic chemistry and analytical applications*, edited by M. A. Deluca and W. D. McElroy, 109–28. New York: Academic Press.

McEnroe, W. D. 1963. The role of the digestive system in the water balance of the two-spotted spider mite. *Adv. Acarol.* 1:225–31.

McGilvery, R. W., and G. W. Goldstein. 1983. *Biochemistry: A functional approach*. Philadelphia: Saunders Company.

McGowan, C. 1979. Selection pressure for high body temperatures: Implications for dinosaurs. *Paleobiology* 5:285–95.

McIntosh, J. R. 1984. Mechanics of mitosis. *Trends Biochem. Sci.* 9:195–98.

McLaughlin, P. A. 1983. Internal anatomy. In *The biology of Crustacea*, vol. 5, *Internal anatomy and physiological regulation*, edited by L. H. Mantel, 1–52. New York: Academic Press.

McMahon, B. R., and J. C. Wilkins. 1983. Ventilation, perfusion and oxygen uptake. In *The biology of Crustacea*, vol. 5, *Internal anatomy and physiology: Regulation*, edited by L. H. Mantel, 289–372. New York: Academic Press.

McMahon, B. R., and W. W. Burggren. 1979. Respiration and adaptation to the terrestrial habit in the land hermit crab *Coenobita clypeatus. J. exp. Biol.* 79:265–81.

McMahon, T. A. 1973. Size and shape in biology. *Science* 179:1201–4.

McNab, B. K. 1971. On the ecological significance of Bergmann's rule. *Ecology* 52:845–54.

McNab, B. K. 1978. The evolution of endothermy in the phylogeny of mammals. *Am. Nat.* 112:1–21.

McNab, B. K. 1983. Ecological and behavioral consequences of adaptation to various food resources. In *Recent advances in the study of mammalian behavior*, edited by J. E. Eisenberg and D. G. Kleiman, *Spec. Publ. Amer. Soc. Mammal.* 7:664–97.

McNabb, R. A. 1974. Urate and cation interactions in the liquid and precipitated fractions of avian urine, and speculations on their physico-chemical state. *Comp. Biochem. Physiol.* 48A:45–54.

McNaught, A. B., and R. Callander. 1975. *Illustrated physiology*. Edinburgh: Churchill Livingstone.

Mechtley, E. A. 1973. *The international system of units: Physical constants and conversion factors*. NASA SP-7012. Washington, National Aeronautics and Space Administration.

Mellanby, K. 1932. The effect of atmospheric humidity on the metabolism of the fasting mealworm (*Tenebrio molitor* L., Coleoptera) *Proc. Roy. Soc. Lond.* 111B:376–90.

Merrill, E. W. 1969. Rheology of blood. *Physiol. Rev.* 49:863–88.

Meves, H., and W. Vogel. 1973. Calcium inward current in internodally perfused giant axons. *J. Physiol.* 235:225–65.

Meyer, D. B. 1986. The avian eye and vision. In *Avian physiology*, edited by P. D. Sturkie, 38–47. New York: Springer-Verlag.

Meyrand, P., J. Simmers, and M. Moulins. 1991. Construction of a pattern-generating circuit with neurons of different networks. *Nature* 351:60–63.

Mill, P. J. 1972. *Respiration in the invertebrates*. London: MacMillan.

Mill, P. J. 1973. Respiration: Aquatic insects. In *The physiology of Insecta*, vol. 6, edited by M. Rockstein, 346–402. New York: Academic Press.

Millard, N. D., B. G. King, and M. J. Showers. 1956. *Human anatomy and physiology*. Philadelphia: Saunders.

Millechia, R., and A. J. Mauro. 1969. The ventral photoreceptor cells of *Limulus*. III. A voltage-clamp study. *J. Gen. Physiol.* 54:331–51.

Miller, P. L. 1966. The function of haemoglobin in relation to the maintenance of neutral buoyancy in *Anisops pellucens* (Notonectidae, Hemiptera). *J. exp. Biol.* 44:529–43.

Miller, P. L. 1973. Respiration—aerial gas transport. In *The physiology of Insecta*, vol. 6, edited by M. Rockstein, 346–402. New York: Academic Press.

Miller, T. A. 1985. Structure and physiology of the circulatory system. In *Comprehensive insect physiology, biochemistry, and pharmacology*, vol. 3, *Integument, respiration and circulation*, edited by G. A. Kerkut and L. I. Gilbert, 289–353. Oxford: Pergamon Press.

Miller, W. 1979. Ocular optical filtering. In *Handbook of sensory physiology*, vol. 7/6A, edited by H. Autrum, 69–143. Berlin: Springer-Verlag.

Miller, W. H., A. R. Mollner, and C. G. Bernhard. 1966. The corneal nipple array. In *The functional organization of the compound eye*, edited by C. G. Bernhard, 21–33. Oxford: Pergamon Press.

Millhorn, H. T., and P. E. Pulley. 1968. A theoretical study of pulmonary capillary gas exchange and venous admixture. *Biophys. J.* 8:337–57.

Milnor, W. R. 1982. *Hemodynamics*. Baltimore: Williams and Wilkins.

Minnich, J. E. 1979. Reptiles. In *Comparative physiology of osmoregulation in animals*, edited by G. M. O. Maloiy, 391–641. New York: Academic Press.

Minnich, J. E. 1982. The use of water. In *Biology of the Reptilia*, vol. 12, *Physiology C Physiological ecology*, edited by C. Gans and F. H. Pough, 325–96. New York: Academic Press.

Mitchel, R. 1972. The tracheae of water mites. *J. Morph.* 136:327–36.

Mitchell, P. 1979. Keilin's respiratory chain concept and its chemiosmotic consequences. *Science* 206:1148–59.

Mitchison, T. J. 1986. The role of microtubule polarity in the movement of kinesin and kinetochores. *J. Cell Sci. Suppl.* 5:121–28.

Moir, R. J. 1968. Ruminant digestion and evolution. In *Handbook of physiology*, vol. 5, sect. 6, edited by C. F. Code, 2673–94. Washington: American Physiological Society.

Mommsen, T. P., and P. J. Walsh. 1989. Evolution of urea synthesis in vertebrates: The piscine connection. *Science* 243:72–75.

Monod, J., J. Wyman, and J. P. Changeaux. 1965. On the nature of allosteric transition: A plausible model. *J. Molec. Biol.* 12:88–118.

Montal, M., R. Anholt, and P. Iabarca. 1986. The reconstituted acetylcholine receptor. In *Ion channel reconstitution*, edited by C. Miller, 157–204. New York: Plenum Press.

Monteith, J. L., and G. S. Campbell. 1980. Diffusion of water vapor through integument—potential confusion. *J. Thermal Biol.* 5:7–9.

Moraczewski, J., A. Czubaj, and J. Bakowska. 1977. Organization and ultrastructure of the nervous system in *Catenulida*. *Zoomorph.* 87:87–95.

Morgareidge, K. R., and F. N. White. 1969. Cutaneous vascular changes during heating and cooling in the Galapagos marine iguana. *Nature* 223:587–91.

Morrison, P. M., and W. J. Tietz. 1957. Cooling and thermal conductivity in three small Alaskan mammals. *J. Mamm.* 38:78–86.

Morrison, S. D. 1968. The constancy of the energy expended by rats on spontaneous activity, and the distribution of activity between feeding and non-feeding. *J. Physiol.* 197:305–23.

Morton, B. 1983. Feeding and digestion in Bivalvia. In *The Mollusca*, vol. 5, *Physiology*, part 2, edited by A. S. M. Saleuddin and K. M. Wilbur, 65–147. New York: Academic Press.

Moyle, V. 1949. Nitrogenous excretion in chelonian reptiles. *Biochem. J.* 44:581–84.

Mullins, D. E. 1985. Chemistry and physiology of the hemolymph. In *Comprehensive insect physiology, biochemistry and pharmacology*, edited by G. A. Kerkut and L. I. Gilbert, vol. 3, 355–400. Oxford: Pergamon Press.

Munro, A. F. 1953. The ammonia and urea excretion of different species of Amphibia during their development and metamorphosis. *Biochem. J.* 54:29–36.

Munshi, J. S. D. 1976. Gross and fine structure of the respiratory organs of air-breathing fishes. In *Respiration of amphibious vertebrates*, edited by G. M. Hughes, 73–104. New York: Academic Press.

Muntz, W. R. A., and U. Raj. 1984. On the visual system of *Nautilus pompilius*. *J. exp. Biol.* 109:253–64.

Murphy, D. J., and S. K. Pierce. 1975. The physiological basis for changes in the freezing tolerance of intertidal molluscs. I. Response to subfreezing temperatures and the influences of salinity and temperature acclimation. *J. Exp. Zool.* 193:313–22.

Murray, R. W. 1960. The response of the ampullae of Lorenzini of elasmobranchs to mechanical stimulation. *J. exp. Biol.* 37:417–24.

Murrish, D. E., and K. Schmidt-Nielsen. 1970. Water transport in the cloaca of lizards: Active or passive transport? *Science* 170:324–26.

Muybridge, E. 1957. *Animals in motion*. New York: Dover.

Nachtigall, W. 1960. Uber kinematik, dynamik und energetik des schwimmens einheimischer dytisciden. *Z. verg. Physiol.* 43:48–118.

Nachtigall, W. 1983. The biophysics of locomotion in water. In *Biophysics*, edited by W. Hoppe, W. Lohmann, H. Markl, and H. Ziegler, 587–600. Berlin: Springer-Verlag.

Nachtigall, W. 1983. The biophysics of locomotion on land. In *Biophysics*, edited by W. Hoppe, W. Lohmann, H. Markel, and H. Ziegler, 580–587. Berlin: Springer-Verlag.

Nagai, M., N. Oshima, and R. Fujii. 1986. A comparative study of melanin-concentrating hormone (MCH) action on teleost melanophores. *Biol. Bull.* 171:360–70.

Nagy, K. A. 1987. Field metabolic rate and food requirement scaling in mammals and birds. *Ecol. Mono.* 57:111–28.

Nagy, K. A., D. K. O'Dell, and R. S. Seymour. 1972. Temperature regulation by the inflorescence of *Philodendron*. *Science* 178:1195–97.

Naitoh, Y., and R. Eckert. 1974. The control of ciliary activity in Protozoa. In *Cilia and flagella*, edited by M. A. Sleigh, 305–52. London: Academic Press.

Naka, K.-I. 1961. Recording of retinal action potentials from single cells in the insect compound eye. *J. Gen. Physiol.* 44:571–84.

Natochin, Y. V., and R. G. Parnova. 1957. Osmolality and electrolyte concentration of hemolymph and the problem of ion and volume regulation in cells of higher insects. *Comp. Biochem. Physiol.* 88A:563–70.

Necker, R. 1981. Thermoreception and temperature regulation of homeothermic vertebrates. In *Progress in sensory physiology*, edited by D. Ottoson, 1–47. Berlin, Springer-Verlag.

Needham, J. 1931. *Chemical embryology.* Cambridge: Cambridge University Press.

Nestler, J. R. 1990. Relationships between respiratory quotient and metabolic rate during entry to and arousal from daily torpor in deer mice (*Peromyscus maniculatus*). *Physiol. Zool.* 63:504–15.

Neuwiler, C. 1983. Echolocation. In *Biophysics*, edited by W. Hoppe, W. Lohmann, H. Markel, and H. Ziegler, 683–97. Berlin: Springer-Verlag.

Newell, P. F., and G. E. Newell. 1968. The eye of the slug *Agriolimax reticulatus* (Mull.). *Symp. Zool. Soc. Lond.* 23:97–112.

Newman, E. A., and P. H. Hartline. 1982. The infrared "vision" of snakes. *Sci. Am.* 246(3):98–107.

Niall, H. D. 1982. The evolution of peptide hormones. *Ann. Rev. Physiol.* 44:615–24.

Nichol, J. A. C. 1948. The giant axons of annelids. *Quart. Rev. Biol.* 23:291–323.

Nichols, J. G., and D. van Essen. 1974. The nervous system of the leech. *Sci. Am.* 230(1):38–48.

Nicholson, G. L. 1976. Transmembrane control of the receptors on normal and tumor cells. I. Cytoplasmic influence over cell surface components. *Biochim. Biophys. Acta* 457:57–108.

Nicklaus, R. B. 1988. The forces that move chromosomes in mitosis. *Ann. Rev. Biophys. Biophys. Chem.* 17:431–50.

Nielsen, B. 1962. On the regulation of respiration in reptiles. II. The effect of hypoxia with and without moderate hypercapnia on the respiration and metabolism of lizards. *J. exp. Biol.* 39:107–17.

Nishi, S., and K. Koketsu. 1960. Electrical properties and activities of single sympathetic neurons in frogs. *J. Cell. Comp. Physiol.* 55:15–30.

Nishiitsutsujo-Uwo, J., and C. S. Pittendrigh. 1968a. Circadian nervous system control of circadian rhythmicity on the cockroach. I. The optic lobes, locus of the driving oscillation? *Z. vergl. Physiol.* 58:14–46.

Nishiitsutsujo-Uwo, J., and C. S. Pittendrigh. 1968b. Circadian nervous system control of circadian rhythmicity on the cockroach. II. The pathway of light signals that entrain the rhythm. *Z. vergl. Physiol.* 58:1–13.

Nishimura, H. 1980. Evolution of the renin-angiotensin system. In *Evolution of vertebrate endocrine systems*, edited by P. K. T. Pang and A. Epple, 373–404. Lubbock: Texas Technical Press.

Noback, C. R. 1967. *The human nervous system.* New York: McGraw-Hill.

Nobel, P. 1983. *Biophysical plant physiology and ecology.* New York: Freeman and Company.

Noble, G. K. 1925. The integumentary, pulmonary and cardiac modifications correlated with increased cutaneous respiration in the Amphibia: A solution of the "hairy frog" problem. *J. Morph. Physiol.* 40:341–416.

Noble, G. K. 1931. *The biology of the Amphibia.* New York: Dover Publications.

Noble-Nesbitt, J. 1977. Active transport of water vapor. In *Transport of ions and water in animals,* edited by B. L. Gupta, R. B. Moreton, J. L. Oschman, and B. J. Wall, 571–98. London: Academic Press.

Noble-Nesbitt, J., and M. Al-Shukur. 1988. Cephalic neuroendocrine regulation of integumentary water loss in the cockroach *Periplaneta americana* L. *J. exp. Biol.* 136:451–59.

Nonnette, G., and R. Kirsch. 1978. Cutaneous respiration in seven sea-water teleosts. *Respir. Physiol.* 35:111–18.

Novikoff, M. M. 1953. Regularity of form in organisms. *Syst. Zool.* 2:57–62.

Nutting, W. L. 1958. A comparative anatomical study of the heart and accessory structures of the orthopteroid insects. *J. Morph.* 89:501–97.

O'Donnell, M. J. 1982a. Hydrophilic cuticle—the basis for water vapour absorption by the desert burrowing cockroach *Arenivaga investigata*. *J. exp. Biol.* 99:43–60.

O'Donnell, M. J. 1982b. Water vapour absorption by the desert burrowing cockroach. *Arenivaga investigata:* Evidence against a solute dependent mechanism. *J. exp. Biol.* 96:251–62.

Oglesby, L. C. 1968. Some osmotic responses of the sipunculid worm *Themiste dyscritum*. *Comp. Biochem. Physiol.* 26:155–77.

Oglesby, L. C. 1978. Salt and water balance. In *Physiology of annelids,* edited by P. J. Mill, 555–658. New York: Academic Press.

Oglesby, L. C. 1981. Volume regulation in aquatic invertebrates. *J. Exp. Zool.* 215:289–301.

Oglesby, L. C. 1982. Salt and water balance in the sipunculan *Phascolopsis gouldii:* Is any animal a "simple osmometer"? *Comp. Biochem. Physiol.* 71A:363–68.

Ohtsuki, I. 1980. Functional organization of the troponin-tropomyosin system. In *Muscle contraction: Its regulatory mechanisms,* edited by S. Ebashi, K. Maruyama, and M. Endo, 237–49. Tokyo: Japan Scientific Societies Press.

Okabe, S., and N. Hirokawa. 1990. Turnover of fluorescently labelled tubulin and actin in the axon. *Nature* 343:479–82.

Oldfield, B. P. 1985. The role of the tympanal membranes and the receptor array in the tuning of auditory receptors in bushcrickets. In *Acoustic and vibrational communication in insects,* edited by K. Kalmring and N. Elsner, 17–24. Berlin: Verlag Paul Parey.

Olive, P. J. W., and R. B. Clark. 1978. Physiology of reproduction. In *Physiology of annelids,* edited by P. J. Mill, 271–368. London: Academic Press.

Olivera, B. M., W. R. Gray, R. Zeikus, J. M. McIntosh, J. Varga, J. River, V. de Santos, and L. J. Cruz. 1985. Peptide neurotoxins from fish-hunting cone snails. *Science* 230:1338–43.

Olsson, R. 1969. Endocrinology of the Agnatha and Protochordata and problems of evolution of vertebrate endocrine systems. *Gen. Comp. Endocrinol. Suppl.* 2:485–99.

O'Mahoney, P. 1977. Respiration and acid-base balance in brachyuran decapod crustaceans: The transition from water to land. Ph.D. thesis, State University of New York at Buffalo.

O'Malley, B. W., and W. T. Schrader. 1976. The receptors of steroid hormones. *Sci. Am.* 234(2):32–43.

Orchard, I., and B. G. Loughton. 1985. Neurosecretion. In *Comprehensive insect physiology biochemistry and pharmacology,* vol. 7, *Endocrinology I,* edited by G. A. Kerkut and L. I. Gilbert, 61–107. Oxford: Pergamon Press.

Orkand, R. K., J. G. Nicholls, and S. W. Kuffler. 1966. The effect of nerve impulses on the membrane potential of glial cells in the central nervous system of amphibia. *J. Neurophysiol.* 29:788–806.

Ottoson, D. 1983. *Physiology of the nervous system.* London: MacMillan Press.

Paganelli, C. V., and H. Rahn. 1987. Diffusion-induced convective gas flow through the pores of the eggshell. *J. Exp. Zool. Suppl.* 1:173–80.

Palmer, J. D. 1976. *An introduction to biological rhythms.* New York: Academic Press.

Pappenheimer, J. R. 1953. Passage of molecules through capillary walls. *Physiol. Rev.* 33:387–423.

Parrish, O. O., and T. W. Putnam. 1977. Equations for the psychrometric determination of humidity from dewpoint and psychrometric data. NASA TN D-8401.

Parry, G. 1960. Excretion. In *The physiology of Crustacea,* vol. 1, *Metabolism and growth,* edited by T. H. Waterman, 341–66. New York: Academic Press.

Parsons, T. S. 1970. The nose and Jacobson's organ. In *Biology of the Reptilia,* vol. 2, *Morphology B.,* edited by C. Gans and T. S. Parsons, 99–191. New York: Academic Press.

Partridge, J. C. 1989. The visual ecology of avian cone oil droplets. *J. Comp. Physiol.* 165A:415–26.

Pasquis, P., A. Lacaisse, and P. Dejours. 1970. Maximal oxygen uptake in four species of small mammals. *Respir. Physiol.* 9:298–309.

Pass, G. 1980. The anatomy and ultrastructure of the antennae circulatory organs in the cockchafer beetle *Melolontha melolontha* L. (Coleoptera, Scarabaeidae). *Zoomorphology* 96:77–89.

Passmore, R., and J. V. G. A. Durnin. 1955. Human energy expenditure. *Physiol. Rev.* 35:801–40.

Patterson, D. J. 1980. Contractile vacuoles and associated structures: Their organization and function. *Biol. Rev.* 55:1–46.

Pattle, R. E. 1976. The lung surfactant in the evolutionary tree. In *Respiration of amphibious vertebrates,* edited by G. M. Hughes, 233–56. New York: Academic Press.

Peaker, M., and J. L. Linzell. 1975. *Salt glands in birds and reptiles.* Cambridge: Cambridge University Press.

Pearse, A. G. E. 1969. The cytochemistry and ultrastructure of polypeptide hormone-producing cells of the APUD series and the embryonologic, physiologic and pathologic implications of the concept. *J. Histochem. Cytochem.* 17:303–13.

Pelli, D. G., and S. C. Chamberlain. 1989. The visibility of 350° C black-body radiation by the shrimp *Rimicaris exoculata* and man. *Nature* 337:460–61.

Penry, D. L., and P. A. Jumars. 1986. Chemical reactor analysis and optimal digestion. *Bioscience* 36:310–15.

Penry, D. L., and P. A. Jumars. 1987. Modeling animal guts as chemical reactors. *American Naturalist* 129:69–96.

Peters, H. 1935. Uber den einfluss des salzgehaltes im aussenmedium auf den bau und die funktion der exkretionsorgane dekapoder crustaceen. *Z. Morph. Okol. Tiere* 30:355–81.

Peters, H. M. 1976. On the mechanism of air ventilation in anabantoids (Pisces: Teleostei). *Zoomorphologie* 89:93–123.

Pettigrew, J. D., J. Wallman, and C. F. Wildsoet. 1990. Saccadic oscillations facilitate ocular perfusion from the avian pecten. *Nature* 343:362–63.

Pfaffman, C. 1975. Phylogenetic origins of sweet sensitivity. In *Olfaction and taste V,* edited by D. A. Denton and J. P. Coghlan, 3–10. New York: Academic Press.

Pfannenstiel, H. D. 1975. Mutual influence on the sexual differentiation in the protandric polychaete *Ophryotrocha puerilis.* In *Intersexuality in the animal kingdom,* edited by R. Reinboth, 48–56. Heidleberg: Springer-Verlag.

Phillips, J. 1981. Comparative physiology of insect renal function. *Am. J. Physiol.* 241:R241–57.

Phillipson, J. 1981. Bioenergetic options and phylogeny. In *Physiological ecology: An evolutionary approach to resource use,* edited by C. R. Townsend and P. Calow, 20–45. Oxford: Blackwell Scientific.

Picken, L. E. R. 1937. The mechanism of urine formation in invertebrates. II. The excretory mechanism in certain Mollusca. *J. exp. Biol.* 14:20–34.

Piiper, J., and P. Scheid. 1982. Models for a comparative functional analysis of gas exchange organs in vertebrates. *J. Appl. Physiol.* 53:1321–29.

Pilgrim, R. L. C. 1954. Waste of carbon and energy in nitrogen excretion. *Nature* 173:491–92.

Pirie, B. J. S., and S. G. George. 1979. Ultrastructure of the heart and excretory system of *Mytilus edulis* L. *J. mar. biol. Assoc. U. K.* 59:819–29.

Pitts, R. F. 1974. *Physiology of the kidney and body fluids*. Chicago: Year Book Medical Publishers.

Place, A. R., and D. A. Powers. 1979. Genetic variation and relative catalytic efficiencies: Lactate dehydrogenase B allozyme of *Fundulus heteroclitus*. *Proc. Natn. Acad. Sci. U.S.A.* 76:2354–58.

Poggio, G. F. 1980. Central neural mechanisms in vision. In *Medical physiology*, edited by V. B. Mountcastle, vol. 5, 544–85. St. Louis: C. V. Mosby.

Poggio, G. F., F. H. Baker, Y. Lamarre, and E. Riva. 1969. Afferent inhibition at input to visual cortex in the cat. *J. Neurophysiol.* 32:892–915.

Poll, M. 1962. Etude sur la structure adultere et la formation des sacs pulmonaires des Prototyperes. *Ann. Mus. R. Afr. centr. Ser. in 8°* 108:129–72.

Porter, K. R. 1976. Introduction: Motility in cells. In *Cell motility*, book A, *Motility, muscle and non-muscle cells*, edited by R. Goldman, T. Pollard, and J. Rosenbaum, 1–28. Cold Spring Harbor Conf. Cell Prolif., vol. 3. Cold Spring Harbor Laboratory.

Potts, W. T. W., and G. Parry. 1964. *Ionic and osmotic regulation in animals*. Oxford: Pergamon Press.

Pough, F. H. 1980. The advantages of ectothermy for tetrapods. *Am. Nat.* 115:92–112.

Powell, F. L., J. Geiser, R. K. Gratz, and P. Scheid. 1981. Airflow in the avian respiratory tract: Variations of O_2 and CO_2 concentrations in the bronchi of the duck. *Respir. Physiol.* 44:195–213.

Powning, R. F., M. F. Day, and H. Irzykiewicz. 1951. Studies on the digestion of wool by insects. II. The properties of some insect proteinases. *Aust. J. Sci. Res.* B4:49–63.

Precht, H. 1958. Concepts of the temperature adaptations of unchanging reaction systems of cold-blooded animals. In *Physiological adaptation*, edited by C. L. Prosser, 50–78. Washington: American Physiological Society.

Preston, R. L., and B. R. Stevens. 1982. Kinetic and thermodynamic aspects of sodium-coupled amino acid transport by marine invertebrates. *Amer. Zool.* 22:709–21.

Pribram, K. H., and D. E. Broadbent. 1970. *Biology of memory*. New York: Academic Press.

Prosser, C. L. 1958. General summary: The nature of physiological adaptation. In *Physiological adaptation*, edited by C. L. Prosser, 167–80. Washington: American Physiological Society.

Prosser, C. L. 1973. *Comparative animal physiology*. Philadelphia: Saunders.

Putnam, R. W. 1979a. The basis for differences in lactic acid content after activity in different species of anuran amphibians. *Physiol. Zool.* 52:509–19.

Putnam, R. W. 1979b. The role of lactic acid accumulation in muscle fatigue of two species of anurans, *Xenopus laevis* and *Rana pipiens*. *J. exp. Biol.* 82:35–51.

Quilliam, T. A., and M. Sato. 1955. The distribution of myelin on nerve fibers from Pacinian corpuscles. *J. Physiol.* 129:167–76.

Quinn, S. J., and G. H. Williams. 1988. Regulation of aldosterone secretion. *Ann. Rev. Physiol.* 50:409–26.

Rahn, H. 1974. Body temperature and acid-base regulation. *Pneumonologie* 151:87–94.

Rahn, H., and B. J. Howell. 1976. Bimodal gas exchange. In *Respiration of amphibious vertebrates*, edited by G. M. Hughes, 271–86. New York: Academic Press.

Rahn, H., and C. V. Paganelli. 1968. Gas exchange in gas gills of diving insects. *Respir. Physiol.* 5:145–64.

Rahn, H., C. V. Paganelli, and A. Ar. 1987. Pores and gas exchange of avian eggs: A review. *J. Exp. Zool. Suppl.* 1:165–72.

Rahn, H., R. B. Reeves, and B. J. Howell. 1975. Hydrogen ion regulation, temperature, and evolution. *Am. Rev. Resp. Disease* 112:165–72.

Rall, W. 1977. Core conductor theory and cable properties of neurons. In *Handbook of physiology*, sect. 1, vol. 1, part 1, *Nervous system*, edited by J. M. Brookhart and V. B. Mountcastle, 39–98. Bethesda: American Physiological Society.

Ralph, M. R., R. G. Foster, F. C. Davis, and M. Menaker. 1990. Transplanted suprachiasmatic nucleus determines circadian period. *Science* 247:975–78.

Ramsay, J. A. 1954. Active transport of water by the Malpighian tubules of the stick insect *Dixippus morosus* (Orthoptera, Phasmisae). *J. exp. Biol.* 31:104–13.

Ramsay, J. A. 1958. Excretion by the Malpighian tubules of the stick insect *Dixippus morosus* (Orthoptera, Phasmida): Amino acids, sugars and urea. *J. exp. Biol.* 35:871–91.

Ramsay, J. A. 1964a. *Physiology of lower animals*. Cambridge: Cambridge University Press.

Ramsay, J. A. 1964b. The rectal complex of the mealworm *Tenebrio molitor* L. (Coleoptera, Tenebrionidae). *Phil. Trans. Royal Soc. Lond.* 248B:279–314.

Randall, D. J., and C. Daxboeck. 1984. Oxygen and carbon dioxide transfer across fish gills. In *Fish physiology*, vol. 10, *Gills*, part A, *Anatomy, gas transfer and acid-base regulation*, edited by W. S. Hoar and D. J. Randall, 263–314. New York: Academic Press.

Randall, D. J., W. W. Burrgren, A. P. Farrell, and M. S. Haswell. 1981. *The evolution of air breathing in vertebrates*. Cambridge: Cambridge University Press.

Rao, K. R. 1985. Pigmentary effectors. In *The biology of Crustacea*, vol. 9, *Integument, pigments and hormonal processes*, edited by D. E. Bliss and L. H. Mantel, 395–462. New York: Academic Press.

Raskin, I., I. M. Turner, and W. R. Melander. 1987. Regulation of heat production in the inflorescences of an Arum lily by endogenous salicylic acid. *Proc. Nat. Acad. Sci. USA* 86:2214–18.

Rayner, J. M. V. 1979. A new approach to animal flight mechanics. *J. exp. Biol.* 80:17–54.

Redfield, A. C., and E. D. Mason. 1928. The respiratory

properties of the blood. III. The acid combining capacity and the dibasic amino-acid content of the hemocyanin of *Limulus polyphenus*. *J. Biol. Chem.* 77:451–57.

Redfield, A. C., and R. Goodkind. 1929. The significance of the Bohr effect in the respiration and asphyxiation of the squid *Loligo pealei*. *J. exp. Biol.* 6:340–49.

Redfield, A. C., T. Coolidge, and A. L. Hurd. 1926. Transport of oxygen and carbon dioxide by some bloods containing hemocyanin. *J. Biol. Chem.* 69:475–509.

Reeves, R. B. 1969. Role of body temperature in determining the acid-base status in vertebrates. *Fed. Proc.* 28:1204–08.

Reid, R. G. B., and F. R. Bernard. 1980. Gutless bivalves. *Science* 208:609–10.

Reppert, S. M., and W. J. Schwartz. 1983. Maternal coordination of the fetal biological clock. *Science* 220:969–71.

Retzlaff, E. 1954. Neurohistological basis for the functioning of paired half-centers. *J. Comp. Neurol.* 101:407–45.

Richards, O. W., and R. G. Davies. 1977. *Imm's general textbook of entomology*. London: Chapman and Hall.

Richter, C. P., and K. K. Rice. 1945. Self-selection studies on coprophagy as a source of vitamin B complex. *Am. J. Physiol.* 143:344–54.

Riegel, J. A. 1972. *Comparative physiology of renal excretion*. New York: Hafner Publishers.

Riggs, A. 1951. The metamorphosis of hemoglobin in the bullfrog. *J. Gen. Physiol.* 35:23–40.

Rimmer, D. W., and W. J. Wiebe. 1987. Fermentative microbial digestion in herbivorous fishes. *J. Fish. Biol.* 31:229–36.

Rios, E., and G. Pizarro. 1988. Voltage sensors and calcium channels of excitation-contraction coupling. *NIPS* 3:223–27.

Rivera, J. A. 1962. *Cilia, ciliated epithelium, and ciliary activity*. New York: Pergamon Press.

Robbins, C. T., S. Mole, A. E. Hagerman, and T. A. Hanley. 1987. Role of tannins in defending plants against ruminants: Reduction in dry matter digestion? *Ecology* 68:1606–15.

Roberts, J. L. 1975. Active branchial and ram gill ventilation in fishes. *Biol. Bull.* 148:85–105.

Robertson, J. D. 1976. Chemical composition of the body fluids and muscle of the hagfish *Myxine glutinosa* and the rabbit-fish *Chimaera monstrosus*. *J. Zool.* 178:261–77.

Robin, E. D., D. M. Travis, P. A. Bromberg, C. E. Forkner, and J. M. Taylor. 1959. Ammonia excretion by mammalian lung. *Science* 129:270–71.

Roeder, K. D. 1948. Organization of the ascending giant fiber system in the cockroach *(Periplaneta americana)*. *J. Exp. Zool.* 108:243–61.

Roeder, K. D. 1966. Auditory system of noctuid moths. *Science* 154:1515–21.

Rogers, E. 1986. *Looking at vertebrates: A practical guide to vertebrate adaptations*. London: Longman.

Rogers, W. P. 1949. On the relative importance of aerobic metabolism in small nematode parasites of the alimentary tract. 2. The utilization of oxygen at low partial pressure by small nematode parasites of the alimentary tract. *Aust. J. Sci. Res.* 2:166–74.

Romer, A. S. 1968. *Notes and comments on vertebrate palaentology*. Chicago: University Chicago Press.

Romer, A. S., and T. S. Parsons. 1986. *The vertebrate body*. Philadelphia: Saunders College Publishing.

Romer, H. 1985. Anatomical representation of frequency and intensity in the auditory system of Orthoptera. In *Acoustic and vibrational communication in insects*, edited by K. Kalmring and N. Elsner, 25–32. Berlin: Verlag Paul Parey.

Romspert, A. P. 1976. Osmoregulation of the African clawed frog *Xenopus laevis* in hyperosmotic media. *Comp. Biochem. Physiol.* 54A:207–10.

Roos, P. J. 1964. Lateral bending in newt locomotion. Koninklijke Nederlandse Akademie van Wetenschappen. *Proc. C.* 67:223–32.

Root, R. W. 1931. The respiratory function of the blood of marine fishes. *Biol. Bull.* 61:427–56.

Root, R. W., L. Irving, and E. C. Black. 1939. The effect of hemolysis upon the combination of oxygen with the blood of some marine fishes. *J. Cell. Comp. Physiol.* 13:303–13.

Rosenthal, J. 1969. Post-tetanic potentiation at the neuromuscular junction of the frog. *J. Physiol.* 203:121–33.

Rosenzweig, M. R. 1970. Evidence for anatomical and chemical changes in the brain during primary learning. In *Biology of memory*, edited by K. H. Pribram and D. E. Broadbent, 69–86. New York: Academic Press.

Ross, G. 1982. *Essentials of human physiology*. Chicago: Year Book Medical Publishers.

Rozin, P. 1965. Adaptive food sampling patterns in vitamin deficient rats. *J. Comp. Physiol. Psychol.* 69:126–32.

Rubenstein, D. I., and M. A. R. Koehl. 1977. The mechanisms of filter feeding: Some theoretical considerations. *Am. Nat.* 111:981–94.

Rubner, M. 1929. *Die gesetze des energieverbrauchs bei der ernahrung*. Leipzig: Deuticke.

Ruch, T. C., and H. D. Patton. 1979. *Physiology and biophysics*. Philadelphia: Saunders.

Rudolph, D. 1982. Site, process and mechanism of active uptake of water vapour from the atmosphere in the Psocoptera. *J. Insect Physiol.* 28:205–12.

Rudolph, D. N., and W. Knulle. 1982. Novel uptake systems for atmospheric water vapor absorption among insects. *J. Exp. Zool.* 222:321–33.

Ruibal, R. 1962. The adaptive value of bladder water in the toad, *Bufo cognatus*. *Physiol. Zool.* 35:218–23.

Ruppert, E. E., and P. R. Smith. 1988. The functional organization of filtration nephridia. *Biol. Rev.* 63:231–58.

Rushton, W. A. H. 1951. A theory of the effects of fibre size in medullated nerve. *J. Physiol.* 115:101–22.

Saffo, M. B. 1988. Nitrogen waste or nitrogen source? Urate degradation in the renal sac of molgulid tunicates. *Biol. Bull.* 175:403–9.

Salt, R. W. 1959. Role of glycerol in the cold-hardening of *Bracon cephi* (Gahan). *Can. J. Zool.* 37:59–69.

Salvini-Plawen, L. v. 1982. On the polyphyletic origin of photoreceptors. In *Visual cells in evolution,* edited by J. A. Westall, 137–54. New York: Raven Press.

Sanders, N. K., and J. J. Childress. 1988. Ion replacement as a buoyancy mechanism in a pelagic deep-sea crustacean. *J. exp. Biol.* 138:333–434.

Sandor, T., and A. Z. Mehdi. 1979. Steroids and evolution. In *Hormones and evolution,* edited by E. J. W. Barrington, 1–72. New York: Academic Press.

Santos, C. A. Z., C. H. S. Penteado, and E. G. Mendes. 1987. The respiratory responses of an amphibious snail *Pomacea lineata* (Spix, 1827) to temperature and oxygen tension variations. *Comp. Biochem. Physiol.* 86A:409–15.

Sargeant, J. R., and R. R. Gatten. 1976. The distribution and metabolism of wax esters in marine invertebrates. *Biochem. Soc. (Lond.) Trans.* 4:431–33.

Satchell, G. H. 1968. A neurological basis for the coordination of swimming with respiration in fish. *Comp. Biochem. Physiol.* 27:836–41.

Satir, P. 1968. Studies on cilia. III. Further studies on the cilium tip and a "sliding filament" model of ciliary motility. *J. Cell. Biol.* 39:77–94.

Saunders, D. S. 1977. Insect clocks. *Comp. Biochem. Physiol.* 56A:1–5.

Saz, H. J. 1981. Energy metabolisms of parasitic helminths. *Ann. Rev. Physiol.* 43:323–41.

Schaller, D. 1978. Antennal sensory system of Periplaneta americana. *Cell Tissue Res.* 191:121–39.

Scharrer, B. 1948. The prothoracic glands of *Leucophaea madeira* (Orthoptera). *Biol. Bull.* 95:186–98.

Scharrer, E. 1959. General and phylogenetic interpretations of neuroendocrine interrelations. In *Comparative endocrinology,* edited by A. Gorbman, 233–49. New York: John Wiley and Sons.

Scheich, H., G. Langner, C. Tidemann, R. B. Coles, and A. Guppy. 1986. Electroreception and electrolocation in platypus. *Nature* 319:401–02.

Scheid, P., and J. Piiper. 1976. Quantitative functional analysis of branchial gas transfer: Theory and application to *Scyliorhinus stellaris* (Elasmobranchii). In *Respiration of amphibious vertebrates,* edited by G. M. Hughes, 17–38. London: Academic Press.

Schick, J. M., A. deZwaan, and A. M. T. deBont. 1983. Anoxic metabolic rate in the mussel *Mytilus edulis* L. estimated by simultaneous direct calorimetry and biochemical analysis. *Physiol. Zool.* 56:56–63.

Schildknecht, H., and K. Holoubek. 1961. Die bombardierkafer und ihre explosionschemia. *Angewandte Chemie* 73:1–7.

Schmidt-Nielsen, B. 1977. Excretion in mammals: Role of the renal pelvis in the modification of the urinary concentration and composition. *Fed. Proc.* 36:2493–2503.

Schmidt-Nielsen, B. 1988. Excretory mechanisms as examples of the principle "The whole is greater than the sum of its parts." *Physiol. Zool.* 61:312–21.

Schmidt-Nielsen, K. 1960. The salt-secreting gland of marine birds. *Circulation* 21:955–67.

Schmidt-Nielsen, K. 1983. Animal physiology: Adaptation and environment. Cambridge: Cambridge University Press.

Schmidt-Nielsen, K., and C. R. Taylor. 1968. Red blood cells: Why or why not? *Science* 162:274–75.

Schmidt-Nielsen, K., B. Schmidt-Nielsen, S. A. Jarnum, and T. R. Houpt. 1957. Body temperature of the camel and its relation to water economy. *Am. J. Physiol.* 188:103–12.

Schmidt-Nielsen, K., C. B. Jorgensen, and H. Osaki. 1958. Extrarenal salt excretion in birds. Am. J. Physiol. 188:103–12.

Schmidt-Nielsen, K., F. R. Hainsworth, and D. E. Murrish. 1970. Countercurrent heat exchange in the respiratory passages: Effect on water and heat balance. *Respir. Physiol.* 9:263–76.

Schmidt-Nielsen, K., R. C. Schroter, and A. Skolnik. 1980. Desaturation of the exhaled air in the camel. *J. Physiol.* 305:74P–75.

Schneiderman, H. A., and L. I. Gilbert. 1964. Control of growth and development of insects. *Science* 143:325–33.

Scholander, P. F. 1940. Experimental investigations on the respiratory function in diving mammals and birds. *Norske Videnskaps-Akad. Oslo Hvalradets Skrifter.* 22:1–131.

Scholander, P. F. 1957. The wonderful net. *Sci. Am.* 196(4):96–107.

Scholander, P. F. 1960. Oxygen transport through hemoglobin solutions. *Science* 131:585–90.

Scholander, P. F. 1964. Animals in aquatic environments; diving mammals and birds. In *Handbook of physiology,* sect. 4, edited by D. B. Dill, E. F. V. Adolph, and C. G. Wilbur, 729–39. Washington: American Physiological Society.

Scholander, P. F., and L. Van Dam. 1954. Secretion of gases against high pressures in the swim-bladder of deep sea fishes. 1. Oxygen dissociation in blood. *Biol. Bull.* 107:247–59.

Scholander, P. F., and W. E. Schevill. 1955. Countercurrent vascular heat exchange in the fins of whales. *J. Appl. Physiol.* 8:279–82.

Scholander, P. F., V. Walters, R. Hock, and L. Irving. 1950. Body insulation of some arctic and tropical mammals and birds. *Biol. Bull.* 99:225–36.

Scholander P. F., W. Flagg, R. J. Hock, and L. Irving. 1953. Studies on the physiology of frozen plants and animals in the arctic. *J. Cell. Comp. Physiol.* 42 Suppl. 1:1–56.

Scholey, J. M. 1990. Multiple microtubule motors. *Nature* 343:118–20.

Schoonhoven, L. M. 1972. Plant recognition by lepidopterous larvae. *Symp. Roy. Ent. Soc. Lond.* 6:87–99.

Schwartz, J. H., L. Bernier, V. F. Castellucci, M. Palazzolo, T. Saitoh, A. Stapelton, and E. R. Kandel. 1983. What molecular steps determine the time course of the memory for short-term sensitization in *Aplysia? Cold Spring Harbor Symp. Quant. Biol.* 48:811–19.

Seegers, W. H., and E. A. Sharp. 1948. Hemostatic agents, with particular reference to thrombin, fibrinogen, and absorbable cellulose. Springfield: C. C. Thomas.

Seeherman, H. J., R. Dmi'el, and T. T. Gleeson. 1983. Oxygen consumption and lactate production in varanid and iguanid lizards: A mammalian relationship. In *Biochemistry of Exercise. int. ser. sport. sci.* 13, edited by H. G. Knuttgen, J. A. Vogel, and J. Poortmans, 421–27. Champaign: Kinetics Publishers.

Sernetz, M., B. Gelleri, and J. Hoffman. 1985. The organism as bioreactor: Interpretation of the reduction law of metabolism in terms of heterogeneous catalysis and fractal structure. *J. Theor. Biol.* 117:209–30.

Serway, R. A. 1983. *Physics for scientists and engineers with modern physics*. Philadelphia: Saunders College Publishing.

Seshadri, C. 1956. Urinary excretion in the Indian house lizard, *Hemidactylus flaviviridis* (Ruppel). *J. Zool. Soc. India* 8:63–78.

Seymour, R. S. 1974. How sea snakes may avoid the bends. *Nature* 250:489–90.

Seymour, R. S., and K. Johansen. 1987. Blood flow uphill and downhill: Does a siphon facilitate circulation above the heart? *Comp. Biochem. Physiol.* 88A:167–70.

Sha'afi, R. I. 1981. Permeability for water and other polar molecules. In *Membrane transport*, edited by S. L. Bonting and J. J. H. H. M. de Pont, 29–60. Amsterdam: Elsevier/North Holland Biomedical Press.

Sherbrooke, W. C., M. E. Hadley, and A. M. del Castrucci. 1988. Melanotropic peptides and receptors: An evolutionary perspective in vertebrate physiological color change. In *The melanotropic peptides*, vol. 2, edited by M. E. Hadley, 175–89. Boca Raton: CRC Press.

Sheridan, P. J., and P. M. Martin. 1987. Autographic localization of steroid hormone receptors. In *Steroid hormone receptors*, edited by C. R. Clark, 188–211. Chichester: Ellis Horwood.

Shoemaker, V. H., and K. Nagy. 1977. Osmoregulation in amphibians and reptiles. *Ann. Rev. Physiol.* 39:449–71.

Shoemaker, V. H., and P. E. Bickler. 1979. Kidney and bladder function in a uricotelic treefrog *(Phyllomedusa sauvagei)*. *J. Comp. Physiol.* 133:211–18.

Shoemaker, V. H., D. Balding, and R. Ruibal. 1972. Uricotelism and low evaporative water loss in a South American frog. *Science* 175:1018–20.

Shoemaker, V. H., K. A. Nagy, and S. D. Bradshaw. Studies on the control of electrolyte secretion by the nasal salt gland of the lizard *Dipsosaurus dorsalis*. *Comp. Biochem. Physiol.* 42A:749–57.

Shoemaker, V. H., L. L. McClanahan, P. C. Withers, S. S. Hillman, and R. C. Drewes. 1987. Thermoregulatory response to heat in the waterproof frogs *Phyllomedusa* and *Chiromantis*. *Physiol. Zool.* 60:365–72.

Shpetner, H. S., and R. B. Vallee. 1989. Identification of dynamin, a novel mechanochemical enzyme that mediates interactions between microtubules. *Cell* 59:421–32.

Sibley, R. M., and K. Simkiss. 1987. Gas diffusion through non-tubular pores. *J. Exp. Zool. Suppl.* 1:187–91.

Sidell, B. D. 1977. Turnover of cytochrome C in skeletal muscle of green sunfish *(Lepomis cyanellus*, R.) during thermal acclimation. *J. Exp. Zool.* 199:233–50.

Simkiss, K. 1986. Eggshell conductance—Fick's or Stefan's law? *Respir. Physiol.* 65:213–22.

Simons, J. R. 1960. The blood vascular system. In *Biology and comparative physiology of birds*, edited by A. J. Marshall, 345–62. New York: Academic Press.

Simpson, E. R., and M. R. Waterman. 1988. Regulation of the synthesis of steroidogenic enzymes in adrenal cortical cells by ACTH. *Ann. Rev. Physiol.* 50:427–40.

Sinensky, M. 1974. Homeoviscous adaptation—a homeostatic process that regulates the viscosity of membrane lipids in *Escherischia coli*. *Proc. Natnl. Acad. Sci. USA* 71:522–25.

Singer, S. J., and G. Nicholson. 1972. The fluid mosaic model of the structure of cell membranes. *Science* 175:720–31.

Skadhauge, E. 1981. *Osmoregulation in birds*. Zoophysiology 12. Berlin: Springer-Verlag.

Skadhauge, E., E. Clemens, and G. M. O. Maloiy. 1980. The effects of dehydration on electrolyte concentrations and water content along the large intestine of a small ruminant: The dik-dik antelope. *J. Comp. Physiol.* 135B:165–73.

Skinner, D. M. 1985. Molting and regeneration. In *The biology of Crustacea*, vol. 9, *Integument, pigments and hormonal processes*, edited by D. E. Bliss and L. H. Mantel, 43–146. New York: Academic Press.

Slama, K. 1976. Insect haemolymph pressure and its determination. *Acta Ent. Bohemoslovaka* 73:65–75.

Sleigh, M. A. 1974a. Metachronism of cilia and flagella. In *Cilia and flagella*, edited by M. A. Sleigh, 287–304. London: Academic Press.

Sleigh, M. A. 1974b. Patterns of movement of cilia and flagella. In *Cilia and flagella*, edited by M. A. Sleigh, 29–92. London: Academic Press.

Slip, D. J., and R. Shine. 1988. Reptilian endothermy: A field study of thermoregulation by brooding diamond pythons. *J. Zool.* 216:367–78.

Smith, F. M., and D. R. Jones. 1982. The effects of changes in blood oxygen-carrying capacity on ventilation volume in the rainbow trout *(Salmo gairdneri)*. *J. exp. Biol.* 97:325–34.

Smith, H. W. 1932. Water regulation and its evolution in the fishes. *Quart. Rev. Biol.* 7:1–26.

Smith, H. W. 1959. *From fish to philosopher*. Boston: Little, Brown and Company.

Smith, K. K., and W. M. Kier. 1989. Trunks, tongues, and tentacles: Moving with skeletons of muscle. *Amer. Sci.* 77:35–38.

Smith, P. J. S. 1981. The role of venous pressure in regulation of output from the heart of the octopus *Eledone cirrhosa* (Lam.). *J. exp. Biol.* 93:243–55.

Smith, R. E., and B. A. Horowitz. 1969. Brown fat and thermogenesis. *Physiol. Rev.* 49:330–425.

Snider, R. S. 1950. Interrelations of cerebellum and brain stem. *Res. Publ. Assoc. Res. Nerv. Ment. Dis.* 30:267–81.

Snodgrass, R. E. 1935. *Principles of insect morphology.* New York: McGraw-Hill.

Snyder, G. K. 1973. Erythrocyte evolution: Significance of the Fahraeus-Lindqvist phenomenon. *Respir. Physiol.* 19:271–78.

Snyder, G. K. 1977. Blood corpuscles and blood hemoglobins: A possible example of coevolution. *Science* 195:412–13.

Somero, G. N., and J. J. Childress. 1980. A violation of the metabolism-size scaling paradigm: Activities of glycolytic enzymes in muscle increase in larger-size fish. *Physiol. Zool.* 53:322–37.

Somero, G. N., and J. J. Childress. 1985. Scaling of oxidative and glycolytic enzyme activities in fish muscle. In *Circulation, respiration, and metabolism,* edited by R. Gilles, 250–62. Berlin: Springer-Verlag.

Somero, G. N., and P. S. Low. 1976. Temperature: A "shaping force" in protein evolution. *Biochem. Soc. Symp.* 41:33–42.

Sommerville, B. A. 1973. The circulatory physiology of *Helix pomatia.* I. Observations on the mechanism by which *Helix* emerges from its shell and the effects of body movement on cardiac function. *J. exp. Biol.* 59:275–82.

Southward, A. J., and E. C. Southward. 1980. The significance of dissolved organic compounds in the nutrition of *Siboglinum ekmani* and other small species of Pogonophora. *J. mar. biol. Assoc. U. K.* 60:1005–34.

Southwick, E. E., and G. Heldmaier. 1987. Temperature control in honeybee colonies. *Bioscience* 37:395–99.

Sparrow, M. P. 1988. Multiple intracellular pathways for calcium activation of contraction in smooth muscle. *Aust. Physiol. Pharm. Soc.* 19:74–81.

Speeg, K. V., and J. W. Campbell. 1968. Purine biosynthesis and excretion in *Otala (= Helix) lactea:* An evaluation of the nitrogen excretory potential. *Comp. Biochem. Physiol.* 26:579–95.

Spotila, J. R., P. W. Lommen, G. S. Bakken, and D. M. Gates. 1973. A mathematical model for body temperature of large reptiles: Implications for dinosaur ecology. *Am. Nat.* 107:391–404.

Spring, K. R., and A. W. Siebens. 1988. Solute transport and epithelial cell volume regulation. *Comp. Biochem. Physiol.* 90A:557–60.

Stahl, W. R. 1967. Scaling of respiratory variables in mammals. *J. Appl. Physiol.* 22:453–60.

Steele, J. E. 1985. Control of metabolic processes. In *Comprehensive insect physiology biochemistry and pharmacology,* vol. 8, *Endocrinology II,* edited by G. A. Kerkut and L. I. Gilbert, 99–145. New York: Pergamon Press.

Steen, J. B. 1971. *Comparative physiology of respiratory mechanisms.* New York: Academic Press.

Steffens, W. 1989. *Principles of fish nutrition.* Chichester: Ellis Horwood.

Stein, W. D. 1981. Permeability for lipophilic molecules. In *Membrane transport,* edited by S. L. Bonting and J. J. H. H. M. de Pont, 1–28. Amsterdam: Elsevier/ North Holland Biomedical Press.

Stent, G. S., W. B. Kristian, W. O. Friesen, C. A. Ort, M. Poon, and R. L. Calabrese. 1978. Neuronal generation of the leech swimming movement. *Science* 200:1348–56.

Stevens, E. D. 1972. Some aspects of gas exchange in tuna. *J. exp. Biol.* 56:809–23.

Stevens, E. D., and G. F. Holeton. 1978. The partitioning of oxygen uptake from air and from water by the large obligate air-breathing teleost piraruca *(Arapaima gigas).* Can. *J. Zool.* 56:974–76.

Stevens, E. D., and W. H. Neill. 1978. Body temperature relations of tuna, especially skipjack. In *Fish physiology,* vol. 7, *Locomotion,* edited by W. S. Hoar and D. J. Randall, 315–424. New York: Academic Press.

Stevens, E. D., G. B. Benion, D. J. Randall, and G. Shelton. 1972. Factors affecting arterial pressure and blood flow from the heart of intact, unrestrained lingcod, *Opiodon elongatus. Comp. Biochem. Physiol.* 43A:681–95.

Stevens, E. D., H. M. Lam, and J. Kendall. 1974. Vascular anatomy of the countercurrent heat exchanger of skipjack tuna. *J. exp. Biol.* 61:145–53.

Stewart, M. G. 1979. Absorption of dissolved organic nutrients by marine invertebrates. *Oceanogr. Mar. Biol. Ann. Rev.* 17:163–92.

Stier, T. J. B., and E. Wolf. 1932. On temperature characteristics for different processes in the same organisms. *J. Gen. Physiol.* 16:367–74.

Stieve, H. 1983. The biophysics of photoreception. In *Biophysics,* edited by W. Hoppe, W. Lohmann, H. Markl, and H. Ziegler, 709–29. Berlin: Springer-Verlag.

Stites, D. P., J. L. Caldwell, and C. S. Pavia. 1980. Phylogeny and ontogeny and reproductive influences. In *Basic and clinical immunology,* edited by H. H. Fudenberg, D. P. Stites, J. L. Caldwell, and J. V. Wells, 156–70. Los Altos: Lange Medical Publishers.

Stoner, L. C., and P. B. Dunham. 1970. Regulation of cellular osmolarity and volume in *Tetrahymena. J. exp. Biol.* 53:391–99.

Storey, K. B. 1986. Freeze tolerance in vertebrates: Biochemical adaptation of terrestrial hibernating frogs. In *Living in the cold: Physiological and biochemical adaptations,* edited by H. C. Heller, X. J. Musacchia, and L. C. H. Wang, 131–38. New York: Elsevier.

Storey, K. B., and J. M. Storey. 1983. Carbohydrate metabolism in cephalopod molluscs. In *The Mollusca,* vol. 1, *Metabolic biochemistry and molecular biomechanics,* edited by P. W. Hochachka, 92–137. New York: Academic Press.

Stormer, L. 1977. Arthropod invasion of land during Late Silurian and Devonian times. *Science* 197:1362–64.

Stossel, T. P. 1990. How cells crawl. *Amer. Zool.* 78:408–23.

Stover, H. 1973. Cold resistance and freezing in *Arianta arbustorum* L. (Pulmonata). In *Effects of temperature on ectothermic organisms: Ecological implications and mechanisms of compensation,* edited by W. Wieser, 281–90. Berlin: Springer-Verlag.

Stride, G. O. 1955. On the respiration of an aquatic

African beetle, *Potamodytes tuberosus* Hinton. *Ann. Ent. Soc. America* 48:344–51.

Stryer, L. 1986. Cyclic GMP cascade of vision. *Ann. Rev. Neurosci.* 9:87–119.

Sturkie, P. D. 1986. *Avian physiology.* New York: Springer-Verlag.

Summers, K. E., and I. R. Gibbons. 1971. Adenosine triphosphate induced sliding of tubules in trypsin-treated flagella of sea urchin sperm. *Proc. Nat. Acad. Sci. U.S.A.* 68:3092–96.

Sundin, U., G. Moore, J. Nedergaard, and B. Cannon. 1987. Thermogenin amount and activity in hamster brown fat mitochondria: Effect of cold acclimation. *Am. J. Physiol.* 252:R822–32.

Sutherland, N. S. 1957. Visual discrimination of orientation and shape by the octopus. *Nature* 179:11–13.

Swan, H. 1972. Comparative metabolism: Surface versus mass solved by hibernators. In *Proc. Int. Symp. Environ. Physiol. Bioenergetics,* edited by R. E. Smith, J. P. Hannon, J. L. Shields, and B. A. Horowitz, 25–31. Bethesda: Federation of American Societies for Experimental Biology.

Szabo, T. 1965. Sense organs of the lateral line system in some electric fish of the Gymnotidae, Mormyridae, and Gymnarchidae. *J. Morph.* 117:229–50.

Takahashi, J. S., and M. Menaker. 1979. Physiology of the avian circadian pacemakers. *Fed. Proc.* 38:2583–88.

Takeda, T. 1989. Cutaneous and gill O_2 uptake in the carp, *Cyprinus carpio,* as a function of ambient pO_2, *Comp. Biochem. Physiol.* 94A:205–08.

Tamm, S. L. 1988. Calcium activation of macrocilia in the ctenophore *Beroë. J. Comp. Physiol.* 163A:23–31.

Tandler, A., and F. W. H. Beamish. 1979. Mechanical and biochemical aspects of apparent specific dynamic action in largemouth bass *Micropterus salmoides* Lacepede. *J. Fish Biol.* 14:343–50.

Taylor, C. R., and C. P. Lyman. 1972. Heat storage in running antelopes: Independence of brain and body temperatures. *Am. J. Physiol.* 222:114–17.

Taylor, C. R., G. M. O Maloiy, E. R. Weibel, V. A. Langman, J. M. Z. Kamau, J. H. Seeherman, and N. C. Heglund. 1981. Design of the mammalian respiratory system. III. Scaling of maximal aerobic capacity to body mass: Wild and domestic animals. *Respir. Physiol.* 44:25–37.

Taylor, C. R., K. Schmidt-Nielsen, and J. L. Raab. 1970. Scaling of the energetic cost of running to body size in mammals. *Am. J. Physiol.* 219:1104–7.

Taylor, H. H., and P. Greenaway. 1979. The structure of the gills and lungs of the arid-zone crab, *Holthuisana (Austrothelpusa) transversa* (Brachyura: Sundathelphusidae) including observations on vessels within the gills. *J. Zool.* 189:359–84.

Taylor, P. M., and M. J. Tyler. 1986. Pepsin in the toad *Bufo marinus. Comp. Biochem. Physiol.* 84A:669–72.

Tazawa, H., M. Mochizuki, and J. Piiper. 1979. Respiratory gas transport by the incompletely separated ventricle in the bullfrog, *Rana catesbeiana. Respir. Physiol.* 36:77–95.

Terwilliger, R. C., N. B. Terwilliger, and E. Schabtad. 1976. Comparison of chlorocruorin and annelid hemoglobin quaternary structures. *Comp. Biochem. Physiol.* 55A:51–55.

Thatch, W. T. 1980. The cerebellum. In *Medical physiology,* edited by V. B. Mountcastle, vol. 1, 837–58. St. Louis: C. V. Mosby.

Theriot, J. A., and T. J. Mitchison. 1991. Actin microfilament dynamics in locomoting cells. *Nature* 352:126–31.

Thillart, G. v. d. 1982. Adaptations of fish energy metabolism for hypoxia and anoxia. *Molec. Physiol.* 2:49–61.

Thillart, G. v. d., V. B. Henegouwen, and F. Kesbeke. 1983. Anaerobic metabolism of goldfish *Carassius auratus* (L.): Ethanol and CO_2 excretion rates and anoxia tolerance at 20, 10 and 5° C *Comp. Biochem. Physiol.* 76A:295–300.

Thomas, R. D. K., and E. C. Olson. 1980. *A cold look at the warm-blooded dinosaurs.* AAAS Selected Symposium 28. Washington: American Association for the Advancement of Science.

Thompson, L. C., and A. W. Pritchard. 1969. Osmoregulatory capacities of *Callianassa* and *Upogebia* (Crustacea: Thalassinidae). *Biol. Bull.* 136:114–29.

Thorson, T. B., C. M. Cowan, and D. E. Watson. 1967. *Potamotrygon spp:* Elasmobranchs with low urea content. *Science* 158:375–77.

Thorson, T. T., and A. Svhila. 1943. Correlation of the habitats of amphibians with their ability to survive the loss of water. *Ecology* 24:374–81.

Thurm, U. 1983a. Biophysics of sensory mechanisms: Fundamentals of transduction mechanisms in sensory cells. In *Biophysics,* edited by W. Hoppe, W. Lohmann, H. Markl, and H. Ziegler, 657–66. Berlin: Springer-Verlag.

Thurm, U. 1983b. Biophysics of sensory mechanisms: Mechano-electric transduction. In *Biophysics,* edited by W. Hoppe, W. Lohmann, H. Markl, and H. Ziegler, 666–71. Berlin: Springer-Verlag.

Toien, O., C. V. Paganelli, H. Rahn, and R. R. Johnson. 1987. Influence of eggshell pore shape on gas diffusion. *J. Exp. Zool. Suppl.* 1:181–86.

Tombes, A. S. 1970. *An introduction to invertebrate endocrinology.* New York: Academic Press.

Torre-Bueno, J. R., J. Geiser, and P. Scheid. 1980. Incomplete gas mixing in air sacs of the duck. *Respir. Physiol.* 42:109–22.

Tortora, G. J. 1983. *Principles of human anatomy.* New York: Harper and Row.

Toulmond, A., and C. Tchernigovtzeff. 1984. Ventilation and respiratory gas exchanges of the lugworm *Arenicola marina* (L.) as functions of ambient pO_2 (20–700 torr). *Resp. Physiol.* 57:349–63.

Toulmond, A., P. Dejours, and J. P. Truchot. 1982. Cutaneous O_2 and CO_2 exchanges in the dogfish, *Scyliorhinus canicula. Respir. Physiol.* 48:169–81.

Towles, D. W., C. P. Mangum, B. A. Johnson, and N. A. Mauro. 1982. The role of the coxal gland in ionic, osmotic, and pH regulation in the horseshoe crab *Limulus polyphemus.* In *Physiology and biology of horseshoe crabs: Studies on normal and environmentally stressed animals,* edited by J. Bonaventura, C. Bonaventura, and S. Tesh, 147–72. New York: Alan Liss.

Tracy, C. R. 1976. A model of the dynamic exchanges of water and energy between a terrestrial amphibian and its environment. *Ecol. Monogr.* 46:293–326.

Treherne, J. E. 1965. The distribution and exchange of inorganic ions in the central nervous system of the stick insect *Carausius morosus. J. exp. Biol.* 42:7–27.

Treherne, J. E., and S. H. P. Maddrell. 1967. Membrane potentials in the central nervous system of a phytophagous insect *(Carausius morosus). J. exp. Biol.* 46:413–21.

Trimm, J. L., G. Salama, and J. J. Abramson. 1986. Sulfhydryl oxidation induces rapid calcium release from sarcoplasmic reticulum vesicles. *J. Biol. Chem.* 261:16092–98.

Truchot, J.-P. 1976. Carbon dioxide combining properties of the blood of the shore crab *Carcinus maenas* (L.): Carbon dioxide solubility coefficient and carbonic acid dissociation constants. *J. exp. Biol.* 64:45–57.

Truchot, J.-P. 1978. Mechanisms of extracellular acid-base regulation as temperature changes in decapod crustaceans. *Respir. Physiol.* 33:161–76.

Truchot, J.-P. 1980. Lactate increases the oxygen affinity of crab hemocyanin. *J. Exp. Zool.* 214:205–08.

Truchot, J.-P. 1983. Regulation of acid-base balance. In *The biology of Crustacea,* vol. 5, *Internal anatomy and physiological regulation,* edited by L. H. Mantel, 431–57. New York: Academic Press.

Truman, J. W. 1985. Hormonal control of ecdysis. In *Comprehensive insect physiology biochemistry and pharmacology,* vol. 8, *Endocrinology II,* edited by G. A. Kerkut and L. I. Gilbert, 413–40. Oxford: Pergamon Press.

Truman, J. W., and L. M. Riddiford. 1970. Neuroendocrine control of ecdysis in silkmoths. *Science* 167:1624–26.

Tse, F. W., and H. L. Atwood. 1986. Presynaptic inhibition at the crustacean neuromuscular junction. *NIPS* 1:47–50.

Tucker, V. A. 1975. The energetic cost of moving about. *Amer. Sci.* 63:413–19.

Turner, C. D. 1966. *General endocrinology.* Philadelphia: Saunders.

Turner, C. D., and J. T. Bagnara. 1976. *General endocrinology.* Philadelphia: Saunders.

Uchiyama, M., T. Murakami, and H. Yoshizawa. 1990. Notes on the development of the crab-eating frog *Rana cancrivora. Zoological Science* 7:73–78.

Ultsch, G. R. 1974. Gas exchange and metabolism in the Sirenidae (Amphibia, Caudata). I. Oxygen consumption of submerged sirenids as a function of body size and respiratory surface area. *Comp. Biochem. Physiol.* 47A:485–98.

Ultsch, G. R., and D. C. Jackson. 1982. Longterm submergence at 3° C of the turtle *Chrysemys picta belli* in normoxic and severely hypoxic water. I. Survival, gas exchange and acid-base balance. *J. exp. Biol.* 96:11–28.

Ultsch, G. R., and G. Gros. 1979. Mucus as a diffusion barrier to oxygen: Possible role in O_2 uptake at low pH in carp *(Cyprinus carpio)* gills. *Comp. Biochem. Physiol.* 62A:685–89.

Unwin, P. N. T., and G. Zampighi. 1980. Structure of the junction between communicating cells. *Nature* 283:545–49.

Ussing, H. H., and K. Zerahn. 1950. Active transport of sodium as the source of electric current in the short-circuited frog skin. *Acta Physiol. Scand.* 23:110–27.

Van Dover, C. L., E. Z. Szuts, S. C. Chamberlain, and J. R. Cann. 1989. A novel eye in "eyeless" shrimp from hydrothermal vents of the mid-Atlantic ridge. *Nature* 337:458–60.

van Holde, K. E., and K. I. Miller. 1982. Haemocyanins. *Quart. Rev. Biophys.* 15:1–129

van Mierop, L. H. S., and S. M. Barnard. 1978. Further observations on thermoregulation in the brooding female *Python molurus bivattatus* (Serpentes: Boidae). *Copeia* 1978:615–21.

Vanfleteren, J. R. 1982. A monophyletic line of evolution? Ciliary-induced photoreceptor membranes. In *Visual cells in evolution,* edited by J. A. Westfall, 107–36. New York: Raven Press.

Veghte, J. H. 1964. Thermal and metabolic responses of the gray jay to cold stress. *Physiol. Zool.* 37:316–28.

Verhoef, H. A., and J. E. Prast. 1989. Effects of dehydration on osmotic and ionic regulation in *Orchesella cincta* (L.) and *Tomocerus minor* (Lubbock) (Collembola) and the role of the coeloduct kidneys. *Comp. Biochem. Physiol.* 93A:691–94.

Vernberg, F. J. 1954. The respiratory metabolism of marine teleosts in relation to activity and body size. *Biol. Bull.* 106:360–70.

Verniani, F. The total mass of the Earth's atmosphere. 1966. *J. Geophys. Res.* 71:385–91.

Vick, R. L. 1984. *Contemporary medical physiology.* Menlo Park: Addison-Wesley.

Villee, C. A., E. P. Solomon, C. E. Martin, D. W. Martin, L. R. Berg, and P. W. Davis. *Biology.* Philadelphia: Saunders College Publishing.

Vinogradov, S. N. 1985. The structure of erythrocruorins and chlorocruorins, the invertebrate extracellular hemoglobins. In *Respiratory pigments in animals: Relation structure-function,* edited by J. Lamy, J.-P. Truchot, and R. Gilles, 9–20. Berlin: Springer-Verlag.

Visschedijk, A. H. J., and H. Rahn. 1983. Replacement of diffusive by convective gas transport in the developing hen's egg. *Respir. Physiol.* 52:137–47.

Vitalis, T. Z., and G. Shelton. 1990. Breathing in *Rana pipiens:* The mechanism of ventilation. *J. exp. Biol.* 154:537–56.

Vogel, S. 1974. Current-induced flow through the sponge, *Halichondria. Biol. Bull.* 147:443–56.

Vogel, S., C. P. Ellington, and D. C. Kilgore. 1973. Wind-induced ventilation of the burrow of the prairie dog *Cyanomys ludovicianus. J. Comp. Physiol.* 84:1–14.

Vonk, H. J. 1937. The specificity and collaboration of digestive enzymes in metazoa. *Biol. Rev.* 12:245–84.

Vreugdenhil, P. K., and J. R. Redmond. 1987. Elastic properties of the aortas of the horseshoe crab, *Limulus polyphemus. Mar. Behav. Physiol.* 13:51–62.

Wachtel, A. W., and R. B. Szamier. 1969. Special cutaneous receptor organs of fish. IV. Ampullary organs of the non-electric catfish *Kryptopterus. J. Morph.* 128:291–308.

Wainwright, S. A. 1982. Structural systems: Hydrostats and frameworks. In *A companion to animal physiology,* edited by C. R. Taylor, K. Johansen, and L. Bolis, 325–38. Cambridge: Cambridge University Press.

Wakeman, J. M., and G. R. Ultsch. 1975. The effects of dissolved O_2 and CO_2 on metabolism and gas-exchange partitioning in aquatic sawlamanders. *Physiol. Zool.* 48:348–59.

Walker, A. M., C. L. Hudson, T. Findley, and A. N. Richards. 1937. The total molecular concentration and the chloride concentration of fluid from different segments of the renal tubule of amphibians. The site of chloride reabsorption. *Am. J. Physiol.* 118:121–29.

Walker, J. C. G. 1977. *Evolution of the atmosphere.* New York: Macmillan.

Walker, J. E. C., R. E. Wells, and E. W. Merrill. 1961. Heat and water exchange in the respiratory tract. *Am. J. Physiol.* 30:259–67.

Wall, B. J. 1971. Local osmotic gradients in the rectal pads of an insect. *Fed. Proc.* 30:42–48.

Wall, B. J., and J. L. Oschman. 1970. Water and solute uptake by rectal pads of *Periplaneta americana. Am. J. Physiol.* 218:1208–15.

Walsberg, G. E., G. S. Campbell, and J. R. King. 1978. Animal coat color and radiative heat gain: A re-evaluation. *J. Comp. Physiol.* 126:211–22.

Walsh, P. J., E. Danulat, and T. P. Mommsen. 1990. Variation in urea excretion in the gulf toadfish *Opsanus beta. Marine Biology* 106:323–28.

Wampler, J. E. 1978. Measurements and physical characteristics of luminescence. In *Bioluminescence in action,* edited by P. J. Herring, 1–48. New York: Academic Press.

Wangenstein, O. D., D. Wilson, and H. Rahn. 1971. Diffusion of gases across the shell of the hen's egg. *Respir. Physiol.* 11:16–30.

Wasserthal, L. T. 1975. The role of butterfly wings in regulation of body temperature. *J. Insect Physiol.* 21:1921–30.

Watanabe, S., and D. J. Hartshorne. 1990. Paramyosin and the catch mechanism. *Comp. Biochem. Physiol.* 96B:639–46.

Waterbury, J. B., C. B. Calloway, and R. D. Turner. 1983. A celluloytic nitrogen-fixing bacterium cultured from the gland of Deshayes in shipworms. *Science* 221:1401–03.

Waterman, A. J., B. E. Frye, K. Johansen, A. G. Kluge, M. L. Moss, C. R. Noback, I. D. Olsen, and G. R. Zug. 1971. *Chordate structure and function.* New York: Macmillan.

Waterman, T. H., and H. R. Fernandez. 1970. E-vector and wavelength discrimination by the retinular cells of the crayfish *Procambarus. Z. vergl. Physiol.* 68:154–74.

Weathers, W. W. 1979. Climatic adaptation in avian standard metabolic rate. *Oecologia* 42:81–89.

Weathers, W. W., and D. F. Caccamise. 1975. Temperature regulation and water requirements of the monk parakeet *Myiopsitta monachus.* Oecologia 18:329–42.

Webb, J. 1989. Neuromast morphology and lateral line trunk canal ontogeny in two species of cichlids: An SEM study. *J. Morph.* 202:53–68.

Weber, R. E. 1978. Respiratory pigments. In *Physiology of annelids,* edited by P. J. Mill, 393–446. New York: Academic Press.

Weber, R. E., and J. Baldwin. 1985. Blood and erythrocruorin of the giant earthworm, *Megascolides australis:* Respiratory characteristics and evidence for CO_2 facilitation of O_2 binding. *Molec. Physiol.* 7:93–106.

Weeds, A. 1982. Actin-binding proteins—regulators of cell architecture and motility. *Nature* 296:811–16.

Weibel, E. R. 1984. *The pathway for oxygen: Structure and function in the mammalian respiratory system.* Cambridge: Harvard University Press.

Weigert, R. G. 1964. Population energetics of meadow spittle bugs (*Philaenus spumarius* L.) as affected by migration and habitat. *Ecol. Mono.* 34:217–41.

Weiner, J., A. V. Lous, D. V. Kimberg, and D. Spiro. 1968. A quantitative description of cortisone-induced alterations in the ultrastructure of rat liver parenchymal cells. *J. Cell Biol.* 37:47–62.

Weis-Fogh, T. 1964a. Diffusion in insect wing muscle, the most active tissue known. *J. exp. Biol.* 41:229–56.

Weis-Fogh, T. 1964b. Functional design of the tracheal system of flying insects as compared with the avian lung. *J. exp. Biol.* 41:207–27.

Weis-Fogh, T. 1976. Energetics and aerodynamics of flapping flight: A synthesis. In *Insect flight,* edited by R. C. Rainey, 48–72. New York: Wiley.

Wells, M. J. 1962. *Brain and behavior in cephalopods.* Stanford: Stanford University Press.

Wells, M. J. 1980. Nervous control of the heartbeat in *Octopus. J. exp. Biol.* 85:111–28.

Wells, M. J. 1983. Circulation in cephalopods. In *The Mollusca,* vol. 5, *Physiology,* part 2, edited by A. S. M. Saleuddin and K. M. Wilbur, 239–90. New York: Academic Press.

Wells, M. J. 1990. The dilemma of the jet set. *New Scientist* 125:30–33.

Wells, M. J., and J. Wells. 1957. The function of the brain of *Octopus* in tactile discrimination. *J. exp. Biol.* 34:131–42.

Wells, M. J., and J. Wells. 1989. Water uptake in a cephalopod and the function of the so-called "pancreas." *J. exp. Biol.* 145:215–26.

Welshons, W. V., and V. C. Jordan. 1987. Heterogeneity of nuclear steroid hormone receptors with an emphasis on unfilled receptor sites. In *Steroid hormone receptors*, edited by C. R. Clark, 128–54. Chichester: Ellis Horwood.

Werblin, F. S., and J. E. Dowling. 1969. Organization of the retina of the mudpuppy *Necturus maculosus*. II. Intracellular recording. *J. Neurophysiol.* 32:339–55.

Werner, G. 1980. The study of sensation in physiology: Psychophysical neurophysiologic correlations. In *Medical physiology*, edited by V. B. Mountcastle, 606–46. St. Louis: C. V. Mosby.

West, B. J., and A. L. Goldberger. 1987. Physiology in fractal dimensions. *Amer. Sci.* 75:354–65.

West, B. J., V. Bhargava, and A. L. Goldberger. 1986. Beyond the principle of similitude: Renormalization in the bronchial tree. *J. Appl. Physiol.* 60:1089–97.

West, G. C. 1965. Shivering and heat production in wild birds. *Physiol. Zool.* 38:111–20.

West, I. C. 1983. *The biochemistry of membrane transport*. London: Chapman and Hall.

West, J. B. 1986. Climbing Mount Everest without oxygen. *NIPS* 1:23–25.

West, N. H., and D. R. Jones. 1975. Breathing movements in the frog *Rana pipiens*. II. The power output and efficiency of breathing. *Can. J. Zool.* 53:345–53.

Westby, G. W. M. 1981. Communication and jamming avoidance in electric fish. *TINS* 4:205–10.

Westerterp, K. 1978. How do rats economize energy losses in starvation? *Physiol. Zool.* 50:331–62.

Wharton, G. W. 1978. Uptake of water by mites and mechanisms utilized by the Acaridei. In *Comparative physiology—water, ions and fluid mechanisms*, edited by K. Schmidt-Nielsen, L. Bolis, and S. H. P. Maddrell, 79–95. Cambridge: Cambridge University Press.

Wharton, G. W., and R. T. Furumizo. 1977. Supracoxal gland secretions as a source of fresh water for Acaridei. *Acaryologia* 19:112–16.

White, D. C. S. 1977. Muscle mechanics. In *Mechanics and energetics of locomotion*, edited by R. McN. Alexander and G. Goldspink, 23–56. London: Chapman and Hall.

White, F. N. 1968. Functional anatomy of the heart of reptiles. *Amer. Zool.* 8:211–19.

White, R. H. 1985. Insect visual pigments and color vision. In *Comprehensive insect physiology, biochemistry and pharmacology*, edited by G. A. Kerkut and L. I. Gilbert, vol. 6, *Nervous system: Sensory*, 431–94. Oxford: Pergamon Press.

Whitten, J. M. 1964. Connective tissue membranes and their apparent role in transporting neurosecretory and other secretory products in insects. *Gen. Comp. Endocrinol.* 4:176–192.

Whittle, A. C., and D. W. Golding. 1974. The infracerebral gland and cerebral neurosecretory system—a probable neuroendocrine complex in phyllodocid polychaetes. *Gen. Comp. Endocrinol.* 24:87–98.

Whittow, G. C. 1973. Evolution of thermoregulation. In *The comparative physiology of thermoregulation*, vol. 3, *Special aspects of thermoregulation*, edited by G. C. Whittow, 201–58. New York: Academic Press.

Wieser, W. 1972. A glutaminase in the body wall of terrestrial isopods. *Nature* 239:288–90.

Wigglesworth, V. B. 1935. *Principles of insect morphology*. London: Chapman and Hall.

Wigglesworth, V. B. 1938. The regulation of osmotic pressure and chloride concentration in the haemolymph of mosquito larvae. *J. exp. Biol.* 15:235–47.

Wigglesworth, V. B. 1945. Transpiration through the cuticle of insects. *J. exp. Biol.* 21:97–114.

Wigglesworth, V. B. 1984. *Insect physiology*. London: Chapman and Hall.

Wightman, J. A. 1981. Why insect energy budgets do not balance. *Oecologia* 50:166–69.

Williamson, S. J. 1973. *Fundamentals of air pollution*. Reading: Addison-Wesley.

Wilson, F. A., and J. M. Dietsky. 1974. The intestinal unstirred layer: Its surface area and effect on active transport kinetics. *Biochim. Biophys. Acta* 363:112–26.

Wilson, T. H. 1962. *Intestinal absorption*. Philadelphia: Saunders.

Wiltschkow, W., and R. Wiltschkow. 1972. Magnetic compass of European robins. *Science* 176:62–64.

Wine, J. J., and F. B. Krasne. 1982. The cellular organization of crayfish escape behavior. In *The biology of Crustacea*, vol. 4, edited by K. Wilbur and A. S. M. Salueddin, 241–92. New York: Academic Press.

Withers, P. C. 1976. Models of diffusion-mediated gas exchange in animal burrows. *Am. Nat.* 112:1101–12.

Withers, P. C. 1977. Metabolism, respiration and haematological adjustments of the little pocket mouse to circadian torpor cycles. *Respir. Physiol.* 31:295–307.

Withers, P. C. 1981a. An aerodynamic analysis of bird wings as fixed aerofoils. *J. exp. Biol.* 90:143–62.

Withers, P. C. 1981b. The effects of ambient air pressure on oxygen consumption of resting and hovering honeybees. *J. Comp. Physiol.* 141:433–37.

Withers, P. C. 1983. Energy, water, and solute balance of the ostrich *Struthio camelus*. *Physiol. Zool.* 56:568–79.

Withers, P. C., and J. D. Campbell. 1985. Effects of environmental cost on thermoregulation in the desert iguana. *Physiol. Zool.* 58:329–39.

Withers, P. C., and S. S. Hillman. 1988. A steady-state model of maximal oxygen and carbon dioxide transport in anuran amphibians. *J. Appl. Physiol.* 64:860–68.

Withers, P. C., S. S. Hillman, and R. C. Drewes. 1984. Evaporative water loss and skin lipids of anuran amphibians. *J. Exp. Zool.* 232:11–17.

Withers, P. C., T. M. Casey, and K. K. Casey. 1979. Allometry of respiratory and haematological parameters of arctic mammals. *Comp. Biochem. Physiol.* 64A:343–50.

Withers, P. C., W. R. Siegfried, and G. N. Louw. 1981. Desert ostrich exhales unsaturated air. *Sth. Afr. J. Sci.* 77:569–70.

Withers, P. C., S. S. Hillman, M. S. Hedrick, and P. B. Kimmel. 1991. Optimal hematocrit theory during activity in the bullfrog (*Rana catesbeiana*). *Comp. Biochem. Physiol.* 99A:55–60.

Withers, P. C., S. S. Hillman, R. C. Drewes, and O. Sokol. 1982. Water loss and nitrogen excretion in sharp-nosed reed frogs *(Hyperolius nasutus:* Anura, Hyperoliidae). *J. exp. Biol.* 97:335–43.

Wittenberg, B. A., J. B. Wittenberg, and P. R. B. Caldwell. 1975. Role of myogloben in the oxygen supply to red skeletal muscle. *J. Biol. Chem.* 250:9038–43.

Wittenberg, J. B. 1958. The secretion of inert gas into the swim-bladder of fish. *J. Gen. Physiol.* 41:783–804.

Wittenberg, J. B. 1963. Facilitated diffusion of oxygen through haemerythrin solutions. *Nature* 199:816–17.

Wodtke, E. 1973. Effects of acclimation temperature on aerobic energy production in eel liver: Oxidative phosphorylation in isolated mitochondria. In *Effects of temperature on ectothermic organisms: Ecological implications and mechanisms of compensation,* edited by W. Wieser, 97–105. Berlin: Springer-Verlag.

Wodtke, E., T. Teichert, and A. Konig. 1986. Control of membrane fluidity in carp upon cold stress: Studies on fatty acid desaturases. In *Living in the cold: Physiological and biochemical adaptations,* edited by H. C. Heller, X. J. Musacchia, and L. C. H. Wang, 35–42. New York: Elsevier.

Wolbarsht, M. L. 1965. Receptor sites in insect chemoreceptors. *Cold Spring Harbor Symp. Quant. Biol.* 30:281–88.

Wolcott, T. G., and D. L. Wolcott. 1985. Extrarenal modification of urine for ion conservation in the ghost crab, *Ocypoda quadrata* (Fabricius). *J. Exp. Marine Biol. Ecol.* 91:93–107.

Wood, D. C. 1970. Parametric studies of the response decrement produced by mechanical stimuli in the protozoan, *Stentor coeruleus. J. Neurobiol.* 1:345–60.

Wood, D. C. 1971. Electrophysiological correlates of the response decrement produced by mechanical stimulation in the protozoan *Stentor. J. Neurobiol.* 2:1–11.

Wood, S. C. 1980. Adaptation of red blood cell function to hypoxia and temperature in ectothermic vertebrates. *Am. Zool.* 20:163–72.

Woodley, J. D. 1980. The biomechanics of ophiuroid tube-feet. In *Echinoderms: Present and past,* edited by M. Jangoux, 293–99. Rotterdam: A. A. Balkema.

Woodring, J. P. 1973. Comparative morphology, functions, and homologies of the coxal glands in orbatid mites (Arachnida: Acari). *J. Morph.* 139:407–30.

Woolley, T. A. 1972. Scanning electron microscopy of the respiratory apparatus of ticks. *Trans. Amer. Micros. Soc.* 91:348–63.

Wright, S. H., and D. T. Manahan. 1989. Integumental nutrient uptake by aquatic organisms. *Ann. Rev. Physiol.* 51:585–600.

Yancey, P. H. 1985. Organic osmotic effectors in cartilaginous fishes. In *Transport processes, iono- and osmoregulation,* edited by R. Gilles and M. Gilles-Baillien, 424–36. Berlin: Springer-Verlag.

Yancey, P. H. 1988. Osmotic effectors in kidneys of xeric and mesic rodents: Corticomedullary distributions and changes with water availability. *J. Comp. Physiol.* 158B:369–80.

Yancey, P. H., and G. N. Somero. 1979. Counteraction of urea destabilization of protein structure by methylamine osmoregulatory compounds of elasmobranch fishes. *Biochem. J.* 183:317–23.

Yancey, P. H., and G. N. Somero. 1980. Methylamine osmoregulatory solutes of elasmobranch fishes counteract urea inhibition of enzymes. *J. Exp. Zool.* 212:205–13.

Yancey, P. H., and G. N. Somero. 1988. Temperature dependence of intracellular pH: Its role in the conservation of pyruvate apparent K_m values of vertebrate lactate dehydrogenases. *J. Comp. Physiol.* 125:129–34.

Yancey, P. H., and M. B. Burg. 1989. Distribution of major organic osmolytes in rabbit kidneys in diuresis and antidiuresis. *Am. J. Physiol.* 257:F602–07.

Yancey, P. H., and M. B. Burg. 1990. Counteracting effects of urea and betaine in mammalian cells in culture. *Am. J. Physiol.* 248:R198–204.

Yang, D., Y. Oyaizu, H. Oyaizu, G. J. Olsen, and C. R. Woese. 1985. Mitochondrial origins. *Proc. Natl. Acad. Sci. USA* 82:4443–47.

Yates, F. E. 1981. Comparative physiology of energy production: Homeotherms and poikilotherms. *Am. J. Physiol.* 240:R1–2.

Yee, H. F., and D. C. Jackson. 1984. The effects of different types of acidosis and extracellular calcium on the mechanical activity of turtle atria. *J. Comp. Physiol.* 154B:385–91.

Yokoe, Y., and I. Yasumasu. 1964. The distribution of cellulase in invertebrates. *Comp. Biochem. Physiol.* 13:323–38.

Yokohari, F., and H. Tateda. 1976. Moist and dry hygroreceptors for relative humidity of the cockroach *Periplaneta americana. J. Comp. Physiol.* 106:137–52.

Yokota, S. D., and S. S. Hillman. 1984. Adrenergic control of the anuran cutaneous hydroosmotic response. *Gen. Comp. Endocrinol.* 53:309–14.

Yokota, S. D., and V. H. Shoemaker. 1981. Xanthine excretion in a desert scorpion, *Paruroctonus mesaensis. J. Comp. Physiol.* 142B:423–28.

Yokota, S. D., S. Benyajati, and W. H. Dantzler. 1985. Comparative aspects of glomerular filtration in vertebrates. *Renal Physiol.* 8:193–221.

Yonge, C. M. 1928. Feeding mechanisms in the invertebrates. *Biol. Rev.* 3:21–76.

Young, J. Z. 1991. Computation in the learning system of cephalopods. *Biol. Bull.* 180:200–08.

Zapol, W. M., G. C. Liggins, R. C. Schneider, J. Qvist, M. T. Snider, R. K. Creasey, and P. W. Hochachka. 1979. Regional blood flow during simulated diving in the conscious Weddell seal. *J. Appl. Physiol.* 47:968–73.

Zdarek, J. 1985. Regulation of pupariation in flies. In *Comprehensive insect physiology biochemistry and pharmacology,* vol. 8, *Endocrinology II,* edited by G. A. Kerkut and L. I. Gilbert, 301–33. New York: Pergamon Press.

Zerbst-Boroffka, I. 1977. The function of nephridia in

annelids. In *Transport of ions and water in animals,* edited by B. L. Gupta, R. B. Moreton, J. L. Oschman, and B. J. Wall, 763–70. New York: Academic Press.

Zeuthen, E. 1953. Oxygen uptake as related to body size in organisms. *Quart. Rev. Biol.* 28:1–12.

Zhao-Xian, W., S. Ning-Zhan, and S. Wen-Fang. 1989. Aquatic respiration in soft-shelled turtles, *Trionyx sinensis. Comp. Biochem. Physiol.* 92A:593–98.

Zinkler, D., and M. Gotze. 1987. Cellulose digestion by the firebrat *Thermobia domestica. Comp. Biochem. Physiol.* 88B:661–66.

Genus Index

The letters 'f', 's' and 't' after a page number indicate a figure, supplement, or table respectively.

Subject Index

The letters 'f', 's' and 't' after a page number indicate a figure, supplement, or table respectively.